T0144096

Human-Centred Computing:

Cognitive, Social and Ergonomic Aspects

Volume 3

Human Factors and Ergonomics
Gavriel Salvendy, Series Editor

Bullinger, H.-J., and Ziegler, J. (Eds.) : Human–Computer Interaction: Ergonomics and User Interfaces, *Volume 1 of the Proceedings of the 8th International Conference on Human–Computer Interaction*

Bullinger, H.-J., and Ziegler, J. (Eds.) : Human–Computer Interaction: Communication, Cooperation, and Application Design, *Volume 2 of the Proceedings of the 8th International Conference on Human–Computer Interaction*

Hollnagel, E. (Ed.) : *Handbook of Cognitive Task Design*

Meister, D. (Ed.) : *Conceptual Foundations of Human Factors Measurement*

Meister, D., and Enderwick, T. (Eds.) : *Human Factors in System Design, Development, and Testing*

Smith, M. J., Salvendy, F., Harris, D., and Koubeck, R. J. (Eds.) : *Usability Evaluation and Interface Design: Cognitive engineering, Intelligent Agents and Virtual Reality*

Smith, M. J., and Salvendy, G. (Eds.) : *Systems, Social and Internationalization Design of Human–Computer Interaction*

Stephanidis, C. (Ed.) : *Universal Access in HCI: Towards an Information Society for All*

Stephanidis, C. (Ed.) : *User Interfaces for All: Concepts, Methods, and Tools*

Ye, N. (Ed.) : *The Handbook of Data Mining*

Jacko, J., and Stephanidis, C. (Eds.) : *Human-Computer Interaction: Theory and Practice (Part I)*

Stephanidis, C., and Jacko, J. (Eds.) : *Human Computer Interaction: Theory and Practice (Part II)*

Harris, D., Duffy, V., Smith, M., and Stephanidis, C. (Eds.) : *Human-Centred Computing: Cognitive, Social and Ergonomic Aspects*

Stephanidis, C. (Ed.) : *Universal Access in HCI: Inclusive Design in the Information Society*

Human-Centred Computing:

Cognitive, Social and Ergonomic Aspects

Volume 3 of the Proceedings of HCI International 2003
10th International Conference on Human - Computer Interaction
Symposium on Human Interface (Japan) 2003
5th International Conference on Engineering Psychology and Cognitive Ergonomics
2nd International Conference on Universal Access in Human - Computer Interaction
22 – 27 June 2003, Crete, Greece

Edited by

Don Harris
Cranfield University

Vincent Duffy
Mississippi State University

Michael Smith
University of Wisconsin-Madison

Constantine Stephanidis
ICS-FORTH and University of Crete

2003 LAWRENCE ERLBAUM ASSOCIATES, PUBLISHERS
Mahwah, New Jersey London

Lawrence Erlbaum Associates, Inc., Publishers
10 Industrial Avenue
Mahwah, New Jersey 07430

Human-Centred Computing : Cognitive, Social and Ergonomic aspects / edited by Don Harris, Vincent Duffy, Michael Smith, and Constantine Stephanidis.

p. cm.

Includes bibliographical references and index.
ISBN 0-8058-4932-7 (cloth : alk. paper) (Volume 3)

ISBN 0-8058-4930-0 (Volume 1)
ISBN 0-8058-4931-9 (Volume 2)
ISBN 0-8058-4933-5 (Volume 4)
ISBN 0-8058-4934-3 (Set)

2003

Books published by Lawrence Erlbaum Associates are printed on acid-free paper, and their bindings are chosen for strength and durability.

10 9 8 7 6 5 4 3 2 1

Preface

The 10th International Conference on Human-Computer Interaction, HCI International 2003, is held in Crete, Greece, 22-27 June 2003, jointly with the Symposium on Human Interface (Japan) 2003, the 5th International Conference on Engineering Psychology and Cognitive Ergonomics, and the 2nd International Conference on Universal Access in Human-Computer Interaction. A total of 2986 individuals from industry, academia, research institutes, and governmental agencies from 59 countries submitted their work for presentation, and only those submittals that were judged to be of high scientific quality were included in the program. These papers address the latest research and development efforts and highlight the human aspects of design and use of computing systems. The papers accepted for presentation thoroughly cover the entire field of human-computer interaction, including the cognitive, social, ergonomic, and health aspects of work with computers. These papers also address major advances in knowledge and effective use of computers in a variety of diversified application areas, including offices, financial institutions, manufacturing, electronic publishing, construction, health care, disabled and elderly people, etc.

We are most grateful to the following cooperating organizations:

- Chinese Academy of Sciences
- Japan Management Association
- Japan Ergonomics Society
- Human Interface Society (Japan)
- Swedish Interdisciplinary Interest group for Human Computer Interaction - STIMDI
- Asociación Interacción Persona Ordenador - AIPO (Spain)
- Gesellschaft für Informatik e.V - GI (Germany)
- European Research Consortium for Information and Mathematics - ERCIM

The 289 papers contributing to this volume (Vol. 3) cover the following areas:

- Ergonomics and Health Aspects
- Cognitive Ergonomics
- Engineering Psychology
- Online Communities, Collaboration and Knowledge
- Applications and Services
- Design & Visualization
- Virtual Environments

The selected papers on other HCI topics are presented in the accompanying three volumes: Volume 1 edited by J. Jacko and C. Stephanidis, Volume 2 by C. Stephanidis and J. Jacko, and Volume 4 by C. Stephanidis.

We wish to thank the Board members, listed below, who so diligently contributed to the overall success of the conference and to the selection of papers constituting the content of the four volumes.

Eduardo Salas, *USA*
Dirk Schaefer, *France*
Neville A. Stanton, *UK*

Universal Access in Human-Computer Interaction
Julio Abascal, *Spain*
Demosthenes Akoumianakis, *Greece*
Elizabeth Andre, *Germany*
David Benyon, *UK*
Noelle Carbonell, *France*
Pier Luigi Emiliani, *Italy*
Michael C. Fairhurst, *UK*
Gerhard Fischer, *USA*
Ephraim Glinert, *USA*
Jon Gunderson, *USA*
Ilias Iakovidis, *EU*
Arthur I. Karshmer, *USA*
Alfred Kobsa, *USA*
Mark Maybury, *USA*
Michael Pieper, *Germany*
Angel R. Puerta, *USA*
Anthony Savidis, *Greece*
Christian Stary, *Austria*
Hirotada Ueda, *Japan*
Jean Vanderdonckt, *Belgium*
Gregg C. Vanderheiden, *USA*
Annika Waern, *Sweden*
Gerhard Weber, *Germany*
Harald Weber, *Germany*
Michael D. Wilson, *UK*
Toshiki Yamaoka, *Japan*

We also wish to thank the following external reviewers:

Chrisoula Alexandraki, *Greece*
Margherita Antona, *Greece*
Ioannis Basdekis, *Greece*
Boris De Ruyter, *Netherlands*
Babak Farschian, *Norway*
Panagiotis Karampelas, *Greece*
Leta Karefilaki, *Greece*
Elizabeth Longmate, *UK*
Fabrizia Mantovani, *Italy*
Panos Markopoulos, *Netherlands*
Yannis Pachoulakis, *Greece*
Zacharias Protogeros, *Greece*
Vassilios Zarikas, *Greece*

This conference could not have been held without the diligent work and outstanding efforts of Stella Vourou, the Registration Chair, Maria Pitsoulaki, the Program Administrator, Maria Papadopoulou, the Conference Administrator, and George Papatzanis, the Student Volunteer Chair. Also, special thanks to Manolis Verigakis, Zacharoula Petoussi, Antonis Natsis, Erasmia Piperaki, Peggy Karaviti and Sifis Klironomos for their help towards the organization of the Conference. Finally recognition and acknowledgement is due to all members of the HCI Laboratory of ICS-FORTH.

Constantine Stephanidis
ICS-FORTH and University of Crete, GREECE

Julie A. Jacko
Georgia Institute of Technology, USA

Don Harris
Cranfield University, UK

Vincent G. Duffy
Mississippi State University, USA

Michael J. Smith
University of Wisconsin-Madison, USA

June 2003

HCI International 2005

The 11th International Conference on Human-Computer Interaction, HCI International 2005, will take place jointly with:

Symposium on Human Interface (Japan) 2005
6th International Conference on Engineering Psychology and Cognitive Ergonomics
3rd International Conference on Universal Access in Human-Computer Interaction
1st International Conference on Virtual Reality
1st International Conference on Usability and Internationalization

The conference will be held in Las Vegas, Nevada, 22-27 July 2005. The conference will cover a broad spectrum of HCI-related themes, including theoretical issues, methods, tools and processes for HCI design, new interface techniques and applications. The conference will offer a pre-conference program with tutorials and workshops, parallel paper sessions, panels, posters and exhibitions. For more information please visit the URL address: http://hcii2005.engr.wisc.edu

General Chair:

Gavriel Salvendy
Purdue University
School of Industrial Engineering
West Lafayette, IN 47907-2023 USA
Telephone: +1 (765)494-5426 Fax: +1 (765) 494-0874
Email: salvendy@ecn.purdue.edu
http://gilbreth.ecn.purdue.edu/~salvendy
and
Department of Industrial Engineering
Tsinghua University, P.R. China

The proceedings will be published by Lawrence Erlbaum and Associates.

Table of Contents

Section 1. Ergonomics and Health Aspects

Section 2. Cognitive Ergonomics

Section 4. Online Communities, Collaboration and Knowledge

Section 5. Applications and Services

Section 6. Design & Visualisation

Section 7. Virtual Environments

Section 1

Ergonomics and Health Aspects

Position of the arm and the musculoskeletal disorders

Arne Aarås

Alcatel A/S
P.O. Box 310, Økern
0511 Oslo
Norway
arne.aaraas@alcatel.no

Gunnar Horgen

Buskerud College
Department of Optometry
P.O. Box 251, Kongsberg
Norway
gunnar.horgen@hibu.no

Abstract

In a laboratory study, postural load during Visual Display Unit (VDU) work was measured by electromyography (EMG). The load on musculus trapezius and rector spina lumbalis (L3 level) was significantly less in sitting with supported forearm on the tabletop compared to sitting without forearm support. In another laboratory study, the muscle load of the extensors of the forearm was significantly less when working with a neutral position of the forearm versus a pronated one, when performing mouse work. This was true for the extensor digitorum communis and extensor carpi ulnaris.

In a prospective field study, two intervention groups of approximately 50 VDU workers were given possibility to support their whole forearms on the tabletop. These groups reported significantly reduction of shoulder pain after two years while only small changes were observed in the control group. Further, after two years the shoulder pain was reported significantly higher in the control group compared with the two intervention groups.

In another prospective field study, one intervention and one control group of approximately 30 VDU workers supported their forearms on the tabletop when performing mouse work. The intervention group who operated the mouse with a neutral position of the forearm reported significantly less pain in the upper part of the body compared with those who worked with a pronated forearm after six months.

1 Introduction

This paper is a summary of several papers published in different journals.

The position of the arm seems important for developing of musculoskeletal disorders. The following three conditions regarding arm position must be considered in order to reduce such discomfort.

1. The position of the upper arm relative to vertical.
2. Supporting the forearm.
3. Position of the forearm.

2 Laboratory studies

2.1 Postural load for various work posture

In a laboratory study, postural load was compared in various work postures for VDU workers. This was done during data entry work with and without possibility to support the forearm on the tabletop. The muscle load on Trapezius and Erector spina lumbalis was significantly less when sitting with supported forearms compared to sitting and standing without support. The static load of the upper trapezius when using keyboard was significantly less when sitting with supported forearms, 0.8 % Maximum Voluntary Contraction (MVC) compared with sitting without support 3.6 % MVC, as group mean values, (Aarås, Fostervold, Ro, Thoresen & Larsen, 1997).

2.2 Postural load of the forearm in neutral and pronated position

Another laboratory study was performed in order to compare the muscle load of the extensors of the forearm when operating two different types of computer mouse. One mouse was operated with a pronated forearm while the other could be operated with the forearm in a neutral position. The muscle load of the forearm was significantly less when using the forearm in a more neutral position compared with a pronated one. Regarding extensor digitorum communis the median values for the group were 4.5 % MVC versus 10.8 % MVC, (Aarås & Ro, 1997).

3 Prospective field studies

3.1 The shoulder moment and the musculoskeletal pain

A field study of the position of the upper arm and the sick leave due to musculoskeletal illness was performed in an assembly plant manufacturing parts for telephone exchanges. Ergonomic improvement of the work stand gave the operators greater flexibility to vary the work posture. This lead to a reduction of trapezius load, measured by EMG from approximately 4.5 % MVC to approximately 2 % MVC as group mean values, when comparing before and after the intervention. Survival statistics of workers employed before and after implementation of the ergonomic adaptations, documented a significant less risk to record a musculoskeletal sick-leave after intervention versus before ($p<0.01$). This significant difference was true for different times after employment until two years, (Aarås & Westgaard, 1987). Figure 1.

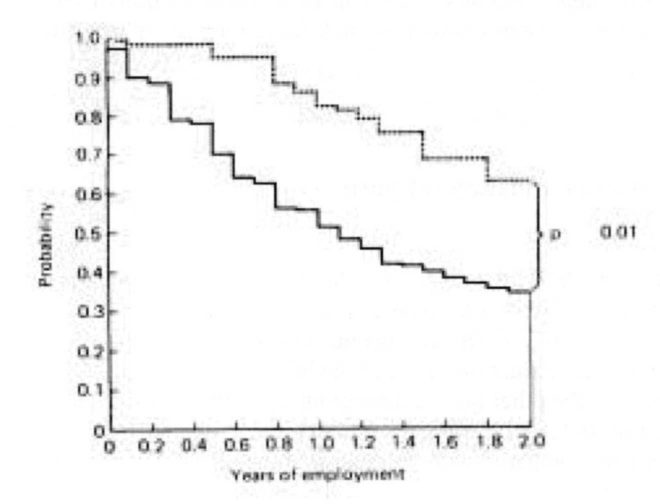

Figure 1: Survival statistics of workers employed before (solid line) and after (broken line) implementation of ergonomic intervention at the STK Kongsvinger plant. The two curves indicate the probability of not recording a musculoskeletal sick leave at different times after employment. Workers employed before the ergonomic intervention, were predominantly workers at the 8 B and cable making systems, after redesign at the 10 C and 11 B systems.

Further, a reduction in postural load was find by calculating the shoulder moment during work. The angles of the upper arm in terms of flexion/extension and abduction/adduction were measured by inclinometers. In addition, video recordings were used for estimating the position of the forearm in order to obtain the shoulder moment. Those groups of workers with a high shoulder moment, 5 Newton meter (NM) and 6.3 NM, had significantly more sick-leave due to musculoskeletal illness compared with those workers with shoulder moment between 1.4 NM and 3.7 NM, (Aarås, 1994). Figure 2.

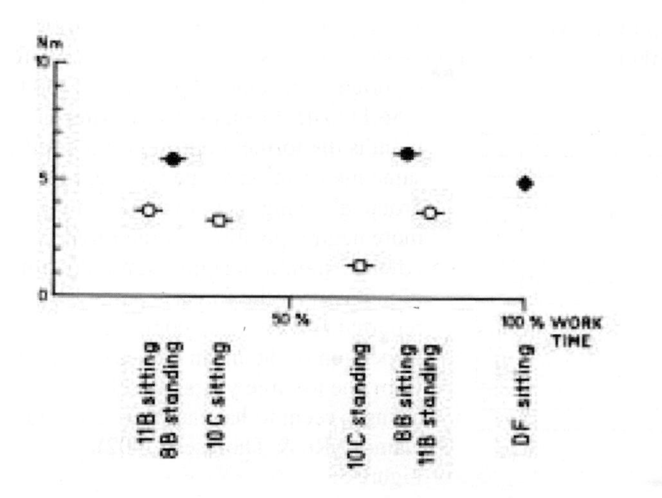

Figure 2: The static shoulder moment is calculated on the basis of the group median value of postural angles of the upper arm and flexion in the elbow joint for different work systems on the y-axis. The time in the sitting and standing position as a percentage of total work time is indicated on the x-axis.

Those workers with a median flexion of the upper arm of 15° and a median arm abduction of 10° with low external load in hand, were beginning to approximate the sick-leave for a group of office workers without continuous work load.

3.2 Supporting the forearm and musculoskeletal pain

Another field study was carried out giving the VDU workers possibility to support their forearms performing software engineering. The same subjects were compared regarding intensity of shoulder pain 2 years after versus before intervention. The average intensity of shoulder pain during the last six months, showed a significant reduction in one of the intervention group ($p=0.02$) and a clear tendency to reduction in the other intervention group ($p=0.08$). No significant changes were found in the control group ($p=0.92$). The two intervention groups reported significantly lower intensity of shoulder pain compared with the control group after two years of the study.

3.3 Supporting the forearm in a neutral position and musculoskeletal pain

In a third field prospective study of VDU workers with pain, the development of pain was evaluated when using a mouse with a more neutral position of the forearm (intervention group) compared with a traditional mouse with a more pronated forearm (control group). After six months, a significant reduction was reported in the intervention group regarding pain intensity of wrist/hand, forearm, shoulder and neck ($p<0.009$). The control group continues with the traditional mouse, reported only small changes in the pain level ($p>0.24$). In the intervention group the forearm pain was reduced as a group mean with 95 % confidence intervals from 52.9 mm (42.7 – 63.0) to 32.8mm (3.6 – 42.0) on a 100 mm Visual Analog Scale (VAS). No such changes were reported in the control group 44.6 (33.0 – 56.1) to 45.3 (34.4 – 56.4). After 6 months the former control group got the same intervention as the intervention group, allowing them to work with a more neutral positions of the forearm. After 12 months a significant reduction of intensity of the forearm pain was reported in the former control group. By inspection of the mean values of the pain from one to three years, no relevant changes seem to have appeared, (Aarås, Dainoff, Ro & Thoresen, 2002). Figure 3.

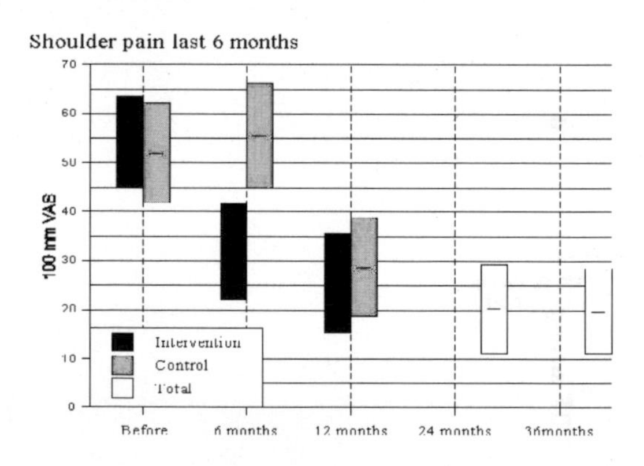

Figure 3: The intensity of average forearm pain during the last 6 months. The values are given as group mean with 95 % confidence interval on VAS.

4 Conclusion

The results from the field studies supported the results from the laboratory studies regarding the importance of working with supported forearm in a neutral position.
Without supporting of the forearm, the position of the upper arm should be as close to the vertical position as possible.

5 References

Aarås, A. (1994). The impact of ergonomic intervention on individual health and corporate prosperity in a telecommunications environment. Ergonomics, vol. 37, no. 10, 1679-1696.
Aarås, A., Fostervold, K. I., Ro, O., Thoresen, M. & Larsen, S. (1997). Postural load during VDU work: a comparison between various work postures. Ergonomics, Vol. 40, No. 11, 1255-1268.
Aarås, A. & Ro, O. (1997). Work load when using "mouse" as input device. A comparison between a new developed "mouse" and a traditional "mouse" design. International Journal Human Computer Interaction, Volume 9, No. 2.
Aarås, A. & Westgaard, R. H. (1987). Further studies of postural load and musculoskeletal injuries of worker at an electro-mechanical assembly plant. Applied Ergonomics 18,3, 211-219.
Aarås, A., Horgen, G., Bjørset, H-H., Ro, O. & Thoresen, M. (1998). Musculoskeletal, visual and psychosocial stress in VDU operators before and after multidisciplinary ergonomic interventions. Applied Ergonomics, 29, 335-354.
Aarås, A., Dainoff, M., Ro, O. & Thoresen, M. (2002). Can a more neutral position of the forearm when operating a computer mouse reduce the pain level for Visual Display Unit operators? A prospective epidemiological intervention study: Part III. International Journal of Industrial Ergonomics 30/4-5, 307-324.

Physical Environments for Human Computer Interaction in Scandinavia

Steen Enrico Andersen

PLH architects, Dampfaergevej 10, DK-2100 Copenhagen Oe, Denmark
sea@plh.dk

Abstract

Clear, innovative solutions for physical work environments, developed in collaboration with the end user, are the basis for a successful design. The user's needs and the building's functional requirements are transformed by the architect into space, material and light. Together with experts in the other building professions, architectural and technical solutions are developed which meet the user's expectations as well as the architect's professional standards.

1 Introduction

The architect's challenge is, simply speaking, a matter of anticipating the needs and requirements of the user; both those which are defined and those which may be unforeseen, and the successful project is one that follows a carefully planned route which has been mapped out in collaboration with the users. Maintaining an open channel of communication between all parties is vital to the successful outcome of the project. Careful detailing and product knowledge are keywords for the architect who continually strives to create buildings that aesthetically and physically withstand the test of time.

2 Workplace Design

Interior consulting examines spatial issues from a different perspective than interior design. The organisation of a company has direct influence on the physical framework and likewise, physical settings have a direct influence on work patterns. An organisation may need support in articulating its physical settings. The aim of space planning is to create work environments that promote flexible work patterns, allow greater interaction and reduce office hierarchy. The design must allow for variations while clarifying internal flow. It is also important to optimise competencies across the organisation by creating a climate for the exchange of knowledge.

Throughout the process the consultant develops tools to reinforce the client's strategic goal. A requirement for a successful result is to establish a collaborative relationship with Management and Human Resources. To assist organisations in identifying their requirements and mapping out their goals, the support process will typically include strategic planning, workshops, analysis of needs and pilot projects. Through this process the specific needs of the organisation and the individual employee are clearly established, and an optimal workplace solution can be achieved.

Figure 1: Space planning diagram for a new facility for Scandinavian Airlines System. Public zones are marked (red), semi-public (beige), private (light blue) and semi-private (dark blue).

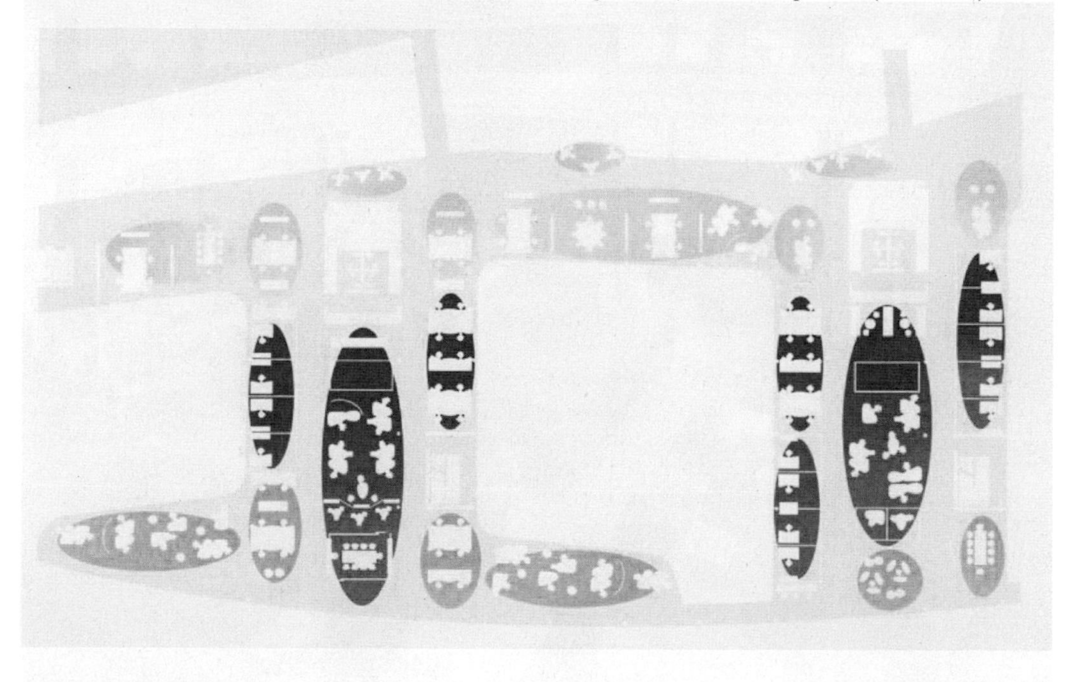

Figure 2: Space planning of a typical "neighbourhood" with access to an internal atrium in Scandinavian Airlines System's new facility. Semi-public zones are marked (beige), private (light blue) and semi-private (dark blue).

Figure 3: Atrium at Deloitte & Touche offices, Copenhagen

3 Deloitte & Touche: "A Resort for Work"

The new offices for Deloitte & Touche in Copenhagen illustrate how an organization's global workplace strategy can be interpreted and reflected in the building design. The 13,500 m^2 office is designed as a 6-storey atrium building, which stands out as a transparent crystalline mass overlooking the harbor of Copenhagen. A glazed roof floods the atrium with diffuse daylight (figure 1). The exterior facade consists of glass louvers providing optimal daylight and thermal regulation to the interior. In contrast to the stringent exterior, the interior is more organically formed with softly curved balconies and timber-clad cores.

The organisation's global space planning paradigms are based on the principles of *The Activity-based Workplace*. The design for the new 13,500 m^2 Danish office reflects Deloitte & Touche's vision of "a resort for work," where the 6-storey atrium offers a variety of work settings available to all employees, and acts as the hub of both the building and the organisation. The design is supported by an advanced use of wireless communication and IT technology (figure 4).

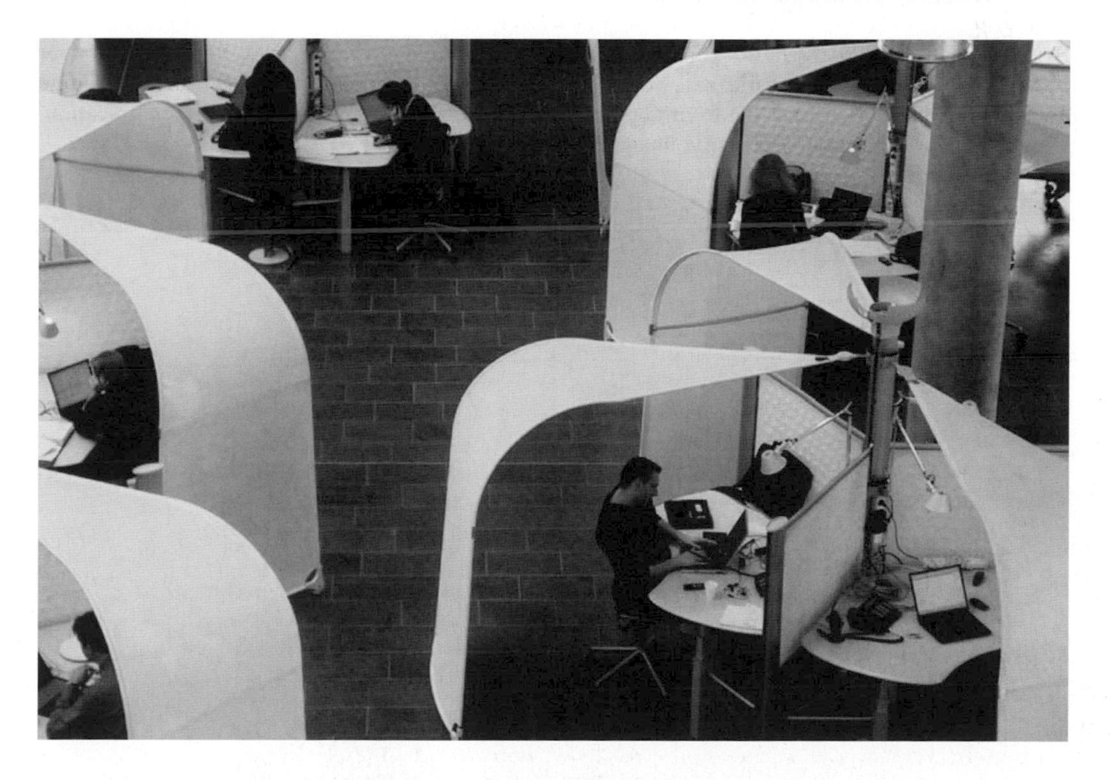

Figure 4: Deloitte & Touche, Copenhagen – Wireless touch down area

4 New Ways of Working Trends in Scandinavian Environments

Workplace design in Scandinavia has undergone dramatic change over the past 10 years – moving from enclosed spaces in narrow buildings to more open, "playground for work" environments, in deeper and square atrium buildings.
Scandinavian management often use a global approach in their cultural interpretation of workplace strategy and the planning of new environments. Priority issues for users in Scandinavia span from advanced use of the latest technology to ergonomics and daylight provision.
The process from vision to implementation is often a top down and bottom up approach, with focus on the balance between organisation, workplace, knowledge and IT within the company.

5 Case Studies

Deloitte & Touche Copenhagen: *A resort for work*: The interpretation of a global workplace strategy and how the project evolved as "Strategy+". Key words from the design brief: "open, fluid, networks, spontaneous and vision."

Scandinavian Airlines System, Copenhagen: *A landside lounge for work*. A design for greater interaction between departments and working groups – 4 zones private, semi-private, semi-public and public areas. Key words from the design brief: "flexibility, openness, communication, visibility, people, lifestyle, diversity, image (figures 1 & 2)."

Engineering Workplaces: Advanced Workplace Concept

Werner Baumeister

DaimlerChrysler AG, Stuttgart, Germany
Werner.Baumeister@daimlerchrysler.com

Abstract

Changes in work environments demand a new approach to workplace design. New ways of organizing work and new processes determine new types of workplaces and require a new method of planning. Advanced workplace concept enables to create process oriented solutions enhancing the efficiency of processes (speed, communication, flexibility, creativity).

1 Background

The work environment is changing more than ever before.
In order to survive and remain competitive in today's global market, businesses have to design their strategies to suit global market conditions.

What does this mean for businesses?

Processes and organizational forms have to be able to reflect market demands.

In the case of automobile development, for instance, this means that companies have to service global markets, that an ever-increasing range of models has to be brought to market in ever shorter cycles, that cost pressures are becoming more and more intense as competitors try to squeeze each other out of the market, and that technical innovations are required at a faster and faster pace.

These are the basic cornerstones of a product development strategy that determines the organizational forms and processes of development work.

This allows us to derive the following development goals :

- Reduction in development times
- Enhancement of development quality
- Reduction in development costs
- Product design geared to manufacturing processes
- More cost-effective production techniques.

2 Features of development work

2.1 Organizational and process transformation

The transformation that has been achieved in this area shows that there has been a move away from strictly hierarchical organizational forms towards project and network-oriented structures. Automobiles, which are now highly complex systems, can only be developed nowadays with the help of networked projects and sub-projects as a means of achieving set goals. A focus on targets and results, a high level of individual responsibility on the part of those involved, rapid feedback processes and intensive self-management are the key features of development work. "Project houses" are an example of these principles in action.

There are increasing demands on the content of development work; the products incorporate increasingly expensive technology, and innovative solutions are in urgent demand as a way of implementing the strategic goal of technological leadership.
In particular, there has been a dramatic increase in the number of electrical and electronic systems in automobiles, and all of these areas need to be mastered by the engineering teams.

As the level of complexity increases, problems can only be solved by using rapid feedback processes across disciplines. Parallel processes lead to simultaneous engineering. There is a transfer of expertise, innovative solutions from one project are transferred to other projects, as for example with the development of electronic or safety systems that can be applied across the board for all types of vehicle (e.g. the Electronic Stability Program (ESP) or air bags).

More and more development partnerships are being entered into, and the extent of development is reduced to the company's own core competency. Internal resources are used to consolidate this core competency. Horizontal development process chains are created. As a consequence of this, management of development processes and the co-ordination of development partners are becoming increasingly significant.

In the core process of development work, we find such function groups as design engineers, test engineers, development engineers and project engineers.

Design engineers provide the digital resources by implementing the project specifications; test engineers turn the digital material into the physical product. Development engineers design complex vehicle systems, and project engineers guide and manage the system partners through all the stages of the development process chain up to the point where a vehicle enters series production.
Service functions such as calculation, testing and inspection support development operations and need to be integrated into the overall process.
As a network of different projects, the organization becomes a dynamic learning system.

2.2 Technological transformation

In much the same way, there has also been a momentous technological transformation. Whereas in the past, most design work was carried out at the drawing board, this has now been almost completely replaced by the use of data processing technology, which has significantly improved the efficiency of the development process. The drawing office, with its drawing boards arranged in rows like school desks, disappeared from development offices many years ago.

Vehicle design is now digital, using CAD systems at workstations. It is now possible to produce a complete virtual representation of a vehicle by digital means, in the form of a digital mock-up. Simulation and calculation functions have shortened or replaced test processes, and have also improved development quality and enhanced the reliability of decision-making.

In this environment, the creation and handling of knowledge, which now has a half-life of just four years, is of crucial importance.
Complete data integration is the precondition for a worldwide data exchange based on integrated IT platforms. Methods and tools to aid an efficient workflow across the value added chain of the product creation process are stored in a Book of Knowledge.
Worldwide communications networks mean that the workplace has become more delocalized, and the system of fixed working hours is also largely a thing of the past.

Worldwide communications networks mean that the workplace has become more delocalized, and the system of fixed working hours is also a thing of the past.

3 Design requirements for development workplaces

This is the background against which we can derive the structural demands for workplaces and the spatial conditions required for engineering.
The key dictum is to create workplace, floor and room structures that meet the demands of the development process. Specifically, this means that process-oriented solutions are required that reflect the development process in spatial terms.

The following design features need to be incorporated:

- Making it possible to experience openness and transparency
- Improving the conditions for communication
- Stimulating speed
- Encouraging spontaneity and creativity
- Enabling self-organization to take place
- Integrating communications technology
- Generating enthusiasm.

It also has to be made clear that "the individual" as a success factor has to be the focus of all design activities.

4 Approaches to design solutions - Advanced Workplace Concept

4.1 Methodical approach to planning

Implementing plans involves more than just looking at the design of an individual workplace or workplace type and looking no further; an integrated view of a process field is essential.
This takes in the full range from the individual, personalized workplace to communication-oriented areas where changing groups of people operate on a project level and in a type of workshop environment.
To meet this challenge at a planning level, we use the "Advanced Workplace Concept", a planning method that we developed ourselves.
This consists of floor modules with design components allowing a wide range of working situations to be represented.

Figure 1: Examples of the "Advanced Workplace Concept"

4.2 Implementation in Engineering

However, user-friendly planning is only possible if we are familiar with the process involved. This is why basic organizational, process, functional, communication and technical details are recorded as part of a process analysis and are then used to produce a design plan.
Our objective here is to plan from the inside outwards.

What results do we obtain for the benefit of Engineering?

The personal workplace for all developer functions requires a flexibly constructed free-form table, the height of which can be adjusted up to eye level using a motor. The linked workplace systems formerly in use proved too inflexible and are no longer practical today in view of a turnover rate of about 40% a year.
Workplaces are provided with up to three data connections. These are needed for the workstation with a 21-inch screen, plus a docking station with laptop, supplemented where necessary by an output device, and a standard telephone connection.
In principle, all workplaces are networked together. The basic requirement is a powerful, comprehensive IT infrastructure, networked beyond the local environment for data exchange.

With more and more computer applications, involving ever-increasing volumes of data, a 100 MB network is required. The option of facilitating transfer amounts in the GB range for new projects is under consideration.

A process-oriented workplace arrangement generally involves about five to eight workplaces at a team level. To support spontaneous personal communication, which is an important element in ensuring process speed and efficiency, each team is assigned a "marketplace", a high, round table that invites spontaneous communication. The short distances involved encourage communication. If work requires full concentration for a time, an appropriate room module offers the opportunity to withdraw from the group environment.

The creation of project areas in the immediate environment, which are also equipped with communication technology to enable the spontaneous organization of co-ordination processes, is becoming more and more important. The range of special areas also includes rooms for digital mock-ups, discussions, virtual reality rooms, or special laboratory areas such as electronic and acoustic laboratories.
There has been a steady increase in the number of areas that do not form part of the personal workplace and which incorporate to a degree the wide range of forms of a communications environment can take.

This makes it possible to achieve a combined "design and testing" approach, in which these two concepts are closely integrated on both a spatial and a process level; in other words, a linking of the real and digital worlds and a corresponding spatial proximity to Engineering and Workshop, (for testing purposes).

The case study in the example shows what the planning solution is like in practice, using an excerpt from the process.

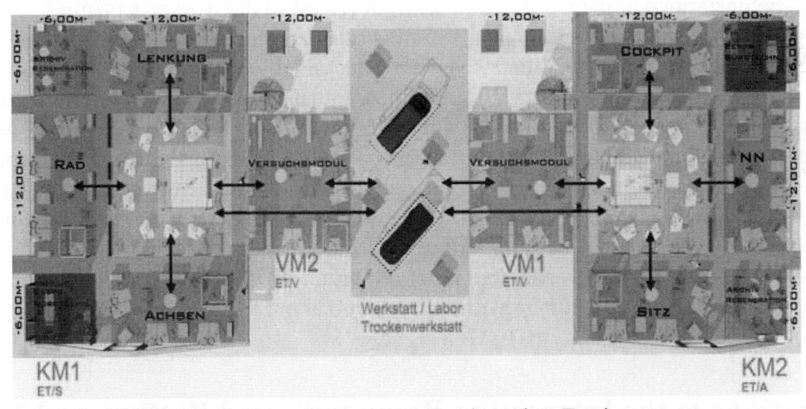

Figure 2: Process Chain in an Engineering Environment

To sum up, the spatial requirements for creative and innovative work have to be met by taking an integrated approach: all aspects must be considered to produce a physical environment in which employees can feel comfortable, and which also satisfies the technological criteria to enable them to develop their abilities and thus contribute to securing the company's competitive position.

Interactions of Visual and Motor Demands on Reaching Actions at Workstations

Marvin J. Dainoff, Leonard S. Mark, Douglas L. Gardner

Miami University
Oxford, Ohio 45056 USA
dainofmj@muohio.edu

Abstract

Research on visually-guided reaching has examined the transitions between various action modes that are used to reach for objects at different distances. People tend to change from one reach mode to another, not at the absolute maximum distance at which one of the modes can be used, but at closer distances. The selection of reach mode depends not only on the reach distance, but also task constraints. Our research has shown that postural stability requirements constrain the reach action, especially when the task has stringent motor and visual requirements. We consider these findings from the perspective that perception and action are not separate modules that operate separately from one another, a view that was advocated by the late James Gibson (1966).

1 Introduction

In designing computer workstations, ergonomists have typically considered the visual demands of the activities to be performed at the workstation without regard to the worker's posture, movements, and goal-directed activities. That vision is often studied as an independent module reflects a belief that our senses are inherently passive detectors of energy and that the processing of that energy is both necessary for constructing our perceptual experience and the province of the nervous system. This perspective does not consider either the functions served by vision or its role in guiding the performance of goal-directed action. The research presented in this paper is grounded in a different tradition (Gibson, 1966), namely that visual perception must be examined in the context of the activities that it serves. However, vision is not merely the servant of action. Action also produces changes in the sensory stimulation that is information about what is out in the world, what is happening in the world, and most importantly, what a person is doing in the world? In short, action provides a person with information about the world that may have further implications for the course of the activity.

2 Applications to the Interaction Between Seated Posture and Task

Research conducted in our laboratory almost 20 years ago showed that computer-based tasks with different constraints (e.g., editing vs. data entry) require different postural configurations (Dainoff & Mark, 1987). Data entry tasks require operators to see small characters. In order to see these characters, a forward leaning posture was found to place the eyes sufficiently close to see the copy. In contrast, the larger text found in editing tasks do not require the worker to be nearly as close to the video terminal. In this situation, a backward leaning posture afforded sufficient stability and comfort for the worker to accomplish the editing task. The outcome of this work points toward the interplay between visual and postural (action) requirements for achieving task goals.

3 Research on Reaching

3.1 A Taxonomy of Reach Actions

More recently, we have examined the activity of visually-guided reaching. Gardner et al. (2001) reported a taxonomy of action (reach) modes, that is, the parts of the body recruited to perform reach actions as reach distance increases (Figure 1): At close distances, people sitting in front of a table might only have to move their arms, an *arm-only reach*, or rotate their torso such that the shoulder of the reach arm extends forward, an *arm-and-shoulder reach*. At somewhat farther distances people tend to lean forward as they extend their arm, an *arm-and-torso reach*. (On rare occasions people *slide forward* in order to avoid standing up.) At still farther distances, leaning will not suffice and people begin to stand, either partially (a *partial standing reach*) or fully (a *full standing reach*). At the farthest distances people have to *locomote* in order to reach the object.

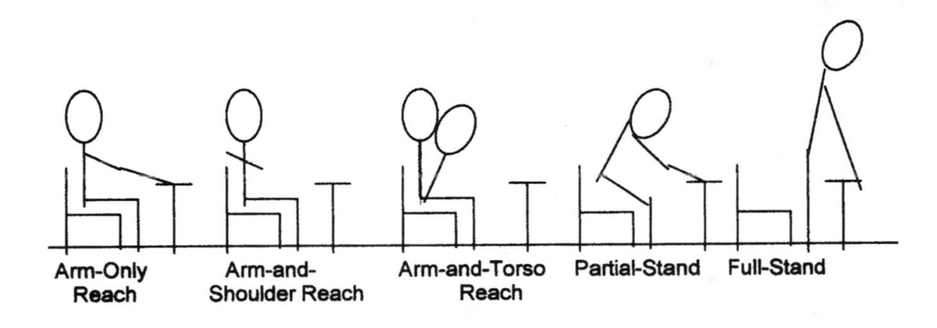

Figure 1: A graphical depiction of the taxonomy presented by Gardner et al. (2001).

3.2 Transitions Between Action Modes

Several studies of these reach actions (Carello et al, 1989; Mark et al., 1997) have shown that people are quite accurate in *perceiving* the absolute limits of their capability to use a particular reach mode, the *absolute critical boundary* (ACB). However, when allowed to reach in whatever manner they want, Mark et al. (1997) found that as reach distance increases, actors change from one reach mode to another at distances closer than the ACB—what Mark et al. referred to as the *preferred critical boundary* (PCB). Figure 2 shows data from Mark et al.'s study of the transition between arm-only and other reaches. What is particularly noteworthy is that subjects' judgments of the comfort rating of arm-only reaches relative to other reach modes changed at the same distance as they changed from using arm-only reaches to more complex reach modes.

3.3 Constraints on the Location Transitions Between Action Modes

Our research has also examined the constraints that determine the location of the PCB between action modes. Gardner et al. (2001) focused on the role of visual and postural task constraints in determining both the location of the PCB between reach modes and whether particular modes would be used at all. In this work, the reach targets were a 3-cm cube, which had to be grasped with the thumb and fingers, and a sewing bead with a 1-mm hole, which had to be skewered with a needle. The target distances were selected so that the closest distance required an arm-and-torso reach (lean forward) and the farthest distance could be accomplished with full-standing reach. Figures 3a (block task) and 3b (bead task) show two results that reveal the importance of task constraints on reaching. The location of PCB was closer to the operator (83% ACB vs. 95 % ACB) for the bead task; (b) a partial-standing reach (crouch) was typically used in the block task, but not in the bead task. We interpret this *mode suppression* in the bead task as indicating that the combination of increased visual and motor demands required a degree of postural stability that could not be achieved using a partial standing reach mode.

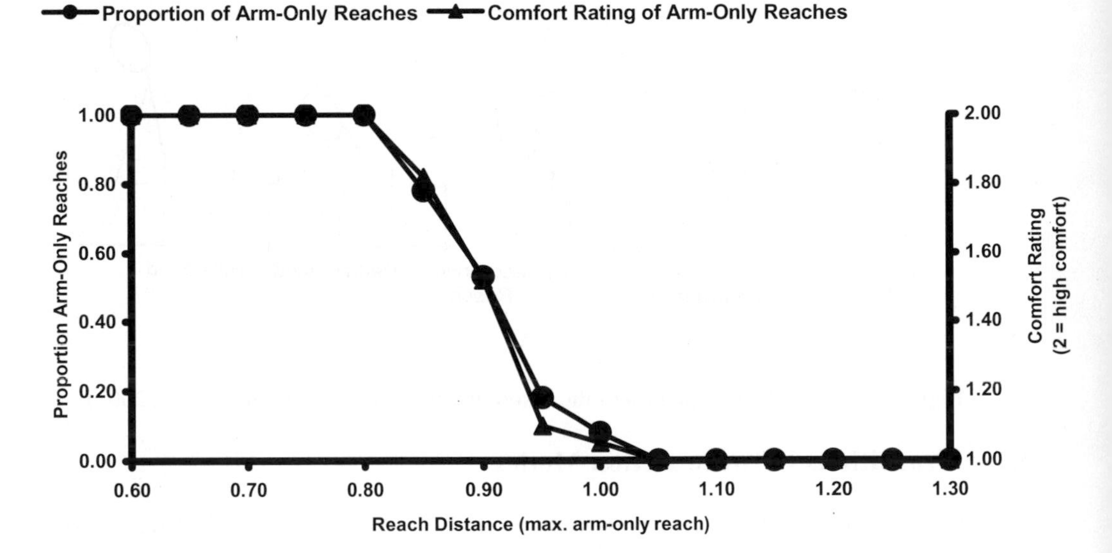

Figure 2: The proportion of arm-only reaches as a function of reach distance scaled with respect to each subject's ACB for arm-only reaches. The PCB between arm-only and more complex reaches occurred at roughly 92% of the distance of the ACB. When the same subjects were asked to judge the comfort associated with an arm-only reach, the location of the PCB coincided with the observed decrease in the comfort associated with arm-only reaches (N=18).

A subsequent study in which bead sizes, diameters, and tasks were varied demonstrated that mode suppression also occurred for a task in which the bead diameter was larger but the skewering tool diameter was close to the bead diameter (high motor demand; low visual demand). However, mode suppression was not found when the task was grasping (as opposed to skewering) a much smaller bead (low motor demand, high visual demand). Gardner et al. (2001) interpreted these results as indicating that the action mode used in either the block or bead task depended on the postural stability needed to satisfy the visual and motor requirements of the particular task. A partial-standing reach, which is similar to a crouch, is effortful to maintain for more than a few seconds. This did not pose a problem for the block task, which could be completed within 1 sec and entailed few postural demands because the block could be picked up easily. However, skewering the bead with a needle entailed a stable postural frame to support the fine motor activity needed to see the hole in the bead and skewer it. The partial standing reach did not afford the necessary stability and thus was avoided.

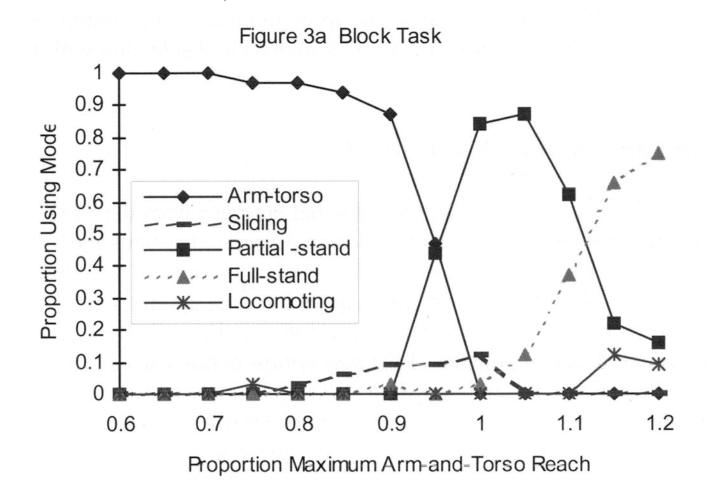

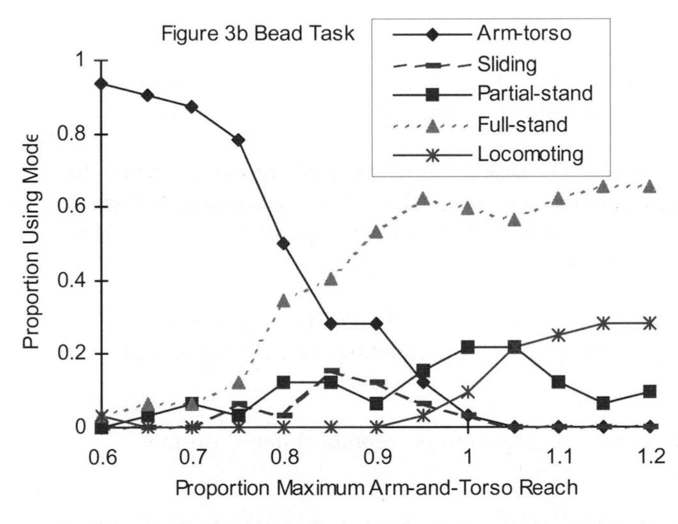

Figure 3: The proportion of reaches using each reach mode for each reach distance for (a) the block pickup and (b) bead task. Reach distance is expressed in units of each subject's maximum arm-and-torso reach.

4 Implications for Workstation Design

These findings have clear implications for consideration of computer workstation with respect to the impact on reach actions of combinations of factors such as optometric correction, monitor location, input device location, and screen/cursor geometry. Designers must be careful not to (inadvertently) constrain reach modes such that fatigue and/or muscle strain results. The extent to which these principles can be extended to other actions is less clear, but worthy of further study. Although the concept of a PCB is general, our study to date has been limited to reach actions in the physical domain. We anticipate that the basic principles will apply to on-screen reaching, but are less sure of the broader applicability to other action taxonomies such as tool selection within a software package.

5 The Study of Perceiving and Acting Are Linked

The relationship between vision, posture and goal directed actions that emerges from the work discussed in this paper reflects interdependencies between the human perceptual system and action. Gibson (1966) proposed that a perceptual system not only entails the sense organs, but also the muscles and limbs entailed in exploratory movements required to pick up information about the world. Within this framework neither actions nor perceptions can be studied in isolation. Rather they must be considered as components of the *perception-action cycle*, which must be considered to be the fundamental unit of analysis. In this analysis, information regarding the world is picked up during the perception phase of the cycle, that information provides the basis for action, the result of which reveals new information. The results of the studies reported in this paper suggest that ergonomic research on the role of vision in computer workstations will benefit from adopting this framework.

6 References

Carello, C., Grosofsky, A. Reichel, F. D., Solomon, H. Y., & Turvey, M. T. (1989). Visually perceiving what is reachable. *Ecological Psychology*, *1*, 27-54.

Dainoff, M.J. & Mark, L.S, (1987). Task the adjustment of ergonomic chairs. In B. Knave and P.-G. Wideback (Eds.), *Work with display units 86: Selected papers from the International Scientific Conference on Word with Display Units, Stockholm, Sweden, May 12-15,* 1986 (pp.294-302). New York: Elsevier.

Gardner, D. L., Mark, L. S., Ward, J., & Edkins, H. (2001). How do task characteristics affect the transitions between seated and standing reaches? *Ecological Psychology*, *13*, 245-274.

Gibson, J. J. (1966). *The senses considered as perceptual systems.* Boston: Houghton-Mifflin.

Mark, L. S., Nemeth, K., Gardner, D. G., Dainoff, M. J., Paasche, J., Duffy, M., & Grandt, K. (1997). Postural dynamics and the preferred critical boundary for visually guided reaching. *Journal of Experimental Psychology: Human Perception and Performance*, *23*, 1365-1379.

Two Field Trials of Brief Rest Breaks to Reduce Musculoskeletal Symptoms

Julia Faucett[1],
jaf@itsa.ucsf.edu

James Meyers[2],
jmmeyers@uclink4.berkeley.edu

John Miles[3],
jamiles@ucdavis.edu

Ira Janowitz[4],
janowitz@earthlink.net

Fadi Fathallah[3],
fathallah@ucdavis.edu

Abstract

Engineering intervention is the control method of choice to reduce exposure to occupational ergonomic hazards, particularly for strenuous work tasks. When employer or task constraints prohibit the use of engineering approaches, administrative intervention may improve working conditions and reduce workers' symptoms of musculoskeletal (MS) disorders. We tested an experimental rest and recovery protocol for its efficacy in reducing symptoms and its effects on productivity during two types of strenuous work task. Both tasks are associated with high ergonomic hazards and frequent MS symptoms, and engineering interventions have proved difficult to identify or implement. The experimental condition consisted adding a 5 minute rest break to every working hour in which there was no other scheduled break (e.g., lunchtime). This resulted in an additional 20 minutes of rest per work day. We tested the intervention in two trials: Trial One compared workers (n=72) randomly assigned to an experimental or a control group during the harvest of commercial strawberries. Trial Two utilized a cross-over design to compare two groups (n=16 pairs of workers) during experimental and control conditions while they inserted bud grafts into 18" high citrus trees. For both trials, workers under the experimental condition reported significantly less severe symptoms than workers under control conditions. The order in which the intervention was given, however, appeared to result in variations in productivity. The introduction of frequent, brief rest breaks every hour resulted in improving symptoms for different groups of workers performing two different work tasks. Effects on worker productivity appeared to be related to more to the trial conditions than the intervention itself, and possibly to work crew competition. We discuss the challenges of implementing these studies in field settings.

1 Introduction

1.1 The Agricultural Ergonomics Research Center of the University of California

The Agricultural Ergonomics Research Center of the University of California develops and tests engineering, administrative and training interventions to reduce the incidence of injury and the severity of symptoms related to musculoskeletal (MS) disorders.

[1] Box 0608-School of Nursing, University of California, San Francisco 94143-0608
[2] School of Public Health, University of California, Berkeley CA 94720
[3] Dept. of Biological and Agricultural Engineering, University of California, Davis CA 95616-5294
[4] UC Ergonomics Lab., 1301 So. 46 St., Bldg. 163 University of California, Richmond CA 94804
Funding: U.S. National Institute of Occupational Safety and Health (PHS-RO1/CCR914508)

Team members include experts in health care, ergonomics, agriculture, education, engineering, and statistics. Although engineering is the most desirable approach for the control of ergonomic hazards, there remain agricultural work tasks for which "off-the shelf" products are not available, and feasible and economic engineering solutions remain elusive. For two such stoop labor tasks, a participative ergonomics approach was implemented in two field settings that resulted in trials to test alternative rest break patterns.

1.2 Intervention

The intervention consisted of a rest and recovery protocol calling for an additional 5 minute break for every working hour in which there was no other scheduled break (e.g., lunchtime). On average, this results in an additional 20 minutes of rest per work shift. Workers are asked to stop their work; they may stand, stretch, walk about or assume a comfortable sitting position.

1.3 Outcome Measures

We have developed and tested a symptom survey that is appropriate for the language, culture and literacy levels of the Mexican immigrant workers who provide the majority of labor in California agriculture.[1] The survey yields a composite measure of symptom severity. We also examine legally required worksite injury logs (OSHA 200); however, underreporting is common in California agriculture where workers have often immigrated recently, and frequently illegally. Additionally, we utilized a bar-code system to measure productivity of each worker (or team) that counts the number of work cycles completed. We also evaluate worker and manager preferences.

2 Trial One

2.1 Strawberry Harvest

The first trial was implemented with a small strawberry production operation (60 acres) specializing in early season fresh market and processing fruit. Strawberry picking involves walking down furrows (approximately 12" wide and 300 yards long) between raised beds (14-18" high) and pushing or carrying a wire cart with strawberry boxes in it. Berries are twisted from vine with the fingers and placed in boxes. Normally pickers remain in a stooped or crouched posture for the length of a row. Workers are paid on incentive basis and move as fast as possible. Extensive research by the engineering team confirmed that there were no available tools or technologies that would alleviate either the stoop or the repetitive hand picking exposures. A survey of workers' symptoms showed that 70% of these workers reported significant pain or discomfort in the mid or lower back at the end of the season end and 30% reported other lower extremity pain. Risk factor exposures include:

- Severe trunk flexion
- Sustained neck extension
- Shoulder and elbow flexion
- Highly repetitive hand picking
- Constant deviation of both hands
- Contact pressure on knee from kneeling on bed.

This was a three day pre-post intervention trial with random assignment to the Experimental Group (Es), who received the intervention or Control Group (Cs), who performed their tasks under their usual work conditions. At the end of Day 1, we gathered baseline data about MS symptoms, fatigue and productivity (n=72). Workers were then randomly assigned to the E and C groups. Symptom and productivity data were gathered at the end of each shift for Days 2 and 3.

2.2 Results

For each day of the rest and recovery trial, Es reported less severe symptoms than Cs, but were also less productive. Nonetheless, after statistically controlling for differences in levels of productivity, Es continued to demonstrate better symptom control. Figure 1 describes difference scores for workers' MS symptoms. Figure 1 was based on the subset of workers who experienced a change in their symptoms. It shows that most of the workers whose symptoms worsened on Days 2 and 3 were Cs. It also shows that those workers whose symptoms improved tended to be Es. A total of 39 indicated either an increase or a decrease in symptoms and 33 workers reported no change in symptoms. Figure 1 graphs the observed frequency that a particular total symptom score difference occurred. Each symptom score difference was calculated by taking the sum of the scores for Days 2 and 3 and subtracting twice the symptom score for Day 1 from the Day 2/Day 3 composite. This variate can be interpreted as twice the difference between, on the one hand, the Day 2/Day 3 average and, the Day 1 score. The Mann U Whitney value is 466.5 ($p<0.04$, two tailed). A negative score value indicates that the average MS symptom score measured during Days 2 and 3 was less than the score measured on Day 1, while a positive score indicates that during these two days symptoms increased. A value of zero indicates no change in symptom severity after the intervention. Productivity for each worker was assessed by counting full boxes of fruit brought to the trucks on each day of the trial. E crews averaged about a 9.4% reduction of productivity as a result of the intervention. This productivity reduction would require a crew of 35 about 29 minutes to make up.

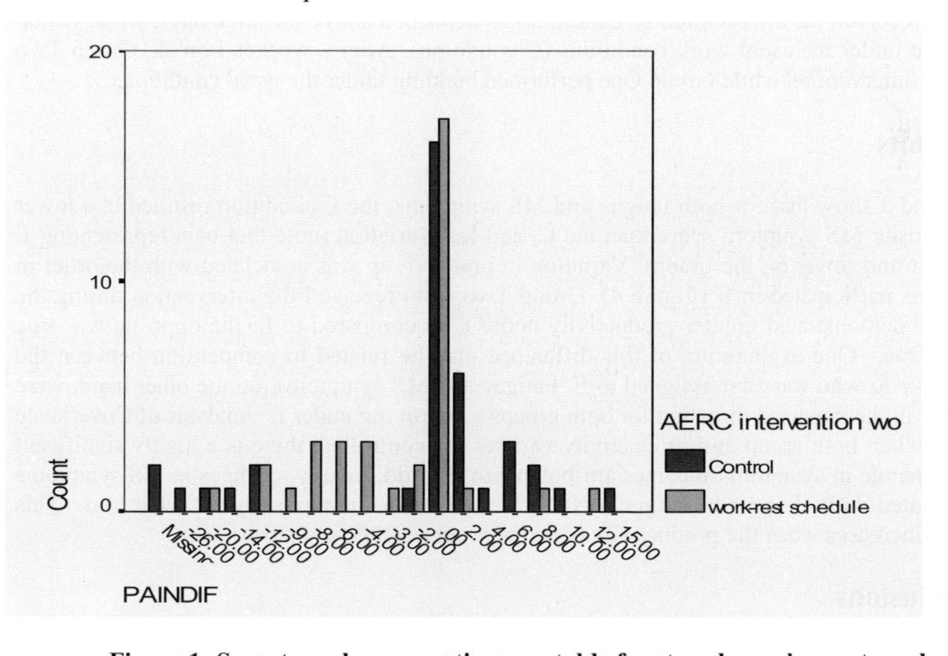

Figure 1: Symptom change contingency table for strawberry harvest workers (n=72).

2 Trial Two

2.1 Tree Nursery

Trial Two took place in a large nursery for orchard trees and focused on the task of grafting buds on young developing citrus trees (~"18 high). Budding and grafting requires a crew of two workers who squat or kneel continuously and crawl from one tree to the next: one worker cuts an exact incision, the second worker inserts the new bud and ties it in place. Productivity was assessed by the number of trees budded by each crew pair each day. Prior to intervention, workers completed over 1600 grafts per shift. Pre-intervention surveys demonstrated that 67% of these workers reported significant pain or discomfort. Over 60% reported pain and symptoms in the lower extremities and 57% reported back pain and symptoms. Risk factor exposures include:

- Sustained kneeling to reach appropriate work area
- Highly repetitive handwork to make cuts, insert buds, and tie off graft (18/min.)
- Trunk flexion up to 80°
- Trunk twisting up to 15°
- Elbow flexion up to 90°
- Pinch grips on both knife and bud
- Static postures of the neck, trunk, and lower extremities for periods of up to two hours.

The trial was designed as a cross-over trial, using 16 pairs of workers randomly assigned to two groups. The intervention was delivered in two trials of three days each; the two trials followed a baseline day of data collection before random assignment into the two groups. In the first trial, Group One received the intervention (E condition) as described above for three days, while Group Two worked under the usual work conditions (C condition). After a weekend break, Group Two received the intervention while Group One performed budding under the usual conditions.

2.2 Results

Figures 2 and 3 show that for both fatigue and MS symptoms, the E condition resulted in a lower mean composite MS symptom score than the C, and less variation (note that bars representing E are narrower and lower on the graph). Variation in productivity was associated with the order in which groups participated in E (Figure 4). Group Two, who received the intervention during the second trial, demonstrated greater productivity under C as compared to E; the opposite was true for Group One. One explanation of this difference may be related to competition between the crews relative to who was first assigned to E. Fatigue and MS symptoms, on the other hand, were consistently in the expected direction for both groups - improving under E. Analysis of Covariance shows that when both group and productivity variates are controlled, there is a highly significant ($p=0.002$) change in symptom outcomes attributable to E. Additionally, changes in MS symptoms were associated with changes in fatigue ($r=0.62$, $p=0.01$) over the three days of each trial. This association increases when the productivity is partially out ($r=0.69$, $p=0.001$).

3 Conclusions

Other investigators, studying workers in diverse industries, have reported the benefits of modifying standard rest periods, in terms of improved symptom reports with modest impacts on productivity.[2-4] Our findings replicate this work in a sample of farm workers doing strenuous stoop labor. Although initially reluctant to implement alterations to work tasks, afterwards workers and managers alike were enthusiastic about wanting to continue to test modified rest interventions - suggesting that the intervention proved promising. There are, however, many challenges to field testing interventions in the occupational environment. Design validity may be affected by social, cultural and political factors such as the labor supply, local labor-management relations, and characteristics of the workforce and management (e.g. ethnicity, gender, immigration status). Injury logs, in this case, were of no use due to apparent under-reporting. Additionally, our attempts to evaluate impact on productivity may have been hampered by inter-group competition. Some of these limitations may be addressed by extending trials over a longer period of time, reversing the intervention or staggering its introduction, adding multiple control groups or data collection intervals, and reducing awareness between groups about the experimental conditions.

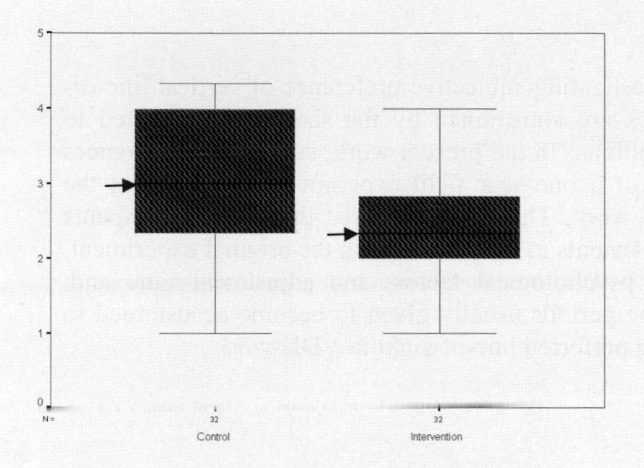

Figure 2: Comparison of fatigue under control and intervention conditions.

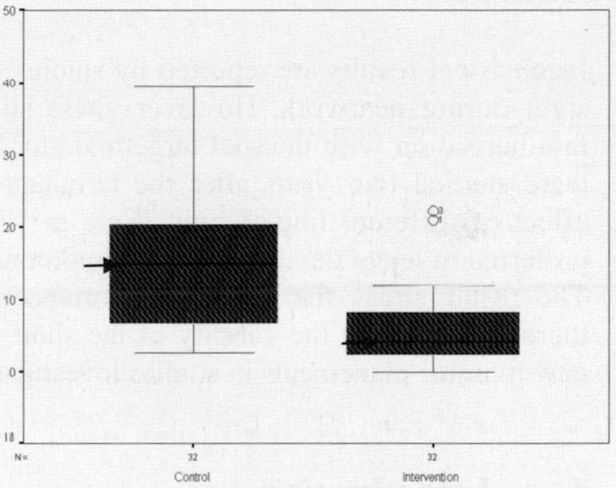

Figure 3: Comparison of musculoskeletal symptoms under control and intervention conditions.

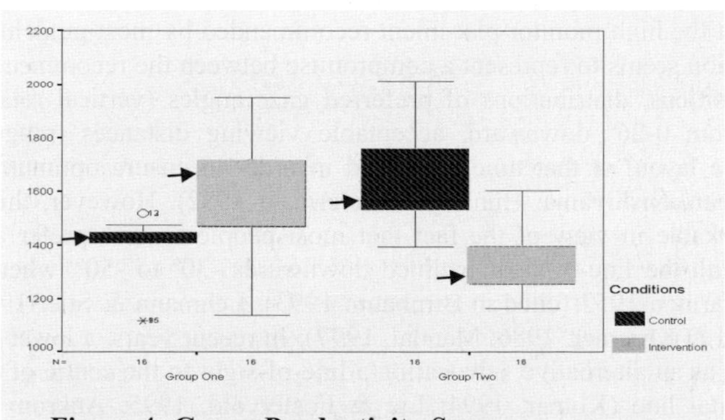

Figure 4: Group Productivity Scores

References

1. Faucett, J., Meyers, J. et al. *Journal of Agricultural Safety & Health 2001;7*(3):185-198.
2. Henning RA, Sauter SL, Salvendy G, Krieg Jr EF. *Ergonomics* 1989;32:855-864.
3. Galinsky TL, Swanson NG, Sauter SL et al. *Ergonomics* 2000;43(5):622-638.
4. Dabenah A, Swanson N, Shell R. *Ergonomics* 2001;44(2):164-174.

VDU-work and the preferred line-of-sight after long term exposure to different monitor placements

Knut Inge Fostervold

Department of Psychology, University of Oslo
P.O.Box 1094, Blindern, N-0317 Oslo, Norway
e-mail: k.i.fostervold@psykologi.uio.no

Abstract

Inconsistent results are reported by studies investigating subjective preference of vertical line-of-sight during nearwork. However, these studies are constrained by the short time allocated to familiarisation with unusual target/monitor locations. In the present work, subjective preferences were studied two years after the termination of a one-year field experiment, investigating the effect of different line-of-sight angle in VDU work. The results showed that most participants preferred to leave the desktop and monitor adjustments as they were set in the original experiment. The result stress the deceive importance of psychological factors and adjustment time and, thereby, questions the validity of the short time periods usually given to become accustomed to new monitor placements in studies investigating preferred line-of-sight in VDU-work.

1 Introduction

Implementation of advanced information technology has totally revolutionised ordinary office work. Due to its frequent use much effort has been invested into the development of ergonomic guidelines in order to prevent symptoms associated with VDU work. However, criticism has lately been raised against the high monitor placement recommended by most guidelines. Originally, the high monitor position seems to represent a compromise between the recommended sitting posture, preferred head positions, distributions of preferred gaze angles (vertical rotation of the eye in orbita) ranging from 0-26° downward, acceptable viewing distances (range 60-90 cm), and prevailing furniture layout at that time, proposed in order to ensure optimum placement of the keyboard (Grandjean, Nishiyama, Hünting & Piderman, 1982). However, this recommendation looks rather remarkable in view of the fact that most people tend to prefer a somewhat flexed sitting position, with the line-of-sight inclined downwards -30° to -50°, when performing other nearwork tasks (Harmon 1952 (cited in Birnbaum 1993); Lehmann & Stier 1961; Çakir, Hart, & Stewart, 1980; Hill & Kroemer, 1986; Mandal, 1997). In recent years, a lower monitor placement has been proposed as an alternative, advocating a line-of-sight to the centre of the screen 30°- 45° below the horizontal line (Kumar, 1994; Lie & Fostervold, 1995; Ankrum & Nemeth, 1995; Burgess-Limerick, Plooy & Ankrum, 1998; Fostervold, 2003).

Several studies have investigated subjective preference of vertical line-of-sight during near-work. With the head fixed in an upright position, Hill & Kroemer (1986) and Kroemer & Hill (1986) observed that the average preferred line-of-sight was declined 28.6°, relative to the Frankfurt plane. Attempting to replicate Hill & Kroemer's results under dark vergence conditions, Heuer, Brüwer, Römer, Kröger & Knapp (1991) reported the average declination, relative to horizontal, to be in the range of -10° to -14°. In real near-work situations, the preferred downward rotation of the eyes is supplemented by head flexion and to some degree back flexion. Thus, the most preferable line-of-sight under ordinary work conditions should be expected to be somewhat more declined. Nevertheless, average declinations in the line-of-sight corresponding to the angles reported by Heuer *et al.* (1991) has been reported for VDU-work (Jaschinski, Heuer & Kylian, 1998; Jaschinski, Heuer & Kylian, 1999). However, the preferred declination is dependent upon the range of free choices offered by the experimental design. In the aforementioned studies the adjustable range of vertical positions was limited to -20° below horizontal. This range falls well outside the range of -30° to -45° that has been derived from near-work conditions in the environment of evolutionary adaptedness (EEA). Studies on preferred vertical line-of-sight show large individual differences with regard to the monitor height chosen to be the most comfortable. Jaschinski et al. (1998) showed that the individually preferred line-of-sight, ranging between horizontal and -20° declined (mean -8°), remained stable for each subject within a period of four weeks, with repeated preference observation. It has been argued that this stability in individual preference reflects individual differences to eyestrain aspects of the oculomotor system, such as resting position of vergence and accommodation (Jaschinski-Kruza, 1991; Best, Littleton, Gramopadhye & Tyrrell, 1996, Jaschinski et al., 1998). When comparing eyestrain in conditions of imposed and freely adjusted screen positions, Jaschinski (1998, 1999) reported that the individually preferred screen position is adjusted to avoid positions that would induce stronger eyestrain or visual discomfort. These results suggest that VDU work place design should provide options for individual adjustment of the height of the screen. A controversial question remains, however. How large a range of inclinations should be covered by these options? Ecological considerations suggest that the range should cover inclinations down to -45°.

A factor that has been given little attention in the discussion of subjective preferences are the transformation process that takes place as subjects adapt to new work conditions. Questions should be asked about how much the short time period subjects usually are given to become accustomed to unfamiliar monitor placement influences the results. To approach this question, preferences were studied in a group of office clerks two years after they participated in a one-year field experiment in Norway. The aim of this field experiment was to study the effect of downward line-of-sight and high line-of-sight in an optimised natural setting.

2 Method

In the original field experiment, the participants were randomly assigned to two different experimental groups, a low monitor group and a high monitor group. A new computer desktop was designed for the low monitor group. The angle of line-of-sight was defined as the intermediate angle lying between a line drawn from the eye to the midpoint of the VDU and a horizontal line drawn at eye level, provided the head-erect sitting posture. Desktop and monitor placement in the high monitor group was optimised according to an angle 15°. The low monitor group was optimised according to a mean angle of 31° below the horizontal line.

Two years after the conclusion of the study a short follow-up questionnaire was distributed to 110 participants that were still appointed at the work place. In addition to background information, the participants were asked about their experience of changes made during the experiment, about changes made after the conclusion of the experiment, and about preferred monitor position. A total of 86 questionnaires (78%), out of the 110 distributed was returned. Two questionnaires were returned unopened, due to ended employment. Out of 84 completed questionnaires, 41 questionnaires were returned from participants originally assigned to the downward line-of-sight group while 42 were returned from participants in the high line-of-sight group.

2.1 Statistics

Person Chi-square was used as statistical method for between subject comparisons. Results were considered significant at p< .05. The phi coefficient (φ) for the chi-square was used as a measures of effect size for between group comparisons.

3 Results

Frequency distributions from the follow-up questionnaire, in relation to line-of-sight group, are shown in table 1.

Table 1: Questions and frequency distribution of the follow up questionnaire, in relation to downward line-of-sight and high line-of-sight

	Downward line-of-sight			High line-of-sight.		
Changes conducted as a part of the study were experienced as:	Improve-ment	Neither nor	Aggra Vation	Improve-ment	Neither nor	Aggra-vation
	28(70%)	11(27.5%)	1(2.5%)	12(29.3%)	29(70.7%)	0
Changed desktop after Study completion:	Yes	No		Yes	No	
	14(35%)	26(65%)		10(24.4%)	31(75.6%)	
Have you changed monitor position after the study?	No	Yes, high	Yes, low	No	Yes high	Yes, low
	30(73,2%)	6(14.6%)	5(12.2)	32(76.2%)	1(2.4%)	8(19.0%)
Why did you monitor position?	Personal wish	Changed desktop	Other Causes	Personal wish	Changed desktop	Other Causes
	7(58.3%)	3(25%)	2(16.7%)	6(66.7%)	1(11.1%)	2(22.2%)
Given the opportunity to choose freely, what monitor position would you prefer?	DLS	SLS	No prefer.	DLS	SLS	No prefer.
	36(87.8%)	5(12.2%)	0	16(39.0%)	12(29.3%)	13(31.7%)

As shown in table 1, a majority (70%) of the participants assigned to the downward line-of-sight group experienced changes conducted as a part of the study as an improvement. In the high line-of-sight group, the majority (70.7%) experienced changes neither as an improvement nor as an aggravation. Only one participant experienced the changes as an aggravation. The "aggravation" category was, therefore, collapsed with the "neither – nor" category in the statistical analysis. The result showed that the interactions between line-of-sight group and the experience of changes conducted as a part of the study was significant ($\chi 2(2)= 13.44$, $p<0.001$, $\varphi = 0.40$).

A majority of participants, 65% in the downward and 75.6% and high line-of-sight group had kept the desktop they used during the study. In addition, a majority reported that they had not changed the monitor position after the conclusion of the study, 73.2% in the downward group and 76.2% in the high line-of-sight group. For participants assigned to the downward line-of-sight group, who reported changes in monitor position, approximately the same number changed to a high position (14.6%) as the number of participants changing to a low position (12.2%). For participants in the high line-of-sight group a higher number of participants changed two a low monitor position (19%) compared to the number of participants who changed to a high position (2.4%). When asked about why they changed monitor position, a majority in both groups reported personal wish as the main cause, 58.3% and 66.7% in the downward and high line-of-sight group, respectively. The statistical analyses revealed no significant interactions between line-of-sight group and any of the above mentioned variables.

In the downward line-of-sight group, a majority of participant (87.8%) preferred downward line-of-sight if given the opportunity to choose freely. The remaining participants in the downward line-of-sight group stated that they preferred the high monitor position (12.2%). In the high line-of-sight group, the choice of preference were approximately evenly divided between downward monitor placement (39%), high monitor placement (29.3%) and no particular preference (31.7%). The statistical analysis revealed that the interaction between line-of-sight group during the study and preferences two years after the conclusion of the study was significant ($\chi2(2)= 23.57$, $p<0.001$, $\varphi = 0.53$).

4 Discussion

Results from the follow-up questionnaire revealed that a staggering high proportion of the participants had chosen to leave the desktop and monitor adjustments as they were set in the study. This outcome could be interpreted as a result of a general contentment with the optimisations accomplished in the study. However, the result is predictable from theoretical considerations and empirical data derived from decision making research. Challenged by the option of deciding between two alternatives were one alternative represents the status quo and the other an unfamiliar situation research have shown that people tend to chose the status quo situation merely because it represents status quo (Beach, 1990). Thus, subjectively the two situations does not hold equal status although they objectively may do so. Moreover, people often feel reluctant towards the option of changing well established routines, as changes entails a transient decrease in comfort and work-efficacy when old structures or systems are transformed into new ones (Venda and Venda, 1995). Paraphrased by the well-known saying from therapeutic research, the result was expected viewed in the light of people's resistance to change. Most likely, the result reflects that a new set of acquired habits has been established among participants working with downward line-of-sight while participants working with high line-of-sight tend to prefer their accustomed practice. Consequently, it was expected that the participants would prefer monitor placements in accordance with monitor placement used during the field experiment. However, the result showed that the preferences were unsymmetrically distributed between the two groups. While, participants in the downward line-of-sight group clearly preferred low monitor placement, the preferences were more evenly distributed in the high line-of-sight group. This difference in preferences indicate that participants in the low monitor group have developed a significantly stronger positive attitude toward the new monitor placement compered to participant in the status quo situation, working with high line-of-sight. Thus, the result stress the deceive importance of psychological factors and adjustment time and, thereby, questions the validity of experimental designs frequently used to study subjectively preferred line-of-sight in VDU-work.

References

Ankrum, D. R. & Nemeth, K. J. (1995). Posture Comfort and Monitor Placement. *Ergonomics in Design,* April, 7-9.

Beach, L. R. (1990). *Image theory: Decision making in personal and organizational contexts.* New York: John Wiley & Sons.

Best, P. S., Littleton, M. H., Gramopadhye, A. K. & Tyrrell, R. A. (1996). Relations between individual differences in oculomotor resting states and visual inspection performance. *Ergonomics,* 39, 35-40.

Birnbaum M. H. (1993). *Optometric Management of Nearpoint Vision Disorders.* Boston: Butterworth-Heinemann, Reed publishing Inc.

Burgess-Limerick, R., Plooy, A. & Ankrum, D. R., 1998. The effect of imposed and self-selected computer monitor height on posture and gaze angle. *Clinical Biomechanics*, 13, 584-592.

Çakir, A., Hart, D. J. & Stewart, T. F. M. (1980). *Visual Display Terminals*. New York: John Wiley & Sons.

Fostervold, K. I. (2003). VDU-work with downward gaze: The emperor's new clothes or scientifically sound. *International Journal of Industrial Ergonomics,* 31, 161-167.

Grandjean, E., Nishiyama, K., Hünting, W. & Piderman, M., 1982. A laboratory study on preferred and imposed settings of a VDT workstation. *Behaviour and information technology*, 1, 289-304.

Heuer, H., Brüwer, M., Römer, T., Kröger, H. & Knapp, H. (1991). Preferred vertical gaze direction and observation distance. *Ergonomics,* 34, 379-392.

Hill, S. G. & Kroemer, K. H. E. (1986). Preferred Declination of the Line of Sight. *Human Factors,* 28, 127-134.

Jaschinski, W., Heuer, H & Kylian, H. (1999). A procedure to determine the individually comfortable position of visual displays relative to the eyes. *Ergonomics,* 42, 535-549.

Jaschinski, W., Heuer, H. & Kylian, H. (1998). Preferred position of visual displays relative to the eyes: a field study of visual strain and individual differences. *Ergonomics,* 41, 1034-1049.

Jaschinski-Kruza, W. (1991). Eyestrain in VDU users: Viewing distance and the resting position of ocular muscles. *Human Factors*, 33, 69-83.

Kroemer, K. H. E. & Hill, S. G. (1986). Preferred line of sight angle. *Ergonomics* 29, 1129-1134.

Kumar, S. (1994). A computer desk for bifocal lens wearers, with special emphasis on selected telecommunication tasks. *Ergonomics*, 37, 1669-1678.

Lehmann, G. & Stier, F. (1961). *Mensch und Gerät. Handbuch der gesamten Arbeitsmedizin.* [Humans and tools. A comprehensive handbook of occupational medicine] Vol.1 (pp. 718-788). Berlin: Urban und Schwartzenbert. (in German).

Lie, I. & Fostervold, K. I. (1995). VDT - Work With Different Gaze Inclination, In: Grieco, A., Molteni, G., Piccoli B. and Occhipinti E. (Eds.), Work with Display Units 94 Selected papers of the Fourth International Scientific Conference on Work with Display Units (pp. 137-142). Amsterdam: Elsevier Science.

Mandal, A. C. (1997). Changing standards for school furniture. *Ergonomics in Design*, 5, 28-31.

Venda, V. F. & Venda, Y. V. (1995). *Dynamics in Ergonomics, Psychology, and Decisions: Introduction to Ergodynamics*. Norwood, NJ: Ablex Publishing Corporation,

Anisotropic characteristics of LCD TFTs and their impact on visual performance: "Everything's superior with TFTs?"

Thomas Groeger, Martina Ziefle & Dietmar Sommer

Psychology Department, Aachen University, Jaegerstr. 17-19, 52056 Aachen
Thomas.Groeger@post.rwth-aachen.de Martina.Ziefle@psych.rwth-aachen.de
Dietmar.Sommer@psych.rwth-aachen.de

Abstract

Two experiments were reported dealing with the impact of anisotropic display characteristics, predominating at LCD TFTs (Liquid Crystal Displays with Thin Film Transistor technique), on visual performance. Anisotropy is defined as the variances of display's luminance and contrast due to different viewing angles. A state of the art LCD TFT (Philips P150x) with two adjustments was examined: one with a background luminance of 100 cd/m^2, one set at its maximum. A detection task was used to measure visual performance at 63 different screen positions. Reaction times (RT) and accuracy were recorded and related to photometric values. Results showed a significant negative correlation between background luminance and visual performance: RT increased with decreasing luminance. Neither the contrast between background and character, nor character luminance was correlated with visual performance. Moreover, if the 63 luminance values were grouped to luminance clusters, it can be proven that visual performance goes parallel with the luminance variations over the screen surface. The anisotropy of LCD-TFTs has therefore to be reassessed as a serious ergonomic factor deteriorating visual performance limiting the appreciation of this display technology.

1 Introduction

Within the last years the market share of flat panel displays, Liquid Crystal Displays (LCDs) with Thin Film Transistor (TFT) technique, has increased continuously compared to Cathode Ray Tubes (CRT). There are multiple reasons for this almost triumphal procession of TFTs, even though still being sold at higher purchase prices than CRTs. Among the visual factors, the by far most important advantage of TFT LCDs is that they are flicker free. As corroborated in several studies, flicker-free displays outperform CRT screens with respect to visual performance and eyestrain. Even in contrast to CRTs running at higher frame rates invisible for the human eye the positive impact of flicker-free TFTs on visual performances remains stable (Ziefle, 2001 a, b, c, 2002). Importantly though, paper still outperforms TFTs (Ziefle, 2001 a, b). Another major difference between TFTs and CRTs refers to the display's different photometric rate of yield. In this context, three photometric indices are crucial: the luminance of the display's background and the luminance of the character/object (both in cd/m^2) as well as the contrast. Contrast is calculated as ratio out of the luminance of the bright background and dark foreground (character)[1]. TFT screens can display information at higher luminance and contrast levels than CRTs. Per contra to the above-mentioned advantages there are still some issues not yet resolved. One major handicap is described as the *anisotropic* (i.e. not equally located) characteristics of the display. According to ISO, anisotropy of a display is given when "the radiation deviates from that of a Lambertian

[1] In the literature different contrast formulas are used, but all relate the luminance of bright and dark areas.

surface by more than 10% at any inclination angle, $\theta<45°$" (ISO 13406-2, 2001, p. 4). Hence, if a TFT display shows a deviation of more than 10 % of its luminance subject to target location or viewing angle, it is called anisotropic.

An extensive research was conducted to determine the impact of photometric parameters on visual performance, thereby mostly using paper or CRTs as experimental displays. The outcomes can be comprised as follows: Some studies suggest that visual performance improves with increasing luminance levels (up to 100 cd/m^2), attributed to an optimization of the eye's adaptation level (e.g. Johnson & Classon, 1995). The majority of studies focus on the display's contrast as the most important factor affecting visual performance (e.g. Plainis & Murray, 2000). Some studies considering the interdependence of contrast *and* luminance hint at a performance decline if certain limits of both measures are exceeded (e.g. Zhu & Wu, 1990). As already mentioned, luminance and contrast are highly interdependent: the contrast as ratio out of two luminance values, neglects to a broad extent the absolute levels of luminance. Thus, it cannot be determined what is more important for visual processing, contrast or luminance. Due to technical restrictions, an independent variation of both photometric indices is not possible in current displays. As an aggravating factor, none of the mentioned studies was conducted with TFT screens. The majority of the TFT-studies deals with flicker effects on performance, excluding anisotropy in their experimental design. A first attempt to quantify the anisotropic impact on performance was undertaken by two recent studies (Hollands & Cassidy, 2000; Hollands, Cassidy, McFadden & Boothby, 2001). Here, the visual performance of TFT and CRT was compared: participants executed a visual reaction task displayed in negative polarity (bright colored symbols on a dark background) in two viewing conditions: a frontal view on the display and an extended viewing angle of 60 degrees. The results showed a CRT advantage over the TFT in the off-axis viewing condition. The TFT inferiority is referred to the anisotropic characteristics predominating if the screen is looked at off-axis. Unfortunately, a detailed specification of luminance values due to different screen locations was not provided in the mentioned study. Moreover, screens displayed in negative polarity do not match every day's situation in the computer workplace.

The present study aims at a detailed insight into anisotropic characteristics of TFT displays: First, it was quantified how much anisotropy weighs for visual performance. Meeting a rather strict benchmark, "best-conditions" were realized: the screen had a small extent of anisotropy compared to industry standards. Users, young and with a high visual acuity performed a purely visual task with no cognitive load. If effects can be found under best conditions, "anisotropy" has to be taken seriously as affecting performance. Second, it was of interests which of the photometric measures account for performance differences: the luminance of bright/ dark areas or the contrast itself.

2 Method

In order to quantify anisotropy, a 15" LCD TFT (Philips P150x, 1024x768 pixels) with two adjustments for brightness and contrast was used in a dark surrounding. In the first experiment, background luminance was set to 100 cd/m², resulting in a contrast ratio of 100:1 in the display's center (P100). In the second experiment, the two display controllers for brightness and contrast[2] were set to a maximum (Pmax). Here, the luminance reached 253 cd/m² with a contrast ratio of 189:1. In order to get exact measurements for each position, the display was divided into 63 (9 x 7) virtual fields (fig. 1 left). To emulate real user viewing conditions, the luminance meter was set to a single central viewing point 60 cm in front of the display. From this point, the luminance values of the bright background and the dark foreground were measured (with a luminance meter

[2] Due to technical restrictions of LCDs, a change of the controller setting for "brightness" or "contrast" does not automatically correspond to a systematic change in the according photometric values.

by Bruel & Kjaer) for all fields (fig. 1 right). Setting the values in the center point as reference, the luminance of the bright background varied from about –30% up to +20%, the black area from -40% up to +50%. The resulting contrast ratio varied from about –40% up to +30% due to the display position.

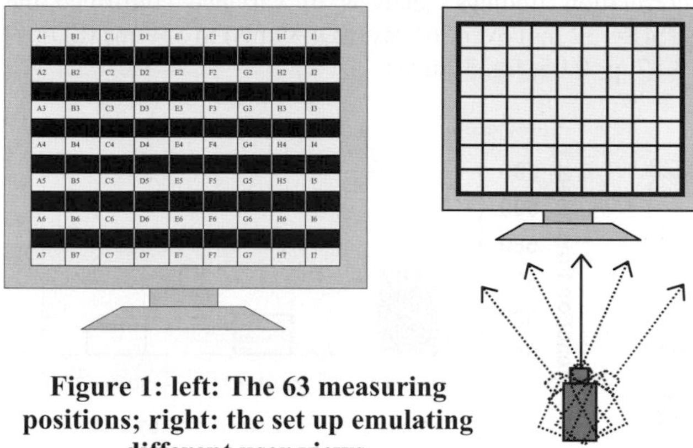

Figure 1: left: The 63 measuring positions; right: the set up emulating different user views.

In order to relate visual performance to photometric measures, a methodology and software was developed specifically for experimental purposes. On each of the 63 fields previously measured regarding to photometry, quadratic Landolt-C's were randomly presented (fig. 2). Participants had to detect the small gaps of these targets. Thus, it was made possible to relate photometric outcomes and performance at multiple screen positions and to reflect visual performance as a function of anisotropic display characteristics. Target location, gap orientation, RTs (in ms) and errors were recorded on a trial basis. 48 participants (m=24 years) with above-average visual acuity (m=1.3) took part in 90 minutes sessions. 24 participants performed the task in the P100 set up (experiment 1), 24 participants in the Pmax condition (experiment 2).

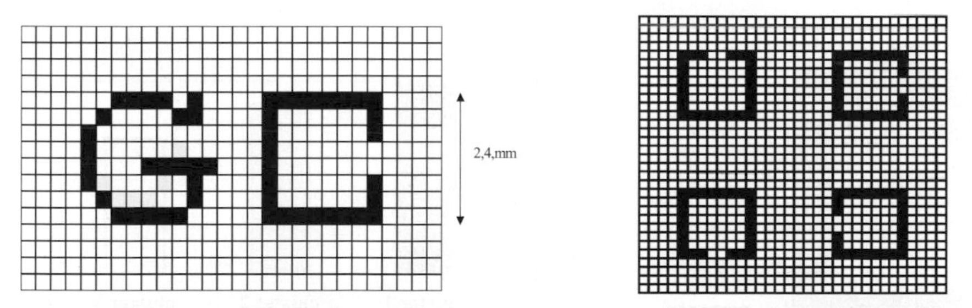

Figure 2: left: The target's height and width subtended 8 pixel, the stroke width 1 pixel and the gap 2 pixels, corresponding to the Capital letter (Courier 12). Right: Landolt C´s with the gaps in the four different orientations.

3 Results

As only very few errors occurred at all (1% in both conditions), and no differences in accuracy were found due to different screen positions, only correct responses were considered for further analyzing. In order to learn which of the photometric indices is responsible for visual outcomes within anisotropic displays, photometric indices were correlated with RTs. In a second step, luminance values were clustered and tested if the different luminance clusters account for significant changes in RTs. Results showed a significant negative correlation between the luminance of the bright background and RTs at both screen set ups (P100, $r=-.45$, $p<.01$ and Pmax, $r=-.53$, $p<.01$), indicating that RTs decreased with increasing luminance. However, RTs showed neither to be correlated with the luminance of dark areas (P100, $r=-.20$, $p>.05$ and Pmax, r=.04, p>.05) nor with the contrast (P100, r=-.08, p>.05 and Pmax, r=-.05, p>.05)[3]. In a next step background luminance values were aggregated to three grades of luminance via cluster analysis. Fig. 3 (left) shows the distribution of luminance clusters over the screen surface for the P100 set up. As can be seen, the luminance variation goes from 104 cd/m² (white) over 91 cd/m² (gray) to 79 cd/m² (black). According to the correlation findings, analysis of variance confirmed the significant decrease of RT from 573 ms (SD=132 ms) over 561 ms (SD=89 ms) to 559 ms (SD=89 ms) with increasing luminance ($F_{(2,46)}$= 3.87, p<0.05, fig. 3, right).

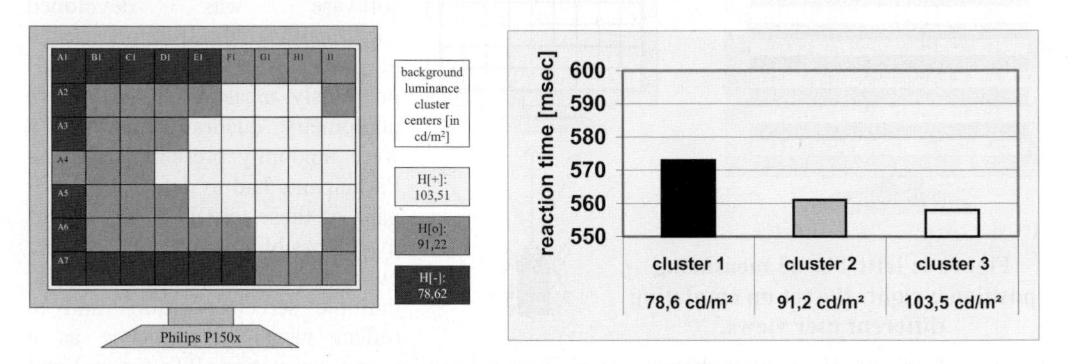

Figure 3: Left: Distribution of background luminance at different screen positions; right: RTs as a function of screen luminance adjusted to 100 cd/m² (P100).

The same result pattern can be seen if the screen is adjusted to its maximum (Pmax), see Fig. 4.

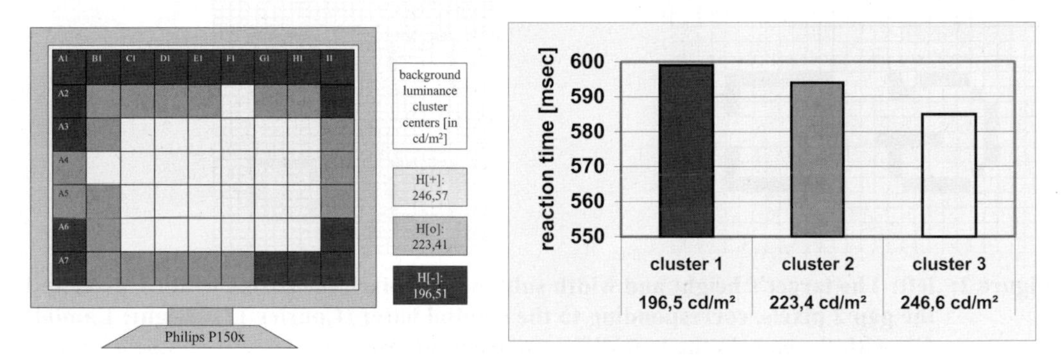

Figure 4: left: Distribution of background luminance at different viewing angles; right: RTs as a function of luminance clusters at the maximum screen set up (Pmax).

Again, three clusters were found with respect to the distribution of luminance over the screen surface (white areas indicate a luminance of 247 cd/m², gray 223 cd/m², black 197 cd/m²). It could be corroborated that the background luminance is responsible for changes in the detection performance. RTs were significantly shorter as a function of luminance ($F_{(2,46)}$= 13.26, p<.05), decreasing from 600 ms (SD=85 ms) over 594 ms (SD=85 ms) to 585 ms (SD=84 ms).

[3] According to the literature the physiological contrast (quotient between the difference of object and background luminance and background luminance) was used. Pmax shows the partial correlation coefficient with controlled background luminance.

4 Discussion

Two experiments were carried out to quantify effects of anisotropic TFT displays on visual performance. The results showed the level of background luminance to be a crucial factor affecting detection performance. Remarkably, even if it is mostly assumed, the display's contrast was not proven as a main photometric source for optimized performance. The results clearly demonstrated that the fluctuations in the display's luminance over the surface have measurable effects on visual detection. Thus, anisotropy must be regarded as a major handicap of the TFT-LC display technology negatively affecting visual performance. One could argue that the effects were rather small and should not be overestimated. However, it has to be taken into consideration that the results refer to best conditions not representing the daily situation in the work place: users of the workforce are older than the tested sample. Further, commonly used displays are less sophisticated with regard to extent of anisotropy. Viewing angles examined here were rather small compared to working situations in which one user has to survey several LCDs or in which more than one user works in front of one screen. Further, working periods on-screen are usually longer and higher demanding tasks than small feature detection have to be accomplished.

The implications of these basic findings will have to be proven in the further studies, i.e. how and how much anisotropy affects visual performance under less optimal conditions. Current experiments deal with the effects of anisotropy at greater viewing angles, with effects of ambient illumination and with the question if negative display polarities may possibly reduce anisotropy. Moreover, as anisotropy is not present within CRT screens, the rat-race between the two display technologies, CRT and TFT, must be renewed, including anisotropy into experimental designs.

References

Hollands, J. G., Cassidy H. A., McFadden S. (2001). LCD versus CRT Displays: Visual Search for Colored Symbols. Proceedings of the Human Factors and Ergonomics Society 45th Annual Meeting (pp. 1553-1557). Santa Monica: Human Factors Society.

Hollands, J. G.; McFadden, S.; Cassidy, H. A., Boothby, R. (2000). Visual search performance on LCD and CRT displays: an experimental comparison. SID 00 Digest, 20.3, 292-295.

Johnson, C. & Classen, E. (1995). Effects of luminance, contrast and blur on visual activity. Optometry and Visual Science, 72 (12), 864-869.

Plainis, S., Murray, I.J. (2000). Neurophysiological interpretation of human visual RTs: effect of contrast, spatial frequency and luminance. Neuropsychologia, 38 (12), 1555-64

Ziefle, M. (2001a). CRT screens or TFT display? A detailed analysis of TFT screens for reading efficiency. In M. Smith, G. Salvendy, D. Harris & R. Koubek (eds.). Usability Evaluation and Interface design (pp. 549-553). Mahwey, Lawrence Erlbaum.

Ziefle, M. (2001b). Aging, visual performance and eyestrain in different screen technologies. Proceedings of the Human Factors and Ergonomics Society 45th Annual Meeting (pp. 262-267). Santa Monica: Human Factors Society.

Ziefle, M. (2001c). User productivity and different screen technologies: CRT screens with high refresh rates vs LCD Displays? The Display Search Monitor, 6 (14), 11-14.

Ziefle, M. (2002). Visual performance in CRT and LCD Displays in different user groups. The Display Search Monitor, 7 (7), 85-88.

Ziefle, M., Groeger, Th. & Summer, D. (under revision). Visual costs of the inhomogeneity of contrast and luminance by viewing TFT-LC-Displays off-axis. International Journal of Occupational Safety and Ergonomics, to be published in 2003.

Zhu, Z. & Wu, J. (1990). On the standardization of VDT's proper and optimal contrast range. Ergonomics, 33 (7), 925-932.

Reduced productivity due to musculoskeletal symptoms: Associations with workplace and individual factors among white collar computer users

Mats Hagberg[1] *, Ewa Wigaeus Tornqvist*[1,2]*, Allan Toomingas*[2]

[1]Department of Occupational and Environmental Medicine, Sahlgrenska Academy at Goteborg University and Sahlgrenska University Hospital, S:t Sigfridsgatan 85, SE-412 66 Goteborg, Sweden.
[2]National Institute for Working Life, SE-112 79 Stockholm, Sweden

Abstract

The aim was to assess whether self-reported reduced productivity occurred in computer users due to musculoskeletal symptoms and the association to workplace, symptom and individual factors.
The study group consisted of 1283 computer users from different occupations, of whom 498 were men and 785 women. Reduced productivity was self-assessed by two questions addressing if and how much productivity was reduced the previous month due to musculoskeletal symptoms. There were 63 women (8.0%) and 42 men (8.4%) of the total study group who reported reduced productivity due to musculoskeletal symptoms. The mean magnitude of the reduction was 15% for women and 13% for men. This outcome was associated with computer mouse position and task for both men and women. For women, work demands, computer problems, and being divorced/separated were also associated with reduced productivity.
These results suggest a variety of interventions that may serve to decrease the impact of reduced productivity due to musculoskeletal disorders in the workplace.

1 Introduction

Most studies of musculoskeletal disorders and workplace factors limit the outcomes measured to disorders and symptoms. Several rehabilitation researchers have addressed sick leave, disablement pension and days lost at work. Recently there has been an increased interest in sickness effects while at work (Hagberg et al. 2002), sometimes referred to as sickness presenteeism (Aronsson et al. 2000).

Musculoskeletal symptoms may affect work performance. Presence at work despite symptoms can be explained by severity that is not sufficient to require complete cessation of work activity, loyalty to the company and co-workers, or economic incentives and personal ethics. This phenomenon limits the utility of lost time at work to estimate employers' cost for these symptoms and disorders (Pransky et al. 1999). The determinants of reduced productivity at work are not necessarily the same as the determinants for symptoms per se or lost time at work. Thus, whether symptoms at work reduce productivity is an important issue in identifying priorities for ergonomic interventions and health promotion. Risk factors for symptom occurrence may be different from risks for decreased performance once symptoms have occurred. The aim of the present study was to assess whether self-rated reduced productivity at work occurred in computer users due to musculoskeletal symptoms, and to identify associations with workplace and individual factors.

2 Study base and methods

Together with the employers and the Occupational Health Care Centers of 46 different work sites, work groups or departments were invited to participate in the study, representing both private companies and public organizations. The work sites differed in size, the smallest including only seven persons and the largest 260. The study group represented both private and public sectors and included a variety of occupations. The study base consisted of 1532 computer users, 636 men and 896 women. The response rate was 78 % for men (n=498) and 88 % for women (n=785), forming a total study group of 1283 (84% overall response rate).

2.1 Questionnaire

The occurrence of musculoskeletal symptoms and occupational exposures during the preceding month and individual factors were assessed by a written questionnaire of 88 questions. Parts of the questionnaire on working conditions had been validated earlier (Karlqvist et al. 1996). In the present study we inquired about working hours, work content (variation of work tasks, hours/week of computer work, work with a non-keyboard input device and data/text entry), physical exposures (duration of sitting/standing and walking, periods of computer work without breaks and position of the non-keyboard input device on the work table) and psychosocial exposures (job demands in relation to competence, job strain and probability of meeting time limits and quality demands, as well as social support and support from a supervisor) and comfort of the work environment. Individual factors included civil status, age, educational level and certain life-style factors. The questions about musculoskeletal symptoms referred to the duration (days) during the preceding month of pain or aches (and numbness in the hands/fingers) in the neck, upper back, lower back, right and left shoulder, shoulder joint/upper arm, elbow/forearm, wrist and hand/fingers, respectively. Reduced productivity was assessed by the question: "Have the symptoms influenced your productivity at computer work during the preceding month?" with a yes and no alternative. This was followed by a second question to those who answered yes: "If your productivity has decreased, please state how much in percent compared to the previous month".

2.2 Statistics

This was a cross-sectional study. Analysis was done using SAS version 8 and JMP. Age-adjusted prevalence ratios were computed according to Mantel-Haenzel in five age classes (18-24, 25-34, 35-44, 45-54, 55-64). A prevalence ratio with a 95 per cent confidence interval not including one was considered statistically significant. Exposure factors were dichotomized according to current standards or quartiles where no obvious standards existed (Table 1). A multivariate analysis was conducted with a proportional hazards approach, the number of risk factors present was used as a simple way of examining the effect of a combination of risk factors.

Table 1. Dichotomized exposure factors

Exposure factors	Exposure definiton
Non optimal computer mouse position	Those who marked their computer mouse position outside a rectangle close to the operator in a workplace layout figure in the questionnaire.
Work demands	The upper quartile of the sum of the scores on 5 questions, each with 5 response levels, concerning work demands (have to work hard, fast etc.) (Karasek and Theorell 1990).
Work control	The lower quartile of the sum of the scores on 2 questions, each with 5 response levels, concerning having good control over work (can decide on how and when work is to be performed) (Karasek and Theorell 1990)
Work management	The lower quartile of the sum of the scores on 8 questions with positive statements about the management (support, encouragement, information, feedback, development etc), each with 5 response levels.
Computer problems	A yes/no question about having computer problems (software or hardware) the last month.
Overweight	Body mass index >25.
Divorced/separated	A question about marital status. Those divorced, separated or widowed were defined as exposed.
Children at home	Exposed were those having children at home.
Smoking	Exposed were those being curren smokers.

3 Results

At least one musculoskeletal symptom was present in most respondents (87% of females and 76% of males). Of these persons with symptoms, 63 women (9.2 %) and 42 men (11.2 %) reported reduced productivity due to their musculoskeletal symptoms. The mean magnitude of the reduction was 15 % for women and 13 % for men, with a median of 10% for both women and men and range of 1% to 75% for women and 1% to 97% for men.

There were only 52 women and 14 men who stated that they had been on sick-leave the previous month in the study group of 785 women and 498 men. Among the 733 women who were **not** on sick-leave (no lost days at work) during the previous month, there were 45 (6.1%) who reported reduced productivity in their work and 40 (8.3%) among the corresponding men.

Non-optimal computer mouse position was a factor weakly associated with reduced productivity among men and women (Table 2). For women, work demands, computer problems and being divorced/separated were also somewhat associated with reduced productivity.
In the multivariate analysis with eight risk factors layout, graphic design work was a strong risk factor for reduced productivity (Table 3). The number of risk factors present increased the prevalence ratio (PR) for reduced productivity; one factor PR=1.13: two factors PR=2.60: three factors PR=3.22.

Table 2. Association between different factors and reduced productivity. Exposed is the number of women and men who had this factor (for definitions see Table 1). Exposed cases is the number of men and women with reduced productivity among the exposed. PRa is the prevalence ratio adjusted for age and 95 CI is the 95th percent confidence interval.

Exposure factors	Women				Men			
	Exposed	Exp cases	PRa	95 CI	Exp osed	Exp cases	PRa	95 CI
Non optimal computer mouse position	591	49	1.13	1.004-1.27	316	33	1.22	1.02-1.47
Work demands	243	25	1.39	1.0007-1.94	113	12	1.25	0.75-2.10
Work control	188	18	1.26	0.83-1.90	164	18	1.28	0.87-1.87
Work management	252	16	0.84	0.55-1.28	127	10	0.91	0.51-1.61
Computer problems	120	14	1.67	1.02-2.74	47	6	1.54	0.70-3.38
Overweight	195	20	1.43	0.97-2.09	227	23	1.23	0.91-1.66
Divorced/separated	146	17	1.60	1.03-2.50	59	5	1.00	0.43-2.37
Children at home	370	27	0.98	0.75-1.29	229	20	0.95	0.70-1.29
Smoking	154	15	1.24	0.77-2.01	62	7	1.34	0.66-2.74

Table 3. Multivariate model of risk factors (proportional hazard regression, 103 event, 1161 censored).. 95 CI is the 95th percent confidence interval.

Risk factors	Prevalence ratio	95% CI
Layout, graphic design	2.15	1.43-3.24
Information searching	1.99	1.23-3.23
Non-optimal computer mouse position	1.70	1.01-2.87
Accounting work	1.49	0.96-2.32
Divorced/separated	1.41	0.87-2.29
Computer problems	1.39	0.83-2.32
Work demands	1.34	0.88-2.05
Data/text input	1.28	0.87-1.90

4 Discussion

Self-reported musculoskeletal symptoms were common among the investigated computer users, and a proportion (9.9%) reported decreased productivity due to these symptoms. Decreased productivity was related to several of the factors ascertained by the questionnaire. The impact of these conditions may be greatest in those performing the most keyboard-intensive activities, compared with occupations where keyboard use is less frequent or intensive, and in those with symptoms of higher persistence.

The decrease was defined as amount of deviation from a set standard of productivity. In a Dutch company (Brouwer et al. 1999), found that 7% of the workers had health problems that reduced productivity (they did not define the type of problem or gender). They estimated that 0.93% of all working hours were lost due to health-related reduced productivity without official sick-leave.

The sick-leave rate due to musculoskeletal symptoms among the participating computer users was low. Staying at work despite musculoskeletal symptoms and reduced productivity could be explained by both loyalty to the employer and co-workers and economic factors. Sick-leave would in Sweden reduce the individual's salary by 20% and the first day on sick-leave would not be compensated. This meant that the employer, without knowing, paid full salary for employees with reduced productivity.
An estimation of the economic impact for the employer can be made. For a man or woman in the study group, the mean loss of productivity per month was 16.8 hours (10.5% x 160h). Since reduced productivity without sick-leave was reported by 8.3% of the men, this could be translated into 1.4 hours lost per month per employee (8.3% x 16.8h). The corresponding figure for women would be 1.0 (6.1% x16.8h) hour lost per month per employee.

5 Conclusion

Although this is a cross-sectional study based on self-report, if any of the association found between reduced productivity and work tasks and individual factors was causal and not consequence, there is a possibility for prevention. Even if only a fraction of the loss is eliminated by investments made in early rehabilitation of musculoskeletal symptoms at the workplace, substantial savings can be made by the employer. Examples of such investments are office ergonomics education and physical training at work. Ergonomic adjustment of work-station design and work technique may reduce the number of operators with poor computer mouse technique.

How can an employer address these issues? Workplace surveillance by questionnaires or interviews and with feedback may be a way of benchmarking and to creating ergonomic or work environment programs. Questions about reduced productivity due to health problems may be an efficient way for the employer to justify ergonomic investments and health promotions and to follow up measures taken.

6 References

Aronsson, G., K. Gustafsson and M. Dallner (2000). "Sick but yet at work. An empirical study of sickness presenteeism." *J Epidemiol Community Health* **54**(7): 502-9.

Brouwer, W., F. Koopmanschap and F. Rutten (1999). "Productivity losses without abscence: measurement validation and empirical evidence." *Health Policy* **48**: 13-27.

Hagberg, M., A. Toomingas and E. Wigaeus Tornqvist (2002). "Self-reported reduced productivity due to musculoskeletal symptoms: Associations with workplace and individual factors among white collar computer users." *J Occup Rehabil* **12**: 151-62.

Karasek, R. and T. Theorell (1990). Healthy Work. New York, Basic Books, Inc., Publishers.

Karlqvist, L., M. Hagberg, M. Köster, M. Wenemark and R. Ånell (1996). "Musculoskeletal symptoms among computer-assisted design (CAD) operators and evaluation of a self-assessment questionnaire." *Int J Occup Environ Health* **2**: 185-94.

Pransky, G., T. Snyder, A. Dembe and J. Himmelstein (1999). "Under-reporting of work-related disorders in the workplace: a case study and review of the literature." *Ergonomics* **42**(1): 171-82.

Support of Creative Knowledge Workers in Flexible Office Environments Through a Positioning System

Udo-Ernst Haner and Alexander Greisle

Institute for Human Factors and Technology Management, University of Stuttgart
Nobelstrasse 12, D-70569 Stuttgart, Germany
udo-ernst.haner@iat.uni-stuttgart.de alexander.greisle@iat.uni-stuttgart.de

Abstract

This paper argues that office positioning systems in flexible office environments are supporting creativity processes through raising awareness regarding the co-presence of colleagues and their availability for communication. Two options for an office positioning systems are presented after giving a brief overview of technology alternatives. Although currently available technologies are still to be optimized, the more urgent issue is organizational and potentially societal agreement on the use of location information, since its collection is potentially privacy infringing.

1 Introduction

In fast-paced economic environments flexible and innovative organizations are more likely to succeed due to a greater adaptability. Organizational flexibility and innovation however are heavily relying on awareness with respect to the situational context as well as on creativity. Both, awareness and creativity of knowledge workers in office environments can be supported through spatial and technological solutions. Positioning systems are such a solution offering awareness in office environments regarding the location of colleagues and – to a certain degree – allowing to infer their activities. A higher degree of communication, interaction and collaboration becomes possible thereby supporting creativity. In the following chapters first the issue and process of creativity is discussed. Then flexible office environments are briefly described and the interrelation with creativity processes discussed. In the second main chapter the range of technology options for positioning systems in office environments are presented, two of which have been tested in the Fraunhofer Office Innovation Center in Stuttgart. They will be described in more detail. Lessons learned include the technological experiences, the impact on creativity and issues to be considered when introducing positioning systems in an organizational context.

2 Creativity in Flexible Office Environments

Creativity defined as the "ability of thinking by oneself or in a team in new and unusual patterns and thereby generating new ideas and new solutions (...) requires the combination of superficially not related facts, which have their origin in individual knowledge, imagination, or information" [Bauer/Haner/Rieck 2001]. This process of combination – respectively that of creativity – is a sequence of four distinguishable phases, which cannot be controlled, but supported. In some phases the process of creativity requires spontaneity and awareness, in others structuring and reflection. In the preparation phase the main activity is (re-)cognition of improvement opportunities and their context. Information gathering and structuring is prevailing. During the incubation phase the issue settles and via (sub-)conscious processes the illumination phase – the spontaneous idea and clue generation – is prepared. The fourth phase is that of verification: the output is contrasted to the initial requirements. These separate phases need different types of spatial and technological support, especially if creativity is to be a team effort, during which personal encounter and face-to-face communication is desired.

Flexible office environments are characterized by people using different settings for carrying out their tasks; e.g. cellular offices or thinking booths when needing concentration, meeting spaces when conferencing, or team spaces when group efforts are needed. Different spatial configurations support certain activities better than others. Knowledge workers are giving up their "own" desk for the benefit of enjoying the more functional alternatives offered. If this non-territorial approach of working is applied by larger groups sharing their environment a possible disadvantage is the loss of awareness regarding the presence and activities of colleagues. The situation is similar to that of mobile working, where the exact location of a person to be addressed is not known upfront. The opening phrase "How are you?" shifts towards "Where are you?".

For re-installing the sense of awareness with respect to co-present colleagues in flexible office environments different alternatives are available. Hotelling systems where knowledge workers can book office space depending on their particular needs for the day can offer awareness through their rather static booking system. As long as a person is not changing the location during the day these systems are suitable. However, in flexible office environments with a range of spatial alternatives this premise is rather inadequate. Alternatively a range of tracking mechanisms are feasible. These are presented in the following chapter, highlighting two different solutions.

3 Technology Options for Positioning in Office Environments

3.1 Overview

Tracking the actual position of office users requires both, technology and organization. In this chapter the focus is on the technological options for positioning systems. Positioning services can be implemented based on a broad range of different technologies depending on the application context, including activities, needs, and physical environment of the user. When considering only wireless alternatives, these can best be categorized based on their coverage areas: global range systems like GPS and local range systems, e.g. WLAN hot-spots. Figure 1 below gives a more detailed overview on the available technology options.

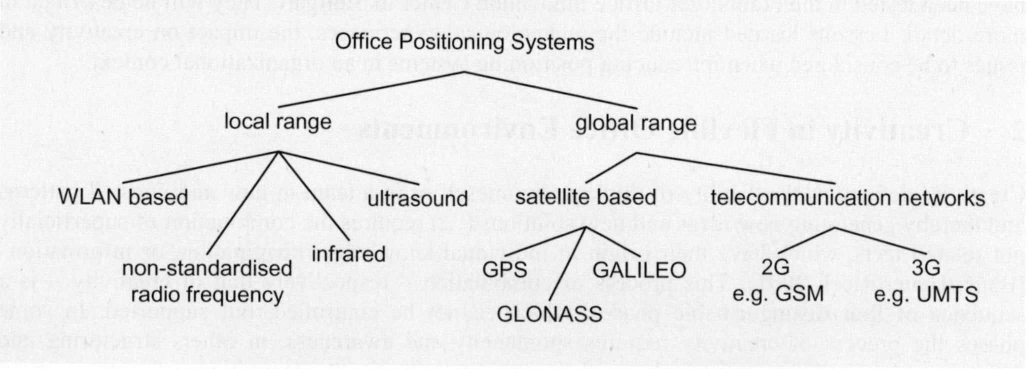

Fig. 1: Classification of different types of positioning systems

Local range systems are limited to provide positioning services only to small predefined areas, like an airport, business premises or an office environment. Global range positioning systems are used to collect and provide position information anywhere – virtually location independent – based on the premise that appropriate infrastructure (e.g. a GSM network) is available. Particularly relevant for positioning in office environments are currently only the local range systems. Global range systems are not providing the necessary accuracy, yet.

To be identified and tracked by any system, users have to carry a special personalized device, be it a tag or an appropriate daily object like a mobile phone or a PDA. These devices communicate with base terrestrial or orbital stations. Depending on the system design the base stations could be proprietary or standard access points that are used for positioning services. Different approaches are then used for locating a personalized device and therefore the user.

Two possible alternatives for the detection of the personalized devices are usual. Signal strength measurement is based on decreasing electro-magnetical fields with increasing distance from a base station. If only one base station is in contact with the tracked device a statement regarding the presence of the device in the coverage area of this base station is possible. If coverage areas of two base stations are overlapping positioning can become ambiguous thereby lowering accuracy. Also, high demands on accuracy cannot be fulfilled as only the presence in the area of a sender can be tracked. The second alternative is to do triangulation based on at least three simultaneously detectable senders. Still, the accuracy provided through signal strength measurement is rather low, since signal strength can be affected through a variety of environmental factors. Unaffected by such factors is triangulation which is based on measuring signal run times from at least three simultaneously detectable base stations. In office positioning accuracy is a main aspect, for it can be necessary not only to identify devices in rooms but also their exact location therein, e.g. third desk on the left in an open-plan office.

In line of sight of satellites the Global Positioning System (and equivalent upcoming systems) could provide the required accuracy (at least when combined with other algorithmic calculations) when a minimum of four satellites are visible [Legat/Lechner 2001]. However in tall buildings this line of sight is not available and therefore these systems are not effective as stand-alone solutions for office positioning. Literally another dimension needs to be considered when office positioning is to be performed in multi-floor office buildings with roughly the same accuracy than in a plan. This aspect is not an issue for local range systems because they can be installed on every floor and handled via the server. To these local range systems belong also GPS- or GSM-based micro solutions, where a reference point helps positioning.

3.2 Systems Tested in Fraunhofer Office Innovation Center

A variety of methods for locating the users in a plan are available. A detailed overview of possible alternatives is given in [Bakke/Bergersen/Haner 2001]. Here, the two systems which have been implemented and tested in the Fraunhofer Office Innovation Center (OIC) are described. The screenshots in Fig. 2 show on how positioning information is communicated to the users.

The interface shown in Fig. 2a implemented as a Java WebStart application, is communicating location information based on Wavetrend hardware that requires the installation of proprietary cabling and the installation of an antenna in every (sub-)unit of space in which location should be possible. The system is using signal strength measurement for locating users. In a fine-grained office environment where partitions are made of permeable and reflecting materials – like the OIC, such a system is not optimal since various opportunities for ambiguous signals are possible. Accuracy is then lower than expected. Spatial conditions are relevant for implementation success.

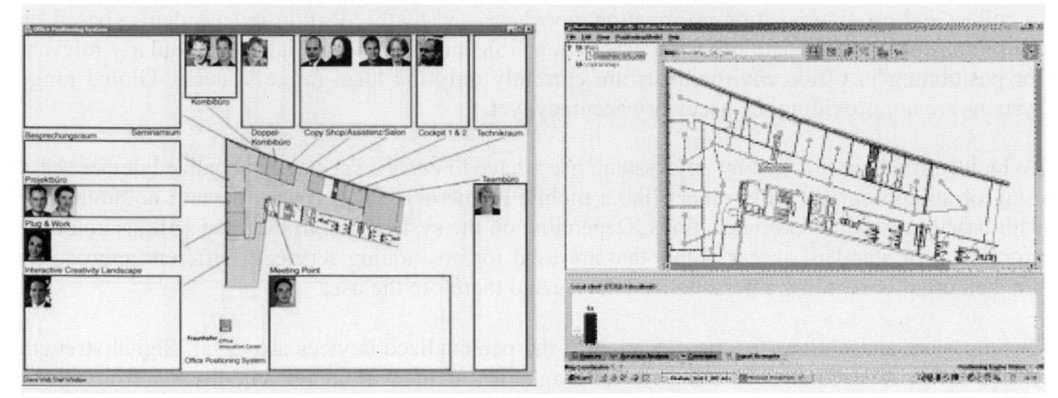

Fig. 2a *Examples of OPS client interfaces* *Fig. 2b*

With the Ekahau systems shown in Fig. 2b a high accuracy due to the WLAN-based triangulation can be achieved. Using the existing WLAN environment leads to a far lower investment of capital and setup time when compared to proprietary systems. Another benefit of this system is the possibility of getting information on the user's movement within the office environment (small arrow in Fig. 2b). This feature is useful for people interested in spontaneous communication with colleagues coming by. One main disadvantage in this usage scenario is that WLAN enabled daily objects like notebooks and/or PDAs are potentially too heavy to carry them permanently along or they disconnect due to power saving. A solution would be a WLAN enabled credit card like device which sustains connection to the network.

The following Tab.1 gives an overview of the systems' differences:

	Wavetrend (Fig. 2a)	**Ekahau (Fig. 2b)**
Tags	Proprietary tags with always on functionality	Standard WLAN enabled tags
Measurement	Signal strength measurement with proprietary antennas	Triangulation with WLAN access points (min. of 3 visible AP for accuracy)
Accuracy	High when little overlap between coverage areas of base stations	Very high (up to 30 cm)
Calibration	Adjustable antenna coverage areas through variable strength of signal detection at antenna side	Manual calibration of reference points at selected positions in office environment
Client	Individual programming possible	Individual programming possible
Vendor	http://www.wavetrenduk.com/	http://www.ekahau.com/

Tab. 1: Comparison of two office positioning systems

3.3 Future Technology Trends

One main aspect on human-computer interaction is the seamless integration of technologies. This implies the need for a single positioning system for both global and local range positioning.

UMTS is one technological option for indoor use and outdoor use. Why not use this system for a seamless positioning in using mobile local based services and office positioning services with one handset? Latest publicly communicated measurements by handset providers show that UMTS achieves an accuracy of 8-12 m under optimal conditions (three or more base stations have to be visible to the handset). In a larger town like Stuttgart only 36 % of the area would be covered by enough base stations for a accurate 2D positioning at the moment [Winkel, et. al. 2000].

GALILEO is a new European satellite based positioning system which is currently under development and is expected to be operational by 2008. The technical planning is not completed yet but the outlook is promising: the system should fulfill the needs for a seamless and integrated 3D positioning system [Tiberius/de Jong 2002; Hein 2003].

4 Lessons learned

Office positioning in flexible office environments can contribute significantly to raising awareness among colleagues about co-presence and availability for communication, inferred from location information. This is particular relevant in creativity processes where co-location, communication and interaction are relevant for the success of a joint effort. Different systems allow for different accuracy but simultaneously have restrictions regarding the scenarios for use. The future trend towards seamless integration will solve most of the technological issues. However there is a more significant aspect to the success of office positioning systems that relates to privacy. Location information is possibly infringing privacy. Therefore according to German law for example any system that potentially keeps track of employees' activities is by default prohibited. Permissions from and acceptance by the employees and their representation is necessary. Organizational and possibly societal agreement on the use of such systems is to be reached.

Acknowledgements

The research leading to this publication has been performed in the context of SANE (Sustainable Accommodation for the New Economy), a two-year project under the Information Society Technologies (IST) Program of the European Union (IST 2000-25257). Special thanks to Omar Arvelo and Nikolay Dreharov, students at the University of Stuttgart, for their technical support.

References

Bakke, J.W./ Bergersen, E./ Haner, U.-E. et al (2001). *D21 – Technology Survey.* Report as part of SANE – Sustainable Accommodation for the New Economy, EU-IST-2000-25257

Bauer, W./ Haner, U.-E./ Rieck, A. (2001): *OFFICE 21 – Inventing an Interactive Creativity Landscape.* Proceedings of the HCI International 2001, 9th International Conference on Human-Computer Interaction, August 5-10, 2001 New Orleans, Vol.3, pp. 658-662

Hein, G. (2003). *Positionierung und Lokalisierung des Anwenders.* Presentation held on the congress „Mobil mit digitalen Diensten", Münchner Kreis, February 6-7, 2003

Legat, K./ Lechner, W. (2001). *Integrated navigation for pedestrians.* Paper presented at GNSS 2001

Tiberius, C./ de Jong, K. (2002). *Developments in Global Navigation Satellite Systems – GPS modernised, Galileo launched.* In: The Hydrographic Journal, No 104, April 2002

Winkel, J./ König, D./ Oehler, V./ Eissfeller, B./ Hein, G. (2000). *Positioning and Navigation Using Mobile Communication: Alternative or Supplement to a Future GNSS-2?* IAIN World Congress/U.S. ION Annual Meeting Receipt, Catamaran Hotel, San Diego, CA, USA, June 26-28, 2000

Use of Electronic Performance Monitoring to Promote Individual and Team-managed Rest Breaks: a Summary of Laboratory Research

Robert A. Henning, Ph.D., CPE

Psychology Department
University of Connecticut
Storrs, Connecticut, USA 06269-1020
Henning@UConnVM.UConn.edu

Abstract

There is growing evidence from both laboratory and field settings that adding frequent short rest breaks to a conventional rest break schedule can reduce the risk of musculoskeletal disorders as well as benefit worker productivity and wellbeing during repetitive computer-mediated tasks. However, there is also evidence that workers performing more complex tasks do not fully comply with these new rest break schedules because the breaks are ill-timed in relation to ongoing task activities. A series of laboratory studies was used to search for an alternative that would promote some degree of worker self-management of the short breaks. Electronic monitoring of the keystroke activity of student typists was used to identify discretionary break periods. The first prototype system administered computer-controlled breaks on a compensatory basis whenever discretionary breaks were insufficient in relation to a target level of break frequency and duration set by the experimenter, but was only minimally effective. The next, more successful prototype incorporated a cybernetic feedback display that showed the typist how discretionary breaks compared to the target level of breaks. In the most successful prototype, feedback information was integrated with the task so that it was not necessary for the typist to monitor a separate feedback display. Two additional prototypes were developed to promote team self-management of breaks. In general, providing cybernetic feedback displays enabled both individuals and teams to self-manage break behavior effectively in a manner that was less disruptive to task activities than a fully-automated or machine-paced rest break administration system.

1 Introduction

1.1 Worker Non-compliance to New Break Schedules at Two Field Sites

In our initial field study of an experimental rest break program (Henning, Jacques, Kissel, Sullivan & Alteras-Webb, 1997), a considerable degree of worker non-compliance was found at two work sites of a large insurance company. Overall, non-compliance among 92 claims processing workers averaged 55 percent based on self-report data from an exit survey.

A number of task constraints could explain why workers did not comply with the new schedule that consisted of four short breaks from computer work each hour (one every 15 min, three 30-s breaks and one 3-min break) in addition to the conventional mid-morning and mid-afternoon rest breaks. For example, some insurance claims required 40 min or more to process, and most VDU (video display unit) operators were organized as teams of 4 to 7 workers that also handled client inquiries via telephone while using the company database. Thus it seems likely that these complex task activities were not as amenable to frequent interruptions as short-cycle, repetitive tasks.

1.2 When Computers Administer Frequent Breaks, Clumsy Automation?

The low rate of compliance at both field sites may be blamed on what has been referred to as "clumsy automation" (Woods, Sarter & Billings, 1997). Rest break administration was automated at the field site with the idea that neither the claims processors nor their supervisors should be burdened with the task of determining when short rest breaks should begin and end throughout the workday. What can be considered clumsy about this fully automated approach to rest break administration is that rest breaks would often occur at inopportune times such as during a critical processing step in the claims processing cycle, phone conversations with clients, team member interactions, or at inappropriate times such as immediately following a worker-initiated discretionary rest break of adequate length.

A fully-automated rest break administration can also be considered clumsy from the standpoint of preventing any local control by workers. In the present case, a machine-paced break system prevented either the claims processors or their supervisors from exerting any control over when rest breaks occurred, even though these workers were strategically positioned within the work organization to determine the most advantageous times for rest breaks given local task constraints. Such constraints were likely to continuously change due to variations in the difficulty level of each claim that was processed, the social dynamics of the team, variations in the length of telephone conversations, and so on. A badly-timed break may also have disrupted cognitive behaviors to such an extent that a worker had to repeat earlier processing steps or risk serious errors when the work was resumed. In sum, fully automating the rest break system in this clumsy manner may have precipitated a host of task-related and psychosocial problems that outweighed the gain in recovery that short rest breaks could offer.

1.3 Worker-managed Rest Breaks as an Alternative to a Fully-automated Break System

Challenging the assumption that workers are unable to control rest breaks in an effective manner during computer-mediated work, a human-centered design effort was initiated to develop a support technology to help VDU operators to self-manage rest break behavior in accordance with rest break regimes that included frequent, short breaks. The basic approach was to give workers control over the timing of short rest breaks but in a way that does not seriously compromise an optimized break schedule. This compromise of an optimized break schedule is justified considering the alternative result in our field research that workers skipped over half of the scheduled short breaks.

The series of laboratory studies reviewed below help clarify the basis for the final successful design guidelines. The design problem was framed as the need for a computer-based support system that would somehow help workers track and adjust their rest break behavior in order to comply with the rest break standard set by management. This decision to develop what can be considered an aided tracking system to help workers self-manage their rest break behavior is consistent with many of the current recommendations for the human-centered design of automation systems, such as retaining workers in the control loop for whatever control process is automated (Woods et al., 1997). An aided tracking system was proposed to help workers track their own work and rest behavior, and thereby extend the workers' ability to self-manage work and rest behavior in a manner that would comply with safe work practices.

Development of prototype systems was guided by cybernetic theory and methodology developed by K.U. Smith and associates (Smith & Smith, 1987), an approach that is compatible with many macroergonomic design principles (Hendrick & Kleiner, 2002). Five prototype systems are described below in the order they were developed and tested with undergraduate student typists as participants. The primary goal was to demonstrate a proof-of-concept that workers could self manage rest breaks, and not to test the effects of frequent rest breaks on worker wellbeing. This helps explain why the specific rest break schedules that were tested did not conform with the rest break schedules recommended by NIOSH (Swanson, Sauter & Chapman, 1989). Possible adverse effects of the prototype system being tested were assessed through analysis of mood state, psychophysiological stress responses, and participant ratings of task difficulty. Strengths and weaknesses of each candidate system are described below to demonstrate the iterative nature of the research and design effort. The last two prototype systems extended the basic approach to teamwork environments.

2 Systems to Promote Self-managed Rest Breaks

2.1 Prototype 1: A Compensatory System (without a feedback display)

A semi-automated support system was developed that allowed typists to substitute discretionary breaks for computer-controlled breaks (Henning, Kissel & Maynard, 1994). It was assumed that the combined effects of task disruption and intrusiveness caused by computer-controlled breaks would be potent motivators for typists to take more discretionary rest breaks, thereby avoiding most computer-controlled breaks. Typists (N=38) entered alphabetical, upper/lower case data for 48 minutes. Typists in the control group only received computer-controlled breaks. Typists in the treatment group received computer-controlled breaks only when discretionary breaks were inadequate in relation to the target standard set by the experimenter (less than 17 s in 5 min). Although an 18% reduction in computer-controlled break time occurred in the treatment group, this was not considered a strong indication that workers had gained control over discretionary rest breaks when using this semi-automated support system. Typists did not appear able to self-monitor their discretionary break behavior effectively in relation to the rest break standard while simultaneously performing computer-mediated work.

2.2 Prototype 2: A Compensatory System with a Separate Feedback Display

In addition to the semi-automated support feature of Prototype 1, feedback about how a typist's discretionary rest break behavior compared with the target level of breaks set by the experimenter was provided in an on-screen graphical display (Henning, Callaghan, Ortega, Kissel, Guttman, & Braun, 1996). Typists (N=31) entered lines of randomized words for about one hour. Typists in the control group received computer-controlled breaks whenever discretionary break time was less than the target standard (30 s every 10 min). Typists in the treatment group also received immediate (i.e., real-time) feedback indicating how their accumulated discretionary break time compared to the target level of break behavior. Results showed that typists in the feedback display condition received virtually no full-length, 30-s computer-administered breaks (M=0.05) compared with typists in the control condition who received at least some (M=0.50) full-length computer-administered breaks. These results indicated that the feedback system did improve the typist's ability to self-manage discretionary breaks. Typists indicated that the feedback display helped them control rest breaks but using it was distracting.

2.3 Prototype 3: A Compensatory System with Task-integrated Feedback

The same feedback information provided in Prototype 2 was imbedded in the on-screen task in an attempt to eliminate the dual-task demands associated with monitoring a separate graphical feedback display while performing the primary work task (Henning et al., 1996). In cybernetic terminology, this was expected to make the feedback more coherent to other sources of task feedback (Smith, Henning & Smith, 1995). Gradual shifts in the vertical position of the current data line were used to indicate the need for more/less discretionary breaks. Typists (N=30) entered lines of randomized words for about one hour and received computer-controlled breaks on a compensatory basis whenever discretionary breaks were less than the break standard set by the experimenter (a total of 30 s every 8 min). Typists in one group received the task-integrated feedback. Typists in the second group received a separate feedback display. The task-integrated feedback was found to be the most effective system for promoting worker-managed breaks. Analysis of the 10-min work periods showed that typists who received task-integrated feedback controlled discretionary rest breaks in a more stable manner, taking neither too much nor too little break time compared with the target break standard in each 8-min work segment. In addition, typists who received task-integrated feedback usually initiated discretionary breaks at the end of data lines, indicating reduced task disruption. Typists also reported less task disruption when using the task-integrated feedback than when using the separate feedback display. No untoward effects on performance, well-being, or user acceptance were found. These results showed that a self-managed rest break system can be used effectively by the typist when task-integrated compliant feedback about break behavior is provided.

2.4 Prototypes 4 & 5: Parallel Team Control Versus Serial Team control

Two additional prototypes were used to investigate fundamentally different approaches for promoting self-management of breaks by two-person teams (Henning, Bopp, Tucker, Knoph & Ahlgren, 1997). The two methods of providing task-integrated feedback information were based on parallel and serial modes of social control identified in social cybernetic theory (Smith, Henning & Smith, 1995). Typists (N=30 2-person work teams) performed a joint typing task that required verbal word exchanges between team members for 46 minutes. As in the earlier prototypes for individual typists, the teams received computer-controlled breaks when the team's discretionary breaks were inadequate. In the parallel control condition, both team members received task-integrated feedback about the discretionary break behavior of the team. In the serial control condition, only one team member received task-integrated feedback. A small sample of team members in a baseline comparison group received no feedback at all. Teams in both the parallel and serial control conditions were much more successful at self-managing discretionary breaks than teams who did not receive feedback. While teams in the parallel mode condition may have outperformed teams in the serial mode condition, there were indications that a greater coordination effort was required by teams in the parallel control condition.

3 Concluding Remarks

It would be ironic if efforts to improve worker wellbeing through the introduction of new rest break systems were subverted by the use of clumsy automation that undermines worker control and wellbeing (Sauter, Hurrell & Cooper, 1989). The present research was able to demonstrate that it is possible for typists to control discretionary rest breaks effectively if the task environment is designed to help them track their own discretionary rest break behavior in relation to the standard of rest break behavior. The final prototypes tested are also likely to meet macroergonomic design needs since it would be possible for management to set break standards at a global level while not undermining worker control at the local level by administering breaks in a machine-paced manner. The surplus of computing power that is now readily available at workstations could and should be used to help promote worker self-management of safe and effective work and rest behaviors.

4 References

Hendrick, H.W. and Kleiner, B.M. (Eds.) (2002). Macroergonomics. New Jersey: Lawrence Erlbaum.

Henning, R.A., Bopp, M.I., Tucker, K.M., Knoph, R.D., and Ahlgren, J. (1997). Team-managed rest breaks during computer-supported cooperative work. *International Journal of Industrial Ergonomics*, 20, 19-29.

Henning, R.A., Jacques, J., Kissel, G.V., Sullivan, A.B., and Alteras-Webb, S.A. (1997). Frequent short rest breaks from computer work: effects on productivity and well-being at two field sites. *Ergonomics*, 40(1), 78-91.

Henning, R.A. Callaghan, E.A., Ortega, A.M., Kissel, G.V., Guttman, J.I., & Braun, H.A. (1996). Continuous feedback to promote self-management of rest breaks during computer use. *International Journal of Industrial Ergonomics*, 18, 71-82.

Henning, R.A., Kissel, G.V, and Maynard, D.C. (1994). Compensatory rest breaks for VDT operators. *International Journal of Industrial Ergonomics*, 14, 243-249.

Sauter, S.L., Hurrell, J.J., and Cooper, C.L. (1989). Job control and worker health. New York: Wiley.

Smith, T.J., Henning, R.A., and Smith, K.U. (1995). Performance of hybrid automated systems - a social cybernetic analysis. *International Journal of Human Factors in Manufacturing*, 5(1), 29-51.

Smith, T.J., and Smith, K.U. (1987). Feedback control mechanisms of human behavior. In G. Salvendy (Ed.), *Handbook of human factors* (pp. 251-293). New York: John Wiley and Sons.

Swanson, N.G., Sauter, S.L., and Chapman, L.J. (1989). The design of rest breaks for video display terminal work: A review of the relevant literature. In A Mital (Ed.), Advances in Industrial Ergonomics and Safety I: Proceedings of the Annual International Industrial Ergonomics and Safety Conference (pp. 895-898). London: Taylor and Francis.

Woods, D.D., Sarter, N., and Billings, C. (1997). Automation surprises. In G. Salvendy (Ed.), *Handbook of human factors* (2nd edition) (pp. 1926-1943). New York: John Wiley and Sons.

Visual Discomfort Among VDU-users wearing Single Vision Lenses compared to VDU-progressive Lenses?

Gunnar Horgen MSc[1]
Arne Aarås MD,PhD[2]

Magne Thoresen MSc[3]

[1] Buskerud College, Department of Optometry, PO.Boks 251, Kongsberg, Norway.
gunnar.horgen@hibu.no
[2] Alcatel/STK AS, P.O.Box 60, Økern, 0508 Oslo, Norway
arne.aaraas@alcatel.no

[3] University of Oslo - Section for Medical Statistics, Oslo, Norway.
magne.thoresen@basalmed.uio.no

Abstract:

Three different types of optometric corrections specially designed for VDU-work and one single vision lens are compared in a prospective field study. The different corrections effects on postural discomfort are evaluated by questionnaires.

1. Introduction

Three different types of optometric corrections specially designed for VDU-work, were compared with single vision lenses regarding postural load in a laboratory study (Horgen *et al.*, 1999). The different corrections' effects on postural load were measured by using electromyography (EMG). Body posture was measured continuously by using three dual axis inclinometers attached to the head, back and upper arm. No significant differences were found between the single vision lenses and the specially designed VDU-lenses regarding muscle load in m.Trapezius and m. Infraspinatus. Small differences were found regarding the head angle. The study concludes that these new lens designs create interesting opportunities for optometrists to optimize the visual conditions for different work tasks. However, the laboratory study did not include any subjective measurements, nor any longitudenal effects on the wearer .

2. Aim of the study

Do specially designed VDU-progressive lenses create a difference in the development of visual discomfort compared to single vision lenses when working on an optimized VDU-workstation? Further, will the VDU-user change his/her assessment of the total visual environment of the workplace when using the VDU-progressive lenses.

3. Study design

The study has a prospective, parallell group design, with four groups of VDU-Workers. Approximately 40 subjects in each group. After careful task analysis with special attention towards the visual angles and distances to the worktasks, the different progressive lenses were given to selected groups of VDU-workers. One control group was fitted with ordinary single vision lenses. The four groups are followed over a time period of 1 year. A questionnaire concerning visual conditons, working conditions, discomfort in different body areas and the status of the subjects'optometric corrections was filled in before the intervention, to establish the baseline situation. A second questionnaire was filled in after six months observation time. In this questionaire some additional questions on the subjects' experience with the new optometric correction was added. A third questionnaire was filled in after one year observation time. No other contact is made with the subjects, except for occasional contacts in cases of moving the workplace, unexpected sick leave etc (Table 1).

4. Methology

4.1 The test lenses

The progressive lenses specially designed for VDU-work available on the Norwegian market were divided into three different design groups. One group is mainly designed for near use, but allowing an increased depth of focus. To represent this design the lens *Interview* (I) from Essilor was chosen. The second group represents lenses that are designed to cover reading, VDU-work and with an increased depth of focus giving sufficient vision to cover most indoor activity (out to approximately 2,5 meters). To represent this design, *Gradal RD* (R) from Zeiss was chosen . The third group represents lenses also meant to cover mainly near visual tasks, but with a possibility to look at infinity, through "a window" placed rather high up in the lens. American Optical's *Technica* (T) was chosen to represents this lens design, see contour plots.

4.2 The subjects

From the total work force at Alcatel Telecom, approximately three hundred and sixty VDU-workers were elected as possible subjects for the study. They were visited at their work site, and a task analysis was performed. Emphasize was put on the ergonomic layout of the workplace, and the different optometric measurements were taken. Each subjects' visual demands were also evaluated, and the subject was either excluded from the study according to the exclusion criteria, or he/she was placed in one of the groups to be randomized afterwards. Based on the task analysis, the subject population was divided into three main samples. One sample that was suitable for all four types of lens designs, a second sample that was unsuitable for single vision lenses, but could use the dataprogressive lenses from the designs 1, 2, or 3, and a the third sample was suitable only for dataprogressive lenses from designs 2 or 3. Based on this procedure, a stratified randomization was performed, and the subjects were divided into four test groups, one Single Vision group (S), one Interview group (I), one Gradal RD group (R), and one Technica Group (T).

Inclusion criteria: -Presbyope, with a minimum of 1,5 dioptres addition for ordinary reading tasks.
-Experienced VDU-users, and (preferably) using single vision correcion at the VDU-workplace.

Exclusion criteria:-Spectacle correction stronger than - 6.00
-Active eye disease or systemic disease with eye complications.
-Taking drugs that might influence either eye functions, or muscle functions.
- Having work tasks that make it impossible to use either of the lenses in the study.
- Subjects suffering from physical handicaps to a degree that makes it difficult to do the measurements.

4.3 Time table

Baseline situasjon	3 Weeks	6 months	6 months
Group "RD"	X	X	X
Group "Int."	X	X	X
Group "Tech."	X	X	X
Eye exam Questionaire Dispensing.	Possible adjustment	Questionaire	Questionaire

Table 1. Time table of the project.

4.4 Visual discomfort and pain in the musculoskeletal system

Pain intensity and duration were assessed on a 10 centimeter Visual Analogue Scale (VAS), for the last month and the last six months period before the intervention, and six and twelve months after the intervention. I has been shown that VAS are both reliable and reproducible (Larsen *et al.*, 1991). The questions used were based on the Standardized Nordic Questionnaires for the analysis of musculoskeletal symptoms (Kuorinka *et al.*, 1987). It has been documented that complaints regarding such symptoms reported by questionnaires, has 80% sensitivity of findings in a clinical examination (Ohlsson *et al.*, 1994).

4.5 Psychological factors

The questionnaire deals with psychological factors both at work and at home. Amount, frequency and duration of VDU-work are considered, and lengths of working periods without breaks, and other work tasks compared to VDU-work. The variation in work tasks as a whole, job control and the opportunity to make contact with colleagues, self-realization in terms of learning, increased skills and utilization of own capabilities and basic need satisfaction (work burden at home etc) was also considered (Westlander, 1987).

4.6 Optometric examination:

All subjects were given a complete optometric examination. The methods and criteria for corrections are described in detail elsewhere (Horgen and Aarås, 1993).

5.0 Results

Only small changes in the development of headache and visual discomfort were registered. However, the subjective evaluation of area of clear vision and overall satisfaction was significantly improved for the I and R lens (P<0,05) (Fig. 4). There were no significant changes for T and SV lenses.

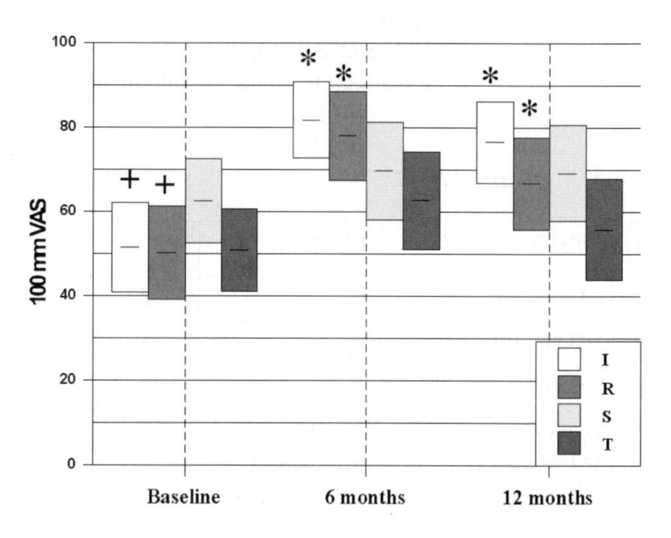

Fig. 4: Degree of satisfaction with the new correction. The values are given as group mean value with 95 % confidence interval. On the vertical axis, 0 means very low satisfaction, 100 means a very high degree of satisfaction. Asterisks denote statistical significance at p<0,05 level. Plus signs indicate from where the significant changes are stemming.

6.0 Conclusion

Lens designs that cover viewing distances from near and out to approximately 2 meters works well compared to SV lenses and lens designs trying to cover greater depth of focus than this.

6. References.

Horgen, G. and Aarås, A. (1993) Optometric Intervention in VDT- Workplaces. In Nielsen, R.A.J., K. (ed.) *Advances in Industrial Ergonomics and Safety V*. Taylor and Francis - London., Copenhagen, pp. 107 - 114.

Horgen, G., Aarås, A., Kaiser, H. and Thoresen, M. (1999) Specially Designed VDU-lenses and Postural Load during VDU-work. *Optometry and Vision Science*. Lippingcott & Wilkins, Seattle USA, Vol. 76, p. 261.

Kuorinka, I., Jonsson, B., Kilbom, Å., Winterberg, H., Bæring-Sørensen, F., Andersson, G. and Jørgensen, K. (1987) Standarized Nordic Questionnaires for Analysis of Musculoskeletal Symptoms. *Ergonomics*, **31**, 735-747.

Larsen, S., Aabakken, L., LIllevold, P.E. and Osnes, M. (1991) Assessing soft data in clinical trials. *Pharmaceutical Medicine*, **5**, 29-36.

Ohlsson, K., Attewell, R.G., Johnsson, B., Ahlm, A. and Skerfving, S. (1994) An assessement of neck and upper extremity disorders by questionnaire and clinical examination. *Ergomics*, **37**, 791-897.

Westlander, G. (1987) How identify organizational factors crucial of VDU-health? A context oriented method approach'. In Knave , B.a.W., P.G. (ed.) *Work With Display Units*. North Holland - Amsterdam, Stockholm - Sweden, pp. 816-821.

Effect of Bezel Reflectance on People Using a Computer Monitor

Claudia M. Hunter and Peter R. Boyce

Lighting Research Center
Rensselaer Polytechnic Institute
Troy, New York USA
gilsoc@rpi.edu,

James H. Watt

Social and Behavior Research Laboratory
Rensselaer Polytechnic Institute
Troy, New York USA
wattj@rpi.edu

Abstract

The effect of a wide range of reflectances of a computer monitor's front frame, or bezel, on several office task performance measures and visual behavior measures was studied. No effects of bezel reflectance were found.

1 Introduction

This study examines the effect of bezel reflectance on office task performance and visual behavior. Different bezel reflectances create different luminance ratios with the monitor screen and the surrounding visual field, which could potentially cause discomfort and reduced visual performance. Studies of the effects of bezel reflectance have not yet been found in the literature.

The potential mechanisms by which different luminance ratios might cause visual discomfort and lower performance are disability glare, discomfort glare, distraction, and possibly misadaptation (Vos, 1985, 1999; Boyce, 1981; Wolska and Switula, 1999). Monitor bezels of low reflectance (i.e., black) cannot be the source in the first two mechanisms, because of the way the visual system works. Low-reflectance bezels could potentially be a source of distraction, and just possibly a cause of misadaptation. However, these mechanisms are not very likely to occur in normal office environments. High-reflectance bezels are potential sources of discomfort glare, disability glare, or distraction. It should be noted that discomfort glare, disability glare, and distraction could potentially occur simultaneously and cause discomfort (Aaras, A., Horgen, G., Bjorset, H., Ro, O., and Walsoe, H., 2001; Wibom and Carlsson, 1986; Boyce, 1986, 1991).

2 Method

Using nineteen paid temporary office workers in a simulated office setting, the study gathered data on task performance, visual acuity, and visual behavior. The simulated office setting (shown in Figure 1) had mid-range reflectances (0.10 to 0.75) on all surfaces except the computer monitors, and office lighting that conforms to standard recommendations, approximately 500 lux on the workplane (IESNA, 2000). A commercially available black monitor, with a bezel reflectance of 0.07, and a pearl white monitor, bezel reflectance of 0.51, were used. A third monitor was specially painted with a high-reflectance matte white finish, 0.87. Against the matte-gray rear wall of the office, the black monitor formed a luminance contrast ratio of approximately 11 to 1; the pearl white monitor formed a ratio of 1.4 to 1, and the high-reflectance "extreme" white monitor formed a ratio of 0.88 to 1.

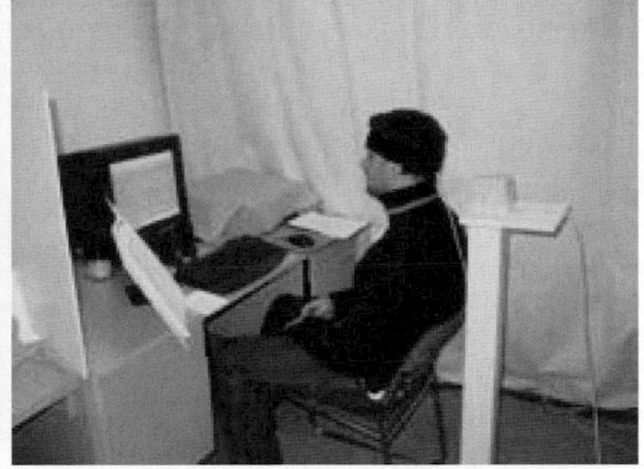

Figure 1: The experimental office setting (left); participant at the eye-gaze tracking station (right)

For an 8-hour workday, participants worked at one of three workstations. Each workstation was equipped with one of the three monitors. Over three days, participants worked with all three monitors. Each day they performed a set of three tasks for approximately one hour each in both morning and afternoon sessions: a data entry task, a numerical verification task, and a naturalistic reading task. The independent variables were: monitor bezel reflectance (black, pearl white, and extreme white), task point size (6- and 12-point type), and time of day (morning or afternoon). The dependent measures were performance on the data entry and numerical verification task. These tasks were designed to ensure the participants' eye movements crossed the bezel frequently or stayed on the screen, respectively. A measure of visual acuity was also made before and after each work session. Eyegaze measures were also collected on one participant in each group of three (these measures are described below).

3 Results

The data from each performance measure were submitted to two 2-way repeated measures analyses of variance. The data were first transformed by removing the inter-subject variation; this transformation and the two analyses were necessary because of the fractional factorial design of the study. The results are shown in Figures 2 and 3.

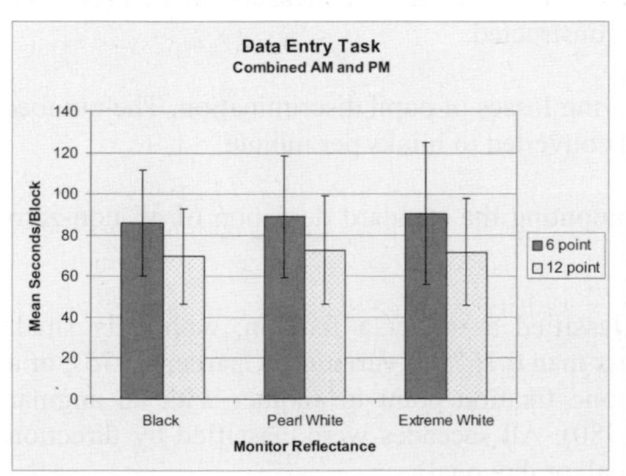

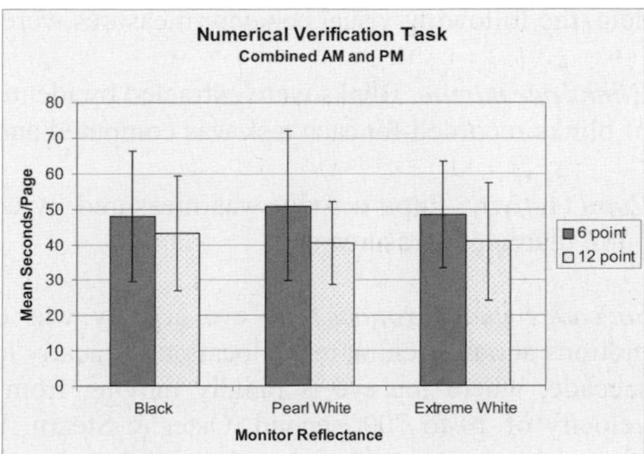

Figure 2: Results for task performance measures

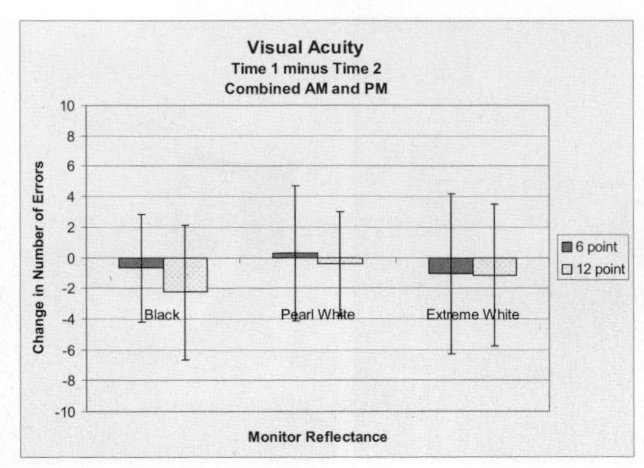

Figure 3: Results for the visual acuity measure

There was no statistically significant main effect of monitor bezel reflectance on any of the measures (data entry measure, $F = 0.47$, $df = 2, 36$, $p = 0.63$; numerical verification measure, $F = 0.34$, $df = 2, 36$, $p = 0.72$; visual acuity measure, $F = 1.53$, $df = 2, 36$, $p = 0.23$), nor any interaction. The results did show predictable statistically significant results of the task point size. Participants performed significantly faster at the 12-point task size than the 6-point size. There was no effect of time of day (morning or afternoon session).

4 Visual Behavior as an Indicator of Fatigue

Eyegaze measurement was undertaken as a "study-within-a-study," to determine if patterns of eyegaze behavior associated with fatigue could be identified, and if so, related to monitor bezel reflectance or type of task being performed.

4.1 Experimental Procedures

The research design was a 3 (bezel reflectance) x 2 (task) x 2 (time of day) repeated measures design. The dependent variables were derived from eyegaze coordinate and pupil size data collected for 3-5 minutes per task at the beginning and end of the 8-hour day, while working on numerical verification and reading tasks. Six participants completed the measurements. From this data, the following visual behavior measures were constructed.

Blinks per minute. Blinks were extracted by identifying losses of pupil discrimination. The number of blinks recorded for each task was computed and converted to blinks per minute.

Pupil Activity. Pupil activity was measured by computing the standard deviation of all non-zero pupil diameter measurements.

Saccades and Fixations. All eye activity was classified as either a fixation, with only small motions about a central mean location (typically less than 0.167° of variation (Gaarder, 1975); or a saccade, where the eye is rapidly moving from one fixation point to another with an angular velocity of 20 to 700°/second (Oster & Stearn, 1980). All saccades were classified by direction (forward or backward horizontal, up or down vertical, or diagonal).

4.2 Results

One-sample t-tests on the differences between the morning and afternoon measurements on the same participant showed two visual behavior measures deviated significantly from zero. The overall saccade rate increased significantly (by 23.7 saccades/min, $p < 0.04$) from morning to afternoon. Most of this was due to an increase in backward horizontal saccade rate (14.8 saccades/min, $p < 0.02$). This indicates a fairly clear fatigue effect on eye motion. No fatigue effect on blink rate or pupil activity was found.

To further investigate these significant changes, a doubly repeated-measures ANOVA was conducted on the multiple dependent visual behavior measures. No effect of monitor bezel reflectance on any measure was found. The numerical verification task showed a significant increase ($F = 8.93$, $df = 1, 5$, $p = 0.03$) in total saccade rate in the afternoon measurements over the morning baseline.

5 Discussion

The results of this study do not support the idea that the reflectance of a computer monitor's bezel, either high or low (0.87 to 0.07), causes visual fatigue or degrades visual performance in a naturalistic office work protocol. Neither the performance measures nor visual acuity worsened as the bezel reflectance was varied. A high-reflectance bezel could still be a problem if the ambient illuminance were high enough, but in that situation the visibility of the whole screen would be diminished. The reasons for the lack of impact of the bezel reflectance are probably the small area of the bezel and the limited range of luminances achievable with any reasonable ambient illuminance.

Saccade rate increased significantly from beginning to end of day. This increase was not related to the monitor bezel reflectance, but was associated with a task that required matching columns of numbers. This change in visual behavior could be due to muscular fatigue or cognitive fatigue. However, the difference in visual behavior between the reading task and the numerical verification task and the overall increase in backward saccades across both tasks places suspicion on cognitive fatigue. The more active "rereading" of items to be matched is consistent with loss of attention and/or shortened short-term memory, both of which are likely outcomes of cognitive fatigue. It is speculated that cognitive fatigue reduces the certainty of the participants' decisions about the similarity or difference of items being matched, producing more saccades between the columns.

Although the study of visual behavior was based on a small subset of participants, the idea that fatigue can be detected by unobtrusively examining eyegaze behavior seems promising as a replacement for simple self-reports of fatigue. Murata, Uetake, Otsuka, and Takasawa (2001) report a strong correlation between pupil diameter and self-reports of visual fatigue in a much shorter experimental task. However, no relationship between pupil size and fatigue was observed in this experiment. Nonetheless, development of reliable unobtrusive measures of computer user fatigue from eyegaze recording may have value in accurately assessing the impact of interface and screen designs, data processing tasks, and other human-computer interactions.

6 Acknowledgments

The authors gratefully acknowledge the financial support of the IBM Corporation for this study. They also wish to thank Vasudha Ramamurthy, Ujjaini Dasgupta, Mike Del Prete, Bryan Watson, and Mike Lynch for their help in running the study.

7 References

Aaras, A., Horgen, G., Bjorset, H., Ro, O., and Walsoe, H. 2001. Musculoskeletal, visual, and psychosocial stress in VDU operators before and after multidisciplinary ergonomic interventions. A 6 years prospective study—Part II. *Applied Ergonomics,* 32, 559-571.

Boyce, P. 1981. *Human Factors in Lighting*. London: Applied Science.

Boyce, P. 1986. Lighting the display or displaying the lighting. In Knave, B. and Wideback, P.-G. (Eds.), *Work with Display Units 86* (pp. 340-349). Amsterdam: Elsevier Science Publishers.

Boyce, P. 1991. Lighting and lighting conditions. In Roufs, J. (Ed.), *The Man-Machine Interface – Vision and Visual Disfunction, Volume 15* (pp. 41-54). London: MacMillan Press.

Gaarder, K.R. 1975. *Eye Movements, Vision and Behavior.* Washington, D.C.: Hemisphere Publishing Co.

Illuminating Engineering Society of North America. 2000. Office lighting. In Rea, M. (Ed.), *The IESNA Lighting Handbook: Reference and Application, 9th ed.* (pp. 11-1 – 11-22). New York: Illuminating Engineering Society of North America.

Murata, A., Uetake, A., Otsuka, M., & Takasawa, Y. 2001. Proposal of an index to evaluate visual fatigue induced during visual display terminal tasks. *International Journal of Human-Computer Interaction*, 13(3), 305-321.

Oster, P.J. & Stern, J.A. 1980. Measurement of eye movement. In Martin, J. & Venables, P. H. (Eds.), *Techniques in Psychophysiology* (pp. 275-308). New York: Wiley.

Vos, J. J. 1985. Disability glare—a state of the art report. *CIE Journal*, 3, 39-53.

Vos, J. J. 1999. Glare today in historical perspective: Towards a new CIE Glare Observer and a new glare nomenclature. In *Proceedings of the 24th Session of the CIE, Warsaw, June 24-30, Vol. 1, Part 1, CIE Publication 133* (pp. 38-42). Vienna: CIE Central Bureau.

Wibom, R., and Carlsson, L. 1986. Work at video display terminals among office employees: Visual ergonomics and lighting. In Knave, B. and Wideback, P.-G. (Eds.), *Work with Display Units 86* (pp. 357-367). Amsterdam: Elsevier Science Publishers.

Wolska, A. and Switula, M. 1999. Luminance distribution on VDT work stands and visual fatigue. In *Proceedings of the 24th Session of the CIE, Warsaw, June 24-30, Vol. 1, Part 1, CIE Publication 133* (pp. 43-47). Vienna: CIE Central Bureau.

Stress in the Office: the Influence of Software-Ergonomic Quality

Jörn Hurtienne

bao – Büro für Arbeits- und Organisationspsychologie GmbH
Kösterstr. 1[B], 14165 Berlin, Germany
Email: j.hurtienne@bao.de

Jochen Prümper

FHTW – Fachhochschule für Technik und Wirtschaft
Treskowallee 8, 10313 Berlin, Germany
Email: j.pruemper@fhtw-berlin.de

Abstract

The transactional stress-model by Lazarus (1999) is used as a theoretical framework to investigate the influence of software-ergonomic quality on irritation, psychosomatic complaints and health problems in computer workers. Two converging studies are discussed, both showing significant effects of software-ergonomic quality. The paper concludes with practical implications drawn from the data.

1 Introduction

Many workers today use computers as tools for accomplishing their daily work. While much research has been done on the specific aspects of *hardware* quality (like specific effects of work with keyboards and screens or the effect of system response times on stress), the effects of *software* quality on workers' experience of stress has rarely been investigated in a systematic fashion.

The aim of this paper is to report two studies on the relationship between the ergonomic quality of a software programme and the experienced stress in computer workers.

2 Theoretical Framework

Theoretical basis of our research is the transactional stress-model by Lazarus (1999) and Lazarus and Folkmann (1984). Rather than assuming a direct relationship between stressors and subjective strain (i.e. consequences of stress), this model proposes a cognitive stance in between: appraisal and coping.

Stressors at the work place are evaluated by the individual. In a primary and secondary appraisal stressors are judged whether they are a threat to the individual and how they can be dealt with. During the primary appraisal the individual categorizes a stimulus as being positive, negative or irrelevant to its own well-being. The recognition of a negative stimulus is usually accompanied by unpleasant emotions or general discomfort. Secondary appraisal then involves a more detailed analysis and the generation of possible coping strategies.

The likelihood of successful coping depends on the available amount of resources. Thus resources can moderate the effect of stressors. Typical resources include decision latitude (control), social support, qualification, knowledge of coping strategies, and social competency. If there are not enough resources available or the individual chooses inadequate coping strategies, continued exposure to stressors can lead to short term stress reactions like physiological changes (increase in heart rate, blood pressure, hormone levels), decreased performance, frustration, anger, or irritation. If stress is continuously present over longer periods of time short-term reactions may transform into long-term stress reactions like psychosomatic complaints, physical health problems, anxiety, depression, burnout, or absenteeism.

We will use the transactional stress model for investigating the relationship between the ergonomic quality of a software and experienced levels of stress in office workers (see figure 1).

Stressors	Appraisal and Coping	Short-term stress reactions	Long-term stress reactions	Demo-graphics
• Time of exposure to a specific software • Software-ergonomic quality	Resources: • Experience with software • Mastery of software • Social Support	• Irritation	• Psychosomatic complaints • Physical health problems	• Sex • Age

Figure 1: Variables used in our two studies according to the transactional stress model by Lazarus (1999), Lazarus & Folkmann (1984)

According to the model we classify 'software-ergonomic quality' and 'time of exposure to that software' as stressors. During primary appraisal the user decides whether the behaviour of the software (e.g. displaying an error message saying that incorrect data has been entered a week ago) imposes a problem on him/her. In the secondary appraisal he/she generates coping strategies (e.g. correcting the data). The chance of removing a problem rises the more experience a user has with this software and the higher his level of mastery of this software is. The availability and degree of social support (e.g. asking colleagues or supervisors for help) are also important resources to the user. Is software-ergonomic quality low and resources are not available for successful coping stress becomes strain. Short term stress reactions like irritation are the consequences and may result in long term-stress reactions like psychosomatic complaints and even physical health problems. The remainder of this paper will look at empirical data to support the hypothesized relationships.

3 Method

3.1 Description of Study 1 and Study 2

We conducted two questionnaire studies to test the hypothesized effects. The participants of our first study were computer workers of an accident prevention and insurance association with local offices in several places of Germany (N = 444). Over two thirds of the participants rated software packages that were developed exclusively for their employer. Other software included mainly Microsoft Office products. All of the variables in figure 1 were measured.

A second study was designed to verify the data of the first study, regarding the relationship of stressors and resources on short term stress reactions. For practical reasons (e.g. privacy issues) we abandoned the measurement of long term stress reactions in this study. 472 workers from different industrial sectors (financial services, car industry, pharmaceutical industry, consulting, IT, telecommunications, trade, public services and others) and different company sizes (small, medium and large) in Germany took part and completed the questionnaire. More than 50% of the participants rated Microsoft Office Products, the remainder included a wide variety of standard-software packages like SAP R/3 and independently developed software.

3.2 Instruments

'Software-ergonomic quality' was measured with the questionnaire ISONORM 9241/10 (Prümper, 1993, 1999). The answers in this questionnaire are coded from –3 to +3. 'Social support' was measured with the subscale 'social support' from the KFZA (Prümper, Hartmannsgruber & Frese, 1995). The answers in this questionnaire are coded from 1 ('do not agree') to 5 ('strongly agree'). 'Irritation' was measured with the irritation scale by Mohr (1986). It contains items like 'I have difficulty relaxing after work' or 'I get irritated easily, although I don't want this to happen' with the answers ranging from 1 ('strongly disagree') to 7 ('strongly agree'). 'Psychosomatic complaints' (like headache, disturbed sleep, sensitive stomach) and 'Health problems' (like high blood pressure, stomach ulcer, bronchitis) were measured by scales after Mohr (1986). The answers in the 'psychosomatic complaints' questionnaire are coded from 1 ('never') to 5 ('almost daily'). The 'health problems' questionnaire score is the count of 'yes' answers to 22 health-problems. All other variables were single items: 'Time of exposure to software' in h/week, 'experience with software' in months and 'mastery of software' (ranging from 1='very bad' to 7='very good').

4 Results

4.1 Study 1

Table 1 shows the descriptive and correlational results of study 1. Pearson correlation coefficients support the hypothesis of an influence of 'software-ergonomic quality' on 'irritation' as an indicator for short-term stress reactions ($r=-.34$, $p<.05$). The second highest influence exerts 'social support' ($r=-.28$, $p<.05$) The 'exposure to the software' in hours per week shows no relation to 'irritation' ($r=.01$, $p>.05$). Neither does 'experience' with the software ($r = -.02$, $p>.05$). Self assessed 'mastery' of the software has an influence on 'irritation' ($r=-.18$, $p<.05$).

Running a regression analysis on these data (see table 3), we again find significant contributions of the stressor 'software-ergonomic quality' and the resource 'social support' to 'irritation' levels, whereas the other variables have no statistical influence. Overall explanation of variance is 20% - an astonishing amount since we did not take other stressors and resources regarding working and general life conditions into account.

The effects of short term stress reactions on long term stress reactions are mirrored by the correlations of 'irritation' with 'psychosomatic complaints' ($r=.60$, $p>.05$) and with 'physical health problems' ($r=.46$, $p>.05$).

Table 1: Descriptives and Correlations of the first study

Variable	Mean	SD	1	2	3	4	5	6	7	8	9	10
1 SW-Exposition, h/week	18.26	13.33	(--)[3]									
2 SW-Ergonomic Quality	-.05	1.25	-.07	(.93)								
3 SW-Experience, months	32.64	27.20	-.04	.10	(--)[3]							
4 SW-Mastery	5.34	1.24	.09	**.29**	**.12**	(--)[3]						
5 Social support	3.32	1.01	-.04	**.22**	-.01	**.13**	(.76)					
6 Irritation	3.31	1.43	.01	**-.34**	-.02	**-.18**	**-.28**	(.93)				
7 Psychosom. complaints	2.30	.81	**.23**	**-.18**	-.02	**-.15**	**-.18**	**.60**	(.93)			
8 Health Problems	4.48	2.72	**.14**	**-.10**	.01	**-.14**	**-.18**	**.46**	**.66**	(.64)		
9 Sex[1]	1.53	.50	**.33**	.06	-.02	-.02	.00	-.03	**.18**	.09	(--)[3]	
10 Age[2]	3.86	.97	**-.14**	**.14**	.10	**-.13**	-.03	.08	**.10**	**.11**	**-.19**	(--)[3]

Note: $387 \leq N \leq 443$. Correlations in bold type are significant at $\alpha = 0.05$. Reliabilities (Cronbach α) in brackets. Footnotes: [1] male = 1, female = 2. [2] < 20 yrs = 1, 20 … 29 yrs = 2, 30 … 39 years = 3, 40 … 49 yrs = 4, > 50 years = 5; correlations with age are Spearman-Rho coefficients. [3] single item scale.

4.2 Study 2

Table 2 shows the descriptive and correlational results of study 2. Again 'software ergonomic quality' and 'mastery' of the software show significant correlations with 'irritation' ($r=-.12$, $p<.05$, and $r=-.13$, $p<.05$, respectively). However, this time the effect is not as pronounced as in the first study. The influence of 'social support' remains ($r=-.21$, $p<.05$).

Table 2: Descriptives and Correlations of the second study

Variable	Mean	SD	1	2	3	4	5	6	7	8
1 SW-Exposition, h/week	22.74	13.47	(--)[3]							
2 SW-Ergonomic Quality	.70	.86	-.05	(.85)						
3 SW-Experience, months	37.76	36.35	**.19**	.02	(--)[3]					
4 SW-Mastery	5.53	1.04	**.16**	**.14**	.07	(--)[3]				
5 Social Support	3.99	.80	.04	**.13**	.04	.06	(.64)			
6 Irritation	3.03	1.13	.03	**-.12**	-.03	**-.13**	**-.21**	(.88)		
7 Sex[1]	1.60	.49	.09	**.11**	.08	-.04	.04	.08	(--)[3]	
8 Age[2]	2.98	1.04	.09	.03	**.27**	**-.16**	-.07	**.11**	.02	(--)[3]

Note: $451 \leq N \leq 472$. Correlations in bold type are significant at $\alpha = 0.05$. Reliabilities (Cronbach α) in brackets. Footnotes: [1] male = 1, female = 2. [2] < 20 yrs = 1, 20 … 29 yrs = 2, 30 … 39 years = 3, 40 … 49 yrs = 4, > 50 years = 5; correlations with age are Spearman-Rho coefficients. [3] single item scale.

The regression analysis (see table 3) again shows significant effects of 'software-ergonomic quality' and 'social support' on 'irritation' scores. However, just 6% of the variance can be explained by the regression model.

Table 3: Regression coefficients for irritation in both studies

Independent variables	β in study 1	β in study 2	Dependant variable
SW-Exposition	n.s.	n.s.	
SW-Ergonomic Quality	-.31	-.10	
SW-Experience	n.s.	n.s.	Irritation
SW-Mastery	n.s.	n.s.	
Social Support	-.22	-.18	

Note: Study1: $R^2 = .20$, Study2: $R^2 = .06$; n.s. = not significant at $\alpha = .05$

5 Conclusions

Using the transactional stress model by Lazarus (1999) as a theoretical guide we could show that software-ergonomic quality is an important and stable factor influencing the experience of stress at the computer workplace. Consequences of poor ergonomic quality include not only short-term issues like higher levels of irritation but also more serious psychosomatic and health problems in the computer workforce. As this is adding to the total costs of ownership of software in a company, resources should be provided to ensure software-ergonomic quality (a) through informed assessment and selection of software prior to purchase; (b) by ensuring user-centred design in internal software-development projects, and (c) through ergonomic customizing of standard software packages at the user site (see Hurtienne, Prümper & Linz, 2002).

Another important aspect is social support to the user. If other employees are competent users of their software they can help each other with solving problems and smooth irritation and anger. The time of exposure to a software and the level of mastery of that software play a role in influencing computer worker's stress levels to a lesser degree. The more people sit in front of their screens the more psychosomatic and health complaints they will express. Here it is desirable to introduce work with various tasks and with a lesser share of time spent at the computer. Eventually an excellent qualification of users to enhance their level of software mastery needs to be considered.

But still, questions for future research remain. Although our findings replicate, they suggest different levels of correlations. Is this due to statistical reasons (decreased variability of the 'software-ergonomic quality' score in the second study) or is it a systematic variation due to different people using different software packages?

References

Hurtienne, J., Prümper, J. & Linz, R. (2002). ERGUSTO: The Ergonomics of Work with SAP R/3. In: H. Luczak, A. E. Çakir & G. Çakir (Eds.), *WWDU 2002 - Work with Display Units: World Wide Work* (pp. 293-295). Berlin: Ergonomic.

Lazarus, R. S. (1999). *Stress and emotion. A new synthesis.* New York: Springer.

Lazarus, R. S. & Folkmann, S. (1984). *Stress, appraisal and coping*. New York: Springer.

Mohr, G. (1986). *Die Erfassung psychischer Befindensbeeinträchtigungen bei Industriearbeitern.* Frankfurt/Main: Lang.

Prümper, J. (1993). Software-Evaluation based upon ISO 9241 Part 10. In: T. Grechenig & M. Tscheligi (Eds.), *Human Computer Interaction* (pp. 255-265). Berlin: Springer.

Prümper, J. (1999). Test IT: ISONORM 9241/10. In: H.J. Bullinger & J. Ziegler (Eds.), *Human-Computer Interaction - Communication, Cooperation, and Application Design* (pp. 1028-1032). Mahwah, New Jersey: Lawrence Erlbaum Associates.

Prümper, J., Hartmannsgruber, K. & Frese, M. (1995). KFZA - Kurzfragebogen zur Arbeitsanalyse, *Zeitschrift für Arbeits- und Organisationspsychologie*, 39, 125-132.

Flexible Working Hours, Stress Factors and Well-being among IT Professionals

Pekka Huuhtanen & Marketta Kivistö

Finnish Institute of Occupational Health, Topeliuksenk. 41aA, FIN-00250 Helsinki, Finland
pekka.huuhtanen@ttl.fi

Abstract

The aim of the study was to investigate the impact of flexible working hours on stress factors and well-being and to analyze possible differences between young and older IT professionals as regards the impact of flexibility. Those working flexible working hours experienced a higher demand for new knowledge, and work strained their private life significantly more, but they had better competence and opportunities for development than those who worked fixed hours, and still better if they worked overtime in addition to fixed hours. After controlling for age, gender and the number of working hours, flexible working hours were still linked to exhaustion, spillover of work to private life, time pressure, low competence, low possibilities to influence, and information overflow. These older IT professionals who worked fixed hours experienced more information overload and poor opportunities for development than the younger ones. This age difference was not found among those with flexible working hours. On the other hand, older age associated with less exhaustion only in the fixed working hours group. The results show that flexible working hours do not merely help workers to plan their time usage, but present also a risk to their well-being.

1 Introduction

A large body of research has revealed that the impact of information technology (IT) on work content and job satisfaction has mainly been positive (Salanova & Schaufeli, 2000). However, high mental job demands together with increased time pressure are a risk for employees' well-being (Schaufeli & Entzman, 1998). A study conducted by the Finnish Institute of Occupational Health (FIOH) among IT -professionals showed that information overflow, time pressure, high mental workload, poor opportunities for development, demand for new knowledge, and lack of competence, were factors lowering the well-being of IT professionals (Kivistö & Kalimo, 2003) .

Flexible working hours are more common among IT professionals than in work life generally in Finland. In information technology (IT) work, according to the study "Knowledge-intensive work" by the Finnish Institute of Occupational Health (FIOH), 87% of the IT professionals conformed to fixed working hours (compared to 94% of all employed people).

2 Aim of the Study

The aim of the study was to investigate the impact of flexible working hours on stress factors and well-being in IT work in Finland. The second aim was to analyze possible differences between young and old IT professionals as regards the impact of flexibility at work.

3 Material and Method

The subjects of the study were IT professionals who worked "free" working hours (called "flexible" here, n=151) and fixed daytime working hours between 6 am - 6 pm (n=850). The data were collected by a questionnaire as a part of a larger study in 2001. 25% of the subjects were women. Free working hours were most common in research and development tasks (39%) and least common in manintenance and assembly tasks (3%).

The following sum scales of stress factors at work were formed: *information overflow:* four items (Cronbach's alpha =.75); *time pressure:* 16 items (.90); *pressure of work to private life:* three items (.71); *opportunities to develop at work*: three items (.79); *demand for new knowledge:* three items (.81); *lack of competence*: two items (.68).

Well-being at work was measured by job satisfaction and the absence of negative symptoms like feelings of stress and exhaustion. Job satisfaction was asked with the question how satisfied one was with his/her present job. Stress was inquired with one question: "Do you feel nervous, anxious, restless or disturbed or do you have sleeping difficulties due to depressing thoughts?" Exhaustion was assessed by five items (Cronbach's alpha=.89) describing fatique, which can not be overcome even by sleeping.

The subjects were divided for the analysis into two age groups: under 35 years (n=424) and over 50 years (n=573) for the reason that the exhaustion was found to be more prevalent among younger workers. The average total weekly working hours in the fixed group were 43.6 hours, and in flexible group 48.0 hours; in the young group 44.5 hours, and in the older group 43.8 hours.

4 Findings

4.1 Flexible working hours, stress factors and well-being

The results show that those working flexible hours had a higher demand for new knowledge, and work strained their private life (working at home; difficulties to get away from work during leisure time; considerable work pressure in private life) significantly more than among those working fixed hours (Table 1). This was even more evident among those working flexible hours if the results of the work had been agreed upon. On the other hand, the flexiworkers had better competence and more opportunities for development than those who had fixed working hours. Flexible working hours were clearly connected with more stress and exhaustion. Interestingly, the groups did not differ regarding information overflow, time pressure, or job satisfaction (Table 1).

Table 1. Stress factors and well-being of IT professionals by the working time schedule, means (standardized scales 1-5), analysis of variance (MANOVA), Tukey's tests (HSD).

Stress factors	fixed	flexible	F	p
Information overflow	3.01	3.03	0.06	.7999
Time pressure	3.59	3.67	1.98	.1597
Work strains private life	**2.37**	**2.63**	**25.29**	**.0001**
Pour opportunities for development	**2.77**	**2.58**	**15.44**	**.0001**
Demand for new knowledge	**3.39**	**3.76**	**48.95**	**.0001**
Lack of competence	**2.02**	**1.87**	**9.88**	**.0017**
Well-being:				
Job satisfaction	3.67	3.76	3.16	.0758
Perceived stress	**2.44**	**2.62**	**7.90**	**.0050**
Exhaustion	**2.21**	**2.49**	**11.95**	**.0006**

4.2 Age, gender and stress factors in flexible work

Flexible working hours were more common among those under 35 years of age (21%) than among those over 50 years of age (11%). The difference was more clear among men. The older age group of those who worked with fixed working hours experienced more information overload and poor opportunities for development than the younger group. This age difference was not found among those with flexible working hours. Work strained the private life of the older flexiworkers more than of both the older and the younger workers in the fixed groups. The demand for new knowledge was greater among the young people with flexible schedules than among those working fixed hours. Young age associated with more exhaustion and stress experiences only in the fixed working hours group (Table 2).

Table 2 Differences in stressors and well-being between young and older IT professionals working fixed and flexible working hours. Analysis of variance (MANOVA)

Working hours:	fixed		flexible		F	p
Stress factors **Age:**	<35	>50	<35	>50		
Information overflow	**2.85**	**3.04**	2.94	2.95	2.81	**.0386**
Time pressure	3.37	3.37	3.31	3.24	0.50	.6843
Work strains private life	2.29	**2.35**	2.52	**2.69**	5.00	**.0019**
Poor opportunities for developing	2.51	**2.93x**	2.56x	**2.64**	21.23	**.0001**
Demand for new knowledge	**3.53**	3.40	**3.84**	3.67	8.16	**.0001**
Lack of competence	1.96	2.01	1.90	1.80	1.81	.1433
Well-being						
Job satisfaction	3.69	3.74	3.70	3.88	1.02	.3842
Perceived stress	2.43	**2.35**	**2.69**	2.36	2.92	**.0330**
Exhaustion	**2.41**	**1.94**	2.59	2.27	12.68	**.0001**

The significant differencies (p<.05) in stress and well-being factors between the groups are indicated in bold text, underlined, or marked with an x.

As regards gender differences, women perceived more conflicts between work and private life than did men, especially when working flexible hours. They mentioned also lack of competence and information overflow more often than men. Men experienced a greater demand for new knowledge. The opportunities for development were better among those men who worked flexible hours than among those who worked fixed hours, but this was not true for women .

4.3 Flexible hours and exhaustion in relation to the number of working hours

In order to test the impact of the volume of working hours on the differences between the groups working flexible and fixed working hours, further comparison was made between short (under 40 h per week) and long (over 50 h) work weeks. Information overflow, time pressure, demand for new knowledge, and the spillover of work to private life, as well as stress experiences and exhaustion were linked to long working hours in both groups. Good opportunities for development were linked to long working hours only in the flexible group, and job satisfaction only in the fixed group. Competence was not linked to the number of working hours.

Finally, stepwise regression analysis showed that after controlling the age, gender and the number of working hours, flexible working hours were still linked to exhaustion, spillover of work to private life, time pressure, low competence, low possibilities to influence, and information overflow *(Table 3)*. A separate analysis of young and older IT professionals showed that after controlling for age, gender and the number of working hours, flexible working hours were linked to exhaustion in the older group only.

Table 3. Regression analysis (stepwise) for Exhaustion (R^2=.40). Age, gender and working hours are standardised.

Stress factors:	Std. beta	Part R^2	F	p
Work strains private life	.28	.22	307.64	**.0001**
Time pressure	.22	.06	90.36	**.0001**
lack of competence	.13	.03	40.38	**.0001**
Pour opportunities for developing	.12	.01	21.47	**.0001**
Information overflow	.10	.01	10.34	**.0013**
Flexible working hours	*.06*	*.003*	*5.24*	***.0223***
Demand for new knowledge	.00	.001	0.76	.3834

5 Conclusions

Good competence was linked to flexible working hours in spite of the number of working hours. Good development possibilities, on the other hand, were linked to flexible and long working hours. This demonstrates that flexible working hours are the privilege of the most capable and ambitious workers, but at the same they are also a burden on their well-being.

The opportunities for development were better among the flexible than fixed group of men, but this was not true for the women. This might reflect the fact that women do not tend to use the flexible schedule for developing themselves, but rather for coping with their private life.

The fact that long working hours were linked to good development possibilities among men with flexible working hours, but not among those with a regular working day, may be due to the differences in work tasks. Flexible working hours were prevalent in work related research, product development, consulting and programming, all jobs in witch development possibilities are considerable, compared to e.g. operating, assembly and teaching tasks, in which very few persons had free working hours.

The demand for new knowledge and the spillover of work to private life, as well as stress experiences and exhaustion, were all greater in the "flexigroup" and they were linked to long working hours in both groups. This confirmed that the stress and well-being factors were due to the working time schedule in spite of the number of working hours.

The spillover from work to private life was linked strongly to flexible working hours, and also explained best strong fatique. This shows that free working hours do not merely help workers to plan their time usage, but present also a risk to their well-being. This in turn seems to build up in the form of prolonged fatique, which is the central and first sympton of burnout.

Well-being at work was lower among those IT professionals who had flexible working hours. This was not associated with longer weekly working hours but may be associated with differences in tasks and responsibilities. Flexible working hours were more common among the young employees, which may mean that "flexiwork" will increase among IT professionals in the near future. The autonomy and demands at work will be analysed further. In mentally demanding IT work with high time pressure, innovative working time solutions should be developed which combine both productivity demands and the individuals' needs in different phases of their work career.

References

Kivistö, M. & Kalimo, R. (2003). IT -ammattilaisen työn henkiset kuormitustekijät ja työuupumus (Mental work load and burnout among IT professionals. Työ ja Ihminen 1/ 2003. English abstract. In press.

Salanova, M. & Schaufeli, W.B. (2000). Exposure to information technologies and its relation to burnout. Behaviour & Information Technology, 19, 385-392.

Schaufeli, W. & Enzmann, D. (1998). The burnout companion to study and practice - a critical analysis. London: Taylor & Francis.

The effect of mental demand on performance and muscle activity during computer use

Jensen B.R.

Inst. of Exercise and Sport Sciences,
University of Copenhagen, Denmark
brjensen@ifi.ku.dk

Laursen B

Nat. Inst. of Public Health,
Copenhagen, Denmark
bla@si-folkesundhed.dk

Garde AH

Nat. Inst. of Occup. Health,
Copenhagen, Denmark
ahg@ami.dk

Jørgensen AH

The IT-University of Copenhagen,
Copenhagen, Denmark
anker@it-c.dk

Abstract

Healthy female subjects performed 4 different computer tasks: a mental demanding task and a reference task, which were performed with keyboard or computer mouse. Performance during the computer tasks, muscle activity in seven forearm, shoulder and neck muscles, and perceived exertion were measured. The subjects performed significantly better when using the keyboard compared to using the computer mouse. The mental demands increased muscle activity of the forearm, shoulder and neck muscles.

1 Introduction

Computer work is known to contribute to the development of work-related musculoskeletal symptoms. However, little is known regarding the specific exposures leading to these symptoms. Especially the computer mouse has been suspected to be one of the contributing factors to these symptoms. It may be hypothesized that there are differences in workload between the use of character-based user interfaces such as keyboard and graphic user interfaces like the computer mouse since use of computer mouse requires extensive hand-eye coordination and therefore may be more difficult to automate. In addition computer work often comprises mental demands. It is hypothesized that motor demands and mental demands may interact resulting in increased muscle load. The aim was to study the influence of mental demands on the performance and the musculoskeletal workload during computer work.

2 Methods

2.1 Subjects

Twelve females with an average age of 32 years participated (1, 3).

All subjects were experienced computer users.

2.2 Work station

The workstation was carefully adjusted to each subject. The table was height adjustable and shaped so that it allowed forearm support. The computer mouse was placed at forearm distance and the sensitivity was set to medium, corresponding to a sensitivity of approximately 6 (VDU curser movement/mouse movement) for slow movements and 12 for fast movements.

2.3 Electromyography (EMG)

Surface electromyography (Ag-AgCl electrodes) was recorded from the forearm muscles (m. extensor carpi radialis (ECR), m. extensor digitorum (ED), m. extensor carpi ulnaris (ECU) and m. flexor carpi radialis (FCR)), the shoulder muscles (m. trapezius (TRAP), right and left) and the right side of the neck (NECK) extensor muscle group. For each task the amplitude probability distribution function of the EMG was determined and from this the static activity level (the level exceeded in 90 % of the time) and the median activity level (the level exceeded in 50 % of the time) were calculated. All results were normalized relative to maximum EMG measured during maximum voluntary contractions.

2.4 Perceived exertion.

The physical exertion was rated during the computer tasks using a 10-graded scale, where increasing numbers indicate increasing exertion. The perceived exertion was rated in six body regions: right hand/wrist, right forearm/elbow, right shoulder, left shoulder, neck and eyes.

2.5 Performance

The performance was measured as the percentage of correct answers and the response time.

2.6 Protocol.

Computer tasks were performed with keyboard (K) or computer mouse (M), and with mental demands (colour word test, CWT) or without (reference, REF). This resulted in four combinations, KCWT, MCWT, KREF and MREF. The duration of each task was 8 minutes (1, 3).

Mental demanding task: Two competing visual stimuli were presented in a computer version of the colour word test. A word designating a colour (e.g. red) was shown on the monitor in another colour (e.g. blue). The subjects were to report the colour with which the word was written. Four different colours were used in random order and also the position on the monitor was randomised. Below the colour word were buttons with the colour words in black. Each stimulus was presented for 0.6-2.0 sec (mean 1.3 s). The subjects were instructed to respond as quickly and correctly as possible. The subjects reported by pushing one of four adjacent keys, which corresponded to the buttons on the monitor (keyboard condition) or with the computer mouse clicking the button on the monitor (mouse condition).
Reference task: The reference task was for the computer mouse condition clicking a single button, which appeared on the monitor in the same way as the colour word in the CWT. For the keyboard condition, the subjects were instructed to pres any of the four keys used in the colour word test

when an asterisk appeared on the monitor. In this way the physical work demand was the same in the reference task as for the colour word task.

3 Results and discussion

3.1 Input device

During the colour word test the subjects performed significantly better when using the keyboard compared to using the computer mouse. Thus, the percentage of correct answers using the keyboard was 82% (SD 9%) compared to the computer mouse condition 54% (SD 12%). Furthermore, the response time was longer 1.1s (SD 0.08s) for the mouse condition compared to the keyboard condition, 0.69s (SD 0.06s). The neurophysiological process during the mouse condition is rather complex and characterized by high demands on the hand-eye coordination, partly due to lack of linearity between mouse and curser displacement when manipulating the mouse. The lower proportion of correct answers in the mouse condition may therefore be explained by the complexity of the motor task during the mouse condition leading to lack of time.

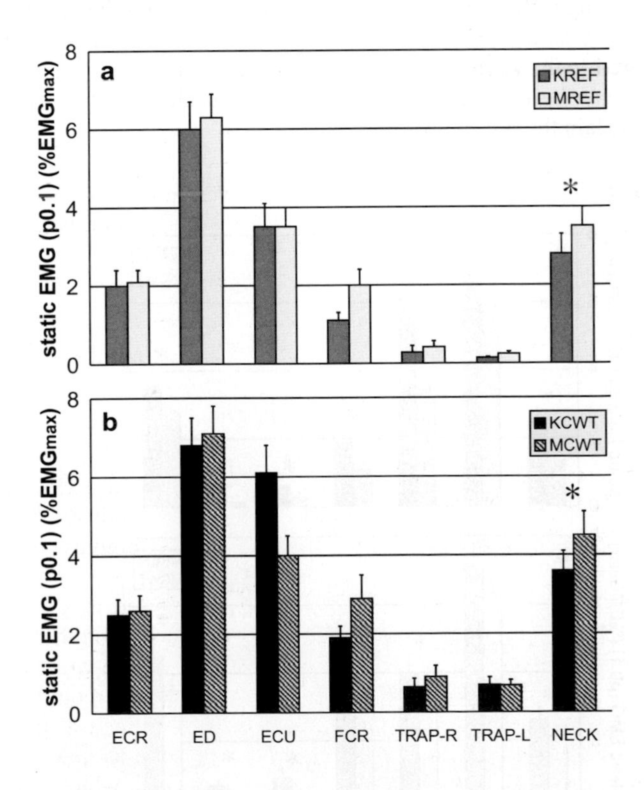

Figure 1. The effect of input device on the static muscle activity for the seven muscles.
a: Keyboard versus computer mouse during the reference task. b: Keyboard versus computer mouse during the colour word test (3). * indicate differences ($p<0.05$).

The perceived physical exertion was significantly higher for the right hand/wrist region in the keyboard condition (mean: 2.7, SD: 0.4) compared to the mouse condition (mean: 2.0, SD: 0.3). Furthermore, there was a tendency to higher values for the perceived exertion for the right forearm/elbow region for the keyboard condition compared to the computer mouse condition whereas for the remaining regions, right and left shoulder, neck and eyes no differences were found. Thus, the perceived exertion seems to be more related to the input mode for body regions close to the input device than for more distant body regions. The higher values of perceived exertion in the keyboard condition may be explained by the relative fixed hand position in the present study where only four keys were used.

The static muscle activity for the seven muscles is presented in figure 1. In general the results showed large differences in the activity level between muscles. The highest muscle activity was found for the forearm extensor muscles (m. extensor digitorum and m. extensor carpi ulnaris), indicating lack of possibility to rest the hand during use of both input devices. The lowest levels of muscle activities were found for the shoulder muscles. This may be explained by efficient support of the forearms. For the arm and shoulder muscles no effect of input device was seen in the present study, neither for the reference condition nor for the colour word test condition (figure 1).

In contrast, for the neck, higher levels of muscle activity were found for the mouse conditions compared to the keyboard condition. This may be due to higher visual demands when using the computer mouse compared to the keyboard. In a previous study it has been shown that a reduction in the eye reaction time can be obtained by increasing the muscle activity of the neck extensor muscles (2). This may explain the present results.

3.2 Mental demand

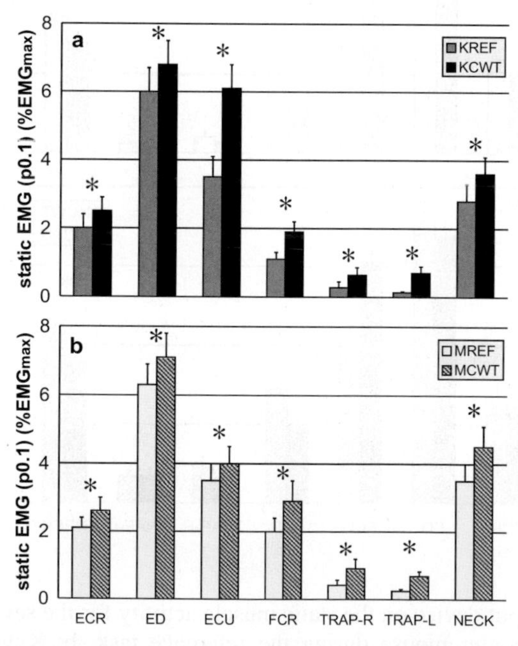

Figure 2. The effect of mental demand on the static muscle activity (seven muscles). a: Colour word test versus reference task during use of keyboard. b: Colour word test versus reference task during use of computer mouse (3). * indicate differences ($p<0.05$).

The colour word test induced high demands on concentration and time pressure.
This was clearly seen as an increase in the mean and the static level of the muscle activity. Thus, the colour word test (KCWT, MCWT) caused a significantly higher mean muscle activity and a higher static muscle activity for all recorded muscles compared to the reference conditions (KREF, MREF) (figure 2).

Previous studies have shown an effect of mental demands on specific muscles (4, 5). The present study showed an effect of mental demands on all the recorded muscles in the forearm, shoulder, and neck, indicating a higher level of co-contraction during the mentally demanding tasks compared to the reference tasks. Thus, the effect of mental demands on muscle activity during computer work seems to be a general effect on the upper extremity muscles and not specifically related to certain muscles. The present findings may be part of the adverse effects of psychosocial work factors on the musculoskeletal system.

4 Conclusion

- Mental demands increased muscle activity of the forearm, shoulder and neck muscles in general
- Increased neck muscle activity was found for the mouse condition compared to the keyboard condition
- Highest levels of muscle activity was found for m. extensor digitorum (forearm muscle) and lowest levels for m. trapezius (shoulder muscle)

5 References

1. Jensen, B.R., Laursen B., Garde, A.H., Jørgensen, A.H. (2001) Effect of mental demand on muscle activity during use of computer mouse and keyboard. *The 2nd PROCID Symposium*, Göteborg, Sweden, pp 105-109.

2. Kunita K, Fujiwara K. Relationship between reaction time of eye movement and activity of the neck extensors. (1996) *Eur J Appl Physiol* 74:553-557.

3. Laursen, B., Jensen, B.R., Garde, A.H., Jørgensen, A. (2002) Effect of mental and physical damands on muscular activity during the use of a computer mouse and a keyboard. *Scand J Work Environ Health* 28(4): 215-221

4. Wærsted, M., Bjørklund, R.A., Westgaard, R.H. (1994) The effect of motivation on shoulder-muscle tension in attention-demanding tasks. *Ergonomics* 37(2): 363-376

5. Wærsted, M., Westgaard, R.H. (1996) Attention related muscle activity in different body regions during VDU work with minimal physical activity. *Ergonomics* 39: 661-676

Information Technology and Moral Stress
How to Avoid Moral Stress and How to Promote Health

Iordanis Kavathatzopoulos, Jenny Persson and Carl Åborg

Uppsala University
Department of Information Technology
Human-Computer Interaction
Box 337, SE-751 05 Uppsala
iordanis@hci.uu.se

Abstract

In the present paper we present data collected before and during the first steps in the implementation of a new information technology system at a large governmental organization. The focus of the study was on the relationships of ethical competence to moral stress and health, to the work environment, and to the use of existing information technology systems and tools. The method used was a survey questionnaire. Moral stress was positively correlated to risky ways of using information technology and to work stress, and negatively correlated to ethical competence and ethical confidence. It was also shown that continuing education, support, information, choice, and participation in the systems development process were correlated to higher ethical competence and confidence.

1 Introduction

Successful construction, implementation and use of information technology systems require many different things. Satisfaction of economical and technological demands used to be (and still are) in focus, but today other aspects, such as user competence, work environment, organizational structure and dynamics, work tasks, and even ethical issues, are afforded a larger role in the functions of a system. In this paper we focus on moral stress and discuss how ethical competence can be achieved and maintained in a highly changing work environment dominated by the use of IT tools. Ethical competence is an important emerging factor in determining the optimal use of information technology systems through the acquisition of greater control of the whole work situation, by the creation of a better work environment, by the reduction of the stress level, and by a more satisfying handling of moral problems (Collste, 2000).

Information technology systems demand a certain kind and amount of knowledge on the part of the user in order to function optimally. According to the psychological theory of Piaget (1980), knowledge or cognitive schemata provide the solutions to certain problems. Such knowledge is adaptive in the sense that it guides action effectively. Information offered as a solution to a concrete problem but not working, is not knowledge but irrelevant information and impossible to learn and use. Persons lacking necessary knowledge are in a state of cognitive disequilibrium. Furthermore, they feel that they cannot satisfactorily control their work situation, and lack of control is a well known stress factor (Karasek & Theorell, 1990).

Thus knowledge and competence have to be there when they are needed, to be able to co-ordinate existing knowledge resources inside a group or organization, or they must be possible to create when missing. If a knowledge gap is observed, persons and organizations try to fill it. Informing, training, and educating people are the most obvious and most frequently used ways of providing the knowledge needed. However, a certain amount of stability is necessary for any educational process to succeed. The construction of introductory courses and other means of assistance for information technology users, including ethical codes and guidelines, presupposes the ability to anticipate future problems in real situations. Solutions to these future problems are the base on which training has to rest. Therefore, in order to be able to construct effective training materials and education programs, or plan and implement needed organizational changes, it is necessary to know what the future will be.

There are many interdependent factors influencing the optimal use of an information technology system; technology changes very rapidly as well. These are two factors that make it very difficult to anticipate future problems and to identify the corresponding knowledge necessary to solve them. This happens even more in the world of ethics. In the fast changing world of today it is extremely difficult to know in advance what will be right and what will be wrong, and therefore it is impossible to offer adaptive and functional answers. Ethical codes and guidelines are constructed in advance and before the problem they target has emerged. Thus the only kind of moral knowledge such guidelines can offer are general principles and not concrete answers to real moral problems.

Persons as well as organizations, however, need moral knowledge. Persons need a psychological problem-solving and decision-making skill to cope with moral problems, that is, ethical competence. In a corresponding way organizations need adaptive processes and routines to handle moral problems satisfactorily. Persons can learn through their own experience but also by training. Organizations can reorganize themselves in order to be able to co-ordinate existing knowledge and expertise. The great advantage with informal cognitive support, above and beyond its applicability, is that knowledge transmitted in this way reaches the target easily, and therefore it is accepted and applied successfully. Both personal skills and organizational processes are general problem-solving methods applicable to an infinite number of problems of the same kind, in this case, moral problems.

Ethical competence is defined here as a psychological process based on ethical autonomy (Piaget, 1932; Kohlberg, 1985; Kavathatzopoulos, in press). An ethically competent person is unconstrained by moral fixations, moral authorities and uncontrolled reactions and is able to start the thought process of considering and analyzing critically and systematically all relevant values in a situation involving a moral problem. An ethically confident person trusts his/her own ability to cope with moral problems and has the emotional strength to implement controversial decisions. Such ethical skills result in better handling of moral problems, achievement of satisfactory solutions, higher control, less moral stress, a better environment, and better health.

2 Aim of the study

The aim of the present study was to investigate the relationship of certain ethical aspects, such as ethical competence, ethical confidence and moral stress, to stress and health aspects in computer work. Data were collected during the development and implementation of a new information technology system. This study was part of a larger investigation program whose overall goal is to describe a base line for the evaluation of the future use of the new system, as well as to come to some conclusions about measures to be taken during the system development process in order to be able to assure the usability of the system.

3 Method: Instrument and participants

For the collection of data a survey questionnaire with fixed alternatives was used. Included were questions dealing with some background information. The dimensions of the questionnaire were about stress and fatigue; about relaxation and recovery (Kjellberg & Wadman, 2002); about health conditions; about work demands, support and control; about work with the computer system; and about ethical competence and confidence. Each dimension was assessed by a single question. The participants were employees at the National Registration Office in Sweden, *Folkbokföring*. 638 persons answered the questionnaire, 568 of whom were women and 53 men. Most of them, 432 persons, were between 40-59 years of age, and they had an average of 15 years of experience at National Registration (SD=10.34). 207 were assistants, 354 were clerks, 36 were junior managers, and 17 were senior managers.

4 Results, discussion and conclusions

The results presented a rather positive picture of work conditions in general at the National Registration Office, and especially of computer work conditions. It was a well functioning work place with a good environment and work climate, satisfying levels of work task control and support from colleagues and superiors, as well as higher levels of interest and positive stimulation, combined with low stress and good health. Participants were seldom confronted with moral problems at work and even less privately. They were not very stressed by moral problems, and it was not difficult to cope with the moral problems that came up. They could explain their decisions convincingly, and they had confidence in their ability to handle moral problems. Their organization had some, but not sufficient, routines to deal with moral problems and support them.

There were, however, some negative aspects reported: (1) Participants were worried that the shortcomings of IT tools or the way those IT tools were used might have serious consequences. (2) They felt that they had little opportunity to influence the construction of IT tools or to select the IT tools they had to use to achieve their work goals. (3) IT tools were complex and difficult to use. (4) There were some worries that computer work might cause health problems. (5) Computer breakdowns occurred rather often. (6) Work tasks were too simple and monotonous but difficult. (7) Too much responsibility at work. (8) Very little training and education for ethical competence.

The results showed that the National Registration Office in Sweden was a rather good work place. There are, however, some important aspects that detracted from that overall positive picture. Despite the fact that there were few moral problems at work, participants showed a high level of concern about what effects the use of their IT tools might have. They were worried that IT tool deficiencies or the possibility to misuse or abuse them might cause harm. At the same time they reported their IT systems often broke down and that using them was complex and difficult. Furthermore, they had few opportunities to choose the IT tool they preferred to use, or to participate in the development and construction process of the new IT systems for their work. These results showed that, even in a well functioning computerized work place, there are serious moral concerns about IT systems. We may therefore assume that in work places where IT systems are flawed a higher level of moral stress would be expected.

Table 1. Correlations between moral stress, ethical competence, ethical confidence, and IT, work environment, stress and health aspects.

	Moral stress	Ethical competence	Ethical confidence
IT tools and the way they are used may cause harm	.32		-.21
Selection of IT tool or to participate in tool construction			.20
Computer work may cause health problems	.21		
Difficulty in fulfilling work task	.28		
Felt worried at work the latest month	.21		
The work is trying	.28		
The work is hard	.25		
Feel relaxed at work during a typical day	-.23		
Active		.23	
Calm	-.21		
Important		.22	
Confirmed		.27	.20
Tense, pressed, worn out	.27		
Stressed	.25		
Irritated	.32		
Angry		-.20	
Upset	.30	-.26	
Exhausted	.23		
Worried	.32	-.27	
Depressed	.23	-.22	
Nervous	.28		
Anxious	.31		
Dejected		-.25	
After work still thinking about work	.22		
Tiredness from work inhibits leisure activities	.24		
Education and competence development opportunities		.22	
Access to information needed at work		.24	.21
Own work is valued and appreciated by colleagues		.31	
Frequency of moral problems at work	.33		
Ethical competence	-.24	1.00	.49
Ethical confidence	-.20	.49	1.00

Note: No. of responses varied from 416 to 576. All correlations significant at $p < .05$.

The results also showed significant correlations (see Table 1). Although the level of correlations was not very high, correlations above the level of .20 were accepted. Scoring on the survey questionnaire was strongly positive regarding almost all aspects of computer work at the National Registration Office. A more heterogeneous result or a more negative result might have produced higher correlation data.

Moral stress was positively correlated to the frequency of moral problems at work, but not to the frequency of moral problems in private life. Despite its low level at that particular work place, moral stress was positively correlated to worries about the possible consequences of actual IT systems, to the systems being difficult to use, and to health concerns related to computer work. On the other hand, participant worries that their own systems might cause harm were negatively correlated to ethical confidence. Ethical confidence was positively correlated to having more opportunities for IT tool selection and participation in IT tool development. Moral stress was consistently correlated to most aspects of work stress and other negative feelings at work. It was also, with the same consistency, negatively correlated to positive feelings and conditions at work. On the other hand, ethical competence and confidence were positively correlated to positive feelings at work (see Table 1).

Ethical competence was also negatively correlated to many negative feelings and conditions at work. Moral stress was positively correlated to the participants' difficulty in shaking off work problems and to their tiredness after work. Moral stress was negatively correlated to ethical competence and ethical confidence. Ethical competence was positively correlated with having the opportunity for continuing education at work, and getting support and encouragement from colleagues. Ethical competence and ethical confidence were correlated with having access to information necessary for their work (see Table 1).

The above results have shown that certain aspects and conditions of work with IT tools are indeed correlated to moral stress and worries about health. Although the correlations do not describe causal relationships, these results, and particularly the correlation results, may give us the support we need in order to formulate a hypothesis on what would be the most fruitful way to prevent moral stress and promote health. It seems that continuing education, support, information, choice, and participation in the process of systems development should be part of any effort to heighten ethical competence and confidence in highly automatized and computerized work places.

5 References

Collste, G. (2000). *Ethics in the age of information technology*. Linköping, Sweden: Centre for Applied Ethics.

Karasek, R. & Theorell, T. (1990). *Healthy work: Stress, productivity and the reconstruction of working life*. New York: Basic Books.

Kavathatzopoulos, I. (in press). Making ethical decisions in professional life. In B. Brehmer, R. Lipshitz & H. Montgomery (Eds.), *How do professionals make decisions?* Mahwah, NJ: Lawrence Erlbaum Associates, Inc.

Kjellberg, A. & Wadman, C. (2002). *Subjectiv stress och dess samband med psykosociala förhållanden och samband [Subjective stress and its correlation to psycho-social conditions and problems]* (No. 2002:12). Stockholm: Arbetslivsinstitutet.

Kohlberg, L. (1985). The Just Community: Approach to moral education in theory and practice. In M. Berkowitz and F. Oser (Eds.), *Moral education: Theory and application* (pp. 27-87). Hillsdale, NJ: Lawrence Erlbaum Associates.

Piaget, J. (1932). *The moral judgment of the child*, Routledge & Kegan Paul, London.

Piaget, J. (1980). *Experiments in contradiction*. Chicago: The University of Chicago Press.

Can Computer Work Retard Aging?

Ülo Kristjuhan

Tallinn Technical University
Ehitajate tee 5, 19086 Tallinn, Estonia
Email ylokris@staff.ttu.ee

Abstract

Life expectancy is increasing in developed countries. Postponement of aging for many years will improve considerably workers' health. It will save billions of dollars in medical costs. It is possible to make this progress more rapid. Optimization of human physical activity is relatively easy in computer work. Using computers is opening new possibilities for creating optimum working environment and prevention. It is easy to inform about avoiding harm in workplaces.

1 Postponing Aging as a Reality

1.1 Save Billions of Dollars in Medical and Social Costs

Everybody wants to enjoy life for more years. Many people "hunger" for more youth. At present most people cannot imagine their life without computers. Therefore, how computer work can influence health and aging is an interesting question. The main causes of mortality are age-related diseases in developed countries at present. When we cure some age-related disease in the middle age, we mainly postpone troubles for this disease to older age. Postponing aging can be more effective for health. It will postpone most diseases and eradicate many diseases. It will prolong youth and middle age. It could diminish the age-related decline of millions of people and their dependency in older years. It could save billions of dollars. Probably computers can help to postpone aging.

1.2 What is Aging?

The concept of aging has been constructed by humans from age – the number of years. It showed the number of years. Afterwards aging was mostly linked to changes in the organism, and less to chronological age. The word aging is not the best:

- Because time does not kill us.
- Using different concepts (adding years and adding biological changes) of the same word can destroy scientific research of aging.
- Aging processes in humans are very individual. No equivalent of the 'aging' concept is found in the nature. Aging is a human "invention".

Aging is a very complicated phenomenon. Aging is a collective name of many processes in the organism; it includes gradual changes in macromolecules, cells, organs, and organ systems and at the organism level. At present there is not even any widely accepted definition of aging in science. There are over several dozens definitions and more than 300 theories of aging. If we speak about aging as a process, we should have criteria that show different stages of this process. Despite of the huge amount of research we do not have good aging biomarkers at present. Every environmental factor can influence this process. It is sometimes thought that there are "programmes of death" in the organism but a huge amount of research has shown that there are mainly "programmes for life". Often aging is determined as a hierarchy of processes determining age-associated loss of functional reserve, or increased probability of death, or progressive decline in the ability to withstand stress damage and disease. Various specialists understand this process differently. For example, a specialist in genetics often considers aging as processes at the gene level and a therapist as processes at the organism level. Their fields of aging knowledge and understanding the phenomenon can be very different. There are no aging processes in some animals, e.g. in the hydra or sea anemone. Aging is neither necessary for the human organism nor inevitable (see Kirkwood, 1999). Peto & Doll wrote in 1997 that such a process as aging does not exist in the human organism as there are not one but many different age-related processes of accumulation of damage that depend on various exogenous – psychological, physical, nutritional, etc. – factors. Age-related changes (also senescence) are a better concept than aging for multiplicity of changes. Physiological age is also a dubious concept. It suggests that all physiological systems "age" at the same rate. Factually most organs remain in a relatively good condition when the organism dies.

1.3 Biological Immortality

Life expectancy is increasing in most countries. Oeppen & Vaupel (2002) write that human mortality experts have repeatedly asserted that life expectancy is close to an ultimate ceiling and these experts have repeatedly been proven wrong. Principally, as there are processes of restoration in the organism, its life expectancy will be much bigger than in case of similar immortal objects. The term "negligible senescence", defined in 1990 by Finch, describes the life history of organisms whose risk of death does not measurably rise, as they get older. Biological immortality, the absence of an increase in rates of mortality as a function of age, is now an established scientific fact. In most organisms, biological immortality arises very late in life (90 years in humans), after a prolonged period of aging. This phenomenon leads to a new perspective on the extension of life. Sometime in the future, researchers might know enough to halt aging earlier, maybe at 45.

1.4 The Present Day Reality

The mortality of elderly is decreasing more rapidly than earlier in developed countries at present. The average 60-years-old is considerably healthier than his or her counterpart 20 years ago. Age-related processes are slower in developed countries at present than dozens of years ago. Data about some concrete person can be sometimes untrue, but data about countries and population subgroups are very informative. Among countries with populations of more than one million, life expectancy is the highest in Japan. The life expectancy of Japanese men jumped by over four months to 78.07 years in 2001 compared with 2000, according to data released by the Japanese Health and Welfare Ministry. Life expectancy for Japanese females increased during one year from 84.60 years to 84.93 years. Some groups of population in the world experience even more dramatic health benefits. In the state of Minnesota (USA) life expectancy at birth is for Asian American females 86.7 years compared to 81 years for total Minnesota females and for Asian American males 79.0 years compared to 74.6 for total Minnesota males (Aging Initiative, 2003).

2 No Dividing Line Between Aging Processes and Diseases

When age-related changes develop, they are connected with unpleasant feelings. On the basis of the World Health Organisation definition of health, we can say that an organism is not healthy.

There is no dividing line between "normal" aging processes and age-related diseases. Atherosclerosis is associated with accelerated cellular aging, with telomere length shortening (Fossel, 1998; Samani, Boultby, Butler, Thompson & Goodall, 2001). Age-related diseases share fundamental and often unappreciated pathology at the cellular and genetic levels, through cell senescence (Fossel, 2002).

3 Predictors of Human Life Expectancy

3.1 Impossibility of Getting Exact Quantitative Data

Many scientists are pessimistic about postponement of aging in human beings. Even retarding the process of aging a little, a few years, seems unrealistic for many specialists in medicine who are fighting for every day of human life. Quantitative elucidation of factors that influence age-related processes in human beings is difficult as it is impossible to organise large human experimental and control groups (hundreds of persons) for dozens of years. Moreover, environment and therapeutic measures are constantly changing. Comparison of workstations with and without computers will inevitably involve confounding factors.

3.2 Individual Accumulation of Damage

Most researchers agree that aging is an individual accumulation of damage. This aspect is important as it points on individual measures to postpone changes due to aging. There are many different processes that can go wrong as we age. These can be affected by an enormous number of inborn genetic variations that modulate how we age. Age-related changes depend on extrinsic factors. Adverse conditions cause the death of the cell. At organism level the brain plays an important role.

3.3 General Predictors of Age-Related Changes

Only approximately a quarter of the variation in lifespan in developed countries can be attributed to genetic factors. Life-style and environmental measures for decreasing risk factors of age-related (cardiovascular, musculoskeletal, etc.) diseases should be effective also for the accumulation of most aging changes (Kristjuhan, 2001) as these diseases are outcome of aging changes. Stress, social support and coping style are significant predictors of developing age-related changes. Many aging theories emphasize the importance of physical, chemical, psychological and social factors.

3.4 Main Risk Factors of Work

As to long-lived persons, studies show that these persons mostly worked regularly almost whole their adult life. At present psychosocial stressors are widespread. Such factors in comparison with others shorten significantly active and total periods of human life. So Patalano (2000) showed that the American jazz musicians died much earlier (57.2 yr) than classical musicians (73.3 yr). Important is low work control. Our studies pointed to an unknown fact that during nonmonotonous work unaccustomed activity is very fatiguing (Kristjuhan, 2000).

4 Computer Workstations as Solutions

4.1 Physical and Psychological Loads and Work Design

There exists a huge amount of information about computer-related musculoskeletal and visual problems. These threaten anyone who uses a computer for long hours. We can avoid these problems using ergonomics. Strenuous exercise is characterised by increased free radical generation and other aging processes (see Dröge, 2002). Using computer technology changes profoundly working conditions. Work can be real pleasure. The work is sometimes so interesting that a worker becomes a workaholic. The work is self-paced. Computer technology also makes work at home a more likely option. A worker can enjoy the comfort of home. It is connected with diminishing stress on roads and economy of time. A topical question is that of optimum physical load. American Centers of Disease Control and Prevention and the American College of Sports Medicine recommend that every U.S. adult accumulate 30 minutes or more of moderate-intensity physical activity (e.g. briskly walking) on most days of the week (see: Pate et al., 1995). Good work design takes into account the need for relatively short physical loads. Organization of favourable body positions and movements and avoiding harmful movements are easy in computerised workstations design. The size of the table is important for optimization of movements. It is possible to create an optimum psychological environment on the screen and in the whole workstation.

4.2 Physical and Chemical Environment Factors

Computers can create a good environment, avoiding its adverse physical and chemical factors. Liquid crystal displays (LCD) are healthier than cathode ray tube displays (CRT) and their prices are decreasing. CRT displays emit harmful radiation, whereas LCD displays do not. To create optimum lighting conditions (using computer lamps) in the workstation is relatively simple. It is possible to have signals for "vision" breaks. The acoustic noise of modern computers is below 40 dB(A) and docs not disturb even mental work. Chemical factors are mostly lacking.

4.3 Computer Assisted Instruction in Everyday Work

Computer workstations enable to use computer-assisted instruction for timely information about details of appropriate work and rest schedules, gymnastic exercises, life-style, nutrition, etc. during the working day and conduct fitness and health assessments. It is possible to connect health and anti-aging databases, and important health sites.

5 Information Through Bodily Sensations

Most researchers use very little internal bodily sensations for avoiding accumulation of damage in the organism. It is often useful to assess fatigue to find the best workstation solutions. We developed a chart of the human body (that divides it into 100 regions) for detailed analysis and quantitative assessment of fatigue using a 10-point scale (10=intensity which disturbs working). Every region has its specific reasons and physiological mechanisms for its fatigue. The chart takes into account the following:

- Spatial expansion of the symptoms of fatigue in different body regions.
- Disposition of anatomical regions.
- Special role of upper extremities in work.

- Symptoms of pathology of the musculoskeletal system.
- Localization of symptoms caused by unfavourable changes in the visceral organs.
- Spatial threshold of discomfort sensations.
- Making a distinction between the regions easy for workers.
- Keeping dimensions of regions as close as possible.

We have used these charts for computer workstations more than ten years. Our studies have shown that it is possible to create healthy computer workstations and postpone aging changes.

References

Aging Initiative. Project 2030. 2003. Retrieved February 13, 2003, from http://www.dhs.state.mn.us/agingint/proj2030/populations/minAsAm.htm

Dröge, W. (2002). Free radicals in the physiological control of cell function. *Physiological Reviews*, 82 (1), 47–95.

Fossel, M. (1998). Telomerase and the aging cell. *JAMA,* 279, 1732–1735.

Fossel, M. (2002). Cell senescence in human aging and disease. *Annals of the New York Academy of Sciences*, 959, 14–23.

Kirkwood, T. (1999). *Time of Our Lives. The Science of Human Ageing.* London: Guernsey Press.

Kristjuhan, Ü. (2000). *Tegevuse optimeerimine*. Tallinn: Tallinna Tehnikaülikool. (Summary in English).

Kristjuhan, Ü. (2001). Solving problems of workforce and human organism ageing. *Tööterviseriskide haldamine ettevõtluses: teeninduses, logistikas, kaubanduses, tööstuses.* Tallinn: Tallinna Tehnikaülikool, pp. 97-101.

Oeppen, J. & Vaupel, J. W. (2002). Broken limits to life expectancy. *Science,* 296 (5570), 1029–1031.

Patalano, F. (2000). Psychosocial stressors and short life spans of legendary jazz musicians. *Perceptual and Motor Skills,* 90, 435–436.

Pate, R. R., Pratt, M., Blair, S. N. et al. (1995). Physical activity and human health. *JAMA,* 273, 402–407.

Peto, R. & Doll, R. (1997). There is no such thing as aging. *British Medical Journal,* 315, 1030–1032.

Samani, N. J., Boultby, R., Butler, R., Thompson, J. R. & Goodall, A. H. (2001). Telomere shortening in atherosclerosis. *Lancet*, 358, 472–473.

Effects of Data System Changes on Job Characteristics and Well-being of Hospital Personnel A Longitudinal Study

Kari Lindström and Merja Turpeinen

Finnish Institute of Occupational Health,
Department of Psychology,
Topeliuksenkatu 41 a A, FIN-00250
Helsinki, Finland,
Email: kari.lindstrom@occuphealth.fi
and merja.turpeinen@occuphealth.fi

Juha Kinnunen

Kuopio University, Institute of Health
Care Administration and Economics,
PL 1627, FIN-70211 Kuopio, Finland,
Email: Juha.Kinnunen@uku.fi

Abstract

Restructuring and mergers in private and public organizations often include also restructuring and changes in the data systems used. In the beginning of the year 2000, three hospital organizations were merged in Finland. The aim of this paper is to compare the changes in the data system as well as other changes that occurred during the transition period, and later after the transition period. In the first questionnaire survey in 2000 altogether 9241. Of those employees, 5681 (61%) participated also in the second survey in 2002. The nature of the data system changed when the organizational change proceeded from the transition phase to the subsequent adjustment phase during the two following years. The poorly managed transition and job insecurity were no longer so clearly related to it as earlier. The data system change was related more to qualitative changes at the workplace, like new types of patients, new work practices and competence demands.

1 Introduction

Mergers are characterized by top-down planning, and the discontinuity of organizational culture and practices. The management's perspective in mergers is at the macrolevel, it is distant and global, and the main actor in the transition is often the decision maker of the change process (Weick and Quinn 1999). In the beginning of the year 2000, three hospital organizations were merged in Finland.

The transition period meant, e.g. that the smallest partner had to adopt the data systems used by the larger partner. Also many other structural changes occurred at the organization level, e.g. subunits with overlapping activities were combined. The data system changes were most frequent among former city hospital personnel who had to leave their own data system, work unit and work practices, and adopt the ones of the biggest partner in the merger. The overall change was especially negatively seen by this partner, and also the changes were perceived to be most comprehensive (Lindström et al. 2002). During the next two years, 2000-2002, the changes continued, but now more within the subunits representing various medical specialities. These dealt mainly with adjustments in the division of labour and patient flow between and within various wards.

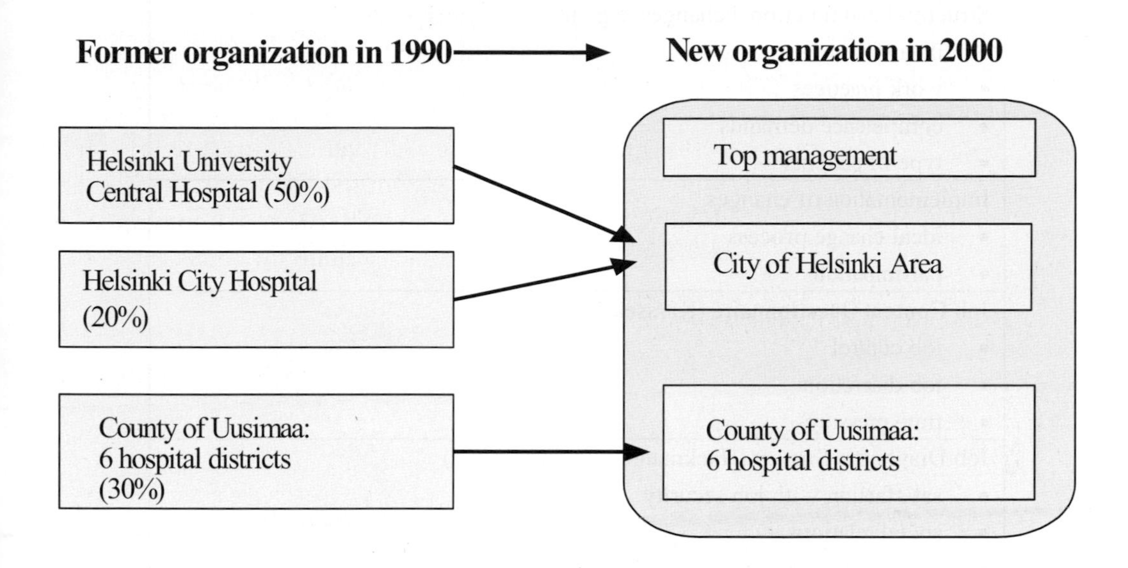

Figure 1: Old and new hospital organization

The aim of this paper is to compare the changes in the data system as well as other changes that occurred during the transition period, and later after the transition period, and to see how these changes were related to perceived job characteristics and well-being.

2 Subjects and Methods

In the first questionnaire survey in 2000, altogether 9241 (58%) employees participated from the total of 15 900. Of those 9241 employees, 5681 (61%) participated also in the second survey in 2002. The target group of this paper is the longitudinal panel group. The biggest occupational groups were qualified nurses, other health care personnel, administrative and office personnel, and physicians. These people came originally from three different hospital organizations; 50% from the University Hospital, 20% from the City Hospital, and 30% from the county hospital. All these organizations represented specialized health care.

The questionnaire included questions about the perceived changes, the transition process, especially about informing, participation, the management of the change process, and the perceived job characteristics at work, as well as the general well-being of the personnel (Table 1).

Table 1: Content of the questionnaire survey in 2000 and 2002

Structural and functional changes, e.g. in
• data system • work practices • competence demands • type of patients
Implementation of changes • ideal change process • participation
Job Content Questionnaire (Karasek 1985) • job control • job discretion • time pressure
Job Diagnostic Survey (Hackmann & Oldham 1975) • satisfaction with job security • social relations • supervisory practices • personal growth
Other scales • management practices • interactional justice • formal procedures
General Health Questionnaire (Banks et al. 1980)

3 Results

In the first study round four months after the transition, the most common changes reported were the new data system (27%), changed competence demands (24%), changed work practices, and a new supervisor (20%). After the transition period during the next two years, the reported changes were more frequent than immediately after the merger. 45% reported changes in competence demands, 38% in work practices, 37% in data systems, and 37% had got a new supervisor (Table 2).

Table 2: Structural and functional changes reported in 2000 and 2002

Main changes reported	**2000**	**2002**
• New competence demands	24%	43%
• New work practices	24%	39%
• Closest supervisor new	20%	39%
• New data system	28%	34%
• New kinds of patients	15%	26%

In the first study round during the transition period, the reported changes in the data system correlated statistically significantly with changes in competence demands, work practices, and a change in the location of the workplace. In the second study round the data system changes were again related to changes competence and in work practices, but also with the new kinds of patients. These reflected the new type of changes that occurred immediately during the transition and later during the adjustment period.

Also the quality of the changes as well the implementation process were seen differently during the transition and in the period following it. The comprehensiveness of the change was perceived to be greater during the follow-up than just after the transition. The management of the change was nevertheless perceived to be better, and the changes less negative later than during the transition.

3.1 Change in Job Characteristics

When the job characteristics just after the transition were compared to those two years later, the time pressure had increased, but the job was perceived as more challenging, job control better, and job security higher. However, the management of the whole organization was seen as less satisfactory in the second study phase, compared to the transition period. The result pointed out the qualitatively different nature of the transition itself and the later adjustment phase.

3.2 Relationships between Data System Changes and Job Characteristics and Well-being

The changes in the data system during the transition period were related to low job security and lower well-being. In the second study round the data system changes were related to low job control and especially to lower well-being. This might indicate that qualitatively the content of the change process in the transition and the consecutive phase differed somewhat.

4 Discussion

The nature of the data system changed when the organizational change proceeded from the transition phase to the subsequent adjustment phase during the two following years. The poorly managed transition and job insecurity were no longer so clearly related to it as earlier (Lindström et al. 2002). The data system change was related more to qualitative changes at the workplace, like new types of patients, new work practices and competence demands. This was understandable, because numerous reorganizations had occurred in various specialities. Patients were relocated to other wards, and also the personnel were relocated to other wards within various medical specialities. However, in both study phases the lowered well-being was related to data system changes. The results of the longitudinal study indicate that the immediate transition after the merger was followed by a rather long adjustment period at work unit level.

5 References

Lindström, K., Kinnunen, J., & Turpeinen, M. (2002). Organizational and data system changes: Psychosocial aspects. Proceedings of the Conference WWDU 2002 World Wide Work, May 22-25, 2002 Berchtesgaden. Eds. by H. Luczak, AE Cakir, G Cakir. (pp. 22-24). Berlin: Ergonomic Institut für Arbeits- und Sozialforschung.

Weick, K. E., & Quinn, B. E. (1996). Organizational change and radical re-engineering: emerging issues in major projects. European Journal of Work and Organizational Psychology, 5 (3), 325-350.

An Observational and Interview Study on Personal Digital Assistant (PDA) Uses by Clinicians in Different Contexts

Yen-Chiao Lu, Yan Xiao
University of Maryland, Baltimore
ylu001,yxiao@umaryland.edu

Andrew Sears
UMBC
asears@umbc.edu

Julie Jacko
Georgia Institute of Technology
jacko@isye.gatech.edu

Abstract

In healthcare settings, mobile computers are frequently being used where environment can be disruptive, such as lighting is inadequate, noise is high, or when the user is moving or walking. In addition, mobile devices often interrupt clinicians' ongoing activities in order to perform computer-based tasks. This preliminary observational and interview study was conducted to begin identifying and documenting the environmental and task factors that contribute to situationally-induced impairments and disabilities (SIID). We investigated how PDAs were used in different contexts by the clinicians. We found that most clinic ians were satisfied with their PDAs and more than half carried the devices with them all the time. A wide range of tasks was performed by the clinicians while they used their devices. Those tasks can be categorized based on the complexity involved: 1) comp eting tasks; 2) prolonged tasks; 3) trivial tasks. In conclusion, how the PDAs are being used by physicians in different situations were identified and examined to advance the development and design or next generation of handheld computers that fit physicians' practice needs.

1 Introduction

Mobile computing has become a way of life for many people for work and leisure. Even for those workers who have offices, they are away from desktop computers in a significant amount of time (Belloti & Bly, 1996). Understanding mobile information access and related human computer interaction issues becomes increasingly important as more mobile computing devices are used and more people are away from regular desktop computers because of the availability (Perry et al, 2001; Kristoffersen & Ljungberg, 1999). In certain information critical fields, such as healthcare, the stakes are high for accessing up-to-date data such as patient medical records and treatment plans. A recent published report by Institute of Medicine (Kohn, Corrigan, & Donaldon, 2000) projected 98,000 deaths every year due to medical errors, whose causes included communication and information access problems. Ubiquitous information access at the point of patient care and other points of information needs can potentially help alleviate problems in communication and information access. To achieve that potential, much understanding is needed about the interaction between uses of mobile computing devices (such as personal digital assistants or PDAs) and work (Kristoffersen & Ljungberg, 1999). Presently in healthcare, a growing number of workers start to use various PDAs, as they take advantage of portable, relatively inexpensive and reliable information management tools to access, retrieve and record data and information anytime and anywhere (Barbash, 2001; Nesbitt, 2002). However, there are published reports on complaints associated with clinical uses of PDAs. Problems such as weight, small screen size, unease use of graffiti and short battery life, were reported (Ammenwerth, Buchauer, Bludau & Haux, 2000; Garvin, Otto & McRae, 2000; Kelly, 2002; Kiél, & Goldblum, 2001; Lapinsky, Weshler, Mehta, Varkul & Hallett, 2001).

Despite these problems, extensive uses of PDAs help clinicians in coordinating and providing care (Moss, Xiao, & Zubaidah, 2002). In many clinical settings, the users may not be in ideal environments for interacting with computing devices as they are designed now. This paper describes a preliminary study to address the questions on factors limiting use of PDAs, especially those beyond the ones reported in the literature on interface problems and organizational support issues. In particular, we conducted an observational and interview study to examine the environmental and tasks factors that might discourage the uses of PDAs in healthcare setting.

2 Method

This preliminary observational and interview study was conducted at a major medical center to begin identifying and documenting the environmental and task factors impeding the use of PDAs. Seven different contexts of PDA usage were studied: the emergency room, shock trauma unit, patient examination room, nursing station, office, ambulance and conference room. First, structured interviews with eight physicians and four paramedics were performed to explore how the clinicians viewed and used PDAs in their practice. The interview was guided by a template with following sections: Scenarios, setting, purpose of using PDAs, environmental characteristics, task characteristics, and comments. Observations were conducted to examine how clinicians actually using the devices in their practice. Both audio recording and photographs were made during the study. The interview data was later transcribed and analyzed.

3 Results

Table 1 is an example of interview results. Our field observations and interviews have shown that PDAs were used widely by clinicians. Applications that clinicians most frequently used included Date Book, Address Book, To Do List, Memo Pad, Calculator, ePocrates and MedCalc.

Table 1: Example of structured interview results.

Scenarios	Environmental Characteristics	Task Characteristics	Comments
In the Urgent Care Physician's Office, a resident was prescribing Keflex for her patient. She wanted to find out the dosage and length of treatment with a PDA to get the needed information from eProcrate.	• Office: Three computers (one of them with high resolution for PAC); not enough desk space. • Lighting: lights are turned off sometimes when physicians need to see more details on radiology images from PAC. • Sound: people walking and talking. • Temperature: room temperature.	• Prescribing medication for patient. • Modifying and printing patient instructions.	• She mentioned that she carries her PDA everywhere. • Most used functions: drug reference (eProcrate), date book (scheduling), address book.

We found that most clinicians were satisfied with their PDAs and more than half carried the devices with them all the time. Two physicians expressed that they "can't live without it". The most frequently used functions were 1) personal information management such as Date Book, Address Book, To Do List, Notepad, Mail and Memo; 2) decision support aids such as medical reference, drug reference and medical calculator; and 3) patient information access such as looking up patient's lab values or medical history.

However, our study showed that there were certain environmental and task factors, as well as problems with physical design of PDA, impeded clinicians to fully utilize their PDAs. Both the environment, in which individuals interact with computers, and the need to attend to multiple tasks, can result in increased demands on the user's cognitive, perceptual, and/or motor skills. When the working conditions, in combination with the user's tasks, result in demands that exceed the user's capabilities, the user experiences what we refer to as situationally-induced impairments & disability (SIID).

Our field observations provide illustration of how SIID occur (Figure 1 and Figure 2). For example, paramedics complete forms while providing medical care during ambulance-based patient transfers. The paramedic's hands, eyes, and even their ears are often busy providing medical care. As a result of the demands created by their primary task, the paramedic experiences visual, motor, and hearing impairments with respect to the secondary form completion task. The paramedic may also experience SIID as a result of the movement of the ambulance and the lack of a stable surface on which to place whatever device is being used to complete forms.

Four categories of the environment characteristics were identified through our observations: 1) *sounds:* in most cases, there were unpredictable background noises. For example, while observing one physician examining a patient at the emergency room, someone was talking loudly in the waiting area; 2) *lighting:* in some cases, the light was dimmed by physicians in order to view slides or x-rays in the office; 3) *nomadic condition*: such as physicians moving from one examine room to another, or a paramedic transferring a patient on a moving ambulance; 4) *temperature*: in the shock trauma unit and operating room are kept cool, the cold temperature might reduce manual dexterity.

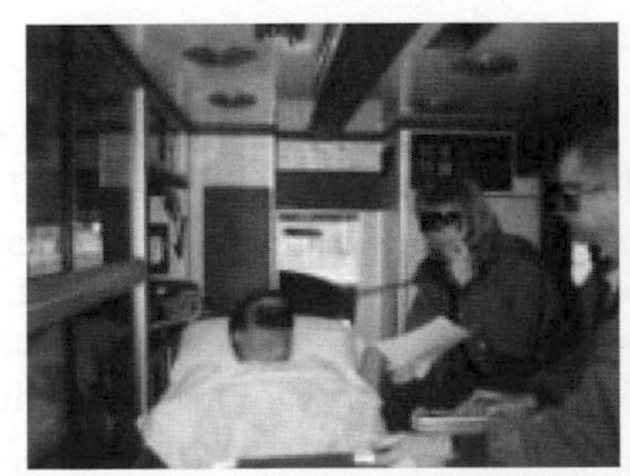

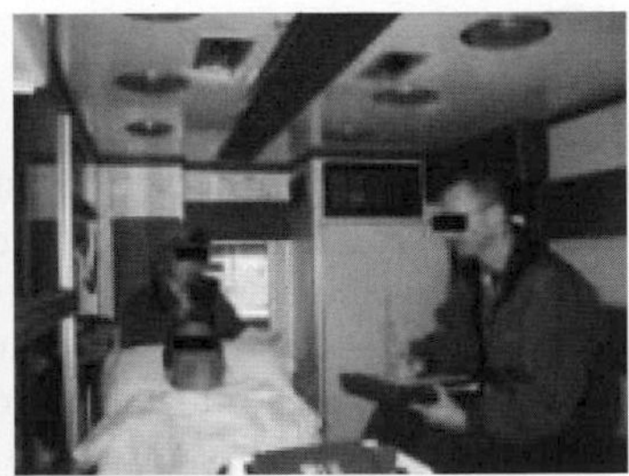

Figure 1. Factors Contributing to Situationally-Induced Impairments & Disability. Left: A nurse inside a moving ambulance is giving a report (via radio) while retrieving information (currently from a paper-based source). Right: A paramedic is recording information (filling a run report) while monitoring the patient.

Our interviews also showed that the clinicians performed a wide range of tasks while they used their devices. Those tasks can be categorized based on the complexity involved: 1) "competing tasks" are tasks that require eyes, ears or hands coordination at the same time: such as doing a physical assessment; 2) "prolonged tasks" are tasks that require extended period of time to complete: such as performing a prolonged surgery; 3) "trivial tasks" are tasks such as attending conference, walking, talking (e.g. communicates with colleague regarding patient treatments), documenting and prescribing medications and/or treatments.

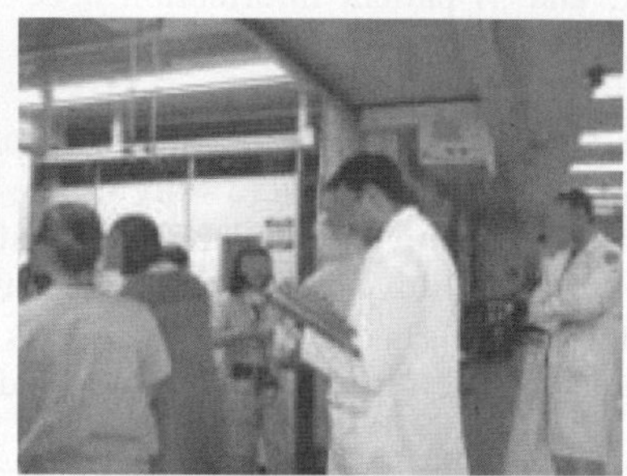
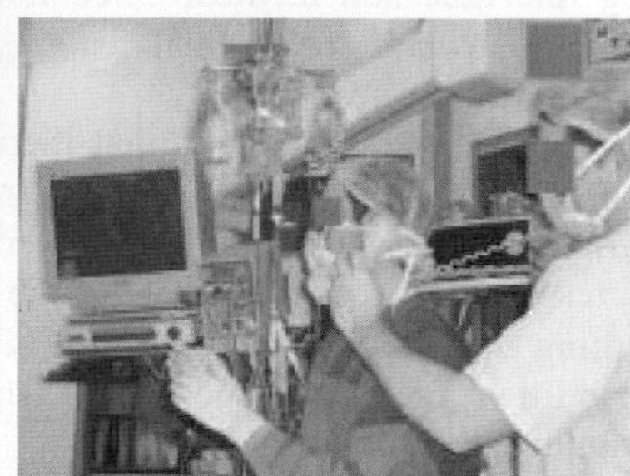

Figure 2. Factors Contributing to Situationally-Induced Impairments & Disability. Left: An anesthesiologist is retrieving information from a computer while his hands are busy taking care the patient. Center: A surgeon is recording events during a trauma patient resuscitation while monitoring the team activities and the patient. Right: A surgeon is in sterile scrub and gloves during a surgery with an assistant helping the surgeon to communicate on the phone.

In addition, clinicians generally preferred using the devices while sitting down at the office or station rather than bedsides with the patients. This might suggest that the devices were not designed to be used by the clinicians while walking, riding in the ambulance or under low light situations.

Last, there were some complaints reported by clinicians regarding the designs of the device. These problems discouraged clinicians to use their PDAs in some circumstances. Reported complaints from this study included: 1) Unease of Data entry with graffiti; 2) Size, weight, and small screen; 3) Delicate devices; 4) short battery life and limited onboard memory. However, as the technology continues to evolve, lighter, faster, durable and more powerful devices with longer battery life and easier data entry mechanism can be developed to meet users' needs.

4 Discussions

As mobile computing becomes more common, SIID will become a significant barrier to the effective use of these systems. Our research has identified factors contributed to SIID in healthcare settings. However, more research is encouraged to further investigate the factors that lead to SIID, the effects SIID have on user interactions, and the similarities between SIID and DII. The results of our study would have a direct impact on how mobile computing devices should be designed for healthcare users. New technologies, in the form of environment-aware mobile devices, should be developed that more effectively address the needs of individuals experiencing SIID.

This study has several limitations. First, only one researcher conducted the observations and interviews; however, the potential for bias was minimized because other researchers had input into the design of the interview questions as well. Second, small sample size and use of convenient samples limited the generalization of the findings. Last, other usage patterns or factors related to SIID might not be identified through the interview and observation methods. However, all of these deficits will be addressed in our next phase of study.

Currently the demands to communicate and distribute clinical information among clinicians are high, PDAs have the potentials and capabilities to streamline the communication process. To advance the development and design or next generation of handheld computers that fit physicians' practice needs, identifying and analyzing how those devices are being used by physicians in different situations is critical.

Acknowledgeme nts

This material is based upon work supported by the National Science Foundation under Grant No. IIS-0121570. Any opinions, findings and conclusions or recommendations expressed in this material are those of the authors and do not necessarily reflect the views of the National Science Foundation (NSF). The authors wish to thank the contribution by Jack Seagull and Paul Regnault.

References

Ammenwerth, E., Buchauer, A., Bludau, B. & Haux, R. (2000). Mobile information and communication tools in the hospital. *International Journal of Medical Informatics, 57*, 21-40.

Barbash, A. (2001). Mobile computing for ambulatory health care: Points of convergence. *Journal of Ambulatory Care Management, 24*(4), 54-66.

Bellotti, V. & Bly, S. (1996). Walking away from the desktop computer: distributed collaboration and mobility in a product design team, Proceedings of the 1996 ACM conference on Computer supported cooperative work, p.209-218.

Garvin, R., Otto, F., & McRae, D. (2000). Using handheld computers to document fami ly practice resident procedure experience. *Family Medicine, 32*(2), 115-118.

Kelly, J. (2002). Going wireless. *Hospitals & Health Networks*, 65-68.

Kiél, J. M., & Goldblum, O. M. (2001). Using personal digital assistant to enhance outcomes. *Journal of Healthcare Information Management, 15*(3), 237-250.

Kohn LT, Corrigan JM, and Donaldson MS, (eds.) (2000). *To Err is Human: Building a Safer Health System.* Washington, D.C.: National Academy Press.

Kristoffersen, S. & Ljungberg, F. (1999). "Making place" to make IT work: empirical explorations of HCI for mobile CSCW. ACM SIGGROUP Conference on Supporting Group Work, 276-285.

Lapinsky, S. E., Weshler, J., Mehta. S., Varkul, M., & Hallett, D. (2001). Handheld computers in critical care. *Critical Care, 5*, 227-231.

Moss, J., Xiao, Y. & Zubaidah, S. (2002). The operating room charge nurse: coordinator and communicator. *Journal of American Medical Informatics Association, 9*(90061), S70-S74.

Nesbittm T. S. (2002). Equipping primary care physicians for the digital age. *The Western Journal of Medicine, 176*, 116-120.

Perry, M., O'hara, K, Sellen, A., Brown, B., & Harper, R. (2001). Dealing with mobility: understanding access anytime, anywhere. ACM Transactions on Computer-Human Interaction, 8(4): 323-347.

Sears, A., & Young, M. (2003). Physical Disabilities and Computing Technologies: An Analysis of Impairments. In J. A. Jacko and A. Sears (Eds.), *The Human-Computer Interaction Handbook* (pp. 482-503).

Task Performance with a Wearable Augmented Reality Interface for Welding

H. Luczak, M. Park, B. Balazs, S. Wiedenmaier, L. Schmidt

Institute of Industrial Engineering and Ergonomics, Aachen University
Bergdriesch 27, D-52062 Aachen, Germany
{ h.luczak, m.park, b.balazs, s.wiedenmaier, l.schmidt }@iaw.rwth-aachen.de

Abstract

This paper describes research on a wearable Augmented Reality (AR) welding support system. "Blind welding" caused by current protective welding helmets is to be replaced by AR supported views of the welding process. In this development the question of 2D to 3D viewing aids arises. For the experiment an integrated helmet with two High-Dynamic-Range cameras (HDRC) and a closed view stereoscopic head-mounted display was used. The experimental task was to position pegs on a pegboard. Alignment angles of the cameras and inter-camera separation distances were varied under the different experimental conditions. It was found that a convergent alignment of the cameras resulted in less mistakes and higher speed. An effect of the inter-camera separation was found in 3D-perception, suggesting the advantage of smaller distances.

1 Introduction

The research described in this paper aims at improving welder's working conditions and, therefore, the quality of the welding process itself. A major problem in the manual welding process is the limited visual information about the work environment. In order to protect the worker from the bright arc and the radiation, the current welding helmets have a darkened eye shield, therefore, the welders performance has to depend upon his other senses than visual. For example, the welder is judging about the quality of the process from the sound generated by the welding torch. However, we usually get very large amount of information from our environment through the visual system, which is actively involved in most of our work tasks. Thus, this project aims at changing the concept of the "blind welding" by creating a new ergonomic wearable visual interface. The main components of the system are two HDRC cameras and a closed view stereoscopic head mounted display integrated within the welding helmet. In the new system the welder has no direct view of his environment, but the high-quality-stereoscopic-image, recorded by the HDRC cameras, is played in real time on the head mounted display. Additionally, some augmentative information, such as the run of the weld seam, machine parameters etc., is superimposed on the images.
The main focus of this research is dedicated to perception of the stereoscopic space and co-ordination ability of the hands, which are important for the quality of task performance. The most important factors influencing human perception of the three-dimensional space are depth cues, which provide information about the spatial location of an object. In an artificially created stereoscopic system some depth cues might be lost due to technological limitations. Therefore, the perception of the location of an object in space is one of the biggest problems with the current

wearable interfaces. This is indispensable for the welding task, since misjudging the spatial location of the objects can be critical for the quality of the weld seam.

2 Related Work

A rather large number of publications deals with stereoscopic systems in telemanipulation tasks. Alignment of cameras and inter-camera-separation distances for stereoscopic systems in teleoperation are described by Rastogi [11]. A number of papers describe advantages of stereoscopic versus monoscopic viewing for teleoperation tasks, such as reduced task execution time, improved accuracy, faster and more accurate perception of the spatial layout [4,9,10]. Skill acquisition and task performance in teleoperation with stereoscopic and monoscopic viewing are described by Drascic [1] and cognitive processes related to human performance with an AR-System for manufacturing and maintenance applications have been discussed by Neumann and Majoros [7]. Perceptual issues involving depth cues in mixed and augmented reality systems are thoroughly described by Drascic and Milgram [2]. Further studies regarding user perception of the distance in AR displays have been carried out by Stephen Ellis and his group at the NASA [3]. The application of AR for manual assembly tasks is described by Luczak et al., Stadler et al. and Wiedenmaier et al. [6,12,14]. The development trends of visual displays are discussed by Luczak et al. and Oehme et al. [5,8].

3 Experimental Design

For the experiment the integrated helmet with the two HDRC cameras and a closed view stereoscopic head-mounted display (DH-4400VPD), and a pegboard (60x48cm) were used. During the experiment subjects had to position pegs on the pegboard while wearing the helmet (compare Tiffin and Asher, Wiedenmaier et al. [13,14]). The view of the environment was recorded by the HDRC cameras and played on the HMD. The frame rate with the current system was 16 frames/s. In order to indicate the hole, in which the peg had to be placed, an arrow appeared in a randomised position on the pegboard (see Figure 1a). The pegs were placed in a box, which was connected to a computer mouse, so that the position of the arrow changed each time when the subject opened the box (see Figure 1b). This type of task has been chosen in order to check the accuracy of hand co-ordination as well as the speed of performance (compare: similar co-ordination of the hands is required during spot welding). 18 volunteers were chosen for the peg-to-the-hole task. No special skill, such as experience with welding or virtual/augmented reality was required. However, the subjects visual ability was tested and those with strong myopia, hyperopia or inability of stereovision were excluded from the experiment. A subject blocking (within subject) design was applied, so that each subject was examined under all conditions.

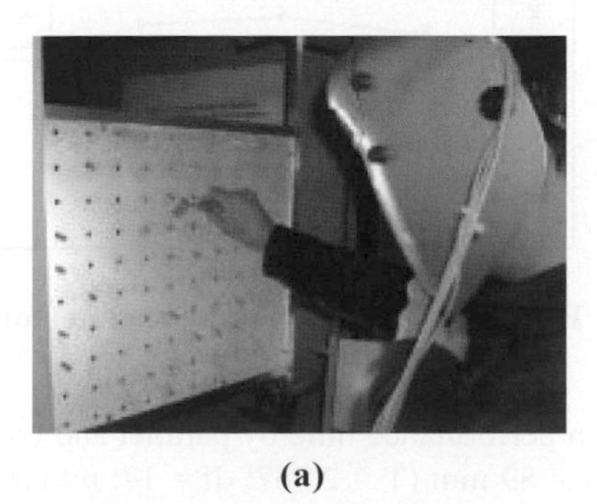

(a)

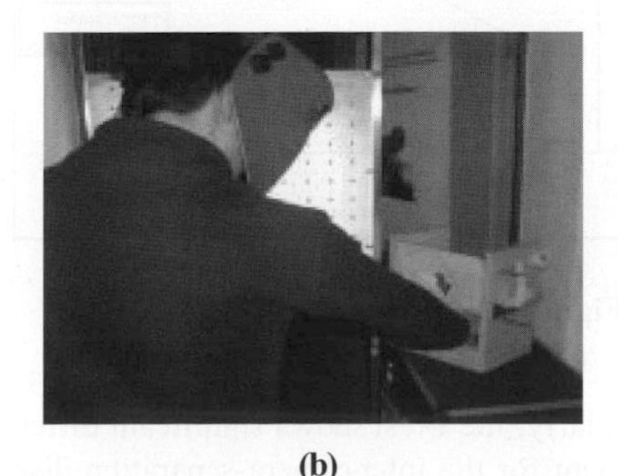

(b)

Figure 1: Task performance with the wearable interface

Six experimental conditions were set up for the experiment, with convergent versus parallel alignment of the cameras and three different inter-camera-separation distances (see Table 1). In order to minimise bias, the sequence of experimental conditions was partially randomised.

Table 1: Experimental settings

a = 45 mm		a = 61 mm		a = 89 mm	
Parallel	Convergent	Parallel	Convergent	Parallel	Convergent
a = 0° b = **8**	a = 6,4° b = 400mm	a = 0° b = **8**	a = 8,7° b = 400mm	a = 0° b = **8**	a = 17,7° b = 400mm

a – inter-camera-separation distances b – distance to the convergence point a – convergence angle

For each inter-camera-separation distance there were two possibilities for camera-alignment – parallel and convergent. The working plane was placed in the distance of less than 400 mm, so that the convergence point was set directly behind the pegboard and the task space was presented as uncrossed parallax on the display. In the condition of parallel alignment, optical axes were assumed to converge in infinity.

For the objective assessment, the time needed to take out the peg of the box and place it into the hole and the accuracy of positioning were measured. Subjects were also asked to fill out a questionnaire regarding their perception of 3D space, visual acuity, and quality of task performance.

4 Results

The time of task performance and the error rate were measured during the test. Each subject had to repeat the peg-to-the hole task (30 pegs) in each of the six experimental conditions (see Table 1). The number of errors and the mean time of task performance were calculated under each condition. Two factors (alignment angles and inter-camera distances) were checked in terms of significance, performing analysis of variance (ANOVA) for repeated measurements and t-Tests. ANOVA indicates significant differences ($F = 14{,}066$; $df = 1$; $p<0{,}05$) in the speed of task performance between convergent and parallel alignment of the cameras. The subjects perform faster when the optical axes of the cameras are aligned with a convergent angle (see Figure 2).

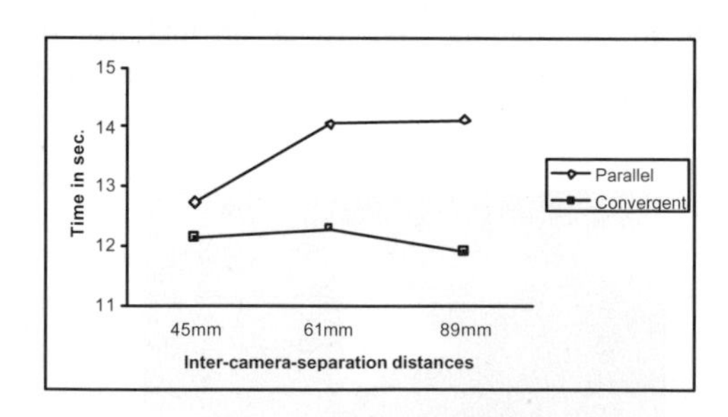

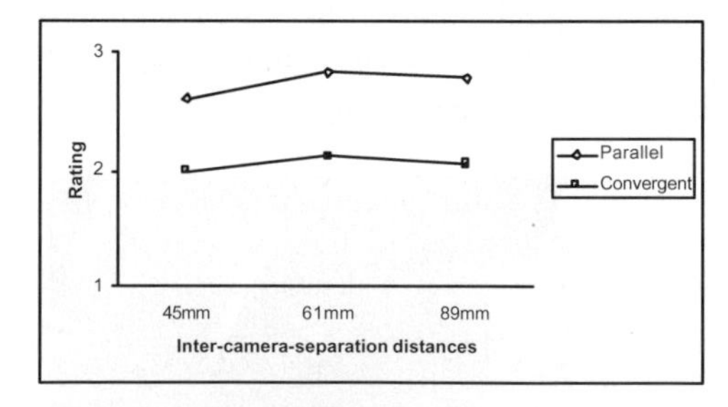

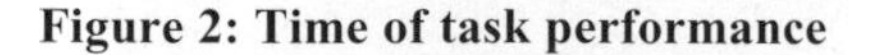

Figure 2: Time of task performance

Figure 3: Perceived quality of performance

Particularly, the t-test shows significant difference in performance time by parallel and convergent alignment for the inter-camera-separation distance a = 89 mm ($T = 2{,}867$; $df = 17$; $p< 0{,}05$) and a

non-significant trend for the inter-camera-separation distance a = 61 mm ($T = 2{,}097$; $df = 17$; $p < 0{,}1$). In the next step the number of errors was calculated for the two factors parallel and convergent alignment across all three inter-camera-separation distances. A T-test comparing means of the two alignments shows significantly less mistakes of the subject with convergent alignment of the cameras ($T = 3{,}684$; $df = 17$; $p<0{,}05$).
The subjects perception of 3D space, visual acuity and quality of performance with different inter-camera-separation distances and alignment angles were measured on a scale from 1 (very good) to 4 (unsatisfying) and also tested. Results of ANOVA show significant differences between convergent and parallel alignment with perception of 3D space ($F = 21{,}25$; $df = 1$; $p<0{,}001$), visual acuity ($F = 36{,}17$; $df = 1$; $p<0{,}001$) and the quality of performance ($F = 26{,}41$; $df = 1$; $p<0{,}001$), whereby the subjects preferred settings with convergent angles (see Figure 3 and 4). Perception of 3D space was significantly better with smaller inter-camera-separation distances ($F = 5{,}75$; $df = 1$; $p<0{,}05$) (see Figure 5). Overall no interaction effects of alignment angles and distance were found.

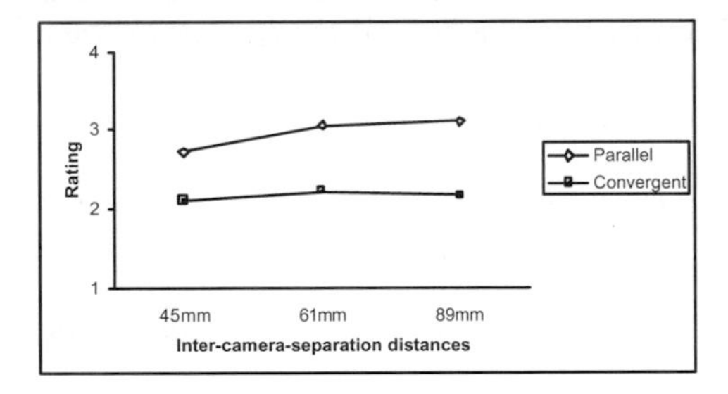

Figure 4: Perceived visual acuity

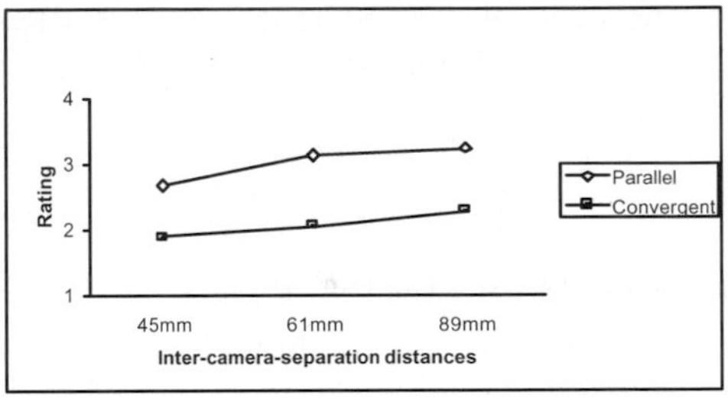

Figure 5: Perceived 3D view

Furthermore, after completing all experimental conditions, the subjects were asked to indicate the best setting in terms of perception of 3D space, visual acuity and the quality of performance. Except for one (a=45mm), all of the subjects chose settings with convergent alignments (a=45mm/ a=6,4°: 38,9%; a=61mm/a=8,7°: 33,3%; a=89mm/a=17,7°: 22,2%). After 1 - 1,5 hour work the subjects reported eye strain (55,6% little; 11,1% much), motion sickness (33,3% little), muscle pain (55,6% little; 11,1% much).

5 Conclusions and future work

In conclusion, significant differences were found between convergent and parallel alignment of the cameras, which indicate that a convergent alignment of the cameras should be used in the further development of the system. People performed faster, made less errors and were more satisfied with the convergent setting across all three different inter-camera distances. The only significant effect of the inter-camera-separation distance was found in subjective perception of 3D space, where smaller distance was preferred. This, however, is an important outcome, since perception of 3D space is one of the success factors for implementation of the system in welding industry. The factor, which was not tested in the experiment, was the effect of the position of the cameras within the integrated helmet. The location of the cameras was not coinciding with the location of the eyes (the cameras were placed above the height of the eyes), which made the system statically more balanced but some discrepancies in the depth perception might have occurred.
Referring to the results obtained, future recommendations for the development of the system would be to keep the inter-camera-separation distance within the range of 45 to 61 mm for tasks

within hand reach and align the cameras with a convergence angle, which should be calculated considering inter-camera-separation distance and distance to the working plane. Further research is needed to optimise the position of the cameras within the helmet and minimise the strain using the system.

6 Acknowledgements

The project is supported by German Ministry of Education and Research (BMBF) under grant number 01IRA07B.

References

1. Drascic, D. (1991). Skill Acquisition and Task Performance in Teleoperation using Monoscopic and Stereoscopic Video Remote Viewing. *Proceedings of the Human Factors Society 35th Annual Meeting*, 1367 – 1371.
2. Drascic, D., & Milgram, P. (1996). Perceptual Issues in Augmented Reality. *Proc. SPIE* (Vol 2653: Stereoscopic Displays and Virtual Reality Systems).
3. Ellis, S. R., & Menges, B. M. Localisation of Objects in the Near Visual Field. *Human Factors*, 40, no. 3, 415 – 431.
4. Kim, W., S., Tendick, F., Start, W., L. (1987). Visual Enhancements in Pick-and-Place Tasks: Human Operators Controlling a Simulated Cylindrical Manipulator. *Proc. IEE Journal of Robotics and Automation*, 5, 418-425.
5. Luczak, H., & Oehme, O. (2002). Visual Displays – Development of the Past, the Present and the Future. *Proc. WWDU 2002 – Work With Display Units – World Wide Work,* 2 – 5.
6. Luczak, H., Wiedenmaier, S. Oehme, O., Schlick, C. (2000). Augmented Reality in Design, Production and Service – requirements and Approach. *Human aspects of Advanced Manufacturing: Agility and Hybrid Automation – III, proc. of the HAAMAHA 2002*, 15 – 20.
7. Neumann, U., & Majoros, A. (1998). Cognitive, Performance, and System Issues for Augmented reality Applications in Manufacturing and Maintenance. *Proc. IEEE* (Virtual Reality Annual International Symposium), 4-11.
8. Oehme, O., Wiedenmaier, S., Schmidt L., Luczak, H. (2002). Comparison between the Strain Indicator HRV of a Head Based Virtual Retinal Display and LC-Head Mounted Displays for Augmented Reality. *Proc. WWDU 2002 - Work With Display Units- World Wide Work,* 387 - 389.
9. Pepper, R. L. & Hightower, J. D. (1984). Research issues in Teleoperator Systems. *Human Factors Society 28th Annual Meeting*, 803-807.
10. Pepper, R., L. (1986). Human Factors in Remote Vehicle Control. *Human Factors Society 30th Annual Meeting.*
11. Rastogi, A. (1996). Design of an Interface for Teleoperation in Unstructured Environments using Augmented reality displays. *MASc Thesis*, University of Toronto.
12. Stadler, A., & Wiedenmaier, S. (2002). Augmented Reality Applications for Effective Manufacturing and Service. *Proc. WWDU 2002 – Work With Display Units – World Wide Work,* 393 – 395.
13. Tiffin, J., Asher, E.J. (1948). The Purdue Pegboard: Norms and studies of reliability and validity. *Journal of Applied Psychology*, 32, 234-247, 1948.
14. Wiedenmaier, S., Oehme, O., Schmidt, L., Luczak, H. (2001). Augmented Reality (AR) for Assembly Processes – An Experimental Evaluation. *Proc. IEEE and ACM International Symposium on Augmented Reality (ISAR)*, 185-186

The Effect of Alternative Keyboards on Musculoskeletal Symptoms and Disorders

J. Steven Moore

School of Rural Public Health
3000 Briarcrest, Suite 300 1266 TAMU
Bryan, TX 77802
jsmoore@srph.tamushsc.edu

Naomi Swanson

NIOSH
4676 Columbia Parkway
Cincinnati, OH 45226
nws3@cdc.gov

Abstract

A two-year longitudinal study was performed to evaluate the effectiveness of two alternative keyboards on musculoskeletal symptoms and disorders. The fixed alternative keyboard was associated with improvement of baseline wrist and carpal tunnel syndrome symptoms. The adjustable alternative keyboard was associated with improvement of baseline elbow and wrist symptoms. The fixed alternative keyboard was also associated with a significant reduction in the incidence of neck symptoms and, among those completely asymptomatic at the start of the study, distal upper extremity symptoms (primarily hand/wrist).

1 Introduction

In 2000, NIOSH initiated a two-year longitudinal study of the effectiveness of alternative keyboards in relieving or preventing musculoskeletal symptoms or disorders. This preliminary analysis examines the impact of the alternative keyboards from two perspectives: (1) improvement of symptoms and disorders among subjects symptomatic at baseline and (2) the incidence of symptoms among those asymptomatic at baseline.

2 Methods

In 2000, 418 individuals from one insurance company with locations throughout the United States volunteered to participate in the study. At baseline, each participant was interviewed and examined by one occupational medicine physician experienced in the evaluation and treatment of work-related musculoskeletal disorders. Individuals were randomly assigned to one of three keyboard conditions – a traditional keyboard (control), a fixed alternative keyboard, in which the keyboard halves were separated and "tented" at fixed angles, and an adjustable split-design alternative keyboard, in which the keyboard halves could be separated and "tented" through a range of angles. Care was taken to ensure that job categories and job demands were evenly distributed across keyboard conditions. During the study, the volunteers completed symptom surveys every three months. Individuals reporting incident musculoskeletal symptoms were contacted by the physician and interviewed. In 2002 (24 months), volunteers were again interviewed and examined by the same physician. The physician was blinded to keyboard condition during the study.

The presence of one or more symptoms reported to the physician during the baseline interview was the case definition for *baseline symptom* cases. If a symptomatic person also had the presence of at least one relevant sign on physical examination, the subject was also identified as a *baseline disorder* case. Individuals who were asymptomatic at baseline, but reported the onset of new potentially work-related symptoms (i.e., symptoms that were not clearly linked with non-work activities) during the observation period were identified as *incident* cases.

Individuals with symptoms or disorders were initially grouped according to affected body part(s) – neck, shoulder, upper arm, elbow, forearm, wrist, hand, and back. Hand/wrist cases were also identified according to the presence of symptoms or disorders of specific conditions, including tendon entrapment of the first dorsal compartment, carpal tunnel syndrome, and the presence of a ganglion cyst.

At 24 months, baseline symptom cases were asked about the outcome of their symptoms. These outcomes were dichotomized as improved versus not improved.

The distributions of baseline cases, baseline cases reporting improvement, and incident cases between subjects using a control keyboard versus a fixed or adjustable alternative keyboard were evaluated using 2x2 tables (chi-square test for homogeneity with $\alpha < 0.05$).

3 Results

Of the 418 initial volunteers, 125 (30%) subjects dropped out of the study during the two-year observation period. The keyboard conditions did not differ in the number of drop-outs, and drop-outs were primarily due to subjects leaving the company or transferring to different jobs. Complete data are available for 289 subjects who remained in the study for the full two years. Four subjects had incomplete follow-up data and are not included in these analyses.

3.1 Prevalence of symptoms and disorders at baseline

The number and prevalence of subjects who were identified as symptom or disorder cases are summarized in Table 1.

Excluding a few instances, all baseline symptom and disorder cases were distributed homogeneously across keyboard treatments.

3.2 Effect of keyboard treatment on baseline symptoms and disorders

Neither the fixed nor the alternative keyboard treatment was associated with improvement of baseline symptoms or disorders affecting the neck, shoulder, upper arm, or back. The fixed alternative keyboard treatment was associated with improvement of wrist symptoms ($p < 0.01$) and carpal tunnel syndrome symptoms (Fisher Exact $p < 0.02$). The adjustable alternative keyboard was associated with improvement of elbow symptoms (Fisher Exact $p = 0.03$) and wrist symptoms (Fisher Exact $p < 0.01$).

3.3 Effect of keyboard treatment on the incidence of symptoms

Two analyses were performed. In the first one, only subjects who were completely asymptomatic at baseline were considered eligible to be incident cases (N = 81). In the second one, all individuals who were asymptomatic for a particular body part at baseline, even though they may have been symptomatic in another body part, were considered at risk.

Table 1. The number and prevalence of baseline cases by keyboard treatment

MS Morbidity	Controls (n = 97)		Fixed Alternative (n = 100)		Adjustable Alternative (n = 92)	
Any Symptom	71	73%	73	73%	64	69%
Any Disorder	24	25%	38	38%	33	36%
Neck Symptom	43	44%	47	47%	44	48%
Neck Disorder	14	14%	22	22%	19	20%
Shoulder Symptom	16	17%	14	14%	14	15%
Shoulder Disorder	4	4%	8	8%	4	4%
Upper Arm Symptom	7	7%	8	8%	6	7%
Upper Arm Disorder	0	0%	1	1%	1	1%
Back Symptom	10	10%	9	9%	6	6%
Back Disorder	1	1%	0	0%	2	2%
Any Baseline DUE Symptom	42	43%	44	44%	44	47%
Any Baseline DUE Disorder	12	12%	18	18%	20	22%
Elbow Symptom	12	12%	12	12%	11	12%
Elbow Disorder	4	4%	4	4%	6	6%
Forearm Symptom	13	13%	20	20%	15	16%
Forearm Disorder	5	5%	10	10%	9	10%
Wrist Symptom	27	28%	29	29%	32	34%
Wrist Disorder	7	7%	11	11%	9	10%
de Quervain's Symptom	3	3%	2	2%	2	1%
de Quervain's Disorder	2	2%	1	1%	2	2%
CTS Symptom	8	8%	6	6%	6	6%
CTS Disorder	3	3%	5	5%	5	5%
Hand Symptom	19	20%	18	18%	19	20%
Hand Disorder	3	3%	1	1%	4	4%

3.3.1 Incident cases among completely asymptomatic subjects

The number and cumulative incidence of cases in this population are summarized in Table 2.

Table 2. The number and cumulative incidence of incident cases among subjects asymptomatic for all body parts at baseline by keyboard treatment.

MS Morbidity	Controls		Fixed Alternative		Adjustable Alternative	
Any Symptom	12	63%	8	35%	14	56%
Neck Symptom	**7**	**33%**	**1**	**4%**	4	15%
Shoulder Symptom	2	9%	0	0%	3	12%
Upper Arm Symptom	2	8%	0	0%	3	11%
Back Symptom	6	26%	3	13%	2	7%
Any DUE Symptom	7	28%	**4**	**15%**	**12**	**43%**
Elbow Symptom	2	8%	0	0%	2	7%
Forearm Symptom	2	8%	0	0%	3	11%
Hand/Wrist Symptom	7	33%	4	15%	11	42%

The fixed alternative keyboard was associated with a reduced number of incident cases of neck symptoms. The odds ratio for incident neck symptoms comparing control keyboards to the fixed alternative keyboard was 0.09 (95% CI: 0.01 – 0.82; Fisher exact $p = 0.02$). The fixed alternative keyboard was not associated with an increased or decreased number of potentially work-related incident cases for the back, shoulders, upper arms, elbows, or hands/wrists.

The adjustable alternative keyboard was not associated with an increased or decreased number of potentially work-related incident cases for the neck, back, shoulders, upper arms, elbows, or hands/wrists.

When compared to the adjustable alternative keyboard, the fixed alternative keyboard had significantly lower number of incident cases affecting the distal upper extremity and a trend toward a reduced number of hand/wrist cases. The odds ratio was 4.3 (95% CI: 1.2 – 15.8; Fisher Exact $p = 0.04$). For the hand/wrist, the odds ratio was 4.0 (95% CI: 1.1 – 15.1; Fisher Exact $p = 0.06$).

3.3.2 Incident cases among cases asymptomatic for specific body parts

The number and cumulative incidence of cases in this population are summarized in Table 3.

Table 3. The number and cumulative incidence of incident cases among subjects asymptomatic for specific body parts at baseline by keyboard treatment

MS Morbidity	Controls		Fixed Alternative		Adjustable Alternative	
Any Symptom	45	52%	35	41%	41	49%
Neck Symptom	**14**	**16%**	**4**	**5%**	7	8%
Shoulder Symptom	4	4%	4	5%	9	10%
Upper Arm Symptom	5	5%	3	3%	4	4%
Back Symptom	11	12%	9	10%	5	6%
Any DUE Symptom	28	29%	21	22%	29	33%
Elbow Symptom	4	4%	7	7%	5	6%
Forearm Symptom	6	6%	6	6%	7	8%
Hand/Wrist Symptom	27	31%	18	20%	27	32%

The fixed alternative keyboard was associated only with a reduced number of incident cases of neck symptoms. The odds ratio for incident neck symptoms comparing control keyboards to the fixed alternative keyboard was 0.36 (95% CI: 0.16 – 0.83; Fisher exact $p = 0.02$). The adjustable alternative keyboard was not associated with an increased or decreased number of potentially work-related incident cases for any body part.

4 Conclusions

Preliminary analyses from this study suggest that the alternative keyboards used in this study may have beneficial, but limited, effects on musculoskeletal disorders among office workers who use computers. In terms of primary prevention, only the fixed alternative keyboard demonstrated a significant effect on the incidence of musculoskeletal symptoms. This effect was definite for the neck and suggestive for the distal upper extremity in general. In terms of an intervention to reduce symptoms, both alternative keyboards were associated with improvement of wrist symptoms. In addition, the fixed adjustable keyboard was also associated with a significant improvement of carpal tunnel syndrome symptoms while the adjustable alternative keyboard was also associated with an improvement of elbow symptoms.

Human Characteristics of Pointing an Object on Small Screen of Personal Digital Assistant by Pen Based Interface

Kazunari Morimoto, Takao Kurokawa and Atsuo Mukae

Graduate School of Science and Technology, Kyoto Institute of Technology
Matsugasaki, Sakyo-ku, Kyoto 606-8585 JAPAN
morix@ipc.kit.a.c.jp, Kurotak@ipc.kit.ac.jp

Abstract

This paper focuses on human characteristics of pointing with a pen in PDA (Personal Digital Assistant) in which the human pointing performance is influenced by the size and position of buttons displayed on small screen. We carried out experiments on pointing in which subjects pointed the square buttons displayed on PDA held on the left hand and a pen held in the right one. The length of the side of square buttons is 1, 3, 5, 7 and 9 mm. One button is represented at one of the 23 positions on the screen for a trial. Twenty subjects who were novice in use of PDA participated in the experiments take a total of 1,150 trials for both sitting and standing posture. The results showed that the object with 9 mm sides was superior in pointing time, error rate and subjective evaluations to the other buttons' size. Pointing performances was not different between sitting and standing posture. Some other human characteristics was revealed that pointing time when target buttons were in the right or left marginal areas was longer of 10% than when they were in the center of the screen.

1 Introduction

PDA with pen interface is widely used in the area of mobile computing. The important characteristics of PDA are that it can be handled with suitable for walking and standing, because it is small and light. However, the pen interface is not considered enough about pointing a small object on small screen. For instance, it doesn't clear whether the size and the shape of a button represented on the small screen are suitable for the input by pen. Some papers report about user interfaces of computer with a tablet of pen input device. Xiangshi (1997) researched on the size of the target and the pen input performances. MacKenzie (1999) examined the software keyboard on screen. On the other hand Shumin Zhai (2000) discussed design techniques for virtual keyboard, and Mukae (2000) evaluated the software keyboard on small screen of PDA. However pointing characteristics of pen interfaces was not discussed thoroughly on the pointing characteristics of novice users. Moreover the postures of both sitting and standing will have board implications for pointing characteristics of PDA held by hand. The aim of this paper is to demonstrate human characteristics of pointing with a pen in PDA in which the human pointing performance is influenced by the size and position of buttons displayed on small screen.

2 Methods

2.1 Size and Location of Objects on Screen

The size of square objects represented on a screen of PDA was determined on investigation of the size of the kana keyboard used for PDA appeared on the market. The side length of the square objects used is 1, 3, 5, 7 and 9 mm. Subjects pointed one object represented on the 23 locations showed in the figure 1. The center of square objects in each size was made the same position. The objects represented on screen were pointed quickly by subjects as possible. The other object displayed on the other position after 1 or 2 seconds from when the subjects pointed an object. One object was represented at one of the 23 locations on the screen for a trial, and each subject made a total of 1,150 trials under the condition of sitting and standing posture. Pointing time and pointing position of a pen were recorded automatically. After the pointing all objects, subjective rating evaluations on usability were carried out about preference of button size, correctness and speed of pointing and ease of use. Pointing movements on the PDA screen with pen were recorded on video to analyze the pointing characteristic.

2.2 Screen and Body Size of PDA

The size of the PDA used in the experiments was the vertical length of 136 mm and the width of 80 mm, which has colour screen of liquid crystal device of TFT. The screen size for displaying the objects was the vertical length of 80 mm and the width of 60 mm that was constructed by many dots of the size of 0.25 mm.

2.3 Participants

Twenty subjects (16 males and 4 females) who were novice in use of PDA participated in the experiments. All of them were right handed. In sitting condition they sat down straight toward the desk of 70 cm height. A distance from the edge of the desk to the PDA screen was about 78 mm. In standing condition they stood up at the fixed position and indicated to hand the PDA with most easy to operate.

3 Results

3.1 Pointing Time

Pointing time was defined that it was the operation time for pointing 115 objects. Figure 2 showed mean pointing time and standard deviation in both conditions of sitting and standing posture. Size 1 to size 5 in this figure indicates the side length of 1 mm to 9 mm respectively. The pointing time of Size 1 was the longest. This figure showed that the bigger the size of objects, the shorter the pointing time. This tendency was not changed in conditions of sitting and standing posture. The statistical difference among mean pointing time of objects was significant.

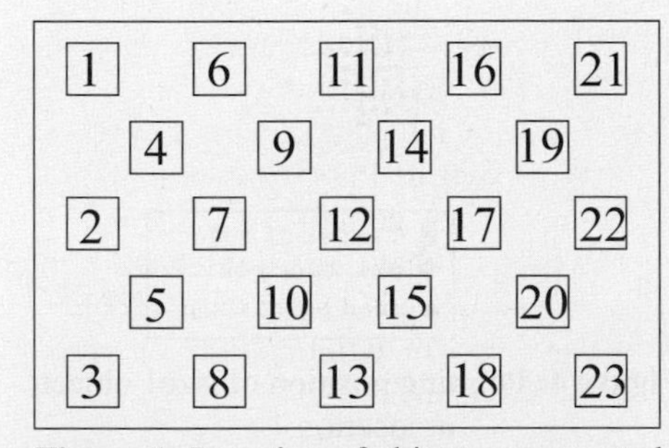

Figure 1: Location of objects represented on screen

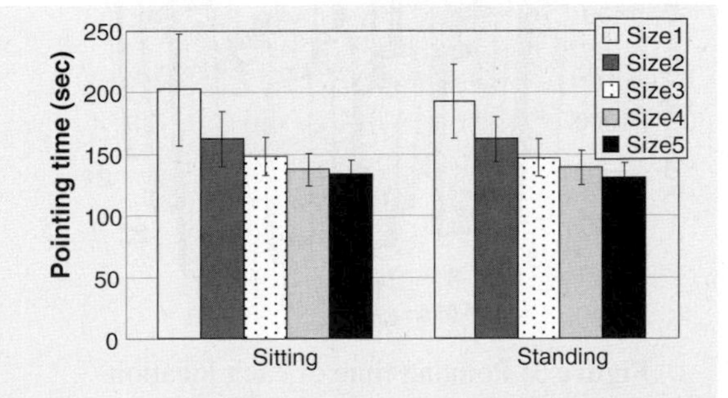

Figure 2: Mean pointing time

pointed 5 each location was shown in figure 3. Pointing time was shorter in the center of screen than in both sides of screen. Subjects pointed 5 times on each object from Size 1 by Size 5 at one position in each condition of sitting and standing posture. Therefore 1000 data were obtained in each location by 5 objects.

To examine the pointing characteristic, pointing positions and the center of objects were demonstrated for each location of object, then calculated mean pointing position of vertical and horizontal axes and their standard deviations. For instance, figure 4 shows the distribution of the pointing position in case of the smallest target that is size 1 object. The deviation was very small, but there are many pointing errors. On the other hand figure 5 shows pointing positions in case of the size 5 target. The deviation of pointing was very big, but almost positions were in the area of the object. χ square tests were done about all the x and y values of pointing position on screen. Pointing position tended to vary in the depth direction from the center of the square rectangle ($p < 0.01$). Subjects tend to point the right side from the centre of the objects in the left side toward the screen. However the objects on the right side of the screen were pointed to the left from the centre of then objects ($p<0.01$).

3.2 Error Rate

Pointing error was defined how many times the subjects pointed outside the objects. Error rate was calculated the times of error by 115 trials. Mean error rates of all subjects and their standard deviations were shown in the figure 6. Error rates of the object of size 1 were the highest in both conditions of sitting and standing posture. On the other hand error rates of the other objects were too small.

3.3 Subjective Rating

Four usability items on object size, the accuracy in pointing, quickness and easiness to point were evaluated by 5 point scale method, and mean rating values and standard deviations about the subjective evaluation values of these items were asked in all subjects. The evaluation value of an object of size 1 was the lowest under the condition of sitting. However the evaluation values of size 1 and size 2 were significantly lower under the condition of standing posture. The evaluation value of accuracy was high for the size 5 in sitting condition and for size 3, size 4 and size 5

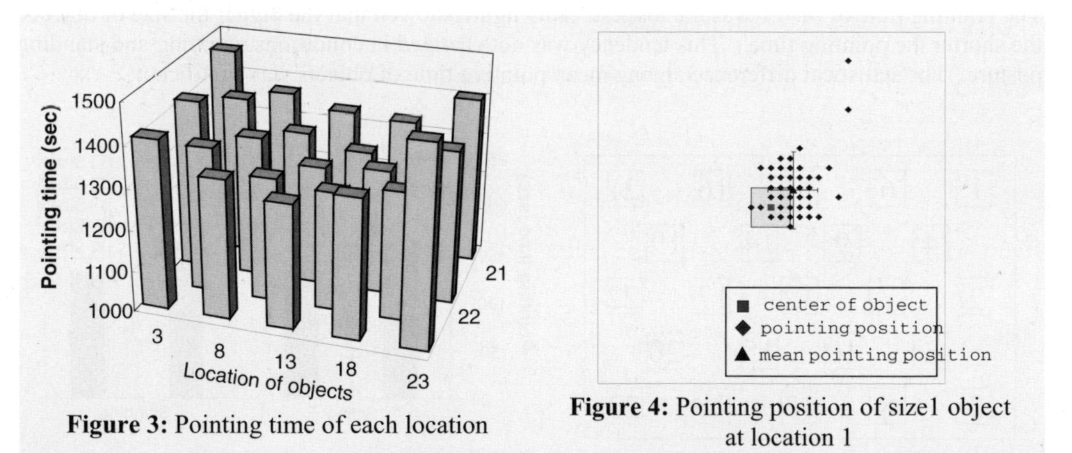

Figure 3: Pointing time of each location

Figure 4: Pointing position of size1 object at location 1

objects in standing condition. The evaluation value about quickness and easiness of pointing was high against the objects of size 3, size 4 and size 5 in both conditions.

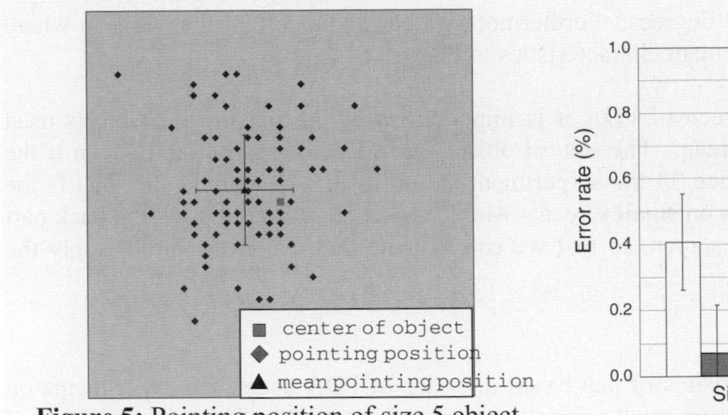

Figure 5: Pointing position of size 5 object at location 23

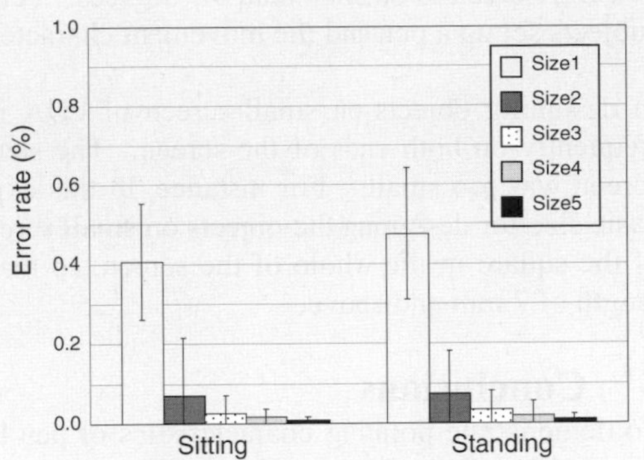

Figure 6: Mean error rate

4 Discussions

4.1 Pointing Time and Error Rates

Pointing time of the objects of size 4 and size 5 was shorter than the other objects in two conditions, therefore it can be said that users can quickly point of the square of above 7 mm in length. Moreover error rates of the objects of size 4 and size 5 were low even in standing posture. We can point out that the performance in pointing time and error rates was high under the condition of the objects used in the experiments. Then we conclude that the minimum object size in pen interface was 7 mm square.

4.2 Pointing Time and Location of Objects

We pointed out that pointing time was to long to point the objects on the both sides of the screen and to short in the central part of the screen by analyzing the pointing time in each location. Pointing time was especially long at the location 1, 2, 3, 4, 5, 6, 21, 22 and 23 on screen in figure 1.

The high correlation with the pointing time and subjective evaluation values was revealed. We examine about this reason. The opinions from 6 to 10 subjects showed that the left end of the screen was far from the position to be holding a pen by the right-handed. There were many subjects who pointed out that it was very hard to point the object at location 23 on the right end of the screen. Ten to sixteen persons beyond the half appealed strongly that it was hard to seek the object at location 23 behind them right hand.

The movement of the hand in pointing was analyzed by using the video recorded the pointing scene. Subjects moved a pen to the left quickly and point an object located on the left end of the screen. Therefore, temporal movement was occurred just before pointing and the pointing speed became slow. On the other hand, many subjects briefly could not notice the object represented at location 23 that were behind the right hand.

4.3 Pointing Position and Location of Objects

Subjects tend to point an upside from the center of the square in all the locations. They point the right part of the center of the square in the left side of the screen. On the other hand, when objects were represented on the right side of the screen, subjects tend to point the left part of the center of the square. We can point out two factors why these tendencies were obtained in these experiments on human pointing characteristic. First factor is that the angle from subjects' eyeball to the corner of PDA screen is smaller than 90 degrees. Ferthermore we can indicate that the position where subjects set up a pen and the movement characteristics of the wrist.

In designing objects on small screen of PDA it is important to avoid locating the objects used frequently on both ends of the screen. The size of objects has to be big as possible, even if the screen was too small. For instance, in this experiment the object of size 4 indicated that is the basic size for designing the objects on small screen. Moreover subjects tend to point the back part of the square in the whole of the screen, so that we can indicate that it is desirable to apply the length of 7 mm and above.

5 Conclusions

To demonstrate pointing characteristics of pen based interface of PDA users, the experiments on size and location of square objects on small screen have carried out by 20. The minimum size of objects for using PDA was induced from pointing time, error rates and subjective evaluations. The side of square object for PDA was at least 7 mm. We obtained some pointing characteristics as follow.

(1) Variation of pointing position expanded as the object size grew larger.
(2) Depth direction of the objects was pointed them represented in all locations.
(3) Pointing time of the objects at the right and left side of screen was longer than that at center of the screen.

Besides the screen was very small, these results exhibited that comfortable use of PDA was restricted to the area of object location. To take high performances using PDA many subjects said that something attachment for supporting a right hand with pen. Therefore it is necessary not only the improvement of the software such as shape and location of objects represented on PDA, but also the development the hardware from viewpoint on usability.

6 References

MacKenzie & Shawn X Zhang (1999) The Design and Evaluation of a High-Performance Soft Keyboard , *CHI*, 25-31.

Mukae A., Yamamoto M., Morimoto K. & Kurokawa T. (2000) An evaluation on Kana Keyboard of Personal Digital Assistant. *Proc. of Kansai branch of JEA*. 28-33.(in Jpn)

Xiangshi Ren & Shinji Moriya (1997). The Best among Six Strategies for Selecting a Minute Maximum Size of the Minute Maximum Size of the Targets on a Pen - Based Computer. *Human-Computer Interaction INTERACT*, 85-92.

Shumin Zhai, Micheael Hunter & Barton A Smith (2000) Metropolis Keyboard - An Exploration of Quantitive Techniques for Virtual Keyboard Design, *Proc. of ACM Symposium on User Interface Software and Technology (UIST 2000)*, 119-128.

Changing Requirements of Laboratory Design

Ina Maria Müller and Christoph Heinekamp

dr. heinekamp Labor- und Institutsplanung
Gaußstr. 12, D-85757 Karlsfeld, Germany
mueller@heinekamp.com heinekamp@heinekamp.com

Abstract

The requirements placed on laboratories have undergone considerable changes over the past years. 50 years ago laboratory workbenches were made from bricks and stone. At this time workbenches and fume cupboards were installed in parallel rows. Nowadays there have been no radical changes in laboratory floorplans but in the configuration of workbenches and fume cupboards.
For reasons of economy laboratory space has to be planned and used efficiently and operation has to be as cost-effective as possible. When designing a new building the floorplan should be based on the functional lab size.
Furthermore, laboratory planning requests a high degree of flexibility both in configuration and technical installation. In the past central supplies provides the laboratories with media and electrical installations. Modern laboratory buildings offer the possibility of supplying different special media according to needs or in different areas by providing them only at places of demand.

1 Lab Design

Looking at the floorplan of laboratory buildings we most often find a collection of small, individual labs with available support space mostly in the 'dark area' - without daylight exposure - of the building. At times we find in industrial research laboratories open-plan laboratories which allow the scientist to share essential equipment.
Transferring the results of the controversial discussion for office buildings about pros and cons of open-plan *vs.* individual offices leads to a laboratory landscape that combines the advantages of both models: defined working areas and centrally positioned infrastructural facilities (see fig. 1). For example you get additional storage space by including the corridor into the lab (see fig. 1 c). Supporting space like rooms for refrigerators and cryogenic freezers are within reach. If required paperwork stations can be located near the window and separated by partition walls.

Equipment for new tasks such as high througput automated biological tools has become a primary driver in the layout of laboratories (U. Lindner, 2001). Especially in biological-medical research there is a need to change lab size to provide flexible usable floor space. Anyhow there is still a need for separated rooms which provide a better cleanliness (e.g. tissue culture rooms) or in which instrumentation like PCR-machines or gas chromatographs or other commonly used instruments are situated.
Separate rooms or closed work cubicles are provided for quiet workstations or noisy machines. The creation of areas/rooms for equipment needs detailed planning concerning required media, power supply and access to LAN (Local Area Network) and notice of malfunction. Occurring influences on room conditions caused by the equipment i.e. thermal loads have to be considered.

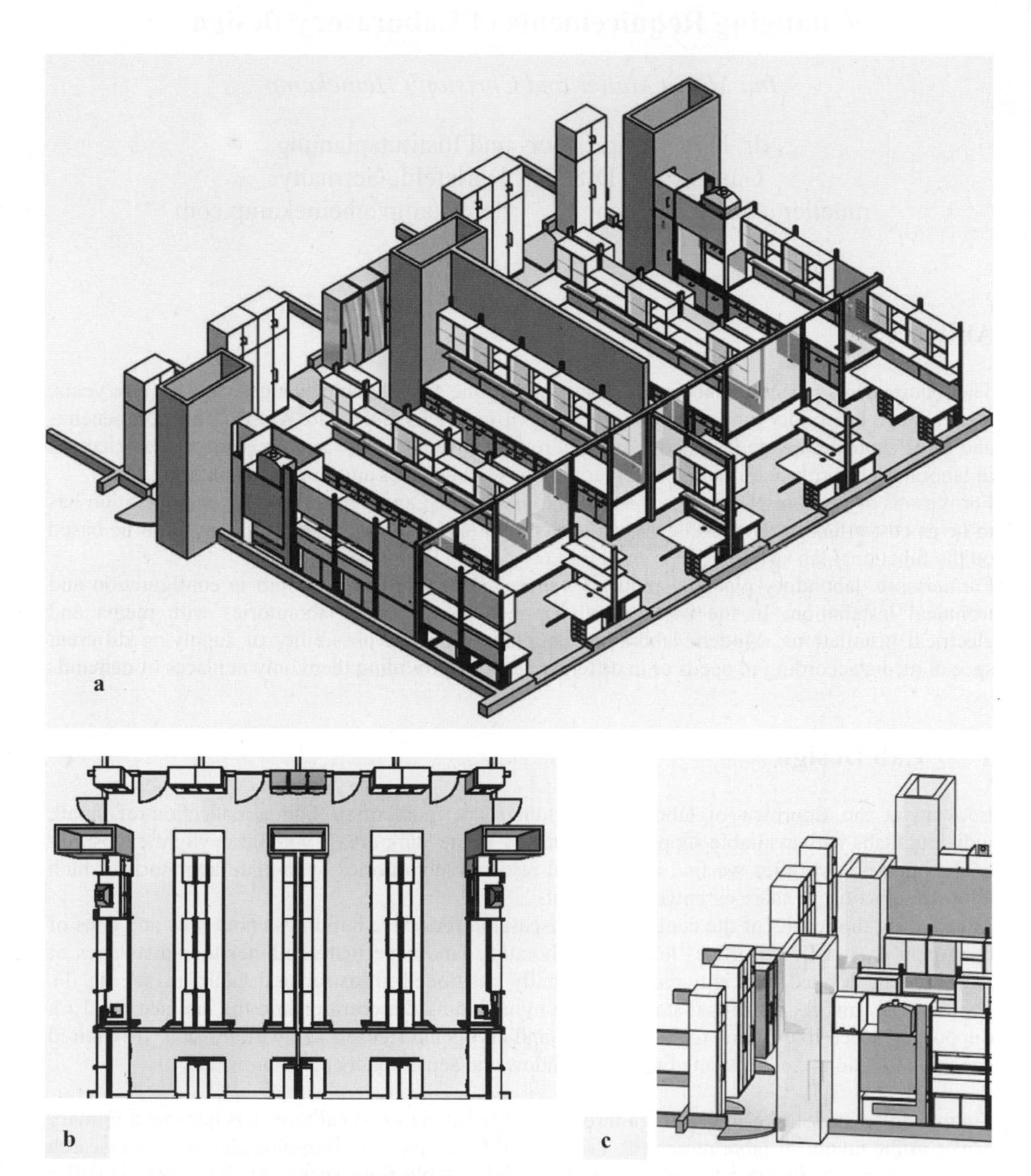

Figure 1: 3d-Floorplan of a laboratory landscape
a) isometry view, b) plan view, c) plan detail: corridor

The flexibility of the lab design is supported by modern laboratory furniture. In the past most of the furniture was bottom fixed, so reconfiguration of the casework was difficult. Another disadvantage of a fixed system is maintenance of the installation: service piping and conduits are not accessible. Because of fixed kneeholes seating in front of bottom mounted casework is only possible at previously fixed positions.

Today laboratory furniture is available which is completely modular: each module – even a fume hood - can be replaced. This system features maximum flexibility but requires a cost intensive installation of power supply and media. Hence this system did not succeed at the market.

Currently systems are popular which are designed as modular systems with integrated media and electrical installations. To receive flexibility of laboratory design it is necessary to avoid installations requiring fixed ventilation to the outside or drainage in the middle of the room. For example, the classic standing bench with a sink at the end should be situated along walls, as far as possible (see fig. 1). Substructures on wheels can easily manhandled by the lab user. These systems allow the adaptation to new tasks which occur – particularly in research and developmental laboratories – in shorter and shorter periods of time.

Figure 2: Ceiling fixed media supply
a) lab view with instruments, b) height-adjustable media supply

The centre of the laboratory should be reserved for integration of large devices as autoanalysers or elaborate analytical technology. Devices resting on the floor need free access and, at best, they are supplied from the ceiling. The height of the supply installations is determined by the devices below them (see fig. 2 a). New developments make it possible to install a height-adjustable media supply, which allows an easy handling (see fig. 2 b).

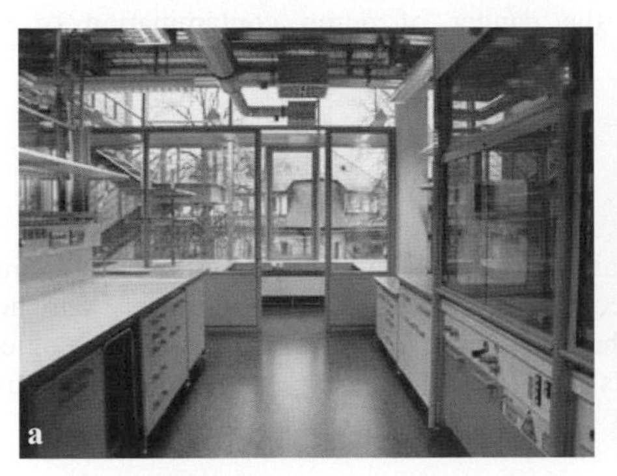

Figure 3: Seated workstation a) view from the lab, b) cabin place

Modern laboratories require, in addition to standing and seated workbenches, seated workstations to set up devices and computers. Office workstations in or near the laboratory are becoming more and more important, since data acquisition and evaluation are now as important as the preparative or analytical work in the laboratory. Therefore it is necessary to create PC-workstations which allow glare-free working. By separating these places with a partition wall (see fig. 1 and fig. 3) you can create a low-noise working area. If required the partition wall can be easily completed with a glass door.

2 Technical installation

Buildings have to be structured systematically regarding building services engineering in order to make laboratories more flexible. Technical devices have to be clearly structured to guarantee subsequent media installations: a variable technical infrastructure is the basis for adaptation processes. Plumbing should be supplied at defined area in the lab to minimize bench marks. Media which are commonly used should be provided by centralised supply.

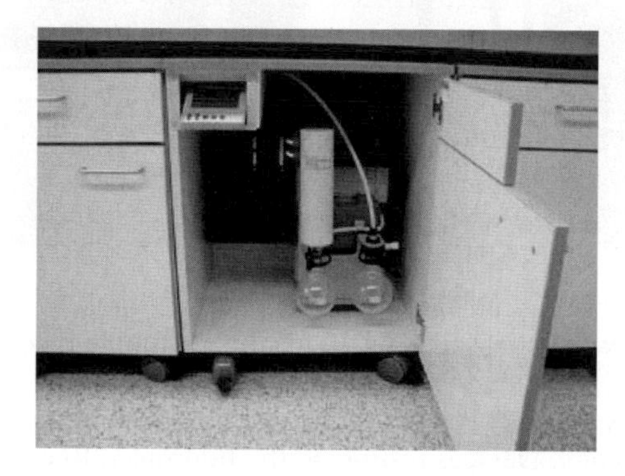

Figure 4: Local vacuum supply

Local media supplies at places of demand ensure cost-effective construction and economical use. For example a local vacuum supply with vacuum pumps placed under the desks provides the user with a high quality vacuum and minimizes the danger of pump contamination or cross contamination between two users (see fig. 4).

3 Safety

Today's safety concepts have to comply with legal requirements and relevant guidelines. Planning of safety equipment i.e. fume hoods achieve compliance with occupational safety and health standards. Harmful substances are exhausted through employment of modern technologies as individual offtakes and ventilated safety cabinets and occupational exposure limits were strictly observed.

Figure 5: Laboratory building with escape balcony a) balcony, b) lab view

When designing the floorplan it is also important that every area of the laboratory has direct access to emergency exits. So open-plan laboratories must have access to an exit corridor rsp. to an emergency staircase.
Ideally the laboratory building is equipped with an escape balcony on which you are able to leave the lab quickly (see fig. 5). A convenient side-effect of this architecture is an enhancement of flexibility in the building.

Up-to-date safety standards make it possible to realize new ventilation concepts: structure and operation of the ventilation installation must be based on the requirements of the fume cupboards and process exhaust systems as well as to continuous ventilation of fire resistant safety cabinets. A workstation-specific ventilation enables the laboratory to be run cost-effectively in line with its needs.

References

Lindner, U. M. (2001). Laboratories in the post-genomic area. Retrieved February 12, 2003, from http://www.ewalab.com/pubs.htm

What is the Most Beneficial Type of Recreation for Computer Operators?

Iiji Ogawa and Takumi Sakamoto

Teikyo University of Science & Technology
2525 Yatsusawa, Uenohara, Kitatsuru, Yamanashi-Ken, Japan 409-0193
iiji@ntu.ac.jp

Abstract

This study was undertaken to assess differences on the performance of a simple data input and a creative task using PC when break periods with free time, a friendly companion animal, and a robotic pet were given. Before/after each performance, subjects' EEGs were measured to investigate their emotional states. Also, during each performance, their heart rates (R-R intervals) were measured to monitor their levels of concentration. Twelve (12) subjects were selected for this study. The results indicated that the break period with the robotic pet AIBO was beneficial for the data input task; on the other hand, in terms of performance for the creative task, the break period with the companion animal poodle or free time was better.

1 Introduction

Many different kinds of recreational activities such as Animal Assisted Activity and Robot Assisted Activity are used in nursing home facilities for therapeutic purposes. Those activities may be beneficial in terms of health and performance for company employees who are engaged in word processing or creative tasks on the job. In fact, a few companies allow employees to bring their pets into the workplace in Japan. Sakamoto and Ogawa (2001) conducted an experiment where balloon volleyball was used as the group recreational activity to find out whether having an individual or a group activity during a break period increased a word processing task performance. They pointed out the importance of the recreational activity and found that a group activity proved more beneficial on a word processing task performance. However, they also noted that individual recreational activities might be more suitable under real working conditions. The objective of this study was to assess differences on the performance of simple repetitive and creative tasks when break periods with free time, a friendly companion animal, and a robotic pet were given.

2 Methods

2.1 Experimental Set-Up

The experiments were carried out in a shield room (W3.5m x D2.6m x H2.4m). The following experimental units were used in the experiments.

2.1.1 PC unit

An IBM APTIVA with a 22-inch monitor, 733MHz (CPU), 194MB (memory), and Windows Me (OS) were used to run the performance tasks.

2.1.2 Electroencephalograms (EEGs)

The Emotion Spectrum Analyzer was used. Using the international 10/20 method, EEGs of 14 mono-polar signals were measured with the ear lobe as a reference. A method suggested by Musha (1996) was applied to quantify four of the emotional states (anger/stress, sadness, joy, and relaxation), based on the cross-correlations between two of the mono-polar signals in terms of θ waves (5 to 8 Hz), α waves (8-13 Hz), and β waves (13 to 20 Hz). In this study, the level of anger/stress was only analysed as an index of subjects' feelings toward the task.

2.1.3 Heart Rate (HR) measuring unit

The 64K byte heart rate memory unit was used to measure the heart rates (R-R intervals). Based on the data collected, each subject's Heart Rate Variability (HRV) was calculated to find out the level of concentration during each task. Many scoring methods of HRV have been suggested in the literature. In this study, the standard deviation of heart rates (R-R intervals) was applied as HRV. When HRV is low, the concentration level can be said to be high (Ogawa & Nishikawa, 1997).

2.2 Performance Tasks

There were two kinds of performance tasks; one was a simple repetitive data input task, i.e., each subject was required to feed many sets of 10 digits from an original manuscript as fast and as accurately as possible, and the other was a creative task which involved designing and building a house under some constraints, using software designed for such purposes. Both of the experimental tasks lasted 20 minutes and were repeated twice. The performance measures of each task were as follows: (1) the total number of 10-digit inputs which the subjects fed correctly into the PC for the data input task, and (2) the completion rate of the house for the creative task as evaluated by an experimenter.

2.3 Subjects

Twelve (12) male university students, between 18 and 24 years of age, participated in the experiments. None of them disliked companion animals.

2.4 Companion Animal

A one-year-old male toy poodle participated in the experiments as the friendly companion animal. The poodle was in good health, well trained, and slightly bigger than the robotic pet used.

2.5 Robotic Pet

The SONY entertainment robot AIBO (ERS-210) was used as the robotic pet. It was programmed in advance in order to be able to play ball with the subjects.

2.6 Experimental Procedure

Before starting the experiments, each subject had a practice period for both performance tasks to minimize the learning effect. The two performance tasks were carried out on the same day for the convenience of the pet owner; however, sufficient rest periods of more than 30 minutes were given between the two tasks. The experimental procedure is shown in Figure 1. For each performance task, a one-minute rest period was set before and after the first and second trials, thus in total, there were four (4) one-minute rest periods in order to measure subjects' EEG at rest. A 5-minute break period with free time, the companion animal, or the robotic pet was placed between the second and third one-minute rest periods. During the 5-minute break period with free time, each subject could do anything except eat, sleep, and engage in exercise. During each performance task, subjects' heart rates (R-R intervals) were measured.

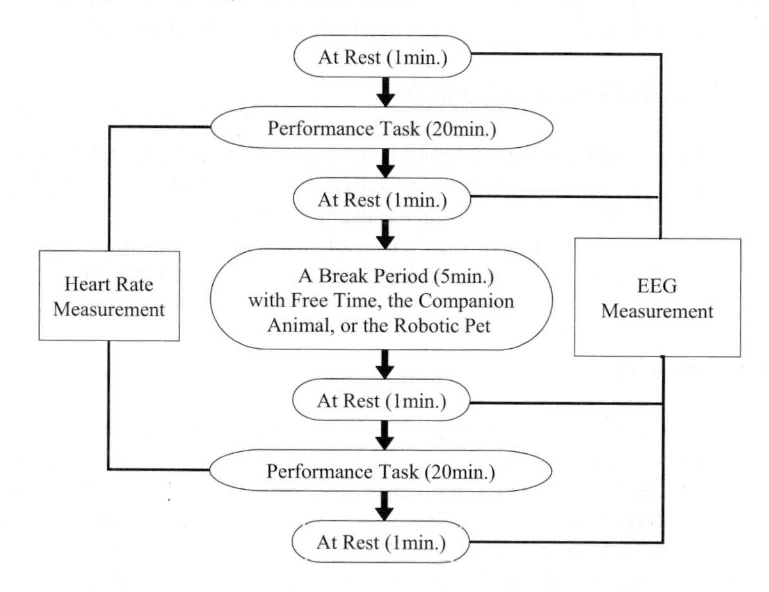

Figure 1: Experimental Procedure for each Performance Task

3 Results and Discussions

3.1 Performance

The Analysis of Variance (ANOVA) for each task was performed separately. The results of ANOVA indicated that individual differences were highly significant ($p < 0.01$) for the data input task and significant ($p < 0.05$) for the creative task. The results of the data input task showed that the break periods were a significant factor on performance ($p < 0.05$). The graphical presentation of the mean of the number of data inputs before/after the break period is shown in Figure 2. The break period with the robotic pet AIBO was found to be the best and statistically different from the other conditions for the data input task performance ($p < 0.05$). As for the creative task, the break periods and the differences between before and after the break period were found to be quite significant factors on performance ($p < 0.01$). Performance was enhanced after the break period with the companion animal poodle and was found to be statistically different from the period with the robotic pet AIBO ($p < 0.05$) as shown in Figure 3.

3.2 EEGs

The EEG data collected at rest were analysed. The results of ANOVA for EEG showed that the break periods were significant ($p < 0.05$) for the data input task and highly significant ($p < 0.01$) for the creative task. For the data input task, subjects exhibited the highest level of anger/stress when they had the break period with free time; on the other hand, their anger/stress levels were the lowest when the break period included the poodle. Both conditions were found to be statistically different ($p < 0.05$). For the creative task, the highest anger/stress level was exhibited when they had the break period with the robotic pet AIBO; on the other hand, the lowest anger/stress was exhibited when the break period included the companion animal poodle. It was also found to be statistically different from the other two break periods ($p < 0.05$).

3.3 HRV

As a result of ANOVA for HRV on both performance tasks, the break periods were found to be a significant factor on HRV ($p < 0.05$). HRV was the highest when the break period included the robotic pet AIBO; however, when the companion animal poodle was in the break period, HRV was the lowest. Both break periods were found to be statistically different ($p < 0.05$). Also, HRV after the break period was higher than before the break and significantly different from each other ($p < 0.05$).

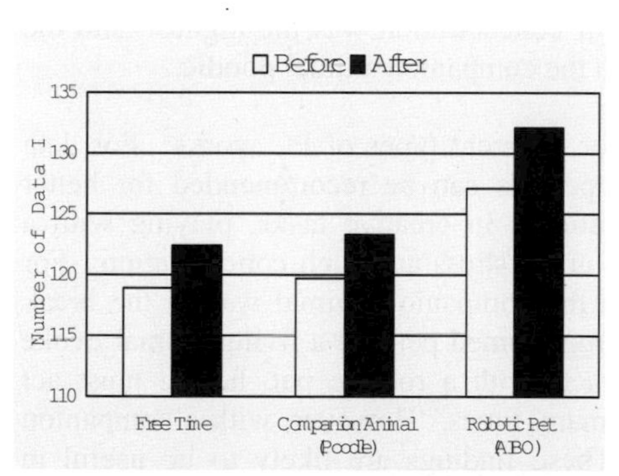

Figure 2: The Mean of the Number of Data Inputs before/after each Break Period with Free Time, Poodle, or AIBO for the Data Input Task

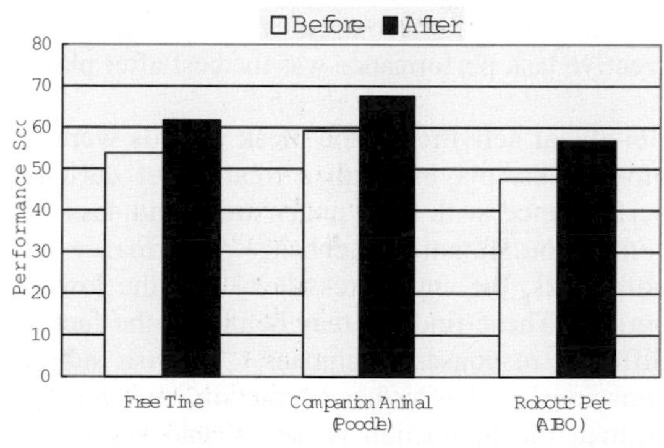

Figure 3: The Mean of the Performance Scores before/after each Break Period with Free Time, Poodle, or AIBO for the Creative Task

4 Conclusions

The following conclusions can be made within the limitation of the experiments.

4.1 For Data Input Task

Subjects demonstrated better performance when they interacted with the robotic pet during the break period. They also demonstrated the highest level of anger/stress when they had the break period with free time and the lowest levels of anger/stress when the break period included the companion animal poodle.

4.2 For Creative Task

Subjects exhibited better performance when they played with the companion animal during the break period. When the robotic pet AIBO was in the break period, their anger/stress levels were the highest for the creative task performance. However, subjects demonstrated the lowest anger/stress level when the companion animal poodle was in the break period. The level of subjects' concentration for creative task before the break period was higher than after the break and significantly different from each other ($p < 0.05$).

4.3 For both Performance Tasks

In spite of the fact that overall performance improved after each break period, the level of concentration was found to be higher before any break period. The level of subjects' concentration for both performance tasks was found to be the highest when the subjects played with the companion animal poodle in the break periods; on the other hand, when playing with the robotic pet AIBO, the level of concentration was the lowest for both tasks. Comparing the task performance with the anger/stress and concentration levels, the level of anger/stress was found to be higher, the level of concentration was the lowest, and the data input task performance was the best after playing with the robotic pet AIBO. Also, for the creative task, the opposite holds true, i.e., the level of anger/stress was the lowest, the level of concentration was the highest, and the creative task performance was the best after playing with the companion animal poodle.

Beneficial activities in the break periods were found for different types of PC works. For data input tasks, playing with a robotic pet during break periods can be recommended for better performance with high anger/stress and less concentration. In creative tasks, playing with a companion animal may enhance performance with less anger/stress and high concentration. For both tasks, the anger/stress level was the lowest when the companion animal was in the break period. These findings may be due to the fact that a programmed pet and a living animal evoke different responses in humans. When a subject interacts with a robotic pet, he/she must act consciously, plan his/her interaction, and think it over many times. However, with a companion animal, the interaction is natural and spontaneous. These findings are likely to be useful in arranging break periods for computer operators. The results reported here are the first in a series of studies planned. One further study aims at investigating the same questions for female and middle-aged operators.

References

Musha, T. (1996). Measure "kokoro." *Nikkei Science Scientific American (Japanese edition)*, 26 (4), 20-29.

Ogawa, I., & Nishikawa, K. (1997). A study of motorcycle riding fatigue. *Proceedings of the 13th Triennial Congress of the International Ergonomics Association*, 6, 400-402.

Sakamoto, T., & Ogawa, I. (2001). The importance of individual/group recreation while working on a word processing task. *the Japanese Journal of Ergonomics*, 37 (supplement), 386-387.

Computer Input with Gesture Recognition: Comfort and Pain Ratings of Hand Postures

David Rempel, Emily Hertzer, Richard Brewer

University of California, San Francisco
1301 South 46th Street, Building 163
Richmond, CA 94804, USA
rempel@itsa.ucsf.edu

Abstract

The design and selection of hand gestures for computer input should consider the discomfort associated with specific gestures when they are performed rapidly and repeatedly. Sign language interpreters perform rapid and prolonged gesturing and experience hand and arm pain with this activity. A questionnaire was administered to 24 professional sign language interpreters to assess pain or discomfort ratings for 60 different hand gestures. Pain ratings differed significantly between specific gestures. The findings of this study can be used to rank order gestures on a scale from most comfortable to most painful. This type of information should be considered in the design of a gesture recognition language.

1 Introduction

Gesture recognition is emerging as a potential method of providing input to computers. The design of hand gestures for computer input primarily follows principles of natural language combined with optimising image recognition. Upper extremity pain and musculoskeletal disorders are well known to be associated with the use of computer input devices (e.g., keyboards and mice), especially with long hours of use and work in awkward postures (Gerr et al. 2001; Marcus et al. 2001). However, little is known about the association of repeated hand gestures to comfort and pain. If specific hand gestures are associated with upper extremity pain, then designers of gesture languages for computer input should minimize the use of those gestures.

Sign language interpreters have extensive experience with hand gestures and also have elevated rates of musculoskeletal disorders associated with signing (Scheuerle, Guilford & Habal, 2000). The purpose of this study was to survey professional sign language interpreters and determine the relative ranking of specific hand gestures according to comfort and pain. The hypothesis was that sign-language interpreters would report different levels of pain with different hand gestures.

2 Methods

Twenty-four experienced sign-language interpreters were recruited from among San Francisco Bay Area professional interpreters [1997-1998 Northern California Registry of Interpreters for the Deaf] and agreed to complete a 50 minute administered questionnaire. The study was approved the UCSF Committee on Human Research.

The questionnaire assessed demographics and history of upper extremity pain and disorders. In addition, information was collected on signing history such as years as a professional interpreter, style of signing, and typical frequency and duration of signing sessions. Pictures of commonly used hand signing gestures and other gestures were shown to the subjects and they rated the comfort or pain associated with that gesture on a 10 point visual analog rating scale. Gestures and postures included specific motions, types of signs, hand locations, size of movements, and speed of movements. Images of shoulder, elbow, forearm, wrist, fingers and thumb postures (Greene & Heckman, 1994) as well as specific signs from American sign language (Butterworth & Flodin 1991) and alphabet and numbers (Gustason, Pfetzing & Zawolkow 1980) were shown.

Median discomfort scores were calculated and Friedman's nonparametric test was used to test for overall differences between treatments (positions or movements) within a given body part. If significant, follow-up tests determined statistically different treatments within each body part.

3 Results

Participant characteristics are presented in Table 1. Twenty of the participants were female with a mean of 16 years of experience as a professional interpreter. Most spent more than 20 hours per week signing. Almost all of the participants experienced symptoms in the hands or arms after long signing sessions, described primarily as discomfort, fatigue, ache or pain. Symptoms were distributed throughout the upper extremity but were primarily located in the right wrist, forearm, elbow, shoulder and upper back. Symptoms occurred on average 11.4 days per month and began during signing sessions but lasted, on average, half a day.

Open-ended questions assessed comfort of hand regions for signing, size of movements and hand motions or postures. The most comfortable location for the hands during signing was between the chest and shoulders with a preference for lower chest area close to the body. The least comfortable area was around the shoulders or face or off to one side. Moderate size movements were the most comfortable and small size movements the least comfortable. The least comfortable hand motions were fast paced, finger spelling, forming a tight fist, or large wrist motions. The most comfortable hand motions were those involving an open hand, straight wrists, and fluid motions.

Table 1: Demographic characteristics of participants (N=24). Values are numbers of subjects in the category and percent of total reporting or average and standard deviation.

Characteristic	n (%)	Mean (S.D.)
Gender (females)	20 (83%)	
Age (years)		40.5 (8)
Years as interpreter		16 (9)
Style of signing		
Am Sign Language	19 (83%)	
PSE	14 (61%)	
Pidgion	2 (9%)	
Signed English	3 (13%)	
Transliterate	2 (9%)	
Other	6 (25%)	
Hours per week signing		
0 to 9	1 (5%)	
10 to 19	2 (10%)	
20 to 29	5 (24%)	
>= 30	13 (62%)	
Mean signing duration (hours)		2.4 (1.3)
Symptoms after a long signing session? (Yes)	23 (96%)	
If "Yes," what type of symptoms?		
Discomfort	17 (77%)	
Fatigue	16 (73%)	
Ache	14 (67%)	
Pain	12 (57%)	
Numbness	10 (50%)	
Burning	9 (45%)	
Occurrences of symptoms per month		11.4 (13.5)
Duration of pain (days)		0.5 (0.4)

Specific, commonly used signs are rank ordered by discomfort. The most comfortable signs were those involving elbow and shoulder motion (e.g., 'announce', 'brave', 'seal', 'nothing') or those involving all fingers moving together (e.g., 'brave', 'cool', 'nothing'). Those involving individual finger motion (e.g., 'selfish', 'quarrel', 'haircut') or wrist motions (e.g., 'how', 'flag') were the most uncomfortable.

Alphabetic characters and number were ranked as most comfortable or uncomfortable to form at the end of a long day of interpreting. Characters and numbers involving a loose fist (e.g., 'a', 'o', 'c', 'm', 'n') or concordant posture of all fingers or just the use of the index finger or thumb (e.g., 'b', '5', '1', or '10') were the most comfortable. Characters or numbers involving wrist flexion or discordant finger postures (e.g., 'q', 'p', 'w', 'f', '6') were the least comfortable.

The most comfortable hand shape was that with some flexion of all fingers while the least comfortable hand shapes were those involving the 'clawed' hand or a hand with the metacarpophalangeal joint flexed to 90 degrees and the interphalangeal joints extended. The most comfortable finger postures were those with the finger joints slightly flexed and the least comfortable involved extending or abducting all fingers or flexing the interphalangeal joints simultaneously. Moving the thumb in opposition to the small finger or in full radial or palm thumb abduction were the least comfortable thumb postures. The neutral wrist posture was the most comfortable while full radial or ulnar deviation or 45 degree wrist extension or flexion were the least comfortable. The most comfortable forearm postures were neutral or 45 degrees pronation, the least comfortable was full supination. Elbow postures of greater than 90 degrees flexion were the most uncomfortable as were shoulder postures of external rotation.

4 Discussion

This study identifies hand gestures that are painful and hand gestures that are comfortable with repeated use by professional sign language interpreters. The findings are generally supported by a previous study of sign language interpreters which identified non-neutral hand and wrist postures as a risk factor for pain along with other factors (Feuerstein & Fitzgerald, 1992). A limitation of the use of sign language interpreters is that their gesturing pattern may differ from gesturing used for computer input. For example, sign language interpreters do not control the pace of their work plus their hand workspace must be visible to an audience.

Based on this study of gestures and hand pain and our understanding of hand physiology and anatomy, we can begin to rank order hand gestures and design gestures for commands for computer input. Gesture instructions that are likely to be performed frequently should be those that are the most comfortable while infrequently performed instructions can be performed with less comfortable gestures. Considering comfort is just one aspect of the complex problem of developing a gesture recognition language. It is important to also understand natural gesture language issues in order to ensure productivity through minimizing errors and learning time. It is also critical to optimise image recognition and instruction interpretation.

References

Butterworth R.R. & Flodin M. (1991) The Perigee visual dictionary of signing : an A-to-Z guide to over 1,250 signs of American sign language, Perigee Books, New York, NY.

Feuerstein M. & Fitzgerald T.E. (1992) Biomechanical factors affecting upper extremity cumulative trauma disorders in sign language interpreters. J Occup Med, 34(3):257-64.

Gerr F., Marcus M., Ensor C., Kleinbaum D., Cohen S., Edwards A., Gentry E., Ortiz D.J., & Monteilh C. (2002) A prospective study of computer users: I. Study design and incidence of musculoskeletal symptoms and disorders. Am J Ind Med, 41:22-235.

Greene, W.B. and Heckman, J.D. (1994) The Clinical Measurement of Joint Motion (1st ed). American Academy of Orthopaedic Surgeons, Chicago, IL

Gustason G., Pfetzing D., & Zawolkow E. (1980) Signing Exact English, Modern Signs Press, Los Alamitos, CA.

Marcus M., Gerr F., Monteilh C., Ortiz D.J., Gentry E., Cohen S., Edwards A., Ensor C., & Kleinbaum D. (2002) A prospective study of computer users: II. Postural risk factors for musculoskeletal symptoms and disorders. Am J Ind Med, 41:236-249.Scheuerle J, Guilford AM, & Habal MB. (2000) Hand/wrist disorders among sign language communicators. Am Ann Deaf. 2000 Mar;145(1):22-5.

Impact of Information Technology on Work Processes and Job Characteristics in the Printing Industry

Pentti Seppälä

Finnish Institute of Occupational Health
Topeliuksenkatu 41 a A, FIN-00250, Helsinki, Finland
E-mail: pentti.seppala@ttl.fi

Abstract

Changes in technology, in work processes and in organizations, and their association with job contents and job characteristics, were studied in four printing companies. Altogether 109 employees from different organizational levels were interviewed, and 453 employees answered the questionnaire. Most profound technological changes had taken place in the pre-press processes. The computer-based integration had offered opportunities to design broad work entities but at the same time this demanded learning a lot of new things. Of the production workers, the pre-press workers had most often experienced that their job had become more variable and interesting.

1 Introduction

The impact of computerization and new technology on job contents and job characteristics has been discussed much during the past two decades (Smith and Sainfort, 1989; Seppälä et al., 1992; Seppälä, 1995; Smith and Carayon, 1995; Smith and Cohen, 1997). It seems that the consequences vary depending on the organization of work, job design principles and management of change. In positive cases, the new technology has increased job variety, autonomy and opportunities to develop new skills at work. Negative examples show the opposite, i.e., more monotony, less demanding job contents and increased stress symptoms.

The printing industry is one of the industrial branches in which advanced information technology has had a profound impact on the production processes, occupations, division of labor and job characteristics. This development was visible already in the beginning of the 1990s when the technology currently being used was just emerging (Seppälä, 1993).

The aim of this study was to analyze what kind of changes had taken place in the production technology, in the work processes and in the work organization in printing companies in the 1990s, and the association of these changes with job characteristics in the production jobs.

2 Material and methods

Four production units of larger printing concerns participated in the study. The products produced in these units were magazines, catalogues, brochures, annual reports, forms and professional journals. The number of personnel ranged from 53 to 330.

The data were collected by interviews, questionnaires and documents received from the companies. Altogether 109 employees from different organizational levels were interviewed, and 453 answered the questionnaire. Only the responses of the production workers from the three major production stages, i.e., pre-press (n=98), printing (n=115), and post-printing (n=98), are dealt with in this paper.

3 Findings

3.1 Changes in technology and organizations in the 1990s

The participating companies started to adopt computer-based text-processing systems already in the 1980s. However, the various parts of the page, i.e., text columns, pictures and photos were prepared and output separately and assembled manually to form a page layout that was copied onto a printing plate. At the end of the 1980s, digital desktop-publishing systems (DTP) emerged, which made possible the handling of texts and photos in electric form within the same system. Currently the entire pre-press process is integrated. The whole page and the entire printing sheet can be processed in electronic form, and by means of the computer-to-plate (CTP) technique transferred directly to the printing plate without the copying phase. Furthermore, the information can be stored on a disk that is inserted into the control system of a printing machine. This speeds up the start of printing.

The pre-press process and the job contents has been greatly affected by the change in the way the material is received in a printing firm. Currently the material from a customer is received mainly in electronic form via line connections, e-mail or a disk. Often the material is almost ready for printing. Only some formatting, demanded by the technology of the printing shop, must be done by the pre-press workers.

Technological development in the printing stage means an increase in computer-based control and pre-setting opportunities in the printing machines, as well as the integration of the printing and some post-printing processes. Also additional personal information (e.g., names, addresses, advertisements, messages, etc.) can be printed digitally by ink-jet equipment parallel to the main printing process.

In the post-printing stage, computer-based automation has increased too. As a result, previously separate production stages, such as cutting, folding, binding, packing and mailing have been integrated and conducted in a large computer-controlled production line. However, the operation of these systems still requires a lot of manual work. In most cases material must still be handled manually in input and output stages in spite of the advanced computer-based control technology of the production lines.

Along with the technological development, the organization of printing shops has become flatter. The number of supervisors has decreased and teamwork has increased. Teamwork, multi-skilling and an autonomous style of working have increased especially in the pre-press departments due to the development of information technology and the increased role of customers in the production process. Pre-press operators now have total responsibility for preparing a page in electronic form so that it is ready for the printing process. For this task they use page making programs by which texts, pictures and photos are integrated into an electronic page ready for printing. The pre-press operators are also directly in contact with the customers when needed. Production teams have been common already for years in printing departments where a group of printers have carried out printing tasks very independently. Nowadays they have also greater autonomy in work scheduling due to the implementation of integrated information management systems.

3.2 Job characteristics experienced by different occupational groups

Technological and organizational changes have influenced occupations, job contents and job characteristics especially in the pre-press stage. Some traditional work tasks, such as composing or entering of texts and copying of films are no longer needed. Broader work entities and a new division of labor can be designed when the material is processed digitally in a computer system. In order to utilize these technological opportunities, a lot of training and motivation to learn new things and assume wider responsibilities are needed. Many pre-press workers had received their vocational training before the era of advanced information technology, which has brought with it constant changes in technology and application programs. Consequently, companies have organized general training in computer use, as well as provided opportunities to take part in special occupational exams.

Table 1: Job characteristics experienced in different occupational groups, %

	Pre-press		Printing		Post-printing			
					Operators		Others	
	a little/ seldom	much/ often	a little/ seldom	much/ often	a little/ seldom	much/ often	a little/ seldom	much/ often
Variety of work	15	55	30	33	26	40	54	19
Can use one's knowledge and skills	16	53	12	56	3	51	50	15
Can learn new things at work	21	36	35	27	42	18	63	8
Opportunities for development at work	49	13	62	14	77	11	82	2
Can decide the order of work tasks	25	42	29	33	37	26	53	19
Haste at work	13	37	32	24	11	26	29	22

The questionnaire study revealed that the experienced job characteristics differed in the different occupational groups. Pre-press workers most often felt that their job was variable and it gives opportunities to use one's skills and knowledge and to learn new things (Table 1). Also over half of the printers were of the opinion that they can use their knowledge and skills to a great extent in their work. The workers in the post-printing stage comprised two distinct groups. The machine operators felt more often than the other workers (mainly women doing manual material handling at machine lines) that their work was variable and gave opportunities to use their skills and knowledge at work. The group of other workers was dissatisfied with most of the job characteristics. Opportunities for personal growth and development were not perceived as good in any of the groups (Table 1).

Table 2: Changes in job characteristics in recent years perceived by different occupational groups

	Pre-press		Printing		Post-printing			
					Operators		Others	
	-	+	-	+	-	+	-	+
Interesting work tasks	26	44	25	30	15	27	49	21
Variety of work	25	45	21	27	9	24	40	24
Opportunities for learning new things	15	63	21	36	18	36	29	27
Opportunities to make independent decisions	15	34	18	35	6	36	29	16
Opportunities to decide how to do one's work	17	20	19	22	15	27	22	21
Quantity of work	22	34	16	42	12	39	18	44
Amount of work tasks and responsibility areas	8	37	4	39	0	41	13	23

1) - = decreased, + = increased, % of respondents

Table 2 shows the extent to which the respondents in the different groups had perceived changes in some job characteristics in recent years. On average, all the groups except the group of other workers, had perceived positive changes more often than negative ones. In all the groups, the majority had felt that the quantity of work and the amount of work tasks had increased or remained unchanged. An increase in the amount of work tasks and responsibility areas was most often felt by the pre-press workers, the printers and the machine operators in the post-printing departments. The pre-press workers felt most often that their work had become more interesting and variable and that it gave opportunities to learn new things. Over a third of the pre-press workers, the printers and the machine operators in the post-printing felt that their opportunities to make independent decisions had increased. On the other hand, it must be noted that many workers in all groups felt that their work had become less interesting and less variable. This finding was supported by the interviews of pre-press workers who said that their work tasks were sometimes very simple when the customers sent almost complete pages to the printing shop.

4 Discussion

The production processes in the printing industry underwent profound changes during the 1990s. The changes were related to the developments in information technology and the companies' strategies to transfer some stages of the production process to customers and subcontractors. This concerns especially the pre-press stage. Information technology has influenced production processes and tools as well as companies' information management systems. The integration of separate functions is an essential feature in both cases.

When one considers these changes from the viewpoint of their impact on job contents and job characteristics, it is evident that they offer possibilities to design meaningful and broad job entities in which employees can learn and develop new skills. On the other hand, some work tasks can become overly simple and monotonous when processes are automated, if the work organization, division of labor and job contents are not redesigned at the same time. For example, if the task of a pre-press worker is only to receive and to forward the material sent by a customer in electronic form, the job will be dull and monotonous. Another possibility is to train the workers to be a pre-press operator who is capable of producing a complete printing sheet. This entity can comprise refinements of typography, pictures and colors of photos as well as preparing the page layout and assembling the final printing sheet. Furthermore, he/she can transfer this entity to the printing plate by means of the modern computer-to-plate (CTP) technique. In addition, direct contacts with customers and responsibility for the quality of the product will increase the meaningfulness of the work.

The companies participating in this study were still in the stage of introducing the most modern technology. Along with the implementation of new technology, also a redesign of jobs and training of the personnel had been started. In many cases, rather broad and variable job contents had already been designed. However, the development had been uneven in different stages of the production process, and many traditional problems still remained. Positive results had been achieved most often in the pre-press stage that was also most affected by the introduction of information technology. The workers in this stage had also experienced their work more positively than the other groups. The jobs in the post-printing stage were divided into two groups, machine operators and others, who had different job contents. The workers in the post-printing stage, whose work was mainly manual material handling at the production lines, were most dissatisfied with their job and the changes in it. A new division of labor and teamwork should be developed in these cases in order to alleviate the problems.

References

Seppälä, P. (1993). Job restructuring in the printing industry. *Nordisk Ergonomi*, 2 (June), 12-15.

Seppälä, P. (1995). Experiences on Computerization in Different Occupational Groups. *International Journal of Human-Computer Interaction*, 7 (4) 315-327.

Seppälä, P., Tuominen, E., & Koskinen, P. (1992). Impact of flexible production philosophy and advanced manufacturing technology on organizations and jobs. *International Journal of Human Factors in Manufacturing,* 2 (2) 177-192.

Smith, M. J., & Carayon, P. (1995). New technology, automation, and work organization: stress problems and improved technology implementation strategies. *International Journal of Human Factors in Manufacturing*, 5 (1) 99-116.

Smith, M. J., & Cohen, W. J. (1997). Design of computer terminal workstations. In G. Salvendy (Ed.), Handbook of Human Factors and Ergonomics, -2nd ed. (1637-1688). New York: John Wiley & Sons, Inc.

Smith, M.J., & Sainfort, P.C. (1989). A balance theory of job design for stress reduction. *International Journal of Industrial Ergonomics,* 4, 67-79.

Situated Interaction with Ambient Information: Facilitating Awareness and Communication in Ubiquitous Work Environments

Norbert Streitz, Carsten Röcker, Thorsten Prante, Richard Stenzel, Daniel van Alphen[1]
AMBIENTE Division, Fraunhofer IPSI
Dolivostr. 15, D – 64293 Darmstadt, Germany
{streitz, roecker, prante, stenzel}@ipsi.fraunhofer.de

Abstract

In this paper, we introduce our approach as well as examples of realizations for situated interaction in the context of future work environments. These environments will be populated with a range of smart artefacts. The artefacts and their mutual interaction are designed to facilitate awareness and notification as well as informal communication. They constitute examples of our approach to develop future work environments going not only beyond traditional PC-based work places but also beyond electronic meeting rooms and roomware components previously developed by us. We address a range of spaces in office buildings including semi-public spaces, e.g., in the hallway, the foyer, and the cafeteria. The approach is not restricted to office buildings but can be extended to other types of buildings and spaces. It is part of our vision that we call "Cooperative Buildings". The artefacts and the software were developed in the EU funded "Disappearing Computer"-project "Ambient Agoras: Dynamic Information Clouds in a Hybrid Worlds".

1 Introduction

"Ambient Agoras" is a project that addresses the office environment as an integrated organisation located in a physical environment and having particular information needs both at the collective level of the organisation, and at the personal level of the worker. This project promotes an approach to designing interactions in physical environments using augmented physical artefacts and corresponding software to support collaboration, social awareness, and to enhance the quality of life in the working environment. Ambient Agoras couples a set of interaction design objectives (ambient displays, mental disappearance of computing devices) with sensing technologies, smart artefacts (walls, tables, mobile devices) and the emerging functionality of artefacts working together. This work is an example of our approach to develop future work environments going beyond traditional PC-based work places and electronic meeting rooms as well as the various roomware components previously developed by us (Streitz et al., 2001).

2 The Role of Informal Communication

In our approach we address issues beyond the traditional PC-based working place and the less traditional but in many places already existing electronic meeting room environments. It is also motivated by the increased awareness of and interest in the role of informal communication in innovative work environments. While people acknowledge for quite some time the importance of

[1] Daniel van Alphen, Productdesign, Hufelandstr. 32, D-10407 Berlin, Germany. E-mail: dva@vanalphen.de

"soft skills" and "social competence", the value of informal communication for the performance and the creativity of an organization tend to be underestimated. So far, there has been little work on computer-supported augmentation of informal communication. Two trends changed the situation: The trend of putting more emphasis on informal communication and the trend in new technology developments where computer-based support is not only tied to the PC workplace anymore but will be mobile and ubiquitous using a wide range of devices. In this context, computer-based support for informal communication becomes an important topic when designing the workspaces of the future in terms of what we have called "Cooperative Buildings" (Streitz el al., 1998). Informal communication and the associated social interactions involve what may be described as "gossip". The way people deal with exchanging "soft facts" influences the general climate and atmosphere of the corporate culture. The increase of temporary project teams and nomadic workers with irregular presence in the office building requires compensating for the loss of continuity. Asynchronous and localized communication may help to overcome the anonymity and alleviate socializing for nomadic workers. Though email and telecommunication cover some of these needs, media that convey atmosphere, rumours, and vague news, etc. have to be developed. The question arises how one can support and augment the exchange of informal communication and atmospheric information using information technology. Our proposal is to develop "smart artefacts" with a focus on ambient displays.

3 Ambient Displays

Ambient displays take a broader view of the notion of "display" usually encountered with conventional graphical user interfaces (GUI) found on PCs, notebooks, PDAs and even on interactive walls or tables. Ambient displays also serve the purpose to communicate information but in a different, usually "implicit" way compared to traditional "explicit" GUI displays. This is achieved by making use of the physical environment around us and conveying information via changes in light, sound, movement of objects, smell, etc. For early examples see Wisneski et al. (1998). In the Ambient Agoras project, we decided to use changes of light, in particular changes in light patterns for conveying information about different states of people and of the physical as well as the virtual environment in an office building. In addition to communicating various types of information, we combined this with the goal to contribute also a decorative atmospheric element - called "GossipWall" (see below). As an example of what can be called "informative art", it serves two purposes at the same time: being a highly decorative spatial artefact and a real time feedback tool for informal processes within a space or a building.

4 Different Zones of Interaction

In addition to developing a new type of ambient display, our goal was also to make the type of information and the way of its communication context-dependent. The service provided by this artefact should be situation-based and depending on the proximity of people passing by. We decided to distinguish among three "zones of interaction" (and their respective modes) dependent on the distance from the GossipWall:

- Ambient Zone
- Notification Zone
- Interactive Zone

This is achieved by integrating sensors into the wall that cover two ranges, which may be adapted according to the surrounding spatial conditions. The sensors allow us to introduce "distance-dependent semantics", implying that the distance of an individual in front of the wall defines the kind of information shown and the interaction offered.

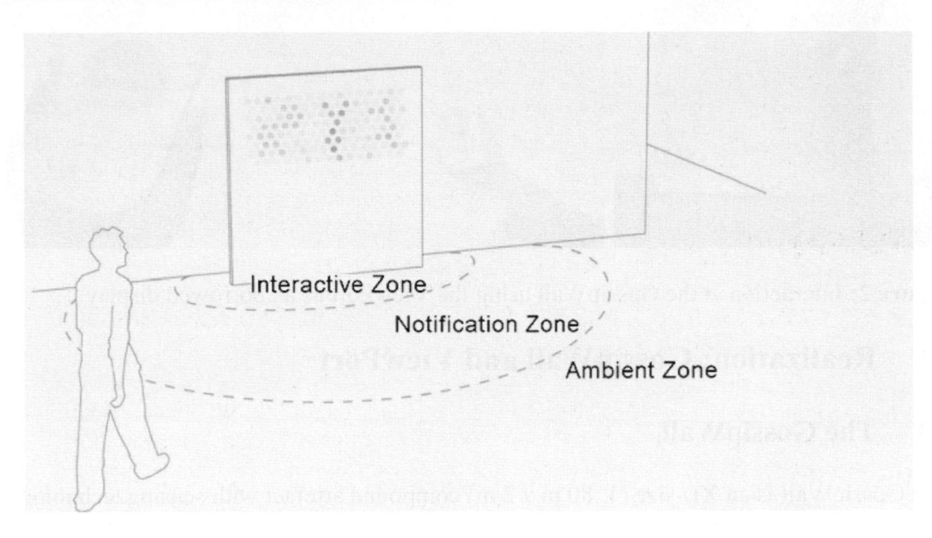

Figure 1: Three Zones of Interaction

Ambient Zone
When people are passing by but are outside the range of the wall's sensors, they experience the "ambient" mode, i.e. the display shows general information that is defined to be shown independent of the presence of a particular person. The parameters chosen to define the atmosphere will be represented as light patterns. Examples are: the number/ percentage of people still in the building, levels of activities, etc.

Notification Zone
If an individual approaches or passes by close to the wall, the person enters the Notification Zone and the wall will react. The GossipWall changes from a stand-by pattern to a notification pattern. This pattern can be a personal one relevant only for that particular person or a group pattern that is shown to all members of that group when passing by. These patterns can be "secret" and only been known to the people that are notified.
While the notification serves already an important purpose, in many cases there is a need to receive more detailed information. This is achieved by combining an ambient, implicit display with another explicit display. We propose a mechanism that we call "*the principle of borrowing a display*". In our realization, we use a mobile device called "ViewPort". The GossipWall borrows the display of the ViewPort and the user has all kind of information "at hand". This includes information about the meaning of the displayed patterns. Depending on the actual application, the user can interact and also enter data, download ("freeze") or browse information. See figure 2.

Interaction Zone
The third zone is active, once the person is very close to the GossipWall. In this case, the person can approach the GossipWall and interact with each single *cell* (= independent interactive pixel). This is able to store and communicate information in parallel in combination with mobile devices. This feature allows playful and narrative interactions, which other media don't supply. There is also a charming element of surprise that may be discovered via single cell interaction.

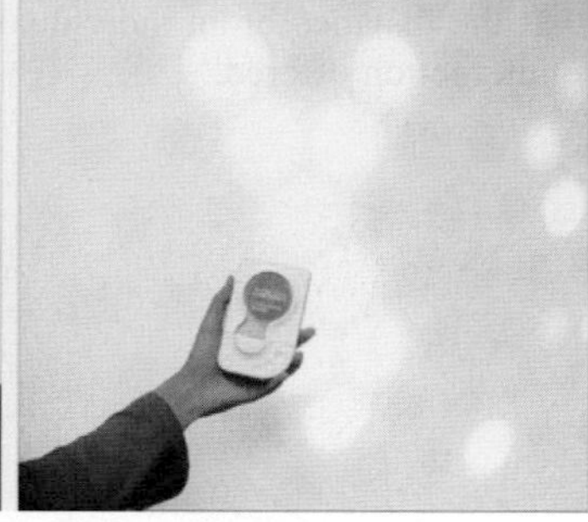

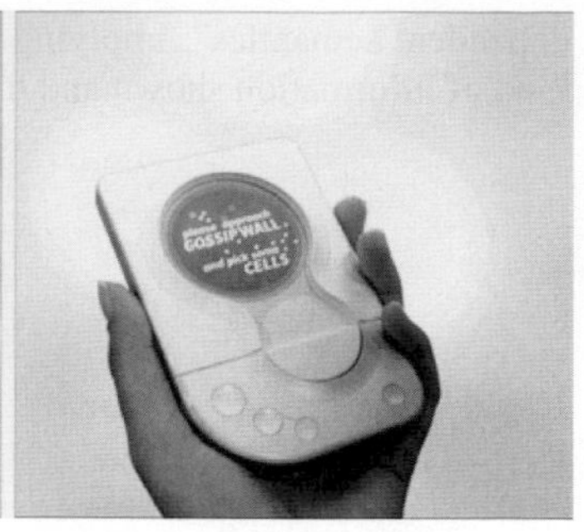

Figure 2: Interaction at the GossipWall using the ViewPort as a „borrowed display".

5 Realization: GossipWall and ViewPort

5.1 The GossipWall

The GossipWall is an XL-size (1, 80 m x 2 m) compound artefact with sensing technology. It does not have a standard type of display but is able to "display" or communicate ambient information, i.e., an "ambient display". It serves the function of "spreading gossip" by providing awareness and notifications to people passing by via patterns, in our case light patterns. Different patterns correspond to different types of information. People can access details via portable M-size artefacts as, e.g., the ViewPort (see below). We call this situation 'the GossipWall is "borrowing" the display of another artefact for explicitly displaying the information'.

We call the device GossipWall, because our focus is on the atmospheric and non-explicit aspects of communication localized within a building. Since individuals passing by can be recognized, there is a range of interaction opportunities including individual information through mobile artefacts as well as anonymous and public communication.

Each of the 124 cells at the GossipWall contains an LED cluster and a short-range transponder. The light intensity of each cell can be controlled. Since dimming of LEDs is not possible, we developed a control unit using pulse width modulation to change the brightness in 256 steps. To support the interaction among the different components we use two independent RFID systems and a wireless LAN network. People within the notification zone are detected via two RFID long-range readers installed in the lower part of the GossipWall.

Once a person is detected, the identification information is sent to the controlling PC for further processing. Depending on the kind of application, data can be transmitted to the ViewPort via the wireless LAN or distinctive light patterns can be displayed for notification.

Within the interaction zone people can access the information "stored" in each cell by reading the cell's ID with the ViewPort's short-range reader. With the received data the ViewPort can access the corresponding information.

5.2 The ViewPort

The ViewPort is a portable M-size compound artefact with a pen-based interactive display and provided with sensing technology. It can be used as a personal, a temporarily personal or public device for creating and visualizing information. It provides also the functionality of visualizing information "transmitted" from other artefacts that do not have displays of their own and are "borrowing" this display as, e.g., the GossipWall.

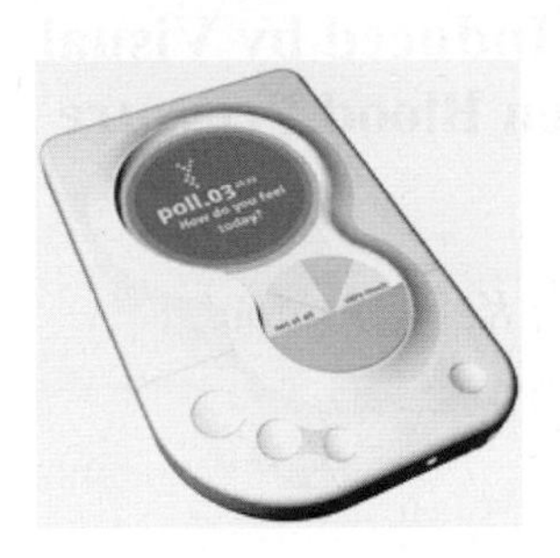

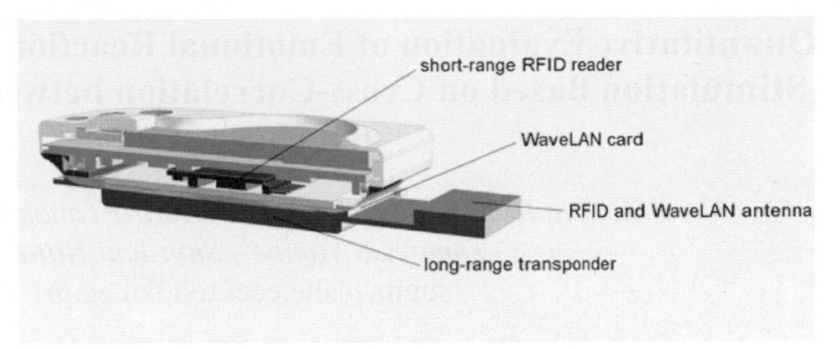

Figure 4: ViewPort with sensing (left) and communication module (right)

The ViewPort is developed on the basis of a Compaq iPAQ 3660 with 32bit RISC Processor, touch-sensitive color display and 64MB RAM. Its functionality is extended by a passive short-range reader unit and a WaveLAN adapter. The integrated RFID reader allows reading ranges up to 100mm. For higher flexibility, the WaveLAN hardware is implemented in a detachable communication module. Additionally, the ViewPort is equipped with a long-range RFID transponder. Thus, the ViewPort can be detected by stationary artefacts, e.g., the GossipWall, while at the same time it can identify nearby artefacts through its own reading unit. The ViewPort is able to offer services that are aware of the context. Sample applications include a GossipWall memory game and a polling functionality that reflects the results of the vote on the GossipWall.

6 Acknowledgements

This work is supported by the European Commission (contract IST–2000-25134) as part of the proactive initiative "The Disappearing Computer" of "Future and Emerging Technology" (FET) (project website: www.ambient-agoras.org). We are especially grateful to Daniela Plewe for her contribution. Thanks are due to members and students of the AMBIENTE division for their contributions to the implementation of hardware and software (www.ipsi.fraunhofer.de/ambiente).

References

1. Streitz, N., Geißler, J., Holmer, T. (1998). Roomware for Cooperative Buildings: Integrated Design of Architectural Spaces and Information Spaces. In: N. Streitz, S. Konomi, H. Burkhardt, H. (Eds.): *Cooperative Buildings - Integrating Information, Organization, and Architecture. Proceedings of CoBuild '98,* Darmstadt, Germany, LNCS Vol. 1370, Heidelberg, Germany, Springer, 1998. pp. 4-21.
2. Streitz, N., Tandler, P., Müller-Tomfelde, C., Konomi, S. (2001). Roomware: Towards the Next Generation of Human-Computer Interaction based on an Integrated Design of Real and Virtual Worlds. In: J. Carroll (Ed.), *Human-Computer Interaction in the New Millennium.* Addison-Wesley, pp. 553-578.
3. Wisneski, C., Ishii, H., Dahley, A., Gorbet, M., Brave, S., Ullmer, B., Yarin, P. (1998). Ambient Displays: Turning Architectural Space into an Interface between People and Digital Information. In: N. Streitz, S. Konomi, H. Burkhardt, H. (Eds.): *Cooperative Buildings - Integrating Information, Organization, and Architecture. Proceedings of CoBuild '98,* Darmstadt, Germany, LNCS Vol. 1370, Heidelberg, Germany, Springer, 1998. pp. 22-32.

Quantitative Evaluation of Emotional Reaction Induced by Visual Stimulation Based on Cross-Correlation between Blood Pressure and Heart Rate

Norihiro Sugita[1], *Makoto Yoshizawa*[2], *Akira Tanaka*[1], *Ken-ichi Abe*[1],
Tomoyuki Yambe[3], *Shin-ichi Nitta*[3]
sugita@abe.ecei.tohoku.ac.jp

Abstract

To evaluate the effect of visual stimulation on the human, the maximum cross-correlation coefficient (ρ_{max}) between blood pressure and heart rate whose frequency components were limited to the Mayer wave-band (0.04-0.15Hz) was employed. The test subjects were classified into two groups: Group A (n=18) whose members were prone to motion sickness and Group B (n=15) whose members were not. In the experiment, each subject was watching the video taken by in-vehicle cameras of a roller coaster in the form of two or three dimensional (2D or 3D) display. The proposed index ρ_{max} decreased at the falling scene of the roller coaster more apparently than conventional indices such as the *LF*/*HF* of heart rate. Moreover, Group B had a significantly higher recovery rate of ρ_{max} from the depression after the falling scene than Group A. This suggests that proneness to motion sickness is closely related to the baroreflex function of the autonomic nervous system.

1 Introduction

Recently, human beings have been exposed to special visual images such as three dimensional images used in virtual reality systems, high-resolution video images with very wide field of vision given by new TV sets, and swaying images of cellular phones used in vehicles. The Pokemon incident which arose in 1997 in Japan indicates that particular video images may have an adverse effect on humans.

Physiological data such as heart rate and blood pressure, which can be measured easily and noninvasively, have been frequently used to evaluate the biological effects of the visual stimulation or mental workloads (Nakagawa & Ohsuga, 1998). However, quantitative and objective methods for processing physiological data to assess the effect of visual stimulation have not been established yet. The purpose of this study is to propose a new index ρ_{max} (Sugita et al., 2002) for this assessment and to ascertain the validity of ρ_{max} by comparing it with the traditional indices obtained from the power spectral density of heart rate. The proposed index ρ_{max} is defined as the maximum cross-correlation coefficient between blood pressure and heart rate whose frequency components are limited to the neighborhood of 0.1Hz (Mayer wave component). On the other hand, the traditional indices are the low frequency component (*LF*; 0.04-0.15Hz) of the heart rate, its high frequency component (*HF*; 0.16-0.45Hz), and the ratio of *LF* to *HF* (*LF*/*HF*). These are well-known indices that may reflect the autonomic nervous activities (Goldstein et al., 1994).

[1] Graduate School of Engineering, Tohoku Univ., Sendai 980-8579, Japan
[2] Information Synergy Center, Tohoku Univ., Sendai 980-8579, Japan
[3] Inst. of Development, Aging and Cancer, Tohoku Univ., Sendai 980-8574, Japan

2 Experiment

2.1 Methods and Protocols

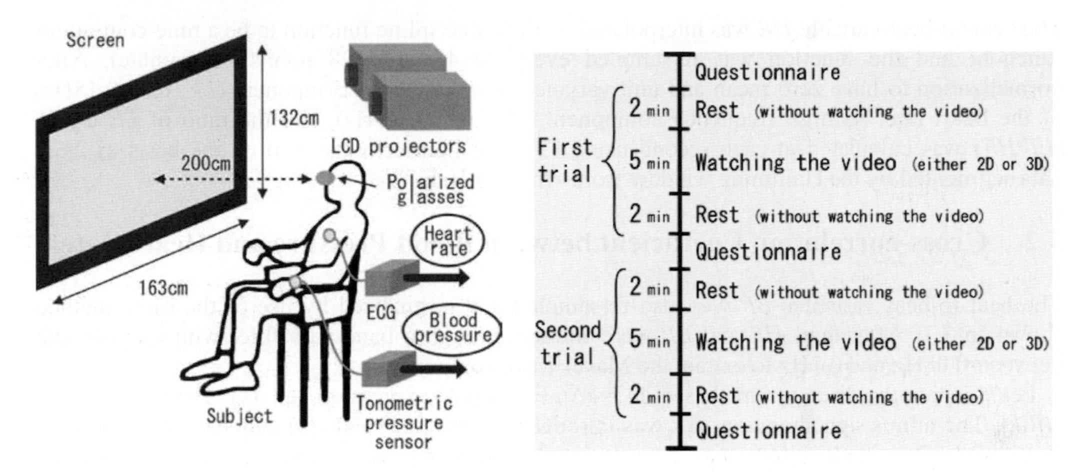

Figure 1: Schematic illustration of experiment.

Figure 2: Protocol of experimental trials for one subject.

Figure 1 shows the schematic illustration of experiment. To give the visual stimulation to the human subject, a video image taken by stereoscopic cameras carried on a roller coaster was used. This kind of image may induce strong emotional reactions such as fear, excitement, unrest, dizziness, nausea and motion sickness, and simultaneously raise physiological responses.

Figure 2 shows the protocol of experimental trials for one subject. One trial run lasted for 9min. In each trial run, the subject watched the image including the same content in the form of either 2D or 3D. In the 3D format, the subject watched the stereoscopic image with polarized glasses. In the 2D format, the subject watched the same content without the glasses. The same subject experienced two trial runs whose display formats, i.e., 2D and 3D, were exchanged for each other. To avoid the order effect, the occupation rates of 2D -> 3D and 3D -> 2D were managed to become close to each other. Questionnaires were charged on the subject to check his physical and mental states before the experiment, after the first trial and after the second trial. The number of test subjects was 33 (21 males and 12 females, aged 18 to 53 year old). They were categorized into two groups on the basis of the previous declaration with respect to easiness to suffer from motion sickness. Group A is the group of 18 subjects who reported to be prone to motion sickness. Group B is the group of 15 subjects who reported to be not prone to motion sickness.

2.2 Measurements

The radial arterial pressure signal acquired by a tonometoric pressure sensor (Nihon Corin; JENTOW 7700) and the ECG signal were stored in a personal computer every 1ms. Mean blood pressure (*BP* [mmHg]) was obtained as the mean value of the radial arterial pressure signal over one heart beat. Heart rate (*HR* [min^{-1}]) was calculated from the reciprocal of the inter-R-wave interval of the ECG signal.

3 Analyses

3.1 Power Spectrum of Heart Rate

The beat-to-beat variable *HR* was interpolated by the cubic spline function to be a time-continuous function, and the function was re-sampled every Δt=469ms (128 points per minute). After normalization to have zero mean and unit variance, low frequency component (*LF*; 0.04-0.15Hz) of the heart rate, its high frequency component (*HF*; 0.16-0.45Hz), and the ratio of *LF* to *HF* (*LF*/*HF*) was calculated at each second using FFT(fast Fourier transform) on the basis of 2min data segmented by the Hamming window from -1min to 1min.

3.2 Cross-correlation Coefficient between Blood Pressure and Heart Rate

The beat-to-beat variables *BP* was also re-sampled and normalized by use of the same method shown in 3.1. After that, *HR* and *BP* were filtered through a band-pass filter with a bandwidth between 0.08Hz and 0.1Hz to extract the Mayer wave component.

Let k denote the discrete time based on $t=k\Delta t$. For simple expression, let $x(k)= BP(k)$ and $y(k)=-HR(k)$. The minus sign shown in -*HR* was introduced so that $x(k)$ and $y(k)$ may become as inphase as possible for simple interpretation in depicted figures. At each second, the cross-correlation coefficient ($\rho_{xy}(\tau)$; the cross-correlation function normalized by mean square values of input and output signals at lag time $\tau=k\Delta t$) from $x(t)$ to $y(t)$ was calculated time-discretely on the basis of 2min data segmented by the Hamming window from -1min to 1min as follows:

$$\rho_{xy}(\tau) = \frac{\varphi_{xy}(\tau)}{\sqrt{\varphi_{xx}(0)\cdot\varphi_{yy}(0)}} \tag{1}$$

where $\varphi_{xx}(\tau)$ and $\varphi_{yy}(\tau)$ are auto-correlation functions of $x(t)$ and $y(t)$, respectively, and $\varphi_{xy}(\tau)$ is the cross-correlation function from $x(t)$ to $y(t)$. Furthermore, the maximum cross-correlation coefficient $\rho_{xy}(\tau)$ for the positive τ was obtained as

$$\rho_{\max} = \max_{0\le\tau\le7} \rho_{xy}(\tau) \tag{2}$$

4 Results and Discussion

4.1 Power Spectrum of Heart Rate

Figure 3 shows the comparison of the variation in *LF* with time between 2D and 3D (n=33 in both cases). This figure indicates that *LF* decreased significantly after the beginning of watching video and recovered after the end of that. This feature can also be observed in the comparison of the variation in *LF* consisting of both 2D and 3D between Group A (prone to motion sickness; n=36; 18 persons times 2 formats) and Group B (not prone to motion sickness; n=30; 15 persons times 2 formats), as shown in Figure 4. Figures 5 and 6 are the same comparisons with respect to *LF*/*HF* as Figures 3 and 4, respectively. The pattern of the variation in *LF*/*HF* is very similar to that in *LF*.

However, in all cases, there was no significant difference between 2D and 3D or between Groups A and B. Past clinical tests showed that *LF*/*HF* of heart rate may reflect the sympathetic nervous activity, then *LF*/*HF* at the exciting scene should have increased. However, the result of the present experiment did not agree with past results.

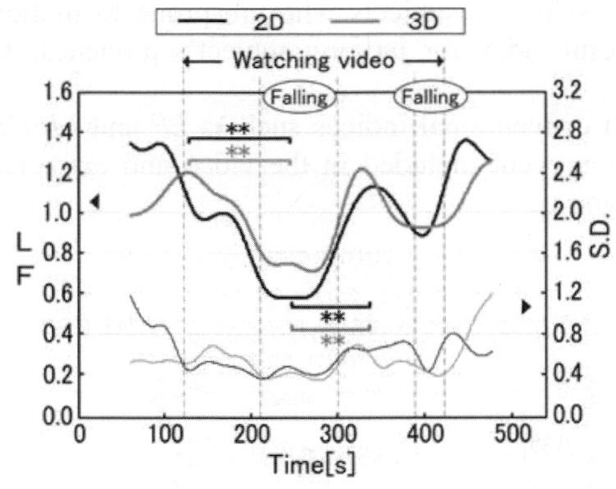

Figure 3: Comparison of *LF* of *HR* averaged over all 33 subjects between 2D and 3D (**: $p<0.01$).

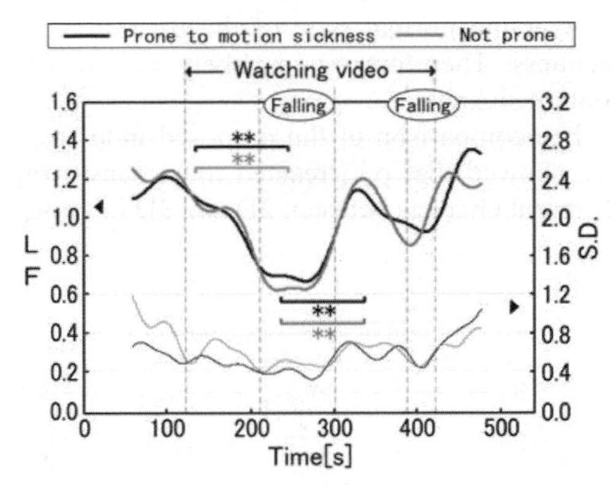

Figure 4: Comparison of *LF* of *HR* consisting of both 2D and 3D between Group A and Group B (**: $p<0.01$).

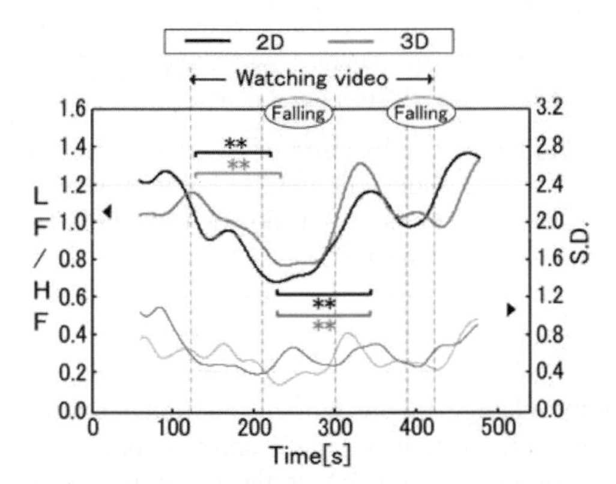

Figure 5: Comparison of *LF*/*HF* of *HR* averaged over all 33 subjects between 2D and 3D (**: $p<0.01$).

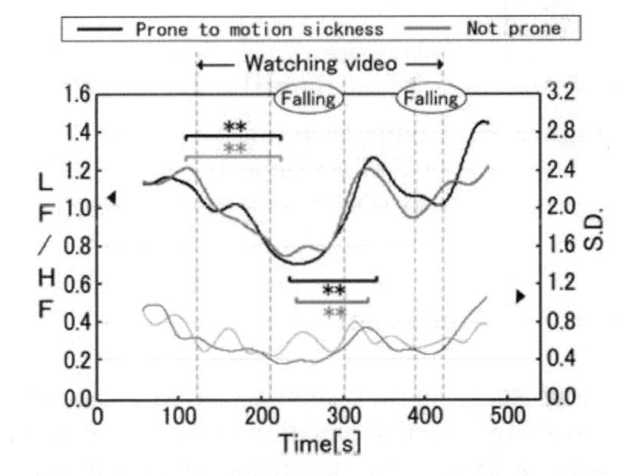

Figure 6: Comparison of *LF*/*HF* of *HR* consisting of both 2D and 3D between Group A and Group B (**: $p<0.01$).

4.2 Cross-Correlation Coefficient between Blood Pressure and Heart Rate

Figure 7 shows the comparison of the variations in ρ_{max} with time between 2D and 3D averaged over 33 subjects. In this figure, before the falling scene of the roller coaster, ρ_{max} of 2D recovered more quickly than that of 3D. This difference may be caused by the difference in intensity of the emotional impact to the subject, i.e., 3D images could give a stronger impact than 2D images. This fact implies that ρ_{max} can reflect the difference in realistic sensation of visual effect. In the falling scene, ρ_{max} decreased significantly in both cases of 2D and 3D. The falling scene disturbed the autonomic nervous system too strongly to reduce the difference between 2D and 3D.

Figure 8 shows the comparison of the variation in ρ_{max} consisting of both 2D and 3D between Groups A and B. This figure shows that Group B had a significantly higher recovery rate of ρ_{max} from the depression after the falling scene than Group A. This result suggests that the capability to recover the normal state of the baroreflex system is low in subjects who are prone to motion sickness. Therefore ρ_{max} is likely to become a useful index for judging subject's proneness to motion sickness.

The comparison of the proposed index ρ_{max} with conventional indices such as *LF* and *LF/HF* has showed that ρ_{max} reacted more sensitively to the event included in the video and extracted different changes between 2D and 3D or Groups A and B.

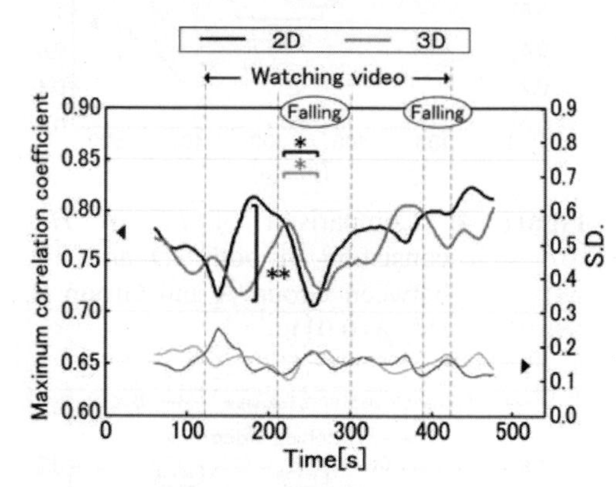

Figure 7: Comparison of ρ_{max} averaged over all 33 subjects between 2D and 3D (**: $p<0.01$, *: $p<0.05$).

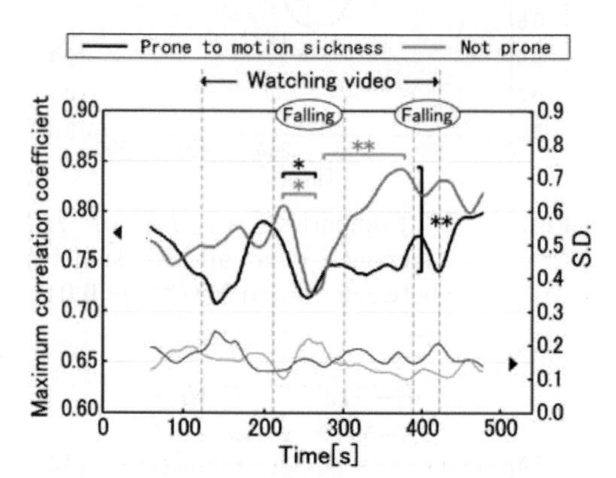

Figure 8: Comparison of ρ_{max} consisting of both 2D and 3D between Group A and Group B (**: $p<0.01$, *: $p<0.05$).

5 Conclusion

In this study, it was shown that ρ_{max} can reflect the change in the human emotional reaction induced by visual stimulation better than conventional indices such as *LF/HF* and may become an index to extract the difference in display format between 2D and 3D or the difference in the constitutional tendency to motion sickness.

In further studies, the proposed method should be improved toward reduction of individual difference and enhancement of reproducibility, and other experiments should be done using other visual sources such as flushing light stimulation that caused the so-called Pokemon incident.

References

Goldstein, B., Woolf, P. D., Deking, D., Delong, D. J., Cox, C., Kempski, M. H.(1994). Heart rate power spectrum and plasma catecholamine levels after postural change and cold pressure test. Pediatr. Res., 36, 358-363

Nakagawa, C., & Ohsuga, M. (1998). The present situation of the studies in VE-sickness and its close field. Transactions of the Virtual Reality Society of Japan, 3(2), 31-39 (in Japanese)

Sugita, N., Yoshizawa, M., Tanaka, A., Abe, K., Yambe, T., & Nitta, S. (2002). Evaluation of effect of visual stimulation on human based on maximum cross-correlation coefficient between blood pressure and heart rate. Journal of Human Interface Society of Japan, 4(4), 39-46 (in Japanese)

World Wide Web and Sustainable Workplaces with Visual Display Units

Hilja Taal

Tallinn Technical University Chair of Work Environment and Safety
Kopli street 101 Tallinn 11712 Estonia
Email: hiljat@staff.ttu.ee

Abstract

At Estonian Tallinn Technical University teaching risk and safety study to engineering students of different specializations (economics, construction, power plant engineering, logistics, etc.) is based mainly on the relevant legislation of Estonia. Therefore legislative acts found on various home pages of Estonian ministries, but also homepages of others Baltic states countries and EU, US, etc. are used. A study conducted at TTU that involved students learning economics showed increased awareness of how health is affected by different factors of work environment, increased knowledge of information available about those factors on the Internet, and that this information influences the choice of products, working habits, way of transportation, etc. The purpose is to avoid work related diseases to reduce the speed of aging processes.
According to this study even one hour a day can cause discomforts, aches and pains if the workstation is not ergonomic.

1 Aging Population and the Use of Computers in Estonia

One of the most notable changes in tommorrow's workforce is its aging. In the European Union in year 2025 about 35% of work force will be 50-64 years of age – double the size to younger workers of 15-24 years of age (17%) (Ilmarinen, 2002). Estonia with a population of less than 1.5 million is characterized by the fact that its population has decreased by a tenth during the last ten years, Estonian birth-rate has dropped significantly, all this causing aging population.
Preventive and developmental measures should prevent the early disability due to work on three levels: individual, enterprise and society level. The purpose is to avoid work related diseases which increase the speed of aging processes.

Computerisation in households has grown rapidly in Estonia. With rapid growth of Internet usage Estonia achieved a leader position among CEEC countries: Internet usage per 100 persons (aged 15-74) in 2000 was 32. The average level in EU countries was 39 in the same year. Personal computer penetration in Estonian households is 28%, penetration rate of Internet in companies in Estonia is 61% (for comparison the average level in European Union is 63%), the number of computers per 100 pupils at primary and secondary level schools is 3,82 (in EU average level is 9.05). The information collection for shopping in Internet is 7%, 2% of people use Internet for shopping, 24% of people use Internet as an information source (Leppik, 2002).

2 Risk Communication

Acceptance of risk is basically a problem of decision making, and it is inevitably influenced by many factors such as type of activity, level of loss, economic, political, and social factors, and confidence in risk estimation.

Communication between people with an increased use of ICT makes it clear to us that communication has different aims: it has a knowledge function, a social function, a control function, and, not the least, an expressive function. The concepts of terms IT and ICT in Europe, surround us more and more often, as well as computer technology, telecommunication technology and media technology. New dimensions in the quality of communications will occur. How to estimate risk criteria? Models are identified as "give sell the public the facts" and "persuade the public to accept the facts", the last one being restricted by a continued societal division between lay people and technical-managerial elite. However there is some precedent for risk consultation on occupational safety and health within workplace and industry communities. (Holmes, 1993)

3 Education Process of Ergonomics

3.1 Methods and Materials

The World-Wide Web (Web) offers an enormous variety of information resources. It is not easy to find the exact answers a task requires. It depends on individual factors such as beliefs and values, motions and motivation, person's ability to think and understand the socio-economic factors, social and cultural effects, micro- and macroergonomics, processes and organization that take place in society. The primary goal is to teach the students by working, the work process and organizational learning.

Teaching risk and safety to engineering students of different specialties (construction engineering, power plant engineering, logistic, etc.) is based on Estonian legislation and the EU directives, on which Estonian legislation is based. Legislative acts found on various home pages of Estonian ministries are being used, consulting also work environment home pages of other Baltic states (www.baltiseaosh.net/index_ee.stm). As Estonia is heading towards the EU (projected to join the union in 2004) it is necessary to become acquainted with prescriptions in other European countries. To get an understanding of problems concerning human factors through Internet we use various homepages of different states and universities. Students make a comparative analysis of the use of different prescriptions of the EU (www.osh.sm.ee/index_ee.stm), ALARA principles, etc. With the help of Internet searches we try to expand problem-based learning in seminars to teach how to avoid occupational diseases and accidents. The use of the Web motivates students, besides, direct pathways (for example on lecturers' own homepages) can be indicated. It is possible to find out about various problems and problem groups in different work environment areas in several countries.

3.2 Use of Risk Assessment

In Estonia continuos explanatory work is required to improve the working habits of all population groups in order to avoid health risks due to work with VDU and the related medical costs.
We are updating our methods of teaching ergonomics and give knowledge through lectures and practical training. One component of practical training is homework (Taal, 1999). The purpose of homework is to fix ergonomics knowledge and make students aware of the actual risks in work

situations through risk assessment. Our aim is to lecture the students about different risks and their consequences, that can occur because of lack of knowledge about the risks in work situations, and to prepare young persons for the right habits and behavior to avoid the rapid acceleration of aging process in different stages of life.

Taking examples from several works (Rantanen, 1995; Westlander, 1993) a questionnaire was compiled, based on ISO/DIS standard 9241 and risk assessment principles. In the checklist for workstation with VDU possible risks are named. Their causes can include technical data of VDU (electromagnetic fields), visual ergonomics, postural ergonomics, indoor climate, workstation layout, mistakes of work organization caused mental stress (group work, training, periodic breaks etc.) and possible health complaints.

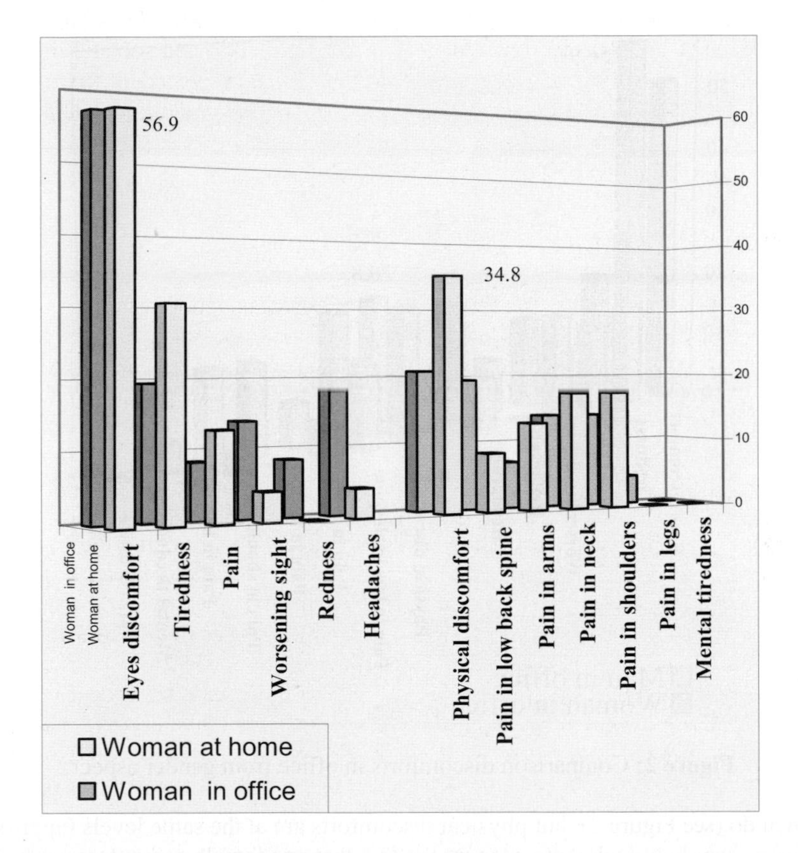

Figure1: Comparison of discomforts by female VDU users working in office and at home

In the first stage of lecturing the students have to clarify the possible risks in doing white-collar work. In the second stage students search in Internet to clarify the complaints and problems of white-collar workers around the world. The third stage of learning is the so-called problem-based analyses, where everyone must estimate the risks working with computer based on a risk assessment checklist find the occurring errors and make suggestions for improvement. For solving problems of postural ergonomics in particular we use Internet (CUergo, 1999).

3.3 Discussion and Overview from Risks Done by Students

Reading homepages is a cause of various eye discomforts: many computer users have computer vision syndrom (CVS). Studies show that 50% – 90% of computer users experience the symptoms of computer vision syndrome (CVS) (Computer, 2000). Some of these symptoms can heal, but some not shows the survey. On Figure1 we see the same level of sight problems that women, working both at home and in office, have. This level reaches 56.9%.

Physical complaints are mentioned by users working at home, comparing to relative low level of complaints by women in office – 34,8% versus 20,8%. Comparing the complaints of women and men working in office shows that women complain on eyes discomforts somewhat

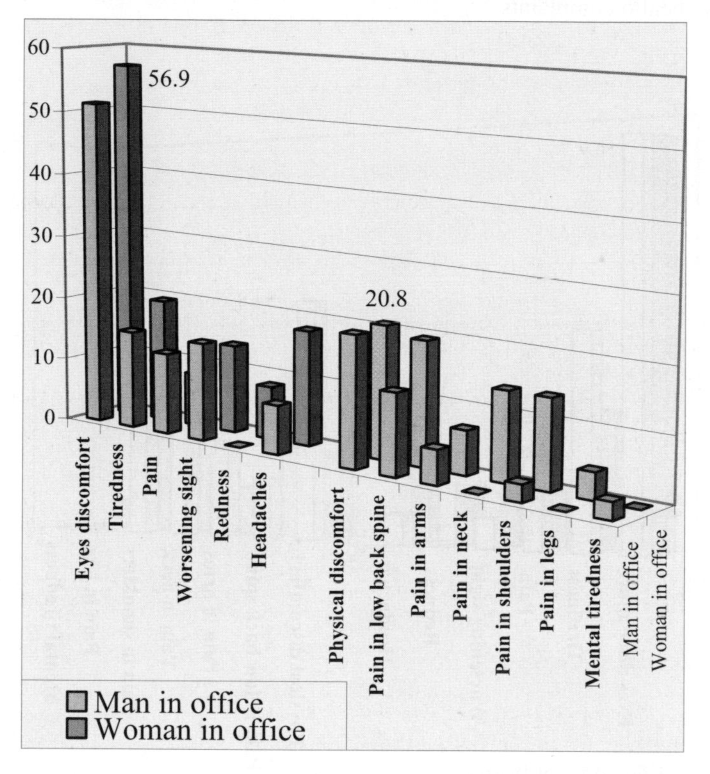

Figure 2: Comparison discomforts in office from gender aspect

more than men do (see Figure 2), but physical discomforts are at the same levels (approximately 21%). However, one should take into account the fact that one female computer user had often 3 complaints, whereas the male users had one or two complaints. From respondents working up to four hours with computer daily for the last 5-10 years one third had no complaints, but two thirds had problems with eyes. From similar respondents with working time 1-4 years 43% had no complaints, 63% had problems with eyes, 20% with spine and 6% had problems with arms. It is not important, whether one works half an hour or more, ignoring sight ergonomics results in health problems. Therefore the new ordinance for working with computers points out that workstations must meet ergonomics requirements independent of hours worked with VDU.

From the respondents 31.5% changed their working place more ergonomic immediately. Most of the changes were done by people, who worked with computer three hours per day – 60% form those working at home and 47% form those working in the office. So, that firstly being acnkowledged about the health risks and ergonomic principles, and secondly changing the workplace healthier, does not demand specific financial expenditures.

3.4 Conclusions

Like all European countries, Estonia as well was at loss due to the phrase "90/270/EEC directive currently applies to employees who habitually use display screen equipment as a significant part of their normal work" in the directive. Therefore the current research was made with 19-24 year old students with minimal health problems before. On the base of this research the Estonian ordinance of 1996 was overviewed and additions were made in year 2000.

References

Holmes, N. (1993). Risk communication in occupational health and safety: scientific slogan, marketing messages or community consultation. *Journal of Occupational Health and Safety*: Australia and New Zealand – 9(4), pp. 339-346.

Ilmarinen, J. (2002). Promotion of work ability during aging. In: Tallinn Technical University Avoiding Aging Catastrophe. Proceedings of the International Symposium 28-29 January 2002, Tallinn, Estonia. (pp. 17-18).

Rantanen, J. (1995). Future perspectives of research - need of multidisciplinary, weaknesses and opportunities. In: 5th International Symposium of the ISSA Research Section: Interdisciplinary research for safety and health protection of work. (pp. 333-338) Bonn: Druckerei Plump, Reinbreitbach.

Taal, H. (1999) Ergonomical Requirements for Working with Display Screen Equipment. In: Proceedings Læring og forandring- vejen til et bedre arbejdsmilj? NES 1999 Nordisk Ergonomiselskab Jubileæumkonferens, Nyborg, Denmark, 8-10 Sept., (pp. 325-330).

Westlander, G. (1993). Some problems and counter-.measures in modern labour. From single causal factors assessment to the analysis of multiple forms of hardship at work. *Journal.Ther. Biol.* 18 (5/6), 659-664.

A web site:
Leppik, L. (2001) Comparision of Estonian and Eurepean ICT sectors' state of affairs and developments. Retrieved September 20, 2002, from http://www.eik.ee/atp/failid/teglik2osa00.pdf

CUergo: Workplace Ergonomics Tools (RULA). Retrieved September 15, 1999, from http://ergo.human.cornell.edu/cutools.html

Computer vision syndrome. Retrieved April 13, 2000, from http://www.doctorergo.com/main.htm

The User-Computer Relation as an Anticipator of Musculoskeletal Strain in VDU Work

Seppo Tuomivaara, Ritva Ketola, Pekka Huuhtanen and Risto Toivonen

Finnish Institute of Occupational Health
Topeliuksenkatu 41 a A, FIN-00250 Helsinki, Finland
Seppo.Tuomivaara@ttl.fi

Abstract

The aim of the study was to analyze the relationship between psychological user-computer relation and musculoskeletal strain among VDU workers. It was assumed that the stressful user-computer relation associates with musculoskeletal strain. It was also assumed that the stressful user-computer relation moderates the success of the ergonomic intervention to decrease the musculoskeletal strain. The subjects (109) were employees working with a VDU for more than 4 hours a week. The data were collected before and 10 months after the intervention. The strain experience decreased from the baseline to the follow-up. The results indicate that the user-computer relation moderated the change in musculoskeletal strain during the intervention. The strain decreased among those physically distressed subjects who did not have problems in their user-computer relation. The interpretations of these results are discussed.

1 Introduction

Psychosocial factors such as monotonous work, high workload, time pressure, and low control and social support are related to the musculoskeletal symptoms of workers (Bonger, de Winter, Kompier & Hildebrandt, 1993). These factors can also be seen as the stressors of the human-computer interaction at work. The new technology has brought along special stressors as well, such as technology breakdowns and slowdowns. (Smith, Conway & Karsh, 1999) The ecological model of musculoskeletal disorders in VDU work by Sauter and Swanson (1996) indicates that work organization factors incluence musculoskeletal outcomes via psychological strain, and individual factors moderate the relationship between work organization and psychological strain.

In this paper the user-computer relation was seen as a potential psychosocial stressor. The cause of psychological strain in the human-computer interaction can be seen as an inconsistency between one's personal capacity and the demands in using computers. The consistency or inconsistency is always also the user's interpretation of his or her capability to use a computer. This kind of user-computer relation reveals the expectations of how one will manage with computers, i.e. how control is perceived. (Tuomivaara, 2000) If the control is perceived as low, the user interprets the situation as arising from his/her own incapability (inconsistency), and will experience anxiety in the use or even the intention to use computers (e.g. Skinner, 1995). Studies have also led to the general conclusion that computer users in less skilled jobs have more stress than those in higher skilled jobs (Smith et al., 1999). The first prediction was that a stressful user-computer relation anticipates the experience of musculoskeletal strain. The relation builds up in anatomical areas which have been found to be under biomechanical strain in VDU work, like the neck-shoulder area, upper extrimities and back (e.g. Punnett & Bergqvist, 1997).

The user-computer relation become more stressful during a technological change or in a situation where technology is a part of the change. The employees in less skilled jobs experienced more stress due to new technology than employees in more skilled jobs (Smith et al., 1999). Stressed employees can also be assumed to perceive their control over computer technology as low. Low perceived control decreases one's motivation to use technology (Tuomivaara, 2000) and take part in the change process. In computerized work tasks also the physiological arousal increases (e.g. Boucsein & Thum, 1997). The second prediction was then that the stressful user-computer relation moderates the success of the ergonomic intervention to decrease the musculoskeletal strain.

2 Material and method

The data in this paper are a part of an ergonomic intervention study carried out in 1998–1999 in three administrative units of a medium-sized city in Finland (Ketola, et al., 2002). The subjects in the entire study were employees working with a VDU for more than 4 hours a week. They were mainly secretaries, technicians, architects, engineers, or draftspersons. For the purpose of this paper, the study population was selected on the basis of participation in the intervention.

2.1 Procedure and subjects

First, in the entire study the subjects filled out a baseline questionnaire (N=416) inquiring about musculoskeletal pain and strain, general health, work environment factors, work with a VDU, and work organizational and psychosocial factors. In the second phase the subjects for the intervention were selected (N=124). Third, experiences related to computer use, workload and the workplace of the selected subjects were evaluated before the intervention (fourth). Fifth, 7 to 10 months after the intervention, selected subjects filled out the same questionnaire as at the baseline. Also 10-months after the intervention, the work places of the subjects were evaluated again by two experts in ergonomics. There were 109 subjects present at the intervention, and 102 in the 10- month follow-up. 58% of the 109 were women, and the mean age of the subjects was 48, (range 27 – 60).

The inclusion criteria for the intervention subjects were reported symptoms in the neck, shoulders, or upper limbs in at last one and, at most, eight anatomical areas (out of 11 areas) during the preceding month, and mouse usage for more than 5% of the work time with a VDU. These subjects were allocated into three groups: ergonomic, education and reference group.

2.2 Measures

Musculoskeletal strain was assessed at the baseline (Time 1) and at the follow-up (Time 2). The strain after the usual workday during the preceding month was assessed by a five-point scale ranging from 1 = 'no strain at all' to 5 = 'very much strain'. The strain was evaluated in different anatomical areas. A manikin was used to define these areas. The strain experience was summarized in to two dimensions: the upper part of the body (excluding shoulders and extremities) (α=0.81 at Time 1 and α=0.88 at Time 2) and the lower part of the body (α=0.62 at Time 1 and α=0.64 at Time 2).

The user-computer relation was measured on three dimensions: computer confidence, difficulties with controlling computers, and anxiety in the use of computers. *Computer confidence* was evaluated by a sum of questions on certainty of using computer programs. There were six programs and the rating was from 1 = 'certain' to 4 = 'uncertain' and 5 = 'not in use' (α=0.85).

The dimension of *difficulties with controlling computers* gave eight statements, with a five-point scale from 1 = 'never' to 4 = 'often' and 5 = 'can not say' (α=0.90). The last dimension, *anxiety in the use of computers* consisted of four things that make the subject feel timid or fear. The statements were answered with yes or no, and the summary dimension was coded 0 = 'no anxiety' and 1 = 'anxiety'.

The level of ergonomics was assessed by two researchers from video-recordings of the subjects doing their usual daily tasks. The rating scale varied from 4 ('poor') to 10 ('excellent'). The mean of these ratings describes the ergonomic level of each workstation. The interobserver repeatability was r = 0.85. *Amount of VDU work* was evaluated by the subjects as an per cent portion of the total working time during the past month. Both evaluations were done at baseline and follow-up.

2.3 Data analysis

A hierachical regression analysis was conducted to examine the anticipation of the user-computer relation at Time 1 regarding the symptoms in the upper and the lower part of the body at Times 1 and 2. The controls are presented in the Results section. The education and reference groups were treated as one reference group because they got no active interference on workstation ergonomics.

3 Results

The musculoskeletal strain at the baseline was rated as moderate (the upper part mean=3.30, the lower part mean=2.67) according to the selection of the subjects. The strain in the upper part of the body (excluding shoulders and extremities) decreased from the baseline to the follow-up measurement (Δ strain=0.37, t =3.69 , p<.001). The strain in the lower part of the body decreased also, but not statistically significant (Δ strain=0.19, t =1.98 , p<.051). The average ergonomic rating at the beginnig of the study was 6.78 and at the follow-up 7.50 (t=-6.92, p<.001). The rise in rating was significant only in the intensive group. The VDU work percentage was 46 at the baseline and 44 at follow-up, the change was not significant. The subjects perceived their confidence in computer use to be between 'quite certain' and 'not very certain'. Difficulties with computers were encountered 'seldom' or 'sometimes'. At least one statement was chosen to be anxiety-provoking in the use of computers by 62% of the participants.

Table 1: Effects of the User-Computer Relation on Musculoskeletal Strain at Time 1

	Upper Part of the Body				Lower Part of the Body			
	ΔR^2	F	df	β	ΔR^2	F	df	β
Step 1	0.06‡	2.41	2,72		0.08†	3.21	2,72	
Age				-0.02				0.30†
Gender				0.24†				0.10
Step 2	0.01	0.40	2,70		0.00	0.16	2,70	
Level of Ergonomics 1				-0.12				-0.01
VDU Work 1				0.04				-0.06
Step 3	0.02	1.51	1,69		0.06†	4.98	1,69	
User-Computer Relation				0.15				0.26†

Note: *p<0.01; †p<0.05; ‡p<0.10. Gender was coded as 1 = female, 0 = male.

According to the first prediction the user-computer relation with experience of low control was linked with musculoskeletal strain in both the upper and lower parts of the body. We tested this prediction at the baseline in two hierarchical regression analyses (Table 1). First for the upper part of the body, and then for the lower part of the body. In both analyses, the variables entered in the first step were gender and age. In the second step, the variables level of ergonomics at Time 1 and the amount of VDU work at Time 1 were added. The standardized sum of three dimensions of the user-computer relation was added in the last and third step. Inspection of the Table 1 reveals that first prediction was supported only in the case of the lower part of the body. The stressful user-computer relation at Time 1 was positively related to the experience of musculoskeletal strain in the lower part of the body at Time 1.

Recall the second prediction that the stressful user-computer relation moderates the success of the ergonomic intervention to decrease musculoskeletal strain. The variables of strain at Time 1, group and changes in ergonomics and VDU work were added to the models in the second step along with the previous variables. The adjustment of strain at Time 1 gave the possibility to predict the change of experienced strain by the user-computer relation. Table 2 indicates that the second prediction was supported as regards both the upper part of the body and the lower part of the body. The stressful user-computer relation was linked positively to the strain in the upper and lower parts of the body at Time 2. This means that the intervention decreased the experience of strain among those physically distressed subjects who did not have problems in their user-computer relation.

Table 2: Effects of the User-Computer Relation on Musculoskeletal Strain at Time 2

	Upper Part of the Body				**Lower Part of the Body**			
	ΔR^2	F	df	β	ΔR^2	F	df	β
Step 1	0.07	2.14	2,61		0.13†	4.66	2,60	
Age				0.13				0.39*
Gender				0.27†				0.12
Step 2	0.36*	5.61	6,55		0.30*	4.73	6,54	
Outcome at Time 1				0.54*				0.53*
Group				0.21‡				-0.07
Level of Ergonomics 1				-0.08				0.15
VDU Work 1				0.25‡				0.02
Change in Ergonomics				0.07				0.06
Change in VDU Work				-0.12				-0.08
Step 3	0.04†	4.12	1,54		0.10*	11.28	1,53	
User-Computer Relation				0.23†				0.37*

Note: *p<0.01; †p<0.05; ‡p<0.10. Gender was coded as 1 = female, 0 = male. Groups were coded as 1 = reference, 0 = ergonomic group.

The difficulties experienced in computer usage were the most powerful dimension of the user-computer relation in the anticipation of strain in the lower part of the body. The experienced anxiety, on the other hand, anticipated strain in the upper part of the body at Time 2.

4 Discussion

The cross-sectional analysis of the data revealed that the relation between the user-computer relation and the strain experience was manifested only in the lower part of the body. Therefore, long-term stable computer use no longer activate the user-computer relation to be connected to the musculoskeletal strain experience in the areas which are under the greatest biomechanical strain in

VDU work. The user-computer relation nevertheless predicted well the strain in both the upper and the lower parts of the body about ten months after the intervention, and after adjusting for the baseline strain, ergonomic rating and amount of VDU work. The best predictor of the change in the strain felt in the upper part of the body was anxiety, and in the lower part of the body, difficulties in the use of computers. The user-computer relation anticipates musculoskeletal strain, especially in situations where the object of change is computer technology and its use.

The results can be interpreted in several ways. The user-computer relation was activated in the intervention and moderated the success of the efforts to decrease musculoskeletal strain. The stressful relation to computers can mediate the musculoskeletal symptoms via three pathways according to Sauter and Swanson (1996): 1) The stressful user-computer relation may increase muscle tension and intensify biomechanical strain, 2) increased psychological stress may affect the awareness and reporting of musculoskeletal strain, or 3) there are rather complex cognitive interpretative processes behind the correlation between strain and the user-computer relation. Here none of these interpretations can be ruled out with certainty. The results nevertheless show that the relation also shapes the reaction to the change by lowering a person's motivation and ability to learn. Consequently, the interventive efforts to decrease the experience of strain remain quite powerless.

Even though more studies are needed to clarify this relationship, the present results helps to take into account the user-computer relation in ergonomic interventions in VDU work. We also need more sophisticated methods which include analysis of the cognitive component (Sauter and Swanson 1996) to evaluate the success of endeavours to decrease musculoskeletal strain.

References

Bonger, P. M., de Winter, C. R., Kompier, M. J., & Hildebrandt, V. H. (1993). Psychosocial factors at work and musculoskeletal disease. *Scandinavian Journal of Work, Environment and Health, 19*(5), 297-312.

Boucsein, W., & Thum, M. (1997). Design of work/rest schedules for computer work based on psychophysiological recovery measures. *Interntional Journal of Individual Ergonomic.*

Ketola, R., Toivonen, R., Häkkänen, M., Luukkonen, R., Takala, E.-P., & Viikari-Juntura, E. (2002). Effects of ergonomic intervention in work with video display units. *Scandinavian Journal of Work, Environment and Health, 28*(1), 18-24.

Punnett, L., & Bergqvist, U. (1997). Visual display unit work and upper extremity musculoskeletal disorders. *Arbete och hälsa, 16*(1-156).

Sauter, S. L., & Swansson, N. G. (1996). An ecological model of musculoskeletal disorders in office work. In S. D. Moon & S. L. Sauter (Eds.), *Beyond Biomechanics; Psychosocial aspects of musculoskeletal disorders in office work.* London: Taylor & Fres Ltd.

Skinner, E. A. (1995). *Perceiced control, motivation & coping.* Thousand Oaks: Sage Publications.

Smith, M. J., Conway, F. T., & Karsh, B.-T. (1999). Occupational stress in human computer interaction. *Industrial Health, 37*, 157-173.

Tuomivaara, S. (2000). *Vapaa-ajan ja työn tietokonesuhteet ja käyttöhalukkuusmallit.* [Human computer relations and psychological models for computer use at work and leisure] Väitöskirja, Tampereen yliopisto, Tampere. [Dissertation. English abstract.]

Lighting of VDT Workstands and Users' Visual Discomfort – Results of an Experimental Study

Agnieszka Wolska

Central Institute for Labour Protection – National Research Institute
00-701 Warszawa, ul. Czerniakowska 16, Poland
agwol@ciop.pl

Abstract

Lighting is one of the main environmental factors which influences users' visual discomfort and well-being. It should ensure that visual work conditions do not result in visual discomfort. The purpose of the study was to model different lighting systems for VDT work and to improve knowledge in the field of choosing best lighting systems for VDT stands with LCD screens, taking into consideration visual discomfort and users' preferences. The results of the study showed that visual fatigue (the smallest changes in visual functions like accommodation and convergence) is lowest for indirect and compound lighting systems. However, considerably bigger intensities of sensitivity to light, redness and heaviness of eyelids were found for compound lighting. On the other hand, in general, direct lighting realized by "dark-light" luminaires is the most preferred lighting system. Some interesting differences related to age, gender and VDT work experience were found.

1 Introduction

Even though computer technology is improved all the time and the quality of screens is incomparably better than even a few years ago, visual discomfort during VDT work is still common. One of the main environmental factors which influences visual fatigue is lighting (Bergqvist, 1984, Dainoff, 1982, Pickett, 1991, Smith, Cohen & Stammerjohn, 1981, Wolska & Switula, 1999). The goal of illuminating engineering is not only to avoid any possible dangers or other negative results, but actually to promote health in the meaning of the World Health Organization definition: "Health is a state of complete physical, mental, and social well-being and not merely the absence of disease or infirmity" (http://www.who.int/about/definition/en/). That means lighting should ensure a luminous environment which is human-friendly and appropriate for the performed visual tasks. Thus, visual strain should be minimized and well-being ensured during visual work under a given lighting system. Visual work is inseparably connected with visual fatigue, but the kind of symptoms and their intensities depend on the different factors. A review of office lighting environment in Poland (Kamienska – Żyła, 1993) and in Japan (Kanaya, 1990) showed that visual working conditions, especially on VDT stands are far from perfect. In many offices introduction of VDT equipment introduction has not been followed by changes in lighting. On the other hand even if there were changes in lighting changes, they have not necessarily led to users' acceptance of the new lighting. Çakir's (1991) longitudinal study (1978–1990) which assessed lighting in offices and VDT workstands in Germany showed that artificial lighting is far from promoting health and well-being in the spirit of the WHO definition of health.

Çakir (1991) concluded that technical improvements in office illumination would not lead, over the years, to an increase in the level of acceptance for artificial lighting.

The purpose of the present study was to model different lighting systems for VDT work and to improve knowledge in the field of choosing best lighting systems for VDT stands with LCD screens, taking into consideration visual discomfort and users' preferences.

2 Method

2.1 Participants

The group consisted of 44 participants (15 women and 29 men) aged 17–37 with a mean age of 22.25 years. The participants were selected according to the criteria of age (under 40 years old), eye state (no known visual defects, visual acuity ranging between 1.0 and 1.5 on Snellen charts for distance – with corrective lenses if needed, spherical refractive errors less than ± 3.5 Dsph, astigmatism less than ± 0.5 Dcyl and no systemic or neurologic diseases) and VDT work experience (novices and professionals). VDT work experience of the participants was evaluated during pre-study group selection on the basis of an interview. The volunteers had to do a list of specially prepared VDT tasks and to fill in a questionnaire. As a result the group was composed of 22 novices and 22 professionals. All participants volunteered to take part in the study and underwent training in VDT work before the experiments. They had to become familiar with the visual task simulated by a computer program

2.2 Visual task

The visual task was simulated by a computer program and consisted of eight Landolt's tests (ie 4 with negative and 4 with positive polarity on the screen). Each test consisted of 352 white or black rings respectively. There were two kinds of Landolt's tests:

- a "step by step" test, in which each of the eight possibilities of gap localization had to be defined for 352 rings on the screen,
- an "option" test, in which rings only with three (indicated in the corner of the screen) gap localizations had to be found (from among 352 rings) and marked. The colour of the marked rings changed and mistakes could not be corrected.

Both kinds of tests were prepared for positive and negative polarity and each of the eight tests had different ring distribution on the screen.

Participants performed the visual task for 1.5 – 2 hours.

2.3 Lighting conditions

Experiments were carried out in laboratory conditions for the following lighting systems suitable for VDT work:

- direct-indirect (DI-L) in the form of "soft-light" fluorescent luminaires ABML 2x58W,
- direct (D-L) in the form of "dark-light" fluorescent luminaires XRD 2x36 W,
- indirect (I-L) in the form of "uplight" fluorescent luminaires TCS 663 2x58 W,
- compound - general and task lighting (C-L) with general lighting in the form of an indirect lighting system and a low-luminance Wacolux 801 desk luminaire (special VDT workplace luminaire).

The main assumption of modelling lighting conditions for each lighting system was to obtain the same illuminance level of about 500 lx on the work surface – a desk and to avoid direct and reflection glare.

2.4 Visual fatigue and lighting assessment

Visual fatigue was evaluated with a visual complaints questionnaire and additionally by simple measurements of two visual functions: the near point of accommodation (NPA) and the near point of convergence (NPC).

The subjective evaluation of visual complaints (asthenopic symptoms) was established by a questionnaire, which was filled in after each experimental session.

The NPA and NPC were measured with an RAF (Clement Clarke, UK) near point rule according to the measurement method described by London (1991a, 1991b). Both parameters were measured before and after each experimental session. The changes in the calculated (mean) values of NPA and NPC before and after the experiment were the measures of visual fatigue.

After each experimental session participants had to fill in a questionnaire on different aspects of their perception of the luminous environment, on the influence of lighting on the participants' well-being and on the occurrence and intensity of asthenopic symptoms. Assessment was performed on a nominal scale and on an ordinal 5-point scale of the degree of strenuousness or intensity.

3 Results

3.1 Visual fatigue

Changes in NPA and NPC were obtained by subtracting "after" values from "before" values whereas changes in the accommodation amplitude (AA) by subtracting "before" values from "after" values for each participant. Those changes were statistically analysed. After the experimental session NPA and NPC moved away from the eyes, which corresponded with the reduction in accommodation and convergence abilities of the eyes. The mean changes of accommodation and convergence after VDT work under different lighting systems are presented in table 1. Although the biggest reduction in accommodation was found for DI-L and the smallest for the C-L lighting system, those changes were not significant. According to ANOVA, convergence changes differ significantly in relation to the lighting system ($F(3)=3.16$, $p=.03$). The biggest convergence reduction was found for the DI-L lighting system, which was significantly bigger than convergence changes for the I-L ($p=.02$) and C-L ($p=.005$) lighting systems.

The mean values of intensity of complaints indicated that all asthenopic symptoms were assessed as small or medium regardless of the lighting system. The biggest intensities of discomfort were found for tired eyes, redness, blurring, sensitivity to light, burning, heaviness of eyelids, lacrimation and itching. It was established (Wilcoxon signed ranks test) that sensitivity to light was significantly bigger for C-L than for D-L ($p=.02$), tiredness of eyes was significantly bigger for DI-L than for either I-L ($p=.05$) or C-L ($p=.04$), heaviness of eyelids was significantly bigger for C-L than for either D-L ($p=.05$) or DI-L ($p=.003$) and redness after the experiment was significantly bigger for both C-L (p=.003) and I-L (p=.01) than for the DI-L lighting system. The most frequently reported complaints were tired eyes, blurring, redness, burning, lacrimation and sensitivity to light.

Table 1: Changes of accommodation and convergence under different lighting systems

Parameter	Lighting system			
	DI-L	**D-L**	**I-L**	**C-L**
NPA, cm				
Mean	0.99	0.71	0.70	0.34
SD	1.46	1.61	1.42	1.64
Variance	3.13	2.60	2.01	2.70
AA, Dptr				
Mean	0.79	0.62	0.59	0.42
SD	1.34	1.34	1.36	1.08
Variance	1.81	1.79	1.84	1.16
NPC, cm				
Mean	1.38	0.89	0.57	0.40
SD	1.69	1.53	1.65	1.49
Variance	2.84	2.33	2.73	2.20

3.2 Lighting assessment

Most participants assessed the D-L system as comfortable for visual tasks and would have liked to work under that lighting system all the time (see figure 1). However the assessments of the considered lighting systems did not differ significantly.

The Friedman nonparametric test showed there was a significant influence of the type of lighting system on mood (p=.011) and on perceiving the room to be too bright (p=.004). The participants' mood changed during the experimental sessions mostly in a positive direction under each lighting system. The room illuminated by the DI-L system was the one most frequently assessed as too bright, which 37% of participants perceived as strenuous (and a possible cause of discomfort glare).

Spearman correlation analysis between the subjective assessment of lighting features and age, gender and experience in VDT work revealed that:

- females more often than males found surfaces in the room excessively bright under the I-L system (r=.54, p<.001), which could be why they less often assessed that system as comfortable for visual work,
- females less often than males chose the D-L (r=-.33, p=.03) and I-L (r=-.47, p=.001) systems for regular work,
- professionals more often than novices chose the I-L (r=.37, p=.013) system for regular work,
- older participants less often than younger ones chose the DI-L (r=-.47, p=.001), D-L (r=-.4, p=.007) and I-L (r=-.39, p=.009) systems for regular work.

4 Conclusions

From the point of view of visual fatigue indirect and compound lighting systems are best. However, a compound lighting system should be used with special care for anti glare realization of task lighting.

The obtained results show that about 1.5 hours of VDT work (with visual attention mainly on the screen) can cause small or medium asthenopic symptoms, regardless of the lighting system. However it seems that direct-indirect and compound lighting systems are the most problematic. They should be used with special attention to their appropriate anti-glare realization.

Generally users prefer a direct lighting system with "dark-light" luminaires but preferences for lighting systems with regard to users' features differ and lighting designers should be provided with some guidance.

Acknowledgements

This study is part of the National Strategic Programme "Occupational Safety and Health Protection in the Working Environment", supported in 1998–2001 by the State Committee for Scientific Research of Poland. The Central Institute for Labour Protection was the Programme's main co-ordinator.

References

Anshel, J. (1998). Visual Ergonomics in the Workplace. London: Taylor & Francis.

Bergqvist, U.O.V. (1984). VDTs and health: A technical and medical appraisal of the state of the art. *Scand J Work Environ Health*, 10 (Suppl.2), 1—87.

Çakir, A.E. (1991). An investigation on state-of-the-art and future prospects of lighting technology in German office environments. Berlin: Ergonomic Institute for Social and Occupational Sciences Research Co., Ltd.

Dainoff M.J. (1982) Visual fatigue in VDT operators. In: E. Grandjean, E. Vigliani (Eds), *Ergonomic aspects of visual display terminals* (pp. 95—99). London: Taylor & Francis,:

Kamienska-Zyla, M. (1993), Ergonomic evaluation of the work of VDT operators in Poland. *Applied Ergonomics*, 24, 432—433.

Kanaya, S. (1990). Vision and visual environment for VDT work. *Ergonomics,* 33, 775—785.

London, R. (1991a), Near point of convergence. In: J.S. Kridge, J.S. Amos, & J.B. Bartlett (Eds), *Clinical procedures in optometry.*(pp. 66—68). Philadelphia: JB Lippincott

London, R. (1991b), Amplitude of accommodation. In: J.S. Kridge, J.S. Amos, & J.B. Bartlett (Eds), *Clinical procedures in optometry.*(pp. 69—71). Philadelphia: JB Lippincott

Pickett C.W.L, Lees R.E.M. A cross-sectional study of health complaints among 79 data entry operators using video display terminals. *Occup Med* 1991; 41: 113—116

Smith M.J, Cohen B.G.F, Stammerjohn L.W (1981). An investigation of health complaints and job stress in video display operations. *Hum Factors,* 23: 387—400

Wolska, A, Switula, M. (1999), Luminance of the surround and visual fatigue of VDT operators. *JOSE*, 5(4), 553—580.

Is the trackball a serious alternative to the mouse? A comparison of trackball and mouse with regard to cursor movement performance in manipulation tasks

Martina Ziefle

Department of Psychology; Aachen University
Jägerstr. 17-19, 52056 Aachen, Germany
Martina.Ziefle@psych.rwth-aachen.de

Abstract

Two experiments were carried out focusing on the motor performance of two input devices, trackball and mouse. In both input devices, as independent variable cursor speed was examined in a low and a high version. A manipulation task had to be carried out. The 30 participants had to move the cursor from a starting point to a target that had to be highlighted precisely. Dependent variables were time and accuracy of cursor movements as well as user judgements due to handling comfort and muscular load. The results show that cursor speed is a crucial factor for motor performance, thus proving the slower cursor speed to result in a faster and more accurate manipulation performance. Contrasting the suitability of both input devices, the trackball´s usability for a quick and precise manipulation seems to be rather limited, showing a considerable lower performance in both, speed and accuracy. Furthermore, the handling comfort was judged to be lower by participants compared to the mouse.

1 Introduction

Since 1964, as the mouse was introduced in the computer work place, many studies (e.g. Card et al., 1978; McKenzie et al., 1991; Sperling & Tullis, 1988) focussed on the usability of the mouse and showed it to be a fast and precise input device. Current health statistics however- and this is a cause for concern- indicate a significantly increased prevalence of different kinds of computer work related muscular disorders (e.g. Aarås & Ro, 1997; Keir et al., 1999). Accordingly, the mouse is strongly discussed as a possible cause for arm, shoulder and neck strains (e.g. Burgess-Limmerick et al., 1999), often referred to as mouse-arm-syndrome. Sustained defective positions of hand, fingers (clicking) and wrists (turning the wrists out- and upwards, i.e. ulnardeviation, and dorsiflexion), are commonly made responsible for the mouse-arm-syndrome. Within the last years, new models of input devices were developed, meeting demands of the market (novelty of products) and demands of ergonomic-clinical needs. Unfortunately, most of these developments are rather concerned with demands of the market. Under an ergonomic perspective, the thumb-controlled trackball comes into question as a promising alternative to the classic mouse. Current thumb-controlled trackballs show to be smooth and easily running. Moreover, using this trackball, the hand is resting on the chassis of the device, thus no wrist deviations occur as the cursor is controlled by the thumb, the strongest digit disposing of an extra muscular supply.
Up to now only very few studies were concerned with a serious evaluation of the performance of modern thumb-controlled trackballs with regard to cursor movement performance and user acceptance. This was undertaken here. Two experiments were carried out focusing on the motor performance of two input devices, the trackball (Logitech Track Man Wheel) and the mouse

(Logitech Mouse Man Wheel). Even if the usability of the trackball is of main experimental interest, the mouse as the benchmark within input devices has been examined as a performance baseline. Thus it may be decided if the trackball can be recommended as a serious alternative to the mouse enabling the users to execute quick and accurate cursor manipulations.

2 Method

2.1 Experimental Variables

The *independent* variable was the cursor speed in a low and a high version. Both cursor speeds were determined in a preliminary study for both input devices by an extensive testing out of software drivers and the maximal cursor speed depending on different controller set ups and input device. The low cursor speed was set at 4751,8 Pixel/s for the mouse and 5151,4 Pixel/s for the trackball. The high mouse cursor speed was set at 5757,9 Pixel/s and at 5844 Pixels/s for the trackball.[1] Each participant worked with the low and the high cursor speed, whereupon the order of both speeds was alternated over participants.

The *dependent* variables were time and accuracy of cursor movement control. The time was defined as the interval between the start of the cursor movement until the object was clicked and highlighted properly. In order to assess movement accuracy, different aspects of motor adjustment were pursued. The clicking error occurs if participants clicked, while the cursor was not adjusted correctly, but outside the target. The highlighting error is present if the button to be pressed is released too early or late. Finally, a new descriptive measure for pointing accuracy has been developed (Sutter and Ziefle a in press, b submitted). This error is given, when the target was not hit correctly at once and the cursor is circling around the target zone. Once, the cursor goes too far, resulting in an overshoot, some other time the cursor subsequently undershoots the target. This impreciseness of fine cursor adjustment was registered, analyzing movements over- and undershoots carried out in the target zone. In order to get precise and valid measurements, a circular raster (virtual butt) was arranged around the target (figure 1). The "perfect" or ideal cursor way was set as the straight interconnection between start and target representing the bull´s eye in the butt. The deviation from the ideal way was descriptively registered, i.e. how often and how far the cursor deviated. As deviations, cursor movements crossing the boundary were considered. Graphically, this can be expressed as the width and amplitude of the executed cursor movements. Here, the distance from the target center was divided into categories, each consisting of 1.2 mm (4 pixel). The radius of the butt covered 12 cm. As for the user judgements, difficulties in the handling comfort was rated on a four point scale (4 = difficult, and 1 = easy). In addition, the extent of the muscular discomfort raised by the use of the input device and its location was indicated by participants.

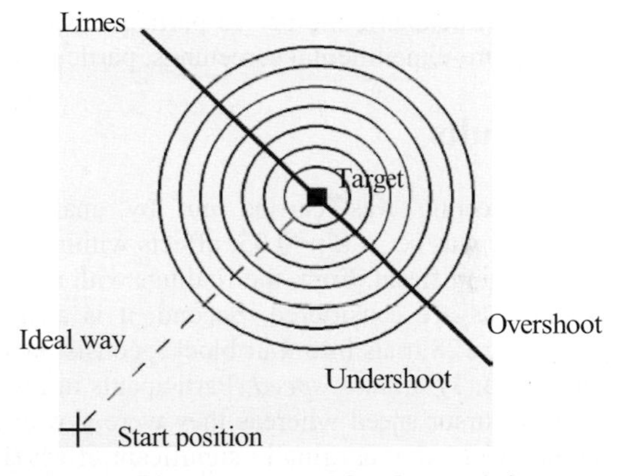

Figure 1: Circular raster for determining impreciseness of cursor movements around the target zone (from Sutter & Ziefle, submitted)

[1] Remarkably, even if the identical software driver (developed by the same company (Logitech) is used in different input devices, the resulting cursor speeds vary distinctly in the different device set ups. Extensively testing provided for a relative good comparability of cursor speeds in both input devices.

2.2 Experimental Task

As experimental task a text manipulation was used. The cursor had to be moved from the start to the center of a letter string where the underlined letters (target) had to be highlighted (figure 2). Each trial began by pressing the space bar and a string appeared on the screen. The cursor was to be moved to the target and if it had been positioned correctly, the black square target changed into green, providing for a visual feedback. Then, the target disappeared and the next string appeared. Participants were instructed to adjust and highlight the cursor as fast and accurate as possible. To exclude confounding effects of different movement directions, the targets appeared in eight directions around the starting point: 45°, 90°, 135°, 180°, 225°, 270°, 315° and 360°. The distance between start and target was 2.5 or 5 cm. At each cursor speed, 128 trials were executed. In total, 256 trials were completed with the mouse (exp. 1) and the trackball (exp. 2) by 15 participants, each. For training purposes, four practice trials were given in the beginning. After participants finished the 128 trials at each cursor speed, the handling comfort and the muscular load raised by the use of the input device were rated. Movement time, accuracy and keystrokes were recorded action-correlated by a software program specifically developed in preliminary studies for similar purposes (Sutter and Ziefle, a in press, b submitted).

Figure 2: Manipulation task with the central target to be highlighted

2.3 Participants

In the first experiment (mouse), 15 participants between 18 and 30 years (M = 25), in the second experiment (trackball), another 15 participants between 18 and 25 years (M = 23.8) took part. As taken from pre-experimental screenings, participants were highly used to computer work.

3 Results

Statistical testing was carried out by analyses of variances for repeated measurements. Significance was set at a $p<0.05$; effects within the less restrictive limit of $p<0.1$ were denoted as marginally significant. First, the findings with respect to movement time and accuracy due to both cursor speeds are considered. Second, it is analysed if the performance improves by training (dividing the 128 trials into four blocks, consisting of 32 trials each).

Mouse (Exp. 1). *Cursor speed*: Participants needed on average 1918.5 ms to highlight the target in the low cursor speed whereas they were slower in the high cursor speed (M = 2085 ms). This difference yielded a marginally significant effect ($F_{(1,14)} = 4.13$; $p = 0.06$). Overall, the frequency of errors was small, yielding no significant effects of cursor speed on accuracy (clicking error: slow cursor speed: M = 0.5; fast cursor speed M = 0.3; highlighting error: slow cursor speed: M = 11.1; fast cursor speed M = 12.7). In addition, a significant *training* effect ($F_{(1,14)} = 6.8$; $P < 0.05$) was present in the low cursor speed. It took on average 2029 ms to highlight the targets in the first 32 trials (block 1) and performance improved by 8% to 1859 ms in the last (trial 97-128). As for the high cursor speed, a marginally significant ($F_{(1,14)} = 3.3$; $p=0.089$ effect of training (almost 10%) was found (block 1: M = 2211 ms; block 4: M = 1994 ms). Though the errors decreased over the 128 trials, accuracy was not significantly affected by training. Figure 3 illustrates an insight into fine motor adjustment in the target area. The target is missed rather often at both cursor speeds, however more often in the fast cursor speed (black bars). Note that these cursor actions are not "necessary" due to a direct and proper highlighting procedure, thus representing extra "costs" lengthening movement times. The distribution is quite asymmetric thus pointing to an overbalance of cursor actions going too far. Apparently, even in the highly trained mouse, the inaccuracy of motor adjustment is present, subtending an area of 4.2 cm around the target center.

Trackball (Exp. 2). Results of cursor speed and effects of training on movement time and accuracy are presented. *Cursor speed*: Again, the low cursor speed (M = 2650 ms) performed better as the high cursor speed (M= 4598 ms), decreasing movement times by more than 40%. The effect of cursor speed on movement times showed to be significant ($F_{(1,14)} = 7.8$; $p < 0.05$). However, cursor speed did not significantly affect accuracy. Concerning fine cursor adjustment around the target zone, the inaccuracy in the trackball was still more distinct, revealing not only a higher frequency of over- and undershoots, but also a wider range (6 cm) as compared to the mouse (figure 4). With respect to the *training*, no significant improvement was identified in the trackball, even if in both cursor speeds the movement time decreases from the first block to the last block (figure 5).

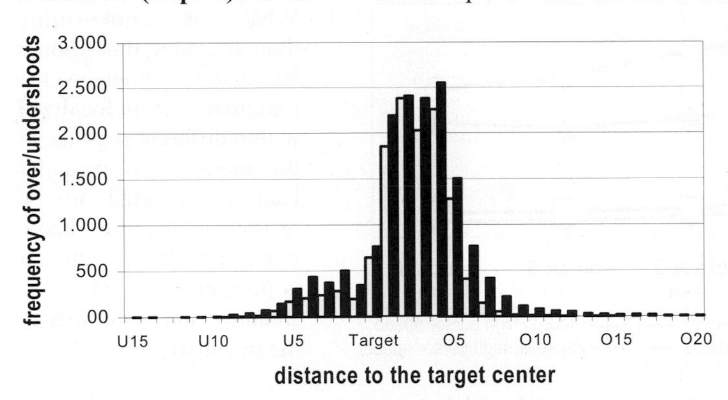

Figure 3: Mouse: Distribution of errors executed in the target zone (O20=Overshoot of 24 mm; U 15=Undershoot of 18 mm). Black bars: fast cursor speed, white bars: slow cursor speed.

Mouse versus trackball: Both input devices differ with respect to the ex ante expertise: per contra to the trackball, the mouse is highly trained. If both input devices are directly compared and the comparison is aimed to be "fair", only the last 32 trials (block 4) should be contrasted. Succinctly, a strong overall advantage of the mouse over the trackball was proven. Analysis of movement times yielded a significant effect of cursor speed ($F_{(1,28)} = 10.2$; $p < 0.05$), indicating that the slow cursor speed showed distinctly better motor performance. Moreover, a significant ($F_{(1,28)} = 5.0$; $p < 0.05$). interaction of cursor speed and input device was found, thus hinting at the high cursor speed to be an aggravating factor if the trackball is used as an input device.

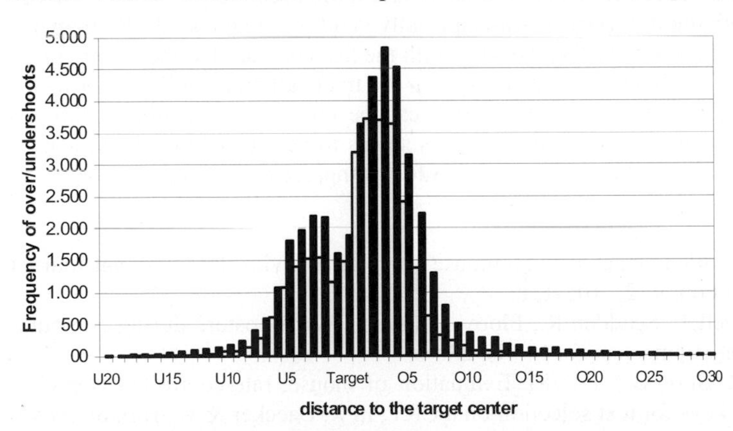

Figure 4: Trackball: Distribution of errors in the target zone (O30=Overshoot of 36 mm; U20=Undershoot of 24 mm).Black bars: fast cursor speed, white bars: slow cursor speed)

A final inspection is concerned with the judged quality of both input devices. *Handling comfort*: No differences in user ratings for both input devices were found with respect to the usage of mouse keys and form. Big differences though were reported due to the easiness of cursor control: in the trackball, the cursor is reported as rather uncontrollable (especially in the high cursor speed) compared to the mouse. *Muscular load:* Only 53% of the trackball and 40% of the mouse users reported muscular load at all. In addition, the overall extent of muscular load is rather low as

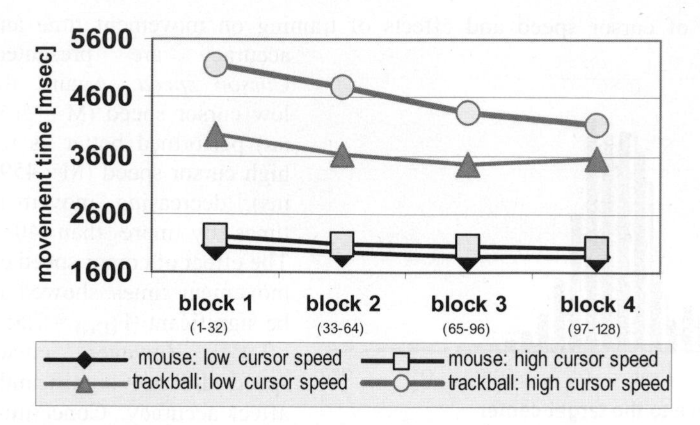

Figure 5: Movement times (means, ms) in both input devices and cursor speeds

young users took part. What is noteworthy though - and this should be pursued further on- the muscular load is localized within different regions: in the mouse, the muscular load is reported to be restricted to the hand, wrist and elbow, while it is primarily located in the shoulder and neck area in the trackball.

4 Discussion and Conclusions

The present study aimed at an ergonomic analysis of the thumb-controlled trackball with respect to the usability of cursor control. Two cursor speeds were tested and time and accuracy of fine cursor adjustment were analysed. The mouse, the benchmark, was taken as a baseline. The outcomes draw a clear picture. First, the cursor speed showed to be a crucial factor for movement times and accuracy, with the slower cursor speed resulting in a faster and more accurate performance. Even if this could be proven for both input devices, it was especially valid for the trackball: Participants showed a significant lower performance (18%) working with the fast compared to the lower cursor speed. Even if a training effect of about 22% was present in the trackball, trackball´s performance was still significantly inferior to the mouse. Over all it can be concluded, that the trackball´s suitability for a quick and precise manipulation of objects seems to be rather limited, showing a considerable lower performance and a lower handling comfort compared to the mouse.

References

Aarås, A. & Ro O. (1997). Workload when using a mouse as an input device. International Journal of Human-Computer Interaction, 9 (2), 105-118.

Burgess-Limerick R., Shemmell J., Scadden R., Plooy A. (1999). Wrist posture during computer pointing device use. Clinical Biomechanics, 14 (4), 280-6.

Card, S.K., English, W.K. & Burr, B.J. (1978). Evaluation of mouse, rate-controlled isometric joystick, step keys and text keys for text selection on a CRT. In R. Baecker & W. Buxton (eds.), Readings in Human-Computer Interaction (pp.386-392). San Mateo, CA: Morgan Kaufmann..

Keir, P., Bach, J. & Rempel, D. (1999). Effects of computer mouse design and task on carpal tunnel pressure. Ergonomics, 42(10), 1350-1360.

McKenzie, I. S., Sellen, A., Buxton, W. (1991). A comparison of input devices in element pointing and dragging tasks. In: Proceedings of the CHI' 91 Conference on Human Factors in Computing Systems. (pp. 161-166). New York: ACM.

Sperling, B. & Tullis, T. (1988): Are you a better mouser or trackballer? A comparison of cursor positioning performance. SIGCHI Bulletin, 19(3), 77-81.

Sutter, C. and Ziefle, M. User's Expertise: A biasing Factor for Performance and Usability of Notebook Input Devices. Submitted for publication in Human Factors.

Sutter, C. and Ziefle, M. (in press). How to handle notebook input devices: An insight in button use strategy. In P. McCabe (ed). Contemporary Ergonomics, in press.

Section 2

Cognitive Ergonomics

Human Performance in Cognitive Tasks Involving Multimodal Speech Interfaces

Azra N. Ali and Philip H Marsden

University of Huddersfield
Huddersfield, England.
a.n.ali@hud.ac.uk p.h.marsden@hud.ac.uk

Abstract

Increasing numbers of human factors orientated studies are focusing on 'text-to-speech' and speech recognition as a modality in speech-based interfaces. Relatively little work, however, has been directed towards the study of the implications of working with multimodal speech interfaces on the quality of human performance. In this experiment audio-visual channel misalignment was investigated to determine how this affects the quality of human-computer interaction. The results indicate that multimodal speech interfaces involving talking head technology can lead to increased workload demands on users when problems arise with the synchrony of visual and audio data. The findings provide a valuable insight into the development of talking head interface (THI) designs and suggests that alignment of the two channels is critical for optimal human-computer performance.

1 Introduction

Increasing numbers of human factors orientated studies focus on 'text-to-speech' and speech recognition as a modality in speech-based interfaces (Oviatt, 1997; Adams, Damper, Harnad, & Hall, 1999). Relatively little attention, however, has been directed towards thc study of the implications of working with multimodal speech interfaces on the quality of human performance. In this study we aim to redress this inbalance by investigating human response to human talking head interfaces (THIs), which are becoming increasingly used in many interface designs, for example, information booths/kiosks, intelligent educational tutors (Miller, 2001, Ward et al, 2003) and as an aid in task learning. One of the reasons THIs are attractive to system designers is that they can help make the design of the interface more natural and convenient for users to exchange information. From a scientific perspective, multimodal speech interfaces are interesting because they involve the overall human cognitive and performance system, in that the interface employs more than one human perceptual motor modality, the auditory and visual channels.

1.1 Multimodal Speech

People use both acoustic and visual modalities to understand speech, although many are not aware of the visual component. Strong evidence for the visual component is found amongst the many people with a hearing impairment who can understand speech by lip-reading (Dodd & Campbell, 1987; Beskow, Dahlquist, Granström, Lundeberg, Spens, & Öhman, 1997). Studies have shown that visual information provided by the movements of a speaker's mouth and face strongly influences what an observer perceives, even when the auditory signal is clear and the observer's hearing is good (Massaro, 1998). When the audio channel is noisy, information from the visual channel significantly improves the accuracy of speech perception, as was demonstrated quantitatively by Sumby and Pollack (1954). Perceiving a place of articulation contrast, such as that between labial /b/ and palatal /d/, is difficult via sound but relatively easy via sight. Thus, congruent audio and video speech channels not only provide two independent sources of information, but do so complementarily: each is strong when the other is weak. Furthermore, the complementarity makes accurate speech perception more resistant to channel noise (Robert-Ribes, Schwartz, Lallouache, & Escudier, 1998). Evidence for strong interaction between audio and visual speech channels in human speech perception is found in the well-known McGurk effect (MacDonald & McGurk, 1976). If humans are presented with temporally aligned but conflicting audio and visual stimuli – now known as 'incongruent stimuli' - the perceived sound may differ from that present in either channel. For example, in 1976, McGurk and MacDonald asked their recording technician to create a videotape with the audio syllable [ba] dubbed onto a visual [ga], most normal adults reported hearing [da] or [tha]. But when the subjects were presented with only one modality (visual or audio) they reported the syllables correctly.

2 Experiment

Previous studies, have concentrated only on participant's responses to congruent and incongruent multimodal speech stimuli in nonsense syllables. The aim of our study is to determine if participant's performance levels are affected when the audio channel is not in synchrony with the visual lips movements using real English monosyllabic words. In addition, we also measured performance levels when noise was added to the stimuli. We used two types of human performance measurements; objective and subjective measurements as detailed below.

2.1 Objective human performance measurements

Objective measurements are important because they provide a true measure of real-time performance of cognitive workload without any user intervention. Our study showed that when the audio channel is not properly synchronised with visual lip movements, it thus increases individual's cognitive workload in trying to determine what the talking head is saying. Extra mental effort is further increased when background noise was introduced. We measured cognitive workload in two different ways:

Decision times and task complexity; our experiments probed the hesitations of participants *via* two measures. One was the total time taken by a participant to select a response to a stimulus from an open list of possible responses, as response execution is regarded as one stage of information processing (Xie, & Salvendy, 2000). The other was the number of replays used by the participant before reaching a decision about the stimulus. Both measures also determined the level of task complexity and were logged automatically by our experimental control software.

Psychophysiology; using psychophysiology measures allows one to monitor subtle changes in individual's physiological functions in relation to variations of a given task (Andressi, 2000). An important aspect of using psychophysiological measures is that it illuminates both the mind and the body of users interacting with multimodal speech interfaces. Using continuous physiological measurement to measure the workload, such as heart rate, blood volume pressure and electrodermal activity, we were able to monitor individual's reactions to multimodal speech stimuli. These data allowed inferences to be drawn about the impact of the McGurk effect on the functioning of participants' levels of autonomic nervous system activity, thus providing an indication of stress levels.

2.2 Subjective human performance measurements

The participants were provided with a report form on which to record 'what they thought the speaker was saying' when receiving an experimental stimulus, thus reflecting on the participant's perceptual motor skills. The report form included the following text-words: one corresponding to the audio channel of the stimulus, one for the video channel, one each for possible results of channel fusion, some random words and a space to write in a word not included on the form. The results showed that where channels were congruent subjects accurately reported the correct response to the stimuli and when channels were not congruent majority of the participants perceived the McGurk fusion. The fusion rate further increased with cocktail party noise.

3 Method

Creating the stimuli and software - video recordings were made of a male (aged 23 years) and a female (aged 22 years), both native speakers English speakers. The video recordings were done inside a quiet, controlled Usability Lab using a standard 8 mm digital Sony Camcorder with built-in microphone for audio. Incongruent video clips were created whereby the audio was manually synchronised (using Adobe Premier) with the visual lip movements of the talking head uttering a different word, example shown in Table 1.

	Original clip	**Incongruent clip**
Audio channel	"map"	"map"
visual channel	"map"	"mat"

Table 1: Showing pairing of incongruent stimulus

The incongruent stimuli, words were grouped into contrasting pairs of consonants (e.g. onsets: *tile/pile, tail/fail* etc, codas: *map/mat, beef/beet,* etc) and vowels (e.g. short: *hod/hid, head/hood,* etc, long: *hoard/hoed, hard/heed,* etc). The incongruent clips, some with 'cocktail party' noise added acoustically and a few natural controls were then saved as *.avi files with a frame rate of 30 per second and frame size of 320mm x 270mm. Software was designed to incorporate all the stimuli clips together. The application had a function whereby it would automatically record the participant's personal details (name, age and gender), duration time and number of replays for each stimulus per participant. The software compiled the data collected into a csv (common separated value) file format, enabling data to be exported into statistical packages for analysis.

Procedure - fifty participants, males and females took part in the experiment with an age range between 21 to 54 years. None had hearing problems and all either had normal vision or wore prescribed corrective lenses. Participants sat about half a metre from the monitor screen and used headphones connected to the computer to listen to the audio. Participants were told that they should report 'what they thought the talking head was saying' when receiving an experimental stimulus on the report forms provided, replaying the clip as many times as they needed to reach a decision. No time limits were set for task completion, and no feedback about the experiment was given to participants. Before commencing the experiment, electrodermal activity and blood volume pressure sensors were attached to the fingers of the participant's non-dominant hand. The first five minutes prior to running the video clips was used as a settling-in period allowing both the subject and sensors to settle down. Participants were prompted when to commence the experiment. Two PCs were used: one to present the multimodal speech stimuli; the other using Datalab 2000 in conjunction with National Instruments BioBench software, to collect the physiological signals from the sensors that were attached to the participant's fingers.

4 Discussion of Results

4.1 Objective results

Exploratory analysis was focused on the total decision times and number of replays per stimulus over all subjects, Figure 1. Both were examined for variation with the state of the two channels (congruent, incongruent without noise and incongruent with noise) as shown graphically. The graph clearly shows that when in incongruent channels (with and without noise), participants took longer to decide in what the talking head was saying, this is echoed also in the number of replays of the stimuli.

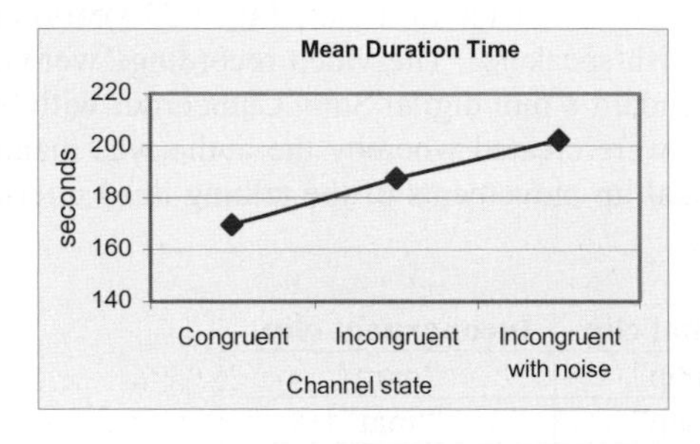

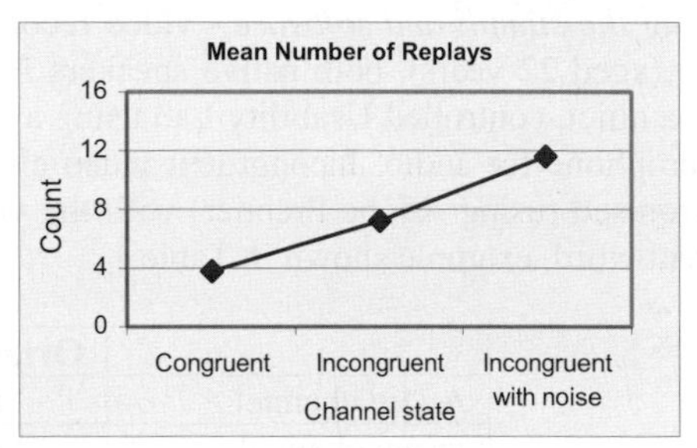

Figure 1: Decision times for various channel states
time to respond in seconds and number of replays of the stimuli

Time to respond: Using ANOVA, with all the stimuli, the null hypothesis that time to respond is independent of channel state was rejected ($F=4.986$, $p<0.01$). With all incongruent stimuli partitioned by types of response (decision for audio, decision for visual, decision for fused), null hypotheses that time to respond are the same for noise and noiseless cases were rejected both for visual and fusion responses ($F=4.979$, $p<0.05$; $F=5.869$, $p<0.01$ respectively) but accepted for audio response at $F=2.637$, $p=0.105$.

Number of replay of stimuli: Using ANOVA, with all the stimuli, the null hypothesis that number of replays is independent of channel state was rejected ($F=23.915$, $p<0.001$). With all incongruent stimuli partitioned by types of response (decision for audio, decision for visual, decision for fused), null hypotheses that time to respond are the same for noise and noiseless cases were rejected both for visual and fusion responses ($F=23.399$, $p<0.001$; $F=23.386$, $p<0.001$ respectively) but accepted for audio response at $F=0.076$, $p=0.783$.

Psychophysiology: data logged from the BioBench software were exported to Microsoft Excel for preliminary analysis. Skin conductance (SC), blood volume pressure (BVP) and pulse rate (BPM) over the second minute of the congruent stimuli were taken as the baseline reading for each participant. Our preliminary results produced the following picture; subjects differ on their baseline values, indicating that each participant has a unique psychophysiological characteristic. BVP – a total change of 68% for realignment of channels with 'cocktail-party' noise from the baseline. SC – increased by 4% for realignment of channels with 'cocktail-party' noise from the baseline. BPM – a slight increase in the pulse rate of 1.5 % but the standard deviation of pulse rate increased considerably from the controlled stimuli to incongruent stimuli and incongruent stimuli with noise.

4.2 Subjective results

Where channels were congruent subjects accurately reported the correct response to the stimuli. However, in incongruent, 42.1% of the participants experienced the McGurk effect fusion and stated accordingly so on the report forms, this increased to 49.8% for incongruent channels with noise stimuli, although more participants this time reported the visual channel rather than the audio channel. Using chi-squared test, differences of fusion rates brought about by noise were tested: the hypothesis of null difference was rejected at $p< 0.01$.This clearly shows under noisy conditions speaker's mouth and face strongly influences what an observer perceives (Sumby & Pollack, 1954).

5 Conclusion

The experiment demonstrates that using multimodal speech interfaces involving talking head technology can lead to increased workload demands on users when problems arise with the synchrony of visual and audio data. This was clearly revealed in the physiological data collected as part of this experiment. In additions, the study also showed performance decrements resulting from a phenomenon known as the McGurk effect. In this situation, the listener misperceives the audio-visual message when the two output channels lack synchrony. The findings therefore provide a valuable insight into the development of THI designs and suggests that alignment of the two channels is critical for optimal human-computer performance. Additional research is planned to investigate the affect of misalignment on short phrases and sentences.

6 References

Adams, L., Damper, R., Harnad, S., and Hall, W., (1999) A system design for human factors studies of speech-enabled Web browsing. In proceedings ESCA Workshop on Interactive dialogue in multi-modal systems, pp. 137-140.

Andreassi, J.L. (2000) Psychophysiology: Human Behaviour and Physiological Response. Lawrence Erlbaum Associates.

Beskow, J., Dahlquist, M., Granström, B., Lundeberg, M., Spens, K-E & Öhman, T. (1997) The teleface project: Multimodal speech communication for the hearing impaired. In proceedings Eurospeech, Rhodes, Greece.

Dodd.,B and Campbell, R (1987) Hearing by Eye: The psychology of lipreading, Lawrence Erlbaum Associates Ltd, U.K

Massaro, D.W., (1998) Perceiving talking faces: From speech perception to a behavioral principle. Cambridge, Mass: MIT Press.

MacDonald, J. and McGurk, H., (1976) Hearing lips and seeing Voices, *Nature* 264, December 23/30: pp. 746-74.

Miller, O. (2001) Talking Heads, The University of Memphis Magazine. Retrieved February 1, 2001, from http://www.memphis.edu/magazine/v19i2/feat4.html.

Oviatt, S (1997) Multimodal interactive maps: Designing for human performance. *Human-Computer Interaction*, 12(1–2), 93–129.

Sumby, W.H and Pollack, I. (1954) Visual contribution to speech intelligibility in noise, *Journal of the Acoustical Society of America*, 26, 212-215.

Robert-Ribes, K., Schwartz, J., Lallouache, T. & Escudier, P.(1998) Complementarity and Synergy in Bimodal Speech, *Journal of Acoustic Society of America*, 103(6), 3677-3689.

Xie, B and Salvendy, G (2000) Prediction of mental workload in single and multiple tasks environments, *International Journal of Cognitive Ergonomics*, 2000, 4(3), 213-242.

Performance on Mobile Phones: Does it Depend on Proper Cognitive Mapping?

Susanne Bay & Martina Ziefle

Psychology Department
Aachen University
Jägerstr. 17-19, 52056 Aachen
Susanne.Bay@psych.rwth-aachen.de
Martina.Ziefle@psych.rwth-aachen.de

Abstract

The present study aims at showing whether performance of novice cellular phone users aged 9 to 16 depends on the quality of the cognitive map of the device's menu structure these users build. The mental representation of the menu structure was surveyed using the card sorting technique after solving 8 tasks on a simulated mobile phone. Moreover, user judgements regarding the phone's usability were collected and their accordance with both, performance and mental representation, was considered. A high interrelation between the performance when solving the relevant task and the correctness of the mental model assessed via card sorting was found: Children who were fast and made few detour steps on the cellular phone structured more cards correctly. On the other hand it was not possible to identify the difficulties the children experienced with the menu structure through their subjective estimations. It may be followed, first, that supporting the process of cognitive mapping, for example by providing maps of the menu organisation might facilitate handling a phone. Secondly, it is concluded that assessment of mental models of technical devices can give deeper insights into usability of a menu, respective the transparency of functions and their structure, than consumer judgements.

1 Introduction

Users' difficulties handling many technological devices including mobile phones are well known and have recently been demonstrated in empirical studies (Ziefle, 2002a, b; Bay & Ziefle, under revision, Ziefle, submitted). However, the specific reasons for cellular phone users' problems are still unclear. One issue may be the particular organization of the phone's functions under superordinate terms in a hierarchical menu tree. It can be assumed that, especially due to the display's inability to show more than a small section of the menu, some users fail forming an appropriate mental representation – a cognitive map – of the menu structure when navigating through the various functions and get lost. This process of building a mental map can be compared to orientation in a natural environment, where three types of spatial knowledge are assumed to be of importance (Thorndyke & Goldin, 1983): *Landmark* knowledge representing salient features on the route, *procedural* knowledge of the sequence of actions required to get from one point to another, and *survey* knowledge which represents an overview of locations and routes in the environment.

Regarding disorientation in hyperspace and hypertext the spatial metaphor is frequently used as explanation to help understanding such problems as "becoming lost" in the menu structure (e.g.

Farris, Jones & Elgin, 2001; Kim & Hirtle, 1995). A number of studies have shown the positive effect of a map of the menu structure (which provides survey knowledge) on users' performance navigating in hyperspace (e.g. Billingsley, 1984; McDonald & Stevenson, 1998). Therefore it is assumed that the quality of the mental map of the menu structure the users themselves build while navigating must also be of importance for the individual efficiency. Farris et al. (2001) examined the cognitive maps acquired while using hypermedia with varying depth. They found, however, that the participants' spatial representation of the website - which was surveyed via drawings - did not reflect the connection structure inherent in the hypermedia. This raises the general question whether proper handling of a cellular phone depends on a well-defined mental map of its menu structure or if the mental representation of the menu's spatial structure is unrelated to performance.

A second issue that is addressed is the usefulness of inferences from mental model assessment on the usability of a device. If the mental model of most users does not correspond to the arrangement of functions in the device, it might be concluded that this device's menu is contraintuitive. But, it has to be shown that a correct mental representation is related to performance, possibly even stronger related than other usability data, namely users' subjective estimations of the ease of use.

2 Method

2.1 Participants

As children are assumed to be open towards new technologies and learn fast, this group was chosen as participants for the study. 21 children between 9 and 16 years participated in the study. The average age was 13 years. All of them had no or very limited experiences using the different functionalities of a cellular phone.

2.2 Procedure

Participants processed four commonly executed operations (calling a person using the internal phonebook, sending a short message, hiding the own phone number when calling and editing an entry in the phonebook) twice consecutively on two mobile phones that were simulated on a touch screen according to two existing models, Siemens C35i and Nokia 3210. The two phones differed regarding menu and navigation keys but were identical regarding all other features. Half of the subjects used the Siemens C35i, half the Nokia 3210. During task processing each action of the user was recorded through logfiles in order to measure time spent on task, detour steps (keystrokes that did not lead directly to the solution of the task) and specific errors.

After completing the tasks the participants were asked for their experienced difficulties handling the phones, specifically, their comprehension of the terms used in the menu, the organization of functions under superordinate terms, and the keys' functionalities. The answers to the questions were to be given referring to smiling (=1), neutral (=2) and sad (=3) faces shown on the questionnaire.

Then they were invited to map their mental model of the organization of the menu functions through card sorting. As the whole menu is too big to be structured by the children and only parts of it are being used when processing the eight tasks, the menu branch they had passed when processing the task of hiding the phone number was chosen. The structure of this menu branch consisted of 22 functions arranged in four levels in both simulated phones. 22 cards, each labelled with one of the function's names, were presented to the participant who had to arrange the cards on a table according to the memorized structure of the menu.

3 Results

To process the task of hiding the phone number when calling the participants spent on average five minutes in the first turn and two minutes in the second. Six children (four using the Siemens C35i, two using the Nokia 3210) did not achieve to solve the task in the first trial, whereas all but one (who used the Nokia) were successful in the second trial. A detailed description of the outcomes of users' performance on the two different cellular phones is being reported elsewhere (Bay & Ziefle, under revision), the focus of this paper is the cognitive mapping of the menu structure and its relation to the performance using the device.

Results of the card sorting task show that 18 out of the 21 participants arranged the cards in a hierarchical tree structure (only those participants are included into further analysis), two laid clusters of two to five items grouped together and one child structured the functions into one row. The children's hierarchical menu trees had on average 3.7 levels thus reflecting the real structure (of four levels) nearly correctly. Yet, as most children did not integrate all functions, the menu branch created through card sorting was not as broad as the actual menu (average of 5.3, 4.6, 3.2, and 2.1 functions on menu levels one to four instead of 9, 8, 3, and 2 in the genuine Siemens branch and average breadth of 7, 4.9, 3.2, and 2 instead of 10, 4, 6, and 3 in the Nokia menu branch). The allocation of the functions to the right superordinate term and the right level was also rather poor: Only 7 out of 18 participants were able to map the complete route of functions that had to be selected to solve the task. The children connected on average 10.2 of the overall 22 functions of the menu branch correctly. As reported elsewhere, kids using the Nokia 3210 processed the tasks significantly faster than with the Siemens C35i (Bay & Ziefle, under revision). The Nokia phone's menu was also mapped better ($t(16) = 2.2$; $p < .05$): While the Siemens C35i users allocated only on average 8.6 of the 22 functions correctly, those participants using the Nokia 3210 mapped 12.3 of 22 cards to the right superordinate term on the right level.

The importance of a correct cognitive map for successful interaction with the device can be seen in three different measures: Firstly, those participants who did not picture a hierarchical menu organization tended to solve on average less tasks ($M = 6$) than the rest ($M = 7.4$) ($t(19) = 2.0$; $p = .055$). Secondly, there is a significant difference in performance in both trials between those kids who mapped the route that had to be passed to solve the task of hiding the own number correctly, and those who did not. In the second trial (where the tasks was solved by 20 of the 21 subject) the seven participants who mapped the route correctly took on average 29 seconds undertaking 13.7 detour steps whereas the others spent 201.5 seconds ($t = 2.73$; $p < .05$) undertaking 157.6 detour steps ($t = 2.4$; $p < .05$), as figure 1 shows. Thirdly, as to be seen in table 1, high correlations were found between the performance in the phone task, especially in the first trial, and the correctness of the mental model of the whole menu branch assessed via card sorting ($r = .61$, $p < .01$ for the first trial, $r = .48$, $p < .05$ for the second). Children who were fast and made few detour steps on the cellular phone thus structured more cards correctly. On the other hand it was not possible to identify the difficulties the children

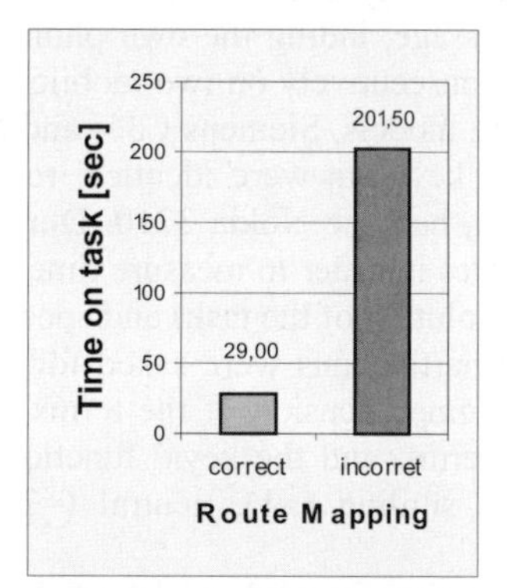

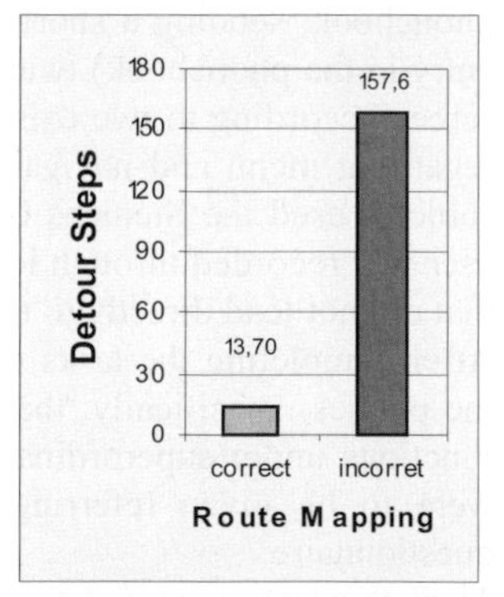

Figure 1. Users' performance hiding the phone number in the second trial depending on mental mapping

experienced with the phone's menu structure through their subjective estimations: The answers to the two questions regarding the comprehensibility of the functions' naming and the felt plausibility of the allocation of functions under superordinate terms did not show any significant correlation with performance measures (see table 1) and cognitive mapping. Yet, the children were able to give differentiated estimations on the experienced difficulties solving the tasks and the comprehensibility of the key functions. The reported overall difficulty using the phone ($r = .61$; $p < .01$) and understanding the phone's navigation keys ($r = .87$; $p < .001$) was significantly correlated with the performance solving tasks on the phone (total time on task).
Moreover, systematic errors participants made in the card sorting tasks can be used to draw inferences about the intuitive allocation of items. For example did three out of ten participants in the Nokia group allocate the function "Call Diverting" under "Phone Settings" whereas it is located on the highest menu level.

Table 1: Correlations between performance on the cellular phone and quality of the mapped menu structure as well as users' estimations of the menu's usability (N=18)

	Time on Task			Detour Steps		
	First Trial	Second Trial	Total (tasks 1-8)	First Trial	Second Trial	Total (tasks 1-8)
Number of correct mappings	**-.61****	**-.48***	**-.48***	**-.54***	**-.43***	**-.42***
Understanding of functions' names	.15	.07	.15	.13	.-.09	.13
Understanding of functions' classification	-.03	-.11	.06	-.06	-.02	.13

**p > .05 **p > .01*

4 Discussion

The study focussed on detecting the relation between a user's efficiency using a cellular phone and his cognitive mapping of the phone's menu structure as well as the usefulness of analysing users' cognitive maps. It was demonstrated that users who have a good cognitive map of a phone's menu structure process tasks on the device faster, making less detour steps and solve more tasks correctly. On the other hand no relation between performance on the phone and subjective estimations of the clarity of the menu could be detected.
Which types of spatial knowledge of the menu structure are of importance for efficient interaction with a cellular phone? Results in this study showed that children who mapped the specific route of functions needed to solve the relevant task correctly, performed better. Moreover, participants that were aware of the menu's tree structure solved more tasks than those who were not. Accordingly, not only landmarks – functions that are recognized as leading to the goal by the users – but also procedural knowledge, cognitive representations of the right path to follow, and survey knowledge of the menu's spatial organization seem to be of importance.
As the present study shows connections between different measures without experimentally varying one of the variables, it can not be clearly identified what causes what. It is not clear whether the correct mental model is necessarily a prerequisite for effective and efficient interaction with the phone or if successful usage of the cellular phone leads automatically to good cognitive mapping of the menu organization. Yet, one could argue that those subjects who spend more time processing tasks undertaking more detours have a better chance of memorizing the functions' names and locations. The opposite was the case, as our results show, corroborating our hypothesis.

Which kind of conclusions can be drawn from the finding that cognitive mapping is presumably of importance for effective interaction with a device like the mobile phone? As the users still had a rather sparse model of the menu structure after solving eight tasks on the cellular phone, but those with a better cognitive map performed better, it might be indicated to include an overview of the whole menu tree into the devices' manual to support the process of cognitive mapping. The effects of this kind of help are already being addressed in further studies (Bay, in press).
It may be argued that participants differed with regard to their spatial abilities and memory and that these differences account for performance differences on the cellular phone as well as in the card sorting task. Further studies will have to show the impact of these abilities on users' efficiency and effectivity solving tasks on the device to provide external validity of the data.
The second interesting finding was the low correlation between users' performance and the subjective estimations regarding the comprehensibility and allocation of functions, even though other measures show that the children were able to give realistic estimations of their experienced difficulties. Other studies confirm that user ratings are less sensitive towards differences in performance than action logging (Ziefle, 2002a,b; Ziefle, submitted). It may be concluded that consumer judgements conventionally solely used by companies for usability assessments are not able to detect all the weaknesses of a technological product. Alternative methods such as performance measures and card sorting to evaluate users' mental models should be considered.

5 References

Bay, S. (in press). Cellular phone manuals: Users' benefit from spatial maps. In *CHI 2003 Extended Abstracts of the Conference on Human Factors in Computing Systems.* New York: ACM.

Bay, S. & Ziefle, M. (under revision). Effects of cellular phone's complexity on children's performance. Manuscript submitted to *Human Factors*.

Billingsley. P. A. (1982). Navigating through hierarchical menu structures: Does it help to have a map? *Proceedings of the Human Factors Society 26th Annual Meeting*, 103-107.

Farris, J. S., Jones, K. S. & Elgin, P. D. (2001). Mental representations of hypermedia: An evaluation of the spatial assumption. *Proceedings of the Human Factors and Ergonomics Society 45th annual meeting,* 1156-1160.

Kim, H., & Hirtle, S. C. (1995). Spatial metaphors and disorientation in hypertext browsing. *Behaviour and Information Technology*, 14 (4), 239-250.

McDonald, S. & Stevenson, R. J. (1998). Navigation in hyperspace: An evaluation of the effects of navigational tools and subject matter expertise on browsing and information retrieval in hypertext. *Interacting with Computers,* 10, 129-142.

Thorndyke, P. W. & Goldin, S. E. (1983). Spatial learning and reasoning skill. In: H. L. Pick & L. P. Acredolo. *Spatial Orientation. Theory, Research, and Application*. New York: Plenum.

Ziefle, M. (2002a). Usability of menu structures and navigation keys in mobile phones: A comparison of the ease of use in three different brands. *Proceedings of the 6th International Scientific Conference on Work with Display Units,* 359-361. Berlin: Ergonomics.

Ziefle, M. (2002b). The influence of user expertise and phone complexity on performance, ease of use and learnability of different mobile phones. *Behaviour and Information Technology*, 21 (5), 303-311.

Ziefle, M. (submitted). Older adults and the man machine interface. Aging effects on the usability of different cellular phones. Manuscript submitted to *Human Factors*.

Developments in the Area of Cognitive Systems: Reducing the Gap Between Production Systems and Naturalistic Decision Making

Nathan G. Brannon

Sandia National Laboratories[1]
Albuquerque, NM 87113-0830
ngbrann@sandia.gov

Richard J. Koubek

Penn State University
University Park, PA 16802
rkoubek@psu.edu

Abstract

Cognitive systems are increasingly being used to support human decision making for complex cognitive tasks. While tools such as expert systems have made contributions to the support of human decision making, systems that adhere to cognitively plausible mechanisms are increasingly being introduced to simulate cognitive performance for addressing tasks of high complexity. This paper includes a brief review of the Adaptive Control of Thought – Rational (ACT-R) system and an architecture based on Klein's Recognition Primed Decision Making model. Finally, implications are discussed leading to a hybrid architecture that seeks to merge existing capabilities in modeling more structured decision making with an instantiation of naturalistic decision making.

1 Introduction

A philosophical approach emerging within research communities, such as those studying human-computer interfaces, has resulted in an effort to improve the automated system's ability to understand the individual utilizing that particular system. Traditionally, automated systems run the same procedures and algorithms regardless of who "logged on." In contrast, it is of interest to design automated systems that conform to a particular user based on factors such as level of expertise, background, and/or individual differences. In theory this would result in a truly tailored and more agile system.

A significant part of understanding a particular user is to better understand and model the manner in which relevant data is being processed. This contention is an important deviation from many decision support systems, which maximize functionality at the exclusion of developing cognitively plausible mechanisms of information processing. For example, a calculator is a decision support system that does not process data in a human-like or biomimetic fashion. While modern and analogous "calculators", such as search engines and automated agents, can manage an increasing level of complexity, it is argued that an improved understanding of specific users and how information is being processed will enhance the capability of man-machine systems. The

[1] Sandia is a multiprogram laboratory operated by Sandia Corporation, a Lockheed Martin Company, for the United States Department of Energy under contract DE-AC04-94AL85000.

automated system may itself be less functional, but the human-machine system can ultimately yield greater capability as application areas continue to increase in complexity.

Models of human performance that utilize cognitively plausible mechanisms have been developed in two traditionally separate research areas, naturalistic decision making (Klein, 1989) and production systems (Brannon & Koubek, 2002). The objective of this paper is to review these two areas and to introduce a conceptual model that merges the respective strengths of each.

2 The Gap Between Models of Structured and Naturalistic Tasks

Production systems traditionally model human performance associated with tasks that are well structured such as arithmetic or logical problems (e.g., Luchins' waterjug problem (Luchins, 1942)). Brannon and Koubek (2002) provide an overview of four different production systems. Naturalistic decision making models often model less structured tasks such as battlefield command and control, fire fighting, or close quarters battle.

Production systems leverage empirical studies and computational tools to construct highly accurate simulations of human cognitive performance. With such accuracy, production systems have been sought for their potential predictive capabilities. Unfortunately this prospect has not been as promising as anticipated. While production systems have made some progress in modelling more real-world tasks, significant discrepancies remain between production system models and the human performance observed in naturalistic settings (Pew & Mavor, 1998).

In contrast to production systems, naturalistic decision making (NDM) models have focused on real-world tasks consisting of more complex and less structured factors (Klein, 1997). In particular NDM models have investigated the characteristics of expert decision making, which involve devoting more resources to gaining situation awareness than deliberating over alternative courses of action. Methods employed to derive such models generally include observation of human performance and semi-structured interviews with subject matter experts. The product is a descriptive model of the behavior including critical cues and heuristics necessary to perform a task. A feature absent from NDM models is a computational representation of the performance depicted in the descriptive model.

Production systems and NDM models each exhibit desirable strengths. However, research has until recently fallen short of generating a computational system that can simulate naturalistic human performance in a cognitively plausible manner.

2.1 Computational Instantiation of Recognition Primed Decision making

An effort to model and simulate naturalistic decision making is being pursued at Sandia National Laboratories (Forsythe & Xavier, 2002). The model is grounded in Klein's Recognition Primed Decision making (RPD (Klein, 1989)) framework and oscillating systems theory (Klimesch, 1996). Much of the development has focused on Level One of Klein's framework, which involves the processing of cues and the resultant course of action. The Sandia model utilizes several modules including the instantiation of episodic memory, emotional processes, and contextual knowledge. Models are populated with expert knowledge unique to particular domains. Simulations have been generated for domains such as close quarters battle, security, and air traffic control.

2.2 ACT-R

Adaptive Control of Thought – Rational (ACT-R (Anderson & Lebiere, 1998)) represents a unified theory of cognition supported by computational instantiations. With early research dating back as far as Anderson and Bower (1972) and ever-expanding applications, ACT-R is arguably the most empirically rich and active computational model of rational cognition. The architecture of ACT-R consists of symbolic and subsymbolic representations of cognition. The symbolic layer captures the processing of production rules while the subsymbolic layer consists of a sophisticated network of cognitively plausible parameters intended to supplement symbolic processing. The subsymbolic parameters provide added flexibility and adaptivity using mechanisms such as partial matching and stochastic functions. Like other production systems, production rules in the symbolic layer are instantiated in the form of condition-action associations. However, among production systems, ACT-R is a unique hybrid system by supporting symbolic processing with neural-like mechanisms in the subsymbolic layer.

2.2.1 Validation

ACT-R has been validated in a variety of tasks such as learning mathematics, programming and factors associated with mental workload. Brannon (2001) developed an ACT-R model of humans performing a resource management task in the Multi-Attribute Task Battery (MATB; Comstock and Arnegard, 1992). While the primary objective of the research was to model mechanisms that could degrade performance, the work included simulations of knowledge acquisition. ACT-R simulation results were compared with subsequently collected human subject data resulting in correlations as high as 0.93.

3 ACT-RPD

The RPD framework is well suited for gaining an understanding of the subtle complexities associated with expert human performance in natural environments (Klein, 1997). In contrast, ACT-R is well suited for capturing human performance associated with well-defined contexts at dynamic levels of expertise. While efforts have been made to augment the complexity with which ACT-R can model human performance in more dynamic tasks (Brannon, 2001), the effort to blend the attributes of RPD and ACT-R appears unprecedented.

Figure 1 depicts the rationale for the sequence of functions necessary to merge the features of RPD and ACT-R. Within a given context, RPD assumes the person being modeled is an expert. A feature of this expertise is the refined nature in which the expert processes information. The expert theoretically can derive necessary actions from one or more cues from the environment (upper left-hand corner of Figure 1). These cues will be fed to perceptual/motor processors comparable to the buffers in Version 5.0 of ACT-R.

The next step is the derivation of an associated situation. This encompasses a significant part of the proposed research. It is possible that the perceived cues are insufficient for situation recognition and therefore the expert has not achieved situation awareness. This phase maps to RPD Level 2 (Diagnosis) where a goal is pushed to decide upon an alternative explanation. If an alternative cannot be derived, more cues will be gathered, otherwise, responsibility returns to pattern recognition.

If a situation is recognized, action must be taken. At this point, the expert has processed a myriad of cues and anti-cues to bring some order to the current context. Once this order has been established, more controlled processing can occur. The means by which ACT-R is initiated is by having RPD designate a goal. An example might be that the smell of smoke in your office is detected and you essentially say, "I *need* to get out of the building."

Once a goal is pushed, ACT-R performs conflict resolution to derive a production that matches the current goal. With the assumption of expert knowledge, it is quite possible a production will be readily available and can fire quickly. This would be consistent with naturalistic decision making. However, for various quantifiable reasons, a production/action may not be available with respect to the designated goal. In this scenario, RPD specifies that the expert conducts mental simulation (RPD Level 3). With respect to the current context, mental simulation will be instantiated by pushing a subgoal to find alternative courses of action. In ACT-R this represents searching alternative productions that match the current goal. If a rule matching the current goal has been found, it will be fired. Otherwise, the expert will need to reassess the situation by processing more cues.

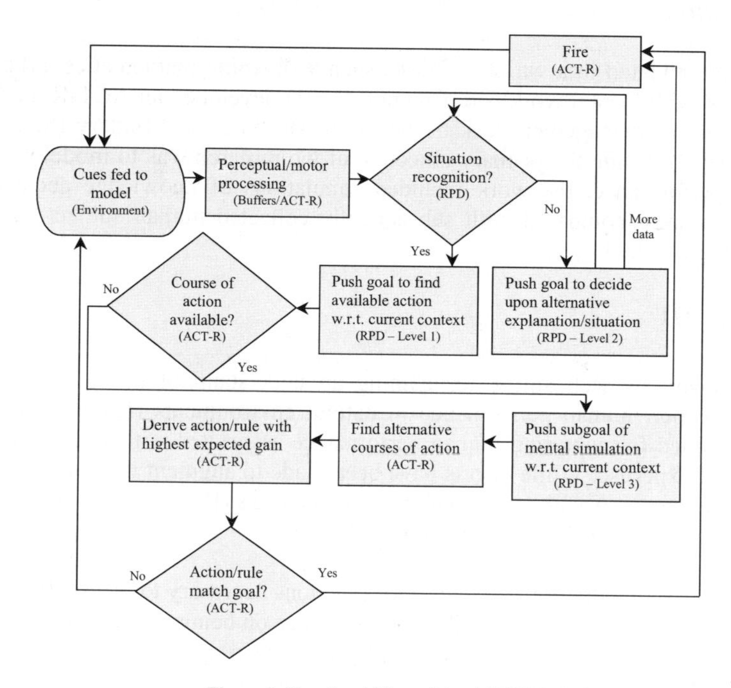

Figure 1: Functional View of the ACT-RPD Hybrid

4 Conclusion

Current research is focusing on refining the dialogue between the LISP based ACT-R system and the C++ version of RPD. A socket interface has been established using Python.

ACT-RPD represents a step closer to merging the proven strengths of production systems and models of expert decision making in naturalistic settings. Furthermore, the work is consistent with more abstract pursuits to better understand particular decision makers and therefore augment the capability of man-machine systems.

References

Anderson, J. R., & Lebiere, C. (1998). *The Atomic Components of Thought*. Mahwah, NJ: Lawrence Earlbaum.

Brannon, N. G. (2001). *Knowledge degradation*. Unpublished doctoral dissertation, Wright State University, Dayton, Ohio.

Brannon, N. G., & Koubek, R. J. (2002). Towards a conceptual model of procedural knowledge degradation. *Theoretical Issues in Ergonomics Science, 2*(4), 317-335.

Comstock, J. R., & Arnegard, R. J. (1992). *Multi-attribute task battery for human operator workload and strategic behavior research* (NASA Technical Memorandum 104174). Hampton, VA: NASA Langley Research Center.

Forsythe, C., & Raybourn, E. (2001). Toward a human emulator: A framework for the comprehensive computational representation of human cognition. In *Proceedings of the Human Factors and Ergonomics Society 45th Annual Meeting* (pp. 537-542). Santa Monica, CA: Human Factors and Ergonomics Society.

Forsythe, C., & Xavier, P. (2002). Human emulation: Progress toward realistic synthetic human agents. *Proceedings of the 11th Conference on Computer-Generated Forces and Behavior Representation*, Orlando, FL., 257-266.

Klein, G. A. (1989). Recognition-primed decisions. In W. B. Rouse (Ed.), *Advances in man-machine systems research* (Vol. 5, pp. 47-92). Greenwich, CT: JAI Press.

Klein, G. A. (1997). An overview of naturalistic decision making applications. In C.E. Zsambok & G. Klein (Eds.), *Naturalistic decision making* (pp. 49-59). Mahwah, NJ: Lawrence Earlbaum.

Klimesch, W. (1996). Memory processes, brain oscillations and EEG synchronization. *International Journal of Psychophysiology*, *24*, 61-100.

Luchins, A. (1942). Mechanization in problem solving. *Psychological Monographs, 54*, 248.

Pew, R. W., & Mavor, A. S. (Eds.). (1998). *Modeling human and organizational behavior*. National Research Council, National Academy Press: Washington DC.

A road-based evaluation of a Head-Up Display for presenting navigation information

Gary E. Burnett

School of Computer Science and IT, University of Nottingham
Jubilee Campus, Wollaton road, Nottingham, UK NG8 1BB
gary.burnett@nottingham.ac.uk

Abstract

This paper reports the findings from a preliminary, road-based evaluation study of a Head-Up Display (HUD) for presenting navigation information. In a repeated measures design, twelve participants drove two routes in an unfamiliar urban environment using a navigation system which presented voice guidance information, plus visual information either on an LCD within the car (Head-Down) or as a virtual symbol outside the car (Head-Up). There was some evidence that the drivers' navigating performance was superior with the HUD compared with the LCD, although it was clear that the HUD led to specific negative behavioural changes. In particular, participants using multi and bi-focal glasses had to repeatedly lower their head to see the HUD image in focus, and several drivers had to adopt awkward driving postures in order to view the HUD image. In addition, participants expressed concerns regarding the paced, omni-present nature of information presented on a HUD. Future work must aim to verify these findings within longitudinal trials, and should address issues relating to the implications of these novel displays on driving performance.

1 Introduction

Head-Up Displays (HUDs) are widely used within the aviation and military fields, but have yet to be implemented on a large-scale within road-based vehicles. HUDs will potentially allow drivers to continue attending to the road ahead whilst taking in secondary information more quickly (Ward & Parkes, 1994). As a consequence, they may be most applicable to situations in which the visual modality is highly loaded (e.g. urban driving), and for older drivers who experience difficulties in rapidly changing accommodation between near and far objects (Burns, 1999).

From a Human Factors perspective, there are clear dangers in simply translating a technology from one context to another, given that vehicle-based HUDs will be used by people of varying perceptual/ cognitive capabilities within an environment where there is a complex, continually changing visual scene. Specifically, there is a fundamental need to understand whether the use of HUDs disrupts distance perception and visual scanning patterns, the potential for cognitive tunnelling, the risk of critical road information being masked and the acceptability of this advanced technology (summarised by Tufano, 1997; and Ward & Parkes, 1994).

In focussing on these research questions, previous empirical studies examining the use of HUDS within vehicles have examined specific objective variables (e.g. reaction times to peripheral targets) within the controlled (but context-limited) environment of the driving simulator (e.g.

James, Eheret & Philips, 1995; Steinfeld & Green, 1998). Inevitably, there is a requirement for 'real-world' research on this topic, and the study reported in this paper aimed to improve understanding on how people use such novel displays in a driving and navigating context, and to establish their opinions on the technology.

2 Method

In a repeated measures design, twelve experienced drivers (half male, half female), aged 40-65 years, drove two routes in an unfamiliar urban environment (the City of Eindhoven, The Netherlands). The two routes were of equivalent complexity, each consisting of 17 decision points and taking approximately 20 minutes to drive.

For each route, participants navigated using a commercially available vehicle navigation system which had been adapted so that it either presented visual information on an LCD as head-down (i.e. in a central high position inside the car) or head-up (i.e. as a green, translucent, virtual symbol focussed immediately in front of the driver, just above the bonnet and approximately three metres outside the car). The visual navigation information was the same for the two display types and consisted of simple arrow-based symbols (left, right and straight on) with some contextual information concerning the layout of junctions including the surrounding roads. Furthermore, in both conditions, simple voice information was provided by the navigation system (e.g. "take the next turning left"). Drivers' exposure to the routes and display conditions were counter-balanced.

At the beginning of the study, and prior to each of the test routes, participants were given a short period of training and familiarisation with the vehicle, the navigation system and the two displays. Whilst the participants were driving and navigating, observational data was collected on comments, questions, errors made and difficulties experienced. Following completion of each route, subjective opinion was captured via a bespoke questionnaire. This addressed issues specific to the displays (e.g. ease of viewing) and general issues such as perceived safety, ease of navigation, and confidence in the information presented. Participants also completed a version of the NASA-RTLX (Raw Task Load Index), a multi-dimensional subjective rating scale for measuring physical and mental workload.

3 Results and Discussion

On the whole, participants made few navigational errors during the test routes for this study (defined as having occurred when they strayed from the route recommended by the navigation system). Nevertheless, on average, participants made less wrong turnings when using the HUD (mean 0.7, SD= 0.778) than when using the in-vehicle LCD (mean 1.1, SD=0.669), a difference which approached significance ($p=0.069$, $df=11$). Figure 1 shows the results for individual participants, and reveals that for 11 of the 12 participants the number of navigational errors made when using the HUD was the same or less than that for the in-vehicle LCD. Although it is tempting to attribute this result to the relative ease by which users of the HUD could assimilate guidance information, observational data indicated a confounding causal factor. In this study, for the navigation system under test, there was a slight delay in the presentation of the voice information in relation to the visual information (up to around two seconds). The viewing angle for the HUD (i.e. within the road-scene) enabled users therefore to access the guidance information quicker than when using the peripheral LCD. In particular situations (e.g. where multiple lane changes were required) this advantage was critical.

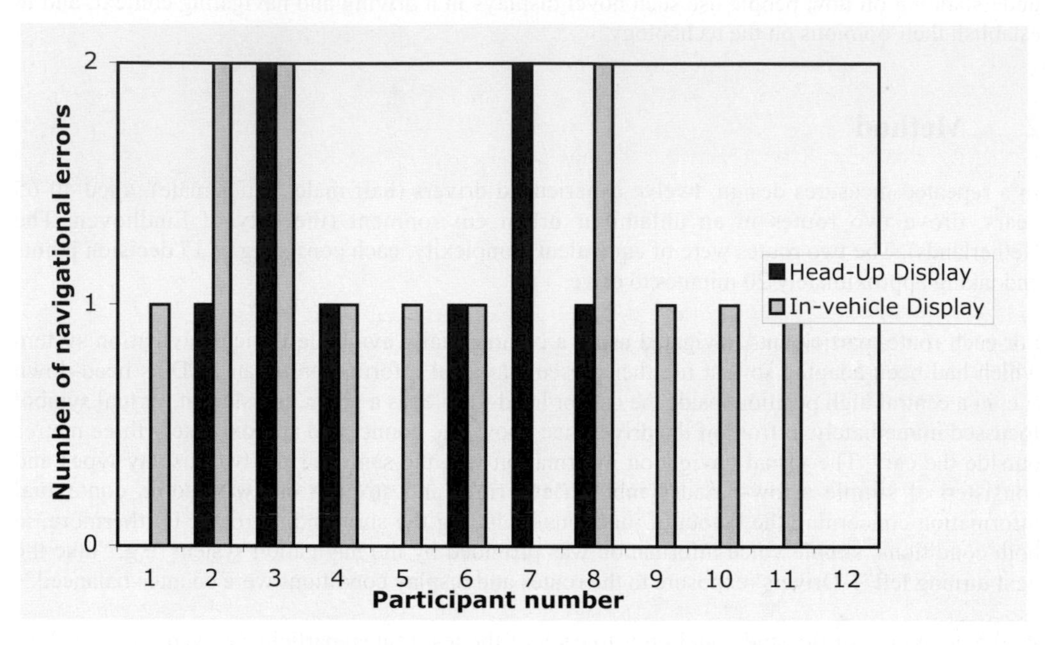

Figure 1: Numbers of navigational errors made by each participant when using navigation system with Head-Up Display or In-vehicle Display

However, this heightened awareness of the timing of presentation of visual navigation information when using the HUD did cause some difficulties. For instance, in situations where the visual information remained on the display for a short period of time (up to three seconds) after the turning had been made, some participants appeared confused in using the HUD and were not sure whether they had to make an additional manoeuvre.

The study revealed two interesting issues regarding the behavioural changes that can occur when drivers are provided with new technology. Six of the participants wore multi or bi-focal glasses to allow them to view the outside and in-vehicle scenes clearly. However, these participants encountered difficulties with the HUD since the image was presented at a low level and so, despite being a 'far' image requiring viewing through the top half of their glasses, was outside the ordinary range of vision for 'far' images with their glasses. This resulted in these participants having to lower their head to see the image in focus, which caused the experiencing of some discomfort in the neck and eye regions. It is predicted that over a longer period, these participants would have experienced a greater degree of fatigue problems caused by the eye and neck muscles being repeatedly strained to view the HUD image.

An important safety question arises from these problems which requires further empirical investigation: if drivers have to lower their head to view a head-up image, causing the road ahead view to effectively lie above their glasses and hence appear out of focus, are they then less able to allocate sufficient attentional resources to the road ahead?

A further behavioural issue related to the eye box position (the site in which the eyes must lie for the image to be seen). In this implementation of the HUD, the range over which the image could be seen was quite small (approximately 5cm). This meant that some participants adopted a variety of awkward sitting positions whilst driving. This limitation in the HUD design proved to be an important factor that greatly influenced the overall preference for a particular display. For instance, eight people found the position of the in-vehicle LCD more satisfactory than the position of the HUD, of which seven also preferred the in-vehicle LCD. The implication of this finding is clear - if an automobile HUD is to be acceptable to the general public, it is vital that it accounts for individual differences in sitting height. To address this anthropometric design issue, consideration needs to be given to the relative benefits of a 'design for extremes' versus a 'design for adjustable range' strategy.

A related point was raised by six participants who remarked that they did not like being confined or forced to a set position in order to see the HUD image, and so could not adopt more relaxed sitting positions. This highlights the dangers of transferring aviation technology into a wholly different environment, in which subjects are not well trained and skilled, and would not usually maintain their eyes in such a fixed position.

Within the questionnaire and comments made during the study, participants expressed a number of negative views regarding the use of the HUD. This is reflected by the key result that although five participants preferred the head-up display overall, seven participants preferred the in-vehicle LCD. In relation to specific questions, five participants commented that they did not like the omnipresent nature of the HUD information, i.e. that they could not choose not to look at the head-up images. The 'forced' visual information was judged to be intrusive, and this is highlighted by the fact that within the NASA-RTLX the HUD was generally perceived to be more distracting (mean = 35.0, SD = 19.758, where 0='Low'; 100='High') than the self-pacing LCD (mean = 23.0, SD = 18.289), a difference which approaches significance (p=0.080, df=11).

As a further point, six participants noted that the road (or the outside scene) was not the correct domain for a guidance image, which should be inside the car and not obstructing the road information. There are two points arising from these comments. Firstly, this appears to be evidence of a powerful expectation factor regarding where driver information should be displayed. Other forms of driver information (warnings, cassette/radio/CD, trip computer data, etc.) have always been presented within the vehicle, and navigation information was seen as just an extension of the information normally encountered. Secondly, the participants believed the HUD image was obscuring some of the road information. From observation of the image (which was transparent and presented at a low level), it is felt unlikely that any important information was actually affected, and the comment arose because once again the participants were not used to an image presented at that location. Such preconceptions could be resolved by long-term exposure to such a novel display.

A few (four) participants remarked they did not feel comfortable with the HUD because it seemed 'strange' or 'unnatural'. This is also a comment largely concerning expectation as regards where to present driver information, but it is felt it may have arisen as a result of two other factors. Firstly, because the HUD image was not projected at optical infinity (the image was focused approximately at the end of the bonnet), it may have induced myopia (i.e. a "pulling" of visual accommodation toward the driver) – Ward & Parkes (1994). Secondly, the HUD image may have disrupted familiar visual scanning patterns, by attracting visual attention towards the image at the expense of important aspects of the road scene. It is difficult to be sure whether these factors influenced the participants' opinions, but it is felt likely that if a driver did experience such problems, they would describe the experience as 'strange' or 'unnatural'. Indeed, when questioned further about these opinions, participants would often refer to a more familiar frame of reference, for example by remarking that the in-vehicle display seemed 'more natural'.

The weather conditions throughout the trials were mainly cloudy with frequent periods of rain. However, four participants experienced the HUD whilst driving in full sunlight, and expressed concern with regard to the reflections on the windscreen they were experiencing. These participants remarked that this increased their rating of distraction caused by the HUD, and is indicative of the problems associated with applying aviation technology to an environment where such widely varying ambient lighting conditions pervade. This study did not aim to address directly the effect of ambient lighting on driver behaviour whilst using a HUD, but it is apparent from such comments that problems might occur. Research is required into the effect that adverse windscreen reflections might have on the perception of both the HUD image and aspects of the road scene.

4 Conclusions

The Human Factors literature is abound with simulator studies investigating specific issues regarding the use of Head-Up displays in road-based vehicles. However, there appears to be a paucity of 'real-world' research on this topic. This study, which took place on public roads in an urban driving environment, provided some evidence that drivers are more effective in the navigation task when using a HUD, as compared to a traditional in-vehicle display. Furthermore, the research generated some guidelines for the design of automobile HUDs, for example concerning the need for consideration of anthropometric data, avoidance of adverse reflections and careful positioning of the HUD image. Nevertheless, fundamental issues remain to be resolved, in particular with respect to the negative behavioural changes that can result from the use of new technology and the considerable mix of opinion as to the potential for HUDs within vehicle navigation systems. Many of the issues uncovered by this preliminary study must be investigated in detailed, larger-scale empirical research, in particular within longitudinal trials.

5 References

Burns, P. (1999) Navigation and the mobility of older drivers. *Journal of Gerontology. Series B, Psychological sciences and social sciences*, 54 (1), 49-55

James, C., Eheret, B., & Philips, B. (1995). Effects of rotation and location on advanced traveler information system displays. In *Proceedings of the Human Factors and Ergonomics Society 39th Annual Meeting*: Vol. 2. (pp. 1077-1081). Santa Monica, CA: Human Factors and Ergonomics Society.

Steinfeld, A., & Green, P. (1998). Driver Responses to Navigation Information on Full-Windshield, Head-Up Displays. *International Journal of Vehicle Design*, 19(2), 135-149.

Tufano, D.R. (1997). Automotive HUDs: The overlooked safety issues. *Human Factors*, 39(2), 303-311.

Ward, N. J., & Parkes, A.M.. (1994). Head-up displays and their automotive application: An overview of human factors issues affecting safety. *Accident Analysis and Prevention*, 26 (6), 703-717.

Analysis and Verification of Human-Automation Interfaces

Asaf Degani

Computational Science Division
NASA Ames Research Center
adegani@mail.arc.nasa.gov

Michael Heymann

Department of Computer Science
Technion, Israel Institute of Technology 1
Heymann@cs.technion.ac.il

Abstract

This paper addresses the problem of verifying and designing human-automation interfaces. The approach focuses on what information is provided to the user (and not on how this information is presented). We describe a formal methodology for verification of interfaces. The methodology is aimed at proving that the information provided to the user via the display (e.g., modes and parameters), enables the user to perform his or her tasks successfully. We assert that a display and corresponding user-manuals are correct if there exist no *error states*, no *blocking states*, and no *augmenting states*. The essentials of the methodology, which can be automated and applied to the verification of large and complex systems, are discussed and illustrated via a simplified automotive example. An extension of this formal approach for generating interfaces and associated user-manuals is briefly discussed.

1 Introduction

Automated control systems such as automotive, medical equipment, and avionics exhibit extremely complex behaviors. These large systems react to external events, internal events, as well as user-initiated events. For the user to be able to monitor the machine and interact with it to achieve a task, the information provided to the user about the machine must, above all, be correct. In principle, correct interaction can always be achieved by providing the user with the full detail of the underlying machine behavior, but in reality, the sheer amount of such detail is generally impossible for the user to absorb and comprehend. Therefore, the machine's interface and related user-manuals are always a reduced, or abstracted, description of the underlying machine behavior. Naturally, we all prefer interfaces that are also simple and straightforward. This, of course, reduces the size of user-manuals, training costs, and perceptual and cognitive burden on the user (Abbott, Slotte, & Stimson, 1996).

In this paper, we will briefly present an approach and methodology for verifying interfaces and user-manuals. The methodology evaluates whether the interface and user-manual information are correct and free of errors, given a description of the machine, the user's task, and the interface (Degani & Heymann, 2002).

2 Formal Aspects of Human-Machine Interaction

In analyzing human automation interaction from a formal perspective, we consider here three major elements: (1) the behavior of the machine (in terms of states and transitions), (2) the user's tasks, and (3) the interface.

2.1 Machine

Computer-based system and automated control systems can be described using a variety of formal models. In this paper, we model a machine as a finite system of states (but note that the methodology described here is general and can be applied to any formalism). Some of the transitions in the system are triggered by the user (e.g., a driver placing the automatic gear in Drive). Other transitions are automatic and are triggered either by the machine's internal dynamics (e.g., automatic transition from 1st to 2nd gear), or by the external environment (e.g., when the car's speed is above 180 miles-per-hour, the transmission prevents the driver from further speed increase). In the models described here, we depict user-triggered transitions by solid lines, while automatic transitions are broken lines. The transitions are labeled by Greek symbols indicating the events under which the machine moves from state to state.

The machine in Figure 1 describes a simplified three-speed transmission system of a vehicle. The transmission has eight states (representing internal torque-levels). These are grouped into three speed modes: LOW, MEDIUM, and HIGH. States L1, L2, and L3 are in the LOW speed mode; M1 and M2 in the MEDIUM speed mode; and H1, H2, H3 in HIGH. The transmission shifts up and down either automatically (based on throttle, engine, and speed values) or manually by pushing a lever (Figure 2). Manual up-shifts are denoted by event β and down-shifts by event ρ. Automatic up-shifts are denoted by event δ, and automatic down-shifts by event γ.

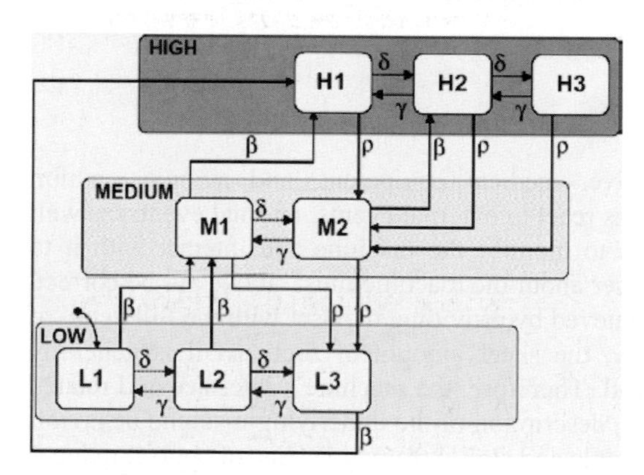

Figure 1. Transmission system.

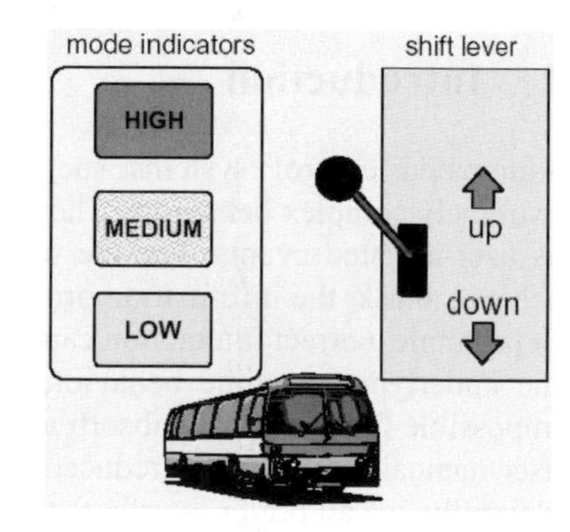

Figure 2. Display and control panel.

2.2 User's Task

The second element is the user's tasks, which in case of this transmission system, consists of tracking the three speed modes unambiguously. In other words, the user must be able to determine the current mode of the machine and predict the next mode of the machine. This requirement is akin to the type of questions users usually ask about automated systems: "What's it doing now?" "What's it going to do next?" and "Why is it doing that?" (Wiener, E. L. personal communication, April 5, 2002). We describe the user's task by partitioning the machine's state-set (the 8 internal states in Figure 1) into three disjoint regions: low, medium and high. Note however that the user is required to track only the modes that correspond to these regions and *not* every individual state of the machine.

2.3 Interface

As discussed earlier, the interface generally provides the user with a simplified view of the machine. In almost any display, especially those for automated systems, many of the machine's internal events and states are hidden from the user -- otherwise, the size of cockpit displays, for example, would be colossal. Hence the display provides only partial, i.e., abstracted, information about the underlying behavior of the machine. Since we want an interface that is correct and succinct, the essence of the interface design problem centres on what information can be safely removed-away, or abstracted, from the display and what information must be presented.

Figure 3 describes one proposed display for the transmission system. The display indicates the three primary modes (LOW, MEDIUM and HIGH), and the driver shifts among modes by pushing up or down on the gear lever. What is also being removed from the interface, user-manual, and consequently from the user's awareness is the automated internal transitions that take place within each mode, or gear. For example, the LOW mode has three possible internal states, L1, L2, and L3. When the user first up-shifts manually into low gear, L1 is the active state. When the driver increases speed, an automatic transition to L2 takes place. This internal transition is not evident to the driver, who is aware only of being in LOW mode.

3 Evaluation of Interfaces

In addition to the display indications, Figure 3 also shows the manual transitions that are needed to move the transmission system from one mode to another (this information is provided in the user-manual). This proposed display is very simple and straightforward: it shows only the three modes (LOW, MEDIUM, and HIGH), all internal states are removed, and all the automatic transitions are suppressed.

Is this an adequate display?

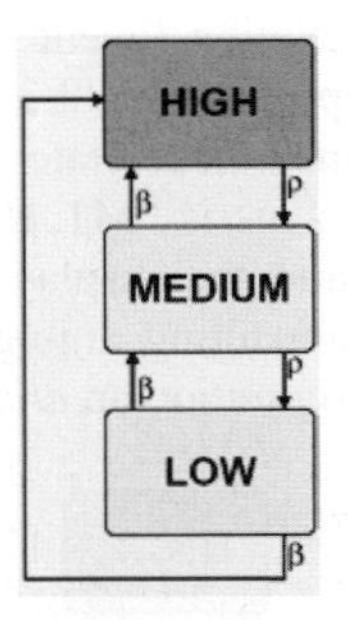

Figure 3. Proposed display.

Intuitively it looks fine, but let's look at it more carefully: The manual shifts from MEDIUM to HIGH or down to LOW, as well as the down- shift from HIGH to MEDIUM, are always predictable – the user will be able to anticipate the next mode of the machine. However, note that the transitions out of LOW depend on the internal states: up-shifts from L1 and L2 take us to MEDIUM, while the up-shift from L3 switches the transmission to HIGH. What we have here is that the same event (β) takes us to two different modes. But since the display hides from us which internal state we are in, we will not be able to predict if the system will transition to MEDIUM or HIGH. Therefore, we must conclude that this display is incorrect and inadequate for the task.

An alternate user-model that may remedy the above problem is depicted in Figure 4. This modified display has two LOW modes (LOW-1, LOW-2). The user-manual further explains that the transitions between LOW-1 and LOW-2 occur automatically, and that upon up-shift from LOW-1, the system transitions to MEDIUM, while on upshift from LOW-2, the system goes to HIGH.

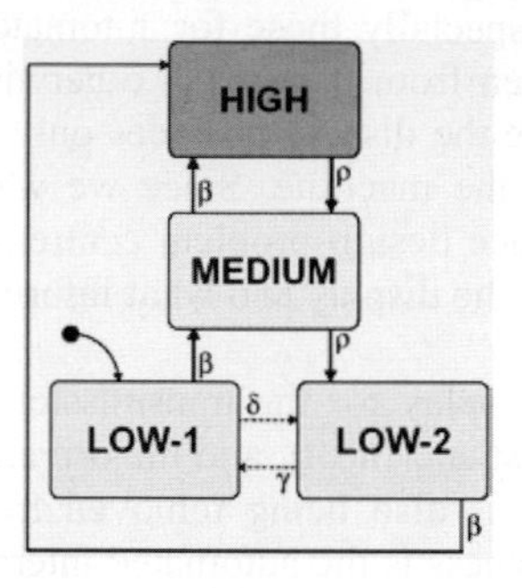

Figure 4. Alternate display.

4 Formal Verification of Interfaces

Again, we ask: is this a good interface?

Well, by intuitive inspection it seems quite reasonable -- we have taken care of the problem with the manual up-shift out of LOW. But let us try to verify this intuition in a more methodological way. The way we do this is by creating a composite model of the display of Figure 4 and the machine model. This is the model presented in Figure 5(c).

We evaluate the composite model by exploring all the composite states. The machine (Figure 5(a)) starts in state L1 and the display (Figure 5(b)) starts in LOW-1. So the first composite state is "L1, low-1." Upon an automatic up-shift transition (event δ), the machine transitions to L2 and the display to low-2, and now we are in composite state "L2, low-2." At this point, the user decides to up-shift manually. The machine will transition to state M1, yet according to Figure 5(b), we are now in HIGH mode. The new composite state is "M1, HIGH," which of course is a contradiction! The user thinks he is in HIGH mode (and the display confirms) where in fact the underlying machine is in MEDIUM (state M1). The resulting ambiguity is a classical mode error (Norman, 1983), and we call such a composite configuration an *error-state*.

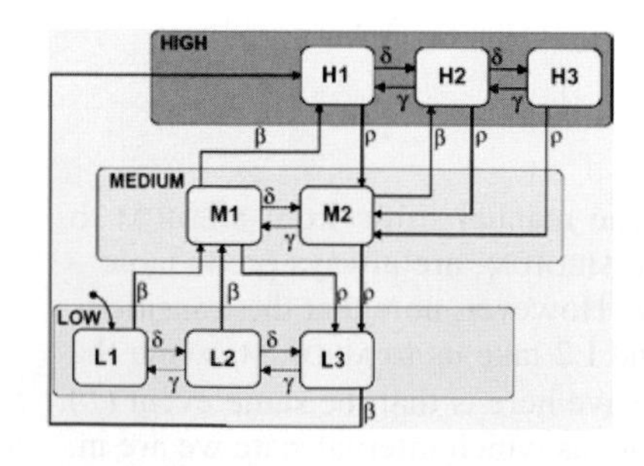

Figure 5(a). Machine model.

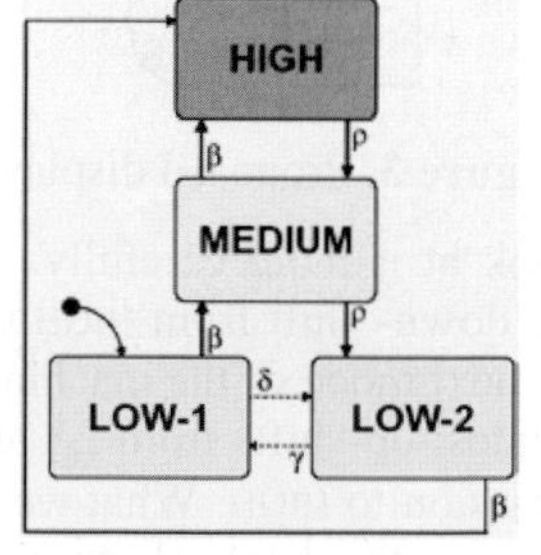

Figure 5(b). Display.

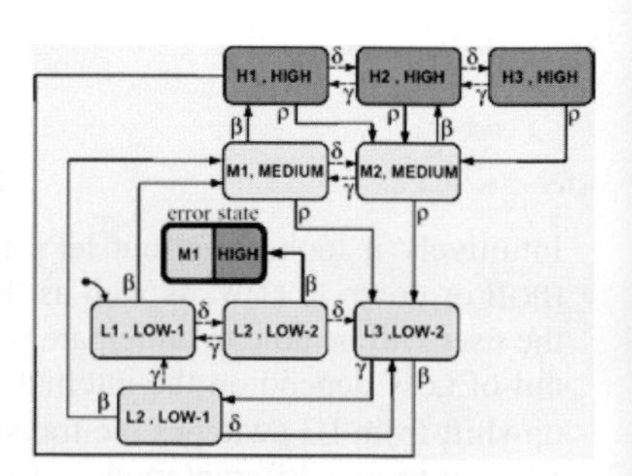

Figure 5(c). Composite model.

5 Conclusions

The objective of the verification methodology is to determine whether a given interface enables the user to operate the machine correctly. The essence of the verification procedure is to check whether the interface "marches" in synchronization with the machine model. We assert that a user-model is correct if there exist no *error states*, no *blocking states*, and no *augmenting states* in the composite model. An error state represents a divergence between the machine and its interface. That is, the display tells the user that the machine is in one mode when in fact the machine is in another. A blocking state represents a situation in which the user can in fact trigger a transition from one mode to another, yet this information is not provided to the user (and when the transition happens, the user is surprised). An augmenting state is a situation in which the user is told that a certain mode change is possible, when in fact it may be the case that the machine will not switch into this mode or sub-mode. The details of this verification methodology and its application to an automated flight control system are provided elsewhere (Degani & Heymann, 2002).

Going back to the transmission system, it is possible to concoct other displays and then iteratively employ the verification procedure to determine their correctness. It turns out that there exist a composite model that exhibits no error states, no blocking states, and no augmenting states. The corresponding display is the one depicted in Figure 6.

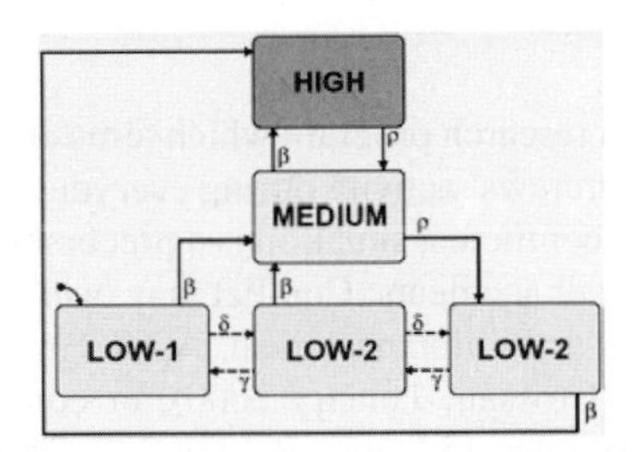

Figure 6. A correct and adequate display.

Finally, while a verification methodology such as the one presented here is quite useful for evaluations of interfaces, it is not a panacea for design. For larger and more complex systems, it may take considerable effort to develop and verify one display after another, with no guarantee of success. Furthermore, even when a correct interface is identified, there is no assurance that it is the simplest possible -- there could be an equally good, or even better abstraction, hiding just around the corner. The development of a methodology and algorithm for generating interfaces that are both correct and succinct is discussed in a recent NASA report (Heymann & Degani, 2002).

6 References

Abbott, K., Slotte, S. M., & Stimson, D. K. (1996). *The interface between flightcrews and modern flight deck systems*. Washington, DC: Federal Aviation Administration.

Degani, A. & Heymann, M. (2002). Formal Verification of Human-Automation Interaction. *Human Factors*. 44 (1), 28-43.

Heymann M., & Degani A. (2002). *On abstractions and simplifications in the design of human-automation interfaces*. NASA Technical Memorandum 2002-211397. Moffett Field, CA.

Norman, D. A. 1983. Design rules based on analysis of human error. *Communications of the ACM*, 26 (4), 254-258.

GHOST: experimenting countermeasures to cure pilots from the perseveration syndrome

Dehais Frédéric
Onera-Cert/Supaéro
2, av E. Belin 31055 Toulouse Cedex 4
dehais@cert.fr

Abstract

In this paper we present Ghost, an experimental environment dedicated to test countermeasures to help pilots breaking up the perseveration syndrome. These countermeasures are based on information removal from the onboard interface, an approach that differs from current system dedicated to warn pilots. The results of an experiment, conducted with 21 private and military pilots in the Ghost environment, are presented.

1 Introduction

ONERA is currently involved in a research program which aims at improving flight safety through a methodology for the analysis of aircrews' activity during everyday flights [1]. The review of civilian and military reports reveals that a conflictual situation is a precursor of loss of the aircrews' situation awareness and is a major cause of air accidents. Conflict may occur through different forms: conflict between the different operators (Air Philippines crash, April 1999), resource conflict (Streamline, May 2000), knowledge conflict (Crossair, January 2000), or conflict between automated systems (TCAS) and human operators (Tupolev 154 and a Boeing 757, July 2002). A first experimentation, dedicated to study the pilot's behaviour facing up conflictual situations, showed that pilots have a trend to persevere in erroneous decisions when they are in conflict [2]. This behaviour, called the "perseveration syndrome", is studied in neuropsychology [3, 4] and psychology [5]: it is known to summon up all the pilot's mental effort toward a unique objective (excessive focus on a single display or focus of the pilot's reasoning on a single task). Once taken up in this logic, the pilot does anything to succeed in his objective, even if it is dangerous in terms of security, and worst, he neglects any kind of information that could question his reasoning (like alarms, data on displays). This behaviour is very dangerous and, according to BEA[1], it is responsible for more than 40 percent of casualties in air crash.

2 Send accurate countermeasures

Therefore the idea is to design countermeasures in order to break this mechanism. The design of the countermeasures is grounded on the following theoretical and empirical results:
• conflictual situations lead pilots to persevere;

[1]The French national institute for air accident analysis. http://www.bea-fr.org

• additional information (alarms, data on displays) designed to warn pilots are often unnoticed when pilots persevere.
Our conjectures are then:
→ with an accurate on-line conflict detection, it is possible to analyze and predict the subject of the pilot's perseveration;
→ instead of adding information (classical alarms), it is more efficient to *remove the information* on which the pilot is focused and which makes him persevere in the wrong way, and instead to display a message to explain his error.
Ex: in the experiment described afterwards, pilots persevere on trying a dangerous landing at Francazal despite the bad visibility, with a particular focus on an instrument called the H.S.I.[2] to locate Francazal airport. The countermeasures will consist in removing the H.S.I. during a few seconds, to display two short messages instead, one after the other ("*Landing is dangerous*"... "*look at central display*") and next to send an explanation of the conflict on the central display ("*Low visibility on Francazal, fly back to Blagnac*").
The idea here is to shock the pilot's attentional mechanisms with the short disappearance of the H.S.I., to introduce a cognitive conflict ("*if I land, I crash*") to affect the pilot's reasoning, and to propose a solution on the central display.

2.1 GHOST

Ghost is an experimental environment designed to test countermeasures to cure the pilot's perseveration. It is composed of:
• Flightgear[3], an open source flight simulator, which means that many modifications can be made, e.g. implementing new displays. Almost all the airports and beacons (NDB and VOR[4]) of the world are modeled;
• Atlas[5], a freeware designed to follow the pilot's route;
• a wizard of Oz interface (see figure1), which we have designed and implemented, which allows a human operator to trigger events (failures, weather alteration and the countermeasures) from an external computer *via* a local connection (TCP/IP protocol, sockets communication).
As far as the countermeasures are concerned, several actions are available to the wizard of Oz:
• *replace* a display on which the pilot is focused by a black one (time of reappearance is adjustable);
• *blink* a display (frequency and time of blinking is adjustable);
• *send a message* to a blinked, removed or faded display;
• *send a message* to the central display for explicit conflict solving.

2.2 Experimental scenarios

As conflict appears when a desired goal cannot be reached [2], we have designed three experimental scenarios where a crucial goal of the mission cannot be achieved:
• scenario 1 is a navigation task from Toulouse-Blagnac airport to Francazal airport including three waypoints (VOR 117.70, NDB 331 and NDB 423). An atmospheric depression is positioned in such a way that the pilot cannot see the landing ground but at the last moment when it is too late to land on it.
• scenario 2 is a navigation task from Toulouse-Blagnac airport to Francazal airport including three

[2]The H.S.I is a display that gives the route to follow and indicates any discrepancy from the selected route.
[3]http://www.flightgear.org/
[4]NDB and VOR are two kind of radionavigation beacons used to guide pilots.
[5]http://atlas.sf.net

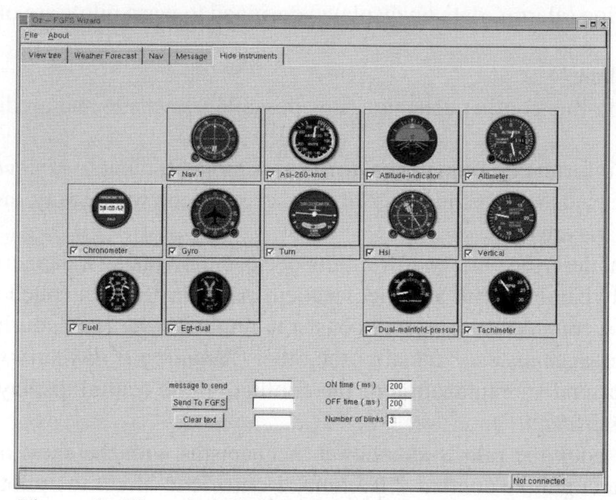

Figure 1: The wizard of Oz interface - countermeasures

waypoints (VOR 117.70, NDB 415 and NDB 423). The visibility is decreased from the runway threshold: from far away, the landing ground is visible but as the pilot gets closer to it, it disappears totally in a thick fog.

• scenario 3 is a navigation from Toulouse-Blagnac back to Toulouse-Blagnac including three waypoints (VOR 117.70, NDB 331 and NDB 423). An atmospheric depression is positioned over waypoint 2 (NDB 331), and as the pilot flies over this waypoint, the left engine fails.

In scenarios 1 and 2 the pilot's conflict can be summarized as "*should I try a dangerous landing at Francazal or should I fly back to Blagnac for a safer landing?*". If our conjectures hold, pilots will persevere and try to land at Francazal.

In scenario 3, the pilot's conflict can be summarized as "*should I go on with the mission despite the failure or should I land at Francazal and therefore abort the mission?*"

2.3 Experimental Results

21 experiments were conducted with Ghost in December 2002. The pilots' experiences ranged from novice (5 hours, small aircraft) to very experienced (3500 hours, military aircraft).

2.3.1 Results for scenarios 1 and 2: "impossible landing"

In these two scenarios, the pilots faced the decision of mission abortion, i.e. not land at Francazal and fly back to Blagnac.

Both scenarios were tested within two different contexts: without countermeasures and with countermeasures.

• Context 1: no countermeasures

7 pilots tested scenario 1 or 2 without any countermeasures (see next table). "Circuits" corresponds to the number of circuits performed by the pilot round Francazal before crashing or landing.

Pilot	Scenario	Circuits	Results
Pilot1	1	3	crash
Pilot2	1	3	crash
Pilot3	1	1	lucky landing
Pilot4	1	1	lucky landing
Pilot5	2	1	crash
Pilot6	2	1	lucky landing
Pilot7	2	1	crash

The results suggest that without any countermeasures, none of the pilots made the right decision (fly back to Blagnac): they all perservered at trying to land at Francazal. Four of them hit the ground, and the three others had a "lucky landing", which means that while they were flying round Francazal, the runway appeared between two fog banks and they succeeded in a quick landing. During the debriefing, all of them admitted they had made an erroneous and dangerous decision.

• Context 2: with countermeasures

12 pilots tested scenario 1 or 2 with countermeasures (see next table). "Circuits" corresponds to the number of circuits performed by the pilot round Francazal before a countermeasure is triggered by the wizard of Oz.

Pilot	Scenario	Circuits	Results
Pilot8	1	3	crash on Francazal
Pilot9	1	2	back to Blagnac
Pilot10	1	2	back to Blagnac
Pilot11	1	2	back to Blagnac
Pilot12	2	3	back to Blagnac
Pilot13	2	2	back to Blagnac
Pilot14	2	2	back to Blagnac
Pilot15	2	2	back to Blagnac
Pilot16	1	2	lucky landing Francazal
Pilot17	1	2	back to Blagnac
Pilot18	1	2	back to Blagnac
Pilot19	1	5	crash on Francazal

The results show the effectiveness of the countermeasures to cure perseveration: 9 pilots out of 12 changed their minds thanks to the countermeasures, and flew back safely to Blagnac. During the debriefing, all the pilots confirmed that the countermeasures were immediately responsible for their change of mind. Moreover, the short disappearance of the data due to the countermeasures did not cause any stress on them. 4 military pilots found that the solutions proposed by the countermeasures were close to what a human co-pilot would have proposed.

The results with Pilot19 suggest that the more a pilot perseveres, the more difficult it is to get him out of perseveration. During the debriefing, Pilot19 told us that he was obsessed by the idea of landing, and that he became more and more stressed as long as he was flying round Francazal. He then declared that he did not notice any countermeasures. Pilot16 and Pilot8 also persevered despite the countermeasures: they declared they knew they were not flying properly and that they had done that on purpose because they wanted to test the flight simulator.

2.3.2 Results for scenario 3: "failure"

Only 2 pilots tested scenario 3. One pilot experimented this scenario without any countermeasure: he did not notice the failure and hit the ground. The second pilot was warned through the countermeasures that he had a failure and that there was a landing ground at vector 80: the pilot immediately performed an emergency landing on this landing ground. Other experiments are currently being conducted with this scenario.

2.3.3 Other results

During the experiments, 9 pilots made errors, e.g. selection of a wrong radio frequency, erroneous altitude, omission to retract flaps and landing gear. To warn them, the wizard of Oz blinked the display on which they were focusing and displayed the error (e.g.: "gear still down"). In each case, the countermeasure was successful: the pilots immediately performed the correct action. During the debriefing, the pilots declared that this kind of alarm was very interesting, and much more efficient and stressless than a classical audio or visual alarm because they could identify at once what the problem was. These experiments have shown also the significance of the message contents on the blinked display. For example, for the case of the erroneous landing at Francazal, the initial message was "Don't land", but the pilots did not take it into account, thinking it was a bug of the simulator. When the message was changed to "Fly back to Blagnac" or "Immediate overshoot" it was understood and taken into account immediately.

3 Conclusion and perspectives

We have presented an approach to detect and cure conflicts in aircraft pilots' activities. An experiment with 21 pilots on a flight simulator has proved the effectiveness of the countermeasures to cure perseveration. Further experiments are to be conducted to go on tuning the countermeasures according to the pilots' feedback. The next step is to design and implement the whole countermeasures closed-loop, i.e to remove the wizard of Oz and perform an automated conflict and perseveration management. This is currently done thanks to :

- the conflict detection tool [2] ;
- Kalmansymbo, a situation tracker and predictor [6] which is based on predicate/transition timed Petri nets for procedure and pilot's activity modeling. It assesses the current situation thanks to the parameters received in real-time from Flightgear and predicts what is likely to happen. Both the current situation and the predictions are inputs for the conflict detection tool;
- a tool for building and sending accurate countermeasures back to the pilot *via* the Flightgear cockpit.

References

[1] L. Chaudron, J.-Y. Grau, P. Le Blaye, and N. Maille. REX : a human factor flght safety research program. ICAS 2000, 2000.

[2] F. Dehais. Modelling cognitive confict in pilot's activity. In *STAIRS 2002: Starting Artificial Intelligence Researchers Symposium*, July 2002.

[3] Henri Cohen, editor. *Neuropsychologie expérimentale et clinique*. Gaetan Morin, 1993.

[4] J. Pastor. Cognitive performance modeling - Assessing reasoning with the EARTH methodology. In *COOP'2000*, Sophia Antipolis, May 2000.

[5] J.L. Beauvois and R.V. Joule. A radical point of view on dissonance theory. In Eddie Harmon-Jones and Judson Mills, editors, *Cognitive Dissonance : Progress on a Pivotal Theory in Social Psychology*. 1999.

[6] C. Tessier. Towards a commonsense estimator for activity tracking. In *AAAI Spring symposium*, Stanford University, Palo Alto CA, USA, March 2003.

Source Recommendation System for Information Search and Retrieval

Narasimha Edala, Lavanya Koppaka, and S. Narayanan

Department of Biomedical, Industrial and Human Factors Engineering
Wright State University
3640 Col. Glenn Hwy
Dayton OH 45435
{nedala, lkoppaka, snarayan}@cs.wright.edu

Don Loritz and Raymond Daley

LexisNexis Alliances and New Technologies
9443 Springboro Pike
Miamisburg OH 45342
{don.loritz, raymond.daley}@lexisnexis.com

Abstract

Today, the Internet is becoming a world-wide knowledge medium – a mechanism for people to collaborate, communicate and share information. While this is good, it also brings the burden of finding contextually relevant information on the user. Recommender Systems are a suite of technologies that help the users identify appropriate sources of information while eliminating the overwhelming amount of information that they must otherwise have to endure. In this article, we will examine a source recommendation system built in the domain of an electronic data-warehouse. We present architectural, implementation and evaluation details of the system.

1 Introduction

In our information oriented society, the web has become pervasive and accessible to all ("A Nation Online", 2002.) The electronic medium has become a critical source for disseminating, collecting, and compiling information. Users are increasingly relying on the World Wide Web to conduct business. In 2001, there were over 8.5 million websites in operation that contain over 8.5 billion web pages ("The Web Characterization Project", 2002.) This metric is an indicator of how increasingly reliant users have become on the WWW for daily information needs. But this overwhelming amount of information also brings the challenge of empowering users with tools and technologies that let them home in on information that rightly addresses a user's need – tools and technologies that understand contextual needs (Lawrence, 2000) of a user (or a user-profile) and customize the content for increased productivity. Recommender Systems are one tool in the suite of technologies that understand user needs. Recommender Systems also benefit information-providers in the process. First, they earn the good-will of clients by providing on-point recommendations. Second, source recommendation systems ensure that clients home in on information quickly to minimize resource usage for the information provider.

In this article, we examine a source recommendation system (SRS) built in the domain of an electronic data warehouse provided by LexisNexis Corporation (LexisNexis Research System, 2003.) SRS is built so as to build a user-profile by analyzing a results-set from an information search and retrieval session and in turn applying this to provide source recommendations for subsequent search and retrieval.

2 Related Research and Approach

Recommender Systems currently employ content-based techniques (Adomavicius & Tuzhilin, 2001, and Mooney & Roy, 1999), collaborative-based filtering techniques (Resnick, Iacovou, Suchak, Bergstrom, & Riedl, 1994, Shardanand & Maes, 1995, and Hill, Stead, Rosenstein, & Furnas, 1995), or a combination thereof to provide recommendations (Melville, Mooney, & Nagarajan, 2001, and Balabanovic & Shoham, 1997). Collaborative filtering techniques find correlations among users and then use these to provide recommendations for other users. Similarly, content-based techniques use sources (or source-histories) being perused by the user as an indicator that other sources similar in nature may also be interesting. In both of these methods, it is critical that user-interests be identified so matching with users of peer group or sources of interest can be performed. However, deducing user-interests is a challenge in itself. Two approaches have been widely employed in the research community to obtain user-interests – explicit user-ratings and implicit user-ratings. The explicit approach requires users to explicitly verbalize or indicate interest on a topic to content providers. This approach faces "sparsity" and "first-rater" problems (Melville et al., 2001). Periodic obstruction of user's information retrieval process in order to explicitly submit interest-ratings is also a major problem with this approach. Implicit approaches (Nichols, 1997, and Oard & Kim, 1998), on the other hand, use automatic percepts such as time spent perusing a page, book-marking, printing, or linking to a page to identify user-interest in the content. Hill, Hollan, Wroblewski, & McCandless (1992) term these indirect metrics as "edit wear" or "read wear". Studies have shown that a strong correlation exists between these implicit percepts and user-interests.

The SRS system employs implicit metrics to record user's contextual information needs during a session. It in turn applies this understanding to provide source recommendations for future search and retrieval purposes. This method provides an unobstructed mechanism through which source recommendations can be provided. This method also provides for overcoming the "first-rater" and "sparsity" problems. We model user-profile as a concept map where individual object-nodes share relational arcs with each other. Mathematically, these objects and relations translate into first order and second order terms respectively in a vector notation. Use of vector space model to encapsulate documents, queries, and user-profiles permits the use of computationally simple and consistent measures such as cosine or fuzzy-min coefficients to compute similarity.

SRS requires an initial estimate of user-interests to provide a first set of recommendations. Direct use of query terms severely limits the vector size. Instead, SRS uses key concepts from the result-set of a first query to form the initial user-profile. Results-sets from subsequent search and retrieval are then used to refine the user-profile. This is done by either including new term-dimensions or adjusting the weights on existing term-dimensions. This refinement is done until an acceptably stable vector representation of the user-profile is formed.

Recall and precision have been traditionally established as work-centered metrics to evaluate IR systems performance. In this article, we employ similar metrics to evaluate performance of the source recommendation system.

3 System Architecture

The LexisNexis knowledge-warehouse contains approximately 50,000 sources. Each source is a compilation of labeled documents with a common attribute such as publisher, topic of publication, industry or geography. These sources are organized into a tree-like hierarchical structure on the LexisNexis web interface, so customers can identify a subset of the source volume to search on. Hierarchical organization of such a large source volume can render the structure too complex for navigation. Novice users or users researching a new topic often face the challenge of identifying a right subset that appropriately covers the user's information needs. SRS is designed to automatically assist users in identifying that right source (or a collection of sources). SRS has two components – the inverted file builder and the source recommender.

3.1 Inverted File Builder

A source in the LexisNexis warehouse is a collection of labeled documents. Documents are tagged with metadata by subject matter experts when they are included into the source volume. This metadata is populated across various segments – segments such as industry, company, geography, subject topic, and so on. Inverted File Builder is an automated module to perform a general search on each source for every source in the 50,000 source volume. Once a sample set of documents from the source is obtained, instances of metadata from these documents are collated to form an inverted index for the source. Inverted File Builder is built in C#, a .NET language, and uses regular expressions, Microsoft HTML object library and COM-.NET interoperability to drive the automatic search and retrieval process.

3.2 Source Recommender

The Source Recommender module is built in C# as an Internet Explorer bar (a browser plug-in). Plug-in design permits users to obtain side-by-side recommendations non-intrusively while performing search and retrieval. Figure 2 shows source recommender providing source recommendations to the user. The Explorer Bar model also permits source recommender to obtain handles to the browser object and the content document window in order to analyze the results from a search and retrieval session. Results-set contains documents, whose metadata segments provide an indication about the user's interests. Source recommender has functionality to collate these metadata segments into a vector form and compare them with the inverted source file created by the inverted file builder. This comparison leads to a ranked list of sources that are similar to user search and retrieval needs.

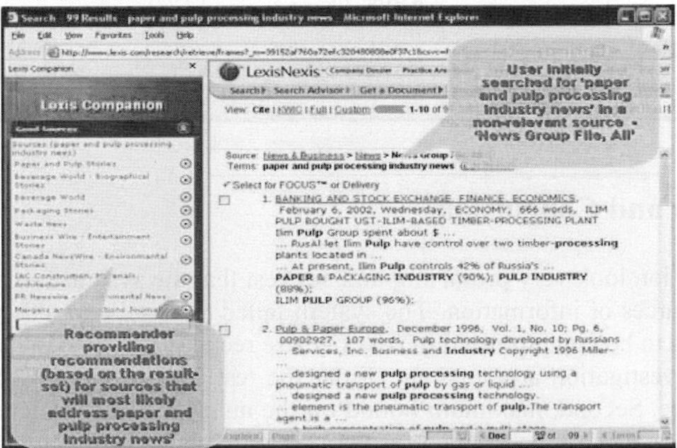

Figure 1. Recommender providing source recommendations to the user.

4 Evaluation of the Source Recommendation System

In order to evaluate the performance of SRS, a prototype testbed of 2500 sources was selected. The SRS system is built to provide a ranked list of up to 25 source recommendations. Since there is a maximum bound on the size of the list, directly evaluating recall is not possible. We instead employ a heuristic procedure that indirectly measures the system recall. A subject matter expert is asked to navigate to a random source within the testbed volume and formulate an "on-point" query that uniquely sets this source apart from the source volume. We will call this the "target source". The subject matter expert is then asked to submit the on-point query to a generic source (with no special topic-focus) such as the "All News Group File". While this search and retrieval is processed on the generic source, source recommender analyzes the results-set and provides the top 25 source recommendations. The rank of the target source is noted if it is recalled within the top 25 recommendations. To evaluate precision of the system, the subject matter expert is asked to use a three-point Likert-scale to rate the relevance of all the 25 recommendations to the test-query.

4.1 Results

A total of 30 test queries (for 30 target sources) were framed by a subject matter expert. The results of the evaluation are presented in Table 1. Results indicate that in 26 (87%) cases, the recommender system correctly recalls the target source within the top 25 recommendations. In the correctly recalled cases, average precision was observed to be 56%. The subject matter expert was neutral about recommendations in about 17% of the cases while disagreeing about 27%. In cases where the recall was not successful, average precision was observed to be only 10%. The subject matter expert was neutral about 27% and disagreed with 63% of recommendations.

Table 1. Results from SRS evaluation

Total Queries	*Recalled target source successfully within top 25 recommendations*		*Rank of the target source*		*Avg. SME Precision Rating*
30	Yes	26 (87%)	Rank 1- 5	17 (57%)	Agree (56%) Neutral (17%) Disagree (27%)
			Rank 6- 10	2 (7%)	
			Rank 11 - 15	4 (13%)	
			Rank 16 - 20	1 (3%)	
			Rank 21 - 25	2 (7%)	
	No	4 (13%)	Rank > 25		Agree (10%) Neutral (27% Disagree (63%)

4.2 Discussion and Conclusion

Results from evaluation look very promising and suggest that the system can effectively guide the users to the right sources of information. The system failed to recall correctly in only a few cases. Precision also seems to be compromised in cases where recall is not successful. A few factors that will need further investigation are apparent. First, the test query may not be precise enough to isolate the test source. Second, maximum bound on the number of recommendations of SRS may not correctly accommodate all the recalled results. Third, terms included into the topic vector were not stemmed or normalized, and a few terms also exhibited polysemy or synonymy which was not compensated for in the vector space model. Future extensions of this work will study all the aforementioned factors, as well as other work-centered factors that assist in SRS scalability and optimization.

References

Adomavicius. G., & Tuzhilin, A. (2001). Extending Recommender Systems: A Multidimensional Approach. IJCAI-01, *Workshop on Intelligent Techniques for Web Personalization (ITWP'2001)*, Seattle, Washington, August 2001 (with A. Tuzhilin).

Balabanovic, M., and Shoham, Y. (1997). Content-based, collaborative recommendation. *Communications of the ACM*, 40(3).

Hill, W., Hollan, J. D., Wroblewski, D., and McCandless, T. (1992). Edit Wear and Read Wear. *Proceedings of ACM CHI-92 Conference*. 3-9. New York: ACM.

Hill, W., Stead, L., Rosenstein, M., and Furnas, G. (1995). Recommending and evaluating choices in a virtual community of use. *Proceedings of CHI-95 Conference*. 194 – 201, Denver, CO.

Lawrence, S. (2000). Context in Web Search, *IEEE Data Engineering Bulletin*, Volume 23, Number 3, 25 – 32.

Melville, P., Mooney, R. J., and Nagarajan, R. (2001). Content-boosted collaborative filtering. *Proceedings of the SIGIR-2001 Workshop on Recommender Systems*, New Orleans, LA.

Mooney, R. J., & Roy, L. (1999). Content-based book recommending using text categorization. *Proceedings of SIGIR-99 Workshop on Recommender Systems: Algorithms and Evaluation*, Berkeley, CA.

National Telecommunications and Information Administration. (Feb 2002). *A Nation Online: How Americans Are Expanding Their Use Of The Internet*. Retrieved Feb 3, 2003 from NTIA publications http://www.ntia.doc.gov/ntiahome/dn/index.html

Nichols, D.M. (1997). Implicit ratings and filtering. Proceedings of Fifth DELOS Workshop on Filtering and Collaborative Filtering. *ERCIM Report ERCIM-98-W001*. 31-36, Budapest.

Oard, D., and Kim, J. (1998). Implicit Feedback for Recommender Systems. *Proceedings of the AAAI Workshop on Recommender Systems*, July 1998.

Resnick, P., Iacovou, N., Suchak, M., Bergstrom, and Riedl, J. (1994). GroupLens: An open architecture for collaborative filtering of netnews. *Proceedings of ACM 1994 Conference on Computer Supported Cooperative Work*, Chapel Hill, NC: ACM, 175-186

Shardanand, U., and Maes, P. (1995). Social information filtering: Algorithms for automating "word of mouth". *Proceedings of CHI'95 - Human Factors in Computing Systems*, 210-217

The Web Characterization Project, Online Computer Library Center, Inc. (2002). *Size and Growth*. Retrieved Nov 15, 2002, from http://wcp.oclc.org/stats/size.html

Symbols, Signs, Messages in Ergonomics of Social Space

Adam Fołtarz

INSTITUTE OF ARCHITECTURE AND TOWN PLANNING
TECHNICAL UNIVERSITY OF LODZ (POLITECHNIKA ŁÓDZKA)
POLAND
LODZ, Al. Politechniki 10
tel. 042 63 135 45
e-mail: foltarz@p.lodz.pl

Abstract

An interaction of receptive stimuli in a sensory system is used to differentiate and organize data. It is one of the basic features of all perceptive sensations. Such an assertion leads to a conclusion that pictures, sound structures, pheromones, odoriferous substances and tactile sensations are formed. These stimuli in short-term memory disappear quickly and must be stimulated to be retained, remembered and interpreted by a nervous system. They are not important. The role of sensitivity - over sensitivity of natural and artificial substitute receptors in receiving sensory stimuli - signals in case of seeing process in a central and peripheral field, localisation, periodicity, a pitch of a tone and a sound, a description of texture of materials, solid bodies and liquids, smells and thermal sensation and emphatic and hostile sensations lead to representation of sensory data stored in long-term memory. The analysis and association of stimuli in a further process causes the active synthesis of signals. It makes us do research on the coherence of language - a code and a content of a message. When and how to look for proper information? Finding proper information in an informational hum is a condition for solving problems of information transmission in a flow of a signal i.e. in the structure of remembering, association of concepts as after - effects of sending a structure and a meaning.
A structure and a free flow of multiple media information are important on different levels of the intellectual and physical activity of participants of interactive social communication. As to psychological aspects of communication in the environment signs and systems of visual, tactile and acoustic information are essential. A semiotic landscape of the surrounding is changing together with the transmitors. Present and constantly multiple media, knowledge about nature and social engineering of communication condition and enable effective and correct research methods and designing, among the others, defining human-factors engineering standards. Urbanised outer spaces, as a necessary substitute of 'human biotope', must help in establishing necessary social bonds.
I conduct practical classes in propaedeutics with first year students in the Institute of Architecture and Town Planning of Technical University in Łódź. The aim of these classes is to make people aware of the fact that social communication is important in town planning on a micro and macro scale. I provide several examples and my own comments.
In conclusion, the issue of social communication is one of the basic problems of the humanization of present day architecture.

Adaptation of the reality for utilitarian purposes goes around existential needs. What determines expectations and functional features? For a Bauhaus constructivist, Marcel Breuer, a chair (il 1) is a clear functional construct with unrevealed literariness. Wawerka, a conceptualist, treated a chair as a symbolic object, which can stand all difficulties of life (il. 2). Compare all four illustrations.

Both contrary aesthetic statements towards a morphology of an object evoke a function through a form (Breuer) and through a symbolic borrowing of shape (Wawerka). Umberto Eco identifies a from with culture. Culture means all human interventions with the environment, transformed in such a way so they could be included in some social relation. According to Ingraden, a from qualifies matter. A functional need of object space is prior, anatomic and intuitive.

Reading of nature's intentions, its binary impressive sphere, visible and hidden constructive structure and meaning (also symbolic) should go on according to one criterion, which has to be multivalue.

Complexity of the problem requires objectivity and intellectual modification. If we add a reflection factor, intuitive measures usually become of a higher level of intellectual activity. A theory and rich experience from Euclides's geometry and discovery of models of surface parts and Gauss's bends and the usage of them in architecture and cosmology could be good examples. It seems that non-Euclides space is exclusively a conceptual construct. Einstein by proving non-expansion of space and time showed its bend of a changeable nature. In a phenomenological philosophy, generally speaking, space is such as we feel it and has meaning to us as it has to our consciousness. In the artistic practice anthropomorphic rules in Greek architecture and intuitive constructs of Gothic cathedral area examples of intellectual reflection in the usage of constructions and also symbolisation of forms and shapes.

For reading out functions and meanings you need multivalue cognitive criteria:

1. from pure practice, through
2. operations of powers and unrevealed phenomena, to
3. signs – formalised symbols in the picture of quasi reality. This reality is a set and a construct. It is a construct – a paradigm of romantic need of fulfilling in the world of shapes and metaphors.

A need of fulfilment in the world of shapes and metaphors is a challenge for architecture and art. Today a truth about "a man as a measure of everything" is particularly carefully analysed and verified when faced with challenges of today's civilisation and culture.
Revived styles want to be called modern. Classical architecture after a fascination of gothic, discovered by Filaret, was considered as *"buona maniera moderna"*.

Following Giorgio Vasari, a term "moderna" was used. It has been accepted to describe a revived style – modernism: a pure style glorifying technology and aesthetics. It turned out to be boring and inhumane. Architecture has to have meaning and a reference point, in renaissance it was Plato's metaphysics[1]. At this stage, let us return to a romantic utopia of an ideal city created by Claude Nicolas Leudoux. Under the influence of morality in the spirit of J. J. Russeau, in the years 1775-79, he made the dreams of encyclopaedia writers about new renaissance come true. The space arrangement designed by him showed the unity of man and the environment. Following this idea he showed the organisation of a city and a symbolism of forms. He subordinates the designation of particular buildings to the idea of the ideal co-existence of dwellers. Ledoux looked for forms full of expression. He used stereometric solids. He associated them with general concepts: surface of sphere as a symbol of eternity and a cube as a symbol of stability.

It may be assumed that a man, an intelligent symbolic creature, perceives the world of symbolic contents and he a subject of symbolism via his creativity and the need of fulfilment.

---

[1] Jencks, Ch. Architektura postmodernistyczna (Postmodernistic Architecture). 1987. Wydawnictwo ARKADY. Warsaw

A principle of binarism defines human perception and also the basic line of symbolic thinking. A world of a human is ordered in the categories of opposites.[2] Duality of forms and contents carries signs of binary opposition of a concept and a thing (signifying and signified). A from – description becomes a construct – a symbolic sign in relation to the contents.

"Artificiality lies in the centre of reality"[3]. An artificial civilisation is a total artificiality – a product of media. Hyper-real poetics of pictures creates metaphysical space. Hyper-real pictures evoke emotions of a momentary nature. Today's man stands in front of a "white screen". Audio-visual measures, complex technology of creating "messages" and contexts (senses and values) conveyed in them achieve a positives cultural meaning when a change in the way of reception – a level of perception takes place (refers to interactive contacts with reality). This contact – a dialogue can't be deprived of usage of an aesthetic criterion. In reference to a relation between a human and the architectonic space, humanisation of the surrounding is vital and filling it up with subjective meanings. A second criterion is to be found in the idea of universal designing. Ewa Kurylowicz gives a number of guidelines for designing a space in the architecture taking into consideration the needs of the disabled. She emphasised the sensual reception of the surrounding and connected with it problem in social communication in the urbanised environment of both healthy and disabled people. In the introduction she says:

"A history of environment adaptation to the abilities of these people shows certain, clearly defined phases: starting with divided enclaves ("parallel worlds") through marking of areas for the disabled, to multi aspect, equal treatment of all people with all space consequences. The latter tendency is based on a view that there is no concept of "standard needs", as a human being can't be treated as "average". On the contrary, under the concept of "a human" a variety of needs and abilities (I would personally add variety of from and techniques of contacts with the surrounding) is to be found."[4]

An interaction of receptive stimuli in a sensory system is used to differentiate and organise data. It is one of the basic features of all perceptive sensations. Such an assertion leads to a conclusion that pictures, sound structures, pheromones, odoriferous substances and tactile sensations are formed. These stimuli in short-term memory disappear quickly and must be stimulated to be retained, remembered and interpreted by a nervous system. They are not important. The role of sensitivity – over sensitivity of natural and artificial substitute receptors in receiving sensory stimuli – signals in case of seeing process in a central and peripheral field, localisation, periodicity, a pitch of a tone and a sound, a description of texture of materials, solid bodies and liquids, smells and thermal sensation and emphatic and hostile sensations lead to representation of sensory data stored in long-term memory. The analysis and association of stimuli in a further process causes the active synthesis of signals. It makes id do research on the coherence of language – a code and a content of a message. When and how to look for solving problems of information transmission in a flow of signal i.e. in the structure of remembering, association of concepts as after-effects of sending a structure and a meaning.

The imaginable picture, coded in the brain, is not objective and is voluntarily associated and is a wonder helpful in the process of association and interpretation, i.e. a wonder of intuitive thinking, imagination and fantasy are used. The above mentioned introvert abilities of the brain fulfil associative functions in the processes of association.

---

[2] Michera, W. Kolory w procesie symbolizacji (Colours in the Process of Symbolisation) in „Symbol i poznanie" (Symbol and Cognition) ed. T. Kostyrko. 1987. PWN: Warsaw

[3] Baudrillard, J. L'echage symbolique et moert. 1976. Paris

[4] Kurylowicz, E. Projektowanie uniwersalne, udostepnienie otoczenia osobom niepelnosprawnym. (Universal Design, making the surrounding available to the disabled). 1996. Centrum Badawczo-Rozwojowe Rehabilitacji Osob Niepelnosprawnych: Warsaw)

Associative ability encourages the use of heuritics in cognitive practices. Especially evoking the habit of directed thinking is the aim.

Operative functions of heuritics:

- evokes a feeling of informational hunger
- favours revealing and association of information
- postulates the need of openness to information
- releases the necessity of creative attitude towards things we have been exposed to (an innovative factor)
- creates a critical attitude towards what necessary and possible
- teaches the skill of seeing the whole and making aware of needs.

This thesis gives three recommendations:

1. Be in touch with what you find around. Let the matter with its unique qualities (without deformations) communicate with you.
2. Hear and listen to unknown messages and at each moment let the quiet direct your thought and sensitivity. Be obedient to the authority of need and be ready to submit to things that does not exist yet.
3. Start with what is present and the most meaningful. Act with existing meanings. Expand activity and think about the language of meanings and act only in the directions which remarkably expand meaning.

- Heuritics puts perception in order.
- Heuritics puts activity in time. It divides activities into separate sequences. It evokes reflection it the following relations:

decision ------------------ activity -------------------- assessment[5]

Heuritics can constitute a universal working technique of a designer and an architect creating languages of meanings and messages which are conveyed in human's surrounding.
Nowadays substitutes of nature are more often created – these are media creatures. A structure and a free flow of multiple media information are important on different levels of the intellectual and physical activity of participants of interactive social communication. As to psychological aspects of communication in the environment signs and systems of visual, tactile and acoustic information are essential. A semiotic landscape of the surrounding is changing together with the transmitors. Present and constantly multiple media, knowledge about nature and social engineering of communication condition and enable effective and correct research methods and designing, among the others, defining human-factors engineering standards. Urbanised outer spaces, as a necessary substitute of 'human biotope', must help in establishing necessary social bonds. It is especially essential in the aspect of uniting Europe in the context of saving "little homelands".
I conduct practical classes in propaedeutics with first year students in the Institute of Architecture and Town Planning of Technical University in Lodz. The aim of these classes is to make people aware of the fact that social communication is important in town planning on a micro and macro scale. I provide several examples and my own comments.
In conclusion, the issue of social communication is one of the basic problems of the humanization of present day architecture.

---

[5] Dietrych, J. System i konstrukcja (System and Construction). 1985. N.T. : Warsaw

1. Bruer Marcel "Krzesło" 1922.

2. Wawerka Stefan "Wojna".

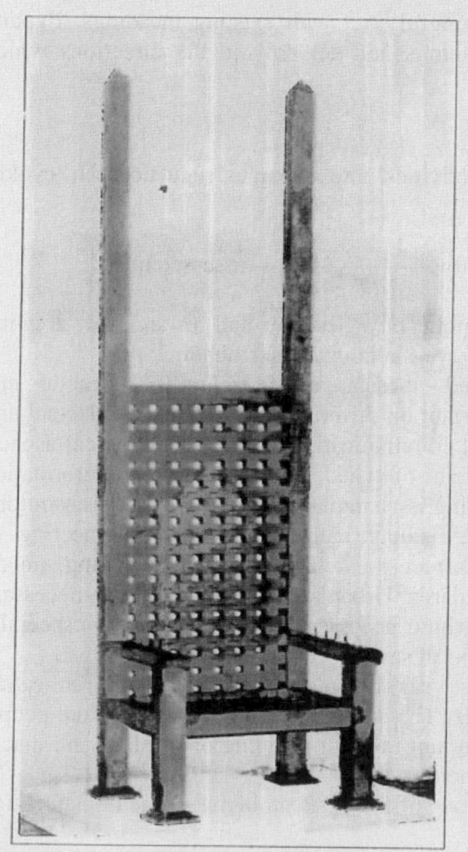

3. Fołtarz Adam tron, "Epilog", proj. scen. do teatru plastycznego

4. Fołtarz Adam siedzisko, ogród jurajski, proj. do patio ogrodowego

The automated measurement of icon complexity; a feasibility study

Forsythe, Alex, Sheehy, Noel, & Sawey, Martin

School of Psychology, Queens University
Belfast, Northern Ireland
a.forsythe@qub.ac.uk

Abstract

Complexity, or the amount of detail or intricacy is one of three important characteristics that designers can use to assess the usability of icons (McDougall et al., 1999). Measuring icon characteristics relies heavily however, on population sampling. This is a costly and slow process and an automated system capable of replicating what users "see" as complex could reduce development costs and time. Measures of icon complexity developed by McDougall et al., (1999) and Garcia et al., (1994) were correlated with automated measures, using Matlab (Mathworks, 2001); software that uses image processing techniques to measure icon properties. The strongest correlates with human judgements of perceived icon complexity (McDougall et al., 1999) were structural variability r_s =.65 and edge information r_s =.64. These findings suggest that image processing techniques can measure icon characteristics in ways that approximate more time consuming measures. The correlations also suggest that ratings of icon complexity may be linked to homogeneity in image structure.

1 Measuring Complexity

Graphical symbols or icons are widely used to convey information efficiently and effectively. Measures of the effectiveness of icons rely heavily on surveys of the perceptions of population samples. Changes to an established set of icons are likely to alter perceptual impact, but to what extent? Currently there are relatively few methods available to the icon designer that can allow them to estimate the likely impact of alterations on users' perceptions. One method involves the iterative testing of icons using samples drawn from populations of potential users. This process is costly and slow, and research into good icon design and usability has been impeded by difficulties in measuring and controlling symbol characteristics. McDougall, Curry & Bruijin (1999) have recently addressed some of these issues by developing the first set of normative values for icons.

McDougall et al identified three important icon characteristics; icon concreteness, distinctiveness and complexity. The characteristics of icon concreteness and distinctiveness are to some degree dynamic icon properties in the sense that the degree to which an icon is judged to be a concrete representation is partly determined by an observer's familiarity with the target object or function. The distinctiveness of an icon will depend not only on its structural features but also on the visual landscape in which it is situated (McDougall et al., 2000). Judgments of icon complexity however may be somewhat less sensitive to factors of user familiarity and visual context and may be computationally tractable using an automated measurement procedure. Garcia, Badre & Stasko (1994) have developed a semi-automated technique to measure icon abstraction. Their technique was based on several icon features including, closed and open figures, and horizontal and vertical lines. They found that icons which are pictorially similar to their real world counterpart are more likely to be judged as complex. In support of this McDougall et al., (1999) reported a high correlation (r_s =.73) between the Garcia metric and the complexity ratings of human observers. However, the Garcia system is painstaking for all but the simplest of icons. The two studies reported here consider whether a fully automated system could be used to quickly produce reliable estimates of perceived icon complexity.

1.1 Method

Matlab's (Mathworks, 2001) Image Processing Toolbox can be used to recover detailed structural information from a black and white, greyscale or colour image. The Toolbox has several image processing functions and three of these were applied to a random sample (n=68) of the McDougall et al (1999) icons. The presence of high correlations between these measures and the metrics developed by Garcia et al., (1994) and McDougall et al., (1999) would indicate similar measurements of icon complexity.

All touching pixels whether horizontally and vertically (but not horizontally) were considered to be connected. This is referred to as a 4-connected neighbourhood, and the difference between this and an 8-connected neighbourhood can be observed in Figure 1b &c. Essentially a 4-connected neighbourhood produces a sharper image.

To test the importance of changes in image structure in the perception of icon complexity, two perimeter detection measures and one structural analysis was performed. If structural variability is important in the perception of icon complexity it would be indicated by high correlations between these measures and the metrics of Garcia and McDougall. Examples of how these metrics treat the appearance of an icon are given in Figure 1.

Figure 1: Automated measures morphs with metric scores.

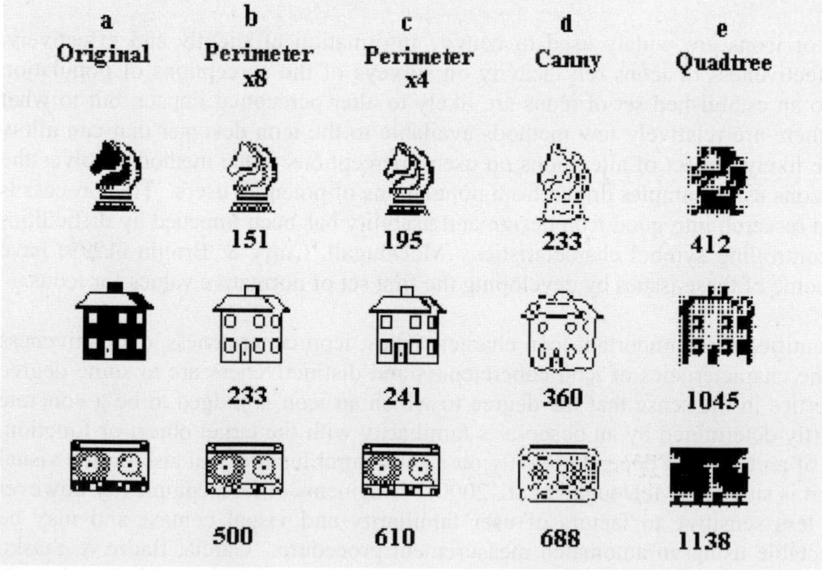

a Original	b Perimeter x8	c Perimeter x4	d Canny	e Quadtree
	151	195	233	412
	233	241	360	1045
	500	610	688	1138

Perimeter detection measures work by examining the changes in intensity that occur at the edges of an image. Edges are located using two criteria which examine areas in the icon were there is a rapid change in image intensity. Either a change in intensity must be larger than a predetermined threshold (edge detection provides a number of estimators which can used to specify sensitivity), or an edge will be detected were the intensity derivative has a zero crossing. For the purposes of this study zero crossings occur at the places were negative and positive pixels are adjacent. For a pixel to be considered an edge pixel it must be activated (on) and it must be connected to at least one non-activated (off) pixel. This is a simplified version of more general edge detectors, such as Canny, which calculate the gradient of intensity values for close-by pixels in colour or greyscale images. The advantage of the Canny method over other edge detection methods is that it works by using two thresholds to detect strong and weak edges and includes the weak edges in the output only if they are connected to strong edges. This means that truly weak edges will be detected in the analysis, but noise – such as shadow or shading- will be ignored. Figure 1d (tape) is a good example of the power of the Canny analysis. Whilst the perimeter measures (Figures 1b & c) retain elements of shading and shadow in their analysis, the Canny analysis retrieves a much more skeletal impression. Furthermore, since Canny is able to deal with greyscale and binary images it is unnecessary to treat the data before processing.

Quadtree decomposition examined homogeneity in the icons. This works by iteratively sub-dividing the icon into quarters and each sub-block is then tested for homogeneity. The sub-division continues until the resulting sub-block is homogeneous, hence a large number of small homogeneous blocks will indicate an image with a lot of structure (Figure 1e). A small number of largish blocks will indicate a more homogeneous image.

Each of these measures was correlated with the metrics developed by Garcia et al., (1994) and McDougall et al., (1999).

1.2 Results

The edge detection and structural variability metrics were correlated with the McDougall et al., (1999) and Garcia et al., (1994) measures of icon complexity. Given the large number of analyses performed, a Bonferroni adjustment was applied and the criterion for statistical significance was set at P<.0008.

Table 1: Spearman Correlations between Icon Attributes.

	McDougall	Garcia	Perimeter	Quadtree
McDougall Complexity	1.00			
Garcia	*.75	1.00		
Perimeter	*.64	*.66	1.00	
Quadtree	*.65	*.65	*.94	1.00
Canny	*.49	*.60	*.88	*84

**Correlations significant p<.0008*

Structural variability (r_s =. 65, p<. 0008) and edge information (r_s =. 64, p<. 0008) correlated strongly with human judgements as reported by McDougall, et al. Thus, perceived complexity appears to be partially accounted for by structural change. The Canny correlations with Garcia (r_s =. 60) and McDougall (r_s =. 49) are smaller but statistically significant. Several of the automated measures correlated significantly with each other. What differences there are between the measures is due to slight differences in the way Matlab implements each of the analyses. However for non-binary images, such as coloured icons or photographs, a Canny analysis could be extremely useful in removing the effects of noise from a calculation.

One probable explanation for these findings is that there are complex neurological response mechanisms that respond to changes in the structure of an image. Structural variations in an image are registered by the brain as changes in intensity, and it is these course and fine changes which provide detail and local information about a stimulus (Beck et al., 1991). Course scales are thought to be processed by the brain as low frequency components and these are processed faster than high frequency components obtained from local information. This difference in processing speed appears to be a function of image complexity. Thus, when an object contains a lot of detail its global attributes are processed faster than its local (Hoeger, 1997; Parker, 1997).

2 Predictive validity

Predictive validity is the extent to which a measure accurately forecasts how a person will think or act, in the future. To what extent can one take a population of icons, alter the same feature in all of them, and use the Quadtree and Perimeter measures to accurately predict how they will be judged by a sample of people?

A simple contrast (black-white) inversion was performed on all the of McDougall et al. icons (n=239) and Quadtree and Perimeter analyses applied to the inverted sample. The same inverted icon set was presented to 30 participants from Queens University Belfast. Following the McDougall et al (1999) method, participants were presented with booklets containing the icons and asked to rate the icons along a 5-point scale for Complexity (i.e. "the amount of detail or intricacy in the icon"). Each participant received the icons in a different order.

The human ratings of icon complexity for the inverted icon set were correlated with the Quadtree and Perimeter measures (cf Table 2). The McDougall et al., (1999) measures of icon complexity were also correlated with the complexity ratings for the new inverted icon set. A Bonferroni adjustment set the criterion for statistical significance at p<.0002.

Table 2: Spearman Correlations

	Inverted Set	Perimeter
Inverted Set	1.00	
Perimeter	*.46	1.00
Quadtree	*.34	*.92

Correlations significant p<.0002

The automated measures of icon complexity correlated .46 (Perimeter) and .34 (Quadtree) respectively with human ratings of perceived icon complexity. The reduction in the size of the correlation coefficient is undoubtedly due to the very large data set. However, the findings indicate that automated measures can reliably predict what humans "see" as complex.

The predicted complexity of the modified icon set correlated .85 p<. 0001 with the original McDougall icon set, indicating that black-white inversion an icons black / white contrasts does not change how human raters "see" complexity in an icon.

2.1 *Conclusion*

These findings suggest that image processing techniques can measure icon characteristics in ways that approximate more time consuming measures. The correlations also suggest that ratings of icon complexity may be linked to the number of edges contained within an icon.

To what extent can designers change the structural properties of icons without changing behavioural responses to them? Decision-support aids based on the image processing measures reported here have the potential to reduce the degree of speculation involved in arriving at an estimate. Some familiarity with Matlab is required for this particular software to be of immediate use to icon designers. Thus, simpler metrics – such as Garcia et al (1994) - may be preferred when calculating the complexity of more straightforward icons. Our measures have however, great flexibility in that they can be applied very quickly to any number of simple or complex icons. They also have the potential to yield estimates of the complexity of completely novel icons because the perception of complexity appears to be partially accounted for by multivariate structural variations that can be quickly measured. The measurement of structural variation may have future use in describing icon information in terms of spatial patterns and intensity change.

3 References

Beck, H., Graham, N., & Sutter, A. (1991). Lightness differences and the perceived segregation of regions and population, *Perception and Psychophysics*. 49(3), 257 269.

Garcia, M., Badre, A.N. & John, T. Stasko. (1994). Development and validation of icons varying in their abstractness, *Interacting with Computers*, 6, 2. 191-211.

Hoeger, R. (1997). Speed of processing and stimulus complexity in low-frequency and high-frequency channels. *Perception,* 26, 1039-1045.

Mathworks. (2001). Image processing toolbox users guide, the Math Works Inc. Boca Raton, CRC Press.

McDougall, S., Curry, M.B., & Bruijn, de, O. (1999). Measuring symbol and icon characteristics: Norms for concreteness, complexity, meaningfulness, familiarity, and semantic distance for 239 symbols. *Behaviour Research Methods Instruments and Computers,* 31,3, 487-519.

Parker, D,M., Lishman, J,R., & Hughes, J .(1997). Integration of spatial information in human vision is temporally anisotropic: evidence from a spatio-temporal discrimination task, *Perception,* 26, 1169-1180.

Supplementary material including a supporting appendix is available at:
http://www.psych.qub.ac.uk/research/projects/aia/

Improving System Usability Through the Automation of User's Routine Intentions: an Image Edition Tool Case Study

Alejandro C. Frery, André R. G. do A. Leitão, André W. B. Furtado, Fernando da C. A. Neto, Fernando da F. de Souza, Gustavo D. de Andrade, José E. de A. Filho

Centro de Informática (CIn) – Universidade Federal de Pernambuco (UFPE)
Av. Prof. Luis Freire s/n, Cidade Universitária, Recife/PE, Brazil, CEP-50740540
{frery,argal,awbf,fcan,fdfd,gda,jeaf}@cin.ufpe.br

Abstract

The concept of user's routine intentions is focused on this paper. A proposal, based on this idea, is made to develop applications with enhanced usability. Its goal is to add new user interface (UI) specification techniques to already known UI description models, in order to make it possible to translate the user's most desired intentions into a discourse of as few as possible intuitive interface interactions. An indication of how to integrate routine intentions in requirement analysis and design is also presented, as well as a case study: the development of a tool for image edition in the frequency domain using the proposed paradigm.

1 Introduction

Usual notations for describing system interfaces, such as transition diagrams and cognitive models, do not lead naturally to the identification of reiterative procedures. Our proposal consists of increasing the semantic content of these (and possibly other) notations, aiming at guiding the design of the human computer interaction towards system usability.

Usable systems provide users with convenient means to accomplish objectives. Such means should help translating the user's intentions into as few as possible intuitive interactions. Good translations are, therefore, prerequisites for the development of usable systems, since the aim of effective interfaces is a concise portrayal of the user's mental map. On the other hand, unsubstantial and confusing translations indicate that the user's intentions have to be dealt by a complex process in order to be executed by the system, indicating a lack of usability.

In most application domains, it can be observed that users repeat certain actions, aiming at satisfying their most frequently desired intentions. Although transition diagrams and cognitive models help improving system usability and describe system interfaces in a clear way, they do not emphasize such routine intentions. Thus, interface design may become less productive, since the option to automatically accomplish monotonous, tedious and even frustrating tasks is not necessarily available, forcing users to spend more time performing their most desired intentions.

This paper formalizes the proposal that user's routine intentions contribute massively to the improvement of human-computer interaction design from the usability viewpoint. As a consequence, it is also suggested that system design processes and frameworks should produce interfaces where the operations that implement routine intentions are as automatic as possible.

A frequency domain image edition tool, through the Fast Fourier Transform (FFT), is used as a case study in this paper. FFT is a powerful signal-processing algorithm, since it allows the definition and an efficient usage of image filters (Jain, 1989). However, as shown here, the manipulation of tools that implement FFT is not an intuitive process, demanding mathematical knowledge on the subject. Usable and intuitive interfaces would allow the FFT potential to be exploited by reluctant lay users, since a key point for a system to be accepted is its usability.

This set of applications was chosen as a case study, also, due to the fact that it is strongly used to combat periodic noise in images. This practice is identified as a routine intention, therefore becoming a major guide for the human-computer interaction design of such kind of tools. An image database, as well as a freeware version of the tool presented in this paper, is available for downloading at www.cin.ufpe.br/~imagens/fourierstudio. A comparison between the tool developed following the aforementioned proposal with other frequency domain image edition tools, in the context of human-computer interaction, is also presented here. This is aimed at confirming the hypothesis that the usability of one application is related to its competence in satisfying routine intentions.

2 Routine Intentions

An intention can be considered as the desire to obtain specific output from an application after some corresponding input is provided. User intentions are strongly related to system functionalities. The bridge between them is the way input and output are interchanged between the user and the system, a major responsibility of the user interface. Some intentions can be fulfilled by simple UI interactions, while others can only be achieved through many complex interface operations. Users, therefore, unconsciously build their own *interface mental map* after dealing several times with an application. Such mental map is a translation set that reflects the effort users have to dispend in order to satisfy each one of their intentions, translating them into UI operations.

Some mental map translations are more commonly performed than others, as they bind user intentions to key system functionalities. These specific intentions are defined in this paper as being *user's routine intentions*. As it will be seen later, it is desired that translations evolving routine intentions and key system functionalities are as simple and automatic as possible.

3 Enhancing Some Common Models of HCI

This section discusses some ideas related to adding semantic content to common models of HCI, in order to make them support the concept of routine intentions. The transition diagram design notation and the GOMS cognitive model will be focused on (Dix, Finlay, Abowd & Beale, 1998).

A transition diagram has a set of nodes that represents system states and a set of links between the nodes that represents possible transitions. Each link is labeled with the user action that selects that link and possibly other computer responses (Shneiderman, 1998). Although transition diagrams are effective for following flow or action and for keeping track of the current state plus current options, they do not emphasize the sets of user actions belonging to key system functionalities, which are related to routine intentions. A simple suggestion to overcome this limitation would be to identify *critical paths* in the transition diagrams related to routine intentions. Some proposals to accomplish this task include painting, with the same color, nodes belonging to the same critical path or labeling links between them as "special routine path links". The interface designer, in a later stage, would be able to collapse nodes belonging to a same critical path into fewer nodes or even into a unique node, bearing in mind the automation of the user actions related to the routine intention. This transition diagram enhancement approach is also applicable to similar notations such as *statecharts*.

GOMS is an acronym for GOALS, OPERATORS, METHODS, and SELECTION RULES. A GOMS model is composed of **methods** that are used to achieve specific **goals**. The **methods** are then composed of **operators** at the lowest level. The **operators** are specific steps that a user performs and are assigned a specific execution time. If a **goal** can be achieved by more than one **method**, then **selection rules** are used to determine the proper **method** (Bodnar, Heagy, Henderson & Seals, 1996). Similarly to transitions diagrams, the GOMS model does not lead naturally to the identification of reiterative procedures: it does not exist a way to treat a goal as a user routine intention. A proposal to enhance the model is to make a slight change to its syntax. Goals would have an additional attribute indicating how recurrent it is. According to the value of this attribute, the interface designer could prioritise goals in order to automate its methods, probably collapsing the operations related to them.

4 The Role of Routine Intentions in UI design

User interface software has traditionally been very difficult to create and maintain. This has been particularly troubling since good user interface design normally relies on an iterative approach based on experience and testing with real users (Hudson, 1994). The proposal presented in this paper states that user's routine intentions play an important role in this iterative process, since its identification in early software requirement analysis stages and its use in UI design can serve as a mean to reducing user frustration. In this section, it is suggested how analysis and design activities related to routine intentions fit in the software development lifecycle. Both Rational Unified Process (Jacobson, Booch & Rumbaugh, 1999) and Microsoft Solutions Framework (Microsoft, 2002), two of the most well succeeded software development frameworks, will be used to illustrate the aforementioned proposal.

An analysis of the Rational Unified Process (RUP) reveals that dealing with routine intentions should start at an early stage, in the *Requirements* workflow of the first *Inception* and *Elaboration* iterations. In this workflow, more specifically at the *Find Actors and Use Cases* activity, the *System Analyst* should identify which use cases represent key functionalities, therefore associated with routine intentions. This will guide the *Use Case Specifier*, in later iterations, to identify use case task sequences that can be automated. The output of this activity will then feed the *User-Interface Designer*, at the *User Interface Modeling* activity, with enough information to model a "routine intention driven" user interface. Similarly, the *Project Designer*, during *Analysis & Design* workflow, will be able to produce meaningful use case designs in order to guide *System Implementers* in their work. Finally, *System Testers*, in the *Test* workflow, should grant that the output produced by the *Implementation* workflow satisfied the *System Analyst*'s understanding of identified user's routine intentions.

Dealing early with routine intentions is also the best strategy for a software development methodology based on the Microsoft Solutions Framework (MSF). As part of clearly understanding user goals and requirements, in the *Envisioning* phase, the person executing the *User Experience* role (or *User Education* role, in previous versions of MSF), should identify user's routine intentions and help the *Product Manager* to introduce them in the *User Profiles* section of the *Vision Document*. During the *Planning* phase, the *User Experience* role should also consider routine intentions when analyzing user needs, creating user performance support strategies and conducting usability testing for all user interface deliverables. This must be reflected in the system *Conceptual*, *Logical* and *Physical Designs*. In the *Developing* phase, the *User Experience*, *Development* and *Testing* roles should work together in order to produce *Internal Releases* compatible to routine intention considerations previously specified. During the *Stabilizing* phase, the focus should be on usage testing to assure routine intentions were properly translated according to the *User Experience* role point of view. User beta testing feedback on this issue is important and must not be ignored. Finally, after the *Deployment* phase and the beginning of a new development cycle, new functionalities are elected and must be checked for correspondence with user intentions. If such a correspondence succeeds, the above process restarts.

Since both frameworks are based on an iterative software development approach in which key functionalities are prioritized, routine intentions can act as a mean for guiding this incremental system building process. Furthermore, some routine user intentions are very complex by nature and, therefore, it is expected that the use of system functionalities related to these intentions demands a large effort (represented by many complex user interface interactions). This can inhibit reluctant lay users to take advantage of these functionalities, since the corresponding mental map subset would not be so encouraging. However, if routine intentions are taken into account during requirements analysis and are properly handled in user interface design, it would be possible for many knowledge-specific applications to be exploited by a larger amount of people. The FourierStudio, presented in next section, is a nice example to illustrate this approach.

5 Case Study: FourierStudio

In order to illustrate and validate the concepts presented in this paper, an image edition tool was developed according to the proposed approach. An instance of the Rational Unified Process was used in its development. However, the Microsoft Solutions Framework (and possibly other frameworks) could be also applied.

FourierStudio is an image edition tool that works in the frequency domain. This domain is an alternative representation of any image, and it is obtained after applying the Fourier Transform in the well-known matrix representation. For computing this transform efficiently, FourierStudio uses the Fast Fourier Transform (FFT) algorithm. Working at the frequency domain is required when dealing with features that are not easily treated in space representation and that can be simply handled in the frequency domain. However, understanding the properties of the frequency domain is demanding, restricting the use of this technique. Though extremely powerful, frequency techniques are not available in pictorial image processing platforms as, for instance, Photoshop. These characteristics guided the development of FourierStudio, an application that combines the power of the Fourier Transform with usability for common users, enabling the use of this image edition technique to new application domains.

During the Requirements workflow of the Inception phase, it was observed that a common use of FFT based image edition tools is to filter periodic noise. This type of problem, illustrated in Figure 1, is typically handled at the frequency domain, where the noise becomes concentrated at clearly defined regions. Therefore, this intention was classified as being a routine intention. During the interface design, this intention was taken into account and its interface operations were automated, attempting to make the user mental map simpler. In order to validate the usability features of FourierStudio, an experience was developed to compare the usage of the tool against the usage of a popular image processing system, ENVI, that offers frequency domain techniques.

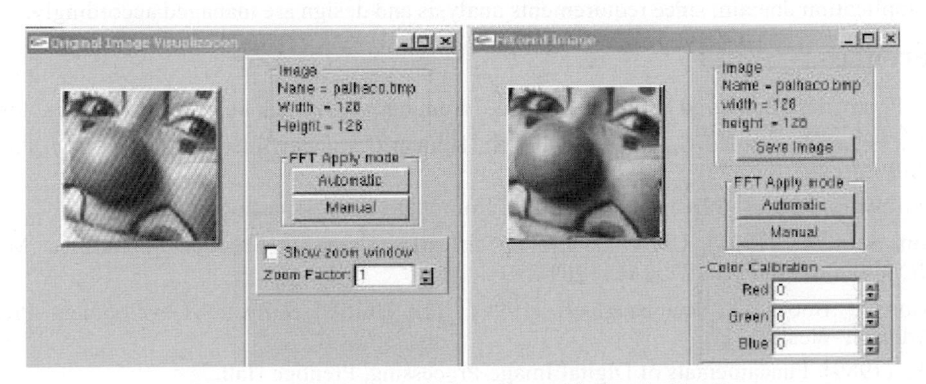

Figure 1: Filtering periodic noise with FourierStudio

ENVI (the Environment for Visualizing Images) is designed to address the numerous and specific needs of those who regularly use satellite and airborne remote sensing data. It uses a powerful structured programming language, the Interactive Data Language (IDL), which offers integrated image processing operations. This characteristic gives enhanced flexibility and capabilities to the application, which is largely used by remote sensing experts. The interface operations needed to achieve the aforementioned routine intention in ENVI contribute to a complex and deep mental map. The process starts at the selection of the image to be edited, using a common windows interface. Then, the Forward FFT option is selected at the main menu, leading to a window where the user must choose the image band in which the operation should be applied. ENVI works with multiple image bands (its internal representation of images), and users always have to manage these bands in memory, keeping track of the results of each operation applied. The Forward FFT operation results in a new band, which the user can visualize and edit. However, to edit this new image, the user must define another image, a mask, that will be combined with the transformed input image. Defining this mask requires accessing a new menu, in which the user must create a filter mask, defining regions and operations associated to them. Then, this mask is sent to memory where it can be multiplied by the FFT result image, leading to a new result band. Finally, the Inverse FFT option can be selected at the main menu window, and selecting the last generated band leads to an edited image band, that can be displayed or saved by the user.

Achieving this routine intention in FourierStudio is performed by a much simpler set of interface operations. The first step is the selection of the image to be edited, using a windows interface like ENVI. Then, the user must click the Edit Button and the application starts the automatic processing throughout this common user intention, ending with the display of the filtered image. Users do not need to manually edit the image frequency domain, neither to visualize it. The application developed using the ideas presented in this paper resulted in a more usable tool, through the automation of interface operations related to routine intentions.

6 Conclusion

Incorporating routine intentions identification and analysis in software development lifecycle is a practice that does not demand too much effort, but can cause a remarkable impact in user satisfaction and system usability. Besides that, encourages reluctant lay users to explore functionalities of more complex systems.

In the presented case study, the technique allowed the power of the Fourier Transform to be used by non-expertise users. In addition, it made possible for users that already know how to work with the FFT to be more productive. This significant achievement is also suitable to any other application domain, since requirements analysis and design are managed accordingly.

References

Adobe Photoshop. Retrieved February 02, 2003, from http://www.adobe.com/products/photoshop/

Dix A., Finlay J., Abowd G., Beale R. (1998). Human-Computer Interaction (2nd ed.). Prentice Hall Europe.

Envi Software. Retrieved December 09, 2002, from http://www.rsinc.com/envi/

Hudson, Scott E., (1998). User Interface Specification Using an Enhaced Spreadsheet Model. *ACM Transactions on Graphics*, 209-239

Jacobson I., Booch G., Rumbaugh J. (1999). The Unified Software Development Process, Addison-Wesley.

Jain, K. (1989). Fundamentals of Digital Image Processing, Prentice Hall.

Shneiderman B. (1998). Designing the User Interface: Strategies for Effective Human-Computer Interaction (3rd ed.). Addison Wesley Longman.

The Microsoft Solutions Framework on TechNet. Retrieved December 19, 2002, from http://www.microsoft.com/msf

Self-Organized Criticality of Color Information of Impressionist's Art Works

Asako Fukumoto   *Dong Sheng Cai*   *Michiaki Yasumura*

Keio University
asako@sfc.keio.ac.jp, yasumura@sfc.keio.ac.jp

Tsukuba University
cai@cs.tsukuba.ac.jp

Abstract

Impressionism is one of the greatest arts beloved by people all over the world. One of the greatest artistic appealing are their specific brightness and color touches, In this research, we use chi-square fitting for color information obtained from the image of their paintings to estimate the number of colors used by its painter as a degree of freedom in images, and then to estimate the fractal dimensions of their color information used in the image. Here we describe an analysis of the color information of impressionist's art works, which shows, first, that, the order of distribution of colors appeared in the paintings tend to obey Zipf's law, and second, both the degree of freedom of the color they used and the fractal dimensions of colors can be useful in characterizing the impressionist works.

Background

Impressionists tried to see the object naturally, focusing on direct and reflected sun light, and tried to record what they perceived with short brush strokes on their canvas, without black and gray, and mixed colors. Unfortunately the special appeal of impressionism is still unclear. In this research, we apply the theory of "Self-Organization-Criticality (SOC)"for the analysis, to show that the color information of the Impressionist's Art works is implicitly structured, the distribution of the color frequency obtained from image, tend to obey zipf's law. And to estimate the number of colors they used in the painting and found that approximately 20 to 40 indexed colors they used in their canvas .We propose that this invisible feature creates the visual appeal of the impressionist's art works and will be useful in characterizing the impressionist works and also color design.

Fig.1: La cathedrale de Rouen, le portrail et la tour Albane, effet du matin;Harmonie blache   Claude Monet
[Impressionist] Musee d'orsay]

1 Theory of Zipf's Law

1-1 SOC : Self Organized Criticality
The behavior of SOC is the tendency of large systems to evolve spontaneously towards a critical state without interference from any outside agent. The process of self-organization takes place over a long transient correlations in both space and time, and is always created by the a long process of evolution. [1] There is the that simple underlying character of SOC, Power-Law

1-2 Power Law
These three kinds of phenomena are so similer, in that they all be expressed as straight line on a double logarithmic plot, which called Power laws.

1) Power spectacle density (1/f noise) $$P(f) \propto \frac{1}{f^{\alpha}} \quad (1)$$

2) Size distribution (Zipf's law) $$N(s) \propto \frac{1}{s^{\gamma}} \quad (2)$$

3) Temporal distribution of Events

(Inter-event-interval distribution) $$N(\tau) \propto \frac{1}{\tau^{\beta}} \quad (3)$$

Since we are focusing on the "distribution of frequency of color information" we apply eg.(2) Size distribution generally called Zipf's law. It is well known for the regularity expressed by the straight line in the logarithmic plot of rank versus frequency of words used in the English language. Zipf also made a number of striking observation of some simple regularities in systems of human origin.[1]

2 Experiment and Features

Since RGB value is continuous, the patterns of dividing the colors of the image is infinity. "The Dividing colors" here we call "degree of freedom", it is quite difficult to define the degree of freedom. To estimate the degree of freedom from the paintings, it is influenced by the subjective view, which means distribution is differ from individual, and hard to estimate the objective value. Here we use the algorism divided method

The Equation(1) obeying the power law, distribution Function N(s) the degree of freedom N, each rank s, Cut off parameter T, and adjustment parameter A

$$N(s) = A\frac{1}{s^{\beta}}e^{\frac{-\tau}{T}} \quad (4)$$

S : rank T :Cut off Parameter , β : slope

A: adjusting Parameter

From this equation, we estimated the "Theoretical Deistribution(Value) "

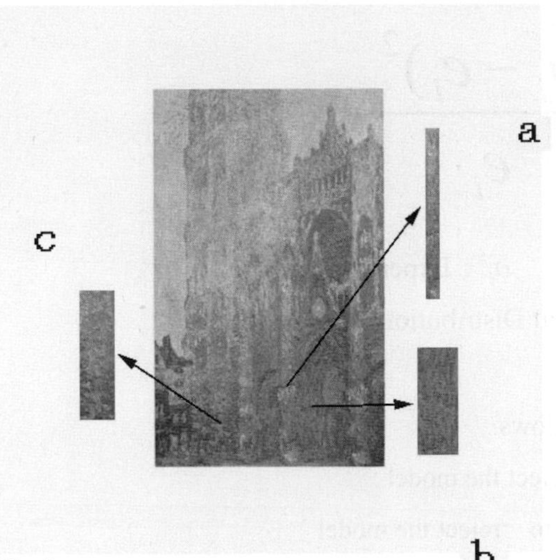

Fig.2
La cathedrale de Rouen, le portrail, soleil,maintinal; Hamonie bluee Claude Monet [Impressionist] Musee d'orsay]

Procedure of Experiment

1. Prepare digital image of Impressionist's art works, here we use the series of "Rouen Cathedral", masterpieces of Claude - Monet ,the representative of Impressionism

2. Transferred the file format to indexed color min 10～max 256(rgb value),using Adobe Photoshop7.0

3. Extracting the histogram data of each indexed colors(1～256) from the image

4. To evaluate the slope value of " Theoretical distribution" applying N(s)(1), evaluate N(s) obtained from the data of procedure3

3 Fcature 1

From Experiment1, we found that "theoretical distribution(Value) obey power law, which means the color Frequency of the image obey Zipf's law.

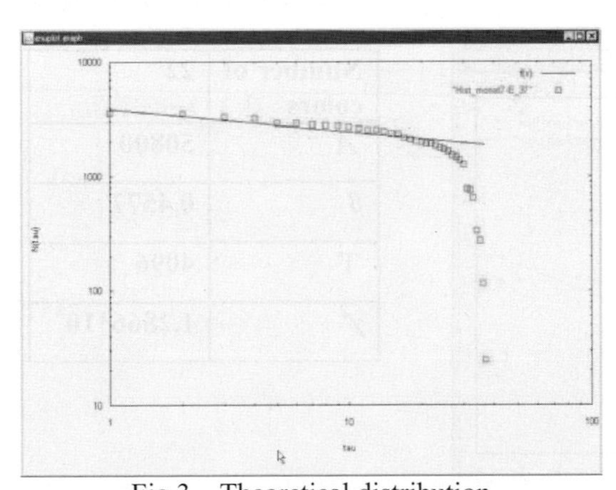

Fig.3 Theoretical distribution

Table.1 Result of the Experiment 1

Adjustment A	**50800**
Slope value β	**0.4577**
Cut off parameter T	**4096**

4 Experiment 2

We are trying to estimate the number of colors artist used in canvas with χ^2 Fitting Test, to investigate which degree of freedom is the smallest inχ^2 Fitting[2]
Here is the equation of χ^2 Fitting applying in this experiment.

$$x^2 = \sum_{i=1}^{k} \frac{(o_i - e_i)^2}{e_i} \quad (5)$$

k : Number of data, o_i : Experimental Data,

e_i : Theoretical Distribution

But there is the problem of χ^2 Fitting Test as follows.

(1) If the size of the sample is large, it tend to reject the model

(2) If the size of the sample is small, it tend not to reject the model

(3) The degree of freedom of χ^2 doesn't depend on the sample, but depend on the value of K

In this experiment, we use the large data. Thus we apply eg.(1) If the size of the sample is large, it tend to reject the model. To solve this problem, we scaled each sample data to 1000.

5 Feature 2

This is the result of the test, and this figure shows that"22 colors" is the most optical divided colors in this image.

Table **2** Result value of χ^2 Fitting Test

Number of colors	**22**
A	**50800**
β	**0.4577**
T	**4096**
χ^2	**1.2866*10³**

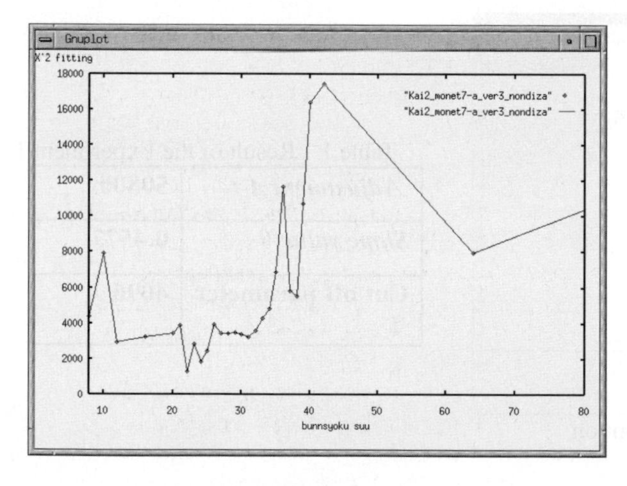

Fig 4 Result of χ^2 Fitting Test

We also examine other painting of Claude MONET, and found that those optical number are from 20-40 colors

6 Conclusions

We proposed an analysis of the color information of Impressionist's Art works, first, that, the order of distribution of colors appeared in the paintings tend to obey Zipf's law, and second, both the degree of freedom of the color they used and the fractal dimensions of colors can be useful in characterizing the impressionist works.

References

[1] Bak, Per, How nature works, Springer-Verlag, 1996.
[2] Mandelbrot, B. Benoit, The fractal geometry of nature, Freeman, 1977.

Cognitive Aspects of Computer Aided Planning Tasks

Maria Giannacourou

University of Piraeus
Karaoli & Dimitriou 80, 18534 Piraeus
email: mgianna@unipi.gr

Lambros Laios

University of Piraeus
Karaoli & Dimitriou 80, 18534 Piraeus
email: llaios@unipi.gr

Abstract

Advanced decision aiding automation raises a number of human performance issues. An empirical investigation was conducted to compare the effectiveness of various levels of strategic planning support, under two conditions of uncertainty, into managerial decision-making performance. Decision process was analyzed into distinct phases and empirical assessment on the effects of a computer system relying on a descriptive decision model are presented describing the benefits of such a system to the decision maker in each phase of strategic planning as well as in the total process.

1 Introduction

Aiding human cognitive performance requires description of application - specific cognitive activities through empirical studies, models of cognitive activities in work domains and through identification of the cognitive activities demanded by the process being controlled and by the structure of the control system design. Many biases have been demonstrated for decisions made under certainty such as the concept of bounded rationality meaning that a decision maker bases decisions on a simplified model of the world. When decisions are difficult, such as decisions under uncertainty, violations of the rational model of decision-making are more systematic and although this procedure reduces the processing load, it can also lead to the elimination of the optimal choice.

In the present study, the comparison between aided vs. unaided decision making seeks to explore some of the shortcomings of human judgment and decision making in order to determine the extent to which computerized decision aids could assist the human judge making better decisions.

2 Method

2.1 Experimental procedure

The laboratory method of experimentation was chosen and a business game was developed based on data from a real enterprise. The subjects were 35 managers or senior executives of small to medium-sized enterprises and they were asked to make decisions on product line, marketing policy, investment in plant and equipment, as well as purchase or sale of stocks in order to maximize profit during the experimental session which represented a three-years period. One forth of the subjects proceeded unaided while the rest used 3 variations of the same strategic planning support system (Strategic Managerial Planning System - SMPS) developed at the University of Piraeus: full support, no support during the environmental scanning stage, no support during the generation of alternatives stage. Two levels of uncertainty were introduced (low - high) in the form of variations in forecasting estimations.

2.1.1 Analysis of data

The analysis of data from the business game revealed 46 dependent variables. Through factor analysis, these variables were reduced to a manageable and meaningful set of 10 variables, using the Varimax procedure, which accounted for 86 percent of the total variance of the original 46 items. The internal consistency of each set was tested using the Cronbach's alpha method. Seven sets of variables had reliability coefficients that ranged from 0.86 to 0.99, after eliminating one item, out of the five items, of the sets and they were included in the consequent analysis. The seven variables were: length of analysis, target sophistication, scanning of strengths, scanning of weaknesses, scanning of opportunities, scanning of threats and decision confidence.

Due to the nature of the data as well as the existence of two independent variables the MANOVA method was selected. The aim of the analysis was to test: the influence of the level of support, the influence of uncertainty, the influence of the interaction between level of support and uncertainty, and the influence of the interaction between dependent variables.

2.1.2 Results

One of the main contentions of this study was that the use of the computer aid would affect the effectiveness of decision-making. Effectiveness was considered as a compound measure, which included economic performance, decision confidence, decision time and quality of decision-making. The analysis of the data revealed that all levels of computer support affected positively economic performance. However, only full support conditions produced statistically significant results (p=0.02). The examination of the data on decision confidence did not reveal any significant differences between users and non-users, all scoring quite high in confidence. Regarding decision time, the data suggest that users took more time for decision making than non-users. However, here again the significant difference is between full support and no support conditions during the first experimental year (p=0.002) and in total (three experimental years, p=0.006). To test decision quality the process of decision making was broken into distinct stages and effectiveness criteria were developed for the decision making process in general such as degree of sophistication (number of stages included in the decision process) as well as for each individual stage of the process which included accuracy and comprehensiveness.

The analysis showed that most significant differences are found in the behavior of users vs. non-users. Thus, computer users followed more sophisticated processes, which included more stages than non-users. The later seemed to pass into action (generation of alternatives) immediately after reading the preliminary information supplied by the business game. However, statistically significant differences were found only for experimental year 1 and 2 (p=0.001 and p=0.01 respectively). Regarding individual planning stages and especially the environmental scanning stage and the generation of alternatives stage, statistically significant relations were again observed between users and non-users. For the target setting stage, the analysis showed that non-users rarely set quantitative targets while computer users, in all experimental conditions, set both quantitative and qualitative targets and thus, significant relations are observed between users and non-users (p=0.001). In the evaluation of alternatives stage no significant differences were observed mainly due to the fact that most subjects, users and non-users, skipped this stage.

Regarding the effects of uncertainty on performance the data revealed significant differences between low and high uncertainty conditions in economic performance (p=0.01), scanning of opportunities (p=0.01), and utility of selected tactics, (p=0.01). Uncertainty and level of support interacted only in initial decision time, (p=0.05). Thus, non-users spent less time than users in reading business game information when uncertainty levels were low while they spent more time when uncertainty levels were high. On the contrary, users spent more time when uncertainty levels were low, except computer users without support during the environmental scanning stage.

3 Discussion

In general, the analysis of the data showed that unaided subjects proceeded in the generation of alternatives immediately after reading the supplied information while aided subjects followed more sophisticated processes. Thus, during the first experimental year aided subjects went through all (five) decision-making stages even when support was missing in certain stages according to the experimental conditions. During the second experimental year subjects in full support conditions went through, again, all five stages while the subjects in the other experimental conditions usually eliminated the stages of the environmental scanning and evaluation of alternatives. During the third experimental year both aided and unaided subjects followed similar decision making processes, generating alternatives immediately after reading the supplied information. These findings suggest that unaided subjects proceed to the solution of the business problem immediately after reading it, without seeking more information. This behavior is also evident when decision time is analyzed. Aided subjects took more time to reach decisions than unaided and this was due to the multi-stage decision processes they followed. Unaided subjects did not ask for more information although they had a catalog with available information but used data and experiences from their own companies to fill possible gaps and justify their choices. Thus, unaided subjects deliberately restricted their search for information. A possible explanation could be either the cognitive demands that the processing of such information entails or the habit to exploit information at the greatest extend due to the difficulty associated with its acquisition (Isenberg, 1986). Thus, the cognitive effort associated with the acquisition and processing of information is responsible for the non-optimal behaviors observed in unaided subjects.

Cognitive theories can also explain for the behavior of subjects without full support. These subjects tend to drop the environmental scanning and evaluation of alternatives stages after the first year. Users' positive evaluation of these stages, as well as utilization of these stages at full support conditions reject incompatibility issues and point, again, towards more cognitive explanations. These two stages are considered as very demanding in terms of cognitive effort due to the multitude of information and the complex processing that is required for their execution. The lack of support during the environmental scanning stage in some of the experimental conditions made the search for and the evaluation of information even harder. Users turned into action (generation of alternative tactics without evaluation) based on incomplete information. The decision making process followed might be considered as quite different from the process proposed by the rational model. However, a closer examination of the process followed by the subjects revealed certain similarities with the rational model, especially in the strategy and target setting stages. Thus, the same subjects who applied heuristics took time and effort to thoroughly set strategies and determine quantitative targets. These findings support more flexible theories for explaining the cognitive strategies that entrepreneurs use during the execution of their tasks. For example, Isenberg (1986) observed that managers generate contingency plans (rational strategy) based on restricted and incomplete information (non rational strategy). In the same direction, Fredrickson (1985) concludes that the strategies used by managers include both logic and intuition. Feng-Yang (1998) observes that such behaviors are more effective in rapidly changing environments. Thus, both literature and study findings agree that the way managers reach decisions is not always the same (rational or non rational) but it adjusts so as to include both rational and intuitive elements since the use of heuristics does not necessarily exclude the use of rational methods. Such behaviors might be more effective in changing and uncertain environments and explain the survival of companies through decisions made this way. The comparison of the results between support conditions showed that the use of rational strategies was more often and faded later in full support conditions. Therefore, it seems that the systematic use of a computer system affects greatly user behavior and might even change user habits. However, further research is necessary before conclusive statements can be made.

Regarding the effect of support on decision time, aided managers spent, in general, more time on decision making than unaided managers although the significant differences are between full support and no support conditions. Similar results have also been observed in the literature, (Mackay et al. 1992). These results can be attributed to the familiarization period which is necessary, when a computer aid is introduced to the decision making process, as well as to the use of more sophisticated decision making methods: more thorough execution of more planning stages than unaided subjects.

On the other hand, managers' subjective evaluation concerning the correctness of their decisions was not affected by computer use and were rather high (reported levels of confidence exceeded 70%). Thus, it seems that when managers expressed their certainty regarding produced results (alternative tactics), they used either very high or very low percentages, which signify acceptance or rejection (Valiris and Laios, 1995). When in doubt, managers choose to reject the alternative tactic and thus, strong confidence levels characterize the remaining ones.

Concerning the effect of uncertainty, subjects under high uncertainty conditions selected low utility tactics, which consequently, led to low economic performance. High uncertainty also seems to affect environmental scanning, as it is evident in the significant reduction of opportunities that subjects scanned. These results support arguments, which attribute low performance to inefficient environmental scanning. Smircich and Stubbard (1985), who studied strategic management procedures, conclude that the environment can be represented as a set of events and relationships between them, which are perceived and understood through certain cognitive strategies. These strategies of information selection and processing, affect strategic decisions and, through them, performance. Thus, the existence of variations in market forecasts, made their evaluation during the environmental scanning stage difficult, leading to the development of unsuccessful cause-effect relations (tactics). In addition, theorists (e.g. Galbraith, 1973) argue that as the level of uncertainty increases more information should be processed in order to make the right decisions. In other words, environmental scanning should increase when uncertainty increases. However, in the present study uncertainty affected negatively the scanning stage and subjects neglected or failed to collect the right information, which could lead them to the correct conclusions, and thus their economic performance deteriorated.

Regarding interaction between level of support and uncertainty the data revealed only one significant relationship in the time variable and especially during the familiarization period (the period in which subjects where getting acquainted with the problem reading the supplied information). Thus, unaided subjects spent less time during the familiarization period under low uncertainty levels and more when uncertainty was high. On the contrary, aided subjects spent more time when uncertainty was low, except those who solved the business problem without support during the environmental scanning stage. A possible explanation might be that the already high information uncertainty was getting greater when subjects tried to solve the business game without support and individuals spent more time during the familiarization period trying to absorb as much information as possible. On the contrary, when uncertainty was low, success prospects made computer users to spent more time trying to formulate the problem in their heads or to discover solutions based on the simulated company's data and on competitors' actions. When uncertainty was high, the use of the computer system might offer reassurance and subjects proceeded in the solution of the business problem, except in those cases where the absence of support during the environmental scanning stage maximized the already existing high uncertainty and subjects spent more time trying to assimilate better the existing information achieving thus, decision times similar to those of the non-users.

4 Concluding Remarks

The results of this study point towards the desirability of computer systems that support each phase of decision making distinctively placing more emphasis on the data gathering and evaluation phase (assessment of SWOTs) and the generation of alternatives (tactics).

A global consideration of human cognitive limits and especially their limited capacity for attending to and working with information as well as the biases suggest that diagnosis is a task for which humans are ill-suited and thus computer support in such tasks as environmental scanning can have positive effects. In order to improve the human component of decision making systems, decision aids can be developed which can help to unburden working memory and un-bias certain tendencies leading thus, to more optimal processes. In addition, decision-aiding systems that encourage rationalization of the decision-making process, imposing structure, promoting comprehensiveness of search for information and extending the scope of solution search, can positively affect planning effectiveness.

References

Feng-Yang, K. (1998). Managerial intuition and the development of executive support systems. *Decision Support Systems*, 24, 89-103.
Fredrickson, J.W. (1985). Effects of decisions motives and organizational performance level on strategic decision processes. *Academy of Management Journal*, 28, 821-843.
Galbraith, J.R. (1973). Designing complex organizations. Reading, MA, Addison-Wesley.
Isenberg, D.J. (1986). Thinking and managing: a verbal protocol analysis of managerial problem solving. *Academy of Management Journal*, 29, 4, 775-788.
Mackay, J.M., Barr, S.H. and Kletke, M.G. (1992). An empirical investigation of the effects of decision aids on problem-solving processes. *Decision Sciences*, 23, 3, 648-672.
Smircich, L. and Stubbard, C. (1985). Strategic management in an enacted world. *Academy of management review*, 10, 724-736.
Valiris, G. and Laios, L. (1995). Designing and validation of knowledge acquisition tools in business domain. *Behavior & Information technology,* 14, 2, 121-131.

Technology as an Equalizer: Can it be Used to Improve Novice Inspection Performance?

Anand K. Gramopadhye[1], *Andrew T. Duchowski*[2], *Joel S. Greenstein*[1], *Sittichai Kaewkuekool*[1], *Mohammad T. Khasawneh*[1], *Shannon R. Bowling*[1], *and Nathan A. Cournia*[2]

[1]Advanced Technology Systems Laboratory
[1]Department of Industrial Engineering
[2]Virtual Reality Eye Tracking Laboratory
[2]Department of Computer Science
Clemson University, Clemson, South Carolina 29634-0920

Abstract

Previous research indicates that as much as 90% of aircraft inspection is visual. Hence, where inspectors look is critical, potentially impacting inspection performance. Moreover, it has been shown that there are vast differences in performance between experienced and novice inspectors. Training on the task leads to superior coverage of the search area, better detection of defects, and improved efficiency and effectiveness. Providing this training, thereby improving the performance of novice inspectors, is the primary focus of this study. The research describes the virtual reality inspection environment and studies currently underway at the Advanced Technology Systems Laboratory (ATSL) at Clemson University in South Carolina. The studies funded by NASA are focused on evaluating alternate feedback and feedforward training strategies in improving aircraft inspection performance.

1 Introduction

Aircraft inspection and maintenance are an essential part of a safe, reliable air transportation system. Training has been identified as the primary intervention strategy in improving in section performance (Gramopadhye, et al., 1998). If training is to be useful, it is clear that inspectors need to be provided with training tools to help enhance their inspection skills. Existing training for inspectors in the aircraft maintenance environment tends to be mostly on-the-job (OJT). Nevertheless, this may not be the best method of instruction (Latorella et al., 1992). For example, in OJT feedback may be infrequent, unmethodical, and/or delayed.

Over the past decade, instructional technologists have developed numerous technology-based devices with the promise of improved efficiency and effectiveness, ushering in a revolution in training. Such training devices are being applied to a variety of technical training applications, including computer-based simulation, interactive videodiscs, and other derivatives of computer-based applications; and technology-based delivery systems such as computer-aided instruction, computer-based multi-media training and intelligent tutoring systems are already being used today. In addition, the compact disc read only memory (CD-ROM) and digital video interactive (DVI) are two technologies that will provide "multi-media" training systems for the future. The use of such technology, specifically computer-based simulators for aircraft inspection maintenance, has a short but rich history (Latorella et al., 1992; Gramopadhye et al., 1997; Blackmon et al., 1996; Nickles et al., 2001), the most advanced and recent example being the Automated System of Self Instruction for Specialized Training (ASSIST), a training program developed using task analytic methodology and featuring a PC-based aircraft inspection simulator (Gramopadhye et al., 2000). The results of the follow-up study conducted to evaluate the usefulness and transfer effects of ASSIST were encouraging with respect to the effectiveness of computer-based inspection training, specifically in improving performance (Gramopadhye et al., 1998).

Despite their advantages, existing multimedia-based technology solutions, including low fidelity simulators like ASSIST, still lack realism as most of these tools use only two-dimensional sectional images of airframe structures and, therefore, do not provide a holistic view of the airframe structure and the complex maintenance/inspection environment. More importantly, the technicians are not immersed in the environment, and, hence, they do not get the same look and feel of inspecting/maintaining a wide-bodied aircraft. To address these limitations, VR technology has been proposed as a solution (Vora et al., 2001).

Using VR, one can more accurately represent the complex aircraft inspection and maintenance situation, enabling students to experience the real hangar-floor environment. The instructor can create various inspection and maintenance scenarios by manipulating various parameters – for example, defect types, defect mix, defect severity, defect location, defect cues -- reflective of those experienced by a mechanic in the aircraft maintenance hangar environment. As a result, students can inspect airframe structure as they would in the real world and initiate appropriate maintenance action based on their knowledge of airframe structures and information resources such as on-line manuals, airworthiness directives, etc. Their performance in tackling these scenarios can be tracked in real-time with the potential for immediate feedback and active learning. Furthermore, instructors can use a progressive-parts approach based on the adaptive needs of the student, or instruction can be delivered asynchronously based on the availability and schedules of the student. The result is an innovative curriculum application, one in which the student will be able to visualize and test the information that is presented, internalizing the lesson. Students will be able to grasp the links between various visual cues presented, the need for specific inspection items and potential maintenance solutions. Repeated exposure to various scenarios along with classroom teaching will help them link theoretical scientific knowledge, for example, physical and chemical characteristics of structures, to various engineering solutions. In response, research efforts at the Advanced Technology System Laboratory (ATSL) at Clemson University have focused on developing the VR simulator for aircraft inspection training.

2 Development of the VR Environment

The development of the VR environment was based on a detailed task analytic methodology (Duchowski et al., 2001). Data on aircraft inspection activity was collected through observations, interviewing, shadowing, and digital recording (using video and still images) techniques. More detail on the task description and task analytical methodology can be found in Duchowski et al. (2001).Various scenarios were developed which were representative of those would occur in the real world environment. A library of defects was developed occurring at various severity and locations. The following defects were modelled: corrosion, crack and broken conduits. By manipulating the type, severity, location and defect mix; experimenters can now create airframe structures that can be used for running controlled studies.

3 The VR Environment

Our operational and deployable VR inspection simulator features a binocular eye tracker built into the system's Head Mounted Display (HMD), which allows the recording of the user's dynamic point of regard within the virtual environment (Figure 1a) (Duchowski et al., 2001). User gaze direction, as well as head position and orientation, are tracked to enable navigation and post-immersive examination of the user's overt spatial-temporal focus of attention while immersed in the environment. Tracking routines deliver helmet position and orientation in real-time, both of which are then used provide updated images to the HMD. User gaze direction is tracked in real-time (Figure 1b), along with calculated gaze/polygon intersections, for subsequent off-line analysis and comparison with stored locations of artificially generated defects in the inspection environment (Duchowski et al., 2002).

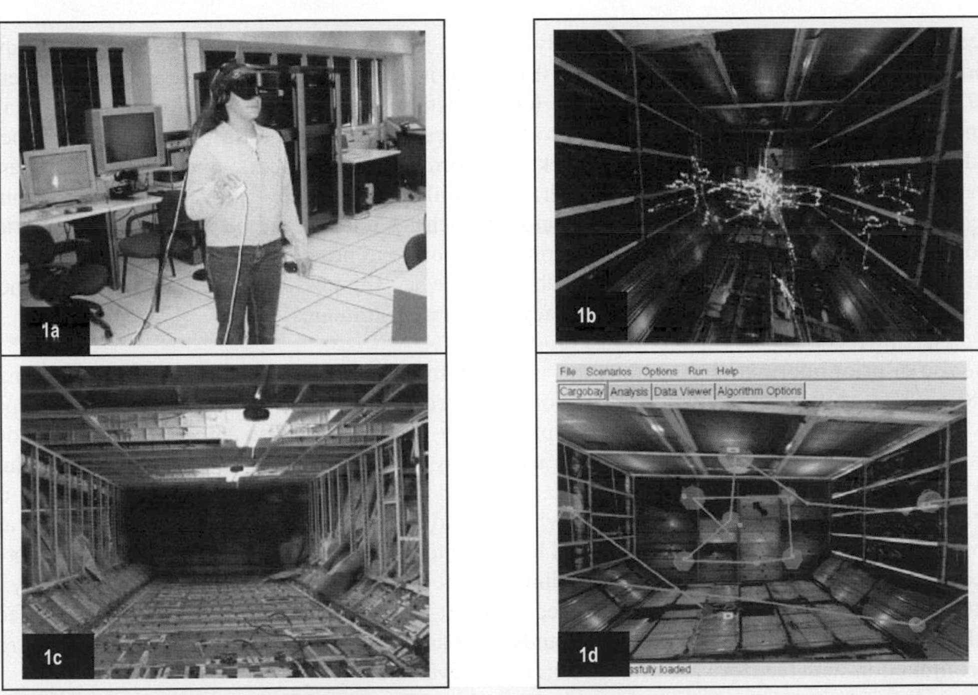

Figure 1: (1a) User immersed in VE, (1b) Real time gaze tracking in VR, (1c) Physical aircraft cargo bay, (1d) Captured fixation in virtual environment

The goal of the construction of the virtual environment is to match the appearance of the physical environment, similar to the example shown in Figure 1c (Vora et al., 2001). Eye movements, following fixation analysis, captured in the existing prototype VR environment are shown in Figure 1d. During immersion, users have access to tools similar to those used by a technician on the aircraft hangar floor (e.g., flashlight, mirror), albeit these are virtual in our environment. The system captures process data (eye movements of the subjects as seen in Figure 1d) and performance data (speed and accuracy) for use as feedback to the subjects. Several defect scenarios have now been developed (e.g., enabling visual search for corrosion, cracks, broken conduits), augmented with user-adjustable parameters. The graphical user interface facilitates on-line user immersion and off-line eye movement analysis.

Since initial development, we have successfully migrated from primary supercomputer rendering engine (a dual-rack, dual-pipe, SGI Onyx2® InfiniteReality™ system with 8 raster managers and 8 MIPS® R10000™ processors) to a personal computer. Our PC platform, running Linux and equipped with an NVidia GeForce4 TI4600 graphics card, attains rendering performance comparable to our former SGI platform at a significant reduction in cost. Multi-modal hardware components include a binocular ISCAN eye tracker mounted within a Virtual Research V8 (high resolution) Head Mounted Display (HMD).

4 Description of Studies

The studies funded by NASA are focused on the evaluation of alternate feedback and feedforward strategies in improving inspection. The following is a description of two experiments underway.

Feedback information has had consistently positive results in all fields of human performance (Gramopadhye et al., 1997), provided that it is given in a timely and appropriate manner. Wiener (1975) has reviewed feedback in training for inspection vigilance and has found it universally beneficial. Traditional feedback provided to the inspector has been performance feedback (speed and accuracy). Another form of feedback is information on process/ measures (e.g., search strategies, eye-movements data). However, it is still unknown how process and performance feedback should be presented to the inspector and what would be the best forms (i.e., statistical and graphical) of presenting this information. Therefore, the primary objective of this study is to evaluate the effectiveness of alternate feedback strategies (process, performance, and combined) on visual search performance: speed, accuracy, and search strategies.

The use of prior information (feedforward) is known to affect inspection performance (McKernan, 1989). This information can consist of knowledge about defect characteristics (types, severity/criticality, and location) and the probability of these defects. Although several studies have been conducted that demonstrate the usefulness of feedforward as a training strategy there are certain research issues that need to be addressed. These issues include: what format should feedforward information be presented in, when should feedforward information be presented, and how much feedforward information should be presented. Hence, this study evaluates the effect of feedforward information in a simulated 3-dimensional environment by the use of virtual reality. A simulated aircraft cargo bay in which inspectors must locate and identify various types of defects is the simulated environment used for the study. The use of job-aiding tools (flashlight and mirror) in conjunction with feedforward information is also evaluated.

5 Discussion and Conclusions

The VR environment developed allows researchers to conduct off-line controlled studies, facilitating the collection of performance (e.g., speed and accuracy) and process measures data (e.g., eye-movements strategies). The results obtained from these studies can be used to understand different aspects of inspection performance. Earlier studies have shown good transfer effects between virtual environments and real-world environments (Vora et al., 2001) in aircraft inspection. The results of these studies will throw new light in the use of feedback and feedforward in improving inspection performance, which will ultimately improve safety. Moreover, using the VR environment, we will be able to train novice inspectors in where to look for defects, thereby bring their performance to that of experienced colleagues in a shorter time span.

Acknowledgement

This research was funded by a grant from NASA (Grant number: NCC 2-1288, Program Manager: Dr. Barbara Kanki) to Drs. Anand K. Gramopadhye and Andrew T. Duchowski.

6 References

Backmon, B., & Gramopadhye, A. K. (1996). Using the aircraft inspectors training system (AITS) to improve the quality of aircraft inspection. *Proceeding of the Industrial Engineering Research Conference* .pp. 447-452. Minneapolis, MN,

Duchowski, A. T., Medlin, E., Gramopadhye, A. K., Melloy, B. J., & Nair, S. (2001). Binocular eye tracking in VR for visual inspection training. *In Virtual Reality Software & Technology (VRST).* Banff, AB Canada, ACM. 1-8.

Duchowski, A. T., Medlin, E., Cournia, N., Gramopadhye, A. K., Melloy, B. J., & Nair, S. (2002). 3D eye movement analysis for visual inspection training. *In Eye Tracking Research & Application (ETRA) Symposium*, ACM. pp 103-110. New Orleans, LA,

Gramopadhye, A. K., Bhagawat, S., Kimbler, D., & Greenstein, J. (1998). The use of advanced technology for visual inspection training. *Applied Ergonomics*, 29 (5), 361-375.

Gramopadhye, A. K., Drury, C. G. & Sharit J. (1997). Feedback strategies for visual search in airframes structural inspection. *International Journal of Industrial Ergonomics.* pp 333-344.

Gramopadhye, A. K., Melloy, B. J., Chen, S., & Bingham, J. (2000). Use of computer based training for aircraft inspectors: Findings and Recommendations. *In Proceedings of the HFES/IEA Annual Meeting*. San Diego, CA

Latorella, K. A., Gramopadhye, A. K., Prabhu, P. V., Drury, C. G., Smith, M.A., & Shanahan, D.E., (1993). Computer-simulated aircraft inspection tasks for off-line experimentation. *In Proceedings of the Human Factors Society Annual Meeting*. pp. 92-96 Santa Monica, CA.

McKernan, K.E. (1989). The benefits of prior information to visual search for multiple faults, *Unpublished Master's Thesis*, SUNY, Buffalo, Department of Industrial Engineering.

Nickles, G., Marshall, J., Gramopadhye, A. K. & Melloy, B. (2001). ASSIST: Training program for inspectors in the aircraft maintenance industry. *International Encyclopedia for Ergonomics and Human Factors*. Taylor and Francis - UK

Vora, J., Nair, S., Medlin, E., Gramopadhye, A. K., Duchowski, A. T., & Melloy. B. J. (2001). Using virtual reality to improve aircraft inspection performance: Presence and Performance Measurement Studies. *The Human Factors and Ergonomics Society Annual Meeting*. Minneapolis, MN.

Wiener, E. L. (1975). Individual and group differences in inspection. *Human Reliability in Quality Control.* By C. Drury and J. Fox (Eds). Taylor and Francis. London.

Measuring Team Situation Awareness by means of Eye Movement Data

Gunnar Hauland

Risø National Laboratory, Denmark, now with Det Norske Veritas, Norway
DNV, P.O.Box 408, N-4002 Stavanger, Norway
Gunnar.Hauland@dnv.com

Abstract

This study suggests that it may be possible to measure Situation Awareness (SA), including the temporal aspect of SA, by means of system operators' visual information acquisition. The results indicate that such measures can be defined both at the individual level and at the team level. SA and Team SA (TSA) can be measured by means of eye movement data as the extent to which operators *distribute their visual attention*, and actively involve themselves in *planning* activities in relation to the available and relevant situation elements.

1 Background and Objectives

The concept of Situation Awareness (SA) is used within human factors research to explain to what extent operators of safety-critical and complex real time systems know what is going on in the system and the environment. SA can be generally described in terms of operators' correct *perception* and *understanding* of a situation. Smith and Hancock (1995) defined SA as *externally directed consciousness*. It is not until the externally defined task is made explicit that the observed behaviour can achieve the status reserved for SA (ibid.). This seems to be in agreement with Endsley's (2000) notion of *objective* SA measures, i.e. measuring SA as the relation between operators' subjective understanding and the objective situation requirements. Furthermore, many definitions emphasise the *temporal* aspect of SA (Shrestha et al. 1995). One might argue that understanding in SA *is* anticipation, i.e. actively seeking information regarding the immediate future of situation elements. In this study, *understanding* was conceptualised as a *current* or *future time reference for thought* relative to the traffic situation.

The first objective of this study was to develop methods for measuring SA that include measures of Team SA (TSA). SA definitions are often limited to address SA of individual operators. Yet, many systems require more than one operator. The second objective was to avoid methodological problems associated with measuring techniques based on the interruption of simulated tasks. Measures during interruptions may not sufficiently capture the relevant and dynamic aspects of awareness. Thus, it was a goal to develop more process-oriented TSA measures. It was assumed that eye movement data can be used to develop such measures (Drøivoldsmo et al. 1998). The complete study is reported in Hauland (2002).

2 Method

The TSA measures in this study were based on Air Traffic Control (ATC) students' visual behaviour, measured by means of eye movement tracking (point-of-gaze), during simulator training at the Danish Civil Aviation Authorities Academy (Copenhagen Airport, Kastrup). Two criteria were used for the validation of measures. First, TSA measures should *predict* ATC system performance. There were multiple system performance measures, e.g. response time measures, number of repeated requests, number of continuous climb and descents given, frequency of radio transmissions and expert ratings of effectivity (Hauland, 2002). Second, TSA measures should be *sensitive* to manipulations in abnormal scenarios (two scenarios with emergency descents). The subjects were ATC students in 35 team combinations of radar and planner controllers. (Between subjects design. Variation in N was due to data loss.)

Three *probe*-events (with a 6 minute time-window) were defined *a priori* and embedded in the simulated tasks (inspired by Dwyer et al. 1997). Probe-events were normal traffic situations designed to require attention and coordination within and between teams, i.e. traffic events that by definition required TSA. Pseudo-pilots made sure that probes occurred as designed. Defining probes was an attempt to ensure *ecologically* valid, TSA *relevant* and *comparable* team samples.

Areas of Interest (AOI) were normatively defined as visual representations of the SA requirements. The subjects' visual information acquisition in relation to AOIs was used as measures of SA/TSA. In this study the visual unit of analysis was *dwells on AOIs*, conceptually defined as *several single fixations within an AOI*. (The operational definition was that *the point of gaze indicator must stop for a minimum of 250 milliseconds inside the border of an AOI. The dwell is terminated when the indicator exits the border of an AOI.)*

Dwells on AOIs, as opposed to single fixations, have the potential for representing a cognitive unit of analysis, i.e. a chunk of information (Hauland, 2002). In SA terminology (Endsley, 2000), AOIs can represent *elements* in the situation, i.e. in accordance with the situation requirements. AOI definitions may have any shape and size as long as the AOI represents a chunk of information. AOIs may be static or dynamic (moving), for example single aircraft and strips (flight information) as well as clusters of aircraft and flight information labels. Furthermore, the AOIs can be defined with properties possibly representing aspects of cognition. Such properties could be *individual* versus *team* related tasks and the *temporal* aspect of the task, i.e. representing *current* versus *future* time reference to traffic information.

The units of analysis were related to *individual competencies*, *team competencies held at the individual level*, and team *competencies held at the team level* (Cannon-Bowers and Salas, 1997). For example, in this study it was considered a team competence measured at the individual level that the radar controller and planner controller monitored each other's tasks. Measurements at the team level consisted of *relations* between two sets of individual behaviours (co-occurrences). Key behaviours were assumed to be the allocation of attention resources within the team.

The *visual attention strategy* variable was hypothesised to measure the perceptual aspect of SA: *Focused* visual attention implied looking at few chunks of information, i.e. AOIs, for long periods of time (more than one second), whereas *distributed* visual attention implied looking at many chunks of information for a short period of time on each AOI (less than one second). The *temporal reference of situation models* (i.e. application of knowledge) was measured by defining *current* or *future* time references to traffic information, represented by the AOIs.

3 Results

Measurements of visual behaviour were *frequencies* and *duration* (seconds) of dwells on AOIs aggregated for each probe time-window (Hauland, 2002).

3.1 Individual Level (attention strategy)

There were significant positive correlations between the planner controller's use of a distributed visual attention strategy and some aspects of ATC system performance (e.g., frequency measure: Kendall's tau = 0.352, $p < 0.01$, two-tailed test, N = 34. Kendall's tau = 0.462, $p < 0.01$, two-tailed test, N = 34. Duration measure: Kendall's tau = 0.407, $p < 0.01$, two-tailed test, N = 34). There was also a significant positive correlation between the duration of the radar controller's use of a focused visual attention strategy and system performance (Kendall's tau = 0.303, $p < 0.05$, two-tailed test, N = 35). This correlation indicates that *focused* visual attention for a longer period of time was less effective in terms of optimal use of the radio frequency.

The duration of the planner controller's use of a distributed visual attention strategy was significantly longer for the normal scenario than for the abnormal scenario (Mann-Whitney U = 17.000, $p < 0.01$, N = 22, two-tailed test). The total duration of the radar controller's use of a focused visual attention strategy in the sector was significantly shorter for the normal scenario than for the abnormal scenario (Mann-Whitney U = 33.000, $p < 0.05$, N = 23, two-tailed test). These results were found only for abnormal scenario number 1.

The operators' use of *visual attention strategies* seems like a valid TSA measure in that the measure can *predict* certain aspects of ATC system performance, although effect sizes were medium. This measure also seems *sensitive* to abnormal developments in the situation.

3.2 Team Level (temporal reference)

The second variable, considered an aspect of the operators' understanding, was the *temporal reference of situation models*, i.e. measuring the extent to which both team members simultaneously were involved in planning activities (active anticipation). That is, the co-occurrence of *future* and *current* time references of situation models in the team.

There was a significant positive correlation between the maximum duration of operator's co-occurring visual information acquisition from AOIs representing future traffic information for both the radar controller and the planner controller (i.e., overlapping) and (the response time measure of) system performance (Kendall's tau = 0.431, $p < 0.05$, two-tailed test, N = 21. Corrected for outlier: Kendall's tau = 0.530, $p < 0.01$, N = 20).

A clear shift from a *future* time reference to a *current* time reference was observed when something abnormal happened (figure1, below).

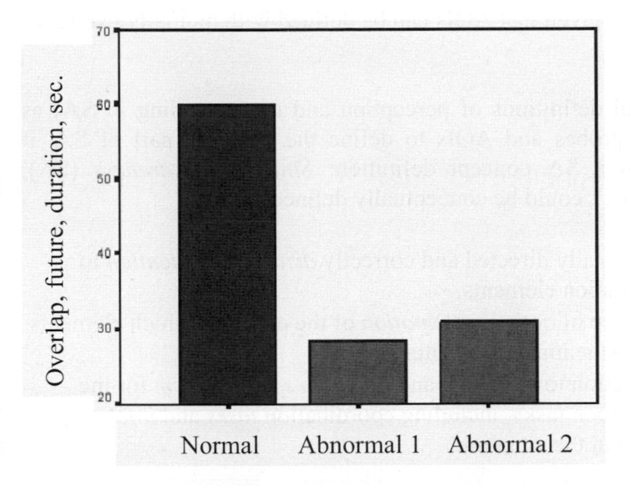

The two controllers' co-occurring visual information acquisition from AOI's representing *future* traffic information for both the controllers had significantly *longer duration* in the *normal* scenario than in the two *abnormal* scenarios / emergency descents (Mann-Whitney $U = 15.000$, $p < 0.01$, $N = 22$, two-tailed test and Mann-Whitney $U = 22.000$, $p < 0.01$, $N = 23$, two-tailed test).

Figure 1: Overlapping and future time references of the team's situation models.

The team goes from an orientation towards the *future* to a more *current* time reference during both the abnormal scenarios (figure 1). This finding was confirmed in that the controllers' co-occurring visual information acquisition from AOI's representing *current* traffic information for both the controllers had significantly *shorter duration* in the *normal* scenario than in the two *abnormal* scenarios (Mann-Whitney $U = 25.000$, $p < 0.05$, $N = 23$, two-tailed test and Mann-Whitney $U = 22.000$, $p < 0.01$, $N = 23$, two-tailed test).

4 Conclusion

The planner controller's frequent use of a *distributed* visual attention strategy may indicate that the perceptual aspect of SA is good. The radar controller's frequent use of a *focused* visual attention strategy may indicate that the perceptual aspect of SA is less good. However, no significant correlations were found for the opposite use of attention strategies, i.e. the planner's use of a *focused* strategy and the radar controller's use of a *distributed* strategy. Abnormal events seemed to cause a shift of visual attention strategies in the team, i.e. away from possibly ideal attention strategies towards less ideal attention strategies. Thus, the distributed and focused visual attention strategies seem like a promising measure of the perceptual aspect of SA.

If both operators simultaneously accessed information regarding future traffic, then system performance seemed to improve for the handling of abnormal events. This correlation was not found for system performance in normal scenarios. Furthermore, the team was less current and more future during normal scenarios, and vice versa during abnormal scenarios. This result indicates that abnormal scenarios cause the student teams to become less oriented towards the future. One might argue that the team ideally should relate as much as possible to different temporal aspects of the task (at all times). Nevertheless, the measure did comply with the validation criteria as defined in this study (Hauland, 2002).

The measure of *temporal reference of situation models* is considered a promising process-oriented measure of the understanding part of SA. When measuring the frequency and duration of the co-occurrence of planning activities, this variable represents a team competence held at the team level, i.e. this variable can not be reduced to aspects of individual operator's behaviour. This measure may be applied in other domains, given that AOIs can be defined with mutually exclusive *current* and *future* properties.

Based on these findings (i.e. operational definitions of perception and understanding in SA) as well as the approach using embedded probes and AOIs to define the situation part of SA, it becomes possible to suggest inputs to a SA concept definition: *Situation Awareness* (SA), including *Team Situation Awareness* (TSA), could be conceptually defined as:

- *Perception*: the operators' externally directed and correctly *distributed attention* to available and task relevant situation elements.
- *Understanding*: sufficient degree of *active anticipation* of the extent to which elements remain and become relevant in the immediate future.
- *Situation Requirements*: in relation to a *domain* and *situation specific norm* for the continuous information acquisition tasks, including coordination *tasks* and *inter-personal relations* as elements in the situation.

In sum, this study suggests that SA and TSA can be measured by means of eye movement data as the extent to which operators *distribute their visual attention*, and actively involve themselves in *planning* activities in relation to the available and relevant situation elements.

5 References

Cannon-Bowers, Janis A. and Salas, Eduardo. (1997). A Framework for Developing Team Performance Measures in Training. In: Michael T. Brannick, Eduardo Salas and Carolyn Prince (Eds.), *Team Performance Assessment and Measurement. Theory, Methods, and Applications*. Lawrence Erlbaum Associates.

Drøivoldsmo, Asgeir; Skraaning jr., Gyrd; Sverrbo, Mona; Dalen, Jørgen; Grimstad, Tone; and Andersen, Gisle. (1998). Continuous Measures of Situation Awareness and Workload. OECD Halden Reactor Project, HWR-539.

Dwyer, Daniel J.; Fowlkes, Jennifer E.; Oser, Randall L.; Salas, Eduardo; and Lane, Norman E. (1997). Team Performance Measurement in Distributed Environments: The TARGETs Methodology. In: Michael T. Brannick, Eduardo Salas and Carolyn Prince (Eds.), *Team Performance Assessment and Measurement. Theory, Methods, and Applications*. Lawrence Erlbaum Associates.

Endsley, Mica R. (2000). Theoretical Underpinnings of Situation Awareness: A Critical Review. In: Mica R. Endsley and Daniel J. Garland (Eds.): *Situation Awareness Analysis and Measurement*. Lawrence Erlbaum Associates.

Hauland, G. (2002). Measuring Team Situation Awareness in Training of En Route Air Traffic Control. Process Oriented Measures for Experimental Studies. Ph.D. thesis, Risø National Laboratory. Risø-R-1343(EN), ISBN 87-550-3074-2 (internet). Available on-line at: http://www.risoe.dk/rispubl/SYS/ris-r-1343.htm

Shrestha, Lisa B.; Prince, Carolyn; Baker, David P.; Salas, Eduardo. (1995). Understanding Situation Awareness: Concepts, Methods and Training. In: Rouse, William B. (Ed.), *Human/Technology Interaction in Complex Systems*. Volume 7, pp. 45-83, JAI Press Inc., Connecticut.

Smith, K. and Hancock P. A. (1995). Situation Awareness is Adaptive, Externally Directed Consciousness. In: Richard D. Gilson (Ed.), *Human Factors*, Vol. 37 (1), 137-148.

Difference Presentation: A Method for Facilitating Users' Adaptation to Software Upgrades

Hiroshi Hayama
The University of Tokyo
3-8-1 Komaba, Meguro-ku, Tokyo, Japan
hiros-h@rogue.co.jp

Kazuhiro Ueda
The University of Tokyo
3-8-1 Komaba, Meguro-ku, Tokyo, Japan
ueda@gregorio.c.u-tokyo.ac.jp

Abstract

We propose and evaluate a new method of designing interfaces, in which the differences between an old version and a new version of a piece of software are presented. When software is upgraded, users can easily adapt to the new version and can reconstruct their mental models by using the method of "difference presentation" explained in this paper. We developed a new interface based on difference-presentation and conducted a usability experiment. The result showed that errors increased just after the upgrade of software in all the groups However, the performance of the subjects shown the difference-presentations was better than that of the control group who were not shown this information. This shows that the difference-presentation is effective in facilitating user's adaptation to a new version and can help to reconstruct the users' mental models. Moreover, our work suggested that two factors are important in facilitating users' adaptation, i.e., the reconstruction of the user's mental model and the users' operation guide. A detailed analysis of the data showed that the difficulty in users' adaptation depends on the type of change or upgrade made to the system: users can more easily adapt to a newly added function than to a function in which two or more old functions are integrated.

1 Introduction

Although many researches have focused on the change in a user's mental model while using adaptive/adaptable user interfaces over extended periods of time, the changes made to the functional spaces of the software itself have seldom been taken into consideration. For example, Thomas (1998) recorded two thousand users' operation logs of the text editor named SAM for seven years, and he analysed how users search for and learn the functions of that text editor. The text editor was not upgraded during the term of research. However, software tends to be upgraded frequently, with the functional space of the software changing continually. It is, therefore, difficult to create adaptive/adaptable user interfaces that have a practical use in the long term without taking subsequent version upgrades into consideration. In this study, we focus on the change in functional spaces over time and propose the "difference-presentation" method in which a user interface clearly displays what and how functions have been changed between versions. It is supposed that the difference-presentation will facilitate users' adaptation to new software versions. To test the effectiveness of the difference-presentation method, we developed programs based on it and conducted a usability experiment.

2 Experimental

2.1 Experimental Software and Procedures

The system used in the experiment was a piece of common schedule management software named csched (see Figure 1).

Figure 1: Screenshot of the csched schedule management software, version 2

The changes made to the functional space were presented by displaying small icons on the left-hand side of the relevant menu items (see Table 1). Before designing the menus of the csched program, we inspected four common programs to reveal how upgrades have been made to commercial programs. The programs were Microsoft Word 97, 2000, 2002, Microsoft Excel 97, 2000, 2002, Paint Shop Pro 5, 6, 7 and Ichitaro 9, 10, 11 (a popular Japanese word processor software). We found the changes between two versions could be classified into six types. Only four of them were implemented in the csched program. The two that were unimplemented were "renamed" and "deleted". We did not implement them in cshed because the structure of cshed's menu would not change even if a menu item were renamed and the test subjects would not be able to select a deleted item in the new version. In addition, only a few menu items were deleted in the new version of software.

Table 1: Icons that Represent the Changes Made to the Functional Space

Icons	Meanings
New!	The item is a new feature of the new version.
Mov!	The item is moved to a different position in the new version
Div!	A single item in the old version was divided into two or more items in the new version.
Int!	Two or more items in the old version were integrated into a single item in the new version.

The experiment was held on every Saturday over the course of six weeks. Subjects were asked to use the old version three times in the first half of the experiment, and then to use the new version three times in the second half. Twenty-four subjects were divided into three groups:

- Control group: Differences were not shown even when the software was upgraded.

- Difference-presenting group: Differences were consistently shown on the menu of the second version.
- Difference-fading group: Differences were shown like in the difference-presenting group, but as the subjects got used to the second version, the differences faded.

The subjects performed six tasks per trial. These tasks include entering plans, copying, searching, changing fonts, etc., and their operations were recorded.

2.2 Results of Experiment

2.2.1 Effectiveness of Difference-Presenting method

The results are shown in Figure 2. The amount of error is defined according to the following formula.

Errors = total steps – steps when operated the most effective way

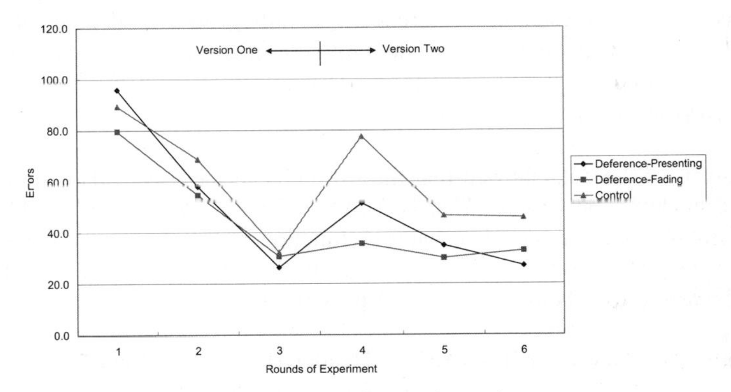

Figure 2: Amount of error in each group

The results show that errors increase in all the groups just after the software upgrade [$F(1,21)=27.311$, $p<0.0001$]. However, the amount of error was relatively smaller in the difference-presenting group and the difference-fading group than in the control group [$F(2,21)=5.482$, $p=0.0356<0.05$]. Also throughout the second half of the experiments, the performances of the difference-presenting group and of the difference-fading group were better than that of the control group [$F(2,21)=3.858$, $p=0.0374<0.05$]. These results show that the difference-presentation is effective in facilitating the user's adaptation to a new version and can help to reconstruct his/her mental model.

2.2.2 About Reconstruction of Mental Models

The above result was acquired from data of subjects who were regarded as having mastered the old versions. When we included data from subjects who had insufficiently learned the old version, the performance of the difference-fading group fell as the difference display disappeared. On the other hand, the performance of the difference-presenting group exceeded those of the previous rounds (see Figure 3).

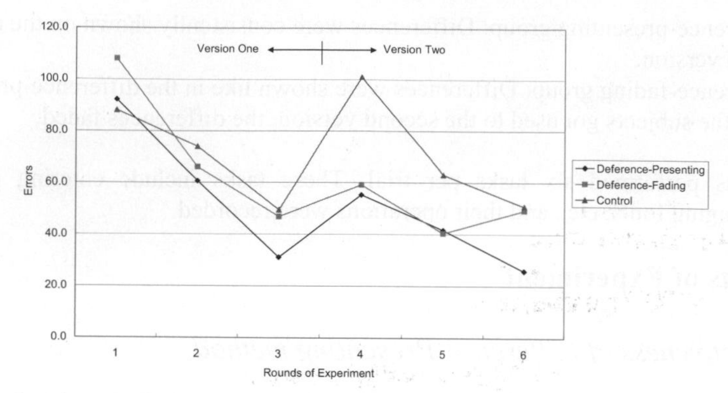

Figure 3: Errors including those of subjects who had insufficiently learned tasks

This result shows that although the difference-presenting group performed well, the reconstruction of their mental models was not sufficient. There are two important factors in facilitating users' adaptation: the reconstruction of the mental model and the users' operation guide. Above result shows the necessity of presenting differences according to the degree of user expertise. Additionally, the users' adaptation must be taken into consideration when we develop a system.

2.2.3 Detailed Results

A detailed analysis of the data showed that the difficulty in the users' adaptation depends on the type of change made to the system. That is, compared with using a newly added function, the subjects' performance was lower when using a function in which two or more of the original functions were integrated. Table 2 shows the mean steps to reach a newly added menu item ([Format(O)]-[Header(H)…]) after finishing the previous task.

Table 2: Mean Steps to Reach the [Format(O)]-[Header(H)…] Menu Item.

Groups	Round 4	Round 5	Round 6
Control (N=4)	12.50	5.25	7.50
Deference-presenting (N=8)	6.25	8.13	5.25
Deference-fading (N=8)	4.38	6.50	3.88

While the inter-group main effect and the inter-round main effect were not significant, the interaction between the groups and the rounds was significant [$F(4,34)=2.695$, $p=0.0471<0.05$]. However, here it is necessary to pay attention to the number of subjects in the control group. Some subjects were so confused that they selected the [Format(O)]-[Header(H)…] menu item twice or more before accomplishing their task. Such data was excluded from the analysis, which means that the actual performance of the control group was worse than the result shown in Table 2. It is reasonable to suppose that the difference-presentation was effective.

Next, let us examine the integrated item. Table 3 shows the mean steps to reach an integrated menu item. The [Format(O)]-[Color(C)…] menu item and the [Format(O)]-[Font(F)…] menu item in version one were integrated into the [Format(O)]-[Cell(E)…] menu item in version two.

Table 3: Mean Steps to Reach the [Format(O)]-[Cell(E)...] Menu Item.

Groups	Round 4	Round 5	Round 6
Control (N=8)	32.60	16.00	6.00
Deference-presenting (N=8)	5.67	9.22	5.22
Deference-fading (N=4)	8.38	8.68	18.13

The inter-group main effect was significant [$F(2,19)=7.400$, $p=0.0041<0.01$], and the interaction between the groups and the rounds also was significant [$F(4,38)=7.331$, $p=0.0002<0.01$]. The result shows that the difference-presentation was effective. In addition, we should note that the difference-fading group's performance fell in round six. These results can be interpreted in the same way as with the two factors of the previous section. The integrated item was relatively more difficult to learn than the newly added item. Subjects in the difference-fading group seemed to not to have been able to reconstruct their mental models sufficiently. In the difference-fading group, the differences between the old version and new version were not shown in round six. They could only use the icons that showed differences as guides for their operations. From software engineering's point of view, we suggest that when developers upgrade the software, they had better consider the consequences of integrating two or more previously defined functions and avoid an easy integration of functions.

3 Discussion and Conclusion

This research showed that the difference-presentation method was effective in helping users to adapt to a new version of familiar software. To develop a more efficient adaptive/adaptable interface, it will be useful to take the changes made to the system into account. Our research suggests that software system development would benefit from a new methodology in which the users' adaptation is incorporated into the life cycle of a system. Note that several researchers point out that even experts do not always use the most efficient function (e.g. Bhavnani & John, 2000), and users continuously develop their skills even after they have one or two years' experience of using certain programs (Nilsen et al., 1993; Doane et al., 1990). Moreover, there is also the "active user's paradox" in which experts have few chances to learn new functions (Carroll & Rossen, 1987). The difference-presentation method should also be helpful to these users to operate programs more effectively.

References

Bhavnani, S. K. & John, B. E. (2000). The strategic use of complex computer systems. *Human Computer Interaction*. Vol. 15, No. 15, 107-137.

Carroll, J.M & Rossen M. B. (1987). The paradox of the active user. In J. M. Carroll (Ed.), *Interfacing Thought: Cognitive Aspects of Human-Computer Interaction* (pp. 80-111). Cambridge, MA: MIT Press.

Doane, S.M., Pellegrino, J. W. & Klatzky, R. L. (1990). Expertise in a computer operating system: Computing and performance. *Human-Computer Interaction*, Vol. 5, 267-304

Thomas, R. C. (1998). Long Term Human-Computer Interaction. An Exploratory Perspective, New York: Springer Verlag.

Nilsen, E., Jong, H., Olson, J. S., Biolsi, K., Rueter, H. & Mutter, S. (1993). The growth of software skill: A longitudinal look at learning and performance. In *Human Factors in Computing Systems INTERCHI '93 Conference Proceedings*, 149-156. New York: ACM.

Generating Insights from Agent-Model Emergent Behavior

Raymond R. Hill
Department of Biomedical, Industrial, and Human Factors Engineering
Wright State University
Dayton OHIO, USA 45435

Abstract

Multi-agent simulations are finding application in an increasing number of areas, in a plethora of disciplines, including most recently military modeling and analysis. Agent simulations are attractive since the simple interaction mechanisms among the agents provide "emergent behavior." Emergent behavior is not the result of specific programming efforts and as such may provide analytical insights not available in legacy modeling approaches. In this paper, we summarize three military-focused, agent-based simulation models and the emergent behavior uncovered and exploited. For each model, we provide a brief summary of the model, its emergent behavior output, and the analytical implications of that output.

1 Introduction

Multi-agent simulations are finding application in an increasing number of areas, in a plethora of disciplines, including most recently military modeling and analysis. Agent simulations are attractive since the simple interaction mechanisms among the agents provide "emergent behavior." Legacy approaches to modeling, particularly military modeling, have had difficulty capturing many of the effects attributed to human decision-making. This is where complex adaptive systems (CAS) theory may help. Instead of modeling top-down—the current paradigm—CAS implements bottom-up modeling. Agents are created in the model and their interactions with each other and their environment drives the system's behavior. Observing the trends in the behavior can help describe past behavior and lend insight into predicting future behavior. The promise of CAS, and agent-based modeling in general, is the ability to analyze model emergent behavior and gain real-world insights not attainable with legacy approaches to modeling. Unfortunately, the methodologies employed to examine emergent behavior are quite immature. In this paper, we examine military-focused agent models and discuss methods for gaining insights by explicitly examining the agent-based model emergent behavior.

2 Background

Legacy modeling methodology advocates a top-down approach to simulation and modeling. CAS approaches are based on model building from the bottom up. This important paradigm shift recognizes that complexity is a property of agent interactions and complex behavior emerges from those interactions. Unfortunately, CAS simulations will not generally follow the same course of action from run to run. Lack of repeatable simulations is a concern among users of CAS. Non-repeatability is a realistic property as life, and especially warfare, is not predictable. The appeal of CAS is that the models may better describe nature. Thus, models created using CAS theory lend themselves to different forms of analyses. Their prescriptive nature allows users to glean new insights despite a lack of classical, statistical predictive power.

Simon (1990) provides an excellent discussion of the issues associated with models used for prescriptive versus predictive purposes. Among the criticisms of CAS theory and agent-based modeling is that experimental repeatability is very difficult. CAS theory advocates indicate CAS approaches are best for the prescriptive abilities provided. Because CAS is based on chaos theory they generally contain basins of attraction in which emergent behavior gravitates to small zones of feasibility. The user can determine with some certainty that reality will fall into some basin of attraction. Thus, knowledge of these basins provides some predictive capabilities. Attention focused on these basins can provide better bounds on potential outcomes. Noting these basins have, to date, required an exploratory modeling approach (Bankes, 1993). Exploratory modeling involves a complete characterization of the feasible space to generate response functions for selected parameters. These response surfaces facilitate the search for local extreme points, information useful when looking for settings to maximize some decision criteria. We use this approach.

3 Examples of Agent-Based Combat Model Analyses

Agent-based modeling is a new approach for the military modeling and simulation community. In the examples that follow, specific agent-based models are introduced and the analysis methodology discussed. Details on each model are excluded due to space considerations.

3.1 Pilot Inventory Complex Adaptive System (PICAS)

Gaupp and Hill (2000) provide an overview of the development and use of a prototype agent-based model focused on examining the behavior of United States Air Force pilot personnel career progression. Figure 1 (a) provides a screen shot of PICAS. PICAS is an artificial life approach to mimic the decisions of actual pilots to remain in the military or to leave the military. Each PICAS agent (dot on the screen) represents a "typical" pilot. Agent color-coding indicates the "age" of the pilot with respect to a pilot's career. Agent behavior is guided with a utility-based decision model, with utility values dynamically altered with sliding scales located on the screen left. These utility curves represent attitudes toward monetary compensation and time-off levels (i.e., quality of life). Environmental factors include: civilian airline job availability, perceptions of the military versus civilian pay gap, the level of military operations tempo, and an individual's relative attitude to time versus money. The user dynamically varies these factors with slider bars located along the screen right.

Analysis with PICAS involves setting personality factors and environmental concerns and then examining the emergent behavior of the agents with the model. PICAS agents have two life decisions: remain in the military or separate into civilian life. An agent's propensities are captured as a lower left screen attractor (separate) and an upper right screen attractor (remain in the military). Figure 1(a) also depicts a typical PICAS output.

Figure 1(b) depicts an analytical scenario of mid-career pilots (e.g., pilots half way to retirement) plotting the propensity to separate as a function of agent attitudes and environmental factors. Separations are the highest when both the environment and the pilots' attitudes are poor, but also increase if either parameter is bad. However, the results indicate less impact due to environmental factors. This information is valuable since it indicates policy makers should concentrate on leveling environmental factors to improve pilot retention.

3.2 Strategic Effects Model (SEM)

The prototype SEM model is a CAS of combatant "agents" in a ground warfare environment (Tighe, 1999). SEM specifically examines a combatant decision cycle and varies how quickly a combatant can make decisions. Military doctrine holds that faster decision making translates into an operational advantage, often termed a "strategic effect."

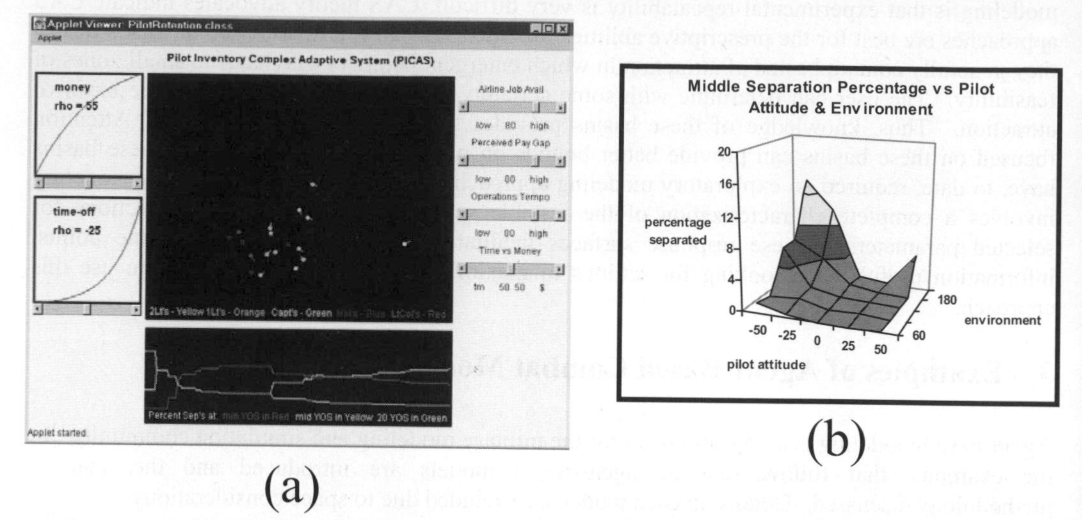

Figure 1: (a) Screen shot of prototype PICAS; (b) Surface Plot of Middle of Career Separators

Figure 2(a) depicts a SEM battle in progress. Each dot represents a combatant agent; Red agents against Blue agents. Lines from an agent indicate a "shot" against an opponent. The objective of each force is to capture the enemy base, indicated by the large colored block.

Research using SEM focused on whether decision speed advantage could overcome traditional force superiorities. One scenario examined involved equally capable forces. The experiment involved varying the decision speed of one side (the Blue side) to determine when a decision speed advantage provides a force enhancing advantage. Figure 2(b) depicts an output landscape varying decision speed advantage up to four times that of an opponent. Review of the surface indicates that when the decision speed advantage rises to approximately twice that of an opponent, military force equality ceases to exist; Blue now has an overwhelming advantage. Similar surfaces were obtained in scenarios involving a numerically superior Red force, a more lethal Red force, and a more mobile Red force. The surfaces differed only in the location of the "knee in the curve" where the decision speed advantage became sufficient to overcome the superior Red force.

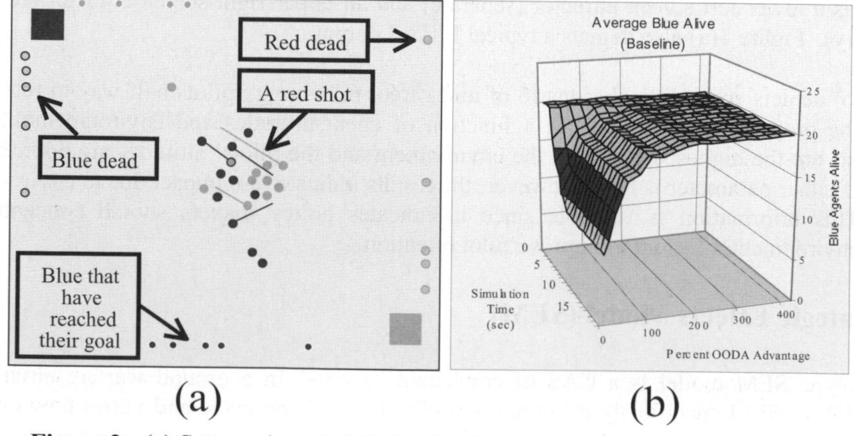

Figure 2: (a) Screen shot of SEM battle; (b) Output surface of SEM scenario

3.3 Hierarchical Interactive Theater Model (HITM)

HITM extends the SEM concept to examine theater-level effects. HITM agents interact horizontally (within forces) and hierarchically (command structures) while capturing reliance on infrastructure assets to prosecute combat actions. HITM is depicted in Figure 3. Each side is equal in terms of territory, assets, and combatants. Agents include pilots, ground combatants and commanders. Agents defend their territory while trying to destroy their opponent's assets, and thus their opponent's ability to prosecute military actions.

A HITM experiment runs the model until one side "wins" the fight. Since both sides are equal in strength, winning a fight is a function of targeting effectiveness against the opponent's ability to prosecute the fight. The stochastic nature of HITM ensures each output differs. The emergent behavior are those actions that lead to military victory.

Czerwinski [1998] suggests nonlinear systems are characterized by three regions and links these regions to a battle. The first region is Equilibrium where damages inflicted by an opponent are local and their effects transient. The second region is Complexity. In this region, the damage inflicted by an opponent requires adaptation in order to overcome the effects. The third region is Chaos where damage inflicted by an opponent propagates and eventually results in destruction. The HITM output in Figure 4 maps a plot of HITM output to these nonlinear regions. The implication is that agent-based modeling might provide a means to objectively examine how theoretical constructs such as Czerwinski's map into targeting doctrine and the uncertainty of warfare. Additional work examined sensitivities of outcomes to specific resource levels.

4. Summary and Conclusions

Agent-based modeling, defined here to encompass CAS approaches, offers the modeling and simulation community new capabilities to capture non-programmed behavior. However, analytical means to examine and infer insights from emergent behavior is still immature. In this paper we discussed three prototype agent models, focused on military issues, in which emergent behavior is observed and exploited for analytical insights.

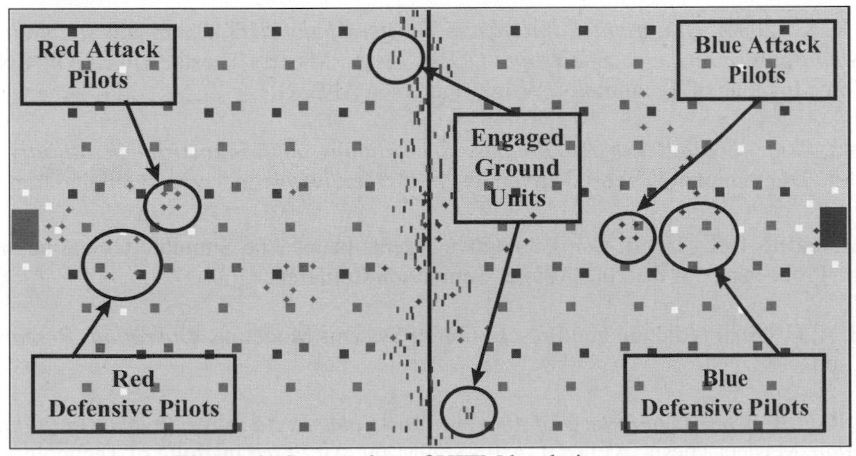

Figure 3: Screen shot of HITM battle in progress

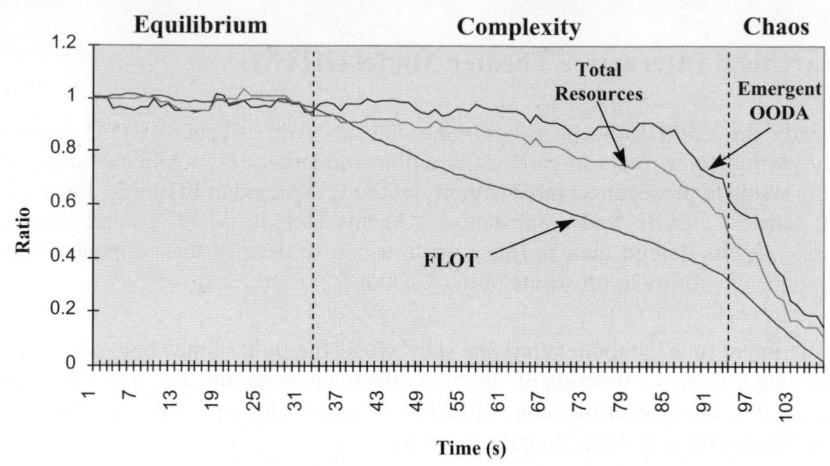

Figure 4: Output of HITM experiment mapped to complex transition points

Future avenues of research abound. For instance, there appears to be a lack of quantitative techniques to summarize the emergent behavior found in agent-based models. To reiterate, the allure of agent modeling is the potential to capture human-based issues prevalent in military operations. By capturing human-based issues via the agents, modeling analysts hope to get better bounds on the range of potential outcomes. To compete in the model-based analytical world, agent models, and their emergent behavior, will require valid, defendable numerical summaries.

There is also the issue of verification and validation of agent models. There is a large literature on verification and validation of models and in particular the output those models generate. One accepted approach is to statistically validate model output with respect to actual observations. The challenge with emergent behavior and military modeling is how to assess the validity of a model output when that model output is compared to a military outcome where the observations may be just a single observation.

5. References

Bankes, S. Exploratory Model. (1993). *Operations Research*, Vol. 41, No. 3.

Bullock, R. K. (2000). *Hierarchical Interactive Theater Model (HITM): An Investigation into the Relationship Between Strategic Effects and OODA Loops*. Masters Thesis, AFIT/GOR/ENS/00M-05, Air Force Institute of Technology, Wright-Patterson AFB, OH.

Czerwinski, Tom. *Coping with the Bounds: Speculations on Nonlinearity in Military Affairs*. Washington, DC: National Defense University, 1998. http://www.dodccrp.org/copind.htm

Gaupp, M. and R. Hill. (1999). Using Adaptive Agents In Java To Simulate U.S. Air Force Pilot Retention. Proceedings of the 2000 Winter Simulation Conference.

Simon, H. A. (1990). Prediction and Prescription in Systems Modeling. *Operations Research*, Vol. 38, No. 1, pp: 7-14.

Tighe, T. R. (1999). *Strategic Effects of Airpower and Complex Adaptive Agents: An Initial Investigation*. Masters Thesis, AFIT/GOA/ENS/99M-09, Air Force Institute of Technology, Wright-Patterson AFB, OH.

A Case Study of two experiences of group-based student projects: Cognitive model vs. Situated learning

Kuo-Hung Huang[1], *Kuohua Wang, S. Y. Chiu*

Department of Education[1]
National Chiayi University, Taiwan
kuohung@mail.ncyu.edu.tw

Graduate School of Science Education
National Chenghua University of Education, Taiwan
kuohua@cc.ncue.edu.tw

Abstract

The purpose of this study is to design two computerized learning environments for the course of "Special-Subject Software Design" and to investigate their effectiveness on the students' learning process and personal development. These two environments are the modeling-oriented learning environment, which is based on the framework of cognitive model and the situating-oriented learning environment, based on the framework of situated learning.

1 Introduction

This article compares two experiences of group-based student projects in the laboratory and in the real world. The researcher adopted group projects from similar motivations, and used available materials to guide them through this process. For the laboratory project, students were asked to build models in the laboratory. They designed artifacts in the world, built the products with Lego blocks, wrote Java programs to control the models, and then tested if the models functioned correctly.

For the situated project, students were asked to attend the authentic activities of the community of computer professionals. They were assigned tasks, which were part of a contracted computer system implementation. Subsequently, the researcher has come to reflect on the experience and to examine the theoretical dimensions of such two approaches in more detail.

2 Related Research

Many cognitive researchers take the 'learning' as 'a process of model building'. When students first come to know some new skills or knowledge, they will form a model in minds. After spending a great deal of time practicing, playing, shaping, restructuring the model, they gradually develop better understanding about the system of the knowledge. Therefore, a project of artifact construction can help them to become effective learners. 'Microworld' is such a learning environment for learners to build their own models (Papert, 1991).

In recent years, the notion of community of practice, proposed by Lave and Wenger (1991), has had great impact on the educational organizations and institutes. Taking the relational views about knowledge and learning, Lave and Wenger claim that "The generality of any form of knowledge always lies in the power to renegotiate the meaning of the past and future in constructing the meaning of present circumstances." Communities of practice relate people with difference characteristics in heterogeneous ways, but "who improvise struggles in situated ways with each other over the value of particular definitions of the situation " (Lave, 1993). Through "Legitimate Peripheral Participation", learners can participate in communities of practitioners and gradually acquire the knowledge and skills during the process of becoming full participants in a sociocultural practice (Lave, 1988; Roth, 1998).

Although there are sound educational and political perspectives to indicate a firm basis for group work being an important component of undergraduate courses, there is still some skepticism as to whether the theoretical advantages are borne out in practice. This article addresses these issues through quantitative and qualitative analysis of the case study with two different projects.

3 Student Projects

The modeling-oriented learning environment utilized the LEGO computerized blocks and the visual software to form a microworld in which the students could learn through interaction with the emerging artifacts. Students used the LEGO blocks to design and build an artifact and implemented the designed functions by controlling the computerized blocks with computer programs. This process provided an opportunity of knowledge integration, problem solving and cognitive development.

The situating-oriented learning environment provided the students with opportunities of participating the community of practice. The students were exposed to the real world of the computer career and the impacts from the workplace inspire their reflection and career planning.

A local hospital had being planned to set up an intranet information system, serving to provide internal information to the hospital staff, but failed to implement it due to the shortage of human resources. After discussion and clarification their expected goal of this system, the researcher decided to take this task as the situated-learning project for the students.

Eight students attended the designed activities in the learning environments respectively for ten months. Meanwhile, the researcher collected and analyzed the data, which consisted of observation, logs of work, interviews and reflections.

4 Discussion

4.1 Subjects

Five female students and four mail students, aged around 19 years old, volunteered to join this project. All of them were first-year computer science majors without any prior work experience. And they did not have strong technical backgrounds. Before joining this project at the beginning of the second semester, four of them failed the course of "computer programming." These students knew one another well and usually formed a group while working on the assignments. They were unconfident in their computer abilities and worried about being rejected when applying for this project.

Five female students form a teamed to take the situated-learning project while the other students took the modeling project.

4.2 Procedure

Since the teachers had changed their roles to the senior co-workers instead of instructors, teachers would intensively participate in their activities but not actively involved in any work. One of the students was elected the project leader and then given the full control over the project. The leader had to negotiate with the other members on the tasks such as visiting the clients, dispatching the duties, scheduling and monitoring the progress of this project.

For the modeling project, students decided to construct an automatic convey machine. By sensing the color of tape on the floor, the machine could know the correct movement to act. They set up a series of stages toward their goal and monitor their progress.

For the situated-learning project, two of the members visited the hospital to interview the clients, gathered materials for publishing, and updated the homepages on the web server once a week. Every Thursday afternoon, a group meeting was held. They discussed the project and identified the difficulties. The teachers only gave them advice and suggestions in the meeting. If necessary, the teacher directly showed them how to use particular software to solve the problems.

4.3 Data Collection

This project started in January, which was the beginning of the second semester. And it ended in July, the end of the second semester. During the period, each student kept his or her own working log. In addition, computer programming tests, laboratory observations, interviews, and project reports became the sources of the qualitative data.

5 Discussion

The results reveal the difference of development of the two groups of students:

- In computer knowledge and skills: The modeling-oriented learning environment emphasizes the staged learning in the specific domain of knowledge, while the situating-oriented learning environment emphasizes the integrated application of variety of knowledge.

- In interpersonal skills: The modeling-oriented learning environment helps the students to recognize the importance of choosing progressive working partners toward task accomplishment, while the situating-oriented learning environment helps the students to recognize the importance of social relationship for the access of resources for the tasks.

- In learning and working style: The modeling-oriented learning environment provides the students with a microworld to learn to think in an assumption-verification perspective, while the situating-oriented learning environment provides the students with a real world to learn to think in multiple perspectives.

- In career planning: The modeling-oriented learning environment helps the students to know their own talents and abilities to decide whether to choose the work, while the situating-oriented learning environment helps the students to know the community of practice to judge whether to participate the community.

References

Lave, J. (1988). *Cognition in practice: Mind, mathematics and culture in everyday life.* New York: Cambridge University Press.

Lave, J. (1993). The practice of learning. In S. Chaiklin & J. Lave (Eds.), *Understanding practice* (pp. 3-32). New York: Cambridge University Press

Lave, J. & Wenger, E. (1991). *Situated learning: Legitimate peripheral participation.* New York: Cambridge University Press.

Papert, S. (1991). Situating Constructionism. In S. Papert & I. Harel (Eds.), *Constructionism* (pp. 1-12). Norwood, NJ: Ablex Publishing Corporation.

Roth, W. (1998). *Designing communities.* Boston: Kluwer Academic Publishers.

Design and Implementation of Steganography based on 2-Tier File Encryption Algorithm

Young-Shil Kim[*], *Young-Mi Kim*[**] *,Sung Gi Min*[***] *, Doo-Kwon Baik****

*Dept of Computer Science & Information, Daelim College
**Dept of R&D CEST CO.LTD,
***College of Information & Communication, Korea University
pewkys@daelim.ac.kr , rose@cest.co.kr , sgmin@korea.ac.kr , sbaik@software.korea.ac.kr

Abstract

Due to the brilliant development of computer science and telecommunication system, all the devices have been popped up for protecting the data. One of them is encryption method. Also various data protecting technologies have intensively been under constructing. Of the technologies, 'Steganography' is the most typical technique. Steganography was designed to get users (not specialized ones) harder to find out the data through hiding data in forms of various materials such as text, image, MPEG, and audio. If some secret message were encrypted, the security level could go higher. Though some attacker might find out the coded secret data, the attacker had to encoding the data. And according to the size of Mask, the size of Cover-data should be decided. Therefore the Mask must be compressed to hide in the Cover-data. At present the most highly developed Steganography is the one with using image technique; the most heated Steganography is the one with using audio technique. But in the conventional method that Mask is stored in Cover-data, the data tends to be larger as the Mask is put only in one of right bits to give an emphasis on the sound quality. For solving this problem, StegoWaveK method is strongly suggested. Also according to those suggested methods, some sound qualities of commercialized products are precisely compared

1 Introduction

Steganography was designed to get users harder to find out the data through hiding data in forms of various materials such as text, image, video, and audio. Currently commercialized Steganography softwares are designed to hide the data in Cover-data after compressing Mask file. But this software is not compressing the Mask file at a regular rate. So when they choose the Cover-data file, that might cause some problem. Each Mask bit stream is uniformly inserted into the last bit, and the listeners could easily do filtering the secreted Mask. Of course, they don't know what data are hidden in the Stego-data (Peter Wayner,2002).
For solving those problems, this study suggests the Steganography which employing 2-Tire File Encryption Algorithm. This Steganography method is basically constructed with the similar method as the conventional commercialized Steganography does. The most generalized Audio Steganography technique is Lowbit Encoding which insert one bit of Mask to the last bit. Attacker has the disadvantage where attack was able to do the Mask which was easily concealed in case of Lowbit Encoding. Also capacity of Stego-data is low. To improve low capacity, we embed more

[*] 526-7 Bisan Dong, Dongan-gu, Anyang-si, Kyungki-do, Korea, 431-715
[**] Hyocheon B/D, 1425-10, Secho-Dong, Secho-Gu, Seoul Korea, 137-070
[***] Anam-Dong, Sungbuk-Gu, Seoul, Korea, 136-075

than one bit in every sixteen bit. But the attacker easily filters Mask when inserted bit is equally bits in every sixteen bits, it is proposed that the Mask should be inserted in forms of sign curve with changing the number of bits. And it is possible to filter Mask form Stego-data made from the type of sign curve, we insert Mask to the area of Cover-data that have greater than threshold value. And It is determined dynamically embedding location according to a property of a file.
2-Tire method is processed through three encryption steps. Based upon this process, we could get Plaintext, and encrypted this. And the Plaintext is used in Steganography which employ wave as Cover-data. As in Steganography, there is almost no difference of data size between Cover-data and Stego-data, average listeners or users could not discern the infromation hiding.

2 Improved File Encryption Algorithm

We propose IFE(Improved File Encryption) algorithm that can improve the problem that showed a specific pattern in addition to encrypt of a file to raise a level of security. The proposed method is compose following steps.
First is the applying stage which employs AES algorithm(Raymond,1999) to enhance over one step level of encryption. AES has variable key length of 128 bits, 192 bits, 256 bits at a variable length block of 128 bits, 192 bits, 256 bits. Therefore, safety of data is improved.
Second is hiding the structure and form of encrypted Ciphertext for removing some particular patterns which could be appeared in encrypted Ciphertext. And it is applied to the MBE(Modyfied Block Encryption) which encrypts Ciphertext using the key based on this after generating random number of Ciphertext blocks. MBE algorithm circulates 255 keys which it has sequential created from image key and carries out each block and XOR operation. The following is MBE algorithm to have applied to IFE algorithm.

Input : $d_{(0)},...d_{(n)}$ created block data from Plaintext
$k_{(0,...255)}$ created key from the table of secret message in numbers
Ouput : $o_{(0)},..o_{(n)}$ result calculated per block
1. $o_{(0)} \leftarrow d_{(0)} \oplus k_{(0)}$
2. For i from 1 to $n-1$
$o_{(i)} \leftarrow d_{(i)} \oplus k_{(i \% 255)} \oplus o_{(i-1)}$

Figure 1: MBE Algorithm

3 Design of StegoWaveK Model

StegoWaveK is a model that uses the 2-Tier file encryption in order to raise a security level of the data which are going to embed the audio Steganography system that can hide Mask of various type file and be encoded.
Commercialized Audio Steganography software has greatly two problems. First, is taking the Low-Bit Encoding way that is the simplest application way of audio Steganography. By listening to a wave file or watching wavelength type simply, users or listeners did not know that information is embedded, but there is the important thing that information embedded by attacker has a problem for a filtering to be able to easily work. Second, we need the Cover-data of 16 times for Mask file arithmetically. It makes difficulties for choice of the Cover-data. Therefore, development of the technology that can embed of large size Mask is necessary.
In order to solve a problem that is able to have been easily analyzed structure and a characteristic of a file because general file encryption algorithm shows a specific pattern and that a filtering can

easily work in attacker, and capacity was low because Commercialized Audio Steganography let the last one bit of 16 bit hiding data, this paper proposes StegoWaveK.
The proposed algorithm in this paper is improved Low-Bit Encoding method, and solves the two types of problems, which is low capacity and easy filtering in Commercialized Audio Steganography. First, in order to insert bigger size of Mask to the limited size of Cover-data, we embed more than one bit in every sixteen bit.
Result from opinion of 100 students listening Stego-data hiding as 1 bit, 2 bit, 4 bit, and 6 bit. The student listened to music using a speaker. All students were not able to distinguish between Stego-data and Cover-data. Therefore, we used headphones for a precision analysis. Most of student do not feel the difference of Cover-data and Stego-data. But if we increase the number of bit to insert unconditionally, then there is significant difference of the two data. Since Cover-data is injured, listener is aware of information embedding.
In order to prove there in no meaningful difference between Cover-data and Stego-data inserted by 1 bit, 2 bit, 4 bit, and 6 bit, we use Cover-data having format of 16 bit PCM Wave file. We transfer 16 bit segment to decimal number and select three thousand of decimal values to analyze. The Correlation Analysis is performed to know what difference is between Cover-data and Stego-data. In the analysis result, since correlation coefficient of relation between Stego-data and Cover-data is close to 1, we know the fact that Stego-data obtains properties of Cover-data. But in the case of 6 bit Stego-data, correlation coefficient is 0.9996 and thus some properties of Cover-data is dropped

```
Procedure Acc234bitsMask_Data_Insertion();
begin
   Cover-data read;
   Calculate a Insertion number of Mask data;
for( i=0 ; i<Insertion_number ; i++)
begin
  Read Insertion bit of Mask data;
  Loaded Mask data overwrite into Cover-data;
end;
end;
```

Figure2: Acc234 Mask Data Insertion Algorithm

The Attacker easily filters Mask when inserted bit is equally 2 bit, 3 bit, or 4 bit. To improve this problem, we propose the method that bits of Mask is inserted in forms of sign curve with changing the number of bits, not inserted by regular rate per 16 bit of Cover-data.
As a result of Correlation Analysis for Cover-data and 1 bit Stego-data, we can know that high correlation relationship appears between Stego-data with a 1 bit sign curve and Cover-data. In insertion of Mask, we use sign curve and can keep a characteristic of Cover-data.
The proposed Mask Data Insertion algorithm improves a problem for information concealed by Attacker to be able to easily become filtering. It is to designate not to insert one bit, but to be able to insert more large size of Mask into Cover-data. This algorithm is inserted bits by accumulated 234bit.

4 Performance Evaluation of StegoWaveK

This chapter discusses the result of comparatively analysis of StegoWaveK, Invisible Secrets 2002(actually commercialized S/W I), and steganos Security Suite 4(commercialized S/W II) with view point of human sight and audition. Especially auditory analysis might be subjective view point. The comparative analysis between Stego-data and Cover-data is precisely carried on four objective result.(J.D.Gordy and L.T.Bruton, 1998). We use the two wave file Kungddarishabara.

4.1 HVS(Human Visible System) Aspect

HVS distinguish between Stego-data and Cover-data using a human visible system. That is, it is that classifies visually a wavelength or frequency in an wave file.

Figure 3: Wavelength Comparison of Commercialized systems and StegoWaveK

It is not easy to discern each one by wave pattern when generated by the two methods, Invisible Secrets 2002 and StegoWaveK which were called from them through CoolEditor, editing tool. This happens because of the distinction of humane HVS

4.2 HAS(Human Auditory System) Aspects

100 students listened to Cover-data and various type of Stego-data after storing different size of 13 plaintexts and 4 wave files of Cover-data as 2 bits, 4 bits, and 234 bits. The students couldn't recognize the difference between the two music files after listening to both of Cover-data and Stego-data. But for easy getting Mask, StegoWaveK method is designed to change the bit to the level which listeners could catch the sound. If the cycle of sign curve is higher, Cover-data is damaged. This is the basic theory of StegoWaveK method. The experiment presented that it is not easy to discern the sound quality or wave form with comparing Cover-data with Stego-data until 4 bit.

4.3 AM(Audio Measurement) Aspect

There are frequency response, gain or loss, harmonic distortion, intermodulation distortion, noise level, phase response, and transient response in audio measurement and analysis method. These valuables are including signal level or phase, and frequency.
For example, SNR(Signal to Noise Ratio) is a level measurement way represented by various forms log, decibel(dB) or ratio. The quality of stego-data ,which conceal secret messages, is measured using SNR in (J.D.Gordy and L.T.Bruton, 1998). SNR represents a relative value rate(Dr. Richard C. Cabot,et al., 2002).

	Commercialized S/W I	Commercialized S/W II	StegoWaveK
PESQ	4.49	2.63	4.43
SNR	99.5	71.3	99.1

Figure 4: SNR and PESQ comparison between cover-data and each stego-data

Figure 4 shows SNR (Signal to Noise Ratio) comparison among cover-data, commercialized audio steganography. It tells that there's no difference in SNR between stego-data generated using suggested system and commercialized audio steganography systems. And PESQ(Perceptual Evaluation of Speech Quality) which evaluates the quality of voice. In fact it is difficult to trust completely the results from the automated voice measurement, but it is reliable enough to use for the way of various tests related

5 Conclusion

The Steganogaphy using wave file in the area of information hiding has the same file size of Cover-data and Stego-data. And then user or listener does not know the fact that information is hided. Also, Human visual system does not recognize the difference between Cover-data and Stego-data with wavelength for wave file by audio edit tools. They couldn't easily discern the wave quality in HVS aspects. Thus we easily transfer the important information.
In the Commercialized Audio Steganography, they insert one bit of Mask to sixteenth digit position of the 16 bit data. It occurs the problem is easy for attacker to do filtering. Because the different type of file, we have the different file size of Cover-data for the same size Mask.
In this paper, proposed StegoWaveK model is first model to solve those problems for Commercialized system. First, we have better function of Cover-data Capacity, and we can easily solve the problem used by Acc234 bit Mask Data Insertion algorithm that Attacker get simply Mask by filtering. In the proposed Acc234 bit algorithm, the low efficiency problem is much improved, and the form of insertion appears as in forms of sign curve type. As a result, though the suggested model uses the Steganography employing Private-Key, this makes it difficult to notice whether there is some hidden information or not as Public-Key Steganography.
Second, to insert Mask, we can readily select Cover-data as providing options for 2 bit, 4 bit, and Acc234 bit. Third, before insertion of Mask, we process encryption procedure not to know the information from Mask and the purpose of encryption is to make it difficult to discern any noticeable features. To reduce the size of Mask, we execute preprocess called compression. 2-Tier file encryption is suggested to solve these problems. And in cyber education field that utilize multi-media contents, this method could be applied to be a solid tool of authentication for it. It can be also used for the blind to keep their information on their own computers concealed from others, because there's no telling the difference the original data and stego-data hiding information.

References

Peter Wayner(2002). Disappearing Cryptogrphy Information Hiding : Steganography and Watermarking(2nd ed.), Morgan Kauffman..

Stefan Katzenbeisser and Fabien A. P. Petitcolas(2000). Information Hiding techniques for steganography and digital watermarking, Artech House.

Raymond G. Kammer(1999). DATA ENCRYPTION STANDARD, Federal Information Processing Standards Publication.

J.D.Gordy and L.T.Bruton(1998) IEEE MWSCAS 2000, "Performance Evaluation of Digital Audio Watermarking Algorithms." , pp. 337-355.

J.Zollner, H.Federrath, H.Klimant, A.Pfitzmann, R.Piotraschke, A.Westfeld, G.Wicke, G.Wolf(1998), "Modeling the security of steganographoc systems", 2nd Workshop on Information Hiding, Portland, LNCS 1525, Springer-Cerlag, pp.345-355, April.

S.K. Pal, P.K. Saxena, S.K. Muttoo(2002), " The Future of Audio Steganography", Pacific Rim Workshop on Digital Steganography 2002 (STEG'02)

Dr. Richard C. Cabot, P.E., Bruce Hofer, Robert Metzler(2002), Chapter 13 Audio Measurement and Analysis

Situation Awareness and Situation Dependent Behaviour Adjustment in the Maritime Work Domain

Thomas Koester
FORCE Technology
Hjortekærsvej 99, 2800 Kongens Lyngby, Denmark
tsk@force.dk

Abstract

Situation awareness is not directly observable in real time in real environments. But the resulting changes in behaviour and communication can be observed. These observations can indicate the presence of situation awareness on level 1, 2 and 3. It is in this paper exemplified how situation awareness can be found among crews on board ferries ply between harbours in Scandinavian waters. The examples include reactive changes of communication indicating situation awareness on level 1 and 2 and proactive changes indicating situation awareness on level 3.

1 Introduction

The ability to perceive elements in the current situation (situation awareness level 1), comprehend the current situation (situation awareness level 2) and make projections of future events (situation awareness level 3) is very important in the everyday routine work on board ships of any type. Together the perception, comprehension and projection form the situation awareness of the operator, the crewmember. The situation awareness is important in the navigation of the vessel, the anti collision work and in the maintenance of a safe and efficient operation of the vessel in any matter. It is possible to find evidence from maritime accident reports, that a lack of situation awareness can actually cause accidents or contribute significantly to the causal development of them. According to a study performed by Grech & Horberry (2002) 71% of all human error types on ships are situation awareness related problems. It is therefore evident, that proper situation awareness has significant effect on the overall safety of any type of maritime vessel and that a lack of situation awareness eventually can cause accidents such as groundings, collisions or structural damage to the vessel. The result of these accidents could be material damage, loss of lives, pollution and extensive expenses to rescue and salvage operations, repair, increased insurance premiums, and loss of income and goodwill.

2 Situation as a concept

The concept of situation awareness is – according to the definition made by Endsley (2000a) – based on the interaction between the human operator and the surrounding working environment. The underlying assumption is that the surrounding working environment contains situations and changes of situations. Situational changes can be classified according to several dimensions such as fast/slow, incremental/significant, obvious (directly observable)/concealed (indirectly observable), repeatedly/single occurrences and planned/unplanned (unplanned can further be divided in prepared/unprepared). Some examples of changes in situation are: Change of watch, arrival into port, change of heading, crossing a traffic separation scheme, arrival into area controlled by Vessel Traffic Service, unexpected fire alarms, risk of collision, calls received by VHF or telephone, safety critical events etc. Common to all situation changes is, that they require appropriate decision making and behaviour. These situation dependent behaviour adjustments are automated through the education, training and experience of the seafarer, but it can also rely on official written procedures appropriate in the given situation. The prerequisite for successful adjustment of the behaviour is a successful perception and comprehension of the situational change and an adequate level of situation awareness.

3 Situation awareness - concept and measurement

The crewmember should – through appropriate situation awareness – be able to adjust and adapt his behaviour including decision-making, communication and performance in accordance with the current situation or change of situation. In the model formulated by Endsley (2000a) situation awareness leads to decision-making, which again leads to performance of actions. These actions will generate feedback in the environment and in the situation awareness closing the feedback loop. The behavioural aspects of situation awareness - the performance of actions - have been given relatively little attention in the literature, which primarily has a focus on the conceptual development of cognitive models of situation awareness and the empirical development of methods and techniques for the measurement and analysis of situation awareness. Situation awareness is an internal cognitive function or process and it is therefore impossible to observe it directly. Situation awareness is furthermore a *complex* cognitive function or process compared to for example attention. This makes psycho-physiological measurement of situation awareness impossible. However, since situation awareness is a concept on an aggregated level it is possible to measure some of the *underlying* cognitive functions (e.g. attention) by psycho-physiological methods. Further, it is possible to measure factors, which have major influence on situation awareness (e.g. stress and workload cf. Grech & Horberry 2002) by means of psycho-physiological methods (van Westrenen 1999). A number of methods for measurement of situation awareness have been suggested in the literature. One of them is the SAGAT methodology (Endsley 2000b). This methodology is based on questionnaires, which is supposed to test the situation awareness in a given moment in time. The work performed by the operator, the pilot, officer on watch etc. has to be interrupted or abandoned when the questionnaires are filled in. This is perfectly possible in simulated environment where the simulation can be momentarily stopped, but it could be a threat to the safety if this method is used in real environments. The method will simply generate a loss of situation awareness. Further, the method is suitable for single measurements of situation awareness, but it is not suitable for continuous monitoring of situation awareness over longer periods of time.

4 Empirical study of situation dependant behaviour adjustments

It is already mentioned that situation awareness generates decisions and further performance of actions (Endsley 2000a). This mechanism can be used for measurement of situation awareness since the performance of actions - behaviour and communication - is directly measurable by means of traditional psychological methods for observation of human behaviour. The assumption is that environmental changes, changes in the situation will generate changes in situation awareness, which again will generate (through decisions made) changes in behaviour and communication. An observable change in behaviour and communication could therefore give valuable indications about the situation awareness. Situation awareness is - as mentioned in the introduction - of vital significance for the overall safety of the vessel. Situation awareness is maintained through for example repeated observations of changes in the environment (the look out function of the crew), through the monitoring of the progress of the vessel according to the plan (the navigational function of the crew) and through communication and exchange of information and observations (Andersen 2001). Maintenance of situation awareness is a basic crew or bridge resource management skill which is gained through the education and training of the seafarer and through experience. Even though situation awareness is not directly observable it is possible to find observations of changes in behaviour or communication, which indicates the presence of situation awareness on different levels. Data from the empirical study gives us examples of *reactive behaviour adjustments* where changes in behaviour and communication are produced as a reaction to unplanned changes of situation such as risk of collision or calls received by radio or telephone. This indicates situation awareness on level 1 and 2. The available data gives further examples of *proactive behaviour adjustments* where changes in behaviour and communication are produced as a consequence of anticipations and projections of future events. This indicates situation awareness on level 3. We would expect the proactive strategy to be used in the case of planned and expected situational changes such as approach to harbour.

4.1 Method

The empirical study contains observations of bridge crew communication from 8 voyages with two different ferries (combined passenger, car and trailer vessels) ply between harbours in Scandinavian waters. The work performed by the bridge crew includes – but is not limited to – the navigation of the vessel, the anti collision work and other tasks related to the maintenance of a safe and efficient operation of the vessel. The data material is collected as online registration (logging or transcription) of the crew communication with automatic time stamping of each logged communication sequence. A communication sequence is defined as a set of sentences with the same focus. If the focus or subject of the communication is changed it indicates a new sequence. Further it is defined that a break between sentences with duration of more than 15 seconds indicates a change in sequence. The communication is categorized according to three different types: Actual, relevant and general. The actual communication is important in the given situation e.g. related to navigation of the vessel, anti-collision work, alarms and management of critical situations. The relevant communication is not important in the given situation, but it is relevant in the overall maintenance of safe and efficient operation of the vessel. The general communication is not important in the given situation, and it is not directly related to the overall maintenance of safe and efficient operation of the vessel, but it can serve the purpose of enhancing the mental workload of the crewmember in order to maintain sufficient attention. Similar recordings could have been made of behaviour other than communication e.g. actions made by the crew, but since communication is easy to register in the form of sentences and since it requires interpretation on a lower level than other types of behaviour it is used for this study. It is assumed, that the crew communication is a (very important) sub-set of the total behaviour of the crew.

4.2 Reactive behaviour adjustments

The following example (see table 1) from the data material illustrates reactive behaviour adjustments in the form of reactive adjustments of communication. The appearance of the alarm indicating failure with a navigation light was unplanned. It is not clear if the alarm was expected or unexpected, but in any case the reaction by the crew to this change of situation is that they focus their communication about the subject: The alarm and the bulb in the navigation light. The behaviour is adjusted according to the changed in situation and as a reaction to the change. This indicates *situation awareness on level 1 and 2*. If the alarm has not been perceived or if the comprehension of the situation created by the alarm had been erroneous, we would not have seen the - in this particular situation - appropriate response from the crew, which is actually found in the data material.

Table 1: Communication on the bridge of the Danish ferry Tycho Brahe

Timestamp	Communication	Comments
16:40	"The gangway has been taken in"	Standard routine communication - actual
16:40	Conversation about the crew	Communication - relevant
16:40	From radio: "loading on car deck finished", "You are now allowed to lift [the ramp] on the car deck"	Standard routine communication - actual
16:41	"Two strokes [on the bell]", "Tycho Brahe leaves Helsingør holding north to westbound traffic"	Standard routine communication - actual
16:41	Alarm (navigation lights)	Unplanned change in situation
16:42	Conversation about the alarm - the bulb should be changed	Reactive change in communication - actual
16:44	Conversation about the surrounding environment followed by conversation about private matters	Communication - general

4.3 Proactive behaviour adjustments

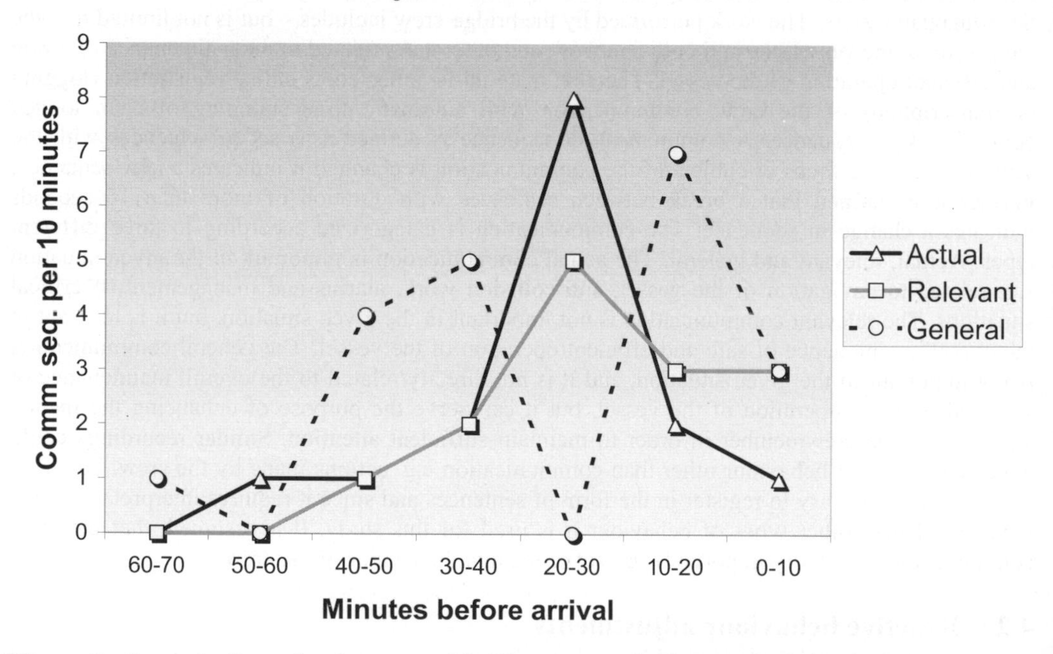

Figure 1: Example from the data material. The communication at the bridge of the vessel is logged, analysed according to the classification in actual, relevant and general communication and counted in intervals of ten minutes.

It is from the data material possible to identify communication patterns, which could be characterized as being proactive, when we look at the communication pattern in the time before major *planned* changes in situation such as arrival in port. The example shown in figure 1 illustrates how the total amount of communication increases until about 20 minutes before arrival in port (the time of entering the breakwaters). It further illustrates a positive correlation between actual and relevant communication and a negative correlation between actual and relevant communication and general communication at the moment when the amount of actual and relevant communication reaches its peak. The change in actual communication reflects the preparation before a change in situation (the approach). But it also reflects the attempt to maintain an appropriate level of situation awareness. Further, the decrease in general communication reflects an adaptation to a potential critical situation (entering the breakwaters) where a maximum of resources for actual and relevant communication is needed. This preparation and anticipation illustrated by the proactive communication pattern is a clear indication of *situation awareness on level 3*.

5 Discussion and conclusions

Results from the study show examples of both reactive and proactive situation dependant behaviour adjustments observed as changes in communication corresponding to unplanned respectively planned situational changes. This indicates the presence of situation awareness on level 1, 2 and 3 where situation awareness on level 3 is shown in the case of a planned change of situation. The best possible situation awareness in any given situation includes all three levels. Navigational students are taught to be ahead during the voyage and pay attention to, anticipate and predict how the situation will develop. It seems that there is a contradiction between the maintenance of situation awareness on all three levels and the fact that the crew is exposed to sudden, unplanned and perhaps even unexpected situational changes. Electronic bridge systems are to day fitted for the purpose of supporting situation awareness on level 1 and 2. Alarms help the crew perceive and comprehend status and situation correctly. But it is possible to design integrated bridge systems which actually also supports situation awareness on level 3 by giving *early warnings* and decision support *way before* the situational changes occurs (Andersen 2001). This is possible by the integration of several unique instruments such as radar, electronic charts, GPS (Global Positioning System) and meteorological instruments with a decision support system, which contains knowledge about the vessel manoeuvrability characteristics given the actual state of sea, wind and current.

6 References

Andersen, P.B. (2001). Maritime Instruments as Media: a Theme in the Elastic Systems Project, Center for Human Machine Interaction. *Proceedings of the First Danish Human-Computer Interaction Research Symposium 27. November 2001*, Aarhus, Denmark.

Endsley, M.R. (2000a). Theoretical Underpinnings of Situation Awareness: A Critical Review. In M.R. Endsley & D.J. Garland (Ed.), *Situation Awareness Analysis and Measurement* (pp. 3-32). Mahwah, New Jersey: Lawrence Erlbaum Associates.

Endsley, M.R. (2000b). Direct Measurement of Situation Awareness: Validity and Use of SAGAT. In M.R. Endsley & D.J. Garland (Ed.), *Situation Awareness Analysis and Measurement* (pp. 147-173). Mahwah, New Jersey: Lawrence Erlbaum Associates.

Grech, M. & Horberry, T. (2002). Human Error in Maritime Operations: Situation Awareness and Accident Reports. *5th International Workshop on Human Error, Safety and Systems Development 17-18 June 2002*, Noah's On The Beach, Newcastle, Australia.

Van Westrenen, F. (1999). The Maritime Pilot at Work. Delft: Uitgeverij Eburon.

Optimizing Text Layout for Small-screens: the Effect of Hyphenation and Centering

Jari Laarni

Center for Knowledge and Innovation
Research, Helsinki School of Economics
HTC-Pinta, Tammasaarenkatu 3
00180 Helsinki, Finland
laarni@hkkk.fi

Abstract

The presentation of Web pages on small-screen interfaces is growing fast. An important usability problem is how best to present text on small screens. Since in many languages, e.g. in Finnish, words are typically quite long, only few words fit on a single line of a small-screen interface. In order to prevent extremely varied right-hand edge of text, it is either necessary to break words at syllable boundaries or center them. The effect of hyphenation and centering was studied in two experiments. The results suggest that not until the number of hyphenated lines is over 50%, hyphenation slows down considerably reading rate. If only deep indents are prevented by hyphenation, the effect of word breaking remains small. The ragged-centered format is a valuable alternative for small screens.

1 Introduction

As small-screen interfaces (e.g. PDAs) become more and more popular, an important usability problem is how best to present text and graphics on small screens. Since only very few words fit on a single line of a PDA screen, there are deep indents at the right hand side if no hyphenation is allowed. It is possible that these deep indents deteriorate reading performance. If, on the other hand, words are hyphenated, reading performance may deteriorate, because hyphenated words have to be recomposed (Nas, 1988). One alternative is to center lines between margins. This kind of ragged-center format may be worth to consider for small screens (Laarni, Kojo & Kärkkäinen, 2002).

What is the optimal text format may depend on reading rate (Dyson & Haselgrove, 2001). For example, when we rapidly scan or skim text, a different format may be the best alternative than when we read at a normal reading speed.

Here three text-presentation formats (flush-left-right-ragged, hyphenated and center-ragged format) were compared. In Experiment 1 participants read texts at two reading speeds. In Experiment 2 it was more carefully studied the effect of hyphenation. There were two hyphenated formats in which either 50% or 80% of the words that do not fit onto a single line were broken at syllable boundaries.

2 Experiment 1

2.1 Method

2.1.1 Participants

Twelve young adults participated in Experiment 1. The participants had normal or corrected-to-normal vision. They were paid for their participation.

2.1.2 Stimuli and apparatus

Six Finnish magazine articles were read. Their lengths vary from 974 words to 1308 words. Each text line contained maximally about 21 characters. In the flush-right format texts were left-justified and right-ragged. In the hyphenated format words that do not fit onto a single line were broken at syllable boundaries. About 50% of the lines were thus hyphenated. Standard Finnish hyphenation was used. In the ragged-centered format lines were centered between margins, and no hyphenation was used. The texts were presented on a Dell Latitude. Display resolution was 1024 x 768 pixels.

2.1.3 Procedure

Three of the texts were read at a normal reading speed, and three of them were read at a rate twice as fast as the normal reading speed. In order to develop a faster reading speed a period of training was carried out (Dyson & Haselgrove, 2001). Participants were asked to read the first article at their normal reading speed. Then they have to speed up their reading and read the next article at twice this rate. If they have not read fast enough, they were asked to read another article.

Text comprehension was measured by five type of questions (title, structure, main factual, incidental and recognition questions). In order to standardize the chance level for different questions the comprehension scores were adjusted (Dyson & Haselgrove, 2001). They were also weighted according the number of questions.

After the participants had read all the articles, they were asked to rate each of the three formats according to their readability and their visual layout on a 7-point Likert scale. They were also asked to give reasons for their preferences.

2.2 Results

On average, the participants read about 60% faster at the faster reading rate (see Table 1). A two-way repeated measures ANOVA (format, reading style) on reading rate (words/min) revealed that the effect of reading style was significant, $F(1,11) = 35.2$, $p < 0.001$, but the effect of format and the interaction between style and format were not ($p > 0.1$). However, as can be seen in Table 1, there was a tendency for the reading rate to be slower for the hyphenated format than for the other two formats.

A two-way repeated measures ANOVA (format, reading style) on comprehension scores showed that the increase of reading rate had no significant effect on comprehension ($p > 0.1$). Neither were there differences in comprehension between the three formats ($p > 0.1$), even though the scores were somewhat higher for the hyphenated format.

Table 1: Mean Reading Speed (Words/min) and Comprehension Rate (% Hits) in Experiment 1 for Three Text Formats When Reading Either with Normal or Fast Speed.

	Ragged-right	*Hyphenated*	*Centered*	*Means*
Normal	251 (64)	231 (76)	235 (73)	239 (71)
Fast	385 (72)	358 (76)	381 (73)	375 (74)
Means	318 (68)	295 (76)	308 (73)	307 (72)

A two-way repeated-measures ANOVA (format, preference type) on preference ratings showed that the interaction between format and preference type (readability vs. layout) was significant, $F(2,22) = 8.2$, $p < 0.01$. This finding suggests that a format that is easy to read is not necessarily aesthetically appealing. In fact, even though the participants thought that the hyphenated format was less readable than the other two formats, the hyphenated format and the centered-ragged format was rated as more aesthetically appealing than the flush-left-right-ragged format.

Hyphenation was thought to be disturbing, because hyphenated words had to be decomposed (cf., Nas, 1988). Some participants preferred the centered format, because it reduced the need to make horizontal eye movements. Adjectives that were used to describe the ragged-right format were cracked, restless, artistic and effective. The hyphenated format was described, for example, as "even", "straight" and "beautiful", and the centered format as "novel", "relaxed" and "smooth".

3 Experiment 2

3.1 Method

3.1.1 Participants

Thirteen volunteers participated in Experiment 2. The participants had normal or corrected-to-normal vision, and they were paid for their participation.

3.1.2 Stimuli, apparatus and procedure

The texts and apparatus were identical to those of Experiment 1. Participants read two articles in each of the three conditions (flush-left-right-ragged, low hyphenation and high hyphenation). In the flush-right format texts were left-justified and right-ragged. In the two hyphenated formats words that do not fit onto a single line were broken at syllable boundaries. In the low-hyphenation condition 50% of the text lines were hyphenated; in the high-hyphenation condition 80% of the lines were hyphenated. Standard Finnish hyphenation was used.

3.2 Results

The main results are presented in Table 2. A two-way repeated measures ANOVA (format, subsession) on reading rate showed that the effect of format was significant, $F(2,24) = 4.7$, $p < 0.05$, but the effect of subsession was not ($p > 0.1$). A post-hoc test showed that there was not a significant difference between the low-hyphenation and the high-hyphenation conditions ($p > 0.1$).

The comprehension scores were adjusted and weighted in a similar way as in Experiment 1. A two-way repeated measures ANOVA (format, subsession) showed that there was no difference in comprehension between the formats ($p > 0.1$).

Table 2: Mean Reading Speed (Words/min) and Comprehension Rate (% Hits) in Experiment 2 for Three Text Formats and Two Subsessions

	Ragged-right	*50% Hyphenation*	*80% Hyphenation*	*Means*
1st	230 (64)	211 (74)	200 (66)	214 (68)
2nd	225 (59)	203 (59)	192 (62)	207 (60)
Means	228 (62)	207 (67)	196 (64)	210 (64)

Only three participants had noticed during reading that there was any difference in layout between formats. The participants thought that the flush-left-right-ragged and the 50%-hyphenated format were more suitable for reading than the 80%-hyphenated format. One-way repeated measures ANOVA on preference ratings showed that the effect of format was significant, $F(2,24) = 10.5$, $p < 0.001$.

4 Discussion

Our results partly support Nas' (1988) findings according to which it takes longer time to read text versions with word divisions than those without. Apparently, hyphenated words have to be recomposed which takes extra time. Even though deep indents at the right-hand side of a line apparently slow reading, they do not deteriorate reading at the same extent as word divisions. There was no difference between the ragged-right and ragged-centered formats for reading rate. For skimming and rapid reading the ragged-centered format may be even recommended provided that the lines are short.

Surprisingly, comprehension was not worse at a fast speed. This finding is inconsistent with several previous findings (Dyson & Haselgrove, 2001; Poulton, 1958). Perhaps medium levels of time pressure are not detrimental for reading comprehension. As Walczyk, Kelly, Meche and Braud (1999) have shown, mild levels of time pressure can even enhance comprehension.

5 Conclusion

The results suggest that not until the number of hyphenated lines is over 50%, hyphenation slows down considerably reading rate. If only deep indents are prevented by hyphenation, the effect of word breaking remains small. The ragged-centered format is a valuable alternative for small screens.

6 References

Dyson, M. C., & Haselgrove, M. (2001). The influence of reading speed and line length on the effectiveness of reading from screen. *International Journal of Human-Computer Studies*, 54, 585 – 612.

Laarni, J., Kojo, I., & Kärkkäinen, L. (2000). Reading and searching information on small display screens. In D. de Waard, K. Brookhuis, J. Moraal, & A. Toffetti (Eds.), *Human Factors in Transportation, Communication, Health, and the Workplace.* (Pp. 505 – 516). Shaker: Maastricht. (On the occasion of the Human Factors and Ergonomics Society Europe Chapter Annual Meeting in Turin, Italy, November 2001).

Nas, G. L. J. (1988). The effect on reading speed of word divisions at the end of a line. In G. C. van der Veer & G. Mulder (Eds.) *Human-Computer Interaction: Psychonomic Aspects.* Berlin: Springer.

Poulton, E. C. (1958). Time for reading and memory. *The British Journal of Psychology*, 49, 230 – 245.

Walczyk, J. J., Kelly, K. E., Meche, S. D., & Braud, H. (1999). Time limitations enhance reading speed. *Contemporary Educational Journal*, 24, 156 – 165.

Evaluating Situation Awareness in Different Levels of Fidelity of Synthetic Environments: Virtual Cockpit Versus Conventional Flight Simulator

Ungul Laptaned, Sarah Nichols, John R. Wilson

University of Nottingham
Virtual Reality Applications Research Team
School of Mechanical, Materials, Manufacturing Engineering and Management
University Park, Nottingham, NG7 2RD, UK
epxul@nottingham.ac.uk

Abstract

The primary purpose of this study is to compare situation awareness (SA) between two levels of fidelity of synthetic environments. Situation awareness was evaluated by objective and subjective measurements. Objective measures included: (a) recall, (b) performance (error rates), and (c) virtual reality situation awareness global assessment technique (VRSAGAT). Subjective measures were: (a) virtual reality situation awareness rating metric (VRSARM), (b) virtual reality situation awareness rating technique (VRSART), (c) sickness, (d) presence, (e) transitions to reality, and (f) usability questionnaire. There were eight subjects with a minimum of a private pilot license plus instrument ratings. The independent variables were: Environment (Virtual Cockpit and Conventional Flight Simulator), and Virtual Environment Experience (Novice and Expert). The results indicated that there was no significant main effect of environment on recall measure and VRSAGAT. However, there was a significant main effect of VE experience on altitude error rates. There were also significant main effects of environment on the situational awareness and immersion subscale of VRSARM. VRSART indicated that there was a significant main effect of environment. A difference main effect of environment on sickness questionnaire was also significant. It was concluded from the results that SA of the pilots who flew CFS was greater than those who flew VC. From the usability point of view, it was concluded that CFS is far superior to VC. The VC, nevertheless, is still in an early stage of development. This experimental research is used to generate design implications for future directions for SA in VC.

1 Introduction

The virtual cockpit (VC) is a virtual simulation for aircraft cockpits with visual, haptic, and audio output to provide the feeling of immersion for a pilot. The cockpit layout is generated from Airbus 340 aircraft design and enables the pilot not only to view the display but also to interact with devices and systems. Pilot system input is dominated by hand input interaction on panels, buttons, dials, and switch and cockpit output devices render the aircraft status to the pilot. Flight guidance displays provide all necessary flight information to the pilot during normal procedures. The VC is a front end with all displays and units that are attached to a conventional flight simulator (CFS). It incorporates and shares a terrain database and a sound simulation with CFS. The VC is different from CFS in many aspects such as it has less field of view, high tracker latency, and lack of force feedback. In the VC use, a pilot sits in a basic mock-up donning with a head mounted display

(HMD) that simulates the cockpit environment surrounding him. In this research, SA is examined and discussed with the advent of cockpit technologies. Endsley (1995b) addressed that "a measure of cognitive components is valuable in the engineering design cycle because it assures that prospective designs provide operators with the critical commodity." This cognitive measure used in explaining performance and decision-making emphasizes the concept of SA in the decision process. Endsley (1988) defined SA as "a person's perception of the elements of the environment within a volume of time and space, the comprehension of their meaning and the projection of their status in the near future." It is an understanding of the state of the environment. SA is a good subjective measure of the quality of any systems, for example, aviation/aerospace system. The user of this system must have knowledge of the current process state at all times, and the ability to use this knowledge effectively in predicting future process states and controlling the process to attain operational goals (Endsley, 1995a). A cockpit virtual reality (VR) is considered to have an ability to promote the feeling of being there. Curry (1999) suggested that the concept of presence deals with the degree to which you feel a part of some virtual space; that the space exists and you are occupying it. In the context of SA, presence might provide the sense for a pilot to perceive the elements, comprehend their meaning, and project their status. In addition, spatially immersive VR systems expressly have been found to have the capability to promote/provoke high SA (Prothero et al., 1995). Therefore, it is essential to determine how the degree of being present could lead to enhance SA. By assessing pilot's SA, it may be able to predict his performance.

2 Method

The experimental design used in this study was a 2×2, factorial design with 1 within and 1 between subject variable. The independent variables were Environment (VC and CFS) and Virtual Environment Experience (Novice and Expert). Pilots in the experiment were recruited from airlines in Germany. 8 pilots volunteered to partake in this study. They attended a test session in one full day (VC and CFS, or vice versa). The input devices were a joystick, throttle, flap, and "n Vision" HMD that used to fly an airplane. Cockpit layouts were generated from Airbus 340 aircraft design with databases that are loaded into the Virtual Design II software package. This software allows inserting, modifying, and replacing geometry, texture, and attached cockpit sounds. With a VC and CFS panel, the key components of display formats consist of Primary Flight Display, Navigation Display, Heads Up Display, and Electronic Centralized Aircraft Monitoring. These components are a basic requirement and standardized for a modern aircraft cockpit. There were a training session, and two tests plan (Level Flight/Speed Change/Pitch-Power-Control and Standard Rate Turn) that need to be completed

3 Measurement of Situation Awareness

SA measurement included objective and subjective measures. Objective measures were: waypoint recalled, performance (error rates), and VRSAGAT (devised by author, 2000). Subjective measures were: VRSARM (author, 2000), sickness (Kennedy, 1993), presence (Witmer and Singer, 1998), VRSART (Kalawsky, 1999), and transitions to reality (Slater and Steed, 2000).

4 Results

Data from 8 pilots was analyzed by using a factorial analysis of variance. Due to space restrictions, only significant results of all questionnaires as well as the total scores of sickness questionnaire, presence questionnaire, and VRSART are reported here.

4.1 Objective Measures

4.1.1 Recall Measure

The means number of waypoint recalled for VC and CFS were 4.25, and 4, respectively. The means number of waypoint recalled for novice pilots was 3.63 and for expert pilots was 4.63. No significant difference or interaction was found. The means name of waypoint recalled for VC and CFS were 2.75, and 2.5, as well as for novice pilots were 2.38 and for expert pilots were 2.88. No significant difference or interaction was found.

4.1.2 Performance Measure

The means altitude error rates (%) for VC and CFS were 20.04, and 18.33, respectively. A significant difference was not found. The means altitude error rates for novice pilots were 29.04 and for expert pilots were 9.33. A significant difference was found, indicating F=5.11, df=1(12), ?<0.041. An omega-square (ω^2) of 0.23 indicates that 23 % of the variability in error rates of altitude is due to VE experience. There was no significant interaction. For the means speed error rates (%) and heading error rates (%) on both environment and VE experience, there were no significant differences or interactions found.

4.1.3 Virtual Reality Situation Awareness Global Assessment Technique

The means response time for VC and CFS were 7.19, and 7.39, respectively. The means response time for novice pilots was 7.22 and expert pilots were 7.35. No significant difference or interaction was found. The means number of correct answers for VC, CFS, novice pilots, and expert pilots were 6.25, 6.13, 6.25, 6.13, respectively. No significant difference or interaction was found.

4.2 Subjective Measures

4.2.1 Virtual Reality Situation Awareness Rating Metric

The means VRSARM scores on situational awareness subscale for VC and CFS were 42.63, and 50.13. A significant difference was found (F=4.91, df=1(12), ?<0.045), and ω^2 of 0.19 indicates that 19 % of the variability in SA factor is due to environment. Therefore, the pilots who exposed to CFS were more situationally aware than those who exposed to VC. The means VRSARM scores of novice pilots were 44 and expert pilots were 48.75. No significant difference or interaction was found. The means VRSARM scores on immersion subscale for VC and CFS were 40.63, and 52.88. A significant difference was found (F=7.07, df=1(12), ?<0.020), and ω^2 of 0.29 indicates that 29 % of the variability in immersion is due to environment. It was concluded that the pilots who exposed to CFS had higher degree of immersion than those who exposed to VC. There was no significant difference on VE experience or interaction found.

4.2.2 Virtual Reality Situation Awareness Rating Technique

The means VRSART scores on total scores for VC and CFS were 91.25, and 107.13, respectively. A significant difference was found (F=8.88, df=1(12), ?<0.011), and ω^2 of 0.36 indicates that 36 %

of the variability in SA with presence is due to environment. Hence, the pilots who felt present and immersed within CFS were more situationally aware than those who exposed to VC. No significant difference on VE experience or interaction was found.

4.2.3 *Presence Questionnaire*

The means presence scores on total scores for VC and CFS were 60.88, and 76.25, respectively. A significant difference was found (F=6.93, df=1(12), ?<0.021), and ω^2 of 0.28 indicates that 28 % of the variability in presence is due to environment. It was concluded that the pilots who exposed to CFS have higher degree of presence than those who exposed to VC. There was no significant difference on VE experience or interaction found.

4.2.4 *Sickness Questionnaire*

The means SSQ scores of total scores during post exposure for VC and CFS were 109.83, and 16.03. A significant difference was found (F=5.74, df=1(12), ?<0.032), and ω^2 of 0.25 indicates that 25 % of the variability in sickness symptom is due to environment. Therefore, the pilots who exposed to VC experienced sickness symptom more than those with CFS. No significant difference on VE experience or interaction was found.

4.2.5 *Transitions to Reality Questionnaire*

(n is number of pilots who gave responses in the corresponding category) - Transitions particularly interesting for Virtual Cockpit are shown in Table 1.

Table 1: Example of Reasons Given for Transitions to Reality

Cause	n	Some Examples
Internal	2	-Absence of sounds -The information given was hard to understand
Experiment	3	-Instructions sounded to me like through a head phone, the noise from outside was not noticed -HMD was too heavy and field of view was low -Light from outside the HMD
Personal	1	-Felt like not belong to the real aircraft
Spontaneous	2	-No apparent reason

5 Discussion

It was found that the recall measure or VRSAGAT did not find any significant differences between VC and CFS, or between novice and expert pilots. This may be due to the low number of subjects attended. Therefore, there should be further experiments with a higher number of pilots to ensure that a significant difference is found. The sickness questionnaire found a difference on the total scores, and concluded that those pilots who flew with VC experienced higher sickness symptom than CFS. It was obvious that the pilots were asked to don a "n Vision" HMD that is in fact the weight of HMD, as commonly used, can promote a high sickness symptom. The presence score was also significantly different between VC and CFS. The total score of presence questionnaire indicated that the pilots who flew CFS had a greater sense of presence than VC. The CFS is capable of promoting a higher level of immersion, which results in a higher sense of presence. There was also an indication of a significant difference for the situational awareness subscale of VRSARM. The result showed that SA of the pilots who flew CFS was higher than those who flew VC. Also, the total score of VRSART on CFS was greater than the total scores on VC, which simply indicated that SA resulting from being present and immersed in CFS was higher than in VC. Results from usability questionnaire (presented elsewhere) indicated that the CFS is easier to use, more naturalistic, and better good control over the interaction within the flight environment, as well as provokes a better degree of presence. It should be noted that VC is in an early stage and would be improved in the future with better display technology and input devices, which are all required for aviation training purposes.

6 Acknowledgements

This work was a part of Ph.D. research carried out at the University of Nottingham. Support from the apparatus use at the Institute of Flight Systems and Controls, Darmstadt Institute of Technology, Darmstadt, Germany, as well as the assistance by Prof. Dr. -Ing Wolfgang Kubbat, Dr. -Ing Jens Schiefele, Kai-Uwe Dörr, Silke Hüsgen, are gratefully acknowledged.

References

Curry, K. M. (1999, May). Supporting Collaborative Awareness in Tele-Immersion. *International Immersive Projection Technology Workshop, Centre of the Fraunhofer Society Stuttgart IZS*, 10 (11), 253-261.

Endsley, M. R. (1988). Situation Awareness Global Assessment Technique (SAGAT). In *Proceedings of the National Aerospace and Electronics Conference* (*NAECON*) (pp. 789-795). New York: IEEE.

Endsley, M. R. (1995a). Toward a Theory of Situation Awareness in Dynamic Systems. *Human Factors*, 37(1), 32-64.

Endsley, M. R. (1995b). Measurement of Situation Awareness in Dynamic Systems. *Human Factors*, 37(1), 65-84.

Kalawsky, R. S. (1999). VRUSE - A Computerised Diagnostic Tool for Usability Evaluation of Virtual/Synthetic Environment Systems. *Applied Ergonomics*, 30, 11-25.

Kennedy, R. S., Lane, N. E., Berbaum, K . S., & Lilienthal, M. G. (1993). Simulator Sickness Questionnaire: An Enhanced Method for Quantifying Simulator Sickness. *The International Journal of Aviation Psychology*, 3(3), 203-220.

Nichols, S. (1999). *Virtual Reality Induced Symptoms and Effects (VRISE): Methodological and Theoretical Issues*. Ph.D.'s Thesis, University of Nottingham, Nottingham, England.

Prothero, J., Parker, D., Furness, T., & Wells, M. (1995). Towards a Robust, Quantitative Measure of Presence. In *Proceedings of the Conference on Experimental Analysis and Measurement of Situational Awareness*, Daytona Beach, Florida, 359-366.

Regal, D. M., Rogers, W. H., & Boucek Jr., G. P. (1978). *Situational Awareness in the Commercial Flight Deck: Definition, Measurement, and Enhancement*. Society of Automotive Engineers, Inc.

Schiefele, J. Dorr, K., Olbert, M., & Kubbat, W. (1999). Stereoscopic Projection Screens and Virtual Cockpit Simulation for Pilot Training. In *Proceeding of 3rd International Immersive Projection Technology Workshop (IPT 99)* (pp. 211-222). Stuttgart, Germany, May, 1999.

Slater, M., & Steed, A. (2000, October). A Virtual Counter. *Presence: Teleoperators and Virtual Environments*. PDF 9.2, In Press.

Witmer, B. G., & Singer, M. J. (1998). Measuring Presence in Virtual Environments: A Presence Questionnaire. *Presence: Teleoperators and Virtual Environments*, 7(3), 225-240.

Developing a Testbed for Studying Human-Robot Interaction in Urban Search and Rescue

Michael Lewis

University of Pittsburgh
Pittsburgh, PA 15260
ml@sis.pitt.edu

Katia Sycara and Illah Nourbakhsh

Carnegie Mellon University
Pittsburgh, PA 15213
katia@cs.cmu.edu illah@cs.cmu.edu

Abstract

We are developing simulations of the National Institute of Standards and Technology (NIST) Reference Test Facility for Autonomous Mobile Robots (Urban Search and Rescue) in order to develop and test our strategies for Robots-Agents-People (RAP) team coordination and control. The NIST USAR Test Facility is a standardized disaster environment consisting of three scenarios of progressive difficulty: Yellow, Orange, and Red arenas. The USAR task focuses on robot behaviors, and physical interaction with standardized but disorderly rubble filled environments. As part of our research effort we are constructing and permanently housing a physical replica of the Orange arena at Carnegie Mellon University. A simulation of the Orange arena was constructed first in order to allow comparisons between simulated and real environments as soon as construction of the physical Orange Arena is completed. We hope to use the simulations to provide a testbed in which to evaluate rapidly prototyped interfaces and control strategies prior to the construction and testing of physical robots. This paper describes our simulation approach based on the use of the Unreal game engine to provide graphics and physics and simplified CAD models textured from digital photographs to model the environment.

1 Introduction

Large-scale coordination tasks in hazardous, uncertain, and time stressed environments are becoming increasingly important. In such domains, e.g. rescue operations in natural and man-made disasters, de-mining, environmental cleanup, civilian and military crisis response, different organizations such as fire fighters, police, and medical assistance come together to cooperate to save lives, protect structural infrastructure and property and evacuate victims to safety. In such environments, human rescuers must make quick decisions under stress, and possibly risk their lives. They must have timely and accurate information on the status of the environment, expected arrival times for additional resources (e.g. medical supplies), and they must coordinate resource allocation and other rescue activities. One solution that addresses the limitations of current manual operations is to introduce enhanced automation that can address the challenges of time stressed, uncertain, dangerous and evolving environments.

To address these challenges requires fundamental research advances in the design of distributed systems that would effectively coordinate with dispersed humans. Our research is founded on three key advances/technical ideas. First, we propose Hybrid Teams of Robots, autonomous Agents, and People (RAPs) consisting of large number of entities that are distributed in space, time, capability, and role. Second, we move away from the traditional human-controlled design of automation, where automated systems are subordinate to their human controllers who give them their goals and tasks and manage task execution. Instead, we advocate a cooperative control (adjustable autonomy) paradigm where current notions of organizational control and system interactions are extended based on adaptive sharing. The various members of RAP groups, be they human, robots or cyber-agents could share common goals, initiative for communication and action, responsibility for coherent group activity, information on the environment, mission, situation an help each other in overcoming barriers to achievement of common goals. Third, we recognize that ad hoc interoperability across different agents, teams and organizations that are brought together "as is", and co-adaptation to each other is key in addressing the challenges present in such large-scale, uncertain coordination domains.

Because our research involves the study of coordination of both simulated and real robots it is essential that we develop robust reference tasks to guide and evaluate our research. To promote the generality and openness we have chosen the NIST Reference Test Facility for Autonomous Mobile Robots (Urban Search and Rescue) (Jacoff, Messina & Evans, 2001) as the reference domain to develop and test our strategies for RAP team coordination and control. The NIST USAR Test Facility is a standardized disaster environment consisting of three scenarios: Yellow, Orange, and Red physical arenas of progressing difficulty shown in Figure 1. The USAR task focuses on robot behaviors, and physical interaction with standardized but disorderly rubble filled environments. As part of our research effort we are constructing and permanently housing a physical replica of the Orange arena of the USAR Test Facility at Carnegie Mellon University to serve as a testbed for RAP research by ourselves and others. We have already completed a virtual version of the Orange Arena and untextured versions of the Yellow and Red Arenas. The simulated arenas are intended to allow comparative studies, experiments controlling large numbers of simulated robots, and hardware in the loop simulations involving both the physical test facility and its simulation.

Figure 1: Orange (near) and Yellow Arenas, NIST photograph

2 Orange Arena Simulation

Our initial simulation of the Orange Arena is keyed to teleoperation just as the initial robots to be deployed to the physical Orange Arena are to be teleoperated. Subsequent simulator development will add: 1) non-imaging sensor models for sound, sonar, laser ranging, and heat sensing for use in inner loop autonomous control and alarms, 2) nonvisible spectra imagery simulating forward looking infrared (FLIR), or greenscale night vision cameras, 3) more exacting kinematic models relating locomotive capabilities to the environment. While these advanced capabilities are essential to simulating semi-autonomous robots, high fidelity simulation of a video feed and an interface for controlling locomotion and camera position are the crucial features for studying human-robot interaction for teleoperation.

2.1 Simulation with Game Engines

Developing simulations from scratch is an expensive and time consuming process. The Robocup Rescue simulation (Kitano et al., 1999) begun following the Kobe earthquake and still under development identifies entities at the level of FireBrigades and AmbulanceTeams, treats the world as a planar graph, and remains predominately text-based. Our initial simulation of the Orange Arena (much smaller area but at higher resolution), by contrast, was complete within two months after data collection and provides realtime interaction with a simulated video feed. This dramatic discrepancy in development times and effort is due to the high quality tools and sophisticated infrastructure created by the commercial video/computer game industry. Collision detection, Newtonian physics, lighting models, and state of the art interactive graphics all come as part of computer games that sell for under fifty dollars.

Because contemporary games are designed so that users can modify both the environment (level programming) and behavior (game programming) of the game, they make powerful, inexpensive tools (Lewis & Jacobson 2002) for researchers who need interactive 3D modeling and graphics. Epic Game's Unreal 2003 game engine used in Unreal Tournament 2003 and America's Army (Zyda et al., 2003) among other games is especially attractive because of the Gamebots (Kaminka et al., 2002) modification written to allow entities within the game to be controlled through standard TCP/IP sockets. Using the Gamebots interface, data from within the simulation such as ranging data showing the robot's distance from walls and objects can be read off the socket while commands are sent back to control the robot within the simulation. Human control can range from a completely autonomous robot using ranging data to avoid walls to a teleoperated one that relays a high fidelity view of its surroundings using the "spectate" function that attaches a viewpoint to a player (in this case the robot she is controlling). Following this strategy we have simulated a teleoperated robot for which we can control locomotion in two degrees of freedom (following the terrain in the third) with separate control of pitch and yaw in the view to simulate a motorized camera.

2.2 Data Collection

A team of ten researchers visited the NIST USAR facility on November 1, 2002 to collect data and photographs for constructing simulations and identifying design requirements for robots to operate in these environments. Data were collected on illumination levels in the three arenas (Orange Arena shown in Figure 2), 80211.b radio reception, and typical arena materials were evaluated for reflectivity and detectability by sensors. Robot level views were video taped from a radio controlled platform in order to provide a comparison standard for developing the simulations and to provide an early indications of factors likely to affect teleoperation in these environments. Extensive digital photographs were taken of the arenas both for additional documentation and for use as textures in the simulations to be developed. Simulation Construction

Arena simulations were based on CAD models of the arenas provided by NIST. These solid models which had been developed using *ProEngineer* software were simplified in *ProEngineer*, exported through *Nugraf* to *3D Studio Max* which converted them to triangular meshes suitable for high frame rate animation, and then re-exported in *Unreal Tournament* format. Textures selected from the collection of digital photographs were applied in the native *Unreal* editor to

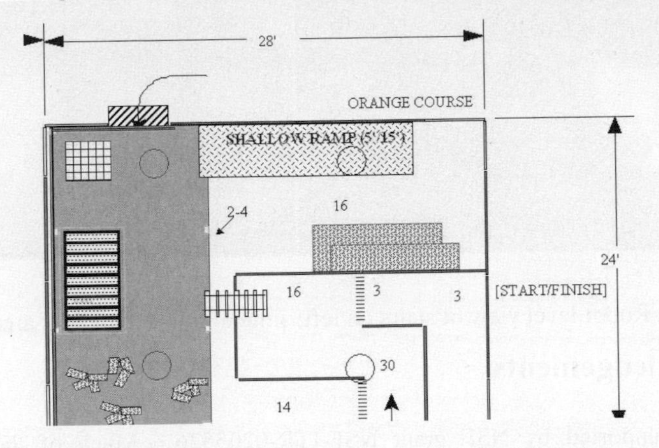

Figure 2: Illumination levels in lux for Orange Arena

faithfully reproduce patterns and colors of the arena's surfaces. Finally, model lighting was adjusted to approximate the extreme (15:1) variations between open areas and shadows under the platform. Special details including mirrors and orange safety fencing were handcrafted and applied to the model. Figure 3 shows elevated views of the actual (left) and simulated (right) Orange Arena. Robot level views are shown in Figure 4.

2.3 Planned Research

The initial research planned for the simulator will involve teleoperation. The first experiment scheduled to begin in late February 2003 will compare off-the- shelf controllers including conventional and inertial joysticks, track balls, game pads, and a 6 DOF SpaceOrb. Participants will perform Fitts law, tracking (path following), and search (treasure hunt) tasks in either the desktop USAR simulation or a simulated Egyptian temple presented in the BNave (Lewis & Jacobson, 2002), a cave-like panoramic display. Subsequent experiments are planned to evaluate teleoperation aiding strategies including simulated tethering, velocity dependent changes in FOV, and aids to situation awareness for conveying platform attitude and camera orientation.

Figure 3: Orange Arena NIST photograph on left, elevated view of simulation on right

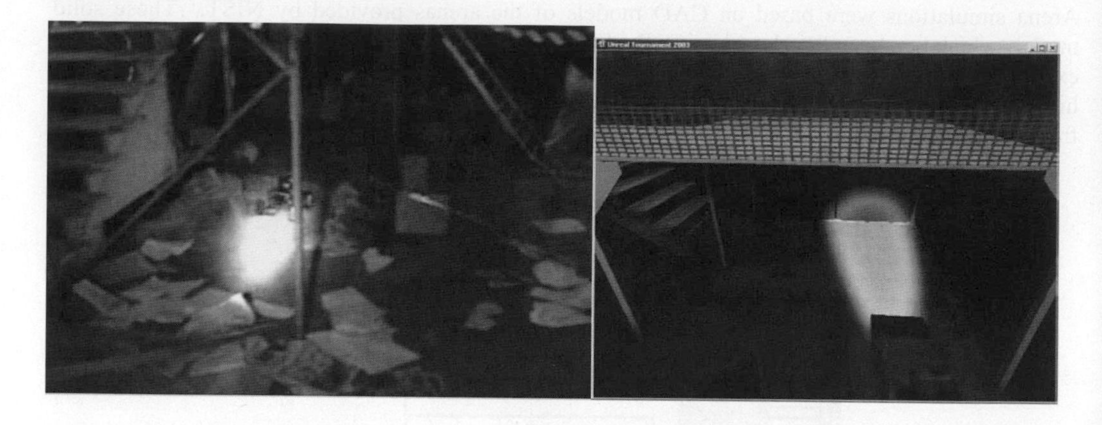

Figure 4: Robot level view of stairs on left, simulation view of stair area on right

3 Acknowledgements

This project is supported by NSF grant NSF-ITR-0205526. Much of the simulation was constructed by Jijun Wang. Mary Berna, Alexander Gutierrez, Terence Keegan, Kevin Oishi, Binoy Shah, Steven Shamlian, Mark Yong, and Josh Young assisted in data collection.

References

Jacoff, A., Messina, E., Evans, J. (2001), Experiences in deploying test arenas for autonomous mobile robots, Proceedings of the 2001 Performance Metrics for Intelligent Systems (PerMIS) Workshop, Mexico City, Mexico.

Kaminka, G., Veloso, M., Schaffer, S., Sollitto, C., Adobbati, R., Marshall, A., Scholer, A., and Tejada, S. (2002). GameBots: A flexible test bed for multiagent team research, *Communications of the Association for Computing Machinery* (CACM), NY: ACM 45(1), 43-45.

Kitano, H., Tadokoro, S., Noda, I., Matsubara, H., Takahashi, T., Shinjoh, A., Shimada, S. (1999). RoboCup Rescue: Search and Rescue in Large-Scale Disasters as a Domain for Autonomous Agents Research, *Proc. 1999 IEEE Intl. Conf. on Systems, Man and Cybernetics*, Vol. VI, pp. 739-743, October, Tokyo, 1999 (SMC 99)

Lewis, M. & Jacobson, J. (2002) Game Engines in Research. *Communications of the Association for Computing Machinery* (CACM), NY: ACM 45(1), 27-48

Zyda, M., Hiles, J., Mayberry, A., Wardynski, C., Capps, M., Osborn, B., Shilling, R., Robaszewski, M., Davis, M. (2003), The MOVES Institute's Army Game Project: Entertainment R&D for defense, *IEEE Computer Graphics and Applications*.

Using symbol size and colour to effect performance confidence

Frederick M. J. Lichacz

Canadian Forces Experimentation Centre
Ottawa, ON, Canada
Lichacz.MJ@forces.gc.ca

Gregory L. Craig

National Research Council-FRL
Ottawa, ON, Canada
Greg.Craig@nrc.ca

Lindsay C. Bridgman

University of Waterloo
Waterloo, ON, Canada
clbridgman@artsmail.uwaterloo.ca

Abstract

This experiment was conducted to examine the relationship between symbol size and colour on visual search and confidence. Using a simulated radar screen, participants were required to identify the most hostile aircraft amongst a display of different aircraft types of varying size and colour and to provide a confidence judgement associated with each response. The findings from this experiment support previous findings insofar as colour and size facilitate detection performance. Also, confidence ratings were observed to be faster and higher when stimuli can be discriminated based on colour or size. Finally, calibration techniques were used to assess the participants' situation awareness.

1 Objective and significance

A vital element of a fighter pilot's arsenal is the ability to quickly identify and verify an enemy aircraft with the highest degree of confidence. Detecting an enemy aircraft as fast and as accurately as possible represents one of the first steps to developing good situation awareness, which in turn should facilitate pilot performance. However, the extent to which detected aircraft are prosecuted depends in large part on the pilot's confidence in his/her situation awareness (Endsley, 1993). Currently, pilots use a radar display with monochrome symbols of identical size as their primary means of target identification and verification. It has been well established that the speed and accuracy with which information is detected is facilitated by both colour (Treisman & Gelade, 1980) and stimulus size (Quinlan & Humphreys, 1987). What is less clear is whether the use of colour and stimulus size can facilitate confidence judgements and the speed with which they are made. Accordingly, the purpose of the present study was to examine the extent to which stimulus colour and size affect visual search and confidence in visual search performance.

Confidence judgements in our beliefs and knowledge systems play an important role in the decision-making processes that guide our everyday activities (Baranski & Petrusic, 1994). This is no less the case in SA research. In the present study, the relationship between confidence and SA makes use of a calibration analyses techniques to determine the participants' under and overconfidence and resolution (Baranski & Petrusic, 2001, 1994). This analysis represents a formal examination of the relationship between confidence and decision-making by looking at the correspondence between a probability assessment, expressed as the occurrence of a particular event and the empirical probability of the occurrence of that event. For example, following a decision or judgement response, the participant provides a subjective probability of the correctness of the response as a confidence judgement (with 0 denoting uncertainty, 50 a guess, and 100 certainty). The relation between the correctness of that response and the corresponding confidence rating is calibrated into a measure of the participant's under and overconfidence where underconfidence (negative scores) is the tendency to underestimate his/her mean percentage correct whereas overconfidence (positive scores) is the tendency to exceed the percentage correct. Perfect calibration occurs when participants provide confidence judgements of 0 and 100 for incorrect and correct answers, respectively, thereby achieving an overall confidence score of zero. In addition to the issue of over/underconfidence, the issue of resolution demonstrates the extent to which a person or group can distinguish an event's occurrence or non-occurrence (Baranski & Petrusic, 1994). This current view of the relationship between confidence and decision-making goes beyond traditional methods, which are typically used in current SA research (i.e., provide a confidence rating to some performance output measure), which can only speak to the monotonic relationship between confidence and adequacy of performance (Adams & Adams, 1961).

The potential of calibrating the relationship between response and confidence into a score of under/overconfidence and utilizing resolution is to obtain a more realistic depiction of an individual's SA, which includes their degree of certainty in their understanding of their environment and propensity to respond in a particular manner than can be achieved by looking at SAGAT responses and confidence ratings in isolation. For example, the extent to which an individual is under/overconfident illustrates the degree of understanding of their environment. It can be argued that the larger the individual's level of under/overconfidence, the less understanding they have about their environment regardless of whether they answer an SA query correctly. Moreover, high levels of under or overconfidence can lead to slow, rational responses or fast, irrational responses, respectively, even when SA queries are correct. Moreover, high and low levels of confidence associated with correct and incorrect SA responses, respectively, confirms that the individual indeed has a good understanding of their environment. What is more, in addition to a deeper understanding of an individual's or group's level SA, knowledge of an individual's level of under/overconfidence and resolution can aid our understanding of disconnects (Endsley, 2000) that are often observed between one's level of SA, as measured by SAGAT queries, and actual response.

2 Method

2.1.1 Participants

Thirty adults were recruited by advertisement from within DRDC Toronto and York University for this study. All had normal or corrected-to-normal vision, and none had any colour-vision deficits as assessed with the AO H-R-R Pseudoisochromatic Plates. Participants were compensated according to DRDC Toronto guidelines.

2.1.2 Apparatus and stimuli

An SVGA colour monitor was used to present a simulated radar screen that displayed various configurations of hostile, unknown, and friendly aircraft symbols. A PC Pentium II computer controlled event sequencing, randomization, and the recording of responses, response times, and confidence reports.

The symbols were of three sizes: small 0.17 x 0.17 inches (20 arc min.), medium 0.26 x 0.26 inches (30 arc min.), and large 0.34 x 0.34 inches (40 arc min). Within each size condition, half of the displays were coloured and half were presented in monochrome green. Half of the displays contained aircraft symbols of heterogeneous sizes and half contained symbols of homogeneous sizes (equal number of small, medium, and large sized aircraft symbols). In the coloured condition, the hostile aircraft were red, unknown aircraft were yellow, and friendly aircraft were blue. In all conditions, different symbol shapes were used for hostile, unknown, and friendly aircraft.

On each trial, the stimulus display consisted of seven aircraft symbols containing one, two, or three hostile targets and six, five, or four unknown and friendly aircraft symbols. The placement of the symbols was determined randomly and counterbalanced across each display.

2.1.3 Design and procedure

Participants were required to indicate as fast and as accurately as possible which quadrant of the radar display contained the most hostile target (i.e., largest, or single hostile target in the display). Following each response, the participants provided a confidence rating ranging from 0% to 100%, as fast and as accurately as possible, indicating how confident they were that their response was correct.

3 Results

For present purposes, only the displays with a single hostile target are reported. All ANOVAs were conducted as 2 (heterogeneous vs. homogeneous symbol presentation) x 2 (colour vs. monochrome display) x 3 (small vs. medium vs. large symbol size) repeated measures designs.

3.1.1 Visual Search

There was a significant effect of symbol presentation ($F(1,29) = 4.33$, $Mse = 37498$, $p < .05$) such that responses were slower when symbols were presented in different (1268 ms) rather than same (1225 ms). A significant effect of display ($F(1,29) = 126.64$, $Mse = 190439$, $p < .000001$) showed that responses were faster when symbols were displayed in colour (990 ms) rather than in monochrome (1504 ms). Finally, there was a significant main effect of symbol size ($F(2,58) = 75.92$, $Mse = 20951$, $p < .000001$) that indicated that responses became faster with increases in symbol size from small (1362 ms) to medium (1247 ms) to large (1132 ms).

There were significant two-way interactions between symbol presentation and display ($F(1,29) = 4.37$, $Mse = 89835$, $p < .05$), symbol presentation and symbol size ($F(2,58) = 32.17$, $Mse = 10896$, $p < .000001$), and display and symbol size ($F(2,58) = 53.93$, $Mse = 18144$, $p < .000001$). These interactions are qualified within a significant Symbol Presentation x Display x Symbol Size ($F(2,58) = 28.27$, $Mse = 12701$, $p < .000001$). Responses did not vary across symbol size for displays containing different and same symbol presentations during the colour display condition. In contrast, in addition to the responses being slower in the monochrome display condition, responses did vary significantly across symbol size during the monochrome display conditions.

3.1.2 *Confidence*

Overall, confidence was higher when the displays were in colour rather than monochrome (98 vs. 95), $\underline{F}(1,28) = 13.38$, $\underline{Mse} = 52$, $\underline{p} < .001$ and increased across target size (96 vs. 97 vs. 97), $\underline{F}(2,56) = 4.75$, $\underline{Mse} = 5$, $\underline{p} < .01$. However, these effects are qualified within a significant three-way interaction between symbol sizes, colour, and target size, ($\underline{F}(2,56) = 3.73$, $\underline{Mse} = 4$, $\underline{p} < .03$) such that there was a significant effect of target size for heterogeneous symbol presentation during the monochrome condition. Post hoc analyses using Tukey's LSD method ($\alpha < .05$) revealed that confidence ratings were significantly different across small, medium, and large target sizes (94 vs. 95 vs. 97, respectively).

There was a significant effect of colour such that confidence ratings were faster during the colour rather than monochrome presentations (924 ms vs. 1168 ms), $\underline{F}(1,28) = 27.02$, $\underline{Mse} = 192388$, $\underline{p} < .00002$). Also, a significant effect of target size revealed that confidence ratings increased in speed across small, medium, and large targets (1079 ms vs. 1038 ms vs. 1022), $\underline{F}(2,56) = 8.44$, $\underline{Mse} = 11729$, $\underline{p} < .001$. These effects were qualified within a significant two-way interaction between colour and target size, $\underline{F}(2,56) = 3.41$, $\underline{Mse} = 17760$, $\underline{p} < .04$. Post hoc analyses using Tukey's LSD method ($\alpha < .05$) revealed that confidence ratings were significantly different across small, medium, and large target sizes (1221 ms vs. 1166 ms vs. 1119) only during monochrome presentation.

Overall, the participants demonstrated good resolution. In addition to being overwhelmingly correct in response to their SA queries (96% correct vs. 4% incorrect), the participants demonstrated significantly higher levels of confidence in their correct than incorrect answers (98% vs. 33%, respectively).

In general, the participants were slightly underconfident in their responses (i.e., their confidence ratings were slightly lower than their percent correct) but well calibrated. There was a significant effect of symbol size such that participants showed greater underconfidence when the symbols were heterogeneous rather than homogeneous (-2.05 vs. –0.99), $\underline{F}(1,28) = 12.07$, $\underline{Mse} = 97$, $\underline{p} < .001$. Furthermore, there was a significant effect of target size such that participants became increasingly less underconfident as the targets increased in size from small, to medium, to large (-2.66, -1.86, and –0.02, respectively), $\underline{F}(1,28) = 17.75$ $\underline{Mse} = 213$, $\underline{p} < .000001$, thus demonstrating increasing calibration or SA. However, these effects are qualified within a significant three-way interaction between symbol size, colour, and target size, $\underline{F}(1,28) = 5.81$, $\underline{Mse} = 60$, $\underline{p} < .005$. From this interaction it was revealed that underconfidence remained stable across both symbol and target size when the display was presented in colour with a mean underconfidence score of -1.6. Similarly, the participants' under/overconfidence scores remained stable across the three target sizes during the monochrome presentations using heterogeneous symbol sizes and across the small and medium size target in the homogeneous symbol condition with a mean underconfidence score of –3.11. However, the participants demonstrated a significant level of overconfidence (3.67) with large target sizes presented in colour with homogeneous symbol sizes.

4 Summary

The present findings replicate previous visual search results insofar as colour and size facilitate detection performance. Moreover, these findings showed that confidence ratings are faster and higher when stimuli can be discriminated based on colour or size. What is more, colour was observed to effectively attenuate the affects of symbol size and symbol presentation on confidence. In the current study, the participants displayed a high degree of SA as evidence by their excellent resolution and low levels of underconfidence. The resolution data showed that participants were well aware when they were right or wrong in their SA query answers via their confidence in their responses: Correct answers, high confidence, incorrect answers, high confidence. Moreover, the low levels of underconfidence revealed that the participants' objective and subjective performance was well calibrated. Together with the resolution data, we would expect the participants to respond in an appropriate manner were they instructed to do so. Indeed, responses would be expected to be in line with their level of SA. Alone, the results of SA queries and basic confidence ratings cannot provide as detailed a view of an individual's SA or propensity to act in a particular manner as can be achieved with resolution and under/overconofidence data. However, as this study represents an initial attempt to study the relationship between calibration techniques and SA, the extent to which these current findings generalize to SA research in general must be made cautiously. Clearly more research within more dynamic and challenging contexts is required and the full range of calibration analyses must be investigated before the full merit of this new way of studying SA can be properly assessed. Nevertheless, the present study has raised important concerns about the adequacy of current SA research and held out the potential for a more fruitful and more accurate manner of studying SA that might better facilitate our understating of this metacognitive concept and its relationship with performance than has heretofore been undertaken.

References

Adams, J. K. & Adams, P. A. (1961). Realism of confidence judgments. Psychological Review, 68, 33-45.

Baranski, J. V., & Petrusic, W. M. (1994). The calibration and resolution of confidence in perceptual judgements. Perception & Psychophysics, 55, 412-428.

Endsley, M. R. (1993). A survey of situation awareness requirements in air-to-air combat fighters. The International Journal of Aviation Psychology, 3, 157-168.

Endsley, M. R. (2000). Theoretical underpinnings of situation awareness: A critical review. In M. R. Endsley & D. J. Garland (Eds.), Situation Awareness Analysis and Measurement. (pp. 3-32). Mahwah, NJ: Lawrence Erlbaum Associates.

Quinlan, P. T. & Humphreys, G. W. (1987). Visual search for targets defined by combination of color, shape, and size: An examination of the task constraints on feature and conjunction searches. Perception & Psychophysics, 41, 455-472.

Treisman, A. M., & Gelade, G. (1980). A feature-integration theory of attention. Cognitive Psychology, 12, 97-136.

User Studies on Tactile Perception of Vibrating Alert

Jukka Linjama

Nokia Mobile Phones
Helsinki, Finland
jukka.linjama@nokia.com

Monika Puhakka & Topi Kaaresoja

Nokia Research Centre
Helsinki, Finland
monika.puhakka@nokia.com
topi.kaaresoja@nokia.com

Abstract

Vibration alert in mobile phones is used to alert the user for incoming call or message, in the situations when loud ringing tones are not acceptable, on not audible in noisy environments. Vibration alert can be enhanced and extended in many ways to be used as a tactile or haptic communication channel. There is not very much relevant information available in the literature of vibration perception in the mobile context. A summary of user studies of vibrotactile perception in different use contexts is presented. Aspects of using vibration patterns in the user interface are discussed.

1 Introduction

1.1 Tactility in mobile phones

Majority of current mobile phones have inherited one tactile feature from pagers -- the vibration alert. Vibration alert is used for an alternative way to alert the user for incoming call or message. It can be used for silent alert in the situations when loud ringing tones are not acceptable, on not audible in noisy environments.

Vibration alert can be enhanced and extended a lot. For example, many Japanese phones have more than one pattern for vibration alert. Panasonic has a service that the user can send email to a mobile phone and attach vibrations to the email (Matsushita, 2000). Vibrations are also used as "shakes" in mobile phone games, and different tactile patterns are progressively being used for indicating user interface feedback.

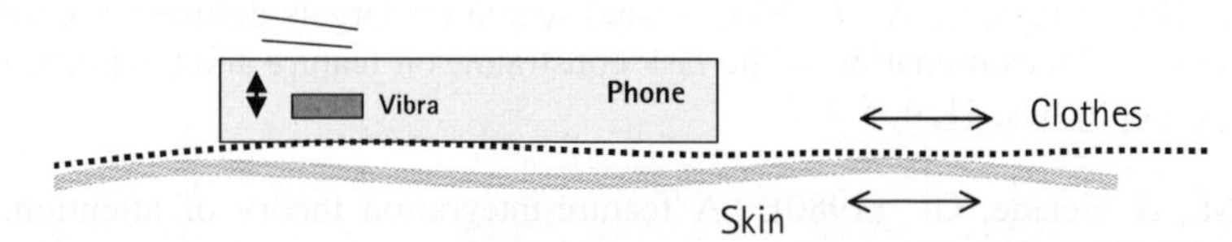

Figure 1. Transmission of vibration from phone to skin in a general case.

Most of information is presented by using auditory and visual stimuli. These sensory channels are often overloaded. This overload may be reduced if we can use as well other channels, like sense of touch, to process presented information (Isdale, 2001, Goldstein, 2002). One of the challenges for utilizing the tactile channel for conveying information in mobile devices is that the characteristics of the tactile contact to body may vary a lot (see Figure 1). In this paper, we focus to the user studies on tactile field, and topics that need to be studied in order to make the existing vibrating alert better and to extend its use towards tactile feedback or haptic applications.

2 Research topics

2.1 Vibration Perception and Physics

Most of the literature of tactile perception is for the hand. Figure 2 shows the detection threshold and sensation level data for continuous sinusoidal vibration (Verillo et al, 1969). It can be seen from Figure 2 that typical responses of mobile phones (also indicated in the figure) lie close to the 40 dB sensation level, i e vibration is clearly perceived in human hand. This is however not the case in a normal alert situation when the phone is in another location (pocket, belt clip…). Other parts of the body are generally less sensitive to vibration than human hand. However, these curves may serve as a guideline and reference for other use cases than handheld situation

Vibration strength of vibration sources can be characterized as a force or motion they produce. In a general case, both of these quantities must be given. Force is best used as blocked force, i e force against a hard and massive structure. Motion strength is best given as a free velocity of the system. Usually, phone is in a relatively loose contact to body, and its movement can be as a first approximation modelled as free motion of a rigid body (Linjama & Kärkkäinen, 2001). In practice, therefore, vibra is felt by movements of the phone, and it is best to use the motion response properties for characterizing the strength of the stimulus caused by the phone.

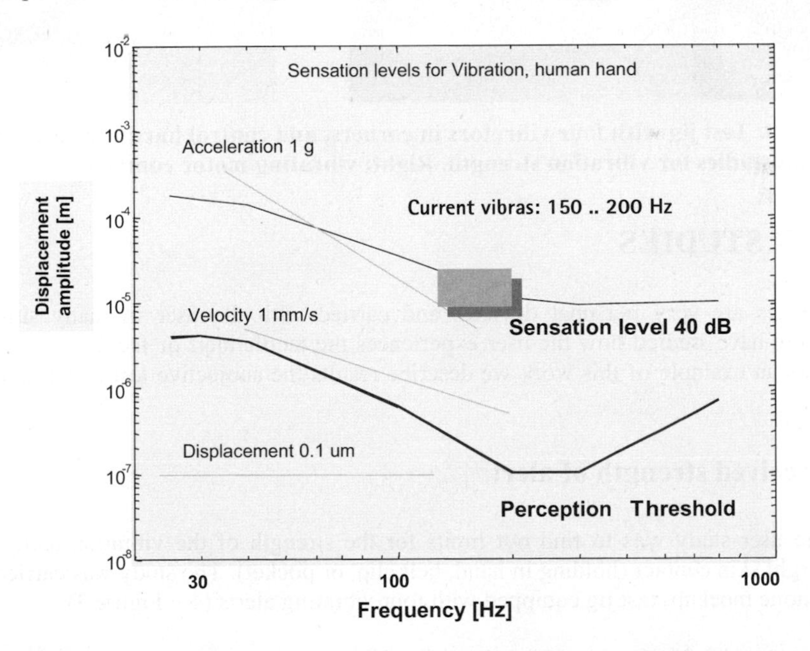

Figure 2. Vibration perception data from literature (Verillo et al 1969). Detection threshold as displacement amplitude for human hand. Also indicated 40 dB equal sensation level curve, and constant velocity (1 mm/s), and acceleration (1 m/s2) levels.

In relevant standards of human exposure to vibration, hand-arm transmitted case gives another guideline how to approach the effects of vibration (ISO, 1990). Suggested weighting of the acceleration in ISO 8041 is essentially band limiting to 8 Hz –1000 Hz and integration to velocity. This suggests that the velocity level would be the best measure for human exposure (and sensation) to vibration. At low frequencies (< 200 Hz) this seems indeed to be the case at moderate vibration amplitudes (sensation level 40 dB) in Figure 2.

2.2 Vibration Technology and Applications

The traditional vibrating component in phones is a tiny rotating motor having an eccentric weight on top of the shaft (see Figure 3). It is cheap and manufacturers have created the lines for mass markets. The feeling of the vibration produced by a rotating motor is very soft. Different solutions are available for applications where sharper tactile patterns might be needed. For instance, in future sound designers might take vibra as a musical part of the ringing tones.

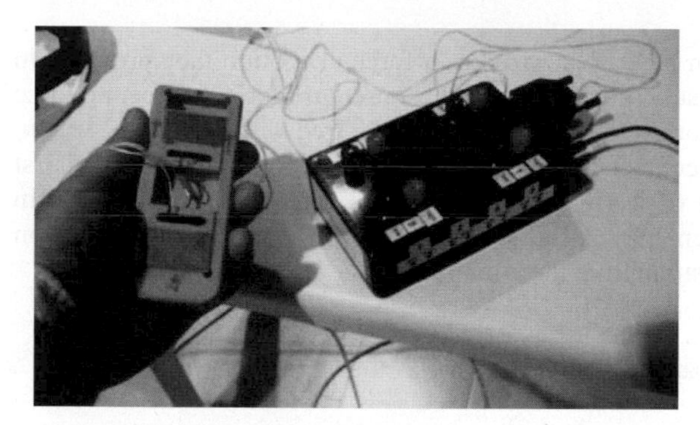

Figure 3. Left: Test jig with four vibrators in corners, and control hardware used in perceptional studies for vibration strength. Right: vibrating motor component.

3 USER STUDIES

Mobile phones are very personal devices, and carried with the user in many situations and locations. We have studied how the user experiences the tactile alert or feedback in this "mobile context". As an example of this work we describe results the subjective test of vibration strength perception.

3.1 Perceived strength of alert

Goal of the user study was to find out limits for the strength of the vibration alert, in case of relatively tight skin contact (holding in hand, belt clip, or pocket). The study was carried out using a mobile phone mockup, test jig equipped with four vibrating alerts (see Figure 3).

Test Setup. Amplitudes (counterweights) of the vibrating alerts were selected to cover relative strengths in scale 1, 2, 4, and 8 (steps of 6 dB). However, the vibration velocity caused by the vibrators is also related to frequency they are driven. By selecting two frequencies (Low - 125 Hz, High – 180 Hz) half-octave apart, the vibration velocity of the stimuli increase in steps of approximately 3 dB, covering the range of 3 to 35 mm/s (rms, in the position of the vibra). The hypothesis was that the stimuli will be ranked to be in a series of increasing strength as 1 low – 1 high – 2 low – 2 high etc. Total of 16 stimuli, and two false (null) stimuli were presented in random order the test persons (N = 20). Four randomized test sequences were used. Task was to rank the strength of sensation in the five-stage subjective rating scale:

0 - Not detectable	1 – Weak	2 – Moderate	3 - Strong	4 - Too strong/ irritating

Before each use case, three sample stimuli (1 high, 4 low, 8 low) were given to the test person as an orientation. Then the stimuli were presented manually in sequence, indicating the number of each stimulus before playing it. Masking noise was played through headphones, so that test persons did not hear the vibra stimuli.

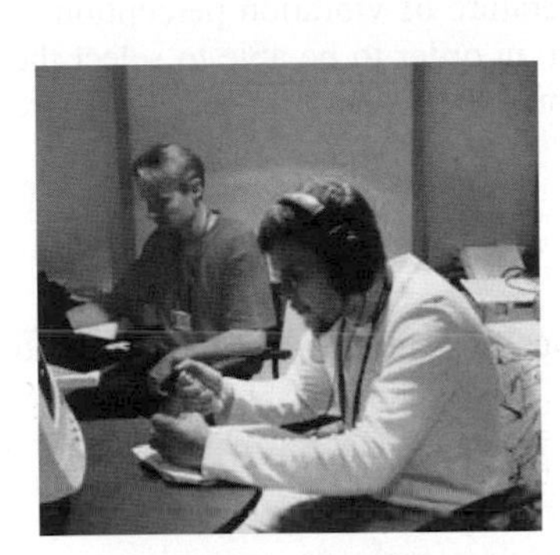

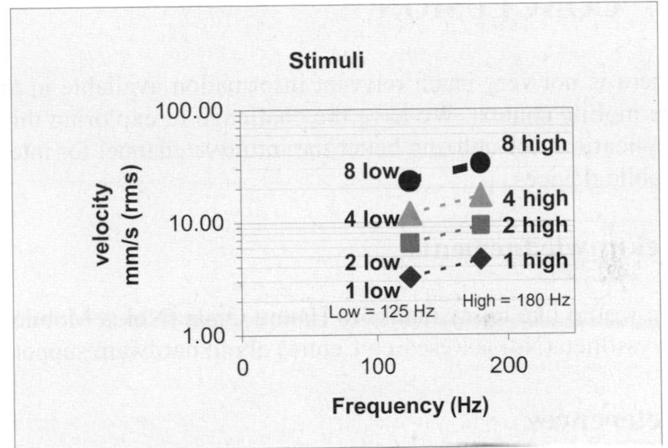

Figure 4. Left: Test case of handheld situation. Test subject on the right with headphones. Right: Stimuli used in strength perception test. Vibration velocity in the location of the vibrator in test jig

Figure 5. Results of subjective strength ratings.

Results and discussion. The results shown in Figure 5 verified the hypothesis of the order of perceived strength of the stimuli. Thus vibration velocity is the best quantity to measure in the mobile device use context, when indicating in one number the perceived strength of vibration.

Another conclusion was that vibration values exceeding approximately 10 mm/s (rms) in mobile phone surface are felt too strong or irritating. As lower intensity vibrations (only 12 dB below this) were often not detected at all in the test, the optimum value range for alert strength is relatively narrow. There was surprisingly little difference in sensations between holding positions of the phone. Obviously all cases tested represent a case of good contact for vibration transmission.

4 CONCLUSION

There is not very much relevant information available in the literature of vibration perception in the mobile context. We have the challenge in exploring this area, in order to be able to select the applications for utilizing better the intuitive channel for interaction – the sense of touch – in future mobile devices.

Acknowledgements

We would like to say thanks to Hannu Ojala (Nokia Mobile Phones), Juhani Tuovinen and Tuomo Nyyssönen (Nokia Research Centre) about hardware support.

References

"BeatMelody," Matsushita Communication Industrial Co, Ltd, 2000, http://www.mci.panasonic.co.jp/pcd/623p/beatmelody/index.html.

Goldstein, E. Bruce, (2002). "Sensation and Perception". Wadsworth Group, a division of Thomson Learning, USA,

Isdale, J. (2001) "Haptics," VRnews, vol. 10, pp. http://www.vrnews.com/issuearchive/vrn1003/vrn1003tech.html#links, 2001.

ISO 8041: 1990 (E). Human response to vibration – Measuring instrumentation. International organization for standardization

Linjama, J & Kärkkäinen, L. (2001). Modelling and characterization of vibrating alerts. Proceedings of Nordic Vibration Research 2001. Stockholm, Sweden, 11-12 May 2001. Scandinavian Vibration Society. 7 p.

Verillo, R T, Fraioli, A J, & Smith, R L. (1969). Sensation magnitude of vibrotactile stimuli. Perception and psychophysics, 6, 366 – 372.

Identifying and Refining the Tasks in a Cockpit Data Link Model

Lynne Martin[1], Savita Verma[1], Amit Jadhav[1], Venkat Raghvan[1], Sandy Lozito[2]

[1]SJSUF, CA, USA
lmartin@mail.arc.nasa.gov, saverma@mail.arc.nasa.gov,
amit_jadhav3@hotmail.com, venkat1044@ yahoo.com

[2]NASA ARC, CA, USA
Sandra.C.Lozito@nasa.gov

Abstract

A modeling effort was undertaken to replicate a mixed-modality voice and data link study by Dunbar, McGann, Mackintosh, and Lozito (2001). The human performance model Air MIDAS was used to represent the tasks associated with data link and voice messages sent by an Air Traffic Controller (ATC) to flight crews. The messages received by the agents were always in pairs – a data link followed by a voice message. Conditions placed on crews in the original study, such as media priorities and intervals between messages, were represented in the model. The model generated output that could be verified against data from the Dunbar et al. (2001) study. The overall correlation between conditions from the study and the modeling conditions was significant. Intra-condition correlations were generally high and positive, except in the voice condition.

1 Introduction

Controller-Pilot Data link Communication (CPDLC), a system for messaging between flight crews and ATC, has been investigated through numerous studies over the last two decades (see Kerns, 1990 for a review). Although it has been implemented on only a few oceanic routes and one terminal area in the U.S. (Fiorino, 2002), future airspace concepts often assume the widespread implementation of CPDLC.

As part of an effort to study the human factors implications of introducing CPDLC into the cockpit, a model of the data link communication process is being constructed using the Air MIDAS (Air Man-Machine Integrated Design and Analysis System) architecture. Once complete, the model will be used primarily as a complement to real-time data link studies, but could also be used as a component of wider-ranging fast-time simulations, such as testing future airspace concepts. The goal of this project is to develop a model with Air MIDAS that has sufficient complexity to enable the testing of activities and procedures. The model would also yield results similar to that of a human-in-the-loop study. However, a model with this level of specificity can only be developed through much iterative work. Thus, the intent of this initial effort was to demonstrate that a simple, working model of the ATC-flight crew communication process could be developed.

1.1 Air MIDAS Architecture

The MIDAS system has evolved over a period of ten years (for a detailed description see Corker

& Smith, 1992) and was originally used for military applications. Air MIDAS development has diverged from Core MIDAS for civilian applications. It has an agent-based architecture and represents both the physical world and the cognitive domain; the former in the form of vehicles, environment, and equipment, and the latter as a human operator model. A human performance profile is generated from the dynamic interplay of task demands; operator characteristics meeting those demands; the functions of the equipment (data link in the present study) with which the operator interacts; and, the operational environment, to create a time course of events.

The human operator performance model is a combination of functionally integrated micro-models of specific cognitive capabilities. The human operator model functions as a closed-loop control model with inputs coming from the world and actions being taken. Some of the human mental constructs represented are verbal and visual working memory, long term memory, and attention. These mental constructs focus, identify, and filter the agent's simulation world information for action and control. Incoming world information is represented by declarative memory structures forming context This is the basis of actions taken by the agent that, in turn, change the external world. The interaction between context and tasks represents human behavior.

2 Method: Scenario and Model Construction

The data link study by Dunbar, McGann, Mackintosh, and Lozito (2001) was selected as a benchmark study. The aim was to replicate their method and a portion of their experimental procedure. Some of the independent variables that Dunbar et al. (2001) included were not considered to simplify the model development. The variables selected highlighted one of the most consistent findings in data link research – that data link transactions take longer to complete than voice transactions.

The scenario was based on a flight from Chicago O'Hare to San Francisco and was taken from the study by Dunbar et al. (2001). All flight parameters, e.g., speed and altitude, were typical for the cruise phase of flight. In the modeling study, the crew, consisting of the pilot-flying and the pilot-not-flying (PNF), received ten messages from ATC. These were sent as five pairs, with the first message in the pair transmitted via data link and the second via voice. The interval between each message in the pair was varied between a 'long interval' of 60 seconds and a 'short interval' of five seconds. This gave a 2 x 2 experimental design – data link or voice by short or long intervals. A shorter interval between messages was assumed to create time pressure to complete the tasks associated with a first message before a second arrives. Assigning different priorities for task completion to the two media (voice and data link) also created pressure. Voice messages were given a higher priority than data link messages in the Dunbar et al. (2001) study and this difference was recreated in the data link model.

The model construction entailed two major tasks – defining data link as equipment within the Air MIDAS architecture and defining procedures for operator-data link interactions (see Figure 1). The functionality of the data link equipment required building new sub-models in Air MIDAS such as dynamic prioritization, and a review log. Only dynamic prioritization will be discussed in this paper. In earlier versions of the model, priorities were fixed for activities. The dynamic prioritization module made activities re-usable by changing their priorities based on message content. Thus, messages containing urgency words like 'expedite' or 'amend' were given higher priorities. It is assumed that crews try to respond more quickly to more urgent messages.

The model's focus was on the interaction between the flight crew and data link. The controller was simply represented as the message sender. The GOMS (Goals-Operator-Method-Selection, Kieras, 1997) task decomposition methodology was used to define the interactions between the agent and equipment. An extended form of GOMS was applied by assigning priorities and resource requirements (visual, auditory, cognitive, and psychomotor) to the tasks. Then, a detailed list of the data link and voice tasks to receive, respond to, and act upon was composed based on the work of McGann, Morrow, Rodvold, and Mackintosh (1998).

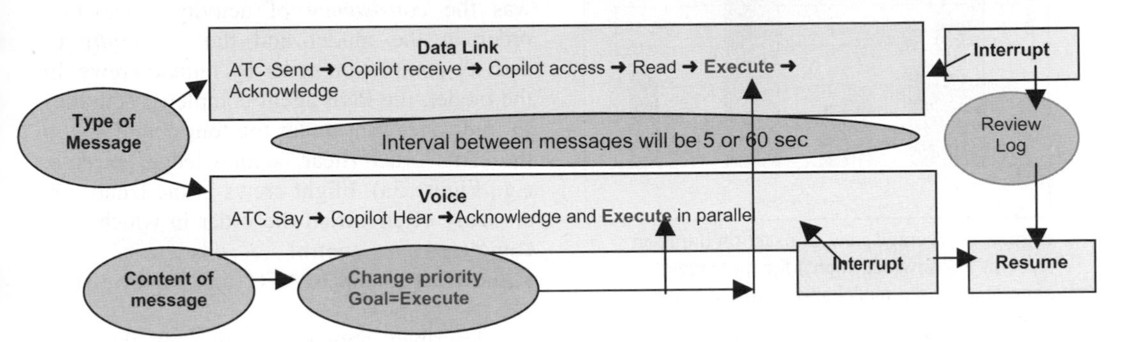

Figure 1: Model depicting data link equipment and flow of information in Air MIDAS

The final step of this work was to run a set of ten trials with the complete model and compare the model output with a data set extracted from the Dunbar et al. (2001) study, to verify whether the model represented the data link task realistically. Ten messages, in five pairs, were sent to the crew-agents during every run. Due to the way the model collected data, the last pair of messages was discarded, leaving four pairs (eight messages) to compare with the Dunbar et al. (2001) data.

3 Results and Discussion

Dependent variables of interest in these preliminary runs were the effects of message interval and task completion time under each medium. It was hypothesized that the effects of message medium would match the findings of Kerns (1990) – voice messages would be completed more quickly but that data link messages could be re-visited and completed after a delay. It was also hypothesized that shorter intervals between messages would overload the agents' resource capabilities, leading to task shedding. In this case, the task would not be completed.

Message completion times were measured by their 'transaction duration', which was defined by Dunbar et al. (2001) as the time between a message being received by the crew and its acknowledgement by them. Average transaction durations are presented in Figure 2 and compared to the Dunbar et al. (2001) results.

A random sample of the model runs was chosen to match the 40 message data set of the Dunbar et al. (2001) study. The transaction durations of these messages were compared. The correlation coefficient of the model transaction durations with those from the earlier study was 0.777 ($df=38$, $p<.01$), indicating that there are parallels between the two sets of data. Messages sent with a long interval had durations that correlated more highly ($r=0.835$, $df=18$, $p<.01$) than those sent with a short duration ($r=0.738$, $df=18$, $p<.01$). Comparing the correlation between messages sent using different media showed that while the data link messages had similar trends ($r=0.545$, $df=18$, $p<.05$), voice messages did not ($r=-0.166$, *ns*).

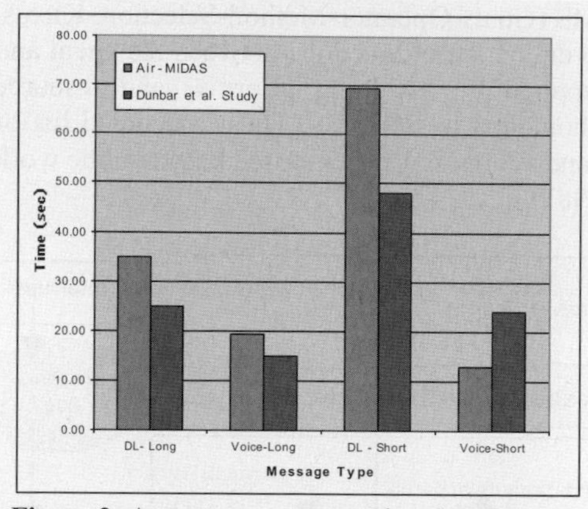

Figure 2: Average crew transaction duration (receipt-acknowledgement) for messages generated by the model and from the Dunbar et al. study (2001)

To investigate why the transaction durations differed for the voice condition between the model and the Dunbar et al. (2001) data, the order of the model's activities was compared to crews' in the original study. The most notable difference between the two sets of data was the *consistency* of activity completion order in the model and the *variability* of activity completion order by human crews. In the model, the PNF agent completed activities for messages sent under the four conditions in the same order (hear, acknowledge, execute, e.g., Figure 3a). Flight crews in the Dunbar et al. study often varied the order in which they completed control changes and acknowledgements to ATC (e.g., Figure 3b).

As described above, selection of the next activity in Air MIDAS was based on the availability of resources and the priorities given to the content and medium of the message. Due to the replicated experimental design, the message medium and its content were identical for both studies and so should not have been the reason for differences observed. However, resources may have differed for two reasons. First, and most likely, the resources allocated to model activities may not have matched those required by crews, making the task scheduling task more straightforward for the model agent. Second, the agents had a quota of resources, which the scheduler managed as efficiently as possible by parallel processing with remaining resources while the primary task was completed. Although Dunbar et al. (2001) did not record participants' workload, it is possible that crews were not as cognizant of scheduling their resources and focused more on completing activities.

No task shedding was observed in the ten model runs completed. The short, five-second interval between messages did not overload agents' resource capabilities to the point where the agent abandoned the task. However, when the PNF agent's activities were listed over time, it showed the agent switching between the two short-interval message tasks. The PNF agent focused all its attention on the incoming voice message (suspending the prior data link message) while receiving and reading back a voice message. The model's activity output shows the PNF-agent tried to

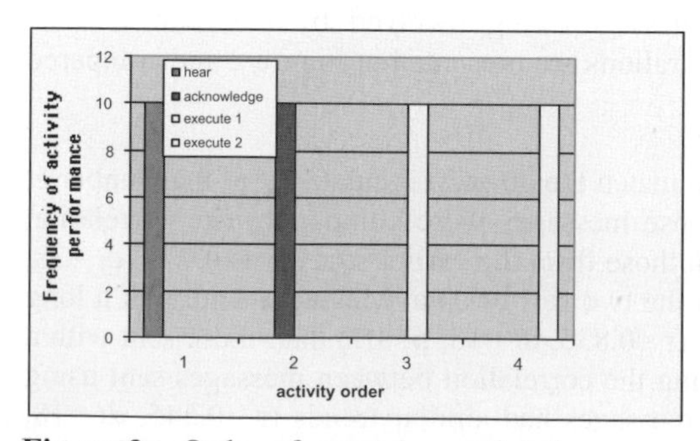

Figure 3a: Order of agents' completed activities for long voice messages

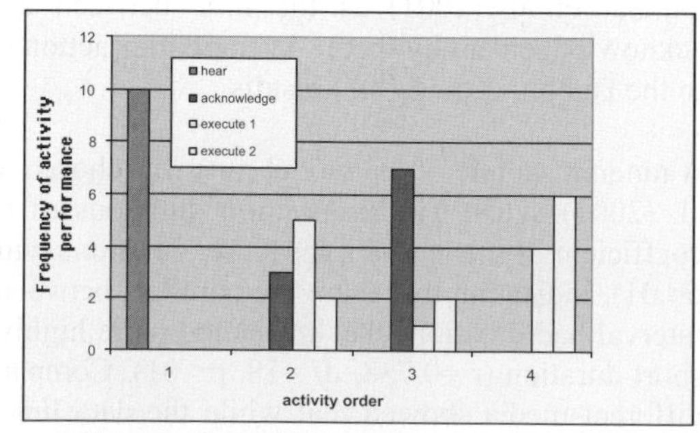

Figure 3b: Order of crews' completed activities for long voice messages

resume data link's 'read aloud' activity but did not succeed until the higher priority voice message was completely handled. The priority of the short-interval voice message over the data link message is also reflected in the shorter transaction duration shown in Figure 2 for short-voice messages.

Figure 2 also shows that the short-interval voice condition is the only instance where the model predicts a quicker message completion time than was observed in the Dunbar et al. (2001) data. As noted already, this may indicate that the model agents used a different strategy than participants. It is possible that in the Dunbar et al. (2001) study, flight crews were able to continue with other (data link) tasks while listening to incoming voice messages, but this may have increased their response time to the voice task.

4 Conclusions

Data gathered from a series of preliminary runs indicated, as expected, that the model needed further calibration and some modification to task parameters. Encouraging findings were that the modeled PNF agent did have to re-schedule tasks during short-interval message sequences, and that data link communication activities sent with a long interval took nearly twice as long to complete as did the voice messages in that pairing. This latter finding has been consistently noted in real time data link studies (Kerns, 1990).

5 References

Corker, K. & Smith, B. (1992). An architecture and model for cognitive engineering simulation analysis: Application to advanced aviation analysis. Presented at *AIAA Conference on Computing in Aerospace*, San Diego, CA.

Dunbar, M., McGann, A., Mackintosh, M., & Lozito, S. (2001). *Re-examination of Mixed Media Communication: The Impact of Voice, Data Link, and Mixed Air Traffic Control Environments on the Flight Deck*, NASA/TM-2001-210919. Moffett Field, CA: NASA Ames Research Center.

Fiorino, F. (2002). Controller-pilot data link goes live in Miami. *Aviation Week & Space Technology*, Article 20021014, October 14 issue, McGraw-Hill.

Kerns, K. (1990). *Data Link Communication Between Controllers and Pilots: A Review and Synthesis of the Simulation Literature*, Report MP-90W00027, Project 1765E. McLean, VA: The MITRE Corporation.

Keiras, D. (1997). A guide to GOMS model usability evaluation using NGOMSL. In M. Helander, T. Landauer, & P. Prabhu (Eds.) *Handbook of Human-Computer Interaction (2nd ed)*. Amsterdam: Elsevier.

McGann, A., Morrow, D., Rodvold, M., & Mackintosh, M. (1998). Mixed-media communication on the flight deck: A comparison of voice, data link, and mixed ATC environments. *The International Journal of Aviation Psychology, 8 (2),* 137-156.

6 Acknowledgments

The authors would like to thank Dr. Kevin Corker, of San José State University, for his support. This work was funded by the National Aeronautics and Space Administration Human Automation Integration Research program, under a cooperative agreement with SJSU (NCC 2-1095).

Use of a Computer Simulation Model to Predict Survival from Metastatic Cancer

Lewis Mehl-Madrona

Program in Integrative Medicine (College of Medicine), and
Center for Frontier Medicine in Biofield Science (Dept. of Psychology)
University of Arizona
mehlmadrona@aol.com

1.0 Introduction

Oncologists commonly predict how long cancer patients will survive. Genetic factors contribute to the pathogenesis of cancer (Burt and Samowitz, 1988; Bussey, 1978), but are not sufficient to explain its onset or its prognosis by themselves. Previously Mehl (1991) used a systems dynamics, mathematical computer simulation modeling approach [Meadows, 1975; Randers, 1981; Bronson & Jaconson, 1989) with the STELLA software package (High Performance Group, 2002) to predict the course of colon cancer. In this paper, the use of a similar approach, building upon the pre-existing model, to predict survival with several types of metastatic cancer is described. Systems dynamics emphasizes the effects of feedback loops in explaining dynamic behavior. Factors which were previously considered in developing the colon model included:

a) Natural history: meaning the most common course of cancer development once cells are present
b) The rate at which normal cells transition to cancer: The seemingly more rapid development of cancerous cells in susceptible populations is a function of the greater pool of abnormal cells present, kept in check by the immune system, whose positive benefits are eventually overcome by the growing numbers of abnormal cells. One genetic alteration (rate of mutation) was sufficient to explain all findings.
c) Prostaglandin and other promoters: Evidence is strong that promotional agents catalyze the conversion to cancerous cells. Prostaglandin is a likely candidate for one major promoter. All promoters under the label "Prostaglandins," referring to all arachidonic acid metabolites. Human malignant cells in organ culture produce substantially more prostaglandin E (PGE) than do normal cells from the same specimen (Bennett, et al, 1977; Jaffe, et al, 1971, 1973). Prostaglandins influence tumor invasion, metastasis, angiogenesis, and osteolysis (Karim, et al, 1976; Levine, 1982; Form, et al, 1982; Rolland, et al, 1980; Form & Auerbach, 1982 Honn, et al, 1981; Bennett, et al, 1975, 1977). Prostaglandins exert a growth enhancing influence on neoplastic tissue *in vivo* (Trosko, et al, 1985) Growth factor (which is also a tumor promoter) stimulates prostaglandin synthesis (Levine, et al, 1982b). Steroids almost completely counteract tumor promotion (Slaga, et al, 1978; Schwartz, 1977) by inhibiting the production of arachidonic acid metabolites through action on phospholipase A2 (Greglewski, 1979; Hong & Levine, 1976). Inhibitors of phospholipase A2 (dibromoacetophenone and other agents) are potent inhibitors of tumor production [46, 49, 50].
d) Effect of non-steroidal anti-inflammatory agents*:* Some non-steroidal anti-inflamatory agents (NSAID's) inhibit cancer growth while the patient or the experimental animal takes the drug.
e) General size and kinetic constraints*:* Colon cancer statistics were used to calculate growth rate for other cancers. The usual nuclear aberation rate is about 0.2 nuclei per crypt. Rat crypts contain about 120 cells per crypt. The colon of rats contains 250,000,000 epithelial cells. A 70 kg man would contain roughly 200 times more cells. 20% of colon epithelial cells are

proliferating at any given time. A Gardner's syndrome patient, by adolescence may have as many as 1000 to 3000 polyps. Most of these are small (10 to 100 mm range). Some may be as large as 4 cm in radius. Assume a patient has 2000 polyps in the following breakdown: 20 large polyps, 300 medium polyps and1700 small polyps. A cell is 1000 cubic microns. A large polyp of 2 cm radius has $4/3\pi r^3$ volume. This gives a content of 35 x 10^6 cells. If we assume a medium polyp has 200 mm radius, it has 10-fold fewer cells. A small polyp of 10 mm radius has 20-fold fewer cells. Then, 20 large polyps will give 700 x 10^6 cells, 300 medium polyps will give 1050 x 10^6cells, and 1700 small polyps will have 300 x 10^6. This gives a total of 2 x 10^9 polyp cells. We will work in the model with a 100th segment of colon which means we will predict 2,000,000 polyp cells by adolescence, on an average, from a population of 500,000,000 normal cells. We assume an infant is born with 50,000,000 normal cells in this segment of colon and no abnormal cells.

2.0 Influence of the mind on cancer

Work with animal models has clearly shown a relationship between psychological determinants and tumor growth (La Barba, 1970: Sklar and Anisman, 1981: Visintainer et all, 1983). Research on the relationship of human personality to cancer has not yielded clear correlations (reviewed by Garssen and Goodkin, 1999). Prospective studies have implicated a number of possible risk factors, the most consistent finding being repression of emotion or depression (Shekelle et al, 1981; Daftore et al,, 1980: Hislop et al. 1987: Jensen, 1987: Kaasa et al.: 1989 Temoshok: 1987; Gross, 1989), a conclusion disputed by Kreitior etc. al (1993) in their more recent review. Other, larger studies have failed to show a connection between cancer and depression (Kaplan and Reynolds, 1988: Hahn and Petifti, 1988; Zonderman et al., 1989). Factors which some have found to be protective include social support (Eiii et al., 1992; Waxier-Morrison et al., 1991; Maunsell et al., 1995) disputed by Funch et al. 1983 Cassileth ct al,. 1988), greater expression of distress (Derogatis et al 1979), smaller numbers of severe or difficult life events (Ramierez 1989), and "fighting spirit" (Pettingale et al, 1985). Yet other investigators failed to find similar relationships (Cassileth et al., 1988. Jamison et al., 1987), and fighting spirit, measured psychometrically, was not protective in a recent analysis of 578 patients done by Watson et al., (1999). Evidence from controlled trials of interventions is mixed. At the time of writing, there are 9 published trials with randomized, case control or sequential cohort designs. Four showed a positive effect of an intervention on survival. In the best-known of these, Spiegel et al., (1 989) demonstrated an 18 month average prolongation of life in women with metastatic breast cancer who attended a weekly support group for up to 1 year. However, the control group in this experiment may have been anomalous, since as pointed out by Fox (1998), its members died more rapidly than similar populations of patients who were not in therapy. Richardson et al., (1990), using a sequential cohort design, found significant effects of a psychological intervention in patients with hematological malignancies. Fawzy et al. (I 993), in a randomized trial, similarly demonstrated a significant survival advantage to patients with malignant melanoma who had taken a brief, group, psycho-educational course 6 years before. Ratcliffe et al., (1995) found a small but marginally significant difference in survival of patients with lymphoma who received training in relaxation compared with randomized controls who did not. All trials designed to test effects on survival have, as indicated earlier, given negative results. The first of these, by Linn et al (1982), was a randomized comparison of the lifespan of patients with a variety of late-stage cancers who either received or did not receive individual counseling (for an unspecified time). The second, by Morganstern et al. (1984), was a case controlled study, which showed a non-significant trend in favor of longer survival in patients in group therapy. Two more recent studies were RCT's using interventions similar to those of Spiegel and of Fawzy, respectively. Cunningham et al., (1988) gave 35 sessions of group supportive and psycho-educational therapy to women with metastatic breast cancer: these subjects did not live significantly longer than controls. Edelman et al. (1 999) used a brief, 11 session, cognitive behavioral intervention which also failed to prolong the life of patents with metastatic breast cancer, by comparison with controls. One further published study, not designed to test for survival and with a weaker design, has also given negative results in a post hoc analysis (Ilnyckyl et al., 1994). One area of study that sheds some light on a possible connection between efforts at mental self help and cancer remission or prolonged survival is the examination of "remarkable survivors' (Ikemi et al, 1975; Kennedy et al, 1976; Achterberg et al, 1977; Roud, 1986; Pennington, 1988; Huebscher, 1922a and b, Berland, 1995). The picture emerging from these studies is consistent with findings from Cunningham's pilot work; patients who lived longer than expected were flexible, self-motivated, and usually reported significant changes in behaviors and attitudes. However, in all these accounts, patients were selected retrospectively, so that there was no way of knowing how many patients showed similar characteristics but failed to survive. All have other serious methodological defects as well: medical documentation was almost nonexistent, and in some cases subjects who had medically curable cancers were included? Most collected data was from a single interview only, usually years after recovery; and only 2 used standard qualitative methods.

3.0 Methods

An existing computer model (Mehl-Madrona, 2002) was modified to include more than just colon cancer and the factors found by Cunningham, et al. (2000) to predict survival with metastatic cancer. A series of feedback loops were conceptualized as diagrammed in Figure 1. The goal of the modeling was to become parameter insensitive, showing that the existence of feedback loops would render the usual statistical task of parameter estimation unnecessary, thereby making the model more immediately and readily generalizable to any patient with metastatic cancer. Equations are available from the author.

Figure 1

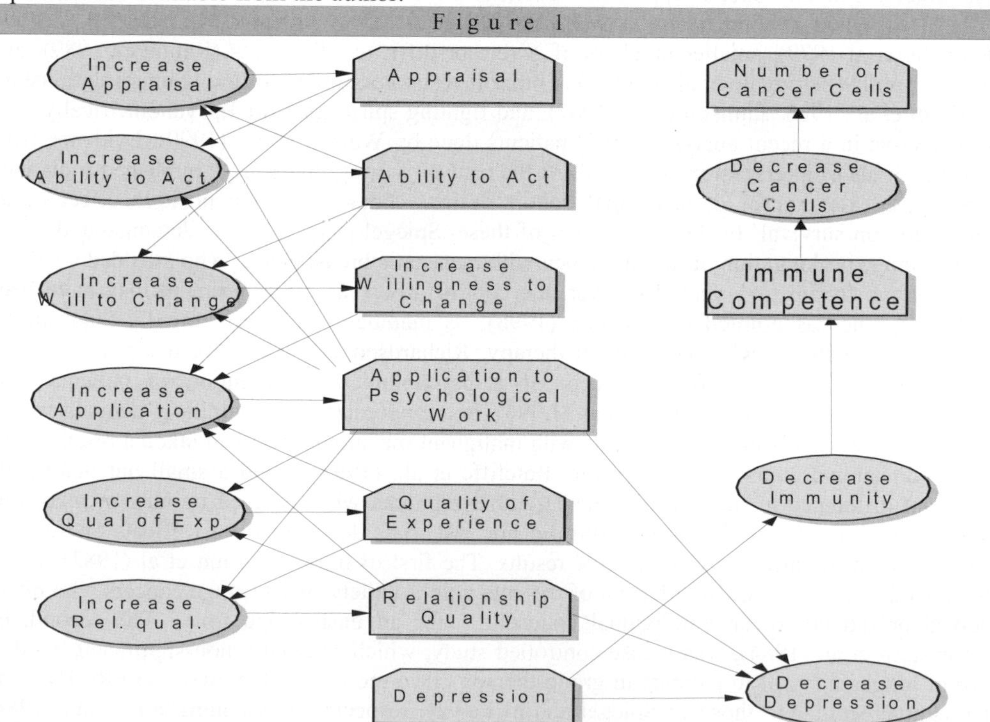

Data from Cunningham, et al, (2000) for each of these 6 dimensions were entered into the resulting computer model. Predictions were made for length of survival and were compared to

Cunningham, et al's patients' actual survival. The strength of the association was measured by pair-wise correlational coefficients between predicted and actual survival, with living patients being adjusted so that the predicted survival matching the current actual survival (4 patients).

4.0 Results

For simplicity, Figure 1 eliminates the influences which serve to diminish the value of the variables. Each variable at the center of the diagram represents a differential equation with incremental increases each dt from those variables which serve to increase its value and incremental decrements each dt from those variables which serve to decrease its value. Appraisal refers to the degree to which patients can accurate appraise the gravity of their situation. Patients who are able to act are more likely to make realistic appraisals. Patients who are in denial are less likely to make accurate appraisals. Application to psychological work is thought to make people more likely to make accurate appraisals (less denial).

In turn, making an accurate appraisal is likely to lead to increased ability to act given. It will also increase the person's willingness to change. Similarly, having the ability to act should increase one's willingness to change and should increase one's application to psychological work. Increased application to psychological work should increase quality of life and quality of relationships. Increased quality of experience should decrease depression and increase relationship quality. Increased relationship quality should also decrease depression. Decreasing depression should improve immunocompetence, which should lengthen survival through slowing the rate of cancer disease progression. Similarly, though not diagrammed decreased depression should feedback to the top of the diagram to increase accuracy of appraisal, willingness to change, ability to act, and application to psychological work. This finishes the web of interconnections that are so typical of systems dynamics models.

Figure 2 is reproduced from Cunningham, et al. (2000) and shows the survival of the patients whose actual survival for the patients that we are trying to predict. The group separates into short-term and long-term survivors. Data were taken from Cunningham, et al. (2000) for scores on the 6 dimensions developed in their qualitative analysis. Standard biological assumptions were made for patients to match the expected cancer progression rates seen for metastatic colon cancer, the condition for which the computer model was originally developed. Further and more accurate medical data is expected to increase the accuracy of predictions made form the psychological data.

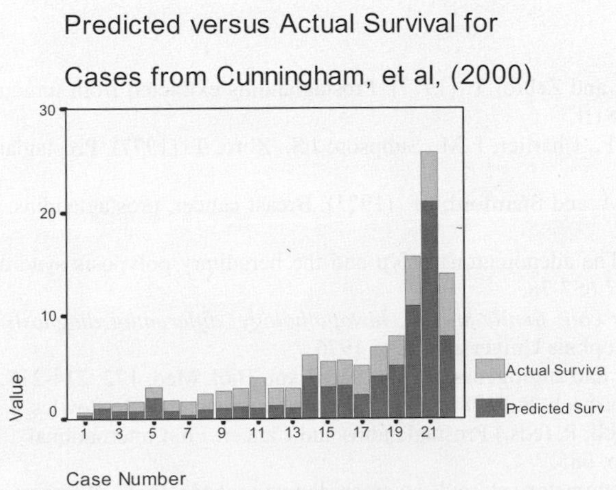

Predicted versus Actual Survival for Cases from Cunningham, et al, (2000)

Figure 2: Predicted versus actual survival for cases (Cunningham et al ,2000)

Figure 2 shows the results of comparing predicted survival to actual survival. The first 11 case numbers were the lower half in survivorship, while the second 11 case numbers come from the upper half of survivors. Only Cases 11 and 14 were reversed representing two incorrect predictions out of 22 cases, for accuracy greater than 90%.Table 1 show that the paired samples correlations between predicted and actual survival was 0.881, which was significant at $p < 0.0009$. The statistical test for paired differences in means was appropriately non-significant.

Table1: Paired Samples Statistics

		Mean	N	Std. Deviation	Std. Error Mean
Pair 1	Predicted Survival	2.1445	22	1.67906	.35798
	Actual Survival	2.1886	22	1.45643	.31051
Pair 2	Predicted Survival	2.1445	22	1.67906	.35798
	Actual Survival	2.1886	22	1.45643	.31051

Table 2: Paired Samples Correlations

		N	Correlation	Sig.
Pair 1	Predicted Survival & Actual Survival	22	.881	.000
Pair 2	Predicted Survival & Actual Survival	22	.881	.000

Table 3: Paired Samples Test

		Paired Differences			95% Confidence Interval of the Difference		t	df	Sig(2tailed)
		Mean	Std. Deviation	Std. Error Mean	Lower	Upper			
Pair 1	Predicted Survival - Actual Survival	-.0441	.79370	.16922	-.3960	.3078	-.261	21	.797

5.0 Discussions

Computer modeling has been growing as a method for understanding data and making predictions based upon that data for the solution of real world problems. Mehl [77] has discussed the development of this approach in detail including the criticisms of the standard statistical approaches which have been applied to the prediction of premature labor without success. Holder and Blose [79] have written about the use of computer modeling in alcohol research. Intriligator [78] has described the problems with conventional statistical analysis of data as has been done in the past for predicting birth complications. The value of this exercise in prediction is to show that the use of feedback loops allows accurate predictions without any parameter estimation. Common statistical models require intricate calculations of parameters for accurate predictions. These parameters are not robust in that they are not often useful for other populations. The systems dynamics model approach can be instantly applied to other populations, for there are no parameters to estimate. The strength of the prediction comes from the accuracy of understanding of existent feedback loops. Future research will test this model on other populations without any further parameter estimation.

6.0 References

Bennett, A., del Tacca, M., Stamford, I.F., and Zebro, T. (1977). Prostaglandins extracted from tumours of human large bowel. *Br. J. Cancer*; 35:881ff.

Bennett, A., McDonald, A.M., Stamford, I.F., Charlier, E.M., Simpson, J.S., Zbro, T. (1977). Prostaglandins and breast cancer. *Lancet* 1: 624-626.

Bennett, A., Simpson, J.S., McDonald, A.M. and Stamford, I.F. (1975). Breast cancer, prostaglandins, bone metastasis. *Lancet* 2: 1218-1220.

Burt, R. W. and Samowitz, W.S. (1988). The adenomatous polyp and the hereditary polyposis syndromes. *Gastroenterol. Clinics North America* 17:657-78.

Bussey, H.J.R., (1975). *Familial polyposis coli: family studies, histopathology, differential diagnosis, and results of treatment.* Baltimore: Johns Hopkins University Press, 1975.

Form,D.M. and Auerbach, R. (1983). PGE2 and angiogenesis. Proc. Soc. Expt. Biol. Med. 172: 214-218.

Form, D,M., Sidky, Y.A., Kubai, L. and Auerbach, R. (1982). PGE2-induced angiogenesis. In Powles, T.J., Bockman, R.S., Honn, K.V., and Ramwell, P. (eds.) Prostaglandins and Cancer: First International Conference, Alan R. Liss, Inc., New York, p. 685.

Gryglewski, R.J. (1979). Effects of anti-inflammatory steroids on arachidonate cascade. In: G. Weissman, et al. (eds.), *Adv. Inflammation Res., Vol. 1*, Raven Press, New York, p.505.

Remainder of References available from the author.

A SPN-Agents based model for Functional Modeling of Brain Regions Interaction

S.Mertoguno, D. Kavraki* and N. Bourbakis*

Wright State University –Information Technology Research Institute, Dayton OH and*AIIS Inc. Centreville OH

Abstract

This paper presents a first stage discussion for the stochastic Petri net graphs (SPNG) based agent modeling of the synergistic operation of the brain regions. By regions of the human brain we mean the ones that control the operation of the human body's parts. Each brain region is considered as an intelligent agent that is connected with the other regions and exchange information for the operation of the brain (making decisions, controlling motion, coordination, operation, etc of parts of the human body, etc.) In addition, each agent is a SPN structure that operates as an autonomous intelligent network itself.

1. Introduction

The study of the human brain has started centuries ago due to human curiosity and fascination around the capabilities and the secrets hidden inside it. In the recent times scientists have developed several methodologies for representing its functional behavior mainly based on observations. Some of these methodologies are stochastic models, graphs, neural nets, automata, etc. [Tsoukalas, Holland]. One of the methodologies for studying complex scientific environments with autonomous functional behavior is the interactive autonomous distributed agents (IDAA) [S.Mertoguno].

A distributed autonomous agents model is particularly good for problem solving problems in complex large scale systems, such planning, scheduling, abstracting, matching, etc.

Moreover, ubiquity, parallelism, associations, mining are involved in many ways on the decision making process for every complex large scale sophisticated system.

A multi-agent system is an efficient way of developing an adaptive brain model. Agents communicate with each other, to match patterns, adapt themselves to new patterns, fire when recognize a pattern, etc.

Another efficient and complementary method for the development of the brain model is the stochastic Petri-net graph (SPNG) [Gattiker, Ligomenides, Silva]. An SPNG is particular flexible in the way of modeling the functionality of either a single brain neuron and/or a brain region that performs certain tasks.

2. The SPNG Model

The SPN graph model is an efficient tool for modeling functional behavior of systems.

Definition: The structure of the SPNG model is based on the stochastic Petri-net graphs. Thus, the formal definition is a 11-tuple:

SPNG = {P,T,I,O,M,L,X,Q,R,F}

where,

> P = {P1, P2, … , Pn} is a finite set of places. Each place Pj represents a particular state of an element Ei, i,jєZ;
> T ={T1, T2, … ,Tm} is a finite set of transitions. A transition Ti represents an action performed on a set of elements at the state S(k,t);
> Ii C[PxT] is the input function;
> Oj C[PxT] is the output function;
> Mi ={mi1,mi2, …, min}, ijєZ, is the vector of markings (token) mij, which represent the status of the places, (for i=0, m0j and j=0,1,2, …, n, denotes the number of tokens in place Pj in the initial marking M0).
> Q = {t1,t2, …, tm} is the vector of time values related with the time required by an action to be performed;
> R = {r1,r2, …, rk} is the set of relationships among elements;
> F = {f1,f2, …, fn} is the set of properties of the elements;
> L = {a1, a2, …, ak} is the set of possibly marking dependent firing rates associated with the transitions T;
> X = {x1, x2, …, xm} is the set of delays associated with the transitions T.

2. 1. The SPNG Neuron Model

A neuron Ni of a brain is represented by the SPNG as an element Ei . Thus, the basic states that a neuron element has are:

P = {Pi = input, Pl = learning, Pm = matching, Pf = firing, Pa = malfunctioning, Pn = new matching, Pd =dead }

The SPNG model of a neuron is graphically given below, fig. 1:

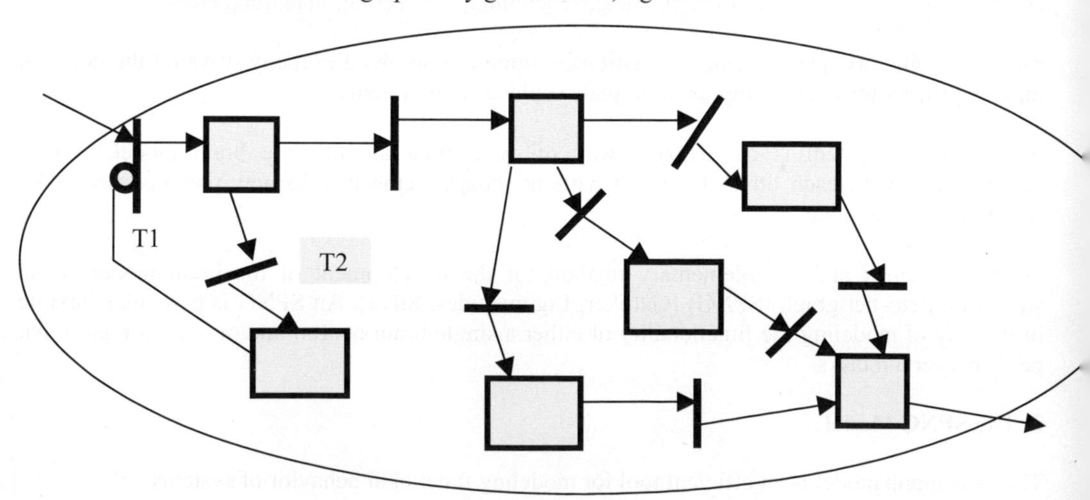

Figure 1. The SPNG neuro model

The brief overall functional behavior of a neuron Ni is as follows:
Ni receives an input from another Nj, if the neuron Ni is dead no input goes through. If it is alive the it moves the input into the learning-training status; Now is this input is a recognizable pattern for this particular neuron, then Ni fires, else it attempts to learn the new pattern. When this new pattern appears again it recognizes it and fires. If however Ni learns a wrong pattern then fires and its behavior becomes abnormal.

3. The SPNG Agent Model

The modeling of the brain based regions that perform certain tasks, such as making decisions, controlling motion, coordination, operation, etc of parts of the human body, etc. The SPNG modeling of the brain is based on the assumption that the brain is partitioned into different regions as shown below, fig. 2. In addition, we have assumed that a region of a certain color consists of neurons wit the same of similar functional behavior. Thus, each color region performs a task, and we represent it as an ellipsis with same color. Now, we can consider that each region (or at the lowest level each neuron) is an autonomous intelligent agent with the capability of learning, performing certain tasks, healing itself, cooperating with other neurons from different regions, etc. Thus, the SPNG agent based modeling is graphically shown in figure 3, where the SPNG crosses the regions by exchanging and processing information. An important problem here is the association of brain regions via the SPNG modeling for the execution of a certain task.

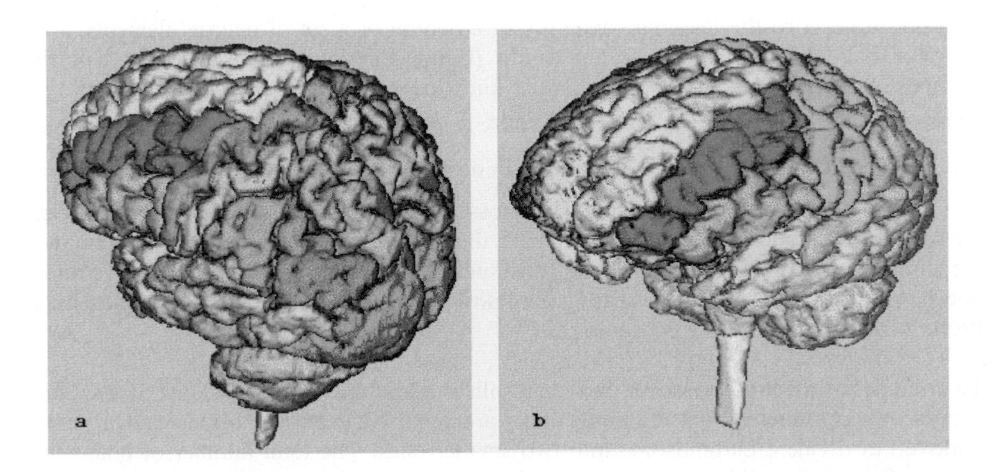

Figure 2: Different color regions for different tasks

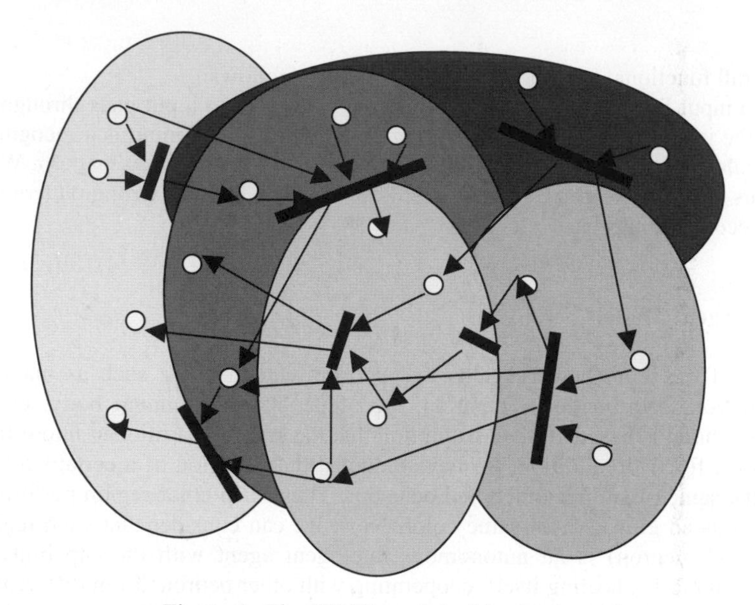

Figure 3: The SPNG model of the brain regions

4. Learning

In order to describe the learning capability of a neuron or a set of neuron elements, a crossover operator is adopted from genetic algorithms [Chung-Reynolds]. This operator will allow two agents exchange information regarding their knowledge graph. From genetic point of view, each gene consists of three nodes, namely N-node, L-node and N-node ($N_i - L_j - N_k$, where N_i and N_k N-nodes and L_j represent Lnode). In order to make learning consistent after crossover, a non-standard point crossover is used. This crossover is defined as one genetic crossover. Due to the length of chromosome could be different, thus the crossover pair may not necessary to be at the same location in chromosome. The constraints of the point crossover are : (1) Prefer at least one N-node are the same in the mating chromosomes. (2) Prefer diversity chromosome pair. (3) Prefer high performance chromosome pair.

To counter the effect of crossover growth, a split is introduced into this model. If a split operation occurs, several dummy links (dummy links are links between an N-node and an L-node) are removed to divide a chromosome into two smaller ones. The total number of nodes in the two agents after a split will be less or equal to the original node before split, since there may be some isolated nodes or small cluster of nodes that were produced when a split occurs. These isolated nodes and small cluster of nodes will be deleted. Let's assume the probability of the occurrence of a split is P_s, and the average number of deleted nodes (isolated nodes & small nodes cluster) is ε, then the growth due to split is $-P_s\varepsilon$. Then the expected individual growth of a graph is $P_c(1+P_2)-P_s\varepsilon$

Assuming that the population before one step evolution P_0 and the population after the evolution step P_1. Similarly, $\overline{\mathrm{N}}_0$ is the average size (in number of nodes) of an individual graph before evolution step, while $\overline{N}_1$ the average size after evolution step. The relation of the above parameters can be stated as: $P_1\overline{N}_1 = P_0\overline{N}_0 + P_0\{P_c(1+P_2) - P_s\varepsilon\}$.

From the fundamental theory point of view, high performance chromosome is forming according to the high performance gene segments (schemata). Due to the dynamic nature of interactions, those schemata could emerge on the temporal learning elements.

Considered the largest ratio of schemata H (S(H)) in a graph (G) is S(H)/G and the survival probability of schemata is P_s. Probability of crossover is P_c. Thus $P_s \geq 1 - P_c \times \frac{S(H)}{G}$. Assume that system average performance is $\overline{f}$. $f(H)$ is the performance of schemata H, where $\frac{f(H)}{\overline{f}} \geq 1$. Assume the mutation disrupt ratio is $o(H)$ and probability of mutation is P_m. The number of schemata m then is increasing to $m(H,t+1) \geq m(H,t) \times \frac{f(H)}{\overline{f}} \times \left[1 - P_c \times \frac{S(H)}{G} - o(H)P_m\right]$.

5. Conclusions.

The SPNG model was described for the functional behavior of a neuron or a region of neurons that share the same or similar task responsibilities. The SPNG model has certain characteristics, such as timing, synchronization, parallelism of events that take place in a brain. The major future goal of this project is to study functional MRI and PET images under certain functional behavior, such as solving a problem, performing simple arithmetic operations, etc, in order to compare functional behavior between SPNG model and real images in the brain [Bourbakis].

REFERENCES

Tsoukalas, L.H., Uhrig, R.E., Fuzzy and Neural Approaches in Engineering, Wiley, New York, 1997.

J.S. Mertoguno and N.G. Bourbakis, "Kydon Vision System: the Adaptive Learning Model" Int. Journal on AIT, vol.4,4,1995, pp.453-470.

J. H. Holland, "Hidden Order, How Adaptation Builds Complexity", Addison-Wesley Publishing Company, 1995.

J. Gattiker and N. Bourbakis, "Functional and Structural Representation of Knowledge Using SPNG", Proceedings Int. Conf. SEKE, MD, 1995.

C.J.Chung and R.Reynolds, "An evolution based tool for real-valued function optimization using cultural algorithms", IJAIT, vol.7, 3, pp. 239-291, 1998

N.Bourbakis, et. al. Bioimaging : Methods and Applications, Kluwer Academic Publisher, 2003

P.Ligomedides, Brian modeling using Petri-nets, IEEE Trans. On SMC, 1987.

M.Siva, L.Recalde, Petri-net and integrality relaxations: a view of continous PN models, IEEE T-SMC, 32, 4, 2002.

Multimodal Interface for Remote Vehicles Command and Control

Shruti Narakesari, S. Narayanan, & Jennie Gallimore
Department of Biomedical, Industrial & Human Factors Engineering
Wright State University
Dayton, OH 45435
snarakes @cs.wright.edu, snarayan@cs.wright.edu & jgalli@cs.wright.edu

Mark Draper
US Air Force Research Laboratory
Crew System Interface Division
WPAFB, OH 45433
Mark.draper@wpafb.af.mil

1 Abstract

Command and Control of Remotely Operated Vehicles is a complex operation, requiring the user to perform a multitude of tasks in a time constrained environment. Conventional mode of communication that uses a single input device for all interactions may sometimes be a hindrance in effective control. One possible solution is to design Multimodal Systems, where a particular action/interaction is a combination of inputs from various sensory units such as voice, body gestures, eye movement and facial expressions. Multimodal systems are gaining increasing popularity as they promise a transparent and more natural means for human-machine interaction.

In this paper we discuss the architecture, and implementation of a Multimodal system for the Command and Control of Remotely Operated Vehicles. The objective of this research is to examine different combinations of input modalities, including voice, keyboard and mouse for an effective communication in time critical situations.

2 Introduction

Multimodal interfaces are systems with capabilities of a transparent and more natural means of communication with the computer. Interaction with the computer can range from a simple combination of keyboard and voice inputs to more complex systems, which incorporate recognition of facial expressions, gestures and handwriting. Advances in the field of new input technologies, algorithms, hardware, speed, distributed computing and spoken language in particular has supported the emergence of these new classes of Multimodal systems. One of the prime application areas of Multimodal interfaces is in the field of Supervisory Control systems.

Supervisory Control Systems (SCS) incorporate intelligent agents that perform most of the lower level functions autonomously while the human operator makes high-level decisions and is responsible for the efficient operation of the system (Narayanan, et al. 2000). Some of the typical application areas that use supervisory control systems include space exploratory missions, undersea oil, nuclear industry tele-surgery, war and military operations and in the field of Remotely Operated Vehicles (ROV). The Remotely Operated Vehicle is capable of performing complex operations and the remote human operator controlling the vehicle is responsible for performing supervisory control tasks such as monitoring, planning, decision making coordinating and troubleshooting (Narayanan, et al. 2000). The communication is via an interface that sends and receives signals to and from the vehicle. The general modes of interaction in the ROV domain have been keyboard, mouse and voice. Research is being conducted to incorporate new interaction techniques such as vision, gesture, handwriting etc into Supervisory Control Systems, however there are drawbacks with each of these individual modes of interaction. An effective solution would be to combine these modes in way that the strengths of each mode can overcome the weakness of the other, giving rise to the concept of Multimodal systems. Further, humans, being multimodal organisms, are capable of communicating in more than one modality at the same time. Thus, incorporating Multimodal interfaces would be a step towards realizing a more natural way of human –machine communication.

2.1 Related research

Uni-modal (where the human controller uses a single modality such as keyboard/mouse or speech alone, to control the system) type of communication inputs in complex systems seem insufficient to interact across all tasks and environments. Further, uni-modal communication does not support effective attention allocation, especially in the context of unexpected changes and could lead to a breakdown in human-machine coordination (Sarter, 2001). A possible solution to this problem could be derived from the basic mode of human-human communication Human interactions generally involve multiple sensory channels. For example, people tend to point naturally while describing the content of a graphic image (Chapman, Smith, Klopfenstein, Jezerinac, Obradovich, & McCoy, 2000), which implies that for certain types of task humans use more than one modality to convey information. Supporting this is the Multiple-Resource Theory (Wickens, 2002) which stresses distribution of tasks and information across various sensory channels of the human. This concept implemented in the field of Human-Machine System is the Multimodal system. Multimodal systems are supposed to increase the bandwidth of information exchanged, give the user flexibility to shift between modalities and combine modalities to convey the appropriate information to the system. Design of Multimodal systems need to be based on the knowledge about how people use a particular modality and switch between modalities or different tasks (Oviatt, et al.,2000). The important steps in development of Multimodal systems involves addressing issues like "content selection" (what are the types of interaction with the system), modality allocation (what type of modalities to use), modality realization (how to implement the particular modality) and modality combination (Maybury, 1994).

Oviatt, et al. (2000) describes the advantages of using speech as input modality: a) it has a high bandwidth of information, b) more natural to use, and 3) it also allows the user to use the hands for other tasks. Speech is the preferred mode for functions like describing objects and events, sets and subsets of objects, out-of view objects, past and future temporal states, as well as commands for actions or iterative actions. Wauchope (1994) describes graphical user interfaces as the most natural paradigm of direct manipulation, in which the user operates directly on visual representation of the domain and control entities using pointing selection and keyboard inputs. In the current system the user views the world through a graphic and performs control actions using a keyboard and/or a mouse.

However neither of these modalities is sufficient on its own. In the current environment the user will be required to perform a multitude of tasks in a short span of time. Graphical interfaces are inflexible and repetitious in use. They are not suitable means of interaction for out-of view objects. On the other hand, using speech-only interactions face the risk of being under specific, vague, ambiguous and leading to misleading commands (Wauchope, 1994). The combination of the two modes can very advantageous as the cognitive processing is complementary in nature, i.e.; humans can effectively divide the task between the Graphical and Speech channels and thus enhance performance (Wickens, 2002). The goal in integrating multiple modalities should be to utilize the strengths of one modality to overcome the weakness of another and more importantly allocate tasks to the modalities for which it is best suited.

3 Design of the Multimodal Command and Control Simulation System

The main challenge to building a robust multimodal architecture is to ensure support for parallel inputs from the users and capability to handle ambiguities from individual modalities. Prior to the design and development of a multimodal architecture for the command and control of remotely operated vehicles we conducted a thorough task analysis of the system and established the need for multiple modes of input to the system. We have identified voice, head mounted displays, haptic device input and graphical manipulation as some of the modalities that would be useful in the current scenario. As a first step we have incorporated voice and graphical manipulation as multimodal inputs, the architecture of the multimodal interface is such that introducing new modalities would require minimal changes to the existing system.

Unmanned Aerial Vehicle Simulation Test bed (UMAST) is a complex military simulation designed to emulate SEAD mission scenarios for Remotely Operated Vehicles. UMAST is a distributed application where multiple users can connect to the simulation server and interact with the system simultaneously (Narayanan, 1999). The human controller acts as a supervisor who monitors the system and makes intelligent decisions based on the deductive capabilities of the machine. The multimodal interface is an Open-Agent type architecture, where the individual modalities are designed as separate reusable components with a plug and play feature.

3.1 Multimodal Components and Information Processing

The multimodal interface consists of two sections

a) The Graphical User Interface: The graphical user interface is designed and developed using a 3D visualization software Vega 3.7.1. The interface consists of three display panels with one panel showing a 3D view of the simulated world, the second one displaying a 2D satellite like image of the same scenario which can be manipulated using the mouse and the third panel displays a 3D " camera view", the area below any of the ROVs. There is also a "control panel" consisting of several graphical objects like panes, buttons, lists etc and is selected using the mouse and the keyboard which is used by the controller to select objects and trigger events in the simulation.

b) The Voice Interface: The voice interface is developed as a separate component using the Nuance 7.0 software. The advantage of Nuance is that it can be developed using many computer languages. Unlike other voice application softwares, the voice recognition engine in Nuance need not be trained to recognize a particular user. Nuance also provides support for many spoken languages and even different versions of the same language. Studies indicate that (Oviatt et al, 2000) action terms or verbs are best expressed by the use of natural language techniques, in our case-the voice stream. Users would find it more natural to use voice to describe commands like "move here" or" fire at target" as opposed to performing a series of button clicking and selection operations on the control panel.

Some multimodal commands are deictic references, while others use voice along with the selection control panel. Dexis is a form of reference in which pointing gestures accompany voice expressions. It is a useful communication technique in which each modality provides minimal information, but in combination with other modalities it can help to mutually disambiguate errors generated by each modality and yield a more accurate combined information (Wauchope, 1994). A command to change the direction of a moving vehicle can be accomplished by using the voice command to say "move here" while clicking on the map to indicate the exact location. In this context we use voice to utter the action nature of the command and the mouse to convey spatial information by clicking. There are other instances when we use the control panel's display to select from a list of available choices for a particular parameter and use voice to describe the action. For example, the task of assigning a type of missile for a "fire" is done by selecting a missile from a list of available choices on the "selection panel" and uttering the action command. By this way we reduce the burden on the user to remember large amounts of information.

3.1.1 Information processing

Having enumerated the different components in the multimodal system, we now discuss the actual process of information flow between these components and the integration of the different modalities.

Integration of the multimodal inputs involves recognizing multiple inputs and combining the information from different sources to form one unified signal. In our system, each input signal is time stamped and processed independently. Voice commands are filtered through the Nuance software which produces text information along with a "confidence level" for the interpretation. The acceptable levels of "confidence" can be set programmatically; if the recognized text is above the confidence level then it is accepted as a recognized result. This recognized text result with the time stamp is sent to the "Integrator". The graphical user interface also performs similar actions, i.e. any change in control parameter and or a mouse click is recorded with a time stamp added to it. Oviatt, et al. (2000) has demonstrated that gesture and speech usually precede or overlap each other within a specific time threshold. The integrator incorporates this concept of temporal unification to combine the two input signals. The time threshold is set based on empirical data collected to identify the approximate range within which both the signals arrive at the integrator. The block diagram representation shown in Figure 1 indicates the process of information flow between the individual modalities.

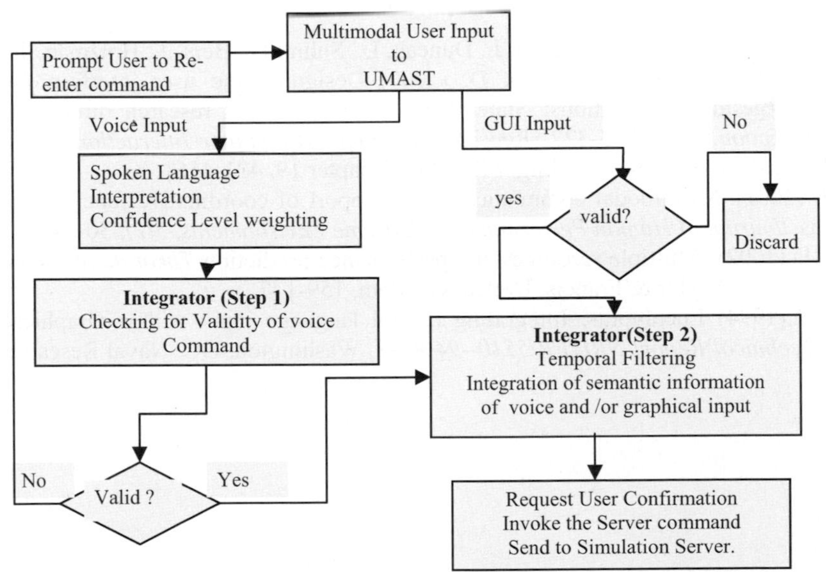

Figure 1: Information flow diagram in the Multimodal Architecture.

4 Conclusion

The multimodal interface developed for the Command and Control of Remotely Operated vehicles is an initial prototype which is aimed at studying the performance of human operators, in supervisory control, with multiple channels of communication. This is an exploratory phase intended to study how the users' actually utilize the available modes of communication, their expectations from the system and identify the shortcomings in performance and possibly use other modalities to make up for it. The idea is to incorporate new modalities in an incremental fashion such that the integration of these inputs is complementary to each other. Eventually we aim to develop a guideline for building efficient multimodal interfaces in Supervisory Control Systems.

5 References

Chapman, R. J., Smith, P.J., Klopfenstein, M., Jezerinac, J., Obradovich, J., and McCoy, C.E. (2000). CSLANT: An asynchronous communication tool to support distributed work in the National Airspace System. *Proceedings of the 2000 Annual Meeting of the IEEE Society on Systems, Man and Cybernetic,*. Nashville, TN, 1069-1074.

Maybury, M.T.(1993), *Intelligent Multimedia Interfaces* Menlo Park, Calif.:AAAI Press ; Cambridge, Mass. : MIT Press

Narayanan, S., N. R. Edala, J. Geist, P. K. Kumar, H. A. Ruff, M. Draper, & M. Haas (1999). UMAST: A web-based architecture for modeling future uninhabited aerial vehicles. *Simulation*, 73 (1), 29-39.

Narayanan, S., Ruff, H. A., Edala, N. R., Geist, J. A., Patchigolla, K., Draper, M., & Haas, M. (2000). Human-integrated supervisory control of uninhabited combat aerial vehicles. Journal of Robotics and Mechatronics, *Special Issue on Intelligent Control in Coming New Generation,*12, (3), 628-639.

Oviatt, S.L., Cohen, P.R., Wu, L.,Vergo, J., Duncan, L., Suhm, B., Bers, J., Holzman, T., Winograd, T., Landay, J., Larson, J. & Ferro, D. (2000) Designing the user interface for multimodal speech and gesture applications: State-of-the-art systems and research directions, *Human Computer Interaction*,15(4), 263-322 [Reprinted in *Human-Computer Interaction in the New Millennium* Addison-Wesley Press, Reading, MA, 2001; chapter 19, 421-456].

Sarter, N.B. (2001). Multimodal communication in support of coordinative functions in human-machine teams. *Journal of Human Performance in Extreme Environments, 5(2),* 50-54.

Wickens, C.D.(2002) Multiple resources and performance prediction *Theoretical Issues in Ergonomics Science*. 3(2), Taylor & Francis, United Kingdom, 159-177.

Wauchope, K.(1994) Eucalyptus: Integrating natural language input with a Graphical User Interface, *NRL Technical Report NRL/FR/5510--94-9711*, Washington, DC: Naval Research Laboratory.

Applying an eye-tracking based process measure for analysing team situation awareness in aviation

Chr. Rud Pedersen

Risø National Laboratory
P.O. Box. 49, 4000 Roskilde, Denmark
rud.pedersen@risoe.dk

Hans H. K. Andersen

Risø National Laboratory
P.O. Box. 49, 4000 Roskilde, Denmark
hans.andersen@risoe.dk

Abstract

This paper discusses the feasibility of applying a Distributed-Focused Visual Attention (DFVA) process measure for analysing team situation awareness in aviation. Distributed Visual Attention (VA) implies looking at several elements of information for a short period of time. Focused visual attention implies looking at few selections of information for longer periods.

The research question is: is the distributed-focused visual attention process measure sensitive to the pilot's behaviour in situations designed to test for Team Situation Awareness (TSA). The measure has been applied to investigate the difference between a "normal" flight and a probe-based flight. The hypothesis is that during the probes with high workload the individual pilots will use a different VA strategy compared to the "normal" work situation. The coding of the distributed-focused process measure is automated using a computer program.

1 Introduction

A loss of situation awareness may lead to human errors, possibly resulting in accidents. A conception of individual situation awareness has been summarised by saying that an operator should be able to (1) recognise the relevant elements in a situation, (2) understand how these elements interact and, on the basis of this understanding, (3) predict system status into the immediate future (Endsley, 1993; 1995). The study of Line Control Rooms on the London Underground (Heath & Luff, 1991) showed how actors maintain fluent reciprocal awareness regarding other actor's activities. In doing so the actors monitor each other's activities by overhearing other actors' radio or telephone conversations.

In Hauland (2002), a measure of TSA within Air Traffic Control (ATC) was validated in terms of TSA measures predicting system performance and being sensitive to changes in the situation. It was found that ATC student's frequent use of a distributed visual attention strategy indicates that the perceptual aspect of TSA is good. The student's frequent use of a focused visual attention strategy indicates that TSA is worse. Abnormal events seem to cause a shift of visual attention strategies in the team, i.e. from possibly ideal attention strategies (distributed) toward less ideal attention strategies (focused). It was argued that, when combined with a probe approach, this measure represents the quality of ATC student's detection of relevant elements in the situation. Regardless of what attention strategies should be considered optimal for the task, the attention strategy is a valid measure of perceptual aspect of TSA in that it is sensitive to abnormal developments in the situation. This is therefore considered a process-oriented TSA measure that can represent both individual competencies and team competencies held at the individual level within the domain of ATC.

Hauland's study profound the background for this study where the DFVA process measure developed within the ATC domain is applied to aviation. The idea is that the DFVA process measure measures the extent to which pilots attend to many relevant aspects of a situation (global/distributed), versus limiting the attention to a few of these relevant elements (local/focused). In this way the measure resembles "scanning entropy" occasionally mentioned in eye-tracking research.

The results from an interview with a flight training chief instructor and a chief pilot indicate that this measure is transferable to the aviation domain in that the instructor and the pilot could confirm that a focused visual behaviour could indicate less good TSA. A focused visual behaviour was referred to as having "tunnel-vision."

1.1 Definitions and analysis methods

Hauland (2002, page 121) uses the following definition for distributed and focused visual attention:

Focused *> 1 second / maximum 2 chunks (+ one quick look)*
Distributed *< 1 second / minimum 3 chunks (+ continue)*

Using this terminology chunks refer to pieces of information that can move around, e.g. aeroplanes on a radar display. In aviation the relevant information for the pilots is found at the instrument panels in the cockpit, which are static and referred to as Area Of Interest (AOI) using eye-tracking terminology.

Hauland had 2 persons make the coding of distributed or focused visual attention manually from video recordings at half speed using dwells as the unit of analysis. Hauland (2002, page 67) defines dwells as several fixations within an area of interest where the Eye Point Of Gaze (EPOG) must be within the area of interest for minimum 250 milliseconds. Regarding reliability of data coding Hauland concludes (page 142) that a 71% agreement between the two observers is regarded sufficient but mentions disagreement related to the timing of the coding of distributed-focused visual attention. Hauland applies the distributed-focused coding to only some of the areas of interest, meaning that there can be "holes" between a coding of distributed VA and a coding of focused VA. In the cockpit the coding of distributed or focused VA is applied to all predefined area of interest; i.e. there are no "holes" between distributed and focused codings.

The definition of dwells from Hauland and their use as the unit of analysis is followed. However, due to the fact that the eye-coordinates of fixations in this study have a minimum duration of 150 milliseconds, the minimum limit for the EPOG within an area of interest is set to 150 milliseconds. From eye-coordinates of fixations and defined AOIs a computer program calculates dwells automatically. These dwells are the basis for an algorithm determining distributed or focused VA. The only omission from the above definitions of distributed and focused VA is the condition that allows a quick look to a third AOI. The benefits of using computer based behavioural coding compared to manual coding are besides time savings a consistent coding.

2 Experimental set-up, data collection and validation

The experiment is part of the Vinthec II European project aiming at develop, assess and evaluate objective measurement methodologies for team situation awareness. The project studies 11 crews in a motion based flight simulator. Each crew flew two flights, one "normal" flight and one with several probes designed to provoke TSA. Each flight lasts for about one hour during which both verbal and visual behaviour were gathered. The visual behaviour was recorded by use of combined head and eye-tracking equipment.

2.1 Eye Data Recordings

Two types of eye-tracking data were obtained: 1) video recordings with a pointer indicating subject's eye point of gaze which requires manually coding of video tapes or 2) automatic data recordings of EPOG in form of eye coordinates for fixations on predefined areas of interest.

Due to the experimental set-up eye-tracking data was not collected for all 22 pilots. As this study compares the "normal" flight with the probe-based, eye-tracking data for the same person for both flights is required. Nine pilots from 6 crews fulfil this and are the basis for the following. For these 9 pilots, eye-tracking data is collected in average 71% of the total experimental time span (standard deviation 11, minimum 45 and maximum 87). Five out of the 18 eye-tracking data files have less than 70% track, which is considered a low tracking percentage. However, in order to be able to compare flights these data are not omitted.

2.1.1 Coding procedures

Manual coding was performed with the default state as focused VA. When the EPOG has move to 3 or more AOI within a time span of 1 second distributed VA was coded. When the EPOG stayed with an AOI for at least 1 second or the EPOG kept moving between 2 AOI focused VA was coded. Notice that different results were found when the default code state was distributed VA and focused VA was coded when ever the EPOG was with in area of interest for more than 1 second or if the EPOG sequentially moved between two and only two AOI. This was not studied further as the first mentioned coding procedure, to our knowledge, is the one closest to the coding procedure used by Hauland (2002).

2.1.2 Validation

Two persons applied the manual coding procedure to a 10-minute time span, each coding this time span 3 times. Coding was done from the scene cameras using Noldus Observer. The following results are calculated using Noldus Observer's reliability analysis with a time-span set to 2 seconds. Within persons concordance was on average 50% and 48%. Between persons using all 6 codings pair wise the average concordance was 42%. The 6 manual codings was compared to the result from the computer-based coding, which pair wise gave 34, 50, 39, 47, 36, 45% concordance (average 42%). Moreover, a time-event plot of the codings showed co-occurrence (see Figure 1). As also mentioned by Hauland, the deviation between codings was due to the timing and the fact that the computer program works with a finer resolution than humans. From these results it was concluded that the computer-based coding was valid.

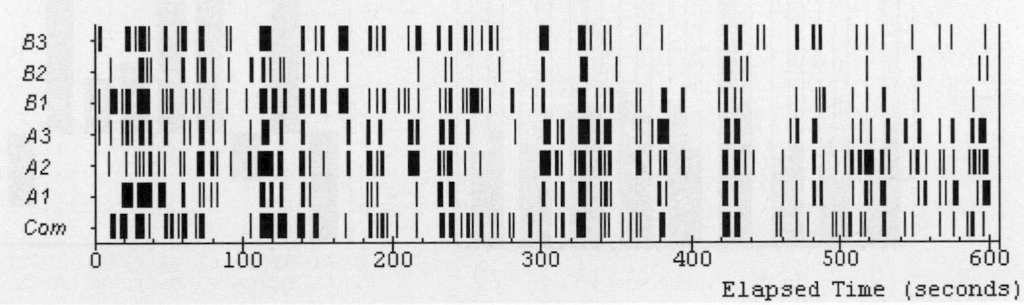

Figure 1. Time event plot of distributed visual attention for crew 02, captain, "normal" flight, starting from time 12:25:00. Com = computer-based coding. A is 3 codings by the same person. B is 3 codings by another person.

2.2 Probes

During the "normal" flight only two probes were inserted during the cruise flight phase to keep the pilots alert. During the second flight several probes were inserted. The probes were designed so one pilot alone could not solve the problem as the objective was to test for TSA. Here the focus is the descent flight phase, i.e. from altitude less than 20000 feet to weight on wheels. The "normal" descent is compared to the abnormal containing these 3 probes: 1) heart attack passenger, 2) primary flight display failure (for the second time in this flight) and 3) glide slope capture failure which required a go-around.

For the "normal" flight the descent phase lasted on average 1074 seconds (standard deviation 133, minimum 904 and maximum 1226). In the probe-based flight only the three above-mentioned probes are studied. The total duration of all 3 probes are on average 490 seconds (standard deviation 148, minimum 338 and maximum 806).

The start of a probe is defined from an event in the simulator log and can therefore be regarded as "objective" compared to the end of a probes, which is defined from manual coding of 1 person determining when then abnormal situation is handled. On average the probes 1, 2 and 3 last accordingly 120, 126 and 168 seconds having removed the outlier of crew 7 for the glide slope failure, which took 678 seconds (go-around twice) in this calculation.

3 Results

The duration of distributed-focused visual attention is calculated and Figure 2 shows in percentage how much distributed and focused visual attention the pilots exhibit during a "normal" and abnormal descent. During the "normal" descent the pilots use a distributed VA strategy in average 28% of the time; the rest of the time a focused VA strategy is used. For the probe-based descent the average of distributed VA was increased to 34%.

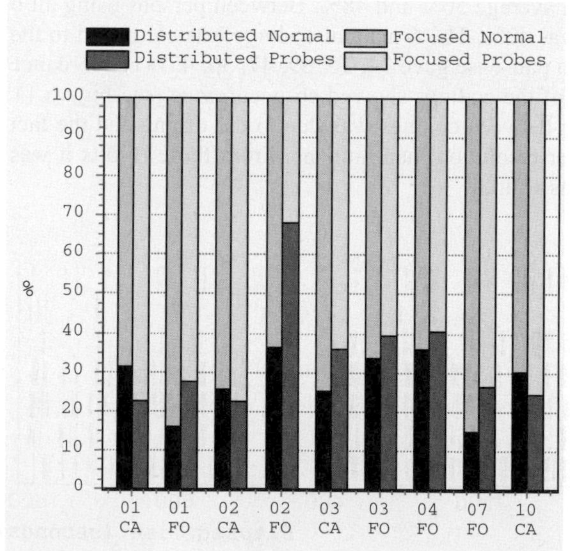

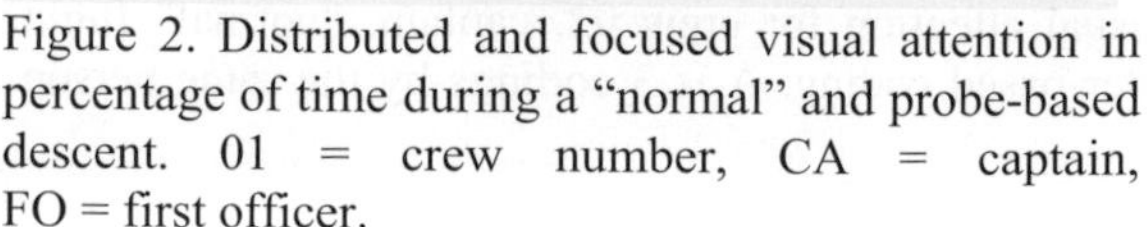
Figure 2. Distributed and focused visual attention in percentage of time during a "normal" and probe-based descent. 01 = crew number, CA = captain, FO = first officer.

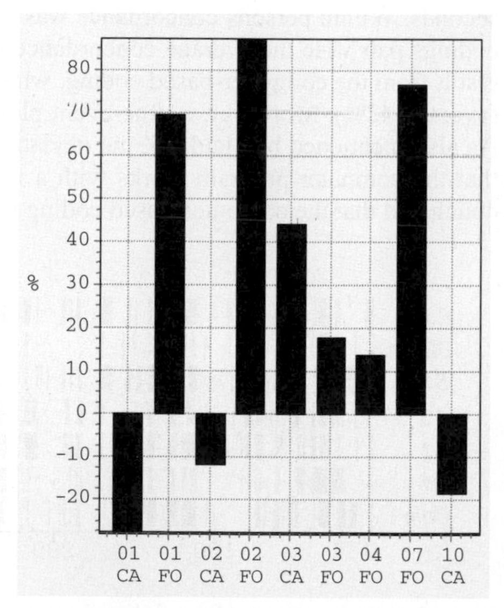

Figure 3. Increase in percentage in distributed visual attention from the "normal" descent to the probe-based.

The increase (in percentage) in distributed VA from the "normal" flight to the abnormal flight has been calculated and is shown in Figure 3. E.g. for crew 01 captain distributed VA decreases from 31% to 23%, which is an increase of 100*(23-31)/31 = -26%.

Three pilots increased their distributed visual attention 70% or more, 3 pilots increased less and other 3 pilot decreased their distributed VA, i.e. they become more focused during the abnormal descent. Notice, however that even with a large increase in distributed VA they are still on average focused 66% percentage of the time during the prove-based descent.

This confirms the result from Hauland (2002) that the DFVA measure is sensitive to changes in the work situation. A comparison to performance measures is outside the scope of this paper; hence whether a distributed or focused VA is preferable during the probe-based descent is still an unanswered question. Moreover, the proper VA strategy might depend on the pilot's role as either pilot flying or pilot not flying. During the three probes analysed the pilots shift roles due to the primary flight display failure probe and therefore the influence of pilot's roles have not been studied.

4 Conclusion

One finding is that the pilots show more focused than distributed visual behaviour in both the "normal" and abnormal flight conditions. Further, there might be a correlation between pilot's visual attention strategy and their workload. Hence, there is some indication that a distributed-focused visual attention measure can be applied to aviation. However, much precautions should be taken as many factors influence the eye-tracking data: 1) tracking percentage, 2) definitions of fixations, area of interest and dwells, 3) definition and coding of distributed and focused visual attention, and 4) definitions and comparison of work situations.

5 Acknowledgement

The work presented in this paper was partly funded by the European Commission and partly by the Vinthec II project partners (contract no. G4RD-CT-2000-00249).

References

Endsley, M.R. (1993). Situation Awareness in Dynamic Human Decision Making. *Proceedings of the 1st International Conference on Situational Awareness in Complex Systems*, Orlando, February 1993.

Endsley, M.R. (1995). Towards a Theory of Situation Awareness. *Human Factors*, 37 (1), 32-64.

Hauland, G. (2002). Measuring team situation awareness in training of en route air traffic control. Process oriented measures for experimental studies. Risø-R-1343(EN) .

Heath, C., Luff, P. (1991). Collaborative Activity and Technological Design: Task Coordination in London Underground Control Rooms. *Proceedings of the Second European Conference on Computer-Supported Cooperative Work*, ed. by L. Bannon, M. Robinson and K. Schmidt, Kluwer Academic Publishers, Amsterdam, 1991, pp. 65-80.

ENHANCING HUMAN-COMPUTER-COOPERATION BY GROUNDING THE ENGINEERING PROCESS ON A UNIFORM, COGNITIVE MODEL

Henrik Putzer, Reiner Onken

Universität der Bundeswehr München
Institut für Systemdynamik und Flugmechanik, D-85577 Neubiberg, Germany
[henrik.putzer|reiner.onken]@unibw-muenchen.de

Abstract

Man-machine cognitive cooperation has already become reality. The development of cognitive systems, however, is still not a well-established engineering process This paper considers the design of artificial cognitive systems as a software development process. A new engineering approach, implemented by a *uniform cognitive system architecture* (COSA), might be a crucial step forward in the sense to achieve engineering standards and more wide-spread application of cognitive systems. - The new approach is based on the Cognitive Process (CP), a new concept of generating cognitive behaviour modelling the human information processing loop. The behaviour of the CP is solely driven by 'a-priori-knowledge' for which the COSA-framework implements a new modelling methodology based on the programming paradigm of *mentalistic notions*. A first application based on COSA is presented along with experiences and results.

1 Introduction

In our days everybody is aware of the increasing complexity of computers and software. With this evolution some capabilities evolved silently: systems turned from 'simple and stupid tools *used* by humans' to 'autonomously acting units *cooperating* with humans' (Onken, 2002).
With this new application of computer systems new requirements evolve, which are hardly covered by conventional engineering approaches. Thus, conventional systems fail in complex environments, especially in the context of human-computer-interaction, because essential skills are missing: *cognition* and genuine *cooperation*. This leads to serious deficiencies resulting in decreased productivity or even decreased security.
To overcome these deficiencies a new approach is needed. On the one hand, this approach should represent a software engineering process to ensure the structured development of complex software. On the other hand, it should be based on a uniform and powerful cognitive model to come up with capabilities like those of human mental processes and behaviour such as cognition and cooperation.
Thus, the first step to take is the creation of a uniform cognitive model - the Cognitive Process - to simulate human mental processes with its evolving capabilities (section 2). The second step is the transformation of this theory into practice by implementing a generic framework for cognitive systems – COSA – which uses the new model as its core component (section 3). A first application – COSY^{flight}, an unmanned, interactive autonomous air vehicle – built with the framework is presented within the context of evaluation (section 4). The paper concludes with results and prospects (section 5).

2 Creating the Uniform Cognitive Model

2.1 The Cognitive Process as a Model

Inspired by cognitive science and artificial intelligence the goal is to let cognitive behaviour evolve from a technical system. Not tying to achieve exact models of the human brain, a paradigm is followed to produce intelligent behaviour that is goal consistent and compatible to human cognition (weak artificial intelligence). Only such an approach will ensure the effective integration of the cognitive process (CP) with a team of human operators (Onken, 2002).
An elaborated recognition-act-cycle is used as the scheme of thought processes (Putzer & Onken, 2003). As shown in figure 1 it is build around an oval part, the *body*, which contains all data in structured symbolic representation. The inner part of the body, slightly darker, contains the *a-priori-knowledge* which is fed into the CP (or learned), before the process starts. This a-priori-knowledge is the (only) origin of application-dependent behaviour of the CP. The outer part of the body contains the *cognitive yield* which is created during runtime. The cognitive yield can be considered to be the situational knowledge of the system because it results from the CP-subprocesses, the *transformators*.
Transformators have access to the whole knowledge in the body and create new knowledge by writing their results into designated areas of the body (at the arrowhead). In figure 1 transformators are located (clockwise) around the body:

- *interpretation* (of the situation) generates the belief concerning the state of the environment, mainly based on a-priori-knowledge about environment models and mainly driven by the input data accessible from sensors and communication interfaces outside the CP,
- *goal determination* (including determination of conflicts and opportunities) generates the actually relevant goals, mainly based on a-priori-knowledge about desires and mainly driven by beliefs,
- *planning* generates a plan for reaching goals mainly based on the a-priori-knowledge about strategy models and driven by determined goals, and
- *plan* realisation generates instructions to execute the plan mainly based on a-priori-knowledge about instructions models and driven by the plan.

The environment represents the *world in real* the CP is interacting with. This includes human operators and other environmental objects.

2.2 CP-Method

The development of knowledge that drives the CP can be seen as software engineering process.

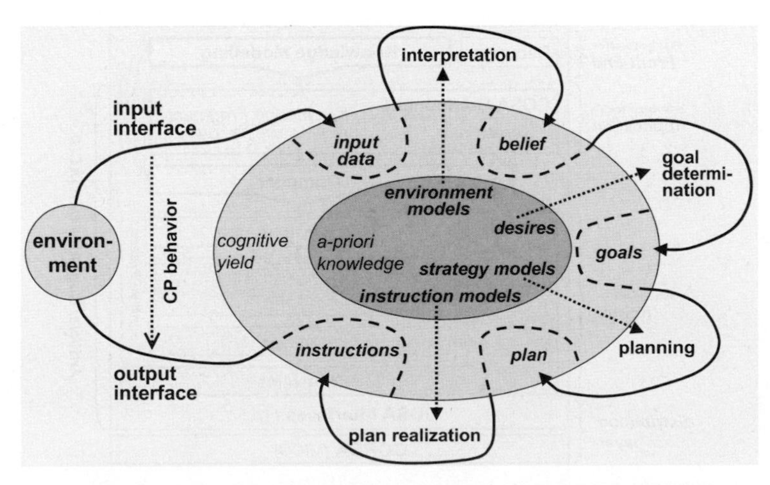

Figure 1: Basic conception of the Cognitive Process (CP)

Thus, there should be an appropriate method for 'programming' knowledge. But what does programming mean to the programmer? Following conventional approaches, one has precisely to formulate <u>how</u> certain things are to be done. Being concerned with many details the programmer must think within the bounds of the processing model of the programming language.
By means of the CP a processing scheme for goal oriented behaviour creation was invented. As a high abstraction, the programmer now specifies <u>what</u> the system has to comply with (the goals) and formulates the knowledge in terms of *mentalistic notions* like desire, goal, belief, etc. This is already a level of abstraction which is used to describe complex human behaviour. Now it can be used to implement cognitive, cooperative systems.
The modelling process consists of five steps (Putzer & Onken, 2003): as the emerging behaviour should be goal oriented it starts with the goals. It continues to model the way how the *goals* are to be achieved (*plans*) and how the plans are to be executed (*instructions*). A model (a-priori-knowledge) of relevant elements from the environment builds the system's *belief.* The underlying *ongoing interaction* defines the micro-behaviour of modelled elements. The combination of all *micro-behaviours* in the CP's body yields the system's *macro-behaviour* (*CP behaviour*).

3 The Framework: From Theory to Practice

3.1 Architecture

The basic idea of the COSA framework is to obtain a reusable platform for the implementation of cognitive systems. This is warranted by its component architecture which is composed of four areas as shown in figure 2 (descriptions on the left hand side):

- The *distribution layer* ensures that components can be distributed across a computer network according to their specific resource requirements. Such resource requirements may consist of computation, memory or peripheral devices like displays or special IO. The COSA framework uses CORBA in its MiCO implementation (Puder & Römer, 1999).
- The *kernel* as the central element integrates all components into one system. After a registration process the controller takes full control. The second responsibility of the controller is the encapsulation of the CP represented by a group of sub-components: The *processor* translates formal knowledge representations into behaviour. SOAR (Newell, 1990) has proved to be a valid platform for implementing cognitive behaviour. Extended by object oriented means SOAR is used as the processor. The *CP-library* implements the scheme of the CP on top of the processor to define structures for the cognitive yield and

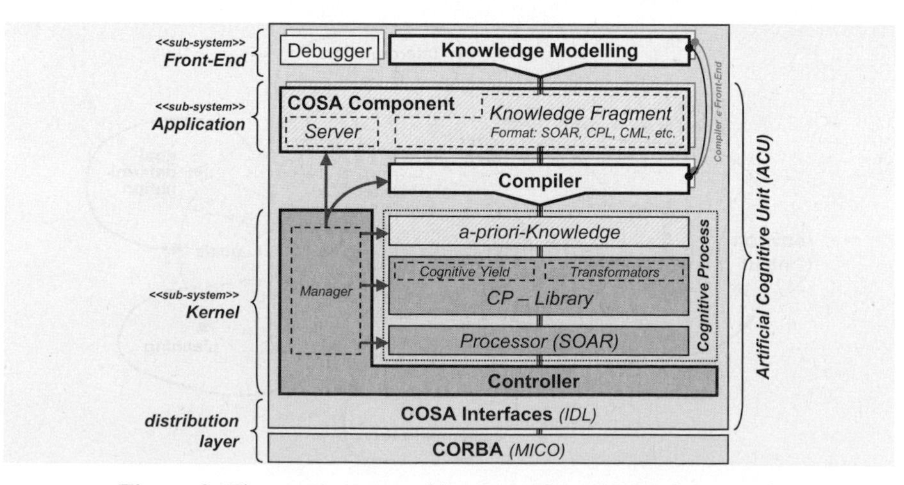

Figure 2: The component architecture of the COSA framework

the transformators. Using this implementation of the CP the component of the *a-priori-knowledge* implements the cognitive behaviour.

- The *application* is formed by a number of reusable *COSA components*. These special components include knowledge fragments which are inserted into the CP before run-time. COSA components may contain *servers* as well. These are functional modules external to the CP. Examples for servers are interfaces to databases, simulation or displays, high frequency feedback control loops or image analysis core functions.
- The *front-end* is the interface for *knowledge modelling* (acquisition and representation). It may support several methods to let the designer use the one which is best suited for his problem. As a fragment this knowledge is directly inserted into COSA components. *Compilers* are used to translate the knowledge into the format of a-priori-knowledge that it can be executed in the processor to let emerge cognitive behaviour from it.

3.2 Modelling Tools

The above described architecture around the processor implements the CP. To support the modelling process given by the CP-method new means like description languages and graphical tools are implemented as a front-end. They are based on the paradigm of mentalistic notions and follow the scheme given by the CP. Thus the underlying language is called 'cognitive programming language' (CPL) (Putzer & Onken, 2003). All tools follow modern software design paradigms as they are supported by the *unified modelling language* (UML).

4 Evaluation

A first and relatively simple application is implemented on the basis of COSA. It's name is 'autonomous cognitive system for the flight-domain' ($COSY^{flight}$). It models an autonomous unmanned air vehicle (UAV) during a military reconnaissance mission. The test bed used is a flight simulator with simulated dynamic environment.

As shown in figure 3, $COSY^{flight}$ has to communicate and cooperate with many other natural and artificial entities during its mission. The first contact after booting is made with the operator, who defines the mission order for the UAV. As far as possible this mission is executed autonomously by $COSY^{flight}$. Confronted with unexpected situations a solution is negotiated with the operator *only* if $COSY^{flight}$ can not decide on its own. $COSY^{flight}$ interacts with further entities like command and control center (CCC), air traffic control (ATC) and other aircraft which are operated by natural or artificial intelligences.

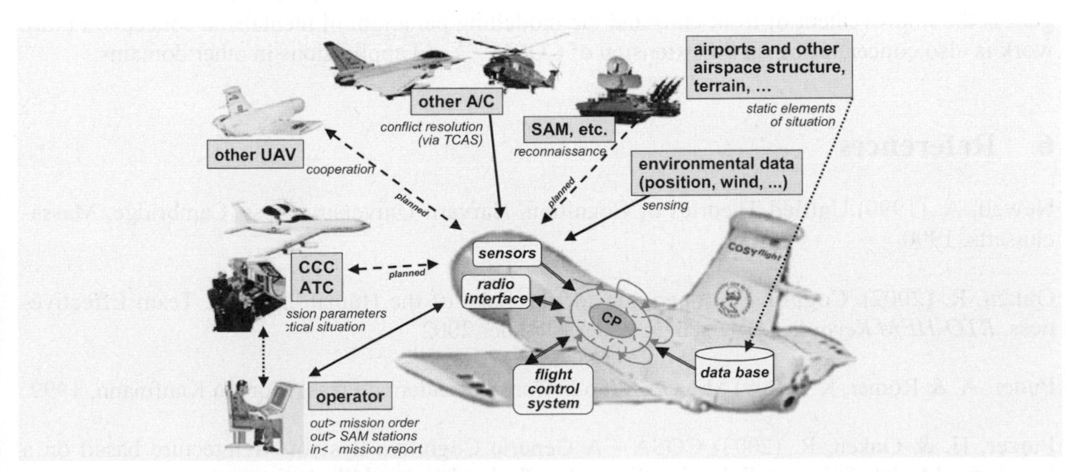

Figure 3: Complex environment of $COSY^{flight}$ with needs for interaction and cooperation

The focus of this application is to verify core features of COSA's architecture to some extent. The focus lies on the autonomous cognitive behaviour of the UAV itself that evolves from the knowledge processed by the CP. As a result goal-based behaviour complying with the actual situation can be observed.

5 Results and Prospects

To overcome deficiencies of current cognitive systems their design is considered as a software engineering process. A new engineering process was created. On one hand, it represents a software engineering process to ensure the structured development of complex software. On the other hand it is based on a uniform and powerful cognitive model, the Cognitive Process. This ensures capabilities like cognition and cooperation, known as human excellencies. These are the essentials for human-computer-team integration.

The Cognitive Process is used as basis of the uniform cognitive system architecture (COSA) framework which is an uniform architecture for cognitive systems. The evaluation of COSA includes an application that models an autonomous unmanned air vehicle (UAV) called COSY^{flight}. The UAV communicates and interacts with its simulated environment to achieve its goals.

Results from the research up to now can be summarized as follows:

- A flexible and usable framework (COSA) on the basis of a new software engineering concept with the CP as its core element.
- The COSA framework eases the creation of cognitive systems, since basic architectural problems are resolved by it.
- The CPL front-end on the basis of the COSA framework can reduce the time for development of cognitive systems through high knowledge abstraction (mentalistic notions).
- Knowledge reuse based on object orientation and knowledge abstraction
- With the application COSY^{flight} the usability of the COSA framework and the CPL front-end is demonstrated.
- In terms of cognitive system modelling COSY^{flight} is based on a-priori knowledge. This and nothing else determines the system's behaviour as a reaction on the external world situation.
- As it is proposed by the CPL method, the behaviour shown by COSY^{flight} is driven by explicit goals which is rarely found in current applications.

For the near future, improvements of COSA are planned in many implementation details, especially concerning the CP library and the interfaces of the processor (SOAR). A research related goal is the improvement of front-ends and the modelling paradigm of mentalistic concepts. Future work is also concentrating on the extension of COSY^{flight} and applications in other domains.

6 References

Newell, A. (1990) Unified Theories of Cognition. Harvard University Press. Cambridge, Massachusetts, 1990

Onken, R. (2002). Cognitive Cooperation for the Sake of the Human-Machine Team Effectiveness. *RTO-HFM Keynote 5*, Warschau, Polen, October 2002

Puder, A. & Römer, K. (1999) MiCO - Mico Is Corba. Academic Press, Morgan Kaufmann, 1999

Putzer, H. & Onken, R. (2003) COSA - A Generic Cognitive System Architecture based on a Cognitive Model of Human Behavior, *Cognition Technology and Work,* (in press)

A Methodology for Reengineering Courses for the Web[1]

Jean-Marc Robert

École Polytechnique de Montréal
Dep. Mathematics and Industrial Engineering
P.O. 6079, Station Centre-ville
Montréal, Québec H3C 3A7
jmrobert@polymtl.ca

Luciano Gamez[2], Walter de Abreu Cybis

Universidade Federal de Santa Catarina,
CEP 88040-900 Florianopolis
Santa Catarina, Brazil
luciano.gamez@polymtl.ca
cybis@inf.ufsc.br

Abstract

This study presents a methodology for reengineering courses for the Web. This methodology provides a framework to set the ground, collect data, discuss issues, make decisions in a structured way, design and evaluate the Web course with the proper information. It starts from the mission, the need for reengineering, and a goal for the project. It uses postulates, principles, requirements, and learning theories as foundations. It comprises a macro-analysis of the current course, a micro-analysis of each course unit, a diagnosis of the course, as well as decisions on what to reuse, modify, add, or suppress. Design and evaluation of the new course are done iteratively. The methodology has been tested and validated on a real project of course reengineering, and is platform-independent.

1. Introduction

With the advent of Internet, more and more colleges, universities, and companies around the world aim at offering several of their courses on the Web. They want to take advantage of this new medium for training or learning, and exploit the power of technology for information search, visualization, and communication, while allowing the students to learn at their own pace, anywhere, and at any time. Hence the great interest in e-learning (Beer, 2000; Driscoll & Reeves, 2002; Lockwood & Gooley, 2001). For a large majority of professors and course designers, offering courses on the Web means reengineering an existing course. That is, decide to re-use, adapt, suppress, and add new material, then design and build the courses on a specific platform. Up so far, there has been a lot of studies on process reengineering in different domains, especially in business (e.g., Davinport & Short 1990; Hammer, 1990). There are also some studies on reengineering in education (e.g., Penrod & Dolence, 1991; Wilkinson 2002). Although the results of such studies are useful to us, there is still a need for a rigorous and detailed methodology that specifically applies to course reengineering for the Web. This paper proposes such a methodology.

2. The approach

---

[1] This research has been partly funded by the Natural Science and Engineering Research Council (NSREC) of Canada, and the Conselho National de Desenvolvimento Científico e Tecnológico (CNPq) of Brazil.

[2] Bolsista do CNPq.

To develop the methodology, we realized the following activities:

- Literature review on process reengineering, course reengineering for the Web, and the use of technology at schools. Let us mention that Hammer (1990) presented 7 key principles for reengineering; Davinport & Short (1990) identified 10 basic steps for reengineering; Penrod & Dolence (1991) discussed various concepts for reengineering higher education; Wilkinson (2002) presented a model of reengineering for education.
- Observation in the classroom and analysis of parts of 3 traditional courses: 2 graduate courses and 1 undergraduate course in Human Factors and Ergonomics. The goal was to collect data on different aspects of the courses: the pedagogical approach, the activities in the classroom, the interaction between the professor and the students, the material, etc.
- Observation and analysis of a WebCT-based course. The goal was to know the characteristics, possibilities and limitations of this specific platform.
- Discussions and decisions about reengineering a course (a 45-hour graduate course on Human-Computer Interaction offered in a presencial mode. Three persons were involved: the professor (a human factors specialist), a software engineer, and a Ph.D. student doing a research on course reengineering.

3. The reengineering process

Figure 1 presents the methodology we propose for reengineering a course for the Web.

Mission. It is essential to start with the high-level mission of the organization wherein the reengineering process will be carried out. In the case of universities, it is clear enough: form competent professionals and specialists through teaching and research, develop the knowledge and technology through research, and provide services to the community.

Recognize the need for reengineering. There is a need to reengineer a course inasmuch as the professor or the course designer wants to reuse (and adapt, where required) existing material for the Web. Then the key question is: Why to develop a Web-based course? The answer generally deals with quality, competitiveness, and success. Offering Web-based courses might be a strategy to reach more students, allow them to study at their own pace, anywhere, and at any time, facilitate their work-and-study life and improve their results, fulfill their requests for interactive and stimulating courses, acquire experience with this new medium, etc. Another key question is: What kind of Web-based course to develop: should it be offered in a presencial mode, in a non-presencial mode, or in both?

Goal. The goal of this project is to reengineer a course for the Web and offer it in a non presencial mode, except for a part of the evaluation, the exams, that will be done in the classroom. Not to confound with the goal of this paper, which is to present a methodology for reengineering courses for the Web.

Background. Here we set the ground for the project by addressing more specific issues.

Postulates. We defined four postulates at the onset of the project: 1) the professors want to re-use as much as possible the material they already have for the course; 2) they want to transfer to the Web things that work fine or they like best in the course; 3) they want to eliminate the flaws of the course, especially when they are in a transition period towards a new medium; 4) they want to have their personal style and pedagogical approach reflected throughout the course.

Principles. Four principles guided our effort to create a stimulating Web-based environment: 1) exploit the Web as much as possible, for information search, visualization, and communication; the challenge is to bring plus value to the course, not simply transfer an existing course to a new medium; 2) create an interactive environment for the students, not only through the use of the interactive computer tool (e.g., click on icons and menus), but also through the manipulation of the course content; 3) allow the students to be active during the course, not passive as it is too often the case in traditional environments; 4) guide the students through the course, that is, propose a path through the mass of information that allows them to stay focused on the content while taking advantage of the new medium.

Requirements. At least three basic requirements must be satisfied to have the professors and course designers accept to use a Web-based course: 1) the course must be easily modifiable since it constantly evolves, and it should be so by the professor him/herself, not by a another person; 2) the professors must be able to follow the students' activities and progression, and provide assistance when it is required; 3) the professor must be continuously aware of the course situation, even though the new medium encourages the students to be autonomous and explore by themselves.

Learning Theories. As professors and course designers, we must have some knowledge (the most the best) about learning theories and adult education in order to make better decisions about the types of learning activities proposed to the students, the dynamics of interaction and communication with others, collaborative learning, etc.

Macro Analysis. The goal here is to get a clear picture of the current course, have an overview, and see the interrelations and the integration between the different aspects of the course. The following aspects were described and analysed: the objectives, content and structure, materials, pedagogical approach, workload and evaluation, media for teaching and learning, students' profile (e.g., previous knowledge, abilities with computers, motivation, availability for teamwork outside the classroom), and professor's profile (e.g., preferences, teaching style, etc.).

Micro Analysis. This step allows one to describe and analyze the details of each course unit (a part of the course that is devoted to a specific topic): the objectives, content and structure, duration, material, pedagogical activities, media, workload and evaluation. This step generates rich and specific information that is essential for diagnosis and design.

Diagnosis. This is a key step for the reengineering process since one will normally want to capitalize on the strengths and the assets of the course, and will try to correct the weaknesses and the problems. The diagnosis may be done by the professor(s), a colleague, a specialist of pedagogy, an observer, a quality control agency, the students, or several of them. It may be done on the basis of different information: the professor's own evaluation, the results of the students, the course evaluation by the students, etc.

Decision. Considering the characteristics and the diagnosis of the course, and in light of the knowledge available on Web-based learning (e.g., the motivation problem of the students), and of the possibilities and constraints of the Web platform, one can decide about different issues of the course: for instance, the pedagogical approach, the types of activities proposed to the students, the participation of the professor, the type of evaluation, the use of different features of the Web platform: the hyperlinks, the discussion forum, the chat, on-line animation, on-line evaluations, the combination of on-line reading vs reading on paper, etc.

Requirements. Considering the diagnosis of the course and the features of the Web platform, one can decide about the parts of the course that will be re-utilized, modified, or suppressed, and the additions to make to the course. Moreover, one has to decide how to modify each course unit, and to create new course material where it is needed. Thus the requirements for the design of the Web course are specified.

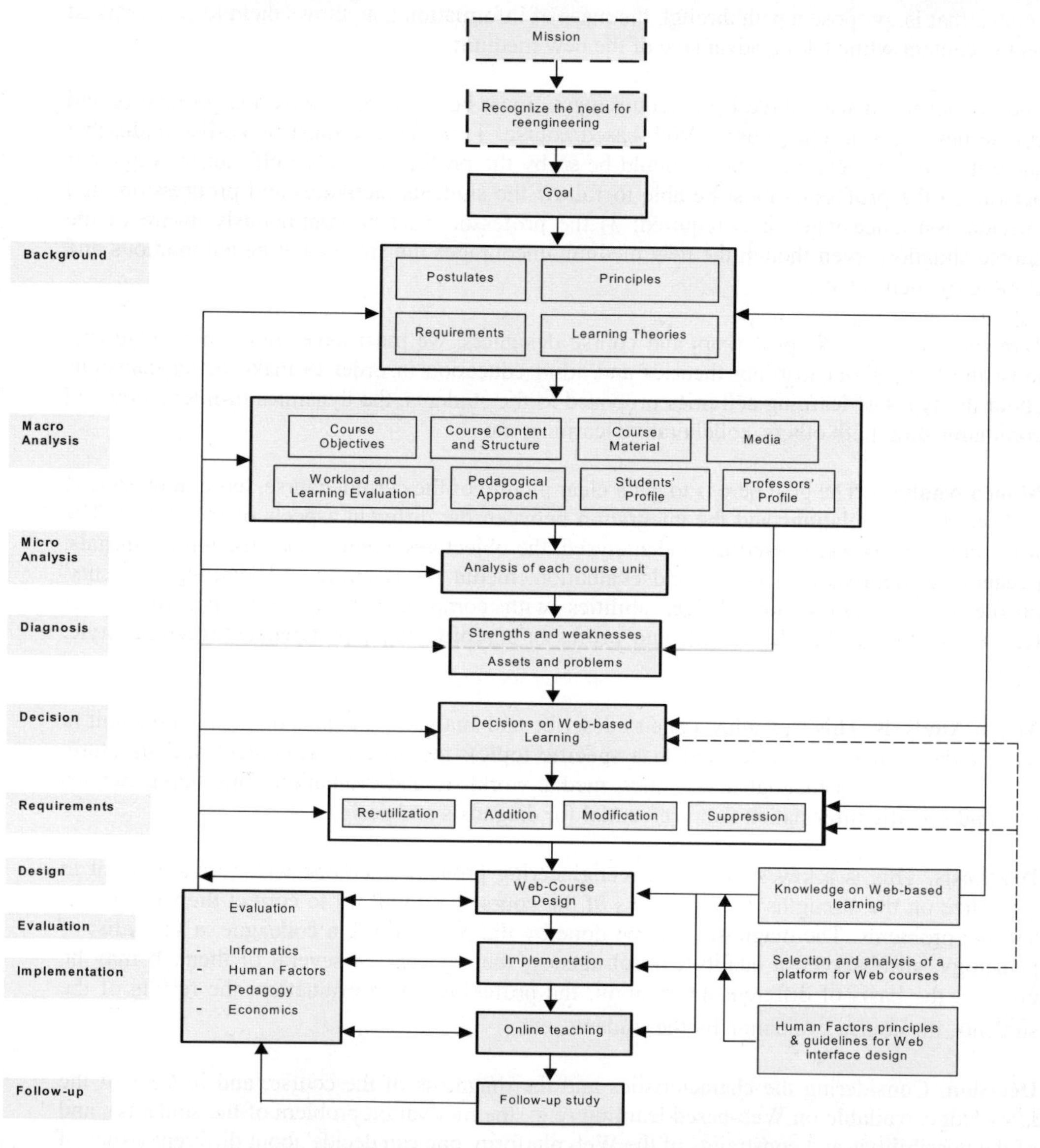

Figure 1: The Methodology for reengineering courses for the Web

Design. The goal of this step is to design the course for the Web, that is, its content and structure, functionalities, tools, interaction styles, user interface, etc. The design process evolves iteratively, that is, through a series of design-evaluation-redesign cycles. It is user-centered in that it takes into account the professors and the students in the process.

Evaluation. Evaluation is essential when creating a new course on the Web. The course has to be evaluated at least under four dimensions: informatics (How is the response time on the network? What amount of information can be displayed on the screen?), pedagogy (Is the course stimulating for the students?), human factors (Is the user interface easy to use?), and economics (How much does it cost?).

Implementation. Once the Web course has been designed, it can be implemented onto a specific Web platform. Our experience indicates that there is a lot of back and forth activities between the design and implementation steps. Because it reveals all kinds of constraints and difficulties, implementation may lead the professors and course designers to change the design.

Online Teaching. Once available, the course can be offered on the Web. There is a lot of knowledge available on how to successfully teach online, i.e. how to motivate students, encourage their participation, improve the quality of learning, etc. (e.g., Beer, 2000).

Follow-up studies. These studies consist in collecting data about the course and its impact after some period of time (e.g., 1, 2, 3 … semesters), in order to improve the course, if necessary. The data collected could be about the students' activities, performance, motivation, and satisfaction, the professors' participation, workload, and satisfaction, the impact on the community and the university, the cost, etc.

4. Conclusions

This paper has presented a methodology for reengineering courses for the Web. This methodology starts from a strategic vision and provides a framework to set the ground, collect data, discuss issues, make decisions, design and evaluate with the proper information. The next step of this research will consist in refining the methodology, and testing and validating it with several other courses.

5. References

BEER, V. (2000). The Web Learning Fieldbook. Using the World Wide Web to build workplace learning environments. Jossey-Bass Pfeiffer, San Francisco.
DRISCOLL, M., REEVES T. C. (2002). Proceedings of E-Learn 2002. World Conference on E-learning in Corporate, Government, Healthcare, & Higher Education. October 15-19, Montreal. Association for the Advancement of Computing in Education, Norfolk.
LOCKWOOD, F., GOOLEY, A. (Eds) (2001). Innovation in open and distance learning. Successful Development of online and Web-based learning. Kogan Page, U.K.
PENROD, J. & DOLENCE, M. (1991). Concepts for Reengineering Higher Education. Cause/Effect, Summer, Vol. 14, No 2, 10-17.
WILKINSON, J. (2002). Reengineering competency-based education through the use of a multimedia CD-ROM. A matter of life or death. Industry And Higher Education, Vol. 16, No 4, 261-265.

Explorations in Modeling Human Decision Making in Dynamic Contexts

Ling Rothrock
The Pennsylvania State University
lrothroc@psu.edu

Alex Kirlik
University of Illinois at Urbana-Champaign
kirlik@uiuc.ed

Abstract: While the lens model equation has been proven useful toward modeling operator behavior that is compensatory (i.e., representative of a linear, weighted strategy), its use in modeling noncompensatory (i.e., formulated as a rule base strategy using AND, OR, and NOT elements) behavior has been less informative. In this paper, we present our hypothesis that a lens model equation counterpart for noncompensatory rule sets, called the Genetics Based Lens Model (GBLM), can be developed. We show some initial results that compare the utility of the regression-based lens model equation in situations where a linear operator strategy is used versus situations where a nonlinear strategy is executed. Likewise, we present preliminary findings comparing the performance of GBLM in modeling a linear operator strategy versus a nonlinear strategy. We conclude the paper with some suggestions for future research.

1 The Lens Model

As researchers attempt to model human decision making in dynamic contexts, an often used criticism is the neglect of naturalistic settings and of ill-structured problems that are prevalent in uncertain, real-world environments (Klein et al., 1993). One method of addressing the problems posed by the naturalistic decision making community was initiated by Brunswik in the early 1950's. Brunswik labeled his approach *probabilistic functionalism* (Brunswik, 1955) to emphasize the relationship between an organism and its environment – a relationship which is mediated through the lens of probabilistic variables in the environment. Figure 1 shows the standard formulation of Brunswik's probabilistic functionalism as constructed by Hammond and his colleagues (Hammond, 1955; Hursch et al., 1964; Tucker, 1964) – more commonly known as the lens model.

The multiple linear regression model of the judge is formulated as $Y_S = \hat{Y}_S + e$ where e is the residual. The correlation between Y_S and $\hat{Y}_S$ is given by R_S and represents the cognitive control of the judge. A corresponding multiple regression model is given for the ecology. In the case of the ecology model, R_e represents the predictability of the criterion. The correlation between $\hat{Y}_S$ and $\hat{Y}_e$, or **G**, has been labeled as linear knowledge to denote the linear correspondence between the judge's decision policy and the optimal model of the criterion. The correlation between the two sets of residuals (Y_S-$\hat{Y}_S$ and Y_e-$\hat{Y}_e$), or **C**, is commonly called unmodeled knowledge. The remaining term, r_a, is the cornerstone of the lens model and its equation is called the Lens Model Equation (Hursch et al., 1964; Tucker, 1964):

$$r_a = GR_eR_s + C\sqrt{(1-R_e^2)}\sqrt{(1-R_s^2)} \quad \textbf{(1)}$$

The LME assumption of a judge whose behavior can be closely approximated as linear does not hold in all cases of decision making. In particular, investigators (Payne, 1976; Wright, 1974; Maule & Hockey, 1993) have found that in contingency situations, decision behavior is noncompensatory.

2 Noncompensatory Lens Model

The noncompensatory lens model used in this research is an extension of work on the Genetics-based Lens Model (GBLM) (Rothrock & Kirlik, In Press). To be consistent with the findings from Gigerenzer and Goldstein (1996), the representation of each rule set in GBLM is of disjunctive normal form (DNF). Hence, GBLM generates sets of rules (called rule sets) that potentially reflect not only fast and frugal heuristics, but also any logical strategy consisting of AND, OR, or NOT operators (Mendelson, 1997). A central element of GBLM is the multi-objective fitness evaluation function. The function evaluates fitness along three dimensions: completeness, specificity, and parsimony. The completeness dimension is based on work by DeJong et al. (1993), and is a measure of how well a rule set matches the entire set of exemplars (i.e., human judgments in a data set). The specificity dimension was first suggested by Holland et al. (1986), and is a measure of how specific a rule set is with respect to the number of wild cards it contains. Therefore, rule sets with less wild characters are classified as more specific. The parsimony dimension is a measure of the goodness of a rule set in terms of the necessity of each rule. Hence, in a parsimonious rule set, there are no unnecessary rules. The ideal rule set, therefore, will match all operator judgments, will be maximally specific, and maximally parsimonious. For more model details, see Rothrock and Kirlik (In Press). For the research reported here, we used multi-objective optimization to explore the Pareto-optimal frontier consisting of non-dominated points in the three dimensions (see Rothrock & Repperger, In Review for another example of using multi-objective optimization in human modeling).

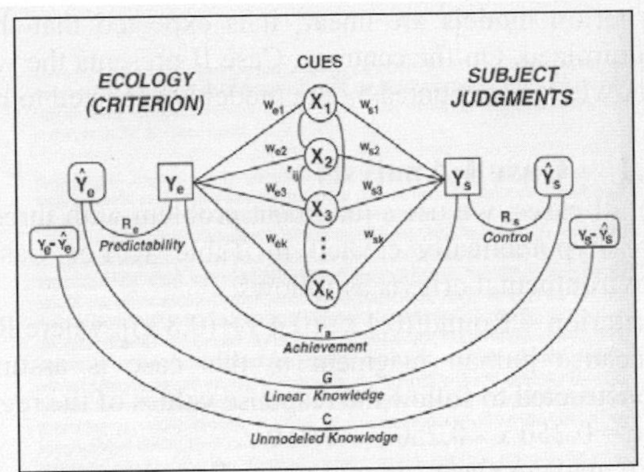

Figure 1. Lens Model with Labeled Statistical Parameters (from Cooksey, 1996)

3 Exploring the Environment Structure and Judgment Strategies

The framework to investigate decision making under different environment cue structures and judgment strategies is shown in Figure 2. In Cases I and IV, the organizing principle of the cues is linear. In contrast, Cases II and III represent nonlinear cue-criterion relationships and are characterized by rule-based descriptions. The analog to the cue-criterion distinctions exists for the judge in terms of compensatory and noncompensatory decision strategies. A compensatory strategy, such as Cases I and III, represents a cue-judge relationship that can be characterized by a linear formulation (Hoffman, 1960). In contrast, a noncompensatory strategy, such as Cases II and IV, is exemplified by conjunctive and disjunctive rules (Einhorn, 1970; Gigerenzer & Goldstein, 1996). For the present study, we focus on Cases I and II. First, Case I is the ideal case for the use of the Lens Model Equation (LME). If both the judgment and

Decision Strategy \ ***Cue Structure***	Linear	Nonlinear
Compensatory	Case I	Case III
Noncompensatory	Case IV	Case II

Figure 2. Framework to Investigate Decision Making Models

criterion models are linear, it is expected that the amount of unmodeled knowledge, or *C*, is minimized. On the contrary, Case II presents the worst case for using the LME. That is, the linear knowledge accounted by the model is expected to be minimal and *C* is maximized.

3.1 Case I Analysis

In all cases, we use a judgment problem with three cues (x, y, and z). The instances of judgment are hypothetically created in Table 1. For Case I, we set the generative equation for the environmental criteria as follows:

Criterion = Round[(0.4 x)-(0.4 y)+(0.8 z)], where Round is the rounding function.

Because human judgment in this case is assumed to be compensatory, the judgments are constructed to follow the response values of the regression equation:

Ye = 0.250 x - 0.250 y + 0.750 z.

We analyze Case I in two ways. First, using conventional regression-based methods, we find the lens model variables and assess model effectiveness. Second, we find the models of the judge ($\hat{Y}_s$) and criterion ($\hat{Y}_e$) *as generated by GBLM* and assess the lens model variables. Results found using multiple linear regression are shown in Table 1. It is evident that the data shown in Table 1 represents the ideal case of the lens model using LME. In addition to possessing good cognitive control, the fact that linear knowledge (*G*) is high and unmodeled knowledge (*C*) is low suggests that our hypothetical subject's decision making behavior is consistent with an effective linear model.

Table 1. Analysis of Case I Using LME.

x	y	z	Ye	Ye_resid	Ye_hat	Ys	Ys_resid	Ys_hat	Re	Rs	G	C	ra
1	0	0	0	-0.25	0.25	0	0.0000000	-0	0.856	1	0.905	0.163	0.775
0	1	0	0	0.25	-0.25	0	0.0000000	-0					
0	0	1	1	0.25	0.75	1	-0.0000000	1					
0	0	0	0	0.00	0.00	0	0.0000000	0					
1	0	1	1	-0.00	1.00	1	0.0000000	1					
1	1	0	0	0.00	-0.00	0	0.0000000	-0					
0	1	1	0	-0.50	0.50	1	-0.0000000	1					
1	1	1	1	0.25	0.75	1	0.0000000	1					

To select rule sets as alternative decision strategies to LME-based models, we use selection criteria that are analogous to the Lens Model Equation. We know the following is true of a LME-based model:

1. Each model must account for all instances of judgment.
2. Each model of judgment must provide a decidable function.

We translates knowledge about LME-based models to the following constraints to define acceptable rule sets:

1. The rule set must account for all instances of judgment (completeness==1).
2. The decision must be specified (0 or 1).
3. Each rule within the rule set must contribute toward the match (parsimony==1)

To further improve the decidedness of the rule set, we apply the rules in order of specificity where the more specified rules are applied first. A sample of the Pareto-optimal rule sets for the criterion model and the subject model are shown in Table 2. Note that two rule sets are needed because the criterion values do not exactly match the hypothetical subject judgment.

Each of the rule sets in Table 2 represents a point along the Pareto-optimal frontier (dimensions to include completeness, specificity, and parsimony). The rule sets are also optimized along completeness and parsimony. The analysis of Case I using the rule sets is show in Tables 3. For Case I, we see that, while *Re* and *Rs* reflect model accuracy, *C* also increased. Hence, the nonlinear aspect of the model is evident through the lens analysis.

Table 2. Sample Pareto-Optimal Case I Rule Sets Using GBLM

Criterion Model Rule Set	Judgment Model Rule Set
$if(x=0 \wedge y=0 \wedge z=1) \rightarrow (criterion=1) \vee$	$if(x=0 \wedge y=0 \wedge z=0) \rightarrow (criterion=0) \vee$
$if(x=0 \wedge y=1 \wedge z=1) \rightarrow (criterion=0) \vee$	$if(x=0 \wedge y=1 \wedge z=0) \rightarrow (criterion=0) \vee$
$if(x=1 \wedge z=1) \rightarrow (criterion=1) \vee$	$if(x=1 \wedge z=0) \rightarrow (criterion=0) \vee$
$if(z=0) \rightarrow (criterion=0)$	$if(z=1) \rightarrow (criterion=1)$

Table 3. Analysis of Case I Using GBLM.

x	y	z	Ye	Ye_resid	Ye_hat	Ys	Ys_resid	Ys_hat	Re	Rs	G	C	ra
1	0	0	0	0	0	0	0	0	1	1	0.775	1	0.775
0	1	0	0	0	0	0	0	0					
0	0	1	1	0	1	1	0	1					
0	0	0	0	0	0	0	0	0					
1	0	1	1	0	1	1	0	1					
1	1	0	0	0	0	0	0	0					
0	1	1	0	0	0	1	0	1					
1	1	1	1	0	1	1	0	1					

3.2 Case II Analysis

For Case II, we set the generative rule set for the environmental criterion as follows:
IF (X=0) AND (Y=0) AND (Z=0) THEN (JUDGMENT=NO) OR
IF (X=1) AND (Y=1) AND (Z=1) THEN (JUDGMENT=NO) OR
IF (ANYTHING) THEN (JUDGMENT=YES)
Because we assume noncompensatory judgments, we set the criterion values as the judgments themselves. Once again, we conduct our analysis in two ways. First, we find the lens model variables based on regression-based analysis. Second, we find models of the judge and criterion as generated by GBLM. Results found using multiple linear regression are shown in Table 4. As expected, the nonlinear component, *C*, of the LME is high. Moreover, based on the values of *Re* and *Rs*, we see that the regression model is not an effective model of decision making for Case II.
As in Case I, we found the Pareto-optimal rule sets for the criterion/subject model in Case II. The assessment of GBLM performance as a substitute for multiple linear regression models in instances of noncompensatory decision strategy use provides an interesting finding. From Table 4, we see that the LME-based model does not provide an accurate account of decision making behavior. When we investigate the performance of GBLM in Case II (Figure 5), we see that *C* remains high. Nevertheless, the models ($\hat{Y}e$ and $\hat{Y}s$) are now able to fully account for the criterion and the judgment.

4 Discussion

The idea that human decision makers use both compensatory and noncompensatory decision strategies is not a novel (Payne, 1976; Einhorn, 1970). However, while some have attempted to extend the Lens Model Equation to account for noncompensatory behavior (Einhorn, 1970; Cooksey, 1996) there does not exist a framework which is able to fully model both compensatory and noncompensatory strategies. This paper presents an effort to use the lens framework, as originally conceived by Brunswik (1955), to explicate both modes of decision making.
From the analysis of the two cases, we demonstrated that GBLM was able to account for cognitive control (R_s) and environmental predictability (R_e) better than its linear counterpart. However, the fact that "unmodeled" knowledge remained high in the Case II GBLM analysis presents a dilemma. Ideally, the unmodeled knowledge for the analysis of GBLM performance in Case II should be low whereas the analysis of LME-based model performance in Case II should reflect a high value. This dilemma offers opportunity for future research.

Table 4. Analysis of Case II Using LME.

x	y	z	Ye	Ye_resid	Ye_hat	Ys	Ys_resid	Ys_hat	Re	Rs	G	C	ra
0	1	1	1	0.25	0.75	1	0.25	0.75	0	0	1	1	1
1	0	1	1	0.25	0.75	1	0.25	0.75					
1	1	0	1	0.25	0.75	1	0.25	0.75					
1	1	1	0	-0.75	0.75	0	-0.75	0.75					
0	0	0	0	-0.75	0.75	0	-0.75	0.75					
0	0	1	1	0.25	0.75	1	0.25	0.75					
0	1	0	1	0.25	0.75	1	0.25	0.75					
1	0	0	1	0.25	0.75	1	0.25	0.75					

Table 5. Analysis of Case II Using GBLM.

x	y	z	Ye	Ye_resid	Ye_hat	Ys	Ys_resid	Ys_hat	Re	Rs	G	C	ra
0	1	1	1	0	1	1	0	1	1	1	1	1	1
1	0	1	1	0	1	1	0	1					
1	1	0	1	0	1	1	0	1					
1	1	1	0	0	0	0	0	0					
0	0	0	0	0	0	0	0	0					
0	0	1	1	0	1	1	0	1					
0	1	0	1	0	1	1	0	1					
1	0	0	1	0	1	1	0	1					

References

Brunswik, E. (1955). Representative Design and Probabilistic Theory in a Functional Psychology. *Psychological Review, 62*(3), 193-217.

Cooksey, R. W. (1996). *Judgment Analysis: Theory, Methods, and Applications*. San Diego: Academic Press, Inc.

DeJong, K. A., Spears, W. M., & Gordon, D. F. (1993). Using Genetic Algorithms for Concept Learning. *Machine Learning, 13*(2-3), 161-188.

Einhorn, H. J. (1970). The Use of Nonlinear, Noncompensatory Models in Decision Making. *Psychological Bulletin, 73*(3), 221-230.

Gigerenzer, G., & Goldstein, D. G. (1996). Reasoning the Fast and Frugal Way: Models of Bounded Rationality. *Psychological Review, 103*, 650-669.

Hammond, K. R. (1955). Probabilistic Functionalism and the Clinical Method. *Psychological Review, 62*, 255-262.

Hoffman, P. J. (1960). The Paramorphic Representation of Clinical Judgment. *Psychological Bulletin, 47*, 116-131.

Holland, J. H., Holyoak, K. F., Nisbett, R. E., & Thagard, P. R. (1986). *Induction*. Cambridge, MA: The MIT Press.

Hursch, C. J., Hammond, K. R., & Hursch, J. L. (1964). Some Methodological Considerations in Multiple-Cue Probability Studies. *Psychological Review, 71*(1), 42-60.

Klein, G. A. (1993). A Recognition-Primed Decision (RPD) Model of Rapid Decision Making. In G. A. Klein & J. Orasanu & R. Calderwood & C. E. Zsambok (Eds.), *Decision Making in Action: Models and Methods* (pp. 138-147). Norwood, NJ: Ablex.

Maule, A. J., & Hockey, G. R. J. (1993). State, Stress and Time Pressure. In O. Svenson & A. J. Maule (Eds.), *Time Pressure and Stress in Human Judgment and Decision Making* (pp. 83-101). New York: Plenum.

Mendelson, E. (1997). *Introduction to Mathematical Logic* (4th ed.). London: Chapman & Hall.

Payne, J. W. (1976). Task Complexity and Contingent Processing in Decision Making: An Information Search and Protocol Analysis. *Organizational Behavior and Human Performance, 16*, 366-387.

Rothrock, L., & Kirlik, A. (In Press). Inferring Rule-based Strategies in Dynamic Judgment Tasks. *IEEE Transactions on Systems, Man, and Cybernetics*.

Rothrock, L., & Repperger, D. W. (In Review). The Application of Multi-Objective Genetic Algorithms to Design and Validate Human-Machine Interfaces in a Manual Control Task. *IEEE Transactions on Systems, Man, and Cybernetics*.

Tucker, L. R. (1964). Suggested Alternative Formulation in the Developments by Hursch, Hammond, and Hursch, and by Hammond, Hursch, and Todd. *Psychological Review, 71*(6), 528-530.

Wright, P. (1974). The Harassed Decision Maker: Time Pressures, Distractions, and the Use of Evidence. Journal of Applied Psychology, 59(5), 555-561.

Analyzing Emotional Human-Computer Interaction as Distributed Cognition: The Affective Resources model

Ioannis Tarnanas, Athanasis Karoulis, and Ioannis Tsoukalas

Dept. of Informatics, Dept. of Psychology – Aristotle University of Thessaloniki
ioannist@psy.auth.gr, karoulis@csd.auth.gr, tsoukala@csd.auth.gr

Abstract

In this paper, we present the results of a research project, concerning virtual therapeutic environments as a new approach to emotional interaction modeling based on the concept of distributed cognition. One new technology for which the WIMP interface is not applicable is Virtual Reality (VR). The term VR was coined in 1989 by Jaron Lanier, although Sutherland (Sutherland, 1965; 1968). Previously, HCI focused mainly on the cognitive aspects of the user and the UI. Slowly however we are becoming aware that computer systems can also have non-cognitive effects on the user that should be taken into account when designing the UI. (Reeves & Nass, 1996) showed that humans have a strong tendency to respond to computers in similar ways as they do to other humans. Through "presence" we see humans responding to virtual scenes as if these were real, at a non-cognitive level. We could say that the computer system no longer only supplies the user with information; it also supplies the user with an experience (Laurel, 1993). We defined as *real-time emotional interaction* a target-oriented situation that demands a complex coping skills inventory of high self-efficacy and emotional decision-making strategies. By providing a model whose concepts are rooted in DC concepts we tried to achieve this visibility inside a virtual scenario. The virtual scenario differed in the number of affective cues, their type and the strategy of interaction. The goal was to investigate how these "affective cues" create a repository that can be used to enhance user performance in a crisis situation without any previous training. Other studies suggest that people have little awareness of their muscle activation during complex work or a crisis situation, so some form of biofeedback training might be necessary to sensitize them to non-optimal movement patterns and unnecessary muscle activation. Our study begins answering some of these questions, by selecting a single application that embodies these new developments. This enabled us to ground our understanding in a real world situation and it developed a test case to test that understanding. The application chosen for this purpose in this study is *Virtual Reality Exposure Therapy* (VRET). The affective resources model described in this paper defined a limited number of affective resource types as abstract information structures that can be used to analyze interaction. These two components of the resources model, information structures and interaction strategies, through the process of co-ordination and integration, provide the link between devices, representations and action that was not so well articulated in the DC literature. DC research identifies *affective resources for action* as central to the interaction between people and technologies, but it stops short of providing a definition of such resources at a level that could be used to analyze emotional interaction. In addition, we tried to provide the foundation for a program of research that extends the DC analysis of single user systems presented here to larger units of analysis more familiar to Collaborative Virtual Environments, Emotional Ergonomics and CSCW research.

1 Introduction

VR systems are used for certain types of tasks for which traditional user interfaces are less suited. Currently, VR is used amongst others in training and education, where users can learn how to operate complicated machines such as airplanes or how to work in dangerous environments such as a building on fire, in entertainment such as videogames, in visualization, for instance allowing people to walk through buildings that haven't been build yet (Brooks et al., 1992) and even in psychological therapy.

In psychological therapy, we can use the fact that VR can affect the user at a non-cognitive level to carefully administer synthetic experiences to the patient necessary to cure his or her mental illness. Particularly in phobia treatment, VR has already been shown to be effective with a great number of case studies and several-controlled studies. At this moment, phobia treatment in VR can be said to be outgrowing the experimental stage and is already in use by a handful of early adopters.

In VRET, the patient is confronted with a fearful situation by displaying a VE with anxiety provoking elements on a display such as a Head-Mounted Display (HMD). During the therapy, the therapist in collaboration with the patient alters the synthetic experience of the patient in such a way that the patient can slowly habituate to more and more fearful situations. For this, the system provides two user interfaces: one for the patient and one for the therapist. The UI for the patient incorporates advanced VR technology since the patient needs to be immersed in the VE, the interface for the therapist is often more conventional. Patient and therapist need to cooperate and coordinate their actions, requiring communication and interaction using these different user interfaces. For systems that are currently in use, these interfaces are very crude and simple, mainly dictated by the limitations of the hardware instead of the needs of the user. Here we see the importance of presence for VRET: If users do not feel present in the fearful situation then they will not experience fear and this fear will therefore not diminish through habituation. We thus need a system that has the ability to produce a sense of presence in the patient. The sense of presence a person experiences in a VE depends on several factors, such as display fidelity, the number of modalities stimulated (e.g. vision, sound, touch) and the individual characteristics of the user. Furthermore, most scholars attribute a special role for the interaction between human and computer in generating a sense of presence. However, the exact role of this HCI in presence and thus in VRET is not yet clearly understood.

2 Design Method

Computer science research studies artificial as opposed to natural phenomena. It deals with human creations and these human artifacts can be both created and studied (March & Smith, 1995). Thus, both natural sciences and the design sciences can be applied to information technology. The natural sciences are concerned with explaining how and why things are, design sciences are concerned with devising artifacts to attain goals. The research described in this dissertation is mainly concerned with the devising of artifacts, requiring a design methodology. This methodology consists of mainly two activities: building and evaluating.

The first question in determining our research approach then is: what should we build? The products one can build in the design sciences are of four types (March & Smith, 1995): constructs, models, methods and implementations. Constructs or concepts form the vocabulary of a domain, a model is a set of propositions or statements expressing the relationships among constructs, a method is a set of steps used to perform a task and an implementation or instantiation is the realization of an artifact in its environment. In this research, the products that we need to build are those that are necessary to facilitate the design of the user interfaces of the class of applications of which VRET is our prime example. Those products are:

1. A means of dealing with the question how to design such systems: a design *method.*
2. An understanding of the relationship between the HCI, presence, fear and effectiveness of the therapy: a *model* describing the relationships between these constructs.
3. An *instantiation* incorporating the former two: the design of the interfaces for both the patient and the therapist for VRET.

Whereas natural science tries to understand reality, design sciences attempt to create things that serve human purposes. Its products are therefore assessed against criteria of value or utility. The second question we need to answer to complete our research approach is: what criteria should be used to evaluate our products? The basic evaluation criteria are: (March & Smith, 1995; Dolby, 1996)

• *Does it work?*
• *Is it an improvement?*

2.1 Existing HCI design Methods

In the usability engineering lifecycle as described by Nielsen (Nielsen, 1993) we find a strong emphasis on focusing on the user and his/her task. The various ways in which Nielsen proposes we do this is by involving the user, also at the earlier stages of development, and by thoroughly analyzing the user and the user's task. The analysis of the user's task is referred to as a Task Analysis, and for several scholars (e.g. Nielsen, 1993; van der Veer & van Welie, 1999) it is basis of HCI design. In general, the HCI design process starts of with such an analysis of the user's current way of working. Based on any usability problems found, a second model is created, this time of the desired way of working, and a design of the system that should support and enable this new way of working. This new design is then evaluated, often through heavy involvement of the users, and possibly redesigned.

2.1.1 VE development process

An investigation of current design practice for VEs (Kaur, 1998) has shown that design is often informal and that an iterative approach is used, partly because most VR-development tools allow quick adjustments of the VE followed by subsequent viewing of the VE by the developers. However, very little user testing is done. In general, the steps taken in the development process is: 1. Requirements-specifications 2. Gathering of reference materials from real world models 3. Structuring of graphical models, sometimes dividing it between designers 4. Building objects and positioning them in the VE and 5. Enhancing the environment with texture, lighting, sound and interaction, and optimizing the environment Most time is spend on creating the appearance of the VE, with little effort spend on usability issues. Important HCI issues such as interaction techniques and current VR application developers do not deal with input devices.

3 Presence Model

In order to design a system that is effective, we will need an understanding of the way in which the HCI affects the effectiveness of the therapy. The simplest way to do this would be to manipulate one HCI variable at a time and observe its effect in real therapy. However, this would be extremely time-consuming and require many therapists and patients before significant results are obtained. We therefore need to divide the problem into smaller problems, using intermediary concepts such as presence and fear. This way we can determine the overall effect of HCI on effectiveness by determining the relationships between these intermediary concepts, which are much easier to investigate and at the same time provide us with a more thorough understanding of the underlying process. We can furthermore make use of the considerable amount of research that has already been performed in this area.

3.1 Evaluation of the presence model

Since this will be the first of such a model attempting to explain the effectiveness of therapy in terms of the HCI, it is irrelevant to ask whether it is an improvement. We will only have to show that it works. This can be tested by showing that predictions made using this model about the UI for VRET are correct. Here, the hypothetico-deductive method is appropriate: we can deduce hypotheses from the model and test these hypotheses using empirical data. These hypotheses should apply to the case of VRET. First of all, we tested the effect of the choice of patient control versus therapist control:
Hypothesis 1: Locomotion controlled by the patient will increase the patient's sense of presence.
Hypothesis 2: Locomotion controlled by the patient will increase the fear a phobic user can experience.

Second, we tested whether, in the case of patient control, there are any differences between the various locomotion techniques and affective cues at the environment. A more natural locomotion technique is one with which the user is closely familiar. Therefore, different locomotion techniques will be natural to different to users. However, the locomotion technique most common to all users is that which we use in everyday reality: walking with our own two feet and looking around by turning our head and body.
Taking this notion of natural in consideration, we can formulate the hypotheses: *Hypothesis 3: A more natural locomotion technique for the patient will increase the presence a user experiences*
Hypothesis 4: A more natural locomotion technique for the patient will increase the fear a phobic user can

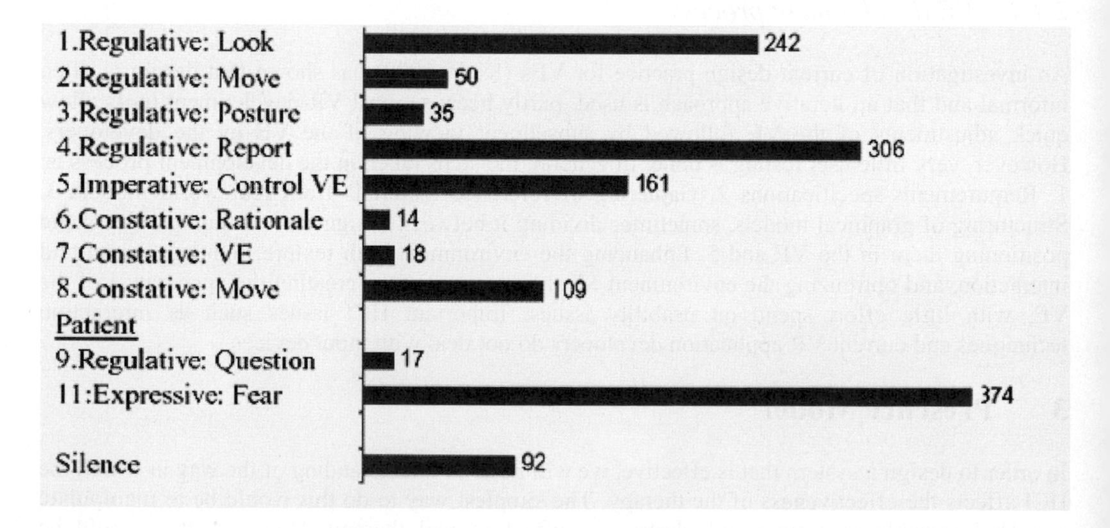

Figure 1. Sample events created using the Affective Resources Model

4 Conclusion

In this study we have discussed the research approach necessary to fulfill our research goal. Since the research goal entails the design of artifacts, we have stated that a design methodology is appropriate. The artifacts that need to be designed are: a design method, a model describing the relationship between the HCI and the effectiveness of VRET therapy using the concept of presence, and the designs of the user interfaces for our system that are a product of the former two artifacts. We classified 1420 affective resources events using the distributed cognition design model described above. A sample of these events can be seen in *figure 1*.

References

Brooks, F. (1992) *FinalAnnual Technical Report: Walkthrough Project*, Report to National Science Foundation, TR88-035NC:UNC-CH Dept. of Computer Science

Dolby, R.G.A (1996) *Uncertain Knowledge: An Image of Science for a Changing World*, Cambridge University Press, Cambridge, England

Kaur, K. (1998) *Designing virtual environments for usability*, PhD thesis, Center for HCI Design, City University, London. June

Laurel, B. (1993) *Computers as theatre*, Addison-Wesley Publishing

March, S.T., Smith, G.F. (1995) Design and natural science research on information technology, *Decision Support Systems*, Vol.15, No.5, December, pp.251-

Nielsen, J. (1993) *Usability Engineering*, AP Professional

Reeves, B., Nass, C. (1996) *The Media Equation: how people treat computers, televisions and new media like real people and places*, Cambridge: Cambridge University Press, 1996

Sutherland, I. (1965) The Ultimate Display, *Proceedings of the International Federation of Information Processing Congress*, Vol.2, pp 506-508

van der Veer, G.C., van Welie, M. (1999) Groupware Task Analysis, Tutorial Notes for CHI99 workshop, 16th May, Pittsburgh PA,

Cognition and Autonomy in Distributed Intelligent Systems

Robert M. Taylor

Dstl Human Sciences
Air Systems, Ively Gate, Ively Road, Farnborough, Hants GU14 0LX, UK
rmtaylor@dstl.gov.uk

Abstract

The aim is to consider the role of humans in distributed and autonomous intelligent systems, with enhanced capability from ubiquitous information, computer and communication technologies. The key human factors (HF) issues are generally believed to be what are appropriate levels of automation for military decision functions, and how to mitigate the associated risks for human understanding, prediction and control of system functioning. This paper develops these issues into a view of human roles through frameworks of military capability, cognition and autonomy.

1 Unattended Cognitive Vehicles

Recent advances in sensors, communications, computing and information technology have provided humans with remote control and supervisory roles for a variety of uninhabited vehicles and platforms. This gives the human operator improved accommodation, particularly for unmanned air vehicle (UAV) control, and the ability to work in relative comfort and safety, located either in a ground control station or a separate supervisory air platform. Increased remote control capability is sought to reduce manning and training costs. A future systems design goal is for one operator to control many platforms (vice the current many controlling one), reducing or eliminating human skill requirements and unwanted performance variability. In the future, humans may have only indirect control of operations, through the initial system specification and design, and through the mission and tasking instructions for supervisory control. Advanced uninhabited vehicles include ideas for Uninhabited Combat Air Vehicles (UCAV) and Unattended Cognitive Underwater Vehicles (UCUV). These are likely to have on-board cognitive automation operating with relatively high levels of autonomy or decision authority. Context sensitive technologies or "intelligent" computer software agents (e.g. Bayesian nets) offer the possibility of being able to control, regulate, direct and adapt system behaviour, within constraint boundaries, in uncertain, novel, and unpredictable situations. The aspiration is to achieve the requisite cognitive agility, precision, reliability and safety of operations with intelligent systems, with the minimum human supervision and human-computer communication.

In looking forward, cognitive automation and autonomous decision systems should be considered in the context of the prevailing military environment and associated developments in defence capability requirements. The military challenge is increasingly unpredictable, opaque and uncertain, and requiring an increasingly network-enabled command and control information infrastructure (CCII). UK MOD strategy for R&D and equipment procurement programs has moved from a platform-centric approach to a capability-based vision enabled by network technology. The UK MOD Defence Capability Framework (DCF) identifies seven key capability dimensions in three axes, namely (1) to command and inform, (2) to sustain and protect, and (3) to prepare, project and operate. Future network web-based information technology and seamless joint intelligence is expected to enable capability and increase battle-space exploitation. This is to be achieved by providing SA and decision/knowledge superiority, balanced information gathering and decision tempo, collaborative planning and command intent, timely integration and synchronisation of components, and optimised tempo and minimised friction. Decision superiority preserves operational flexibility, enables courses of action to be developed ahead of the adversary and maintains the initiative in the battle-space.

2 Distributed Cognition

It is generally believed that human involvement in critical decisions on the use of military force is paramount. Humans should decide the objectives, strategy, influences and intended effects of military force. Cognition provides humans with essential mental abilities for judging and deciding the appropriate and effective use of military force. This includes the ability to appreciate novel, complex and dynamic situations involving uncertainty and time pressure, to visualise complex new patterns, relationships and interactions in situations, and to interpret unstructured and incomplete information. Cognition enables humans to hypothesise and draw inferences about possible consequences of courses of action, to predict patterns, and to identify opportunities and alternatives. Cognitive skills provide the ability to anticipate consequences, to make decisions at the right tempo, and visualise how situations might evolve over time.

Information processing frameworks for cognition based on structural descriptions of fundamental stages of internal mental processes, free of context. An alternative class of pragmatic frameworks considers cognition in context and natural situations. This approach recognises that human performance is constrained by the conditions under which it takes place, and focuses on the variety of human performance and what cognition does, rather than what cognition is and the internal mechanisms for achieving it. Hollnagel (2002) summarises this approach in the following terms:

- Cognition is distributed across multiple natural and artificial cognitive systems and not confined to an individual cognitive agent
- It is part of a stream of activity not confined to a short moment in response to an event
- Sets of active cognitive systems are embedded in a social context which constrains their activities and provides resources
- The level of activity has transitions and evolutions and is not constant
- Most of the activity is supported by something or someone beyond the individual cognitive agent i.e. by an artefact or another agent

Hollnagel's Contextual Control Model (COCOM) considers the functions necessary to explain the orderliness of human action, with levels of cognitive control. It is intended to be applicable to a range of systems, including individuals, joint cognitive systems and complex socio-technical systems. In the Extended Control Model (ECOM), Hollnagel (2002) recognises that performance can take place at several levels simultaneously, or as multiple concurrent control loops; some are closed-loop and reactive, some are open-loop and pro-active. Four levels of activity or control loops are currently distinguished, namely controlling, regulating, monitoring, and targeting or goal-setting. The COCOM/ECOM framework is intended to support understanding of human roles and collaborative functioning with intelligent automation and joint cognitive systems.

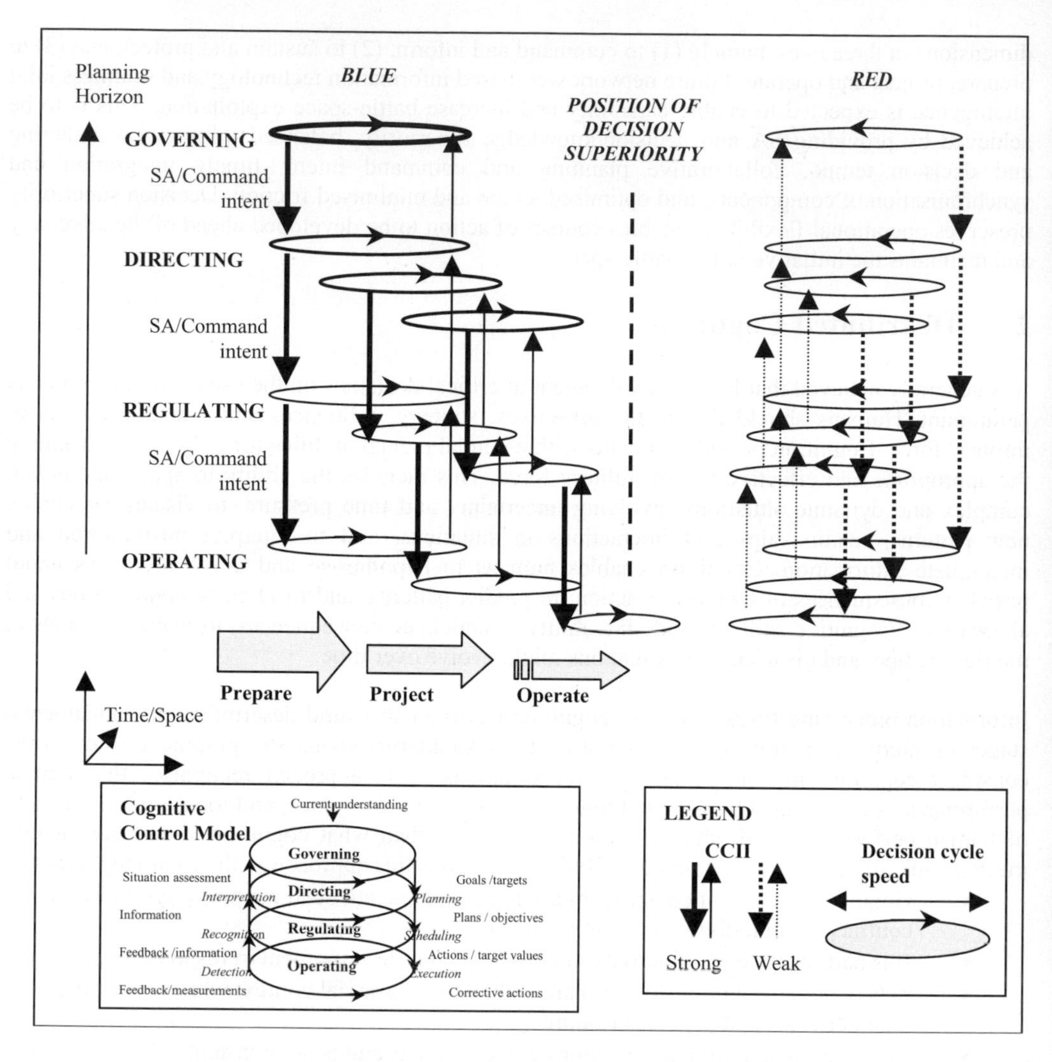

Figure 1. Decision agility through cognitive command and control

There are several conjunctions between the concepts of COCOM/ECOM theory of cognitive control and the UK MOD Defence Capability Framework. Shared situation appreciation, collaborative planning and command intent and the enabling of decision agility and combat identification in operations are important for a balance of tactical and opportunistic cognitive control modes. The coupling of detailed preparation and planning coupled with high situation awareness and command intent, are the foundation of decision agility and superiority in novel situations, and this makes sense in terms of cognitive control requirements. Preparation and planning enable predictable decisions to be made ahead of time and frees cognitive resources to control the situation both tactically and efficiently. In Figure 1, an ECOM model (governing, directing, regulating, operating) is used to illustrate the provision decision agility through strong CCII, command intent, planning and situation awareness

3 Multi-Agent Adjustable Autonomy

Adjustable contractual autonomy levels (At Call, Advisory, In Support, Direct Support), developed for cockpit decision support (PACT) have been used in UK MOD work on the management of multiple UAVs from manned cockpits and ground stations, and underwater autonomy (Taylor, 2001). Recent DARPA work indicates similar autonomy solutions e.g. ICAV Intelligent Control of Unmanned Air Vehicles project on mixed initiative distributed intelligence architecture. Future envisioned UCAV operations, involving real-time, multiple (group) collaborating autonomous vehicles in joint operations with manned platforms, seem likely that autonomous control levels will need extending beyond human commanded computer support, to cover autonomous complimentary, co-ordinated and co-operative planning and interactions.

Autonomy issues and implementation solutions have been addressed in work on multi-agent intelligent systems for problem solving in complex dynamic environments (Barber et al 2000). Mixed-initiative systems, dynamic adaptive autonomy and adjustable autonomy have been proposed to enable multi-agent systems to perform effectively with adaptability and flexibility. In the context of single-agent to human-user interaction, autonomy has generally been viewed as freedom from human influence. But for multi-agent systems, where the human user may be remote from operations, autonomy becomes a matter of the agents self-direction and goals, and the capability to dynamically form, modify or dissolve the agent organisation into goal-oriented, problem-solving groups. The degree of autonomy is considered to be implicit or explicitly linked to individual goals, and focuses on the decision making process used to determine how a goal is pursued free from intervention, oversight, or control by another agent (technical or human). Autonomy with respect to goals can be considered to be on a variable scale:

- Consensus or distributed control through consensus (working as team member, sharing decision-making control equally with other decision-making agents, all with equal authority),
- Master control (makes decisions alone, may communicate/order other agents with authority)
- Locally autonomous (makes decisions alone, only agent with authority).
- Command-driven or centralised control (makes no decisions about how to pursue goals, has authority, but must obey orders given by another agent),

Table 1 couples these agent autonomy levels with the PACT levels to provide a summary of the responsibilities in cognitive control model terms (governing, directing, regulating, operating).

Adjustable autonomy gives the agent architecture the ability to adapt their problem-solving to situations particularly in domains with unreliable communications and the possibility of agent failure, high degrees of uncertainty and resource contention needing distribution of tasks and co-ordinated planning to resolve conflicts. Distributed problem solving structures are generally thought to perform faster for complex tasks, when operating under uncertainty and changes in the environment, when few resources are shared, and when communication is unreliable. Centralised structures perform faster for simple tasks, when many resources are shared, when communication is reliable, and when there is no requirement to negotiate. Autonomy level agreements and communication protocols, joint intentions, and employing conventions for explicit commitment to specific interaction styles are considered necessary to establish reliability and trust. A central problem in adjustable autonomy is the transfers of control. The transfer from agent to human is believed to require a balancing of the costs of interrupting a human user with the benefits for highest quality decision making when the human has superior decision-making expertise. Transfer could be based on the expected utility of transfer, high uncertainty, incorrectness and significant harm, lack of decision capability, or thresholds of learnt rules. In multi-agent applications, cognitive strategies are needed for reasoning with adjustable autonomy in the operating context (situated autonomy) to provide the correct co-ordination, reordering, scheduling and to balance the costs, benefits, uncertainty and implications within the multi-agent group.

Table 1. Adjustable autonomy levels for intelligent multi-agent systems

AUTONOMY	GOVERNING	DIRECTING	REGULATING	OPERATING
Consensus Autonomy	Multiple intelligent computer agents	Multiple intelligent computer agents	Multiple intelligent computer agents	Multiple intelligent computer agents
Master Autonomy	Intelligent computer agent	Intelligent computer agent	Intelligent computer agent + Authorised support agents	Intelligent computer agent + Authorised support agents
Local Autonomy	Intelligent computer agent	Intelligent computer agent	Intelligent computer agent	Intelligent computer agent
Automatic/ Commanded Autonomy	Operator	Computer agent performing some interpretation & planning + Operator interrupt	Computer agent performing recognition & scheduling + Operator interrupt	Computer/intelligent agent performing detection & execution agent + Operator interrupt
Assisted Direct Support	Operator	Operator authorising + Computer agent performing some interpretation & planning	Operator authorising + Computer agent performing recognition, & scheduling	Operator authorising + Computer agent performing detection & execution
Assisted In Support	Operator	Operator performing + Optional computer agent advising & performing some interpretation & planning	Operator performing + Optional computer agent advising & performing recognition & scheduling	Operator performing + Optional computer agent advising & performing detection & execution
Assisted Advisory	Operator	Operator performing + Computer agent advising interpretation & planning	Operator performing + Computer agent advising recognition & scheduling	Operator performing + Computer agent advising detection & execution
Assisted At Call	Operator	Operator + Optional computer agent	Operator + Optional computer agent	Operator + Optional computer agent
Command	Operator	Operator	Operator	Operator

There is considerable potential for read-across for control architectures from cognition and joint cognitive systems for the control of distributed multi-agent systems. They use decision resources efficiency and enable the decision agility and adaptiveness needed for the manouevrist approach to military problem-solving. The use of cognitive control models will increase the transparency of control architectures and control authority for human user appreciation of the planning and interaction situation during collaborative problem-solving.

References

Hollnagel E. 2002. Cognition as Control: A Pragmatic Approach to the Modelling of Joint Cognitive Systems. Special Issue of IEEE Transactions on Systems, Man and Cybernetics A: Systems and Humans – "Model-Based Cognitive Engineering in Complex Systems (In Press) http://www.ida.liu.se/~eriho/

Taylor R.M. 2001. Cognitive Cockpit Systems Engineering: Pilot Authorisation and Control of Tasks. In R. Onken (Ed), CSAPC'01. Proceedings of the 8th Conferences on Cognitive Sciences Approaches to process Control, Neubiberg, Germany, September 2001. University of the German Armed Forces, Neubiberg, Germany.

Barber, K., Goel, A. and Martin C. 2000. Dynamic Adaptive Autonomy for Multi-agent Systems. Journal of Experimental and Theoretical Artificial Intelligence. Vol 12. Part 2, pp 129-148.

Bring Out Creativity!

Adi Tedjasaputra

translate-easy.com
Gading Elok Timur 6/BO2 -23
Kelapa Gading Permai
14240 Jakarta Utara
Indonesia
adi@translate-easy.com

Eunice Ratna Sari

translate-easy.com
Skovvej 20 -141
6400 Sønderborg
Denmark
sari@translate-easy.com

Abstract

Games are widely used as means of edutainment. One of the main reasons is the belief that games can help learning processes while entertaining the players. Whilst there are still some sceptical reactions on the issue, the study that we conducted asserts the belief.
For the purpose of the study, we have designed and developed a board game that can help intercultural and multidisciplinary designers to learn, share and work together while having fun. The results, conclusions, and observations during the study were parts of the design and development process of the game itself.

1 Introduction

When one looks at the sky, it is usually associated with: thinking, looking for inspiration or vision, daydreaming, reflecting, or contemplating. In these occasions, sky as a metaphor of inspiration, vision, dream, or idea has triggered a cognitive process.

On the other side of the coin, sky is also a part of nature. The shape of the clouds, the air movements, the starlight (including sun and its reflection on the moon), and sometimes the air pollution determine the characteristics of the sky in a complex and systematic manner.

Inspired by these facts, we developed a game that facilitates the transformation of creativity into ideas, solutions, and actions through trivial and perceptible seasonal phenomena. (see Figure 1)

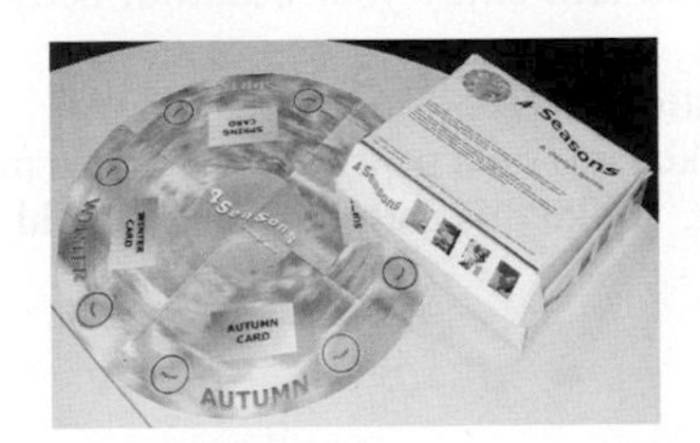

Figure 1: The “4 seasons” game

2 The game

The design and development of the "4 Seasons" game involved prospective users during the process. It consisted of five user testing sessions, five discussion and focus group sessions, a couple of card sorting sessions, a couple of word association sessions, some brainstorming sessions, and some iterative refinements. Four of the user testing and discussion sessions were conducted during the design and development process. The fifth user testing session and discussion group was done after the final game board was produced. In total, it involved 17 design students and 5 design institute staffs.

Initially, we wanted to design a game that simulates an interaction between a designer and a client. In the simulation, a designer has to learn some problems and some needs of the clients through some hints given by the clients. Collaboratively, they communicate with provided design materials. The nature of communication can be in the form of drawing, sketching, prototyping, role-playing/acting, or story telling.

Evolving from this, we blended the inspiration from nature with the designer-client concept, and eventually chose the four annual seasons as the theme of the game.

In the "4 Seasons" game, a participant is confronted with natural phenomena of a season and their effects. Collaboration and co-design with another participant to generate one or more scenarios of problems, ideas, and solutions is the core of the game. Acting, talking, role-playing, sketching/drawing or some other possible activities are encouraged during the collaboration.

This is an example of a conversation in the game:

P1 is the player who got a season card and P2 is the player who sat on his left hand side.

P1 got: **Summer card: Sun::Burn**

P1: I always go to the beach in Sønderborg and do sunbathing on summer. I hate when the sun burns my beautiful skin.
P2: Well.., you can use a sun screen then. They are available in any supermarket.
P1: Do you mean the sunscreen cream?
P2: Yes, you can use that too.
P1: Hmm.., but I don't like using sunscreen cream.
P2: Why?
P1: Because my skin is very sensitive and I get skin allergic easily.
P2: If that's the case, why don't we design a "transparent jelly plastic" that you can wear when you do sunbathing? This way, people can still enjoy your beautiful body and skin but you don't have to worry about getting sunburn.
P1: Ummhhh, I don't know. Will it be sticky?
P2: This is what I have in mind… (P2 sketches on a paper what is on his mind)
P1: P1 points to the sketch while saying, "I like this part, but can't we add a color to this part?"
P2: …

3 Results and Discussion

3.1 Perceptible and Concrete

Nature is a source of inspiration. Some creative ideas and solutions are the results of mimicking nature (Benyus, 1997). In particular, natural phenomena and their effects have a couple of significant characteristics: concrete and perceptible. They can be perceived through senses.

During the study, we observed that perceptibility is important. The perceptibility did not only ease the process of association within the game, but it also smoothened the transformation of creativity into scenarios, problems, and solutions. This is illustrated in the following cases:

In the third user testing session, an architect picked up a card related to "winter", "electrostatic", and "hurt". In collaboration with another participant, they hardly produced ideas or solutions. Two out of three other participants did not even understand "electrostatic".

In another collaboration, a participant who picked a card related to "autumn", "garden", "dirty" started out the association process by acting out a scenario of cleaning a bunch of dirty leaves with a broom and a backache problem during autumn. Because of his act, the other participant had a better understanding of the problem in hand and thus articulated inspiring, interesting, and innovative ideas and solutions together.

From a total of eight rounds in the fourth user testing session, there was only one occasion when a participant could not create a scenario and it was because the participant did not understand the instruction (see Table 1).

Table 1: The Results of User Testing Sessions

User testing session	Number of participants	Number of rounds	Results
1	4	2	Participants could create creative ideas, problems, and solutions
		2	Participants fumbled and mumbled
		4	Participants were stunned and could not come up with an idea.
2	3	6	Participants could articulate ideas, problems, and solutions
3	5	3	Participants faced difficulties in creating scenarios of problems
		12	Participants communicated creative ideas, problems, and solutions
4	4	1	Participants faced difficulties in creating scenarios of problems
		7	Participants communicated creative ideas, problems, and solutions

3.2 Roles Defining

Confrontation with problems can only stimulate creativity if it is followed by an activity of identifying and eliciting problems. It is in accord with what Franken argues about delineating problems (Franken, 2002).

However, we observed that strictly defined roles could also inhibit creativity. We observed this issue in two different user testing sessions. In the second user testing session (see Table 1), we implemented the game with role-based playing: designer-client and in the third session emphasized the importance of collaboration with no strictly defined roles.

In the second session, the participants were trapped in the paradigm of designer-client and unconsciously restricted their potentials in creating creative scenarios and solutions. They were limited by the role they had at a particular round. A general pattern was that a client would only tell or complain about their problems and asked a designer to give some ideas or solutions for them, whereas the designer responded with plausible solutions and answers. However, in the third session, the participants could play interchanging roles unconsciously, which resulted in interesting collaborative processes and outcomes.

In the third session, there was an interesting situation. During the game, suddenly a participant mentioned that he was the designer and his mate was the client. Suddenly the situation changed drastically. The participant who was called the client suddenly lost her ability to articulate her ideas. She only mentioned her problems to her "designer" mate and asked him for some solutions. This was repeated in the next round. However, in a subsequent round, one of the participants realised that the game instruction did not ask them to perform designer-client role and then the situation changed again to a non-strictly defined role situation. After this, during the rest of the session, the participants could communicate creative problems, ideas and solutions while having fun.

3.3 The Judging System

The use of judging system was inspired by the idea of punishment and reward. Eisenberg and Selbst argue that appropriate reward will enhance creativity in a subsequent task, but improper reward will have a reverse impact (Eisenberg & Selbst, 1994). Langer argues that premature judgement is a characteristic of mindlessness (Langer, 1989). Strickland and Franken argue that judgement will inhibit creativity (Strickland, 1989; Franken, 2002). Based on these thoughts, we were interested in understanding the impact of a judging system on creativity in our game.

We did some experiments to figure out a judging system that has three characteristics: involves punishments and rewards directly or indirectly, applied in timely manner, and brings out creativity.

For these purposes, the first experiment adopted an open democratic judging system with Danish grading system. This was chosen with an assumption that the participants are familiar with the local grading scale. However, in a discussion session after the first user testing session, all participants uttered that they faced a difficulty in using the Danish Grading System because they had not known how to measure creativity with this grading system and they were too many numbers to be considered in a very short time. The second problem was that an open system created inconvenience. A participant felt strongly inconvenient to give marks for the other participants and at the same time the other participants knew his mark. Three out of four persons felt very bad, embarrassed and discouraged when they got low marks and they were aware that the other participants knew their marks.

In the second user testing session, we applied close democratic judging system. A player did not know the marks of the other participants and how the other participants judge a particular participant. Since Danish grading system was considered unsuitable to measure creativity, we experimented with a grading scale of 0 to 9 with an assumption that this grading system was reasonably universal (10-scale grading system) and thus understandable. During the game, the body language of the participants showed that they were suspicious and curious about their marks. Subconsciously, this situation distracted the flow of the game. The participants found that the scale was easier to understand, but there were still too many numbers to be considered in a very short time.

Based on this situation, we created a scale of 0 to 3 and instructed the participants to have discussions among themselves, but to give marks individually. The feedbacks of the participants from last three user testing sessions showed that this judging system was satisfactory for the following reasons: it was simple and understandable, it stimulated strategic decision-making and negotiation, and it also encouraged collaboration.

However, in the last user testing session, there was an interesting situation where one group forgot to use the judging system. This group said that they had had a lot of fun because each participant produced interesting, novel, and wild ideas, while the rest responded enthusiastically and discussed the ideas together.

4 Reflection

This paper reminds us that an interesting game should be playful, interesting and without pressure. The 4 Seasons was initially made to facilitate creativity transformation, but along the way, it evolved into a game that can develop designer's creative thinking and collaborative skills by creating a context that subsequently triggers problems that can challenge and engage designers in positive and productive manner while strategically planning, negotiating, and making decisions.

References

Benyus, J. (1997). Biomimicry: Innovation inspired by nature. USA: William Morrow and Company, Inc.

Franken, R.E. (2002). Human Motivation. California: Wadsworth/Thomson Learning.

Goldenberg, J.,& Mazursky, D. (2002). Creativity in product innovation. UK: Cambridge University Press.

Eisenberger, R.,& Selbst, M. (1994). Does reward increase or decrease creativity? *Journal of Personality and Social Psychology*, 66, 1116-1127.

Langer, E.J., (1989). Mindfulness. Reading, MA: Addison Wesley.

Strickland, B.R., (1989). Internal-external control expectancies: From contingency to creativity. *American Psychologist*, 44, 1-7.

An Aircraft Preference Study on the Application of Vector Maps in U.S. Navy Tactical Aircraft

Michael E. Trenchard, Maura C. Lohrenz, Stephanie S. Edwards

Naval Research Laboratory
Mapping, Charting, and Geodesy Branch
1005 Balch Blvd.
Stennis Space Center, MS USA 39529-5004
trenchard@nrlssc.navy.mil, mlohrenz@nrlssc.navy.mil,
sedwards@nrlssc.navy.mil

Abstract

Cockpit digital map displays have long been considered a good situational awareness (SA) tool for the pilot. However, most cockpit map displays have been limited to display of digitized paper charts and imagery due primarily to limited computational capabilities in tactical aircraft. One significant problem encountered by tactical aircraft pilots is map display clutter. Important mission planning and real-time overlays are often rendered over the map display during flight. The clutter resulting from the combination of these overlays with the underlying map display can lead to reduced situational awareness. Unfortunately, the feature content of digitized paper charts and imagery cannot be altered in these map displays to help alleviate the clutter. Next-generation cockpit map displays will not have the computational limitations of previous systems. Vector map displays can provide for added customization and embedded information within the map and address the problems associated with cluttered map displays. Of course, there are tradeoffs to be made between the potential of added flexibility in the cockpit versus a potentially higher pilot workload. A web-based Vector Map survey was developed to gather Navy and Marine Corps aircrew preference data to evaluate functional aspects of vector maps in the cockpit and in mission planning. The aircrew that responded to the Vector Map survey represented a full cross-section of Navy and Marine Corps Tactical and Rotary Wing platforms, and their associated missions, who currently have or are expected to have a requirement for cockpit Moving Map functionality.

1 Background

Today's U.S. Navy and Marine Corps cockpit digital moving map is primarily driven by three geospatial databases: CADRG (Compressed ARC Digitized Raster Graphics), CIB (Controlled Image Base), and DTED (Digital Terrain Elevation Data). CADRG is the U.S. Department of Defence (DoD) standard digitized paper chart map product. CADRG is a simple scanned, digital representation of a paper chart that encompasses a wide range of specific chart series and equivalent map scales. CIB is the U.S. DoD standard digital imagery product available in 10-, 5-, and 1-meter resolutions. Both CADRG and CIB are "raster" databases (i.e. represent a simple pixel-by-pixel reproduction of a picture). DTED is a similar database in that each element represents an elevation value. There are several fundamental problems associated with using these raster-scanned databases as base-maps for aircraft moving-map systems:

- The source paper charts originally were designed to support a wide variety of users. As such, they are general-purpose, with excessive detail irrelevant to many operations.
- The source charts were never intended to be used in a digital context; scanning, transforming, and compressing chart data can render them nearly illegible.
- The source charts were designed to be stand-alone, and overlaying new mission planning symbols on them often results in a distracting, cluttered display.
- There is inconsistency in colors, symbols, and text among the charts (e.g., foreign chart producers, varying chart ages and standards, etc.). The cartographic variability causes disruptive changes in appearance when crossing chart boundaries or changing between chart series.
- Raster chart data are inflexible. Each pixel is a simple RGB (red, green, blue) value with no spatial information, so it is impossible to either customize or query the map.

Vector maps and, in particular, the Vector Product Format (VPF) produced and distributed by the National Imagery and Mapping Agency (NIMA, 1996), are geospatial databases comprised of point, line, and area features that can be queried and displayed by geospatial location and thematic content (e.g. transportation, vegetation, industry). These databases are fully attributed and conform to international DIGEST standards. The ability to query a map on thematic content can provide the pilot with the ability to selectively add or remove detail from a map based on particular mission needs. Vector maps can solve the problems that are inherently associated with typical raster charts. However, while there is value in the added flexibility to design and customize a vector map, there is also the potential for increased aircrew workload to manage the level of flexibility provided in the cockpit and in mission planning. Customization and workload issues with vector maps were first identified in an NRL study conducted in 1995 to help define the baseline mapping requirements for the Tactical Aircraft Moving Map Capability (TAMMAC) digital map system (Lohrenz, Trenchard, & Myrick, 1997). This survey is a next step in helping to define which types of vector map functions are most desired by aircrew who either currently have or soon will have an integrated cockpit moving map display system.

2 Approach

The Naval Research Laboratory (NRL) developed an internet-based, on-line Vector Map survey to gather U.S. Navy and Marine Corps aircrew preference data to evaluate functional aspects of vector maps in the cockpit and in mission planning. NRL used a similar internet-based survey technique as a lower-cost and less intrusive alternative to one-on-one pilot interviews in a pilot-centered study of aircrew mapping needs for MCM (Mine Counter Measures) and ASW (Anti-Submarine Warfare) missions (Trenchard, et al., 2000). The aircrew that responded to the Vector Map survey represented a full cross-section of U.S. Navy and Marine Corps Tactical and Rotary Wing platforms and their associated missions, who currently have or are expected to soon have a requirement for a cockpit moving map capability.

After a short introduction page to familiarize participants with the purpose and scope of the vector map survey, each participant was instructed to complete a registration page that collected information on familiarity with moving map systems, flight experience, platform and mission information. Information from the registration page was used to develop an aircrew profile by both platform and mission type.

The survey was comprised of five functional vector map demonstration pages and a final rankings page. The demonstration pages allowed each participant to view a preset demo of a particular vector map function and/or interact with the vector map display. Once the function had been demonstrated, the participant was then asked to rate (Table 1) the function for use in both the cockpit and mission planning. In addition to rating each vector map function, a "comments" section was included on each page to allow the participant to add qualitative feedback on the vector map function shown. Finally, the participant was asked to rank the five vector map demonstrations in order of priority for implementation.

Table 1: Vector map ratings scale

5	Extremely Useful
4	Of Considerable Use
3	Of Use
2	Of Limited Use
1	Of No Use

A second part to the survey was comprised of the evaluation of six geospatial content vector map pages. However, this paper will only focus on the vector map functional aspects of the survey.

3 Results

3.1 Aircrew Profile

The platforms and missions that were identified by the aircrew participants are shown in Table 2.

Table 2: Vector map survey participants

Platform	Missions	# Participants
AV-8B	Primary - Air-to-Ground Combat (AGC); Secondary - Air-to-Air Combat (AAC)	25
F/A-18 C/D/E/F	Primary / Secondary – AAC; AGC (tied)	27
H-1 (AH-1W, UH-1N)	Primary – AGC Secondary – Forward Air Controller-Airborne	19
H-60 (SH-60B, SH-60F, HH-60H)	Primary – Anti-Submarine Warfare (ASW) Secondary - Search and Rescue (SAR)	16
Other	Various	7
Total		**94**

3.2 Vector Map Functional Analyses

The following vector map functions were evaluated:

1. Customize Detail of Map Features [Declutter]
2. Reorder Vector Layers [Reorder]

3. Upright Text [Upright Text]
4. View Map Metadata [Metadata]
5. Database Query [DB Query]

Declutter demonstrated the ability to customize the detail of map features by thematic and selective feature content. Reorder demonstrated the ability to de-conflict map features by re-ordering the feature content. Upright Text demonstrated the ability to separate layers of vector information to allow text information to remain in a north-up orientation regardless of the orientation of rest of the map, thus providing better text readability of the map. Metadata demonstrated the ability to access detailed map feature attribution (e.g., runway lengths, composition) and data quality (e.g. source, accuracy). DB Query demonstrated an advanced method to quickly declutter a map based upon pre-defined queries on the attribution of map features to customize a vector map display. Figure 1 shows that Declutter attained the highest overall ratings and weighted priority rankings for cockpit implementation.

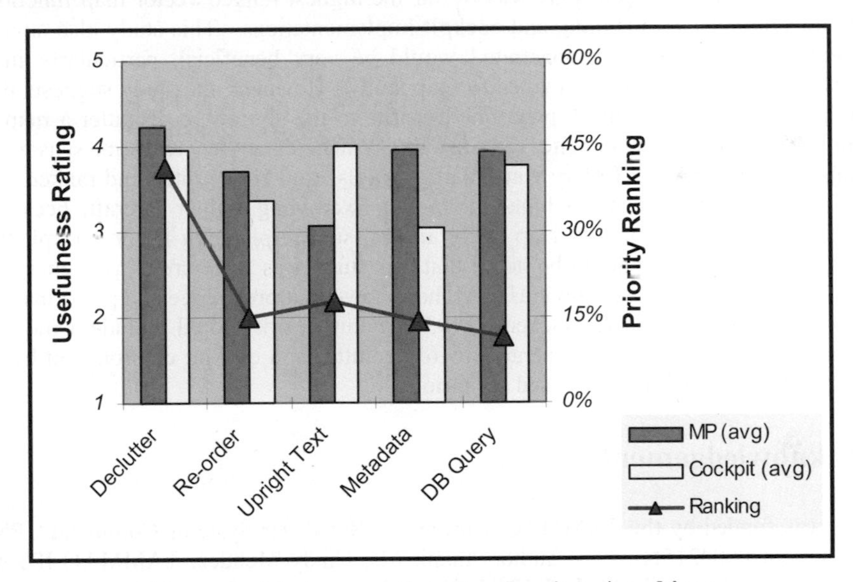

Figure 1: Vector map functional ratings / rankings

Both the AV-8B Harrier and F/A-18 Hornet communities addressed concerns with task prioritization and workload issues in the cockpit relative to the Vector Map functions outlined in the survey. Customizing details of map features and viewing metadata functions would allow them to tailor the map display to minimize task saturation and access specific data if necessary. There was also concern about potential of the upright text function to adversely affect situational awareness. The re-order vector layers and database query functions were seen as potentially task intensive, but the ability to set aircrew profiles during mission planning was highly desired and could mitigate a potential increase in task workload. The ability to interact with the moving map during mission execution was especially desired by the H-1 and H-60 communities. The high priority assigned to customizing map features, reordering vector layers and keeping text upright are consistent with the high priority of map utilization for threat avoidance, terrain avoidance and navigation. Detailed map study during mission planning would reduce the priority for additional map information in flight via the metadata and database query functions. Among all the participants, but particularly those with less flight experience, the database query function was seen as a high priority to assist with determination of available emergency landing airfields. T-test analysis revealed a significant ratings preference toward the customization of map features (declutter) in cockpit implementation for both the AV-8B and H-1 platforms. In mission planning implementation, T-test analyses revealed significant ratings preference toward declutter and database query for both the AV-8B and the H-60, and declutter for the F/A-18. Forced rankings indicated a high preference for declutter for all platforms that participated in the survey. In all demonstrations except the Upright Text demonstration page, all platforms indicated a higher rating preference for implementation in mission planning versus the cockpit. However, all vector map functions had average ratings of at least 3 (of use) or higher for both mission planning and cockpit implementation overall.

4 Conclusions and Recommendations

The ability to declutter a map display was by far the highest ranked vector map function for all platforms for both mission planning and cockpit implementations. This study also suggests that all the vector map functions demonstrated would be very beneficial, particularly in mission planning where more time can be devoted to map study. However, the study suggests there may be benefits to providing a reduced, predefined profile to the aircrew to declutter a map quickly. For example, a pilot may wish to better visualize divert airfields or view real-time sensor, imagery, or mission-specific overlays. Rotary wing platforms (H-1 and H-60) rated and ranked secondary functions of reorder and metadata higher than did fixed-wing fighter aircraft, because rotary aircrew could devote more time to map study and mission re-planning in the cockpit than their fixed wing counterparts. It should be noted that this study was an aircrew preference study and not a performance-based study. Previous work has shown that preference and performance do not necessarily correlate (Merwin & Wickens, 1993). Future work should include a more detailed demonstration of the possible implementations of declutter in a cockpit environment by platform and mission type in a performance-based scenario.

5 Acknowledgements

This work was funded by the TAMMAC team at the Naval Air Systems Command (PMA 209; program element 0604215N). The authors thank Ms. Cindy Mendez, TAMMAC IPT lead and Mr. Dan Shannon, Deputy TAMMAC IPT lead for their support of this work. Finally, we would like to thank the participants who contributed their time, expertise, and enthusiasm to this study.

References

National Imagery and Mapping Agency (April 1996). Interface Standard for Vector Product Format, MIL-STD-2407.

Lohrenz, M.C., Trenchard, M.E., Myrick, S.A. (October 1997). Digital Map Requirements Study in Support of Advanced Cockpit Moving Map Displays. NRL/FR/7441—96-9652.

Trenchard, M.E., Lohrenz, M.C., Myrick, S.A., Gendron, M.L (October 2000). A Two-part Study on the Use of Bathymetric and Nautical Chart Information in a Moving-Map Display. *Human Performance, Situational Awareness & Automation Conference* (Savannah, GA).

Merwin, D., and Wickens, C. (1993). Comparison of eight color and gray scales for displaying continuous 2D data. *Proceedings of the Human Factors and Ergonomics Society 37th Annual Meeting.*

Task Decomposition: Why do Some Novice Users Have Difficulties in Manipulating the User-interfaces of Daily Electronic Appliances?

Kazuhiro Ueda, Masaki Endo

University of Tokyo
3-8-1 Komaba, Meguro-ku, Tokyo,
153-8902 Japan
{ueda, endo}@cs.c.u-tokyo.ac.jp

Hiroaki Suzuki

Aoyama-Gakuin University
4-4-25 Shibuya-ku, Tokyo,
150-8366 Japan
susan@ri.aoyama.ac.jp

Abstract

A series of experiments were conducted to explore a source of difficulties in users' manipulating daily electronic appliances and to propose a method for supporting such technologically inept users as finding difficulties in using the appliances. Users are required to decompose their goals into a set of sub-goals or subtasks in a specific way when using these appliances. We made a preliminary experiment, using a copier, to derive a hypothesis that the users who had extreme difficulties in using these appliances failed to decompose a task or that their decompositions were different from the one that the designers had assumed. In order to test this task decomposition hypothesis, we compared the performance of those who had been instructed the general idea of task decomposition with that of those who had not. The result of the experiment showed that the experimental group achieved faster with fewer errors than the control group, which indicated that the hypothesis was right. Based on this view, we then proposed a method to help users understand "task decomposition" by providing all operating sub-functions in a tree structure. We made two experiments to find the effectiveness and robustness of the method.

1 Introduction

The purpose of the study is to explore a source of difficulties in technologically inept users' manipulating daily electronic appliances and to propose a method for supporting these users.

People use various electronic appliances on a daily basis. Many appliances have become highly multi-functional to meet various users' demands. The complex operation of such appliances, as a result, has made them hard to be used so that many people find it hard to understand the user-interfaces. To improve the usability of these appliances, some methods of user-interfaces have been proposed: Desktop icons, pull-down menus, etc [Norman, 1988]. Despite the improvements, however, some people (the so-called technophobes) still have trouble using the appliances.

Why so? For example, an icon or a pull-down menu never provides a user with the whole structure and meanings of functions because it represents only a small part of a whole task or function. When we use an appliance to carry out a task, on the other hand, we must decompose the task into some subtasks that correspond to the functions equipped with the appliance: We must carry out *task decomposition*. The interfaces that give no information about the whole structure of functions, however, will not facilitate users' task decomposition. This is thought the reason why some users still have trouble using electronic appliances.

So we assume that some novice users, by nature, do not clearly conceive the idea of task decomposition. Here task decomposition means that every task can be decomposed into subtasks and that, by completing each subtask, the original goal can be accomplished. This task decomposition hypothesis was derived from the result of a preliminary experiment by using a copier: Three undergraduate students, who had used copiers only to make copies piece by piece, were asked to make five double-sided copies of three pages of A4-sized, one-sided manuscript and to sort them for the given number of persons. One of the three subjects, who could not complete the task without referring to the manual, tried to carry out the task, or to push the start button, without finishing the settings of all the necessary sub-functions. This was assumed to be caused by her failure in task decomposition because she never missed the buttons necessary for the task.

This task decomposition hypothesis will be tested, using a copier, in Experiment 1. We will also investigate how such technologically inept users can be supported, based on the hypothesis, when they carry out tasks with using appliances. For this purpose, we will propose the method of *visualizing all the sub-functions* of an appliance and test its usefulness in Experiment 2 & 3.

2 Experiment 1

We conducted an experiment, with a copy simulator, to test the "task decomposition hypothesis."

2.1 Method

Subjects: Ten undergraduate students, who found it hard to manipulate daily electronic appliances, participated in the experiment. The choice of the subjects was based on a questionnaire: The questionnaire consisted of 11 questions about experiences of using daily electronic appliances and computer software. From the results of the questionnaire, we counted the score of each candidate and selected, as our subjects, the 10% candidates from the lowest. All the ten subjects were divided into two groups so that the average scores of the two groups were the same: In one group, subjects were taught, in advance, the general and abstract idea of task decomposition (experimental group) while, in the other group, subjects were not (control group). We hypothesized that the experimental group would show better performance than the control group.
Tasks: We asked the subjects to carry out the six tasks, each of which consisted of three sub-tasks. For example, the first task required each subject to make three double-sided copies of four pages of an A4 sized, one-sided manuscript and to group them for a given number of persons.
Device: We did not use an actual copier but a simulator, because the subjects' operation needed to be recorded in detail and a simulator would not be disabled by paper jam during the experiment. The simulator used is shown in the upper part of Figure 1: The subjects needed not to set manuscripts on ADF but to operate clicking a mouse, instead of pushing buttons. They could confirm, on the display of the simulator, whether they could carry out the tasks accurately or not.
Procedure: All the subjects were first asked to carry out three practical tasks to get the feel for operating the simulator. In addition, the subjects only of the experimental group were taught both a daily example of and the abstract idea of task decomposition. All the subjects were then asked to carry out the six tasks. The simulator displayed these tasks one by one and was designed so that the subjects could proceed to the next task only when they completed a task without errors. They were asked to achieve the tasks by themselves: We did not answer their any question.

2.2 Results

First we focused on the performance of achieving the first task. The subjects of the experimental group took only 100 sec. on average while those of the control group did 400 sec. In addition, the subjects of the experimental group made only 0.4 errors on average while those of the control

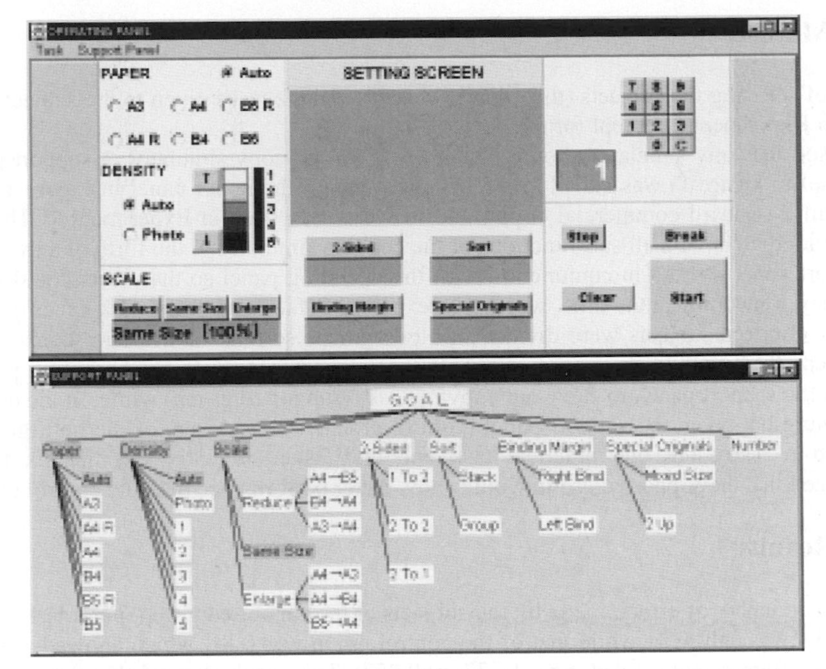

Figure 1: Simulator for Experiment 1 (the upper panel only) and Experiment 2

group did 1.8 errors. T-tests revealed that the experimental group completed the first task faster and with fewer errors than the control group (both the differences were significant, $p < 0.01$).

We then examined the average time and number of errors per one task through all the six tasks and, in this respect, compared between the performances of the two groups. So we could find that the subjects of the experimental group took only 66.3 sec. on average while those of the control group did 161.3 sec, which showed a marginally significant difference between the two groups (The two groups showed no significant difference in the average number of errors).

The above results showed that the subjects of the control group learned to understand the functional structure of the simulator. However, they took no less than 14 min. to do the learning. We can, therefore, conclude that the subjects of this experiment, namely some of the technologically inept users, did not have the idea of task decomposition or use it explicitly and that they could improve their performances given some instruction about task decomposition.

3 Experiment 2

The above result indicates that the idea of task decomposition serves as a map to help users to search their ways in the task space. If so, the task decomposition needs not always be internalized as in Experiment 1: Instead, it can be externalized in the user-interface itself. We then propose the method of visualizing all the sub-functions of an appliance to help the users, who have difficulties using machines by the lack of the idea of task decomposition, to make copies accurately and smoothly. Based on this method, we built a copy simulator with an operating panel (i.e. the upper panel of Figure 1) and a corresponding support panel (i.e. the lower panel of Figure 1), the latter of which visualized the structure of task (or functional) decomposition in the user interface. To test the usefulness of the method, we then investigated whether technologically inept users could perform well in making copies using the proposed support panel of the simulator.

3.1 Method

The way of selecting the subjects (the number was 14) and the tasks given to the subjects were the same as in Experiment 2, except for the device used.

We used the copy simulator shown in Figure 1. On the copy simulator, a support panel (the lower panel in Figure 1) was added to the operating panel (the upper panel in Figure 1) that was provided in a standard commercial copier and that was also used in Experiment 1. This support panel provided users with all sub-functions of the copying simulator in the form of a tree structure. The support panel worked in conjunction with the operating panel so that users could see which subtask they were doing at the time, which was expected to facilitate users' task decomposition.

All the fourteen subjects were divided into two groups so that the average scores of the two groups were the same: In one group, subjects were asked to operate the operating panel, with looking at the support panel, to carry out copying tasks (supported group) while, in the other group, subjects were not given the support panel (control group). The subjects of the both groups were first asked to carry out one practical task and then asked to carry out the six tasks. We hypothesized that the supported group would show better performance than the control group.

3.2 Results

The average number of errors, made by the subjects of the supported group, per a task through all the six tasks was only 0.31 while that of the control group was 0.81, which showed a marginally significant difference (one-sided, $t(8)=1.777$, $p=0.057$). This result showed that the support panel was useful for the novices to use the interface smoothly and to facilitate their task decomposition.

4 Experiment 3

It is true that the proposed support panel of Figure 1 is effective for some novices to use the copier interface smoothly. This may be, however, caused by the relatively understandable structure of functions in the operating panel. We need, therefore, clarify whether our proposed method can intrinsically facilitate users' task decomposition, whatever the design of a corresponding operating panel is. So we will make the same experiment as Experiment 2, using another operating panel.

4.1 Method

The way of selecting the subjects (the number was 14), the tasks given to the subjects and the procedure were the same as in Experiment 2, except for the device used.

In this experiment, we used a copy simulator (Figure 2) with a different design from the one in Figure 1 to determine the robustness of the proposed supporting method. This operating panel is different from that in Figure 1, in that the former has some unspecific buttons such as "Basic" and "Added", which might prevent users' understanding the interface.

4.2 Results

The result was almost the same as that of Experiment 2. The average number (0.07) of the errors per a task made by the subjects of the supported group was significantly fewer than those (0.86) made by the subjects of the control group (one-sided, $t(7) = 3.586$, $p = 0.004 < 0.01$). This result shows that the effectiveness of the supporting method dose not depend on the design of operating panel, which means that this supporting method can intrinsically support some novice users.

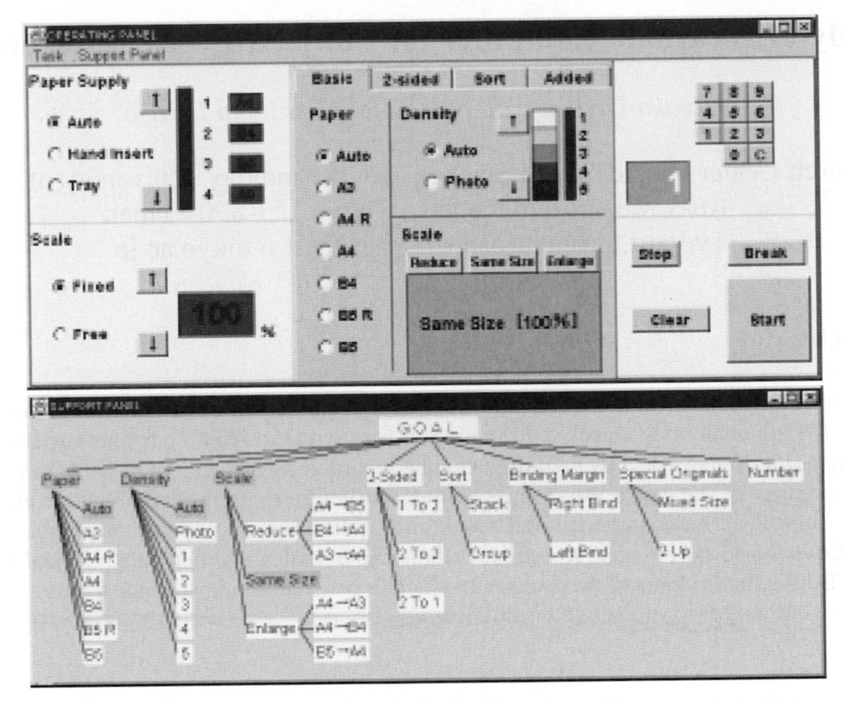

Figure 2: Simulator for Experiment 3

5 Discussions and Conclusion

Experiment 1 showed that one of the reasons why some users find it hard to use appliances was their failure in task decomposition. Some previous studies pointed out the importance of building "mental models in manipulating machines" [Gentner & Stevens, 1983], but they gave no answer how to build a mental model. We believe that the task decomposition contributes to this problem.

To help some novice users, we proposed the method of visualizing all the sub-functions of an appliance. Experiment 2 and 3 showed the usefulness and robustness of the method. Our proposal is totally opposite to the "less is more" principle [Nielsen, 1993]. Appropriately reducing the number of functions displayed, which is necessary in the "less is more" way, is a tough problem. However, our approach, displaying all the functions, can skip this problem. Our idea is also different from the "information appliance" [Norman, 1998] because a functional structure of an appliance is "visible" in our method, which enables users to leverage "flexibility and power" of a versatile appliance, while it should be "invisible" in the Norman's idea. Our approach is rather similar to GOMS model [Card et al., 1983]: The both are, however, different from each other because our method is for facilitating users' manipulation while GOMS is a user-model itself.

References

Card, S. K., Moran, T. P. & Newell, A. (1983). *The Psychology of Human-Computer Interaction*. Hillsdale, NJ: Laurence Erlbaum Associates.
Gentner, D. & Stevens, A. L. (1983). *Mental Models*. Hillsdale, NJ: Laurence Erlbaum Associates.
Nielsen, J. (1993). *Usability Engineering*. San Diego, CA: Academic Press.
Norman, D. A. (1988). *The Psychology of Everyday Things*. New York: Basic Books.
Norman, D. A. (1998). *The Invisible Computer*. Cambridge, MA: MIT Press.

Study of Wearable Computer for Subjective Visual Recording

Ryoko Ueoka, Koichi Hirota, Michitaka Hirose

Research Center for Advanced Science and Technology ,University of Tokyo
4-6-1 Meguro-Ku Komaba Tokyo 153-8904 Japan
{yogurt,hirota,hirose}@cyber.rcast.u-tokyo.ac.jp

Abstract

This paper describes the wearable computer for subjective experience recording. As the wearable computer fits a human body closely all the time, it is possible to record personal experience from egocentric point of view. And by reconfiguring the recorded data, it will be possible to reconstruct subjective reality of the world.We discuss memory structure to define the objective elements of recording for subjective experience and focus on the discussion to record subjective vision.We conducted two basic experiments to estimate the characteristics of a person's physical behaviour in space. Taking the findouts of these experiments into consideration, we made a prototype system of wearable computer for subjective visual recording.

1 Introduction

Today, the wearable computer is becoming a reality due to the significant downsizing of computer devices.[1] And it has received great attention mainly due to the effectiveness of outdoor information display devices.However, they also hold great potential for scanning and recording real time information about the immediate surroundings and ambience around the user because a wearable computer fits the human body closely.

It can be said that recording individual experience is one of the most important and practical applications of wearable computers. The desire to record one's own experiences is so strong that the trend of downsizing recording devices such as cameras and video cameras has been constant. Not only is recorded history is worthwhile to allow individuals to recall their experiences, but also it will become a valuable record of the period for the future. Taking the characteristcis of wearable computer into account, our final goal of this study is to construct a time machine to be able to go back to the past where a person has once experienced. In order to achieve this goal, we need to clarify what elements should be recorded as well as how much the resolution and quality of the experience record is necessary for inducing realistic sensation when to be played back later.

2 Episodic Memory and Experience Record

For defining the elements of experience record, we refer to the definition of " episodic memory" which was proposed by Endel Tulving.[2] According to his classification, episodic memory fully depends on individual experience received by its senses. And there seems to be little source based on social concensus but most of the credibility is determined by oneself. In this sense, one's experiences are comprised of self-centered information. Therefore even if various people are in the same situation, each one of them has different impression of their experiences. In order to record experience objectively, we focus on recording five senses, according to Tulving's classification, which is a basic sourceof episodic memory.

Figure 1 shows five categories of elements for subjective experience recording. These are based on human beings' five senses.(Visual, Aural,Haptic,Gustatory,Olfactory senses) These senses seem to be the inputs of experiences affecting internal and external condition of a person.

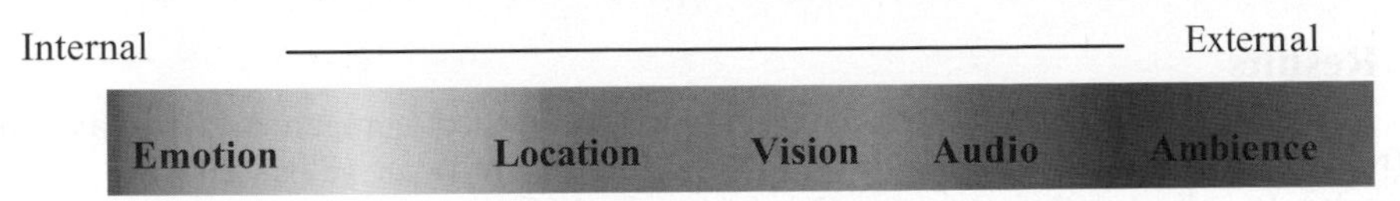

Figure 1: Hypothesis of five elements of experience recording

3 Subjective Visual Recording

It is generally said that 80 to 90% of information that human beings process is composed of visual information.Therefore to record precise subjective vision is one of the most important tasks for this study.

Study of context awareness wearable computer has been known for using a camera as the input of physical environment.[3] In this study, the main purpose of recording is to achieve advanced human-computer interaction so that required range and quality of vision is determined based on computer vision algorithm.[4][5] It tends to adopt omnidirectional camera for recording omnidirectional physical environment of the wearer so as to estimate the situation from the wide range of environment. This recorded data is used for vision processing for estimating the behaviour in the specific location or situation, so the image quality is not adequate for recalling the experience later by viewing them or the equipment is relatively heavy for long time use. There is also a study concerning wearable and video retrieval method of the past events.[6] The combination of head mounted camera and physical sensors for retrieval algorithm and real time video image matching with the database are known for the study. As these studies are focusing on retrieval algorithm, there is no discussion about subjective recording for inducing reality when being retrieved.

In this study, we focus on finding out the proper method for recording subjective experience with approximating the field of vision in everyday situation. We conducted two basic experiments to find out the characteristics of eye movement in space and reliability of peripheral field of vision when a person performs an ordinary behaviour. According to the results, we made a prototype wearable system for subjective visual recording.

4 Experiment 1 Eye movement experiment

The purpose of this experiment is to analyze eye movement profile in space.

4.1 System

We use ISCANCo.Ltd. Cap Mounted Eye Camera(System 4000/C monocular type eye tracking system) for recording eye tracking. The eye camera records eye image reflected by embedded infra-red ray in the inner side of the cap. After adapting corneal reflecting method for calibrating eye camera and augen, the output of the tracking data is converted to two dimensional pixel unit(x,y) accordance with the relative eye position on the recorded scene video. The data acquisition frequency is 60Hz per second.

4.2 Method

A subject wears the cap system and runs two tasks. One is to read a book for two minutes (distance from a book to head is about 30.0cm.)and the other is to look around a room(3.0m x

3.0m) for two minutes. This is to evaluate the difference of eye movement while to perform an eye gazing task of reading and looking around a space. Three subjects(two males and one female) conducted an experiment.

4.3 Results

There is no significance neither means and variances when compared to the axis of two tasks.($t(4)=0.17$ n.s.,$t(4)=0.03$, $F(2,2)=2.51$ n.s, $F(2,2)=2.96$ mean values and SD of x and y repectively) It indicates that the overall behaviour of eye movement is almost same in both different type of tasks. Though the eye traces tend to become vertically when to read (a book written in Japanese is vertical.) and horizontally when to look around. The significant difference of two tasks is the ratio of which eye tracking was not recorded within the angle of scene video.(Out of edge ratio:0.061 for task 1 and 0.427 for task 2. 1.0=whole recorded data) This implies that the range of eye looking in space is not able to stay within the angle of scene video.

5 Experiment 2 Peripheral Vision Experiment

5.1 Method

A subject walks the predetermined route in school building wearing a head tracking sensor,CCD camera and visual control mask. There are three kinds of masks. Each mask is set different restriction of visual angle, 60,90 and 120 degrees. At the beginning of the experiment, there is a test walk to instruct the route. Then a subject wears a standard equipment, those with three kinds of masks randomly. The tracking sensor records a subject's three dimensional motion (roll,pitch and yaw axis) in 10 Hz and CCD camera records a subject's visual information with 60 degree angle in AVI file format. Four subjects(males) conducted the experiment. All of them get accustomed to the location.

5.2 Results

We examine the difference of head motion of each task to evaluate the influence of peripheral vision in ordinary behaviour.Figure 2 shows mean and standard deviation of pitch angle of each task. By restricting the angle, people tend to move head frequently to confirm their current situation. And The result implies that people tend to see peripheral information in space so as to acquire environmental information unconsciously.

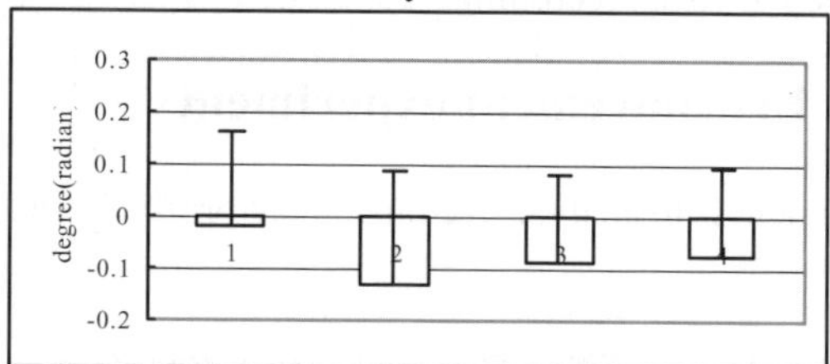

Figure 2: Mean and standard deviation of pitch angle of each task
(negative-head decline, positive-head up)

6 Wearable Prototype System

Taking the previous two experiments into consideration, we made a prototype system of recording system regarding to record approximate peripheral vision of human being. We fix the multiple camera system on the front of body for achieving horizontally wide range of visual recording from the viewpoint of a wearer.With the prototype system, we evaluate the approximation of recording of subjective vision.

6.1 System Configuration

In order to record wide range of vision with relatively high resolution,we used three progressive CCD camera (Horizontal:50degree,Vertical:40degree)array 30-degree spokewise worn on the front of the body.In order to achieve high transfer speed of image data taken by these three cameras so as to record them in one PC, IEEE1394 connection cameras are used. The camera attached wear is made not to shift the position of camera caused by a wear's swing. We fix the camera on the position where the rib is becoming the stabler not to move them. Figure 3 shows system configuration of a prototype.The system records images taken by three cameras with 320 x 240 pixel jpeg format file and head and body tracking data with accordance with GPS data as a text file. All data are recorded every 1 second.

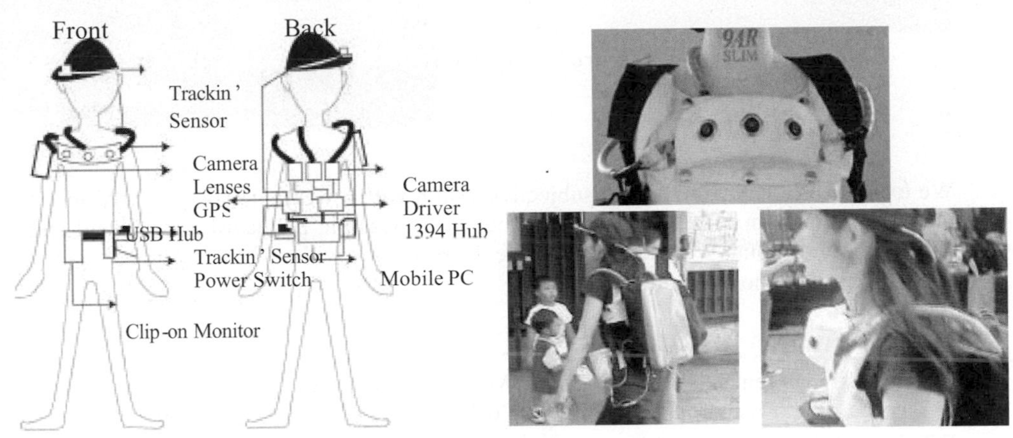

Figure 3: System configuration

Figure 4: Camera array(top) Wearer(bottom)

6.2 Results

Out of two sources of tracking data, we calculate the difference of Yaw and Pitch angle between body and head. Figure 5 and 6 show the histogram of the results respectively. We calculate interval estimate of mean value ;

$$5° \le \mu_{yaw} \le 9°$$

$$-10° \le \mu_{pitch} \le -9°$$

(* $p<0.05$ By setting the rejection region using Masuyama rejection test[8]($p<0.05$) , 6 points out of total data are rejected.)

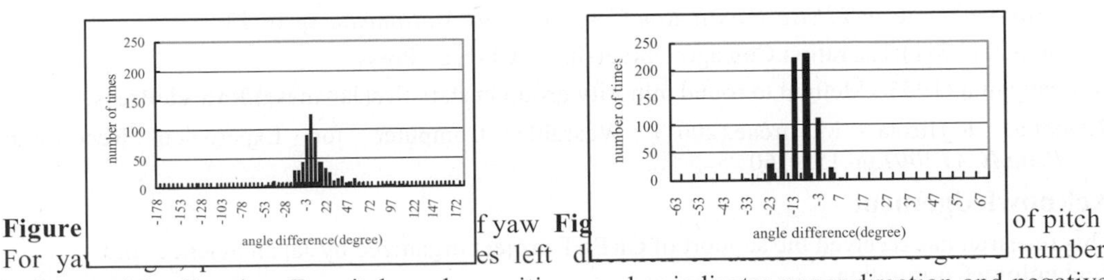

Figure ... f yaw **Fig** ... of pitch

For ya... es left d... number indicates right direction. For pitch angle, positive number indicates upper direction and negative number indicates lower direction.This estimated mean value indicates that average recording region of visual angle is about 101 degree to 105 degree in horizontal angle and 30 to 31 degree in vertical angle.

Figure 7: Sequentia

7 Conclusion

We examined a method to record subjective vision with a wearable computer.By adopting a front body as a platform of camera arrays, it becomes possible to record wide visual angle from a subjective view point. We need to integrate the prototype with other recording elements discussed [9] to construct memory recalling system.

References

S.Mann.(1998). Humanistic Intelligence:WearComp as a new framework for Intelligent Signal Processing. *Proc.of the IEEE*, vol.86,No.11,pp. 2123-2151

E.Tulving.(1985). Episodic Memory(Japanese).Kyoiku Press.

A.Dey and G. Abowd.(1999). Towards a Better Understanding of Context and Context-Awareness,

VU Technical Report.GIT-GVU-99-22.

T.Starner,B. Schiele,And A. Pentland. (1998).Visual Contexual Awareness in Wearable Computing. *Proc. of the Second Internatinal Symposium on Wearable Computers,*IEEE PRESS,pp50-58.

W.Runsgasarityotin and T.H. Starner.(2000).Finding location using omnidirectional video on a wearable computing platform. *Proceedings of the Fourth Internatinal Symposium on Wearable Computers*,IEEE Press,pp61.68.

K.Aizawa,K.Ishijima,andM.Shiina.(2001).Automatic Summarization of Wearable Video.

Proc.vof theSecond IEEE Pacific Rim Conference on Multimedia,pp.16-23

Polyak.S.L.(1941).The ratina.Chicago:University of Chicago Press.

G.Masuyama.(1953). Method to round minority group in statistics(Japanese).KawadePress.

R.Ueoka, K.Hirota, M.Hirose.(2001) Wearable Computer for Experience Recording. *Proc.ICAT2001*,pp.155-160.

Acknowledgement

This research has received the support of CREST project organized by Japan Science and Technology Corporation.

Intelligent Agents as Cognitive Team Members

P. Urlings [1], J. Tweedale [1], C. Sioutis [2], N. Ichalkaranje [2] and L. Jain [2]

[1] Defence Science and Technology Organisation
[2] University of South Australia, Australia
pierre.urlings@defence.gov.au lakhmi.jain@unisa.edu.au

Abstract

Advances in automation and artificial intelligence, especially in the area of intelligent (machine) agents, have enabled the formation of rather unique teams with human and machine members. This paper describes initial research into intelligent agents using a Beliefs-Desires-Intentions (BDI) architecture in a human-machine teaming environment. The potential for teaming applications of intelligent agent technologies based on cognitive principles will be examined.

Intelligent agents using the BDI-reasoning model can be used to provide a situation awareness capability to a human-machine team dealing with a military hostile environment. The implementation described in this paper uses JACK and Unreal Tournament (UT). JACK is an intelligent agent platform while UT is a fast paced interactive game within a 3D-graphical environment. Each UT game is scenario based and displays the actions of a number of opponents engaged in adversarial roles. Opponents can be humans or agents interconnected via the UT games server. Specific research goal is the development of human-agent teaming concepts. Cognitive agent behaviour such as communication and learning will also be addressed.

1 Introduction

During recent decades we have witnessed not only the introduction of automation into the work environment but have also seen a dramatic change in how automation has influenced the conditions of work. While some years ago the addition of a computer was considered only for routine and boring tasks in support of human, the balance has dramatically shifted to the computer being able to perform almost any task a human is willing to delegate. Advances in automation and especially in artificial intelligence have enabled the formation of a rather unique team with human and machine members. The rapid pace of developments in microprocessor technology has been the main driver, producing impressive achievements in many supporting technologies, including information processing, intelligent agents, and methods for cognitive and complex knowledge engineering.

2 Intelligent Agents and Cognitive Engineering

An emerging candidate technology for realisation of complex human-machine systems is Distributed Problem Solving (DPS) and in particular its multi-intelligent-agent concept. DPS provides a natural transition from the functional framework of a human-machine team into a multi-agent system architecture where the inherent distribution and modularity of functions is preserved and allocated to the distributed problem solving agent members of the team. A multi-agent system consists of heterogeneous agents that have a range of expertise or functionality, normally attributed to humans and their intelligent behaviour. These intelligent agents have the potential to function stand-alone but are also able to cooperate with other agents. This provides the basis for a human-machine teaming architecture, consisting of multiple cooperative agents, where each agent implements an assistant function. Each agent has his own local data, belief sets, knowledge, operations and control that are relevant for the problem or task domain of the function. BDI-agents, which implement a Beliefs-Desires-Intentions architecture, are of particular interest for teaming applications. The *beliefs* of these agents are its model of the world and its *desires* contain the goals it wants to achieve. The *intentions* are the actions that the BDI-agent has decided to implement in order to fulfil its desires (Rao & Georgeff, 1995).

The cognitive classification has been proposed as a promising approach to overcome the deficiencies of traditional automation (Rasmussen, 1986). Traditional automation is considered not being capable of compensating for the natural human cognitive deficiencies, which results in automation concepts in which the human either becomes overloaded or is kept out-of-the-loop. The core element of the cognitive approach is the natural human information-processing loop, which has the following elements: monitoring of the situation (perception and interpretation), diagnosis of the situation (including goal activation and evaluation), planning and decision-making, and plan execution. Most current automation research has adopted the cognitive approach. Major contribution to this success is the fact that the human problem solving strategy and the cognitive process as initially proposed by Rasmussen have been translated into mature and practical engineering frameworks. The approach of "Cognitive Work Analysis" (Vicente, 1999) has already been successfully introduced into the design of intelligent BDI-agents (Lui & Watson, 2003). The BDI-architecture allows the implementation of the cognitive loop within the agent, which is required to stay in line with the human cyclic cognitive process.

3 Research Challenges

This paper describes initial research of using intelligent agents to enhance situational awareness of a human-agent team in a military environment. Although there are several methods of achieving situational awareness, this research will concentrate on intelligent agents using a BDI architecture and reasoning approach. Major research aim is to improve the team-process of gathering situational awareness by *an order of magnitude* greater than that of a human playing the game alone. The proposed increase should result from reorganising the human-machine interaction, by imposing a paradigm shift as shown in Figure 1. The development of intelligent agents has been driven in the past by simulations, operations analysis and training applications, where the modelling and inclusion of human-like intelligence and decision-making behaviour was required. The development of intelligent agents has also been driven by automation and the need for decision support to human operators in critical or high-workload situations. This resulted in the development of machine assistants.

The aim of the proposed research is to achieve a paradigm shift that realises an improved teaming environment in which humans, assistant agents and cooperative agents are complementary team members. In formulating such teams, the main change required from a simulation point of view is that humans and machines are not *interchangeable*, they are *complementary*. The main change from an automation point of view is a move away from a comparison of tasks in which a human or machine is superior. Human and agents are not *comparable*; rather they are *complementary*. It is expected that the human will still remain in charge of the human-agent team but intensive management and coordination between the different team members will be required. Agent communication will therefore be a main focus of the proposed research. Aspects of communication will include control and management, coordination, self-learning, performance monitoring, warning, and assertive behaviour.

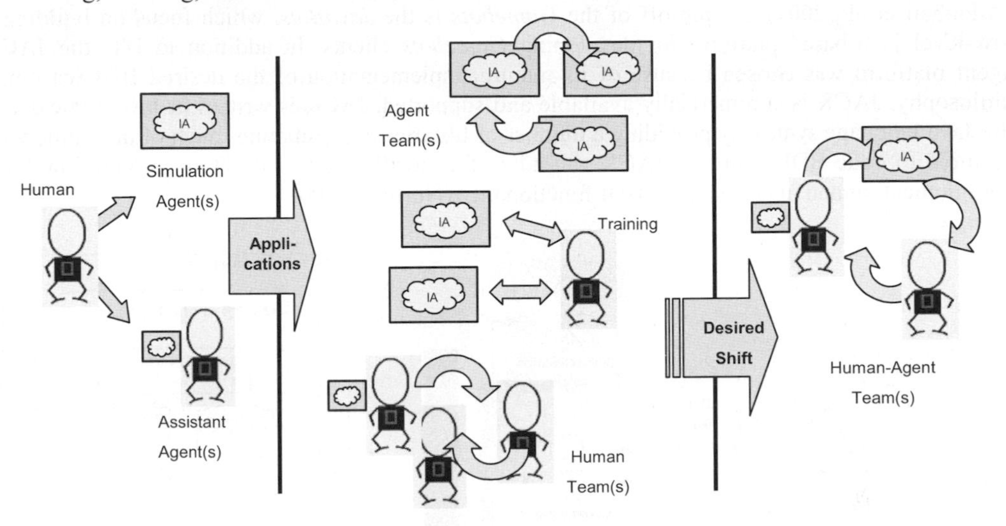

Figure 1: Human-Agent teaming

Another research aim is to introduce aspects of learning to the existing BDI philosophy. Most of the functionality of the agents can be implemented using existing syntax, but a problem arises from the fact that BDI agents exhibit open-loop decision-making. That is, a BDI agent requires an *event* from the environment to activate its BDI reasoning, which is completed by selection and execution of an appropriate plan. In addition, BDI-agents are optimised for specific a scenario, which means that an agent written for a particular environment will not easily translate to another (Norling, 2000). These shortcomings could be removed by introducing learning into the agent. The feedback caused by learning introduces a closed loop decision process analogous to the Boyd's Observe-Orient-Decide-Act (OODA). The decision process can proceed more rapidly, and decisions can be evaluated before they are committed. The side with the faster OODA loop triumphs in a battle (Clemens, 1997).

4 The problem of Situational Awareness

Significant research interest in automated situational awareness (SA) and the use of intelligent agents exists in military operational environments with the incentive to embody and include elements of cognitive behaviour (Ichalkaranje et al., 2003). Situational awareness can also be related to non-military scenarios such as accident sites, bush fires and civil emergencies. For research purposes even the environment of popular computer games like "RoboCup Rescue" and "Quake" are of interest. These games have already been widely used for research into a wide range of intelligent agent technologies.

Several publicly available game platforms were evaluated for use in the research described in this paper and "Unreal Tournament" (UT) was found to be most suitable. UT is a fast paced interactive game within a 3D graphical environment. It is a co-development by Epic Games, InfoGrames and Digital Entertainment. The UT game-engine has great flexibility for development or customisation. Earlier research resulted in a UT extension of *Gamebots* that allows bots in the game to be controlled via external network socket connections. The "thinking" of the bots can therefore be implemented on any machine and in any language that supports TCP networking (Adobbati et al., 2001). A spin-off of the *Gamebots* is the *JavaBots,* which focus on building a low-level java-based platform for developing *Gamebots* clients. In addition to UT, the JACK agent platform was chosen because of its mature implementation of the desired BDI reasoning philosophy. JACK is commercially available and supported. JACK is written in Java and extends the Java language syntax by providing a number of libraries encapsulating much of the framework required to code BDI solutions. JACK can adapt the JavaBots class of UT to provide low-level communication and implement the BDI functionality (Norling, 2000).

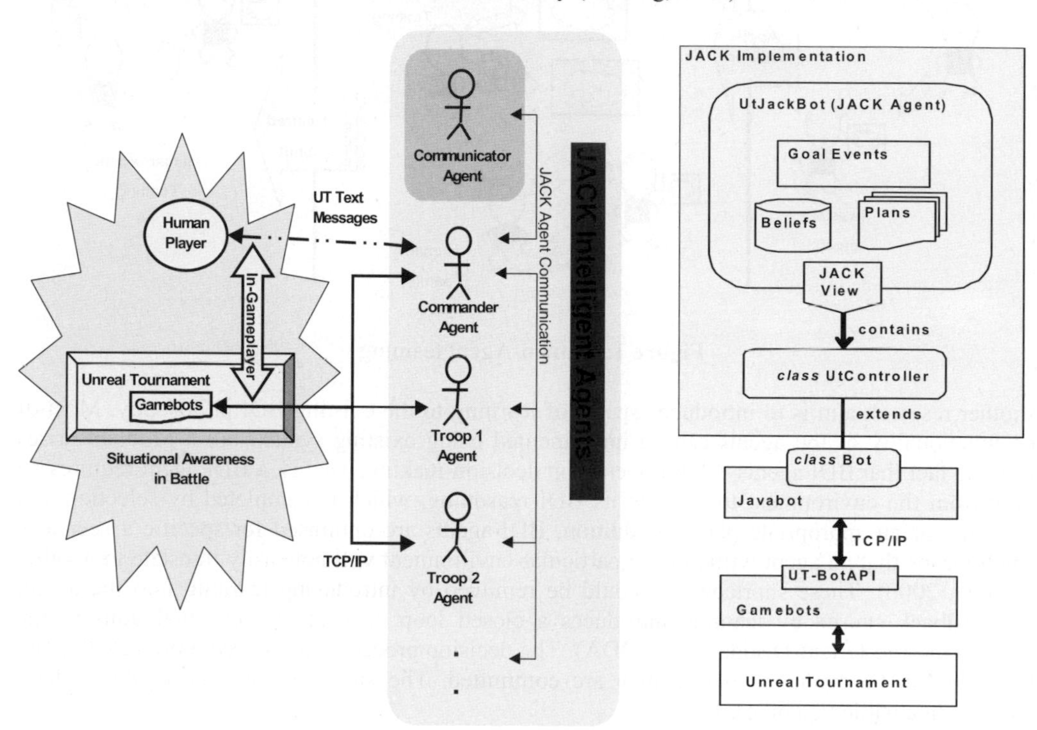

Figure 2: The research environment

The resulting research environment is shown in Figure 2. Unreal Tournament is used to visualise the scenario as it develops. A federation of agents is controlled by JACK code via a Communicator agent that is tailored to conduct all suitable protocol conversions. There are in total three types of agents. The Commander agent manages and coordinates the troops under his command and also informs the human player about what is happening and suggests what to do. The human can also give orders to the Commander agent, which in turn controls the team directly. The Communicator agent does not connect to UT, but facilitates the communication between the Commander and the human by analysing and understanding the text messages sent to/from the human. Several Troop agents explore the environment, provide environmental feedback and carry out orders received from the Commander agent. The existing UT environment will be used for the core implementation. A modification to GameBots adds RemoteBot entries in the UT's game menu. When a RemoteBot game is selected the GameBots server is started and it listens for TCP connections. JavaBots are then used to set up the TCP connection and to handle low-level communications to the GameBots server. To this purpose the bot-class within JavaBot needs to be extended with an UtConroller. UtController provides a streamlined connection interface and it parses incoming messages as meaningful Java objects. The JACK environment sits on top of all the above.

5 Conclusions

This paper argues the case for the use of intelligent agents as cognitive team members in a human-machine team environment. This team will be still supervised by human(s) but will be further composed of machine members that provide complementary capabilities. The Beliefs-Desires-Intentions (BDI) type of agent is expected to exhibit the cognitive behaviour that is required to align with the cyclic character of the knowledge based information process of the human.

Research is proposed in using intelligent agents that are able to enhance situation-awareness in a hostile environment. To this purpose the environment of Unreal Tournament (UT) has been selected because of its mature development tools and its large research following. The JACK intelligent agent shell was chosen because of its BDI-reasoning philosophy and its compatibility with UT. Through the use of UT text message communication, the human player will have an improved awareness of what is happening in UT during the game. The human player will be able to coordinate with the agent bots as a team in order to win the game. The purpose is to provide him with a situational awareness by an order of magnitude greater than that of playing alone. In addition to improved teaming and communication, the research will focus on agent learning and the addition of learning to the BDI-reasoning concept.

References

Adobbati, R., Marshall, A.N., Scholer, A., Tejada, S., Kaminka, G., Shaffer, S., & Sollitto, C. (2001). Gamebots: a 3D virtual world test-bed for multi-agent research. Information Sciences Institute & Carnegie Mellon University.

Clemens, S.M. (1997). The one with the most information wins? The quest for information superiority. In: Graduate school of logistics and acquisition management: Air Force Institute of Technology. 1997, pp. 128.

Ichalkaranje, N., Sioutis, C., Tweedale, J., Urlings, P., & Jain L. (2003). Intelligent Agents for situational awareness in a hostile environment. In: Proceedings of the Conference on Knowledge-based intelligent Electronic Systems, KES-03, September 2003, Oxford-UK

Lui, F., & Watson, M. (2003). Modelling Command & Control and decision-making using Cognitive Work Analysis approach for Intelligent Agents in land operations. In: Proceedings of the Conference on Knowledge-based intelligent Electronic Systems, KES-03, September 2003, Oxford-UK.

Norling, E. (2000). Flexible, reusable agents for modelling human operators. Workshop on Defence Human Factors Special Interest Group, HF-SIG, Melbourne-Australia.

Rasmussen, J. (1986). Information processing and human-machine interaction: an approach to cognitive engineering. New York: North Holland.

Rao, A.S., & Georgeff, M.P. (1995). BDI Agents: from theory to practice. In: Proceedings of the 1st International Conference on Multi-Agents Systems", ICMAS-95, June 1995, San Francisco-USA. pp. 312-319.

Vicente, K.J. (1999). Cognitive work analysis: toward safe, productive, and healthy computer-based work. Mahwah-New Jersey: Lawrence Erlbaum Associates.

The Influence of Colour Coding on Information Extraction from Computer-Presented Tables

Darren Van Laar, Kim Chapman & Mark Turner

Department of Psychology, University of Portsmouth,
King Henry Building, Portsmouth, PO1 2DY, United Kingdom.
Darren.Van.Laar@port.ac.uk
Kim.Chapman@port.ac.uk
Mark.Turner@port.ac.uk

Abstract

This study investigates whether different types of colour coding may be used to enhance the retrieval of information from computer-presented colour tables. Forty-eight participants were asked to perform searches for specific cell values in a 12 by 12 table of numerical values. Eight different table formats were investigated in which colour was used to group information in separate rows and columns. Colour-coded tables were also compared to a simple monochrome table format. Search times for tabular information differed as a function of the colour coding used. The results however were complex with no consistent evidence that either row-coded or column-coded table formats led to faster reading times than simple monochrome tables. Despite this, colour-coded formats were consistently rated as being more attractive, easier to use and requiring less mental effort than monochrome table formats. The observed results may be complicated by differences in the search strategies adopted by participants, the reported difficulty of the search task or by restricted exposure to each colour format.

1 Introduction

Current advances in computer technology have provided new means of communicating large amounts of information to visual display users. However, there is no consensus as to the way in which displays should be designed to present data in the most efficient manner. The effectiveness of any display will be dependent upon the requirements of the task and the capabilities of the user although tables have been shown to have a systematic advantage over some graphic displays with respect to the speed and accuracy of reading and comprehension (Meyer, Shinar & Leiser, 1997).

International Standard 9241 (British Standards Institution, 1998) supports the principle that colour coding can enhance the visual and cognitive processing of information in displays, for example by separating closely spaced data, reducing 'visual clutter' and improving user performance. Colour coding may enhance the perceptual processing of tabular displays by highlighting or grouping specific table elements and by helping to locate items during search tasks (Christ, 1975). Colour has been found to be consistently more effective than using flashing or boxed target stimuli as a means of emphasizing information (Fisher & Tan, 1989). Sidorsky (1980) found that colour could substantially reduce processing times for complex information in battle displays, whilst Brawn and Snowden (1999) showed that items in a display could be attended to separately if they are of a different colour. Structuring tables so that two colours of similar luminance are used to code alternate lines of information may also provide a 'horizontal guide' for the eye (Reynolds, 1982).

Other authors have suggested the use of colour in tables may constitute an aesthetically pleasing but redundant attribute. Too many colours on a display may be distracting and detrimental to performance (Van Nes, 1986). The use of colour may also induce visual fatigue or confusion by making data seem to be related in instances where they are not (Filley, 1982). Wickens and Andre (1990) argue that since the perception of colour-coding is not pre-attentive there may be no a priori reason why it should lead to faster search times than monochrome displays, although the use of colour may be justified when considering the benefits to reading accuracy. Foster and Bruce (1982) examined the effect of using 4 different methods of colour coding when searching for information in a 13 (row) by 6 (column) tabular display. Participants who were asked to answer a series of written multiple choice questions based on the content of the display were found to do so fastest when the table was coded with alternate coloured rows (yellow and green). This format was significantly faster than a row banded format (where every three rows were either yellow or green) or a column-coded format (first column cyan, last column white, other columns alternated yellow and green), but was not significantly faster than a simple monochrome format (all data in yellow).

The ambiguity of findings regarding the use of colour to enhance tabular information may arise from the arbitrary use and limited availability of colours in early computer displays, differences in the search tasks examined or differences in environmental conditions between previous studies. More recent advances in display technology and colour coding methods (e.g. Van Laar, 2001) now allows greater sophistication in the manner with which colour may be applied to displays. This fact, coupled with the increasing use of coloured tables to display information (e.g. via the Internet) suggests that there may be a need to revaluate previous research on which current human factors practice is based.

The purpose of this project was to investigate whether different types of colour coding may be used to enhance participants' retrieval of information from computer-presented colour tables. It was hypothesised that the time taken to retrieve information from tabular displays, the number of reading errors made and the perceived difficulty of information retrieval would be lower for tables that used colour to group information compared to simple uncoloured tables.

2 Method

2.1 Table formats

A repeated-measures design was used in which participants answered questions using eight different colour-coded table formats. The table formats investigated included three row-coded formats (groups of one, two or three rows the same colour), three column-coded formats (groups of one, two or three columns the same colour) and a combined-coding format (using joint row and column coding). All colour formats were compared with a simple monochrome format. All non-monochrome tables used two colours: 'green' and 'orange'. The luminance and saturation of both colours were made equal, and the specific hues used were chosen to ensure there was still high colour contrast between the colours. Unsaturated versions of these colours were used in addition to the original colours in the combined-coding format. Whilst visually more complex than any of the other tables, the combined-coding format was designed using a perceptual layering technique (Van Laar, 2001) which potentially allowed participants to attend to the row coding or the column coding as one layer of data. All text in all table formats was black.

2.2 Procedure

Experimental sessions lasted approximately 40 minutes during which time each participant completed a total of 88 test trials (11 per table format), in addition to practice trials. On each trial, participants were required to indicate whether a prompt question was true or false by searching a 12 by 12 table of two-digit numerical values. Two alternative question structures were used to encourage either row scanning or column scanning search strategies. The order in which participants completed each table format was counterbalanced whilst the question structure and whether a true or false response was required were randomised for each participant. The location of the target cell was controlled such that each row and column was used only once within each colour format. The target cell value was randomly chosen and appeared only in the target cell and not elsewhere in the table. Reaction times and error responses were measured for each trial. Mental workload (the NASA Task Load Index (Hart & Staveland, 1988)) and user satisfaction (an adapted version of the Questionnaire of User Interaction Satisfaction (Shneiderman, 1998)) were measured after viewing each table format. Additional data was also gathered on participant age and previous computing experience (subjectively rated using a 5-point scale: low to high). A total of 48 participants (15 males and 33 females) took part in the study. All were volunteers with normal colour vision who were paid £5 for their participation in the experiment.

3 Results

3.1 Errors and response times

Error rates during the experiment were low (approximately 1% of trials completed). The majority of participants made no errors (n=27) or only one error (n=13) and no difference in the total number of errors made was evident between different table formats. The mean time taken to respond to a question was 8.35s, (SD, 1.37s; range 5.67s to 11.54s). Response times were positively correlated with participant age ($r(n=48)= .36$, $p<.01$) and negatively correlated with self-rated computing experience ($r(n=48)= -.29$, $p<.05$) suggesting older participants and those who reported being less able computer users took longer to respond. To examine the unique effect on response times of colour coding format, question structure and required response (true or false), an 8x2x2 repeated measures analysis of covariance was performed. A significant difference in response times between colour-coding formats was observed when participant age and previous computing experience were used as covariates ($F(7, 315)=2.34$, $p<.05$). For row-coded tables, response times were fastest when alternate coloured rows were used (1 row), whilst for column-coded tables response times were slowest when alternate coloured columns were used (1 column, Table 1). However, no significant difference in response times were found between row-coded or column-coded tables compared to the monochrome format or combined format. Response times were significantly longer when answering questions that required a false (mean, 8.62s) as opposed to a true response (mean, 8.08s) for all eight formats ($F(1,47)=30.8$, $p<.001$). No effect of question structure was found and no significant interaction effects were found.

Table 1: Mean Response Times (s) and Standard Deviations (s) for each colour-coding format.

	Colour coding format							
	1 row	2 rows	3 rows	1 column	2 columns	3 columns	combined	mono-chrome
Mean	8.35	8.44	8.37	8.50	8.30	8.14	8.40	8.28
SD	1.56	1.68	1.95	1.70	1.64	1.68	1.68	1.41

3.2 Subjective workload and user satisfaction

NASA Task Load Index data (range 0-100) gathered following each of the eight colour coding formats was used to assess subjective impressions of the mental workload required to complete the visual search tasks (Table 2). Mean workload scores differed significantly between colour coding formats ($F(7, 329)=5.5$, $p<.001$). Pairwise comparisons showed mean workload scores for the monochrome format to be significantly greater ($p<.001$) than for all other colour formats, indicating the monochrome format was more difficult to use and required more mental effort. No statistically significant differences in workload were found for comparisons between row-coded, column-coded or combined-coded table formats. A similar trend was found for subjective satisfaction ratings ($F(7, 329)= 7.1$, $p<.001$). Mean satisfaction scores in the range 0 (low satisfaction) to 130 (high satisfaction) for each colour format are shown in Table 2. Participant's were found to rate the monochrome table format more negatively, suggesting it was significantly more frustrating, less stimulating and more difficult to read than all other colour table formats, although no subjective differences in satisfaction were evident between coloured table formats.

Table 2: Mean NASA-TLX weight-workload (WWL) score and subjective satisfaction ratings for each colour-coding format (standard deviations in parentheses).

	Colour coding format							
	1 row	2 rows	3 rows	1 column	2 columns	3 columns	combined	mono-chrome
Workload (WWL)	42.3 (16.2)	43.0 (13.9)	46.9 (17.8)	41.6 (17.6)	43.5 (15.0)	43.2 (16.3)	43.4 (17.9)	53.1 (16.2)
Overall Satisfaction	75.4 (16.0)	72.8 (14.2)	74.3 (18.9)	75.1 (17.3)	76.7 (16.8)	72.7 (18.2)	72.8 (19.2)	60.1 (17.2)

4 Conclusion

Search times for tabular information differed as a function of the colour coding used. The results however were complex with no consistent evidence that either row-coded or column-coded table formats led to faster reading times than simple monochrome tables. No evidence was found to suggest a combined row- and column-coded format produced better user performance or satisfaction than other formats. Despite this, all colour coded formats were consistently rated as being more attractive, easier to use and requiring less mental effort than monochrome table formats. Of the colour coded formats investigated, multiple column-coded formats tended to yield marginally faster reading times than row-coded formats of similar bandwidth, in contrast with Foster and Bruce (1982) who found row-coded formats to be superior. This may partially be explained by differences in table size: in the present study cell widths were slightly greater than cell heights, as with modern spreadsheet applications. This gave rise to a landscape-style table compared to the 13 rows by 6 columns, portrait-style table employed by Foster and Bruce (1982). Searching across a row may then have been relatively easier for participants in Foster and Bruce's study since there were fewer cell values to evaluate.

The use of colour coding in this study highlights the interaction between perceptual factors and task requirements when extracting information from tables. The high cognitive effort required to perform table-reading tasks, combined with participant learning over the small number of trials completed with each format, may have offset any relative performance advantage inherent in each colour format. Although not directly quantified in this research, differences in user search strategies were evident and these may have also influenced performance on the table formats. Such factors illustrate the complexity of the user's task when searching for information in a spreadsheet-like table. When faced with this task, differences in where scanning commences or in which part of the table the required value is situated can become major predictors of overall task performance. Future research should therefore not only attempt to quantify the interaction between search task and colour but should also consider the development of strategies as users become more practiced with specific table formats. In this way, not only will improved guidance for the colour coding of information be generated, but also a better understanding of the relationship between cognitive and perceptual processes for such typical office type tasks may be reached.

Acknowledgments

This research was made possible by a Nuffield Foundation research bursary. The authors are grateful to the Nuffield Foundation for their support.

References

Brawn, P., & Snowden, R. J. (1999). Can one pay attention to a particular colour? *Perception and Psychophysics*, 61, 860-873.

British Standards Institution (1998). BS EN ISO 9241-8: Ergonomics requirements for office work with visual display terminal (VDTs). Part 8. Requirements for displayed colours. London: British Standards Institution, 1998, 1-27.

Christ, R. E. (1975). Review and analysis of colour coding research for visual displays. *Human Factors*, 17, 542-570.

Filley, R. D. (1982). Opening the door to communication through graphics. *IEEE Transactions on Professional Communication*, PC-25, 91-94.

Fisher, D.L., & Tan, K.C. (1989). Visual displays: The highlighting paradox. *Human Factors*, 31, 17-30.

Foster, J. J., & Bruce, M. (1982). Looking for entries in videotex tables: a comparison of four colour formats. *Journal of Applied Psychology*, 67, 611-615.

Hart, S.G., & Staveland, L.E. (1988). Development of NASA-TLX: Results of empirical and theoretical workload. In P.A. Hancock and N. Meshkayi (Eds), *Human Mental Workload* (pp.139-183). Amsterdam: North-Holland.

Meyer, J., Shinar, D., & Leiser, D. (1997). Multiple factors that determine performance with tables and graphs. *Human Factors*, 39, 268-286.

Reynolds, L. (1982). Display problems for Teletext. In D. Jonassen (Ed.), *The Technology Of Text* (pp. 415-437). Englewood Cliffs, NJ: Educational Technology Publications.

Shneiderman, B. (1997). Designing the user interface: Strategies for effective HCI (3rd Edition). Wokingham: Addison-Wesley.

Sidorsky, R.C. (1980). Colour-coding in tactical displays: Help or hindrance? US Army Research Institute technical paper. Alexandria, VA: United States Army.

Van Nes, F. L. (1986). Space, colour and typography on visual display terminals. *Behaviour and Information Technology*, 5, 99-118.

Van Laar, D. L. (2001). Psychological and cartographic principles for the production of visual layering effects in computer displays. *Displays*, 22, 125-135.

Wickens, C. D., & Andre, A. D. (1990). Proximity compatibility and information display: Effects of color, space and objectness on information integration. *Human Factors*, 32, 61-77.

Emergence of Shared Mental Models During Distributed Teamwork: Integration of Distributed Cognition Traces

Rita M. Vick[1], *Martha E. Crosby*[2], *Brent Auernheimer*[3], *Marie K. Iding*[4]

University of Hawaii, USA[1, 2, 4], California State University, Fresno, USA[3]
{vick, crosby, miding}@hawaii.edu[1, 2, 4], brent@CSUFresno.edu[3]

Abstract

Team decision-making becomes more complex when problem specifications and availability of resources change over time. Time constraints may also amplify complexity. Ad hoc team response to this type of change can be studied over time by analyzing distributed cognitive traces and emergence of resource coordination response patterns during the team decision process. Knowledge of team work style patterns gained through such analysis provides criteria for design of adaptive collaborative work systems and development of more effective use patterns.

1 Motivation for the Study

Research efforts have focused variously on individually held abstract concepts becoming shared (Cannon-Bowers & Salas, 2001; Langan-Fox, Code, Langfield-Smith, 2000; Levesque, Wilson, & Wholey, 2001) and distributed cognition as an approach to group problem solving (Zhang, 1998). A framework has been suggested for use of distributed cognition as a way to describe interaction (Wright, Fields, & Harrison, 2000). However, I know of no other work attempting to demonstrate, as is the case here, how ad hoc coordination of resources, supervised by a team's emergent shared mental model, is facilitated by synthesis of emergent distributed cognitive traces.

The post-interaction residue left throughout a work or learning environment is a set of cognitive fragments. A transcript of the interaction that occurred over a period of time makes up a set of traces. Traces, in the form of digital artifacts, are footprints that define the path of interaction and resource coordination taken by a team. The emergence of patterns can be tracked over time as teams structure a decision model in response to changing requirements, complexity, time constraints, and availability of resources. The objective of this paper is to demonstrate how distributed cognitive traces stored in environmental artifacts are joined through discourse to form sufficient shared cognition to negotiate production of a satisfactory decision model.

2 Distributed Cognitive Information Traces (DCITs)

Distributed computer-mediated group decision-making requires coordination of multiple resources and persistent situation assessment in a multitasking environment. If outcome requirements change without commensurate extension of time-to-completion, task complexity increases due to time constraint and the additional cognitive load induced by the need for cognitive remodeling as teams re-evaluate goal, criteria, and alternatives. Increased mental load can be reduced by using artifacts in the work environment as intermediate storage for as yet incomplete information and resources can be leveraged to create shortcuts. What might be termed "distributed cognitive information traces (DCITs)" are left in the work environment. "Traces" may take the form of suggested ideas (captured by audio, a chat archive, a meeting support system report archive), referenced URLs (left in a browser history menu or a list of bookmarks), e-mail, or whiteboard drawings. Initially partial or ambiguous information may emerge as meaningful elements in the decision process. Information left in the environment can be used to trace the path of the decision-making process as it transpired during a given team session. This residue, even though omitted from the final outcome, serves as a valuable record of ad hoc interim choices that can later be used to explain the rationale for decisions that may have resulted in unanticipated or undesirable events.

3 Resource Coordination Response (RCR) Patterns

Patterns of idea acceptance and rejection, developed during the decision process, reveal how consensus emerged over time. Analysis of these patterns reveals a group's unique temporal pattern for development of shared understanding of the problem and potential ways to manage a solution. More than that, pattern analysis provides a narrative of how a group managed available human, interactive, and archival resources. Assuming a team engages in regular decision making activity, the team develops a pattern for resource use that is perceived to be effective. Because the kind of team work studied here involves synchronous, time-constrained, complex multitasking, the cognitive load resulting from the problem-solving aspect of the work can cause the team's resource use pattern to default to a habituated mode where the resource coordination function becomes an ineffective "resource coordination response (RCR)" pattern. Because the coordination function tends to become reflexive, effective use of available resources does not occur.

4 Adapting Software to Individual Conceptual Model Convergence

Knowledge of group coordination patterns indicates process points where intervention in the work process can facilitate improved outcomes. For example, alerting a group to the need for more frequent situation assessment may avoid time consuming focal divergence through a built-in affordance enabling teams to view their own concurrent performance by presenting an on-screen visualization of emergent team performance patterns. Successively more obtrusive stages of embedded decision team guidance would be: (1) real-time visual representation of resource availability and usage patterns enabling teams to oversee and adjust their own performance "on the fly," (2) built-in facilitative agency to mitigate emergence of, for example, ineffective RCR patterns and related divergent interaction patterns (DIPs), and (3) comprehensive support that would include aggregate user models of team work style and problem-solving patterns.

At the abstract level, group members maintain cognitive representations of decision goals, criteria, and alternatives as well as plans for achieving the goal and possible mediating relationships among criteria and alternatives. Members of a decision-making group draw on their individual conceptual models of goal, criteria, and alternatives while assessing resource appropriateness and availability during the decision process. Individually held components are modified during interaction with other team members and may be externalized, emerging as available resources through representation as physical, digital, or linguistic artifacts. Once available in the immediate (virtual) environment (e.g., printed textual information), via storage and transfer media (e.g., online data stores, e-mail, the Web), or within the human-computer interface (e.g., archived digital notes, text chat statements), individual conceptual information becomes a tangible resource. As a team interacts to use or archive information, individual conceptual models converge to form a single conceptual model that is represented externally in the interface (by the completed decision model in the present study) and is archived as the team's representation of the problem solution.

5 Method

A system design decision problem was given to zero-history teams of three to five undergraduate senior-level computer science students using TeamEC™ (http://www.expertchoice.com/) decision support software. Teams were placed in a computer laboratory setting where they communicated only via text chat supported by NetMeeting (http://www.microsoft.com/windows/netmeeting/) to simulate synchronous distributed computer-mediated work. TeamEC™ was run using NetMeeting for application sharing, file transfer and shared digital whiteboard (used to graphically demonstrate ideas and as a form of team memory). Transcripts of team discourse, whiteboard, and decision model files created within the specified timeframe were archived. Teams were provided with a written job request, although end-product design requirements changed over time. The task was to complete a decision model that defined and prioritized relationships among the factors the team believed significant to successful implementation of their solution. This required accurate specification of an appropriate overall goal covering design, development, and/or marketing roles for production of a collaborative software package. Student teams assumed each role in a series of rotations over a ten-week period. Critical sub-tasks were: determine the criteria required to support the goal; specify alternate ways to meet the criteria; prioritize relationships among all elements; define all model elements; justify or eliminate inconsistencies. Chat transcripts were analyzed for situation assessment, resource coordination, idea generation, and model structuring activity intervals and for external information resource use (e.g., the Web, instructor feedback via e-mail, scenario source material) are shown as they occurred (see Figure 1).

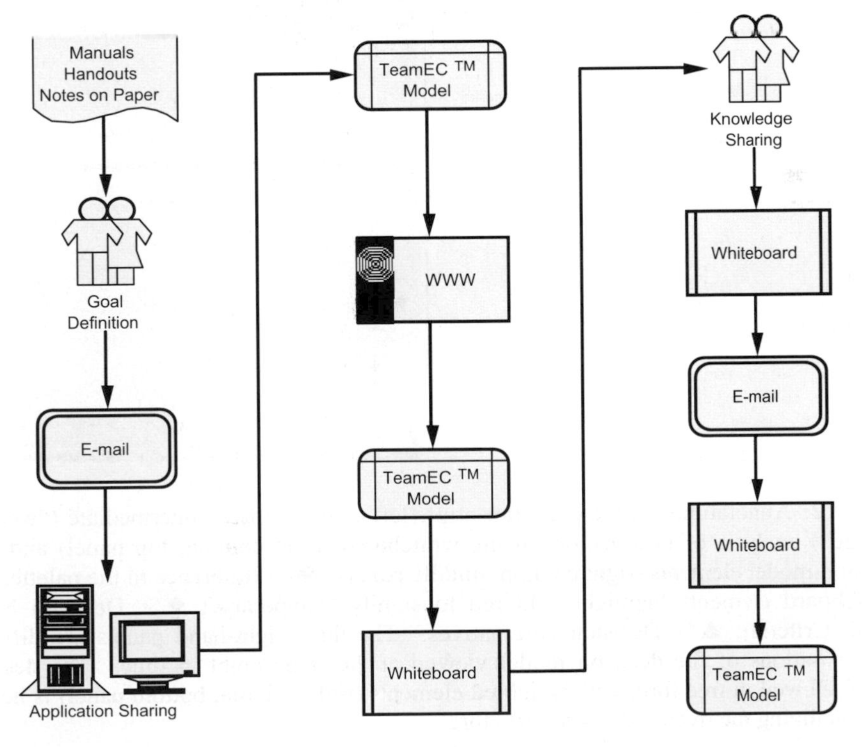

Figure 1: Team workflow showing alternating intervals of different types of information gathering, synthesis, and processing.

Analysis of chat transcripts showed that teams left "distributed cognition traces" in the form of fully and partially expressed ideas that supported definition and interrelationship of goal, criteria, and alternatives spread throughout the digital capture facilities of their work environment. Traces appeared in chat statements and on the whiteboard. Selected trace fragments accumulated to structure the final decision model as demonstrated in Figure 2.

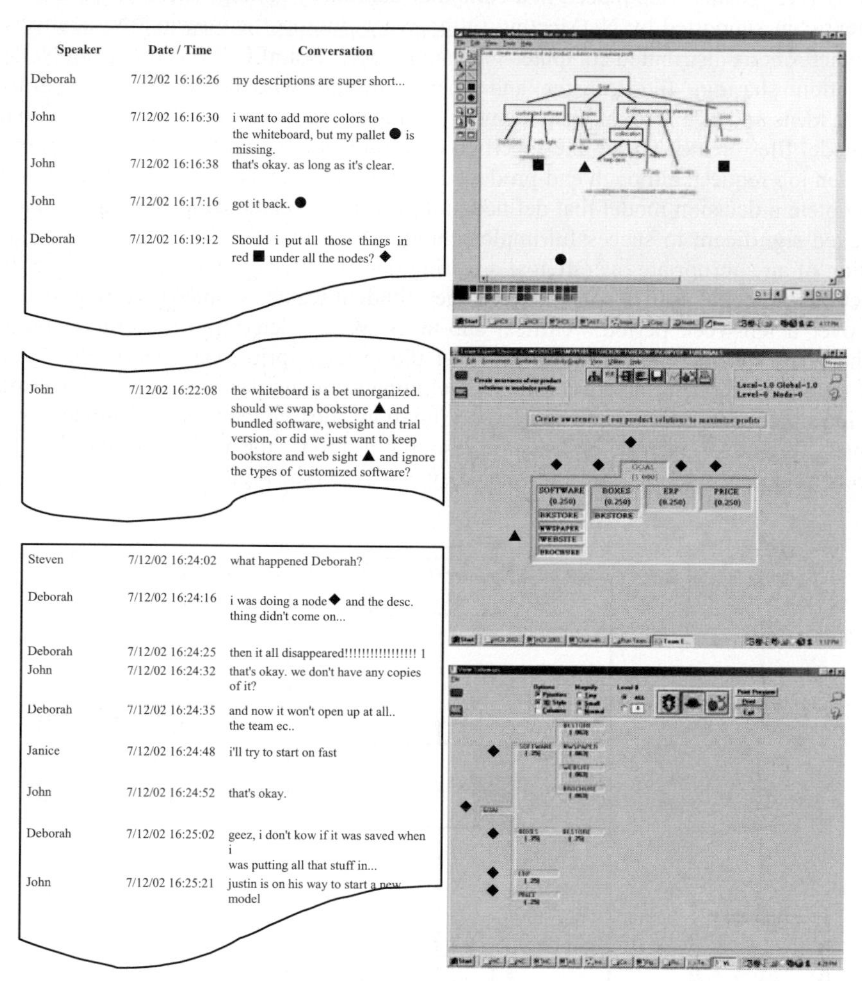

Figure 2: Annotations on the chat transcript (left column) match intermediate ("working storage") analysis of information on the whiteboard (right column, top panel) and final decision model elements (right column, middle panel). ● = Reference to the palette; ■ = Whiteboard elements highlighted in red to signify "temporary"; ◆ = Decision Nodes (Goal, Criteria); ▲ = Decision Alternatives. The three right-hand panels are different representations of the decision model viewed at the same point in time. The decision model viewed in tree form with weighted elements (right column, bottom panel) is helpful for visualizing the overall decision structure.

The overall workflow from multiple decision sessions was analyzed to differentiate sequences of artifact usage. The post-work-process environment revealed evidence of how and when emergence of decision elements occurred and so provided a narrative of knowledge sharing and dissemination of cognition through the physical work environment and among team members. Synthesis and structuring of conceptual relationships during interactive discussion revealed emergence of new understanding based on team-generated ideas as well as from external resources. The form and content of final team decision models emerged collaboratively over time and throughout the discourse. The completed decision model instantiates the team's shared understanding of the solution to the system design decision problem.

6 Discussion

Realistic constraints (e.g., communication limited to synchronous text chat, specification changes, time limitations) were designed into the problem to provide a naturalistic decision-making environment. Given a high rate of environmental change, decision teams may need to implement mid-stream changes in decision mode (e.g., a preference-based decision may, depending on unforeseeable economic factors, become cost-based). Ability to quickly restructure decision models is especially desirable when the consequences of a decision impact multiple components of a workflow. Embedded decision team guidance can assist maintenance of team productivity despite resource limitation and internal or external environmental change. This work is based on the assumption that distributed team decision making is best done synchronously. While versions of the same technology used in this study might be used asynchronously for less critical or less time-sensitive decision-making problems, sustaining sufficient persistence over time to enable consistent and effective decision modeling would prove challenging. Finally, this work indicates that while text chat is not a rich medium, teams grounded in this work style can effectively and inexpensively perform synchronous distributed decision-making.

7 Acknowledgements

This research was supported in part by DARPA grant NBCH1020004 and ONR grant no. N00014970578 awarded to Martha E. Crosby.

8 References

Cannon-Bowers, J. A., & Salas, E. (2001). Reflections on shared cognition. *Journal of Organizational Behavior*, 22, 195-202.

Langan-Fox, J., Code, S., & Langfield-Smith, K. (2000). Team mental models: Techniques, methods, and analytic approaches. *Human Factors*, 42(2), 242-271.

Levesque, L. L., Wilson, J. M., & Wholey, D. R. (2001). Cognitive divergence and shared mental models in software development project teams. *Journal of Organizational Behavior*, 22, 135-144.

Wright, P. C., Fields, R. E., & Harrison, M. D. (2000). Analyzing human-computer interaction as distributed cognition: The resources model. *Human-Computer Interaction,* 15, 1-41.

Zhang, J. (1998). A distributed representation approach to group problem solving. *Journal of the American Society for Information Science,* 49(9), 801-809.

Learning and Forgetting Aspects in Student Models of Educational Software

Maria Virvou, Konstantinos Manos

Department of Informatics
University of Piraeus,
Piraeus 18534, Greece
mvirvou@unipi.gr; konstantinos@kman.gr

Abstract

In this paper we describe the part of the student modelling process of an educational application that keeps track of the students' learning and remembering facts that are taught to him/her. For the purposes of this process we have adapted and incorporated principles of cognitive psychology into the overlay technique that is used in the system for student modelling. As a result, the educational application takes into account the time that has passed since the learning of a fact has occurred and combines this information with evidence of the user knowing or not knowing something. This process gives the system an insight of what is known and remembered and what needs to be revised and when. In this way the system may dynamically adapt the content of each lesson to the specific needs of each individual student at a particular time.

1 Introduction

Since Information Technology has been so widely spread and offered its services to many domains and disciplines, it has also been widely acknowledged that it can be very useful for assisting education. To this end many educational software researchers employ several means of computer technology to improve the presentation, the structure and reasoning abilities of educational software. One important goal has been the improvement of attractiveness and aesthetics of educational applications based on multimedia, virtual reality etc. However, on the other hand, multimedia educational products are often criticised that they do not support the learner well nor exploit the capability of the medium (Laurillard 1995, Montgomery 1997, Moore 2000). From this point of view, the incorporation of a student modelling component into educational software may be quite important for rendering the system more adaptive to the student's learning needs, abilities, weaknesses and knowledge.

This paper describes the enhancement of a Virtual Reality educational game, by adding a module that could measure-simulate the way students learn and possibly forget through the whole process of a game-lesson. The virtual reality game is called VR-ENGAGE (Virvou et al. 2002) and teaches geography. In particular, we are going to describe a student modeling process that keeps track of what a student is being taught and will actually remember after the end of the lesson. This is achieved by the adaptation and application of models of cognitive psychology to the particular circumstances of the educational software application.

2 Student Modelling in the Educational Software Application

The educational software application aims at teaching students in a motivating way. Therefore teaching and testing takes place in the environment of a virtual reality game. Indeed, recently researchers in educational software point out the virtues of computer games relating to children and adolescents' education. For example, Muntaz (2001) notes that a range of cognitive skills are practised in computer game playing given the sheer number of decisions children make as they weave their way through various games.

In VR-ENGAGE the ultimate goal of a player is to navigate through a virtual world and find the book of wisdom, which is hidden. To achieve the ultimate goal, the student-players have to be able to go through all the passages of the virtual world and to obtain a score of points, which is higher than a predefined threshold. While students navigate through the virtual world they meet several animated agents that either show them parts of the lesson theory or ask them questions that the students have to answer. The total score of a player-student is the sum of the points that the player has obtained by answering questions.

The system also has the ability to adapt teaching to the specific needs of each student so that it may maximise the amount of knowledge that the student learns. For example it may dynamically select which part of the theory the student is going to see and when. For this purpose there is a student modelling component that keeps track of what the student has already seen (and when s/he saw it), what the student seems to have learnt given his/her answers to questions and what the student is likely to remember by the end of the lesson. This information is stored in the long term student model to be used in subsequent lessons.

Student modelling in VR-ENGAGE is based on the overlay technique. The overlay model was invented by Stansfield, Carr and Goldstein (1976) and has been used in many early user modelling systems (Goldstein, 1982) and more recent systems (Matthews et al. 2000). The main assumption underlying the overlay model is that a user may have incomplete knowledge of the domain. Therefore, the user model may be constructed as a subset of the domain knowledge. This subset represents the user's partial knowledge of a domain and thus the system may know which parts of the theory the user knows and which s/he does not know. However, as Rivers points out (1989), overlay models are inadequate for sophisticated modelling because they do not take into account the way users make inferences, how they integrate new knowledge with knowledge they already have or how their own representational structures change with learning. One additional problem with the overlay technique is that it assumes for the student an "all or nothing" knowledge of each part of the domain (either a student knows something or not).

The overlay technique has to be used in conjunction with inference mechanisms about the students' knowledge. The inference mechanisms that have been employed so far have been mainly based on students' actions in assessment tests that show evidence of the students' knowing or not knowing something. However, even in cases where the student shows evidence of knowing something at a particular time, s/he may forget it after a while. Therefore in our research we take into account what parts of the theory the student has been shown, how often this has happened and what s/he is likely to remember. Therefore the overlay technique is extended to include degrees of knowledge for each fact. Each degree represents the possibility of a student knowing and remembering something given the time this was learnt. For this purpose, we use a forgetting model.

There are two popular views of forgetting (Anderson, 2000). One, the decay theory holds that memory traces simply fade with time if they are not "called up" now and then. The second view states that once a material is learned, it remains forever in one's mental library, but for various reasons it may be difficult to retrieve. These theories may seem to be "conflicting", but when someone has "forgotten" something, there is really no way for us to tell whether it has been completely removed from his/her mental library or it is very (almost impossibly) difficult for him/her to retrieve it. For our study, both theories have practically the same meaning; if a student finds it hard to remember a fact that s/he has learnt (either due to memory fading or difficulty of retrieval) then the learning process was not that good and should be modified.

A classical approach on how people forget is based on research conducted by Herman Ebbinghaus and appears in a reprinted form in (Ebbinghaus, 1998). Ebbinghaus worked for a period of one month and showed that memory loss was rapid soon after initial learning and then tapered off. In particular, Ebbinghaus' empirical research led him to the creation of a mathematical formula which calculates an approximation of how much may be remembered by an individual in relation to the time from the end of learning (Equation 1).

$$b = \frac{100 * k}{(\log t)^c + k} \qquad (1)$$

In Equation 1:

- t: is the time in minutes counting from one minute before the end of the learning
- b: the equivalent of the amount remembered from the first learning. As it is evident from the logarithmic nature of the formula, b lowers greatly at the beginning and starts to stabilize after time passes on.
- c and k : two constants with the following calculated values: $k = 1.84$ and $c = 1.25$

Linton (1979) also conducted research on retention of knowledge and worked for a period of six years. Linton's results were similar to Ebbinghaus' results. Finally, Klatzky (1980) also reports the results of a study that consisted of experiments on retention. These experiments involved repetitions of a memorised list of words after a pre-specified break length, typically up to few days. This study showed that memory decay is a power function of the break length. For example, subjects forget 55% of the words within a six hour break time and 80% percent within 72 hours. However, these results are very close to Ebbinghaus results. Indeed, if Ebbinghaus' formula was used, one would find that subjects forget 60% of the words within a six hour break and 75% within 72 hours. Such differences in the results have little importance for the purposes of the incorporation of a forgetting model in the educational application. Therefore Ebbinghaus' mathematical formula has been used in VR-ENGAGE to give the system an insight about the students' learning and forgetfulness.

In our model there is a database that simulates the mental library of the student. Each fact a student encounters during the game-lesson is stored in this database as a record. In addition to the fact, the database also stores the date and time when the fact was last used along with a numerical factor describing the likelihood of the student's recalling the given fact. The smaller the factor the less likely it will be for the pupil to remember the fact after the end of the game-lesson.

Our research goal is to render the educational game more effective in teaching the student. This will happen if after the course the student actually ends up with many facts with high factors in his/her mental library. To model that, we assume that the student has a blank mental library on the subject being taught; meaning, that during the first lesson there is nothing in the mental library of the student to be retrieved.

3 Learning a new fact

While the student plays the educational game s/he encounters a "tutor" that provides him/her with a piece of information to be taught (Virvou et al., 2002). This is the first encounter of that information so it is added to the memory database. The data saved in the database are:

- **ID**: It is a string ID of the fact being taught
- **TeachDate**: It is the date and time of the first occurrence
- **RetentionFactor(RF)**: a number showing how likely it is for the student to actually remember the given fact after the end of a "game lesson"

When a fact is inserted in the database, the TeachDate is set to the current date and time, while the RF is set to a base number equal to 95. Taking as a fact that any RF bellow 70 corresponds to a "forgotten" fact, using the Ebbinghaus' power function we may calculate the "lifespan" of any given fact. The RF stored in the "mental" database for each fact is the one representing the student's memory state at the time showed by the TeachDate field. Thus, whenever we need to know the current percentage of the fact that a student remembers, the equation 2 is used.

$$X\% = \frac{b}{100} * RF \quad (2)$$

Where:

- b: is the Ebbinghaus' power function result (Equation 1), setting t=Now-TeachDate
- RF: is the Retention Factor stored in our database

4 Recalling – Using a fact

During the game, the student may also face a question-riddle (which needs the "recall" of some fact to be correctly answered). In that case the fact's factor is updated in accordance to the student's answer. For this modification an additional factor, the Response Quality (RQ) factor, is used. This factor ranges from 0 to 3 and reflects the student's answer's "quality" in the way illustrated in Table 2.

Table 2: Response Quality Factor reflecting the student's answer's quality

RQ	Description	Modification
0	No memory of the fact	RF' = X – 10 TeachDate = Now
1	Incorrect response; but the student was "close" to the answer	RF' = X – 5 TeachDate = Now
2	Correct response; but the student hesitated	RF' = RF + 10
3	Perfect response	RF' = RF + 15

Table 2 illustrates the formulae used to modify the RF of the correspondent fact, depending on the RQ value. In the formulae used:

- X is the value calculated using Equation 2
- RF\` is the new Retention Factor value
- RF is the old Retention Factor value

In the cases where the RQ is 0 or 1, the student has a difficulty remembering a fact that has already been taught to him/her. Thus, in these cases the TeachDate is reset so that the Ebbinghaus' power function is restarted. This will generate a rapid loss of memory in the beginning and will stabilise later on. By lowering the RF (by ten or five) even for a student with very strong memory, it gives the fact a "lifespan" of a couple of minutes, thus it is necessary for the game to repeat the teaching process of the given fact if it wishes the student to remember it.

5 Conclusions

In this paper, we have shown how principles of cognitive psychology have been used by the student modelling process of an educational application. This is done so that the system may gain an insight of what students remember from the course material that has already been taught to them. For this purpose, the system takes into account how much may be remembered by an individual in relation to the time from the end of learning. In addition, it takes into account what the student has been able to remember from the material taught as this has been recorded in his/her performance in tests. This information is used by the system to adapt the teaching process accordingly. Depending on what a student remembers or not the system proceeds in presenting new course material or repeats certain parts of the course material that has already been taught. In this way, the educational software application may become more personalized and adaptive by responding appropriately to each individual student's needs regarding the way the course material is being taught to him/her.

References

Anderson J.R. (2000). Learning and Memory: An Integrated Approach (2nd ed.). John Wiley & Sons, Inc.

Virvou M., Manos C., Katsionis G., Tourtoglou K., VR-ENGAGE: A Virtual Reality Educational Game that Incorporates Intelligence, IEEE International Conference on Advanced Learning Technologies (2002), pp.

Ebbinghaus, H. (1998) "Classics in Psychology, 1885: Vol. 20, Memory", R.H. Wozniak (Ed.), Thoemmes Press, 1998

Klatzky, R.L. (1980) "Human memory – Structure and Processes", W.H. Freedman and Co., San Francisco.

Laurillard D. (1995) 'Multimedia and the Changing Experience of the Learner', *British Journal of Educational Technology* 26(3), pp. 179-189.

Linton, M. (1979) "Real-world memory after 6 years- Invivo study of very long-term-memory" *Bulletin of the British Psychological Society* 32 (Feb): 80-80 1979

Montgomery M. (1997) Developing a Laurillardian CAL Design Method, *Proceedings of ED-MEDIA/ED-TELECOM World Conferences On Educational Multimedia, Hypermedia and Educational Telecommunications*, Vol. 2, pp. 1322-1323.

Goldstein, I. (1982). The genetic graph: A representation for the evolution of procedural knowledge. In D. Sleeman & L. Brown (Eds.), Intelligent Tutoring Systems. London: Academic Press.

Matthews, M., Pharr, W., Biswas G. & Neelakandan, (2000). "USCSH: An Active Intelligent Assistance System," *Artificial Intelligence Review* 14, pp. 121-141.

Moore, D. (2000) "A framework for using multimedia within argumentation systems". *Journal of Educational Multimedia and Hypermedia* 9(2), pp. 83-98.

Muntaz, S. (2001) "Children's enjoyment and perception of computer use in the home and the school". *Computers & Education* 36, pp. 347-362.

Rivers, R. (1989) Embedded user models – where next? *Interacting with Computers* 1, pp. 14-30.

Stansfield, J.C., Carr, B., & Goldstein, I.P. (1976) Wumpus advisor I: a first implementation of a program that tutors logical and probabilistic reasoning skills. At Lab Memo 381. Massachusetts Institute of Technology, Cambridge, Massachusetts.

Biologically Inspired Analysis of Complex Systems: Back to Nature

M. Wheatly[1], S. Narayanan[1], R. Koubek[5], C. Harvey[1], L. Rothrock[5], P. Smith[2], M. Haas[3], and W. Nanry[4]

Wright State University[1], Dayton, OH, The Ohio State University[2], Columbus, OH, Air Force Research Laboratory[3], Dayton, OH, Air Force Institute of Technology[4], Dayton, OH. Penn State University, State College,PA
michele.wheatly@wright.edu

Abstract

Researchers are turning to the natural sciences for new paradigms in seeking solutions to engineering applications. Examples include using swarm intelligence to address business optimisation problems and the six-legged robot RHex that was inspired by cockroach locomotion. Our research uses a multidisciplinary approach to develop a biologically inspired adaptive interface that could potentially impact industrial and military applications. The basis of the work is that biological organisms perform potentially at levels many orders of magnitude better than silicon-based systems. As such, biological models ranging from sub-cellular to system/population level can inspire algorithms, software, and complex systems. Effective metaphors of information processing in biological systems can inform information technology applications.
In this paper, we describe the use of the immune system as a paradigm for designing an operational control system that is self-evolving. This article summarizes the overall approach and presents details of the analysis of the battle field scenario and its mapping to immune systems.

1 Introduction to Biological Inspiration

Nature has provided, over millions of years of evolution, a cornucopia of paradigms that could be used to inspire solutions to engineering and business problems. For example swarm intelligence has been used to address business optimisation problems (Bonabeau, Dorigo & Theraulaz, 2000) and insect locomotion has been used to design multi-legged robots (Altendorfer et al., 2001). Natural selection of adaptive characters is considered to be the determining mechanism of evolution. Adaptation to the environment is a feature of living organisms that provides the basis upon which natural selection can act over evolutionary time. Adaptive systems, be they biological or artificial, have two characteristics in common: they sense and respond to a changing environment. Biological systems are inherently decentralized where complexity emerges from simplicity. Biological inspiration seeks to inspire not mimic nature. Evolution works on building just-good-enough rather than optimal solutions. Humans can typically build better designs than nature (planes being a good example). Further, organisms are constrained by their history, development and genes so that their adaptations are far short of designer perfection. Examples in humans would be the unwanted junction of the food-conveying oesophagus with the air-conveying trachea that poses the ever-present danger of choking on food. Other design flaws might be the inability to manufacture vitamin C, the absence of a second heart and a host of problems that accompany upright bipedalism.

2 Complexity Emerging from Adaptive Systems

Holland and his colleagues at the Santa Fe Institute (Holland, 1995) have studied complex adaptive systems (biological, social and artificial) and concluded that they possess four characteristics (aggregation, nonlinearity, flows and diversity) and three driving mechanisms (labeling, internal models and building blocks). Complexity emerging from adaptation is a tenet of complex adaptive systems, just as it is the credo of the theory of evolution. A conceptual framework can be constructed to enable promising biological paradigms to be archived and retrieved for application to artificial systems. Natural systems involve living organisms and the environment. Living organisms exhibit a hierarchical organization from single cells through multicellular organisms to population and ecosystem biology. The environment is complex and changes over space and time. It is renewable and sustainable, yet limited in resources. Both environment and living organisms have increased in diversity over evolutionary time. Consistent with the theory of evolution, organisms adapt to their environment. The environment can select for certain characteristics (survival of the fittest) and thereby alter organisms; yet organisms can alter the environment. Central to all biological models and their interaction with the environment is that information must be acquired, processed, distributed and used.

3 Biological Information Technology

The focus of the present study was to use biological inspiration to model an adaptive computer interface. An NSF workshop recently identified the natural sciences as holding the most promising solutions for IT research (Hickman, Darema, & Adrion, 2001). Within Biological Information Technology there are three foci for future work. In silico systems will be developed for fundamental understanding of cellular information processing devices and architectures (an example would be using genomics information in drug discovery). Hybrid systems (bio-silicon devices such as solid-state microelectrodes, implants and other transduction methodologies) will assist our understanding of information transfer between cells and tissues (an example would be neuronal networks). Systems biology will explore communication among (intra) and between (inter) organisms. The report identifies a need to develop better simulation tools to analyse large-scale modelling of hierarchically organized biological systems along with metrics to evaluate when biology exhibits a "better" way of doing things.

4 A Biologically Inspired Adaptive Computer Interface

The long-term goal is to develop and test a rapidly customisable interface that adapts to operator state in real time and thus maximizes personnel and technology investment. Fixed interfaces are poorly suited to dynamic/fluctuating work environments, since they produce sub-optimal performance outside the design envelope. An *adaptive* interface is defined as one where the appearance, function or content of the interface can be autonomously changed *by the interface* (or underlying application itself) in response to the current goals and abilities of the user (Rothrock et al., 2002) by monitoring user status, the system task, and the current situation. The adaptive computer interface ideally would: incorporate representations of operator states; represent interface features that are adaptable; represent cognitive processing in a computational framework; provide measures of merit for matches between input state variables and interface adaptation permutations; operate in real-time mode; and incorporate self-evolving mechanisms.

The empirical (top down) approach to adaptive interface development has focused on features of the interface/environment that can change or adapt as well as on state variables of the user and/or task that are monitored to trigger the adaptation. Under sponsorship of the Air Force Office of Scientific Research and the Air Force Research Lab, the Synthesized Immersion Research Environment (SIRE) lab has been using several novel autonomous research stations that can be pulled together to form a multi-friendly/adversary flight simulation.

The Dayton Area Graduate Studies Institute/Air Force Research Laboratory recently funded a multidisciplinary team to model (bottom up) a biologically inspired self-evolving adaptive interface. The team included academic researchers and government scientists from the disparate areas of biology, collaboration, operations research, human factors, biomedical engineering, simulation, HCI, control theory, cognitive modeling, optimization and empirical approach. The field was critically reviewed (Rothrock et al., 2002). Goals were: to develop the theoretical paradigm; to instantiate the new mechanism in a computational environment; and to validate the mechanism in the SIRE lab.

5 Application: Close Air Support (CAS) Mission

The application selected was the high threat close air support (CAS) mission involving ground and air forces. CAS entities include the Forward Air Controller (FAC), who controls the air forces, and pilots, who follow directions of the FAC, protect friendly forces, and eliminate enemy units. In high threat situations, there is reduced time for positive identification, decreased communication with the FAC, and a more circuitous battlefield chain of command.

6 Biological Inspiration: The Immune System Metaphor

The immune system (IS) was selected as a potential metaphor for the battlefield mission due to commonality of purpose: to protect against threats through identification and destruction of enemy (in this case pathogens) using "detectors" and "effectors" respectively. Further the immune system is highly adaptive, "learns" from experience and has novel modes of communication (Hofmeyr, 2001). In the simplest terms immune detection involves the ability to distinguish harmful nonself from everything else. Elimination involves choosing the right effectors for a specific pathogen. The architecture of the IS is hierarchical with defenses on several levels. Physical (skin) and chemical (pH) barriers prevent entry. Thereafter, the first line of defense is the non-specific arm of the IS that involves complement (chemical alarm signal) and macrophages (roaming scavenger generalist cells). Chemical signals from cytokines induce the inflammatory response characterized by increased blood flow to the infection site. These responses allow time for the adaptive or specific branch of the IS to activate to deal with a specific pathogen through deployment of lymphocytes. Receptors on the surface of an immune cell recognize epitopes (3D chemical structures) on the surface of the pathogen and evaluate the intensity (affinity) of the threat. The IS must have sufficient diversity of lymphocyte receptors to react to a range of pathogens (10^{16} different epitope varieties). The IS uses genetic recombination to produce enough different receptors. Activation of a complementary lymphocyte is followed by proliferation (clonal selection) and then differentiation into effectors (such as plasma B cells that produce antibodies) and memory B cells (that effect a more rapid secondary response). Autoimmunity is prevented through T helper cells that detect a combination of self and non-self advertised as a warning signal by antigen presenting cells (formerly macrophages). Selection of appropriate effectors is determined by chemical signaling (viruses are eliminated by T killer cells, extracellular bacteria are eliminated by macrophages or complement, production of specific antibodies).

7 Mapping of CAS and IS

Unified Modeling Language (UML, Fowler & Scott, 2000) was employed as a representation model for communicating syntax to assign attributes and behaviors to each class within both CAS and IS (Table 1).

Table 1: Mapping Between CAS Mission and IS Metaclasses

Metaclass Properties	CAS Mission	IS
Foreign, hostile	Enemy troops	Pathogen
Target ID	FAC	Antigen presenting cell
Target recognition, attacks	Pilot	T Helper, B Plasma
Destroys enemy	Weaponry	Antibodies, cytotoxic T cells
Environment	Battlefield	Vertebrate body

8 Incorporating Unique Solutions of IS to Inform CAS Mission

Attention has focused on how the pathogen evades detection by the IS or subverts host defences. Like human enemies, pathogens are capable of using disguise (decrease expression of MHC 11), altering their identity (antigenic drift/shift), using camouflage (acquiring a surface layer of host "self" antigen), sleeping underground (encystment, latency), adopting armour (cuticular protection) or seeking refuge (residing protected area). Tactics to destroy host defences include: disabling weaponry (proteases that cleave antibodies), suppressing defences (stimulate prostaglandins to suppress inflammation), creating a smoke screen (stimulate mass release of antigens), exhausting weaponry (clonal exhaustion of lymphocytes) or disarming opponents (secrete chemicals to repel neutrophils). Hijacking and hostage taking are also observed in the IS.

An alternative strategy has been to focus attention away from the adaptive (specific) response and toward the critical non-specific first response. Metaphors for combat might include border patrol (limited access to body by epithelial tissues), border guards (macrophages and complement elicit a generic inflammatory response), detainment (holds infection at bay), and alerting appropriate specialists (activates the specific IS). Successful immunity relies upon constant surveillance, information from internal and external sources, and communication through the environment.

Decentralized control is also a unique feature of the IS. Control is exerted from multiple agents: the antigen (enemy), the antigen presenting cell (FAC), antibodies (weaponry), lymphocytes (pilot), the brain (central command) and genotype (country of origin). Security can be breached from the homeland (intracellular pathogens) or via the adjoining oceans/airspace (pathogens spread in blood and lymphatics). Defences within the IS are strategically placed to optimise target recognition (APCs converge in lymphoid tissue where T cells patrol).

Target recognition within the IS has prompted considerable attention since a finite genome is capable of recognising infinite numbers of different antigens. Memory is conferred on the IS through residual antigen that can reactivate the specific IS through memory cells. Thus the bottom line is that the enemy is never completely eliminated! Communication in the IS takes the form of a range of chemicals (cytokines/integrins/selectins) that are broadcast into the environment. Entities of the IS recognize and respond to appropriate messages.

9 Synergistic Activities

A graduate course (Advanced Topics in Human-Computer Interactions: Biological Inspired Model Based Adaptive Interfaces) was offered in Fall of 2001 through IVN for students at WSU, OSU and AFIT. The course reviewed HCI paradigms, contextual electronic information retrieval, multiobjective adaptive interfaces, operations research methods, interactive critiquing as a form or decision support, biological paradigms, physiological systems and collaborative systems.

10 Future Perspectives

The initial study has yielded two IS metaphors that are promising for design of new technology to better inform the battlefield: communication and target identification. An alternative approach will be to define problems in the CAS mission (fratricide, delayed communication or information retrieval, elimination of a CAS element, avoiding detection) and then extrapolate a solution from the IS. A simulation of the abstracted IS in Arena is ongoing and will allow us to explore potential emergent behaviors that may inspire development of the interface. An adaptive interface for simple task (multi attribute task battery) will be tested and validated in the SIRE laboratory.

The team proposes to explore other life science paradigms that could be extrapolated more broadly to inform how humans communicate and interface with complex systems in areas such as decision-making and risk management, complexity, IT and collaboration. This will involve organising interdisciplinary symposia/workshops and appropriately training graduate students.

11 References

Altendorfer, A., Moore, N., Komsuoglu, H., Buehler, M., Brown Jr., H. B., McMordie, D., Saranli, U., Full, R., & Koditschek, D. E. (2001). RHex: A biologically inspired hexapod runner. *Journal of Autonomous Robots,* 11, 207-213.

Bonabeau, E., Dorigo, M., & Theraulaz, G. (2000) Inspiration for optimization from social insect behaviour. *Nature,* 406, 39-42.

Fowler, M., & Scott, K. (2000). UML distilled: A brief guide to the standard object modeling language (2nd ed.). Reading, MA: Addison-Wesley.

Hickman, J.J., Darema, F., & Adrion, W. R. (2001). NSF report of the workshop on: Biological computation: how does biology do information technology?

Hofmeyr, S. A. (2001). An interpretative introduction to the immune system, In L. A. Segel & I. R. Cohen. (Eds.), *Design principles for the immune system and other distributed autonomous systems* (pp. 3-26). New York : Oxford University Press.

Holland, J. H. (1995). Hidden order: How adaptation builds complexity. Reading, MA: Helix Books.

Rothrock, L., Koubek, R., Fuchs, F., Haas, M., & Salvendy, G. (2002). Review and reappraisal of adaptive interfaces: toward biologically-inspired paradigms. *Theoretical Issues in Ergonomics Science,* 3(1), 47-84.

Location of the Titles Matters in Performance with Tables and Graphs

Li Zhang, Xiaolan Fu, Yuming Xuan

Institute of Psychology, Chinese Academy of Sciences, Beijing, China
zhangl@psych.ac.cn
fuxl@psych.ac.cn
xuanym@psych.ac.cn

Xiaowei Yuan

ISAR User Interface Design, Beijing, China
xiaowei_yuan@isaruid.com

Abstract

There has been considerable debate in recent years about multiple factors that determine performance with tables and graphs. The present experiment attempted to assess the relative efficiency of line-graphs and tables by studying the effects of the location of the titles, the complexity of the presented data, and the difficulty of the tasks. The results showed that table group extracted information faster than line-graph group, but line-graph group achieved higher recall accuracy than table group. For both tables and graphs, the title on the top led to better performance than the title down below.

1 Introduction

Tables and graphs are very commonly used for presenting data. Although computers facilitate the use of tables and graphs for information presentation, valuable information still depends largely on a person's ability to obtain from the data displayed. Human abilities to recognize information are highly sensitive to the exact presentation, which refers to the forms of presentation that make human search and hold the information easily (Larkin & Simon, 1987). A set of data presented in the exact presentation will help people enhance the speed and the quality in data mining for valuable information.

Summarizing the literatures on table and graph, Paling (1999) found that a major question in this field focused on which was better for a given situation, task, or population. Some empirical studies reported the superiority of graphic presentation (e.g., Benbasat & Dexter, 1985). But some other studies reported the reverse (e.g., Remus, 1984). Other researchers found either mixed results or no difference (Smith, Best, Stubbs, Archibald, & Roberson-Nay, 2002). Meyer (2000) developed Visual Search Model (VSM) to predict the performance of various tasks with tables and graphs of different levels of complexity. The results showed that tables had an initial advantage over graphs for all tasks, and there were complex interactions between the variables.

Researchers have found multiple variables that affect the relative efficiency of using tables and graphs, e.g., the display's graphic properties, the presentation conditions, the complexity of the data, the task, the users' characteristics, and the criterion for optimal performance. The multitude of relevant variables and the existence of complex interactions between them prohibit the empirical study of all combinations of the relevant variables (Meyer et al., 1997). Therefore, no exhaustive research-based set of guidelines can specify the optimal display for every possible condition (Meyer, 2000).

Major questions of research on the display of quantitative information included whether any differences exist between tabular and graphic displays and whether the organization of elements in a display affects its usability and user's performance. The present experiment attempted to assess the relative efficiency of line-graphs and tables by studying the effects of the location of the titles, the complexity of the presented data, and the difficulty of the tasks.

2 Methods

2.1 Participants

Forty-nine 19-23 years old undergraduate students (26 males and 23 females) of China Agricultural University and Beijing Forestry University served as paid participants.

2.2 Design and Materials

A 2×2×2×2 design was used in this experiment. The between-subjects factor was display format (tables vs. line graphs). The within-subjects factors were the title location of tables and line-graphs (top vs. down), the complexity of the presented data that was manipulated by changing the number of points in a data series and the number of data series (2×5 vs. 4×7), and the difficulty of the information extraction tasks (easy vs. difficult). The easy information extraction task was to read an exact value, while the difficult task was to read an exact value and then did an addition or a subtraction.

Forty tables and 40 line-graphs from the same 40 sets of data were used in the experiment.

2.3 Procedure

The experiment was conducted using 17 inches color display with stimulus presentation and data recording controlled by an E-prime program developed for this experiment.

Participants were randomly assigned to one of two groups, namely table group and line-graph group.

At the beginning of the experiment, participants received a brief explanation of the experimental task and completed four practice trials. After they indicated that they understood the task requirements they began to perform information extraction tasks. Each trial began with a question appeared at the top of the screen for 4 seconds. Then the three optional answers and a table (or a line-graph) were shown simultaneously below the question, while the question remained visible. The participants pushed the responding keys on computer keyboard for making selection and were urged to respond as quickly and accurately as possible. Totally they had to complete 40 trials that were randomized by computer.

After the information extraction task, participants did continuous subtraction calculation on the paper for three minutes as a distraction task.

Finally, participants received a recalling task in which they were asked to answer the 40 questions in the previous information extraction task again one by one as accurately as possible. But this time they had to try to retrieve the correct choices without either table or line-graph. In each trial, a question and the three optional answers were presented simultaneously that remained on the screen until the participants made responses by pressing the corresponding keys. Again, the order of the 40 questions was randomized.

3 Results

The two major performance measures in this study were response time and accuracy. Three (2 males and 1 female) participants' experimental data were eliminated as extremum, therefore, the effective number of participants was 46.

3.1 Information extraction task

Means of response time and accuracy of information extraction task are reported in Table 1. A 2×2 ×2×2 analysis of variance (ANOVA) revealed a significant main effect of display format on response time, $F\ (1, 44) = 22.08$, $p < .001$. Table group (8.12s) responded significantly faster than line-graph group (11.39s). The main effect of tables and line-graphs' titles was also significant, $F\ (1, 44) = 27.64$, $p < .001$. The title on the top led to faster responses (9.20s) for both tables and line-graphs than the title down below (10.31s). Participants responded more quickly when doing easy task (6.72s) than difficult task (12.79s), $F\ (1, 44) = 330.36$, $p < .001$. But the main effect of the complexity of the presented data was not significant, $F\ (1, 44) = 3.61$, $p > .05$. The interaction between display format and the difficulty of the tasks was significant, $F\ (1, 44) = 6.13$, $p < .05$. The interaction between the complexity of the presented data and the location of titles was also significant, $F\ (1, 44) = 11.80$, $p < .01$. And the interaction among display format, the complexity of the presented data, and the location of titles was significant, $F\ (1, 44) = 12.45$, $p < .01$.
A 2×2×2×2 ANOVA revealed a significant main effect of location of titles on accuracy, $F\ (1, 44) = 18.60$, $p < .001$. The title on the top led to higher accuracy (0.97) than the title down below (0.94). Low complexity of data led to significant higher accuracy (0.97) than high complexity of data (0.94), $F\ (1, 44) = 5.79$, $p < .05$. Easy information extraction task led to significant higher accuracy (0.98) than difficulty task (0.93), $F\ (1, 44) = 24.11$, $p < .001$. But the main effect of display format was not significant, $F\ (1, 44) = 0.08$, $p > .05$. Both table and line-graph groups obtained higher accuracy (0.96 vs. 0.95).

Table 1: The Response Time and Accuracy of Information Extraction Task *(N=46)*

			Easy Task				Difficulty Task			
			Title on the top		Title down below		Title on the top		Title down below	
	Display Format	Complexity of Data	*M*	*SD*	*M*	*SD*	*M*	*SD*	*M*	*SD*
Response time (s)	Table	2×5	5.28	2.06	5.55	1.76	9.18	3.17	11.14	3.81
		4×7	5.01	1.63	6.16	2.01	10.85	4.87	11.82	4.71
	Line-graph	2×5	8.39	2.24	7.51	1.88	14.92	4.37	14.23	4.12
		4×7	7.47	2.29	8.40	2.73	12.51	2.08	17.71	5.83
Accuracy	Table	2×5	0.99	0.04	0.98	0.07	0.96	0.10	0.95	0.11
		4×7	0.99	0.04	0.96	0.08	0.98	0.06	0.84	0.17
	Line-graph	2×5	0.99	0.04	0.98	0.06	0.94	0.10	0.94	0.11
		4×7	0.99	0.04	0.98	0.06	0.94	0.11	0.86	0.16

3.2 Recalling task

Means of response time and accuracy of recalling task are reported in Table 2. A 2×2×2×2 ANOVA revealed a significant main effect of difficulty of the tasks on response time, F (1, 44) = 45.64, $p < .001$. Participants recalled the answers of difficulty questions faster (5.01s) than easy questions (7.05s). But there were not any significant main effects of display format, location of titles, and the complexity of presented data, Fs (1, 44) < 0.07, $ps > .05$.
A 2×2×2×2 ANOVA revealed a significant main effect of display format on accuracy, F (1, 44) = 4.41, $p < .05$. The line-graph group obtained significant higher accuracy (0.60) than the table group (0.53). It was interesting that the title on the top also led to significant higher accuracy (0.61) for both tables and graphs than the title below down (0.52), F (1, 44) = 17.40, $p < .001$. Participants achieved significant higher accuracy on simple data (0.60) than on complex data (0.54), F (1, 44) = 8.69, $p < .01$. But it is surprising that participants acquired significant higher accuracy on the items of difficulty task (0.63) than on the items of easy task (0.50), F (1, 44) = 35.57, $p < .001$. The interaction between location of titles and the complexity of the presented data is significant, F (1, 44) = 21.24, $p < .001$.

Table 2: The Response Time and Accuracy of Recalling Task *(N= 46)*

			Easy Task				Difficulty Task			
			Title on the top		Title down below		Title on the top		Title down below	
	Display Format	Complexity of Data	*M*	*SD*	*M*	*SD*	*M*	*SD*	*M*	*SD*
Response Time (s)	Table	2×5	4.19	1.95	6.56	4.24	7.02	4.05	7.41	3.37
		4×7	5.36	4.46	4.38	2.96	7.26	3.72	6.49	3.26
	Line-graph	2×5	4.56	2.81	4.27	1.36	6.26	2.70	7.14	3.16
		4×7	4.87	2.3	5.68	6.72	7.37	2.97	6.45	2.52
Accuracy	Table	2×5	0.62	0.20	0.39	0.17	0.68	0.19	0.56	0.24
		4×7	0.43	0.21	0.46	0.19	0.60	0.30	0.54	0.19
	Line-graph	2×5	0.75	0.19	0.34	0.19	0.73	0.21	0.70	0.22
		4×7	0.46	0.24	0.55	0.24	0.63	0.24	0.62	0.19

4 Discussions

4.1 Information extraction performance

The results of the information extraction task showed that tables led to faster responses. It indicates that tables were better for retrieving specific numbers of facts, which is accordance with the results of past research (J. Coll & R. Coll, 1993). Low complexity of data led to faster responses and higher accuracy. It may be that simple data acquires little memory burden than complex data. The easier information extraction task led to faster responses and higher accuracy too. It is easy to understand that participants spend much more time on difficult task than to easy task. Surprisingly, the title on the top led to faster responses and higher accuracy for both tables and graphs. This indicates the location of titles is an important factor to affect the usability of tables and graphs and user's performance. The results showed that the title on the top is better for people to extract information from both tables and graphs.

4.2 Recalling performance

The results of the recalling task showed that line-graph group achieved higher accuracy than table group. We think that, because of the salient strongpoint of graph in comparison with table, such as diverse, clear, straight, beautiful, interesting in reading, multi-dimensional, and continuous, graphs give deep impressions to people. Participants performed better on the items of the difficult task than on the items of easy task. A possible explanation is that comparing to the easy task, people spend more time on the difficult task when completing information extraction task and hence lead to a better memory to the related items. Surprisingly again, the title on the top also led to higher accuracy of recalling for both tables and graphs.

4.3 Summary

The results of the information extraction task showed that tables led to faster responses, but tables had equally high accuracy as graphs. Simple data led to faster responses and higher accuracy and so did the easier information extraction task. The title on the top led to faster responses and higher accuracy for both tables and graphs. The results of the recalling task showed that line-graph group achieved higher accuracy than table group. Participants remembered more items of the difficult task. The title on the top also led to higher accuracy of recalling for both tables and graphs.
To conclude, our findings demonstrate that the efficiencies of line-graphs and tables are different and that the organization of elements in a display affects user's performance. The location of the titles matters in user's performance with tables and graphs.

References

Benbasat, I. & Dexter, A. (1985). An experimental evaluation of graphical and color-enhanced information presentation. *Management Sciences, 31(1),* 1348-1364.

Coll, J. H. & Coll, R. (1993). Tables and graphs: A classification scheme for display presentation variables and a framework for research in this area. *Information Processing & Management, 29,* 745-750.

Larkin, J. H., & Simon, H. A. (1987). Why a diagram is (sometimes) worth ten thousand words? *Cognitive Science, 11,* 65-100.

Meyer, J. (2000). Performance with tables and graphs: Effects of training and a visual search model. *Ergonomics, 43,* 1840-1965.

Meyer, J., Shamo M. K., & Gopher, D. (1999). Information structure and the relative efficacy of tables and graphs. *Human Factors and Ergonomics Society, 41,* 570-587.

Paling, S. (1999). Summary of tables-related literature. Retrieved May, 2002, from http://istweb.syr.edu/~tables/biblio3.doc

Remus, W. (1984). An empirical investigation of the impact of graphical and tabular data presentations on decision making. *Management Sciences, 30(5),* 533-541.

Smith, L. D., Best, L. A., Stubbs, D. A., Archibald, A. B., & Roberson-Nay, R. (2002). Constructing knowledge: The role of graphs and tables in hard and soft psychology. *American Psychologist, 57(10),* 749-761.

Section 3

Engineering Psychology

Assessment and Training Using a Low Cost Driving Simulator

R. Wade Allen*, Theodore J. Rosenthal*, George Park*, Dary Fiorentino**, Erik Viirre†

*Systems Technology, Inc., Hawthorne, CA 90254, USA, rwallen@systemstech.com
**Southern California Research Institute, Los Angeles, CA 90066, dary@adelphia.net
†University of California at San Diego, La Jolla, CA, eviirre@popmail.ucsd.edu

This paper describes a low cost PC platform, including a driving simulation, that is designed to provide orientation, subject record keeping, training, performance measurement and evaluation of driver behavior. The system is based on ordinary, low cost PC technology, and can be run on laptop or desktop computers. The heart of the platform is a driving simulator system that includes an easily programmable scenario definition language for designing driving courses and providing performance measurement. This driving simulation and its application have been described elsewhere. Two applications of the PC platform are summarized herein. The first application involves training novice drivers. The second application concerns the assessment of elderly driver capability.

1. Introduction

The application of driving simulation in research typically requires a controlled laboratory environment and experienced investigators to run subjects. The purpose of the platform described in this paper is to permit reasonably automated subject registration, briefing training and evaluation of driving behavior in classrooms and other relatively uncontrolled environments with minimal supervision. To accomplish this requires functions that 1) allow a new subject to log-in to a database, 2) brief the subject on the background and objectives of the training/evaluation, 3) familiarize the subject with the driving simulation, 4) administer the required training and/or evaluation, 5) log data and 6) provide performance assessment and give feedback to the subject. Simulator performance measurement includes assessment of vehicle motions, driver control responses, and relative motions with respect to other vehicles and pedestrians. Typical measures include accidents, violations, speed and lane deviations, time to collision, use of turn indicators, reaction time, etc. This paper includes a description of the training, evaluation and measurement platform, and a summary of two current applications, one involving training novice driver skills, and the second regarding the evaluation of older driver skills.

2. Platform

The training/evaluation platform includes features for subject registration, orientation and familiarization, automated administration of simulator runs, data logging, and performance assessment and feedback to the subject. The platform is a software shell that administers a simple database and can launch applications such as MS Power Point® for orientation briefings, and administer the driving simulator for familiarization and training/evaluation runs. The software platform also allows simple text based introductions and summary statements to be placed before and after the administration of each application. The platform elements are as follows:

Registration - Subjects are issued a floppy disk that will uniquely identify them, store their data and track their progress (in larger networked systems a common hard disk database server could be used). New subjects are asked to enter their name and a unique ID (identification) number. The platform then advances the subject to the orientation.

Orientation - The platform can play MS Power Point® files including associated narration (sound or wave) files to give desired orientation information and instructions to subjects. MS Power Point® allows the presentation to be prepared in a range of languages and account for regional conditions and application objectives. The narration wave files can be easily prepared using a Windows utility (e.g., Sound Recorder). The orientation for a USA based project starts with background information to motivate the subject, and then presents important information on the roadway environment. Typical example orientation slides are shown in Figure 1 for a novice driver training project that instruct students about turning safety and situation awareness concepts with accompanying audio narration. The orientation ends with a description of the simulator configuration that the subjects will be experiencing, including an overview of the displays and controls. Subjects are assumed to be involved in a more broadly based driver's education curriculum, and the computer orientation is designed to give the information essential for understanding and coping with the roadway environment that will be presented subsequently in the driving simulator training.

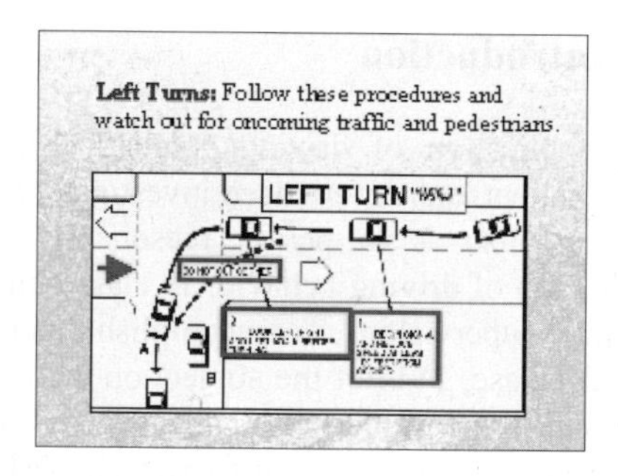

Figure 1. Novice driver orientation slides

Driving Simulation - After the orientation the platform will automatically launch the driving simulator. The first exposure is a familiarization run which slowly introduces the subject to steering and speed control, then intersections with traffic control devices, then finally traffic and pedestrian conflicts. After the familiarization run the platform will present the subject with standardized training and/or evaluation scenarios. Typical driving scenes are illustrated in Figure 2. The subject can be presented a fixed number of driving scenarios during which performance can be evaluated. Performance criteria can be designated for assessing subject capability.

In training situations subjects can be graduated when they have achieved a certain performance level. In evaluation applications, subjects can be assessed as to the adequacy of their driving capability, and given additional trials to improve or rehabilitate their skills. The simulator has

Figure 2. Simulation display scenes

previously been described in some detail elsewhere (Allen, Rosenthal, et al., 1999; Allen, Rosenthal, et al., 2001).

3. Applications

Two applications of the above assessment and training platform are included here that relate to problems with young and old drivers as suggested in the Figure 3 accident rates for the State of California, USA (Aizenberg and McKenzie, 1997). The youngest drivers suffer from lack of experience in real, hazardous driving environments. Driving simulation provides a means for training young drivers situation awareness and hazard recognition and avoidance in a controlled, concentrated manner. Elderly drivers suffer from deterioration of physical and mental skills that allow them to recognize and avoid hazards. The platform provides a means for assessing elder driver skills, and possibly providing some rehabilitation, or suggestions for controlling their exposure (e.g. avoid busy streets, take care at intersections, or stop driving in the extreme).

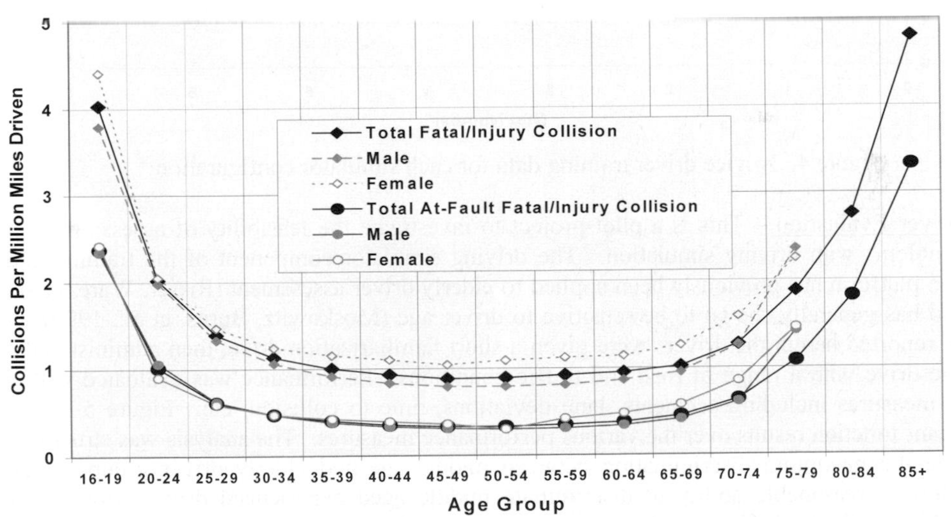

Figure 3. Accident rates by age group in the state of California, USA

Novice Driver Training – This is the second phase of a project to develop a low cost simulator for training novice drivers. The objective of this application is to train novice drivers on complex, hazardous driving situations that will reduce their high accident rate in the first few years of driving. The first phase of this project focused on the feasibility of a low cost, driver-training simulator (Allen, Cook, et al., 2000). In the current second phase of this project the focus is on training a large number of novice drivers in the State of California, USA and following their accident records through the California Department of Motor Vehicles. This project involves several levels of simulation, from desktop single screen versions to full cab, wide-angle display systems as illustrated in Figure 3. Figure 4 shows preliminary learning curves for road edge excursions and accident rates for each of the simulator configuarations. ANOV was performed on each measure and included run as a within subject variable and gender and site as between group variables. Run was significant for both accidents ($p<.05$) and road edge excursions ($p<.001$). Configuration was significant for accidents ($p<.001$). Gender was significant for road edge excursions ($p<.01$). Several interactions were also significant for accidents. The configuration differences and interactions will be investigated when the training of the entire subject population (n=600) is complete. The learning trends probably represent the drivers learning to deal with the complexity of the roadway and traffic environment in general.

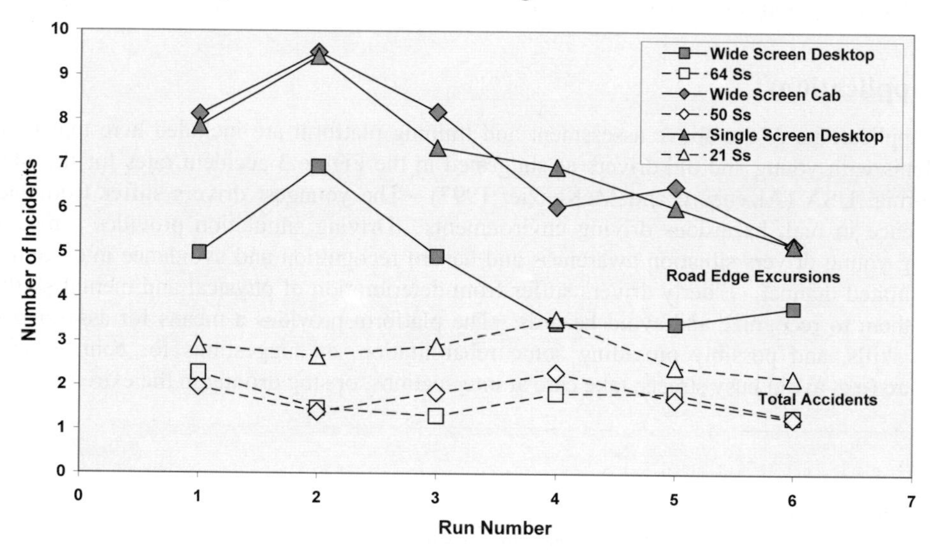

Figure 4. Novice driver training data for each simulator configuration

Older Driver Evaluation – This is a pilot project to investigate the feasibility of assessing older driver problems with driving simulation. The driving simulator component of the training and evaluation platform has previously been applied to elderly driver assessment (Risser, Ware, et al., 2000) and has generally shown to be sensitive to driver age (Moskowitz, Burns, et al., 1999). In the work reported herein the drivers were given a short familiarization drive, then administered a 15 minute drive with a range of road and traffic conditions. Performance was evaluated with a range of measures including accidents, lane deviations, time to collision, etc. Figure 5 shows discriminant function results over the various performance measures. The analysis was structured to provide the maximum discrimination between middle age and elderly driver groups. The results show a reasonable ability to discriminate middle aged experienced drivers from older drivers. Three subjects in the younger group overlap with three subjects in the older group. The range of performance of the older drivers is not surprising as we know the effects of aging are very ideosyncratic. The range of middle aged driver performance is a little more surprising.

4. Concluding Remarks

The objective of this research was to develop and demonstrate a PC based driver evaluation and training platform that can be routinely run outside of the research laboratory by non-research personnel (e.g. therapists, trainers, teachers). The driving scenarios and performance measures have proven to be sensitive to training and age. The PC-based platform described herein has proven to be easy to administer in school classrooms as well as in clinics and laboratories. PC technology will continue to improve for the foreseeable future in terms of computing power and visual and auditory display. The potential capability of PC-based training and evaluation platforms will also advance, limited primarily by software development.

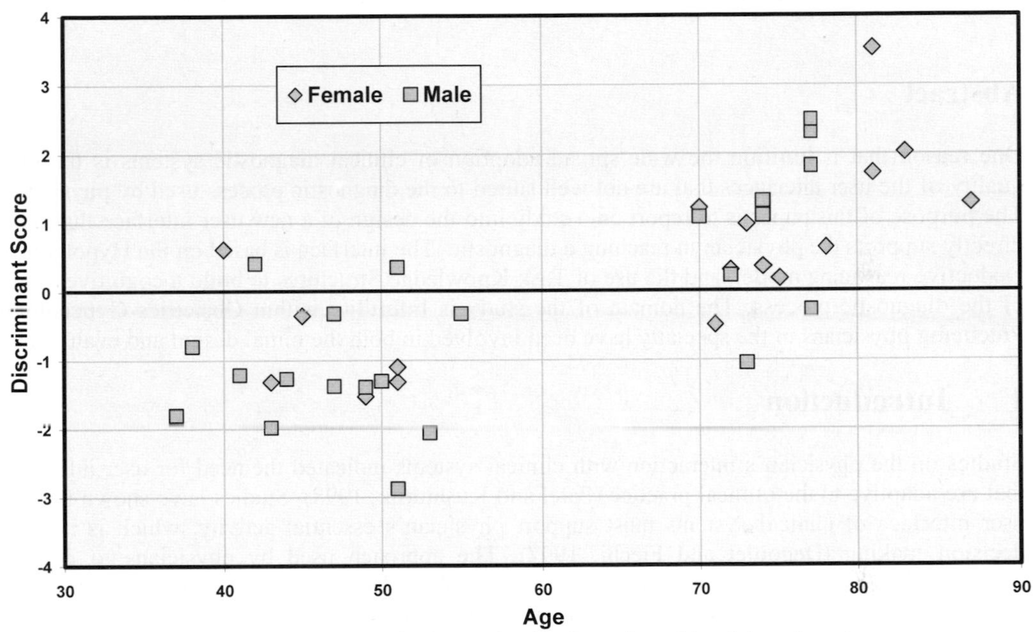

Figure 5. Discriminant analysis results for detecting older driver impairment

5. References

Aizenberg, R. and McKenzie, D.M., (1997) "Teen and Senior Drivers," California Department of Motor Vehicles Report No. CAL-DMV-RSS-97-168, December 1997.

Allen, R.W., Rosenthal, T.J., et al., (1999), Low Cost, Pc-Based Techniques For Driving Simulation Implementation, DSC1999, Paris, France.

Allen, R.W., Cook, M.L., et al., (2001) "Low Cost PC Simulation Technology Applied To Novice Driver Training", Proceedings of the International Driving Symposium on Human Factors in Driver Assessment, Training, and Vehicle Design, Snowmass Village, CO, USA.

Allen, R.W., Rosenthal, T.J., et al., (2001) "A Scenario Definition Language for Developing Driving Simulator Courses," DSC2001, Sofia Antipolis (Nice), France, September 2001.

Moskowitz, H., Burns, M., Fiorentino, D., Smiley, A., and Zador, P. (1999), *Driver Characteristics and Impairment at Various BACs.* Final Report, NHTSA, U.S. Dept. of Transportation, Washington, D.C.

Risser, M. R., J. C. Ware, et al. (2000) "Driving Simulation Performance in the Elderly with Mild Cognitive Impairment." Sleep 23 (Abstract Supplement #2).

Clinical System User Interface derived from cognitive task analysis of the physicians' diagnostic process

Nawal H AlShebel *Peter M Dew*

King Saud University
P O Box 7805 Riyadh 11472
nshebel@ksu.edu.sa

University of Leeds
Leeds LS2 9 JT
Dew@comp.leeds.ac.uk

Abstract

One reason that is limiting the wide spread adoption of clinical diagnostic systems is the poor quality of the user interfaces that are not well suited to the diagnostic process used by physicians. The purpose of this paper is to report on a study into the design of a new user interface that more directly supports the physician in reaching a diagnostic. The interface is based on the Hypothetico-Deductive reasoning model, and the use of Task Knowledge Structures to build a cognitive model of the diagnostic process. The domain of the study is Infertility within Obstetrics Gynecology. Practicing physicians in the specialty have been involved in both the initial design and evaluations.

1 Introduction

Studies on the physician's interaction with clinical systems indicated the need for user interfaces that are adaptive to the clinical practice (Patel and Kushniruk, 1998). Studies have shown that the user interface of clinical systems must support physician's essential activity, which is medical decision making (Degoulet and Fiechi, 1997). The approach used by physicians to reach a diagnosis is described by (Shortliffe and Barnet, 2001) as "hypothetic-deductive reasoning" model. This paper reports on the use of this model together with a cognitive task analysis of the clinical process to derive a cognitive user interface model. This model has been based on the user requirements obtained from users/physicians at King Khalid University Hospital in Riyadh. The procedural and declarative knowledge of the diagnostic process has been developed using Task Knowledge Structures (Johnson and Johnson, 2000). The model was used to design a user interface prototype of a clinical system. The knowledge domain of the study was Infertility within Obstetrics/Gynaecology. To model the knowledge domain interviews were conducted with Gynaecology consultants to capture the knowledge component of the user interface and undertake early evaluation studies.

2 Hypothetico-Deductive Reasoning Process

Strategies used during clinical decision making have been identified and critically reviewed in literature by Croskerry, 2002. This review have shown that the Hypothetico-Deductive model is the most widely accepted model of clinical decision making (Kovacs and Croskerry, 1999). However, little progress has been made to conceive new technological methods improve the medical diagnostic process (Patel, Kaufman and Arocha, 2002).

The use of the Hypothetico-Deductive Reasoning process by physicians is described by (Shortliffe and Barnet, 2001). Briefly, the process is "sequential , staged data collection, followed by data interpretation and the generation of the hypothesis, leading to hypothesis directed selection of the next most appropriate data to be collected". (Chapman, 2002), identifies four stages for the process: 1. Hypothesis Generation by listing potential diagnosis triggered by initial patient cues; 2. Hypothesis evaluation by data gathering to confirm/exclude original diagnostic hypothesis; 3. Hypothesis refinement by adding new diagnostic hypothesis; and 4. Hypothesis verification by confirming diagnostic Hypothesis with data gathered. This model forms the basis of the user interface design discussed below.

3 Clinical Diagnostic Task Model

A task Knowledge structure (TKS) is a summary representation of the different types of knowledge that are recruited and used in task behaviour (Johnson and Johnson, 2000). In TKS, the task knowledge is represented in terms of roles, goals, objects and actions. Two types of knowledge are represented; 1) Procedural Knowledge represented by goal structure of actions and 2) declarative knowledge represented by taxonomic structure of objects and attributes used to reach the goals. TKS is used in this paper to design a cognitive model of a user interface to support the medical diagnostic process. For example a sample of a goal structure TKS model for a general diagnostic process is shown in Figure-1 below:

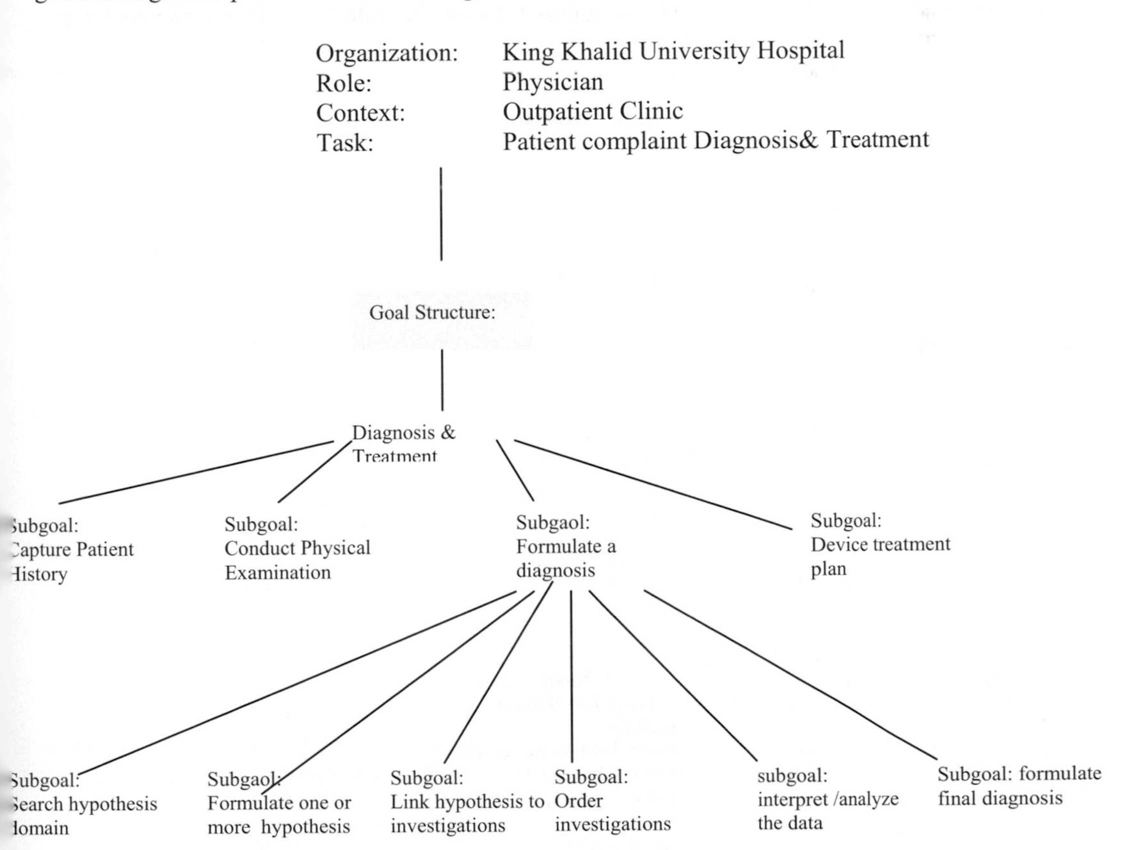

Figure-1 A sample goal structure of TKS diagnostic model

Physicians' main goal (diagnosis and treatment) is represented by a number of sub-goal structures that specify the actions to be conducted by physicians to reach the main goal. At the next level, further breakdown of the sub-goal "formulate diagnosis"; It has been found that TKS has helpful in identifying the actions to be conducted by the physician to reach a final diagnosis. These actions determine the functionality of the user interface. Following this , the taxonomic structure (objects and attributes) associated with such goals/actions and used it (taxonomic structure) to model the knowledge representation used in the user interface.
The gaol and sub-goal structure of the diagnostic process is generic to all medical specialties, because it adopts the clinical diagnostic Hypothetic-Deductive reasoning approach. The object attributes in the taxonomic structure is specific for the medical specialty knowledge domain. It can be seen that TKS enables us to build a cognitive model of the user interface from which it is possible to realise higher quality user interfaces. The next stage is to build to model the domain.

3.1 Infertility Knowledge domain Taxonomic Structure

For this study the knowledge domain was Infertility within Obstetrics/Gynaecology. The Taxonomic structure (represented by objects and attributes used in the infertility diagnostic process) is shown in Figure-2 , this is referred to as the Infertility diagnostic declarative knowledge.

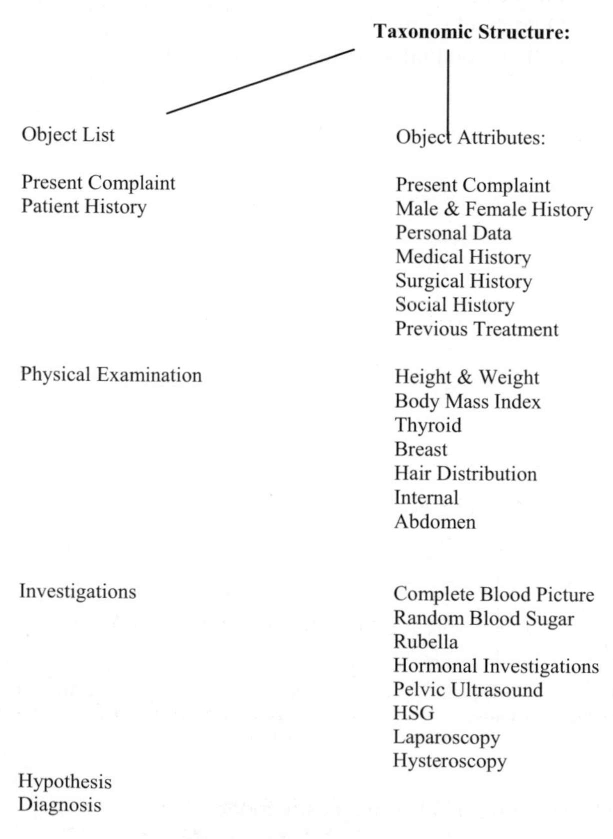

Figure-2 A sample of Taxonomic Structure Infertility Diagnostic Process

This model was derived from interviews conducted with Gynaecology consultants at King Khalid University Hospital. TKS helped us analyzing and collecting data within the context of the user/physician domain and organization. The goals/actions are used to define the functionality of the user interface, while the taxonomic structure is being used to provide the visible representation of the user interface.

4 User Interface Prototype

The user interface concept was derived from the Task Knowledge Structures and the Hypothetico-Deductive reasoning process. An outline of the conceptual user interface is shown below in Figure-3:

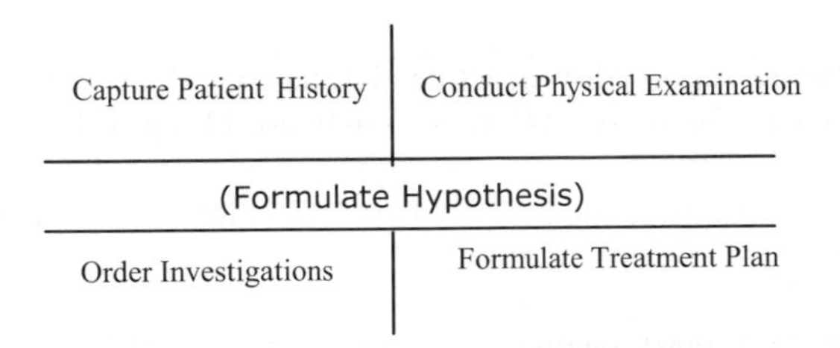

Figure-3: User Interface Concept

The main screen has five main functions .The representational components and their association to the functions are mapped from the Task Knowledge. Formulate Hypothesis function is the core element of the user interface, providing the knowledge support for physician main activity to reach a final diagnosis. The supporting knowledge functions were derived from TKS models and the domain specific Infertility Literature Knowledge. The following is a display of the main screen design (Figure-4):

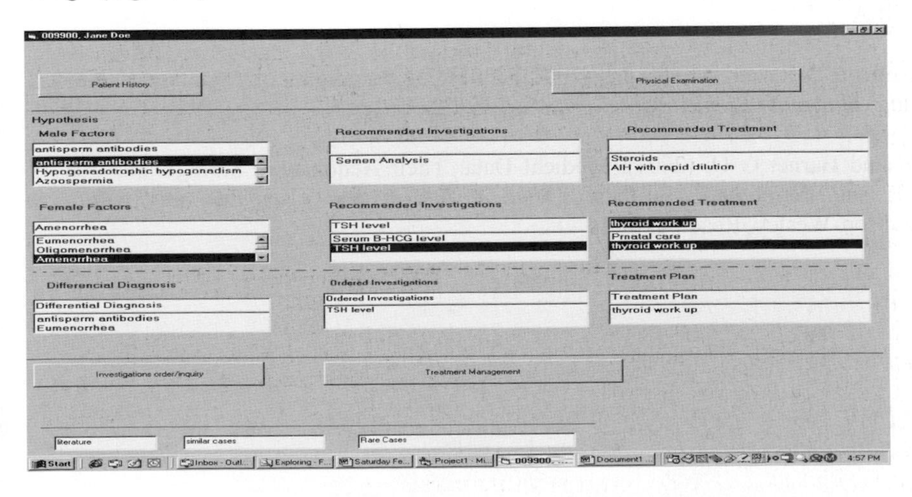

Figure-4 User Interface Prototype design – main screen

5 Conclusion

This paper has shown that TKS can be used to model the physicians' diagnostic deductive reasoning process. Both a generic and a domain specific Task Knowledge Structure model of the diagnostic process have been derived. The TKS model was used to derive a clinical user interface prototype, which was evaluated by Gynaecology consultants.

References

Chapman DM, Char DM, Aubin CD,(2002), Clinical Decision Making, In Marx, Hockberger and Walls (Eds.), Rosen's Emergency Medicine: Concepts and Clinical Practice, 5th ed., pp. 107-115, Mosby Bpublishing.

Croskerry P, (2002), Achieving Quality in Clinical Decision Making: Cognitive Strategies and Detection of Bias, *Academic Emergency Medicine*, vol.9, no. 11, pp. 1184-1204.

Degoulet P and Fieschi M, (1997), Medical Software Development in: *Introduction to Clinical Informatics: Computers in Healthcare*, Springer.

Johnson P and Johnson H, (2000), Getting the Knowledge into HCI: Theoretical and Practical Aspects of Task Knowledge Structures. In Schraagen, Chipman and Shalin (eds.),*Cognitive Task Analysis* , pp. 201 – 214.

Kovacs G and Croskerry P,(1999), Clinical Decision Making An Emergency Medicine Prospective, *Academic Emergency Medicine*, vol. 6, no. 9., pp. 947 – 952.

Patel V L and Kushniruk A W (1998), Interface Design for Health Care Environments: the Role of Cognitive Science. In Christopher G. Chute (ed.), *Proceedings of the 1998 AMIA Annual Symposium*, AMIA

Patel V L, Kaufman D R, and Arocha J F,(2002), Emerging Paradigms of cognition in medical decision-making, *Journal of Biomedical Informatics*, 35, 52-75.

Shortliffe E H and Barnet G O, (2001), Medical Data: Their Acquisition, Storage and Use. In Shortliffe E H and Perreault L E, (eds.), *Medical Informatics: Computer applications in healthcare*, Addison-Wesley, Reading, Mass

Task Analysis Method of Advertising Design Process Using Computer Media Based on Cognitive Behaviour Description and Eye Tracking Technique

Hirotaka Aoki

Tokyo Institute of Technology
2-12-1 Oh-okayama, Meguro-ku, Tokyo 152-8552, Japan
aoki@ie.me.titech.ac.jp

Abstract

This paper illustrates a task analysis methodology of designing print advertisements based on behaviour, verbal protocol and eye-tracking data. The methodology reveals transitions of tasks performed as well as information acquisition processes. We apply the methodology to data obtained in a series of experiment in which two practicing designers participated. Based on the analysis results, inferences concerning factors affecting the designers' behaviours are discussed.

1 Introduction

Visual advertisements (ads.) are designed by the experts, so-called graphic designers, having much skill on fine arts and relating fields. However, it is recognised that the design processes carried out by these experts are intangible because these processes are strongly connected with each expert's implicit know-how. In this context, this paper illustrates a series of task analysis methodology of printed ad. design processes in the concept embodiment stage. In this study, the concept embodiment stage is referred to as a series of tasks in which designers try to reduce differences between design goal and given design elements by arranging them, by colouring them, etc. In this stage, designers are required to choose good design elements, layouts, and so forth, by which the concept of advertised product/service can be conveyed to consumers. Therefore, it seems that the concept embodiment stage plays a crucial role in design of effective print-ads.

We use the following three kinds of data as the basis of analysis: Designers' verbal protocol data, behavioural data, and eye tracking data that are obtained during ads. design tasks. The proposed analysis method is applied to data observed in the experiments (experiments 1 and 2), in each of which one of two practicing advertising designers participated. In the experiment 1, the following two factors' influences on design activity are focused: Level of design difficulty and sketching. In addition, the analysis method is also applied to data obtained in experiment 2, in which the experimental environment was set as almost the same as the practical/daily situation. We discuss the influences of the aforementioned two factors on design processes based on analysis results. We also discuss the designers' individual differences by comparing between analysis results of experiments 1 and 2.

2 Task Analysis Method

2.1 Transcription of Behaviour and Protocol Data

In this study, tasks conducted in concept embodiment stage are analysed based on a designer's behaviour, verbal protocol, and eye movement data. Basically, verbal protocol data are collected by use of concurrent verbal protocol method [1]. A designer's behaviour and protocol data are transcribed onto a paper in timeline style as shown in Fig. 1. This figure shows these data obtained during designing a car ad., which consists of design elements of woman, car, texts groups (texts 1~2), rectangle, brand-logo, and so forth.

2.2 Task Transition Analysis: A Macro-Perspective Analysis

2.2.1 Tasks Included in Design Concept Embodiment Stage

In design concept embodiment stage, many types of tasks are carried out. We characterise these tasks by the following three aspects: (1) Action, (2) objects being acted upon, and (3) information involved in objects. Each aspect is classified into sub-classes as follows:

- Action (Control of shape/Control of position/Control of shape and position)
- Object(s) being acted upon (Element/ Element-group/Whole)

- Information involved in object(s)
  - Information format (Pictorial /Textual)
  - Information content (Product-related/Emotion-related/Purchase-related)

These aspects are identified based on characteristics of behaviour and protocol data. In Fig.1, for example, it can be found that this designer successively manipulates all the design elements from 11 sec. to 1 min. This characteristic of behaviour data seem to show that the designer tries to change shapes of positions of the whole design elements around this moment. Therefore, we can identify this period as a duration in which Task 1 was performed. We also understand that an action carried out in Task 1 is control of position and shape for whole design elements, and that these elements include product-, emotion-, and purchase-related information represented in pictorial and textual formats. Directing attention to the verbal protocol obtained at 1 min. 42 sec. we can interpret that the designer change his/her task to a new phase. In this way, we infer tasks (i.e. action and objects) from behaviour and protocol data.

2.2.2 *Extraction of Task Transition*

To analyse task transitions, an analytical diagram called task-characteristics-based transition network is proposed. The task-characteristics based transition network enables us to obtain visually represented designer's time allocation spent at each task characteristic as well as a transition frequency between all the combinations of task characteristics.

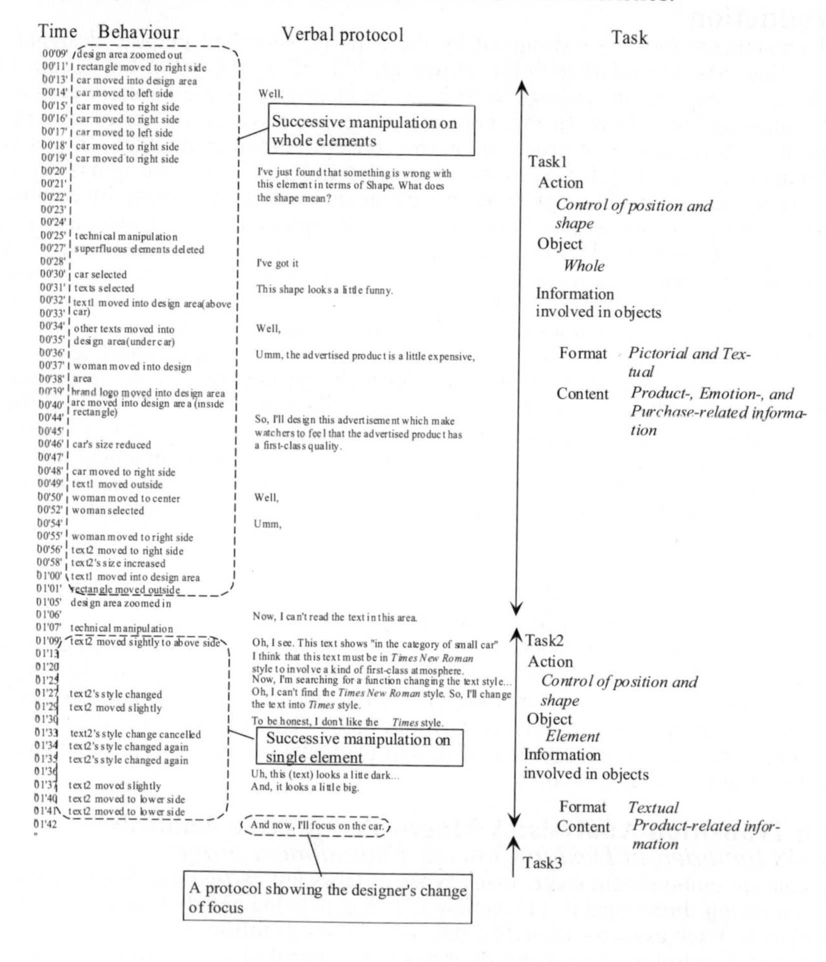

Fig. 1 Data transcription

2.3 Eye-Tracking Analysis Combined with Behaviour and Protocol Data: A Cognitive-Perspective Analysis

It is often said that eye-tracking technique holds a potential ability for analysing human cognitive processes performed unconsciously (ex. [2]). Another analytical diagram called attention allocation network is described at every task to obtain interpretations of a designer's cognitive processes. Based on these network forms, we estimate designer's watching patterns by considering areas of interests and the most frequently observed adjacency pairs of design elements gazed at. These aspects are related to how a designer acquires information from design elements at each task.

3 Experiments

To obtain data during advertising design, we performed the following two experiments (experiments 1 and 2). Two practicing designers who have perfect design skills participated in each of the experiments. Both designers were paid about 150 dollars for each ad.

In experiment 1, a designer (Designer A), who has over 10 years' design experience, was asked to design two car ads., and was asked to design two cosmetics ads. The designer was instructed that he had to design car ads. by arranging the following pictorial/textual design elements into design area: Visual images of advertised car, woman, arc-shaped sign, rectangle-shaped sign, brand-logo, and so forth, and textual elements such as brand-name, prices, catch-copy and so forth. As to the cosmetics ads., we ask him to design ads. by use of the same design elements as the car ads. except images of advertised products. This instruction is motivated by the intention of controlling the level of design difficulty by changing the similarity between product concepts and design elements. This control resulted in making the design of car ads. relatively more difficult than that of cosmetics ads. In addition, a subject was permitted to generate sketches while thinking of design ideas of each one of cars and cosmetics, and was asked only to think design ideas without sketches in design of the rest of the ads. The choice of the with/without sketch condition was motivated by interests in estimation of effects of sketch on design productivity (ex. [3]). In each ad., a subject was asked to design one printed ad in A4 size magazines. At the beginning of the experiment, the subject was shown a short brochure of the advertised product, and asked to design A4 size ad. by only using design elements which can be found in the brochure. After this instruction, a subject thought design ideas under with/without sketch condition, and design ads. using software Adobe Illustrator 10. The same twenty minutes is provided to the designer in order to generate design idea for both with/without sketch condition. During designing ads., visual image on display was recorded by video recorder with subjects' verbal protocols. Subjects' eye movement data were also recorded by use of an ASL 4000 eye tracking system. The subject was frequently asked to talk aloud as he worked on the design task.

Another designer (Designer B) having over 7 years' design experience participated in experiment 2. In this experiment, the designer was asked to design the two cosmetics ads. (same as experiment 1). The equipments provided in this experiment were the same as those in experiment 1. However, any limitations, such as means for generating ideas, were not given. In addition, we did not ask the designer to "thinking-aloud" during designing. This was motivated by the intention of corresponding the experimental environment with the designer's daily work situation. Therefore, the amount of protocol data obtained in the experiment 2 was much smaller compared to experiment 1.

4 Results

4.1 Task Transition Analysis

Fig. 2 and 3 show task-characteristics-based transition networks described based on data obtained in experiments 1 and 2. In these figures, both dimensions in each table indicate "Action" and "Object(s) being acted upon" which are included in task characteristics. For example, the upper-left section in each table shows a task in which a designer manipulates a single design element's shape and position. The circle in each section represents that the task is carried out, and the size of these circles corresponds to the total duration of the task. The arcs between these circles show task transition paths, and the numbers represent the observed frequency. In the following sections, we focus on data of Designer A obtained in experiment 1, and discuss the tendencies found.

4.1.1 *Influence of Sketching*

Comparing the total task duration of "with sketching" and that of "w/o sketching" within the same level of difficulty level by using data obtained in experiment 1, we can observe that more time is necessary to finish designing ads. if sketching is not carried out in idea generating phase (1,350 sec. vs. 972 sec. for cosmetics ads., and 1,782 sec. vs. 888 sec. for cars ads.). This difference is mainly caused by the duration of tasks in which design element group are manipulated in both cosmetics ads. ($\chi_0^2(\varphi=9)=450.1$, $p<0.01$).

It seems that a designer manipulating design element group has put some semantic interpretation on these categorised design elements. In other words, it seems that a designer has formed in mind some semantic relationships between design elements, which a design element group is composed of, just before she/he starts manipulating the design group. From the result obtained and the above-mentioned inference, it may be conjectured that the sketching performed in idea generation phase plays a crucial role in the following concept embodiment stage in terms of constructing/generating semantic relationships among design elements.

4.1.2 *Influence of Design Difficulty*

No significant difference is identified between both difficulty conditions (difficult and easy) in terms of total task duration. As can be understood from Fig. 2, however, tasks in which the shapes of objects are changed are found during designing of cars ads. This may suggest that this designer perceived some differences between the design concepts of cars' ads. and each design element /element group, and struggled to reduce the differences by changing shape. As to the cosmetics ads., it seems that this kind of tasks are not necessary since the differences mentioned above do not exist.

4.2 Eye-Movements Differences Between Designers

Fig. 4 indicates attention allocation networks for Designers A and B during designing of the same cosmetic product (i.e., Cosmetics Ad 1). In these networks, every node represents a fixated area, and its size corresponds to the fixation duration. Arcs between nodes represent scan-paths of a designer, and its thickness shows the movement frequency. In Fig. 4, fixated areas for which the fixation duration ratio is less than 5% in total time are omitted. In addition, the scan-paths having lower frequency ratio than 0.5% in total movements are also eliminated. As a remarkable

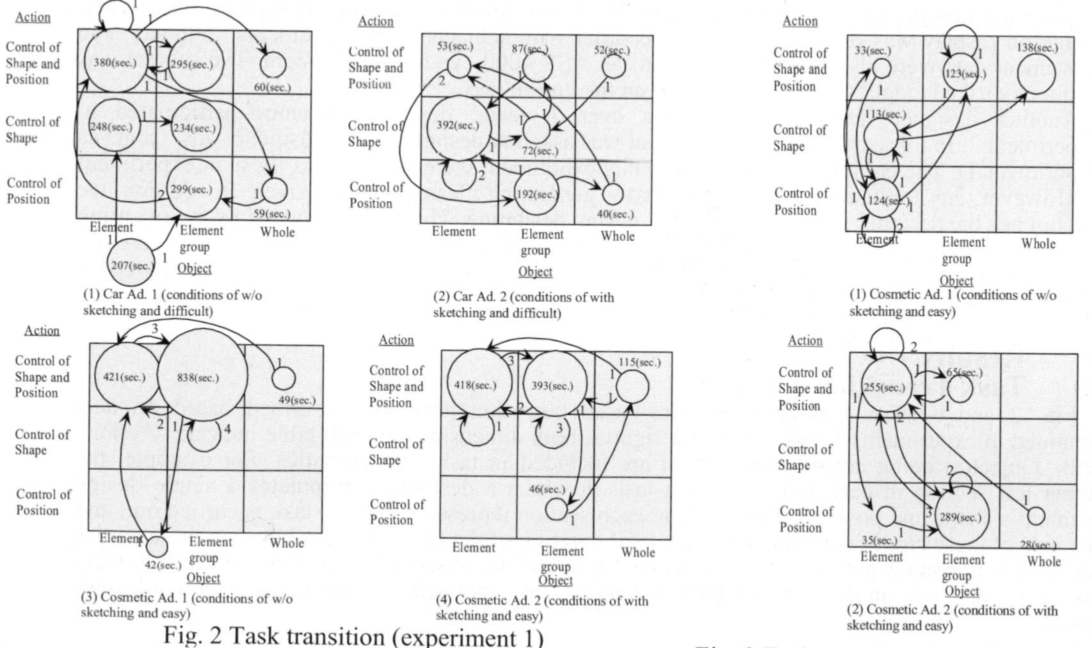

Fig. 2 Task transition (experiment 1)

Fig. 3 Task transition (experiment 2)

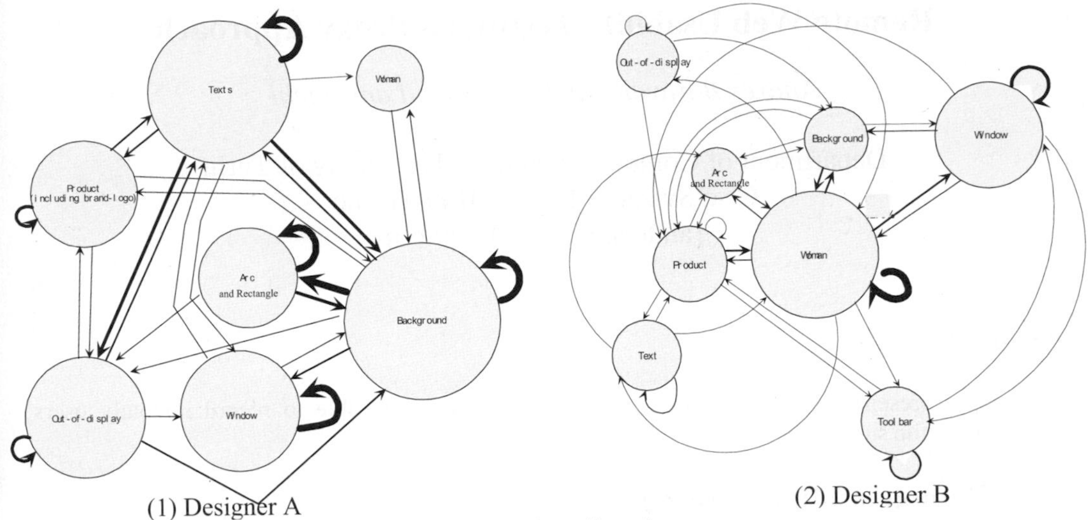

(1) Designer A (2) Designer B

Fig. 4 Attention allocations

difference observed in Fig. 4, more attention is paid for *background* for Designer A. In this study, the term *background* means the design area that doesn't include any design elements. Therefore, it can be estimated that a designer monitors relatively wide area around fixated position in the design area during fixating on *background* since he/she keeps high arousal level all through designing. Considering this estimation, we put an interpretation on the observed difference between two designers as follows: Designer A tends to acquire information from relatively wider area (including more amount of design elements) than Designer B do. Focusing on the individual attributes of designers, we can find a difference of experiences on ads. design. On one hand, Designer A mainly designs print ads. in his daily working life. On the other hand, Designer B conducts design of ads. on www, and sometimes designs print ads. From these considerations, the observed differences of information acquisition tendency between the designers may be caused by the designers' experiences on ads. design as well as the experimental environment.

5 Conclusions

In the present paper, we illustrate an task analysis methodology for concept embodiment stage in print ads. design based on behaviour, verbal protocol and eye-tracking data. Applying the methodology, we could infer factors affecting designers' behaviours from analysis results. However, the number of designers and designed ads. is too small to discuss the feasibility of the task analysis methodology as well as the observed phenomena. Therefore, we plan to collect more amounts of data, and to elaborate the illustrated analysis methodology in near future.

References

[1] Hansen, J. P. and Itoh, K. (1995). Building a cognitive model of dynamic ship navigation on basis of verbal protocols and eye-movement data. In Norros, L. (ed.): 5th *European conference on cognitive science approaches to process control*, Espoo, Finland, 325-337.

[2] Itoh, K., Hansen, J. P. and Nielsen, F. R. (1998). Cognitive modelling of a ship navigator based on protocols and eye-movement analysis, *Le Travail Humain*, 61(2), 99-127.

[3] Schon, D. A. and Wiggins, G. (1992). Kinds of seeing and their functions in designing, *Design Studies*, 13(2), 135-156.

Acknowledgements

This research was partly supported by Grant-in-Aid for Young Scientists, No. 14780349, the Japan Society for the Promotion of Science, and by a research fund of the Rikohgaku-shinkohkai, Tokyo, Japan. We would like to acknowledge the following excellent designers: Kaori Miyayama, Sho Ishii, and Yoji Takemura, not only for their participation in this research, but also for their providing great amount of useful suggestions to this research. We are also indebted to Prof. Kenji Itoh, Tokyo Institute of Technology, who always encouraged this research.

Remote Web Usability Testing: a Proxy Approach

Andres Baravalle and Vitaveska Lanfranchi

Department of Computer Science - University of Turin
C.so Svizzera 185 – 10149 Torino
{andres, vita}@di.unito.it

Abstract

This paper presents OpenWebSurvey, an open-source software able to record remotely users behaviour while surfing the Internet.

1 Remote usability testing

Remote usability testing allows researchers to evaluate the usability of web sites gathering information from remote users. The main advantage of this approach is that it facilitates evaluation in a real world environment, but without the need for costly technologies.
In a previous research (Perkins, 2002), Perkins defines three basic types of remote evaluation: attended, automated and instrumented.
Attended methods require the researcher to inspect visually the user behaviour. This kind of evaluation generally can be performed using common remote viewing software, as Norton Pc Anywhere or Microsoft NetMeeting.
Automated methods collect and automatically analyse data. In this category we suggest to distinguish two main approaches: the spider approach and the user logging approach. The first approach is used mainly to analyse the compliance of a web interface to standards or guidelines, like WAI (W3C, 1999) or U.S. Section 508 (U.S. Congress, 1999), and to find scripting errors in the code. The second approach focuses on automatically analysing the user behaviour during a normal web navigation session. This approach is normally based on log files analysis.
Instrumented methods collect user behaviour using special software applications to perform a test and log the user behaviour.
We present a brief survey of methods for remote web usability testing and we then introduce OpenWebSurvey, its architecture and capabilities. A case study is then presented.

2 A classification of tools for remote web usability testing

We classify the tools for instrumented evaluation of web usability based on the kind of application that logs the user behaviour: a client web browser, the web site under investigation and a different site. We call these approaches respectively "web browser", "web site" and "proxy" approaches.
The web browser approach is based on the development of a specific client side application that the user will use to surf the web. That application is usually a customisation of a specific browser, commonly Microsoft Internet Explorer, but it can as well be a different web browser at all. The main disadvantages are that the users are forced to install and use a particular client, some hardware and/or software platform, and that they will have to install the client by theirself, that it is not always a fast and easy task. This kind of approach is adopted, amongst others, by Usable Tools (2001).

The web site approach is based on a web application that is inserted in the web site under investigation and that monitors the user behaviour. Usually it requires to have full access to the web site code and so it can be used only in few cases, as most of the web sites will not allow external researchers to modify their source. This kind of approach is adopted, amongst others, by Paternò, Paganelli and Santoro (2001) and by Trausan-Matu, Marhan, Iosif and Juvina (2002).
Trausan-Matu, Marhan, Iosif and Juvina (2002) are developing an e-learning system based on personalization of the information and of the interaction. They adopted the "web site" approach to record the behaviour of students browsing e-learning web pages.
The proxy approach is based on an application that acts as a proxy between the client and the server monitoring the navigation. The main advantage of this approach is that the users do not have to install anything on their computer and that the researchers do not need to have access to the web site code. This kind of approach is adopted, amongst others, by WebQuilt (2001), a logging and visualization system that captures user actions on the web and analyzes the collected data.
The main difference between our approach and WebQuilt is that WebQuilt is a logging tool that can be used for usability researches only combined with external surveys tools. OpenWebSurvey instead combines user logging and online surveys, presenting the subjects some tasks to be accomplished while surfing, in order to analyze their reactions to specific stimuli.

3 OpenWebSurvey

OpenWebSurvey is a web based tool, based on the proxy approach, able to record, store, share and process data for web usability analysis (see Figure 1).

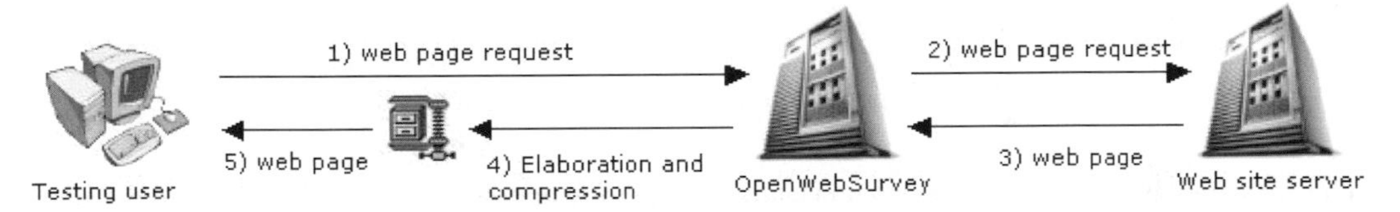

Figure 1. OpenWebSurvey architecture

It does not require any client or server installation, since the user can answer the survey questions and navigate the web trough a normal web browser. The subject connects to OpenWebSurvey site and finds some questions to be answered while surfing some sites. Usually the subject is presented some tasks to be accomplished and some questions to answer, in order to analyse behaviour responding to specific stimuli.
The main advantage of presenting questions during navigation is that the subjects have to react immediately and to find the information requested in a pre-established (by the researchers) context. The researcher can choose a domain for the investigation not only by the tasks but also by the sites that they decide to link to every question. Besides the subjects do not know the question or the task to be accomplished before starting navigation and so they cannot establish in advance their navigation path or search strategies. The collected data can then be analysed to better understand the cognitive and emotional factors that can influence search strategies and web navigation.
The testing interface is a double framed browser window: the upper frame contains the survey questions and the input fields for the answers, while the lower frame contains the site under investigation (see Figure 2).

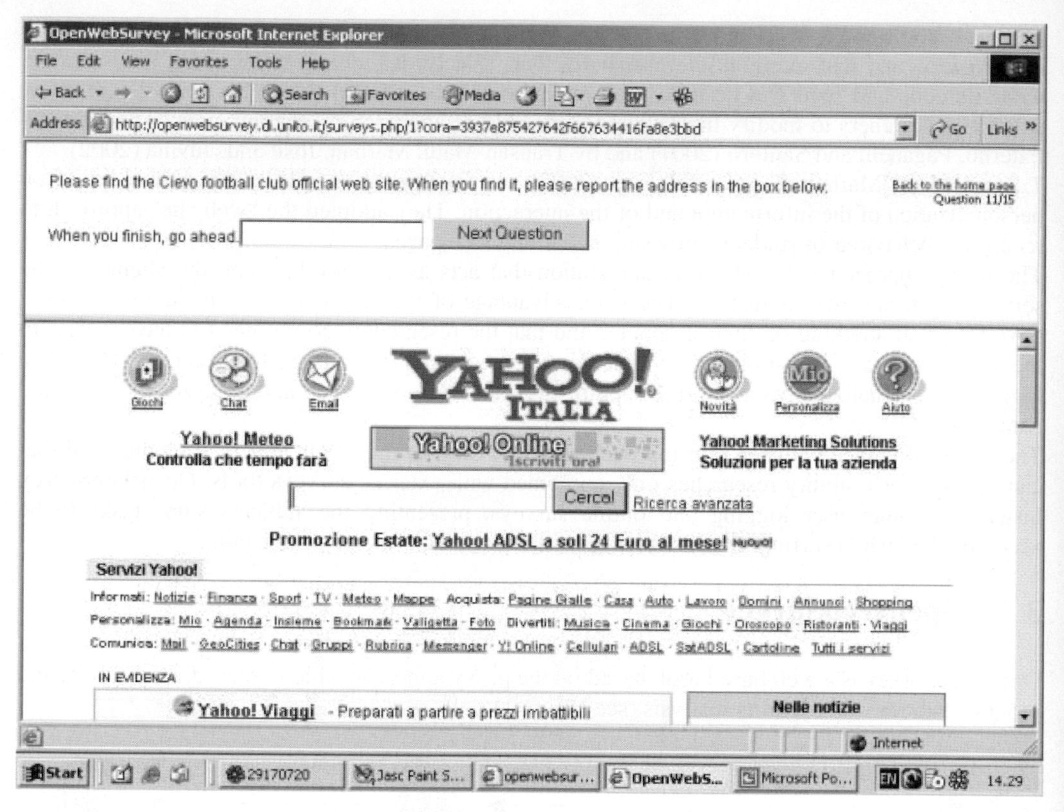

Figure 2. OpenWebSurvey interface

The web sites under investigation are visualized in the lower frame: their code has been rewritten by OpenWebSurvey server for monitoring web navigation. The layout and the navigation are left unmodified: the user can follow any link, play sounds, watch movies and so on. Normally an average user does not realize that the site has been modified. OpenWebSurvey stores data about visited pages (load time and some client side actions), about the entire session (visited pages, total visit time, page visit time, general information about the user system) and about survey answers.The navigation path of the subjects is rebuilt paying attention to the link followed to reach the page, to the keywords used during the search activities and to the number of times the subject came back to the same page and so on. We keep track of every site page visited by the subject, recording date and time of request and creation of the modified page. Moreover, for every question we keep track of the user answer(s), of the keywords the user inserted, of date and time of request and of answer and, if applicable, of the pages visited to find the answer (see Table 1).
OpenWebSurvey is able to process HTML pages with any kind of standard object, as sounds, images or movies and even JavaScript code, if it finds full URLs (e.g.: http://www.mydomain.com/mypage.html). OpenWebSurvey is not able to process links in binary data, as objects managed by plugins (e.g.: Flash or Director movies, Java applets), or JavaScript URLs built with concatenation techniques.

Table 1. Survey analysis

Question number	Answer	Start time	Answer time	Visited pages
3	3	12/01/2003 16:44:33	12/01/2003 16:45:37	0 http://junior.virgilio.it/
4	implicit	12/01/2003 16:44:37	12/01/2003 16:45:52	0 http://junior.virgilio.it/ 1 http://junior.virgilio.it/directory/cgi/dir.cgi?ccat=46901 2 http://junior.virgilio.it/directory/cgi/dir.cgi?ccat=55941

4 Case study

OpenWebSurvey was tested on a usability survey of web sites for children, in collaboration with the Department of Psychology. The subject groups were some classes of a school in Turin (Italy), where the children were already able to use a computer and surf on Internet. The children were asked to perform some tasks, in order to understand the strategies that they use to find information. The site used in the experiment was a search engine for children, junior.virgilio.it.

We defined two major kinds of search strategies: category based and keyword based. Category based search is when subject search the information requested using the categories pre-established by the search engine. Some of the categories used by the site *junior.virgilio.it.* were: *animals*, *tales*, *cartoons*, *science* and *television* (just to mention some of them). Keyword based search is when subjects type in the input fields some words relevant for the information they would like to find.

The first task was, simply, to find something about Harry Potter on the site, since we wanted to analyse what children do if we do not provide further hints about the way they should search. Then we showed the subjects how to perform category searches and keyword searches, and we asked them to perform more searches. Some questions forced the user to find the answer with a category-based search, others forced the user to search using keywords. At the end the subjects were asked to perform some searches in order to investigate which search strategy they would choose.

The main goal was to test if OpenWebSurvey was efficient for research purposes, while the analysis of children responses in order to understand better their behaviour during the navigation was only a complementary goal.

5 Preliminary results and discussion

The main problems we encountered were due to the intrinsic characteristic of our approach: some subjects did not like the pre-established web site of navigation and wrote a different address on the navigation bar and then left the survey. Also the sequence of questions bored some children who left the survey. In some interesting cases children loved excessively a task and tried to perform it repeatedly. When the question changed they pressed the "back" button to go back to the preceding question.

OpenWebSurvey log data showed that most of the children performed a category based search, since it was more intuitive and easy to understand. Preliminary results showed that most of the children didn't have much trouble in finding pages related to objects or themes they knew very well, for example *Pokemon* cartoons; they performed much worse searching for non-common themes.

Currently we are planning to extend this survey to different age groups, with the aim to understand if there is a mutual relation between search preferences and aging. We are also working on several improvements in the administrative part of the interface in order to increase the software analysis capabilities. The next step will be to choose a metric to evaluate browsing behaviour.

References

Hong, J.I., & Landay, J.A. (2001). *WebQuilt: A Framework for Capturing and Visualizing the Web Experience*. WWW10, May 1-5, 2001, Hong Kong.

Paternò, F., Paganelli, L. & Santoro, C. (2001). *Models, Tools and Transformations for Design and Evaluation of Interactive Applications* [Electronic version]. Invited presentation at PC-HCI 2001, Patras, Greece, December 2001.

Perkins, R. (2002). *Remote Usability Evaluations Using the Internet*. The 16th British HCI Group Annual Conference, South Bank University, London (England), September 2002.

Trausan-Matu, S., Marhan, A., Iosif, G., & Juvina, I. (2003). *Generation of Cognitive Ergonomic Dynamic Hypertext for E-Learning*. This volume.

U.S. Congress (1998). *Summary of Section 508 Standards*. Retrieved September 1, 2002 from Section 508 web site: http://www.section508.gov/index.cfm?FuseAction=Content&ID=11

Usable Tools (2001). *Usability Browser*. Retrieved August 1, 2002 from Usable Tools web site: http://www.usabletools.com/ub.htm

W3C (1999). *Web Content Accessibility Guidelines 1.0*. Retrieved September 1, 2002 from W3C web site: http://www.w3.org/TR/WCAG10/

Development of an Error Management Taxonomy in ATC

Thomas Bove

Risø National Laboratory
Roskilde, Denmark
Thomas.bove@risoe.dk

Henning Boje Andersen

Risø National Laboratory
Roskilde, Denmark
Henning.b.andersen@risoe.dk

Abstract

It is well known that human errors contribute substantially to the risk of incidents and accidents in safety critical areas such as air traffic control (ATC). At the same time, it is also widely acknowledged that it is impossible to completely eliminate the occurrence of errors. Nonetheless, in many situations the consequences of errors may largely be controlled by timely and effective recoveries. Therefore, in recent years there has been an increased focus on error recovery as a means to reduce the consequences of errors. In this paper a conceptual model and an associated taxonomy for analysing error management events in ATC is presented. The framework can be used as a foundation for enhancing the understanding of how errors normally are handled before *or* when they lead to adverse consequences. By applying the taxonomy it becomes possible to identity not only the weaknesses of a human-machine system but also its strengths.

1 Introduction

The error management framework presented in this paper has been developed on the basis of an extensive literature review of error and error management frameworks. Each of the frameworks that were reviewed was evaluated on the basis of a series of pre-determined criteria (Wiegmann & Shappell, 2002): Reliability, comprehensiveness, diagnosticity and usability. *Reliability* concerns whether the framework can produce robust results across repeated trials. *Comprehensiveness* is the extent to which the framework covers all of the relevant variables that it purports to cover. *Diagnosticity* refers to whether the framework can provide insight into the phenomenon of interest. *Usability* is related to whether it be applied to practical settings. All of these requirements are of crucial importance when considering the utility of a given conceptual framework.

2 Literature review

2.1 Human error

Some of the more prominent and influential human error frameworks are the traditional Information Processing Model (see e.g. Wickens 1992), The Skills-Rules-Knowledge model (see e.g. Rasmussen 1982), The Model of Unsafe Acts (Reason, 1990) and the HERA Taxonomy (Isaac, Shorrock, Kirwan, Kennedy, Andersen & Bove, 2000). Few attempts have been made to apply these taxonomies to the analysis of errors in complex domains and it is therefore difficult to make firm conclusions about their relative usefulness. Nonetheless, studies indicate that these taxonomies can accommodate a large part of the observed errors and that reliable classifications can be obtained with these taxonomies. In spite of the fact that the reviewed taxonomies have been relatively successful on the quantitative level (being able to describe most of the identified errors in the reliable manner) there may be some reasons why an information processing model – such as the one found in Wickens' framework and the HERA technique - would be the most appropriate framework to analyse human errors in ATC. In particular, the stages in the information-processing models are frequently mentioned in the ATC literature. This may be a reflection of the fact that errors that occur in ATC, such as hearback errors, visual misidentifications or decision/planning errors, seem to be most compatible with the information-processing framework. This compatibility is also reflected in the fact that only the reviewed information processing models have been applied to the domain of ATC. In particular, the HERA technique has been specifically adapted to the ATC environment – and been thoroughly validated within this environment - and therefore seems to be a good platform on which to base error analyses.

2.2 Threat and error management

To be able to develop a classification system of the error management process it is useful to the have a model that can be used as an organising principle. Currently few models are available to describe the generic structure of the error management process. Some of the most promising frameworks are to be found in Helmreich, Klinect & Wilhelm (1999) and Kanse & Van der Schaaf (2000). In the present context Helmreich et al.'s model of threat and error management seems to contain several important benefits. Its chief benefit is that it provides a description of the main stages of threats, errors and error management. In this manner error management is placed within a larger context of human behaviour. The model is originally derived empirically from observations of flight crew behaviour in line operations (e.g. Klinect, Wilhelm, & Helmreich, 1999), but has also been applied to the analysis of incidents and accidents (e.g. Jones & Tesmer, 1999). In this manner the framework has been useful in the study of human errors and their management in both normal and abnormal conditions. This means that the framework is able to deal with both successful and unsuccessful behaviour. The model is unique insofar as it is the only model that incorporates threats as an integrated part of the model. This is an issue that has not previously been emphasised in any other model of error and error management. Finally, the model provides an intuitively logical structure to understand the error management process and the concepts do not require any theoretical background and should therefore be easy to understand.

2.3 Performance shaping factors

In the discussion of the nature of human error it has been argued that it is misleading to view accidents as *just* happening as the result of human errors (or failed error recoveries). Rather, on a more accurate view accidents are the results of human-task mismatches (Rasmussen, 1982). A logical consequence of this insight is that features of the context and the work situation should be taken into account when analysing the chain of events in critical scenarios. Hence, the literature review has included taxonomies pertaining to the contextual influence on human performance or, in short, Performance Shaping Factors (PSFs). The PSFs are based primarily on TRACEr (Shorrock & Kirwan, 1998), HERA (Isaac et al., 2000), ADREP2000 (Cacciabue, 2001), ASAP (Helmreich & Merritt, 2000), BASIS (O'Leary, 1999) and research on Recovery Influencing Factors (Kanse & Van der Schaaf, 2001).

3 A framework of threat and error management

The framework presented in this section has been developed on the basis of the previously described literature review and it has been further refined and tested on the basis of incident reports, critical incident interviews and simulator studies. The framework is based on a model of error management and provides an opportunity to analyse both the cognitive and behavioural activities of the individual actors involved in the error management process as well as the influence (positive and/or negative) of a series of contextual factors.

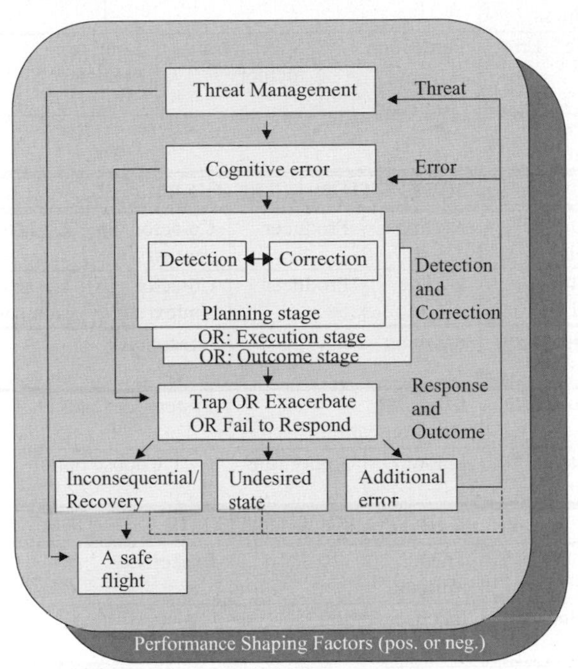

Figure 1: A model of error management

The model of error management starts out with potential threats (such as a thunder storm, a problem pilot or a military exercise) from the environment that should be dealt with. The controller may try to avoid threats leading to problems and errors resulting in a continued safe flight. If the threat is not discovered or adequately handled it may result in an error. The error can be analysed on the basis of the cognitive mechanisms underlying its occurrence. The error might not be detected, but if it is the detection and the recovery may happen at different stages in the evolution of the error and on the basis of different kinds of information sources. Different kinds of responses might be produced and the result may vary from being inconsequential to an undesired state or a new error. In the case where the outcome is an additional error a new error analysis can begin. Even if the outcome is inconsequential or leads to an undesired state new errors may still occur in the event sequence. In addition to these individual activities a list of PSFs constitutes an integrated part of the framework. These factors can be used to expand the analysis beyond the individual level to include team and organisational factors that are relevant to gain a comprehensive understanding of why the event occurred and how it was prevented from developing into an even more serious situation.

Based on the model presented above an error management taxonomy has been developed. The dimensions and classifications associated with the individual actions as well as the main groups of PSFS are shown below. The PSFs can have both a positive and a negative influence on the course of events.

Table 1: The taxonomy

THREAT & ERROR						
Threat & Error	Threat Preparedness	No Anticipation		Anticipation		
	Cognitive Error Type	Perception	Short-term memory	Long-term memory	Decision	Response
	Procedural violation	Yes		No		
DETECTION & RECOVERY						
Who: Actor	Error/state detected by	No one	Producer	Co-actor in context	Co-actor outside context	System
	Error/state corrected by	No one	Producer	Co-actor in context	Co-actor outside context	System
When: Processes	Detection Stage	Planning	Execution	Outcome		
How: Processes	Detection source	External communication	System feedback	Internal feedback		
	Error/state correction	Ignore	Apply rule	Choose option	Create solution	
RESPONSE & OUTCOME						
What: Behaviour & outcome	Error/state Response	Trap/ Mitigate	Exacerbate	Fail to respond		
	Error Outcomes	Inconsequential/ Recovery	Undesired state	Additional error		
PERPFORMANCE SHAPING FACTORS						
Why: Influence of contextual factors (main groups)	1. Traffic, airport and airspace	*Pos.*	*Neg.*			
	2. Procedures and documentation	*Pos.*	*Neg.*			
	3. Workplace design, HMI and equipment factors	*Pos.*	*Neg.*			
	4. Training and experience	*Pos.*	*Neg.*			
	5. Person related factors	*Pos.*	*Neg.*			
	6. Social and team factors	*Pos.*	*Neg.*			
	7. Company, management and regulatory factors	*Pos.*	*Neg.*			

4 Conclusion

The purpose of this study was to develop an error management framework. To achieve this goal an extensive literature review was conducted. This included both a review of existing error and error management taxonomies as well as performance shaping factors that may influence the whole process from error production to error recovery. The framework described in this paper has been validated on the basis of different kinds of data material (for a comprehensive review of the results please refer to Bove, 2002). First of all, the framework has been applied to error events found in critical incidents and in a simulator study. Secondly, the framework has been evaluated by a series of human factors experts who have been involved in research that is highly relevant in relation to the current project. The results indicated that the framework performed successfully on the basis of the previously described evaluation criteria. For example, it was demonstrated that the framework contained a high level of diagnosticity insofar as 12 out of the 13 a priori defined hypotheses - based on theoretical expectations and previous research – were confirmed by using it on the data material from the empirical studies. Furthermore, the results showed that it was both possible to achieve fairly robust analyses on the basis of the framework and that experts find it highly relevant in relation to the study of error management. Hence, the results indicate that the framework is useful for error management studies.

References

Bove, T. (2002). Development and Validation of a Human Error Management Taxonomy in Air Traffic Control. Ph.D. dissertation. Risø-R-1378 (EN). Risoe National Laboratory, Roskilde, Denmark.

Cacciabue, P.C. (2001). Human factors insight and data from incident reports: the case of ADREP-2000 for aviation safety assessment. *Engineering Psychology and Cognitive Ergonomics.* Vol 5. Harris, D. (ed.). Ashgate. England.

Helmreich, R.L., Klinect, J.R. & Wilhelm, J.A. (1999). Models of threat, error, and CRM in flight operations". *Paper from the 10th International Symposium on Aviation Psychology.* Columbus, Ohio May 3-6, 1999.

Helmreich, R.L., & Merritt, A.C. (2000). Safety and error management: The role of Crew Resource Management. In B.J. Hayward & A.R. Lowe (Eds.), *Aviation Resource Management* (pp. 107-119). Aldershot, UK: Ashgate. (UTHFRP Pub250)

Isaac, A., Shorrock, S.T., Kirwan, B, Kennedy, R, Andersen, H.B., & Bove, T. (2000). Learning from the past to protect the future – the HERA Approach. *Paper for 24th conference of European Association For Aviation Psychology (EAAP) 2000.*

Jones, S.G., & Tesmer, B. (1999). A new tool for investigating and tracking human factors issues in incidents. In Proceedings of the *Tenth International Symposium on Aviation Psychology* (pp. 696-701). Columbus, OH: The Ohio State University.

Kanse, L. & Van der Schaaf, T.W. (2000). Recovery of Failures in the Chemical Process Industry. *3rd International Conference on Engineering Psychology and Cognitive Ergonomics.*

Kanse, L., & Van der Schaaf, T.W. (2001). Factors influencing recovery from failures. *CSAPC '01. 8th Conference on Cognitive Science Approaches to Process Control*, 24-26, September 2001. Universitat der Bundeswehr, Neubiberg, Germany.

Klinect, J.R.; Wilhelm, J.A. & Helmreich, R.L. (1999): "Threat and Error Management: Data from Line Operations Safety Audits". *Paper from the 10th International Symposium on Aviation Psychology.* Columbus, Ohio May 3-6, 1999.

O'Leary, M. (1999). The British Airways Human Factors Reporting Programme. *3rd Workshop on human error, safety, and system development (HESSD '99)*, Liège (BE), 7-8 Jun 1999.

Rasmussen, J. (1982). Human Errors. A taxonomy for describing human malfunction in industrial installations. *Journal of Occupational Accidents, 4*, 311-333.

Reason, J. (1990). Human Error. Cambridge University Press. Cambridge.

Shorrock, S.T. & Kirwan, B. (1998). The development of TRACEr: a technique for the retrospective analysis of cognitive errors in ATC. *Paper presented at the 2nd Conference on Engineering Psychology and Cognitive Ergonomics.* Oxford: 28 - 30 October.

Wickens, C.D. (1992). Engineering Psychology and Human Performance. Harper Collins Publishers. 2nd Edition. New York.

Wiegmann, D.A. & Shappell, S.A. (2002). Human error perspectives in aviation. *The International Journal of Aviation Psychology, 11(4)*, 341-357.

Use of a Train Signal Control Simulator to Develop a Valid Measure of Shared Mental Models

N. D. Bristol & S. Nichols

University of Nottingham
Institute for Occupational Ergonomics
University of Nottingham
University Park
Nottingham NG7 2RD, UK

epxndb@nottingham.ac.uk

Abstract

Arguments are presented for a measure of team interaction to accompany the more commonly administered measure of 'concept relatedness rating', when investigating the presence of shared mental models. The performance of teams holding shared mental models (as detected by the measures described) were compared to that of teams not thought to hold shared mental models, or to be using 'groupthink', another teamwork style, during a train signal control simulation task. Communication was restricted during two of the six tasks performed, in an attempt to imitate the limitations typical of an emergency situation, in which the benefits of shared mental models are thought to be realised.

1 Shared Mental Models

Johnson-Laird (1983) describes a mental model as a means of representing an external or real-world feature internally, and in such a manner that it can be manipulated and investigated mentally, without the cost and time required by that of the external feature. When a whole team of workers are faced with a problematic situation, each team member will form a mental model of the situation and subsequently a problem-solving strategy. Having similar or *shared* mental models is thought to approximate to 'being on the same wavelength', 'speaking the same language' or 'being able to relate' and is likely to be a very desirable feature of teamwork.

Cannon-Bowers & Salas (1992) describe the existence of shared mental models within a team as enabling members to draw on common knowledge bases that can be used to form accurate and similar expectations about their own or another's performance or the situation generally. These expectations allow team members to coordinate effectively without the need for extensive overt strategising. Serfaty et al (1998) suggest that shared mental models provide a team with a common problem space in which team members can explain the behaviour of fellow team members and predict their information needs. This makes it possible to engage in implicit coordination - particularly useful in high workload or emergency situations where time is of the essence.

Certain teams will develop shared mental models naturally, especially if members are carrying out similar types of interdependent tasks in a shared workspace. However, it is often the case that team members are scattered about in remote locations and sometimes even in different time zones. For such *virtual* teams, it is highly unlikely that the benefits of shared mental models will be realised naturally, without some sort of intervention. Indeed, some researchers have explored the possibility for shared mental model team training (Converse et al, 1991; Rouse et al, 1992; McCann et al,2000). However, without first establishing a sound measure for detecting the degree of shared mental model presence, the design of training may be considered premature.

2 Development of Shared Mental Model Measures

Following an initial exploration of a variety of measures within the field of railway signal control, two different yet complimentary measures were selected for the task of shared MM measurement: causal relation ratings and a desirable communication checklist.

2.1 Causal Relation Ratings

Causal relation ratings are a process, whereby the MMs of individual team members are compared. This is carried out through a series of ratings, which question the extent of causal effect between pairs of concepts: 'Does concept A have an effect on concept B and, if so, how much?'. Although this process might not include every pair of concepts contained in each team member's mental model, it should cover a substantial proportion of it. A more general approach to the measurement of concept relations had been adopted in a previous study ('are concepts A and B related and, if so, how much?'), but this method was found to be unreliable; in many instances, two team members gave similar relatedness ratings for a particular pair of concepts, yet during informal interviews gave different reasons for this relatedness. It was felt that increased specificity in the questions of relatedness- such as one-way causal effect- would minimise this problem.

2.2 Desirable Communication Checklist

Without a measure of team process, one cannot be certain that any similarity in team members' causal relation ratings has resulted from the presence of shared MMs, rather than another team process such as 'Groupthink'.

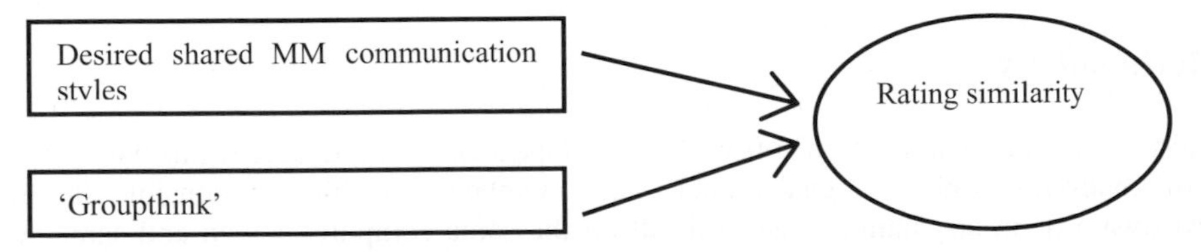

Figure 1: The two routes to rating similarity

As shown in figure 1, the team members may show similar ratings of concept relationship and therefore have a shared understanding of a problem, but unless they have approached the problem via a similar route, such a similarity may be a product of 'Groupthink', whereby this is simply a result of members striving for unanimity, which may override their motivation to realistically appraise alternative courses for action (Janis, 1972). Admittedly this could be a successful means of achieving coordination and assimilated thought processes, but it is doubtful whether such a method will lead to accurate understandings or optimal decision-making. It is therefore important that the process leading to the shared understanding is identified and recognised as a fundamental part of the shared MM measurement procedure.

Using recordings of the conversation between team members, a series of tallies are kept for every time a particular style of communication takes place. The tally chart comprises various aspects of team interaction- some proposed by researchers to support the development of shared MMs and others proposed by researchers to reflect processes of 'Groupthink':
Shared MMs:

- Discussion prior to or during a task about what might be expected
- Specific procedural statements (thinking aloud)
- Verbalisation of reasoning behind one's choice of action or behind a suggestion
- General strategy statements for future performance
- General strategy statements relating to past performance
- Statements about one's own needs or situation
- Statements about other's needs or situation
- Questioning (seeking opinion) of other about course of action to take
- Constructive discussion at end of task: how the team performed, what could be improved

Groupthink:

- The debating of issues raised by players- disagreement voiced
- Observation of silent disagreement to decisions: tense movements, gestures suggesting anger
- Failure to consider arguments raised fairly/ putting pressure on other to drop argument
- Stereotype comments about 'outgroups'

3 Validating the Shared Mental Model Measure in a Controlled Environment

In order to discover the validity of these techniques as shared mental model measures, various hypotheses concerning their behaviour in response to other variables such as performance levels, workload and communication restrictions needed to be tested. The work of railway signal controllers is safety critical and it would therefore be inappropriate to manipulate the variables required in these tests in the real life situation. Instead, the controllable environment of a computer simulated signal control game was opted for.

3.1 Methodology

Using a railway signalling simulation software (Train Dispatcher 2.0), dyadic teams were each set the task of controlling railway signals. Each team member was either responsible for trains running westward or trains running eastward, all on the same computer screen and same set of tracks. In addition, participants were required to share a single computer mouse, which was the only control needed for the task. All trains ran along a single bi-directional track and therefore, coordination between eastward and westward trains was essential for smooth operation and consequently a high performance score. Teams were required to carry out six ten-minute sessions, each consisting of the same train schedules and the same set of tracks. Sessions one, two, four and five were carried out under condition A, which meant that they were free to communicate in any way they felt appropriate, while sessions three and six were carried out under condition B, limiting participants to non-verbal communication only. This condition is included to represent emergency situations where time limitations restrict the amount of explicit interaction that can occur and in which shared MM benefits are thought to be realised.

Following the six sessions, all participants completed a set of twenty causal relation rating questions. All communication during sessions one, two, four and five was recorded for later analysis using the desirable communication checklist.

3.2 Findings

The first prediction was that members of each team displaying the desired communication styles would give similar causal relation ratings, since the communication styles enforce a level of shared understanding. The second prediction is that members of teams giving similar causal relation ratings will not however necessarily all have adopted the desired communication styles, since it is possible to produce shared understanding through other teamwork approaches such as 'groupthink'. Indeed, the results showed that all teams displaying the desired communication styles also showed similarity in ratings amongst their members, but some teams with similar ratings did not adopt the desired communication styles.

The third prediction is that those teams thought to hold shared MMs (those displaying the desired communications and rating similarity) will show improvement in sessions three and six- despite communication restrictions, since they will be following the same train of thought and have less need for explicit communication, whereas teams without shared MMs will not continue to improve in sessions three and six and may even show a decline in performance level.

Table 1: Sessions 1-3 performance curves for teams holding *shared MMs* (N=6):

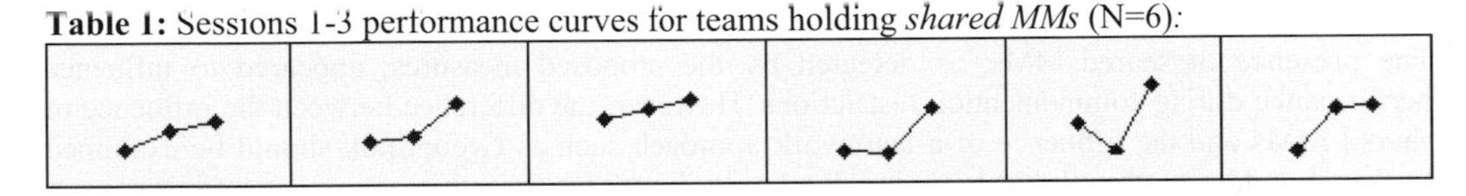

Table 2: Sessions 1-3 performance curves for teams using *Groupthink* (N=7):

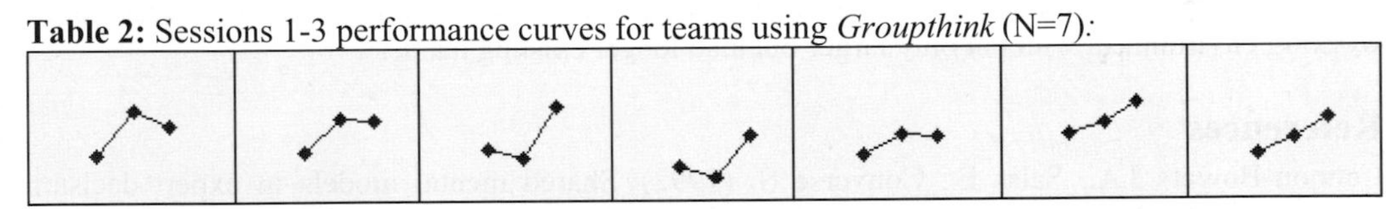

Table 3: Sessions 1-3 perf. curves for teams *with neither shared MMs nor Groupthink* (N=4):

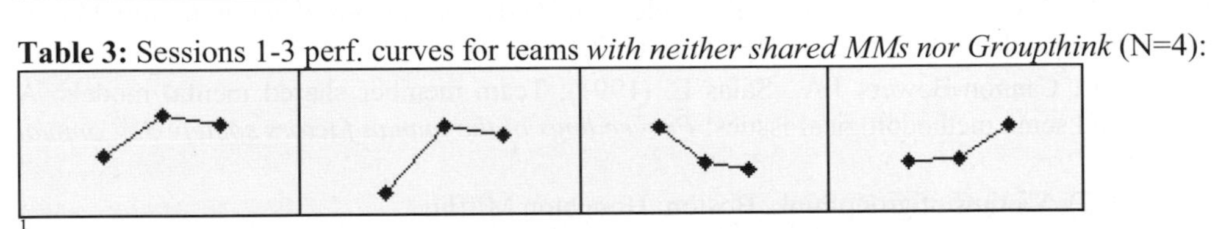

Table 4: Sessions 4-6 performance curves for teams holding *shared MMs* (N=7[1]):

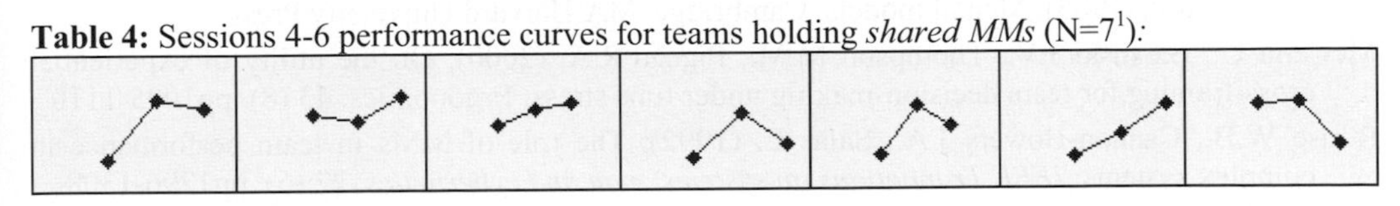

[1] The number of teams in each group changed; some developed a different interaction style in sessions 4-6.

Table 5: Sessions 4-6 performance curves for teams using *Groupthink* (N=5):

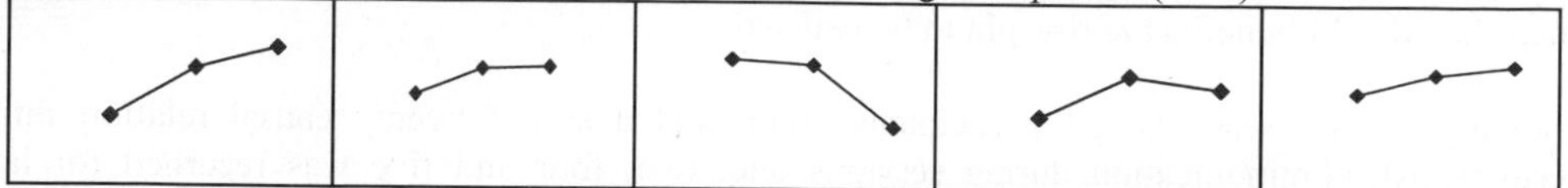

Table 6: Sessions 4-6 perf. curves for teams with *neither shared MMs nor Groupthink* (N=4):

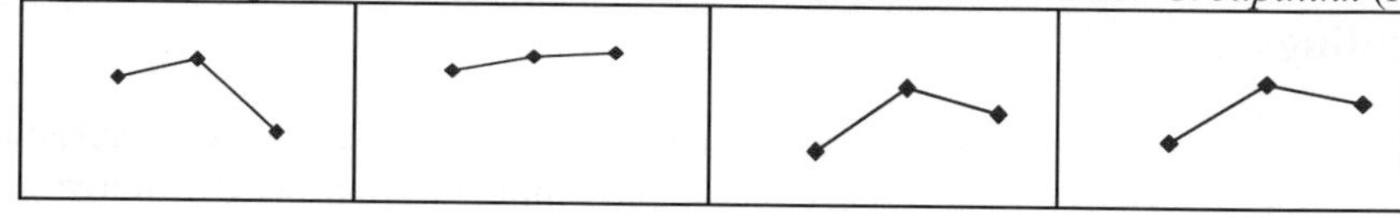

All of the teams thought to hold shared MMs showed improvement in session three despite communication restrictions (table 1), in contrast to only four of the seven teams thought to use a 'Groupthink' approach (table 2) and only one of the four teams not thought to use either technique (table 3). In session six, only three of the seven teams thought to hold shared MMs showed improvement (table 4), no different to the three of five teams thought to use a Groupthink approach (table 5) and one of the four teams not thought to use either technique (table 6). This result could have been affected by a limit in achievable scores. A few of the teams displaying shared MM characteristics had already reached some of the highest achieved scores by session four, leaving little room for improvement in sessions five and six.

4 Implications

The presence of shared MMs, as detected by the proposed measures, appeared to influence performance during communication restrictions. However, the difference between the influence of shared MMs and the influence of a teamwork approach such as Groupthink should be examined further. In addition, the current trials involved only dyadic teams within a simplistic environment. The administration of the measures now needs to be examined within a more complex 'real world' working environment with not only larger but also longer existing teams.

References

Cannon-Bowers J.A., Salas E., Converse S. (1992); Shared mental models in expert decision making; In G.A.Klein, J.Orasanu, & R..Calderwood (Eds) *Decision-making in Action: Models and methods*; (chapter 12, pp221-245) Norwood NJ: Ablex

Converse S.A., Cannon-Bowers J.A., Salas E. (1991); Team member shared mental models: A theory and some methodological issues; *Proceedings of the human factors society 35th annual meeting.*

Janis I.L.(1972); Victims of groupthink. Boston: Houghton Mifflin

Johnson-Laird P. (1983); Mental models. Cambridge, MA:Harvard University Press

McCann C., Baranski J.V., Thompson M.M., Pigeau R.A. (2000); On the utility of experiential cross-training for team decision-making under time stress; Ergonomics, 43 (8), pp1095-1110

Rouse W.B., Cannon-Bowers J.A., Salas E. (1992); The role of MMs in team performance in complex systems; *IEEE Tranactions on systems, man and cybernetics*, 22 (6), pp1296-1308

Serfaty D., Entin E., Johnston J.H. (1998); Team coordination training; In J.A. Cannon-Bowers & E.Salas (eds) *Making decisions under stress;(*pp221-245) Washington DC: American Psychological Association Press.

Scenario Development for Testing Safety Devices in Automotive Environments

P.C.Cacciabue, M.Martinetto

European Commission
Joint Research Centre
Via E.Fermi,1
I – 21020, Ispra
pietro.cacciabue@jrc.it

S.Montagna, A.Re

University of Turin
Department ofPsychology,
Via Po, 14
I – 10123, Turin
Re@psych.unito.it

Abstract

Driver behaviour is one of the main causes in road accidents, contributing in over 90 per cent of accidents. The investigation on the human factor can not be taken out of the traffic context. The focus of the study presented in this paper is on traffic scenarios, as detailed specification of the accident scene by mean of all the contributing variables and their relationships. The variables taken into account refer to actors involved, setting and accident dynamic. Such approach offers (in a retrospective view) an overall overview and comprehension of the road accident phenomenon and proposes (in a prospective view) the scenarios specification as a practical tool for generating experimental conditions to test any safety device, both in simulated and real road environment.

1 Introduction

Every year Europe counts around 45.000 deaths and 1.5 millions injured people: this means that one person out of every 200 European citizens is injured in a traffic accident each year (European Parliament, 1998). This implies a huge cost for society that has to face both human costs and economical costs (medical treatments, invalidity pensions etc.).

The problem of road safety has to be approached from different points of view, involving various disciplines. The traffic system includes a variety of factors that have to be treated in a systemic approach. The model proposed as reference (Figure 1) results from the integration of two of the most effective paradigms used in transportation domains:

1. The Driver-Environment-Vehicle by Della Valle and Tartaro (1999), where the basic elements are the *vehicle* (including reliability, maintenance, etc.); the *driver* (physical, psychological conditions, etc.) and; the *environment* (traffic flow, weather, infrastructures, etc.); and
2. The SHELL model (Edwards 1972; Hawkins, 1987), where a more comprehensive model is taken into account for the interaction between the human being and all the external factors. The central *Liveware* (L) includes all the human resources and features of the subject; the *Environment* (E) refers to the social, cultural, physical environment; the *Hardware* (H) concerns any physical component, equipment, vehicles, tools and so on; the other *Liveware* (L) include all the people interacting with the subject, in particular the passengers; the *Software* (S) refers to the norms, procedures and rules, that define the interactions among the other components of the model.

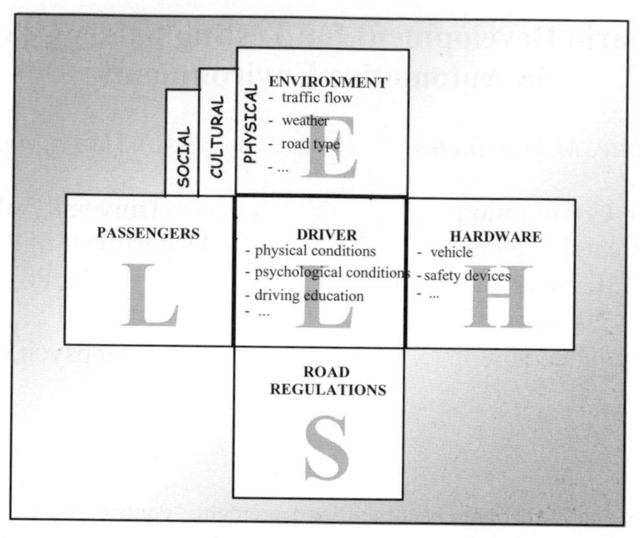

Figure 1: Integrated Model of Driver-Environment-Vehicle System.

The integrated model has been adapted to the specific aims of the European Commission co-funded research project EUCLIDE (*Enhanced human machine interface for on vehicle integrated driving support system*, GRD-2000-26801), that are to develop and test a new anti-collision device for supporting drivers in bad visibility conditions (Polychronopoulos et al., 2003). The system detects the obstacles on the road, provides warnings, but delegates any recovery action to the driver.

An adequate test bed has to be developed in order to ensure that technical functionality and human factor requirements (efficiency, effectiveness and user friendliness) are properly satisfied (ISO/FDIS, 1999). One of the most crucial safety critical aspects consists in the specification of dynamic road accident scenarios to be implemented both in simulators and in field tests. In the case of EUCLIDE, particular attention was given to bad weather conditions and darkness, when the main benefits of the new system are expected. The work carried out for definition and selection of most significant scenarios is based on the integrated model described above. In the following sections the actual process of definition of the scenarios studied and considered for testing is described in detail.

2 Road Accident Scenarios

In the domain of road safety it is difficult to find a standard and commonly accepted methodology for accident investigation. In particular, a universal definition of "accident scenarios" and a general methodology for scenario identification are not available. For this reason, the EUCLIDE project has spent efforts in defining the most critical scenarios particularly related to the use of the anti-collision system under design.

2.1 Road Accidents Investigation

The development of road accident scenarios started from the investigation of the main European (EUROSTAT, CEMT, OECD) and, in particular, Italian (ISTAT) road accident databases. This first analysis allowed the identification of general accident trends and causes all over Europe. Problems in international data comparability were encountered. In spite of the effort spent so far to standardise European data, many discrepancies still remain in conceptual definitions, taxonomies and methodologies for data collection. Sources analysed definitely contain a large amount of data about accidents, but the information provided is not sufficient to accurately identify scenarios. Data collected are reported only in the aggregated level, so that it is impossible to discriminate among single accidents and define factors contributing to accident scenarios. Some alternative sources have been identified in insurance companies, car manufacturers and police databases. Unfortunately these sources turned out to be inaccessible, because of the high level of confidentiality of the information contained therein.

Finally, newspaper articles offered the most useful information for reconstructing accident scenarios. The research now focused on the Italian situation: the data were collected from the thematic articles of two main Italian newspapers. A total number of 300 articles were selected among those published in the year 2000. The selection focused on the richness and detail of the information provided: particular attention was dedicated to the presence of data on weather conditions, traffic conditions, seriousness of accident consequences and accuracy of accident dynamic description. The use of newspaper articles, even if with some inconveniences (e.g. almost only the road accident with most serious consequences are reported), showed to be powerful for longitudinal studies. The reference to a whole year and to all subsequent articles on a same road accident allowed the updating of data on causes and seriousness of consequences.

2.2 Data Analysis and Scenarios Identification

A Content Analysis was performed to organise the data collected. The analysis allowed the specification of the articles' subject; the main reported concepts; the translation of these concepts into few general categories concerning driver, vehicle, environment and accident dynamic; the identification of all the variables referred to each category. For each article a schedule was compiled with all the information provided.

The quantitative analysis included the following levels of detail:

1. *Frequency Analysis*. On a descriptive level the frequencies of variables have been investigated to describe the road accident phenomenon. This analysis enabled to check external validity of data: the articles data are in line with the Italian official databases. This means that the selected sample adequately represents the whole group of accidents.
2. *Contingency Tables*. On an interpretation level the Contingency analysis allowed the identification of the most significant (Chi-Square) relations among variables. The outcomes provided further information in order to define the most risky scenarios.
3. *Logistic Regression*. On a predictive level the multivariate analysis checked the likely contemporary occurrences of different variables in road accident scenarios. The outcome was the identification of sub-systems of variables that increase the probability or the seriousness of a particular accident.

This analysis process led to the final identification of a set of accident scenarios, guided by the specific requirements of testing the new anti-collision system. A number of 32 accident scenarios were sorted out accordingly to the following criteria:

- The high frequency of the accident dynamic in the sample.
- The presence at least of one person dead or two seriously injured.
- The presence of bad weather conditions.

- The presence of more estimated causes for the accident.
- The presence of information about the human factor.

The specific scenario structure adopted in this research is in accordance with the general definition given by Carroll (2000). Each road accident scenario has been described by three sets of variables:

1. *Actors* involved in the accident: This category includes driver related variables, i.e., age, sex, and psycho-physiological condition immediately before the accident, and vehicles related variables, i.e., number of vehicles involved, and vehicle power.
2. *Setting*: The accident scene includes a number of environmental variables, i.e., type of road, route, paving type and maintenance condition, and infrastructures.
3. *Accident dynamic*: It contains events sequence and outcomes of the accident. Actions performed by the actors and external events are considered within this category.

The scenarios reconstruction has allowed the identification of the main variables contributing to the event and its consequences. Table 1 presents an example of the schedule produced for each accident. The analysis of all the accident schedules led to some considerations and possible directions for future work: these are reported in the following section.

Table 1: Example of the scenario structure.

SCENARIO	*VARIABLES*	*CONSEQUENCES*
• *Accident Dynamic*: Collision between two vehicles, one with medium and one with high-power engines. • *Actors*: Young male driver (high-power engine) coming back from a party, overtaking a car near to a bend, is involved in a frontal crash. • *Setting*: Collision occurs at the end of a straight road, continually lined, lightly uphill, tree-lined, and with a damaged surface.	⇒ Familiar road ⇒ Night ⇒ High speed ⇒ Forbidden manoeuvre ⇒ Passengers on-board ⇒ Drunkenness	Frontal collision with 2 deaths and one seriously injured person.

3 Discussion and Perspective

The description of critical scenarios has allowed the detection of some factors that are not usually taken into account in the most known studies on road accidents and to supply a powerful experimental tool. The knowledge of real situations on the road is the basis for a realistic simulation and implementation of scenarios with different objectives. In particular, the scenario is a useful tool for the evaluation of the new safety devices that need to be tested within the real context of use.

Table 2 reports the three integrated sets of variables proposed for accident scenario description. These variables should guide accident analysis into the creation of an integrated and exhaustive road accident database. The small database purposely created for this research has allowed the specification of the set of scenarios to be implemented in the EUCLIDE system test plan.

Table 2: Scenario variables.

Actors Variables	*Setting Variables*	*Accident Dynamic Variables*
Age	Traffic condition	Manoeuvre type
Gender	Rain	Obstacles
Driving experience	Fog	Place
Nationality	Snow/ice	Hour
Vehicle type	Wind	Vehicles involved
Car power	Clear	Injured people
Safety devices	Artificial lighting	Seriousness of injuries
Speed	Day time	Deaths
Passengers	Road type	Road rules violation
	Route	Foregoing situation
		Province
		Official cause
		Additional cause
		Accident type
		Short description

The scenario tool can be adapted also to different experimental situations: while not all the contextual variables can be part of the experimental plan, the most relevant ones can be sorted out accordingly to the specific features of the device to be tested. This study just offers a methodological framework and a flexible tool for future researches aimed at evaluating the benefits of new safety systems.

References

Carroll, J. (2000). Making use – Scenario-based design of human-computer interactions. MIT Press Cambridge, Massachusetts.

Della Valle, G., and Tartaro, D. (1999). L'analisi degli incidenti stradali: metodologia, strumenti, risultati. In Seconda Giornata di *Studio La sicurezza stradale e la mobilità sostenibile*, Milano.

Edwards, E. (1972). Man and machine: systems for safety. In *Proceedings of British Airline Pilots Association Technical Symposium* (pp.21-36). British Airline Pilots Association, London.

European Parliament, Directorate-General for Research (1998). The European Community and Road Safety, Working Paper, Transport Series, TRAN 103 EN, from http://www.europarl.eu.int/workingpapers/tran/pdf/103_en.pdf

Hawkins, F.H. (1987). Human factors in flight. Gower Technical Press.

ISO/FDIS 13407 (1999) (E). *Human-centred design processes for interactive systems*.

Polychronopoulos A., Kempf D., Martinetto M., Amditis A., Cacciabue P. C., Andreone L. (2003). Warning Strategies Adaptation in a collision avoidance/vision enhancement system. HCI International 2003, Crete, Greece, June 22-27.

Designing a Pleasurable Web Pad User Interface with the Participatory Function Analysis

Chien-Hsiung Chen

Graduate School of Design
National Taiwan University of Science and Technology
43 Keelung Road, Section 4
Taipei, 106 TAIWAN
cchen@mail.ntust.edu.tw

Hong-Tien Wang[1] *Hung-Liang Hsu*[2]

Center for Research of Advanced Information Technologies
Tatung Company
22 Chungshan North Road, Section 3
Taipei, 104 TAIWAN
[1]avinw@tatung.com
[2]dannyh@tatung.com

Abstract

The purpose of this study is to explore the users' perceptions of a Web Pad's interface pleasures based on the technique of participatory function analysis. It is a direct and powerful tool that emphasizes users' collaborative activities in an iterative interaction design process. In the experiment, both static user interface elements shown on the Web Pad's desktop screen and users' dynamic interaction with the interface functions were evaluated based on this technique. The results generated from this study reveal that it is impossible for an interaction designer to create user interfaces that guarantee every user's pleasure feels. However, some trends regarding interface pleasure still could be found. In addition, potential interaction problems can be found by using the technique of participatory function analysis, which can significantly improve the Web Pad's interface usability.

1 Introduction

Humans are pleasure-seeking animals. We do a variety of leisure activities just for the purpose of entertaining ourselves. This implies that most of us want our lives to be filled with more fun and pleasures even while we are working. Jordan (1999, 2000, 2002) points out that all the products surrounding us are potential sources of pleasure. He argued if a user wants to achieve the sense of pleasure, a product should be equipped with appropriate functionality. Then an interaction design can enhance the product's usability based on the functionality. Only after a product's usability has been achieved can the interaction design take the pleasure factors into account to make it more fun to operate. The process for pleasure considerations in the context of user interface is the same. In fact, the term "interface" can be viewed as either a concrete or an abstract medium to be used to facilitate the communication between a user and a product. That is, a concrete interface can promote tangible interaction by focusing on the design of physical interfaces (e.g., a keyboard, a mouse, or any other type of input and output device) with ergonomic considerations. An abstract interface can facilitate intangible interaction by incorporating users' psychological understandings (e.g., users' mental models) in the design process to emphasize the design of user-friendly interfaces (Chen, 1998). An interaction designer can incorporate pleasures into both types of mediums.

Tiger (1992) identifies four levels of pleasures: physiological pleasure, social pleasure, psychological pleasure, and ideological pleasure. Tiger mentioned that the physiological pleasure is mainly sensory pleasure, such as taste and smell. The social pleasure can be achieved if an individual has good social interactions with other members of the same society. The psychological pleasure can be obtained based on an individual's achievements, such as after spending some efforts to accomplish a task. The ideological pleasure is related to an individual's spiritual feels, such as after reading a book, appreciating an artists' work, or listening to favourite music. To many interaction designers, their goal is to design easy-to-use user interfaces to fulfill users' needs. However, merely reaching this stage is still not enough. What users want are even more than the degree of easiness. They want to have fun during their task processes.

In this study, the authors emphasize that designing pleasurable interactions are the bases for realizing pleasurable user interface design. In the context of Web Pad user interface design, the last three types of pleasures (i.e., social, psychological, and ideological pleasures) mentioned above can be incorporated in the iterative interaction design process. This is because a Web Pad can be used for wireless communication among users for social pleasure, such as e-mail communication and Internet navigation. Its embedded applications can help users accomplish various types interaction of tasks to achieve psychological pleasure, such as personal information management. In addition, the media player provided in the Web Pad can be used to play music and view video for ideological pleasure. In order to improve the Web Pad's interface pleasure, the technique of participatory function analysis was used in this study. It is a powerful and empirical tool to help elicit potential users' internal feelings regarding a Web Pad's functions. This paper aims at applying the technique of participatory function analysis to investigate users' perceptions and understandings of a Web Pad's user interface design in terms of interface pleasure.

2 Participatory Function Analysis

Participatory design concepts often used in a user-driven design process in a collaborative manner. It is generally viewed as Scandinavian approach (Greenbaum & Kyng, 1991; Kyng, 1994). In this study, the proposed participatory function analysis is a technique derived from participatory design concepts. It is a design and evaluation technique that helps interaction designers characterize users' Web pad interaction activities. It is a qualitative-based approach that emphasizes involving users in the iterative interaction design process. More specifically, participants are required to express their views verbally during the experiment regarding how they perceive pleasures while interacting with the Web Pad's interface functions. Their protocols were further analysed after the experiment. A list of pleasure and displeasure word phases regarding the design factors that affect users' perceptions of Web Pad interface pleasures were then generated. Based on these summarized phases, the authors can modify the Web Pad's interface functions to make it more pleasurable to interact with.

3 Experiment

In this research study, two innovative Web Pad user interfaces designed with pleasure considerations were first created for the experimental purpose. Both Web Pad user interfaces were equipped with a digital photo album on the desktop screen to enhance their pleasure feels. In addition, one user interface was created based on the "gardening" metaphor (see Figure 1). That is, all the interface elements shown on the Web Pad desktop were designed with the images of gardening tools. Nonetheless, the other user interface was designed without applying any metaphor (see Figure 2). The interface elements (e.g., screen layout, desktop buttons, and color scheme) shown on these two desktop systems were also different. In order to investigate which type of user interface can be better perceived by the potential users as with more pleasure to interact with, the technique of participatory function analysis were used to investigate users' pleasure perceptions regarding these two Web Pad's user interface designs.

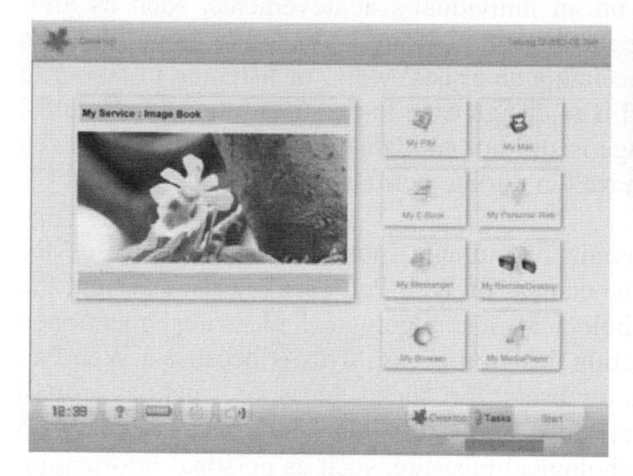

Figure 1: A Web Pad desktop designed with "gardening" metaphor

Figure 2: A Web Pad desktop designed without any metaphor use

3.1 Participants

A group of 18 student volunteers from the Department of Industrial and Commercial Design at National Taiwan University of Science and Technology were recruited as participants. Their ages range from 18 to 25 years old. In this experiment, no participant was paid to participate in the study. However, some students received class credit for taking part in the experiment. All participants were asked to interact with both Web Pads in the experiment. However, half of them were randomly selected to interact with the metaphorical Web Pad user interface first, and then proceed to the non-metaphorical Web Pad. The other half of participants were randomly asked to interact with the non-metaphorical Web Pad user interface first, and then to the metaphorical Web Pad. By so doing will balance participants' pleasure feels influenced from the preceding Web Pad to the second Web Pad user interface.

3.2 Methods

In the experiment, participants were asked to explore the Web Pad's interface functions and express their pleasure feels during the process. That is, participants were asked to perform various interaction tasks by walking through the Web Pad's wireless functions, such as Web browsing, e-mail sending/receiving, and multimedia downloading and playing. Potential interface usability problems were also found in this stage.

The think-aloud method was applied to help collect participants' word phases regarding their feels about interface pleasures. That is, when interacting with the Web Pad user interfaces, all the participants were asked to speak out loud how they felt about the interface design, and indicated their pleasure or displeasure feels. Participants' verbal data were recorded to allow post task analyses.

3.3 Results and Discussions

After the participants' verbal data were collected, they were classified and coded based on two categories, i.e., static user interface elements and dynamic interaction styles. The features of static user interface elements include screen layout, desktop buttons, photo album, and user overall impression. A table was created to show how participants felt about the static user interface elements regarding their pleasure feels. See Table 1 for a brief summary.

Table 1: Brief Summary of Static User Interface Elements regarding Users' Pleasure Feels

Metaphorical Web Pad	Pleasure feels	Reasons	Frequency	Non-metaphorical Web Pad	Pleasure feels	Reasons	Frequency
Screen layout	+	Good unity	2	Screen layout	+	Dynamic feels	2
	-	Boring	2		-	Too bright	2
Desktop buttons	+	Good metaphor	5	Desktop buttons	+	Clear indications	6
	+	Personal touch	3		-	Large size	7
	-	Small text	2		-	No user customization	4
Digital photo album	+	Good color scheme	3	Digital photo album	-	Small size	6
	-	Small size	2		-	Rigid	4
Overall impressions	+	Friendly feels	6	Overall impressions	+	Good color scheme	9
	+	Warm and fragrant	6		+	Simplicity	6
	+	Cherish the past	5		+	Contemporary	4
	+	Romantic feels	3		+	Clarity	4
	+	Elegant feels	2		+	Rational feels	4
	-	Old and soft	2		-	Cold	2

Note: + means positive pleasure feels; - means negative pleasure feels.

From Table 1, participants' express both pleasure and displeasure feels towards static user interface elements. In fact, it is impossible for an interaction designer to create user interfaces that guarantee every user's pleasure feels. For example, some participants possessed pleasure feels towards the metaphorical Web Pad because its screen layout was designed with good unity. Nonetheless, some participants thought the screen layout was too boring. Similarly, some participants felt pleased because the screen layout of the non-metaphorical Web Pad is designed with dynamic feels. However, some participants thought it was too bright. By comparing the participants overall impressions on these two Web Pad user interfaces, the authors found that most participants felt the metaphorical Web Pad user interface pleasurable because it was designed with warm and fragrant gardening style that made participants feel romantic and elegant. It provided participants with friendly feels. Most female participants preferred to interact with this type of user interface. On the contrary, the non-metaphorical Web Pad user interface gave participants contemporary feels because of its color scheme and simplicity. Participants felt the interface pleasurable because of its clarity and rational feels. Most male participants expressed their favors in interacting with this type of user interface.

In addition, the participants' dynamic interaction styles generated based on interacting with various Web Pad functions were also analysed. Potential Web Pad interaction problems were found. These problems were corrected after the experiment to improve the Web Pad interface usability. For example, the tapping of application buttons on metaphorical Web Pad user interface did not provide instant feedbacks to the participants. The pop up windows could not be resized which might intervene with the participants' view on the screen. However, some interesting suggestions made by the participants were not implemented immediately because of the time and budget constraints. For instance, participants would like to have options to select their favourite desktop metaphor. Therefore, more metaphorical desktop skins (e.g., a living room or other home settings.) need to be created to satisfy their particular needs. Similarly, when interacting with the non-metaphorical Web Pad user interface, participants pointed out several potential interaction problems. For example, the hand writing area seemed too small for the participants to write text freely. The pop-up window covered the application buttons. Also similar to the metaphorical Web Pad user interface, participants would like to have opportunities to select their favourite desktop skins or customize button icons. By customizing the Web Pad user interface, users can incorporate their personal identities into the system, which may make their interactions with the Web Pad user interface a more pleasurable task.

4 Conclusions

This study was intended to explore the users' perceptions of a Web Pad's interface pleasures based on the technique of participatory function analysis. In the experiment, participants express both pleasure and displeasure feels towards static user interface elements. However, some trends pertaining to interface pleasure still could be found. For example, in this study, female participants preferred to interact with the metaphorical user interface because of its warm and fragrant gardening image while male participants liked to use the non-metaphorical user interface because of its contemporary and simplicity feels. In addition, potential dynamic interaction problems were also found during the experimental process, which led to modifications after the experiment. The authors have successfully demonstrated that the application of participatory function analysis technique can be used to elicit users' perceptions towards interface pleasure, and at the same time to improve the Web Pad's interface usability.

5 Acknowledgements

The authors would like to acknowledge our great gratitude to Dr. Alan Pan, Mr. Ming-Che Weng, Ms. Chiao-Tsu Chiang, Ms. Yueh-Chi Wang, and Ms. Chieh-Yu Chan for providing invaluable design suggestions. The authors also want to express special thanks to Mr. Bo-Ching Chiou and Ms. Chia-Ying Tsai for assisting the design of simulation and the execution of experiment.

6 References

Chen, C.-H. (1998). *Color in Human-Computer Interaction*. Doctoral dissertation. The University of Kansas, Lawrence, Kansas, USA.

Greenbaum, J., & Kyng, M. (1991). *Design at Work: Cooperative Design of Computer Systems*. Hillsdale, NJ: Lawrence Erlbaum Associates.

Jordan, P. W. (1999). Pleasure with products: Human factors for body, mind and soul. In W. S. Green & P. W. Jordan (Eds.), *Human Factors in Product Design: Current Practice and Future Trend*. London: Taylor & Francis.

Jordan, P. W. (2000). Inclusive Design: A Holistic Approach. *Proceedings of the IEA2000/HFES2000 Congress*. San Diego, California, USA, 6-917~6-920.

Jordan, P. W., & Green, W. S. (2002). *Pleasure with the Use of Products*. London: Taylor and Francis.

Kyng, M. (1994). Scandinavian design: Users in product development. *Proceedings of the SIGCHI Conference on human factors in computing systems*. Boston, Massachusetts, USA, 3-9.

Tiger, L. (1992). *The Pursuit of Pleasure*. Boston, MA: Little, Brown and Company.

Incorporating Cognitive Usability into Software Design Processes

Michael Feary
NASA
Ames Research Center, CA, USA
michael.s.feary@nasa.gov

Lance Sherry
Athena Technologies, Inc.
Manassas, VA, USA
lsherry@athenati.com

Peter Polson
University of Colorado
Department of Psychology
Boulder, CO, USA
ppolson@psych.colorado.edu

Karl Fennell
United Airlines
UA Training Center
Denver, CO, USA
Karl.Fennell@UAL.com

Abstract

The goal of the research is to provide a usability inspection method that is useful in many aspects of the initial phases of the design process. The method presented is task oriented, and provides categories which are broad enough to capture a wide range of aircraft automation activities but do not require detailed explanations of human performance. The expectation is that such a structured method should allow designers to rapidly assess prototype designs, and provide guidance for the design of automation functionality.

1 The Role of Usability Inspection Methods in the Design of Functionality

While design teams acknowledge that task oriented usability evaluation is an important part of the design process, modern day techniques are perceived as an effort-intensive activity for a limited return on investment (Ivory and Hearst, 2001). While future development in automated usability methods may eventually supply a more comprehensive solution to providing easily obtainable, objective usability measurement, design teams have a need today for methods which are less expensive to use, but still provide some aid for making HCI related design decisions. The motivation for the work described in this paper is to propose a method which attempts reduce the cost and increase the usefulness of usability inspection for the design of software functionality.

Surveys reveal that many usability inspection methods focus on usability problems in user interfaces (Nielsen and Mack, 1994), however, the stage at which the user interface prototype is available may already be relatively late in the design process. As a result, modifications to the functionality are more costly than if improvements had been suggested earlier, and the results of the evaluation are deemed less useful. If a usability inspection method could be used to examine potential usability problems during the requirements specification and function allocation phases of design, then changes would be less costly and more useful for design teams.

Additionally, task decomposition methods such as the Goal, Operators, Methods, Selection Rules (GOMS) methodology (Card, Moran, & Newell, 1983), have been successfully used to define improvements to training and interfaces in aviation (Irving, et al., 1994), but are also too time intensive to allow use in fast paced design efforts. The motivation for the development of the method described below is to increase the usefulness of usability inspection techniques and task decomposition methods by enabling the methods to be used very early in the design process, and reducing the time required to complete an analysis and provide input to the design of functionality.

The 'RAFIV' method follows in the tradition of the Cockpit Cognitive Walkthrough (Polson and Smith, 1999), of trying to reduce the amount of time required to conduct analyses by defining categories of activities which are applicable to the domain. What separates the RAFIV method is the addition of task reformulation, verification, and monitoring stages which broaden the range of aircraft automation tasks which can be captured, and. In addition, the RAFIV method focuses on evaluating the ease of learning and ease of use (Wharton et al., 1994) that improves the efficiency of the method by focusing on the match between the task and the functionality instead of attempting to provide detailed explanations of human cognitive and perceptual activities for each task.

2 A Method for Introducing Cognitive Usability Engineering Early in the Design Process

The RAFIV method is a step of a design process which uses all available information to develop as comprehensive list of tasks as possible, then uses a formal software engineering method, such as the Situation – Goal – Behavior (SGB) method (Sherry, 1995) to design functionality which accomplishes the tasks in the least complex fashion possible. In concert with these engineering methods, the RAFIV method is used to evaluate the functionality against the task list.

2.1 The RAFIV Method

The method that has been developed to provide an analysis of the cognitive usability of an automation – interface system is referred to as the Reformulate – Access – Format – Insert – Verify and Monitor (RAFIV) method (Sherry et al., 2002). The method begins with the specification of a set of tasks that are representative of the environment in which the new design will be used.

In this method **Task** refers a higher level goal which must be accomplished regardless of whether or not automation is employed. An example of this in the aviation environment is the Air Traffic Control (ATC) clearance "climb and maintain (altitude)", which can be accomplished, manually, or through one of several aircraft automation functions (see Table 1). At this level, tasks may be derived from the environment, for example, many tasks in the aviation domain take the form of ATC clearances. Additional tasks may be derived from training and operations manuals of existing aircraft, as well as the task data elicited from Subject Matter Experts (SMEs). This reduces the time required of the SMEs by generating a large number of tasks which the SMEs can then evaluate and modify.

Once the set of tasks has been agreed upon, evaluators can subdivide the elements to accomplish the tasks into one of the five RAFIV categories.

The **Reformulate** stage of the method first determines if there is an automation feature which supports the task, and then examines the amount of translation required to allow the function to successfully accomplish the task. The mission task must be translated into a definition of the function (or feature) of the automation that will be used, and the data that must be entered (Palmer, Hutchins, Ritter, & van Cleemput, 1992). The analysis for measuring the level of difficulty for completing this stage of a task focuses on the answers to questions about whether the functionality of the device supports completion of the task, whether there are subtasks required to complete the tasks, and whether data manipulation is required. In the aviation example given, two aircraft automation functions that could be used to accomplish the "climb and maintain" are listed (see column 2, table 1). The example highlights one of the unique aspects of the RAFIV method, that of differentiating between tasks and functions which may support a task. One of the difficulties facing airline training departments today is providing guidance on which of the many aircraft automation features to use for different situations in the complex, dynamic aviation environment. Training departments have attempted to resolve this problem by developing "automation philosophies", but the solution to this problem needs to be addressed in design.

Once the user's task has been reformulated into a description of what is being requested of the automation, the user must transfer the description to the automation via a sequence of actions. These actions have been divided into three steps by (Polson, Irving, and Irving, 1994), following the GOMS description of Operators (Card et al., 1983).

The **Access** stage examines the interface actions required to navigate the user interface from which he or she can give a recognizable task command to the automation.

The **Format** stage examines the amount of data manipulation required for give recognizable task commands to the automation, including correct ordering, spelling, grammar, etc. of information to be entered.

The **Insert** stage examines the efficiency of giving the executing task commands to the automation.

After the user has transferred the task description to the automation, the user must then determine whether the user's goals match that of the automation.

The **Verify** stage examines the matching between the user's task description and the description of automation function performance via the automation interface (see column 6, table 1). The verify stage assesses feedback for all functions which support tasks, an area which is deficient in many modern aircraft (Sherry et al., 2001). Included in this stage is an additional assessment of feedback for **Monitoring** the automation's progression towards completion of the task. This can be a very difficult requirement as some automation functions may be armed, meaning that the commencement of the requirement to monitor the function may not occur until long after the other stages have been completed.

Task	Reformulate	Access	Format	Insert	Verify and Monitor
Climb and Maintain (altitude (feet))	MCP/FCU climb function Requires altitude value	MCP/FCU	None required, dial altitude value	Pull altitude select knob	Verify climb to appropriate altitude on PFD, monitor performance and timely level off of aircraft
	FMC/MCDU VNAV climb function	MCDU Climb page MCP/FCU VNAV/Man	If required to enter altitude, "/" must be	If required, press execute button /insert LSK Press MCP	Verify climb to appropriate altitude on PFD, monitor performance and

	requires altitude value	aged climb function	entered first	VNAV button, or pull FCU altitude select knob	timely level off of aircraft

Table 1. Example of RAFIV method

2.2 Increasing Utility in Aviation Design

Once a set of high-level tasks have been agreed upon, the steps of the RAFIV process can be accomplished quickly, depending on the designer/evaluator's knowledge of the proposed functionality. The **Reformulate** evaluation will require understanding of applicable functionality for solving a particular task, but does not require SME involvement during the generation of the reformulate elements. The **Access**, **Format** and **Insert** task elements can be evaluated with a prototype of the interface, or a description of the proposed interface objects and interface object behavior. Finally, analysis of the **Verify** and **Monitor** task elements requires knowledge of the proposed interface feedback, although it can also be used to generate feedback which matches the task description. Once the generation of the RAFIV steps has been completed on the first pass, SMEs can evaluate the already developed breakdown of the task steps.

Currently the RAFIV method is in use in the design of future aircraft automation, and elements of the method have been used successfully in conjunction with formal systems engineering methods (Sherry et al., 2003). Analysis measures begin with a count of the number of steps required to accomplish and analysis of the means of reducing the count. Beyond these simple measures, HCI usability characteristics can be examined. An example of this is the "label following" tendency (Polson and Lewis, 1990). The label following tendency is supported by evidence (Kitijima and Polson, 1997; Wharton et al., 1994) that users will take action on user interface objects that most closely match the user's goal. For example if in the "Climb and Maintain" aviation example given above, there was a button labelled "climb and maintain", then users would be very likely to select that interface object, regardless of whether or not it was the correct action to accomplish the task.

Although use of this paradigm exclusively would lead to an explosion of interface objects to match the number of tasks, another aspect of a RAFIV analysis which mitigates this is an assessment of the frequency of tasks. For example, it is less costly to train users to recall an "Access" element which is used very frequently because the risk of memory decay of that element is less than an "Access" element which is used infrequently. This will of course lead to tradeoff decisions in the design process, but the information gained by conducting the RAFIV analysis should lead to more informed tradeoff decisions.

3 Discussion

Introducing a method for usability inspection that addresses perceived problems by reducing the 'costs' of the process at the same time as increasing the benefits it offers may encourage design teams to make further use of usability inspection methods and perhaps more general HCI principles in design. The RAFIV method (Sherry et al., 2003), is intended to provide a step towards utilizing domain expertise more efficiently.

As a method that allows task analysis information to be used repeatedly in the design and evaluation process, RAFIV could provide motivation for organizations to invest the necessary time and resources to create explicit task analyses.

The methods described are not intended to provide a complete set of tools for evaluating HCI systems, but are intended to provide tools that designers can use in the early stages of the design process to rapidly evaluate and modify designs to improve cognitive usability. There are many other aspects of usability which these methods do not address, but providing designers with assistance in areas where there is a known lack of tools will begin to close the gaps, and could begin to motivate additional attention to developing methods which address automation – interaction issues in design.

References

Card, S., Moran, T., & Newell, A. (1983). *The Psychology of Human-Computer Interaction.* Hillsdale, NJ: Erlbaum.

Ivory, M.Y., & Hearst, M.A. (2001). The State of the Art in Automating Usability Evaluation Methods. *ACM Computing Surveys, Vol. 33*, No. 4, pp. 470-516.

Kitijima, M., & Polson (1997). A Comprehension-based model of exploration. *Human-Computer Interaction, Vol. 12.*

Nielsen, J. & Mack, R. (1994). *Usability Inspection Methods.* New York: John Wiley & Sons.

Palmer, E., Hutchins, E., Ritter, R.D., VanCleemput, I. (1992). *Altitude Deviations: Breakdowns of an Error Tolerant System.* NASA/TM DOT/FAA/RD-92/7, Moffett Field, CA, USA.

Irving, J., Polson, P. G.,. Irving, J. (1994). Applications of Formal methods of Human Computer Interaction to Training and Use of the Control And Display Unit. Tech Report 94-08, University of Colorado. Boulder, CO, USA.

Polson, P. G., & Lewis, C. (1990) Theory-based design for easily learned interfaces. *Human-Computer Interaction, Vol. 5*, No. 2 & 3, pp. 191-220.

Polson, P., and Smith, N. (1999). The Cockpit Cognitive Walkthrough. In the proceedings of the *Tenth International Symposium on Aviation Psychology*, Columbus, OH, USA.

Sherry, L. (1995). A Formalism for the Specification of Operationally Embedded Reactive Systems. In proceedings of the *International Council of System Engineering*, St. Louis, MO, USA.

Sherry, L., Feary, M., Polson, P., Palmer, E., and Fennell, K. (2003). Drinking from the Fire Hose: Why the FMS/CDU can be hard to train and difficult to use. NASA Technical Memorandum (in press) Moffett Field, CA, USA.

Sherry, L., Feary, M., Polson, P., Mumaw, R., and Palmer, E. (2001). A Cognitive Engineering Analysis of the Vertical Navigation (VNAV) Function. NASA Technical Memorandum 210915.

Wharton, C., Rieman, J., Lewis, C., & Polson, P. (1994). The Cognitive Walkthrough Method: A Practitioner's Guide. In J. Nielsen and R. Mack (Eds.) *Usability Inspection Methods.* New York: John Wiley & Sons.

Usability Laboratories - Quantitative and Qualitative Approaches

Regine Freitag, Wolfgang Dzida, Barbara Majonica, Karsten Nebe, Natalie Wole

Fraunhofer Institute AIS
Sankt Augustin, Germany
regine.freitag@ais.fhg.de
wolfgang.dzida@ais.fhg.de

C-LAB
Paderborn, Germany
barbara.majonica@c-lab.de
karsten.nebe@c-lab.de
natalie.woletz@c-lab.de

Abstract

Since a usability laboratory is a cost factor, it is important to know which investigations can be performed cost-efficiently and how the results can be used effectively in the design process. Taking into account the today's advanced laboratory equipment there are several aspects which should be considered when running an inhouse lab. Shall the laboratory be equipped for stationary, portable or remote investigations, regular or occasional usage, summative or formative testing, mainly recording critical events or also providing design support? Quantitative as well as qualitative usability investigations can be supported by a usability lab helping to reveal critical features at the user interface early and to find the sources of a hidden mismatch. Only when we understand what causes a usage problem we can achieve a more usable product.

1 Introduction

Best practices of usability engineering are increasingly adopted in software development projects. Consequently, there is a shift from testing products at the end of a project towards early and continuous usability testing in the development process. Many problems with using a product can now be detected rather early, since most of the usability requirements are getting specified during the design phase by explorative prototyping.

But is it really necessary to utilise the expensive equipment of a usability laboratory for usability prototyping and testing? Heuristic evaluation (Nielsen, 1994) and cognitive walkthroughs (Lewis & Wharton, 1997) are much cheaper approaches to evaluate a prototype for usability. In favour of the walkthrough and evaluation experts one could say that there are almost always obvious defects in design proposals an expert can uncover by inspection without consulting users. Moreover, when designing and evaluating the first prototype it may suffice for the expert to envisage hypothetical users. However, in favour of the questioned usability laboratory one could also say: "Don't trust the experts. Listen to the users." Only users can give authentic judgement, as to whether a design proposal meets the usability requirements.

There is a growing awareness within development teams that user involvement during the usability engineering process is a significant prerequisite for the success of a product. Depending on the status of the project and the design issues to be investigated users can give substantial support in developing a usable system. Usability tests within a usability laboratory can help the usability engineers to better understand how users want to perform their tasks with a system and why they have troubles with design proposals. To decide for the appropriate use of a lab we should consider a sequence of decision steps before observing and analysing user performance:

- Users have problems in performing their tasks with a prototype or a piece of software and it is necessary to demonstrate and document the complex but short living situations encountered.
- The usage problems need to be clearly understood in order to analyse their impact on user and task performance.
- If the test results reveal a need for changing the design the re-design rationale has to be specified in order for a proposal to get improved and tested subsequently.

In a usability laboratory applying a state-of-the-art equipment can support all of these steps.

2 Documenting critical situations of system use

A usability lab can help to demonstrate and document critical usage situations. "A usability lab system produces an amalgamation of perspectives, and provides a communication breakthrough" (Johnson & Connolly, 1999, p. 1125). Showing the designers a video with users struggling with a prototype is extremely valuable, especially, when it is not possible for the designer to take part in the user observation personally. Edited videotapes enriched with structuring title screens and explanatory voiceovers can help to communicate the goals and experiences of user studies to other people as well (Roe Purvis, Czerwinski & Weiler, 1994).

Typically, when users work on complex tasks, it may be difficult for the usability assessor to thoroughly understand the task and the context of use and to reliably interpret the user's trouble. An understanding can be improved by an on-site analysis. However, many short living observational situations cannot be studied sufficiently. Users may get nervous about the elaborate way of understanding a trouble. A recording of the observational data allows to analyse the situation afterwards on the assessor's own speed.

If it is important to explore how different users deal with a proposed design solution a usability lab can be used to record the variety of ways chosen by the users. For the users there is obviously no "one best way" of performing a task. Individual differences in approaching a task may give the designer valuable feedback on whether a design proposal fits more or less well with the users' intentions.

During usability prototyping it often suffices to pursue the user's flow of interaction with the system and document the tricky usage situation appropriately. If it is evident what causes the problem, there is no need to spend further effort in a well-equipped usability lab, which helps deliver the same result but at much higher costs. "Often, there is no real reason to expend resources of gathering statistically significant data on a user interface that is *known* to contain major usability problems and has to be changed anyway." (Nielsen, 1994, p. 7).

3 Identifying problems with system use

When features of the system need to be scrutinised for their impact on user performance a *quantitative approach* is a cost-efficient way. Quantitative data achieved by counting and measuring system activities and user actions can help to detect critical functions, misleading features or dysfunctions in terms of user performance.

The equipment of a usability laboratory allows capturing defined events whenever they occur during the user session. Events become directly observable as mouse movement, mouse clicks, triggering of a function name, etc. or they represent generalised categories, e.g., begin and end of a certain task, "user errors". Based on the recording of events data like time needed for a certain task, error-handling time, or frequency of events can be calculated. The focus of such investigations is not on just demonstrating and documenting the usage of a prototype but rather to locate failures and shortcomings of a system systematically on a descriptive level.

When evaluating systems addressed to a broad user group remote usability testing can supplement a traditional usability lab. By using Internet facilities in the conduct of a prototyping session multiple users can be included more easily having them perform tasks at their original workplaces. For a comparative study of remote and stationary testing see (Chen, Mitsock, Coronado & Salvendy, 1999). This can be an additional cost-efficient way to collect quantitative data in order to identify problems, which can be examined more thoroughly with only a few users.

Additionally, statistical data give interesting hints with respect to the applicability of innovative features. Comparative usability tests help to estimate if criteria like user performance, error rate or satisfaction are influenced by a new technology compared to a conventional one. Comparative testing is also a means to "quantify the benefits of the new interface" (Salasoo, White, Dayton, Burkhart & Root, 1994, p. 511). Comparative studies can as well be used to estimate the pros and cons of different design proposals by opposing e.g. the duration of task performance, the number of "user errors" or the number of navigational steps. The results contribute to coming to a suitable design rationale. The data collected during the design process cannot be interpreted as "real" data, because a usability prototype is usually optimised to illustrate a (part of a) scenario, which is not necessarily realistic in performance and functionality. Still, a formative evaluation conducted during the design process can give valuable hints to problems and pitfalls of the coming system.

Often system characteristics are evaluated in an advanced state of the development process. In this case the use of the system can be investigated under more realistic circumstances than with early prototypes. The results help to demonstrate if the usability principles and the usability goals are met (Majonica, 1996) and to detect still existing deficiencies.

However, any attempt to identify valid usability problems can fail when the evaluator get drunk with masses of observational data captured, for instance, by video or logfile recording. The ability to focus on relevant usage data (e.g. snapshots) is crucial for cost and success.

A qualitative approach can also help to isolate usability problems. When in the same situation different users complain about different problems this can be an indication of so-called opportunistic user behaviour. Opportunistic use of a system is characterised by making use of seemingly suitable opportunities to reach the intended goal because there is no straightforward way of performing the current task (Rosson & Carroll, 2002, p. 163). Since it can hardly be foreseen when this situation will happen, the laboratory helps to capture the whole session. Afterwards, all interesting observational data can be focused and rapidly extracted in terms of snapshots for enabling repeated analyses of critical cases.

4 Identifying a mismatch

In our laboratories of Fraunhofer AiS, TÜV Secure iT and C-LAB we learned that recording and reproducing critical situations may not suffice to uncover and understand usage problems.

A qualitative approach is required when a usage problem has been identified but it is difficult to understand what causes the trouble. Though the design solution fits well with the needs of some users, others complain that the way to do a task does not make sense. This situation typically occurs with various user target groups and significant differences in the contexts of use.

At the joint laboratory of Fraunhofer AiS and TÜV Secure iT (Dzida & Freitag, 2001) we set up a general coding scheme intended to be applied in different settings and projects. The Task Intention Code (TIC) is based on a general task model (Dzida & Valder, 1984) and offers a specific template for coding intentions in a semi-formal way (Dzida, 2003). First we analyse the user's task flow in terms of a sequential task model by separating the task flow into four phases: evaluating the current state, preparing a task, executing a functional command and evaluating the result. If the user's interaction with the system reveals an opportunistic behaviour or even an error recovery, we trace the dialog in order to find the source of what we call a mismatch. TIC helps to capture the critical snapshot within the mass of observational data.

Once having identified the critical situation, the user's task is classified according to the coding schema of TIC. From this classification the analyser can generate a hypothesis about what the user might have intended to achieve. Having validated the user's intention, it can be compared with the designer's intention. Dissenting intentions indicate the source of a mismatch, which can be reconciled by improving the design proposal. By the notion of mismatch we want to avoid blaming the user or the designer. For clarifying the mismatch we use the elementary task model of TIC, thereby dividing a task into several components each of which to be offered in the computer dialogue. A mismatch occurs, for instance, if a tool is missing on the display or cannot be identified by the user. The notion of mismatch implies that the user's intention to do something does not meet the designer's intention to support the user. A mismatch between intentions can be precisely and completely identified. The application of TIC, a general coding scheme based on intentions, supplies results which can be directly translated into design by clarifying what elements of a task are missing at the user interface or need more affordance or guidance.

5 Conclusion

Often a summative evaluation at the end of the development process is requested, e.g. as a benchmark with similar systems or earlier versions. Although descriptive data can demonstrate the advantages of the system in question, they neither unveil the causes of usability problems nor the intentions of the users or their mental models. While critical usage situations can be infered from quantitative data, a qualitative method helps to find the hidden causes of usability problems. Thus quantitative and qualitative methods can be seen as complementary approaches each with their own strengths and weaknesses. However, a combination of both in order to measure the overall quality of the system implies methodical constraints (Wottawa & Thierau, 1990), (Shackel, 1991).

Within the usability laboratories of today third generation data collection and analysis tools (Hoiem & Sullivan, 1994) support quantitative and qualitative evaluations. Different kind of data can be collected and analysed. However, getting useful results from a lab investigation is still a complex and time-consuming task. The case of an investigation should be clarified very carefully in advance so that the equipment of a lab can be tailored to the problem at hand.

A general classification schema of user behaviour should be one of the facilities of a usability lab. Discussing different coding schemes ("SIG Usability Laboratory", 2002) usability experts came to the result that we should bundle our efforts to construct an ontology of user behaviour which can be used as a common basis for performing qualitative evaluations within a usability laboratory. In our experience a general scheme such as TIC supports the classification and explanation of critical incidents. Still the usability assessor's skill in exploiting the laboratory tools is a key issue when applying a usability laboratory.

6 References

Chen, B., Mitsock, M., Coronado, J., & Salvendy, G. (1999). Remote Usability testing through the Internet. In H.-J. Bullinger & J. Ziegler (Eds.), *Human-Computer Interaction: Ergonomics and User Interfaces* (pp. 1108-1113). Mahwah, NJ: Lawrence Erlbaum.

Dzida, W. (2003). TIC - Task Intention Code. *This volume.*

Dzida, W., & Freitag, R. (2001). The Usability-Engineering Laboratory at GMD. *ERCIM News*, No. 46, July 2001. 64.

Dzida, W., & Valder, W. (1984). Application Domain Modelling by Knowledge Engineering Techniques. In: Shackel, Brian: *INTERACT'84*, 320-327.

Hoiem, D. E., & Sullivan, K. D. (1994). Designing and using integrated data collection and analysis tools: challenges and considerations. *Behaviour & Information Technology, Special Issue Usability Laboratories*, 13 (1 and 2), 160-170.

Johnson, R., & Connolly, E. (1999). Mobile and Stationary Usability Labs: Technical Issues, Trends and Perspectives. In H.-J. Bullinger & J. Ziegler (Eds.), *Human-Computer Interaction: Ergonomics and User Interfaces* (pp. 1123-1127). Mahwah, NJ: Lawrence Erlbaum.

Lewis, C., & Wharton, C. (1997). Cognitive Walkthroughs. In M. Helander, T. K. Landauer, P. Prabhu (Eds.): *Handbook of Human-Computer Interaction* (pp. 717-732). Amsterdam: Elsevier.

Majonica, B. (1996). *Evaluation eines Informations-Systems für die Unterstützung von Instandhaltungsaufgaben.* Münster: Waxmann.

Nielsen, J. (1994). Usability laboratories. *Behaviour & Information Technology, Special Issue: Usability Laboratories*, 13 (1 and 2), 3-8.

Roe Purvis, C. J., Czerwinski, M., & Weiler, P. (1994). The Human Factors Group at Compaq Computer Corporation. In M. E. Wiklund (Ed.), *Usability in Practice - How Companies Develop User-Friendly Products* (pp. 111-145). Boston: Academic Press

Salasoo, A., White, E. A., Dayton, T, Burkhart, B. J., & Root, R. W. (1994). Bellcore's User-Centered Design Approach. In M. E. Wiklund (Ed.), *Usability in Practice - How Companies Develop User-Friendly Products* (pp. 489-515). Boston: Academic Press

Shackel, B. (1991). Usability – context, framework, definition, design and evaluation. In B. Shackel & S. J. Richardson (Eds.), *Human Factors for Informatics Usability* (pp. 21 – 37). Cambridge, University Press.

Rosson, M. B., & Carroll, J. M. (2002). Usability Engineering – Scenario-Based Development of Human-Computer Interaction. San Francisco: Morgan Kaufmann, 163.

SIG Usability Laboratory (2002). Performing usability investigations with a Usability Laboratory. Measuring Behavior 2002, Amsterdam, discussion.

Wottawa, H. & Thierau, H. (1990). *Evaluation.* Bern, Huber.

Perceptual Distributed Multimedia Quality: A Cognitive Style Perspective

Gheorghita Ghinea *Sherry Y. Chen*

Department of Information Systems and Computing
Brunel University, Uxbridge
Middlesex
UB8 3 PH
{George.Ghinea; Sherry.Chen}@brunel.ac.uk

Abstract

This paper examines the relationships between cognitive styles and Quality of Perception, a user-centric measure of distributed multimedia quality. 76 users took part in an experiment with two treatments of video clips. Cognitive Style Analysis was applied to identify users' cognitive styles as Field Independent, Intermediate, and Field Dependent. Results showed that colour treatment is an important factor to Intermediate user in the use of video clips. Moreover, types of audio and levels of video significantly impacted on the three cognitive style groups' Quality of Perception.

1 Introduction

The ultimate measure of effectiveness of a distributed or standalone multimedia presentation is the user experience in terms of enjoyment and information assimilation. Key to this is the issue of the *quality* of the multimedia presentation. Quality, in our perspective, has two main facets in a distributed multimedia environment: *of service* and *of perception*. The former illustrates the technical side of computer networking and represents the performance properties that the underlying network is able to provide. The latter characterises the perceptual experience of the user when interacting with multimedia applications and forms the focus of this paper.

2 Background

The networking foundation on which current distributed multimedia applications are built either do not specify Quality of Service (QoS) parameters, or specify them in terms of traffic engineering parameters such as delay, jitter, and loss or error rates, which do not convey application-specific needs such as the influence of clip content and informational load on the user multimedia experience.

On a technical level, the QoS impacts upon the perceived multimedia quality in distributed systems. However, previous work examining the influence of varying QoS on user perceptions of quality has almost totally neglected multimedia's infotainment duality, and has concentrated primarily on the perceived entertainment value of presentations displayed with varying QoS parameters. Accordingly, Apteker et al. (1995) examined the influence that varying frame rates have on user satisfaction with multimedia quality, whilst Steinmetz (1996) presents the perceived effect of synchronisation skews between media and Wijesekera et al. (1999) investigated the impact of media loss on users. However, research has largely ignored the influence that the user's psychology factors, especially cognitive styles, have on the perceived quality of distributed multimedia.

Cognitive style is an individual's consistent approach to organising and processing information. Riding and Rayer (1998) defined cognitive style as "an individual preferred and habitual approach to organising and representing information" (Riding and Rayner, 1998, p.25). Among a variety of cognitive styles, previous research indicated that the dimensions of Field Dependence/-Independence have significant effects on individuals' preferences to multimedia information systems. Several studies suggested that Field Independent individuals could particularly benefit from the control of media choice. A study by Chuang (1999) produced four-courseware versions: animation+text, animation+voice, animation+text+voice, and free choice. The result showed that Field Independent subjects in the animation+text+voice group or in the free choice group scored significantly higher than those did in the other two groups. No significant presentation effect was found for the Field Dependent subjects. Conversely, Lee (1994) indicated that auditory cues are important to Field Dependent learners.

Consequently, empirical investigation of examining the relationships between the provision of network-level QoS and the requirements of different cognitive style groups in distributed multimedia applications become paramount, because such evaluation can help use network resources efficiently and results in better end-to-end performance, which in turn has a direct positive impact on the user experience of multimedia. To address this issue, this paper presents an empirical study, which examines how cognitive styles influence users' perception of distributed multimedia information systems with respect to different levels of QoS.

3 Methods

Users were presented with two multimedia clips in MPEG-2 format. One of the clips depicted a high school band playing a Dixieland/jazz tune against a background of multicoloured and changing lights, whilst the other was an advertisement for a washing liquid. In the Band clip, although audio is important (for enjoyment purposes), it caries little significance as a conveyor of information, since the video stream is dominant from this point of view. In the Commercial clip however, there is a more balanced distribution of information among the media components. Thus, in this clip, the qualities of the product are praised in four ways - by the unseen narrator, both audio and visually by the couple being shown in the commercial, and textually, through a slogan display

Each clip was shown with a specific pair of QoS parameters. In our experiments we varied frame rate (5, 15 and 25 frames per second), and colour depth (full 24-bit colour and black and white). After each clip, the user was asked a series of questions (ranging from 10 to 12) based on what had just been seen and the experimenter duly noted the answers. Lastly, the user was asked to rate the quality of the clip that had just been seen on a scale of 1 - 6 (with scores of 1 and 6 representing the worst and, respectively, best perceived qualities possible). Users' cognitive styles were identified by using Riding's (1991) Cognitive Styles Analysis (CSA) in this study because Wholist/Analytic (WA) measured in the CSA is equivalent to Field Dependence/-Independence (Ford, 1995), and the CSA offers computerised administration and scoring.

3.1 Participants

This study was conducted at Brunel University's Department of Information Systems and Computing. 76 subjects participated in this study. Despite the fact that the participants volunteered to take part in the experiment, they were extremely evenly distributed in terms of cognitive styles, including 25 Field Independent users, 25 Intermediate users, and 26 Field Dependent users. All of them were inexperienced in the content domain of two clips. Tables 1 to 3 illustrate the distribution of the sample in different content treatments.

Table 1: The distribution of the sample within Field Dependent Group

	Frame Rates		
Colour Treatment	*5*	*15*	*25*
Black/White	5	7	4
Colour	2	6	2

Table 2: The distribution of the sample within Intermediate Group

	Frame Rates		
Colour Treatment	*5*	*15*	*25*
Black/White	3	4	3
Colour	4	6	2

Table 3: The distribution of the sample within Field Independent Group

	Frame Rates		
Colour Treatment	*5*	*15*	*25*
Black/White	5	6	3
Colour	2	7	2

3.2 Cognitive Styles Analysis

The cognitive style dimension investigated in this study was the level of Field Dependence. A number of instruments have been developed to measure Field Dependence, including the Group Embedded Figures Test (GEFT) by Witkin et al. (1971) and the Cognitive Styles Analysis (CSA) by Riding (1991). The GEFT derives scores for Field Independence by requiring subjects to locate simple shapes embedded in more complex geometrical patterns. However, a criticism of this approach is that levels of Field Dependence are inferred from poor Field Independence performance (Ford and Chen, 2001).

The CSA differs from the GEFT in that it includes two sub-tests. The first presents items containing pairs of complex geometrical figures that the individual is required to judge as either the same or different. The second presents items each comprising a simple geometrical shape, such as a square or a triangle, and a complex geometrical figure, as in the GEFT, and the individual is asked to indicate whether or not the simple shape is contained in a complex one by pressing one of two marked response keys (Riding and Grimley, 1999).

It seems that these two sub-tests have different purposes. The first sub-test is a task requiring Field Dependent capacity. Conversely, the second sub-test requires the disembedding capacity associated with Field Independence. In this way, the CSA overcomes the GEFT limitation that affects the measures of Field Dependence and Field Independence, because Field Dependent competence is positively measured rather than being inferred from poor Field Independent capability (Ford and Chen, 2001). In addition, the CSA offers computerized administration and scoring. Therefore, the CSA was selected as the measurement instrument for Field Dependence in this study.

The CSA measures what the authors refer to as a Wholist/Analytic (WA) dimension, noting that this is equivalent to Field Dependence/Independence (Riding and Sadler-Smith, 1992). Riding's (1991) recommendations are that scores below 1.03 denote Field Dependent individuals; scores of 1.36 and above denote Field Independent individuals; students scoring between 1.03 and 1.35 are classed as Intermediate. In this study, categorisations were based on these recommendations.

4 Results

4.1 Frame Rates

The results showed that frame rate did not influence users' performance and enjoyment, regardless of their cognitive styles and colour treatment. The reason may be that this study took a small sample in the treatment of each frame rate, which is a limitation of this study. Future research should conduct such type of the experiment with a larger sample. However, if our results can be replicated on a larger scale, they would confirm previous research (Apteker et al. 1995), thus highlighting that significant bandwidth reductions can be obtained if one takes perceptual considerations into account when transmitting multimedia.

4.2 Colour Treatment

In the video clip with the support of audio, colour treatments play an influential factor for Intermediate users to the use of multimedia information systems without regard to the frame rates. Intermediate users in subject content with multiple colours significantly performed better than those in the subject content with black/white colours did (t= -2.256, p=.041). However, this factor had significant effect on Field Independent and Field Dependent users. There is a need to conduct further research that examine why a multiple-colour treatment is important to Intermediate users who possesses a more versatile repertoire of information processing approaches.

4.3 Types of Audio

At first glance, it might appear surprising that the types of audio influence users' performance in the use of video clips, regardless of their cognitive styles. All of three cognitive style groups obtained better scores in the Band video clip than those did in the Commercial video clip with human's narration (p=0.000). Table 4 illustrates their mean scores:

Table 4: Impact of Audio on Cognitive Styles in the Band and Commercial Clips

	QoP (%)		
Video Clips	**Field Independent**	**Intermediate**	**Field Dependent**
Band Clip	76.00	84.00	76.92
Commercial Clip	26.67	18.67	33.33

However, we believe that the reason behind this finding is the fact, highlighted above, that the audio component in the Band clip, although important for enjoyment purposes, is not so much for informational ones. In contrast, the audio component of the Commercial clip conveys a much higher informational load.

4.4 Levels of Video

Band and Commercial Clips present different levels of informational video, which in our experiments have been found to also influence users performance, regardless of their cognitive styles. Thus, the three cognitive styles had a better performance in the Commercial clip than those did in the Band clips. Table 5 illustrates the significant differences of their mean scores.

Table 5: Impact of Video on Cognitive Styles in the Band and Commercial Clips

	QoP (%)		
Video Clips	**Field Independent**	**Intermediate**	**Field Dependent**
Band Clip	54.00	50.40	52.63
Commercial Clip	65.71	73.71	66.73

5 Conclusion

We have presented results which link persons' cognitive styles to Quality of Perception, a measure of perceptual quality in distributed multimedia systems. Whilst our study is a small scale study, both in terms of content matter and sample size, our ultimate aim is to build an integrated solution for the delivery of multimedia data, which takes into account not only a user's subjective preferences but also his/her cognitive style.

6 References

Apteker, R.T., Fisher, J.A., Kisimov, V.S. & Neishlos, H. (1995). Video Acceptability and Frame Rate. *IEEE Multimedia*, 2(3), 32-40.

Chuang, Y-R. (1999). Teaching in a Multimedia Computer Environment: A study of effects of learning style, gender, and math achievement. Retrieved August 8, 2000, from http://imej.wfu.edu./articles/1999/1/10/index/asp

Ford, N. (1995). Levels and types of mediation in instructional systems: an individual differences approach. *International Journal of Human-Computer Studies*, 43, 241-259, 1995.

Jackson, M., Anderson, A.H., McEwan, R. & Mullin, J. (2000). Impact of video frame rate on communicative behaviour in two and four party groups. *Proceeding of the ACM 2000 Conference on Computer supported cooperative work*, 11-20, Philadelphia, United States.

Lee, C. H. (1994) *The Effects of Auditory Cues in Interactive Multimedia and Cognitive Style on Reading Skills of Third Graders,* Unpublished Ed.D. Dissertation, University of Pittsburgh, USA.

Riding, R. J. *Cognitive Styles Analysis*, Birmingham: Learning and Training Technology, 1991.

Riding, R.J. & Rayner, S.G. *Cognitive Styles and Learning Strategies,* London: David Fulton Publisher, 1998.

Steinmetz, R. (1996). Human Perception of Jitter and Media Synchronisation. *IEEE Journal on Selected Areas in Communications*, 14(1), 61 –72.

Wijesekera, D., Srivastava, J., Nerode, A., & Foresti, M. (1999). Experimental Evaluation of Loss Perception in Continuous Media, *Multimedia Systems*, 7(6), 486-499.

Definition of a Common Work Space

Benoît GUIOST, Serge DEBERNARD, Patrick MILLOT

LAMIH, UMR CNRS 8530
Université de Valenciennes et du Hainaut-Cambrésis
Le Mont Houy, 59313 Valenciennes Cedex 9, France
Email: {Benoit.Guiost, Serge.Debernard, Patrick.Millot}@univ-valenciennes.fr

Abstract

This paper takes up the human-machine cooperation into the framework of the air-traffic control. It presents a study, which had made in order to design a new support tool for air-traffic controllers and results that are obtained. This study consists in a modeling of the human-human cooperation thanks to an encoding of cognitive' activities.

1 Introduction

The Air Traffic Control (ATC) is a complex system with a low degree of automation where the place of human operators is preponderant. To keep the air-traffic safety, in a high increase context, it's necessary to provide assistance tools with controllers.
The French ATC is a public service in charge of the flight security and the regulation of air traffic. Their main task consists in detecting and preventing collisions between the aircrafts. These situations are called conflicts and controllers must avoid them. Our studies concern the "en route" control (i.e. the air traffic management except for takeoffs and landings).
Airspace is divided into several sectors of control. Two controllers supervise each of them: a planning controller (PC) and a radar controller (RC). Each controller has a radar screen that displays information of each flight (flight level, trajectory, destination...). The experimental platform, called SPECTRA V2, was given a support tool to the human operators and a mini common work space (CWS) (Lemoine et al., 96).
The work presented in this paper is the continuity of this platform. Specifications of the new cooperative system are based at CWS' knowledge and the three cooperative forms of Schmidt (Schmidt, 91). In order to confirm these specifications, we have tried to model human-machine cooperation by studies the human-human cooperation. To make them, air-traffic controllers had been put into a realistic control situation. The post of the RC had been divided in two. It's the cooperation between the two RC, which is presented in this paper. In a first time, we will introduce concepts that are used.

2 Common Work Space: support of the human-machine cooperation

2.1 Human-Machine Cooperation

The cooperation between two agents can be defined like that (Hoc, 96):
" Two agents are in a cooperative situation if they meet two minimal conditions:

Each one strives towards goals and can interfere with the others on goals, resources, procedures, etc...

Each one tries to manage the interference to facilitate the individual activities and/or the common task when it exists.

The symmetric nature of this definition can be only partly satisfied"
The goals that are quote in this definition are not the same those are fixed by the supervision and the control of a process but the goals for achieving a particular task.
Interferences are interactions between the activities of several agents. Their nature can be positive or negative. Positives interactions refer to normal interaction between agents and negative interactions refer to conflicts between agents.
The role of the human-machine cooperation is to minimize negative interactions. The cooperation can be seen by two approaches, structural and functional.
At the structural level, human-machine cooperation can be seen by two forms (Millot, 88) (Millot, 97):

Vertical structure:
In this type of structure, support tools can't ever act of the system. The operator one has this capacity. However, support tools have same information of the process that the human operator and can also give him advice.
Horizontal structure:
In this type of structure, human operator and support tools are connected to the process droved and/or supervised. Therefore, the support tool must have reasoning capacity in real time. The two decision-makers are at the same hierarchic level. Different tasks are allowed between agents to regulate the workload of human operators.

At the functional level, human-machine cooperation can be view of three forms (Schmidt, 91): augmentative, debative and integrative:

Augmentative form: the agents have the same abilities and must perform one task that is too extensive for an operator alone. This task is divided in several sub-tasks that are shared between agents.
Debative form: the agents have the same abilities too but they compare their results to obtain the best solution.
Integrative form: the different agent abilities are supplementary. The task is divided in sub-tasks that are allowed between operators in function of their abilities. Each operator contribution is integrated to achieve the task.

2.2 The common frame of reference (COFOR)

When a human operator must achieve a task, he builds a frame of reference himself. This frame of reference permits at the operator to build him a representation of the process and of the process state. Thanks to this, human operator can plan their actions and detected abnormal evolutions of the process (Lemoine, 98).
In order to identify contents of this frame of reference we base ourselves in the Rasmussen's simplified model of problem resolution (Rasmussen, 83). This model contains four activities: information elaboration, identification, decision-making and action. Indeed, the operator's frame of reference is composed of different items (Pacaux-Lemoine et Debernard, 01): Information, problems, strategies, solutions and commands.
In the framework of cooperation between human operators (human-human cooperation), agents swap them our data of our frame of referential in order to build a common frame of reference. They used our COFOR to achieve tasks when goals or sub-goals are linked. Therefore, the COFOR is a reference between operators and contained items that are defined on top.
In order to minimize and optimise exchanges (number, duration and contents), which are verbal, agents can build for themselves an interpretation of the COFOR of their colleagues. However, the representation of a colleague's COFOR can be wrong. Therefore, to improve the cooperation between human operators and the support tools, an implementation of this COFOR seems necessary (Debernard, 02). This implantation is called Common Work Space (CWS).

Debernard and Hoc propose to define the structure and functions of the support tools by notions of CWS and the three forms of cooperation of Schmidt (Schmidt, 91). The three forms are implemented like that (Debernard S., Hoc J-M., 01):

Debative form: agents add data, that judge significant in the CWS. When interference appears, they must negotiate to erase them.

Integrative form: only an agent adds data in the CWS. Others agents take information, which are necessary to achieve our task.

Augmentative form: The shared of tasks and method of data addition are decided by agents themselves.

In order to verify the coherence of their proposals, a complementary cognitive study had been performed at the experimental platform AMANDA V1 (Automation and MAN-machine Delegation of Action Version 1). We will introduce you the results of it.

3 Experimentations

3.1 The experimental platform

The used experimental platform permit to put professional controllers into realist situations. It is composed of three work stations (1 PC & 2 RC). Each station is composed of two screens. At the left screen, controllers have a "radar view" that displays different controlled aircrafts. At the right screen, controllers have a "strip view". It models the board of strips habitually used by air-traffic controllers. In reality, strips are in paper. It contains the flight plan of each aircraft, I.E. the flight goal (trajectory and flight level).

3.2 The experimental protocol

In order to identify the contents of the RC' COFOR, the experimental platform was composed of one PC and two RC which have the same traffic to control. Therefore, aircrafts are shared into the two RC and a RC can't act of an aircraft who was not belong to him. In addition, all aircrafts of a conflict was not allowed at the same controller. However, controllers must implicitly cooperate to solve conflicts.

In France, air-traffic controllers control on average twenty-five aircrafts per hour. In order to controllers cooperate together, they had been put into realist situation but overload. Used scenarios were imposed a control of fifty aircrafts per hour.

The experimental protocol was composed of a training period for each controller. This period must help controllers to learn the functioning of graphical interfaces and input device (keyboard & mouse). Therefore, controllers must make two exercises on the platform and a full training scenario. Five trinomials of controllers had been participated of this experiment. They had been followed the entire protocol and the experimental scenario was the same for all.

3.3 Data collect

Data who are studied come from two records sources. The first source is the platform itself. It contains a function, which permits to record all the controllers' action of the system (ex: order in cap to a flight). Next, this function permits to re-play at the screen everything that the controllers have made during the experiment. Indeed, we can link up different system states with voiced records. This voiced' record is the second source. All the operators had said is recorded to a mini-disc. Next, records are transcribed in a text file, called verbal exchanges. They are encoded and analysis to extract items of COFOR.

3.4 Encoding and exploiting of data

In order to exploit verbal exchanges, it's necessary to encode each of their exchanges. The used method is the same of Lemoine's method used to encode the air-traffic controllers cognitive activities (Lemoine, 98).
This encoding is based on the used of predicate that correspond at different activities of COFOR build and update and then at detecting and solving of interference. An item and a variable are attributed to each predicate in function of the content of verbal exchange. Item corresponds at the translation of information that was defined on top. For example, "turn AFR1627 behind BAW436" is a strategy; "look at AFR and MON at south" is an information of the place of a problem.
Variables permit to characterize the used item in order to know exactly data that are used to build and update the COFOR. Finally, each predicate is dated in function of verbal exchanges appeared.
Items are "data", "problem", "strategy", "solution" and "implementation".
Variables are "raw" or "way" for the item data; "aircraft", "flight level", "constrain" or "place" for the item "problem"; "pass behind", "direct", etc for the item strategy; instructions for items solution and implementation.
Four predicates are used for encoding verbal exchanges: SIMPLE_SUPPLY, SIMPLE_REQUEST, INTERFERENCE_DETECTING and INTERFERENCE_SOLUTION. The two first predicates translate activities of building and updating of the CWS. INTERFERENCE_DETECTING: correspond at the detecting of interference between the frame of reference of a controller and the COFOR. Finally, INTERFERENCE_SOLUTION: correspond to activities which reduce difference between one or several items between the frame of reference of a controller and the COFOR.
All of those predicates contain four arguments: controller, problem number, item and variable. In addition, INTERFERENCE_SOLUTION contain the type of used method to erase the interference: ACCEPTANCE, IMPOSITION and NEGOTIATION.

4 Study results

As the encoding of verbal exchanges had been defined to analyse a precise activity, all of exchanges haven't be encoded. Indeed, twenty percents of verbal exchanges can't be used. Therefore, the rest, eighty percents of verbal exchanges have been encoding as COFOR' activities and confirm the existence of them. Seventy-three percents of activities are simple_supply and simple_request: operators perform tasks of information of our frame of reference. This singularity means activities of build and update of the COFOR. Beyond different activities that concern the COFOR, this study had permitted to give prominence to items of controllers work more them. There more used items had been selected to build future CWS of support tool. Clusters (representation of a problem to the graphical interface of the support tool) will contain signature of conflict aircrafts (flight number), the flight level, the strategy built by controllers and solutions found by the support tool. Number of simple_supply (58% of verbal exchanges) and there contents prove that controllers have a strong activity of problem updating. Indeed, interfaces of support tool must be transmitted to controllers problem' evolutions in real time.
In addition, it appears that interference_detection are made by a majority of problem data (28%). Indeed, the interfaces had been built to permit controllers take in new data easily. For example, all of new information add by system is highlight while controllers have not carried out them. Interference_solution are made also of problem data. This situation confirms that the build of problem representation is preponderant to solve conflicts.
Finally, as interference_solution are made also of solution, interfaces must permit controllers to build together their solutions. We'll give them also tools to test and improve their solutions. In order to build this function, they will be able to view our solutions thanks to a trajectory simulator tool. Controllers will verify our solutions and also solutions of system. If it's necessary, they will

be able to change instructions before the conflict expiration. This option may avoid urgency interference_detection and improve solution quality which are achieved.

5 Conclusion

We have presented concepts of human-machine cooperation used to the design of support tools for the air traffic control. These concepts are at the base of the study that has been made in order to define functions of a new support tool. Firstly, these works had permitted to confirm precedent proposals of the integration of a Common Work Space into a support tool. Secondly, this study had permitted to specify the content and the functioning of human-machine interfaces that carry out the CWS between human operators and the support tool.

The whole of data had permitted to build the experimental platform AMANDA V2 who integrate the CWS and a support tool which is able to solve a conflict from strategies build by controllers. This platform is actually in test with professional controllers.

References

DEBERNARD S., HOC J-M., (2001), Designing Dynamic Human-Machine Task Allocation in Air Traffic Control: Lessons Drawn From a Multidisciplinary Collaboration. In M.J. Smith, G. Salvendy, D. Harris, R. Koubek (Ed.), Usability evaluation and Interface design: Cognitive Engineering, Intelligent Agents and Virtual Reality, volume 1. (pp. 1440-1444). London: Lawrence Erlbaum Associate Publishers.

DEBERNARD S., CATHELAIN S., CRÉVITS I., POULAIN T., (2002), "AMANDA Project: Delegation of tasks in the air-traffic control domain". In M. Blay-Fornarino, A-M. Pinna-Dery, K. Schmidt, P. Zaraté (Ed.), Cooperative Systems design. pp 173-190. IOS Press, 2002. (COOP'2002)

HOC J. M., (1996), Supervision et contrôle de processus - La cognition en situation dynamique, University' press of Grenoble, France.

LEMOINE M-P., DEBERNARD S., CREVITS I., MILLOT P., (1996), « Cooperation between humans and machines : first results of an experimentation of a multi-level cooperative organization in air traffic control ». Cooperative Supported Cooperative Work: the journal of collaborative computing; Vol. **5**, N°2-3,pp.229-321; December 1996

PACAUX LEMOINE M. P., (1998), Coopération Hommes-Machines dans les procédés complexes: modèles techniques et cognitifs pour le contrôle de trafic aérien. Thesis. University of Valenciennes and Hainaut-Cambrésis, France, 1996

MILLOT P., (1988), Supervision des procédés automatisés et ergonomie, Traité des nouvelles technologies Série automatique, Editions HERMES. Paris, Décembre, 1988.

MILLOT P., (1997), La supervision et la coopération Homme-Machine dans les grands systèmes industriels ou de transport, Colloque national GIS Sciences de la cognition : Sécurité et cognition, Paris, Septembre 1997.

PACAUX-LEMOINE M.P., DEBERNARD S., (2002), Common work space for human-machine cooperation in air traffic control. *Control Engineering Practice.* **10** (5), 2002, pp. 571-576.

RASMUSSEN J., (1983), Skills, rules, and knowledge; signals, signs and symbols, and other distinctions in human performance models, *IEEE Transactions on systems, man cybernetics*, SMC13, n°3, May/June 1983.

SCHMIDT K., (1991), Cooperative Work: a conceptual framework. In J. Rasmussen, B. Brehmer, & J.Leplat (Eds), Distributed decision-making: cognitive models for cooperative work. pp 75-110. Chichester, UK: John Willey and Sons.

A Multidimensional Scale for Road Vehicle Handling Qualities

Don Harris, Jamie Chan-Pensley and Shona McGarry

Human Factors Group, Cranfield University
Bedford MK43 0AL, UK
d.harris@cranfield.ac.uk

Abstract

Advances in technology will provide many options for optimizing the ride/handling compromise of modern cars. With increased computerization and suspensions not dependent upon steel springs there will be many ways to refine a car's behavior. This paper describes the development of a multidimensional scale to assess road vehicle dynamic behavior based upon a technique developed to evaluate aircraft handling qualities. Scale dimensions were extracted and verified from analyses of driver opinion. This was followed by an assessment of scale sensitivity and diagnosticity. The results suggest that the scale developed is sensitive and has content and construct validity.

1 Introduction

Technology will give automotive engineers many more options to choose from when optimizing the ride/handling qualities compromise of future motor vehicles. With increased computerization and suspension systems not dependent upon springs, there will be many ways available to refine their behavior. Computer-controlled hydro-pneumatic suspensions will allow effective spring rates; damping characteristics; roll characteristics; and anti-dive/anti-squat characteristics to be optimized through software changes. With the introduction of 'drive-by-wire' systems, other characteristics of the driver-vehicle interface will also be capable of modification via software changes (e.g. steering gain and feel, and throttle response). With such flexibility becoming available to determine the on-the-road behavior of future vehicles, considerably more refined tools will be required to describe and evaluate these cars. However, such a metric is not currently available. At present, a vehicle's dynamic qualities are evaluated by the comments of highly skilled test drivers and by the measurement of engineering parameters related to ride and handling.

Several years ago a similar situation was faced by the aerospace industry with the advent of 'fly-by-wire' systems. These systems allowed new methods to refine an aircraft's handling qualities. At this time the scale most commonly used to evaluate handling qualities was the one-dimensional Cooper-Harper scale (Cooper & Harper, 1969). A fundamental problem this scale though, was that the complexity of the aircraft's behavior was not reflected in the structure of the scale. It also did not describe the interaction between an aircraft's dynamic qualities and the flight task. Harris, Gautrey, Payne & Bailey, (2000) developed a multidimensional aircraft handling qualities scale (Cranfield Aircraft Handling Qualities Rating Scale – CAHQRS). This built upon concepts from the NASA–TLX (National Aeronautics & Space Administration–Task Load Index) workload scale (Hart & Staveland, 1988) and the Cooper-Harper scale. This took into account the interaction between handling qualities and the task. Trials using the CAHQRS showed it to have greater test-re-test reliability than the Cooper-Harper scale and greater diagnosticity (Harris *et al.* 2000).

In many ways the requirements for the evaluation of vehicle dynamic behavior are similar to those of modern aircraft, although in this case there is an interaction between the type of car and its desirable road behaviour. What may be a desirable characteristic in one type of vehicle (*e.g.* a soft ride in an executive car) may not be so desirable in another category of vehicle (*e.g.* a sports car). This paper describes the development and initial validation of a multidimensional scale to assess vehicle dynamic qualities. The method is based on that used to develop the CAHQRS. The scale requirements were to exhibit good content validity, construct validity and construct reliability.

2 Method: Overview

The method to develop the scale progressed in three phases. Phase 1 elicited the basic dimensions of vehicle ride, feel, performance and handling. Phase 2 replicated the first phase in an independent sample to ensure that these dimensions were stable (*i.e.* they had construct validity). Phase 3 transformed these dimensions into a rating scale to assess the dynamic behavior of road vehicles and also evaluated the scale's sensitivity and discriminant validity.

2.1 Phase 1: Elicitation of Dimensions

Over 400 adjectives to describe the dynamic qualities of cars were collected from road tests published in the motoring press. From these adjectives the 52 most frequently used formed the basis of a series of bi-polar adjective pairs. These were presented to 50 participants who were asked to evaluate their meaningfulness. After initial parsing, 33 bi-polar pairs remained which were made into a short self-completion questionnaire. Each item used a 1-5, (desirable to undesirable) rating scale. Three hundred questionnaires were distributed to motorists. Respondents were asked them to assess the ride and handling qualities of the vehicle they used most frequently.

Table 1: PCA summary statistics for the dimensions of vehicle dynamic qualities (n=233)

Component number	Component Name	Eigenvalue	% of variance post-rotation	Cronbach's alpha
1	Steering Qualities	11.83	15.7	0.91
2	Performance	2.87	14.3	0.91
3	Ride Composure	2.01	10.5	0.80
4	Handling Qualities	1.82	8.8	0.78
5	Ride Comfort	1.28	8.0	0.77
6	Grip	1.20	4.7	0.83
7	Unnamed	1.06	3.7	N/A

Data from the returned instruments were subject to principal components analysis (PCA). Seven components with an Eigenvalue in excess of unity were extracted accounting for 65% of the sample variance. These components were subject to a Varimax rotation. A summary of the post rotation PCA solution, including dimension names, is presented in Table 1. The values for Cronbach's Alpha were all over the minimum acceptable level of 0.7 for internal reliability. As the seventh component extracted was composed of only a single item, this dimension was deleted.

2.2 Phase 2: Verification of Dimensions

The objective of the Phase 2 was to cross-validate the latent structure elicited in the Phase 1 PCA to assess its stability, reliability (internal consistency) and construct validity. All the items describing each of the six dimensions of vehicle behavior were formed into a second questionnaire using an identical response format to that described previously. These were distributed to another independent sample of UK drivers in the same manner as the first questionnaire.

A total of 224 questionnaires were returned in time for analysis. The data were subject to a confirmatory factor analysis (CFA) using maximum likelihood extraction. A slightly modified model of indicator variables from that in the initial PCA produced a model with an extremely good fit to the hypothesized latent structure of vehicle behavior (Chi^2 minimum = 150.97; df=120; p<0.03: adjusted Chi^2 = 1.26: goodness of fit = 0.93: adjusted goodness of fit = 0.90). The structure from the CFA is described in figure 1.

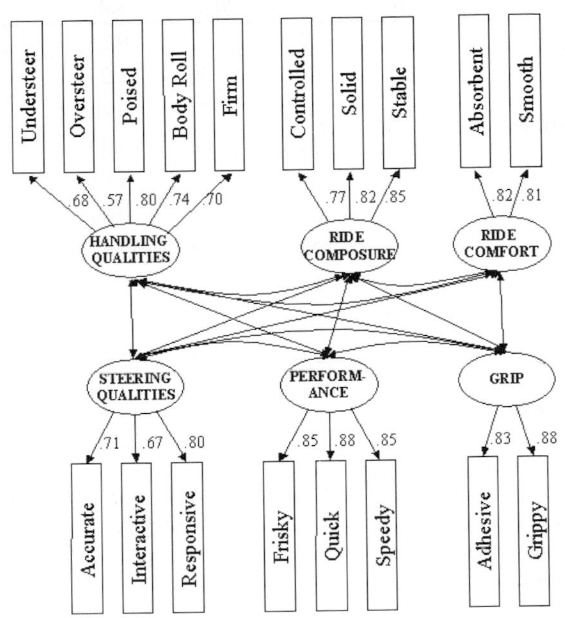

Figure 1: Final CFA factor structure to describe vehicle dynamic qualities. The numbers alongside the arrows are standardized regression weights between latent and indicator variables.

2.3 Phase 3: Scale Construction and Validation

The instrument was comprised of two basic components, one scale rating each dimension of vehicle behavior for its adequacy and a further ranking the importance of each of these aspects with respect to the type of the vehicle (*c.f.* NASA-TLX weighting index). Scale scores were calculated from an average of the scores on items making up each sub-scale. After completing these ratings respondents ranked the importance of each dimension of the vehicle's road behavior with respect to the role/category of the vehicle. These importance rankings ran from 1 (most important) to 6 (least important). To test the sensitivity and validity of the scale it was distributed to drivers of 'superminis'; small hatchbacks; medium-sized cars; small executive cars and small, sporting coupes. Within each of these sub-samples it was ensured that there was a large sample of at least two models, including the class-leading vehicle (as defined in Autocar magazine).

3 Scale Performance

For the instrument to be valid it must be able to discriminate meaningfully between vehicles and the scales must also be sensitive. Scales that discriminate only between disparate categories of vehicle are of limited utility. To test sensitivity, the ratings of dynamic behaviors of at least one other model within each category were compared to the class-leading vehicle. To illustrate, a comparison was made of the European class-leading small hatchback (car A) against the best selling vehicle in this class (car B). These results are shown in Table 2.

Table 2: Comparison of two leading small hatchbacks in terms of their rated dynamic qualities. Note: low figures (range of 1-5) indicate superior dynamic behavior.

	Car	n	Mean	s.d.	t	p
Performance	A	24	1.92	0.53	-2.27	0.03
	B	24	2.46	1.04		
Steering Qualities	A	24	1.71	0.49	-1.74	0.08
	B	24	2.04	0.80		
Grip	A	24	1.75	0.63	-0.88	0.38
	B	24	1.92	0.69		
Ride Comfort	A	24	2.02	0.67	-1.27	0.21
	B	24	2.31	0.91		
Ride Composure	A	24	1.64	0.43	-1.84	0.07
	B	24	1.99	0.82		
Handling Qualities	A	24	1.97	0.50	-2.50	0.02
	B	24	2.48	0.86		

Looking across the whole sample, an analysis of the of the importance rankings indicated that drivers of small, sporting coupes regarded handling, steering and grip as being more important than ride quality. Conversely, drivers of executive cars emphasized ride composure and comfort over performance and handling qualities.

However, the greatest power of multidimensional scales lies in the graphical presentation of scores. When presenting the data in this manner, scale ratings of dynamic qualities were plotted on the y-axis and importance rankings on the x-axis. Short histogram bars indicate more desirable ratings of that aspect of a vehicle's dynamic behavior, and narrow bars indicate higher levels of importance. This emphasizes the interaction between the dynamic qualities and the category of the vehicle. An example is given in Figure 2. In this case, a poorly rated sports coupe (car X) is contrasted with a highly rated executive saloon (car Y). The width of the bars indicate ride comfort and ride composure are regarded as being more important by drivers of executive saloons, whereas drivers of sporting coupes place greater emphasis on grip, steering and handling qualities. The drivers of car Y rated all aspects of their vehicle's dynamic properties superior to those of car X, as indicated by the shorter histogram bars.

4 Conclusions

Using the same approach as in the development of the NASA-TLX and the CAHQRS, a scale has been developed which allows everyday drivers to assess meaningfully the dynamic behavior of their vehicles. The manner by which the scale was developed means that the its content validity should be assured. The replicability of the underlying factor structure is also indicative of good construct validity. The use of a NASA-TLX type of scale format once again shows the advantages of being able to quantify and display the interactive nature of scale ratings within their context. In this case the interaction is between the category of motor vehicle and its dynamic behavior. Drawing conclusions from simple ratings of a vehicle's behavior on each of the sub-scale dimension overlooks the complex interaction with the category of the vehicle and hence the relative importance of each aspect of its dynamic qualities.

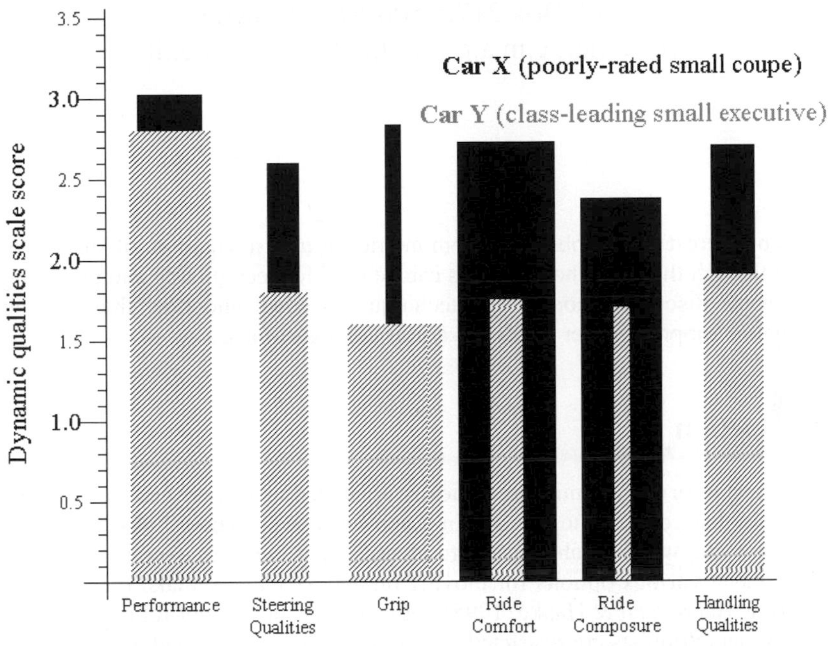

Figure 2: Graphical presentation of the multidimensional vehicle dynamic qualities scale. Low ratings indicate desirable dynamic behaviors. Narrow histograms indicate greater importance.

Further trials using the scale are planned to evaluate its criterion validity and assess its test-re-test reliability. These will use both an engineering simulator and real vehicles in a series of test-track trials. These trials will further assess the scale's sensitivity. However, initial results suggest that there is a sound basis on which to continue further scale development.

5 References

Cooper, G.E. & Harper, R.P. (1969). *The use of pilot rating in the evaluation of aircraft handling qualities.* Report number NASA TN-D-5153. Moffett Field, CA; NASA Ames Research Center.

Harris, D., Gautrey, J., Payne, K. & Bailey, R. (2000). The Cranfield Aircraft Handling Qualities Rating Scale: a multidimensional approach to the assessment of aircraft handling qualities, *The Aeronautical Journal,* 104, 191-198.

Hart, S.G. & Staveland, L.E. (1988). Development of the NASA task load index TLX: Results of empirical and theoretical research. In P.A. Hancock and N. Meshkati (Eds.) *Human Mental Workload* (139-183). Amsterdam: North-Holland.

From Browsing Behavior to Usability Matters

Eelco Herder & Betsy van Dijk

University of Twente, The Netherlands
P.O. Box 217, 7500 AE Enschede
{e.herder, e.m.a.g.vandijk}@cs.utwente.nl

Abstract

Many features of interest can be observed from metrics of user navigation, of site structure and of paths followed through the site. These metrics can be used for recognizing and predicting usability matters. This paper discusses models and techniques needed, and how they can be used for adapting navigation support to user needs in various contexts of use

1 Introduction

Online services and information units are typically organized in a hypermedia structure, consisting of pages and links that connect those pages in a non-linear way. These links form the *navigation infrastructure* of a site, which enables users to identify their current position in the site structure, to distinguish among various options for moving on and to actually make their way through the web site (Thüring, Hanneman & Haake, 1995). Obviously, users cannot follow non-existing links. In a similar way, automobilists are restricted to following the existing road infrastructure.

Highways are built to have people reach their destination as fast as possible. Local roads are usually more time-consuming, but more interesting. When one is in a hurry, it is frustrating if there is no option but a scenic route. Correspondingly, when you need to carry out online banking activities in limited time, you want to have immediate access to those services, instead of having to wade through a sea full of interesting, but currently irrelevant information.

When driving an old Volkswagen Beetle with uncomfortable seats, a blurred windshield and malfunctioning handles, you might want to reach your destination as fast as possible. Ironically, using this car prevents you from doing so. In a similar way, mobile devices provide limited means of interaction, and they are typically used while on the move and while having to deal with time constraints. Studies have shown that mobile users display a more conservative link exploration behavior (Buyukkokten, Garcia-Molina, Paepcke & Winograd, 2000). This behavior, though it may seem efficient at first sight, is typically less effective than regularly returning to navigational landmarks (McEneaney, 2001).

Adapting the navigation infrastructure to the user and to the context of use is seen as a proper way of providing adequate navigation support for varying user needs (Brusilovsky, 2001). However, most web applications are designed for a prototypic user, accessing the service with a browser on a high-resolution monitor, using sophisticated means of interaction, ignoring the fact that mobile internet access is increasing (Menkhaus, 2001). The only way of providing personalized navigation support for these applications is by adapting the existing link structure, without requiring or expecting an author's effort for link annotation or the like.

In this paper we present an intermediary-based link adaptation strategy, making use of site structure and navigation paths. We explain the concept of *lostness in hyperspace*, which covers many important usability matters, and how these matters can be inferred from user navigation through a web site. We conclude with a brief description of an agent that acts as an intermediary between information providers and users

2 Lostness in Hyperspace

Being lost is considered as one of the fundamental difficulties which users experience when interacting with hypermedia systems. Therefore, much hypermedia research has been devoted to this issue (Otter & Johnson, 2001). At some point users may not know where they are, how they got there or where they should go next – or, alternatively, they know *where* they want to go, but not *how* to get there (Thüring et al., 1995). These problems may be caused by *cognitive limitations* – such as low spatial ability and memory capacity (Neerincx, Lindenberg & Pemberton, 2001), by *unfamiliarity* with the conceptual structure of a site (Otter & Johnson, 2001), by problems related to the *user context* – such as device characteristics, time constraints, task-set switching, by *improper orientation clues or navigation facilities* within an information domain (Thüring et al, 1995) or by a combination of these factors.

When lost, navigation behavior usually changes. Users might arrive at a particular point and forget what was to be done there, they might neglect to return from a digression, or they might miss some pages that contain relevant information (Otter & Johnson, 2001). As a result, navigation performance and user satisfaction will drop dramatically (Smith, 1996). Therefore, although the term 'lostness' applies to the interaction *process*, it is clearly related to the interaction's *outcome* (Otter & Johnson, 2001).

The problem of lostness is even worse for mobile devices. Because of the small screen size, users can be given less orientation clues than in a desktop setting. To make things worse, information displayed on small screens needs to be split in small chunks, thus requiring the user to select and scroll more intensively, which only adds to the problem. As a result, users may not have confidence in being able to use the resources to achieve their goals if it involves tedious and error-prone navigation (Jones, Buchanan, Marsden & Pazzani, 2001).

3 Metrics of Browsing Behavior

From user navigation many interesting features can be inferred, which indicate navigation strategies chosen and usability problems encountered. However, this is not feasible without proper models of user navigation and quantitative representation of its characteristics, which are presented shortly in this paragraph. A more complete overview is given in (Herder, 2002).

The node-and-link structure of online content lends itself to be modeled as a graph – the *site graph*. User navigation can be seen as an overlay of the site graph, consisting of only the pages visited and the links followed – the *navigation graph*. This navigation graph can contain information about one user session, about one particular user or about an entire user group. Each edge/link can be assigned a weight, which indicates the distance between the source node and the destination node. Several notions of distance can be applied, such as textual similarity and transition frequency – which expresses the probability a user will follow the link (Herder, 2002). The graph structure reveals many features which have their impact on user navigation.

On a detailed level, different types of pages can be recognized by the linkage to and from a page. Pirolli (Pirolli, Pitkow & Rao, 1996) classified pages in four main categories (home pages, index pages, reference pages and content pages). Index pages, for example, are usually quite small and have a large number of outgoing links. Kleinberg (Kleinberg, 1999) deploys the link structure for recognizing so-called *authorities* and *hubs* – pages that are seen as important and pages that link to many important pages respectively. Although links within a site can hardly be seen as reliable indicators of the significance of a page, they do convey semantic relationships between pages. Following such a link can be regarded as recognizing the relationship as interesting.

Seen from a higher level, pages can be grouped into *clusters*, richly interconnected parts of the site graph dedicated to a topic with but a few links to other topics (Pirolli et al., 1996). Global characteristics of each cluster, or of each site as a whole, can be calculated based on the link structure. One of the most obvious measures is the size of a cluster or a site. *Cluster size* and the *average connected distance* (Botafogo, Riflin & Shneiderman, 1992) are indicators of the average navigation effort required to reach a certain page. For each individual page its *depth*, the distance from a navigational landmark, indicates the effort required to reach this page. The average *net density*, the ratio between the number of pages and the number of links, and the *distribution* of the links indicate the amount of freedom in navigation offered. The more links, the more complex an interaction will be. Complexity can be reduced by making a site more *linear* (Botafogo et al., 1992). Besides global characteristics, interesting spots within the graph can be located as well – such as groupings of navigation landmarks, cycles and central and peripheral areas.

Users have been shown to display different navigation strategies, varying from goal-directed to explorative (Shneiderman, 1997). We expect these strategies can be recognized from global characteristics of user navigation. These characteristics include *path length*, *path linearity* and *path density* (McEneaney, 2001) – the latter two measures indicate the amount of backtracking a user performs. Global characteristics as well as other interesting phenomena – such as a user navigating in cycles – can be discovered from the navigation graph itself. On the other hand, one needs to know the infrastructure in which a user is navigating to be able to successfully interpret the navigation. Did a user return to a navigational landmark or to a content page? Furthermore, it would be interesting to know the correlations between navigation behavior and site structure. Do more densely linked sites invite less linear navigation strategies? Does site linearity shorten navigation paths? Do link junctions or dead-ends lead to usability problems? Modeling user navigation as an overlay of site structure facilitates dealing with these issues, as it allows for comparison.

4 From Metrics to Usability Matters

As pointed out earlier, lostness is seen as a fundamental usability problem in hypermedia research. However, the term covers a wide range of user problems, which are unlikely to be captured by a single measure. Otter (Otter & Johnson, 2001) suggests that we could make use of a 'battery of measures' instead, which correlate well with one another and which have been shown through empirical studies and in the literature to measure lostness to some degree.

Unfortunately, contradictory results can be seen from the literature. Smith (Smith, 1996) bases her measures for lostness on the assumption that users who cannot locate information they require and which exists in the system, can be regarded as lost. She proposes a lostness measure based on degradation of user performance, which is indicated by an increase of page revisits and by a large number of navigation steps compared to the optimal route. The usefulness of this measure is said to be confirmed in an experimental setting. McEneaney (McEneaney, 2001), on the other hand, directly compares *patterns* of navigation and search outcome measures, therewith taking into account spatial and temporal features of user paths. From his experiments it turned out that less successful users adopted a "page turning" strategy, while subjects who did well on the search task often revisited navigational landmarks, such as a table of contents. These contradictory results cannot be explained by system goals, since both authors used a system that provided teenagers study advices.

Several explanations for these contradictory results can be thought of. Most importantly, neither McEneaney nor Smith related user navigation to the site structure. As argued above, navigation infrastructure directly influences users' navigation strategies. McEneaney mentioned the presence of a reading sequence using 'next' and 'previous' buttons in his system. Further, we do not know to what types of pages users returned, or how much time they spent on each page.

The metrics described in the previous paragraph provide data on user navigation within a site structure. From these data user navigation strategies and usability problems can be inferred. As mentioned before, users display various kinds of navigation strategies, varying from goal-directed activities to explorative browsing. These various navigation types call for different types of navigation assistance. Strongly hierarchically structured sites appear to be more suitable for goal-directed activities (Modjeska & Marsh, 1997), as they provide better structured orientation clues (Thüring et al., 1995). Cross-references do increase navigational efficiency and provide more freedom for exploration, but at the cost of some disorientation (Modjeska & Marsh, 1997).

Taking the strategy employed into account, problems related to lostness – as described in the second paragraph – can be predicted from the site structure (for example, too many cross-references) or observed from the navigation path. Insight in relationships between user path metrics, site metrics and interaction success measures is expected to be obtained from the results of an experiment we are planning to conduct. Since we are using numerous metrics for predicting usability matters from site structure and user navigation, machine learning techniques – such as Bayesian Learning (Zukerman & Albrecht, 2001) – appear to be useful for this purpose. The actual design of the learning algorithms will depend highly on the conclusions drawn from our experiment. Once the usability matters can be inferred reliably enough, one can seek suitable adaptation strategies to solve them.

5 From Usability Matters to Adaptation Strategies

In this paper we discussed how usability matters can be inferred from metrics of user browsing behavior in a site structure. This diagnostic task is the first step to be taken for adapting the navigation infrastructure to user needs in various user contexts, and to recognize or predict usability problems related to lostness in hyperspace. The next step is to find suitable adaptation strategies for the usability matters found. On page level, these adaptations can be created using common adaptive hypermedia techniques, such as link ordering, link hiding, link highlighting, link annotation and direct guidance (Brusilovsky, 2001). On site level, these adaptations can be seen as a personalized sub graph imposed on the site graph, which can be evaluated by comparing user navigation and the personalized site structure once again (Herder, 2002).

In the PALS project[1] (Lindenberg, Nagata & Neerincx, 2003) we are designing an intermediary agent that provides this functionality. The agent employs site crawling and web logging techniques to gather information on site structure and on user navigation. We opted for an intermediary-based approach for two reasons. First, as mentioned in the introduction, we constrained ourselves not to make use of server-side adjustments. Second, since we are looking for means to enhance user experience in both desktop and mobile contexts, methods that do not require intensive processing on the client side are preferred, as it might be a device with limited

[1] PALS stands for Personal Assistant for online Services; the project is supported by the Dutch Innovative Research Program IOP-MMI. Our research partners are TNO Human Factors and the University of Utrecht.

processing power (Buyukkokten et al., 2000). Moreover, an intermediary-based approach facilitates the sharing of knowledge on a user's general browsing behavior, so that it will be easier to recognize and to deal with usability matters on various devices and in various contexts of use.

6 References

Botafogo, R.A., Rivlin, E. & Shneiderman, B. (1992). Structural Analysis of Hypertexts: Identifying Hierarchies and Useful Metrics. *ACM Transactions on Information Systems* 10 (2), 142-180.

Brusilovsky, P. (2001). Adaptive Hypermedia. *User Modeling and User-Adapted Interaction* 11, 87-110.

Buyukkokten, O., Garcia-Molina, Paepcke, A & Winograd, T. (2000). Power Browser: Efficient Web Browsing for PDAs. *Proceedings of the CHI 2000 Conference*, The Hague, 430-437.

Herder, E. (2002). Metrics for the Adaptation of Site Structure. *Proc. of the German Workshop on Adaptivity and User Modeling in Interactive Systems, ABIS02*, Hannover, 22-26.

Jones, M., Buchanan, G., Marsden, G. & Pazzani, M. (2001). Improving Mobile Internet Usability. *Proc. of the 10th Intl. WWW Conference*, Hong Kong, 673-680.

Kleinberg, J. M. (1999). Authoritative Sources in a Hyperlinked Environment. *Journal of the ACM* 46 (5), 604-632.

Lindenberg, J., Nagata, S. & Neerincx, M. (2003). Personal Assistant for Online Services: Addressing Attention, Emotion and Navigation. *Workshop on Human Information Processing and Web Navigation, HCI 2003*, Crete (to appear).

McEneaney, J. E. (2001). Graphic and numerical methods to assess navigation in hypertext. *Intl. Journal of Human-Computer Studies* 55, 761-786.

Menkhaus, G. (2001). Architecture for client-independent Web-based applications. *IEEE Proceedings of the TOOLS-Europe Conference 2001*, Zürich.

Modjeska, D. & Marsh, A. (1997). Structure and Memorability of Web Sites. *Working Paper in the Department of Industrial Engineering,* University of Toronto.

Neerincx, M.A., Lindenberg, J. & Pemberton, S. (2001). Support Concepts for Web Navigation: a Cognitive Approach. *Proc. Of the 10th Intl. WWW Conference*, Hong Kong, 119-128.

Otter, M. & Johnson, H. (2001). Lost in hyperspace: metrics and mental models. *Interacting with Computers* 13 (1), 1-40.

Pirolli, P., Pitkow, J. & Rao, R. (1996). Silk from a Sow's Ear: Extracting Usable Structures from the Web. *Proceedings of the CHI 1996,* Vancouver, 118-125.

Shahabi, C., Zarkesh, A., Adibi, J. & Shah, V (1997). Knowledge Discovery from Users Web-page Navigation. *Proc. of 7th Int. Workshop on Research Issues in Data Eng. on High Performance Database Management for Large-Scale Applications*, Birmingham, UK.

Shneiderman, B. (1997). Designing Information-Abundant Websites: Issues and Recommendations. *International Journal of Human-Computer Studies* 47 (1).

Smith, P.A. (1996). Towards a practical measure of hypertext usability. *Interacting with Computers* 8 (4), 365-381.

Thüring, M., Hanneman, J. & Haake, J.M. (1995). Hypermedia and Cognition: Designing for Comprehension. *Communications of the ACM* 38 (8), 57-66.

Zukerman, I. & Albrecht, D.W. (2001). Predictive Statistical Models for User Modeling. *User Modeling and User-Adapted Interaction* 11, 5-18.

A Formal Method for Analysing Field Data and Setting the Design Requirements for Scheduling Tools

Peter G. Higgins

School of Engineering and Science
Swinburne University of Technology
Hawthorn, Australia, 3122
phiggins@swin.edu.au

Abstract

A methodology is presented that can be used to analyse scheduling behaviour by formally structuring field data collected from the manufacturing environment. The method links the tasks used in controlling production to the environmental factors to various levels of abstraction in regard to ends and means. It is an extension of Cognitive Work Analysis (CWA), which incorporates two different types of analysis: Work Domain Analysis (WDA) and Control Task Analysis (CTA). WDA uses an abstraction hierarchy (AH) - a generic framework for describing goal-oriented systems - to describe a system in a way that distinguishes its purposive and physical aspects. WDA is event independent and is quite separate from Control Task Analysis (CTA), which is a subsequent event dependent analysis of the activity that takes place within a work domain. The discussion leads to the expected benefits for tool designers and tool users from of using this methodology.

1 Introduction

In intermittent job shops or industries where there is inherent uncertainty or where human judgment is necessary, production scheduling is generally acknowledged to be a skilled craft practised by experienced human schedulers (McKay & Wiers, 1999). They are practical persons who understand the capabilities of machines and work practices within their domain. Their decision-making behaviour is mostly rational and goal directed (Higgins, 1999). Their decision strategies are often complex and hinge on their awareness of the subtle relationships between factors that comes from an intimate knowledge of the plant, products, and processes (Higgins, 2001). Knowledge and intuition, gained through years of first-hand experience, are the principal tools they employ to generate and maintain satisfactory schedules (Rodammer & White, 1988). They produce realistic schedules by balancing many competing and conflicting goals. Numerous constraints imposed by *environmental factors* restrict the degrees of freedom on decision choices. Their goals and the means they use to assess performance depend upon context. Various *cognitive factors* also influence decision-making activities. Their behaviour is situated activity embedded in the particular work environment (Suchman, 1987). They have to make effective decisions in circumstances where there is no clear prediction of the state of the system, within an environment in which information regarding jobs, materials and resources are ill defined and scheduling goals are diverse. In Suchman's terms, there is an "irremediable incompleteness" of instructions. They act pragmatically. They don't generate alternative schedules and then compare their strengths and weaknesses. Instead, they recognise typical situations and ways to respond. Predisposed towards actions that require little expenditure of time and cognitive effort, they behave like Klein's (1989) proficient decision maker, who evaluates possible responses one at a time. He argues that proficient decision makers try to anticipate what would happen if they carry out a specific action, by imagining its execution in the specific working environment. For simple cases, they easily recognise the situation and know straight away how to act. Klein calls this recognition-based reaction. It is similar to Rasmussen's (1986) rule-based level of performance. For more complex cases, they consciously evaluate feasible choices.

To support their decision-making behaviour, schedulers need software tools that help them seek patterns within data, recognise familiar work situations, and explore different decision-making strategies under novel circumstances (Higgins, 1999). How can designers develop software tools to support schedulers, if human behaviour in job-shop scheduling is too complex for complete specification by rules? Make the design focus the support of situated activity. Preserve their initiative to evaluate with utmost control: place them at the centre of the decision architecture.

2 Design Requirements via Modelling Decision Making

Development of the design requirements of software to support scheduling as practised depends upon an analysis of the socio-technical system: the manufacturing system and the problem-solving operations of human decision-makers. The analysis is posited on a systems-thinking context in which production schedulers, whose decisions are perceived to be rational and goal-directed. It uses a formal language that encompasses both the engineering system and the problem-solving operations of the human decision-maker. It has a framework that acts as a template on which to plot the information used in the making of decisions. In forming structural models of scheduling activities on which a human-computer scheduling system can be designed, it is not necessary to detail the actual mental processes engaged by schedulers. A model has to support activities associated with the conceptual models held by domain experts, without necessarily preposing their mental models (Wilson and Rutherford, 1989).

Higgins (1999) proposed a formal methodology for representing the sources and forms of information used by decision-makers. His method extends Cognitive Work Analysis (CWA), developed by Rasmussen (1986) and further elaborated by Sanderson (1998) and Vicente (1999). CWA is a pragmatic systems-based approach to the analysis, design, and evaluation of human-computer interactive systems. It is a structural model of the activities decision makers may use (Rasmussen, 1986). It is a mechanism for generating descriptions of system purpose and form, explanations of system functioning and observed system states, and predictions of future system states. It incorporates two different types of analysis: Work Domain Analysis (WDA) and Control Task Analysis (CTA). Because scheduling is an intentional process, goals vary. Higgins (1999, 2001), therefore, added a goal structure as a third component to the analyses.

WDA describes a system of resources in a way that distinguishes its purposive and physical aspects; its concern is "what" is being acted upon. The description is at different levels of abstraction (see Figure 1). The manufacturing processes are expressed in different languages at the levels of *physical function* and *physical device*. The construction of a WDA can begin at any level. An expedient starting point is the level of physical function. The generic physical functions required for processing production orders are depicted diagrammatically. Under this is a depiction of the manufacturing process in terms of physical devices. Arcs drawn from the level of function to device show the physical means for achieving the functional ends. These physical functions can be associated with physical devices. Each job has its own mapping between these levels, as job shops are characterised by diversity of product. The purpose-related function level depicts the purpose of the production facility. It sets the mapping between functions and devices to the job's specific requirements. The priority/values are the criteria used to measure the performance of the system in regard to its functional purpose. The hierarchy is one of means and ends. For a particular level, the level below depicts the means for achieving its ends.

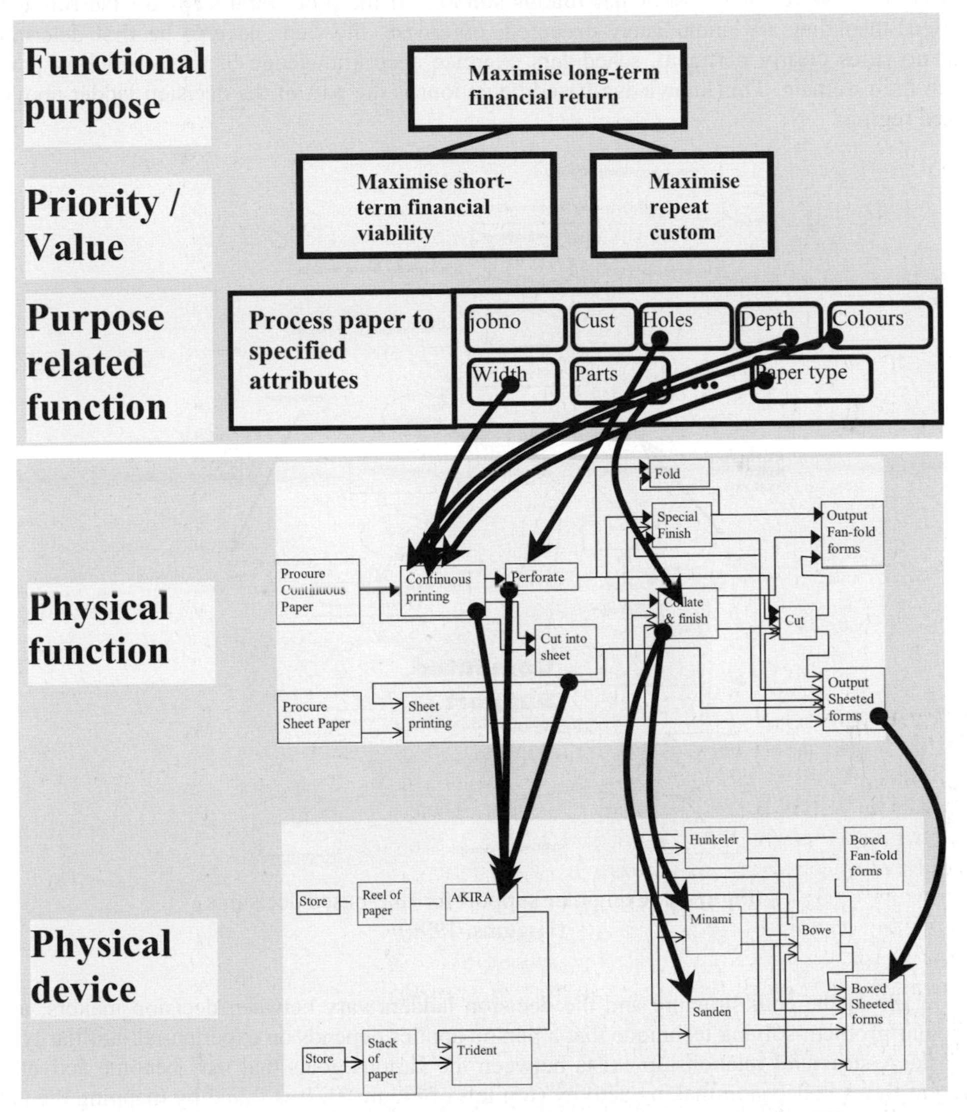

Figure 1: Work Domain Analysis for scheduling printing presses showing feasible means-ends links for a particular job specification

CTA is an event *dependent* analysis that shows how activities are directed towards specific goals. It has a structure of recognition-action cycles that provides a framework for locating information used in decision-making. Using concepts of skills, rules and knowledge, decision-making processes can be structured in a framework made up of recognition-action cycles: known as Rasmussen's decision ladder. It focuses on the user's mental context. For each specific goal, a 'decision ladder' is used as a template to represent the control activities associated with decision behaviour that is directed towards the goal. Support for rule-based decisions relates to the shaded region in Figure 2 (Higgins, 1999). On observing cue-patterns in the data, a decision maker (person or software) sees a particular rule as suitable. If the procedural steps for the rule can be recalled, then they are immediately executed, otherwise, the steps have to be first determined. With no rules clearly pertinent, schedulers exercise deep knowledge of the scheduling practice within their domain. This knowledge-based behaviour is the part of the decision ladder above the shaded region.

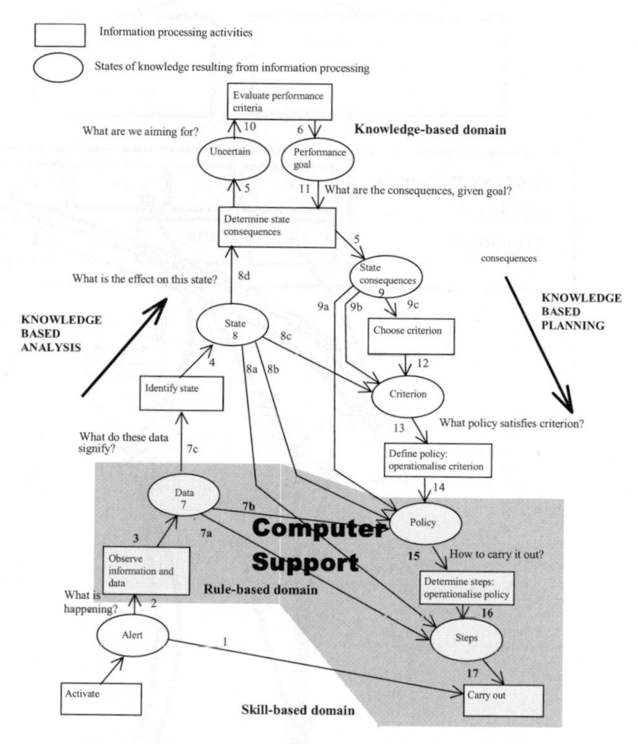

Figure 2: Computer support for rule-based decisions (Higgins, 1999)

The details in the goal structure and the decision ladders vary between decision makers, as the particular problem-solving technique that a person applies depends on experiential familiarity with the task. A structural relationship exists between the various goals that may become activated at different times in decision-making activity (top left of Figure 3). It is found by mapping the actual operational objectives, which form the 'ultimate goals' in the various decision ladders, to goals at higher levels of abstraction. The higher a goal is up the hierarchy, the less directly it relates to immediate operational activity. High-level goals tend to be attained through satisfaction of low-level goals, rather than being directly linked to 'ultimate goals' of decision makers. The relationship between a decision ladder, the goal structure and the abstraction hierarchy is shown in Figure 3. There are links between the high-level goals and the functional purpose and the priority/values in the 'means-ends' abstraction hierarchy. The apex of the goal structure coincides with the functional purpose level of the abstraction hierarchy, and the level immediately below coincides with the highest-level priorities in the abstraction hierarchy.

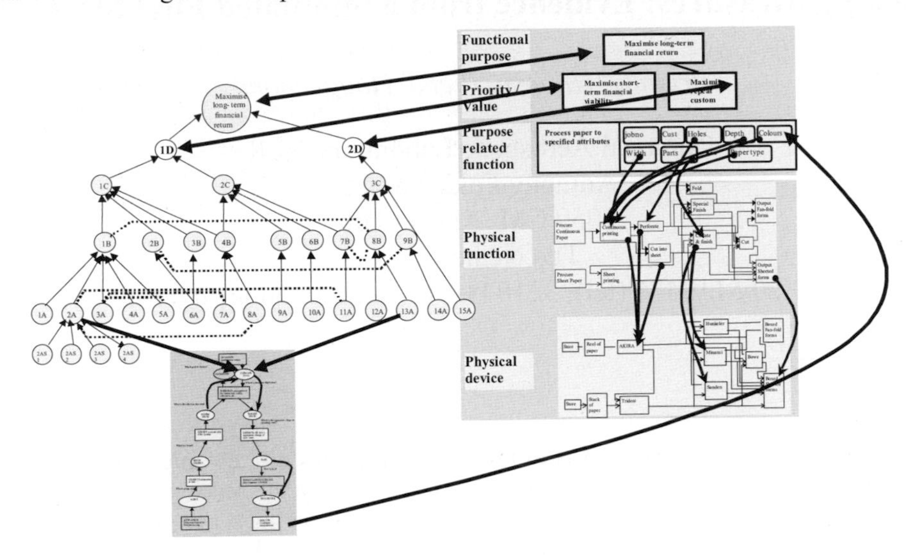

Figure 3: The relationship between the goal structure, decision ladder and abstraction hierarchy.

References

Higgins, P. G. (1999). *Job Shop Scheduling: Hybrid Intelligent Human-Computer Paradigm.* Ph.D. Thesis, The University of Melbourne, Australia.

Higgins, P. G. (2001). Architecture and Interface Aspects of Scheduling Decision Support. In B. MacCarthy and John Wilson (Eds.) *Human Performance in Planning and Scheduling* (pp. 245-281). London: Taylor & Francis.

Klein, G. (1989) Recognition-primed decisions. In W. B. Rouse (Ed.) *Advances in Man-Machine Systems Research, Vol. 5* (pp. 47-92). Greenwich, Connecticut: JAI Press.

McKay, K.N., & Wiers, V.C.S. (1999). Unifying the Theory and Practice of Production Scheduling. *Journal of Manufacturing Systems,* 18 (4), 241–255.

Rasmussen, J. (1986). Information Processing and Human Machine Interaction: An Approach to Cognitive Engineering. New York: North-Holland.

Sanderson, P. M. (1998). Cognitive work analysis and the analysis, design, and evaluation of human-computer interactive systems. In P. Calder and B. Thomas (Eds.) *Proceedings 1998 Australian Computer Human Interaction Conference, OzCHI'98* (pp. 220-227). IEEE.

Suchman, L. A. (1987). Plans and Situated Actions: The Problem of Human-Machine Communication. Cambridge UK: Cambridge University Press.

Wilson, J. R., & Rutherford, A. (1989). Mental Models: Theory and Application in Human Factors. *Human Factors*, 31 (6), 617-634.

Vicente, K. J. (1999). *Cognitive Work Analysis: Towards Safe, Productive, and Healthy Computer-based Work*, Hillsdale, NJ: Lawrence Erlbaum Associates.

Evaluating Crew Interaction via Task Performance and Eye Tracking Measures: Evidence from a Simulated Flightdeck Task

Brian G. Hilburn, Piet J. Hoogeboom

National Aerospace Laboratory NLR
Anthony Fokkerweg 2
1059 CM Amsterdam
The Netherlands
+31 20 5113476 (tel); 3210 (fax); {hilburn, hoog}@nlr.nl

Abstract

The chief aim of the present study was to investigate measures of crew interactivity. This study was conducted in the context of a larger research project aimed at the development of measures for aircrew shared situation awareness (sSA). Pairs of participants took part in a PC-based flight simulation that involved the coordinated performance of several flight-related tasks. Taskload and Interactivity were varied within participants. Dependent measures included aspects of overt performance, as well as eye tracking behaviour. Whereas taskload manipulations were reflected in performance measures (as well as in pupil diameter and blink rate aspects of eye tracking), interactivity effects were apparent only via eye tracking measures. Although crew interaction and shared situation awareness appear important elements of task performance in many settings, these results suggest that overt performance measures might not necessarily be the best means to assess them.

1. Introduction

Situation Awareness (SA) is a currently popular construct used to describe an operator's comprehension of complex and dynamically changing system dynamics. The term grew out of the tactical fighter domain, in which a pilot's ability to continuously comprehend various system characteristics (vehicle energy state, opponent's location, etc.) could mean the difference between life and death. The prominent role of SA in current human factors research is due, in large part, to the increasingly cognitive nature of human-machine tasks.

The current lack of a consensus definition of SA is reflected in the range of measurement techniques used to assess it. Techniques have ranged from self-report measures, to over-the-shoulder evaluations of overt task performance, to various "screen-freeze" procedures (e.g. requiring an air traffic controller to reconstruct the traffic pattern from memory (Sollenberger & Stein, 1995)), to the use of psychophysiological measures such as P300 (Endsley, 1995) or eye point of gaze (Smolensky, 1993, Durso et al., 1995).

Until now, most theoretical and empirical research into SA has tended to focus on the single operator context. The present study was, first, an experimental investigation into the assessment of *shared* Situation Awareness (sSA) on the civil flightdeck. This was done by manipulating team interactivity, as an indirect manipulation of sSA. Second, the study sought to evaluate the potential utility of eye tracking measures in this regard.

2. Method

A total of 24 non-pilot participants, ranging in age from 20 to 38 ($\bar{x}$ =25.1), took part in this study. Twelve two-person crews were randomly created, with team members randomly assigned to the position of either Captain or First Officer.

The Multiple Attribute Task (*MAT* (Comstock and Arnegard, 1992)) battery was run on a single desktop PC. A video splitter permitted the simultaneous display of the MAT on two side-by-side 17-inch colour monitors. The MAT, as displayed in Figure 1, presents four subtasks, each of which taps a different aspect of cognitive-motor performance: Compensatory joystick tracking; System monitoring; Radio communications; and Resource (i.e. fuel) management. The First Officer made inputs via a single joystick (for the tracking subtask), and several keyboard keys (for the monitoring and resource management subtasks). The Captain made inputs (for the Resource Management subtask) via the mouse.

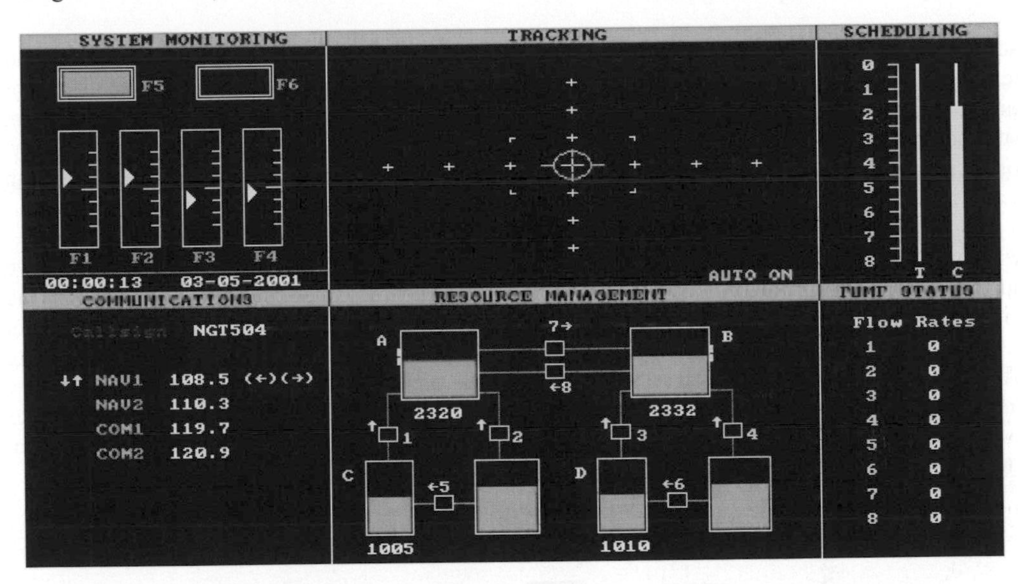

Figure 1. The Multi-Attribute Task (MAT) Battery, as displayed onscreen.

Eye tracking measures were collected using NLR's GazeTracker system. GazeTracker is a free head system that makes use of corneal reflection video recording, and a magnetic head tracking system, to provide a theoretical eye point of gaze resolution of 0.5 degrees.

A 2x2 within-subjects repeated measures design was used. *Interactivity* (Co-operative versus Solitary) was manipulated by presenting participants one of two alternative sets of instructions, the first designed to encourage overall team performance, the second individual subtask performance. *Taskload* (Low versus High) was manipulated via scripted changes in task difficulty (both frequency of system monitoring events, and amplitude of the tracking subtask forcing function). Dependent variables consisted of four performance measures (Tracking RMS error; Monitoring response time, hit rate and false alarm count; and Resource Management mean RMS error), as well as various eye tracking related measures (e.g. fixation time, pupil diameter, blink rate).

3. Results

3.1 Performance measures

A two-way repeated measures ANOVA showed a significant main effect of Taskload (Low versus High) on performance for all three subtasks. High taskload was associated with worsened performance in terms of Mean Tracking RMS error number ($F(1,11) = 70.18$, $p < .05$)), Fuel management RMS error ($F(1,11) = 39.01$, $p < .05$), and System Monitoring false alarm count ($F(1,11) = 16.44$, $p < .05$)). These effects are portrayed in figures 2 through 4, respectively.

The effect of Interactivity was not significant for any of the dependent measures, although a trend was found toward decreased tracking error and increased monitoring hit rate under co-operative work.

3.2 Eye tracking measures

The Taskload manipulation turned out to be less apparent in the eye tracking measures than had been expected. Pupil diameter was the only measure that showed a significant main effect of Taskload ($F(1,15)=8.09$, $p<.05$). Figure 5 shows the average pupil diameters (standardised within participant) for both the Captain and First Officer, as a function of both Taskload and Interactivity. Notice in this figure that increased pupil diameters (i.e. larger data values) are associated with greater indicated workload.

Results revealed a significant effect of Team Interactivity on a number of parameters of the eye tracking data: dwell time, fixation duration and number of fixations of the captain on the Monitoring window, $F(1,10) = 16.72$, $p <.05$; $F(1,10) = 16.72$, $p<.05$; and $F(1,10) = 17.56$, $p<.05$, respectively. Table 2 shows that different modes of team interaction led to corresponding differences in number of fixations and average dwell time of the captain on his own window (Resource management) and on the windows of the co-pilot (Monitoring and Tracking). Further there was a significant main effect of Team Interactivity on the transition rate of the captain, $F(1,10) = 19.76$, $p<.05$.

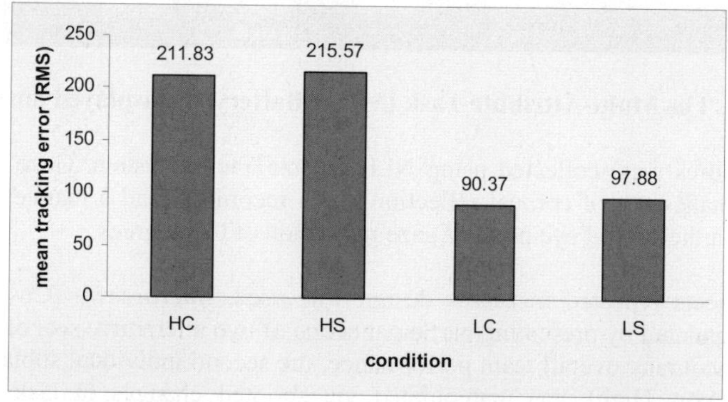

Figure 2. Mean Tracking RMS Error, by Taskload and Interactivity (H/L= High/Low Taskload; C/S=Co-operative/Solitary).

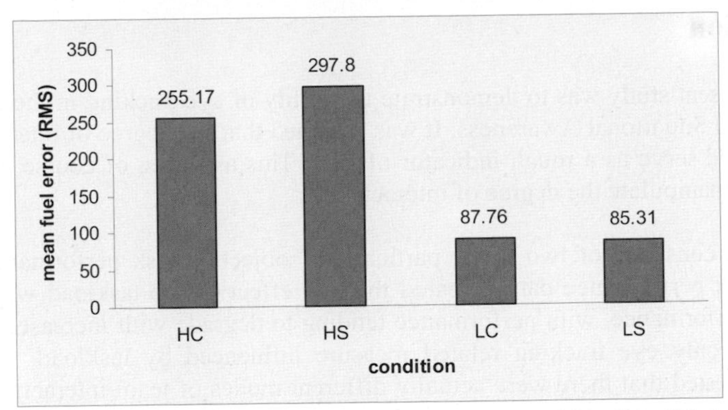

Figure 3. Mean Fuel Management RMS Error, by Taskload and Interactivity (H/L= High/Low Taskload; C/S=Co-operative/Solitary).

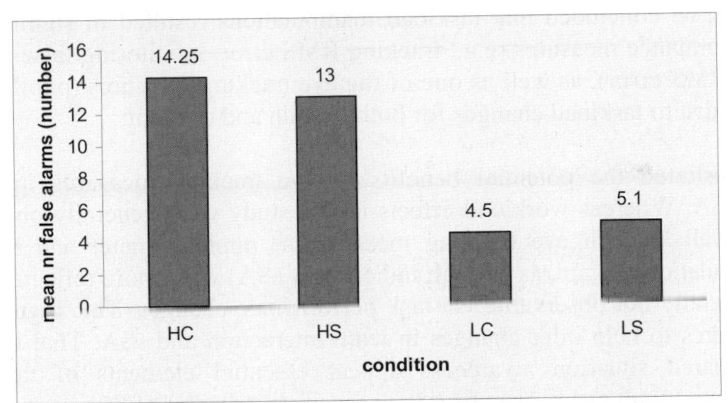

Figure 4. Mean System Monitoring False Alarm Count, by Taskload and Interactivity (H/L= High/Low Taskload; C/S=Co-operative/Solitary).

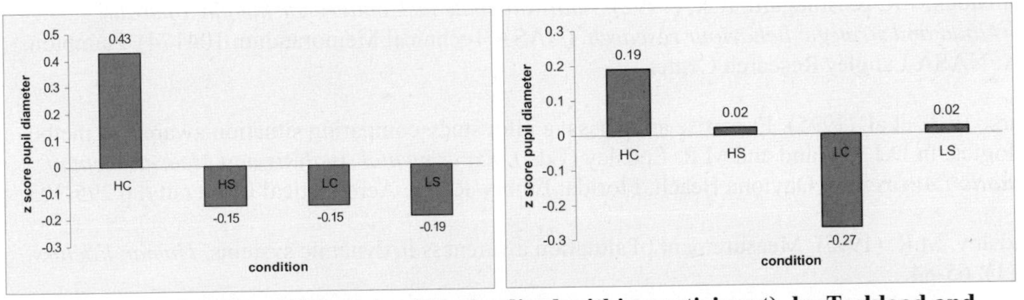

Figure 5. Mean Pupil Diameter (standardised within participant), by Taskload and Interactivity
(note that greater pupil diameters represent higher indicated workload)
(H/L= High/Low Taskload; C/S=Co-operative/Solitary).

4. Discussion

The aim of the present study was to demonstrate the utility of eye tracking in the measurement of flight crews' shared Situational Awareness. It was assumed that the degree of interactivity between crewmembers could serve as a rough indicator of sSA. This assumes, of course, that instructions were sufficient to manipulate the degree of interactivity.

The obtained data consisted of two parts: participants' objective task performance, and the eye tracking data. Task performance data revealed that the effect of the taskload was significant on most aspects of performance, with performance tending to degrade with increased taskload. Pupil diameter was the only eye tracking related measure influenced by taskload. Further, the eye tracking data suggested that there were actually different modes of team interactivity. Under high interactivity (i.e. the Co-operative condition), there were significantly more and longer fixations, and a longer dwell time of the Captain on the First Officer's displays, compared to the solitary (low interactivity) condition.

In summary, it can be concluded that taskload manipulations resulted in significant changes in objective task performance measures (e.g., tracking RMS error, monitoring false alarm count, and fuel management RMS error), as well as one of the eye tracking measures (pupil diameter). Pupil diameter was sensitive to taskload changes for both captain and co-pilot.

This study demonstrated the potential benefits of eye tracking measures in assessing team interactivity and sSA. Whereas workload effects in this study were generally observable via task performance (as well as such eye tracking measures as pupil diameter and blink rate), team interactivity manipulations (again, as a rough indicator of sSA) were more reflected in eye tracking measures, and generally not observable via task performance changes. This argues for the use of eye tracking measures to help infer changes in team interaction and sSA. That is, although team interaction and shared situation awareness appear essential elements of task performance, performance itself might not necessarily be a very sensitive indicator of these constructs.

5. References

Comstock, J.R. & Arnegard, R.J., (1992). *Multi-attribute task battery for human operator workload and strategic behaviour research.* (NASA Technical Memorandum 104174). Hampton, VA: NASA Langley Research Center.

Durso, F.T. et al. (1995). Expertise and chess: a pilot study comparing situation awareness methodologies. In D.J. Garland and M.R. Endsley (Eds.*), Experimental Analysis and Measurement of Situation Awareness*. Daytona Beach, Florida: Embry-Riddle Aeronautical University, p 295-303.

Endsley, M.R. (1995). Measurement of situation awareness in dynamic systems. *Human Factors*, 37(1), 65-84.

Smolensky, M.W. (1993). Toward the physiological measurement of situation awareness: the case for eye movement measurements. In *Proceedings of the Human Factors and Ergonomics Society 37th Annual Meeting* (p 41). Santa Monica, California: Human Factors and Ergonomics Society.

Sollenberger, R.L. & Stein, E.S. (1995). A simulation study of air traffic controller situational awareness. In D.J. Garland and M.R. Endsley (Eds.*), Experimental Analysis and Measurement of Situation Awareness*. Daytona Beach, Florida: Embry-Riddle Aeronautical University, p 211-217.

Input Requirements to a Performance Monitoring System

Erik Hollnagel

CSELAB
Institute of Computer and Information Science, University of Linköping, SE-581 83 Linköping, Sweden
eriho@ida.liu.se

Yuji Niwa

Institute of Nuclear Safety System
64, Sata, Mihama-cho, Makata-gun, Fukui 919-1205, Japan
niwa@inss.co.jp

Abstract

One of the critical issues in building performance monitoring systems is the range of performance data. Many systems require access to specific types of information, either from sensors or synchronised simulations of internal user states. In practice it is necessary to make very conservative assumptions about the availability of such specific information. An important research problem is therefore whether it is possible to build a performance monitoring system, taking into account both that the data collection must be unobtrusive and reliable, and that the system must be robust and produce neither too many false alarms nor too many misses. The paper describes the input requirements to a performance monitoring system, reflecting the above considerations

1 Introduction

Because the complexity of human-machine systems and demands to their performance both continue to grow, the scope of system design should change from focusing on the interface and interaction to focusing the functioning of the joint system. Significant practical and theoretical efforts have been dedicated to improve the information exchange between humans and machines, particularly in the areas of visual information presentation or display design. The interaction between human and machine is, however, not an end in itself but only meaningful from the perspective of the overall system. This is most easily seen in the control of dynamic industrial processes and transportation systems, where the success criterion is how well the joint human-machine system is able to keep the process safe and efficient.

Effective control is functionally equivalent to managing performance variability. It is therefore important to consider the possible sources of this variability and how they can be monitored and regulated – in particular the variability of the humans who are part of the system. The classical solution has been either to reduce the variability by letting automation take over part of the human's functions, or to constrain its range, e.g., via training or procedures. Since neither solution has been able to solve the problem of providing effective control, other possibilities must be explored. One of these is to monitor the operators' performance and to intervene in a suitable fashion when the variability becomes unacceptably large. Proposals for performance monitoring, ranging from plan recognition to adaptive user interfaces, have been around since the late 1980s (e.g. Alty & Hollnagel, 2000; Kautz, 1987). One of the critical issues in building performance monitoring systems is the access to data, where it is essential that only actually available data types are required.

A Progress Monitoring System (PMS) should keep track of all relevant events. These comprise changes in plant parameters, input/output via the operator-machine interface, communication with others in the control room or in the plant, access to non-instrumented information, etc. In terms of the frequency of events, there are clear differences. Typically, events that involve the operator – such as input/output actions and acts of communication – are relatively infrequent, at least during normal operation. Events that refer to the plant, such as changes in parameter values, alarms and other binary indications, change in plant states, etc., occur with a very high frequency. It therefore makes sense to consider two parallel processes for a PMS: one for information about process changes in either analogue or binary form, and one for information about operator activities. The latter is the subject of the work reported here.

2 Specification of Plant Information

The specification of the information needed by a PMS can be derived from an analysis of the activities that are to be monitored. This paper considers part of the Steam Generator Tube Rupture (SGTR) Emergency Operating Procedure (EOP), which represents in a structured form the activities that the operating team must carry out in order to keep control of the situation. Since the actual form of the SGTR EOP may vary between countries and utilities, it is practical to render the procedure in a generic format that can support the development of a PMS, such as the template shown in Table 1.

Table 1: Generic template for action recognition.

Type	Contents	Description
Pre-condition	Shows if a precondition has been specified	Defines precondition as a state.
	Component	Identifies component(s) referred to by precondition
	Status	Defines required status of component(s).
Post-condition	Shows if a post-condition has been specified	Defines post-condition as a state.
	Component	Identifies component(s) referred to by post-condition
	Status	Defines required status of component(s).
Action Type [X]: Control Communication Information Confirmation Decision making	Shows if a type [X] action has been specified	Defines action to be carried out.
	Component	Identifies component(s) referred to by action
	Status	Defines required status of component(s).

Practically every step of the SGTR EOP refers to one or more source of information, usually components or plant sub-system. In the implementation of a PSM it is necessary that all information sources be identified, so that valid and reliable information can be obtained. For the current report, it is sufficient to consider the information sources that are used for the various action types, since these also include the information sources used by pre-conditions and post-conditions. These are shown in Table 2.

For the technical components, it may be assumed that the required information is available and that it is valid and reliable. For the use of CRTS and other interface components, it is reasonable to assume that such interaction can be reliably registered and identified, at least for modern or advanced control rooms. For the three remaining categories goes it is necessary to make more conservative assumptions. While some of the data can be provided by the use of various types of sensors, cameras and microphones, there are technical, organisational and ethical problems in introducing surveillance equipment at work.

Table 2: Components used by various action types in the SGTR EOP.

Action type	Tech	CRT I/O	Verbal com	Telephone	Field ops.	None
Control	YES				YES	
Communication		YES	YES	YES		
Information		YES	YES			
Confirmation	YES					
Decision-making						YES

One solution to this problem is to complement direct monitoring with pseudo-monitoring and indirect monitoring. Pseudo-monitoring achieves some of the objectives of direct monitoring by taking advantage of pre-defined relations between action types and equipment. For instance, the determination of whether an operator communicates with the radiation protection unit can be made either by analysing the conversation (using speech recognition) or by noting that the operator has dialled the number to the radiation protection unit at the right time. If it is possible to define criteria for pseudo-monitoring events, then this is clearly a simple and robust way of monitoring at least some of the operators' actions. Intermediate monitoring makes use of plant indicators and instrumentation to infer operator actions. Changes in process parameter values, in particular trends, are usually evidence of specific operator actions. Most action types are amenable to intermediate monitoring, which obviates the need of specialised equipment. On the other hand, to implement intermediate monitoring in an actual control room may well require several changes to the instrumentation and control (I&C) to make the recognition both easier and more certain.

2.1 Action Recognition Principles

Of the two parallel input streams to a PMS, the most difficult is the monitoring of specified operator actions. The template described in Table 1 is based on the principle that an action may be characterised three ways: (1) by the pre-conditions, (2) by the action itself, and (3) by the post-conditions or outcome. One advantage of basing the recognition on three constituents rather than one is that it can be made more certain; in addition it will be possible to have a partial recognition, which is better than having none at all.

The logical principle for action recognition can be represented by an event tree, which describes how the different outcomes are produced. The input is the operator action, which is then analysed by considering whether the pre-condition, the action, and/or the post-condition can be recognised. If all three are defined and if all three can be recognised, then the outcome of the recognition is certain. If none can be recognised, then it is impossible to determine what the operator did. In between there are six other cases, where the recognition can take place with different degrees of certainty.

The first test is if the action is part of a pre-defined set. Much of the effort in developing a PMS must therefore go into providing a clear, operational definition of the actions, since the validity of the recognition depends on the validity of the action recognition. Following that three types of recognition can take place with the possible outcomes shown in Table 3.

Recognition Of Pre-Conditions. Pre-conditions can refer to the use of a CRT, verbal communication, telephone communication, or technical components, where the latter are in majority. Pre-condition rules assume that it is always possible to determine whether a pre-condition has been achieved, i.e., the outcome will be TRUE or FALSE. The only exception is for pre-conditions that are defined in terms of verbal communication where the state is defined as UNKNOWN.

Recognition of Actions. Actions can refer to technical components, the use of CRT and displays, verbal communication, telephone communication, and field operations as described in Table 2, which also shows how the different types of actions refer to different types of information. The principles used are the same as for a pre-condition with the addition of Action:Control that refers to a field operation and Action:DecisionMaking that has no reference. The former case is treated in analogy with telephone communication. The latter case is treated in analogy with verbal communication.

Recognition Of Post-Conditions. Post-conditions can refer to technical components or telephone communication, and the rules are completely analogous to the detection of pre-conditions. There are, however, only few procedure steps that contain post-conditions.

Table 3: Outcome from action recognition

	TRUE	**FALSE**	**UNKNOWN**	**Not Defined**
Pre-conditions	Pre-condition is recognised and been met	Pre-condition is recognised but has not been met	It cannot be determined whether pre-condition has been met	No pre-condition defined for this procedure step
Actions	Action is recognised and has been carried out	Action is recognised but has not been carried out	It cannot be determined whether action has been carried out	
Post-conditions	Post-condition is recognised and has been achieved	Post-condition is recognised but has not been achieved		No post-condition defined for this procedure step

Since there can be four outcomes from the recognition of pre-conditions, and three for each of the recognition of actions and post-conditions, it follows that there will be altogether 4*3*3=36 different outcomes. These fall into the following logical groups: in five cases the action is recognised; in seven cases it is partly recognised; in nine cases it is uncertainly recognised; and in fifteen cases it is not recognised.

3 Example: Pressure Equalisation – FRAME-23

The SGTR-EOP altogether consists of 85 different knowledge frames, of which Frame-23 is illustrated here. Frame-23 is part of the activities needed to ensure pressure equalisation, and comprises the following actions: "Cool down rapidly reactor coolant to under the target hot leg temperature (zero power temperature-27.8 C) using main steam relief valves of intact loops". A detailed analysis of the components referred to by frame-23 shows that only the monitoring of the effects on the secondary system, and the opening of the Power Operated Relief Valve (PORV) can be recognised by the PMS.

Since the status of the pre-condition remains UNKNOWN (Table 4), it follows that the only possible outcomes of the recognition are "action partly recognised", "action uncertainly recognised", and "action not recognised". Which outcome will actually be the case depends on te state of the system when the recognition is made.

Table 4: Recognisable features of Frame-23.

TYPE	YES/NO	DESCRIPTION
Pre-condition	YES	All isolations have been made
	Component	
	Status	
Post-condition	YES	Indication lamp [green]=>[red] **AND** Hot leg temperature of reactor coolant: TR-410: 411C, 421C, 431C: TI-410P, 420P, 430P [decreased]
	Component	Indication lamp **AND** TR-410: 411C, 421C, 431C: TI-410P, 420P, 430P
	Status	[green]=>[red] **AND** [decreased]
Action: Control	YES	Cool down rapidly reactor coolant to under the target hot leg temperature (zero power temperature-27.8 C) by opening MSRV of intact loops
	Component	Controller
	Status	Opened (manually)

4 Conclusions

The analysis of all the actions defined for this subsequence showed that only a small number of them could be recognised. This is not surprising since it reflects the fact that the quality of the recognition depends on the input available. In an actual control room, many sources of input are unavailable such as operator-to-operator communication, decision-making, etc. In general, it is only possible to get reliable data from actions or events that involve manifest changes to components of the technological system, such as changes in displays, activations of controls, changes in indicator status alarms), etc. For the pressure equalisation subsequence as a whole, it was found that 31.25 % of the actions could be partly or completely recognised while the remaining 68.75 % could not be recognised.

Although more extensive analyses are needed to see if this is representative for EOPs as a whole, the results do suggest that the design of PMSs should carefully specify the input requirements. By realising in advance the needs of performance monitoring, it may be relatively easy to incorporate changes to the interface and the S&C, which makes a PMS feasible. However, the limited coverage in terms of functions means that the output from a PMS should not be applied for critical purposes, and certainly not for error detection of operators. A simpler and more useful purpose is to use a PMS to produce input to other systems, for instance computerised procedures. Here a PMS, using the principles of pseudo-monitoring, can keep track of how the operators carry out control actions and verifications. This can be used as input to, for instance, automated scrolling of the procedures (Niwa et al., 1996). In addition, a PMS may also be able to detect clear violations of the procedures, and generate alerts to the operators.

5 References

Alty, J. L. & Hollnagel, E. (2000). Adapting the Operators Needs to Abnormal Situations. *Proceedings of the European Annual Conference on Human Decision Making and Manual Control*, Alty, J.L., (ed), pp 145 - 155.

Kautz, H. A. (1987). *A formal theory of plan recognition*. PhD thesis, University of Rochester, Rochester, N.Y.

Niwa, Y., Hollnagel, E. & Green, M. (1996). Guidelines for computerised presentation of emergency operating procedures. *Nuclear Engineering and Design, 167,* 113-127.

Requirements Analysis and Task Design in Dynamic Environments

Johan F. Hoorn *Gerrit C. van der Veer*

Vrije Universiteit, Faculty of Sciences, Department of Computer Science, Section Information Management and Software Engineering, subsection Human Computer Interaction, Multimedia and Culture
De Boelelaan 1081a
{jfhoorn, gerrit}@cs.vu.nl

Abstract

Design of complex interactive systems is never an isolated project. In the case of large companies or public administration, the business goals, and, consequently, business processes develop. Market influences and political situations force an organization to adjust. Design projects, in these circumstances, are at risk to start from requirements that lose validity during system development. State of the art task analysis approaches like Groupware Task Analysis (GTA) and the related design approach DUTCH make an explicit distinction between current and envisioned future task worlds. However, even these approaches do not cover the issue of evolving requirements. In our analysis, we expand the task analysis approach of GTA and DUTCH to encompass business models and their development over time during the design process. To relate the extended conceptual framework to empirical assessment, we present a model of requirements change in terms of emotional relevance and valence of a system to the stakeholders' personal and business goals and work processes.

1 Design Approach and Task Analysis

The DUTCH design approach is a general method to design complex interactive systems. DUTCH takes into account, for instance, the work situation, the task world, and the system users in a dynamic way, so that the approach renders possibilities to get to some form of requirements management. The analytical part of the DUTCH approach (Figure 1, left) is called Groupware Task Analysis (GTA) (Figure 1, right) (Van Welie & Van der Veer, 2000). Instead of only analyzing the individual at work, GTA also recognizes the impact of the work group, business, or organization. To allow for more agent, work, and situational aspects, two task models are discerned, one that describes the state of affairs of the task environment on the work floor (Task Model 1) and one that envisages the desired situation after the system has been built and implemented (Task Model 2). Because Task Model 2 (TM2) describes a more-or-less ideal situation, it is limited by practical constraints such as technical possibilities, the client's budget, and legislative directives. During and after analysis, specifications of the requirements are prepared, guiding detailed decisions about 'what the user actually wants.'

Because longitudinal knowledge about situational change is important for requirements management and because TM2 helps forming the agreed requirements, the elements that make up a task world like TM2 should be thoroughly understood. An event is a change in the task world's condition at a given point in time. Changes may reflect shifts of attribute values of internal concepts such as object, task, agent or role. They could also reflect shifts of external concepts such as political climate or water supply. Events affect the order of task performance by triggering tasks. DUTCH provides for situational changes by means of task models. Events, tasks, and agents impinge upon their environment and New User Action Notation or NUAN models in relation with TM2 resemble the requirements specification as laid down in the agreed requirements (cf. Kotonya & Sommerville, 1998). However, information about business processes is needed to make this resemblance near perfect.

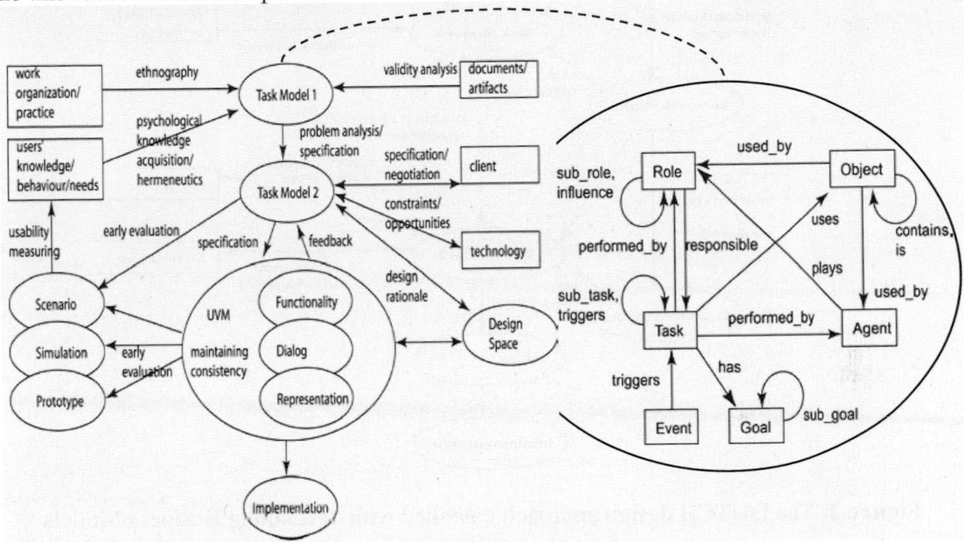

Figure 1: DUTCH design (left) with Groupware Task Analysis as its core (ellipse right)

2 Integration with Business Models

In Figure 2, the adapted DUTCH approach shows that the task hierarchy of Business Model 1 (BM1) corresponds with the task or process hierarchy in TM1 and of BM2 with the task or process hierarchy in TM2. BM1 relates business requirements to TM1 (goals, problems, Critical Success Factors or CSFs). The single-headed arrow from TM1 (work floor) to BM1 (executive office) could actually be double-headed in that the work floor should have an understanding of the business plans as well. Analysis of BM1 and TM1 leads to TM2. The task hierarchy of TM2 is used to build BM2, which relates the business requirements to the desired situation. BM2 illustrates how business goals, CSFs, and problems affect TM2 and serves as a justification of TM2 to that extent. A double-headed arrow between BM2 and TM2 should be common practice but it demands a client who cooperates strongly with the management to help design the future business model. Business goals may have a strong impact on task hierarchy. Business goals probably affect the task hierarchy on a higher level (closer to the root) than user requirements do. If such a business goal is at issue, it limits the use of a detailed investigation of the current tasks at a lower level in TM1. For this reason, the design team should not wait with obtaining business requirements until TM1 is finished.

To represent (dis)agreement in system requirements we interlaced the general requirements engineering approach with DUTCH and DUTCH with business models. Next, we want to turn to the motivational aspects, the emotional halo around the business and task models to argue that the reasons why requirements change probably are fixed and that those may serve as stable indicators of the direction of change. A positive emotional bias of stakeholders towards the system may be one of the CSFs that are crucial for goal achievement and a business problem when that bias inverses over time.

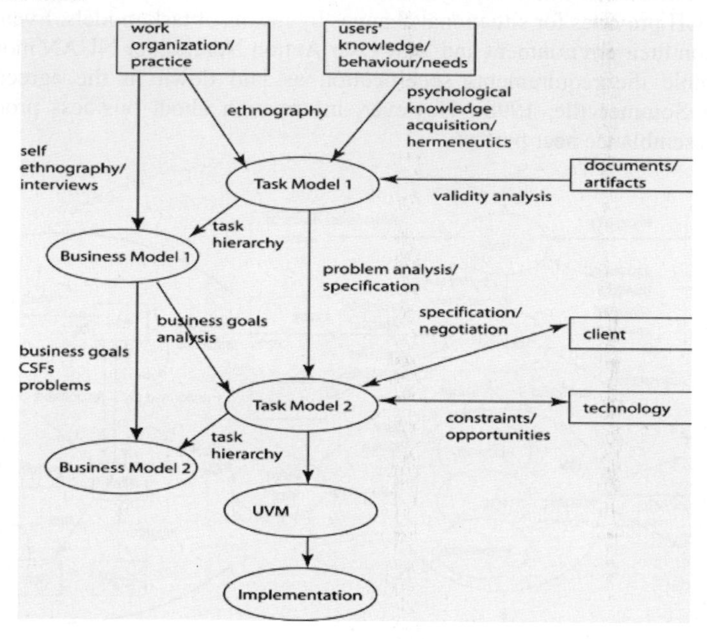

Figure 2: The DUTCH design approach extended with developing Business Models

3 Model of Requirements Change

The change of system requirements may lie in a switch of focus between business and personal goals. The focus switch can also happen *within* the business or personal goals. A classic business goal is maximizing profit. For example, accelerating production speed only to raise more money is an egotistic goal. However, in the nonprofit sector the mission is more philanthropic. Improved accuracy in artery surgery, for example, is a more altruistic goal. Yet, both commercial businesses and public services have commercial as well as ideological goals albeit in different numbers with different importance. The same distinction between egotism and altruism can be made also in the personal goals of employees (whether laborers or managers). On the one hand, they work to improve their own circumstances (the pursuit of money and/or happiness). On the other, employees show loyalty to the group (e.g., help the 3rd world by joining a fair trade company). In that case, they subscribe to the business goals. Perhaps *egotism* and *altruism* are the fixed components of business goals and obviously, there is a trade-off between them that may follow the twists of changing situations, therefore resulting in a different demand on the system requirements.

Archetypal for business processes is that they should be efficient, meaning that they are accurate while fast. Usability of a system that manages processes also should be efficient while arousing satisfaction (Nielsen, 1993). However, if processes go extremely fast, they push the borders of the user's limited capacity of information processing (cf. [3]). There lacks the time to think and users feel out of control. If processes go too slowly, users may get fed up with the delay of goal achievement while they have to reach targets. If a process is extremely accurate, it may also be extremely vulnerable to any change in its environment, because it requires that every element is always at the expected place at the expected time. Overspecialization of processes not only makes bad adaptive systems but also bad adaptive users. This may suffocate the users and make them feel there is no room for their own ingenuity. Hassenzahl, Platz, Burmester, and Lehner (2000) explicitly connect usability of and satisfaction with a system (fun, challenge) with its efficiency. Their upshot is that the most effective interface can actually be boring. If a process is sloppy, on the other hand, users may become irritated because their successfulness in achieving goals is hampered by errors. Perhaps, then, the two fixed components of processes are *speed* (fast vs. slow) and *accuracy* (accurate vs. inaccurate) and here too the effect of their trade-off on the attitude towards the system (i.e. the user's satisfaction) might account for changing system requirements of the stakeholders (Figure 3).

The underlying emotional mechanism of these trade-offs, we suppose, is that critical events change the business situation or task environment (Figure 2), which make stakeholders change their attribution of relevance (important-unimportant) to goals and processes and subsequently redirect the valence (positive-negative) towards goals and processes. The hypothesis of *egotism-altruism trade-off of goals* is based upon four assumptions. On the side of the business goals there is the *maximizing-profit* assumption (egotistic business goal) versus the *bring-good-to-the-people* assumption (altruistic business goal). On the personal side we construed a *work-my-way-up* assumption (egotistic personal goal) versus the *employalty* assumption (altruistic personal goal). The hypothesis of *speed-accuracy trade-off of processes* also relies on four assumptions: The *efficiency-autocracy* assumption states that processes may be extremely fast and accurate but that the users feel they work under a tyrant controlling them in stead of vice versa. The *quick-and-dirty* assumption is about processes that are fast even at the cost of being error prone. Users may feel such a process is a ruthless implementation of reaching maximum profit. The *safe-at-all-costs* assumption relates to processes that are slow and accurate. The slowness may invoke boredom while extreme accuracy (to always be on the safe side) may be suffocating. The *adaptive-but-inefficient* assumption posits that processes that are slow and inaccurate may be inefficient but there is plenty of time to adapt to change and there are hardly any restrictions on reorganizing structure and/or contents.

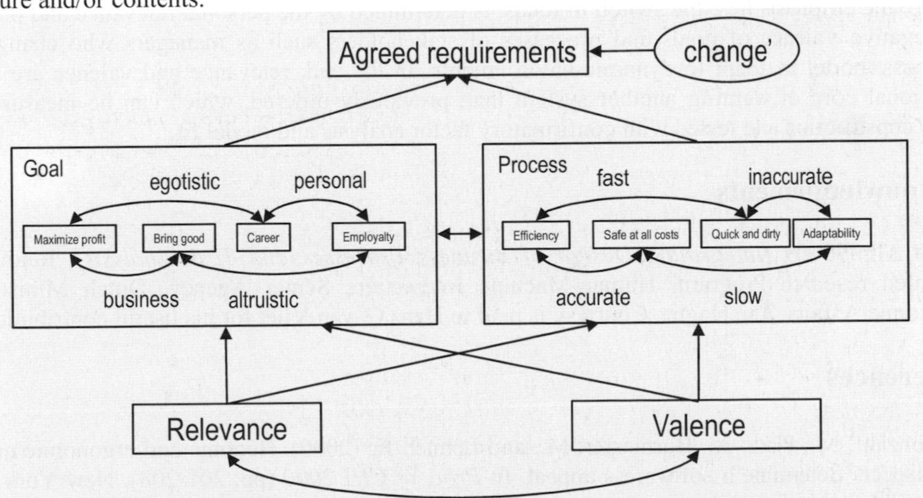

Figure 3: Emotional values of goals and processes for stakeholders

Figure 3 explains what happens in the mind of a stakeholder or client after an event (e.g., the attack on the World Trade Center) has taken on meaning for him or her. In the perspective of this event – which can be inside or outside the stakeholder's head – relevance and valence are estimated of the goals and processes. Shifts in relevance and valence attribution may change goals and processes, which change the system requirements. All these constructs are in rectangular boxes, indicating that they are observable variables, whether registered by structured questionnaires, reaction-time experiments, or electronic document search. However, the prediction of requirement change will not be perfect, so that a latent variable *change'* is devised as an unobserved variable (in the ellipse) to soak up the random variation in the *Agreed requirements* scores (agree-disagree). The single-headed arrows should be read as regression weights and the double-headed arrows as covariances. Given this path model, future research can be directed to causally model the latent structures in (changing) business processes and goals with longitudinal observations in multiple user groups (e.g., managers vs. laborers). This last suggestion implies that the dynamics described in Figure 3 is not limited to an individual client or stakeholder but can be seen also as a description of average (group) behavior.

4 Conclusions

To explain why system requirements change during development, we joined up requirements engineering, design, and task analysis and placed them within a wider social or business context. The design approach DUTCH accounts for variation in situations by employing task analysis models. The task world ontology is affected by its environment (i.e. events and human agents) and together with NUAN models the *envisioned* task world (TM2) provides about the same information as a requirements specification. Business models are needed to make this resemblance near perfect. Changes in business goals and processes may influence a task hierarchy more so than user requirements do.

Business models formulate directives on business goals and work processes. Goals can be decomposed into personal and business goals with an egotistic or altruistic thrust. People want to make a career but also deliver quality work. Processes can be described in terms of speed and accuracy, in which slowness correlates with higher accuracy and fastness with higher inaccuracy. Where the emphasis lies, the switch in focus, is determined by the personal relevance and positive or negative valence of goals and processes of stakeholders such as managers who change the business model to adapt to dynamic environments. In the end, relevance and valence are at the emotional core of wanting another system than previously ordered, which can be measured by scale construction and tested with confirmatory factor analysis and model fit.

Acknowledgements

Grant Mmi99009 *(Integrating Design of Business Processes and Task Analysis)*, Innovation Oriented research Program, Human-Machine Interaction, Senter Agency, Dutch Ministry of Economic Affairs, The Hague. Courtesy is paid to Hans C. van Vliet for his useful contributions.

References

Hassenzahl, M., Platz, A., Burmester, M., and Lehner, K. (2000). Hedonic and ergonomic quality aspects determine a software's appeal. In *Proc. of CHI 2000* (pp. 201-208). New York, NY: ACM Press.

Kotonya, G., & Sommerville, I. (1998). *Requirements engineering: Processes and techniques.* New York, NY: John Wiley.

Nielsen, J. (1993). *Usability engineering.* Boston, MA: Academic Press Professional.

Van Welie, M., & Van der Veer, G. C. (2000). Task based groupware design: Putting theory into practice. In *Proc. on Designing interactive systems* (pp. 326-337). New York: ACM Press.

Human Factors in Web-assisted Personal Finance

Ion Juvina and Herre van Oostendorp

Institute of Information and Computing Sciences, Utrecht University
Padualaan 14, Utrecht, The Netherlands
{ion, herre}@cs.uu.nl

Abstract

This paper aims at grounding a series of experimental studies in the expanding area of human factors in using Web applications. Based on a preliminary experimental study, literature review and exploratory task analysis, a hypothetical model of factors relevant for Web navigation is developed in order to be experimentally tested.

1 Introduction

Web navigation allows the user to approach an information space in a rather natural manner, basically in the same way as orientation in physical space or seeking for food (Pirolli & Card, 1995). The research literature (e.g. Chen & Rada, 1996) presents general information-processing mechanisms (related mainly to the dynamic nature of mental models) and individual differences (cognitive, affective, and conative[1]) that influence user performance and satisfaction in navigating through hyperspace. However, only individual studies concerning specific factors are reported and there is still a need to investigate a coherent set of factors together in order to identify their mutual relationships and their relative influence on performance and satisfaction. In other words, an integrative predictive model of web navigation is required. We address the following research questions: What are the relevant factors of a navigation model? What is the predictive power of the model and of each particular factor? What are the interactions between factors in predicting users' performance, satisfaction and reliability?

This paper presents the preliminary part of a research project that aims at building a psychometric model that predicts and explains human performance in web-assisted tasks. First, a preliminary experimental study is presented as background for an exploratory task analysis. A hypothetical model built according to task analysis results is further presented. The paper ends with conclusions.

2 Pilot study

The pilot study addressed the following questions: Which variables are important to users' satisfaction with a web site? What is the *relative* importance of these variables? The practical goal of the research was to enhance the frequency of visits and revisits to the site. Therefore, user satisfaction was selected as a dependent variable. We used some indicators of user performance (e.g. search time) as predictors of satisfaction, for this reason performance was not used as a dependent variable. Performance is presumed to have an effect on satisfaction and not vice-versa.

[1] Conation refers to the intentional and personal motivation of behaviour (Bandura, 1997)

2.1 Background

Usability "in the strict sense" of a web site has been measured based on the heuristics of Nielsen (2000) and the suggestions of Sullivan (1998): ease of navigation, consistency, feedback, controllability, lay-out, content reliability, interactivity, interestingness (of provided information). Perceived usability has been measured by collecting judgments from end users based on a questionnaire containing rating scales. Besides these 'cognitive ergonomic' (usability) qualities of a web site, other independent variables include: Cognitive abilities of visitors as expressed by search time, correctness of answers, and goal directness; Experience with the Internet; Attitude towards the Internet; and Informational match (the extent to which presented information meets specific information needs of the user).

2.2 Method

The participants (n=31) were recruited on the basis of an invitation published on the web site that was the object of research. The participant applications were screened and a sample representative of the Dutch visitors of the site was selected. The web site examined was that of the Dutch Ministry of Traffic and Public affairs. It contains more than 40.000 public information documents. Participants were situated in a quite room and searched the website to answer a number of questions, while being videotaped. Search time, number of visited pages to reach the needed information and number of correct answers were registered. After completing the website search, participants filled out a questionnaire, which measured the variables Usability, Experience with the Internet, Attitude towards the Internet, Informational match and Satisfaction based on judging of statements with rating scales.

2.3 Results of the pilot study

In a first multiple linear regression analysis (with the 7 main factors - Usability, Search time, Goal directness, Correctness of answers, Experience with the Internet, Attitude towards the Internet and Informational match - as predictors) Satisfaction was significantly related to Usability. The *Beta* coefficient of Usability is .74; the other factors were not significantly related to satisfaction. The multiple R^2 is .55; $F(1,29)=34.68$, $p <.001$. The total amount of variance explained by usability is thus 55%.
A secondary analysis with the 8 usability aspects (Ease of navigation, Consistency, Feedback, Controllability, Lay-out, Content reliability, Interactivity and Interestingness of provided information) now *also* separately included in the analysis, showed that Satisfaction is only significantly related to the usability aspects Interestingness (*Beta* = .62) and Ease of navigation (*Beta* =.28). These two usability aspects explain 59% of the variance. The multiple R^2 is .59; $F(2,28)=19.42$, $p <.001$.

2.4 Conclusion of the pilot study

Unexpectedly, on the basis of multiple linear regression analysis it appeared that only the usability of the web site is a strong predictor (55% explained variance) of the users' satisfaction with the website. Particularly the usability dimensions *ease of navigation* and *interestingness* of presented information are strongly related to satisfaction. The other factors such as Experience with the Internet or Cognitive ability appeared to be weakly and not significantly related to satisfaction. We can conclude that in this particular case (this site, these subjects with the task as used), usability is a strong predictor of users' satisfaction with the site, particularly the usability dimensions

Interestingness (how interesting is the provided information in itself, independent of users' own interest) and Ease of navigation (how easy it is to navigate through the web site) are strong factors. It is important to mention a number of restrictions to the generalization of the present study. (1) The sample was initially self-selected: the subjects who participated in the study had previously visited the website and then volunteered for the study. With a more representative sample (e.g. random selection of participants from the general population), one might expect to obtain also significant effects for some of the other factors that were weakly and not significantly related to satisfaction in this study (e.g. 'Experience with the Internet'). (2) The scope of navigation was restricted to one site and the task was simple, containing only an information search component without other components as problem solving or decision making. (3) Only a limited number of factors were included in the multiple-regression analysis. One might expect that other factors might also be relevant, as for instance 'prior domain knowledge'. (4) Objective measures are necessary to complement the subjective judgments of usability and satisfaction (e.g. experts rating usability and performance indicators along with satisfaction as dependent measures). (5) A more complex statistical analysis (LISREL) is necessary to reveal the mutual relations and the direction of influence between the variables of the model.
The results of the pilot study revealed the importance of studying a set of factors together in order to estimate their relative importance and directed relationships between them. The study was also an informative way to define methodological requirements and constraints for such research.

3 Task analysis

To overcome the restrictions of the pilot study presented above, an explorative task analysis process was considered appropriate, for the following reasons: (1) to achieve an initial insight regarding what factors are relevant to be included in a navigation model; at a later time these factors will be included in a larger research investigation to determine their interrelationships; (2) to better characterize the domain on which the research findings could be generalized since it is expected that some of the factors are domain specific (e.g. spatial ability, Sjolinder, 2002); (3) to collect realistic task instances.

3.1 The task domain

The model is intended to be ecologically valid, first by being relevant to the domain of web-assisted Personal Finance and next by being extendable to other kinds of web-based applications. By web-assisted Personal Finance we refer to using the Internet to perform activities like: setting financial goals, saving and investing, financial decisions, transactions, etc. We have selected this domain because of its broad availability and utilization, shown by the increasing presence of web sites that provide Personal Finance assistance (Yahoo Personal Finance, CNN Personal Finance, MSN Money, FinanCenter, ThisIsMoney etc.). Most of our task analysis activities were conducted using FinanCenter (www.financenter.com), a website ranked by Forbes magazine as having the best personal finance calculators on the Internet.

3.2 Tasks

In order to ensure ecological validity, our experimental tasks have to be realistic and consequently rather complex. They require not only information-seeking activities but also involve problem-solving and decision-making processes. The tasks that were used for task analysis included personal life planning components (e.g. "Establish a personal budget"), information retrieval components (e.g. "Find the definition of financial goal"), problem solving components (e.g. "How much do you need to save in order to buy a car in 4 years"), and personal decision-making components (e.g. "What kind of car can you afford").

3.3 Technique

The main goal of the task analysis process was to acquire a qualitative insight in both task domain and navigation behavior. Besides observation and semi-structured interviews, a knowledge elicitation technique - thinking aloud protocols – was used to collect information to answer the following questions: How do people navigate? More specifically, what kind of knowledge and skills do they use in navigation? And what are the possible individual differences in this regard? This method was also used to collect information about technology and task related factors involved in Web navigation.

3.4 Results of task analysis

Based on interpretation of task analysis protocols, we conceived a hypothetical model of Web navigation. The inclusion of a specific factor in the model was based on observations and verbalizations of the subjects during task execution as well as on subsequent interviews. For instance, the inclusion of 'spatial ability' was suggested by an utterance as "I need a site-map" in the thinking-aloud protocol followed by specific interview questions. Inevitably, some of the potential factors were left out. For instance, 'attention' appeared not to be relevant as a separate variable but it is functioning rather as an implicit component in working-memory and self-regulation. When asked about attention, subjects reported that there were no specific requirements. As they were sufficiently motivated and willing to achieve the task, attention was probably automatically triggered.

The model (fig.1) consists of 6 categories of predictive factors (independent variables) and 3 categories of predicted factors (output/dependent variables). Predictive factors are further divided in personal factors (cognitive, affective, conative and demographic) and environmental factors (technology-interface and task-context). Predicted factors are grouped in 3 categories (performance, satisfaction and reliability[2]). Obviously this model consists of a more comprehensive set of factors as compared with the one investigated in the pilot study presented above.

At the moment, the model contains only the factors that are considered relevant for web navigation, based on task analysis and literature review. Further on, relationships between these factors (with associated weights and directions) and within the categories of factors will be included in the model, based on a series of experimental studies. Some of the relationships can be already inferred from the literature. For example, one possible association is the following: The degree of user control made possible by the interface is expected to be mediated by 'locus of control' (as cognitive style) in its effect on performance (Anderson 2001). Thus 'internalists' would benefit from having control over the interface and task structure while 'externalists' would benefit more from being directed by the interface and having a given task structure. Such inferences will be tested in our experiments and possibly integrated in our model.

---

[2] Reliability refers to the measure at which the undesirable effects of task execution are avoided or minimized.

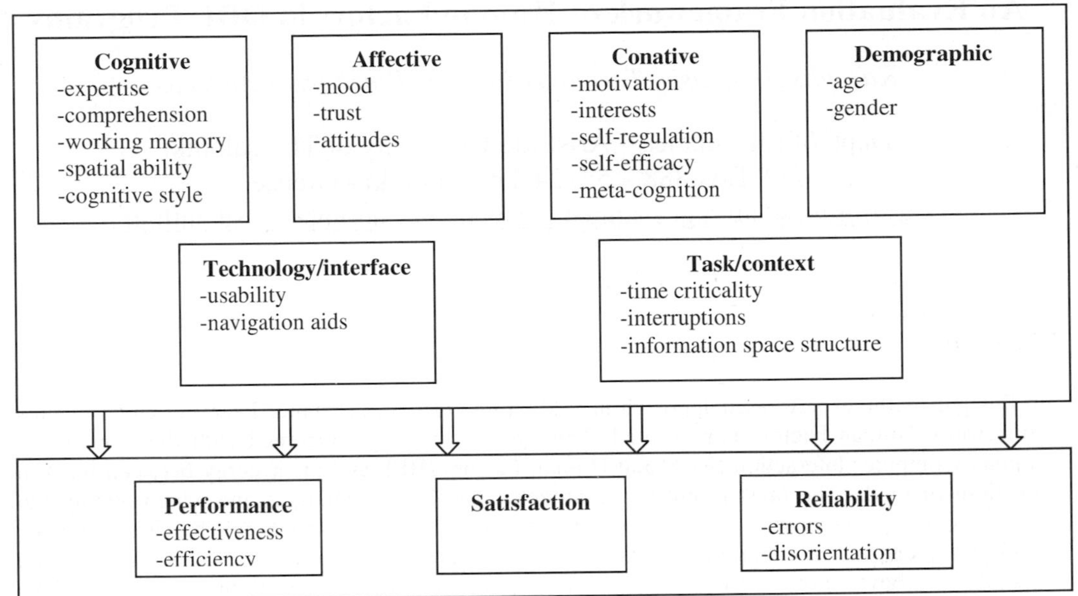

Figure 1: The hypothetical model of web navigation

4 Further developments

In the next phase of the research the model will be tested in a series of quantitative studies. When we will have acquired a clear understanding of the predictive power of our model and of each factor, we will subsequently manipulate some of the factors and test them in a real-life application in order to answer the following question: what manipulations (human and technical) can improve the user performance, satisfaction and reliability in web assisted personal finance? The predictive power of factors will determine which manipulations should come first in order to maximally improve the desired effects.

References

Anderson, M.D. (2001). Individual Characteristics and Web-based Courses. In C.R. Wolfe (Ed.). *Learning and Teaching on the World Wide Web*. (pp. 45-72). London: Academic Press.

Bandura, A. (1997). *Self-efficacy: The exercise of control*. New York: W. H. Freeman.

Chen, C., & Rada, R. (1996). Interacting with hypertext: A meta-analysis of experimental studies. *Human-Computer Interaction*, 11(2), 125-156.

Kieras, D. E., & Meyer, D.E. (2000). The Role of Cognitive Task Analysis in the Application of Predictive Models of Human Performance. In J. M. Schraagen, S.F. Chipman, V.L. Shalin, *Cognitive Task Analysis*. (pp. 237-260). Mahwah, NJ: Lawrence Erlbaum Associates.

Nielsen, J. (2000). *Designing Web Usability: The Practice of Simplicity*. Indianapolis, ID: New Rider Publishers.

Pirolli, P., & Card, S. (1995). Information foraging in information access environments. *Conf. proceedings on Human factors in computing systems*. (pp. 51-58). Denver, CO: ACM Press.

Sjolinder, M. (2002). *Individual differences in spatial cognition and hypermedia navigation*. Retrieved October 24, 2002 from http://www.sics.se/humle.

Sullivan, T. (1998). *User testing techniques. A reader friendliness checklist*. Retrieved September 9, 2002 from http://www.pantos.org/atw.35317.html.

An Evaluation Framework of Human Factors in ODL Programs

Karoulis Athanasis, Tarnanas Ioannis, Pombortsis Andreas

Dept. of Informatics – Aristotle University of Thessaloniki
PO. Box 888 – 54124 Thessaloniki - Greece
karoulis@csd.auth.gr ioannist@psy.auth.gr apombo@csd.auth.gr

Abstract

In this paper the expert-based approach in evaluating Open and Distance Learning environments concerning Human Factors is presented. This study is mainly concerned with the domains of Human-Computer Interaction (HCI) and Human Factors (HF), and the synergy between them in the field of ODL. The most promising approach to evaluating ODL, is one which attempts to frame the ideas from distributed cognition research in a way that is more usable by HCI designers. Though the distributed cognition framework acknowledges a vast majority of cases suited to HCI designers and evaluators, it has never been tried before as a framework for the evaluation of the cognitive work that is *distributed* among people, between persons and artifacts, across time and between abstract resources of information in the ODL domain. So, the objectives of this paper are, on one hand to present contemporary research on the domain, and, on the other hand to pinpoint the main concerns and to propose trends for further research on this complicated field that combines ODL, HCI and HF in such a holistic manner.

1 Introduction

Most contemporary organizations spend enormous amounts on education and training, in order to enhance the abilities of their human resources. A clear tendency in this direction is to utilize some alternative educational approaches, like Computer Based Training (CBT), distance learning or another technology enhanced variation that better suits the needs of the particular organization.
Contemporary research on Human Factors has been concerned with participation and skill in the design and use of computer-based systems. Collaboration between researchers and users on this theme, starting with the pioneering work of Donald Norman (Norman, 1986) and Alan Newel (Card et al., 1983), has created a shift from the idea that Human Factors is a passive observation of users to the idea of Human Actors, where you create models for the interactions between actors, artifacts and the settings in which interaction occurs (Preece et al., 1994).

2 The Educational Evaluation

Many writers have expressed their hope that constructivism will lead to better educational software and better learning (e.g. Brown et al., 1989). They stress the need for open-ended exploratory authentic learning environments in which learners can develop personally meaningful and transferable knowledge and understanding. The lead provided by these writers has resulted in the proposition of guidelines and criteria for the development of constructivist software and the identification of new pedagogies. A recurrent theme of these guidelines, software developments and suggestions for use is that learning should be authentic, on a cognitive and contextual level. A

tenet of constructivism is that learning is a personal, idiosyncratic process, characterised by individuals developing knowledge and understanding, by forming and refining, which finally leads to the five main socio-constructive learning criteria (Squires & Preece, 1999) that must be met in order to characterize an educational piece as socio-constructive: credibility, complexity, ownership, collaboration and curriculum.
Human Factors research in the educational domain is investigating the visual search of computer menus and screen layouts. These models provide detailed empirically-validated explanations of the perceptual, cognitive, and motor processing involved (Kieras, 1988). However, there are certain known difficulties in the evaluation of educational environments in general, such as the difficulty to anticipate the instructional path every student will follow during the use of the software (Tselios et al., 2002), or the fact that such environments usually expand as they are used by the students (Hoyles, 1993). Nevertheless, in general, it is acclaimed to study the usability of an educational piece in relation to its educational value (Squires & Preece, 1999; Tselios et al., 2002).
However, the problem of evaluating educational software in terms of Human Factors is one shared by many HCI practitioners. It is important to know if a design is being used as it was intended. When a design is being used differently than intended, then it is important to discover if that use is compatible with current domain practice. The most promising approach to evaluating educational HF (Human Factors), is one which attempts to frame the ideas from DC (Distributed Cognition) research in a way that is more usable by HCI designers. This approach is described as the Distributed Information Resources Model (or Resources Model for short). The Resources Model seriously takes the idea introduced to HCI by Suchman (1987), that various types of information can serve as resources for action and a set of abstract information structures can be distributed between people and technological artifacts. The Resources Model also introduces the concept of interaction strategy and describes the way in which different interaction strategies exploit different information structures as resources for action. In this sense the Resources Model is an HCI model in the DC tradition. The aim of the continuing work on the domain is to apply the Resources Model to collaborative distance educational settings where resources are shared between individuals. In this way it is hoped to explore the value of the Resources Model as a framework for bridging between HCI, ODL and CSCW research.

3 Expert-based vs. User-based Evaluations

The most commonly applied evaluation methodologies are the expert-based and the empirical (user-based) evaluation ones, according to the taxonomy of most researchers on the field (eg. Preece et al., 1994; Lewis & Rieman, 1994). Expert-based evaluation is a relatively cheap and efficient formative evaluation method applied even on system prototypes or design specifications up to the almost ready-to-ship product. The main idea is to present the tasks supported by the interface to an interdisciplinary group of experts who will play as professional users and try to identify possible deficiencies in the interface design.
However, according to Lewis & Rieman (1994) «you can't really tell how good or bad your interface is going to be without getting people to use it». This phrase expresses the broad belief that user-testing is inevitable in order to assess an interface. Why then, don't we just use empirical evaluations and continue to research expert-based approaches as well? This is because the efficiency of these methods is strongly diminished by the required resources and by various psychological restrictions of the participating subjects. On the other hand expert-based approaches have meanwhile matured enough to provide a good alternative. Nowadays, expert-based evaluation approaches are well established in the domain of HCI. Cognitive Walkthrough (Polson et al., 1992) and its variations (Rowley & Rhoades, 1992; Karoulis et al., 2000), Heuristic Evaluation and its variations (Nielsen, 1993) and Formal Usability Inspection (Kahn & Prail, 1994) are the most encountered approaches in the field.

An expert based approach to Human Factors evaluation for educational software is primarily performed by using a distributed cognition approach and a multidisciplinary team of experts. A lot of the drawbacks to the application of this theory are shared with many other evaluation techniques used in HCI, particularly those relying on heuristics. One advantage of distributed cognition with respect to too much data, is that the theory helps focus on where to look in the data by the emphasis on domain expertise and task relevant representational state. A second problem is that this method is time consuming, as with all analysis. Again, this is a problem that many analysis methods face, even those not video based. It is a problem encountered in virtually all educational industries using HCI, where the pressure to "get the product out" is essential.

4 From Human Factors to Human Actors

Computational cognitive modeling is emerging as an effective means to construct predictive engineering models of people interacting with computer systems (Preece, et al., 1994). Cognitive models permit aspects of user interfaces to be evaluated for usability by making predictions based on task analysis and established principles of human performance. Cognitive models can predict aspects of performance, such as task execution time, based on a specification of the interface and task. Cognitive modeling can thus reduce the need for user testing early in the development cycle. Cognitive models can also reveal underlying processing that people use to accomplish a task, which can help designers to build interfaces and interaction techniques that better complement the actual processing that humans apply to a task. When built with an architecture such as EPIC (executive process-interactive control, Kieras, 1988), ACT-R/PM (atomic components of thought with rational analysis and perceptual/motor enhancements, Anderson & Garisson, 1995), or EPIC-Soar (the Soar architecture with enhancements from EPIC), cognitive models also contribute to Allen Newell's grand vision of a unified theory of cognition (Card et al., 1983). Methodologies for applying cognitive architectures to predict aspects of human performance and learnability are still evolving (Kirschenbaum et al., 1996). On the other hand, Heuristic Evaluation and Formal Usability Inspection are criteria- or heuristic-based methodologies. So, the next point of concern is the appropriate list of criteria or heuristics needed to assess the environment. As already stated, a good starting point provides the socio-constructivist view of instruction. Some studies in the field (eg. Tselios et al., 2001) are based on the constructivist approach for open learning environments, sometimes also known as microworlds.

5 Concerns

The main question that remains unanswered with regards to the learnability of the environment, is whether the experts are really able to predict the problems experienced by the learners and the cognitive domain of the environment. Some studies in the field (eg. Tselios et al., 2001) argue that an expert evaluator cannot predict the students' performance, although he/she can assess heuristically the learnability of the environment, with mediocre results.

Another issue that needs further research is the protest stated by Spoole and Schröder (2001) regarding the optimal number of evaluators. The authors presented the results of some recent research that brings into question the long-standing rule of thumb put forth by Jakob Nielsen concerning his description of Heuristic Evaluation. The point of contention is whether one can truly find the majority of usability problems in complex web sites by employing a very small number of testers. The Nielsen rule of thumb, referred to by the authors as the "parabola of optimism," states that five testers can find 85% of the problems in a web site regardless of its size. In contrast, Spoole's and Schröder's research indicates that the number of testers needed increases linearly with the size of the web site. There are some objections to this statement, mainly because the heuristic methodology is not task based and because simple test users have been proved not to

perform well in heuristic approaches. However, the claim of Spoole and Schröder (2001) becomes too important to be ignored, because educational sites can expand greatly, users mainly perform tasks in such sites and ODL includes de facto the cognitive parameter of inexperienced users.
The expert-based approach, as presented in this paper does not yet seem suitable for a holistic evaluation of the environment. We are entering the age of ubiquitous computing in which our environment is evolving to contain computing technology in a variety of forms. Such environments, also called "media spaces" (Stults, 1988) are developments in ubiquitous computing technology, involving a combination of audio, video and computer networking. They are the focus of an increasing amount of research and industrial interest into support for distributed collaborative work.
These concerns are of interest for ODL ubiquitous environments, as large amount of educational and private data is circulating between the participating entities. The assessment of such an environment is a challenge for expert-based evaluation of Human Factors.

6 Conclusion

The two roles of expert based evaluation in Human Computer Interaction (HCI) are to aid design and analysis. Distributed Cognition is a theoretical framework that differs from mainstream cognitive science by not concentrating on the individual human actor as the unit of analysis. Distributed Cognition acknowledges that in a vast majority of cases cognitive work is not being done in isolation inside our heads but is *distributed* among people, between persons and artifacts, and across time. This has a natural fit for ODL, where the behavior we are interested in is the interaction of all the involved entities through the communication channel, the system of people and artifacts. What makes a system cognitive is the presence of processes applied to representational states that result in cognitive work. Tracking the representational states can uncover the specific cognitive processes being employed. In a system, these representational states can be directly observable. While the movement of the boundary of the unit of analysis from individual to system facilitates observation, there is still the difficult task of deciding which observed representations are actually relevant representational states for a particular cognitive activity. Theory and domain expertise work together during observation to aid an analyst in determining these task relevant representational states. However, the method differs from these other fields by the pervasive influence of the distributed cognition theory.
Cognitive theory is far from sacrosanct. Indeed, in recent years the dynamicism of mainstream cognitive theory has been shown by its adaptation and incorporation of the connectionist challenge from below and its recent response to the challenge of situated action from above. We firmly believe that applied cognition must be based upon cognitive theory.

7 References

Anderson, T.D., and Garisson, D.R. (1995) Critical Thinking in Distance Education: Developing Critical Communities in an Audio Teleconferencing Context. *Higher Education,* 29, 183-199.

Brown, J.S., Collins, A. and Duguid, P. (1989) Situated Cognition and the Culture of Learning. *Educational Researcher*, 18 (1), 32-42

Card, S.K., Moran, T.P. and Newell, A. (1983) *The Psychology of Human-Computer Interaction*. Lawrence Erlbaum Associates. Hillsdale: New Jersey.

Hoyles, C. (1993). Microworlds/schoolworlds: The transformation of an innovation. In C. Keitel and K. Ruthven (Eds), *Learning from computers: Mathematics Educational Technology*, Berlin: Springer-Verlag, 1 -7.

Kahn, M., and Prail, A. (1994). Formal Usability Inspections. In Nielsen, J. and Mack, R.L. (edts) *Usability Inspection Methods*. New York: John Wiley & Sons, 141-171.

Karoulis, A., Demetriades, S., Pombortsis, A. (2000) The Cognitive Graphical Jogthrough – An Evaluation Method with Assessment Capabilities. *Applied Informatics 2000 Conference Proceedings*, February 2000, Innsbruck, Austria, 369-373. Anaheim, CA: IASTED/ACTA

Kieras, D. E. (1988). Towards a practical GOMS model methodology for user interface design. In M. Helander (Ed.), *The handbook of human-computer interaction*. 135-158. Amsterdam: North-Holland.

Kirschenbaum, S. A., Gray, W. D., & Young (1996). Cognitive architectures for human-computer interaction. *SIGCHI Bulletin*.

Lewis, C. and Rieman, J. (1994). *Task-centered User Interface Design - A practical introduction*, Retrieved from ftp.cs.colorado.edu/pub/cs/distribs/HCI-Design-Book

Nielsen, J. (1993). *Usability Engineering*. San Diego: Academic Press.

Norman, D.A. (1986). Cognitive engineering. In Norman, D. & Draper, S. (eds) *User Centered System Design* Hillsdale, NJ: Lawrence Erlbaum Associates.

Polson, P.G., Lewis, C., Rieman, J. and Warton, C. (1992) Cognitive Walkthroughs: a Method for Theory-based Evaluation of User Interfaces. *International Journal of Man-Machine Studies*, 36, 741-773.

Preece, J., Rogers, Y., Sharp, H., Benyon, D., Holland, S., Carey, T., (1994) *Human-Computer Interaction*, Addison-Wesley Publ. Company.

Rowley, D. & Rhoades, D. (1992 May). The Cognitive Jogthrough: A Fast-Paced User Interface Evaluation Procedure. *Proceedings of ACM CHI '92,* Monterey, California, 389-395.

Spoole, J., and Schröder, W. (2001). Testing Web Sites: Five Users Is Nowhere Near Enough. *Proc. of ACM - CHI 2001*

Squires, D. and Preece, J. (1999) Predicting Quality in Educational Software: Evaluating for Learning, Usability, and the Synergy Between Them. *Interacting with Computers,* 11(5) 467-483

Stults, R. (1988). The Experimental Use of Video to Support Design Activities. *Xerox PARC Technical Report SSL-89-19*, Palo Alto, California.

Suchman, L.A. (1987) *Plans and situated actions: The problem of human computer interaction*. Cambridge: Cambridge University Press.

Tselios, N.K., Avouris, N.M., and Kordaki, M. (2001). A tool to model user interaction in open problem solving environments. In N. Avouris, N. Fakotakis (ed.) *Advances in Human-Computer Interaction*, 91-95, Typorama Publ., Patras, Greece.

Tselios, N., Avouris, N and Kordaki, M. (2002). Student Task Modeling in design and evaluation of open problem-solving environments, Education and Information Technologies, 7(1) 19-42.

Making Instructions 'Visible' on the Interface: An Instructional Approach to the Acquisition and Retention of Fault-finding Skills

Tom Kontogiannis and Nadia Linou

Dept. of Production Engineering & Management,
Technical Univerisity of Crete, Chania, Greece,
Email: konto@dpem.tuc.gr; linou@safety.tuc.gr

Abstract

This study explored how information technology can be used within an instructional approach in the acquisition and retention of fault-finding skills. Instructions are made *visible* on the user interface by presenting trainees with a set of *tell tale* signs to help them discover diagnostic rules. A group of subjects was trained in using this interface while discovery was guided by providing them with a model of the plant. Another group received the same model but practised on a conventional interface while a third group was trained to apply a set of heuristics with the support of the plant model. The new user interface enabled trainees to achieve higher diagnostic scores as well as retain their fault-finding skills better on a retention task. Making instructions *visible* seemed to reduce the workload in remembering instructions and allowed trainees to impose their own organisation of knowledge on the instructions provided by the training method.

1 Introduction

Over the last twenty years, several instructional methods have been developed and evaluated, such as teaching plant theory, heuristics, and decision flow charts. However, having to learn and follow a specific method may increase the workload of operators in remembering instructions. Teaching a mental model of the system, for instance, has proved to be a good basis for fault-diagnosis skills (Patrick & Haines, 1988) but increases the workload of trainees in remembering relationships between process parameters and translating causal relationships into diagnostic search strategies. Other training methods (e.g., diagnostic heuristics) may restrict operators into a single strategy, hence, denying opportunities for exploring alternative solution paths. There is a need, therefore, that operators are equipped with efficient strategies and, at the same time, are allowed some flexibility in adapting instructions to develop their own strategies.

An approach to overcome such problems is to design the interface in ways that causal links and constraints of a technical system are made transparent to the users. Several studies have used this approach by 'externalising' a user model of causal links of parameters (Heuer, Ali, & Hollender , 1995) or by making the constraints of the system transparent on the interface (Vicente, Christoffersen & Pereklita, 1995). A concern with these studies is that interface design has not been integrated with training, hence giving rise to problems of user misconceptions and motivation loss due to the difficulties involved in understanding principles of operation without the benefit of prior training.

The present study aims to explore how to blend system representations with verbal training information so that trainees develop comprehensible mental models and acquire versatile fault-finding skills. An earlier study by Kontogiannis & Linou (2001) showed that fault-finding skills are better supported when a set of instructions and principles of operation are made visible or transparent on the interface. These *tell tale* signs are supplemented with verbal instructions about plant theory so that trainees discover for themselves a set of heuristics for fault diagnosis. The proposed approach has been investigated in the context of a distillation plant. A new interface was developed portraying several *tell tale* signs designed to prompt trainees discover a set of diagnostic heuristics. Since fault-finding skills may not be exercised for long time intervals, the issue of skill retention has also received a lot of attention in this study. Our hypothesis has been that, by making causal links and diagnostic rules visible on the interface, trainees would develop a better understanding of the system that may also provide the basis for the retention of skills. Hence, trainees may be able to recall better and reconstruct aspects of fault-finding that have not been practised for long time intervals.

2 Blending *Tell Tale* Signs and Training Instructions

In this study, a set of heuristics was developed on the basis of a fault-symptom matrix where a large number of plant failures were compared in terms of their symptoms. On the basis of this comparison and the experience of the authors, a set of generic heuristic was developed, providing the basis for the design of a new control panel. Each diagnostic heuristic was represented on the new interface as a set of *tell tale* signs, exploiting the graphic facilities of information technology. The principle of functional grouping was applied to design multiple-trace recorders showing the sequence and pattern of change of process parameters. Figure 1 shows the sequence of change of the bottom, middle and top level temperatures (TI2, TICreb, TI3) of the distillation column which can prompt trainees to discover heuristic # 1 (i.e., the temperature that changes first, indicates the general location of the fault – '*the half split rule*').

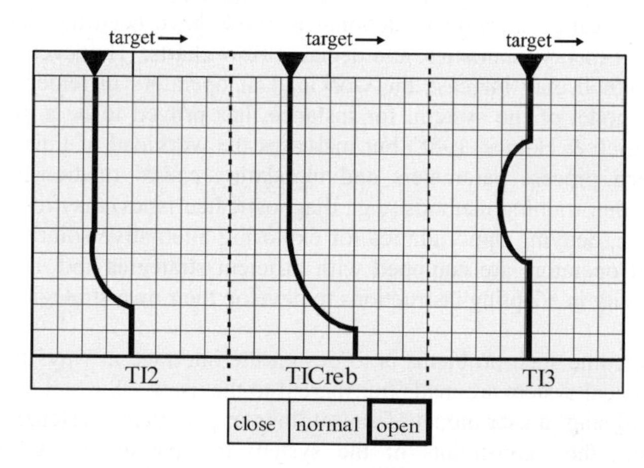

Figure 1: Tell tale signs showing temporal sequence patterns (heuristic #1)

The panel sector in Figure 2 was designed in order to facilitate the diagnosis of controller failures and control valve failures (i.e., valves stuck in position). In level-controllers, parameters and valves should change in the same direction in contrast to temperature-controllers where they should change in opposite directions (heuristic # 2). The arrows in the last column remind trainees of the correct direction of change. In Figure 2, the valve of the automatic level controller LIC2 has been stuck, failing to prevent the increase of the level in the bottoms of the column.

Controller	var.	valve	Temp
TICheat	–	+	
TICreb	–	+	
Controller	**var.**	**valve**	**Level**
LIC2	+	0	
LIC3	–	–	
FIC3	–	–	

Figure 2: Tell tale signs for identifying controller failures (heuristic #2)

The sector in Figure 3 was designed in order to both explain the causal relationships between parameters and prompt subjects to develop diagnostic rules. The two bars may prompt subjects to remember that changes of the column level (LIC2) can be due to changes of the temperature of the feed (TICheat) and the temperature of the reboiler (TICreb). In turn, changes of the temperature of the feed (TICheat) can be due to changes of the flow rate of the feed (FIf). A careful examination of Figure 3 shows that the middle level temperature (TICreb) failed to return to its target value, an indication of a potential under-heating problem of the reboiler (i.e, heuristic #3). This information is not easily detected on the recorders (Figure 1) but it has high diagnostic value.

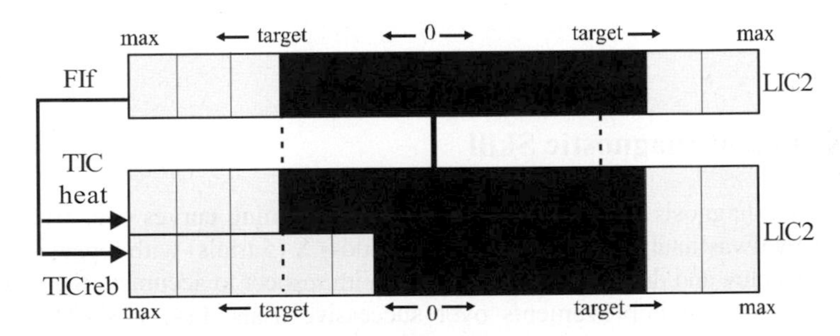

Figure 3: Tell tale signs for deriving principles of operation and discovering heuristic # 3

3 Method

Three experimental groups took part in a series of fault diagnosis tests and trained as follows:

- **Condition T(old)** was trained in plant theory to perform a set of diagnostic tasks and tested on the traditional panel.
- **Condition T(new)** received the same plant theory but practised and tested on the new control panel.
- **Condition T+H** received a set of diagnostic heuristics and a similar plant theory while tested on the traditional panel.

In our hypothesis, we anticipated that the T(new) group would achieve better diagnostic scores than the T(old) group. However, we would not expect any significant differences between the T+H and T(new) groups since both groups received the same plant theory and were instructed or prompted to develop similar diagnostic strategies. Approximately 30 subjects were assigned to the three groups and received a training course for about 12 hours.

The traditional control panel consisted of indicators (i.e., bar graphs) and recorders showing the evolution of parameters over time. Hence, the progression of the scenario could be seen only for the parameters displayed on the recorders. The new interface was equivalent to the traditional panel in the sense that the same variables were displayed but in different formats.

The sequence of training was as follows:

- **Training** – subjects were trained in fault diagnosis according to their respective methods and received aided practice on 8 failures.
- **Acquisition Test** – subjects took a series of five tests each consisting of 8 faults. An effort was made to make tests equivalent in terms of novel elements and difficulty.
- **Retention Test** – subjects were tested six weeks later in a target fault set, comprising eight faults that had been practised twice in the acquisition phase.

Three performance measures were used in the retention phase: (i) a *retention score* (i.e., success in finding the faults of the target set regardless of whether they were found or not in the acquisition phase), (ii) a *recall score* (i.e., success in diagnosing faults that were correctly diagnosed in the acquisition phase), and (iii) a *reconstruction score* (i.e., success in discovering faults that were not diagnosed in the acquisition phase).

4 Results

4.1 Acquisition of Diagnostic Skill

Differences in fault diagnosis were analysed in terms of learning curves over five trials with the original task. A two-way analysis of variance (3 methods) X (5 trials) with repeated measures was carried out for accuracy and duration of diagnosis. With respect to accuracy, the results indicated that there were significant improvements over successive trials, $F(4, 128)=21.8$, $p<.001$. All groups improved over trials to the extent that accuracy in trials 4 and 5 was greater than in earlier trials. Methods had also significantly different effects, $F(2, 32)=6.44$, $p<.01$, that were analysed separately for each trial because of an interaction between trials and methods. A Tukey test found that the T(new) group was better in accuracy than the T(old) and T+H groups ($p<.01$) in trials 4 and 5. With respect to task duration, no significant differences were found between methods. The new interface enabled the T(new) group to diagnose 85% of the faults which exceeded the T(old) and T+H groups by 25% in accuracy.

The study also looked into the acquisition scores for the eight faults of the target set. A one-way between-subjects ANOVA showed that there were significant differences between methods in the acquisition scores for the target fault set, $F(2, 32)=6.03$, $p<.01$. The Tukey′s post multiple comparisons test showed that the T(new) group was also better than the other two groups ($p<.02$).

4.2 Retention of Diagnostic Skill

A one-way between-subject ANOVA found significant differences in the retention scores of training groups, $F(42.32)=10.3$, $p<.001$. The Tukey test showed that the T(new) group was better in accuracy than the T(old) and T+H groups ($p<.01$). However, this result could be due to the degree of original learning rather than ability to remember. It is important, therefore, to examine group differences in terms of their recall and reconstruction scores which are independent of diagnostic performance in the acquisition phase. A one-way between-subject ANOVA found significant differences in the recall scores between training methods, $F(2,32)=6,9$, $p<.01$. According to the Tukey test, this could be attributed to the fact that the T(new) group recalled more faults of the target set than the T(old) group ($p<.02$). A comparison in terms of reconstruction scores failed to find significant differences.

5 Discussion of Results

In summary, the results showed that making instructions visible has helped the T(new) group to perform more accurate than the T(old) and T+H groups in the original task. In addition, the T(new) group was better in retaining their skills as shown in the superior retention and recall scores they achieved. Making instructions *visible* reduces the workload in remembering instructions, encourages discovery of diagnostic rules, and allows trainees to impose their own organization of knowledge. These learning mechanisms may provide a better basis for training in the acquisition and retention of skills.

Making instructions visible is only a means of communicating 'knowledge in the world' to the users. The training designer should make sure that this heuristic knowledge covers a wide range of events and make explicit the main assumptions underlying the specified knowledge. Any events or faults that alter structural features of the plant may render some heuristics invalid which can be a major source of diagnostic errors. The challenge for training design is to build a reliable body of knowledge on the basis of operational experience, system reliability studies and process simulation models.

References

Heuer, J., Ali, S., & Hollender, M. (1995). Mediation of mental models in process control through a hypermedia man-machine interface. In T.S. Sheridan (Ed.). *Analysis, design and evaluation of man-machine systems* (pp. 499-504). IFAC Symposium. Oxford, England: Pergamon.

Kontogiannis, T., & Linou, N. (2001). Making instructions visible on the interface: an approach to learning fault diagnosis skills through guided discovery. *International Journal of Human Computer Studies*, 54, 53-79.

Patrick, J., & Haines, H. (1988). Training and transfer of fault-finding skills. *Ergonomics*, 31, 193-210.

Vicente, K., Christoffersen, K., & Pereklita, A. (1995). Supporting operator problem-solving through ecological interface design. *IEEE Transactions on Systems, Man and Cybernetics*, SMC-25, 529-545.

Reading News from a Pocket Computer: an Eye-movement Study

Jari Laarni
Ilpo Kojo
Lari Kärkkäinen

Center for Knowledge and Innovation Research, Helsinki School of Economics
HTC-Pinta, Tammasaarenkatu 3
00180 Helsinki, Finland
laarni@hkkk.fi

Pekka Isotalus

Department of Phonetics, University of Helsinki
Vironkatu 1 B,
00014 University of Helsinki, Finland
pekka.isotalus@helsinki.fi

Abstract

We measured participants' eye movements when they chose news items and read online news stories from a pocket computer. The news site contained personally-relevant information and an animated interface agent that, for example, commented the news items. All the readers noticed personally-relevant information. They also tended to read first the animated agent's comment. The layout of the front page had an effect on the order in which the news stories were chosen when participants read all the stories. When the participants had to read only two stories, they tended to choose them according to the content of headlines.

1 Introduction

A growing number of people read news online. There are, at least, three trends that characterize the use of www news services. First, since pocket computers provide easy, wireless access to the Internet, it is more and more common that people use pocket computers to access news services. Another trend is to tailor news services for each visitor's personal needs and preferences. Personal information that can be used includes information the user has explicitly given, for example, by filling out a questionnaire and information based on his/her previous visits to the Web site. The third trend that helps to make the site more comfortable to the user is to use an animated interface agent that exhibits human-like behaviour, such as expresses emotions and motives and talks with the user.

There are several interesting questions concerning the development of personalized news services for small-screen interfaces. Some of them will be studied here. The first question is whether a reader notices personally-relevant information, e.g., his/her own name, when he/she visits a news site. Studies of selective attention have shown that, for example, a person's own name tends to capture attention both in vision and audition (Mack & Rock, 1998). Thus, a good guess is that, the reader immediately notices his/her own name when he/she first encounters it at a news site.

Animated interface agents are thought to make a computer system more life-like, engaging and motivating. The results of empirical studies are, however, ambiguous (Dehn & van Mulken,

2000). For example, it is questionable whether agents have a positive effect on learning (André, Rist & Müller, 1999). In order to produce positive effects, one prerequisite, of course, is that the user notices the presence of an animated agent. Thus, the second question we studied is, to what degree users notice the animated agent and its comments.

There are, at least, two different ways to choose news items. A news story can be chosen from a simple list of items. Another possibility is to choose a story from a front page which contains headlines and a couple of photos. The third question we are interested in is whether there are differences in performance between readers who choose the news items from the list and those who choose them from the front page. For example, it is possible that the stories are read in a different order in these two cases and there are differences in free recall.

The fourth question is whether text or pictorial information is first noticed and which one is viewed longer. There is some evidence that photos and images have not the special status they have been thought to have when people surf the Web (Poynter Institute, 2000). On the other hand, concrete interactive images and photos may facilitate recall and memory of facts (Lutz & Lutz, 1977; Waddill & McDaniel, 1992).

The fifth question we are interested in is whether the layout of a front page and the properties of headlines have any effect on the order in which the stories are chosen. For example, people may notice a headline but not choose it, or they may totally fail to notice the headline. Our aim is to study what particular features cause people to notice a headline, whether people view headlines/choose news stories in any particular order and how viewing time varies as a function of particular features of the photos and the headlines.

The above-mentioned questions were studied in two experiments in which users' eye movements were measured. To our knowledge, there is little research that has addressed the characteristics of eye movements when people inspect front pages of online news sites. Especially, there is little research on reading online news from pocket computers.

2 Method

Totally 24 participants volunteered in the two experiments. All the participants were young adults. They were all unaware of the purpose of the experiment. The participants were paid for their participation.

The stimuli were displayed on a pocket computer (Casio Cassiopeia). Display resolution was 240 x 320 pixels. A chin rest was used to stabilise a participant's head. A book stand held the computer.

The stimulus material included a welcome page, a front page, a headlines page, a couple of news stories (six stories in Experiment 1 and twelve stories in Experiment 2) that were chosen from the online edition of a Finnish afternoon paper and a farewell page. The headlines were on the front page and on the headlines page. The front page was a typical front page of an afternoon paper with two photos and several headlines of news stories. In the headlines page there was a simple listing of the headlines. By clicking on the headline the user saw the whole news story.

The agent was a still picture of a young woman designed to resemble an attractive news anchor. The anchor was shown on the welcome and farewell page. Moreover, she appeared with three new

Figure 1: The welcome page of the site. The black dot shows that the reader is just viewing the forehead of the agent.

stories in which she expressed a short comment. For example, she could say: "This is news about physical exercise".

In the welcome page there was a balloon, a face figure and a small triangle (link to the next page). The text of the balloon said: "Hello X (= the first name of the participant)! You are welcome to read Iltalehti news stories." The farewell page was quite similar. There was a photo of the anchor and a balloon. For example, the text of the balloon could say: "Thank you for using Iltalehti news service! Goodbye and welcome back!"

Text appeared on the screen about 40 cm from the reader's eyes. The angle between gaze and a page was about 105 deg in all conditions. The luminance of the text was about 1.0 cd/m^2 against a background of about 60 cd/m^2.

The participants' task was to read all the six news stories in Experiment 1. They read them in an order they liked. Half of them chose the news items from a list of headlines, the other half chose them from the front page. The participants used the stylus to advance to the next page and to scroll the text. In Experiment 2 the participants chose two news stories from a selection of twelve stories and read them. They also had time to skim a couple of other stories.

The participants' eye movements were recorded using a head-mounted gaze tracking system (SMI iView). A participant's right eye was monitored with a miniature infra-red camera while one infra-red LED (light-emitting diode) illuminated the eye. The eye tracking system was controlled by a PC computer. In Figure 1 the position of the black dot indicates the position the reader is viewing. Video images of the pupil and corneal reflections were captured at 50 Hz by the eye tracker. The eye movement system was calibrated using a set of 5 screen locations. iView software was used to detect fixations and calculate their durations. To be considered a fixation, a gaze point had to fall within a spatial area between about 2 x 2 deg, and had to have a minimum duration of 100 ms.

We used here an observational method to analyse eye-movement data. The data were first recorded to video clips, and then an observer analysed the clips. We also gathered information about reading speed and memory for the stories. In addition, the participants completed a questionnaire assessing the behaviour of the agent.

3 Results and discussion

3.1 Experiment 1

On the welcome page, 88% of the participants looked first at the balloon text. All the participants noticed their own name in the balloon. 12% of the participants did not look at the photo at all. Concerning the farewell page, the results were similar: 75% of the participants looked first at the balloon text. When the participants first opened the front page, 63% of them looked first at photos. Those participants that chose the news items from the front page returned to this page later in order to choose the next item. In 52% of these cases they looked first at the news item list. The photos were looked first only at 26% of the cases. It seems to be that the readers looked at the photos when they first encountered them. Later the photos were ignored, as was all the information that was not relevant for the reader's goal.

There seems to be no correlation between what was looked first and which story was chosen first when the readers first encountered the front page. In the later visits, the correlation was, however, clearly higher: when the readers later returned to the front page to choose the next story, they looked straight to the location of the headline they were going to choose.

The layout of the front page had a clear effect on the order in which the news stories were chosen. Those who chose the stories from the front page read the highlighted stories either first or last. The news stories which were listed in a box were read in the order of their presentation. Also those who chose the texts from the list read them in the order of their presentation.

The average time the participants spent reading was 12.5 minutes. By which way texts were chosen had no effect on total reading time. The time duration that was spent in choosing news items from the front page and from the list was also comparable. The readers spent viewing the front page longer in the first visit when they chose the items from the front page than when they chose them from the list. Those who chose the news items from the front page seemed to remember somewhat more items than those who chose the items from the list.

3.2 Experiment 2

One problem with Experiment 1 was that, since the reader had to read all the six news stories, he/she needed not choose anything and leave something out. Therefore, in Experiment 2 only two stories had to be completely read. Totally, the participants chose ten stories, and 50% of them chose the most popular story. Those who saw first the front page typically chose either one story from the front page and the other one from the list or both stories from the list. Those who chose the news items from the list spent quite a lot of time in scrolling the list back and forth. In this case, the content of headlines determined which stories were read. That is, it seems to be that, contrary to Experiment 1, the layout of the front page and the order in which the news items were listed had only a small effect on the order in which the items were chosen.

The agent and its comments were also here well noticed. However, the participants thought that the agent was not very useful. Their comments were either neutral or mildly negative. For example, even though the smiling of the agent was noticed, the agent was thought to be quite humorless and impersonal.

4 Conclusions

The readers tended to notice well the personally-relevant information. The agent and its comments were also well noticed, even though the agent was here a still picture. Thus, there is a good chance that an agent makes a computer system more engaging and motivating provided that the agent is interpreted as friendly and jolly. To what degree animated interface agents increase computer immediacy, i.e., the feelings of closeness with the computer, remains, however, to be seen (Isotalus & Muukkonen, 2002; LaRose & Whitten, 2000).

The layout of the front page had a clear effect on the order in which the news stories were chosen only when all the news stories had to be read. Since the readers seemed to choose news stories according to the content of their headlines, it is important which information the headlines give. Headlines with a short summary might help readers. The choice would also be easier if the news stories have been divided in categories.

5 Acknowledgements

This work has been undertaken in the framework of the E-TRACKING project (IST-2001-32323) which is partly funded by the European Commission under the Information Society Technologies 5th Framework Programme. The partners of the project are: Università degli Studi di Pavia, Université de Nice Sophia Antipolis, Giunti Ricerca, Fraunhofer Institute for Applied Information Technology, Telefónica Investigación y Desarrollo, ARDEMI.

6 References

André, E., Rist, T., & Müller, J. (1999). Employing AI methods to control the behavior of animated agents. *Applied Artificial Intelligence,* 13, 415-448.

Dehn, D. M., & van Mulken, S. (2000). The impact of animated interface agents: A review of empirical research. *International Journal of Human-Computer Studies,* 52, 1-22.

Isotalus, P. M., & Muukkonen, H. (2002). Animated agent immediacy and news services with handheld computers. *Communication Quarterly*, 50, 78-92.

LaRose, R., & Whitten, P. (2000). Re-thinking instructional immediacy for web-courses: A social cognitive exploration. *Communication Education,* 49, 320-338.

Lutz, K. A., & Lutz, R. J. (1977). Effects of interactive imagery on learning. Application to advertising. *Journal of Applied Psychology,* 62, 493-498.

Poynter Institute (2000). Definitely not your father's newspaper/Surprise! All eyes on text. Retrieved November 3, 2000, from http://www.poynter.org/centerpiece/050300.htm

Rock, I. (1998). Inattentional Blindness. Cambridge, MA: MIT Press.

Waddill, P. J., & McDaniel, M. A. (1992). Pictorial enhancement of text memory: Limitations imposed by picture type and comprehension skill. *Memory & Cognition,* 20, 472-482.

Personal Assistant for onLine Services: Addressing human factors

Jasper Lindenberg

TNO Human Factors
The Netherlands
lindenberg@tm.tno.nl

Stacey F. Nagata

Utrecht University
The Netherlands
nagata@cs.uu.nl

Mark A. Neerincx

TNO Human Factors
The Netherlands
neerincx@tm.tno.nl

Abstract

The Personal Assistant for onLine Services (PALS) project aims at substantially improving the user experience of mobile internet services. It focuses on a generic solution: a personal assistant, which attunes the interaction to the momentary user needs and use context (e.g. adjusting the information, presentation and navigation support to the current context, device and interests of the user). Based on scenario analysis and literature research the initial outcomes of the PALS project provide high level user requirements and a research agenda for establishing a well-founded adaptation theory with corresponding user-interface concepts.

1 Introduction

In contrast with the PC-based Internet, the mobile Internet has mainly provided poor user experiences and has not met expected levels of performance by end users (Ramsey & Nielsen, 2000). To enhance the user experience substantially and to utilise the (envisioned) possibilities of mobile services, we propose a Personal Assistant for OnLine Services (PALS, Figure 1).

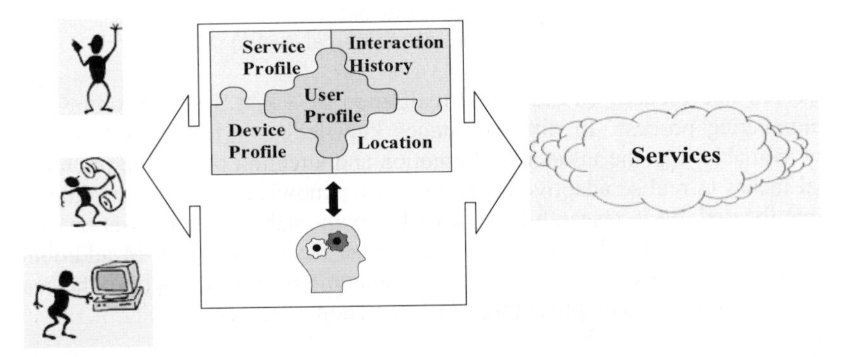

Figure 1: The personal assistant attunes the interaction to (mobile) Internet services.

PALS mediates between the user and the service by adapting the user interface to the individual user's interests, capacities, usage history, device and use context (social, environmental and technical). PALS will be developed using a cognitive engineering (CE) method that provides theoretically and empirically founded concepts for adaptation (cf. Neerincx, in press).

2 Method

To guide the PALS design process a cognitive engineering method is used (Figure 2). CE originated in the 80's to improve computer-supported task performance (e.g., Norman, 1996) and emerged from the fields of cognitive science and artificial intelligence. CE aims at generating new or enhanced human-computer interactions by increasing insight in the cognitive factors of human performance (e.g. Neerincx, in press).

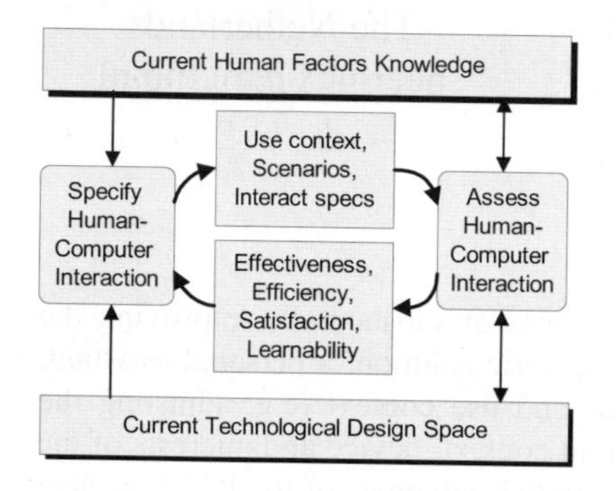

Figure 2: The cognitive engineering method

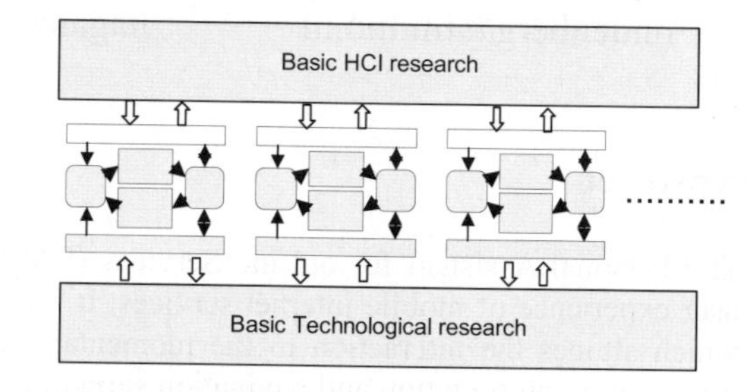

Figure 3: Combining basic research and CE-cycles

Cognitive engineering provides techniques to guide the iterative development of PALS. In the first iteration PALS is conceptualised through scenario based design. The subsequent iterations are used to narrow and specify details of the PALS concept through prototype design and testing of the concept. However, the goal of the project is not only to realise an effective PALS but also to generate fundamental Human Computer Interaction (HCI) and technological knowledge. Therefore, three research lines can be distinguished within the PALS project:

1. *PALS creation:* using a cognitive engineering approach.
2. *Basic HCI research*: extending the HCI knowledge base.
3. *Basic technological research:* extending the technological design space.

The first research line focuses on the actual realisation of a PALS demonstrator guided by the cognitive engineering process. In different stages knowledge or technology is needed that is currently not available, e.g. the influence of emotion and attention on mobile interaction, and the technological means to realise adaptive navigation. This knowledge will be developed within the two, discipline focused, basic research lines of PALS, both enabling the realisation of an effective PALS and extending the HCI and technical knowledge base (Figure 3). In addition to the CE generated questions that 'feed' the basic research, autonomous processes within the basic research line will 'feed' the CE process by providing new interaction concepts.

3 PALS Creation

In the *PALS Creation* research line scenarios were developed to refine the original PALS concept and establish high-level user requirements. Based on PALS and user characteristics which were identified from the scenario analysis and literature review, four general user requirements were distinguished:

- Adaptation of dialogue (AD), adapt the information presentation, dialogue structure, and modality to the current context of use, available cognitive resources and emotional state of the user.
- Adaptation of content (AC), adapt information selection based on long-term and momentary interests of the user and the current context of use.
- Pro-active scheduler (PS), provide suggested actions based on scheduling information and current context of use.
- Continuity of interaction (CI), maintain a sense of continues use of services over devices and usage sessions.

The following example scenario illustrates the application of the user requirements.
Henk is travelling to a business meeting via train. In the train he consults his route planner to find the exact location of this meeting. Last night, Henk planned the route on his desktop PC. When Henk accesses the route planning service, his pre-planned route is immediately presented (CI) in a suitable format on his PDA (AD) including his current location (AC). On arrival at the train station Henk puts his PDA in his pocket. At the busy and noisy station PALS suddenly "beeps" and "vibrates" hard (AD). Henk needs an international train ticket for the coming weekend and PALS suggests buying a ticket now knowing there is enough spare time (PS). During his transaction at the service desk PALS holds back incoming, non urgent, messages (AD).

The current HCI knowledge base and technological design space will enable the realisation of PALS to a certain extent. However, HCI knowledge and enabling technologies are lacking to fully realise the identified user requirements. These issues will be dealt with in the two basic research lines of the PALS project. The basic HCI research is described in this paper, information on the basic technological research which focuses on adaptive navigation and dynamic user profiles can be found in Herder & van Dijk (to appear).

4 Basic HCI Research

To meet the identified user requirements fundamental knowledge is needed. Theories on cognitive capabilities (e.g. attention) and emotion (e.g. arousal) are needed to realise an adequate adaptation of dialogue. These research topics are addressed in the PALS basic HCI research line by focusing on attention and cognitive load for multitasking activities in a mobile context (defined by task-set switching and time pressure), and the role of emotion (defined by valence and arousal level).
The results of the basic research will refine the user requirements, enhance the empirical and theoretical foundation of PALS and enable the realisation of PALS by providing essential knowledge to the CE development cycle.

4.1 Attention

The mobile context is a fluid environment that contributes to the complexity of using a mobile device to conduct a task. Literature and scenario analysis have identified attention and cognitive load as important factors for efficient and satisfactory interaction in the mobile environment. Research on attention in the mobile context is required to establish an effective adaptation of dialogue.
Particularly, high rates of task interruption, low amounts of attention dedicated to a task, time constraints and flexibility of task completion are important aspects of the mobile context (cf. Vaananen Vainio–Matilla & Ruuska, 2000). These aspects play a role in usage issues with web services, resulting in lowered performance on tasks and navigation, on a mobile device. The focus

of the study is to investigate user performance of an activity (transaction, information search) on a device (e.g. mobile device, desktop) while task switching (duration of time to switch between tasks) due to interruptions (computing or external) and experiencing time constraints. The hypotheses suggest that interruptions from the computing environment (e.g. Instant Message, email etc.) are more disruptive to performance on web tasks when compared to interruptions from external tasks (e.g. cell phone calls, pagers etc.) Defined tasks with a step-by-step process (e.g. purchasing transaction) are less vulnerable to interruptions and declines in performance than less well defined tasks (e.g. information search). Time constraints will produce significant declines in performance for tasks with interruptions compared to tasks without time limits.
User requirements for communication with the personal assistant (AD) will be addressed by findings on how users handle interruptions in different dialogue modes (e.g. auditory, visual). For example, support for the hypothesis that internal computing interruptions are more disruptive to performance than external interruptions, suggests that PALS should mediate internal computing interruptions. Results on task switching and interruptions may infer a design guideline to present a mobile device user with assistant dialogue (e.g. visual) for a reminder (messages unrelated to current task) during natural pauses in a task or between tasks

4.2 Emotion

Emotion plays an essential role in human problem-solving behaviour in general and, consequently, proves to affect the user experience and performance of human-computer interaction substantially (e.g., Picard; 1997). For example, Klein, Moon & Picard (2002) showed that "emotional support", consisting of sympathy and compassion expression, leads to less frustration and more prolonged interaction with a computer game. Emotional states, caused by an internal or external stimulus, are defined on two dimensions: the arousal level—low versus high—and the valence—positive versus negative (Scheirer et al., 2002). The personalisation of mobile services should take into account the user's states and possible effects of the human-computer interaction on these states.
For the user requirement "adaptation of dialogue" (AD), we therefore aim at attuning the user interface appearance (including a possible presentation of a PALS component) and moment of information presentation to the user's state. The first emotion experiments will investigate the effects of device type (iPAQ versus laptop) and emotional state on user's behaviour and trust. User interface appearances, personalisation elements and interaction aspects prove to affect trust (Karvonen, 2000) and we assume that emotion plays a role in these effects. Trust depends on persistent, competent behaviour of a system that has a purpose of serving the user. It develops as a function of experience, and is influenced by errors from the system, interface characteristics and individual differences on part of the user (Muir, 1994). The experiments will test if the user's locus of control, a general belief of the control one exerts over external events (Lazarus, 1991), can explain some of the individual differences. In sum, this part of our study will provide insight in the relations between emotion, device, trust and performance, in order to develop a personalisation mechanism that establishes adequate user behaviour and an appropriate sense of trust for different interaction devices and emotional states.
Subsequently, we will focus on the requirement "adaptation of content" (AC) that should result in attuning the content (e.g. a specific set of games) to the user's state. Experiments will provide insight in the interaction effects between user's emotional state, preferences and information filters. The "pro-active scheduler" requirement, e.g. suggest to do no risky transactions in a specific state, will be further explored in the CE traject (figure 2 and 3) to derive the basic research questions.

5 Results and future work

After the first year of PALS the original concept is refined by defining four high level user requirements and a research agenda has been set for establishing a well-founded adaptation theory with corresponding user-interface concepts. During the conference we will present a first version of the PALS demonstrator and initial results from experiments on emotion and attention that test the theoretical and technical foundation of PALS.

Acknowledgement

The PALS project is funded by IOP MMI Senter. We would like to thank Jan Willem Streefkerk for his contribution to the emotion section of this paper and the other PALS members for their contribution.

References

Herder, E., van Dijk, E.M.A.G. (to appear). From Browsing Behavior to Usability Matters. Proceedings of the 10th International Conference on Human-Computer Interaction, Crete, 2003.

Karvonen, K. (2000). The beauty of simplicity. Proceedings of the 2000 ACM Conference on Universal Usability, Arlington, VA, 85-90.

Klein, J., Moon, Y., & Picard, R. (2002). This computer responds to user frustration: Theory, design and results. *Interacting with Computers, 14, 119-140.*

Lazarus, R. (1991). Emotion and adaptation. New York, NY: Oxford University Press.

Muir, B. (1994). Trust in automation: Part I. Theoretical issues in the study of trust and human intervention in automated systems. *Ergonomics*, Vol. 37, 1905-1922.

Neerincx, M.A. (in press). Cognitive task load design: model, methods and examples. In: E. Hollnagel (ed.), *Handbook of Cognitive Task Design*. Chapter 13. Mahwah, NJ: Erlbaum.

Neerincx, M.A., Van Doorne, H. & Ruijsendaal, M. (2000). Attuning computer-supported work to human knowledge and processing capacities in ship control centres, In: Schraagen, J.M.C., Chipman, S.E. & Shalin, V.L. (Eds.), *Cognitive Task Analysis*. Mahwah, NJ: Erlbaum.

Norman, D.A. (1996). Cognitive Engineering. In: Norman, D.A. & Draper S.W. (Eds.) *User-Centered System Design: New perspectives on human-computer interaction.* Hillsdale, NJ: Erlbaum.

Picard, R. (1997). Affective Computing. Cambridge, MA: The MIT Press.

Ramsey, M. & Nielsen, J. (2000) WAP Usability Déjà vu: 1994 All Over Again, Report from a Field Study in London, Fall 2000. Nielsen Norman Group. pp 2-91.

Scheirer, J., Fernandez, R., Klein, J., & Picard, R. (2002). Frustrating the user on purpose: a step toward building an affective computer. *Interacting with Computers*, Vol. 14, 93-118.

Vaananen-Vainio-Mattila, K. & Ruuska, S. (2000). Designing mobile phones and communicators for consumers'needs at Nokia. In: Bergman, E. (Ed.) *Information appliances and beyond: interaction design for consumer products* (pp. 169-204). San Francisco, CA: Morgan Kaufman.

The Use of Haptic Cues Within a Control Interface

S. M. Lomas, G.E. Burnett

University of Nottingham
Jubilee Campus, Nottingham, NG9 1BB
sml@cs.nott.ac.uk
geb@cs.nott.uk

J. M. Porter & S. J. Summerskill

Loughborough University
Leicestershire, LE11 3TU
j.m.porter@lboro.ac.uk
s.j.summerskill2@lboro.ac.uk

Abstract

This paper investigates the use of three different types of haptic coding (location size and shape) used either singularly or in combination with each other. Ten blindfolded right hand dominant participants were asked to feel ten items on each of seven boards, which incorporated the different combinations of coding. Location coding was found to be superior compared to size and shape coding in terms of the time taken to find the correct item. Both size and shape coding were useful but relied on participants sequentially feeling each item until they found the correct one. Results from the verbal protocol revealed features of the shapes, which could be incorporated into the interface design of products where non-visual use would be desirable such as vehicle dashboard controls, telephones and television remote controls.

1 Introduction

In the field of computing most user interfaces have traditionally relied upon the user acquiring information from the system using their visual, and to a lesser extent, auditory senses. As the diversity of environments that computers are used in increases, the use of visual and auditory cues is not always possible or ideal. For example, in a car environment, controls for computers, such as navigation systems, should minimize the visual demand, allowing the driver to concentrate upon the driving task. As such, there is a requirement for alternative control interfaces that emphasize the use of non-visual cues (Burnett & Porter, 2001).

The human hand can sense a wide variety of haptic (tactile and kinaesthetic) features e.g. edges, textures, contours and force feedback (Prynne, 1995), thus it is possible to convey information regarding a control's function, mode of operation and current status without using the visual or auditory systems. This is of particular benefit to the elderly population as studies have shown that, unlike the visual and auditory systems, the tactile sense does not deteriorate to the same extent (Steenbekkers and van Beijsterveldt, 1998).

There has been an increasing interest in the field of Human Computer Interaction (HCI) for new interfaces that use haptic information. Indeed, studies have been performed in a variety of applications, for example in Virtual Reality to increase the user's sense of presence, and in aviation to warn the pilot that the plane is about to stall. In automotive applications, collision avoidance systems have been the main source of investigation. For example, Faerber, Faerber, Godhelp and Schuman, 1991 describe an active steering wheel which applies a torque in the direction the driver should turn whereas Bittner, Lloyd, Norwak and Wilson, 1999 have investigated the use of haptic braking, through the brake pedal.

In the area of human factors there have been many studies investigating tactual coding methods for manual controls to assess how a user can differentiate between two or more controls by the sense of touch. These have resulted in many guidelines regarding control coding (e.g. Jenkins, 1947; Bradley, 1969; Chapanis and Kinkade, 1972; Boff and Lincoln, 1988). In reality the use of one type of control coding on its own for a control (e.g. shape, size, location) is not always feasible. For example, the use of a *single* type of control coding mechanism for an interface containing a high number of controls in a limited space (e.g. in a car or mobile phone) may make differentiation of the controls difficult. In contrast, the use of a range of coding formats in *combination* with each other (e.g. shape and location) will increase the number of possible variations in control coding, potentially allowing greater differentiation between controls. This study was performed to compare the use of singular and combination coding cues for controls, which facilitate their non-visual use.

2 Procedure

Three types of control coding were investigated: location, shape and size. Ten right-hand dominant people (five male, five female), aged 22-34, participated in this study. Participants were blindfolded and seated at a desk. Seven boards containing ten items each were presented randomly, in turn, to each of the participants (Figure 1).

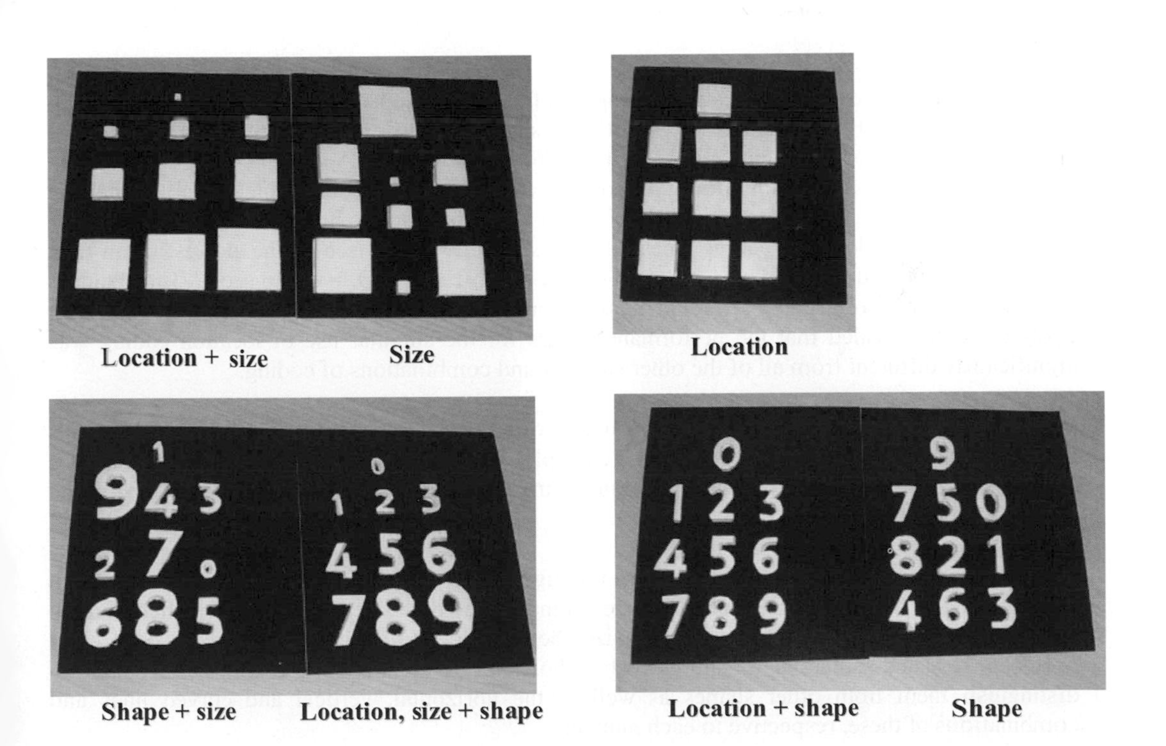

Figure 1: Boards Containing Combinations of Coding

The items where either the numbers 0-9, the items being shaped as these numbers, for the coding which involved shape, or squares for the coding which did not involve shape. Tiresias font typeface was used for its simplicity (Perera, 2002). Squares and numbers were used as all the participants were familiar with these features and they contained a range of features applicable to a wide range of potential interfaces. The size of the numbers and squares was either constant or varied depending upon whether size coding was involved. The items on the boards were arranged so that one item was in the first row in the centre, the remaining rows were located below this in a 3x3 matrix. The boards contained individual and combinations of haptic coding as shown in Figure 1.

Participants were asked to tactually find each of the items in turn randomly, using their left hand only, until they had attempted to identify all of the items on each board. They were required to verbalise what they were feeling and searching for as they completed the task. The participants' performance was recorded by video camera to establish the amount of time it took them to both initially touch the correct item, confirm the item and complete the task overall.

Participants completed a practice session before attempting the actual study. This session involved them attempting to identify ten shapes by touch and allowed them to become familiar with the procedure. Once all the data had been collected the videotapes were analysed to establish the time taken to find the correct item.

3 Results

Of the coding types evaluated location coding was found to be superior to shape and size coding in terms of mean task times (Figure 2). Statistically significant differences were found between the different combinations of coding ($F (6, 570) = 4.86$; $p<0.01$). When location coding was present, either alone or in combination with other coding types, participants were able to find the correct item far quicker than when it wasn't. The fastest time taken to hit the correct item was for the board that just had location coding (mean = 1.52 sec, S.D. = 1.60 sec). The slowest mean time was when shape and size coding were used (mean = 10.91 sec, S.D. = 16.76 sec). The standard deviations were also larger for finding items when location coding was not present (Figure 2). Post-hoc tests revealed that the performance times for the singular use of location coding was significantly different from all of the other singular and combinations of coding.

When individuals had formed a mental model of where the items were located, certain locations on the boards appeared quicker to find. For example, items located in the 'corner' positions; items '1', '3', '7' and '9', for boards with location coding, were found faster than items located in the middle of rows.

Both size and shape coding were helpful in finding items but relied on sequential feeling of each item in order to find the correct one. For size codin, finding the extreme sizes, the smallest and the largest items, was the easiest, the middle sizes being harder to differentiate (Figure 3). Verbal protocols revealed some shape features such as holes in the numbers 8, 0 and 6 which helped to distinguish them from other shapes as well as the horizontal, vertical and curved lines and combinations of these, respective to each number.

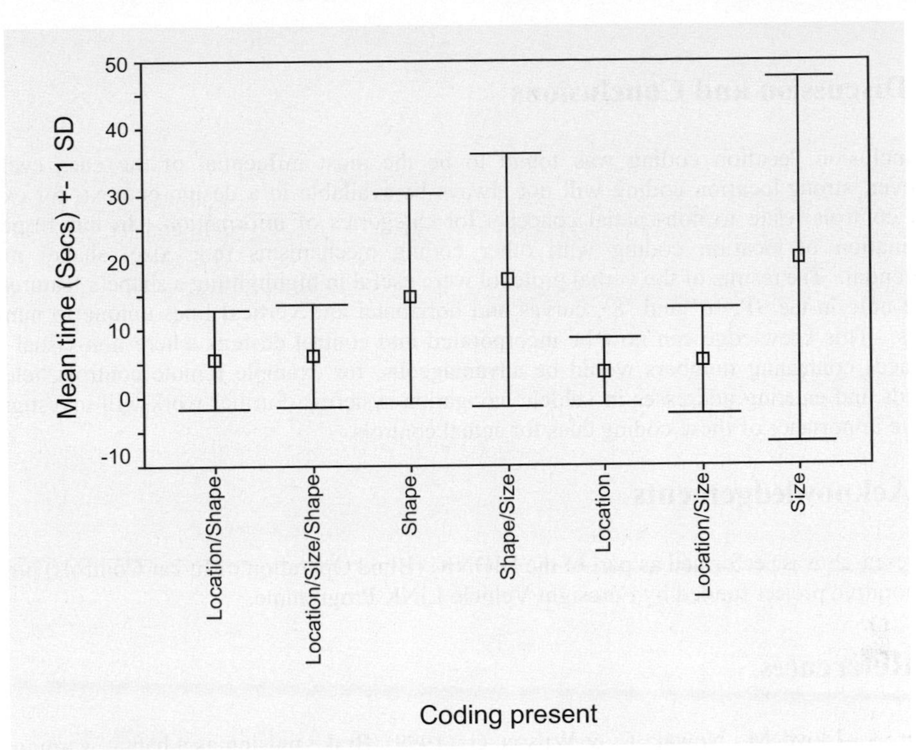

Figure 2: Mean Time Taken To Find The Correct Items per coding type

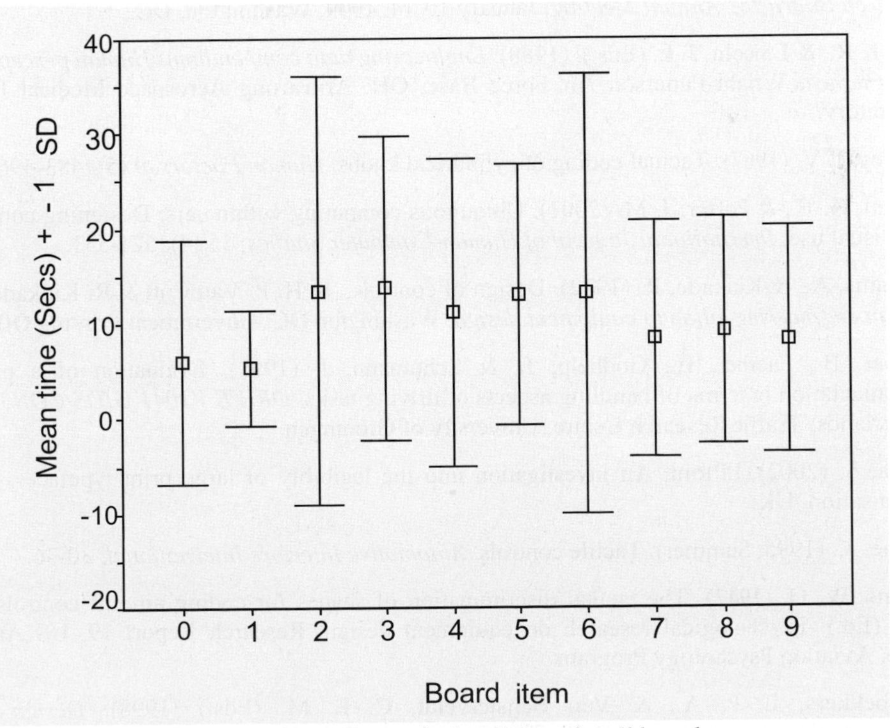

Figure 3: Time taken to find each item for all boards

4 Discussion and Conclusions

In conclusion, location coding was found to be the most influential of the cues evaluated. However, strong location coding will not always be available in a design context, for example when controls relate to non-spatial concepts for categories of information. In this respect the combination of location coding with other coding mechanisms (e.g. size, shape) may be preferential. The results of the verbal protocol were useful in highlighting a shape's features, such as the hole in the '0', '6' and '8', curves and horizontal and vertical lines unique to numerical shapes. This knowledge can now be incorporated into control designs where non-visual use of interfaces containing numbers would be advantageous, for example remote controls, telephone keypads and entering addresses in vehicle navigation systems. Further work will investigate the relative importance of these coding cues for actual controls.

5 Acknowledgements

This research was performed as part of the BIONIC (Blind Operation of In-car Controls) project, a collaborative project funded by Foresight Vehicle LINK Programme.

6 References

Bittner, A., Lloyd, M., Nowak, C. & Wilson, G. (1999). Brake pulsing as a haptic warning for an intersection collision avoidance (ICA) countermeasure, *Proceedings of the Transportation Research Board 78th Annual Meeting*, January 10-14, 1999, Washington, DC.

Boff, K.R., & Lincoln, J. E. (Eds.). (1988). *Engineering data compendium: Human perception and performance*, Wright-Patterson Air Force Base, OH: Armstrong Aerospace Medical Research Laboratory.

Bradley, J. V. (1967). Tactual coding of cylindrical knobs, *Human Factors*, 9 (5), 483-496.

Burnett, G. E., & Porter, J. M. (2001). Ubiquitous computing within cars: Designing controls for non-visual use, *International Journal of Human-Computer Studies*, 55 (4), 521-531.

Chapanis, A., & Kinkade, R. (1972). Design of controls. In H. P. Van Cott & R. Kinkade (Ed.s), *Human engineering guide to equipment design*, Washington DC: Government Printing Office.

Faerber, B., Faerber, B., Godhelp, J., & Schumann, J. (1991). Evaluation of a prototype implementation in terms of handling aspects of driving tasks, *DRIVE V1041 GIDS/CON3*, Haren, Netherlands: Traffic Research Centre, University of Groningen.

Perera, S. (2002), LPfont: An investigation into the legibility of large print typefaces, Tiresias Organisation, UK.

Prynne, K. (1995, Summer). Tactile controls. *Automotive Interiors International*, 30-36.

Jenkins, W. O. (1947). The tactual discrimination of shapes for coding aircraft controls. In P. Fitts, (Ed.), Psychological research on equipment design, Research Report 19, US Army Air Force, Aviation Psychology Program.

Steenbekkers, L. P. A., & Van Beijsterveldt, C. E. M. (Eds.) (1998). *Design-relevant characteristics of ageing users*. Delft, The Netherlands: Delft University Press.

Hazard Perception training in the BSM Driver-Training Simulator

Susan McCormack

British School of Motoring
1, Forest Road, Feltham, Middlesex, TW13 7RR, UK
smccormack49@aol.com

Abstract

There are many advantages to using a simulator for driver training. As a driver-training organisation, one of the main advantages for BSM is that we are able to monitor driving behaviour and give feedback to trainees. The BSM driver-training simulator replicates the driving task extremely effectively to allow driver training to take place. The BSM simulator was originally designed to develop physical control skills whilst driving, but it has recently been identified as an effective tool for the development of hazard perception skills as well.

It is well known that novice drivers are more likely to be involved in an accident during the first 2 years of passing their driving test due to a lack of driving experience and in particular, decreased perception of unfolding hazards. In order to train hazard perception skills, a Simulator Hazard Perception Programme (SHPP) has been designed, based on the Hazard Perception Test. The SHPP has been designed by driving instructors, giving credibility to the educational value of the events depicted. In the BSM simulator, the SHPP allows the trainee to be exposed to different hazards and practise appropriate responses. With repeated practise, responses become learned and incorporated into cognitive schemas allowing novice drivers to develop effective cognitive skills in driving. It is proposed that hazard perception training in the simulator will better prepare them for hazard perception training on the road, where situations cannot be contrived or replicated and experience cannot be guaranteed.

This paper sets out how and why BSM has decided that hazard perception training through cognitive skills development in a simulator is an effective way of subsequently reducing accident liability amongst newly qualified drivers. By recreating risky situations in the simulator we are adding to the learner driver's repertoire of driving experience; by practising manoeuvres in a risky context we are giving them experience to which they can relate when they experience a similar event in the real world.

1 Introduction

BSM is the largest driver-training organisation of its kind in the world. With 107 customer centres and 2,500 franchised instructors BSM has a 20% market share of learner drivers. Established in 1910 BSM has consistently been at the forefront of change and innovation, working closely with the Driving Standards Agency and other motoring organisations to improve road safety and promote high driving standards.

BSM is therefore creating a simulator-based hazard perception programme with the aim of reducing the accident liability of new drivers. Learners pass their driving test with very little experience of the sort of decisions and judgements they are required to make in everyday driving situations: the lights are on green – will they change shortly? How sharp is the bend? How much speed do I need to lose? When should I start braking? How does the rain affect my stopping distance – do I know how long it takes to stop the car in the rain – do I know how big a gap to leave between myself and the driver in front? [DETR, 2000] Whilst these issues may have been covered in the theory component of the driving test, novice drivers may have had little opportunity to practise manoeuvres, especially under risk of accident.

Faced with a new situation an inexperienced driver has little to fall back on to help them make a safe judgement. Moreover, novice drivers are often lulled into a false sense of security because the driving task appears easy (e.g. they can now change gear and find the brake with ease and apply the indicator) and they become over confident and even complacent because they have little appreciation of the risks associated with driving.

The simulator is 'low-cost' and uses standard technology. Historically, it was believed that the best way to teach in a simulator was to replicate the task as accurately as possible. Yet research using flight simulators shows that this is not always necessary for effective training to take place [Dennis & Harris, 1998]

2 A Description of the BSM Driver-training simulator

The Faros Company in France purchased the driving simulator activity of Codes Rousseau at the end of 1994. Since then, Faros have continued the development of simulator products now intended for different markets such as initial driver training, further training, functional rehabilitation of people with reduced mobility and security awareness. Faros also developed a partnership with BSM to design specific training lessons for the British market.

In 1997 BSM took delivery of its first 10 Faros driving simulators. These were situated in Edinburgh, Newcastle, Leeds, Cambridge, Birmingham, Manchester, Bristol, Cardiff and Baker Street, London. Each year the number of simulators increases so that by the end of 2003 there will be 97 simulators in the UK, so that 91% of BSM retail outlets will have a simulator.

2.1 Hardware Architecture

The simulators run on a PC platform using Microsoft Windows (9x or 2000) OS. Standard 3D graphic boards generate images and driver interaction is achieved by interfacing real Vauxhall car equipment.

A single screen simulator is built around a unique PC computer, which integrates the visual system, the sound system and a custom I/O board. This same computer runs the simulation and training software.

The I/O board handles all the driver's interfaces:

- Analogue and digital inputs for the pedals, the gearbox, the steering wheel, the buttons and switches (signals, ignition, horn, etc).
- Analogue and digital outputs for the dashboard, with a frequency controlled output for the speedometer.

The steering wheel is linked to an active force feedback system, controlled by the software according to the speed of the car, the road surface, and the kind of power steering. The pedals passive force feedback mechanism reproduces the feeling of a real car clutch and brake. There are also three screen versions of the simulator. The three screens simulator needs a second computer to generate left and right displays. In this case, the two computers communicate via an Ethernet network using TCP and UDP transfer protocol. They are synchronised by a common system clock, and frame interpolation is done if the transfer time is too long.

2.2 Software Architecture

The simulator is designed as the combination of simulation software and training software. The simulation software includes the following specifications: automatic traffic generation, vehicle dynamics, visual and sound generation and I/O controls. The training software contains two levels:

- At the highest level is the syllabus manager: the pupil can go through the different menus (the syllabus tree), and select an exercise (if the syllabus manager validates this choice, depending on the log file and the level reached by the pupil).
- At the internal level is the exercise manager, which implements the scenarios. This module controls and interacts with the simulation module to handle traffic, car behaviour, and events.

Initially, BSM translated the existing driver training syllabus in the Faros simulator into English and the system was reprogrammed to allow for left-hand driving. We introduced the programme into the learner driver's lessons. As a result of qualitative research based on this novice programme and a two hour motorway lesson, a new 5 hour programme was introduced. This programme incorporated several key features to aid learning, in particular, automatic pilot.

One of the biggest stumbling blocks to teaching people to drive is that tasks cannot be taught in isolation of each other. For example, a gear change can only be executed once novice drivers have successfully learned to move away. Similarly, novice drivers must also understand how to stop the vehicle alongside being able to make it move. It is almost impossible to repeatedly practise isolated tasks. To address this issue, the new 5 hour programme therefore presents 'automatic pilot' at the start of an exercise. Automatic pilot allows learners to practise moving away but not how to stop in the first exercise and the screen simply fades away. When they come on to learn about stopping a vehicle, the simulator will move them away automatically and their task is to only stop.

Other key features of the 5 hour simulator programme are :

- Video demonstrations allow learners to learn at their own pace.
- The simulator is built from the dynamics of a Vauxhall Corsa (tuition vehicle) and teaches all the controls of the car and how to use them in preparation for on-road driver training.
- Within about 10 minutes a complete beginner with no previous experience can be moving off in the simulator.
- Throughout the exercises there is balanced feedback to the learner, which identifies strengths and weaknesses of their performance and is linked to the specific actions of the learner.
- As learners progress through the exercise the amount of help they are given is determined by the success of the previous actions.

The 5 hour simulator is presented as a 5x1 hour package. Typically learners spend five hours in the simulator, over 2 or 3 days, and the fifth hour is followed immediately by the first hour with the driving instructor on the road. In this way, the simulator complements the driving instructor-based training and the learner has a smooth progression from the simulated environment to the real environment.

3 Hazard Perception

The new Hazard Perception Test (HPT) was introduced to the UK in November 2002. The HPT is a computer-based test of 14 real-life scenarios each containing one or more scoreable developing

hazards. The Test was introduced because of the high accident rate amongst new drivers. Previous research has shown that training improves hazard perception score [McKenna & Crick, 1994] The HPT score of learner drivers after 3-4 hours training is similar to the average score of experienced drivers. The focus of the training in preparation for the HPT is on effective scanning techniques, keeping a safe separation distance; the correct use of speed and adequate anticipation and forward planning.
Since the introduction of the HPT BSM has introduced a Hazard Perception Training package to prepare learner drivers for the HPT. This has also fuelled the project for the development of hazard perception skills in the simulator.
All the clips used in the HPT were rigorously tested to ensure that they distinguish between inexperienced and experienced drivers. Trials were run where groups of learners, novices and experienced drivers responded to the clips in the bank and the clips that did not discriminate between experienced and inexperienced drivers were discarded.

4 Simulator Hazard Perception Programme (SHPP)

As a result of the success of the BSM simulator, BSM decided to incorporate Hazard Perception rather than focus solely on psychomotor skills development. Clearly, there is a place for the simulator to develop cognitive skills, which complement the learner's on-road tuition in hazard perception and risk management skills.
In order to develop the aforementioned effective hazard scanning techniques, the visual display must be as clear as possible. Therefore, all the simulators are being upgraded to the same specification using Pentium IV 2 GHz processors with GeForce4 graphic cards with 128Mb DDR memory. The overall memory is 256Mb and the hard disk is 80GB.
We can recreate risky situations in a simulated environment to accelerate the learning process for the inexperienced driver so that they too can develop coping strategies. By training the components of hazard perception we can identify to the driver exactly which aspect of hazard perception needs to be developed.
We are proposing a one-hour simulator-based programme, which will begin with a short video introduction explaining what hazard perception is and information about the exercises they are going to complete and how to navigate their way around the programme. This will be followed by a 10-minute assessment. It is anticipated that participants in this part of the simulator programme will be able to control and drive a car to a certain extent. There will be the facility to switch onto an automatic gearbox rather than manual if they prefer. The assessment will consist of a series of scenarios, each one monitored to specifically assess 1, 2 or 3 of the 8 cognitive abilities measured in the programme.
The cognitive abilities in the programme are:

- Assessment of speed and distance,
- Avoiding Risk,
- Changing Plans,
- Divided Attention,
- Focus,
- Obeying traffic regulations,
- Reaction Time,
- Visual Scanning.

The feedback at the end of the assessment will highlight the 4 weakest cognitive abilities. The driver will be advised that they are required to improve performance for these four areas. For each of these development exercises, there will be a short video explanation describing the cognitive ability in terms of a driving skill showing real life footage of when it is crucial to apply the skill. A 3D reproduced accident, describing what could happen if the cognitive skill was not applied, will be shown. For example, if a learner showed decrement in judging space and distance it is possible that they may drive into a parked car on the real road. The simulator will reproduce this and show the accident from the driver's, bird's eye and front points of view. Finally, the development exercise will have a practice session where the driver will spend time practising with voice-over instruction. Taking visual scanning as an example, the driver notices an ambulance on the left-hand side of the road. As the driver's attention is drawn to this, a child steps out from between the parked cars on the right hand side of the road. How soon the driver reacts by taking the foot off the gas will be monitored. Learning to remember to keep scanning the road even when your attention has been distracted is valuable experience and is a critical skill for real driving.

A final assessment will consist of the same route as driven in the initial exercise with the same scenarios. The four weakest areas will be monitored again but all controls and monitoring will be tolerant of errors in recognition that it is possible for some people to do worse in the final assessment compared with the first because of information overload. However, the benefit will be evident in their on-road driving thereafter. Feedback after the final assessment will be based on the summary, stating these are the areas the learner has been focusing on and congratulating them on completing the programme and advising them to practise hazard perception in their own driving.

5 Conclusion

BSM is now able to develop a revolutionary project, which will track hazard perception through cognitive training in the simulator, thus better preparing learner drivers to identify and deal with the dangers on the road with the aim of passing their hazard perception test and driving test and ultimately reducing their accident liability.

References

DETR : Tomorrow's Roads – Safer for Everyone, *The Governments Road Safety Strategy and casualty Reduction Targets for 2010. DETR, London, UK.*

McKenna, F.P., & Crick, J.L. (1994). Hazard perception in drivers : A methodology for testing and training. TRL Contractor Reprot 313. Crowthorne, UK : Transport Research Laboratory.

Dennis, K.A., & Harris, D. (1998). Computer-based simulation on an adjunct to ab initio flight training. The international journal of Aviation Psychology, 8 (3), 261 – 176.

Scenario-based Drama as a Tool for Investigating User Requirements with Application to Home Monitoring for Elderly People

Fran Marquis-Faulkes, Stephen J. McKenna, Peter Gregor & Alan Newell

Division of Applied Computing, University of Dundee, Dundee, DD1 4HN, UK.

fran@computing.dundee.ac.uk, stephen@computing.dundee.ac.uk
pgregor@computing.dundee.ac.uk, afn@computing.dundee.ac.uk

Abstract

Drama on video is being used as a tool to investigate user requirements for a fall detector within the context of a monitoring system based on visual tracking. The system is being designed to have the ability to provide passive monitoring in the homes of older people so that in the case of a fall being detected, the emergency services are called automatically. As with any such system, it is important that its presence is seen as supportive and not invasive. Drama has been found to be an effective way to focus potential users on discussions about the usage of the system, prior to important design decisions being made. This paper reports on the first stage of an iterative process and the fall detector design decisions made as a result of discussions with potential user groups. Issues to be considered in future work are discussed briefly.

1 Introduction

A system providing fall detection and movement monitoring to support elderly people living at home is being developed using computer vision technology (Nait-Charif & McKenna, 2003). Demographics and the costs of care clearly indicate the need for such systems. They must be sensitively designed with user involvement to ensure that the system presence is experienced as supportive and not invasive.

The objectives of the study reported here were two-fold: (i) to establish whether drama on video would work successfully in terms of focussing discussion amongst older potential users of a home monitoring system and (ii) to gain information from potential users (including sheltered housing wardens) about issues surrounding the use of current community alarms, contexts of use and what users would wish for future systems. This study is significant because the use of drama allows elderly potential users to be involved effectively in the process of design at the pre-prototyping stage. Theatre has been used only occasionally in the context of new product design (Sato and Salvador, 1999; Howard et al, 2002). There has been related research recently on the use of scenarios (Carroll, 2000), narrative (Benyon & Macaulay, 2002) and how stories capture interactions (Imaz & Benyon, 1999).

2 Methods

Four scenarios were developed based on information from focus groups and user stories. These were performed by a local theatre group, Foxtrot, who specialise in interactive theatre that raises topical issues and follows on from the pioneering work of Boal (Boal, 1995). The scenarios were filmed and shown to three groups of older people (with 7-15 participants) and one group of sheltered housing wardens (7 participants). Design constraints and user requirements for a fall detection and activity monitoring system were explored. Each scenario was developed to provoke discussion of a slightly different aspect of system design: (i) participants experience and anxieties about falling, (ii) the way in which information about a fall would be sent to and received by carers, (iii) the information the "faller" wants and needs about the system, and (iv) issues related to activity monitoring rather than fall detection. Groups were encouraged to discuss issues that occurred to them following each separate scenario and these discussions were recorded on video. More details of scenarios used and results are provided in (Marquis-Faulkes et al., 2003). A qualitative analysis was performed in which themes were identified and important points documented. Unresolved issues suitable for exploration using drama in a second iteration of the method were also identified.

3 Results

The first objective was clearly met: drama on video was used successfully to focus discussions amongst elderly people on design at a pre-prototyping stage. The second objective: "to gain information from potential users about issues surrounding the use of current community alarms, contexts of use and what users would wish for future systems" was also met and the information obtained is summarised below under four headings.

3.1 Design of the Monitoring System For Fall Detection

There was a perceived need for a passive fall detection system and its design can be broadly specified from the requirements of potential users.

- Potential users were content to have a monitoring device based on visual tracking provided that only a computer analyzes the output. None of the participants wanted visual monitoring that allows other people to actually watch them in their homes.
- Older people, especially those who still live in their own homes, were not enthusiastic about carers having too much information about their activities except for the purposes of helping them in case of a fall.
- Participants have experience with the current community alarms where a voice comes through the system to the flat occupant to ask if they need help; from this experience, a voice connection with any new system is expected and preferred.
- Sheltered housing wardens, in particular, felt that the new monitoring system would need to be linked with the existing community alarm system, i.e. using same call centre as is used at present. It would be too confusing for tenants and carers if there were two systems in place.

- The most likely places to fall were reported as the bedroom and the bathroom with going to the bathroom in the night being a particularly risky event for older people. The lobby was also said to be an area of risk.
- Careful consideration needs to be given to the relationship of the new monitoring system to other technology such as the door lock and the security system, so that in the case of a fall being detected in a private house that carers or emergency services have access.
- In aesthetic terms the potential users would rather have something which is unobtrusive, that looks like a smoke alarm for example, so that others do not know that they have such a system in place.

3.2 Communication between the System and the Faller

Discussions have given some insight into the important area of communication between a fall detection system and the faller and between the system and the carer or emergency services. Results so far are as follows.

- Participants said that the faller needs to know that the system has "seen them and what it is doing to raise help".
- Wardens commented that they had information about tenants (e.g. medical history and personality) that enabled them to prioritise care depending on likely urgency of need. This led to the suggestion that a monitoring system should have different settings: to be more sensitive if, for example, someone had just come out of hospital.
- All the groups said that they preferred human rather than computer-based systems. It was stated that the current personal connection and verbal communication between faller and warden/call centre provided an important and reassuring link.
- It was important that a potential faller could not turn off the fall detection system completely. Users, however, do need to be able to "clear the system" in the case of a false alarm to prevent help being called unnecessarily.
- Users need to be able to press a button to call for help if necessary so that an active system is integrated with a passive monitoring system.

3.3 Errors in the System

There is the issue of errors in the system: what happens when falls are missed or when the system mistakenly reports a fall (a false alarm).

- An anxiety was voiced that, if carers are relying on the system and therefore perhaps ringing up or visiting less, it is very important that it work reliably.
- Those who had experience of the current community alarm system, confirmed how important it was that they could clear a false alarm.
- The issue of missed falls is more complex. Potentially ambiguous situations, such as falling asleep in a chair or having a stroke while sitting down, were discussed as problem areas for an automatic visual monitoring system. While it was not intended that a fall detector recognise such situations a carer might expect to be alerted and a monitoring system could be designed to cope with this by, for example, responding to prolonged inactivity.
- Research is required to explore further how the dangers of missed falls and the inconvenience of false alarms can be avoided and more scenarios are planned to explore these issues.

3.4 Activity Monitoring

There are issues and anxieties concerning the monitoring of activity, rather than simply fall detection.

- A non-invasive fall detector was felt to be useful, but there was reluctance, expressed most strongly by those living in their own homes, to have any activity monitoring.
- Potentially, a monitoring system could build up a profile of the person's normal activity patterns and then alert a carer if the person was moving around less than normal, or using the kitchen less, or the bathroom more, than normal. Whether this was acceptable was explored in one scenario and generally it was felt to be too invasive. There are situations and circumstances where it might be felt that the reduction in privacy was worth the safety issues. For example, where the monitoring system could offset a move to sheltered housing or allow a more rapid return from hospital. This needs further exploration.

4 Conclusions and Future Work

The user requirements gathering methodology of provoking discussion by using scenario-based theatre has proved of particular value with audiences of elderly potential users, enabling them to focus on the details of a monitoring system at the pre-prototyping stages. Older people when presented with new technological possibilities in an appropriate form are very capable of considering and discussing desired functionality. Furthermore, they provided useful and interesting suggestions about what would make such a system really useful to them. They also had strong views about what they did not want of a monitoring system. For example, they particularly did not want any images of themselves at home to be accessible to anyone and could think of no circumstances where this would be necessary or acceptable. The discussion has led to areas of further work being pinpointed and future drama being designed to explore these areas. A series of new scenarios has been filmed. The plan is that this research is iterative: the issues mentioned by potential users in the first series of scenarios have been used to make provisional decisions about the design of the monitoring system. For example, it has been decided that:

- no video images will be transmitted to carers
- a new system will be integrated with current community alarm systems
- there will be voice communication with the monitoring system's user as there is with current community alarms
- the sensor will have an appearance which is unobtrusive, like a smoke alarm for example, so that others will not know that such a system is in place.

It was clear from participants' responses to other issues that more detailed investigations were required. This was particularly the case for the following areas.

- The details of the voice communication between the monitoring system and the faller
- What happens when there is a false alarm?
- What are the subtle conditions which a human carer would immediately recognise as a potential emergency but to which a visual tracking system would not respond? How can the monitoring system be designed to minimise the likelihood of missed falls?

Involving potential users in this way allows them to play a real part in the design of a product being developed to support them to live independently; drama used in this way engages older users and focuses discussion on the functioning of fairly complex technology in an enjoyable context using relevant media. Discussions were fairly open and unstructured to allow participants to raise issues of importance to them. This meant that unexpected topics were raised that the researchers had not predicted.

Acknowledgements

This work was funded by UK EPSRC EQUAL grant number GR/R27419/01. The authors would like to thank Foxtrot Theatre especially the Director, Maggie Morgan. Many thanks to all the potential users (older people, carers and wardens) who gave their time to the evaluation process.

References

Benyon, D. & Macaulay, C., (2002) Scenarios and the HCI-SE design problem. Interacting with Computers 14, 397-405

Boal, A. (1995) The rainbow of desire, Routledge, London

Carroll, J. M. (2000) Making use of scenario-based design of human computer interactions. MIT Press, Cambridge, MA

Howard, S., Carroll J., Murphy J., Peck, J. & Vetere, F. (2002) Provoking innovation: acting-out in contextual scenarios, People and Computers XVI Human Computer Interaction Conference

Imaz M. & Benyon D. (1999) How stories capture interactions, Human Computer Interaction, Interact 99, IOS Press.

Marquis-Faulkes, F., McKenna, S. J., Newell, A. F. & Grcgor, P. (2003) Using drama and video with older people in the requirements gathering for a fall monitor, Submitted to: Technology and Disability, IOS Press.

Nait-Charif, H. & McKenna, S. J. (2003) Improved particle filtering for tracking poorly modelled motion, Submitted to: IEEE Conference on Computer Vision and Pattern Recognition

Sato, S. & Salvador, T. (1999) Playacting and focus troupes: theatre techniques for creating quick, intensive, immersive and engaging focus group sessions, Interactions, Sept-Oct, 35-41

A Human Interface for In-Vehicle Information Space Using Drivers' Eye Movements

Hirohioko Mori

Dept. of Systems Information Eng., Musashi Institute of Technology
1-28-1, Tamazutsumi, Setagaya, Tokyo, 158-8557, JAPAN
mori@si.musashi-tech.ac.jp

Abstract

In this paper, we propose an in-vehicle human interface system, using driver's eye movements, to save drivers from cognitive overload caused by the enormous amount of information delivered by Intelligent Transport Systems services. In our system, a driver can obtain various kinds of information just by gazing at physical cues distributed in the external environment in a subtle manner. If the cue takes up the driver's attention, the system refers to the information engaged to the objects and displays it so that it is superimposed on the physical objects using the head-up display. The experiments showed that traditional salient displays were not only disruptive and decreased the safety of cruising but also made hard for the drivers to understand and utilize the delivered information. On the other hand, in our system, drivers can be saved from cognitive overload and to be able to utilize the delivered information effectively, because capturing drivers' eye movement made it possible for drivers to select and access only their necessary information actively and to maintain the relationship between the virtual information space and the drivers' view of the physical world.

1 Introduction

One of the major objectives of the ITS (Intelligent Transportation Systems) is to enhance driver safety by various information services about the situation of the road, traffic, and so on. According to Highway Industry Development Organization Japan (1998), more than 30 information services are currently under consideration in ITS. Each of them is indeed significant for drivers and useful to maintain their safety. However, if all the services were to be delivered consistently to the driver, the enormous amount of information would become disruptive. Not only would this frustrate her/him and decrease the efficiency with which she/he can drive a car safely, but also put her/him into some danger, as the driver will suffer from cognitive overload if ITS information grabs too much attention.

This paper proposes a new human interface system that reduces such cognitive loads by utilizing the driver's gaze information. In this system, various physical marks (which are hanged or painted on physical objects) are distributed in the environment such as on the roads, on a wall, on cars, on signboards, and used as a cue to access the information. In order to obtain necessary information, a driver gazes at a mark that is attached to the related object. The information is then displayed on the head-up display in a way that overlays the related objects. In this manner, by using the driver's gaze information, this system allows the driver to actively select and access only his/her necessary information among the enormous range presented.

The proposed system aims to build a calm information space inside the car by accessing the information actively in response to the driver's intention. It makes an enormous amount of information subtle and discreet by distributing information cues in the environment thus reducing the possibility of cognitive overload. By overlaying the information on the related physical objects, real world information and virtual world information are fused and the real world is visually enhanced and augmented.

2 Some Usage Scenarios

In this section, we provide an overview of our system and highlight some of its features by taking examples of situations in which the system works efficiently.

2.1 Making the Invisible Visible

The first example of a scenario in which our system works effectively is the situation that inspired us to design and develop our system. Consider that there is a twist on the road with soundproof walls along the twist and the tail-end of the traffic jam begins just after the twist. In such situation, drivers will not have noticed the traffic jam because the soundproof wall shuts out the view of the situation they are approaching. To resolve this problem, we considered that if we could view the forward dangerous situation behind the wall just as though it was transparent, the traffic accidents in such areas could be reduced.

The combination of our proposed system and the concept of the "ubiquitous video" [3] proposed in the computer science field makes this possible. In ubiquitous video environments, video cameras are distributed everywhere, and we can easily access real world information in real time through them from a distant place.

In this scenario, a video camera is mounted at the location where the situation of the exit point of the twist can be monitored, and the mark engaged to the video camera is stuck on the entrance part of the soundproof wall. When a driver enters the twist on the road and she/he watches the mark on

(a) Gazing at the Mark

(b) Picture from Video Camera Appears

Figure 1. Seeing the view ahead through the Wall.

the wall (Figure 1(a)), the real time situations obtained though the video camera are superimposed on the wall and the driver can feel as if she/he sees the traffic situation ahead as though the wall is transparent (Figure 1(b)).

Some similar scenarios can be considered. Car drivers will know of experiencing feelings of frustration when the forward traffic situation is obscured by a big truck is driving ahead. In our system, when we watch the mark on the back of the truck, we can see the traffic situation ahead of it through the video camera mounted on the truck

Another scenario is the parked car on the road. Parked cars cause a dangerous situation when people are trying to cross the street or children are playing behind them. In such situations, our system allows drivers to be able to see and check situations behind parked cars, by watching the appropriate mark, they can see via the video camera mounted on the parked car.

2.2 Making Public Signboard Private

The scenarios mentioned above concern a combination of the "ubiquitous video" infrastructure and seeing real time video information through our system. Our system, of course, works even without the ubiquitous video infrastructure, as it can be used not only in displaying video information, but also to provide many types of information such as still pictures and texts.

For example, assume a driver passes an ETC (Electric Tall Collection) gate. In such a situation, the information about the road fare is planned to be provided. Some drivers, however, are not interested in the information because they already know how much the charge is. In our system, only the drivers who want to know the charge get the information about it. These drivers watch the mark attached on the ETC gate to obtain information (Figure 3), other drivers can ignore the mark. Indeed, it should be natural for the drivers to watch the gate if they want obtain the information connected with it.

One of the great advantages of our system in this scenario is that it allows drivers to decide whether to get information on the signboard in the traditional manner, while at the same time the information can change corresponding to each driver. In the above scenario, the road charge for each driver is assumed to be different. In this situation, it is impossible to display the private and particular information such as the specific road charge on the public electric signboard, not only because of the numbers of cars passing through the gate simultaneously, but also because of the problem of privacy. Our system allows the driver to access private information in the same manner of traditional public signboards.

3 Experiments

We have conducted an experiment to evaluate our system from the following question: do the manner of active access to them reduce the driver's cognitive load?

3.1 Methods

In this experiment, all participants are involved in three conditions. One is the condition where many pieces of information delivered in ITS services are displayed continuously on the head-up display using texts. In the second condition, many information cues are embedded in the scenes and if the participants watch a cue, the information engaged in the cue is displayed on the head-up display in the form of text. In the last condition, information is accessed in the same manner as the second condition, but the displayed information is video pictures of the real time situation. We did not add the condition of auditory notifications because some subjects abandoned the experiment in this form in the pilot experiment, because of annoyance of too much information. In three conditions, the travelling scenes are all the same and it is the typical traffic situations in the center area of Tokyo: the streets are wide, relatively crowded, and many cars were parked on the streets.

The primary task the participants should perform is to step on the brake when the small red light on the screen is on. This task mimics a driver's reaction to the brake light of the car just ahead. All participants are instructed to react as soon as possible to maintain a safe drive. In addition to the primary task, the participants are instructed to pay attention to the information concerning the parked car in the next lane (this information is given among other information). If the participants receive and notice warning information that someone is standing or playing behind the parked car, they are required to step the brake as soon as possible.

Both reaction times of the primary and secondary tasks were measured and, after the experiments, some brief interviews were held. In addition to the above three conditions, participants performed only the primary task without any provided information as the control condition.

3.2 Results

Figure 2 shows the average reaction times to the warning notifications. This figure indicates that there is a significant difference between the salient condition and two other conditions. These results suggest that providing information continuously and passively is disruptive, both frustrating drivers and decreasing safety, such as avoiding emergent people and obstacles while driving a car. By contrast, drivers can effectively use the information obtained actively by intention during their driving tasks.

The results of the subjective evaluation of the participants support these results. Figure 3(a) shows the averages of the subjective ratings in each condition about the question of "what utility did the notification of information have in helping you to perform your tasks?" with a range of answers at 5 levels. Figure 3(b) also shows the average ratings of the question of "To what extent did the notifications information disturb your driving?" Looking at this result, we can easily see that the notifications under the salient condition are more annoying and more harmful to the driving task.

The differences between the embedded-text and embedded-video conditions are smaller than the comparison with the salient condition, but still significant. Indeed, during the brief interviews after the experiment, some subjects claimed that with the notification in the form of the video pictures, dangerous situations could be more easily understood and could be managed more easily than in text form. These results suggest that drivers can utilize visual

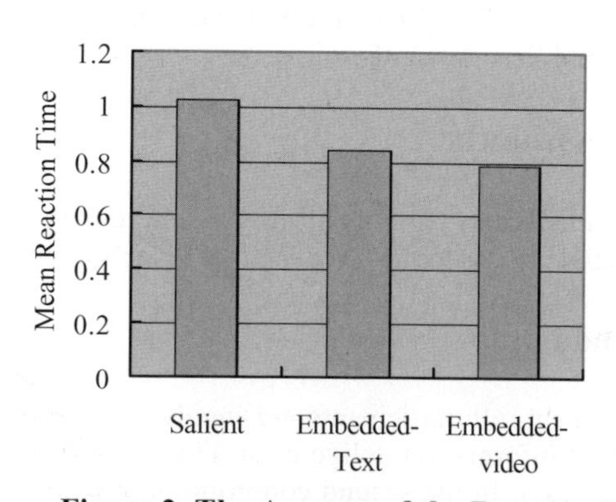

Figure 2. The Average of the Reaction Times

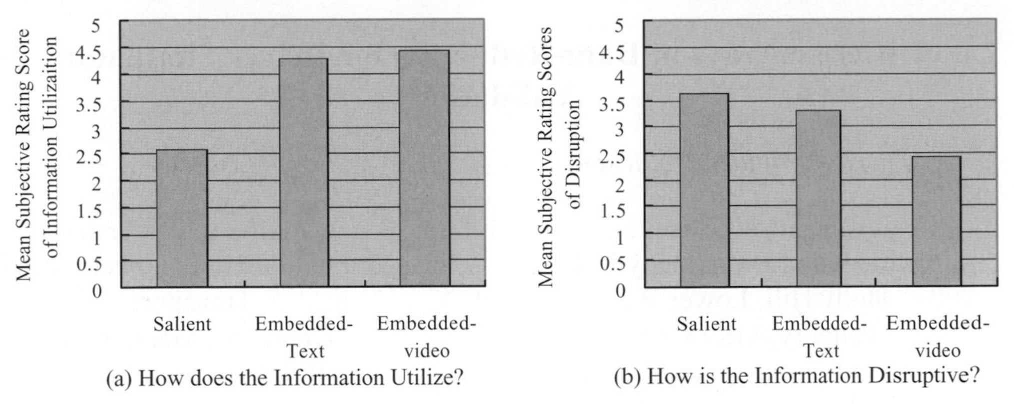

Figure 3. The Mean of Subjective Rating Scores.

information and can transform the knowledge into action by the human motor system using less human cognitive resources than is needed with text information.

To conclude, the results indicate that displaying information in the salient manner not only annoys drivers, but may also cause them cognitive overloads and push them into dangerous situations. Embedding the information in the environmental context in the ambient form can save the driver from such situations, and the active access to the information caused by the driver's intentions can be more effective in utilizing the delivered information.

4 Conclusion

In this paper, we propose a human interface system to save a driver from cognitive overload caused by the enormous amount of information delivered by ITS services. This system allows the driver to obtain information actively by gazing at physical cues distributed in the external environment. In our system, the information is displayed in a subtle manner that protects drivers from cognitive overload. The experiments showed that enormous information displayed in salient manner give the bad effects on the safe cruising and our system improves such situation. They especially showed that accessing information distributed in the road environment in subtle manner actively and maintaining the relationship between the virtual information space and the drivers' view of the physical world allows drivers to reduce the drivers' cognitive loads and to utilize the delivered information effectively. Though lowering the cognitive load simply equals improved safety, too much information delivered beyond drivers' cognitive capacities put them in the dangerous situation it is important, as our experiments shows, to keep their cognitive loads in appropriate levels within their capacities and to save drivers from such situation with good human interface designs.

References

Highway Industry Development Organization, Japan (1998). ITS Handbook Japan. (Highway Industry Development Organization, Japan).

Buxton W. (1997), Living in Augmented Reality: Ubiquitous Media and Reactive Environments. In *Video Mediated Communication* (pp.363-384), Hillsdale, N.J., Erlbaum.

Different Ways of Data-Reduction for Driver Simulator Validation

Thera K. Mulder-Helliesen

Independent Consultant
Revolution E Company Ltd
Holly Hill, Lower Way
Padbury, MK18 2AX
Email: info@reveco.co.uk
Web: www.reveco.co.uk

Lisa Dorn

Cranfield University
Department of Human Factors and Air Transport
Cranfield, MK43 0AL
Email: l.dorn@cranfield.ac.uk

Abstract

A key problem with simulator validation studies, whether used for research or training purposes, lies within the chosen methods of statistical analysis. Data on a typical validation study are often arranged in at least a four-way data matrix. Generally, statistical analyses on simulator-car data are comparisons of means (paired *t*-tests) and correlations. However, means and/or standard deviations can be computed over different ways and each method of averaging means sacrificing potential information on sources of variance in the data. This might lead to a poorly validated driving simulator, with disastrous consequences for research or training applications.

This paper reports a simulator validation study that was conducted for a low-cost, fixed-base driving simulator. Sixty-one participants were instructed to drive a stretch of road in their own cars accompanied by two observers, and drive the same, simulated road on the driving simulator. Data were collected on speed, lane-position and steering for both conditions on 12 (distance) measurement points, as well as subjective evaluations of the simulator's realism (face validity).

Participants' face validity of the simulator is quite low. Results of the *t*-tests and correlations on the different ways of averaging are not very convincing in terms of simulator validity. Only speed and steering show a very modest relative validity. The scatter plots for the means of simulator-car data though, illustrate that analysis per measurement point results in a much higher proportion of explained variance of the simulator data than when analysing per participant. Results are discussed with reference to the use of more appropriate methods of analysis for simulator validation studies.

1 Introduction

Currently, the three areas for the application of driving simulators are system and road design, training of driving skills but primarily driving behaviour research. However, to obtain any scientific confidence about the results of driving simulator studies in any of these areas, one first has to establish whether it is actually a trustworthy tool for its intended purpose. In order to do so, one has at least to allow for the transfer of basic driving skills, the reliability of the data collected and for this paper in particular, whether the measures of driving performance taken from the driving simulator are truly a valid representation of real-life driving.

Validity of a test or other measurement device is defined as the extent to which it measures what it is supposed or designed to measure (Gleitman, 1991; Korteling & Sluimer, 1999). As Sanders (1991) points out: "the primary aim of validating a simulator is of course to establish how well the

simulation actually reflects 'reality' " (p.1011). However, according to Blana (1996), it has proven extremely difficult to apply the psychological definition of validity to driving simulators. The human-machine-environment interaction of a simulator is extremely complicated and the major problem in defining its validity is that human performance is persistently confounded with system performance. Among simulator researchers though, the concept of validity is often seen as a single, independent feature. The problem then arises that they are frequently over-generalizing in their validation studies of driving simulators (Korteling & Sluimer, 1999), which has potentially serious consequences for the use of driving simulators for research but particularly for training purposes. In simulator validation research, physical validity (Blana, 1996) is described as the physical correspondence between a driving simulator and a real car, such as accelerating/braking, steering and the driving environment. Subjective evaluation of the realism of the driving simulator is generally called face validity. Behavioural validity (Blana, 1996) is described as the extent to which the behaviour (perceptual, cognitive and motor processes) of a participant on the simulator corresponds to his or her behaviour on the real task under the same conditions. This type of validity is often divided into absolute and relative validity (Blaauw, 1982; Blana, 1996; Harms, 1996; Kaptein et al., 1995).
Although modern driving simulators may be very realistic, interpretation of the data always requires careful consideration. The artificial environment of a driving simulator differs in many respects from that of real-life driving, which may influence participants' driving performance (Harms, 1996). At the same time, the majority of researchers often *assume* that drivers will behave in the simulator as they behave on the real road when driving under similar conditions (Blana, 1996). Generally, however, these conditions are anything but similar. Most of previous validation studies compare driving on a simulator along a simulated road, to driving in an instrumented (that is, not the participants' own) vehicle, along a completely different stretch of a real road, or even on test tracks without any interacting traffic. Obviously, these dissimilarities may affect participants' behaviour on both systems differently and one should therefore be very cautious about generalizing the results from these studies to the driving population in the real world.

For this paper, a key problem with simulator validation studies lies with not only study design but also the chosen methods of analysis. Data on one-sample validation studies are usually arranged in at least a four-way data matrix: participants, methods (simulator-car), dependent variables and measurement points (distance or time intervals). The main objective for validating a driving simulator is comparing participants' driving behaviour for both methods on the dependent variables and the measurement points. This is typically preformed by disproving differences in means of participants' scores between simulator and car (paired *t*-tests), or by proving relationships between participants' scores for both methods (correlations). However, analyses such as these use means or standard deviations and whether these will be computed over participants and/or measurement points is a decision for the researcher. Most commonly used are the 'overall' means (of all participants' scores on all measurement points) and the means of all measurement points (in order to look at individual differences between drivers). Comparing the means of all participants per measurement point is rarely conducted. Yet, using this method of analysis, one gains an insight of how all participants, on average, are behaving on speed limits, curves, overtaking, straight driving etc. at the specific measurement points.

2 Validation Study

This paper reports on a validation study that was conducted on a low-cost, fixed-base driving simulator (STISIM, Systems Technology Inc., 1999). An existing stretch of road was reconstructed on the driving simulator in exactly the same way as the real road and also included an average amount of traffic and parked cars according to photographs, videos and daily observations. The route was a single-lane rural road, about 2.5 miles long with curves and bends and if one kept to the speed limits (30-60 mph) it would take roughly five minutes to drive one way.

2.1 Method

Sixty-one participants (age: mean=31.7, std=11.5), 30 males and 31 females, all car-owners and with a full British driving licence, were instructed to drive the same road in the simulator as well as in their own cars. Half of the sample performed on the simulator component first and then the real road, and half performed on the real road first, followed by the simulated drive. In the controlled real-life driving condition, two observers accompanied the participants. Data were collected on the dependent variables speed, lane-position and steering, for the simulator as well as the real car, for every 0.2 miles (12 measurement points). Additionally, data were collected on the subjective comparison between simulator and car as a measurement of face validity.

2.2 Statistical Analyses

The four-way data matrix consists of 61 participants, 2 methods, 3 dependent variables and 12 measurement points. By means of paired *t*-tests and Pearson's correlations, the two methods (simulator versus car) are compared on the dependent variables, based on 'overall' means (for 61×12 observations), means of the 12 measurement points and means of 61 participants.
Before looking at the results, it is imperative to point out that when testing for differences between simulator-car means the actual null hypothesis, that is, the hypothesis of 'no difference', for a validation study is in fact the experimental hypothesis. The object of this method of analysis, such as the paired samples *t*-tests, is therefore to *accept the null hypothesis as tenable*. As the sample size (*N*=61) is rather small, the two-tailed alpha level, that is the probability of *rejecting* the null hypothesis when it is *true*, is set higher than normal for this kind of analysis: $\alpha = .10$.

3 Results

3.1 Face Validity

As a measurement of face validity, participants completed a questionnaire on a five-point scale on the realism of the simulator for six variables: steering, speed, accelerator and brake pedal, and road-scenery. Face validity for this sample is not very favourable: apart from the road-scenery, the majority of participants rated the simulator performance rather low, indicating that driving on the simulator did not feel very realistic to them.

3.2 Paired *t*-tests

Based on the 'overall' means, the results are not very good: only the variable steering indicates similar average behaviour on both methods. When comparing the means of the measurement points, it reveals that only 28% of the participants drove the same average speed, while respectively 74% and 90% of the participants show similar average behaviour on the variables lane-position and steering on both systems. The outcomes based on the means of participants are particularly poor for the variables speed and lane-position: apart from one measurement point, the means between simulator and car differ significantly. The variable steering shows similar average behaviour on seven points. However, despite the differences, the plots in Figure 1 do reveal a strong similarity between the simulated driving condition and real-life driving condition for speed choice and steering behaviour of the average participant.

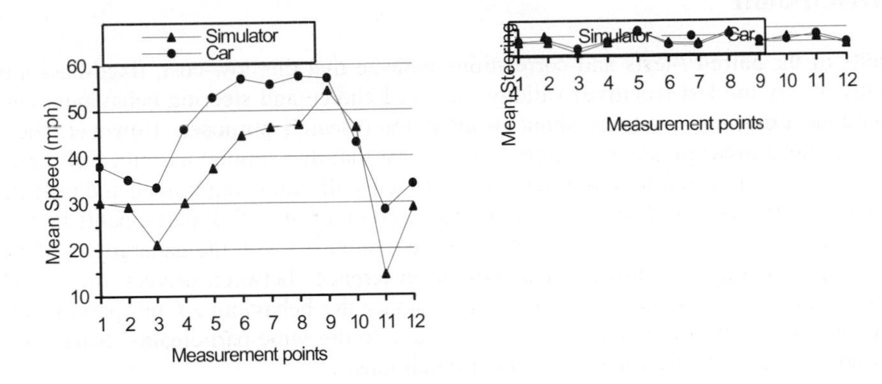

Figure 1: Mean speed and mean steering for simulator and car per measurement point

3.3 Correlations

None of the correlations between simulator-car data is very remarkable. Only the variable speed reveals a reasonable positive relationship: based on the 'overall' means $r = .62$, ($p < .01$) while based on means of measurement points, 84% of the participants show a significant association ($p < .05$) between simulator and car on this variable.

3.4 Scatter plots

Figure 2 presents the scatter plots for the variable speed, based on the means of participants (left) and means of measurement points (right). Comparing the two, it is apparent that the analysis based on the means of participants offers a much better prediction of the simulator values than when looking at individual differences. Moreover, the proportion of explained variance (R^2) for the simulator values is much higher when analysing per measurement point than when analysing per participant. The same is true for the variables lane-position and steering as well.

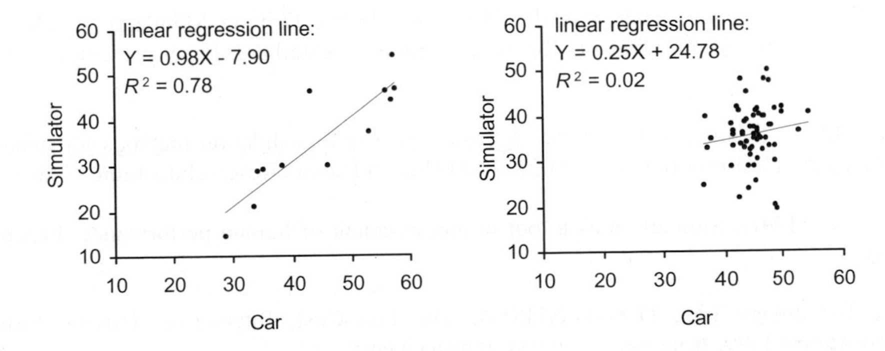

Figure 2: Scatter plots for mean speed (mph) car versus simulator for 12 measurement points (left) and 61 participants (right)

4 Discussion

The results of the paired *t*-tests and correlations indicate that this low-cost, fixed-base simulator shows only a very modest (relative) validity on speed choice and steering behaviour. Therefore, one should be extremely cautious about using it for research purposes. However, the similar patterns of simulator-car means in Figure 1 do suggest that this simulator could well be used for training, assuming one carefully describes exactly which skills are being trained and/or learned.
The scatter plots (Figure 2) illustrate that analysis based on means of all participants improves the way the validity of a driving simulator is established, compared with the generally used means of the measurement points. Regardless of individual differences between drivers, the validity of a driving simulator should be based on assessing whether the behaviour of any participant on the different variables in the simulator is truly comparable to the same participants' behaviour on the same variables in real-life driving under similar conditions.
In applying statistical methods such as paired *t*-tests and correlations, one should realise that information is being 'averaged out', thereby losing potential vital sources of variance in the data. For validation purposes, it might be more appropriate to use multi-dimensional methods, which might provide an improved insight in all sources of variance in the data, as well as possible interactions between the dependent variables, such as speed choice and steering behaviour.

References

Blaauw, G.T. (1982). Driving Experience and Task Demands in Simulator and Instrumented Car: A Validation Study. Human Factors, 24, 473-486.

Blana, E. (1996). Driving Simulator Validation Studies: a Literature Review. Working Paper 480, Institute for Transport Studies, University of Leeds.

Gleitman, H. (1991). Psychology. Third Edition. W.W. Norton and Company, New York-London.

Harms, L. (1996). Driving Performance on a Real Road and in a Driving Simulator: Results of a Validation Study. Vision in Vehicles-V, Elsevier Science B.V.

Kaptein, N.A., Theeuwes, J. & Van der Horst, R. (1995). Driving Simulator Validity: Some Considerations. TNO Human Factors Research Institute, Soesterberg. Paper presented at the TRB 75th Annual Meeting.

Korteling, J.E. & Sluimer, R.R. (1999). A critical review of validation methods for man-in-the-loop simulators. TNO-report TM-99-A023, TNO Human Factors Research Institute, Soesterberg.

Sanders, A.F. (1991). Simulation as a tool in measurement of human performance. Ergonomics, 34, 995-1025.

Systems Technology Inc., (1999). STISIM, The Low-Cost, Interactive, Driving Simulator. Retrieved August 1999, from http://www.systemstech.com.

The Development of a Bus Simulator for Bus Driver Training

Helen Muncie and Lisa Dorn

Cranfield University
Cranfield, Bedfordshire, MK43 0AL
h.muncie@cranfield.ac.uk

Cranfield University
Cranfield, Bedfordshire, MK43 0AL
l.dorn@cranfield.ac.uk

Abstract

This paper describes the development of the UK's first bus simulator for bus driver training. The design and construction of the ARRIVA bus simulator (ABS) is based on training needs and is described in detail here. The ABS is a fixed-base moderate fidelity system built to present a wide field of view and provide high resolution to facilitate the visual detail required for bus driver training. Future validation studies will focus on face and behavioural validity of the ABS and the design of driver training methods to improve bus driver safety.

1 Introduction

Simulators have been successfully employed for training in the military and commercial sectors for several decades. While most simulation technology has been developed to satisfy aviation training requirements, driving simulators are not often used for professional driving training, especially in the UK. A particular group of professional drivers that may benefit from the advantages of simulator-based training are bus drivers. At present, the average new bus driver receives two weeks instruction in a driving school and approximately 2-3 weeks on-the road experience with a mentor driver. However the traffic environment is increasingly complex and there is little control over the events trainees are confronted with in the real world. A simulator offers the possibility of training in various traffic conditions, including emergency situations that are rarely encountered in traditional in-vehicle training, but are events that may lead to bus incidents (Dorn et al, 2002). ARRIVA Passenger Services Ltd recognised that a bus simulator has great potential as a time saving and cost-effective method of developing a driver's experience with decision-making in traffic and took the decision to construct one for this purpose in 2002.
Training simulators serve two main functions: to present information that is required in training and to incorporate features to facilitate and enhance practice and learning (Flexman and Stark, 1987). Specific training objectives should drive the design of any simulator intended for use as a training device. One of the major issues to be addressed is the evidence that drivers typically prefer a high fidelity system rather than the one best suited for training (e.g. Andre and Wickens, 1994). Yet, it is the level of fidelity needed to support learning that should determine the level of fidelity built into the simulator rather than the preferences of the driver. If the intention is to train complex skills then the simulator needs to closely replicate the task environment so that all relevant cues and elements of the tasks are available to the driver to allow the transference of skills. According to Lintern (1991), intentional departures from reality between the simulator and real world environments can focus attention on task related perceptual invariants and reduce

distractions that may prevent the driver from recognising key perceptual cues. To exemplify, a novice student may gain more from a low fidelity simulation in which relevant cues are emphasized and irrelevant cues are reduced to enhance initial learning. If the fidelity of the simulation is too low, transfer to the real environment may be impeded. However, it is also recognised that a moderate fidelity simulation is also important to motivate behaviour that is consistent with that experienced in the real world if we are to positively affect driver's acceptance of the simulator in training (Salas, Bowers and Rhodenizer, 1998).

2 The Display System

An analysis of ARRIVA's incident database revealed that most bus incidents occur at junctions and bus stops. As these events require the driver to sample from a wide field of view (FOV) the simulator's visual display needed to present 180° FOV for these events to be recreated accurately. This decision is further supported by the work of Kaptein, Horst and Hoekstra (1996) in a comparison of braking performance across simulated and real world trials. The results suggested that a wide FOV is essential for validity purposes. Given that braking manoeuvres is also a key feature of bus driving behaviour at bus stops and junctions it was clear that a wide FOV would be necessary for training purposes.

Other important features to consider in the design of a bus simulator are high-resolution depth cues, critical for many driving manoeuvres (Kemeny, 1999). High visual update and refresh rates are necessary to display close proximity moving models, such as other vehicles (Stoner, 1995). It is particularly important to have a good visual resolution for simulated bus driving for accurate detection of events, especially in the distance. Although, high scene detail contributes to the realism of the visual environment, the amount of interactive objects that can be programmed is limited because the update rate of the visual scene slows down when there are many complex 3D models to build in rapid succession. The complexity of the simulated environment then is a compromise between an optimal update rate and a high fidelity interactive environment. The guiding principle then in the selection of the visual display was to achieve the highest resolution and a wide FOV.

With regards to the use of a motion platform we drew evidence from the aviation literature showing that complex cockpit motion systems have no measurable training benefit (O'Hare and Roscoe, 1990; Waag, 1981). Manoeuvres, procedures and flight scenarios can all be effectively trained on a fixed-base flight simulator. There is insufficient research on the use of motion platforms in driving simulators to conclude that there would be any added benefit for the additional expense incurred.

3 Description of the ARRIVA Bus Simulator (ABS)

The design and construction of the ABS is based on the training and fidelity needs described. The basic construction of the ABS is fixed-base and uses the STIsim PC-based interactive driving simulator model 400 (Systems Technology Incorporated). The simulation software includes a simple vehicle dynamics model, a simple power train model, visual and auditory feedback and a performance measurement system. Driving tasks and events are programmable within a unique Scenario Definition Language (SDL), which allows the user to specify sequences of tasks, events and performance measurement intervals (Allen, Rosenthal, Aponso, Harmsen and Markham, 1999).

The STIsim software runs on a Windows 2000 operating system, which allows networking for increased computational capability (Netgear sport 10/100 Mbps Fast Ethernet switch Model FS105.) Three AMD Athlon ™ XP 1800+ AT/AT COMPATIBLE Pentium computers with 262 MB RAM compute complex vehicle dynamic responses to the human operators control input with an adequate update rate to satisfy visual, proprioceptive and auditory cueing requirements. PC processors are fitted with PCIM-DDAO6/16 analog output and digital I/O board, V266B Motherboard, PCI-QUAD04 Four channel quadrature encoder input board, NVidia Riva TNT2 Model 64 128-bit 3D Processor graphics card provides visual cueing and Soundblaster sound processor card provides auditory cueing. A Philips multimedia speaker system (20W RMS) provides audio feedback to the driver. Four Samsung SyncMaster 753DFX colour monitors with 17 inch display and 1024 x 768 resolution support three driving displays (right, left, centre) to give a 180° field of view plus an operator's display. The virtual environment is presented to the driver by projecting the three driving displays onto a 180° curved screen, which is 6 metres in diameter and 2.75 metres high through 3 Panasonic LCD projectors mounted on the top rear of a 360° chrome framework. The boundaries between the three projector images are blended and geometrically corrected by an additional graphics processing system. The driver sits in an actual Volvo bus cab situated in the centre of the screen so that the driver's immediate environment is as realistic as possible. A modular steering unit with speed sensitive feel provided by a computer controlled torque motor through a full-size steering wheel (+/- 360 degree steering capability) and a modular accelerator and brake pedal unit have been adapted so that a normal bus steering wheel, brake pedal, accelerator pedal and speedometer are available. Figure 1 depicts how the simulator components are connected.

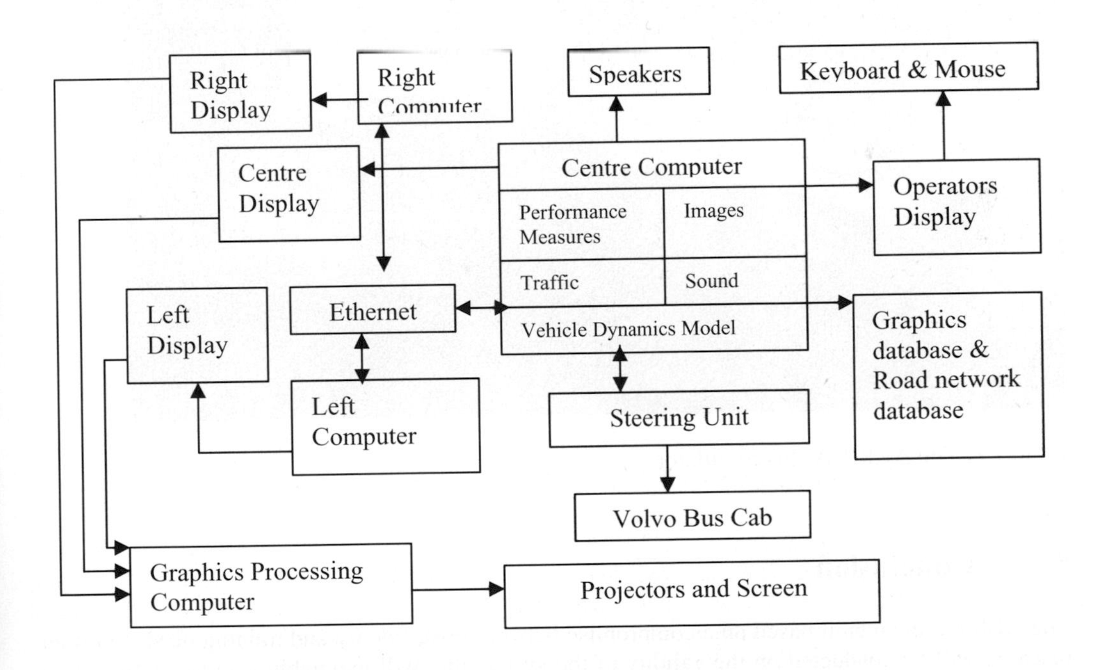

Figure 1: Functional Diagram of the ARRIVA Bus Simulator

4 The Validation Phase

Now that the ABS is built the next step is to develop the simulated scenarios for validation and training purposes. The scenarios have been designed based on an analysis of over 20,000 bus incidents to improve ecological validity during training. Findings indicated that most incidents occurred at junctions and bus stops so the scenarios incorporate many high-risk situations that bus drivers may encounter in both rural and urban areas. We are preparing to conduct validity studies to assess the acceptability of both the hardware and software components of the ABS. To be an effective training device the bus simulator must be comparable to driving in the real world. The first study concerns the degree of physical similarity between the simulator and its real world counterpart and emphasizes the components, layout, and dynamic characteristics of the simulator. Secondly, behavioural validity studies will be undertaken to assess whether the bus driver behaves in a similar way in the simulator as they do in a real bus. However, even if the ABS is carefully validated, the ultimate value of the simulator will be determined by its success in training.

Figure 2: The ARRIVA Bus Simulator

5 Conclusion

The ABS has been built based on a compromise between cost, fidelity and training needs. Further research will be conducted on the validity of the system that will inevitably lead to configuration and engineering changes in the future. Issues of fidelity and validity in the design of driver training simulators offer significant challenges for future research, but they are critical issues if we are ever to fully appreciate the value of driver training simulators. It is vital to proceed in a systematic way in the introduction of this new technology for driver training, rather than develop a quick-fix simulator that may not be fit for purpose. The next stage would be to design a simulator-based driver-training program and investigate transfer effectiveness to real driving. The questions that need to be answered are; what skills can be effectively transferred from the simulator to the real bus? And what retards or enhances their transfer?

References

Allen, R.W., Rosenthal, T., Aponso, B., Harmsen A., and Markham, S. (1999). Low Cost, PC-based techniques for driving simulation implementation. *Proceedings of the DSC'99*, France, July 1999, p 31-44.

Dorn, L., Garwood, L., and Muncie, H. (2002). The accidents and behaviours of bus drivers. *Behavioural Research in Road Safety*: 12th Seminar, Department for Transport, Dublin, HMS.

Kaptein, N. A., Horst, A. R., and Hoekstra, W. (1996). *Effect of Field-of-View and Scene Detail on the Validity of a Driving Simulator for Behavioural Research.* TNO Report TNO-TM-96-AO22.

Kemeny, A. (1999). Simulation and perception. *Proceedings of the DSC'99*, France, July 1999, p 13-28.

Lintern, G. (1991) Instructional Strategies. In Morrison J. E (Ed). *Training for Performance: Principles of Applied Human Learning*. Wiley Press.

Roscoe, S. N. (1990) Simulator qualification: Just as phoney as it can be. *International Journal of Aviation Psychology*, 1(4), pp 335-339.

Salas, E., Bowers,. G.A., and Rhodenizer, L. (1998). It is not how much you use it but how you use it: Toward a rational use of simulation to support aviation training. *International Journal of Aviation Psychology*. 8 (2). p 197-208.

Stoner, J.W (1995). Developing a high fidelity ground vehicle simulator for human factors studies. In, L. Hartley (Ed). *Fatigue and Driving*. (p 207-217). Taylor and Francis.

Waag, W. L. (1981).*Training effectiveness of visual and motion simulation* (AFHRL-TR-79-72) USAF Human Resources Laboratory, Operations Training Division, William Air Force Base.

Wickens, C.D. and Andre, A.D. (1994). Performance-preference dissociations. *Society for Information Display Digest* (pp. 369-370). Playa del Rey, CA: Society for Information Display.

An Analysis of Potential of Human Error in Hospital Work

Yusaku Okada

Keio University
3-14-1 Hiyoshi Kohoku-Ku Yokohama Japan
okada@ae.keio.ac.jp

Abstract

This paper discusses the human error management in general hospital. In human error management one of the most important requirements is exact extraction of the causes of human error, which are performance shaping factors (PSFs). Therefore, we produced PSF keyword table so that the hospital risk manager can analyze the PSFs exactly and easily. The strategies that prevent human error should be planned from the viewpoint of PSFs. So we examined how to estimate the possibility of human error occurrences, in order to decide the execution priority of the strategies. The method was applied to several medical treatments by nurses, and PSFs and the possibility of human error occurrences were obtained. From these results, the human error prevention strategies were planned and executed. Moreover, the validity of the strategy was confirmed by reduction of the incident number of cases.

1 Introduction

As not a few incidents has been reported in general hospital, hospital manager have to aim the establishment of human error management system. Though some plans to examine human error executed in Japanese many general hospitals, the effect is a little. Therefore, this paper intends to propose the procedures to make effective strategies to prevent human error.
In order to solve the mechanism of human error occurrences, causes of human error and a flow of performances should be obtained exactly. For that purpose, the detailed information about occurrence condition and content of operation must be indicated by incident report. However, incident reports of many Japanese hospitals do not satisfy the requirements. Therefore in many cases, the risk manager judged that the cause of human error is careless mistake, and he did not examine the human error prevention plan. So the similar human error will occur in the future.
So this paper intends to discuss the methods to produce the effective strategy to prevent human error. Furthermore, we apply the methods to the practical medical treatments in general hospital, and investigate the validity of the method.

2 Analysis of Incidents by Human Error in Hospital

In order to propose an effective strategy to prevent human error, the causes of human error occurrences must be obtained in detail. However, it is not easy for risk manager of hospital to extract the causes of human error occurrences exactly. So we propose the measure to support the extraction of the causes in this chapter.

2.1 Performance Shaping Factors (PSFs)

The causes of human error are called as performance shaping factors (PSFs). PSFs are work environments, manuals of operations, design of instruments/interfaces, physical stressor, psychological stressor, skill of operator, operator's character, and so on. There are many kinds of method that extracts the PSFs. For example, Variation Tree Analysis, Event Tree Analysis, Why-Why Analysis, are. If the hospital risk manager uses the methods, the immediate factors which caused human error directly will be extracted. But it is difficult for hospital worker without most knowledge about human factors to detect the latent factors which exist in background of human. Immediate factors should be removed instantly. However, the various human errors would not be prevented by the only improvement of the immediate factors. The origin of human error occurrences should be removed so that the potential of human error generating will decrease.

2.2 Categorized PSFs on Human Cognitive Performance Model

So we improved the design of human error report, and reinvestigated the human error incidents for past three years in a general hospital of Tokyo, Japan. As a result, about 250 PSF items were obtained. In order to classify the PSFs, we used a simple model for human performance on operator's protocols (see Figure 1). The model has four steps; 1) receive information on interfaces, 2) derive knowledge from long-term memory resource (LTM), 3) interpret the knowledge creates dynamic image of performance, 4) generate as practice action. PSF items were categorized into five; AFFORDANCE on Receiving Information, Context of information, Ability of operator, AFFORDANCE on Action, Environment (included factors on management).

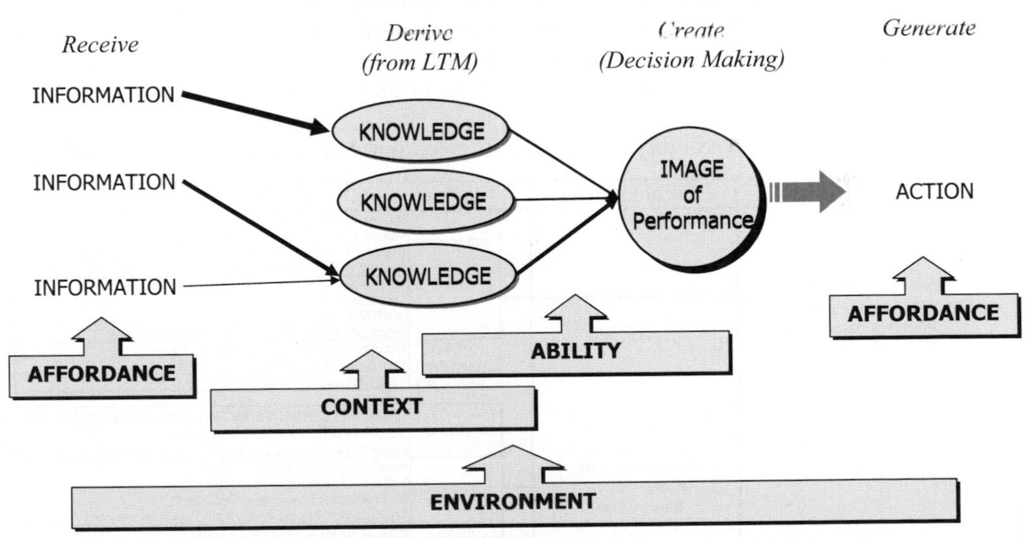

Figure 1: Human Cognitive Performance Model

2.3 PSF Keyword Table for Hospital works

Moreover, the characteristics of PSF items in each category were summarized as PSF keywords table (see Table 1). If the hospital risk manager refer to the PSF keyword table, the extraction of latent PSFs will become easier.

Table 1: PSF Keyword Table

AFFORDANCE	CONTEXT	ABILITY	ENVIRONMENT
Existence of Information	Checkup	Knowledge	Installation of Machine
Quantity of Information	Checkup by Others	Experience	Maintenance of Machine
Quality of Information	Work Sharing	Skill	Members of Workers
Affordance	Indication (Oral/Paper/Board/Others)	Team Work	Schedule (Time Pressure)
Information Sharing	Procedures (Usual / Interruption / Emergency)	Recovering	Environmental Condition
Information Processing	Manual Documents	Estimation of Risk	Shift
Usability	Communication	Custom	Management of Patient
Readability	Flow of Operation (Dependence)		

3 Estimation of Possibility of Human Error Occurrence

This chapter discusses the method to estimate the potentiality of human error in usual operations of hospital. This method is applied to several nursing works, but this paper shows the results in case of operations in dialysis room.

3.1 Methods

First, the usual sequence of the operation is drawn on a diagram. Next, the PSFs are extracted by using of the PSF key word tables, and are written down the diagram. (see Figure 2)

The strategies that prevent human error should be planned from the viewpoint of the PSFs. However, all of strategies cannot execute in practice since there are several restrictions on management. So we intend to produce the execution priority by the possibility of human error occurrences.

Here, we assume that the possibility of human error occurrences is determined by the condition of

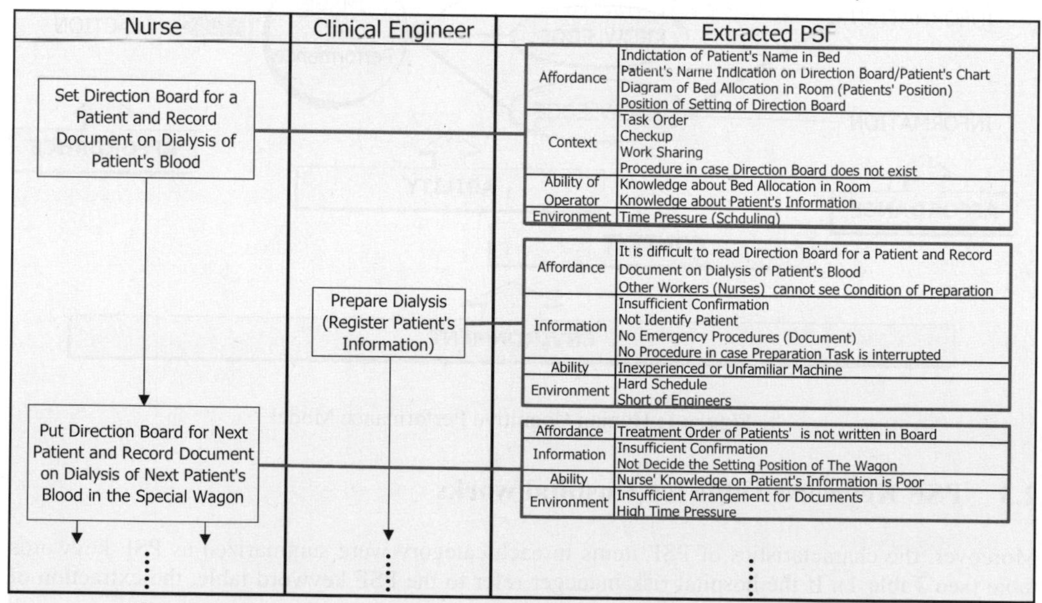

Figure 2 Diagram for Ana;ysis of PSFs in Usual Opration (in Dialysis Room)

PSFs and the weight of the effect of PSFs to human error occurrences. That is, Pi = Ci Wi where Pi is the possibility of human error occurred by PSF-i., Ci is the insufficiency of condition of PSF-i., and Wi is the weight of effect of PSF-i to human error. In this study, Ci is decided based upon the result of subjective evaluation by a risk manager and workers of hospital. Wi is determined in consideration of both the direct and indirect effects of PSFs on human error. The direct effects of PSFs are decided with reference to the past incident reports. The indirect influences of PSFs are determined by the matrix that is given by one-pair comparison between PSFs. Moreover, the potential of human error generating in the operation (**HEP**) is estimated by

$$HEP = \sum_i Pi = \sum_i CiWi$$

The human error prevention strategies are planed and executed according to the result of HEP value.

3.2 Case Study: A operation in dialysis room

A part of results which evaluated the conventional operation in dialysis room is shown in Table 2. Observing the HEP data in Table 3-1, the PSFs that should be removed were obtained. And the human error prevention strategies were planned as the improved operation. For example of the strategies, change of the design of the direction board is. As a result, it is obtained that HEP is decreased in improved operation (see Table 3-2).
Furthermore, the number of incidental cases is also decreased. Average of incidental cases was about 8 before improvement, but now average is less than 2.

4 Conclusions

As a result of analysis for several operations in hospital works, many problems (PSFs) on human error are found out. Though practice hospital workers were conscious of the problems, the potential of human error occurrence could not be estimated. So the priority of strategy to remove the problems was not obtained. Unless the problems (PSF) cause accidents, the strategy for the problems is not examined and is not executed. It is reason why the number of accidental cases did not decrease in spite of execution for risk management.
Human error prevention should be not only to prevent the occurrence of the same accident but also to remove the factors that would cause human error in the future. These two aspects would be requirements on human error management.
As Event Sequence Diagram with PSF keyword table supports the analysis of causes of incident,

Table 2 Evaluation Value of Indices and Possibility of Human Error Occurred by PSFs

Operator	Step	Operation	Performance Shaping Factors				
			Category	Item	Insufficiency Ci	Weight Wi	Possibility of Human Error Occurred by PSFs Pi
Nurse	1	Set Direction Board for a Patient and Record Document on Dialysis of Patient's Blood	Affordance	Indication of Patient's Name in Bed	0.8	1	0.80
				Patient's Name Indication on Direction Board/Patient's C	0.6	0.4	0.24
				Diagram of Bed Allocation in Room (Patients' Position)	0.6	0.8	0.48
				Position of Setting of Direction Board	0.4	0.6	0.24
			Context	Task Order	0.8	0.6	0.48
				Checkup (Confirm)	0.8	1	0.80
				Work Sharing	0.8	0.8	0.64
				Procedure in case Direction Board does not exist	0.8	0.8	0.64
			Ability of Operator	Knowledge about Bed Allocation in Room	0.6	0.4	0.24
				Knowledge about Patient's Information	0.6	0.4	0.24
			Environment	Time Pressure (Scheduling)	0.6	0.8	0.48

Table 3-1 Possibility of Human Error Occurrence in Conventional Operation

Step	Conventional Operation					
	Potential of Human Error Generating (HEP)	Number of PSF Items	Human Error Occurrence Possibility on Each PSF			
			High	Middle	Low	Negligible
1	5.28	11	2	5	4	0
2	4.80	10	1	4	5	0
3	2.80	7	1	2	3	1
4	3.04	8	0	2	5	1
5	3.48	5	3	2	0	0
6	4.88	9	1	6	2	0
7	3.12	6	0	4	2	0
8	3.08	6	0	5	1	0
9	5.80	10	1	7	2	0
10	5.60	10	0	9	1	0
11	3.88	6	2	3	1	0
12	4.88	10	1	5	4	0
13	4.68	10	1	5	4	0
14	2.48	5	0	3	2	0
15	4.48	8	2	5	0	1
16	1.76	4	0	3	0	1
17	4.44	8	2	4	2	0
18	5.44	9	2	6	1	0
19	1.24	3	0	1	2	0
20	2.24	4	0	4	0	0
21	5.24	11	2	5	4	0
22	4.80	10	1	4	5	0
Total	87.44	170	22	94	50	4
Ave	3.97		1.00	4.27	2.27	0.18

Table 3-2 Possibility of Human Error Occurrence in Improved Operation

Step	Improvement Operation					
	Potential of Human Error Generating (HEP)	Number of PSF Items	Human Error Occurrence Possibility on Each PSF			
			High	Middle	Low	Negligible
1	2.80	1	0	0	1	0
2	3.32	1	0	0	1	0
3	2.08	0	0	0	0	0
4	2.56	1	0	0	0	1
5	0.96	1	0	0	0	1
6	2.80	0	0	0	0	0
7	2.16	1	0	0	1	0
8	1.76	1	0	0	1	0
9	2.64	1	0	1	0	0
10	3.84	1	0	1	0	0
11	2.44	1	0	0	1	0
12	3.96	8	0	3	2	3
13	3.28	1	0	0	1	0
14	1.36	1	1	0	0	0
15	2.76	1	0	0	1	0
16	1.76	1	0	1	0	0
17	3.12	1	0	0	1	0
18	3.92	1	0	0	1	0
19	0.92	1	0	0	1	0
20	1.92	0	0	0	0	0
21	2.80	0	0	0	0	0
22	3.32	0	0	0	0	0
Total	56.48	24	1	6	12	5
Ave	2.57		0.05	0.27	0.55	0.23

background of incident and mechanism of human error occurrence would be obtained more detail. Therefore, the appropriate strategy that prevents the reoccurrencc of same human error would be proposed. .

The method to estimate the possibility of human error occurrence can be able to remove the factors that would cause human error in the future. But the quantity of incidental data investigated in this paper is not enough to define the quotation to estimate the possibility (HEP). So we should be reconsidered the quotations, the variables, and the coefficients on many hospital incident data. Finally, we aim to establish the human error management system in hospitals.

References

Embrey, D. (1984).SLIM-MAUD: An approach to assessing human error probabilities using structured expert judgment.
NUREG/CR-3518, USNRC.

Gertman, D.I., et al. (1992).
INTENT: A method for estimating human error probabilities for decision-based errors.
Reliability Engineering and System Safety, 35, 127-136

Hollnagel, E, (1998). Cognitive reliability and error analysis methodology, Elsevier

Kirwan, B. (1992). Human error identification in human reliability assessment, part 1: overview of approaches. Applied Ergonomics, 23, 299-318.

Reducing Interaction Style Errors in Task-Switching

Antti Oulasvirta

Helsinki Institute for
Information Technology
P.O. Box 9800, FIN-02015 HUT,
Finland
oulasvir@hiit.fi

Hannu Kuoppala

Helsinki University of Technology
Usability Group
P.O. Box 5400, FIN-02015 HUT,
Finland
kuoppa@cs.hut.fi

Abstract

Task-switching is commonplace in modern human-computer interaction. When a user switches between two tasks that each require a different interaction style (IS), there is a possibility that the active IS is not updated to correspond with the upcoming task. One experiment tested the idea that making the user interfaces (UIs) visually dissimilar would reduce such errors. Participants switched between two tasks that differed in IS (point-and-click vs. drag-and-drop). Participants in the internal cue group had to internally remind themselves of the to-be-activated IS, because the task interfaces were identical, whereas in the external cue group, a visual cue (presence/absence of a red frame) was available to reduce working memory load. An external cue enhanced performance in the main task, but caused IS errors in switches back to the other task. The paper closes by discussing underlying cognitive factors and implications for design.

1 Introduction

The hallmark of modern operating systems is the support for user-driven multi-tasking. It was already estimated in 1987 that a task-switch occurs every 15 minutes during use (Card & Henderson, 1987). Upon every switch between two tasks, the user must deactivate the old interaction style (IS) and activate the correct one. IS switching is prone to errors, especially when the two user interfaces (UIs) are similar. Consider, for example, a common situation: an applet is loaded within a browser window, and while interacting with the browser, the user has performed link navigation by means of single clicking; However, the applet may require double clicking links to navigate. Little is known about UI factors that advance the correct selection of IS. We therefore begin the investigation with a simple and specific question: If the switched-to UI is visually dissimilar to the UI of the preceding task, will users more easily activate the correct IS? For example, will changing the background color of a dialog box to blue suffice if the competing dialog box is gray? The question relates to the question of in what conditions *consistency* should be strived for in design (Nielsen, 1989).

Related questions have been addressed, however, in cognitive psychology. The central topic has been the *task switch cost* (Jersild, 1927)—that is, the decrease in reaction time (RT) and increase in errors upon switching. Task switch cost is a result of perceptual and central capacity limitations. Core factors, such as number of task repetitions, stimulus-response mapping similarity between A and B, task preparation time, and preparation information have been studied in detail (see Pashler, Johnston, & Ruthruff, 2001). To benefit from the vast amount of research done, the approach taken here is to modify the *task-switching paradigm* more into a human-computer interaction related setting.

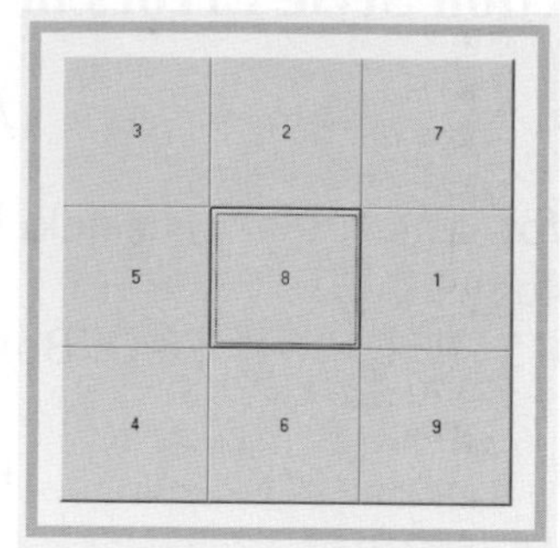

Figure 1: The digit matrix task. Participants were instructed to arrange digits to ascending order. The (red) frame around the matrix was provided only in the external cue condition.

The present experiment (N=6) tests the idea that making the UIs of the switched-to and switched-from tasks visually dissimilar would reduce IS errors. Participants switch between two tasks that differed in ISs (point-and-click vs. drag-and-drop). Participants in the internally cued group must internally remind (i.e., covertly rehearse the IS of the forthcoming task) themselves of the to-be-activated IS; there are no visual cues available in the display to help in selecting the correct IS because the UIs of the two tasks are identical. In contrast, in the externally cued group, a distinctive visual cue (a red frame) is available on the display to help distinguish between the ISs.

The main hypotheses are as follows. First, the externally cued group is expected to outperform the internally cued group, because maintaining a reminder in working memory slows down or restricts allocation of attentional resources to processing of the task itself. Second, the 164 experimental trials are expected to suffice to learn to associate the external cue with the IS. Third, learning the association should decrease errors in IS selection, speed up RT, and enhance overall performance in the main task.

2 Method

Six undergraduates from the Helsinki University of Technology volunteered for the present experiment. Participants in the two experimental groups (internal vs. external cue) were matched by age and sex. All participants were experienced in using computers.

The experimental task consisted of arranging randomly placed digits in a 3x3 matrix in ascending order (see Figure 1). Participants were instructed to avoid using a wrong IS and to arrange as many digits as possible during the given 7 seconds. Two ISs (T1 and T2 henceforth) were used to arrange digits. T1 was always *precued* by digit '1' and T2 digit by '2'. The precue was shown on screen for 400 ms before the task began. In T1, digits changed places by clicking two of them (point-and-click), whereas T2 required drag-and-drop. Prior to experimental trials, the two ISs were introduced to participants by demonstration. Participants were told that point-and-click was always associated to pre-cue '1' and drag-and-drop to '2'.

Cue Type (external vs. internal) was manipulated as a between-subjects factor. In the external cue group, the UI in T2 had a visual cue, a red frame surrounding it (see Figure 1). In the internal cue group, T1 and T2 were visually identical. The externally cued group were not told that the red frame would indicate T2; instead, this association had to be learned in the course of the experiment.

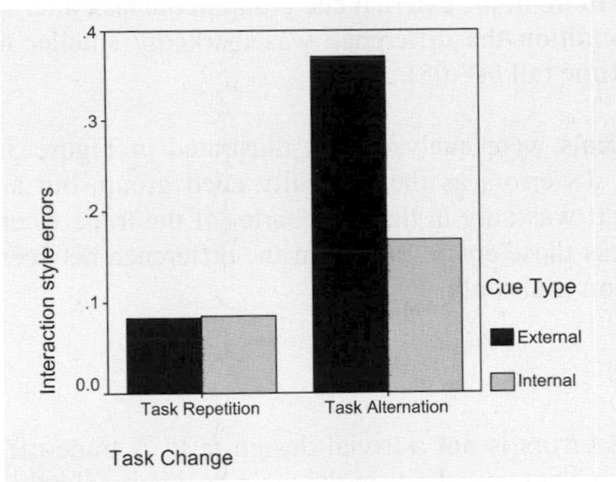

Figure 2: Interaction style errors in T1 as a function of Task Change and Cue Type. Switching from T2 to T1 was more error prone in externally than internally cued group.

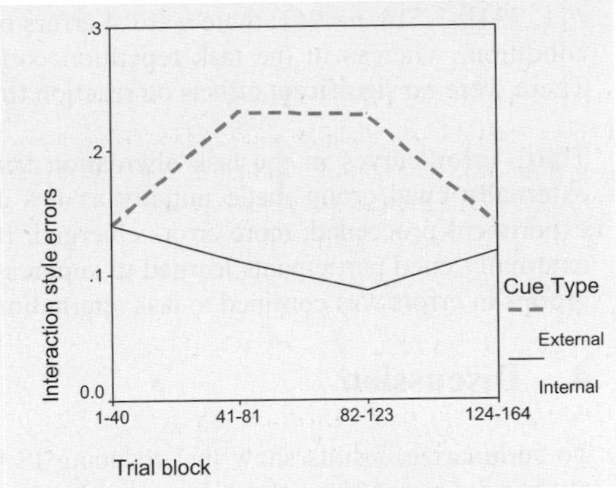

Figure 3: Mean interaction style errors per trial in task alternation condition in T1 as a function of trial block and Cue Type.

T1 and T2 appeared in blocks of four where T2 appeared always in one of the first three positions; hence, within each block, T2 was always preceded by 1, 2, or 3 T2s. Each participant performed 164 trials altogether. Four rest periods were equally spaced across trials.

3 Results

One-way univariate analyses of variance (ANOVA) were separately performed for the number of correct moves, IS errors (e.g., using point-and-click instead of drag-and-drop), arrangement errors (switching two digits to descending instead of ascending order), and reaction time (time needed to make the first move). Only those results relevant for the presented research questions are reported here.

First, the externally cued group made significantly fewer arrangement errors in T2, $F[1, 980]=6.613$, $p<.01$. Mean difference in the number of arrangement errors in T2 between the groups was .27. No differences were found between the groups neither in the number of correct moves nor in reaction time in T2, which supports the assumption that performance between the two groups was on the same level. However, in a direct contrast to the hypotheses, there was no significant difference in the number of IS errors ($p>.1$) in T2.

Second, a situation where T1 had been preceded by T2 (*task alternation*) was compared to the situation where T1 had been preceded by one or more T1's (*task repetition*). An unexpected disadvantage for the external cue group in switches from T2 to T1 was observed. A two-way univariate ANOVA yielded a significant interaction effect between Cue Type (internal vs. external) and *Task Change* (task alternation vs. task repetition) in the number of IS errors, $F[1, 734]=6.715$, $p<.05$. In the task alternation condition, there were .2 IS errors more in the internally than in the externally cued group. In contrast, in the task repetition condition, the difference was minimal (.001) (see Figure 2). Interaction effect between Cue Type and Task Change was marginally significant in the number of correct moves, $F[1, 734]=8.880$, $p=.054$. There were .4 more correct moves in the task alternation condition for the internally cued group. Again, this difference was diminished in the task repetition condition to only .08. In addition, a similar pattern was found in the number of digit arrangement errors. The interaction was marginally significant,

$F[1, 734]=3.516$, $p=.061$; there were .4 errors more in the external cue group in the task alternation condition, whereas in the task repetition condition the difference was markedly smaller (.05). There were no significant effects on reaction time (all *ps*>.05).

Third, error curves in the task alternation trials were analyzed. As illustrated in Figure 3, the externally cued group made initially as few IS errors as the internally cued group, but as the experiment proceeded, more errors emerged. It was only in the last quarter of the trials when the externally cued participants learned to suppress these errors. Note that the difference between the groups in errors was confined to task alternation trials only.

4 Discussion

To summarize, results show that reducing IS errors is not a trivial design task. A trade-off was observed: On one hand, the red frame helped in arranging digits in the main task immediately after the switch (while not helping in choosing the right IS); On the other hand, it actually increased the number of IS errors when returning back to the no-frame task. This negative effect was, however, eliminated with practice, but only after approximately 120 trials.

Explanations for the findings are proposed here. First, it is suggested that performance in the internally cued group was more prone to digit arrangement errors (in the frame-cued task T2) because of the precue reserving capacity from working memory (WM). It may be that participants maintained the precue in WM some time after the switch. This extra load distracted the main task, which required storing and comparing two perceived digits at a time to make a decision. Previous work has revealed a specific mechanism, *inner speech*, utilized in internal reminding upon task-switching (Baddeley, Chincotta, & Adlam, 2001). Inner speech is sequential in nature, limited in capacity, immediately accessible to consciousness, highly practiced for most adults, robust, and relatively well-protected from interference from other cognitive processes (Emerson & Miyake, in-press). These properties make it an effective means for reminding oneself of the upcoming task. However, reserving its limited capacity may hamper performance in the upcoming main task. It is thus suggested that designers should ensure that the WM requirements of the task-switch do not interfere with the main task.

Second, it is be explained why the number of IS errors was actually increased in the externally cued group when returning back to the no-frame task. The explanation forwarded here is that participants did not notice the offset of the red frame because it was *peripherally located*; that is, while attention was focused to the main task (i.e., arranging digits), the disappearance of the surrounding frame was not registered (for the plausibility of this assumption, see Mack & Rock, 1998), thus leading to the observed errors of perseverance. In the current experiment, noticing the disappearing of the red frame may require shifting attention away from the task, which explains the found deficit. It can be speculated that if the external cue were instead *centrally located*, integrated to the currently attended features of the UI, participants would more easily notice it. In addition to being centrally located, the cue could have provided *affordance*. For example, the cells in the matrix could have "afforded" being drag-and-dropable or point-and-clickable. Another, although potentially more disturbing, way to ensure that distinctive features are noticed could be to use short flashes or abrupt *on*sets of the crucial features upon switch to capture attention. It is important to ensure that distinctive features of the UI are attended to upon task-switching.

The third finding was that the disadvantage for the externally cued group disappeared in the last quarter of the experimental trials. This suggests that whatever underlies the observed disadvantage for the externally cued group in task alternation trials can be suppressed with practice. Additionally, it is known that switches from "stronger" (i.e., more repetitions) to "weaker" tasks are easier than vice versa (Monsell, Yeung, & Azuma, 2000). These notions imply that special attention must be paid for the design of cues of UIs that are rarely used.

To conclude the paper, we want to propose the possibility of using the presented experimental paradigm for studying affordance. In usability, "affordance" means how effectively a UI element activates the correct action—an aspect that is well captured in the dependent variables of the present experimental paradigm. On the other hand, in task-swithing research, it is known that "uncertainty" over the response required in the upcoming task increases task switch cost (Fagot, 1994; see also Duncan, 1984). The two concepts, uncertainty and affordance, are obviously related. A further advantage of the present paradigm is that it affords studying speed-accuracy trade-offs in IS selection. Future work by the authors explores these possibilities.

Acknowledgements

We thank Jari Laarni for valuable comments. This work has been supported by Alma Media, Elisa Communications, the Finnish Work Environment Fund, Nokia, Sonera, and SWelcom.

References

Baddeley, A., Chincotta, D., & Adlam, A. (2001). Working memory and the control of action: evidence from task switching. *Journal of Experimental Psychology: General,* 130 (4), 641–657.

Card, S.K., & Henderson, A. Jr. (1987). A multiple, virtual-workspace interface to support user task switching. *Procs. of ACM CHI'87– Human Factors in Computing Systems,* 53–59.

Dunbar, K., & Sussman, D. (1995). Toward a cognitive account of frontal lobe function: Simulating frontal lobe deficits in normal subjects. In J. Grafman, K.J. Holyoak, & F. Boller (Eds.), *Structure and functions of the human prefrontal cortex,* 289–304. NY: New York Academy of Sciences.

Duncan, J. (1984). Selective attention and the organization of visual information. *Journal of Experimental Psychology: General,* 113, 501–517.

Fagot, C. (1994). *Chronometric investigations of task switching.* Ph.D. dissertation, University of California, San Diego.

Emerson, M.J., & Miyake, A. (in press). The role of inner speech in task switching: A dual-task investigation. To appear in *Journal of Memory and Language.*

Jersild, A.T. (1927). Mental set and shift. *Archives of Psychology,* whole number 89.

Mack, A. & Rock, I. (1998). *Inattentional Blindness.* Psyche, Cambridge, MA:MIT Press.

Monsell, S., Yeung, N., & Azuma, R. (2000). Reconfiguration of task-set: Is it easier to switch to the weaker task? *Psychological Research,* 63 (3–4), 250–264.

Nielsen, J. (Ed.) (1989). *Coordinating User Interfaces for Consistency.* Boston, MA: Academic Press.

Pashler, H., Johnston, J.C., & Ruthruff, E. (2001). Attention and performance. *Annual Review of Psychology,* 52, 629–651.

Rogers, R.D., & Monsell, S. (1995). Costs of a predictable switch between simple cognitive tasks. *Journal of Experimental Psychology: General,* 124, 207–231.

Pupil Dilation as an Indicator of Cognitive Workload in Human-Computer Interaction

Marc Pomplun and Sindhura Sunkara

Department of Computer Science, University of Massachusetts at Boston
100 Morrissey Boulevard, Boston, MA 02125-3393, U.S.A.
Email: marc@cs.umb.edu, ssunkara@cs.umb.edu

Abstract

Pupil dilation is known to quickly respond to changes in the brightness in the visual field and a person's cognitive workload while performing a visual task. Pupil dilation is rarely analyzed in usability studies although it can be measured by most video-based eye-tracking systems and yields highly relevant workload information. This is mainly due to two problems: First, the variety of factors that can influence pupil dilation, and second, the distortion of pupil-size data by eye movements: The size of the pupil as seen by the eye-tracker camera depends on the person's gaze angle. In the present study, we developed and implemented a neural-network based calibration interface for eye-tracking systems, which is capable of almost completely eliminating the geometry-based distortion of pupil-size data for any human subject. Moreover, we compared the effects of cognitive workload and display brightness on pupil dilation and investigated the interaction of these two factors. The results of our study considerably facilitate the use of pupil dilation as a quick and reliable indicator of a person's cognitive workload.

1 Introduction

In the evaluation of human-computer interfaces, an increasing number of researchers conduct analyses of users' eye movements during task completion (e.g., Goldberg & Kotval, 1999). Gaze trajectories can indicate difficulties that users encounter with certain parts of the interface and point out inappropriate spatial arrangement of interface components. However, when performing such studies, scientists often neglect the analysis of another variable that they receive as a "byproduct" of video-based eye tracking, namely the size of the user's pupil.

It is well known from a variety of studies that participants' pupils dilate with increasing cognitive workload being imposed (see Kahneman, 1973). This effect has been demonstrated for tasks such as mental arithmetic (Hess, 1965), sentence comprehension (Just & Carpenter, 1993), and letter matching (Beatty & Wagoner, 1978). Besides cognitive workload, the intensity of ambient illumination is the other major factor determining the size of a person's pupil. Changes in illumination can therefore interfere with the use of pupil size as a measure of cognitive workload (Kramer, 1991). To reliably measure workload, we have to compensate for such changes in illumination (Nakayama, Yasuike & Shimizu, 1990; Porter, Troscianko & Gilchrist 2002). Furthermore, scientists face a technical problem: Since participants move their eyes during experiments, their pupils assume different angles and distances towards the monitoring camera of the eye tracker. This, in turn, means that the size of the pupil as measured by the system - the number of pixels that belong to the pupil in the camera image – varies with the participant's gaze angle. This effect is especially strong if the camera is located below the eye (see Figure 1).

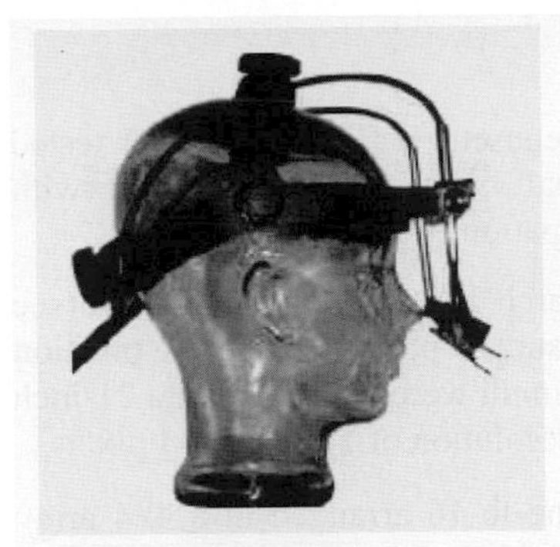 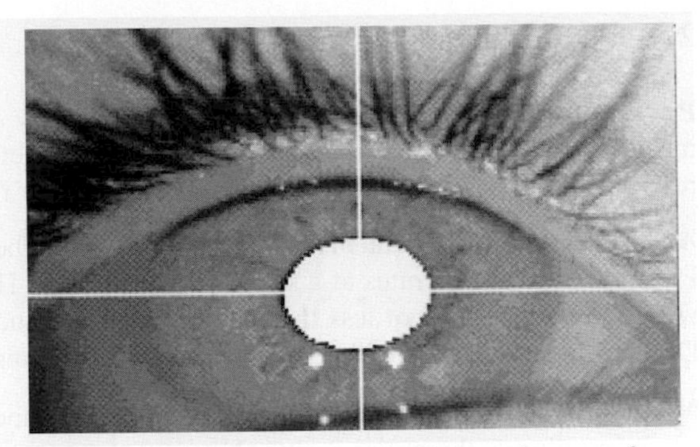

Figure 1: Left panel: The headset of the EyeLink-II system; Right panel: Camera image of a participant's left eye with the pupil area recognized by the system (white)

In order to reduce the noise in pupil size measurement caused by eye movements, we implemented a neural-network based calibration interface for video-based eye trackers and evaluated it empirically in Experiment 1. Using the increased precision achieved by the new interface, in Experiment 2 we investigated, from a practical perspective, in which way the brightness of the screen in a human-computer interaction task interferes with the measurement of cognitive workload as indicated by pupil size and whether this interference can be substantially reduced.

2 Experiment 1: A Pupil Calibration Interface and its Evaluation

Since the setup of the eye tracker - that is, the camera position and orientation relative to the participant's eye – is different for every experimental session, it is not feasible to use a fixed geometric calculation for correcting the measured pupil size. Instead, we introduced a pupil calibration procedure prior to the experiment to determine the relative size of the pupil as a function of the participant's gaze position. Participants were asked to fixate on each point in a 3×3 array four times to collect pupil size data for these 3×3 gaze positions. We chose to use only nine calibration points to make the calibration procedure as quick and little disruptive as possible. Given the continuous small changes in pupil size and the resulting variance, additional calibration points would not have led to a substantial improvement of the calibration.

Obviously, interpolation is necessary to estimate, based on the calibration data, the change in the measured pupil size as a function of the current gaze position. For such interpolation tasks, a type of artificial neural network called Parametrized Self-Organizing Map (PSOM) has proven to be well-suited (Pomplun, Velichkovsky & Ritter, 1994). PSOMs are a variant of the Self-Organizing Maps (Kohonen, 1990), but learn much more rapidly than the latter ones and are capable of representing continuous, highly non-linear functions. In the present context we used a PSOM with nine neurons and fed it with the calibration data, that is, the measured average size of the pupil at the nine calibration points, divided by the pupil size measured while looking at the center of the screen. During the subsequent experiment, by interpolating the calibration data, the PSOM estimated the factor by which the measured pupil size differed from the one that would have been measured if the subject had looked at the center of the screen. Then the currently measured pupil size was divided by the PSOM's output and thereby standardized, which we assumed to strongly reduce the variance in pupil size data that is due to eye movements. We conducted Experiment 1 in order to test the effectiveness of our calibration interface at improving the signal-to-noise ratio when measuring the effect on pupil size exerted by changes in display brightness.

2.1 Method

Participants. Ten students from the University of Massachusetts at Boston were tested individually. All participants had normal or corrected-to-normal vision. They were naïve with respect to the purpose of the study and were paid for their participation.

Apparatus. Eye movements were recorded with the SR Research Ltd. EyeLink-II system (see Figure 1), which operates at a sampling rate of 500 Hz and measures a participant's gaze position with an average error of less than 0.5 degrees of visual angle. Stimuli were presented on a 21-inch Dell Trinitron monitor with a refresh rate of 85 Hz and a screen resolution of 1152 by 864 pixels.

Materials. The stimulus displays showed the numbers from one to 16 arranged in a 4×4 array spanning almost the entire screen. None of the 16 positions coincided with any of the nine target positions used for calibration. Two different displays were created: One showing white numbers on a black background (luminance < 1 cd/m^2) and another one presenting black numbers on a white background (luminance 82.4 cd/m^2).

Procedure. Each participant was sequentially presented with the two stimulus displays. The order of presentation was counterbalanced across participants. They were asked to find the numbers in ascending order and read them out loud. Subsequently, participants were asked to repeat the task, but this time in descending order.

2.2 Results

All pupil size data, both the uncorrected and the corrected ones, were separated into 16 groups based on the participant's gaze position during their measurement. For this purpose, the screen area was divided into four by four equally large rectangular parts. A two-way analysis of variance (ANOVA) with the factors background color (two levels: black and white) and gaze position (16 levels) revealed significant effects by background color, $F(1; 9) = 32.82$, $p < 0.001$, and gaze position, $F(15; 135) = 22.30$, $p < 0.001$, as well as a significant interaction between the two factors, $F(15; 135) = 2.00$, $p < 0.05$. The gaze-position and interaction effects demonstrate that, as predicted, the measured pupil area is systematically influenced by the participant's gaze position. Figure 2 (left) illustrates this finding.

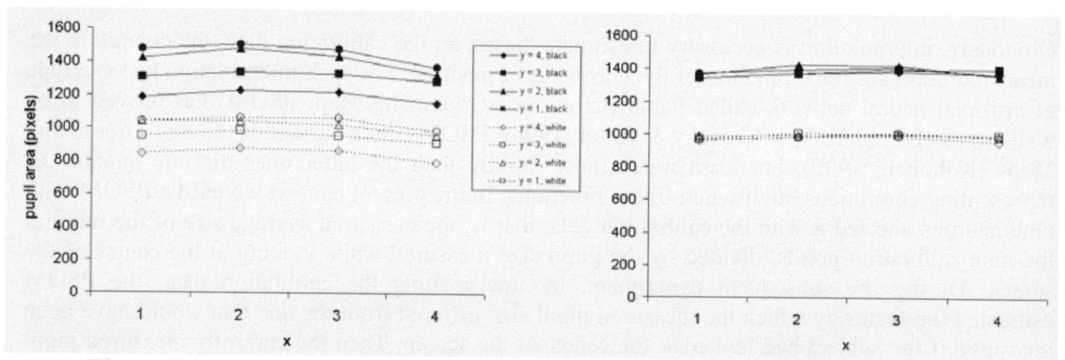

Figure 2: Measured average pupil size by gaze position (x = 1, …, 4; y = 1, …, 4) and background color before correction by the PSOM (left panel) and afterwards (right panel).

An analogous ANOVA for the corrected pupil size data also showed a significant effect by the factor background color, $F(1; 9) = 33.21$, $p < 0.001$, but no significant effect by gaze position,

$F(15; 135) < 1$, and no interaction effect, $F(15; 135) < 1$. This indicates that our calibration interface greatly reduced the systematic influence of the gaze position on the pupil size measurement (see Figure 2, right).

3 Experiment 2: Workload and Brightness Effects on Pupil Dilation

To investigate brightness and cognitive workload effects on pupil size in human-computer interaction, we devised a gaze-controlled human-computer interaction task that ran in three different speeds, thereby creating three different levels of task difficulty and, assumedly, cognitive workload.

3.1 Method

Participants. The same ten participants from Experiment 1 also participated in Experiment 2.

Apparatus. The apparatus was the same as in Experiment 1.

Materials. The stimulus displays showed a grid of 4×3 cells (see Figure 3, left). At the beginning of a trial, all cells were empty. Then, in each cell, one of four possible items could appear: a red square, a red circle, a blue square, or a blue circle. These items then increased in size twice before they disappeared. The participants' task was to avoid any blue circles from attaining their maximum size. To achieve this, they could look at any growing blue circle and press a designated button at the same time to eliminate that item. Any failure caused a loud buzzer sound to be played. In the "easy" condition, every second one cell was randomly chosen to be updated, that is, if it contained an item, this item would grow (or disappear if already fully-grown), otherwise a new, small item of random type would be placed in the cell. In the "medium" and "hard" conditions, the updating interval was reduced to 200 and 75 milliseconds, respectively.

Procedure. Each of the three levels of task difficulty was combined with two levels of background brightness (black and white, as in Experiment 1), resulting in six different trial types. Each type was presented to each participant four times. Before the experiment, participants were instructed not to let any blue circle reach its full size. The experiment started with an easy practice trial whose data were not analyzed, followed by the 24 experimental trials in random order. Each trial lasted 30 seconds.

3.2 Results and General Discussion

A two-way ANOVA revealed that the (corrected) pupil size was significantly influenced by the factor task difficulty (levels easy, medium, and hard), $F(2;18) = 35.13$, $p < 0.001$, and the factor background color (levels black and white), $F(1; 9) = 41.08$, $p < 0.001$, while there was no interaction between the two factors, $F(2; 18) < 1$. Figure 3 (right) illustrates how the increase in pupil area induced by higher task demands was almost identical for black backgrounds (1231, 1315, and 1441 pixels) and white backgrounds (872, 961, and 1102 pixels).

This finding suggests a method for accurate cognitive workload measurement even in situations where the display brightness cannot be kept constant. The idea is to perform an additional calibration procedure in which the display brightness - in the same display that will be used in the subsequent experiment - is systematically varied to determine the participant's pupil size as a function of brightness. During the following experiment, by subtracting the calibration value for the current display brightness from the currently measured pupil size, the amount of pupil dilation induced by cognitive workload can be computed.

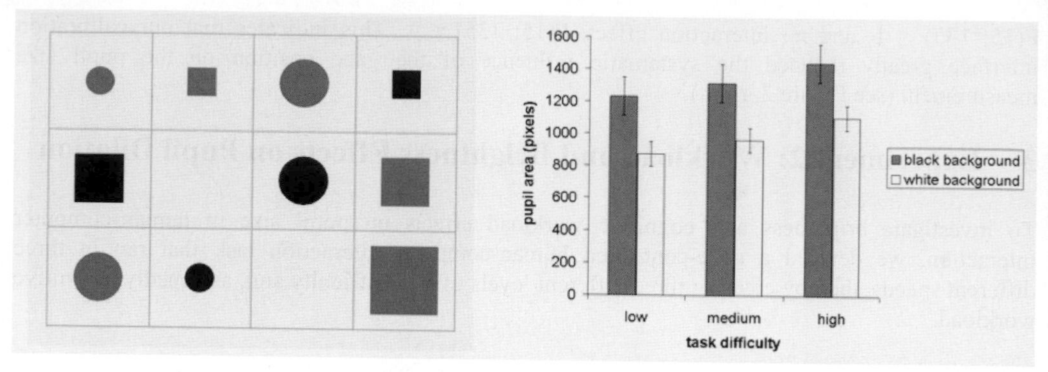

Figure 3: Left panel: Screenshot of Experiment 2 - red objects are shown in light gray, blue ones in dark gray. Right panel: Results of Experiment 2.

In summary, we have presented a technique for substantially reducing the eye-movement induced variance in video-based pupil dilation measurement. Our proposed calibration procedure takes only about 30 seconds and strongly and reliably improves measurement precision. Moreover, we have pointed out how to separate brightness effects from workload effects on pupil size. All in all, the present study can be considered a small but significant advance in using pupil dilation for the analysis of cognitive workload in human-computer interaction.

Acknowledgements. The authors would like to thank Naing Naing Maw for her valuable help with programming the experiments for the present study.

4 References

Beatty, J. & Wagoner, B.L. (1978). Pupillometric signs of brain activation vary with level of cognitive processing. *Science, 199*, 1216-1218.

Goldberg, J.H. & Kotval, X.P. (1999). Computer interface evaluation using eye movements: Methods and constructs. *International Journal of Industrial Ergonomics, 24*, 631-645.

Hess, E. H. (1965). Attitude and pupil size. *Scientific American, 212*, 46-54.

Just, M.A. & Carpenter, P.A. (1993). The intensity dimension of thought: Pupillometric indices of sentence processing. *Canadian Journal of Experimental Psychology, 47*, 310-339.

Kahneman, D. (1973). *Attention and effort.* New Jersey: Prentice Hall.

Kohonen, T. (1990). The self-organizing map. *Proceedings of IEEE, 78,* 1464-1480.

Kramer, A.F. (1991). Physiological metrics of mental workload: A review of recent progress. In D.L. Damos (Ed.), *Multiple-task performance.* (pp. 279-328). London: Taylor & Francis.

Nakayama, M., Yasuike, I. & Shimizu, Y. (1990). Pupil size changing by pattern brightness and pattern contents. *The Journal of the Institute of Television Engineers of Japan, 44*, 288-293.

Pomplun, M., Velichkovsky, B.M. & Ritter, H. (1994). An artificial neural network for high precision eye movement tracking. In Nebel, B. & Dreschler-Fischer, L. (Eds.), *Lecture notes in artificial intelligence: AI-94 Proceedings* (pp. 63-69). Berlin: Springer Verlag.

Porter, G., Troscianko, T. & Gilchrist, I.D. (2002). Pupil size as a measure of task difficulty in vision. *Perception 31S,* 170.

Locating Relevant Categories in Web Menus: Effects of Menu Structure, Aging and Task Complexity

Jean-Francois Rouet, Christine Ros, Guillaume Jégou and Sabine Metta

CNRS and University of Poitiers, France
Laboratoire Langage et Cognition
99 avenue du Recteur Pineau
86022 Poitiers Cedex
France
jean-francois.rouet@univ-poitiers.fr

Abstract

Fifty younger, intermediate and older adults were asked to perform a series of search tasks using a 400-item hierarchical Web menu. Menu structure (deep, broad-alphabetic and broad-categorized) and question length and implicitness were manipulated in a within-subject design. Older adults tended to show poorer performance especially with deep menus. Longer and implicit questions also lead to poorer performance. We discuss possible implications of these results for individual and task-related factors of information search performance.

1 Introduction

Most Web portals and other online services use menu structures in order for users to access their contents. In some portals, menus provide access to hundreds of categories in potentially all areas of interest to the general public. Menu design varies a lot from one portal to another. Some portals offer deep menus with only a few options available at each level, while others offer broad menus with many categories available at each level. In some cases, the items are listed alphabetically, while in other cases the items are grouped by semantic categories (e.g., finance, travel...). Given the increasing importance of using the Web for everyday activities, it is important to assess the effects of these design strategies on users' information-seeking performance.

1.1 Menu Structure, Search Performance and User Preference

Past research has found that selection in a menu can be facilitated if items are grouped according to semantic categories (Giroux, Bergeron, & Lamarche, 1987) and if the depth/breadth ratio is optimal (Parkinson, Sisson, & Snowberry, 1985). More recently, Yu and Roh (2002) compared three types of menu presentations on university students' performance and evaluation of a virtual shopping mall. They found that a pull-down menu that maximized visibility of intermediate categories was the most effective in terms of searching speed for both specific and more global search tasks. Students' evaluation of design quality and disorientation, however, did not vary across menu types.

1.2 Issues of Task Specificity and User Characteristics

Most of the studies conducted so far involved university students as participants. It is unclear if the results of these experiments would replicate in naturalistic web-based information search tasks performed by individuals from the general public. University students constitute a homogeneous population with reading and reasoning skills higher than average. In real-life Web usage, individual variables such as the user's age and experience with the Web may interfere with design options. An important issue is to find out how younger vs. older Web users react to different menu structures. So far, studies tended to show a negative main effect of age on search performance (e.g., Westerman, Davies, Glendon, Stammers & Matthews, 1995). However, the relationship between age, search strategy and menu design has been seldom investigated so far.

Another characteristic of laboratory menu experiments is that they often use simple, explicit search objectives. When searching the Web for real-life purposes, the phrasing of the question may be implicit, ambiguous, and it may not correspond to the menu labels. The extra effort needed to infer which categories are relevant given a specific search task may also interfere with design options.

2 Method

2.1 Participants

The participants were 50 members of a panel of 100 adult volunteers representative of the population of a mid-size French town. The group included 9 male and 7 female younger adults (age range 24-36 years); 5 male and 14 female intermediate adults (37-53 years); and 8 male and 7 female older adults (54-80 years). All the participants had been regular users of the Internet for over 18 months at the time of the experiment.

2.2 Materials

A 400-item menu structure was designed after existing Web-based portals. The menu structure presented a hierarchy of general interest categories and subcategories (e.g., education, travel, jobs, sports and so forth). Three versions were developed: broad-categorized, broad-alphabetic and deep. The broad menu versions involved a larger number of items per page (i.e., a maximum of 42), but only two levels of selection (main menu-submenus). The deep menu structure involved a maximum of only 6 items per page, but four levels of selection. In the categorized version of the broad menu, items were grouped according to semantic categories (e.g., "education", "travel", "jobs"), whereas they were ordered alphabetically in the other two versions. The number and wording of target categories were identical across versions. Twelve search questions were written based on keywords from the menu hierarchy, so as to be compatible with all three menu structures. For instance, the question "Find accommodation in a palace hotel" was written based on the hierarchy "tourism>accomodation>hotel>palace". Each question was written in four different versions in order to manipulate explicitness and length. Explicitness was manipulated by replacing the original content words by synonyms (e.g., "Find lodging in a luxury inn"). Length was manipulated by adding 2 or 3 content words to each question (e.g., "find temporary accommodation in a nice palace hotel in Paris"). Although consistent with the meaning of the question, the additional content words did not match with any category label. Short questions comprised 2 to 4 content words, while long questions comprised 5 to 8 words. Each question corresponded to a unique subcategory in the menu structure. A login database was programmed in order to control remote access to the Web menus and to deliver questions and versions according to a balanced within-subject experimental design.

2.3 Procedure

The participants were visited at their homes upon appointment. They were interviewed on their knowledge of the Internet using a standard written protocol, and they were asked to perform a series of search tasks using the list of probes and menu versions. The equipment available at each participant's home was a standard PC with a 14-inch screen. The connection was made with the Web server through a standard 56K modem. The experimenter assisted the participants with connection and login to the Web site. The experimenter gave standard explanations and directions for the execution of the search tasks. The participants performed the 12 search tasks using each question presented in one of the four versions, and one of the three versions of the menu. For each question, the participant had to locate and select the relevant subcategory using a check box. They could re-read the question while searching and could also give up the search using a "give up" button. The participants were encouraged to search as accurately as they could. Nevertheless, if the search remained unsuccessful after 3 minutes, the experimenter prompted the participant to move on to the next question. The categories selected along with the selection delays were automatically recorded and stored in the database.

Based on previous research, we expected older adults to be slower and less accurate than the two other age groups. Furthermore, we expected that the difficulty of locating target categories would increase with menu depth and question length and explicitness. We also expected the difference between younger and older adults to increase with menu depth, due to age-related cognitive limitations.

3 Results

Search success, search time and question lookbacks were used as dependent measures. The data were analyzed using mixed factorial designs, with age level as a between-subject factor (three levels) and question wording, length and menu type as within-subject factors (where applicable).

3.1 Search Success

The average success rate was 52.5%. On 9% of the occasions, the participants selected another category in the correct submenu. On 38.5% of the occasions, they selected another category or failed to provide any answer. Several main effects can be seen. First, performance decreased slightly with age, with an average success rate of 55%, 54% and 48% in age groups 1, 2 and 3, respectively. Second, The broad categorized menu had a success rate (57%) slightly higher than both the deep and the broad alphabetic menus (50%). Third, both question explicitness and length had an impact on search success. Short explicit questions were answered in 67% of the cases, as opposed to 57%, 45% and 41% for long explicit, short implicit and long implicit questions, respectively. To examine the possibility that menu type interacted with age, we performed a two-way ANOVA with age group as a between factor and menu type as a within-subject factor, on the percentage of questions correctly answered by each participant for each menu structure (regardless of question type).

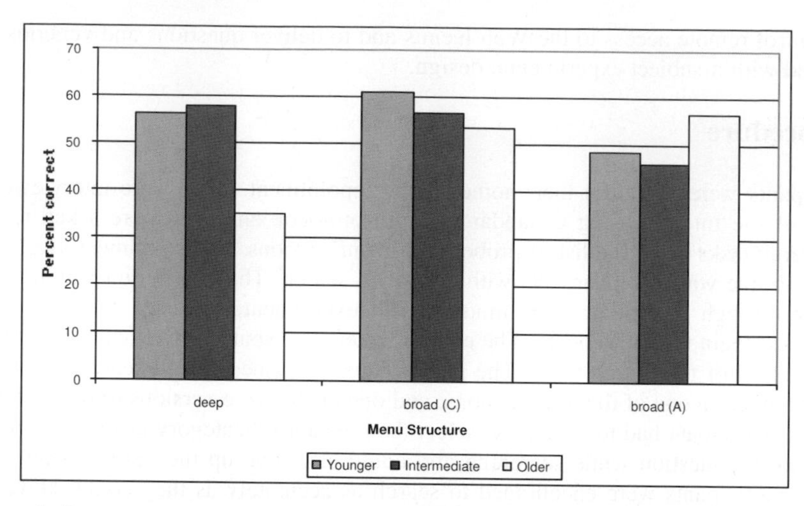

Figure 1: Percentage of correct answers as a function of age group and menu structure.

As can be seen on Figure 1, the age group factor tended to interact with menu type. The percentage of correct answers was lower for older adults in the deep menu condition. Even though the interaction failed to reach statistical significance ($\underline{F}(4,94)=1,81$, NS), a partial comparison indicated an effect of the age level on performance using the deep menu ($\underline{F}(2, 47) = 3.83$, $\underline{p}<.05$). The average performance of the younger, intermediate and older adults was 56.3, 57.9 and 35%, respectively. In the two other menu conditions, age did not significantly affect performance.

3.2 Search Time

We analyzed average search time for correct answers only. Because of the high rate of failed trials, data were collapsed across questions and menu conditions (Table 1). The data from one participant in the older adult group had to be discarded because the log file was incomplete.

Table 1: Average search time for correct answers as a function of age group

	Younger	Intermediate	Older
N	16	19	14
Mean *(SD)*	42,59 *(14,85)*	48,87 *(14,39)*	64,57 *(28,20)*

A one-way ANOVA with age group as a between-subject factor showed that search time increased with age ($\underline{F}(2, 46)=5,01$, $\underline{p}<.05$). Younger and intermediate adults were faster than older adults, but did not differ from each other.

3.3 Question Lookbacks

We analyzed the percentage of questions re-read as a function of age group and menu type. A two-way ANOVA showed a main effect of menu structure on the rate of re-reading. Re-readings were less frequent with the broad categorized menu than in the other two conditions ($\underline{F}(2, 94)= 3,72$, $\underline{p}<.05$). Older adults tended to re-read less questions than the two other groups, especially when searching the deep and broad-categorized menus. Again, the two-way interaction failed to reach significance ($\underline{F}(4, 94)=1,74$, NS), but a partial comparison showed that older adults re-read significantly less questions when using the broad categorized versions than when using the other versions ($\underline{F}(2,28=3.90, \underline{p}<.05)$. Finally, longer and implicit questions were re-read more often than short and explicit questions (29.3, 38.7, 44.7 and 50% for short-explicit, short-implicit, long-explicit and long-implicit questions, respectively). This pattern suggests that linguistic factors directly impacted the difficulty of remembering the question while searching.

4 Discussion

This experiment provided additional evidence that searching a hierarchical menu structure is far from a trivial task for general public adults. Even though we made sure that each question corresponded to a category in the menu hierarchy, the success rate was just slightly over 50%. We observed that any complication of the question, either due to the use of synonyms, or to the intrusion of extra-words caused a decrease in search success. This finding suggests that adults tend to search based on lexical similarity rather than on a deeper semantic representation of the question content.

Older adults were slower than younger and intermediate adults. Their poorer performance with deep menus is consistent with previous findings and suggests that older adults have more trouble dealing with longer series of subgoals (in this case, a longer series of selections), due to decreased working memory capacity. Besides, the relatively small impact of menu depth on younger users contradicts previous results, but maybe interpreted in terms of the high visual complexity of the broad menu versions used in this experiment.

Other factors may also differentially affect younger and older users' performance. Older adults have more negative attitudes toward technology (Czaja & Sharit, 1998, and they tend to underestimate their computer knowledge (Marquié Jourdan-Boddaert & Huet, 2002). These and other non-cognitive factors may be responsible for the effects observed in the present study. For example, the fact that older adults showed a pattern of question lookbacks different from younger and intermediate adults suggests a shift in strategy rather than a plain effect of cognitive capacity. Re-reading the question is a deliberate, strategic action that probably depends on one's representation of task and environmental constraints. Although we lack independent data to conclude on this issue, we may tentatively suggest that reliance on the environment's affordances (here, the external representation of information) may account for at least some of the observed differences. Put together, the results emphasize the high cognitive load associated with menu search, and the importance of explicit categories and search probes.

References

Czaja, S.J., & Sharit, J. (1998). Age difference in attitudes toward computers. *Journal of Gerontology : Psychological Sciences*, 53(5), 329-340.

Giroux, L., Bergeron, G., & Lamarche, J.-P. (1987). Organisation sémantique des menus dans les banques de données, *Le Travail Humain*, 50(2), 97-107.

Marquié, J.C., Jourdan-Boddaert, L., & Huet, N. (2002). Do older adults underestimate their actual computer knowledge ? *Behaviour and Information Technology*, in press.

Parkinson, S.R., Sisson, N., & Snowberry, K. (1985). Organization of broad computer menu displays. *International Journal of Man-Machine Studies*, 23, 289-297.

Westerman, S.J., Davies, D.R., Glendon, A.I., Stammers, R.B., & Matthews, G. (1995). Age and cognitive ability as predictors of computerized information retrieval. *Behaviour and Information Technology*, 14, 313-326.

Yu, B.-M., & Roh, S.-Z. (2002). The effects of menu-design on information seeking performance and users' attitude on the World Wide Web. *JASIST*, 53(11), 923-933.

Measuring Driver Fatigue and Establishing Kolintang Music Treatment for Decreasing Fatigue in Driver: A Preliminary Study to Develop Smart Sensor of Fatigue for Car Driver

Rozmi Ismail, Yohan Kurniawan, Ismail Maakip,
Mohd. Salleh Abd . Ghani, Mohd. Jailani Mohd Nor & Daud Sulaiman

Accident and Safety Research Group, School of Psychology and Human Development, National University of Malaysia, MALAYSIA

Abstract: The objective of the project described in this paper is to develop methods for studying the relationships between physiological status of the driver and the effectiveness of music treatment to reduce fatigue. Previous findings have indicated that music could somehow reduce fatigue when driving. Evidence has showed that the listening to radio/music can reduce fatigue crash-related accident (42.6%). In this study we used wooden music because it is a slow music and give a calm feeling. For several reasons such as safety, reproducibility, and low cost, a laboratory-based driving simulator is being used for the project experiments. Initial experiments were conducted with a cohort of 5 male and 4 female drivers aged between 20-30 years under carefully controlled conditions. In this experiment, subjects were drove continuously for 2 X 2 hours. Physiological measurement includes EEG & EMG was used before and after treatment. Video camera also used to record body movement of the driver under experimental and treatment conditions. In the treatment condition, we used instrumental music (wooden music namely Kolintang (an Indonesia's wooden music)). This experiment is divided into 2 parts; in the first two hours subjects were drive under no treatment condition, whereas in the second part, the subjects were drive for two hours under treatment condition (music). Results of the study showed that in the future the research is going to develop smart sensor to alert fatigue and automatic compatible music treatment when a driver feels fatigue and drowsy. This research is funded by IRPA project number 03-02-02-0016-SR0003/07-02.

Introduction

The meaning of fatigue extends all the way from human feelings of being tired to a weakening of inanimate objects such as steel girders. Even when limited to human applications, the definition of fatigue varies from physiological impairment to subjective feelings to performance decrement. However, fatigue may be viewed quite broadly in a context analogous to psychological activation theory. If a person is overloaded, they suffer from an exhaustive type of fatigue. If they are insufficiently loaded, he suffers from a type of fatigue known as boredom or monotony. Excessive fatigue of either type will affect his/her task performance. A driver who is fatigue when driving can be drowsy. Drowsy driving is a serious problem that leads to thousands of automobile crashes each year. In 1995, NHTSA (The National Highway Traffic Safety Administration) began employing the Crashworthiness Data System (CDS) to obtain more in-depth information on driver inattention-related crash causes, including drowsiness and many forms of distraction. Driver drowsiness is one specific form of human error that has been well studied. Loss of driver alertness is always preceded by psychophysiology. Fatigue is more dangerous than speeding. Over 39.000 serious injuries and nearly 3.500 deaths occurred on roads in the UK last year (according to DTLR statistic). The three major forms of driver inattention and their percent involvement in 1995 CDS crashes are: distraction (13.3 %), looked but did not see (9.7%), and sleepy/fell asleep (2.6%). In Western Australia about 30% of all fatal vehicle crashes, and 1 in 4 fatal truck crashes are caused by fatigue. About 1 in 3 drivers have felt close to falling asleep at the wheel in the past year and 13% have fallen asleep at least once. Thirteen percent of WA truck drivers reported occasional nodding off at the wheel in past six months and 10% report fatigue is often or always a problem for them. In 1985 the American Automotive Association examined 221 truck crashes where the truck was towed away. Crashes were determined to be due to fatigue if the driver had worked more than 16 hours and if the characteristics of the crash were consistent with falling sleep such as running off the road or on to the wrong side of the road. It was estimated that fatigue was the primary cause in 40% and a contributory cause in 60% of crashes. One study concluded that in articulated vehicles the incidence of fatigue crashes ranges between 5 and 10% of all crashes, about 20-30% of injury crashes and about 25-35% of fatal crashes. Studies of car crashes where the crash had similar characteristics of running off the road or a head on collision when not over taking show much the same prevalence. A U.K. Study revealed 20% of freeway crashes were attributed to fatigue by car drivers. Usually fatigue is related with sleepiness because when someone feels tired, they will be feeling sleepiness. One cause someone sleepless is loss sleep. Sleep is a neurobiological need with predictable patterns of sleepiness and wakefulness. It is an episodic phase characterized by reduced awareness and responsiveness to both internal and external stimuli, and by motor inhibition.

Objectives and Hypotheses

The main objective of this research is to develop methods for studying the relationships between physiological testing driver states and music treatment for reducing fatigue to driver. This research aims to gather data of fatigue criteria to driver and knowing the effect of wooden music to reduce fatigue when driving. We use wooden music because it is a slow music and give a calm feeling. Evidence showed that listening\to radio or music can reduce fatigue and crash among drivers. We hypothesized that there should be a relationship between feeling of fatigue and errors while driving. We also predicted that under music condition the level of fatigue will be reduced therefore less errors reported.

Laboratory experiment (driving simulator)

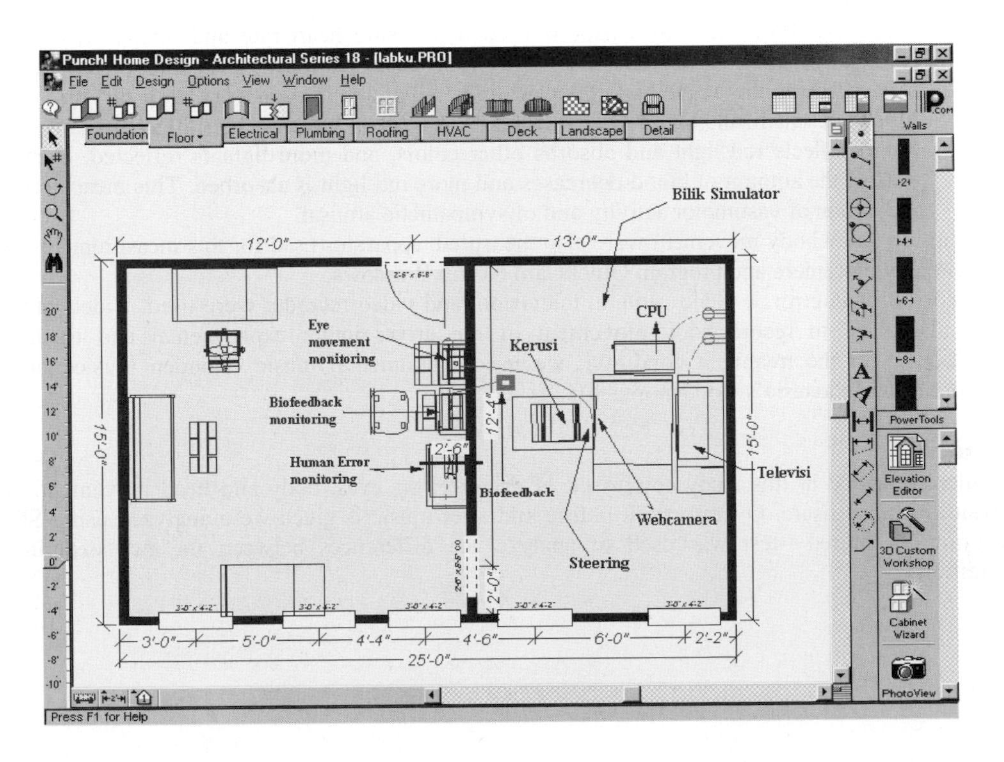

Method

For reasons of safety, reproducibility, and low cost, a laboratory-based driving simulator is being used for the project experiments. (See appendix 1). The experiment laboratory were built according to driving condition. The driving simulator completed with in car simulator such as steering, accelerator and so on. The driving games software (like video arcade game) were modified to suit with the experiment condition to measure driver behavior as well as driver status when driving under prolonged condition.

Subjects:

Initially 9 subjects were involved in (5 male and 4 female Malaysian drivers) aged between 20-30 years. All subjects have had a driving licensed and had driving experience at least 2 years.

Design and Procedure:

In this experiment, subjects were continuously driven for 2 X 2 hours. This experiment is divided into 2 parts; in the first two hours subjects were drive under no treatment condition, whereas in the second part, the subjects were drive for two hours under treatment condition (wooden music).

Apparatus:

For the measurement of physiological changes, Biofeedback (Procompt+) is used. This software is normally used to measure physiological status of subject or patient under treatment. The measurement consists of:

a. EMG: The MyoScan-Pro is the surface EMG (sEMG) or Electro Myogram. It measure muscle activities. EMG is recorded from a sensor that is placed on the skin's surface. In this research, 400 W (wide bandwidth) position is used because the sensor will be sensitive to the full 20-500 Hz bandwidth.
b. EEG: EEG sensor detects and amplifies the small electrical voltages. The frequencies most commonly looked at, for EEG, are between 1 and 40 Hz, amplitude between 0.1 and about 200 micro-Volts. The frequencies above 40 Hz are interpreted as EMG noise from neighboring muscles.
c. BVP: The BVP-Flex/Pro sensor is used to measure heart rate and inter-beat interval. BVP, also called photoplethysmography, bounces infra-red light against a skin surface and measures the amount of reflected light. This amount will vary with the amount of blood present in the skin. At each heart beat (pulse), there is more blood in the skin – blood reflects red light and absorbs other colors, and more light is reflected. Between pulses, the amount of blood decreases and more red light is absorbed. This measure is an indication of vasomotor activity and of sympathetic arousal.

Eyes blinking and body movement were also measured, apparatus used for this measurements was a Logitech web camera and program QuickCam 6.0 for Windows.

To record human error, a video camera, television, and video recorder were used. Video camera were also used to record body movement of the driver under experimental and treatment conditions. In the treatment condition, we used instrumental music (wooden music namely *Kolintang* (an Indonesia's wooden music)).

Results

Results presented in this study comprised of driver error; eyes, body and head movement; and physiological measure. Overall result before and after music is given were analyzed using SPSS program. A paired t-test was used to analyze the differences between the two experiment conditions.

A. Driver Error

	Paired Differences						
			95% Confidence Interval of the difference				
	Mean	SD	Lower	Upper	T	Df	Sig.(2-tailed)
Total error before music treatment and after music treatment	139	1070.69	-684	962	.389	8	.707

Result: No significant difference between error before and after music treatment.

From the result showed that there is no significant difference of driving error before and after the music treatment were given. This indicates that treatment (music) does not effect or improve drivers fatigue level. Errors that occur when driving however not decrease after music is given differed from opinion that music can reduce fatigue and sleepiness related accident.

B. Eyes, head, and body movement

	Paired Difference						
			95% Confidence Interval of the Difference				
	Mean	SD	Lower	Upper	T	Df	Sig.(2-tailed)
Eye, head, and body movement before music treatment and after music treatment	139	1071	-684	962	.389	8	.707

Result: No significant difference between eye, head, and body movement before and after music treatment.

Eyes, head and body movement were measured to find out the level of fatigue and sleepiness among drivers. The more the eyes, head and body move the more fatigue and sleepy the driver is. The total of eyes, head and body movement before and after treatment were recorded and the data were analyzed. From the result showed that there is no significant difference between the total eyes, head and body movement before and after music were given. Although music was given, the level of fatigue or sleepiness is still the same.

C. Physiological Measurement

Results of Physiological measurement were also showed no significant differences between before and after treatment. The EEG, EMG and BVP indicators showed no significant which means that driver status has not change very much before and after treatment was given.

	Paired Difference						
			95% Confidence Interval of the Difference				
	Mean	SD	Lower	Upper	T	df	Sig.(2-tailed)
EEG before – after music treatment	-36.74	1564.05	-1238.97	1165.50	-.07	8	.946
EMG before – after music treatment	1.11	19.54	-13.90	16.13	.17	8	.868
BVP before – after music treatment	42.54	142.37	-66.90	151.97	.87	8	.396
Physiology before – after music treatment	6.91	1593.16	-1217.70	1231.53	.01	8	.990

Discussions and Conclusion

So far we have developed a laboratory simulator to measure driver status under experimental and control conditions. Results of the study showed that driver status during prolonged driving can be monitored physiologically or physically. Driver status as indicated by EEG, BVP and EMG graph showed that drivers do experienced fatigue and sleepiness as the number of errors made during driving significantly correlated with the number of errors. This indicates that under fatigued and sleepy condition drivers unable to detect traffic changes, thus more errors were made such as driving off the road or hitting something. Numbers of eyes blink also increased as drivers session goes on. However when the drivers' performance were compared with before and after music treatment were given the results showed no significant difference. According to this results one would argue that music is not a good approach to combat tiredness when driving. It is too early to conclude that the music treatment can reduce fatigue, as the experiment showed the opposite which remain a question whether certain type of music could reduce fatigue. The *kolintang* wooden music used in the experiment may consider as a sentimental music and did not help the driver to reduce fatigue level. Research is underway to implement aroma therapy as an alternative method to reduce fatigue level instead of using sentimental music.

References

Corfitsen M.T. 1994. *Tiredness and visual reaction time among young male night time drivers: A roadside survey*. Accident, Analysis and Prevention. Vol 26(5): 617- 624.

Fell D.L., and Black B. 1997. *Driver fatigue in the city*. Accident, Analysis and Prevention. vol 29(4): 463-469.

Feyer A.M, Williamson A. and Friswell R. 1997. *Balancing work and rest to combat driver fatigue: An investigation on two-up driving in Australia*. Accident, Analysis and Prevention. Vol 29(4): 541-553.

Hartley Laurence.1995. *Fatigue and driving: Driver impairment, driver fatigue and driving simulation*. Australia: Taylor and Francis (http://www-nrd.nhtsa.dot.gov/depertments/nrd-13/driver-distraction/PDF/1.PDF)

Hughes P.K and Cole B.L. 1986. *What attract attention when driving?* Ergonomics. Vol 29(3): 377-391.

Lenne M.G, Triggs T.J and Redman J.R.1997. *Time of day variation in driving performance*. Accident, Analysis and Prevention. Vol 29(4): 431-437.

Lovsund P., Hedin A. and Tornros J. 1991. *Effect on driving performance of visual field defects: A driving simulator study*. Accident Analysis and Prevention. Vol 23(4): 331-342.

Macdonald W. A and Hoffmann E.R. 1984. *Driving awareness of traffic sign information*. Australian Road Research Board Internal Report AIR : 382-391.

Maycock G. 1997. *Sleepiness and driving: The experience of UK drivers*. Accident, Analysis and Prevention. vol 29(4): 453-462.

McKnight, A. J., Shinar, D. & Hilburn, B. 1991. *The visual and driving performance of monocular and binocular heavy duty truck drivers*. Accident Analysis and Prevention. August; Vol 23 (4) : 225-237.

Pack A.I, Pack A.M, Rodgman E, Cucchiara A, Finges D.F, & Schwab C.W.1995. *Characteristic of crashes attributed to the driver having fallen asleep*. Accident Analysis and Prevention. Vol.27(6): 769-775.

Recarte, M.A.& Nunes, L.M. 2000. *High percentage of car accidents due to inattention rather than lack of driving ability*. http://www.shpm.com/articles/ sports/caraccidents.html.

Renge K. 1980. *The effect of driving experience on a driver visual attention. An analysis of object looked at: Using the 'verbal report' method*. International Association of Traffic Safety Science Research. 4: 95-106.

Summala H., Nieminen T., and Punto M. 1996. *Maintaning lane position with peripheral vision during in-vehicle tasks*. Human Factors. Sept 1996 v38(3): 442- 452.

Testin F.J and Dewar R.E. 1981. *Devided attention in a reaction time index of traffic sign perception*.Ergonomics.Vol24(2):111-124.Appendix 1

Designing for Psychological Effects: Towards Mind-Based Media and Communications Technologies.

Timo Saari

M.I.N.D. Lab / CKIR / Helsinki School of Economics
Tammasaarenkatu 3
00180 Helsinki, Finland
saari@hkkk.fi

Abstract

Mind-Based Media and Communications Technologies are technologies that can produce immediate emotional and cognitive effects in their individual users as they are consuming multimodal information. With personalization technologies one may vary the content and way of presenting information to create targeted psychological effects. Implications to HCI are discussed.

1 Introduction

When perceiving information via media and communications technologies, the mind is psychologically transported into a quasi-natural experience of the events described. This is called presence. In presence, information becomes the focused object of perception, while the immediate, external context, including the technological device, fades into the background (Lombard and Ditton, 1997). Various empirical studies show that information experienced in presence has real psychological effects on perceivers, such emotion based on the events described or cognition of making sense of the events and learning about them (Reeves and Nass, 1996).

Communication systems may be considered as consisting of three layers (Benkler, 2000). At the bottom is a physical layer that includes the physical technological device and the connection channel that is used to transmit communication signals. In the middle is a code layer that consists of the protocols and software that make the physical layer run. At the top is a content layer, which consists of multimodal information. The content layer includes both the substance and the form of multimodal information (Billmann, 1998; Saari, 2001). Substance refers to the core message of the information. Form implies aesthetic and expressive ways of organizing the substance, such as using different modalities and structures of information (Saari, 2001). Technologies may be considered Mind-Based because they take into account the characteristics of different segments of users and alter information presented to them in a systematic manner to create emotional and cognitive effects. For instance one may vary the form of information per user profile, which may more or less systematically produce, amplify, or shade different psychological effects.

2 Model of emotional and cognitive effects

When a user of an information system is interpreting information, a complex set of interrelated "gateway variables" may influence emotional and cognitive effects. These gateway variables may be clustered as Mind (individual differences and social similarities of users), Content (information substance and form embedded in technology with certain ways of interaction) and Context (social and physical context of reception) (Saari, 1998; Saari, 2001). If one is able to predict which types of variations of the gateway variables (Mind, Content and Context) and their sub-variables produce which types of consequent psychological effects this may result in "design-rules" for psychological effects for certain individuals or segments of individuals. The key idea is that the gateway variables have some values just before a user receives certain information or interacts with a system in a session. It means that they constitute some preliminary probabilities and directions of the possible emotional and cognitive effects that are about to occur during and immediately after consuming the particular information. As this is rather complex it may be sensible to consider only the key variables and their interactions with selected individual differences and consequent most probable and most intense psychological effects within a certain task.

It is obvious that the substance of information, i.e. what is said, influences psychological effects. However, there is evidence in literature that varying the form of information creates emotional and cognitive effects. In media studies it has been found that different modalities, such as visual and auditory, may lead to different kinds of psychological influences and the valence of a preceding subliminal stimulus influences the subsequent evaluation of a person evaluated (Cuperfrain and Clarke, 1985; Krosnick et al, 1992). In educational studies it has been shown that different ways of processing information influence learning and emotion of stimuli with certain modality (Riding and Rayner, 1998). Research concerning emotional influences on the cognitive processing of information has often concentrated on how different emotions related to information change the way users pay attention to, evaluate and remember the mediated message. This research has results on the influence of emotional information as increasing the user´s self-reported emotion (Lang et al, 1996); attention (physiological and self-reported) (Lang et al, 1995) and memory for mediated messages, particularly arousing messages (Lang, 1990; Lang et al 1995; Lang et al, 1996). Studies in experimental psychology have shown that recognition and memory can be influenced or even enhanced by previous exposure to subliminal visual or auditory images of which the subjects are not consciously aware (Kihlström et al, 1992).

3 Personalization of information

Personalization technologies introduce a new set of tools that enable automatic and rule-based variations of information for individual users (Riecken, 2000). Based on the principle of variability, many potential versions of the same content or service may be available for different users (Manovich, 2001). One may discuss the packaging of information, which means how the different elements of information are put together into a certain type of package, including form and substance. The content can be selected and organized in different ways and the presentation of content can be tailored to suit the needs and preferences of the individual. This may include personal preferences for layouts or colour schemes.

Personalization can be explicitly controlled by the user or it can be done adaptively based on user profiles. With the possibility of real-time adaptation of information for different users it is hypothesized that one may vary the form of information within some limits per the same substance of information. For instance, the same substance can be expressed in different modalities based on the available selection of alternate modalities of the same or roughly the same substance. Table 1 adresses the key factors which may influence emotional and cognitive effects of information. For instance, if a user has a profile indicating that he processes text more fluently than audio within a given communications device the system may try to select the information modalities accordingly.

Naturally one may vary also the components of the code layer, i.e. ways of interaction and visual-functional aspects of user controls when producing psychological effects. This has the advantage of being less dependent on a particular substance. The same approach partly applies to visual layouts and structure as subcomponents of form of content with certain substance in a given modality.

Table 1. Key factors influencing psychological effects of technology adapted from (Saari, 2001).

Layer of technology	Key factors
Physical	**Hardware** - large or small vs. human scale - portable or non-portable - close or far from body (intimate-personal-social distance) - user changable covers or additional devices to alter appearance or functionality
Code	**Interaction** - degree of user vs. system control and proactivity through user interface
	Visual-functional aspects - way of presenting controls in an interface visually, temporally and functionally
Content	**Substance** - the essence of the event described - type of substance (factual/imaginary; genre, other) - narrative techniques used by authors
	Form *1. Modalities* - text, video, audio, graphics, animation, etc. *2. Visual layout* - ways of presenting various shapes, colours, font types, groupings and other relationships or expressive properties of visual representations - ways of integrating modalities into the user interface *3. Structure* - ways of presenting modalities, visual layout and other elements of form and their relationships over time (linear and/or non-linear structure)

4 Emotion Media

Mind-Based Media and Communications Technologies are created via real-time variations of i) substance, ii) form and iii) code layer (interaction and controls) within a certain technological device per certain user profiles. These elements interact in complex ways when producing psychological effects. The role of hardware should not be neglected. A device with a large screen or a portable device with smaller screen with user-changeable covers may also influence the emerging effects. Examples of such technologies include Knowledge Media (Saari, 1998) that would enhance in-depth learning and Emotion Media (Saari, 2001) that would produce certain types of emotions. One may also think of Presence Media (Saari, 2002) that may produce desired types of presence. To realize various types of Mind-Based Media and Communication Technologies, such as Emotion Media, one needs to i) profile users, ii) create a database for design-rules for psychological effects and iii) utilize a personalization engine which conducts the variations of information based on desired psychological effects. A technique for such operationalization has been proposed as Psychological Customization (Saari and Turpeinen, 2003).

Emotion Media may be used as a concrete example of Mind-Based Technologies. Emotion may be considered a reaction to events ranked as relevant by a particular individual; and emotions have physiological, affective, cognitive and behavioral components (Brave and Nass, 2002). One may focus on "primitive" emotional responses or emotions requiring more extensive cognitive appraisal and processing. Both of these types of emotions can be linked to various psychological consequenses. For instance, emotions may be the basis of mood, a general and longer-term emotional state not linked to a particular emotional response (Frijda, 1994). In accordance with particular emotional reactions or moods attention may be increased, memory may be influenced, performance in problem solving may be enhanced and judgement and decision making may be influenced (Clore and Gasper, 2000; Reeves and Nass, 1996; Isen, 2000). Consequently, with Emotion Media one may focus on i) creating immediate and primitive emotional responses, ii) creating mood and iii) indirectly influencing secondary effects of emotion and mood, such as attention, memory, performance and judgement.

5 Implications for HCI

Mind-Based Media and Communications Technologies as a research area has some overlap with usability studies and design studies that also research user experiences. For instance, in usability studies the pleasantness and the aspect of having fun with interfaces have been addressed (Monk and Frochlich, 1999). Affective computing has been developed in the area of computers and emotion (Picard, 1997). Accordingly, in design-related research there has been for some time discussion about emotion and design (Hirsch et al, 2000). However, Mind-Based Media and Communication Technologies pose a possible change in thc perspective to technology. One may view technology as a source for creating added value for a user, such as enabling desired emotional and cognitive effects. In contrast, in usability studies technology is often viewed as a tool to be used efficiently. This approach sometimes lacks focus on the depth of user experience outside the efficient operation of machinery and motivation to use an actual "media" or service. In studies of emotion and design there are gaps in explaining and predicting i) which component of the design influences which type of emotional effect and ii) how different user profiles interact with the design-components and psychological effects. Consequently, this type of information may currently not be robust enough to act as basis of design-rule databases to produce systematic psychological effects.

To be able to realize various archetypes of Mind-Based Media and Communication Technologies one may have to conduct a number of experimental studies in which different applications with certain tasks are tested in laboratory and field conditions. Laboratory methods, such as psychophysiology, may be used for indicating emotional effects in addition to self-report methods. However, the real challenge comes from conducting research with more ecological validity than laboratory studies. This brings into focus the nature of the task of the user in real life. Naturally, in accordance with Psychological Customization techniques, one should develop content management technologies to utilize the design rules acquired from user studies. Considering the large number of experiments and elaboration needed to research the feasibility of various Mind-Based Media and Communications Technologies it seems evident that collaboration and input from various research groups will be necessary on this potential new area of HCI.

6 References

Benkler, Y. (2000). From Consumers to Users: Shifting the Deeper Structures of Regulation. *Federal Communications Law Journal* 52, 561-63.

Brave S. and Nass, C. (2003). Emotion in human-computer interaction. In Jacko, J.A. and Sears, A. (Ed.), *The Human-Computer Interaction Handbook. Fundamentals, Evolving Technologies and Emerging Applications*. (pp. 81-96). London : Lawrence Erlbaum Associates.

Billmann, D. (1998). Representations. In Bechtel, W. and Graham, G. (Ed.) *A companion to cognitive science* (pp. 649-659). Malden, MA: Blackwell publishers.

Clore, G. C. and Gasper, K. (2000). Feeling is believing. Some affective influences on belief. In Frijda, N.H., Manstead, A. S. R. and Bem, S. (Ed.), *Emotions and beliefs: How feelings influence thoughts* (pp. 10-44). Paris/Cambridge: Editions de la Maison des Sciences de l'Homme and Cambridge University Press.

Cuperfain, R. and Clarke, T. K. (1985) A new perspective on subliminal perception. *Journal of Advertising*, 14, 36-41.

Frijda, N.H. (1994). Varities of affect: Emotions and episodes, moods and sentiments. In Ekman, P. and Davidson, R.J. (Ed.), *The nature of emotion* (pp. 59-67). New York: Oxford University Press.

Hirsch, T., Forlizzi, J., Hyder, E., Goetz, J., Stroback, J., and Kurtz, C. (2000). The ELDeR Project: Social and Emotional Factors in the Design of Eldercare Technologies. *Conference on Universal Usability*, 2000, 72-80.

Isen, A. M. (2000). Positive affect and decision making. In Lewis, M. and Haviland-Jones, J. M. (Ed.), *Handbook of emotions* (2nd ed.) (pp. 417-435). New York: Guilford Press.

Kihlström, J. F., Barnhardt, T. M. and Tataryn, D. J. (1992). Implicit perception. In Bornstein, R. F. and Pittmann, T. S. (Ed.), *Perception without awareness. Cognitive, clinical and social perspectives* (pp. 17-54). New York: Guilford Press.

Krosnick, J. A. , Betz, A. L., Jussim, J. L. and Lynn, A. R. (1992). Subliminal conditioning of attitudes. *Personality and Social Psychology Bulletin*, 18, 152-162.

Lang, A. (1990). Involuntary attention and physiological arousal evoked by structural features and mild emotion in TV commercials. *Communication Research,* 17 (3), 275-299.

Lang, A., Dhillon, P. and Dong, Q. (1995). Arousal, emotion and memory for television messages. *Journal of Broadcasting and Electronic Media*, 38, 1-15.

Lang, A., Newhagen, J. and Reeves. B. (1996). Negative video as structure: Emotion, attention, capacity and memory. *Journal of Broadcasting and Electronic Media*, 40, 460-477.

Lombard, M. and Ditton, T. (1997). At the heart of it all: The concept of presence. *Journal of Computer Mediated Communication*, 3 (2).

Manovich, L. (2001). The language of new media. Cambridge, MA, London, England: The MIT Press.

Monk, A.F. and Frohlich, D. (1999). Computers and Fun, *Personal Technology*, 3, 91.

Picard, R. W. (1997). Affective Computing. MIT Press, Cambridge, 1997.

Reeves, B. and Nass, C. (1996). The media equation. How people treat computers, television and new media like real people and places. Stanford: Cambridge University Press.

Riecken, D. (2000). Personalized views on personalization. *Communications of the ACM,* V. 43, 8, 27-28.

Riding, R. J. and Rayner, S. (1998). Cognitive styles and learning strategies. Understanding style differences in learning and behavior. London: David Fulton Publishers.

Saari, T. (1998). Knowledge creation and the production of individual autonomy. How news influences subjective reality. Reports from the department of teacher education in Tampere university. A15/1998.

Saari, T. (2001). Mind-Based Media and Communications Technologies. How the Form of Information Influences Felt Meaning. Acta Universitatis Tamperensis 834. Tampere: Tampere University Press.

Saari, T. (2002). Designing Mind-Based Media and Communications Technologies. *Proceedings of Presence 2002 Conference*, Porto, Portugal, 79-87.

Saari, T. and Turpeinen, M. (2003). Towards Psychological Customization of Information for Individuals and Social Groups. Manusript submitted *to Interact 2003.*

Effects of perceptual and semantic grouping on the acquisition of hypertext conceptual models

Ladislao Salmerón

Department of Experimental Psychology
University of Granada
Campus Cartuja. 18071 Granada - Spain
salmero@fedro.ugr.es

José J. Cañas

Department of Experimental Psychology
University of Granada
Campus Cartuja. 18071 Granada - Spain
delagado@ugr.es

Inmaculada Fajardo

Department of Computer Science
University of the Basque Country
Lardizábal 1. 20018 Donostia - Spain
acbfabri@si.ehu.es

Miguel Gea

Department of Computer Science
University of Granada
Saucedo Aranda. 18071 Granada - Spain
mgea@ugr.es

Abstract

Research conducted on hypertext systems had not provided solid results about the process involved in the acquisition of hypertext conceptual models (Farris, Jones and Elgin, 2002). In this paper, we propose a model to assess this process, and some predictions following this model are tested in an experiment. Finally, we present a preliminary model of how users acquire hypertext conceptual models and a guideline based on the results obtained.

1 Introduction

Users form and use internal models while interacting with hypertext systems. Understanding the model held by users is important in order to avoid problems such as to get disoriented in hypertext systems (Otter and Johnson, 2000). Researchers have traditionally measured user models in order to map expert models into the hypertext structure (Jonassen, 1990). This approach is based on the mapping assumption that states that if the hypertext reflects the semantic structure of the expert, a novice user will apprehend more easily the expert knowledge. However, research following this idea has not always provided successful results. For example, Jonassen (1993) conducted several experiments to test the mapping assumption, and found that merely making visible content structures in the interface was not sufficient for helping novice users acquire expert structure. Rather, he found that what were determinant for this acquisition was the nature of the processing task and goals for learning while interacting with a hypertext.

We believe that research following the mapping assumption has not provided robust results because of the lack of a consolidated theory of how users acquire hypertext conceptual models (Farris, Jones and Elgin, 2002). Most researchers follow the theory of the cognitive map, which was first proposed for explaining navigation in physical environments. This theory states that the process by which navigational knowledge of physical places is acquired develops at several stages

(Wickens, 1992). It begins with the identification of landmarks, followed by a development of route knowledge, ending with the acquisition of a survey type cognitive map which allows persons to plan journeys along routes not previously travelled. The researchers following this theory consider that the user internal representation is similar to a cognitive map of the hypertext, where the "landmarks" are the nodes and the "routes" are the links (Dillon, McKnight & Richardson, 1993). However, this approach has served more as a metaphor of how users navigate through the system than as a cognitive theory of the mental processes involved in user interaction with hypertexts (Broechler, 2001; Farris et.al, 2002). Therefore, we are still in need of a theory of how novice users acquire conceptual models of hypertext.

As an alternative to the cognitive map theory, we propose to start from considering an empirical model of how a mental representation of the system (a mental model) is constructed (Cañas, Antolí & Quesada, 2001). These authors assume that this mental representation is formed through combining in Working Memory (WM) information stored in Long-Term Memory (LTM) and the characteristics extracted from the task. This combination would demand a varying amount of cognitive resources. Then, the next steps would be to consider a theory of Cognitive Load to make predictions about how and when the demands of cognitive resources would affect the construction of the mental representation acquired after navigating in a hypertext system. In doing that, we can follow Tripp and Roby (1990) who proposed that if the cognitive load associated to navigation is reduced, the users would have more free resources to learn the structure of the hypertext.

First, we would assume that cognitive load could come from the user attempt to maintain and consider unrelated concepts in WM (Sweller, 1988). Therefore, a well structured hypertext system could help to reduce user's cognitive load by allowing her/him to chunk meaningful data. We could call this prediction the "Semantic Grouping – SG hypothesis". Second, cognitive load could come from the lack of perceptual clues of the interface. For example, let us think in an interface menu with 10 items. A none perceptually grouped menu had to be processed item by item. The limited capacity of WM to 7+-2 chunks would difficult this process. On the contrary, the same menu can be designed forming two or more perceptual groups. In this case, the user only has to process 2 chunks. We could call this prediction the "Perceptual Grouping - PG" hypothesis.

We have conducted an experiment to test both SG and PG hypothesis on the acquisition of hypertext conceptual models. Both SG and PG hypotheses are necessarily related. For example, in order to chunk meaningful data (SG) it is necessary to group this information perceptually (PG). Therefore, we have tried to isolate the effects of both grouping factors by using three different type of web menus: 1- Semantically Grouped menu (SGM), grouping related items using the same color for related concepts; 2- Arbitrary Grouped menu (AGM), using the same pattern of color used in (1) but distributing items without any semantic relation; and 3- Arbitrary Ungrouped menu (AUM), using the same distribution of items of (2) but without color grouping.

2 Experiment

Participants. Participants were 42 undergraduate students at the University of Granada, 36 women and 6 men. Average age was 21.6 years old. They received course-credits for their participation.

Materials. A web site of a Scientific Workshop was designed for this experiment. It was chosen because it contained information that is not familiar to our participants (pre-grade university students). Then, we could expect that they did not have form a conceptual model of this kind of web site. The site consisted of 11 pages hierarchically organized. The main page contained links to the rest of the pages. Each link was surrounded by a coloured rectangle (Figure 1). Three menu interfaces were created, as stated above: 1- Semantically Grouped menu (SGM); 2- Arbitrary Grouped menu (AGM); and 3- Arbitrary Ungrouped menu (AUM). Groups of the SGM were created according to an expert solution. Four experts performed a card sorting task with the items of the main menu. The two grouping solution was chosen because it had the better intra-expert agreement. The experts agreed in labelling the two groups as "Bureaucratic" and "Content info".

Figure 1: The main pages of the SGM (left), AGM (centre) and AUM (right) condition.

Procedure. Participants performed a 20 items search task through the web site. The presentation of the items was divided in two stages. In stage 1 each of the first 10 items referred to information contained in one of the 10 different content-pages. In stage 2, each of the 10 items referred also to one of the different content-pages. Participants had 90 seconds in order to find each item. After the search task participants performed a judgment relation task. The concepts of this task consisted of the combination of the 10 links of the menu, which resulted in 45 pairs. Participants had to evaluate the relation between each pair of links on a scale from 1 (not related) to 6 (much related).

Design. The design was a 3 x 2 Mixed Factorial. Main page (SGM, AGM and AUM) was the between-participants variable and Stage (1 and 2) was the within-participant variable. Performance on the search task was measured by two dependent variables: response time and lostness. Lostness was defined as: $L = (N/S - 1)^2 + (R/N - 1)^2$ (Smith, 1996), were N is the number of different nodes visited while performing the search task; S is the total number of nodes visited while performing the search task; and R is the number of nodes needed in order to accomplish the search task. The greater the values, the greater the lostness. The third dependent variable was the judgment ratings.

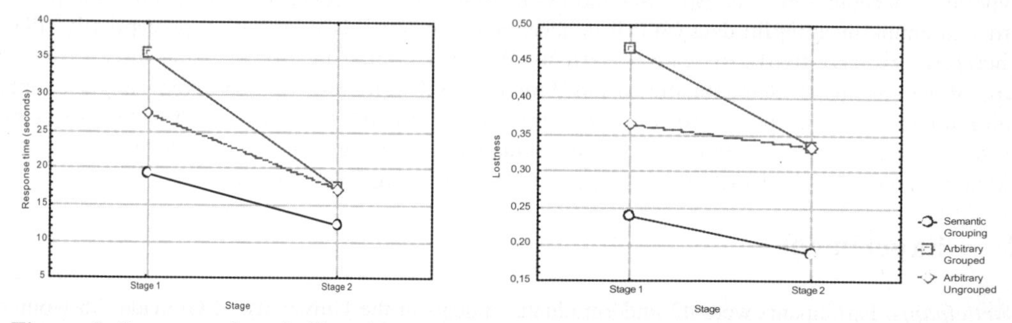

Figure 2: Response time (left) and Lostness index (right) as a function of Main page and Stage.

Results. The construction model of Cañas et. al. (2001) and the Cognitive load theory predicted that users would acquire a better conceptual model and perform better when menu items are chunked. The results of the present experiment showed that this prediction is valid for the SGM, but it is opposed (interference instead of facilitation) for the AGM. Time response and Lostness data showed that participants of the semantically organized page were faster and get lost less than participants of the two arbitrary pages (Figure 2). In addition, data suggest that the effect of semantically grouping persists even with learning, because these differences were observed in both Stage 1 and 2. Participants of the AGM performed worst than those of the AUM at Stage 1, although differences disappear at Stage 2. Judgment ratings data revealed that participants of the semantically grouped menu acquired a better conceptual model of the system structure than those of both arbitrary pages, although the results were at best marginally significant (Figure 3).

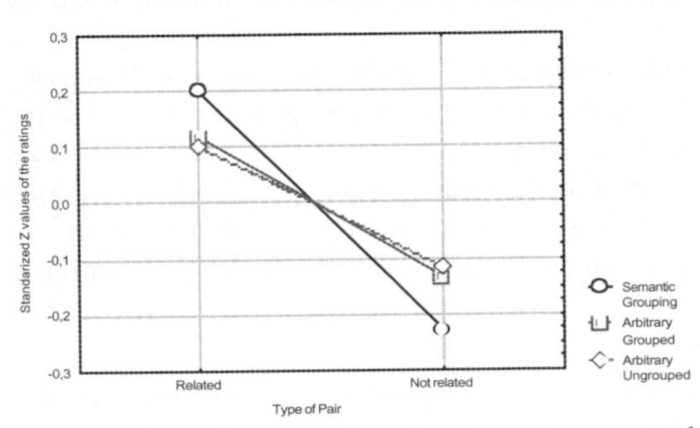

Figure 3: Standardized Z values as a function of Main page and Type of pair.

3 Conclusions

Considering results from all three conditions, we proposed a preliminary model of how users acquire conceptual models. It is of special relevance here the distinction of Cañas et. al. (2001) of considering that the mental model of a system can be constructed with information from the interface or / and from the LTM. This distinction corresponds to that made by Waern (1990). She considers that mental models can be acquired following a bottom-up or / and a top-down process.

When users try to find information on a new hypertext, they seem to engage on a bottom-up process of finding the relationship between menu items. In a first stage, they can consider that items that are grouped on the menu are related on the system structure. Then, users can interact with the hypertext following a representation of the system constructed from that interface information. This process can be inferred from performance data of the stage 1. Users of the SGM could form a good conceptual model from the semantic menu and then performed better than those of the AGM. On the contrary, users of this hypertext seemed to follow the wrong conceptual model revealed from their arbitrary menu. If users are not provided with information about the structure, they would have to interact following the external representation of the menu item by item in isolation. As the cognitive load proposed, these users would spend many resources in navigation, and it would interfere the learning of the system. This seems to happen with participants of the AUM at stage 1, that performed worst than those that could form a good system conceptual model (SGM), but better than those that followed a wrong model (AGM).

In addition, it is proposed that users only form a permanent representation of the system with the interface data if they can interpret the chunks perceived (a top-down process). Although users of our hypertext system had no experience with scientific workshops, they could interpret the contents by their knowledge about the world. As stated above, in our case the semantic grouping was following the expert distinction between "Bureaucratic" and "Content info", which can be observed in other domains common for our users, like when they engage in a course registration. If users can not interpret the chunks provided by the interface, they can return to the previous bottom-up process and interact with the system following only interface data and ignoring the previous chunking information. This new state is similar to when the interface do not provide information about the structure. This process can be inferred from performance at stage 2 and judgment ratings data. On the one hand, participants of the SGM continue following the system conceptual model provided by the interface, and then performed better than the other conditions. Also, judgment ratings data showed that they had learnt a more accurate conceptual model of the hypertext. On the other hand, participants of the AGM seemed to reject the representation inferred by the arbitrary chunks, and passed to perform equally than those of the AUM.

Finally, following the results of our experiment we propose a preliminary guideline. As the results showed especially high differences between conditions at stage 1 (after 10 search tasks), it will be specially suited for hypertext designed for incidental interaction. Concretely, if a semantic grouping can be done, we proposed grouping menu items related by means of Gestalt laws (e.g. similitude, proximity...). If a semantic grouping can not be done, it is better to eliminate information on the interface that could induce to group items (e.g. similar color or form).

Acknowledgements

This research was partially founded by a grant of the Spanish Education and University State Secretariat to the first author.

References

Boechler, P.M. (2001). How spatial is hyperspace? Interacting with hypertext documents: Cognitive processes and concepts. *Cyberpsychology & Behavior, 4,* 23-46.

Cañas, J.J., Antolí, A, and Quesada, J.F. (2001). The role of working memory on measuring mental models of physical systems. *Psicologica, 22,* 25-42.

Dillon, A., McKnight, C., and Richardson, J. (1993). Space -The final chapter or why physical representations are not semantic representations. In C. McKnight, A. Dillon, and J. Richardson (Eds.), *Hypertext: A psychological perspective*. New York: Ellis Horwood.

Farris, J. S., Jones, K.S., and Elgin, P. D. (2002). User's schemata of hypermedia: what is so 'spatial' about a web site? *Interacting with Computers, 14*, 487-502.

Jonassen, D. H. (1990). Semantic network elicitation: tools for structuring hypertext. In C. Green and R. McAleese (eds.) *Hypertext: State of the Art*. Oxford: Intellect.

Jonassen, D. H. (1993). Effects Of Semantically Structured Hypertext Knowledge Bases on Users' Knowledge Structures. In C. McKnight, A. Dillon, and J. Richardson (Eds.), *Hypertext: A psychological perspective* (pp. 169-191). New York: Ellis Horwood.

Otter, M, H. and Johnson, H. (2000). Lost in hyperspace: metrics and mental models. *Interacting with Computers, 13*, 1-40.

Smith, P.A. (1996). Towards a practical measure of hypertext usability. *Interacting with Computers 4*, 365-381.

Sweller, J. (1988). Cognitive load during problem solving: effects on learning. *Cognitive Science, 12*, 257–285.

Tripp, S.D. and Roby, W. (1990). Orientation and disorientation in a hypertext lexicon. *Journal of computer based instruction, 17*, 120–124.

Waern, Y (1990). On the dynamics of mental models. In *Mental models and human-computer interaction 1*. Ackermann, D and Tauber, M.J. (Eds). Elsevier Science. Amsterdam.

Wickens, C.D. (1992). *Engineering Psychology and Human Performance*. NY: Harper Collins.

Predicting Design Induced Pilot Error: A comparison of SHERPA, Human Error HAZOP, HEIST and HET, a newly developed aviation specific HEI method

Paul M. Salmon, Neville A. Stanton, Mark S. Young, Don Harris, Jason Demagalski, Andrew Marshall, Thomas Waldmann and Sidney Dekker

Brunel University, Department of Design, Egham, Surrey, TW20 0JZ, paul.salmon@brunel.ac.uk

Abstract

At the moment, there appears to be no human error identification (HEI) technique developed specifically for use in the aviation domain. Similarly, there appears to be very little research into the prediction of potential design induced pilot error in the cockpit. As part of a DTI/EUREKA! funded project, the authors have developed a novel HEI methodology specifically aimed at predicting design induced pilot error on civil flightdecks. It is proposed that, in line with recent FAA/JAR recommendations, this method will be used in the certification of civil flight decks. The purpose of this study is to compare the newly developed aviation HEI method against three contemporary HEI methods, SHERPA, Human Error HAZOP and HEIST when used to predict potential pilot error on an aviation landing task using the auto-land system. The study aims to demonstrate that the newly developed method will be more accurate at predicting design induced pilot error than the existing HEI methods.

1 Introduction to HEI

Human Error in high risk, complex systems is a problem of great concern to human factor's professionals. When committed by commercial airline pilots or control room operators in nuclear power plants, hundreds of lives can potentially be put at great risk. In response to a number of high profile, high fatality catastrophes attributed in part to human or operator error, such as Three Mile Island, Bhopal and Chernobyl, the prediction of potential operator or human error in complex systems has been investigated extensively over the past three decades. As a result of this research, an abundance of human error identification (HEI) or human reliability analysis (HRA) techniques were developed throughout the 1980's and 1990's. Techniques such as THERP, Human Error HAZOP, SHERPA, PHECA and CADA were developed specifically to identify potential human error in high risk, complex systems such as nuclear power plants and chemical processing plants.

2 HEI in Aviation

It is apparent that the major cause of all aviation accidents is pilot or human error (McFadden and Towell, 1999). Estimates vary, but recent research suggests that human error has been identified as the source of at least 60% of the incidents that occur in commercial aviation (McFadden and Towell, 1999). Furthermore, studies also suggest that 70% of all aviation accidents are classified as pilot error (McFadden, 1993). Also evident is a growing number of high profile aviation catastrophes involving a large loss of life that are attributed to pilot error, such as the Mont St Odile disaster in Strasbourg, 1992, which claimed the lives of 87 people. This catastrophe involved an A320-111impacting into the side of a mountain. The crash was attributed to pilot error caused by a faulty design which led the flight crew to inadvertently select a 3,300 feet per minute descent rate instead of the required 3.3° flight path angle on the aircrafts approach to Strasbourg airport. It is surprising then, considering the continued incidence of design induced pilot error in civil flight decks, that there appears to be very little in the way of published research concerning human error identification or prediction in aviation. Even more surprising, perhaps, is the apparent lack of proven HEI methods developed specifically for the aviation domain. Indeed, it appears that, valid and reliable methods for predicting errors on modern day flight decks simply do not exist. As part of a DTI/EUREKA! funded project entitled , "Prediction of Human Errors on Civil Flightdecks", the authors have developed a novel HEI methodology to be used specifically for predicting design induced pilot error on civil flight decks. From a recent (1996) FAA Human Factors Team Report on the interfaces between flightcrews and modern flightdeck systems (FAA, 1996) it was recommended that the FAA should require the evaluation of flight deck designs for their susceptibility to design induced pilot error as part of the certification process (Demagalski, 2002). It is therefore proposed that, in line with recent FAA/JAR recommendations, this method will be used in the certification of civil flight decks. The purpose of the current study is to compare the newly developed HEI method against the three contemporary methods, SHERPA, Human Error HAZOP and HEIST. The study aims to demonstrate that the newly developed method will be more accurate at predicting pilot error than existing HEI methods.

3 Methodology

A total of 37 subjects were trained in one of the HEI methods (8 trained in HET, 9 trained in SHERPA and HAZOP, 11 trained in HEIST). All of the subjects used were undergraduate design students with no prior experience of HEI techniques. Once trained sufficiently, subjects were then required apply their methodology to a hierarchical task analysis (Annett, Duncan and Stammers, 1971) of the landing task, 'land at New Orleans using the autoland system', in order to predict any potential design induced pilot error. The predicted errors were then compared to actual error data reported by pilots using the autoland system, which was obtained via questionnaire by Cranfield University (Demaglaski, 2002). Validity statistics for each method were then computed. A brief description of each of the four methods used is given below:

4.1 Human Error Template (HET)

HET is a novel HEI methodology, developed by the authors, aimed specifically at predicting design induced pilot error on civil flight decks. The method comes in the form of an error template and is applied to each bottom level step in a HTA of the task under analysis. The HET technique works by indicating which of the HET error modes are credible for each task step, based upon the judgement of the analyst. The analyst simply applies each of the HET error modes to the task step in question and determines whether any of the modes produce any credible errors or not. The twelve HET error modes are shown below:

- Fail to execute
- Task execution incomplete
- Task executed in the wrong direction
- Wrong task executed
- Task repeated
- Task executed on the wrong interface element
- Task executed too early
- Task executed too late
- Task executed too much
- Task executed too little
- Misread Information
- Other

For each credible error (i.e. those judged by the analyst to be possible) the analyst should give a description of the form that the error would take, such as, 'pilot dials in the airspeed value using the wrong knob'. Next, the analyst has to determine the outcome or consequence associated with the error. Finally, the analyst then has to determine the likelihood of the error (Low, medium or high) and the criticality of the error (Low, medium or high). If the error is given a high rating for both likelihood and criticality, the aspect of the interface involved in the task step is then rated as a 'fail', meaning that it is not suitable for certification.

4.2 Systematic Human Error Reduction and Prediction Approach (SHERPA)

SHERPA (Embrey, 1986) uses hierarchical Task Analysis (HTA) (Annett, Duncan, and Stammers 1971) together with an error taxonomy to identify credible errors associated with a sequence of human activity. The SHERPA technique works by indicating which error modes are credible for each task step in turn, based upon an analysis of work activity. This indication is based upon the judgement of the analyst. SHERPA is conducted on each bottom level task step taken from the HTA. Using subjective judgement, the analyst uses the SHERPA human error taxonomy to classify each task step into one of the five following behaviour types, Action, Retrieval, Check, Selection and Information communication. The analyst then uses the taxonomy and domain expertise to determine any credible error modes for the task in question. For each credible error (i.e. those judged by the analyst to be possible) the analyst should give a description of the form that the error would take, such as, 'pilot dials in wrong airspeed'. Next, the analyst has to determine any consequences associated with the error and any error recovery steps that would need to be taken in event of the error. Finally, ordinal probability (Low, medium or high), criticality (Low, medium or high) and any potential design remedies (i.e. how the interface design could be modified to eradicate the error) are recorded.

4.3 Human Error Hazard and Operability Study (HAZOP)

HAZOP (Kletz 1974) is a well-established engineering approach that was developed in the late 1960's by ICI (Swann and Preston 1995) for use in process design audit and engineering risk assessment (Kirwan 1992a). Originally applied to engineering diagrams (Kirwan and Ainsworth 1992) the HAZOP technique involves the analyst applying guidewords, such as Not done, More than or Later than, to each step in a process in order to identify potential problems that may occur. A more hu-man factors orientated version emerged in the form of the Human Error HAZOP, aimed at dealing with human error issues (Kirwan and Ainsworth 1992). In the development of another HEI tool Whalley (1988) also created a new set of guidewords, which are more applicable to human error. These Human Error guidewords are Not done, Repeated, Less than, More than, Sooner than, Later than, As well as, Mis-ordered, Other than and Part of. The guidewords are applied to each step in the HTA to determine any credible errors (i.e. those judged by the subject matter expert to be possible). Once the analyst has recorded a description of the error, the consequences, cause and recovery path of the error are also recorded. Finally, the analyst then records any design improvements to remedy the error.

4.4 Human Error Identification in Systems Tool (HEIST)

HEIST (Kirwan, 1994) is a technique that has similarities to a number of traditional HEI techniques such as SRK, SHERPA and HRMS (Kirwan, 1994). The technique forms part of the HERA methodology (Kirwan, 1998b). HIEST can be used by the analyst to identify external error modes via using the HEIST tables which contain various error prompt questions which are designed to prompt the analyst for potential errors. An example of a HEIST error identifier prompt would be, "Could the operator fail to carry out the act in time?" There are eight tables in total, under the headings of Activation/Detection, Observation/Data collection, Identification of system state, Interpretation, Evaluation, Goal selection/Task definition, Procedure selection and Procedure execution. The analyst applies each table to each task step from the HTA and determines whether any errors are credible or not. For each credible error, the analyst then records the system cause or psychological error mechanism and error reduction guidelines (both of which are provided in the HEIST tables) and also the error consequence.

5 Results

To compute validity statistics, the error predictions made by each subject were compared with error incidence data reported by pilots using the autoland system for the same flight task, supplied by Cranfield University. . The signal detection paradigm was used as it has been found to provide a useful framework for testing the power of HEI techniques and has been used effectively for this purpose in the past (Stanton and Stevenage 2000). The signal detection paradigm uses the following four categories:

1) Hit – Predicted errors that actually have occurred
2) Miss – Failure to predict errors that have occurred
3) False Alarm – Predicted errors that have not occurred
4) Correct rejections – Correctly rejected errors that have not occurred

The signal detection paradigm can be used to calculate the sensitivity index (SI). This provides a value between 0 and 1, the closer that SI is to 1, the more accurate the techniques predictions are. The SI scores for each of the subjects is shown in figure 1 below.

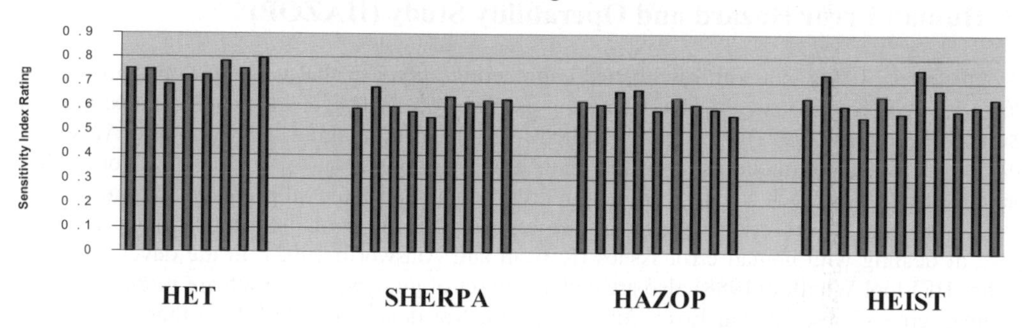

Analysis of the data revealed that subjects using the HET methodology achieved the highest SI scores, with an average SI score of 0.74. Subjects using the HEIST methodology achieved an average SI score of 0.63, whilst both SHERPA and Human Error HAZOP subjects achieved an average SI score of 0.61. The results imply that of the four methods, the HET methodology was the most accurate in terms of error prediction. The results also indicate, however, that all four of the methods were moderately successful in their error predictions.

6 Conclusions

In conclusion, the results demonstrate that of the four techniques, the HET methodology performed the best when used to predict potential pilot error on a landing task using the autoland system. It can therefore also be tentatively concluded that of the four methods, the HET methodology is the most suited for use in aviation. The main reason for this result is that the Error mode taxonomy used by the HET methodology is clearly the most suited to the types of task that exist in civil aviation. The HET error mode taxonomy was developed specifically for the civil aviation domain, whereas SHERPA, Human Error HAZOP and HEIST's error taxonomies were developed for the nuclear power domain. As a result of this, some of the reported error incidences cannot be predicted by the techniques as their taxonomies do not allow it. For example, the guidewords used by the Human Error HAZOP methodology do not allow the analyst to predict an error such as, "Pilot enters airspeed using the heading knob instead of the Speed/MACH kniob", i.e. pilot presses the wrong button. The HET error mode taxonomy however prompts the analyst to predict this error, with the error mode, 'Right action on wrong interface element'. The results obtained also suggest that analysts can be trained successfully in the HET methodology, and also that analysts with no prior experience of the aviation domain can achieve moderately successful accuracy in predicting potential pilot errors for a given flight task. It is conceivable that the SI scores reported will increase significantly when HEI/HRA aviation experts are used as subjects.

References

Annett, J.; Duncan, K. D.; Stammers, R. B & Gray, M. J. (1971) Task Analysis. Training Information No.6. HMSO: London.

BASE (1997) Boeing Airplane Safety Engineering, Statistical summary of Commercial Jet Airplane Accidents – Worldwide Operations, 1959-1996. In K. L McFadden & E. R Towell (eds.) *Aviation human factors: a framework for the new millennium.* Journal of Air Transport Management, 5, 177-184.

Demagalski, J; Harris, D; Salmon, P; Stanton, N; Marshall, A;Waldmann, T; Dekker, S; (2002) Design Induced Errors on the Modern flightdeck during Approach and Landing. In S. Chatty, J. Hansman, G. Boy (eds) *Proceedings of the International Conference on Human-Computer Interaction in Aroenautics,* 173 – 180.

Embrey, D. E. (1986) SHERPA: A systematic human error reduction and prediction approach. Paper presented at the *International Meeting on Advances in Nuclear Power Systems*, Knoxville, Tennessee.

Kirwan, B. & Ainsworth, L. K. (1992) A Guide to Task Analysis, Taylor and Francis, London.

Kirwan, B. (1992a) Human Error Identification in Human Reliability Assessment. Part 1: Overview of approaches. Applied Ergonomics, 23 pp. 299-318.

Kirwan, B. (1994) A Guide to Practical Human Reliability Assessment, Taylor and Francis, London.

Kirwan, B. (1998b) Human Error Identification Techniques for Risk Assessment of High Risk Systems– Part 2: Towards a Framework Approach. *Applied Ergonomics*. 29 (5), 299-318.

Kletz, T. (1974) HAZOP and HAZAN: Notes on the identification and assessment of hazards. In C. D. Swann & M. L Preston (eds.) *Twenty five years of HAZOPs. Journal of loss prevention in the Process Industries*, vol 8 (6)

McFadden, K. L. (1993) An empirical investigation of the relationship between alcohol and drug related motor vehicle convictions and pilot flying performance. In K. L McFadden & E. R Towell (eds.) *Aviation human factors: a framework for the new millennium.* Journal of Air Transport Management, 5, 177-184.

McFadden, K. L. & Towell, E. R. (1999) Aviation human factors: a framework for the new millennium. Journal of Air Transport Management, 5, 177-184.

Stanton, N.A & Stevenage, S. V. (2000) Learning to predict human error: issues of acceptability, reliability and validity. In J. Annett & N. A. Stanton (eds.) *Task Analysis*. Taylor and Francis, London.

Swann, C. D. & Preston, M. L. (1995) Twenty five years of HAZOPs. Journal of loss prevention in the Process Industries, vol 8 (6), pp. 349-353.

Whalley (1988) Minimising the cause of human error. In B. Kirwan & L. K. Ainsworth (eds.) *A Guide to Task Analysis*. Taylor and Francis, London.

The Presentation of Conflict Resolution Advisories to Air Traffic Controllers - A Human Factors Perspective

Dirk Schaefer, Mary Flynn

EUROCONTROL Experimental Centre
BP15, F-91222 Brétigny cedex, France
firstname.lastname@eurocontrol.int

Gyrd Skraaning

OECD Halden Reactor Project
P.O.Box 173, 1751 Halden, Norway
gyrds@hrp.no

Abstract

This paper presents the results of the evaluation of different modes of presenting conflict resolution advisories to the air traffic control (ATC) officer. The Conflict Resolution Assistant system (CORA) presently being developed at EUROCONTROL aims at supporting the ATC controller through the provision of information about conflicts between pairs of aircraft as well as between aircraft and airspace and a list of resolution advisories for each conflict. Different options of presenting these advisories can be imagined and it is not obvious which one is preferable in terms from the human operator's point of view. Specifically the depth of automation, the resolution timeline and the sorting principle can be implemented in different fashions. Experiments in the EUROCONTROL Human Factors Laboratory were performed to provide some evidence on the preferred solution. The different design philosophies were implemented on a prototyping facility and presented to air traffic controllers during experiments using short traffic scenarios ('vignettes'). Subsequently subjective ratings of human-automation cooperation, mental workload, and situation awareness were elicited. Team debriefings at the end of each day were concerned with the usability of the design solutions and the operators' preferences.

1 Introduction

Air traffic control (ATC) services are responsible for the safe separation between aircraft in most parts of the world in which commercial aviation is performed. In order to meet the society's demand for safe and economic flights, ATC faces the challenge of significantly increasing air traffic capacity whilst maintaining, or even better, increasing today's levels of safety. This places a tremendous demand on the air traffic controller team and a variety of user support systems have been introduced to alleviate this situation.

The air traffic controllers primary tasks includes the observation of the traffic situation, the detection and resolution of conflicts[1] and the communication with the flight crew within the sector of their responsibility. Monitoring functions and the detection of conflicts may be supported through semi-automated systems. The short-term conflict alert (STCA) is a safety net that detects, if there is an imminent risk of collision and alerts controllers accordingly. STCA calculates the future separation between any two aircraft based on actual position and speed vector. If a separation infringement is likely to occur based on this analysis, a visual warning is presented to

[1] The term conflict refers to a predicted violation of the minimum separation between two aircraft, typically 5-6 nautical miles horizontally and 1000 feet vertically.

the controller. STCA is operational in the majority of ATC centers in Europe. It is widely considered a useful tool despite a certain number of false alerts. One limitation of STCA is that fact that it is entirely based on the aircraft state vector and does not have access to data about the flight plan, ATC clearances and the planned trajectory. Predictions of future positions and hence the detection of potential conflicts between aircraft can thus only be obtained, with reasonable accuracy, for a very short span in the future and STCA is commonly limited to a timeline 30-90 seconds.

The medium-term conflict detection system (MTCD) tries to avoid these limitations by considering the aircraft's trajectory, i.e. its planned flight trajectory in space and time. The trajectory is commonly calculated by a ground-based trajectory prediction system but may also be the result of the aircraft down-linking its planned route. MTCD is still under development at EUROCONTROL and operational evaluations in shadow mode trials are presently ongoing in three ATC centers in Europe[2]. MTCD predicts conflicts, i.e. separation violations between any pair of aircraft within a timeframe of typically 20 minutes into the future. Unlike STCA which is primarily used by the executive controller in order to avoid imminent conflicts, the timeframe of MTCD permits more strategic use through the planning controller who can plan new flight paths for the aircraft involved and, if need be, coordinate them with the controllers of adjacent sector. As a consequence, MTCD will help to assign aircraft safer and more efficient trajectories.

2 The Conflict Resolution Assistant System

After a conflict has been detected, the air traffic controller identifies a suitable maneuver to avoid the conflict. The next step would be to support the controller in the identification of a suitable conflict resolution. The Conflict Resolution Assistant system (CORA) aims at supporting the air traffic controller, more specifically the planning controller, through the provision of conflict resolution advice. Upon detection of a conflict CORA would typically generate one or more conflict resolution advisories, i.e. maneuvers that one or both aircraft would have to execute in order to resolve the conflict, and display this information to the controller. Different ways of displaying these advisories can be imagined, and it requires some clarification, which of these is preferable in terms of operator performance and satisfaction.

On the one hand the calculation of conflict resolution maneuvers may be supported very efficiently by computer-based systems since it involves arithmetic that can be performed much more accurate by computers. This may result in reduced fuel burn and flight time and increased safety; it might also greatly reduce the human operator's workload.

On the other hand the human operator has a wealth of information and expertise inaccessible to any automated system and it is very desirable that the human operator has the final authority over the maneuver to be implemented. Apart from safety reasons and 'out-of-the-loop' problems (poor monitoring performance, skill degradation, and complacency) another aspect needs careful consideration: assigning flight paths and resolving potential conflicts between aircraft is the core of controller work and much job satisfaction results from it. Assigning this task, at least partially, to automated systems requires careful evaluation. A situation in which the controller's responsibility would be entirely reduced to pushing a button once resolution has been identified is clearly unacceptable. If the calculated resolutions are optimal and reliable under all circumstances

---

[2] Malmoe, Rome, and Maastricht.

then there might be little argument for involving a human operator at all. Since that this is not the case nor will be in the near future, an active involvement of the air traffic controller is crucial.

Different degrees of involvement of the human operator can be imagined for CORA, ranging from a more user-driven approach where the controller deliberately activates the calculation of resolution advisories should he deem it helpful and then selects one or refuses all in favor of his own resolution; up to a more automated version in which CORA calculates and implements the optimum resolution unless the controller intervenes. The user-driven approach may be preferable in terms of transparency and user-involvement, the automated approach may be preferable in terms of reduction of operator workload[3]. The optimum may lie between the two extremes.

3 Experimental Evaluation of Different Design Solutions for CORA

Three different design principles were defined for CORA:

- *User-driven:* Conflicts are displayed in the CORA conflict display and the controller can trigger the CORA system to calculate conflict resolution advisories. The advisories are then displayed in a list, typically one advisory per category (climb, descend, turn left, turn right, speed) together with a quality/cost index for each advisory.
- *Automatic:* Upon detection of a conflict, CORA calculates a resolution advisory; the 'best' (in terms of cost index) is then automatically displayed to the controller. If the controller does not reject the resolution within a certain time span, the resolution is implemented automatically. The controller can, on request, obtain conflict information in the CORA conflict display.
- *Collaborative:* Upon detection of a conflict, CORA calculates various resolution advisories and flags the availability of resolution advisories. If invoked by the controller, the conflict resolution display shows a list of advisories, typically one per category (climb, descend, turn left, turn right, speed) together with the quality/cost index of each advisory.

Apart from the design principle, two other aspects required evaluation: the appropriate timeline (how many minutes prior to the occurrence of the conflict the resolutions should be calculated and displayed) and the sorting principle (display of resolutions in a fixed order according to type of maneuver or ranked according to quality of resolution expressed by a calculated quality/cost index).

3.1 Method

The three design principles plus three timelines (5, 10, 15 minutes) and two sorting principles (fixed order, ranked) were implemented on a prototyping facility in EUROCONTROL's Human Factors Laboratory. Consequently, a within-subject design with three independent variables (3*3*2) was applied and 15[4] traffic 'vignettes' of about 10-12 minutes duration were implemented. The controllers worked as planning controllers, thus not advising clearances to the aircraft but observing and planning the traffic. Upon detection of a conflict CORA was activated

[3] A further downside of the automated option is the higher degree of dependence on the availability and accuracy of the resolution calculation.

[4] For analysis purposes a 3*3 design plus a 2*3*2 design was used since the sorting option is not sensitive to the automatic design philosophy (only one resolution displayed). Since the scenarios are partly overlapping a total of 15 simulations per user were performed.

and the controller would select one of the resolution advisories[5]. Upon completion of the vignette a short e-questionnaire was completed, concerned with human-automation cooperation aspects, mental workload and situation awareness. After all exercises were completed, an end-of-day questionnaire was filled in which was predominately concerned with the operators evaluation of the concepts and their individual preferences.

A total of ten air traffic controllers participated in the experiment, nine of which were active controllers from different control centers in Europe. Figure 1 shows the CWP with CORA as used during the experiment – in this case a conflict between AZA574 and RAM323 has been detected. The resolutions are presented according to the user-driven principle and in fixed order.

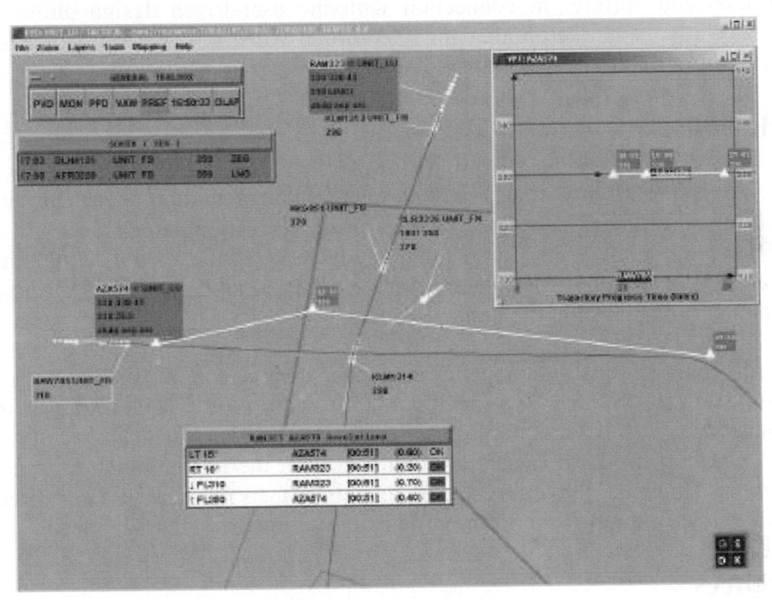

Figure 1: Controller Working Position with CORA.

3.2 Results

Design principle: 7 out of 10 controllers preferred the *collaborative* design philosophy whilst 3 controllers preferred the *user-driven* design philosophy. The analysis of mental workload showed an interaction effect between design philosophy and timeline: for the automatic design philosophy workload seems to be highest for the longer timelines (10 and 15 min) whilst for the user-driven and collaborative design philosophy workload is higher for the shorter timeline (5 min).

Timeline: Seven out of the ten controllers preferred the 10-minute timeline, while three expressed that the 15-minute timeline supported their work most effectively. The 5-minute timeline was viewed as too short, one reason being that the resolution alternatives would time out before the

[5] The concept of CORA foresees that the resolution would be communicated to and implemented by the executive controller. Teamwork and communication aspects, however crucial, were not within the scope of this study.

controller had time to go through it. Mental workload was significantly higher during scenarios with a 5-minute timeline when compared to scenarios with a 10- or 15-minute timeline; no significant difference existed between 10 and 15 minutes. Human-automation cooperation was rated significantly higher during scenarios with 10- or 15-minute timeline as opposed to scenarios with 5-minute timeline; no significant difference was observed between 10 and 15 minutes.

Sorting principle: All but one controller expressed a preference for the ranked presentation of the resolution advisories (in order of quality/cost index). One controller preferred the fixed order and questioned the usefulness of the quality/cost index since he feared it would exert a certain pressure on the controller to implement the resolution with the lowest cost index even if it violated his own judgement. The analysis of human-automation cooperation revealed an interaction effect between design-philosophy and sorting: in connection with the user-driven design philosophy human-automation cooperation was rated higher for fixed order whilst it was rated higher for ranking in connection with the collaborative design philosophy.

4 Conclusions

The high degree of flexibility that a prototyping platform and a dedicated human factors experimentation facility offers is of great value for the evaluation of concepts and early design prototypes. The presentation of conflict resolution advisories to the air traffic planning controller was investigated with regard to the design principle, the timeline and the sorting principle in which they should be displayed to the human operator.

The analysis of human-automation cooperation, mental workload, and situation awareness did not reveal systematic effects for all independent variables which may be due to the small number of test participants which was beyond the experimenters control. The applicability of some of these concepts on a situation in which dynamic interaction with the traffic situation is restricted may also be questioned. However, the observed experimental effects point in the same direction and are in agreement with the findings from the end-of-day questionnaire and the debriefings.

5 References

Hilburn, B. & Flynn, M. (2000) CORA2: Air Traffic Controller Attitudes toward Future Automation Concepts. EUROCONTROL ASA.01.CORA.2.DEL02-A.RS and B.RS

Kirwan. B. & Flynn, M. (2002). CORA2: Investigating Air Traffic Controller Conflict Resolution Strategies. EUROCONTROL, ASA.01.CORA.2.DEL04-B.RS

Prototype implementation of the CORA HF Experiment downloadable from http://www.eurocontrol.fr/projects/edep/

Schaefer, D. & Smith, D. (in Press): Human-in-the-Loop Studies in the EUROCONTROL Human Factors Laboratory. *Transactions of the Society for Modeling and Simulation International; Special Issue "Simulation for Air Traffic"*

Skraaning G. & Heimdal J. (2003). CORA HF Analysis Results. EUROCONTROL EEC, France

Skraaning, G. & Heimdal. J. (2002). CORA HF Experimental Plan and Method. EUROCONTROL EEC, France.

The Man without a Face, and other Stories about Human-Centred Automation in Nuclear Process Control

Gyrd Skraaning Jr., Ann Britt Miberg Skjerve

OECD Halden Reactor Project
PO Box 173, N-1751 Halden, Norway
gyrds@hrp.no, annbm@hrp.no

Abstract

The automatic agent is like a man without a face in the nuclear power plant control room. His analytical skills go beyond human capabilities, but he is cognitively inflexible and cannot operate without human support. Even though humans and automation cooperate to achieve common goals, it is generally accepted that human operators should indirectly infer the activities of the automatic system from process events and/or changes in the system state. This conventional design approach makes the automatic agent a silent partner and a poor team player. The OECD Halden Reactor Project conducted two simulator experiments in order to investigate the impact of explicit feedback from the automatic system on the quality of human-machine interaction. Both studies provided the control room operators with immediate visual and verbal information about the behavior of the automatic system. The experiments demonstrated that explicit automation feedback improves operator performance and human automation communication during nuclear power plant disturbances.

1 Introduction

During the last fifty years, the use of automation has markedly increased within high-risk industrial production systems, and automatic systems have gained in autonomy and authority. The driving force behind this development is probably that the allocation of functions to machines improves safety and productivity, because computers perform procedural tasks more reliable, precise and effective than humans. Automation can also reduce operator workload and fatigue, and may therefore provide a better utilization of human resources. These advantages of automation have given us the power to control industrial production systems of increasing complexity. Automatic systems are reliable, precise and effective because they are algorithmic, i.e., they follow infallible step-by-step recipes for obtaining a pre-specified result (Haugeland, 1986). This formalism is also the fundamental weakness of automation, since algorithms are incapable of inductive reasoning, innovative thinking, and often lack context sensitivity. Humans possess these flexible cognitive qualities that may be critical when the operating task is unpredictable and a priori uncontrollable (Jordan, 1963). Many nuclear power plant disturbances, and most accidents, can be characterized as *dynamic* and *ill-defined* problems, where the content of the problem changes as a function of the human-machine interaction itself (Frensch & Funke, 1995), and the problem has a vague initial state, diffuse operational possibilities, and/or an unclear final state (Reitman, 1964). Such situations cannot be handled by automatic algorithms alone. Consequently, the interaction between humans and automation is not a transitory stage, but a permanent situation.

Technological improvements that lead to safer and more effective execution of some functions in a human-machine system, can negatively affect other functions within the same system. Herein lie the unwanted side effects of automation, since the introduction of automatic agents has a tendency to deteriorate human performance. When automatic systems take over crucial functions:

- It becomes harder for the operators to understand incoming information.
- It is more difficult to predict the future development of the system.
- The role of the operator changes from an active contributor to a passive monitor.
- Manual control skills degrade and make it difficult to handle automation failures.
- It is problematic for the operators to know when to trust or mistrust automation.
- New tasks and cognitive demands are introduced.
- The operational complexity is increased.

To overcome these problems, one may regulate the level of automation and allocate functions strategically between humans and automatic systems. This conventional solution defines optimal task distributions for a given system configuration, but can probably not identify the underlying factors that shape an effective human-automation partnership. Hence, there is a need for an alternative approach that can improve the way human operators and automation work together.

It will be presumed here, that the cooperation between humans and automation in existing control rooms is suboptimal, because operators are not adequately informed about the activity of the automatic system due to inadequate human-machine interface design (Endsley, 1996; Norman, 1990; Sarter and Woods, 1995, 1997; Woods, 1996). Adhering to the broad human-centred design perspective, Billings (1991, p. 7) suggests that the concept of Human-Centred Automation (HCA) should be defined as, "... automation designed to work cooperatively with the human operators in the pursuit of stated objectives." Hence, the automatic system should be designed as a tool to support humans in achieving the operational goal. The HCA perspective makes it necessary to specify how human-automation communication should be promoted. In this respect, the OECD Halden Reactor Project performed two closely related experiments in a full-scale nuclear power plant simulator, using licensed operators as participants (Skjerve et al., 2001a, 2001b, 2002). The two experiments offered an opportunity to test whether the quality of the human-machine interaction was enhanced by the introduction of an experimental user interface designed to improve the team player capacity of the automatic agent. It was hypothesized that explicit feedback from the automatic system would improve operator performance and facilitate the cooperation between humans and automation.

2 Method

With respect to the experimental design and the selection of performance indicators, the HCA-2001 experiment was a replication of the HCA-2000 experiment. Both studies were performed in the Halden Man-Machine Laboratory (HAMMLAB), which was established by the OECD Halden Reactor Project in 1983. The laboratory has three full-scale nuclear power plant simulators that are used during highly realistic experimental studies of human-machine interaction in the nuclear control room. The participants were licensed operators from the simulated nuclear power plant (Loviisa in Finland). All participants were male, and their age ranged from 28 to 60 years. Each participant was given a dedicated role within an operating crew. There were eight crews in the HCA-2000 study and six crews in the HCA-2001 study. Before the experiments, the operators received relevant training on special characteristics of the simulator process model, the user interface, and the data collection procedure.

A 2x2 within-subject design was employed in the experiments. The first treatment was the type of human-machine interface with two levels of manipulation, i.e., an *experimental* interface that gave explicit feedback about the activities of the automatic system, and a *conventional* interface where the operators had to infer the activities of the automatic system from process events and/or changes in the system state.

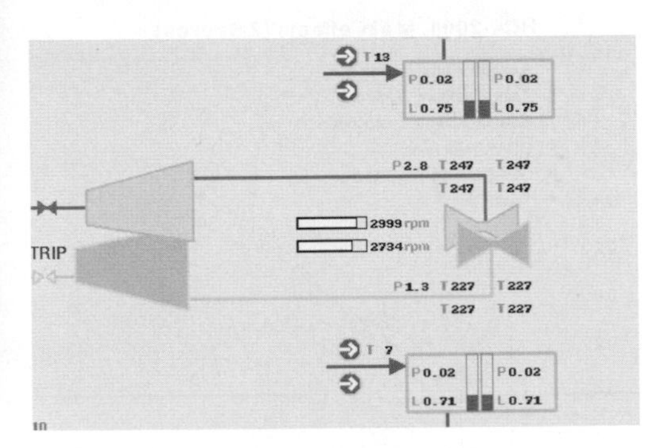

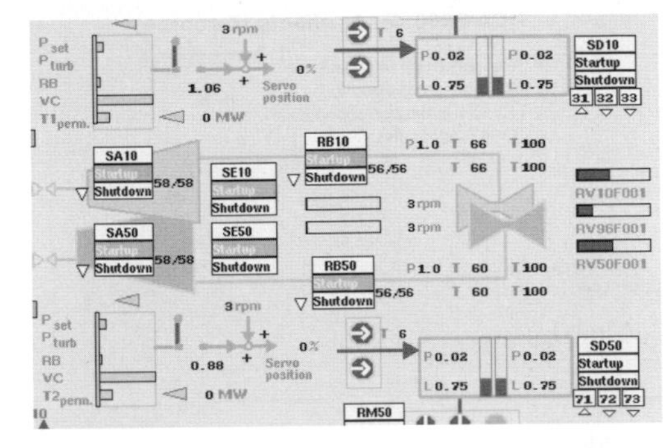

Figure 1. Extract from the conventional interface (left), and the experimental interface (right).

The experimental interface had several unique properties (see Figure 1): (a) key automatic devices were represented in the process overview display, (b) pre-recorded verbal feedback associated with the activity of the automatic devices was announced through loudspeakers (e.g. "controller out of range, compensating for low water levels"), (c) there were dedicated controller displays, and (d) computer-based logic diagrams allowed for remote tracking of automation sequences. The conventional interface had none of these features, but was identical to the experimental interface in all other respects. Thus, the conventional interface resembled the human-automation design solutions that can be found in existing control rooms. The second treatment compared two types of ill-defined nuclear power plant disturbances that challenged the cooperation between human operators and automation. Different scenarios were used in the two experiments, and there were two scenario variants per scenario type to avoid learning effects. The outcome of this second experimental manipulation will not be reported here. To eliminate order effects, the presentation order of experimental conditions and scenarios were counterbalanced.

The Operator Performance Assessment System (OPAS) was used to measure the extent to which the operators were able to detect information in the user interface, and carry out important intervening actions (Skraaning, 2003). In addition, operator response times were calculated for critical actions (ibid.). The quality of the human-automation communication was estimated by three self-rating instruments: (a) the Halden Cooperation Scale (Skjerve, 2002), (b) the Halden Trust Scale (Strand, 2001), and (c) the Halden Task Complexity Index (Braarud, 1998).

3 Results

In the HCA-2000 experiment, there were main effects of interface type in the hypothesized direction for: operator response time ($F(1,7)= 13.42$, $p=0.01$, $\omega 2=0.28$), OPAS detections ($F(1,7)= 5.61$, $p=0.05$, $\omega 2=0.13$), task complexity ($F(1,7)= 10.94$, $p=0.01$, $\omega 2=0.24$), and cooperation ($F(1,7)= 6.90$, $p= 0.03$, $\omega 2=0.15$). In the HCA-2001 experiment there were main effects of interface type in the hypothesized direction for: operator response time ($F(1,5)= 26.56$, $p= 0.01$, $\omega 2=0.52$), OPAS intervening actions ($F(1,5)= 6.86$, $p= 0.05$, $\omega 2=0.20$), OPAS detections ($F(1,5)=$

15.89, p= 0.01, ω2=0.38), subjective task complexity (F(1,5)= 8.75, p=0.03, ω2=0.24), trust (F(1,5)= 21.84, p=0.01, ω2=0.46), and cooperation (F(1,5)= 21.10, p= 0.01, ω2=0.46). Disordinal interaction effects between the treatments were not observed in any of the experiments. The effect of interface type therefore had a straightforward interpretation (Howell, 1997). Figure 2 illustrates the statistically significant main effects that were uncovered by the two experiments.

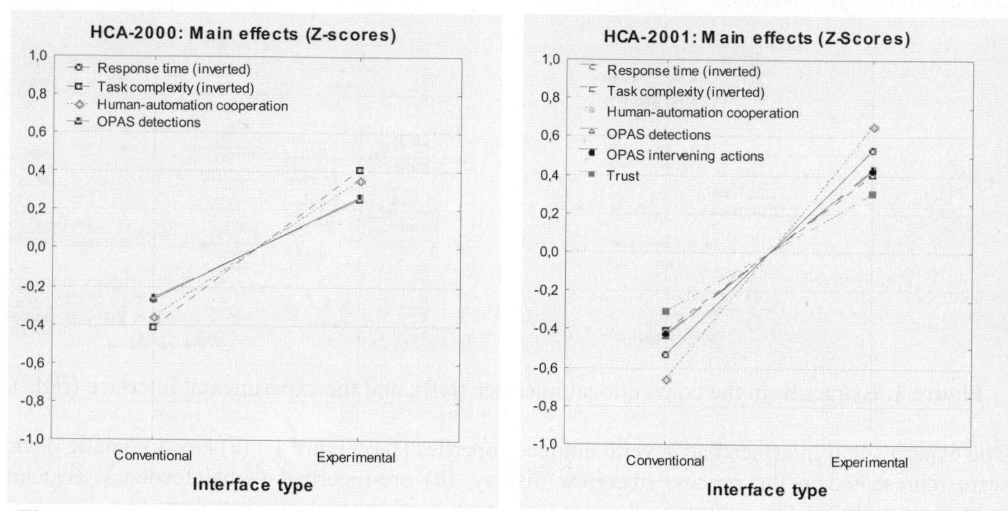

Figure 2. Statistically significant main effects in the HCA-2000 and HCA 2001 experiment.

4 Discussion

As hypothesized, the experimental interface enhanced operator performance and human-automation communication across a wide range of task conditions in the HCA experiments. This is interesting, since the operators only had limited exposure to the experimental interface before the data collection. One may therefore speculate what the long-term consequences of explicit automation feedback would mean for the safety and efficiency of complex industrial production systems. It should also be noted that the experimental interface was far more complex than the conventional interface (see Figure 1), and added to an already heavy information load on the operators. Still, the operators benefited from the extra information.

The magnitude of the experimental effects (measured by partial omega squared, ω^2) was higher, and there were statistically significant effects on more dependent variables in the HCA-2001 experiment, as compared to the HCA-2000 experiment. This was expected, since the experimental interface was developed further and improved before the second study.

The empirical evidence reported here, suggests that it is time for the man without a face to step out of the shadows of the nuclear power plant control room. In fact, the automatic agents of the future should probably be designed to concurrently inform the operators about automation activities. The question may even be raised of whether new technology should be used to humanize the cooperative aspects of automation far beyond simple feedback displays. This investment would be worthwhile, since human operators seem to have a permanent role in the control room. As a first step in this direction, it is proposed that the design and evaluation of cooperative machine agents should be established as a prioritized research area.

5 References

Billings, C. E., 1991. Human-Centered Aircraft Automation: A Concept and Guidelines, NASA Technical Memorandum 103885, NASA Ames Research Center. Moffett Field, CA.Endsley, M. R. (1996). Automation and Situation Awareness. In Parasuraman, R. & Mouloua, M. (Eds.), *Automation and Human Performance. Theory and applications* (pp. 161-165). Mahwah, NJ: Lawrence Erlbaum Associates. Frensch, P. A., & Funke, J. (1995). *Complex problem solving: the European perspective.* Hillsdale, NJ: Lawrence Erlbaum Ass.

Braarud, P. Ø. (1998). *Complexity factors and prediction of crew performance.* OECD Halden Reactor Project, HWR-521, Halden, Norway.

Haugeland, J. (1986). *Artificial intelligence: The very idea.* Cambridge, MA: MIT press.

Howell, D. C. (1997). *Statistical methods for psychology* (4th ed.). CA: Wadsworth.

Jordan, N. (1963). Allocation of functions between man and machines in automated systems. *Journal of Applied Psychology*, 47, 161-165.

Norman, D. A., 1990. The 'problem' with automation: inappropriate feedback and interaction, not 'over-automation,' in: Broadbent, D.E., Baddeley, A., Reason, J.T. (Eds.), *Human Factors in hazardous situations,* Clarendon Press/Oxford University Press, New York, pp. 137-145. Hillsdale, NJ: Lawrence Erlbaum Associates.

Reitman, W. (1964). Heuristic decision procedures, open constraints, and the structure of ill-defined problems. In M. W. Shelley & G. L. Bryan (Eds.), *Human judgements and optimality.* New York: Wiley.

Sarter, N. B., Woods, D. D., 1995, How in the World Did We Ever Get into That Mode? Mode Error and Awareness in Supervisory Control. *Human Factors*, 37, 5-19.

Sarter, N. B., Woods, D. D., 1997. Team Play with a Powerful and Independent Agent: Operational Experiences and Automation Surprises on the Airbus A-320. *Human Factors*, 39, 553-569.

Skjerve, A. B., Andresen, G., Skraaning, G., & Saarni, R. (2001a). *The influence of automation malfunctions on operator performance. Study plan for the Human-Centred Automation 2000 experiment.* OECD Halden Reactor Project, HWR-659, Halden, Norway

Skjerve, A. B., Andresen, G., Skraaning, G., Saarni, R., & Brevig, L. H. (2001b). *Human-Centered Automation 2000 Experiment. Preliminary Results.* OECD Halden Reactor Project, HWR-660, Halden, Norway.

Skjerve, A. B., Strand, S., Saarni, R., & Skraaning, G. (2002). *The influence of automation malfunctions and interface design on operator performance. Study plan and preliminary results of the HCA-2001 experiment.* OECD Halden Reactor Project, HWR-686, Halden, Norway.

Skjerve, A. B. M., 2002. *The Halden Co-operation Scale. Human-Automation Co-operation in Control Room Settings.* HWR-685, OECD Halden Reactor Project, Halden, Norway.

Skraaning, G. (2003). *Experimental control versus realism: Methodological solutions for simulator experiments in complex operating environments.* Working document. OECD Halden Reactor Project, Halden, Norway.

Strand, S. (2001). *Trust and automation: The influence of automation malfunctions and system feedback on operator trust.* OECD Halden Reactor Project, HWR-643, Halden, Norway.

Woods, D. D. (1996). Decomposing automation: Apparent simplicity, real complexity. In Parasuraman, & Mouloua, M. (Eds.), *Automation and Human Performance. Theory and applications,* 37-65. Mahwah, NJ: Lawrence Erlbaum Associates

Situational Awareness Displays in Driving

Neville A. Stanton

Department of Design, Brunel University, Egham, Surrey, TW20 0JZ, UK

Abstract: This paper reports on the evaluation of in-car displays used to support Stop & Go Adaptive Cruise Control. Stop & Go Adaptive Cruise Control is an extension of Adaptive cruise Control, as it is able to control to car to a complete standstill. Previous versions of Adaptive Cruise Control have only operated above 30 kph. The greatest concern for these technologies is the appropriateness of the driver's response in any given scenario. Three different driver interfaces were proposed to support the situational awareness of the driver: an iconic display, a flashing iconic display, and a representation of the radar. The results show that drivers correctly identified more changes detected by the system with the radar display than with the other displays, but this increased detection was accompanied by the higher levels of workload. It was suggested that the radar display might be most useful in training drivers about the limitations of the system.

Situational awareness in driving

The concept of situational awareness offers an explanation of how the driver manages to combine high-level goals (such as driving to a destination) with low-level goals (such as avoiding collisions) in real-time (Sukthankar, 1997). Drivers are required to keep track of a number of critical variables in a dynamic and changeable environment, such as: their route, the position, their speed, the position and speed of other vehicles, road and weather conditions, and the behaviour of their own vehicle. Drivers also need to be able to predict how these variables will change in the near future, in order to anticipate how to adapt their own driving (Gugerty, 1997). Research has also shown that in situations of information overload, drivers cope by focusing attention on the position of hazards. Research has suggested that poor situational awareness is a greater cause of accidents than excessive speed or improper driving technique (Gugerty, 1997). Previous research has established the importance of the driver maintaining a good level of situational awareness (Stanton & Young, 2001; Stanton et al, 2001). Review of research on situational awareness by Endsley (1995) has identified three facets that may be important to drivers: mode awareness, space awareness and time awareness. Stop & Go Adaptive Cruise Control (ACC) represents an extension of Adaptive cruise Control, as it is not only able to maintain set speed and brake automatically to maintain a safe gap between vehicles, it is also able to bring the car to a complete standstill For the driver of a car with Stop & Go ACC, a higher level of situational awareness will mean that they are able to readily identify the current operational mode of the vehicle (e.g., cruising or following), identify the spatial relevance of other vehicles (e.g., in-path target or not), and identify the temporal relevance of other vehicles (e.g., impending contact or not). Integration of all this information should ensure that the driver responds appropriately to the dynamic road-vehicle environment. The driver receives information about changes in the environment directly via visual, auditory and tactile senses. In addition, the driver interface will be an important source of information regarding the status of the Stop & Go ACC system. Ideally, the driver should be able to integrate the information presented by the Stop & Go interfaces together with the information presented directly from the environment in a timely manner. The driver should also be able to determine if any intervention on their part is required. Thus the experimental study set out to assess the objective and subjective levels of drivers' situational awareness. It is also important that the driver interface should not place too much cognitive demand on the driver. In this respect it should be perceived to be largely intuitive and easy to use. Thus the study also explored the issues of driver workload and interface usability.

Experimental method

The following experimental methods were employed in the testing of the three interfaces.

Participants: Six male and six female participants were recruited for this study. All were experienced drivers of manual cars.

Design: The experimental design was a completed repeated, within-subjects, factorial design. This means that all participants experienced all of the experimental trials. There were two independent variables in the study, one called ‘interface type’ and the other called 'task type'. The five dependent variables are measures of situational awareness, driver workload, host vehicle driver reaction time, an interface usability rating, and driver intervention. In addition, a static test was performed at the end of the trials.

Equipment: A car equipped with Stop-and-Go ACC was used as the experimental vehicle. The three interfaces tested in the study were based on the original ACC system, a development from the original system, and an attempt to map the radar representation onto the Stop & Go system. The original system simply presents an amber follow icon when the vehicle enters follow mode and the icon is extinguished when the vehicle leaves follow mode. This is the simplest interface. A development of this interface was to indicate the presence of a new in-path target (e.g., a new vehicle) by flashing the icon red at first, before assuming steady state of the amber icon. The third interface represented a departure from the follow icon design. This interface encapsulated the driver requirements on temporal, spatial and mode information, by mapping the in-path target data onto a representation of the radar display. This offered a direct mapping between the position of the in-path target in the world and its representation on the driver interface. Accompanying the visual information on the Stop & Go interface is vehicle proprioceptive information on vehicle braking or accelerating, the LCD message display, auditory tone, standard vehicle instruments, engine noise and visual information outside the windscreen. Three questionnaires were used in the study: the NASA-TLX (a workload questionnaire), SART (a situational awareness questionnaire), and a specially developed usability questionnaire.

Driving tasks: Five driving tasks were performed in the study: lead car cut-in, lead car braking, cornering, stop and start, following at slow speeds. In addition, a static ‘target identification’ test was undertaken to determine if drivers could correctly detect the objects that the radar had identified as the leading vehicle.

Procedure: On agreeing to participate in the study, biographical data was sought to ensure that the gender groups matched. Then each participant spent one hour driving the experimental car to become acclimatised with Adaptive Cruise Control, one week prior to the study. At this point they were assigned to their experimental condition, to ensure matching of gender. On the day of the study, the participant drove to the test track to take part. The order of events is listed below.

(i) Participants were briefed on nature of study
(ii) Asked to sign a participant consent form
(iii) Participants then read through the ACC manual.
(iv) Participants then re-acquainted themselves with the car.
(v) Then they secured the seatbelt and attached the microphone.
(vi) The experimenter pointed out the three interfaces in the car.
(vii) Participants then had a practice circuit.
(viii) Then participants started the experimental study.
(ix) At the end of each trial, the participant answer the questionnaires.
(x) After the final condition was completed, the static test was undertaken.
(xi) Finally, the participant was debriefed on the nature of the study.

Results

The results section is divided into seven subsections to reflect the different analyses undertaken on the dependent variables. The dependent variables were driver verbal protocol response times, static test, driver workload, interface usability, and subjective situational awareness. Analysis of variance revealed no statistical differences between different interface designs for the verbal response times of drivers for any of the tasks. The static test revealed statistically significant difference between the hit rates of drivers for the three interface designs ($\chi^2_2 = 11.619$, $p<0.005$). Post-hoc, pair-wise, comparisons of the different interfaces reveal statistically significant differences between the standard icon condition and the radar display condition ($Z=-2.494$, $p<0.05$), and between the flashing icon condition and the radar display condition ($Z=-2.666$, $p<0.01$). There were no statistically significant differences between the standard icon and flashing icon displays ($Z=-0.578$, p=ns).

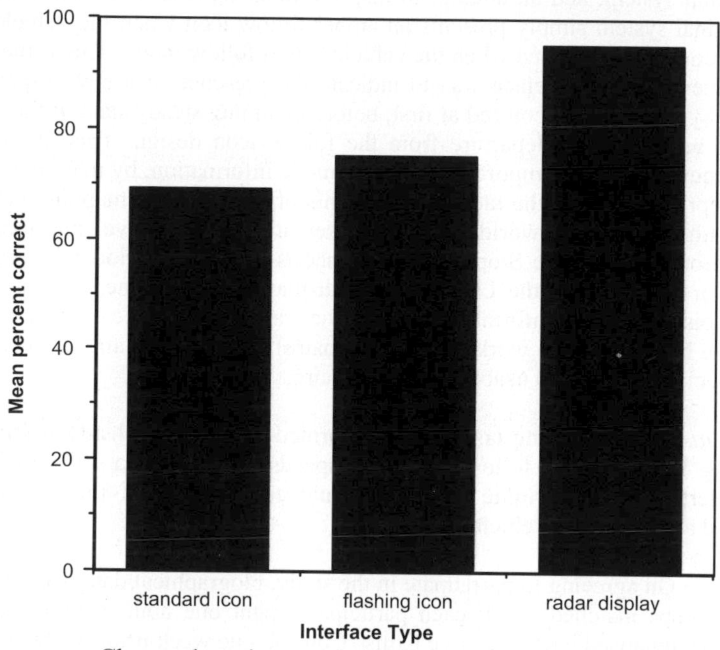

Figure one. Change detection rates with the three interfaces.

The mental workload scale revealed statistically significant difference between the rated load for the three interface designs ($\chi^2_2 = 15.073$, $p<0.001$). Post-hoc, pair-wise, comparisons of the different interfaces reveals statistically significant differences between the standard icon condition and the radar display condition ($Z=-2.964$, $p<0.005$), and between the flashing icon condition and the radar display condition ($Z=-2.673$, $p<0.01$). There were no statistically significant differences between the standard icon and flashing icon displays ($Z=-0.973$, p=ns). The physical workload scale revealed statistically significant difference between the rated load for the three interface designs ($\chi^2_2 = 6.513$, $p<0.05$). Post-hoc, pair-wise, comparisons of the different interfaces reveals statistically significant differences between the standard icon condition and the radar display condition ($Z=-2.323$, $p<0.05$), and between the flashing icon condition and the radar display condition ($Z=-2.324$, $p<0.05$). There were no statistically significant differences between the standard icon and flashing icon displays (Z=-0.512, p=ns). The temporal workload scale revealed statistically significant difference between the rated load for the three interface designs (χ^2_2 = 13.543, $p<0.001$). Post-hoc, pair-wise, comparisons of the different interfaces reveals statistically significant differences between the standard icon condition and the radar display condition (Z=-2.67, $p<0.01$), and between the flashing icon condition and the radar display condition (Z=-2.527, $p<0.05$). There were no statistically significant differences between the standard icon and flashing icon displays (Z=-1.334, p=ns). The performance scale revealed no statistically significant differences between the rated performance for the three interface designs (χ^2_2 = 1.442, p=ns). The effort scale revealed a statistically significant difference between the rated effort for the three interface designs (χ^2_2 = 7.35 $p<0.05$). Post-hoc, pair-wise, comparisons of the different interfaces reveals statistically significant differences between the standard icon condition and the radar display condition (Z=-2.849, $p<0.005$). There were no statistically significant differences between the flashing icon condition and the radar display condition (Z=-1.94, p=ns), nor between the standard icon and flashing icon displays (Z=-1.334, p=ns). The frustration scale revealed statistically significant difference between the self-rated frustration for the three interface designs (χ^2_2 = 2.905, p=ns). The workload results are shown in figure two.

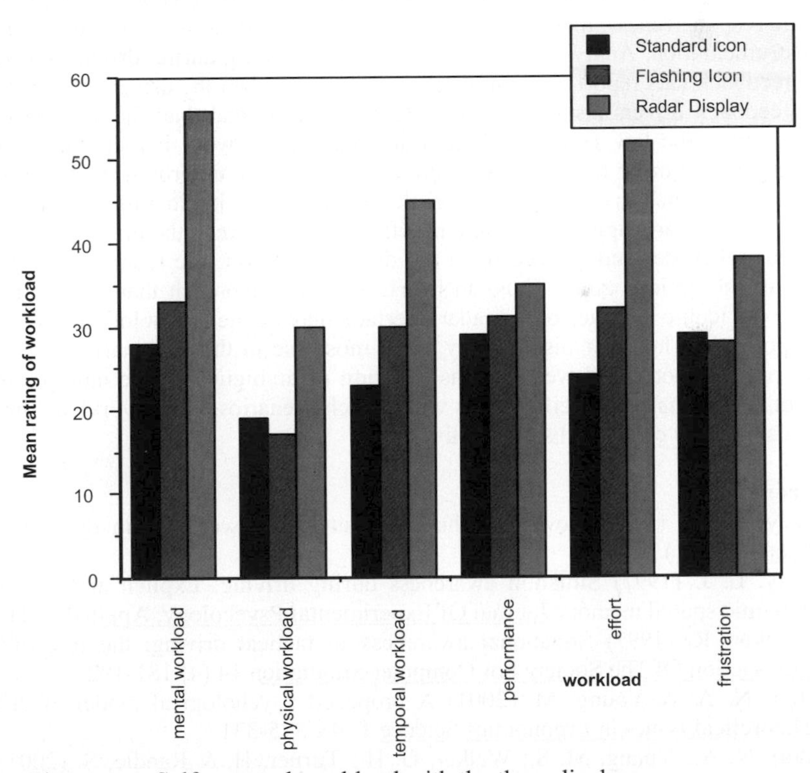

Figure two. Self-reported workload with the three displays

Analysis of the subjective ratings of interface usability showed that there were no statistically significant differences between the three conditions (χ^2_2 = 2.13, p=ns). Analysis of the subjective ratings of situational awareness showed that there were no statistically significant differences between the three conditions (χ^2_2 = 1.66, p=ns).

Conclusions

The radar display put a fair demand on drivers. This demand was due, in part, to the additional information and, in part, to the sighting of the display. In prototype form, the radar display required the driver to look right and down (as it was sited in the centre-middle of a left-hand driver car, covering the navigation display). A location more central to the driver's vision might reduce the demand placed on the driver. The radar display could have benefits for training drivers to have a realistic appreciation of how the Stop & Go ACC works. The radar display shows, in an unambiguous manner, what the radar has detected and what it can and cannot see. Once the driver had mastered the operation of the system, then the interface could be switched off, leaving the iconic display. The radar display could be selected by drivers who prefer the additional information, particularly in situations where multiple in-path targets are likely to be present. The radar display revealed its benefits in the static test, where participants showed that they could more accurately identify the in-path target detected by the Stop & Go system. Thus the radar display is of most benefit in ambiguous situations, where two or more possible in-path targets are presented. In simple situations, the iconic interface is superior. Recent research on manually operated cars has shown that SA is probably the most important psychological variable in optimising driver performance. Walker et al (2001) conducted an on-road trial of drivers with high and low feedback cars. High feedback vehicles were characterised by: higher degrees of vehicle responsiveness, more sophisticated drive train and chassis configurations, and richer levels of instrumentation. Analysis of verbal protocols recording during driving showed that drivers of high feedback cars report more at all of the SA levels than the drivers of low feedback cars. The high feedback drivers also report lower levels of workload than the low feedback drivers. This suggests that the low feedback drivers are required to work harder, than their high feedback counterparts, in order to extract SA information from the environment. Even then, their level of SA does not match that of driver's of high feedback cars. In situations where the demand placed upon the driver are higher, the iconic interface may help keep the driver within the optimal zone. Whereas when demands placed upon the driver are lower, the radar display may help keep the driver in the optimal zone. Thus, an interface design approach that enables the driver to either sample the iconic interface or the radar interface may be the best solution. There is a paradox with this approach. The radar display may bc of most use in those scenarios where most demand is being placed upon the driver, such as situation of ambiguity where multiple in-path targets are present. There may be specific points within such scenarios where workload dips sufficiently that the radar display can be utilised effectively.

References

Endlsey, M. R. (1995) Toward a theory of situation awareness in dynamic systems. Human Factors 37 (1) 32-64

Gugerty, L. J. (1997) Situation awareness during driving: Explicit and implicit knowledge in dynamic spatial memory Journal Of Experimental Psychology: Applied 3 (1) 42-66

Sukthankar, R. (1997) Situational awareness in tactical driving: the role of simulation tools. Transaction Of The Society For Computer Simulation 14 (4) 181-192

Stanton, N. A. & Young, M. (2001) A proposed psychological model of driving automation. Theoretical Issues in Ergonomics Science 1 (4), 315-331.

Stanton, N. A.; Young, M. S.; Walker, G. H.; Turner, H, & Randle, S. (2001) Automating the driver's control tasks. International Journal of Cognitive Ergonomics 5 (3), 221-236.

Walker, G. H., Stanton, N. A., & Young, M. S. (2001). An on-road investigation of vehicle feedback and its role in driver cognition: Implications for cognitive ergonomics. International Journal of Cognitive Ergonomics, 5, (4), 421-444.

Predicting Pilot Error: Assessing the Performance of SHERPA

Neville A. Stanton[1], Paul Salmon[1], Don Harris[2], Jason Demagalski[2], Andrew Marshall[3], Thomas Waldmann[4] and Sidney Dekker[5]

[1]Brunel University, Egham, UK
[2]Cranfield University, UK
[3]Marshall Associates, UK
[4]University of Limerick, Ireland
[5]Linkoping Institute of Technology, Sweden

Abstract
This paper introduces SHERPA (Systematic Human Error Reduction and Prediction Approach) as a means for predicting pilot error. SHERPA was initially developed for predicting human error in the nuclear industry about 15 years ago. Since that time validation studies to support the continued use of SHERPA have been encouraging. Most research shows that SHERPA is amongst the best human error prediction tools available. Yet there is little research in the open literature of error prediction for cockpit tasks. This study attempts to provide some evidence for the reliability and validity of SHERPA in aviation domain.

Introduction

Human error is an emotive topic. Psychologists and Ergonomists have been investigating the origins and causes of human error since the dawn of the discipline (Reason, 1990). Traditional approaches suggested that error was an individual phenomenon, the individual who appears responsible for the error. Indeed, so-called 'Freudian slips' were treated as the unwitting revelation of intention: an error revealed what a person was really thinking but did not wish to disclose. More recently, error research in the cognitive tradition has concentrated upon classifying errors within taxonomies and determining underlying psychological mechanisms (Senders & Moray, 1991). The taxonomic approaches by Norman (1988) and Reason (1990) have led to the classification of errors into different forms, e.g. capture errors, description errors, data driven errors, association activation errors and loss of activation errors. Reason (1990) and Wickens (1992) identify psychological mechanisms implied in error causation, for example the failure of memory retrieval mechanisms in lapses, poor perception and decision-making in mistakes and motor execution problems in slips. Taxonomies offer an explanation of what has happened, whereas consideration of psychological mechanisms offer an explanation of why it has happened. Reason (1990), in particular, has argued that we need to consider the activities of the individual if we are able to consider what may go wrong. This approach does not conceive of errors as unpredictable events, rather as wholly predictable based upon an analysis of an individual's activities. Since the late 1970's much effort has been put into the development of techniques to predict human error based upon the fortunes, and misfortunes, of the nuclear industry. Despite this development, many techniques are poorly documented and there is little in the way of validation studies in the published literature.

Validating Human Error Prediction

Whilst there are very few reports of validation studies on ergonomics methods in general (Stanton and Young, 1999a), the few validation studies that have been conducted on HEI are quite optimistic (e.g. Kirwan, 1992a, b; Stanton and Baber, 1996). It is encouraging that in recent years the number of validation studies has gradually increased. Empirical evidence of a method's worth should be one of the first requirements for acceptance of the approach by the ergonomics and human factors community. Stanton and Stevenage (1998) suggest that ergonomics should adopt similar criteria to the standards set by the psychometric community, i.e. research evidence of reliability and validity before the method is widely used. It may be that the ergonomics community is largely unaware of the lack of data (Stanton and Young, 1998) or assumes that the methods provide their own validity (Stanton and Young, 1999b). The development of HEI techniques could benefit from the approaches used in establishing psychometric techniques as two recent reviews demonstrate (Bartram et al, 1992, Bartram et al, 1995). The methodological concerns may be applied to the entire field of ergonomics methods. There are a number of issues that need to be addressed in the analysis of human error identification techniques. Some of the judgments for these criteria developed by Kirwan (1992, b) could be deceptive justifications of a technique's effectiveness, as they may be based upon:

- User opinion
- Face validity
- Utilisation of the technique.

User opinion is suspect because of three main reasons. First it assumes that the user is a good judge of what makes an effective technique. Second, user opinion is based on previous experience, and unless there is a high degree of homogeneity of experience, opinions may vary widely. Third, judgments may be obtained from an unrepresentative sample. Both Kirwan (1992, b) and Baber & Stanton's (1996) studies used very small samples. Face validity is suspect because a HEI technique might not be able to predict errors just because it looks as though it might, which is certainly true in the domain of psychometrics (Cook, 1988). Finally, utilisation of one particular technique over another might be more to do with familiarity of the analyst than representing greater confidence in the predictive validity of the technique. Therefore more rigorous criteria need to be developed. Shackel (1990) proposed a definition of *usability* comprising effectiveness (i.e. level of performance: in the case of HEI techniques this could be measured in terms of reliability and validity), learnability (i.e. the amount of training and time taken to achieve the defined level of effectiveness) and attitude (i.e. the associated costs and satisfaction). These criteria together with those from Kirwan (1992, b: i.e., comprehensiveness, accuracy, consistency, theoretical validity, usefulness and acceptability) and the field of psychometrics (Cronbach, 1984; Aiken, 1985) could be used to assess HEI techniques (and other ergonomics methods) in a systematic and quantifiable manner.

Systematic Human Error Reduction and Prediction Approach (SHERPA)

SHERPA (Embrey, 1986) uses Hierarchical Task Analysis (HTA: Annett *et al.* 1971) together with an error taxonomy to identify credible errors associated with a sequence of human activity. In essence the SHERPA technique works by indicating which error modes are credible for each task step in turn, based upon an analysis of work activity. This indication is based upon the judgement of the analyst, and requires input from a subject matters expert to be realistic. The process begins with the analysis of work activities, using Hierarchical Task Analysis. HTA is based upon the notion that task performance can be expressed in terms of a hierarchy of goals (what the person is seeking to achieve), operations (the activities executed to achieve the goals) and plans (the sequence in which the operations are executed). Then each task step from the bottom level of the analysis is taken in turn. First each task step is classified into a type from the taxonomy, into one of the following types:

- Action (e.g. pressing a button, pulling a switch, opening a door)
- Retrieval (e.g. getting information from a screen or manual)
- Checking (e.g. conducting a procedural check)

- Selection (e.g. choosing one alternative over another)
- Information communication (e.g. talking to another party)

This classification of the task step then leads the analyst to consider credible error modes associated with that activity. From this classification the associated error modes are considered. For each credible error (i.e. those judged by a subject matter expert to be possible) a description of the form that the error would take is given. The consequence of the error on the system needs to be determined next, as this has implications for the criticality of the error. The last four steps consider the possibility for error recovery, the ordinal probability of the error, its criticality and potential remedies.

Studies of SHERPA

Kirwan (1992b) conducted a comparative study of six potential HEI techniques. For this study he developed eight criteria on which to compare the approaches. In his study, Kirwan recruited 15 HEI analysts (three per technique, excluding group discussion). Four genuine incidents from the nuclear industry were used as a problem to focus the analysts' effort. This is the main strength of the study, providing a high level of ecological or face validity. The aim of the was to see if the analysts could have predicted the incidents if the techniques had been used. All the analysts took less than two hours to complete the study. Kirwan presented the results for the performance of the techniques as both subjective judgments (i.e.: low, medium and high) and rankings (i.e. worst and best). No statistical analysis was reported in the study, this is likely to be due to methodological limitations of the study (i.e. the small number of participants employed in the study). From the available techniques, SHERPA achieved the highest overall rankings and Kirwan recommends a combination of expert judgement together with the SHERPA technique as the best approach. A study by Baber & Stanton (1996) aimed to test the hypothesis that the SHERPA technique made valid predictions of human errors in a more rigorous manner. In order to do this, Baber & Stanton compared predictions made by an expert user of SHERPA with errors reported by an observer. The strength of this latter study over Kirwan's is that it reports the use of the method in detail as well as the error predictions made using SHERPA. Baber & Stanton's study focuses upon errors made during ticket purchasing on the London Underground, for which they sampled over 300 transactions during a non-continuous 24-hour period. Baber and Stanton argue that the sample was large enough as 90% of the error types were observed within 20 transactions and after 75 transactions no new error types were observed. From the study, SHERPA produced 12 error types associated with ticket purchase, nine of which were observed to occur. Baber & Stanton used a formula based upon Signal Detection Theory (Macmillan & Creelman, 1991) to determine the sensitivity of SHERPA in predicting errors. Their analysis indicated that SHERPA produces an acceptable level of validity when used by a expert analyst. There are, however, a two main criticisms that could be aimed at this study. First, the number of participants in the study was very low; in fact only two SHERPA analysts were used. Second, the analysts were experts in the use of the technique; no attempt was made to study performance whilst acquiring expertise in the use of the technique. Stanton & Stevenage (1998) conducted two experimental studies to test the learnability of SHERPA with novice participants. In the first study, the error predictions of 36 participants were compared to those who had no formal error methodology, to see if people using SHERPA performed better than heuristic judgement. Similar to the Baber & Stanton (1996) study, these predictions were made on the task that required people to make a purchase from a vending machine. Participants using the SHERPA technique correctly predicted more errors and missed fewer errors than those using the heuristics. However, they also appeared to incorrectly predict more errors. There appears to be a trade-off in terms of training such that a more sensitive human error identification is achieved, at the cost of a greater number of false positives. This is probably a conservative estimate, as no doubt if the observation period was extended indefinitely, more error types would be observed eventually. In the second study, 25 participants applied SHERPA to the vending task on three separate occasions. The data reported by Stanton & Stevenage show that there is very little change over time in the frequency of hits and misses however, the frequency of false alarms appears to fall over time and consequently, the frequency of correct rejections appears to increase. In terms of the overall sensitivity of error prediction, this shows remarkable consistency over time.

Predicting Pilot Errors in the Autoland Task Using SHERPA

The purpose of this study was to evaluate the SHERPA methodology applied to the analysis of the flight deck for the autoland task. There are many limitations on this study. For starters, there is no attempt to evaluate the dialogue between the pilot, the co-pilot and air traffic control. It is already assumed that autoland will be used. There are also limitations with regard to the collection of error data from pilots, which largely relied upon self-report to a questionnaire survey. Nevertheless, within these limitations, some insight into the success with which SHERPA can be applied to an aviation domain can be gleaned. Eight graduate engineering participants aged between 22 and 55 years took part in this study. All participants were trained in the SHERPA methodology. The training comprised an introduction to the key stages in the method and a demonstration of the approach using a non-aviation example, using an in-car task from Stanton & Young (1999a). Participants were then required to apply the method to another non-aviation task with guidance from the instructors from a public technology task from Stanton & Stevenage (1998). The purpose of this was to ensure that they had understood the workings of the SHERPA method. A debriefing followed, were participants could share their understanding with each other. When the instructors were satisfied that the training was completed, the main experimental task was introduced. This required participants to make predictions of the errors that pilots could make in the autoland task. To make their error predictions, participants were given a HTA of the autoland task developed by the authors (comprising some 22 subtasks under the main headings: setting up for approach, lining up for the runway, and preparing the aircraft for landing), a demonstration of autoland via Microsoft flight simulator, the SHERPA error taxonomy, and colour photographs of: the autopilot panel; levers for flaps, landing gear and speed brake; the primary flight displays; and an overview of the cockpit. Participants were required to make predictions of the pilot errors on two separate occasions, separated by a period of four weeks. This enabled intra-analyst reliability statistics to be computed. The predictions were compared with error data reported by pilots using autoland. This enabled validity statistics to be computed. The signal detection paradigm provides a useful framework for testing the power of HEI techniques. In particular, it identifies type I errors (a miss: when the error analyst predicts the error will not occur and it does) and type II errors (a false alarm: when the error analyst predicts that there will be an error and there is not) in the judgement of the analyst. Analysis of the data revealed the mean reliability of analysts between time one and time two using SHERPA as approximately 0.7 and mean validity, expressed as an mean of the hit and false alarm rates as approximately 0.6. These values are moderate, but it should be noted that this was the first time the participants had applied the SHERPA method in anger and that they were not aviation experts. The pooled error predictions are compared to the errors reported by pilots in table one. If the error predictions are pooled, the validity statistic rises to approximately 0.9 which is very good indeed.

Conclusions

In conclusion, the results are promising for the use of SHERPA in predicting pilot error. Whilst more studies are needed to investigate different tasks, the current study shows that novices were able to acquire the approaches with relative ease and reach acceptable levels of performance within a reasonable amount of time. This supports the investigation by Stanton & Stevenage and is quite encouraging. The study also shows that HEI techniques can be evaluated quantitatively. Human error is a complex phenomenon, and is certainly far from being completely understood. Yet in attempting to predict the forms in which these complex behaviours will manifest themselves armed only with a classification systems and a description of the human and machine activities it is amazing what can be achieved. Despite the gaps in our knowledge and the simplicity of the techniques, the performance of the analysts appears surprisingly good. This offers an optimistic view of the future for human error identification techniques. There are a number of other criticisms that need to be addressed, however. Stanton and Stevenage (1998) propose that clearer documentation on the methodologies needs to be provided, and that cross validation studies should be undertaken.

Table 1. Pooled error data

		Errors Observed	
		Yes	No
Errors Predicted	Yes	Hits = 52	F. A. = 4
	No	Misses = 5	C. R. = 179

Acknowledgement

This research is supported by a grant from the Department of Trade and Industry as part of the European EUREKA! research programme

References

Aitkin, L. R. (1985) Psychological Testing and Assessment. Allyn & Bacon: Boston.

Annett, J.; Duncan, K. D.; Stammers, R. B. & Gray, M. J. (1971) Task Analysis. Training Information No. 6. HMSO: London.

Baber, C. & Stanton, N. A. (1996) Human error identification techniques applied to public technology: predictions compared with observed use. Applied Ergonomics. 27 (2) 119-131.

Bartram, D.; Lindley, P.; Foster, J. & Marshall, L. (1992) Review of Psychometric Tests (Level A) for Assessment in Vocational Training. BPS Books: Leicester.

Bartram, D.; Anderson, N.; Kellett, D.; Lindley, P. & Robertson, I. (1995) Review of Personality Assessment Instruments (Level B) for use in Occupational Settings. BPS Books: Leicester.

Cook, M. (1988) Personnel Selection and Productivity. Wiley: Chichester.

Cronbach, L. J. (1984) Essentials of Psychological Testing. Harper & Row: New York.

Embrey, D. E. (1986) SHERPA: A systematic human error reduction and prediction approach. Paper presented at the International Meeting on Advances in Nuclear Power Systems, Knoxville, Tennessee.

Kirwan, B. (1992a) Human error identification in human reliability assessment. Part 1: overview of approaches. Applied Ergonomics, 23 pp. 299-318.

Kirwan, B. (1992b) Human error identification in human reliability assessment. Part 2: detailed comparison of techniques. Applied Ergonomics, 23 pp. 371-381.

Macmillan, N. A. & Creelman, C. D. (1991) Detection Theory: a user's guide. Cambridge University Press: Cambridge.

Norman, D. A. (1988) The Psychology of Everyday Things. Basic Books: New York.

Reason, J. (1990) Human Error. Cambridge University Press: Cambridge.

Senders, J. W. & Moray, N. P. (1991) Human Error. LEA: Hillsdale, NJ.

Shackel, B. (1990) Human factors and usability. In: Preece, J. & Keller, L. (eds) Human-Computer Interaction. Prentice-Hall: Hemel Hempstead.

Stanton, N. A. & Baber, C. (1996) A systems approach to human error identification. Safety Science, 22, pp. 215-228.

Stanton, N. A. & Stevenage (1998) Learning to predict human error: issues of reliability, validity and acceptability. Ergonomics 41 (11), 1737-1756

Stanton, N. A. & Young, M. (1998) Is utility in the mind of the beholder? A review of ergonomics methods. Applied Ergonomics. 29 (1) 41-54

Stanton, N. A. & Young, M. (1999a) A Guide to Methodology in Ergonomics: Designing for Human Use. Taylor & Francis: London.

Stanton, N. A. & Young, M. (1999b) What price ergonomics? Nature 399, 197-198

Wickens, C. D. (1992) Engineering Psychology and Human Performance. Harper Collins: New York.

Assessing the Effect of New Technology on Driver Behavior – a Theoretical Model

Rebecca Stewart and Don Harris

Human Factors Group, Cranfield University
Bedford MK43 0AL, UK
r.j.stewart@cranfield.ac.uk and d.harris@cranfield.ac.uk

Abstract

European Human Vehicle Interface (HVI) standards are being developed for the automotive industry in which support, psychological comfort and safety will be assessed, particularly in the context of new in-vehicle technology. Drivers' behavior will be evaluated to ascertain the effects of this technology. By doing this it is assumed that the performance of the driver reflects the quality of the HVI, however this may be an unwarranted assumption. A system model is proposed illustrating this conundrum. When new technology is placed in a car it is tested by observable driver behavior (e.g. speed). The drivers' internal cognitive states (e.g. risk, workload, situation awareness) are then inferred from this behavior but this may not be valid.

1. Upcoming Technology in the Motor Vehicle

In the next 25 years it is likely that the driver interface in the modern car is going to undergo a dramatic change. A great deal of new technology is going to be introduced into road vehicles. Up until relatively recently, the primary task of the driver has simply been the manual control of the vehicle which has been exercised via inputs to the steering wheel, gear lever and pedals. Using the typology proposed by Stanton & Young (2001) to describe in-vehicle automation, some new in-vehicle technologies will be implemented in the category of primary control of the vehicle, where control, to some degree, is delegated to the vehicle. Technologies in this area range from simple cruise control, to more advanced cruise control systems that will keep station with the vehicle in front, to highly automated systems that will also keep the car centered in the lane of the carriageway. Some equipment already being installed in modern motorcars has already successfully automated many secondary control functions (e.g. automatic gearboxes; anti-lock brakes or; traction control). The automation of primary control functions needs to be undertaken equally as successfully to maintain (or enhance) the overall level of safety.

The second major task often imposed on the driver is that of navigation. At the moment navigation usually involves the use of paper maps, notes on pieces of paper, road signs and the driver's memory. This is already beginning to change in some vehicles with the installation of sophisticated, in-car navigation systems that guide the driver by providing prompts at appropriate points during a journey. This is an example of the second major area of in-vehicle automation identified by Stanton & Young (2001), driver cognitive support. As a further example, the driving task may be made easier at night with the implementation of night vision systems that project images from a vision enhancement system onto a head-up display (HUD). These systems may also be able to identify hazards in the environment and highlight them on the HUD. The HUD may also pass information from the navigation system to provide navigation information to the driver. The car may also be connected to the world-wide web, passing information to the driver concerning the location of services in the immediate vicinity that s/he will require or information about local traffic congestion. The communication interface in modern vehicles also needs to be considered. In short, the driving task may begin to change away from one of vehicle control and evolve in the direction of information/system management.

This change in the role of the driver reflects the change in the role of the commercial pilot that began about 30 years ago. The role of the pilot is now one of a flight deck information systems manager and system monitor rather than as a 'hands on' controller. Many lessons can be learned from the evolution of the flight deck and the concomitant changes in the pilot's tasks with respect to the changes that can be expected in the behavior of the driver. Harris and Smith (1997) describe some of the parallels in technology (and pitfalls) when transferring technological principles from the aviation to the automotive environment. There are important differences in the two application areas that limit the ability to generalize from one environment to the other. The most critical (and perhaps obvious) conclusion that can be drawn, though, is that new technology will change the behavior of the driver. It is essential that in the broadest sense, the behavior of the driver must change for the better.

2. Assessing the Introduction of New Technology on the Driver

As part of the European Union-funded ROADSENSE (Road Awareness for Driving via a Strategy that Evaluates Numerous SystEms) program the driver's responses to the implementation of new technologies will be evaluated in the broad categories of safety, comfort and efficiency. Part of this program also aims to establish a new, European quality standard for the design and approval of the Human-Vehicle Interface (HVI) for motor vehicles. The objectives are that these new technologies should have a high level of efficiency of interaction (usability); they should increase driver comfort; and that they should not degrade (and should preferably enhance) the overall level of road safety (i.e. not just that of the driver and passengers in the vehicle) (RoadSense, 2000). The problem is how do you evaluate these goals?

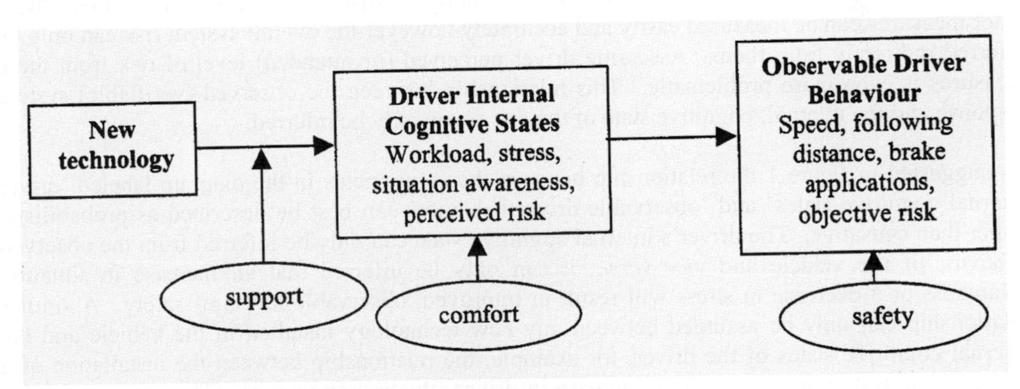

Figure 1. Simple system model of driver responses to new technology. A hypothesized mechanism for drivers' responses with respect to the introduction of new technology into the motor vehicle.

There is an implicit assumption that new interface technologies (irrespective of their function) will change driver behavior in one or more of the broad categories described in the ROADSENSE program. However, any changes in driver behavior as a result of the se technologies may not be either (a) directly attributable to the equipment and/or (b) may not be directly measurable. A simple conceptual model describing the relationship between new driver interface technologies and behavior is proposed in figure 1. It can be seen that the three different general categories of measure proposed in the ROADSENSE program address different stages in the model.

Support (or usability) is a measure of the quality of the user interface. As a measure it addresses neither the interface on the equipment nor the human information processing system. Measures in this category assess the conceptual 'fit' between the representation of the information on the equipment or the quality of the control interface, and the human perception of what is on the interface. This 'fit' is also known as the human machine interaction (HMI). If the driver's perception of what is on the equipment interface closely corresponds to what is actually on the interface and the driver fully understands the information that the system designer actually intended to transmit, then the interface can be considered to be a high quality interface. An interface that is easy to use (perhaps in terms of requiring few control inputs) and is also well defended against errors can also be regarded as a 'usable' interface. Usability has many dimensions; ease of use; satisfaction, whereby the system is not frustrating to use; learnability; effectiveness; efficiency, few physical resources are needed and errors, meaning that few mistakes are made (Marshall, 2001). If an interface encompasses all these factors then the driver's internal cognitive states are less likely to be overloaded. Comfort addresses the psychological comfort of the driver rather than the physical comfort, for example through the improvement in the design of the vehicle's seats. Increasing the driver's situation awareness, for example by allowing them to see further ahead in darkness by equipping the vehicle with a night vision system may increase the driver's psychological comfort level by reducing their uncertainty of elements in their surrounding environment. High levels of uncertainty or workload may lead to high levels of driver stress. Conversely a reduction in driver stress may lead to an increase in driver psychological comfort. Safety is something that can only be inferred and never measured directly, however many objective measures can be made that are highly related to overall system safety. For example, overall system risk (of an accident) may be highly related to the free speed of vehicles, the lateral acceleration of the vehicle when cornering or their following distance from the car in front. These latter measures can be measured easily and accurately however the overall system risk can only be inferred indirectly from them. Assessing driver perceived (or intended) level of risk from these measures is even more problematic. This relationship between the observed (verifiable) system response and the internal, cognitive state of the driver can only be inferred.

As suggested in figure 1 the relationship between the components in the diagram labeled 'driver internal cognitive states' and 'observable driver behavior' can best be described as probabilistic rather than causative. The driver's internal cognitive state can only be inferred from the observed behavior of the vehicle and *vice versa*, it can only be inferred that an increase in situation awareness or a decrease in stress will result in improved, observable levels of safety. A similar relationship can only be assumed between any new technology installed in the vehicle and the internal cognitive states of the driver, for example, the relationship between the installation of a night vision system in a car and an increase in driver situation awareness. The outcome of this argument is that it is actually quite difficult to establish if the new technologies actually result in an increase in overall levels of system safety (or alternatively compromise it), as all effects of the technology are mediated, to a lesser or greater extent, by the driver. There is no direct link between observed driver behavior and the new equipment in the vehicle. It is for this reason that it is necessary to assess the quality of the user interface and driver psychological comfort in addition to the observable measures of performance, which are assumed to be related to safety.

3. Reliability and Validity

Reliability and validity are fundamental measurement concepts. Reliability is the ability of a measurement instrument to produce the same value on two (or more) occasions when all other aspects are the same. Validity concerns whether the measurement instrument evaluates the parameter that it purports to measure. Figure 1 begins to describe a framework for the evaluation of new driver technology by proposing measures taken at three different levels in the driver-vehicle system. Some problems begin to become apparent, though, when issues in the reliability and validity of the measurement metrics falling into each of these categories are considered. Some of the issues concerned with the validity of inferring that changes in driver behavior are directly attributable to the introduction of new technology have already been alluded to at the end of the previous section. However, the issues of the reliability of measurement also need to be considered. Although the science of Ergonomics has been around for about half a century and has begun to contribute to major improvements in equipment interfaces, there has recently been criticism of many of the measurement techniques utilized (Stanton and Young, 2000). Many measurement techniques employed by human factors professionals to evaluate interfaces have never been evaluated for their reliability and validity. Some subsequent evaluations of the measurement metrics have shown them to be very poor, well below the standards regarded as being acceptable in other areas of psychometric measurement, for example aptitude or personality testing. With a little further consideration a conundrum begins to become apparent. The most reliable measures that can be taken of driver performance are related to the behavior of the vehicle. However, when considered in terms of the model proposed in figure 1, it can also be seen that these measures are perhaps the least valid as measures of the effects of new technology installed in a car. They cannot be directly related to the new technology as they are all mediated by the internal cognitive states of the driver. The measures most closely related to the new equipment (the usability of the interface and the internal cognitive states of the driver) are more valid as measures of the effects of the new technology, however, because of the nature of these measures, they are also the least reliable.

4. Measures

The relationship between validity of the measures taken to assess drivers' responses to new technology and the context of the study (e.g. driving simulator, test track or 'on-the-road') is a complex and intimate one. Some methods and measures (e.g. collision avoidance studies) may only be safely undertaken in a simulator or on a test track. This allows for tighter control of the experimental situation and offers a greater amount of flexibility in the choice of measures employed, thereby enhancing the internal validity of the study, unfortunately, at the expense of the ecological (external) validity of the study. Will these behaviors generalize to driver behavior on the road? Studies of driver behavior performed on the road have greater ecological validity but sacrifice internal validity. Furthermore, in this case the range of measures that may be employed is considerably restricted as a result of practical, safety and other ethical issues. As an objective for the HVI standards, any new technology that is introduced into vehicles will support the driver, should increase the drivers' psychological comfort and should not degrade safety. To assess the new technology, behavioral measures will be used, but how do we know if they measure what they are supposed to measure? E.g., speed measures speed, and steering wheel position variance measures the variance of the steering wheel position. The problem arises with the assumption that a change in, for example speed, is attributable to risk. Are observable driver behaviors valid and reliable measures of the drivers' internal cognitive states? The concomitant measures of comfort, support and safety would lend itself to a path analysis to help work out any relationship.

From researching the literature there seems to be little evidence that the validity and reliability of these measures have been assessed. This is confirmed by a survey carried out by Stanton & Young (1998). A hypothetical model depicting future areas of research into reliability and validity is shown in figure 2.

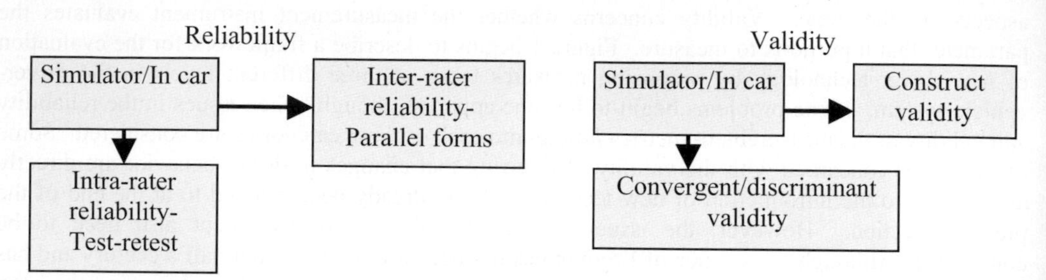

Figure 2. A simple model depicting possible methods of testing

5. Conclusions

There will be many changes and advances in technology in the near future that will be implemented in cars. Many of these new systems are aimed at helping the driver by way of supporting them whilst driving, increasing their safety and by increasing their psychological comfort. The measures used to assess these are being used based on assumptions that haven't been fully justified, e.g. when a drivers' observable behaviour changes, for example their speed or following distance, it is attributed to the drivers' internal cognitive states, for example an increase in risk, stress etc. If HVI standards are to be introduced into the automotive industry these assumptions need to be carefully considered and tested to assess that they are measuring what they are assumed to measure and that they do this on more than two occasions. The next step therefore is to assess the validity and reliability in the simulator and in-car of these measures.

6. References

Harris, D., & Smith, F. J. (1997).What can be done versus what should be done: a critical evaluation of the transfer of human engineering solutions between application domains. In D. Harris (Ed.), *Engineering psychology and cognitive ergonomics* (pp339-346). Aldershot,: Ashgate.

Marshall, C. J. (2001). *A usability scale for interactive television.* Unpublished masters thesis, Cranfield University, Bedford, UK.

RoadSense (2000). Annex 1 – Description of Work, V 1.0. RoadSense GRD1-2000-25572.

Stanton, N. A., & Young, A. S. (1998). Is utility in the mind of the beholder? A study of ergonomics methods. *Applied ergonomics, 29*, (1), 41-54.

Stanton, N. A., & Young, M. S. (2000). A proposed psychological model of driving automation. *Theoretical issues in ergonomics Science, 1* (4), 315-331.

Stanton, N. A., & Young, M. S. (2001). Psychological factors of using adaptive cruise control. In D. Harris (Ed.), *Engineering psychology and cognitive ergonomics* (pp. 399-405). Aldershot, UK: Ashgate.

Modeling of Knowledge Structure Transformation with First-Order Clauses in Dynamic Systems

Kuo-Hao Tang

Feng Chia University
100 Wenhwa Rd., Taichung, Taiwan
khtang@fcu.edu.tw

Abstract

The qualitative difference between novices and experts has long been an interesting research area and theories such as knowledge structure have been proposed to describe this difference. To this point, it is not clearly understood how knowledge structures are generated and transformed during training. In order to depict the knowledge structure transforming processes, an aggregation production simulation model was established and the first order clauses in predicate formats are used to represent knowledge structure in this study. A single clause can be considered as at a given timing where a participant believed an action was needed, the snapshot of the system status, the actions taken and the consequent outcome of the system. By rearranging these collected first order clauses, the cognitive model of a participant was then developed by grouping these clauses into chunks and the graphs representing a participant's cognitive model can be generated according to appropriate sequencing of these chunks. Thus, the changes of the graphs of a particular participant from beginning to a later learning stage represent the change of knowledge structure. Qualitative analysis from the chunked knowledge structures shows that more structured frameworks appear after learning, and the timing of making a strategic decision becomes more precise as well. Statistical analysis also shows that performance and mental workload on running the simulation model improve significantly during learning.

1 Introduction

In virtually performing every cognitive intensive task, novices and experts differ in how they process information and solve problems. The qualitative difference between novices and experts has long been an interesting research area and theories such as knowledge structure have been proposed to describe this difference. Knowledge structure refers to the manner in which knowledge or information is organized within individuals. As learning occurs, increasing well-structured and qualitatively different knowledge developed. These structures enable individuals to build a representation or mental model that guides problem solving process and further learning. Knowledge structure theory also proposes that a significant determinant of performance is the manner in which humans structure their knowledge about the domain under consideration. Early research (de Groot, 1965; Chase and Simon, 1973) focused on chess players; experts were able to reproduce board configurations more efficiently than novices by using patterns and chunks. Later research by Adelson (1984) found differences between novice and expert computer programmers. Novices focused on how the program operated, while experts focused on what the program did. Adelson (1984) considered the novices to have a concrete representation of the domain, while experts had a more abstract representation, or knowledge structure. Research in this area by

Koubek and Salvendy (1991) proposed three levels of knowledge structures: novice, expert, and super-expert, with the levels corresponding to the levels of operator skill: surface feature, task specific, and abstract/hierarchical, respectively. The knowledge structure approach implies while some people are able to develop a high level of conceptual understanding of domain knowledge, others simply are not.

The importance of knowledge structure assessment has also been addressed and recognized (e.g., Glaser et al., 1987; Lane, 1991), it is still the case that "cognitive theories about knowledge structures have progressed far ahead of research on methods for their assessment" (Snow & Lohman, 1989, p. 304). In recent years a number of methods have been developed to elicit an individual's mental model for a set of concepts. These techniques have used a simplified model to represent structural knowledge as a network of interconnected nodes. In the model's basic form, each node has represented a concept, and the meaning is determined by its connections to other concepts (nodes) in the network. Using this model, these techniques elicit knowledge structure representations for a set of concepts by requiring a subject to directly manipulate the nodes representing the concepts, and the connections representing the relationships among them (Rogan, 1988; Hegarty-Hazel and Prosser, 1991). While these methods have been used to elicit knowledge structure representations in different areas, the knowledge acquisition processes are usually not embedded during the process of performing a task. And the concept (node) is usually represented by a keyword in the domain knowledge. Therefore, it is difficult to compare the processes of problem solving between a novice and an expert. In order to depict the knowledge structure transforming processes, first order clauses in predicate formats are used to represent knowledge structure in this study, and an aggregation production simulation model was established as a test bed.

2 Method

The aggregation production plan is a production plan at a middle level. It adjusts policy variables such as amount of labor according to external variables such as demands and internal variables such as inventory and sales levels. The decision variables in the simulation model include the amount of overtime, the amount of hiring and layoff, product price, and desired inventory level to build anticipation inventory for future demand. The internal variables include current labor force, production and sales quantity, inventory level, inventory cost, overtime cost, material cost, production cost, and finally, hiring and firing cost. The external variables include market demand and interest rate. During simulation, the external variables change with time to present the randomness and cyclic patterns of real world. The simulator was based on system dynamics model with time delay and the underling rules controlling these variables follow economics law. The bottom half of the screen shows the current production quantity, sales quantity, inventory level, market demand and profit in accordance to time in line format as shown on Figure 1. The simulator was developed by using Visual Basic and system dynamics concepts were adopted in developing the rules underlying the simulator.

A participant monitors the pattern change on the screen and stops the simulator when the participant believes some actions need to be taken. After the participant intervenes the simulator, the parameters for the decision variables can be readjusted and the simulator continues again using new parameters. The simulator then records data input by the participant and the system status as well as system outcomes and rearrange them into first order clauses in predicate formats. 12 subjects participated the experiment, each participate performed three one-hour sessions. All clauses and training performances were recorded. NASA-TLX test measuring mental workload was conducted at the end of each hour.

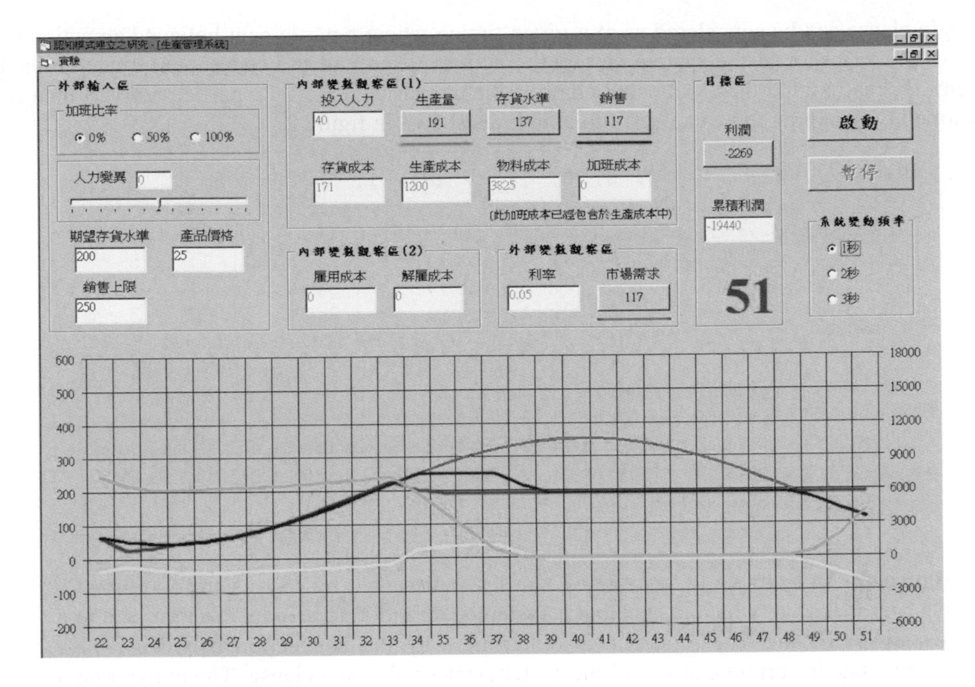

Figure 1: An example of the simulator's screen printout.

In order to capture structured knowledge reflecting if a participant successfully making correct and necessary adjusts, the first order clauses are defined as follows:

Define the following three sets:

X = {all internal and external variables}

Y = {all decision variables}

Z = {profit}

Given that $x \subset X$, $y \subset Y$, and $z \subset Z$, three predicates can be defined as:

Sys_state(x): current system status represented by current values of internal and external variables.

Action(y): a participant's adjustments represented by the adjustments to the decision variables.

Goal(z): the system outcome in terms of profit change.

The following patterns then collected during participant's interventions to acquire structured knowledge:

IF {Sys_state(x) and Action(y)} Then Sys_state(new_x)

IF {Sys_state(x) and Action(y)} Then Goal(z)

A sample of data collected is listed below:

10:01:16 PM, counter: 854
Sys_state(0, 1,-1,-1):- Sys_state(0, 0, 0,-1), Action(0, 0, 0, 0, 1)
10:01:25 PM, counter: 857
Sys_state(-1,-1,-1,-1):- Sys_state(0, 1,-1,-1), Action(-1, 0, 0, 0,-1)
10:01:39 PM, counter: 864
Sys_state(0, 1, 1, 1):- Sys_state(0, 1,-1,-1), Action(0, 0, 0, 0,-1)
10:01:56 PM, counter: 874
Sys_state(-1,-1,-1,-1):- Sys_state(-1, 1,-1,-1), Action(0, 0,-1, 0, 0)

A single clause can be considered as at a given timing where a participant believed an action was needed, the snapshot of the system status, the actions taken and the consequent outcome of the system. By rearranging these collected clauses, the cognitive model of a participant was then first developed by grouping these clauses into chunks as shown in figure 2.

```
( 0, 1,-1,-1)+ ( 0, 0, 0, 0,-1) -> (-1, 1,-1,-1)   8
  (-1, 1,-1,-1)+ ( 0, 0, 0, 0,-1) -> (-1,-1,-1,-1)   4
    (-1,-1,-1,-1)+ ( 0, 0, 0, 0,-1) -> ( 0, 0, 0, 0)   4
      ( 0, 0, 0, 0)+ ( 0, 0, 0, 0,-1) -> ( 1,-1, 1, 1)   5
      ( 0, 0, 0, 0)+ ( 0, 0, 0, 0,-1) -> ( 0,-1, 1, 1)   3
      ( 0, 0, 0, 0)+ ( 0, 0, 1, 0, 0) -> ( 1, 1, 1, 1)   3
        ( 1, 1, 1, 1)+ ( 0, 0, 0, 0, 1) -> (-1,-1,-1,-1)   4
          (-1,-1,-1,-1)+ ( 0, 0, 0, 0,-1) -> (-1, 0,-1,-1)   3
            (-1, 0,-1,-1)+ ( 0,-1, 0, 0, 0) -> ( 0, 0, 0, 0)   3
            (-1, 0,-1,-1)+ ( 0, 0, 0, 0, 1) -> (-1,-1,-1,-1)   3
            (-1, 0,-1,-1)+ ( 0, 0, 0, 0,-1) -> ( 1, 0, 1, 1)   3
            (-1, 0,-1,-1)+ ( 0, 0, 0, 0,-1) -> ( 0, 0, 0, 0)   4
        ( 1, 1, 1, 1)+ ( 0, 0, 0, 0, 1) -> ( 0, 0, 0, 0)   3
  (-1, 1,-1,-1)+ ( 0, 0, 0, 0,-1) -> ( 0,-1, 0, 0)   3
```

Figure 2: An example of rearranged clauses represented by "Sys_state + Action -> Sys_state" format, which means that a system state and a participant's action lead to a new system state. The sequence is determined by the connection between a new system state in current clause and the system state in the next clause. The number next to the clause is the frequencies of that clause.

3 Results

From the rearranged clause as shown on Figure 2, graphs representing each participant's knowledge structure or mental model to the system can be built using system status as node and action as arc as shown on Figure 3. Figure 3 shows the graph representations of knowledge structure developed during three sessions for a specific participant. The changes of the graphs of a particular subject from beginning to a later learning stage represent the change of knowledge structure. It is clear that the knowledge structure of this participant became more complex from the first session to the last session. Qualitative analysis from the chunked knowledge structures shows that more structured frameworks appear after learning. Participants learned more strategies and were more capable of handling fluctuating external environment. At later stages, participants also showed better control of the timing of making a strategic decision and became more precise and resourceful in selecting appropriate strategies. These similar patterns of changing knowledge structure appear in all 12 participants across the three sessions. Given each clause as a potential strategy, correlation analysis revealed that as participants used different strategies across three sessions ($p<0.01$). If divided participants into two groups according to the final performance as higher achievers and lower achievers, correlation analysis with a Fisher's test ($\alpha=0.05$) revealed that at the third session, higher achievers had more consistent strategies than the lower achievers. ANOVA analysis also shows that performances and the mental workloads measured from NASA-TLX on running the simulation model improve significantly during the three sessions, both have a p-value less then 0.001. Overall, the experimental results suggest how the knowledge structure changes as a learner becomes more familiar with the domain knowledge. The results also point out that learners' knowledge structures become more stable at later stages and higher achievers share more similar strategies than others.

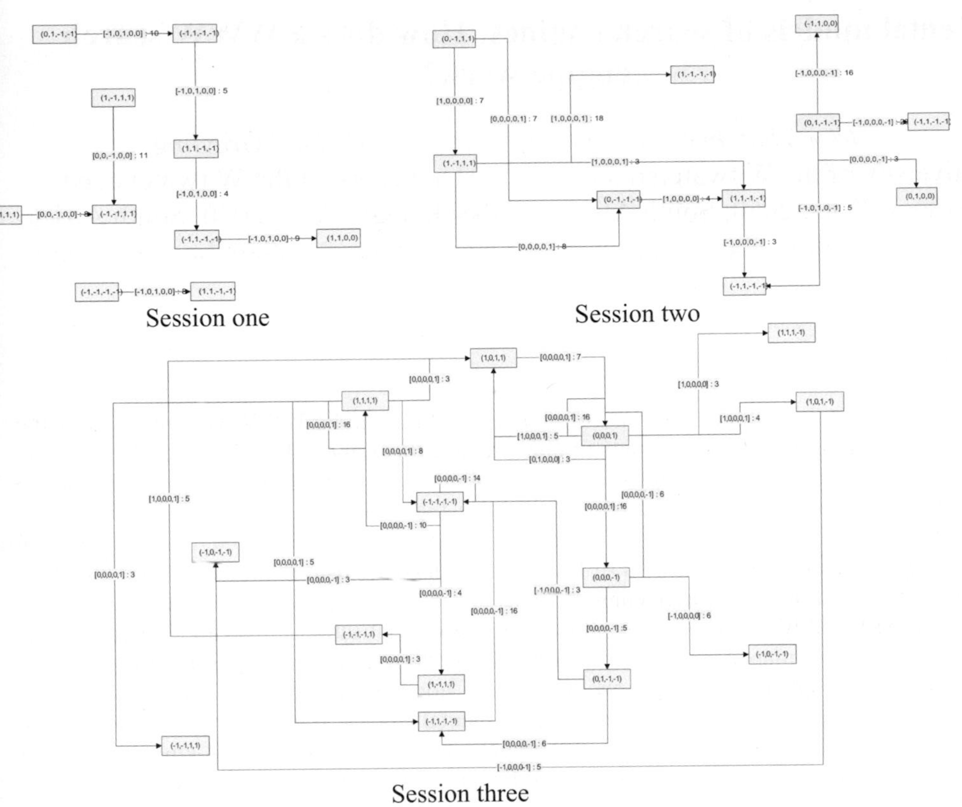

Figure 3: An example of a participant's knowledge structure across three sessions.

4 References

Adelson, B. (1984). When Novices Surpass Experts: the Difficulty of a Task may Increase with Expertise. *Journal of Experimental Psychology*, 10(3), 483-495.

Chase, W. G. & Simon, H. A. (1973). Perception in chess. Cognitive Psychology, 4, 55-81.

de Groot, A. D. (1965). *Thought and Choice in Chess*. The Hague: Mouton.

Glaser, R., Lesgold, A., & Lajoie, S. (1987). Toward a cognitive theory for the measurement of achievement. In R. R. Ronning J. Glover, J. C. Conoley, & J. C. Witt (Eds.), The influence of cognitive psychology on testing (pp. 41-85). Hillsdale, NJ: Lawrence Erlbaum Associates.

Hegarty-Hazel, E., & Prosser, M. (1991). Relationship between students' conceptual knowledge and study strategies: Part 1: Student learning in physics. *International Journal of Science Education, 13,* 303-312.

Koubek, R. J. and Salvendy, G. (1991). Cognitive Performance of Super-Expert on Computer Program Modification Tasks. *Ergonomics*, 34(8), 1095-1112.

Lane, S. (1991). Implications of cognitive psychology for measurement and testing: Assessing students' knowledge structures. Educational Measurement: Issues and Practice, 10(1), 31-33.

Rogan, J. M. (1988). Conceptual mapping as a diagnostic aid. *School Science and Mathematics, 88*(1), 50-59.

Snow, R. E., & Lohman, D. F. (1989). Implications of cognitive psychology for educational measurement. In R. L. Linn (Ed.), Educational measurement (3rd ed., pp. 263-331). New York: ACE/Macmillan.

Mental models of search engines: How does a WWW search engine work?

Andrew Thatcher
University of the Witwatersrand
Psychology, Wits, 2050, South Africa
thatchera@umthombo.wits.ac.za

Michael Greyling
University of the Witwatersrand
Psychology, Wits, 2050, South Africa
greylingm@umthombo.wits.ac.za

Abstract

A number of authors have suggested that information seeking on the WWW is dominated by the use of search engines. This study attempts to unearth the general mental models that WWW users have about the search engines that they use. Eighty volunteer subjects completed two tasks, one to establish their mental model task and one to establish the effectiveness of their mental model. From an analysis of the identification of fifteen salient features of search engines, four mental model clusters were established. In two of these mental model clusters, subjects had quite naïve mental conceptualisations of search engines, whereas in the other two mental model clusters subjects demonstrated slightly more sophisticated conceptualisations of search engines. The results across the performance measures suggest small performance advantages across the mental model clusters, although these are largely statistically non-significant. This would suggest that while many WWW users have quite unsophisticated mental models of search engines they are still able to use them relatively effectively. It would appear that the design of search engine interfaces obscures many of the salient features necessary for users to form more accurate and complete mental models.

1 Introduction

Searching in many closed hypertext systems is usually dominated by following hyperlinks whereas the sheer volume of information and webpages on the WWW would make this searching strategy ineffective for certain types of searches. A number of researchers have argued that searching on the WWW is dominated by search engine use (e.g. Spink et al, 2001), proceeded by following links within or between websites to locate more specifically relevant information content. It is also apparent that most users utilise many different search engines in the course of their interactions with the WWW and sometimes many different search engines within the same information-seeking task (e.g. Hodkinson et al, 2000). We were therefore interested in investigating what sort of mental models users would develop of search engines in general. The term 'mental model' has been used to describe an image or mental picture, an analogy, a qualitative simulation, a representational format, and an abstract mapping of properties and relations. In short, mental models are used in describing, explaining and predicting events in the system domain (Rouse & Morris, 1986). Mental models are also used in predicting future events, finding causes of observed events, and determining appropriate actions to cause changes in the system. Many authors have suggested that mental models of problem domains in complex computer systems are generally inaccurate, inconsistent, or incomplete (e.g. Staggers & Norcio, 1993). Mental models are acquired through interaction with a system or by observing others interacting with the system. The mental model emerges by comparing the existing prior knowledge (or a metaphor) with the new system by means of an analogical model. But a complete and transparent mapping of the new system may be difficult if the elements of the existing metaphor conflict with the functionality of the new system. There are a small number of studies that have referred (usually just in passing) to the users' mental models of search engines. These are studies restricted to investigations of a single search engine (e.g. Moukdad & Large, 2001; Spink et al, 2001) and where the mental model is indirectly inferred from users' queries (e.g. Schacter et al, 1998). To date there has been no systematic investigation of users' general mental models of search engines on the WWW. Given that each search engine has a different method of collecting and categorising information from the WWW (e.g. Sullivan, 2002) we were interested in seeing whether users have developed general cognitive models to understand search engine functioning and whether cognitive conceptualisations of this type would help information seekers to effectively use search engines. Indeed, evidence from single search engine studies suggests that some users mistakenly apply incorrect metaphors that they may have gained from using another search engine interface (e.g. Moukdad & Large, 2001).

2 Methodology

In order to uncover users' cognitive models of WWW search engines we had 80 volunteer subjects complete two WWW search tasks defined by the researchers. One task was designed to explore users' mental models of search engine use (this mental model task was a general purpose browsing task) and the second task was a directed search task (the performance task) used to determine the effectiveness of these cognitive representations. Subjects were allowed to search for the required information using their 'usual' searching strategies and had the freedom to select search engines of their own choice. The ordering of the two tasks was randomly assigned to control for any learning effects between the mental model task and the performance task. While the subjects were completing the tasks, client-side log files were collected together with a record of all on-screen actions using Lotus ScreenCam. After the subjects had completed each task, the ScreenCam recordings were played back to the subjects and the researcher asked the subjects to explain each of the decisions and choices that were made, as a form of retrospective protocol. After both tasks were completed the subjects were each asked to draw and then explain in words how they thought a search engine worked. Prior to the search tasks, subjects filled out a biographical questionnaire (where their experience with WWW and with computers was also assessed). The subjects were from range of different occupations (including students, scholars, accountants, and IT professionals) and ages (Range: 16-35 years). Fifty of the subjects were male and 30 were female. The majority of the subjects spoke English as a first language (63%), although all subjects were competent in speaking English. A composite measure for WWW experience was established from 7 questions aimed at determining the depth and breadth of knowledge and experience (a similar composite measure was established for computer experience) (e.g. Fisher, 1991). These composite measures of experience showed a great deal of variability (e.g. from subjects who had only used the WWW once previously to subjects who used the WWW every day for a variety of purposes). The process of deriving the mental model categories was based on two phases. Firstly a full list of the salient general features of search engines was developed based on the drawings and written descriptions and augmented with information from the retrospective protocols from the mental model task. Secondly, these salient features were assigned to each subject on the basis of an inspection of this data. Two raters working independently assigned the features. The inter-rater agreement coefficient indicated that there was a sufficiently high degree of agreement between the raters in their assignment of the features (91% agreement, Cohen's Kappa = 72%). Where there was disagreement, consensus was reached through discussion between the raters. The fifteen salient features included understanding that a search engine is a database collection, that different search engines use different 'collecting' algorithms, that there are algorithms to 'rank' results, etc. Next, a Ward's cluster analysis (Ward, 1963) was performed on the salient features in order to determine clusters of features that would describe possible mental model categories. A number of different clusters were established from the cluster analysis, with each cluster representing a different 'mental model' category. Next we looked at the influence of computer and WWW experience as antecedents to these clusters. Following this process the categories were compared against the subjects' performance on the directed search task. Performance was operationalised as the time taken to find the answer to the search question, and whether the search answer was correct or not. The patterns of salient features in the mental model categories enable us to establish which features of users' understanding of search engine functioning are important in determining search performance.

3 Results

Firstly, it was obvious that the majority of subjects identified relatively few of the salient features of search engines, suggesting that they held rather incomplete mental models. At best only one subject indicated 10 of the 15 salient features and that subject did not overtly recognise that search engine results are ranked. The majority of subjects only identified fewer than 5 of these salient features). There were two additional categories, identified as salient features by the researchers, which could not be identified in any of the subjects' data (e.g. that different search engines have different algorithms to rank results and that search results allow users to find similar results). In the analysis to determine the mental model clusters, the number of clusters used was determined by examining the Pseudo T^2 statistic, the cubic clustering criterion (CCC) and the dendogram (Milligan & Cooper, 1985). The Pseudo T^2 showed distinct slope changes at 2 and 4 clusters while CCC showed a slope change at 4/5 clusters. The dendogram indicated two very distinct clusters, each of which divided into two more clusters. Three of the resultant clusters were very distinct while the fourth cluster could be further divided into two clusters. A careful examination of the results for the different permutations indicated that 4 clusters gave the most meaningful and parsimonious description of the data.

3.1 Identifying the clusters

In order to identify the key characteristics of the clusters both descriptive statistics and a CHAID analysis (Kass, 1980) were performed. The CHAID was performed using the cluster identity as the dependent variable and the 15 salient features as the independent variables. The clusters were roughly of equal size (a typical consequence of Ward's method) ranging from N=16 in cluster 2 to N=26 in cluster 4. Clusters 1 and 2 tended to have the simplest mental models and were primarily identified by their lack of recognition of the salient features. In particular no subjects in these two clusters indicated an awareness of multiple search engines and only one subject indicated an awareness of the idea that search engine examine the search engine's database, rather than searching the entire WWW. No subjects in these two clusters recognised that the search engine results were ranked and some 45% of subjects made no mention of the WWW in their description of a search engine. The key distinction between these two clusters was that those subjects in cluster 1 tended to place some importance on the process of entering key words as a method of searching for information and the consequent matching of these keywords to the available information on the WWW, while those subjects in cluster 2 emphasised searching for information by means of browser-defined categories. In contrast, subjects in clusters 3 and 4 showed more detailed mental models endorsing many more of the salient features including recognising the presence of multiple search engines, the notion of the search engines examining a database generated from the WWW, and the principle of refining search terms to improve the outcome. All subjects in cluster 3 highlighted the presence, or made use of, multiple search engines and nearly half the subjects indicated that different search engines used different algorithms for matching to produce a different set of results. Significantly fewer subjects in cluster 4 endorsed this view. In contrast to subjects in cluster 3, many more subjects in cluster 4 understood that search engines derived their database from the WWW. All the subjects in cluster 4 emphasised the process of matching keywords to the database and nearly half the subjects described a process of ranking the search engine results based on the quality of the match.

3.2 The predictive value of the mental model clusters

Clusters were then related to the degree of prior experience and the task performance. Although the mean trend appeared to indicate that subjects in clusters 3 and 4 showed higher composite computer and WWW experience, no statistically significant differences were found. This suggests that experience in the use of the WWW (and with computers in general) does not necessarily improve the depth of the mental models of search engines produced. Two measures of task performance were used. Firstly, the ultimate success of the task was noted (i.e. task completed successfully with the correct answer being located) and secondly the time taken to complete the task. An analysis of the 'correctness' variable indicated no difference between the clusters and the range of performance (from 50% success rate for subjects in cluster 2 to 65% success rate for subjects in cluster 4) being concomitantly small. In order to analyse the time variable the Kruskal-Wallis test was used assuming that uncompleted tasks took longer than any completed task. The results indicated a significant difference between the clusters ($\chi^2_{(3)} = 12.46$, $p<0.01$). Post hoc tests indicated that those subjects in cluster 2 took significantly longer than subjects in the other three clusters. Subjects in cluster 4 showed the shortest completion times while subjects in clusters 1 and 3 had very similar completion times. However, these differences were not statistically significant. As the time to completion could be considered censored data, the analysis was confirmed by means of a Cox proportional hazards model (Cox, 1972). Results again indicated that subjects in clusters 1, 3 and 4 outperformed subjects in cluster 2, but only subjects in cluster 4 did so significantly.

4 Discussion.

Similarly to Staggers and Norcio (1993) it would appear that the majority of subjects in this sample had mental models that were neither complete nor accurate. A large proportion of the subjects did not even recognise that there were a number of different search engines and that different search engines might provide different sets of search results. Spink et al (2001), who looked specifically at 'relevance feedback' (which is one of the overt expressions of the search engine's ranking mechanism), found that users of their search engine rarely considered this feedback. These results might suggest that search engine interfaces keep many of their relevant operational features hidden from users. The results of the performance measures suggested small performance advantages across the mental model clusters but these performance effects were not widespread. This might suggest that each of these mental model clusters was equally poor (or equally good) at allowing subjects to find relevant information. The only salient feature of the poorest performing cluster was the reliance on the use of search engine categories to find information, rather than the use of keywords. It is not surprising that this more indirect approach takes longer on average than the use of a keyword search. One feature of the two more 'advanced' mental model clusters was the recognition that search engines apply search algorithms to collected databases. On the whole, the lack of significance in the various outcome measures examined, given the relatively large sample size, suggests that whatever advantages there are to be gained from a more comprehensive mental model are relatively small. In addition the lack of significance in the computer and web experience measures appears to suggest that the repeated use of the WWW (and therefore also possibly of search engines) does not necessarily enhance the complexity of users' mental models, over and above the simple understanding of the interface itself. It may be interesting in future studies to probe the user's understanding of where the information comes from and how it is generated. This was not possible in the open-ended tasks set out for the participants in this study. The differences in performance may however be masked by the relatively crude classification of the users' mental models as well as measurement error in the task performance (i.e. some subjects had longer performance times due to WWW download speeds). A more comprehensive battery of search tasks, which control for noise (e.g. WWW download speeds, subject-area knowledge, etc.), may well produce more significant results. It was hoped that this analysis would provide search engine designers with a set of salient design features that would enable users to effectively utilise their search engine. From the results presented here it is only possible to suggest that search engine interface designers make a number of elements clear to users (specifically the nature and composition of the database, and the ranking mechanism of the search results requires some attention)

References

Cox, D.R. (1972). Regression models and life-tables (with discussion). Journal of the Royal Statistical Society, Series B, 34, 187 -220

Fisher, J. (1991). Defining the novice user. Behaviour and Information Technology, 21, 437-441

Hodkinson, C., Kiel, G. & McColl-Kennedy, J.R. (2000). Consumer web search behaviour: diagrammatic illustration of wayfinding on the web. International Journal of Human-Computer Studies, 52, 805-830

Kass, G. V. (1980). An exploratory technique for investigating large quantities of categorical data. Applied Statistics, 29, 119-127.

Milligan, G.W. & Cooper, M.C. (1985). An examination of procedures for determining the number of clusters in a data set. Psychometrika, 50, 159 -179.

Moukdad, H. & Large, A. (2001). User's perceptions of the web as revealed by transaction log analysis. Online Information Review, 25, 349-358

Rouse, W.B. & Morris, M.M. (1986). On looking into the black box: prospects and limits in the search for mental models. Psychological Bulletin, 100, 349-363

Schacter, J., Chung, G.K.W.K. & Dorr, A. (1998). Children's Internet searching and complex problems: performance and process analyses. Journal of the American Society for Information Science, 49, 840-849

Spink, A., Wolfram, D., Jansen, B.J. & Saracevic, T. (2001). Searching the web: the public and their queries. Journal of the American Society for Information Science and Technology, 52, 226-234

Staggers, N. & Norcio, A.F. (1993). Mental models: concepts for human-computer interaction research. International Journal of Man-Machine Studies, 38, 587-605

Sullivan, D. (2002). How search engines work. Retrieved from the WWW, 14 October 2002: http://www.searchenginewatch.com/webmasters/work.html

Ward, J.H. (1963). Hierarchical grouping to optimize an objective function. Journal of the American Statistical Association, 58, 236 -244.

Generation of Cognitive Ergonomic Dynamic Hypertext for E-Learning

Stefan Trausan-Matu
Computer Science Dept.,
Politehnica Univ. & ICIA
313, Spl. Independentei,
Bucharest, Romania
trausan@cs.pub.ro

Alina Marhan, Gheorghe Iosif
Institute of Philosophy and Psychology
13, Calea 13 Septembrie,
Bucharest, Romania
{amarhan, giosif}@racai.ro

Ion Juvina
Institute of Inf. and Computing Sciences, Utrecht University
14, Padualaan,
Utrecht, The Netherlands
ion@cs.uu.nl

Abstract

This paper presents the theoretical and empirical foundations of a project that aims at developing an e-learning system ("SINTEC") used as a complementary tool in the education of students in computer science at the Polytechnic University of Bucharest. User (learner) models are used for the dynamic generation of personalized web pages that conform to the cognitive ergonomic requirement to maintain a holistic view in the learner's mind.

1 Introduction

In the SINTEC project (a former name was MidWeb, see Cristea, Tapus & Trausan-Matu, 2002) we use the results from two previous projects in which members of the present research team were involved. One of the previous projects studied the particularities of teaching and learning in the domain of computer science and their relevance for designing intelligent tutoring systems in this domain. Some results of this previous project that are relevant for the present project are:

- Recommendations for the organization of content to be learned: a) Sequential (prerequisite-new) and hierarchical (topics-units-elements) organization of knowledge; b) Differentiation of knowledge presentation based on the type of knowledge (declarative vs. procedural); c) Diversification of presentation modes (texts, formulas, images, animations) according to students' preferences and styles;
- User modeling tools (knowledge tests, personality questionnaires, and user diagnosis based on navigation patterns);
- Pedagogical strategies: a) Learner-centered approach (personalization); b) Maintaining a coherent organization of content (holistic character) during the whole learning process.

A second previous project (an INCO Copernicus project called LarFLaST, see http://www-it.fmi.uni-sofia.bg/larflast/) has developed a set of web tools that support learning foreign terminology in finance. In this project, the first author experimented with the idea of mapping the semantic network of the domain ontology on a network of personalized web pages automatically generated.

In the following sections our approach in designing an e-learning system is presented. Section 2 presents some of the user characteristics that we want to include in the user model and section 3 describes our approach to personalization of the learner-system interaction in SINTEC.

2 User Modeling

User modeling is a set of techniques used in developing interactive systems. A user model is a representation of those characteristics of the user that are relevant for the interaction. The main advantage of user modeling consists in reducing the cognitive workload of the user by providing information relevant to the task at hand, to the assumed previous knowledge of the user, to user preferences etc.

Based on literature review and on our own previous research, we consider relevant to include in the user model of an e-learning system the following characteristics:

Learner Knowledge. Diagnosis of learner knowledge is essential in an instructional interaction. The system could become aware of the level of student's knowledge by the aid of classical knowledge tests. Moreover, specific knowledge that is intended to be achieved can be assessed directly by questions or puzzles, or indirectly from user history, navigation paths etc. The actual diagnosis is done by comparing student's knowledge with domain/expert knowledge or rule-based inferences (e.g. if you passed a last test then you know more or less the knowledge that was tested). Particularly of interest for instructional systems is also erroneous knowledge, since it can be used to improve the instruction outputs (Shute & Psotka, 1996). Errors are identified by comparing incorrect or unusual answers with a database of errors maintained in the system.

In the context of an e-learning system it is important to consider not only domain knowledge but also knowledge and skills (expertise) about how to navigate through the hypermedia structure of the system. Navigation expertise could be assessed indirectly based on its proved correlations with other indicators. For example, people with high computer use expertise show a higher number of mouse movements and dynamic navigation patterns (Antonietti & Calcaterra, 2001). And novices tend to make use of a linear structure in hypermedia systems when it is made available, while experts tend to navigate non-linearly (Eveland & Dunwoody, 1998).

Cognitive style refers to an expression of psychological differentiation within characteristic modes of information processing. One of the most investigated cognitive styles in relation to instructional hypermedia systems is *Field in/dependence*. This style differentiates between subjects based on their preferred way to process a stimulus in relation to its context. Field independent subjects tend to separate the stimulus from the context while field dependent subjects regard the stimulus as being inherently connected with its context. It has been proved that field dependent users prefer to be guided by the system in navigation and task completion. They use the *home* or *back/forward* keys, analogical maps and video content more frequently than field independent users. Their probability to feel disoriented or get lost in hyperspace is higher than that of field independent users (Chen & Macredie, 2002). Field independent users prefer free navigation and non-linear task structure, want to have control in the interaction and make use more frequently of indexes, search engines, *find* option, and URLs (Chen & Macredie, 2002). These research results are important aids in assessing cognitive style. Usually, field independency could be measured by a dedicated psychological test (Group Embedded Figures Tests). However when the test is not available or not applicable, the system could infer the cognitive stile by looking, for example, at use frequency of indexes, search engines, the *find* option, and URLs (Chen & Macredie, 2002).

In intelligent tutoring systems (e.g. LarFLaST – Trausan-Matu & al., 2002), a user model (or learner model, because the main users are students) is dynamically built and monitored from the answers at drills (Angelova & al., 2002). The learner models typically contain at least lists of known or unknown concepts. Such models are further used to adapt the instruction strategies to learner characteristics. They may contain also general facts about the user (age, gender, test results etc.) and inferred characteristics of the user (for example, the cognitive style or "intelligent" if test scores are above a certain value). Inferences based on this model can be further made automatically, based on rules provided by experts.

3 Dynamical generation of cognitive ergonomic web pages

One of the main goals of a learning process is to progressively construct in the learner's mind a representation of the declarative knowledge available in the domain considered (a "model" of the domain). The skeleton of this model may be considered as a semantic network of the main concepts (the "ontology") of the domain. Our first idea was to map this semantic structure into a net of dynamically generated web pages, each concept having a page and each relation a link. The learning process consists of several cycles of browsing and reading the generated web pages, answering to questions posed by the system, and reading and browsing the newly generated (according to the learner model inferred from the answers to questions) web pages (see Figure 1).

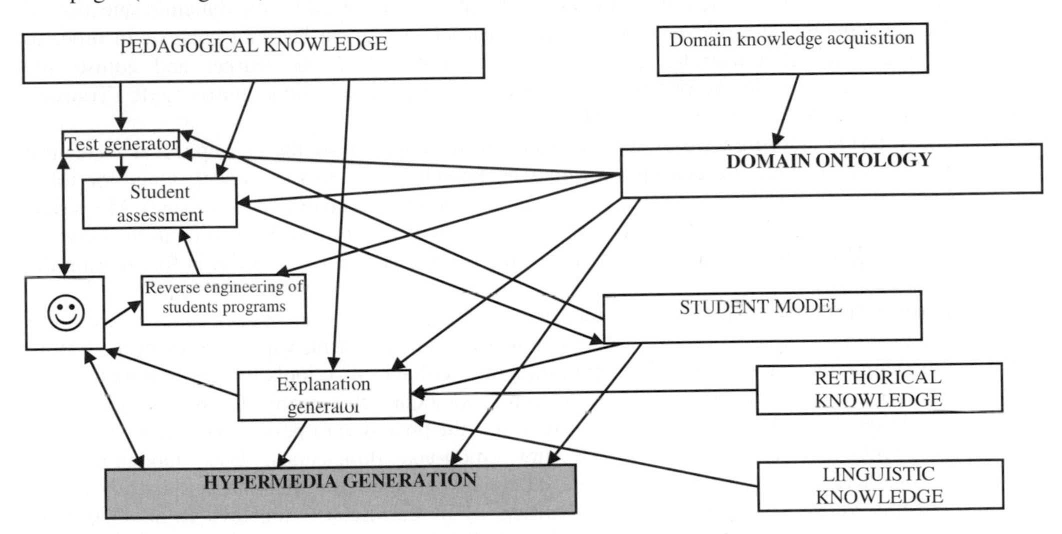

Figure 1. Synopsis of personalization: a student model is developed and upgraded by student assessment; a relevant selection of the domain ontology is provided to the student; and a set of pedagogical knowledge and rules guides the student-system interaction.

Some major problems with e-learning systems that use the web navigation in a collection of web pages are the cognitive overload and the "lost in hyperspace" (Herder & Van Dijk, 2003) situation. Moreover, if the web pages are generated dynamically, their number and hypertext structure might easily get out of any control.

To tackle the complexity of hypertexts on the web and their speed of change, the assuring of a holistic character of the knowledge model to be constructed in the learner's mind is very important. From previous research, we have concluded that one solution might be to design the learning process in order to induce and keep the sense of the whole in the learner's mind (Trausan-Matu & al., 2002). This simplifies the understanding of complex sets of concepts, being in consonance with cognitive-ergonomic and pedagogical principles.

Our approach was to support the construction of the learner's own knowledge (mental) model by incrementally presenting him, (hyper)textually and graphically, a filtered subset of the domain's concept ontology (taxonomy) at each step of the learning process (Trausan-Matu & al., 2002). To assure the holistic, easy understandable character of the structure, instead of general types of semantic nets, we decided to use simple ontologies, which may be seen as semantic nets with a very limited number of relations between concepts (super-concept, sub-concept, and similar-concept). Such ontologies, which may be considered "theories of the domain", are presented to the learner graphically, in a pseudo-tree (taxonomic) structure, generated dynamically in a web page, at each step of the learning process. The tree structure

reflects the super-concept relation (Trausan-Matu & al., 2002). The dimension of this taxonomy is limited in size to avoid cognitive overload. However, a minimal structure is kept for making an understandable whole (a theory for the given stage of the learner).
Some concepts from the ontology web page have links to other concept web pages (also dynamically generated). Each such web page has a textual, specific content, with links to other web pages for concepts, each link corresponding to the super-, sub- or similar- concepts. In these pages also contextual cues (what are the learning goals and contexts) and meta-cognitive knowledge (what s/he does or doesn't know, what is her/his learning style etc.) are provided in order to try to prevent learner's disorientation in the complex hyperspace on the domain ontology.
The learner uses as learning material the network of personalised web pages automatically generated from the filtered ontology. Personalization is achieved by the dynamic selection of the concepts that are relevant for a specific learner in a given learning context. In order to acquire this, the e-learning system maintains a model of the learner and adjusts the pedagogical strategies according to learner's knowledge level and cognitive style (Trausan-Matu & al., 2002, Angelova & al., 2002).
The filtering algorithm we used was based on the selection of the concepts that were in a given relation with the concepts recalled as known in the learner model. Only for these concepts web pages are generated (and links included in the taxonomy web page). Moreover, some concepts will not be filtered if their absence will make unclear some relations between remaining concepts. In addition, the size of the taxonomy is another criterion for filtering the concepts.
Starting from the experience in the two previous projects mentioned above, we have conceived and implemented the SINTEC e-learning environment, which is now in the testing and evaluation phase. This environment integrates collaborative tools for learning on the web (whiteboard, shared display, shared electronic notebook, chat room, discussion forum etc.) with knowledge-based techniques. A main aim is the provision of personalized instruction that takes into account also the social and emotional dimensions. User modelling and personalisation of instruction are achieved by knowledge-based automatic diagnosis of the concepts known or unknown by the learner, as in LarFLaST (Angelova & al., 2002). In addition to LarFLaST, SINTEC keeps a log of all actions performed by the user. This log can be used for several purposes: for abstracting users' profiles, or for studying learners' behavioural patterns.

4 Conclusions

The LarFLaST dynamic generation of web pages was successfully tested in real conditions of learning processes. The students appreciated the graphical, filtered presentation of the pages in the context of the domain's taxonomy. In SINTEC, a system designed for the usage of a large number of students, we want to write and evaluate new algorithms for filtering the concepts and for introducing also some socio-emotional (pedagogical) rules in the personalization of the learning process.
There are other approaches (e.g. deBra & al., 1999) that apply user modelling and dynamically generated hypertext to achieve personalization through filtering of links to be seen. We consider this approach not best suited for e-learning because it is locally oriented, with easy potential of destroying the holistic character of knowledge model to be induced in the learner's mind.
Another approach (Herder & Van Dijk, 2003) intends to use an intermediary agent that performs automatic user diagnosis and then restructures the navigation infrastructure based on this diagnosis. In addition, we use an automatic web page generator that recreates not only the infrastructure of links but also the content of the pages. The use of some metrics, as mentioned in (Herder & Van Dijk, 2003) will be considered in the near future also in SINTEC. We consider that the remote web testing approach (Baravalle & Lanfranchi, 2003) could be integrated with our approach.

Acknowledgments

We wish to thank Herre van Oostendorp, Andres Baravalle, and Jean-Francois Rouet for their helpful comments. Part of this work was performed under the EU COPERNICUS project LARFLAST, FF-POIROT project, ICIA research program, and INFOSOC project SINTEC.

5 References

Angelova, G., Boytcheva, S., Kalaydjiev, Ognyan., Trausan-Matu, St., Nakov, P., & Strupchanska, A. (2002). Adaptivity in a web-based CALL system. In F. van Harmelen (Ed.) *Proceedings of the European Conf. on AI.* (pp. 445-449). Amsterdam: IOS Press.

Antonietti, A., & Calcaterra, A. (2001). *Relationships between thinking style and net surfing style.* XII Conference of the European Society of Cognitive Psychology. Edinburgh.

Baravalle, A., & Lanfranchi, V. (2003). *Remote Web Usability Testing: A Proxy Approach.* This volume.

Chen, S. Y., & Macredie, R.D. (2002). Cognitive Styles and Hypermedia Navigation: Development of a Learning Model. *Journal of the American Society for Information Science and Technology* 53(1), 3-15.

De Bra, P., Brusilovsky, P., & Housen, G. (1999). Adaptive Hypermedia: From Systems to Framework. *ACM Computing Surveys* 31(4).

Cristea, V., Tapus, N., & Trausan-Matu, St. (2002). MidWeb - Towards a Knowledge and Web Based Virtual Environment for Training in High Performance Computing. In *Proceedings of First RoEduNet Conference.* (pp. 21-26). Cluj-Napoca: Mediamira Publishers.

Eveland Jr., W.P., & Dunwoody, S. (1998). Users and navigation patterns of a science World Wide Web site for the public. *Public Understand. Sci.* (7): 285-311.

Herder, E., & Van Dijk, B. (2003). *From Browsing Behaviour to Usability Issues.* This volume.

Shute, V.J., & Psotka, J. (1996). Intelligent tutoring systems: Past, Present and Future, In D. Jonassen (Ed.) *The Handbook of Research for Educational Communications Technology.* New York: Simon & Schuster MacMillan.

Trausan-Matu, St., Maraschi, D., & Cerri, S. (2002). Ontology-Centered Personalized Presentation of Knowledge Extracted From the Web. In S.Cerri, G.Gouarderes (Eds.). *Intelligent Tutoring Systems.* (pp 259-269). Springer, Lecture Notes in Computer Science 2363.

Operational barrier to control human error

F. Vanderhaegen

Laboratoire d'Automatique et de Mécanique Industrielles et Humaines (LAMIH)
UMR 8530 CNRS - Le Mont Houy - 59313 Valenciennes Cedex 9 - FRANCE
frederic.vanderhaegen@univ- valenciennes.fr

Abstract

An operational barrier is a tool to prevent a system from the occurrence of human errors, to recover them or to limit their consequences. This paper presents three examples of the impact of operational barriers on human behaviour. The SPECTRA system is based on a dynamic air traffic control task allocation between humans and an operational barrier. The MPV system proposes five operational barriers to control the diagnosis process of phone troubleshooting. The TRANSPAL includes several barriers to control the traffic flow on a railway process. The results based on a multi-criteria performance analysis, on a barrier activation rate analysis and on a conditional probability analysis are discussed in order to determine the interests and the biases of the use of such operational barriers. The main conclusion leads to design future operational barriers not only as human error tolerant systems but also as violation tolerant systems.

1 Introduction

The main present challenge when designing future human-machine system is to focus on the human behaviour study to define more reliable system taking into account two important compromises:

- Human as a source of reliability. Despite the possibility to design an entire automated system, human operators may be maintained into the control and supervisory loop of the process in order to avoid a loss of human expertise when the automated system failed.
- Human as a source of error. Despite a permanent presence of human operators into the control and supervisory loop of the process, they may be fallible in case of the occurrences of particular control situations, such as overloaded, urgent or degraded situations.

In order to reduce the occurrence of a human error or to limit error propagation or to protect the human operator from technical failures, designers take measures using barriers. Generally, these barriers are provided for enhancing safety and are situational and operational means:

- Situational means such as the definition of safety procedures, of rules, of norms, as the specification of training or situation awareness programs, or as the ergonomic improvement of working positions.
- Operational means such as automated tools to support the decisions of human operators in order to prevent the occurrence of their errors, to regulate their activities or to recover the consequences of their errors.

This paper focuses on the study of the impact of operational barriers to support error recovery and error prevention.

2 Operational barrier and human reliability

Human reliability has to be defined as the probability that humans perform correctly their allocated tasks in given conditions, and that they do not assume any additional tasks which may degrade the human-machine system (Swain and Guttmann, 1983). Human error is the opposite concept but relates much more to the probability that an deviation occurs when performing a task regarding a prescription without considering any possible additional tasks! An internal human error depends on the occurrence on external and internal events that may affect human performance. An internal event relates to human behavioural factors whereas external event depends on other factors such as environmental or technical factors. An internal error appears when such an event occurs combined with an intention to act on the process (Reason, 1990). An action on the process and the presence of an internal error may lead to an external error. A external error differs from an additional task such as a violation: the consequence or the occurrence of an external error is unintentional whereas the occurrence or the consequence of a violation is intentional. The genesis of a internal error, and the occurrence or the consequence of an external error can be controlled either by human operators or by technical defences called barriers. Such barriers can then be designed to support human error prevention or recovery, taking into account each step of a given problem solving from the process state perception to the real action on the process. Hollnagel (1999) defines a barrier as a means that may either (1) prevent an action from being carried out or a situation to occur, or (2) prevent or lessen the severity of negative consequences. The author distinguishes four classes of barriers:

- Material barriers (e.g., grid of protection, safety belt) physically prevent an action from occurring or limit the negative consequences of a situation.
- Functional barriers (e.g., sensors, keys) logically or temporally link the occurrence of actions with events.
- Symbolic barriers (e.g., panels, signals) require interpretation.
- Immaterial barriers (e.g., rules, procedures) are not physically in the work situation.

Material and functional barriers are operational means whereas symbolic and immaterial ones are situational. A barrier can belong to more than one of these classes. Serial defences can be built in order to control human error genesis. For instance, they relate to barriers to support the detection of event that may degrade human performance, barriers to declare human intention at each step of a decision making, and barriers to control the quality of an action. Three examples of operational barrier have been developed at the University of Valenciennes and an experimental protocol was defined for each one in order to study the impact of barriers on the human behaviour:

- The SPECTRA platform (French acronym for Experimental System for Air Traffic Control Task Allocation) consists in sharing dynamically tasks between human air controllers and an operational barrier. SPECTRA involved 15 human controllers in the course of 3 experiments: one without any barrier, one in a explicit mode during which the operational barrier is activated or deactivated by human operator and one in a implicit mode during which the task allocation is controlled by the operational barrier integrating a task demand estimator (Vanderhaegen, 1999). The shared task is the supervision of conflicts between planes, including the conflict detection verifying if planes may transgress minimum separation norms, the conflict solving by sending an adequate order to planes in order to avoid conflicts, and the problem solving in order to verify if a conflict is over and to orient deviated planes to their initial way. The operational barrier is able to detect all conflicts but to solve and control only those between two planes.
- The MPV platform (French acronym for Multi-Point of View System) consists in providing human operator with a series of barriers based on points of view on a same list of failures to

be diagnosed: a structure based viewpoint, a function based viewpoint, a frequency based viewpoint, a similarity based viewpoint and a test based viewpoint. The MPV platform aims at studying the phone troubleshooting diagnosis made jointly by a human operator and these operational barriers. 3 human experts have operated on MPV both with these barriers and without any barrier during 15 scenarios (Jouglet, Piechowiak, Vanderhaegen, 2003). The shared task is a part of a failure diagnosis process: the location of failures.

- The TRANSPAL platform (French acronym for Train Transformation System) proposes a series of barriers that can be removed (Vanderhaegen, Zhang, Polet, Wieringa, 2002). It consists in controlling trains from a departure depot to an arrival one, crossing and stopping at transformation areas on which human operators have to operate on the products located into the trains. 20 human operators have used TRANSPAL during two experiments: one integrating all the operational barriers that are signals at depot, at switching device, at transformation areas and one with the possibility to remove some of them.

3 Presentation of the results

Table 1, Table 2 and Table 3 give the impact of each studied operational barrier on performance criteria, the barrier activation rate and the conditional reliability rate. All rates integrates the average difference between results (e.g. standard deviations). Regarding the results on performance criteria, a positive value is an improvement and a negative one is a degradation. The values of the performance criteria and the activation rate relate to a comparison between the results obtained when the barriers are used with those obtained without any barrier. On Table 3, these values are assessed comparing the results obtained with barrier removal with those obtained without any removal, i.e. with all the proposed barriers. For the SPECTRA results (see Table 1), the safety criterion is based on the conflicts for which aircrafts have transgressed the minimum separation norms and the economical criterion is based on the aircraft consumption. SPECTRA improves the air traffic safety and does not penalize the economical criteria. Despite the importance value of the standard deviation of the safety related results, the implicit mode is more efficient than the explicit mode in term of safety improvement regarding the experiments made without barrier.

Table 1: Results of the SPECTRA experiments

	With explicit / without barrier		With implicit / without barrier
Safety improvement rate	0.46 ± 0.46		0.78 ± 0.25
Economical improvement rate	0.02 ± 0.03		0.01 ± 0.03
Barrier activation rate	0.39 ± 0.16		0.44 ± 0.10
	Without barrier (for 270 conflicts)	Explicit mode (for 165 conflicts)	Implicit mode (for 152 conflicts)
Detection rate	0.89 ± 0.10	0.95 ± 0.08	0.97 ± 0.04
Safety rate upon detected conflicts	0.78 ± 0.10	0.85 ± 0.10	0.94 ± 0.07
Reliability rate	0.70 ± 0.15	0.81 ± 0.11	0.91 ± 0.08

The activation of the operational barriers is more important in implicit mode than in explicit one. The allocation policy managed by the operational barrier forces human controllers to take in charge of fewer tasks in implicit mode. This implicit mode is also the more efficient one in terms of the detection of the conflicts allocated to the human controllers and the safety level of the detected conflicts. Nevertheless, the task demand estimator gave sometimes erroneous assessment because it did not take into account an additional charge related to the supervision by the human operators of the conflicts allocated to the operational barrier event they could not take them back or because human controllers solved conflicts before the operational barrier detect them. In explicit mode, the allocation policy is not homogeneous and some human controllers used the SPECTRA platform as a detection support tool instead of an action support tool allocating conflicts to themselves after having analysed the solution proposed by the operational barrier.

For the MPV results (see Table 2), the quality relates to the diagnosis that can be wrong or correct and the second criterion is the response time of the human operators to make the diagnosis. The barriers of the MPV platform improve the quality of the diagnosis but without facilitating the human behaviour in term of response time. The function based point of view is the more useful barrier. The structure and similarity based points of view facilitate the filtering of some suspected failures. The test point of view is efficient to solve the doubts on these failures. The frequency based barrier is useless: it relates to the occurrence probability of each failure whereas human operators would prefer a point of view on the occurrence probability of a given failure for a given customer.

Table 2: Results of the MPV experiments

Quality improvement rate	0.28 ± 0.05	
Time answer improvement rate	-0.29 ± 0.16	
Function based barrier activation rate	0.47 ± 0.08	
Structure based barrier activation rate	0.20 ± 0.16	
Similarity based barrier activation rate	0.20 ± 0.02	
Test based barrier activation rate	0.13 ± 0.13	
Frequency based barrier activation rate	0.01 ± 0.00	
	Without barrier (for 45 scenarios)	With barrier (for 45 scenarios)
Quality rate	0.64 ± 0.03	0.82 ± 0.03
Unrecovery rate upon correct diagnosis	0.93 ± 0.05	0.84 ± 0.06
Reliability rate	0.60 ± 0.04	0.69 ± 0.07

The analysis of the quality criterion is refined considering the unrecovery rate of the diagnosis process, i.e. assessing the correct diagnosis without any recovery action. The barriers of the MPV platform remain efficient when considering the quality of the human diagnosis made without recovering. Nevertheless, the recovery process is sometimes required and this is also supported by the barriers of MPV to detect the incoherencies on the human reasoning. The motivation to locate failures increases when the human operators use the operational barriers because they prefer to improve the service quality given to customers instead of their response time because this service is free.

For the TRANSPAL results (see Table 3), four criteria are assessed: the traffic safety in terms of collision, derailment and possible accident due to a delay of an announcement of train movement at transformation stations; the quality related to the respect of the timetable; the production related to the percentage of product treated at the stations; and the human workload related to the occupational rate. Despite important differences between the human operators' results, the removal of barriers improves the quality and the workload criteria but degrades the safety and the production ones. Barrier removal can free human operators from constraints that may penalize performance criteria. This confronts the problem of the erroneous perception because some human operators perceive an improvement of all the performance criteria when they remove barriers while it is not the case objectively! The errors on the announcement of train moving at transformation areas have penalized the safety criterion. Considering the rate of the announces that were realized correctly, even if the experiments with barrier removals reduce the omission of these announcements, they increase the amplitude of their delay.

Table 3: Results of the TRANSPAL experiments

Safety improvement rate	-2.73 ± 4.64	
Quality improvement rate	0.27 ± 0.92	
Production improvement rate	-0.11 ± 0.64	
Workload improvement rate	0.21 ± 0.07	
Barrier activation rate	0.73 ± 0.15	
	With barrier removal (for 440 announces)	With all barriers (for 440 announces)
Performed announce rate	0.94 ± 0.06	0.90 ± 0.08
Safety rate upon realized announces	0.85 ± 0.11	0.85 ± 0.11
Reliability rate	0.80 ± 0.10	0.77 ± 0.14

4 Conclusion

The operational barriers studied on the paper are efficient to increase the performance depending on the human behaviour. The SPECTRA's barrier facilitates the workload regulation and reduces the occurrence of unsafe events. The MPV's barriers facilitates both the human error prevention and recovery when diagnosing failures. The activation of the TRANSPAL's barriers is useful to avoid unsafe events and to degrade the production quantity, but penalizes the human workload and the traffic quality. Nevertheless, on the one hand, human operators sometimes reduce the role of an operational barrier by removing or inhibiting or bypassing it in order to free them from constraints. On the other hand, this barrier removal can be erroneous when for example human operators do not perceive correctly these constraints. The studied operational barriers facilitate the control of human error in terms of consequences regarding predefined objectives. However, they do not take into account the possible occurrence of additional tasks that may degrade the human-machine system such as barrier removals that can be assimilated to particular violations. Future operational barriers have then to be not only tolerant to human errors but also to violations and to free the human operator from errors of violation, i.e. errors of perception that motivates them to remove a barrier or errors of removing, considering possible combination of inter-individual preferences and organisational objectives.

References

Hollnagel, E. (1999). *Accident and barriers.* Proceedings of the 7th European Conference on Cognitive Science Approaches to Process Control, Villeneuve d'Ascq, France, pp. 175-180.

Jouglet, D., Piechowiak, S., & Vanderhaegen, F. (2003). A shared workspace to support man-machine reasoning: application to cooperative distant diagnosis. *Cognition, Technology & Work*, Manuscript: 0108, In press.

Reason, J. (1990). Human Error. Cambridge University Press, Cambridge, UK.

Swain, A. D. & Guttmann, H. E. (1983). Handbook of Reliability Analysis with emphasis on Nuclear Plant Applications. NUclear REGulatory Commission, NUREG/CR-1278, Washington D.C.

Vanderhaegen, F. (1999). Cooperative system organization and task allocation: illustration of task allocation in air traffic control. *Le Travail Humain*, 62 (3), 197-222.

Vanderhaegen, F., Polet, P., Zhang, Z., & Wieringa, P. (2002). *Barrier removing study in a railway simulation*. Proceedings of the 6th International Conference on Probabilistic Safety Assessment and Management, June 23-28, San Juan, Puerto Rico, USA.

E-TRACKING: eye tracking analysis in the evaluation of e-learning systems

Daniela Zambarbieri

University of Pavia
Via Ferrata 1 – 27100 Pavia – Italy
dani@unipv.it

Abstract

The analysis of eye movements during visual exploration can provide information on what a subject is observing and for how long. Therefore it can integrate the classical tests in the field of usability evaluation by providing quantitative and objective data.
The aim of this paper is to describe ETRACKING, a research project partly funded by the European Commission under the Information Society Technologies 5th Framework Programme. One of the main objectives of the project is the definition and validation of a quantitative method for the evaluation of the level of functionality, usability and acceptability of e-learning systems. This methodology integrates the classical qualitative tests with the analysis of exploration eye movements.

1 Introduction

Clear vision of an object is guaranteed only when its image falls on the fovea, that is the central part of the retina. In order to explore a visual scene, the eyes have to move in such a way as to bring the image of an object of interest onto the fovea and to maintain it stable. It follows that, when a subject is exploring a visual scene, eye movements can tell us exactly where the subject is looking, what the subject is observing and for how long.
Starting from these statements it becomes immediately obvious that the study of eye movements represents a powerful tool in all studies dealing with visual attention and exploration. A specific situation in which visual exploration is performed is that of a subject interacting with a computer display. The study of how a user interacts with the displayed information has become the major topic in the field of human-computer interaction (Nielsen, 1993).
At present, the usability of an information tool is assessed by means of methods such as expert evaluation, user's interview and task execution (Dix et al., 1993; Garzotto & Matera, 1997; Rubin, 1994) that are mainly qualitative and subjective methods.
The analysis of exploratory eye movements can integrate the classical evaluation tests by providing information that is quantitative and objective.
The aim of this paper is that of describing a research project on the application of eye movement analysis for the evaluation of e-learning systems, the E-TRACKING Project IST-2001-32323.

2 E-learning

E-learning represents one among the different applications of information technology that can provide a very powerful tool in different contexts of everyday life.

E-learning systems are normally provided through hypertext architecture and by using multimedia tools. The development of both multimedia technology and hypertext structures has reached a very high level. In particular, hypertexts can also be adaptive and able to modify themselves according to the user goals and intentions (Brusilovsky, 1996). Therefore from a technological point of view, no limits exists to the implementation of appealing and high-tech systems. But in order to be effective, an e-learning system must be able of actually transferring the knowledge to the user, and this ability is highly influenced by its level of functionality, usability and acceptability.

When a subject is facing an e-learning system, his exploration behaviour can influence significantly the level of learning and in return tell us what would be the best location on the interface to put important information. If a part of the screen is not explored, the information that this part contains are not acquired by the subject.

Detailed analysis of eye movements on a text can provide rich information on both fixations and re-fixations that can be considered as quantitative indicators of the level of difficulty of the text. But of course, also the attention level of the subject must be considered, since fixation duration per se is not enough to guarantee that the subject is actually acquiring information. Therefore, other cognitive parameters are evaluated together with the movements of the eyes.

3 Eye Movements and Recording Techniques

Saccades are the fastest eye movements produced by the human oculomotor system and they are aimed at shifting the subject's line of sight. The need to reduce the execution time of a saccade is due to the lack of vision during movement execution caused by the image of the scene slipping on the retina. Saccades are produced by a pulse activation of the extraocular muscles, and the maintenance of the lateral position is guaranteed by tonic activation of the same muscles. Saccades velocity depends on the amplitude of the movement and can reach a peak value of 500°/s in humans.

Fixation is the time between two successive saccadic eye movements and it represents the period of time during which visual information can be acquired. Visual exploration is therefore composed by alternating saccades and fixations. The two-dimensional movement executed by the eyes during the exploration of a scene is normally called "scanpath" (Norton & Stark, 1971).

Since visual information is acquired by the central nervous system only during fixations whereas saccades are used to shift the gaze from one point to the other, it is reasonable to infer that the analysis of the scanpath is a powerful tool for the study of exploration strategies and the underlying cognitive processes.

Several techniques are currently available to detect the rotation of the eyes in the orbit (Young & Sheena, 1975). The electro-oculographic technique uses surface electrodes placed at the canthi of each eye to measure the variation of the electrical potential produced by rotation of the eyeball. In fact, the eye can be considered as an electrical dipole with the cornea positively charged with respect to the retina.

The infrared technique is based upon the reflection of infrared (IR) light by the area on both sides of the edge between the white sclera and the darker iris. Special glasses support IR light emitting diodes and IR light sensitive phototransistors.

Video-oculography represents an evolution of the previous technique since it still uses IR emitting sources to produce corneal reflexes, but the measure of eye position is made through the processing of the eye image taken by a video camera. In this way eye movement can be detected simultaneously in two dimensions.

Finally the scleral search coil technique is based upon the magnetic induction of a small coil. The induction coil is embedded in a flexible ring of silicone rubber that is placed on the eye like a contact lens. The subject's head is positioned inside a magnetic field.

Each one of these techniques can be characterised according to some parameters such as: precision, stability, linearity, field of measurement, and finally, subject's comfort. For instance, electrooculography is simple to use and has a field of measurement up to ± 40°, but it is noisy and unstable. On the other hand, the search coil technique is very precise, linear and stable, but lens application is not very comfortable for the subject. Therefore the most suitable techniques for recording two dimensional visual scene exploration during interaction with the computer are those relying on video camera based tracking.

4 The E-TRACKING Project

E-TRACKING is a research project partly funded by the European Commission under the Information Society Technologies 5th Framework Programme. The partners of the project are: Università degli Studi di Pavia, Université de Nice Sophia Antipolis, Giunti Ricerca, Fraunhofer Institute for Applied Information Technology, Telefónica Investigación y Desarrollo, ARDEMI.
The main objectives of the project are: i) to define and validate a quantitative method for the evaluation of the level of functionality, usability and acceptability of e-learning systems, ii) to identify guidelines for the development of functional, usable and acceptable e-learning systems, iii) to develop "optimised" prototypes of e-learning systems based on the identified guidelines.
The methodology that is defined and validated during the project integrates the classical qualitative tests with the analysis of exploration eye movements.
Inside the project, acquisition of subject's eye movement during the exploration of the graphical interface is made by using a videooculographic device: EyeGaze, LC Technologies Inc. (Fairfax, Virginia). The EyeGaze (Figure 1) is a "remote" system composed of a video camera mounted below the computer screen, and a video monitor allowing to control that the subject's eye is focused and centered within the camera field.

Figure 1: the EyeGaze system for eye movement recording

The system makes use of the Pupil-Centre/Corneal-Reflection method to determine gaze direction. A small infrared light emitting diode illuminates the eye and generates a very bright reflection on

the cornea. This effect enhances the camera's image of the eye and is used by the image processing algorithms to locate the centre of the pupil. The software calculates the co-ordinates of gaze position based on the relative positions of the pupil centre and cornea reflection within the video image of the eye. The power of the light emitting diode is low enough not to give discomfort to the subject such as smarting eyes and weeping. The basic feature of the system is that no contact exists between the recording device and the subject and therefore no discomfort is caused even by very long acquisition sessions, and subject's behaviour is not influenced neither modified by the presence of the device.
The acquisition software developed within the E-TRACKING project records both eye movements and any interaction of the subject with the computer, such as mouse click and navigation. Then, the software reconstructs the 2D trajectory of the eyes on the display (scanpath). For each page it is possible to define regions of interest, that is specific parts of the document that can provide detailed information on the allocation and shift of subject's visual attention. Then the SW computes a number of quantitative parameters such as the number of accesses to each region, the duration of each access, the sequence of accesses among the regions, and so on.
Figure 2 gives an example of a scanpath (red trace) of a subject exploring the home page of the E-TRACKING project Web site.

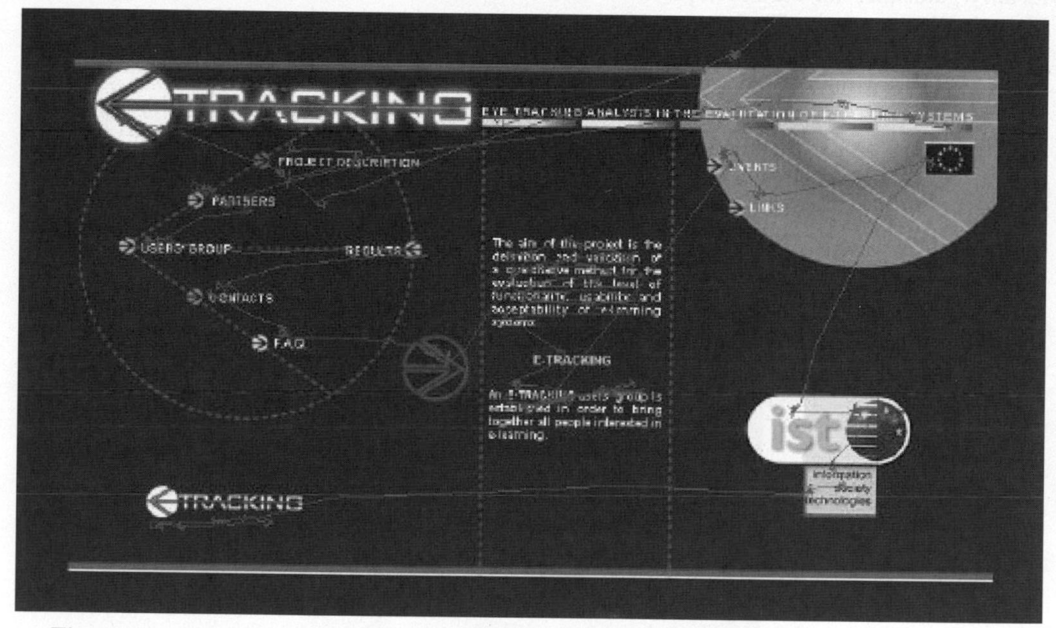

Figure 2: an example of scanpath of a subject exploring the E-TRACKING project web site

5 Conclusions

Eye movements analysis during visual exploration of a graphical interface represents a powerful tool to investigate subject behaviour. Therefore they can be used to integrate the classical usability evaluation methods.
This method can be applied to any situation that involve the interaction of a subject with the computer such as e-learning, web navigation, use of application software and so on.
The quantitative data provided by eye movements analysis will be interpreted in relation to the specific content in which the graphical interface is applied.

6 References

Brusilovsky, P. (1996). Methods and techniques of adaptive hypermedia. *User Modelling and User-Adapted Interaction* , 6 (2-3), 87-129.
Dix, A., Finlay, J., Abowd, G., & Beale R. (1993). Human-computer iteraction. Pentice Hall
Garzotto, F., & Matera M. (1997). A systematic method for hypermedia usability inspection. *The New Review of Hypermedia and Multimedia*, 3.
Nielsen, J. (1993). Usability engineering. San Francisco: Morgan Kaufman Publ.
Norton, D., & Stark L. (1971) Scanpaths in saccadic eye movements while viewing and recognizing patterns. *Vision Res.* 11, 929-942
Rubin, J. (1994) Handbook of usability testing: how to plan, design, and conduct effective tests. New York: John Wiley & Sons.
Young, L.R., & Sheena, D. (1975) Survey of eye-movement recording methods. *Behav. Res. Met. & Instrum.* 7, 397-429.

7 Acknowledgment

This work has been undertaken in the framework of the E-TRACKING project (IST-2001-32323) which is partly funded by the European Commission under the Information Society Technologies 5th Framework Programme. The partners of the project are: Università degli Studi di Pavia, Université de Nice Sophia Antipolis, Giunti Ricerca, Fraunhofer Institute for Applied Information Technology, Telefónica Investigación y Desarrollo, ARDEMI.

A Systematic Barrier Removal Methodology: Application for Transportation System

Z. Zhang, F. Vanderhaegen, P. Polet

Laboratoire d'Automatique, de Mécanique et d'Informatique industrielles et Humaines (LAMIH), UMR 8530 CNRS
University of Valenciennes - Le Mont Houy
59313 Valenciennes Cedex 9 – France
{zzhang; vanderhaegen; ppolet}@univ-valenciennes.fr

Abstract

The paper presents a systematic Barrier Removal (BR) methodology with emphasis on the urban-guided traffic applications. A BR is presented as a particular human violation, and then Erroneous BR (EBR) is discussed. By integrating the EBR concept, the proposed methodology includes six main steps. Application in Urban Guided Transport Management Systems (UGTMS) is illustrated.

1 Violation

Human errors have received close attention from psychologists and others for well over a century. The study of violations is still at an early stage (Reason, 1995). The importance of violations in industrial safety increased after the Chernobyl accident. According Reason (1987, quoted by Hudson et al., 1998), of the 7 human actions that led directly to the accident, 5 were deliberate deviations from written rules and instructions rather than slips, lapses or mistakes. It's the similar situation in the railway industry: an investigation of railway accidents in Britain between 1989 and 1992 shown that violations play a considerable role in accidents to staff, e.g. personal injury and fatalities (Free, 1994). Although the violations have been mentioned in a wide variety of contexts including car driving, plane piloting, nuclear industry (waiver review, Zhang et al. 2002b), computer programming, bureaucratic environments (Besnard, 2002), the research on violations is still insignificant in comparison with what is known about slips, lapses, and rule & knowledge-based mistakes (Villera, 1999).

Violations can be defined as any deliberate deviations from the rules, procedures, instructions or regulations introduced for the safe or efficient operation and maintenance of equipment (Hudson et al., 1998; Mason, 1997). The Health and Safety Executive (HSE, 1995) classified violations under four categories: routine, situational, exceptional and optimising; Mason defines the factors that influence the likelihood of violations as: Direct Motivators and Behavior Modifiers; Besnard et al. (2002) demonstrate by two examples that violations can generate very different situations (contributing to or impairing system safety). Whatever the merits of the above approaches of safety violation, there are a number of key observations that can be made of all of them. Table 1 tries to compare the different violation methodologies by providing their features.

The researches on Barrier Removal (Vanderhaegen et al./Polet et al./ Zhang et al. 2002) focus relatively on the violation contributing to system safety. The present study is designed to explore the prospective analysis of these forms of violation, called Barrier Removal (BR) with emphasis on the urban-guided traffic applications.

Table 1: Comparison of violation-related methodologies[1]

Methodologies	Violation forms	Comments
Mason, 1997	Routine, Situational, Exceptional, Optimizing	Pertinent framework of violation classification; not always easy to list exhaustively all items in the questionnaire
Hudson et al. 1998	Routine, Situational, Exceptional, Optimizing	It depicts exhaustively all violation forms
Besnard et al. 2002	Exceptional	Dealing with exceptional violations; Well illustration of mutual status of violation
Vanderhaegen et al. 2002	Routine, Situational, Optimizing	Violation as removal of barrier in terms of benefit, cost & potential deficit

2 A systematic Barrier Removal methodology

2.1 Erroneous Barrier Removal (EBR)

The barrier concept is common in the field of nuclear safety, but also in other areas (Harms-Ringdahl, 1998). An AEB model was developed for barrier analysis (Svenson, 1991), and an extensive study with this model has been conducted to analyze the reliability of the existing barrier functions in a given system. But the common criteria for determining when adequate risk reduction has been achieved need to be defined (*Kecklund et al, 1996*). There are several another studies on the barrier analysis, e.g., barrier analysis related to the MORT (*DOE, 1992*), analysis and explanation of organizational accidents (*Reason, 1995*), etc. Hollnagel (1999a/b) distinguished four classes of barriers in term of prevention (occurrence of hazardous event) and protection (severity of hazardous event). Based on this barrier classification, a retrospective application (accident analysis) has been conducted.

The motivation to remove a barrier, i.e. to make a violation, can be erroneous, e.g. difference between the perception of the benefit, cost & potential deficit and the real benefit, real cost & real potential deficit (Zhang et al. 2003). Therefore, there is an error of BR or difference between the viewpoint of several references such as designers and users. It is named as EBR. Violations do not solely lead to undesired events. They are actions that intentionally break procedures (Parker *et al.*, 1995; Reason, 1987), e.g. aiming at easing the execution of a given task. When they are coupled with a valid mental model, they can ensure or even increase the safety level of a system (Besnard, 2002). Bearing in mind this fact, the EBR could be distinguished from the correct BR through comparing the performance variation (between prior-removal and posterior-removal) of human operator. The EBR has several basic features:

- Removability. The EBR exists so long as the barrier is removable and removed.
- Variation between the subjective judgment and the objective sources. If there is an improvement according to subjective judgment, the operator may accept to remove a barrier. But facing this removed barrier, when there is a degradation regarding the objective sources, it is considered as an EBR.
- EBR should be identified in terms of different performance criteria.

And the causes of EBR can be an error of evaluation and/or an error of perception of human operator. By integrating the EBR concept, a systematic Barrier Removal methodology is discussed in following section.

[1] Intended sabotage is not a subject of the common HRA methods. It is always assumed that the operator is willing and tries to avert damage from the system.

2.2 Main steps of the systematic BR methodology

The proposed systematic BR methodology encompasses several main steps (Figure 1). The 1st step is the **Preliminary Risk Analysis (PRA)**. It aims at analyzing and then defining the functions, specification associated to the given complex system. Then, **functional analysis** is implemented in the 2nd step. This can be conducted mainly based on the principles proposed in APRECIH model (Vanderhaegen, 2001). Identification of **associated barriers** from function list is the 3rd step. A practicable and operable barrier definition and classification should be then established. Barrier classification mentioned in subsection 2.1 may be referred. Noted that barriers could be identified through analyzing the function list. It could be, if possible, identified from the procedural and contextual analysis as well. The procedure & context in normal operation, near miss, incidents, breakdowns or accidents situations etc. can be involved.

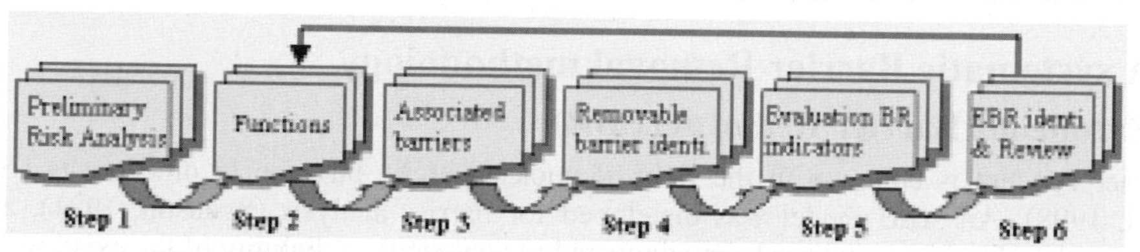

Figure 1: Main step illustration of systematic BR methodology

The 4th step is **identification of removable barriers,** i.e. filtering from associated barriers of the step 3. The theory of safety-related violations of system barrier (*Polet et al., 2002*) could be taken into account. A barrier removal is associated to an operational risk which constitutes a combination of cost: the cost of removing the barrier, the benefit (negative cost) immediately after removing the barrier, and the potential deficit (critical cost) due to the exposure to hazardous conditions that are created after the barrier has been removed.

Based on the identification of removable barriers, **evaluation of indicators on BR** is then implemented in the 5th step. The similarity between the removals of barrier for a given complex system can be found out and then memorized in a connectionist network in terms of mono-performance mode and multi-performance mode (Zhang et al. 2002a/b). Based on BR indicator data in terms of different performance modes and the relative statistics, probabilistic prediction of the removal of a changed/new barrier on a simulator is accomplished (Zhang et al. 2003).

Moreover, the evaluation of indicators on BR can be implemented by different groups, e.g. EBR analyst, operator group, designer group etc. EBR identification in a group e.g. operator group, Zhang et al. (2003) gives an example. Note that the above-mentioned predictive analysis model is usable in each group's evaluation. The comparison between the evaluations of different groups allows us to verify the identified result. **EBR identification** is undertaken such that those barriers that were often erroneously removed can be taken into account in the next function analysis (Cf. Figure 1). Here, the next function can be a changed/new function mode for the designers; it can be also for the utilities/operators. In the 6th step, the internal **review** of complete study by a group separate from the actual performers. The process between step 5& 6 may be repeated if necessary.

3 Illustration of a field study

In order to study the human reliability in the safety and conformity assessment on the transportation system, the proposed systemic BR methodology is used for human factor impact analysis on Functional Requirements Specification (FRS) of an European transport project. It names UGTMS founded by the European Commission (EC) under the transport R&D Programme of the 5th framework Programme (*contract n° GRD2-2000-30090*) (European Commission, 2002).

Table 2: Partial results derived from the step 2 & 4 (F: functional; I: Immaterial; M: Material)

FRS in D1	Removable Barriers
1.3. Manual mode	Speed limit system (F)
1.6. Operation principles	Temporary speed restriction rule (I)
……	……
2.1.1.1. Modify train pathway	Timetable (I)
2.1.4. Manage Traffic	Respecting the direction of movement (I)
2.1.5. Provide Interface with OCC MMI	Symbolic information for train driver on the interface (I)
	Procedure of movement supervision (I)
2.3.7.1. Enforce Procedural Sequence when intentionally passing a Signal at Danger	Train entering a section, which could be occupied, is not allowed to use the information intended for the train occupying the track ahead (I)
	Train speed supervision sys (M)
2.3.7.2. Enforce Procedural Sequence when accidentally passing a Signal at Danger	Train speed supervision sys (M)
2.4.1. Authorize Movements	Anti-collision (F)
	Safe headway (F)
	Route setting (F)
	……

Several matrixes are established with the proposed methodology: the 1st matrix for PRA/FRS Functions; the 2nd matrix for Functions/Associated Barriers; the 3rd matrix for Barrier/Removable Barriers; the 4th matrix of evaluation on removal indicators (includes 3 sub-matrix: Analysts preliminary viewpoint, UGTMS consortium, Operators). Table 2 shows an example of partial results derived from the step 2 & 4. D1 in Table 2 means a Deliverable identification number for the project.

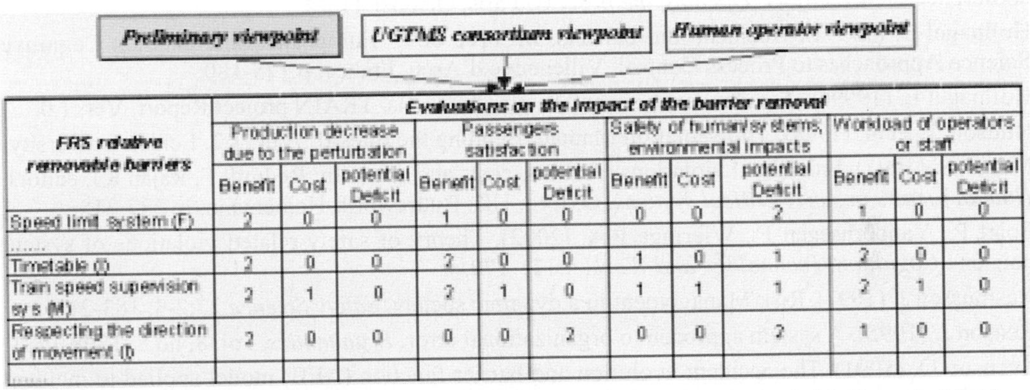

Preliminary viewpoint — *UGTMS consortium viewpoint* — *Human operator viewpoint*

FRS relative removable barriers	Evaluations on the impact of the barrier removal											
	Production decrease due to the perturbation			Passengers satisfaction			Safety of human/systems, environmental impacts			Workload of operators or staff		
	Benefit	Cost	potential Deficit	Benefit	Cost	potential Deficit	Benefit	Cost	potential Deficit	Benefit	Cost	potential Deficit
Speed limit system (F)	2	0	0	1	0	0	0	0	2	1	0	0
……												
Timetable (I)	2	0	0	2	0	0	1	0	1	2	0	0
Train speed supervision sys (M)	2	1	0	2	1	0	1	1	1	2	1	0
Respecting the direction of movement (I)	2	0	0	0	0	2	0	0	2	1	0	0

Figure 2: An example of evaluation between different groups (0: Low; 1: Medium; 2: High)

Regarding to the step 5 (Cf. Figure 1), an example of analysts' preliminary evaluation is illustrated in Figure 2. During the evaluation, not only the safety-based criterion is used, other criteria should be introduced as well. For instance, productivity/economic based criteria (e.g. cost for railway company due to the perturbation), or quality based criteria (e.g. implication for passengers satisfaction), or workload related criteria (e.g. workload of operators or staff) are introduced into evaluation on BR indicators. The research on UGTMS application is still ongoing. The ultimate goal is, in each period of defense-in-depth, to reduce the probability of occurrence of EBR by reducing the benefit of removal of a barrier, increasing the cost and the potential deficit; by making the human operator's perception of the benefit low, the cost high and the potential deficit high, e.g. improving the human-machine interface, reducing the perception error; by surveillance or monitoring of states of the barriers in terms of benefit, cost and potential deficit; and by protection and mitigation measures for EBR.

4 Conclusions

A systematic BR methodology is presented with emphasis on the railway industry, particularly the urban-guided traffic applications. Several matrixes are finalized along with the analysis flowchart of the methodology. The feasibility is illustrated with UGTMS project supported by the EC.

The EBR comes from the error of evaluation and/or error of perception of BR. EBR analysis can be undertaken for a given complex system such that those barriers that were often removed erroneously can be taken into account. In the future, fuzzy logic can be performed for the treatment of different group's evaluations on impacts of BR. Data collection allows quantifying a probability of EBR based on the frequency of occurrence and some indicators on EBR. It aims at trying to integrate probability of EBR into the general Probabilistic Hazards Analysis (PHA) in railway system, guiding the specification of violation prevention/protection support tools and finally reaching the design of "violation-free" system.

References

European Commission (2002). Deliverable 1, First report for a preliminary definition of UGTMS. GROUTH GRD2-2000-30090, Ver. 1.0, 28 June.

Free R. (1994). The role of procedural violations in railway accidents, Ph.D. Thesis, University of Manchester.

Harms-Ringdahl L. (1998). Proceedings Society for Risk Analysis - Europe. The 1998 annual conference: *Risk analysis: Opening the process*. Paris, October 11-14.

Hollnagel E. (1999a). Accident and barriers. In: Proc of 7th European Conference on Cognitive Science Approaches to Process Control, Villeneuve d'Ascq, France, p.175-180.

Hollnagel E. (1999b). Accident analysis and barrier functions. TRAIN project Report, Ver. 1.0.

Hudson, P. et al. (1998). The Violation Manual, Bending the rules II , Ver. 1.2, Leiden University.

Mason S. (1997). Procedural violations – causes, costs and cures. In: Redmill F, Rajan KJ, editors. *Human factors in safety critical systems*. Oxford, UK: Buttreworth-Heinemann, p. 287-318.

Polet P., Vanderhaegen F., Wieringa P.A. (2002). Theory of safety-related violations of system barriers. *Cognition, Technology and Work*, 4:171-179.

Rasmussen J. (1997). Risk Management in a dynamic society. *Safety Science* 27:2-3, 183-213.

Reason J. (1995). A system approach to organizational error. *Ergonomics*, vol38, no.8, 1708-1721.

Svenson O. (1991). The accident evolution and barrier function (AEB) model applied to incident analysis in the processing industries. *Risk Analysis* Vol. 11, No 3, (499-507)

U.S. Department of Energy (1992). DOE-NE-STD-1004-92, Root Cause Analysis Guidance.

Vanderhaegen F. (2001). A non-probabilistic prospective and retrospective human reliability analysis method–application to railway system. *RESS* 71(1):1-13.

Vanderhaegen F., Polet P., Zhang Z., Wieringa P. A. (2002). Barrier removal study in railway simulation, *PSAM 6*, Puerto Rico, USA, June.

Villera S., Bowersb J., Roddena T. (1999). Human factors in requirements engineering, *Interacting with Computers* 11, 665–698.

Zhang Z., Vanderhaegen F. (2002a). A method integrating Self-Organizing Maps to predict the probability of Barrier Removal. Warren Neel Conference on Data Mining, Knoxville, USA, June.

Zhang Z., Polet P., Vanderhaegen F., Millot P. (2002b). Towards a method to analyze the problematic level of Barrier Crossing. In: Proc of lm13/Esrel2002, Lyon, France. p. 71-80.

Zhang Z., Polet P., Vanderhaegen F. (2003). Towards a qualitative predictive model of violation in transportation industry, Esrel2003, Maastricht, The Netherlands, June.

Section 4

On-line Communities, Collaboration and Knowledge

Web Interfaces between users and a centralized MAS for the technological watch

Emmanuel ADAM, Mélanie LECOMTE

LAMIH - UMR CNRS 8530
Université de Valenciennes et du Hainaut Cambrésis,
F - 59313 Valenciennes Cedex 9
emmanuel.adam@univ-valenciennes.fr
melanie_lecomte@hotmail.com

Abstract

Web retrieval becomes more and more important for the knowledge management area, and we think that multiagent systems are a good answer to this problem. We propose, in this paper, a centralized Information multiagent system to help actors of technological watch cells (called CIMASTEWA). This system is an extension of a previous distributed multiagent system and is set-up in within a n-tiers architecture. In order to securitize the search, notably concerning the survey by spies, we have added particular search strategics. And, in order to encourage the users to share their results, we propose to organize groups of users according to their center of interests. This functionality is one of our main perspectives concerning this information multiagent system that has been developed to answer to demands from technological watch cells.

1 Introduction

The boom in Internet technology and company networks has contributed to completely changing a good number of habits, which have been well established in companies for several decades. Enterprises are now set off on a race for information: being the first to find the good information becomes an essential objective to competitive enterprises. So, it is important to own a fast tool of information search and distribution. Admittedly, tools have already been suggested such as: search engines, meta-engines, tools for automatic search (which search at determined regular intervals), and, more recently, information agents, capable of searching information, of sorting and of filtering it.
The problem of these solutions is that they do not taking the human factors, such as the notion of the group or even the man-machine co-operation into account. We have previously developed a method to design multiagent systems for helping the actors of complex administrative systems (Adam, 1999). So we have reused this method to develop a multiagent system for helping cooperative information management within a team of technological watch. A MultiAgent System (a MAS) was firstly designed; it was composed of a set of information mutliagent system called CIASTEWA (for Cooperative Information Agents System for TEchnological WAtch). Each of these CIASTEWA was dedicated to a watchman and was located on his/her computer. This distributed architecture allowed a larger flexibility at system level than a centralized system, but

was not enough securitized for a use in a technological watch context. So we propose a more centralized system, taking into account both the cooperation between actors and the needs of security in information search. Indeed, requests made by the agents to the Internet can be tracked by hackers or survey systems (like Echelon, Carnivore for instance). That is why we have defined some information search strategies in order to foil the spies.
This application has been developed in a small data-processing company, for technological watch cells, and we are currently adding more complex search strategies, which use the dynamic characteristics of the MAS, for our own needs. This article presents firstly the context of our approach, which is the technological watch context. Next, the core of the information MAS that we have designed during a previous project and a presentation of our n-tier application for securitized information search are proposed. In a third part, we present the set of interfaces that allow users to interact with this distributed system, and finally, our current and future works regarding this system.

2 A multi-agent System for the technological watch

An information multi-agent system (IMAS) is composed of information agents that search, on the basis of requests that are sent to them (directly or indirectly through of a database) information from databases (local or distributed) or from the Internet sites.
Generally, activities of these agents are coordinated by coordinator agents, which own knowledge on them (such as address or their search domain for instance). Coordinator agents can: send requests to information agents either in targeted way (if they own knowledge on information agents' competence) or in a general way (by broadcast techniques); gather information found by information agents, to check it, or to filter it.
Most of information multi-agents systems are directly in touch with the user, upstream (to integrate new requests) and/or downstream (during the presentation of search results). In order to have an interface reactive and distributed (to be accessible for all users), some IMAS propose the use of interface agents acting as an interface between the users and the system.
So, in order to propose a IMAS to help actors of technological watch cells, which takes into account the notion of group, we have designed a MAS, called CIASCOTEWA (for Cooperative Information Agents System for COoperative TEchnological WAtch), which is composed of several IMAS, called CIASTEWA whose the architecture is shown in figure 1.
Each of our CIASTEWA is made up of:

- a local database that contains the requests, their results and information on the user,
- an interface agent that assists the users to express their requests and allows to them the interaction with the results, provided by information agents, or with the other users of the group,
- a coordinator agent that has the task to coordinate actions of the other agents,
- an information responsible agent, which distributes the request that are recorded in the local database to the information agents according to a search strategy.
- information agents that have to find information on the Internet.

Each CIASCOTEWA helps the user, to who it is dedicated, to search relevant information and to communicate with other actors. In order to maintain or create the feeling of community or group in the actors, which is often forgotten with the use of new technologies (the individuals are isolated at their workstation), we have proposed to develop self-organizing capacities in order to generate communities of CIASTEWA, which have to answer at the same kind of requests. This reorganization is indicated to users in order to encourage them to cooperate, if they want to, with other users having the same centers of interests.

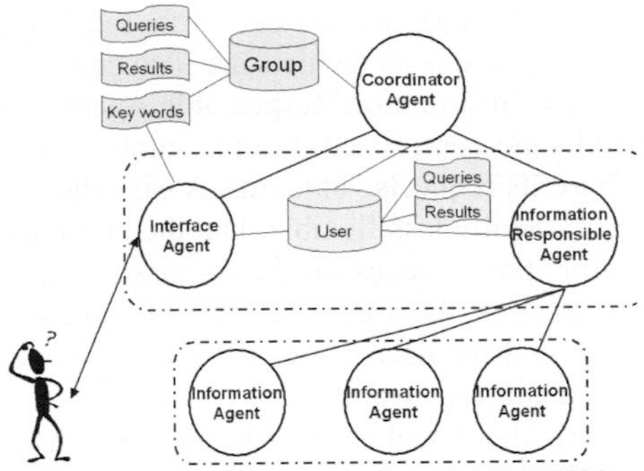

Figure 1. Architecture of a CIASTEWA

In fact, works on the generation or identification of communities in IMAS are appeared recently, like in (Helmy et al., 2002) where the agentification of Web server needs the creation of agents communities. And, we think that "finding the right person" who could know where are located the answers is the best way to find the good information. Some works are carried out in this direction by (Jie, 2000) and (Kanawati, 2002). For example, in a large laboratory, it is frequent that searchers have momentary the same center of interest without knowing it. In our system, they are informed of this and so encouraged to exchange their information.

However, information search, in technological watch domain, implies to develop securitized search strategies, such as adding 'noise' around the requests and centralizing the databases in a main database on a server with access control.

3 A centralized MAS for the technological watch

In order to answer to particular needs of actors of technological watch cells, we have proposed to move our system from a totally distributed architecture to a more centralized one. And we designed a MAS, called CIMASTEWA (for Centralized IMAS for TEchnological WAtch). The data relating to the users, their queries and the results gathered are then recorded in a centralized database (fig. 2). In this new architecture, we have only one IMAS, which accesses both to the database and to the Internet. In fact, this IMAS is a CIASTEWA whose the interface agent has been suppressed.

The users have access to this IMAS through their browsers by calling dynamic web pages (jsp or asp pages for example) and dynamic web components (like java bean for instance). They record their queries in the centralized database that the coordinator agent checks periodically to detect if a request has to be done. In this case, the request is sent to the information responsible agent, which distributes it to the information agents according to a search strategy. The information agents send back their results to the information responsible agent, which filters and merges them to record them into the database.

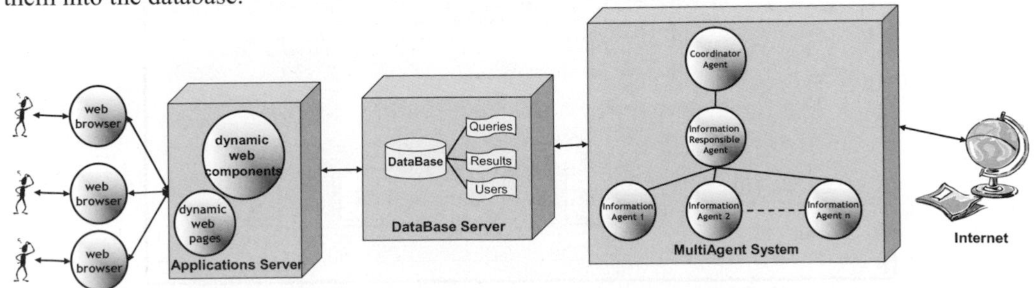

Figure 2. Architecture of the CIMASTEWA

Regarding the notion of security, if a user wishes one of its requests to be scrambled, this one is then decomposed into elementary requests, which are launched with false requests. The real results are then recomposed by the Information Responsible Agent with a particular process.
Since all the requests and results are centralized into the same database, it is possible to a user to choose if he / she wants to share its requests. This makes easy the communication of information between users. This system is not really less flexible than the previous one regarding the adding of user; we just have to give to him / her an access to the database.
According to the user needs, it could be interesting to have different search strategies for the Information Agents. For example: if the user wants to have results rapidly, an invitation to tender can be launch to these agents; if the user wants to have all the possible results, then the requests are sent to all of them by a broadcast technique. On the other hand, according to the queries, different search strategies can be also chosen. For example: if the words used in the request correspond to specialties of information agents, then the search is organized by specialty; if some requests are subsets of other requests, then information agents are structured hierarchically.

4 Application

The CIMASTEWA has been developed, mainly in java, in a small data-processing company to answer to the demand of technological watch cells. The coordinator agent, the information responsible agent and the information agents have been developed in java, using the Magique multiagent platform, which is particularly well adapted to the hierarchical structure of our MAS. Currently, the search strategy used by our information agent is a broadcast strategy: the information responsible agent sends the request that has to be launched to information agents, which are each dedicated to a search engine. So our system is able to search information on 5 classical search engines and two search engines that are specialized in magazines and documents.
The application server that we use is Tomcat, two databases have been developed (in Progress® and in Ms Access®). The MAS has not particular location, it only needs to have access to the database and to the Internet, in order to execute the searches.

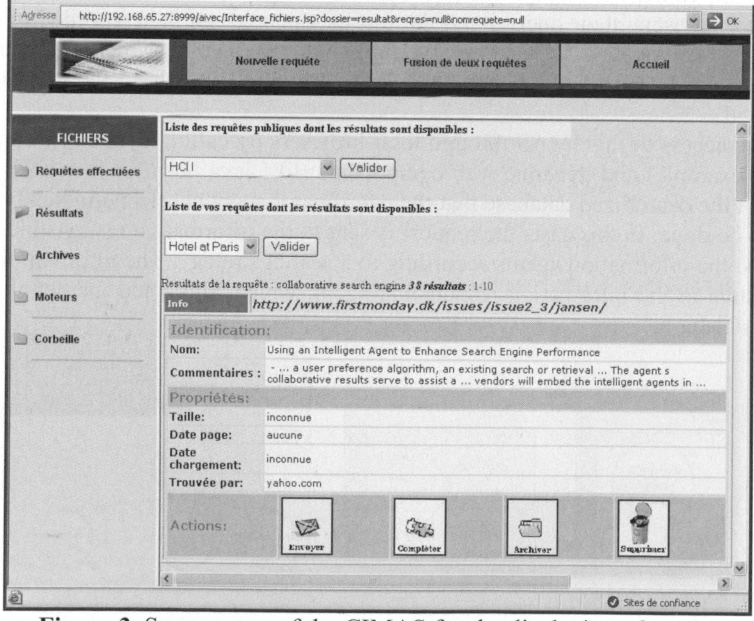

Figure 3. Screen copy of the CIMAS for the displaying of results

The human machine interfaces have been developed in web pages (using jsp, javascript and css processing language). Through these pages, the users can: add, modify, remove requests; consult previous results; choose to share or not theirs requests or results; separate a request in two parts; merge results of two requests (in fact, this duplication is the first strategy of scrambling).
The figure 3 presents a screen copy of the web page allowing to the user the consulting of personal or public results: to send them to another actor (through the user mail service), to add some information, to archive them and/or to delete them. The results proposed by the CIMASTEWA of the figure 3 come from a request on 'collaborative search engine'.
The functionality concerning the identification of interest communities of the users has not been yet added to this new IMAS. Currently, we are developing the functionality of self-organizing in the CIMASTEWA, which will allow it to choose the best search strategy according to the users or to the requests.

5 Conclusion

We have developed a Centralized IMAS to help actors of technological watch cells, or searchers in a laboratory, that have to search information, to share it in a cooperative aim. Our system proposes a search strategy in order to hide the actual searches asked by the users. Sharing the queries and their results allows to the user to keep in mind the notion of group, which is important, and allows them to win time in their searches. This system is currently used in our laboratory by a few searchers of our team. We plan to set-up this system in a larger way in our laboratory in short term. Indeed, we carry out researches and developments: to self-organize the information agents in order to have the more relevant results; and to identify automatically communities of users. So, we could measure impact of our MAS on the behaviour of the searchers.

References

Adam, E. (1999), "Specifications of intelligent human-machine interfaces for helping cooperation in human-organizations", In H.J. Bullinger, J. Ziegler (Ed.), Ergonomics and user interfaces, vol. 1, London: Lawrence Erlbaum Associates, 311-315.

Helmy T., Amamiya S., Mine T., Amamiya M. (2002). "An Agent-Oriented Personalized Web Searching System" , in Giorgini P., Lespérance Y., Wagner G., Yu E. (Ed.), Proceedings of the Fourth International Bi-Conference Workshop on Agent-Oriented Information Systems (AOIS-2002 at AAMAS*02), Bologna (Italy), July 16, 2002.

Jie M., Karlapalem K., Lochovsky F. (2000). "A Multi-agent framework for expertise location", in Wagner G., Lesperance Y. and Yu. E. (Ed.), *Agent-Oriented Information Systems 2000*, iCue Publishing, Berlin, June 2000.

Jonnequin, L., Adam, E., Kolski, C., Mandiau, R. (2002). "Co-operative agent design in a technological watch context". In C. Kolski, J. Vanderdonckt (Ed.), *Computer-Aided Design of User Interfaces III*. (pp. 357-366). Dordrecht: Kluwer Academic Publishers.

Kanawati R., Malek M. (2002). "A Multiagent for collaborative Bookmarking", in Giorgini P., Lespérance Y., Wagner G., Yu E. (Ed.), *Proceedings of the Fourth International Bi-Conference Workshop on Agent-Oriented Information Systems (AOIS-2002 at AAMAS*02)*, Bologna (Italy), July 16, 2002.

Acknowledgments

The authors thank SOLVAY S.A. and ALCOR SPRL for their support, and also the Region Nord-Pas de Calais and the FEDER supporting our current reflections concerning new interactions between people and organizations (Project TACT NIPO).

Semantically Enhanced Hypermedia: A First Step

I. Alfaro, M. Zancanaro, A. Cappelletti, M. Nardon, A. Guerzoni

ITC-irst
Via Sommarive 18, Povo – TN 38050, Italy
{alfaro, zancana, cappelle, nardon, annaguer}@itc.it

Abstract

The paper introduces a framework to automatically build hypermedia links from a semantically annotated repository of multimedia data. The system architecture is based on a relational database accessible through XML queries on an HTTP connection.
A shallow semantic representation is encoded as a set of key words that correspond to entities in the domain. This representation is used to annotate the texts and the images in the database. Communicative strategies are then employed by the graphical interface to dynamically produce links to the data.

1 Introduction

Typically, hypermedia systems are manually prepared by content experts and technical editors. The preparation phase consists of—besides preparing the content material—deciding how different pieces of information should be linked together. These links are then explicitly inserted in the texts or multimedia content. One of the main drawbacks with this approach is that as the hypermedia grows, it becomes increasingly difficult to add new material as this needs to be manually linked to that already existing in the system. In this paper, we describe a framework for creating a multimedia database semantically annotated that can be used to automatically build hypermedia links. The main disadvantage of our approach is the need for explicitly annotated data with respect to a semantic model. The research trend on semantic web (Berners-Lee et al., 2001), however, has the potential to provide a suitable infrastructure to create these kinds of repositories.
In the following section, we will describe the system architecture and we will briefly discuss the semantic model employed together with the annotation format used. In section 3, we will present a walk-through of a hypermedia developed on a museum setting. Section 4 will discuss the set of semantic strategies used for automatically building hypermedia from a semantic multimedia database.

2 Semantically Annotated Multimedia Database

The Multimedia Database (MMDB) allows the storing of a repository of multimedia information (i.e. text, images, speech, audio, etc.) and its annotation with respect to a shallow representation of its meaning. It allows the retrieval of data chunks with respect to topics (e.g. all the texts with a given topic) and to make some limited semantic inferences on topics (e.g. all the texts whose topic belongs to a given class). The MMDB is intended as a module in a system to build adaptive presentations of information; it is both a data repository and a shallow knowledge base system.

2.1 Topic Classification

The knowledge about the domain, in our case life in the Middle Ages, is encoded as a set of key words—called topics—representing entities, such as characters and animals, and processes, such as hunting and leisure activities. At this phase of the work, only one relation can be defined between topics; that is, the *member-of* relation that denotes that a given topic belongs to a given class. For example, the topic *fox_hunting* is in a *member-of* relation with the topic *hunting*, which means that *fox_hunting* is a form of *hunting*.

2.2 Text Annotation

Since the system was designed to navigate through paragraphs rather than entire documents, paragraphs within each document were explicitly marked. A title was assigned to each paragraph in order to use it as a text anchor for the links to be displayed on the interface.

The meaning of each paragraph was then attributed assuming that it discussed a single "concept" already represented by a given topic in the knowledge base. This semantic annotation, though simple, allows the classification of text according to what is being discussed. For example, it is possible to search the database for all the paragraphs that discuss, in one way or another, hunting. This is possible simply by looking for paragraphs for which (a) the main concept is the topic hunting or (b) the main concept is one of the topics that are in a *member-of* relation with hunting.

A second type of semantic annotation attributed to each paragraph assumes that aside the single concept, a paragraph also introduces other entities that, although not the central meaning of the text, can nevertheless be used to provide smooth shifts toward new arguments. These entities, still represented as topics in the knowledge base, are called the paragraph's "Potential Focus List" (PFL). Figure 1 illustrates the semantic annotation for one of the texts used in the hypermedia introduced in section 3 below.

```
<paragraph title="Fox Hunting" concept="fox_hunting" pfl="fox,badger">
On the right there are two hunters coming through the snow, each with two dogs on a lead. One of the hunter's dogs is making for a bush in which a badger is hiding The whiteness of the snow-covered countryside is broken by the low bushes and the fir trees under which there are some foxes. </paragraph>
```

Figure 1. Semantic annotation of paragraph

2.3 Image Annotation

As with the document-paragraph situation, the semantically relevant element was often not the image itself but the scenes it contained. The images, thus, are sub-divided into details that could be expressed in a single topic, such as a hunting or a harvesting scene. A detail is represented by its bounding box; that is, the region occupied by the scene in the image coordinates. Depending on the complexity of the information, the details can be embedded inside other details.

2.4 System Architecture

The back-end of the system is organized as a relational database (implemented in mySQL). In order to provide a way to connect the database to any type of graphical user interface, a query engine has been developed to retrieve data from the MMDB through a standard HTTP connection. The engine, implemented using Java Servlets, transforms the request in a list of SQL queries and returns the data in XML format (figure 2). The MMDB is able not only to provide the actual data but also to perform semantic queries, for example, retrieving all the texts that have a concept that belongs to a certain class, thus making the implementation of the semantic strategies of the interface (section 4) easier.

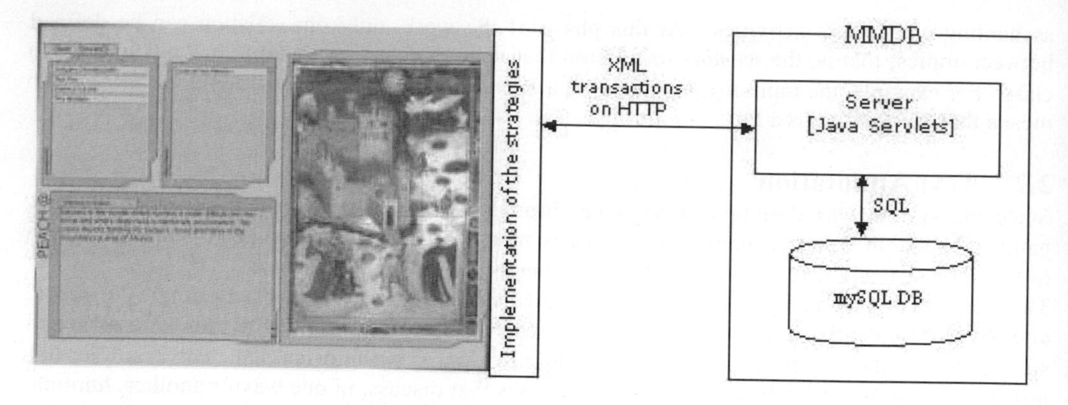

Figure 2: System Architecture

3 The "Torre Aquila" Hypermedia: A Walk-through

Our system has been conceived as a universal tool that can be applied to any multimedia domain. As a case study, we have chosen to work with documentation about the Cycle of the Months of Torre Aquila at the Buonconsiglio Castle in the city of Trento, Italy. This fresco is composed of eleven panels (each one representing a month) painted during the 1400s and illustrates the activities of aristocrats and peasants throughout the year. The fresco introduces a number of characters as well as many different activities, from falconry to wine harvesting.

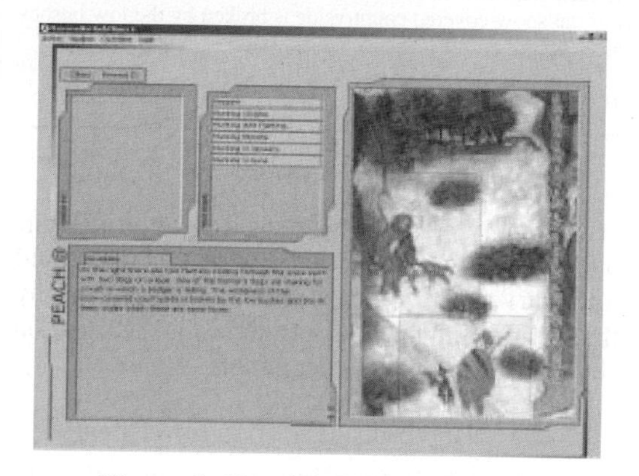

Figure 3: Hunting detail in January

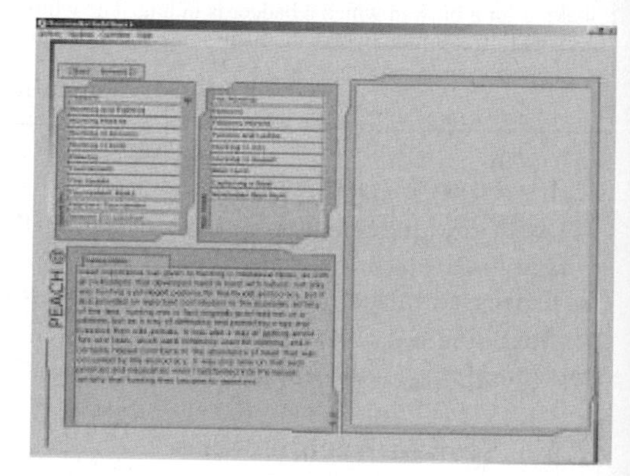

Figure 4: General presentation about hunting

Figures 3 and 4 represent a snapshot of an interaction with our system. In figure 3, an image of a detail from the month of January is seen along with the text (lower left box) that describes the two hunters chasing a fox and a badger. In the two boxes on the top left side of the screen, the system generates links to related information. Given that the text describing the hunting scene explicitly mentions foxes and badgers, informative texts about foxes and badgers in the Middle Ages are generated as link proposals. Similarly, as the image represents a hunting beat, a general text about hunting as an activity is also suggested (i.e. new link generated inside one of the two top boxes).
Figure 4 illustrates the interface that would appear when the user clicks on this last suggestion. On the right side of the interface, the detail of the month of January has disappeared and the new text is displayed in the box on the bottom left, while on the top left boxes, new navigating options

have been automatically generated. These are links to details about hunting beats represented throughout in the Cycle of the Months; namely, Fox and Badger in January, Falconry in July, August and September, and Bear Hunting in November.
This example demonstrates three different communicative strategies. The first one (links to informative texts about foxes and badgers in the Middle Ages, Figure 3) employs information about potential foci in the current text in order to propose links to other texts that elaborate one or more of these foci. The second strategy (a general text describing hunting, Figure 3) uses the idea that the "hunting beat" depicted in the month of January is a particular example of hunting as an activity and thus proposes a link to another text that describes this activity in general terms. The third strategy (Figure 4) recognizes that when the current text describes a general activity or idea, it would be appropriate to propose links to specific examples that may be found in the database about the current topic.
By using the three communicative strategies just described, the system is thus able to propose new information to the user following a semantic approach. In other words, the system "navigates" the information space and makes consequent decisions about what new information could be interesting to the user by utilizing the "knowledge" of what the user is currently exploring on the interface.

4 Implementation of the Strategies

A communicative strategy is a mechanism to build links between relevant information. Each strategy is composed of two steps. First, a rule that defines how to find a new topic given the current topic, and second, a query in the database to retrieve all the information, i.e. paragraphs and image details, indexed by the new topic. For example, the three strategies introduced in Section 1 can be defined as follows:

a) **"Follow the PFL" Strategy:** This strategy chooses the new topic among those of the potential focus list of the text currently being displayed on the interface. The database is then queried to retrieve all the paragraphs that have been annotated with that topic as the main concept. For each paragraph found, a new link is created on the interface in one of the two top boxes seen in figures 3 and 4.

b) **Generalization Strategy:** This strategy chooses the new topic as the class to which the main concept of the text currently being displayed on the interface belongs. As with the above strategy, the database is then queried to retrieve all the paragraphs that have been annotated with that topic as the main concept and for each paragraph a new link is created.

c) **Class members Strategy**: This strategy is applied when the main concept of the text currently being displayed on the interface is a class. In this case, the new topic is chosen among the members of that class. The database is then queried to retrieve all the paragraphs that have been annotated with that topic as the main concept. For each paragraph found, a new link is created on the interface.

When a link is clicked on the interface, the corresponding paragraph is displayed and the topic that represents the main concept is used to query the database to search for a new image detail to propose. If that detail contains other embedded details the picture is transformed into an image map where details become clickable areas. If one of the details is clicked, the corresponding topic is used to search the database for a new paragraph (and possibly a new detail) to propose to the user, thus the three rules above are re-applied to build new links to other relevant information.

5 Conclusion

In this paper, an architecture to dynamically link multimedia content previously annotated with respect to a shallow semantic representation has been presented. The advantage over traditional hypermedia lies primarily in the ease with which the system can be updated since new data can be automatically connected to the rest of the information.

Similar works are being conducted by other research groups. In particular, the Multimedia and Human-Computer Interaction group at CWI in Amsterdam is working on a tool to automatically infer semantic relations between media objects and then map them onto spatial and temporal relationships within SMILE files (Little et al., 2002). They, too, have an agenda to exploit semantic web technology for hypermedia in the automation of some tasks related to hypermedia creation and management (van Ossenbruggen, 2002). While in both theirs and our system semantic model are a crucial part of the strategies used for the automatic generation of the hypermedia, CWI focuses on the generation of the interface layout for the hypermedia, rather than on link construction; as is the case with our system.

The research field of adaptive hypermedia is also relevant to our approach in that it studies the use of semantic relations for tailoring the hypermedia to a specific user; in particular, works by (Kobsa et al., 2001) and (Brusilovsky, 2001) have investigated many relevant issues. By the same token, our previous work done on Macronodes (Not and Zancanaro, 2000) provided an important stepping stone to our research in that it analyzes the structure and annotation of texts for a multimedia presentation.

6 Acknowledgments

This work has been conducted in the context of the PEACH project (Personal Experience with Active Cultural Heritage, Stock and Zancanaro, 2002) funded by the Autonomous Province of Trento. More information on PEACH can be found at http://peach.itc.it/.

References

Berners-Lee, T., Hendler, J., & Lassila, O. (2001). The Semantic Web. *Scientific American*, May.

Brusilovsky, P. (2001). Adaptive hypermedia. *User Modeling and User Adapted Interaction*, 11(1/2), 87-110.

Kobsa, A., Koenemann, J., & Pohl, W. (2001). Personalized Hypermedia Presentation Techniques for Improving Online Customer Relationships. *The Knowledge Engineering Review* 16(2), 111-155.

Little, S., Geurts, J., & Hunter, J. (2002). Dynamic Generation of Intelligent Multimedia Presentations through Semantic Inferencing, In *Proceedings of the 6th European Conference on Research and Advanced Technology for Digital Libraries*, September.

Not, E., & Zancanaro, M. (2000). The MacroNode Approach: mediating between adaptive and dynamic hypermedia. In *Proceedings of International Conference on Adaptive Hypermedia and Adaptive Web-based Systems*, Trento, Itay, August.

Stock O., & Zancanaro M. (2002). Intelligent Interactive Information Presentation for Cultural Tourism. Invited talk. In *Proceedings of the International Workshop on Natural, Intelligent and Effective Interaction in Multimodal Dialogue Systems*. Copenhagen, June.

van Ossenbruggen, J., Hardman, L. & Rutledge, L. (2002). Hypermedia and the Semantic Web: A Research Agenda. *Journal of Digital Information* 3(1), August.

A Methodology for the Administration of a Web-Based Questionnaire

Maria João Antunes, Eduardo Anselmo Castro, Óscar Mealha

Universidade de Aveiro
Campus Universitário de Santiago, 3810 193 Aveiro, Portugal
mjoao@ca.ua.pt, ecastro@ua.pt, oem@ca.ua.pt

Abstract

In recent years there was an increase in the number of Internet surveys, in particular those that use email and Web services, exploring their ability to reach many potential respondents, with reduced effort and low costs. In spite of increasing sophistication, it has been considered that the strategies and approaches implemented were not the most efficient. The low response rates in questionnaires show that the target group needs an extra stimulus to collaborate in these surveys. Taking this information into account a carefully planned strategy was developed, offering a new approach, that attracts target population attention, exploring Web potentialities. This paper presents the questionnaire methodology used in this survey.

1 Introduction

The revolution in information and communication technologies had a profound impact in the major sectors of society and, above all, in the way people communicate and access information. However, the arrival of Internet services generated the idea that the spatial dimension of communication becomes unimportant: interactions no longer need to be structured around places. However, only a limited number of interactions are independent of spatial proximity and personal ties. The greater the involvement of individuals in interaction networks, embedded in places, the greater their attractiveness to the global networks (Castro & Butler, 2003). Concerning the subject briefly introduced, present research has the objective of answering to the following questions: i) Is Internet supporting new forms of relationship, enabling the connection of people from outside circles, or simply keeping contacts based on existing strong ties (Granovetter, 1973)?; ii) How membership of networks, embedded in geographical places, affects peoples' capacity to interact in wider networks?

2 Presenting the Strategy Used

2.1 Questionnaire

Looking at different methods of data collection in social research, it was decided to choose the questionnaire, as the best option to obtain the answers, of a large number of persons, to the questions raised. The survey was designed using a computer self-administered questionnaire, in an open web site. Compared to other main methods of data collection, like mail questionnaires and interviews, web surveys present several advantages: low cost of diffusing and processing, close control over the order in which questions are answered, check on incomplete responses, less time consuming to conduct the study and to process the data, opportunity for passing on the questionnaires to other, and ability to reach a widely dispersed population. The questionnaire, directed to the Portuguese community of Internet users, older than 17 years, was divided into four main parts: i) Personal data; ii) Use of Internet to communicate with relatives and friends tied by strong relationships; iii) Use of Internet to develop new relationships; iv) Use of Internet to electronic commerce.

2.2 Questionnaire Respondents and Motivation Strategy

The questionnaire was diffused using two major sampling sources: invitations and snowballing. In the first case, collaborations from outside organizations, like portals, universities, technical universities, and a regional section of a major Portuguese public service, were used. The portals exhibit, during a certain period, a banner invitation. The other partners delivered, in their internal emails lists, a message, composed by the research team, encouraging participation in the study. This email had embedded the URL of the questionnaire site. After respondents' submission of data, they have the option to pass on to acquaintances and friends an invitation to collaborate in the study. This invitation was integrated in the site; respondents only need to provide other electronic addresses and to present their own name and email.

To motivate people to go to the web site, and respond to questions, researchers need to develop a carefully planned strategy. In this sense, it is considered that the risk of obtaining a low response rate is reduced if the questionnaire gives some stimulus to collaborate in the survey. As a consequence the strategy developed exploits Web potentialities, to attract potential respondents and to make the questionnaire user friendly and interesting. Four key aspects were taken into account to design the message: i) focus on research objectives; ii) clear presentation of the questionnaire objectives; iii) interaction strategy and iv) Web instrument usability.

2.2.1 Focus on Research Objectives

In the process of questionnaire build up, researchers need to ensure that every question raised has a strong link with the conceptual framework of the study (Oppenheim, 1999). In the present survey all questions included were designed to maximize the efficiency in terms of responses to the objectives of the overall research.

2.2.2 Giving Clear Presentation to Respondents of Questionnaire Objectives

Internet questionnaires are usually delivered as an email attachment or as an email with a hypertext link to a simple web based questionnaire. In this survey the inquiry is part of a web site that provides, in a simple manner, full information about objectives and the framework of the research. The web site has four links: "start", "more about", "credits" and "contacts". The home page (link "start") provides, using simple language, some important information about inquiry. A brief explanation about the general objective of the research reinforce respondents anonymity, presents an estimate of the time required to complete the questionnaire, and remembers that there are no difficult or embarrassing questions. In this page respondents can also find the links to the inquiry and a description about the best monitor and browser options to visualise the questionnaire and the entire site (see Figure 1).

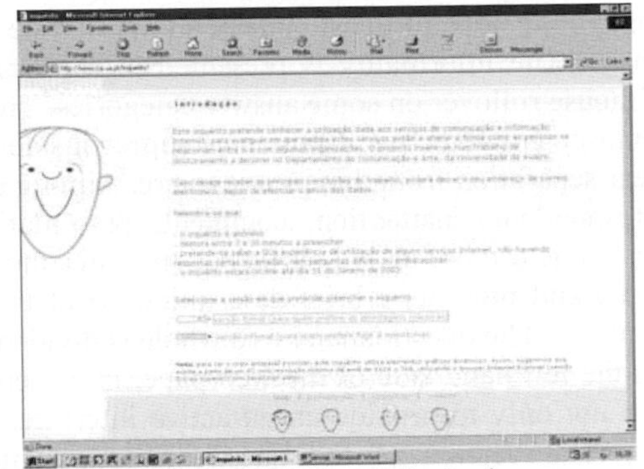

Figure 1: Questionnaire web site home page

Some authors suggested that presentation of research objectives could result in a respondents' tendency to help or obstruct the study, able to invalidate questionnaire results. On the other hand, it is also considered that when researchers do not present the rationale of the study respondents are tempted to guess what is wished (Foddy, 1996). Our option was to present the major objectives and strategy used, to collect the information needed, using the link "more about".

2.2.3 Interaction Strategy

A research conducted through a questionnaire raises great challenges to obtain respondents co-operation. In the particular case of a web-based questionnaire it is believed that the interface, and the interaction strategy used, play a major role. In the present survey, the web site was centered on an interface agent, Columbus, a figure represented only by his face, whose expression shows happiness and informality. The navigation on the web site was designed to be easy and intuitive. Light-coloured greens and blues, in a white background, and black font and drawings were considered a balanced and appealing graphic design expected to keep respondents at the web site. Aiming to introduce some innovative features in questionnaire design, and to motivate respondents' collaboration, it was possible to choose between two versions for filling the questionnaire: a formal or an informal version. However, both versions have the same questionnaire content and structure (see Figures 2 and 3).

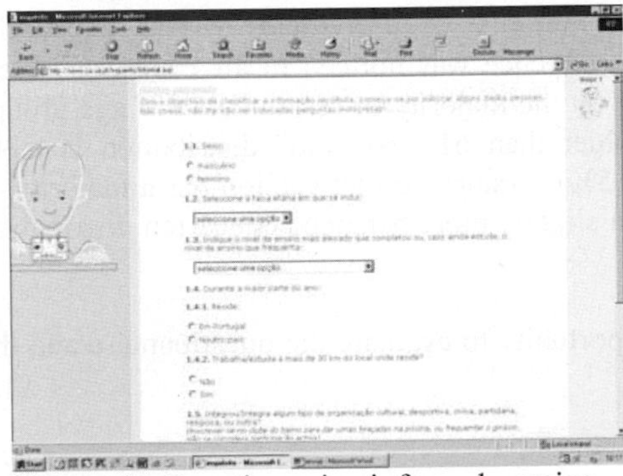

Figure 2: Questionnaire informal version example page

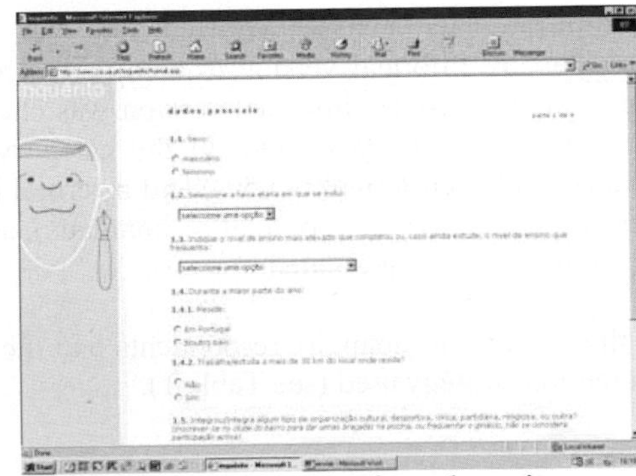

Figure 3: Questionnaire formal version example page

A comparison between the two versions of questionnaire shows that de informal one is friendlier and strongly based on image. The informality is presented in short and funny statements that appear when respondent's mouse rollover on some answer categories. The comments are related to the answers and have an amusement objective. Comments appear inside a yellow box, with a blue font, as a way to keep a clear separation from the questionnaire. Aiming to obtain a surprise effect, and as a way to avoid respondents inattention, comments presentation are sporadic. At the informal version Columbus gains a more active participation. After the selection of this version, Columbus introduces himself and presents a brief description about the structure of the inquiry and the interaction strategy used. The questionnaire as a whole is divided into eight forms. In each form Columbus appears in the left hand side of the web page, in a scene that invokes the theme being asked. Image is used not only to provide an attractive interface, but also to reinforce the context questioned. The formal version of the questionnaire assumes a more conservative appearance and Columbus image is the same throughout the forms.

2.2.4 Web Instrument Usability

According to A. Oppenheim (1999), the attractiveness of questionnaire and answering process is also an important element in maintaining respondent's cooperation. Questionnaire design follows a respondent funnel approach. Each part starts with a broad filter question and it is followed by specific ones (Oppenheim, 1999). The use of filters enables exclusion of some respondents from certain questions (Oppenheim, 1999) and guides them to another appropriate sequence. In present survey the use of DHTML and Javascript made possible, in a transparent manner, to present to respondents only questions related with their own situation, decreasing information per web page and tailoring dynamically the questionnaire to the respondents' universe.

3 Results Obtained and Conclusions

The data for the study were collected between November 2002 and January of 2003 from 2930 respondents, older than 17. 54.6% of respondents are women and 45.4% men. 80.8% are ages between 18 and 35, and 12.4% between 36-45. Only 6.8% are older than 45. Concerning education, 46.2% of respondents attended or attend a degree level, 16.9% attended or attend high school, 10.9% and 9.2% attended or attend, respectively,.a master or PhD level.

Preliminary data analyses showed that the vast majority of individuals (85%) chose the questionnaire informal version to complete. This preference was noticed in all education levels and in all age groups. Informal version was chosen by the majority of respondents of age groups between 18 and 60 years old. In the category "older than 61 years old" distribution is more balanced between formal (43.5%) and informal (56.5%) version. Formal version has a majority of men respondents (61.1% of total respondents), and a slightly more balanced education distribution between categories presented.

At the end of questionnaire respondents had the opportunity to evaluate the questionnaire and the interaction strategy used (see Table 1).

Table 1: Categories to Evaluate Questionnaire Interaction Strategy and Results Collected

Answer Categories	Informal Version	Formal Version
Rejection of questionnaire	3.8%	5.5%
Regrets about the choice of the version	5.9%	11.1%
Sympathy of the layout but uneasy contents	2.8%	6.6%
Sympathy of layout and contents	57.5%	22.7%
"Love at first sight"	30.1%	54.1%

From presenting data, it seems that interaction strategy used in informal version improved in fact respondents' questionnaire completion, and/or their perception about their positive contribution to the answering process. Concerning the formal version, and in spite the percentage of formal version respondents that "Regrets about the choice of the version" (11.1%), a huge majority supported the questionnaire strategy. Among individuals who chose formal version, there are a clear distinction about strategy used related to age. Individuals older than 51 express more positive opinions, individuals under 25 years old, have a more critic evaluation. It seems that younger respondents need in fact an extra stimulus to appreciate the task.

The opportunity to encourage friends and acquaintances to collaborate in the survey is a successful strategy: 20.5% of responses come from snowballing whilst banner exhibition on portals counts 27.4% of total responses and messages delivered by regional portals and other institutions to its mailing lists count 15.4% and 29.1% respectively. Other forms of diffusing the questionnaire count 7.5%.

In the context of a questioning survey there are some factors that contribute to response rates, probably the most important of all is the topic being asked, and its degree of interest to respondents. Concerning this aspect the research team has little to do, once issues to be investigated were primary defined, but, with respect to all the other aspects involved, it is believed that there were a lot of things that could be done. The results obtained show that the strategy presented in this paper was successful, inducing positive effects on respondent's co-operation.

Acknowledgements

Special thanks are due to Pedro Monteiro (Mental Factory, Lda) for his valuable contribution in the design and development of the survey web site, and to Leonel Neves for the overall database development and programming. This research had the support of various portals and institutions: Clix, campus.sapo.pt, Aveiro Cidade Digital, Alentejo Digital, CRSS Aveiro, ESEL; ESEnf. Viana do Castelo; ESTG Viana do Castelo; IPC; IPG; IPP; IPS; IPV; ISCA Coimbra; ISCA Porto; ISEG; Universidade de Aveiro; Universidade da Madeira.

References

Castro, E. and C. Jensen-Butler (2003). Network externalities telematics and regional economic development. Forthcoming in *Papers of Regional Science*.

Foddy, W. (1996). Como perguntar: teoria e prática da construção de perguntas em entrevistas e questionários. Oeiras: Celta Editora.

Granovetter, M. S. (1973). The strength of weak ties. *American Journal of Sociology*, 78 (6), 1360-1380.

Oppenheim, A. N. (1999). Questionnaire design, interviewing and attitude measurement (New ed.). London: Pinter.

Properties of Controlling Models for Expressions Linked to Visual Knowledge

Eiichi Bamba
Faculty of Science and Engineering, Kinki University,
bamba@mec.kindai.ac.jp

Abstract: Human makes the visual information like the pictures on the computer display conscious and receptive as stimuli if necessary. When necessary knowledge to execute the demand operations is included in received information, then human tends to rake up more information as possible and enlarge knowledge. As a result, intellectual properties grow larger and behaviors accompany with emotions. Therefore, this paper considers the management process (the expressive process) for the dispatching control of stimuli information to expressive aide to attach importance to emotion. In this paper, the introduced psychogenic particle system driven by consciousness forms the quantized system with the sequence of the conscious information as the input event and the driven information of emotion as the output respectively. The probabilistic properties of the psychogenic particle system are expanded into semi-Markov process, and the psychogenic particle system makes the dispatches of the conscious information to the emotion driven factors correspond to one-to-one. Therefore, the autonomous vehicle(AV) considered in this paper makes sure of the intellectual property, and the dynamics of the inner states of AV conform to semi-Markov process.

1 The System Control Model for Conscious Information from Stimuli

If the conscious space can be regarded as a kind of the system, then stimuli are fit for with the input and expressions for the output respectively. Hence, the following four correspondences exists between stimuli, conscious space and expressions. — (1) the states that comply with stimuli and expression, *e.g.*, the awareness states, (2) the states that comply with stimuli but not with expressions, *e.g.*, the arousal states, (3) the states that don't comply with stimuli but expressions, *e.g.*, the conditioning(pre-conscious) states, (4) the states that neither comply with stimuli nor expressions based on consciousness, *e.g.*, the unconscious states. From the viewpoint of the system control models, the case (1) is identified with the dynamics of C/Os., the case (2) with C/UOs., the case (3) with UC/Os., and the case (4) with UC/UOs.. On the input side of the noticed system, receptors distribute to each of functional regions in cerebrum to process intermittently the accepted environmental information. Mechanisms of the receptor in this case are realized by twofold gates, *i.e.* arousal gates and information gates. A role of those equipped gates is like to a role of Glia cell in living body. Those schemes are shown in Fig.1.

2 The Proposition of the Psychogenic Particle System and Its System Model

Emotion on the emotional processes are expressed either concurrently or time sequentially by way of some processes. Thus, the mechanisms to perform those processes are required. On the other hand, it is interpreted from the various doctrines with the mentally psychological emotions that those mechanisms are due to the mental movements, but the quantitative descriptions of a mental are hardship intrinsically. Hence, it is in the course of nature that the psychogenic particle based on Cognitive System Driven Concepts are organized. The inputs into the psychogenic particle system are conscious information, and the output are emotion driven factors, *i.e.* Cognitive Driven Factors, Perception Driven Factors and Request Driven Factors. Therefore, inputs and outputs on the psychogenic particle system are the sequence of the events and described by the continuous—variable continuous—time system named as the quantized system as shown on Fig. 2.
The set of consciousness is dividable into the three subsets of Recognition, Pre-consciousness and Unconsciousness. Moreover, the information in each of those subsets is variable with the acceptance of stimuli momentarily. The event of consciousness is generally described by the non-numerical terms, and the recognized consciousness is the discrete events. Therefore, the outputs or the state variables of the four subsystems in the system control model(Fig.1) construct the sequence of the discrete control actions to the psychogenic particle system. Hence, we obtain the following proposition.

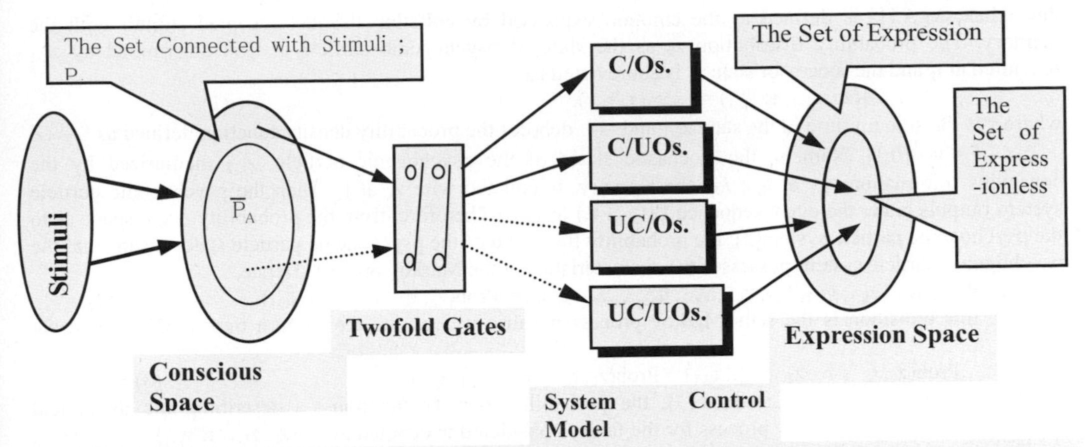

Fig.1 : Set Relations between Stimuli, Conscious Space, System Control Model and Expression

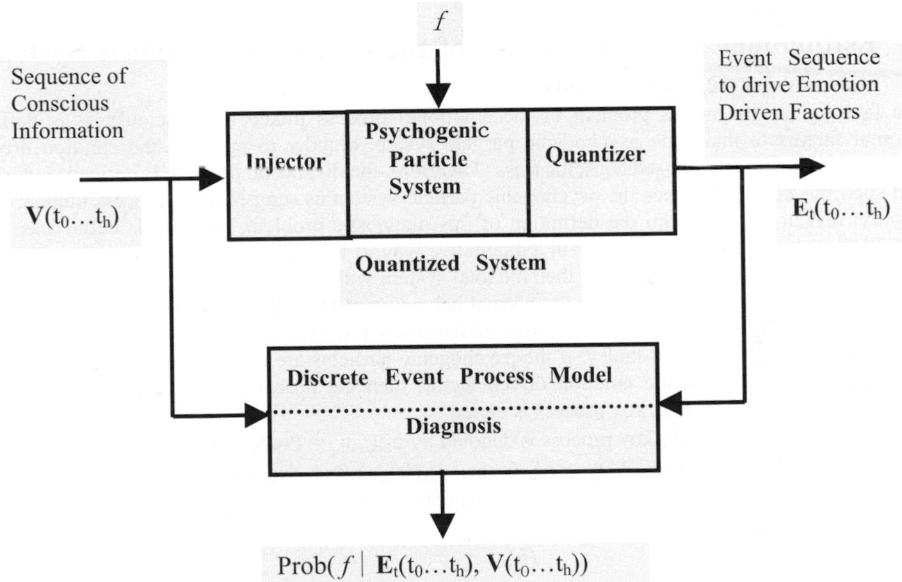

Fig.2 : The System Structure of the Quantized System

Proposition 1 : Consciousness composes the timed discrete events $\mathbf{V}(t_0 \ldots t_{hj})$, where t_{hj} denotes the duration arousing the j-th consciousness (j=1,…,3) i.e., Recognition, Pre-consciousness, and Unconsciousness. In the quantized system for the psychogenic particle system, the quantizer(Fig.2) maps the state $\Box^n$ onto a finite set $\Box_x = \{0,1,\ldots,4\}$ of the qualitative values, and introduces a partition of $\Box^n$ into the finite number of the disjoint sets $Q_{\mathbf{x}}(i)$ to denote the set of the state $\mathbf{x} \Box \Box^n$ with the same qualitative value i. The numbers of the disjoint sets $Q_{\mathbf{x}}(i)$ correspond to those of the subsystems in the system control model for consciousness. The dynamics of psychogenic particle system are definable as the following, because this state-space model is usually applied to the control models.

Definition 1 : The psychogenic particle system is the family and the dynamics of the particles are described by $d\mathbf{x}/dt = f(\mathbf{x}(t),\mathbf{u}(t),f)$, $\mathbf{x}(t_0)=\mathbf{x}_0$, where $\mathbf{x}(t) \Box \Box^n$, $\mathbf{u}(t) \Box \Box^m$, and f denotes the fault. The fault, f is defined as the external factors to make the psychogenic particle system blunder to motivate the emotion driven factors corresponding with the aroused consciousness. The schema-triggered affect proposed by Fiske, S.T. and

Pavelchak, M.A.[1] is defined as the emotion expressed by collating the just accepted stimuli with the memory. The probability distribution z_k as the state of psychogenic particle at t_k is determined by z_{k-1} remained at t_k and the successor state z_k is abbreviated as

$$\square_T(z_k, z_{k-1}, \tau, \nu_{k-1}) = f_{zkzk-1}(\tau, \nu_{k-1}), \quad (1)$$

where τ is the sojourn time in the state z_{k-1} and $\square_T$ denotes the probability density function defined as $\square_T : Z \times Z \times \square^+ \times \nu \rightarrow [0,1]$. Namely, the stochastic states of the psychogenic particle, z_{k-1} memorized by the conscious information ν_{k-1} at t_{k-1}, *i.e.* the memory, is collated with z_k at t_k. Then the psychogenic particle system outputs either the event sequence $\mathbf{E}(t_0,\ldots,t_h)$ or not. Therefore, from the probability event space as to the psychogenic particle system[2], the probability transition of the psychogenic particle states means that the psychogenic particle system possesses the characteristics of the Markov process. Hence,

$$\mathrm{Prob}(z_k, t_k \,\square\, z_{k-1}, \nu_{k-1}, t_{k-1}, z_{k-2}, \nu_{k-2}, t_{k-2},\ldots,z_0, \nu_0, t_0) = \mathrm{Prob}(z_k, t_k \,\square z_{k-1}, \nu_{k-1}, t_{k-1}). \quad (2)$$

Further, this transition is the semi-Markov process because we assume the sojourn time τ of z_{k-1}, and we obtain

$$\mathrm{Prob}(z_k, t_k \,\square\, \nu_k, z_{k-1}, \nu_{k-1}, t_{k-1}) = \mathrm{Prob}(z_k, t_k \,\square\, z_{k-1}, \nu_{k-1}, t_{k-1}). \quad (3)$$

Abiding by the proves of Lunze, J.[3], the probability density function $\square_T$ describing the dynamical properties of the semi-Markov process for the fault f considered is denoted as $\square_T(z_k, z_{k-1}, \tau, \nu_{k-1}) = f_{zkzk-1}(\tau, \nu, f)$. Therefore, the set of the output event for semi-Markov Process, *i.e.* the psychogenic particle system, is given by

$$M_t(e_0, \mathbf{V}(t_0,\ldots,t_h), f) = \{(e_0, t_0; e_1, t_1; \ldots; e_H, t_H) \,\square\, f_{ek-1ek}(\tau, \nu_{k-1}, f) > 0, k=1,\ldots,H\}, \quad (4)$$

where e_j, $j=1,2,\ldots H$ denotes the event generated by the psychogenic particle system at the time tj.

3 Estimations and Verifications of the Psychogenic Particle System Based on the System Diagnosis

The fault in the diagnostic problem of the emotional process linked to consciousness is defined as the external factors to make the psychogenic particle system blunder to motivate the emotion driven factors corresponding with the aroused consciousness. Therefore, the diagnostic problem is defined as the problem to find such the fault that makes the psychogenic particle system as semi-Markov process blunder to drive the emotion driven factors. From the definition of this diagnostic problem, the fault is defined as the external factors changing the system behavior in a qualitative way. If the aroused consciousness doesn't motivate the corresponding emotion driven factor, then the total system with stimuli and emotions, *i.e.* an individual, loses normality. Therefore, this means that the normal behaviors of the individual system are changed. Hence, the following theorem is provable by being based on the above point of view.

Theorem 1 : By the external fault f, if the psychogenic particles are made to be unsuccessful in linking the conscious information on the emotion driven factor, then the probability distribution of the psychogenic particle states as the semi-Markov process is varied.

From this theorem, semi-Markov process is denoted by $p_M(f, t_h) = \mathrm{Prob}(f \mid \mathbf{E}_t(t_0,\ldots,t_h), \mathbf{V}(t_0,\ldots,t_h))$ so that the relations, if $\mathbf{E}_t(t_0,\ldots,t_h) \in M_t(e_0, \mathbf{V}(t_0,\ldots,t_h), f)$, then $p_M(f, t_h) > 0$ or else, $p_M(f, t_h) = 0$ hold. From the conscious dispatches of the psychogenic particles, $p_M(f, t_h)$ is obtainable as

$$p_M(f, t_h) = \frac{p_a(f, t_h)}{\sum_f p_a(f, t_h)} \quad (5)$$

by the auxiliary function, $p_a(f, t_h) = f_{eH+1eH}(t_h - t_H, \nu_H, f)p_M(f, t_H)$.

The case of $p_M(f, t_h) = 0$ means that the conscious information accepted from stimuli is adequately dispatched to its corresponding driven factor or emotions by the psychogenic particle. Otherwise, $p_M(f, t_h) > 0$ means that the psychogenic particle blunder to dispatch to the right driven factors of emotions by the fault f, but the emotion driven factors are also driven by the proper conscious information. Therefore, if we obtain the result of the positive $p_M(f, t_h)$, then it can deduce that the individuals don't stay the normal mental state from the theorem 1. Hence, if we can find the existence of such f as $p_M(f, t_h) > 0$, then it becomes clear whether the individuals are normal for the expressed emotion or not. Therefore, by finding out the solution of the diagnostic problem and removing the fault, we can make the emotion system linked to consciousness robust to the fault. This paper proposes to make the emotion system linked to consciousness robust to the fault. Hence, if it is proved that the psychogenic particle system is semi-Markov process, then the psychogenic particle system makes the dispatches of the conscious information to the emotion driven factors one-to-one.

4 Application Example

This example carries out the computer simulation of system control models for consciousness in the processes to emotions from stimuli, and attends to the intelligent navigation of Autonomous Vehicle(AV) loaded up by such the task as obstacles avoidance. Contents of the task are that AV navigates from the start point **A** to the end **B** on the replete surfaces with obstacles, and the appearing situations of obstacles are conformed to the table of random numbers. Movements of AV are executed by C/Os.. If any, C・UOs. detects the obstacles, then the state variables of C/UOs. are translated to UC/Os.. The system structures are depicted as Fig.3. In the avoiding behaviors for the suddenly random appearances of obstacles, it is proposed that AV expresses a hatred and puzzlement[2]. This expressions mean that UC/Os. of the psychogenic particle system on AV fails to dispatch the conscious information to Cognitive Driven Factors.

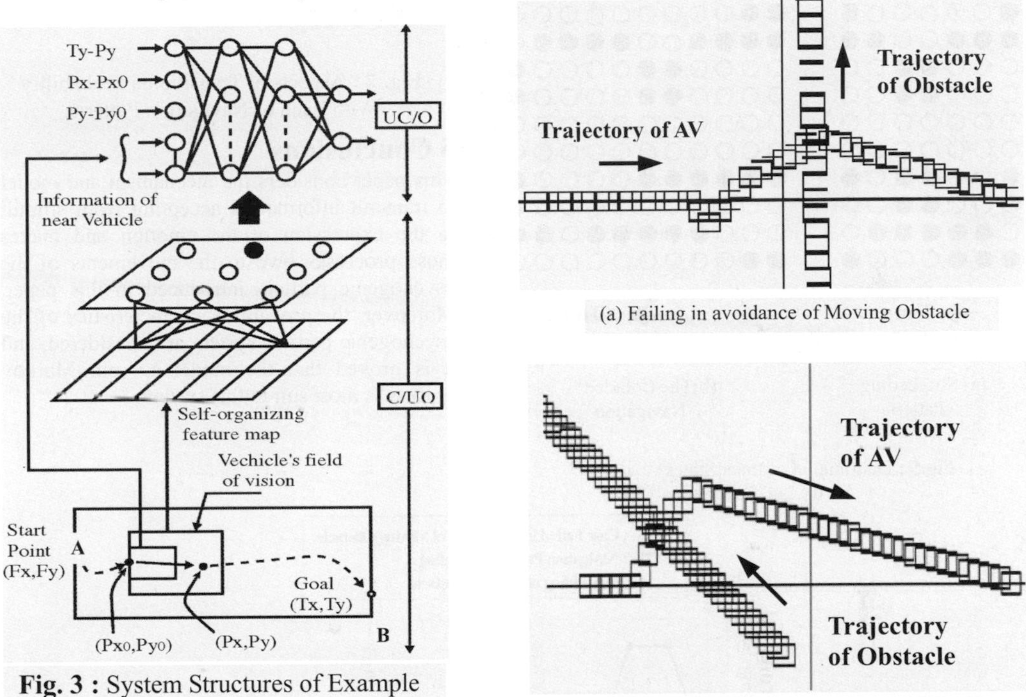

Fig. 3 : System Structures of Example

(a) Failing in avoidance of Moving Obstacle

(b) Succeeding in Avoiding of Moving Obstacle

Fig. 4 : The Navigation Pattern

This navigation system to avoid obstacles is controlled by neural networks of Kohonen type with 4 layers viewed from Fig. 3. AV may collide with the obstacle by the moving scheme of obstacles viewed on Fig.4. Fig. 4 (a) shows the case failing in avoidance of the moving obstacle and (b) the case succeeding in avoiding of the obstacle. If cells constructing each layer are regarded as the psychogenic particles introduced in this paper, then the firing situations of those cells at the 2nd layer are given by Fig. 5 for each navigation pattern. In Fig.5,●denotes the fired cells. The ratio with the fired neurons is shown on Fig. 6, and this figure presents that the characteristic curve for the navigation patter succeeding in avoiding of the moving obstacle indicates the trend to converge. Fig. 7 is the comparison between the values at t_k and t_{k-1}($k = 1,2,\ldots,11$)of the function to describe the fired neurons by the difference equation, and denotes those transitions. In the case succeeding in avoiding obstacles, some probability rules *e.g.* ergodic process or Markov process, with the transitions of the fired neurons are observable from this figure. Therefore, information to recognize a obstacle is dispatched to the behaviors of avoidance, and the probability of those dispatches have some statistic regularity. Hence, it is made sure that the dispatcher as the psychogenic particle exists.

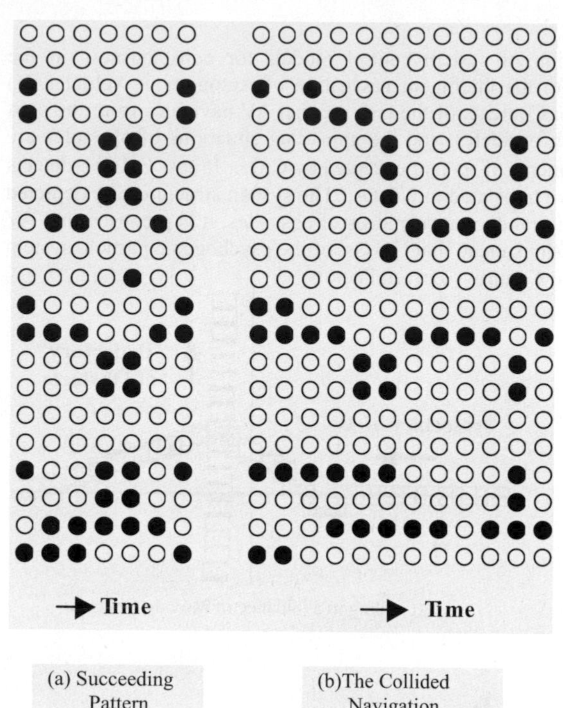

(a) Succeeding Pattern

(b)The Collided Navigation

Fig.5 : Distribution of Inner States on the 2nd Layer

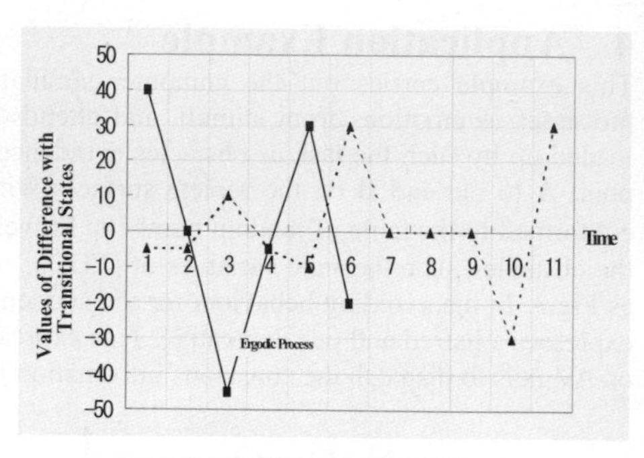

Fig. 7 : Aspects of Transitional Probability With Each of Navigation Patterns

5 Conclusions

This paper considers the mechanism and model to transmit information accepting from stimuli to the expression of the emotion and makes those processes owe to the movements of the psychogenic particle introduced in this paper. Moreover, the probability characteristics of the psychogenic particle system are considered, and it is proved that the model as semi-Markov process is most suit to this system.

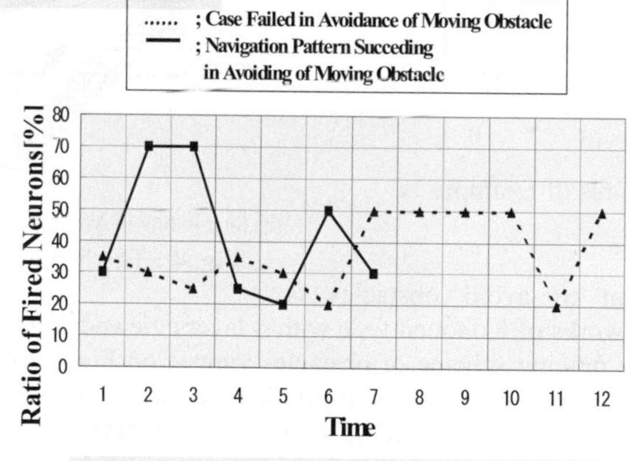

Fig.6 : Ratio of Fired Neurons on the 2nd Layer for Each of Navigation Patterns

6 References

[1] Fiske, S.T. & Pavelchak, M.A.(1986). Category-based vs. piecemeal-based affective response: development in scheme- triggered affect. In R.M.Sorrentio & E.T.Higgins (Eds.), Handbook of motivation and cognition(pp.167-203). Guilford

[2] Bamba, E. and Nakazato, K.(2000). Fuzzy Theoretical Interaction between Consciousness and Emotions. *Proc. of 9th IEEE International Workshop on robot and Human communication 2000*, 219-223

[3] Lunze, J.(2000). Diagnosis of Quantized Systems Based on a Time Discrete-Event Model. *IEEE Trans. on Systems, Man, and Cybernetics —Part A : System and Humans*, 30(3), 322 335

Are Users Ready for Electronic Prescription Processing?

D Bell
Huddersfield University
Queensgate
Huddersfield
d.bell@hud.ac.uk

P Marsden
Huddersfield University
Queensgate
Huddersfield
p.h.marsden@hud.ac.uk

M Kirby
Huddersfield University
Queensgate
Huddersfield
m.a.r.kirby@hud.ac.uk

Abstract

This paper describes preliminary results of a stakeholder analysis aimed at investigating user attitudes, preferences and beliefs to the proposed implementation of a national electronic prescription processing (EPP) system. The stakeholders discussed in the paper are primarily professionals such as GPs and Pharmacists but the opinions of patients – the recipient of the proposed service - have also been taken into consideration. Findings show that whilst most stakeholders accept that EPP is inevitable and some are quite enthusiastic about the opportunities that a nationally integrated system will provide, there is little evidence to suggest that healthcare professionals are ready for adoption of new work practices implicit in the proposed change. They do not appear to have adequate computational skills and the majority does not have appropriate technology or the means to obtain it before the expected implementation date in 2004.

1 Introduction

The present Labour Government (1994→) has expressed their intention of introducing an Electronic Prescription Processing system (EPP) by 2004 [1]. It is their goal to convert all paper based prescription services to an electronic transference system suggesting that the benefits to the patients will include fewer trips to the GP surgery to collect repeat prescriptions, and an end to illegible and incomplete prescriptions [2]. There is the added suggestion that the new system could reduce fraud which currently costs the NHS c£100m a year, mostly through patients evading the prescription charge but fraud by contractors is estimated to constitute c£10m a year [3]. The research described in this short paper has been sponsored by the EPSRC; the intention being to conduct a socio-technical study to help identify the factors which could impact on the usability of an EPP system. Our research began in January 2001 and will conclude in July 2002. The main objectives of the study were to: -

- Study human factors issues implicit in converting a large paper based confidential information system into a highly secure distributed electronic system that uses public data networks for its transmission

- To understand the attitudes, perceptions and working practices of the stakeholders.

2 Background

In it's current form the largely paper based processes of prescribing dispensing and settlement are time consuming, inefficient, inaccurate and open to fraud. Some errors in paper based prescribing are due to clinical handwriting, wrong or illegible drug doses or drug names, some of which could have proved fatal without the Pharmacist's intervention [4]. In addition, reliable and current information about prescribing trends and costs are not available, which makes forward planning difficult for the NHS. Finally hospitals cannot get patient records to find out what has been prescribed (or vice versa) and there is no record of when or if a prescription has actually been dispensed. Most if not all of the problems could be overcome to a greater or lesser extent via the deployment of a well-designed distributed information system.

The proposed shift from paper to electronic systems can provide significant advantages to an organisation, for example, electronic systems can be designed to provide almost instantaneous audit data to highlight discrepancies that might otherwise compromise the overall integrity of a system. They facilitate rapid sharing of data to enable an organisation to respond quickly to changing market conditions. On the negative side, however, inappropriate transmission of confidential data can leave the system open to abuse. In addition, in the absence of feelings-of-ownership, users can frequently adopt an overly conservative perspective. It has been well documented that user resistance can function to obstruct an otherwise successful deployment of the technology [5]

This paper studies the conversion of a paper-based process (that of prescribing and dispensing prescription medicines) into a securely distributed one, and from this draws valuable lessons that can be applied to other similar conversions. A secure distributed electronic prescribing application needs to address many and varied issues, including scalability, user friendliness, security, and ease of administration. These multi-faceted problems often have mutually exclusive solutions - for example high security usually means inconvenience to the user, which is in direct contradiction to ease of use. Therefore the final application is often a compromise of several competing factors, and it is this interplay of issues which, when better understood, can subsequently be applied to the building of secure electronic applications in other problem domains.

3 Methods

There were three main stages to the data collection in this study, the first of which was a large-scale postal survey encompassing two geographic health authorities. Participation was requested from a total of 1,700 possible stakeholders of which 417 were GPs, 162 were pharmacists, 319 were dentists, 796 were reception staff attached to the disciplines outlined and the remaining 120 were patients. The overall response rate was 31.12% (n=499).
From the survey study participants were recruited to take part in an observational study (pharmacists n=6) and a qualitative interview stage of the investigation (GPs n= 5, patients n=10, pharmacists n =6)
This paper is based predominantly on the quantitative findings of the first stage of the study but will draw upon the qualitative findings where appropriate.

The qualitative data has been analysed using SPSS v.10 a statistical package for social scientists. The quantitative materials collected have been analysed with the assistance of NUD*IST and thematic analysis [6]

4 Current System

The first stage of this study investigated the Governments statements regarding incomplete prescriptions and the effects these had on the current service. It was discovered that the majority of prescriptions are computer generated with 97% of GP practices being computerised, they are dispensing between 15 and 230 prescriptions per GP per day increasing the total amount for the study group to between 1,815–27,830 prescriptions. Of the prescriptions dispensed 39.6% (n=48) GPs stated that they are handwriting between 3-12 prescriptions daily. Of the pharmacists participating in the study all stated that they have cause to contact the GP to clarify prescriptions. Figure 1 below indicates a cross section of the types of errors requiring clarification as suggested by the pharmacists and GPs combined.

The 'y' axis indicates the percentage frequency of the error mentioned by the participating groups, **not** the frequency of the problem occurring.

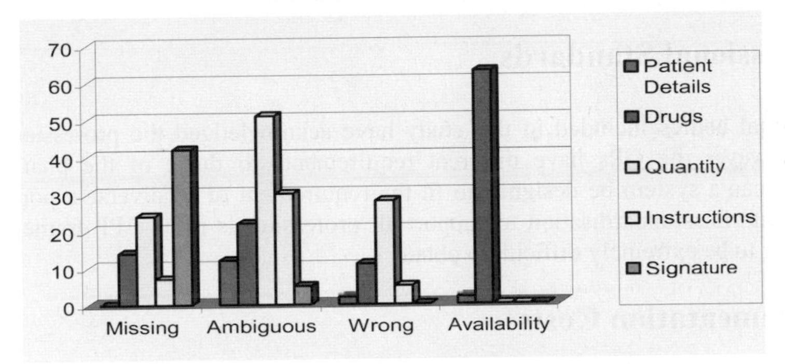

Figure1. Occurrences of Prescription Errors

The occurrence of missing signatures on prescriptions is a fundamental error, which can result in significant delays for patients and inconvenience for both the GP and pharmacists. The availability of drugs also constitutes a major problem for the patient especially when they have traveled to the pharmacy. These data indicate that there are significant problems associated with the current prescription processing system and that there is an urgent need for action. It is envisaged that many of the problems reported above could disappear with the introduction on an EPP system. However, experience suggests that automating the prescription processing process might introduce new forms of error and for this reason it is vital that members of the stakeholder community are invited to participate in the design process to help identify potential hazards and threats to the integrity of the proposed system. In the following section we some will be compatible and others will crash under the pressure.

5 Existing Working Methods

In the current paper based prescription system the professionals employ various tricks of the trade to make the system work efficiently for them. They also endeavour to provide an acceptable quality of service for their customers. There are a number of systems, which the pharmacists and GPs will use, which may not be accessible with the introduction of an EPP system. For example, at present a patient may attend a pharmacy and request the loan of a small quantity of their drugs until the prescription is written and delivered later in the day. Currently a pharmacy may loan the drugs to the patient, if they have a record of the patient and their medication. They are concerned that this behaviour would be curtailed if there were any delays in the electronic transfer of the prescription, or this service may have to be completely abolished restricting the services they currently provide for their customers.

6 Business Processes

The current prescription system allows the individual stakeholder groups to function independently of each other, they communicate when there is a problem with a prescription but the majority of the time they have autonomous systems. The separate professions conduct their business with divergent techniques and in order for an EPP system to work the individuals will be forced to come together and develop standardised working methods, which will unify practice. The problems that may arise from this unification could result in limiting the services they provide to the patients. The fear of practice being restricted by EPP was expressed by 55.6% (n=20) of the pharmacists and 20.7% (n=25) of the GPs were in agreement

7 Professional Standards

The professional bodies included in this study have acknowledged the professionalism of their colleagues however the GPs have different requirements to those of the pharmacists or the dentists. How can a system be designed to fit the requirement of so diverse a population? There are requirements for standardisation to support all professionals in an EPP framework, however these are going to be extremely difficult to obtain.

8 Implementation Costs

At this stage in the investigation there are several concerns regarding the cost of implementing another new system. Initially there will be a cost involved when acquiring the necessary new technology whether in the form of new computers or software. The GP will receive financial backing from the Government but the pharmacists are expected to supply their own systems. In addition to the initial costs there are concerns that the system will change over time and that there will be additional hidden costs needed to keep the system working efficiently for all involved. There will be a cost involved in the connectivity issues when taking on a new system. Currently the practices and hospitals work independent of each other but the EPP system will force the individual groups to work together highlighting system communication inadequacies. The professional stakeholders are concerned about who will meet these costs and who will provide the financial support in keeping the systems functional. Although pharmacist 4 indicated agreement that the proposed system may be beneficial in many ways there was concern that it *'may just increase the costs in the system'*.

9 Patients Perceptions

Up to this stage in the investigation the analysis has concentrated specifically on the views of the health professionals but the importance of the patients in this survey has been incorporated as a significant factor in the acceptability analysis.

The patients do not have strong opinions about the change in systems but this may be due to the lack of information in the public domain when the survey began. The opinion from the patients, which seemed most relevant, was that they believe the new system will be more convenient for them, that it will save them time travelling to a GP surgery and possibly having to travel then to a pharmacy to collect their drugs. Of the patients interviewed the general view was that computerisation was the way forward and 6 of the 7 people stated that they had access to a computer however only 2 of which had any significant experience of using them. The others had only bought them because *'it seemed like the thing to do.'* The 1 remaining patient was opposed to any computerisation and was unhappy about having to use ATM machines to access earnings.

10 Conclusions

The results of these investigations – which will be described at length elsewhere - are instructive in relation to providing an answer to the general question: Are stakeholders prepared for the adoption of a national EPP system. Based on the evidence provided in this study it seems highly unlikely that an electronic prescription processing system can be introduced effectively without major problems being encountered, owing to the state of legacy systems currently in use and the reluctance of the stakeholders to invest time and money into bringing the new systems on-line. Although most stakeholders are accepting of the inevitability of EPP they do not appear to be attitudinally prepared for changes in technological and have little awareness as to how the proposed changes will affect current work practices.

It appears that the current system can limp along satisfactorily for the foreseeable future. However, with the Governments intention to implement a new electronic system by 2004 the pressure is on for change causing anxiety and uncertainty within the stakeholder population. This stakeholder analysis has been useful in identifying problems in the current system and foreseeable problems with the proposed system. With further investigation and adaptation this technique could prove extremely useful in the future development of electronic transfer technology from the stakeholder perspectives. Further work on this study will involve a complete CUSTOM stakeholder analysis and task analysis of the observational studies, which will be available in the near future.

11 References

[1] The NHS Plan. (2000) *A Plan for Investment; A Plan for Reform.* http://www.nhs.uk/nationalplan/nhsplan.htm

[2] Department of Health (2001) *Electronic Transmission of Prescriptions (ETP) Consultation of amendments to POM order.* http://www.doh.gov.uk/pharmacy/etp.htm

[3] Warden. J. (1998) *UK Aims to Deter Prescription Fraud.* BMJ 1998; 316:167-172 (17th January)

[4] Marsden, P. & Green, M. (1994) Procedures in Manufacturing Systems. Some common problems and human factor solutions. In F Aghazadeh (ed) *Advances in Industrial Ergonomics and Safety VI.* London; Taylor and Francis.

[5] British Medical Journal (1990) pp986-990

[6] Crabtree, B. F., and Miller, W. L. (1992). A template approach to text analysis: Developing and using codebooks. In B. F. Crabtree and W. L. Miller, (eds.), *Doing Qualitative Research.* Newbury Park, California: Sage.

Assuring Information Quality in Industrial Enterprises: Experiments in an ERP Environment

Thomas Bellocci[1], Mark R. Lehto[2], Shimon Y. Nof[2]

[1]Saint-Gobain Glass
France
Thomas.Bellocci@saint-gobain.com

[2]School of Industrial Engineering
Purdue University
West Lafayette, IN
lehto@ecn.purdue.edu; nof@purdue.edu

Abstract

This work addresses the problem of information quality and decision quality in distributed information systems. A TQM-based definition of information assurance is introduced to fit the needs of inter-networked enterprises that rely on information and decision support for the fulfillment of their objectives. Information failure types and impact are investigated in distributed systems like ERP (Enterprise Resource Planning).

1 Introduction

The evolution of information systems from a centralized to a distributed organization has enabled the development of inter-networked enterprises, where information systems not only support business functions but are also integral parts of the business operation. For example, ERP systems (Enterprise Resource Planning) are essential for organizations and their supply chains. Some companies completely rely on their information system for the execution and coordination of daily business operations. An effective information system will deliver valuable, timely information and correct data between different departments of the company. Unfortunately, the distribution of information sources and the high speed of data transfer have increased the vulnerability of companies to information failures. Incorrect information in ERP systems can have serious consequences for inter-networked companies that become even more critical when a company tries to manage its supply chain. As a consequence, inter-networked companies need to consider both the security and quality aspects of the data they use for managing their decisions and operations.

Computer scientists were the first to address the topic of information assurance and security. Their approach of information management focuses on information security from internal and external threats (e.g., Longley and Shain, 1986; Shirey, 1995; Voas, 1999). A clear distinction is made between information security and information assurance (e.g., Dobry and Schanken, 1994; Jelen and Williams, 1998). Information security is a feature of the functional components of a product or system, whereas information assurance "is a measure of confidence that the security features and architecture of an automated information system accurately mediate and enforce the security policy" (Longley and Shain, 1986).

Today's networked enterprises are, however, often primarily concerned with the quality aspects of their information for the purpose of achieving their performance goals, and with reaching a global improvement in the trustworthiness and value-addition of information. For example, insurance companies and government agencies may be concerned about the quality of the data they collect and store (Wellman, et al., 2003), or the quality of the responses they provide to clients seeking help (McGlothlin, et al., 2003). Companies are now applying new approaches to improve the administration of distributed information systems (e.g., Schwartz and Zalewski, 1999; Steinitz, 1998). A Total Quality Management (TQM) approach has been advocated by several researchers. In particular, the user-centered approach of Wang (1998) identifies the following quality dimensions of information: intrinsic quality, accessibility, contextual value, and representational value. The central challenge to the developer of distributed information systems is determining how important the particular dimensions of information quality are for the particular application of concern. Without clear knowledge of the true needs for information assurance, a company may employ local, specialized solutions that are either too restrictive, or not comprehensive enough.

A list of all the requirements a company must fulfill if it wants to assure the quality of its information according to the TQM-based definition, along with a non-exhaustive list of measures, for each category, was developed and presented by Bellocci and Nof (2001a). These comprehensive requirements were derived from the RACF parameters (Resource Access Control Facility) developed by IBM and presented by Schwartz and Zalewski (1999), and from the TDQM parameters described by Wang (1998).

This paper presents some initial results, regarding the importance of particular information quality assurance requirements, based on laboratory experiments and an Industry survey addressing the impact of information failures on ERP performance. It was expected that this work would ultimately lead to a more refined set of assurance requirements for ERP applications.

2 MICSS lab experiments

The initial step in refining the assurance requirements derived from the literature, and showing the variable needs in information assurance, involved a series of experiments conducted with an ERP software simulator called MICSS (Management Interactive Case Study Simulator) [http://www.mbe-simulations.com/, June 2001]. MICSS was used to simulate the functioning of a company with a team-oriented view. The four views considered in the simulator are those of Marketing, Production, Purchasing and Finance. Each of these views can combine to operate the company to be profitable.

In an ERP system, a company enters an operational policy P to ease and automate some of the basic functions of the business, such as production planning. A policy can be described as a k-tuple recording the value d_i of each data item D_i composing the policy; e.g. $P = (d_1, d_2, \ldots, d_k)$. Often the company follows a baseline policy BP, recognized to provide good performance results regarding profits p and due-date-performance DDP.
As a consequence, three typical scenarios regarding information can be encountered in an ERP system. When the company wishes to input its baseline policy BP, the value d_i of a data item D_i can be one of the following:

1. Correct, defined as BP
2. Correct but delayed, defined in this research as $D_4(D_i)$ or $D_8(D_i)$
3. Wrong, defined as $W_d(D_i)$ or $W_h(D_i)$.

A set of experiments was run with MICSS to simulate failures in information exchange and analyze the potential consequences of failures on the performance of the company. A complete description of the experimental process is described by Bellocci and Nof (2001a). The detailed statistical analysis is described by Bellocci et al. (2001).

The main conclusions from the lab experiments were as follows:

1) The experiments showed that information failures had a significant impact on the performance of a company only under specific conditions.
2) Profit was very sensitive to information failures. Due Date Performance reacted more slowly and was impacted significantly only after long lasting and large errors.
3) The impact of information failure depended greatly on the Data item concerned with the failure. For instance, the consequences of a problem concerning Price were usually much more serious and long lasting than when the error concerned Batch Size.
4) Data items had different characteristics that made them more sensitive to a specific type of failure. For instance, a delay of 8 months had a large impact on Profit when it concerned Price, but no real impact when it concerned Batch Size.
5) A difference in the length of delay influenced the performance of the company only when Price was concerned with the error.
6) The impact of an information failure depended on the error size, except when the error concerned Order Level.
7) Each Data item had strong particularities as to which performance metric of the company it affected, and the type of failures that it was most affected by.

It should be remembered that the lab experiments were constrained by the fact that the MICSS simulator does not model open-market competition, and all the decisions were carried out within the scope of the controlled experimental system. Hence, some of the observed effects cannot represent real industry performance. On the other hand, the lab experiments do indicate the relative importance of information quality variables and their impact on decision quality.

3 Industry survey

Based on the lab experiments, an industry survey was designed to assess information assurance requirements in the corporate world. Two questionnaires were developed. One was sent to the information system manager of a given company, and the other one to the department managers of the same company (e.g., production manager, marketing manager…). The objective of the first survey questionnaire was to understand the general approach of companies regarding information security and assurance. The second survey questionnaire was designed to study the actual information assurance problems encountered by users of the company's information system.

A detailed description of the survey is given in Ray, et al. (2001a). The questionnaires, and the detailed analysis of companies' answers are available in Ray, et al. (2001b). The questionnaires were sent to approximately 50 companies in the United States, Europe and Asia. The analysis was based on the 9 questionnaires returned by information system managers, and the 10 questionnaires returned by department managers. The main conclusions from the survey were as follows:
1) Respondents thought that information assurance failures had a significant impact on company performance.
2) Respondents were more concerned by information significance than information security or integrity in their information systems.
3) Profit and Due Date Performance (the reputation of the company) were the parameters most affected by information assurance failures.
4) System Authorizations, Firewalls and Antivirus programs were the most popular preventive measures that companies used to assure the quality of their information and decisions. This observation shows that companies are equipped to handle information security and integrity problems, but not yet to handle information significance problems.

5) Companies introduce flexibility in their information systems mainly using user groups having access to different resources using passwords.
6) The process of assuring the data is too time-consuming for information system users. In a decision-making process, information system users spend more time on acquiring the necessary information, and arguing about its accuracy than using the data.
7) Most companies disregard information due to the fact that it may come from unreliable sources or be inaccurate.
8) If information is missing at the time of changing the policy in their ERP system, most of the companies can wait, but not for very long. If they have to wait longer, they go ahead and change their strategies.
9) Users have difficulty changing processes because of consequential damages due to tight integration.

4 Summary and Discussion

The development of inter-networked enterprises created a new computing environment in which information assurance is critical. This work investigated the information assurance needs of today's companies. A new definition of information assurance was introduced following the TQM approach to better fit the needs of inter-networked enterprises. A list of requirements for information assurance was developed. The critical aspects of information assurance failures in ERP systems were also investigated using MICSS lab experiments and an industry survey.

The lab experiments showed large variability in the impact of information failures, depending on the failure type and the data item concerned by the failure. As a consequence, variable information assurance should be introduced in information systems. The industry survey demonstrated that information significance and its impact on decisions and performance are the true concern of information systems users in inter-networked companies. It also helped in developing a list of examples of assurance tasks that should be automated in distributed information systems.

The following directions can be recommended for future research:
1) It has been assumed in the simulation models that the assurance policy level was fixed over time. The influence of assurance policy level variation over time could be investigated.
2) The simulation models focused on the sequence of optional assurance tasks followed by production tasks. In such a case, the Combined assurance model appeared to be the most advantageous. The development of assurance protocols to distinguish between the processing of requests that need parallel assurance tasks or the ones that need serial assurance tasks could be investigated.
3) The variable assurance approach presented in this research work showed that significant resources can be saved by adjusting the assurance tasks to the request and the context. However, additional resources can be saved if the assurance tasks that are performed on concurrent requests are taken into account, as they increase indirectly the assurance level of the given request. Negotiation-based variable assurance protocols could be investigated to solve this research problem.

5 Acknowledgments

This research was supported in part by the CERIAS (Center for Education and Research in Information Assurance and Security) at Purdue University and by the PRISM Center for Production, Robotics, and Integration Software for Manufacturing & Management.

6 References

Bellocci, T., Planning Variable Information Assurance in Agent-Based Workflow Systems, Master's Thesis, School of Industrial Engineering Purdue University, December 2001.

Bellocci, T., and Nof, S.Y., Context of Information Assurance in Inter-Networked Enterprises, Research Memorandum No. 01-XX, School of Industrial Engineering Purdue University, December 2001a.

Bellocci, T., and Nof, S.Y., Agents and Protocols for Variable Information Assurance in Workflow Systems, Research Memorandum No. 01-XX, School of Industrial Engineering Purdue University, December 2001b.

Dobry, R., and Schanken, M., Security Concerns for Distributed Systems, Annual Computer Security Applications Conference, 1994, 12-20.

Jelen, G., and Williams, J., A Practical Approach to Measuring Assurance, 14th Annual Computer Security Applications Conference, Phoenix, AZ, Dec 1998.

Longley, D., and Shain, M., Data & Computer Security – Dictionary of Standards Concepts and Terms, Stockton Press, 1986.

McGlothlin, J. and Putz-Anderson, V., Advantages and Disadvantages of Quality Assurance in the National Institute for Occupational Safety and Health (NIOSH) Publications Office, Tenth International Conference on Human-Computer Interaction, Crete, Greece, June 22-26, 2003.

MICSS (Management Interactive Case Study Simulator) [http://www.mbe-simulations.com, June 2001].

Ray, P., Bellocci, T., and Nof, S.Y., Information Assurance in Networked Enterprises: MICSS Class Experiments and Industry Survey Conclusions, CERIAS Technical Report 2001-37, Research Memorandum No. 01-08, School of Industrial Engineering Purdue University, June 2001a.

Ray, P., Bellocci, T., and Nof, S.Y., Information Assurance in Networked Enterprises: MICSS Class Experiments and Industry Survey Analysis, CERIAS Technical Report 2001-38, Research Memorandum No. 01-09, School of Industrial Engineering Purdue University, June 2001b.

Schwartz, A.P., and Zalewski, M.A., Assuring Data Security Integrity at Ford Motor Company, Information Systems Security, 1999, 18-26.

Shirey, R., Security Requirements for Network Management Data, Computer Standards & Interfaces, v 17 n 4, September 1995, 321-331.

Steinitz, D., Information Security Management at British Airways: Implementing a Strategic Security Program, 15th World Conference on Computer Security, November 1998.

Voas, J., Protecting Against What? The Achilles Heel of Information Assurance, IEEE Software, January 1999, 28-29.

Wang, R.Y., Total Data Quality Management, Communication of the ACM, v 41, n 2, February 1998.

Wellman, H., Sorock, G., Lehto, M.R., Automated Identification and Correction of Coding Errors in an Accident Narrative Database, Tenth International Conference on Human-Computer Interaction, Crete, Greece, June 22-26, 2003.

Visualization Techniques for Personal Tasks on Mobile Computers

Gerald Bieber

Fraunhofer Institute for Computer Graphics, Rostock
Joachim-Jungius-Straße 11,
D-18059 Rostock, Germany
gerald.Bieber@rostock.igd.fhg.de

Christian Tominski

University of Rostock,
Institute for Computer Graphics
Albert-Einstein-Straße 21,
D-18059 Rostock, Germany
ct@informatik.uni-rostock.de

Abstract

Mobile computers support users to achieve their personal goals. An important task for digital assistance is to represent a user's personal agenda. This has to be done in an intuitive way. In this paper we describe how interactive visualization can be used to reach this goal. Therefore, we describe how spatial-temporal aspects of personal tasks can be represented visually considering the limitations of mobile devices. Furthermore, the personal task management and visualization system *eGuide* is introduced.

1 Introduction

The ubiquitous digital assistance (e.g. PDAs, PalmPilots, PocketPCs, Smartphones etc.) helps users to manage and optimize activity sequences for personal tasks. However, there is a large amount of personal data a user somehow must be informed of. This has to be done in an intuitive way to make it easy for users to get all relevant information from their data.
An adequate way to display a personal agenda is the use of visualization techniques, i.e. mapping abstract data to visually perceptible representations. The capabilities of the human visual system allow the simultaneous recognition of an enormous amount of visual information. Thus, much information can be encoded in one single picture and therefore, visualization is an adequate way for representing a user's personal agenda.
In the field of mobile computing limited resources (especially limited display capabilities) of mobile devices must be taken into account. On the other hand, mobile devices can be used to collect information of the current situation (e.g. location tracking, current time, bio sensors etc.). In doing so, situation aware visualizations can be realized. Thus, the amount of data to be displayed can be reduced and rare display-space can be saved.
In this paper we describe how visualization can be used for personal task management. Section 2 addresses visualization of personal tasks under the limitations of mobile devices. Furthermore, we present a system for personal task management and visualization developed for CeBIT in section 3 and give a conclusion and a perspective on future work in the final section.

2 Personal task visualization on mobile computers

A personal task management system running on an ultra mobile device (e.g. PalmPilot or PocketPC) supports users to optimize execution of their tasks and activities by providing an interactive agenda. It needs input of user tasks, the present situation, and user preferences before an analyzing unit can calculate a schedule. Furthermore, an efficient visualization engine is needed as well. Basic techniques for visualizing personal activities can be classified as follows:

- representation of spatial aspects to visualize where activity occurs,
- representation of temporal aspects to visualize when tasks have to be performed,
- alphanumerical representations like list, text, and hierarchy views,
- representation of task priority, e.g. by varying intensity of colors of graphical objects,
- other representations.

Nowadays, personal tasks are often represented alphanumerically. Most applications only provide text views and simple list views; in few cases there is limited support for hierarchically structured data. A user has to scroll through the data to find relevant information. Thus, searching for information often takes longer than completing tasks. A few special techniques address the visualization of personal tasks on mobile computers. One example is DateLens (Benderson, Czerwinski, & Robertson, 2002). It is based on Table Lens (Rao & Card, 1994) and provides a calendar interface for representation of a schedule. The technique DateLens makes use of the Focus & Context approach (Keahey, 1998) and thus, is also eligible for large data sets. DateLens shows that adaptation of known visualization techniques is a successful way for creating lightweight visualizations for mobile computers.

Although several visualization techniques exist, they are all designed for one specific purpose only. Thus, it was our aim to integrate techniques that address temporal, spatial and personal aspects of tasks in a new visual personal task management system. An additional design goal was to create a system where all information representations are linked together.

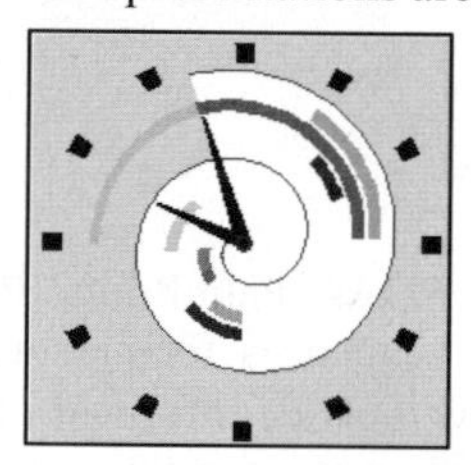

Figure 1: SpiraClock's watch hands are representing the current time. Tasks are visualized by curved shapes placed at the spiral according to their beginning and finishing time. Tasks of the current hour are at the outer cycle of the spiral; farther tasks are placed at the inner cycles. The status of task performance is represented by a change of color intensity.

One technique we found eligible for visualizing temporal aspects of personal data is SpiraClock (Dragicevic and Huo, 2002); but the support for spatial aspects is limited. Though SpiraClock is already implemented, we had to re-implement it. Our implementation of SpiraClock (see Figure 1) considers the requirements of mobile devices. Thus, graphical effects have been omitted and the interface now meets the special needs of pen-based input. Furthermore, it had to be designed as a modular component to allow an easy integration into our system. In order to realize linking, tasks are represented as curved shapes within the shape of the spiral. These task shapes are input sensitive and can be clicked to open a selection of possible links (e.g. to get further information about a task or the get a map view centered at the location where the task has to be performed).

As mentioned before, by using SpiraClock it is difficult to visualize a larger number of locations. Thus, we needed a technique to represent spatial aspects of the personal agenda. Maps are commonly used for this purpose. There are two possibilities when using maps:

- raster maps, based on raster images or
- vector maps, based on a geometrical description of locations.

The advantage of raster images is that they are easy to handle and fast to display. However, they are not as flexible to meet the requirements of a dynamic task management system. Thus, we decided to use a vector map. An eligible base for the development of a map visualization engine has been found in the Scalable Vector Graphics (SVG) specification of the World Wide Web Consortium, more exactly, in the SVG Mobile Profiles: SVG Tiny and SVG Basic (W3C, 2003). An advantage of SVG is that it is a standardized file format based on XML grammar. Thus, it is easier to acquire and integrate customer specific map content, and even the automatic generation of map content from databases is possible by using new technologies like XSLT. Furthermore, SVG content can be dynamically changed when the Document Object Model is used as internal data structure. By doing so, we can react to arbitrary changes of a user's environment by adapting the SVG data structure. This can be realized by changing attributes of SVG nodes. Furthermore, using SVG gives us the opportunity to realize transformation of the map content (e.g. for rotating the map according to a user's heading), as well as rendering the map content to different regions of the display (e.g. for realizing an overview map). Another argument for the use of SVG is that the desired linking capabilities are an essential part of the SVG specification. Since our system is intended to link all information, we implemented an SVG viewer for SVG Tiny that fulfills our special needs. Thus, we are able to display graphical map content. Additionally a linkage of graphical map objects with further object specific information can be realized by simply clicking an object.

Nevertheless, we especially had to consider the limited display space of mobile devices. Personal task data depends on resources like location and time. To give an example, one might imagine a manager at a trade fair listening to a presentation. After finishing the presentation the manager's next task will mainly depend on where he is located and the current time. Fixed dates of the manager's schedule must be considered too. Thus, our intention was to show a user relevant information only and to focus the visualization on possible next task, i.e. spatial-temporal nearby information such as presentations starting during the next minutes, booths of companies saved in the managers contact file and so forth.

We found the approach of situation awareness suitable for our needs. One important task for a situation aware application is location tracking, i.e. determining where a user is located. According to the availability of location information we differentiate between:

- continuous location tracking and
- discrete location tracking.

While continuous location tracking permanently provides a system with data about a user's coordinates in space (e.g. by calculating WLAN field strength), discrete location tracking means getting information about the location of a user at specific positions in space only.

Nearly every today's mobile computer is equipped with an IrDA sensor. Thus, we are using special IrDA beacons to realize a discrete location tracking. These beacons can be easily mounted at doors, information booths or other places of interest. The positions where the beacons are located are stored in a small database on the mobile computer. Thus, when getting a signal from a specific beacon, we know the position of the user. Furthermore, the internal clock of mobile computers provides us with current timestamps. Aware of current location and time we are able to search within an activity database for relevant information, e.g. presentations shortly starting behind a nearby door. Additionally, with the information we found, we are able to dynamically calculate a route through a user's tasks. In connection with the route a representation of the relevant information is presented. Now the capabilities of our map engine get involved. By altering the transformation matrix of the map view a zoom and pan according to the users location is realized. By adding and removing nodes of the SVG data structure relevant information becomes visible while irrelevant information is hided. Another important feature of the map engine is the intuitive representation of the personal route. Since, the route is calculated not only by considering space, but also time, we can provide a user with temporal aspects of the route as well. Therefore, we developed a color scaled route representation that can be enhanced by placing

annotations at certain points of the route (see Figure 2). In doing so, users are always informed about where they have to go and when they will be at a desired location. This eases a user's personal task planning significantly. Additionally, priority of tasks and status of task performance can be visualized be changing color intensity or saturation. This can be realized again by altering attributes of nodes of the SVG data structure.
For future development the integration of further sensors is planned; new sensor technologies for location tracking have been already investigated (e.g. WLAN navigation and motion sensors). Interesting sensors might be acceleration or velocity sensors to dynamically change font size while a user is in motion, compass sensors to realize an automatic map rotation according to a user's heading and biosensors to track current personal body situation.

3 eGuide – task visualization for the CeBIT

A lot of fair guides do already exist; Fraunhofer IGD implemented the first official mobile electronic guide for the computer show CeBIT (Bieber & Giersich, 2001). The problem is that task management is missing in the known solutions and visualization techniques are rarely used. The scope of our research is dynamic personal assistance for visitors of trade shows or conventions under consideration of a users current situation integrating task management and spatial-temporal aspects of visualization. Thus, a new system called *eGuide* has been created for such in-house scenarios considering the stated aspects. To realize a platform independent system we are developing our application according to the PersonalJava™ Application Environment Specification (Sun Inc., 2000).

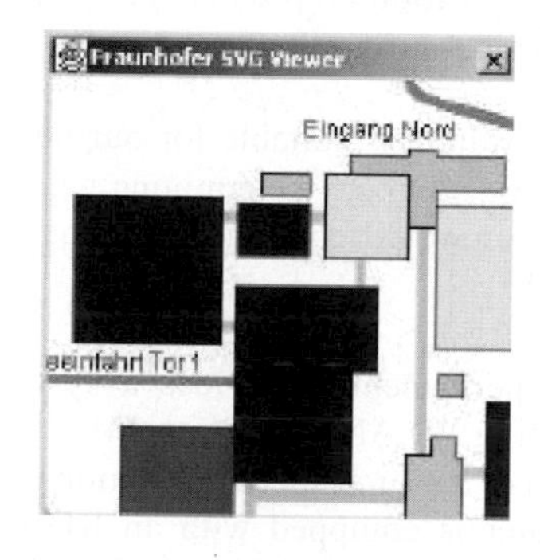

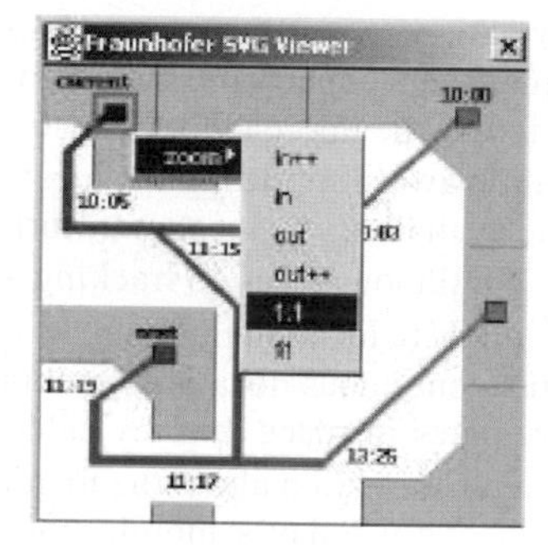

Figure 2: Map view including importance information encoded by using color intensity (left); SVG map, simple but meaningful figures of convention halls (graphical abstraction) (middle); Personal tour planning and representation using the SVG map engine (right).

Following our aim to integrate intuitive representations, the system *eGuide* provides list views, text views and hierarchical views, as well as an agenda view (e.g. SpiraClock, see Figure 1), and a map view (adapted SVG map engine, see Figure 2). An important feature of our representations is that they are all linked together. In order to realize the linking concept a user's preferences (e.g. the aim to visit specific exhibitors or to listen to certain presentations), as well as present situation of a convention (e.g. hall and booth locations, opening hours, schedule of presentations etc.) are considered. Thus, if the map shows a booth of an exhibitor the user is interested in, via click on the booth's representation a link to further information is opened. Additionally, by selecting an exhibitor form an offered list a user can be linked to detailed company information in an extra text view or to the map view, while the map is centered at the exhibitor's booth.

The intention for scheduling is to dynamically calculate a personal and individual plan (for the complete duration of a convention) and to create a visual representation of this plan a user can interact with. Since we have to find an optimized way through the different aims, tasks, and user preferences, a fast heuristic task-scheduling algorithm has been implemented. In doing so, not only a personal route through the exhibition can be calculated, but also an estimation of length and duration of the entire journey can be given. The use of color scales to represent task priority and task performance status allows a user to detemine the possible next task easily. This is even increased by using the color scaled route representation.
The approach of the Fraunhofer Institute is to represent personal tasks in an intuitive way. To reach this goal, we use an iterative process of application development. Requirements for this iterative approach are the constant use of the application and it's easily adaptable software architecture; both requirements are met by *eGuide*. Within our research we plan to evaluate the results of personal activity scheduling and graphical representation at the booth of Fraunhofer at CeBIT 2003, at International Workshop on Mobile Computing (IMC, 2003) in Rostock and other related events. The user feedback enables us to realize the iterative approach and thus, to successively design a more and more user friendly and intuitive application.

4 Conclusions and future work

A convention or trade fair visitor has to perform various different tasks under the consideration of space, time, and personal preferences. We described how a system that integrates personal task management and visualization can be used to help users optimizing their tasks and thus, achieving a better navigation and support within personal schedules. We presented how a dynamic spatial-temporal representation of the personal agenda can be realized in connection with input of several sensory components. Furthermore, we described the electronic trade fair and convention guide *eGuide* and that an iterative user feedback based approach for application development can be used to create a user-friendly application.
Further research will address the improvement of the used data structures and scheduling algorithms and the enhancement of the visualization to realize an even more user-friendly application. One of our main interests is the integration of further sensory components.

References

Bieber, G. and Giersich, M., (2001). Personal mobile navigational systems - design considerations and experiences. In Selected Readings in Computer & Graphics 2001, (pp. 505-514). Stuttgart: IRB Verlag.

Benderson, B.B., Czerwinski, M.P., and Robertson, G.R. (2002). A Fisheye Calendar Interface for PDAs: Providing Overviews for Small Displays. HCI-Lab Technical Report #2002-09, University of Maryland.

Dragicevic, P. and Huot, S. (2002). SpiraClock: a continuous and non-intrusive display for upcoming events. In Proceedings of the ACM Conference on Human Factors in Computing Systems (pp. 604-605). New York: ACM Press.

Keahey, T. A. (1998). The Generalized Detail-In-Context Problem. In Proceedings of the IEEE Symposium on Information Visualization (pp. 44-51). Los Alamitos: IEEE Computer Society.

Rao, R., and Card, S. K. (1994). The Table Lens: Merging Graphical and Symbolic representations in an Interactive Focus+Context Visualization for Tabular Information. In Proceedings of Human Factors in Computing Systems (pp. 318-322). New York: ACM Press.

Sun Inc. (2000). PersonalJavaTM Application Environment Specification. Sun Microsystems, Inc.

W3C (2003). Mobile SVG Profiles: SVG Tiny and SVG Basic. www.w3.org/TR/SVGMobile/

Agent-Based User Interface Customization in a System-Mediated Collaboration Environment

Holger Brocks, Ulrich Thiel & Adelheit Stein

Fraunhofer IPSI
Dolivostr. 15, 64293 Darmstadt, Germany
{brocks,thiel,stein}@ipsi.fraunhofer.de

Abstract

Within a collaborative information environment, the system acts as a mediator between remotely located users. The user interface of such an information system has not only to provide support for the task to be performed, but also to reflect the ongoing discourse context. The MACIS Framework has been developed to support the design and implementation of collaborative, interactive information systems. Using a multi-agent approach, the whole information system, including its user interface, is decomposed in terms of various classes of cooperating agents. Since the users are also modeled as part of the multi-agent organization, interactions between all constituent agents are described in a uniform way.

1 Introduction

Intelligent information systems are characterized as interactive systems which support various tasks using knowledge about the users and their task-related actions. To support collaboration between users imposes additional requirements on the underlying technical infrastructure. The MACIS Framework (Multiple Agents for Collaborative Information Systems) is designed to support the development of complex, collaborative information systems exhibiting intelligent, pro-active behavior. The system, including its user interface, is described in terms of various classes of cooperating agents. Since users are also modeled as part of the multi-agent society, collaboration between users is seamlessly incorporated as a special form of inter-agent communication.
The user interface comprises dedicated user interface agents (UI agents), which serve as intelligent mediators between the users and the system. Hence, user input is no longer regarded as some method invocation, but as a communicative act expressing the current discourse goal of the user. Since users and agents are situated within a shared context, their interdependencies and interrelations are organized according to their roles within the information system.

2 Modeling User Interfaces for Collaborative Systems

Direct manipulation user interfaces, allowing for multi-threaded interaction and immediate semantic feedback, help to provide a comprehensible, predictable, and controllable environment for the user. Unfortunately, direct manipulation (see Shneiderman, 1997) does not take aspects of collaborative work into account. Within a collaborative environment, there is no de-facto control over the actions of other users. Hence, in the context of collaboration we have to shift from direct manipulation towards delegation (see Maes, 1994) in order to provide an adequate, non-intrusive representation of the interaction process. This approach regards the interaction with a collaborative

system as formal communication between the user, his remote human partners as well as components of the user interface which are constructed as software agents.
Wooldridge and Jennings define an agent as a software system which is characterized by autonomy, social ability, reactivity, and pro-activeness (Wooldridge & Jennings, 1995). Multi-agent systems can be classified as a society of agents which cooperate to achieve common objectives. Situated within a shared context, the dependencies and relationships between those agents are organized according to the roles they represent in the system.
Taking an agent-based perspective we have to extend the notion of an intelligent information system as a technical infrastructure to a more holistic level, which not only takes the system but also its users into account. Consequently, the concept of inter-agent communication has to also be applied to cover the communication required for human users to collaborate with each other. This requires the interactions between the user and the system (via UI agents) as well as corresponding communication acts between cooperating agents located in distributed peers to be modeled in a uniform way. In the following, we present a holistic approach towards development environment for collaborative information systems, which incorporates the users as integral parts of the multi-agent organization.

2.1 The User Interface as a Multi-Agent Society

The strong resemblance between the models for inter-agent communication and human-computer interaction enables us to seamlessly integrate the users as part of the multi-agent community. To be more specific, we propose to model interface components as autonomous software agents, which serve as intermediaries for the user to interact with the system. Hence, user input is not regarded as some method invocation, but instead as a communicative act expressing a discourse goal of the user, which allows for rich social interactions between both user and information system, and (possibly remotely located) cooperating system agents (Jennings, 2000).
User interface agents (UI agents) specialized in human-computer interaction accept user input and display data, i.e. they provide their own specific set of user interface functions. (Presentation-Abstraction-Control, Buschmann et al., 1996). UI agents, as building blocks of the user interface, are thought of as an organization of active components in that they can be dynamically instantiated and reconfigured, they are capable of forming relationships, and they are able provide active assistance for the sub-task they support. Service agents constitute the data model of the application. Their responsibilities range from data acquisition and management, operations on data structures, coordination activities towards distributed problem solving, i.e. they represent the functional core (see Coutaz et al., 1995) of an interactive system.
And finally, the users are to be considered as part of the multi-agent community in that human-computer interaction can be interpreted as a form of agent communication. However, they need UI agents as intermediaries in order to communicate with the information system. Users and UI agents interact to perform a common task, i.e. UI agents mediate between the system and the user.
Hence, an intelligent information system can be defined as a multi-agent environment, where various types of agents (users, service agents, and user interface agents) cooperate to achieve shared goals. Given some formal representation of the tasks to be performed, the overall complexity of the information system can be decomposed in terms of the corresponding sub-tasks, and the agents required supporting them.

2.2 The MACIS Framework

Without any inherent support for collaboration functions, standard software engineering techniques alone fail to provide an adequate methodology for developing collaborative information systems. Using a multi-agent approach, the MACIS Framework[1] introduces collaborative aspects in a natural way. Dynamic, system-mediated negotiations between users and inter-agent communication are modeled in a uniform way.

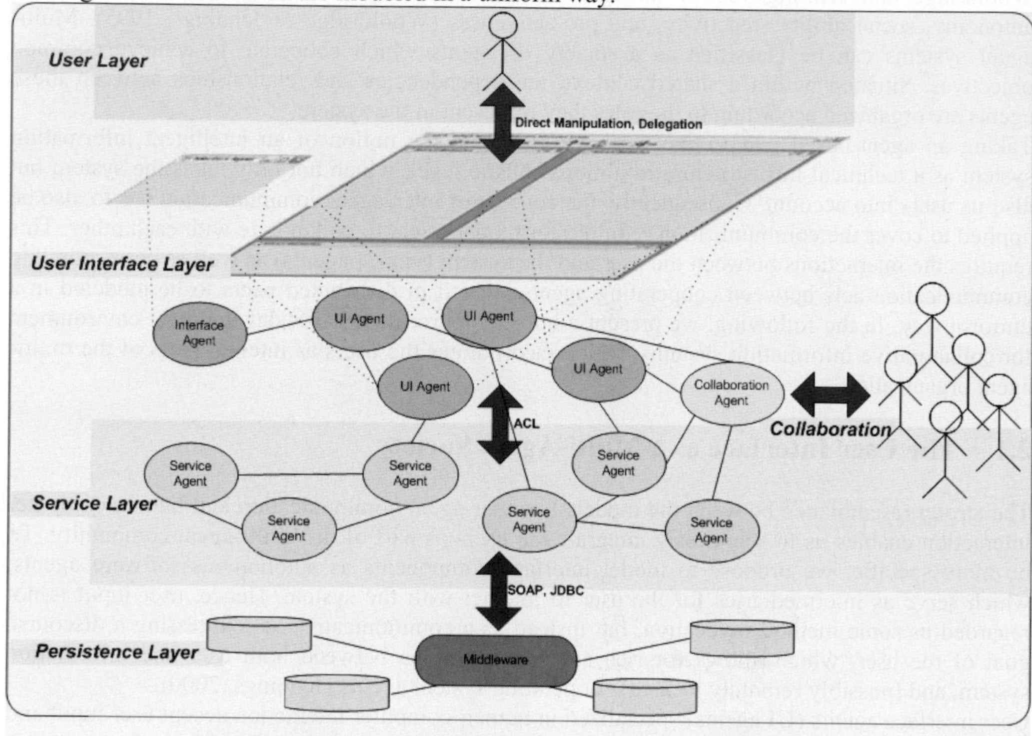

Figure 1: The MACIS Framework – Conceptual Overview.

2.2.1 User Layer

The user layer provides a representation of the current user in terms of the tasks to be performed, her role in those tasks, and the communication acts required to model her interactions with the system.

2.2.2 User Interface Layer

The actual user interface is composed of UI agents, which may form organizational relationships to support more complex tasks. The user interacts with UI agents in a standard way, e.g., using direct manipulation techniques.

2.2.3 Service Layer

The service layer is comprised of those agents which do not provide a user interface. They interact with other agents (including UI agents) to achieve their design objectives. Service agents can also be grouped as complex semantic agent entities to perform more extensive tasks.

2.2.4 Persistence Layer

The persistence layer is responsible for the storage and retrieval of the information objects constituting the application domain.

---

[1] The MACIS Framework is under development. JADE (Java Agent Development Framework, [JADE, http://sharon.cselt.it/projects/jade/]) serves as an underlying FIPA-compliant multi-agent platform, which provides the messaging infrastructure required for inter-agent communication.

2.2.5 *Collaboration*

The MACIS Framework incorporates asynchronous, system-mediated collaboration between users working in distributed peers by forming dynamic organizations of users, UI agents, collaboration agents and service agents.

2.3 Case-Study: System-Mediated Collaboration in COLLATE

The European-funded COLLATE project aims to design a platform-independent and cross-organizational collaborative information space for the Humanities. Historical film documentation serves as example domain for the setup of a technical environment for shared access to a digital repository comprising digitized multi-format documents on several thousand European films of the early 20th century.

Set in the domain of historical film documentation, the main goal is to support collaborative, interpretative work on digitized source documents. In the course of conducting scientific analyses and interpretations, users are enabled to explicitly request assistance from other users for accomplishing their tasks. The recipients of those requests are notified about its arrival and react to it via a corresponding UI agent. Thus, collaborating users can engage in lively discussions about certain interpretations, reference other cases of precedence, or simply argue certain statements.

Assume *User A* is looking for censorship documents referring to the film "M" by Fritz Lang. From the document list shown, a Czech censorship decision is of particular interest to her. She requests the document to be retrieved and views the annotations already associated to it. On page 2, she spots a document passage, which has been marked-up and annotated by *User B*.

The annotation by *User B* contains a brief summarization (and translation) of the application procedure in the Czech Republic for the film "M". Since *User A* knows this film very well, she elaborates on the subject and provides some additional information about the censorship of the film "M" in Germany before and after the Nazi regime took over. Nevertheless, she isn't completely confident about some of those facts (there were contradicting documents), therefore she asks her colleagues to check her statements.

But which colleagues should be asked? *User B* would be a likely candidate, other colleagues are known personally, but since she doesn't know all users of the system it not clear who would be best qualified to verify her statements. Meanwhile, a collaboration agent has already analyzed the current working context, i.e. the active document set, and requests (from another agent) to retrieve the IDs of similar documents. Given this set of potentially related documents and the active document set, it contacts other remotely located collaboration agents and asks whether any of their users is or has recently been working on some of those documents. Those collaboration agents, whose document set (partially) matches the request then decide, whether they are willing to cooperate and comply with the request. Based on the retrieved set of positive answers, the collaboration agent of *User B* prepares a ranked list of possible collaborators, and requests the UI agent responsible for annotations to incorporate that list as a menu in its interface. From the list displayed, *User A* chooses the persons she would like to ask for some feedback on her comment. The cooperating collaboration agents receiving the incoming request then assess its relevance with respect to the current discourse context, decide about an appropriate visualization in the user interface (e.g., pop-up a window), and might pro-actively instruct a service agent to retrieve the related data items. If the user chooses to process the request, the document, including its associated metadata, is accessed via corresponding UI agents without any further delay.

Unlike interface agents described in (Maes, 1994 or Rich & Sidner, 1996), who are designed to assists human-computer collaboration; collaboration agents act as intelligent mediators between human users and the ongoing discourse situation. Collaboration agents instruct the UI agents to reconfigure their presentation and their choice of appropriate collaboration functions. Furthermore, they pro-actively initiate communication with other collaboration agents located in remote peers in order to, e.g. seek out users working on similar documents, and notify the user about potential collaborators.

3 Conclusions and Future Work

We have demonstrated that a multi-agent approach using the MACIS Framework is ideally suited to support virtual teams discussing the interpretation of documents. System-mediated, task-level support for human collaboration can be provided by dynamic, situation-adequate instantiations of corresponding UI agents which adapt and react to their environment in a flexible way, i.e. the user interface actively reflects the influences evoked by collaborating users.
UI agents enable an elegant, non-intrusive visualization of the status of an ongoing collaborative discourse in that they are capable, at least at a syntactic level, of dynamically adapting their appearance according to the surrounding discourse situation. Depending on the current user role and the actual task, UI agents act as active mediators in the discourse context, i.e., they offer a situation-adequate choice of collaboration functions to the user.
Hence, a complex intelligent information system demonstrating this level of autonomy in its constituent modules has to maintain a conceptual model of the collaborative discourses being established, i.e. the various actors (users, UI agents, system agents) and the communicative acts contributed as part of their cooperative negotiation for achieving shared objectives.
A prototype implementation of the MACIS framework has been used for designing the client application in COLLATE, a European-funded project aimed at establishing a platform-independent and cross-organizational collaborative information space for the Humanities.

4 References

Wooldridge, M.J. & Jennings, N.R. (1995). Intelligent Agents: Theory and Practice. Knowledge Engineering Review, 10(2), 1995.

Shneiderman, B. (1997). Direct Manipulation for Comprehensible, Predictable and Controllable User Interfaces. Proceedings of the 2nd international conference on Intelligent User Interfaces (IUI'97), 1997.

Buschmann, F.; Meunier, R.; Rohnert, H.; Sommerlad, P. & Stal, M. (1996). Pattern-oriented Software Architecture, Jon Wiley and Sons Ltd, 1996.

Coutaz, J.; Nigay, L. & Salber, D: (1995). Agent-Based Architecture Modelling for Interactive Systems. Critical Issues in User Interface Engineering, Springer Verlag, 1995, 191-209.

Jennings, N.R. (2000). On Agent-Based Software Engineering. Artificial Intelligence, 117(2), 2000, pp. 277-296.

Maes, P. (1994). Agents that Reduce Work and Information Overload. Communications of the ACM, 37(7), 1994.

Rich, C. & Sidner, C. (1996). Adding a collaborative agent to graphical user interfaces. Ninth Annual Symposium on User Interface Software and Technology, 1996, pp. 21-30.

Organisational Learning: An Investigation of Response to Rapid Change in a Traditional Environment

Kathy Buckner and Elisabeth Davenport

School of Computing, Napier University,
Merchiston Campus, 10 Colinton Road, Edinburgh EH10 5DT
k.buckner@napier.ac.uk; e.davenport@napier.ac.uk

Abstract

This paper is a case study of 'collective' HCI in a 'top-down' systems project – a very large computing center in a new university in the United Kingdom. The authors present the findings of a small empirical study of computer supported teaching and learning in the new environment. Using activity theory, they discuss individual and organizational responses to this initiative in its first 18 months of operation.

1 Introduction

In September 2001, a very large computing centre (VLCC) was opened at an institute of higher education. There had been little prior consultation with instructors and students, and the centre did not conform to existing perceptions of a teaching space. This led to initial difficulties in relation to acceptance and adoption of good practice. An earlier report focused on observations and findings in the actual teaching space through covert participative observation (Buckner & Davenport, 2002). This paper presents the next phase of a longitudinal study of classroom transformation in the VLCC; it focuses on organisational impact.

2 The context: 'user-free' design

Traditionally, spaces designed for teaching with computers in UK higher education institutes have been comprised of laboratory or workshop facilities containing between 20 and 50 computers. The computers are normally positioned either around the walls of the room or in parallel rows. In either case there is a clear 'front' of class from where the teacher can instruct if they wish; identifiable boundaries (in the form of classroom walls) and a semi-secure entrance which can be closed if the instructor desires. Normally there is a computer linked to a dedicated projector from which the instructor can demonstrate particular activities and other teaching artifacts such as overhead projectors and whiteboards. The use of the space can be controlled by the instructor (eg s/he can choose to allow student not in their class to use the spare capacity or not) and the attention of the students can be gained and held through clearly identifiable lines of sight and lines of communication.

An efficiency drive in the institution that hosts the VLCC has led to a number of modifications in the use of space. The VLCC which was set up during the summer of 2001 was inspired by a vision of economies of scale. It offers a high-density open plan teaching environment for computing students, consisting of 18 clusters or 'pods' of 24 tightly packed machines, back to back on two rows of twelve. The facility is open 24 hours a day, and at full capacity can house 500 individuals. Unlike the traditional teaching spaces, each pair of pods is bounded only by a waist high surround, and access to them is by means of a system of intersecting aisles. There are no walls or other form of physical boundaries between pods although some 'virtual boundaries' are formed where upper and lower tiers are separated by height and there are no physical barriers at the entrance to each pod.

Teaching staff returning from summer vacation in September 2001 were faced with a radically new environment in which to teach with little time to acclimatise or rethink teaching strategies prior to the start of Semester. The authors saw this as an opportunity to explore emerging work practice, and account for reactions and strategies to what was in effect a 'user-free' design and implementation exercise.

3 Theoretical background

We consider the VLCC to be a collective activity system in which the study of activity focuses on the actions of and interaction between individuals, their relationship and interaction with systems and artifacts and the continually changing organisational setting (Wells & Harrison, 2000). Emerging literature on social computing (Dourish, 2000) and more established perspectives such as activity theory (Engeström, 2000) have informed our investigation of 'CHCI', or collective human computer interaction.

Using Dourish's (2000) approach to the study of embodied interaction we have focused our attention on mundane practices that occur within the specific context of the learning environment of the VLCC, to see if what might start as somewhat 'chaotic' practice becomes more orderly through adaptive behaviour.

To further aid our understanding of what and how changes in practice occur, we consider the notion of organisational learning. Argyris and Schön (1996 p16) suggest that "organisational learning occurs when individuals within an organisation experience a problematic situation and inquire into it on the organisation's behalf." The 'problematic situation' in this case is 'how to support student learning effectively in a non traditional teaching space'. A particular focus of the investigation is the way in which tacit knowledge about effective teaching methods is externalised and the extent to which it has or has not become embedded in organisational practice. To date (February 2003) we have completed two phases of the study (Buckner & Davenport, 2002). These have involved 'covert' observation: the first author enrolled as a part-time student; the second author 'picked up' information at staff meetings and other informal gatherings.

4 Empirical study

In Phase One: 'The Phantom Wall Syndrome' we observed modifications in habits and perceptions when work practice was forced into the new habitat of the VLCC. Unsurprisingly, previous habits, practices and expectations of behaviour were carried over and then adapted for use in a new environment during the initial stages of change. Again, unsurprisingly, some practitioners, in particular those that could be referred to as the early adopters - adapted more readily than others to the change (Rogers, 1983). Discussion among teaching staff (formal and informal) in the early stages focused on problem definition rather than problem solution with emphasis on negative aspects of the change.

In Phase Two: 'Adaptive Practice' there was some evidence that practice had been modified to meet some challenges of the new environment. At both individual and organisational level, we observed changes in ownership and occupation strategies at the level of individual learning spaces or pods, and there were observable changes in the ergonomics of getting, seeking and maintaining attention. These were to some extent due to the failure of management to respond rapidly to initial requests in the first semester by "shocked" instructors for teaching artifacts to be provided. As a result when screens and projectors appeared in the second semester they were under used. Interpersonal practice, or the conduct of face-to-face teaching, was already undergoing a transformation as part of a national change in the focus of higher education that emphasised individual learning. In the VLCC this was evident in a shift away from whole group interaction to individual or small group interaction supported by electronically mediated scaffolding. Overall although some instructors perceived the VLCC as a rather brutal intervention others could see how it fitted with a process of transformation that was already underway.

We have just completed a third phase of the study, that we have labelled 'Partial Adaptive Practice'. Given our findings in the 2nd phase, we anticipated that emerging shifts in practice would result in prescriptive organisational intervention in the form of mentoring and/or provision of guidelines for new staff. In fact, there has been relatively limited organisational activity of this kind, and it is only in the past month that colleagues have organised a survey (a little over half of those involved in VLCC teaching responded) to assess perceptions of the VLCC experience. Though reports on specific problems have been raised in staff meetings since the opening of the VLCC this is the first attempt to present consolidated evidence to the appropriate committees. We are intrigued by the time lag between identification of problems and intervention at the organisational level and have turned to activity theory (Engeström, 2000) to account for the apparent organisational inertia.

5 Analysis and discussion

We initially constructed activity diagrams for two groups: learners and instructors. Student transformation by learning (the object of activity) is influenced by complex interactions between the subject (ie learner or instructor), tools (eg computers, projectors; signage, learning resources etc), rules (eg speed of learning; control of space), community (eg instructors; students; technical support; management), division of labour (by type of learning or instruction). We used the diagrams to distinguish between the learner (L) experience and instructor (I) experience. (These are not, of course, homogeneous groups, and we concede that we have used the diagrams as rather blunt instruments). There is some divergence in the resulting 'profiles', notably in the areas of tools, rules and division of labour. But within the 'I' group, there are further 'primary' contradictions (see Turner and Turner, 2001). We have captured these by using the terms Instructor-Adopters (I-A) and Instructor-Resisters (I-R): these can be modelled in two separate activity diagrams capture differences in their respective activity systems. (In reality these are not homogenous groups and instructors are positioned somewhere along the continuum from adopter to resister).

On the basis of interviews with key informants, our perception of the goal or object of the instructors is that while both are focused on the student, I-As are more focused on the outcome of facilitating effective student learning. There are however two categories of resister, some I-Rs (I-R/L) may be similarly focussed on facilitating student learning but are resistant to change because they believe that other types of classroom better support student learning while other I-Rs (I-R/T) are more focused on transforming the student through effective teaching. This fundamental difference has a significant impact on the extent and manner in which affordances in the physical and virtual teaching and learning environment are exploited. It also affects the way in which different instructors adapt to the space. Students appear to have embraced the new environment, and thus the instructor who focuses on 'student learning' is more likely to embrace the new space.

I-As are most likely to see useful tools as those which support 'individual' self paced student learning. These may include web-based resources and workbooks which the student can use either individually or with colleagues in small group activities. I-As do not normally require external props such as whiteboards and PC projectors as their interaction with students is on a one-to-one or small group basis. I-R/Ts on the other hand prefer to lead whole group activities (at least for a part of the time allotted for the class) which often require the use of tools such as PC-projectors and whiteboards. Tensions arise when these artefacts, previously accepted as standard, are no longer readily available or easily accessible.

Division of labour, differs according to type of instructor. I-As allow the student to lead the learning process, and proactively support the student according to individual need. I-R/Ts however prefer to lead the students as a class or group through the learning process and consequently find an environment which is more conducive to independent learning more difficult. Some managers have suggested that a decrease in the level of visible presence of instructors in the VLCC indicates that they are simply abandoning classes, however our observations suggest that I-As and I-R/Ls have altered the balance of visible and invisible work by 'entering' the machine. They do this by supporting learning through the development and use of online learning materials. They may still be present in the classroom though demonstrators may assume some of their support role there, and they are not as highly visible as traditional up-front 'teachers'.

Engeström's 'subject/rules/object' complex has allowed us to unravel a major nexus of contradictory practice in the instructor group. How to 'set the pace' is a major concern. I-As generally work in a facilitating manner in which the student will set the pace of learning. I-As ensure that each student is progressing satisfactorily but they not overly concerned if students approach the problem from different perspectives. I-R/Ts, however, prefer a more didactic approach in which they set the pace. They are more comfortable if all the students in a class are working on the same problems each week, and find it more difficult to adapt to individual learning styles, speed and approach. This is less well suited to the VLCC where it is more difficult to lead and control whole group activities. I-R/Ts thus prefer to have more exclusive control and ownership of the space in which they teach – which as we have seen in previous phases is extremely difficult with regular 'incursions' from students who are not members of the class. I-As, in contrast, are content with less exclusive control and are more tolerant of incursions provided that the interlopers do not disturb the class.

6 Conclusion

We have found that the community of practitioners who regularly instruct in the VLCC discuss their practices with others. New instructors discuss teaching approaches with co-instructors. Established instructors discuss practices in informal situations eg at coffee, in the corridor and more formally in discussing teaching approaches for the next semester. In this way an environment is engendered in which knowledge, values and assumptions are shared. The tacit knowledge of 'how to' teach in this particular space becomes externalised and available for adoption by others. Significantly, it does not rely on formalised training sessions.

To date, no organisational policy on teaching and learning in the VLCC has emerged. We have thus not found evidence of explicit organisational learning in relation to the VLCC in its first 18 months of operation. This was contrary to our expectations, as the department that hosts the VLCC has well-established structures to promote organisational learning by means of formal and informal programmes for staff development and for good practice through the teaching and learning group. The VLCC appears to have been a 'no go' area as far as institutional response is concerned, though, as we note above, individual instances of difficulties, and specific requests for equipment were presented to staff meetings. Our (limited) analysis with activity theory has provided us with insights into this paradox. It suggests that the issues are complex (Fleck and Howells, 2001), and that a naïve institutional response (in the form of workshops, for example, to promote good practice) might do more harm than good. The VLCC touches on fundamental issues of differences in teaching philosophy and personal sense-making that will take time to resolve. What we describe as 'organisational inertia' in an earlier section may in fact be 'organisational caution'.

7 Acknowledgements

We gratefully acknowledge the support of staff and students who have contributed to this study.

8 References

Argyris, C.& Schön, D.A. (1996) Organizational Learning II: Theory, method and practice. Reading MA: Addison Wesley.

Buckner, K. and Davenport, E. (2002) *Teaching and learning in the VLCC: actions, reactions and emerging practice in a very large computing centre.* In S. Bagnara, S. Pozzi, A.Rizzo and P. Wright (Eds.) Proceedings of the 11th European conference on cognitive ergonomics. Instituto di Scienze e Tecnologie della Cognizione Consiglio Nazionale delle Ricerche. 355-360.

Dourish, P. (2000). Where the action is. MIT Press.

Engeström, Y. (2000). Activity theory as a framework for analysing and redesigning work. *Ergonomics* 43(7) 960-974.

Fleck, J. and Howells, J. (2001) Technology, the technology complex and the paradox of technological determinism. *Technology Analysis and Strategic Management*, 13 (4), 523-532.

Rogers, E. (1983) Diffusion of Innovations (3rd ed.) New York, Free Press.

Turner, P. and Turner, S. (2001). Describing team work with activity theory. *Cognition, Technology and Work* 6, 127-139.

Wells M. & Harrison R. (2000) *Emergent Behaviour of Human Activity Systems.* Workshop on Distributed Cognition and Distributed Knowledge: Key issues in design for e-Commerce and e-Government 14 - 15 June 2000 in Schärding, Austria http://falcon.ifs.uni-linz.ac.at/workshop/wells.doc last accessed 23 October 2002.

Explication and Legitimisation of Arguments and Outcomes in Sense-making and Innovation by Groups: Some Implications for Group Support Systems

Petrie Coetzee

Faculty of ICT
Technikon Pretoria
Private Bag X680
Pretoria 0001
South Africa
Petrie@techpta.ac.za

Jackie Phahlamohlaka

Department of Informatics
University of Pretoria
Hatfield
Pretoria 0002
South Africa
jphahla@hakuna.up.ac.za

Abstract

We argue the need for explicitly facilitating group support system (GSS) processes in those type of activities focused on sense-making by groups and particularly for making arguments explicit and ensuring legitimacy of arguments and outcomes. To this end we consider some relevant views and concepts from literature on design of inquiry systems. Certain principles bearing on debate and particularly on legitimacy and explication of arguments and outcomes are surfaced. Two proposals are offered for facilitating legitimisation and explication of arguments and outcomes. Implications are derived for use and design of GSS in view of these suggestions.

1 Introduction

We are interested in providing users of group support systems (GSS) with conceptual frameworks for using GSS so that they may more adequately achieve communal purposes. These interests lead us into research focused on design and development of such frameworks. We are particularly interested in such frameworks for application in the context of use of GSS for problem solving, sense-making and innovation. These areas of application of GSS have this in common: they can invoke intense debate amongst participants in the GSS process. We do not here wish to present or discuss any particular conceptual frameworks, since we are presently in the very process of experimenting with possible frameworks. We wish to report in due course on the outcomes of those experiments. We do however wish to focus on merely two particular aspects here. One is the aspect of explication of arguments in the process of sense-making and the other is that of legitimisation of outcomes of sense-making. In this paper we merely wish to argue for the sensibility of being concerned with these particular two aspects and for the sensibility of formally incorporating means for addressing these concerns in GSS processes.

2 Explication of Arguments

If sense-making or innovation are activities aimed at solving or resolving problematic situations, they need to lead to a decision on a proper or adequate way of dealing with the situation. This implies that alternatives need to be generated for dealing with the situation. This requires multiple perspectives to be brought to bear. Generation of such alternatives might require creativity, which may need to be fostered under conditions of divergent thinking. Yet convergence to one or more solutions requires convergence to some solution. We suggest that one means of obtaining clarity on arguments presented in favour of alternative solutions is by requiring explication of arguments.

Ulrich (1983) points out that although contemporary practical philosophers, such as Jürgen Habermas, have developed "ideal" models of practical discourse which give us essential insights into the conditions that would allow us to explicate/justify disputed validity claims, these models are impractical (that is, not realisable), precisely because they are ideal. Ulrich dealt with this problem in two ways: firstly, he developed an approach called Critical Systems Heuristics. Central to this approach is making of boundary judgements by those involved, the demonstration that such boundary judgements cannot be justified rationally, and the translation of their own subjective ways of being affected by the boundary judgements in question into rational, cogent argumentation (Flood & Jackson, 1991, p.112). Ulrich argues that although we *cannot know* the totality of relevant conditions behind any phenomenon concerning a human cultural system we are investigating, we nevertheless *need to think* such a totality because of our moral obligations, and that this implies that we need to define boundaries. Such boundaries cannot be rationally determined but are determined by our interests in the system. Ulrich argues that we need to always be critically self-reflective, which also holds for boundary judgments. This means that discussants need always to be acutely aware of their boundary judgments being tentative (Ulrich, 1983, pp.227-228). Ulrich also proposed a program of research aimed at developing a conceptual framework, which, amongst other functions, would embed conventional "hard" and "soft" systems tools within well-defined institutional and procedural arrangements for rational rebate (Flood & Jackson, 1991, p. 247).

3 Legitimisation of Outcomes

Legitimisation of outcomes, over and above explication of arguments, is a further consideration in GSS contexts that involve sense-making. We take legitimacy of an act or action as referring to the extent to which such an act is accepted and institutionalised within a community. We furthermore take legitimisability of a proposal for action as the extent to which such a proposal has propensity for generating legitimacy within a community or, when applied to a party in deliberations concerning what action to take, within such a party. This implies that legitimisability refers to propensity for generating acceptance of a proposal and commitment to a proposal. It also implies that legitimisability is intersubjectively assessable by a party, concerning its own interests and concerns.

Dealing with problematic situations using GSS is comparable with doing systems design using systems methodologies, such as those expounded in Flood & Jackson (1991). We will here consider aspects of methodologies of Ulrich (1983) and Flood & Romm (1996), taking them as appropriate and adequate for our present purposes.

Ulrich (1983, pp.265-267) considers the "problem of rational practical discourse" and conceives of a "critical" solution to it. He articulates three principles to this end, namely the principles of dialectics, of the polemical employment of reason, and the democratic principle.

Ulrich (1983, pp.268-276) advocates dialectical reasoning and forms of debate and insists that dialectical reasoning entails critical self-reflection, even in a radical sense, when applied to own thought processes. He furthermore advocates the polemical employment of reason, entailing that it is quite unnecessary to prove or even to pretend that a polemical statement may not be false or merely subjective as follows (Ulrich, 1983, p.305). The democratic principle states "the sovereignty and equality of all citizens, be they involved or 'merely' affected by planning" It insists that those who are to be affected by a systems change be able to articulate or to have their subjective concerns and interests represented and to see these accounted for and incorporated in systems design, since it is they who carry the authority of legitimisation of the system (Ulrich, 1983, p. 160, p. 266, pp. 311-312).

Flood & Romm (1996) concern themselves with emancipatory practice while being concerned with developing a meta-methodology for coping with diversity in types of model, methodology, and theory in problem solving. This concern with emancipatory practice leads them to focus on, amongst others, "might-right" issues. They ask: "Is rightness buttressed by mightiness and/or mightiness buttressed by rightness"? "Mightiness", for them, refers to "might is right" or the force of tactic, while "rightness" refers to "right is might" or the force of reason (also "the force of better argument" as Habermas phrases it). They argue, following postmodernists such as Foucault, and contrary to what Habermas preaches, that debate or argumentation ought in principle *not* to be resolved through consensus (Flood & Romm, 1996, pp. 48-49). They argue that neither of the two mentalities, namely of "might is right" (that is, might as tactic) and "right is might" (that is, right as argument), must dominate argumentation. Their position emphasises the variety generation role of politics in the process of generation of knowledge. The focus is on exploring new modes of relationship and transforming relations between discussants, debaters, or disputants rather than on them reaching agreement on action. Disputants, in this view, must be empowered to transform their relations. (Flood & Romm, p. 74). The essential notion here has the intent of enabling sustained debate by two means: by recognising that conclusions from argument must be mostly looked upon as essentially being tentative and that power, whether political or economic or whether emanating from some other sphere, must be looked upon as not inviolable. They place particular emphasis on deliberations and decisions having to be merely locally relevant and focused, albeit that decisions need to be widely informed (Flood & Romm, 1996, pp. 11-12) rather than having to be globally relevant and applicable. They furthermore question the value of polemical forms of argument (Flood and Romm, 1996, p. 233).

4 Analysis

From consideration of the expositions given above and accepting qualified postmodernist points of departure (in the spirit of Flood & Romm) we come to the following assessment:

- Debate needs to be dialectical and critical self-reflection needs to apply for all discussants (following Ulrich).
- The principle of democracy needs to apply, entailing involvement of those to be affected by actions flowing from decisions following deliberation and debate (following Ulrich).
- Validity or truth of the outcomes of debate cannot be guaranteed objectively but depend on interests and positions of discussants and disputants (following Ulrich and Flood & Romm). Explication of the various viewpoints, arguments and boundary judgments becomes helpful, even necessary during debate. It becomes means for discussants to come to better understanding of their various approaches to issues at hand in the process of searching for

common ground. If common understanding is reached on handling of issues, the explication of that common understanding and preceding explications of positions and arguments can serve as minutes of what agreement was reached, on what the basis of the agreement was, as well as providing a record of various positions that preceded and led to the accord.

- Acceptance of outcomes needs to be based on considerations of legitimacy (following Ulrich). Legitimacy is rooted in mutual interests of discussants and in mutual respect, understanding and acceptance of each other (following Flood & Romm). Boundaries need to be defined, with such boundaries being determined by the interests of participants within well-defined institutional and procedural arrangements for rational debate (following Ulrich).
- Considerations and arguments need to be only locally focused (that is, bear on local concerns and concerned with local implications) (following Flood & Romm). ("Local" refers to both space and time.)
- Conclusions from arguments need to be taken as tentative, since critical reflection can bring new insights that may have significant value (following Ulrich and Flood & Romm). Focus should be on exploration of new modes of relationship and transforming relations between discussants rather than on forever binding agreements. (This follows from Flood & Romm.)

On the basis of this assessment we now wish to argue for two suggestions.

Legitimacy is central and essential for acceptance of outcomes of argumentation. Even though outcomes may be but tentative, they need to have legitimacy to be of value. Agreement is required for legitimacy, whereas, 'reasonableness' (as grounded in and demonstrated by good reasons) is required for cogent argument and its explication. Because demonstrations of legitimacy may not be practically efficacious and arguments supported by 'good reasons' may not necessarily be accepted by the individuals concerned, forceful impositions of requirements may be needed. D'Agostino, writing on "public justification" in Zalta (1997), remarks that perhaps the Foucauldians and post-modernists are right in claiming that notions of legitimacy are inherently and inescapably themselves instruments of power, rather than 'rational' alternatives to force. We take this to mean that any group claim can be argued as being legitimate since the conception of legitimacy is itself controversial, and hence can be imposed only by force - not by inducements of 'reason'.

We suggest that legitimacy may be enhanced through agreement prior to debate on articles or principles of understanding to be followed during debate. Such articles/principles, in our view, need to be focused on governing the process of ensuring mutual respect, understanding and acceptance of each other among the discussants. The articles must constitute a frame of reference that can re-orient debate towards mutual understanding when needs be. Compilation of such articles/principles amounts to making boundary judgments, and particularly on the meta-level of the debate.

We furthermore argue that once aspects of power have been surfaced at the start of a debate and commitment to the process to be followed has been obtained, the explication of arguments becomes critical. We maintain that explication of argument can induce some 'reasonableness' into arguments. It can be an aid for better understanding of considerations and their interrelations that lead to particular conclusions. Whereas articles of understanding can serve to aid guidance of relations between discussants towards mutual respect and understanding, explications of arguments can serve to aid 'reasonableness' in argumentation.. Awareness of both, we maintain, can clarify both in-process and post-process misunderstandings of particular understandings reached during debate. This can possibly, we suggest, enhance internal legitimacy – that is, legitimacy amongst discussants – and also serve as an aid in establishing external legitimacy – that is, legitimacy amongst the affected, namely those who do not directly participate in the debate.

One of the classic examples as to how the explication of arguments could be pursued is given by Toulmin et al. (1979). With such an explication framework, groups could meaningfully and innovatively engage in the sense-making process in pursuing intended outcomes of their arguments.

5 Implications for Group Support Systems

We derive the following implications for use and design of group support systems if the two suggestions requiring a set of articles of understanding as well as explication of arguments are accepted:

- Groups that are to use GSS need to enter any session with a frame of mind that is attuned to the necessity and the intent of both having to derive and use an articles-of-understanding and provide explication of argumentation. This implies that they must learn about these requirements beforehand. They must not merely have an understanding of these requirements but need to enter any session with a commitment concerning their realisation. The same applies to the facilitator.
- Design of software for GSS can incorporate functions that facilitate learning and realisation of these two requirements for both discussants and facilitator

References

Boland, R.J., & Tenkasi, R.V. (1995). Perspective making and perspective taking in communities of knowing. *Organisational Science* 6(4), 350-372.

Zalta, E.N. (Ed.). (1997). Stanford Encyclopaedia of Philosophy. Palo Alto, CA.: Metaphysics Research Laboratory, Centre for the Study of Language and Information, Stanford University.

Flood, R.L., & Jackson, M.C. (1991). Creative problem solving. Chichester: John Wiley & Sons.

Flood, R.L., Romm, R.A. (1996). Diversity management: Triple loop learning. Chichester: John Wiley & Sons.

Hayek, F.A. (1973). Law, legislation and liberty (Vol. I: Rules and order). University of Chicago Press.

Toulmin, S., Rieke, R., & Janik, A. (1979). An introduction to reasoning. New York: Macmillan.

Ulrich, W. (1983). Critical heuristics of social planning. Berne: Paul Haupt.

Users driven optimization for a web-based university Management Information System

J. Del Río, J. A. Taboada, J. Flores, M. V. Gómez-Sobradelo

Universidad de Santiago de Compostela, Dep. electrónica y Computación,
Santiago de Compostela, España, 15706
{sandman, eljose, eljulian, elbilly }@usc.es

Abstract

This paper presents an Management Information System called SID of the University of Santiago de Compostela. It was developed using web architecture and the management of its is fully allowed using web interface. Interface design, aggregation levels and functionalities system has been adapted taking into account users considerations.

1 Introduction

University of Santiago de Compostela (USC) needs to obtain a set of tables and indicators with an specific information about different areas. These data will be incorporate of the National Plan of Quality (PNC) whose main aims are: to establish a favourable framework to allow comparison among Universities serving as common element of reference, discussion and evaluation; to provide an assembly of criteria that can be used by Universities for auto evaluation.

Academic, Economic and Research areas are most important for the PNC. One of our main problems is that most users can not analyse the high amount data required for the PNC. Furthermore, information is stored in isolated databases with different desing criteria that makes impossible combine all of them in one data warehouse. So, we need to obtain a set of indicators based on grouped and combined information whose help users to get a global vision for the USC. Due to high computational cost for make these indicators, we need to reduce it making an optimal intermediate scheme.

We present in this article a web solution that allows to get information requested by PNC and can be used to make another analysis as historical ones for USC internal intentions.

2 Objectives

Indicators demanded by the PNC is characterized to be dynamic as much in the information used as in the data showed. So, our main objective is to have a system that can answer to PNC requirements. In the system design we consider that information proceeds from different databases that were created in an isolated form and with different objectives. In the practice this does not allow the creation of a database that combines the available information without assuming incoherence problems. Furthermore, PNC dynamic character reflects in a design that has to be flexible to allow different types of analysis, and open, to be able to include new structures not expounded *a priori*. As specification of the system it was imposed that it should be web based so users don't need specific training since they are very familiarized with the web environment.

Due to SID's characteristics and objectives, minimize the respond time became a main objective. In this sense, the retard in data visualization was determined by the connection type and not for the nucleus of our application. Also, the simplicity in the interfaces and in the navigation are primordial criterion in our application whose objective is to reduce users learning times and improve transmission times. We incorporated a complete administration and management module that allows carrying out all the essential tasks.

Lastly, information security it is a critical aspect that should be treated carefully. The system has to be capable to guarantee data privacy without it supposes an auto limitation. Anytime the system carry out security checks to guarantee that those authorized users can only make the requested tasks. However, this debit should never result in a decrease of the answer speed or in an increment in the complexity of the interfaces.

3 Description

The system is oriented to allow high level analysis of disaggregated information in several sources. Also our system will grant the access security and easy modification of data and user privileges through reports and graphical interface. Among its features we can emphasize:

- Access to the information related with the Quality Plan, maintaining the security levels needed.
- Visualization of historical data with possibility to select the number of years to be shown.
- Graphical analysis of current and historical data. In both cases the user can choose the type of graphic to be displayed.
- The SID is provided with a search and filtering engine in order to select the relevant information to the user.
- The SID allows the elaboration of user reports by combining all necessary information in one single window.
- Management of privileges

4 SID Analysis

SID has been developed in three stages architecture: lower one manages information and assured data consistency; second layer establish and verify the operation security in such a way that the privacy and the veracity of the stored information was assured. To ease the management of the security and administration of the users, this had been included in the system. Finally, the third stage handled the communication with the user and the efficiency of data transmission. In this stage techniques of usability that improve the answer of the system to the user have been included adapting the interface to users needs and demands.

4.1 Information Management

The first task that the SID performed is the management of information to adapt itself to the demands of the Quality Plan. To achieve this, it was needed to combine data from all organisms in the administration of the USC. This required the combination of data which have been elaborated with different objectives and that they do not usually share features that permit their interrelation. Although it is possible to design a database that eliminates problems of incoherence associated to the different sources, the transfer of information of existing data would need a manual processing to adapt them to the new program (this is nonviable task due to database size). Using the relational model of data bases, it has been possible to establish a structure that can relate all the elements implied in the Quality Plan admitting possible incoherencies between them. Despite this, it is necessary to be careful with the consults such as the incoherencies that exist in the data do not reflect in the results that are shown to the user. In this section it is included the diagram entity-relation used and also its characteristics are detailed. To illustrate the incoherence problems we consider the situation shown in the figure 1.

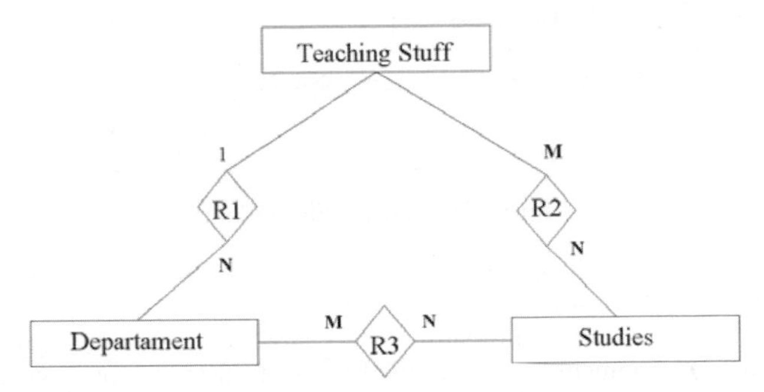

Figure 1: Relationship between teaching stuff, departament and studies.

We can observe how teaching stuff data are related with the department and studies (R1 and R2) that correspond, respectively, with databases of investigation and teaching activity's areas. In the investigation database, one professor can only be assigned to a knowledge area . On the contrary, in the teaching activity database a professor can appear tied to more than one area depending on the subjects that he imparts. When a simultaneous evaluation of research hours and teaching activity per knowledge area is requested, we find a teachers reallocation problem. As it was already commented, to solve this type of problems it is necessary a research and teaching activities database readapting to a new model and an exhaustive revision of already introduced historical data. Since we can't do this we resort to implement a database that allows to pick up these incoherencies and eliminate them in the data post-processing. The design complexity can make that the time used in processing the information represents a problem and cause a bottleneck in SID´s speed. To solve this, the nested queries have been limited to a maximum of two depth levels as well as using other habitual techniques to improve the efficiency of a database.

4.2 Information Security

It is evident that you cannot approach a project of these characteristics without making security considerations. As this information has an strategic character, it is very important that those properly authorized people can only consent to the available data in the system. It is also necessary to guarantee that the ones who upgrade the information are the system managers and no other user that are no validated (Martorells 2000). We have imposed the administrator figure to negotiate the user administration and to take charge to grant and refuse permissions to the users. In the figure 2 is shown the flow chart that an user follows, always transparently to the same one, to execute an action like visualize or upgrade some data.

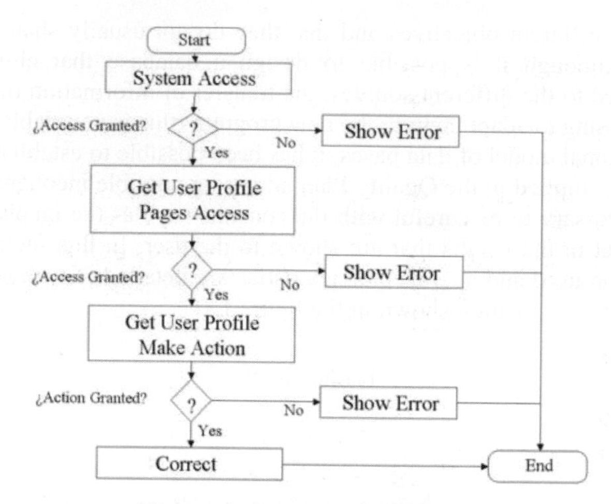

Figure 2: Security Flow Chart

The system shows a different error message to indicate the user the cause for which he cannot continue with the action that he is carrying out. As a complement to the security module the system uses a complete user pursuit that indicates us who aceed to our system, when and what work he is doing in each moment. As additional steps, an user connection is blocked if this is already connected, and the maximum time of inactivity is limited for not leaving residual connections. When an error takes place, one can make a historical analysis of the recent actions to discover it's origin.

4.3 Information display

The task of our system last layer is taking charge or the user communication and showing him the requested information so that it is more useful and easily to interpret. Apart from guaranteeing the information reliability assuring that it is free of incoherencies and processing errors, our main goal it is to minimize answer times. This task doesn't only affect to this stage, but it's the one in which it's reflected. The pursued objective is to show the information in a time limit of ten seconds to avoid losing the user attention . This threshold comes marked by different usability studies that state that user attention practically disappears in that time (Nielsen 2000).

The Web documents design has been guided to get this objective by minimizing the use of drawings, logos and other ornamental elements and appealing, whenever it is possible, to standard HTML elements . A secondary objective that is pursued with this strategy is to be able to visualize the information in the maximum type of navigators, avoiding the dependences in this sense. On the other hand, the simplicity of the pages allows us to increase the number of concurrent connections since the server´s processing time per connection is minimized. Other way to win speed in the system it is minimize the number of intermediate web pages to pass an user to reach the objective that he's looking for. To solve this the information tree has been limited to a maximum of two levels. It has been designed to simply consent to another web page with two connections. With this imposition in the design we made a very simple system that the user doesn't have to memorize the form he gives to reach a certain information and it is never lost, without knowing how to retrieve the system start web pages. Lastly the system was adapted to facilitate the impression of the reports since, in many cases, the written support was required for some analysis.

Figure 3 sample a screen corresponding to a report of the economic module. In this case it we tried to visualize the SID information related with the selected chapter, indicating the historical series values. The main characteristic is that the number of lines and rows of the table was adapted to the number of the historical ones. To guarantee the correct data visualization we have included error check algorithms that not depend on database processing.

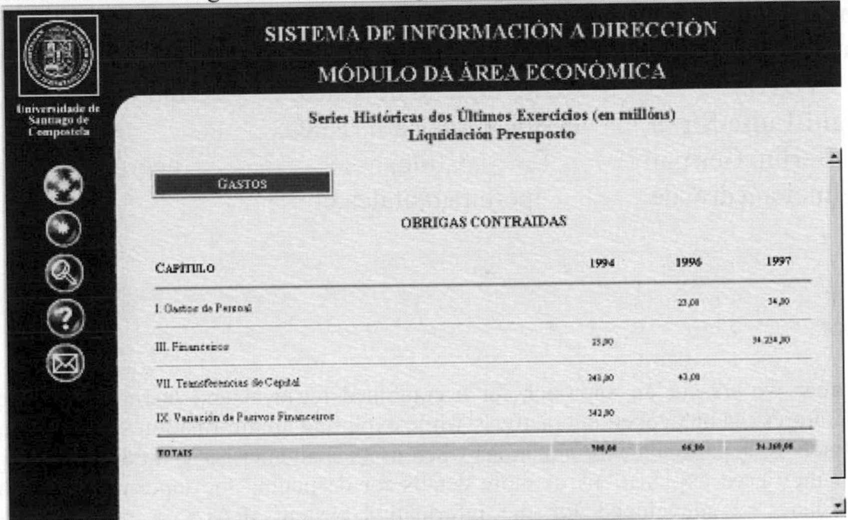

Figure 3: Financial system screen capture

5 Conclussions

The main SID guarantee is that it's working in a uninterrupted way since 1998. During this time, the implicated users have included the system like another tool in their habitual work. In this period the SID acceptation has increased the functionalities demand and the information showed to users.

Along this project we have been able to verify that it is possible to carry out an administration system that leans on in a database in which information consistency is not guaranteed. This is specially important when we have to combine very different sources with different origins whose objective is, in some cases, antagonistic.

Also, the system analysis demonstrates us that the users prefers the speed and the simplicity versus the design and the graphic ornamentation. The users were very satisfied with the obtained answer times and they didn't show in any case discourages for the delay in the use of graphics.

6 References

Kendall & Kendall (1997). Análisis y diseño de sistemas, Ed. Prentice Hall
Pons Martorell, Manuel (2000). Control de accesos, Ed. Escuela Universitaria Politécnica
Korth, Silberschatz (1998). Fundamentos de bases de datos (3rd ed). Ed. McGraw-Hill
Nielsen, Jakob (2000). Usabilidad, Diseño de sitios Web, Ed. Prentice Hall

A learning companion - design of personal assistance in an adaptive information and learning ambience

Kurt Englmeier

German Institute for Economic Research (DIW)
Königin-Luise-Str. 5, 14195 Berlin, Germany
kenglmeier@diw.de

Javier Pereira

Departamento de Informática de Gestión, Universidad de Talca, Avda. Lircay s/n, Talca, Chile,
jpereira@utalca.cl

Narciso Cerpa

Facultad de Ingeniería, Universidad de Talca, Avda. Lircay s/n, Talca, Chile,
ncerpa@utalca.cl

Abstract

In this paper we present an approach for a combined retrieval and learning ambience where learning features can be evoked interactively while exploring an information space. These features lend themselves, in particular, for situations when users encounter issues in the retrieved material for which they need explanations or more details for deepening the topic in focus. The model presented here was developed for the information system IRAIA, a portal for economic information from huge data collections of Economic Research Institutes (ERI) and National Statistical Institutions (NSI). The approach draws on IRAIA's model of context-oriented retrieval and observe, record and analyze the users' search and navigation by applying concepts of the respective taxonomies. In the following we outline the principles of retrieval in IRAIA, the actual integration of learning and guidance, and our model of a more situation-sensitive design of the learning features.

1 Learning features in an information retrieval environment

Traditional text- and guidebooks carefully guide users through the topics of a course or a content domain in a specific order the authors have deemed most appropriate. Integrated into a retrieval environment they can be pretty helpful to guide the users to explanations or more details on a specific topic. They can broaden a certain topic or even compensate for missing knowledge by adding explanations to the pages a reader visits. If the user, for instance, is not so familiar with the concept of "business climate" used as an indicator measuring business activities, she or he can switch from the retrieval section to a document or a set of documents that explains this concept, and come back afterwards to continue retrieval. Actually the user simply marks a certain passage of text and prompts the system to provide her or him with documents from a certain stock of explaining texts. This is quite a straightforward technique, but has the crucial disadvantage of being too "short-sighted" in some situations. Even though user evaluation showed that this feature is highly welcomed it also tuned out that restricting the selection of adequate documents solely on a selected passage of text quite often ignores important aspects that appear in the recent past of the user's retrieval and are implicitly known to her or him. An ambience more sensitive to an actual retrieval situation has to extend the thematic shade towards the retrieval track in an attempt to determine what the user's preferences, problems, and interests are.

2 Situation-sensitive information spaces

The design of a situation-sensitive information ambience starts with a concise, coherent, and comprehensive abstraction of content that in the end can also reflect a particular retrieval situation. The abstraction of content arises from data when they are combined, arranged, and presented accordingly on a suitable level of abstraction. Only an appropriate combination of texts, images, graphs and the like is in the position to convey the information that is contained in these usually separate and otherwise not fully perceptible components.

IRAIA's rationale of context-oriented retrieval is derived from related models of retrieval environments for interacting with large data collections [Agosti et al. 92]. A retrieval environment as envisaged here lets users explore data collections within a semantic co-ordinate system derived from taxonomies of the respective information domain. These taxonomies exist for a variety of application areas. They are a solid basis for a controlled and structured vocabulary and therefore and most appropriately for the content semantics which define domain-related content.

Based on a number of successful approaches in adaptive hypermedia and web systems [see Yang et al. 99, for instance], our design of a situation-sensitive information space is intertwined with a strong linguistic layer that raises significantly the expressiveness of the interaction mode. This layer provides an abstraction of content that endows the users with a concise as well as comprehensive vocabulary. Arranged hierarchically and grouped along major content facets, this vocabulary acts as a stable co-ordinate system easy to comprehend and memorise. The users are thus put in the position to localise without much effort. Successfully searching and navigating now means guided travelling from information to information just by changing the semantic co-ordinates, i.e. by pointing to relevant concepts. This structure on the other hand allows pinpointing the semantic location of any kind of information. It also supports the identification of the right retrieval or learning strategies that sees content searching and navigating becoming content-assisted retrieval.

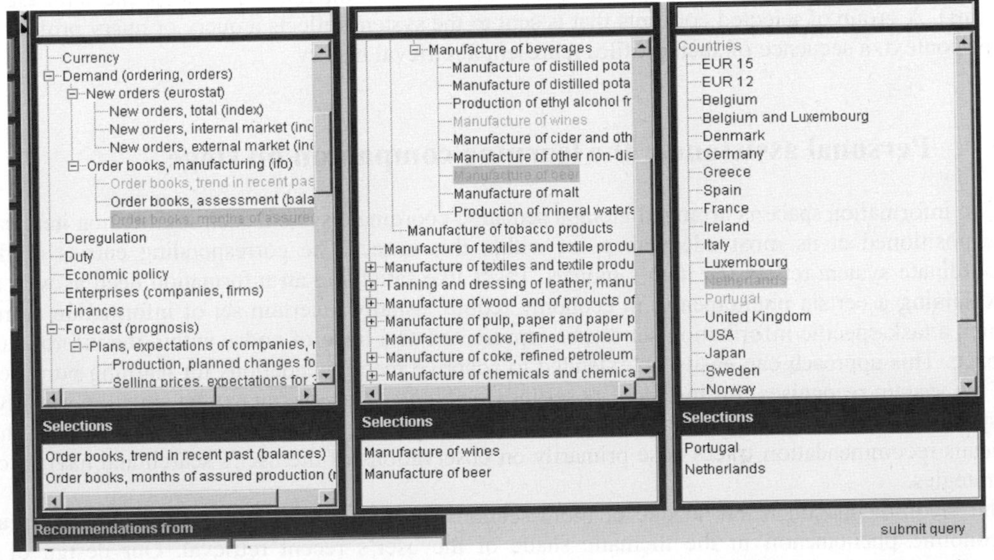

Figure 1. Searching and navigation in a semantic coordinate system. Selected concepts make up the initial query profile. While realising his retrieval strategy the user usually performs iterative steps of defining a query and analysing the retrieved results. In IRAIA, the documents are annotated solely with entries from the hierarchies.

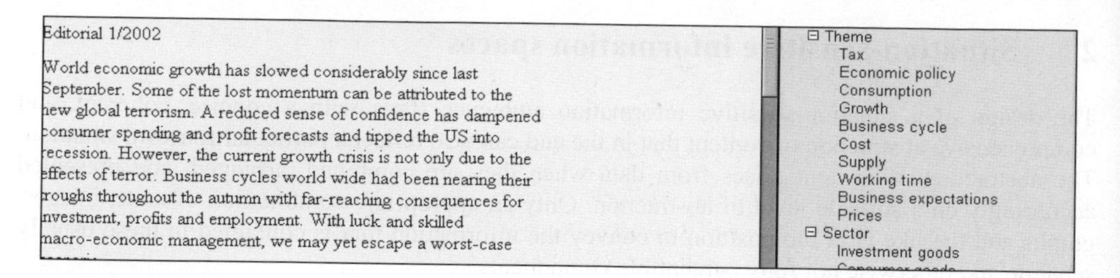

Figure 2. While viewing a document (a text or time series) the interface shows also the annotated concepts (similar to the screen above) that are thus already familiar from the initial query formulation. Modifying the set of annotated concepts is thus tantamount with repetitive query formulation.

Content can be rendered simply and efficiently on a suitable abstraction level by concept hierarchies displayed in adjacent windows. For IRAIA we produced a set of three concept hierarchies that merges two of the most important structures in this field: eurostat's NACE nomenclature and the industry systematic of the IFO institute for economic research. The unified taxonomy creates a semantic coordinate system that enables exact and automatic positioning of coherent documents even if they are of different types. The positioning process compares, roughly speaking, the content of the concept hierarchies with those of the documents and decides which tree nodes are the most prominent ones for this document. The entries of each hierarchy are ranked according to the belief that those are relevant to the document. This annotation process serves to create references from the document to the hierarchies of one or more thematic domains. Thus, a document's descriptor consists of a set of references to nodes of the concept hierarchies.
The coordinate system also provides users with the necessary orientation while exploring the information space. Like in using languages it helps users as a passive vocabulary to identify the topics of their information problem. The user just pinpoints to the relevant concept (i.e. a phrase of terms). A group of selected concepts that is sent to the system reflects a query or query profile. In this context, a sequence of such profiles represent a retrieval history.

3 Personal assistance: the learning companion on stage

If an information space is arranged along a semantic coordinate system each information item can be positioned at its most adequate point within this space. The corresponding entries of the coordinate system reflect the item's indices. Tasks like satisfying an information need as well as explaining a certain phenomenon in economy require usually a certain set of information items. Such a task-specific information set can be represented as a trace of nodes within the information space. This approach can easily be extended to adaptive recommendations for learning purposes. If the system recognises, for instance, a user moving along a trace it can extract situation-sensitive aspects from this trace or – strictly speaking from its representation on the abstraction level. This means recommendation traces base primarily on observations of the user's search and navigation strategies.
The recommendation refers to one or more sections of a data collection suitable to explain an economic phenomenon in the thematic shade of the user's recent retrieval. Our design of a recommendation companion results from combining the approaches for agents as personalised companions [André & Rist 02, Billsus & Pazzani 99], agents for automatic text analysis [Wermter 00, for an overview see Mladenic 99] and for developing interaction strategies [Durfee 01, Englmeier et al. 01].

3.1 A framework for situation-sensitive recommendations

Let us define $O=\{o_1,o_2,\ldots,o_n\}$ as the n-abstractions set of a user's actual (observed) retrieval track and $T_j=\{t_{j1},t_{j2},\ldots,t_{jm_j},\ldots,t_{jn_j}\}$ as the n_j-abstractions set of a guidebook with a sequence of sections t_{ji} $(i=1..\ n_j)$. Remember that the abstractions render nodes of the concept hierarchies. Concepts of the hierarchies constitute both query profiles and content profiles of the documents including those that were visited. An important role has the recommendation profile, when after a number of navigation steps the users request additional material that explains a phenomenon, encountered at a time series visited for instance. At this point k the user ask for recommendations for additional situation-sensitive material. We assume that a guidebook (or parts of it) is adequate for a user's retrieval if it contains sections that have profiles similar to those of the user's recent retrieval and, at the same time, only a few different ones.
We assume that an information problem triggering and steering the user's retrieval can be divided into a smaller number of sub-problems that can be observed frequently over a number of more complex problems. Even if two users have the same information problem they satisfy their need in a different way resulting in different sequences of profiles. Certain clusters among these sequences, however, are very similar. We therefore apply our analysis to these "atomic" problems. At the same time it can be assumed that the user request additional material addressing the most recent retrieval step. It usually coincides more with such an "atomic" problem rather than a larger sequence of past retrieval steps. Let $O_c=\{o_{k-L},\cdots,o_{k-1},o_k\}\in O$ be a sequence of profiles in the actual retrieval and $T_{jc}=\{t_{jm_j-L},\cdots,t_{jm_j-1},t_{jm_j}\}\in T_j$ the part to be compared from an abstraction of a guidebook.

3.2 The agent's decision model

In the following, we present a slightly simplified model the agent resorts to when measuring the similarity between an actual retrieval profile and profiles of a guidebook's sections.
Let us assume that I_{O_c} and $I_{T_{jc}}$ are the indexes of concepts in O_c and T_{jc} respectively, w_i the weight of a concept i, C the set of indexes of concepts in O_c also contained in T_{jc}, and D the set of indexes of concepts in O_c not contained in T_{jc}.

Then, we define the rate of common concepts as $\phi=\dfrac{\sum_{i\in C} w_i}{\sum_{I_{o_c}} w_i}$. In the same way we define the rate of distinct concepts as $\delta=\dfrac{\sum_{i\in D} w_i}{\sum_{I_{T_{jc}}} w_i}$. Therefore, a sufficient similarity is expressed here by $\phi\geq 0.75$ and $\delta\leq 0.25$. If two profiles $t_v\in T_{jc}$ and $o_u\in O_c$ fulfil this criteria we say both are similar or $t_v\approx o_u$. Thus, we can define the following similarity function: $\sigma=\phi(1-\delta)$.
If O_c and T_{jc} contain exactly the same concepts holds $\sigma=1$, whereas it holds $\sigma=0$ if they are completely different. If half of concepts are similar then we have $\sigma=1/4$. In other words, the

existence of distinct concepts impose a penalty to the similarity aggregated function. The more the distinct concepts in T_{jc} are important, the bigger this penalty.

The sections obtaining the highest similarity score are then chosen to be recommended for the user's further reading.

4 Conclusion

Our paper presented features for agents supporting learning and guidance. The features' model is developed around a semantic use model that represents user interests and goals as well as learning tasks. In general, the model can be regarded as an architecture of learning and guidance that determines the lay-out of the semantic traces within an information space. In a further step, this model can be enhanced by additional adaptability taking into account different levels of user knowledge. While providing users with explanatory material it could present more detailed information or hide sections of the guidebook for the more advanced users. In addition, it can allow users to specify exactly how the system should be different, for example, tailoring a certain set of traces to provide information within a certain thematic context. This kind of personalisation could speed up learning as well as exploring a guide or textbook.

References

[Agosti et al. 92] Agosti, M.; Gradegnio, G.; Marchetti, P. "A hypertext environment for interacting with large databases." Information processing and management, 28 (1992), 371-387.

[Billsus & Pazzani 99]. Billsus, D.; Pazzani, M. "A personal news agent that talks, learns and explains." In: Etzioni, O. et al (eds), Proceedings of the third annual conference on autonomous agents. Seattle (USA). 1999. 268-275.

[Durfee 01]. Durfee, E. "Scaling up agent coordination strategies." Computer. 34/7 (2001) 39-45.

[Englmeier et al. 01]. Englmeier, K.; Mothe, J.; Pauer, B. "Users bootstrap searching the Web through interactive agents supporting best practice sharing." In: M.J. Smith; G. Salvendy; D. Harris; R.J. Koubek. Usability Evaluation and Interface Design: Cognitive Engineering, Intelligent Agents and Virtual Reality. Mahwah, USA (2001). 923-927.

[Mladenic 99]. Mladenic, D. "Text-learning and related intelligent agents: a survey." IEEE INTELLIGENT SYSTEMS, 14/4 (1999), 44-54.

[Wermter 00]. Wermter, S. "Neural network agents for learning semantic text classification." Information Retrieval, 3/2 (2000), 87-103

[Yang et al. 99]. Yang, Y.; Carbonell, J.; Brown, R.; Pierce, T.; Archibald, B.; Liu, X. "Learning Approaches for Detecting and Tracking News Events." IEEE INTELLIGENT SYSTEMS, 14/4 (1999), 32-43.

Acknowledgement

Research outlined in this paper is part of the project IRAIA that were supported by the European Commission under the Fifth Framework Programme (IST-1999-10602). However views expressed herein are ours and do not necessarily correspond to the IRAIA consortium.

Design of ICT support for Communities of Practice: Case study of a Trade union information system

Leni Ericson, Niklas Hallberg and Toomas Timpka

Department of Computer and Information Science
Linköping University, Linköping, Sweden
lener@ida.liu.se

Abstract

Informal, social learning systems, such as Communities of practice, are arenas for distributed learning and building collective knowledge. A case study of the development of an information system aimed at supporting trade-union shop stewards is used to present system features that support the creation and maintenance of Communities of practice. In the design process, Participatory Design methods were used. It is argued that information and communication technology can be configured to advance these communities with, for instance, case databases for experience-based narratives and seamless connections between applications.

1 Introduction

The present rapid changes in the social and economic environment forces most type of organizations to adjust to meet the changes. These adjustments effect organizational structures and work procedures. Today, assignments in formal and informal organizations often are highly specialized with no connections to learning environments for developing knowledge and professional identity. In particular, this problem is significant in non-profit organizations with scarce resources. Here, the concept of the *third sector* is used to denote non-profit non-government enterprises in a society. The concept encompasses both formal and informal organizations, where Trade unions represent one of the largest and most influential organizations. Recent decentralization of Swedish Trade unions has increased the responsibilities and workload of local shop stewards, and has increased expectations on performance. Shop stewards, in most cases, perform their assignments on a voluntary basis in their spare time and have insufficient contact with other representatives, often in shifting environments with little or no preparatory training or studies (Pilemalm, 2002). Subsequently, they are dependent on experience-based knowledge from their trade and support from peers and experts to gain skills and develop organizational identity. To be able to participate in the community and culture of the practice is thus of vital importance.

Between 1997 and 2001, a national project, *Distance Learning for Local Knowledge Needs* (DLK), was carried out in collaboration between the Swedish Trade Union Confederation (LO), its 18 affiliations, and Linköping University. The overall objective of DLK was to advance trade union use of information and communication technology (ICT) and to support union studies. The project was aimed at integrating ICT into the day-to-day work of 225.000 shop stewards of the Swedish blue-collar trade unions. As a part of the project, a web-based information system was designed over a period of two years. The development of the design was based on Participatory Design (Pilemalm, 2002).

Informal communities of individuals performing similar tasks or with the same professions have been recognized as *Communities of practice*—that is, systems in which information and knowledge are distributed and created (Lave & Wenger, 1991). Experienced-based knowledge is gained in practical enterprises and in communications between peers and through apprenticeships and mentorship. Such knowledge-development systems have been described as *social learning systems* where members are able to gain knowledge, build social capital, and develop organizational identity (Wenger, 1998). The possibility to use ICT for community development has been widely recognized (Johnson, 2001; Preece, 2000). Notwithstanding, few concrete examples have been given on how ICT should be configured to meet the needs of Communities of practice.

This study is based on case study methods. It is argued that ICT can be configured to advance Communities of practice with, for instance, case databases for experience-based narratives and seamless connections between applications. The features used to analyze the construct *Communities of practice* are derived from the definition of Lave and Wenger (1991). The specific aim is to explore how selected features of the concept are supported in the design of the information system developed during the DLK project.

2 Background

Theoretically, the development of knowledge in a *community of practice* occurs through situated learning and distributed cognition where learning occurs in the space between personal experience and knowledge in the community (Wenger, 1998). This learning takes place in activity systems, in which individual, social and material issues are interdependent and where mutual reciprocity is expected in social exchange between community members (Thibault & Kelley, 1959). Thus, knowledge and identity are a social construction and shaped in an ongoing process. Participation in Communities of practice provides opportunities for individuals to take part of and contribute to the collective knowledge of the community. Communities of practice thus form arenas for gaining information about tacit, experience-based knowledge, where novices learn from peers and experts and have access to the total sum of collective knowledge.

Communities of practice are characterized by information being freely distributed, knowledge about ways of performing work tasks and assignments is provided and membership is free and shifting (Brown & Duguid, 1991). Other significant features are that knowledge is gained through distributed cognition, knowledge can be acquired about how to be a respected member in the community and problem-solving is performed in situated learning (Davenport, 2001).

Johnson (2001) stresses the need of different levels of expertise simultaneously present in the community and a progression from being a novice to an expert. Brown and Duguid (1991) underlined the importance of a free flow of narratives between peers and experts. Lave and Wenger (1991) define Communities of practices as personal social networks at work, or systems in which information and knowledge are distributed and created. Learning is situated in the culture of the community and from experiences of problem solving in practical work. Progress of learning is connected with building an identity as a community member. To learn is to learn is to learn the culture of the community, with its social structure and power relations.

2.1 Case study

Trade union representatives are a heterogeneous group with different commissions. Nevertheless, their union work has common characteristics, such as information mediation, conciliation, active negotiation, and support to union members (Pilemalm, 2002).
As a means to facilitate local shop stewards, a prototype of a supporting web-based information system was developed in the DLK project. A participatory-design method, Action Design, was used to develop the prototype (Pilemalm, 2002). System content and structure were designed according to findings concerning shop stewards' preferences and needs. Requirements for the information system were targeted to support *information seeking and sharing*, *communication*, *gaining knowledge* and *information management* and *information material.*

3 Results

It was found that an information system developed to support a Community of practice should provide one or more of the following features:

- Support for building of relations between members on several levels in the community, peer to peer and with other members
- Participation in the culture of the practice
- Access to information about the community and practice
- Access to information about what needs to be learned to be a full practitioner or respected member
- Support for a free flow of information, narratives and experience-based knowledge

The information system developed in the DLK-project contains communication tools such as conferencing systems, chat, e-mail, and sections with information about laws, agreements, general labor union information, and web search tools. The system also contains on-line courses and information about courses in union matters. Further, there are databases containing frequently asked questions, labor-union projects and cases with experience-based narratives. The design of the system features *case database* and *horizontal navigation* are described below in detail.

3.1 The Case database

The Case database is an experienced-based repository, where shop stewards can submit and retrieve information about how other representatives have solved problematic situations. It makes experience-based knowledge available to peers and other representatives. The Case database is accessible via the Internet and a web-browser. To perform a search in the Case database, search strings can be entered in a free-text search box, and/or a term chosen from drop-down menus in one or more of the categories *affiliation*, *assignment*, *topic*, *solution* or *proposals for solution*. The visitor selects a narrative, which is available in a pop-up window. Representatives are encouraged to share their own experiences. From the database interface they can proceed to a page and write down their own narrative in a text box and send it to the case-database moderator. It is of importance to the narrative that it includes descriptions of the type of problem, if and how it was solved and opinions of possible steps to prohibit the problem from occuring in the future.
The representative is supported and guided in writing the narrative, optionally supported by a set of about twenty help questions. For example, how did the problem situation occur? What type of information/knowledge did you need and from where did you get it?

Narratives can be edited or rejected by the moderator if individuals or work places can be identified. The moderator is also responsible for indexing narratives.

3.2 Horizontal navigation

Navigation in the system is both hierarchical and horizontal. A navigation bar is always visible on the left-hand side of the browser window and supports navigation. The bar contains an expandable menu where each section of the system can be reached. Sections have a depth of one to three levels. In some cases, a pop-up window shows the information required. Thus, the user is not forced to make unnecessary moves in the system. To support and enhance users' domain and system knowledge, system web pages are cross-referenced internally. Figure 1 shows an overview of the page linkage. The internally cross-references presuppose an editor who supervises and updates links in the internal web pages. Subsequently, navigation is possible between pages in the same sections and between pages from different sections and affiliations. Pages in the system, such as discussion groups about certain topics and pages with information are linked to other pages in the system, connecting case-database narratives, discussion groups, e-mail, sections with frequently asked questions and sections with data information.

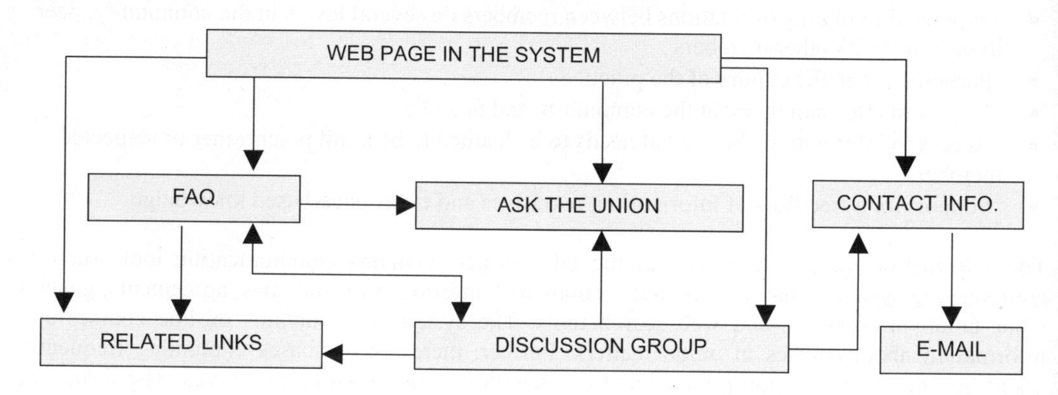

Figure 1. An example of horizontal linking in the system.

A web page can contain information on a specific topic, information about relevant law, links to other sites covering the topic, and other types of related information that enhance the users' knowledge.

4 Discussion

Formerly, apprentices were socialized into professions by participating in social learning communities such as Communities of practice (Lave & Wenger, 1991). They observed the work of masters, exchanged information with peers and incorporated experiences in situated learning. In modern society, an individual can have a unique assignment in their work team and have little contact with peers or more experienced professionals. Professionals, who work alone with modest contact with peers, need opportunities to exchange information and knowledge with others in the same profession (Uncapher, 1999). A trade-union shop steward, for example, may be the only representative at a workplace and have limited possibilities for support or contact with other representatives when experiencing problematic situations. Information and communication technology has the ability to support work groups and Communities of practice (Johnson, 2001).

These learning communities have proven to be forerunners of organizational innovations and developments (Brown, 1998). Nevertheless, informal, invisible networks such as Communities of practice do not get the same attention and support as formal structural settings and networks, even if they are as important for organizational development and learning (Westerberg, 1999).

5 Summary and conclusion

In this study, an information system was designed with the aim to support shop stewards' creation and maintenance of Communities of practice. The use of a participatory-design method was a key element for identifying the essential system requirements for supporting the shop stewards' Community of practice. For example, the design of the case database and the horizontal-navigation feature are two aspects of the implementations of these requirements. A central finding was that a repository of narratives from shop stewards' everyday work situations, a *Case Database,* supports users by making experience-based knowledge available. In the process of writing narratives, representatives have access to help questions to support reflection and formulation of experiences. Other findings are that *horizontal linking* in the system design, connecting case database narratives, discussion groups, e-mail, sections with frequently asked questions and sections with data information, supports the free flow of information, narratives and experience-based knowledge. Horizontal linking, connecting shop stewards active in different sections of the system, also facilitates distributed learning and relationship building. Finally, this study illustrates that participatory-design methods, such as Action Design, are well suited for developing support systems for Communities of practice.

References

Brown, J. S. & Duguid, P. (1991). Organizational learning and Communities-of-Practice: Toward a unified view of working, learning, and innovation. *Organization Science,* 2 (1), 40-47.

Brown, J.S. (1998). Internet technology in support of the concept of 'Communities in Practice'. *Accounting Management and Information Technologies,* 8, 227-236.

Davenport, E. (2001). Knowledge management issues for online organizations: 'Communities of Practice' as an exploratory framework. *Journal of Documentation*, 57 (1), 61-75.

Johnson, C. M. (2001). A survey of current research on online Communities of Practice. *The Internet and Higher Education,* 4, 45-60.

Lave, J. & Wenger, E. (1991). Situated learning: Legitimate peripheral participation. Cambridge: Cambridge University Press.

Pilemalm, S. (2002). Information Technology for Non-Profit Organizations. Dissertation No. 749, Linköping University: Linköping Studies in Science and Technology.

Preece, J. (2000). Online communities: Designing usability, supporting sociability. Chichester: Wiley.

Thibaut, J. & Kelley, H. (1959). The social psychology of groups. New York: Wiley.

Uncapher, Willard (1999). Electronic homesteading on the rural frontier. In M. Smith and P. Kollock (Eds.), *Communities in Cyberspace* (pp. 264-290). London: Routledge.

Wenger, Etienne (1998). Communities of Practice and social learning systems. *Organization,* 7 (2), 225-246.

Westerberg, K. (1999). Collaborative networks among female middle managers in a hierarchical organization. *Computer Supported Cooperative Work (CSCW)*, 8 (1), 95-114.

Human Computer Interaction and Cooperative Learning in Mobile Environments

Bernd Eßmann and Thorsten Hampel

Heinz Nixdorf Institute
University of Paderborn
Fürstenallee 11
33102 Paderborn
{bernd.essmann, hampel}@uni-paderborn.de

Abstract

Starting out on network supported cooperative learning environments for classical learning scenarios, this article discusses their possible enhancements with respect to new standards of mobile learning scenarios. Exemplary, the pursued adaptation will be described by means of the learning platform sTeam.

1 Introduction

Through mobile devices like laptops and PDAs new forms and possibilities of computer supported cooperative learning are arising. Computer supported learning is no longer bounded on to definite locations where stationary Personal Computers are available. This independence of the learner's location is even enhanced by networks based on GSM, Bluetooth or WLAN (WiFi). On the one hand mobile clients allow access to classical network infrastructures; on the other hand they allow a spontaneous networking among themselves.
Besides the structuring of documents the mutual awareness and finding of other learners is a new challenge within mobile environments. Within these mobile scenarios the computer is able, e.g. to transfer finding of one another from the virtual to the real learning environment. For this purpose one's own position may be transmitted to the computers of co-learners, so that also real meetings may be possible. Furthermore, learners in such scenarios may communicate through virtual learning environments and may exchange learning materials without depending on classical network infrastructures. To develop learning environments for the mentioned learning scenarios, technical solutions are required which guarantee *spontaneous networking* and *communication between the clients*. With regard to this aspect only unsatisfactory solutions are at hand up to now, which require by far too extensive technical knowledge on the side of the user.
In addition, conceptual questions concerning the domain *mobile learning environments* and user interfaces of clients regarding these environments need to be answered. Questions like, e.g. how to *tailor desktop applications to small devices*, like Personal Digital Assistants (PDAs) and questions of the development of *client applications especially adapted to these devices*. Spontaneous networking requires the development of concepts which allow the *mapping of cooperative actions* and the *awareness of one another through these spontaneous networks*. This is even more important as co-learners are much harder to locate in a system of spontaneous structures. With respect to the conceptual design of learning environments for the mentioned structures, questions about the *access to shared documents and about their persistence* arise. The approach of group forming and

cooperative knowledge areas must be reconsidered again as mobile learning networks more likely produce *spontaneous groups* of learners, that might be disbanded after one meeting.
It is an important goal to *collect the results of spontaneous meetings* and to make them available for the future. Also the spontaneous learning groups themselves, or parts of them, must be transferred to a server in the form of permanent learning groups with their accompanying persistent knowledge areas. The technical potential of spontaneous networking additionally offers completely new possibilities to adapt the learning environment to the user's context.
This article first concentrates on concepts of linking Peer-2-Peer (P2P) networks with classical client-server topologies and furthermore discusses the design of specialized user interfaces for accessing knowledge areas.

2 Goals and approaches

To test these and other concepts in a real life situation, we follow the approach of a step by step integration of mobile devices into the existing infrastructures. By this way, learners may profit instantly from the new environment. Furthermore classical CSCL/CSCW-scenarios are still the rule at nowadays universities. These scenarios may not be neglected with respect to the development of learning environments for mobile learning. We utilize the existing learning environment sTeam[1], which is designed for classical network infrastructures. On the basis of a strong orientation towards flexible learning objects with changeable attributes, the basic architecture of sTeam may be adapted to any learning scenario. Thus it is our goal to design an advanced sTeam for both, mobile and classical, learning scenarios.
The user interfaces of existing clients must be scaled to PDAs to guarantee a good usability. The network may be accessed through so called Access Points (APs) within classical network infrastructures[2]. Web browser, email-, news- and chat-clients may be used as first interfaces between PDA and sTeam as their protocols are already integrated into the sTeam server. Additionally, the existing synchronous Java clients are ported to the Java run time environment for PDAs. In its course whiteboards and synchronous awareness components are now available for these devices. To use the full potential of mobile networking sTeam is going to be prepared for mobile P2P scenarios. The developments concerning this area are accompanied by interdisciplinary exchange with the lately established graduate college "Automatic configuration in open systems" at the HNI in Paderborn[3].
At last, P2P networks are planned, where *the establishing of spontaneous working groups* as well as the *data reconciliation between the group members* and *between eventually existing servers* can be realized. Besides for a spontaneous networking between learners such wireless networks offer the possibility of *contextual awareness*, where the learning environment integrates into different

---

[1] sTeam *[sti:m]* is developed at the University of Paderborn, it supports cooperative learning through the net. Learners are enabled to form groups and establish common knowledge areas in which they exchange learning materials as well as work on them. An asynchronous web interface and synchronous Java-Clients, especially the remarkable "Shared Whiteboard", are available clients for sTeam. For further information about sTeam please see (Hampel 2001).
[2] Because a lot of universities are already equipped with a thorough net of APs, the required infrastructure already exists.
[3] Within the mentioned graduate college systems are developed, which by self configuration, integrate themselves into a certain environment. The research field ranges from hardware components to network protocols and routing problems up to the domain of applications. Our research attempts may be addressed to the last mentioned domain. Through vertical knowledge transfer between the different areas of the graduate course, research results have been positively influenced. (http://wwwhni.uni-paderborn.de/gk/index_e.php3)

contexts of the learners and correspondingly the required learning resources are provided for. According to the chosen current context any learner may become a member of a learning group which is related to the context or may get access to the related knowledge area. For example, persons who are present in a certain lecture hall are automatically integrated into a group which corresponds with the attended lecture. The lecturing materials in this particular case are offered by mobile devices.

3 Integration of Client-Server and Peer-2-Peer topology

One advantage of learning environments based on P2P technology is their independence of existing network infrastructures. As an disadvantage of P2P networks it may be stated that the availability of learning materials can not be guaranteed within these networks. They are in a certain topology distributed throughout the network. If one user is leaving the learning network, the resources provided by him or her will no longer be accessible. The server centered solution of sTeam guarantees the permanent access to learning materials within classical network infrastructures, whereas the P2P topology allows the establishment of networks which allow cooperative learning without implementing such structures. Thus as a medium-range solution the *fusion of both topologies* is an important research goal.

sTeam offers a good client-server solution, which can be supplemented by any desirable P2P capabilities. Originally, sTeam needs a permanent connection for the communication via COAL protocol[4]. As in mobile scenarios persistent network connections can not be guaranteed (cell changing, fault signals), the Application Programming Interfaces (APIs)[5] of the clients must be designed to tolerate faults with respect to the server connections. Additionally, there are increasing demands on APIs to facilitate the spontaneous networks between the clients requesting the adaptation to a steadily changing topology with no longer guaranteed availability of a main server. In this case, a P2P architecture is demanded where the clients themselves provide for the distribution of learning objects throughout the network. Furthermore, synchronization mechanisms are integrated which allow to keep the results of different learning scenarios persistently on the main server.

This proceeding still allows the support of classical learning scenarios as well as the integration of mobile devices without denying the benefit of spontaneous learning networks. In the first step, existing clients will be prepared for mobile devices.

4 A Web User Interfaces for sTeam on PDAs

Thanks to integrated or upgradeable network interfaces, PDAs, technically speaking, may be used as client devices of a network based learning environment such as sTeam. However, they have special demands on the developer with respect to the design of user interfaces due to their small display and their pen control.

Among others an HTML4.0-able web-browser and a POP/IMAP-able email-client are integrated into the PDA used by us[6]. Furthermore news- and chat-clients are available through third parties. These clients allow the PDA to access the learning environment without special adaptations. In

---

[4] COAL is a communication protocol especially designed for sTeam that allows the manipulation of learning objects on the server.

[5] The sTeam client APIs administrate the connections to the server and provide for necessary functions. They also allow to execute a certain function directly on the server (Remote Method Invocation).

[6] We use Sharp SL5500G, also called Zaurus. With a StrongARM 206 MHz CPU, 16 MB ROM, 64 MB RAM and a touch sensitive 3,5" display with a resolution of 240x320 pixels, Sharp SL5500G is a typical representative of the current PDAs.

contrary to pure communication clients the presentation of the web interface is generated by the learning environment. Initially this HTML based client is designed for 15" displays or larger. It can be viewed on the PDA but because of the small display a lot of scrolling is required within the presentation. Thus a page layout especially tailored to small displays is desirable (Jones et al., 1999). One possible solution could be a re-authoring system which transforms the web pages in order to be viewed with a PDA (Brickmore & Girgenson, 1998). Since sTeam facilitates XSL-stylesheets for the generation of web pages[7] another possible solution would be to refer to different display resolutions within the stylesheet, which would be more tailored to the PDAs capabilities. By doing so, the confusing structure of the stylesheets would be hard to maintain. That is why sTeam chooses another approach.

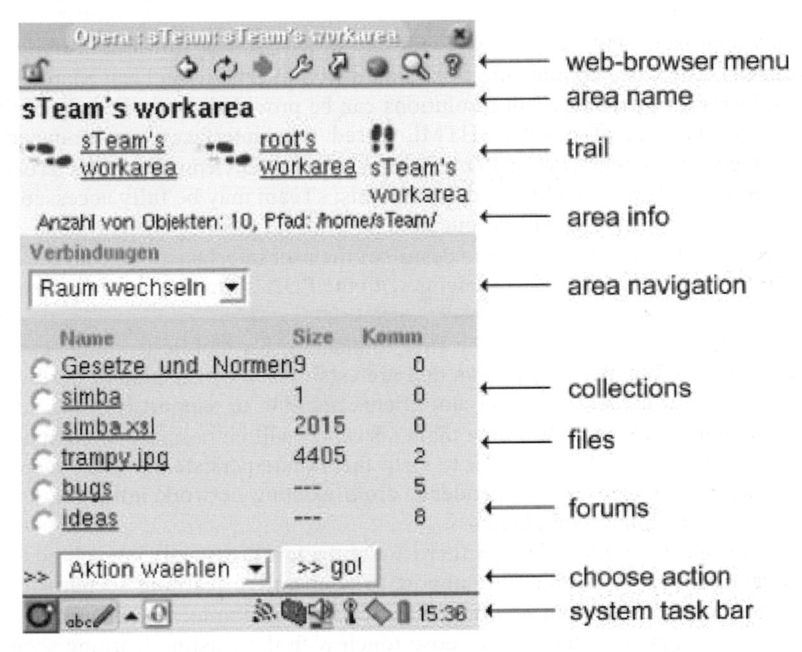

Figure 1: The sTeam web-client with a page layout tailored to PDAs.

If the browser asks to view a certain object, sTeam looks up which stylesheet has to be chosen for the generation of its HTML presentation. sTeam first tries to use the browser identification code which is passed on by the web-browser to recognize the resolution of the browser[8]. For each supported browser resolution sTeam-objects possess an attribute which allows the classification of the appropriate stylesheet[9]. Furthermore, the web interface allows assigning a certain stylesheet to a

[7] Basically for the presentation of knowledge areas over web interfaces sTeam uses XSL-stylesheets which are applicable to XML data generated by the server. The XML file is attached to each stylesheet. It determines which functions may apply to the ordered object to generate the XML data-flow required for the presentation of the information asked for. A XSLT-parser applies the XSL-stylesheet to the data-flow to generate the HTML-Code for the browser presentation.

[8] At present sTeam acknowledges two client resolutions. The first is called QVGA (Quarter Video Graphics Array) with 320x240 pixels, which is usual for current PDAs, and the second is called XGA (eXtended Graphics Array) with 1024x768 pixels which is used as minimal resolution for all other clients.

[9] For the standard client the appropriate attribute is called "xsl:content", for PDAs "xsl:qvga: content". These attributes are registered for all objects with the help of so called sTeam-Factories.

certain group. By this way each user group considering its role is supported with a specific view on knowledge areas. Each resolution is connected to a certain group? stylesheet classification. The combination of browser resolution and active group determine the stylesheet, which is used to display the data. This mechanism allows the flexible generation of views on objects of the learning environment. The resulting view is optimized with respect to the client device as well as to the role of the active user group. Because PDAs have only small displays for the presentation at their user interface, the concerning stylesheets are designed more compact and they are sparing with graphical elements (see figure 1).

5 Conclusions

Our previous approach may be acknowledged to be a first step towards mobile, cooperative knowledge areas. The basic architecture of sTeam´s web interface has been adapted in a way that flexible web pages for different client resolutions can be provided. With the help of the mentioned technology first sketches of compact HTML based user interfaces are produced. They have proofed that the useful integration of PDAs within learning environments like sTeam is possible. Even by means of mail-, news-, chat- and web clients, sTeam may be fully accessed. After providing for asynchronous clients, it will be our goal to port the existing synchronous Java-Clients to the PDA. Here it will be necessary for the design of the user interface to take all the characteristics of a PDA into account. Furthermore, problems with the PDA's run time environments[10] have to be solved.

By supplementing the COAL-protocol the first foundation for mobile learning sessions has been laid. Building up on spontaneous networks that are established by the help of hardware-close protocols (IEEE802.11, Bluetooth, etc.), sTeam clients are able to support learning sessions among themselves without being dependent on a main server. It will be possible to compare the learning results of and on with those on the server to keep them there persistently. This scenario proves to be a good trade-off between the independence from existing network infrastructures and the dependability on server-centered topologies.

Although, with respect to mobile devices there are still a few previously not solved technical questions existing, PDAs are a meaningful support to cooperative learning scenarios. The resulting independence of the learner from a certain location thus guarantees a new quality of computer supported learning. It is important not to loose touch with the existing learning scenarios, so that the potential progress with respect to structuring persistent knowledge spaces which has been made is not given away to an isolated application for mobile devices.

References

Brickmore, T., & Girgenson, A. (1998). Web Page Filtering and Re-Authoring for Mobile Users. *The Computer Journal*, 42, 534-546.

Hampel, T. (2002). Virtuelle Wissensräume – Ein Ansatz für die kooperative Wissensorganisation. Paderborn, Germany, Department of Mathematics / Computer Science, PhD Thesis, December 2001

Jones, M., Marsden, G., Mohd-Nasir, N., Boone, K., Buchanan, G. (1999). Improving Web Interaction on Small Displays. *Computer Networks*, 31, 1129-1137.

[10] SL5500G is equipped with a PersonalJava run time environment, which is based on Java 1.1.8. sTeam Java API is based on functions of J2SE. The corresponding micro edition J2ME 1.2.2. presently has the status of a technology study.

Modeling Business Information in Virtual Environment

Dimitris Folinas, Vicky Manthou, Maro Vlachopoulou

University of Macedonia, Department of Applied Informatics
156 Egnatias str, 540 06 Thessaloniki, Greece
+30310891866, +30310891893, +30310891867
folinas@uom.gr, manthou@uom.gr, mavla@uom.gr

Abstract
As organizations confront a dynamic global environment, they are increasingly forced to create virtual collaborations so as to remain competitive. As a result, they come to possess abundant data and information, jointly developed products or services, concluded contracts, shared promotional plans and negotiated prices, terms and demand expectations. These information-intensive cooperative activities comprise the core elements of every successful business and are greatly affected by the Internet. In this paper, a web-based integrated platform and the adequate mechanisms for modeling dynamically business processes and data to enable e-partnerships, is developed. Based on the proposed model sequential tasks are supported for the collaboration of partners, information exchanges and processes.

1 Introduction
A Virtual Organization (VO) is a temporary or permanent association of geographically dispersed organizations communicating with each other, fulfilling customer requests and / or triggering services that carry out some complex workflow of transactions (Camarina-Matos, & Afsarmanesh, 1998, Van Aken, 1998). The participant actors are chosen according to their competencies and special abilities. The basic characteristics of a VO are lack of geographical boundaries, absence of informational barriers, form fluidity, co-operative and instant partnerships capability, exceptional speed and agility and a unity of appearance. The advantageous properties of a VO depend to a large extent on the modeling of business information and this constitutes a major challenge but also one of the biggest problems. The modeling of business information in a virtual environment needs a secure, extendable and reliable web inter-organizational platform capable of handling high volumes data / information sharing, high transaction complexities and capable of supporting high levels of automation and integration (Folinas, Manthou & Vlachopoulou, 2001).
In order to enable organizations to set up and manage quickly and reliably virtual collaborations, an Internet-based platform is recommended to satisfy the demand for coordinated inter-organizational business processes and to establish standards of data and processes. Standards play an important role to open business coordination, since they define the form of the interpretation of messages, documents, processes and services passing through the business partners. A business process can be defined as a set of one or more linked procedures or activities, which collectively realize a business objective or policy goal. There is a need to automate processes across all of these partners and ensure that transactions flow quickly. Every transaction is determined as the sequential exchange of business documents / messages, resulting in the transition to a new state for the partners. The objective is the development and use of common business vocabularies that describe the structure and the semantics of the exchanged messages. In this paper a web-based integrated platform and the adequate mechanisms for modeling dynamically business processes and data to enable e-partnerships, is developed. Based on the proposed model sequential tasks are supported for the collaboration of partners, information exchanges and processes. The

specification of the collaborative objectives and shared processes is the initial concern of the model. The determination and analysis of data / information requirements related to each sharing process are followed. The direct and dynamic development and adoption among involved partners over the Internet is necessary for the establishment of business information standards. eXtensible Markup Language (XML) Schemas are used to encode information needed in agreeing on specific exchanges such as, sales terms, prices, terms of payment and other contractual information. The collaboratively design and development of business vocabularies aims to define the required business information for each instance of a business message and its structure for the virtual communication with the involved entities. The acquisition of vocabularies and required modules by all partners enables the implementation and utilization of the corresponding XML instant business documents. Innovative integration technologies for the management of information across the involved parties in a virtual environment are exploited in this model. Unified Modeling Language (UML) diagrams and XML Schemas are used for modeling business integration and information requirements. Shared processes and business vocabularies are designed and developed in a totally visual manner requiring mostly business than technical background by users. The proposed model endows partners in a virtual environment with a real and innovative web-based solution for improving their interoperability and competitiveness.

2 Business Information Modeling

The dynamic creation and operation of VO demands clear-cut standards between partners, specifying the information to be exchanged and the steps to be taken in this exchange of data. The proposed model of dynamic network presents a framework for modeling business information in a virtual environment. The objective is two-fold: to ensure that the exchange of business information will take place in real time and without human intervention and that all the enterprises among VO will have access and use the same information. In order to achieve the above objective five steps must be followed: *definition and establishment of Information and Communication Technological (ICT) standards of data transmission, modeling of business information, development, acceptance and application of business vocabularies* (see Figure 1).

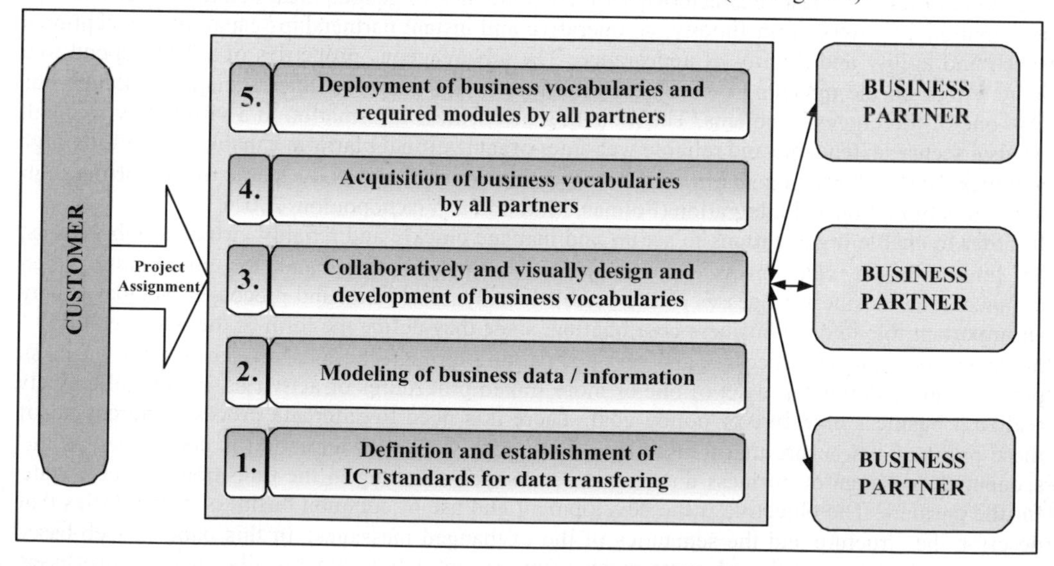

Figure 1: A model for business data and process integration in a virtual environment

2.1 Definition and establishment of ICT Standards

Technical and communications standards must be determined in the first step for data transmission between two or more VO nodes as well as safety mechanisms to ensure availability, confidentiality and integrity. This task consists of the selection of transfer protocols as TCP/IP, FTP, HTTP, SMTP etc. and the mechanisms for management of messages. In both cases the basic idea is the connection of systems in a synchronous or asynchronous manner. Synchronous communication requires a private channel installation between sender and receiver. It is used in real time systems and demands availability of both nodes. Examples of such technologies are the Distributed Computing Environment Remote Procedure Call (DCE-RPC) and the Java Remote Method Invocation (RMI). The asynchronous communication allows a more loosely connection between sender and receiver without requiring continuous connection between them (i.e. wireless communication). Message queuing mechanisms operate as intermediaries administrating and auditing data flows. Basic examples of this approach are IBM MQSeries, Progress SonicMQ and MSMQ.

2.2 Modeling of business data / information

The process of business information modeling in a virtual environment consists of four phases: *determination, analysis, design of business information requirements* and finally the *creation of business vocabularies*. First, the problem must be described and the requirements must be registered for the accomplishment of the various business transactions. This phase consists of the specification of collaborative objectives and common processes determining the various roles (business entities or systems) and their relationships (Carlson, 2001). For this purpose, Use Case diagrams are used to model the interdependence between them and the roles of involved business entities. For each Use Case diagram, an Activity diagram will be created on demand that describes the assigned functions. In the second phase the analysis of requirements is performed. It can be accomplished with the determination and the initial planning of the business objects or classes. The integration requirements according to the proposed model are: the integration of data / processes with the adoption or creation of messages, and the modeling of process flows that includes a series of exchanged messages. Hence, interactions of objects (exchanged messages) must be constructed using Activity or Sequence diagrams that demonstrate the interaction of objects in specific period of time. The main objective of design is to implement a realistic system of static models in practice, which were created during analysis. In this phase, the vocabularies of business messages are developed with the use of Class Diagrams, which represent the basic elements, their relationships and the restrictions of the system. Furthermore, in this phase the creation of State diagrams is required in order to represent the state of every object and the various events that lead to every objects state. During the final phase the business vocabularies (XML Schemas) from UML diagrams are developed. These XML Schemas are used for the creation and the validation of the messages. The transformation from diagrams to schemas is the research subject of many projects (Booch, Christerson, Fuchs, & Koistinen, 1998, Conrad, Scheffner & Freytag, 2000, Routledge, Bird & Goodchild, 2002). Practically an intermediate stage between the conceptual models (corresponding classes) and natural models of XML Schemas is applied to all of the approaches. This can be achieved by extending UML diagrams with the use of Stereotypes, which can describe more complicated business rules. At the end of this phase apart from the corresponding diagrams the following are created: the required vocabularies with the form of XML Schemas, which must be human readable and must describe the content with precision. The above deliverables are created in a totally visual way. In fact the user constructs the diagrams and designs the business vocabularies determining the metadata that describes the structure, syntax, data types and the business rules. All the above will be stored in a central repository that one business entity must be responsible for its management.

2.3 Design and development of business vocabularies

The previous step dealt with business information modeling in order to fulfill a particular business goal. Furthermore, another critical point to consider is the adoption of business vocabularies by all partners. The requirement for a specific message initializes searching in existing schemas data repository. These schemas are pre-defined messages that are established either by organizations / institutions or are outcomes of older modeling projects. Pre-defined messages can be used unchanged or modified in order to meet the particular requirements of a specific business problem. There is also the possibility for the creation of one business vocabulary from scratch. The basic characteristic of the above procedures is their collaboratively execution.

2.4 Acquisition of business vocabularies by all VO partners

The next step starts after the development of business vocabularies as it was described in the previous section. However, before the initiation of VO's life cycle, various partners have already submitted their own business profile (business rules and constraints, financial data, internal records updating process etc.). The above information should be based on business objects, which enterprises use for process implementation. These business vocabularies will be delivered electronically for control and evaluation to the various members. Partners must first accept the sufficiency of vocabularies and suggest afterwards additional issues of content. In case of disagreement on the appropriateness of vocabularies, modifications must be applied until business vocabularies meet the particular enterprises needs.

2.5 Deployment of business vocabularies

The main operation is the validation process of the messages, which are the instances of business vocabularies. A common business message (XML document) consists of two parts, the header with information that concerns the sender, receiver, time, type, subjects of routing, safety and management of errors, and the body that encircles the content. Another point of interest focuses on the syntax / content of business messages that are examined in order to confirm the rules of the corresponding vocabulary which was created in the previous steps. This validation process is crucial due to the fact that the input data must satisfy the requirements of business systems or applications. Thus, it includes two additional operations: conformity to business rules that each VO member has set, and transformation of messages content to another form if it is required.
In the pre-mentioned cases, control is executed in the application level with the use of Extensible Stylesheet Language for Transformation (XSLT) technology that was first presented as a means of XML document transformation in various forms like pdf, doc, etc, and for various devices such as Personal Digital Assistant (PDA), mobile phones etc. Today XSLT has a wider use as a transformation language of XML documents between two different business vocabularies. Finally, the whole process of business exchanges between VO members must be unanimously agreed (i.e. main functions and transmission protocols of message management, application functionality, and various legal / financial subjects).

3 Model Evaluation

The proposed web-based platform emphasizes the collaborative effort in the process of modeling business information. The objective is to ensure that all partners in a virtual environment have access and use the same information. The following profits are anticipated: more realistic identification of requirements of the business processes, time and cost reduction in the XML Schemas design and development process because of the user-friendly interface, ability to reuse business objects, processes and vocabularies for similar scenarios, risk reduction due to collaborative effort and more effective and centralized control and management by a specialized business entity.

There are many points that need to be considered for the effective design and operation of the proposed model. Particularly, the acceptance of the business vocabulary from VO members is not sufficient enough for the initiation and maintenance of the exchange messages / business documents process. EDI experience proves that business vocabularies need to be extendable (Folinas et al., 2001). They must have the ability to adapt easily and directly in order to cover the specific needs of a potential business partner or environmental changes. Probably a transformation will be required from one vocabulary to another. Also, system integration among partners is not implemented only with the modeling of business information. Advanced applications must be developed for the full integration of enterprises' information systems. The proposed methodology can set the base for the creation of appropriate Web Services since it covers the most important phases in analysis and design of various business processes. The development of the above services can be implemented with the use of innovative technologies such as J2EE, .Net etc.

4 Conclusions

In a virtual environment the concept of service management emphasizes cooperative relationships, integration of processes and information systems, and inter-organizational problem solving. The field of VO represents a very challenging research and application area for collaboration networks approaches. VO aims to achieve a high level of structural flexibility by dynamically reconfiguring the whole network. The proposed model attempts to address the above challenges and suggests a cooperation platform that will support the development of this virtual environment in regard to modeling business information. A web-based platform is designed to cope with the new dynamic networked environment based on a series of specific tasks. Considerable value, such as faster time-to-market and lower costs should derive from leveraging this platform. Further research can be carried out for developing vocabularies in specific business domains. The main objective is the establishment of an open environment that encourages collaborative relationships and is more responsive to individual customer needs.

References

Booch, G., Christerson, M., Fuchs, & M., Koistinen, J. UML for XML Schema Mapping Specification. Received October 21, 2002, from http://www.rational.com/media/uml/resources/media/uml_xml schema33.pdf.

Camarina-Matos, L., Afsarmanesh, H. (1999). Virtual Enterprises: Life cycle supporting tools and technologies. European Esprit PRODNET II project.

Carlson, D. (2001). Modeling XML Applications with UML, Practical e-business Applications. Object Technology Series. Addison-Wesley.

Conrad, R., Scheffner, D., & Freytag, J. (2000). XML Conceptual Modeling Using UML. Proceedings International Conceptual Modeling Conference, Salt Lake City, USA, 309-322, Springer Verlag.

Folinas D., Manthou V., & Vlachopoulou M. (2001). Logistics Service Management in Virtual Organization Environment: Modeling Systems and Procedures. Proceedings of 17th International Logistics Congress: Logistics from a to O, Thessaloniki, 127-139.

Routledge, N., Bird, L., & Goodchild, A., (2002). UML and XML Schema. Thirteenth Australasian Database Conference (ADC2002), Melbourne, Australia. Conferences in Research and Practice in Information Technology, Vol.5 Xiaofang Zhou, Ed.

Van Aken, E. (1998). The Virtual Organization: a special mode of strong inter-organizational cooperation. Managing Strategically in an Interconnected World. Chichester: John Wiley & Sons, 301-320.

Cyber Crime Advisory Tool - C*CAT: a holistic approach to electronic evidence processing

Sandra Frings [*], *Mirjana Stanisic-Petrovic* [*], *Jürgen Falkner* [*]
Robin Urry [#], *Neil Mitchison* [#]

[*] Fraunhofer IAO, Nobelstrasse 12, D-70569 Stuttgart, Germany
[#] Institute for the Protection and Security of the Citizen, JRC, I-21020 Ispra (VA), Italy
Sandra.Frings@iao.fhg.de, Mirjana.Stanisic@iao.fhg.de, Juergen.Falkner@iao.fhg.de,
Robin.Urry@jrc.it, Neil.Mitchison@jrc.it

Abstract

For people to have confidence in computer systems, it is important to be able to offer clear proof of what has happened in an electronic event, whether this be an intrusion, a disputed purchase or order, a manipulation of important data, or the explosion of a "logic bomb". The increasing migration of business activity to the Internet makes this proof even more important - since there are more opportunities for misdeeds of one sort or another - and also increases the complexity of transactions and interactions, for which recourse to a paper audit trail may simply be impossible. Moreover, in many cases suspicious events are investigated first by company personnel who have no specific training in the legal requirements for handling evidence to be brought before a court; if these personnel "taint" the electronic evidence they handle, a court - especially a criminal court with its high standards of proof - may be unable to use the evidence subsequently.

For this reason the project "Cyber Tools On-Line Search for Evidence" (CTOSE), http://www.ctose.org, partially funded by the European Commission (IST Programme), was initiated. The project focuses on a holistic approach - the CTOSE process model and methodology - to the handling of potential electronic evidence. This includes the sub-processes of identification, collection, tamper-free storage, restricted and controlled access, analysis, judicial presentation and documentation of electronic evidence arising from high-tech crime or disputed electronic transactions, while taking into consideration the demands of security and privacy. This approach requires certain pre-conditions, in particular a good IT security infrastructure with adequately secured logs and records; given these, it offers a methodology for the handling of electronic evidence aimed at ensuring its legality and admissibility in a judicial or quasi-judicial process.

Introduction

The growing shift of business activities into the Internet requires the citizen and the company to have confidence in the accuracy of records kept of computer-based events. For this, it is necessary to be able to offer clear proof of what actually happened in an electronic transaction or interaction. Therefore it is not only necessary to set up a good IT security infrastructure to be prepared for the worst but also to have an efficient investigative methodology, including appropriate guidelines, methods and tools, for processing electronic evidence, designed to ensure the chain of custody of electronic evidence from the first investigation on.

For this reason the project "Cyber Tools On-Line Search for Evidence" (CTOSE), http://www.ctose.org, partially funded by the European Commission (IST Programme), was initiated to develop a common methodology for electronic investigations and investigators. The CTOSE project is led by the Joint Research Centre of the European Commission (based at Ispra in Italy), with the participation of the *Institute for Human Factors and Technology Management IAT of the University of Stuttgart*, (in close cooperation with the Fraunhofer Institute of Industrial Engineering IAO), the *Centre de Recherche en Informatique et Droit of the Université de Namur* (Belgium), the Department of Management of the University of St. Andrews (United Kingdom), and the industrial partners: Alcatel CIT (France) and QinetiQ plc (United Kingdom - formerly DERA).

1 Background

Rules of evidence are well established within the European Union Member States' statute law, case law and precedent. These rules in general have their most exacting formulation in the requirements for evidence to be put before criminal courts; as such they have been incorporated into "standard operating procedures" for law enforcement agencies as best practice guidelines for evidence collection and investigation procedures (e.g. Internet Society, Network Working Group 2002). This practice has been modified over time to accommodate changing needs driven, inter alia, by technological developments. The need to handle electronic evidence is one of the changes which has required new rules of evidence collection.

In this paper high-tech crime is defined as follows:

> The use of information and communications technology to commit or further a criminal act, against persons, property, organisations or a networked computer system.

Guidelines exist both nationally and internationally for the seizure of electronic equipment and computers, the processes to follow for warrants and court orders and the process to follow for electronic copies of evidence. A history of cases and best practice has been developed and incorporated within law enforcement training courses at local level. However there is consensus that better coordination and integration is required both for training, prevention and electronic evidence investigation procedures. At present this best practice is almost totally confined to the law enforcement (LE) community. For the vital chain of custody to be preserved, this law enforcement best practice has to be supported by all electronic investigators, including those from the civilian community. Such a holistic approach does not exist anywhere in Europe.

2 Objectives of CTOSE

Taking the above into consideration, the project's objectives focus on the whole process of handling potential electronic evidence, in a holistic approach. This includes the identification, collection, tamper-free storage, restricted and controlled access, analysis, judicial presentation and documentation of electronic evidence arising from high-tech crime, computer intrusion, or a disputed electronic transaction over the Internet. (In this paper such an event is called an "incident". It should be noted that in practice it is likely that only incidents of a certain severity will in fact justify what is likely to be a time-consuming and fairly expensive investigation.). The approach also involves paying special attention to the demands of security, privacy, and legal admissibility of evidence.

The project aims to develop the following results by May 2003:

- an electronic guideline (C*CAT – Cyber Crime Advisory Tool) to be used in case of an incident to give relevant information on actions and decisions (excluding legal advice),
- a legal expert system, directly connected to C*CAT, to cover the legal aspects when handling an incident,
- a simulation environment (CTOSE demonstrator) to be used both to validate the underlying process model, and as an educational awareness and validation tool to explain the holistic approach and C*CAT,

all including organisational and technical aspects.

3 The CTOSE Process, its Sub-processes and Phases

The project started with a survey of the needs of users, both users of evidence - i.e. investigators and those involved in legal and judicial processes - and users of the analysis provided by electronic evidence, i.e. owners and suppliers of electronic systems. According to the results of the survey and additional research the project defined three phases:

- Running Phase – This is the "steady-state" position where electronic transactions and interactions are occurring and no disputes or suspicion of improper activities have arisen.
- Assessment Phase – This occurs when some prima facie evidence is found (by a person, an automatic detection device, manual analysis of monitored data, etc.), or a dispute is initiated (trigger), that requires a more focused analysis and assessment of the business risk arising from some suspicious event. The system may or may not continue to run normally in this phase; in either case it is expected that any investigative actions will be taken by the normal system management.
- Investigation Phase – This starts when there is a decision to proceed with a formal investigation. Normally this will involve a senior management decision, bearing in mind the interest of the company (e.g. to get the system running again, to minimise the financial loss, etc.). The investigative activity will be carried out by a specialist investigator, external expert, system administrator, etc. with the goal of finding out what was done and by whom, and of course how to prevent it from happening again. At a certain point in this process, law enforcement bodies may be called in.

The CTOSE process starts in the running phase and ends after the investigation phase with a phase which could be called "post-event", where the information and experience gained during the first three phases is fed back into the model to develop it further.

The main sub-processes during handling electronic evidence are the identification, collection, analysis, storage, access, presentation, and documentation of information which might be used as electronic evidence. These sub-processes all deal with managing information. They have to be performed under very exacting constraints, often at considerable speed. Therefore both the human-computer interface and the interfaces between the various people involved are of vital importance.

4 The Electronic Guideline and CTOSE Results

The aim of the project is to give organisational, technical and legal advice in the running phase to prepare for preventing any kinds of incidents, and in the assessment and investigation phase to support investigative actions and decisions.

In the project a number of "components" in the process model were worked out, such as actions, decisions, roles, skills, advice, documents, etc. which define the structure of the process model and the sequence of actions to be taken.

The electronic CTOSE process model will be the interface between the information to be processed and the people involved in investigating a computer-related crime. The more this model can cover of the real world, the more support the people involved have available to solve a case. These people might be IT specialists, police, lawyers, judges or mangers of businesses.

For the end of the project we aim to have a holistic process model for investigating computer related events which will give sufficient detail on which activities and decisions have to be carried out, who is participating in them and who is responsible for decisions (definition of roles and skills), and which constraints and documents are to be expected and considered (legal issues – especially privacy and data protection, time, standard operating procedures, policies, strategies, etc.).

This process model will be available in electronic form, and will act as an electronic guideline in case of an incident ("C*CAT – Cyber Crime Advisory Tool" connected to the legal expert system). It will give information on what has to be done for stakeholders to be prepared for the event, what has to be done during the event and how the electronic evidence collected has to be handled to ensure a legally correct and admissible chain of custody for use in court or other quasi-judicial settings.

The electronic process model will be easy to use, by first allowing the user to define the situation (by selection among different choices). Then the necessary actions and decisions will be presented. In each case the user is able to ask for more advice and hints. Since the integrity of the chain of custody is critical, following correct procedures is of vital importance. At the end of the process, the user will be able to give feedback concerning the usage of the model to further optimise it.

The future aim of CTOSE is to spread its methodology as much as possible. Therefore, CTOSE is developing a simulation environment (CTOSE demonstrator) to be used as an educational awareness and validation tool. The demonstrator will explain the process model using several different scenarios. Each scenario will provide the user in a clear and understandable way with *do's* and *don'ts* when handling electronic evidence for each step of the holistic process, which starts with forensic readiness in the running phase and may end up in a civil or criminal court.

5 Advantages resulting from CTOSE

The persons involved in the process will have a checklist of what to do in case an incident occurs. They will know, to a certain level of detail, what to do, whom to contact, what to consider and what information is needed by external investigators. Their reports will have an uniform structure. Furthermore, a common process model and architecture will protect the chain of custody, and ensure that best practice which has been developed through time by law enforcement experts is followed by all investigators.

If the documentation prepared for and by prosecutors is based on the CTOSE process and guideline, the result will be a familiar structure leading to better understanding and therefore less participation of experts in court will be needed. Further benefits of the project, in the form of efficient training of personnel, standardising procedures and making processes transparent will help organisations and law enforcement fight and prevent high-tech crime.

6 Conclusions

Why is there a need for research like CTOSE?

Because as a society - the Information Society - we have become massively dependent upon computers and electronic devices, which all have memories and are the repository of our society's records and information. This being the case, in any form of dispute or investigation, the finding, storing, analyzing and presentation of electronic evidence plays a major role.

The project team agreed upon the need to have a common approach to using and preserving this chain of custody in a process model similar to those used in the current physical world. It was necessary to understand the users' needs, the technological constraints, the legal boundaries and the requirements for protection of the citizen, while enabling effective investigations into crimes and disputes, all this without jeopardising the deployment of efficient and secure IT systems.

The operational co-ordination of electronic investigations is being carried out at national law enforcement levels - with international coordination by Europol and Interpol, as well as internally within organizations, and by expert external firms. A variety of excellent commercial tools exist for many of the actions of an electronic investigation. The legal profession is building its expertise and case law in this new activity, while the legal framework itself is being developed by proposals such as those in the Council of Europe's Convention on Cybercrime (Convention on Cybercrime 2001). The CTOSE Program is unique in accommodating all of these disparate activities and making tools and processes available to integrate them both technically and procedurally.

7 Future Work

CTOSE will deliver the CTOSE demonstrator and the electronic guidelines (C*CAT and legal expert system) by the end of May 2003. This will be the launching platform for further development, diffusion, and the production of commercial products, to be carried out by a group made up from the partners and Special Interest Group members to further develop a commercial product and services. If this is achieved successfully, the CTOSE methodology should come to be widely diffused throughout Europe and the rest of the technologically developed world. As it becomes more and more widely used, and as a wide range of commercial products conformant to the CTOSE methodology becomes available world-wide, we expect that judicial processes will come to expect that electronic evidence is presented in accordance with the agreed methodology. There is a delicate balance to respect here: on the one hand CTOSE must be publicly available at reasonable cost, while on the other hand there must be enough commercial interest to ensure continuing investment in development of CTOSE-compatible products.

References

Convention on Cybercrime 2001. Retrieved January 29, 2003, from http://conventions.coe.int/Treaty/EN/projets/FinalCybercrime.htm

Internet Society - Network Working Group (2002). Guidelines for Evidence Collection and Archiving. Retrieved January 31, 2003, from ftp://ftp.ietf.org/rfc/rfc3227.txt

Discussion over a shared file system

Younosuke Furui *Katsuya Matsunaga* *Kazunori Shidoji*

Division of Cognitive Science,
Department of Intelligent Systems,
Graduate School of Information Science and Electronic Engineering,
Kyushu University
6-10-1 Hakozaki, Higashi-ku,
Fukuoka 812-8581 JAPAN

furuiy@brain.is.kyushu-u.ac.jp matsnaga@brain.is.kyushu-u.ac.jp shidoji@brain.is.kyushu-u.ac.jp

Abstract

This paper describes a new way of computer-mediated collaboration based on two-way linking between discussion and file sharing. We have been developing a system named *Pilgrim* that links together a Web-based discussion service and an SMB-based file-sharing service. This system allows users to have asynchronous discussions on any types of files via networks. It also helps users to be aware of the discussions and reach them easily, by providing backlinks from the files to the discussions. We believe the two-way linking feature will be beneficial to the discussions.

1 Introduction

This paper introduces a new way of computer-mediated collaboration based on two-way linking between a Web-based discussion service and an SMB-based file sharing service. We have been developing a system named *Pilgrim* that aims at helping small-sized or medium-sized groups to collaborate via networks. It provides a discussion service, a file-sharing service, and a two-way linking feature for the two services. With this system, users can exchange messages asynchronously, and refer to any type of files by including hyperlinks to the files in the messages. Furthermore, this system provides backlinks from the files to the messages. When the users navigate the files, the backlinks allow them not only to be aware of the discussions but also to traverse among the messages and files in any direction. In this way, this system integrates the two services, discussion and file sharing.
The rest of this paper is organized as follows: Section 2 explains several technical terms and concepts as the background of our work; Section 3 describes the problematic issues and our solutions; Section 4 describes the system design and implementation; Section 5 describes the current status of our work; Section 6 summarizes this paper and describes our future work.

2 Background

This section explains several terms and concepts about today's computing infrastructure and computer-mediated collaboration.
Today, the World-Wide Web (WWW, or simply *Web*) is an information infrastructure that is essential to our daily work (W3C, 2001). *Discussion*, or the electric bulletin board system (BBS), is one of the most popular services on the Web. It allows two or more people to post and/or read

messages asynchronously by means of Web browsers. A typical Web-based discussion service allows URL's to be included as clickable hyperlinks in the messages.

File sharing also is one of the most popular services for collaboration. A file server provides this type of service via a network protocol such as Server Message Block (*SMB*) or Network File System (NFS). *Samba* is a file server that runs on Unix or Unix-like operating systems such as Linux, and supports SMB (Hertel et al., 2001). With this protocol, a Windows-based PC can mount one of the server's directories on the PC's file systems as a drive, and the user can access remote files in the same manner as accessing local files.

A Web server can service on a wide area network, and a typical Web server provides a read-only file sharing service, while SMB realizes a read/write file sharing service on a local area. Therefore, these two are complementary.

Internet Shortcut (or simply *shortcut* in this paper) is a file type defined by the Microsoft Windows operating system. The extension part of the file name is "url". This type of file is supposed to include text data as the following example:

```
[InternetShortcut]
URL=http://brain.is.kyushu-u.ac.jp/
```

When a user opens the file, for example, by double-clicking on it, a Web browser starts running and acquires the Web page from the specified URL. Therefore, it is possible to utilize this type of file as a kind of link from the user's local file system to the Web.

3 Issues and our solution

3.1 Issues in combining Web and SMB

Since the Web and SMB are complementary as mentioned in Section 2, simply combining a Web server and an SMB-based file server would be useful. For example, users could share files, both in the read-only mode on a wide area network and in the read/write mode on a local area network.

This paper focuses on the case where the Web server also provides a discussion service. In this case, users can have online discussions by means of Web browsers; the discussion messages may include hyperlinks for referring to some of the shared files, each of which has a unique URL given by the Web server. Such a hyperlink correlates the message and a shared file. However, since the hyperlink is one-way, there is no user interface for traversing from the shared file to the message. This raises the following issues:

- *Usability*: When users access shared files via SMB, they cannot reach discussions correlated to the files by taking advantage of the "click-and-browse" usability on the Web. This prevents discussion and file sharing from being integrated more tightly.
- *Awareness*: When the users access the files via SMB, they could be even unaware of the correlated discussions, because there is no user interface provided for notifying the users of the discussions. We believe awareness could make these discussions more collaborative.

3.2 Our solution: shortcut files for backlinking

A possible solution to these issues is to present a *backlink*, which is a link from a shared file to a discussion message that includes a hyperlink to the file. With this solution, a user who is accessing the shared file being discussed would find the backlink, get aware of the discussion, and be able to traverse to the Web by clicking on the backlink. This realizes a two-way linking feature between the Web-based discussion service and the SMB-based file service. This would also make it possible for two or more files to have a relationship among them via a discussion. In order to realize this, we considered the following two strategies:

- Present a backlink as a kind of file nearby the file being discussed. The Internet Shortcut file, which was mentioned in Section 2, would be useful in this case.
- Provide a utility program that helps users to find a relationship between the discussion and the shared file. For example, a small pop-up window, a message at a corner of the utility window, and a beeping sound, would be possible functions of the program.

We adopted the first strategy, which has three advantages in comparison with the second one:

First, since the shortcut is a file, the user can use the conventional file utilities to navigate and find the shortcut in the same manner as accessing ordinary files. All she or he has to pay attention to is that the shortcut file has the read-only permission, because this file must be under the control of a backlinking mechanism.

Second, since the shortcut file can be copied, it is portable and preservable. Suppose that a directory has several files, one of which is referred to by a discussion on the Web. Even if the whole directory was compressed into a file, it would still include the backlink to the discussion.

Third, since the file type of the Internet Shortcut is defined by the Windows operating system, it is unnecessary to install new software in the user's PC. All she or he has to do in order to follow the backlink is to double-click on the shortcut file.

4 System design and architecture

4.1 Design of the backlinking mechanism

According to the strategy described in Section 3, we designed and implemented a backlinking mechanism as a part of the Pilgrim system. In this section, we describe our concerns and determination on the design, which is shown in Figure 1.

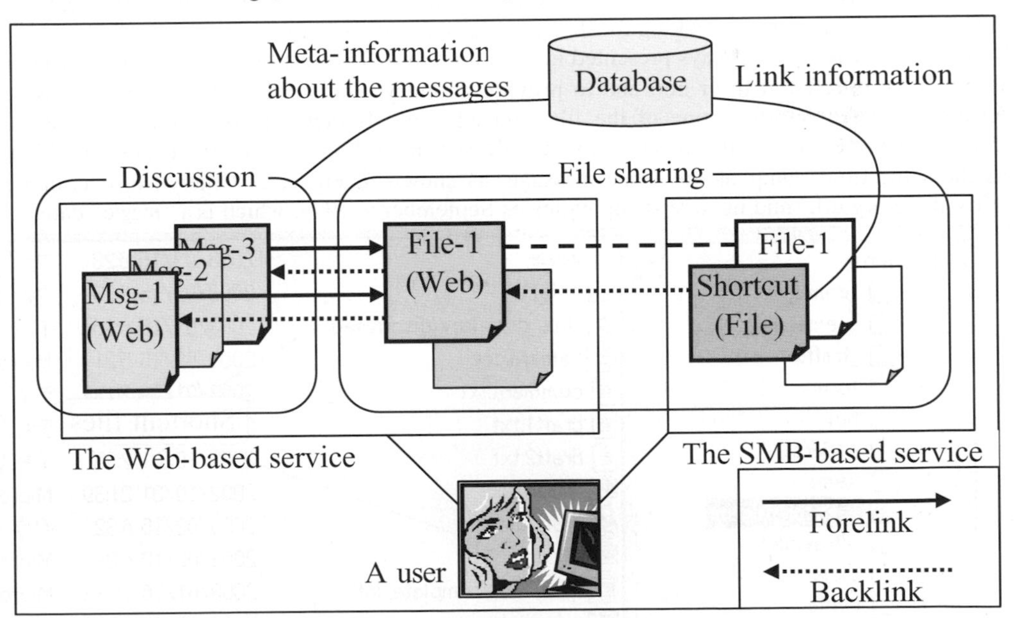

Figure 1: An overview of the system design

The Web-based discussion server realizes both the discussion and file sharing on the Web. It allows users to share files in the read-only mode on the Internet by giving a unique URL to every file. Users can browse the files as a Web page. If browsed is not a file but a directory, the Web page also includes the listings of all files and subdirectories in it. The users can also have discussions by writing, posting and reading messages, which may include the URL's of the files. The inclusion of a URL correlates the discussion message and the discussed file. When the users visit the message's Web page, the URL is shown as a hyperlink.

The file server makes the files accessible in the read/write mode via SMB on a local area network. The users can take advantage of their familiar file utilities to navigate and search the files. In addition, the file server is connected to the discussion server. When a user scans a directory, the file server determines which file in the directory is correlated to which discussion message, in order to generate shortcut files.

A virtual shortcut file for backlinking appears in the directory via SMB (as shown in Figure 2), if there is a discussed file. The shortcut file does not really exist, but has been temporarily generated by the file server according to the correlation, and includes the URL of the discussed file's Web page. When the user double-clicks on the shortcut, she or he would reach the Web page, which accompanies the hyperlinks for backlinking to the correlated messages. Thus, all the user has to do to reach one of the messages is one double-click and one single-click. In this way, every correlation between a file and a message appears as a two-way link at the both ends.

The name of a shortcut file is that of the discussed file with an extension ".url" at the end. For example, "draft3.doc.url" is a shortcut to the Web page of "draft3.doc". (Since the Windows standard shell does not show the extension of the shortcut file's name, users will find two "draft3.doc" as shown in Figure 2.) Therefore, when the files are listed in alphabetical order, the shortcut file appears nearby the discussed file. This makes it easier for users to find the shortcut file and become aware of the discussion messages.

The time stamp of the shortcut file indicates when a discussion referred to the file for the last time. This helps the users to find whether the file is discussed recently or not. In addition, they can also find out such discussions by looking for shortcut files that have late time stamps.

A special shortcut file is always presented in each directory. This has two roles: to show whether the directory is referred to or not, and to provide an entry point for initiating a new discussion concerning the directory or one of the files placed there. If there are one or more messages referring to the directory, the special shortcut file is named "_this_directory_discussed.url" and has the same time stamp as the latest message, as shown in Figure 2. Otherwise, it is named "_this_directory.url," and its time stamp indicates September 9, 1999, which is a "magic" date.

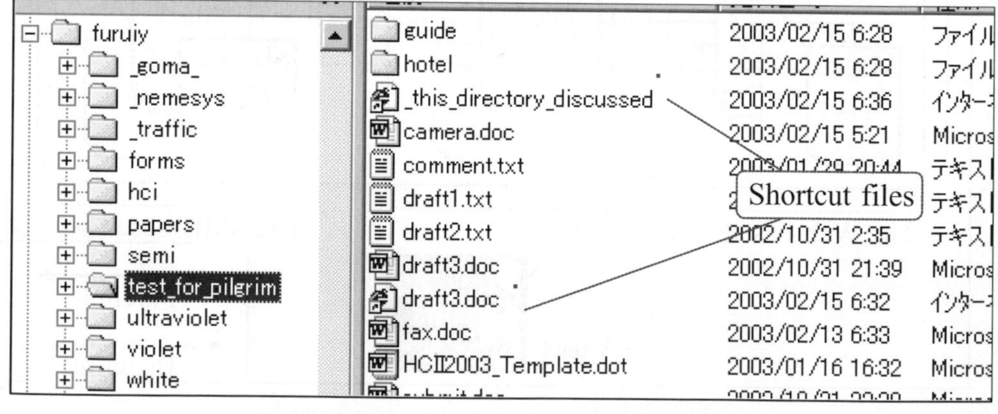

Figure 2: Shortcut files for backlinking to the Web

4.2 Architectural overview of the Pilgrim system

We constructed the Pilgrim system on top of RedHat Linux 7.2, Apache Tomcat 4.0.3 and PostgreSQL 7.1, by using Java 2 SDK 1.4 Standard Edition. The discussion server is a JSP/Servlet application. When a user posts a message by filling in a Web form, the server writes it in a text file, and stores meta-information on the message into the PostgreSQL database. Such meta-information contains the date, title, author, and the URL's that the message includes in it (if it does). The message is displayed on a Web page that the server makes up on the basis of the text file and meta-information. The server also allows the user to browse a shared file, by providing a Web page that shows the file and the listings of all messages referring to the file (if they exist). The listings are based on the meta-information stored in the database.
We adopted Samba 2.2.4[ja] as the SMB-based file server, and customized it in order to allow it to connect to the PostgreSQL database. When a user navigates a directory shared via SMB, the file server acquires the meta-information from the database, determines which messages include hyperlinks to the files in the directory, and makes virtual shortcut files appear in the directory.

5 Current status and problems

We have realized the two-way linking feature as a part of the Pilgrim system, but it still has a performance problem, which is caused by the overhead for acquiring the meta-information. With a Linux server that has a Pentium II processor (450 MHz), it takes approximately fifteen seconds for the file server to show a directory that has eighteen subdirectories, one file, and one shortcut file, while it takes only a second for the original Samba server. Due to this problem, we have not conducted a long-term experiment yet. However, we think that it is possible to shorten the time by improving the caching function, which is a part of the customized Samba server and reduces the number of database queries.
We have been running the system for nine months without the SMB-to-Web backlinking, for the purpose of the internal discussions of our laboratory members. Currently, it is possible to access the shared files, have discussions on our local area network as well as on the Internet, and take advantage of the Web-to-Web backlinking.

6 Summary and future work

This paper introduces a new technique of computer-mediated collaboration that enables two-way linking between a Web-based discussion service and an SMB-based file sharing service. It resolves issues concerning usability and awareness. The Pilgrim system allows users to share files and have discussions, which may refer to the files. It also provides backlinks from the files to the discussions. The users can access the backlinks by using their familiar file utilities, get aware of the discussions, and traverse among the discussions and the files in any direction.
Our future work will include a Web-based file updating function, which realizes read/write access on the Web, performance improvement, and a long-term experiment for evaluating this technique.

References

World-Wide Web Consortium (2001). About The World-Wide Web. Retrieved January 20, 2003, from http://www.w3.org/WWW/

Hertel, C., Samba Team, & jCIFS Team (2001). Samba: An Introduction. Retrieved February 15, 2003, from http://us1.samba.org/samba/docs/SambaIntro.html

The Competence Card – A Tool to improve Service

Walter Ganz
Fraunhofer Institute for Industrial Engineering
Nobelstrasse 12, 70569 Stuttgart, Germany
Walter.Ganz@iao.fhg.de

Anne-Sophie Tombeil
Fraunhofer Institute for Industrial Engineering
Nobelstrasse 12, 70569 Stuttgart, Germany
Anne-Sophie.Tombeil@iao.fhg.de

Performance Measuring and Scorecard Development as a Scientific Challenge.

The objective of assisting service enterprises in Germany in improving their performance in significant fields of growth in the service economy, and in keeping pace with international developments, is the focus of the »Fit for Service« research and development project. The development and testing of benchmarking instruments specific to the service economy and the definition of crucial success factors are intended to enable enterprises to learn from best practices and thus improve their ability to implement their service strategies and compete on (international) service markets.

The partner enterprises and the research organizations involved in »Fit for Service« are developing the methodology of the Competence Card (CC) as a key instrument for implementing service benchmarking and are filling it with sample contents. The salient points of the discussion surrounding the Balanced Scorecard as a strategic performance management system have been taken up in developing the Competence Card and elements of the EFQM model have been given consideration in addition. The objective of this way of proceeding is to combine the strengths of these two management systems and render them usable for the strategic design of service benchmarking with regard to content. The Competence Card therefore provides an instrument which assists organizations or organization networks, and actors involved in shaping the sociopolitical environment in which these organizations act, to consider from a strategic point of view those content relevant aspects which should be explored for the purposes of benchmarking in learning from the best by means of comparison (cf. Ganz/Holzschuh/Tombeil 2001; Ganz/Tombeil 2000). Whereas the Competence Card is intended to throw light on the structure, selection and implementation of contents for benchmarking in the run-up to and in the course of benchmarking processes, these processes can themselves be supported by other instruments such as the DIN PAS 1014 which has been especially developed for service providers.

The results of the project to date show that the Competence Card has proven its value as an instrument for initiating and supporting benchmarking processes. The logic of the Competence Card assists decisionmakers in enterprises as well as researchers in structuring the discussion surrounding catch-up and lead strategies for service providers in Germany, in prioritizing and operationalizing strategic objectives in the respective fields of growth and in identifying and operationalizing those competencies which would appear necessary to achieve these objectives. On the basis of the methods developed so far, sample contents are being drawn up step-by-step in the »Fit for Service« benchmarking clubs and the associated service research organizations around which enterprises intent on succeeding in the service economy can orientate their activities.

Over and above these successes on the cooperative project level, service research, of which »Fit for Service« is a part and to which a contribution is to be made, has a vested interest in examining how the results produced here over the course of a two-years period fall in line with international research and developments in measuring and evaluating performance by comparison as a basis for improved performance. The discussions in the Performance Measurement and Performance Management subject areas were spotlighted in order to venture a positioning of the Competence Card in the international state of the art.

On closer scrutiny of the international discussion, it becomes apparent that the issue of adequately measuring and evaluating performances – be they of a personal or organizational nature – under the headings Performance Measurement (PMe) and Performance Management (PMa) is considered a highly relevant subject both from the scientific and corporate viewpoints. The conviction that only that which can be measured and evaluated can be successfully controlled and managed would appear to remain unbroken in scientific and corporate practice. Great efforts are thus being made to develop methods and instruments with which all sorts of performances can be evaluated and improved. Even a good decade since its first publication by Kaplan and Norton (1992), the discussion surrounding PMe and PMa is still dominated by the concept of the Balanced Scorecard (BSC) which seems to be one of the most popular and widespread management concepts so neverthe less a noticeable disillusionment with its actual uses regarding implementation in tangible terms can be ascertained (cf. Cobbold/Lawrie 2002).

Both the popularity of the BSC concept and disillusionment in regard of its implementation and effect have led to various further developments of the original concept since the early 1990's. In an informative article concerning the state of the discussion (Cobbold/Lawrie 2002), Cobbold and Lawrie identify three generations of Balanced Scorecard. The development of the BSC can be summarized very briefly as follows. In the BSCs of the first generation, which were topical from the early to mid-1990's, the issue of selecting specific parameters for inclusion in the BSC and clustering them along the four BSC perspectives predominated. Conceptual further developments since the mid-1990's focussed on facilitating the selection of suitable parameters by introducing strategic objectives. Furthermore, attempts were made to portray the links between individual strategic objectives with the procedure introduced in the literature under the term »Strategic Linkage Model«, and thus take the first steps away from a system of measurement towards a strategic management system (cf. Kaplan/Norton 1996, 2000).

Reflection on the development of the Competence Card in »Fit for Service« against this backdrop reveals that the instrument falls in line with the international discussion well and that central elements of discussion have been taken up in the course of its development and have been conceptualized. The Competence Card developed in »Fit for Service« can thus be seen to latch onto the »Destination Statement« element at two points. Firstly, the research and development work is oriented around the common vision of all the partners in the association, namely successful positioning in the service economy by high service performance. Secondly, the strategic implementation instrument of the CC in the »Fit for Service« clubs at least implicitly supports the development of visions for specific fields of growth.

In addition the Competence Card in »Fit for Service« attempts to pave the way in the area of performance measurement and performance management by means of three elements, with special emphasis on the benchmarking concept. Firstly, the element of »competencies« is explicitly anchored in the methodology of the CC. In this context, competencies describe the ability of an organization to achieve, maintain and surpass its strategic objectives. This means that not only the »what«, i.e. which objectives an organization desires to achieve, is brought into discussion, but also the »how«, i.e. with the assistance of which competencies these objectives can be achieved. Whereas the BSCs of the third generation concentrate on strategic linking of objectives and the associated fields of perspective, the Competence Card focuses on establishing a link between objectives and competencies in addition to the synergetic definition of objectives. Practical in-company experience and scientific insight are to serve as a basis for formulating cause-and-effect relationships between competencies and objectives. The relevance of these interdependencies is examined in the course of the project and they are reformulated if necessary.

Secondly, the four fields of perspective familiar from the Balanced Scorecard are omitted in the Competence Card concept. Foregoing the fields of perspective defined by the conventional BSC is based on the assumption that for contexts, such as »Fit for Service«, in which the use of the Competence Card does not aim at, or remain restricted to, a specific enterprise, but where the card is to be used to enhance the management of central elements with regard to content in entire fields of growth, it is important to allow as much flexibility in goal definition as possible, rather than introducing fixed causal relationships – a frequent reproach leveled at the one-way chain of the fields of perspective in the BSC (cf. Brignall 2002). On the contrary, the actors in the development process of the Competence Card are to be allowed to decide for themselves which dimensions of control and organization are relevant to the field of growth and are to be addressed, and which mutual and multidimensional interdependencies between objectives and goal achievement are to be monitored and examined.

The disadvantage of omitting the four BSC fields of perspective in the instrument of the Competence Card may lie in the fact that a tendency to imbalanced selection of those objectives relevant to enterprises may manifest itself in the absence of prescribed guidelines concerning which fields are to be given consideration. The objection may be raised at this point that it is also hardly possible to speak of a true balance in implementation when working with Balanced Scorecards, at least in the versions of the first and second generation, and other Performance Measurement Systems as well. It would rather more appear that financial aspects have dominated strongly in previous application of the BSC despite the introduction of four perspectives.

In addition, an imbalanced preference for certain objectives from the strategic management standpoint may even be desired. Namely when the intention is to see structurally or institutionally weaker stakeholder groups adequately represented and imbalances of power equalized between interest groups – which are represented by the four fields of perspective of the BSC (cf. Brignall 2002). Finally, it can be said, at least on the basis of empirical evidence in »Fit for Service«, that enterprises are actually quite well equipped to recognize and address those stakeholders which are important to them. It has been shown that aspects from the employee sector, the customer sector and specific business processes are addressed in the Competence Cards of the »Fit for Service« clubs both in the areas of strategic and operative objectives and also in the competencies of the enterprises.

Thirdly, the Competence Card introduces a three-level model in formulating the operative goal. This anchors the concept that the definition of objectives which are to be achieved on the way to higher service performance and success in the service economy are formulated not only with regard to the enterprise level, but also with regard to the market and policy, politics and polity. With respect to a central criticism of the BSC concept which complains that »...they neglect several other important stakeholders whose need for performance-related information is worthy of recognition« (Brignall 2002) and which makes demands for more space in Pme and Pma systems for influential factors from the outer spheres, the Competence Card promotes a wider perspective in performance measurement and management activities, not to forget benchmarking processes, by anchoring the three-level model in concept. In addition to the enterprise level, developments and development conditions in existing and future service markets must be addressed and the area of social and political processes such as demographic developments, changing values or issues of state regulation must also be afforded due consideration. The philosophy and singular characteristics of the Competence Card are summarized in the **figure 1** below.

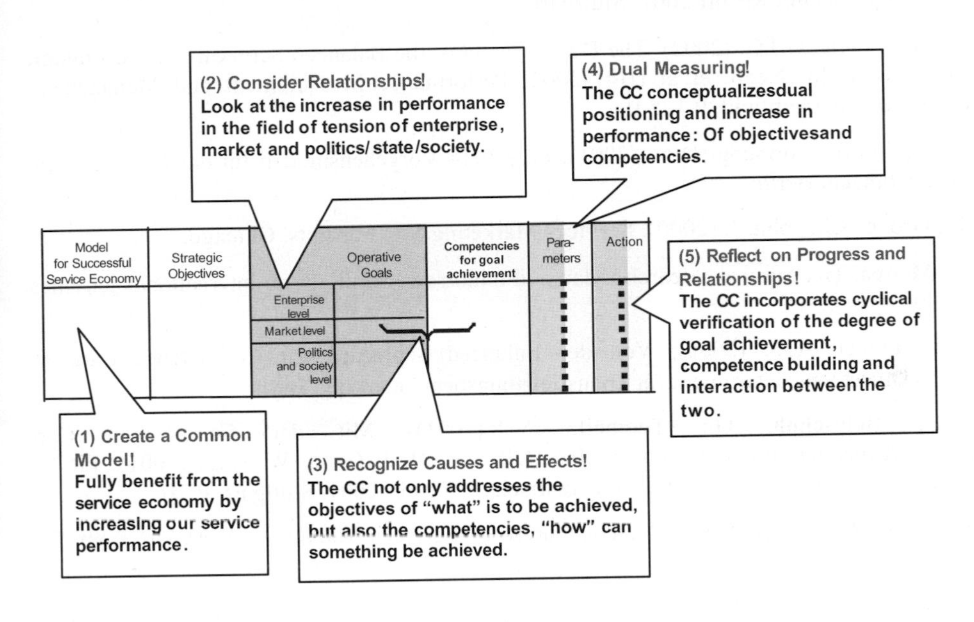

Figure 1: The singular characteristics of the Competence Card

From a general point of view, it can be maintained that the instrument of the Competence Card developed in »Fit for Service« falls very much in line with the state of the international discussion on performance measurement and management, and scorecard development. The concept addresses the main criticisms of existing BSC versions and offers new solutions on the basis of tried-and-tested elements from existing systems. Major focuses of further development are the conceptual anchoring of competencies and the explicit representation of interdependencies between these competencies and the set objectives, dual measurement, namely of the degree of goal achievement and the degree of competence building, and the introduction of a three-level model which positions performance measurement and performance management in the field of tension of enterprises, market and conditions in the sphere of policy, politics and polity. The attempt made in »Fit for Service« to organize and support benchmarking with regard to strategy and content using a scorecard may also lend interesting impetus to the international discussion. It can be stated that further elaboration of the Competence Cards will bear fruit – whereby operationalization of competencies and objectives and examination of the results in initial benchmarking cycles are now at the center of attention – and is capable of making a relevant contribution to the state of the art of service research.

References

Brignall, S. (2002): The Unbalanced Scorecard: A social and environmental Critique. In: Neely et al (Hg) (2002): Performance Measurement and Management: Research and Action. Cranfield, P.85 ff..

Bullinger, H.-J; Ganz, W. (Hg) (2001): »Fit for Service« Benchmarking für die Dienstleistungswirtschaft. Report 2001. Stuttgart.

Cobbold, I.M; Lawrie, G.J.G. (2002): The Development of the Balanced Scorecard As A Strategic Management Tool. In: Neely et al (Hg) 2002: Performance Measurement and Management: Research and Action. Cranfield, P.125 ff.

Deutsches Institut für Normung (DIN) (2001): PAS 1014 Vorgehensmodell für das Benchmarking von Dienstleistungen. Berlin.

Fisk, R.P.; Grove, S.J.; John, J. (2000): Services Marketing Self Portraits. Chicago.

Ganz, W.; Meiren, Th.: Service research today and tomorrow. Spotlight on international activities. Stuttgart 2002

Ganz, W.; Tombeil A.-S. (2000): Von der Balanced Scorecard zur Competence Card als strategisches Umsetzungsinstrument für Dienstleistungsbenchmarking. Berlin.

Ganz, W.; Holzschuh, G.; Tombeil, A.-S.(2001): Mit der Competence Card Dienstleistungsbenchmarking unterstützen. In: Bullinger, H.-J; Ganz, W. (Hg) (2001): Fit for Service Benchmarking für die Dienstleistungswirtschaft. Report 2001. Stuttgart, P.7ff..

Grove S.J.; Fisk R.P., John, J. (2002): The Future of Services Marketing, Forecasts from Ten Service Experts. New Orleans.

Kaplan, R.S, Norton, D.P. (1992): The Balanced Scorecard – Measures That Drive Performance. In: Harvard Busines Review, Vol 70, Jan-Feb.

Kaplan, R.S, Norton, D.P. (1996): Translating Strategy into Action. Boston.

Kaplan, R.S, Norton, D.P. (2001): Die Strategie - Fokussierte Organisation. Stuttgart.

Rust, R.T.; Zahorik, A.J.; Keiningham, T.L. (1996): Service Marketing. New York.

A Group Development System for Improving Motivation, Performance and Team Climate in Virtual Teams

Susanne Geister, Udo Konradt & Guido Hertel

University of Kiel, Department of Psychology
24098 Kiel, Germany
geister@psychologie.uni-kiel.de

Abstract

With the spreading of decentralized or "virtual" teams, group development systems and tools become essential that consider the characteristics of dispersed work. Compared to traditional face-to-face teams, virtual teams are shown to be more vulnerable to motivation losses and show lower levels of trust, cohesion, and satisfaction. Particularly, the lack of feedback and information about team processes within the virtual teams, which are due to the de-location and reduction of social contacts among team members, has been repeatedly mentioned as a major problem. Based on a theoretical model of motivational processes within groups, a system for group feedback was developed in order to assist virtual teams to cope with reduced team-related information. The Online-Feedback-System (OFS) is a web-based tool, which supports team members to collect crucial indicators of the ongoing team processes on a regular base. Basically, the OFS provides three aspects of team feedback: (1) motivational feedback, (2) task-related feedback in relation to MBO, and (3) feedback about the team climate regarding satisfaction, team identity, and conflict management.

After describing the features and functions of the OFS, two studies will be presented that addressed two major goals: First, effects of the OFS were evaluated by comparing teams that used the OFS with controls that did not use the OFS. Second, characteristics of successful virtual teams were investigated by exploring the relation between motivational variables, collective strategies, and team performance. The results show that the OFS was able to enhance performance of virtual team members. Moreover, high performing teams were characterized by a higher motivation of the members, a better task-related cooperation, and more non-task-related communication within the team. Together, the results show positive effects of the feedback system and demonstrate that information and feedback are crucial in virtual teams.

1 Introduction

Virtual teams are a new and challenging form of work units. By using new information technologies they facilitate cooperative work independent of time and space. Virtual teams offer several strategic advantages, e.g. connecting competent employees for a job regardless of their location, providing greater flexibility to individuals, and saving expenses for traveling and office-equipment. However, apart from these advantages there are several challenges due to the isolation and the reduction of social contacts among the team members. Compared to traditional face-to-face teams, virtual teams are shown to be more vulnerable to motivation losses, and usually have lower levels of trust, cohesion, and satisfaction (Hertel, Konradt & Orlikowski, under review; Rocco, Finholt, Hofer & Herbsleb, 2000; Warkentin, Sayeed & Hightower, 1997). Because strategies of direct influence and control on team members are restricted, team leaders need to employ more indirect and delegative forms of leadership. A management concept that has been recommended for remote work is Management by Objectives (MBO). MBO has an emphasis on three components: (1) goal setting, (2) participation of employees in planning and decision-making, and (3) feedback during task fulfillment. However, for successful virtual teamwork not only performance-related feedback as in MBO is needed, but also motivational and socio-emotional feedback. These feedback aspects are particularly difficult to exchange by electronic communication, even though they are essential for team-building related to cohesion, trust and team identity. An Online-Feedback-System (OFS) was developed in order to provide virtual teams with additional and reliable information about team processes. The OFS is based on a theoretical model of motivational processes in groups and is expected to work as a team-building tool, enhancing motivation, performance, and satisfaction in virtual teams.

2 Theoretical and Empirical Background

2.1 Group Support Systems

Group Support Systems (GSS) can increase group work performance in terms of decision quality or effectiveness, number of creative ideas generated, and equality of group participation (Fjermestad & Hiltz, 1999). However, some studies reported that GSS did not increase but sometimes even decreases group satisfaction (Fjermestad & Hiltz, 1999). These different effects of GSS are especially important for virtual teams. As outlined above, virtual teams often show lower levels of motivation, cohesion, trust, and team identity compared to face-to-face teams. Virtual teams work more task-oriented and exchange less relationship-oriented information compared to face-to-face teams (Chidambaram, 1996). Consequently, GSS in virtual teams should devote special attention not only to performance indicators but also to climate indicators such as motivation and satisfaction of virtual team members. Huang, Kwok-Kee and Lim (2002) criticize that most GSS research focuses on face-to-face teamwork; few studies were conducted in supporting virtual teamwork and team-building. Two notable exceptions are the following studies: Huang and Lai (2001) showed that a GSS embedded within a team-building framework was able to enhance performance *and* satisfaction of virtual teams. Zumbach et al. (2002) showed that socio-emotional feedback in virtual learning communities lead to an increase of motivation and quality of interaction processes. The Online-Feedback-System provides motivational, task-related and relationship-related feedback in order to support motivation, performance and satisfaction of virtual teams. Next, the design and objectives of the OFS will be described in more detail.

2.2 Online-Feedback-System

The Online-Feedback-System (OFS) is based on the VIST-Model (Hertel, 2002; see also Hertel, Niedner & Herrmann, in press) that postulates four central motivational variables as determinants of team members motivation and related performance: *Valence* (subjective importance of team goals for team members), *Instrumentality* (perceived indispensability of own contributions), *Self-Efficacy* (perceived capability to fulfill the required tasks) and *Trust* (perceived trust in other team members and the electronic support system).

The OFS is a web-based tool, which supports team members to collect the status of different crucial indicators of the ongoing team processes on a regularly base (e.g. once a week). It provides basically three aspects: (1) motivational feedback about the VIST-components, (2) task-related feedback in the context of MBO, and (3) relationship-related feedback regarding satisfaction, team identity, and conflict management. Feedback is given to each member on an accumulated team level. Team members' ratings are collected on a 7-point-scale ranging from "strongly disagree" to "strongly agree". All statements are worded in the same way: Agreements are equivalent to positive evaluations and are colored green. On the opposite, disagreements indicate a negative evaluation of team aspects and therefore are colored red. By using colored scales a plain feedback is supported. Thus, team members can detect in a simple way which team aspects are rated as satisfying and which aspects are rated as improvable.

3 Research Questions and Methodology

3.1 Research Questions

Firstly, the *effectiveness of the OFS* was explored in a longitudinal study. It was assumed that by using the OFS, employees concentrate on relevant motivational aspects. An implicit goal setting process is triggered, which leads to an increase of motivation by optimising motivational components and their conditions. Furthermore, information and feedback about current team processes is spread out through the team, which leads to an improved team well-being. Increased motivation and satisfaction in turn, mediate an increase of performance.

Secondly, *characteristics of successful virtual teams* were explored. It was expected that a high motivation (i.e. high scores in the VIST-variables) and a well-functioning communication process are related to a high team performance.

3.2 Participants and Procedure

Student teams: The first study examined 53 student teams from different German universities. Students were on average about 24 years old (M = 24.1, SD = 2.7) and were on average in their third year (60% studying psychology, 40% business administration). Two third of the participants were female, one third male. Team members communicated only by electronic communication tools (e.g. E-mail, phone) and did not meet personally. Each team consisted of two same-gender team members from different cities. Motivation for participation were ensured by embedding the project in student courses, giving course credit for their participation and offering a reward for every 10th team (drawn by a lottery).

The cover story of the study explained that each team is supposed to be a "consulting company" that has to develop two problem solutions for two clients. Before the teams started, they were asked to exchange basic personal information, to name their "consulting company", and to play an "ice-breaker" game. During the study, participants answered an evaluation questionnaire two times: First at the beginning of the project (before using the OFS) and secondly at the end of the project. Teams were randomly assigned to the OFS-teams (26 teams) vs. the Non-OFS teams (27 teams). Each team was given three weeks for solving the problems. Thus, each of the OFS-teams used the OFS tool for five weeks in total.

Business teams: In the second study, 45 virtual teams of a large Internet provider company were studied. The task of the teams was to advise and supervise chat rooms, boards and quizzes. The teams consisted on average of 13 team members (M = 12.9, SD = 9.7), existed about 30 weeks (M = 31.3, SD = 11.1), and had no restriction in their time span. Team members were on average about 30 years old (M = 33.6, SD = 10.5). Gender was distributed equally. Although part of a commercial Internet provider, these teams did unpaid work and cooperated voluntarily in their free time. Accordingly, they worked mostly from their home PC, which resulted in a high level of "virtualization". Since team members worked without salary, the company tried to maintain motivation of team members by non-monetary rewards, e.g. informal parties and raffles.

As in the first study participants answered the evaluation questionnaire two times: First at the start of the project (before using the OFS) and secondly at the end of the project. Teams were randomly assigned to the OFS-teams (22 teams) vs. Non-OFS teams (23 teams). OFS-teams used the OFS seven weeks in total. Each week a mail was sent for remembering team members to complete the OFS.

3.3 Measures of the Dependent Variables

Process variables measured motivation and collective strategies. Motivation was measured according to the VIST-model, with items for valence, instrumentality, self-efficacy and trust (Hertel et al., in press). Collective strategies were measured regarding cooperation (task-related communication) and communication (relationship-oriented communication).

Outcome variables consisted of performance and satisfaction measures. Performance was measured by asking participants to rate team performance (overall performance, quality, quantity, motivation, and compliance to deadlines) in percent (range from 0 – 100%).

Additional items: Demographic measures (e.g. age, sex, field of study, team size, distribution of team members) were also included in the evaluation questionnaires. At the end of the project, the OFS users were asked to rate the comprehensibility of the OFS objectives and the usability of the OFS. All items (except performance ratings and demographic data) were answered on 5-point scales ranging from "disagree strongly" (1) to "agree strongly" (5).

4 Results

Return rate dropped from 95% to 87% for student teams, and from 64% to 23% for business teams. The use of the OFS was also different in Study 1 and 2. One-hundred percent of the student teams used the OFS at least once, compared to 86% of business teams. Measured on an individual level, 98% percent of students used the OFS at least once, compared to 30% of business team members. The data of the evaluation questionnaire were examined as follows: First, scales were composed according to exploratory factor analyses and scale reliability analyses. Second, individual data were aggregated on team level, which was justified by different indicators of within team agreement (r_{wg}, ICC). Please note that the following results are preliminary.

Effects of the OFS: In the first study with student teams, an ANOVA with repeated measures with perceived performance of teams as dependent variable revealed a significant interaction effect ($F(1,50) = 6{,}11$; $p < .02$). No such interaction effect regarding performance occurred in the second study for the business teams. Although the interaction effects for motivational variables and satisfaction were not significant, means for valence, trust and satisfaction showed the expected pattern.

The vast majority evaluated the OFS itself as positive: 66% of students and 62% of business team members explained that the objectives of the OFS were comprehensible. 90% of students and 81% of business team members evaluated the usability of the OFS as positive.

Characteristics of successful teams

Student Teams: Motivational variables correlated significantly with perceived team performance (valence $r = .58^{**}$, instrumentality $r = .24+$, self-efficacy $r = .30^{*}$, and trust $r = .63^{**}$). Also cooperation ($r = .65^{**}$) and communication ($r = .43^{**}$) correlated significantly with perceived team performance. A linear regression analysis with perceived team performance as dependent variable revealed significant effects of the valence ($\beta = .37^{**}$) and trust ($\beta = .43^{**}$) components. Regarding the collective strategies only cooperation contributed significantly to performance ($\beta = .61^{**}$).

Business Teams: Two of the motivational variables correlated significantly with the perceived performance of teams (instrumentality $r = .42^{**}$, trust $r = .78^{**}$, valence $r = -.02$ ns, and self-efficacy $r = .07$ ns). Also cooperation ($r = .31^{*}$) and communication ($r = .47^{**}$) correlated significantly with perceived performance. A linear regression analysis revealed that particularly trust ($\beta = .74^{**}$) regarding the motivation and communication ($\beta = .42^{**}$) regarding collective strategies contributed significantly to perceived team performance.

(*Note:* $+p < .1$, $^{*}p < .05$, $^{**}p < .01$, two-tailed)

5 Discussion and Implementations

Results show that the OFS was able to enhance performance of virtual team members in one of the two studies. Student teams who used the OFS rated their performance higher compared to student teams who did not use the OFS. No such effect was found for the business teams. However, the vast majority of users in both studies rated the OFS itself as positive. Diverging results might be due to the different samples. Compared to student teams, business teams were larger, more heterogeneous (regarding age, sex and education), worked voluntary in their free time and did not use the OFS as frequently as the student teams. Although the OFS is assumed to support a wide range of virtual teams there might be moderating effects of heterogeneity or team size. Moreover, in further studies frequent use of the OFS should be ensured by discussions of goals and functions of the Group Development System before the implementation. To summarize, at least the results of the first study with student teams suggest that the OFS is indeed able to enhance performance of virtual teams. However, contrary to the expectations, no mediating effects were found regarding motivation and satisfaction, although the means of motivational variables and satisfaction show the expected pattern. Further research is needed to explore mediating processes in more detail.

The analyses of characteristics of successful teams showed that the VIST-variables were related to perceived performance of the virtual teams. Thus, the VIST-model might provide helpful information in order to maintain and enhance motivation in virtual teams. Especially trust was perceived as essential for performance of virtual teams. Together, the results show overall positive effects of the OFS, demonstrating that information and feedback also about climate issues can be crucial in order to improve the performance of virtual teams.

References

Chidambaram, L. (1996). Relational development in computer-supported groups. *MIS Quarterly*, 20(2), 143-163.

Fjermestad, J. & Hiltz, S.R. (1999). An assessment of group support systems experimental research: Methodology and results. *Journal of Management Information Systems*, 15(3), 7-15.

Hertel, G. (2002). Management virtueller Teams auf der Basis sozialpsychologischer Theorien: Das VIST Modell [Managing virtual teams based on models from social psychology: The VIST model]. In E.H. Witte (Ed.), *Sozialpsychologie wirtschaftlicher Prozesse* (172-202). Lengerich, Germany: Pabst Publishers.

Hertel, G., Konradt, U., & Orlikowski, B. (under review). *Managing Distance by Interdependence: Goal Setting, Task Interdependence, and Team-based Rewards in Virtual Teams.*

Hertel, G., Niedner, S. & Herrmann, S. (in press). Motivation of software developers in open source projects: An internet-based survey of contributors to the Linux kernel. *Research Policy.*

Huang, W.W., Kwok-Kee, W. & John, L. (2003). Using a GSS to support Virtual Team-building: A theoretical framework. *Journal of Global Information Management*, 11,18-72.

Huang, W.W. & Lai, V.S. (2001). *Can GSS Groups Make Better Decisions and Feel Good at the Same Time? A Longitudinal Study of Asynchronous GSS Groups.* Proceedings of the 34th Hawaii International Conference on System Sciences.

Rocco, E., Finholt, T.A., Hofer, E.C., & Herbsleb, J.D. (2000). *Out of sight, short of trust.* Proposal of the Collaboratory for Research on Electronic Work, University of Michigan, USA.

Warkentin, M.E., Sayeed, L., & Hightower, R. (1997). Virtual Teams versus Face-to-Face Teams: An Exploratory Study of a Web-based Conference System. *Decision Science*, 28, 975-996.

Zumbach, J., Muehlenbrock, M., Jansen, M., Reimann, P. & Hoppe, H.-U. (2002). Multidimensional Tracking in Virtual Learning Teams. In Gerry Stahl (Ed.), *Computer Support for Collaborative Learning: Foundations for a CSCL community* (650-651). Hillsdale, NJ: Erlbaum.

Intercultural virtual cooperation: Psychological challenges for coordination

Elizabeth R. Grant, Hartmut Schulze, Siegmar Haasis

DaimlerChrysler Research and Technology
RIC/EP, P.O. Box 2360, 89013 Ulm, Germany
Elizabeth.Grant@daimlerchrysler.com

Abstract

Process coordination is critical for avoiding delays in intercultural engineering projects. We present initial findings from a 6-month field study of intercultural (German-American) production planning in the automobile industry applying concepts from Wehner, Clases, & Bachmann's (2000) interorganizational cooperation model. Our findings suggest that intercultural cooperation creates significant hurdles for identifying when partners' activities in the cooperation are uncoordinated. We call this blindness to unexpected events, problems, and crises *illusory coordinatedness,* in contrast to the state of initial coordinatedness in the Wehner et al. (2000) model. We propose mechanisms at the individual, team, and organizational levels that make perceiving lack of coordinatedness so difficult in intercultural cooperation.

1 Motivation

Intercultural cooperation is increasing in the automobile industry as global networks between producers and suppliers grow. These cooperations include joint development of IT and documentation systems, as well as joint product development and production planning. One critical goal for these cooperations is to coordinate activities successfully, to minimize the costly delays attributed to process loss. If coordination is, as Wehner et al. write, based on "an implicit mutual understanding that develops in everyday practice over time," (p. 989), then we should see difficulties maintaining a state of coordination in intercultural distributed work teams, who share little of their everyday practice and do not share implicit, or often even explicit, mutual understandings or conventions (Mark, 2002). This should be particularly true for cooperative work with low formalizability (Ziegler, 2002) like production planning.

In this paper, we apply Wehner et al.'s (2000) model of cooperation at work to a 6-month field study of an intercultural production planning project in the automobile industry. We will describe the theoretical cooperation model and then, using data from interviews, surveys, and observations in the field, introduce what we understand as the psychological mechanisms that hinder perceiving coordinatedness in intercultural projects. In contrast to what Wehner et al. (2000) describe as initial coordinatedness between cooperation partners, this is often, in intercultural projects, *illusory* coordinatedness, the state of believing activities are coordinated when they are not.

2 Cooperation Model from Wehner, Clases & Bachmann (2000)

Wehner et al. (2000) have proposed a model of the process of coordinating joint activity in inter-organizational cooperation. In this model, a state of initial coordinatedness is considered to be the precondition and product of joint activity. When this coordinatedness of the partners' activity is disrupted by an unexpected event, several options for subsequent interactions exist: a) corrective cooperation, in which the participants regulate their own behavior without questioning their coordinatedness, or, when the unexpected event is perceived as a crisis, b) expansive cooperation, in which the structural aspects of coordinatedness are brought into question, c) co-construction, in which mediators are brought in and time is set aside to "generate organizational solutions that transcend single cases," (p. 990), and, completing the loop back to a state of renewed coordinatedness, d) remediative coordination.

This model, defining modes of "co-operation as goal-directed and process-related joint activity" (Wehner et al., 2000, p. 984), is a rich source of hypotheses for empirical investigation in real cooperations. As the mechanism motivating choice of cooperation mode, the perception of a lack of coordinatedness is a critical aspect of this model that needs further test and elaboration.

2.1 Perception: The missing mechanism in the model

Our research has confirmed that a critical aspect of interorganizational cooperation is the timely perception that an ongoing process or joint activity is not coordinated, that process gears are slipping, expectations are not being met, and delays that could be costly are preventing the cooperation partners from reaching their goals. These "problems" or "unexpected events", in our field study over six months in a production planning project were rarely identified in a timely way, and were not always identified as equally serious by different cooperation partners. Periods of coordinatedness were hard to identify. Much of the time, the partners continued working under the illusion that their activities were coordinated, only to find out later that they were not, that events had not progressed as they assumed. Quite serious information (e.g., that a subsupplier had stopped work and sent their workers on forced vacation) took two months to reach customer management and be openly addressed with the supplier. This was not an exception. The failure to identify the lack of coordinatedness resulted in serious project delays.

What keeps people and teams from realizing that they are not coordinated? What constitutes for individuals and organizations an "unexpected event"? Further, at what point do they identify a situation as a "problem" and regulate individual behavior versus identifying a situation as a "crisis" and calling in mediators to support expansive cooperation? We propose that this perception is particularly difficult for intercultural virtual teams, and, using data from a production planning project, present several hypotheses about psychological factors which interact with social and organizational factors to create illusory coordinatedness.

3 Field Study: Intercultural Production Planning

Over a period of six months, we completed the initial stages of a field study of the cooperation between three partners in an intercultural production planning project. A German production-planning team was working together with an American plant and an American supplier to plan the production process for a new car model. We completed two rounds of interviews with all three teams (first wave n=17, second wave n=22) and, during the second wave, conducted a survey (n=25) of factors related to the cooperation. Here we present initial findings.

Planning a production process for cars consists to a large degree in communicating with product designers, with the supplier designing the production tools, and with the factory, an internal cooperation partner who provides the financial resources for the project and runs the plant for the life of the car model. As the production planners see themselves as ultimately responsible for the success of the production process, they direct the production planning. However, this must be accomplished within the physical and legal constraints of the plant. Areas of negotiation between partners include culture-based differences in production philosophies and the technical solutions created by the two internal partners together with the tool supplier (who has subsuppliers for robots, weld guns, etc.).

Thus, production planning consists mostly of communicative and problem solving acts in which cultural differences in communication and thinking highly influence cooperation. The partners have high interdependency but their joint activities have low formalization. There are not clear states of coordinatedness but rather degrees of coordinatedness. We propose several mechanisms that lead to illusory coordination.

4 Mechanisms leading to illusory coordinatedness

4.1 Different experience-based process scripts

Employees in Germany and America showed different expectations for the process of cooperation and planning based on their experiences performing these activities in their own cultures. Psychologically, these expectations take the form of cognitive scripts stored in memory and activated by the current activity. This is a common cognitive risk for problem solving (Grant & Spivey, in press): activated experience-based knowledge in the form of associations or analogies (Mark, 2002), can lead to misleading intuitive solution attempts. The lack of shared expectations is also a risk for lowering trust in interorganizational cooperations (Lane, 1998), which in turn can lower the likelihood of information sharing and problem identification.

At the beginning of the project, employees reported that they expected German and American planning practices to be similar, that they would perform their planning tasks as usual. However, the process scripts that each team followed in its own country involved different steps, different timing, different artefacts and norms (e.g., for which specification documents take priority), different cultural symbols and language in process documentation. Thus, the German and American teams started out relying on their own experience of prior planning projects. No one completed an intercultural training, although some Germans had made attempts to improve their English (no Americans, including those residing in Germany, learned German).

4.2 Assumption of similarity

Misperceptions are also caused by the tendency in intercultural projects for people to believe that their partners from another culture will behave in a similar way to themselves (Adler, 1997). In our case, supporting evidence for this comes from interviews—no one expected a German-American relationship to be so different. Data show that 1) they didn't expect differences in work practices or processes, 2) that it initially took 6 months to identify differences, 3) and that the problems that arose took longer to solve.

4.3 Expectation adjustment over time

Six months into the project, survey results showed that 92% of Germans and 80% of Americans (mean of both teams) reported that, unlike they expected, work practices of production planners are different in both cultures. The most common description we heard about this cooperation was that "it's a normal project, but then again, it's not a normal project." This statement reveals the confusion about whether or not the cooperation was running as expected. After the initial period of assuming similarity and then reaching a crisis when one team escalated what they considered a problem to top management (the other team did not perceive a problem), it had become hard for planners to identify what was normal and what was not normal, to tell which standards they should use to judge whether or not their activities were coordinated. For example, when an email reply failed to arrive or the planners did not receive an updated plant layout form the supplier, both German and American teams had difficulty interpreting whether or not it was an "unexpected event" to be taken seriously. Different employees developed different perceptions and harbored fantasies about the reasons email replies came late.

Over time, the team members had learned not to rely on their own experience. Unable to trust their tacit knowledge about work processes and social interactions, they lowered their expectations, making it yet harder to tell whether or not activities were coordinated. A typical response to what they would have earlier thought was an unexpected event became, "Well, it's normal for *them*, maybe it's normal in America, and actually we sometimes have problems with German suppliers too. Maybe it's just the normal supplier-customer relationship." This confusion about standards against which to judge an event made it difficult to identify a situation as a problem. In fact, employees began to accept dysfunctional and uncooperative behavior as normal, which created even less motivation to expansively cooperate or develop norms for future interactions. This correlated with increased statements of mistrust after one year of the project compared to initial interviews six months into the project, which in turn created a high risk for the project: employees expressed resignation about the possibility of identifying or escalating problems. When one side had identified a problem or tried to escalate it, the other side had not understood the seriousness of the issue, resulting in general demotivation to identify and escalate problems. The supplier did not want to draw attention to problems or even questions about tool variants, because of their fear that the customer would "go behind a post and laugh at them" or think they were "stupid."

Managers and technical planners who did not have a counterpart at the other company showed lower levels of trust and a higher pervasive sense of lack of coordinatedness than did planners who worked on a daily basis with a counterpart overseas. Yet the managers still held the illusion of coordinatedness at the work process level. They showed surprise when they learned after the fact that specific activities were not happening (e.g., that the weld gun designs had been on hold for 2 months). That is, in spite of general mistrust, managers held a global belief in coordinatedness in the face of disconfirming evidence. Planners showed higher levels of trust, but their reports about process coordination varied within a single interview. Again, it was clear that it was hard for them to judge what constituted a coordination breakdown, much less which events showed a lack of cooperation or goodwill on their counterpart's side.

4.4 Organizational Factors

Assumptions of similarity when work processes actually differed also took place at the organizational level, which reinforced individuals in their illusion of coordinatedness. As the German and Ameircan cultures were assumed to be similar, no formal intercultural training or systematic preparation for intercultural factors took place. There was no kick-off that included technical planners in the start phase or discussions between organizations about differences in planning philosophies or work processes. We discovered significant differences in production philosophies between the two cultures (e.g., quality vs. lean manufacturing) as well as work processes (e.g., product freezes vs. continuous product changes, labor and maintenance planning), which were not addressed by the group. The social and organizational factors that provide the context for illusory coordinatedness at the individual level should not be underestimated.

5 Conclusion

We have provided initial evidence from a field study of intercultural production planning in the automobile industry that intercultural cooperation creates difficulties for perceiving when joint activities are not coordinated. We call this phenomenon *illusory coordinatedness*. By allowing process failures to go undetected, this is a significant risk for intercultural engineering projects.

The German and American cooperation partners under study often held a belief that their processes were coordinated when, in fact, they were not. Uncoordinated acts were not identified as problems and escalated. We propose that different experience-based process scripts and the assumption of similarity at the individual and organizational levels led to illusory coordinatedness. As difference was experienced over time, employees became disoriented about the standards against which they should judge events in the cooperation, which made it more difficult to identify a lack of coordination. This correlated with mistrust, especially at the management level.

Further research should address how people and organizations identify when coordination is real or not so they can participate in cooperative acts. Wehner et al.'s (2000) descriptive model of cooperation can be elaborated to include the social and organizational mechanisms that drive identifying and discriminating between degrees of coordinatedness that motivate cooperative behavior, as well as the model's outcomes and relevance for intercultural and virtual cooperation.

References

Adler, N. J. (1997). *International dimensions of organizational behavior. (Third Edition).* Cincinnati, OH: South-Western College Publishing.

Grant & Spivey (in press) Eye movements and problem solving: Guiding attention guides thought. *Psychological Science*.

Lane, C. (1998). Introduction: Theories and issues in the study of trust. In C. Lane & R. Bachmann (Eds.), *Trust within and between organizations*, pp. 298-322. Oxford: Oxford University Press.

Mark, G. (2002). Conventions for coordinating electronic distributed work: A longitudinal study of groupware use. In P. Hinds & S. Kiesler (Eds.), *Distributed work*, pp. 259-282. Cambridge, MA: MIT Press.

Wehner, T., Clases, C., & Bachmann, R. (2000). Co-operation at work: a process-oriented perspective on joint activity in inter-organizational relations. *Ergonomics, 43,* (7), 983-997.

Ziegler, J. (2002). Modeling cooperative work processes: A multiple perspectives framework. *International Journal of Human-Computer Interaction, 14(2),* 139-157.

Self-Administered Cooperative Knowledge Areas
- Evaluation of the WWW Interface in Terms of Software Ergonomics -

Thorsten Hampel, Bernd Eßmann
Computer Science, Heinz Nixdorf Institut
Fürstenallee 11, 33102 Paderborn, Germany
E-Mail: {hampel|bernd.essmann}@upb.de

Abstract
The paper describes our experience with the use of cooperative knowledge areas in teaching practice and gives a critical evaluation of the approach. The evaluation and the proposed improvements focus on the web-based user interface.

1 Introduction
Using cooperative knowledge organization methods is by no means an easy undertaking. Besides developing and adapting suitable platforms and tools, we are confronted with numerous questions concerning the appropriate integration of such tools into the daily practice of classroom teaching.[1]
As part of the Paderborn open-source initiative sTeam[2], a technical platform is being developed to allow web-based cooperative work and learning using materials. In addition, the tools and environments produced – based on sound theoretical foundations – are being tested in everyday practice, critically evaluated and continuously adapted.[3] The results of this multilayered approach have already been presented in a number of publications.[4]
Despite the largely positive experience gained so far with the practical application of our approach, both teachers and learners have encountered a number of problems and uncertainties arising specifically from *cooperative* knowledge construction scenarios.
These include problems with navigation and orientation in a virtual environment. At the same time, a cooperative knowledge organization environment offers functions and options that are untypical of individually used applications, e.g. annotating material, as well as options that are unknown in author-centred learning and work environments, e.g. inserting one's own documents in a work area. To this extent, many users must first be encouraged to develop an understanding of cooperative knowledge construction and a feeling for cooperative use scenarios.
All of the problem areas mentioned are closely tied up with the design ergonomics of the software interfaces used. A specific problem here is that of suitably mapping different action options on

---

[1] The term Blended Learning is currently used in the literature to refer to the integration of distance learning scenarios in classroom teaching.
[2] The [open]sTeam project was funded by the *DFN-Verein* (German Research Network Association) for a period of two years.
[3] Parallel to this, theoretical concepts for the media-based evaluation of cooperative knowledge construction functions are being developed. This has resulted in a tentative understanding of the evaluability of technology in terms of cooperative actions on materials in the form of elementary media functions.
[4] For more on our experience with the development of cooperative tools, in particular the sTeam system architecture and the theoretical media function approach, the reader is referred to (Hampel, 2002) and (Hampel & Keil-Slawik, 2002). The concept of cooperative knowledge areas is given detailed treatment in (Hampel, 2002, pp. 107ff.). Its practical application in a number of regular university courses has been evaluated and documented (cf. Hampel, 2003).

materials (media functions) in a web-based user interface,[5] but the awareness of other co-learners/users must also be suitably embedded in the virtual knowledge area.
This paper reports on a experiment in which participants in the course *Praxis der Systemgestaltung* evaluated and improved the existing ^{open}sTeam system in terms of software ergonomics. As part of the course tutorials, the students (approx. 400 active participants) were asked to evaluate a scenario, which involved using the system to work cooperatively, and to make suggestions for improving various masks in terms of their functionality and software ergonomics. Questions ranged from the user convenience and task suitability of the dialogues to navigability and orientation in cooperative knowledge areas. Also being tested were the implementation of the chosen virtual-knowledge-construction metaphors and the area itself along with its documents and gates to other areas, as well as the rucksack as an important instrument for transporting material between different points of the environment. For technical or conceptual reasons, not all of the suggestions and criticisms made appeared directly practicable; some of them, however, were able to be implemented while the experiment was going on.[6]

2 The Experiment – A Cooperative Work Scenario

Initially, the evaluation of the sTeam system's user interface and our concept of cooperative knowledge areas was deliberately focused on a core area of the overall functionality. It would certainly have made little sense to call for an evaluation of all areas of the web interface or to allow the users complete freedom of choice in their evaluation of potential interfaces. What was needed here was a middle course that both allowed users a degree of freedom in selecting the appropriate functionality and use options and essentially took into account the key ideas underlying cooperative work. To this end, a scenario from the area of joint evaluation and analysis of sources (web pages) was developed and recommended as a model to the students:

I. *Creating a shared knowledge area:* As part of tutorial work, the students are organized in working groups. With a view to obtaining smaller learning groups to work on the individual assignments, a new group must first be created and then 3-4 cooperation partners allocated to it. – The creation of a working group causes a new area to be created.
II. *Setting authorizations in an access rights dialogue:* The next step is to adjust the access rights within an area such that each member of the working group is given full read and write access rights to the work area. (If the above created area is chosen, the rights are, in principle, already correctly preset for the group; only if an existing area is used do the rights have to be adjusted accordingly.) The access rights dialogue should be used in this way either to explicitly define the rights for the group or to effect inheritance of the rights from the "superarea".
III. *Linking knowledge areas by means of gates:* Gates must be created from the individual participants' home areas to the new shared work area. These are designed to support not only swift navigation between cooperative knowledge areas but also mutual awareness among users of a shared work area.
IV. *Creating reference objects and creating/inserting a text file:* The real aim of group work is to cooperatively evaluate a number of web pages on different course topics (ergonomics guidelines, etc.). Here, each group member first looks at a reference, then creates a link from the shared knowledge area to the web page and evaluates it in the form of a shared text file.

---

[5] The World Wide Web has become the de facto standard for the use of Internet technologies. However, with its author-centred architecture, it is not well-suited for supporting cooperative knowledge construction scenarios such as mutually annotating texts or jointly creating and structuring them. Nevertheless, given the availability of the WWW at all learning locations, tools and environments must necessarily be based on existing browser technologies and protocols and provide compatible interfaces to the WWW.
[6] This was possible because the sTeam's open and modular architecture, supporting design of a web interface using XML, XSLT and various scripting technologies, makes it very well-suited for implementing different approaches, but also because parallel to the evaluation a developer was constantly engaged in improving and extending the interfaces.

V. *Mutual criticism/discussion by annotating material:* Once each group member has viewed the complete text file, the evaluations/reviews of the other cooperation partners are attached to the respective reference objects in the form of annotations.

Parallel to the above scenario, the user interface dialogues employed were to be tested in terms of their functionality and ergonomics. The users were to specify which dialogues provided good and which poor support for the above scenario. In particular they were to identify functions that either inadequately supported or speeded up the respective work steps.
For each point criticized, a concrete description was to be given of the problems caused and, if possible, suggestions made for remedying them. A second task was either to assess an sTeam access rights dialogue according to the software ergonomics criteria on dialogue design discussed in the course, or to evaluate and redesign an arbitrary dialogue outside the above scenario.
All suggestions for improvements were in turn deposited in the knowledge areas used and discussed and annotated by the supervisors/tutors.
Here, a shared whiteboard was used for graphical annotation of the screen masks.

3 Results of the User Survey – Implementing a Scenario

The range of suggestions and ideas developed by the users was almost as broad as the problem complex of cooperative knowledge organization.[7] A great deal of experience with the use of cooperative knowledge areas in teaching had already been gained in earlier practical applications of the system (cf. Hampel, 2003). One of the most important results that emerged here was the key aim of reducing the required motoric effort for successfully completing a task/activity (cf. Hampel, 2003).[8] This paper focuses on the results obtained directly from the above experiment. Here, previous experience and results were largely corroborated.[9] Unfortunately, the following selection of results and examples can only reflect a small portion of the findings and ideas – the examples given are those in which suggestions by the users led to concrete improvements.

3.1 Awareness of Cooperative Action Options

The students' evaluation is based not only on the dialogue design criteria presented in the course, but also on the seven criteria for dialogue design[10] specified in EN ISO 9241 Part 10.
In terms of functionality, the mapping of the above scenario made it possible to identify a whole series of weaknesses and problems. These largely relate to finding/identifying specific functions in the user interface or awareness of the options provided by the environment. Another problem was that feedback on cooperative actions was frequently poor.[11]

---

[7] The design of cooperative knowledge areas involves addressing issues encompassing coordination, cooperation support, mutual awareness and the embedment of cooperative actions on material, i.e. cooperative document management.
[8] An example of this is the integration of tools and methods into the virtual cooperative knowledge area, which is called for at all times and should be as smooth as possible. For instance, users wish to use their own email client for sending and receiving messages or retain their accustomed tools and methods for creating and working on material.
[9] These included, for example, the finding that good performance and constant availability of user interfaces and the system as a whole is a prerequisite for acceptance by the users.
[10] Suitability for the task, self-descriptiveness, controllability, conformity with user expectations, error tolerance, suitability for individualization, suitability for learning.
[11] Here, a problem that needs addressing within the knowledge area is that of the user's own action options, but also that of group and activity awareness. A typical criticism with respect to the first point was the difficulty in determining how to create a new area or a gate to another area. Similarly, it was difficult to identify the cooperative action options in the rucksack metaphor.

The study's findings provide tentative solutions for avoiding/reducing the effect of the above weaknesses by showing *action options directly at the points where cooperative actions are performed.* For instance, besides the list of gates to other areas, an icon for creating a new gate to an area is added. Similarly, an icon/template for creating a subfolder (collection) or a subarea can be added to the mapped material available within an area. Also, the option of annotating any desired material can be highlighted by the simple prompt "It is possible to comment on/annotate this object" or an appropriate icon can be added. The evaluation yielded a whole series of such concrete suggestions for improving the awareness of cooperative action options.
Similarly, suggestions were made for facilitating cooperative actions by providing agents/wizards. For instance, a wizard could guide users through a typical cooperative action (workflows), like creating a new area, inviting potential cooperation partners and setting the necessary access rights. Another guided dialogue could support the actions of creating a gallery and depositing slides (for a seminar, talk or lecture). There are a number of possibilities for facilitating cooperative actions in this way.

3.2 Navigability, Trails, Orientation

A second key problem complex in cooperative knowledge organization is navigation and orientation in a virtual environment of linked knowledge areas. Even a small course with a manageable number of tutorial and student learning groups results in a large number of different knowledge areas: lecture areas, tutorial group areas and student learning group areas. Users participating in the survey frequently expressed the wish for improved individual orientation and navigation between knowledge areas. The list of suggestions for improvements also points in this direction.[12]
For instance, it was suggested that an individual's path through different knowledge areas be visualized in the form of a trail and utilized as a navigation aid. The prototypical implementation and successful testing of this suggestion was possible while the evaluation was still in progress.
Parallel to this, the attempt was made to visualize the current environment of an area in tree form. The essential criterion here would appear to be making the visualization specifically confinable to neighbouring – i.e. those accessible by gates – and super-/subordinate areas of a given depth. – A Java navigation aid, already available as a prototype, is currently being integrated into the sTeam system web interface. (cf. Eßmann, 2002)
Other suggestions concerning navigability in virtual knowledge areas relate to the implementation of extended search options. For example, the explicit suggestion was made to provide a specifically restrictable search function within knowledge areas and their content as a key navigation instrument. The displayed results of a search must be directly usable as a basis for further navigation (users' paths through material).

3.3 Ergonomic Redesign of Software, Scalability of Functionality

Most of the described suggestions for improvements relate to the structuring, controllability and conformity with user expectations of the graphical interface/web interface. Particularly with regard to the setting of access rights, a whole series of improvement options were proposed.[13] Most of the ideas proposed relate to scalability of the functionality provided. Take the dialogue for setting access rights: here a specific restriction of options was initially called for, e.g. setting access rights for the *individual's own* user group or groups belonging to an identical parent group. Also typical

---

[12] Interestingly, the mentioned suggestions for improvements – and some of those that have been elaborated in terms of design – relate to simple implementation ideas, but ones which may be all the more effective because of their simplicity.
[13] Here, technical implementability was not always taken into account, possibly limiting feasibility.

combinations of access rights (roles) should be available. Only in a second step should the whole range of options be made visible. Similarly, an improved access rights dialogue should make clearer the consequences of a particular combination of access rights, especially in conjunction with the inheritance of access rights.[14]
A final important demand, clearly voiced by the users participating in the survey, was the option of internationalizing user interfaces. – This could be implemented as part of current development efforts.

4 Summary/Outlook

Given the large number of ideas and suggestions for improvements put forward during the experiment, it is difficult to address them all in sufficient detail. At the same time, though, the very wealth and diversity of aspects relating to cooperative knowledge construction means that the design and use of cooperative systems must be a continuous, mutually stimulating process. The success of an approach that creates new opportunities for working and learning cooperatively with material depends on the extent to which it proves possible to practically implement the developed systems and the ideas underlying them.
In terms of ergonomics and human-computer interaction, a whole series of sound ideas and prototypical implementations are already available, but so far few constructive criteria have been developed for the design of cooperative user interfaces. An exciting future area of research would be, for example, to explore the extent to which cooperative user interfaces should – and can usefully – be coupled and structured. All the more reason, then, to pursue further practical experiments and research in this area. As an open-source platform, the sTeam system is not only ideally suited to the prototypical implementation and evaluation of cooperative user interfaces but also to the development of everyday practical solutions. The system's full potential will become evident when it comes to integrating the new opportunities offered by spontaneous networking or mobile terminals (mobile phones, tablet PCs, etc.).
This article has attempted to demonstrate the great potential offered by the active involvement of users in system design.[15] The chosen approach would appear to be a worthwhile investment of resources, particularly given its open-source philosophy.

5 References

Eßmann, B (2002). *Semantische Strukturierung virtueller Räume – Konzeption und Umsetzung eines Navigationswerkzeuges für das sTeam-System*, Master's Thesis, University of Paderborn, Mai 2002.

Hampel, T., (2003). Our Experience With Web-Based Computer-Supported Cooperative Learning – Self-Administered Virtual Knowledge Spaces in Higher Education. In: *Site 2003 International Conference.* Charlottesville (Va.), USA: ACM, in preparation.

Hampel, T. (2002). Virtuelle Wissensräume. – Ein Ansatz für die kooperative Wissensorganisation, University of Paderborn, PhD Thesis.

Hampel, T. & Keil-Slawik, R. (2002). sTeam: Structuring Information in a Team - Distributed Knowledge Management in Cooperative Learning Environments. *ACM Journal of Educational Resources in Computing* 1(2).

[14] If, for example, access rights to material are initially derived from the access rights set for a specific area or container – which makes sense for most cooperation situations – it must be clearly evident from the respective material what effect the explicit setting of access rights has.
[15] The involvement and surveying of some 400 students of computer science, business informatics and engineering definitely provided a good basis for critical evaluation and (re)design of the system.

Changing Technology – Equal Opportunities?

Niina Helminen

M.Sc. (Psychology), Researcher, TAI Research Centre
Helsinki University of Technology, P.O.B. 9555, FIN-02015 HUT
niina.helminen@hut.fi

Abstract

The study examined equality in a technological implementation as a micro level phenomenon. Equality issues were identified in five dimensions: between gender, age and work groups, between professions and between employment relations. Inequalities arose from managerial and co-worker values, and were manifested in providing training and equipment, and allocating tasks.

1 Introduction

All industries are governed by rapid and profound technological changes, which alter and disclose implicit organizational status hierarchies. 'The Lifecycle of Competencies' project examined the impacts of digitalization in radio and television production on job-descriptions, competencies, well-being and equality. This article focuses on experiences of equality among personnel. The main research question is: How are technological implementations experienced in terms of equality? Although subjective experiences of equality are essential in determining for example change resistance they have been overlooked in the extant literature. The research focus has been on the impacts of implementations on macro structures but here the aim is to describe micro level equality issues, which are more directly related to individual and organizational well-being.

Previous studies indicate, that inequalities may be experienced between gender, age, status and professional groups. Firstly, technology has created inequalities between genders. The status of women is lower than that of men both in work life generally (Korvajärvi & Lehto 2000, 207) and in technological sectors particularly. Inequality is caused by past discrimination of women in labor markets, gendering of jobs, hierarchical blockages, status-related work arrangements and women themselves (Brew & Garavan 1995). Technology discriminated against women in the 1990's: it created new jobs, but they were occupied by men, and automation further diminished employment of women (Lehto & Sutela 1999, 149). Today equality is evaluated as better in jobs using technology (Pekkola 2002), which implies that technology is no longer gender bound. Yet, there are differences in experiencing technological features (Teo & Lim 2000), computer anxiety (Qureshi & Hoppel 1995), training (Martinsons & Cheung 2001) and implementations (Venkatesh & Morris 2000), which suggest that gender affects technology adoptions and usage.

Secondly, technology creates inequalities between ages. Generally, technology has been seen to favor young employees and discriminate against the aged. This can, however, result more from differences in foreknowledge and technological expertise than age (Freudenthal 2001). There are strong myths about the effects of aging on capabilities, even though aging decreases mental competencies only slightly and actually increases social competencies (Tuomi 1999). Differences have been found in work attitudes with young employees valuing opportunities for self-development, middle-aged employees being concerned with mental stress, and the aged emphasizing social relationships, good adjustment and security of work (Lindström, Lehto & Kandolin 1989, 71), the latter even excessively as a negative age-pattern for initiative in educational activities has been found (Warr & Fay 2001). Moreover, technology adoption differs between ages: productivity-oriented factors determine technology usage for young whereas social pressure factors are more essential to the aged (Morris & Venkatesh 2000). Training needs of young employees are more related to the object of work, whereas the aged require training of the functioning and managing work tools (Tikkanen 2002).

Thirdly, technology affects equality also between professions and status roles. Technical staff can experience inequality, as they have a lower status in material capital (resources are targeted to content producers), in social capital (their freedom to move to non-technical tasks is limited), and in symbolic capital (they are undervalued) than content producers (Zachry, Cargile, Cook, Faber & Clark 2001). Inequality can also be linked to organizational hierarchies. According to Agnew, Forrester, Hassard & Procter (1997), implementations may degrade the jobs of the lower status employees simultaneously when they improve the jobs of higher status employees. These findings are connected to the finding, that the negative effects of the implementation of technology are higher in automated and routine information work than in the demanding and challenging knowledge work (Aronsson, Dallner & Åborg 1994; Korunka, Weiss, Huemer & Karetta 1995).

2 Material and Methods

'The Lifecycle of Competencies' -study was conducted in the Finnish Broadcasting Company, which is a large media enterprise with almost 4000 employees. Approximately 52,6 % men and 47,4 % women work in journalistic and editorial professions. The average age is fairly high with 37,3% of employees 50-59 years of age. Three cases participated in the study: Case1 was from radio documentary production, Case2 from television news production and Case3 from television documentary production. All units were digitalized at the time of the study. Due to digitalization, journalists begun to gather and edit material independently with digital editing software.

At first, researchers observed work processes, conducted expert interviews and studied organizational documents. The data was gathered by conducting 32 thematic interviews to mainly journalists and media journalists, but also to a few sound editors and producers, to one news editor, web editor, production planner and announcer. Interview themes dealt with changes in job-descriptions and competencies as well as well-being and equality. The research question related to equality was: How does personnel experience digitalization in equality? Moreover, two sub questions were formed: Which employees/employee groups are experienced to have an unequal status in digitalization? How does inequality between employees/employee groups manifest itself? Equality was defined inductively by the interviewees. The within-case and cross-case analyses were performed by using inductive and deductive qualitative content analysis approaches.

3 Results

Equality issues were experienced in five dimensions during digitalization: between genders, ages, employment relations, professions and work groups. Firstly, equality issues came about between genders. The personnel of the two television channels, in Case2 and Case3, saw that equal opportunities had been offered to all employees regardless of their gender. There again, some employees in Case1 experienced, that men had been offered better opportunities to succeed in digitalization, which was manifested in the attitudes of management and in providing men more versatile tasks and technology-related training. Secondly, equality issues were experienced between age groups. In all the cases some interviewees felt that employees of different ages had been supported equally and that individual traits had a stronger effect on coping with digitalization than age. There again, in all the cases also negative experiences came about, according to which the young employees had better prospects in digitalization than the aged. It was viewed that managers and young employees had doubts about the learning capabilities of the aged employees, which induced stress among the aged, who themselves felt capable to overcome digitalization. Thirdly, equality issues came about between different employment statuses. In all the cases there were views that permanent and temporary employees had been treated equally. Yet, in all the cases it was experienced that the status of temporary employees was lower than that of permanent ones. This was experienced by the freelancers, but also by some interviewees of the permanent staff. It was seen that freelancers did not get as much training as the permanent employees and did not have the same opportunities to attend training when working for example in a weekly program production. Fourthly, equality issues occurred between professions, especially between journalists and sound editors in Case1. Due to digitalization, journalists begun edit sound, which the sound editors experienced as a threat. The interviewees revealed, that cameramen and editors in television production had similar experiences. The necessity for editing professionals in digital production had been questioned in the organization. The interviewees further stated, that there were no plans of training sound editors to journalistic tasks. Finally, equality issues manifested between work groups. In Case2, some interviewees experienced that the online news unit (responsible for teletext and Internet news production) was eclipsed by the television news unit when the digital television news channel was launched. Moreover, task allocation was considered unequal as online journalists had to work also in the television unit, but for television journalists working in the online unit was not obligatory. It was also experienced that topical programs were valued more than cultural programs: employees in topical programs got more training, and the training and digital tools were offered them earlier than employees in cultural programs.

The results are summarized in figure 1. The dimensions of inequality are illustrated by rectangles, the sources of inequality by ovals and the manifestations of inequality by dash lined ovals. The arrows represent possible causal relationships with the different elements.

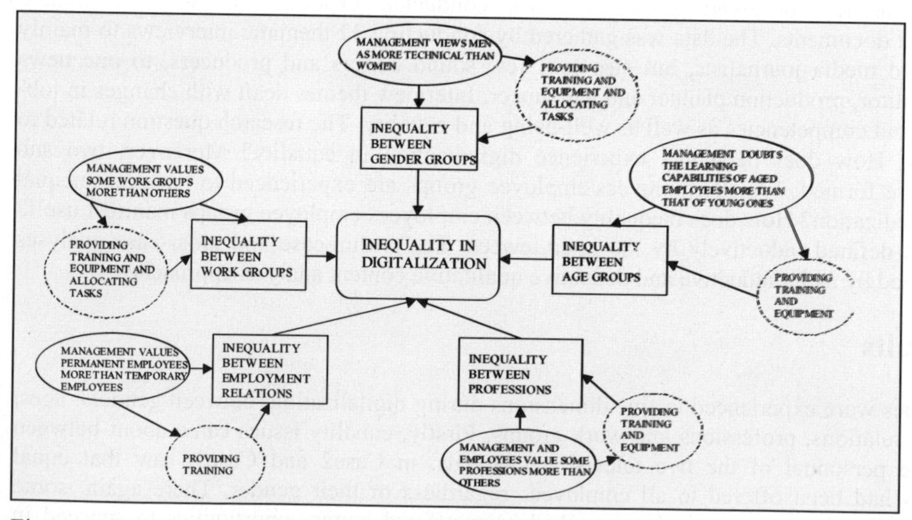

Figure 1. Equality issues, their sources and manifestations.

4 Conclusions

Equality issues were experienced between gender groups, age groups, professions, employment relations and work groups. Firstly, it was experienced that women had a lower status in digitalization than men, which was manifested in providing training and allocating tasks. This notion contradicts with findings by Pekkola (2002), that technology promotes equality between genders. Secondly, aged employees had a lower status in digitalization than the young ones, which was manifested in managerial and co-worker doubts about the learning capabilities of aged employees. These doubts were unnecessary in the light of previous research (Tuomi 1999). It was also seen that all employees had unique backgrounds regardless of age, which is confirmed by Freudenthal (2001). Overall, experienced equality might be more important to aged employees, who emphasize the security of work (Tuomi 1999). Thirdly, temporary employees had a lower status in digitalization than the permanent ones, which was manifested in providing training. Equality may be of special importance to temporary employees, as Elman and O'Rand (2002) argue that employees without employee benefits are more concerned about job insecurity. Fourthly, sound editors were experienced to have a lower status in the organization, which was manifested in digitalization in providing training and equipment, and in questioning the need for the professional group. It was experienced, that sound editors had a lower status in all the three levels Zachry et al. (2001) suggest: the resources (training and equipment) were targeted to content production (journalists), the work of sound editors was undervalued and the freedom of sound editors to move to other tasks within the organization was limited. Also, the work of sound editors can be viewed as information work, in which the negative influences of implementations are higher than in knowledge work, such as journalistic work (Aronsson et al. 1994; Korunka et al. 1995). Fifthly, equality issues were identified between work groups in allocating tasks and providing training and equipment to different work units. This inequality is connected to inequality between professions, as some work groups felt undervalued.

The contribution of the study is in clarifying ways in which equality is manifested (providing training and equipment, allocating tasks) and in identifying employees/employee groups who are at risk for inequalities (women, the aged, technical professionals, some work groups, temporary employees), according to which managerial actions to reduce inequalities can be taken. Although new information was obtained, the concept of equality is still fairly abstract and undefined. Further research is needed, particularly about experiencing equality in different change situations.

References

Agnew, A., Forrester, P., Hassard, J. & Procter, S. (1997). Deskilling and Reskilling within the Labour Process: The Case of Computer Integrated Manufacturing. International Journal of Production Economics, 52, 317-324.

Aronsson, G., Dallner, M. & Åborg, C. (1994). Winners and Losers from Computerization: A Study of the Psychosocial Work Conditions and Health of Swedish State Employees. International Journal of Human-Computer Interaction, 6, 17-35.

Brew, K. & Garavan, T.N. (1995a). Eliminating Inequality: Women-Only Training, Part 2. Journal of European Industrial Training, 19 (9), 28-35.

Elman, C. & O'Rand, A.M. (2002). Perceived Job Insecurity and Entry into Work-Related Education and Training among Adult Workers. Social Science Research, 31, 49-76

Freudenthal, D. (2001). The Role of Age, Foreknowledge and Complexity in Learning to Operate a Complex Device. Behaviour & Information Technology, 20 (1), 23-35.

Korunka, C., Weiss, A., Huemer, K.-H. & Karetta, B. (1995). The Effect of New Technologies on Job Satisfaction and Psychosomatic Complaints. Applied Psychology: An International Review, 44, 123-142.

Korvajärvi, P. & Lehto, A.-M. (2000). Knowledge Intensive Work and Gender (in Finnish). Työ ja ihminen, 2000, 14 (1), 206-215.

Lehto, A.-M. & Sutela, H. (1999). Equality in Work (in Finnish). Työmarkkinat 1999:19. Helsinki: Tilastokeskus.

Lindström, K., Lehto, A.-M. & Kandolin, I. 1989. Age and Work. The Work Conditions and Work Attitudes of Employees of Different Ages (in Finnsh). Rep. No. 154. Helsinki: Tilastokeskus.

Martinsons, M.G. & Cheung, C. (2001). The Impact of Emerging Practices on IS Specialists: Perceptions, Attitudes and Role Changes in Hong Kong. Information & Management, 38, 167-183.

Morris, M.G. & Venkatesh, V. (2000). Age differences in Technology Adoption Decisions: Implications for a Changing Work Force. Personnel Psychology, 53 (2), 375-403.

Pekkola, J. (2002). Telework, Knowledge Work and the Quality of Work Life in Finland (in Finnish). In M. Härmä & T. Nupponen (toim.) Työn muutos ja hyvinvointi tietoyhteiskunnassa. Rep. No. 22. Helsinki: Sitra, 23-34.

Qureshi, S. & Hoppel, C. (1995). Profiling Computer Dispositions. Journal of Professional Service Marketing, 12 (1), 73-79. .

Teo, T.S.H. & Lim, V.K.G. (2000). Gender Differences in Internet Usage and Task Preferences. Behaviour & Information Technology, 19 (4), 283-295.

Tikkanen, T. (2002). Learning at Work in Technology Intensive Environments. Journal of Workplace Learning, 14 (3), 89-97.

Tuomi, K. (1999). Ageing Workers and Age Mix at the Workplace. Työterveiset 3.

Warr, P. & Fay, D. (2001). Short Report: Age and Personal Initiative at Work. European Journal of Work and Organizational Psychology, 10 (3), 343-353.

Venkatesh, V. & Morris, M.G. (2000) Why Don't Men Ever Stop To Ask Directions? Gender, Social Influence, and Their Role in Technology Acceptance and Usage Behavior. MIS Quarterly, 24 (1), 115-139.

Venkatesh, V., Morris, M.G. & Ackerman, P.L. (2000). A Longitudinal Field Investigation of Gender Differences in Individual Technology Adoption Decision-Making Processes. Organizational Behavior and Human Decision Processes, 83 (1), 33-60.

Zachry, M., Cargile Cook, K., Faber, B. D. & Clark, D. (2001). The Changing Face of Technical Communication: New Directions for the Field in a New Millennium. Annual ACM Conference on Systems Documentation, 2001, Santa Fe, New Mexico, USA.

Group Knowledge Acquisition System Using Two or More Domain Knowledge

Yoshinori Hijikata   Toshihiro Takenaka   Yukitaka Kusumura   Shogo Nisida

Department of Systems and Human Science
Graduate School of Engineering Science, Osaka University
1-3 Machikaneyama, Toyonaka, Osaka 560-8531, JAPAN
hijikata@sys.es.osaka-u.ac.jp

Abstract

This paper proposes a group knowledge acquisition system, which makes consistent of the knowledge constructed from cases and every member's domain knowledge. The system compares the above sets of the knowledge, and detects the flaw between them. Finally the system makes questions to each member according to the flaw. By answering these questions, the members can get rid of the flaw, and improve the reliability of the knowledge.

1 Introduction

In an organization, it is important for its members to share their knowledge. In this research, we construct a group knowledge acquisition system, which makes consistent of the knowledge constructed from cases and every member's domain knowledge. The "case" means a description of an actual matter and a situation. The "domain knowledge" means a set of rules, which a member describes from his own experiential knowledge. KAISER (Tsujino & Nishida, 1986) is a knowledge from the cases and the user's domain knowledge. However this system does not support multi users so that it is not practically feasible in real organizations. The feature of our system is to compare the knowledge constructed from cases, one user's domain knowledge and another user's domain knowledge, and asks questions to the appropriate user for improving the knowledge. To make the knowledge consistent, it is necessary to clarify the difference between the elements of knowledge.

In this research, we arranged the following three points:

- Patterns of the difference between the elements of knowledge
- Questions according to the above patterns
- Priority level of the above patterns

Based on the above-mentioned points, we propose a method for providing questions to the users by detecting the differences between the elements of knowledge.

2 System outline

This system needs three inputs: one user's domain knowledge, another user's domain knowledge and some cases. And the knowledge based on the cases is obtained by doing inductive learning from the cases. The system compares the above three sets of knowledge, and detects the flaw between them. The "flaw" means a contradiction (difference), which happens because of the error by doing inductive learning from cases, input error of the expert, the expert's misunderstanding. Finally the system makes questions to each user from the flaw. By answering those questions, the users can get rid of the flaw, and improve the reliability of the knowledge.

3 Knowledge description

This system covers a classified problem. This problem is described by class, attribute, and attribute value. The followings are description form of the case and the domain knowledge:
"Case": all attribute values and class
[(attribute : value)???(attribute : value) (class)]
"Domain Knowledge": if then rule
[if (attribute : value)???(attribute : value) then (class)]

The system obtains the knowledge based on the cases by inductive learning. We use a decision tree for inductive learn ing from the cases. The decision tree is constructed by ID3 (Quinlan, 1986). Description form of the decision tree is as follows:
"Decision tree": if then rule
[if (attribute : value)???(attribute : value) then (class)]

4 Method for detecting flaws and questions for knowledge improvement

4.1 Definition of flaw

We define the following three types as flaws:

- Type 1 (contradiction):
  The conditions of two rules are same, but the judgments are different.
- Type 2 (lack or excess of the clause of the condition):
  The conditions of two rules are different. But the conditions become same, if we add or remove a certain clause of the condition. This flaw is classified into two kinds on whether the judgments are same or different. When the judgements are same, we call it Type 2-a. When the judgments are different, we call it Type 2-b.
- Type 3 (lack of rule):
  The same rule exists in two. But this rule does not exist in another one.

4.2 Procedure for detecting flaws

Firstly, three sets of rules (the domain knowledge of two users, and the decision tree) are compared by the round robin with each other. After comparing rules, the system checks whether or not the flaw exists according to the patterns of conditions and judgments (explained later). The detected flaw is ordered by the priority (defined later). The system outputs questions (explained later) according to the flaw.

4.3 Combination of condition and judgment

Because there are three sets of rules in comparison, there are the following four kinds of classifications in their conditions:
(1) The three conditions are corresponding.

(2) The conditions of the two user's domain knowledge are corresponding.
(3) The conditions of one user's (User A or User B) domain knowledge and the decision tree are corresponding.
(4) The three conditions are not corresponding.

We arranged all combinations of above the classifications of conditions. Judgement can also be classified into four kinds in the above-mentioned way. We arranged the flaw according to the combination. We also arranged the questions according to the flaw. For example, when three conditions are corresponding, the flaw type and the question are arranged in Table 1. When the conditions of the two user's domain knowledge are corresponding, the flaw type and the question are arranged in Table 2. In some combinations, the system has to consider whether the clauses of the condition are lacking or excessive (see Table 2). We also arranged all patterns of the lacks and excesses of the conditions (see Table 3).

Table 1: Combinations of the classifications of conditions (1) and the classifications of judgments

Pattern No.	Condition If	Judgment Then	Flaws or not. [question to users]
	A B Tree	A B Tree	
1-1	X X X	X X X	Not flaw (ideal pattern) [No question]
1-2	X X X	X X Y	Type1 [Cases may be wrong.]
1-3	X X X	X Y X	Type1 [Judgement of user A may be wrong.]
1-4	X X X	Y X X	Type1 [Judgement of user B may be wrong.]
1-5	X X X	X Y Z	Type1 [Both users should reconfirm Knowledge]

There is no meaning in these signs (X Y Z), but we use them as a symbol, which shows the combination.

4.4 Priority level of flaws

The priority level of the flaw must consider the following three points: (1) The conditions of the rules are same, but the judgments are different. (2) There are many numbers of rules with the same condition. (3) There is a possibility of Flaw Type 2 or not. The priority levels of the flaw are as follows:
(1) The conditions of three rules are same, but the judgments are different.
(2) The conditions of two rules are same, but the judgments are different. And Type 2 exists in another rule.
(3) The conditions of two rules are same, but the judgments are different. And Type 2 does not exist in another rule.
(4) Two rules are same. And Type 2 exists in another rule.
(5) Two rules are same. And Type 2 does not exist in another rule.
(6) There is no rule of same condition, but Type 2 exists.

Table 2: Combinations of the classifications of conditions (2) and the classifications of judgments

Pattern No.	Condition If	Judgment Then	Flaws or not. [question to users]
	A B Tree	A B Tree	
2-1	X X Y	X X X	Possibility of Type 2-a and 3 - If this rule of decision tree has Type2-a. [Question of Table 3.] (see Table 3) - If the rules of decision tree has Type3. [Input this pattern's case.]
2-2	X X Y	X X Y	Possibility of Type 2-b and 3 - If this rule of decision tree has Type2-b. [Question of Table 3.] (see Table 3) - If the rules of decision tree has Type3. [Input this pattern's case.]
2-3	X X Y	X Y X	Type1 [Judgement of users A and B are contradictory.] Possibility of Type2-a and Type2-b - If this rule of decision tree has Type2. [Question of Table 3.] (see Table 3)
2-4	X X Y	Y X X	Type1 [Judgement of users A and B are contradictory.] Possibility of Type2-a and Type2-b - If this rule of decision tree has Type2. [Question of Table 3.] (see Table 3)
2-5	X X Y	X Y Z	Type1 [Judgement of users A and B are contradictory.] Possibility of Type2-b - If this rule of decision tree has Type2. [Question of Table 3.] (see Table 3)

Table 3: All patterns of the lacks and excesses of the conditions

	Lack or excess of the clause of the condition	Same or different of the clause of the judgement	For example: Relation to [if (A a1)(B b1) then X]	Question for users
1	Lack	Same	if (A a1) then X	Can't the conditions of one of rules be added or deleted?
2	Lack	Different	if (A a1) then Y	Isn't the rule generalized or specialized?
3	Excess	Same	if (A a1)(B b1)(C c1) then X	Can't the conditions of one of rules be added or deleted?
4	Excess	Different	if (A a1)(B b1)(C c1) then Y	Isn't the rule generalized or specialized?
5	Lack and Excess	Same	if (B b1)(C c1) then X	No question
6	Lack and Excess	Different	if (B b1)(C c1) then Y	No question
7	No overlap	Same	if (C c1)(D d1) then X	No question
8	No overlap	Different	if (C c1)(D d1) then Y	No question

5 Implementation

We implemented our system on Windows2000 with graphical user interface by Microsoft VC++6.0. A screenshots of this system is Figure 1.

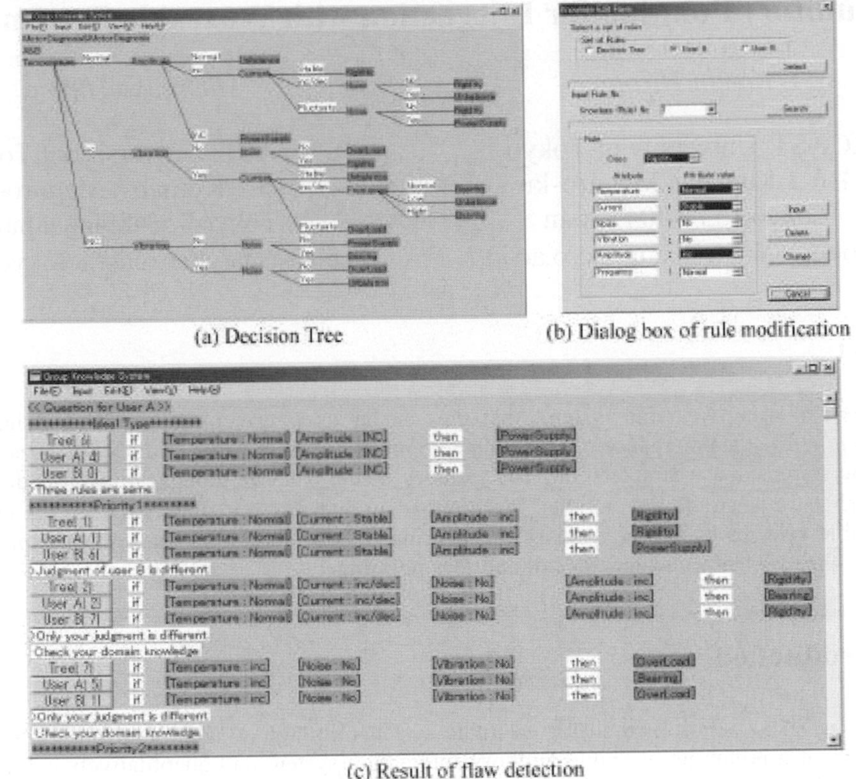

(a) Decision Tree (b) Dialog box of rule modification

(c) Result of flaw detection

Figure 1: Screenshot of system

When the user inputted a text file of the case, the system draws the decision tree. The user can input his/her domain knowledge by a text editor or the dialog box of this system. In this dialog box, the user can select classes and attribute values from a list box. After the users inputted a text file of the cases and their domain knowledge, the system detects the flaws and shows them. On the screen in which the result of the detecting flaws is displayed, the users can change their domain knowledge by pushing a button.

6 Conclusions

This paper proposes a group knowledge acquisition system, which makes consistent of the knowledge constructed from cases and every member's domain knowledge. We implemented the prototype system and shows how the system works. In the future we plan to evaluate our system by an experiment with actual users.

7 References

Tsujino, K., & Nishida, S. (1992). A Knowledge Acquisition System driven by Inductive Learning and Deductive Explanation: KAISER. *Japanese Society for Artificial Intelligence*, 7 (1), 149-159.

Quinlan, J. R. (1986). Induction of Decision Trees. *Machine Learning*, 1, 81-106.

Auditory Pointing for Interaction with Wearable Systems

Koichi Hirota

RCAST, University of Tokyo
4-6-1 Komaba, Meguro-ku
Tokyo 153-8904 Japan
hirota@cyber.rcast.u-tokyo.ac.jp

Michitaka Hirose

RCAST, University of Tokyo
4-6-1 Komaba, Meguro-ku
Tokyo 153-8904 Japan
hirose@cyber.rcast.u-tokyo.ac.jp

Abstract

It is a merit of using wearable systems that they enable us to use computers while doing other tasks. Even before the emergence of wearable computers, we were using auditory devices such as headphone stereos in a wearable style. The auditory media is hence thought to be a media that is most suitable for wearable interactions. We propose an interaction method between the user and the wearable systems based on auditory localization. In this paper, we discuss our experiments examining the efficiency and accuracy of auditory pointing.

1 Introduction

We propose an interaction method for wearable systems through an auditory pointing interface. We implement a prototype wearable auditory interaction system and quantitatively evaluate the time required for auditory pointing operations.

Many studies have investigated the use of sound in HCI. In the earliest study, Gaver submitted the concept of the Auditory Icon (Gaver, 1989). With this concept, he was attempting to introduce sound into the GUI. The Mercator Project (Mynatt & Edwards, 1992) aimed at the development of an auditory interface for the blind, by transforming the GUI into an interactive auditory interface. The technology of auditory localization using HRTF (head-related transfer function) was established by about 1990 (Wenzel et al., 1990), and has been used in virtual reality systems and applied to communication in virtual environments (Aoki et al., 1994). Recently, the use of mobile computers in the real world has been an important research topic and new concepts such as the wearable computer (Mann, 1997) have been submitted and various new applications have been investigated (Ueoka & Hirose, 2001).

We propose an approach to interaction with wearable systems through a pointing and selection mechanism using an auditory pointer and icons; the pointer and icons are represented around the user as localized sound sources. The features of such an auditory interface are that it does not disturb the use of sight of the user and can present information even for the orientation that is outside of the user's visual range.

2 Wearable Auditory Localization System

We implemented a wearable auditory interaction system by combining an auditory device, a notebook PC, and two motion sensors (Figure 1). The auditory device is capable of localizing four sound sources independently by simulating the effect of HRTF at 16bit 48kHz precision in real time. We used HRTF data that was measured at the MIT Media Lab Machine Listening Group using a dummy head (KEMAR head) and is available to the public (Gardner & Martin, 1994). The HRTF data consists of the coefficients of 711 orientations that cover an entire sphere except the area below -45deg elevation, or head tilt. Since the sampling rate of the data was 44.1kHz, we re-sampled them at 48kHz. In our module, the localizing orientation is changed discretely by switching these coefficients, and the maximum discretization error is 6deg. Also, we do not treat the localization according to the distance from the sound source.

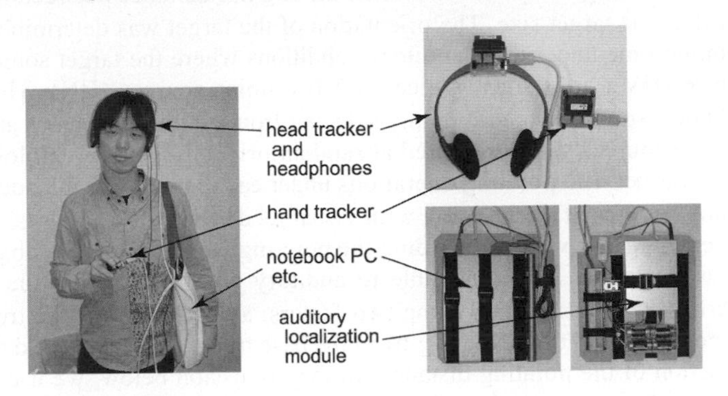

Figure 1: Wearable Auditory Localization System

The motion sensor (MDP-A3U7, Tokin) measures its own orientation in three degrees of freedom, by integrating terrestrial magnetism, gravitation, and gyro sensors; the maximum update rate is 125Hz, or 8ms in interval time, and the RMS error in the worst case is 15 deg.; the two motion sensors are used for tracking the motions of the user's head and the pointing device, respectively.
The notebook PC controls the auditory module and motion sensors and sets tasks for experiments; the PC receives the sensor data, selects the HRTF coefficients according to the motion of the user, and sends out control commands to the auditory module. The computation time for each cycle of the process is approximately 25ms, and the time required to send out a command to the module is approximately 8ms.
In the experiments, the entire device is used while being worn by the user. The time delay through the entire device is thought to be approximately 41ms, and the weight of the entire device is approximately 3.8kg, including 170g for the head tracking sensor and the headphones and 100g for the hand tracking sensor.

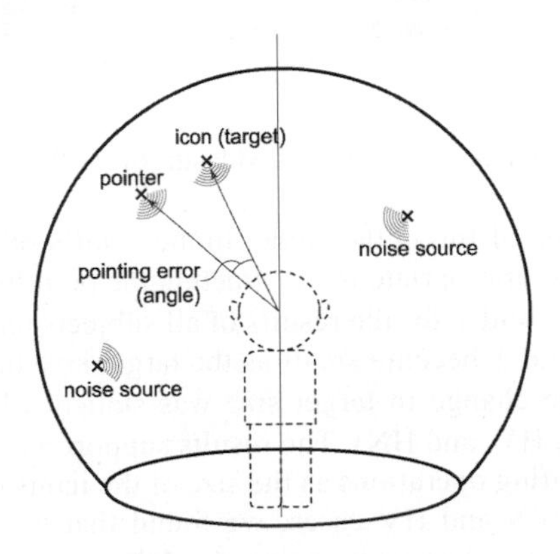

Figure 2: Auditory Pointing

3 Auditory Pointing

We attempted to evaluate the time required for pointing in relation to the size of the target and the angular distance to the target (Figure 2). The size of the target was varied to 6, 12, and 24deg, where we consider that a target is pointed at when the angular distance between the target and the pointer is smaller than the target size. The orientation of the target was determined randomly. We measured the pointing time under three auditory conditions where the target sound is white noise (WN), human voice (HV), and human voice with two noise sources (HN). The sounds of the human voice and noise sources were randomly selected from among the verbal articulations of 'a' to 'z'. Also, the noise sources were presented at random orientations. We employed 10 subjects, and each subject performed 100 pointing operations under each target size and auditory condition. Although, the characteristics of the pointing with visual feedback follow the Fits' law (MacKenzie, 1992), there is no existing knowledge regarding the pointing with auditory feedback, and we have not confirmed that the law is also applicable to auditory pointing. This is the reason why we attempted to describe the pointing time using two factors: a base trend (or the trend of minimum time required for pointing) and the scattering from the base trend. We formulated the base trend by fitting a linear function of the pointing distance. In the discussion below, we use the value of the base trend function when the pointing distance is 60deg (C_{60}) as an index. We obtained the scattering component (or scattering time) of each pointing operation by subtracting the base trend time from the pointing time. The scattering factor was modeled by an exponential distribution function where the frequency of occurrence decreases according to the increase in scattering time, and obtained the time constant of the distribution (T). If we consider the nature of the exponential distribution, the scattering time is below T in 63% of pointing operations.

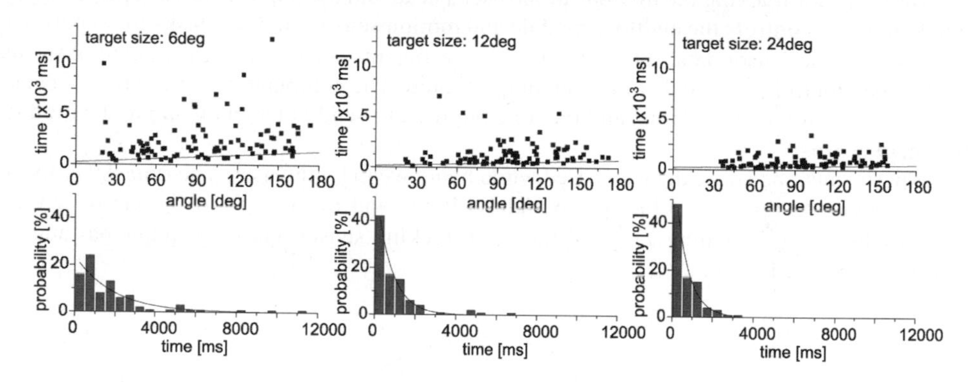

Figure 3: Examples of Measurement Results

Figure 3 shows some examples of fitting the model to the results of a subject, which suggest that the parameterizing algorithm is appropriate for abstracting the pointing characteristic.

We computed the values of C_{60} and T for the results of all subjects under each condition (Figure 4). Generally, the values of C_{60} and T become small as the target size becomes large. The difference of those values caused by the change in target size was statistically significant (α=0.02) under each auditory condition (WN, HV, and HN). The results support our experimental finding that we can perform more certain pointing operations as the size of the icons increases.

Although by comparing the WN and HV cases, we found that both C_{60} and T slightly increase when we use a human voice rather than white noise, the difference was not statistically significant (α=0.02).

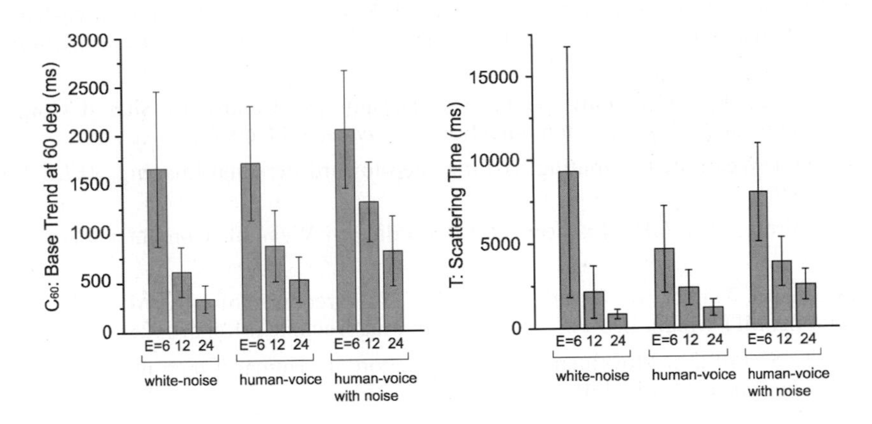

Figure 4: Effect of Conditions on C_{60} and T

Also, the value of T is increased by the presentation of noise sources. The increase in the value under the HN condition relative to the HV condition was statistically significant (α=0.02) under the condition of each target size. According to comments from the subjects, it takes more time to find the target orientation when noise exists. Also, they report that certain vocalized sounds, such as 'c', 'p', and 't', or 'f' and 'n', are confusing because they have similar pronunciations.
Among the experimental conditions, the pointing operation was most efficient when the target size was 24deg under the WN condition, where the averages of C_{60} and T were 0.34s and 0.83s, respectively; pointing operations were the least efficient when the target size was 6deg under the HN condition, where the averages of C_{60} and T were 2.05s and 7.97s, respectively.

4 Conclusion

We found a relatively large variation among individuals in terms of pointing time. In our considerations, such variations originate from the individual difference in the degree of matching to HRTF data, and we expect that the user performance would be improved if we were able to use HRTF specifically tuned for each user. Through the experiments, we clarified the usability of the auditory pointing interface under some simple auditory conditions, and consider that the auditory interface is applicable for wearable systems.
In our future work, we intend to investigate the application of wearable auditory interaction method. We expect that navigation while walking along streets will be a most promising application area; if we can operate the navigation system through auditory pointing, users will be able to keep eyes free from the operation and to continue to pay attention to the surrounding environment; also, we will be able to provide a way of indicating directions intuitively by means of the localization of sound sources.

References

Gaver W. (1989). The SonicFinder: An Interface That Uses Auditory Icons. Human Computer Interaction, 4, 67-94.

Mynatt, E. & Edwards, W. K. (1992). Mapping GUIs to Auditory Interfaces. Proc. ACM UIST'92, 61-70.

Wenzel, E. M., Stone, P. K., Fisher, S. S. & Foster, S. H. (1990). A System for Three-Dimensional Acoustic 'Visualization' in a Virtual Environment Workstation. *Proc. Visualization'90*, 329-337.

Aoki, S., Cohen, M. & Koizumi, N. (1994). Design and Control of Shared Conferencing Environments for Audio Telecommunication. *Presence*, 3(1), 60-72.

Mann, S. (1997). Wearable Computing: A First Step toward Personal Imaging. *IEEE Computer*, 30(3), 25-29.

Ueoka, R. & Hirose, M. (2001). Experimental Recording by Wearable Computer. *Proc. HCII2001*, 753-757.

Gardner, W. G. & Martin, K. D. (1994). HRTF Measurements of a KEMAR Dummy Head Microphone. MIT Media Lab Perceptual Computing Technical Report #280.

MacKenzie, I. S. (1992). Movement Time Prediction in Human-Computer Interfaces. *Proc. Graphics Interface`92*, 140-150.

Audience of Local Online Newspapers in Sweden, Slovakia and Spain - a comparative study

Carina Ihlström[1], *Jonas Lundberg*[2] *and Ferran Perdrix*[3]

[1]School of Information Science, Computer and Electrical Engineering, Halmstad University, P.O. Box 823, S-301 18 Halmstad, Sweden
Carina.Ihlstrom@ide.hh.se
[2]Department of Computer and Information Science, Linköping University
S-581 83 Linköping, Sweden
jonas.lundberg@ida.liu.se
[3]Department of Computer and Industrial Engineering. GRIHO, University of Lleida. Jaume II nº69 E-25007 Lleida, Spain
ferran@griho.net

Abstract

Since a new online audience for local newspapers has emerged during the last years, in response to the growth of the Internet, we need to know who they are, what their reading habits are, and what their view on emerging technologies are, to be able to design good online newspapers. We have conducted a study using online questionnaires at three local online newspapers in three different countries: Sweden, Slovakia and Spain. The objective of this paper is to describe the differences and similarities between the three countries regarding audience profiles, scenarios of use, opinions of current and future issues and to discuss design implications.

1 Introduction

Online newspapers can currently be characterized as a variant genre, not yet a fully developed genre of its own. Computer mediation is seen by Shepherd and Watters (1998) as a force towards change, where a genre moving from one medium to another first becomes replicated trying preserve content, structure and purpose, then changes to a variant, utilizing the medium more. Then it may change dramatically into an emergent genre, where its roots as a replicated genre are not obvious. Today, most online newspapers consist of a set of genre elements. Eriksson and Ihlström (2000) identify the news stream that presents recent stories ordered by publishing time and the headlines that are used to present stories of highest general interest together with the archives. Other elements included in a repertoire of genre elements together with the above are navigation elements, landmarks and advertisements (Ihlström and Lundberg, 2003).

As newspapers launched their online editions, a first effect was the formation of an online audience. Apparently, to get a sizeable audience is important, since the competition has grown with the expansion of the Internet (Aikat, 1998). This audience has for example been found to have different habits depending on whether or not they subscribe to the printed edition in a Swedish study of online audiences (Ihlström & Lundberg, 2002). The different habits can be described as scenarios of use (e.g. derived from the opinions of the users). The detected scenarios of use can later be used as input in a design process (Carroll, 2002). Since these scenarios differ between audience groups, these differences can't be detected by server logs alone. Therefore, in our study, which is a part of the Electronic Newspaper Initiative (ELIN-IST-2000-30188), we have used an online questionnaire approach to get data about audience groups in three different countries.

Cyber Dialogue reported that although 49% of online U.S. adults visited a local newspaper site on a monthly basis, only 7 percent did so on a daily basis (Runett, 2000). Bellman, Lohse and Johnson (1999) reported that out of their survey with 10.180 respondents, 19,1% used the Internet at home to read news. Readers of online editions of local papers tend to be readers of that paper, but online editions of national papers reach people who don't read the print edition (Chyi & Lasorsa, 1999). In this paper we present a comparative European study between Sweden, Slovakia and Spain, using online questionnaires at the Östgöta Correspondenten (www.corren.se), Korzar (www.korzar.sk), and Diari Segre (www.diarisegre.com) online newspapers, respectively. All three newspapers have a printed edition of the newspaper as well.

The objective of this paper is to describe the differences and similarities between the three countries regarding audience profiles, scenarios of use, opinions of current and future issues and to discuss design implications.

2 Method

The study was conducted in December/January 2001/2002. At three newspapers, the survey was presented for approximately one week. The questionnaire contained 19 questions, no open question included. Having corrected for item non-respondents for the demographic questions (1-5), a total of 2.864 answers were collected. Of these, 2.311 were from Östgöta Correspondenten, 159 were from Diari Segre and 358 were from Korzar. The Diari Segre and Korzar audiences are smaller than the audience of Ösgöta Correspondenten.

The survey was been presented as a pop-up questionnaire. A one-page design was used, which is faster for the respondent than a multi-page design, although the item non-response is higher. There was no use of advanced graphics, which gives less partial non-response from modem users (Manfreda, Batagelj & Vehovar, 2002). Only respondents who pushed the submit button were included in these surveys. In all surveys, there were some multiple-choice questions.

3 Results

3.1 Audience profiles

The Swedish Audience has the largest proportion of senior readers (11.4% over 55 years), and also the largest proportion of subscribers to the printed edition, almost 50%. 53.1% are in the ages between 16 and 35. They are fairly well educated with 48.4% university educated and 41.2% with comprehensive school. They are mostly full time workers (61%), students (16.4%) or part time workers (6.6%). The audience consist of approximately 2/3 male, and only 1/3 female readers, in all three countries.

The Spanish Audience has the largest proportion of young readers (29.3% less than 25 years) but the smallest amount of senior readers (2.6% more than 55 years old). They have the highest percentage of university educated readers with 62.6% and 23% have comprehensive school. Only

1/5 of the readers are subscribers to the printed edition. The also have the highest amount of students (18.5%) but the lowest amount of full time workers (57.4%).

The Slovak Audience does not subscribe to the printed edition (only 3.6%). Almost 57% of the readers are between 26 and 45 years old. They are the most educated readers with either university educated (57.8%) or comprehensive school (40.2%). The have the highest amount of full time workers (74.4%) but the lowest amount of students (10.6%).

3.2 Scenarios of use

Main scenario: More than a third of the respondents of the three countries read the online editions once daily or several times a week at a minimum. The information in the online newspaper that the audience read is similar, primarily the most recent news (average 67.1%) and the local news (average 52.8%) of the three papers. In all three countries, men also read sports news to a much greater extent than the women.

The Swedish Scenario: The audience reads the online newspapers mostly on a daily basis (27% once a day and 14.4% several times a day), 29.1% reads sometimes each week while 18.5% reads several times a week. They read it at work (45%) and at home (38.7%). More than 40% reads the online edition before lunch and 23.3% reads it during the evening. They primarily read the most recent news (70.8%) and local news (56.2%). The Swedish audience is the most interested in sports (35.2%) but they do not read much culture news (11%).

The Spanish Scenario: A large proportion (33.3%) of the Spanish audience reads the online edition several times each day (7.7% once a day). 10.3% reads several times each week while 15.4% reads some times each week. They read it from their home (33.8%) or from work (29.7%). Mostly, they read the online edition in the evening which differs from the other countries. They read the most recent news (64.6%) and the local news (44.1%), while not being very interested in economy (12.3%). The women are more interested in culture news (45.5%) than the men are (19.8%).

The Slovak Scenario: In Slovakia the audience mostly reads the online edition once a day (36%) or several times a week (33.8%). Only 8.1% reads several times a day. A clear majority reads from work (62.9%), which is the highest amount of the three countries. Only 22.3% reads from home. More than 37% reads during the early morning and only 16.8% in the evening. They read the most recent news (65.9%) and the local news (58.1%). The Slovak audience reads the economy news (21.8%) more than the audience from the other countries as well as the national news (36.3%). The women are more interested in culture news (20.8%) than the men are (8.8%).

3.3 Current and future issues

In Slovakia, 37.7% of the audience were positive to micro payment for more and faster local news in the future, compared to 8.3% in Spain and 10.5% in Sweden. Regarding attitudes towards advertisements, some differences between the three countries were found. In Slovakia 34.1% were positive to advertisements compared to 21.1% in Sweden and 11.8% in Spain. Most people were neutral (average 42%) in all three countries. Concerning alternative media for receiving news, 37.1% (average) preferred to use their TV, whereas few people wanted news in their handheld computer (average 13.8%) or mobile phone (average 17.1%). No differences between the countries regarding alternative media were found. The preferred alternative forms for news was radio news in all countries (Spain 37.4%, Sweden 22.9% and Slovakia 31.5%), SMS news was mostly preferred by the Slovak audience (24.6%) and the Swedish audience (21%) preferred video news more than the other audiences. Regarding moving images and sound over broadband connections, as much as 55.8% of the Swedish audience thought it would be more interesting (Slovakia 45.3%, Spain 39.5%). As regards personalisation, a majority in Sweden (56.6%) would not like a fully personalized paper compared to 22.9% in Slovakia and 33.3% in Spain. The most positive were the Slovak audience with 40%.

4 Discussion

In our study, we found results which differ from those of Runett (2000), namely that our respondents read the paper more often, with an average of 42.2% in the three countries reading once, or more than once daily, compared to the 7% reported by Runett. Furthermore, in our study, 36.1% read from their home, which is almost twice as many compared to the results of Bellman et al (1999). Compared to the results of Chy and Larosa (1999) this study did show national differences regarding subscribing and non-subscribing readers. Only Sweden was close to their results, with around 50% subscribing readers.

The *main differences* between countries found were; ***a)*** The high amount of subscribing readers in Sweden (48.1%) compared to Slovakia (3.6%). ***b)*** More senior readers in Sweden (11.4%) compared to Spain (2.6%). ***c)*** In Slovakia the audience mostly read the online newspaper from work (62.9%) and only 22.3% reads from home whereas in Sweden 38.7% reads from home and 44.9% from work. ***d)*** The Slovak audience was almost four times more interested in micropayment for more local updated news than the other countries. ***e)*** The Swedish audience was the most negative to full personalization (56.6%) compared to the Slovak audience with a positive attitude of 40%. ***f)*** The attitudes towards advertisements were most positive in Slovakia (34.%) compared to the Spanish audience with only 11.8%. These differences could be due to cultural differences and/or differences in infrastructure. Further research is needed to explain these differences.

The main similarities found are used to propose *design implications* that are more general and could be used at most online newspapers. The motivation to read online news is principally to read the most recent news, almost 70% of the audience use the online newspaper for that reason. The news stream element with its timestamps is a good form for presenting a selection of the most recent news items (Ihlström & Lundberg, 2003). In Sweden the high interest in sports news, in Spain the culture news and in Slovakia the national news could be a reason to give that a prominent position on the first page. However, there is a difference between male and female readers regarding sport and culture news, which could for example be resolved through personalization, or by presenting the additional sports/cultural news items in a specialized news stream. Interestingly, the Slovak audience reads the most recent news to about the same degree as the other countries in spite of not having a news stream on the front page of the online newspaper, and they are four times as positive to micropayment for more and faster local news in the future. We believe that micropayment could be a good means for paying to have such a news stream added to a personalized edition, in particular since the Slovak audience was the one most positive towards personalization. The headlines section should primarily present local or domestic daily news, since this is an important use of the online edition. However, since more than half of the audience members visit the newspaper less than once daily, we suggest that the headline section is extended to include a clearly marked part where the most important news from the past days is included. This could be clearly labeled and/or the date could be used as a mark on all articles in the headlines section. Taking together the preference for video and audio, and the desire to use the television set as a media terminal, future online news sites will have to present content in new ways. This is an issue for future research, such as is being conducted in the ELIN project, where future media production and content forms are explored.

As an instrument for design, these implications are not complete, but could be interpreted as specific for the current audience. A genre analysis together with analyzing server logs, would show the impact on overall audience size and reading habits, of design alterations. Focus groups could be used to gain insights in what design, news and new services that could interest the potential audience and to develop new forms for news presentation and interaction.

5 Conclusion

We have found both similarities and differences between the three countries leading to some design implications. The results suggest that it is hard to create one design for all countries, since the preferred contents, alternate terminals, willingness to pay etc differ. The online newspaper genre is currently evolving into a genre of its own. With the use of future media terminals, personalization and more audio and video over broadband connections the genre might evolve to an emergent genre (Shepherd & Watters, 1998). This indicates that the media companies have to rethink the way to present the news in the future.

References

Aikat, D. (1998). News on the Web - Usage Trends of an On-line newspaper. *Convergence* 4 (4): 94-110.

Bellman, S., Lohse, G. L. & Johnson, E. J. (1999). Predictors of Online Buying Behavior. *Communications of the ACM*, 42 (12), 32-38.

Carroll, J. M. (2002). Making use: scenario-based design of human-computer interactions. Cambridge, Mass., MIT Press.

Chyi, H. I. & Lasorsa, D. (1999). Access, use and preferences for online newspapers. *Newspaper Research Journal*, 20 (4), 2-13.

Eriksen, L. B. & Ihlström, C. (2000). Evolution of the Web News Genre - The Slow Move Beyond the Print Metaphor. In Proceedings of 33' Hawaii International Conference on Systems Science. IEEE Press. CD-ROM.

Ihlström, C. & Lundberg, J. (2002). The Audience of Swedish Online Local Newspapers – a Longitudinal Study. In Proceedings of ELPUB2002 in Chemnitz, 92-102.

Ihlström, C. & Lundberg, J. (2003). The Online News Genre Through the User Perspective. In Proceedings of 36' Hawaii International Conference on Systems Science. IEEE Press. CD-ROM.

Manfreda, K. L, Batagelj, Z. & Vehovar, V. (2002). Design of Web Survey Questionnaires: Three basic Experiments. *Journal of Computer-Mediated Communication* 7(3). Available at: http://www.ascusc.org/jcmc/vol7/issue3/vehovar.html

Runett, R. (2000). Ride the Momentum. Presstime Magazine. Available at http://www.naa.org/presstime/0011/nmc.html

Shepherd, M. & Watters, C. (1998). The Evolution of Cybergenres. In Proceedings of the Thirty-First Annual Hawaii International Conference on Systems Science, 2, 97-109.

*i*FlashBack: A Wearable System for Reinforcing Memorization Using Interaction Records

Yasushi Ikei, Yoji Hirose

Tokyo Metropolitan Institute of Technology
6-6 Asahigaoka, Hino, Tokyo 1910065
ikei@tmit.ac.jp,
yoji@krmgiks5.tmit.ac.jp

Koichi Hirota, Michitaka Hirose

Research Center for Advanced Science and Technology,
The University of Tokyo
4-6-1 Komaba, Meguro, Tokyo 1538904
{hirota, hirose}@cyber.rcast.u-tokyo.ac.jp

Abstract

This paper describes a basic development of a wearable system, *i*FlashBack, that reinforces the user's memorization regarding interaction to real-world objects. The goal of the system is to exploit infinite capacity of human brain by assisting memorization using recorded real-worold information. The enhancement of possibility of successful rccall about the user's action and the object helps the user access vast materials distributed in the real-world for intellectual production works. Memorization reinforcement is achieved by drawing attention, promoting rehearsal by replaying a video of interaction, and organization of memory by providing relevant information. The action of the user is captured by a hat-mounted miniature camera, an RFID reader, and arm posture sensors attached to the user. Video clips are played back for rehearsal (video-aided rehearsal) with relevant information immediately after the interaction and also at appropriate times for the user's context. We conducted a preliminary experiment to confirm our idea. A model task to move cups between divided areas (slots) was performed by three subjects, which suggested the *i*FlashBack system could reinforce memorization of interaction in line with our expectations.

1 Introduction

Despite remarkable advancement of computer and software technologies, it is not yet easy for a computer to manage enormous versatile information which comes out from the objects in the real environment the user deals with in the workplace. Even though the specifically tailored sensor and computing agent might handle the individual aspect of real-world processing, indeed collecting and transferring it to the user in a comprehensive way involve human interface challenges. Human performs intellectual tasks using various kinds of information materials distributed over the world as the occasion demands. If that is the case, to maintain the meta-information (such as the category of the topic) and the location of an individual information source within the user's memory (brain) as much as possible is the most efficient measure since the information is eventually used by the user him/herself. Although the research of context-aware computing is gradually incorporating the user's status to establish intelligent agents to manage distributed information, to exploit further the large human capacity in memorizing information is considered as a right complement of the alternative approach that effectively works in conjunction with the computer agent-based approach.

Machine memory aid has been discussed in various ways among which the episodic memory aid is most relevant to this study. In the Forget-me-not system (Lamming & Flynn, 1994) the user's context of action was automatically recorded to provide the user who forgot past actions with easy retrieval interface that worked on fragments of related items in his/her episodic memory. The location, people around, events, and time stamps constitute the database for the later retrieval. The proactive presentation of related information based on the context was discussed in the wearable remembrance agent (Rhodes, 1997). These systems are designed to provide a substitution of human memory or a recall aid by presenting episodic information in the database reflecting the user's context. Their goals are not necessarily in fortifying the memory trace itself, but in helping the retrieval of relevant item by the cues easily memorable for human.

In this paper we introduce the *i*FlashBack, a wearable system that aims to reinforce human memorization of the action to interact with objects that we use in everyday work environment. The objects include books, magazines, bulletin reports, paper materials, stationeries, information devices, appliances, etc. that hold information related to our intellectual activities. We refer and arrange them in the 3D real environment so that we can access them physically and cognitively effortless under sometimes strict spatial constraint. If the total amount of objects we own grows enormous, to memorize them turns into a difficult task limiting our activity of efficient intellectual production. The *i*FlashBack provides memorization aid by drawing attention and promoting rehearsal with a replay of recorded events. The interaction of the user is captured by using a camera, RFID readers and arm posture sensors attached to the user and the environment. The video clips are played back with relevant information at appropriate times to the user for effective rehearsal (video-aided rehearsal) to reinforce memorization of the action. We built a prototype system using a miniature camera, RFID systems and an arm posture sensor for detecting, recording and replying interaction events.

2 Memorization Aid Based on Interaction Records

We focus on the aspects of memorization enhancement of drawing attention, rehearsal, and organization of human memory. When the user manipulates the object, the *i*FlashBack presents messages to draw attention in an appropriate manner. The messages are built primarily based on the captured interaction records such as video images and sounds the user saw and heard with the objects. The property data and related links of objects stored in the database are also involved in the message. The object is labeled and identified by an RFID tag that is read by a tag reader the user wears. For promoting rehearsal of the interaction event, the messages are presented to the user immediately after the user's interaction ended or at the adequate time when the user is in the resting state to prefer reviewing the action. The *i*FlashBack system has three core functions—detection of interaction, recording of interaction, and presentation of recoded cues.

2.1 Detection

State changes of the user's interaction are detected by wearable and environmental sensors. The state changes include the start of contact to an object (picking up a real-world object), the end of contact (releasing the object), and the user's resting state. The detection is performed by the RFID system—small RFID tags are affixed to objects, and RFID readers are put on the user's hands. Arm posture is measured by acceleration sensors that indicate the progress of interaction.

2.2 Recording

During the user's interaction, the scene and sound are captured by a camera attached to the brim of a hat that covers the work space of the user's hand. The camera view involves a part of the user's body not only of the hands but the front side of torso in order to increase the sense of engagement of the user in the video. In addition, the state-change time stamp, the object's ID data, and the location are recorded.

2.3 Presentation for Reinforcing Memorization

The presentation is performed in three phases of those

a) at state changes, for drawing the user's attention to the change occurred in the interaction,
b) during the interaction, for promotion of memorization by showing relevant information, and
c) after the interaction, for promoting rehearsal at appropriate timing by adequate quantity.

The presentation procedures are as follows:

a) State changes are informed by a short sound, a screen blink, a color change, or tactile cues.
b) Relevant information such as the title of related objects and events as well as the target (held) object is presented on the wearable display while the user holds the object.
c) A video clip that recorded the process of interaction is shortened and retouched, then played back at the wearable display for appropriate times immediately after the interaction is terminated.

The merits of using the memorization aid are:

- To help build a correct index map of information media in the real environment,
- To reconfirm the interaction, or to modify and/or reestablish memory trace in case attention was insufficiently paid to that interaction,
- To support a later search in the captured interaction data, plus to provide more cues for the search, which reduces unwanted extended conversation with the database,
- To intensify consciousness of engagement by projecting a part of the user's figure in the captured video, and
- To organize the memory of interaction fragments with existing consistent schema by providing relevant data even if the fragments were not properly related.

The problems to be addressed are as follows:

- Presentation may cause a nuisance when voluntary attention is directed to a different task.
- Excessive repetition may increase a cognitive load of the user.
- Too many similar images may decrease the uniqueness of each cue for recall.
- Context behind the user's action needs to be estimated accurately to optimize timing for appropriate presentation.

3 Model Experiment of Video-Aided Memorization

We conducted a preliminary model experiment to investigate the effectiveness of using a video clip that recorded the user's interaction to objects for enhancing memorization. The model task performed by the subjects was to move an object, a paper cup, to a different place within an area provided for the experiment. The *i*FlashBack system presented the video clip of the user's motion immediately after each placement. A free recall of five consecutive trials (cup shiftings) was called for after the five trials, and the performance was compared with that without flash-backing the action images.

3.1 Experimental Settings

3.1.1 Target objects

The objects were twenty paper cups in four colors: red, blue, green, yellow, and white; five for each color. An RFID tag was attached at the bottom. The cups were placed obverted for stable reading by an RFID reader.

3.1.2 Target location

Divided areas for placing a cup were set up on a table and the opposite bookshelf. Each division was a square 12 cm on a side. A 6×6-divided area was drawn on a table. Thirty-three divisions in three racks were set up at the bookshelf. Each division was indicated by small boundary marks.

3.1.3 Apparatus worn and placed

The subject wore a cap that implemented a small CMOS image sensor camera (Kanebo Ltd.) with a wide-angle lens (120-degree) for capturing a video of the user's movement transferring a cup by the hand. An RFID tag reader (V700 Series, Omron Corp.) was attached to the user's palm to detect a cup when it was held and also to catch when it was released on the table or the rack. A small VGA display (CO-3, MicroOptical Corp.) was attached to the bow of glasses for providing the user with a video clip. A computer was placed nearby on the table to relieve the load to the subject.

3.1.4 Procedure

Twenty cups in four colors were randomly placed in 69 divisions set up on both the table and the bookshelf. The subject was asked to take up any cup and place it in one of the vacant divisions. When the subject held a cup, a cue sound (Windows' ding.wav sound) was presented on a speaker for attracting user's attention. Simultaneously, capturing the video was started to record the scene in front of the subject involving the target cup, other cups placed around, the movement of the user's hand, and the location the cup was released. Releasing the cup produced a cue sound (Windows' notify.wav sound) again for confirmation of the end of a single pick-and-place task. Video recording lasted 0.2 seconds after the release. Then within about one second, the recorded video was replayed with the same cue sounds. The playback was repeated three times during that time the subject was asked to watch them carefully to memorize the scene—the color of the cup moved, the divisions from and to. During the playbacks, the cup moved was returned to the original division by the experimenter, which concluded one trial. The subject performed additional four trials immediately after the playbacks—establishing one session by five trials. At the end of the session, the subject was asked to recall freely the five trials and repeat them by moving the cups. The response was recorded.

Then the subject was asked to perform, for five minutes, two-digit arithmetic operations of adding and subtracting presented on a CRT screen of a desktop PC. This procedure was inserted to prevent the mental preparation (such as building memorization strategy) for the next session. Then the subject performed another session in which the video playback was not provided. Instead, the subject was supposed to do rehearsal of the action (taking up and placing a cup) in his/her own way. In the both sessions, the interval of trials was 30 seconds unless the video playbacks exceeded the interval due to the slow action. In that case, the next trial was started immediately after the end of the playbacks. The subject performed this pair-session five times in straight five days.

3.1.5 Subjects

Three volunteer subjects from the institute, the graduate students 24.3-year-old in average age, participated in the experiment.

3.2 Results

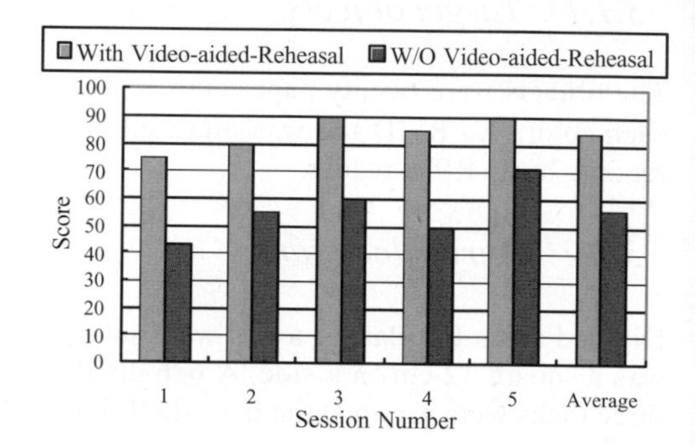

Figure 1: Recall accuracy of object transfer task

Figure 1 shows the mean score of recall accuracy over three subjects. One hundred marks mean that the subject could precisely recall and repeat 25-time transfers without fault. A near positioning (placed an object on the neighboring division to the right one) scored a half, and a one-end right answer (correct start/end division) did a quarter. The result indicated definite improvement of free recall ratio primarily related to the short term memory. A typical serial-position curve showing recall performance dependency on the position in a repetition was observed within a session——primacy and recency effects. Subjective evaluations reported in the interview claimed that the video was quite effective, and both the number of repetition and the timing to start a clip were appropriate. However, sound cueing might not be important in this experimental situation.

4 Conclusion and Future Work

The *i*FlashBack system detects the user's interaction to physical objects in the real workspace to present the digest video of the interaction from the user's viewpoint as if he/she reflects based on his/her own visual memory, but more vividly and accurately to aid efficient rehearsal. The video itself was quite impressive since the action unconsciously performed was recaptured even with a slightly distorted first-person viewpoint. From our preliminary observation, it is expected that reinforced memorization will be possible by flash-backing to an eye of a scene and the apparent user's engagement to the objects, and it would be more effective with relevant information involved in the presentation. To investigate availability of this method to the long-term memory retention is one of future works. In addition, presentation timing control——to flash or not, and when——based on the importance of objects assessed at the specific user's context might be a challenging aspect of this research to make the *i*FlashBack system of practical utility.

5 References

Lamming, M., Flynn, M. (1994). Forget-me-not: Intimate Computing in Support of Human Memory. *Proc. FRIEND21: Int. Sympo. on Next Generation Human Interface*, 125–128.

Rhodes, R. J. (1997). The Wearable Remembrance Agent: a System for Augmented Memory. *Proc. ISWC'97*, 123–128.

Retrieval System for CAD Data on the Internet

Hyun-Seok Jung, Byung-Gun Lee, Cheol-Min Joo

Department of IE, Dongseo University,
San 69-1, Jurye-dong, Sasang-gu, Busan, 617-716, KOREA
e-mail : hsjung@dongseo.ac.kr, bglee@dongseo.ac.kr, cmjoo@dongseo.ac.kr

Abstract

Nowadays, various computerized management technologies are developed and introduced for effective enterprise management. Especially, design and drawing automation technology is essential to the design stage in the aspect of drawing productivity and reducing the product developing time. Accordingly, makers have to purchase many copies of expensive CAD software. In some aspect, this flow makes the management costs of a maker higher.

In this research, we try to construct a software which can etrieve and manage drawings information without commercial CAD software, for example, AutoCAD. As a result of this research, In case of a department that is not concerned with design and drawing task(for example, as purchase department, QC department, inspection department, especially, many subcontract companies that work based on received drawings), there is no need to purchase so many CAD software.

1 Introduction

In the field of design and drawing, various kinds of CAD software are introduced for the purpose of ensuring the competitiveness, productivity and quality. As a result, the maker can keep the design and drawing task in high quality. Especially, the design task of a maker in Machinery and Construction field use so many computerized tools, and can make a good design process(Lee, 1999, Lee, 1997, Yoon, 1996).

But drawing document management system, such as drawing file management, distribution system, confirm system etc, is based on manual work. Actually, drawing information is important so many parts of a company, such as purchasing department, quality control department, inspection department and so on. Many subcontract companies, which have not their own design department, may use the given completed drawings. In this case, they must buy the CAD software to utilize distributed CAD files. Because of the high buying cost of CAD software, almost all of these concerned departments control the CAD files as printed drawings with drawing register.

In both case of CAD file and printed drawing management, the problems are (1) With only the drawing file name, one cannot understand the contents of a drawing file, (2) It is difficult to classify the drawings, (3) It is difficult to find out a drawing to be revised, (4) A batch process for drawings in the same category is difficult, (5) Most important problem is if one wants to use drawing retrieval system, he/she must buy one of commercial systems. But because these retrieval systems are mainly based on CAD system, they must also buy one of these CAD systems. It may be the economic pressure to the company.

To overcome these problems, there is a need to develop a retrieval system that is independent to CAD software and such a system must provide drawing registration, drawing classification and key word retrieval functions. In this research, we develop this kind of retrieval system that can be used on the internet.

2 Methodology

The basic approaching method of this research is (1) to survey the reality of drawing management of several makers, (2) to numerate the results of this survey, (3) to set up the relation between these gathered information, (4) to classify these information into drawing input information and retrieval information and (5) to develop the retrieval software. In this process, ISO9000 Quality Management System offered a good reference. The requirements of this system concerned with design and drawing file management is very clear, easy to understand and easy to represent as a computer program. Through this process, we can make a kind of standard for managing drawings.

3 The Structure and Functions of Developed Software

3.1 Software and Database Structure

The development tools are MiscroSoft Visual BASIC 6.0, ASP Programming Language, NAMO Web Editor and MicroSoft Access as a Data Base(Joo, 1998, Norton, 1998, Kim, 1999, Yoo, 2000, Kim, 2000, Daerim, 2001, Cho, 2002).

Developed software is based on Windows system(98/NT), and the entire structure of this software is shown in figure 1. This system needs a database control server and many client computers. These computers communicate through developed application software.

The database consists of (1) User Management DB with 5 fields, which controls user information and (2) Drawings Information DB with 15 fields, which controls basic drawing information and detail information.

Figure 2 shows the relation between these two DBs and three modules, such as, User Management Module, Drawing Management Module and Drawing Retrieval Module.

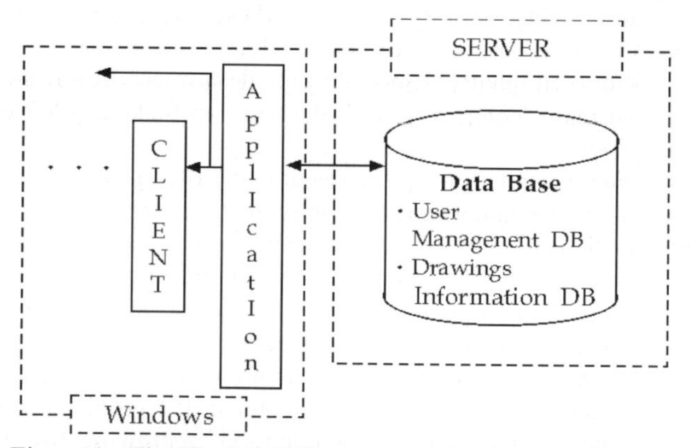

Figure 1 : The Hardware Structure of the Retrieval System

3.2 Modules and Functions

Main Page are shown in figure 3. There are 4 menus, such as User Management, Drawing Registration, Drawing Retrieval and Exit.

The User Management Module controls the registration, update and deletion of user authority, user ID and password.

The Drawings Information Management Module works for managing the basic information about drawings. In case of drawing file registration, 2 kinds of registration method are prepared. One is file registration and the other is file group registration. Using file group registration function, registration time can be effectively shortened. Figure 4 shows the registration page and figure 5 shows the concept of file group registration.

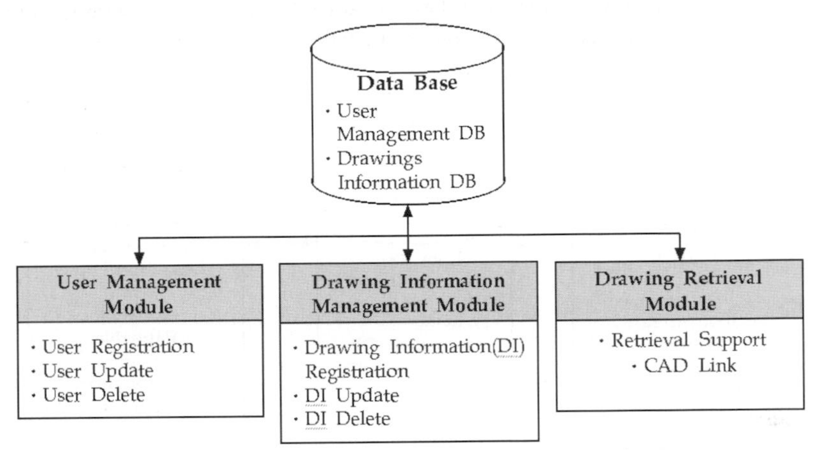

Figure 2 : The Relation Between Modules and DB

At the time of drawing retrieval, one can use Drawing Retrieval Module(Figure 6). In this module, one can use 4 steps retrieval priority. If this system can find out any matching drawing, then empty grids are filled with the information about that drawing. One can update this information easily. If one double click this selected file, then system shows the preview page shown in figure 7.

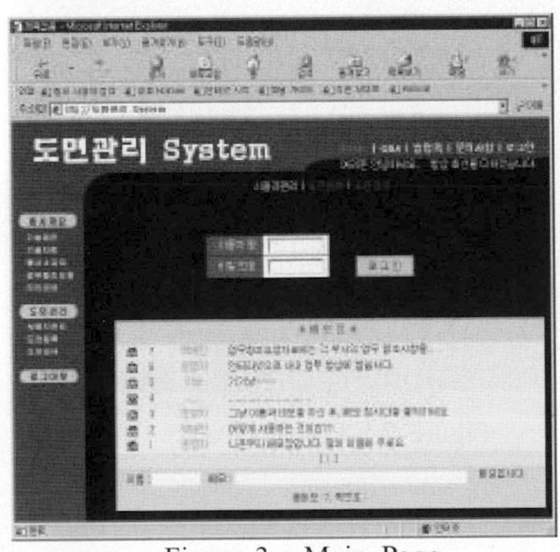

Figure 3 : Main Page

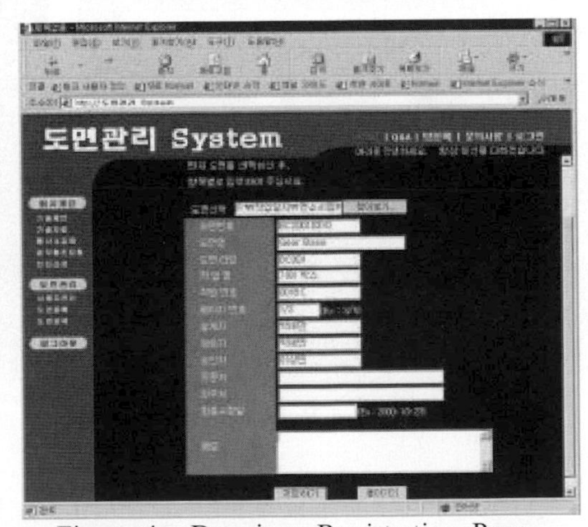

Figure 4 : Drawings Registration Page

Preview page can show the drawing without expensive CAD software. At the time of updating the drawing, one click the "Open" button, then AutoCAD program run and one can update the drawing(Figure 8).

4 Results

One of the problems of a maker that controls many drawings is how to manage these drawings efficiently. Through this research, (1) drawing management and drawing retrieval functions are reinforced, (2) one can preview drawing files without CAD software, (3) through the result (2), many department can use drawing files and they can save the space for drawings, and (4) through the result (2), inter-department communication can be taken easily.

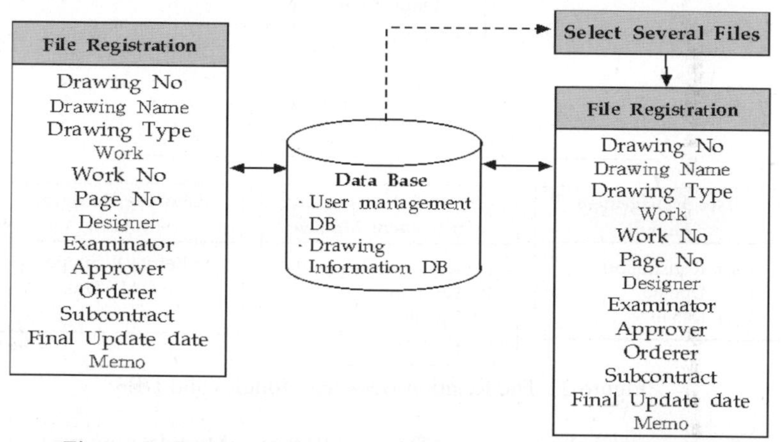

Figure 5 : Drawing Information Management Module Concept

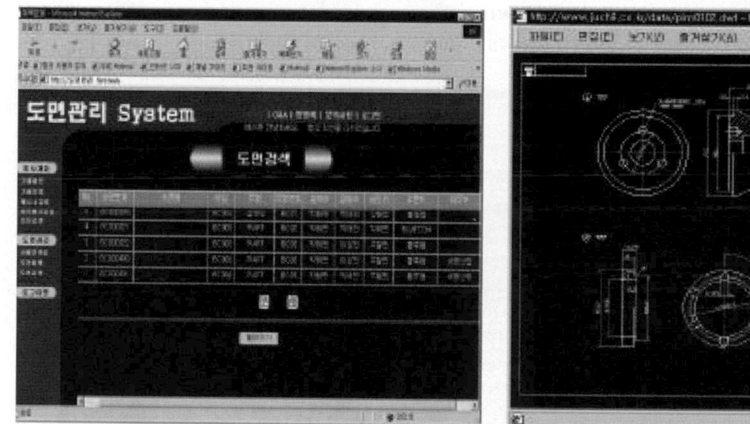

Figure 6 : Drawings Retrieval Pages

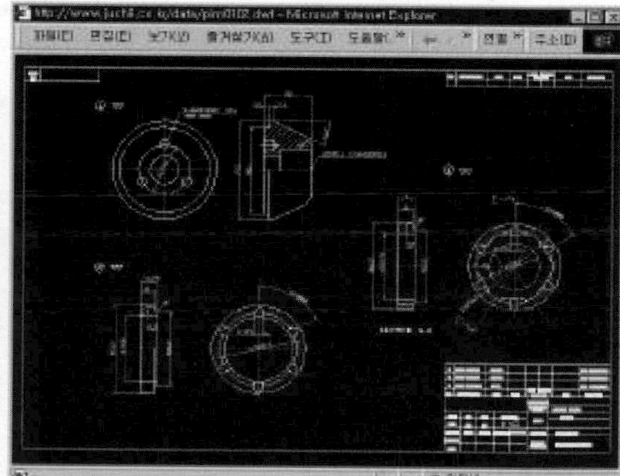

Figure 7 : Preview Page

5 References

Lee, K. W.(1999), A Study on the Process of Computer-Aided Extraction of Morphological Information from Architectural Drawings, Engineering Research, Ulsan Univ., Vol 30, No 1, pp 539-549

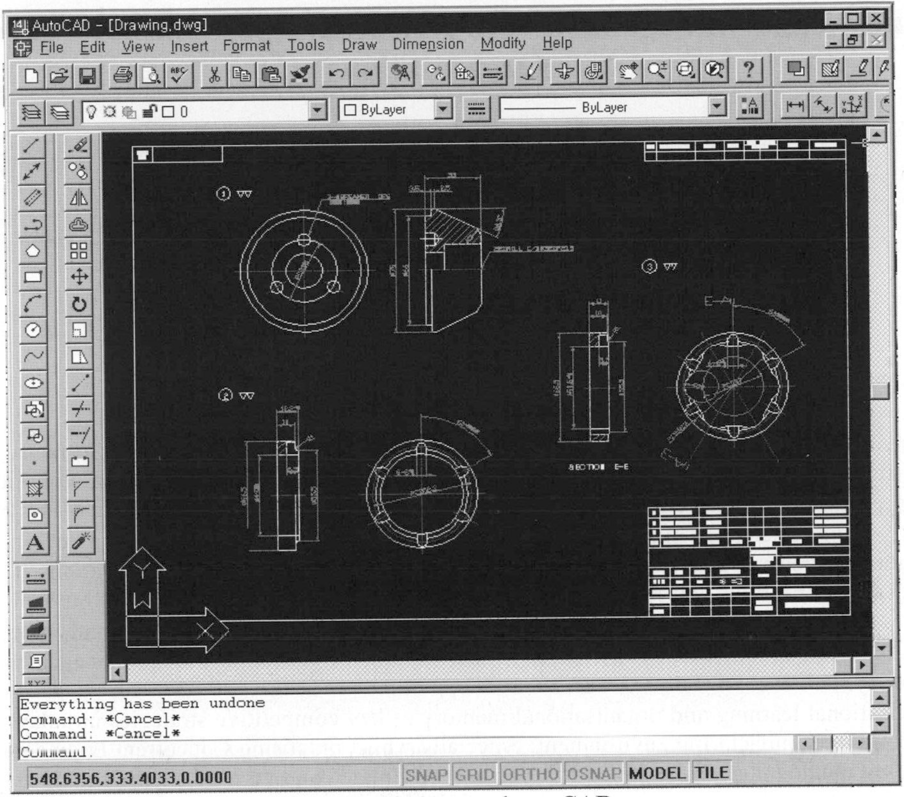

Figure 8 : Update mode on CAD

Lee, S. S.(1997), Computer Aided Design Verfication Systems of Drawings by CAD Systems, Kunkook Engineering Research, Kunkook Univ., Vol 21, No 1, pp 89-106

Yoon, K. B.(1996), A Research for Managing Archiectural Drawings Using Information Technology, Wonkwang Univ. Research Journal, Vol 16, No 1, pp 627-638

Joo, K. M.(1998), Visual Basic Programming Bible, Youngjin Press

Daerim(2001), Microsoft SQL Server 7.0, Daerim Press

Norton, P(1998), Korean Window NT Server 4, Information & Culture Press

Yoo, H. S.(2000), IIS4 & Proxy Server 2, Hyejiwon

Kim, H. M.(2000), Application of ASP on Window NT, Cyber Press

Cho, H. M.(2002), ASP Web Programming, Kame Press

Kim, K. S.(1999), Microsoft Visual Basic Database Programming, Samyang Press

Knowledge Management: An Essential Ingredient for Learning Organisations

Kayis, B., Ahmed, A., Reidsema, C., Webster, O.

The University of New South Wales
Sydney, 2052 NSW Australia
b.kayis@unsw.edu.au

Abstract

In the late 1990's Knowledge Management (KM) emerged as a popular topic in management, engineering and information systems literature. While information is often considered as interpreted data, knowledge is considered as action-oriented information which is available, usually subjectively, to improve the quality and success of decisions made in organisations. Knowledge creation and management are essential inputs for learning organisations. Learning organisations need an environment in which alternatives and opportunities can be explored by continuously increasing the pool of knowledge that is structurally gathered and documented through previous individual experiences. Thus, experience-based knowledge management is the art of capitalising on failures and missed opportunities (Jarke, 2002). This paper proposes organisational learning and organisational memory as key competitive strategies in today's highly competitive manufacturing environment, especially whilst practising Concurrent Engineering. The proposed methodology is explained by referring to validated and verified results in industry.

1 Introduction

It follows then, that manufacturing companies practising Concurrent Engineering need to be "learning organisations". Learning organisations occur if decentralised experiences; positive or negative are *shared* and *documented* across the boundaries of not only organisational departments but also all the way along the customer-supplier chain in manufacturing. Sharing increases the communication overhead and has thus traditionally limited the degree of virtualisation, and helped form unity, minimise risks and shouldering responsibility between different partners in the organisation.

The proposed research addresses organisational learning and organisational memory as key competitive strategies in today's highly competitive manufacturing environment, especially whilst practising Concurrent Engineering. This paper first presents a detailed overview of the topic which is followed by a description of the implementation methodology for a knowledge management solution approach to an existing problem within the Aerospace manufacturing industry. The conceptual approach, data and results acquired during a collaborative research project with an industrial partner is validated and verified.

Concurrent Engineering (CE) is a product development approach aimed at improving product quality and decreasing cost and time-to-market through participation of team members from various functional areas simultaneously. Tasks are carried out together by various team members from different departments at the same time. One of the more critical aspects of CE from a new product design and manufacturing perspective is that of knowledge creation and the subsequent management of that knowledge.

In the late 1990's Knowledge Management (KM) emerged as a popular topic in management, engineering and information systems literature. While information is often considered as interpreted data, knowledge is considered as action-oriented information which is available, usually subjectively, to improve the quality and success of decisions made in organisations. Knowledge creation and management are essential inputs for learning organisations. Learning organisations need an environment in which alternatives and opportunities can be explored by continuously increasing the pool of knowledge that is structurally gathered and documented through previous individual experiences. Thus, experience-based knowledge management is the art of capitalising on failures and missed opportunities (Jarke, 2002).

2 Methodology

The Knowledge Management approach used in the research will cover New product Development Process, by including the following stages:

1. Construct Database

- Database as knowledge base
- Database as case history (previous experiences from several other projects) repository

2. Development of Knowledge-Based System

The Database as Knowledge-Base was developed by acquiring a large question set that will be utilised by a Knowledge-Based System. This is the precursor step to identifying risks within a Concurrent Engineering product development environment. These questions were sorted into their appropriate knowledge categories.

The need for, and difficulty in developing computer based tools to support design process planning, has been shown to be highly influenced by the complexity of coordinating knowledge transfer to support decision making utilising incomplete information within a design process which is ill-structured in nature. The intelligent exploitation of knowledge that presently resides within computerised workgroups as uncoordinated and often unconnected islands of information is seen as a precondition to maximising the benefits that may be achieved through planning the design process in a CE environment (Alex et al., 1993).

An expert system is predominantly utilised within current research regarding intelligent decision support, and is one of the most common approaches to developing knowledge-based decision support systems to support designers and other participants in a CE environment. An expert system architecture is designed to allow systems developers to develop applications that can emulate the problem-solving behaviour of a human expert within a particular discipline. An expert system is a technique that is able to capture the knowledge and reasoning of an experienced expert for re-use in assisting the less experienced in making decisions. Several tools and methods including risk mangement decision support systems available for multi-site engineering projects are summarised in Mingwei et al., 2002.

Risk assessment is a process that attempts to answer three questions (Larson 1996):

1. What can go wrong?
2. What is the likelihood that it will go wrong?
3. What are the consequences?

One of the more common ways of managing risks in a project is through the application of a risk management procedure given by Figure 1.

The approach used in this project is to utilise the questions derived from analysis and interview of the product development domain of the industrial partner as knowledge and rules.

Although expert systems are capable of delivering quantitative information, determined from models that have been developed through extensive research of a particular problem, their strength lies in their ability to utilise these heuristics or rules of thumb in interpreting qualitatively derived values where there is a lack of quantitative information. Consequently, one of the most important features of the expert system is that it is capable of reasoning in the absence of certainty and with incomplete information through the use of confidence values. In establishing the risk context in this research, the system interrogates the user to determine from the responses, a qualitative assessment of the risk profile of the organisation, the project and the product and process. Confidence values are input by the user in response to questions posed.

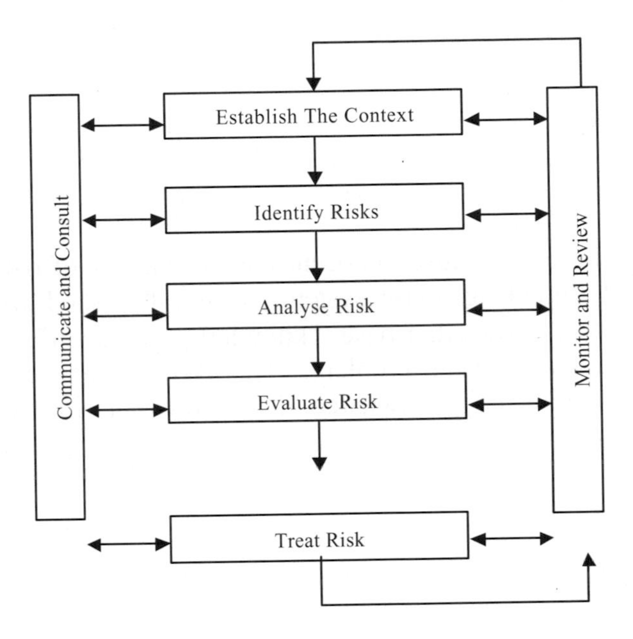

Figure 1. Risk Management Procedure

The Database as a Case History Repository covers the "lessons learned" from previous projects based on several individual experiences of team members and managers. The Database as Activities Repository is aimed to be used for constructing a process network for each project completed. The activities of each project was stored in a structured manner within a database, which then became an input to associate several possible risk factors with each activity type to yield a risk profile from a process/activity viewpoint.

The second stage which is the development of Knowledge-Based System would incorporate the results of Stage 1 and support the users to share, utilise and manage the previous and current knowledge on several projects; thus enabling a new generation of knowledge management strategies in today's competitive manufacturing environment (Kayis et al, 2002).

3 Results

This research helps provide a tool for the managers to test number of possible outcomes for specific decision choices by using a knowledge base, especially considering the following decisions:

- Technical
- Strategic
- Interpersonal
- Interdisciplinary
- Commercial

Each of the above decisions aim to minimise the occurrence of variety of risks in a project environment whilst knowledge-base system to be established would allow the user to search its own knowledge base, "knowledge reservoir", for a possible existing solution to a given design problem. The users are expected to fill in the knowledge reservoir with their knowledge, data and experience which would assist and support the establishment of "a learning organisation".

References

Jarke M., 2002, Experience-based Knowledge Management, Control Engineering Practice, Vol. 10, pp. 561-569.

Kayis B., Amornsawadwatana P., Ahmed A., Kaebernick H., 2002, A Risk Assessment Approach for Concurrent Engineering Projects in Multiple Sites, Proceedings of the 3rd Asia-Pacific Conference on Systems Integrity and Maintenance, pp. 36-41.

Duffy, Alex H.B., Andreasen M.M., MacCallum K.J. and Reijers, L.N., 1993, Design Coordination for Concurrent Engineering. Journal of Engineering Design. Vol. 4, No. 4, pp 251-265.

Larson N. and Kusiak A., 1996, Managing Design Processes: A Risk Assessment Approach. IEEE Transactions on System, Man, and Cybernetics-Part A: Systems and Humans, Vol. 26, No. 6, November, pp. 749-759

Zhou, M., Nemes, L, Reidsema, C., Ahmed, A., Kayis, B., 2002, Tools and Methods for Risk Management in Multi-Site Engineering Projects, Proceedings of International DIISM Conference, Japan.

Approaching Online Self-Representation in a Community of Practice[1]

Cecília Kremer[*], *Judith Ramey*[**], *Clarisse Sieckenius de Souza*[*]

Computer Science Department[*]
Pontifícia Universidade Católica
R. Marquês de São Vicente 225
Rio de Janeiro – RJ – 22453-900
Brasil
ceciliak@tecgraf.puc-rio.br
clarisse@inf.puc-rio.br

Technical Communication Department[**]
College of Engineering
University of Washington
14 Loew Hall – Box 352195
Seattle – WA – 98195-2195
USA
jramey@u.washington.edu

Abstract

Although research in online communities is still in its early stages, we believe that the issue of online self-representation deserves more attention. Self-representation is crucial to form relationships, and relationships are central to communities and the building of social capital. This work presents a case study that illustrates our approach to systematic analysis of community members' self-representations and points out some implications of designing such information. The approach was taken during the contextual inquiry and analysis stage in the design of a communication tool for a medical community of practice. Our approach is rooted in the work by Erving Goffman and its main outcome is the proposal of dimensions to be considered in the design of users' online self-representation and how they relate to the issue of trust.

1 Introduction

The concept of community has had different defining criteria across time (Wellman, 1982) that evolved from physical features, such as size and location, to the social features of strength and type of relationships among community members (Preece, 2000). According to such social features, a community can now be defined as a set of relationships where people interact socially for mutual benefit (Andrews, 2002). Therefore, it is crucial to the formation and maintenance of a community to offer its members resources to establish relationships among themselves.

The concept of relationship involves at least two participants and the fact that each plays a role in it. In our daily lives, in order to build a relationship with others, we may catch ourselves trying to express the role we are willing to play and to assess whether our interlocutors are willing to play a counterpart role and engage in the relationship we are proposing. As explained in Goffman's work (Goffman, 1959), the way we present ourselves to others is central to the definition of the roles we are willing to play and consequently of which relationships we are willing to engage in.

[1] This work was developed under the grant of Stewart, Brent K., Principal Investigator, "Patient-centric Tools for Regional Collaborative Cancer Care Using the Next Generation Internet", National Library of Medicine, 1999-2002.

The importance of self-presentation to the development of relationships and the importance of relationships to the development of communities were the motivation for our work. Although we understand that research in online communities is still in its early stages (Preece, 2000), we believe that the issue of self-representation should be treated in more depth, as it is crucial for community support and the building of social capital (Putnam, 1995).

In this paper, we propose an approach to help online community designers think about community members' self-representation and how to support it. We provide dimensions of the presentation of self in an organized fashion, hoping they can be of use to other designers and researchers. We would like to make clear that our goal is not to mimic face-to-face interaction in online environments, but to build on existing Sociology works to inform their design.

Our approach was developed for a community of practice, that is, a community defined by "a common disciplinary background, similar work activities and tools, and shared stories, contexts, and values" (Millen, Fontaine & Muller, 2002). In the following section we will describe the community in focus. Then we will explain the proposed dimensions for online self-representation and how we used them in the design of an online community.

2 The Community of Practice

This work was done in the early stages of the design of a communication tool for a medical community of practice, which holds weekly face-to-face meetings to discuss complex and unresolved cancer care cases. A subgroup was moving to a remote location and our task was to design an online environment to enable the meeting to continue, supporting the existing communication and relationship patterns.

This community has a core and steady set of members that works in the same building, meets face-to-face and uses email mostly to confirm information about the forthcoming meeting (discussion is held face-to-face or over the phone). This core group is primarily made up of medical doctors with different specialties who lead the meetings and are very participative. The larger part of the community is formed by periodic or sporadic participants, including medical residents and students, other medical personnel, and sometimes few drug companies' representatives. They attend meetings to learn and are not expected to play an active role.

We observed that this community has a strong hierarchy and that communication practice takes it into account. Thus, to facilitate online communication, we have enriched our design to capture the dimensions that represent the community members' relative standing.

3 Dimensions for Online Self-representation

We relied on work by the sociologist Erving Goffman (Goffman, 1959) to enrich our design. Although he discussed society in a general manner, his ideas were a useful analytical resource for the design of our communication support tool.

Based on his concepts, we organized the design space of self-representation as shown in Figure 1. Our focus will be on the *front dimension* and its *personal front* part. The front dimension involves information about an individual, mainly "the expressive equipment of a standard kind intentionally or unwittingly employed by the individual" (Goffman, 1959, p. 22). The personal front part allows more directly for self-representation and identification of an individual community member.

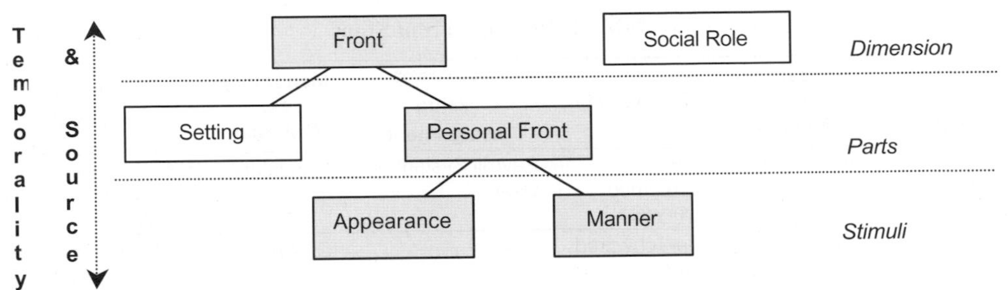

Figure 1: Self-representation design space

We are interested in the expressive equipment the designer will provide online community members for self-representation. In our approach, the personal fronts should be organic, naturally occurring in the domain, and the signs identified as stimuli for them should be necessary and sufficient to support appropriate choices for social interaction. Note that the set of signs should not be exhaustive. As described by Goffman, "a significant characteristic of the information conveyed by front is its abstractness and generality" (Goffman, 1959, p. 26). When people communicate, they use stereotypical thinking to frame a "pattern of expectation and responsive treatment" and "orient themselves in a wide variety of situations" (Goffman, 1959, p. 26). Therefore, designers should observe the target community and find out what guides members to make their communication productive. If too much information can overburden them, making it difficult to know one another, too little may also be harmful, since it may put members in awkward positions or require efforts to gather missing information necessary to a comfortable communication.

Before presenting examples of personal front signs we considered relevant in our context, we should note that these signs can be divided into two kinds: *appearance* and *manner*. Appearance is the set of signs that tells us about a member's social status and current ritual state (e.g. whether s/he is engaged in formal or informal social activity). Users may use appearance signs to declare their office rank, age, looks (a photo or iconic self-representation), etc. Manner is the set of signs that conveys the interaction role the performer is expecting to play. It tends to be produced through behavior along time and therefore to be unwittingly or less intentionally generated.

We have observed community members and found personal front signs conveyed by appearance: Professional Maturity (e.g. doctor) and Academic Title, and by manner: Active Participation in Meetings and Importance of Contributions. The different natures of these signs imply different representations. Appearance signs are usually declarative and require input by a user or external source. Manner signs require behavioral pattern observation by the system or users, as in reputation systems (Jensen, Davis & Farnham, 2002). The first two columns in Table 1 show examples of personal front signs, indicating if they are conveyed through appearance or manner.

A self-representation design should also prevent community members from inadvertently conveying destructive personal front information, such as outdated or inconsistent information. This would disrupt or hinder communication, lead to inconvenient behavior and harm social capital building. For instance, a resident who became a doctor but kept an old online personal front could be treated as a student and not as someone who is expected to be more active in meetings. To avoid outdated information, we represent the temporality of the individual personal front sign values (Table 1, third column). Thus, the designer could set the system to periodically provide updates or remind users of their current self-representation, and ask them to re-validate it. For instance, an annual reminder for office rank re-validation would be adequate for our community members since it does not change often and system interruption overloading should be avoided.

Table 1: Personal Front Signs

Personal Front Sign	Appearance vs. Manner	Temporality	Source of value assignment	Mandatory vs. Optional	Trust Level
Professional Maturity (e.g. resident)	Appearance (declarative)	Low (confirmed every year)	Payroll System or User	Mandatory	High if source is system Otherwise trust depends on trust on user
Academic Title (e.g. professor)	Appearance (declarative)	Low (checked every year for new titles)	User	Optional	Trust depends on trust on user. Can be checked w/ institutions
Frequency of meeting attendance	Manner (behavioral)	High (updated every meeting)	User or System (depends on the technology)	Mandatory (derived by system)	High if source is system Otherwise depends on trust on user. Can be checked w/ users
Participation level in meetings	Manner (behavioral)	No need to be confirmed: source = system	System generated = number of times the user contributed	Mandatory (system derived)	High if source is system Otherwise depends on trust on user. Can be checked w/ users
Importance of contributions	Manner (behavioral)	No need to be confirmed: source = users	Generated by users who evaluate the contribution	Optional	High: derived from other users More users = more trust
Credibility	Manner (behavioral)	No need to be confirmed: source= system or users	System: overall contributions' evaluation Users: evaluation	Optional (depends on users' informing)	High: derived from other users More users = more trust

To avoid inconsistent information, we represent two qualitatively different relationships among personal front signs and sign values. A Restrictive Relationship may state: "someone whose Professional Maturity is *resident* cannot inform Academic Involvement as *medical fellow*"; whereas an Expectation Relationship may state: "a person whose Professional Maturity is *medical doctor* and Academic Involvement is *professor* is expected to provide highly important Contributions and have high Credibility". As pointed by Goffman, we often expect consistency between appearance and manner. In an online community it may be helpful to inform a member about what others expect from a conveyed (self)personal front, motivating reflection about it as well as better positioning towards others, which is vital to social relationships.

As shown in Table 1's fourth column, we also represent the source of personal front sign values. For instance, Professional Maturity can be informed by the user or the hospital payroll system, Participation in Meetings can be automatically generated from users' behavioral patterns, Importance of Contributions can be informed by users and Credibility may be derived from an overall evaluation of the user's different Importance of Contributions. The source of the different Personal Front sign values is important because it is directly related to trust.

Several aspects influence trust in online communities (Andrews, 2002). Goffman argues that people check "governable" upon "non-governable" signs (Goffman, 1959, p. 7). The former, mainly verbal assertions, can be easily manipulated by the person using them, while having little concern or control of the latter. When someone says a food tastes good, one may check whether s/he is chewing it with pleasure (Goffman, 1959, p. 7). When applying for a job, one's résumé states how good one is, but employers look for referra ls, prizes, etc. (see also Donath, 1996).

Trust building in online communities is hindered in part because users are provided with fewer non-governable signs than in real life. In our approach, the source of personal front sign values allows designers to evaluate support for trust. For instance, if all personal front sign values are input by users, the designer may add non-governable signs. The shift from governable to non-governable signs, or their combination, can benefit trust development in an online community.

4 Concluding Remarks and Future Works

Previous work on identity and deception in virtual communities (Donath, 1996) employ concepts which can be paralleled to governable and non-governable signs. While that work took an ethnographic approach to analyze existing Usenet communities, ours has studied an offline community to design an online environment. Despite the importance of trust, the main focus of our work is self-representation, how it influences relationships among community members, and how this affects social capital building. This approach values personal signs that are necessary and sufficient to support self-representation and appropriate choices for social interaction.

This work has provided a case study that illustrates an approach to a systematic analysis of the type of presentation of self that a given community of practice may require, and has pointed out some design implications of these findings. Our next step is to design how community members would be enabled to explicitly present these dimensions of themselves through the online community interface. In the future, we plan to explore how this approach can fit other communities. Our long-term goal is to produce a method to help designers of online communities to support community member self-representation and relationship building.

5 References

Andrews, D. C. (2002). Audience specific online communities. *Communications of the ACM*, 45 (4), 64-68.

Donath, J. S. (1996). Identity and deception in the virtual community. Retrieved February 7, 2003, from http://smg.media.mit.edu/people/Judith/Identity/IdentityDeception.html. Final draft prepared for Kollock, P. & Smith M. A. (1998). Communities in Cyberspace. London: Routledge.

Goffman, E. (1959). The presentation of self in everyday life. Doubleday, New York: Anchor Books.

Jensen, C., Davis, J. & Farnham, S. (2002). Finding others online: reputation systems for social online spaces. *CHI 2002 Proceedings*, 4 (1) 447-454.

Millen, D. R., Fontaine, M. A. & Muller, M. J. (2002). Understanding the benefit and costs of communities of practice. *Communications of the ACM*, 45 (4), 69-73.

Preece, J. (2000). Online communities: designing usability, supporting sociability. West Sussex, England: John Wiley & Sons.

Putnam, R. D. (1995). Bowling alone: America's declining social capital. *Journal of Democracy*, 6 (1), 65-78.

Wellman, B. (1982). Studying personal communities. In P. M. N. Lin (Ed.), *Social structure and network analysis*. Beverly Hills, CA: Sage.

Management of Information and Knowledge in Human Computer Interaction using System for Cusp Surface Analysis

Yasufumi Kume, Chung-Yong Liu and Loren Cobb

Kinki University, Industrial Technology Research Institute and Aetheling Consultant
3-4-1 Kowakae Higashiosaka Osaka 577-8502 Japan
kume@im.kindai.ac.jp

Abstract

This paper describes the system for Cusp Surface Analysis and the application of this system to the management of information and knowledge in human computer interaction. Cusp surface is the quantitative cusp catastrophe model in our previous papers. This paper focuses on the quantitative cusp catastrophe model for creative process comparing with the qualitative catastrophe model proposed by King. In this model, two control factors are coping characteristics and psychological stress, and a state variable is creativity of self-actualization. The relationship between self-actualization and coping characteristics, and the relationship between self-actualization and psychological stress are clarified in detail. And the usefulness of this system is shown by other examples.

1 Introduction

Organizational aspects are the balancing and fusing of people, technology and organization. It depends on the development speed of information communication technology. The speedy behaviour is more required in research and development process than production processes. This is the research and development and new product development including marketing, design and shipping. In the point at the theoretical issues of agile manufacturing, organization is focused on. In the point at issues of present production system, the system is not possible to correspond to need of market. Agile manufacturing is not speedy production but the speed including the organization problem. Interactive human computer communication is unexpected relation. Fundamental one is the employment for the talent people.

Stress at work, especially psychological stress becomes important problem. As automation of production system progress, workers have been needed creativity namely ability of human. If psychological stress exceeds the limit, it brings to inappropriate conditions. It is important to grasp how workers are changed environment, that is, the relationship between stressor, stress and coping. Creative process may be defined as "the transform or emergence of new ideas or artifacts with result from a series of interconnected actions or events". Wallas proposed the creative process as a series of more or less discrete phases or stages [1]. He termed as preparation, incubation, illumination and verification. These terms are still commonly employed in the field of creative studies. In actuality it would be better to conceive of creative thinking in more holistic terms, as a total pattern of various processes process that overlap and interweave between the occurrence of the original stimulus and formation of final product. A more complete and satisfactory way of looking at creative process is to regard it as a type of "whole field phenomenon". This leads us to a more complete and holistic picture that recognizes the immense variety to be found in approaches to creativity by different individuals. Note that the system is "contained" in a field of mental activity divided into two regions of principle forms of thinking, convergent and divergent. The ability of the creative individual to use both types with equal and combinatory dexterity is essential to the development of original thinking, as each stage of the process is attended by a particular set of mental condition. This paper focuses on the cusp catastrophe model for creative process.

2 King's Catastrophe Model for Creative Process

This section examines the development of a single structural model of creative process and explores the implications for graphics interface design. The need for an open systems framework for studying the dynamics of graphics interfaces is identified. A single structural model of creative behavior in the form of topological "map" is presented using catastrophe theory as a tool for the analysis of discontinuous behavior in open systems. Catastrophe theory was developed by Rene Thom, whose classic work on the subject, "Structural Stability and Morphogenesis, was first published in 1972. Thom's thesis is cantered on the dynamic mapping of the morphological changes in biological structures by use of topological surfaces. The theory deals with nature and dynamics of continuous and discontinuous changes in systems behavior and is ideally suited to an analysis of creative process that, by its nature, is characterized by a discontinuity, the phenomenon of illumination or intuitive insight. The rigorous mathematical foundations of the theory are not used in this particular application. A catastrophe is "any discontinuous transition that occurs when a system can have more than one table state or follow more than one stable pathway ". Five important properties of a cusp catastrophe have been identified viz, bimodality, divergence, sudden transition, hysteresis and inaccessibility. These forms will be describe in general terms, followed by a specific application to the analysis of creative process. A creative behavioural topology may be proposed in the form illustrated [1)]. Two control parameters have been chosen convergent and divergent thinking, these having been identified as characteristic of creative activity. The four phases of creative process, i.e., preparation, incubation, illumination and verification, are shown in their respective positions governed by the relative amounts of convergent/divergent thinking evident during each phase. Preparation, characterized by highly divergent thinking, is shown as the bottom surface. Incubation, the relatively inaccessible phase, appears as the behavioural characteristic of that portion between the two surfaces. Illumination is shown as the convergent factor that causes the system to move from divergent search to convergent association (the sudden transition). Verification appears as the upper surface of the topology. Given this initial structure, we may now examine the possible dynamic conditions of creative process in terms of a single catastrophe model.

3 Catastrophe Model for Creative Process Considering Psychological Stress

In King's model, control factors are divergent thinking and convergent thinking, and state variable is creativity. In other words, the information for creative activity is collected by divergent thinking and unified a logic system by convergent thinking. The creativity becomes catastrophe-jumping phenomenon. Conventional logic system is dissolved and the information that is the materials for building a new logic system is collected. It seems that the condition of complication that result in operating divergent thinking is the condition of psychological stress. The information that cannot be unified by the conventional logic system is unified to a new logic system by operating the convergent thinking. This is equivalent to the operation of coping characteristics by which a new logic system is formed in the view point of coping for psychological stressor. In the new logical system, one does not cling to conventional logic system of strain of activity due to define function, namely suppression or illogical thinking, but the information under the condition of complications can be unified and comprehended by the logic system. A catastrophe model for creative processes is shown in our previous paper [2]. Control factors are psychological stress and coping characteristics was proposed. In order to have generality of self-actualisation in the basis of creativity, the level of creativity may be estimated by psychological stress and coping characteristics in the new model. The effect of psychological stress on creativity becomes clear. King's model may be used as the management of the information which diverges in creative activity and the new model as the management of human factors in creativity.

3.1.1 Cusp Catastrophe Model of Creative Process

Moslow, A.H. divided the desire of human into five stages. They are psychological desire, dignified desire, and sel-actualization desires etc.. Self-actualisation desire is the desire of wanting to realize the highest figure, which he should have, while demonstrating a possibility, to the maximum extent. This desire can be seen in creative work. The workers in creative work are influenced by surrounding various environments. They are the influence by the external environment of technical change, and influence by inner environment of such as human and organization. In this environment, workers are in a high strain state. In the state of this strain, it is evaluating these effects for themselves. If this evaluation exceeds its own tolerance level, workers will lapse into the psychological stress under the much extension. Therefore, in this paper, the cusp catastrophe model of the creative process shown in our previous paper [2] was proposed. Two control factors are coping characteristics and psychological stress, and state variable is creativity of self-actualization.

3.1.2 Experiments and Its Result

In this paper, the experiment was conducted to 30 experimental subjects. The S-A creativity inspection by Onda was used for evaluation of creativity of self-actualization. This evaluates creativity by fluency, originality and flexibility as the thinking characteristics of creativity. It asked for the score of individual creativity of self-actualisation based on this appraisal method. Investigation which asks for self-actualisation of how much to adopt 19 clusters of the coping behavior by Sakata was conducted in evaluation of the coping characteristics. At this time, frequency was carried out to from 0 to 100. What evaluated the results of investigation by the geometric average was considered as the score of individual coping characteristics. The stress appraisal method for the labourers by Natsume was used for evaluation of psychological stress. This is investigation, which asks for self-consultation how much stress is felt in the ranges from 0 to 100 about each item.

3.1.3 Cusp Surface Analysis and Its Results

In cusp surface analysis, it is assumed that the variable expressed with $x_1, x_2, \ldots, x_n$ exists in relation to parameter α, β, γ [3] . Then formula is changeable into the form of the following formula.

$$0 = \alpha(\underline{x}) + \beta(\underline{x})\{y - \gamma(\underline{x})\} - \delta\{y - \gamma(\underline{x})\}$$
$$\alpha(\underline{x}) = \alpha_0 + \alpha_2 x_1 + \alpha_2 x_2$$
$$\beta(\underline{x}) = \beta_0 + \beta_2 x_1 + \beta_2 x_2$$
$$\gamma(\underline{x}) = \gamma_0 + \gamma_1 x_1 + \gamma_2 x_2$$

, where ä is a constant.

In cusp surface analysis, x_1 is the score of coping characteristics, x_2 is the score of psychological stress and y is the score of creativity of self-actualisation. Each parameter was presumed as follows by maximum likelihood method.

$$0 = \alpha(\underline{x}) + \beta(\underline{x})\{y - \gamma(\underline{x})\} - 1.032\{y - \gamma(\underline{x})\}$$
$$\alpha(\underline{x}) = -0.205 + 0.347x_1 - 0.642x_2$$
$$\beta(\underline{x}) = 0.282 - 0.349x_1 - 1.761x_2$$
$$\gamma(\underline{x}) = -0.056 - 0.292x_1 + 0.35x_2$$

Fig 1.is the cross section in the proposed method with the mean value x_2 = 44.9 of psychological stress. Where the score of coping characteristics is low, the level of creativity of self-actualization is also low in this figure. Moreover, catastrophe phenomenon occurs in the place of x_1(coping characteristics) = 50.0, and the creativity of self-actualisation is changing on the comparatively high level after that.

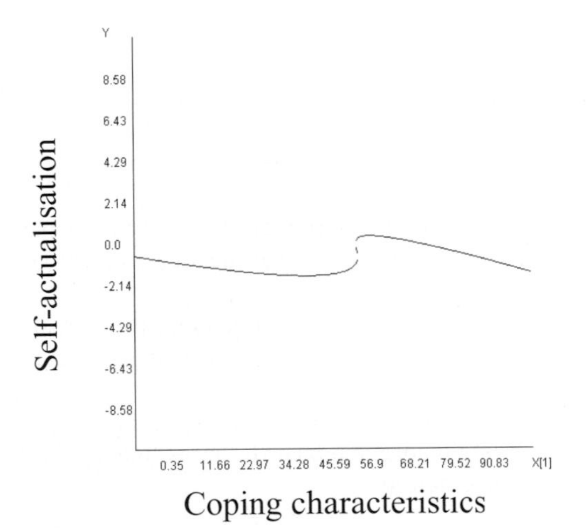

Fig.1 Relationship between self-actualisation and coping characteristics

Fig.2 is the cross section in the proposed model with the mean value x_1 = 45.6 of coping characteristics. In this figure, catastrophe phenomenon has occurred even in the place about x_2 (psychological stress) = 40.0. In other words, the level of creativity of self-actualisation has caused a change rapid between them, since the solution (anti-mode) which is right in the middle may have a low possibility of existing compared with other solutions and may exist in either a lower solution (mode) or the upper solution (mode).

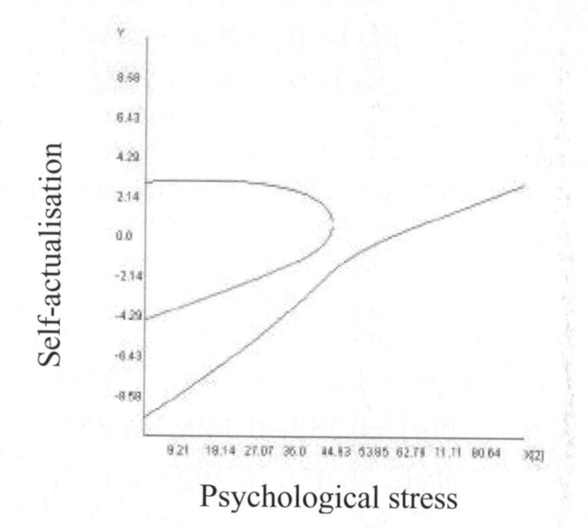

Fig.2 Relationship between self-actualisation and psychological stress

From Fig.1 and Fig.2, it is clarified that management of information and knowledge in human computer interaction is the management of human stress and coping characteristics in creative activity for human computer interaction.

4 Conclusion

The geometric feature of stochastic cusp catastrophe model in the discontinuous phenomenon of the creative process governed by irregular nature was clarified including the human factors, and the model was quantified by cusp surface analysis. The result obtained by this paper is summarized as follows.

1) It was shown by stochastic cusp catastrophe model proposed by this paper from the level of coping characteristics and psychological stress that the level of creativity of self-actualisation can be presumed.
2) In cusp surface analysis, it became possible to treat quantitatively creativity of self-actualisation, the coping characteristics and psychological stress, and it was able to clarify the action.
3) It is clarified that catastrophe model for creative processes for talent people considering psychological stress becomes the theoretical basis for stress management in the creative activity for human computer interaction.

References

1) Kume, Y. &Sato, N. (2000). Cusp Catastrophe Model for Creative Process in Agile Manufacturing. Manufacturing Agility and Hybrid Automation ? ,117-122.
2) Kume, Y., Yamamto, N & Cobb, L. (2002). Proc. of Japan-USA Symposium on Flexible Automation, 687-690.
3) Cobb, L. (1992). An Introduction to Cusp Surface Analysis.

Contextualizing Search Results in Networked Directories

Christoph Kunz, Veit Botsch, Jürgen Ziegler, Dieter Spath

Fraunhofer Institute for Industrial Engineering
Nobelstrasse 12
70569 Stuttgart, Germany
{christoph.kunz, veit.botsch, juergen.ziegler, diether.spath}@iao.fraunhofer.de

Abstract

This paper presents a novel approach for the representation and contextualization of search results in a thematic category network, which can be present in the form of web directories like the Google category system or the Open Directory Project. An interactive matrix display is used for showing relations between concepts and concept hierarchies of the directory, which are displayed along the two axes of a matrix. The interface allows pure browsing of the thematic structure as well as querying the underlying search engine and exploring the hits on a thematic level. The results of a search are not only shown as the commonly used, but effective plain text lists, but are classified in the category system. In this way, the tasks of browsing the categories and searching the information space are combined in one interface.

1 Introduction

The exponentially growing amount of information available on the internet, in an intranet or a file system increases the interest in the task of retrieving information of interest. Search engines usually return thousands of results per query, which are displayed in a plain text list with only a short abstract describing the context of the given query term in which it appears. The list-based structure of search results offers little to assist the user in browsing large sets of pages with respect to a thematic category. The user must process the results linearly using a single fixed ordering scheme, which is based on some ranking algorithms, in order to identify relevant pages. This processing is additionally impeded by the small amount of hits (~20) that can be displayed in a browser window at one time thus forcing the user to scroll to the next hit page, having to wait for the server response.
On the other hand, in the middle of the 1990s yahoo started to categorize web pages by subject into a hierarchical organized web directory. This work has been followed by Google, Altavista and recently the Open Directory Project (www.dmoz.org). By a web directory a thematic navigational structure is given for drilling down from root level to subjects of interest. Also, the web links contained in a web directory are of high quality because of being selected by hand. While this is useful for searching links relating to a specific topic, the interface support for browsing subjects is inadequate. For example, browsing in Google´s directory from top level to the topic "Google" takes six steps. In each step the user is confronted with choosing the next subtopic from a list out of about thirty topics. This step by step procedure hampers the process of getting an overview of the whole topic structure. Also, the tasks of searching and browsing are separated in most of the web search engines. The information located in the hierarchy of subjects is not used to classify and to display the search results in a holistic interface.

For designing appropriate interfaces it is essential to understand the users search problem (Figure 1, from Nonaka & Takeuchi, 1997). First, the user has a specific information shortcoming which should be compensated by the search system in an efficient way: on the one hand the result set of a query should contain all relevant information for the need, on the other hand it should not contain unnecessary information. The information shortcoming is divided in a objective and a subjective part. The objective part is that demand, which is needed for solving a particular task. The demand the user believes she needs for solving the task is called the subjective information need. Objective and subjective information needs are often very different, especially if the user´s task is hard to specify or if the information space and its interrelation is unknown.

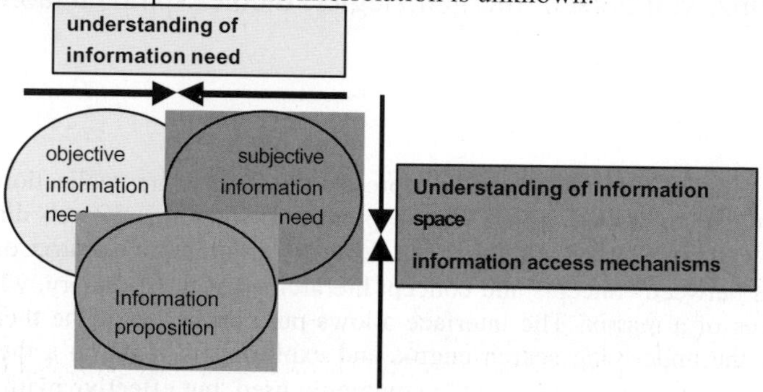

Figure 1: The search problem (Nonaka & Takeuchi, 1997)

In general, the main goals of search and retrieval interfaces is to get the three sets in Figure 1 to completely overlap. That is supporting the user´s understanding of her own information needs and providing an overview on the structure and relations of the information space as well as effective mechanisms for accessing information within a specific context. A graphical contextualization of the search results could assist in communicating these types of information. A visualization can provide a broad, concise representation of the results which the user can quickly scan to understand why and how the results are related to the query and to thematic subjects. Ideally, users can then interact with the visualization to obtain sites and pages of interest.
Against this background, the work reported here aims at improving the users understanding of both her own information needs and the logical structure of the information space as well as providing a context of resulting pages and valuable access mechanisms.

2 Related Work

A number of systems have been built to visualize hypertext systems such as the Web (Andrews; 1995, Ayers & Stasko 1995, Hasan et al., 1995, Mukherjea & Foley, 1995). These systems have primarily focused on representing the pages and links between them as an aid for navigation and for understanding the structure of the space.
For category systems, the conventional display approach is to map all the hierarchy into a region that is larger than the screen and then uses scrolling to move around the region. Other, more sophisticated approaches, were chosen in the hyperbolic tree (Lamping & Rao, 1994) or in cone trees (Robertson et al., 1991), where the whole hierarchy is visualized by distorted views or put into three dimensional space.
Whereas visualization kits like The Brain (http://www.thebrain.com), Antarctica (http://maps.map.net) (both in Patience & Chalmers, 2002) or the search engine "kartoo" (http://www.kartoo.com) try to put search results into context of thematic topics to which they

belong (see figure 2). The Brain uses an interactive graph display to visualize the topic hierarchy. As a result of a given query term, The Brain returns these web pages which were categorized into the topic, that relates best to the query term. Directly related topics are then depicted around the best fitting one. Consequently, the query is not matched against relevant pages but to relevant topics, yielding in an incomplete or partly non relevant result space.
This drawback is overcome by Antarctica, which uses a geographic map metaphor for linking websites and topic. Topics and corresponding pages are placed on a region in the map. A given query results in zooming into that region the super topic belongs to. However, due to space limitations only a few most relevant pages can be visualized.
"Kartoo" visualizes search results in a network, where specific pages are related to each other by some key words, which are contained in this pages. The user then can decide if he wants to add these keywords to his query or to remove them. Yet, the system is not giving some information, why the resulting nodules are related to the query or about the content of a page. Also, due to space limitations only a few pages can be depicted at one time.

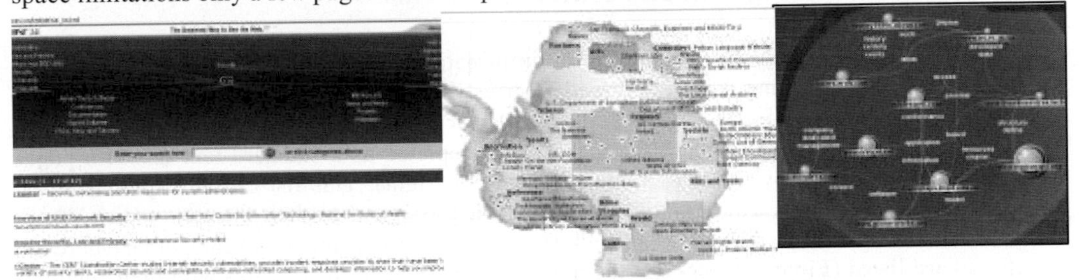

Figure 2: The Brain, Antarctica and kartoo

Sebrechts.and his team (Sebrechts et al., 1999) compared the value of search interfaces in plain text, 2D and 3D space with the result that plain text lists perform best with respect to users' computer skills and experience.
According to these results, none of the approaches introduced supports all of the interface tasks like providing insight in own information needs, the structure of the information space or returning all relevant information in a given context.

3 Visualizing networked web directories with MatrixBrowser

A networked web directory can be seen as a big concept graph where all the web pages are instances of corresponding concepts. In order to visualize a web directory, the central idea of the MatrixBrowser (Ziegler et. al, 2002) approach and prototype is to map the underlying graph structure to a highly interactive adjacency matrix (Figure 4). Adjacency matrices are a well-understood alternative graph representation where the nodes of the network are shown along the horizontal and vertical axes of a matrix. Both the direction of an association (for directed graphs) as well as different types of associations can be visualized by using arrows and graphical symbols shown inside the cells. This can be done in conjunction with other techniques such as tool tip descriptions of different association types.
The second main feature is that MatrixBrowser provides mechanisms for presenting hierarchical information structures directly as interactive tree widgets. This "Windows Explorer"-like technique is widely known and intuitive to use. In MatrixBrowser, the user can flexibly place hierarchical structures of the directory along both axes and explore them with the familiar expand/collapse procedure. In this way, the information shown in the matrix can be better

structured and the amount of visually displayed material reduced. Also filter mechanisms are provided to reduce the amount a displayed nodes.
As a result of using tree widgets as axes of the matrix, not all concepts and relations are visible all the time. MatrixBrowser allows using expand/collapse functions not only for the trees but also for the cells. If the explicit relations in the net are not visible because their super ordinate concepts are in a collapsed state, an interactive symbol is shown, that can be clicked for expanding/collapsing the associated trees. With these techniques, the user can flexibly drill-down into the network or condense parts of it.
A net-like interactive visualization in the upper left corner of the matrix shows all of the neighbours in the net of a node, which is selected in one of the two hierarchies. So not only the context of a node within the hierarchies but also in the net can be seen easily.

4 Contextualizing search results

Beside the exploratory way of opening up the information space, the search interface provides mechanisms for performing keyword based queries and offering then navigational structure providing a context and effective access of the result space.
As an effect of performing a query, the result set is depicted as a standard text list with pages descriptions as returned by the search engine (Figure 3, right). This list can be explored in the familiar way of scrolling. The resulting pages are then assigned to their thematic topics. These partitions of the directory are then visualized offering a thematic context. All categories where hits are found are listed (Figure 3, left) and the most relevant concept hierarchies with their relations are displayed in the matrix (Figure 3, middle). The amount of hits per concept is visualized with a bar. If similar hits are found in concepts on both axes of the matrix, additional relations are drawn. The size of each relation is corresponding to the amount of similar hits in the linked concepts.

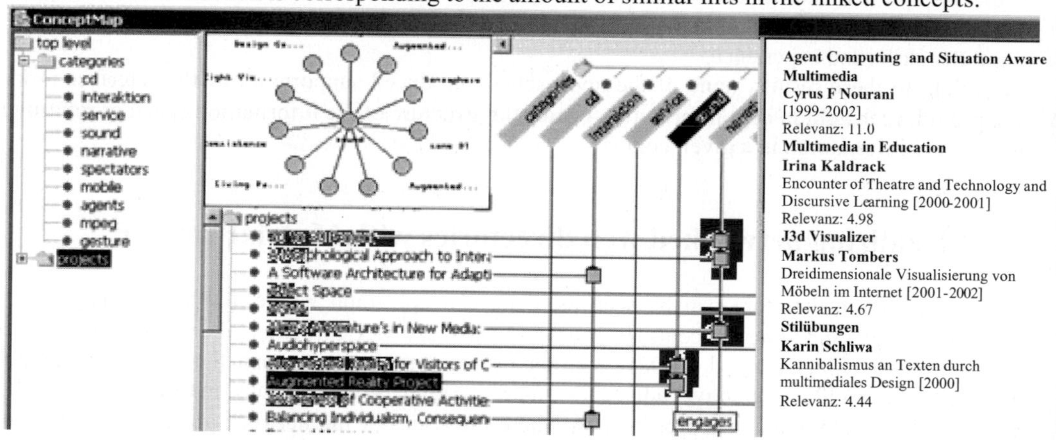

Figure 3: The search interface

By selecting specific concepts or relations, the textual hit list shows only the intersection of the set of pages corresponding to the topic and the set of resulting pages. In case of the selection being a relation, the intersection of the sets of both concepts and the result set is shown.
All listed concept hierarchies can be placed on the axes by drag&drop mechanisms, offering a flexible way to explore the concept network of the search engine and the result set of a query

visually in the matrix and by interactively drilling down from high-level concepts to lower-level nodes and relations.

5 Conclusion and future work

We have presented a new graphical user interface and prototype for exploring web directories and contextualizing search results in the category network of these directories, that allow user to navigate and explore the results in a familiar and intuitive way. Hierarchical category systems are displayed in interactive tree widgets, so the user can increase or reduce the amount of information displayed and refine his query without input of any textual data. The two tasks of searching and browsing are combined in one graphical interface.

Future work will concentrate on combining the interface with more detailed metadata structures than web directories like ontologies. By introducing richer metadata possibilities are explored, how semantic queries can be formulated in a graphical and interactive way in the sense of "give me all documents of topic "information visualization", which focus on "visualizing search results" of conference "HCI".

In addition to these functional enhancements, the graphical user interface is being evaluated in usability studies.

References

Andrews, K. (1995). Visualizing Cyberspace: Information Visualization in the Harmony Internet Browser, *InfoVis'95*, IEEE Press, Atlanta, pp. 97-104.

Ayers, E.Z., Stasko, J.T. (1995). Using Graphic History in Browsing the World Wide Web, *Proceedings of the Fourth International World Wide Web Conference*, Boston.

Hasan, M., Mendelzon, A., Vista, D. (1995), Visual Web Surfing with Hy+, *CASCON'95*.

Lamping, J.; Rao, R. (1994): Laying Out and Visualizing Large Trees Using a Hyperbolic Space. *Proceedings of the ACM Symposium on User Interface Software and Technology*, p. 13 – 14.

Mukherjea, S., Foley, J. (1995).Visualizing the World Wide Web with the Navigational View Builder. *Proceedings of the Third International World Wide Web Conference*, Germany.

Nonaka, I., Takeuchi, H. (1995). The knowledge creating company, *Oxford University Press*, London.

Patience, N., Chalmers, R. (2002).Unstructured Data Management: the elephant in the corner (guest or customer access required), *the451 Report*.

Robertson, G.., Mackinlay, J., and Card, S. (1991): Cone trees: Animated 3D visualizations of hierarchical information. In *Proceedings of the ACM SIGCHI Conference on Human Factors in Computing Systems*, p. 189-194.

Sarkar, M. et al. (1993). Stretching the Rubber Sheet: A Metaphor for Viewing Large Layouts on Small Screens, *ACM Symposium on User Interface Software and Technology (UIST)*

Ziegler, J.; Kunz, C.; Botsch, V.; Schneeberger, J. (2002). Matrix Browser as a New Interactive Visualization for Large Networked Information Spaces, *Proceedings Information Visualization '02*, London, UK.

Methods for Estimating the Person's Busyness as Awareness Information in the Medium-sized Laboratory Environment

Itaru Kuramoto, Yu Shibuya, Tomonori Takeuchi and Yoshihiro Tsujino

Kyoto Institute of Technology
Goshokaido-cho, Matsugasaki, Sakyo-ku, Kyoto 606-8585 JAPAN
{kuramoto, shibuya, takeuchi, tsujino}@hit.dj.kit.ac.jp

Abstract

We propose the "busyness" as one of the awareness information for smooth communication. It indicates whether the user allows others to disturb him/her or not. His/her busyness is useful to avoid disturbing him/her and distracting his/her work eventually, and therefore a communication with him/her is smooth. We introduce the Busyness Estimation System for detecting the user's busyness automatically using his/her body activities. As the result of experiments, the proposed system is helpful to estimate "very busy" and "not busy" situation.

1 Introduction

The awareness information is important to collaborate with people. It includes a certain person's situation, like whether he/she is busy, where he/she is, and so on. It is important to know such situation in order to talk to him/her without disturbing his/her current activity. The goal of our current project is to construct a communication support system using the awareness information (Dourish & Bly, 1992; Reynolds & Picard, 2001).
In this paper, we try to make a smooth communication among the people with the awareness information. First, we propose a new measurement of the awareness information, "busyness". When a person knows the other's busyness, he/she can decide easily whether to start communication or not. Second, we introduce Busyness Estimation System (BES), which calculates user's busyness automatically using his/her use of PC and his/her body activities. In order to prevent the system from interfering his/her work, it needs no special operation for estimating his/her busyness (Gellersen & Beigl, 1999).

2 Busyness

2.1 Definition

One of the awareness information for a smooth communication is "busyness". The word "busyness" has rather ambiguous meanings, so in this paper we propose the level of the busyness which represents whether the person allows other people to disturb him/her or not. We classify the busyness into three categories. The level 1, "not busy state" is the lowest level of the busyness. It means that a person in this level is not working so hard and that he/she has time to talk with other people. The level 2, "busy state" is the middle level. It means that a person is working hard in this level. He/she does not want to be disturbed unless someone strongly wants to make a communication. The highest level 3, "very busy state" means that a person strongly denies any interruption from other people. He/she is too busy to talk with someone else.

2.2 Human Activities

In order to estimate the busyness of a person, his/her activities are detected and used. We classify the human activity into three categories in the viewpoint of the flow of information through a human. They are retrieving, thinking, and representing, as shown in Figure 1.
When a person is retrieving the information, the level of the busyness depends on the dynamics of the information that he/she is getting. If the information is fixed or static, like as a paper document, the person can pause his/her retrieving when someone requests to begin communication with him/her. In this case, after the communication, he/she can easily resume the retrieval because the same information remains there, and his/her busyness is not so high. On the other hand, if the information is dynamically changing, it might be difficult to resume the retrieval at the break point. In this case, he/she is estimated as very busy.
In the representation case, we consider that the person is very busy. If he/she pauses to represent some information in the middle of the task, it is very hard to resume his/her task again because it is very difficult to remember everything that he/she was thinking before pausing the task. Consequently, representations make a person very busy.
When a person is talking with other people, he/she is not only retrieving some information but also representing. As mentioned above, he/she is in "very busy state" (Whittaker et al., 1994).
When a person is thinking or concentrating about some issue, he/she is also in "very busy state". However, it is very difficult to detect whether he/she is thinking or not. In this case, he/she might have to show his/her state explicitly.
From above consideration, we treat two out of three categories, retrieving and representing some information. Thus, we try to estimate the busyness of the person by detecting his/her activities.

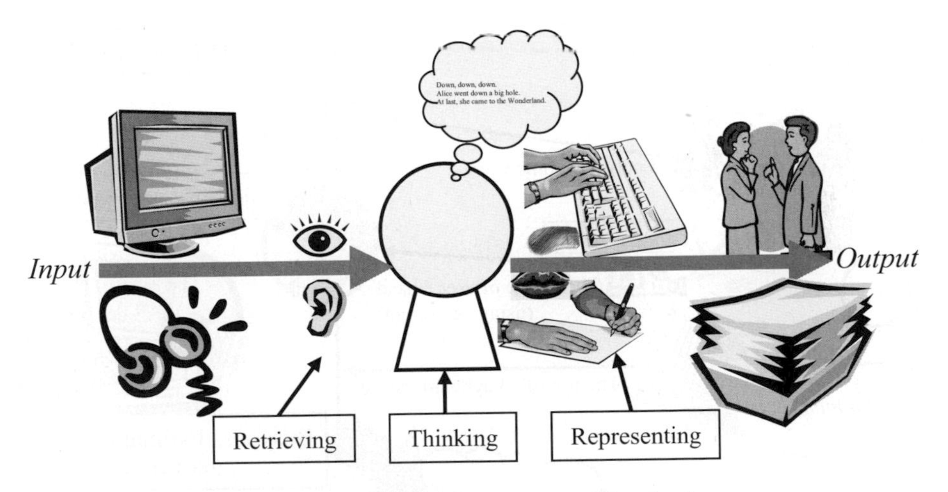

Figure 1: The flow of the information through a human

3 Busyness Estimation System

3.1 User Activity Elements

As we assume a user is in the laboratory environment, he/she is expected to work using his/her own PC and/or a pen. In order to estimate his/her busyness, it is a common idea to monitor the use of a keyboard and a mouse. When he/she is using the keyboard, we consider he/she is representing something like making documents by a word processor. Using the mouse means either representing or retrieving. For example, scrolling a text area is one of the retrieving activities using it. The use of the mouse is less strongly concerned with the human activities than other elements. In addition, we add two more elements for his/her busyness calculation. One is the use of a pen as one of body activities. He/she is representing information or thinking deeply when he/she hands a pen and uses it. That is, he/she is in the "very busy state". The other is person's conversation. Talking with other people means not only retrieving information but also representing it. It is considered that he/she is in the "very busy state". We summarize the human activity elements for detecting busyness in Table 1.

Table 1: The collection of the human activity elements

	Keyboard	Mouse	Pen	Conversation
Retrieval	-	**C** (static)	-	**SC** (dynamic)
Representation	**SC**	**C**	**SC**	**SC**

(**SC**: Strongly Concerned, **C**: Concerned)

3.2 System Overview

Figure 2 shows the overview of BES. It includes two subsystems. One is the Sound Detection Subsystem (SDS), which detects the user's conversation. It uses a microphone connected to his/her PC for getting sounds. The other is the Video Processing Subsystem (VPS), which detects the user's use of his/her pen. A small video camera is connected to the PC for getting his/her use of the pen from the side of him/her. Using them, Busyness Estimation Algorithm running on his/her PC calculates his/her busyness.

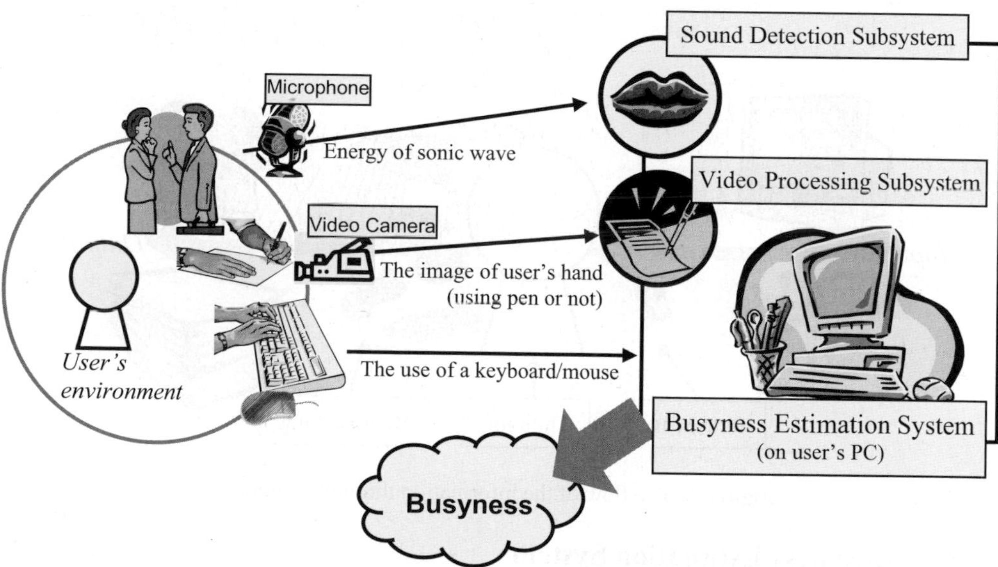

Figure 2: The Overview of the Busyness Estimation System

3.3 Busyness Estimation Algorithm

We introduce Busyness Estimation Algorithm for estimating the busyness from the awareness information. Figure 3 shows the flow of it.

First, it judges whether the person is talking or not. When a microphone gains the sounds, his/her busyness goes up to the level 3. The SDS outputs only whether he/she is talking or not.

It does not collect the contents of the conversation because of prevention of the privacy problem. The SDS judges he/she is talking not only when the microphone gets sounds but also while he/she pauses talking for a short while. The threshold of pausing is 0.5 seconds.
Second, when a conversation is not detected, the algorithm evaluates the use of a pen. If he/she uses a pen, his/her busyness level increases to 3. In order to recognize his/her use of the pen, we put a color mark on a tail of the pen. The VPS decides his/her use of the pen in case of the color mark is on the upper half of the captured image. Figure 4 shows the image. This method can be implemented with only a common video camera and an ordinary pen.
Third, when the use of a pen is not detected either, the algorithm calculates the busyness by the amount of keystrokes and/or the distance of mouse movements. There are two thresholds of the busyness level, the threshold of the level 2 (T_2) and that of the level 3 (T_3). When the user types more keys than T_2 in the level 1, the level goes up to 2. However, When he/she in the level 3 stops typing and the number of keystrokes falls less than T_2, the level does not falls to 1 directly, but goes down to 2 because it is assumed that the user's busyness is not going down drastically. Figure 5 shows the state transition diagram of the busyness for using a keyboard/mouse.

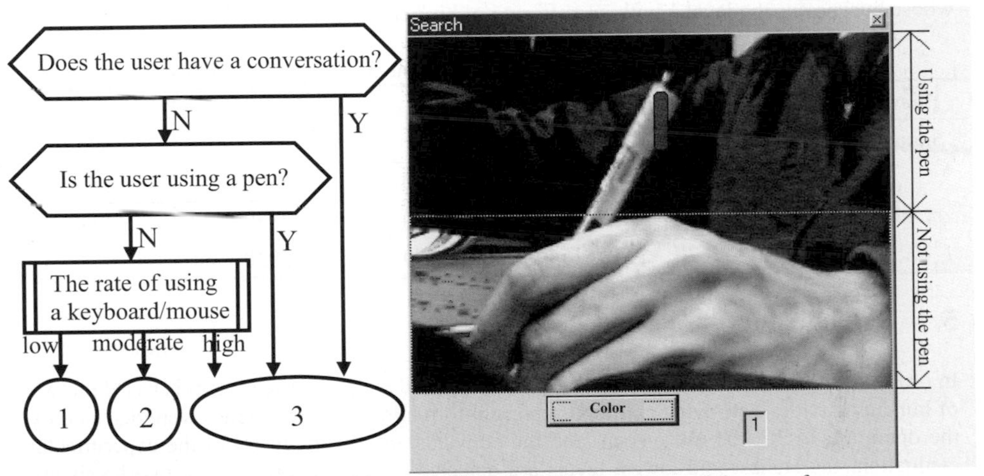

Figure 3: The flow of the algorithm **Figure 4:** The detection of the use of a pen

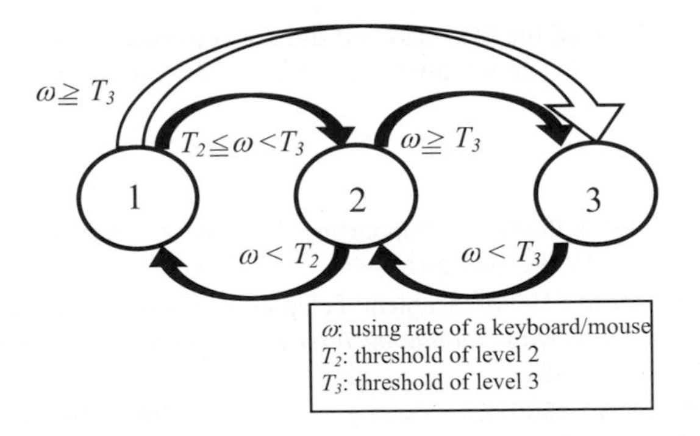

Figure 5: The state transition diagram of the busyness from a keyboard or a mouse

4 Evaluation

Experimental evaluation was held to find out whether the proposed system could detect user's busyness correctly. There were three subjects, who are the members of our laboratory. We implemented the proposed system on their PCs and recorded their busyness. Furthermore, their activity was video-recorded for the later analysis. During the experiment, the subjects performed their ordinary works (using PCs) and an additional task of annotating to some documents (using their pens). After two hours' experiment, subjects were asked to remind their busyness with the recorded video in order to verify whether their subjective busyness fit the one our system provided.
Table 2 shows the result of the estimation of the busyness. The correct estimation rate of the busyness was 45% for level 3, 29% for level 2, and 63% for level 1. The level 2 is the most difficult to detect but it is not so serious problem. People are usually interested in making a communication with others without disturbing them. So, it is important to know the good timing to make a communication, that is, level 1, or to know the non-acceptable duration, that is, level 3. Our busyness estimation system will helpful to detect such important levels. In addition, the level 1 is more important to start conversation because we can talk with a person whose busyness is the level 1 without suffering of his/her distraction.

Table 2: The result of the estimation of the busyness (by second)

Busyness Level	System Estimation Level			Summation	Correction Ratio (%)
Subjective Level	3	2	1		
3	**3360**	2350	1790	7500	44.8
2	1860	**2140**	3270	7270	29.4
1	320	1540	**3170**	5030	63.0

5 Conclusion

In order to estimate the busyness of people, we proposed the BES. The system is based on two of human activities, retrieving and representing information. These activities were detected by the operation of PC, the use of pen, and the conversation. As the result of the experimental evaluation, the highest and lowest levels are detected rather correctly. These two levels are helpful enough to make a smooth communication, especially level 1 in the case of beginning it.
In this paper, we only detected the busyness and did not consider the presenting method of it to another people. The busyness presentation system is also needed for the practical use.

6 References

Dourish, P. & Bly, S. (1992). Portholes: Supporting Awareness in a Distributed Work Group, *Proceedings of ACM CHI '92*, 541-547.

Gellersen, H-W. & Beigl, M (1999). Ambient Telepresence: Colleague Awareness in Smart Environment, *Springer Verlag: Managing Interactions in Smart Environments, P.Nixon, G.Lacey, S.Dobson ed.* 80-88 .

Reynolds, C. & Picard, R.W. (2001). Designing for Affective Interactions, *Proceedings from the 9th International Conference on Human-Computer Interaction* 499-503.

Whittaker, S., Frohlich, D. & Daly-Jones, O. (1994). Informal Workspace Communication: What Is it Like And How Might We Support It, *Proceedings of ACM CHI '94*, 131-137.

A Design that Meets User's Goals Creates Usable Security

Antti Latva-Koivisto and Yki Kortesniemi

Helsinki Institute for Information Technology
P.O.Box 9800, 02015 HUT
Antti.Latva-Koivisto@hut.fi, Yki.Kortesniemi@hiit.fi

Abstract

Poor usability of security-related software has received attention lately. Some have claimed that usability and security are inherently incompatible, others that security is such a demanding field that a new area of usability, usability of security is required. We beg to differ. Based on our analysis of existing work, we suggest that the key missing point in the field is focus on user's goals and the link from goals to system and interface design.

1 Introduction

The increase in viruses and other threats has forced significant security-motivated changes on email and other familiar technologies. The results, however, have been anything but usable (Whitten & Tygar, 1999). Usability of a system means the extent to which users can achieve their goals with it effectively, efficiently and with satisfaction (ISO, 1998). It can therefore be measured by running a usability test where the assignments given to the test users correspond very closely to users' actual goals. A goal-based test assignment may tell the user that she is an executive salesperson at SAP, which is taking part in a bid contest to sell a large CRM software system to Sony. She needs to send her highly confidential offer to Ms. Kathryn Taylor at kathryn.taylor @sony.com by Friday noon. A regrettably common kind of test task such as "send an email offer encrypted with your private key to unrealistic@email.address.com" does not correspond to any user's goal and, hence, cannot tell us anything relevant about the system's usability.

Whitten and Tygar (1999) ran a usability test on Eudora email client and PGP 5.0, where the test assignments were fairly well based on users' actual goals, and did not contain feature-based tasks, such as encryption or signing. The results were drastic: few users managed to complete the assignments, and hence the system was unable to meet even the basic usability requirement, let alone be efficient or provide satisfaction. In this paper we introduce the design rationale of goal-derived design (Laakso 2003) in the context of security software. The interface design is derived from prioritized users' goals, and the interface design is finished before any technical design or implementation begins. We then reflect on previous work in the light of goal-derived design and give examples on the radical implications of applying goal-derived design to security software.

2 Goal-Derived Design

Since a system's usability can be verified by testing it with user's goals, it should be obvious that we can get the best results if we consider those same goals from the very beginning of system design. The system design must hence start with uncovering the goals and designing the user interface. Here, we conform to Raskin's (2000) and Cooper's (1999) view. If the design is not based on goals, it should not come as a surprise if at that the verification stage the users are not able to achieve their goals using the system.

Goals answer the question why a user is doing something. Ideally, the interface design should be *derived* from the goals. To do that, we need to enrich the goals with contextual data describing the user, the user's decision criteria and other data that is obtained from user observations and contextual interviews (Beyer & Holtzblatt, 1998). What should be noted is that users' current tasks are not goals (Cooper, 1995). For example, Whitten and Tygar (1999) seem to accept that the current tasks of encrypting, decrypting, publishing public keys, signing and verifying signatures are necessary for the user to carry out. However, performing those tasks will not advance the user directly toward the goals. Those public key operations are technology, and using any technology is no user's goal. Motivating users to learn or understand that technology is always the wrong solution, since it does not match their goals. From goal-based point of view, this is akin to claiming that the solution to creating usable cars is to motivate the users to acquaint themselves with the intricacies of ignition.

To do goal-derived design, we must prioritise goals and concentrate only on a small number of important and frequently occurring goals. If all goals – common and rare – get the same treatment, everything will be equally easy to use, i.e. equally difficult, and common goals will be far too difficult to reach. It's better to concentrate on few common goals and make achieving them exceedingly straightforward. It is acceptable to have infrequently occurring goals harder to reach. Designing, for instance, an email software on the basis of just about five different but typical goals should be quite sufficient to produce a very effective, efficient and deeply satisfactory system.

This goal-based approach ensures that the software designers are answering the right question. Only then can they decide on how to answer the question. Technology-based security leads to a situation where wrong questions are answered in a well, sophisticated manner. Check-list-based usability, similarly, leads to beautiful, consistent and standard-adhering systems, such as PGP 5.0, that solve irrelevant problems, such as how to motivate the user to use the software in a secure manner or how to present public and private keys in the user interface.

2.1 Designing Secure Software

From the point of view of interface design, the objective of the system's designers is to allow the user to reach his goals with the system in as direct and straightforward manner as possible. To achieve this, the designer must carry the responsibility of the security of the whole system. Security, however, brings about an unpleasant temptation: the easy way to allow the user to achieve his goals and complete his job is to give him all possible permissions, which, however, brings about severe security problems. Saltzer's and Schroeder's (1975) security principle of *least privilege* (programs and users should operate using the least set of privileges necessary to complete the job) brings about the responsibility of the designers: they must work out the user's goals in order to allow him to always achieve goals directly, without giving him or his programs too many privileges.

The designer also must not dump setting up the security to the user. It does not help if the user "can" configure the system to match his needs - it's laziness on the designer's behalf. The end-user's goal is never to configure the system to make it secure. Certificates, signatures, public and private keys should be like electricity: the user does not need to see it or understand it in order to benefit from it and use it. Whitten (1999) claims that we cannot know what the user wants, which, however, is false. We can and we should. We can quite certainly know in advance that the user will *not* want anyone else to be able to read the confidential offer she sent Ms. Kathryn Taylor.

2.2 Method Description

The collection of these goals takes place in the form of user observations and contextual interviews. The raw data of these methods produces scenarios, from which goals need to be extracted by figuring out why the users do what they currently do. If scenarios as such are used to drive the design, we must acknowledge the fact that the scenario already imposes some sequence of actions and fixes current tasks that could be automated, and hence influences the design with adverse effects. We need to ask "why" enough many times to truly get outside the system we are designing. In case of email software, we need to get concrete situations in which email is used or could be used. "Sending an email" is no-ones actual goal, "arranging a date with a lover" is. To drive the design, we need to supplement the goal with a concrete description of the situation at the time the goal becomes active. Such descriptions are goal-based use cases.

Once we know a set of goal-based use cases, we need to prioritize them. The most typical or the market-wise most important goal is used to draft a design that supports this one and only goal as directly as possible. No compromises need to be made for the first goal – only few features are needed to support it, and only they should be included in the design. Everything should easily fit on one main screen. Then we take the second-most important goal and supplement the design for the first goal to support also the second one, again as directly as possible, with a minimum number of steps. Some compromise that makes the first goal slightly more awkward to achieve might be necessary, but usually not. Gradually, more goals are introduced into the design, but all interface elements are kept on one main window to minimize navigation or "excise" tasks (Cooper, 1995). Eventually, compromises must be made such that the low-priority goals are more difficult to achieve than the high-priority goals. Support for them is then moved to some secondary screens or tabs, instead of the main screen. The design is constantly simulated to ascertain that the support for the top-priority goals stays very direct. (Laakso, 2003)

At all times, the features should support the goals as directly as possible and there must be no features that do not correspond with any goal. If the designers suspect that some feature is necessary, the goal for which the feature is needed must be uncovered and prioritized. Only then can the designers decide what the best way to support that goal is, taking into account the priority of the goal. The result of this interface design process is a user interface that is very simple and yet very effective, and which deeply satisfies the user's needs. The system might not support all *possible* goals, but then again, it should not. A system that addresses all possible goals cannot support the most important ones in any significant way, and hence lacks the potential for success.

All the steps the user has to take need to have productive results from the user's point of view. There must be no unnecessary "excise tasks" just to satisfy the needs of the software (Cooper, 1995). No technology-driven features are presented to the user, such as "saving a file" (Raskin, 2000). Coming back to our email example, it follows from the goal-based approach that the interface must not contain a feature labelled "encrypting an email" or "sign an email using a private key". All such security-related must be automated, as already Saltzer and Schroeder (1975) noted. At this point, the technologist will out-cry that the email software cannot automate this since the necessary PKI infrastructure is missing. Quite correct, but from the user's point of view, the system, as a whole, must be designed to meet the users' goals and the goal-derived design will then also have an effect on the design of email-architecture and email standards.

3 Analysis of Related Work

A big problem in usability of security is that from a goal perspective, wrong questions are answered. Cranor (2003) has designed an interface for a Platform for Privacy Preferences (P3P) user agent, using which a user can configure options regarding privacy preferences. However, no user's goal is *configuring preferences*. No organisation of configuration options can come even close to the usability of a goal-derived solution where configuration is never needed. Answering wrong question is the result of not appreciating user's goals and starting the interface design too late. In the case of P3P, interface design cannot fix a flawed specification afterwards. Goal-derived design needs to take place early: already during the development of standards and specifications. Apparently, P3P was developed with little regard to user's actual goals, work flows, or even the way that users would actually see and configure said privacy preferences.

A popular approach to improving usability of security-related software includes teaching the users the technology and the tasks that spring from it to let the users achieve their goals. According to Whitten and Tygar (1999), a criterion for usable security software is that users are reliably aware of the security tasks they need to perform. With a goal-derived interface, the users would not need to perform any such tasks. Instead, all the steps they need to take would be directly related to their goal, so there would not be any extra tasks they should be aware of. The writers also say that security software should communicate an accurate conceptual model of the security to the user quickly. However, a better solution for the user is if such conceptual model is not needed. Or perhaps just a very simply model: "This software is secure." In their more recent work, Whitten and Tygar (2003) teach the users security mechanisms through a technique of "safe staging". Although they demonstrate success in teaching the users, their example, key certification, is just another feature of a secure email system that should be completely automated for a vast majority of users, since key certification does not match any end-user's goal.

Yee (2002) has identified several principles necessary for designing interfaces for secure systems. The first principle, Path of Least Resistance, corresponds well to letting the user achieve a goal as directly as possible and at the same time, securely. Problems with Yee's principles are that goals are not explicitly mentioned even though they should be used to generate the design. The principles, on the other hand, are not generative. Rather, they present criteria to a solution but don't help much in creating it, similarly to many other check-list usability heuristics (e.g. Nielsen, 1994). Other principles, such as revocability and expected ability, are just new wordings for familiar interface design principles, namely, making all user's actions reversible (Cooper, 1995) or preventing errors.

Brown and Snow (1999) have taken the right approach of automating the encryption and signing of all email by building a proxy system that creates an automated security layer under existing email software. Their shortcoming is that security and usability cannot be added to a system afterwards. Both must be accounted for by the designer at the requirements analysis and design phase. Nevertheless, their approach shows that automation will be feasible if it is designed in and the email system is built from the ground up.

The most recent approaches to improving usability of secure software are demonstrating similar ideas to the one in this paper. According to Smith (2003), building PKI on top of existing systems cannot be done successfully, but rather applications should be rethought from the ground up and cryptography should be integrated into them. Similarly, Grinter and Smetters (2003) assert that better interfaces on top of existing systems won't work, and instead, we need to start from the user. Rethinking should indeed take place on the foundation of the goals. However, both Smith and Grinter and Smetters also suggests that users should be made aware of the use of cryptography, or that they should be fed back information about the security state of the system, thus unnecessarily imposing underlying implementation model to the user.

4 Conclusions

The usability of secure software stems from the same principles as all usability: fulfilling the users' goals. In this paper we have discussed a general design method that derives the user interface from goals and, thus, helps us answer the right questions and produce imminently more usable systems.

References

Beyer, H. and Holtzblatt, K. (1998). Contextual Design. Defining Customer-Centered Systems. San Francisco, CA: Morgan Kaufman Publishers.

Brown, I. and Snow, C.R. (1999). A proxy approach to e-mail security. Software – Practice and Experience, 29(12) 1049-1060, October 1999.

Cooper, A. (1995). About Face. The Essentials of User Interface Design. Foster City, CA: IDG Books.

Cooper, A. (1999). The Inmates Are Running the Asylum. Indianapolis, IN: Sams Publishing.

Cranor, L. (2003). Designing a Privacy Preference Specification Interface: A Case Study. Workshop on Human-Computer Interaction and Security Systems, CHI 2003, April 5-10, 2003, Fort Lauderdale, Florida. Retrieved February 15, 2003 from http://www.andrewpatrick.ca/CHI2003/HCISEC/index.html

Grinter, R.E. and Smetters, D.K. (2003). Three Challenges for Embedding Security into Applications. Workshop on Human-Computer Interaction and Security Systems, CHI 2003, April 5-10, 2003, Fort Lauderdale, Florida. Retrieved February 15, 2003 from http://www.andrewpatrick.ca/CHI2003/HCISEC/index.html

ISO (1998). ISO 9241-11:1998 Ergonomic requirements for office work with visual display terminals (VDTs) - Part 11: Guidance on usability.

Laakso, S.A. and Laakso, K.-P. (2003). Hyvän käyttöliittymän varmistaminen GUIDe-projektimallilla. University of Helsinki, Department of Computer Science and Interacta Design Oy. Unpublished article. February 12, 2003. Retrieved February 15, 2003, from http://www.cs.helsinki.fi/u/salaakso/papers/GUIDe.html

Nielsen, J. (1994). Enhancing the explanatory power of usability heuristics. Proceedings of the CHI '94 Conference, ACM Press, 1994, pp. 152-158.

Raskin, J. (2000). The Humane Interface. Addison-Wesley.

Saltzer, J. H. and Schroeder, M. D. (1975):. The Protection of Information in Computer Systems. In Proceedings of the IEEE, vol. 63, no. 9, September 1975, pp. 1278-1308. Retrieved February 15, 2003 from http://web.mit.edu/Saltzer/www/publications/protection/

Smith, S. (2003). Position Paper: Effective PKI Requires Effective HCI. Workshop on Human-Computer Interaction and Security Systems, CHI 2003, April 5-10, 2003, Fort Lauderdale, Florida. Retrieved February 15, 2003 from http://www.andrewpatrick.ca/CHI2003/HCISEC/index.html

Whitten, A. and Tygar, J.D. (1999). Why Johnny can't encrypt: A usability evaluation of PGP 5.0. Proceedings of the 8th USENIX Security Symposium, August 1999, Washington.

Whitten, A. and Tygar, J.D. (2003). Safe Staging for Computer Security. Workshop on Human-Computer Interaction and Security Systems, CHI 2003, April 5-10, 2003, Fort Lauderdale, Florida. Retrieved February 15, 2003 from http://www.andrewpatrick.ca/CHI2003/HCISEC/index.html

Yee, K.-P. (2002). User Interaction Design for Secure Systems. In Proceedings of the 4th International Conference on Information and Communications Security, Singapore, 2002.

Integrating and Evolving a Mob: The Growth of a Smart Mob into a Wireless Community of Practice

John Lester

Massachusetts General Hospital
Harvard Medical School
Boston, MA USA
JL@Tmail.com

1 Introduction

The concepts of "smart mobs" and "communities of practice" are of growing interest in social science and computer science (1, 2). With the introduction of communication tools such as wireless data communication devices and sophisticated cellular telephones, individuals are able to be extremely mobile yet stay constantly in touch with each other. These technologies are enhancing and creating new kinds of bonds between human beings. Patterns of smart mob behavior have been observed in various cultures, particularly Japan, and these patterns are beginning to be seen in the United States and Europe (2).

This paper examines the successful evolution of a smart mob into a wireless community of practice. It begins with an examination of a popular wireless blogging website "Hiptop Nation" (http://hiptop.bedope.com). Hiptop Nation acts as a central blogging site for owners of the T-Mobile Sidekick device, a wireless handheld data communications device recently introduced by Danger (http://danger.com). See Figure 1. The Sidekick supports wireless AOL Instant Messaging, email, SMS text messages, and web access. Users of the Sidekick can post wireless public blogs on Hiptop Nation via their Sidekick device and upload photographs from the Sidekick's digital camera.

On Halloween, October 31 2002, Hiptop Nation sponsored a photo-scavenger hunt competition across the US. Participants were users of the Hiptop Nation blog site who were placed into competing teams, coordinated their actions as well as acquiring and uploading photographs from across the US exclusively via their Sidekick devices. The hunt lasted for 24 hours. Participants in the hunt moved from behaving like a classic smart mob to behaving like a highly focused community of practice. This paper will review some of the strategies and tools observed in this particular evolution, and propose some ideas about the phenomenon of smart mobs growing or "crystallizing" into mobile communities of practice.

Figure 1: The T-Mobile Sidekick device

2 Hiptop Nation: Common Goals and Experiences

The Hiptop Nation website was created by Mike Popovic, a Webmaster in southern Maine, soon after the introduction of the T-Mobile Sidekick device last fall. While similar to other communal blog sites such as Metafillter (http://www.metafilter.com), Hiptop Nation is unique in a couple important ways. It is a communal blog website designed to be used exclusively by owners of the T-Mobile Sidekick device. Furthermore, the primary focus of Hiptop Nation is to share photographs taken with the Sidekick. Because of these factors, Hiptop Nation has a user base of individuals who all share some basic common goals and experiences. Namely, they all use Sidekicks and like to post interesting photographs with commentaries while being mobile.

Two weeks before Halloween, Mr. Popovic announced that Hiptop Nation would be holding a "Halloween Photo-Scavenger Hunt" (http://hiptop.bedope.com/halloween.html). The rules were similar to those of conventional scavenger hunts. Five teams of participants with eleven members each were created, and each team was given the ability to post to a team blog. The author of this paper was a member of "Team Raven" (each team had Halloween-themed names), and the basic goal of the hunt was for each competing team to capture photos of identified scavenger items using their Sidekick devices. These photos would then be uploaded to the participant's team blog.

On the morning of Halloween, Mr. Popovic posted the list of scavenger items along with the number of points to be awarded for each item (http://hiptop.bedope.com/hunt/the_list.html). At that point, the URLs of the various team blogs were publically posted on Hiptop Nation so that everyone on the Internet could monitor the progress of the 5 teams in real time. At the end of the hunt, which lasted for the 24 hrs of Halloween day, Team Raven was identified as having the most points and declared the winner.

3 Moving from Smart Mob to Mobile Community of Practice

The members of Team Raven had all known each other before the scavenger hunt and were all regular contributors to Hiptop Nation. However, their methods of interpersonal communication and organizational strategies changed significantly immediately before and during the scavenger hunt.

Previous to the scavenger hunt, team members posted to Hiptop Nation in an informal and friendly, communal fashion. Sharing interesting photographs and commenting on each other's posts, team members often referred to Hiptop Nation as "their home" where they could be creative and work together to create a website that was an interesting, poetic collection of images and thoughts from many different people's daily lives.

During the scavenger hunt, members of Team Raven underwent a rapid and smooth transformation. Team members were still friendly with each other, but participants quickly realized the need for specialization and organization. A week before the hunt began, a Team Captain was immediately elected. A "Team Member Availability" spreadsheet was created that outlined the availability and geographic location of each team member during the day of the hunt, and another "Found/Outstanding Items" spreadsheet keeping track of which hunt items were found and which were still outstanding was created the day of the hunt. A MUD (multi-user dungeon) was created to allow group chats, and an interface to AOL Instant Messaging was created to allow team members in the field to communicate using the AIM client on their Sidekicks. Team members who were unavailable to be in the field "hunting" for items (typically because they had to be at work sitting at their desks) were put on "triage" detail, actively monitoring the MUD and updating the web-based "Found/Outstanding Items" spreadsheet listing found/outstanding hunt items. An additional HTML-based chatroom was created as a backup in case the MUD/AIM interface failed.

During the hunt, Team Raven naturally established a workflow somewhat similar to that of a very efficient taxi company. Team members in the field used the web browser on the Sidekick to look at the "Found/Outstanding Items" list. Then they used AIM or email to communicate with team members who were on triage detail and "bid" on items. Once bid on, an item was marked in the "Found/Outstanding Items" spreadsheet as "pending" to avoid wasted duplicate efforts (e.g., multiple people trying to take photos of the same thing at the same time). If multiple people bid on an item, the person who was geographically closest was given the bid in order to guarantee that whoever could find and take a picture of an item in the shortest time always had the opportunity to do so. At one point, the AIM/MUD interface crashed, and the team quickly moved to the HTML-based chatroom as a backup. At no point did the communication system completely break down.

The members of Team Raven, as individuals using Hiptop Nation, initially behaved like a classic smart mob. Smart mobs naturally emerge when communication and computing technologies amplify human talents for cooperation. However, smart mobs typically lack specific, complex goals, and they do not typically grow into more sophisticated communities of practice. Why and how did Team Raven evolve so quickly and easily from a smart mob into a mobile community of practice?

The answer lies in the fact that Team Raven was formed by members who participated in a successful communal blog. By being active members of this communal blog, they had three crucial elements that a successful community of practice needs. Namely, the establishment of social capital (3), the creation of multiple weak ties between individuals (4), and a strong sense of place that the Hiptop Nation gave to all its members (5). Social capital is critical in creating trust between people, an essential element of collaboration. Particularly important is the fact that weak ties have been shown to be essential resources for new ideas and innovative thinking (4), and Team Raven could not have succeeded without a great deal of creative thinking and planning. The final and fourth element was the introduction of a specific and challenging goal, the scavenger hunt itself. These four elements came together to create a group with all the aspects of true community of practice (6, 7, 8).

Team Raven created a very goal-oriented community of practice because all of the social and technological elements were eminently available. After the hunt was over, the members of Team Raven returned to using Hiptop Nation as before, and the mobile community of practice dissolved. However, there was a good amount background chatter about "how much we'd love to do something like that again" and "wow, I feel like our team could accomplish anything!" It seemed that Team Raven was (and still is to this day) primed to reform a mobile community of practice. All that is needed is a new complex goal.

4 Conclusion

Based on experiences with Hiptop Nation, it appears that by having ubiquitous mobile data communication devices and a successful communal blog, it is possible to create an ideal environment within which a smart mob can grow into a goal-oriented mobile community of practice. Communal blogs play a critical role in the creation of three essential elements of community: the establishment of social capital, the creation of weak ties that foster creativity, and the formation of a sense of "place" within which everything can happen (3, 4, 5). The final crucial ingredient is 4) a complex goal.

In these special circumstances, a smart mob can not only quickly change into a mobile community of practice, but once its goals have been achieved it can just as quickly "dissolve" back into a smart mob. This is metaphorically similar to the way certain liquid solutions can quickly crystallize, dissolve back into liquid, and then recrystallize based on external influences. By adding an external influence, namely a specific shared goal, one can "precipitate" the crystallization of these smart mobs into powerful mobile communities of practice. After the goals have been achieved, during which participants have gained expertise in their particular domains, the group can dissolve back into a smart mob and be ready to rapidly recrystallize whenever new goals are introduced.

The increasing popularity of communal blogs, coupled with more sophisticated ubiquitous mobile communication devices (9, 10, 11), will most likely make this interesting social phenomenon more common in the future. A future opportunity will be the deliberate cultivation of this phenomenon, as it has the ability to create incredibly effective and creative goal-oriented teams of mobile individuals.

References

1. Wenger E, McDermott R, and Snyder W. Cultivating Communities of Practice: A Guide to Managing Knowledge. Harvard Business School Press, Boston, MA, 2002. p. 4-9.

2. Rheingold H. Smart Mobs: The Next Social Revolution. Perseus Publishing, Cambridge, MA, 2002.

3. Preece J. Supporting community and building social capital. Communications of the ACM, April 2002, Vol. 45, No. 4, p 37-39.3.

4. Teigland R, Wasko MM. Creative ties and ties that bind: examining the impact of weak ties on individual performance. Proceedings of the twenty first international conference on Information systems, Brisbane, Queensland, Australia, 2000, p. 313-328.

5. Harrison S, Dourish P. Re-Place-ing Space: The role of Place and Space in Collaborative Systems. Proceedings of the Conference on Computer Supported Cooperative Work, September 17-20, Boston, Massachusetts, 1996, ACM Press, p. 67-76.

6. Brown JS, Duguid P. Organizational Learning and Communities of Practice: Toward a Unified View of Working, Learning, and Innovation. Organization Science, 2:2, 2992, 40-57.

7. Wenger E. Communities of Practice: Learning, Meaning and Identity. Cambridge University Press, Cambridge, England, 1998.

8. Lave J, Wenger E. Situated Learning: legitimate Peripheral Participation. Cambridge University Press, Cambridge, UK, 1991.

9. Luff P, Heath C. Mobility in Collaboration. Proceedings of ACM Conference on Computer Supported Cooperative Work, Seattle, Washington, November 14-18, 1998, ACM Press, p. 305-314.

10. Palen L, Salzman M, Youngs Ed. Going wireless: behavior and practice of new mobile phone users. Proceedings of ACM Conference on Computer Supported Cooperative Work, Philadelphia, Pennsylvania, 2000, ACM Press, p. 201-210.

11. Vanaanen-Vainio-Mattila K, Ruuska S. User needs for mobile communication devices. First Workshop on HCI for Mobile Devices, Glasgow, Scotland, May, 1998.

Fruitful Collaborations: Integrating Research and Practice

Michael D. Levi

U.S. Bureau of Labor Statistics
2 Massachusetts Ave., NE
Washington, DC 20212
USA
levi_m@bls.gov

Gary Marchionini

University of North Carolina
School of Information and Library Science
Chapel Hill, NC 27599
USA
march@ils.unc.edu

Abstract

Successful partnerships enrich one another and evolve together in unexpected directions. Such has been the case during a six year long collaboration between information and computer scientists from a number of U.S. universities who have been working with staff from the U.S. Bureau of Labor Statistics and other U.S. Federal statistical agencies.

This paper treats the collaboration as a case study. It sets the context by summarizing specific ways that academic and government participants worked together toward their individual and shared goals, enumerates the concrete benefits each side gained, discusses some of the ongoing tensions that surfaced during the work, and ends with the authors' reflections on the factors that led to their mutual satisfaction.

1 Introduction

There is a gulf between research and practice. This is as true in the world of Human-Computer Interaction as it is in all other technological environments. Academics are interested in developing and validating theoretical models that have broad applicability. Project managers need to get high-quality systems out the door as quickly and cheaply as possible. Academics are rewarded if they publish papers in peer-reviewed journals. Project managers are rewarded if their systems increase profits or expand the customer base. Academics are primarily concerned with the abstract. Project managers focus on the concrete.

These are natural, healthy, and inevitable differences in perspective which can illuminate one another. All too frequently, however, efforts to transfer insights from the lab to the cubicle or *vice versa* end in frustration. Either the business needs of the practitioner community are not adequately met or the scholarly interests of the researchers are short-changed.

By contrast, successful partnerships enrich one another and evolve together in unexpected directions. This has been true for the following case study: a six year long collaboration between information and computer scientists from a number of U.S. universities who have been working with staff from the U.S. Bureau of Labor Statistics and other Federal statistical agencies to broaden public access to government statistics.

2 Context

In 1996, Gary Marchionini and Carol Hert, two academic information scientists, began working with the U.S. Bureau of Labor Statistics (BLS) to improve BLS outreach to constituent groups by means of the World Wide Web. Over the ensuing six years other researchers and Federal statistical agencies joined the project[1].

Early on the researchers and Federal staff agreed on three general goals:

- To gain a better understanding of how non-specialists think about, access, and use statistical data.
- To investigate and document how federal statistical agencies can adopt and adapt technologies to better serve the needs of diverse constituencies.
- To develop effective user interfaces tailored to this task domain and user community.

Researchers interviewed agency staff, intermediaries, and end users; analyzed Web transaction logs; content-analyzed samples of e-mail sent to the BLS help desk; collected and categorized reports on BLS statistics from the popular press; and prototyped and tested a variety of novel user interfaces.

Out of these efforts the researchers created user and task taxonomies, methodologics for assessing end-user behavior, transaction log techniques that included path analysis, an email content coding scheme, interface guidelines and recommendations, linguistic crosswalks from general usage to the specialized vocabulary employed by BLS economists, and a number of prototype applications including a "relation browser" (Brunk & Marchionini, 2001) and an electronic “table browser” (Marchionini & Mu, in press.)

BLS, in turn, reworked its Web site to adopt an information-intensive design, created a new glossary and index for the site, reviewed and changed much of the vocabulary employed, and is beginning to develop and deploy the infrastructure required to support electronic tables. BLS staff organized two ACM CHI workshops (Levi & Conrad, 1997; Levi & Conrad, 1999) and co-authored several papers with the university team members.

3 Research Benefits

The researchers found that their involvement in these various projects has been quite rewarding. Some of the specific benefits to the researchers included:

- Exposure to a new and challenging task domain in government statistics that allowed them to test ideas and forced them to think beyond text and multimedia contexts to consider statistics as a highly compressed medium in its own right.
- Access to broad and deep primary data sets; the BLS Web site alone has tens of thousands of pages and provides ad-hoc access to a database of more than 90 million observations.
- Access to a substantial base of usage data: Web logs, search terms, and help desk inquiries.

---

[1] There have been a number of reports and papers detailing results from this collaboration. Many of them are available at http://www.ils.unc.edu/govstat/bls.html

- Long-term observations of changing organizational behavior that could be documented from multiple sources of evidence ranging from interviews with staff over time to changes in public usage patterns in the Web logs.
- A test bed to validate theories of interface design.
- Funding to support researcher time and graduate student support.

4 Government Benefits

Federal staff at BLS found their involvement equally rewarding. Among the concrete benefits accruing to BLS were:

- Specific work products. Many of the insights from reports, design recommendations, and system prototypes have found their way directly into the BLS public Web site and are a core driver of ongoing work.
- HCI Education. Though the principles of Human-Computer Interaction were familiar to BLS analysts prior to this collaboration, the partnership gave Federal staff a deeper and more firmly established understanding of usability engineering as an ongoing component of software development.
- Inspiration. The importance of motivation can not be overemphasized. Close collaboration with academically-oriented researchers gave BLS developers the impetus for imaginative thinking and the courage to take creative risks.
- New information useful in internal politics. One unexpected outcome of the collaborative work was that empirical research results proved surprisingly helpful in marketing changes to higher levels of management and overcoming organizational roadblocks.

5 Inherent Tensions

The past six years have not been entirely conflict-free. Several important areas of tension have recurred in various forms.

The first area is one of focus. Researchers have a specific interest and mandate to explore or derive broad principles and subsequently to rigorously validate those theories. Their work with BLS is only one piece of a professional life filled with teaching duties and other research projects. BLS developers, for their part, are often frustrated by general principles which appear too broad and open to interpretation to give immediate guidance on a particular sticky point. Developers need to make specific systems work effectively and efficiently. Usability is only one of many competing demands on staff resources and project budgets.

The second area in which tensions arose is that of timing. Researchers often work within the framework of academic calendars. Graduate assistants, in particular, may need to finish certain projects within semester boundaries. Researchers would like to see their insights bear visible fruit as soon as results are available. Federal staff, by contrast, often work within a slow-moving culture which has intermittent short windows for systemic change. Thus research results may come too early or too late to be put to immediate use.

Finally, there is the issue of prototypes *vs.* working systems. Researchers build operational prototypes, are justly proud of them, and want to see their labor put to good use. Federal staff, however, generally are looking for ideas rather than code. Concerned with long-term maintenance and adherence to organizational standards, BLS developers have been reluctant to take on a code base that was not created in-house and may not fit organizational policies and practices.

6 Factors Leading to Success

Despite the conflicts listed above, all parties in this collaboration agree that our joint efforts have been hugely valuable. The authors have identified the following factors as being critical to such an outcome:

Recognize that the interaction works both ways. Though most writing on technology transfer assumes that it is a uni-directional process in which results are passed from a research organization to a development group, this collaboration assumed that the interaction would be bi-directional. Research results certainly provide direction for practice, but equally important is that practice provides guidance for useful research. Mutual respect has allowed us to overcome numerous obstacles. In many senses this partnership instantiates the mutually-beneficial theoretical model of two cooperating 'learning organizations' (Senge, 1990).

Pick the right projects. Especially at the beginning it was important to choose topics that were narrow enough to show measurable results relatively quickly. This allowed us to assess the value of our collaboration fairly early. Projects were targeted to address specific questions from practitioners but also to provide incremental steps in a long-term research agenda.

Remain patient. The conflict between research and practice timetables has been addressed earlier. Sometimes researchers were too fast, sometimes too slow. The same held true for practitioners. Only patience could bridge this gulf.

Pick the right people. As in most human activities, personalities were key. Staff active in a true collaboration must be interested, motivated, and have a basic understanding of the aims, strengths, and limitations of their partners.

Communicate frequently. This partnership was characterized by geographic separation, with many of the researchers working hundreds of miles distant from BLS headquarters in Washington, DC. E-mail proved to be an effective medium for frequent communication. On-line correspondence, however, had to be supplemented by regular face to face conversations. The team held biannual meetings at BLS and took advantage of *ad-hoc* opportunities for discussion at conferences such as ACM CHI, ASIST, and ASA.

Participants found that joint papers and presentations offered a unique opportunity for collaborative reflection in the course of which all parties learned more about the others. The willingness on both sides to participate in conferences beyond the familiar (e.g., the academic team members participated in statistical and government service conferences and the BLS team took part in information science conferences) was an important factor in sustaining the work.

Make meaningful commitments. Our experience shows that all parties must be clear on the time each will invest in the effort. The time allocated must be realistic and sufficient to accomplish the goals. Adequate funding must be made available and both sides must make commitments in terms of time and attention. Most of all, partners must be willing to engage with one another on matters large and small.

7 Conclusion

Based on the collaboration in this project there was a progression of BLS Web site design from a primarily agency-centric presentation of labor statistics to an information-intensive, user-centered presentation that supports user interaction and initiative. The vocabulary has become more user-oriented, there is attention to diverse user populations with prominence given to commonly sought data and services, the design is information-intensive with basic data on the home page (e.g., the unemployment rate), alternative content categories (e.g., geographic and topical), as well as extensive links organized according to user's requests. This design aims to minimize mouse clicks and provides "look aheads" through mouse hovers. On a more abstract level, the collaboration has led to a model of organizational interface in which the user interface serves to link the public to the corporate culture of BLS (Marchionini, 2002). Thus there have been both practical and theoretical advances based upon this collaboration.

In the authors' experience, two characteristics are essential to effective collaboration between activities as disparate as research and practice. The first is flexibility, the second is trust. Neither one is trivial, neither is automatic, neither comes without conscious effort. To flourish, participants from both communities must be willing to look beyond immediate operational or career demands and accept that plans and products are malleable but that continuing a long-term conversation is worth the intermittent annoyances and frustrations.

It is only through a series of *quid pro quo* exchanges of flexibilities and commitments that trust builds over time. Shared values can provide the basis for this. In the case study we have described, that shared valuc was a strong dedication by all parties to facilitating citizen access to government information, and thus to improving the human experience.

References

Brunk, B. & Marchionini, G. (2001). *Toward an Agile Views WWW Sitemap Kit: The Generalized Relation Browser*, January 2001, SILS Technical Report 2001-06, UNC-Chapel Hill. http://ils.unc.edu/ils/research/reports/relation_browser.pdf

Levi, M. & Conrad, F. (1997). Usability Testing of World Wide Web Sites: A CHI 97 Workshop. SIGCHI Bulletin, 29(4). http://www.acm.org/sigchi/bulletin/1997.4/levi.html

Levi, M. & Conrad, F. (1999). Interacting with Statistics: A CHI 99 Workshop. SIGCHI Bulletin, 31(4). http://www.acm.org/sigchi/bulletin/1999.4a/levi.pdf

Marchionini, G. (2002). Co-evolution of user and organizational interfaces: A longitudinal case study of WWW dissemination of national statistics. Journal of the American Society for Information Science, 53(14), 1192-1209.

Marchionini, G. & Mu, X. (in press). User Studies Informing E-Table Interfaces. Information Processing & Management.

Senge, P. 1990. The fifth discipline: The art and practice of the learning organization. NY: Doubleday.

Practices of KM for high-tech industry: Empirical study in Taiwan's industries

[1'3]*Chung-Yong Liu*, [1'2]*Ta-Hsien Lo*, [3]*Yasufumi Kume and* [2]*Benjamin J.C. Yuan*

[1]Industrial Technology Research Institute, [2]Chiao Tung University and [3]Kinki University
Oak 2 Bldg, 4F, 1-15-6 Ebisu, Shibuya-Ku, Tokyo 150-0013, Japan
cyliu@mb.kcom.ne.jp

Abstract: The importance of sound knowledge management for high-tech companies needs not to be emphasized. Otherwise the risks of running business would be high. Hence we need to address the purpose of it and understand the appropriate methods. The defining of the objectives and goals of it involves decision-making process using some reasonable rules for multiple objectives and criteria. This research surveys two approaches for applying the Multiple Criteria Decision Making (MCDM) method and Evidential Dominance Rule (EDR) to examine the KM trends and strategies of high-tech R&D organizations and enterprises in Taiwan with some empirical studies. The study concludes with the findings for the key points of setting up a knowledge management related projects or an organizational decision-making.

1. Introduction. A brief survey of knowledge management literature and our aim of research

Around the '50s of 20th century statisticians like W. E. Deming seriously thinking marketing and production systems in terms of Theory of Knowledge . These efforts were extended and applied in the quality and production management revolutionary movements in '80s and 90s. In the meanwhile, the rising of information technology and related sciences enhanced the capability of people and organizations in the areas of planning, designing and management of knowledge. The diffusion of the concept of "Knowledge Management"(KM) in thc industry and society make its meaning as in a jungle words, hence it is difficult for the people in the modern high-tech industry to select its objectives, strategies and tactics in the implementations of KM. Since our scope of this paper will focus on the applications of KM in some organizations and companies of Taiwanese major information industry which is basically a manufacturing- cum-service country, we'll make the decision process of KM as our topics. For a research and knowledge-based service organization like ITRI, the term KM is defined as an integral managerial process combines the functions of IT (or "MIS"). In addition to this,, KM is expected to provide a framework for the innovation process and the application and diffusion of new technology. For a business enterprise, the term of KM is more for profit-driving and of cost-effectiveness consciousness. [1]. That is, the main purpose of KM is the gaining of a profitable knowledge base. Corporations might also try to upgrade their performance through teamwork-building and alliances activities. This might create a cooperative atmosphere for knowledge sharing hence make the costs of business running lower. [2]. This research is based on our two published papers, one is to study the objectives and strategy of KM within the high-tech industry and to rank the strategy of KM in considering with multiple criteria and objectives. The second half of this paper is an application of evidential dominance rules to the analysis of questionnaires in order to construct a constitutive approach for communication technology and group decision making in human network system[3].

2. Build up a hierarchical strategic model for evaluation of KM practices

We built up a model with common objectives and general criteria to describe and measure the goal and strategies of KM. They are described as follows. The criteria are listed: A). Technology promotion ; B). An increase of profit level ; C). An increase of managerial effectiveness ; D). Operational talent promotion ; E). Adding values. According to other related researches the strategy of KM for high-tech industry can be summarized and represented by the following five descriptions: 1). Utilization of IT tools ; 2). Founding a KM organization ; 3). The launch of a KM project ; 4). Leveraging the know-how of others; 5). KM-related incentives for staff. Promotions, the offering of prizes, and bonuses are often the incentives encouraging management development. In this research we build up a hierarchical strategy model re 1 to show the relevance system of KM hierarchy, which also may be analyzed by applying the MCDM method to rank the strategies.

3. Evaluating the hirerarchical system

This study applied MCDM (a fuzzy AHP questionnaire) [4] to evaluate hierarchy strategies. The process of evaluating includes three steps: First, evaluating the weights for the hierarchy relevance system by AHP. The AHP weighting is mainly determined by the evaluators who conduct pair-wise comparisons to reveal the comparative importance of two criteria. The relative importance derived from these pair-wise comparisons allows a certain degree of inconsistency within a domain. Saaty used the principal eigenvector of pair-wise comparison matrix derived from the scaling ratio to find the comparative weight among the criteria of the hierarchy system.

$$(\boldsymbol{A}\text{-}\lambda_{\max} I)\ \boldsymbol{W}=\mathbf{0} \qquad (1)$$

And

$$C.I.=\frac{\lambda_{\max}-n}{n-1}$$ where if $C.I.\leq 0.1$, then en we may satisfy with the consistency.

B. Getting the performance value by meas + uring the fuzzy number $\mu_{\tilde{A}}(x)$ from evaluators (linguistic variable) and defuzzying their fuzzy numbers by the following calculation. $\mu_{\tilde{A}}(x)$

$$=\begin{cases}(x-L)/(M-L) & L\leq M\leq M\\(U-x)/(U-M) & M\leq x\leq U\\0 & otherwise\end{cases} \qquad (2)$$

The linguistic variable in equation 2 can also be expressed by a triangular fuzzy number (TFN). And, each linguistic variable can be indicated by a TFN within a range of 0-100 to show its degree of importance (see below).

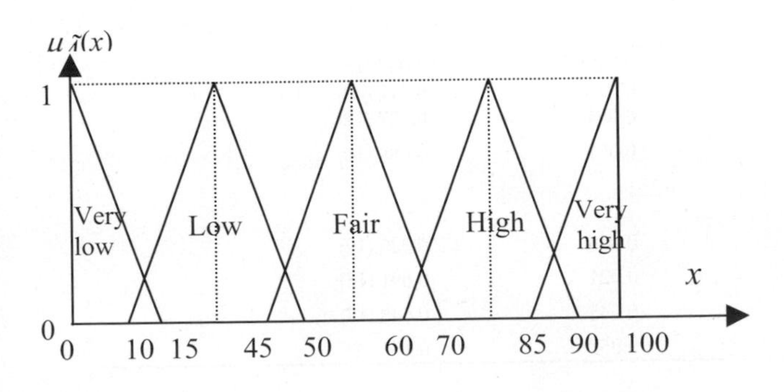

The membership function of the five levels of linguistic variable is applied to measure the achievement of the performance value for each objectives and criteria. And then, the credit point of each strategy can be measured. Each linguistic variable can be indicated by a TFN. Alternatively, the evaluators could subjectively assign their personal weights to the linguistic variables. Assume there are m evaluators, the fuzzy performance value of evaluator k toward strategy i under objective/criteria j is calculated by the equations of 3 to 8. The sign ⊙ denotes fuzzy multiplication, and the sign ⊕ denotes fuzzy addition.

$$E_{ij}^{k} = (LE_{ij}^{k}, ME_{ij}^{k}, UE_{ij}^{k}), j \in S \quad (3)$$

$$E_{ij} = (1/m) \odot (E_{ij}^{1} \oplus E_{ij}^{2} \ldots \oplus E_{ij}^{m}) \quad (4)$$

$$E_{ij} = (LE_{ij}, ME_{ij}, UE_{ij}) \quad (5)$$

$$LE_{ij} = (1/m) \odot (\sum_{k=1}^{m} LE_{ij}^{k}) \quad (6)$$

$$ME_{ij} = (1/m) \odot (\sum_{k=1}^{m} ME_{ij}^{k}) \quad (7)$$

$$UE_{ij} = (1/m) \odot (\sum_{k=1}^{m} UE_{ij}^{k}) \quad (8)$$

The defuzzification is to locate the Best Nonfuzzy Performance (BNP) value. And, the defuzzification of BNP is shown in equation 9.

$$BNP_i = [(UE_i - LE_i) + (ME_i - LE_i)] / 3 + LE_i, \forall i \quad (9)$$

4. An Empirical study

We did an empirical study for ITRI, TSMC and Winbond, as an example to show the practicability and usefulness of the proposed method. The processes and results of evaluating the KM strategy of high-tech companies are shown as follows:

A. Evaluating the objectives / criteria weights:

Objectives/Criteria	Weights	Total Weights(w_j)
a. Technology promotion	0.224	
1.integrate & systemize tech	0.047	0.141 (10)
2.implementation of tech	0.054	0.164 (5)
3.the quantity & quality of IPR	0.060	0.192 (2)
4.solution tank of technology	0.063	0.193 (1)
b. An increase of profit level	0.219	
5.decreasing time to the market	0.050	0.093 (14)
6.production and yield	0.059	0.132 (11)
7.service/ sale & profit	0.053	0.157 (6)
8.decreasing production cost	0.057	0.143 (9)
c. Managerial Effectiveness	0.230	
9.policy effectiveness	0.048	0.099 (13)
10.reflection of problems	0.061	0.187 (4)
11.solution to the problem	0.064	0.189 (3)
12. performance evaluation	0.057	0.154 (7)
d. Operational Talent	0.176	
13.forecast of market & tech	0.049	0.047 (18)
14.objectives of operation	0.022	0.011 (20)
15.forecast of risk & failure	0.054	0.127 (12)
16.response to emergencies	0.051	0.088 (16)
	0.151	
e. Adding Values		
7.leverage blend name	0.028	0.026 (19)
18.valuation promotion	0.021	0.091 (15)
19.boost of stock price	0.045	0.048 (17)
20.long-term competitiveness	0.057	0.151 (8)

B. Estimating the performance matrix:

Evaluator	Very low	Low	Fair	High	Very high
1	(0,10,15)	(10,15,30)	(30,50,70)	(70,80,90)	(80,90,100)
2	(0,10,20)	(20,30,40)	(40,50,60)	(60,70,80)	(80,90,100)
…	…	…	…	…	…
m	(0, 0,10)	(10,25,40)	(40,55,70)	(70,80,90)	(90,95,100)

C. Ranking the KM strategies for high-tech company:

Judging from above evaluation the ranking and BNP value of five strategies is C(68.31) > B(66.76) > A(63.21) > E(58.39) > D(55.49). And C.I.=0.05.

D. Discuss the ranking of strategies :.

Judging from above analysis, we may conclude that the overall industrial trend of KM is promoting technology and increasing managerial effectiveness. The characteristic of strategy B and C can be described as following three cases.

ITRI (A Leading R&D Organization in Taiwan), TSMC (A World Leading IC Foundry Company) and Winbond (A Leading DRAM Company in Taiwan). The comparison of KM strategy among ITRI, TSMC and Winbond is shown in Table1.

Table 1. The KM strategy of ITRI; TSMC and Winbond

	Strategy	Responsible	Activities	Title
R&D Org. ITRI	C	Volunteers Staff: 300+	KM website & Task forces	Six Leading Projects
Foundry TSMC	C	IT Dept. Staff: 10	Research team & Experts' list	KM Project
DRAM Winbond	B	KM center Staff: 20	Tech Committee DM center	KM Center

5. A Study based on evidential dominance rules for CHINA INVESTMENT REGIONS

We proposed an application of evidential dominance rules to the analysis of questionnaires prepared by Taiwan Electrical and Electronic Manufacturers' Association (TEEMA) . The effectiveness of our proposed method is shown by application of evidential dominance rules to analysis of the questionnaires. It is clear that evidential dominance is effective for converging alternatives. Thus, it is possible to grow the agility of decision-making [3]. We give an example of application of our method as a sample to enhance understanding of our approach. The weighting allocation for five Indices for Investment Environment are: natural environment with 10% weighting, infrastructure 20%,public facilities 10%,social environment 10% and legal and political environment, 50%. The first three indices are concerned with natural conditions and buildings/ facilities, while the last two indices are for the artifact elements. The formula for Weighted Satisfaction is as the following,: Weighted Satisfaction = 10% Natural Environment + 20% Infrastructure + 10% Public Facilities + 10% Social Environment + 50% Legal and Political Environment. Next step of the analysis is to apply the rules of Evidential Dominance to analyze the data. First, we use the survey data by segment of areas, and expressed them in intervals with minimum value and maximum value. The Investment Environment of the basic assignments include A:natural environment with the weighting of 10%, B: infrastructure 20%, C:public facilities 10%,D:social environment 10% and E:legal and political environment, 50%. By applying the evidential dominance rules, the results of the Investment Risks Analysis are summarized as follows. Yangtze River Delta Area is selected as the best area for investment environment. The highest risks area by the rule of max. value evidential dominances are Southeast Costal Area. These results are consistent with the survey study of TEEMA. We can come to the conclusion that the method of interference by the interval with maximum and minimum values and the focused elements is valid. In the result of this example, considering to select alternatives by evidential dominance effectively, decision maker wants to obtain the level of information is not preferred with high level alternatives.only .It is effective to focus on just one phase of alternatives with well-balance information. Also, the more the number of alternatives, the more reduction rate decreases. Decision-maker can investigate preferred alternatives effectively. Therefore, it is clarified that evidential dominance is effective to focus on just one phase. Agile manufacturing is brought about by the integration of organization, people and technology into a coordinated independent system. In agile manufacturing, the priority relation of linkage for humans is focused on one phase of the linkage. The most optimal linkage is cleared. It is very important to make a database of linkage for human about the knowledge and information needed. After all, someones should decide the priority relation or they should perform decision-making themself.

Table 2 Mainland China investment environment analysis for regions

Region	A	B	C	D	E	F	G
1.Southeast Coastal Area	613	3.62	3.16	2.92	2.87	2.68	2.91
2.Yangtze River Delta Area	519	3.82	3.51	3.41	3.60	3.18	3.37
3.Five Central Province Area	67	3.55	3.28	2.95	2.99	2.90	3.05
4.Southwest Region Area	89	3.60	3.16	3.04	3.15	2.80	3.01
5.Bohai Bay Area	117	3.48	3.39	3.35	3.44	3.02	3.21
6.Northeast Region Area	44	3.41	3.38	3.34	3.35	2.98	3.17
7.Northwest Region Area	19	3.33	3.20	2.92	3.14	2.98	3.06
Total Sample Size	1,468	3.67	3.32	3.15	3.22	2.92	3.12

6. Conclusion

For organizations in high-tech sector or information technology industry, to take advantage of the opportunities in Greater China economic integration, the best strategy is to sponsor KM projects or set up a professional organizational for it. In addition to screen for a better region for new investment, we can utilize various IT tools and related incentives for staffs. Furthermore, management effectiveness and technology promotion are two of the key measures for performance and benefits of a KM projects. The trend of KM activity in the year 2003 is likely to be in the areas of IPR management and the implementations of solution-based service and think tank for technology.. Finally, both Multiple Criteria Decision Making (MCDM) method and Evidential Dominance Rule (EDR) are proven here to be reliable methods to solve some problems with multiple criteria for decision-making.. present the conclusions of this study.

Reference

[1] Christopher .M. L., Lois S. P. (2001). Developing competencies and capabilities through knowledge management: a contingent perspective PICMET, Technology in the knowledge era, 257.

[2] Leonard D., Sensiper, S. (1998). The role of tacit knowledge in group innovation, Cal. Management Rev., 40, 112.

[3] Kume Y., Liu C.Y. & Kainouchi H. (2002). Constitutive Approach for Communication Technology and Group Decision Making in Human Network System, 2002 APIEMS's Proceedings, 1061-1066

[4] Lee Z.Y, Tzeng H.C., Yu H.C. (2001). Fuzzy MCDM approach for IC company's strategy in the semiconductor industry, PICMET, Technology in the knowledge era, pp 776-787.

Quality Assurance in the National Institute for Occupational Safety and Health (NIOSH) Publications Office.

James D. McGlothlin[1]

Vern P. Anderson[2]

[1]Purdue University
550 Stadium Mall Drive
West Lafayette, IN 47906
Jdm3@purdue.edu

[2]NIOSH
4676 Columbia Parkway
Cincinnati, OH 45226
Vep1@cdc.gov

Abstract

Establishing information quality is critical for United States Government Agencies, especially for agencies such as the National Institute for Occupational Safety and Health (NIOSH) that promote and protect the safety and health of the nation's workforce. In 2002, the Education and Information Division (EID) responded to more that 117,000 phone calls for specific information related to health and safety, and to 63,000 requests for NIOSH publications. Since the events of September 11, 2001, NIOSH has seen an increase in requests for documents related to the World Trade Center, anthrax, bioterrorism, and emergency preparedness. Since much of this information can be accessed through the NIOSH website: http://www.cdc.gov/niosh/homepage.html it is critical that this information is accurate, consistent, and reliable. This paper discusses two projects managed by EID that provides insight into how NIOSH serves its customers and maintains a standard of excellence in information quality.

1 Introduction

Quality assurance at the Federal level is in part driven by the "Government Performance and Results Act (GPRA) that was enacted by the U.S. Congress in 1993 because of public concerns of, among other things "waste and inefficiency in Federal programs undermine the confidence of the American people in the Government and reduces the Federal Government's ability to address adequately vital public needs." (U.S. Whitehouse OMB website, Feb. 13, 2003) The goal of GPRA was, in part, to systematically hold Federal agencies accountable for achieving program results by helping Federal managers improve service delivery through program objectives and by providing program results and service quality. The latest amendment specified that beginning March 31, 2000, and each year thereafter, the head of each agency shall prepare and submit to the President and the Congress, a report on program performance for the previous fiscal year; a kind of quality assurance that presents strategic goals met. GPRA is managed by the Office of Management and Budget (OMB) from the Executive Office of the President of the United States.

The information used to develop this paper is from NIOSH's Education and Information Division (EID) and focuses on two projects where the summaries of such projects were developed to address the reporting requirements of GPRA. The first project is the ***Technical Information Inquiry Service,*** and the second project is the ***NIOSH Document Dissemination: Serving the Public.*** Both projects support NIOSH's Strategic Goal 4 which states: "Provide workers, employers, the public, and the occupational safety and health community with information, training, and capacity to prevent occupational diseases and injuries." In order to deliver in this goal it is imperative that the quality of the information being delivered is the best scientific information available.

2 NIOSH's Technical Information Inquiry Service

Technical Information Inquiry Service is managed by the Education and Information Division (EID) of NIOSH and provides responses to inquiries from the general public that are received via the NIOSH 800-number and the Internet. The NIOSH 800-number served more than 117,500 callers during FY2002. Of these, more than 25,863 callers spoke directly with a technical information specialist, a 5% increase over Fiscal Year (FY) 2001. This increase is due to the number of callers requesting information about their Office of Compensation Analysis and Support (OCAS) claims. During Fiscal Year (from October 1 – September 30) 2002, the information requested most often from NIOSH technical information specialists related to (1) OSHA, MSHA, EPA, State or local regulations; (2) OCAS; (3) chemical hazards; (4) NIOSH resources or programs; and (5) safety/injuries. Figure 1 shows the distribution of 800-number calls by State for FY2002. The States with the greatest number of callers included California, Ohio, Texas, Florida, and Michigan. During FY2002, more than 10,000 fax documents were transmitted to callers by the automated fax-on-demand 800-number service. Internet technical inquiries continue to increase, demonstrating that more users are requesting assistance through the NIOSH Web site. EID responded to 3,796 such inquiries (a 29% increase over FY2001), 1,039 of which were from outside the United States. During FY2002, a large number of information resource files were incorporated into the information database system. The ability to provide electronic information transmission directly from the database provides convenient real-time access to material for our customers. For FY2003, the major focus of the database has been on adding files relating to the World Trade Center, anthrax, bioterrorism, and emergency preparedness. Also, a routing system to answer calls in Spanish was added to the NIOSH 800-number; 157 Spanish-language calls have been received since December 2001.

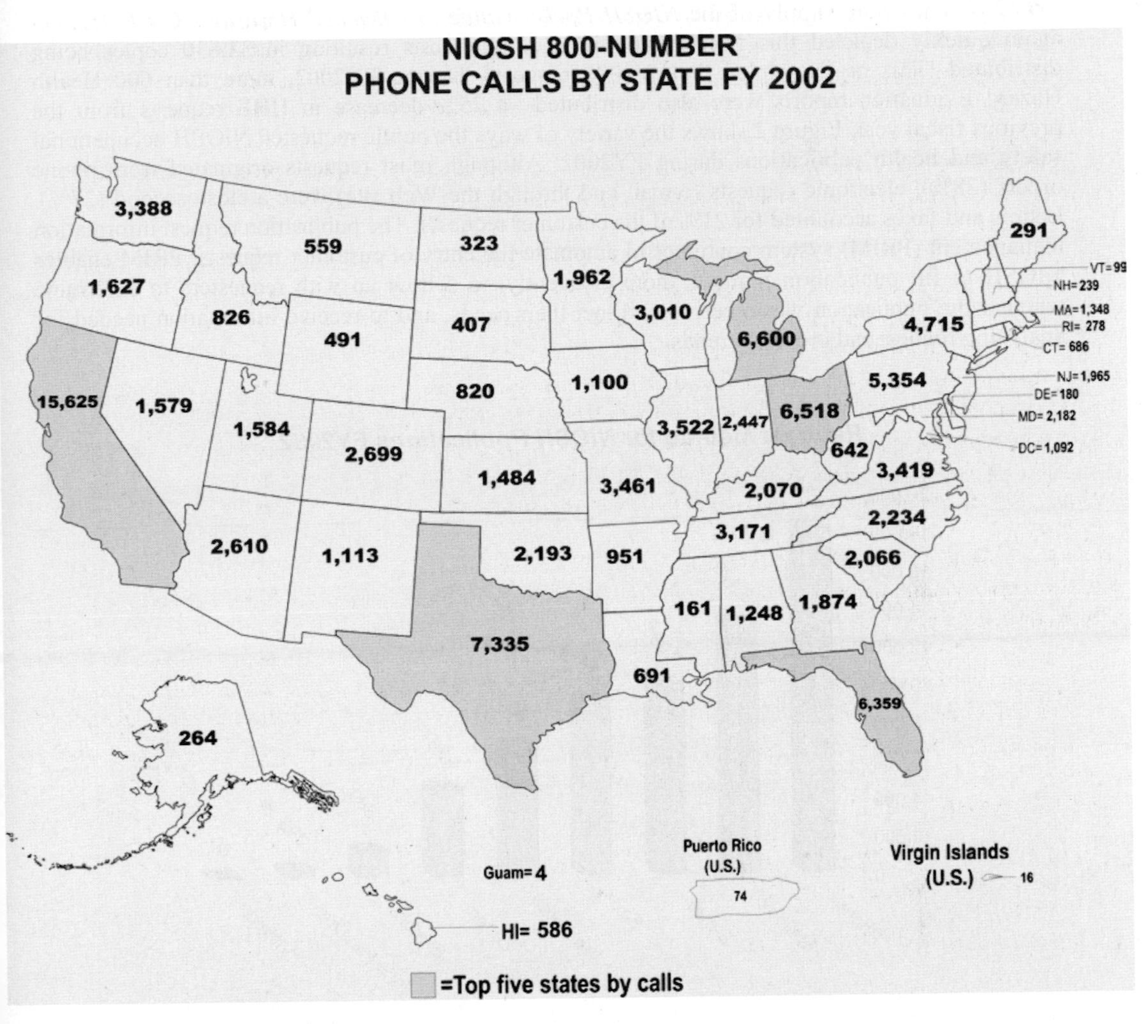

Figure 1: The number of phone calls by state for 2001-2002.

3. NIOSH Document Dissemination: Serving the Public

NIOSH Document Dissemination: Serving the Public is a NIOSH Publications Clearinghouse responded to more than 63,000 customer requests for NIOSH publications in FY2002, distributing approximately 1.9 million publications. The Publications Clearinghouse stocks more than 4,200 titles of NIOSH publications. Of the 1.9 million publications distributed, 1.8 million were distributed in 5,290 separate bulk (greater than 10 copies) mailings. The top five publications in total distribution for FY2002 included (1) *What Every Worker Should Know - How to Protect Yourself from Needlesticks* (280,509 copies distributed); (2) *Latex Allergies, A Prevention Guide* (114,646); (3) *Exposure to Blood: What Health-Care Workers Need to Know* (66,036); (4) *Alert: Preventing Allergic Reaction to Natural Latex* (63,612); and (5) *Are You a Working Teen?*

(59,124). The entire supply of the *NIOSH Pocket Guide to Chemical Hazards - CD-ROM* was again quickly depleted this fiscal year with 9,732 requests resulting in 33,830 copies being distributed. This nearly tripled the FY2001 requests. During FY2002, more than 600 Health Hazard Evaluation reports were also distributed—a 25% decrease in HHE requests from the previous fiscal year. Figure 2 shows the variety of ways the public requested NIOSH occupational safety and health publications during FY2002. Although most requests originated from phone orders (28%), electronic requests (e-mail and through the Web site) were a close second (27%). Letters and faxes accounted for 21% of the customer requests. The publication request information management (PRIM) system continues to automate the entry of customer requests. PRIM enables NIOSH to fill publication requests more efficiently, to follow up with requesters to determine whether the publication was received and met their needs, and to receive information needed for analyzing request and ordering trends.

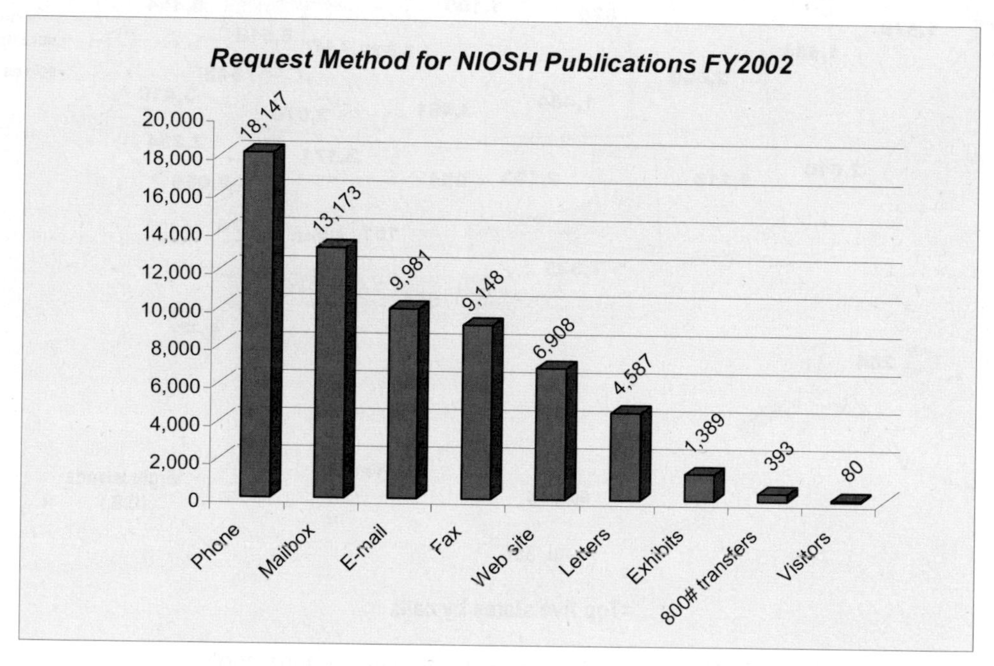

Figure 2: Variety of ways NIOSH clients ordered publications in 2002

4. Summary and Discussion

Maintaining information quality is a top priority for U.S. Government agencies such as NIOSH. In addition, annual reports to the Office of Management and Budget (OMB) as part of GPRA offer assurance that these agencies address vital public needs. What is interesting about the EID projects mentioned above is the increasing demand for its information services and documents not only in the United States, but worldwide. For example, there was a 30% increase in NIOSH website internet technical inquires from 2001 to 2002. What is more interesting, is that 27% (1,039/3796) of the internet technical inquires came from outside the United States. This growing trend of internet requests shows the reputation of NIOSH as an information quality resource on an international scale. In particular, the quality of the responses NIOSH provides to its clients seeking help is critical in maintaining its role as a pre-eminent national and international resource in occupational safety and health.

Within the last year, significant improvements have made to the NIOSH web site to improve internet access by its clients to technical information. This is consistent with the NIOSH strategic goal (as defined by GPRA) for both projects discussed above (NIOSH, EID, FY 2002 Major Accomplishments Report to OMB). Other government internet entities also are concerned about the quality of the data they collect and store (Wellman, et al., 2003), especially in the post September 11, 2001 environment. The changing needs of the public to access information using the internet has resulted in NIOSH developing reliable information systems to provide electronic safety and health information through a database in real-time directly to its clients. NIOSH will continue to use its resources to improve the administration of its information systems to deliver on the Nation's promise: *"Safety and health at work for all people through research and prevention."*

5. References

1. NIOSH, Education and Information Division, Fiscal Year 2002, Major Accomplishments Report to the OMB. October, 2002.

2. The Government Performance Results Act (GPRA) 1993. Office of Management and Budget, Executive Office of the President of the United States. http://www.whitehouse.gov/omb/mgmt-gpra/gplaw2m.html#h1

3. Wellman, H., Sorock, G., Lehto, M.R., Automated Identification and Correction of Coding Errors in an Accident Narrative Database, Tenth International Conference on Human-Computer Interaction, Crete, Greece, June 22-26, 2003.

Human oriented Intelligence Image Processing System for Integrated Visual Inspection

Masao NAKAGAWA

Shiga University
1-1-1 Bamba, Hikone, Shiga, Japan
mnaka@biwako.shiga-u.ac.jp

Hidetoshi NAKAYASU

Konan University
8-9-1 Okamoto, Kobe, Japan
nakayasu@konan-u.ac.jp

Abstract

This paper deals with the automated visual inspection method for advanced manufacturing in order to perform management of information and knowledge in agile manufacturing for improved management of product flexibility and product quality. The defective picture sampling process based on the proposed algorithm is performed for original picture image for a check of the industrial product. In the proposed algorithm, three kinds of inspection information such as location, size and level of defects are treated for clustering image data. In this paper, the proposed method developed by the authors previously, has been improved as a knowledge-based processing in line with the inspection policy such as classification information on color, brightness and contrast of image pixels and on position and shape property of the defects detected from the image data.

1 Introduction

The advance which recent information processing technology is remarkable has done large contribution in the field of the image processing where it required a great deals of expensive and large equipment until now. Recently, several kinds of image processing method has been applying to the automated visual inspection system for defects on the product surface (Chin, et. al., 1982, Huang, et. al., 1992). One of the aims of this research is also in the development of the heuristic and simple method which is used in the judgment process in the automated visual inspection system instead of inspector. A sort capacity by human vision is extremely high-performance, therefore such a soft information processing to a sort of images has been regarded unfit on the computer which is good at digital information processing.

On the other hand, a visual inspection process has been holding the problem in productivity, since the performance of a precision and a speed will be degraded by fatigue of the inspector. In order to meet these problems, some research works were tried for the standardization of an operation time of visual inspection (Morawski, 1992, Arani, 1984, Drury, 1972, Spitz, 1978), though it is not reached to the place which fixes a good evaluation measure, and research of productivity of the production system which consists of a process including such a human being has left lots of problems unresolved.

This paper deals with the automated visual inspection method for advanced manufacturing in order to enhance human skill with machine performance for improved management of product flexibility and product quality. Because of automation of visual inspection, the defective picture sampling process based on the proposed algorithm is performed for original picture image for a check of the industrial product. In the proposed algorithm, three kinds of inspection information such as location, size and level of defects are treated for clustering image data. In this paper, the emergency algorithm developed by the authors previously (Nakagawa, et. al., 1998), has been improved as a knowledge-based processing in line with the inspection policy such as classification information and position information of the defect in the inspection image. An algorithm is proposed intending for external view image processing of FRP panel in visual inspection process within total panel production system. The verification of validity of the proposed algorithm was also made for FRP panel.

2 Analyze Shape for FRP Defects on Visual Inspection

2.1 Shape Property

It is necessary to recognize the shape for the feature extraction of the image, since the information on the location of the object is important information in the pattern recognition of the image.
The shape property is calculated by Elliptic approximation based on shape moment.

2.2 Shape Moment

The moment means principal axis of inertia. $p+q$ order moment M_{pq} of binary digit image $f(x, y)$ with pixel can be expressed by next equation.

$$M_{pq} = \sum_{x \in A_N} \sum_{y \in A_M} (x - x_g)^p (y - y_g)^q f(x, y) \qquad (A_N = \{1,2,...,N\},\ A_M = \{1,2,...,M\}) \qquad \text{eq.1}$$

The area size of the defect is shown on 0-order moment M_{00} in the binary digit image. Coordinate $G(x_g, y_g)$ of center of gravity is obtained, when first-order moment of M_{10} and M_{01} is normalized at M_{00}. The moment of center of gravity is called the Central Moment.

2.3 Elliptic Approximation

The feature of the algorithm is an approximation by ellipse for the shape for the feature extraction. The details of this elliptic approximation are written in the followings.

Eigenvalue λ_1, λ_2 of matrix $\begin{bmatrix} M_{20} & M_{11} \\ M_{11} & M_{02} \end{bmatrix}$ is required by following procedures.

$$\lambda_1 = \frac{M_{20} + M_{02} + \sqrt{(M_{20} + M_{02})^2 - 4(M_{20}M_{02} - M_{11}^2)}}{2}$$

$$, \lambda_2 = \frac{M_{20} + M_{02} - \sqrt{(M_{20} + M_{02})^2 - 4(M_{20}M_{02} - M_{11}^2)}}{2}$$

eq.2

The direction which agrees with the eigenvector of λ_1 is called the principal axis of inertia. By using asked eigenvalue of λ_1, λ_2 and area size A of defects in binary digit image $f(x,y)$, principal axis a and minor axis b of the ellipse can calculate it in next equations.

$$a = \sqrt{\frac{A/\pi}{\lambda_2/\lambda_1}}, \quad b = a \times \frac{\lambda_2}{\lambda_1} \qquad \text{eq.3}$$

The aspect ratio of the shape as following. $r = b / a$

The angle required like the following is an angle θ of principal axis and X-axis of the approximate ellipse. Angle is required like the following by positive and negative of M_{11}.

$$\theta=\tan^{-1}\left(\frac{-C+\sqrt{C^2+4}}{2}\right) \because M_{11}>0,\quad \theta=\tan^{-1}\left(\frac{-C-\sqrt{C^2+4}}{2}\right) \because M_{11}<0 \qquad \text{eq.4}$$

3 Classification by Rule Table method

That is a matter of great importance to industrial manufacturing process what like the property. In this paper, we are studying classification method for the purpose of classification into 4 categories (heavy, middle, light and non-defect categories). We make the Rule Table (RT) method for construct method imported knowledge and skills.

RT method has two stages, it is 3x3 matrix, shows Figure 1. The first stage's valuation basis is area size and brightness of defect, the second stage's valuation basis is length and aspect ratio of defect. The kind of value in the tables is 3 (heavy), 2 (middle) and 1 (light). The tables are created which based on knowledge and skills by inspector. Because some inspectors have pattern of behavior that they find large defects in the first stage after find small defects. Applied Rule Table is shown in Figure 2.

Criterion for evaluation of the first judge table are area size and brightness, and the second judge table are length and aspect ratio.

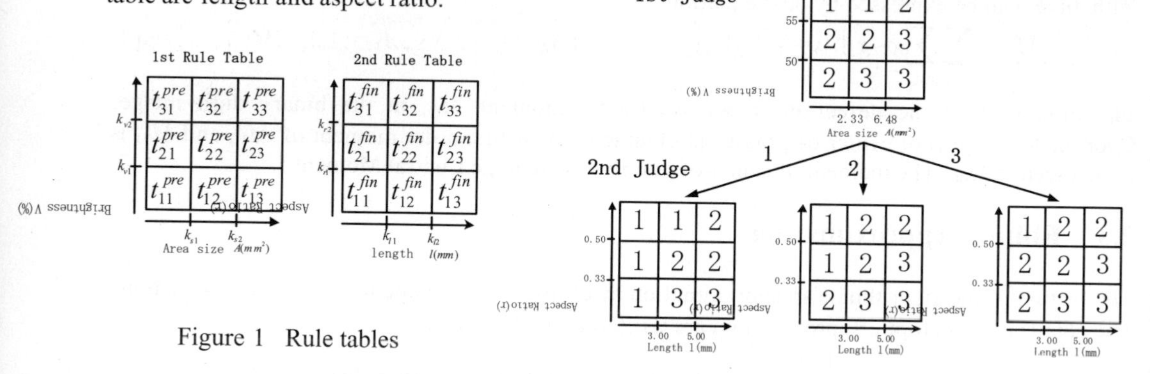

Figure 1 Rule tables

Figure 2 Applied Rule tables

4 Application to FRP products

For the evaluation of the performance of the proposed algorithm, an examination of automated visual inspection is performed for the specimen of FRP panel products. The material for inspection is the MAT FRP with polyester matrix and glass fiber with volume fraction 30%.

The rule table which shows Figure 2 was made by the inspector. The second judge tables was separated each categories from the first judge table.

The image processing is carried out for FRP products, and the feature quantity which is included for the image is extracted. Figure 3 is an image captured by CCD camera. The execution result of image processing after the interval-thresholding processing is shown in Figure 4. This figure is a result of extracted object for the image of Figure 3.

The proposal technique was also able to extract not only size but also feature quantity of angle and aspect ratio of the defect. The Elliptic Approximation on the surface of the FRP panel gives useful information for inspection. The execution result is shown in Figure 5.

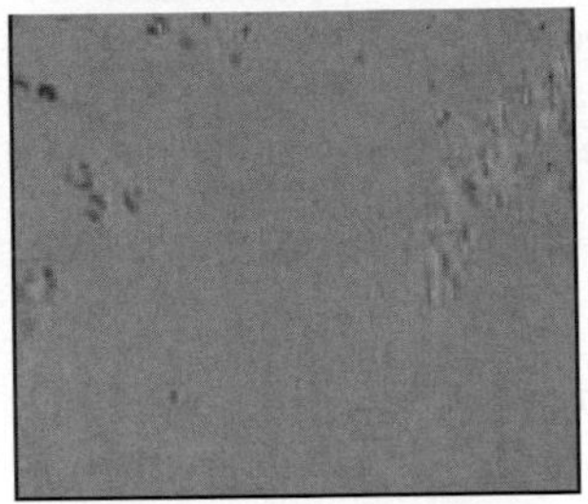
Figure 3 Original Image

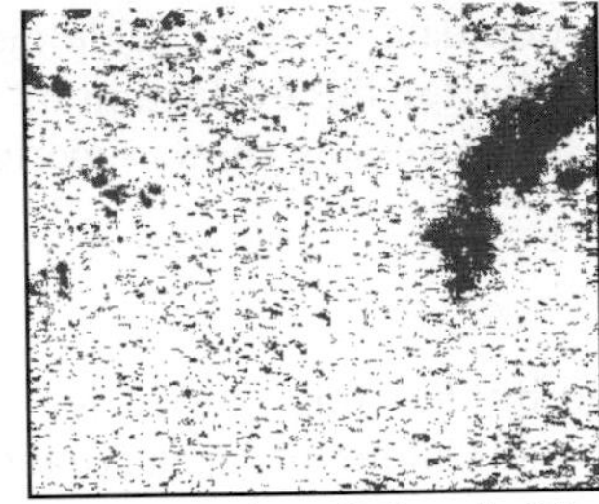
Figure 4 Results of Thresholding

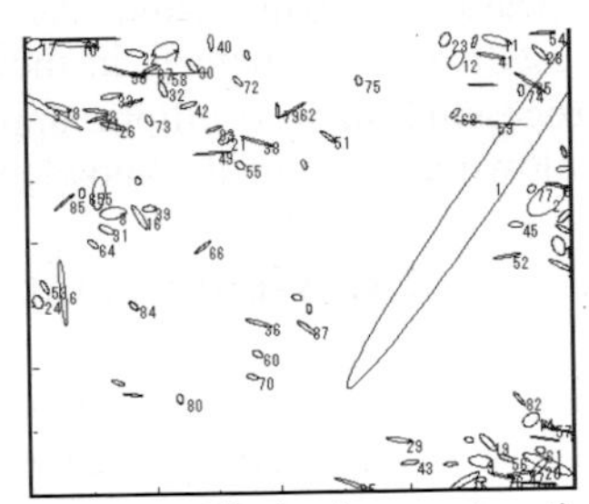
Figure 5 Elliptic Approximation

5 Measurement method of the visual threshold value of human

It explained it to be able to classify the result that was processed the digital image with ellipse approximation by Rule Table Method. The visual threshold value of the human to the attribute of the shape of an ellipse is investigated, to explain the basis of Rule Table that a human bound. The method is called Method of constant stimuli. This method is the degree that shows a certain value and reference value to a human and is able to recognize the difference is investigated. We analyzed the result with a maximum likelihood solution method, and calculated the psychometric curve in Figure 6.

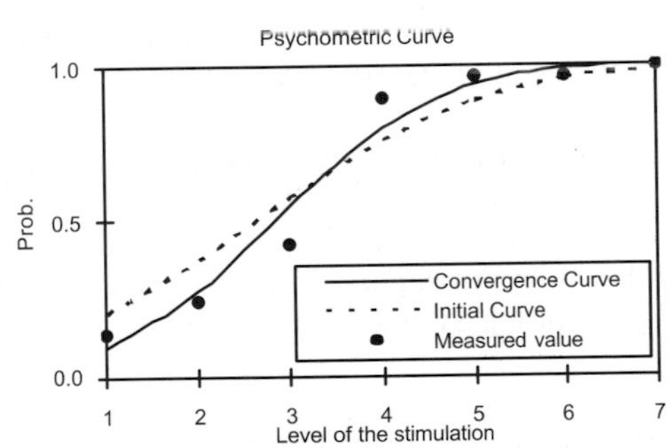

Figure 6 Psychometric Curve

The panelist compares and judges standard stimulation R_0 and change stimulation R_j. If the probability that judges that R_j is bigger than R_0 is defined with P, the rule nature it is admitted as the function of R_j. This function is known that it becomes Normal Distribution empirically from ago, follows next equation.

$$P = \frac{1}{\sqrt{2\pi}\,\sigma}\int_{-\infty}^{R} \exp\left(-\frac{(R - R_j)^2}{2\sigma^2}\right) dR \qquad \text{eq.5}$$

We are obtained to this method with the sensory threshold value of the human. It is able to expect to be made Rule Table to a more accurate thing by continuing this measurement.

6 Conclusion

The proposed image processing algorithm consists of three levels of threshold method, and elliptic approximation. The automated interval-threshold method is based on maximum variance method. The elliptic approximation generates the simple shape in stead of the original defects from the appearance image data. As the practical application to FRP panel, the performance of classification of defects in the inspection data by Rule Table method was well. Another image processing such as elliptic approximation for the evaluation of the area of defects, and border following based on the knowledge-based processing shows superior performance for practical use.

7 Acknowledgement

This w\ork was financially supported by the Grant-in-Aid for Scientific Research Fund of the Ministry of Education, Science, sports and Culture of Japan (Grant No. 11650109).

8 References

Arani, T.T., Karwan, M.H. and Drury, C.G., 1984, A variable -memory model of visual search, Human Factors 26, 680-688.

Chin, R.T., et al., 1982, Automated vis ual inspection: A survey, IEEE Transactions on Pattern Analysis and Machine Intelligence 4(6), 557-573.

Drury, C.G., 1972, The effect of speed of working on industrial inspection accuracy, Applied Ergonomics 4, 2-7.

Huang, C., Cheng, T. and Chen, C., 1992, Color images' segmentation using scale space filter and markov random field, Pattern Recognition 25(10), 1217-1229.

Morawski T.B., Drury C.G. and Karwan M.H., 1992, The optimum speed of visual inspection using a random search strategy, IIE Transactions 24(5), pp.122-133.

Nakagawa M. and Nakayasu H., 1998, An Algorithm with Semantic Information on Visual Inspection for Advanced Manufacturing System, Proc. 6th Int. Conf. HAAMAHA, Hong Kong, pp.151-154.

Spitz, G. and Drury, C.G., 1978, Inspection of sheet materials: Test of model prediction, Human Factors 20, 521-528.

TV Viewing and Internet Use: Experiences from a Large-Scale Broadband Field Trial in Norway

Siri Johanne Nilsen, Kari Hamnes, Kristin Thrane, Rich Ling & Marianne Jensen

Telenor R&D
Snarøyveien 30, N-1331 Fornebu, Norway
Siri-johanne.nilsen@telenor.com

Abstract

In this paper, we present a field trial of several interactive broadband services involving 900 pilot users situated in two geographic locations in Norway. Results from this study indicate that regardless of gender, high speed Internet via the PC is the broadband service that is the most appreciated by the users. We argue the existence of a mental divide between TV viewing and Internet use, compounded by related usability problems, that negatively influences the possible convergence of TV- and PC-centric services.

1 Introduction

The Telenor Hybrid Broadband @ccess (HB@) corporate project (Ims et al., 2001) has during 2000 and 2001 tested how different access solutions may be combined in order to provide next generation interactive broadband services in cities, townships and rural areas. In this paper, we present a field trial of several interactive broadband services involving 900 pilot users situated in two geographic locations in Norway, Stavanger and Oslo. The paper reports on the domestic use of TV and Internet services. It describes the pilot users' use of the services, and looks for answers to the following questions:

- What are the usage patterns of the various TV - and PC-centric services?
- Which elements of the TV- and PC-centric interactive broadband service offering were the most popular?
- What kind of usability problems did the users experience in the use of TV- and PC-centric services?

2 Field trials in Stavanger and Oslo

In *Stavanger* a VDSL (Very High Speed Digital Subscriber line) trial was conducted with 750 users. The VDSL technology is designed to carry high-speed data traffic over normal twisted pair copper telephone lines that are already in place. The set-top boxes in the homes provided three parallel TV channels and it also offered a 2,5 Mbit/s Ethernet Internet connection for a PC. The set-top box had a radio-based remote control such that users could control several TVs placed in different locations within the home from the same set-top box. In addition, the Stavanger pilot also included a separate box (NetGem) in order to offer Web surfing and e-mail on the TV.

The *Oslo* service offering included xDSL[1] to 104 users, and wireless LMDS (Local Multipoint Distribution System) systems to 47 users. The set-top box used in the Oslo trial included software for Web surfing and e-mail on the TV via a TV portal maintained by the HB@ project.

The services offered can be categorised as TV-centric or PC-centric. For TV-centric services (e.g. digital TV channels, Video on Demand, Surf on TV, Electronic Program Guide), the TV-screen is the main focus of attention and interaction for users. Equivalently, PC-centric services require the user to interact with a PC in order to access the services.

In this field trial, data collection regarding the user investigations employed both qualitative and quantitative research methods. There have been telephone interviews with all 900 trial users, several focus groups, 28 in-depth interviews in users' homes, a willingness-to-pay investigation, traffic logging, Call Centre logging and additional analysis like conjoint analysis. The combination of qualitative and quantitative methods gives complementary insights into important information about user habits, usage patterns and the drivers spearheading broadband use.

3 Findings

3.1 Use of TV Centric Services

Trial users saw *Video on Demand* (VoD) as a complement to regular video rental, and not as a substitute for broadcast material. This was one of the most important reasons why the trial users did not try the VoD-system, or that they did not use it frequently. In the qualitative interviews families would talk about evenings with non-interesting programs on the TV, but they did not see VoD as an option on normal evenings.

The *Electronic Program Guide* (EPG) offers a potential for flexibility, control and overview of the existing broadcast transmissions in addition to some extra services. The opinion according to the qualitative interviews in subjects' homes seemed to be that the EPG was user friendly and indeed quite popular. Younger members of the household were generally more inclined to take advantage of the additional functionality of the EPG. Some used the EPG's title search, and some used it to pre-program their TV watching schedule. Many thought that the additional functionality (title search, pre-programming etc) was quite good, but that the fiddling involved was just not worthwhile. As a general rule of thumb, in a TV context, most people are adverse to functionality that does not function near-instantly and would rather settle for status quo than make an extra effort, the reason being that people are used to the TV's perceived ease of use.

Trial users in Stavanger and Oslo (LMDS and VDSL) had the possibility to both send and receive *E-mail on TV* through their TV-portal offer. Both pilots had wireless keyboards and in the Oslo pilots they also had a wireless mouse. Most homes with the NetGem box (Stavanger) had at least tried to access their mail account.

The E-mail on TV option was used only to a limited extent. One reason for this may be that the trial users already had an always-on option in their PC-centric service offering. This means that the barrier to checking e-mail via that PC was lower than in homes where one would need to boot up a computer or connect to a network in order to use this service.

---

[1] The major portion of the users (65) had VDSL. There were also 39 ADSL users in this trial.

In Stavanger, the users had the ability to *Surf on TV* via the NetGem box. In the qualitative interviews, most of the users shared the opinion that this function was user-friendly and stylishly designed, even though very few had started to use it on a regular basis. Seen from a more critical perspective the term "cute" may well sum up the verdict of the users regarding the Surf on TV concept.

3.2 Use of PC Centric Services

The HB@ trial users quoted improvement in Internet speed as the greatest asset in the PC-centric service offering. This result comes out both in the qualitative and in the quantitative data. The reason for this is very simple. There is already an established range of services, applications and content that call for broadband solutions that the customers have been aware of for some time. However, these services have been outside their reach until now due to bandwidth limitations.

Internet use was not evenly distributed among the family members. The children and teenagers were the most avid and regular users, but often had a narrower range of use than adults. This somehow contradicts popular belief that the youngsters are the digital pioneers. It is our impression that interest for the Internet depends on age and belonging to a social network (work, friends) where regular access and some use of the Internet is a matter of course.

Table 1: PC-centric usage preferences for different user groups.

	Children	Teens	Mother	Father
Games	Popular	Heavy		Light
Simple web surfing	Popular	Popular	Light	Popular
Downloading/ftp		Heavy		Popular
Mail	Light	Popular	Light	Heavy
News (usenet)		Light		Popular
Chat	Popular	Popular	Light	Light
E-commerce/Transactions			Light	Heavy
Education	Light	Light	Popular	Popular
Own content/Other	Light	Popular		Light
Intranet			Light	Heavy

Table 1 is a rudimentary way of summing up the dominant use patterns for PC-centric services, as derived from the interviews. Though not statistically valid, we nevertheless feel that it provides a useful summary. The term "Light" refers to use on an irregular basis. "Popular" refers to widespread, regular and fairly time-consuming use. "Heavy" refers to widespread, very frequent and very time-consuming use.

3.3 Popularity of services

In this section we discuss the results of so-called conjoint analysis. The point of the analysis was to reveal the elements in a multi service package that excited the most interest among users. We carried out such an analysis in the HB@ project based on five elements. These included 1) speed of the Internet connection, 2) number of TV channels available on the system, 3) number of parallel TV streams 4) access to VoD and 5) access to an EPG.

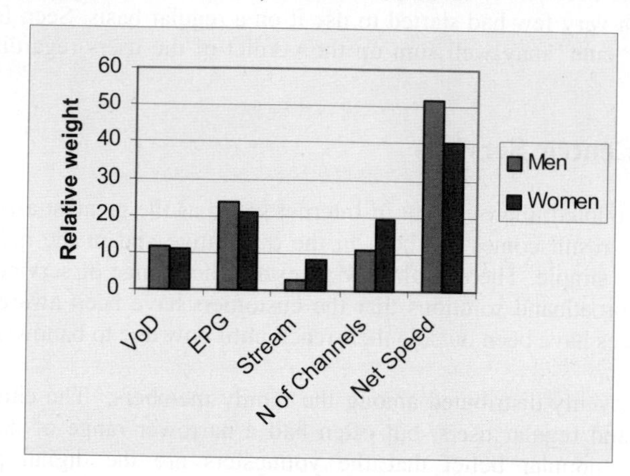

Figure 1: Relative interest in various elements of an eventual broadband service offering by gender (the weightings sum to 100)

The results of the analysis (Figure 1) show that, in general, the speed of the Internet in a package offering is the dominant element. If all elements sum up to 100, access to high speed Internet can be seen as having a weight of almost 50%. Interestingly, access to an EPG is the second most important element for the users with a weight of 23%. The final three elements, in order of diminishing importance are the number of channels available (13%), access to VoD (11%) and access to several parallel TV streams (4%). The number of TVs in the household or the presence of satellite dishes only marginally influenced the results.

There were, however, some gender differences in the ranking of the services. In general, the men were more focused on high speed Internet while the women put more emphasis on the number of television channels and television streams available. This may indicate that men often dominate the central television device that may have a broad range of television channels, whereas women and children are relegated to using secondary TVs that may not have the same capabilities. Women's interest in many channels and streams may in part be due to this. However, access to high speed Internet is still an important issue for women as well.

3.4 Usability Problems

The verdict of the users is that the perceived utility and versatility of Surf on TV was limited. It had limitations in terms of interface design and functionality (i.e. scrolling, text readability, no possibility to download and store material etc.). Some trial users felt that surfing on the TV was slower than their Internet offer from HB@.

The users reported usability problems related in particular to the use of interactive services on the TV that required the use of keyboard and/or mouse (e.g. E-mail on TV). There was some confusion as to which input device to use in certain settings (remote control, mouse or keyboard), and the presence of mouse and keyboard in itself was an unwelcome addition to the TV viewing setting. Other problems like long response times, instability in the system and poor visibility of materials on the TV screen were also reported.

It turns out that many of the new interactive services have technical weaknesses – first and foremost instability – that turns out to be major disadvantages in a leisure setting. The users' benchmarks for home entertainment technology are established technologies like the TV set, the stereo rack and the radio. These technologies have over the years been honed into exceptionally trustworthy technologies in stark contrast to many new interactive services. Timeliness, ease-of use and convenience are important parameters in a home setting. During the interviews we were met by a very low tolerance for technical errors and time-delay on the TV platform as opposed to services on the PC platform. People tend to be far more forgiving towards unstable services on the PC platform than on the TV platform.

4 Concluding Remarks

Although most subjects in our surveys and interviews had a vague notion of how the HB@ TV channels were digital and integrated in the digital broadband solution, a clear distinction between broadcast and Internet services still seems to exist. To put it differently, there is little evidence of convergence in people's minds. Our impression is that this puts "hybrid" services like Surf on TV at a disadvantage.

Users have very different requirements for stability and service quality depending on whether they are using a TV- or PC-centric service: TV's that will need time to "boot" will not be tolerated.

Current PC-centric services usually include more interaction than traditional TV centric services, and require different and functionally more advanced interaction devices (keyboard, mouse) than those currently used in the TV-centric setting (remote control).

ICT and in particular the TV, can be defined as artefacts of beauty and importance around which other furniture and indeed whole rooms are organized (Silverstone, 1995). The introduction of "PC-style" interaction devices was generally not appreciated in the more relaxed TV viewing "sofa" setting, and may in part have contributed to the lack of popularity of Internet services via the TV (E-mail on TV, Surf on TV) in the HB@ project.

To achieve convergence of TV-centric and PC-centric services, and the success of such services, service providers will need to address issues like those raised in this paper.

References

Ims, L.A., Loktu, H., Haga, K., Johannessen, O., Meinich, F., Thrane, K., Ling, R., Andersson, K. & Myrvold, Ø. (2002). Towards deployment of VDSL for interactive broadband services: Experiences from a large-scale consumer market trial. In Proceedings of ISSLS 2002. Seoul, Korea, April 2002.

Silverstone, R. (1995). Media, communication, information and the 'revolution' of everyday life. In S. Emmott (Ed.), *Information superhighways: Multimedia Users and Futures* (pp. 61-77). London: Academic Press.

Web LogVisualizer: a Tool for Communication and Information Management

José Nunes[1] jnunes@ca.ua.pt
Florin Zamfir[2,3] florinzamfir@hotmail.com
Óscar Mealha[1] oem@ca.ua.pt
Beatriz Sousa Santos[3] bss@ieeta.pt

[1]Dep. of Communication and Art, University of Aveiro, Portugal
[2]Dep. of Computer Science, University of Criova, Romania
[3]Dep. of Electronics and Telecommunications/IEETA, University of Aveiro, Portugal

Abstract

This paper describes an instrument intended to help a usability specialist or organization communication and information manager to analyze an institutional web site information flow and communication activity. Details like the site map, page information distribution and hotspot location on the pages, for the general site view, but also information like site statistics and user session selection and analysis are offered to the user. In the single session analysis, if more information (like mouse or eye tracking information) is available, it can be added to the information extracted from a log file and used appropriately.

1 Introduction

Technologically mediated scenarios are becoming more and more pervasive in the day-to-day activity of a growing number of individuals and institutions. Specifically, internet/web technologies and services have a strong presence in institutions worldwide. Internal web sites (intranets) are developed in conformance with internal communication strategies, and reflecting internal information and workflow. An emerging problem concerns the management of these constantly growing intranets/extranets. Organizational information and communication specialists lack the efficient tools to analyze activity and behavioural patterns, and understand what is really going on inside the institutions. Answers for what's wrong or right and why, usually are not delivered by traditional web analysis instruments (Tauscher & Greenberg, 1997), (Bieber, Vitali, Ashman, Balasubramanian & Oinas-Kukkonen, 1997). In fact, these instruments tend to be biased by classical technical metrics, in most situations, for technical tuning and not for organizational communication and information analysis.

Efficient analysis and diagnostic tools must be designed to cope with these sophisticated infrastructures. The problem lies in identifying user-system mismatch at the human-computer communication level, which must be thoroughly identified, and problems pinpointed to the design team. The system must serve the organization, adapt perfectly to its internal communication strategies, and sustain efficiently its information and workflow patterns. Good feedback instruments on problem identification are fundamental.

The human-computer communication process is very complex and difficult to evaluate. When evaluation of the information and workflow patterns is needed, there are new challenges to face, and an extra set of variables to evaluate and analyze.

Users must be able to use the internal/external organizational information and communication system in a "minimal cognitive effort" state of mind.
This work describes some typical solutions presented by commercial products for web log analysis and presents some of their limitations, in terms of human-computer interaction evaluation. Besides this, some proposals to complement this typical approach are also presented alongside the prototype conceptualization.
In section 2, the common approach of available commercial products will be discussed. Then, in section 3, our proposal and its prototype conceptualization details will presented.

2 Common Commercial Approach

Common commercial products deliver many solutions for Web Log Analysis and there are many implementations on the market. In fact, every implementation presented by each company has its own view of the problem and its own approach to analyse the information stored at web server logs.
However if we take a look to some products on the market, like the ones from 10-Strike Software[1], Alentum Software[2], eIQnetworks[3], Flowerfire[4], Leech Software[5], Mach5 Enterprises[6], NetIQ Corporation[7], SurfStats International[8], ZY Computing[9] among many others, we realize that they use, in most situations, the same information and present their results in a similar way.
Some of the more common information used by web log analysers are related to web server log tags: date, time, client IP address, username, client URL request, client URL request parameters, request status, bytes sent, client user-agent, client cookies, and client referer. Note that, some tags may be absent on some server or client configurations.
This information is subject to some statistical processing and the results are usually presented using tables and / or graphics like the examples presented on Figure 1and Figure 2.

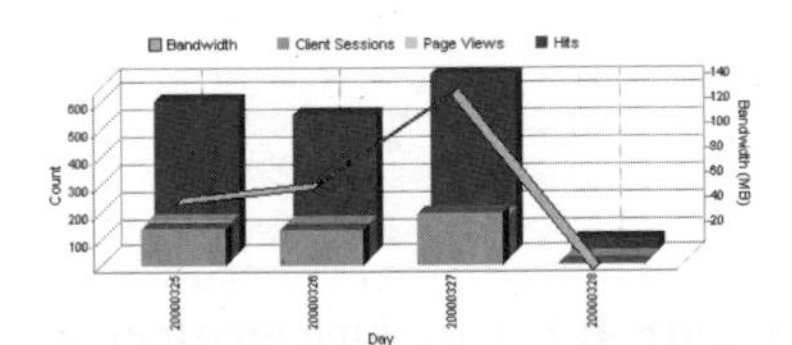

Figure 1: Mixed bar chart / graph example[8].

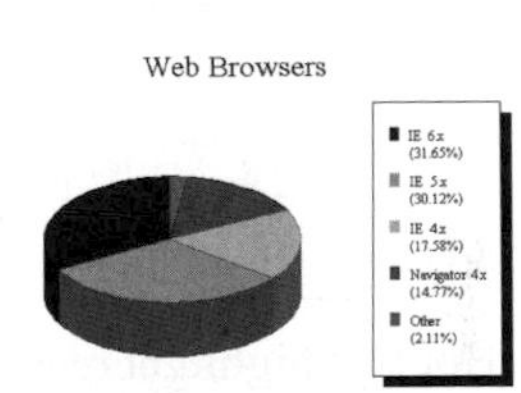

Figure 2: A pie chart[2].

3 A Prototype: Conceptual Model

The proposed prototype sustains its analysis on a usability perspective and on details concerning web site usage. The main target is an organizational web site comprising information and synchronous/asynchronous communication services. Inherent to the analysis process, and besides

---

[1] 10-Strike Software (2003). 10-Strike Log-Analyzer. http://www.10-strike.com/logan/
[2] Alentum Software, Inc. (2003). WebLog Expert. http://www.weblogexpert.com/
[3] eIQnetworks, Inc. (2003). eIQ LogAnalyzer. http://www.eiqnetworks.com/products/webanalytics.shtml
[4] Flowerfire (2003). Sawmill.Net. http://www.sawmill.net/
[5] Leech Software, Inc. (2003). OpenWebScope. http://www.openwebscope.com/
[6] Mach5 Enterprises (2003). Mach5 FastStats Analyzer. http://mach5.com/products/analyzer/index.php
[7] NetIQ Corporation (2003). WebTrends. http://www.netiq.com/webtrends/default.asp
[8] SurfStats International Ltd. (2003). SurfStats Log Analyzer. http://www.surfstats.com/
[9] ZY Computing, Inc. (2003). 123LogAnalyzer. http://www.123loganalyzer.com/index.html

site architecture, are user behavioral issues. It is intended to be possible to explore in at least the following different perspectives:

- Site structure/architecture, allowing exploration/browsing on a goal based orientation – flexible probing/visualization schemes can integrate an overall, holistic site view with a very detailed hotspot location on a page (Figure 3);
- Statistical information collection and analysis based on 2D/3D visually formatted outputs – intended to give statistical details about the usage of the site (why/how people use the web site, which pages are visited and in what form?);
- Identification of user session for individual and multi-session selection and detailed analysis of the selected session – intended to identify the limits of a user session, so it is possible to study the behavior of that user when using the site or how s/he performs a certain task or attains a specific goal (Figure 5 and Figure 6).
- This instrument can also be used to make some analysis on a user session from a controlled experiment, where it could be possible to capture and analyze added information like mouse movement/event tracking, eye movement/gaze tracking (Figure 4) or even other biometric information (merging this information with the information extracted from the site LOG file) – allowing to identify how a specific web user is interacting with the page.

The final goal is to producing a complementary tool to empower decision-making concerning system design problems or organizational technologically mediated communication and information management.

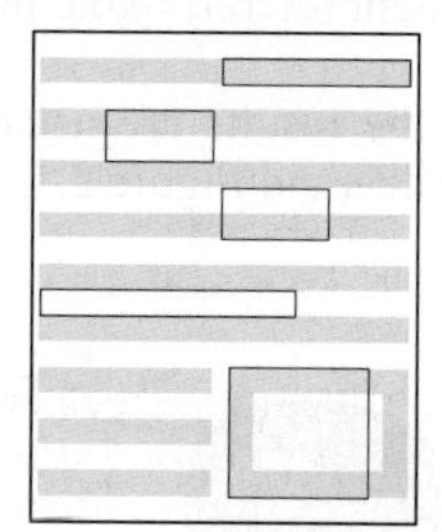

Figure 3: Hot spot highlight representation.

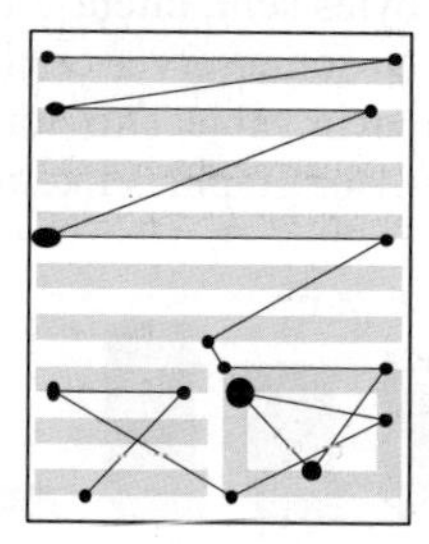

Figure 4: Eye tracking registration.

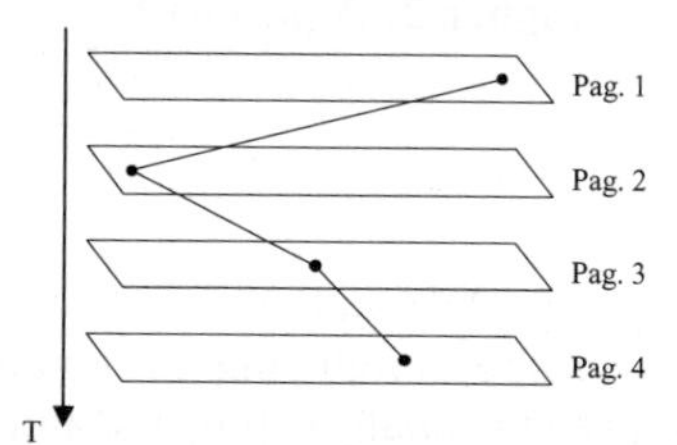

Figure 5: User selected links represented on a timeline– 3D view.

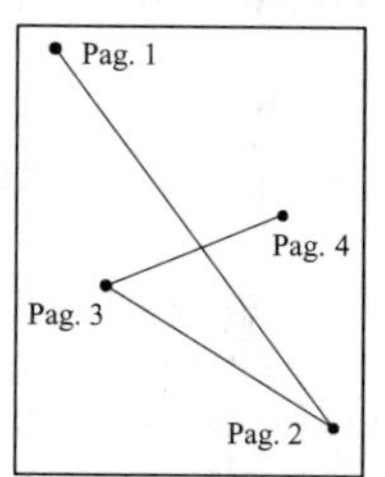

Figure 6: User selected links – projected on 2D plane (visual field).

The use of direct manipulation techniques in the interaction procedures could be particularly useful when associated to a statistical visually formatted feedback, used to inspect and analyze individual institutional behavior and organizational communication and information flow patterns. In the sequence of this framework some user interface design guidelines can be enumerated:

- The prototype must present a coherent and integrated environment for the minimization of context loss through the synchronization of different views of the information;
- The manipulation and analysis procedures should use a direct manipulation approach;
- The various functions and tools must be designed according to the correlated visual inspection strategies;
- Statistical output must, preferably, be done in an integrated visual environment, using color code schemes, coded 2D/3D objects and be based on spatial occupation and distribution rules and schemes from traditional information visualization techniques;
- The analysis and visualization schemes adopted must be easily used to support decisions;
- Complementary data can be obtained from specific instruments (e.g. eye movement/gaze, mouse tracking, etc), and can be used to enrich session analysis (Figure 7);

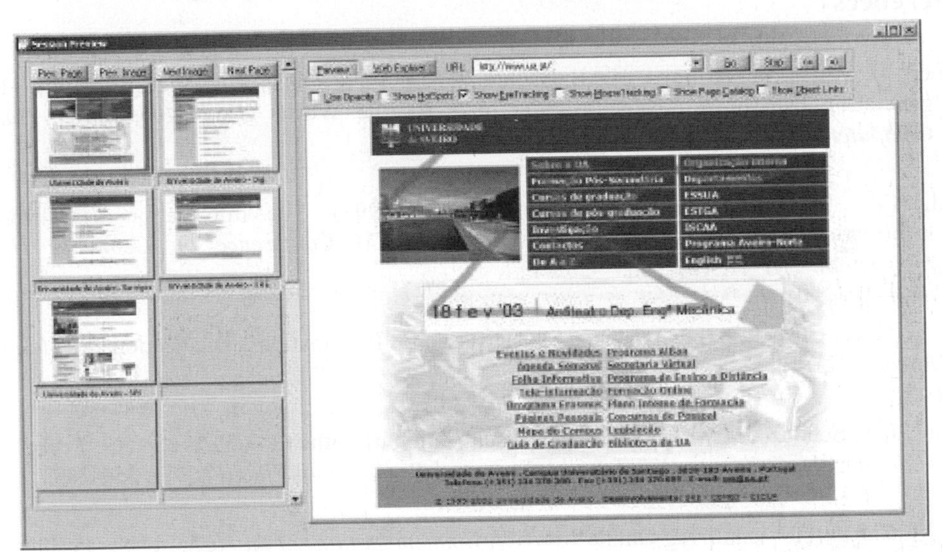

Figure 7: Session preview correlated with eye tracking information.

4 Conclusion and Future Work

With the use of the prototype briefly described in this paper, it has already been possible to identify, on one hand, some aspects that result easier with this application then with traditional approaches and, on the other hand, some areas where improvements will be necessary.

Technologically mediated communication scenarios can be analyzed and understood more easily with the help of specifically designed tools similar to the one conceptualized in this paper, since it allows different observation schemes based on the data obtained from a web server log file. The possibility to mix this information with the structure of the site and user-computer interaction monitoring data seems to be a big step toward a deeper understanding of the communication process, impossible with most available applications.

We can already identify as future work the following enhancements/research for the presented instrument:

- Site Diagram Views: site representations, like tree maps, are not suitable for all kinds of sites and further investigation is needed. Alternative 2D and 3D approaches would be developed and evaluated. The main goal will be to get a deeper understanding on the site contents, its distribution and correlation on each page and among pages.

- Statistical Analysis: it would be interesting to have more detailed information about session characteristics and user behavior during site navigation. 3D representations will be developed and evaluated for analysis of user navigation data such as different paths used and time spent to achieve a predefined goal. These schemes should augment a holistic view of navigational maps without loss of detaile in interaction and service network context.
- Session Analysis: it could be interesting to produce an integrated view of more than one session, to compare different users or the same user in different occasions. Once again, new 2D and 3D views could be interesting to get a better understanding of the relation between the information distribution inside a page/pages and user behavior.

References

Bieber, M., Vitali, F., Ashman, H., Balasubramanian, V. & Oinas-Kukkonen, H. (1997). Forth generation hypermedia: some missing links for the World Wide Web. *Int. J. Human-Computer Studies*, 47, 31-65.

Card, S.K., Robertson, G.G. & Mackinlay, J.D. (1993). The Information Visualizer, an Information Workspace. *Proceedings of the ACM Conference on Computer Human Interaction, CHI'91*, 181-188.

Catarci, T. &. Cruz, I. F (eds.) (1996). Special Issue on Information Visualization. *ACM Sigmod Records*, 25(4).

Cugini, J. & Scholtz, J. (1999). VISIP: 3D visualization of paths through Web sites. *Proceedings of the Int.l Workshop on Web-based Information Visualization (WebVis'99)*, 259-263.

Eick, S. G. (2001). Visualizing Online Activity. *Communications of the ACM*, 44(8), 45-50.

Mealha, O., Nunes, J. & Santos, B.S., Modelo de Análise da Comunicação Humano-Computador: uma proposta. *I Congresso Ibérico das Ciências da Comunicação.*

North, C. & Schneiderman, B., Snap-Together Visualization (2000): Can Users Construct and Operate Coordinated Views?. *Int. J. Human-Computer Studies*, 53, 851-866.

Robertson, G.G., Card, S.K. & Mackinlay, J.D. (1998) Information Visualization Using 3D Interactive Visualization. *Communications of the ACM*, 26(4), 56-71.

Schneiderman, B. (1998). Designing the user Interface: Strategies for Effective Human-Computer Interaction, Addison Wesley Longman.

Tauscher, L. & Greenberg, S. (1997). How people revisit web pages: empirical findings and implications for the design of history systems. *Int. J. Human-Computer Studies*, 47, 97-137.

Tufte, E.R. (1994). Envisioning information. Cheshire, Connecticut: Graphics Press.

Innovative UI Concepts for Mobile Devices

Birgit Otto

Siemens AG
Corporate Technology
Otto-Hahn-Ring 6
D-81730 Munich, Germany
birgit.otto@siemens.com

Fritjof Kaiser

Siemens AG
Corporate Technology
Otto-Hahn-Ring 6
D-81730 Munich, Germany
fritjof.kaiser@siemens.com

Abstract

Today's mobile devices already provide enough computing power and network connectivity for complex enterprise applications. New application domains like wireless ad hoc services and location based services arise. However, this becomes a challenge when such an amount of functionality shall still be accessible and usable by the end user on small and limited devices. In this paper a new UI-concept for mobile devices, the so called Rotating Spaces, is presented. It allows to handle a complex amount of applications on a small display. In every state of the UI all applications are available (no hidden information) and the user is aware of running applications. The organization of applications and information is easy and efficient.

As a proof of concept for the general approach the Rotating Spaces have been realized and usability tested in a project for mobile wireless enterprise services in a corporate campus.

1 Problems

User interfaces of mobile applications have several completely different constraints than PC-based applications: small screen sizes, limited input-and output mechanisms, low performance and new usage paradigms. This is a challenge for designing highly attractive and usable UIs. Due to the very small screen size every application fills nearly the whole screen. Therefore the desktop-metaphor that is used for PCs and that offers icons distributed on the screen to open applications is not adaptable for a small device. Menus seem to allow a space saving navigation to open applications but in a lot of cases they become too long so that scrolling is needed. Existing Pocket PC Systems solve this problem by offering different possibilities to open or close applications, each of them following another approach: via the start menu, via a programs-window (icons), shortcuts to the last used applications, a list to open/close masked running applications and a document-based access to applications (note, date, contact, etc.). This variety does not seem to be the best solution. Testing the usability of Pocket PCs we have seen that also experienced PC users initially have difficulty to find and manage applications and information of a Pocket PC. Furthermore for a user it is not really clear if an application is closed or only masked. This information is especially important for applications that waste a lot of power as well as for applications that use cost generating connections to the internet or to other servers.

The goal of the project presented here was to find a metaphor and new concepts for the organization of and the interaction between applications, documents and all kinds of information of a PDA.

2 Information Architecture

2.1 Dividing the Information Space into Object Classes

The above described problems led to a systematically reorganization of all objects (e.g. programs, documents, devices, etc.) of the whole information space of the PDA. The idea was to divide the information space into different classes of objects. The more classes exist the more difficult the significant differentiation of them is. This might lead to a difficulty for the user to search an object in the right class. On the other hand the interaction between objects should take place only between objects of different classes. The challenge was to identify as few classes as possible and as much classes as needed. At the end three classes were identified: applications, documents and persons (see Figure 1). One new result is the identification of persons as equitable objects to documents and applications. A person object contains all person-relevant information as name, address, phone numbers, etc.

2.2 Interaction between Objects

As mentioned above interaction takes place between objects of different classes. Most interactions can be initiated from both objects. For example, in the case of showing a document with the video projector the user can first open the document and afterwards search the projector or vice versa. Similar interactions are possible between documents and printers and so on. There exist also interactions that involve three objects from three spaces, for example write an email to a person and attach a document. Also interactions with three objects can be initiated from different objects like sending a document to a person via email. In the realization of the object classes we took into account that every object of every class can start an interaction.

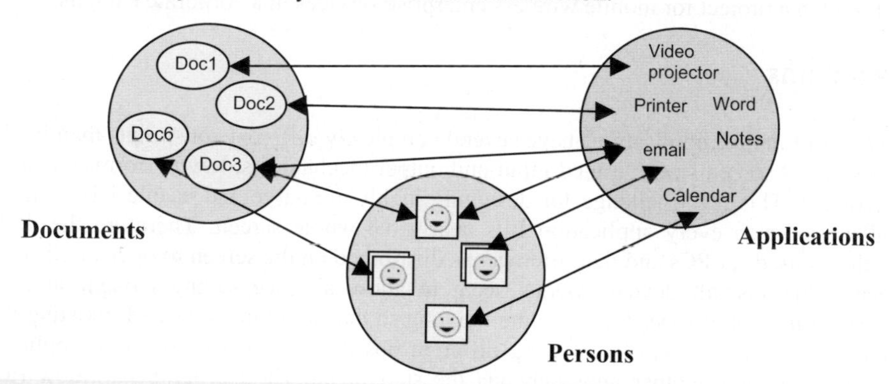

Figure 1: Identified classes with some exemplary objects and some possible interactions.

3 Graphical Realization – The Spaces Concept

To graphically realize the object classes defined above an abstraction of the room-metaphor (Berger et al., 1997) led to the concept of the so called ***"Rotating Spaces".*** Every object class is realized as one space. For navigation between the spaces the metaphor of a roll is used: all three spaces are placed on the roll in such a way that one space is prominent in the middle of the screen, the others are positioned above and below on the margin. With a click on one of the margin spaces a rotation of the roll brings the desired space into the middle of the screen and the others onto the

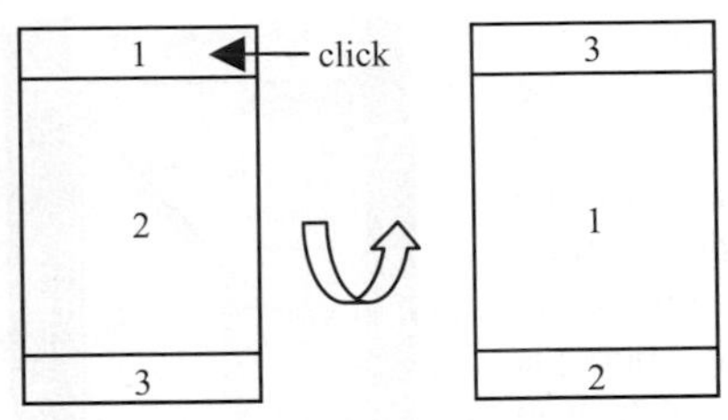

Figure 2: Behavior of the Rotating Spaces

margin (see Figure 2). A different color for each space provides a visual orientation helping to distinguish and to recognize the spaces easily. This concept allows a very easy navigation between the different spaces, each space filling the screen as much as possible. Opened applications or documents are placed as icons on the margin of their original space in order to view and access them in any state of the GUI.

The Rotating Spaces have been realized within a project called Enterprise on Air (EoA) in which a system for enterprise services for mobile users in a corporate campus has been developed (see Figure 3).

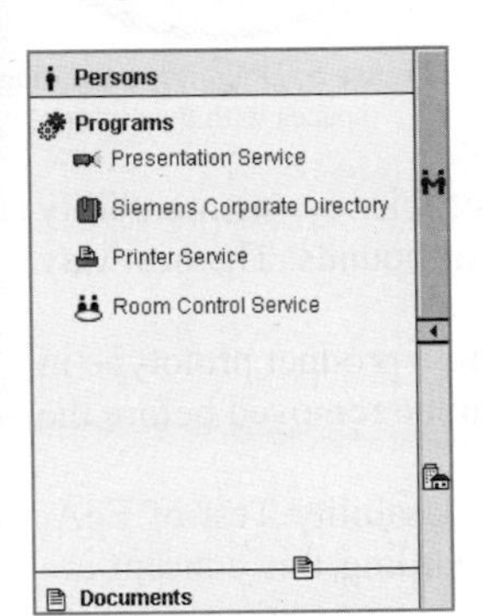

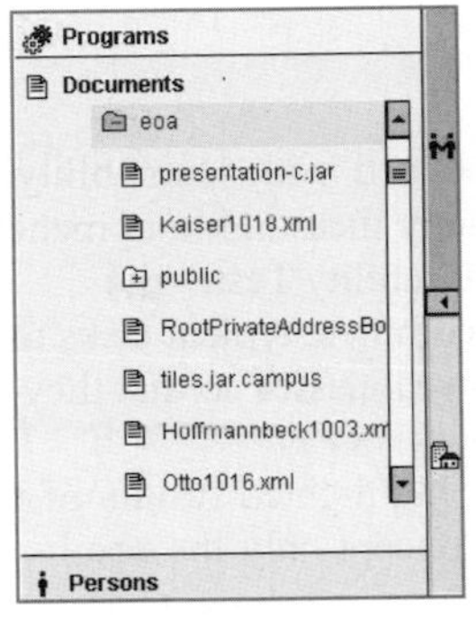

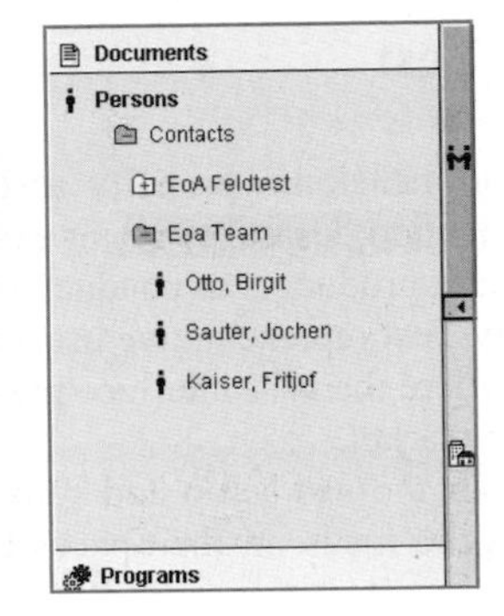

Figure 3: The Rotating Spaces "Program Space", "Person Space" and Document Space" in the EoA project.

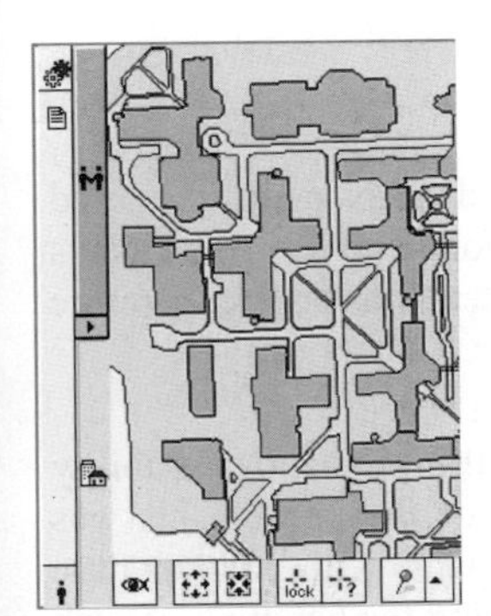

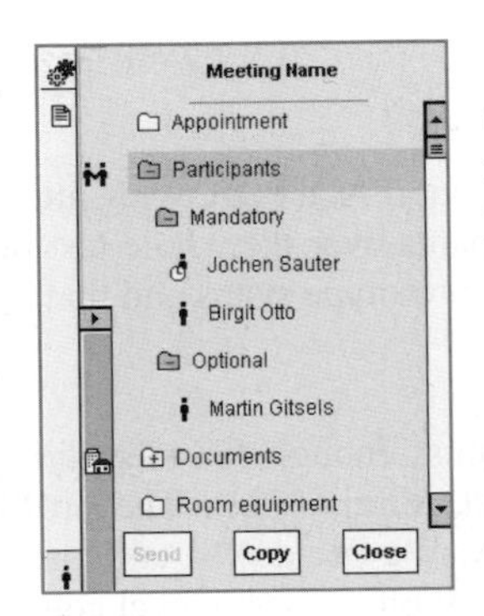

It offers navigation to and information about points of interest and wireless meeting room services such as printing, presentation of slides and room control functionality. The analysis of the application scenarios led to the additional requirement of providing some kind of shortcut to special applications, the so called ***"Sliding Spaces"*** implemented here as "Campus Space" and "Meeting Space" (see Figure 4).

Figure 4: The Sliding Spaces "Campus Space" and "Meeting Space"

When the Sliding Spaces are closed, they are placed on the right border of the display and have the aspect of tabulators. They can be opened by clicking on the tabulator or via drag & drop (see Figure 5). The Sliding Spaces slide into the screen from right to left. Horizontally they fill the whole screen, in vertical direction they can reside in every position on the screen apart from a margin on the left side where the Rotating Spaces are visible and a margin on the right side where their tabs are shown. In fact, it is important for the Spaces Concept that all inactive spaces are visible on the margin.

Trees have been chosen for the visualization of the information inside a space. Trees are a very space efficient way to navigate and to represent information. In order to increase attractiveness and usability small icons are placed at the beginning of each item of the trees. Direct interaction between objects of different spaces is possible via drag & drop, performed with the respective

icon. If an object is dragged onto the margin of another space this space scrolls automatically into the middle of the screen and the object can be dragged onto an object of the new space (for example drag a document onto a printer in order to print it). Also objects that are placed on the margin of a space can be used as a drag & drop source or destination. Objects can also be dragged into the sliding spaces to be available there. For instance, in order to configure a new meeting, documents, persons and applications can be dragged into the Meeting Space. Drag & drop between different spaces generates a link to the source object. Drag & drop within the same space signifies a move of the source object.
The Spaces Concept has been implemented in Java on a Pocket PC.

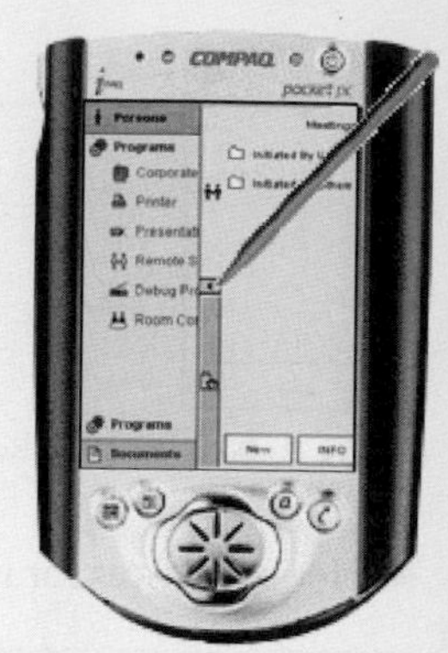

Figure 5: Dragging the sliding spaces with the stylus

4 Usability Test

Usability is a multidimensional property associated with learnability, efficiency, memorability robustness and satisfaction. Usability is not exactly measurable in inches or pounds. The best way to test the usability of a product is to conduct a Usability Test.
In a typical Usability Test representative users complete typical tasks using a product prototype in order to determine where the user interface has weaknesses so that they can be removed before the product is put on the market.
This chapter describes the test setup and some highlighted results of the Usability Test of EoA. Since this paper has the focus on the Spaces Concept only the results regarding this concept are reported.

4.1 Test setup

4.1.1 Test item

EoA was tested while running on a Compaq iPAQ PDA with attached LAN card. The used version of EoA was the same for all participants over the whole test period. The test participants were informed that the test object is still in a prototype status and that system crashes are therefore possible.

4.1.2 Test participants

The test was conducted with five participants, enough for identifying the substantial usability problems of an UI concept (Nielsen, 1994a, Virzi, 1992). The profile of the participants was equivalent to the defined user group of EoA. We decided to take participants with none or little knowledge about PDAs in order to put them all on the same level concerning the handling of such devices. We also decided to deny persons with programming experience as test participants, because programmers often tend to ignore usability problems unconsciously.

4.1.3 Tasks

The Usability-Test consisted of nine tasks. The main focus of the tasks was set on the EoA elements Rotating and Sliding Spaces, integrated services (e.g. presentation service) and Meeting Space.

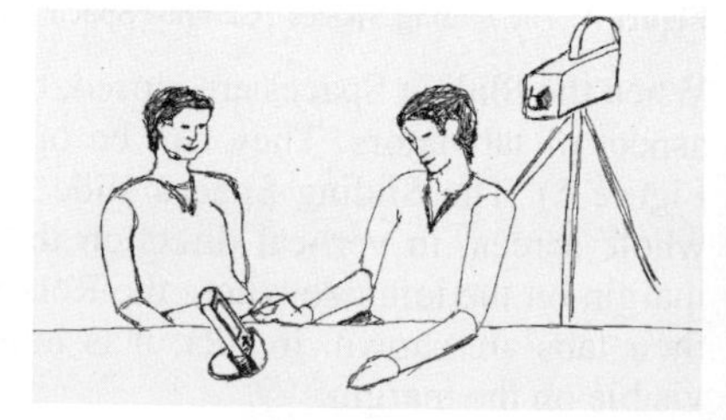

Figure 6: Sketch of field trial set up

4.1.4 Test Environment and procedure

The test participant and the conductor were sitting side by side in front of the PDA. The PDA was fixed on a cradle. The screen of the PDA was recorded by a video camera that was located out of

sight of the test participant, although the test participant was aware that he or she is filmed (see Figure 6). As generally conventional (Nielsen, 1994b, Dumas & Redish, 1994) the comments of the test participant and the conductor were recorded.
During the Usability Test the "thinking aloud" technique (Nielsen, 1994) was used, whereby the user is requested to speak his thoughts out loud while working on the task.
A test protocol was created for each test participant, featuring the main problems of the user interface as well as suggestions for improvements.

4.2 Usability Test results regarding the Spaces Concept

The new UI concept of three vertical Rotating Spaces was recognized and accepted by all test participants. The navigation between these spaces has shown no usability problems, either. The Sliding Space concept was not recognized at the first time, possibly due to a too small icon that triggers the Sliding Space from the right to the left side (in the first version tested the Sliding Spaces did not open by clicking on their margin tabulator but only by clicking onto the small arrow icon). Once the concept of sliding spaces was recognized the participants had no problems to navigate from Sliding Space to Rotating Space and vice versa. The drag & drop facility was used by some of the participants, but the performance of the prototype was too slow to give the user the feeling of controllability over this process.
Conclusion: The Usability Test of EoA has shown only minor usability problems in the areas that where covered by the tasks. These problems can be easily removed by changing the wording of buttons and labels or using improved icons. By increasing the performance of EoA, drag & drop solutions are expected to be usable, too.
After a field trial with an improved implementation of the system (18 participants, 3 weeks) users pointed out that they enjoyed especially the use of the Sliding Spaces.

5 Outlook

The concept of "Rotating Spaces" has been successfully deployed for a UMTS application realizing chat, instant messaging and buddy lists as the three Rotating Spaces. Thereby the margins of the spaces have been used to visualize the state, for example incoming messages in chat areas or new messages in the messaging box. This project showed impressively that the concept of the Rotating Spaces is highly transferable to special application domains.

6 References

Berger, M., Hohl, H., Jarczyk, A., Otto, B., Schneider, M. & Völksen, G. (1997). CoNus: Workspace-Based Intuitive Collaboration in Virtual Enterprises, Proceedings of the WET ICE '97, Workshop on Information Infrastructure for Global and Virtual Enterprises, IEEE Computer Society Press, Los Alamitos, CA, 1997

Dumas, J.S., Redish, J.C. (1994). A practical guide to usability testing. Norwood, N.J.: Ablex Publishing Corporation, second printing.

Nielsen, J. (1994a). Estimating the number of subjects needed for a thinking aloud test. Journal of Human-Computer Studies, 41, 385-397.

Nielsen, J. (1994b). Usability Engineering. Boston: Academic Press.

Virzi, R. (1992). Refining the test phase of usability evaluation: How many subjects is enough? Human Factors, 34, 457-468.

Observations from the Introduction of an Awareness Tool into a Workplace, and from the Use of its 'Status'-field

Samuli Pekkola[+*], *Niina Kaarilahti*[*] *and Pasi Pohjola*[*]

[+] Department of Information Systems, Agder University College, Serviceboks 422, 4604 Kristiansand, Norway

[*] Department of Computer Science and Information Systems, University of Jyväskylä, PO Box 35 (Agora), 40014 University of Jyväskylä, Finland

{samuli, nkkaaril, ppohjola}@cc.jyu.fi

Abstract

Awareness tools and instant messaging systems have become popular during the last couple of years – among teenagers. In this paper, instead on focusing leisure time use, we study how an awareness tool is adopted and used in a real workplace to assist employees work processes. Especially we focus on the use of the 'status'-field, which is an indicator of availability displaying whether the person is agreeable to be contacted by other people. The study shows that the 'status'-field, as such, is seldom used as intended. Instead, the users have merely accommodated to a general awareness information, that one has logged-in to the system, and considered it sufficient.

1 Introduction

Teenagers' use of instant messaging has increased rapidly during the last couple of years. The use ranges from informal conversations to event planning and schoolwork collaboration (Grinter & Palen, 2002). There are also a few studies about using instant messaging within workplaces (Nardi, Whittaker & Bradner, 2000; Isaacs, Walendowski, Whittaker, Schiano & Kamm, 2002). There the use conceptualises mainly to four functions. They include quick questions and clarifications, co-ordination and scheduling work tasks, co-ordination impromptu social meetings, and keeping in touch with friends and family (Nardi et al., 2000) – being in line with the teenagers' use of instant messaging. However, in this paper, we do not focus merely on instant messaging but take a broader scope and analyse how awareness tools, or, in other words, extended instant messaging systems, are introduced into a workplace.

What are the extended instant messaging systems then? Awareness of other users and their activities has been identified essential for co-ordinated activities in non-computerised environments (e.g. Goodwin & Goodwin, 1996; Heath & Luff, 1996). And when using computers, activities take place both within and through them (e.g. Bowers, O'Brian & Pycock, 1996). The awareness information needs to be explicitly available at the same magnitude. Instant messaging usually addresses the work within computers, although they are used for co-ordinating other activities as well. However, they do not generally support outside-of-computers-awareness consequently being insufficient to support the full spectrum of activities of the workplace. Therefore, to distinguish this divergence, such systems are regarded to as extended instant messaging systems, or as awareness tools.

In the real life, to be able to be aware of others in a workplace, one needs to be present there obviously. It is often regarded more appropriate behaviour to state one's presence and not to be hidden and invisible. Within computerised worlds, the same directive can be considered applicable. It is considered more appropriate to reveal and show one's presence to others. The concepts of awareness and presence are therefore mutual, and both parties need to approve their representations within both virtual and real environments.

In this paper, we shall discuss experiences gained from a group of users using an awareness tool for communication and collaboration with their colleagues, as a part of their daily work processes. The prototype, *de facto,* supports the awareness of other users and their activities in a workplace. One of its most prominent features is a 'status'-field, indicator of availability, which illustrates person's presence by the computer, and his or her willingness to interact with the others. Here we shall study some implications of introducing the system into a workplace and how the 'status'-field was utilised.

First, the prototype is briefly introduced. Second, the case organisation and some central findings from the case study are presented and discussed. Finally some conclusions are drawn and generalisations are made.

2 The prototype

The VIVA prototype was designed to support synchronous and distributed co-operative work. It combines multiple media (audio, text chat, instant messages (tiny textual messages), file transfer, email, shared whiteboard, and co-authoring) into an aggregate so that the users may utilise the medium (or the set of media) they find the most appropriate (Pekkola, Robinson, Korhonen, Hujala & Toivonen 2000). For example, occasionally users may use audio and whiteboard, sometimes an instant message is enough, and once in a while they may need the combination of audio, text-chat and file transfer. This makes it impossible for systems designers to make any assumptions of which media are incorporated at any moment (c.f. Ehrlich & Cash 1999, Reder & Schwab 1990). Such aggregate also allows better integration of media so that the awareness information can be presented across different media. This means that users working through the whiteboard connection are aware of others having an audio conversation, and vice versa.

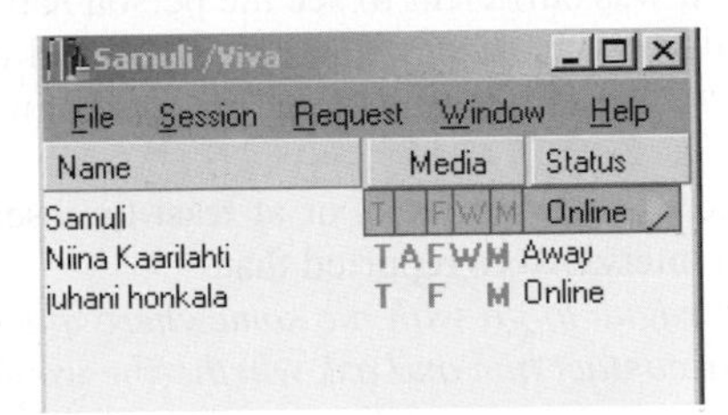

Figure 1. The awareness view of the VIVA prototype.

The awareness view of the VIVA prototype is illustrated in Figure 1. It shows the users online; the medium they are using, or are willing or able to use; and what is their availability, the 'status'. As this paper focuses on the 'status', let us have a detailed analysis of it. From Figure 1, there are three users logged-in to the system. Two of them are explicitly 'Online', i.e. by their computer, and one states to be 'Away'. Other alternatives are 'Busy', 'Idle' and a free form text. The 'Idle'-status illustrates that a user might be online, but has not touched a mouse for a while, or is 'Away', but has forgotten to change an appropriate status. The 'Idle'-status changes automatically to 'Online' immediately the mouse is activated.

The status of availability is central to this paper, so descriptions of the prototype, multiple media, and the support for awareness across different media are not considered in this paper (c.f. Pekkola et al., 2000).

3 Case study and central findings

The VIVA prototype was introduced into a metal industry company, and its' software and process development, and maintenance and administration groups, counting up 18 people. The prototype and the concept were new to the users, so they did not have many expectations. The system was used between three to seven months, during and after which the users were interviewed and their log-files were analysed. Although both groups worked with orgnisation's internal and external stakeholders, the VIVA prototype was used only for in-group communication and collaboration, not when communicating with external partners. This denotes the user community existed before the introduction of an awareness tool.

The VIVA prototype was designed to start automatically when the computer was switched on. This was intended to lower the threshold of use, of which it, as a matter of fact, did, and the users spent considerably long times online. For example, the average VIVA session (from log-in to log-out) lasted up to 277min (4h37min), although the average deviation was also great; 244min. Some people simply logged out immediately after the log-in. Automated start-up had also other implications. Since the system was always 'on', the 'status'-field illustrated whether a person was online and nearby the computer, so that visit to one's office, or interaction through VIVA, were worth to make. Yet, as the log-file analysis revealed, the users did not change their statuses actively. There were only 114 occurrences (for 10 users) when it had been changed from 'Online' to something else. The reasons for minimal conscious activations could be numerous. First, the users might have simply forgot to switch the status to appropriate position, or they were inexperienced to use it. Third, maybe they did not see it beneficial for themselves (paralleling Grudin's (1994) classical finding of disparity between the work and benefits), or were concerned about the privacy issues (that others see them and what they were doing). Or finally, and the most interestingly, saw the mere 'Online'-status adequate. It was sufficient to see the person had switched his or her computer on and was somewhere around in the building. There was no need for more precise information. All these reasons entangle around the same results; 'status' was not used very actively.

Nevertheless, some had learned a practice to use, or at least to observe the statuses actively, and utilise them in their work. Two interviewees reported that:

> "*When I need any colleague to go with me somewhere quickly, VIVA shows me who are present, so I can easily contact him and ask whether he would like to join with me.*" (MT)
>
> "*Some people use VIVA to indicate their presence. It is useful. When, for instance, I tried to make a phone call, I noticed that [the receiver] has switched his status to 'away': I didn't need to call.*" (PH)

The latter quote also points out some implications of availability and accessibility. The phone call was purposeless as the receiver was not in his office and not accessible by the phone. The accessibility information (whether the medium was in use) was supported in the VIVA prototype, although the users did not report that they had used it extensively or that it anyhow influenced their work. However, the feature was designed to support peripheral, i.e. background, awareness, which, as such, is difficult to recognise even in non-computerised environments. Hence, although any usage of the accessibility information was not reported (it cannot be recognised from the log-files), one can surely assume it influenced people's work – as the latter quote illustrates.

4 Discussion and conclusions

The VIVA prototype, and awareness tools in general, were new to the organisation and to its employees. This surely had its implications to the study, since users had to learn to use new tools and adapt new work practices. This takes time, which is evidently longer than in this study, and requires motivated users, which, in turn, can be obtained by making the benefits of the system as obvious to users as possible (Orlikowski, 1992; Prinz, Mark & Pankoke-Babatz, 1998). Also, Orlikowski (1992) noted that "early use of Notes in the office has proven more valuable for facilitating individual productivity than for collective productivity". Similarly, individual users have utilised the VIVA prototype, but this time the benefits are more difficult to recognise, since other users must be logged-in and be online *at the same time*.

Awareness tools require that the users are logged in to the system constantly – as default. The log-file analysis revealed that approximately 17% of all sessions (141 out of 839) were between ten seconds to ten minutes, so chances to accidentally meet an interesting person and spontaneously discuss the topic are minimal. Too often people logged in and then immediately out when a person they were looking for was not logged in. This is exactly the same process as to check new email, but the process is not adaptable, applicable or acceptable in this context. The longer the person is online, the greater the chances for interactions are, and further, the greater the benefits of an awareness tool are. Random short visits are just not appropriate. However, as identified, learning the new work practices took much more time than it was assumed.

In addition to the learning aspect, there might be other reasons for such a practice of random visits. One of the interviewees said that "*there is only* [name removed] *online, and I don't want to communicate with him just now*" (JR), so the person logged out immediately even though they were working in the same team. In other words, the interviewee did not want to belong to the same 'community' with the other (c.f. Harper, 1996). This sets the claim by Dourish and Bly (1992), that awareness tools increase community building, to a strange light. The awareness tool, in this case, did not increase the involvement to a community, but hampered the use of such a tool since there were no people who were considered interesting online. The evaluation of a person was based on external knowledge and the awareness tool did not change it at all. And also, when the interviewee did not want to reveal his presence to the other, he simultaneously refused to gain the potential benefits of being online. He preferred not to be seen by the 'uninteresting person' or to see others either. Feeling of being watched (by a colleague) overwhelmed personal, individually bound benefits.

The adoption of awareness tools into a workplace resembles the introduction of Lotus Notes in organisations back in early 90's. Although the users now have more experiences and expectations on the use of computers in general, they are still not familiar with the idea of revealing their presence in a virtual environment. They may leave their office door open to anyone, but they do not want to do the same with awareness tools. This is most likely caused partly by the lack of training and experiences, partly by organisational cultures and unsettled use of new tools. We believe these issues need to be seriously considered when introducing awareness tools into workplaces.

The conscious and intended use of 'status'-field was observed minimal. Users just did not change the value often. Instead they leant on the log-in information and considered it sufficient. There was no need for more detailed information, although its' utilisation was occasionally observed – as seen from the quote by PH above. This leads us to make the second observation; there is no need to build different variations of 'statuses'. What the users want is mere 'online' information that one's personal computer has been switched on. Other alternatives were not considered important – not in our small sample at least.

In this paper, some observations from the introduction of an awareness tool and from the use of the "status"-field were discussed. Instead of using the features as they were designed, users adopted much simpler practice (c.f. Grudin, 1994) and used the logging-in information for one's availability. One can correctly question whether the "status"-field is gratuitous. In this context it mostly was one, although profound training might have changed the situation.

5 References

Bowers, J., O'Brian, J., & Pycock, J. (1996). Practically Accomplishing Immersion: Cooperation in and for Virtual Environments. In: *Proceedings of CSCW'96*. ACM Press, 380-389.

Dourish, P., & Bly, S. (1992). Portholes: Supporting Awareness in a Distributed Work Group. In: *Proceedings of CSCW'92*. ACM Press, 541-547.

Ehrlich, K. & Cash, D. (1999). The Invisible World of Intermediaries: A Cautionary Tale. *Computer Supported Cooperative Work: The Journal of Collaborative Computing*, 8, 147-167.

Goodwin, C., & Goodwin, M. (1996). Formulating Planes: Seeing as a situated activity. In: Y. Engeström and D. Middleton (Eds.): *Cognition and Communication at Work*. Campbridge University Press, 61-95.

Grinter, R. E., & Palen, L. (2002). Instant messaging in teen life. In: *Proceedings of CSCW'02*. ACM Press, 21-30.

Grudin, J. (1994). Groupware and Social Dynamics: Eight Challenges for Developers. *Communications of ACM* 37(1), 92-105

Harper, R. H. R. (1996). Why People Do and Don't Wear Active Badge: A Case Study. *Computer Supported Cooperative Work: The Journal of Collaborative Computing*, 4, 297-318.

Heath, C. & Luff, P. (1996). Convergent activities: Line control and passenger information on the London Underground. In: Y. Engeström and D. Middleton (Eds.), *Communication and Cognition at Work*. New York, Cambridge University Press, 96-129.

Isaacs, E., Walendowski, A., Whittaker, S., Schiano, D. J., & Kamm, C. (2002). The character, functions, and styles of instant messaging in the workplace. In: *Proceedings of CSCW'02*. ACM Press, 11-20.

Nardi, B. A., Whittaker, S., & Bradner, E. (2000). Interaction and outeraction: instant messaging in action. In: *Proceedings of CSCW'00*. ACM Press, 79-88.

Orlikowski, W. J. (1992). Learning from Notes: Organizational Issues in Groupware Implementation. In: *Proceedings of CSCW'92*. ACM Press, 362-369

Pekkola, S., Robinson, M., Korhonen, J., Hujala, S., & Toivonen, T. (2000). Multimedia Application to Support Distance Learning and Other Social Interactions in Real-time. *Journal of Network and Computer Applications*. 23 (4), 381-399

Prinz, W., Mark, G., & Pankoke-Babatz, U. (1998). Designing Groupware for Congruency in Use. In: *Proceedings of CSCW'98*. ACM Press, 373-382.

Reder, S. & Schwab, R. G. (1990). The temporal structure of cooperative activity. In: *Proceedings of CSCW'90*. ACM Press, 303-316.

Making Privacy Protocols Usable for Mobile Internet Environments

John Sören Pettersson, Claes Thorén, Simone Fischer-Hübner

Centre for HumanIT, Karlstad University
651 88 Karlstad, Sweden
[john_soren.pettersson, claes.thoren, simone.fischer-huebner]@kau.se

Abstract

Privacy and identity management for the mobile Internet is still in its infancy. In this study we investigated some fundamental aspects of user interface design for P3P-enabled WAP browsers on mobile phones. We first analysed the non-English laymen's understanding of privacy vocabulary in order to choose appropriate terminology for the UI. We furthermore adapted a concept for early user-testing of multimedia products to the small displays of mobile phones. Thereby we were able to make adaptations of the user interface design for P3P-enabled WAP browsers in Mobile Internet environments.

1 Introduction

The Platform for Privacy Preferences Protocol (P3P, [W3C 2002]) allows user agents to come to a semi-automated agreement with websites about the privacy practices for personal data processing by those sites. P3P can enhance the user's privacy in the Mobile Internet, because it allows the user to control under which circumstances it wants to reveal what kind of personal data to web or WAP (Wireless Internet Protocol) sites. However, P3P also poses problems for user interface designers. According to European privacy legislation, users have to be well informed about the consequences for releasing personal information to web/WAP servers. Nevertheless, users should not be confronted with extensive information that might be anticipated as bothersome, so that they might even decide not the use the privacy features at all. Ensuring both user-friendliness and informativeness is even more a challenge for the mobile Internet with restricted devices.

One of the most prominent and used P3P user agents for ordinary web browsers is the Privacy Bird from AT&T. Cranor (2002) describes its development through a succession of prototypes. Our work has been inspired by the Privacy Bird design. However, the obstacles when trying to inform a user through the small screens of ordinary WAP phones are numerous. Moreover, informing users is not only a question of display size. The privacy terminology as used within the P3P project has been driven by experts and is not readily comprehensible to the everyday user (Cranor, 2002, p. 255). Differences between languages may furthermore pose problems for user interface designers trying to express privacy policies in the language preferred by the user.

In this presentation, we report about a pilot study encompassing both a quick survey of non-native English speakers understanding of common privacy terminology (sec. 2) as well as the subsequent development (sec. 3) of a user interface proposal (sec. 4). The latter phase was informed by the survey and by user testing at various stages. We will also report about the usability lab system employed (sec. 3.2). This system has been constructed to facilitate the development of interactive interfaces. Especially, we explain how we are currently using it for developing interfaces for P3P-enabled WAP browsers for various handheld devices with small displays. Suggestions for user interface design will be presented in the concluding section (sec. 5).

2 Vocabulary test

The vocabulary test was performed with twenty-four Swedish first-year multimedia students a few weeks after the academic year had started before they were thoroughly familiar with IT terminology (Thorén 2003). One part of the questionnaire consisted of words and terms in English and the assignment was to explain what the words meant. Some words were selected based on the Privacy Bird interface while some further relevant terms were added like *location data*. Below we give a few illuminating examples of the low frequencies of correct answers.

The term *personally identifiable information* was given correct interpretations by only 25 percent of the students. The word *preferences* is important for the P3P protocol, as users have to specify their privacy preferences. However, only 37.5 percent gave a satisfactory answer.

Even though the sensitivity of information is dependent on its purpose of use, some kind of information, such as health data can be per-se highly sensitive for users. For the students' explanations of *sensitive information*, it is notable that many gave examples which could be said to illustrate some sensitive information, whereas 50 percent of the answers were classified as wrong. Implication for user interface design is then to exemplify 'sensitive' information or let the user define himself what sensitive information is for him. (see Fig. 3.b).*Location data* is of particular relevance in a mobile setting, as the possibility of tracking the user's location is a severe privacy threat in the mobile Internet. However, 71% answered that they did not know.

In conclusion, some of the problems that the Swedish respondents had stemmed from their unfamiliarity with the technology and the conceptual sphere it belongs to. However, there is also a discernible barrier provided by the English language. A presentation of privacy policy and preference statements in users' own language might not only be desirable but also necessary.

3 Initial designs and tests

In this study the focus has been on textual rather than voice presentation. We cannot assume that users will use earpieces or phones with free speech capabilities to be able to hear alerts or to listen to the information of the privacy preference setting pages. Certainly, other kinds of alerts are possible when using an ordinary mobile phone, especially using the ringer signal, vibration and the LED indicator. However, using a ringer signal makes it impossible to use WAP at many public places and shared work places. Vibrations and LED indicators are not present on every model of mobile phones. LED indicators might furthermore not be immediately visible for persons using the WAP function of their phones. Concentrating on on-screen information also made it possible to compare the merits of different display sizes.

We have used the displays of Ericsson 520m and R380 with screen size 100 x 80 and 368 x 120 pixels respectively, in these pilot experiments. The Ericsson R380 has a fold-out display which makes the screen wider and somewhat longer and this increases the capabilities of the interface. It also has graphical icons that due to the size of the display become large and clear. The experiments have not been based on real telephones but emulators appearing on the screens of an ordinary computer monitor. This we consider enough for comparative tests: if one layout works better on these emulators than another layout, it would probably work better also on a real appliance.

Two lines of prototypes were thus planned for. The prototype A series using images of the telephone with the small screen, and the prototype B series using the larger screen.

The layout was inspired by AT&T's Privacy Bird (Cranor, 2002). This included finding icons to alert the user if a web/WAP site's privacy policy does not conform to the user's privacy preferences, as well as a preference setting menu where the user after installing the P3P user agent can define his privacy preferences. Alerting needs three modes:

- The site's privacy policy is in agreement with user's privacy preferences;
- The site does not have a privacy policy;
- Disagreement between the site's policy and the user's preferences.

AT&T's Privacy Bird has different colours and different symbols: a double-note, a "?", and a "!" respectively. The note being appropriate for a bird, we chose a rather more general symbol to indicate that all is well: a smiling face (a so-called smiley), while "!" was used for warning. Initially, these two alert icons were placed in the lower right corner of the displays of the prototypes because especially in prototype A, the title bar was too short to house both a rubric and an alert icon. Setting preferences was done according to a low-medium-high scale, just like in the Privacy Bird. However, depending on the screen size, different ways of implementing icons and preference-settings were used for prototypes A and B.

Only one test with one test user was performed with prototype A (see below for how the user tests were conducted). It was obvious that this person simply did not notice when the alert symbols were switched on, or the icon got in the way for the ordinary text (or disappeared). The whole case of alerting will have to be rethought to possibly including ringer and vibration. Since the alert function did not work, the privacy preference setting was not elaborated, even if the user had problems to understand the brief explanations that were in fact included in Prototype A. However, considering the limited size of the screen, it is obvious that the right place to inform the user about possible preference settings might not be the display of the device itself.

Prototype B left more room for general information as well as alerting. Since the placement of the icons in prototype A was not felicitous at all and because the page title bar is longer in the Ericsson R380, the title bar was chosen. More explicit indications where also chosen, as shown in Fig. 1, resulting in pages like the one in Fig. 4.

Figure 1: The mobile privacy icons.

3.1 User tests

The test subjects were brought into the room one by one and were informed that they were going to do two things: read a newspaper and book a hotel room; half of the subjects were also told to set the privacy preferences first. When this was clear and understood the main menu of the cellular phone was presented to them and then they were on their own. They were instructed to think aloud at all times so that their trail of thought could be used to understand the results of the test. Every test took between 10 and 20 minutes. After a test was completed, the test subjects commented on their experience and gave suggestions. In all, three different stages of prototype B were tested. Also privacy and one English language expert user went through such sessions.

3.2 Experimental technique

In spite of the fact that we have not considered informing the test subject via voice, we have used an experimental set-up often employed in user testing in natural language processing projects. It is called Wizard of Oz because just like the wizard in the story with the same name, the test leader controls the output after hearing or reading the inputs of the test subject. In this way no real natural-language decoder is needed, and the developers can gain experience on what vocabulary and commands users like to employ. (Kelley 1984; Dahlbäck et al. 1993)

Naturally, such a faked test makes it possible to test interfaces of systems that are not yet implemented. However, when the feedback from the 'computer' is more than merely voice, some preparations of texts or graphics sequences are needed. To meet this demand, we have developed a system at Karlstad University named Ozlab. The Ozlab system is meant to facilitate the setting up of Wizard-of-Oz experiments, especially experiments on human-computer interaction based on graphical output and input (with 'graphical input' we mean drag-and-drop manipulations of things on the screen). During a test-run, the wizard has access to various functions that enables him to simulate the user interface of a computer program.

The Ozlab has been initially developed for other projects. Nevertheless, since it allows for very easy set-up of user interface prototypes, including interface-in-interfaces as in the above case where on-screen emulators of the Ericsson telephones contain the screen of the mobile devices, it has been very suitable for developing and testing the P3P-enabled WAP browser user interfaces for the two different screen-sizes. A simplified version consist of the full Ozlab software but run on two adjecent computers – a set-up we call 'mini-Ozlab'.

With a mini-Ozlab we can easily communicate around interactive solutions and make quick look-and-feel demonstrations. In Fig. 2, Claes Thorén makes an introductory interactive demonstration for professor Fischer-Hübner; she gains an immediate understanding of the user interface. The appearance of an interactive product is always hard to predict. Wizard-of-Oz tests make it possible for the expert-as-user as well as for the designer-as-wizard to get a deeper understanding of the interactivity of a specific design proposal. Even inexperienced designers have used Ozlab in multimedia prototyping (for an example, see Pettersson 2002).

Figure 2: Testing interactive design *as* interactive design

The draw-back of a ready-made Wizard-of-Oz system (or at least of the 2002 version of this system) is that it does not allow for an automatic run-time inclusion of web or WAP pages on the test subject's screen. However, the wizard-controlled output makes it possible to use material from other sites than P3P-enabled sites in contrast to Cranor's usability studies for the P3P Privacy-bird where specially prepared sites were used for the tests (Cranor 2002). (However, in our tests with inexperienced users mentioned in section 3.1 above, material from two already P3P-enabled sites were used.)

For the future, we are planning to make Ozlab tests with mobile devices, but new issues concerning the privacy of test persons will appear when monitoring users' real use of WAP.

4 A design proposal for small displays

As the vocabulary test have indicated that many users are not familiar with privacy technology, we have decided to use a design with hyperlinks, where links can be clicked by the users for getting a definition or explanation for specific terms with which many users are unfamiliar. Fig. 3.a shows such a design with layered GUIs. If the user clicks on "sensitive information", the screen shown in Fig. 3.b will appear which allows the user to define what sensitive information is for him/her.

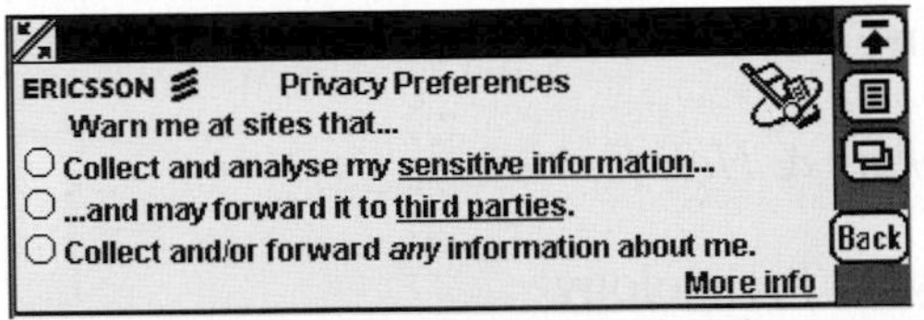

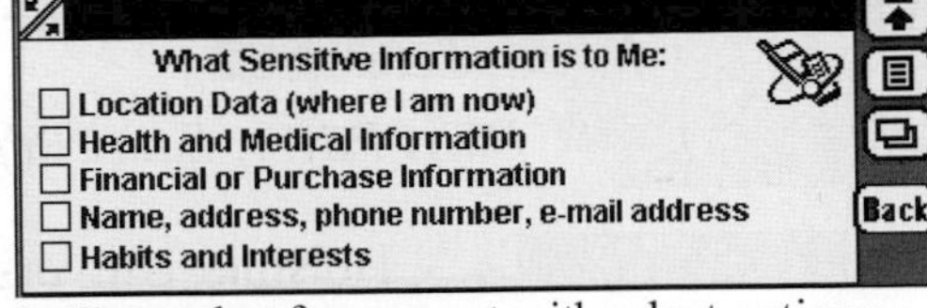

Figure 3: a. Preference menu with links **b.** Example of a concept with select options

The figures present design solutions that are the outcome of a series of usability tests in Ozlab. Fig. 4 shows a screen shot with a privacy icon (smiley face) in the title bar, which symbolises that the user's privacy preferences are consistent the WAP site's privacy policy.

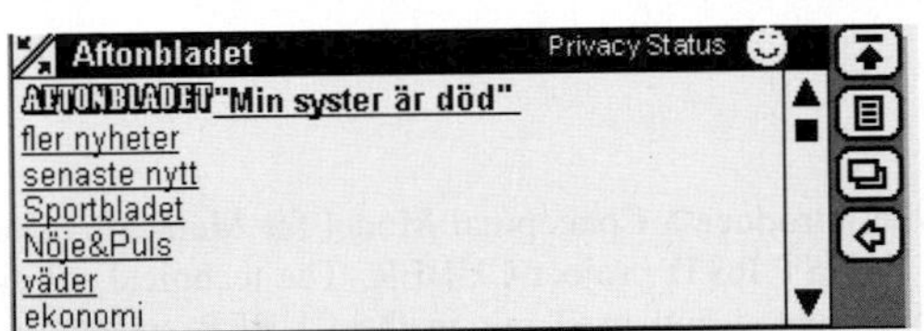

Figure 4: Privacy Status indication by a smiley

5 Conclusions

This study suggests that the unfamiliarity with privacy terminology among ordinary users is probably best met with: 1. User interfaces in native languages (more research is needed!), 2. The replacement of technical terms by descriptive words from ordinary language with examples, 3. Multiple-layered texts with significant words linking to lower levels.

The problems stemming from the small screens have highlighted the need to regard two purposes of user interfaces for privacy enhancing technologies: 1. Teaching (users have to understand the alternatives when setting their preferences), 2. Alerting (when browsing).

For the first kind, small-display phones like the Ericsson R520, are not suitable even if setting the privacy preferences could possible be done by someone familiar with the available options. For the second kind, small displays will definitely be difficult, because easily observable alerts in the display mean intrusive alerting, which will make WAP browsing a rather annoying experience. Ringer or vibration may be an option to select by the user.

On the system development as a process, we venture to claim the inherent multi-disciplinary nature of the question of PET. Exactly where the different experts fit in a design development cycle could need a comment. With a quicker and cheaper prototyping technique, we can make user-tested design proposals long before implementation. Every sort of stakeholder will thus have an earlier say in any design work.

References

Cranor, L.F. (2002) *Web Privacy with P3P*, Chapter 14: "User Interfaces", O'Reilly.

Dahlbäck, N., A. Jönsson & L. Ahrenberg (1993) "Wizard of Oz Studies – Why and How." *Knowledge-Based Systems*, 6 (4), 258-266.

Kelley, J.F. (1984) "An iterative design methodology for user-friendly natural language office information applications", *ACM Transactions on Office Information Systems*, 2(1), 26-41.

Pettersson, J.S. (2002) "Visualising interactive graphics design for testing with users". *Digital Creativity*, 13 (3), 143-155.

Thorén, C. (2003) Finding a usable vocabulary for privacy technology. Poster subm. to CHII2003.

W3C, World Wide Web Consortium (2002), "The Platform for Privacy Preferences 1.0 (P3P1.0) Specification", W3C Recommendation 16 April 2002, http://www.w3.org/TR/P3P/

The Conceptual Model for E-Learning Meta-Data Structure

Päivi Pöyry, Lauri Repokari & Heli Kautonen

Helsinki University of Technology
Software Business and Engineering Institute
P.O.Box 9600, FIN-02015 HUT, Finland
ppoyry@soberit.hut.fi

Abstract

The purpose of this paper is to introduce a Conceptual Model for Metadata information in E-Learning based on the research done within the IST R&D project CUBER. The technical solution based on this conceptual model for metadata was implemented and tried out in the CUBER system. A special user interface was constructed for the input of the metadata. The users of this "metadata authoring interface" were interviewed for evaluating the usefulness of the conceptual model. The first experiences and evaluations of the usefulness of this system and the conceptual model will be reported in this paper.

1 Introduction

In the recent years the e-learning sector has developed rapidly, and the amount of learning resources available on the Internet has grown exponentially. At the same time it has become overly troublesome to find the needed learning resources due to the fact that although there is a vast amount of information available, the information is in most cases unstructured and poorly classified (Sampson et al. 2001). Problems arise because the information is highly dispersed, and heterogeneous semantics are being used. Firstly, the potential learner searching for learning resources must carry out several searches with different search engines or course brokers instead of using one, centralized course information broker. Secondly, the varying concepts and semantics used in the different applications may confuse both the learner and the educational organization. Same concepts are being referred to with different vocabularies and same vocabularies are being used for describing totally different concepts. The situation gets even worse when the e-learning resources or institutions get international: the differences between the national educational systems and concepts get highlighted and conflicts may arise. It is extremely difficult to find information effectively if there is no consensus about the use of concepts.

Due to these problems the field of e-learning is facing a real challenge: how to organize and classify educational information in an effective manner in order to facilitate the information retrieval. As a possible solution we introduce a conceptual model for e-learning metadata and the use of standardized metadata for organizing the educational information. However, metadata alone does not solve the problem, with a highly organized set of standardized metadata the learning resources can be described, organized and classified effectively, which in turn enables effective information searching and retrieval.

The purpose of this paper is to introduce a Conceptual Model for Metadata information in E-Learning based on the research done within the IST R&D project CUBER. The expressive power of the emerging standard Learning Object Metadata (LOM) has been found rather limited for some areas, e.g. higher education (Simon 2001). In such cases where the requirements for metadata are very strictly tied to the context area, the LOM standard can serve only as a backbone to the application specific metadata model, and an Application Profile based on the LOM standard can be created in order to meet the context and application specific requirements. This research and development project concentrated on developing a Conceptual model for E-Learning with aggregation levels and other extension elements. The model developed in CUBER is conformant to LOM, i.e. it is an application profile of LOM. The CUBER aggregations Material, Course, Package, Programme will be introduced, since they form the backbone of the CUBER metadata. The innovation of the metadata work lies in the use of different metadata for different aggregation levels; the level of learning resources description varies according to the aggregation level in question. The CUBER aggregation levels aim to represent the "reality" in higher education in Europe, which is in fact a compromise between different and partly conflicting educational systems. Because there are extension elements that are not described in LOM, CUBER has developed a mapping for some metadata elements in order to ensure interoperability with other systems using the LOM standard.

2 E-learning and Metadata

2.1 E-Learning

E-learning can be seen as a ICT-enabled form of education, especially related to utilizing the Internet as the primary tool and channel of learning and teaching. The major difference between traditional distance education and e-learning is the enabling technology that helps to overcome the barriers of time, place and distance (Krämer 2000). E-learning has been referred to as a new form of studying that is free from the constrains of time and place. E-learning has served as a tool of knowledge management and competence development for many knowledge intensive companies. (Learningcircuits 2002) However, e-learning is not restricted to business life and corporate training. Instead, e-learning has started to gain increasing interest at the traditional universities that have faced the growing challenge of going on-line with their learning and teaching resources. The rise of e-learning sector has given the universities a chance to start building virtual universities that operate mainly on the Internet as virtual organizations formed by one or several educational institutions.

2.2 Metadata and educational metadata

Put simply, *metadata* is descriptive and classifying *information about an object.* It describes certain important characteristics of its target in a compact form. Metadata plays a central role in improving searching and categorising objects within a defined context of use. In order to be able to use metadata efficiently across different contexts and systems, the metadata scheme should be standardised. (Jokela 2001; Wason 2002.) There is a growing interest in using metadata in the field of education (Britain & Liber 1998), and at the moment metadata becoming increasingly important when digital government and e-commerce are emerging (Gilliland-Swetland 2000).

Metadata can be categorised in many ways (Jokela 2001). *Descriptive metadata* can be further divided in two sub-categories: *contextual metadata* and content-based *semantic metadata.* With contextual metadata we mean the conditions and the environment in which the metadata is created. Semantic metadata refers to the semantic characteristics of the object (Jokela 2001). Semantic metadata is very much domain specific, which means that the nature of semantic metadata is highly dependent on the concepts and semantic structures of the specific field (Jokela 2001), e.g. higher education. *Educational metadata*, which is semantic metadata, may describe any kinds of educational objects, e.g. study courses. The pedagogical features of the course, the contents, special target groups, and the technical requirements of the study course can be described with educational metadata (Lamminaho 2000.) Using existing standards offers possibilities for enhancing interoperability and reuse of data between different information systems, but the metadata models developed for different projects may not suffice because they may lack the information fields necessary for the special application area. In addition there is the problem of lacking consensus about the use of metadata and terms. The highly specialized vocabularies cause difficulties in the interoperability between the information systems (McClelland et al. 2002).

3 The Research and Development project CUBER

CUBER was a research and development project funded by European Commission's 5th Framework Programme. The project focused on creating the infrastructure and grounds for a Virtual University. The CUBER project's goal was to develop novel broker middleware combining a specific search engine and a knowledge base of standardised metadata (Krämer 2000). This kind of broker service enables the universities to offer their study resources for a larger audience in the Internet. For the learners the CUBER system provides the possibility to search for study resources with an efficient search engine, to compare the offers of various universities and to combine the study resources from many universities into individually tailored entities. (CUBER 2002)

4 LOM and CUBER metadata

Dublin Core Metadata Initiative (DCMI) and IEEE's Learning Object Metadata Standard (LOM) have formulated shared principles for the creation of new metadata sets (Duval et al. 2002). According to Duval et al., *application profiles* may be created in order to construct domain specific metadata sets that have selected elements from one or more metadata schemas and combined them into a one coherent metadata schema. This expresses the principles of modularity and extensibility required from a good metadata set. By using application profiles it is possible to create metadata sets tailored to the requirements of the particular application/system (Duval et al. 2002). CUBER metadata has been designed to be an application profile of LOM standard.

4.1 Overview of the Metadata model

The metadata model of CUBER consists of nine categories that contain metadata elements. The metadata categories function only as rubrics under which the related metadata elements are gathered. The metadata elements, for their part, carry information as values. They can also include sub-elements that carry more detailed information about the study elements described. (IEEE/LTSC 2002.) The metadata schema is organised hierarchically in the form of a tree. The metadata categories are placed in the top of the hierarchy, and the metadata elements are right below them. One category can have several metadata elements. The metadata sub-elements are subordinated to the metadata elements, and one metadata element can have one or several sub-elements. (IEEE/LTSC 2002.)

4.2 The CUBER Metadata Extensions and Aggregations

The CUBER extensions have been added to the LOM Base Scheme as independent metadata elements and sub-elements in order to be able to describe all the characteristics of the study elements and to be able to provide the functions of the CUBER system. All metadata categories of LOM have been used in CUBER as such. The greatest number of changes was made within LOM Category 'Educational', because of the special need of CUBER to describe the educational characteristics of the study elements. First of all, the vocabularies of selected metadata elements, such as 'Educational.Context', were altered. The teaching activities, examination, ECTS-credits, study guidance, enrolment, dedication to studying, pre-requisites, and related official degrees can be described with the additional CUBER metadata elements.

CUBER_Aggregation is introduced as a central extension element. Firstly, the level of abstraction is more detailed in LOM than in CUBER. LOM aggregations are much more atomic, when compared to the aggregations of CUBER. Second, the content and context of the CUBER_Aggregations is different from those of the aggregations of LOM. CUBER needs to describe its Learning objects or Study elements in a wider context that enables the description of larger study elements, such as study packages and programmes. Furthermore, the main functionality of the CUBER system is based on the aggregation levels defined in CUBER, and it would be very problematic to integrate the aggregations of LOM into CUBER.

The aggregation levels – study *material*, study *course*, study *package* and study *programme* – are used to describe the relationships and the differences between the study elements. Material is the smallest unit followed in the hierarchy by course. After package, the largest unit is the programme. The *descriptions* of the study elements in the different aggregation levels vary according to the individual aggregation level. The metadata schema offers a possibility to use free-text descriptions in some of its elements, and the level of abstraction gradually moves from concrete and specific towards abstract and general according to the level of aggregation.

5 Evaluation of the model and discussion

Before adopting and further developing the CUBER system, its ability to describe content providers' educational offers had to be examined. The ideal moment for starting the evaluation was after the release of the final prototype, at the same time when the partners started providing their information, i.e. uploading their data to the system. The purpose of the study was not to validate the quality of the system but to explore the acceptance of the conceptual model, to detect possible conflicts and to discover further requirements. Therefore, the study concentrated on capturing the local concepts to be fit into the CUBER system, revealing gaps between the local conventions and the model offered by CUBER, and detecting outcomes and effects of possible conflicts to the use of the system.

The study focused on users of the content provider interface, since they were considered better experienced in educational concepts and educational systems, as well as more experienced in the CUBER information model. The study was conducted in two phases. It included an orientational questionnaire that was sent to the respondents in advance and an interview that was based on the questionnaire. Even though the number of respondents was considerably small, the responses were carefully analysed on every detail and they were adequate for exposing the desired information. The results showed how well the information provided by an individual respondent organisation matched with the model offered by CUBER.

The results form the feedback and the interview can be summed up to the following findings: The study element aggregations are efficient for describing the study elements of the provider organisations, although the supply of an individual organisation did not match with the model in full. These mismatches were not considered to cause any sort of defect or obstacle to the use of the system. The idea of displaying data for aggregations on a different abstraction level was welcomed, since different information, indeed, is needed of different study elements. There are several metadata elements, whose use and options need to be reconsidered. There may also be need to find better-accepted definitions for some individual elements. Nevertheless, there should be more examples of all elements and their usage. The results indicate a need to further define, model, and finally visualise the relations between study elements and other studies. Although the system is now perceived as acceptable, better functions on this concept may eventually increase the usefulness and thus usability of the system.

As the final conclusion the metadata model was found suitable for e-learning solutions, even though further developments are needed. The preliminary results indicate that this model is functional. Further research will focus on the expressive power of this metadata model.

References

Britain, Sandy & Liber, Oleg. 1998. *A Framework for Pedagogical Evaluation of Virtual Learning Environments.* JISC Technology Application Programme. <http://www.jisc.ac.uk/jtap/htm/jtap-041.html>

CUBER 2002. The official Web-site of the CUBER project.<http://www.cuber.net>

Duval, E., Hodgins, W., Sutton, S. & Weibel, S.L. 2002. Metadata Principles and Practicalities. D-Lib Magazine April 2002. <http://www.dlib.org/dlib/april02/weibel/04weibel.html>

Gilliland-Swetland, A.J. 2000. Introduction to Metadata, Setting the Stage. <http://www.getty.edu/research/institute/standards/intrometadata/>

IEEE/LTSC 2001. IEEE Learning Technology Standards Committee (LTSC)/IEEE P1484.12 Learning Object Metadata Working Group. 2001. *Draft Standard for Learning Object Metadata.* <http://ltsc.ieee.org/doc/wg12/LOM_WD6-1_1.doc>

Jokela, Sami. 2001. *Metadata Enhanced Content Management in Media Companies.* Acta Polytechnica Scandinavica. Mathematics and Computing Series No. 114. Helsinki University of Technology: Doctoral thesis.

Krämer, Bernd J. 2000. *Forming a Federated Virtual University Through Course Broker Middleware.* LearnTEC 2000, Karlsruhe, February 2000 < http://www.fernuni-hagen.de/D VT/Publikationen/Papers/LearnTEC2000.pdf >

Lamminaho, Virva. 2000. *Metadata specification: Forms, Menus for Description of Courses and All Other Objects.* CUBER project: Deliverable D3.1.

Learningcircuits 2002. <http://www.learningcircuits.org/glossary.html>

LOM 2002. Learning Object Metadata standard (LOM) of IEEE. <http://ltsc.ieee.org/wg12/index.html>
McClelland, M., McArthur, D. Giersch, S. & Geisler, G. 2002. Challenges for Service Providers When Importing Metadata in Digital Libraries. D-Lib Magazine April 2002. <http://www.dlib.org/dlib/april02/mcclelland/04mcclelland.html>

McClelland, M., McArthur, D. Giersch, S. & Geisler, G. 2002. Challenges for Service Providers When Importing Metadata in Digital Libraries. D-Lib Magazine April 2002. <http://www.dlib.org/dlib/april02/mcclelland/04mcclelland.html>

Sampson, D., Kargiannidis, C., Karadimitriou, P. & Papageourgiou, A. 2001. EM^2 - an Educational Meta-data Management tool. Proceedings of ED-MEDIA 2001 World Conference on Educational Multimedia, Hypermedia & Telecommunications. AACE –Association for the Advancement of Computing in Education.

Simon, Bernd. 2001. Do e-learning standards meet their challenges? UNIVERSAL Project. <http://www.ist-universal.org>

Wason, Tom. 2001. *Dr. Tom's Metadata Guide.* IMS Global Learning Consortium. <http://www.imsproject.org/drtommeta.html>

Col•lecció: Collective Bookmark Discussion Applying Social Navigation

Rodríguez Henrry
IPLab/KTH/NADA
henrry@nada.kth.se

Noël Sylvie
CRC Industry Canada
sylvie.noel@crc.ca

Abstract: Collecting Web-objects for different purposes is a common task for teams that use the WWW as infrastructure for working. The team might need to discuss these objects in particular or in general. We present Col•lecció, a simple Web-based tool that allows users to include URLs in a shared space and to link comments to them. The users studies show that the system supports a social navigation approach based mainly in the interaction in the shared space.

1 Introduction

The exponential growth of the World Wide Web (WWW) has made retrieving any particular Web page difficult, and people need better ways of searching than what is possible with present information retrieval technology (Hardin, 2002). One increasingly popular solution is to use other people's suggestions in order to find things on the Web (Wexelbalt, 1998). This is known as social navigation (Benyon & Höök, 1997). For an individual, the usual way to access a Web object is through its URL (Uniform Resource Locator). Almost all Web browsers let users manage their URLs via the bookmarks feature. Generally speaking, bookmarks let users retrieve later those objects included in the bookmark. In the bookmarks, URLs can be organized or annotated according to the user's preference and they are searchable. However, bookmarks have an individualistic approach, that is, they are directly accessible only by the person using the computer on which the browser is situated. Groups that use the WWW as a social infrastructure for knowledge-oriented work (teams, see Severinsson et al. 2003) may find it more difficult to share information found on the Web. Making sure that everyone has access to the Web objects is problematic. We present Col•lecció, a prototype aimed at collecting URLs that members of a team wish to share and discuss. By using a social navigation approach and centring the interaction on the dialogue, this system presents a potential alternative for organizing information from the WWW that a team may need in its activities.

2 Related Work

In 1993, Mosaic, a browser developed by NCSA, let its users make annotations to a document found on the Web and was capable of handling both personal and group annotations. Today, surprisingly, none of the currently most popular Web browsers offer the same sort of annotation feature. As a result documents on the Web can only be read passively by third parties. However, some systems have been developed to allow Web annotations, among others *CritLink* (Yee, 1998) and *Annotea* (Kahan & Koivunen, 2001). Ovsiannikov, Arbib, and McNeill (1999) presented an empirical study of annotation on paper and demonstrated that electronic annotations could be synchronized in real time. Col•lecció allows users to link annotations to the Web sites that have been included in the system. Some systems try to create communities by searching in users' bookmark files to find similarities and then sharing the URLs. For example, *CoWing* (Kanawati et al. 2001) uses assistant agents called *WINGS* to collect and organize URLs. *TopicShop* (Amento et al., 2000) helps users evaluate and identify high quality sites using site profiles and a work area presenting thumbnail images, annotations, and grouping techniques. In order to benefit from using bookmark files to create a collection of links, algorithms are required for clustering and categorizing web pages. One problem with this approach is that empirical studies indicate that many users bookmark "just in case" and not many of these bookmarks are frequently used (Abrams et al., 1998). Another approach to supporting social navigation is to allow users to put their bookmarks on a remote server, such as like http:// mylunx.com, http://www.mybookmarks.com, or http://www.backflip.com. These links can be shared though users are required to set access rules to define privacy. A different approach to using agents to grasp users' bookmarks files or putting them in a server is used by *WebStickers* (Ljungstrand et al. 2000). This system couples Web pages with physical representations, or tokens. These tokens are used to access Web pages by scanning an attached barcode. These tokens can be handed over to another users. An original way of collecting Web pages is offered by *Hunter Gatherer* (Schraefel et al. 2002). This system, aimed at a single user, collects, represents, and can edit components from within Web pages. The idea here is to collect specific content inside a Web page and not the Web page as a whole. In other words, a *collage* of Web pages can be made. *WebTagger* (Keller et al. 1997) provides both individuals and groups users with a shared space of URLs that can be organized and that is searchable. Users can supply feedback on the utility of the URLs that is mainly used for ranking. None of the systems mentioned here dedicate special attention to the need that users might have for discussing either the "collection" globally or any particular item. Our prototype addresses these issues and also provides social interaction that helps users navigate in the collection that the team has created.

3 Col•lecció

Col•lecció is a Web-based tool that runs in an ordinary Web browser supporting JavaScript. It is based on the Web browser's bookmark feature but adds a collective aspect to it. Col•lecció lets users add Web objects in a shared space. Though there are many types of Web objects, focus has been placed on HTML and XML files during the design of this tool. Col•lecció also allows users to link comments to these Web objects (Web document hereafter). These comments are shared within the group and can be used to establish a dialogue among the members. Dialogue is a key component in collaboration. By adding URLs to the system, the team organizes a sub-set of the WWW that might be of interest to the team's activities. The general idea is that members of a group form a collection of Web documents. Members can make comments about the corresponding Web document. The set of comments could thus evolve into a dialogue between participants, aligned with the document, and serves as a communication channel throughout the reviewing process, for example (see Rodr"guez, 1999). The Web browser window is divided into two vertical frames. The left frame is used as an index of contents. The URLs are hypertext links labelled with descriptive titles, which can be grouped under a topic label or by the name of the user who added the link. The right frame is divided into three horizontal frames. The Content frame shows the Web document. The Comment frame shows the comments that have been attached to that Web document. The Command frame is filled with four button commands: *Add URL, Add Comment, Delete URL,* and *Show last events*. When the Add URL button is clicked, a window pops up, presenting a form where the member can write the URL (or just cut-and-paste) and a descriptive title for the link. The Add Comment button calls up a separate window with a form where the member can make a comment. The Delete URL button deletes the current item from the Index of contents that was shown in the Content frame. These actions are immediately implemented, thus making the updated Index of contents or new comments available when another user enters the system. When the "Add comment" button is clicked, the Add-Comment Window (ADW) of the system pops up. This window is divided into two vertical frames. The left frame shows the Web document that the user has selected from the index of contents followed by its comments, if any. This lets the user view the comments in case he/she needs to refer to a previous comment. In this way the system supports a dual context for the annotations, both to the Web document and to the past comments. The right frame presents a text box where the user can write comments. When a comment is submitted, an notification email is sent to the person who added the Web document and to those who have participated in the discussion so far. This is used as an awareness mechanism (Dourish and Bellotti, 1992). Another way of promoting awareness is through the "*comment counter*" that tells the users how many comments have been made on a document. A more general form of awareness is the *Show last event* button. It presents an activity report which shows the changes that the shared space has gone through, including when and who made the changes. Furthermore, the system can show the Web document in an independent window bigger than the Content frame.

4 Case Studies

We present here three recent case studies of the use of Col•lecció. The URLs collected and the interaction among participants (the comments) were analysed. We also asked the 20 students to evaluate the system via a Web-based survey.

4.1 Case study 1: Collecting e-journal Web sites

Six graduate students distributed in Europe, Asia, and Latin America with at least four years of experience using the WWW and Web design wanted to establish a new research journal Web site (e-journal). This work was done to fulfil personal goals and not for professional reasons. Five of the students were familiar with Col•lecció and had participated in the system's graphic design. The group was encouraged to use Col•lecció to gather some examples of e-journals, the intention being to collect models that could be used for the design of the new e-journal. The users were also asked to hold a preliminary discussion of the collected e-journals using the system. The group collected 11 e-journals during a period of one week. These e-journals were discussed over a two-week period. The discussion consisted of 69 comments containing approximately 5,500 words. The URLs were grouped by the name of the member who included it.

4.2 Case study 2: Collecting Web-based collaborative writing tools

Four researchers in HCI located in North America and Europe were encouraged to use Col•lecció to develop a discussion on a topic of mutual interest (Web-based collaborative writing tools). One of the researchers had used the system before. Again, this project was done for personal, not professional, reasons. The participants used the system for a one-year period. However, only two of the members used the system regularly. Nineteen URLs were added to the system. Two main topics were created: "Web-based collaborative writing tools" which held more than 47% of the URLs collected; and "Readings on collaborative writing". As well, the researchers could add links under their own name. There were 72 comments, containing 6,866 words.

4.3 Case study 3: Collecting articles from digital newspapers

Twenty students in a course given in a university in Sweden were asked to use the system to collect articles (a maximum of three per student) from newspapers' digital versions. All the students had at least two years of experience using the WWW. The class was once a week. Only five of the students knew someone else in the class before the course started and this was their first lab. All the articles were grouped under the students' names, not under topics. To fulfil the course requirements, students had to collect the articles in the first week and were to start discussing the articles in the second week. A total of 39 articles were included from 17 different domains and 191 comments were posted holding approximately 12,500 words. The average of comments per document was 4.8 (sd=2.7). This is the only study in which the use of the tool was mandatory. We collected 19 responses to the survey.

5 Discussion

Some teams' work may need to be done around a collection of Web objects. Our case studies show that certain tasks do not need a complex system in which access rights need to be set up or analysis of the users' bookmark files be made. Col•lecció supports discussion of a very specific, selected sub-set of the WWW created by the user.

5.1 Supporting Social Navigation

Bookmarks are a "personal web information space" to help users navigate on the WWW (Abrams et al.1998). Pitkow, as cited by Abrams et al., reports that a great number of users rely on bookmarks as a strategy for locating information and that bookmarks were used slightly more than other navigation strategies such as a search engine. A collective bookmark implies a "collective web information space" that can be used by members of a group. The agent here is the same user; that is a human. Comments can contain relevant information about the web page that could help the user make decisions as to whether or not to read the page, as in the following example[1] from the case study 2

Comment # 2 Date: 02/04/05 Time: 17:13:37
Laura Craig:
Ok, about EquiText.
First it's in Portuguese, so I haven't been able to try it out and see :-). You can write on-line within your Web browser.
Comment # 3 Date: 02/04/05 Time: 17:20:28
Charles White:
I will check this system later. I do not speak Portuguese but knowing Spanish I can read and understand

The students in case study 3 were asked about their reading strategies. Not surprisingly, the article's title was indicated as an important factor in selecting a particular article to read. Students mainly used the original article's title for the title included in Col•lecció. However, some formulated their own titles. We observed that the titles formulated by the students were related to a common ground or to the task itself. For example, one of the titles[2] was named using the topic during a discussion held in class. The original title of the Web site might be an inadequate descriptor of the site content (Abrams et al. 1998) and even if it is adequate we consider that the user could give a more informative name to the site in relation to the task in which they are involved. In fact, the most commented article (12 comments) was one of the few articles in which the users formulated its title and adapted it to the context of the task. Furthermore, this article was the most selected by students when they were asked to select three of the articles that they liked most in the survey. This suggests that adapting the document's title to the group's goal helps users understand how that document is relevant to the group. Another navigation strategy mentioned by many students was the use of the *comment counter* presented in every item of the index of contents. For example, one students said *"I read those articles that had few comments and made a comment to it as I think that everybody should get a comment, then I read the articles that had more comments as I thought that they should be interesting"*. In some cases, navigation could also be based on how interesting students judged a particular comment: *"If I came across a very interesting comment, I also checked what articles that person had added."* Also the interaction among the users influenced the navigation through the articles. As one student wrote *"I read all the comments my article got. Furthermore, I got more interested in reading the articles that were included by those who had commented on my article."* Only one student reported being familiar with all the Web domains from where the articles were taken and 62% had bookmarked at least two of the articles in his/her personal bookmarks. All this means that the interaction of the group was influencing students' navigation patterns in the shared space and in the WWW.

5.2 Supporting collaboration

Collecting, sharing and discussing Web pages can be very important. For example, in case study 1, each URL collected by the participants was an example of the object that they wanted to design. The role of examples in any work is essential. When a participant added an e-journal Web site, he/she indicated the interesting aspects of the site for him/her. Doing this the users (e.g. those who had experience in publishing) helped other users focus on particular details that could be overlooked by the others for different reasons. Some comments suggested strategies that the participants thought the rest should follow in order to accomplish the task. In study 3, we asked the students to choose at least three of the students to work in a project based on this experience. As expected, in general, they selected those with whom they interchanged comments. As well, students that were more prolific proved to be the most popular. One student wrote in a comment *"I think that I have learned to know some of the students a little bit better. It is interesting to see that one can know a person by the way they write and by the topic they select to comment"*. This suggests that users who were not familiar with the other members of the group could learn to know each other better by the URLs they added, by the topics they selected to comment, as well as by the content of what they wrote. Even more, the user might expect others to estimate him/ her through his/her comments and choice of URLs. One student sent an email when looking for a partner for another lab in the same course and said *" So now you can look at my articles or article comments, and see if I'm a worthwhile lab partner"*. This is an important issue in collaborative work because trust in the group is based on these interactions. In fact, researchers can also use these criteria to determine who to work with: the two most prolific participants in case study 2 later decided

---

1 The examples are fragment of the comments that were recorded in ColïlecciÛ. The names have been changed.

2 The original title was "Playing Counter-Strike" and the formulated title by the student was "Regarding games as a research platform for CSCW studies"

to work together on an article.

6 Other aspects and future work

Users display a feeling of ownership for those links that they included in the system. In case study three, for example, several times students referred in their comments to the article they included as *"my article"* even though the student was not its author. When a comment is made to a particular link, the system sends an email notification to the person who included it. This might support this ownership feeling shown by the authors. One of the students stated, *"I don't care who put the article but I care about the topic. So I don't need to see the name of the people. Later on, I was looking for the articles of the ones that added comments to my articles."* The system supports both these tasks to some extent. It seems it would be useful to be able to switch between different modes of organizing the index of contents. There is potential for improving Col•lecció's user interface. For example, when URLs are grouped by topic, the system does not show who added each URL. While adding such a label would be simple, it could potentially overload the information presented in the index of content. Another improvement could be to add a URL and its title automatically with a simple click. There exists a potential legal problem with Col•lecció, if Web site subscribers choose to publicly share documents for which the site owners require paid access.

7 Reference

Abrams, D.; Baecker, R. and Chignell, M. (1998) Information archiving with bookmarks: personal Web space construction and organization, Conference proceedings on Human factors in computing systems, p.41-48, April 18-23, 1998, Los Angeles, California, United States. ISBN:0-201-30987-4

Amento, B.; Terveen, L.; Hill, W. and Hix, D. (2000), TopicShop: enhanced support for evaluating and organizing collections of Web sites, Proceedings of the 13th annual ACM symposium on User interface software and technology, p.201-209. ISBN:1-58113-212-3

Benyon, D. and Höök, K. (1997) Navigation in information spaces: Supporting the individual, In Proc. of INTERACT'97, Sydney, 14-18 July 1997, 39-46.

Dourish, P., & Bellotti, V. (1992) Awareness and coordination in shared workspaces. In Proceedings of ACM CSCW'92 Conference on Computer-Supported Cooperative Work, (pp. 107-114). New York: ACM Press.

Hardin, S. (2002). James Hendler and Ben Shneiderman on the Next Generation of Interfaces. Bulletin of the American Society for Information Science and Technology Vol. 28, No. 3 February / March 2002

Kahan, J. and Koivunen, M-R. (2001) Annotea: an open RDF infrastructure for shared Web annotations, Proceedings of the tenth international conference on World Wide Web, p.623-632, May 01-05, 2001, Hong Kong, Hong Kong

Kanawati, R. and Malek, M. (2001) CoWing:A collaborative bookmark management system, in Cooperative Information Agents V, M. Klusch, F. Zambonelli (Eds.). Lecture notes in Artificial Intelligent 2182, © Springger-Verlag Berlin Heidelberg, pp 38-43

Keller, R.; Wolfe, S.; Chen, J.; Rabinowitz, J. and Mathe, N. (1997) A Bookmarking Service for Organizing and Sharing URLs. Proceedings of the 6th Intl. WWW Conference

Ljungstrand, P.; Redström J.; Holmquist L. (2000) WebStickers: Using Physical Tokens to Access, Manage and Share Bookmarks to the Web, in proceedings of DARE 2000, April, 2000 Elsinore, Denmark. Pp 23-31. ISBN 1-58113-367-7/00/04

Ovsiannikov, I,. Arbib, M., McNeill, T, (1999) Annotation Technology. International Journal of Human-Computer Studies 1999 v.50 n.4 p.329-36

Rodriguez, H. (1999). The Domain Help System. Technical report TRITA-NA-P9912, CID-56, NADA, The Royal Institute of Technology of Sweden. ISSN 1403-0721

Schraefel M.; Zhu, Y.; Modjeska, D.; Wigdor, D. and Zhao, S. (2002) Hunter gatherer: interaction support for the creation and management of within-web-page collections, Proceedings of the eleventh international conference on World Wide Web 2002, Honolulu, Hawaii, USA, pp 172 ñ 181. ISBN:1-58113-449-5

Severinson-Eklundh, K., Lantz, A., Groth, K. Hedman, A., Rodriguez, H. & Salln‰s, E-L. (2003) The World Wide Web as a social infrastructure for knowledge-oriented work, in Oostendorp H. van (Ed) *Cognition in a digital world* ISBN 0-8058-3507-5, pp 97-126

Wexelblat, A. (1998). History-rich tools for social navigation. Proceedings of the CHI 98. page 381

Yee K. CritLink Mediator http://crit.org/critlink.html

Cognitive Strategies and the Process of Teaching and Learning

Júnia C.A. Silva *Vânia P. de Almeida* *Rafael G. Orbolato*

Federal University of Sao Carlos
Rod. Washington Luis, km 235
C.Postal 676 13565-905
Sao Carlos, SP, Brazil

junia@dc.ufscar.br vania@dc.ufscar.br orbolato@dc.ufscar.br

Abstract

E-learning has been spread to a large number of people and each day even more students face the challenge of getting knowledge using computers in the spite of teachers presence. Design quality educational material for e-learning is a challenge that teachers should be concerned in overcoming. In the charge of crating material that can facilitate learning, a pedagogical approach should be considerate. We implemented an editor – Cognitor, which contains cognitive activities to help teachers in their task of creating material that promotes active learning, reducing the knowledge acquisition complexity.

1 Introduction

We are in front of a new way of teaching and learning and it is important to realize that to promote e-learning is not just to make non-electronic material ready to be delivered on line.
Teachers should have in mind that it is necessary to make the material as clear as possible. So, comes the question: How to design a web educational material in a way to facilitate the knowledge acquisition?
Considering usability as the needed effort to use the software, and if we consider the material as an interface to students, so if we make the material even easier to be understood, we are reducing the students effort and so increasing this material usability.
In this work, we defend that to have a well prepared material for e-learning it is necessary to consider some pedagogical principles on the design of the material. Some pedagogical research areas were studied and the Cognitivism was chosen.
Analyzing some available computer environments, we realized that just a few of them offer to teachers a tool to design the material. Trying to supply this absence of e-learning editors, especially those supported by pedagogical principles, we computer implemented an editor provided by what we called "cognitive activities" that are items based on the cognitive strategies that when used can promote the active learning.

2 Cognitivism and E-learning

The Cognitivism studies how human beings assimilate information from the environment they are in. Computers and cognitive principles have much in common, as observed Neisser (Neisser, 1976), "... computer activities seem like cognitive processes in some aspects. Computers get information, manipulate symbols, store items in memory and search then back, classify inputs, recognize patterns …". Gardner (Gardner, 1995), years late, saw in computers a way to confirm the cognitive principles: "It was necessary the computers advent and the birth of the Information Theory to confirm legitimate to the cognitive studies."

In this way, being the Cognitive Science really interested in the information proceedings and being computers machines what, effectively, facilitate this process; it is natural the use of computers to help the learning process supported by cognitive paradigms.

This study about Cognitivism, took us to Gagné (Gagné, 1974), who dealt with the learning internal processes by items that were called "dominions". One of these dominions is made by cognitive strategies that according to Gagné are capacities internally organized, that learners use to guide their own attention, learning, memory and thinking processes. The student uses a cognitive strategy, for instance, to identify several characteristics of what he is reading, to select and codifies what he learns, to retrieve the information and so on. Gagné relates these strategies with the concepts of "learning how to learn" and "learning how to think".

It is important to realize that these concepts have to be considered when talking about e-learning, because frequently learners face these two challenges in the tutors' absence and so efforts to edit adequate material to facilitate this process should be considered.

Liebman (Liebman, 1998) thought about how to apply the cognitive strategies principles in her classes at University of Illinois and suggested the following strategies to promote active learning:

- *Advance Organizers*: remarks by teachers to help students move to new topics.
- *Concept Maps*: diagrams used to express temporal, categorical, causal, hierarchical and other relationships.
- *Framing*: visual displays of the underlying structure of related information.
- *Metaphors and Analogies*.
- *Organizing*: includes applying taxonomies, listing similarities or differences, analyzing form and function and itemizing advantages and disadvantages.
- *Rehearsals*: strategies to keep the material being processed in the students' working memories long enough that they can more firmly establish it in their memories.

Inspired by these strategies proposed and successfully used by Liebman, and knowing about the inherent difficulties of the educational material edition, we developed an edition tool that, by way of what we called cognitive activities implementation (Liebman's strategies interpretations) and class structures (defined before the editing process) that we called "document organization", can help tutors in their task of editing quality material and facilitate the knowledge acquisition.

3 An Educational Material Editor for E-learning

The Cognitor offers common edition insertion options like text and images and the insertion options of the cognitive activities that are the main focus of this work.

The main interface of Cognitor is shown in Figure 1 and is composed by four important areas:

- *A:* Document Organization Area. Using the options of this area, teachers can define a new organization, choose one previously defined or even change one.
- *B:* Cognitive Activities and Web Pages Edition Options.
- *C:* Actual Page Edition Area;
- *D:* Pages Control Area. The options of this area allowed inserting, removing and moving between the material pages.

To create the material, teachers may choose one document organization previously defined (that can be modified) or create a new document organization to be used in the material. After, selecting a cognitive strategy, the respective module is activated and will help them during the necessary edition steps of this strategy.

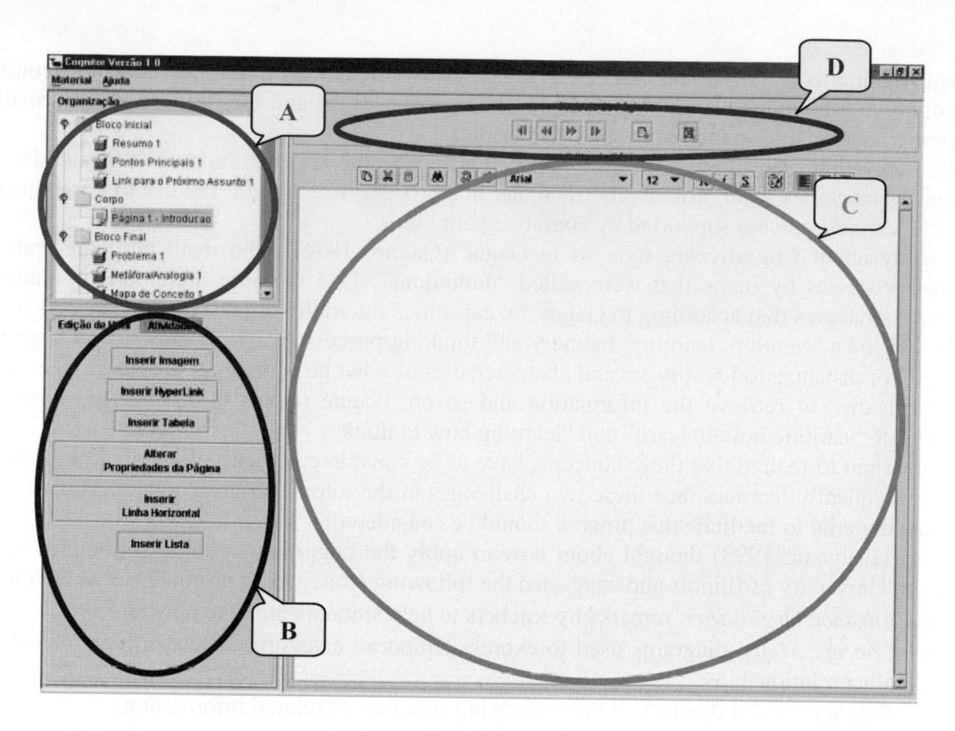

Figure 1: Cognitor Main Screen

After choosing the organization, it is shown to teachers an advice with some tips about how to edit material for e-learning. Also, when choosing a Cognitive Activity other advices are shown with some tips about the related strategy and also how to edit the activity. One of these advices is shown in Figure 2.

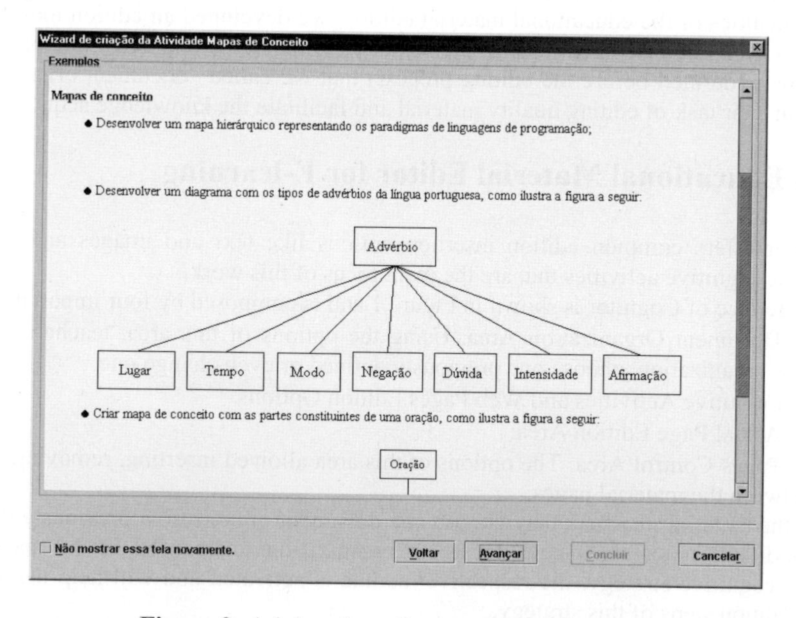

Figure 2: Advice about Concept Maps Diagrams

After the advice, the Activity module is displayed. Figure 3 shows the module that represents the Concept Map strategy. This module allows teachers to edit many kinds of diagrams.

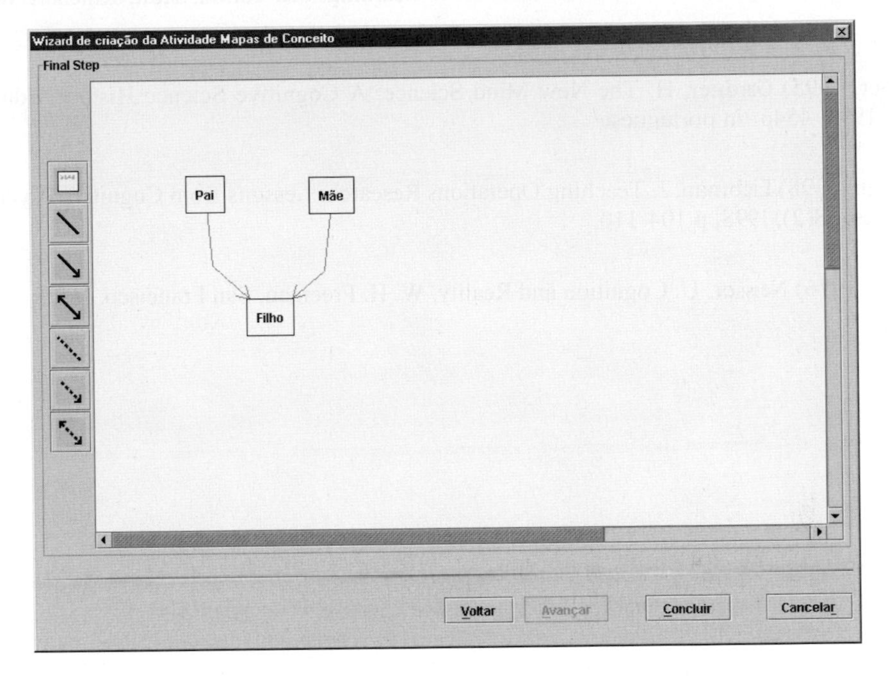

Figure 3: Concept Map Module

After creating all the material pages, teachers should choose the first and the last pages and save the material (group of related web pages).

4 Final Considerations

It is important to realize that e-learning environments must be adaptable to teachers and students of varied fields, and that pedagogical support in the creation of material is needed. Some cognitive strategies presented were chosen as a base to support teachers in the creation of instructional material. Although these strategies suggested seem to be appropriated for this propose, as they have already been tested with success in conventional classrooms, this will only be confirmed with the results of the usability evaluation that will be made. Some methods are being studied to the editor evaluation as software and also to the material, looking to validate the use of the Cognitive Strategies in the charge of reducing the knowledge acquisition in e-learning context.

References

(Gagné, 1974) Gagné, R. M. The Conditions of Learning. 3rd editon. Holt, Rinehart Winston, 1974

(Gardner, 1995) Gardner, H. The New Mind Science: A Cognitive Science History. Edusp, SP, Brazil, 1995, 454p. /in portuguese/

(Liebman, 1998) Liebman, J. Teaching Operations Research: Lessons from Cognitive Psychology. Interfaces, 28(2), 1998, p.104-110.

(Neisser, 1976) Neisser, U. Cognition and Reality. W. H. Freeman, San Francisco, 1976.

Towards an understanding of Common Information Spaces in Distributed and Mobile Work

Gabriella Spinelli and Jacqueline Brodie

Department of Information Systems and Computing, Brunel University
Uxbridge, Middlesex UB8 3PH, United Kingdom
gabriella.spinelli@ brunel.ac.uk, jacqueline.brodie@brunel.ac.uk

Abstract

Collaborative tasks necessitate the pro-active role of users in interpreting the nature of occurring events and in evaluating the coordinated use of resources in shared spaces. Through fieldwork observational studies, this research, aims to depict a picture of diverse instances of Common Information Spaces (CIS) and reveals the impact of communication and information technologies on collaborative work practices.

1 Introduction

The study described in this paper aims to unpack the socio-organisational domain, where collaborative activities are performed, as well as comprehend the nature of the overall resources employed to support them. We argue that collaborative tasks are not just a collection of objects and events but that they also require the user to interpret actively the shared context where collaboration takes place and the resources through which collaboration is achieved. This study reveals two aspects relevant to the understanding of collaborative work: i) the creation and management of Collaborative Information Spaces [Bannon and Bødker, 1997] and ii) the coordination of mobile work in collaborative work practices.

2 Methodology

Through the use of fieldwork an ethnographic oriented methodology was used to document users' activities, their context of work and the artefacts they employed. The data collection covered a period of time of approximately eight months, in a variety of working environments, and employed techniques such as digital video recording, contextual interviews and participatory user data reviews. The latter were particularly relevant for the outcome of this research. Informants participated in collaborative sessions where they reviewed some of the observational data and provided useful insight into the understanding of critical collaborative scenarios that highlighted the disruptions that can be caused by the use of technology in collaboration.

We tried to frame the study within a consistent domain of observation: therefore, three organisations were selected on the bases of the activities they performed. The three teams that were shadowed were all involved in design activities of different types, as listed below:

- the conceptual design of an information appliance;

- the engineering design of an innovative public building;
- the design of a new set of national standards in construction procedures

3 Observations: nature of CIS

One of the core idea within Computer Supported Cooperative Work (CSCW) is that cooperative activities require a communication - or shared informational - space as a common ground [Moran and Anderson, 1990] by which properties such as mobility and awareness become relevant to the overall collaborative performance. Such a notion has been extended and formalized by Bannon and Bødker [Bannon and Bødker, 1997] emphasizing the role of the collaborative informational space, CIS, as the set of available resources enabling creative decision making processes.
Three diverse instances of collaborative work emerged from the observations:

- a physically-centred collaborative space (the project space), a dedicated environment where a group of professional designers collected and manipulated information in order to support their activities;
- a virtual space, resulting from the combination of web application and video conferencing technologies for the collection, retrieval and storage of organisational knowledge to support problem solving activities;
- a distributed space arising from the collected use of several digital devices (mobile phones, faxes etc.) and protocols of communication (circulation of the people, email, snail mail etc.) in order to overcome the obstacles imposed by remote collaboration.

All the instances of collaborative space observed in this study do not find counterparts just in the physical world. They resemble more a collection of established organisational practices and technologies used to achieve collaborative tasks. This observation led us to postulate that we cannot rigidly define collaborative space by simply considering its physical boundaries. This consideration thus directed our research towards:

- the identification of those tasks that make up the dimensions of collaborative work such as collective brainstorming, displaying, capturing, collecting-storing-retrieving information and task distribution;
- the understanding of how co-ordination can be achieved by the employment of resources in situations such as meetings, (physical and virtual *containers* for the coordination of the team members' tasks) and to link and manage the streams of individual and collaborative work;
- the analysis of the impact of individual's mobility on collaborative spaces considering how work on the move often stretches the boundaries of the collaborative dimension, violating the social rules of the work environment and requiring instantaneous re-arrangement of the modalities of work to avoid *breakdowns*.

4 Findings

This section provides an overview of the main points of interest emerging from our field data. Firstly, we illustrate how using physical space and resources impact on collaboration at an individual level of granularity. We briefly describe the combined use of the devices within the space to facilitate the work practice. Secondly we introduce the repercussions of physically centred work practices on a collaborative dimension.

4.1 A Physically centred space for the performance of collaborative work

Physical space is currently the most effective way to support collaborative activities due to the natural interaction that individuals are able to establish with their environmental resources. From an evolutionary point of view human kind has learnt how to structure their environment in order to have the best chance of success. This cognitive strategy is often referred to as *'structural coupling'* (Kirsh, 1995) and it highlights how space cohabitants and structures evolve simultaneously in an interwoven *ecology* that resembles biological systems. The parallelisms between ecology and the office have been drawn already [Kirsh, 2001] and it seems in this context to be appropriate to describe the advantages that the physical work environment provides at an individual and cooperative level. The project space that we observed was a dedicated room without any PC or land phone connection. Beside the personal stationery that each individual brought into the room, paper-based artefacts were mainly available: foam boards, whiteboards, flip-charts and post-it notes. The consistent availability of the project space together with the flexibility offered by the combined use of resources constituted the means through which individuals were able to support the design process.

At an individual level the advantages offered by the physical space (in our interpretation also encompassing artefacts) can be listed as follows:

- *To simplify choice.*

Information relevant to the design process was collected as prints out, hand-written notes, sketches, photos etc. pinned down to foam boards. Several foam boards were placed around the perimeter of the room and over time constituted a layered structure resembling an onion ring. In order to highlight only the information strictly necessary for the task the relevant foam boards were put on the foreground thereby channelling the attention of the team. The unneeded information was hidden away, in storage, for later use.

- *To simplify perception.*

In the design review process, design concepts were sketched on cards and tested with the users. The users' evaluations were summarised into good and bad points and transcribed on to post-it notes of different colours. The post-it notes were then stacked at the bottom of the cards that were progressively positioned on a new board, which collected the concepts the teams dealt with. The symbolic positioning of the cards and the clustering strategies used to add the users' evaluations to the concepts allowed the team members to detect, with a quick glance, what stage of the process they were at and to plan future activities accordingly. Also this structural strategy enhanced awareness in those team members that were not present during the revision by providing a tangible representation of the performed task that they could effortlessly access.

- *To simplify internal computation.*

Toward the end of the project, the design team started to prepare a brochure to be presented to the client. At that stage the design concepts considered most promising were expanded and each of them occupied a whole foam board. They were also laid-out in A4 pages. The project manager had the task of leading the creation of a suitable arrangement of the concepts according to a structure that made sense for the brochure. The manager wrote the names of the different concepts down initially and then attempted a possible order sequence on a sheet of paper. Next she placed the A4 sheets on the table according to the written sequence. Once she had all the A4 papers spread on the table she moved them around until she achieved a more satisfactory arrangement, which was then mirrored in the project space by the positioning of the foam boards in identical sequence. The support provided by the external representations of the concepts on the A4 sheets allowed the individual to delegate to the environment cognitive tasks that she would have had to perform internally otherwise. The manipulations were operated on physical objects that embedded the information processes and retained memory of it in the represented structure. Moreover by mirroring in the environment the order of the concepts, the whole team was made aware of the ongoing task and the team could comment on this since it was effortlessly available.

From the points illustrated above it is also possible to envision how the creation of a physical project space can enhance features that are fundamental for collaboration [Kirsh, 1995], for example, by providing:

- Peripheral awareness of co-workers that could aid keeping track of the overall team activity;
- Joint monitoring of the devices, present in the room, that embedded the history of the collective work practice;
- Broadcast communications without additional effort, since the distribution of the tasks is embedded in the environment and easily accessible.

Working in the same environment where the information is organized offers overall the benefit of using the physical space to back up any potential disruption caused by human or technological factors. However, a physical collaborative space is not the most common manifestation of a CIS because collaboration often occurs in a distributed way; and organisations seldom support or even envision a project space as beneficial to the work.

4.2 Distributed and virtual collaborative places

The teams that did not benefit from the support of a physical collaborative space configured their work practices around the limitation and the assistance that the available technology provided them with. However when technology fails, in order to secure the team's overall performance, demanding and overwhelming work strategies need to be adopted at the expense of the individual. The virtual CIS observed displayed only short-term advantages. In the longer term, when the complexity of the activities increased, the team members needed someone to mediate and manage the shared space that they had created. Moreover the lack of connection between web based technologies and the teleconferencing system to simulate a synthesized virtual environment caused disruption between the creation of knowledge during the decision making process (on-line meetings) and the updating of that information on the web (repository of the team knowledge). Virtual and distributed spaces, in general support only a few aspects of collaborative work when compared to the richness offered by a physical collaborative space with its accompanying advantage of situated awareness. Lack of awareness in the virtual and distributed CIS's led to impoverished interpretations of the objects and events in the shared space by the actors involved, ultimately culminating in a proliferation of breakdowns.
The observation of remote collaboration also opened up this study to the investigation of the relationship between mobile work and the creation and maintenance of collaborative spaces.

4.3. Stretching the boundaries of collaboration: mobility

Individuals working on the move need to order their activities taking into account the deprivation they will experience because the majority of the resources available in fixed collaborative environments are missing. Using the idea of place in collaboration (emerging from space and accompanying structural resources) [Harrison and Dourish, 1996], we observed that for mobile workers, the collaborative workplace consisted primarily of the communication that the mobile phone was able to support. We observed that mobile workers have to focus their activities on the information transmitted and on the space for collaboration created by mobile phones because no other artefacts used on the move were as capable of supporting such *heterogeneous* activities. Mobile phones were used by the majority of mobile workers because they are extremely flexible artefacts: facilitating an immediate response to events while allowing the sharing of attention across other cognitive activities such as walking, dealing with travelling procedures and so forth. However, although mobile phones were the most popular resources for those on the move they constrained the patterns of collaboration possible to the user because of the limited nature of the communication they can currently support.

Although the use of the mobile phone allowed individuals to still reach colleagues and to establish basic forms of communication and interaction, the resulting communication was often extremely impoverished and unable to support an individual's desire to access a rich amount of collaborative knowledge on the move. Such information could be partly accessed by combining the use of several artefacts together - but this arrangement more often than not, generated uncomfortable modalities of work because the technologies in use were predominantly desktop oriented. The co-ordination of multiple artefacts concurrently also necessitated a great deal of planning *a priori*. This inevitably forced individual's to seek a tabletop or flat-surface to work on while also demanding an almost exclusive focus of attention - thereby depleting mobile work from some of it's primary characteristics, i.e. flexibility and spontaneity.

5 Conclusions

In conclusion, a common observation across our fieldwork was the disruption generated by the current digital technology used in emergent workplaces. These disruptions are often avoided or reduced through expensive cognitive behaviours that individuals employ. However, these alternative strategies for keeping collaboration alive often resulted in inefficient working practices and in impoverished cooperation. Also on the basis of our research findings we stress the inadequacies of current mobile technologies to support access to collaborative knowledge on the move. Users seek to experience at least *engagement* in their virtual communities through the use of the mobile phone but the demands of the devices are essentially disruptive and impoverished. As we have illustrated above users are left with the need to plan in advance to accomplish work that should instead be achieved from more flexible and spontaneous activities.

6 References

Bannon L., Bødker S. 1997. Constructing Common Information Spaces. *Proceedings of the 5th European CSCW Conference*. Dordrecht: Kluwer Academic Publishers.

Harrison S. and Dourish P. 1996. Re-place-ing space: the roles of place and space in collaborative systems. *Proceedings of CSCW 1996*. ACM Press New York, NY, USA

Kirsh D. 1995. The intelligent use of space. *Artificial Intelligence*. 73, 1-2 (Feb. 1995) 31-68.

Kirsh D. 2001 The context of work. *Human-Computer Interaction*, vol. X, 2001.

Moran T.P. and Anderson R.J. 1990. The workday world as a Paradigm for CSCW Design. *Proceedings of CSCW 1990*. Los Angeles, CA, USA.

Adaptive and Context-Aware Information Environments based on ODIN – Using Semantic and Task Knowledge for User Interface Adaptation in Information Systems

Maximilian Stempfhuber

Social Science Information Centre
Lennéstr. 30, 53113 Bonn, Germany
st@iz-soz.de

Abstract

In the context of Digital Libraries, where a large set of heterogeneous and distributed information is accessible to users with different backgrounds and expertise, designing a "user interface for all" becomes a complex task. The result is often a tendency to either fall back to the least common denominator – and neglect the demands of many users – or to put everything in a single, very complex user interface. In both cases, the limitations of the user interface will have a negative influence on user satisfaction. With ODIN we present a framework for designing object-oriented, dynamic user interfaces which has been developed in the context of information systems for text documents, time-series data and geographic data. Systems based on ODIN use a model of the semantic dependencies of the application domain to dynamically adapt the user interface to optimize screen layout and information density. The adaptation is subject to the principles of tight coupling and loose coupling, which describe the dependencies between elements of the user interface and the data. Design patterns are used to define both principles and to guide selection and implementation of the principles. We extensively use visual formalisms for displaying semantic relationships in the data, for interactive exploration of query results and for query refinement. The elements of ODIN have already been tested in the domains of text retrieval and urban planning and are currently re-designed.

1 Introduction

Current developments in Digital Libraries aim at the integration of different types of data (e.g. text documents, references to literature, statistical data, specimen collections, survey data, maps, images, video, audio) into portals on the Internet. This gives a large and often vague user group access to distributed and heterogeneous information from a multitude of sources and domains. Along with the wealth of information come many different knowledge structures for indexing and organizing the information (e.g. thesauri, classifications or ontologies), different metadata standards for structuring the data (e.g. Dublin Core[1]), and different types of visualizations commonly used depending on the type of data. The integration of this different features of the data is mostly done on the technical level in the form of query or metadata harvesting protocols and by using form entry query interfaces or common visualization techniques, like a hyperbolic tree browser (Lamping & Rao, 1994).

[1] Dublin Core Metadata Initiative, see http://www.dublincore.org

This means for the user interface designer, that there is no sharply defined, homogeneous user group, whose requirements can be determined and modeled. Often, the least common denominator is used, which leads to Google-like interfaces where novices expect to easily find only relevant results and experts can not comprehend how the query result was achieved. In other cases, domain specific interfaces, like complex ontology browsers, satisfy an expert's needs but complicate a novice's initial access to the system.

To create more generic, but adaptive user interfaces, many of the problems caused by the heterogeneity of the data have to be solved on the retrieval layer of an information system before adequate interfaces can be designed. (Hellweg et al., 2001) describe an approach which tries to build up knowledge structures that integrate different types of information and at the same time preserve their structural and semantic richness. This allows to build an overall view of the information space and to enhance it with structural or semantic details depending on the user and the context.

2 ODIN – A framework for dynamic user interfaces

ODIN (Stempfhuber, 2003) is a framework for dynamic, object-oriented user interfaces which integrates user interface design solutions with design patterns from software engineering. It was developed in the context of research for information systems in the domains of market research (Stempfhuber, Hellweg & Schaefer, 2002), scientific reference databases for literature, and urban planning (Stempfhuber, Hermes, Demicheli & Lavalle, 2001). Common to all three projects was the complexity of the underlying data and the user's information needs, which led to restrictive constraints on screen layout, information density and interaction with the system. The broad scope of users – ranging from novices to domain experts with more than 20 years of experience – rendered previous heuristics useless, and would have led in the majority of cases to sub-optimal solutions. Furthermore, the goals not to implement different interfaces for different user groups and not to introduce artificial sequences of interaction made it necessary to find alternatives to static screen layout. The ODIN framework consists of:

- An object-oriented concept for the overall design of the information environment.
- A model which represents the data sources with all their structural and semantic features and relationships, as far as relevant for the retrieval process.
- A set of adaptable user interface elements for query formulation and exploration of results.
- A set of actions the user interface elements carry out as a response to a user action.
- A set of rules which couples the semantic model and the user interface elements.

The concept for the overall design of the user's workspace is based on the notion of object-oriented user interfaces (IBM, 1992; Mandel, 1994). They use objects as standalone, self contained entities which are represented by icons and are manipulated directly (drag&drop) or with context menus. Indirect manipulation, where the user first selects an object with the mouse and then applies a menu function on the object, is omitted. In addition, the concepts of inheritance (similar objects show similar behaviour), containment (an object may contain only other objects of specific types) and instantiation (objects are created from prototypes or blueprints) are introduced. Furthermore, properties of an object can be changed without influencing other objects. In ODIN, this allows a user to focus on problem-specific interaction, where all relevant concepts, like different search and visualization tools, documents, containers for retrieved documents, and system components (e.g. printer or trashcan) are separate objects which interact using drag&drop.

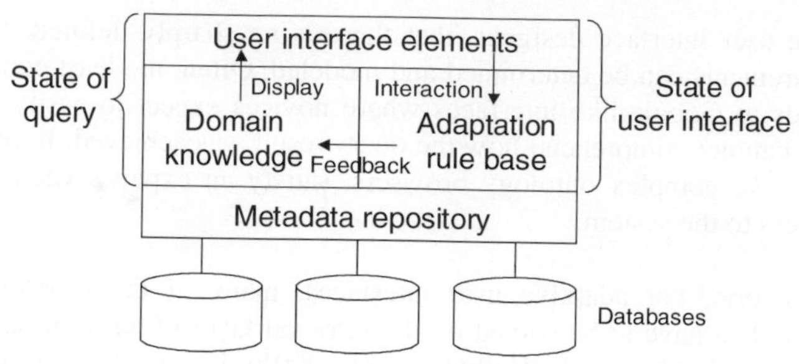

Figure 1: The ODIN architecture

Figure 1 shows the architecture of ODIN. A metadata repository contains information about all data sources, including searchable fields, index data of searchable fields and the technical mapping from user interface elements to fields in the databases. The metadata is also used to dynamically construct the user interface. The domain knowledge consists of the index data, which can be used for searching, together with relationships between different indexing vocabularies of different data sources, which have been extracted from the metadata or are directly modelled in the index data. This allows the system to dynamically determine which indexing vocabularies are needed in the query process and to automatically map query terms between data sources. The user interface elements display the domain knowledge and allow for user interaction. All interactions are reported back to the system, where a rule base is used to evaluate user input, system state and active domain knowledge. The rule base may activate or inactivate domain knowledge, which in turn influences the state of the user interface elements.

2.1 Primary user interface elements of ODIN

The user interface elements of ODIN are based on visual formalisms (Nardi & Zarmer, 1993), dynamically change size and content, and are adaptable to the user's specific needs. The basic visual formalism is an interactive table, whose cells are either empty or occupied by a two-state checkbox. The checkbox reflects an existing semantic relationship between both dimensions of the table and can be activated or deactivated by the user. It can be used for query formulation and exploration of result sets at the same time (output-as-input principle).

Databases ▽
☒ Database 1 (DB1)
☒ Database 2 (DB2)
☒ Database 3 (DB3)

Document types ▽	DB1	DB2	DB3
Monographs	☒	☒	
Journal articles	☒	☒	☒
Gray literature		☐	☐

Document languages ▽	DB1	DB2	DB3
English	☒		☒
French		☐	
German	☒	☒	

Figure 2: Primary and secondary filters

Figure 2 shows a primary ("databases") and two secondary filters ("document types", "document languages") in ODIN. Filters can be opened and closed to save space on the screen and are used for settings which remain unchanged over a longer period of time. It may be seen as a limitation to have only two dimensions for combining attributes because many more attributes are involved in query formulation, but user interviews showed that there is a natural hierarchy which determines a primary attribute (e.g. the databases to be searched) which is then combined pairwise with additional attributes. Here, a combination of document type and language is only hypothetical.

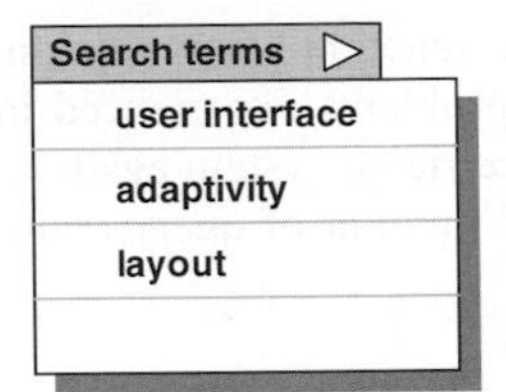

Search terms ◁	DB1	DB2	DB3	
user interface	☒ 34	☒ 56	☐ 104	90
∨ adaptivity	☒ 12	☒ 71		83
¬ layout				
	46	127	0	23

Figure 3: Query formulation

The tabular display is also used for query formulation, so the user faces the same visualization and interaction principles. Figure 3 shows a control for entering search terms in its compressed and extended form. For searches with Boolean logic, the Boolean operators, the number of hits per search term and database, the total of hits per search term, and the total of hits per database can be displayed. For other types of databases, graphical displays or direct manipulation controls (e.g. sliders) could be used. The information about database hits for each term and database can be stored locally or retrieved online from the server. Once retrieved from the server, all further calculations are carried out at the client, which allows fast updates of the display if a user explores the query by selecting or deselecting checkboxes or changing operators.

In addition, controls for selection lists, AND/OR grids for Boolean queries, entry forms, dynamic status displays and displays for result presentation are available in ODIN. Again, the table is used as a general means to visualize the dependencies in the data.

2.2 Design patterns in ODIN

Two important principles in ODIN are that controls are dynamically resized or moved to optimise space usage, and that one control may adapt its content in response to changes in some other control. In the past it proofed difficult to formally decide, when to use which principle and to formally describe the effects on other controls. Design patterns (Gamma, Helm, Johnson & Vlissides, 1995) are more and more being used in the HCI design process (Roberts, Berry, Isensee & Mullaly, 1998; Borchers, 2001). We found that the observer pattern (Gamma et al., 1995, p. 293) exactly describes the tight coupling between "synchronized" controls, and that the mediator pattern (Gamma et al., 1995, p.273) corresponds to loose coupling, where a control may watch other control but will not respond to every change in state.

3 Conclusion

ODIN has been tested in the context of the retrieval of text documents and geographic information with 10 subjects (Petrick, 2003). The goal was to test the users' comprehension of the table display and to get information about the level of complexity which can be handled by novice users. For both systems, nearly all of the users understood the table display and the majority used the checkboxes for query re-formulation. Interestingly, in cases where information was encoded twice, e.g. with checkboxes and with colour coding (vertical bar to the left of the search terms in the system with geographic information), most subjects used the checkboxes for exploring the data. The text retrieval system used the maximum information from figure 3, which could not be interpreted correctly by nearly half of the users. From the test it was not clear if this was caused by limitations in the prototype (e.g. very small checkbox and font) or by the complexity of the Boolean queries.

Currently, the ODIN user interface is being completely re-designed in a cooperation with graphics designers and will then be reimplemented to eliminate the problems experienced in the first user test. In a new cycle, we will then test the ODIN-based text retrieval system again with novice and expert users to get a better understanding about the optimal amount of query-related information for different user groups.

References

Borchers, J. (2001). A Pattern Approach to Interaction Design. Chichester et al.: Wiley

Gamma, E., Helm, R., Johnson, R., & Vlissides, J. (1995). Design Patterns. Elements of Reusable Object-Oriented Software. Reading, Massachusetts: Addison-Wesley.

Hellweg, H., Krause, J., Mandl, Th., Marx, J., Müller, M. N.O., Mutschke, P., & Strötgen, R. (2001). Treatment of Semantic Heterogeneity in Information Retrieval. IZ Working paper 23. Bonn: IZ Sozialwissenschaften (http://www.gesis.org/en/publications/reports/iz_working_papers/index.htm)

IBM (1992). Object-Oriented Interface Design. IBM Common User Access Guidelines. Carmel: QUE.

Lamping, J., Rao, R. (1994). Laying Out and Visualizing Large Trees Using a Hyperbolic Space. In Proceedings of the 7th annual ACM Symposium on User Interface Software and Technology, 13-14.

Mandel, Th. (1994). Thc GUI-OOUI War. Windows vs. OS/2. The Designer's Guide to Human-Computer Interfaces. New York. Van Nostrand Reinhold.

Nardi, B. A., & Zarmer, C. L. (1993). Beyond Models and Metaphors: Visual Formalisms in User Interface Design. *Journal of Visual Languages and Computing*, 4, 5-33.

Petrick, M. (2003). Evaluierung von grafischen Benutzungsoberflächen auf der Basis von ODIN. IZ Working paper 30. Bonn: IZ Sozialwissenschaften (http://www.gesis.org/en/publications/reports/iz_working_papers/index.htm) (to apear).

Roberts, D., Berry, D., Isensee, S., & Mullaly, J. (1998). Designing for the User with OVID: Bridging User Interface Design and Software Engineering. Indianapolis: Macmillan.

Stempfhuber, M. (2003). Objektorientierte Dynamische Benutzungsoberflächen – ODIN. Behandlung semantischer und struktureller Heterogenität in Informationssystemen mit den Mitteln der Softwareergonomie. Forschungsbericht 6. Bonn: IZ Sozialwissenschaften (to apear).

Stempfhuber, M., Hermes, B., Demicheli, L., & Lavalle, C. (2001). Enhancing Dynamic Queries and Query Previews: Integrating Retrieval and Review of Results within one Visualization. In Oberquelle, H., Oppermann, R., & Krause, J. (Eds.), *Mensch & Computer 2001: 1. Fachübergreifende Konferenz* (317-326). Stuttgart: Teubner. (Berichte des German Chapter of the ACM; Vol. 55).

Stempfhuber, M., Hellweg, H., & Schaefer, A. (2002). ELVIRA: User Friendly Retrieval of Heterogeneous Data in Market Research. In Callaos, N., Harnandez-Encinas, L., & Yetim, F. (Eds.), Proceedings of SCI 2002, 6th World Multiconference on Systemics, Cybernetics and Informatics; July 14-18, 2002, Orlando, USA; Vol. I: *Information Systems Development* (299-304). Orlando: TPA Publishers.

Use of the Kansei Engineering Approach in a Decision Support System for the Improvement of Medium-sized Supermarket Chains

Yumiko Taguchi

Department of Commerce and Business Administration, Shohoku College
428 Nurumizu, Atsugi-shi, Kanagawa-ken, 243-8501 Japan
E-mail: taguchi@shohoku.ac.jp

Tsutomu Tabe

Department of Industrial and Systems Engineering, College of Science and Engineering, Aoyama Gakuin University
6-16-1, Chitosedai, Setakagaya-ku, Tokyo, 157-8572 Japan
E-mail: tabe@ise.aoyama.ac.jp

Abstract

This study describes the adoption of the kansei engineering approach for the development of a support system that the presidents of medium-sized Japanese supermarket chains can use to improve their existing shops.

1 Introduction

During an interview, the president of a medium-sized supermarket chain was eager to improve the existing shops in his chain, in order to increase their low sales volumes. When this president applied the strategic decision process (Mintzberg et al., 1976) to determine his measures for improvement, we identified the following steps in his methodology: “decision recognition,” “diagnosis,” “design,” and “evaluation-choice”. Breaking down those steps into further detail, we observed the following eight processes.

1) Recognizing the need to improve an existing shop that has been suffering from low sales or declines in sales volumes (decision recognition). 2) Analyzing the existing shop condition: grasping how the customers evaluate the shop by sending out questionnaires about the shop, and grasping the trends in sales, profit, and costs through analysis (diagnosis). 3) Grasping the problems: locating the concrete problems based on process 2) (diagnosing the shop’s weakness). 4) Designing improvements: the president designs improvements to resolve one or more problems. For example, if the customers are dissatisfied with the “comfort” of the existing shop compared to a nearby competitor, the president might be able to gain an advantage over the competitor shops by improving the “comfort” of his own shop. Accordingly, he designs improvements inside of the shop (e.g. changing the color of the lighting and installing new lighting appliances) that increase the customer’s comfort level inside the shop (design). 5) Evaluating and choosing improvements: the president evaluates the desirability of each designed improvement and then chooses the improvement(s) deemed to the most desirable (evaluation-choice). 6) The president repeats processes 4) and the 5) for each of the other problems. 7) Making the improvements: the president makes the improvements chosen in the above processes. 8) Confirming the improvement effect: a certain period of time after making the improvements, the president measures how sales have changed and confirms if the improvements have been effective or not.

In developing a decision support system for improving existing shops, we earlier proposed a method for providing information to support processes 2) (analyzing the existing shop condition) and 3) (grasping the problems). This system provides the degree of customer satisfaction corresponding to process 2), and perceptual maps representing the position of the president's existing shop in relation to competitor shops corresponding to process 3) (Taguchi & Tabe, 2001). However, our trials in developing this system neglected the support of processes 4) (designing improvements) and 5) (evaluating and choosing improvements), particularly the design of concrete improvement in process 4). Accordingly we recently designed a method to support these latter two processes, in particular the design of improvements, based on the kansei engineering approach. Having done this, we clarify a method for developing an overall support system for the actions to be taken to improve existing shops.

2 Approach

2.1 Support policies

Here we list the support policies corresponding to each process to be supported.

2.1.1 Analyzing the existing shop condition and grasping the problems

Support for the processes "analyzing the existing shop condition," and "grasping the problem" is based on the earlier method (Taguchi & Tabe, 2001).

- Analyzing the existing shop condition
  - Providing perceptual maps representing the position of the president's existing shop in relation to competitor shops so that the president can easily understand how customers evaluate the existing shop.
  - Providing the relative degrees of customer satisfaction in the features of the existing shop and competitor shops so that the president can easily understand the weak points of the existing shop in relation to competitor shops.
- Grasping the problem
  - The system provides the results of customer questionnaires so that the president can identify the concrete problems responsible for the weak points of his shop, based on the degree of customer satisfaction.

2.1.2 Designing improvements

When the president suspects that the customer responses on subjective items (the items influenced by the customer's subjective feelings, for example, "comfortable") among the questionnaire items could be a problem, the system provides photographs of shop design items (e.g., color of the floor, lighting appliances used) for each item together with an evaluation of each design item (the degree to which the customer finds it "desirable"), so that the president can consider how the shop design should be improved. Three steps are required to achieve this: changing the verbal expressions used in the questionnaire items to subjective ones, extracting the factors of shop design, and linking the items with shop design factors. This study proceeds as follows, based on kansei engineering approach.

Step 1: Based on literature surveys, we extract subjective items among the factors in relation to the degree of customer satisfaction and establish the shop design items (we call them 'design items'). As a result, subjective items are underlined, as shown in figure 1. In addition, each design item is broken down into the design categories shown in figure 2. These categories are essentially concrete design features of the design items (e.g., the design item "Presence of net" in figure 2 is broken down into the design categories "draping net around produce area" and "not draping").

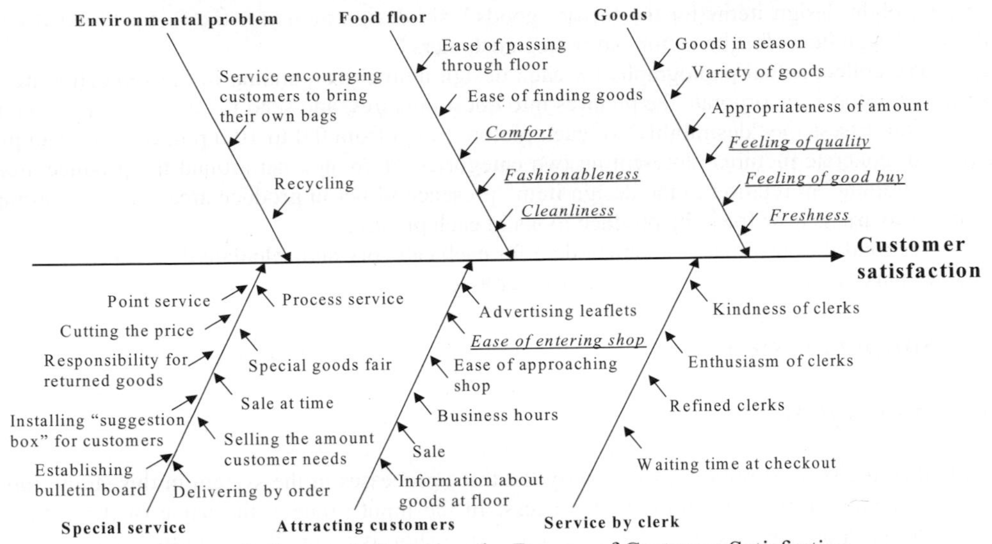

Figure 1: Factors Influencing the Degree of Customer Satisfaction

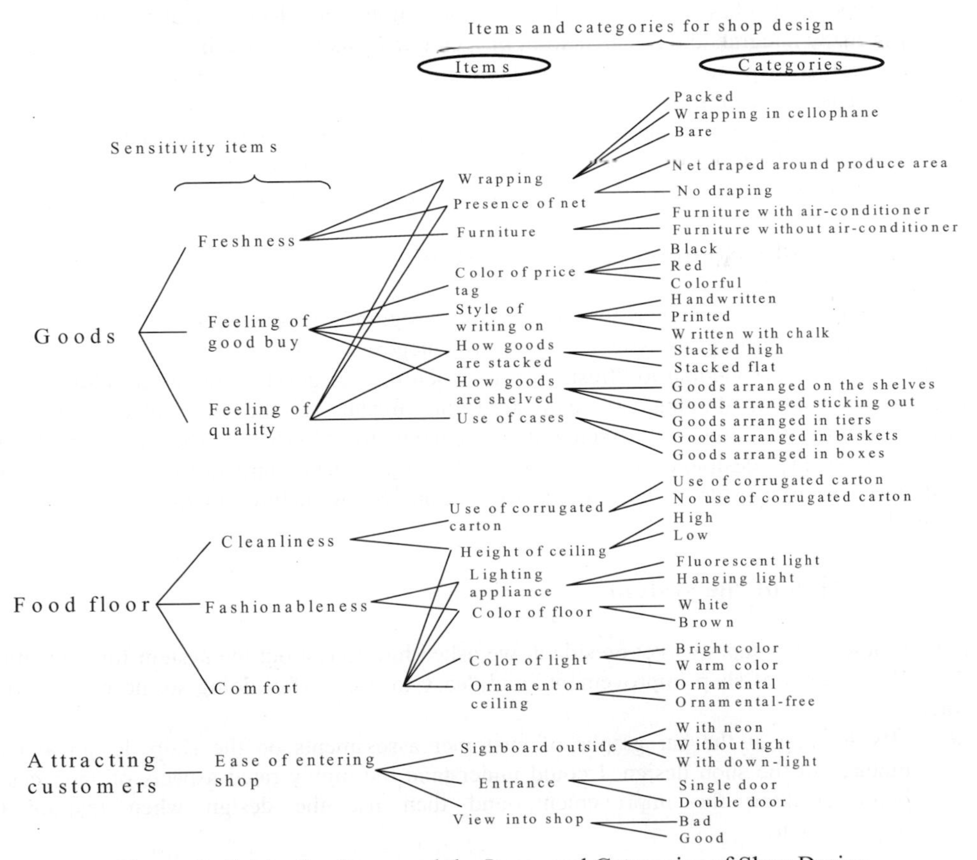

Figure 2: Subjective Items and the Items and Categories of Shop Design

There are eight design items for the factor "goods," six design items for the factor "food floor," and three design items for the factor "attracting customers."
Step 2: We collect actual photographs for each design item corresponding to the subjective items described in Step 1, incorporate the pictures into questionnaires, and send the questionnaires out to customers to assess the "desirability" of each picture (rated from 0.0 to 10.0 points). For example, we present concrete pictures representing two categories "draping a net around the produce area" and "not draping" in relation to the design item "presence of net in produce area", and we prompt customers to assess how fresh the produce looks in each picture.
Step 3: We total the customer assessment data for each category and calculate the average for each category. This average is considered the customer assessment.

3 Support system

3.1 System flow

To actualize the support policies, we employ the three processes in the system in this study: input process, calculation process, and output process. In the input process, the name of the data file storing the questionnaire response data on the existing shop is input. There are two types of data: 1) data on the customer's expectation, and 2) data on the customer's feeling of sufficiency. The calculation process also has two types of data: 1) calculation of the degree of customer satisfaction by applying fuzzy integral, and 2) calculation of factor analysis for extracting factors representing axes of perceptual maps using SPSS. The output process outputs five types of information: 1) perceptual maps, 2) graphs of the degree of customer satisfaction for the existing shop and competitor shops, 3) graphs of questionnaire response data that was used for calculating the degree of customer satisfaction, 4) bar graphs of the customer's assessment of the categories for each shop design item, 5) pictures of the categories of each shop design item.

3.2 Example of system screen in operation

Figure 3 shows two system screens for the process "designing improvements." The screen on the upper left shows the customer's assessments of two categories of the design item "presence of net in produce area" corresponding to "fresh." Thus, when the categories "draping a net around the produce area" and "no draping" are compared for "freshness," the latter receives a higher assessment. When the president designs improvements for "fresh," he/she can consider the category "no draping" desirable for improvement. If the president wants to look at the picture in this category, he/she can do so by clicking the button "Show picture" in the lower right-hand corner of the screen.

3.3 Validation of the system

In an interview with a supermarket president, we asked him to test out the system for evaluations of "the effectiveness of shop improvement" and "easy to use." After doing so, he made several comments.

- "By looking at the bar graphs of customer assessments on the shop design and the pictures of the shop design, I could understand the highly rated aspects of shop design before making the improvement, and then use the design when making the improvement."

- "It's good that I can understand the flow of support provided, by looking at the graphs on questionnaire response data and the bar graphs on the customer assessments of the shop design."
- "The customer assessments of the shop design seem to be influenced by the customers' preference for the pictures shown to them. I would recommend adopting the shop designs that attract customers for use in the system."

On the basis of his comments, this supermarket president seems to have approved of the system in this study.

4 Conclusion

Among the processes that the presidents of medium-sized Japanese supermarket chains use to improve their shops, we applied the kansei engineering approach to clarify one of the developments for the support system, with a close focus on "designing improvements." In addition, we validated this development. In later studies, we hope to expand the scope of this development system to cover three dimensions, as a means of helping supermarket presidents understand the spatial aspects of shop design.

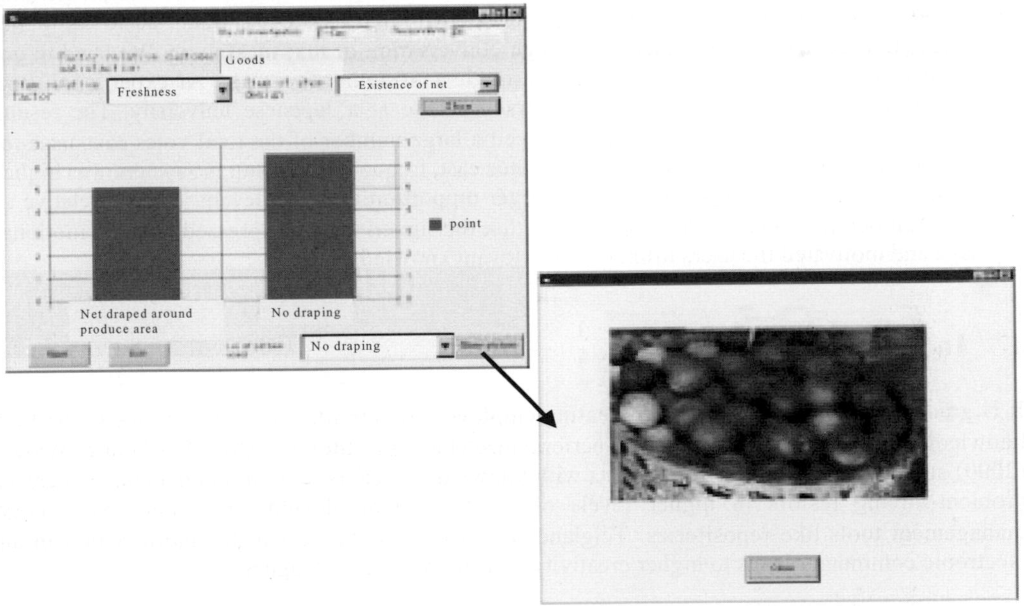

Figure 3: System Screen for the Process "Designing Improvements"

References

Mintzberg, H., Raisinghani, D. and Thèorêt, A., (1976). The Structure of Unstructured Decision Processes. *Administrative Science Quarterly*, 21, 246-275.

TAGUCHI, Y., TABE, T., (2001). Developing of Decision Support Tool for Deciding Whether to Open a New Shop and How to Improve Existing Shops, *Abridged Proceedings of the Ninth International Conference on Human-Computer Interaction,* 251-253.

A Simple Representation of Socio-emotional Interactions to Promote On-line Community Involvement for Knowledge Sharing

Shinji Takao

NTT Advanced Technology Corporation
HIT Center
Higashi-Totsuka-West Bldg., 9F, 90-6,
Kawakami-cho, Totsuka-ku, Yokohama-shi, Kanagawa 244-0805, Japan
takao@hit.ntt-at.co.jp

Morio Nagata

Keio University
Faculty of Science and Technology
3-14-1, Hiyoshi, Kouhoku-ku,
Yokohama-shi, Kanagawa 223-8522, Japan
nagata@ae.keio.ac.jp

Abstract

This study aims to demonstrate that we can increase task relevant ideas exchanges in the electronic community for the purpose of knowledge sharing by providing a simple means of implementing socio-emotional communication with simple signs. We assumed that socio-emotional communication need not necessarily be through conversation or text messaging. We tried to get community members to express gratitude through a system of voting. An experiment was conducted in three classes of a computer literacy course at a Japanese university. The results showed that a system with signed voting received a large number of the total votes cast, users of the system expressed higher satisfaction with votes cast, frequency of using the system was higher than other systems and the system yielded a larger proportion of task relevant writings relative to the total number of messages. We can thus say that the signed voting expressed a socio-emotional message and motivated the users to share task relevant knowledge.

1 Introduction

It is gradually being recognized that creating employee communities is important for promoting knowledge management and enterprise performance (Lesser & Storck, 2001). Teigland & Wasko (2000) suggested that connecting experts with knowledge-seekers and the mutual engagement in problem-solving results in higher levels of creativity than developing "static" knowledge management tools like repositories. Teigland & Wasko also suggested that participation in an electronic community leads to higher creativity than collocating colleagues.

In such communities, socio-emotional communication, as well as task related communication, have been promoted. Some researchers examined a bulletin board system used by doctors in the United States and reported that the proportion of socio-emotional messages to all messages exchanged was about 30% (Rice & Love, 1987). Socio-emotional content has little or no relevance to task knowledge, so that it has been considered that such communication should be limited, in order to make task-related communication more efficient. However, socio-emotional communication plays an important role in making communication smooth, motivating people to communicate with others (Kelly & Bostrom, 1995) and mitigating stress (House, 1981). Few CSCW studies focused on how to handle or support task irrelevant socio-emotional communication efficiently in asynchronous text-based communication.

2 Hypothesis and Design

Some researchers have recently tried to support synchronous socio-emotional interaction without conversation for families whose members live apart (Mynatt, Rowan, Jacobs & Craighill, 2001; Itoh, Miyajima & Watanabe, 2002). These studies suggest that socio-emotional communication does not have to be by conversation or text message. Therefore, we hypothesized that simple and textless signs may be useful for efficient socio-emotional communication. We employed a system of voting to represent such interaction.

We set up an electronic conferencing system for knowledge sharing (figure 1). The system was based on the WWW so that users could use it through WWW browsers. The users could vote on each written communication. The number of votes cast and the names of voters were shown with each written communication. We tried to promote socio-emotional interactions by expressing member support or thanks with signed voting for each message.

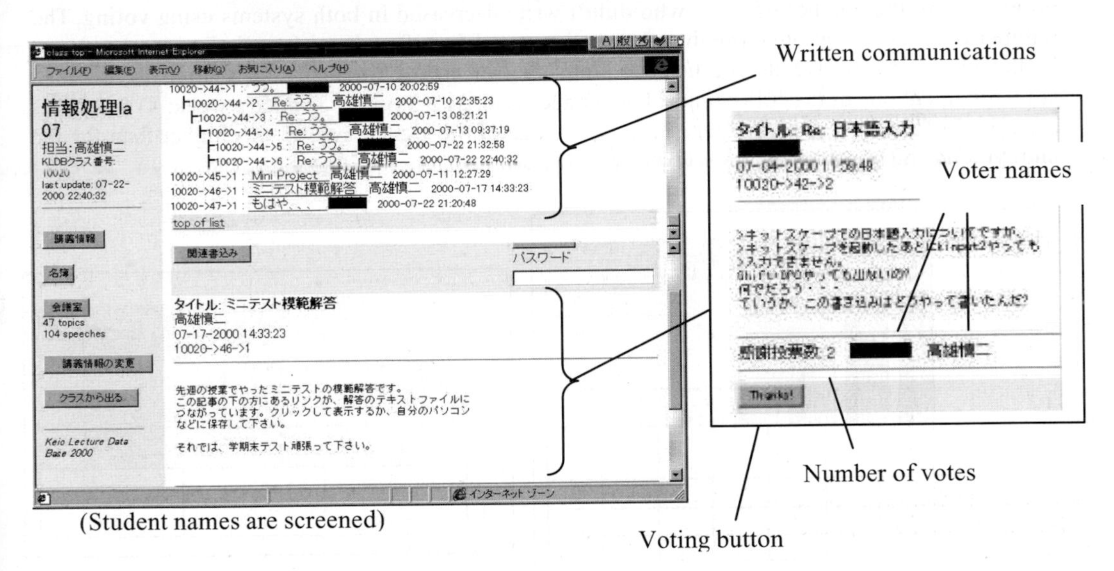

Figure 1: Electronic conferencing system

3 Method

We conducted an experiment in three practice classes in an information literacy course at a Japanese University. The course aimed to educate all the freshmen in basic computing skills: email, editing HTML, elementary programming, and so on. The promotion of knowledge sharing among students worked well, as some students were already highly skilled in these areas whereas there were others who knew very little. Such a situation was somewhat similar to the situation that occurs in the workplace where there are experienced workers and inexperienced newcomers.

We deployed three types of electronic conferencing systems: the system with the signed voting mentioned above, a system with anonymous voting which showed only the number of votes cast with each written communication, and a system where there was a rule that each message would receive equal points and nothing was shown with any written communication.

We asked the students to use the systems over a period of one term (3 months). In the classes using the anonymous voting and the signed voting, we announced that students would get extra points in proportion to their written communications and votes gained, while, in the class with the equal points rule, we announced that students writing questions on the system would earn 1 point per question and those writing answers would earn 2 points per answer. These classes were following the same curriculum, using the same textbook and teaching materials, and were composed of almost the same proportion of male and female students, aged around 18-20. Teachers were male, aged around 20-30. The same teacher was in charge of the class with the equal points rule and the class with signed voting system. We did a questionnaire survey at the end of the term.

4 Results

Figure 2 shows the number of users classified by frequency of writing. Compared to the equal point system, the number of users who didn't write decreased in both systems using voting. The number of users who wrote more than 7 times increased from 0 to 4 and 6 (including teachers). In the questionnaire, we determined the frequency of access (reading and writing) by having students select one of the following: 1) never accessed, 2) 2-3 times, 3) once a week, 4) 2-3 times a week, 5) almost everyday (Figure 3). The average of the signed system was higher than for the anonymous one (significant at 5% level by LSD test).

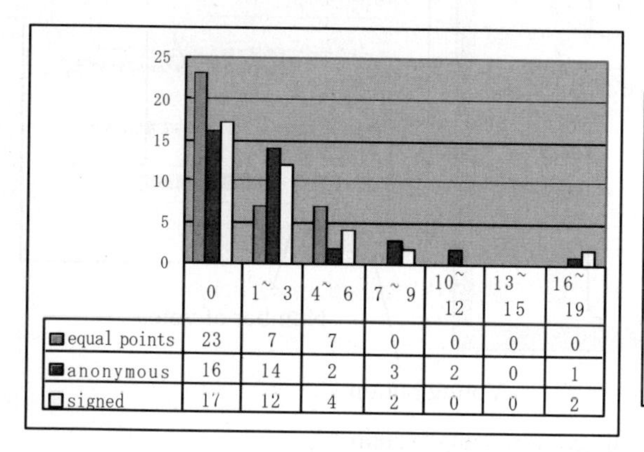

Figure 2: Number of users classified by frequency of writings

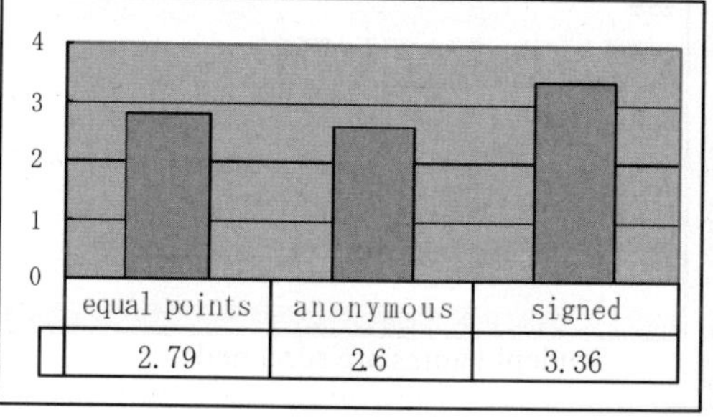

Figure 3: Comparison of frequency of access

Table 1: Comparison Between Anonymous Voting and Signed Voting

	Total writings	Total votes	"Do you feel happy when voted for?"	"Do you feel sad when not voted for"
anonymous	87	7	3.5	2
signed	92	126	4.0	2.5

Table 1 shows comparisons of the anonymous system and the signed system with regard to the total numbers of writings, votes, and the averages of answers to the 5-level questions; "Do you feel happy when voted for?" and "Do you feel sad when not voted for?" (5=Yes, very. 1=No, not at all). There was little difference in the number of writings, but an enormous difference in votes. Both sadness and happiness were a little higher in the signed system.

We asked them "Why did you vote?" with 5 possible answers. The answer "To signify I read" was exceedingly high in the signed system users (figure 4). It is suggested that the signed voting worked somewhat as greeting-like messaging. The fact that students voted not only for task relevant writings but also for socio-emotional text messages also supported this finding.

We classified the contents of writings in 2 categories of "task relevant" and "task irrelevant". Task relevant writings included questions and answers about lectures, exercises and use of the system. Task irrelevant writings include chat, conversations about school life and lectures from other courses. Examples of these writings are shown in table 2. Figure 5 shows that there were more task relevant writings in the signed system than in the anonymous one, while there were more task irrelevant writings in the anonymous one than in the signed one.

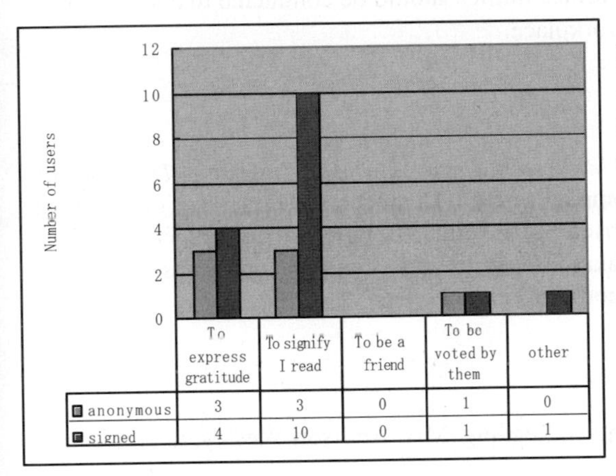

Figure 4: Reasons to vote

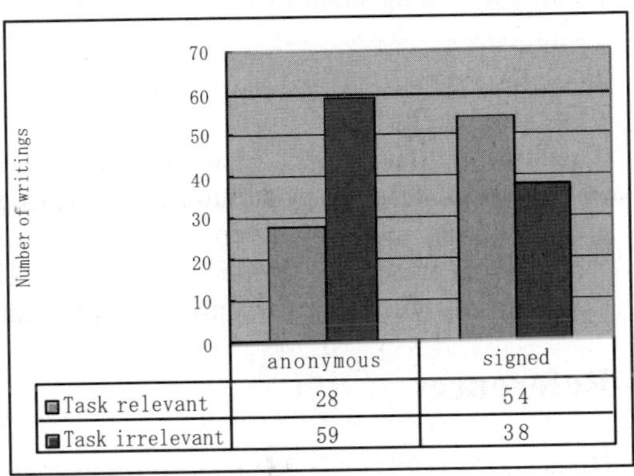

Figure 5: Comparison of classified writings

Table 2: Examples of Writings

Category	Sub category	Example
Task relevant	Class work	Everybody seems to enjoy handling their picture … You can process your picture with this command: %gimp
	System use	Would anyone teach me how to change the password?
Task irrevant	Community	Hello, those who live alone. I'm afraid your diet is out of shape. The co-op will hold a diet consultation…
	Chat	Hello, my name is Chiiko. I'd love to be friends with all of our class…

5 Discussion

The system with signed voting received a large number of the total votes cast while the system with anonymous voting received few votes. Those people using the system with signed voting expressed higher satisfaction with votes cast, and frequency of use (including read only) of the system with signed voting was higher than other systems. According to answers to the questionnaire, they used the signed vote to signify that they read the writings. These results indicate that signed voting worked as a mean of socio-emotional interaction such as expressing support or gratitude. Furthermore, the system with signed voting yielded a larger proportion of task relevant writings relative to the total number of messages. This can be explained by the effect of the signed voting. It worked as a socio-emotional interaction so that students felt less need to write socio-emotional text messages, while it motivated them to share task relevant knowledge. On the other hand, anonymous voting and the equal points rule did not work as a means of socio-emotional interaction because of the anonymity.

6 Limitations and Directions for Further Research

Due to the nature of the computer literacy course, we were able to control a number of factors in a real setting. However, it is always difficult to control group dynamics. Differences of character in teachers and students remained uncontrolled. Further studies should be conducted to confirm these results, including studies of employees in the workplace.

Acknowledgements

This study is based on a result of *the Keio University Multimedia Model Campus Deployment Project,* which has been conducted by Telecommunications Advancement Organization of Japan since October 1997. The experiment was carried out through the good offices of Prof. Nobuo Saito, vice president of Keio University, Takashi Hattori and Manabu Omae, instructor and assistant of Faculty of Environmental Information of Keio University.

References

House, J. S. (1981). Work stress and social support. Reading, MA: Addison-Wesley.

Itoh, Y., Miyajima, A., & Watanabe, T. (2002). 'TSUNAGARI' communication: fostering a feeling of connection between family members. *Extended abstract of CHI '02*, 810-811.

Kelly, G. G., & Bostrom, R. P. (1995). Facilitating the socio-emotional dimension in group support systems environments. *Proceedings of the 1995 ACM SIGCPR conference on Supporting teams, groups, and learning inside and outside the IS function reinventing IS*, 10-23.

Lesser, E. L., & Storck, J. (2001). Communities of practice and organizational performance. *IBM System Journal*, 40 (4), 831-841.

Mynatt, E. D., Rowan, J., Jacobs, A., & Craighill, S. (2001). Digital family portraits: supporting peace of mind for extended family members, *Proceedings of CHI '01*, 333-340.

Rice, R E., & Love, G (1987). Electronic emotion: socio-emotional content in a computer-mediated communication network. *Communication Research*, 14, 85-108.

Teigland, R., & Wasko, M. M. (2000). Creative ties and ties that bind: examining the impact of weak ties on individual performance. *Proceedings of the twenty first international conference on Information systems,* 313–328.

Mobile communication, image messaging and photo sharing: A preliminary comparison of Japanese and Finnish teenagers

Sakari Tamminen, Salla Hari and Kalle Toiskallio

Information Ergonomics Research Group, SoberIT,
Helsinki University of Technology
Metsänneidonkuja 10, 02150 Espoo
[sakari.tamminen, salla.hari, kalle.toiskallio]@hut.fi

Abstract

In the era of globalization cross cultural research is becoming more and more relevant in the area of HCI. This preliminary comparison between the Japanese and Finnish teenage girls implies that cultural differences are to be found e.g. in the attitudes and motives relating to mobile communication and photographing. These cultural differences in the social practices should be carefully considered when designing new consumer products for global markets.

1 Introduction

Globalization means simultaneous mixture of worldwide economy and marketing, and increasing exposure to specific cultural phenomena. The latter is not a mere curiosity, since cultures organize our lives by guiding how to interpret and value different things. One of the best-known theory of cultural differences is based on the empirical work of Geert Hofstede from the late 70's to its revised form in the 90's. He conducted large numbers of interviews and through statistical analysis determined patterns of similarities and differences among the studied cultures. He noticed that most of the differences could be accountable by five dimensions, or factors, of cultures. These are power distance, the level of individualism, masculinity/feminity, uncertainty avoidance and long / short term orientation of action (e.g. Hofstede 1991).

Cultural differences bring also new problems to product design. One product and its design could be well received in one country, but rejected by consumers in another because of cultural effects. Colours, symbols and the styles of communication differ between different cultures, to name but a few possible concrete factors of UI-design related to cultural differences. The problems embedded in cross cultural interface design have been noticed within the HCI community for several years (e.g. Nielsen 1990, Simon 2001, Marcus 2002, Duncker 2002, Russo & Boor 1993).

However, there have not yet been many empirical studies on how mobile communication technologies are assigned differing meanings in different cultures. This preliminary study is focused on cultural effects among Japanese and Finnish teenage girls on their mobile communication practises. The study sheds some light on Japanese and Finnish teenage girls' mobile communication, image messaging and photo sharing cultures.

2 The setting

In this study, 8 teenage girls (15-19 yrs, 4 Japanese, 4 Finnish) were interviewed by semi-structured interviews about their mobile communication behavior, mobile messaging and interpretations about new mobile technologies. In addition, a light contextual inquiry of teenagers' image usage was carried out by analyzing the subjects' photo albums.

The reason for selecting only girls to the interviews lies in the fact that there exists a wide body of research literature (e.g. Brosnan 1998, King 2001, Simon 2001, Wajcman 1991) about the differences between the genders in the expression of technical competence, know-how and preference related to technical artifacts. Thus, by interviewing only girls we tried to eliminate the possibility of different gender effects in our preliminary data. The differences between the responses are thus explained by cultural differences related to the interpretations of mobile messaging and photo sharing.

All interviews were recorded and transcribed for qualitative analysis, which was done by categorizing emergent themes and attitudes in the respondents' discourse. The analysis method of these interviews of thus close to the discourse analytical perspective (e.g. Silverman 2000), where the focus is on recognizing different sets of discourses and the meaning embebbed to them. The focus on our analysis was on the different meanings assigned to various communication practices and how they relate to the social relations of the interviewees. Our preliminary study has two main results. First concerns the use of mobile communication devices, typical use practices and attitudes towards these technologies. Second themes are related to traditional photographing, the use of photo albums and mobile imaging.

The analysis of current photograph use practices was done by comparing the subjects' photographs. All the interviewees were asked to show their photo albums to the researchers and to explain the reason why certain photos were arranged in the album and some were left out. We focused especially on the different kinds of marking and manipulation strategies of emotionally meaningful and " important" photos taken by Japanese and Finnish girls.

3 Differences

One clear difference between the cultures can be inferred from the level of technical know-how in the discourse of Japanese girls. In their talk the mobile phone was positively interpreted both as a technical and social object. The Japanese girls spontaneously described different kinds of technical details of their phones, such as the amount of memory (in bytes) in the phone and the different types of possible ringing tones (monotonic/polyphonic). This is an interesting finding, especially since it has been documented in many international studies that girls' technical competence is lower than boys' (Brosnan 1998, King 2001). It thus seems that the technical know-how is be related, besides gender, to also to the culture of the respondents. Compared to Japanese girls, Finnish girls of our data were rather indifferent with new mobile devices, even stating explicitly that they are not interested in new mobile technologies. However, it is socially convenient to use the devices. For example, by not having a phone girls feel that they would quickly found themselves outside of the circle of their friends. The emphasis in their talking was on the social interpretation of the mobile phone, whereas the technical interpretation was given negative connotations.

There were also differences in evoking emotions with messaging. Japanese girls use voice messages, audio clips and pictures to personalize their messages. They mix all these elements effectively in everyday communication. Finnish girls don't find this interesting, and image messages are used only in special occasions (e.g. birthdays). In addition, differences in photographing culture were discovered. The photographs taken by the Japanese girls are subjected to systematic manipulation before they are arranged in photo albums. They shape the pictures by either tearing or cutting the originals for a "nicer" look. Other forms of manipulating the pictures include the overwriting and framing of pictures with comments.

Finnish girls, in their turn, do not manipulate their original pictures in the same way. If comments are made, they are written next to the photos – original picture are highly valuated, thus experienced as something untouchable. The photos in the albums were also arranged in straight columns, opposite to the Japanese style of artistically "layering" the pictures one over another without any major horizontal or vertical justification. This observation could be interpreted so that perhaps Japanese girls have more iconographic and Finnish more logos centric cultures of image perception and manipulation. This kind of difference is derives from the fact that the history of visual culture in Japan and Finland varies. Japan has thousand years' tradition in painting, whereas Finland has been a Lutheran country in which visual culture has been valued only in very restricted areas of public life.

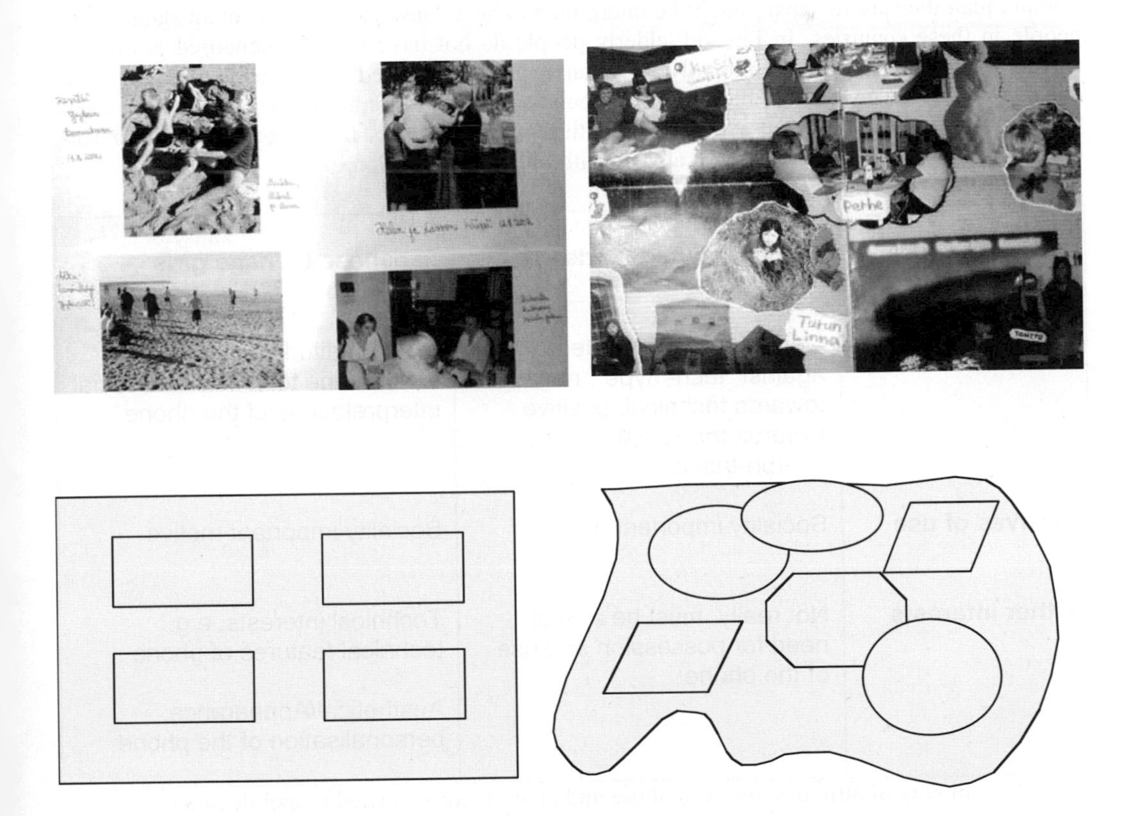

Figure 1. The differences between the Finnish and Japanese arrangement and manipulation of photos.

4 Similarities

One striking similarity in the communication patterns was the nature of the communication with the elders. Both groups are short worded with their parents. The style of communication is hierarchical, which sets limits to the contents. Communication is seen best as "reporting" to the parental inquiries about their life. Compared to the former, communication with friends is highly democratic - sent images are used to reinforce the social relations within a small circle. Another similarity is the objects of taken photos. Both Japanese and Finnish teenagers take their photos from "funny" social events, which are intuitively relevant for both to the sender and receiver.

These findings refer of course to very general phenomena – sociality – that has many universalities. However, they might be also consequences of similarities in cultural changes, which Finland and Japan have been going through last 5 decades. They share late and fast transition to industrialisation and general modernization after the Second World War, and beginning of digital era of consumer society from the 90' s. Furthermore, in both societies there is a rather sharp generation gap in consuming of digital devices.

However, the seeming phenomenon of elderly people being more familiar with mobile devices in Finland than they are in Japan, might be interpreted to be a consequence of different status of old people in these countries. In Finland, elderly people do not have such an honoured position in society as they have had traditionally in Japan (e.g. Japan scored higher on Hofsted's Power distance scale). If they want to be part of any social activities, they must actively participate them. One of the strongest practical symbols to do this in these days is the usage of digital devices, especially mobile phone, and perhaps also pc with Internet connection.

Use of mobile phone	**Finnish teenage girls**	**Japanese teenage girls**
Attitudes	General blasé attitudes against "tech- hype": negative towards technical, positive towards the social interpretations	Playful attitudes: positive towards the technical and social interpretations of the phone
Motives of use	Sociality important motive	Sociality important motive
Other interests	Not really, must be a "real" need for possession and use of the phone	Technical interests, e.g. technical features of phone Aesthetical/Appearance, personalisation of the phone

Table 1: Summary of attitudes, motives of use and other interests related to mobile phone.

5 Conclusions and Future Work

The results of this preliminary study show that sociality is an important motive for both Japanese and Finnish teenagers according their use of mobile phones. In addition Japanese girls were more technically oriented and aesthetical personalization of their phones was considered as an important factor. Finnish girls were indifferent with technical issues but reported more concern of the aims and reasons of their communication, especially with mobile pictures. Japanese teenagers seem also to be more playful with pictures, which manifests in the heavy manipulation of the photos. These results imply that when designing mobile devices and services, cultural differences e.g. in the amount of technical features and aesthetical values of the target groups, should be carefully considered. Our findings point out several topics to be explored further in cross-cultural research on the product design of mobile devices & services. Culture affects not just on how people perceive interfaces, but also to interpretations and valuations of mobile devices (technical/social device, attitudes), communication patterns and image/picture cultures. The cultural adaptation of a product is related to its social acceptability within the selected target groups, and also to how well the new offered techniques could reinforce the social identity of the group members.

Our future work consists of studies on the cultural differences with a bigger sample (of Japanese and Finnish girls) on how teenagers manipulate their photos taken with a mobile phone camera, and how they use these photos to reinforce their social communication.

References

Brosnan, M.J. (1998). Technophobia. The psychological Impact of Information Technology. Routledge Publishers.

Duncker, E. (2002). Cross-Cultural Usability of the Library Metaphor. Proceedings of *JCDL'02,* July 13-17, 2002, Portland, Oregon, USA, 233-240.

Hofstede, G. (1991). Cultures and Organizations: Software of the Mind. New York: McGraw & Hill.

Marcus, A. & West Gould, E. (2002). Crosscurrents. Cultural Dimensions and Global Web User-Interface Design. Interactions, 32-44.

Nielsen, J. (1990). Designing User Interfaces for International Use. New York: Elsevier Science.

Russo, P. & Boor, S. (1993). How Fluent is Your Interface? Designing for International Users. Proceedings of InterCHI 1993. 242-347.

Silverman, D. (2000). Doing Qualitative Research: A Practical Handbook. London: Sage

Simon, J.S. (2001) The Impact of Culture and Gender on Web Sites. An Empirical Study. The DATA BASE for Advances in Information Systems. 32 (1), 18-37.

Wajcman, J. (1991). Feminism confronts technology. UK: Polity Press.

Designing and Evaluating Government Websites Within the Context of the Electronic Democracy

A.G. van der Vyver

Monash, South Africa
Private Bag X60, Roodepoort, 1725
braam.vandervyver@infotech.monash.edu

Abstract

The number of government websites on the Web has grown dramatically over the last two years. The quality of many of the sites has proved to be of a questionable nature. The mere fact that many of those that access government websites are restricted to low technology facilities offers great challenges to designers who venture into this field. This paper offers a number of guidelines and suggestions that can be used to design and evaluate government websites.

1 Introduction

The information revolution has paved the way for the design and the implementation of various facets of the electronic democracy. The creation of electronic communication networks to facilitates communication between government and the citizen has called for the principles of representative democracy to be revised. Schalken, Depla & Tops (1996) formulated a scenario of organized electronic political and democratic processes with many participants replacing representative democracy with its parliamentary system. To them representative democracy is only a surrogate for direct democracy.

The digital divide that earmarks many developing countries has, however, been a major barrier in the way of the establishment of workable electronic democracies all over the world. This digital divide is not only a technological phenomena that describes the gap between the have's and the have not's but also a widening knowledge gap. Outside of the developed world, Internet access is only enjoyed by a selected few whilst millions of others have to rely on internet kiosks and other public facilities to overcome this growing draw-back. Low literacy levels, further, complicates the issue.

The design of a government website requires that people of all walks of life should be taken into account when it is designed. Readability of text will always be an important aspect. Volume of text on the informational pages should be restricted. The citizen of today is a lazy reader. Hot links, should, however, be used to white papers, draft legislation and legislation recently passed. Cultural aspects may also come into play when decisions on logo's and graphics are considered. A professional, user-friendly end-product, that is accessible to a broad section of the population, would satisfy most of the stakeholders.

2 The Creation of the Electronic Democracy

The American I.T. pioneer, Jim Clarke Warren made the following statement in 1976 with regard to the information revolution: "The more I see of it, the more I slowly come to believe that massive information processing power which has traditionally been available only to the rich and powerful in government and large corporations will truly become available to the general public. And I see that as having a tremendous democratizing potential, for most assuredly, information - data and the ability to organize and process it. It is an exciting vision to me" (Browning, 2002, p. 15)

Less than twenty years later, in October 1994, the first E-democracy project was launched in Minnesota. Whilst the two political candidates exchanged comments, 600 members of the voting public participated in the debate by exchanging comments and commentary from their computers. The foundation for the establishment of a virtual democracy was laid.

Hague & Loader (1999, p. 3) stated that "at present, the notion of digital democracy can refer to a wide range of technological applications including televised 'people's parliaments', or citizens' juries, e-mail access to electronic discussion groups, and public information kiosks ..."

The key word to democracy is dialogue. The airing of different viewpoints and the opportunity to influence decisionmakers and voters are at the heart of a democracy. Browning (2002, p. 5) pointed out that the Internet's greatest strength is its ability to support simultaneous interactive communications among many people. "Unlike the telephone, which primarily supports, one-to-one communications, or radio and television, where information flows in only one direction, from a single source to an audience that can only listen passively, the Net allows information to flow back and forth among millions of sources at practically the same time."

The instant interactive element of communication is not the only benefit of the Net. The distribution potential of the Internet is enormous. The practice of "forwarding" enables the author of an e-mail message to reach the operators of mailing lists with one communication act. These operators can then post the message to their subscribers. In some cases these mailing lists contain the addresses of thousands of subscribers. The multiplier effect with regard to message distribution becomes huge.

In this paper the emphasis will be on government websites as tools of the electronic democracy.

3. Government Websites: Important Tools of the Electronic Democracy

One of the most important implementation tools of the electronic democracy is government websites. The design of such websites poses major challenges to the webmasters taking charge of such websites. The following aspects will require sound decisionmaking:

3.1 Visibility of the system status.

This principle requires that the users are always informed of what is going on. Mayhew (1992, p. 53) state that it is important to provide plenty of context information on screens ... to help users keep track of where they are in the dialog, how they got there, and how to get elsewhere."

Instone (1997, p. 1) state that the two most important things that a user need to know is:

- Where am I?
- Where can I go next?

He emphasizes that each page should be branded in that it should show to which section it belongs. This requires that links to other pages should be clearly marked. Instone (1997, p. 1) issues the following warning: "Since users could be jumping to any part of your site from somewhere else, you need to include the status in every page."

Text and sound could for instance be combined to convey the message that an e-mail inquiry has successfully been sent to the web administrator of a government website (Dix et al. (1998, p. 24).

3.2 Match between system and real world

This heuristic encourages the use language, phrases and concepts from the user's world. Dix et al. (1998, p. 413) warns against the use of system-specific engineering terms. Instone (1997, p. 2) recommends that real-world conventions should be followed. Information should appear in a natural and logical order.

Mayhew (1992, p. 52) adds the following guidelines to obtain clarity:
- take the knowledge and understanding of the listener into account
- use vocabulary that will be familiar to the listener
- make sure that the order of statements or arguments are consistent with the intended effect

It could become immensely complicated to interpret this principle. Their message is very seldom targeted to a very specific clearly-defined target market. Exceptions include messages to the owners of motor vehicles or people who are HIV-positive.

The general public, who is the target market for most of the messages that are generated in the governmental domain, represents people coming form all walks of life, representing a myriad of cultural mixes

3.3 User control and freedom

The broad principle dictates that the user should always get the feeling that he/she is in command of their own navigation. Instone (1997, p. 2) give a very restricted interpretation of the principle. He explains that "users often choose system functions by mistake. In order to escape from this dilemma "... they will need a clearly marked "emergency exit" to leave the unwanted state without having to go through extensive dialogue." He recommends a "home button" on every page as the ultimate escape option. "Do and undo" functions should also be supported in the design.

User control is of paramount psychological importance when a user engages into communication with a government website. The user will already entertain some images of the governmental body he/she is.

3.4 Hints to secure a reasonable level of usability

The following hints can be offered to designers of government websites:

- Content of the site: The content should be kept in digestible format without over-simplifying important concepts. Hot links can be used to guide users to explanatory copy. Long texts of proposed legislation and policies should also be hotlinked.
- Simple menus and indexes: It should be used to facilitate navigation by the user.
- Graphics: Graphics should be restricted to a minimum to accommodate users with limited bandwidth at their disposal.
- Interactivity: Opportunities for interactive participation should be created. This may include invitations to comment on proposed legislation and policies. E-mail addresses of politicians and bureaucrats should be hot-linked.
- Bulletin boards: They should be hot-linked to enable citizens to participate in continuous debate on major issues.
- Buttons should be used to make provision for referenda.
- Checkboxes should be used to compile a condensed demographic profile of the participating citizen, i.e. gender, age, occupation etc.
- Quantitative tools like sliders and spin buttons should preferable be avoided.
- Forms should be carefully designed to avoid unnecessary scrolling and paging. Fields should be large enough to accommodate answers. The basic lay-out of the form should not be distorted when an answer exceeds the size of the field.
- Information should be presented in an objective and clear manner. All forms of propaganda and unethical persuasion should be avoided.
- The latest version of legislation and policy should appear on the website. Legal matters should be explained in layman's language.
- Answers to frequently asked questions should be included, provided that this concept is clearly explained.
- Political and polemical issues should be avoided. Content should be restricted to informational and educational topics.
- Supervisors, preferably form the ranks of the communication officers should supervise members of the public when they access the Internet at public access points.

4. Conclusion

The Internet in its present format has been with us for less than ten years. Much progress has been made in the field of website design. The uniqueness of the medium has been captured in its own sets of design principles. The design of government websites has not been at the top of the research agenda. In this paper and attempt was made to provide guidelines to the designers of government websites on how to empower the citizens that are going to visit these websites in the future.

References

Browning, G. (2002). Electronic Democracy. Using the Internet to Transform American Politics. Cyber Age Books: Medford: New Jersey.

Cooper, A. (1995). About Face. IDG: Foster City, CA

Depla, P.F.G., Schalken, C.A.T. & Tops, P.W. (1996). Technology and Modernization of Local Government in Emerging Electronic Highways by V. Bekkers, Koops, B-J, & Nouwt, S. Kluwer Law: The Hague.

Dix, A., Finlay, J., Abowd, G. & Beale, R. (1998). Human-Computer Interaction. (Second Edition). Pearson Education: Harlow, Essex.

Grossman, L.K. (1995). The Electronic Republic. Penguin: New York.

Hague, B.N & Loader, B.D. (1999). Digital Democracy. Routledge: London.

Instone, K. (1997). Site Usability Heuristics for the Web. Retrieved February 1, 2001, from http://www.webreview.com/1997/10_10/strategists/10_10_97_2.shtml

Mayhew, D. (1992). Principles and Guidelines in Software User Interface Design. Prentice Hall: Englewood Cliffs, N.J.

Van de Donck, W.B.H.J., Snellen, I, Th. M. & Tops, P.W. (1995). Orwell in Athens. IOS: Amsterdam.

The Use and Usefulness of Communication, Collaboration and Knowledge Management Tools in Virtual Organizations

Matti Vartiainen, Marko Hakonen & Niina Kokko

Laboratory of Work Psychology and Leadership, Department of Industrial Engineering and Management, Helsinki University of Technology
P.O.Box 9500, FIN-02015 HUT, Finland
Email: matti.vartiainen@hut.fi

Abstract

The use and usefulness of communication and collaboration tools are studied. The data was collected by interviews and a questionnaire from three companies. The study shows that traditional tools are used to communicate such as emails, one-to-one calls, meetings, faxes and mails. For collaboration, shared databases, joint folders in intranet, and project management software are used. The study underlines the need to develop better tools for virtual collaboration.

1 Purpose and research questions

Information and communication technologies have made it possible to work flexibly anywhere in the world where the appropriate infrastructure exists and employees are competent to use it. The amount of virtual work increases. The results of the Emergence 18-country employer study[1] show that there are over 9 million eWorkers in Europe. eWork was defined as any work, which is carried out away from an establishment and managed from that establishment using information technology and a telecommunications link for receipt or delivery of the work.

The purpose of this paper is to describe and evaluate the use and usefulness of tools used in virtual teams and projects for communication, collaboration and knowledge management. The challenges and improvement needs of the tools are also investigated. The research questions of the study are:

- How often members of virtual teams and projects communicate and collaborate, and what media they prefer?
- To what degree communication tools support the production and social functions of teamwork?

2 Background

The targets of this study are virtual organizational units, i.e., virtual teams or projects, which are regarded as basic units of dispersed organizations. Although dispersed working and the use of electronic communication and collaboration tools characterize well virtual organizations, they consist of several additional features that must be taken into account. The relationships of any

[1] The EMERGENCE project is funded by the European Commission's Information Society Technologies (IST) programme. Altogether 7268 employees in 18 European countries answered in 2000 a survey concerning eWork in Europe.

actors being them individuals, dyads, teams, projects, organizations or networks can be characterized with four dimensions and their sub-dimensions (Fig. 1): *space* (same place vs. different, dispersed place; fixed vs. mobile), *time* (same, synchronous vs. different, asynchronous; permanent vs. temporary), *mode of interaction* (face-to-face vs. mediated), and *diversity* of actors (similar vs. different). Even the simplest comb ination of these four dimensions generates sixteen types of organizations, which describes the variety of demands for organizational design and development that virtual organizations provide. Depending on the dimensions and their combination, we can speak, for example, about co-located or multi-site teams, permanent or temporary teams, etc. A fully virtual organization is a specific, "extreme" constellation of the four dimensions.

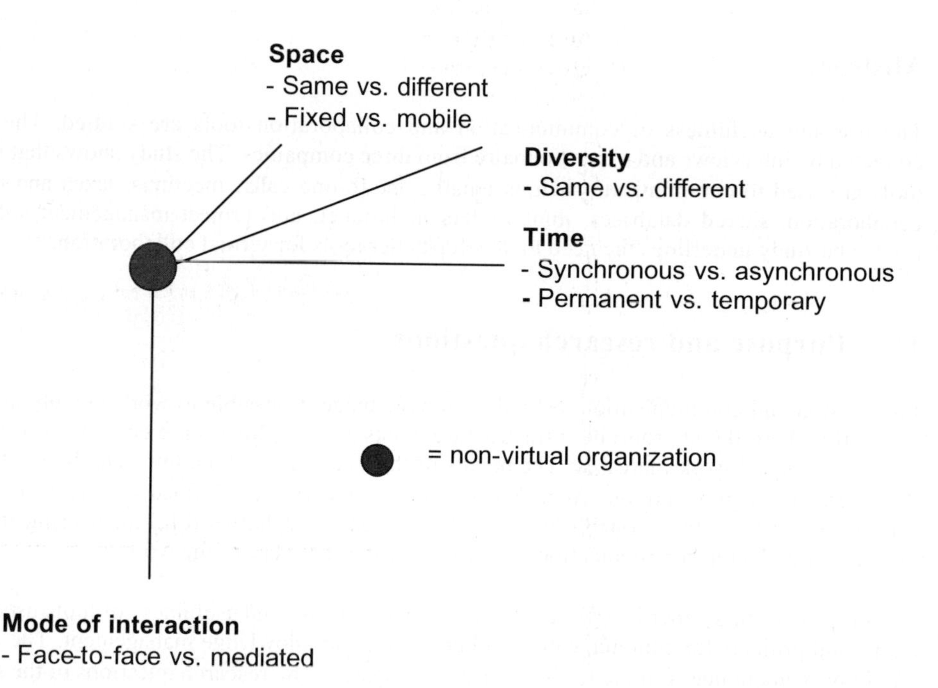

Figure 1: Dimensions of virtual organizations (Modified from Gristock 1997, 8)

For most organizations, being virtual is a matter of degree. DeSanctis, Staudenmayer and Wong (1999) highlight that the virtual organization is not so much a pure form as a continuum for describing a range of relationships along the dimensions of space, time, culture and organizational boundary. According to them, relative to the traditional organization, relationships in the virtual organization are more geographically distributed, more asynchronous, more mult icultural, and more likely to extent outside the firm. People involved in a team may speak several languages and have divers e cultural, educational and vocational backgrounds. The degree nature of the dimensions defining virtuality implicates that there have always been organizations like that, for example, the past Chinese and Venice merchants worked around the globe as a distributed network based on trustful relationships. They worked dispersed, asynchronously and communicated through messengers and letters. The really new thing is the technology used for communication and collaboration. Therefore, virtual organizations are mo stly identified as organizations, where employees work apart for a common goal and communicate via ICT.

How effective virtual organizations are? McGrath (1991) suggested that groups should fulfil three functions to be effective: production function, member-support function, and well-being function. *Production function* implies that team performance meets or exceeds the performance standard set by clients. *Member-support function* requires that working in a team results in satisfaction, learning, etc. *Well-being function* is related to the degree to which the attractiveness and vitality of a team is strengthened. Based on McGrath, Duarte and Snyder (2001) divide the stages of virtual team development into the realms of task and social dynamics. The goal of well-managed *task dynamics* is productivity. The goal of well-managed *social dynamics* is a feeling of team unity. In this research, the frequency of use and usefulness of collaboration technology (Andriessen, 2002) are studied. The collaboration technologies are divided roughly into communication and collaboration tools. The usefulness of communication tools for teamwork performance ('production function') and getting to know each other ('social function') are studied in more detail in addition to the frequency of use and usefulness of collaboration tools.

3 Data and Methods

The data was collected in three companies. Two of them form a joint kraft mill process design project involving several sub-contractors and consultants, altogether over a hundred people. The third company provides expert services for design and construction of roads. The employees of the companies work in several locations all over Finland. First, a context analysis was made in each company by collecting documents and by interviewing company management. Second, a target team or project in each company was selected for detailed analysis. Their members (in all, n = 16) were interviewed. Third, a questionnaire was delivered to the members (n=71) of virtual teams and projects, which were accessible in each company. The interview consists of an open-ended part and a semi-structured part. Interviews aimed at giving in-depth understanding of the phenomena, whereas the questionnaire provides with statistically generalizable data.

4 Results and Discussion

4.1 Use and usefulness of communication and collaboration tools

The results show that the *communication toolbox* of a virtual worker (Table 1) consists of emails, one-to-one calls, scheduled and informal face-to-face meetings, faxes and mails. Also company intranet and text messages are widely in use, whereas chat, discussion lists, video- and teleconferences are seldom available. Emails and telephone calls are used daily, informal meetings and company intranet weekly, although there seem to be a large variation in intranet use. Scheduled face-to-face meetings are arranged monthly, as well as faxes and mails are sent. Surprisingly, it can be seen that even if text messages, teleconference, discussion list, videoconference and chat are available, they are practically never used for work-related communication.

The most useful communication *tools for work performance* are: emails, scheduled and informal face-to-face meetings, one-to-one calls, faxes, and to some extent also company intranet and usual mail. Teleconference and text messages are somewhat useful, discussion list and chat only slightly. Videoconference was considered not at all useful!

The very useful communication *tools in getting to know each other* are scheduled and informal face-to-face meetings. One-to-one calls and emails were considered clearly useful. Company intranet, fax (!), mail, and teleconference were regarded as somewhat useful, whereas videoconference, text messages, discussion list and chat were only of little use in getting to know other team members.

Table 1: Frequency of use, and usefulness of communication and collaboration tools in virtual teams and projects

	Frequency of use					Usefulness - performance			Usefulness – to know each other		
COMMUNICATION TOOLS AVAILABLE	In use			Not in use							
	N	Mean*	SD	N	(%)	N	Mean**	SD	N	Mean**	SD
Scheduled F-t-F meetings	68	3.3	1.0	1	1.4	68	4.5	0.7	67	4.8	0.5
Informal F-t-F meetings	68	3.9	1.1	1	1.4	67	4.4	0.9	67	4.8	0.5
Videoconference	23	1.0	0.0	46	65	8	1.3	0.5	8	2.1	1.1
Telephone (one-to-one)	69	4.4	0.8	-	-	69	4.5	0.7	67	4.0	1.0
Teleconference	36	1.3	0.6	33	47	24	2.7	1.3	19	2.6	1.2
Special discussion lists	23	1.3	1.0	46	65	13	1.9	1.1	8	1.6	1.1
Company intranet	54	3.7	1.7	15	21	53	3.9	1.3	49	2.8	1.4
Chat on the Internet	22	1.1	0.3	47	66	12	1.5	1.0	9	1.4	1.0
E-mail	69	4.7	0.5	-	-	68	4.8	0.4	67	3.5	1.3
Text messages (SMS)	53	2.0	1.2	15	21	45	2.5	1.3	40	2.1	1.1
Fax	65	3.2	1.2	4	6	66	4.0	1.2	64	2.7	1.4
Mail	65	2.9	1.2	2	3	61	3.7	1.2	61	2.7	1.4
COLLABORATION TOOLS AVAILABLE	In use			Not in use							
	N	Mean*	SD	N	(%)	N	Mean**	SD	N	Mean**	SD
Shared folder in intranet	49	3.3	1.4	19	27	41	3.5	1.2	-	-	-
Shared databases	60	3.8	1.3	8	11	55	4.1	1.1	-	-	-
Team's website	24	2.5	1.3	44	62	20	2.9	1.2	-	-	-
Project management software	48	2.4	1.4	20	28	35	3.4	1.2	-	-	-
Data conferencing (e.g. NetMeeting)	20	1.2	0.5	48	68	8	2.0	1.3	-	-	-
Databases to find experts (e.g. 'Yellow pages')	41	2.4	1.3	27	38	31	2.8	1.2	-	-	-
Group decision support system	14	1.2	0.6	53	75	8	2.3	1.3	-	-	-
Group calendar	33	2.7	1.6	35	49	28	3.5	1.1	-	-	-

* 1=never, 2=less than monthly, 3=monthly, 4=weekly, 5=daily
** 1=not at all, 2=slightly, 3=somewhat, 4=clearly, 5=very

The *collaboration toolbox* of a virtual worker consists of shared databases, joint folders in intranet, and project management software often including group calendar. Group decision support systems, data conferencing, and team's joint website are seldom available. Shared databases are considered clearly useful for team's performance, as well as shared folders, group calendar and project management software.

Interviews shed some light on why some tools were preferred for good job performance. Various tools seem to have different functionality. Emails were favored due to their chronological order and possibilities of re-use, although many interviewees complained about excessive information by too many messages. F-t-F meetings were considered necessary for building a 'big picture', creativity, innovativeness and decision-making. One-to-one calls were seen necessary when quick responses were needed in high-priority affairs.

4.2 Discussion

The survey shows that the e-mail was the most used communication tool, which is quite a predictable result in virtual settings. The fact that respondents considered e-mail as the most useful tool in terms of team performance, and the fourth best in getting to know each other, is somewhat surprising. As e-mail is usually not regarded as very rich media, one could predict that it could not fulfil all task and hardly any social functions in a virtual team. However, as face-to-face meetings are quite rare, the virtual workers have to rely on minimal cues even in social functions like trust formation. This may be problematic, since misunderstandings cannot be avoided.

The main question related to the communication and collaboration technologies is their ability to replace immediate social proximity by supporting informal, rich-content communication. In co-located work places, people meet each other in formal meetings and occasionally face-to-face during café breaks and lunch times or just in an elevator, and get to know each other and even gain important pieces of knowledge for their work. The media richness model (Daft & Lengel 1984; Picot, Reichwald & Wigand 2001) relates the richness of information content to the complexity of tasks: the more complex the task, the 'richer' media is needed, and the more structured the task is, the more effective the 'poor (or simple)' media is. According to the model, the effective communication is to be found by combining different media to meet the demands of the tasks and by paying attention to disturbances of excessive and barriers of inadequate information. Kraut et al. (1990a, 287) also claim "that the more spontaneous and informal the communication, the less well it was supported by communication technology." In the near future, the number of eWorkers will be manifold from now. There is a burning need to develop better tools for virtual collaboration.

References

Andriessen, J.H.E. (2003). *Working with groupware. Understanding and evaluating collaboration technology*. London: Springer Verlag.

DeSanctis, G., Staudenmayer, N. & Wong, S.S. (1999). Interdependence in virtual organizations. In C.L. Cooper & D.M. Rousseau (Eds.), *The virtual organization*. Trends in organizational behavior, Vol. 6, (pp. 81-104). Chichester: John Wiley & Sons.

Daft, R.L. & Lengel, R.H. (1984). Information richness: a new approach to managerial behaviour and organization design. *Research in Organizational Behavior*, 6, 191-233.

Duarte, D.L & Snyder, N.T. (2001). *Mastering virtual teams. Strategies, tools, and techniques that succeed*. San Francisco: Jossey-Bass.

Gristock, J. (1997). Communications and organizational virtuality. *VoNet-Newsletter*, 1 (5), 6-11.

Kraut, R.E., Fish, R.S., Root, RW. & Chalfonte, B.L. (1990). Informal communication in organizations: from, function, and technology. In S. Ostkamp & S. Spacapan (Eds.), *People's reactions to technology in factories, offices, and aerospace*. The Claremont Symposium on Applied Social Psychology (pp. 145-199). Sage.

McGrath, J.E. (1991). Time, interaction, and performance (TIP). A theory of groups. *Small Group Research*, 22 (2), 147-174.

Picot, A., Reichwald, R. & Wigand, R.T. (2001). Die grenzenlose Unternehmung. Wiesbaden: Gabler.

Putting Order to Episodic and Semantic Learning Memories: The Case for KLeOS

Giasemi N. Vavoula and Mike Sharples
University of Birmingham, UK
{g.vavoula, m.sharples}@bham.ac.uk

Abstract

KLeOS is a Knowledge and Learning Organisation System which reflects the hierarchical organisation of learning into activities, episodes and projects (Tough, 1971; Vavoula & Sharples, 2002), and allows the user to organize and manage their learning experiences and resources as a visual timeline, while at the same time visualising their episodic learning memories. The prototype demonstrates functionality at three different levels, allowing the user to (a) manage their learning projects; (b) monitor the learning episodes they complete and associate them with projects where applicable; and (c) perform learning activities within episodes. In addition, it incorporates a knowledge map, which the user updates as they progress through their learning experiences and which reflects their semantic learning memories. The learning episodes (episodic learning memories) are interlinked with the relevant knowledge nodes in the map (semantic learning memories) allowing for browsing of past learning experiences and knowledge. KLeOS has been evaluated to assess its (a) usability and desirability, and (b) its effectiveness as a knowledge retrieval tool, against R-KLeOS, a reduced version of the software which does not support the interlinking between episodic and semantic learning memories. Although the users identified some shortcomings in the interface design of KLeOS, overall it was rated as a usable and useful tool that they would be willing to adopt. No significant difference was found between the effectiveness of KLeOS and R-KLeOS as knowledge retrieval tools, however there was some evidence that more prolonged use with real-world meaningful learning tasks might favour KLeOS.

1 Introduction

A distinction has been made between episodic and semantic memory (Tulving, 1983). Episodic memory is involved in the recording and subsequent retrieval of memories of personal happenings and doings, whereas semantic memory relates to knowledge of the world that is independent of a person's identity and past (Tulving, 1983). Electronic records of people's episodic memories have been used in the past for information retrieval (Bovey, 1996; Lamming & Newman, 1991; Lansdale & Edmonds, 1992). The system presented in this paper, KLeOS, maintains an interlinked record of the user's semantic as well as episodic learning memories.

KLeOS is an aid for the recording, organisation and recall of learning. It was designed to reflect the three-level organisation of the learning practice into activities, episodes and projects (Tough, 1971; Vavoula & Sharples, 2002) by supporting: learning activities, i.e. discrete learning acts like reading and discussing; learning episodes, i.e. collections of activities that are performed in a given time interval; and learning projects, i.e. collections of episodes related thematically or by purpose. Projects, episodes and activities are displayed as segments on a timeline, and together they represent the user's episodic learning memories.

In addition, the system incorporates a map of the user's acquired knowledge in the form of notes made on a concept map, which represents the user's semantic learning memories. Each node in the map represents a note, and is tagged with information about the relevant activity, episode and project. Navigation mechanisms are provided for the user to explore their knowledge map, to traverse their learning episodes on the timeline, and to move from the map to the timeline and vice versa.

KLeOS provides a device for testing the effectiveness of the combination of episodic and semantic learning memories as a knowledge retrieval aid. A reduced version of the software, R-KLeOS, was implemented that was identical to KLeOS except for the interlinking mechanism: there was no direct connection between an episodic memory and related semantic memories, or vice versa. Two groups of seven people each performed a (pre-determined) learning task relating to learning about earthquakes using the two versions of the software. Measurements were made of the time taken to retrieve knowledge in order to complete a quiz immediately after the learning task and a second quiz two weeks later. The same people participated to a questionnaire-based usability evaluation after the second quiz. Section 2 of this paper describes the two versions of the software in more detail, and section 3 briefly presents the results of the usability study and the learning task experiment.

2 KLeOS: A Knowledge and Learning Organisation System

Project Timelines: A timeline is a common graphical representation of events (Plaisant et al., 1998; Plaisant, Rose, Rubloff, Salter, & Shneiderman, 1999). Time is depicted on a line and events of a specific type are shown on the respective time points. In KLeOS, learning projects are represented as lines parallel to a timeline (see fig. 1a). The timeline itself extends indefinitely into the past and future, whereas the project lines start and end at the moments specified by the user and their learning activities. The projects timeline offers zooming facilities, as well as a facility for moving forwards and backwards in time. Four different views (daily, weekly, monthly and yearly) are supported in the current version. The vertical arrangement of the project lines codifies information about the order of creating the project lines: the closer a project line lies to the timeline, the less recently it was created. The width of a project line codifies information about the project's importance: the wider the line, the more important the project.

Episodes Interface: In KLeOS episodes are shown as blue marks at the appropriate location (i.e. time) on a project line (see fig. 1a). KLeOS allows the user to carry out and monitor learning episodes and review them at a later time.

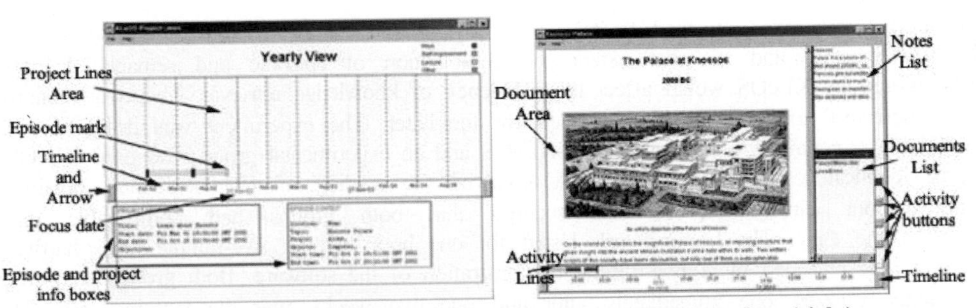

Figure 1: (a) Project timelines (left), (b) Episodes interface (right)

A recorded episode can be reviewed by locating its mark on the project line and clicking on it. A new window opens that displays the information for that episode. A Notes List and a Documents List are displayed in this window. The Documents List contains all the documents that were used during the episode, in chronological order, with repeated entries where necessary. The manipulation of documents is done in the course of performing learning activities: an activity involves the manipulation of a document of some type, for example a report for reading, a transcript for a discussion, etc. At the bottom of the window, appears a timeline against which are drawn Activity Lines, i.e. lines that correspond to the activities for the respective documents in the

Documents List, and signify how long an activity lasted. Clicking on a document in the Documents List, or on an Activity Line, will take the user back to the document and its respective activity. The relevant document is opened in the Document Area (fig. 1b), and the Notes List displays all the notes that were made during that activity. The four buttons "R, W, D and S" represent the different activities that the user can perform (reading, writing, discussing and searching) In the current version, only "R" is supported. The button "P" stands for Pause, and is used whenever the user wants/has to pause their activity for any reason. In review mode the buttons are disabled: the user cannot alter the structure of a past episode while revisiting it by performing new activities. Reviewing is also not implemented as a learning activity itself; however, this is considered for implementation in future versions.

Basic Concept Mapping Tool: Notes in KLeOS are made using the Basic Concept Mapping Tool, on a knowledge map. A concept map (Buzan, 1995), or knowledge map, is a network of nodes, which represent concepts, facts, or ideas, and links between those nodes representing relations between them. The Basic Concept Mapping Tool provides this kind of functionality as a method for making notes. In the current version the navigation mechanism implemented in the concept map itself is restricted to visually following links and scrolling. However, the map is also navigable through the Notes List: if the user wishes to review a note in the map that appears in the Notes List, they can click on the list item and they will be transferred to the map, with the relevant note highlighted. The notes in the map are tagged with information about the episode/activity during which they were created. Thus, while in the map, the user can revisit the relevant episode/activity of a node by activating a pop-up menu. This functionality allows the user to navigate between their semantic (knowledge map) learning memories and their episodic (episodes interface) learning memories, using combinations of cues from both to retrieve past knowledge.

R-KLeOS: The reduced version of KLeOS, R-KLeOS, is identical to the original software except that the episodes interface does not provide the Notes List; instead, a button is available for the user to access their concept map whenever needed. Similarly, the concept mapping tool in R-KLeOS does not have the feature of accessing a node's context, i.e. the user cannot move from a node in the map to the relevant episode/activity of that node.

3 Evaluating KLeOS

An evaluation session was designed to (a) assess KLeOS's usability, usefulness and desirability and (b) test whether and to what extent the combination of episodic and semantic learning memories offered by KLeOS would affect the efficiency of knowledge retrieval. The two versions of KLeOS described in section two were used for the latter. The experiment was designed with two conditions, a control group who used R-KLeOS and an experimental group who used KLeOS, to perform identical learning tasks. The task involved the reading of 15 documents containing information about earthquakes. Pre-tests secured that both groups had comparable prior knowledge of the topic. The learning task lasted for one hour, with a five-minute break halfway through, and was preceded by a five-minute demonstration of the software. Both groups answered two knowledge quizzes, one immediately after and one two weeks after the learning task. The participants were timed during the two quizzes, to compare response times. 14 people participated in total, seven in each group. After the second quiz, they were asked to use the Product Reaction Cards from the Desirability Toolkit (Benedek & Miner, 2002) and to fill in a questionnaire with questions about the usability of KLeOS and the general appeal of the concept. Those using R-KLeOS were first given a demonstration of KLeOS, with data recorded during their own session.

3.1 Usability and desirability

The software scored well in learnability (4.13 out of 5), controllability (3.74) and suitability for learning (3.69). Despite the slowness of the interface and the occasional instability of the software, most participants (10 out of 14) said that they would use it as a learning tool. The greatest advantages of using the software were thought to be (a) the ability to keep a record of one's learning and revisit that record (mentioned 7 times), (b) the ability to organise learning resources and knowledge and relationships between them (mentioned 6 times), and (c) the simple and easy to use interface (mentioned 3 times). The greatest disadvantages of using the software were thought to be (a) that it is time consuming to maintain, (b) it is not flexible enough (e.g. to work in cases where the relevant material is not on the computer, or to integrate with other tools), and (c) it is lacking a text search facility. The questionnaire also examined which features of KLeOS were most likeable as well as their utility as recall aids. Among the most likeable features the Basic Concept Mapping Tool was the most common, followed by the linking (between the documents, the activities and the relevant notes) and the Documents List. The least likeable features related mostly to aspects of the interface, like limited navigation and manipulation of concept map objects, the colours, shapes and general appearance of the GUI, the slowness of screen refreshing, etc. The Documents List and the Concept Map scored higher than the participants' own memory as recall aids. The Notes List did not score as high although it was identified as a useful feature. The Activity Timelines were not thought to be of help as recall aids. In fact, the participants did not make any use of this feature.

3.2 Effectiveness of Knowledge Retrieval

The two groups achieved comparable scores in the two post-task quizzes. In the first quiz, the control group achieved an average score of 48.57 and the experimental group 47.14. In the second quiz, the control group achieved 73.21 and the experimental 71.43. Both groups did significantly better in the second quiz, however, the comparable results between the two groups in the two tests suggest that this improvement in performance is likely to be due to different difficulty levels of the two quizzes rather than the use of the software.

The overall time it took to complete a quiz was, on average, slightly greater for the experimental group for both quizzes by approximately one minute. The times that were measured were:

(a) *The time it took to produce a non-answer:* this equates to the time it took a participant to conclude that they had not previously accessed the information needed to answer a question. This time increased for both groups in the second test in relation to the first.

(b) *The time it took to produce a correct answer, without accessing the (previously read) documents with the relevant information:* this equates to the time it took a participant to produce a correct answer based on memory. These times were comparable between the two groups in both tests, and no difference between the two tests was observed.

(c) *The time it took to produce a correct answer while reviewing the (previously read) documents with the relevant information:* this equates to the time it took a participant to produce a correct answer when they could not remember the answer. These times were significantly reduced for both groups in the second test in relation to the first. However, between the groups the results were similar.

(d) *The time it took to access a (previously read) document that contains information about the current question.* In the case of questions that referred to more than one documents, the time to access the first document was measured. Both groups were slightly faster in the second test, and the experimental group was slightly faster than the control group in both tests; however, no significant difference was observed.

Overall, the times taken by the two groups to answer the two tests were comparable, with only minor trends supporting the hypothesis that the linking between semantic and episodic learning memories would make knowledge retrieval more efficient.

Visiting documents was the feature that was most used by both groups, during both tests. The control group visited an average of 2.09 documents per question in the first test, and 2.32 in the second test. This is an increase of 11%. The experimental group visited an average of 2.02 documents in the first test, and 1.57 in the second test. This is a decrease of 22%.

The linking was used by the experimental group an average of 0.20 times per question in the first test, and 0.34 times in the second test. The low use of the linking feature in relation to direct document access could be attributed to the newness of such a facility.

Although no significant difference in the use of the system between the two groups in the two tests was found, the above constitute some evidence that the linking facility gets more used the more time has elapsed since the original learning episode; and the full version of the software which includes this facility allows the user to reach the previously read relevant documents after less navigation within the documents pool, in less time. Further testing, with longer time intervals between tests and in naturalistic settings where the users perform everyday, personally meaningful learning tasks and retrievals, is necessary before this can be positively confirmed.

4 Conclusions

KLeOS provides the user with a graphical, timeline-based representation of their episodic learning memories, by depicting learning episodes and activities against a timeline, in the context of a learning project. It also allows them to associate their learning activities with relevant notes that they make using a concept-mapping tool. In an initial evaluation, users found this as a useful representation and considered it as an easy to use, desirable, and novel learning tool. A comparison of KLeOS with a reduced version of it, which did not provide the interlinking between the timelines and the knowledge map, did not find any substantial difference in the effectiveness of the two systems in knowledge retrieval.

5 References

Benedek, J., & Miner, T. (2002). *Measuring Desirability: New methods for evaluating desirability in a usability lab setting.* In Proceedings of Usability Professionals' Association.

Bovey, J. D. (1996). Event-based personal retrieval. *Journal of Information Science, 22*(5), 357-366.

Buzan, T. (1995). *The Mindmap Book* (2nd ed.): UK:BBC Books.

Lamming, M. G., & Newman, W. M. (1991). *Activity-based Information Retrieval: Technology in Support of Personal Memory* (Technical Report EPC-1991-103). Cambridge: Rank Xerox Research Centre.

Lansdale, M., & Edmonds, E. (1992). Using memory for events in the design of personal filing systems. *International Journal of Man-Machine Studies, 36*, 97-126.

Plaisant, C., Mushlin, R., Snyder, A., Li, J., Heller, D., & Shneiderman, B. (1998). *LifeLines: Using Visualization to Enhance Navigation and Analysis of Patient Records* (HCIL Technical Report No. 98-08).

Plaisant, C., Rose, A., Rubloff, G., Salter, R., & Shneiderman, B. (1999). *The Design of History Mechanisms and their Use in Collaborative Educational Simulations* (HCIL Technical Report No. 99-11).

Tough, A. (1971). *The Adult's Learning Projects: A Fresh Approach to Theory and Practice in Adult Learning.* Toronto: Ontario Institute for Studies in Education.

Tulving, E. (1983). *Elements of Episodic Memory.* Oxford:UK: Oxford University Press.

Vavoula, G. N., & Sharples, M. (2002). KLeOS: A personal, mobile, Knowledge and Learning Organisation System, *Proceedings of IEEE International Workshop On Wireless and Mobile Technologies in Education,* pp. 152-156.

Automated Identification and Correction of Coding Errors in an Accident Narrative Database.

Helen Wellman[1], Mark Lehto[2], Gary Sorock[1]

[1]Liberty Mutual Research Institute for Safety

71 Frankland Rd, Hopkinton MA
Helen.Wellman@LibertyMutual.com
GarySorock@LibertyMutual.com

[2]School of Industrial Engineering
Purdue University
West Lafayette, IN
lehto@ecn.purdue.edu

Abstract

Narrative information contained in accident databases is often extracted and classified by human coders into cause-of injury groups for meaningful analysis. The coding process is time consuming and tedious and different interpretations of coding protocols may result in disagreement between coders. In this study, we compared human-assigned codes with computer-assigned codes for a set of accident narratives with high probability prediction strengths. We found that a computer algorithm operating at a high accuracy threshold was able to flag difficult narratives for review that humans coded inconsistently.

1 Introduction

Insurance companies, governmental agencies, private industry and universities use information contained in accident databases for a variety of purposes, some of which include surveillance and identification of unique or emerging safety problems, monitoring of program effectiveness, and actuarial purposes (Sorock et al, 1997). The quality of the conclusions drawn during such analyses depends in part on how accurately the data are entered into the accident database. A trained professional often codes the accident narrative into cause-of-injury groups and enters the narrative into a free-text field. The information contained in the free-text portion of safety databases has been used for a variety of purposes. In particular, in some of our past work, we were able to predict accident codes with considerable confidence by analyzing the free-text narratives contained in a large insurer's database for construction work zone accidents (Lehto et al.,1996). In a follow-on effort, we were able to predict cause-of-injury codes assigned by an analyst for the National Health Interview Survey (NHIS) by processing the free-text narratives (Wellman et al., 2003).

These earlier applications support the conclusion that the free-text narratives can be used to identify, and perhaps even correct errors in the way accidents are classified by human coders in existing accident databases. Even after receiving training, human coders often disagree with each other and may make coding errors. In this paper we examine the possibility of identifying and correcting such problems using the same approach followed in our earlier work where we used a computer tool to classify accidents by processing the free-text narrative portion of an accident database.

2 Methods

Four experienced coders assigned 13,487 workers' compensation accident narratives to 17 broad level (2 digit) event categories according to the Bureau of Labor Statistics Occupational Injury and Illness Classification coding system (BLS, 1992). The narratives were extracted from workers' compensation claims occurring in 1999 and filed with one large U.S. workers compensation insurer. Approximately 56% of the narratives assigned by claims handlers as injuries caused by "manual material handling" or "mechanical material handling" had been previously extracted and manually coded for a different study. Six thousand additional narratives that were not "manual material handling" or "mechanical material handling" claims were randomly extracted and manually coded for this study.

The 13,487 narratives and codes described above were then used to train a computer to code accident narratives into 17 event categories using a machine-learning approach based on Fuzzy Bayes logic. The keyword list used by the Fuzzy Bayes program included single words and combinations of up to four words (multiple words). The program was then rerun on the narratives blinded to codes to independently classify the accident narratives. The Fuzzy Bayes classifier chose the word or word combination in each narrative with the highest probability of association with an event category. The resultant maximum probability is defined as the prediction strength of the computer assigned code. Details of the methods are described elsewhere (Wellman et al., 2003).

Following the independent assignment of codes by the computer, narratives which were coded differently by the computer and the human coder with: 1) greater than or equal to a prediction strength of 0.95 (n=64) and 2) greater than or equal to a prediction strength of 0.90 but less than 0.95 (n=120) were filtered out for investigation. Two human coders recoded the 184 narratives blinded to the previous human- or computer-assigned codes. Codes were compared and if there was disagreement, the two coders worked together to decide on the best code. These were considered the "gold standard" codes. These codes were then compared to the original codes assigned by the computer and human coders for the same subset of 184 narratives.

The data were split into two groups for further analysis since the coders felt that it was more difficult to assign some codes than others based on the limited information contained in the narratives. The first group (n=62, 34%) included events leading to sudden-onset injuries with an external source of injury (Traumatic Group, Table 1). The second group (n=109, 66%) included events leading to an overexertion, repetitive motion or bodily reaction injury with less well documented external source of injury (Bodily Reaction Group, Table 1). The remaining 13 narratives were unclassifiable. An example of the former and later categories are respectively:

1. "Employee grabbed dock ladder which was not attached and he fell to the ground" (BLS code 13, "Fall on same level", assigned to Traumatic group)

2. “When moving boxes of product, was continuously bending and lifting, felt gradual pain in back.” (BLS code 23, “Repetitive motion”, assigned to Bodily Reaction group)

Table 1: Assignment of BLS codes into Traumatic and Bodily Reaction Groups:

BLS OIICS Codes[1]	Description of BLS OIICS codes	Group Assignment[2]
01, 02, 03, 06, 11, 13, 40, 41, 42, 43.	Struck against object; struck by object; caught in or compressed by equip; rubbed, abraded or jarred by vibration; fall to lower level; jump to lower level; fall on same level; transportation accidents including non-highway	Traumatic
20, 21, 22, 23	Repetitive motion; Bodily reaction; Overexertion	Bodily reaction

[1]Bureau of Labor Statistics Occupational Injury and Illness Classification 2 digit event codes
[2]Events leading to sudden-onset injuries with an external source of injury were assigned to the “Traumatic” group. Events leading to an overexertion, repetitive motion or bodily reaction injury with less documented external source of injury were assigned to the “Bodily reaction” group

3 Results

For the narratives in which the computer used a prediction strength greater than or equal to 0.95, 28/64 (43.8%) were coded correctly by the program following the new review (Table 2). This compares with 24/64 (37.5%) which were coded correctly by the original human coder following the new review. For the remaining 12/64 (18.8%) narratives, the gold standard code did not match either the computer or the original human code. For five of these cases, we could not decide between two codes based on the limited information available in the narrative (23, repetitive motion or 22 overexertion), so a broader code was used (20, bodily reaction unspecified).

For the narratives in which the computer used a prediction strength greater than or equal to 0.90 but less than 0.95, 45/120 (37.5%) were coded correctly by the computer and 57/120 (47.5%) were coded correctly by the original human coders following the new review. Eighteen cases (15%) remained in which the gold standard code did not match either the computer or the original human code. In summary, when the computer is more certain of a correct classification, it was more often correct than the human coder. When less certain, the human was correct more often.

When dividing the data into traumatic vs bodily reaction groups for analysis, the human coder was more often correct with the traumatic group than the bodily reaction group. The program had more correct codes than the human coder for the bodily reaction group (50.0% vs 29.2% at the 0.95 threshold level); whereas the human coders were more often correct for the traumatic narratives (61.5% vs 30.8% at the 0.95 threshold level). Results were similar for the lower threshold level.

Table 2: Accuracy of computer (Comp) vs human coder (HC)

	Threshold strength .95 to <1.0 (N=64)				Threshold strength .90 to .<.95 (N=120)			
	Human Coder correct	Human Coder incorrect	Total Comp	% Overall Total	Human Coder correct	Human Coder incorrect	Total Comp	% Overall Total
Computer correct	0	28	28	43.8	0	45	45	37.5
Computer incorrect	24	12	36	56.3	57	18	75	62.5
Total HC	24	40	64	100.0	57	63	120	100.0
% Overall Total	37.5	62.5	100.0		47.5	52.5	100.0	

4 Discussion

In this study, we compared human-assigned codes with computer-assigned codes for a set of narratives with high probability prediction strengths. We found that a computer algorithm operating at the highest threshold level was able to flag for review difficult narratives that human coders are coding inconsistently.

This study demonstrated that a computer program can identify coding errors particularly when it is highly certain of the code. When the computer was less certain, the human was more likely to be correct. In addition, when humans are faced with coding ambiguous accident groups like bodily reaction or other non-traumatic events, the computer may be relied upon more heavily for the correct code. The computer was about as likely to be correct as the human coder for the more ambiguous narratives. It is important however to manually review a random sample of conflicting and non-conflicting predictions.

5 References

Burean of Labor Statistics, Occupational Injury and Illness Classification Manual, U.S. Department of Labor Bureau of Labor Statistics, December 1992.

Chan C, Luis B, Chow C, Cheng J, Wong T, Chan K, Chui S. Validating narrative data on residential childhood injury. J Safety Res, 2001;32:377-389.

Lehto, M., Sorock, G. Machine learning of motor vehicle accident categories from narrative data. Methods of Information in Medicine, 1996;35,:309-316.

Sorock, G., Smith, G., Reeve, G., Dement, J., Stout, N., Layne, L., and Pastula, S. Three perspectives on work-related injury surveillance systems. American Journal of Industrial Medicine. 1997;32:116-128.

Wellman, H., Lehto M, Sorock G, Smith G. Computerized Coding of Injury Narrative Data from the National Health Interview Survey. Acc Ann Prev 2003 (in press)

Estimation of Useful Field of View on the Situation of Driving Work

Kimihiro Yamanaka
Konan University
Okamoto, Higashinada-ku,
Kobe, 658-8501, Japan
kimihiro@ hcc1.bai.ne.jp

Hidetoshi Nakayasu
Konan University
Okamoto, Higashinada-ku,
Kobe, 658-8501, Japan
nakayasu@konan-u.ac.jp

Kazuaki Maeda
Konan University
Okamoto, Higashinada-ku
Kobe, 658-8501, Japan
kmaeda@konan-u.ac.jp

Abstract

It is examined by the experiment in order to estimate the useful field of view on driving situation of load vehicle. There are four kinds of experiments carried out. The first is to investigate the relationship between the response time and the factors of experiments. The second is to evaluate the region of the useful field of view for the experimental factors. In the third experiment, the event related potential P300 was measured to verify the relationship of the response time and useful field of view. Some trials for the estimation of useful field of view were examined by the fourth experiments to the combination of driving situation.
Key words: Useful field of view, Response time, Event related potentials (ERP) P300, Driving a load vehicle

1 Introduction

The human error is one of the important factors why the structural safety and reliability have not been improved until now (Miura, 1985., Rassmussen, 1983., Reason, 1990). In this paper, the human response is studied when one recognizes visual stimulus in order to clarify how to occur and what the principle factors are (Hollnagel, 1993., Rassmussen, 1986). It is examined by the four kinds of experiments to evaluate the relation between the useful field of view and physiological signal in driving work when the driver recognizes a target mark as a visual stimulus under background driving scene. In the experiment, the region of useful field of view, response time and ERP (Event Related Potential) P300 on Cz are measured to evaluate a human behaviour.

2 Experimental System for Useful Field

2.1 Experimental Equipments and Environments

The schematic aspects of experimental devices are shown in Figure 1. The response time and physiological signal of examinee are measured when one can recognize a visual stimulus in Figure 2 in the electromagnetic shield room. A visual stimulus consists of foreground scene and background scene. In the foreground, the target mark appears at random spatially and temporally programmed. The foreground is superimposed on the background by imposer (IDK, DSC06d-HR). The output data of NTSC signal from imposer is converted into RGB signal by the up-scan converter in Figure 1. The background scenes were transformed into 3D movies by NuVision's stereo display system (MacNaugton Inc, NuVision 21SX) consisted of liquid crystal modulator shutter, synchronizer, and stereo glasses. The shutter is a liquid crystal modulator that works in conjunction with stereo glasses to present the appropriate image to the left and right eye. The synchronizer accepts many video, input configurations and automatically synchronizes the shutter with the display of stereo image on CRT.

The two kinds of buttons were given to the examinee in one's hands as shown in Figure 2. The examinee must push the button of right side as soon as possible, when one can detect or recognize

the target mark S_2. On the other hand, the examinee must put the mark of S_1 on by one's left side button when S_1 disappears in order to fix the visual attention into the centre of CRT.
Three kinds of scenes were used as the background scene such as the unique grey scene, the 2D movies and the 3D movies of driving an automobile. Physiological signal was also measured by NeuroFax EEG-1518 (Nihon Kohden Inc.) such as EEG (electroencephalograph). In this experiment, 10 university students from 20 to 24 years old was selected as candidates of examinees who have regular class automobile licenses with normal vision or corrected normal vision.

2.2 Foreground and Background Scenes

There are two kinds of experiments for useful field were carried out called as "the experiment of a detectable" and "the experiment of a recognizable" as shown in Table 1 and Figure 2. The former experiment is to measure the useful field that is the range to detect a simple circle target of S_2. The region measured by this experiment is called as "detectable field of view". The latter experiment is to measure the region of the useful field that is the range to be able to recognize a specific shape in Snallen's chart as a target mark among the standard marks with similar shape. The region measured by the latter experiment is called as "recognizable field of view".
In the foreground scene, two kinds of marks are used as signals to measure the region of useful field of view as shown in Table 1. S_1 is used to keep the visual attention of examinee on the centre of CRT. S_1 is a simple white circle with the size from 0.38° to 0.48°. The duration between S_1 and S_2 is 1700msec. In this experiment, S_1 appears or disappears during the specified duration so that the examinee must put on S_1 by one's left side button when S_1 disappears. This additional task T to recover the disappearance of S_1 is also the factor of the experiment. These levels of T are from 15 to 53 cycles/min in Table 2.
Three kinds of background scenes such as a unique grey scene, 2D and 3D driving scenes are used for the disturbance effect on the visual attention. The driving scenes are recorded in practical road by 2D and 3D video camera KS55Z (Kastam, KS55Z). The patterns of driving scenes are listed in Table 2 where V and D mean the factors of the speed and traffic demand. The recorded scenes consist of 2 kinds of level for factors V (velocity of automobile: V_1=40km/h, V_2=80km/h) and D (demand of traffic: D_1=non-crowded, D_2=crowded) as shown in Table 2.

3 Experiments and Discussions

3.1 The First Experiments -Response Time and Factors under Unique Background

3.1.1 The experiment of a detectable

This experiment was examined to determine the experimental conditions so that a simple white mark and an unique grey scene were selected as a target mark S_2 and a background scene. Several candidates for the experimental factors are listed in Table 3. The experiments were performed for two kinds of appearance region of S_2 for Z_3 and Z_5 with the combinations of the factors in Table 3. It is seen that the larger response time is when the smaller the size of S_2 is in spite of the combinations of the differences of luminosity of target mark and background. These results are independent on the individual of examinee such as male and female. It is statistically found in Table 4 that there are no dependencies on male and female for the response time under all combination of size of S_2. Table 4 also shows that the response time is independent on the differences of luminosity between S_2 and background under all combination of sizes of S_2. Especially, it is noted that the differences of luminosity are independent on the experiment condition whether the difference of luminosity of S_2 is positive or negative to the background.

3.1.2 The experiment of a recognizable

In the experiment of a detectable, S_2 is displayed 48 times per experiment. In the experiment of a recognizable, S_2 is displayed 256 per experiment, the number of standard mark of S_2 is 208 times and the target is 48 times. In the experiment, the mark S_2 appears the same times per zone, as shown in Figure 2.This experiment was examined for two kinds of appearance zones of S_2 such as Z_3 and Z_5. From the results of previous experiment by 3.1.1, the difference of luminosity of S_2 to background was determined as 10lx. In Figure 3, the average relation for 10 examinees between response time and the size of S_2 is drawn. It is seen that the response time is dependent on the size of S_2 for every cases of appearance zone Z_3 and Z_5. Especially, it is noted that the discrepancies of the dependencies on the size of S_2 become to be remarkable when the appearance zone is in Z_5.

3.2 The Second Experiment -Useful Field and Experimental Factors

The aim of this experiment is to evaluate the useful field and experimental factors under the 3D driving scene with fixed condition V_2D_2 in Table 2. Therefore the experiments of a recognizable were examined for the three kinds of appearance zones of S_2 such as Z_4 (5.85° ~7.80°), Z_5 (7.80° ~9.75°) and Z_6 (9.75° ~11.70°) with eight kinds of directions from fixation point S_1. As the experimental factors, two kinds of size of S_2 such as 0.33° and 0.38° and three kinds of level of T are selected such as T_1=15 cycles/min, T_2=34 cycles/min, and T_3=53 cycles/min.
It is seen that the factor of additional tasks are independent on the useful field. Figure 4 show the maximum and minimum region of useful field of a recognizable. The maximum region means the region where one can recognize no marks of S_2 outside the region. On the other hand, the minimum region means region where one can recognize all marks of S_2 inside the region. It is found that there are no differences of maximum region on the size of S_2. However it is noted that the maximum region for small size of S_2 is strongly narrower than that for big size of S_2.

3.3 The Third Experiment –ERP (P300), Response Time and Useful Field

Another recognizable experiments were performed for three kinds of background scenes such as unique, 2D and 3D driving scenes for three kinds of appearance zones. For this experiment, EEG was also measured with response time and useful field for fixed conditions such as 0.38° sizes of S_2, 10lx differences of luminosity and T_3. From the ANOVA for response time, it is found that the factor of background scene is the most sensitive factor to the results. It is seen from Figure 5(a) that the response time is dependent on the situation of background scene. It is found from Figure 5(b) and Figure 6 that there are big differences for appearance region and kinds of background the amplitudes of the event related potential P300. It is well known that P300 shows the degree of concentration for visual task or visual attention. From this investigation, the tendency for the amplitude of P300 corresponds to those for response time in Figure 5(a).

3.4 The Fourth Experiment –Useful Field on The Driving

The effect of driving conditions on useful field of view and ERP (P300) was also investigated in the fourth experiment. In this experiment, "detectable" and "recognizable" field of view was measured with response time and ERP (P300) on Cz for all combinations of driving situations on automobile speed V and traffic condition D. This experiment was carried under the fixed experimental conditions such as 0.48° size of S_2, 53 cycles/min additional task t_3 and 10lx differences of luminosity. Table 5 shows the summary of ANOVA on effectiveness of factors. It is seen that T is more effective factor and D and T×V are the effective factors in the experiment of a detectable. On the other hand, in the experiment of a recognizable, D and D×V are more effective factors. These results agree with Miura (Miura, 1985., Miura, Shinohara & Kanda, 1998.) who performed with practical driving situation from the standpoint of psychological experiments. The results of response time and region of useful field of view are summarised in Figure 7 for the conditions of V and D. It was interesting that the distinction on the response time between the experiment of a detectable and a recognizable were from 230 to 290 (msec) and width of visual field of a detectable and a recognizable were from 1.50 ° to 3.15 °. The behaviour of P300 in Figures 8 and 9 show two kinds of typical features. The first is concerned with the effect of factors V and D. The seconds is concerned with the effect of appearance zone of target marks. In the former feature, it is found that there are big differences of the amplitude of P300 in comparison of (a) with (b) or of (c) with (d). This fact means that the effect of traffic demand D is strongly dependent on the depth of visual processing. On the other hand, it is found that there are little differences of the amplitude of P300 in comparison of (a) with (c) or of (b) with (d). This shows that the factor V is little dependent on the depth of visual processing.

ACKNOWLEDGEMENTS

The authors express great thanks to the Grand-in Aid for Scientific Research Found of the Ministry of Education, Science Sports and Culture of Japan (Grant No. 11650109) for their financial support and the Hirao Taro Foundation of the Konan University Association for Academic Research for their financial support.

REFERENCES

Hollnagel, E. (1993) "Human Reliability Analysis -Context and Control," Academic Press, pp.147-202.

Miura, T. (1985) "What is the Narrowing of Visual Field with the License of Increase of Speed?" Proc. of the 10th Congress of the Int. Association for Accident and Traffic Medicine, pp.HF2.1-2.4.

Miura, T., Shinohara, K. & Kanda, K. (1998) "Visual Attention in Automobile Driving: From Eye Movement Study to Depth Attention Study", Proceedings of the 2nd International Conference on Psychophysiology in Ergonomics, Symposium Presentation, pp.7-8.

Rassmussen, J. (1986) Information Processing and Human Computer Interaction – An approach to Cognitive Engineering-, North-Holland.

Rassmussen, J. (1983) "Skills, Rules, Knowledge: Signals, Signs and Symbols And Other Distinctions in Human Performance Models," *IEEE Trans. On SMC*, SMC-13, pp.257-267.

Reason, J. (1990) *Human Error*, Cambridge Univ. Press.

Table 1: Experimental Conditions

	DETECTABLE		RECOGNIZABLE	
	SIZE OF SIGNAL	DISAPPEARANCE RATE	SHAPE OF SIGNAL	DISAPPEARANCE RATE
S_1	● 0.38~0.48°	15~53 cycles/min	● 0.38~0.48°	15~53 cycles/min
S_2	● 0.31~0.54°	---	TARGET STANDARD 0.31~0.54°	---

Table 2: Conditions of Factors

FACTOR	LEVEL OF FACTOR		SCENE
T :TASK	15~53 cycles/min		FOREGROUND SCENE
V :AUTOMOBILE SPEED	V_1: 40 km/h	V_2: 80 km/h	BACKGROUND SCENE
D :TRAFFIC CONDITION	D_1: NON-CROWDED	D_2: CROWDED	

Table 3: Experimental Factors

SIZE OF TARGET		APPEARANCE REGION		DIFFERENCE OF LUMINOSITY BETWEEN TARGET AND BACKGROUND	
m_1	0.31°	z_1	0°~1.95°	l_1	5lx
m_2	0.33°	z_2	1.96°~3.90°	l_2	10lx
m_3	0.38°	z_3	3.91°~5.85°	l_3	15lx
m_4	0.47°	z_4	5.86°~7.80°		
m_5	0.54°	z_5	7.81°~9.75°		
		z_6	9.76°~11.70		

Table 4: Results of Statistical Test

SIZE OF TARGET (Deg.)		0.31°	0.33°	0.38°	0.47°	0.54°
TEST ON MALE AND FEMAIL	μ MALE	388.25	371.09	367.28	385.89	370.03
	μ FEMALE	397.26	403.79	365.01	376.99	373.88
	|μ MALE – μ FEMALE|	9.01	32.70	2.28	8.90	3.85
	JUDGE BY t-DISTRIBUTION	**	**	**	**	**
TEST ON LUMINOSITY	μ POSITIVE	380.10	366.29	370.41	353.22	360.27
	μ NEGATIVE	363.69	370.62	374.53	358.31	363.69
	|μ POSITIVE – μ NEGATIVE |	16.41	4.33	4.11	5.09	3.42
	JUDGE BY t-DISTRIBUTION	**	**	**	**	**

Table 5: Results of ANOVA

FACTORS	T	V	D	V multiple D	T multiple D	T multiple V
DETECTABLE	E*	N*	E	N	N	E
RECOGNIZABLE	---	N	E*	N	---	---

E*: MORE EFFECTIVE,E : EFFECTIVE,N : LESS- EFFECTIVE, N*: NON- EFFECTIVE

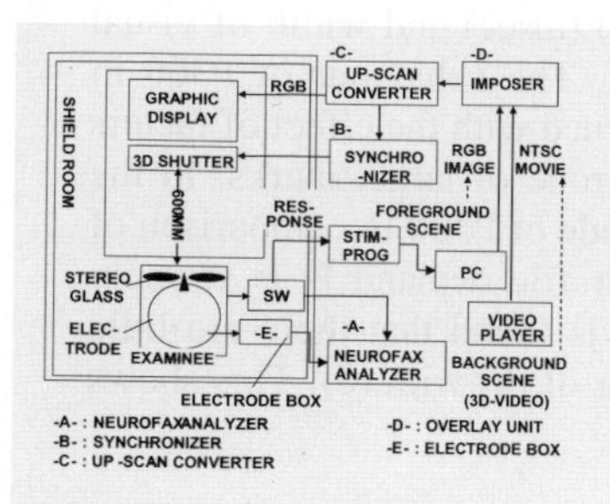

Figure 1: Schematic Aspect of Experiment

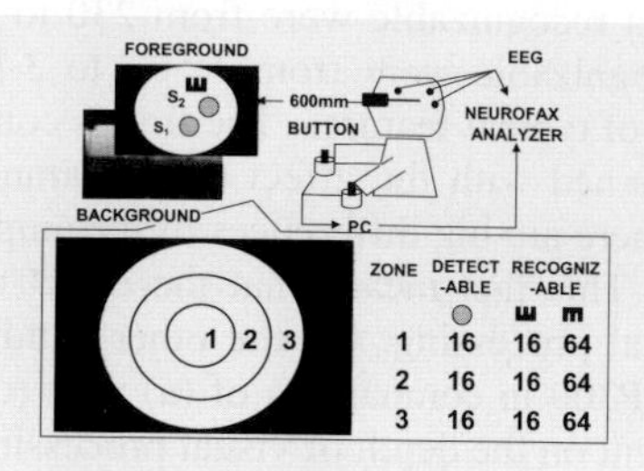

Figure 2: Experimental Situation

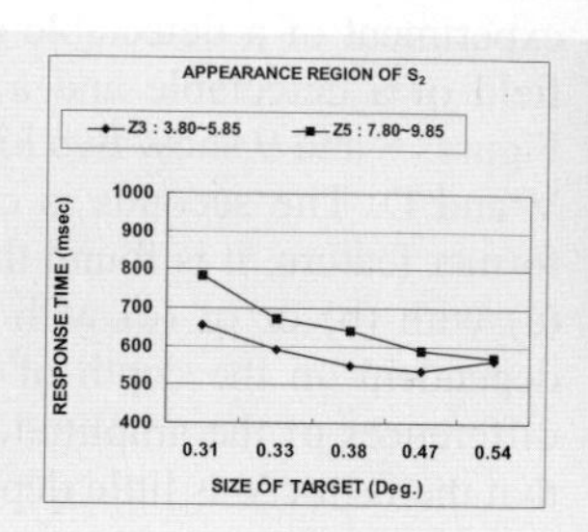

Figure 3: Response Time and Size of S_2

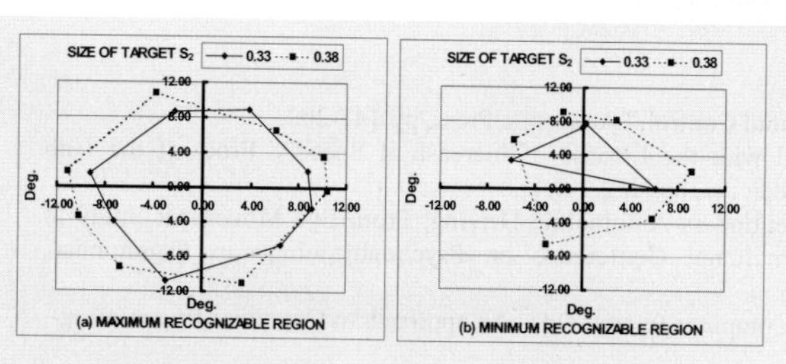

Figure 4: Recognizable Region

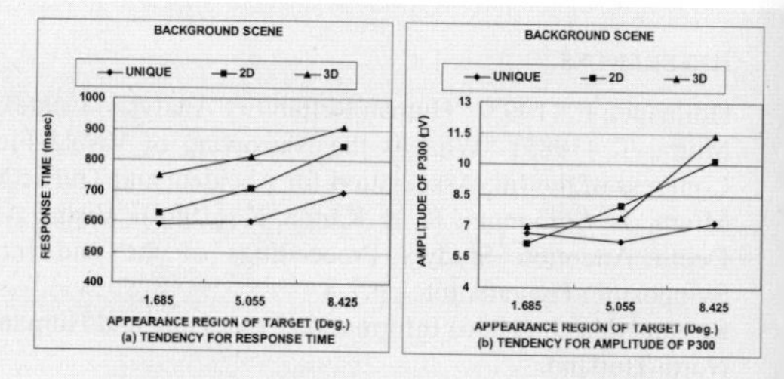

Figure 5: Response Time and P300

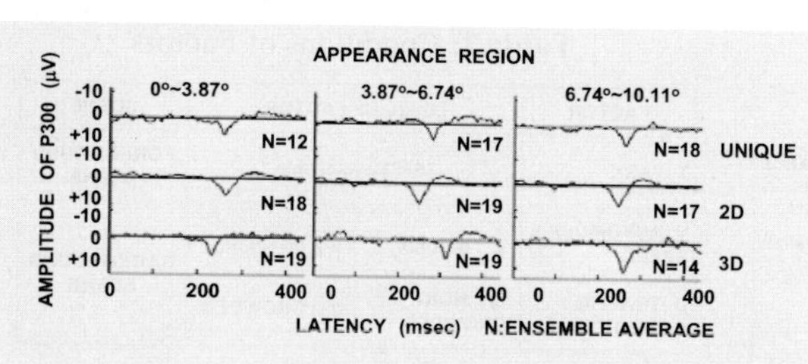

Figure 6: P300 for Background Scenes

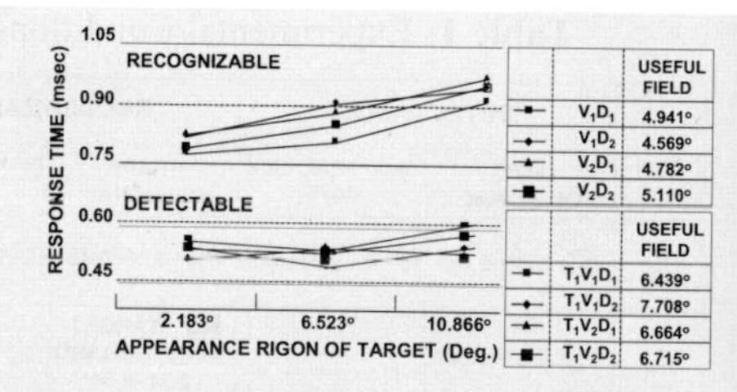

Figure 7: Response Time and Useful Field

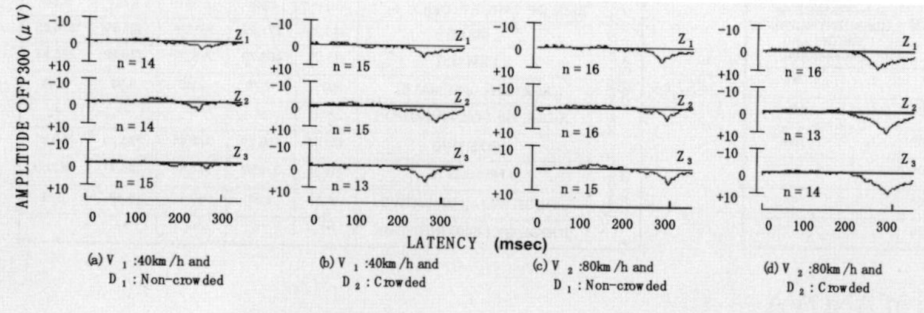

Figure 8: P300 for Driving Situation

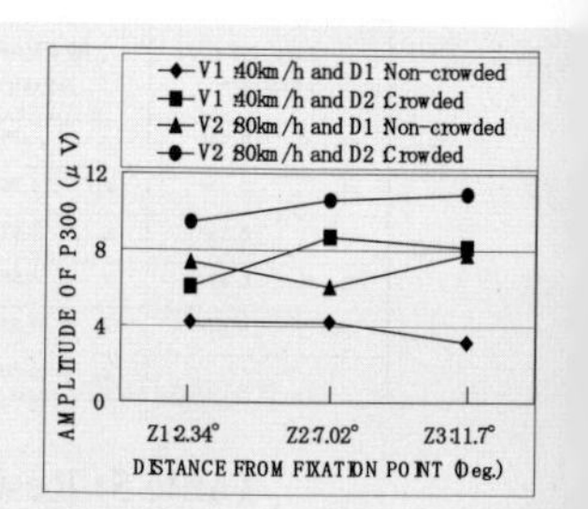

Figure 9: P300 for Situation

Section 5

Applications and Services

An automated system for studying brain function and brain connectivity in a clinical setting

Konstantinos Arfanakis †, Ian A.Heaton ‡

Department of Biomedical Engineering, Illinois Institute of Technology †
10 W. 32nd Street, E1 116
Chicago, Il 60616
arfanakis@iit.edu

Department of Radiology, Miami Children's Hospital ‡
3100 S.W. 62nd Avenue
Miami, FL 33155
hian@nova.edu

Abstract

The latest advances in magnetic resonance imaging (MRI) research have produced unique tools to probe brain activity and structure. However, only a small percentage of clinics around the world routinely use these techniques due to their complexity. These methods require a specialized user who often makes decisions on challenging statistical issues and provides input to computationally intensive algorithms. Therefore, the first goal of this project was to develop an automated system that will allow physicians in any clinic with access to an MRI scanner, to perform studies of brain function and connectivity. This system makes use of two MRI techniques: functional MRI (fMRI), to detect brain activity following stimulation, and diffusion tensor imaging (DTI) to study white matter fiber structure. This automated system does not sacrifice quality, or accuracy, for ease of use. Several predetermined protocols that cover a spectrum of possible clinical procedures are implemented and followed depending on the application. Thus, no step of the process is compromised and the user is not overwhelmed. Additionally, the fact that all users follow the same predetermined protocols makes data sharing and data comparisons meaningful. Therefore, the second goal of this project is to connect each clinical site to a central database where all results will be transferred. This database may serve as a teaching tool and as a forum for interaction between clinicians on ongoing studies.

1 Introduction

Over the last two decades, magnetic resonance imaging (MRI) has become an invaluable tool for physicians around the world. MRI, with its high spatial resolution and sensitivity to variations in tissue characteristics caused by pathology or differences in tissue morphology, offers diagnostic information that cannot be acquired with any other currently available imaging modality. In addition, two MRI techniques that have been recently developed, functional MRI (fMRI) [1] and diffusion tensor imaging (DTI) [2], have revolutionized studies of brain function and brain connectivity respectively. In the past, brain function has been studied with the induction of lesions, with electroencephalograph (EEG), with positron emission tomography (PET), and with electrodes attached to the surface of the cortex. The first method was abandoned for ethical reasons. EEG is still used in clinical and research settings and it offers excellent temporal but poor spatial resolution. PET is also used for studies of brain function and is characterized by acceptable spatial but poor temporal resolution. It also involves administration of radioactive materials to the subject. Finally, the use of surface electrodes is too invasive and nowadays is performed only during brain surgery. On the other hand, fMRI is not invasive, and provides sufficient spatial and temporal resolution. For these reasons, fMRI has been an excellent method to study brain function in normal subjects, in patients with neurological and developmental disorders, brain injury, brain lesions and tumors. The most popular application of fMRI in the clinic has been in pre-surgical planning, where functionally significant brain regions are mapped in order to be excluded from resection during surgery. In regards to studying white matter fiber structure and brain connectivity, DTI is the first imaging, non-invasive technique to provide such information. Until recently, the only information about the structure of white matter fibers and the connections between different brain regions arose from pathological studies in excised brain tissue.

Although the advantages in the use of fMRI and DTI for studying brain function and brain connectivity are obvious and significant, only a small percentage of clinical sites routinely utilize these techniques. This is mainly due to the complexity of the post-processing methods that need to be applied on the raw MRI data in order to get clinically useful results, and the lack of specialized MRI personnel in most non-academic institutions. Therefore, the first goal of this study was to develop an automated system that performs all the post-processing steps that are involved in studying brain function and connectivity with fMRI and DTI respectively. This system retrieves the raw MRI data, performs all necessary post-processing, and outputs the final results for use by clinicians. The only responsibility of the user is to follow predetermined scanning protocols for acquisition of the raw data, depending on the clinical application. Therefore, no post-processing step is compromised, and expertise on fMRI or DTI is not a prerequisite for successful completion of the procedure. In addition, the fact that all users follow the same predetermined protocols means that all acquisition parameters and post-processing methods are the same for all users. This makes data sharing and data comparisons between clinical sites meaningful. Thus, the second goal of this study is to connect all systems to a central database where all final results (and no original data) will be transferred. This database will serve as a training tool for new clinicians as well as a forum for interaction between collaborating physicians.

2 Background

2.1 FMRI

In fMRI, the subject is asked to perform a task that is relevant to the brain region under investigation. Typical fMRI tasks involve: finger tapping for studies of the motor cortex, looking at a changing pattern for studies of the visual cortex, listening to a story for studies of the auditory cortex etc. During the performance of each task oxygenation is increased in the corresponding brain regions. This in turn causes an increase in MRI signal [3]. This increase is less than 5% of the resting state signal (in 1.5Tesla MRI scanners, most available scanners). To increase detection power of the activated brain regions, the stimulus is turned on and off several times within the same data acquisition (Fig.1). This induces a similar variation in the MRI signal only in regions responsible for the performance of the task. Then, the signal from all voxels in the brain is correlated with the expected hemodynamic response, and statistical maps are created that contain high probability values for only those areas that were activated (Fig.2).

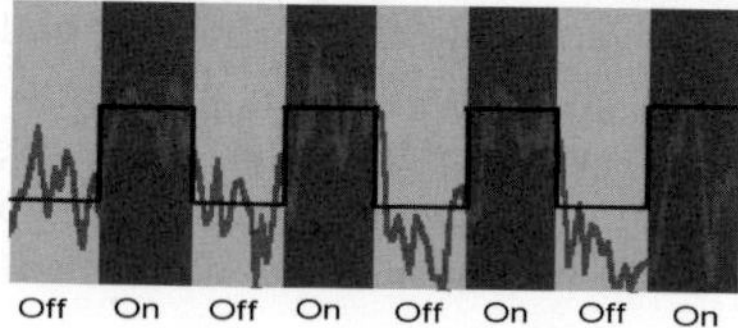

Figure 1. The black boxcars, as well as the changes of background color from light blue to dark blue represent the time intervals during which the task is off and on respectively. The red line is an example of the variation of the MRI signal throughout the performance of the task.

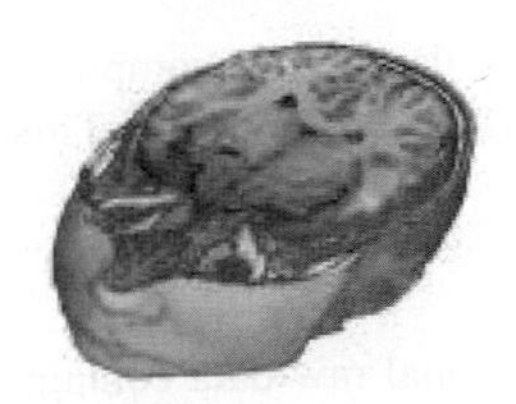

Figure 2. 3D image of an fMRI statistical map overlaid on MRI anatomical images. Activation is shown (orange color) in the auditory cortex due to the performance of an auditory task.

2.2 DTI

Diffusion tensor imaging (DTI), as implemented in MRI [2, 4], is a noninvasive imaging technique that can be used to probe, *in vivo,* the intrinsic diffusion properties of deep tissues. DTI characterizes diffusive transport of water by an effective diffusion tensor **D**. This symmetric 3x3 tensor is of great importance since it contains useful structural information about the tissue. The eigenvalues of **D** are the three principal diffusivities and the eigenvectors define the local fiber tract direction field [2]. Moreover, one can derive from **D** rotationally invariant scalar quantities that describe the intrinsic diffusion properties of the tissue. The most commonly used are the trace of the tensor [2, 5, 6], which measures mean diffusivity, and Fractional Anisotropy (FA)[5, 6, 7], which characterizes the anisotropy of the fiber structure. In each voxel of the brain, the eigenvector that corresponds to the largest eigenvalue reveals the primary diffusion direction of protons. This direction is thought to match the orientation of white matter fiber tracts that cross through the voxel. Therefore, DTI may provide information about the direction of fiber bundles in each voxel of the brain. Following the vectors provided by DTI from one voxel to the next and connecting these voxels provides 3D images of fiber tracts [8, 9] (Fig.3). This method can be used to study white matter integrity and brain connectivity.

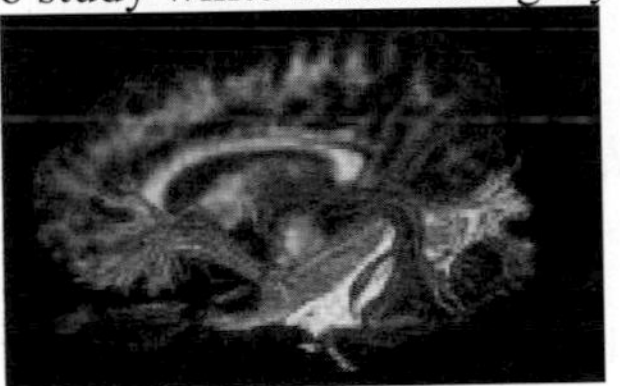

Figure 3. 3D image of 3 white matter fiber bundles traced with DTI (superior longitudinal fasciculus in blue, fronto occipital fasciculus and uncinate fasciculus in orange, inferior longitudinal fasciculus in green and yellow. In the background (grayscale image), a sagittal slice of an FA map reveals the location of other large white matter fibers tracts.

3 System Design

3.1 Brain function

3.1.1 Data acquisition

In any fMRI exam, the user has to select a task which will be performed by the subject and which will most likely activate the regions under investigation. In this system, several pre-packaged tasks are provided to the user. These tasks are previously published and tested by several researchers, and are proven to activate a wide variety of neural systems (motor, sensory, auditory, visual, language etc.). Also, these tasks are provided in several forms depending on the application. They could be audio files with narration of a story (auditory task), video files of an alternating checkerboard (visual task), audio files with commands for the subject to start or stop finger tapping (motor task) etc. These predetermined tasks ensure that data acquisition will produce raw data in the format expected by the post-processing part of the system.

3.1.2 Data transfer

The user determines the location of the anatomical and functional raw data in the computer of the MRI scanner and the system transfers the data with DICOM protocols to a workstation. This process is independent of the MRI scanner vendor.

3.1.3 Motion correction

Since most fMRI exams last a few minutes it is expected that the functional raw data contain motion. Artifacts due to motion will produce false activation and therefore need to be corrected for. Motion correction algorithms are incorporated in this system in order to co-register functional raw data acquired at different time points during the exam.

3.1.4 Statistical analysis

The result of motion correction is used as input in the statistical analysis. Since all task parameters are predetermined, this system includes pre-specified statistical models to perform accurate statistical analysis for each application. The user is only required to select, from a list, the task that was applied during acquisition.

3.1.5 Co-registration of statistical maps and anatomical images

In order to produce clinically useful results, statistical maps must be overlaid on anatomical images for reference. Therefore, the system co-registers functional raw data with anatomical images. Since statistical maps are in the same coordinate space as the functional raw data, the transformation created during co-registration is applied to the statistical maps. Finally, statistical maps are overlaid on anatomical images and the results are presented to the user.

3.2 Brain connectivity

3.2.1 Data acquisition, data transfer, co-registration with fMRI results

In all MRI exams the user is required to select certain acquisition parameters. In order to study brain connectivity with this system the user is required to select pre-specified DTI acquisition parameters. These are ones that are widely accepted by the research community. Similar to the prepackaged fMRI tasks, predetermined DTI acquisition parameters ensure that raw data will have the form expected by the post-processing part of the system. The data is then transferred to a workstation as described in section 3.1.2. DTI raw data is co-registered with fMRI results.

3.2.2 Calculation of diffusion tensors and fiber tracking

Once the DTI raw data are in the same space as fMRI results the diffusion tensors in each voxel of the brain are calculated. Then, the eigenvalues and eigenvectors of each tensor are estimated to obtain the white matter fiber tract orientation in each voxel. Regions of the brain with high probability to be related to the fMRI task performed by the subject are identified from the fMRI results. Finally, the system uses these regions as starting points for the fiber tracking algorithm. The results include 3D images of fiber tracts that originate from, or are spatially related with, the activated regions.

4 Ongoing work

The output of the functional and connectivity part of the system can be transferred to a central database for access by all users. In order for this collection of clinical cases to be informative, certain steps need to be taken before transferring the results to this database. Each case has to be categorized. Therefore, the user will be required to make selections from a number of lists, that best describe the specific case. These lists will include age, gender, weight, diagnosis (tumor, stroke, epilepsy, normal, multiple sclerosis, Alzheimer's, etc.). Patient names and patient ID numbers will be removed prior to integration in the database. However, a universal ID will be issued for each study. This will allow other users to communicate with the site that performed a specific study in order to exchange diagnoses, or even make a request for the raw data.

Thus this system will not only automate studies of brain function and brain connectivity but it will also serve as a training tool for clinicians, it will promote this type of advanced imaging studies, and will improve clinical practice through not only the results themselves but also through interaction of clinicians on different cases.

References

1. Bandettini PA, Jesmanowicz A, Wong EC, Hyde JS. Processing strategies for time-course data sets in functional MRI of the human brain. Magn Reson Med 1993;30:161-173.
2. Basser PJ, Mattiello J, Le Bihan D. MR diffusion tensor spectroscopy and imaging. Biophys J 1994;66:259-267.
3. Bandettini PA, Wong EC, Hinks RS, Tikofsky RS, Hyde JS. Time course EPI of human brain function during task activation. Magn Reson Med 1992;25:390-397.
4. Basser PJ, Mattiello J, Le Bihan D. Estimation of the effective self-diffusion tensor from the NMR spin echo. J Magn Reson B 1994;103:247-254.
5. Pierpaoli C, Jezzard P, Basser PJ, Barnett A, Di Chiro G. Diffusion tensor MR imaging of the human brain. Radiology 1996;201:637-648.
6. Basser PJ. Inferring microstructural features and the physiological state of tissues from diffusion-weighted images. NMR in Biomedicine 1995;8:333-344.
7. Basser PJ, Pierpaoli C. Microstructural and physiological features of tissues elucidated by quantitative-diffusion-tensor MRI. J Magn Reson B 1996;111:209-219.
8. Mori S, Frederiksen K, van Zijl PC, Stieltjes B, Kraut MA, Solaiyappan M, Pomper MG. Brain white matter anatomy of tumor patients evaluated with diffusion tensor imaging. Ann Neurol 2002;51:377-380.
9. Witwer BP, Moftakhar R, Hasan KM, Deshmukh P, Haughton VM, Field AS, Arfanakis K, Noyes J, Moritz CH, Meyerand ME, Rowley HA, Alexander AL, Badie B. Diffusion tensor imaging of white matter tracts in patients with cerebral neoplasms. Journal of Neurosurgery 2002;97:568-575.

The Clinical Perspective of Large Scale Projects: A Case Study with Pediatric Brain Tumors & Multiparametric MR Imaging

L. Astrakas[1], *A. Aria Tzika*[1], *F. Makedon*[2], *S. Kapidakis*[3], *S. Ye*[2], *J. Ford*[2]

Abstract

A multiparametric MR approach (e.g., conventional imaging, hemodynamic MRI, diffusion weighted MRI and MR spectroscopic imaging) enhances our ability to evaluate and treat pediatric brain tumor patients. The development of a local database will provide an infrastructure for better data collection and management as well as more power and flexibility in detailed data analysis. A well-organized database system will have significant implication for the clinical management and research in every brain tumor tracking project. Finally a large-scale database that will share data between different sites in order to be effective has to be easily accessible and overcome the different data formats. We describe a novel negotiation approach to building such large projects.

1 Introduction

Recent progress in medical technology coupled with increases in computing capabilities have created a new landscape in the hospital environment. Automated diagnostic medical procedures (examination protocols) have "industrialised" the acquisition and storage of medical data and have created both opportunities and problems regarding their management and exploitation. Perhaps the most representative example of this trend is the Picture Archiving and Communications System (PACS), which combined with the Radiology Information System (RIS) has created filmless radiology departments and revolutionised access to diagnostic images. Digital organization schemes in modern hospitals have given the opportunity, both to the clinician and the researcher, to combine data from different sources (imaging data, pathology data, clinical data, etc) without significant effort. As we will show in the case of pediatric brain tumors, this multi-source approach helps the clinical evaluation of a patient and the production of scientific knowledge.

Very often, multi-center cooperative research organisations are created to combine resources from different institutions and communities. The Pediatric Brain Tumor Consortium (PBTS) is an example in the area of central nervous system (CNS) tumors in childhood. Such large-scale organizations usually have as their mission the improvement of treatment strategies, clinical protocols, and data analysis procedures, and generally the understanding and cure of a disease. However, most of them lack a complete, secure, and effective way to share data and methodologies among their members. In the following we will discuss as important steps toward this goal using the paradigm of pediatric brain tumors, based on our clinical experience.

2 Mutiparametric MR assessment of pediatric brain tumors

Magnetic Resonance Imaging (MRI) is widely used in the diagnosis and follow-up of pediatric patients with brain tumors because of its ability to provide anatomic detail. However, conventional

[1] NMR Surgical Laboratory, Massachusetts General Hospital,Boston, MA 02114, atzika@partners.org
[2] Dartmouth College, 6211 Sudikoff Laboratory, Hanover, NH 03755, makedon@cs.dartmouth.edu
[3] Archive and Library Sciences Dept., Ionian U.,Corfu 49100, Greece, sarantos@ionio.gr

MR imaging alone does not give information about tissue biochemistry, it is not as good for classification of tumors (degree of malignancy) and the interpretation of conventional MR images may lead to poor estimation of the extent of active tumor. MR contrast enhancement assists in defining tumor borders, but is not reliable in the determination of malignancy. For these reasons we have extended the conventional MRI protocol to include hemodynamic MR imaging (HMRI), proton MR spectroscopic imaging (MRSI) & diffusion-weighted MR imaging (DWMRI).

Proton MR Spectroscopy shows the presence and amount of hydrogen protons attached to different cerebral molecular compounds. A generated spectrum corresponds to a scale of resonant frequencies vs amplitude (concentration). Molecular compounds identified within cerebral tissue include N-acetyl-aspartate (NAA, a neuronal marker), choline (a cell membrane marker), creatinine & phosphocreatinine (energy metabolites), and lipids/lactate.

Perfusion MR shows the microscopic vascular proliferation ("neovascularization") associated with tumor growth. Cerebral tissue perfusion can be assessed following a dynamic injection of Gadolinium. During the first pass transit through the cerebrovascular bed (which lasts only 5 to 15 seconds), gadolinium is restricted to the intravascular space. The restricted intravascular presence of highly paramagnetic contrast molecules (gadolinium) creates microscopic field gradients around the cerebral microvasculature, resulting in a change (shortening) of T2 relaxation and signal loss. From the amount of signal loss, the concentration of gadolinium in each pixel can be calculated, and a pixel by pixel relative estimate of blood volume can be inferred. Maps of cerebral blood volume (CBV) and cerebral blood flow (CBF) can be generated.

DWMRI demonstrates the water movement in different areas of tissue. It produces Apparent Diffusion Coefficient (ADC) maps that describe the mobility of the water molecules in the brain. Hyperintense ADC areas correspond to free water molecules (e.g. cysts) and hypointense ADC areas correspond to restricted water molecules (e.g. necrotic areas).

We applied *multiparametric MR* (including conventional imaging, HMRI, DWMRI and MRSI), to examine the relationship between MR imaging, MRSI parameters and to correlate them with histopathology indices and with clinical evaluation data. The hypothesis is that these relationships reflect the functional and biochemical status of tumors because they are influenced by the tumor's physiologic status and relate to tumor histopathology. We have proven that multiparametric MR studies, (a) predict histologic findings, (b) follow serial changes in tumor grade, (c) corroborate responses to chemotherapy and/or radiation, and (d) are accurate in differentiating tumor from radiation necrosis, normal tissue and other structural abnormalities (Tzika, Zurakowski et al. 2001; Tzika, Cheng et al. 2002; Tzika, Zarifi et al. 2002; Tzika, Astrakas et al. 2003).

Figure 1 is an example of the multiparametric approach in a 7 year-old with a brain stem glioma.

2.1 Retrieving and pre-processing the data.

So far all the stages of retrieval and analysis of our imaging and spectroscopy data have been done in a manual or semi-automatic way. However, the procedures of identification and retrieval of the interesting raw data and their subsequent pre-processing have to become fully automated and unsupervised. This has clear advantages especially in the case of manipulation of large amounts of raw data because (a) the human error is minimised (or eliminated) and (b) an unsupervised analysis can reveal important hidden parameters that may escape a human biased approach.

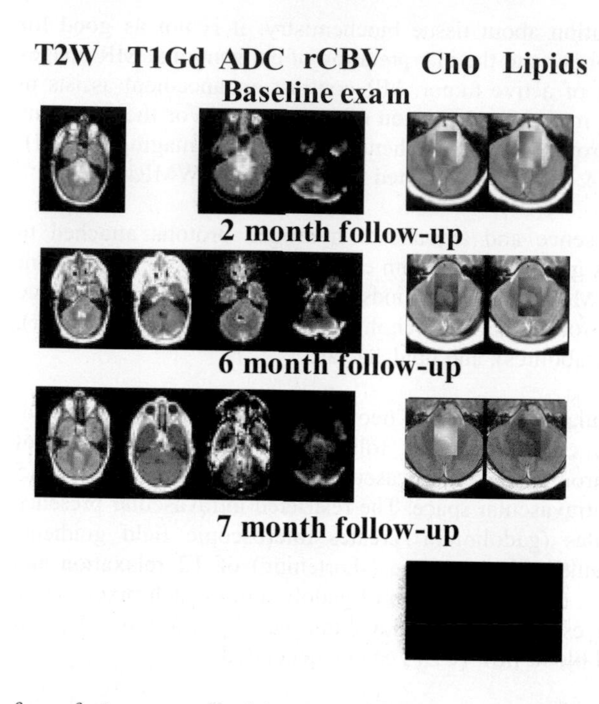

Figure 1: Conventional T2-weighted MR images (T2W), Gd-enhanced T1-weighted (T1Gd), perfusion MR imaging (rCBV image), diffusion-weighted MR imaging (ADC image) and MR spectroscopic imaging with metabolite mapping, within the posterior fossa of a 7 year-old boy with a brain stem glioma. The baseline MR imaging exam (prior to therapy) showed a large mass in the pons, compressing the 4th ventricle of high T2 signal on T2W, with no significant enhancement on T1Gd ima ges, but with high ADC and low perfusion on rCBV. Metabolite images (Cho images) showed increased Cho in the region of T2 signal abnormality, compared to normal surroundings. After 2 months, resolution of both T2 and ADC hyperintensities was detected and the Cho image showed decreased intensity, & perfusions remained low on rCBV. Lipid metabolite images showed increased lipids implying apoptotic cell death by 7 months. These findings were consistent with stable/ regressing disease.

The need for unsupervised data pre-processing is evident by the way that MRS data analysis has progressed from single voxel/spectrum to 3D grids (8X8X8) of voxels. But the amount of 3D spectra generated prevents manual processing and signal processing manipulations can not be performed subjectively since they are unknown. This has led to robust, black-box methods of spectra quantification, Bayesian estimation or prior knowledge (Vanhamme, Sundin et al. 2001).

Similar automated algorithms are needed for the perfusion imaging. Currently human guidance is required for the identification of the middle cerebral artery. Another problem in the case of perfusion is that the original images usually have small spatial resolution and the rCBV maps have even smaller since the algorithm that produces them fails for many pixels that we later omit. On the contrary, ADC maps are much easier to be produced.

Different imaging modalities in multiparametric approaches pose additional difficulties to automated data analysis (e.g., differences in contrast and resolution, displacements or rotations of the subject). More complex is the intra-subject registration in a common-brain template-atlas. Although many robust algorithms exist (rigid or affine transformations and non-rigid transformations) the field of image registration is still active (Maintz and Viergever 1998).

2.2 Analysing the data.

After the preprocessing of data, the role of the clinician/researcher becomes important because she/he defines the scientific criteria according to which the analysis will be performed. In pediatric brain tumors, we used spectroscopic, hemodynamic and diffusion imaging and less contrast enhanced or T2 imaging. From the many hemodynamic parameters (rCBV, rCBF, MTT, etc) we chose the rCBV because we could see changes in vascularization more clearly. When we saw that the highest choline regions are of particular importance for tumor monitoring, we used them. The above shows that human intervention is fundamental in the design of the analysis of medical data.

After the goals and hypotheses of the analysis have been established, a variety of tools are used in an automated fashion (volumetric, feature extraction, statistical analysis, neuronal networks, clustering algorithms, etc. - we used statistical tools (ANOVA, linear regressions, correlation, etc).

The development of a database is essential for better data collection and management, more power and flexibility in detailed data analysis and for research in every brain tumor tracking project. The database should contain the raw data, the parametric maps generated by the post-processing of the raw data as well as metadata. For pediatric brain tumors, raw data are MR images, histopathology and clinical treatment data (e.g., response to treatment, radiation dose, chemotherapy dose, etc). Parametric maps should be the ADC map from DWI, rCBV map from HMRI and metabolite maps (Cho, tCr, NAA, L) from MRSI. *Metadata* can describe important features of the raw and processed data and are used for the final analysis. The choice of metadata is crucial for success.

3 Need for large scientific databases to integrate data

In many cases the local database of a hospital is not adequate for scientific research because of (a) small volume of patients, especially in rare diseases, (b) limited human/hardware resources (preventing hindering robust analysis). These problems can be solved creating *collaborative schemes* where knowledge and data are shared between different groups. In the field of genomics this happened many years ago mainly due to the huge amount of data needed to be organized.

In the medical field, although there are many disease-oriented separate organization efforts, there is no universal way to access medical data, possibly due to the sensitive nature of these data as well as their large inhomogeneity. Digital Imaging and Communication in Medicine (DICOM) standard is a universal method for the transmission of medical images and their associated information. DICOM continuously evolves and now describes not only raw images but parametric maps too. Similar standards are needed in other areas of medical information to enable sharing (e.g. for histopathology data). As a solution, multi-center collaborations, instead of sharing the data among their members, gather them in a central archive and create a large-scale database, not always accessible to everyone. We propose a flexible front end to these databases, to make them more usable by : (a) translating different local database formats into an accepted formalism (b) providing access to more groups with different permissions through a main internet portal.

4 The design and implementation of SCENS: A negotiation system

SCENS[4] is designed to support sharing in scientific research, by providing incentives specific to a research community (Ye, Makedon et al, 2003). It is a trusted third party software infrastructure enabling independent entities to interact and conduct multiple forms of negotiation. In the design and implementation of SCENS, we focus on the flexibility and scalability. The SCENS services implemented include user authentication, data quality assessment, web based negotiation, web service based negotiation, usage tracking, user evaluation and other services. We also developed synthetic tools to simulate the negotiation activities and help evaluate the whole system.

---

[4] Secure Content Exchange Negotiation System

SCENS has a flexible 3-layer service structure, which provides different level of negotiation services for different type of users. **Layer 1** behaves as a traditional web based negotiation support system for human beings. Users can customize the negotiation agents provided by SCENS through multiple parameters. **Layer 2** provides support for negotiation strategy customization for users, especially negotiation agents, which are treated as web service consumers, through web services. **Figure 2** shows the interactivity between Layer 1 and Layer 2. **Layer 3** is designed to provide an open automatic negotiation environment. DAML+OIL (Hendler and McGuinness 2000; Joint US/EU ad hoc Agent Markup Language Committee 2001), a language with which to create ontologies and to markup information, is used in Layer 3 to define a negotiation ontology, which allows agents to acquire knowledge about how to conduct negotiations. This knowledge includes negotiation protocols, negotiation proposals and conditions, etc. The agents communicating with Layer 3 can be used in any negotiation activities given the proper negotiation ontology. In Layer 2, in contrast, the knowledge about negotiation rules is actually hand-coded into the agents.

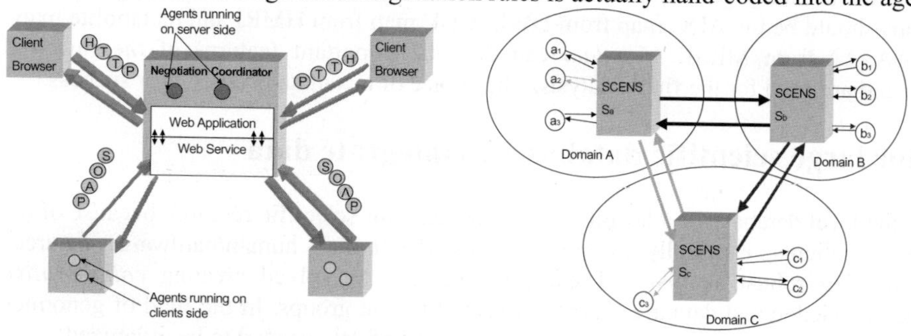

Figure 2 Interactivity between Layer 1 and 2

Figure 3: a2 and c3 are negotiating

References

Hendler, J. and D. L. McGuinness (2000). "The DARPA Agent Markup Language." IEEE Intelligent Systems 16(6): 67–73.

Joint US/EU ad hoc Agent Markup Language Committee (2001). DAML+OIL language specification, http://www.daml.org/2001/03/daml+oil-index.

Maintz, J. B. and M. A. Viergever (1998). "A survey of medical image registration." Med Image Anal **2**(1): 1-36.

Tzika, A. A., L. G. Astrakas, et al. (2003). "Multiparametric MR assessment of pediatric brain tumors." Neuroradiology **45**(1): 1-10.

Tzika, A. A., L. L. Cheng, et al. (2002). "Biochemical characterization of pediatric brain tumors by using in vivo and ex vivo magnetic resonance spectroscopy." J Neurosurg **96**(6): 1023-31.

Tzika, A. A., M. K. Zarifi, et al. (2002). "Neuroimaging in pediatric brain tumors: Gd-DTPA-enhanced, hemodynamic, and diffusion MR imaging compared with MR spectroscopic imaging." AJNR Am J Neuroradiol **23**(2): 322-33.

Tzika, A. A., D. Zurakowski, et al. (2001). "Proton magnetic spectroscopic imaging of the child's brain: the response of tumors to treatment." Neuroradiology **43**(2): 169-77.

Vanhamme, L., T. Sundin, et al. (2001). "MR spectroscopy quantitation: a review of time-domain methods." NMR Biomed **14**(4): 233-46.

Ye, S., F. Makedon, et al. (2003). SCENS: A system for the mediated sharing of sensitive information. In submission to the Third ACM/IEEE Joint Conference on Digital Libraries.

Theories on the Impact of Information and Communication Technology and Psychosocial Life Environment

Bradley, G.

IT university, Institute for Microelectronics and IT
Royal Institute of Technology, Stockholm, Sweden,
Bradley@imit.kth.se

Abstract

An interdisciplinary research program on "Computer technology and work life" was initiated and led by the author at Stockholm University from 1974 to 1988. The program inspired many other research programs in Sweden in the field. A theoretical framework was developed including two theoretical models, one more general and one where the concepts and their interrelationships were specified. The models were tested empirically in three large work organisations in Sweden, representing three main historical periods of computer technology. It was also used as a model in discussing what might be desirable goals in the information society. The present fourth period, the "Network period", is characterised by a *convergence* of three main technologies: computer technology, telecommunication technology, and media technology (ICT). ICT is used in almost every activity and embedded in many things around us. The author proposes a superimposed theoretical model reflecting "ICT and psychosocial life environment", a revised model of her initial models. Sociological theories of Information Society are discussed. Finally, future research is addressed with reference to theoretical models revised, and conclusions address major psychosocial processes, psychosocial life environments and a call for synthesis.

1 Theoretical model on the computer technology and work life

The RAM research programme on "Computer technology and work life" was an interdisciplinary research programme initiated and led by Bradley at Stockholm University from 1974 to 1988. RAM referred to the Swedish expression for "Rationalisation" and "Work Environment". A theoretical framework was developed by G. Bradley entitled "Computer Technology and Working Environment" (first published in 1977). The framework included two theoretical models, one more general (figure 1) and one model where the concepts and their interrelationships were specified (Bradley 1977, 1989). The models were empirically tested in three large work organisations in Sweden, representing three main historical periods of computer technology – from systems with batch processing to microcomputerisation. The psychosocial work environment was considered in terms of the following perspectives: 1.Three levels of analyses (individual, organisational, and societal) 2.objective and subjective work environments 3. interplay between levels 4. interplay between objective and subjective work environments 5.interplay between working life and private life 6. a life-cycle perspective.

The content of some of the concepts in the models may be summarised as follows. The *objective work environment* refers to areas of work that are germane to large groups of employees. The *subjective work environment* consists of perceptions and attitudes related to corresponding sets of factors in the objective work environment. The subjective work environment is closely linked to the concept of job satisfaction, which could be seen as a summarising concept.

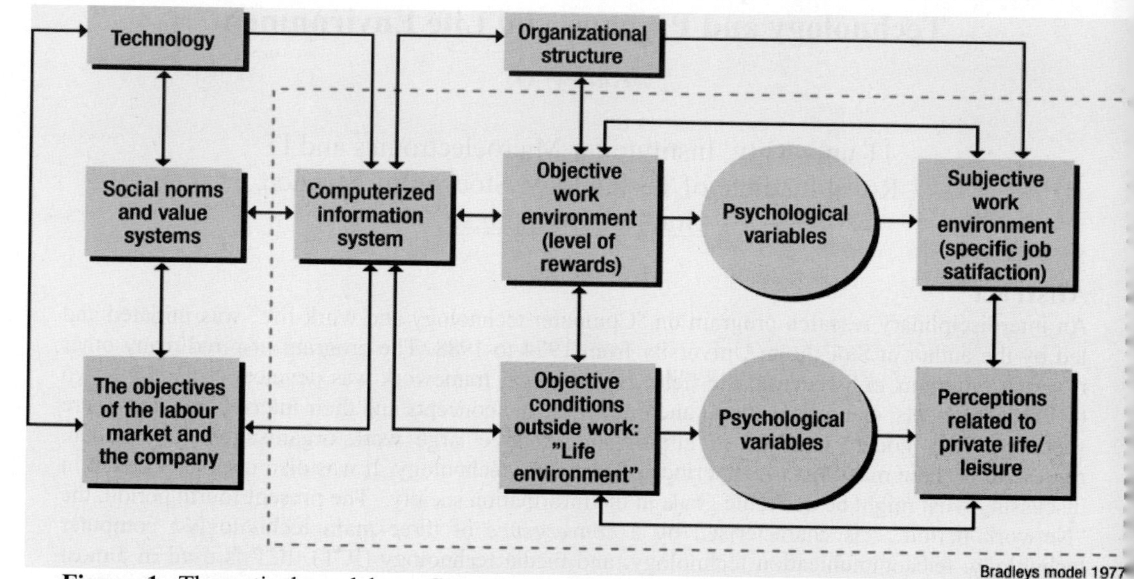

Figure 1. Theoretical model on Computer technology and work environment (Bradley 1977, 1989)

Psychological variables is a general term covering a number of intermediate, psychologically relevant variables such as the *level of aspiration* and the *weight* attached to specific work-environment areas. The concept of *psychosocial* refers to *the process involving the interaction* between the objective environment and the subjective one. *Essential concepts within the psychosocial work environment* include factors such as contact patterns and communication, organisational structure and design, work content and workload, participation in decision-making, promotional and development patterns, salary conditions and working hours.

Objective conditions outside work refers to behaviour, consumption and conditions that prevail during the hours spent away from work, according the traditional definition of work. The theoretical models used in the RAM program were used as models in discussing what *structure* a computerised society should have (see the two-way arrows in the figure) and what might be desirable *goals*. This was also described in a special chapter on actions strategies in Bradley (1986) and also in my chapter in the book Computers and Society (Beardon & Whitehouse,1993).

An extensive research strategy was applied with qualitative stages/methods and quantitative stages/methods for collection and analysis of data. Indices were created through multivariate analysis, and they corresponded well to the theories in the field of work and organisational theory). Theories, methods and results from the RAM programme are summarised in "Computers and the psychosocial work environment" (Bradley 1989). These measures and tools are still relevant for studies of the social and organisational impact of ICT.

2. Theoretical models revisited

In later projects a fourth period in the evolution of computer technology has been explored, both the psychosocial and the societal impact of ICT has been in focus, best referred to in our ongoing research programme entitled "Interplay ICT – Humans – Society". The fourth and present period of computerisation I would like to refer to as the "Network period", very much based on *the convergence* and integration of three main technologies; computer technology, tele technology and media technology. ICT is more and more being used in almost every activity and *embedded* in most of things around us. The graphical representations in my theoretical models have been changed, converging circles better reflect the ongoing process.

Both *Convergence* and *Interactions* are important features in the model. Convergence means a move towards a common point. Interaction means that technology interact with the social world with values and believes, there is an ongoing interaction between the clusters of circles.

- A convergence of computer technology, telecommunication technology and media is occurring to become ICT
- Professional Role (Work Life) and Private Role (Private Life) and Citizen's Role (Public Life) converge to become a Life Role
- Work Environment and Home Environment are converging to a Life Environment
- Effects on the Individual become more multi faceted and complex. This is valid both regarding the psychological and the physical effects (Effects on Humans)
- Technology, Economy, Norms/Values and Labour Market interact and converge and is entitled Globalisation
- *A new emphasis* on certain dimensions in the psychosocial environment occur
- New *dimensions* are appearing in the psychosocial environment. Openness for unforeseen implications is required.

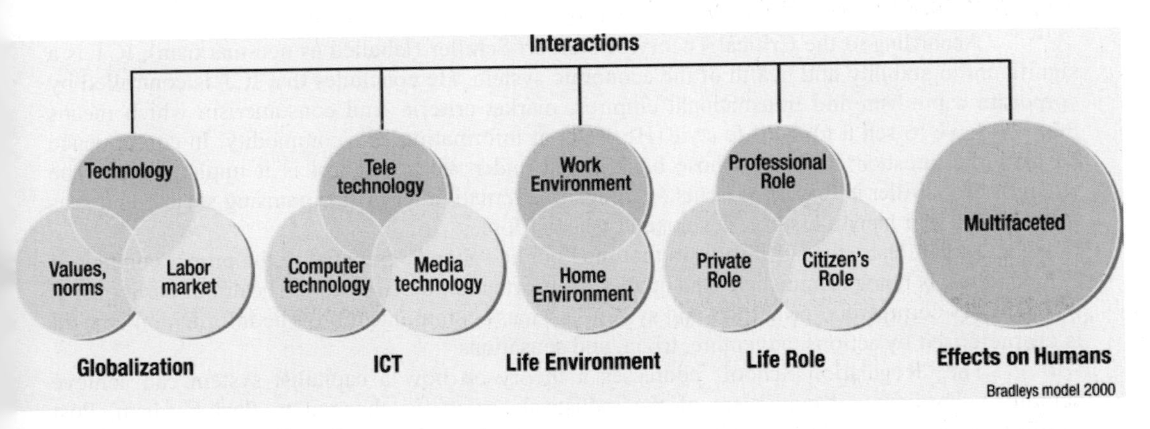

Figure 2. Convergence Model – ICT and Psychosocial Life Environment (Source:Bradley 2001)

Within informatics a discussion of focus is taking place: both analysis and design need to address not only the work process and management connected to the sphere of production life, but also *people's life environment.* Not only professional roles but also our roles as citizens and private persons are crucial. Community research in a broad sense comes to the fore, with respect to both physical and virtual communities. Analysis and design of *ICT and societal systems* both at local level and globally become important. There is also a need for new and additional actors at the deeper and broader integration of ICT in the society (children, elderly, and consumer organisations).

3. How do these theories relate to theories of the Information Society?

Frank Webster in "Theories of the Information Society" (Webster 1995) provides a point of departure of an interrogative and sceptical view of the concept of an "Information Society" (IS). His approach is to start form contemporary social theories instead of social impact approaches.

The following are categorised as Pro IS theories: Post-industrialism; postmodernism; flexible specialisation; the information mode of development. The following are Against IS theories: neo-Marxism; regulation theory; flexible accumulation; the nation state and violence; the public sphere. Webster brings forward five definitions of IS which represent criteria for the new society, by its own or combined: Technological; economical; occupational; spatial and cultural.
I choose to briefly describe the main content of five theories.

Anthony Giddens (1990) means that the origins of today´s "information societies" are to be found in surveillance activities that are largely driven by the requirements of a world organised into nation states. Modern world are nations, which are "information societies" and have always been so but need ever more to maintain the following: 1. allocative resources as planning, administration 2. authoritative resources as power an control 3. information which is the core of modern military affaires. Giddens argues that information today has a great significance but is a continuity. In "Runaway World" (2000) Giddens focuses "globalization": The capital flow is the big issue; power is taken away from local and nations but is also brought downwards; immediate electronic communication enriches as well as trivialises our lives; concerning the potential for democracy both an expansion of democracy and visibility of the limits of democracy.

According to the Critical Theory of Herbert Schiller (labelled as neo-marxism), ICT is a significant to stability and health of the economic system. He concludes that ICT is controlled by corporate capitalism and transnational empires, market criteria, and consumerism which means that you have to sell a global life style. He looks at information as a commodity. In consequence he asks the question: ICT for whose benefit and under whose control is it implemented? The strength of Schiller is that he presents openings to alternative ways of organising society and sees that IS has a real human history, developed by social forces.

Jurgen Habermas (1989) fears that the "public sphere", essential to the proper conduct of democracies, is being diminished. The quality of information determines the health of participants. ICT stresses commercial principles and systems of mass communication. The information content is characterised by actions, adventure, trivia, and sensations.

The "Regulation School" addresses a theory on how a capitalist system can achieve stability. Fordism – Keynianism of the industrial society is changed to Post Fordism. Post Fordism is characterised by: 1. Globalisation of market, production, finance, and communication 2. Corporate restructuring e. g. downsizing, outsourcing, ICT infrastructure, less mass production 3. Flexible specialisation 4. Effect on the labour e g flexibility of employees, wage flexibility, and time flexibility.

The most quoted research during the latest years has been Manuell Castells books on the Network Society. He talks about networks of suppliers, producers, customers, standard coalisions and technology cooperation networks and argues that a multifaceted virtual culture is appearing. In his latest book (Castells 2001) he analyses the digital divide in a global perspective

Webster asks if the there is a break or continuity that dominates and his answer is that there is still features of capitalist continuity, but some shift in orientation with some novel form of work organisation and some change in occupational pattern. But there is no system break witnessed so far (Webster 1995). However a group of theorists categorised as postmodernists (Barthes, Baudrillard, Vattimo, Lyotard) point out that the information society contains a quite new type of society – *a paradigm shift*. A British – Nordic collaboration some years ago tried to form a coherent theory of the ICT society, supported by the National Boards for Research in Humanities and Sciences.

4. Some conclusions

My conclusion from the Convergence theory (figure 2) and the more macro oriented theories above, is that the introduction and use of ICT in the new life environment having moved into our homes, should not be left to the stearing factors that have been present in work life over the years. A crossdisciplinary research and action program is crucial and should be integrated with full scale models for various applications and "reflection". Basic human needs and a people push technology not a technology push should be the leading principal.

The big challenge in the near future is the home in a broad sense, as many human roles are converging to *one life role* and the home is more and more understood in terms of *virtual space as well as physical.* Driving forces are converging and embedded technologies. The following trends are enforcing the home as a *communication sphere:* The home as the extended family center; a care center; a multimedia center; a center for democratic dialogue; a market place; a learning center; an entertainment center. A "smart home" should be service and not product oriented. A "smart home" should help the individual to *good health, safety and joy.* A "smart home" should enable us to deepen *human qualities* and provide humans with psychological and physical strength to change society in a humane direction.. The design should be based on *human needs, as well as human abilities and preferences*. Important human needs are the need for having a safe and secure life, the need to influence our life conditions, the need for social belonging, the need for learning and developing oneself, and finally our need for meaningful life content. In parallel it is important to *prevent various stress phenomena* in our life environment.

References

Beardon, C. & Whitehouse, D. (Eds.) (1993). *Computers and Society.* Oxford: Intellect.

Bradley, G. (1977*). Computer Technology, Work Life, and Communication*. The Swedish delegation for long term research. FRN. Stockholm: Liber (In Swedish).

Bradley, G. (1979). *Computerization and some Psychosocial Factors in the Work Environment.* Proceedings of the conference Reducing Occupational Stress, New York, 1977. NIOSH Publication No. 78-140, p 30-40. U.S. Department of Health, Education, and Welfare.

Bradley, G. (1989) (publ. in Swedish 1986). *Computers and the Psychosocial Work Environment.* London/Philadelphia: Taylor & Francis.

Bradley, L., Andersson, N., Bradley, G. (2000). *Home of the Future - Information and Communication Technology (ICT) - changes in society and human behavior patterns in the net era.* FSCN research report R00-1. http://www.mh.se/fscn

Bradley, G. (2001). *Humans on the Net. Information and Communication Technology (ICT) Work Organization and Human Beings*. Stockholm: Prevent. ISBN 91-7522-701-0.

Castells M. (2001*). The Internet Galaxy*. Oxford Oxford University Press.

Giddens, A. (1990). *The consequences of Modernity*. Cambridge: Polity.

Giddens, A. (2000). *Runaway World.* London: Routledge.

Habermas, J. (1989). *The Structural Transformation of the Public Sphere*. Cambridge: Polity.

Schiller, H. (1993*). Public Way of Private Road?* The Nation, 12, July pp. 64-66.

Webster, F. (1995). *Theories of The Information Society*. London: Routledge.

"Stressors of organizational conditions" – a new design-oriented work analysis instrument

Markus Buch

Institute of industrial science and ergonomics, University of Kassel
D-34109 Kassel, Germany
buch@ifa.uni-kassel.de

Abstract

"Stressors of organizational conditions" is a new work analysis instrument which was developed on the basis of the Job Analysis Inventory (Tätigkeitsanalyseinventar, Frieling, 1999). Its purpose is to assess the stress potential of a job and to deduce stress reducing work redesign interventions if necessary. The reliability and validity data are summarized.

1 Introduction

Results of European studies (e.g. European Foundation, 2001) suggest that rapidly changing working conditions come along with an increase in stress. In order to reduce stress it is necessary to fall back on design-oriented stressors. The question which characteristics qualify stressors as design-oriented remains open. Correspondingly there is a lack of instruments serving as a basis of job (re-)design, despite the multitude of instruments aiming on the economic assessment of jobs' stress potential. The paper presents a new job analysis instrument that meets these requirements: "stressors of organizational conditions". This instrument pursues the objective to assess the necessity of job redesign and furthermore to deduce job redesign interventions. It is intended to include different sectors, professional groups and activity classes in the application area. First of all the question "What makes a stressor design-oriented?" is discussed, then the instrument is presented.

2 Design orientation

Which features are necessary for stressors to meet the demands of job redesign? Undoubtedly test theory criteria are fundamental in this area. To redesign working conditions the following characteristics are helpful:

- Person-unspecific work analysis. Design-oriented work analysis is concerned with the working conditions resulting from the work context and the work content. Nevertheless the job or position as a interaction of the incumbent and the work environment is the unit of analysis. This takes into account that with increasing degrees of freedom in actions the incumbent is able to define his working conditions.
- Threshold limit values. In the application field of job redesign work analysis is to be expanded to work evaluation. Designer of work systems or jobs need information whether interventions are indicated. Therefore threshold limit values are mandatory.

- Concrete and detailed items. When job redesign is indicated it is required to have information where to start. This information is obtained when the stressor level can be traced back to separate changeable features.

These requirements serve as the starting point in the development of the work analysis instrument "stressors of organizational conditions" of the TAI.

3 Theory

The construction is based on an extension of Kannheiser's (1984) activity-oriented stress concept. Results, concepts and models of stress research are integrated in this view. In essence these are:

- activity latitude and control (e.g. Karasek, & Theorell, 1990),
- action theory-oriented stress concepts (e.g. Semmer & Dunckel, 1991),
- social support (e.g. Viswesyaran, Sanchez, & Fisher, 1999),
- role stress (e.g. Kahn & Byosiere, 1992),
- effort/reward imbalance model (Siegrist, 1996).

This conceptual basis leads to the deduction of the following seven stressors:

1. Dependency on supervisors,
2. Dependency on organizational conditions,
3. Low communication and support,
4. Standardization,
5. Uncertainty / unsecurity,
6. Time-, performance- and competition pressure,
7. Low participation.

4 Structure and evaluation

The module consists of 144 items derived of four TAI subchapters describing the organizational conditions. These subchapters are: 3.1 general conditions of the job, 3.2 organizational structure and work process organization, 3.3 features of execution and, 3.4 technical and organizational causes of fault. Figure 1 contains the assignment of item groups to the subchapters.

subchapters	groups of items
3.1 general conditions	*contract conditions*: labour relations, formal specifications of the job
	working time arrangements: rhythm, daily/ weekly working time, shift work
3.2 organizational structure and work process organisation	*organizational structure*: single/ group work center, number of team members, span of control
	work process organization: material flow, information flow, human interaction
3.3 features of execution	*planning time*: planning horizon, percentage of preparatory functions
	target values: kind, achievement
	overlap of orders: frequency and degrees of freedom
	mistakes: results and consequences
	feedback: frequency, kind
3.4 technical and organizational causes of fault	*technical causes of fault*: frequencies and coping
	organizational causes of fault: frequencies and coping

Figure 1: Relevant subchapters of the TAI with groups of items

The classification of the items is based mainly on standard rating scales like frequency, duration, yes/no. These classifications lead to an evaluation at the stressor-scale (fig. 2) guided by normative considerations.

Stressor-level	
0	no necessity for job redesign identified
1	job redesign desirable
2	job redesign necessary
3	job redesign urgently necessary

Figure 2: Stressor scale

The three levels of interpretation are 1) conditions 2) triggers and 3) stressors. Triggers can be interpreted as the source of stressor level. Stressors consist of three to six triggers. The trigger with the highest level defines the level of the stressor. The underlying assumption is that redesign implying working conditions can not be compensated by functional conditions in other domains. Figure 3 presents the assignment of triggers to stressors.

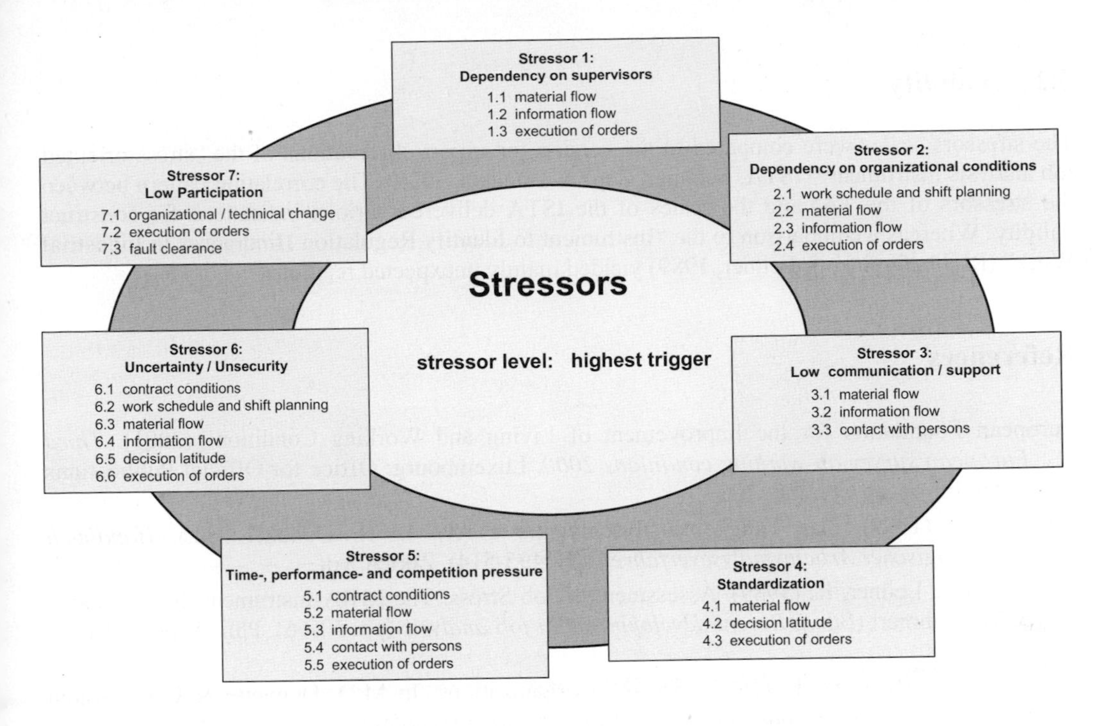

Figure 3: Assignment of triggers to stressors

5 Test criteria

To take the peculiarities of job analysis instruments into account the assessment of the test criteria was guided by suggestions of Oesterreich and Bortz (1994).

5.1 Reliability

The design model of "independent double-analysis" was realized. It considers three types of error that could reduce reliability: 1) the study factor, 2) the time factor and 3) the person factor. This is achieved by two different raters analysing two different incumbents doing the same job. The industrial sample consisted of 40 incumbents working in the mechanical engineering or the automotive industry. Jobs of assembly workers, assembly line workers, production workers, work group speakers, foremen and master craftsmen were analysed by altogether seven raters.

On the stressor-level the reliability is good: Cohen's weighted kappa values range from .71 to .86. The kappa statistic for 19 of the 27 triggers also shows good concordance (>=.7). Only one trigger is below weighted kappa=.5.

5.2 Validity

The stressors scales were compared to the version for expert observations of the "stress-oriented job analysis instrument" (ISTA, Semmer, Zapf & Dunckel, 1999). The correlation pattern between the stressors of the TAI and the scales of the ISTA delivered various indications for construct validity. Whereas a comparison to the "Instrument to Identify Regulation Hindrances in Industrial Work" (RHIA, Greiner & Leitner, 1989) yielded mainly unexpected relations.

References

European Foundation for the Improvement of Living and Working Conditions (2001). *Third European survey on working conditions 2000*. Luxembourg: Office for Official Publications of the European Communities.

Frieling, E. (1999). Das Tätigkeitsanalyseinventar (TAI). In H. Dunckel (Ed.), *Handbuch psychologischer Arbeitsanalyseverfahren* (pp. 495-514). Zürich: vdf.

Greiner, B. & Leitner, K. (1989). Assessment of Job Stress: The RHIA-Instrument. In K. Landau & W. Rohmert (Eds.), *Recent developments in job analysis* (pp. 53-66). Philadelphia: Taylor & Francis.

Kahn, R.L. & Byosiere, P. (1992). Stress in organizations. In M.D. Dunnette & L.M. Hough (Eds.), *Handbook of industrial and organizational psychology*, Vol. 3 (pp. 571-650). Palo Alto: Consulting Psychologists Press.

Kannheiser, W. (1984). Theorie der Tätigkeit als Grundlage eines Modells von Arbeitsstress. *Zeitschrift für Arbeits- und Organisationspsychologie*, 27, 102-110.

Karasek, R.A. & Theorell, T. (1990). *Healthy work. Stress, productivity, and the construction of working life*. New York: Basic Books.

Oesterreich, R. & Bortz, J. (1994). Zur Ermittlung der testtheoretischen Güte von Arbeitsanalyseverfahren. *ABO aktuell*, 3, 2-8.

Semmer, N. & Dunckel, H. (1991). Stressbezogene Arbeitsanalyse. In S. Greif, E. Bamberg & N. Semmer (Eds.), *Psychischer Stress am Arbeitsplatz* (pp. 57-90). Göttingen: Hogrefe.

Semmer, N., Zapf, D. & Dunckel, H. (1999). Instrument zur Stressbezogenen Tätigkeitsanalyse (ISTA). In H. Dunckel (Ed.), *Handbuch psychologischer Arbeitsanalyseverfahren* (pp. 179-204). Zürich: vdf.

Siegrist, J. (1996). *Soziale Krisen und Gesundheit*. Göttingen: Hogrefe.

Viswesyaran, C., Sanchez, J.I. & Fisher, J. (1999). The role of social support in the process of workstress: A meta-analysis. *Journal of Vocational Behavior*, 54, 314-334.

Breaking New Ground in Interactive Configuration of Production Environments by the Use of Intelligent Computer Tool

Frank Butke, Thomas Rist and Wilfried Sihn

Fraunhofer-Institute
Manufacturing Engineering and Automation IPA
Stuttgart, Germany

Abstract

In the face of an increasing demand for tailor-made and highly customized products, the design and development of its production systems became a very important task and critical success factor. Especially because production systems themselves are highly customized and exceedingly tailored to the specific demands of the end product manufactures and their specific products.
This paper presents two different solutions of how the mechanical engineering industry can use configuration tools to assist its distribution staff in the sales process. The main support for the technical salesmen is in having the possibility to present the future production system to their customers in a very realistic way. By that they are not any longer forced to convince their customers by columns and arrays of numbers or long part lists and static drawings of the production system.
The presented computer tools offer the distribution staff a totally new quality of abilities in presenting the results of the joint design process with their customers. At the end of that joint meeting the customer can see "his" future production system in an operating and running state on screen or projection. Additionally he gets qualified and highly descriptive charts and figures for the evaluation of the production systems performance.

1 Introduction

Traditionally, the strength of the German capital goods industry lies in its highly innovative and sophisticated products, its top quality and ability to tailor products to the specific needs of customers. In the past, these factors were sufficient to ward off low-cost competitors. However, the globalization of markets, an increasing cost and pricing pressure, and a growing individualization of customer needs are changing the market conditions and increasing competitiveness (Westkämper et al. 2001; Lay, Schneider 2001). As a result, the mechanical engineering industry had to undergo a radical change, mainly caused by the need to integrate new information, communication and multimedia technologies. The new technologies provide support and serve as catalyst, posing both a challenge and an opportunity.
In the past, the wish to satisfy individual customer needs collided with the actual striving for efficient manufacturing methods, causing a conflict that could not be solved. Custom engineering and design activities often required lots of reworking and adaptation. This variety resulted in the costly production of special one-of-a-kind machines. Most companies have come to recognize this trend. They try to counteract this disparity with building block solutions. The use of modules and components makes it possible to provide a great variety of machines at lower costs.
To be able to present the customer a wide range of systems, the equipment components must be designed as modules which are part of a building-block system. During the configuration process, the modules are combined into an individual system.

2 Configuration

Configuration systems make it possible to design and display products from electronic product catalogues on the basis of defined rules. Configurators allow to slash the time from initial customer contact to order placement. A market survey is given in (Gronau 2001).

The sales personnel uses the configurator to quickly specify the demands that customers place on a production system. The desired system features are entered via user interface. The configuration process is supported by multimedia contents, CAD drawings, etc., which are generated by the configuration systems from the company's BOMs (bills of material). The basic idea behind this procedure is to store the knowledge required for the design process in the configurator, and support the customer with automated proposals. Hence, configuration systems enable even field workers without expert knowledge to provide the customer with details about feasibility and price. At the end of a sales talk, a binding offer can be submitted, which may also include multimedia components. The more the result of the configuration process looks alike the real product, the more the customer is able to decide weather he likes to buy or not. Configuration systems improve the information flow between customer and manufacturer, and shorten the processing times between initial contact and the preparation of tenders.

Four things are necessary for a configurator. First, all the modules which are available have to be constituted, all the rules and restriction that define which selection is possible and which combination of modules does not work has to be build up in a user interface which assist the operator in configuration process. Last but not least the representation of the outcome of this configuration has to be selected.
The result of a product configuration could be a data sheet which describes the details of the configured resource respectively machine or it could be a simulation and 3D animation of a complete production system.
To give an example, on the web side of nearly every automobile producer there is an configurator to create the individual car. The representation of the output is a picture of the car in the selected color and a data sheet which describes the details of the selection. The evaluation of this result is static. In the data sheet the price of the selection or the power of the engine could be used to evaluate the configuration.
The configuration result could be evaluated in a static or in a dynamic way. Static in the terms of data which could be compared with other variants. Dynamically evaluation could be done by simulation.

3 Field of applications

3.1 Overview

In the following chapters examples will be given how configuration is used to assist purchasing staff to get better and faster results in the sales talk with the customer.
In both cases a software, called "I-Plant", developed at the Fraunhofer IPA, is used to configure the products of a mechanical engineering company individually with the customers. The software is a Virtual Reality layout planning tool designed to be extremely easy to handle. It is part of a entire planning tool for a team-based planning processes (see Figure 1).

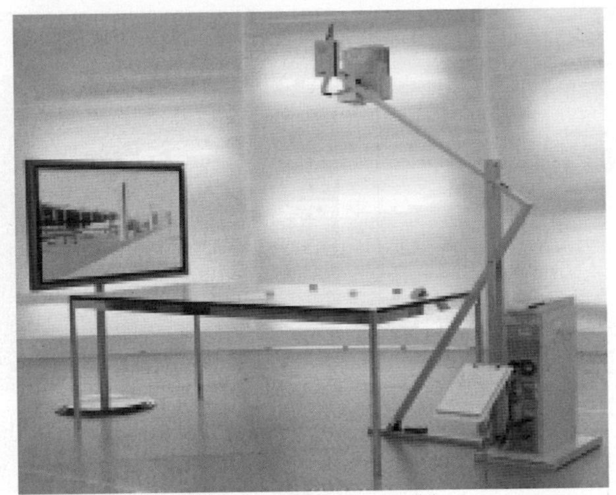

Figure 1: Planning Table

The backbone of the software is a database which is used to store all the available modules as well as the alphanumerical information of the configuration (e.g. location in the layout, item number of the module). In the software the user can plan changes by a simple grasp, shift and insertion of the modules. This software provides the graphical and the configuration user interface as well as the library of all the available modules.

3.2 Resource configuration and design of production system layout

In this case the configuration task was to provide the distribution staff with a tool which can be handled very easily. It is able to demonstrate the customer which possibilities he has, concerning the special circumstances to integrate the new machine or system in his shop floor and how it would look like. The representation of the result is a three dimensional static model. The evaluation is mainly by eye and based on the expertise of the salesman and the customer. Objects of evaluation are e.g. the size and the spatial needs of the system modules, the integration into the vacant space of the production hall, the design of the material handling, etc.
The products of the mechanical engineering company usually consists of one main machine and many components which could be chosen either optional. Some of the components are necessary but the location relatively to the main machine could be different. In some cases there is also a combination of machines for building up a more complex production system necessary.
For the definition of the producer specific rules and restrictions for the configuration there is a special script language available. This script language was also developed at FhG-IPA. By using the script language and building up a concrete producer specific script you define the rules, restrictions and degrees of freedom for combining the modules in the configuration process. The emerging script file also defines the design of the window menus which leads the user through the configuration process. The operator only sees the window menus and makes the choices of the components (see Figure 2).

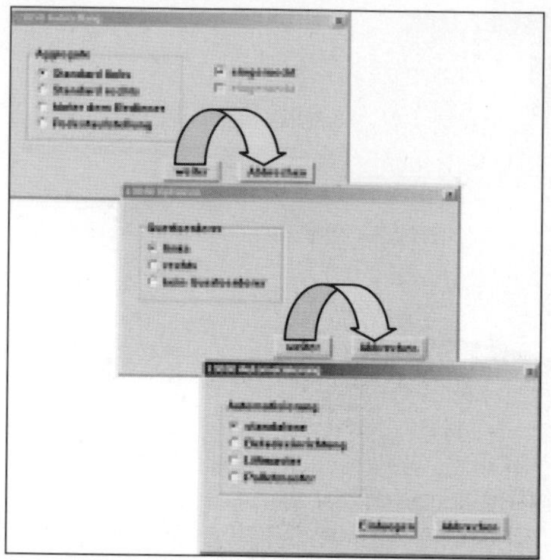

Figure 2: Configuration masks

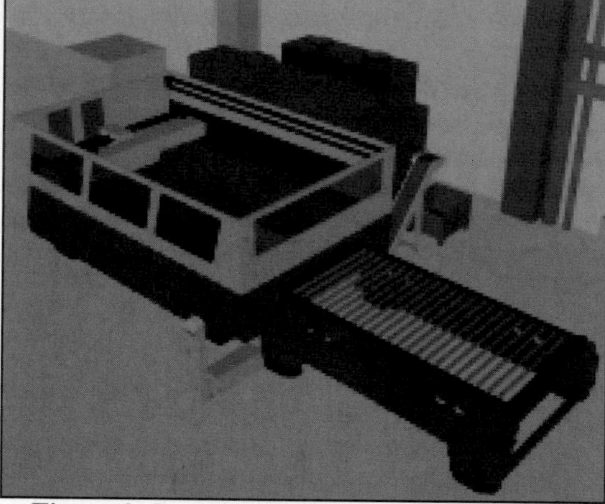

Figure 3: Location of the machine in the layout

The operator starts the I-Plant software, selects the machine type (main machine) the customer likes and works through the menus which opens by selecting the machine type. The result of the configuration is a 3D-object of the machine with all its components which is shown in he I-Plant software. The user now moves the machine to the optimal location in the layout (see Figure 3).

By taking the shop floor layout of the customer it is possible to select the right installation of the machine. On the one hand it could be checked if the selected configuration fits to the layout and on the other hand the customer sees the product he wants to buy and gets more convinced in deciding right.

3.3 Process configuration and simulation

In the second case the software I-Plant is used to configure a complete and complex production system. The operator builds up a production line with single modules which are stored in the library. For every module exists a simulation model. Both the I-Plant software and the generic simulation use the same database. Description of the system and system load are stored in a database, which is only read and interpreted when the individual simulation model is generated. This process is controlled by the generic simulation model.
The representation in this case is in the first step also a static three dimensional module and the alphanumerical data in the database. In the second step the production system will be simulated. The representation in this step is a three dimensional real time animation and different report charts (see Figure 4) which are created from the simulation.

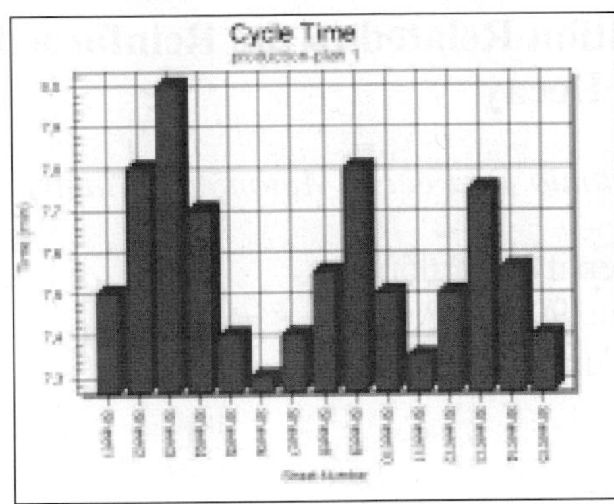

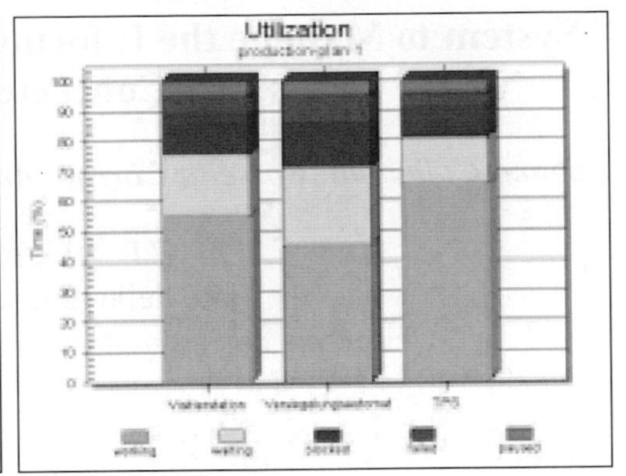

Figure 4: Report charts of the simulation

In the sales talk different variants can be made and best configuration for the customer can be found.

4 Conclusion and Outlook

The paper describes two different levels of configuration of a production system.
In order to offer the customers a large variety of individual production systems, modular components are an absolute necessity. During a configuration process, these modular components are combined to create a customer-tailored system. The subsequent simulation helps to dynamically assess the total operative performance. The 2D/3D visualization of the system could be either in the I-Plant software or in a simulation tool. With the help of different configuration options, the system provider can demonstrate the different performance measures a system is able to achieve to his customer. The use of configurators helps the system manufacturer to distinguish himself from his competitors in soliciting business. The potential buyer benefits from the visualization as it gives him a good impression of „his own“ system. The simulation-based performance measures also provides additional certainty concerning the decision to be made.
Configurators have proven a fair opportunity to reduce time in the sales process. They enhance the reliability to a result of a planning process. The visualization and the simulation of the configuration makes it easy to evaluate a planning result.

References

E. Westkämper, J. Niemann, M. Stolz. 2001. Advanced Life Cycle Management in Digital and Virtual Structures. In: CIRP 34th International Seminar on Manufacturing Systems, 16-18 May, 2001, Athens, Greece. Athens, Greece, S. 1-5.

G. Lay, R. Schneider. 2001. Wenn Hersteller zu Servicedienstleistern werden. In: HARVARD BUSINESS manager Nr. 2, S. 16-24.

N. Gronau, E. Weber. 2001. Marktüberblick Konfiguratoren in PPS-Systemen. In: PPS-Management 6 Nr. 4, S.59-75.

W. Sihn, T.-D. Graupner, T. Kuhlmann, H. Richter. 2002. Internetbasierte Konfiguration und Simulation von Produktionssystemen. In: Proc. "Simulation und Visualisierung 2002", Otto-von-Guericke-Universität Magdeburg, (Hrsg: Schulze, Schlechtweg, Hinz), 225-235.

A System to Manage the Information Related to the Reinforced Concrete Decay

Simona Colajanni, Rossella Corrao, Antonio De Vecchi, Antonietta Giammanco

D.P.C.E. – University of Palermo
Viale delle Scienze, 90128, Palermo
[simcola, corrao]@dpce.ing.unipa.it; devecchi@unipa.it

Abstract

The paper reports on the first results of a research carried out at the Department of Project and Civil Engineering (D.P.C.E.) of the Faculty of Building Engineering in Palermo.
The aim of the research is to create a system able to allow different kind of users to have easily access to the information related to the reinforced concrete decay pathologies and to the actions that is possible to carry out for their repairs.
Nowadays the recovery of technical elements made by reinforced concrete is a topical subject because of the widespread of survey even if it is not much investigated and popularized. The information on this subject are not much systematized and computerized: for this reason is not always possible to establish, for each technical elements, a connection between the decay pathologies and the recovery actions.
It is easy to understand that it is important to have a tool that can guide the users to retrieve these information and to compare them each other with the aim to select the best action for the recovery of a specific technical element affected by a particular decay pathology.

1 Introduction

Starting from the technical elements made of reinforced concrete (Foundation structures, Pillars, Beams, Floors, Balconies and projecting parts, Stairs, Protection elements, Ornamental elements, Retaining wall) the information related to the causes of decay have been selected and put in connection with the phenomena and the recovery actions that is possible to adopt with the aim to repair the buildings built with this material. Moreover the system allow users to have access to a survey that shows both different kind of actions, that are adopted to resolve real situations of decay, and the specifications of products on the market that are used in those particular situations.

2 Design Strategies and User interface

The system pages have been created by using Microsoft FrontPage with the aim to carry out an hypertext that is possible to place even on the Net.
Each page consists of three parts in which there are (starting from the left):

- the buttons that are related to the technical elements. These buttons are always active on the pages of the system.
- the list of the decay pathologies. The items of the list are active with regard to the technical element that has been selected by the user.

- the information field, in which text and images are displayed. The contents of this field depends on the user requests; the information are displayed on a light color field when a subject are completely described. When the information are related to an investigation that the user have activated by clicking on a key word the field are colored with a blue color. The user can review the information displayed by the system by using the navigation arrows that allow the sequential access to the information. When the display of the information needs of more fields the user can select them by using the numbers linked to each field. Navigation arrows and numbers are placed in the information field.

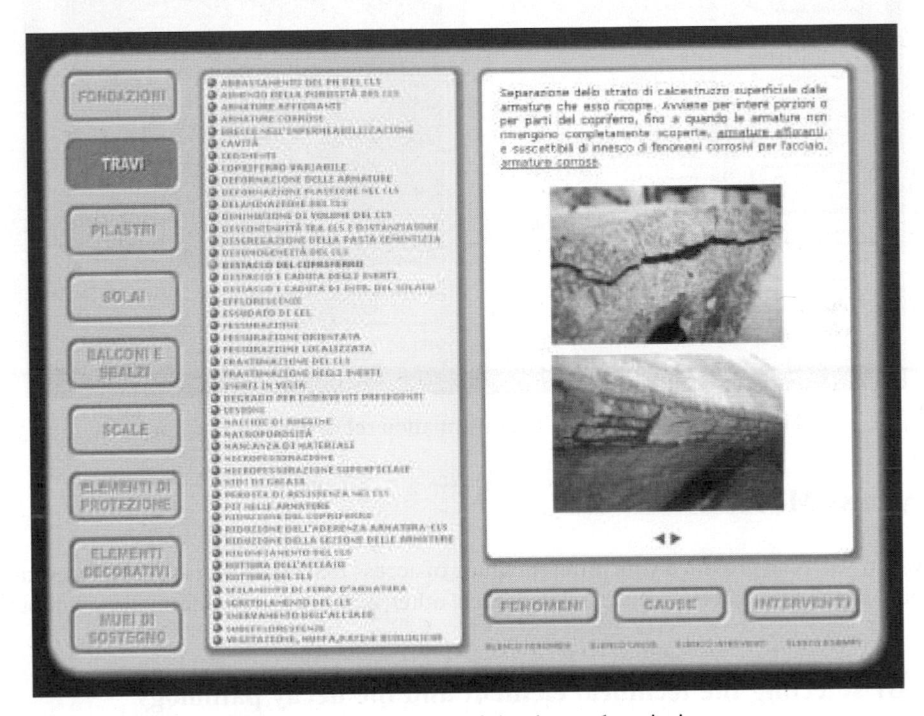

Figure 1: The display of the decay description

The system pages are equipped with specific research tools (Phenomena, Causes, Action buttons)[1] that can be used to investigate particular matters and to put in relation different information of the system. Moreover there are four different links related to the lists of Phenomena, Cause, Actions and Real cases[2] that the user can select with the aim to have directly access to specific information.
The display of the information in the system has been managed without opening different pages, actually the first page displayed by the system as a result of the user requests (by the selection of the technical element and the decay pathology) are simply modified through the changes of information field attributes.

---

[1] In the figures: Fenomeni, Cause, Interventi.
[2] In the figures: Elenco Fenomeni, Elenco Cause, Elenco Interventi, Elenco Esempi.

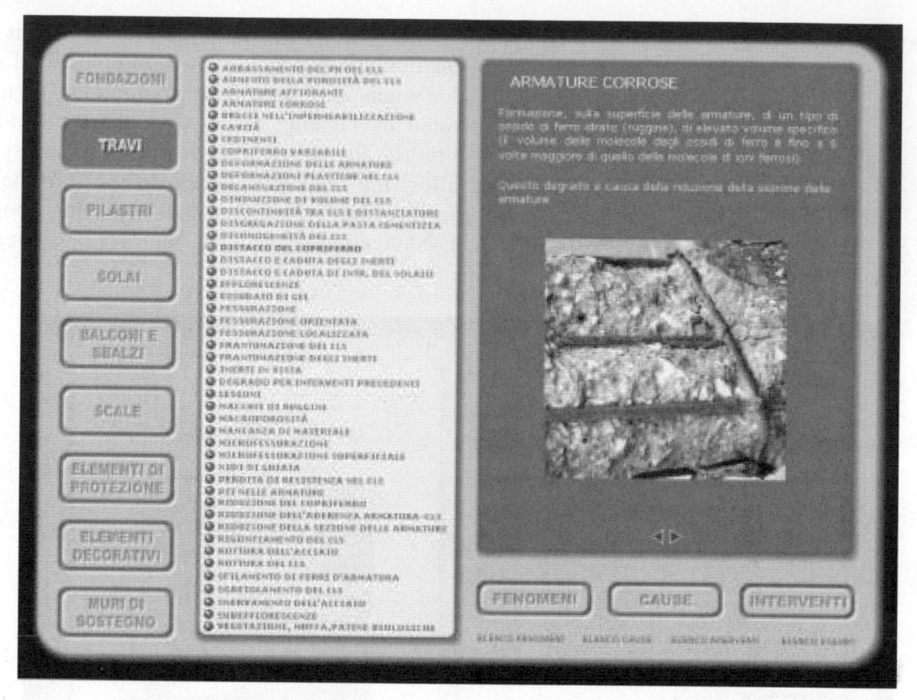

Figure 2: The display of the information related to the key words

3 Access Modes

The system has been provided with different kinds of access that allow users to research structured information about subjects that are related each other or to research information without any connection.

3.1 By selecting the technical element and the decay pathology

The user can have access to the system through the home page that displays the buttons related to the technical elements.
By clicking on the button related to the technical element the system displays a list of all the decay pathologies. Only the items related to the specific element are active and the user can click on them according to his specific needs. In this way is possible to have access to the information that describe the decay. The system displays this information in the field placed on the right of the page. The button related to the technical element and the item of the list related to the decay pathologies remain lighted on the page in the way to stress the user research path.
This kind of operation has been adopted with the aim to help user in the spotting of the decay pathologies related to the specific technical element, actually the other pathologies are always visible even if they can't be clicked.
The description of the decay pathologies is composed by text and images. The text can have key words linked to other information that the system displays in the same information field.
This kind of access allow user to put in relation the information about the decay pathologies with the related phenomena and the actions that is possible to carry out for the recovery of the specific technical element by using the Phenomena and the Actions buttons.

3.1.1 *The Phenomena button*

By clicking the Phenomena button the system displays a list of Phenomena related to the decay that the user has been analyzing. Each items of the list allow user to have access to the related description that the system displays on the information field in which the list is displayed. When the user navigates in the section related to the Phenomena the system makes active another research tool related to the Causes button. By using this tool is possible to have access to the causes that produce the decay. For each cause the system can display the description.

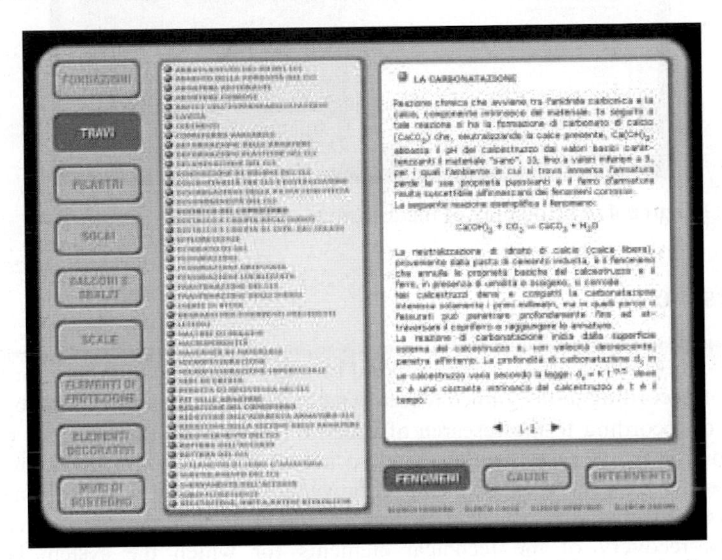

Figure 3: The display of the information related to the Phenomena button

3.1.2 *The Actions button*

By clicking the Actions button the user can have access to a list of actions that is possible to carry out for the recovery of the technical element affected from the decay.
By clicking on each item is possible to have access to the description of the actions and the different phases that are required for its realization.
The information are displayed in the same part of the page in which there is the information field. In this case the system allow user to study in deep the methodologies, the techniques, the real cases and the products by using the key words.
The information related to the products, in particular, are linked to the web sites of the firm of suppliers.

3.2 By selecting the Lists

During the navigation in the system the user can have access to the lists related to the Phenomena, the Causes, the Actions and the Real Cases, arranged on the base of specific classification. From each item in the list is possible to activate a field in which the information (text and images) are displayed.

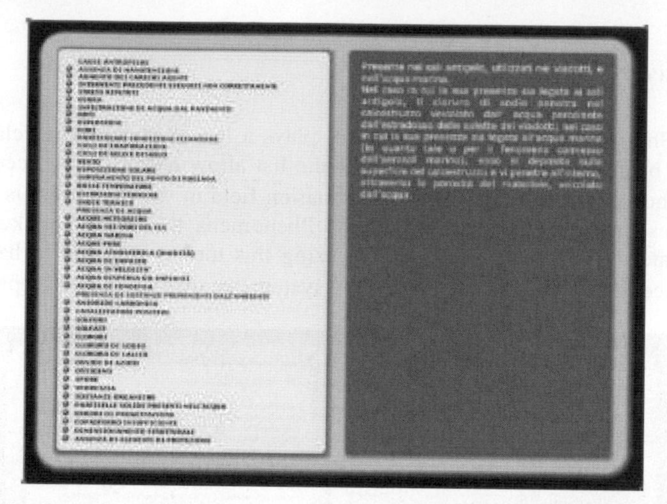

Figure 4: The display of the information related to thematic lists

4 Conclusions

The system described above allows different kind of users to have access to information that are structured or not, according to the guided research path, that the user can activate by using specific research tools, or according to the research of specific information that the user can activate by using the thematic Lists. These different kind of access allow users to perform study activity related to the reinforced concrete decay both at university level and at professional level. In the last case the system allows expert users to select and study in deep the actions that is possible to perform for the recovery of the technical elements for which the system is able to give information. Moreover, the expert users can have access to the specifications of products by using the specific research tools or by activating the Internet connection and by connecting to the suppliers web sites.

At present, the system has been structured with the aim to use it off line, even if the software used to its development allow to create web pages; future developments of the system does not rule out its placement on the Net.

5 References

Corrao, R., Fulantelli, G. (1998). Cognitive accessibility to information on the Web: insights from a system for teaching and learning Architecture through the Net. In AA VV, *Towards an Accesible Web, Proceedings of the 4th ERCIM Workshop "User Interfaces for All",* Långholmen-Stockolm. Web reference: http://ui4all.ics.forth.gr/UI4ALL-98/corrao.pdf

Corrao, R., De Vecchi, A., Colajanni, S. (2001), Easily Access to Technical Information of the Old Construction Handbooks by a Wbi System. M.I.C.R.A. - Manuale Informatizzato per la Codifica della Regola d'Arte. In Stephanidis, C. (ed.), *Universal Access in Human-Computer Interaction (HCI) - Towards an Information Society for All,* Vol. 3, (pp. 793-797), New Orleans, Louisiana, Mahwah, New Jersey, USA, LEA -Lawrence Erlbaum Associates, Inc., Publishers, ISBN 0 8058 3609 8.

Young urban knowledge workers – Relationship between ICT and psychosocial life environment

Ulrika Danielsson

Department of Information Technology and Media
Mid Sweden University, 831 25 Östersund, Sweden
ulrika.danielsson@mh.se

Abstract

The overall purpose of the research project "Home of the Future" concerns changes in the future living and working conditions with emphasis on the impact of Information and Communication Technology (ICT). Four pilot studies have been performed, one in USA and one in South East Asia and two in Sweden. This paper presents the fourth study and the objective of the study is to find a relationship between ICT and the life roles of the people in the occupational group "young urban knowledge workers in the IT-business" in a big city. The study was performed during the autumn of 2000 and the spring of 2001, and a follow up study was carried out in the autumn of 2002. Ten people were interviewed, five women and five men. The study comprises the professional role, the private role and the citizen role. The study regards the life environment, life role, work environment and psychosocial effects. The results from the interviews show that the private role and the professional role merge together into one life role and that the boundaries between the home and the workplace become increasingly hard to define.

1 Introduction

Since the home is considered both a physical and a virtual space its value and importance has been the focus of many discussions about how man will live and learn in the future. Today we don't only talk about working from home, but also about the possibility of carrying out private tasks and errands from your workplace, "Homing from work". In recent years telecompanies have presented us with different technical solutions that allows people to work from other places than the premisis of the company and in this way become more mobile. Employees, mainly within the IT-sector, can for example choose to work from home at any time of the day and decide for themselves when to have their spare time allocated. This way of working means that there are new demands on the employee to be able to handle the boundary between work - spare time and between private role – professional role. To obtain an in depth understanding of the change of peoples' life roles in the Net Era, analyses of human and organizational behavior related to the professional role, private role and citizen's role are performed.

The overall purpose of the research project "Home of the Future" (www.mh.se/fscn) concerns the changes in the future living and working conditions, with emphasis on the impact of Information and Communication Technology (ICT). Four pilot studies have been performed, one in USA (Bradley, Andersson & Bradley, 2000) and one in South East Asia (Bradley & Andersson, 2001). The third pilot study adopted a rural perspective regarding the behaviors, attitudes and values associated with the interaction between humans ICT in the IT-industry (Nörler & Samuelsson, 2001). The fourth study, which is presented in this paper, was conducted in a city environment in Stockholm. A detailed report of the results is available, published in Swedish (Danielsson, 2002).

1.1 Purpose

The objective of the fourth study is to find a relationship between ICT and the life roles of people in the occupational group "young urban knowledge workers in the IT-business " in a big city. The study comprises the professional role, the private role and the citizen role. The study also concerns people's life environment and health.

1.2 Problems

The problem area of this study concerns the following three roles, which are reflected in the structure of the question instrument:

- Professional role (work content and different aspects of psycho-social work environment, physical and virtual meeting places etc.).
- Private role (the home as work environment and private environment, services on the net etc.)
- Relation between work and spare time (establishing boundaries between working time and spare time)

Examples of questions tied to the roles:

The Professional role

- What problems can arise in the interplay between psychosocial and organizational factors and ICT that are relevant for work in and from the home?
- How do the IT-people of today define the terms "home" and "workplace"?
- Are there any physical boundaries?

The Private role

- How do the IT-people of today define their spare time?
- Are there any boundaries between work time and spare time?

The Citizen role

- To what extent do the IT-people of today take an active interest in societal issues, e.g. political issues, and are active within voluntary, non-profit organizations?

1.3 Method

The interviews were carried out throughout October 2000 to January 2001. At the implementation of this study semi-structured interviews were used and some questions had set multiple answering alternatives. The group of interviewees was defined as "young urban knowledge workers in the new IT business" in the environment of a big city, Stockholm the capital of Sweden.
Criteria for determination of the study group:

- The people should work at an IT-firm, as an employee or / and co-owner.
- Single or co-habiting (they have moved away from their parents, but not yet started a family, i.e. had children).

- Educational background with IT-related academic education (graduated alt. not completed)
- 50% men and 50% women.
- The people should be interested in thinking and expressing themselves within the framework of the presented problem of this study.

2 Interaction between ICT and psychosocial life environment

2.1 Objective and subjective work environment

When preparing for this study I acquainted myself with the theoretical models of Prof. Gunilla Bradley, which show relations between computer technique and psycho-social work environment, where also the interplay between objective – subjective work environment is described (Bradley, 1986). Since work environment has both a physical and psychological side, it is of great importance to assume a broad perspective of the work environment problems; in this way one obtains an overall picture of the connections between society, company and the individual. There is an abundance of factors behind a person's stance and attitude towards their work environment (Bradley, Bergström & Sundberg, 1979, 1984). These factors might for example be a difference in background, gender, age, needs, wishes, experience, and expectations. In this study I have, with a concrete selection, focused on " young urban knowledge workers in the IT-business ". For an overall picture of their total environment I have included both the home environment and work environment of the interviewees. These environments converge to form a "life environment".

2.2 The convergence model

In an attempt to understand the interplay between ICT, working life and private life, different roles and effects on the individual, I make use of Bradley's model of the interaction between ICT and the psychosocial life environment (Bradley, 2001). The model is a development of an earlier model, which shows the connection between computer technology and work environment (Bradley 1986). On the basis of the convergence model I have analyzed "Young IT-people's" different roles so as to try to understand how they converge into one life role. From this ground an analysis is made of what factors influence these individuals in their usage of ICT in their life environment.

3 Results

3.1 The professional role

Strong social networks - Workmates and friends often turned out to be the same people. This meant that the interviewees considered themselves to have a strong social network with a strong confidence in their friends and thus also in their workmates. When recruiting it was not necessary to check previous references since they often employed friends from their social network. The solidarity among the workmates was very good and as a result of this the interviewees felt very safe in their professional role.

Flexibility concerning spare time - The interviewees did not consider working more than 40 hours per week as a problem, irrespective of time and place. The responsibility of their professional role gave them freedom and they allowed working time to dominate over spare-time rather than the opposite. The problem with leisure time appeared when they planed for their vacation, since it was hard to predetermine the workload at the planned point of time. Since they had chosen their jobs because of their interest in it and because it was both fun and interesting, it became hard for them to define a boundary between work-time and spare-time.

Expert in the area (necessary knowledge) - The professional role was described as a role you grew into and you were responsible for your own professional development yourself. The necessary knowledge and qualities which they desired in a new employee were an ability to cope with stress and to manage yourself, engagement in the task, a certain technical knowledge and an ability to work well in a group.

Own responsibility - The interviewees considered it to be up to each individual to be responsible for the development of both their professional and private role. Also, the employees must themselves be able take responsibility for the allocation of time and place of work, they thought.

3.2 The private role (and citizen role)

The home - All except for one of the interviewees lived in central Stockholm. The living situation mainly consisted of one or two-room flats with a kitchen or kitchenette. Most of them had their meals outside their home. The fact that they chose to meet their friends outside the home was partly due to the cramped living space and partly since restaurants and entertainment often are present in a big city. All of the interviewees found it difficult to define a line between home and workplace.

Social relations - The most central thing in life for the interviewees were their social relations. The continuous usage of mobile phones meant that they were always connected to work.

The citizen role - None of the interviewees were politically engaged or involved in neighborhood issues.

Psychological effects - Both men and women experienced positive and negative stress, depending on interest in and the time allocated to their work. But it was mainly stress, fatigue and anxiousness they related to a poor work environment. Many of the interviewees experienced worries about their work and had problems sleeping and they had to engage their body in some kind of physical activity that demanded deep concentration to experience complete relaxation from work.

4 Conclusion

In this paper I have presented results showing how the professional and private role merge together into one life role. The results from the report show how the boundaries between the home and the workplace become increasingly blurred and hard to define. The study with a rural perspective also indicated that the private role and the professional role are converging.

How will the continuing interplay between people and ICT develop? As the professional role and the private role converge, how will you be able to distinguish a dividing line? Are boundaries necessary? In that case, how should they be drawn up? Could this new lack of boundaries affect our health? What happens to the health of those who cannot keep these places apart and continuously carry out private tasks and work tasks independent of time and place? Independence of time and place and a shift between our roles seem to signify our time. Finally I want to state that these results show a greater need for studies within this area. A follow-up study has been carried out during the autumn of 2002 and will be presented in 2003.

References

Bradley, G., Bergström, C. & Sundberg, L (1979). *Yrkesroller – Livsmiljö. Psykosociala aspekter på tjänstemännens arbetsvillkor och arbetsmiljö*. Wahlström & Widstrand.

Bradley, G., Bergström, C. & Sundberg, L (1984) *Sekreterarrollen och ord- och textbehandling – kontorsyrke i förändring*. Stockholm: Arbetsskyddsfonden.

Bradley, G. (1986). *Psykosocial arbetsmiljö och datorer*. Stockholm. Akademilitteratur.

Bradley, G. (2001). *Humans on the Net. Information and Communication Technology (ICT) Work Organization and Human Beings*. Stockholm: Prevent.

Bradley, L., Andersson, N. & Bradley, G. (2000). *Home of the Future – Information and Communication Technology (ICT) –changes in society and human behavior patterns in the net era*. FSCN-report R00-1. http://www.mh.se/fscn

Bradley L. & Andersson, N. (2001). *Singapore and Malaysia going smart with broaband*. FSCN-report R-01-12. http://www.mh.se/fscn

Danielsson, U. (2002). *Young urban knowledge workers – in the IT-business*. FSCN- report R-02-34. http://www.mh.se/fscn.

Nörler, R. & Samuelsson, J. (2001). *Samspelet mellan IT och människa* – En studie i Jämtland. M.Sc. thesis in Informatics at Mid Sweden University.

Combining Virtual Reality with an Easy to Use and Learn Interface in a tool for Planning and Simulating Interventions in Radiologically Controlled Areas

Angélica de Antonio, Xavier Ferré, Jaime Ramírez

Facultad de Informática, Universidad Politécnica de Madrid
28660 Boadilla del Monte, Madrid, Spain
{angelica, xavier, jramirez}@fi.upm.es

Abstract

This paper describes the mechanisms that have been devised to facilitate the use of a tool called HeSPI. This tool has been developed within a project funded by the European Commission and is intended to be used by maintenance operators of a nuclear power plant in order to plan and simulate a future intervention. The tool provides a 3D model of the scenario for the intervention and a 3D human mannequin that represents operators. The combined use of a graphical user interface and a voice recognition package, together with other design mechanisms such as predefined actions and semantic information, have been evaluated with real users and the results are also discussed.

1 Introduction

This paper presents some of the results obtained in the VRIMOR project, funded by the European Commission under the 5th Framework Programme, running for two years, having commenced in February 2001. The consortium was led by NNC Ltd (UK) and included Tecnatom (Spain), Z+F (UK), CIEMAT (Spain), SCK-CEN (Belgium), and Universidad Politécnica de Madrid (UPM) (Spain). The general aim of the project was to integrate environmental scanning technologies with human modeling and radiological dose estimation tools and to deliver an intuitive and cost-effective system for use by operators involved with human interventions in radiologically controlled areas (Lee et at., 2001). The usability level of the resulting products was one the main success criteria of the project.

As a part of this project, the UPM team has developed a planning and simulation tool, called HeSPI, based on a combination of a graphical user interface and a voice interface. The goal of HeSPI in the VRIMOR project is to help the personnel of a nuclear power plant (NPP) in the design of a maintenance intervention.

The expected users of the tool are the specialist in maintenance in the NPP and, secondarily, the specialist in radiological protection. The design of an intervention imp lies making decisions about how many operators will participate in the intervention, the actions to be performed by all of them, the paths to be followed by each of them through the controlled environment, the time required for each action and the necessary interaction among operators.

HeSPI takes as input the geometrical model of the environment where the intervention is going to take place. This geometrical model, in VRIMOR, has been produced with the laser scanner provided by Z+F but, in general, any model in the VRML language can be imported into HeSPI.

The tool provides the designer of an intervention with a humanoid 3D model or mannequins that can be loaded into the desired environment and will be used by the designer as if he or she was manipulating a puppet.
Once the user has loaded a geometrical model of the environment in VRML format, he or she must design the intervention by specifying the individual subtasks to be executed by each virtual operator. In order to do this, the user should be able to make the mannequin move around the environment and perform different kinds of actions, adopting varied postures, interacting with the objects in the environment and manipulating tools and equipment.
The 3D humanoid model that is used by the HeSPI tool is called Jack, and is commercially distributed by EDS. Jack is a very complex and powerful human mannequin, with many degrees of freedom, inverse kinematics, and ergonomic constraints. However, it requires a long training for users to be able to design simulations by Jack's environment. The main challenge of HeSPI was to encapsulate Jack into an easy to use and learn user interface, taking into account that the intended users are not expected to have any experience in the usage of three-dimensional applications, and they are not willing to spend a long time learning to use tools such as Jack. It is important to notice that the goal of HeSPI is not to provide NPP personnel with a complete human simulation tool. Many of the functionalities provided by Jack are in fact hidden to the user of HeSPI. We have completely substituted the user interface provided by Jack with a new user interface that only gives the user access to the functionalities that are required for their needs.
Another aim of the HeSPI tool was to minimize the amount of extra hardware equipment to be used in order to increase the feasibility of adoption by NPPs. Our hypothesis was that a combination of a graphical user interface (GUI) and a voice recognition system would offer good enough interaction possibilities for the success in the objectives of the project, without need of any specialised input/output device (such as a virtual reality helmet or a joystick).
In the design of an intervention it is also important to minimize exposure to radiation, both to individual operators and collectively. This goal is achieved by exporting from HeSPI the trajectories that have been followed by each of the virtual operators in the simulation of the designed operation. These trajectories can then be imported into the VISIPLAN tool, provided by SCK·CEN, that allows the analysis of the designed intervention from the radiological point of view. Test trajectories have been successfully exported to and analyzed by the VISIPLAN tool.

2 Design Approach

The design of the HeSPI tool was strongly oriented towards the requirements and characteristics of the final user of this tool. Our previous experiences in the development of systems based on the use of virtual environments for this kind of users (we had already developed some training applications for NPP personnel using 3D environments) indicated that NPP personnel are not used to interact with computer programs in general, and most of them have never used a 3D interface before.
Therefore, usability became our main design criterion and success measure. We have followed a usability-oriented development process in which a usability expert was deeply involved and supervised the technical developments at each stage of the process. Continuous feedback from real users would have been desirable but the availability of NPP personnel in this project was limited to the final evaluation session. In the VRIMOR project the Spanish NPP "Almaraz" helped to define user requirements and evaluate the results.
Simplification of the user interaction with the mannequin is mainly achieved in HeSPI through three mechanisms:

- A voice recognition package

- A pre-defined set of generic and configurable mannequin actions from which the user can select the desired ones.
- A mechanism to facilitate the interaction of the virtual operator with the objects in the scenario.

2.1 Voice recognition

In the design of HeSPI we considered that issuing voice commands was a natural way of composing a new intervention (Kamm and Helander, 1997). We though of the scenario as if it was a kind of theatre stage and the mannequins were virtual actors. In this setting, the maintenance specialist would act as the director, indicating each actor what to do at every moment. Voice then arises as very intuitive way of interacting with those virtual actors.
To enforce this idea, we decided to restrict the use of voice to the composition of the operation. The rest of the user interface is a conventional GUI based on windows, buttons, etc. We also decided that the voice recognition could be activated or de-activated at wish by the operators. We have assumed in our design that the user who activates the voice recognition seeks for simplicity, and is willing to sacrifice some of the flexibility provided by the tool. Therefore, when using voice recognition, some of the parameters available for the configuration of the subtasks are set automatically to the default value. The result is that the use of the tool is simpler and the design of the operation is quicker, but the precision and aesthetics in the execution of some subtasks by the human mannequin can be lower. Our hypothesis was that for most of the applications of this tool, the required precision would not be very high. The results of the evaluation seem to confirm that our intuition was correct. The Almaraz operators clearly preferred simplicity over precision.

2.2 Predefined generic and configurable actions

As we are aware of the big complexity inherent to the interaction with 3D humanoid models, we have devised a way to reduce this comp lexity by "reducing the degrees of freedom". The idea is to provide the user with a set of predefined actions, such as *walk*, *climb a ladder*, *pick up an object*, etc. When the designer selects an action, he will only have to set values for some parameters. For instance, to make the mannequin walk, the user will have to select the destination point; to make the mannequin rotate, the user will have to specify the rotation angle; to make the mannequin pick up an object, the user will have to indicate the object to be picked up. The choice of this mechanism was based on the hypothesis that most of the actions that maintenance operators have to perform in different scenarios are very similar. Therefore, by incorporating into the tool an extensive set of these actions, the design of a new intervention would consist mainly on the selection, configuration and sequencing of the appropriate predefined actions. Only a small number of actions will have to be built from scratch. Our internal evaluation has shown that considerable reductions in modelling time can be achieved through this mechanism.
The particular subtasks involved in different interventions in a NPP have been studied and we have abstracted from them the set of general actions that are currently predefined in HeSPI.
Building actions from scratch is also possible in the HeSPI tool by using the facilities provided for the definition of new postures. The capability of Jack to automatically animate the mannequin from one posture to another one in a realistic way is then used for the construction of a new action as the transition from one posture to another one.
The definition of new postures in HeSPI is one of its most complex functionalities, and it is performed through a set of dialog windows provided by Jack. We have experimented with the use of voice recognition for the implementation of this functionality but the results were not satisfactory, so we decided to maintain Jack's interface, which is usable enough for our purpose.

In this way, new predefined actions can be added to the HeSPI tool. Therefore, if the tool is applied on a regular basis for the planning of interventions and the new required actions are added to the tool, the percentage of actions to be designed from scratch will be decreasing over time and the user efficiency in the use of the tool will be increasing. This advantage of the tool has not been directly assessed on the evaluation session in Almaraz, due to limitations on the availability of the final users; but we believe that it will be very much appreciated by a regular user of the tool.

2.3 Interaction with virtual objects

One of the most complex kind of actions to be modelled in an intervention is the interaction with the objects in the environment. We wanted to provide the user with mechanisms that were, at the same time, generic to be applicable to any possible environment, but simple to use. We have devised a pre-processing step for new environments in which the important objects are associated with semantic information, and some invisible geometrical figures (what we call "the cubes") are attached to these objects to facilitate manipulation. The attachment of the cubes to the objects of the environment is performed automatically by the tool during the pre-processing step, once the user tells the system that the mannequin could interact with a given object, and provides a name for this object (the name is necessary to allow for the identification of the objects when using voice recognition. When using the GUI the objects can be clicked on with the mouse). Ten small invisible cubes will be attached to the object. Six of these cubes will act as handles, and our mannequin will have the built-in capability to grasp an object by any of these cubes. The remaining four cubes will act as position markers, and our mannequin will have the built-in capability to walk towards the object and stop in front of any of these cubes.

This mechanism has its roots in our search for flexibility combined with easiness of use. It allows the user to issue commands of the type "walk to the box" or "grasp the box", without having to go into a low level and detailed design of how the animation of the mannequin should be, but nevertheless obtaining a quite precise animation.

We have designed a functionality of the mannequin that allows it to determine which is the cube that is closest to its current position. If the operator, for example, issues the command "grasp the box", the mannequin will automatically grasp the object with one hand using the closest handle cube. Using this option, the operator does not have to deal with cubes at all, unless the resulting animation is not totally satisfactory for him. In this case he or she can make the cubes visible and select the cube to be used by himself to make the animation more precise. It is also possible to change the position of any cube for a better accuracy. The detailed and complex option is always available, but most of the times the more general and simpler option will be preferable.

3 Evaluation

The evaluation of the HeSPI tool has adopted two forms:

- A technical evaluation of the correctness, reliability, efficiency and robustness quality factors of the tool has been performed by the UPM's staff prior to the user evaluation.
- A user evaluation at Almaraz mainly centred on the usability quality factor. For this purpose we run usability tests with four participants (2 maintenance specialists, 1 radioprotection specialist and 1 computer technician) and we passed questionnaires to them and to other potential users that observed the testing session. For the design of the questionnaires we mainly based on (Schneiderman, 1998).

The results of the evaluation of HeSPI at Almaraz have been quite satisfactory. We can conclude that the required functionality has been successfully incorporated into the tool and that real users are able to plan and design quite complex interventions easily and quickly.

Most of the comments of the operators were quite positive, showing their interest on the methodology and recognizing its usefulness. The NPP personnel stated that if they had the set of technologies provided by VRIMOR they would use them, without any doubt, because they consider them useful. This is the most important conclusion that can be drawn from the evaluation session regarding the applicability of these tools.

The most remarkable drawback noted by the operators was the perceived complexity of the applications. This, in a way, confirms the validity of our motivation to build simplified interfaces to allow the use of three-dimensional human simulation tools by NPP operators. These simulation tools (such as Jack) are too complex to be used and require a high level of training and skills. We have invested our efforts in the design of simpler interfaces, and we can say that, although the complexity perceived by the operators is still high, they in fact were able to design not-so-simple operations in only a few hours of training and use.

The evaluation at Almaraz demonstrated that our main hypotheses were correct. The evaluators (or test participants, from the point of view of usability testing) felt comfortable with the completeness and precision of the set of predefined actions provided by HeSPI. Regarding the use of a combination of a graphical user interface (GUI) and voice recognition, the answer to the question "The possibility to use voice commands makes the system easier to use", was "Quite agree", and several answers to the open questions in the questionnaire mentioned that the use of the voice interface made the use of the tool simpler and quicker.

For a more detailed description of the user evaluation and its results see ('Technological and User', 2003).

4 Applicability

There are two main areas of application of the human simulation tools developed in VRIMOR (specifically HeSPI) within the context of maintenance interventions in a NPP: planning of the intervention and training of the operators that will perform the intervention. Since the Almaraz evaluation centred on the planning of the intervention, the use of HeSPI for training has not been evaluated with real users yet, although the evaluators commented on the clear usefulness of these tools for this task. The amount of details required in the animations to be useful for training is still to be determined. Therefore, the validity of our hypotheses and design mechanisms, for this purpose, is still to be assessed. A balance among the effort required on the part of the designer and the quality and precision of the resulting simulation has to be found. Training is one of the application areas that UPM plans to pursue for the future exploitation of HeSPI, and we intend to study how this effort-precision trade-off can be managed and incorporated into our design.

References

Lee, D., Salve, R., de Antonio, A., Herrero, P., Pérez, J.M., Vermeersch, F., & Dalton, G. (2001). Virtual Reality for Inspection, Maintenance, Operation and Repair of Nuclear Power Plant, FISA 2001, EU Research in Reactor Safety mid-term symposium on shared-cost and concerted actions, Luxembourg, November 2001

Kamm, C., & Helander, M. (1997), Design Issues for Interfaces using Voice Input. In M. Helander et al. (Eds.), *Handbook of Human-Computer Interaction,*(2nd edition). Elsevier Science, Amsterdam (The Netherlands).

Shneiderman, B. (1998). Designing the User Interface: Strategies for Effective Human-Computer Interaction. Addison-Wesley, Reading (MA).

Technological and User Perspective Review Report (2003). Vrimor Project Deliverable D5

Empowering the User in Product Design with Virtual Reality

Oya Demirbilek & Aybuke Aurum***

*University of New South Wales
UNSW, Faculty of the Built Environment, Sydney, Australia
o.demirbilek@unsw.edu.au

**University of New South Wales
UNSW, School of Information Systems, Technology and Management, Sydney, Australia
Aybuke@unsw.edu.au

Abstract

Positive empowerment of users through their simultaneous involvement in decision-making whilst a product is being created is a novel approach that has promising potentials. This involves the use of collaborative design combined with emerging interactive virtual environment technologies. The present paper illustrates a process model that addresses the development of an end-user involvement model. The objective of this paper is to describe a study that investigates the involvement of end-users in product design in order to improve the overall effectiveness of the process of designing, and aims to develop tools that allow collaborative interactive 3D designing capabilities for all stakeholders (i.e. end-users, professional designers, and product retail companies). An overview of the theoretical framework of the study will be presented in the hope to widen the debate as to the merit of such research and possible means to improve the research methodology.

1 Introduction

Industrial design (or product design) refers to the process of designing objects for human use, including the designing of products, packaging, furniture, appliances, transportation, clothing, and any other imaginable things used in daily life (Norman, 1988). The primary goal of industrial design is to produce products that satisfy the needs of the users; other important goals are ease of manufacturing, and acceptable aesthetics. Present industrial design practice typically includes users in the early stages to identify requirements, and then again, to evaluate prototypes of designs. Users are not involved in the actual design process itself. This lack of user involvement can lead to designs that do not quite satisfy the users' intended requirements (IDFORUM, 2001). This paper investigates a model that will enhance users' involvement in the design process in order to improve the overall effectiveness of industrial design. The aim is to develop a virtual computer-based system that enables users to customize and design products at an acceptable level, using three dimensional graphical interaction and interfaces.

The research aims include understanding and subsequently providing techniques, design solutions and tools to support the virtual collaborative activity of designing in a virtual interactive 3D environment for end-users, professional designers, and product retail companies. This study will help ensure such software environment and tools address actual real user needs, and that the design & development process are being appropriately guided by these needs. It will also lead to an improved relationship between end-users/designers/manufactures/retail companies (Aurum, & Martin, 1999; Booth, et al., 2001). The main aim of this study is to allow end-users to participate actively from their homes or their work environment, either to design their own customized objects and order them, or to be involved in bigger design projects, such as public design projects, where they will contribute their daily life experience and their own creativity.

2 Background

In industrial design, a product that has been designed according to the needs of users is not guaranteed to end up corresponding to their real requirements. It all depends on 'how' people are represented in the design process. In addition to this, due to a lack of connection between design practice and design research, a substantial number of designers, in the UK, are making use of their colleagues and themselves to 'represent' end-users instead of using more representative sample spaces of users (Hasdogan, 1996). The premise is that, involving users in the design process assists in reaching better design solutions to existing or hidden design problems, and direct useful feedback. Furthermore, Balu, (2000) argues that users can create knowledge, if they are allowed to do so. The best product designs have been reported to occur when designers are involved in collecting and interpreting customer data (Holtzblatt, & Beyer, 1993). User-centered techniques involve either having the designer participating in the user's world, or the user participating in design activities. Both approaches are useful, as long as the user can be as effective as possible in both roles. The designer plays a key role in providing any material that might help and facilitate this transfer of experience. Traditional market research methods, through focus groups, interviews, and questionnaires, have been focusing more on what people say and think while being efficient and cost effective (Sanders, 1999). Information on what people do can be collected in real life context places where people live or work. This approach combines psychology and design, and focuses on what end-users make or can create, in an attempt to help designers to learn from end-users memories, current experiences and ideal experiences (Sanders, 2000).

The many advantages of end-users' direct involvement into the process of industrial design can be summarized as follows: users notice things that tool researchers don't (Pancake, 1997); users help "sell" the research. Users do get involved in the sense of making a real commitment to a research project. This is called "the spontaneous supporter phenomenon", developing spontaneously as users can follow the results of their ideas and criticism involved in later iterations of the design (Pancake, 1997). Furthermore, users can bring direct real feedback; they are experts related to their daily life and the objects around them; they bring in important personal daily life knowledge; they are excellent at reacting to suggested designs; and saying what is wrong with it. Ultimately, users can generate more ideas and points of views to be explored; the knowledge collected is otherwise inaccessible to designers and other stakeholders (Demirbilek, 1999).

Considering the previous advantages, we believe that the development of an interactive 3D collaborative design software based on an expert system, specialized towards letting end-users achieve professional results will be of great use in providing a direct and ongoing feedback on users' preferences, needs, and trends to companies (design consultancies or producing companies). The stakeholders involved are as follows: the end-user – designers (product designer & software designer) –engineers (industrial & software) – manufacturers – and marketing people. The active participation of real people in the design process involves collaboration between these people and professional designers (Burdman, 1999). Furthermore, a de-specialization of many professions is foreseen (Jones, 1999). The industrial design profession has come a long way and is facing serious changes with the rapid advancement of technology and changing user demands. According to Sanders, the role of designers will change, allowing them to become more involved in creating the tools for the end-users to express their real needs and dreams. Psychology will be an important attribute and designers will be translators and interpreters of visual expressions created by end-users. Sanders (1999) adds that designers will use this translated and interpreted information as a source of inspiration in design.

3 Significance

Over the last several years there has been a huge growth in 3D CAD software oriented towards use by professional designers. With the onset of this growth, many product retail companies have paralleled this trend by investing in the creation of interactive Web interfaces, allowing their customer to customize their products, in a limited way. Virtual Reality is also an emergent technology, which has the capability of impacting the product design process. It is a high-end user interface that involves real-time simulation and interactions consisting of the usual auditory and tactile features. Thus, the user (or designer) feels submerged as if they were actually in the three dimensional space (Gatarski, & Pontecorvo, 1999; Kerttula, & Tokkonen, 2001).

Despite this rapid growth, there is often very little understanding of how this technology can be most effectively utilized to suit larger population samples and how it impacts on consumer behaviour and end-user feedback (Dahan, & Srinivasan, 1998). Commercial examples of such attempts can be found on web pages where end-users can customize, their "own" shoes online, for example, using a palette of different basic shoe types, different colors, textures and materials (Nike.com). Non-commercial sites demonstrate user customization with interactive small games (such as sodaplay.com, tcm.org). The development of these web sites is closely related to marketing strategies, allowing end-users to purchase their 'own' designs. All of these examples provide an indirect interactivity that does not allow for the benefits of real users inputs.

3.1 The Proposed Model: Virtual Collaborative Design (VCD) Model

The virtual collaborative design (VCD) model (see Figure 1) between end-user and designer has been adapted from a previous study (Demirbilek, 1999). This design model links end-user requirements and needs with various designing activities. In the VCD model, the contribution of end-users and designers to the various stages of the industrial product design process is explained schematically. There are 8 main stages and the intensity of the contribution is shown with thin and thicker lines (thicker lines denoting substantial contribution). The contribution discussed here is happening through online virtual collaboration. The 8 stages of the VCD Model are as follows:

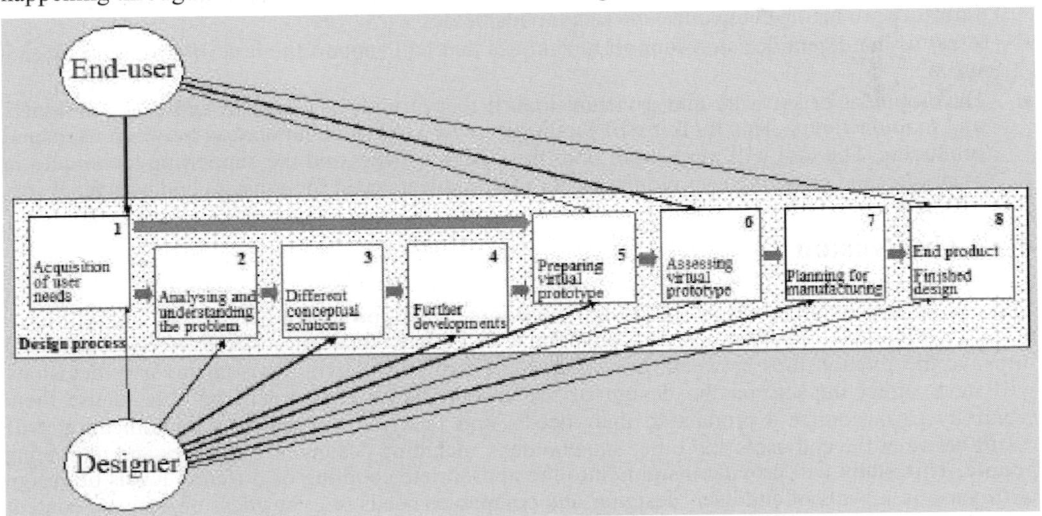

Figure 1. Virtual Collaborative Design (VCD) Model.

Stage 1 of the VCD model is the acquisition of user needs, related to the object to be designed – the input from end-users is substantial; stage 2 is the analysis of the problem(s) related to the object to be designed; stage 3 is the creation of different conceptual potential solutions; stage 4 involves further development where one of the concepts is developed further; stage 5 presents a virtual prototype of the designed object to be seen in three dimension on the screen – end-users can help here by giving feedback; stage 6 is the assessment of the previous 3D virtual prototype – end-users can again asses that stage and give feedback; stage 7 is the re-evaluation of all the previous stages and plans for the final manufacturing of the designed product; stage 8 is the final design. This model needs to be refined and validated using results from user and designer surveys and user usability testing experiments. This requires the development of an Interactive virtual 3D software system that will apply the VCD Model and allow non-professional end-users to be involved into "designing" products in a foolproof manner, as well as learn from the interaction. End-users will be able to simultaneously participate in stages 1 - 5 - 6 – 8 of the process,

3.2 Methodology

The research methodology combines hypothesis-based theory building with an empirical evaluation and refinement. Supporting all the above components will be a series of user/designer surveys and SUMI questionnaire, and usability testing experiments. Experiments will be conducted and the software system will be tested through extensive use of the Human Computer Interaction (HCI) labs. This will be followed by case studies involving furniture companies to test the model and the virtual software system, allowing end-users to be involved in the different stages of the furniture design process. The aim is to provide an environment where the consumer can determine and modify the foreseen functions on one product as well as deciding on the following: a) the form, b) dimensions and proportions, c) colors, d) materials and textures, e) finishes, and finally f) cost.

The foreseen key outcomes of the present study are as follow:

- An improved understanding (represented as a model) of the boundaries and limitations of end-users' involvement in the process of virtual collaborative product design.
- A documented and validated development process that addressed end-users' requirements and involvements in order to facilitate more cost-effective development of internet-enabled designing and purchasing systems. The process will support its rapid adoption by online furniture producing companies and related businesses.
- A set of intelligent decision support techniques that will support the task of designing by end-users.
- Development of software and a virtual toolkit that provides support to end-users, designers and manufacturers. This toolkit will facilitate a new kind of collaboration between users and producers. The tool will implement both the design process and the supporting techniques in managing and resolving user requirements and producers need for constant end user feedback.

4 Conclusion

The planed collaborative virtual 3D design environment will provide a benchmark and will have many development opportunities for commercial and educational purposes. Furthermore, it will improve the relationships between the stakeholders for the following reasons: a) user decisions will have direct impacts on the design of the end-product; b) users will be able to use their creativity to customize a product to their needs; and finally, c) improved communication will result between the end-users and other stakeholders, including designers, engineers and marketing people. This study will provide insights into the appropriate coupling of different levels of design with various aspects of end-user, designer and companies needs (e.g. business objectives, content requirements, functional and non-functional requirements). The development of the Virtual 3D design toolkit will benefit to consumers, as well as design education, design practitioners and manufacturers as it will open new directions for designing products, learning to design and buying products online. An effective development process will allow real users to interact in a direct manner, helping in the design process of the product, learning from it, and providing direct feedback to the involved companies. Successful companies or organizations will probably be the ones that can attract end-users to their web pages with user-friendly interface designs, and encourage them to participate in ongoing design projects with the help of carefully designed virtual toolkits. The study builds upon earlier research into user involvement in design (Demirbilek, 1999), and decision support (Aurum, et al., 2001; Aurum, & Martin, 1999; Booth et al., 2001). The non-professional user involvement in the design characterization model provides a requirements framework identifying the key elements that should be specified and designed, as well as linkages between these elements.

5 References

Aurum, A., Handzic, M., Cross, J., & Van Toorn, C. (2001). Software Support for Creative Problem Solving. *IEEE Int. Conf. On Advanced Learning Technologies, ICALT'01*, 160-163, Madison, Wisconsin, USA.

Aurum, A., & Martin, E. (1999). Managing both Individual and Collective participation in Software Requirement Elicitation Process. *Proceedings of the 14th International Symposium on Computer and Information Sciences ISCIS'99*, 124-131.

Balu, R. (2000). Design Vision. *Fast Company*, (36), 362.

Booth, R., Regnell, B., Aurum, A., Jeffery, R., & Natt och Dag, J. (2001): 'Market-Driven Requirements Engineering Challenges: An Industrial Case Study of a Process Performance Declination'. *6th Australian Workshop on Requirements Engineering, AWRE'01*, 41-47, Sydney, Australia.

Burdman, J. (1999). Collaborative Web Development. Addison-Wesley

Demirbilek, O., 2001. Users as designers. *Include 2001 Conference Proceedings*. London: Royal College of Art, 18-20.

Dahan, E, & Srinivasan, V. (1998). The Predictive Power of Internet-Based Product Concept Testing Using Visual Depiction and Animation. Working paper, Retrieved May 11, 2001, from http://web.mit.edu/edahan/www/

Gatarski, R., & Pontecorvo, M. S. (1999). Breed Better Designs: the generative approach. *DesignJournalen*. Sweden: SVID.

Hasdogan, G. (1996). The Role of User Models in Product Design for Assessment of User Needs, *Design Studies* 17:1, 19-24.

Holtzblatt, K., & Beyer, H. (1993). Making Customer-Centered Design Work for Teams, Communications of the ACM, October. Retrieved March 12, 2001, from http://www.incent.com/index.html

IDFORUM. (2001). Mailing-list, Subject: Design your own product. Retrieved March 18, 2001, from http://interaction.brunel.ac.uk/idforum/

Jones, J.C. (1999). The future of ergonomics (and everything!): a theory of despecialisation. *2nd International Virtual Conference on the Internet on Ergonomics CybErg' 99*: Increasing the Accessibility, Diversity and Quality of International Ergonomics Interaction Special Populations, Perth, Retrieved January 10, 2000, from http://www.curtin.edu.au/conference/cyberg.

Kerttula, M., & Tokkonen, T. (2001). Virtual Design of Multiengineering Electronics Systems.

Norman, D.A. (1988). The Psychology of Everyday Things. New York: Basic Books.

Pancake, C. M. (1997). Can Users Play an Effective Role in Parallel Tools Research? *International Journal of Supercomputing and HPC*, 11 (1), 84-94.

Sanders, E. B. N. (1999). Postdesign and Participatory Culture. Useful and Critical: The Position of Research in Design. Tuusula: Finland. University of Art and Design Helsinki (UIAH)

Sanders, E. B. N. (2000). How we do it: Understand Consumers. Sonicrim. Retrieved March 16, 2001, from http://www.sonicrim.com/red/how/under.html

Userfit Tools, (1996). Tools & Techniques. Userfit Design Handbook. Retrieved June 21, 2001, from http://www.lboro.ac.uk/research/husat/include/1-4.html

A Strategy for Formal Service Product Model Specification

Klaus-Peter Fähnrich, Sören Auer

Universität Leipzig
Institut für Informatik, Abteilung Anwendungsspezifische Informationssysteme
{faehnrich|auer}@informatik.uni-leipzig.de

Abstract

A formal product model contains all information to systematically reproduce a specific product (as economic asset). There exist several approaches for formalizing product model information in old economy (e.g. discrete parts manufacturing). The service sector until now evolved to the most important sector in all developed economies. Especially knowledge plays a more and more crucial role for delivering many services. For complex, IT based service products, high in variants existing approaches are not suitable but formalization is desired (e.g. it allows easier export or trade of such products). The paper elicits a possible strategy for defining formal product models for knowledge-based services.

1 Services

In practice there is mostly no need for a clear definition of the concept "service". Services are mostly demarcated to real assets ("Services are everything which can not alight on one's feet"). There is no commonly accepted service definition in science. A service definition accentuating the relation between client and provider gives DIN 9004/2: "A service is result of activities performed at the interface between client and service provider and internal activities of the service provider resulting in fulfillment of client requirements".

contact intensity	variant variety: low	variant variety: high
high	**Customer Integrativ Services (Mass Service)** **Strategies:** customer orientation brand creation total quality management Customer centric.	**Knowledge Intensive Services (Professional Service)** **Strategies:** customer management knowledge management virtualization Flexibility centric.
low	**Particular Services (Service Factory)** **Strategies:** process optimizing automation economy of scale Process centric.	**Variant Rich Services (Service Shop)** **Strategies:** variant management complex product models order oriented resource -allocation Product centric.

The manifoldness of services makes a typology desirable. There exist many different typologies for services using characteristics as "customer interaction", "product customization" or "labor intensity" etc. for classification. A recent one empiric rendered and based on the "customer contact intensity" and "variant variety" is by Barth, Hertweck, and Meiren and shown above.
Using this typology we are particularly interested in the service shop sector. Service providers here have to cope with enormous product complexity especially caused by varieties of variants. Furthermore resources to render a service will be allocated order specific.
Why Formal Product Models? It Works Without! First of all it should be originated from the object of desire (services) for more efficiency. Services are until now mostly developed 'ad hoc' – a determined strategy is needed. The information technology support in the added value chain gets more and more the deciding competitive factor (not only in the service industry). Services are responsible for 65% of world gross product but only for 25% of world trade. Standardization will have positive effects on trade and export of services. Service products mostly exist in a variety of variations. Formal Product Models may help to cope with product variations, to sustain product flexibility / reduce product complexity and to extend product lifespan. Furthermore only formal product models will enable a useful product comparison.

2 Case Study: Collective Insurances

In the following case study the need and possible advantages of formal product models in the collective insurance business will be shown. In the following scheme collective life insurance services are characterized by several important service attributes:

product type	custom made product	modular product	standard product
main input factors	manpower	machines, equipment	information systems
main service object	customer	tangible objects	intangible objects
product extend	single service		service bundel
product type	customer oriented		business oriented
duration of supply	short (< 1 day)	medium (< 1 month)	long (> 1 month)
point of interaction	offer oriented	demand oriented	separate location
role of the customer	actor	observer	without direct participation

Service description: company pension fond for employees, volume: 10.000 companies with 480.000 employees, turnover: 50.000 € per client.
Collective insurances are subject to a whole string of dependencies in their business model. On law adjustments by the legislator or policy changes of the reinsurance for instance lots of products have to be updated. The following business model scheme shows the most important dependencies:

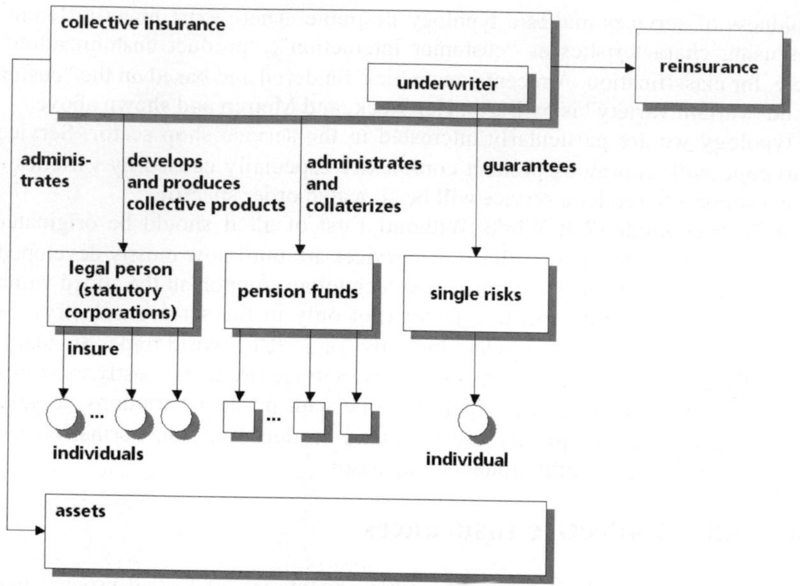

Collective insurance products are defined on different levels resulting in a product tree. With a skeleton agreement on the basis of a product group a concern / organization could supply his employees with specialized insurance products. Condition adjustments on a particular level in the product tree (e.g. product family or product group) should effect in changes on all underlying product instances. In a collective insurance the product lifecycle stages product development, order processing and production are highly toothed.

3 Existing Standards used in Discrete Parts Product Modeling

The vision that services will be eventually "constructed" (as tangible products now already) leads to the desire to transfer methods and concepts or to learn from standards already used in discrete parts manufacturing. A concept widely used in product modeling is constraint programming. Constraints are factors limiting a system. For arrays of products it is mostly easier to define product entities and attributes on which the products are based and to specify limiting factors or rules between those than to construct every product individually. There exists an extensive theory on constraints and constraint solving. For interoperability and flexibility multi layer technology in the definition of the information structure is commonly used. Combination of these concepts can result in the specification of a formal modeling language.

- *STEP* (Standard for the Exchange of Product Model Data) – ISO 10303 [ISO-94], is a family of standards defining a robust and time-tested methodology for describing product data throughout the lifecycle of the product. STEP is widely used in Computer Aided Design (CAD) and Product Data Management (PDM) systems. Major aerospace and automotive companies have proven the value of STEP through production implementations. Product data models in STEP are specified in EXPRESS (ISO 10303-11), a modeling language combining ideas from the entity-attribute-relationship family of modeling languages with object modeling concepts.
- *VDI Guidelines for Product Models* – There exist VDI guidelines for construction methodology of products in machine building, process engineering or fine mechanics [Gra+93]. VDI guideline 2210 e.g. deals with electronic data processing in construction and development. It distinguishes the following stages of product development: functional

analysis and formulation of principles (which result in a functional and an entity structure of the product), furthermore design and detailing.

- *EDIFACT* – EDIFACT, defined and maintained by United Nations (UN) bodies serves as standardization at the electronic exchange of commercial documents and business transactions. All EDIFACT messages are based on ISO9735 that contains a detailed description of all syntax units. In the standards rules are defined which describe how to create EDIFACT messages. These rules describe structure and form of the content of the particular message thus allowing the exchange between different application systems.

4 Product Model Requirements for IT Based Services

The case study showed the complexity of product modeling for IT based services. Here we want to collect some requirements for a product model for IT based services:

- *Modularization* – support of sensible fine grained modularisation; enabling whole (possibly incomplete) arrays of products either by explicit specification or by description of required attributes; extensibility, for creation of new products as well as in case of changes during product lifecycle
- *Interoperability* – use of open standards as far as possible (services are rendered at the interface between client and service provider); implementation independence; rapid prototyping – development and production of services are sometimes not precisely separated; gradual learning curve; tight relation to the object oriented paradigm (widely used in software development).
- *Knowledge* – propagation of constraints (as SLAs - Service Level Agreements); support different views on the products / cover the complete product lifecycle (R&D, marketing, production, management); minimal description complexity.

5 Concepts Useful to Sustain the Knowledge Character of IT Based Services

The fact that development and delivery of variant rich services (as e.g. insurances) is a very complex business (as e.g. shown in the case study) leads to the question if simple transfer of tangible product modeling methods as presented in the last chapter is sufficient or if it is needed to apply results from actual "knowledge processing" research?

- *Semantic Specification by means of ontology* – The main purpose of an ontology is to enable communication between computer systems in a way that is independent of the individual system technologies, information architectures and application domain. The key ingredients that make up an ontology are a vocabulary of basic terms and a precise specification of what those terms mean. First approaches exist in application of ontology theory for product modeling [Strobl98].
- *Formal Concept Analysis* – is a branch of applied mathematics. Based on a mathematization of *concept* and *concept hierarchy* it activates mathematical methods for conceptual data analysis and knowledge processing [GaWi99]. Formal Concept Analysis can be used for clustering data as well as for supporting the work with ontology (merging, comparing). A relatively new direction in Formal Concept Analysis research is analyzing implications between attributes of objects.
- *Generative Programming (GP)* – is about designing and implementing software modules which can be combined to generate specialized and optimized systems fulfilling specific requirements. The goals are to decrease the conceptual gap between program code and domain concepts, high reusability/adaptability, simplify managing variants of a component.

To meet these goals, GP deploys e.g. the principles: separation of concerns, parameterization of differences, analysis and modeling of dependencies and interactions, separating problem space from solution space [Cza+98].

6 Future Working Plan

Goal is the formulation of widely usable Formal Product Models for IT based, variant rich, knowledge intensive services. The following components / concepts have to be developed:

- An *information infrastructure* for knowledge-based services should combine structure models, attribute models and resource models. How far could formal concept analysis help to identify basic entities?
- A plug-in infrastructure of for *defining relations between common entities* has to be developed. Possible relations are ontological relations (may be used to define relations between products) and constraints (suited for representation of incomplete information). Can varieties of products can be described by constrains? How can the consistency of formal product models and the implicit relations be guaranteed?
- *Selection or specification of a language for representation of formal product models for IT based services* - requirements (as modularization, reusability etc.) should be fulfilled on all layers of data storage (schema, data model, file format). Interoperability could be guaranteed by making use of technologies in the XML context (e.g. Web Services). Is EXPRESS (the STEP representation language) suitable as representation language for knowledge based services? How to model the different relation types between entities?
- *Product Data Management System* – PDM systems enable product data sharing. Furthermore a PDM system must support methods for analyzing, visualizing, restructuring of stored product information. With XML represented product models can standard XML databases be used as PDM systems? Which requirements do they have to support additionally?

7 Literature

[AndTri00] Anderl, R.; Trippner, D. (Hrsg.): STEP Standard for the Exchange of Product Model Data. B. G. Teubner, 2000.

[Cza+98] Czarnecki, K.; Eisenecker, U.; Glück, R.; Vandevoorde, D.; Veldhuizen, T.: Generative Programming and Active libraries. In Proceedings of the Dagstuhl Seminar on Generic Programming. LNCS 1766, Springer-Verlag, 2000.

[GaWi99] Ganter, B; Wille, R: Formal Concept Analysis - Mathematical Foundations. Springer 1999.

[GlaMei02] Glanz, W.; Meiren, T. (eds.): Service research today and tomorrow. IAO Stuttgart, 2002.

[Gra+93] Grabowski, H.; Anderl, R., Polly, A.: Integriertes Produktmodel. DIN Deutsches Institut für Normung e.V., Berlin, Wien, Zürich 1993.

[ISO-94] ISO 10303; Industrial Automation Systems and Integration - Product Data Representation and Exchange. 1994.

[Lu02] Lubel, J.: From Model to Markup – XML Representation of Product Data. XML Conference Proceedings, 2002.

[MeiLie02] Meiren, T., Liestmann, V. (Hrsg.): Service Enineering in der Praxis. IAO Stuttgart, 2002.

[Strobl98] Strobl, G.: Entwicklung und Wiederverwendung wissensbasierter Produktmodelle auf der Grundlage formaler Ontologien. Herbert Utz Verlag, 1998.

State of the Art in Service Engineering and Management

Klaus-Peter Fähnrich, Walter Ganz, Thomas Meiren

Fraunhofer Institute for Industrial Engineering
Nobelstrasse 12, 70569 Stuttgart, Germany
Klaus-Peter.Faehnrich@iao.fhg.de;
Walter.Ganz@iao.fhg.de;Thomas.Meiren@iao.fhg.de

Abstract   Design of the study

The main focus of the study presented in this paper is the investigation and analysis of the current state of art in the field of "Service Engineering and Management". A key aim of the study which was undertaken by staff from the Fraunhofer Institute for Industrial Engineering during the period from July to August 2002 was to identify long-term trends and to reveal deficits in current service research.

In order to obtain the most comprehensive view of the different approaches and topics of current service research, the study began by elaborating an model which addresses the four broad topics »basic research«, »applied research«, »mega trends« and »growth sectors« (see Appendix Table 1) which taken together, account for numerous contrasting aspects of current service research.

The actual appraisal activities basically consisted of a series of expert interviews, backed up by a study of international journals, international conferences, and internet communities. For the interviews a total of 25 internationally renowned experts in service research were selected. Care was taken to ensure that the chosen interviewees were not only able to offer expertise from the broadest possible range of topics but also that they represented the most important regions engaged in service research (Europe, North America, South-East Asia and Australia). Methodologically the expert interviews were supported by a semi-standardised interviewer guide.

Service research – paving the way for a paradigm shift?

Is the persistently dominant mindset in the academic world which continues to regard economic activity chiefly in terms of »manufacturing«, i.e., the production of industrial goods, and thus perceives research activity as coupled to this paradigm, still adequate in today's world? Should we not - especially in disciplines which take an ever-broadening approach to service research which goes beyond the latter's sub-disciplinary aspects – seriously question our understanding of what we mean when we talk about »economic activity«? Have we not reached a stage at which academic efforts need to be concentrated on developing and establishing new conceptual and procedural assumptions for the analysis of the economy and society as a whole which reflect the current economic weight and the future potential of services? These questions address a central aspect of the current discussion of service research in the countries surveyed.

A significant number of experts believe that it is imperative to work on the continuing development of the theoretical underpinning and foundations of »services« as a research object. The decisive task must be to identify the quintessential nature of services, to »understand the logic of services« both generally and in specific fields. In order to achieve this, theoretical and empirical research work is required which provides descriptions of a service perspective which are backed up by real-life examples.

Researchers in the USA, in particular, are currently engaged in theoretical and conceptual work which represents a departure from previous assumptions – firstly, by the IHIP model (Intangibility, Heterogeneity, Inseparability, Perishability[1]) and, secondly, by calls for the critical scrutiny – or more radically in some cases, the complete rejection – of the previously widely accepted dichotomy of services and goods. This dichotomous view of economic activity in terms of the production of goods and services should, it is argued, be replaced by a holistic perspective and integrative theories and models.

Summing up the different aspects mentioned - like the need of the creation of a »holistic service perspective« perceving economic activity as »totality of intangibles and tangibles«, the emphasize on the relationship between an offering and the structures, processes, resources and partners involved in delivering is, or the fact, that services above all are »people work« thus »human aspects« and social-interactive components are of considerably greater importance than in the production of goods and special attention must therefore be given to people as customers and employees in the service economy and thus as central business »assets« - it is clear that the theoretical underpinning of »services« as a research object is currently a high priority topic in current debate. One of the most eminent authorities in the field regards the apparent deficit in terms of a common theoretical understanding and a lack of fundamental models and systems as one of the key problems confronting the development and management of services. Most observers identify the challenge as accelerating the creation of generally applicable theory relating to both generic aspects of services and – given the breadth and heterogeneity of the field – of sub-theories and concepts applicable to both individual service sectors and aspects of these sectors. However, rethinking the essence of what constitutes economic activity will not take place overnight – on the contrary, this process in turn can only take place as part of long-term perspective which also supersedes the current persistence of a (sub)disciplinary pigeonholing attitude.

Service research – on the threshold to a multidiscipline?

The international survey revealed that service research today has (still) not managed to establish itself as an autonomous academic discipline confident of its own institutional legitimacy as is the case, for example, with engineering or economic science. Service research activities are often highly fragmented, carried out in isolation from one another and, bearing this in mind, a multidisciplinary approach promises to become a central factor in the success of research in this field fulfilling a dual function. Firstly, the creation of conscious links between research in various (sub)disciplines may help to reach a »critical mass« on the academic agenda which the fragmented approach adopted to date has not been able to achieve, or at least not visibly. Secondly, given the relevance – demonstrated in the previous section – accruing from the formation of meta theory as a common point of reference for future research in various (sub)disciplines, and the creation of a new service perspective, multidisciplinary research endeavours would be warmly welcomed by many of the interviewed experts.

It is important to emphasise both the broadly-shared view that it will no longer be possible to pack the knowledge of the future into traditionally neat »boxes« and the conviction that it will only be

---

[1] Cf. amongst others Grove, Fisk and John (2002): The Future of Services Marketing, Forecasts from Ten Services Experts.

possible to generate robust ideas in the future in the context of an international exchange of research work. In contrast to current scientific practise – particularly in the USA, but also in Europe – future research needs to be organised around the object of research – in this case services – and not within the bounds of established academic tradition.

Summing up it is clear that – owing to its multi-facetted nature – the field of service research is perhaps more than many other areas of academic endeavour not only a particularly suitable candidate for multidisciplinary strategies, it is in fact practically dependent upon such approaches if it is to continue to generate new insights. Researchers and those responsible for supporting institutional esearch both have a role to play in working towards greater mu ltidisciplinary approaches and for ensuring that the recognition of the need for and benefits of cooperative research translates into concrete joint projects and cross-functional transfers.

Lessons to learn – the need for research and action

Differentiated statements regarding the status and future of selected research fields were garnered from the interviews with international experts and opinion leaders. Six central fields of research development emerged emerged from the plethora of views expressed **(refer to figure)**.

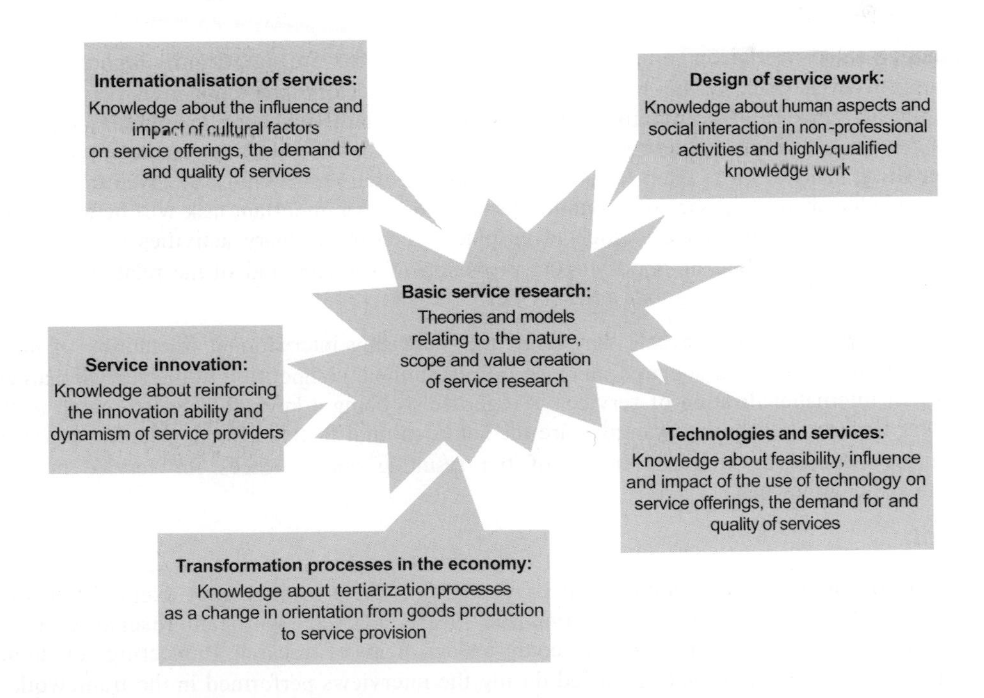

Figure: R&D needs as reflected in the assessments of international experts

The project "Service research today and tomorrow" is funded by the German Federal Ministry of Education and Research (BMBF). My acknowledgement of the gratitude belongs to the BMBF

Let's point in a nutshell to the most interesting findings from our point of view.

Almost all the experts identified a need for more research into the attractive design of service work. There is an intense awareness that, more than in the case of any activities in the manufacturing sector, service work is »people work« and that there is too little knowledge about the human aspects of both the provider and customer side of the service equation. Key concepts here include excessive or too few demands on employees, staff satisfaction and loyalty, subjective perceptions of quality and customer retention as well as the management of interactive company-employee-customer relationships. This revaluation of the human aspects of service research is accompanied by a need for further research and development in the area of adequate »performance management systems«. The challenge is now to take account of the specific nature of services, particularly in the areas of productivity, quality and social interaction, and to map the service-specific components commensurately in systems for evaluating and measuring performance.

In the view of the experts, technologies and services represent a cross-sectional topic of enduring interest. Particular attention needs to be paid to information and communication technologies as well as to service automation technologies. A need for more research is also identified with regard, in particular, to the role of technology in the provision, support, brokerage and sale of services. Research into service excellence in settings heavily dependent on technological support and the creation of high levels of service quality via the deployment of technology are also regarded as particularly relevant.

Even though value creation and employment in the service sector are significantly higher in all the countries surveyed than in the primary and secondary sectors, the collective perception of »economic activity« or of »economy and society« is still largely shaped by familiar industrial models and images. Because these models and images seldom do justice to the requirements of service reality, many of those interviewed believe that research priority must be given to obtaining more knowledge about processes of economic tertiarization. An important task will be to improve understanding both of the change which takes place when the primary activities of companies switch from the production of goods to the provision of services and of the relationships and increasing interaction between goods and services.

Last but not least the study reveals that given the increasing international orientation of many service companies and the associated increase in international competition in the service markets, the topic of internationalisation of services is regarded as being a key subject of research in the future. Particularly noteworthy examples are the socio-cultural factors affecting service offerings and service demand as well as studies of the resulting consequences for service markets, companies and employees.

Outlook

The overall quantitative values and qualitative responses of the presented study exemplify that the themes around which the survey was organised clearly elicited significant resonance among international experts from a wide range of countries and fields of research. Pioneering new trends or entirely novel topics were not identified during the interviews performed in the framework of this appraisal. For an point of view from German service researcher it may therefore, with cautious optimism, be possible to deduce that the fundamental direction and current approaches adopted by service research in Germany do, in general, reflect the international state of the art. Researchers in Germany would be able to respond to international activities and to use these to define the direction of their own activities, nevertheless exporting the successes and results of service research performed in Germany has to date met with very little success.

To get Germany into international networks of service research(ers) rather than following in the research footsteps of other countries in some of their strongest fields of research, such as service marketing, it would perhaps be wiser to combine an increasingly international approach with greater emphasis on our own domestic areas of research and to contribute the results of such work to international discussion in order to build a reputation for Germany as a country in which original research work is undertaken. One priority field in which Germany not only evinces strengths in relation to new service research but also has traditional research and development strengths is the interrelationship between technology development, management concepts and ergonomic work design.

To summ it up, the success of a more emphatically multidisciplinary and internationally-oriented approach to service research is critically dependent on learning with and from each other. If we are to generate the knowledge we will need in the future, it should be paramount to recognize differing paths towards the development of service economy and service research as well as invest in international research programms and knowledge exchange.

References

Ganz, W; T. Meiren (Ed.2002): Service research today and tomorrow; Stuttgart
Grove, Fisk and John (2002): The Future of Services Marketing, Forecasts from Ten Services Experts.

Fisk, Dorsch and Grove (2002): A Retrospective on the Frontiers in Services Conference: Ten Years of Contribution to Service Knowledge, presentation given at the 11th international Frontiers in Services conference in Maastricht, 27 - 29 June 2002.

Holistic development of new services

Mike Freitag[a], *Thomas Meiren*[a], *Hans Wurps*[b]

[a]Fraunhofer Institute for Industrial Engineering
Nobelstr. 12, 70569 Stuttgart, Germany
Mike.Freitag@iao.fhg.de; Thomas.Meiren@iao.fhg.de

[b]Océ Printing Systems GmbH
Siemensallee 2, 85581 Poing, Germany
Hans.Wurps@ops.de

Abstract

In recent years, the service sector has seen a steady change in market structures and competitive situations, and hence an increase in dynamic innovation. This change has been brought about by liberalisation and deregulation as well as by the increased efforts of companies in respect of globalisation. The leading edge in service markets is no longer secured solely through cost, quality and technology premiership, but above all through the systematic development of innovative services. Many enterprises, however, currently lack both suitable procedures and the appropriate organisational framework. This paper presents ways in which services can be designed, gradually refined and ultimately implemented. A case study demonstrates the success factors and critical aspects that can influence their installation in real companies.

1 Introduction

In increasingly dynamic markets, service enterprises no longer secure a leading edge through cost, quality and technology premiership alone. On the contrary, differentiation through innovative services has emerged as an important concomitant factor and in many cases is developing into a crucial unique feature for companies to set themselves apart from their competitors. The central challenges facing enterprises above all compel them to offer continuously improved and new services in the marketplace, to always remain one step ahead of the competition and at the same time to comply exactly with customer needs and expectations.

Although far greater attention has been accorded in the past few years to questions of service development (Ramaswamy 1996, Cooper/Edgett 1999, Fitzsimmons/Fitzsimmons 2000), many companies still exhibit surprising uncertainty when it comes to identifying ideas for new services and translating them systematically into offerings that can be marketed. The development of services as encountered in practice generally takes the following form: new services are "commands" handed down by the company management, they frequently miss the development target when viewed within the context of market and company situations, customer needs and expectations are not analysed or this work is only undertaken at too late a stage and employees' requirements are not adequately taken into account. In addition to this, separate management structures for the development of services can only be found in a few companies.

2 Conceptualisation of service products

Constitutive approaches founded on three dimensions have proved to be suitable for the conceptualisation of services on the basis of various definitions (Donabedian 1980, Kleinaltenkamp 1998). These three dimensions are structure, process and outcome. Concepts and models have to be elaborated for all three, in order to be able to develop service products. The outcomes of the development of service products are thus resource models, process models and product models (Fähnrich et al 1999, Meiren 1999). Figure 1 shows the relationships.

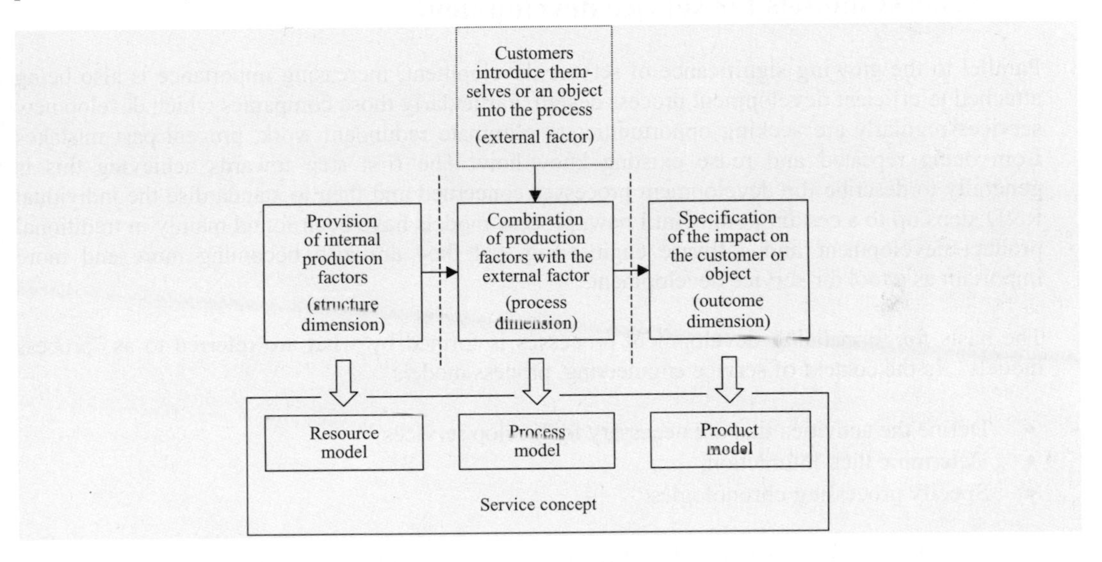

***Figure**. 1: Derivation of a methodology for service concepts*

The product model describes the service product to be developed. All the service contents and the structure which links them must be specified in a product model. Ideally, therefore, the sub-services that make up extensive service products should be modular, so that the overall product structure can be kept as simple and as transparent as possible. Specific customer requirements can then be taken into account by combining different modules ("customising") and satisfied efficiently by the offering company thanks to their clearly defined structure.

The product models, which describe *what* a service does, must be followed by process models, which specify *how* the service is provided. It is important to ensure that all the process steps are specified and the associated interfaces defined. Above all, the processes must be represented transparently and lucidly, because they need to be understood by every single employee. Process optimisation should be a further priority from the outset. Subsequent modifications, which are likely to be costly, can then be avoided. Process steps which add no value should be eliminated along with unnecessary media discontinuities, while parallel, time-saving handling of sub-processes should be actively encouraged (Bullinger/Meiren 2001).

The role of resource models is to plan the deployment of resources – an essential precondition of a service's efficient provision. They cover planning activities relating not only to personnel requirements but also to the use of operating facilities as well as information and communication technologies. Personnel requirements planning is particularly important (Freitag 2002).

The service concept described here is generic in nature and can hence be applied to almost any service product. Concrete methods and procedures must now be elaborated for practical situations. The following section describes which methods and procedures are already implemented in companies today.

3 Process models for service development

Parallel to the growing significance of service development, increasing importance is also being attached to efficient development process design. Particularly those companies which develop new services regularly are seeking opportunities to eliminate redundant work, prevent past mistakes from being repeated and reuse existing know-how. The first step towards achieving this is generally to describe the development processes concerned and then to standardise the individual R&D steps up to a certain point. Until now, process models have been found mainly in traditional product development and software engineering, yet they are also becoming more and more important as a tool for service development.

The basis for formalising development processes is formed by what are referred to as "process models". In the context of service engineering, process models

- Define the activities that are necessary to develop services,
- Determine their interaction,
- Specify processing chronologies.

Process models include detailed documentation of project flows, project structures and project responsibilities, enabling them to support project planning, project steering and project monitoring.

The following main process model types are recommended in the literature for service development:

- Phase models,
- Spiral models,
- Prototyping models.

Sequentially structured phase models are the most commonly encountered type, because they are suitable for a very wide range of tasks (Scheuing/Johnson 1989, Edvardsson/Olsson 1996, Ramaswamy 1996).

Spiral models run through the development cycle several times, enabling the speed of the R&D process to be adapted to each specific problem and meaningful interim results to be obtained at an early stage. Medium to high-volume, investment-intensive customer support services, for instance call centres, help desks or hotlines, are good examples (Shostak/Kingman-Brundage 1991).

Approaches for service prototyping and rapid service development are meanwhile also the subject of fledgling discussions in the service sector. The services developed using these methods, however, are almost exclusively intended for extremely dynamic competitive environments or in cases where the involvement of the customer as a co-developer or consultant is critical.

In addition to the basic models mentioned above, several other process models are also conceivable, though in most cases they have yet to be validated in practice.

4 Case Study

Océ Printing Systems is a copying and printing specialist offering a full portfolio of products that cover the entire spectrum of printing production needs. Its role as a vendor of integrated solutions demands, in particular, a wide range of both accompanying and independent services to support its core business alongside its actual products.

In contrast with the development process for new products, however, services have in the past tended to originate spontaneously, in other words new services have usually been the outcome of personal initiatives while separate R&D management functions for services have been non-existent. Océ Printing Systems has now taken up this challenge within the framework of a project sponsored by the German Ministry for Education and Research, and designed and implemented an R&D process for services that is capable of describing the complete flow from the generation and appraisal of the original idea through its conceptualisation and implementation to the ultimate launch of the new services in the marketplace. The project focused especially on maximum integration of the customer view as well as on the formulation of a holistic approach, in other words one which takes account of technical, organisational and employee-specific aspects, to develop new services. Figure 2 shows the most important process steps together with a detail of the conceptualisation phase.

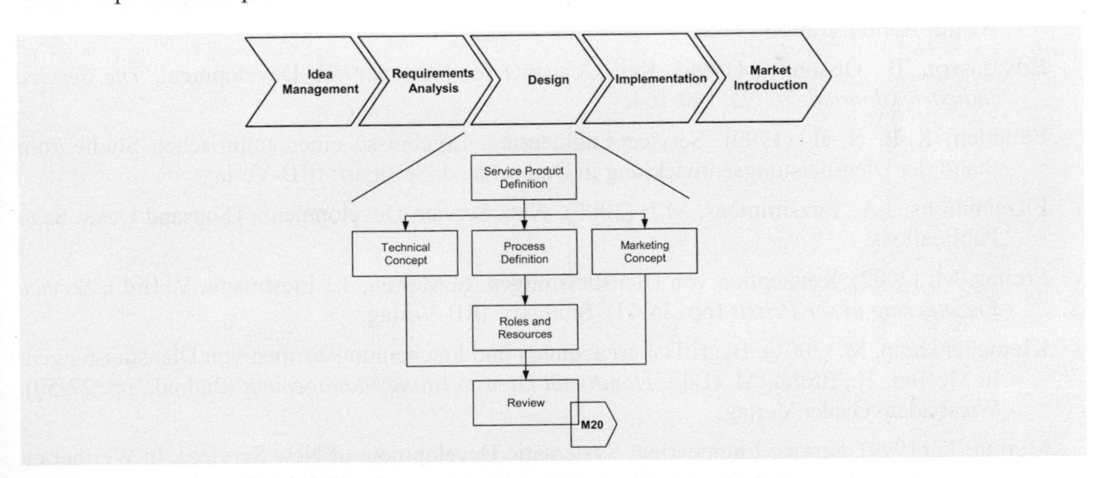

***Figure 2**: Service Development Process of Océ*

Parallel to the definition of the R&D process, the individual process steps and the related activities were also described in detail in an R&D manual, which additionally specifies roles and competencies. Suitable methods and tools were selected and checklists generated to ensure that the process is handled in line with actual needs.

The newly created process is currently being implemented in the company and is put to the test each time a new service is developed. Océ Printing Systems is thus one of the first enterprises to have introduced a separate service development process, as it makes a successful transition from manufacturing firm to vendor of innovative services.

5 Outlook

With the increasing "industrialisation" of services growing importance is likely to be attached during the next few years above all to efficient service development. There are already clear signs today in many companies that basic product strategies such as standardisation, modularisation and customising are in future set to penetrate more and more into services, often crucially influencing competitive success. The leading edge will in all probability be claimed by those enterprises that create service development structures and processes early on, and that consequently avail of the capability to develop and offer new service products which are in line with customers' needs - yet at the same time financially successful - in minimal time.

6 References

Bullinger, H.-J., Meiren, T. (2001). Service Engineering. In Bruhn et. al. (Ed.), Handbuch *Dienstleistungsmanagemen*t (pp. 149-175). Wiesbaden: Gabler Verlag.

Cooper, R.G., Edgett, S.J (1999). Product Development for the Service Sector. Cambridge: Perseus Books.

Donabedian, A. (1980). The Definition of Quality and Approaches to its Assessment. Ann Arbor: Health Administration Press.

Edvardsson, B., Olsson, J. (1996). Key Concepts for New Service Development. *The Service Industries Journal,* 16 (2), 140-164.

Fähnrich, K.-P. et. al. (1999). Service Engineering. Ergebnisse einer empirischen Studie zum Stand der Dienstleistungsentwicklung in Deutschland. Stuttgart: IRB-Verlag.

Fitzsimmons, J.A., Fitzsimmons, M.J. (2000). New Service Development. Thousand Oaks: Sage Publications.

Freitag, M. (2002). Konzeption von Dienstleistungen. In Meiren, T., Liestmann, V. (Ed.), *Service Engineering in der Praxis* (pp. 34-43). Stuttgart: IRB-Verlag.

Kleinaltenkamp, M. (2001). Begriffsabgrenzungen und Erscheinungsformen von Dienstleistungen. In Meffert, H., Bruhn, M. (Ed.), *Handbuch Dienstleistungsmanagement* (2nd ed., pp. 27-50). Wiesbaden: Gabler Verlag.

Meiren, T. (1999). Service Engineering: Systematic Development of New Services. In Werther et. al. (Ed.): *Productivity & Quality Management Frontiers*. Bradford: MCB University 1999.

Ramaswamy, R. (1996). Design and Management of Service Processes. Reading: Addison-Wesley.

Scheuing, E.E., Johnson, E.M. (1989). A Proposed Model for New Service Development. *The Journal of Services Marketing*, 3 (2), 25-34.

Shostak, L., Kingman-Brundage, J. (1991). How to Design a Service. In Congram, C. (Ed.), *The AMA Handbook of Marketing for the Service Industries*. New York: Amacom.

"Job redesign" – still between work organization and work rationalization

Ekkehart Frieling & Sascha Störmer

Institute of industrial science and ergonomics, University of Kassel
D-34109 Kassel, Germany
frieling@ifa.uni-kassel.de

Abstract

With a revised failure-recording sheet and a systematic failure recording it was possible to generate a reliable error statistic. The results were examined more closely with the action-slip cause-analysis and some configuration measures deduced.

1 Problem

In an automotive company, which is manufacturing premium vehicles, it was complained about assembly errors in final assembly although everything was done in order to avoid mistakes from technical view. Production engineers were convinced that the errors were due to human failure and this again was due to personal deficits.

2 Description of the research area

At the assembly area of the examined plant vehicles are being assembled in a two-shift system between 5 a.m. and 10 p.m. (shift change is at 13:30). In the analyzed area the customer specific harness is partially mounted, additionally single electrical modules are built in, plug connections are made and the parts of the interior lining are built in.

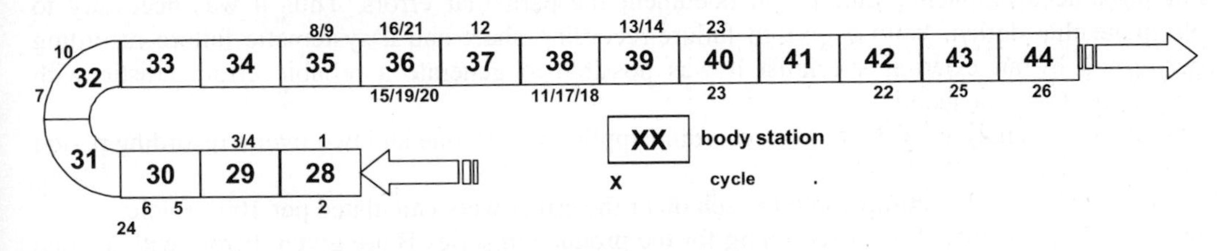

Figure 1: Assignment of the work cycles to the body stations in running direction of the steel plate

The following table partially shows single cycles (the cycle length is approximately 70 sec.) that are allocated to the assembly sectors 28-44. The individual cycle length differs between 3 and 5 minutes.

cycle	production series	rough assembly description
cycle 1 position 28 R	A	Lay customer-specific harness in luggage compartment right, assemble relay luggage compartment;
	B	Acoustic insulation right luggage compartment; installation central locking-motor; removal of customer-specific harness thread help left
cycle 2 position 28 L	B	Relocation customer-specific harness from left to right; attach customer-specific harness to holder left; attachment parts (3 cable ties, 5 clips); 2 cable ties assembled to battery mounting; Band assembled
cycle 3/4 position 29 R	A	Initialization of basic module completed; thread hose line trough front hood
	B	Instrument mounting complete; installation customer-specific harness center-console right; instrument mounting complete; mounting windshield heater

Table 1: Rough work content of the cycles and allocation to the body stations

On the assembly line two different vehicle versions from diverse production series (A = simple automobile and B = complex automobile) are assembled. The employees are organized in groups. The group size differs between 15 and 25 employees. Besides a group speaker which is elected, a foreman exists who coordinates the group, partly acts as a spare man and represents the craft masters' interests. He helps when problems occur and partially takes over the initial training of new employees. On the average group dialogs take place once a month for half an hour. During this time the assembly line is stopped. When a production shortage occurs the group dialogue is cancelled. A crafts master oversees two groups. Per group an additional spare man exists.
The employees are mainly male, skilled workers and averagely between 32-35 years of age.
The employees' utilization ratio is 95 % according to information provided by the plant. The vehicle to be assembled is on a steel plate conveyor, which can be adjusted in height according to the requirements. The worker rides along on the assembly line.

3 Failure recording / data acquisition

To narrow down the cause of error it is necessary to record the errors systematically concerning the selected area. As a more detailed analysis of existing failure recording statistics shows it was not possible to explicitly define and document the particular errors. Thus it was necessary to eliminate this deficit. With a revised failure-recording sheet and a systematic failure recording performed by an external specialist it was possible to generate a reliable error statistic with employees being included.
The following analysis of the failure recording applies to shift one and two over a recording period of four weeks.
To make the error data comparable to each other the errors were calculated per 100 vehicles.
In figure 2 the results of error recording for the production series B are given. Errors with a value smaller then 0.1 error per 100 vehicles are not being considered.
Here the error „grommet left not okay" proved to be the most frequent error (1,39 errors/100 vehicles). Since this error occurred in both shifts with almost the same frequency (shift 1: 1,29 errors/100 vehicles; shift 2: 1,48 errors/100 vehicles), it was examined more closely with the action-slip cause-analysis (Algedri & Frieling, 2000).

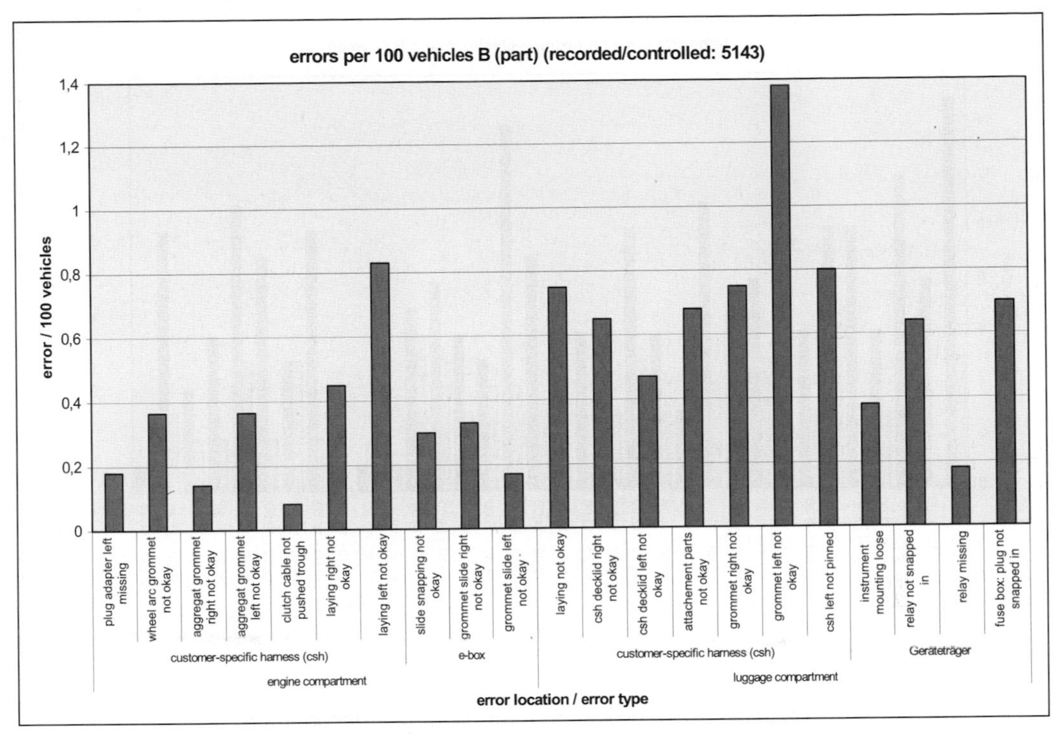

Figure 2: Analysis of the failure-recording sheet for vehicle type B (part)

The failure analysis of the simpler vehicle (A) results in a different pattern. This type of vehicle is already assembled for a longer period (approx. 3 years) and the employees have fewer problems with the wiring harness.
In the next step the errors that occurred in the individual vehicle areas in the analysis period were summarized and compared to the planned time of the corresponding vehicle area. In production series B the greatest part of errors (54,9 %) by far belongs to the luggage compartment, which takes the greatest part of assembly time from the total planning time with 35,3 %. The second largest share of errors with 20,6 % concerns the engine compartment with a planning time quota of 13,8 %.
With production line A a similar pattern occurs. Here the engine compartment with an error rate of 47,1 % and a planning time quota of 29,3 % is the most error-prone area. But concerning the luggage compartment a blatant difference between error rate (35,8 %) and planning time quota (10,1 %) can be detected.
The largest error potential exists in the area of the luggage compartment. Assembly employees assume that the reason for a higher occurrence rate of errors in the luggage compartment is mainly a bad illumination situation.
From the analysis of the error frequency depending on the time of occurrence (cf. figure 3) it becomes clear that certain time effects do exist. The errors accumulate at the beginning of a shift, at about lunch break and just before the end of the shift.

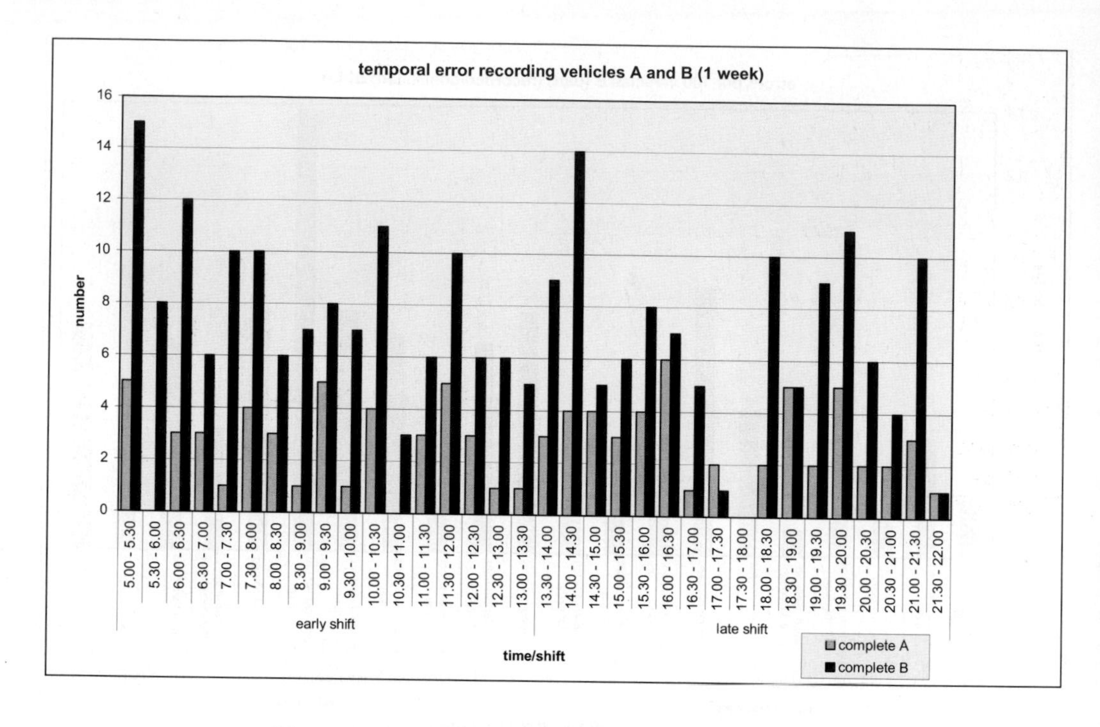

Figure 3: Error distribution for the time of recordation

4 Application of the action-slip cause-analysis to the error „grommet left not okay"

Since the production error „grommet left not okay" was the most frequent product defect per 100 vehicles and occurred frequently in both shifts, it was examined more closely with the action-slip cause-analysis. First the individual actions of the employees were observed and a selection of critical actions was undertaken.
As universally valid causes for errors the bad illumination situation in the luggage compartment and the wheel arc can be named as well as the unfavorable working position when securing the handling points in the wheel arc (figure 4). In addition the employees can only see the working area poorly because of the very low working height.

In addition the individual assembly operations don't follow an inner logic but are rather orientated at the best utilization of the employees. Thus a fragmentation of the workflow results and as a consequence single work-steps are left out by accident.
As another cause the poorly understandable cycle information comes into consideration. Because of time the employees have no possibility to get informed about the complete scope of assembly in their work cycle, when a cycle change occurs they are instructed about the added operations by the foreman. Especially concerning operations which exceed the usual scope the eventuality for missing out this operation after a certain period of time exists.

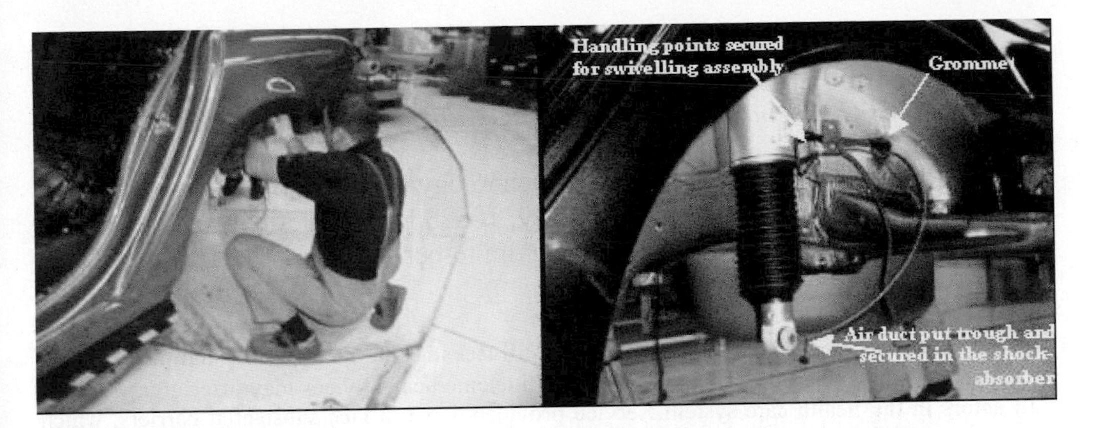

Figure 4: working position when securing the handling points (right) and pulled in as well as secured grommet with handling points and air duct

5 Configuration measures

In order to improve the lighting situation new fluorescent tubes shall be installed in a part of the craft master area in cooperation with the department responsible for illumination. The distance of the fluorescent tubes to the ground has to be decreased to enhance the luminous intensity.
As far as possible the assembly steps which are to be accomplished in the wheel arc should be relocated to a „higher“ cycle, i.e. they should be carried out in a cycle with the car body in assembly being in a high (the highest) position. The employees then can view the working area much better. Additionally the illuminance in the wheel arc will be increased by the lateral light incidence.
From the work organizational point of view only few configuration scopes exist concerning the present flow line assembly. As basic starting-points for configuration the following aspects were derived:

- Reduction of the assembly line speed at the beginning of the early shift;
- A stronger integration of the employees concerning the cycle;
- Design of work cycles which represent a coherent sub-process;
- Reactivation of group work and assignment of planning tasks (planning of free shifts and vacation, job-rotation, qualification) to the group;
- Reduction of group size to approx. 8-10 persons;
- Systematic error recording by group members;
- Reduction of preparatory work for not to affect the ergonomic working conditions;
- Reduction of beer consumption by massive campaigns.

References

Algedri, J. & Frieling, E. (2001). *Human-FMEA. Menschliche Handlungsfehler erkennen und vermeiden*. München: Hanser.

Frieling, E. (Hrsg.). (1997). *Automobilmontage in Europa*. Frankfurt/Main: Campus.

Successful Business Models of Telemedical Services

Hans Georg Gemuenden, Carsten Schultz, Katrin Salomo & Soeren Salomo

Technical University Berlin, Chair for Technology and Innovation Management
Hardenbergstr. 4 - 5, HAD 29, 10623 Berlin, Germany
hans.gemuenden@tim.tu-berlin.de

Abstract

The diffusion of telemedical services in Germany still lacks behind its potentials and the diffusion rate experienced in some other countries. Although telemedical services may offer great benefits to all actors in the health care system, service providers have to face substantial barriers, which prohibit or endanger successful development and marketing of the service. Based on a systematic analysis of these barriers we develop an understanding of business models as a system of success factors which address these barriers. We identify complex customer definitions and the difficult transformation from customer value to customer specific functions as critical barriers to successful telemedical services. Thus, we need to adapt our general concept of business models to the specific situation of telemedical service offerings. In particular, we suggest customer relationship portfolio management to be a central part of successful business models in the telemedical service industry.

1 Potentials and barriers of telemedical services

Based on the definitions of Field (Field, 1996) and of the WHO, we use the following definition of telemedical services:

Telemedical services are characterized by the use of information and communication technologies to exchange valid information for diagnosis, therapy and prevention of diseases and injuries in order to overcome distances in space and time between the involved medical staff and patients.

The implementation of telemedical applications is believed to have several positive effects on efficiency and effectiveness of medical processes. Primarily they improve the information exchange between two or more co-working physicians ('doc to doc') or between physicians and patients. The improvement of information exchange offers the possibility to increase the flexibility of the service production. Telemedicine enables service providers to react faster to the needs of the patients. Additionally, medical capacities can be optimized. A further important advantage of the use of telemedicine is related to improved process control, which is also discussed in the context of Disease Management Programs and Managed Care. Beyond that, telemedicine creates the possibility to extend the service offerings by new components through the possibilities of. The medical services may be extended to new diseases and existing treatments may be prolonged (Eswaran & Gallini, 1996).

Despite these positive performance effects, practical experiences from diverse telemedicine projects as well as numerous studies in this area point out the substantial barriers to successful telemedical services. These barriers can be attributed less to technical problems, but rather to specific factors relating to the medical services environment. Restraining factors can be grouped into input, throughput and output related problems. In terms of input problems cooperative service production, requirement of highly qualified staff and generally complex leadership structures can be identified. The difficulty of the customer definition also increases complexity (Ginter & Duncan, 2000). Variable, complex, and highly specialized service production defines process barriers. Compared to other industrial service offerings medical services have constantly to tackle crisis situations – i.e. medical services is 'standardized crisis management' and, thus, constantly exposed to time pressure and uncertainty together with small error tolerance. Additionally, output

definition and measurement are difficult. Hence, controlling the service process is a further challenge for the management.
Apart from the restrictions based on the medical task of the service offering other barriers related to the health care system, to the production process and to the adoption of these telemedical services can be identified. The German healthcare system is vertically and horizontally differentiated. Strong regulation further increases complexity, thus demanding telemedical services to address health care system related restrictions. Production process barriers result from the integration of information technology into medical services. Adoption barriers generally depend on the newness of the telemedical services – with more innovative services, adoption risks increase. As telemedical services include patients, insurance companies and physicians as potential customers, management of adoption is even more difficult. (Hensel, Schultz, & Gemuenden, 2002). Furthermore, telemedical service providers are exposed to regulative, technological and economic uncertainties. The economic uncertainty is based on unknown behavior of patients and competitors. The technological uncertainty results from future developments of telematic and medical technology. The regulative uncertainty is caused by ongoing reforms of the German medical system.
In order to be successful, telemedical service providers have to meet these challenges to overcome the described barriers and to minimize the uncertainties. Successful telemedical services are characterized by specific attributes concerning the development and operation of their business. The identification of critical success factors and their combination to possible business models is the aim of the present study "Successful business models of telemedical services" at the Technical University of Berlin, which is financed by the German Ministry of Education and Research.

2 A business model concept

A business model is a set of substantial characteristics, which describe an overall business. Along these characteristics successful and unsuccessful businesses can be distinguished these characteristics, thus, form critical success factors (Hauschildt, 1993). In order to classify these characteristics of business models both a resource and a market orientated perspective is relevant. Dynamic development of these characteristics needs to be assessed as well.
Prior research has suggested different critical elements of business models and their internal and external relationships. Among others, customer value proposition, internal organization, information flow and external relations to other actors are suggested as important characteristics of business models. See Gemuenden & Schultz, 2003 for an overview of the actual literature. However, these concepts of business models show some limitations with respect to:

- customer equity, i.e. how and to what degree customers contribute to the company value,
- customer focus, i.e. other relevant market players like actual and potential competitors are not within the focus of these concepts,
- dynamic perspectives, i.e. business models need to be assessed not only in a static perspective, but, must include dynamic evolution of these models, and
- structure and interdependencies of the business model characteristics, i.e. we need a more refined understanding of business model elements and their relationships.

Based on a review of the success factor and strategic management literature, and as a reaction to the above outlined limitations, we define business models as followed:
„A business model combines systematically the value proposition for the costumers and the company as well as all strategic success factors influencing the internal resources and external relations. Furthermore it accounts for existing dynamic interdependencies."
Figure 1 shows the significant aspects of business models.
Business models are in general developed for individual companies, addressing very company specific issues. Nevertheless, individual business models can be compared between different

companies after controlling for relevant market characteristics. It is, thus, possible to analyze and compare business models pertaining to one specific market like medical services. Along certain characteristics individual business models can be clustered into different "strategic groups" (Dranove, Peteraf, & Shanley, 1998). This analysis will focus the core value model, as we expect these characteristics to differ strongly between different groups of telemedical service providers.

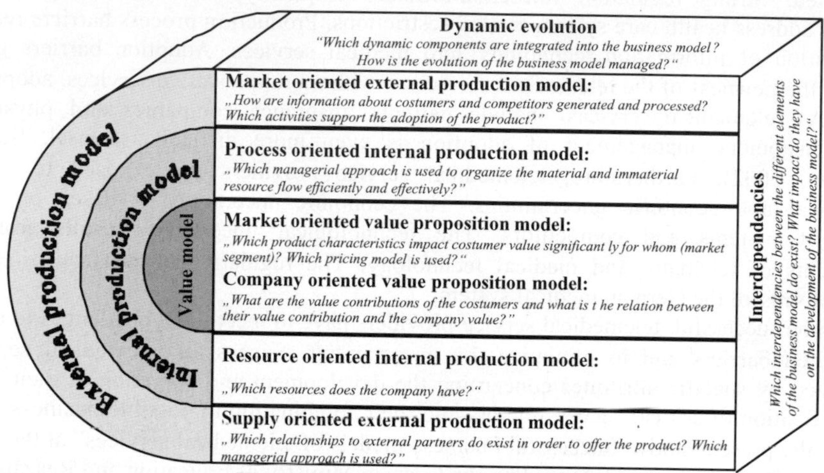

Figure 1: business model concept

3 Value model of telemedical services

Telemedical services can be differentiated into the relations between two or more physicians (D2D) as well as between physicians and patients (D2P). Applications for teledocumentation – famous examples are electronic patient records – form the basis of almost all telemedical services. Generally, the telemedical market can be distinguished with respect to the central customer definition. On the one hand the customer definition may focus patients, on the other hand physicians may be the central customers. This work concentrates on D2P telemedical services, which offer a medical service directly to patients as central customers (Hensel et al., 2002).However, apart from this central customer a medical service offering has to deal with additional types of costumers. In most cases, although the patient receives the medical treatment, this service is paid directly by insurance companies without further involvement of the patients. Furthermore, medical services are 'classical' experience products, i.e. the patient is only to a limited extend able to assess the quality of the offering. Hence, the patient has to rely on the judgments of the respective medical personnel. Typical customer activities as assessment, adoption and payment for the received service, normally performed by one single subject, are here consequently separated into three different actor groups. Patients, insurance companies and external medical personnel have to be integrated into our customer definition. Each of these 'customer components' have to be assessed with respect to their customer value and customer equity.

Within the framework of the telemedicine D2P sector (see figure 2) activities can be distinguished into primary (direct treatment) and secondary (treatment support) services. Telemonitoring aims at generating medical data and their 'just in time' delivery to the telemedicine service provider. The service provider edits the medical data and offers it to both the patient and the medical personnel in charge of performing the treatment. Telemonitoring is of specific relevance in the home care sector. In case telemedical data is directly used for diagnostic purpose, it is called telediagnostic. If telediagnostic is supplemented by a telematic based treatment, this area is named teletherapy. A further important characteristic of telemedical services is their relation to specific diseases.

Treatment specific services are developed and operated with respect to individual diseases. Which have to possess a high rate of digitizeable data. Treatment spanning services monitor and control the integrated treatment processes.

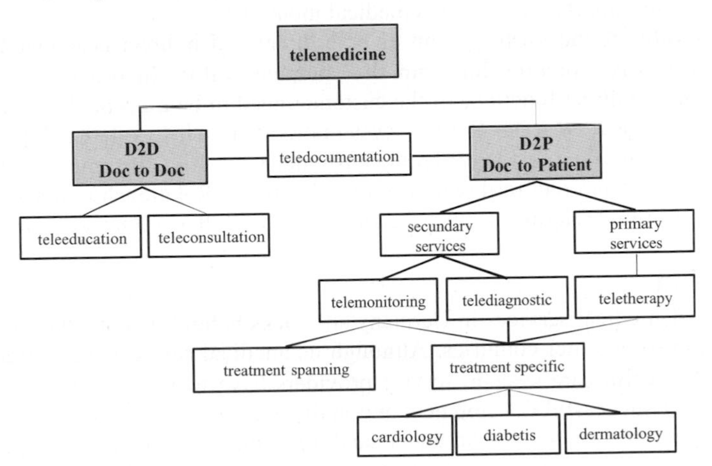

Figure 2: Telemedicine framework

The offered telemedical services have to enclose specific product characteristics, in order to create a costumer value for patients, insurance companies and external medical personnel. Among the most important characteristics affecting the customer value are the relative advantage and the compatibility of telemedical services. The relative advantage depends on increased efficiency and effectiveness offered by the service. The compatibility describes whether telemedical services are able to integrate into existing treatment processes or not.

Customer value in itself is not sufficient to market the service. The service has to meet the legal requirements of the German health care system. Restrictions result in particular from prohibition of exclusive remote treatments and the imperative of a personal medical service offering by physicians. These legal restrictions contribute to the fact that a separate compensation of telemedical services through the public health insurance companies only occurs to a limited extend so far in Germany. Actually, the financial compensation of telemedicine must be based almost exclusively on efficiency improvements of existing treatments. Beyond that, it is only possible to offer the service directly to the patient, who then will not receive compensation from their insurance or to offer the service only to patients with private insurance. Since the large majority of the medical services are still financed by the public health insurance companies, also telemedical services need to fulfill the public insurances requirements in order to generate substantial revenues. However, in the course of current reform of the German health care system further payment possibilities will be created. Different potential payment models can e.g. include monthly fixed amount payments, variable payments according to individual treatments, head-based or case-based lump sum payments or combinations hereof.

The price model is the critical link between the costumer value and the company value. In this context the service provider has to balance the profit function and the remaining functions of the customer within the business relation. Functions of customer relationships can basically be distinguished into direct and indirect functions (Walter, Ritter, & Gemuenden, 2001). Direct functions combine the profit realized with the customer, the efficiency improvements through extended customer base as well as compensation effects from loyal customers, reducing the level of fluctuations from economic or seasonal cycles. The service provider may benefit from indirect customer functions as the costumer may help the service provider to develop the service, as the

service provider may experience a reputation spill-over and as the service provider may more easily generate information about actual and latent market needs. Additionally the service provider may be able to accelerate regulative permission and contract procedures through customer support – a benefit of specific importance in the telemedical industry.
Hence, customer value is the starting point of both direct and indirect customer functions. Both functions have a positive relationship with the company value. In order to generate positive company value, first, indirect functions of the customer need to be translated into direct functions, which again increase positive cash flow. A systematic approach to relationship management is believed to moderate the relations between the customer value and customer functions as well as the transformation from indirect to direct customer functions. Relationship management includes management of individual relations to specific customers as well as an active approach to portfolio management.

4 Conclusion

The diffusion of telemedical services in Germany still lacks behind its potentials and the diffusion rate experienced in some other countries. Although telemedical services may offer great benefits to all actors in the health care system, service providers have to face substantial barriers, which prohibit or endanger successful development and marketing of the service. Based on a systematic analysis of these barriers we develop an understanding of business models as a system of success factors which address these barriers. We identify complex customer definitions and the difficult transformation from customer value to customer specific functions as critical barriers to successful telemedical services. Thus, it is necessary to adapt our general concept of business models to the specific situation of telemedical service offerings. In particularly we suggest, these business models to comprise a systematic approach to customer relationship portfolio management. Our research project „Successful Business Models of Telemedical Services" sponsored by the German Ministry of Education and Research will further investigate this interesting research topic.

5 References

Dranove, D., Peteraf, M., & Shanley, M. (1998). Do strategic groups exists? An economic framework for analysis. *Strategic Management Journal, 19*(3), 1029 - 1044.

Eswaran, M., & Gallini, N. (1996). Patent policy and the direction of technologial change. *Rand Journal of Economics, 27*(4), 722 - 746.

Field, M. J. (1996). *Telemedicine: A Guide to Assessing Telecommunications in Health Care.* Washington, D.C.

Gemuenden, H. G., & Schultz, C. (2003). Entwicklung eines Geschäftsmodellkonzepts und erste Anwendung auf den Bereich telemedizinischer Dienstleistungen, *Organisationsdynamik in Theorie und Praxis - Festschrift zum 65. Geburtstag von Oskar Grün* (will be published in September 2003). Stuttgart: Schaeffer-Poeschel.

Ginter, P. M., & Duncan, W. J. (2000). The content of health care strategy. In J. D. Blair & M. D. Fottler & G. T. Savage (Eds.), *Advances in Health Care Management* (Vol. 1, pp. 35 - 65). Amsterdam et al.: Elsevier Science.

Hauschildt, J. (1993). Innovationsmanagement - Determinanten des Innovationserfolges. In J. Hauschidt & O. Grün (Eds.), *Ergebnisse empirischer betriebswirtschaftlicher Forschung: Zu einer Realtheorie der Unternehmung* (pp. 295 - 326). Stuttgart.

Hensel, K., Schultz, C., & Gemuenden, H. G. (2002). Markteintrittsstrategien und Netzwerkmanagement als kritische Erfolgsfaktoren telemedizinischer Dienstleistungen - erste empirische Bestätigung. In A. Jäckel (Ed.), *Telemedizinführer Deutschland* (Vol. Ausgabe 2003, pp. 30 - 35). Ober-Mörlen: Medizin Forum AG.

Walter, A., Ritter, T., & Gemuenden, H. G. (2001). Value Creation in Buyer–Seller Relationships - Theoretical Considerations and Empirical Results from a Supplier's Perspective. *Industrial Marketing Management, 30*, 365 - 377.

Visualization of Interaction Patterns in Collaborative Knowledge Networks for Medical Applications

Peter A. Gloor
Rob Laubacher
MIT CCS
Cambridge, MA USA
(pgloor, rjl}@mit.edu

Scott Dynes
Dartmouth Tuck CDS
Hanover, NH USA
sdynes@acm.org

Yan.Zhao
Dartmouth College
Hanover, NH USA
Yan.Zhao@dartmouth.edu

Abstract

By combining virtual communities with Internet portal and content management technologies, Collaborative Knowledge Networks (CKNs) (Gloor, 2002) share, access and extend the tacit and explicit knowledge within and across organizations. CKNs are a special kind of web-enabled communities of practice, where like-minded people collaborate and work together towards a common goal, sharing the same vision and values. CKNs are highly relevant also for biomedical Web-based communities to share knowledge and collaborate. In this paper we describe a system for the semi-automatic localization of CKNs in organizations. We identify structural properties and parameters of successful CKNs, based on automated analysis of e-mail archives. We then outline applications of CKNs in the medical field, looking at research collaboration such as new drug development, educational communities such as patient communities, and diagnosis and treatment communities such as clinical trial communities.

1 Introduction – What are CKNs

CKNs are nothing new. In fact, one of the most successful CKNs was started around 2000 years ago in the town of Jerusalem by a carpenter. But the fundamental principles are still the same: a truly innovative idea, which goes against conventional wisdom, is sold by a charismatic leader and a small group of dedicated disciples to an initially skeptical audience. The idea then catches on, and changes the way that the environment behaves.

For an eminent example, look at the way how the Web itself arose as a CKN, driven by visionary leaders: The community of early Web developers exhibited all the characteristics of a successful CKN at work, forming an intrinsically committed, dedicated community. Members joined out of their free will and collaborated not for immediate monetary gain, but because they shared the same values and beliefs. In the meantime the Web has become one of the main drivers of change for our economy, creating billions of dollars of wealth during this process. Even today thousands of dedicated volunteers work together in numerous CKNs to further drive the development of the Web.

Collaborative Knowledge Networks are a concept for entities that have always existed in and across organizations: groups of self-motivated individuals driven by the idea of something new and exciting - a way to greatly improve an existing business practice, or a new product or service for which they see a real need.
CKNs have obtained a great boost from the Internet by globally extending communication and collaboration facilities with instantaneous reach to anybody on the Internet. By combining virtual communities with Internet portal and content management technologies, CKNs can share, access and extend the tacit and explicit knowledge within and across organizations. In this way CKNs are

like communities of practice (Wenger, McDermott & Snyder, 2002), where people collaborate and work together towards a common goal. CKNs are distinct from communities of practice and other forms of online collaboration and virtual teams in both the informal nature of their work, and in the motivation for individuals to join and participate. In CKNs the motivation is internal – they do it because they are driven to do it, as opposed to participation being required as part of work or learning assignments as is often the case in virtual teams and communities of practice.

Due to this internal motivation CKNs cannot be mandated into action; in fact, ordering a CKN into existence is against the very foundation of how CKNs operate. People join and work in CKNs not because they have been told to do so by their superiors, but because they are personally motivated by and convinced of the vision and goals of the CKN community. CKNs are often spawned within an organization, but they quickly break organizational boundaries to include members from outside, bringing in new insights and knowledge otherwise not available within the organization. As noted above, CKNs involve individuals not necessarily related in terms of the corporate hierarchy.

CKNs are an important, unrecognized source of innovation within organizations; identifying and supporting CKNs should be a primary goal of every organization.

2 How to find CKNs

Just as Google is very effective at finding pertinent documents based on viewing patterns, we believe analysis of e-mail and other interaction logs of organizations will enable one to determine communities and core contributors. From the patterns and content of these interactions we will be able to create an index of the CKNs that exist within the organizations, as well as identify agents (individuals or groups) who are sources of expertise and also users of different classes of knowledge.

We are proposing a new methodology to trace the emergence of CKNs and their development over time, and to compute metrics of a CKN's efficiency and its implications for an organization's performance by mining computer logs such as email archives. Our proposed system computes and visualizes the structure of existing CKNs in organizations. We validate this methodology using information on the CKN collaboration structure obtained by interviews with CKN members and other assessments. To address the issues of the potential organizational impact of CKNs we will then compare the different types of CKNs with other corporate performance metrics such as new product output, share price, or profitability. This will allow us to draw conclusive evidence of the impact of CKNs on the performance of the organization.

3 Architecture of the CKN Visualization System

We have implemented a flexible three-level architecture (Figure 1). In the first step, the e-mail messages are parsed and stored in decomposed format in a SQL database. In the second step the database can be queried to select messages sent and/or received by a certain group in a certain time period. In the third step the selected communication flows can be visualized using SNA visualization tools such as Pajek (Batagelj & Mrvar, 1998) and ucinet (Borgatti, Everett & Freeman, 1992).

This architecture provides an optimal testbed of high scalability and flexibility: the number of messages to be analyzed is only limited by the size of the database, and temporal queries can be run in an ad hoc way. We will also be able to experiment with different visualizations of the retrieved structure, identifying graphical representations that adequately reflect the temporal nature of the social networks.

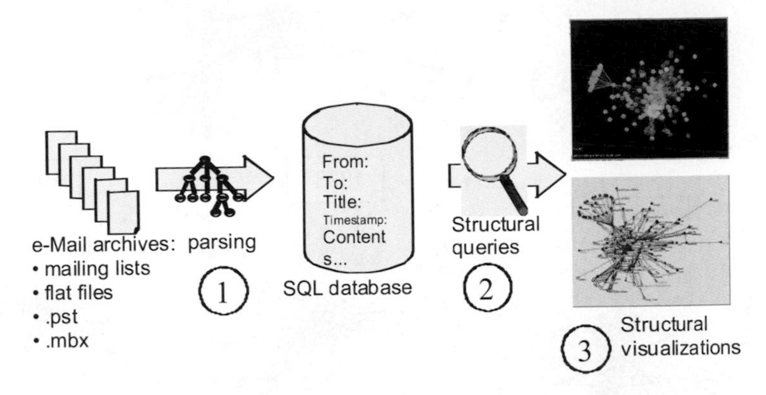

Figure 1: CKN e-mail analysis system architecture

4 First Results of CKN Visualization: KIF Community

As a proof of concept we have analyzed the e-Mail temporal communication flow of 555 e-mail messages contained in the http://www.ksl.Stanford.EDU/email-archives/ of three mailing lists on shared reusable knowledge bases, the KIF knowledge interchange format, and on reusable ontologies. The messages come from about 220 active members. The "knowledge representation" CKN was active in the period between 1990 and 1994. For this initial analysis we examined the temporal flow of emails as a function of time. Figure 2 shows the email interactions for each year of the five years covered by the archive (1990 until 1994).

The goal of this initial analysis was to determine the feasibility of identifying the different roles of the community members of the CKN over time. The initial results are encouraging: in some years there are clearly individuals who are the main senders of emails, and others who are major recipients. It is interesting to note that these individuals change with time, and that the activity level of members of the CKN differs from year to year. This is particularly evident in 1990 and 1991, where there is very little overlap in the group of active members.

There is a rich literature about small group interactions, and we were curious as to whether the results of our initial analysis would conform to patterns identified in this literature. Comparing the activity over the lifetime of the "knowledge representation" CKN with Tuckman's five classic stages of group development (Tuckmann, 1965) (Forming, Storming, Norming, Performing, Adjourning), we can comfortably fit the activity patterns into this framework: 1990 is the forming year, where one active individual is initiating the CKN, while a second stream of scattered information is bypassing this most active member.

Storming involves a core team actively involved in determining the direction of the group; 1991 shows the storming phase with a core team interacting closely. 1992 and 1993 show the greatest activity of the CKN, bystanders of previous years now also become active contributors. This is consistent with Norming (selling the goals of the group to a wider audience and Performing (working towards the goals of the group). The last year, 1995, shows a decrease in the activity of the CKN, the activity level of the most active members degrades dramatically. This corresponds to the Adjourning stage.

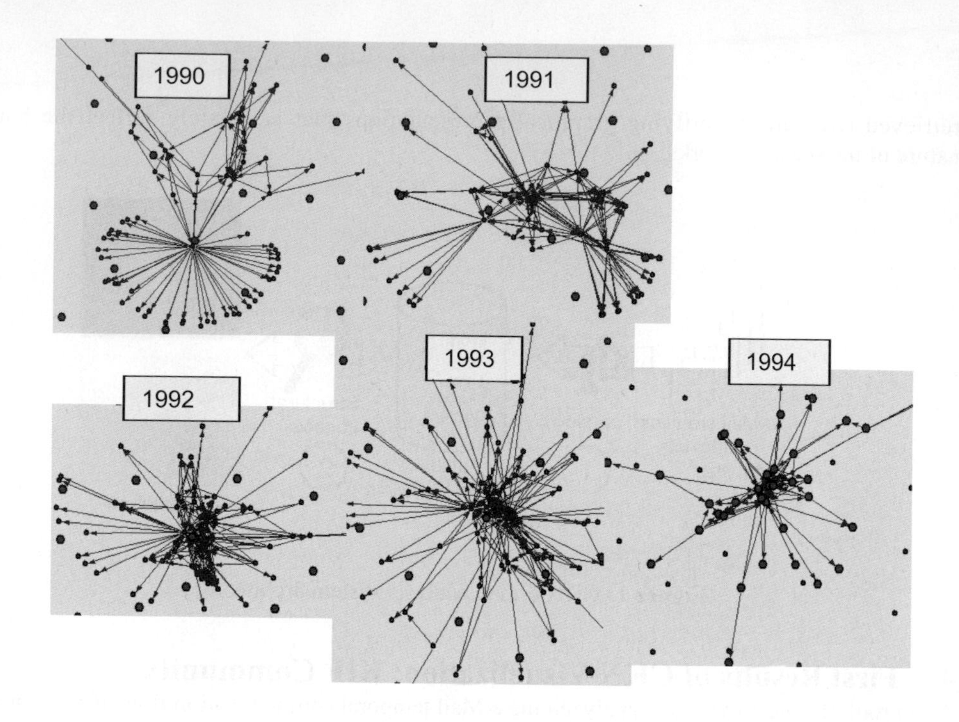

Figure 2: Email interactions as a function of time displayed in pajek. Each member is depicted as a dot. An arrow depicts email interaction between a pair of members.

5 Medical Applications of CKN

CKN concepts are of high practical relevance in the medical field. For example, they have been applied to analyze and improve the new drug development process for pharmaceutical companies (Gloor, 2002). Pharmaceutical companies are refocusing on core competencies and exploring new, more collaborative organizational models, as soaring R&D complexities and stricter regulatory standards have tripled R&D costs within the last 20 years. Companies are increasing and strengthening their collaborative relationships with external organizations for new product opportunities. Also, pharmaceutical decision-making now involves a network of influencers such as governments, insurers, HMOs, etc, although the main area of focus remains the physicians.

With growing levels of wealth and education, patients are becoming more health conscious and increase their influence in the drug buying process by turning to online shopping. For example, http://www.healthtalkinteractive.com/ provides patients with interactive advice by experts and the opportunity to learn from the experience of their own patient community.

Applying CKN concepts internally leads to many advantages for pharmaceutical companies. First, they learn about innovations, which are happening in the company although senior management is not aware of it. This leads to better and more focused allocation of resources and to the identification of new business opportunities. Second, the company becomes more efficient in working together – thus dramatically cutting time to market for new inventions. Third, localization of CKNs helps to find knowledge faster. Visualization of CKNs allows to streamline communication processes, and to locate inefficiencies. Fourth, subject matter experts who might not be high up in the corporate hierarchy can more easily be identified, allowing to better reward key contributors.

6 Conclusions

Social network analysis (SNA) researchers (Wassermann & Faust, 1994) have looked at knowledge networks for some time, but mostly by interviewing people and producing a snapshot view. (Girvan & Newman, 2001) computed the community structure of various popular SNA test data sets using a hierarchical clustering algorithm. (Guimera et al., 2002) computed the overall community structure of a university network by automatically analyzing the e-mail log. (Ebel, Mielsch & Bornholdt, 2002) analyzed the e-mail logs of Kiel Unversity and found a small world networking structure with scale-free behavior (Barabasi, 2002). In our work we automatically analyze the e-mail logs of large organizations in order to extract and cluster the CKN sub-communities. We investigate large-scale networks over time, thus providing a unique temporal view on social networks of thousand of people.

Our research will further the understanding of how virtual teams are formed, how they function, and how they die. The knowledge and tools developed will allow medical researchers, managers, and politicians to better understand how to find CKNs, how they function, what drives members to join, what the crucial roles in CKNs are, and how their success can be measured.

7 Acknowledgements

We are grateful to Tom Malone, Hans Brechbuehl, John Quimby, Tom Allen, M. Eric Johnson, JoAnne Yates, Fillia Makedon, and Jue Wang for their help and encouragement. The Dartmouth DEVLAB provided the IT infrastructure.

8 References

Barabasi, (2002). A. Linked: the new science of networks. Perseus, Cambridge, MA.

Batagelj , V. Mrvar, A. (1998). Pajek - Program for Large Network Analysis. Connections 21, 2, 47-57.

Borgatti, S., Everett, M. Freeman, L.C. (1992). UCINET IV, Version 1.0, Columbia: Analytic Technologies.

Ebel, H. Mielsch, L. Bornholdt, (2002). S. Scale-free topology of e-mail networks. arXiv:cond-mat/0201476v2 12 Feb 2002.

Girvan, M. Newman, M.E.J. (2001). Community structure in social and biological networks. arXiv:cond-mat/0112110v1, 7 Dec 2001.

Gloor, P. (2002). Collaborative Knowledge Networks. eJETA the electronic Journal for e-Commerce Tools and Applications, vol 1, no. 2, November. www.ejeta.org.

Guimera, R., Danon, L., Diaz-Guilera, A. Giralt, F., Arenas, A. (2002). Self-similar community structure in organizations. ArXiv:cond-mat/0211498 v1, 22 Nov 2002.

Tuckman, B.W. (1965). Developmental Sequence in Small Groups, Psychological Bulletin, 63, 384-99.

Wasserman , S., Faust, K. (1994). *Social Network Analysis : Methods and Applications.* Cambridge University Press.

Wenger, E. McDermott, R. Snyder, W. (2002). *Cultivating Communities of Practice.* Harvard Business School Press.

The usage of CRM system at modelling quality of products (CRM FQ - Customer Relationship Management for Quality)

Marek Goliński
Joanna Kałkowska

Poznan University of Technology
Faculty of Computing and Management
Institute of Management Engineering
Strzelecka 11 Str., 60-965 Poznan, Poland
Marek.Golinski@put.poznan.pl, Joanna.Kalkowska@put.poznan.pl

Abstract

The paper describes the possibilities to connect marketing activities with quality. Marketing activities should involve all company activities, especially those that have an influence on the competitiveness. However, quality concerns mostly manufacturing process. Nowadays, economy market requires to apply a new methods of management to gain the economic goals. It is possible by creating the proper offer with taking into consideration market requirements beginning from the phase of product design. To achieve this goal it is necessary to manage the information to join marketing with quality. Such a functions can be realised by CRM FQ *(Customer Relationship Managment for Quality)* system which determines the connection CRM *(Customer Relationship Managment)* system with QFD (*Quality Function Deployment)*. The paper presents also some problems connected with use of CRM systems for modelling quality of goods.

1. Introduction

In today's highly competitive global economy, the demand for high quality products manufactured at low costs with shorter cycle times has forced a number of manufacturing industries to consider various new product design, manufacturing, and management strategies. Companies try to implement a new techniques or methods which support management to be more competitiveness. One of such a method seems to be use of CRM (Customer Relationship Management) systems as well as QFD (Quality Function Deployment).

The necessity of adaptation to the market requirements is connected with the use of effectiveness management tools, techniques and efficient information system. A management information system is any system that provides information for the management activities carried out within an organization. Nowadays, the term is almost exclusively reserved for computerized systems. That consists of hardware and software that accept data and store, process and retrieve information. This information is selected and presented in a form suitable for managerial decision making and for the planning and monitoring of the organization's activities [Curtis, 2002].

All the aspect of management as well as marketing are closely connected with the information. Recent developments in providing business with the information and processes necessary to understand and track customers behaviour has been termed customer relationship management (CRM). Analysis techniques and sophisticated software tools have been developed to exploit the potential information contained in databases of customer details and activity. CRM is often refined into customer profiling, developing categories of customer and attempting to predict their behaviour.

2. CRM FQ systems as a connection quality with marketing areas.

To assess the usefulness of implementing management information systems (including CRM) in companies, there were realised marketing researches on the 40 companies from Wielkopolska area in Poland. The companies were chosen by the selected aim method, where the size of the company and activity area was taken into consideration. The influence of information from environment to the process of modelling quality of products were being analysed. The questions of the researches checked if correctly management information contributes to reach the economic assumptions. The researches confirmed that about 90% questionnaires agreed that the correctly information management has an influence for the realized strategies of company. They also noticed a lack of information flow between functional activities (marketing with production or marketing with quality) as well as lack of comprehensive management information system. The necessity of integration activities from quality management as well as marketing management field leaded creation of tools to make possible simultaneous realisation tasks in this two mentioned areas. For the last few years, commonly applied tools supported marketing are CRM systems. It makes possible permanent and direct contact with actual and potential customers.

CRM is comprehensive approach which provides integration of every area of business that touches the customer – marketing, sales, customer service and field support – through the integration of people, process and technology, taking advantage of the revolutionary impact of the Internet. CRM creates a mutually beneficial relationship with customers. The task of implementing and maintaining a CRM solution can seem overwhelming. But when the goal of cheaper, faster, better customer service is considered, the results far outweigh the challenges [Goldenberg., Bajarin & Selland, 2003].

The use advanced information technology techniques improves also activities of quality management area. One of such a method is QFD (Quality Function Deployment). Thanks to application this method it is possible to take into consideration possibly huge number of factors which has influence for quality of products at the all stages of designing. QFD is a systematic process for translating the voice of the customer into product and process design, and than onto the operating systems for assurance that those design are faithfully reproduced in manufacturing. The "customer's voice" is satisfied by transforming the various needs into defined design requirements, which are in turn deployed through further QFD stages of cooperation – part design, process planning and production control – to ensure that the critical customer – satisfying features are indeed produced. This whole integrating procedure uses a series of matrix charts as the tool for carrying the information [Mill, 1994]. Similarly CRM basic task is to identify the benefits the customers expect about the product and use of the data in stage of designing and manufacturing of the product. At creating CRM FQ information is the resource which makes possible simultaneous realisation QFD and CRM. Many definitions can be proposed for the term information. Information is a data processed for a purpose. Information is an essential commodity for the operation and development of a modern business organization. It is distinguished from data in that information has been processed in order to serve some purpose in the business environment. Information is used in planning, in monitoring and control of business activities and in decision making [Curtis, 2002]. About the value of information decides its usefulness in a concrete case in a given time. It is connected also with the features of information. The most often are distinguished the following features of information:

- usefulness – usefulness for decision making needs in a company
- measurability benefits – expression in a amount (money) substantiated benefits of having information
- accuracy – reliable description of reality
- topicality – delivery information in a proper time
- complete – exhaustive number of information

Because of the complexity activities of quality and marketing, the information which supply CRM FQ determines system that will include a collection of elements realised given information functions and interactions between this elements. CRM FQ is a kind of bridge connecting CRM with quality.

The essential function of CRM FQ is to keep the band with customer and the continuous research of the environment to support the decision – makers with information required to successfully create the quality of products. To realise that goal it is needed to build loyalty of a customer, to get a new customers, to make a service available 24 hours, etc. Such a system which meet all this expectations can be representing CRM FQ elements and interactions among them.

3. Modelling CRM FQ system

The basis of application of CRM FQ system is understanding that company's success is reached by the acceptance its offer on the market. System CRM FQ is an example of practical usage of information-decision making processes at creating quality of products. To achieve effective results of application CRM FQ system it is necessary to realise five steps as following:

1. The initial phase of modelling CRM FQ (see Fig. 1) system is the description the goals of company in a creating quality of products at strategic and tactical area. On the basis on that information the decision problems will be connected with the finance resources, the present and target trade estimation, trends analysis in technology, researches of the expectations and opinions of the users and potential customers with the researches for the current economic activity.
2. From the proper description of decision needs – ably asked a question concerning the quality depends the creation of the offer which will satisfy the customers. On that stage of model construction it can be useful such a methods like: value analysis, brainstorming, Pareto analysis, analysis of construction and technological parameters. The results of that research will helps in precising realisation of marketing researches.
3. The modelling requires to choosing the method which can support information management in CRM FQ. The tool which can be usefull in modelling processes, its analysis and optimalisation is software called ARIS (Architecture of Integrated Information System, IDS Scheer). ARIS is the tool for those professionals just getting started in business process optimisation. ARIS is the ideal solution for enterprise-wide utilisation. ARIS is an organisational tool for the design of business processes that is easy to use for both beginners and professionals. The target group for ARIS with its modelling, presentation, and reporting functions, are the employees in engineering departments and occasional users who document their knowledge in the form of graphic models.

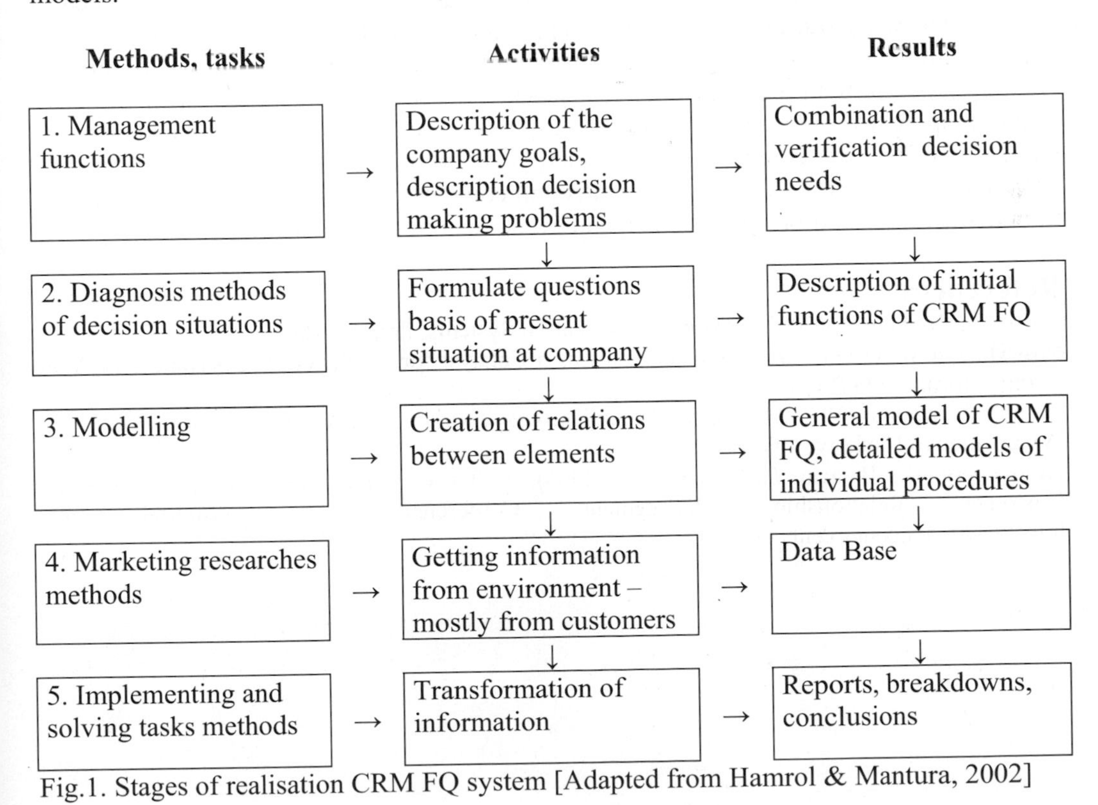

Fig.1. Stages of realisation CRM FQ system [Adapted from Hamrol & Mantura, 2002]

4. The next stage is connected with winning the knowledge. On the basis on a data from marketing researches there are build bases concerns data bases of customers with their opinion and expectations, list of potential customers, information concerning competitors, description of the technology of substitute and complementarity products. The detailed datas about the customer expectations are used to define required features of the product. The features are ranged according to their importance. Next correlation among particular technical parameters of the product are searched. For this subject a useful methods depend on interwieving both the customer and engineering staff. Gathering such information proceeds during the direct and indirect opinions. The interview, Focus Group and Conjoin Analysis are the most often applied research techniques.
5. The last stage is connected with the transformation of information to make them available according to the decision - makers needs. To make this information useful it's necessary to determine procedures concerns: sorting, evaluating, classification, updating, analysis, interpretation, archiving, and dissemination. All the activities connected with the information transformation should subject to parameters and get established changeability range like mentioned above information evaluation.

4. Conclusions

Creating CRM FQ system in company that needs a commitment significant means, systematical work during designing and application. The benefits which company can gain by integrating marketing field with quality should compensate the difficulty of creating system with parallel cost reduction. The savings connected with effective functioning CRM FQ system gave measurable benefits concerning increase the competitiveness of company and reduction overhead costs from marketing and production area. CRM FQ is a set of skills and competencies that will enable a company to better leverage and profit from each and every customer relationship.

References

Curtis G., Cobham D. (2002). Business Information Systems, Prentice Hall.
Mill H., Ion W. (1994). Development Tools as a Catalyst for Teamworking, Advances in Agile Manufacturing, IOS Press.
Hamrol A., Mantura W. (2002). Zarządzanie jakością. Teoria i praktyka. Warszawa-Poznań Wydawnictwo Naukowe PWN.
Goldenberg B., Bajarin T., Selland C. (2003). DCI's Customer Relationship Management Customer Relationship Management, Conference & Exposiotion from http://www.dci.com/events/crm

Designing a Data Management System for Monitoring Camera in Emergency

Yoshinori Hijikata, Yiqun Wang and Shogo Nishida
Graduate School of Engineering Science, Department of Systems and Human Science Osaka University
1-3 Machikameyama, Toyonaka, Osaka 560-8531, JAPAN
hijikata@sys.es.osaka-u.ac.jp

Abstract

In case of emergency, we need to grasp the situation and make correct assessment quickly. The video data taken from monitoring cameras are important information in the emergent situation. In our research, we attempt to build a support system for identifying the video data of the monitoring camera by dealing with video data as spatio-temporal data. We proposed a method, which displays those video data in the city on one virtual large wall, which is a virtually built in the city. For this presentation, the search key in our system is the location and the size of this large screen and from which direction to look at. We also developed a general algorithm to index spatio-temporal data to get an idea for developing the algorithm for this search key.

1 Introduction

Recently, the research on the advanced disaster prevention system becomes increasingly important. When a large-scale disaster happens, the video data taken from monitoring cameras are one of the information for us to grasp the situation and to make correct assessment. In many countries, there is a local disaster prevention center, which is established by the government to prepare for a large-scale disaster. In the case of emergency, the video data, taken from the disaster scene, are carried to the local disaster prevention center, and are displayed in some monitors. In the future, with the spread of net-cameras, many private cameras will connect to the local disaster prevention center through the Internet, when a disaster takes place.

It is necessary for us to check these video data to find problems in the disaster-scene, or to investigate the cause of disaster. However, if there are a lot of video data, the above task will become hard for people to do. Hence, the work to search for video data, should be done by computer. In fact, there are the following three methods to search video data: (1) Image recognition. (2) Using the index of video contents. (3) Using the information of cameras. Here, the information of the camera means the position, the recorded time and the direction of the camera. These information can be gotten by using some sensors, hence the last method is easier than the other two types methods. These data can be considered as spatio-temporal data.

In general, two kinds of search-query, which are range-search and nearest-neighbor search, are often used in spatio-temporal data search. However, the search-query should be more intuitive and simpler for the operator of the local disaster prevention center to grasp the scene situation quickly. One of such search query is the search query like "look at a certain place in a certain direction". For example, the operator may want to look at the building of the other side or want to look at the city from the sea side to know the situation of the disaster. The above search query is useful to perform these demands.

In our research, we deal with the video data as spatio-temporal data. And we attempt to build a system that supports the above search-query for the fast search of video data of monitoring cameras, which records video of a place seen from a certain place during some period and display them in a simple way.

In this paper, we consider the system outline. Before implementing the above system, we proposed XAT structure (a spatio-temporal data structure), and evaluated the performance of the algorithm. The results will be used for considering the detail of the data structure for our system. Concretely, in this paper, we will explain the system's outline, the method of search key transformation and the method to display video data and the algorithm of a spatio-temporal data structure and the computer experiments.

2 Basic Idea

2.1 Spatio-temporal data and video data

By using sensors, we can record some information about the camera's status, while the camera is taking video data. The information includes,

- Temporal information: The video data's recorded time.
- Spatial information: The camera's position.
- The other information: The camera's direction and zoom rate.

The video data can be considered a spatio-temporal data. Therefore, we use spatio-temporal data structure to manage the IDs to manage video data indirectly.

2.2 Problem setting

In our research, we choose the spatial query type, which consists of the location and the length of the projective plane and the direction to look at (Figure 1 (a)). However, no general spatio-temporal data structures support this query type. It is necessary for us to find a method, which can transfer the query type to be spatio-temporal range search (Figure 1 (b)).

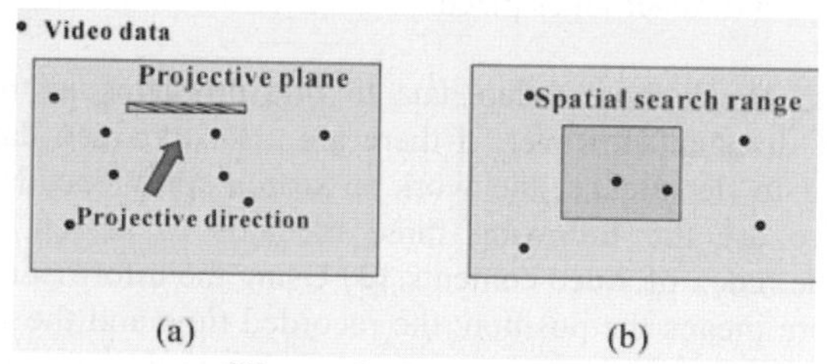

Figure 1: Query Types

When a large-scale disaster happens, we will get a lot of video data in a short time. According to the user's search request, sometime there are too much video data in the search result. Hence, It is necessary to show them in a simple way for understanding.

2.3 Approach of our research

According to the above problem, our research approach is:

1. Not only the spatio-temporal information of the camera, but also the camera's direction is used for indexing video data
2. The search key of the operator will be transformed to the search key for the data structure that manages the video data.
3. To show the result, at first the system will display the video data taken farther from the projective plane, which is easy to grasp the whole scene, then show the closer video data interactively according to the user's operation
4. The search key transformation (described in 2) will also be phased in for realizing the function described in 3. (Feedback of the user's operation to the search key transformation)

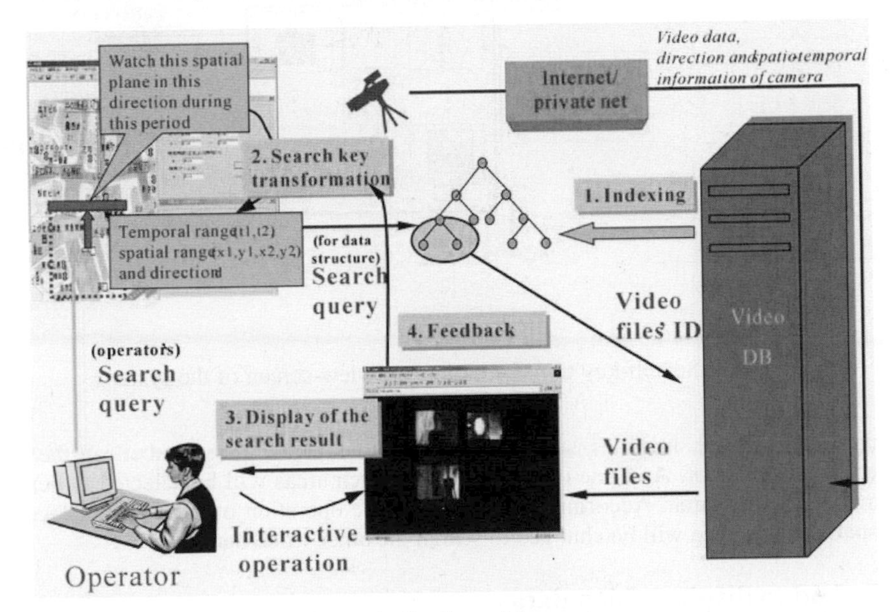

Figure 2: Outline of the system

The outline of the system is shown in Figure 2.When a large-scale disaster happens, the video data and the information of cameras will be sent to the local disaster prevention center through the Internet, which are recorded by many net-cameras. These video data will be stored in the video database, and indexed by the direction and the spatio-temporal information of the camera. These indexes of video data will be managed by a spatio-temporal data structure. For searching the video data, the operator gives the operator's search-key on the map window, which is a part of GUI of the system to display map. Then the operator's search key will be transformed to a spatio-temporal search-key, which is the search key of the data structure. Based on the search results, the video data are requested from the database, and displayed in the view-screen, which is also a part of GUI of the system. The search query can be changed with the operator's interactive operation on the view-screen.

2.4 Method of Search-key transformation

Based on the search-query (Figure 1 (a)), we defined the operator's search-key as the temporal search range, the projective plane and the projective direction. Before the system performs the search through the spatio-temporal data structure, the operator's search-key will be transformed to the temporal search range, the spatial search range and the camera's direction index. Figure 3 (a) shows the transformed spatial search areas of the operator's search-query (Figure 1 (a)).

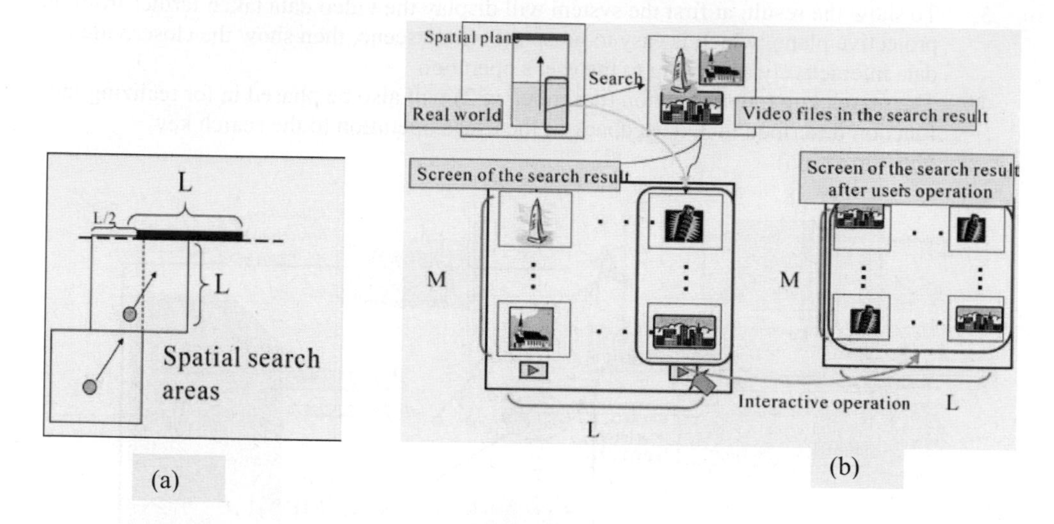

Figure 3: Search-key transformation and view-screen of the system

The spatial search range includes several spatial search areas, the number of the areas is determined by the operator. And one of these spatial search areas will be selected to perform the spatio temporal range search. According to the interactive operation of the operator (see Section 2.5), the spatial search area will be changed to search the other video data.

2.5 Method to display video data

For understanding the situation, the operator needs to check much video data, which are the results of spatio-temporal search. Hence, it is important for the system to display a limited number of video files to the operator at a time, and provide a function to narrow down the area to display video data according to operator's request. In our research, the system shows the search result by using the method, which is shown in Figure 3(b).

In this method, the view-screen has M × L areas to display video data. In each area, a video-data player is embedded for displaying video data. Because of the limitation of the view-screen's size, the number to divide view-screen should be also limited. If the positions of the cameras are close to each other and in the same direction, the contents of video data will be similar. In the system, the areas in the side direction in the view-screen are equivalent to the areas in the actual space. Before showing the search result to the operator, all the video data will be sorted by the cameras' coordinate at first. Then according to the number L, determined by the operator, video data is grouped into L groups. Instead of showing all the video data to the operator, it will select M video data in random from each group.
On the view-screen, the system provides the following functions.

1. Request for the other video data, which is not shown in the current view-screen.
2. Request for nearer (farther) cameras' video data.

(3) Narrow down the video data in the vertical direction to the projective screen.

In Figure 3(b), the operator performs Function (3) to narrow down video data, which is taken from the cameras existed in the right side of the spatial search area. In this way, the operator can get the rough image of the situation from some representative video data.

3 Spatio-temporal Data Structure

In our research, we proposed XAT (extended Adaptive Tree) structure, a data structure of spatio-temporal data, for fast search for moving objects. The object's moving track is managed as the collection of piecewise linear segments, which can be approximated for the curvy line (1). The basic idea of XAT structure is to construct two tree structures, spatial data structures and temporal data structures, instead of using only one tree that manages the spatio-temporal data. One of these trees, which is stricter in search condition, is selected by comparing the size between the temporal search-range and the spatial search-range.

The search process in XAT structure is divided into two steps to improve the search speed. The first step roughly narrows down the potential solutions (moving objects) according to the given search range. The last step fixes the real solution by checking the object's moving tracks. Furthermore, the following devices were done in XAT structure.

- Moving object's temporal data, usually treated as a segment in the time coordinate direction, is transformed into a point in the occurrence / extinction plane.
- Prepare a spatial tree structure and a temporal tree structure to manage the object's moving track segments to make the check process fast.

4 Conclusions and Future Works

In this paper, we proposed a data management system for monitoring cameras in emergency. The basic idea is to treat the video data as spatio-temporal data and by using a general spatio-temporal data structures to manage them indirectly. And we explained the method of search-key transformation, the method to display video data. We also explained the outline of XAT structure, which is a spatio-temporal data structure for moving objects. The next steps of our research are to construct the video data DB to complete the whole system.

Acknowledgments

This research was partly supported by the Japan Society for the Promotion of Science under Grand-in-Aid for Scientific Research (No. 13GS0018).

5 References

1. M. Nabil, A.H.H. Nuu & J. Shepherd (1996). Modelling Moving Objects in Multimedia Databases. *Proc. of the 5th International Conference on Database Systems for Advanced Application*, 67-75.

New Interaction Concept toward Reestablishing the Human Bonds in Daily Life

Naotake Hirasawa

Otaru University of Commerce
Midori 3-5-21, Otaru, Japan 047-8501
hirasawa@res.otaru-uc.ac.jp

Abstract

New research tasks of HCI in dairy life have emerged by innovation of ICT (Information and communication technology). One of them is how the human bonds should be reestablished by using the technology. Recent researches in Japan could give clues about new way of human communication based on ubiquitous computing. This report discusses the communication and reveals some guidelines for designing the human communication system in dairy life.

1 Introduction

The innovation of ICT has caused huge changes in not only business environment but also individual dairy life. Consequently our way of life is getting convenient and can be expected to be more comfortable. On the other hands, our feeling in dairy life is not always filled with contentment. For instance, social problems such divorce, suicide, school refusal, child abuse and so on are unsolved, if anything, are increasing.

Under these circumstances ICT is entering a new phase. The remarkable theme is ubiquitous computing. By the technology the relation between our dairy life and computer is getting much closer. Especially integrating it into communication technology like Internet could bring new and various possibilities of human communication.

Regarding the social problem described above, although the causes are very complicated, crucial problem could be caused by dividing the human bonds of various aspects. The main concern of HCI research has been relation between human and computer system. As a result of the researches the relation has been improved. Meanwhile human relations through the computer, especially in dairy life have never occupied the attention of HCI researchers.

In this paper, from the aspects of human communication, perspectives of HCI research based on ubiquitous computing would be discussed to direct our concerns to re-establish human bonds.

2 Ubiquitous computing research in dairy life

In Japan manufacturers of electric appliances expect strongly information appliances with ubiquitous computing because electric appliances market has already matured. That is, information appliances market is supposed to create a new and huge market. Consequently several manufactures or research institutes have begun to conduct research on ubiquitous computing in dairy life.

Some manufactures have set up organizations for ubiquitous computing researches; others have established exhibition halls to demonstrate future ubiquitous computing society[1].

Regarding university projects, "The Smart Space Laboratory (SSlab) project" [2] at Keio University is implemented on large scale. The project investigates the possibility and feasibility of information appliances in living conditions. As other projects "TRON" or "Stone Room" in Tokyo University might be well-known in Japan.

On the other hand, ubiquitous computing research projects for enhancement of daily life have already been conducted in the world. For example, "the Aware Home Research Initiative (AHRI)" at Georgia Tech[3] is widely noticed for new challenges of an intellectual home environment.

These projects might aim at verifying the possibility of their technical feasibility; however, very few are tested in real daily life. The following researches are validated in real daily life.

3 Some field test cases

3.1 Meeting Pot

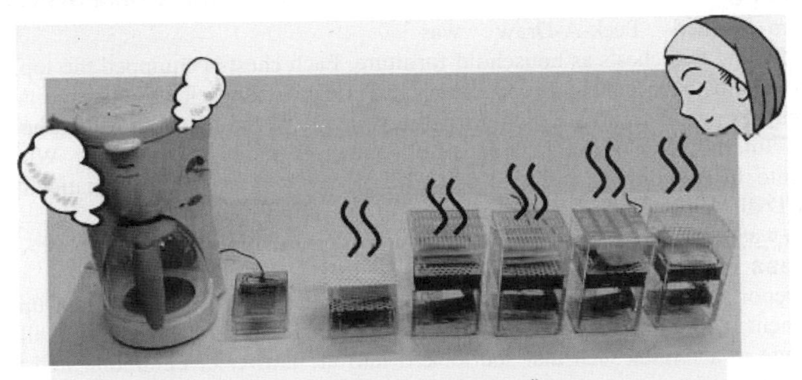

Fig. 1 Meeting Pot System [4]

Aiming

Siio etc.[5] envisioned a system to encourage people who work at isolated office room to have informal communication. The system was installed in newly constituted university where each faculty staff knew little. They were encouraged to come to coffee room as common space.

System outline

The system is called "Meeting Pot" that is a networked coffee maker. When someone tries to make coffee, it informs the event to staff at their office by coffee aroma and/or email. They can know it by browsing also. It transmits through network that coffee will be ready when the pot's switch is on. The receiver unit which is installed in each staff room starts to generate coffee aroma, and displays that people are gathering to the coffee room.

Effectiveness

It is reported that people come to the coffee room as a result of the installation. The evaluations in the daily office activities, for instance difference of information effectiveness between coffee aroma and email, are being carried on.

3.2 Peek a Drawer

Aiming

The trend toward the nuclear family has been natural. Consequently it is often difficult to live in co-located extended families and we find ourselves living across the country from members of our extended family. Even in the situation most of grandparents still want to share a part of their living space and their lives with their children and/or grandchildren. Siio etc.[7] envisioned system that made them possible to know about their grandchildren's favourite toys, their artwork, schoolwork and all those things that would be naturally shared if they lived close to one another.

Fig. 2 Peek-A-Draw System[6]

System outline

The system called Peek-A-Draw was implemented using two chests as household furniture. Each chest is equipped the top drawer with a digital camera, halogen lamps, a reed switch, LCD display and electronic circuits. The digital camera faces downward to take a picture of the contents of the upper drawer. The pictures are shared over Internet connection between the computers in the two chests. When one puts something into his/her drawer and closes it, the picture will be sent automatically to another drawer's LCD display.

The system was installed for grandmothers and grandchildren's room 400km apart.

Effectiveness

Siio etc. is reported that grandmother could not understand it well before setting up the system. After instalment, Siio found that she knew it easy to use and begin to use. Additionally he pointed out her change in activities such that she looked into the drawer or communicated with a paper written in letters.

3.3 Family Planter

Aiming

Miyajima etc. [9] constructed a new concept to feel the bond between family members when they live apart. A service of the concept for enhance the feeling were designed. It was called "TSUNAGARI Communication". The service emphasised not on intentional communication like conversation but on cue information like member's existence. By communicating the cue, they hypothesize that people felt the bond and a feeling of ease.

System outline

They developed electronic planter networked by Internet that was called Family Planter. The main functions are three. First function is to sense the existence of people and let another planter to lighten. Second is to sense the movement of people and let another planter to go round.

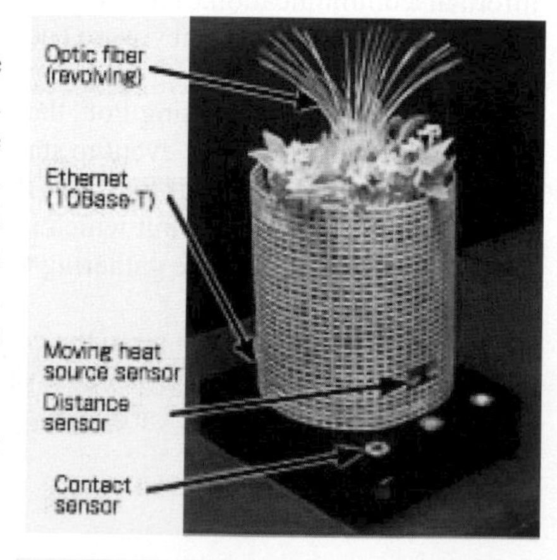

Fig.3 Family Planter System [8]

Final is to sense the touch of people and let another plant to ring. The systems were installed for four families who had members apart.

Effectiveness

Miyajima stated that family members could feel stronger bond, remind members on the other side more than before and ease their anxiety. Consequently she reported distinct internal changes of the members were admitted. Furthermore she pointed out that human relation as background was important because there were distinct differences of feeling between child and child-in-law.

4 Implications from these cases

These cases can be regarded as new communication styles based on ubiquitous computing. Three systems are built in computer into coffee maker, chest and planter which are all household commodities. From the field tests with these systems some suggestions for designing computer mediated communication system could be revealed.

Feeling and imagination of other existences

In dairy life we communicate each others not only consciously through verbal or non-verbal interaction but also unconsciously through social cues, such as footsteps, sound of cooking, sound of bathroom, etc. Feeling the cues promotes to imagine others existences. Filled with the imagination we could feel co-existence with other people intrinsically and have a secure feeling. Concern about co-existence issues are growing in Japan [10]. To encourage the feeling of co-existence in daily life the following guidelines would be needed.

Mutual communication

Information between users should be communicated mutually. Some systems are proposed to watch whether partners apart live safely. But the systems send the only information from one side and people who have a watched component often feel inferiority to opposite people.

Privacy

Privacy policy of system depends on the relationship between people who use it. Based on the policy information contents, quality, etc. should be designed.

Information access

How to access cue information should be designed properly. That is, it is needed to decide whether people should access the information consciously or get it naturally from environment; in the latter case, how people should control it.

Combination with lifestyle

The procedures to access the information should be minimized and combined with daily lifestyle. In particular for elder people the simple procedures are more important.

Multi-modality

Various multi-modals such auditory, tactile, etc. information should be considered. Furthermore the proper combination between them should be designed properly.

5 Toward re-establishing the human bonds

Some researches has suggested that main goal of ubiquitous computing is to connect one artifact with another artifact and let them to talk each other. Therefore that makes our daily life more convenient. But thinking about our daily life, it has already been convenient enough. Ubiquitous computing should be tackled from the point of view that it can create various possibilities for enhancing human communications. Because establishing secure human relationships between family, friends, colleagues etc. is urgent and crucial problem in our dairy life.

Acknowledgment

I would like to thank Professor Siio who was willing to receive my interview about his recent research projects.

References

1) http://www.panasonic-center.com/en/main.html

2) http://www.ht.sfc.keio.ac.jp/SSLab/

3) http://www.cc.gatech.edu/fce/ahri/

4) http://siio.ele.eng.tamagawa.ac.jp/projects/pot/index.html

5) Siio, I., Mima, N., "Meeting Pot", Interaction 2001, Symposium series of IPSJ, Vol. 2001, No. 5, pp.163-164, 2001

6) http://siio.ele.eng.tamagawa.ac.jp/projects/decor/indexj.html

7) Siio, I., Rawan, J., Mynatt, E., "Peek-a-drawer communication by furniture", Conference Extended Abstracts on Human Factors in Computer Systems (ACM CHI 2002), pp. 582-583, 2002.

8) http://kankyo.lelab.ecl.ntt.co.jp/eng/taisei_eco_inn.htm

9) Miyajima, M., Ito, Y., Watanabe, T., “TSUNAGARI Communication – A field Test for Family Members Living Apart”, Human Interface 2002, p565-568, 2002

10) Shimizu, H., Kume, T., Miwa, Y., Miyake, Y., Ba and Co-creation, NTT publication, 2000

The Computer Human Interface as a Partner in the Doctor Patient Relationship

Daniel B. Hoch [1,2]
Stephanie Prady [1]
Yolanda Finegan[1]
Lisa Daly [1]
John Lester [1,2]

Massachusetts General Hospital
Harvard Medical School
Boston, MA USA
dhoch@partners.org

Abstract

A variety of pressures on the provision of heath care have changed the provider-patient relationship and the ways we track that relationship. Once one on one with pencil and paper, the relationship is now strained by ever less time to manage more and more complex medical information. Record keeping systems, the "patient record", must now be viewed as a part of the therapeutic relationship. In this theoretical paper, we explore the pressures on the relationship between provider, patient and medical record. We then propose some qualities of the patient record and the interface to it that may improve medical care. We present a model in which the computer interface to the patient record is a partner with the healthcare provider and patient.

1 Introduction

The traditional medical encounter includes the patient, provider and the patient's medical record. The patient record includes clinical, billing and documentation components and is a medical legal document, subject to a variety of regulations, the most recent of which in the United States is the Healthcare Insurance Portability and Accountability Act (HIPAA)(1). That legislation, and prior efforts, have been aimed at assuring the confidentiality, accuracy, and quality of the medical record as a communication tool with the ability to appropriately monitor reimbursements and costs, and in so far as is possible, creating a tool that can be used to further our knowledge about medical care.

Since the medical record serves so many functions it can also be considered a partner in the clinical encounter. Therefore, the human computer interface in this setting is of critical importance. In this paper, we develop the idea that a well-designed interface to a shared medical record/knowledge base is a critical part of the future "doctor's appointment".

2 Reality vs. Possibility

Information technology and computers have made extremely slow inroads into the practice of health care for several reasons. Firstly, physicians are by nature somewhat cautious and tend to adopt new strategies or technologies only when there is proven benefit. Secondly, unlike financial information for example, most healthcare information is almost entirely textual. Further, the sheer volume of this text creates problems for developers designing storage and retrieval solutions. Finally, legacy systems often confound efforts to rethink information systems design in hospitals or doctors offices.

The ideal medical record must satisfy many needs. To name just a few, the application must; 1) contain all relevant data pertaining to an individual's health, 2) be accessible and readily interpretable by all persons involved in care of an individual throughout his life, 3) be protected from illicit review and use, 4) be easily updated and modified, 5) be easily shared with providers and other people important to a patients health, 6) be accurate and up to date, 7) facilitate communication and 8) foster new discoveries and research without compromising privacy.

In reality, medical records are, non-integrated, non-relational, fragmented and incomplete data sets, created in disparate geographical locations on paper, cards and computers. Although individual records tend to be more standardized and organized with the advent of the SOAP structure (notes categorized as Subjective data, Objective findings, an Assessment and Plan) (2), medical records are generally repositories more likely to be WORN (Write Once Read Never), than WORM (Write Once Read Many) (3). We might even propose the interactive ideal is even more organic, as in WMRM, (write many read many)

A major roadblock is getting the necessarily large volume of data into the computer in the first place. Data entry and retrieval during the clinical encounter has been a challenge since computers were first introduced onto clinicians' desks, often against their wishes. Advances in software and hardware such as speech recognition, natural language processing and touch-screen pens have streamlined the process, however, issues of computer literacy, perceived onus, redundancy and interview style continue to challenge acceptance. Many of these problems are a function of the interface. Interface design, whether for desktop, laptop or handheld computing device, must harness the power of the computer but be at worst benign to the clinical encounter.

3 Forces of Change

The present medical environment makes overwhelming demands on human knowledge and availability. The need for information technology to help meet these demands, through the use of an electronic medical record is now well established in the field of primary-care (4). An accurate electronic medical record must become a partner in the therapeutic relationship. In this view, the computer, and electronic medical record, is no longer an optional, peripheral tool, but a central necessity.

Technology is not the only force bringing the computer into the exam room. Increasing financial constraints in medicine are also driving ventures to streamline the process of care and make it more efficient. This, in turn, changes the dynamic between data, healthcare provider, patients and information technology. The current model seeks to empower patients so they might take active roles in their own care. As a result, a whole new realm of data is introduced into medicine - patient information. Traditionally limited to brief verbal diatribes from the physician on a need-to-know basis, patients are now being encouraged to acquire health knowledge for themselves. This new field, consumer health informatics, emphasizes the relationship between the empowered consumer and health care provider. That relationship can benefit from the increased access to knowledge, improved communication and even “intelligence” offered by the tools of information technology, if the interface is designed to allow it.

Interestingly, these new models come at a time when medical knowledge is accruing exponentially and it is impossible for any one physician to be well versed in all aspects of a subspecialty, let alone all of medicine. People are living longer with more chronic diseases, with the problems of fragmented care, multiple providers and increased medication use. This complexity has stretched our present methods for ensuring quality and patient safety to the limits. A well-designed, medical information and record-keeping interface can improve safety and enhance quality. For example, an up-to-date, shared system could be “policed” by intelligent agents working in the background to improve safety and quality.

4 The Ideal Interface and a Call to Innovate

The features of the ideal interface fall into several broad categories. We have arbitrarily identified these categories as; 1) hardware and the physical environment, 2) software affecting the presentation of information, 3) clinical decision support, 4) patient education and support and 5) support for research.

The hardware should be unobtrusive and should facilitate the one-on-one, group, or telemedical appointment. A small computer screen and keyboard, facing away from the people in the room, is sure to distract the user from eye contact, the observation of body language and other gestures that make a face-to-face encounter effective. A number of existing technologies could easily improve the physical interface. A large screen interface, such as could be created using "electronic paper" (E Ink, Cambridge, MA) could be shared by the provider and patient. Alternatively, a tablet PC could be used to scratch out notes while still making eye contact with the patient. But data entry could be facilitated even further by voice recognition. Sadly always about two years from perfection, this technology could be used to free up the clinicians hands and increase the speed with which text is collected.

Visual presentation of important clinical information is another area that needs far more attention. The interface should optimally display the quantitative and semi-quantitative data that is central to clinical decision-making. The needs of both the patient and provider must be taken into consideration in design of this representation. For example, a “clinical snapshot” that quickly allows both patient and provider to review the impact of a new medication on the patients health, e.g. a chart of blood pressure measures, medications prescribed and dosage, would be worth many thousands of words (3, 5). The effective representation of clinical data can make it much easier for the layperson to understand and adhere to a therapy as well as improve decision-making on the part of the health-care provider.

Other tools to support clinical decision-making have been shown to decrease the cost of care and decrease medical errors (6, 7). Automatic reminders, such as a reminder to the provider to perform a depression inventory during a mental health office visit, can appropriately increase the frequency with which the inventory is administered (8) In that setting, documentation also improves. Such documentation can also be used to retrospectively examine adherence to established clinical guidelines (9). However, post hoc reviews can add to the perception that "big brother" is watching and the impact on workflow and representation of the guideline to the clinician and patient must help overcome barriers to adoption, not create them (10). From the clinician's perspective, the customary physician "computer illiteracy" may limit the use of these powerful tools (11). On the other hand, those developed for the consumer are variable in quality and may not appeal to the clinician (12)

Education has already been mentioned as an important component of the interface but warrants its own separate discussion. The impact of patient education on satisfaction with health care, compliance with treatment, cost and even treatment outcomes has been well presented by Lorig and her colleagues (13, 14); much as treatment guidelines can be used to facilitate appropriate clinician behaviour, educational materials can improve patient behaviours like compliance. Continuity between the information systems used in the office, during the encounter, and systems that foster ongoing communication with the patient, is essential. Plans initiated in the office can include reminders sent later by telephone (15), cellular phone or two-way pager (16). In an overview of these types of interventions presented by Revere *et al* (17), 34 of 37 intervention trials reported improved outcomes.

The final and perhaps most interesting aspect of the medical interface must be its ability to support ongoing research. Sir William Osler encouraged us to learn from each clinical encounter. To facilitate such learning, it is again critical that the design take into account the needs of both clinician and patient. Expert patients bring much to the table (18). Face to face patient "support" groups have already begun taking an extremely active role in research design and data gathering. For example, a support group of parents of children with gastroesophageal reflux was instrumental in helping researchers identify a gene that causes the condition (19). Research on the efficacy of thrombolysis in stroke has also benefited from intensive "consumer" input (20).

5 Conclusion

In an environment that makes vast demands on human knowledge and interpersonal communication, the computer becomes a partner. The design of the interface and knowledge bank must take into account the diverse needs of the providers and patients. The roles that medical computing must support in a seamless, temporally and spatially integrated way include patient and provider empowerment, patient-patient, patient-provider and provider-provider communication, access to medical knowledge, integration of patient-centered clinical research and tools that apply that knowledge. The interface itself can do much to move medical information technology from its present status as a necessary evil to that of a valuable instrument for the shared accumulation and use of knowledge as part and parcel of ongoing patient care.

References

1. Starmer CF. Hitting a Moving Target: Toward a Compliance-driven Patient Record. J Am Med Inform Assoc. 2002;9(6):659-660.
2. Weed LL. Medical records that guide and teach. N Engl J Med . 1968;278(12):652-7 concl.
3. Powsner SM, Tufte ER. Graphical summary of patient status. Lancet. 1994;344(8919):386-389.
4. Bates DW, Ebell M, Gotlieb E, Zapp J, Mullins HC. A proposal for electronic medical

records in u.s. Primary care. J Am Med Inform Assoc . 2003;10(1):1-10.
5. Powsner SM, Tufte ER. Summarizing clinical psychiatric data. Psychiatr Serv . 1997;48(11):1458-1461.
6. Doolan DF, Bates DW, James BC. The use of computers for clinical care: a case series of advanced u.s. Sites. J Am Med Inform Assoc . 2003;10(1):94-107.
7. Committee on Quality of Healthcare in America, Institute of Medicine. Crossing the Quality Chasm. A New Health System for the 21st Century. Washington, DC: National Academy Press; 2001. 337 p.
8. Cannon DS, Allen SN. A comparison of the effects of computer and manual reminders on compliance with a mental health clinical practice guideline. J Am Med Inform Assoc . 2000;7(2):196-203.
9. Maviglia SM, Teich JM, Fiskio J, Bates DW. Using an electronic medical record to identify opportunities to improve compliance with cholesterol guidelines. J Gen Intern Med . 2001;16(8):531-537.
10. Maviglia S, Zielstorff R. D, Paterno M, Teich J. M, Bates D. W, Kuperman G. J. Automating Complex Guidelines for Chronic Disease: Lessons Learned. J Am Med Inform Assoc . 2003;:M1181.
11. Turner S, Iliffe S, Downs M, Bryans M, Wilcock J, Austin T. Decision support software for dementia diagnosis and management in primary care: relevance and potential. Aging Ment Health . 2003;7(1):28-33.
12. Schwitzer G. A Review of Features in Internet Consumer Health Decision-support Tools. J Med Internet Res . 2002;4(2):E11.
13. Lorig KR, Laurent DD, Deyo RA, Marnell ME, Minor MA, Ritter PL. Can a Back Pain E-mail Discussion Group Improve Health Status and Lower Health Care Costs?: A Randomized Study. Arch Intern Med. 2002;162(7):792-796.
14. Lorig KR, Ritter P, Stewart AL, Sobel DS, Brown BW,Jr., Bandura A, Gonzalez VM, Laurent DD, Holman HR. Chronic disease self-management program: 2-year health status and health care utilization outcomes. Med Care. 2001;39(11):1217-1223.
15. Friedman RH, Kazis LE, Jette A, Smith MB, Stollerman J, Torgerson J, Carey K. A telecommunications system for monitoring and counseling patients with hypertension. Impact on medication adherence and blood pressure control. Am J Hypertens . 1996;9(4 Pt 1):285-292.
16. Dunbar P. J, Madigan D, Grohskopf L. A, Revere D, Woodward J, Minstrell J, Frick P. A, Simoni J. M, Hooton T. M. A Two-way Messaging System to Enhance Antiretroviral Adherence. J Am Med Inform Assoc . 2003;10(1):11-15.
17. Revere D, Dunbar P. J. Review of Computer-generated Outpatient Health Behavior Interventions: Clinical Encounters "in Absentia". J Am Med Inform Assoc . 2001;8(1):62-79.
18. Lorig K. Partnerships between expert patients and physicians. Lancet. 002;359(9309):814-815.
19. Hu FZ, Preston RA, Post JC, et al. Mapping of a gene for severe pediatric gastroesophageal reflux to chromosome 13q14. JAMA. 2000;284(3):325-334.
20. Koops L, Lindley RI. Thrombolysis for acute ischaemic stroke: consumer involvement in design of new randomised controlled trial. BMJ. 2002;325(7361):415.

Communicating the Company Brand in the Investor Market: The Collocational Analysis of the Case Company's Quarterly Reports

Järvi P., Vanharanta H.

Pori School of Technology and Economics
Technical University of Tampere
P.O. Box 300, 28101 Pori, Finland
pentti.jarvi@tase.jyu.fi

Magnusson,C., Arppe A

University of Helsinki
Department of General Linguistics
P.O. Box 9, 00014 Helsinki, Finland
camilla.magnusson@helsinki.fi

Abstract

Investors are an important stakeholder group of a company. They invest capital in a company and hope to receive e.g. dividends, profitable share issues, security and rises in the stock market prices. Maintaining and developing good corporate investor relations is a company's core goal. Annual and quarterly reports are an important written tool in creating trust in investor relations and other stakeholder relations, as well as in supplier and subcontractor relations (Bogart 1986, Tuominen 1995).

This study addresses analysing quarterly reports from a communications strategic, a brand theoretical and positioning strategic viewpoint. The study addresses the issue through a method which introduces both a quantitative tool based on linguistic theory and qualitative decisions of the researchers. The empirical research objects of this study are the six quarterly reports of the case company, Sandvik (1/2001-2/2002). The method used is a collocational network. The crucial question is how the collocational networks will be interpreted from viewpoints mentioned before.

1 Aims of Study

Osborne et al. (2001) state that the text in annual reports reflects the strategic thinking of the management of a company. The specific aims of this study are, what exactly do companies say in their quarterly reports, what do they emphasise, how do they do this, and especially how the brand strategic and communication strategic thinking seen in the contents of their messages? The longitudinal research gives an opportunity to find changes in strategic thinking. The target audience in this case is supposed to be investors or stockholders, and thus potential and actual shareholders, and also other stakeholders. The real and conscious goals of the company were not studied.

Whereas many brand studies concern customer behaviour (cf. Aaker 1996, Blackston 1993, Keller 1993), investors and shareholders have not been common objects of studies. It is probable that in this context the company name is the main brand. Collocational networks can be used as a method for studying what concepts in quarterly report texts are linked to the company name, interpreted here as a brand name. The second essential feature of brand thinking is that a brand is based on the long-term strategic decisions, e.g. what kind of values a brand must reflect at all times and how to keep its consistency (Chevron 1998, Low & Fullerton 1994).

2 Brand Theoretical Aspects

The company has also to differentiate its communication on one hand to its customers and on the other hand to its investors or stockholders. Alden et al. (1999) state that themes in communications to investors may include more features of professionalism compared to themes in customer communications.

Keller (1993; cf. also Davis 1995 and Lassar et al. 1995) divide motives of brand into financial and strategy-based, consumer-oriented motives. Shareholders and owners have financial expectations towards the company. The brand is seen as an important asset for them among other assets of the company. "Brand equity is the incremental utility or value added to a product by its brand name" (Boonghee et al. 2000) and brand equity from the financial aspect is seen as market value of a brand or a firm's stock. The images are constituencies for brand equity (Biel 1993). Wrong brand strategic decisions and communication may cause negative images and reactions on the market (Light 1997).

The method in this study differs from traditional methods used in the brand context. This study aims to introduce a linguistic point of view to the study, and the main concept in this method is a collocational network. So-called collocational networks have been used by Magnusson (2002) to study and visualise linguistically the conceptual structure of quarterly reports. This study aims at giving a brand and positioning strategic interpretation of such network diagrams.

3 Methods

Collocational networks have been presented by Williams (1998), based on a theoretical framework laid out by Phillips (1985). Their purpose is to extract from a text or a group of texts the most central concepts, interpreted as individual words, and the most significant links between these concepts. This is attained by using a statistical measure to rank the significance of co-occurrences of all collocates of all words in the text as compared to the occurrences of the words in the text(s) as a whole. A collocate is a word appearing within a specific window, i.e. distance of a particular word, the latter which is often dubbed the node of the collocate. The nodes and their collocates need not be adjacent, but can be separated by any number of words based on the size of the window.

Based on the chosen statistical measure, a cut-off point is used to select which words are incorporated into the network. The network is finally constructed by combining and arranging all the selected node-collocate word pairs into one diagram. If some word appears in several word pairs, it forms a nuclear node. To a nuclear node can be attached either individual leaves, i.e. words which appear only once in and in conjunction to the node in question, or other nuclear nodes.

4 Analysis of Sandvik's Quarterly Reports 1/2001 - 2/2002

Sandvik has four business areas (invoiced sales in parenthesis, million SEK in 2001): Sandvik Tooling (16561), Sandvik Specialty Steels (14528), Sandvik Mining and Construction (13501), and Seco Tools (4259).

There are only two larger networks in Sandvik's first quarterly report 1/2001. 'Sandvik' is the most central word in the first network. It occurs 38 times in the text connected with concepts of its business areas as 'mining', 'tooling' and 'steel', later also the word 'construction'. In the second network are no central words. All words ('rates', 'units', 'exchange', 'comparable', 'fixed') are financial concepts. The same network was found in all quarterly reports in 1/2001-2/2002. Other networks in 1/2001 consist of pairs of words referring to business itself ('operating' and 'profit', 'sales' and 'invoiced, 'order and 'intake') or they are time related ('preceding' and 'year', 'first' and 'quarter). The structure of networks is quite same in the later reports.

The collocational network maps of 2/2001 and 3/2001 reminds the map of 1/2001. There is a new network in 2/2001 without any central word. The words in it, as 'items', 'affecting' and 'comparability', are present also in networks of 3/2001 - 4/2001, but then the words disappear.

In 4/2001 'sandvik' is the central word and its frequency is 75. The largest network consists of 11 words, as e.g. 'year', 'quarter', 'amounted', 'operating' and 'profit'. The same network is seen in 1/2002 and 2/2002, but with 9 words. Compared to all other quarterly reports there are in 4/2001 lots of words and lots of smaller networks (figure 1). The new words in 4/2001 - 2/2002 are 'employed', 'number', 'employees' which reflects good employment in the company. New words are business concepts as 'capital', 'return', 'gain', 'share(s)' and 'earnings', but these words disappear later.

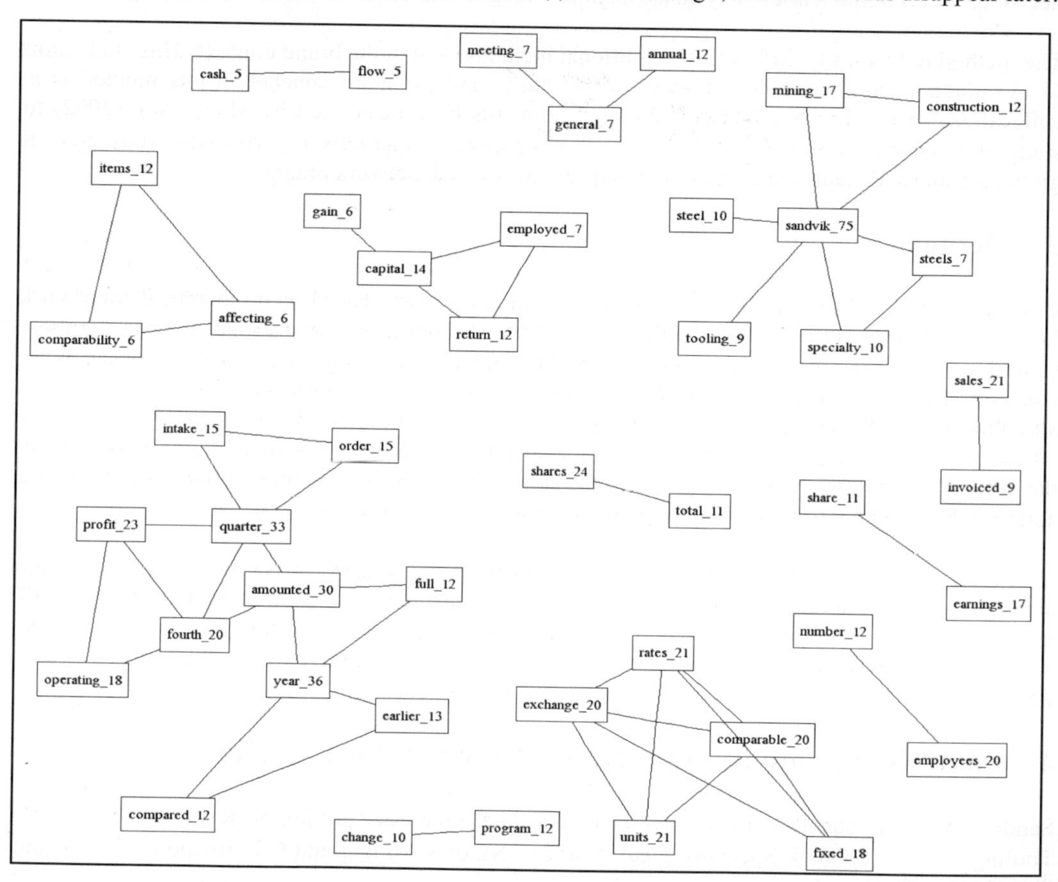

Figure 1: An example of collocational networks (Sandvik 4/2001)

There are not dramatic changes in the contents and structures of collocational networks in quarterly reports 1/2001 - 2/2002. 'Sandvik' is the only central word in all reports. A little change concerning 'sandvik' happens after 4/2001. When in 4/2001 the frequency of 'sandvik' is 75, it is in 1/2002 only

28 and in 2/2002 only 26. 'Sandvik' has only a few links to its business areas. Sandvik is still a central word in 2/2002, but it has only two links to words 'specialty' and 'mining'. Sandvik has emphasised the word 'specialty' in many reports (1/2001, 4/2001-2/2002).

5 Discussion and conclusions

We have used in this paper the method of collocational networks in analysing quarterly reports of Sandvik. The networks show the concepts that are central in the reports and the links between these concepts. Changes in the reports lead to changes in networks and words used in them. New words are introduced, others disappear. This does not mean, that the company is doing short-term brand strategy, because the company brand is frequently used and it is always the most central word in networks. Single or new product names or sub-brands are not used in the case company. The company brand's links to different concepts remain quite same in the research period. The company brand was tied quite strictly to the company's business areas or industries. The six quarterly reports show stability, consistency and professionalism.

The appearance of new words in the networks indicates that these words just have been introduced in the reports, or have become more frequent in the reports, or occur together more often with other central words. The new words represent concepts, which the company now wishes to emphasise. The networks and links also show the essential associations and images the company wishes to create in the minds of investors or stockholders and other stakeholders. Were these associations the real and conscious goals of the company and its managers, was not studied.

Different categories of words are emphasised in different networks. The analysis of the case company produced five different categories of words. There are general business words (e.g. 'sales', 'order', 'share(s)', 'profit'), industry specific words (e.g. 'mining', 'tooling', 'steel(s)', 'construction'), the company brand itself ('sandvik'), financial words (e.g. 'rates', 'units') and time-related words ('quarter', 'earlier', 'year'). The company does not emphasise words of market (e.g. 'demand', 'competition', 'market area') or words referring to customers, i.e. other companies.

The company uses financial and stakeholder-oriented words appealing to financial motives of actual and potential investors, shareholders, analysts and other stakeholders. These words are however; very different from those used in consumer communication.

The challenge for the future research is how to use collocational method better as a tool to analyse, plan and control a company's communication strategy. It may be also interesting to study how stock markets have really reacted after publishing quarterly reports, and using also other financial performance indicators.

References

Aaker, D. A. (1996). Building strong brands. New York: The Free Press.

Alden, D. L., Steenkamp, J.-B. E. M., & Batra, R. (1999). Brand Positioning Through Advertising in Asia, North America, and Europe: The Role of Global Consumer Culture. *Journal of Marketing*, 63, 75-87.

Biel, A. L. (1993). Converting Image into Equity. In D. A. Aaker & A. L. Biel (Eds.), *Brand Equity & Advertising* (pp. 70-78). New Jersey: LEA: Hillsdale.

Blackston, M. (1993). Beyond Brand Personality: Building Brand Relationships. In D. A. Aaker & A. L. Biel (Eds.), *Brand Equity & Advertising* (pp. 113-124). New Jersey: LEA: Hillsdale.

Boonghee, Y., Donthu, N., & Lee, S. (2000). An examination of selected marketing mix elements and brand equity. *Academy of Marketing Science,* Spring 2000. Retrieved April 27, 2000, from http://proquest.umi.com/pqdweb.

Bogart, J. (1986) Preparing Annual Report. In: J. L. Di Gaetani (Ed.), *The Handbook of Executive Communication*, (pp. 136-158). Dow Jones-Irwin: Homewood.

Chevron, J. R. (1998) The Delphi Process: a strategic branding methodology. *Journal of Consumer Marketing,* 15, (3), 254-264.

Davis, S. (1995) A vision for the year 2000: brand asset management. *Journal of Consumer Marketing*, 12, (4). Retrieved June 6, 2000 from http://www.emerald-library.com/

Keller, K. L. (1993). Conceptualizing, Measuring, and Managing Customer-Based Brand Equity. *Journal of Marketing*, 57, 1-22.

Lassar, W., Mittal, B., & Sharma, A. (1995). Measuring customer-based brand equity. *Journal of Consumer Marketing*, 12, (4). Retrieved June 6, 2000, from http://www.emerald-library.com/

Light, L. (1997). Brand loyalty management: The basis for enduring profitable growth. *Direct Marketing*, March, 36-43. Retrieved March 10, 2001 from http://proquest.umi.com/

Low, G. S., & Fullerton, R. A. (1994). Brands, brand management, and the brand manager system: a critical-historical evaluation. *Journal of Marketing Research*. Retrieved December 14, 2000, from wysiwyg://bodyframe.130/http://eho…

Magnusson, C. (2002). Analysing the Language of Quarterly Reports Using Collocational Networks. Unpublished Master's Thesis, Department of General Linguistics, University of Helsinki.

Osborne, J. D., Stubbart, C. I., & Ramaprasad, A. (2001). Strategic groups and competitive enactment: a study of dynamic relationships between mental models and performance. *Strategic Management Journal,* 22, 435-454.

Phillips, M. (1985). Aspects of Text Structure. An Investigation of the Lexical Organization of Text. Amsterdam: North-Holland.

Tuominen, P. (1995). Relationship marketing - a new potential for managing corporate investor relations. In J. Näsi (Ed.), *Understanding stakeholder thinking* (pp. 165-182). Jyväskylä: LSR-Publications.

Williams, G. C. (1998). Collocational networks: Interlocking patterns of lexis in a corpus of plant biology research articles. *International Journal of Corpus Linguistics,* 3, 151-171.

The Development of a CAD System for Carbon Fishing Rod

Cheol Min Joo, Hyun Seok Jung*, Byung Gun Lee**, Yasufumi Kume****

* Division of Information System Engineering, Dongseo University, Busan, Korea

** Division of Leisure & Sports Science, Dongseo University, Busan, Korea

*** Division of Industrial Engineering, Faculty of Science and Engineering, Kinki University, Osaka, Japan

Abstract

In this paper, a study is introduced on the development of the computer aided design system for carbon fishing rod which is designed to aid the fishing rod designer offering all the information needed in making rod or assembling items with graphic representation. In a multi-pieces carbon fishing rod design, the joint part of any two consecutive pieces is very important for the quality of the rod. In general, additional carbon patterns are used to strengthen or fit the joint part and their sizes are decided by experiment or by trial and error. The genetic algorithm is used to decide the sizes of additional carbon patterns for the joint part and included in the computer aided design system. The design system is developed to play under graphic user interface to maximize the efficiency and convenience in fishing rod design.

1. Introduction

A carbon fishing rod is joined with several pieces by the method of telescopic, taking a piece put in, or taking a piece put over, etc. The pieces of a rod are made to roll up the carbon sheet on the prepared circular molding (called mandrel). Sometimes new set of mandrels is made on the time when the new rod item is designed, but the common pre-made mandrels are re-used, in general. So, the selection of the proper set of mandrels is very important for a good action, robust, and light carbon rod. The computer aided design system for carbon fishing rod introduced in this paper helps the rod designers to select a proper set of mandrels simply by quarry the on hand mandrel information from the database. The most important thing in designing a good action, robust and light carbon rod is the selection of the proper kind of the carbon sheet and the decision about the size of the patterns. There are three types of patterns in a piece for a rod, main pattern and two sub patterns. The main patterns are the essential element of the piece, the one sub patterns are needed for strengthen the joint part (called strengthen sub pattern), and the other sub patterns are needed to join the two consecutive pieces smoothly (called joint sub pattern). In general, the sizes of the patterns are decided by experiment or by trial and error. In our design system, the proper kind of carbon sheet for the pattern of a piece is selected simply by quarry the carbon sheet information from the database, and the size of the pattern is decided to utilize the graphic representation of that pattern in the viewer composed with the information about inner and outer diameter of the piece which made from the decided patterns, location and size of the decided patterns, and the difference between the diameters of the piece and the joined pieces. The information is reset immediately on the time when a new pattern is added or the size or location of the decided patterns is modified. Especially, the size of a joint sub pattern can be decided automatically using the genetic algorithm included in the design system. The genetic algorithm is developed to find the most suitable size of a joint sub pattern to join the two consecutive pieces smoothly. The design of a carbon fishing rod is completed with assembling the assemble items, handle, guide, reel seat and cap, etc. In the design system, the proper kind of assemble items for a rod is selected simply by quarry the assemble item information from the database, and the assemble location of a assemble item is decided to utilize the graphic representation of the whole rod in the viewer composed with the information about the shape and location of the decided assemble items. The standard length and weight, the possible number of patterns derived from a unit carbon sheet, the manufacturing cost of a rod, and all other information needed to design the rod are calculated automatically during the design of the rod, and can be viewed any time in a graphic form or a table form. All the information for cutting the pattern, making the rod, assembling the assemble items, standard length and weight of the rod, and the manufacturing cost can be printed out on a paper in the form of manufacture ordering sheet with the graphic representation. The computer aided design system for carbon fishing rod introduced in this paper is developed to play under graphic user interface to maximize the efficiency and convenience in fishing rod design.

2. Structure of the system

The system is composed with modules for user management, elementary information management, rod design, assembly, and cost calculation (refer to figure 1). The elementary information needed for rod design such as rod items, mandrels, carbon sheets, and assembly items are all managed in the module for elementary information management. In the module for rod design and assembly, the kind, the location or the size of mandrels, carbon sheets, patterns, and assemble items for design a carbon rod is decided to utilize the information from the database and the graphic representation of the whole rod in the viewer composed with all the information needed for design such as inner and outer diameter. The production cost with carbon sheets of a designed rod is calculated automatically in the system.

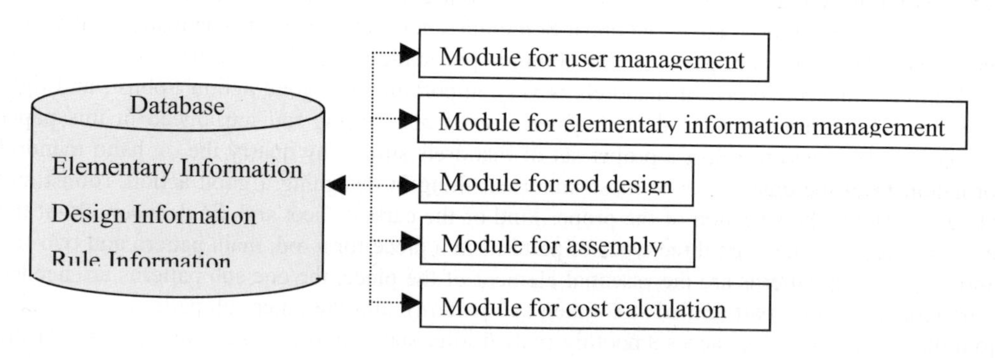

Figure 1. Structure of the system

3. Embodiment of the system

The core of the system is the module for design, assembly, and cost (Joo (1999,2000)). The modules are designed to aid the fishing rod designer offering all the information needed in making a rod with various graphic representation on screen (refer to figure 2).

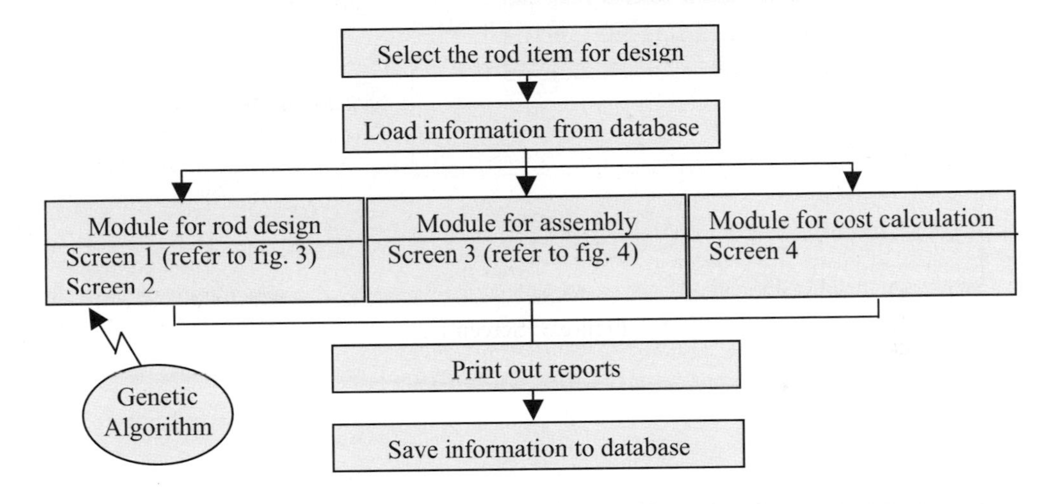

Figure 2. Core of the system

The screen 1 of the module for rod design is composed with two parts. The upper part shows information about selected mandrel, selected carbon sheet, and the size of patterns of all pieces of the rod, and the selection or editing the information is possible simply by quarry from the database on additional screen. The lower part shows information of the selected piece about inner and outer diameter made from the decided patterns, location and size of the decided patterns, and the difference between the diameters of the piece and the joined pieces. The screen 2 of the module for rod design is also composed with two parts. The left part shows information about cutting of the selected pattern, and the right part shows the information about possible number of patterns derived from a unit carbon sheet. Mouse working on the screen 1 and 2 can change the related information directly. The screen 3 is for the module of assembly. The upper part shows information about kind, size, and location of the selected assemble items such as handle, guide, reel seat and cap, and the selection or editing the information is possible simply by quarry from the database on additional screen. The result is represented by a scale-down graphic representation of the whole rod on the lower part with the shapes and locations of the decided assemble items. The production cost with carbon sheets of a designed rod is calculated automatically in the system, and can be viewed any time in a table form on the screen 4.

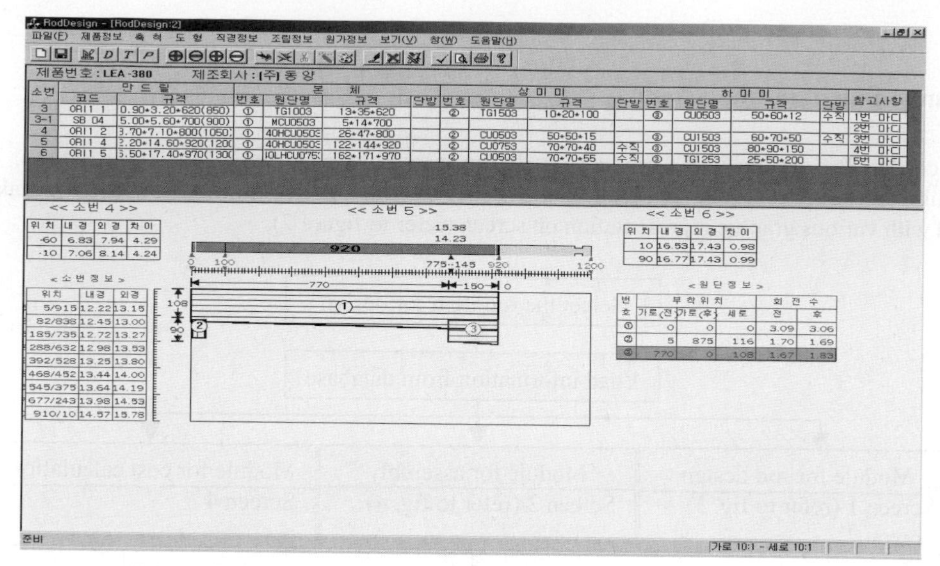

Figure 3. Screen 1

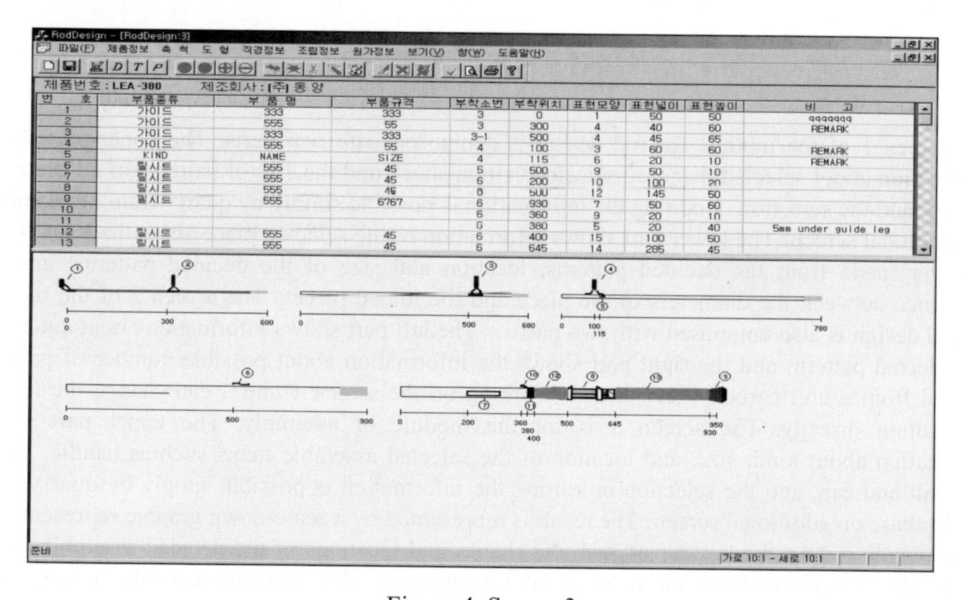

Figure 4. Screen 3

4. Conclusions

In general, the design of a good action, robust, and light carbon rod is dependent on the experience and a number of trial and error of an expert, because the operations for rod design is not standardize. This is the reason to increase the price of fishing rods and the percentage of bad items in production. The system for carbon fishing rod introduced in this paper is developed to approach systematically and scientifically on the operations for rod design, and aid the fishing rod designer offering all the information needed in making rod or assembling items with graphic representation. The period for design and develop the rod item is expected to reduce and the quality is expected to improve using the system. As a result, the computer aided design system contributes largely to the price competitive power of the fishing rods.

References

[1] Joo, C.M., Lee, B.G., and Jung, H.S. (1999). A Study on the Development of a CAD System for Fishing Rod Design, Technical Report, Dongseo University, Korea.

[2] Joo, C.M., Lee, B.G., and Jung, H.S. (2000). Development of a Software for Fishing Rod Design, Technical Report, Small and Medium Business Administration, Korea.

Virtual museum of informatics history in Siberia

V.N. Kasyanov, G.P. Nesgovorova, T.A. Volyanskaya

A.P.Ershov Institute of Informatics Systems
Lavrentiev pr. 6, Novosibirsk, 630090, Russia
kvn@iis.nsk.su

Abstract

In the paper the virtual museum of Siberian informatics history being under development at A.P. Ershov Institute of Informatics Systems in Novosibirsk is presented.

1 Introduction

The history of informatics (computer science), like the history of any other science, is an important and inseparable part of this science. During previous years the teaching of the history of computer science was introduced into the computing curriculum of many Western universities. At the same time, informatics history of the Eastern Europe and the USSR was practically unknown in the Western Europe, although some works on this problem have been published (Ershov, 1975; Ershov & Shura-Bura, 1980), and in 1996, the IEEE Computer Society, in connection with the 50th anniversary of its founding, presented the Computer Pioneer Award to 16 scientists from Central and Eastern Europe countries, including outstanding Russian scientist, academician Alexej A. Lyapunov, who "developed the first theory of operator methods for abstract programming and founded Soviet cybernetics and programming" ("CS Recognizes", 1998).

Research in programming in Siberia has been started after A.A. Lyapunov and his disciple Andrei P. Ershov had arrived to the Novosibirsk Academgorodok (at the beginning of the 60th of the last century). Academician A.P. Ershov and his disciples have founded the Siberian School of programming and informatics that was the third one in the USSR, after Moscow and Kiev. And now, many years after its founder A.P. Ershov died (Bjørner & Kotov, 1991), it keeps on playing an important role in spite of all these difficulties endured by the Russian science and education. This gives us an opportunity to independently investigate formation and development of informatics in Siberia, namely, in the Novosibirsk Scientific Centre, against the background of the Russian and world processes.

In the paper our project SVM of the virtual museum of informatics history in Siberia is described (Kasyanov, Nesgovorova, & Volyanskaya, 2002). The SVM museum is intended for being used by different categories of users, and museum visitors with different preferences, goals, knowledge and interests may need different information and may use different ways for navigation. Therefore, we give a particular attention to adaptation problems in our project.

The project is supported by the Russian Foundation for the Humanities (grant N 02-05-12010) and is based on informatics history pages of the Web-system SIMICS (Kasyanov, 2000).

2 Structure of the virtual museum and its content

Data bases (DBs) of our Web-based museum SVM provide storage and processing of the information about following objects: publications, archival documents, projects, data about scientists in informatics, scientific teams, various events concerning informatics history, conferences, and computers. All the above objects are the *exhibits* of our virtual museum. Every exhibit has the following attributes: a Unique Universal Identifier (UUID) of an object, a name, sometimes a date, a brief description (or an annotation), a full description (or a file), a name of a person who added this exhibit, the date of its addition, the possibility of its modification and participation in exhibitions and modification permissions.

A set of exhibits united according to the thematic, chronological or typological criteria can be represented as an *excursion* or an *exhibition*. Both an exhibition and an excursion have the following attributes: UUID, a name, a name of a person who made it, a brief description, and a reference(s) to the file(s), representing it contents. An excursion is composed of one section (a file), while an exhibition can consist of a few sections (exhibitions or sub-exhibitions). Another difference between an exhibition and an excursion is the following. An excursion is a story about the museum (elapsing in the time) followed by demonstration of its exhibits in a definite order. An excursion, for example, may be a clip or a presentation for MS PowerPoint and may be conducted for users not only in the on-line mode but sometimes off-line. In contrast to an excursion, an exhibition consists of exhibits which a visitor is looking over himself and only on-line. Usually, several ways of navigation, including a free movement among exhibits, are provided. All exhibitions (and excursions) are divided to *permanent* and *temporary* ones. A *hall of exhibitions* and a *hall of excursions* are composed as open ones, i.e. accessible to all users of the museum.

There are also *reserve halls* in our museum: the library, the archives, the chronicle, the halls devoted to scientists in informatics, scientific teams, projects, computers, conferences, the hall of new exhibits and the hall of preparation of exhibitions and excursions.

The *library* consists of books, articles and so on. In addition to the general exhibit attributes each library exhibit has the list of authors and etc. An *archive* consists of text, graphic, audio and video materials. A *chronicle of events* contains description of the most remarkable events of informatics history in Siberia. A *hall of scientists* in informatics presents information about the prominent scientists in informatics, such as the biography, the main published works and achievements, photos and so on. Besides the general information it is provided the following data about scientists: their education, scientific degrees, titles and posts, scientific interests, the text of the biography, photos, the main publications and projects. The *hall of scientific teams* presents information about groups, laboratories and institutes. Except for the general attributes each team has the address and etc. In the *hall of projects* information about informatics projects is placed; apart from the general information, a project has the dates of its beginning and finishing. The *hall of computers* presents computers, which were used and created in the Siberian Division of the Russian Academy of Sciences. In addition, each exhibit has the name of the designer, and the photo. The *hall of conferences* contains the following information about each scientific event: where and when it was held, its status, and the general exhibit information.

New entries to the museum (adding by users) are placed in the *hall of new exhibits*. Exhibitions and excursions creating by users of museum are being composed in the *hall of preparation of exhibitions and excursions.*

3 Users

All users of our Web-based museum are divided in two main categories: unregistered users (visitors) and registered ones (specialists) with different access level to information resources. *Visitors* have access only to the part of museum information that is opened for public access (for example, in the form of excursions and exhibitions). In this case all resources are accessible only for review and search. Visitors are divided in two subcategories depending on their knowledge level of subject domain: beginners and experts. *Beginners* have an opportunity to look over excursions, and *experts* — exhibitions and electronic conferences of users. *Specialists* have access to reviewing all information resources of our museum, including reserve halls, closed for public access; they can also take part in electronic conferences and write in a visitors' book.

All specialists are divided in two main groups depending on their access level to resources: a group of *simple specialists* working only in the hall of new exhibits and a group of *museum employees.* Volunteers, excursion guides and expositors are selected from a group of simple specialists. *Volunteers* have permissions to add new exhibits of any type. *Excursion guides* may create their own excursions, and *expositors* — the exhibitions. Objects made up or added by them are at first situated in the hall of new exhibits, and then administrators of the corresponding resources (for example, chief excursion guide or chief expositor) make a decision on including them in the museum resource. Volunteers, excursion guides and expositors have no permissions to modify the museum DBs.

A group of museum employees can be presented as a hierarchical structure, at the very top of which is a *director* (or a chief administrator). He has all permissions to administrate the museum DBs, including DB of museum users. The second level of the hierarchy consists of *administrators* of corresponding museum resources. They are appointed by the director: the chief expositor, the chief excursion guide, the chief librarian, the chief archivist, the chief chronologist, the chief biographer, the chief specialist in scientific teams, the chief projector, the chief engineer, the chief secretary. Resource administrators have all permissions to administrate DBs of the corresponding resource types. They also control specialists working with DBs of corresponding types of resources. The third level of the hierarchical structure includes museum employees appointing by administrators of the corresponding types of resources: librarians, archivists, chronologists, biographers, specialists in scientific teams, projectors, engineers, and secretaries. They have limited permissions to change DBs of the corresponding resource types.

4 User interface

Current implementation of the DBs of our virtual museum provides storage and processing of data about the following objects: publications, archival documents, projects, events, scientific teams, scientists in informatics, computers, and arrangements. At present, the hypermedia interface of the DB for information filling of our virtual museum has been designed and implemented: for review, search, insert and update of data on above objects and for linking them together. An interface for registration and authentication of the museum users and for holding user electronic conferences has been implemented. The following main functions are supported by the interface.

The registration of a new visitor consists in filling a special form. It contains required (a user name, e-mail, a password for entry to a system) and additional (a country, an index, an address) fields. If the user wants to insert information to the museum (to add exhibits, make up excursions or exhibitions), he should mark the corresponding points in the registration form. A login (a user name for entry to the system) is generated automatically by a special algorithm and sent to user's e-mail address. The director or a corresponding administrator does the registration of a new specialist.

A search mechanism by key criteria is implemented; there is an opportunity to carry out a pattern search over all DBs. It is possible to choose the form in which the search results are represented. Brief information about objects and a reference to complete one is put out as a search result.

An interface for insert and update is implemented for all resource types of our museum. Data input is realised with filling of the corresponding forms depending on resource type: forms for input general and additional information, and forms for linking objects together. UUID is generated automatically and assigned for every new added object. The following information about added object is automatically put on the DB: the name of a person who added an object, the date of addition, possibility of modification and taking part in exhibitions, modification permissions. An interface for data modification is implemented via data edition in the corresponding fields of the form.

An interface for interconnection between objects linking them together is implemented when information inserting or updating. Connection of objects is implemented via a choice of corresponding objects (that are needed to link together with this object) from a list of all possible objects (for each object type). Information links between objects are generated as hyperlinks from given object to objects connected with it.

An interface for holding a user electronic conference is implemented. Unregistered users can only view the conference information, while registered ones can send information to the conference. This interface supports all standard functions of electronic conferences: sending a new message, receiving an answer and search for messages according to some key criteria.

5 Adaptive hypermedia and interface adaptation

Adaptive hypermedia (Brusilovsky, 2001) is an alternative to the traditional approach of the hypermedia systems development. It is intended to increase the functionality of hypermedia by making it personalized. Adaptive hypermedia systems are all hypermedia systems, which reflect some specific features of the user such as preferences, knowledge and interests in the user model and use this model for adaptation of different visible system aspects to the user.

The most of museums presented in WWW now are traditional hypermedia systems and give the same information and navigation to all users. At the same time, our virtual museum SVM is intended for being used by different categories of users, and museum visitors with different preferences, goals, knowledge and interests may need different information and may use different ways for navigation. Therefore, we give a particular attention to adaptation problems in our project (Kasyanov, Nesgovorova, & Volyanskaya, 2002). For adaptive information presentation in our museum SVM it is supposed to use the adaptive presentation methods, such as additional and prerequisite explanations and sorting, and the adaptive navigation support methods, such as direct guidance, link sorting, hiding, annotation and generation. It is supposed that the model of a registered user of the SVM will consist of tree parts: the model of categories, the model of knowledge and the model of preferences.

The model of categories is supported for all registered users, the model of knowledge and preferences — for all categories of users except the group of museum employees. The model of categories represents the access permissions to the museum information resources. It is implemented as a static stereotype model (a set of attribute-value pairs). The names of types of the DB resources are used as attributes of the model; the access permissions to these resources (view, insert, modify and their combinations) are the attribute values. A stereotype of the same name, which is characterised with specified attribute values, corresponds to each user category.

The model of knowledge is used to model the user domain knowledge. It is supposed to implement a model of knowledge as an overlay model based on the structural domain model. The structural model is used for representation of presented by the museum information as a structure of interconnected concepts and relations between them (acyclic graph). The overlay model is intended to present the user knowledge as an overlay of the domain model. The overlay model for a user is a table structure. This structure determines the values of the following attributes: knowledge of concept (studied, not studied), reading (read, not read), ready for reading (ready, not ready) for each domain concept. The overlay model is a dynamic one: it is automatically updated when a user reviews the information.

The model of preferences represents different user preferences, in particular, a method of information presentation (using only a text, graphics, audio, video and so on). It is implemented by a static stereotype model. The attributes of this model are the methods of information presentation mentioned above, and their values are true or false.

References

Bjørner, D., & Kotov V. (Ed.). (1991). Images of Programming. Dedicated to the Memory of A.P. Ershov. Amsterdam: North-Holland.

Brusilovsky P. (2001). Adaptive hypermedia. *User Modelling and User-Adapted Interaction, 11* (1), 87-110.

CS Recognizes Pioneers in Central and Eastern Europe, (1998). *IEEE Computer, 6*, 79-84

Ershov, A.P. (1975). A history of computing in the USSR, *Datamation, 21* (9), 80-88.

Ershov, A.P., & Shura-Bura, M.R. (1980). The early development of programming in the USSR. In *A History of Computing in the Twentith Century* (pp. 137-196). New York: Acad. Press.

Kasyanov, V.N. (2000). SIMICS: information system on informatics history, In *Proc. Intern. Conf. on Educational Uses of Information and Communication Technologies. 16th IFIP World Computer Congress 2000* (p.168). Beijing: PHEI.

Kasyanov, V.N., Nesgovorova G.P., Volyanskaya T.A. (2002). Virtual museum of informatics history in Siberia. In *Modern problems of program construction* (pp.169-181). Novosibirsk: IIS SD RAS. (In Russian).

The Impact of Flexibility Management on Total Chain of Manufacturing

Kayis, B., Kara, S., Skutalakul, K.

The University of New South Wales
Sydney, 2052 NSW Australia
b.kayis@unsw.edu.au

Abstract

As the business environment has grown more and more competitive, gaining competitive advantage is highly related with the ability to meet customer's demand. However, there has been an increasing trend in the need of customer over the past decades; the years to come will belong to "flexibility". It is with this flexibility that the company would be able to supply customised products, cope with immediate design change and fluctuating order sizes, satisfy the varying delivery time and price levels and meet the demand for different quality levels. This research was carried out to examine the organisational process of some Australian manufacturing companies, regarding their organisational structure, manufacturing practices, supplier's contribution and customer's contribution. The analytical evaluation and interpretation of survey data allows for the formulation of relationships involving different parameters related to the total chain of acquisitions, processing and customer. Such relationships could be used to predict the contribution of flexibility that is being practiced by the Australian manufacturing industry.

1 Introduction

The current thinking amongst those driving manufacturing is that industry is moving away from mass production towards mass customisation. The response to the market tends to be dramatic to cope with the ever-changing demand of customers in order to be or remain competitive. Nowadays, all the efforts of restructuring, re-engineering, new management approaches are simply to make companies flexible enough to respond to the fast-changing customer needs, as flexibility, another competitive weapon used in today's competitive markets, is defined as "the ability to respond effectively to the ever-changing and increasing needs of the customer" (Mendelbaum, 1978, Sethi and Sethi, 1990).

The flexibility at acquisition-processing-distribution stages can be achieved if companies use flexibility mechanisms/parameters between and within each stage to enhance their existing capabilities and strategies (Aggarwal, 1997). At acquisition stage, the relationship and participation with suppliers is critical in relation to supplier flexibility. The participation could be incorporated in terms of different aspects within the industries, such as design, equipment, maintenance, policy planning etc. To achieve flexibility at the supplier interface, firms should look at some of their major suppliers as partners, and where possible encourage them to take part in strategic activities such as product and process design and development. In many cases, suppliers should have complete responsibility of component testing and quality control (Kamath and Liker, 1994).

The companies' competitive priorities are what the company strategy demands from manufacturing. The overall picture in Australia, as it has emerged, is that the Australian manufacturers seem to be striving for higher quality but not primarily seeking lower costs and flexibility. Australian manufacturers are aiming at overcoming what they perceive to be the relative deficiencies of their manufacturing compared to their competitors, whereas in the near future the challenge will be low cost - flexible manufacturing with exceptionally high quality. The successful adaptation of low cost-flexible manufacturing lies on the enhancement of existing manufacturing capabilities to develop competition advantage (Kara et al., 2002).

2 Formation of Relationships

Flexibility of elements could be linked at acquisition-processing-distribution stages of manufacturing. Relationship and correlation of data gathered in this research would enhance the observations on flexibility mechanisms at different stages, including the supplier-customer flexibility and its impact on the total chain of manufacturing. The relations formed are used to support the overall flexibility assessment of Australian industries. Accordingly, the relationships are divided into cases of further analysis as follows:

Case 1 Effect of supplier's participation in the manufacturing problems.
Case 2 Effect of customer's participation on the manufacturing problems.
Case 3 Dependency of manufacturing objectives on the system bottleneck.
Case 4 Process choice and manufacturing job's context/depth.
Case 5 Dependency of manufacturing objectives on demand for variation.
Case 6 Demand for the variation and the ability to respond to such demand.
Case 7 Level of suppliers and level of customers.
Case 8 Effect of ownership on manufacturing objectives.
Case 9 Manufacturing objectives and product market spread.

In the paper, mainly Cases 2 and 4 are covered in detail. Logistic regression models and significance tests are used to analyse the cases.

The summary of the findings related to Cases 2 and 4 are as follows.

2.1 Analysis of Case 2

Out of the 9 models formed, only 4 models are proved to have relevant relationships. Then, the independent variables under those models are tested for their significance effect. At last, the models would be comprised of just the significant variables that were proven significant.

The relevant models with their significant parameters have the dependent variables of manufacturing problems, namely Product/process design (Y3), Material Handling (Y4), Policy/planning (Y7), and Equipment/technology/maintenance (Y8).

2.1.1 Findings

- Co-ordination with customers in terms of product design, delivery time criteria, and raw material used helped reduce the problem of product/process design. This could be done in the form of voice of customer (VOC), a technique to gather customer's requirements to improve the design process. This way, changes could be made during the design phase where cost of such changes is minimal, especially that emphasis on product design, material used, and delivery time could make the design of the product and its process simpler accordingly.
- The material handling problem is affected considerably by the product/process design, raw material to be transferred, and equipment/technology/maintenance through the participation of customers. Once the industries have paid attention regarding the mentioned criteria, their material handling system tends to be simpler and more satisfying.
- The existing policy/planning with the customers usually draws attention to the product design, timely delivery commitment, and establishment of detailed policies.
- The equipment/technology/maintenance planning of the industries requires the consultation on product/process design, policy, technical aspects, and marketing/sales from the customers. This tends to be acceptable due to the fact that once customers have agreed on thorough procedures of such criteria, conflicts regarding the selection of equipment used to correspond to the process, the level of technology required to satisfy the customer's needs, and the maintenance of such technology would be reduced.

2.2 Analysis of Case 4

It is very important for the companies to have the process choice decision made in accordance to the type of product, market requirements, competitive positioning, production volumes, product variety, and the job's context and depth. Out of the factors that could affect the determination of process choice, the study has chosen the manufacturing job's context/depth to find the type of relationship with the determination of process choice. Keeping in mind that there should be a link between the two, the study would be able to examine the suitability of current Australian industry practices in terms of process choice. The process choice would generally be categorised as job shop, projects, mass production, or continuous flow, whilst the scale of the job's context and depth being classified as 1 (Narrow) to 5 (Broad).

The simple linear regression model is used to fit the relationship between the job's context/depth and the process choice. The model is further tested with ANOVA (Analysis of Variance), comprising of the F-test for testing the significance of regression. Also, the test on the significance of coefficients was introduced.

The linear regression model is further interpreted, where the positive relationship could be seen. As the job's context and depth varies from narrow to broad, the process choice corresponds from continuous flow to job shop, respectively.

2.2.1 Findings

- The relationship conforms to the theory. As the context of the jobs get broader, the flexibility in manufacturing process should increase to respond to such changes.
- Job shops are matched with broaden context of the jobs, since the knowledge, skills, and equipment are capable of performing different jobs, which may be quite different in size and type. Equipment that are mostly general purpose and flexible allow the skilled workers to work on varying specifications.
- On the other hand, continuous flow process is generally automated process with minimal labour resources. The process line is then dedicated to the high volume production with low flexibility in varying the job type.
- As the job's context narrows down, the justification for investing in processes that are dedicated to make that specified products increases. Generally, the process starts of with the one that is flexible yet not very costly and tends to move to the low cost production by the dedicated repetitive production, which may be capital intensive.
- The current determination of Australian industries on right process choice to match the job's context/depth is found to be satisfying. Choosing the appropriate process choice could be a good start for manufacturing objectives in the long run.

3 Conclusion

This research serves as a baseline model on how customer-supplier relationship could have an impact on manufacturing flexibility as well as examining the current flexibility practices of Australian industries.

As different elements under manufacturing flexibility have suggested, the manufacturing flexibility of Australian industries as affected by customer-supplier participation is "Medium", as Table 1 below suggests. The level of flexibility for several flexibility elements in the research have been rated as "High", "Medium" and "Low" for simplicity of understanding.

FLEXIBILITY ELEMENTS	FLEXIBILITY LEVELS
Supplier Flexibility	Medium
Customer Flexibility	Medium
Manufacturing Flexibility Process Flexibility Design Flexibility Human Resources Flexibility Policy Flexibility	Medium

Table 1: Flexibility Elements and Related Flexibility Levels

Even though the flexibility level under different criteria is evaluated to be at Medium level according to the survey data, it does not mean that the performance of Australian industries do not need improvement. As flexibility is defined as the ability to keep up with the ever-changing and increasing demand, the industries need to keep in mind, the continuous improvement methodology including the aspects of flexibility.

References

Aggarwal, S.I., 1997, Flexibility Management. The Ultimate Strategy, Industrial Management, Vol. 39, No.: 1, Jan-Feb, 31-35.

Energuc, S., Simpson, N.C., Vakharia, A.J., 1999, Integrated Production Distribution Planning in Supply Chain: An Invited Review, European Journal of Operations Research: 115, 219-236.

Kamath, R., Liker, J.K., 1994, A Second Look at Japanese Product Development, Harvard Business Review, 154-170.

Kara, S., Kayis, B., O'Kane, S., 2002, The Role of Human Factors in Flexibility Management: A Survey, human Factors and Ergonomics in Manufacturing, Vol. 12, No.: 1, 75-119.

Mendelbaum, M., 1978, Flexibility in Decision Making: An Exploration and Unification, PhD Thesis, Department of Industrial Engineering, University of Toronto, Toronto, Canada.

Narasimhan, R., Das, A., 1999, Manufacturing Ability and Supply Chain Management Practices, Production and Inventory Management Journal, Vol. 40, 1, 4-10.

Sethi, A.K. Sethi, S.P., 1990, Flexibility in Manufacturing: A Survey, International Journal of Flexible Manufacturing Systems, Vol. 2, 289-328.

Vickery, S., Calantone, R., Droge, C., 1999, Supply Chain Flexibility: An Empirical Study, The Journal of Supply Chain Management, National Association of Purchasing Management, August/Summer, 2, 27-33.

Interaction in a Relaxed Attitude

Yosuke Kinoe

Faculty of Intercultural Communication,
Hosei University,
2-17-1, Fujimi, Chiyoda-ku,
Tokyo 102-8160, Japan
kinoe@i.hosei.ac.jp

Toshiyuki Hama

Tokyo Research Laboratory,
IBM Japan, Ltd.
1623-14, Shimotsuruma, Yamato,
Kanagawa 242-8502, Japan
hama@jp.ibm.com

Abstract

We focus on non goal-oriented process in a relaxed situation of our everyday lives. We proposed interaction design methodology for supporting non goal-oriented processes by applying situation analysis used in Ethnographic study. Based on the methodology, we developed a prototype system using Augmented Reality framework for supporting people in a relaxed attitude.

1 Introduction

Non goal-oriented process plays an important role in our everyday lives. In a relaxed situation, people do not have explicit goals, for example, while watching TV program, listening to music in a living-room, talking with their family in a dining, walking in a park, and browsing a magazine in a study. People in a relaxed situation do not pursue productivity but enjoy *void* space and time of everyday lives. Beyond *ease-of-use*, different design goals are required for a non goal-oriented type of interaction.

In this paper, we focus on interaction supporting for non goal-oriented portion of our everyday lives. We propose a methodology based on situation analysis to provide HCI design principles for designing non goal-oriented type of interaction in a relaxed situation. As a case study, we analyzed a situation where an individual user relax on a sofa and look around a web page fragments projected on a wall in a living room. We developed a prototype system for supporting the situation by applying the methodology.

2 HCI for Non Goal-Oriented Processes of Everyday Life

We analyze class of activation levels of mental attitude in our everyday processes and focus on people in a passive mode. To analyze interaction between people and situations is important for understanding their everyday processes. We propose an HCI design methodology based on situation analysis.

2.1 Activation Levels of Mental Attitudes

We classified our everyday processes into three modes, according to activation/relaxation levels of mental attitude; *active, passive*, and *sleep* mode. Goal-oriented type of task-accomplishing process can be characterized as "active." On the other hand, people in a passive mode lack an active attitude for achieving their concrete goals and they also lack strong motivation to interact with a system. Most of non goal-oriented process can be characterized as "passive."

People in passive mode easily terminate their non goal-oriented processes if they encounter trivial "ease-of-use" problems in the interaction. Non goal-oriented process is very sensitive from a viewpoint of "ease-of-use" requirements. HCI design strategy need be modified according to activation levels of users' mental attitude.

2.2 Understanding Actor's Situation

People are apt to play a specific social role bound by a situation. Space and architecture act upon people (Le Corbusier, 1923) as a medium (Bollnow, 1963, chapter 5). The course of everyday actions depends in essential ways on its material and social circumstances (Suchman, 1987, p.50). Therefore, the design of environment for supporting everyday processes also should contain constraints determined by its situation of a target process. "Situation consciousness" is an important aspect for supporting non goal-oriented everyday processes.

2.2.1 Situation Analysis: A Methodology

Situation analysis is a very useful tool in the design of support environment for everyday processes. We propose a methodology of situation analysis based on a framework of Ethnographic study used in cultural anthropology. In our method, a situation is characterized with a combination of attributes by analyzing from the following analysis viewpoints:

a) ***actor(s)*** involved in a target process, e.g. individual, family, friends or community,
b) ***activities,*** e.g. reading news, watching TV program and talking with intimate friends,
c) ***place and location,*** e.g. private, public or ritual space; closed or open; narrow or extended; light or gloomy; in a living room, study, workshop or shopping mall,
d) ***devices and tools,*** e.g. real- and virtual-world objects available in a target process,
e) ***activation levels of actors' mental attitudes,*** e.g. active, passive or sleep mode,
f) ***postures and behaviours,*** e.g. standing, walking, sitting on a sofa or relaxing in a bathtub,
g) ***interests and preferences,*** e.g. like or dislike a particular topic such as music, politics, sports or scientific discovery,
h) ***importance of temporal continuity*** with previous and next sub-process (Kinoe & Mori, 1993; Kinoe, 2001), e.g. requirements to avoid unexpected interruptions while enjoying entertainment programs, and
i) ***importance of spatial continuity*** with periphery, e.g. requirements to maintain spatial continuity while doing a series of related activities in a target process (Rekimoto & Saitoh, 1999).

The analysis viewpoints for situation analysis consist of static and dynamic elements. Dynamic elements of a situation including *(e) actors' mental attitudes, (f) postures and behaviour* and *(g) interests and preferences* may change during a target process. On the other hand, static elements including *(a) actors, (b) activities, (c) place and location, (d) devices and tools,* and *(h)(i) importance of temporal and spatial continuity* are assumed stable during a process.

2.2.2 Links to HCI Design Principles

Design strategy of HCI for everyday processes can be derived from the analysis results of actor's situation. For example, based on the result of situation analysis, people use material and social circumstances to achieve intended actions in everyday lives (Suchman, 1987). In order to support everyday processes, HCI design framework using Augmented Reality can be re-considered as an effective approach which allows actors to achieve their situated actions by utilizing actors' everyday behaviours and real-world tools and devices included in its circumstances.

3 Designing Interaction in Relaxed Attitude: A Case Study

We focus on a situation where an individual user browse web page fragments projected on a wall in a living-room. High level of HCI design strategy can be determined by the analysis from the static elements of a situation such as *actors, activities, place and locations,* and *devices and tools.* On the other hand, low level of HCI design strategy can be considered by the analysis from the dynamic elements such as *postures and behaviours, interests and preferences,* and *temporal and spatial continuity.*

We further break down the design principles for an interactive system in this specific situation in accordance with the situation analysis in the previous section. We also propose general system framework of the application of this kind.

3.1 Breakdown of Design Principles

We give shape to the design principle of an interactive system in a relaxed mode based on the situation analysis of the previous section. We consider a situation where an individual user is browsing multimedia information contents without any specific purpose. Background conditions of the system design are that *actor* is an individual, that *place* is private (living room), and that *tools and devices* should not impose any intentional usage to a user. In the situation, a user can freely browse whatever information he/she is interested in, and no communication with other actors and no censorship of the information contents do not have to be considered in designing the system. On the other hand, user's attitudes, user's interest, and behaviours are explicitly modelled in the system, and they are used to navigate the information content retrieval.

3.2 Framework

We propose a system framework of this interactive system (Figure 1). The system consists of the following six components: (1) sensor devices embedded in the environment, (2) user physical model, (3) user metal model, (4) information model, (5) presentation model, and (6) physical world model.

Sensor devices are attached to user's body as well as real objects in the environment. Thus a user is located in an Augmented Reality environment, and can interact with the system both intentionally using real objects in the environment and unintentionally by the system's interpretation of his/her body movements and biological metrics.

User physical model component tracks user's posture and behaviour by way of sensor devices. One of the roles of this component is to reconstruct user's local view in the system and map its local coordination to the global coordination of the physical world. The other role is to give an interpretation of user's postures and behaviours as his/her attitude and interest change, and to notify the changes to user mental model component.

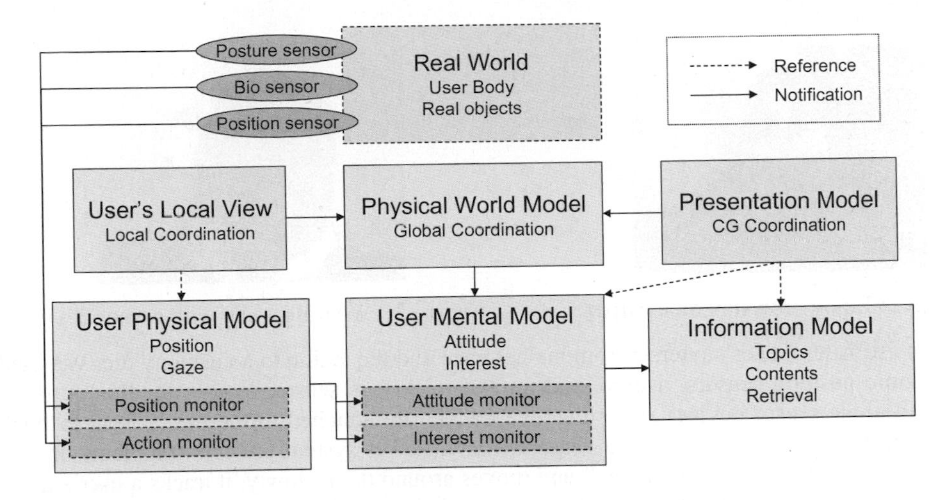

Figure 1: System Design

User mental model component plays a central role in terms of direct reflection of our situation analysis. The model maintains interests and attitude of a user. User's attitude is modelled as state transition among active/passive/sleep mode, and the transition is triggered by the notification from user physical model component.

Information model component maintains information contents a user looks like being interested in, and retrieves appropriate contents from internet if necessary. The contents are visualized in the environment by the presentation model component in accordance with the user's attitude. In an active attitude, full contents are displayed with explicit interaction devices for a user to explore further details of the contents as an ordinary goal-oriented process. In a passive attitude, several fragments of the contents are displayed according to the changes of user's interests, which are indirectly estimated by the system through user's interaction with the environment.

Finally world physical world model component deals with the global coordination of all the participants in the environment, including user's local view, sensor attached real objects, and projected information contents, in order to enable Augment Reality environment.

4 Experimental Prototype

We developed an experimental system based on the system framework proposed in the previous section. The system is designed mainly to support a user in a passive mode to browse web pages without specific purpose in a living room. In this mode, few explicit interactions to the environment are expected for a user to enjoy the system (Figure 2). On the other hand, the system itself works rather actively and sensitively to the behaviours of the user so as to estimate user's interests and preference.

Spatial and temporal continuity are more important aspects of system design in an implementation level than in general framework level. While Augmented Reality technique seems to be generally appropriate for interactions with this sort of system, these aspects should be evaluated in terms of actual implementation of real objects and interpretation of sensor data from the devices attached to the objects and a user's body.

Figure 2: Enjoy browsing on a carpet.

Figure 3: Wearing a cap with sensor devices.

The system estimates a user's interest from his/her gaze and the action to a cushiony die. We used two supersonic position sensors, a gyro, and an eye tracker as sensor devices for this purpose. Position and angle sensor devices as well as eye tracker are attached to the user's head to track user's gaze in the physical environment (Figure 3). While the system projects fragments of web pages related to a certain topic on the wall and moves around them slowly, it tracks a user's gaze. If a user's gaze is constantly tracking a certain moving fragment of web pages, the system assumes a user is interested in the topic of the fragment, and retrieves more contents related to the topic by way of an internet search engine. Otherwise the system changes the topic of web pages slowly and randomly. Another position sensor device is attached to the cushiony die next to the user (Figure 2). If the current topic is boring to a user, he/she can naturally roll the die. Then the system retrieves completely different topics at random.

More intentional action is required for the transition to an *active* attitude. Eventually a user may find very interesting information, and want to explore the contents further. In this case, a user can notify the system by raising his/her arm. The system will change the presentation mode, and display the whole contents of the web page. If there is no explicit interaction meanwhile in an *active* mode, the system naturally goes back to a *passive* mode.

5 Concluding Remarks

We proposed a basic idea of HCI design methodology based on situation analysis for supporting non goal-oriented processes in a relaxed situation. We also proposed a generic system framework and developed an experimental system for supporting a situation where an individual user in relaxed situation in a living room. Evaluation through various case studies in typical everyday situations is essential for improving the methodology.

References

Ballnow, O. F. (1963). Mensch und Raum. Stuttgart: Kohlhammer.

Le Corbusier. (1923). Towards a New Architecture. Dover Pubns.

Kinoe, Y. & Mori, H (1993). Mutual Harmony and Temporal Continuity: A Perspective from the Japanese Garden. ACM SIG-CHI Bulletin, 25(1), 10-13.

Kinoe, Y. (2001). Cross-Devices Temporal and Spatial Continuity: Essential Aspects for Designing Everyday Digital Life. Proceedings of the HCI'2001 Conference, 285-287.

Rekimoto, J. & Saitoh, M. (1999). Augmented Surfaces: A Spatially Continuous Work Space for Hybrid Computing Environments. Proceedings of the ACM CHI'99 Conference, 378-385.

Suchman, L. A. (1987). Plans and Situated Actions. Cambridge University Press.

Comprehension-Based Approach to HCI for Designing Interaction in Information Space

Muneo Kitajima

National Institute of Advanced Industrial Science and Technology (AIST)
1-1-1 Higashi, Tsukuba, Ibaraki 305-8566 Japan
kitajima@ni.aist.go.jp

Abstract

The traditional view of HCI has been based on a cognitive psychological analysis, in which people have to translate their intentions into the language of the computer and have to interpret the computer's response in terms of how successful they were in achieving their goals. However, with the ubiquity of information appliances, this view becomes inadequate. People are no longer simply interacting with a computer they are interacting with but they use various combinations of information appliances and media available for them to accomplish their tasks defined in information space. This opens up the possibility that any single task can be accomplished in various ways. This paper argues that such users' interaction processes can be better viewed as the processes consisting of comprehension of the current situation formed by integrating various sources of available information, followed by selection of an appropriate action based on the comprehension. A set of implications for interaction design for information space is derived based on this comprehension-based view of HCI.

1 Introduction

As the integration of communication and computer advances, our computerized tasks place more emphasis on interaction with information than interaction with computers. Interaction with devices based on familiar interface conventions becomes less focused but interaction with information becomes more important. There are devices that aim to support activities in information apace. These devices are called information appliances[1] or internet appliances[2], providing interface for accessing and manipulating information.

The purpose of this paper is to define an appropriate viewpoint to understand users' activities in information space, and then derive implications for designing interaction for information space. This paper starts by showing that *comprehension* should be one of the fundamental cognitive skills

---

[1] A device for accessing or manipulating information, special-purpose in contrast to a general-purpose computer. The idea is a machine with computing power but designed and used like other consumer electronics, such as stereos, TVs, and toasters. It serves a limited function, enabling it to have a simplified user interface and to fit its intended task more perfectly than a general-purpose machine. (Adapted from http://www.usabilityfirst.com/glossary/)

[2] A device designed to simplify use of the internet and simplify setup compared to a general-purpose computer. Buttons and other controls are minimized and built in to the hardware. The application may permit any kind of web browsing or may be limited to a very restricted functionality such as an email reader, a picture display, or a coffee pot that can be activated over the internet. (Adapted from http://www.usabilityfirst.com/glossary/)

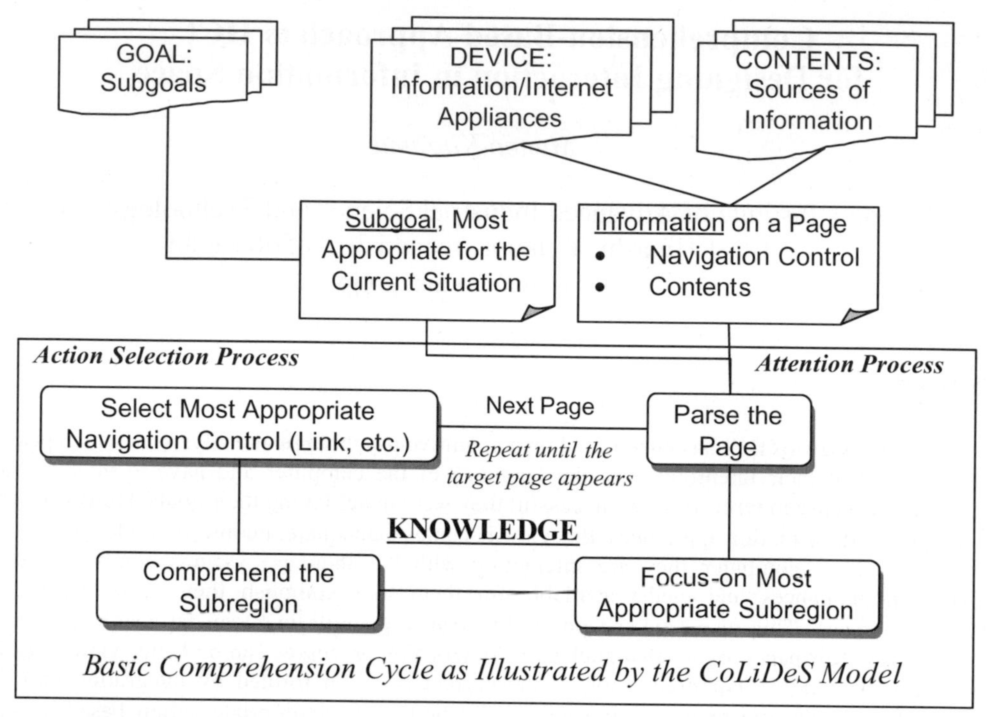

Figure 1: A comprehension-based view of HCI in information space.

that the users should apply in order to deal with interaction in information space, followed by brief description of comprehension-based cognitive models my colleagues and I have developed. It concludes with key design questions for designing interaction for information space.

2 Comprehension as a Key Cognitive Skill

Interacting with information space is different from interacting with a computer for performing traditional computerized tasks in two important ways. First, there are multiple ways to interact with information space. A task in information space can be accomplished by using different information appliances or internet appliances, which would have different physical interfaces. And thus actual physical action sequences would differ device by device. Second, comprehension of a situation in which a task is carried out can vary according to the recruited knowledge by the individual performing the task. Because people have diverse background knowledge, the same physical data provided by the devices can be comprehended various ways. These would result in the observation of multiple paths for accomplishing the task. They are different from each other not only at the level of physical actions but also at the level of mental operations.

Actual physical action sequences observed in users' interaction in information space can be changeable, and thus they cannot be considered as good indexes for understanding people's activities in information space. They cannot be used for evaluating usability as is done, e.g., by the GOMS modeling techniques (Card, Moran, & Newell, 1983). Then what can be qualified as the fundamental cognitive processes that organize people's activities in information space, which should provide useful insights for designing interaction in information space?

I suggest that interaction in information space should be viewed as *goal directed activities driven by comprehension process* in which information on the interface is comprehended by the user with his/her background knowledge in the context of the intention of accomplishing the current goal. Many of our activities are purposeful because we interact with our environment to achieve specific goals. This is true too when we are in information space. What we actually do at a given moment, however, is determined not only by the goals and the environment – contents in the information space – but also by the knowledge utilized to comprehend the situation. In order to select what to do next, we integrate these sources of information: goals, information from the environment, and knowledge relevant to the current situation – the underlined elements in Figure 1.

The process of comprehension, typically observed in text comprehension, is a highly automated collection of cognitive processes that make use of massive amounts of knowledge stored in long-term memory. In text comprehension, for example, readers activate knowledge from long-term memory relevant to the current reading goal and integrate this knowledge with the current goal and representation of text. Conflict among activated knowledge elements may exist which necessitates an integration process to arbitrate this conflict within an appropriate time frame. I suggest this same skill is important in interacting with information abundant environment, where people have to derive appropriate meaning of the situation by recruiting appropriate knowledge for the particular situation. This skill is different from the skill required for performing routine tasks where people just execute precompiled methods for accomplishing tasks, as seen in the typical view of HCI which has been based on a cognitive psychological analysis; people have to translate their intentions into the language of the computer and have to interpret the computer's response in terms of how successful they were in achieving their goals (Norman, 1986).

3 Comprehension-Based Models

3.1 The LICAI Model – a Cognitive Model of Performing by Exploration

The author and his colleagues have developed a series of comprehension-based computational models that deal with cognitive processes of how computer-literate users perform various office automation tasks by using graphical user interfaces, such as word processing, spreadsheet, and graphing (Kitajima and Polson, 1995, 1997). These models are based on the construction-integration architecture, originated from the cognitive models of text comprehension (Kintsch, 1997). Kitajima and Polson (1995) proposes a model of action planning and error by experienced users, and Kitajima and Polson (1997) has extended it to a situation where novice users discover correct actions by exploration. The latter model is called LICAI, which stands for LInked model of Comprehension-based Action planning and Instruction taking.

LICAI models novice users' activities by the following processes, each places emphasis on comprehension process that utilizes knowledge in LTM and creates a coherent understanding of the situation:

1) *Goal Formation Process* – comprehends the task for the purpose of transforming it into workable subtasks, for example, the goal of purchasing an item at an online store is decomposed into a set of subgoals, including browsing a catalogue, selecting an item to purchase, paying for the selected item, and so on.
2) *Goal Selection Process* – comprehends the interface to sequence the subtasks properly.
3) *Action Selection Process* – selects an object-action pair to execute, consisting of the following sub-processes;

a) *Object Selection Process* – comprehends the interface to select the right object.
b) *Action Selection Process* – comprehends the interface to select the right action, for example, click, drag, and type.

3.2 The CoLiDeS Model – a Cognitive Model of Web Navigation

Web navigation can be considered as a typical example of activities in information space; it consists of a series of link selection, each selection leads to a new web page, where next selection is carried out. In each selection, the user compares the representation of the current page with the representation of the task goal by using general knowledge about the page and the goal, and selects a link that would contribute most to accomplishing the goal.

Kitajima, Blackmon, and Polson (2000) have developed a comprehension-based model of web navigation by extending the LICAI model. The model is named CoLiDeS, standing for Comprehension-based Linked model of Deliberate Search. Figure 1 illustrates CoLiDeS schematically. CoLiDeS adds the attention process to the LICAI model in order to model interactions with web which is richer in information and less formatted than those with office applications. The CoLiDeS model is used to develop CWW – the Cognitive Walkthrough for the Web – a usability inspection method for detecting and repairing usability problems, mostly in navigational problems in informational sites (Blackmon, Polson, Kitajima, and Lewis, 2002; Blackmon, Kitajima, and Polson, 2003).

The processes in the CoLiDeS model are defined as follows:

1) *Forming goal and selecting a subgoal*: These processes correspond to the instruction taking process and the goal selection process of the LICAI model, respectively.

2) *Parse the page and focus-on one sub-region:* On encountering a new web page, the user first parses the page into meaningful units. The user then selects one of the units, schematic object, by comprehending the parsed page. This process is called the attention process. This process results in making available the contents in the schematic object to the user.

3) *Comprehend the selected sub-region and select link:* Finally, the user comprehends the contents and selects one link. This process corresponds to the action selection process of the LICAI model.

Parse and focus-on are important addition to the LICAI model and play crucial role in attention management for processing new contents on the pages.

4 Implications for Designing Interactions for Information Space

By looking at the processes defined in the CoLiDeS model from the viewpoint of knowledge use, we could derive key questions at each stage of interaction to be answered affirmatively when designing interfaces for tasks performed in information space. Table 1 summarizes the features of knowledge use at each stage of interaction process as defined by the CoLiDeS model and the third column holds key issues for designing *usable* interface for information space.

One of the serious problems in the information abundant environment is that people cannot reach the desired information that *does exist* in the environment by successively following links. This

Table 1: Cognitive processes necessary to perform tasks in information space and their implications for designing usable user interfaces.

Stage	Feature of the Cognitive Processes	Key Issues for Designing Usable Interface for Tasks in Information Space
Goal Formation	Comprehend the task in order to transform it into workable subtasks	Is the task representation appropriate for activating critical knowledge, such as schema, script, etc.?
Goal Selection	Comprehend the interface to sequence the subtasks properly	Is the interface representation appropriate for activating the right subgoal?
Action Selection	Comprehend the interface to attend to appropriate portion of the interface	Is the interface representation appropriate for activating critical knowledge for parsing the interface?
	Comprehend the interface to select the right object	Is the representation of the interface object appropriate for activating knowledge necessary to relate it with the current subgoal?
	Comprehend the interface to select the right action	Is the representation of the interface object appropriate for activating eligible action for the current subgoal?

problem would be resolved, not fully but partially though, by considering the key issues listed in Table 1. Think first how people comprehend the environment. This is the key to designing usable, if not effective and efficient, interface for supporting activities in information space.

5 References

Blackmon, M. H., Polson, P. G., Kitajima, M., & Lewis, C. (2002). Cognitive Walkthrough for the Web. In Proc. of CHI'2002 (pp. 463-470). ACM Press.

Blackmon, M. H., Kitajima, M., & Polson, P. G. (2003). Repairing Usability Problems Identified by the Cognitive Walkthrough for the Web. In Proc. of CHI'2003 conference.

Card, S. K., Moran, T. P., & Newell, A. (1983). *The Psychology of Human-Computer Interaction*. Laurence Erlbaum Associates.

Kintsch, W. (1997). Comprehension: A Paradigm for Cognition. Cambridge University Press.

Kitajima, M. & Polson, P. G. (1995). A comprehension-based model of correct performance and errors in skilled, display-based human-computer interaction. *International Journal of Human-Computer Systems*, 43, 65-99.

Kitajima, M. & Polson, P. G. (1997). A comprehension-based model of exploration. *Human-Computer Interaction*, 12, 4, 345-389.

Kitajima, M., Blackmon, M. H., & Polson, P. G. (2000). A Comprehension-based Model of Web Navigation and Its Application to Web Usability Analysis. In S. McDonald, Y. Waern & G. Cockton (Eds.), *People and Computers XIV - Usability or Else!* (pp.357-373). Springer.

Norman, D. A. (1986). Cognitive engineering. In Norman, D. A. and Draper, S. W. (Eds.), *User Centered System Design* (pp. 31-61). Lawrence Erlbaum Associates, Hillsdale, NJ.

The Concept of New Interface Design for Elder Persons

Daiji Kobayashi and Sakae Yamamoto

Dept. of Management Science, Faculty of Engineering
1-3 Kagurazaka, Shinjyuku-ku, Tokyo, 162-8601 Japan
daiji@hci.ms.kagu.tus.ac.jp / sakae@hci.ms.kagu.tus.ac.jp

Abstract

Computers and computer networks would drastically progress, and it would be expected that the progress change our daily life. As computer technology advances, the shape of human interfaces would transform in the future. To begin with, the computer and the computer network users would be required to use the human interfaces efficiently. However, now, many Japanese elders are not able to handle the human interfaces. In consideration of this problem, in this paper, a new interface design concept that would support elder persons is discussed based on some simulation experiments.

1 Introduction

In Japan, recently, it is the serious problem that there are many persons who are not able to get information using personal computers (PCs) and the World Wide Web (WWW). Namely, there are many people who can not get information that they need using PCs and the WWW. According to looking on the Japanese elders' situation, we inferred that the operation of the human interfaces have not been taken into account the Japanese elders' characteristics, but the large party's convenience. This is the typical problem for elder persons in Japan. To give an example of input devices, there are many elderly people who have to do using the keyboard to operate the PCs, although ordinary middle and elder persons are not used to the type-writer in daily life. In other words, there are high barrier for elders against using keyboards with PCs. For these problems, recently, the new idea called "accessibility" is suggested. Accessibility means to guarantees using information devices to the elder and the disabled, hence accessibility is needed to develop the new type of human interfaces. Now, some recommendations for elder and disabled considered or established as the international standards[1, 2, 3]. According to these recommendations, it is inferred that a requirement of elders is the same as young persons need. However, we have some doubts about these recommendations for applying to the human interface design, and we considered whether these recommendations should be applied to the human interface design for elders and disabled and other people who need to get information using PCs and the WWW.

---

[1] ISO/TS 16071 (2001), *Ergonomics of human system interaction —Guidance on accessibility for human computer interfaces.*

[2] ISO/IEC GUIDE 71 (2001), *Guidelines for standards developers to address the needs of older persons and persons with disabilities.*

[3] Section 508 of the Rehabilitation Amendment Act (1998)

2 Method

In order to suggest new design concepts, we have done the experimental approach. In these experiments, 39 Japanese elder persons who are 64 to 87 years old reported what they think, feel, or how they tried to do searching specific information using a WWW browser. The each experimental websites that we built was structured as a simple menu selection system. In the experiments, the subjects' verbal protocols and their behaviour were recorded by video camera, and we asked the subject's opinion about mouse operation for information search after. Based on these experimental data, we considered their manner for searching information by WWW browser and the characteristics of information search using mouse devices. In the consequences, the elder persons' behavioural model and their mental process model were tried to illustrate with charts. After trying to estimating these elder's models, we also research the young groups in the same way as the reference condition.

3 Experiments

3.1 Hierarchical Structure of Websites

Firstly, the influences of their characteristics on the access of menu selection were considered experimentally. In order to investigate their characteristics on web-surfing, two types of simple websites are built. These websites are consisted of the different numbers of hierarchical structure's layer, although they represent the same information about a community. The one type named "1-layer" is composed of a top page and additional pages, and the other named "4-layers" is consisted four layers of hierarchical structure. The 1-layer's top page shows almost the site's information; the font size of the text is consequently smaller than another type's. The document on the top page has a few keyword linked the additional pages which shows supplemental information about keyword. The 4-layers shows some titles of categories of the information on the top page; the second page shows some concrete title about selected category on the top page; the third page shows the outline about selected concrete title on the second page; finally the last page shows the details of the information about selected outline. Using these websites, 19 elder persons (65 to 83 years old) and 9 young persons (22 to 27 years old) browse information they were fascinated with. However, 9 elder persons have the experience of using PC application software except WWW browsers before experiments.

As the results of the experiments, some characteristics of the subjects' behaviour were observed as the followings:

- In case of using 4-layers, it was observed that all inexperienced elder persons were bewildered and did not access positively, meanwhile, experienced elder persons and young persons accessed some layers; however, the 4 of 9 experienced persons did not access the last page.
- In case of using 1-layer, all of the elder persons were not embarrassed and got most of the information. This tendency among elder persons was the same as young persons' tendency, although the young persons complain about the compressed layout of top page, and they preferred 4-layers.

According to these behavioural characteristics, the behaviour of the elder persons' were summarized as the model of Figure 1 (a), and the model of the young persons were also summarized as the Figure 1 (b) in order to compare these. These models expressed how difficult the elder persons were to understand the hierarchical structure of websites, and to select information they need on the website.

By the way, in these experiments, some subjects pointed out the usual problems about each screen's layout, font size, and so on; however, it would be necessary for the elders the viewpoint of selecting information that is different from these usual viewpoints.

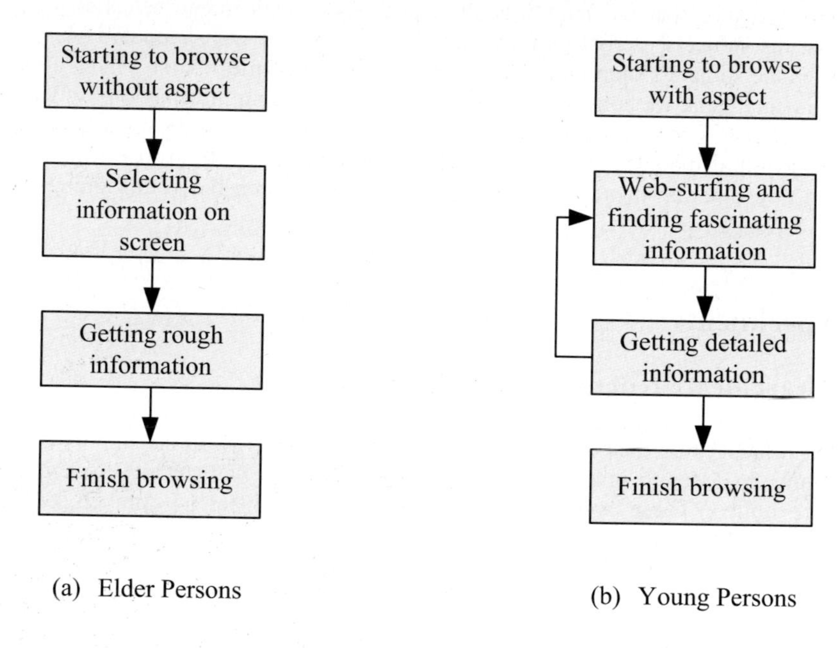

Figure 1: Behavioural models of elder and young persons.

3.2 Arrangement of Information on a Webpage

Secondly, the relation between elders' characteristics and the arrangement of information on a webpage were considered. To investigate the relation between elders' manner and the arrangement of information, 2 types of gift shopping websites were built and some simulation experiments were performed. These shopping websites displayed the same gifts by different arrangement in a page. One type of websites called "Scroll type" displayed the detailed information of each gift, and these gifts were arranged lengthways. Therefore, the users need to scroll the page in order to get all gifts' information. Another type called "Scroll-less type" displayed the rough information of each gift, and all gifts were arranged in a page. And, the users could get all gifts' information without scroll operation. The subjects using "Scroll type" were 10 elder persons (64 to 87 years old) and 7 young persons (20 to 23 years old). And, the subjects using "Scroll-less type" were other 10 elder persons (65 to 73 years old) and 8 young persons (21 to 23 years old). All of them were tried to buy some gifts for their friends within the designated budget, but all the elder persons did not have experiences of using PC; therefore a supporter often followed the elder's object and operated the mouse.

As the result of the experiments using "Scroll type", 8 of 10 elder persons selected gifts from displayed gifts on screen without scrolling, namely, they selected each gift from a few choices. When there were not displayed gifts they wanted, they took their budget into account and changed their choice. Meanwhile, the young persons selected gift from all choices with scrolling, and they selected the satisfying gifts. According to these results, the subjects' mental processes were summarized as the Figure 2. The Figure 2 (a) expressed especially the elder's following two characteristics of the manner: One is to get information more passively than young persons; and the other is to fit their intention to less information. The elder's manner in searching the expected information would indicate that "Scroll type" webpage is not preferable.

In case of using "Scroll-less type," 8 of 10 elder persons select the gifts which were not displayed without scrolling in "Scroll type," and the elders preferred this type of webpage.

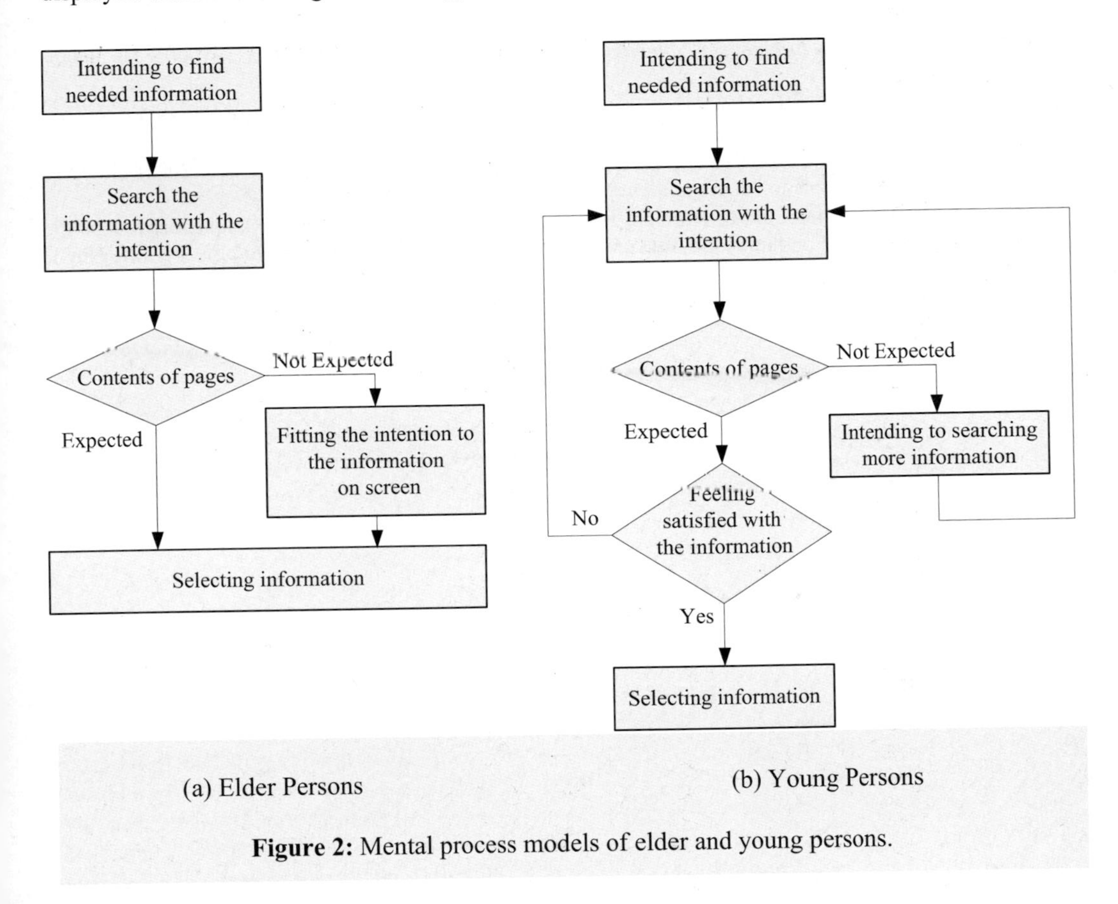

Figure 2: Mental process models of elder and young persons.

4 Discussion

The recommendation (ISO/TS 16071, 2001) state that the concept of accessibility is not limited to disability issues. However, the above experimental results indicate the different conclusions from the recommendations. For example, the experimental results suggest that the requirements of arrangement of web pages for elders are different from the young users' requirements. From the view of their manner, the recommendations for the interface design of the WWW would be not the same between them. With regard to this there are not human interfaces that are made by the conventional concept and satisfy all users. Therefore, the every particular approach to interface design would be needed. This concept means that human interfaces should accommodate individually. That is to say, the human interfaces should be customized to suit them to the each user's requirements. In order to make the human interfaces, the bottom up approach for removing the each user's barrier is rather requirement. Accordingly, the method of evaluating the achievement of accessibility would be required.

References

(1) ISO/TS 16071 (2001). *Ergonomics of human system interaction —Guidance on accessibility for human computer interfaces.*

(2) ISO/IEC GUIDE 71 (2001). *Guidelines for standards developers to address the needs of older persons and persons with disabilities.*

(3) Eghtesadi, K., Kaye, S., O'Hare, M., Pierce, J., Uslan, M., & Duslig, Y. K. (2002). Making Photocopying Accessible to All. *Ergonomics in Design,* Summer 2002, 17-22.

(4) Zaphiris, P., Shneiderman, B., & Norman, K. L. (2002). Expandable Indexes vs. Sequential Menus for Searching Hierarchies on the World Wide Web. *Behaviour & Information Technology,* Vol. 21, No. 3, 201-207.

Managing industrial service portfolios using a platform approach[1]

Johannes R. Kuster *Volker Liestmann* *Volker Stich*

Research Institute for Operations Management,
Aachen University of Technology
Pontdriesch 14-16, 52062 Aachen, Germany

ku@fir.rwth-aachen.de lm@fir.rwth-aachen.de st@fir.rwth-aachen.de

Abstract

Industrial service companies offer an increasing range of product variants in order to satisfy several customer segments characterized by different requirements and needs. However, increasing service variety might not only lead to more attracted customers but can also lead to disproportional higher complexity. In order to be successful, service companies have to seek for an optimum within their product structures with regard to the opposing influences of variety and complexity. This paper aims to present the results of a research study about the possibilities of applying platform approaches to industrial services. Service managers, consultants and researchers were interviewed on their experiences of balancing the variety and complexity of service portfolios. Within a process framework based on existing work in the fields of platform development of physical products and service development the three main platform design principles, i.e. modularization, standardization and integration, are presented with regard to their applicability on services and verified in several industrial service firms.

1 Introduction

Numerous companies of different industrial sectors offer a great variety of services and service variants that cannot be justified considering the purely economical factor. This is mostly resulting from an uncoordinated planning process for service programs. In addition, customer orientation often accompanies the extension of the service portfolio in times of expansion. In phases of recession companies try to differentiate from their competitors by new service variants. On the other hand, companies fail to regularly adjust their service spectrum. This problem also occurs in the goods-producing sector (Meyer and Lehnerd 1997; Robertson and Ulrich 1998). In the field of industrial services this trend, however, needs detailed examination.

Normally the described increase of product variants leads to a disproportionate increase of internal complexity in the goods-producing sector and thereby to an increase of production costs (Schuh 1989). Due to this fact, methods have been used for several years now, which contribute to the solution of this conflict. These very often base on the principle of reusability, which allows the internal standardization as well as the individualization to the outside. A successful approach is the design of product families according to the platform principle.

---

[1] The research reported in this document was made possible in part by the Bundesministerium für Bildung und Forschung (BMBF), Number 01HR0019.

Meyer and Lehnerd (1997) define a platform as "a set of subsystems and interfaces that form a common structure from which a stream of derivative products can be efficiently developed and produced". The applicability of the platform approach is, for this, not explicitly restricted to the goods-producing sector. In the service area the platform approach is not yet widely used (Sunbo 1994; Meyer et al. 1998). However, if an initial situation comparable to goods-producing sector would be present in the service sector, the principles of platform design could also be considered there as a solution. It is obvious that compared to physical goods the specific features of services make a direct transfer and application of the existing methods impossible. For that reason it is necessary to check in detail in how far similarities exist between the problems in the goods-producing and the service sector.

2 Research Setting

An exploration was conducted in Germany in 2002 which addresses this idea. The aim of the study was to determinate if the conditions for applying the platform approach are actually given. For this purpose we formulated five theses. Thirty service experts were interviewed in this study:

2.1 Thesis 1: There is an increase in customer demands towards industrial service providers independent of the type they belong to

In this research we differentiate two types of service providers in the industrial context. The first group consists of service providers that place their service in the direct sense at the customer's disposal; i.e. the service is their main offer. They normally differ from both other groups in their objective to generate maximum profit by their services solely. For the second group of service providers the service itself only plays a complementary role (maintenance etc.) compared to the actual main task (e.g. production of machines). The objective of these services is thus normally to support the core business.

2.2 Thesis 2: Increasing customer demands lead to an increasing number of services respectively variants offered

To be able to fulfill the demands of the customers, the service portfolio is subject to a "constant" change. Customers demand an increasing range of services, i.e. completely new services (e.g. financial services combined with the purchase of a machine), a higher service-level (e.g. decreasing response times in a case of machine failure) or even the participation of the service provider on its own risk, as it is given in the case of BOT models (Build-Own-Transfer). The increasing and changing customer demands thus produce a permanent necessity of service variation and service innovation.

2.3 Thesis 3: The offer of new services and service variants has a negative influence on the objectives of service providers

As a framework for the objectives we took the aspects of the Service Profit Chain according to Heskett et al. 1997 (Figure 1). The SPC shows the connection between internal, mainly personnel-related factors (motivation, training, etc) and external objectives. The increasing complexity due to a growing service offering leads to an increasing complexity of the service production and finally decreasing profitability.

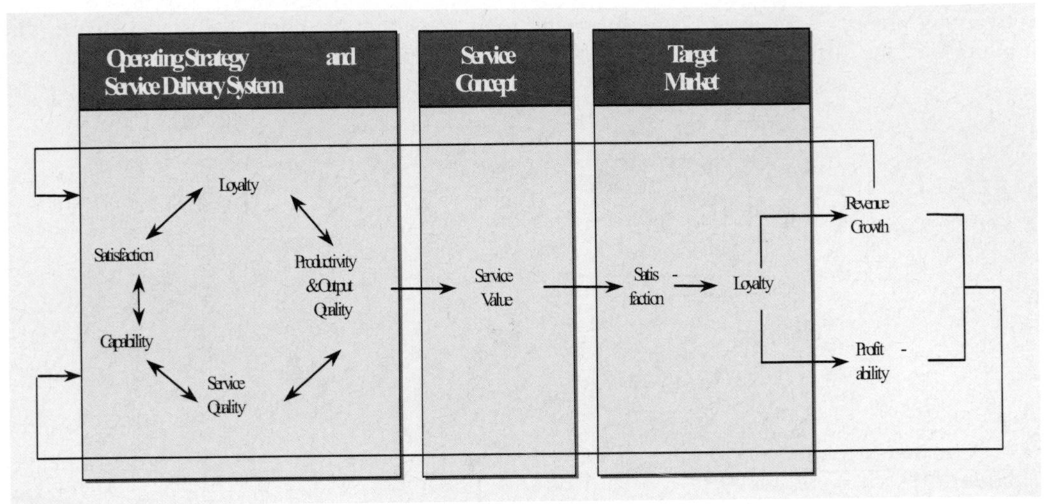

Figure 1: The Service Profit Chain

2.4 Thesis 4: There is a difference between variation and innovation

Innovations, thus completely new services in the portfolio, as well as variations, i.e. changes in the offered service-levels, have an effect on the complexity and with this an effect on the objectives of the company. Nevertheless, these consequences on the objectives differ because different effects occur and consequently the company has to take different measures.

2.5 Thesis 5: No general concepts exist for the control of the conflict between variety and complexity

In order to remain successful, service companies have to aim for professionalism and find concepts how they can control the variety of their services. For this, it is not the aim to eliminate complexity for example through a reduction of the service spectrum. Rather, concepts are meant how the organization can be shaped to maintain a service spectrum as broad as possible and simultaneously work efficiently. In practice most companies try to take up measures, to get rid of single variety-related complexity problems. We understood this thesis, however, in that way not yet to speak of a comprehensive method, which is firmly anchored in the companies.

Summarizing, it should be noted that an increase in the number of services and higher service-levels lead the companies towards a problem of complexity. This is especially to be seen as a management problem since, on the one hand, missing capabilities and, on the other hand, the concern over a lack of service quality leads to enormous decrease of expected profits.

3 Process Framework for Service Platform Planning

To explain the principles of service platform design a framework must be laid out beforehand. Robertson and Ulrich (1998) give an established process framework for platform planning. Their approach provides an open information platform that helps integrating different planning activities. Furthermore, it leaves room for creative problem solving processes through its iterative converging process framework. Three planning tools constitute the core of this framework: the product plan, the differentiation plan, and the commonality plan (Figure 2).

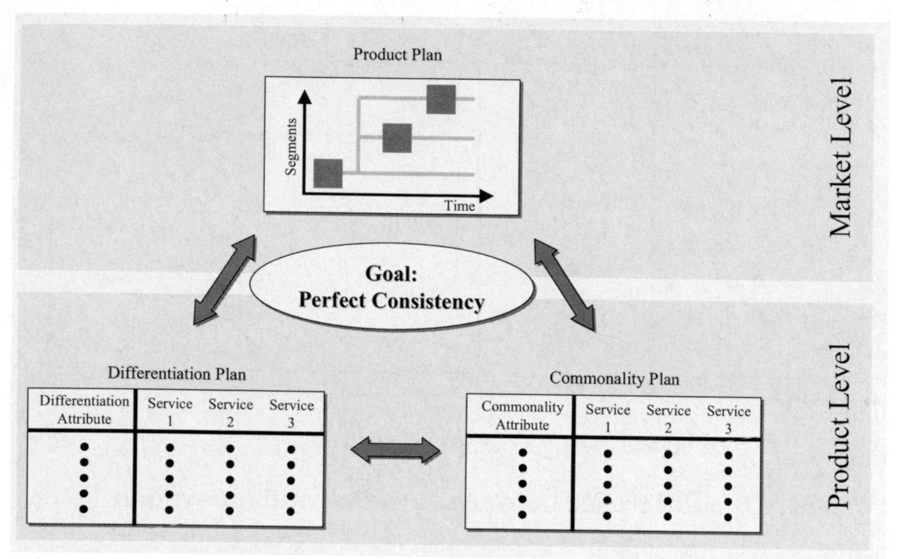

Figure 2: Process Framework for Service Platform Planning

4 Design Principles for Service Platform Design

The following principles support the balance at the product level as represented in figure 2. They originate from the designing of physical products and are here interpreted against the background of service development (Koller 1994; Pahl and Beitz 1996).

Modularization: The principle of modularization supports the process of building subsystems from an overall system. According to this, subsystems are defined under consideration of an easy reconfiguration of the system. Therefore modules ought to be comparable on the same level and it needs to be ensured that uniform interfaces are being defined. The model of the prerequisites of the service from Edvardsson and Olsson (1997) can be taken as a basis for modularization; i.e. the modules are defined either under view of the service concept, service process or service system. The result of the modularization are corresponding partial service modules that are constituted by partial concepts, partial processes or partial structure elements.

Standardization: This design principle refers to reducing the variety of several equal or similar subsystems to one single element that is used in all the different cases of use likewise. Usually, this measure comes along with a reduction of functionality. But using standardized subsystems can also exploit the potentials of scale effects to a large degree. In the service sector these effects are less economies of scale or scope as observed in the procurement and production of goods but rather "real" experience in performing the service processes. The objective is therefore to enhance the capabilities of the personnel with regard to robust processes. In the same way, the personnel could be trained to improve the company's capabilities, and utilize the positive effects of the SPC. Hence it can be assumed that by applying the principle of standardization the major potential for reducing the complexity inherent in service programs can be exploited.

Integration: The third platform design principle is "integration". It aims at reducing the number of interfaces by merging two or more partial services into one. In the service sector most interface problems occur as a loss of information. A high degree of integration mainly prevents this kind of failures. However, this principle rather shifts complexity than it reduces it, which can also lead to an increase of complexity at the subsystem level. In many cases in the goods sector it has been proven that a multifunctional design, which is often not required by the customer, can be less complex or cheaper than providing several unifunctional variants due to economies of scale. These kinds of economies however have not been recognized in the service sector.

The development team should iteratively apply the three principles modularization, standardization and integration at all levels of the service platform system. The more skillful they define the elementary modular components the better they can integrate and standardize them in order to align the need of differentiation from the customer's point of view and the companies aim to reduce complexity. The balance of variety and complexity becomes thus more manageable.

5 Conclusion

In summary, it can be stated that in principal platform approaches can be applied in the service sector in order to balance variety and complexity. As a prerequisite for this the actual problem of an increasing variety in the service sector was identified and fundamental principles for platform development were described. The approaches from the goods sector can surely not be transferred in all their facets. The research in this field however is only at its beginning. Thus, service platforms and their underlying principles and views provide a valuable input for the development of service portfolios.

References

Edvardsson B. and Olsson J., (1996), "Key Concept for New Service Development", *The Service Industries Journal*, vol 16, no 2, pp 140-164.

Heskett J.L., Sasser W.E. and Schlesinger L.A., (1997), The Service Profit Chain, The Free Press, New York.

Koller R., (1994), Konstruktionslehre für den Maschinenbau. Grundlagen zur Neu- und Weiterentwicklung technischer Produkte mit Beispielen, vol 3, Springer Verlag, Berlin.

Meyer M. H., DeTore A. and Walter J., (1998), Product Development for Services. Working Paper, Northeastern University, Boston.

Meyer M. H. and Lehnerd A. P., (1997), The Power of Product Platforms – Building Value and Cost Leadership, The Free Press, New York.

Pahl G. and Beitz W., (1996), Engineering Design - A Systematic Approach, Springer Verlag, New York.

Robertson D. and Ulrich K., (1998), "Planning for Product Platforms", *Sloan Management Review*, vol 39, no 4, pp 19-31.

Schuh G., (1989), *Gestaltung und Bewertung von Produktvarianten. Ein Beitrag zur systematischen Planung von Serienprodukten*, VDI-Verlag, Düsseldorf.

Sundbo J., (1994), "Modulization of Service Production and a Thesis of Convergence between Manufacturing Organizations", *Scandinavian Journal of Management*, vol 10, no 3, pp 245-266.

Yu J. S., Gonzales-Zugasti J. P., and Otto K. N., (1998), "Product Architecture Definition Based upon Customer Demands", *Proceedings of The 1998 ASME Design Engineering Technical Conferences,* September, Atlanta.

Enhancing Remote Control Performance: Enabling Tele-Presence Via a 3D Stereoscopic Display

K. Y. Lim and Roy S. M. Quek

Centre for Human Factors & Ergonomics
Nanyang Technological University, School of MPE,
50, Nanyang Avenue, Singapore 639798
Email: mkylim@ntu.edu.sg, royquek@cyberway.com.sg

Abstract

To achieve a competitive edge and hold its position as the world's busiest port, a freight company in Singapore has developed and implemented the world's first semi-automated camera-based remote control crane system to pick-up and land freight containers. A centralised computer commands the crane to move a container between locations, while the operator performs the skilled task of vertically landing/picking up the container. However, the 2 dimensional (2D) camera-based system compromises depth perception required for the container landing/pick-up task. Thus, the operator may experience difficulty in judging container height relative to the prime mover chassis. As a result, the operator's ability to control container landing/pick up speed and throughput may be affected. Similarly, container landing impact may be controlled poorly. To address these problems, depth perception could be restored through tele-presence achieved via a desk-top three-dimensional (3D) stereoscopic display system. Subject tests with a scale model have shown that the 3D stereoscopic display, can enhance operator performance of container landing impact ($P<0.01$). Thus, it may be concluded that the display has shown promise.

1. Background

A local freight company implemented a new remote controlled camera-based system for container handling. This semi-automated system incorporates a computer that performs the task of shuttling containers between designated locations, before passing the job to one of a number of available operators. The computer removes from the operator the task of trolleying the crane to a designated location. The operator now performs only the skilled component of manipulating the spreader to land or hoist up a container. To perform these tasks, an operator needs to adjust the position of the spreader with or without a container, and occasionally dampen its sway before lowering it. By removing the operator from being physically in the crane (with its attendant poor working posture and vibrations) to a centralised air-conditioned remote control room (see Figure 1), the new system improves the working conditions of the operators and enhances their efficiency and productivity. To enable remote operation, a quad-screen is used to display to the operator, camera views of the four corners of a container/spreader (see Figure 2). This quad plan view is relayed from cameras mounted directly above the spreader. However, such a 2-D camera-based remote control system deprives the operator of the depth perception required for efficient pick-up and landing of containers.

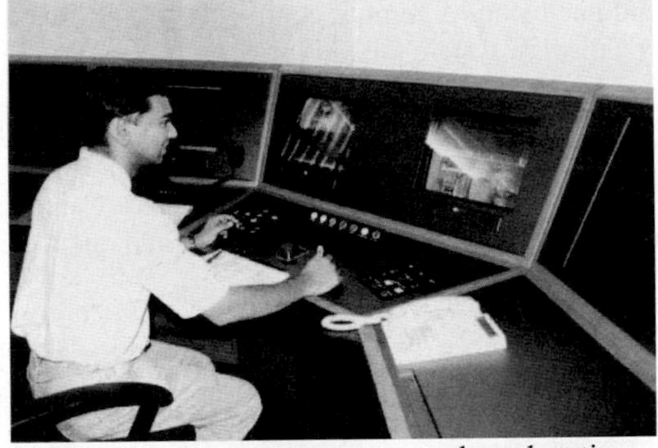

Figure 1: Remote crane control workstation

Figure 2: A quad view 2D display of four corners of a container

Previous studies by Ng et al (2000) have compared the efficacy of various 2D display formats, in enhancing remote task performance of lateral positioning and vertical landing of a container. 2.5D perspective displays have also been assessed. The results revealed that a 2D binary view of two corners of a container could enable the same level of performance as a 2D quad view display. However, the 2D quad-view display continues to be used commercially, as it is already implemented.

Although the quad-screen display may be useful for positioning the container above the prime mover chassis, the operator may find it difficult to judge the speed of container landing and the distance of the container to the prime mover chassis or to another container. The operator needs to monitor container height and speed, to maximise throughput and avoid damaging either the container contents or the prime mover chassis, due to an excessively heavy landing impact. However, a 2D quad display is unable support well the depth dependent container landing task, as onsite stereoscopic depth cues are missing.

In an attempt to restore depth cues to the operator, a desktop virtual reality 3D stereoscopic display system is developed. Although such a display may provide some physiological depth perception cues necessary for the container landing/pick-up task, it is unclear to what extent it could enhance operator performance over the existing 2D quad view display system. This is because both 2D and 3D displays provide some psychological depth cues, such as relative size, motion parallax, shadow and occlusion. A remote control test rig is thus constructed to assess the efficacy of a 3D stereoscopic display system (see Figure 3).

The test rig constructed is shown in Figure 4. The container is controlled along 3 orthogonal axes, via a 4-channel digital proportional radio controller. The container is thus able to travel in the y-axis (front & back), x-axis (left & right) and z-axis (up & down). Cameras are mounted onto the test rig to generate 3 different display designs, namely 2D, 3D and quad-view displays.

Figure 3: Desktop 3D stereoscopic display

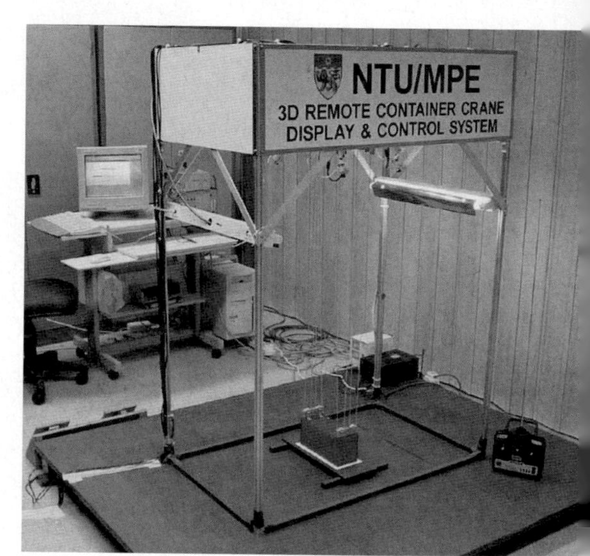

Figure 4: Remote control crane model

2. Experiment

48 subjects (gender balanced) were recruited for the experiment. The subjects were university students without prior experience of container handling. Each subject was required to repeat the test ten times for each of the display designs. The display designs tested comprised the following: a 2D quad-view display of four corners of a corner, a 2D plan view of the container and finally a stereoscopic plan view of the container. The display design tests were balanced to avoid order effects. Before testing, subjects were allowed five repetitions as practice.

Test subjects were instructed to land a container weighing 2.1kg onto a target area as quickly as possible and with the least impact. The target container landing area comprised a force platform coloured matt dark gray to resemble the colour of bitumen roads found in container yards. The landing force and time data were recorded by a computer using the bundled Bioware software. The data were plotted graphically as shown in Figures 6 and 7. Peak force and time data were then extracted and subjected to ANOVA.

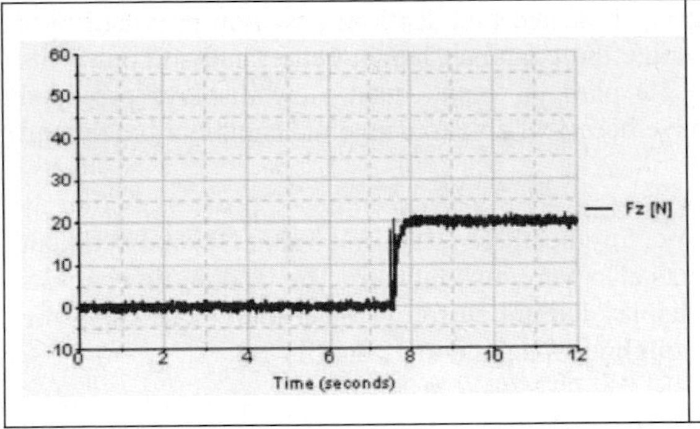

Figure 6: Force-time graph illustrating a low impact landing

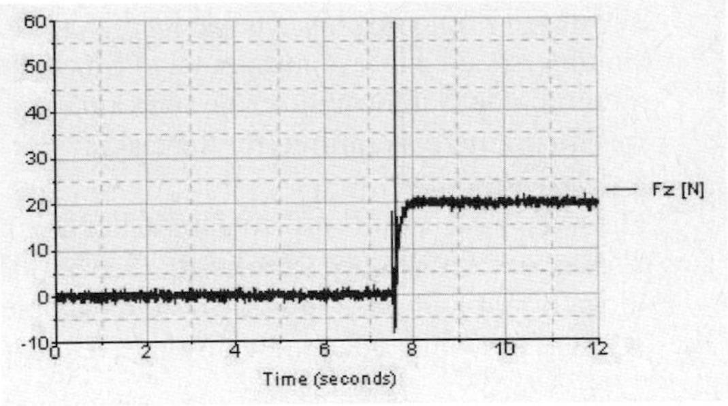

Figure 7: Force-time graph illustrating a high impact landing (indicated by the spike)

3. Results and Discussion

The results in Table 1 show that display design was a significant factor ($P<0.01$). Next, the results were subjected to a Tukey test (Table 2) to ascertain the source of significant differences. The results revealed that there were no significant performance differences between the 2D and quad-view displays. In contrast, significant differences were found between these displays and the 3D display. The ANOVA results revealed that there was no significant gender effect.

Table 1: ANOVA table for container landing performance (impact)

Source	SS	DoF	MS	F	F 5%	F 1%	Sig	Ho
A (gender)	111.5	1	111.5	1.6	4.09	-	-	Retain
Bet Subj	3129.4	46	66.6					
B (Display)	1056.3	2	528.2	**26.8**	**2.6**	**3.12**	**P<0.01**	**Reject**
AxB	121.6	2	60.8	3.1	2.6	3.12	-	Retain
within subj	1814.3	92	19.9					
Total	6233.1	143						

Table 2: Tukey test for display design effects on container landing performance (impact)

		3D Plan View	**2D Plan View**	**Quad-View**
Minimum Pairwise Difference d = 2.59 (P<0.01)	μ =	23.6	29.1	29.5
	3D Plan View		5.5	5.9
	2D Plan View			**0.4**
	Quad-View			

The ANOVA results for container landing impact indicated that depth information provided by a 3D stereoscopic display, enabled subjects to gauge the container height better, and so control its landing speed more appropriately. Since the 2D plan view and quad view displays provided minimal depth information, the subjects may have been less able to gauge the container height and speed parameters required for the task.

The ANOVA results of container landing time for the three displays (Table 3), showed that display design was not significant. Gender effects also were insignificant. The landing time results showed that subjects using a 3D stereoscopic display did not perform slower than those using the 2D plan view and quad view displays. Thus, throughput appeared to be unaffected.

Table 3: ANOVA table for container landing performance (time)

Source	SS	DoF	MS	F	F 5%	Sig	Ho
A (gender)	1.0	1	1.0	0.3	4.09	-	Retain
Bet Subj	176.5	46	3.8				
B (Display)	1.8	2	0.9	1.4	2.6	-	Retain
AxB	0.3	2	0.2	0.2	2.6	-	Retain
within subj	61.1	92	0.7				
Total	240.7	143					

4. Conclusion

The experiment results showed that a 3D stereoscopic display improved subject performance in terms of container landing impact. The display did not affect container landing time performance when compared with 2D plan view and quad-view displays. Thus, it was concluded that a 3D stereoscopic display showed promise in enhancing operator performance of container landing.

However, the benefits would have to be verified with an actual size crane, as scale-up effects might arise. Nevertheless, according to guidelines reported by Boff and Lincoln (1988) concerning stereoscopic viewing (effective up to 65m for central vision), it may be expected that the efficacy of the stereoscopic display would hold. Furthermore, performance time benefits might also be revealed when the system is tested at a greater container height. Tests on an actual size crane are planned and the results will be reported at a later date.

References

Ng M.C. and Lim K.Y., (2000). An Assessment of Various Two-Dimensional Display Designs for a Camera Based Remote Freight Handling System, In Lim K.Y. et al, Proceedings of the Joint Conference of APCHI 2000 & ASEAN Ergonomics 2000, Singapore, Elsevier Science, pp. 68-74.

Boff K.R. & Lincoln J.E., (1988). *Engineering Data Compendium: Human Perception and Performance.* AAMRL, Wright-Patterson Air Force Base, Ohio.

Intelligent Human Interface for Road Tunnel Fire Ventilation Control System

Kazuo Maeda and Ichiro Nakahori

Sohatsu Systems Laboratory Inc.
Kobe Industrial Promotion Center 708
1-8-4 Higashi Kawasaki chou, Chuo-ku,
Kobe,650-0044 Japan
maeda@sohatsu.com , nakahori@sohatsu.com

Abstract

We recently faced fire disasters, and their terrifying consequences – Tuern tunnel in '99 and Gotthard tunnel in '01. Skilled operators could not respond immediately to the lorry fire accident of Mont Blanc tunnel in '99. Thirty nine people were killed then. Because the information was too complicated for operators to understand properly problems involved. To prevent such desasters, we would like to propose a new intelligent human interface for the road tunnel fire ventilation control system that uses (a) a new simulation technology about the precise fire spreads with high speed, (b) a car navigation technology integrated with geographic information system, and (c) a web technology based on intelligent agent model.

1 Introduction

Following the frequent and tragic large-scale fire accidents in road tunnels in Europe, the radical review of design criteria concerning the avoidance of road tunnel fires, has begun in Japan.
As facilities in road tunnels become large in scale and wide in the area they have to cover, so their monitoring and control systems must also become large and complex. In this context it is said that more attention should be given to the function of the systems to be focussed on emergencies, that are rarely experienced but which force an immediate response, and which make it absolutely essential for operators to obtain comprehensive information immediately in order to judge and operate the safety facilities with the maximum speed and efficiency.
On the other hand, the brilliant innovations in information technology, high-speed simulation technology and new sensor technology, such as heat distribution sensors, in recent years have improved the emergency monitoring and control systems of road tunnel facilities. We can now use the new communication technology, such as the car navigation systems, the cellular phone and the Web technology, in addition to conventional emergency phones and information boards.
In the conventional emergency system like Human-Computer Interaction, the contact point between operators and equipment, including monitors and devices, has played a main roll. Here, we introduce a new computer interface such as the Human-Human Interaction and Human-Society Interaction in addition to the conventional Human-Computer Interaction, as stated above.[1] The HHI interface helps operators, fire fighters, police officers and drivers to communicate. The HSI helps to give information to public authorities and rescue services and mass media. Operators, fire fighters, police officers and passengers can all play an important collaborative role in improving the chances of surviving an emergency. This paper describes the outline of a model based road tunnel emergency ventilation control support system and its new user interfaces such as HCI, HHI

and HSI that make use of IT technology, which is developing rapidly, and high-speed tunnel ventilation simulation technology. [2)]

2 A model based road tunnel emergency ventilation control support system

2.1 The outline of the fire ventilation control support system.

Road tunnel fires are a very dangerous accident that threatens human life. It is necessary to secure the evacuation of passengers and the rescue operation in extreme heat and with little visibility. We need (a) the rapid detection of traffic accidents and fire accidents by checking the fire alarm or emergency phone or by monitoring ITV, (b) to provide fire information to the related public authorities, (c) evacuation instructions for passengers by emergency broadcasting or an information board, (d) emergency ventilation control with lower wind velocity for securing an evacuation environment and (e) traffic control by preventing the entry of vehicles into the tunnel. It is very important to give the most suitable evacuation instructions depending on the position and conditions where each person is found. In other words, their evacuation environments differ depending on their positions relative to a fire, the smoke concentration and the airflow which varies locally. The basic structure of the proposed system is shown in figure 1.

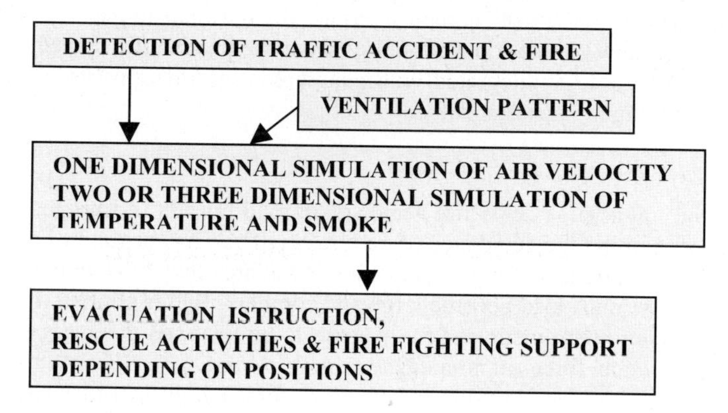

Figure 1 The basic structure of the proposed system

2.2 The road tunnel fire ventilation control support system and a typical example.

Our system efficiently supports a series of tasks from fire detection and its recognition by operators, to fire fighting activities, providing fire information to passengers, police, fire stations and the tunnel equipment administrator, by a model based approach. We can select the optimum fire ventilation control through the high precision numerical simulator of the road tunnel fire in consideration of the human factor and the running condition of the vehicles.

Firstly, we calculate the air velocity and the smoke concentration along the longitudinal direction by a one dimensional fire ventilation simulator, and calculate more detailed smoke concentration distribution and air velocity in the transverse direction by two or three-dimensional fire ventilation simulators giving a boundary condition from former results of air velocity, and finally we support evacuation instructions and water sprinkling control. A typical example is shown in figure2.

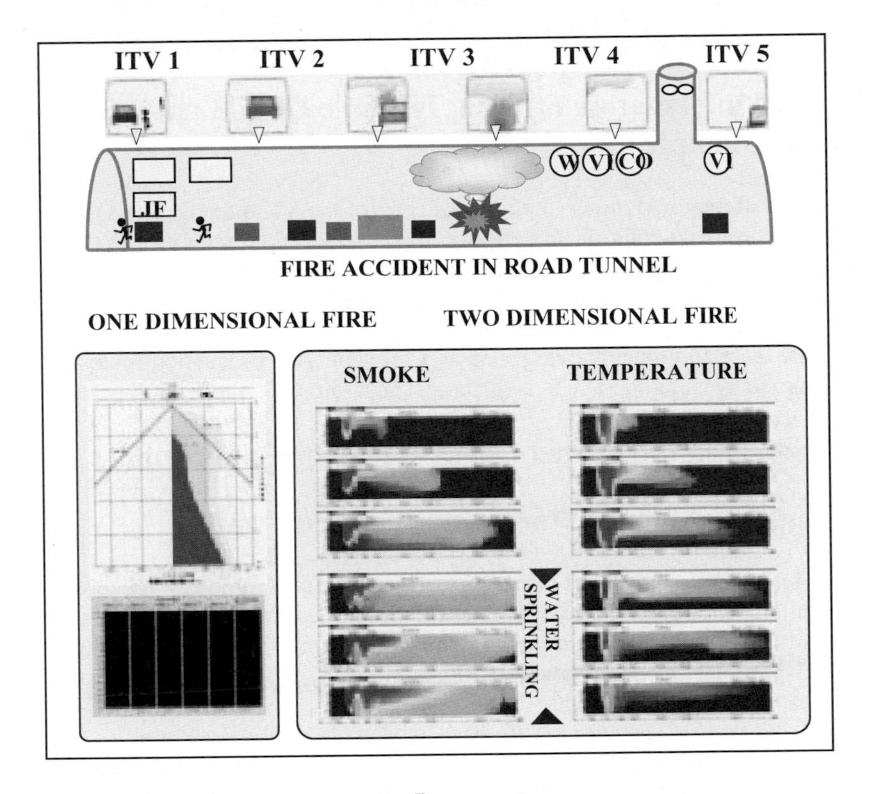

Figure 2 A typical example of the fire operation support system

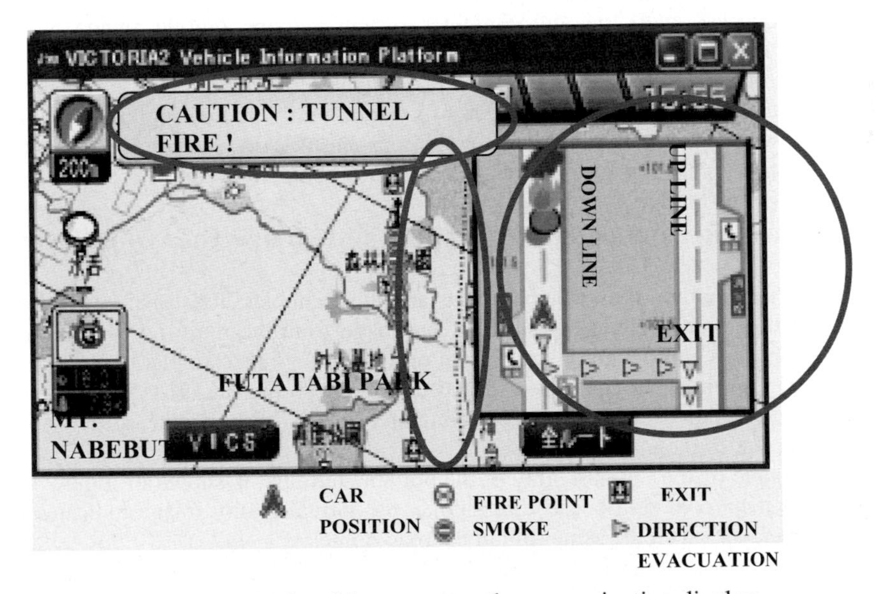

Figure 3. A typical example of the screen on the car navigation display, in the fire information offering system

3 The fire information offering system by VICS navigation and the typical example

Fire information offering system use the infrastructure of the VICS navigation system and support evacuation activities depending on the position of the vehicles to the fire. We can receive fire information which the FM multiplex system broadcasts to the navigation system. Drivers receive a message of "go straight on" downstream of the fire point, and "Stop and leave by the emergency exit" with a screen image such as evacuation route upstream of the fire point. We superimpose the simple figure of the road tunnel drawing such as a fire point, the evacuation route and facilities on the navigation map by using the VICS navigation. This example is shown in Figure 3.

4 The wide area integrated fire management and support system

4.1 The outline of the road tunnel WEB system

The mobile agents move, judge and take necessary actions autonomously on computer networks by using the intelligent agent technology. First of all, our system sends the mobile agents to the monitoring and control information server and the equipment information server at the same time. Next, it retrieves all the fire information and integrates the on-line ventilation plant data and the equipment's information at high speed.

In the conventional systems, the public Road Corporation, road maintenance companies and the other related public authorities, such as the police and the fire station, don't communicate with each other because of the lack of mutual protocol.
Recent technology can work on any computer system and the mobile agents on the virtual machine gather information on the various kinds of fires on many computers in many companies and the public sections as a user's agent. A mobile agent combines the on-line monitoring and control data, map information and equipment information easily. It also supports each person who requests their own information, such as the conditions of suffering, the rescue conditions and the restoration conditions, by moving one road maintenance company to another public section, plus passengers and rescue activities.

4.2 Wide area fire management support system and the typical example

We can get the information of on-line plant data and the tunnel facilities easily with Web browser in the form of text data, voice, image and screen image from the multi-purpose PC of the road maintenance company and the portable terminal of motorists. For example, the co-operation type mobile agent of the wide area control center, co-operates with the fire crisis control agent, judges and takes urgent action automatically on the computer servers of the police, a fire station and the maintenance section of the manufacturer.
The typical example of the virtual reality of the portable terminal is shown in figure 4. When the emergency exit is hard to find because of smoke, the virtual reality image indicates the visible facilities for emergencies on the same spot in the road tunnel.

(a) An actual tunnel image of the fire,

(b) A virtual reality of the fire

Figure 4. Typical example of three-dimensional tunnel virtual reality

5 Summary

We propose the framework of a model based road tunnel emergency ventilation control support system and its new user interfaces such as the HHI and HSI in addition to the conventional HCI by using IT technology, which is developing rapidly, and high-speed tunnel ventilation simulation technology.

All of the operators, fire fighters, police officers and passengers can play an important collaborative role to greatly increase the chance of surviving an emergency.

The safety environment is dependent upon its position relative to a fire. Smoke concentration and airflow vary from place to place. Our system can guide passengers and rescue people accordingly.

References

1) Shogo Nishida, Yutaka Saeki, "Human Conputer Interaction Tecnology", Omu-sha, (1991)

2) Kazuo Maeda,etc.,"Tranjent Analysis of Ventilated Tunnels with Junctions Using Graph Theory, 11th International Symposium on Aerodynamics & Ventilation Of Vehicle Tunnels (2003) (To be appeared)

Multi-Functional Data Collection Interfaces for Biomedical Research Collaboration

Fillia Makedon[1], *Tilmann Steinberg*[1], *Laurence G. Rahme*[2], *Aria Tzika*[3], *Heather Wishart*[4], *Yuhang Wang*[1]

Abstract

This paper describes data collection interfaces for research collaboration in biomedical applications where there is need for secure sharing of sensitive data. These interfaces are multi-functional because they (a) are structured templates for entering experiments and tools within a given domain; (b) provide hierarchical entry/presentation of data; (c) offer security via control of access by user type for each level of hierarchy; (d) provide tracking of data usage; (e) combine workflow management of tasks, thus enabling collaboration; (f) facilitate interoperability for heterogeneous data via common metadata representations; (g) enable advanced searching capabilities allowing multiple parameters; (h) offer assessment services evaluating the quality of entered data; and (i) support intelligent data management and services (*e.g.* notification and risk analysis).

1 Introduction

A fundamental challenge in biomedical research is providing access to data collections to facilitate early detection or discovery. One major technical barrier is the non-interoperability of data due to different methods (formats) of collection or representation (Roland, Svensson *et al.* 2001). This paper describes the Catalog system, an interface built to be open, flexible, user-centered and modular. Catalog is self-standing but can also serve as the front end of any collaboration system where different researchers must collect and share data within a given domain. The system is currently being developed at the Dartmouth Experimental Visualization Laboratory (DEVLAB) for different target applications in neuroscience, molecular biology and heart dynamics research. It collects data for use in *Negotiation Based Sharing (NBS)* (Ye, Makedon et al. 2003), and is extensible and amenable to intra- and inter-domain data sharing.

NBS is based on two simple connected principles: (a) the use of extracted metadata to represent primary data and publicize ongoing research and (b) the use of a negotiation mechanism (SCENS) to provide incentives for users to share their data. The Catalog system collects metadata rather than primary data. The Catalog system must provide ease of use, security, scalability and sustainability (Makedon, Ford *et al.* 2002). Since maintaining the rights of information owners and providing incentives for participation are key to the SCENS framework (**Figure 1**), the data owners (A, B, C) submit metadata of their datasets and policies for their use via the Catalog system. Catalog enables the owner to define the conditions of sharing: when, how, by whom, and for how long. A subscriber-user (A) can query the metadata for qualifying sets, and then enter negotiation mode for access to the original data. Once the data owners' requirements have bee n met, SCENS sends a

[1] Dartmouth College, 6211 Sudikoff Laboratory, Hanover, NH 03755, makedon@cs.dartmouth.edu
[2] Department of Surgery, 50 Blossom Street, Massachusetts General Hospital, Boston, MA 02114
[3] NMR Surgical Laboratory, 51 Blossom Street, Massachusetts General Hospital, Boston, MA 02114
[4] Dartmouth Medical School, Dartmouth-Hitchcock Medical Center, Lebanon, NH 03756

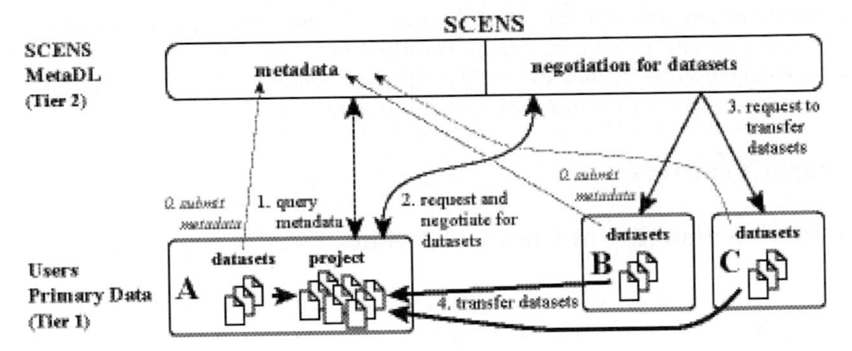

Figure 1. The SCENS framework. *Dashed lines: metadata transfer; thin solid lines: negotiation traffic; thick solid lines: actual data transfer. The* ***Catalog System*** *resides in Tier 2, negotiation.*

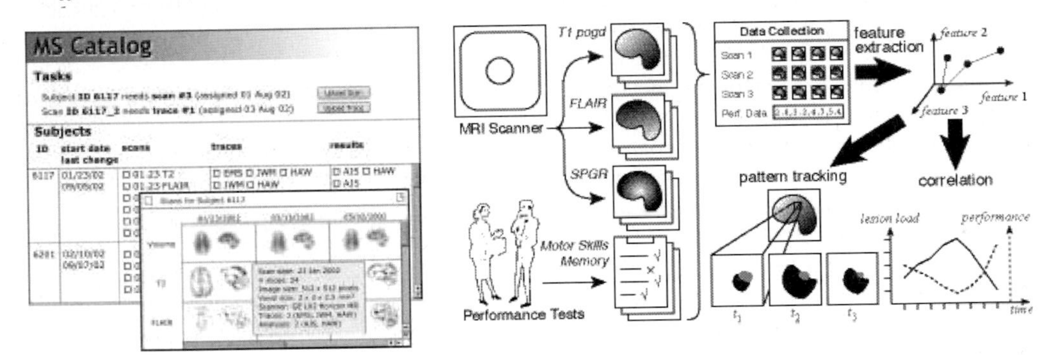

Figure 2. Catalog Interface for MS **Figure 3.** Data acquisition, integration, & analysis

request to the primary data owners, B and C, to transfer the datasets A has requested.

This paper reports on the development of the Catalog system for spatiotemporal multi-modal data streams (**Figure 2** shows an example for MS). The Catalog system helps to "fuse" these data streams and arrive at high-level patterns for early diagnosis. The initial goal is to create local metadata that are later uploaded to a central SCENS server for sharing with other Catalog users.

2 Related Work

Metadata-based collections are used, among others, in the **BIRN** project (Marx 2002), **GenBank** (National Institutes of Health), the **European Computerized Human Brain Database (ECHBD)** (ECHBD), the **fMRI Data Center** (Grethe, Van Horn et al. 2001), and the **Open Archives Initiative** (Open Archives Initiative). Similar ideas also exist in the **Common Data Model** (Gardner, Knuth et al. 2001); **BrainMap** (Fox and Lancaster 2002); and the **BioImage Database Project** (Carazo and Stelzer 1999). In all these, metadata are used to link to data that have not been themselves integrated into the system and do not contain Catalog facilities (Makedon, Ford et al. 2002) but are centralized indexes of distributed data sources. Two cataloging tools are **Axiope** (Axiope project), which combines independent local metadata into a federated system; and **NeuroSys** (Pittendrigh and Jacobs), which provides metadata based on domain ontologies.

The Catalog system is suited for federated systems where there is need to integrate different schemas from different domains and data models using a common metadata standard. Our system builds on existing metadata standards (OAI, **Dublin Core Metadata Initiative** (DCMI)). Other influential metadata projects include the **METAe** project (METAe) and **BrainML** (Weill Medical College of Cornell University Laboratory of Neuroinformatics).

3 System Features

3.1 Multi-Functionality of Data Collection

The first challenge is to enhance the data collection process so that it combines **collection**, **integration**, and **analysis** effectively (see **Figure 3**), by integrating multi-modal data, workflow management, tracking data origins, and analysis tools (*e.g.* consistency checking) at multiple time points. This requires the development of an effective common representation of imaging, clinical, and treatment data. To date, no generally available system exists for the simultaneous consideration of multiple imaging modalities and none that is systematic, scaleable and automated.

A second challenge is to automate metadata extraction for multidimensional data. This facilitates stream fusion and pattern detection, which enables researchers to identify (and be notified of) key events in the course of the disease progress (*e.g.*, when the ratio of lesion volume exceeds a certain percentage). This process is based on expert knowledge and patterns learned from training data.

A third challenge is how to integrate data sharing in the data collection process (Ye, Makedon et al. 2002). To address a seamless integration at an early stage in the data collection stage, the Catalog system must (a) prompt for and encode particular data sharing or usage conditions set by data owners, and conditions imposed by laws and institutional policies; and (b) offer multi-level access and security features. Presenting high-level information of the usage of the local data by others can aid in policy making.

Finally, a fourth challenge is to create a *domain-independent* system. By comparing system development in different target applications, one can determine common needs, critical information, and how to develop evaluation criteria, *e.g.* user satisfaction, that promote system use.

3.2 Data Collection, Analysis and Metadata Extraction in Biomedicine

The applications described below have different types of data, complexity, and clinical goals. Yet there are similarities: they both require effective summarization of changes with metadata; in both, early detection of critical events requires utilization of prior knowledge and analysis that extracts features of substance from the original data; and correlation with patient performance before or after (drug) treatment over time assumes objective quantification of these data as well. **Figure 4** gives an example of the type of data collection, integration, and analysis facilities needed: the original data and some "private"metadata are for research only and must be secured from outside access. Metadata extraction research involves (a) finding a minimal set of expressive features for further analysis, and (b) codifying security and usage policies to safeguard (manage) the public metadata and ensuring that it is followed.

We are studying and developing new algorithms and techniques for metadata extraction, including multiresolution description of the shape of various anatomical structures. An important benefit of these multiresolution deformable models (Kakadiaris, Papadakis et al. 2002) is that they support both global deformation parameters, which efficiently represent the gross shape features of an

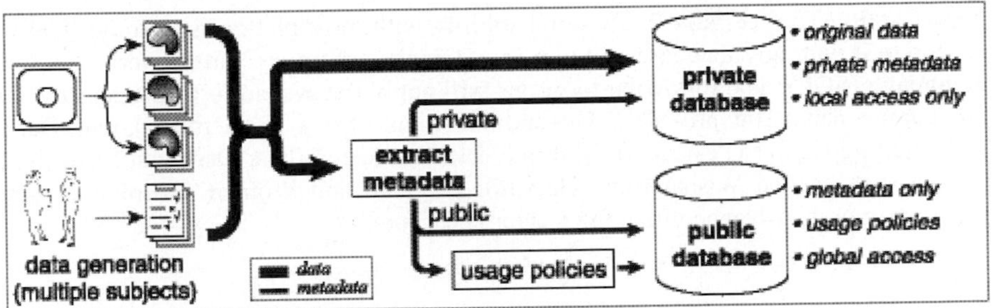

Figure 4. Data acquisition and dissemination.

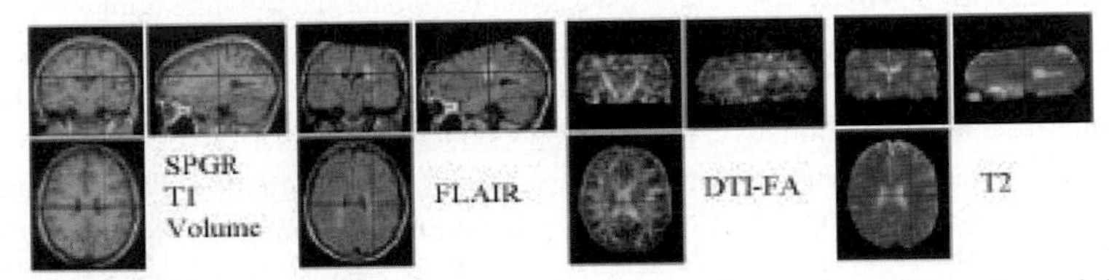

Figure 5. MR image sets of same brain locations taken using different modalities.

object, and local parameters, which capture shape details. Identifying the dimensionality of the wavelet feature vector that best describes each structure, based on statistical shape variability, allows its use as an efficient and compact metadata descriptor.

3.3 Examples in Biomedical Research

Spatiotemporal Multimodal Streams (MS Lesion Analysis): Understanding MS activity requires monitoring neurobiological changes with magnetic resonance imaging (MRI) (Lee, Smith et al. 1999). Most MS-related research and clinical practice has depended on conventional MRI imaging techniques (T2, FLAIR; see **Figure 5**) to detect the lesions that are characteristic of the disease, but there is increasing MRI-based evidence (Filippi and Grossman 2002) of additional more diffuse pathology in normal-appearing brain tissue (NABT) that likely also contributes to symptoms. These diffuse changes in NABT are visible only on advanced types of MR scans, such as diffusion tensor imaging (DTI). However, no one scan type can image all the specific pathological processes or abnormalities. We are developing a data-collation paradigm useful for all modalities, and for any new imaging technology, that will make it easier to integrate data and examine the full extent of pathology in the central nervous system, allowing for more accurate monitoring of disease progression and more informed treatment decisions (Mainero, De Stefano et al. 2001).

Multiparametric Data Tracking (Brain Tumor Tracking): Multiparametric MR imaging using MRSI, HMRI and DWMRI can provide an enhanced assessment of neuroepithelial brain tumors, potentially allowing the distinction between higher-grade and lower-grade tumors, which becomes clinically important especially when the tumors are inoperable. An initial database infrastructure for better data collection and management allows for more power and flexibility in detailed data analysis. Using metadata and secure negotiation services (SCENS) will make it possible to carry out clinical research on a larger scale and share work.

4 Ongoing Work

The described system is currently under development at the DEVLAB. We are using Java for deploying the Catalog Tool (**Figure 2**) for compatibility with most platforms, and the JDBC API to integrate this tool with any database solution (*e.g.* Oracle, Informix) that supports the Structured Query Language (SQL). The following locations will apply the system to their respective areas of research. Tumor data: Harvard-Mass General Hospital (Tzika & Astrakas) and Dartmouth-Advanced Imaging Center (Pearlman). Multiple sclerosis lesion data: Dartmouth Hospital, Brain Imaging Lab (Saykin and Wishart) and Harvard, Brigham and Womens' Hospital (Guttmann). Heart data: Dartmouth-Advanced Imaging Center (Pearlman).

References

Axiope project Data Sharing White Paper, http://www.axiope.org/datasharingwhite.html.

Carazo, J. M. and E. H. K. Stelzer (1999). "The BioImage Database Project: Organizing Multidimensional Biological Images in an Object-Relational Database." Journal of Structural Biology **125**: 97?02.

DCMI The Dublin Core Metadata Initiative, http://dublincore.org/.

ECHBD European Computerised Human Brain Database, http://fornix.neuro.ki.se/ECHBD/Database/.

Filippi, M. and R. I. Grossman (2002). "MRI techniques to monitor MS evolution." Neurology **58:** 1147-1153.

Fox, P. T. and J. L. Lancaster (2002). "Mapping context and content: the BrainMap model." Nature Reviews Neuroscience 3(4): 319?21.

Gardner, D., K. H. Knuth, et al. (2001). "Common data model for neuroscience data and data model exchange." Journal of the American Medical Informatics Association 8(**1**): 103?04.

Grethe, J. S., J. D. Van Horn, et al. (2001). "The fMRI data center: An introduction." NeuroImage 13(**6**): S135.

Kakadiaris, I. A., E. Papadakis, et al. (2002). g-HDAF Multiresolution Deformable Models For Shape Modeling and Reconstruction. British Machine Vision Conference, Cardiff, UK.

Lee, M. A., S. Smith, et al. (1999). "Spatial mapping of T2 and gadolinium-enhancing T1 lesion volumes in multiple sclerosis: evidence for distinct mechanisms of lesion genesis? [see comments]." Brain 122(**Pt** 7): 1261-70.

Mainero, C., N. De Stefano, et al. (2001). "Correlates of MS disability assessed in vivo using aggregates of MR quantities." Neurology. 56(**10**): 1331-4.

Makedon, F., J. C. Ford, et al. (2002). MetaDL: A Digital Library of Metadata for Sensitive or Complex Research Data. European Conference on Digital Libraries (ECDL2002), Rome, Italy.

Marx, V. (2002). "Beautiful Bioimages for the Eyes of Many Beholders." Science 297(**55**78): 39?0.

METAe The Metadata Engine Project, http://meta-e.uibk.ac.at/.

National Institutes of Health GenBank, http://www.ncbi.nlm.nih.gov/Genbank/.

Open Archives Initiative OAI Home Page, http://www.openarchives.org/.

Pittendrigh, S. and G. Jacobs NeuroSys: An Electronic Laboratory Notebook and Semi-structured Database, http://www.nervana.montana.edu/~sandy/paper.doc.

Roland, P., G. Svensson, et al. (2001). "A database generator for human brain imaging." Trends in Neuroscience 24(1**0**): 562?64.

Weill Medical College of Cornell University Laboratory of Neuroinformatics BrainML functional ontology for neuroscience, http://brainml.org/.

Ye, S., F. Makedon, et al. (2002). A Negotiation Framework for Secure Data Sharing. Hanover, NH, Dartmouth College Computer Science Department.

Ye, S., F. Makedon, et al. (2003). SCENS: A system for the mediated sharing of sensitive information. In submission to the Third ACM/IEEE Joint Conference on Digital Libraries.

Maintenance Support of Corporate Directories with Social-filtering

Kazuo Misue

Fujitsu Laboratories Ltd.
4-1-1 Kamikodanaka, Nakahara-ku,
Kawasaki City, Kanagawa 211-8588,
JAPAN
misue.kazuo@jp.fujitsu.com

Takanori Ugai

Fujitsu Laboratories Ltd.
4-1-1 Kamikodanaka, Nakahara-ku,
Kawasaki City, Kanagawa 211-8588,
JAPAN
ugai@jp.fujitsu.com

Abstract

While classified directories provide useful documents with an easy user interface, their maintenance cost is extremely high. This paper proposes to apply social filtering to support the maintenance of corporate directories, such as the collection of documents (URLs) and their classification into categories. Preliminary experimental results indicate that such a process can be effectively supported.

1 Introduction

Many corporations now engage in knowledge management, and the sharing of information within the corporation is a key part of knowledge management. For the purpose of information sharing, search engines and classified directories (also known as catalogues) are widely used. Classified directories are convenient for users because they organize a set of useful documents (or URLs) into a hierarchy of thematic categories.

To provide this convenience, classified directories require continuous maintenance of the category hierarchies and collected URLs. Content maintenance, which we focus on in this paper, consists of collecting useful URLs and classifying them into categories; the cost of doing this manually is extremely high.

The purpose of our research is to decrease the cost of maintaining corporate directories (classified directories in a corporate intranet). We previously developed a directory management system for corporate directories (Ugai, 1999). This system provides facilities to collect useful URLs and to classify them into hierarchical categories (Ugai, Katayama & Tsuda 2001). It classifies URLs based on their contents by applying machine-learning methods.

We tested several classification methods in earlier experiments, but could not obtain sufficient accuracy for some categories. The poor accuracy was due to not having enough training documents (leaf categories have only a few URLs), the difficulty of choosing adequate feature words, and so on.

We propose the use of social filtering to cope with such problems. As far as we know, no attempts to apply social filtering to the maintenance of a corporate directory have been reported. Thus, the main subject of this paper is our investigation of the effectiveness of social filtering for corporate directory maintenance.

2 Preliminaries

2.1 Basic Idea

We propose that social filtering be used for URL recommendation to support the collection and classification of useful URLs. While we cannot expect well-balanced categories in personal bookmarks, these bookmarks could provide up-to-date and useful URLs for some specific topics. Our basic idea, for a category in a corporate directory, is to (1) find similar categories in bookmarks, (2) if the corporate directory lacks any URLs from the bookmarks, propose the missing URLs as new entries. This is similar to the approach of (Rucker & Polanco, 1997). However, we orient the recommendation to corporate directories and investigate the feasibility of applying social filtering for this purpose.

Special mechanisms are needed to refer to personal bookmarks on personal computers (Mori & Yamada, 2000). We currently provide an online bookmark management service on our corporate intranet. Users of the online bookmark service gain the benefit of bookmark portability (they can access the same bookmarks from any computers connected to the intranet). A similar service is available for the Internet (Blink), but using it may allow other users to know URLs within the corporation.

We apply the social filtering model used by (Rucker & Polanco, 1997). The similarity is defined between two categories, and popularity is defined for each URL. For a target category T, some similar category S is found. If some URLs in S are highly popular but are not in T, these are recommended to category T.

2.2 Basic Data

In this section, we describe the numerical data regarding a practical corporate-directory and personal bookmarks that we used in our investigation.

Table 1 gives information about the size (the number of included URLs) of the corporate directory and bookmarks. Table 2 shows how many URLs are shared between the directory and the bookmarks.

Table 1: Size of the corporate directory and bookmarks

	Corporate directory	Bookmarks
# of categories	210	369
# of URLs (entries)	1537	3096
# of URLs (unique)	1318	2257
# of URLs / category	1 ~ 45	1 ~ 117
Average	7.31	8.39
Standard Deviation	6.78	10.03

For each category in the directory, we also checked the number of categories in the bookmarks having at least one common URL. There were 249 pairs of categories sharing at least one URL.

74 categories of 210 in the directory had corresponding categories in the bookmarks; the average number of categories in the bookmarks sharing URLs was 3.36 (the standard deviation was 3.15). Table 3 shows the top ten categories in the directory in terms of the largest number of corresponding categories in the bookmarks.

Table 2: The number of shared / unshared URLs

	# of URLs	percentage
Belong to only the directory	2166	62.2%
Belong to only bookmarks	1227	35.2%
Belong to both the directory and bookmarks	91	2.6%
total	3484	100.0%

Table 3: Top ten categories sharing URLs with many bookmark categories

Directory category	# of categories sharing URL(s)
Information for the whole company	15
Headquarters	11
Personnel affairs	10
Internal procedures & Formats	10
Newly-arrived information	10
Dictionaries & Search engines	9
BBS in the Intranet	9
Telephone directories	8
Local information	8
Information processing group	8

2.3 First Impression from the Basic Data

Category size: The average category size (the number of URLs) was almost the same for both the corporate directory and the bookmarks. However, the category sizes in the bookmarks varied more widely than the sizes in the directory. The corporate directory was well organized for the convenience of employees in general, while the personal bookmarks were more likely to be organized according to personal preferences.

Shared URLs: Only 91 (2.6%) of the URLs were shared between the directory and the bookmarks. Useful recommendations, though, cannot be given without careful control of the similarity values. However, from the other point of view, the bookmarks contained 1127 URLs which the directory did not include. Thus, if we can find adequate categories to use as recommendation targets, we can expect considerable beneficial effects.

Similarity: The ten categories shown in Table 3 have a high probability of being recommended because of the correlation between the number of shared URLs and the similarity values. The results from our earlier experiments showed that contents-based automatic classification does not work well for many categories in the top ten directory categories because of the difficulty of extracting adequate feature words. Social filtering might be able to overcome this weakness of content-based classification techniques.

3 Design of Social-filtering

In this section, we analyze the basic data described in Section 2 to formalize the similarity and popularity for social filtering.

3.1 Formalization of Similarity

The similarity is used to find categories that include URLs recommended to a target category. We considered the following two formalizations of the similarity.

(s1) The number of shared URLs (independent of the category sizes)
(s2) The number of shared URLs / the size of the union between the two categories (dependent on the category sizes)

Candidate (s1) is the same as in (Rucker & Polanco, 1997). In this formalization, for example, for a category $\{a, b, c\}$, two categories $\{a, b, d\}$ and $\{a, b, d, e, f\}$ are given the same similarity (here, let a, b, ..., f be all the different URLs). We thought some adjustment should be made to cope with a wide variety of category sizes in the bookmarks, though, as in candidate (s2), and thus employed (s2).

3.2 Formalization of Popularity

The popularity is used to give priority to URLs in categories with a high similarity. We considered the following two formalizations of the popularity.

(p1) Frequency of occurrence in all categories (independent of a target category)
(p2) Frequency in the categories with positive similarity (dependent on a target category)

Candidate (p1) offers the merit of easy computation, but irrelevant URLs are more likely to be recommended. Using candidate (p2) might decrease the number of irrelevant recommendations. To evaluate the difference, we compared these two formalizations by applying all 210 categories in the corporate directory. We found 636 pairs of URLs which were ordered candidates of a recommendation. Only 34 pairs (5.31%) changed the order between candidates (p1) and (p2). We thought this difference was small, so we used (p1).

4 Experiment

Using a prototype system, we evaluated how many useful URLs would be recommended when the social filtering technique was used.

Our first application of the social filtering generated 22 URLs as recommendations for 17 of the 210 categories in the corporate directory. One of the authors manually checked the 22 URLs and judged that 18 of these URLs should be accepted as useful. Table 4 shows the number of recommended URLs and the number of accepted URLs for some of the categories.

We cannot draw significant results based on the limited amount of data obtained through the experiment, and further work is needed. However, we can make some general comments based on the data.

Table 4: Effectiveness of recommendation by using the social filtering

Category	# of recommended URLs	# of accepted URLs
Internal procedures & Format	1	1
Information for the whole company	4	3
Directories & Search engines	1	0
Local information	1	1
Other 13 Categories	15	13
Total	22	18 (82%)

While 22 URLs were recommended, 18 of these were judged to be useful; that is, the first recommendation was 82% accurate. This accuracy is much higher than that realized when using content-based categorization techniques, which provide only about 30% accuracy (Ugai & Misue, 2003).

The content-based categorization techniques did not work well for categories like "Internal procedures," "Information for the whole company," or "Local information." Table 4 shows that social filtering appeared to work well for such categories in that it provided useful recommendations.

5 Concluding Remarks

Our prototype system using social filtering provided better recommendations for some categories than has been achieved with content-based categorization even though the data scale of hundreds of categories is too small for effective social filtering. This result suggests that social filtering may be applicable to support corporate directory maintenance, and might help to alleviate the problem of the high cost of managing a corporate intranet.

6 References

BLINK. http://blink.com/

Mori, M., & Yamada, S. (2000). Bookmark-Agent: Sharing of Bookmarks for Search Assists, *IEICE Transactions*, J83-D-I (5), 487-494 (in Japanese).

Rucker, J., & Polanco, M. J. (1997). SiteSeer: Personalized Navigation for the Web, *Communication of the ACM*, 40 (3), 73-75.

Ugai, T. (1999). Browseable Directory Management System for Intranet, *Information Processing Society Japan, Research Report*, 99-GW-33-13 (in Japanese).

Ugai, T., Katayama, Y., & Tsuda, H. (2001). Usage of automatic acquiring and classification for directory maintenance, *The 62nd IPSJ Annual Conference Special Track I*, 249-252 (in Japanese).

Ugai, T., & Misue, K. (2003). Interaction between a large directory and bookmarks, *Proc. of Interaction 2003* (in Japanese).

SeL-Mixer: A Music Authoring Environment Fusing Virtual and Physical Activities

Hirohiko Mori

Dept. of Systems Information Eng.,
Musashi Institute of Technology
1-28-1, Tamazutsumi, Segagaya,
Tokyo, 158-8557, JAPAN
mori@si.musashi-tech.ac.jp

Kazunobu Azuma

Dept. of Systems Information Eng.,
Musashi Institute of Technology
1-28-1, Tamazutsumi, Segagaya,
Tokyo, 158-8557, JAPAN
linus@leo.bekkoame.ne.jp

Abstract

Tangible User Interface (TUI) is one of the expected methodologies to achieve the "Calm Technology" in our everyday life. TUI aims to providing the mutual harmony between the physical world and the virtual world by utilizing physical objects around us to manipulate computers. Though many researches have been done in the field of TUI, in most of them, one computational function is assigned to certain physical object statically in advance.
This paper describes a TUI system on the music authoring task which allows user to assign computational functions on physical objects dynamically. In our system, several kinds of physical elements for the music authoring, e.g. sliders, jog dials, push bottoms, are provided for users. Users can easily build their own music mixer, so as to fit for their desired operations, just by layouting the elements on the desk and assigning the computational functions on them that are filtered according to the characteristics of the elements. Users can also change the layout and the assignments dynamically in each scene during the task. This system allows users to customize their working spaces under the TUI environment as flexibly as the GUI ones

1 Introduction

In music authoring tasks, high costs for using professional equipments in the mixing studio afford even the professional technicians to utilize personal computers to perform their tasks. Some professional technicians, however, sometimes claimed that there are some usability problems in performing the mixing tasks with the personal computers. Though, for example, simultaneous multiple operations on some sliders with a few fingers or with double hands provide very important feelings for adjusting some parameters, only one parameter can be allowed to adjusted at once because of using the mouse. Using the mouse also provide no force feedback though adequate force feedback is very important to adjust them. To solve such usability problem, adopting tangible user interface (TUI) technique should be effective.

Tangible user interface (TUI) is one of the expected methodologies under the ubiquitous computing environment and aims to maintain a mutual harmony between the physical and virtual world by utilizing physical objects to manipulate computers. Though many researches have been done in the field of TUI, in most of them, one computational function is assigned to certain physical object statically in advance (e.g., Gorbet, et. al(1998)). Therefore, such approaches sometimes cause the lack of the potential flexibility of the virtual world.

In this paper, we propose SeL-Mixer that is a music mixing system that aims to put both the advantages of the TUI and the flexibility of the GUI to account. In SeL-Mixer, various kinds of physical components for the music mixing tasks are provided for users. Users can create and customize their digital mixer as just by putting the components on the physical desk. Users can also dynamically change their assignment of computational functions by selecting from automatically filtered functions according to the characteristics of the components. In SeL-Mixer, therefore, users can create and change their workspace according to their task in any time with the physical objects.

2 SeL-Mixer

SeL-Mixer allows users not only to create their workspace according to their task but also choose their favourite components to adjust audio parameters.

In the SeL-Mixer, many kinds of physical components, such as sliders, jog dials, and push buttons, are prepared for the users. To create their own mixing system, the user repeats two processes: one is to select and put a physical component in an adequate position on the desk, and the other is to assign a function to it.

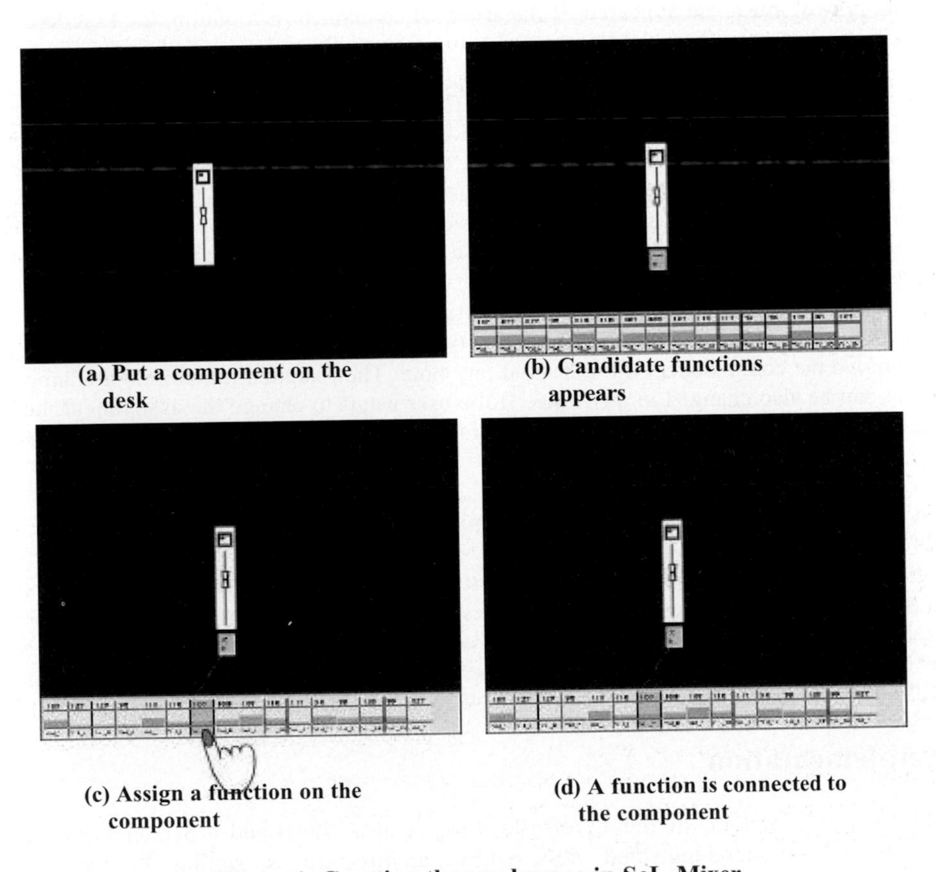

(a) Put a component on the desk

(b) Candidate functions appears

(c) Assign a function on the component

(d) A function is connected to the component

Figure 1. Creating the workspace in SeL-Mixer

Figure 2. Arranging physical components on the LiveDesk

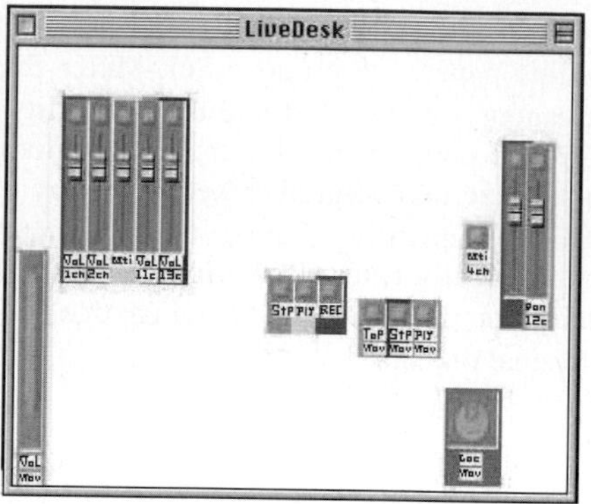

Figure 3. Created Virtual Mixer on the screen

First, the user selects his/her favourite component to control his/her target parameter (for example, the volume level of channel 1) and put it on the desk (Figure 1(a)). Then, the list of the candidate functions that is available to control with the placed component is appeared on the desk (Figure 1(b)). These candidates are determined based on the characteristics of the parameter. If a jog dial, for example, is put on the desk, only the functions that can control a continuous parameter such as the stepping play of music are appeared. If the arranged component is a button, the functions that can control discrete parameter such as normal play function come up.

Then, the user can assign a function from the list. This can be done just by pointing one function on the list and the component with the finger. Then, the function and the physical components are connected with a line. Repeating these processes, the user can create their customized digital mixer and can control the music parameters with the physical components (Figure 2.) The layout of the physical components arranged on the physical desk is also displayed on the computer display unit in front of the user. He/she can also use it as the traditional mixing software with the mouse (Figure 3.)

The user can change the layout dynamically during the task. To change the layout, he/she can just move or remove the components on the desk at any time. The assigned function to a component on the desk can be also changed in any time. If the user wants to change the assignment, he/she pushes the button on the component. Then, the list of candidates appears on the desk, and he/she points a parameter by finger.

The technicians sometimes want to control multiple parameters synchronously. SeL-Mixer also supports to assign multiple functions on a physical object. To add another function on a component on the desk, user should push the button on the component twice just like double clicking of the mouse. Then, the function list comes up on the desk and, by selecting the function from the list, the extra function is assigned on the component. When, for example, the functions to control the volume level of channel 1 and channel 2 are assigned on a slider component, the both level s can become higher or lower simultaneously in keeping their relative relationship.

3 Implementation

The SeL-Mixer is composed of the LiveDesk(Mori, et al. (1999)) and a personal computer. LiveDesk is a computer-augmented desk whose architecture is similar to the Digital

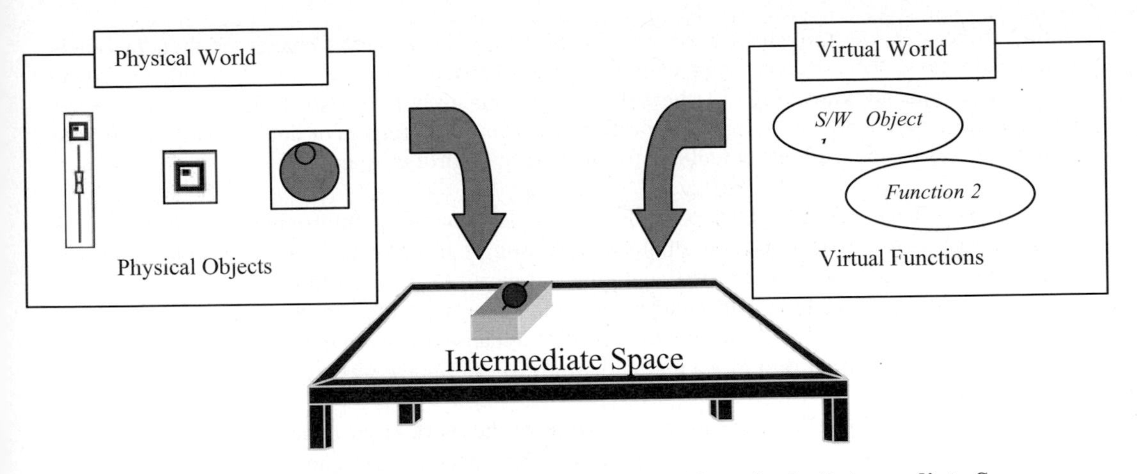

Figure 4. Virtual Functions Meet Physical Objects in the Intermediate Space

Desk(Newman & Wellner(1992)). It has a computer screen on its surface and one video camera is attached on above it to recognize the physical objects and hand gestures on the desk.

In the SeL-Mixer, LiveDesk monitors what types of components and where they are arranged on the surface of the desk continuously. It also detects the movement of the users' fingers. These detected data are always transferred to the personal computer.

The personal computer is used for the processing of the audio mixing task. By receiving the types and axis of the component on the desk, it creates the digital mixer. The physical components are connected to this computer. All parameters are controlled in MIDI signals.

4 Features of SeL-Mixer

SeL-Mixer has the both advantages of GUI and TUI. One merit of the TUI is to provide intuitive controls for the computer by utilizing the physical objects. On the other hand, too many physical objects sometimes cause complication to find the desired functions. GUI can provide the potential high flexibility beyond such constraints of physical world by creating the virtual world inside the computer.

Kinoe & Mori (1993) claimed that one of the important design principles in Human-Computer Interaction in everyday life is that both virtual and physical worlds should be separated from each other, rather than one world is taken into the other and encloses it (for example, in the desktop metaphor methodology the physical world is taken into the virtual world and enclose it). They also claimed that some intermediate space such as "en", Japanese traditional veranda in Japanese architectures, helps to create a mutual harmony between these heterogeneous worlds. This contributes not to take the constraints of one world into the other.

Considering the audio mixing task, while the traditional physical mixer sometimes provides the force feedback and the intuitive controls of a number of parameter, some usability problems, such as the difficulty of the multiple and simultaneous operations among the distant controllers, are caused because their layout has been fixed in advance. When some new functions, furthermore, are developed, whole the equipment should be replaced to adopt new functions. Such problems are caused by the constraints that the physical world potentially has. On the other hands, while the digital mixers of the software can solve such problems because the change of their layout is technically easy, the different types of problems are caused. For example, changing the layout requires the special knowledge of the software for the user. It is also difficult to provide the adequate force feedback or multiple operations by the multiple fingers or double hands that is very important to give some adequate feelings for the user to control the parameters.

To overcome these difficulties, the SeL-Mixer provides a working environment that has both the flexibility of virtual world and the intuitive operation of the physical world by employing the intermediate space between the virtual and physical world (Figure 4.) In the SeL-Mixer, physical components that do not have particular roles in advance meet the virtual functions and charge various functions on the desk. They change its roles dynamically even in performing one task, and eventually, the SeL-Mixer provides a mutual harmony between the virtual and physical world.

SeL-Mixer, eventually, allows the users to customize the work environment very intuitively and operate the multiple controllers simultaneously. SeL-Mixer also can update easily just by downloading them without requiring the user to learn the way to operate the new environment.

5 Conclusion

In this paper, we propose a working environment which have both advantages of TUI and GUI in the field of music authoring tasks. We believe that it is important to fuse the advantages of GUI and TUI and to compensate the disadvantages of each other for the future computing environments. In the near future, we will develop fused computational environments of further advantages of both approaches.

References

Gorbet, M., Orth, M. & Ishii, H.(1998). Triangles: Tangible Interface for Manipulation and Exploration of Digital Information Topography, In Proceedings of the ACM CHI '98(pp.49-56). New York: ACM.

Kinoe, Y., Mori, H. (1993), Mutual Harmony and Temporal Continuity: A Perspective from the Japanese Garden, SIGCHI bulletin Vol. 25, No. 1, (pp. 10-13), ACM Press.

Mori, H. Kozawa, T., Sasamoto, E., Oku, Y. (1999). A Computer-Augmented Office Environment: Integrating Virtual and Real World Objects and behavior, In H. Bullinger & J. Ziegler (Ed.), Human – Computer Interaction Vol.2 (pp.1605-1069). Lawrence Erlbaum Publishers.

Newman, W., and Wellner, P.(1992). A Desk Supporting Computer-based Interaction with Paper Documents. In Proceedings of the ACM CHI 92 (pp.587-592). New York: ACM.

Coordinated Interfaces for Real-time Decision Making in Hierarchical Structures

Mie Nakatani, Shinobu Yamazaki and Shogo Nishida

Graduate School of Engineering Science, Osaka University
Toyonaka, Osaka 560-8531 JAPAN
mie@sys.es.osaka-u.ac.jp

Abstract

This paper deals with a design of visual interfaces for real-time decision making in hierarchical structures. Hierarchical structures are often adopted to control or manage large scale real-time systems, where the whole problem is divided into small portions of sub-problems which can be managed by each member. In this paper, real-time problem solving in hierarchical structures is analyzed first, and its process model is proposed. Then the visual interfaces are designed to support real-time decision making in hierarchical structures on the basis of the process model. The visual interfaces are applied to the management of the fire systems, and its results are evaluated.

1 Introduction

Here we focus on real-time problem solving in hierarchical structures, which is typically observed in control and operation of large scale systems such as fire systems or transportation systems. The structure of an organization should be suitably designed for the type of problem which might arise in it (Malone, 1987). In hierarchical structures, the whole problem is divided into small portions of sub-problems which are assigned to the members with suitable specialties. Each member deals with the assigned sub-problem according to his/her specialty and position. Hierarchical structures cause problems entirely different from those which are dealt with in the GDSS (Nunamaker, 1989) when they are used for real-time problem solving. We believe that special-purpose interfaces are necessary to cope with these problems.

2 Process Model for Problem Solving in Hierarchical Structures

The process of real-time problem solving in hierarchical structures is investigated. First of all, we interviewed fire people, and asked how they make decisions and how they change their decisions. Two important factors, that is, "intention" and "situation", were induced from the analysis of the interview. Both of them are defined as follows.

Intention : Intention is what the operator wants to do on his/her problem. Intention is composed of "target", which is the goal to be reached, and "strategy" which is the path to reach the target. It is very important that each operator understands intentions of other operators at different levels.

Situation : Situation is the state of the system or subsystem to be managed by the operator. The situation changes every moment, and it is important that the situation of the system is expressed for the operators to be understood intuitively and easily.

Operators at each level of hierarchy have their own intentions depending on the situation. In a two-layer model of hierarchy, the upper level manages the whole system and the lower levels manage each area. The intention at the upper level, which is composed of its target and strategy, is decided on the basis of the situation of the whole system. Then orders in perspective are sent to the lower levels based on the intention. At each lower level, intention is formed from the orders from the upper level and the situation in its area, and concrete operations are conducted based on its intention. The changes of the situations are caused by these operations at the lower levels, and in some cases the changes of intentions are caused as a result. This process continues until the problem is solved totally.
The proposed model was traced for the real problem solving process observed in the fire systems, and it was confirmed that the concepts of intention and situation were suitable to express the real-time problem solving process.

3 Design Concept of Visual Interfaces for Hierarchical Decision Making

Coordination between different levels in a hierarchical structure is very important in real-time problem solving. The following functions are proposed based on the process model in the previous section.

(a) Visualization of intention
Showing the intentions at other levels in concrete visible form is considered to be useful to enhance mutual understanding between operators. It will be useful for the lower situations. On the other hand, at the lower layer, fire people are trying to go to the fire points as fast as possible to extinguish lower level operators if they can see the intention of the upper level operator together with the issued orders concretely.
(b) Visualization of situation
Situation is the current state of the system which is dealt with at some levels of hierarchy. It is considered to be useful for the operators to express the situation in the space defined by a small number of important indices. If the situation in each area is expressed by a few important indices, it will be useful for the operators at each level.
(c) Visualization of the gap between intention and situation
Another important function for the operators is to show the gap between intention and situation at each level. This gap shows the difference between the intended operation and its result, that is, the gap becomes zero if the operation works as he intends, and the gap becomes large if the operation does not work as he intends. Direct expression of this gap will be useful to detect the problem in the current situation and will give suggestions to the related operators.

The following steps indicate how to build the interfaces.
(1) N important indices are selected to express the situation of the system. These indices are calculated from the measured data in the system.
(2) Target, which is the goal to be reached, and strategy, which is the path to reach the target, are plotted in the space defined by the selected indices. Figure 1 shows an example of the expression of target and strategy when N equals two. In this case, the decided intention is expressed on the 2-dimensional plane, where the initial situation and target are expressed as a point respectively and strategy as the path between them. Different strategies are expressed as different curves as shown in Figure 1.
(3) Situation at each time is also mapped into the N-dimensional space defined by the selected indices. Then the changes of the situation are expressed on the same plane as the intention. Figure

2 shows an example. If the results of the operations are the same as the intended one, the changes of the situation become the same as the curve of the strategy. On the other hand, if there exists a difference between intention and situation of the system, the plot of the situation becomes different from the strategy curve as shown in Figure 2. The difference shows that there exists a gap between intention and situation, and also gives a clue to think of the reason why such a gap is caused and the method to change the situation (Details are shown in Koiso, 1999).

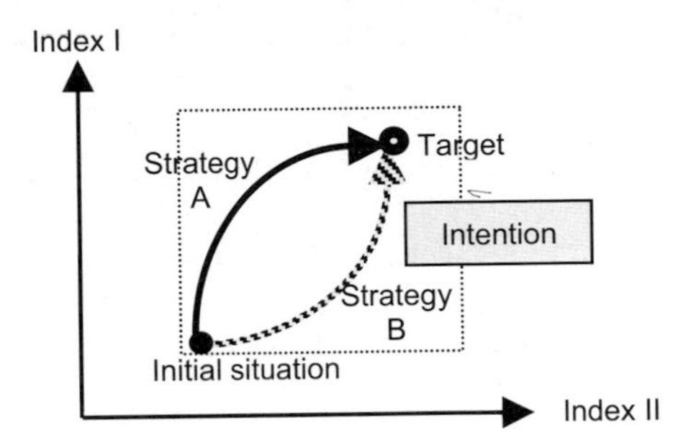

Figure 1: Expression of intension and situation on the same plane (two indices case)

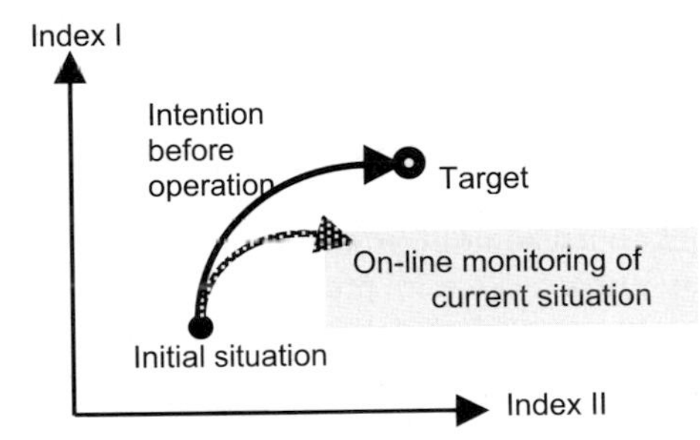

Figure 2: Expression of the gap between intention and situation (two indices case)

4 Implementation and Evaluation of the Visual Interfaces for Management of Fire Systems

The proposed interfaces were implemented on the personal computer for the fire system management by JAVA language. Fire systems are usually composed of two or three layers. Here we suppose one upper layer (U) and two lower layers (L and R) as shown in Figure 3. At the upper layer, the total strategy for the priority order of indices is decided by considering the whole situation. At the lower layers, the distribution of fire engines and fire people is decided based on the upper's strategy. The following indices are picked up to express the situations. The following indices are picked up to express the situations.

(1) Risk Index (RI)
RI shows how high the risk such as the danger of explosion etc. is. This value becomes large when the risk is high.

(2) Burning Index (BI)
BI shows how fast the fire is predicted to extend. If the fire is in wooden building area, the value is high. On the other hand, the value becomes low if the fire is in the fire-proof building area.

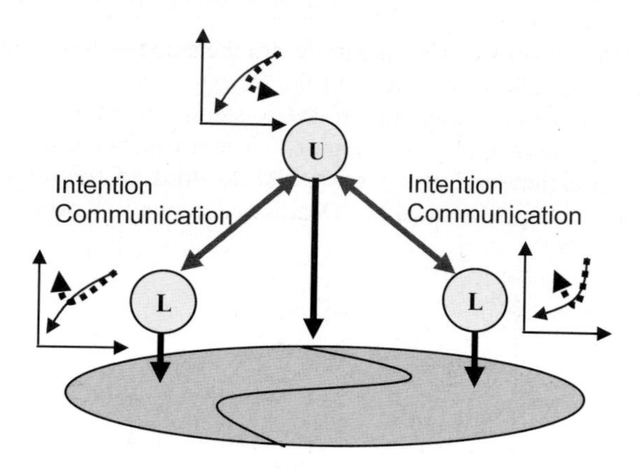

Figure 3: Hierarchical structure

(3) Current Burning Area (CBA)
CBA shows the burning area currently. When the fire is extinguished, CBA becomes zero.
(4) Water Supply Margin (WSM)
WSM is the index to show whether the reserved water is sufficient to extinguish the fire or not.

Figure 4 shows concrete interfaces of the proposed system. The interfaces were consisted of three area; an input area of strategy, a fire map, a visualization area of he intention and the gap. We evaluated the interfaces from a viewpoint of problem solving time and quality of communication using a fire system simulator. Three tasks were prepared by giving various types of fires in the fire map. Table1 shows the average of problem solving time by ten groups (three subjects in each). The proposed interfaces are useful to reduce time to extinguish all fires in Task 1 and Task 3. Almost of subjects could find the gap between intention and situation at a glance with the interfaces and could change the strategy quickly. About Task 2, although the subjects could find

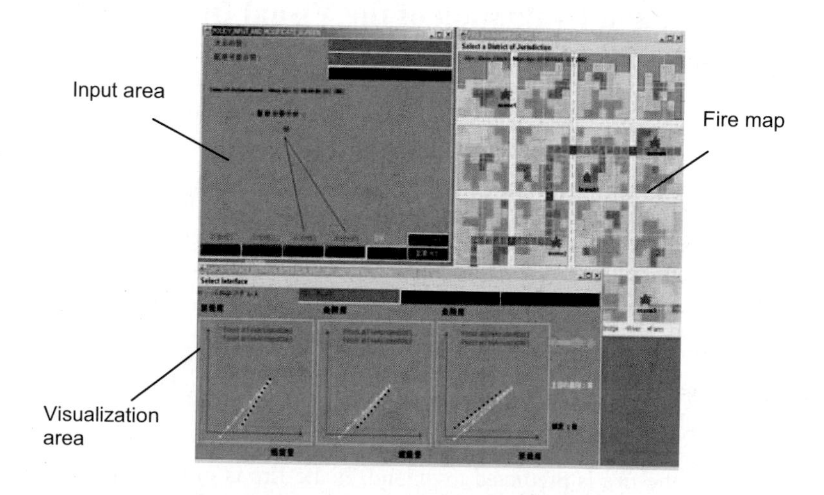

Figure 4: Example of interfaces

Table1: the result of experiment (sec.)

	With IF	Without IF
Task 1	127.4	136.6
Task 2	123.0	118.0
Task 3	113.8	151.6

the gap, it took too much time for only one group to make a new strategy. So the average time of "With IF" was longer than "Without IF". After the experiments, the subjects were interviewed and the following opinions were obtained.

- Representation of intention in concrete form is very useful for mutual understanding.
- Visualization of the gap between intention and situation is a good interface, and it is convenient to catch the image of what is going on in the system.
- Just expressing the gap is not sufficient and more concrete suggestions are desirable to reduce the gap.

The results of the experiments showed that the coordinated interfaces are basically effective for the hierarchical system management.

5 Conclusions

In this paper, we proposed visual interfaces for real-time decision making in hierarchical structure. First, real-time problem solving in hierarchical structure was analysed. Then the interfaces to visualize the gap between intention and situation were proposed and a prototype system with GUI was developed on PC. The interfaces were evaluated by the experiments with the fire system. The results showed that the system is effective for mutual understanding.

6 Acknowledgements

This work was partially supported by the Japan Society for the Promotion of Science under Grant-in- Aid for Creative Scientific Research (Project No. 13S0018).

References

Nunamaker, J.F. editor. (1989). Special issue on GDSS, Decision Support Systems, 5(2).

Malone, T.W. (1987). Modeling coordination in organizations and markets, Management Science, 33(10), 1317-1332.

Koiso, T. and Nishida, S. (1998). Concept of Visual Interfaces to Coordinate Real-time Decision Making in Hierarchical Structures, International Journal of Cybernetics and Systems, 29(1), 47-57.

Nunamaker, J.F. editor. (1989). Special issue on GDSS, Decision Support Systems, 5(2).

Malone, T.W. (1987). Modeling coordination in organizations and markets, Management Science, 33(10), 1317-1332.

Koiso, T. and Nishida, S. (1998). Concept of Visual Interfaces to Coordinate Real-time Decision Making in Hierarchical Structures, International Journal of Cybernetics and Systems, 29(1), 47-57.

Future Trends of Human Interfaces for Public Facilities in Japan

Yasuhiro Nishikawa

Mitsubishi Electric Corp.
Amagasaki, Hyogo, Japan
Nishikawa.Yasuhiro@wrc.melco.co.jp

Shogo Nishida

Osaka University
Toyonaka, Osaka, Japan
nishida@sys.es.osaka-u.ac.jp

Members[1] of the HI in Public Systems Committee in IEEJ

Abstract

In recent years, it is becoming more difficult for an operator to handle public facilities which cover broad and large areas. Therefore, to support the operator's tasks, new concepts of human interfaces must be introduced. From the viewpoint above, we discuss about the human interfaces and define their future trends for public facilities in Japan.

1 Introduction

IEEJ (Institute of Electrical Engineers of Japan) had started a small committee to research around human interfaces for public facilities in 2000. The committee consists of members[1] from municipal waterworks bureaus, consulting companies and manufacturing corporations. Many discussions were held about the human interfaces of the system already installed or proposed by the companies. In 2001 the committee sent a questionnaire to over a hundred plants for user survey about present conditions of human interfaces and in 2002 the committee had finished and published its report[1)] about the problems and the solutions on the human interfaces.
In this paper, the word "Public facilities" are used to mean the plants for water supply or sewer/rainfall collection. The focused systems are the supervisory control systems using CRT where the committee members were working. The following discussions are upon the result of the committee about human interfaces for public facilities in Japan.

2 Supervisory Control Systems for Public Facilities

2.1 Requirements

The public facilities are controlling widespread lifeline equipments and obliged to keep them in good condition. For that purpose, they had the following requirements.

- Necessity of long running continuance
- Necessity of high reliability and long lifetime
- Distributed facilities in large area
- Secureness
- Robustness for load fluctuation

---

[1] The members are Shogo NISHIDA (Chairperson; Osaka University), Hiroshi YAMASHITA (Secretary; Mitsubishi Electric), Yasuhiro NISHIKAWA (Assistant Secretary; Mitsubishi Electric), Chikao SHIMONAGAYOSHI (Nihon Suido Consultants), Eiji NANBA (Yokogawa Electric), Hiroshi SUWA (Japan Sewage Works Agency), Kazuhiko KOBAYASHI (Fuji Electric), Kazuyuki YOKOYAMA (Meidensha), Ken HATABE (Mitsubishi Electric), Kunio FUKUDA (Kyoto Municipal Waterworks Bureau), Mariko SUZUKI (Hitachi), Yoshihiko MIYAYAMA (Osaka Municipal Waterworks Bureau), Yoshihiro NAKAMOTO (Kobe City Waterworks Bureau), Yukio HATSUSHIKA (Toshiba)

The systems have to be designed under considerations of these requirements and the operators are on the duties to maintain them. In a broad sense the systems are operated under two different conditions, one is "Normal condition" and the other is "Emergency condition." These two conditions require operators to take different actions and the requirements of human interfaces should be different. To support the operators, good interfaces for each condition are expected.
The distributed facilities may cause a time delay to control. Short delay makes the system works better. But it is always limited by the hardware specs and so on. Secureness issues are very important for the customer. Especially for water supplier, the operators also have to observe the facilities by video cameras against trespassers. The problems of load fluctuations are not avoidable. They are depending on the human life styles, economic activities and many other factors.

2.2 Human Interfaces

The human interface technologies are always focusing on the users characteristics and making effort to adapt the systems to this point. From this viewpoint, human interfaces can be categorized into the following three ones.

- Physical Interfaces
- Cognitive Interfaces
- Cooperative Interfaces

The physical interfaces are pointed toward human kinematics to make matching with machines to reduce human stress. In this term, "human" means a well-trained person dealing with the system.
After the physical interfaces, cognitive interfaces are spotlighted. The innovations of computer technologies made computers more familiar and it is getting more important to evaluate the systems at new viewpoint on how easily the novice operators can control the system. On this trend, knowledge is introduced from cognitive science and as the great result computer mice and icon/window based systems are developed.
Next trends are brought by advanced network and multimedia technologies. Cooperative interface technologies are focused on how to support cooperative work between plural persons. Society science is introduced and basic technologies for groupware have been developed.

In 1980s, computer controlled systems were widely introduced in public plants in Japan and those are utilized for business efficiency and automation. At that time the operators can only receive information about plants and transmit the sequence of controls. But now supervisory control systems become to cover large areas and deal with ten thousands of telemeters and controllers. To handle such large system, many advanced (difficult) technologies are utilized in the system, those are artificial intelligence, fuzzy theory, modern control theory and etc. The new technologies enlarge the coverage area of the system, but it is necessary to confirm that the technologies really make the system easy to understand and to operate for the operators.

Figure 1 shows an example of web-based supervisory system. This system requires only web browser to view the information and doesn't matter what OS is used. This system has an advantage when a novice operator needs to consult an expert of the plant. The expert can understand the situation of the plant and make consultation from different place. The expert operators are free from operation room to serve another work for themselves. In the same sense, web-based video systems are being introduced.
The other topic is PDAs (Portable Digital Assistants) which are portable computers to be carried to everywhere. Tele-communication technologies enable a PDA to keep it connected with the network to supervise and control the plant from everywhere. This type of PDA systems is normally used for maintenance and inspection. At a serious emergency scene like big earthquakes,

a PDA system plays a special role with its portability on restoration. From control center, managers can control and manage the workers with PDAs. This makes the restoration process more efficient and short-timed. These PDA systems support cooperation of their works.
For supporting cooperation in a control center, large display systems are being introduced. This system has a large display over 100 inches set up in front of operators and the operators can control a window system in the display. Each operator also has their personal monitors and some of such systems enable the windows in the display to be exchanged between the personal monitor and the large display. Figure 2 depicts a screen image of a large display system.

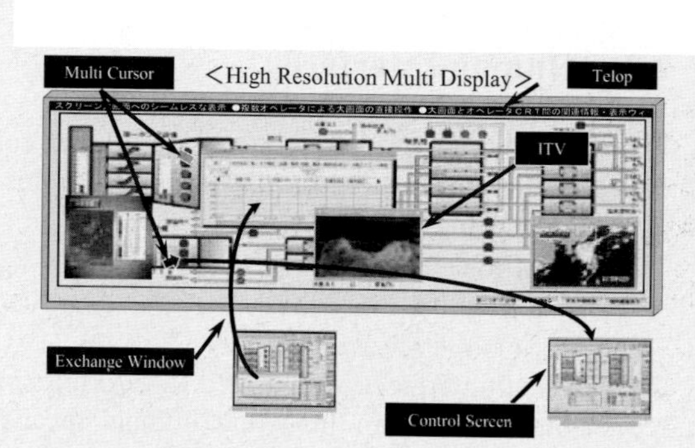

Figure 2: Large Display System

3 Survey

In 2001 questionnaire papers were sent to the public plants to research about satisfaction for their systems and collect the user-side opinions. Organizing the questionnaire answers some facts become obvious.
Figure 3 summarizes the level of satisfaction for wide-area supervisory control system. 46% of the operators have some complaints, the reasons of which are "bad response speed", "bad usability" and a few opinions. Serious problems are in response speed and system usability.
Figure 4 shows the frequency feeling tiresome for experts and Figure 5 shows the frequency of unclearness in operation for novices. Over 60% of both expert and novice operators sometimes feel a sense of incompatibility to operate. To satisfy both types of operators there is a way that a system has some user selectable interfaces such as using both menus and keyboard shortcuts. Another answer in the questionnaire shows that 80% operators use their personal computers. This also mentions another solution to adopt the same interfaces of well-used operating system.

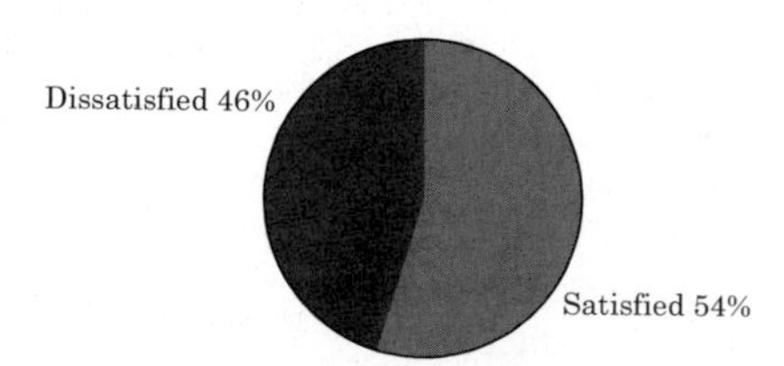

Figure 3: Satisfaction of Supervisory System

Figure 6 shows appearance of cryptic terms in the user interfaces. 47% of operators don't understand all the messages in the system. Figure 7 shows the degree of satisfaction for system's manuals. 65% of operators have complaints about the manuals. These two results show that there might be no way to understand the cryptic terms for the operators. In the public facilities, complete and enriched useful manuals are required urgently because they have some personnel changes in relatively short time.

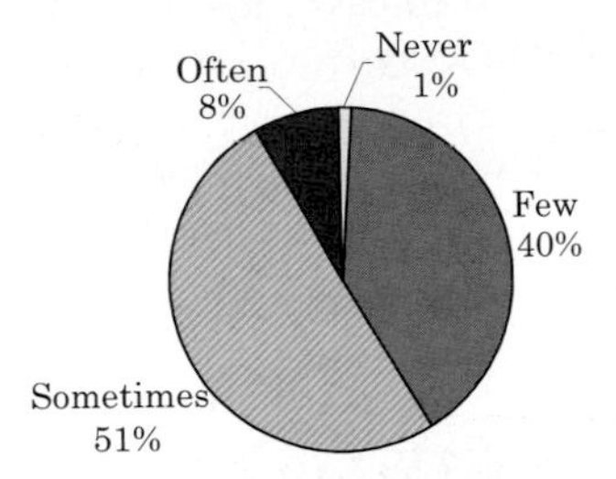

Figure 4: Frequency of feeling tiresome

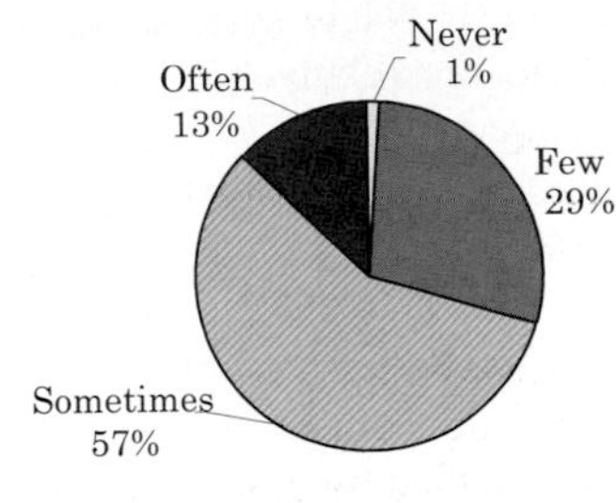

Figure 5: Frequency of feeling unclearness

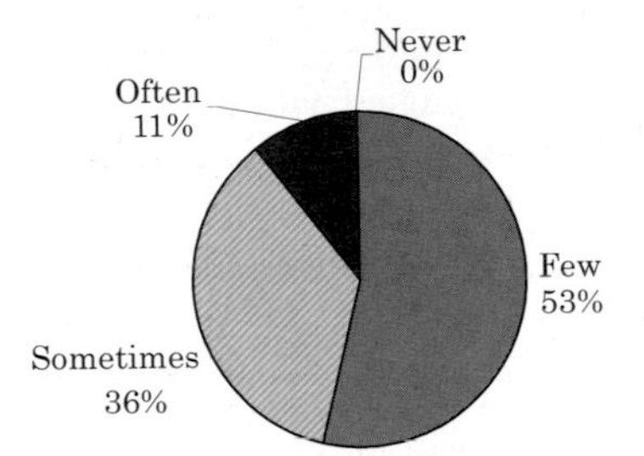

Figure 6: Cryptic Terms Appearance

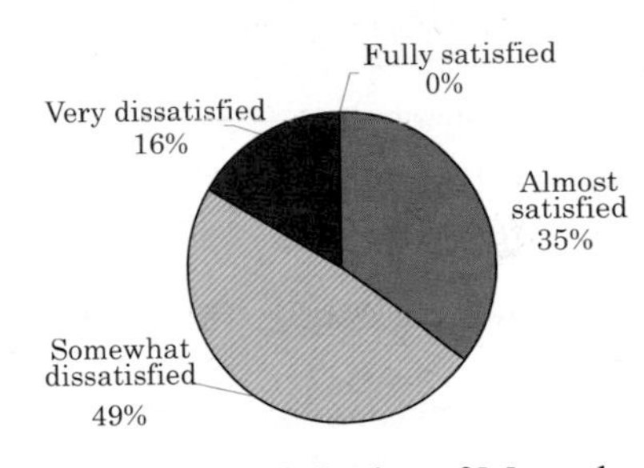

Figure 7: Satisfaction of Manuals

Next topic is system usability in abnormal situations. 31% of operators are dissatisfied with usability in that situation (Figure 8) and the reasons are "difficult usage", "takes time to operate" and "insufficient information informed to operate." Another question was asked about satisfaction with obtaining information through the system and its result shows that 34% of operators are dissatisfied with it. The reasons are shown in Figure 9. In an abnormal situation, operators need different information from a normal situation and system has to display them in proper way. The answer "Inadequate arrangement" mentions that a problem is around how to display the information. Alarms and messages are so many generated in abnormal situation that adequate collecting, filtering and arranging technologies are required.

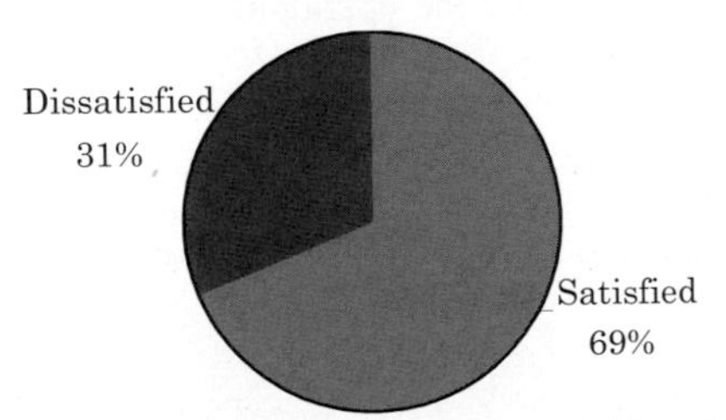

Figure 8: Satisfaction in Abnormal Situations

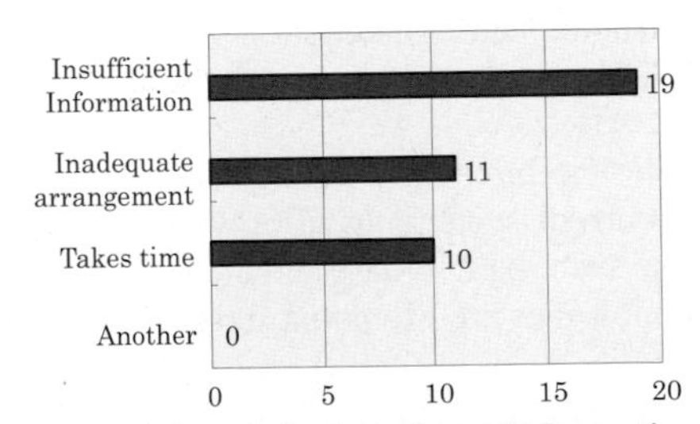

Figure 9: Dissatisfaction about Information

At last the expectations for new technologies were asked. The targets are "Web-based system", "Mobile (PDA) system", "Digital video system" and "Large display system."

"Web-based system" was already installed by 12% and planned by 33%, "Mobile system" was already installed by 4% and planned by 24%, "Digital video system" was already installed by 0% and planned by 17%, and "Large display system" was already installed by 24% and planned by 19%. A common negative opinion is "Current System is enough" (or high costs.) A common positive opinion is "New system enables what was impossible." Web-based systems and large display systems are relatively more expected than the others. These systems support operators with cooperative interfaces.
This questionnaire survey revealed some problems and user needs for human interfaces.

- Response speed
- Usability (for both experts and novices)
- Manuals and understandable interfaces
- Adequate arrangement of information

These are to be settled by new technologies and system producers' efforts.

4 Conclusion

It is introduced about the past systems and the current systems in public facilities. The problems in current systems become revealed by the user questionnaire to over 100 plants. This field has still many problems and more advance of technology is expected in the future.

The future trends are signaled by these current systems and user needs. The following technologies are desired in the future.

a) Quantitative Evaluation of Usability
Typical index is response speed. Many users evaluate the system on the speed and complain about it. This has been discussed for long years and operators aren't satisfied yet. This may be changed by new evaluation method from the viewpoint of cognitive interfaces because the operators might need speeds for improvement of other factors.

b) Adaptive Interfaces
Because the systems need to support many situations, conditions and operators, the human interfaces have to adapt to each of them. It is required two technologies to realize such interfaces, those are "User modeling" and "User adaptation." Estimation of average user is not enough for new systems. The information collection and filtering technologies are effective to support operators in many situations.

c) Communication Support Technology
The significant progress of computer and network technologies realize communication across distance and collaboration works. This technology is expected by the operators. Many systems will be proposed and installed.

Public facilities have strong relations to the society they belong to for their mission. Nevertheless the current systems have only the interfaces between plant systems and operators. In the future, those systems will have the interfaces between plants, operators and their inhabitants. The network technologies are promoting this vision.

5 References

1) "Report on Human Interfaces in Public Systems" (in Japanese), IEEJ technical report No.901 (2002)

Visualizing Medical Imagery in Situ: Augmented Reality as X-Ray Vision

Charles B. Owen and Arthur Tang

METLAB, M.I.N.D. Lab, Michigan State University
East Lansing, MI USA
cbowen@cse.msu.edu

Abstract

Optical and video see-through stereoscopic displays allow medical imagery to be viewed registered onto or inside of patients, effectively providing the physician with X-ray vision. However, there are a large number of technical and psychological issues than remain to be solved for such systems to be widely adopted and truly effective. Indeed, augmented reality, though highly promising, has not always been shown to be as effective in user studies as anticipated. Part of the blame can be assigned to the underlying technologies currently in use; the limits of calibration, latency, tracking, display quality, and image registration can lead to inappropriate and unexpected use behavior. And, the fundamental goal of many of these systems, the viewing of structure in place inside a patient, is at odds with the well-developed knowledge that our eyes are not able to see inside objects and people. This paper examines technical hurdles in achieving x-ray vision, means of eradicating them, and how we can exploit these perceptions to open up a new human-computer interface ability.

1 Introduction

One of the most exciting features of augmented reality is the capability of endowing the user with what appears to be X-ray or see-through vision. See-through vision in augmented reality can be accomplished by superimposing and aligning 3-dimensional data so that they appear to be locating within a physical object. The data can come from an existing database, or from a real time scanner (such as echocardiograph, CAT or MRI scanner). This capability is of particular interest to medical professionals as so much of their interest is in visualizing what occurs inside the human body. Indeed, most medical imagery technologies are designed to visualize the interior of the body in some form or another, from simple x-rays to 3-D reconstructions of MRI brain imagery.

As an example, surgeons want to minimize invasion into the patient's body so as to reduce or avoid surgical complications, anesthesia, hospital stays, pain, and scars to the patient. These goals can be achieved by the latest medical procedures such as endoscopic biopsy and laparoscopic surgery to avoid open surgical intervention to the patient. However, these procedures take away the natural hand-eye coordination of traditional surgical operation, and require special training on the procedures. By spatially positioning the data inside the patient's body using augmented reality, surgeons can retain their natural hand-eye coordination during these procedures [1].

Preliminary experiences suggest that humans do not perceive virtual objects behind real objects in augmented reality. The concept of X-ray vision goes against the well-learned concept that the human body (and other objects) is not transparent. The paper describes results of an experimental pilot study of depth perception of see-through vision in augmented reality.

2 Depth Perception in the Human Vision System

Depth perception in the human vision system is one of the oldest problems in cognitive science. It is a complex perceptual process and there is no precise model of how the human brain calculates the depth of an object. Table 1 lists some of the most important factors in the perception of depth for the human vision system.

Accommodation and convergence	Accommodation refers to the change of focus of the lenses in the eyes, and convergence is the movement of the eye balls. Rene Descartes is one of the first to link accommodation and convergence to depth perception [2]. Accommodation and convergence are factors within a meter or two and change very little beyond two meters.
Pictorial cues	Pictorial depth cues are the impression of depth created by a monocular image. Pictures are 2D, but we still offer a perception of depth. Likewise, individuals with vision in only one eye are still capable of significant depth perception.
Interposition	Interposition is the fact that closer objects occlude or obscure farther objects. The impression of depth created by interposition alone is weak. But when the depth range of two objects is small, it becomes an important cue in creating depth.
Relative size	The more distant an object, the smaller the image on the retina. The relative size of an object in a picture gives the viewer an impression of how close or how far the object is relative to other objects in the picture.
Linear perspective	In linear perspective, parallel lines are converging closer to the vanishing point when recede in distance. Leon Battista Alberti, a Humanist Art theorist, first described a linear perspective construction method for constructing an architecturally correct image of a check board pavement [3].
Texture gradients	Depth information can also be generated by the texture of an object. Texture gradient refers to the density of texture on the surface of an object gets progressively denser when it recedes in distance.
Motion cues	Nearby objects move faster than distant objects in the scene when the location of the objects are in motion. *Motion perspective* refers to the motion cue when the object is in motion, while *motion parallax* refers to the motion cue when the viewpoint is in motion.
Stereopsis	Stereopsis is the depth cue generated by the disparity of two eyes. Our two eyes have a slightly different view to the world due to its difference in spatial location, and our brain uses this slight difference to calculate the perception of depth in a scene. Leonardo da Vinci discovered the differences between monocular vs. binocular vision [4].

Table 1-Depth Perception in the Human Vision System

3 Problems with See-through Vision in Augmented Reality

In real environments, depth cues are consistent and seldom conflict. When adding a virtual environment in augmented reality, contradictions between the real and the virtual may be introduced. These contradictions confuse the human vision system and affect the perception of depth in the scene. Many technical barriers, some of which are listed in Table 2, must be overcome by improved implementations of both the hardware and software processes for AR.

Rosenthal, et al. report a study using AR see-through vision for guiding needle biopsies and report improved results, but only relative to conventional display technologies [5]. Drascic *et al.* addressed the perceptual issues in augmented reality environment, including depth perception [6]. They specifically contrast the use of visualization methods in isolation with mixed reality concepts. As an example, visualization of a real environment through stereoscopic displays (video see-through) can effect depth perception, but the cues tend to be consistent so users adapt quite well. It is only when multiple display modes are combined that depth perception suffers. They provide an extensive list of problems that can limit depth perception in mixed reality environments, irregardless of any occlusion cues. They separate that category, which they call Interposition Failures, into a class referred to as "the hard problems." Rolland *et al.* conducted an experiment in quantiatively measure the depth perception in a see-through head-mounted display [7]. Furmansk, Azuma, and Daily have coined the term Obscured Information Visualization (OIV) as a name for visualization of objects that would normally be hidden or obscured [8]. They conducted studies that demonstrated the subjects tend to visualize virtual elements as ahead or occluding content, irregardless of the element depth.

Calibration errors	Calibration of mixed reality displays is complicated since it is the human perception of registration that the calibration seeks to capture. For some display systems, (optical see-through displays), calibration is difficult to achieve [9]. Calibration errors can change the perception of depth for virtual objects greatly.
Calibration mismatches	Even when correctly calibrated, it is essential that projection modes between the real and virtual environments correspond. It is often the case that graphics are rendered using a pin-hole camera model, which may not accurately model radial distortion in a camera system or the distortions introduced by the eye/display combination.
Stereoscopic spacing	Humans have adapted stereoscopic vision based on a known interpupillary distance. Cameras (real and virtual) in mixed reality environments are not constrained to the spacing of human eyes and often have larger or smaller spacing, contributing to perceptual errors.
Registration errors	In creating virtual environments, it is surprisingly difficult to exactly match a real environment. As an example, we have observed that lighting differences between the real and virtual environments are quickly recognized by users and contribute to confusion. Also, many real-time systems suffer from a fixed latency, which causes dynamic registration errors during motion.
Rendering quality	Real-time computer graphics are very much a compromise, trading limited quality for speed. As an example, flat textures may be applied to polygons to mimic surface characteristics. Facetted objects are used to simulate smoothly shaped objects. At no time is a quality difference more obvious then when blended with reality, which the scene generation is a complex physical process not likely to be emulated in real-time in the foreseeable future.
Vision limitations	Augmented reality displays are a new technology and subject to many limitations.
Fixed focus and field of view	Existing display technologies present virtual content at a fixed focus depth and a limited field of view. (e.g., the Sony Glasstron head-mounted display presents virtual content with a visual field of view of 22° and a fixed focus depth) This is in contract to real environments.

Table 2 - Technical barriers to see-through vision

4 A Depth Perception Experiment

We conducted a simple experiment using 4 male undergraduate or graduate students to investigate the perception of depth in see-through vision using augmented reality. All had previous experience in augmented reality environments. The experimental setup consisted of a Polhemus 6 degree-of-freedom magnetic tracker, Dell workstation with dual Pentium III Xeon 800MHz processors and Wildcat II 4210 graphic accelerator. A Sony Glasstron LDI100 optical see-through head-mounted display was used to augment the vision field of the users. Real time computer graphics used the ImageTclAR Toolkit developed by the Media and Entertainment Technologies Lab.

Figure 1 illustrates the setup. A virtual Duplo block was presented to subjects, who were asked to place a real Duplo block at the side of the virtual block so that the two blocks are at the same range to the subject. The range of the real block was then recorded and compared to the position of the virtual block. There were 8 predefined ranges on the workspace for the virtual block and two conditions for each location of the block when it is being presented: normal condition and see-through vision. Duplo blocks were chosen because they are familiar to users, easily manipulated, and can be unambiguously placed. In the see-through vision condition (Figure 2), a box was placed on top of the workspace so that the virtual object is inside the box.

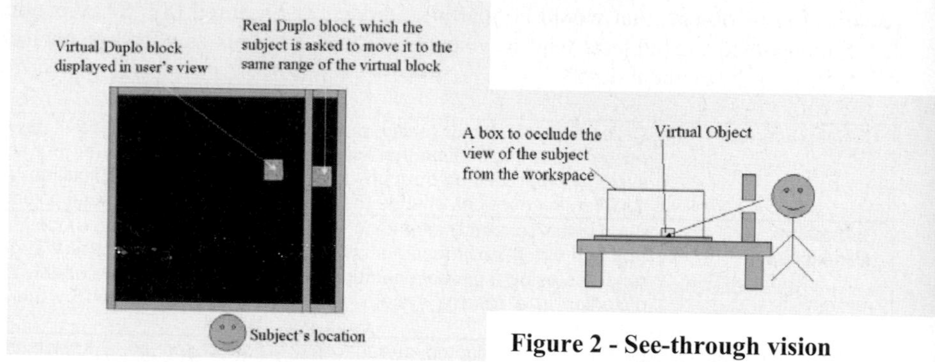

Figure 1 - Experimental setup

Figure 2 - See-through vision configuration

Participants were briefed about the experimental procedure and then performed a display calibration procedure. Stimulus materials were presented to the participants sequentially, and participants were asked to place the real block at the same range as the virtual block they perceived. The location of the real block was recorded before the next stimulus was presented

Figure 3 illustrates the perceived depth for the two conditions. The error was relatively small for closer ranges and varied little between the normal and see-through condition. Figure 4 shows just the depth perception error as a percentage of the actual depth. The error clearly increases both absolutely and as a percentage as the distance to the block increases.

In general, depth perception is more accurate in the closer range than in the farther range in both normal and see-through conditions. Subjects tend to perceive the virtual object closer to the viewpoint than its actual location. One explanation is that attention is diverted to the surface of the physical block, and the depth perception is averaged out to the surface of the box, so the perceived depth of the virtual object in the see-through vision condition is always shorter than the normal condition. This is in contrast to Rosenthal, et al., who found that users nearly always perceived the virtual object as in front of the occluding surface [5]. Part of the difference can be attributed to the monocular monitor-based presentation in their experiment, which eliminated stereo cues, and the smaller relative difference between the virtual object depth and the depth of the occluding surface. Also, participants were aware that objects in the see-through condition were inside the box. Clearly, a surgical user of X-ray vision is aware that the virtual element is intended to be inside an object or person, irregardless of conflicting visual cues.

One subject mentioned that he used the edge of the workspace at the far side as a reference to determine the position of the virtual block in the normal condition when the virtual object is far away from the view point. This subject had results significantly better than the other subjects, indicating that some subjects may consciously seek new cues in mixed reality environments. There is also discussion about whether it is valid to use the physical location of the block to compare with the perceived location of the virtual block in both conditions. This is a rather compelling argument since there is always a difference between perception and reality. The experiment can be improved by adding another condition for displaying a real Duplo block instead of using the predefined location and to compare that data with the normal and see-through vision conditions.

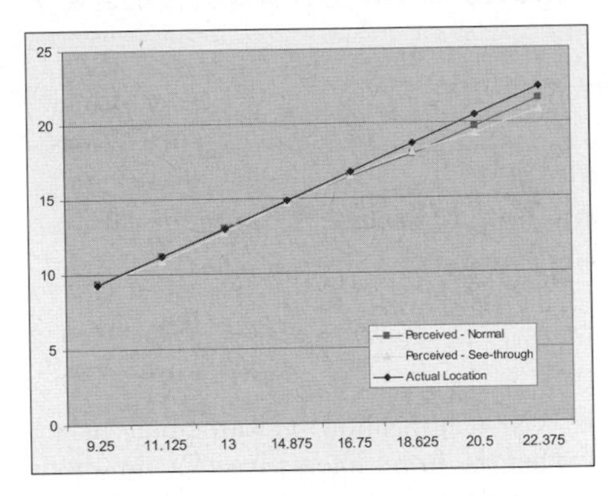

Figure 3 - Perceived depth for two conditions

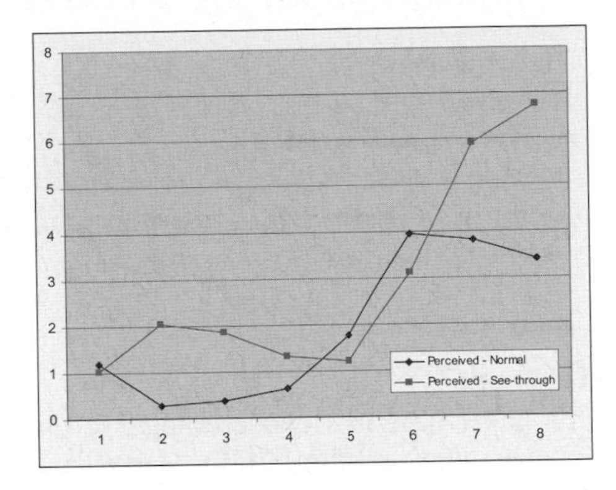

Figure 4 - Relative error for two conditions (percent)

In this small pilot study we do not have enough statistical power to draw any significant conclusions. But we can identify tendencies and design an experiment to further explorer depth perception in see-through vision in the future. Clearly, X-ray vision is a desired application; we need to clarify the obstacles to that application so they can be removed with better designs.

5 References

1. Holden, J.G. and J.M. Flach. *Hand-eye coordination in an endoscopic surgery simulation.* in *3rd Symposium on Human Interaction with Complex Systems (HICS '96).* 1996. Dayton, OH.
2. Descartes, R., *Discourse on Method, Optics, Geometry and Meteorology.* 1637, Indianapolis: Bobbs-Merrill.
3. Alberti, L.B., *On Painting.* 1435, London: Senex and Taylor.
4. Wade, N., H. Ohn, and L. Lillakas, *Leonardo da Vinci's Struggles with Representations of Reality.* Leonardo, 2001. **34**(3): p. 231-235.
5. Rosenthal, M., et al., *Augmented reality guidance for needle biopsies: An initial randomized, controlled trial in phantoms.* Medical Image Analysis, 2000. **6**: p. 313-320.
6. Drascic, D. and P. Milgram, *Perceptual issues in augmented reality*, in *SPIE: stereoscopic displays and virtual reality systems III*, J. Merritt, Editor. 1996: San Jose, CA, USA. p. 123-134.
7. Rolland, J.P., D. Ariely, and W. Gibson, *Towards quantifying depth and size perception in virtual environments.* Presence: Teleoperators and Virtual Environments, 1995. **4**(1): p. 24-49.
8. Furmanski, C., R. Azuma, and M. Daily. *Augmented-reality visualizations guided by cognition: Perceptual heuristics for combining visible and obscured information.* in *ISMAR 2002 IEEE and ACM International Symposium on Mixed and Augmented Reality.* 2002. Darmstadt, Germany.
9. McGarrity, E., et al., *A New System for Online Quantitative Evaluation of Optical See-Through Augmentation*, in *Proceedings of the International Symposium on Augmented Reality (ISAR'01).* 2001: New York, NY. p. 157--168.

Implementation Studies with GUIs utilizing Fisheye Lens
- Application for CCTV based Surveillance and Interactive TV -

Nobuyuki Ozaki

Toshiba Corporation
3-22 Katamachi, Fuchu-shi, Tokyo, Japan
nobuyuki.ozaki@toshiba.co.jp

Abstract

This paper describes two examples of omnidirectional vision utilization, in particular, fisheye cameras. One is from CCTV surveillance and the other is from interactive TV in digital broadcasting. It first discusses the concept of dewarping fisheye projection into normal projection and estimates suitable original image resolution for a fisheye image in terms of the image quality. From this result, we selected image capture devices for each case. According to their applications, some GUIs are presented with snapshots.

1 Introduction

For human interface, it is important to have abundant interactivity in handling realistic images. Although computer graphics technology can provide realistic images with an enormous manual procedure and time-consuming computations, images captured by cameras, which are essentially realistic, have great advantage provided they can be interacted. Interactivity is achieved by showing viewers a part of the broader or panoramic view by translating the part. This view can be created in two ways: by image stitching among multiple normal perspective cameras or by means of non-perspective omnidirectional cameras.

A multiple camera system [FouthView][EyeVision] is huge and complex to arrange for shooting. This makes it very expensive, and not cost effective for easy use. However, an omnidirectional camera system may be practical in the field.

There are two categories of omnidirectional camera: the fisheye [Zimmermann] type and the catadioptric type. These types originate a robot's sensing device for environment understanding [Cao][Yamazawa]. Recently, many research activities have focused on the catadioptric type camera [OMNIVIS'00] [OMNIVIS'02], as this has the advantage that it can simultaneously cover 360 degrees horizontally.

When we consider several applications in which the gaze point is the center of the view, the disadvantage of the catadioptric type is significant, because it cannot project images onto the center area of the view. This is caused by the mirror structure. As our usage requires the center area for projection, the fisheye camera is best suited for our purpose.

Chapter 2 explains the characteristics of the fisheye lens and the concept of dewarping the fisheye image. Chapter 3 describes the practical application of the CCTV (Closed Circuit Television) surveillance system. Chapter 4 explains the second application, used in interactive TV in digital broadcasting,. The advantages of both types are addressed.

2 Fisheye Camera

2.1 Projection Model

A fisheye lens covers a 180-degree field of view: a hemisphere. Equidistance projection, modeled as shown in Figure 1, is most commonly used. It is based on the following equation:

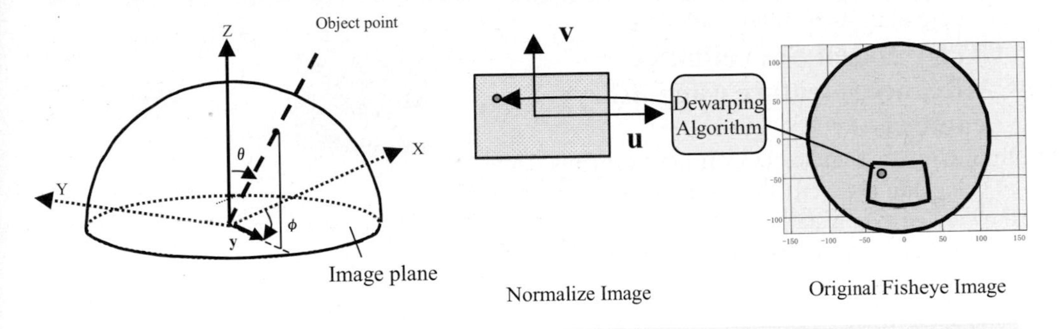

Figure 1: Projection Model

Figure 2: Dewarping

$$y = f \cdot \theta$$

where θ is the zenith angle and y is the radial distance. The projected shape grows planiform based on θ/sin(θ) when the zenith becomes larger: nearer an object point is projected to the edge of the image plane.

The concept of dewarping the original fisheye image into a perspective image is shown in Figure 2 by first defining the area in the original image to be dewarped. As this defined area is smaller than the whole image, translating and enlarging this area assumes that the camera has virtually the functions of paning, tilting, and zooming. Pixel-wise calculation is performed during the dewarping: to determine the intensity of the specific pixel on the dewarped image by locating the corresponding original position on the fisheye image and placing its R, G, and B color components as the new intensity.

During the calculation, the located pixel position is normally not an integer, so that the new intensity can be interpolated by the neighbor pixels. However, according to the comparison test between non-interpolated and interpolated calculation, the interpolated way has little distinctive improvement in resolution even with the additional time consuming calculation. As a result, the located position is simply rounded off to become an integer.

2.2 Determining Original Image Size Based on Quality Evaluation

The size of the original image affects the quality of image; image quality is affected by image blur; image blur is affected by placing the same pixel intensity near during the dewarping process. The probability of the neighborhood pixels falling into the same position in the original image becomes very high when the original size becomes smaller. Once the size of the dewarped image is defined, it requires the appropriate original image size to maintain the image quality.

The quality evaluation cannot be directly extracted on the basis of the projection model as the model is strongly non-linear. The following numeric simulation shown in Figure 3 gives us the proper original image size for a given 400*300 size of dewarped image. During the dewarping process, we counted the duplicated pixels that fall into the same pixel location in the original. If this duplication occurs frequently, the image is blurred. The vertical axis in the figure is the number of duplicated pixels; it shows the magnitude of blur. Therefore, the numeric simulation suggests that it is preferable to keep more than 1000*1000 size in the original image.

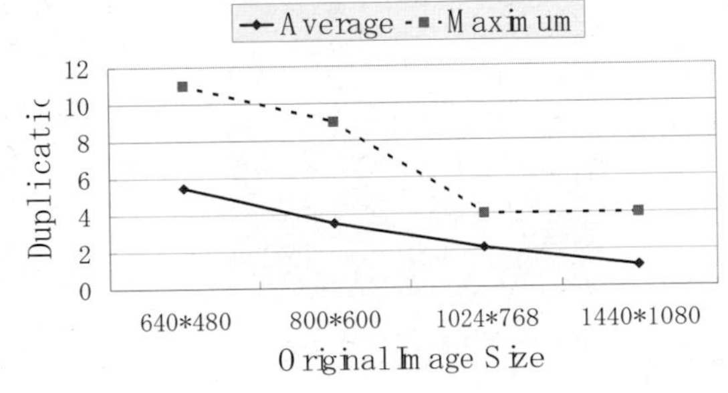

Figure 3: Quality Evaluation

For the following applications, we need to use a camera that has more than 1000*1000 resolution.

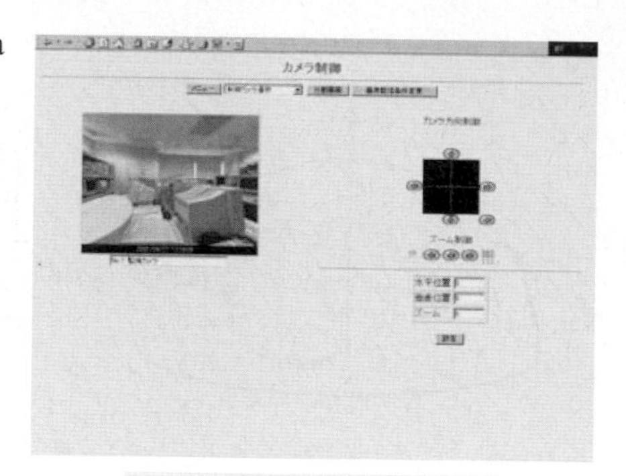

Figure 4: Unified GUI

3 CCTV based Surveillance

3.1 Advantage feature using fisheye cameras

The following are advantages of fisheye cameras.

- Their setup is exactly the same as normal ones so they are easy to install.
- Their system can capture a 180-degree field of view with one shot.
- Their system has virtually similar functions of conventional paning, tilting, and zooming equipment.
- Their system can handle each request concurrently with the same camera by showing the designated views, while conventional camera control equipment can handle only one request at a same time after appropriate priority scheduling of multiple concurrent requests.

3.2 Component for a CCTV Digital Surveillance System

The author has been developing a CCTV digital system [Ozaki], which encodes CCTV analog videos into digital images, or JPEGs (Joint Photographic Coding Expert Group), with a single frame step function. A server system gathers all the digitalized images and a user, accessing the server, can see views from CCTV and also control camera equipment from a client terminal's internet browser. This system can handle several types of encoders and different communication protocols of camera control equipment. With the fisheye camera being installed, the system can control its orientation as well without extra information, or a communication protocol for camera control.
The following are examples of the fisheye feature incorporated into the CCTV surveillance system.

3.2.1 Unified GUI

Figure 4 shows one GUI. As shown, the user can see both types of cameras, normal cameras with control equipment and fisheye cameras, without noticing which is which. The left image in Figure 4 shows the result of dewarping. Camera control is performed from the GUI; the black square area on the right side is the clickable area for panning and tilting and also the feedback indicator of the present status. Another button is used for zooming. With the exact same operation, the user can control cameras of both types.

3.2.2 Narrowing the region of interest after detecting an unusual situation

We propose a procedure for narrowing the region of interest using a fisheye camera. Normally the

Figure 5: Object Detection

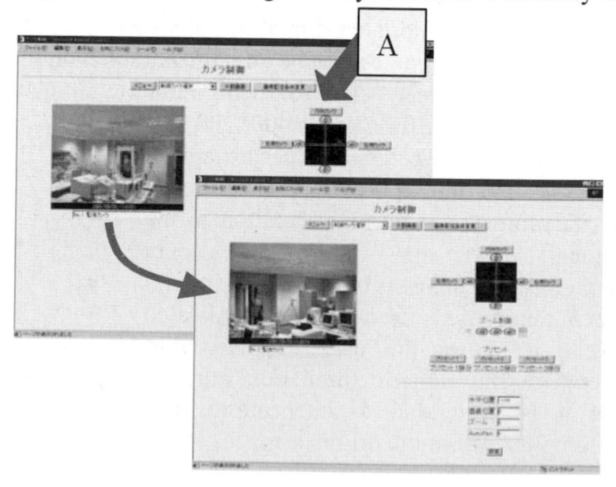

Figure 6: Camera switching

system supervises in a macroscopic way with fisheye cameras so that an unusual situation can be detected without moving the camera's orientation. Once the situation change is detected, users need to scrutinize the situation in detail. For this purpose, the camera system can focus on the region of interest by virtually zooming, panning, and tilting or switching into a different camera that has a good look if multiple cameras are installed.

Figure 7: Shooting Scenes

The narrowing procedure is shown on the condition of the rectangle room: four cameras are installed in each wall and one is a fisheye camera that can almost cover the room.

- The system supervises the room with the fisheye camera. Image processing technique enables the system to detect situation change automatically. The system then notifies users that something is happening in the room by displaying the frame with the green circle attached as the object being detected as in Figure 5.
- Users check with the dewarped image shown on the left side of Figure 6. To scrutinize the object from a better angle, users click the button, shown as A in the figure, that can switch the camera placed at the opposite side.
- After selecting the opposite camera, users select the best angle by the appropriate panning, tilting, and zooming and scrutinize the object.

4 Interactive TV

4.1 Advantages using Fisheye Cameras

The following are the advantages of fisheye cameras.

- A wide area of view enables interactivity in displaying images.
- As the gaze point is still the center of the view during the shooting, it has a smaller gap for a film crew to shoot scenes. This is different when using the catadioptric type.

4.2 Assumed Service and Prototype Demonstration Systems

Interactive TV would be attractive to TV viewers because of its interactivity in displaying visual information and could become a promising future service for digital broadcasting. To verify its visual interactivity, a light prototype system was developed on a stand-alone PC focusing GUIs that suit a data broadcasting service. We also applied it to the internet video distribution service as a second prototype system.

4.3 Procedure from Shooting Scenes to Authoring

Figure 7 shows snapshots during shooting. A fisheye lens that resembles a big sole eye is mounted on a camera.

The image quality condition discussed in 2.2 requires a high vision density camera for shooting. Since devices that encode and decode high vision density video are very expensive, we conducted cost effective approaches. We captured frames as still images and displayed them continuously on the prototype system. We captured frames and encoded them with software into MPEG4 (Moving Picture Expert Group 4) and decoded them with software for internet video distribution.

4.4 Interactive TV GUIs

Figure 8 shows the GUI aimed at data broadcasting services. The upper left video in the figure shows the main video, which is encoded by MPEG2, and the other area indicates service by data broadcasting. The upper right image is the display of the interactive TV provided with four arrow-like buttons that make the view shift up, down, right, and left. The orientation can be

confirmed by the lower left two human animations: the upper human shows panning and the lower shows the tilting angle.
With this GUI, the viewer would feel as if the viewer were walking with one's head turning to look around. This interactive TV was popular with test subjects.
Figure 9 shows the second prototype of internet video distribution: the MPEG4 video streaming. Dewarping algorithm is implemented after decoding into frames.

5 Conclusion

This paper points out the advantages of utilizing fisheye cameras for specific applications. Based on a dewarping algorithm, image quality is discussed with a view to selecting suitable camera devices. Two cases are presented as applications: CCTV based surveillance and interactive TV. A fisheye camera is very handy to use. The author believes that this is a promising device for applying interactivity in images. The creation of the contents for interactive TV was funded by TAO, Telecommunications Advancement Organization of Japan.

References

EyeVision, from http://www.pvi-inc.com/eyevision/eye_out.html

FourthView, from http://www.sony.net/Products/fourthview/

Zimmermann, S., & Martin, H.L. (1991). An Electronic Pan/Tilt/Zoom Camera System. NASA Conference Publication, 261-270.

Cao, Z.L., Oh, S.J. & Hall, E.L.(1986). Dynamic Omnidirectional Vision for Mobile Robots, Journal of Robotics System, 3(1), 5-17.

Yamazawa, K., Yagi, Y. & Yachida, M.(1996). Visual Navigation with Omnidirectional Image Sensor HyperOmni Vision, IEICE Transaction, D-2, Vol1.J79-D-2, No.5, 698-707.

OMNIVIS'00, Proceedings of the Workshop on Omnidirectional Vision (OMNIVIS'00)

OMNIVIS'02, Proceedings of the Third Workshop on Omnidirectional Vision (OMNIVIS'02)

Ozaki, N.(2002). Visually enhanced CCTV digital surveillance utilizing Intranet and Internet, ISA Transactions, Vol. 41, No.3.

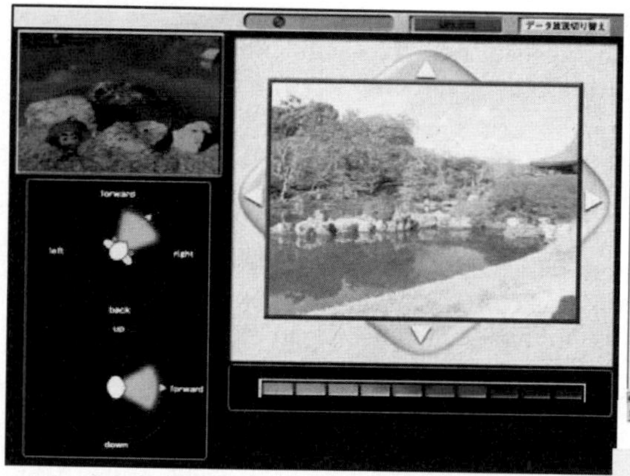

Figure 8: Interactive TV

Figure 9: MPEG4 Video Distribution

New Heuristics for Improving Heuristic Evaluation

Peter J. Patsula

Division of Information Studies
School of Communication and Information
Nanyang Technological University
patsula@pmail.ntu.edu.sg or patsula@usefo.com

Abstract

In recent years, a growing backlash aimed at heuristics and heuristic evaluation is gaining momentum. This paper proposes four new heuristics to improve heuristic evaluation drawing upon Signal detection theory, Miller's "Magical Number Seven," Fitts's Law, the "Principle of Truth," and the Mental Model Imprinting theoretical framework.

1 Backlash Against Heuristics and Heuristic Evaluation

Simple, yet powerful, Nielsen's (1994) heuristics and heuristic evaluation (HE) techniques have received considerable attention over the years. However, a growing backlash aimed at HE and other discount usability inspection methods (UIMs) from both academic and commercial fields is gaining momentum. Cocton and Woolrych (2002) argue that discount methods are "too risky" to justify their "low-cost" and "may actually backfire and end up discrediting the field" (p. 13). Although they concede that "there will probably always be a place for discount methods," the challenge remains "to improve all HCI methods, so that discounted and 'full-strength' methods can be applied in more contexts" (p. 17). Other HE problems researchers have reported include: biases in the current mindset of the evaluators (Nielsen & Molich, 1990); variability of techniques used from one practitioner to the next (Gray, Atwood, Fisher, Nielsen, Carroll & Long, 1995); too many low-priority problems detected (Jeffries, Miller, Wharton & Uyeda 1991); and too many "false positives" (Sears, 1997). Contributors to *Usability News* also argue that:

- standards have been advocated prematurely (Light, 2002a)
- the dominance of one key thinker means that other important voices in the field are being "ignored, underestimated, or overridden" (Light, 2002b)
- Nielsen's focus on "rational left-brain activities" ignores the "emotional subjective side of human beings" (Olsen, 2002)

While Olsen (2002) admits that part of the "Nielsen backlash" may be due to "envy," to a greater extent it can be attributed to "a larger backlash against problems within the HCI and usability fields." Nielsen, with his HE approach, has really "just been a lightning rod" for a growing sense of dissatisfaction with usability design approaches.

One way to improve HE is to rethink how the heuristics themselves are applied. Tversky and Kahneman (1974) suggest that decision makers should be taught to encode events in terms of probabilities rather than frequencies since probabilities account for both events that do and do not occur. In applying this concept to HE, evaluators could be taught to encode heuristics as having both positive outcomes that improve usability, and negative consequences or risks that adversely affect usability. Aspects of this approach have already been explored under claims analysis (Carroll & Rosson, 1992), where *claims* have been proposed as a means of describing design tradeoffs and to record HCI knowledge (Sutcliffe & Carroll, 1999).

2 Proposal for Four New Heuristics

This paper proposes four heuristics to address current weaknesses in HE. These heuristics provide designers and evaluators with new insights into *how* HE might be better structured and *what* kinds of understandings and criteria need to be considered to improve performance.

2.1 Probability Processing Bias

Signal detection theory (SDT) assumes four possible outcomes when a human operator attempts to detect a *signal* amidst *noise*: hit, miss, false alarm, or correct rejection. Bailey (2001), reviewing recent research on what heuristic evaluators thought usability problems would be compared with problems users actually had, summarized that 36% of all the identified problems were usability problems (hits), 43% of the identified problems were not problems at all (false alarms), and 21% of the identified problems users had were originally missed by the evaluators (misses). If this research is true, it can be summarized that more than half of the problems identified by HE will not be problems at all. These results seriously question the usefulness of HE.

Evaluation metrics such as thoroughness, validity, and effectiveness, along with the concepts of hits, misses, false alarms, and correct rejections, have been examined in the context of usability evaluation method outputs (Sears, 1997; Gray & Salzman, 1998; Harston, Andre & Williges, 2001) and heuristic evaluation outputs (Cocton & Woolrych, 2002). To improve HE, these same concepts could also be applied to a single heuristic or even categories of heuristics. In applying SDT to a single heuristic, it is being assumed that no heuristic or evaluator is perfect (i.e., correct 100% of the time), and thus within a certain measure of accuracy, any heuristic being applied will detect some problems, miss some problems, and also give off false alarms. For any heuristic, the "signal" can be considered the application of the heuristic to a design or decision problem, while the "noise" can be considered the uncertainty as to whether the heuristic will actually work or whether it is relevant. The total number of usability problems discovered by a heuristic (i.e., the number of "hits_{sh}"), as well as the total number of "false alarms_{sh}," the "validity_{sh}" of a single heuristic ($_{sh}$), could be assessed as follows:

$$\textbf{validity}_{sh} = \Sigma\ \text{hits}_{sh} / \Sigma\ (\text{hits}_{sh} + \text{false alarms}_{sh})$$

Similarly, thoroughness and severity could also be measured, and effectiveness calculated:

$$\textbf{thoroughness}_{sh} = \Sigma\ \text{hits}_{sh} / \Sigma\ (\text{total number of actual problems in the design being evaluated})$$
$$\textbf{severity}_{sh} = \Sigma\ \text{severe problems}_{sh} / \Sigma\ (\text{severe problems}_{sh} + \text{nuisance problems}_{sh} + \text{minor problems}_{sh})$$
$$\textbf{effectiveness}_{sh} = \text{validity}_{sh} \times \text{thoroughness}_{sh} \times \text{severity}_{sh}$$

The effectiveness formula assumes that validity_{sh}, thoroughness_{sh}, and severity_{sh} are equally important in determining the overall performance potential of a single heuristic. In applying this formula, heuristics that detect the largest percentage of the most severe usability problems, while also giving the least number of false alarms, will have the highest effectiveness rating.

Heuristic #1: *Probability Processing Bias.* Because heuristics are probabilistic in nature, meaning that sometimes they work, sometimes they do not, and almost always, there are tradeoffs, they should be applied with the understanding that they may not always provide the best solution for unique or complex problems. Evaluators must factor in their tendency to overestimate the usefulness of a heuristic and underestimate the severity of its tradeoffs.

2.2 Consistency Processing Bias

The mental-model theory of deductive reasoning (Johnson-Laird, 2001) proposes that "reasoners use the meanings of assertions together with general knowledge to construct mental models of the possibilities compatible with the premises" (p. 434). A phenomenon predicted by one of its core assumptions—"the principle of truth"—is that because reasoners normally represent only what is true or consistent with what they already know by default in their models of the premises, and not what is false, they often make "illusory inferences," that is, inferences that are untrue (Johnson-Laird, 2001). This tendency is similar to the confirmation bias which has been described as a tendency for people "to seek information and cues that *confirm* [their] tentatively held hypothesis or belief, and not seek (or discount) those that support an opposite conclusion or belief" (Wickens & Hollands, 2000, 312). This bias produces a sort of "cognitive tunnel vision" (Woods & Cook,

1999) in which operators fail to encode or process information contradictory to or inconsistent with their initially formulated hypothesis. Along similar lines, the Mental Model Imprinting (MMI) theoretical framework (Patsula, 2002)—which hypothesizes two main channels of information processing in mental model acquisition called "consistency processing" and "usability processing"—predicts that new information sources consistent with what is already known are more easily processed. As a result, reasoners usually default to consistency processing when encountering new information due to their tendency to process more quickly what Norman (1988) refers to as "knowledge in the head" rather than "knowledge in the world."

In applying "the principle of truth" and MMI's "consistency processing bias" to HE, a case can be made that evaluators, if left to their own inclinations, will tend to ignore tradeoffs. To prevent this, it is necessary to incorporate procedures into HE that allow analysts to view the application of heuristics more objectively. Interestingly enough, recent studies have noted improved performance in "illusory inference" experiments when subjects have had the opportunity to develop strategies to handle falsity (Johnson Laird, 2001). Considering this, tradeoff heuristics for single heuristics or categories of heuristics could be developed to give evaluators a more balanced understanding of risks involved.

Heuristic #2: *Consistency Processing Bias.* Because evaluators have a tendency to consider and remember only what is "true," "successful," or "consistent" in the application of a heuristic to a design problem, they often ignore tradeoffs and other contradictory information and may succumb to "illusory inferences." To prevent this, the analysis and assessment of both positive outcomes and negative consequences needs to be built in to HE.

2.3 Cognitive Load

Signal detection theory tells us that all evaluation methods and heuristics, including those used in HE, are fundamentally limited because with increasing "noise" there is an increase in the probability that any particular heuristic will be applied incorrectly (a false alarm). However, to make matters worse, because the working memory is limited, under the right "bad" conditions, the accuracy of HE can degrade even further. The performance of a heuristic also depends on its impact on the human cognitive processing system. Miller (1956) tells us that our capacity for processing information is generally limited to "seven, plus or minus two" alternatives. In information theory, this is equivalent to 2.5 to 3 bits of information. What this means is that as the number of choices increases (i.e., the number of heuristics being applied), at a certain point depending on the nature of the information being processed, the limits of the mind to handle new information will be reached. This limit can be referred to as an "information bottleneck." Miller (1956) further argues that by organizing stimulus inputs simultaneously into several dimensions and successively into a sequence or chunks, human information processors can manage to break (or at least stretch) this "information bottleneck."

Kurosu, Matsuura, and Sugizaki (1997) have taken information theory to heart to propose a new type of HE, called the "categorical inspection method – structured heuristic evaluation" (sHEM). They maintain that because of the limitations of human memory and attention, evaluators may find difficulty in maintaining all the usability heuristics active in working memory. Thus, there is a tradeoff between the number of the activated usability heuristics in the working memory of the evaluator and the number of problems that can be detected. To solve this tradeoff, they introduced structure into a set of usability heuristics using the following methodology: (1) increase the number of guidelines used in the session; (2) divide guidelines into several sub-categories; (3) divide the whole session into several sub-sessions based on the sub-categories; and (4) limit the number of guidelines for each sub-session to within the evaluator's working memory. Experimental results confirmed better performance of this method compared with Nielsen's HE method, cognitive walkthroughs, and user testing (Kurosu, Sugizaki & Matsuura, 1999).

Heuristic #3: *Cognitive Load.* For any design or decision process, the heuristics used should be limited to manageable chunks of six to eight principles, otherwise the interactions within working memory can become too complex and tradeoffs will go unrecognised. When a shorter list of heuristics is not possible, heuristics should be divided and structured into more manageable chunks to ease the working memory load.

2.4 Cognitive Distance

Fitts (1954) investigated the relationship between the variables of time, accuracy, and distance, and discovered that the time required to move the hand or stylus from a starting point to a target obeyed the basic principles of the speed-accuracy trade-off as pointed out by the Hick-Hyman Law (Hick, 1952). Mathematically, Fitts's Law can be described as: $MT = a + b[\log_2(2A/W)]$. In practical terms, this means that the farther a subject has to move a mouse to click on an object (A), the more effort it will take, and the smaller the object is (W), the harder it will be to click on. In other words, the easiest objects to target are those that are closest and the biggest. The speed-accuracy connection between Fitts's Law and the Hick-Hyman Law suggest similarities between mental schema and motor schema. The "availability heuristic," proposed by Tversky and Kahneman (1974), further supports this connection by predicting that "the ease with which instances or occurrences [of a hypothesis or decision choice] can be brought to mind" depends upon factors such as hypothesis simplicity and frequency of the instance or occurrence (p. 1127). Wickens and Hollands (2000) note that "frequent events or conditions in the world generally *are* recalled more easily"(p. 309), and thus, "a hypothesis that is easy to represent in memory … will be entertained more easily than one that places greater demands on working memory" (p. 310).

Conceptually speaking, Fitts's Law could be applied to mental schema as follows: Size "W" could equal the probability of success or the likelihood of a heuristic working as noted by its percentage of "hits" or "$\text{effectiveness}_{sh}$" score. In more abstract terms, size "W" could also be conceptualised as the size of the mental model or neuronal network that can be brought into play to solve a design problem. On the other hand, distance "A" could equal the load on working memory, which could be quantified as the total bits of information that need to be processed. In more abstract terms, distance "A" could be conceptualised as the efficiency or speed of the mental model or neuronal network at recognizing and processing a heuristic along with its tradeoffs. Such tradeoffs include the negative consequences of the application of the heuristic in competition with other heuristics, along with the tendency for the heuristic to be misapplied (i.e., its "false alarm" rate) or overlap and be confused with other heuristics. Practically speaking, what this means is that in applying Fitts's Law and the Availability Heuristic to the selection of heuristics: (1) Heuristics should be easily understood by target users. A heuristic that is easily apparent or "available" means that there is a large existing mental model or neuronal network to process it quickly. (2) Heuristics should also have a tendency to avoid generating numerous and complicated tradeoffs. Heuristics with fewer tradeoffs can be processed and applied more quickly and accurately.

Heuristic #4: *Cognitive Distance.* Effective heuristics reduce the load on the target user's working memory when applied alone or in combination. Heuristics selected should be easy enough to understand and apply, general enough to be widely applicable, and specific enough not to overlap.

3 Conclusions: Applying the "New Heuristics" to HE

No studies have yet been conducted to evaluate the above four heuristics, although a structured heuristic evaluation approach with tradeoffs analysis (sHEwTA) is under development. The essence of sHEwTA is to maintain the simplicity and low-cost of Nielsen's (1994) HE, while at the same time improving upon its methodology by structuring in tradeoffs analysis. Following the conceptualisation of sHEwTA, the next logical phase would be to evaluate its performance compared with Nielsen's HE and other usability evaluation methods, as well as assess and compare its individual heuristics using validity, thoroughness, severity, and effectiveness metrics.

4 Acknowledgements

I am very thankful to Yin Leng Theng (Nanyang Technological University, School of Communication and Information), for her guidance in narrowing the focus of this paper, as well as her helpful comments, proofreading, and recommendations. Special thanks also to Elisabeth Logan (Nanyang Technological University, School of Communication and Information) for her support and guidance, and Kee Yong Lim and Martin Helander (Nanyang Technological University, School of Mechanical and Production Engineering) for their thoughtful suggestions.

References

Bailey, B. (2001). Heuristic evaluations vs. usability testing. Retrieved January 8, 2003, from http://www.humanfactors.com/downloads/jan01.asp

Carroll, J. M., & Rosson, M. B. (1992). Getting around the task-artifact cycle: How to make claims and design by scenario. *ACM Transactions on Information Systems*, 10(2), 181–212.

Cocton, G., & Woolrych, A. (2002). Sale must end: should discount methods be cleared off HCI's shelves? *ACM Interactions,* 9(5), September, 13-22.

Fitts, P. M. (1954). The information capacity of the human motor system in controlling the amplitude of movement. *Journal of Experimental Psychology*, 47, 381-391.

Gray, W. D., & Salzman, M. C. (1998). Damaged merchandise? A review of experiments that compare usability evaluation methods. *Human-Computer Interaction*, 13(3), 203-261.

Hartson, H. R., Andre, T. S., & Williges, R. C. (2001). Evaluating usability evaluation methods. *International Journal of Human-Computer Interaction*, 13(4), 373-410.

Hick, W. E. (1952). On the rate of gain of information. *Quarterly Journal of Experimental Psychology*, 4, 11-26.

Jeffries, R., Miller, J. R., Wharton, C., & Uyeda, K. M. (1991). User interface evaluation in the real world: A comparison of four techniques. In *CHI '91 Conference Proceedings* (pp. 119-124). New York: ACM Press.

Johnson-Laird, P. N. (2001). Mental models and deduction. *Trends in Cognitive Science,* 5(10), 434-442.

Kurosu M., Sugizaki M., & Matsuura, S. (1999). A comparative study of sHEM (structured heuristic evaluation method). In *Human-Computer Interaction: Ergonomics and User Interfaces. Proceedings of HCI International '99, 8th International Conference on Human-Computer Interaction*, (pp. 938-942). Mahwah, NJ: Lawrence Erlbaum Associates.

Kurosu, M., Matsuura, S., & Sugizaki, M. (1997). Categorical inspection method-structured heuristic evaluation (sHEM). In *1997 IEEE International Conference on Systems, Man, and Cybernetics* (pp. 2613-2618). Piscataway, NJ: IEEE.

Light, A. (2002a). Prodding at the limits of user-centred design? Retrieved January 8, 2003, http://www.usabilitynews.com/news/article481.asp

Light, A. (2002b). Ann's Rant: Stop, or Dr Nielsen gets it! – the backlash in usability? Retrieved January 8, 2003, http://www.usabilitynews.com/news/article493.asp

Miller, G. A. (1956). The magical number seven, plus or minus two: Some limits on our capacity for processing information. *The Psychological Review*, 63, 81-97.

Nielsen, J. (1994). Enhancing the explanatory power of usability heuristics. In *CHI '94 Conference Proceedings* (pp. 152-158). New York: ACM Press.

Nielsen, J., & Molich, R. (1990). Heuristic evaluation of user interfaces. Proceedings of ACM CHI'90 (Seattle, WA, 1-5 April 1990), 249-256.

Norman, D. A. (1988). *The psychology of everyday things*. New York: Harper & Row.

Olson, G. (2002). Response: The backlash against Jakob Nielsen and what it teaches us. Retrieved January 8, 2003, http://www.usabilitynews.com/news/article603.asp

Patsula, P. J. (2002). Mental model imprinting: A theoretical framework for interface and website design. Submitted to the *7th SEAES and 4th Malaysian Ergonomics Conference*. Available http://www.usefo.com/usableword/paper2002_mental_model_imprinting.pdf

Sears, A. (1997). Heuristic Walkthroughs: Finding the problems without the noise. *International Journal of Human-Computer Interaction*, 9, 3, pp. 213-234.

Sutcliffe, A. G., & Carroll, J. M. (1999). Designing claims for reuse in interactive systems design. *International Journal of Human-Computer Interaction*, 50, 213-241.

Tversky, A., & Kahneman, D. (1974). Judgment under uncertainty: Heuristics and biases. *Science*, 185, 1124-1131.

Wickens, C. D., & Hollands, J, G. (2000). *Engineering psychology and human performance*, (3rd ed.). Upper Saddle River, New Jersey: Prentice Hall.

Woods, S.D., & Cook, R. (1999). Perspectives on human error. In F. Durso (Ed.), *Handbook of applied cognition.* West Sussex UK: Cambridge University Press.

Towards a Unified Model of Simple Physical and Virtual Environments

Thomas Pederson

Department of Computing Science, Umeå university
SE-90187 Umeå, Sweden
top@cs.umu.se

Abstract

This paper presents a general modeling approach intended to facilitate design of physical-virtual environments. Although the model is based on elements found in typical office environments, certain care has been taken to open for the modeling of more diverse settings with minimal ontological changes. The design approach finds inspiration in the technology-driven areas of Ubiquitous/Pervasive Computing (Weiser, 1991) and Graspable/Tangible User Interfaces (Fitzmaurice, Ishii & Buxton, 1995) as well as more empirical and theoretical research on Knowledge Work (Drucker, 1973; Kidd, 1994), office organisation (Malone, 1983), and Distributed Cognition (Hollan, Hutchins & Kirsh, 2000).

1 Motivation

The purpose of our work is to investigate the possibilities of modeling physical and virtual environments as one environment, centered around a specific human activity. This idea is motivated by the observation that physical environments (e.g. an office, a shop floor, a sports stadium, or a house) and virtual environments (e.g. the desktop environments offered by personal computers (PCs), digital assistants (PDAs), and cellular phones) are not viewed as completely separate entities by human agents themselves when performing modern information technology-supported activities because objects and processes tend to have representations in both worlds. The hypothesis is that by taking on a joint physical-virtual design perspective, it would be possible to design physical-virtual environments that better support these increasingly common, increasingly intertwined, physical- virtual human activities. Specifically, we believe that objects that have representations in many locations (including physical and virtual places) would be more easily handled taking on this stance. The work presented in this paper is an attempt to move towards a greater understanding of the possibilities and challenges involved in bridging the physical and the virtual worlds.

2 Modeling Approach

The fundamental differences between the physical and virtual worlds forces any unifying modeling effort such as the one briefly described in this paper into a series of challenging design trade-offs between the preservation of typical characteristics of one of the worlds at the cost of losing modeling power when describing the other. The modeling approach described in this paper is centred around the concept of *containment* in the physical and the virtual world and the related concepts of human *intra-* and *extra-manipulation* of objects related to each other through containment relationships.

3 Modeling Focus: Manipulation and Organisation of Objects

In order to keep our model simple, we limit ourselves to activities that, more or less, a) have a clear meaning, b) are observable by a human agent, and c) are observable by an artificial agent. Although this narrows the scope of the model significantly (it leaves out for instance pure cognitive or social processes) we believe that for our purposes, the gain in modeling power compensates for it. A distinction is made between physical, virtual, and physical-virtual activities. The notion of activity is furthermore divided into operations, actions and activities based on what abstraction level the phenomena takes place. ("Activity" is used whenever no distinction is necessary.) The model leaves for the designer to categorise the studied activity and sub-activities along these dimensions although agent domain knowledge and potential breakdowns (Bødker, 1989) should influence the decision.

Objects are categorised by the analyst/designer as belonging to the group of domain objects, tools, containers, or agents. (see table 1 for an example categorisation) in a given activity context. It is important to note that the same physical or virtual object can be viewed as for instance a domain object in one situation and as a container in another, depending on the activity and on the interest of the analyst/designer. E.g. an office room might be viewed as a container when a person is performing a knowledge work activity in it and as a domain object in the context of constructing a building.

Container objects play an important structural role. The proposed framework tries to capture two important activity-supporting functions of containers: 1) to provide more or less structured space for long-term storage of "cold" objects, and 2) to provide a) space, and b) tools for manipulation of "warm" domain objects. The notion of "warm" and "cold" objects is borrowed from the empirical work of Sellen & Harper (2002) used by them to distinguish between objects assumed to be relevant for a currently ongoing activity, and those not. Although some container objects frequently play both roles, some are more tuned towards the storage function and others towards supporting the manipulation of domain objects. Whenever a distinction is needed, we denote them "storage objects" and "workshop objects" respectively.

The (re-)organisation of objects is an important part of Knowledge Work activities (Malone, 1983; Kirsh, 1995; Sellen & Harper, 2002). If we concentrate on the physical and virtual organisation activities, they are both possible to observe and have a relatively clear purpose and thus qualify as activities suitable for modeling using our framework.

Table 1: An example categorisation of objects in the physical and virtual worlds

		physical	virtual
domain objects		a book on a bookshelf in an office	a web page in the context of a search on the Internet
tools		a screwdriver when mending a car	a clipboard
containers	**workshop**	a desktop on which you can find pens, a stapler, etc.	a word processor application window
	storage	a refrigerator	a folder in a file hierarchy
agents		a human	a reminder-application

4 Containment Hierarchies

Both physical and virtual environments can be modeled as hierarchies based on the objects situated in them and containment relationships between those objects. Because the physical and virtual worlds typically differ in their structure, "containment" cannot mean exactly the same thing in both worlds. Furthermore, structural constraints ensures a very regular tree structure for the physical world while in the virtual world, cheap "cloning" of objects on the one hand, and independency from laws of nature such as having only three spatial dimensions on the other, opens up for a more irregular structure in the virtual world where for instance the same object can appear at more than one place. Thus, virtual containment hierarchies belong to the class of hierarchies called semi-lattices (Hirtle, 1995). The actual definitions of physical and virtual containment has been omitted here for space reasons but examples can be inferred from table 2.

4.1 Intra- and Extra-Manipulation

The effects of manipulating an object is propagated upwards and downwards in containment hierarchies according to mechanisms decribed as intra- and extra-manipulation. An extra-manipulation of an object on one hierarchical level is identical to an intra-manipulation of that object's parent-object one level above. Further, a container object's internal state is equal to the set of external and internal states of that object's children who in turn depend on the states of their children and so on. Since practically all objects can act as containers (e.g. when interested in the spatial relationship between fibers in an apple, the apple can act as container) and thus nesting is unavoidable, one object has to be chosen to act as a reference object (RO) whenever an analysis is to be done, to avoid confusion. Extra-manipulation of a RO changes the relationship between the RO and its surroundings (technically, its sibling and parent objects in the containment hierarchy). An intra-manipulation of a RO changes the internal structure of the RO itself (technically, the relationship inbetween RO's children as well as the relationship between them and RO).

4.2 Short- and Long-Term Manipulation

While it is sometimes hard in the physical-world to clearly distinguish between short-term and long-term lasting object manipulations, such a difference is more evident in the virtual environment offered by the WIMP paradigm: Extra-manipulation of containers of the type "window" (see table 2) do seldom last for long and are seldom part of the result of the activity but instead motivated by the temporary management of screen real-estate. We denote such manipulations short-term manipulations. Long-term manipulation of objects are manipulations that have a long-term effect.

4.3 Conditions for Object Manipulation in the Physical and Virtual Worlds

Table 2 illustrates the differences between physical and virtual objects with regard to how they afford short- and long-term intra- and extra-manipulation operations. For reasons of space, only domain and workshop objects (see table 1) are included. Furthermore, the virtual environment modeled in table 2 is the one presented by the widely spread WIMP (Windows, Icons, Menus and Pointing device) interaction paradigm currently dominating the area of personal computers. Virtual environments offered by other interaction paradigms would look different. The WIMP paradigm allows the manifestation of virtual objects in three distinct forms which complicates the modeling since each object form allows for different manipulation opportunities, in contrast to for instance the physical world whose objects rarely take on dramatically different shapes.

Table 2: Examples of intra - and extra -manipulation afforded by physical and virtual objects [1]

			short-term		long-term	
			intra-manipulation	**extra-manipulation**	**intra-manipu-lation**	**extra-manipulation**
Virtual World (WIMP environment)	Window RO	workshop	- DM-spatial translation of children objectsW (domain objects, tools) within the spatial boundaries of RO - to include an object as a child ("open"-menu item, "show tool x"-menu item)	- DM-spatial translation of RO within the spatial boundaries of the parent (the desktop) - DM-spatial resizing of RO - hide RO ("minimize"-button) - de-activate RO ("close"-button)	- to adjust workshop preferences settings such as simple/ advanced menus; picas or points, etc.	- ?
Virtual World (WIMP environment)	Window RO	domain object	- DM-spatial translation of children objects (text etc.) within the spatial boundaries of RO (e.g. scrollbar, PgUp/Dn, zoom)	- DM-spatial translation of RO within the spatial boundaries of the parent (the workshop object) - DM-spatial resizing of RO - transform RO from W to Wm ("minimize"-button) - de-activate RO ("close"-button)	- to change the content (the children) of a domain object (e.g. the text in a document) or the content structure (e.g. the location of files in a file system)	- ?
Virtual World (WIMP environment)	Icon RO		- ?	- to select/deselect RO (by clicking (DM)) - to rename RO	- to change file properties (pop-up menu)	- DM- cross-storage-container translation (or duplication) of RO
Virtual World (WIMP environment)	Minimized Window RO	workshop	- ?	- hide W clone of RO ("minimize"-item in pop-up menu) - show W clone of RO (single-click on RO) - de-activate RO ("close"-item in pop-up menu)	- ?	- ?
Virtual World (WIMP environment)	Minimized Window RO	domain object	- ?	- DM-spatial translation of RO within the spatial boundaries of the parent - transform RO from Wm to W ("restore"/"maximize"-button) - deactivate RO ("close"-button)	- ?	- ?
Physical World	Any RO	workshop	- DM-spatial translation of children objects within the spatial boundaries of the RO (e.g. to move a bottle of wine across the dining table) - to include an object as a RO child (e.g. to put a book on the desktop)	- DM-spatial translation of RO (e.g. to put your work bag (the RO) containing pens, papers and laptop on the seat beside you on the morning train)	- to repaint the walls of a living room (the RO)	- to move a desk (the RO) from one room to another
Physical World	Any RO	domain object	- DM-spatial translation of children objects (e.g. book pages) within the spatial boundaries of RO (e.g. to turn the pages) - to write things on a blackboard (the RO)	- DM-spatial translation of RO within the spatial boundaries of the parent (e.g. to move a pawn (the RO) forward in chess) - DM-spatial resizing of RO (e.g. to roll-up the blinds of a window)	- to overline lines of text in a text document using a highlighter pen	- DM-translation of RO to a storage container .(e.g. to move a book from a bag to a shelf) - DM-spatial resizing of RO (e.g. to crumple up a piece of paper)

[1] Legend: RO = Reference Object; DM = Direct Manipulation; object postfixes W, I, Wm = Window, Icon and Minimized Window manifestation respectively.

5 Conclusions & Future Work

Although space restrictions has limited the presentation to only the corner stones of the framework, we believe to have shown that the proposed unified terminology and the selected structural characteristics of physical and virtual environments together has the potential to enable the modeling of simple physical and virtual environments as joint single environments. The presented concepts have been proven useful in the design and analysis of the physical-virtual prototype system Magic Touch (Pederson, 2001).

Future work includes application of the model onto common physical-virtual environments such as offices, industrial shopfloors, building construction sets, home environments and entertainment settings. The result of each application effort is expected to become a mix of mutually connected, containment-based, physical and virtual hierarchies that together represent structural affordances & constraints within the specific physical-virtual environment in the light of specific activities.

6 Acknowledgements

I would like to thank Lars-Erik Janlert and the members of the Cognitive Computing Lab at Umeå Center for Interaction Technology (UCIT) for many useful reflections and ideas.

7 References

Bødker, S. (1989). A Human Activity Approach to User Interfaces, In Human-Computer Interaction, **4**:171-195.

Drucker, P. F. (1973). Managment: Tasks, Responsibility and Practices. New York: Harper & Row.

Fitzmaurice, G.W., Ishii, H. and Buxton, W. (1995). Bricks: Laying the Foundations for Graspable User Interfaces. In *Proceedings of CHI'95*, ACM Press, 442-449.

Hirtle, S. C. (1995). Representational Structures for Cognitive Space: Ordered Trees and Semi-Lattices. In A. V. Frank and W. Kuhn (Eds.), *Spatial information theory: A theoretical basis for GIS*. Berlin: Springer-Verlag.

Hollan, J., Hutchins, E. & Kirsh, D. (2000). Distributed Cognition: Toward a New Foundation for Human-Computer Interaction Research, in *ACM Transactions on Computer-Human Interaction (TOCHI)*, ACM Press, 2000.

Kidd, A. (1994). The Marks are on the Knowledge Worker. In *Proceedings of CHI '94*, Boston, ACM Press.

Malone, T. (1983). How Do People Organize Their Desks? Implications for the design of office information systems, in *ACM Transactions on Office Information Systems*, 1(1), 99-112, January 1983.

Pederson, T. (2001) Magic Touch: A Simple Object Location Tracking System Enabling the Development of Physical-Virtual Artefacts in Office Environments. Short paper for the Workshop on Situated Interaction in Ubiquitous Computing, CHI2000, April 3, 2000, The Hague, Netherlands. Slightly edited version published in *Journal of Personal and Ubiquitous Computing* (2001) 5:54-57. Springer Verlag, February 2001.

Sellen, A. J. & Harper, R. H. R. (2002). The myth of the paperless office. MIT Press. ISBN 0-262-19464-3.

Weiser, M. (1991). The Computer for the 21st Century. In *Scientific American*, September 1991, pp. 933-940.

Monitoring and Control of Systems by Interactive Virtual Environments

Jochen Manfred Quick

Centre for Advanced Media Technology
Nanyang Technological University
Nanyang Avenue, Singapore 639798
quick@camtech.ntu.edu.sg

Abstract

In this paper we introduce an application, which visualizes systems in a virtual environment and allows the user to control the system by interacting with the virtual scene. The use of a virtual environment as an interface for system control allows the realistic presentation of simulated systems, enhanced representation of real systems with additional information and remote control of systems at distributed or hazardous locations.

1 Introduction

Through technological advances, mainly in the field of IT, many systems and processes become increasingly fast and complex. This is true for diverse areas such as financial markets, logistic supply chains, workflow management systems, and collaborative design processes. The increase of speed and complexity is largely due to automation of processes that otherwise had to be conducted by humans, who are by nature prone to making mistakes, are relatively slow in their perception of information and can concurrently handle only a limited amount of it at a time.
However, key functions in decis ion-making, monitoring, and control of systems and processes still have to be realized by human beings since artificial intelligent machines cannot fully substitute their experience and lack the ability to conduct certain complicated decisions yet.
To prevent humans from being the weakest link or the bottleneck in a highly dynamic and complex system, their perception of information can be increased by realistic visual presentations, because unlike machines, humans can better perceive information through images that resemble their natural environment than through rows of numbers and figures. In these visualizations animations will represent the changes in the states of the dynamic systems. Animated visualizations are also suitable to enhance incoming data with additional information in order to support the user in the decision making process.
In this article we introduce a system for the visualization of simulation results and the control and monitoring of manufacturing processes in the area of manufacturing and logistics. Our special focus is directed on the human machine interface aspects of the visualization system.

2 Related Work

In (Luckas 2000) a system framework founded on animation elements is described. The animation elements could be use to visualize discrete events, e.g. discrete simulation results (Luckas & Broll

97). (Dörner, Elcacho & Luckas 1998) discusses methods for rapid generation of animation elements. Animation Elements encompass geometric description of its visual appearance and adaptable animation behaviour. Animation Elements had an object-oriented design and emphasis was put on reusability on overloading of behaviour methods. (Dörner & Grimm 01) introduces the concept of 3D components and 3D frameworks. Instead of conventional software objects software components where now used to encapsulate visual appearance and animation behaviour. The advantage was, that the animation components could now be visually composed into larger scenes. In (Quick 2002) the concept of component-based visualization of simulation results was introduced. A tool was described, which could analyse simulation results and visualize them. No interaction with the simulation system was possible and the application could not visualize real time data.

3 System Architecture

The visualization system discussed in this paper was implemented in Java™. Besides its platform independence, Java™ provides a good component framework (Herold, 1998) supporting visual building tools. It has good communication APIs like JDBC™ (Hamilton, Cattell & Fisher, 2002). Further more the 3D Graphics capabilities of Java 3D™ (Sowizral, Rushforth & Deering, 2000) proved to be sufficient for the visualization of systems and processes in a virtual environment.

3.1 Visual Building Tools

In order to make the wide use of virtual environments for monitoring and control of processes possible, a framework has to be developed which supports the fast implementation of reliable process monitoring and control tools using virtual reality technology. Since the realistic visualization of even simple events in a virtual environment can already require complex 3D models and animation methods, reuse of models and animation behaviour is highly desirable so as to keep development costs down and reduce sources of errors. At the same time we have to keep our framework flexible enough to be able to visualize processes from different application areas and to communicate with them in often proprietary data formats. To reach these conflicting goals of reuse and flexibility, we built a component framework (Weinreich & Sametinger, 2001), which can be adapted to specific tasks through a visual programming tool. Visual programming tools can be seen as a compromise between a system architecture, which can easily be configured just by setting of parameters, but is not flexible enough for our purpose and a system, which is flexible by allowing the use of program code in a general purpose language for configuration, but is too complex to handle for getting fast and reliable results.

3.2 Scene modelling and generation

Before a system can be visualized in a virtual environment its geometry has to be created and its animation behaviour has to be configured. In order to make is easier and faster to build a virtual model of the system, it is composed from reusable animation objects and animation behaviours. They are stored in an animation component library and can be customized with visual programming tools to meet the requirements of new system visualizations. If there is a component for a new system missing, it can be modelled or implemented and added to the animation library. Better still it will be composed from existing animation objects and animation behaviours.
The framework supports automatic generation of the virtual environment from data that defines the process. This helps to prevent inconsistencies between the system process and its virtual

representation in the common case where parts of the process are changed and reduces maintenance costs in keeping the visualization system up to date.
Since in most cases there is no sufficient computer readable specification of the system available, the virtual model of the system partially has to be composed by the user. However, data that is available can be used for automatic system configuration.

3.3 Animation Framework

The animation behaviors of the animated objects in the virtual environment have to be able to visualize system processes in real time. This can be a challenge since a typical animation sequence is defined by a start event and an end event. To visualize a process in real-time the visualization system cannot wait for the end event to begin the animation, but has to start the animation as soon as it receives the start event. To make this possible the developed animation behavior framework is implemented in a way that allows the control of an animation by additional events after it has already been started.
Each animation object consists of object models and animation behaviour objects. The behaviour objects receive animation events, which result from state changes in the controlled system, and generate animation interpolators, which drive the animation of the virtual system model.
While in some cases we would like to observe the system in real time, in others it is useful to replay an animation, probably from another perspective or in slow motion. To be able to choose the starting point of an animation randomly, the state of the virtual system model, which is computed from the beginning state plus the accumulation of all following animation events, has to be recorded. To achieve this, animation interpolators will be stored and reused in the event of an animation sequence replay.

3.4 Visualization and Interaction

To help the user in making decisions, our system has to support the user with additional information. It provides the option to select single objects of the visualized system. A popup menu opens where the user can choose from a list of available information descriptions. The selected information will be displayed in a separate information window until it is deselected. Due to the component oriented concept of the visualization framework, the content as well as the representation of additional object information is not predefined by the visualization system, but depends entirely on the object itself. The object provides the system with the list of selectable information and chooses the information display component too. These components can either be system provided, e.g. standard displays of tables or bar charts, or they come together with the animation object definition from the animation object library. Display components will mostly consist of standard 2D GUI elements, but it is also possible for them to provide new 3D views.
The generation of control messages for the visualized system can be accomplished by the use of these GUI components as well. Through user interactions the components make changes of parameters or create instruction events, which will be passed on through the related animation object to the external system.
Besides the use of additional GUI components, direct interaction with the animation objects in the virtual scene is possible. Animation objects that provide listener interfaces for certain interaction event types will automatically be called when an event, e.g. picking and dragging of an object, occurs. The animation object's behaviour itself is then responsible for the creation of a control message and visual feedback in the virtual scene. Any kind of input device can be used for interaction between user and animation object, as long as its is supported by the visualization application.

3.5 Data Exchange with controlled Systems

In order to control and monitor various systems with our application suitable communication interfaces have to be utilized by the application to observe and steer the external system. So far IO-Streams and database connections are used to receive and send data.
Fields in a database can be easily bound to properties of Animation Objects or Behaviour Objects. This means, that changes of either property will automatically trigger the update of the property bound to it. This form of communication is mostly used to configure the scene model with parameters of the observed system at start time or to change system configurations through interaction with the virtual scene.
IO-Streams provide better responsiveness and are thus better suited for immediate real time messages between the observed system and the control application. Messages send from the system to the application reflect changes in the state of the system, while messages in the other direction encapsulate control instructions for the system. Another advantage of IO-Streams is that they require no special data format and new are thus a simple way to connect new systems to our control application. However, this also means, that the data has to be translated into an application internal format in order to be able to use it with library objects and behaviours. To accomplish this, the incoming byte stream will first be converted into generic events, e.g. location change events or action events. Then, translation rule component objects convert these input events into animation events. Configuration wizards can adapt the translation rule components to system specific requirements. Finally, the translation components will be connected to animation objects, which will receive the generated animation events.

4 Application Example

The introduced application has been tested for the visualization of simulation result of a manufacturing process. The production process of several assembly lines was visualized. As also described in (Rohrer 2000), the visualization of simulation results as a virtual system model helps to verify that the computer simulation model reflects the conceptual model and to validate whether the simulation model reflects reality. The visualization helps the simulation expert to understand complex dependencies within the simulated system. The simulation outcome can be better communicated and helps to build the confidence of the customer, who is not an expert in computer simulation, in the results. Changes in the simulation model, like the placement of machinery or storage areas, can be done within the virtual model, helping the simulation expert to optimise the assembly process.

5 Summary and Future Work

We have introduced an application for the visualization and control of system processes, which is easy to adapt to new systems. Its component based software architecture supports the extension of the application with new object models and new object behaviour. The visualization application can be used in system simulation, user training and system control. Application areas are manufacturing and logistic systems, project management of construction sites or education on virtual models.
Future research work will focus on advanced visualization techniques, which support users to detect and observe events of interest in large virtual environments. Other future research goals are advanced integration methods for knowledge databases providing additional information for

decision making and introducing new advanced interaction devices and interaction techniques to the control application.

References

Dörner, R., Elcacho, C., & Luckas, V. (1998). Behavior Authoring for VRML Applications in Industry. Technical Notes of Eurographics'98. Lisbon.

Dörner, R., & Grimm, P. (2001). Building 3D Applications with 3D Components and 3D Frameworks, Workshop Structured Design of Virtual Environments. At Web3D. Paderborn.

Hamilton, G., Cattell, R., & Fisher, M. (2002). JDBC Database Access with Java: A Tutorial and Annotated Reference. Addison-Wesley.

Herold, E.R. (1998). JavaBeans™. IDG Books Worldwide.

Luckas, V., & Broll, T. (1997). CASUS - An Object-Oriented Three-dimensional Animation System for Event-Oriented Simulators. Proceedings of Computer Animation'97, 144-150. Geneva.

Luckas, V. (2000). Elementbasierte, effiziente und schnelle Generierung von 3D Visualisierungen und 3D Animationen. Dissertation at the School of Computer Engineering, Darmstadt University of Technology. Darmstadt.

Quick, J. M. (2002). Component Based 3D Visualization of Simulation Results. 2002 Advanced Simulation Technologies Conference. San Diego, California.

Rohrer, M. W. (2000). Seeing is believing: The Importance of Visualization in Manufacturing Simulation. WSC2000

Sowizral, H., Rushforth, K., & Deering, M. (2000). The Java 3D™ API Specification (2nd ed.). Addison-Wesley.

Weinreich, R., & Sametinger, J. (2001). Component Models and Component Services: Concept and Principles. In Component-Base Software Engineering. Addison-Wesley.

Web-based Toolkits for the Management of Customer Integrated Innovation

Ralf Reichwald, Sascha Seifert, Dominik Walcher

Technische Universität München, Germany
Department for General and Industrial Management
Prof. Dr. Dr. h.c. Ralf Reichwald
Leopoldstraße 139, 80804 München, Germany
Tel.: +49-89-289-24800
reichwald@wi.tum.de, seifert@wi.tum.de, walcher@wi.tum.de

Abstract

To master the increasing pressure of competition it's crucial for companies to invent and promote new products, respectively services efficiently. Within a customer centric production strategy of customized products and services web-based Toolkits could be an appropriate instrument to integrate customers into the innovation process. In this paper we will discuss how web-based Toolkits could be used in business-to-consumer industries.

1 Introduction

Successful innovations are necessary to ensure the long-term survival of firms.[1] These innovations mostly cause enormous investments, thus a failure in terms of an agglomeration of flops could easily jeopardize the continuity of the whole company.[2]

A market-orientated alignment of companies with a special focus on customer needs is presented very often as the only loophole. Customer orientation becomes more and more the critical guideline. Listening to the customers voice demands an integration of the customers and accordingly their knowledge into the value chain. Research shows that the knowledge of customers as a crucial resource for innovations is not identified and exploited sufficiently:[3]

Results of the study "Service Excellence in Germany" show that 90% of the researched companies do not integrate internal and external customers systematically in the innovation process of new products and/or service. This finding is especially crucial as talking with the customer has been identified as the most important source of new product ideas – and the single best factor distinguishing successful from unsuccessful companies in new product development.[4] This is especially true when customer integration is viewed as a continuous integral process which

---

[1] Barclay, I.; Benson, M. H. (1987), pp.103-112.
[2] Lüthje, C. (2000), pp. 1-2.
[3] Gales, L.; Mansour-Cole, D. (1995), pp. 79-85.
[4] Meyer A., Sperl U. (1998), pp. 337-338.

enables steady organizational learning from the customer – following the principles of a learning organization.[5]
Although the idea of customer integration in the innovation process of an enterprise is nothing new only little effort has been taken to explore its potential in business-to-consumer industries. Pioneering applications can be found mostly in business-to-business markets, such as the development of custom integrated circuits.

To meet and understand the expectations and requirements of customers, structures and processes within the company has to be developed to completely integrate the customer. The guideline of customer-orientation has to be evolved into the concept of customer-integration.[6]

2 Web-based Toolkits and customer integration

Customer-integration refers to the systematical collection and preparation of information to generate innovations, modifications or service specifications, which totally meet the customers requirements.[7] In this way, the customer is integrated into the value creation of the supplier. Consumers take part in activities and processes which used to be seen as the domain of the companies. The customer becomes a "co-producer" respectively "prosumer". While this view is not new[8], it is only today that we see a broader application of this principle in practice. However, as the main part of the interaction with the customer takes place during the configuration and therefore the design of a customer specific product, it seems appropriate to call the customer rather a co-designer than a co-producer.

Current market research methods for inquiring custom tailored needs start at the end of the product developing phase (e.g. surveys or reclamation-analysis) reflecting experiences in usage and consumption. At this stage changes and adaptations in function or form of a product respectively the performance of a service are combined with high costs and can cast doubt on the profitability of the product.
We will present a new way of integrating the customer in the innovation process through web-based Toolkits, a concept which allows a much earlier customer integration than conventional instruments.

Web-based Toolkits (Toolkits customers can access and use over the Internet) are a new way of facilitating new product development. They are able to attenuate classic innovation conflicts between time-to-market, product quality, and development costs and increase the likelihood of innovation success. Through the usage of human interfaces relevant information of customers can be generated faster, more frequent, cheaper and real-time.
Instead of the costly approach of screening the market for new product needs, which are converted into novel or adapted products and the time-consuming iterations within the process of developing,[9] Toolkits allow producers to outsource certain design tasks to customers. A supplier provides customers with tools so that they can design and develop the application-speci?c part of a product on their own. That is, users can create a preliminary design, simulate or prototype it, evaluate its functioning in their own use environment, and then iteratively improve it until

[5] Picot, A., Reichwald R., Wigand R.T. (2003).
[6] Kleinaltenkamp, M. (1995), pp. 77-78.
[7] Meyer, A.; Blümelhuber, Ch.; Pfeiffer, M. (1999), pp. 4-5.
[8] Toffler, A. (1970).
[9] von Hippel E. (2001).

satisfied.[10] As the concept is evolving, Toolkits guide the user to insure that the completed design can be produced on the intended production system without change. In this way, products can be developed much more quickly and at lower cost. Customers, in turn, get exactly what they want – a custom product that suits their individual needs precisely.[11]
If a manufacturer outsources design tasks to users, it must also make sure that users have the information they need to carry out those tasks effectively. Five important objectives can be detected to characterise Toolkits:

- *Trial-and-Error*: It is crucial that user Toolkits for innovation enable users to go through complete trial and error cycles as they create their designs: Research into problem-solving has shown that trial-and-error is the way that problem-solving – including learning by doing – is done.[12]

- *Solution Space*: Economical production of custom products and services is only achievable when a custom design falls within the pre-existing capability and degrees of freedom built into a given manufacturer's production system. The scope and potential of this "solution space" can reach from limited configuration of variations to completely new innovations.

- *User-Friendliness*: User Toolkits for innovation are most effective and successful when they are made "user friendly" by enabling users to use the skills they already have and work in their own customary and well-practiced design language without engaging in much additional training.

- *Module Libraries*: Custom designs are seldom novel in all their parts. Therefore, libraries of standard modules that will frequently be useful elements in custom designs are a valuable part of a toolkit for user innovation. Provision of such standard modules enables users to focus their creative work on those aspects of their design that are truly novel.

- *Translation*: The "language" of a toolkit for user innovation must be convertible without error into the "language" of the intended production system at the conclusion of the user design work. If this is not so, then the entire purpose of the toolkit is lost – because a manufacturer receiving a user design essentially has to "do the design over again."

To profit from web-based Toolkits the construction and administration of information and communication technology infrastructure is essential. It encompasses the combination of appropriate computer architectures, operating systems, networks, as wells as appropriate software. The following example shows how Toolkits can be used as a powerful tool to integrate the customer in the innovation process of business-to-consumer industries.

[10] Thomke S., von Hippel E. (2002).
[11] Franke N., Schreier M. (2002), pp.2-10.
[12] von Hippel, E. and Marcie T. (1995), pp. 1-12.

3 Example for web-based Toolkits

The German company Hyve, specialised in consulting and developing customized innovations, has created a Toolkit for "Swarovski" an Austria based Jewelry Manufacturer.[13] For the "Crystal Tattoo Collection" new designs should have been developed. Similar to pearl necklets these pieces of jewelry can be attached on the skin with the help of a special glue. Within a online design competition, which took place from 8th Jan. to 8th Feb. 2002 in English as well as German, Hyve developed a Toolkit, that enabled the visitor to create completely new types of skin tattoos. The incentive for the best three designs was 300€, 200€ and 100€. After passing a survey asking the shopper questions about style, design, preferences and pice-expectations (average 14,3 min.) pearls with different sizes and colors could be freely arranged on a Java based assembly grid (average 18,47 min.).

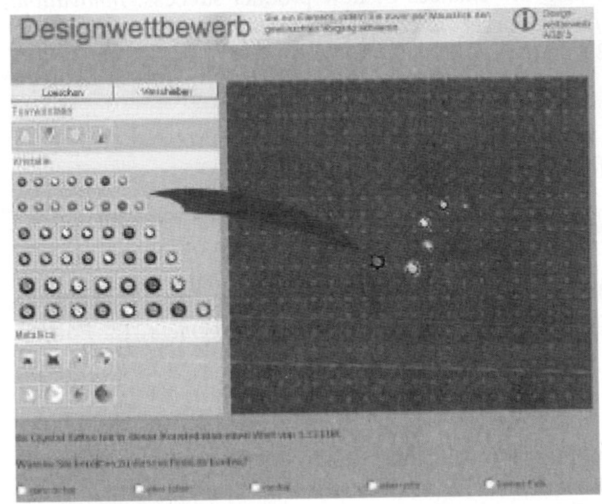

Figure: Hyve Toolkit

The number of pearls, which are used for the design, was not limited. With a simple "drag and drop" technique arrangements can be completed, changed or discarded, thus the "trial and error" – objective is satisfied. Contrary to product configurators, which can be found more and more at online shops, not only variations can be created, but completely new forms. Only a few restrictions, as the size of the grid and the impossibility of overlapping pearls, are given in that "solution space." These restrictions result from the objective, that designs must be transferrable into the production system without error (Translation objective). Moreover using the toolkit is very easy and the visitor doesn't need any additional training (User Friendliness).
306 visitors used the Toolkit and 263 motives were created. A professionel jury assesed the creations and gave awards. Analysing all motives helped to identify new trends and find new patterns, symbols and ornaments. Some designs were revised and produced for the mass-market. With the results from the survey information about product- and pricepreferences could be detected and a tailored marketing campaign was designed.[14]

---

[13] http://www.hyve.de.
[14] www.hyve.de, also Interview with Johann Fueller, director of Hyve AG, 15. Jan. 2003

This example of a web-based Toolkit shows that an early integration of the customer in the innovation process enables a firm to generate competitive advantages through customer centric strategies.
Based on the needs of current innovation processes and the experiences from successful examples the use of web-based Toolkits for integrating the customer in new product developments can possibly be transferred to other branches of consumer goods. The team of Prof. Reichwald at the Technische Universität München currently researches and develops specific applications in the sports product industry (shoes, textiles, equipment). We will present first results and possible web-based Toolkit applications at the HCI International 2003.

References

Barclay, I.; Benson, M. H. (1987). Improving the chances of new product success. *Innovation: Adaptation and growth, international conference on product innovation management*, v. Rothwell R. u. Bessant J., Amsterdam, pp. 103–112.

Barron J. (1988). Thinking and Deciding. New York, Cambridge University Press.
also von Hippel, E. and Marcie T. (1995). How "Learning by Doing" is Done: Problem Identification in Novel Process Equipment. *Research Policy 24* pp. 1-12.

Franke N., Schreier M. (2002). Entrepreneurial Opportunities with Toolkits for User Innovation and Design. *The International Journal on Media Management*, Vol. 4 – No. 4.

Gales, L.; Mansour-Cole, D. (1995). User involvement in innovation projects: Towards an information processing model. *Journal of Engineering and Technology Management*, 12. Edition, pp. 79-85.

Hyve AG. http://www.hyve.de

Lüthje, C. (2000). Kundenorientierung im Innovationsprozess: Eine Untersuchung der Kunden-Hersteller-Interaktion in Konsumgütermärkten, Wiesbaden, pp. 1-2.

Kleinaltenkamp, M. (1995). Kundenorientierung und mehr - Customer Integration. *Absatzwirtschaft*, 8/1995, pp. 77-78.

Meyer A., Sperl U. (1998). Service Excellence in Germany – A study of service practice and performance in Germany. *Quality Management in Services VIII*, Ingolstadt pp. 337-338.

Meyer, A.; Blümelhuber, Ch.; Pfeiffer, M. (1999). Dienstleistungs-Innovation. *Handbuch Dienstleistungsmarketing*, pp. 4-5.

Picot, A., Reichwald R., Wigand R.T. (2003). Die grenzenlose Unternehmung, 5. Auflage, Gabler.

Thomke S., von Hippel E. (2002). Customers as Innovators: A New Way to Create Value. *HBR*, April 2002.

Toffler, A. (1970). Future Shock, New York: Bantam Books.
von Hippel E. (2001). Perspective: User Toolkits for Innovation. *Journal of Product Innovation Management*.

Work and Off-the-Job Acticities: An Important New Field of Work Analysis

Marianne Resch

Universität Flensburg, Internationales Institut für Management
Munketoft 3b
D- 24937 Flensburg
Germany
m.resch@uni-flensburg.de

Abstract

There is (still) a need for work analysis methods allowing a careful analysis of off-the-job-activities. The discussion on "Work Life Balance" shows that the issue of coordinating work and off-the-job activities is one affecting both women and men. The technological advances and the increasing number of workers taking work home with them points to similar problems. The AVAH method (the German acronym stands for "Analysis of Work in Household") was developed with a view to enabling industrial psychologists to analyze and evaluate off-the-job activities and work activities in the household.

1 Old and New Issues in Work Analysis

1.1 Dual Burden and Work-Life Balance

The dual burden of family and work and its potential negative impact on health has long been the subject of discussion. Today, interest is refocusing on this old issue, not least because of recent conspicuous changes in women's behaviour in this area. There has been a marked increase in the number of studies on work, family and health. At the same time, the discussion centring on the new buzz word "Work-Life Balance" shows that the issue of coordinating work and off-the-job activities is one not only affecting working mothers. Attention is focusing on interaction effects between characteristics of paid work and family situation. Many authors (e.g. Barnett & Marshall, 1991) take the view that consideration of such interactions is central to understanding the relationship between stress and health. In the Health Benefits Model, reference is made to the potential resources of the so-called "Dual Life Model". The Role Expansion Model goes further, emphasizing the fact that assuming a variety of roles not only involves conflicts in terms of organizing one's time, but also offers opportunities for building self-confidence, developing social support and acquiring skills (on this discussion, cf. e.g. Sorensen & Verbrugge, 1987). In fact, overview articles on the relationship between work and health show that there is no clear evidence to support the sweeping assumption that working mothers are subject to excessive overload. So far, however, there has been little attempt to collect proper data on work within the family and activities outside the area of paid work. In the absence of suitable methods to obtain such data, many studies give only a very rough characterization of family work, for instance merely asking about the number of children or people living in the household (e.g. Lennon & Rosenfield, 1992).

1.2 Remote Work and Border-Crossing Work

Another "old" issue is the study of remote work or telework. There has always been this type of work, but it has in the past been confined to specific – generally low-skill – activities. Technological advances are making it increasingly possible to free skilled office work from constraints of time and space. The respective costs and benefits of this development for the individual workers, the companies employing them and society as a whole is the subject of fierce debate (on this, cf. e.g. Ulich, 2001, p. 375ff.). One of the risks seen to exist here is that traditional role patterns tend to be reinforced and problems arise in terms of coordinating family routines and needs.

The discussion on new working-hours models, which dispense with traditional clock on/off systems in favour of a system relying purely on results monitoring, points to similar problems. One consequence of such working-hours models is the tendency to break down the barriers between working life and private life. Workers take work home with them, make telephone calls from home or download their emails. This new issue too poses the question of how to reconcile these different areas of activity. A recent study, for example, points out that one of the key issues here is not the quantity but the quality of the time parents spend with their children outside their paid work (Fredriksen-Goldsen & Scharlach, 2001). One of the problems identified by the study – and one of which children are, the study claims, particularly conscious – is parents' psychological absence at home, e.g. when, despite the change in location, parents have their minds on their professional work or actually perform professional duties at home.

Studies on the above problems are rare, and of the available studies few contain careful activity analyses and observations. This means that we lack both the data to make an psychological evaluation of work activities performed at home and the means to assess the impact of the information and communication technologies be increasingly used in the private sphere. What is more, there are practically no findings available on the processes of coordinating and delimiting the different activities.

1.3 Work Analysis Tasks

"Work-Life Balance" and "Border-Crossing Work" are current buzz words in the discussion about changing the organization of work. The issues addressed are not entirely new. It can be shown that there is (still) a need for work analysis methods allowing a careful analysis of off-the-job activities. Here, theoretical and methodological standards must be similar to those applied as in the analysis, evaluation and design of paid work.

When developing analysis methods to meet this need, a number of special factors must be taken into consideration. For instance, the activities being studied are performed in the private sphere, and their scale and content would appear to be largely the result of individual decisions. Which parts of these activities are to be considered work rather than leisure is in itself a difficult question. In the past, there have been neither solutions to such a question nor any awareness that a problem existed, it being accepted for the area of paid work – on which most studies focused – that the beginning and end of a work activity, as well as the way it was performed and the demands it involved, were largely formalized and prescribed by the employer organization. There is thus scarcely any difficulty in defining paid work or in distinguishing it from, say, a private telephone call made at the workplace. In the private sphere, however, there are no comparable formalizations.

A further difficulty facing work analysis concerns the content of the work activity. Particularly in family households, numerous social activities are performed. Both theoretical models and

investigation methods have so far been mainly developed for studying industrial work and office and administrative work activities. They must therefore be reviewed and adapted.
The AVAH method (the German acronym stands for "Analysis of Work in the Household") was developed with a view to enabling industrial psychologists to analyze and evaluate off-the-job and work activities in the household (cf. Resch, 1999). Developing this method is a first step towards extending work analysis to the study of paid *and* unpaid work.

2 The AVAH Method

2.1 General Characterization

The AVAH method is a sound and methodologically tested analysis instrument based on action theory. It allows the study of activities outside the sphere of paid work and a differentiation between family work and leisure activities. The activities are also evaluated according to human-work criteria, which in industrial psychology are considered to be of key importance for psychosocial health, well-being and personal development. The characteristics captured by the AVAH method relate to the decision and planning processes connected with performance of the activity, the forms of cooperation between household members, the temporal flexibility of the activities and the degree to which the action is bound to another person's activity flow ("care intensity").

2.2 Methodological Approach

Analysis using the AVAH method involves conducting two interviews (each lasting approx. 1.5 hours). The first interview is designed to determine the off the job activities. Subjects are asked to name activities they perform at least once a month. They are asked about the location, duration, form and social embedment of the activities. This information forms the basis for subdividing the flow of action into analysis units.
The second interview is then used to collect such information on the individual activities as allows a characterization in terms of the evaluation criteria.
The conduct and evaluation of the interviews is based on an investigation manual containing definitions and commentaries as well as questions – some of them standardized. These questions are directed to the investigators of the studies. The information needed to answer them should be obtained in a conversation with the subject that is only roughly guided by the manual. The questions of the investigators should be worded so as to refer directly to the concrete conditions of the investigated person, not in the abstract manner of the investigation manual. This approach has proved successful with the work analysis methods considered here (cf. e.g. Oesterreich, Leitner & Resch, 2000).

2.3 Formal Characteristics

The method's reliability has been tested in a number of different studies. So-called "dependent double investigations" were conducted, in which tape-recorded analyses were re-evaluated by other investigators. In addition, "independent repeated investigations" were carried out, in which the action of several persons was repeatedly studied by different investigators. On the whole, satisfactory to good concurrences were obtained.
Furthermore, the study yielded results that appear to confirm the criteria-related validity of the method. For example, it was, as expected, possible to show a correlation between the decision

requirements made on a person through their paid work and the level of the decision processes of the off-the-job activities determined using the AVAH method (cf. Resch, 1999).

2.4 Extensions

In the discussion on career and family, the time taken up by (work) activities – or more precisely: the duration and flexibility of such activities – is seen as a key factor. Shorter and more flexible working hours are viewed positively here. But the question is: Under what circumstances do a shorter duration and variable arrangement of activities really guarantee an improved ability to coordinate the whole system of activities?
We are currently working on extensions of the AVAH method relating to the coordination of different activities/activity areas. The aim is to enable the coordination of activity areas to be studied. A computer program is now available that allows computer-supported analysis of time patterns. The current extension is designed to characterize the individual activities in terms of attributes that are relevant for the temporal coordination of the whole system of activities.

3 Outlook

The AVAH method is a first attempt at developing an instrument to extend work analyses to activities outside the realm of paid work. Appropriate application areas for the AVAH method are the work organization chosen in a specific household and the resulting division of labour between women and men. The aim is to study the gendered division of labour in quantitative *and* qualitative terms (e.g.: "Do women take on tasks offering less temporal flexibility more often than men?"). Another prospective area of research is the relationship between family work and paid work, e.g. the question as to the interaction between stress and resources in the respective areas. Comparative analyses of the two forms of work are possible (e.g.: "Are there correlations between the forms of cooperation in family work and the characteristics of paid work?"). Another example is studying the skill-building potential of household work and the question as to how skills acquired outside the realm of paid work can be recognized for career purposes or on re-entering professional life.
There are practical applications for the results obtained using the method. One conceivable application is to use the method or parts of it when advising working parents. It could help to provide information on individuals' current work and life organization and at the same time furnish a basis for anticipative evaluation of career- and/or family-related decisions. The AVAH method can also play an important role in the development and testing of employer-initiated measures to improve the reconcilability of career and family (e.g. "What impact do measures taken by employers to offer more flexible working hours have on the temporal arrangement and coordination of work and off-the-job activities?").
The development of methods like AVAH is promoting theoretical and methodological debate on what are in some cases new issues. These include extending the concept of work or developing models to describe the coordination processes in off-the-job activities. Such methods can also be used to tackle other issues more extensively and intensively, e.g. to study the time patterns of recreation phases in the course of the day (cf. the concept of "total work load", Frankenhaeuser, 1991). Another important field of research is assessing the impact of the constantly growing use of technology in this area. Desirable here is the development of other approaches in this new research area concerned with off-the-job activities.

References

Barnett, R.C. & Marshall, N.L. (1991). "The Relationship between Women's Work and Family Roles and Their Subjective Well-Being and Psychological Distress". In M. Frankenhaeuser, U. Lundberg & M. Chesney (Ed.), *Women, Work and Health. Stress and Oportunities* (pp. 111-136). New York: Plenum Press.

Frankenhaeuser, M. (1991). The Psychophysiology of Sex Differences as Related to Occupational Status. In M. Frankenhaeuser, U. Lundberg, M. Chesney (Ed.), *Women, Work and Health. Stress and Oportunities* (pp. 39-61). New York: Plenum Press.

Frederiksen-Goldsen, K.I. & Scharlach, A.E. (2001). Families and Work. New Direction in the Twenty-First Century. New York: Oxford University Press.

Lennon, M. C. & Rosenfield, S. (1992). Women and Mental Health: The Interaction of Job and Family Conditions. *Journal of Health and Social Behavior*, 33, 316-327.

Oesterreich, R., Leitner, K. & Resch, M. (2000). Analyse psychischer Arbeitsanforderungen und Belastungen in der Produktionsarbeit. Das Verfahren RHIA/VERA- Produktion. Göttingen: Hogrefe.

Resch, M. (1999). Arbeitsanalyse im Haushalt. Erhebung der Bewertung von Tätigkeiten außerhalb der Erwerbsarbeit mit dem AVAH-Verfahren. Zürich: vdf.

Sorensen, G. & Verbrugge, L.M. (1987). Women, Work, and Health. *Annual Review of Public Health*, 8, 235-251.

Ulich, E. (2001). Arbeitspsychologie (5th ed.). Zürich: vdf.

New technology driven processes for the construction sector – the research project ViBaL

Alexander Rieck

Fraunhofer Institute for Industrial Engineering
Nobelstrasse 12, 70569 Stuttgart, Germany
Alexander.Rieck@iao.fhg.de

Wilhelm Bauer

Fraunhofer Institute for Industrial Engineering
Nobelstrasse 12, 70569 Stuttgart, Germany
Wilhelm.Bauer@iao.fhg.de

Abstract

It is important to understand the basic specific problems of usability, process integration, human interaction and networking to achieve a fundamental change in the Architectural, Engineering and Construction (AEC) sectors –from the actual preindustrial process to knowledge based, better, faster, more profitable und human oriented AEC sectors. New technologies like Virtual Reality, Mobile Augmented Reality or Portal Technologies will play a central part in this new process.

1 Introduction

ViBaL's intention is to optimise the information flow in the AEC sectors to build qualitatively better houses and buildings faster and more favourable. Unlike other industrial sectors (i.E. automotive, aeronautical, pharmacy) the AEC sectors could not benefit from the revolutionary development of information and communication technology. The manifold reasons are not only based on the obvious lack of functioning and generally accepted interfaces between the different used technologies, but also on the processes, which originate from the preindustrial time. To achieve a fundamental change of the whole AEC sectors one has to change the traditional rules and implement a new robust technology through all links of the process chain.

2 Project objectives

ViBaL's focus is the development of a Digital Mock Up (DMU) with fundamentally new features specific for the needs of the construction and building industry. By the end of the project a demonstrator will be developed to show the possibilities of a DMU in an integrated communication environment.

In the current research project ViBaL three Fraunhofer Institutes (which are FHG-ISST, FHG-IAO, FHG-IMK) co-work on new solutions for the construction industry in the fields of exactly these issues.

Based on a previous research project at a major construction company, where a future scenario has been developed, ViBaL is mainly divided into four major work packages:

- analysis of the current constructional process and the development of new integrated processes
- design of a new parallel realtime rendering technique.
- description of an experimental augmented reality system
- development of a web-based portal technology

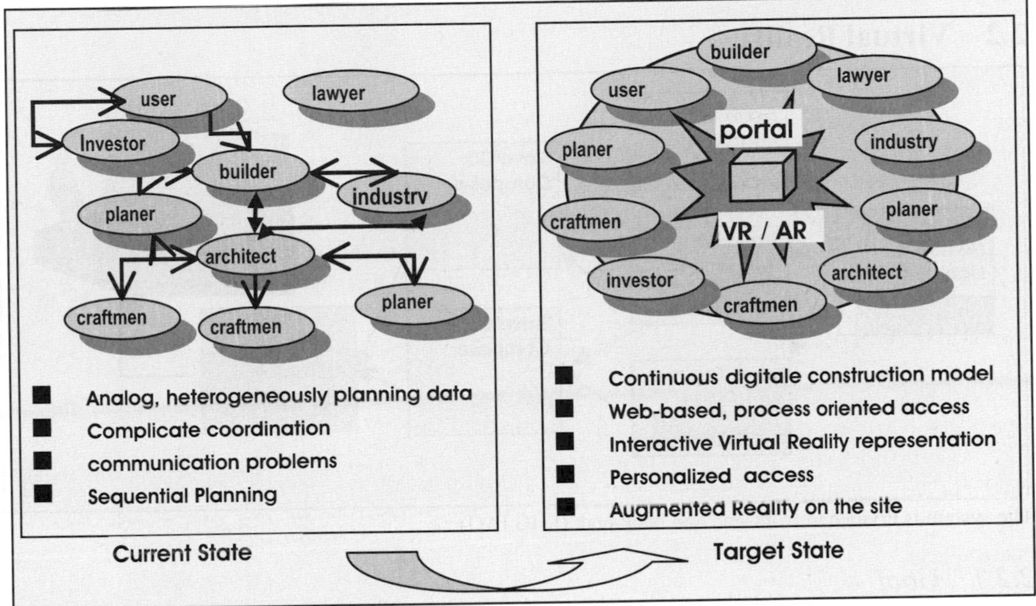

2.1 Constructional Process

2.1.1 *Current State*

The current construction processes are not standing abreast with changes of the information age. Because the new technologies (IT, VR etc.) are usually being used only isolated, the data cannot be constantly passed in a whole system.

The creation of value is shaped by sequential and separated actions of the individual working fields. However, constructional processes often begin before the actual planning is completed. Usually information are still handled over by paper. Meanwhile important 3D information of complex spaces con not be shown in regular 2D plans. Changes in the planning or other problems of the construction are not automatically delivered to all partners in real time. The huge amount of paper based information finally leads to a lost of knowledge.

2.1.2 *Target State*

All information needed in the constructional process are available at any time, everywhere and updated for all partners of the constructional project. All planning tools used are feed by the same information pool and all partners put their information automatically in the same pool. These data can be used for both numerical and visual simulation.

With the beginning of the planning all geometrical data are in 3D. All other information like 2D sketches, excel charts and manuals are linked to the 3D data. On the constructional site the 3D data will be augmented through a head mounted display of the construction worker or will directly be linked to a prefabrication (cnc machines).

2.2 Virtual Reality

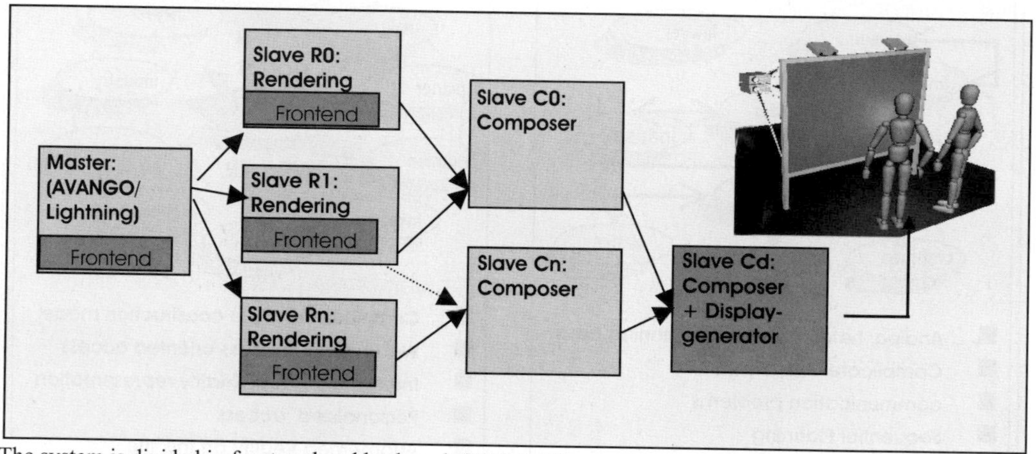

The system is divided in front-end and back-end. (FHG IAO)

2.2.1 Goal

For real time visualisation in Virtual Reality (VR) and Augmented Reality (AR) special graphic workstations with particular graphic subsystems (pipelines) are usually being used. The architecture of the graphic workstations possesses sufficient reserves regarding internal ranges and multiprocessor capability, in order to ensure that a certain number of graphic pipelines is used with full bandwith (for example for CAVEs). In the PC assortment graphic subsystems are already available which offer similar power reserves as the classical graphics workstations. However, these systems do not offer a large scalability. According to the maximum data flow of the graphic pipeline the geometrical complexity used in interactive frame rates is limited. The power reserves of current graphic hardware are not sufficient for visualisation resulting from the available project. complete external building views can be already today visualised interactively; however, usually substantial geometrical simplifications of the original drafts are required . Further constructional components, like reinforcing, electricity installation, etc. increase the complexity substantially.

2.2.2 Current Application

2.2.2.1 Parallelism

The ViBaL visualisation system supports both Sort-first and Sort-last procedure. The first implementation is based on the Sort-first procedure, Sort-last and hybrid technologies are being developed at present. Special attention is put to the utilisation of the existing resources like the use of Threads, which bring advantages on two-processor computers. It has been examined how in the case of Sort-last technology the net range could be used more effectively with an hierarchic depth buffer instead of a traditional Sort-load technology.

2.2.2.2 Distribution of the scene data

The distribution of the scene data on the individual rendering knots can take place in the immediate mode or in the retained mode. In the immediate mode all primitive are conveyed to the rendering knots for each new picture. The needed net range depends directly on the scene complexity. In the retained mode the scene data are locally stored on each computer knot and transfer only the change per frame. The required net range depends on the substantial changes and is usually smaller. Therefore this strategy is used first.

2.3 Augmented Reality

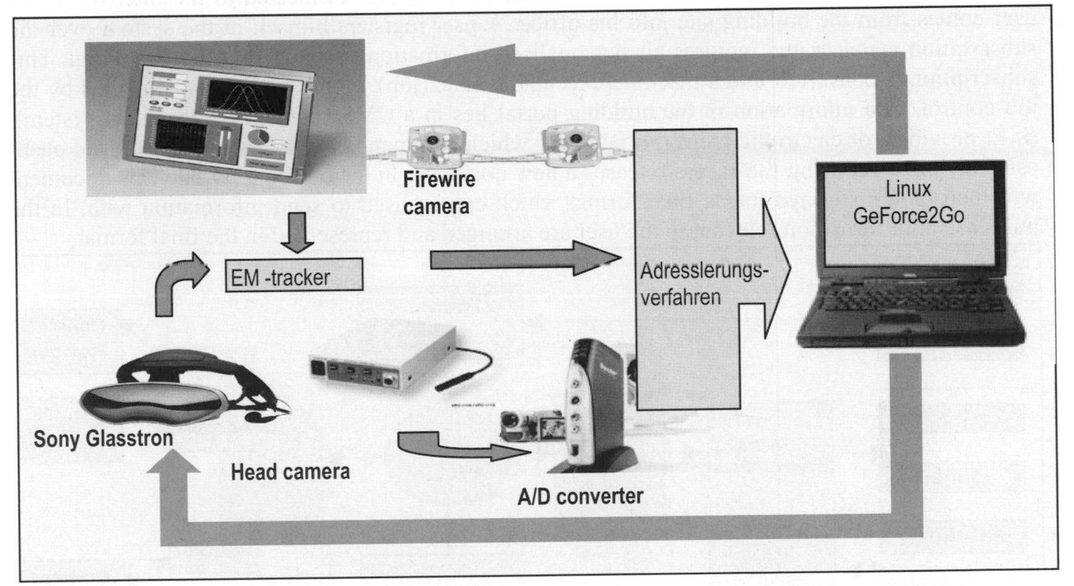

(FHG IMK)

One of the major breaks in the information chain on the construction site is the current use of paper based construction information. Once the 2D plan is printed out, all 3D and additional information from a state of the art CAD system is lost.

2.3.1 Goal

The aim of the project is to develop augmented reality system at the actual construction site. An experimental augmented reality system implemented at the Fraunhofer IMK Lab is currently being described as an example to be enhanced in the future.
The application creates an augmented reality that shows users portion of the structures hidden behind architectural or constructional finishes respective objects supposed to be in the surrounding space and allows them to see additional information about the objects. Fraunhofer's augmented

reality system uses a see-trough display to overlay the human natural view with a graphic. As the person moves around the position of his/her head is tracked allowing the overlaid material to remain tied to the physical world.

2.4 Portal Technologies

The various participants involved in the different parts of the building process need specific information according to the current process step and the task they have to fulfil. Depending on the current process step and the role of the user context-relevant information is provided. Access takes place over different terminals (Smartpad, PDA, Handy, fax devices). For this purpose the heterogeneous data structures must be transformed and set into different target formats. The information loss caused by the transformation must be kept as small as possible. An appropriate architectural concept is prototypical implemented. The interaction with the building portal over different (mobile) terminals can be done by a multimodal dialogue control. I. E. an order can be started via a WAP Handy and be led via a stationary computer connected to the InterNet, if the user comes from the building site into his office. A user registers himself at the system over the subscription manager and requires his personalised information needs of the building portal. This subscription is converted into a machine-executable code (job), which is then implemented by the job control. The information in the building portal lies in a CMS (content management system), (i. E. provider for electronic maps, portals over which commands can be given etc.). In the push-case the job supervising the access releases a new document in the building portal. This document will then be transformed into a final format which can be used to send information with. In the pull-case all information relevant to the user are arranged and represented in the final format.

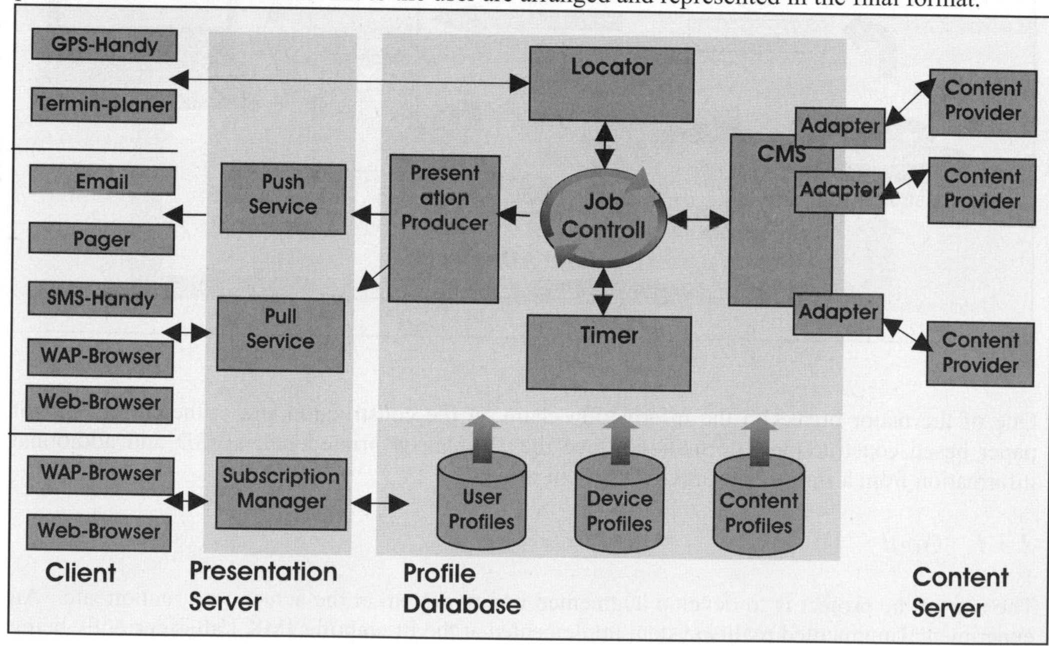

(Fraunhofer ISST)

Training and Assistance to Maintenance in an Augmented Reality environment

Bernd Schwald

ZGDV e.V. Computer Graphics Center
Fraunhoferstr. 5,
64283 Darmstadt, Germany
Bernd.Schwald@zgdv.de

Blandine de Laval

Thales Optronique SA
Rue Guynemer,
78283 Guyancourt, France
blandine.de-laval@fr.thalesgroup.com

Abstract

The background of this paper is an industrial environment, where complex mechanical tasks have to be performed. In this context, for instance power plants or aircraft manufacturers, maintenance tasks require either a lot of documentation and training or a big amount of experience. No matter, if documentation is available in paper form or in electronic form, e.g. on a laptop, the worker cannot simply focus on his work, but has to concentrate on the separated documentation as well. This paper presents the concept of a system, using Augmented Reality for an enhanced 3D workspace, which integrates the documentation and further help into user's environment and especially in his view. Interaction with the system is performed in a multi-modal way, combining speech and a virtual pointing device.

1 Introduction

The need for extensive documentation for a certain task is the starting point for the work presented. The focus is on tasks in the industrial context, i.e. automobile, aircraft or power plant industry, for those objects as well as their maintenance are quite expensive, such that the implementation of an Augmented Reality (AR) system appears profitable. Nevertheless, this approach is adaptable to other domains.
AR has its roots in the Virtual Reality (VR) research. The basic idea of AR is the combination of a real scene, i.e. the view of a user, and a virtual one, generated by a computer. This provides the user with an augmented view and can be used here to generate an enhanced 3D workspace. The user gets some basic information, how to perform a task step by step and can get additional help through plans, pictures or videos on demand.
More background on AR and related ideas can be found in section 2 and a brief overview over the hardware and the technical details can be found in section 3. Further information on system components, tracking and calibration procedures can be found in (Schwald & de Laval, 2003). The design, implementation and configuration of the 3D workspace of this AR system is one of the main aspects of this paper and described in section 4.
While the easy access to documentation is one important request of the addressed complex tasks, having hands free for working is another one. In order to fulfil this demand, the system allows controlling by speech commands as well as selection of objects with a virtual pointing device. The concept of the user interaction is outlined in section 5.

2 Augmented Reality in Industry

The combination of a real scene with a computer generated virtual scene is the main idea of AR. In many cases, the user has to wear a kind of goggles, a so-called Head Mounted Display (HMD), where a video stream from the real scene, captured by cameras mounted on the display, is augmented with virtual objects and shown in the goggles, e.g. in the domain of cultural heritage (Gleue & Dähne, 2001). The system presented here, uses an optical see-through HMD. The user is looking through a half-silvered mirror that combines the real view with the projection from small displays inside the HMD.
Implementations of AR demonstrators for industrial applications, date back to the early 1990's (Caudell & Mizell, 1992) and were advanced in the recent years, e.g. repair of copier machines (Feiner & Seligmann, 1993), wire bundle assembly in airplanes (Curtis & Mizell, 1998) or the insertion of a lock into a car door (Reiners et al., 1998), using different technologies. A case study of AR systems for the maintenance of power plants can be found in (Klinker et al., 2001). Recent work was done within the Arvika project (Wiedenmayer & Oehme, 2001), (Alt et al., 2002). Apart from industrial applications, AR is a growing field for current research in domains such as medicine, engineering design, cultural heritage and others.

Figure 1: Stand, mechanical element and user equipment

3 User Equipment

The augmentations superimposing the real scene can be placed in different ways, as described in section 4.1. One possibility is to overlay parts of the mechanical element, e.g. to highlight a screw, that has to be unscrewed. To allow such an overlay in a correct way, i.e. independent from the viewpoint of the user, position and orientation of the head (eyes) of the user, and the mechanical element has to be known. The determination of position and orientation in real-time is performed by two six degree of freedom (DOF) tracking systems, an electromagnetic and an optical infrared tracking system. These tracking systems are mounted on a stand, as shown in Figure 1. Furthermore the user is wearing a lightweight helmet, integrating the HMD, a six DOF tracking sensor, headphones, a microphone for speech control and a video camera for demo purposes. The lightweight helmet is connected to a transmission unit on user's belt and the transmission unit is connected to the system's PC. More detailed information on user equipment, the stand and tracking systems can be found in (Schwald & de Laval, 2003).

4 The 3D workspace

Information is provided to the user in form of a 3D virtual workspace, meaning that 3D information, thanks to the HMD, surrounds the user so as to help him in his tasks.
In the two subsections below, we describe the contents of this interface and its visual aspect.

4.1 Maintenance procedures

A maintenance procedure consists in a scenario divided into steps, again divided into actions. An action is a very basic task, such as removal of a screw for instance.
The creation of scenarios and of the corresponding database of information is very similar to the procedure of creation of classical user paper or electronic documentation. The objective is to build in parallel documentation for this new system and reuse as much as possible existing information databases. For this purpose, an editing tool was developed to help scenario authors. Its simple interface enables to easily connect a database with a maintenance procedure and its visualisation. Thereby it also ensures that the database and the scenario are appropriate for the system.

Figure 2: An augmented view: scenario instructions

At launching of the system, users can choose equipment and a maintenance procedure related to this equipment. While performing their tasks, they receive at each step of the procedure all the information needed to perform this task (see Figure 2). This information depends on the user's level of competence, as a beginner would not have the same needs in terms of information as someone who is more experienced. In this purpose, information displayed can be default, meaning displayed to all users, or alternative, meaning displayed only to user's request. The list of alternative augmentations available in a given step is indicated by a list of symbols as illustrated in Figure 2.

4.2 Visual aspect of the interface

The interactive 3D-augmented workspace, created around the user, can contain different types of information. The most frequently used are 3D models of pieces of equipment and text instructions. But users have also access to any kind of plans, pictures, symbols, videos and help menu.
To be able to visualise all these augmentations (=information data), several registration modes are proposed:

- *Headset registration:* Information registered to the headset (HMD) is always visible in front of the user's eyes, wherever he moves. Typical headset-registered information is instructions, symbols, and warnings. They indicate the contents of the task as well as the important safety notions.
- *User registration*: information, such as a complex plan, can be placed next to the user at the same place at any time. A window placed for example on the left-hand side of the user will always remain at that position, for user's position and orientation is known.
- *World registration*: Information is placed in a fixed place of the world, as if it was positioned there in reality. Typically a picture showing the list of tools needed during the procedure is placed above the tools table.
- *Equipment registration*: 3D models of real pieces of equipment or arrows pointing on these elements are overlaid on their real counterparts. They enable to illustrate on the equipment itself its functioning, if necessary and more generally to focus the attention of the user on the equipment's parts of interest.

This way of working with 3D "augmented" elements enables each user to feel that the equipment and its documentation as well as the maintenance procedures are all integrated in one single environment.

5 User interaction

The requirement for the user interaction is to be simple and intuitive to make the user willing to use the system. Users mostly need their both hands to perform their maintenance tasks, such that an interaction tool like a mouse was not envisaged. Therefore simple speech commands are used to give users the possibility to navigate through the maintenance scenarios. Thanks to these speech commands, users can access any available information, at any time.
To make the system simple to use but flexible as well, some tools are provided to interact with 3D-workspace and modify its configuration:

- *Look&feel module:* it enables to define the visual aspect of the 3D workspace in terms of windows size, colour, position, mainly depending on the application. Users are proposed to choose the look&feel profile most suitable for them.
- *Virtual Pointing Device:* In combination with speech commands, the VPD enables users to designate augmentations to modify their characteristics, such as the position. In a

future version of the system, this tool could be used to retrieve information about pointed elements (such as pieces of equipment).

An extension of the documentation related to equipment worked at can also be performed online: the "annotations" functionality enables users to record any comments using the microphone. Saving this information in the database enables to improve and update the information available in real time. Users will then be able to access to this information for future use. In a future version of the system, it could be envisaged to record as well information thanks to the video camera fixed on the HMD.

6 Results and Conclusions

The system was implemented within the Starmate project, funded by the EU. The project is close to its end and the system is almost fully implemented: the feasibility of an AR-based system, dedicated to assistance to maintenance is now proved.
The follow-up of the project is currently being discussed. Indeed the prototype developed would have to be quite significantly improved to have a system usable in real work environment. The main improvements would be related to user equipment: the objectives are to have a lighter system that would be more easily installed and moved. Another idea would be to join Starmate to telemaintenance so as to benefit both from the system itself in terms of assistance and from a distant expert, capable of giving instructions as well. Contacts we have had up to now with potential users & customers of the system are quite encouraging and lead us to think a bit more seriously about possible commercialization of the system.

References

Alt, T., & Edelmann, M., et al. (2002). Augmented Reality for Industrial Applications – A New Approach to Increase Productivity. *Proceedings of the 6th International Scientific Conference on Work With Display Units.*

Caudell, T. P., & Mizell, D. W. (1992). Augmented reality: An application of head-ups display technology to manual manufacturing processes. *Proceedings of Hawaii International Conference on System Sciences*, Vol II, pp 659-669.

Feiner, S., & Seligmann, D. (1993). Knowledge-based augmented reality. *Communications of the ACM (CACM)*, 30(7), pp. 53-62.

Gleue, T., & Dähne, P. (2001). Design and Implementation of a Mobile Device for Outdoor Augmented Reality in the ARCHEOGUIDE Project. *VAST 2001 – Virtual Reality, Archeology, and Cultural Heritage International Symposium,* Athens.

Klinker, G., & Creighton, O.,et al. (2001). Augmented maintenance of powerplants: A prototyping case study of a mobile AR system. *Proceedings of ISAR '01 - The Second IEEE and ACM International Symposium on Augmented Reality*, New York, NY.

Reiners, D., Stricker, D., et al. (1998). Augmented Reality for Construction Tasks: Doorlock Assembly. *IWAR, 98 - 1rst International Workshop on Augmented Reality*. SanFancisco.

Schwald, B., & de Laval, B. (2003). An Augmented Reality System for Training and Assistance to Maintenance in the Industrial Context. *Journal of WSCG'2003*, 11 (3), 425-432.

Wiedenmayer, S., & Oehme, O. (2001). Augmented Reality (AR) for Assembly Processes. *Proceedings of ISAR '01 - The Second IEEE and ACM International Symposium on Augmented Reality*, New York, NY.

The Internet in the home: Changing the domestic landscape

Andy Sloane

CoNTACT Research Group
School of Computing and IT, University of Wolverhampton, Lichfield Street, Wolverhampton, U.K.
A.Sloane@wlv.ac.uk

Abstract

This paper outlines some of ways that the Internet has influenced the home information landscape in the last few years. The introduction of information and communication technology (ICT) into the home has had profound effects on the home, its inhabitants and on the ways in which homes interact with each other and other organizations in society. What is clear is that the changes that have happened in recent years are not the end of this transition but there is a need for a better understanding of the way in which information in the home is used, processed and produced in order to better serve the needs of homes and the people in them.

1 Introduction

The introduction of the computer into homes in the 1980s and 90s had an effect on the way people interacted with information and communication technology (ICT) and consequently on how they interacted with other people in the home. Mostly this was caused by two aspects of ICT: the use of computers as a work tool and the use of computers as a leisure device. However, in recent times the changes to the application of computers in the home have meant that they are now more capable of fulfilling a greater role in this environment. The increased capabilities of modern ICT allows many tasks that have been outside the scope of computers to be performed easily in the home and the consequent impact on the home and its occupants will be significant.
Since the late 1990s the Internet has added to the effects and created a new structure for homes that has produced a more profound change in the delivery of information and in the inter-personal relationships of people within a single home situation and between people in different homes. It is mainly these aspects of change that will be discussed in this paper.

2 The place of the home in the growth of the Internet

Homes, and the ICT users situated within them, have been responsible for part of the massive growth of the Internet in recent years. The home users' requirements are diverse and this has fuelled a growth in areas that were not a priority for the research and business-led Internet before home users became interested, e.g. genealogy. The growth has also fuelled Internet companies that specialise in providing services to home users although most of these do not solely rely on the home user for their existence.
The home user in the United States now accounts for about half of Internet traffic[Nielsen-Netratings, 2003] with twice as many active users than work-based users, but who are using the Internet for fewer hours. The home user elsewhere in the world is, occasionally, proportionally

more active but it is difficult to compare figures since the proportion of businesses using the Internet is very different.
In general, the statistics show that the home user has a significant impact on Internet usage statistics and the increasing penetration of the Internet into the home will see further growth in the near future. Additionally, the changes to the home and mobile computing environment that are envisaged (see Lyytinen and Yoo, 2002 and Sloane 2002) will inevitably increase the use of the Internet by home users.
What is more significant for home users in the long term is not the place of the home user on the Internet but the place of the Internet in the home environment where it will have direct effects on the everyday lives of people. This is discussed further in section 4.

3 The applications and information that have contributed

The various applications that have, so far, contributed to the growth of the Internet have been derived from business applications – being those of email and the web browser. This is understandable as any new technology will use existing aspects to "colonise" new spaces. The change in emphasis to the needs and requirements of the home user will occur as more users become aware of the possibilities and their needs start to become met by software and hardware providers. There is some evidence of this happening already and some research has been done on speculative applications that would require a more embedded version of the Internet in homes for them to become a widespread reality (Mynatt and Rowan 2000, Go et al 2000). When the Internet and this pervasive computing milieu become more widespread users will require new forms of interaction to allow them better access to the processing power of these systems.
At present, the use of the Internet is restricted to a few applications that mainly provide information and communication and there is no great degree of user interactivity beyond the keyboard and screen. In future these standard applications will become less important as the applications that promote information use and communication will become an integral part of the home user's system.
To find some direction for these technologies it is necessary to look at the use of information in the home and to examine trends of use. These will increasingly be dependent on the embedding of systems in the home and influence the information landscape of the home.

4 The information landscape of the home

The home is a very diverse environment with any one "home" being different from the many other types and styles of living that are encountered around the world, or even within a single cultural group. The Internet is a technology that can augment the home and create virtual and distributed extensions to the home environment that will require new design methods and new requirements analysis techniques to elicit meaningful data from naïve users.
The introduction of the Internet into the home has changed the relationships that exist within homes and between people in different homes. The obvious case of the use of the chat forum and other extended discussion boards shows how the technology can quickly become invasive and embedded within the home milieu, sometimes as a beneficial influence and sometimes the opposite. What is now becoming apparent is that the Internet will play a big part in everyday life in the future. ICT in general, and the Internet in particular, supports communication and enhances home life. The information used and generated forms the information landscape of the home. What is now crucial to new applications is to understand this information landscape and to use it to inform application development.

The applications that have dominated the television era have been entertainment-led and it is clear that this will continue to play a big part in users requirements. However, users will increasingly want to develop other avenues that become available, particularly inter-personal communication and information sharing that would reduce the real, physical space between them. The place of the Internet in the home is to support and nurture this individual requirement and to provide new avenues of interaction that are not possible without it. To a certain extent this is already happening, and it is also evident in the mobile communications area. Both the fixed communication links and their mobile additions will lead to new applications and requirements from users that are, as yet, not fully understood. What will determine the direction of application development is a better understanding of home user requirements. This will, however, need to be elicited with a great deal of care as the home is a difficult environment in which to carry out experiments and find useful data

5 The users and activities of the home

As pointed out, studying the users of homes is not easy; there are many pitfalls in the area of user requirements gathering and analysis and in general methodology that need careful design to allow for the effects of the home environment in any experiment. This is crucial to the design of new hardware and software for the Internet age. A comprehensive methodology for elicitation is required that will not bias the data. The twin areas of interest are the users themselves and the activities they perform in the home.

The users: Homes accommodate people of every conceivable opinion, desire and need. This makes design a very difficult process. Users are too diverse to get a consensus of how a piece of technology should perform. There is, therefore, a need to make the end product as configurable as possible – this, in turn, leads to added complication with too many alternatives confusing the user. Design in this type of environment is considerably more complex than in a more heterogeneous environment such as in the workplace.

Home activities: The activities that take place in the home are also difficult to categorise and understand from a traditional computing perspective. The constraints of a work environment disappear at home and the user is free to do as much or as little as they need. There is no compulsion to use the Internet for email; users do it for fun and enjoyment. The same is true of other ICT developments – there is no requirement for them to be used; they need to be accessible, enjoyable and usable. However, the diversity of users also poses a problem since some will view some activities as essential and interesting whilst others will requires applications that provide the minimum interaction for the same ends.

This diversity of users and their interests is one of the critical design points of the home environment and design for ICT in the home is at a very early stage of development compared with that from the work environment. It is this problem that leads to the need for further study in the home environment.

6 The need for study to provide further applications

The final area that will be considered is the need for study of the home and it's users to inform the user requirements of future applications. The home is a complex area and study of specific homes does not give general results. Therefore, with careful design and wide ranging testing new applications could be developed that will not only meet user requirements for the home Internet but also be commercially viable to their creators.

One of the main influences on the design of new home applications or appliances will be the degree of pervasiveness and embeddedness in the home environment. As devices become part of the "home-scape" the design problems change. For traditional computer interactions the study of the human-computer interface has been a worthwhile avenue but as the devices become more pervasive the design issues change to become more diverse. It is more likely that the problems will be centred on how the communication is made with the device rather than how it is structured in the first instance. After that the interface requirements will again be relevant but the interfaces are likely to be very different from today's screen and keyboard standard.

7 Conclusions

The home is a key area for pervasive computing technology; with it's diversity of users and varying needs requirements it has a complexity that is not always appreciated. The users in the home are becoming more sophisticated and learning what ICT can offer the home. When the design of appliances, devices and systems takes full account of home user requirements there will be a new dimension to ICT in the home and it should be able to support the activities that home users want. However, until there is a comprehensive understanding of the home, it's occupants and their interactions that contains a multi-disciplinary approach both the users and the suppliers of ICT will be losing out. The users are a complex group with incredible diversity and the plethora of home styles and types needs serious consideration. To enforce strict conformance to standard solutions is a recipe for future problems. Home diversity is a characteristic of human everyday life and surrendering this diversity is not a likely choice for the information age.

References

Lyytinen K and Yoo Y(2002), "Issues and challenges of ubiquitous computing", Communications ACM, **45**,12, 63-65

Nielsen-Netratings (2003) Internet usage statistics. Retrieved 4th February 2003 from http://pm.netratings.com/nnpm/owa/NRPublicReports.Usages Accessed 04/02/03

Sloane (2002) "The Internet in the home environment" Proc. IADIS Internet/WWW 2002 conference, Lisbon, Portugal, November 13-15 2002, ISBN 972-9027-53-6, pp481-484.

Go K, Carroll J and Imamiya A (2000), "Familyware", in Sloane A and Van Rijn F Eds. (2000) "Home Informatics and Telematics: Information, Technology and Society", Kluwer Academic publishers, Boston MA., ISBN 0-7923-7867-9, pp125-140

Mynatt E and Rowan J (2000), "Cross-generation communication via digital picture frames" in Sloane A and Van Rijn F Eds. (2000) "Home Informatics and Telematics: Information, Technology and Society - Volume 2", IFIP, ISBN 3-901882-12-X, pp77-84.

Framing the Flightdeck of the Future: Human Factors Issues in Freeflight and Datalink

Alex W Stedmon[Ψ], *Sarah C Nichols*[Ψ], *Gemma Cox*[Ψ], *Helen Neale*[Ψ], *Sarah Jackson, John R Wilson*[Ψ] *& Tracey Milne*[γ]

[Ψ]VR Applications Research Team,
University of Nottingham,
Nottingham, NG7 2RD, UK
alex.stedmon@nottingham.ac.uk

[γ]QinetiQ,
Cody Technology Park
Farnborough GU14 0LX, UK
tjmilne@qinetiq.com

Abstract

Most flightdeck automation has taken place within the immediate cockpit environment, however, the modern 'flightdeck system' is a distributed, collaborative, decision-making network, which encompasses the complex integration of other parties, practices and procedures. In order to anticipate the potential human factors requirements for the flightdeck of the future it is necessary to describe the flightdeck system in its current form and predict how it might change with the introduction of different initiatives such as freeflight and datalink. This paper presents an initial outline of a framework that will be developed to classify different stages of flight where the potential use and impact of flightdeck IT may vary, and consider the impact of variations from a planned flight-plan and associated communication requirements. With so much automation predicted for the future, an understanding of the relationships between different parties, levels of task delegation (freeflight) and mode of communication (datalink) is needed, in order to establish the extent to which these initiatives support or detract from safe operations in the flightdeck system of the future.

1 Introduction

With aircraft levels set to double in the next 15 years some degree of automation is required to support safe increases in air traffic capacity (Kirwan & Rothaug, 2001). Most flightdeck automation has taken place within the immediate cockpit environment, however, in addition to pilots and their cockpit displays, the modern 'flightdeck system' encompasses the complex integration of other aircrew, air traffic controllers (ATCOs), ground crew, and auxillary agencies (such as airline companies and service staff) and their related practices and procedures.

This system represents a working team characterised by 'trust in the system, functionality of team members, communication within the team, and where authority should be invested in the team' (Taylor & Selcon, 1990). As such, a distributed, collaborative, decision-making network exists whereby the goals of safety and efficiency are mutual but the preferred tactics/procedures used by each part of the team may be different. From this perspective, a joint cognitive system emerges incorporating a number of operators and a number of systems (Hollnagel, 2001). For example, during a typical flight, a pilot will communicate with other members of flightcrew and different ATCOs; the pilot will receive information from flightdeck instruments and displays; and through 'eavesdropping' other radio communications between other aircraft and ATCOs may develop an awareness of other activities occurring in nearby airspace (Midkiff & Hansman, 1993). These sources of information contribute to both the pilot's and ATCO's attention demands, mental workload and situation awareness (SA); and may affect subsequent communications and/or behaviour.

2 The FACE project

The Flightdeck and Air Traffic Control Collaboration Evaluation (FACE) project, funded by the Engineering and Physical Sciences Research Council, takes a systems perspective to investigate the human factors requirements for the flightdeck of the future. The aim of the FACE project is to anticipate the potential impact on the overall flightdeck system of two specific elements of flightdeck IT: datalink and freeflight. In order to achieve this, it is necessary to describe the flightdeck system in its current form and predict how it might change with the introduction of different technologies. Initial methods, such as a literature review, expert interviews and field studies, have investigated aspects of the flightdeck system in its current form. From these approaches, specific scenarios can be used to investigate the impact of new technological initiatives through experimental trials, simulator studies and real-world observation techniques. This paper presents an initial outline of a framework that will be developed to classify different stages of flight where the potential use and impact of flightdeck IT may vary, and considers the impact of variations from the planned flight-plan and associated communication requirements.

3 Framing the flightdeck of the future

The basis of the research approach within this project is the concept of interacting HCI systems distributed over space and time. With the constant change in the nature of the relationships between the parties and tasks being conducted, the focus of the approach highlights social and organisational factors, as well as more traditional factors within HCI and cognitive ergonomics. With a number of complex components to accommodate within the framework, it is important to capture and represent the various relationships within the joint cognitive system. Figure 1 below illustrates four components of the flightdeck system that need to be incorporated into a holistic understanding of the flightdeck of the future. These form the focus of the FACE project.

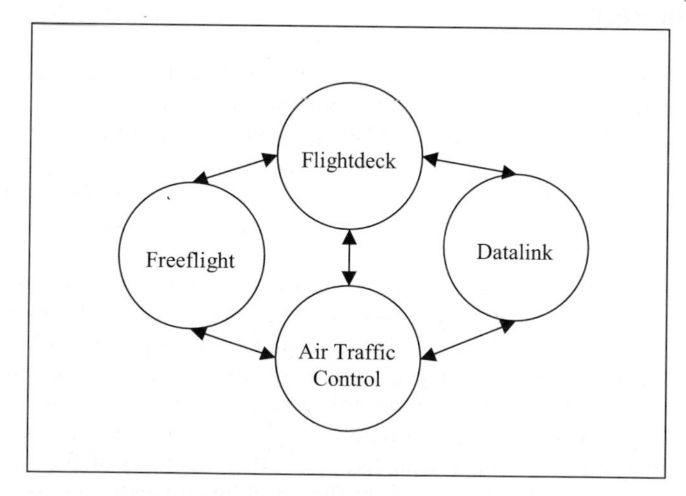

Figure 1: Components of flightdeck of the future

This schematic diagram illustrates the separate components of the flightdeck system and the inter-relationships between them. To concentrate on only one component at the expense of others would only provide a partial understanding of the wider system. As such, any framework that supports an understanding of the wider system, needs to accommodate all the individual components as well as provide an integrated understanding of the relationships of the components. It should be noted that Figure 1 does not attempt to delineate the human and technical input for communications, but illustrates the potential lines of communication between systems.

4 Towards an initial framework

Key stages of flight activity have been identified and this provides a basis for the investigating commonalities between different stages, associated communications, and underlying human factors issues.

The key stages are summarised as follows:

- **On stand (departure)** – final preparations for flight, weather checks and allocation of take off slot from ATC;
- **Taxi** - pilot receives instructions from ATC regarding which routes to take to runway, and may also be required to confirm position with ATCO in cases of poor visibility;
- **Take off** - pilot conducts planned take off procedure. The flight-plan starts at this point;
- **SID** - standard instrument departure, pilot may be instructed by ATC to deviate from this soon after leaving ground other air traffic is in the vicinity;
- **Climb** - with ATC permission, the pilot proceeds with flight-plan and climbs to cruising height;
- **Cruise** - typical communications would be position confirmation, height change, or radio frequency change;
- **Descent** - with ATC permission the pilot prepares for arrival by descending to safe height for landing;
- **STAR** - standard terminal arrivals, pilot receives instructions regarding which runway to land on, and their exact cue to begin their landing procedure;
- **Approach** - pilot conducts planned approach but may be instructed by ATC to deviate from this if there are any obstructions on the runway;
- **Land** - pilot conducts planned landing procedure. The flight-plan finishes once the plane has landed;
- **Taxi** - the pilot is instructed as to where in the airport the plane should be taken.
- **On stand** - pilot completes flight and communications with ATC.

The outline of flight activities is based on the notion of utilising a generic flight-plan as a blueprint for sub-dividing flight activities. In addition to this, generic pre- and post- flight activities are also included in order to capture all activities from start to finish of the process. All flights are organised according to a prescribed flight-plan from the outset. If changes occur then contingencies are generally accepted procedures (such as alterations to flight levels due to air-traffic or weather conditions).

It is important that deviations from a prescribed flight-plan are considered and included in the framework. Under these conditions, changes to the joint cognitive system become particularly important. However, as it is reasonably common that there will often be standard deviations from the flight-plan, it is important to distinguish these from more unusual and unexpected incidents. Level 1 (standard) circumstances may be quite common alterations to the general flight-plan (or pre-flight/postflight activities), such as a missed departure slot, or a change in altitude request to attempt to make up time. Level 2 (non-standard) circumstances may be more severe situations that require dramatic alterations to the flight-plan, such as an engine failure, bird-strike, or severe weather.

5 The need for a structured approach

Using the above outline of typical flight activities, it is possible to identify key types of communications that may be of interest, and human factors issues associated with freeflight and datalink.

5.1 Freeflight

Freeflight refers to the concept of pilot mediated air traffic control where future control vested in the pilot and changes in the degree of delegation (or responsibility) in controller-pilot task distribution will occur (Wickens, Mavor, McGee, 1997). With reference to the proposed classification system above, potential areas of impact for freeflight can be considered. For example, freeflight technology might only be suitable for routine activities or within particular stages of flight, where workload may be lower and allow the pilot to maintain an awareness of flight activity in nearby airspace. This highlights another issue related to the joint cognitive system whereby the ATCO would no longer have explicit responsibility for keeping all pilots informed of nearby activity. At present pilots may use information 'eavesdropped' from other pilot-ATCO communications to maintain an implicit understanding of events around them.

Freeflight automation must therefore be supported with suitable tools in the air and on the ground, with careful thought for the quantity/format of the information/advice presented to different users. Whatever the degree of freeflight, the human operator should have enough knowledge of the system and appropriate skill sets to optimise ATM performance and where current levels of error detection and recovery are maintained (Kirwan & Rothaug, 2001).

5.2 Datalink

The development of datalink technology has focused on reducing the burden placed on R/T channels and enhancing the overall effectiveness of the communications, surveillance and navigation network (Harris & Lamoureux, 2000). However, the handling of datalink needs careful study, information must be presented in the right form, and an appropriate balance between direct voice and datalink communication must be established. Options for datalink protocols and communications media require further research to understand which provide solutions capable of meeting long term bandwidth, signal integrity and cost requirements. The total content of the data transmitted to and from the aircraft needs to be considered in an integrated way rather than examining each requirement separately.

With reference to the classification presented, a clear distinction is evident between the impact of datalink in expected and unexpected events and their associated communications. It is likely that in expected communications, such as a request for change in height as stated on the flight-plan, the use of datalink could avoid errors that may occur due to mishearing, low radio quality, or perceptual confusion between similar flight numbers. However, in the case of non routine situations, such as a pilot running low on fuel, the potential impact of datalink could be more critical, as a text based mode of communication could mask urgency that would be evident in a spoken communication. Furthermore, the impact of the loss of radio communications on pilot 'eaves-dropping' other pilot-ATCO communications, and using the information to maintain SA, cannot be ignored.

6 Conclusion

A flexible flightdeck system is a strategic target, with increased freedom for management shared between service providers on the ground and in the air. With so much automation predicted for the future, a better understanding is needed of the relationship between the pilot and ATC, and how this interaction might affect levels of task delegation (freeflight) and mode of communication (datalink). It is crucial, therefore, when considering these aspects to establish the extent to which datalink, along with different levels of task delegation, support or detract from safe operations in the flightdeck system of the future.

7 Acknowledgement

The work presented in this paper is supported by the Engineering and Physical Sciences Research Council, Project GR/R86898/01: Flightdeck and Air Traffic Control Evaluation.

8 References

Harris, S., & Lamoureux, T. (2000). The Future Implementation of Datalink Technology: The Controller-Pilot Perspective. In, M. Hanson (ed). *Contemporary Ergonomics 2000.* Taylor & Francis Ltd, London.

Hollnagel, E. (2001). Cognition as Control: A Pragmatic Approach to theMmodelling of Joint Cognitive Systems. Special issue of IEEE Transactions on Systems, Man and Cybernetics A: Systems and Humans.

Kirwan, B., & Rothaug, J. (2001). Finding Ways to Fit the Automation to the Air Traffic Controller. In, M. Hanson (ed). *Contemporary Ergonomics 2001.* Taylor & Francis Ltd, London.

Midkiff, A. H. & Hansman, R. J. Jr. (1993). Identification of Important Party Line Information Elements and Implications for Situational Awareness in the Datalink Environment. *Air Traffic Control Quarterly*, Vol. 1 (1) 5-30.

Taylor, R.M., & Selcon, S.J. (1990). Psychological Principles of Human-Electronic Crew Teamwork. In, T.J. Emerson, M. Reinecke, J.M. Reising, and R.M. Taylor (eds.). *The Human Electronic Crew: Is the Team Maturing?* Proceedings of the 2nd joint GAF/USAF/ RAF Workshop. RAF Institute of Aviation Medicine, PD-DR-P5.

Wickens, C.D, Mavor, A.S, McGee, J.P. (1997). *Flight to the Future : Human Factors of Air Traffic Control*. National Academy Press, Washington, D.C.

Asymmetric Communication Mode to Realize Collaboration between Remotely Located Participants for Dealing with Failures

Tadashi Tanaka

Hitachi Ltd , Systems Development Laboratory
292, Yoshida-cho, Totsuka-ku, Yokohama-shi, Kanagawa-ken, 244-0817 Japan
ta-tana@sdl.hitachi.co.jp

Hiroshi Yajima

Hitachi Ltd , Research & Development Group
1-5-1, Marunouchi, Chiyoda-ku, Tokyo, 100-8220 Japan
yajima@gm.hqrd.hitachi.co.jp

Abstract

Compared to face-to-face communication, conventional telecommunication takes more time, and it is very difficult for specialists to take the initiative during the consultation, especially in an emergency situation. This is because in a tele-consultation, specialists can only obtain limited information and they cannot interrupt the caller any time they want to, disrupting their ability to handle the nature of the call in a timely manner. In this study, we propose a conversation method that combines multiple modes, such as voice and text. Callers ask questions to specialists by voice, and specialists respond by text. This prevents the collisions that occur when the caller and specialist try to speak at the same time. A specialist can interrupt easily by using our suggested mode. We verified the effect of our new conversation method by an experiment in a situation where a user consults with a specialist about appropriate response to failures.

1 Introduction

Studies in the field of social psychology have pointed out that physiological arousal is an important factor that hinders human information processing in emergency situations. It is known that arousal automatically happens, and that we cannot control it. As a result the information processing abilities are restricted.

We propose that we support people facing emergency situations by giving them a script with information corresponding to emergencies in order to aid understanding of a given situation or to provoke action as the form of external communication. This enables ordinary people without specialist knowledge to cope effectively with emergency situations.

We proposed asymmetric communication mode (Koizumi et al., 1998), in which customers ask for specialists' advice by voice, and specialists respond with their query by text. Using these different modes, we think that specialists can avoid voice collisions and interruptions. As a result, specialists are able to get information about situations quickly and efficiently. We examined the effect of the asymmetric communication mode by an experiment in a situation where a customer consults with a specialist about appropriate response to failures.

2 Problems with Tele-consulting in Emergency Situations

2.1 Positioning of an Emergency Situation

Emergency situations have following characteristics: frequency of occurrence is lower than ordinary events, it's difficult to anticipate the situation, and there is usually little time to decide which strategy we should adopt to cope with the situation. If we anticipate the situation in advance, we can plan coping strategies and deal with the situation adequately. In cases of unexpected situations, if we have enough time, we ponder the matter and respond appropriately. When events occur frequently, we act on customary practice, and are less likely to make a mistake in dealing with the situation. On the other hand, emergency situations occur infrequently, and there is little time to consider our options, so we can't take the above approaches.

Table 1: Positioning of an emergency situation

	requires no sense of urgency	**Requires sense of urgency**
Frequently occurs	Act according to plan	Act according to habit
Infrequently occurs	Act according to due consideration	Emergency situation

2.2 Model of Human Information Processing in Emergency Situations

To consider an environment that supports humans in an emergency situation, we analyze the information process that occurs when a person faces an emergency situation. We suppose that the information process under such a situation consists of the following steps:

[a] Perception of information about the situation
[b] Cognition of the situation
[c] Deciding a coping action
[d] Executing a coping action

Studies in the field of social psychology have pointed out that physiological arousal is an important factor that hinders human information processing in emergency situations. It is known that arousal automatically happens, and that we cannot control it. Our attention concentrates on inner stress in the state of over-arousal. As a result, our capacity to process information from the outside world decreases, causing us to heed less external information (Eysenck, 1982). In other words, the attention we need to process step [a] shortens, and the information processing abilities in steps [b] and [c] are restricted.

3 Asymmetric Communication Mode

3.1 Approach

Several mechanisms have been proposed to reduce the burden on information processing and enable us to act appropriately in a matter of minutes, including script (Abelson, 1981) and heuristics (Kahneman, et al. 1982). These involve specialist and empirical knowledge that is effective in each particular situation. Specialists possessing such knowledge can automatically execute a series of actions without oppressive information processing because the script becomes active when the person perceives some information about the situation.

For example, a specialist in a specific computer system has a script for understanding emergencies that might occur in that system, so the specialist is able to understand the system's situation in shorter time than ordinary people. Likewise, the purpose of evacuation drill is to make people acquire the script for evacuation procedures by repetitive practice.

We propose that we support people facing emergency situations by giving them a script with information corresponding to emergencies in order to aid understanding of a given situation or to provoke action as the form of external communication. This enables ordinary people without specialist knowledge to cope effectively with emergency situations.

3.2 Method of External Communication from Specialists

People in remote locations use mutual voice communication in a general way. Because of a lack of non-verbal information, however, some problems exist with remote voice communication. For example, exchanges between speakers don't always work smoothly; collision of utterances can arise and people cannot hear each other clearly. This causes reduction in communication efficiency. In addition, in emergency situations over-arousal can reduce the capacity of attention for external communication, and consequently, the problems mentioned above become more serious.

In the asymmetric communication mode environment we propose, specialists give instructions by text-based panels displayed on a customer terminal. This can avoid failure of hearing due to lack of attention, and communication efficiency reduction due to collision of utterances.

3.3 Proposed Asymmetric Communication Mode

In this study, we propose asymmetric communication mode, in which customers ask for specialists' advice by voice, and specialists respond with their query by text. Using these different modes, we think that specialists can avoid voice collisions and interruptions.

Both the specialist and customer have a PC terminal. In addition, the specialist has a headphone system and the customer has a microphone. On the customer's monitor are displayed questions and instructions are shown in the HTML format. They can also see one button by which they respond to specialists. In the meantime, a specialist can see the customers' responses and links that are linked to questions and interruptions in the HTML format. They can hear the customer's voice and respond via text messages.

In this system, it's very important for a specialist to interrupt arbitrarily and at will. Hence, the specialist can work at his or her own pace and shorten the amount of time that they spend on each inquiry.

4 Evaluation of Asymmetric Communication Mode

We analyzed the smoothness of consultation under the following conditions:

- Tele-conference condition,
- Asymmetric Communication Mode condition

The trial completion time was used as the measure of consultation smoothness. In addition, verbal data were recorded during the trials for later analysis.

4.1 Method

4.1.1 Experimental conditions

We tested each pair of subjects under the following conditions.

- Condition 1: Tele-conference condition,
  Subjects consult with each other using a terminal that provides only a videophone function.
- Condition 2: Asymmetric Communication Mode condition
  Subjects consult with each other using the asymmetric communication mode function.

4.1.2 Subjects

Six subjects, including two computer specialists and four customers took part in the experiment. All subjects (aged 26-35) had experience in using PCs. The specialist subjects were trained to use the staff terminal and perform various tasks. On the other hand, the customer-side subjects had no experience in using the terminal or tasks involving voice recording.

4.1.3 Task

The task was to engage in tele-consultation about computer system failures. Customers were given instruction sheets about a system failure and had to explain it to specialists via videophone. Specialists gave instructions to customers via a text-based instruction panel. The specialists' goal was to accurately convey the solution to the problem given to them.

4.1.4 Design

A pair of subjects conducted one trial per condition with a total of two trials per session. A specialist member took part in both sessions with a different customer for each session. The order of conditions was counterbalanced between pairs of subjects.

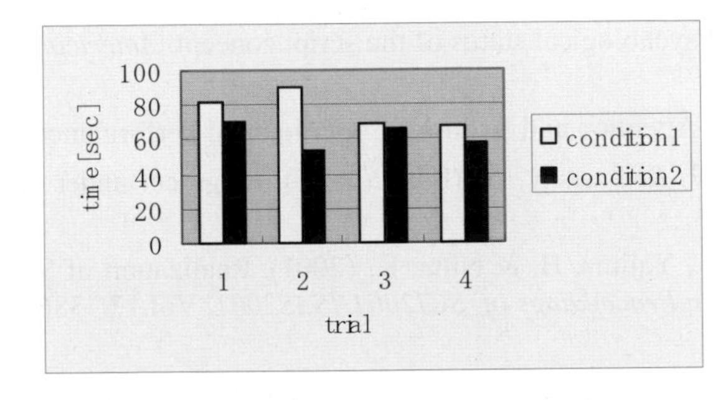

Figure 1: The total completion time.

4.2 Result

For each trial, we measured the time needed to complete the consultation task. We show the result in Figure 1. It is clear that the total completion time under the asymmetric communication mode condition was shorter than under the tele-conference condition.

4.3 Discussion

Smoothness was considered to be the time needed to complete each trial. We found that the average completion time under the tele-conferencing condition was significantly longer than under the asymmetric communication mode condition. This result shows that the asymmetric communication mode is effective when one of the participants is in an emergency situation and has imperfect knowledge about the system. We think this is because a specialist can control the flow of dialogue effectively using a text-based instruction panel.

This time, the subjects took part in the experiment in a relatively sober atmosphere. We assume that the subjects would come under a state of arousal in a real emergency situation, but this experiment wasn't carried out under such a condition. In the case that the subjects were in a higher state of arousal, the possibility of collision of utterances would have been higher, and as a result, the asymmetric communication mode would have been even more effective. Conversely, under a very stressful situation, the subject might not have recognised the instructions on the text-based panel due to lack of attention. We need to conduct further research to confirm whether the proposed communication mode is effective in more realistic situations.

5 Conclusion

We proposed asymmetric communication mode for consultations in emergency situations. The evaluation experiment we conducted showed that proposed communication mode was effective when one of the participants was in an emergency situation and had imperfect knowledge about the system. We think this is because a specialist can control the flow of dialogue effectively using a text-based instruction panel. We need to conduct further research to confirm whether the proposed communication mode is effective in more realistic situations.

References

Abelson, R. P. (1981). Psychological status of the script concept. *American Psychologist*, 36, 715-729.

Eysenck, M. W. (1982). Attention and Arousal: Cognition and Performance. Springer Verlag.

Kahneman, D., Slovic, P. & Tversky, A. (Eds.) (1982). Judgment under uncertainty. Cambridge University Press.

Koizumi, Y., Tanaka, T., Yajima, H. & Nitta, K. (2001). Realization of Specialist's Self-Pace in Tele-Consultation. *in Proceedings of SCI2001/ISAS2001*, Vol.17, 386-390.

Interrogating Search Engine Design using Claims Analysis and General Design Heuristics

Yin-Leng Theng

Division of Information Studies
School of Communication and Information
Nanyang Technological University
31 Nanyang Link, Singapore 637718
tyltheng@ntu.edu.sg

Abstract

Using Google as a concrete example of better search engines, this paper attempts to make explicit the use of Carroll's Claims Analysis and Nielsen's Web design heuristics to detect potential usability problems, and hence propose recommendations for new heuristics to ensure effective search engine design.

1 Introduction

One of the biggest challenges for Web search engine designers and developers is to help users find what they want given the astronomical amount of information available on the Web. Earlier studies focus on enabling users with efficient browse, search and linking facilities. Campagnoni and Ehrlich (1989) conclude from their study that most test subjects using the Sun Help Viewer on-line help system for the Sun386i preferred a browsing search strategy over the index. However, once the size of the Web exceeds browsable proportions, sophisticated search facilities such as keyword search, content search and fuzzy (inexact) search are indispensable for finding specific information (Maurer, 1996). Allison and Hammond (1989) recommend using metaphors such as index for more directed search, and a guided tour facility for exploratory search. Later studies include providing more accurate, faster and more efficient search and linking facilities on the Web such as automating indexes (e.g. Web robots or spiders) to walk the entire server tree, text compression techniques, and machine learning techniques (Maurer, 1996). More recent work focus on understanding users' search and retrieval behaviour using transaction logs (Jones et. al., 1998), user models to generate summaries adapted to users' information (Lopez et. al., 1999), and interaction frameworks to detect usability problems encountered by users (Blandford et. al., 2001).

2 The Study

Despite much work done on improving Web accessibility, Nielsen (1996; 1999b; 2002) commented that Web usability is still poor, and is worsened with new Web technology and applications. In this paper, we describe an initial study using Carroll's Claims Analysis (Carroll, 2000) to interrogate search engine design, building upon prior work employing scenarios and Claims Analysis to improve the usability of digital libraries (Theng et. al., 2002).

Beyond summarizing positive and negative consequences for effective search engine design, a main contribution of the paper is making explicit the use of Carroll's Claims Analysis and

Nielsen's Web design heuristics that inspired recommendations of new heuristics for effective search engine design.

Protocol

The study was conducted during a one-hour lecture on "Human-Computer Interaction (HCI)" with two separate Masters of Information Studies classes of sixty-three students at Nanyang Technological University (Singapore). The students were each given screen dumps of the Google search engine with the search query "interface design guidelines" entered and the associated pages of results returned for the query. Google was selected because it represents a better example of available search engines.

The students were divided into five groups. They performed iterative walkthroughs of the Google search engine using Carroll's Claims Analysis. Claims Analysis was developed to enlarge the scope and ambition of scenario-based design to provide for more detailed and focused reasoning. A claim is a hypothesis about the effect of the features on user activities, that is, the positive outcomes and negative consequences or risks that may adversely affect usability. Norman's influential model of interaction is used as a framework in Claims Analysis for questioning user's stages of action when interacting with a system. The stages of actions are broadly divided into three phases (Carroll, 2000) :

- *Before executing an action*. This phase intends to prompt claims on the design before users perform an action. Two stages of users' actions that address formation of goals (Stage 1a) and planning (Stage 1b) are involved.

- *When executing an action*. This phase (Stage 2) obtains claims by questioning users on how well the system helps them to perform the action.

- *After executing an action*. Two stages (Stages 3a and 3b) prompt users to interpret system's response and evaluate the system's effectiveness in helping to complete a goal.

The students were given forty-five minutes to formulate positive and negative consequences of the Google search engine design with their respective groups before congregating to discuss their findings with the rest of the class.

3 Results and Analysis : An Illustration

As an illustration, Table 1 shows students' comments in response to the questions asked in the Goal Stage (Stage 1a) for the Google search engine. Columns 2 and 3 record claims highlighting positive consequences or negative consequences/risks. The rest of the responses for the other stages were constructed in this manner. We made the following assumptions : comments with positive consequences suggest compliance with well-accepted general Web design heuristics (see Table 1, Column 2); while comments with negative consequences/risks indicate violation of these design heuristics (see Table 1, Column 3). The well-accepted general Web design heuristics are based on Nielsen's ten good deeds (Nielsen, 1999a) and twenty top mistakes in Web design (1999b).

Figure 1 shows some negative consequences detected when interrogating Google search engine during the Goal Stage (Stage 1a).

Table 1 : Goal Stage (Stage 1a) for Google Search Engine

Stage	Positive Consequences	Negative Consequences
Questions to prompt: 1. How does the search engine evoke goals in the user? 2. How does the search engine encourage people to import pre-existing goals? 3. How does the search engine suggest a particular task is : - appropriate or inappropriate? - simple or difficult? - basic or advanced? - risky or safe?	Comments: Simple, clean design Compliance - Feature: Minimalist design. Comments: Provide advanced search, catering to experienced users. Compliance - Feature: Control and freedom for users. Comments: Provide simple search with search box occupying the centre of page, users not distracted from the main task. Compliance - Feature: Minimalist design; Good affordance to type in query and button "Google Search" to submit query. Comments: Other goals displayed. Compliance - Feature: Control and freedom for users.	Comments: Not sure what "I'm feeling lucky" means. Violation - Feature: Not speaking users' language. (see Fig. 1, 1a). Comments: Might click the wrong buttons. Too close. Violation - Feature: Buttons/links too close and/or too small. (see Fig. 1, 1b). Comments: Help is not provided. Violation - Feature: No help available. (see Fig. 1, 1c).

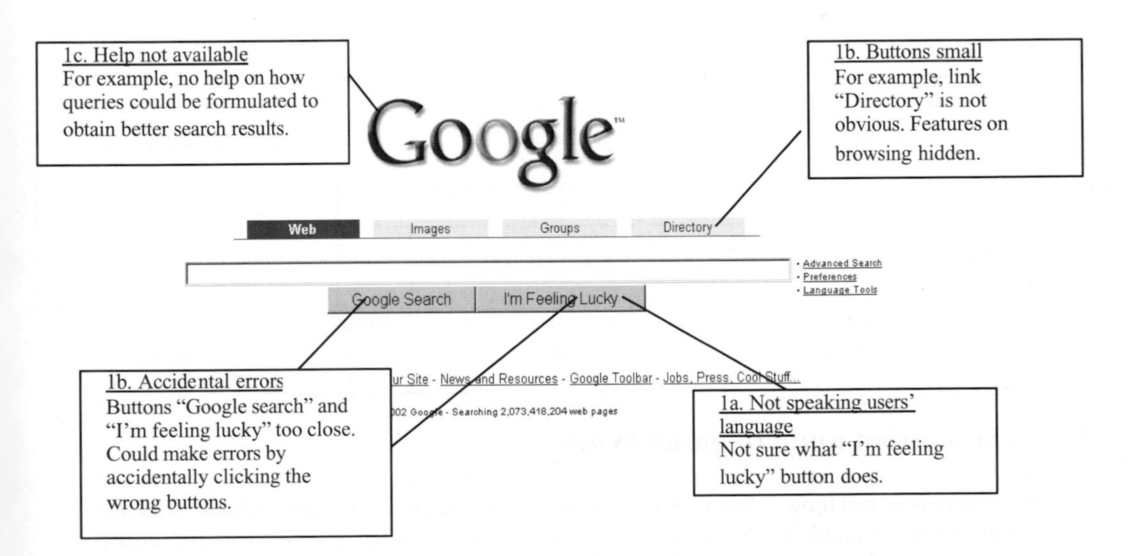

Figure 1 : Interrogating Google Search Engine for negative consequences (some examples) in the Goal Stage (Stage 1a)

4 Recommendations for New Heuristics

By categorizing students' comments in tables similar to Table 1, a list of claims with positive outcomes for the five stages in relation to design heuristics was generated. Besides the heuristics proposed by Nielsen (1996; 1999a; 1999b; 2002), the students' comments were collated from *positive consequences* to suggest recommendations for new heuristics such as : (1) displaying hierarchy of search results; (2) indicating the number of search results; (3) providing "search within search results"; (4) asking for user feedback and satisfaction; (5) providing abstract of articles; (6) giving suggestions to refine search; (7) reminding search goal by making query term(s) remain in textbox; (8) making query terms bold in search results; (9) providing status of search results; (10) providing information indicating date of publication; and (11) providing links to related categories.

Unless properly dealt with, negative consequences/risks could potentially affect usability of a system (Theng et. al., 2002). Table 2 identifies new heuristics (see Column 2) resulting from *negative consequences* (see Columns 1, 3) detected during the 5 stages such as : (i) making features visible; (ii) making buttons of reasonable size; (iii) ensuring clarity and avoiding ambiguity of terms used; (iv) limiting reasonable set of search results; (v) providing help to novices for query formulation; and (vi) making labels/names meaningful to avoid memory overload.

Table 2 : New heuristics from negative consequences of Google search engine design

Design mistakes identified by subjects	New design heuristics	Negative consequences (in Stages S1 – S5)				
		S1	S2	S3	S4	S5
Features "hidden"	Make features visible.			X	X	X
Buttons too small	Make buttons of reasonable size.			X	X	X
Buttons "misleading"	Ensure clarity and avoid ambiguity of terms used.	Novices confused over 2 buttons : "Google search" and "I'm feeling lucky".				
Too many search results returned	Limit reasonable set of search results. Allow the users to specify the number.				X	X
Query formulation difficult for novice users	Provide help to novice users to formulate and re-formulate queries.				X	X
Buttons/categories/directories not meaningfully labelled	Make labels/names meaningful to avoid memory overload.	Directory not immediately evident. Does not encourage users to import pre-existing goal of "browsing"				

5 Conclusion and On-going Work

In our experience teaching Claims Analysis to undergraduate and postgraduate students in universities in the United Kingdom and Singapore, students often found the initial concepts difficult to grasp. However, once understood, students discovered Claims Analysis useful in providing them with a framework to justify their design rationales. As in most HCI techniques, Claims Analysis required a competent level of "craft skills" to administer well, thus making it difficult to use in practice. This paper attempted to make explicit the use of Carroll's Claims

Analysis and Nielsen's Web heuristics to detect usability problems, and suggest recommendations of new heuristics for effective search engine design. This is initial work. Certainly, more work needs to be done to further refine, test and use them in real-world situations before they emerge as principles for search engine design. On-going work involves evaluating different search engines with different groups of users and designers.

6 Acknowledgements

The author would like to thank 2001/2002 and 2002/2003 students from the Masters of Information Studies at Nanyang Technological University (Singapore) for their feedback.

References

Allison, M. and Hammond, N. (1989). A learning support environment : The Hitchhikers' Guide. *Hypertext : Theory into practice. Intellect Books* (pp. 62-74).

Blandford, A., Stelmaszewska, H. and Bryan-Kinns, N. (2001). Use of multiple digital libraries : A case study. *Proceedings of First ACM and IEEE Joint Conference in Digital Libraries*, Ronaoke (Virginia), 179-188.

Campagnoni, F. and Ehrlick, K. (1989). Information retrieval using a hypertext-based machine. *Communications of the ACM*, 31(7), 271 – 291.

Carroll, J.(2000). Making Use: Scenario-based Design of Human-Computer Interactions. The MIT Press.

Jones, S., Cunningham, S. and McNab, R.J. (1998). An analysis of usage of a digital library. *Lecture Notes in Computer Science: Research and Advanced Technology for Digital Libraries, Second European Conference ECDL'98*, 261 – 277. Springer-Verlag.

Lopez, M., Rodriguez, M. and Hidalgo (1999). Using and evaluating user directed summaries to improve information access. *Lecture Notes in Computer Science: Research and Advanced Technology for Digital Libraries, Third European Conference ECDL'99*, 198 – 214. Springer-Verlag.

Maurer, H. (1996). HyperWave : The Next Generation WEB Solution. Addison-Wesley.

Nielsen, J. (2002). Top Ten Web-Design Mistakes of 2002, Alertbox, December 23, 2002. Retrieved February 5, 2003, from http://www.useit.com/alertbox/990503.html

Nielsen, J. (1999a). Ten Good Deeds in Web Design. Alertbox, October 3, 1999. Retrieved February 5, 2003, from http://www.useit.com/alertbox/991003.html

Nielsen, J. (1999b). The Top Twenty Mistakes of Web Design, Alertbox, May 30, 1999. Retrieved February 5, 2003, from http://www.useit.com/alertbox/990503.html

Nielsen, J. (1996). Top Ten Mistakes in Web Design. Alertbox, May, 1996. Retrieved February 5, 2003, from http://www.useit.com/alertbox/9605.html

Theng, Y.L., Goh, H.L., Lim, E.P., Liu, Z., Pang, L.S., Wong, B.B. and Chua, L.H. (2002). Intergenerational Partnerships in the Design of a Digital Library of Geography Examination Resources, *5th International Conference on Asian Digital Libraries, ICADL2002*, 427-439, LNCS 2555, Springer-Verlag.

Overview of Process Trend Analysis

Slim Triki

LAMIH, University of Valenciennes
Slim.triki@univ-valenciennes.fr

Bernard Riera

LAM, University of Reims
bernard.riera@univ-reims.fr

Abstract

In modern process plants controlled by distributed control systems, the role of operators has changed from being primarily concerned with control to a broader supervisory responsibility: analysing operational data, identifying unusual conditions as they develop and responding rapidly and effectively by taking corrective actions. This is a challenging task because of the amount of data operators have to deal with. In recent years there has been a significant progress in applying intelligent systems based on trend analysis method for process monitoring and diagnosis. However, it is recognised that in such process dynamic trend signals are often more important than variable value at the current sampling instant, indeed the qualitative behaviour of the plant is of interest rather than sequence of measured data of high sampling rate. Computer based processing of dynamic trend signal is aimed at noise removal and dimension reduction using minimum data points to capture the features characterising the trend signals. Data are reduced by removing redundant components while preserving, in some optional sense, information which is crucial for pattern discrimination. In addition, process trend analysis softwares have to be seen as tools for operators. The idea is to supply useful information to the operator to facilitate his perception of process state. This paper surveys trend analysis methods and their applications in process industry. Basic principles of several methods are presented.

1 Introduction

For slow process, temporal reasoning is a very valuable tool to diagnose and control the process. Human process supervision relies heavily on visual monitoring of characteristic shapes of changes in process variables, especially their trends. Although humans are very good at visually detecting such patterns, for control system software it is a difficult problem. Much of the utility of collecting data, comes from the ability of humans to visualise the shape of the (suitably plotted) data, and classify it. Unfortunately, the sheer volume of data collected means that only a small fraction of data can ever be viewed. Industrial plants have therefore been called, "data rich, but information poor". The problem of extracting valuable information from times series data has received considerable attention in other fields as indicated by growing interest in data mining and knowledge discovery problems (Apté, 1997; Agrawal et al. 1998). Several high level representations of time series have been proposed, including piecewise linear representation (PLR), wavelet, symbolic mappings, etc. The aim of this paper is to present basic principles of systems that are able to detect meaningful temporal shapes from process data, analyse them and use this information in process monitoring, diagnosis and control. Researchers with different background, for example from pattern recognition, digital signal processing and data mining, have contributed to the trend analysis development and have put emphasis on different aspects of field. At least the following issues are characteristic to the process control viewpoint employed here:

- Methods operate on process measurements or calculated values in time series which are not necessarily sampled at constant intervals.

- Trends have meaning to human experts.
- Application area is in process monitoring, diagnosis and control.

In the following section, the most promising methods are presented based on an extensive literature research.

2 Trend Analysis Methods and Applications

2.1 Hierarchical Piecewise Linear Representation (HPLR)

Piecewise Linear Representation (PLR) of time series data is a good technique to reduce the complexity of the raw data. Obviously this is an approximation of the original data, so it should try to minimise the approximation error. The segmentation problem can be framed in several ways (Koegh et al., 2001) :

- Given a time series S, produce the best representation using only K segments.
- Given a time series S, produce the best representation such that the maximum error for any segment does not exceed some user-specified threshold, *max_error*.
- Given a time series S, produce the best representation such that the combined error of all segments is less than some user-specified threshold, *total_max_error*.

In figure 1 we give an example of hierarchical piecewise linear representation of time series data *S*. *S* is broken into two "linear" segments, the upward A_1 and the downward A_2. In principle, we can use A_1 and A_2 to approximate S. The approximation error may be large, but A_1 and A_2 give us a general idea of the shape of S.

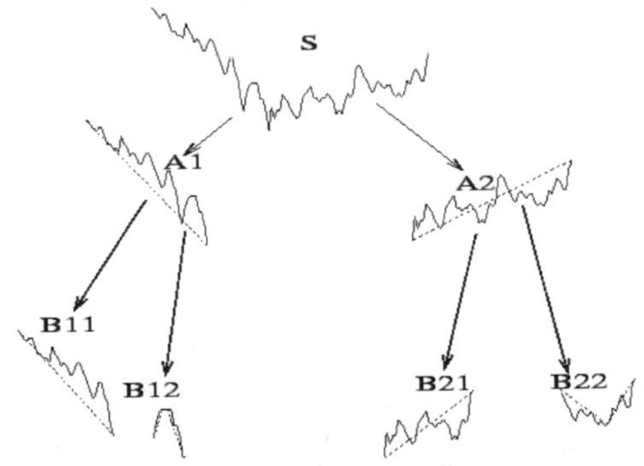

Figure 1: Hierarchical piecewise linear representation of time series data.

To give more detailed description, we can further break A_1 and A_2; method followed here consists of breaking each segment into two "linear" segments and this process is repeated until the desired "precision" is achieved. PLR allows explicit information of significant information of a trend. Qualitative features like "increasing" or "decreasing" could be directly expressed by the type of matched slope shape. (S. Charbonnier et al., 2002) applied an on-line segmentation algorithm developed to pre-process continuously monitored data in Intensive Care Units, in a purpose of alarm filtering.

2.2 Qualitative trend analysis and triangular episodic presentation

Qualitative trend analysis utilises the trend information present in sensor measurements. There are two basic steps: identification of trends in measurements and interpretation of trends in terms of fault scenarios. The identification procedure should be robust to momentary variations (due to noise) in the signal and should capture only important variations. Filtering might distort the essential qualitative character of the signal. (Janusz & Venkatasubramanian, 1991) developed an episode approach that uses nine primitives to represent any plots of function. Each primitive consist of the signs and the first and second derivatives of the function. Therefore, each primitive gets the information about whether the function is positive or negative, increasing, decreasing, or not changing and the concavity. An episode is an interval described by only one primitive and the time interval the episode spans. A trend is a series of episodes that when grouped together can completely describe the dynamic feature. The approach automatically converts on-line sensor data to qualitative classification trees. (Cheung & Stephanopoulos, 1990) developed a slightly different approach called triangular episode that uses seven triangle components to describe a dynamic trend. (Bakshi & Stephanopoulos, 1996) used wavelet decomposition of functions in deferent scales and zero-crossing of wavelet derivatives to find the inflections of decomposition. In this way, episode can be identified automatically by computers. Based on episode analysis, dynamic trends can be interpreted as symbolic representations.

2.3 Dynamic Time Warping

Dynamic time warping (DTW) uses dynamic programming to align the time series under study and a given template so that some total distance measure is minimised (Rabiner & Juang, 1993). This technique has been widely used in the area of speech recognition. DTW algorithm aligns the time axis of a template *Y* with the time axis of observation sequence *X*: $X = x_1, x_2, ..., x_i, ...x_n$ and $Y = y_1, y_2, ..., y_j, ...y_m$. A warping path, *W*, aligns the elements of *X* and *Y*, such that the selected distance measure between them is minimised. $W = w_1, w_2, ...w_p$, is a sequence of grid points, where each w_k corresponds to a point $(i, j)_k$. (Kassidas *et a*l., 1998a) have applied DTW to batch process trajectories synchronisation in order to reconcile timing differences among them. They used data from an industrial emulsion polymerisation reactor to demonstrate the method. The same group has also studied off-line diagnosis of faults in continuous processes using DTW.

2.4 Wavelet

Recently, wavelet based signal processing methods have gained popularity for example in denoising, data compression and feature extraction. Wavelet analysis preserves both the time domain and the frequency domain information of the signal. (Vedam & Venkatasubramanian, 1997) have developed a wavelet theory based nonlinear adaptive system for identification of trends from sensor data named W-ASTRA. It identifies primitive from sensor data and use them as input to a knowledge base to perform fault diagnosis. They demonstrate the application of the system on a simulated fluidised catalytic cracking unit (FCCU). The results showed that W-ASTRA could perform accurate diagnosis of all the fault cases included in historical data base and can notify the occurrence of unknown faults. (Flehming et al., 1998) have applied wavelet based approach to identification and localisation of polynomial trends in noisy measurements. Their method yields the least-squares polynomial for the identified intervals and a quantitative measure for their goodness of fit. Subsequent to a wavelet transformation and denoising of the measurement, candidate intervals are identified by a hierarchical search in the time-frequency plane. The coefficients of the trend polynomials are determined by least squares approximation in the time-frequency representation. The authors conclude that the algorithm is efficient enough to operate with moving time windows in on-line applications.

2.5 Qualitative temporal shape analysis

(Konstantinov & Yoshida, 1992) have proposed a generic methodology for qualitative analysis of the temporal shapes of process variables. Their procedure consists of three phases: analytical approximation of the process variable, its transformation into symbolic form based on the signs of the first and second derivatives of an analytical approximation function and degree of certainty calculation. At the first step process variable $x_j(t)$ is approximated by a polynomial $x_j^*(t) = c_0 t^0 + c_1 t^1 + ... + c_m t^m$, $t \in [t_1, t_2]$; where m is the order of the polynomial and c_k, $k = 0,...,m$ are the unknown coefficients. To speed up the procedure in real-time environment, the approximation equation: $F^T.F.\{c\} = F^T.\{x\}$ (1), was modified. Matrix F can be simplified to a form where it can be calculated in advance taking into account the polynomial order and the length of the discrete time interval. Consequently, polynomial coefficients can be solved from $\{c\} = (F^T.F)^{-1}.F^T.\{x\} = Q.\{x\}$ (2), where Q is a constant matrix. At the second step feature strings are extracted from the analytical approximation function. The extraction of a sequence of the derivative signs is formally described by the operators $SD1[x_j(t)] = sd1 = (+,-,...)$ and $SD2[x_j(t)] = sd2 = (+,-,...)$, $t \in [t_1, t_2]$. Some simple patterns can adequately de presented only by $sd1$. The qualitative shape of process variable is represented by combining these strings: $qshape[x_j(t)] = SD1[x_j(t)]$; $SD2[x_j(t)] = (+,-,...)$; $(+,-,...)$, $t \in [t_1, t_2]$

The third step, degree of compatibility calculation is formulated as follows:

$$dc = cmp1(sd1, sd1^L).\left(1 - k_1 \frac{cmp2(sd2, sd2^L)}{R} - k_2 \frac{dev}{dev_{max}}\right), \text{where}: cmp1(sd1, sd1^L) = \begin{cases} 0 \text{ if } sd1 \neq sd1^L \\ 1 \text{ if } sd1 = sd1^L \end{cases} \quad (3)$$

Where $cmp2(sd2, sd2^L)$ gives the relative number of the symbols in the second derivative string that do not match. Parameters k_1 and k_2 are user adjustable weights and dev_{max} sets limit to the maximal allowed deviation. From the equation above it can be seen that the first derivative string must match perfectly to give a degree of certainty above zero. In this way it is given a higher priority than to the second derivative. The shape analyser was implemented as a server which runs after receiving a request message from the inference engine. The message contains information about the time interval, shape descriptor and variable. The shape analyser replies with the evaluated degree of certainty. The authors have applied the shape analysing scheme to the supervision of a recombinant amino acid production in laboratory scale. They tested the system in handling of process phase transitions, detection of foaming, automatic termination of the process and monitoring of instrumental failures. Their main conclusion was that the use of temporal shapes together with current values of process variables and their derivatives in a rule based supervisory system provides for a natural representation of process dynamics by eliminating nonessential quantitative details and allows for more abstract and robust knowledge representation.

3 Concluding remarks

This paper has been devoted to presenting some promising approaches to process trend analysis. Although the methodological background varies considerably in the presented approaches, certain common features can be found:

- Most of the authors refer to the emulation of human perception and reasoning capabilities in motivating their work.
- Trend analysis operates as a complementing, not as stand alone system and normally provides information to some reasoning mechanism.

- Computational efficiency issues play an essential role in development towards on-line systems.
- No fine design methodologies can be found.

One perspective is to integrate trend analysis with temporal reasoning systems. Another direction for further research could be the investigation of possibilities of trend information in supervisory control. Indeed, we note that today SCADA systems do not integrate advanced tools for trend analysis. In addition, there are very few papers about evaluation with operators. However, there are a lot of works about "ecological" interfaces enabling theoretically a direct perception of process state by HO. We would like in the future to make the link between these two domains.
In conclusion: Trend analysis has found its place in process monitoring, diagnosis and control. A lot of work has been performed in the definition and the improvement of methods and algorithms. Now, it seems necessary to propose and to evaluate (with operators) advanced supervisory systems integrating these new tools.

References

Agrawal, R., P. Stoloroz and G. Piatetsky-Shapiro (eds.). *Proc. 4th Int. Conf. on Knowledge Discovery and Data Mining*. AAAI Press, Menlo Park, CA (1998).

Apté, C. Data Mining: An Industrial Research Perspective. *IEEE Trans. Computational Sci. Eng.*, 4(2), 6–9 (1997).

Bakshi, B. R., & Stephanopoulos, G. (1996). Reasoning in time: modelling, analysis and pattern recognition of temporal process trends. In G. Stephanopoulos, & C. Han, *Intelligent Systems in Process Engineering-Paradigms from Design to Operations* (pp. 487–549).

Cheung, J. T. Y., & Stephanopoulos, G. (1990). Representation of process trends-1: a formal representation framework. *Computers and chemical Engineering*, 14, 495–510.

Flehming, F.;Watzdorf R. V.; Marquardt,W., 1998, "Identification of Trends in Process Measurements Using theWavelet Transform", Computers & Chemical Engineering, Vol. 22, pp. S491–S496.

Janusz, M. E., & Venkatasubramanian, V. (1991). Automatic generation of qualitative descriptions of process trends for fault detection and diagnosis. *Engineering Applications Artificial Intelligence.*, 4, 329–339.

Kassidas, Athanassios; MacGregor, John F.; Taylor, Paul A., 1998, "Synchronization of Batch TrajectoriesUsing Dynamic Time Warping", AIChE Journal, Vol. 44, No. 4, pp. 864–875.

Keogh, E., Chu, S., Hart, D., Pazzani, M. (2001). An Online Algorithm for Segmenting Time Series. The 2001 IEEE International conference on Data Mining.

Konstantinov, Konstantin B.; Yoshida, Toshiomi, 1992, "Real-Time Qualitative Analysis of the Temporal Shapes of (Bio)process Variables", AIChE Journal, Vol. 38, No. 11, pp. 1703–1715.

Rabiner, Lawrence; Juang, Biing-Hwang, 1993, "Fundamentals of Speech Recognition", Englewood Cliffs/NJ, USA. Representations for Large Data Sequences. *Technical Report cs-95-03*, Department of Computer Science, Brown University.

Vedam, Hiranmayee; Venkatasubramanian, Venkat, 1997, "AWavelet Theory-Based daptive Trend Analysis System for Process Monitoring and Diagnosis", 1997 American Control conference, Piscataway/NJ, USA, pp. 309–313.

Charbonnier S., Becq G., Biot L., Carry P., Perdrix JP Segmentation algorithm for ICU continuously monitored clinical data 15th World IFAC congress, July 21-26 2002, Spain.

A Study of a Southern California Wired Community: Where Technology Meets Social Utopianism

Alladi Venkatesh, Steven Chen, and Victor M. Gonzales

Center for Information Technology (CRITO)
University of California, Irvine, CA 92697
avenkate@uci.edu

Abstract

The purpose of the study is to examine the role of Information and Communication Technology (ICT) in the creation, organization, and functioning of a networked community. The reference community is Ladera Ranch located in Orange County, South of Los Angeles.

1 Introduction

In the last few years, there has been a growth of networked or wired communities in some selective parts of the US, and also in the UK and Europe (Dohoney-Farina 1999, Horan 2000). Both from a technological and social point of view, such communities have operated with varying degrees of success. In order for these communities to function as viable social settings and derive the benefits from new technologies, several factors come into play: A well understood concept of family and community networking, the technological readiness of families, the quality of services, advanced technical support of developers, high bandwidth into the homes, sustaining financial backing, continuous user interest, updating of technology as new systems appear, and a host of other social and technical factors. Experience shows that some wired communities may have failed because the standards of performance on these various factors may not have been met.

One wired community that is attempting to succeed at all costs is Ladera Ranch, located in South Orange County, halfway between Los Angeles and San Diego.

2 Study Purpose

The purpose of the study is to examine the role of Information and Communication Technology (ICT) in the creation, organization, and functioning of a networked community.

3 Background Description of the Wired Community and Key Issues

Ladera Ranch is a planned/wired community that came into existence in 1999 and the first set of residents began to move into the new housing in early 2000. There are currently 2000+ families living in Ladera Ranch which is expected to grow to 8,000 homes within the next 5/7 years. All the homes in Ladera Ranch are completely hard-wired for a variety of communication devices. Practically every room in the home is Internet/cable ready. All the communication infrastructure and services are provided by COX Cable, a major player in their industry.

Here is how Ladera Ranch promotion material introduces the community to prospective dwellers on their website. (http://www.laderaranch.com/homes)

"With the rugged beauty of protected Rancho Mission Viejo lands as a backdrop for the best new homes in Orange County, life at Ladera Ranch reflects the independent spirit of the American West while nurturing the diversity of the Southern California Style."

The concept driving the planned community is inscribed in "Ladera Life" which is the marketing slogan. The developers of Ladera Ranch have apparently learned from the successes and failures of similar start-up communities nationally and have meticulously planned their own version of a community where the idea is a) the latest technology should be available to the residents, b) the technology should be user friendly and the residents should be able to master it in a very short time, and c) no matter how advanced or sophisticated the technology is the ultimate success of the community lies in achieving its community oriented goals. Over the last two years, as the community has begun to expand and homes have been susceptible to great demand and appreciated in value, the Ladera Ranch management has been putting less and less emphasis on technology component as a marketing tool while still paying attention to it from an infrastructure and amenities standpoint. This community orientation represents some sort of social utopianism of the suburbia that dates back to Leavittown more than fifty years ago, and more recently to California's experiments with various planned communities.

The research question facing us is how is Ladera able to blend the promise of new technology with social utopianism of sorts. What are the perceptions and behaviors of the residents in regard to their participation in a technology driven networked community? How successful is this community both from a technological point of view and from a social angle?

4 Research Method and Analysis

This is an in-depth study of 25 households and their use of computers and communication technologies in their homes. The study is a work in progress as more families are being added to the sample. Research methods include long interviews, ethnographic observation and monitoring of an online community forum. Community ethnographies date back to the pioneering work of Herbert Gans (1967) in the New Jersey town of Levittown. In terms of wired communities, some recent work has been reported by Carroll and Rosson (1996) in Blacksburg, Virginia, and Hampton (2001) in Toronto, Canada.

4.1 Findings

In this short space, we can only highlight some of the important findings. In the following, we limit our discussion to the community Intranet. (A more complete document and research report can be obtained from the first author.)

4.1.1 Reasons for moving into a wired community

Different families have moved into the wired for different reasons. One important theme that emerges from our study is that technology is only one aspect of the wired community life. People move into the community also because of its physical attraction, its location near a mountainous region and road access to the neighborhood/city attractions and workplaces. The community is perceived as children friendly because of all the amenities and the freedom and the safety with which children can run around with minimum supervision within the immediate neighborhood. This is indeed a major concern of families living in the urban regions of America. What the developers have successfully done is to market the community on the basis of the latest technology but by promising traditional community values and living patterns.

Thus many of our respondents felt that the park like environment, the scenic background, the walking trails, play grounds for children, and the ability to take their dog for walk are the major reasons for moving into the area. The technology is simply a bonus.

4.1.2 Intranet as the Centre-Piece of the Community

The center-piece of the wired community is the Intranet. Each household has access to the community Intranet via a dedicated password. The intranet serves different purposes. First, it creates a public forum where members can exchange their view and ideas and offer suggestions and critique on various community related issues. These issues can range anywhere from such mundane topics as trash collection to more serious problems concerning traffic congestion to the planning of the next shopping mall. A second purpose of the Intranet is to permit members to start their own interest groups and social clubs. Thus there are wine testing clubs, moms with infants, bridge groups and so on. Some of them are more active than others.

We observed various behavior patterns in terms of the Intranet use. We found that there are some community regulars who are always on the Intranet and dominate the discussions. Some families complained that those people who dominate the discussion boards may be the reason why there is not a more even handed expression of views. There are others who go on the Intranet only when the topic is of interest to them and when they are directly affected by related developments. So, for example, people in the immediate vicinity of a shopping area were more vocal about some of the plans that affected them directly by management's decisions concerning those matters. A large number of the community members are interested in following the Intranet material instead of actively participating in the discussions. They feel a need to be aware of the happenings and follow the Intranet mostly for the news value and sometimes for the entertainment value when discussions get heated.

The public Intranet is also used for announcing garage sales and exchange of medical information and announcements from the local schools and the shopping areas.

In general, we found that women tended to read or directly participate in the public forums than men. Thus there is a gender phenomenon here. This is particularly true of stay-at-home mothers who are tending to their young children. This is also true of working women. When it comes older retirees, there does not seem to be much gender variation.

4.1.3 Special Interest Groups and Social Clubs

There are many social clubs and interests groups that are active. This is a very important part of the community life. What is interesting is that the technology plays a key role in keeping community members in contact with each other. For instance, MC's (a member in our sample) participation in the Ladera Life Wine Club has exposed him to a handful of new contacts. Correspondences with these people occur over the community website, and if not for the community message boards, MC would never have met this group of people. The wine club contacts can be classified as weak relationships, but they are important in showing the role of technology in a network system. Because MC had access to the community board, he was exposed to a whole new group of people. MC's family/ social network was expanded and shaped by his participation in the community board.

4.1.4 Some Emerging Dimensions of Communication

Based on our study of the wired community we have identified the following dimensions of the Intranet as an emerging communication technology.

- Distant vs. Local
- Family vs Community (Social)
- Urgent vs. Non urgent
- Socialize vs. Informative
- Engaged Dialogue vs. Terse Dialogue
- Personal vs. Casual
- Formal vs. Informal
- Archaic vs. Technical
- Convenience vs. Hassle
- Easy vs. Hard
- Faster vs. Slower
- Protected/ Secure vs. Unprotected/ Insecure
- Vulnerable vs. Invulnerable
- Intimate vs. Non-Intimate

4.1.5 Network Issues – Some General Observations

From the social dimensions and emerging trends, certain patterns begin to coalesce. Three large patterns of technology and communication were identified: Network Growth, Network Management, and Negotiation and Empowerment.

4.1.5.1 Network Growth

Communication technology and the human decision to employ it expand the social network in two ways: by expanding the emotional web of social and by making it more feasible for new relationships to enter the network. With the onset of computer-based communications, individuals can send an e-mail, nice and quick, and maintain a relationship without the time and emotional investment of writing a handwritten letter or making a phone call. Technology has offered new ways to maintain relationships, ways that have new properties that are not afforded by just one form of communication. An individual can now maintain a relationship with someone, be it a strong or weak relationship with forms of communication that offer low and high intimacy. Thus, technology has provided individuals with different levels of bonding (e.g. high and low levels of bonding) in their social networks; the emotional web of an individual's network has grown.

4.1.5.2 Network Management

The Intranet as a communication technology lends itself to Network Management in two ways: by improving the efficacy of scheduling events, and by giving families more security measures.

Having a breadth of communication technology provides more options for families to manage family and community events. For instance, the convenience of scheduling a family party is not more apparent than in an Intranet communication scenario where other families can be contacted through one correspondence. Similarly, participating in community events, like CB's (one of our subject families) children's athletics team, is made more efficient via the intranet. Without the Intranet, families would still do what they do, but there would be a lack of this convenience. The Intranet provides extra security for families. When an individual goes out of the home, he or she can check up on the rest of the family in by firing off a quick cellular phone call, as in the case with JD.

4.1.6 Negotiation and Empowerment

People are empowered by the negotiation process of communication technology via the Intranet. The tech-enabled network allows its members to choose a mode of communication that suits their experiential needs.

Take this scenario: DG needed to contact her neighbor to notify her of something. However, DG was fearful of being "trapped" on the phone , because the neighbor is an absolute "chatterbox." So DG chose to used the Intranet. DG's experiential needs in this case were quickness, a measure of protection from a long, affected conversation. The nature of her correspondence was informative—not a social call. Where the DG-example shows how people can "push-off" on a correspondence, a second example will show the opposite side of the coin—when communication scenarios call for stronger impressions. Of course, having a bevy of communication technologies give people a choice, and having a choice is part of an empowered state of being.

5 References

Carroll, J., & Rossen, M.B (1996), Developing the B;lacksburg Electronic Village, Communications of the ACM, 39 (12), 69-74.

Dohoney-Farina, S (1999). The Wired Neighborhood, Yale University Press.

Hampton, K. N (2000). Living the Wired Life in the Wired Suburb: Netville, Glocalization and Civic Society, Unpublished PhD Dissertation, University of Toronto.

Horan, T (2000). Digital Places:Building Our City of Bits, Urban Land Institute, Washington D.C.

Gans, H (1967). The Levittowners: Ways of Life and Politics in a New Suburban Community, New York: Columbia University Press.

Collaborative Architectural Design Supported by an Information and Knowledge Pump

Johan Verbeke

Hogeschool voor Wetenschap & Kunst
Sint-Lucas Architecture
Paleizenstraat 65-67, B - 1030 Brussels
Belgium
jverbeke@archb.sintlucas.wenk.be

Marnix Stellingwerff

Delft University of Technology
Faculty of Architecture
Postbus 5043, 2600 GA Delft
the Netherlands
m.c.stellingwerff@bk.tudelft.nl

Abstract

This paper describes a theoretical framework for Collaborative Architectural Design, pointing out the main issues requiring attention in future developments. Architectural Design has become a complex collaborative work with a strong relation to place and time. Moreover, collaborative work is not limited to the building phase but nowadays extends to the full lifespan of the building. Buildings get value and meaning by the use, the environment and their capacity to remain useful over time. The paper takes into account knowledge management issues, translates these issues to the domain of architecture and discusses how an 'information pump' may help to support a design team. The paper discusses how the system may be developed and where the focus for next research efforts is placed.

1 Collaborative Architectural Design

Architectural Design, as a complex field of study, is often viewed and researched in abstract and simplified ways. A common understanding of the design process comes from (Zeisel, 1984). In *'Inquiry by design: Tools for environment-behaviour research',* Zeisel presents a 'design development spiral' consisting of empirical knowledge which is refined in cycles of *image-*, *present-* and *test-actions*. The cycles become fruitful if they lead to distinct conclusions and decisions. However, in time, the character of the design spiral changes. Different 'stages' of the design process are mentioned (Cross, 1989), (Goel, 1995): analysis of the problem, concept generation, preliminary design and detailed design. A general investigation of the diverse publications in the field reveals three distinct phases where the focus is on: design in early stages, collaborative design and completion of the design specifications (see numbers 1,2,3 in figure 1).

During the early design stages several explorative steps are to be set. Problems of a design task are initially 'ill-defined'; more then one solution is possible. The first ideas, sketches and possible solutions are freely concurring with each other. The architect, or in most cases a whole team of different experts, works (metaphorically) in a 'solution space' that develops from initially 'dream-like' ideas and unresolved visions into a more definite set of documents, which describe the design 'object' as precise as possible. This solution space can be defined as the addition of all thoughts and wishes in the minds of the design team, the client and other people involved, including the set of expressed sketches, models and schemes.

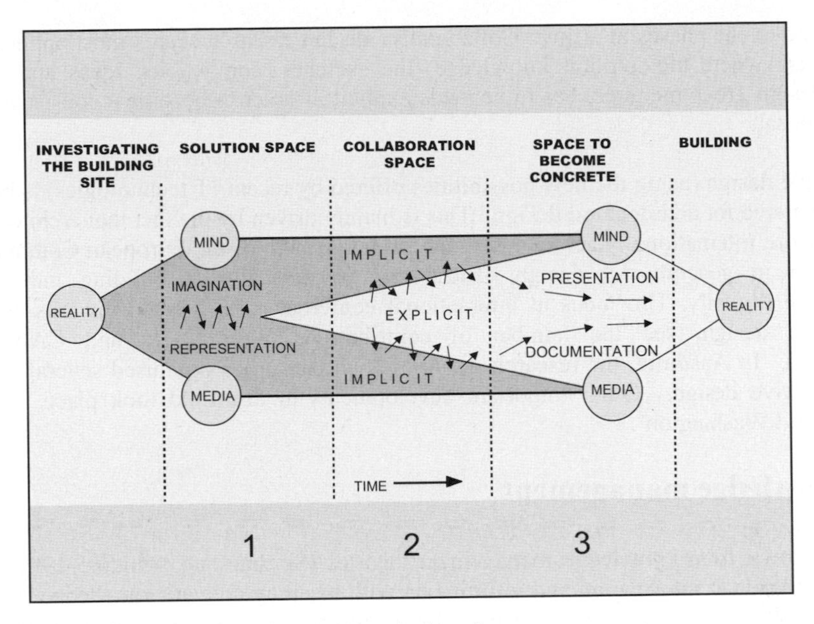

Figure 1: design phases.

At the start of the design process the solution-space seems unlimited. Later on several fruitful solutions might be distinguished, other initial ideas and visions have to be rejected. At a certain moment a design concept can become 'generative' and solutions seem to fit according to the 'rules' of the design itself. In this respect, two types of solution-moves or transformations are identified (Goel, 1995): lateral and vertical transformations. Lateral transformations are real changes from one idea to a different one (from a to b), vertical transformations are developments of one single idea (from a to a+). In general, it is beneficial to the design process if lateral transformations are made before the vertical transformations take place; first exploration, then concretisation.

In the early stages of design, the design representations are still open to debate, they should be produced and exchanged in an open way without prejudgements and without too many constrains. Later on, a phase of collaboration can take place. Each member of a design team influences the ideas with personal knowledge. The most fruitful alternatives can be selected. The design representations in this phase become loaded with domain specific information (e.g. from an engineer or a building cost calculator). After selection and adjustments, the implicit aspects of the design become explicit.

In the end of the design process, the ideas must (sometimes literally) become concrete, after all, the design has to be built. The focus is on presentation of definite images and clear documentation for the builders. The final phase of the design process is interesting for research that is focussed on practical issues like standardisation of documents and optimisation of the collaboration processes between building partners. This phase can benefit from research that helps to streamline and control the refinement of design information.

Thus in each phase of the process the roles of architects and their collaborators change. Likewise, it can be assumed that the tools and design-media to create and manage the design should adapt to the changing needs of the designers. Distinct research for design media and working methods can focus on one of the phases at a time. Collaborative design research seems most applicable in the second phase where the implicit knowledge (the sketches, convictions, ideas and wishes) of different design group members has to be made explicit in order to become a consistent and clear design proposal.

Collaborative design (using the new possibilities offered by recent IT technologies) is becoming a key research area for architectural design. This is mainly driven by the fact that architects become more and more internationally active due to the further growth of the European Communities and globalisation in general. All design offices and partners in the building industry try to communicate digitally. The focus at international conferences is directed towards architectural collaborative design (see the number of contributions at eCAADe and CAAD Futures Conferences). In Australia, the research-group of John Gero[1] has organised several conferences on collaborative design. Other important developments in this field took place in Brussels[2], Coventry[3] and Washington[4].

2 knowledge management

It is well known from knowledge management theories (Skyrme and Amidon, 1997), that data with context leads to information and information with meaning delivers knowledge (figure 2). In a collaborative architectural design process a lot of information is exchanged. The information is based on large data sets relating to the design task and the characteristics of a specific building site. In several cases, there is exchange and transfer of knowledge between the partners involved. However, the situation is far from ideal. Most design efforts are isolated processes; mistakes and solutions appear in similar ways in each new design project. Gained knowledge leads to one new building. After completion, the building partners split up and the knowledge is dispersed.

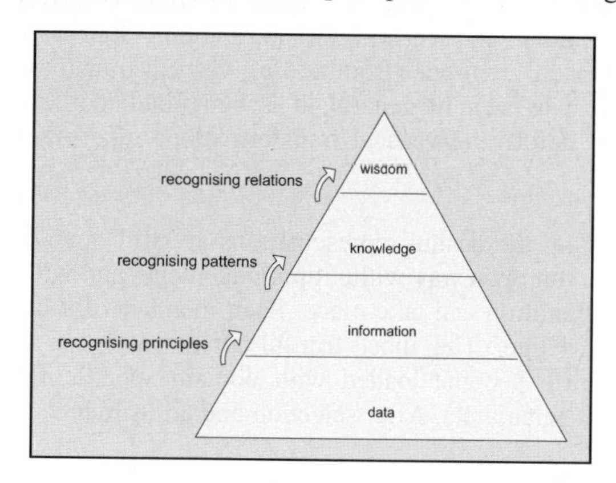

Figure 2: data development pyramid.

[1] http://www.arch.usyd.edu.au/kcdc/ The Key Centre of Design Computing and Cognition, University of Sydney.

[2] European Workshop: {ACCOLADE} Architecture-Collaboration-Design, 28/8-1/9/2000, at Sint-Lucas Brussels.

[3] CoDesigning 2000, S.A.R. Scrivener, L.J. Ball and A. Woodcock, Coventry University, U.K., 11-13/9/2000.

[4] Technology in architectural practice PIA Spring Conference - Six Degrees of Collaboration, April 5-6, 2002, AIA National Headquarters, Washington D.C. http://www.aia.org/pia/tap/conference

An issue of utmost importance, is the distinction between tacit (Polanyi, 1958) and explicit knowledge. While knowledge in each separate design process needs to undergo a development from implicit ideas to concrete explicit documents, there is also a fundamental lack of explicit knowledge in the field of architectural design. The design processes have no 'long term memory'. It is as if the information of a project loses meaning when the building is built and when the next project comes in view. Knowledge has to be situated for new demands, over and over again. Nevertheless, a lot of tacit knowledge is available in practice. SME's, such as design and engineering offices and other partners in the building proccss, such as clients and local governments, have gained powerful 'own knowledge' in subsequent projects. Some explicit knowledge related to Computer Aided Architectural Design has been created (see e.g. ACADIA, CAAD Futures and eCAADe Conference proceedings). Knowledge in the field is growing, but is in need for an integrated approach.

Little attention is paid to the exchange of implicit knowledge and how developments of implicit and explicit knowledge enforce each other in innovation. In terms of Nonaka and Takeuchi (1995), the knowledge spiral is not operational as the interaction between the explicit and implicit creation of new knowledge in the field of architectural collaborative design is still lacking. The proposal of a knowledge pump in this paper intends to react to this situation. Newly established research consortia, such as the {ACCOLADE} group (see Footnote 2), need to bring together competent players, each holding an important piece of knowledge, in order to take a step forward in the field.

3 An information and knowledge pump

We argue that the field of architecture is in need for an information and knowledge pump (van Heijst & van der Spek, 1998). The information pump is seen as supporting the different parties in the building process with a system that handles objective and subjective information and knowledge and also acts in a supportive way. The idea is to have a database in which it is possible to find interesting information while, at the same time, it is possible to comment (annotate) and change that information or to add other information. The 'information pump' is the most difficult database application to start and to maintain, as it depends on the commitment of active creator- and user groups. Ideally, the information gets richer, more reliable and more refined as it is judged and altered by many people. Data in an information pump no longer has one single author. Ideas merge and change during the process. The database functions as a medium that stores, organizes and reflects ideas. The content in an information pump is vivid but the database can keep track of changes in the process. The insights of the users and possible solutions to specific topics develop in an iterative way. Activity (giving, commenting, rearranging, listening) and subjectivity are the fuel to let the information pump do its work.

3.1 Specifications for an information and knowledge pump

It is difficult to describe a new application from scratch, therefore we have undertaken several experiments and made sub-prototypes for distinct features of a possible system. Many of those prototypes originate from experiments during a PhD project: e.g. the exploration of 3D sketching (Stellingwerff, 1999), adaptable context representations (Stellingwerff, 2002), and the enhancement of avatars and knowledge agents (Stellingwerff, 2001). Other inspirations come from the {ACCOLADE} workshop and from several preparatory meetings for a succeeding research project.

In general terms, the information and knowledge pump needs to bridge the gap between the 'world of architectural experience' and the 'world of computation'. The architectural context, explicated knowledge and the design partners can be interactively represented in a Communication Environment (CE). This implies the development of new Human-Computer-Interfaces, the use of CAVE systems or probably an enhanced desk-CAVE and standardized and programmable VR software (e.g. VRML / X3D or Java3D). Besides the technological aspects, there is the need for a thorough understanding of human factors and user scenarios need to be developed. Our ongoing research focuses on Communication Language (CL), Communication Behaviour (CB), Goals and Roles (GR) and Education (E) for Computer Supported Collaborative Work.

3.2 State of the project

The general outline and design for a first prototype knowledge pump has been made. We currently investigate the opportunities of generally available web interfaces to make connections to a MySQL database. Such interfaces become conceivable trough special functions in PHP scripts. See e.g. for an interface between MySQL – PHP – VRML: http://web3d.vapourtech.com/workshop/php-vrml/ and for an interface between MySQL – PHP – Flash: (http://www.flash-db.com/Ex1/.

4 References

Cross N. (1989). Engineering design methods, Strategies for Product Design, Chichester.

Goel V. (1995). Sketches of Thought, MIT Press, Cambridge, MA.

Heijst van, G., van der Spek, R. & Kruizinga, E. (1998). 'The lessons learned cycle'. In Borghoff, U.M. and Pareschi, R. (eds) *Information Technology for Knowledge Management*, Berlin, Springer-Verlag, pp. 17-34.

Nonaka, I., & Takeuchi, H. (1995). The Knowledge-Creating Company: How Japanese Companies create the Dynamics of Innovation, New York, Oxford university Press.

Polanyi, M. (1958). Personal Knowledge: Towards a Post-Critical Philosophy, Chicago, University of Chicago Press.

Skyrme, D.J. & Amidon, D.M. (1997). Creating the Knowledge-Based Business, Wimbledon, Business Intelligence ltd.

Stellingwerff, M.C. (1999). SketchBoX. In: A. Brown, M. Knight & P. Berridge (eds.); *Architectural Computing from Turing to 2000.* 17th eCAADe Conference, Liverpool.

Stellingwerff, M.C. (2001). The concept of carrying in collaborative virtual environments. In: M.C. Stellingwerff, J. Verbeke (eds.); *{Accolade} Architecture-Collaboration-Design.* European Workshop, DUP Science, Delft, the Netherlands.

Stellingwerff, M.C. (2002). Architects' Visual Literacy. In: *New Impulses in Planning Processes, digital / analogue environmental simulation*, Schmidt, Schlömer (editors), proceedings of the 5th conference of the EAEA, Essen, Germany.

Zeisel, J. (1984), Inquiry by Design: Tools for Environment-Behaviour Research, Monterey, Cal: Brooks/Cole, Cambridge University Press.

Communicating in the Home: A Research Agenda for the Emerging Area of Home Informatics

Vivian Vimarlund and Sture Hägglund

Department of Computer and Information Science
Linköping University
vivvi@ida.liu.se; stuha@ida.liu.se

Abstract

Home Informatics is a multidisciplinary research area concerned with everyday IT services and communication in the home. From an economic point of view it is imperative that consumers' preferences are taken into consideration to develop cost-effective technology for home communication. Examples presented illustrate the challenges we are confronting when developing IT-based services and products for work from home, everyday interaction with appliances and services and for home healthcare.

1 Introduction

There is today considerable interest in the use of technologies for networked communities and for the use of IT[1] for communicating at home or in domestic environments (Miles 1987; Venkatesh 1996). Technology vendors and research groups are competing to suggest how information infrastructures and terminals can be built to provide services at home. Studies both in the USA and in Europe has reported that the interest for developing technologies permitting a more interactive role for the domestic user in performing tasks ranging from shopping, to banking, to communication with the work place and with the production of healthcare services has increased during the last two years (Kuhn 2001). In Europe, for instance, the EU has funded hundreds of IT projects that could improve, and perhaps change significantly how services will be provided in the near future. However, the area is still diffuse and it still remains to be understood how to build cost-effective technology that satisfies the needs of individuals and the society as a whole.

2 Home Informatics

Presently, many forms of information carriers are used in the home from simple handwritten notes, through printed documents to full audio-video information in the form of television broadcast (Sloane 2000). The research reports in this area have often been focused on the private home life and its "electronics". But rather than being concentrated on how or what IT applications are used or demanded to facilitate everyday tasks, the discussion has concentrated on what kind of state-of-the-art IT can be implemented and used in future homes. Furthermore, it is often assumed that individuals have at least a computer at home, and that information structures are used at all

[1] The term IT is used to denote all kinds of information systems and computer technology that have the capacity to collect, store, process and report data from various sources to provide information necessary for management and decision-making.

levels to move from public spheres (e.g. the work place) to more private ones (the homes). The term Home Informatics has consequently been used to refer to and for study of the applications of state-of-the-art IT products that are emerging for use by individuals of private households (Miles 1986). It covers not only items of hardware like home computers and new consumer electronic goods, but also both the software that programmes this equipment, and the services – like online information systems – that may be used with the hardware and software. Hereby, more than often, current research in the area of home informatics has mainly studied one side of the equation, the market space perspective and thus the supply side. However, the real challenge in the area lies in the consumer perspective, understanding factors that promote user acceptance and economic feasibility.

3 Selected Topics in Home Informatics Research

In our studies related to the area of Home Informatics at Linköping University we have emphasized the everyday use of information technology and appliances for computing, discussing issues as what kind of IT innovations individuals prefer for communication at home, why households will or not will use IT for communication at their private residences, and the effects of the use of IT for home communication. In this sense, the area is contrasted with Informatics in general, which has had an emphasis on work-life systems and professional use of IT at workplaces. For the purpose of this paper, we have selected examples from on-going research that help to illustrate the potential of the area of home informatics, the importance to understand the ecology and market forces for everyday IT products and finally, the importance of the identification of factors that influence the acceptation of technology for use at home.

3.1 Working from Home through Tele-working

In this area the research performed has been directed to study how home workers use their time and what usually characterises a "tele-worker". The main findings from studies related to this area indicate that tele-working, despite the fact that it has become increasingly popular, often is performed during half days, and therefore does not substitute the neither the frequency of travels nor diminish the quantity of cars on the roads, but rather dislocates it to less congested times of the day. Another interesting result is that tele-working usually is practiced during a time-limited period, often once or twice a week, and it is restricted to a number of "white colour" works, and thus more extensive in big cities. The common tele-worker is usually characterized as a highly educated individual with a high degree of responsibility in work tasks, with high income and is often a male. It is however, important to note that an indirect effect as a consequence of working from home has been observed, namely the increase of service trips, shopping trips, or leisure trips during working time at home. This finding would suggest that working at home implies being available in the home environment and thus a number of "home activities" are expected to be performed for the individual who "stay" at home. Another important indirect effects identified has been the fact that "tele-workers" buy cars to a higher degree than regular workers. Households that work from home even tend to own more than one car, increasing the possibility that other members of the family who normally use public transport or bicycles, actively use the " second" car.

The results obtained in this area seem to indicate that the effects of working from home have given raise to spill-over effects such as for instance the increase of home-related activities and increase of "non-working activities" during working time, as a consequence of being able to spend the entire day as pleased. However, before drawing any conclusion regarding the cost-effectiveness of tele-working, further studies have to be performed in order to study the real benefits and costs of working at home, as well as the benefits of being a "digital company". It is possible that the benefits are nevertheless higher than the costs when individuals have the possibility to make choices in accordance to their preferences (Lancaster 1996)

3.2 Home Healthcare

Our research in this area focus on how to use IT to support home healthcare and medical services for the elderly, with an emphasis on the empowered citizen/patient. Here we are running field studies involving mobile healthcare teams and IT support for supervision of health status and chronic disease symptoms from the home, where security, trust and accountability constitute important issues. An interesting trend in the area is the potential for integration of consumer products for communication and information processing with professional medical equipment.

However, four major issues requiring improvements for an efficient use of IT for home healthcare have been identified. a) systems failures and disconnection, b) complicated interfaces, c), lack of specific knowledge of using central functions when IT innovations are implemented, and d) environmental issues. Systems failures and disconnection caused by e.g., missed alarms or by interference, as well as complicated interfaces, are today considered a major problem in home healthcare. This is because the consequences of the use and/or implementation of a non-effective IT are often loss of confidence, anxiety, suffering from confusion and stress both for healthcare personnel and patients, thus diminishing of life quality for the patients who choose to receive healthcare services at home.

Issues related to the lack of specific knowledge on how to manage IT have also been noticed because they can raise unexpected effects and need consequently to be solved to avoid decrease in available time to assist patients or decrease in the service production. Experiences from research on the effects of end-users participation in the development of health information systems has shown that involving end-users during the whole IT development process, stimulate a learning by doing process and thus induce to a better knowledge and management of the final system when it is implemented (Vimarlund &Timpka 2002), Moreover, the acquisition of practical knowledge and skills in IT has been proven to be the key to reducing errors when dealing with IT innovations, increasing the individual work productivity and allowing more adaptability to new challenges that IT-innovations require. More difficult to solve are however environmental risks factors, for instance when adapting the home space to deliver home healthcare. This is due to the fact that solving this issue presupposes a well-working infrastructure and especially the integration of the social space with the technological space (Venkatesh, 1996). The future of home healthcare is thus apparently dependent on the well-functioning interaction between technology and a number of organizational and social dimensions, such as individuals preferences regarding IT, an appropriate interface, the level of sophistication in the use of IT and special environmental factors that have to be managed to ensure growth of this area.

3.3 Everyday Interaction with Appliances and Services

The combination of interaction modalities for optimal convenience, robustness and ease-of-use when manoeuvring home appliances and information services is of paramount importance. Issues studied in our research span from language technology and novel input devices, to support for a connected dialogue and empirical studies of user preferences, situated with respect to various usages and user contexts.

For the future, there are several possible paradigms for how we will interact with external information resources and services in the home. There is a competition today between the PC, the phone and the interactive television paradigms. In our research, we have also studied attitudes and preferences when a multitude of services are offered though the TV connected to a settop box terminal. Speech interaction can be an alternative interaction technique for interactive and digital television (Ibrahim & Johansson 2002). Our investigations indicate that multi-modal input can provide convenient and robust interaction by combining natural language, spoken input and direct manipulation. In the interactive TV experiments we found that the output should be displayed visually rather than audible and that a three-entity model recognizing the distinction between three entities: the user, the system, and the domain data was useful.

More and more people are now interacting with computers using a language that is not their mother tongue. For instance, large numbers of non-native English speakers use English versions of software. But will the fact that these users are proficient enough to understand and express themselves in the English used in these situations make the cultural differences disappear? (Dahlbäck et al 2001) The goal with our research in this area is also to see how the end user receives interaction with intelligent applications; especially the help that might be needed to understand complex system. In summary, our research clearly indicates that multi-modal interaction provides important opportunities for convenient home communication. To design such systems it is necessary to develop methods and techniques that allow the combination of end-user participation and early tests of new products. To model and predict optimal and cost-effective interactive home services is an exciting challenge for the area of multi-modal interaction.

4 Discussion

Consumer behaviour and its implications for the supply side have been studied from rather diverse perspectives in several sub-disciplines of economic theory e.g., managerial economics contributing the results to important strategies for marketing, product development and information policies. The Home informatics market does not differ from other markets and therefore the success or failure of IT applications will depend, as in other market spaces, on the utility consumers derive from consumption of services offered for two different periods, the near and distant future respectively. In the near future, the consumption of IT for working at home or to receive home healthcare, will always be influenced by the price effect. Individuals will, as a consequence of a price effect, allocate more time to use of IT when available technology becomes cheaper. In the more distant future, however, the consumption, often called latent, will depend on increased incomes. Potential consumers of home robots, interactive TV, IT-supported home healthcare services, or IT for work at home, will be those that put a high value on information services for improving of everyday life and communication at home.

To identify groups of consumers that bear some similarities e.g., preferences regarding latent consumption patterns, is crucial for the area of home informatics. This is due to the fact that if consumers have positive attitudes to some IT-based products and services, they are more likely to accept, use and buy the products when they are finally produced and available on the market. However, it is also important to differentiate between early and late adopters even in this area. Almost all research today is directed to study how early adopters, who are usually more educated people and thus with more economic resources, use and adopt innovations at home. If home informatics products will sell, it is crucial to concentrate on later adopters, their willingness to accept and their willingness to pay for IT for home communication.

5 Summary and conclusions

Home Informatics can be understood as a multidisciplinary research area concerned with systems and services delivered in the home (in a wide sense, including also other everyday localities and whereabouts, also mobile, inhabited by individuals and their relatives) and as such, is has to be studied from both the supplier (services providers) and the demand (services consumers) perspective. The development of IT/IS for home communication is therefore not any longer a prioritized interest for only HCI designers or IT developers. Our examples show that the increasing interest in home communication addresses many of the concerns of interest in areas as for instance interactive systems design, security, and/or management of the information, or to issues related to market economy and to consumer behavior.

In the society of today, consumers more frequently make deliberate choices rather than being passive victims following the dictates of producer's marketing efforts. To involve end-users to influence the design of services and user interfaces before industrial production is of crucial importance. Home informatics should take inspiration from experience in the areas of consumer behavior, health informatics and user-oriented approaches to developing IT products and systems for home communication. This is due to the fact that experiences obtained in these areas have shown the need to include measures of users' individual preferences for successful systems development and implementation process (Vimarlund & Timpka 2002).

References

Dahlbäck, N, Swaamy, S, Nass, C, Arvidsson, F & Skågeby, J (2001) Spoken interaction with a computer in native or non-native language – same or different. *Proceedings of INTERACT 2001, Tokyo, Japan, July 2001.*

Ibrahim, A. & Johansson, P. (2002) Multimodal dialogue systems for interactive TV applications. Proceedings of the 4th *IEEE International* Conference on Multimodal Interfaces, Pittsburgh, USA, 2002.

Kuhn, LG (2001) Telehealth and Global Health Network. From Homecare to Public Health Informatics. *Computer Methods and Programs in Biomedicine*; 74:155-175.

Lancaster,K.J (1996) A New Approach to Consumer Theory. *The Journal of Political Economy*, 74(2): 132-157.

Miles, I (1987) *Home Informatics* A Report to The Six Countries Programme on Aspects of Government Policies towards Technical Innovation in Industry. Science Policy Research Unit. University of Sussex Falmer, Brighton, U.K.

Sloane, A; Van Rijn F (2000) *Home Informatics and Telematics: Information, Technology and Society*. Kluver, Boston.

Venkatesk, A (1996) Computers and other Interactive technologies for the Home. *Communication of the ACM. Vol.39;No12.*

Vimarlund,V; Timpka T (2002). Design Participation as an Insurance Risk-Management and End-User Participation in the Development of Information Systems in Healthcare Organizations, *Methods of Information in Medicine*, 41:76-81.

Knowledge Management Systems: Issues concerning collaboration

G.A.Vouros

Department of Information and Communication Systems Engineering
School of Science, University of the Aegean, Karlovassi 83200, Samos, Greece
georgev@aegean.gr

Abstract

Knowledge management activities are inherently complex and distributed. They involve human-machine, as well as human-to-human interaction, driving people and machines to form emergent organizational structures for achieving specific goals in specific contexts collaboratively. Interaction in these contexts is driven by human capacities, capabilities, intentional states, available information and interface information presentation capabilities In this paper we are identifying emergent issues for empowering both humans and machines to achieving their goals for effective management of knowledge.

1 Introduction

The goal of knowledge management is to leverage the performance of any organization member involved in any task requiring knowledge, by providing active help to her towards achieving her goals. Knowledge management activities include the identification, acquisition, development, exploitation and preservation of organizational knowledge (Abecker et al, 1998a) (Abecker et al., 1998b) (Staab et al, 2001). The aim is to get "knowledge-powered organizations", i.e, organizations where knowledge management happens in the background as part of the day-to-day job (Smith et al., 2000). Towards this aim, people must get the right information at the right time and at the right form to perform specific tasks, get connected with colleagues that may provide solutions or hints towards solving problems, form groups of people with different areas of expertise and/or different competencies to achieve a shared goal, be equipped with the necessary applications and data to fulfil their tasks, and form decisions in real time. More than that, people must be able to provide feedback and share their knowledge, which must be actively and constantly captured, stored, and organized in the background, so as to be exploited in tasks performance and be disseminated to interested colleagues.
A vital aspect for achieving this aim is to empower both, humans and machines, to collaborate towards achieving their joint goals in the context of organizations' goals. Interaction among the collaborated parties is driven by human capacities, capabilities, intentional states, machines' capabilities, available information and interface information presentation capabilities.
This paper aims to identifying some of the emergent issues for empowering both humans and machines to achieving their goals for effective management of knowledge. We approach this target from the standpoint of human-centered computing.
Section 1 of this paper describes the notion of human-centred computing and how this can contribute to achieving our aims in the context of Communities of Practice and Communities of Interest. Section 2 describes research issues towards realizing our aims. The paper concludes by summarizing the main points.

2 Appliances for managing organizational knowledge: Principles & Context

To maximize the value of computing to society in the age of human and machine symbiosis (Shafto & Hoffman, 2002), we must aim at computational tools that fit and further develop existing practices of specific humans into specific contexts. In other words, we must aim at developing human-centered "computing appliances" for amplifying and extending the cognitive and learning abilities of humans in specific contexts. As it is argued in (Hoffman, Hayes & Ford, 2002), for the development of such tools we must take the triple *people-machine-context* as the unit of analysis. It involves studying people capacities, capabilities and goals, computational mechanisms and interface capabilities, and context. Context comprises "activities conceived as identities of participation". It involves people organizations, roles that people play within these organizations, *intentional activities* that people follow by participating in *social practices* in the context of these organizations, *norms* and *constraints* that are inherent in these activities (Clancey 1997), policies, procedures, as well as devices and media that people use for communicating and doing their work.
To make the above points more concrete, W.Clancey (Clancey 1997) provides an example by contrasting between the classical machine oriented approach for systems' development and the human-centered one that takes into account activities within an organization:

"In modeling medical diagnosis (Buchanan and Shortcliffe, 1984) we choose the physician's activity of examining a patient. We even viewed this narrowly, focusing on the interview of the patient, diagnosis and treatment recommendation, ignoring the physical exam. But the physician is also in the activity "of working at the outpatient clinic". We ignored the context of patients coming and going, nurses collecting the vital signs, nurses administering immunizations, parents asking questions about siblings or a spouse at home at home etc. In designing medical systems like Mycin, we chose one activity and left out the life of the clinician.... Indeed, when we viewed medical diagnosis as a task to be modeled, we ignored most of the activity of a health maintenance organization! Consequently, we developed a tool that neither fit into the physician's schedule, nor solved the everyday problems he encountered."

The present article discusses issues concerning the human-centered design of computing appliances for the management of knowledge in the context of emergent knowledge management organizational units: *Communities of practice (CoP)* and *Communities of interest (CoI)*. Communities of practice are the fundamental organizational units for managing knowledge within an organization. These are small groups of people who work together towards performing shared activities. Individuals within a CoP share the same context in terms of practice and domain of interest. They perform joint activities, use the same media/devices for interacting with the world, share the same viewpoints, values and constraints that are inherent in the activities they perform and share nomenclature, policies and procedures. CoP provides the means for newcomers to learn about practices and for their members to share knowledge.
Communities of practice consist of agents with specific interrelated roles, where each role has responsibilities, obligations and permissions. Permissions define the "social laws" of the community, defining the situations in which agents performing a role are prohibited to perform an action. Although communities of practice have a stable social structure (in terms of roles, roles interrelations, agents performing roles, and social laws), it is true that such structures can also emerge within an organization.
A community of interest (CoI) is an emergent organizational structure, which consists of people that share the same problem for a short time period and may come from different backgrounds, bringing different perspectives to the problem at hand (Fisher & Ostwald, 2001). Bridging the different perspectives and establishing a shared context is a challenge for every CoI. CoIs involve agents with specific roles. Each agent has certain responsibilities and obligations, and imposes

his/her own constraints to the group for achieving community's goal, according to its own physical, social and intentional context, abilities, capacities and preferences.
Both types of communities interact with the world by doing *real work,* using their knowledge as a tool. This type of interaction in is called "*knowing*" (Cook & Brown, 1999). Knowing comprises interaction with the world as well as interaction among communities' members. New knowledge emerges as part of these interactions. While individual knowledge mostly emerges through practice, interaction among community members results in exchanging and creating group knowledge conversationally.

3 Appliances for the management of knowledge: Challenging issues

Effective participation within communities demands for collaborative appliances, rather than mere interactive ones.
Designing appliances that merely interact with humans, as well as among themselves is not enough. Members of communities of practice need to participate in communities' activities coherently and consistently, with respect to their physical, social and intentional context(s). Such members, and this is true especially for newcomers, must get advice and be guided for practicing their knowledge in the context of the organization, always respecting the norms, policies and getting constant information on their tasks to be completed.
The development of collaborative, rather than mere interactive applications is considered to be a key issue for the deployment of knowledge management appliances. Collaborative applications perform as active partners, recognize and adopt to users' preferences and contexts of activities, share the goals of users and exhibit helpful behaviour when this is needed for achieving a shared goal.
Although designing collaborative interfaces is necessary for the effective deployment of systems, we must also focus on designing systems at the level of roles and activities involved, and provide systems with social deliberative abilities (i.e. abilities for reconciling participation of users' in many different contexts, and guiding users to participate effectively in joint activities).
In (Partsakoulakis & Vouros, 2003) we have stress the importance of roles for social deliberation and we have also proposed formal frameworks for developing social deliberating agents (Vouros, 2003), as well as an overall architecture for incorporating social deliberating agents in organizations (Kourakos-Mavromichalis & Vouros, 2001), (Vouros, Kourakos-Mavromichalis & Partsakoulakis, 2003) servicing communities members to achieve their joint goals.
Collaborative interfaces is a vital aspect that must be considered. This should be not be done separately from the larger system, but must also incorporate knowledge about roles, activities and the contexts of activities. Collaborative interfaces must exploit tasks, policies and get knowledge on constraints agents face or impose during their activities.

Effective sharing of knowledge demands for agreed ontologies that shape communities' information space.
Ontologies explicate conceptualizations that are exploited by humans during their real work (i.e. *practice*) and shape their information space, i.e. the way they interact with systems, among themselves, retrieving and organizing information. Being part of knowledge that people possess, ontologies evolve in CoP and CoI as part of *knowing*. Knowledge that ontologies comprise can be either individual or group knowledge. While group knowledge represents shared and commonly agreed conceptualizations, individual knowledge represents a specific perspective for (at least) some of the domain aspects and is not shared by all community members.

For information integration and collaboration between different applications, as well as for the effective communication between organizational memories and workers, shared and commonly agreed, "plug-and-exploit" ontologies are very important. To reduce the costs of building such ontologies, it is important for them to be reusable and multifunctional. This means that ontologies must be human as well as machine exploitable via the appropriate application interfaces, with respect to standards. Intelligent search and retrieval, information presentation, participation in dialogues, adaptation of applications and interfaces, effective information browsing and querying, organization and indexing of information, are among the several tasks that ontologies must support.
Ontology management in the context of CoP involves the development, evaluation and exploitation of conceptualizations that emerge as part of *knowing*. We conjecture that important issues that researchers need to resolve in order to deploy ontology management tools in organizations arise due the lack of human-centered design of these tools (Kotis & Vouros, 2003). In particular, conventional tools for the management of ontologies have not viewed people activities of managing conceptualizations of domains within the context of organization structures. Therefore, the research community has focus on object-centered representation formalisms and on technological issues, and has not put special emphasis on the physical way of interacting with these conceptualizations and on the way conceptualizations arc formed by means of people interacting among themselves and practicing.

4 Conclusions

The challenge is great. Not only for designing and developing the key technologies, but also for devising the interplay between the technologies and for supporting the collaboration between the applications themselves, in the context of an active knowledge management system. A paradigm for such interplay of technologies and systems' collaboration is the development of collaborative tools for ontology construction. Such tools involve a number of agents that collaborate among themselves towards the construction of a commonly agreed ontology. Each human agent is equipped with a knowledge editor that facilitates collaboration with other humans. The knowledge editor via its collaborative human-interface realizes ontology presentations based on the user profile, competencies, preferences and pragmatic constraints. For instance, a user may get a diagrammatic display of the ontology, while another may get a natural language description of the concepts definitions. A similar example involving applications' collaboration is the one between an annotation tool for multimedia documents and a knowledge editor for ontology construction.

5 References

Abecker A., A.Bernardi, K.Hinkelmann, O.Kuhn, M.Sintek., (1998a). Techniques for Organizational Memory Information Systems, DFKI Document D-98-02, 1998.

Abecker A., A. Bernardi, K. Hinkelmann, O. Kuhn, and M. Sintek. (1998b). Towards a Technology for Organizational Memories. IEEE Intelligent Systems, May/June 1998, pp. 40-48.

Clancey, W. J. (1997). "The Conceptual Nature of Knowledge, Situations, and Activity", In: P. Feltovich, R. Hoffman & K. Ford. (eds). *Human and Machine Expertise in Context,* Menlo Park, CA: The AAAI Press. 247–291.

Cook, S.D.N., Brown, J.S., (1999) "Bridging Epistemologies: The Generative Dance Between Organizational Knowledge and Organizational Knowing", Organizational Science, Vol. 10, No 4. Jul/Aug 1999, pp. 381-400.

Fisher, G., Ostwald, J. (2001) "Knowledge Management: Problems, Promises, Realities, and Challenges", IEEE Intelligent Systems, Vol 16. No.1, Jan/Feb 2001, pp 60-72.

Hoffman, R., Hayes, P., Ford, K.M. (2002). The Triples Rule. *IEEE Intelligent Systems, May/June 2002*, 62-65.

Kotis, K., Vouros, G., (2003). Human Centered Management of Ontologies, Department of Information and Communication Systems Engineering, University of the Aegean Technical report, also, submitted to IEEE Intelligent Systems, Special Issue on Human Centered Computing, R.Hoffman (ed), 2003.

Kourakos-Mavromichalis, E., Vouros, G., (2001). "Balancing Between Reactivity and Deliberation in the ICAGENT Framework", In "Balancing Between Reactivity and Social Deliberation", LNAI 2103 Springer Verlag, M.Hannebauer et al (ed), pp 53-75, 2001,.

Partsakoulakis, I., Vouros, G., (2003). "Roles in MAS: Managing the Complexity of Tasks and Environments", to appear in T.Wagner (Ed.) "Multi-Agent Systems: An Application Science", Kluwer Academic, 2003

Shafto, M., Hoffman, R. (2002). Human-Centered Computing at NASA. *IEEE Intelligent Systems, September/October 2002*, 10-14.

Smith R.G., Farquhar A. (2000) The Road Ahead for Knowledge Management: An AI Perspective. AI Magazine, Vol. 21, No.4, Winter 2000, pp.17-40.

Staab, S., R.Studer, H.P. Schnurr, Y.Sure, (2001). Knowledge Process and Ontologies, IEEE Intelligent Systems, Jan/Feb. 2001, pp 26-34.

Vouros, G.A. (2003). Role-Oriented Social Deliberation, *Submitted for publication.*

Usability Testing in Chinese Industries

Xiaowei Yuan

ISAR User Interface Design, Beijing, China
xiaowei_yuan@isaruid.com

Xiaolan Fu

Institute of Psychology, Chinese Academy of Sciences, Beijing, China
fuxl@psych.ac.cn

Abstract

Nowadays, usability testing has become one of the most important and popular methods to test the usability and acceptability of products. Usability testing came from the West, and spread to China at the end of the 20th century. Since then usability testing has made effect on Chinese industries. This paper will mainly introduce the research and application of usability testing in Chinese industries.

1 Introduction

Usability testing can be best described as putting a product on a trial-market. In the test, participants represent the target group of product users and are confronted with carefully designed scenarios that occur in the actual use of a product or program. For example, project engineers configure parameters for chemical plants using a new version of a programming tool, or doctors make clinical diagnoses with the help of a prototype of a computerized tomography scanner. Because these scenarios are very similar to the users' real-life situations, the test offers the opportunity to discover the problems of products and to further improve product at the early deployment stages. Therefore, usability testing enables planners to optimize the product specifications.

A usability test delivers detailed recommendations for improvement and further creative development of the products. In fact, by employing sophisticated psychological observation and communication methods in a professionally organized test environment, few users (e.g., 5 or 6 users) are sufficient to reveal 75% of the improvement potential.

The sophisticated methodological evaluation of the critical acceptance factors also enables people to strategically enhance the market attractiveness of their products. From the reports of usability testing, managers, marketing consultants, and development specialists obtain the experiences of the customers' reactions to products, therefore, they can better develop product strategies and improvement ideas.

2 Application of Usability Testing in China

Usability testing as a research method was introduced to China from the West at the end of the 20th century. It has now been used in Chinese industries and has achieved great success.

2.1 Application in Foreign Enterprises

In Chinese industries, usability testing has mainly been introduced by the foreign enterprises. In the Chinese branches of large-scale international enterprises, such as Siemens and Microsoft, laboratories of usability testing have been established. They conduct usability testing research and carry out actual tests by using their research achievement.

2.2 Application in Chinese Domestic Enterprises

Being influenced by the western modernized enterprises, Chinese domestic large-scale enterprises, including Haier and Legend, has begun to pay attention to usability testing and its roles. They tried to use the method of usability testing in the industrial practice. For example, usability testing was applied in the compilation of the user's manual of "Tianjiao and Tianrui double-mode computer" produced by Legend Group. By usability testing, problems were found in the compilation of the user's manual, and a better manual for the product was finally produced.

2.3 The Birth of a Professional Company

When the Chinese industries were increasing their understanding of usability testing, with professional technology and experience, Prof. Dr. Yuan, who was the leader of User Interface Design of Siemens China, established the first Chinese professional usability testing company in Beijing, ISAR User Interface Design Co. Ltd, in 2002. The birth of the professional company means that the operation research of Chinese industries has stepped into a new phase.
ISAR User Interface Design is an industry oriented support center for usability engineering in China, including user analysis, user interface design, and usability inspection. It devotes to increasing the awareness of usability engineering, conducting application oriented research on human-computer interaction, and providing services for its applications in Chinese industry.
ISAR User Interface Design has successfully conducted several HCI projects derived from Chinese industry. It has also helped Chinese domestic big enterprises, such as Legend and Haier, establish their own Usability Centers. These centers play important roles in evaluating the usability of products.

3 Usability Testing of Mobile Phone

At present, in the design of information technology product, it is a new trend to shift from technique design to usability design. In the competitive cell phone market in China, the usability of cell phone products is becoming the most important criteria of customer selection. Therefore, usability is a critical factor to develop cell phone's application technologies and increase cell phone's marketing competition. It is highly required to do a systematic study on cell phone's usability and its' potential deign development.

3.1 Purposes of Testing

For benchmarking, the usability testing has been done on Xiaxin A8, Motorola V60, and Samsung A308 for the following two purposes: 1) To do comparative usability analysis on three type cell phone products, and find their weak points in terms of usability and give them improvement expertise; 2) To standardize basic usability testing routine.

3.2 Designing the Test Scenarios

The design of the scenarios is the script of the usability testing. In order to make the testers perform well and to reflect what consumers are really thinking of when using the product, this design of script should also contain the process of the task.
According to the manual of the three brands of cell phones as well as the engineering design of all parts and the function keys of the machine, we divided the whole design of the scenarios into four parts: 1) the installation of battery, 2) the settings of the mobile phone, 3) the settings of the communication, and 4) the settings of the short message service. In the designed scenarios, participants were asked to complete all the tasks in a relaxing atmosphere.
A two-part questionnaire was designed too. The first part is about personal background, such as sex, age, educational level, profession, the experience of former use of mobile phone, and the brand used. The second part is about the satisfaction degree of each operation.

3.3 Participant Selection and Test

In the aspects such as the knowledge about the tested cell phones and the usage skills, the participants were consistent with the consumer characteristic. In other words, they were embody the ordinary consumers. In this testing, seven end users were recruited as participants.
After completing the planning, there was the rehearsal of the test. That is "the test of the test", which is a part of the whole usage test. It would help us to evaluate whether the scenarios design is reasonable and whether the time is sufficient.
The method of Thinking Aloud was used in all tests.

3.4 Evaluation and Analysis

3.4.1 Description of Errors

After the test, we made statistical analysis with the problems occurred in the test, and finally classified the errors into 5 types:

1) Project faults (PF): User's target is always erroneous when completing the tasks.
2) General faults (GF): Design violates users' general knowledge and experiences.
3) Operation faults (OF): User has the right intention, but always makes operational error.
4) Sense faults (SF): Misconception due to incorrect translation/interoperation of the information or symbols (e.g., color, form, size, position etc.).
5) Interpretative faults (IF): Incorrect interpretation of information.

The error degree was calculated according to the number of the faults founded in the six formal tests. The error degrees were different from A to D: the error degree was A if it occurred in 4 or more tests, B if in 3 tests, C if in 2 tests, and D if in 1 test.
Besides, we made some detailed analysis mainly with Error A and Error B in five aspects of the applicable principle: easy-learning, efficiency, consistency, fault toleration, and satisfaction. The examples of Error A were showed in Table 1.

3.4.2 Satisfaction of End Users

A 10-point scale was used in the questionnaire to test the satisfaction of end users to the cell phone products at the following 16 aspects:

1) Battery Installation

2）Rings Setting

3）Calling Forward

Table 1: The Examples of Error A

Error Grades or Categories	Tasks	Issue Description	User Comments / or Suggestions	Expertise	Issue Type	Source
Motorola V60: Rings Setting						
A	Ring Setting	Disconcertion of setting functions of "My rings"		Change to "Set My Rings"	IF	1, 2, 3, 5
Samsung A308: Rings Setting						
A	Mute Setting 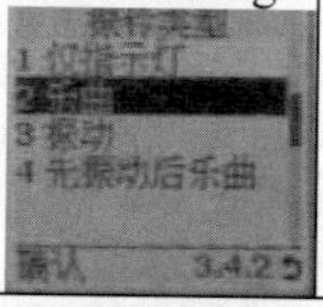	Do not know "LED only" means mute.		Change to common "Mute" symbol	IF	1, 2, 3, 4, 5, 6
Xiaxin A8: Forward Setting						
A	Menu Setting 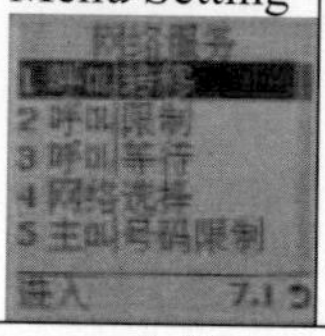	Assume it in phone settings, however, it is network service setting	List it in phone settings	list it in phone settings	PF	1, 2, 3, 4, 5, 6
Xiaxin A8: Time and Alarm Setting						
A	"Universal time" and "Time"	"Universal time" is in toolkit, but not the right one		Change "Universal time" to "Time Zone"	IF	1, 3, 4, 5, 6
Motorola V60: Time and Alarm Setting						
A	Alarm Setting	Participant 1 and 5 assume this function available Participant 2 and 1 can reach the right one only by exclusive selection.		Put it in toolkit, not in calendar or schedule.	PF、IF	1, 2, 3, 4, 5, 6

4）Time Setting

5）Alarm Setting

6）Random Key and Open Cover Lid to Receive

7) Calling Waiting and Shifting
8) Save Phone Number
9) Directory Lookup
10) Input Method Shifting
11) Punctuation mark and Symbol Shifting
12) Text Input
13) Short Message Viewing
14) Number Extracting from Short Message
15) Key Operations
16) Hand Satisfaction degree

Figure 1 showed the users' satisfaction to three cell phone products, Xiaxin A8, Motorola V60, and Samsung A308. Obviously, Samung A308 got the highest satisfaction.
Finally, based on the analysis, we also provided our suggestions on the structure and interactive design.

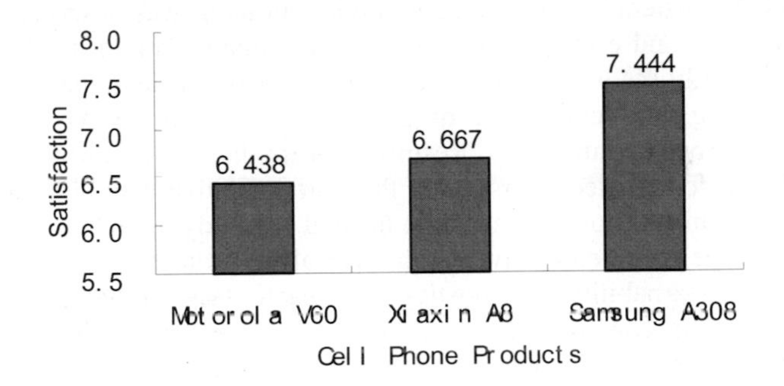

Figure 1: Users' Satisfaction to the Three Cell Phone Products

4 Summary

Usability testing has become one of the most important and popular methods for testing the usability and acceptability of products in China. ISAR User Interface Design is the first industry oriented company for usability engineering in China. ISAR has successfully conducted several HCI projects derived from Chinese industry. In a case study, 3 brands of cellular phone, Samsung, Motorola, and Xiaxin, were selected. Seven participants were required to use these three cellular phones to perform several tasks. The results of the test showed the shortcomings of the three cellular phones on usability as well as the user's satisfaction degree of every function. The improvement suggestion was provided finally.

References

Mayhew, D. J. (1999). The usability engineering lifecycle. Morgen Kaufmann Publisher, Inc.
Mandel, T. (1997). The elements of user interface design. John Wiley & Song, Inc.
Nielsen , J. (1993). Usability engineering. Morgen Kaufmann Publisher, Inc.
Rubin, J. (1994). Handbook of usability testing. John Wiley & Song, Inc.
Siemens CT IC 7. (1999). Leitfaden zur organisation und durchführung von usabilitytests. Munich.
Wiklund, M. E. (1994). Usability in practice. Academic Press, Inc.

Reliability Analysis and Design in Computer-Assisted Surgery

A. Zimolong[1,2], K. Radermacher[2], M. Stockheim[3], B. Zimolong[4], G. Rau[2]

[1]synaix-healthcare, Aachen, Germany
[2]Institut für Biomedizinische Technologien,
Helmholtz-Institut Aachen, Aachen, Germany
[3]St. Josef-Hospital, Universitätsklinik, Bochum, Germany
[4]Institut für Arbeits- und Organisationspsychologie
Ruhr-Universität Bochum, Bochum, Germany
Zimolong@gmx.de

Abstract

Interaction with a CAS system by 25 expert and novice surgeons was observed and assessed using a model of failure mode and effect analysis applied to human error. In total 152 incidents were observed, of which 46 (30 %) were found to be critical for the patient, and 99 (65 %) critical for the process. Comparing interaction time of users with different levels of system experience allowed for quantification of learnability. Results indicate that the preoperative planning is more complicated to learn for novice users than the intraoperative handling of the system. By calculating the ratio of interaction time vs. time needed for studying help material and comparing this value to an average obtained by averaging over all tasks to be performed with the system resulted in specific learnability diagnostics for each task. Finally, a multidimensional questionnaire was used to compare perceived error tolerance, learnability and user satisfaction between novice and experts users. Novices rated learnability significantly *better* than experts, even though they encountered severe problems when interacting with the system. In contrast, experts' rating closer matched findings obtained by observation, which indicates that ratings of system quality by novice users might by less reliable.

1 Introduction

Clinical relevance and improved surgical outcome of systems for computer assisted surgery (CAS systems) have been demonstrated in past studies. With introduction of these systems into clinical routine however additional factors may jeopardize outcomes: studies analyzing adverse events involving the use of technical equipment in medicine found that in 77,2 % of all events investigated the equipment was functioning according to specifications, but users had problems in correct application (Leape, 1994). Handling of technical equipment thus constitutes a hazard for the patient, which especially applies when introducing new systems or complex tasks (Weingart, Wilson, Gibberd & Harrison, 2000). To avoid these errors it is necessary not only to assess fulfillment of technical specifications, but as well to assess the system's usability and reliability in interaction and clinical use.
In doing so, it is necessary to consider that different operators commit the same errors in comparable situations. Mishaps tend to fall into recurrent patterns, an accident proneness of individuals can't be maintained as a valid concept (Hacker, 1998). Instead of blaming the actor for negligence, the error should be understood as the final result of a sequence of events occurring under specific set of circumstances. Thus work place and equipment design insufficiently adapted to clinical needs may constitute the trigger for human error, as investigations on critical incidents in medicine already discovered (Hyman, 1994), (Cook & Woods, 1994).

Accordingly, it is necessary to differentiate between active failures committed by the operator and latent conditions within the system (Reason, 1990). While links between active failures and adverse events usually can be established easily, latent conditions have been introduced into the system by engineers, designers and managers separated in space and time from the event. Reducing the chances of error and increasing reliability thus needs to be achieved by reducing the number or the effects of latent conditions already during system design or introduction into the clinical environment. Minimizing occurrence of latent conditions however is not sufficient, as human operators tend to test system limits to minimize work load or when learning handling of the equipment (Rasmussen, Pejtersen & Goodstein, 1994). Systems need to be able to tolerate the occurrence of errors and contain their damaging effects. With error tolerant systems problem-based and explorative becomes possible, which in comparison to other methods of learning is most effective (Hacker, 1994). Thus the system's reliability use is closely related to it's error tolerance, learnability and extend of latent error provoking conditions.

2 Methods

Work analysis of system's reliability in a clinical context was conducted in two different clinical centers, involving a total number of 25 clinicians with various levels of expertise in conventional surgery and in handling of the CAS system. All participants were asked to solve a number of standardized tasks which were covering all steps of the preoperative and intraoperative procedure involving the system. For simulating intraoperative steps an artificial model of the hip was used. Regardless of the test user's expertise all received standardized information on background knowledge necessary to solve the task, e.g. definitions of landmarks used for planning, information on the registration procedure, and on the type of imaging data used. Next the goals to achieve were presented and explained. These goals were described in a way that subjects were able to assess their own task fulfillment. After introduction users were asked to solve the given tasks by themselves without any additional help from the evaluators. During the whole time the task goals were displayed on a separate computer screen and subjects were able to recall the background information on push of a button. Detailed step-by-step information on how to solve the task was not given orally to the user, but could be recalled from the tutoring system. After subjects had started to solve the given problem, the role of the evaluators was restricted to observation. If however users got stuck with the task, after a defined period of time (2 minutes) evaluators intervened, restored the last stable system state and/or provided additional information on how to solve the problem. After the task had been solved, depending on the result evaluators also interacted with the system to assure comparable starting conditions for all users. System's reliability was determined on multiple dimensions, according to the theoretical framework presented above:

Adverse events: Criticalness of events and their frequency of occurrence are important factors to consider when targeting redesign of the system under evaluation. Investigating error tolerance and uncovering latent system failures was done by observing interaction problems and errors. Problems the users encountered were analyzed to identify failure-modes and effects (FMEA-method). Depending on its effects problems were classified into 5 different groups: May the error go unnoticed, so it affects the final result? Or gets the task sequence interrupted, so the user is not able to proceed? Is it necessary to correct the error, or may one compensate its effects? Or did the user deviate from the optimal solution, but still accomplished the task? Criticalness of the problem was then rated in dependence of its impact on the planning process and on the final result: At what point of the process does the consequence of the error occur and how hard / how likely is it to correct the problem? If it is possible to correct the error, how hard / how likely is it to detect this? Is it possible to compensate the consequences of the error, hence error correction is not necessary? For assessment of the error, aspects concerning difficultness and likelihood were rated on a scale from 0 to 10 and subsequently analyzed for exceeding a threshold set for 20 % of the scale.

Learnability and intuitiveness: For quantification of learnability two different approaches were used. Firstly, interaction and learning time of users interacting with the system was determined from the time each user activated the on-line help available on the separate computer. This time was determined for each task to solve. As task complexity varied, learning time was related to the time need for interacting with the system. This ratio was assumed to be constant, so tasks with a smaller than average ratio of learning vs. interaction time were regarded to be easy to understand but hard to perform, whereas tasks with a ration above average are hard to understand but easy and quick to perform. Secondly, for quantification of the overall learnability of the system interaction time of subjects was compared depending on previous experience with the system. Data was fitted to an exponential curve and curve parameters of the pre- and intraoperative software modules were compared.
Subjective assessment by questionnaire: A questionnaire was assembled which covers multiple dimensions to assess error tolerance, learnability and user satisfaction. From the data received, a first validation of the questionnaire was performed by cross-correlation of questions. Furthermore, ratings given by expert and novice users were compared.

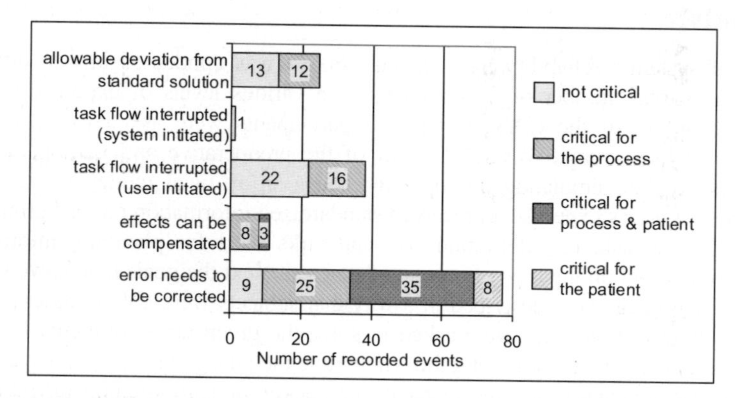

Fig. 1: Classification of adverse events into 5 groups depending on type of consequences and effects on the task sequence. Adverse events of each group are grouped by criticalness.

3 Results

In total 152 incidents were observed, of which 46 (30 %) were found to be critical for the patient, 99 (65 %) critical for the process, and 45 (30 %) were not critical. The majority of critical events occurred when errors needed to be corrected or compensated, while in cases of tolerable deviation from the standard solution or if the task sequence got interrupted the majority of adverse events were not critical (cp. Fig. 1). It should be noted that even though 30 % non-critical events were recorded, this does not allow any estimation of error tolerance of the system. If the user can't proceed because he doesn't know how to do the next step, this event might be rated to be non-critical because the average users find the solution in an acceptable time. System initiated interruption of the task sequence in case of error is an indicator, but in case of functioning error tolerance the events would go unnoticed from the observers.
Overall learnability and intuitiveness of the system was determined by analyzing interaction time of each participant in relation to his/her experience with the system. As the CAS-system features two software packages of different design, overall learnability was determined for each (cp. Fig. 2). Comparison of results for both modules yields a significant difference: for the preoperative planning, novice users start with interaction times almost 5x higher (485 %) than expert level. In contrast, for intraoperative handling of the system novices start with interaction times which are about 2x (212 %) higher. Users with an experience of 10 sessions will reach approx. 226 % or 186 % of expert time with the preoperative or intraoperative modules, respectively.

Next to assessment of overall learnability and intuitiveness, these aspects were also analyzed for each task. This was done by calculating the ration of expert vs. novice interaction time for each task and comparing this ration to the time users spent with studying help information (Fig. 3). Due to varying complexity of each task both values are changing over tasks, however a highly significant (α=1 %) correlation between both could be found. This means that in average a constant factor links both values: the more time novice users needed in comparison to experts, the more time they invested to study help information. Yet for some tasks this factor is smaller or higher than average, which was further analyzed. Results indicate that for tasks with a higher-than-average study-time (*hard to learn but quickly to perform* tasks) these involved new concepts for the users, mental compatibility with known concepts was low. Also the opposite could be observed: easy to learn, but time consuming to perform. The task concerned required repeated activities, where however novices' users were unsure about the quality criteria to assess their outcome by.

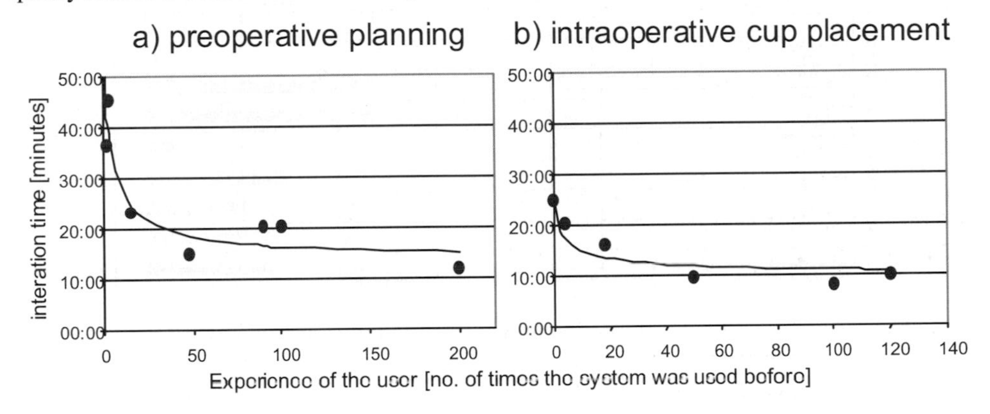

Fig. 2: comparison of overall system learnability for both software module of the system

Finally participants were asked to rate the system by questionnaire. Assessments by novice and expert users were analyzed separately and compared, as different ratings between both groups were expected. In fact, different ratings of system learnability could be found: novice users had a more positive feeling than expert users (p<0,6, Wilcoxon test). Assessment of error tolerance also revealed differences between both groups (p<0,1 U-test). Occurrence of hopeless situations was less frequently admitted by novices than by expert users, despite adverse events they encountered during the trial. System's support by error messages on the other hand received better numbers by experts than novices hence overall error tolerance was rated fairly equal. No differences could also be found in user satisfaction, which involved questions on system's suitability for the task and affect. The same holds for assessment of mental workload.

4 Discussion and summary

In the study presented, interaction with a CAS system by 25 expert and novice surgeons was observed and assessed using a model of failure mode and effect analysis applied to human error. In total 152 incidents were observed, of which 46 (30 %) were found to be critical for the patient, and 99 (65 %) critical for the process. Comparing interaction time of users with different levels of system experience allowed for quantification of learnability. Results indicate that the preoperative planning is more complicated to learn for novice users than the intraoperative handling of the system. These findings are supported by statements of the users. By calculating the ratio of interaction time vs. time needed for studying help material and comparing this value to an average obtained by averaging over all tasks to be performed with the system allows for specific learnability estimation for each task. This way tasks were identified which are easy to understand but hard to perform, as well as task which are hard to understand but easy to perform. For the latter redesign should concentrate on facilitating the underlying concepts or improving knowledge transfer of theses concepts, while for the first the interface design or system functionality needs to be improved.

Finally a multidimensional questionnaire was used to compare perceived error tolerance, learnability and user satisfaction between novice and experts users. Interestingly, novices rated learnability significantly *better* than experts, even though they encountered severe problems when learning how to use the system. Specifically, hopeless situations were not attributed to the system, but novices put the blame on themselves. In contrast, experts' rating closer matched findings obtained by observation, which might indicate that ratings of system quality by novice users is less reliable.

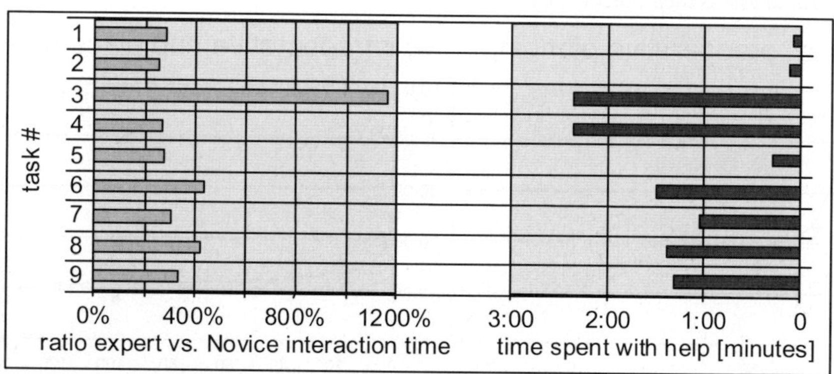

Fig. 3: assessment of learnability for each task: ratio of expert vs. novice interaction time, compared to time novice users spent with help information.

Acknowledgements

The authors wish to express gratefulness to Mrs. A. Marks (Ruhr-Universität Bochum), who contributed significantly to the success of this work, and to all the participants at the St-Joseph-Hospital Bochum and the BG-Unfallklinik Ludwigshafen.

Literature

Cook, R.I., Woods, D.D. (1994). Operating at the Sharp End: The Complexity of Human Error. In: M.S. Bogner (Ed.), *Human Error in Medicine* (pp. 255-310). Hillsdale NJ: Erlbaum Publishers.

Hacker, W. (1994). Arbeits- und organisationspsychologische Grundlagen der Software-Ergonomie". In E. Eberleh, H. Oberquelle, R. Oppermann, (Eds.), *Einführung in die Software-Ergonomie* (pp. 53-94). Berlin: Gruyter.

Hacker, W. (1998). *Allgemeine Arbeitspsychologie*. Bern: Huber

Hyman, W.A. (1994). Errors in the Use of Medical Equipment. In: M.S. Bogner (Ed.), *Human Error in Medicine* (pp. 327-348). Hillsdale NJ: Erlbaum Publishers.

Leape, L.L. (1994). The Preventability of Medical Injury. In: M.S. Bogner (Ed), *Human Error in Medicine* (pp. 13-26). Hillsdale NJ, Erlbaum Publishers.

Rasmussen, J., Pejtersen, A.M., Goodstein, L.P. (1994). *Cognitive Systems engineering*. Wiley.

Reason, J. (1990). *Human Error*. Cambridge: Cambridge University Press.

Weingart, S.N., Wilson, R.M., Gibberd, R.W., Harrison, B. (2000). Epidemiology of medical error. *BMJ*. 18, 774-777.

Constraint-Based Teamwork Analysis in the Software Industry

Bernhard Zimolong, Thorsten Uhle, Stephan Kolominski, Patrick Wiederhake

Ruhr-University of Bochum
Work and Organizational Psychology
Bernhard.Zimolong@ruhr-uni-bochum.de

Abstract

A framework of contextual and organizational factors which may impinge on the selection and staffing of project teams was developed. In five software development teams contextual team task analyses have been carried out. We found substantial differences in the ratings between the firms.

1 Introduction

Teamwork behaviours, resulting from team member interactions, need to be differentiated from taskwork behaviours, which are the position-specific behaviours of individual team members. Teamwork performance is embedded into an inter- and intraorganizational setting, i.e. into a physical, social and economical environment. The contextual factors as well as teamwork and taskwork factors contribute to the overall performance of the teams.

Teamwork behaviours focus on coordination requirements between team members, such as backup, mutual performance monitoring/error correction, and information flow. Individual taskwork behaviours refer to functional requirements to perform individual tasks and meet individual responsibilities. Examples from the teamwork test (Stevens & Campion, 1999) are planning and task coordination, and goal setting and performance management. The specific competencies that enable teams to coordinate, communicate, and synchronise task relevant information to accomplish their goals and missions may be summarized into three primary categories: Knowledge, Skill, and Ability (KSA, Stevens & Campion, 1999). Knowledge includes among others task specific team-mate characteristics, shared mental models, task sequencing, team role interaction patterns, teamwork skills, and team orientation. Some of the skills consist of adaptability, shared situational awareness, mutual performance monitoring, communication, decision-making, assertiveness, interpersonal coordination, and conflict resolution (Adolph, 2000). Attitudes embody motivation, shared vision, team cohesion, mutual trust, collective orientation, and importance of teamwork.

Prerequisite to the measurement of team performance is the team task analysis. A team task analysis provides information about team selection and staffing, team learning objectives and team competencies needed. It enables the understanding of the nature of task interdependency in teamwork and to distinguish collective team tasks from individual tasks. According to the distinction between teamwork and taskwork behaviour, the measures for assessing team skills may be grouped into team KSA and individual KSA requirements.

Not yet available is a sound, validated and systematic methodology for analysing team tasks and the contextual factors, in which the activities are, embedded (Paris et al., 2000). Furthermore, there is a gap between work analysis and design. Descriptive procedures analyze the actual state of the system and show how the environment supports and constrains work. However, no new, innovative design can be derived from descriptive work analyses alone. The ecological framework of Vicente (2002) consists of descriptive and deductive procedures for designing human-computer interfaces for complex sociotechnical systems. It identifies the constraints that an environment imposes on work and provides instruments how to get to an improved interface design solution. The concept is less suitable for the design of social systems that is for teamwork or project groups. The Contextual Design (CD) approach (Beyer & Holtzblatt, 1998) also uses a combination of descriptive (inductive) and deductive procedures. They help finding out how people work and lead the team through the process of discovering design alternatives for the product, the work practice, and work systems. It was developed as a practical framework for the design of customer-centred systems in the software industry, however, not for the right composition of teams.

Traditional staffing procedures in project teams, which normally determine position requirements, recruit applicants, then assess and select those most qualified to perform the job, are not entirely adequate for the composition of teams. Teams require special staffing considerations. Team performance depends not only on KSA required for individual task performance, but also on competencies of individual team members that facilitate team interaction and functioning. One would need to know how staffing requirements might vary according to team type and function, such as production teams or customer service teams. Additionally, contextual factors may play an important role. Consider the following factors: market strategy, budget constraints, duration of project, availability of qualified workforce, type of product (mass vs. customized), and customer demands. All of these factors might influence the decisions on selecting the appropriate applicants. If stable individual characteristics associated with superior abilities for team coordination and performance can be identified, then steps to select the right people under the constraints of market and company strategies can be made. Before this can be done, however, one must first develop a framework of contextual and organisational factors, which may impinge on team performance, and conduct a team task analysis to evaluate these factors.

2 Study

We developed a framework of contextual and organizational factors which may impinge on the selection and staffing of project teams in the software industry (see table 1). The content domain was identified through an extensive review of several major bodies of literature on groups, organizational psychology, and work analysis and design. A team task analysis has been carried out in five software development teams of five different companies during 2002/2003. The size of the companies ranges from small to large, i.e., from 8 to 4.200 employees.

Structured interviews were conducted with the first level managers and project team managers. Up to date, three teams counting 15 members returned their questionnaires. The data base of this report involves 19 questionnaires, 15 from team members and four from project managers. Due to spatial constraints, we report on team and task factors that facilitate team interaction and functioning and on individual and teamwork KSA critical to teamwork performance.

3 Results and Discussion

The results of the significance ratings of team- and task work factors by Project Managers (PM) and Team Members (TM) are captured in table 2. Most important team factors are Support by management, Team management, Communication, and Team cohesion. PM and TM agree upon the significance of these factors and put them at the first four positions. The least significant factor is KSA homogeneity. The overall agreement between PM and TM is r=.41. Obviously both groups do not share the same view of the significance of teamwork and taskwork factors. Remarkable differences (≥ 1 scale unit) can be seen between the ratings of the firm's project managers with respect to Team cohesion, Mutual relief, KSA and

Table 1: Taxonomy of factors influencing selection and staffing of project teams

Factors	Descriptors
Contextual factors	Extra organizational factors:
- Market	Share of market, market conditions, business fluctuations
- Labour market	Skilled labour demand, functional and social competencies
- Customer	Customer profile, customer demands
Organizational factors	Strategic targets, type of products (mass/customized), service orientation, culture, leadership style, qualification systems, payment system
Structural factors	Work organization, project teams Job contract, regulations of working hours
Team design factors	Division of labour, scope of team responsibilities, composition of teams, team size, individual skills and traits, function of key people, project management
Team process factors	Process management, controlling , role of customer in the design process, quality control, documentation
Contingency factors	Technical and financial resources, autonomy of project team

Adapted from Paris, Salas, & Cannon-Bowers, 2000, p. 1058

Table 2: Significance ratings (means) of team- and task work factors that facilitate team interaction and functioning

Team/Task[1] factors	Firm A (1/ 5)		Firm B (1/2)		Firm C (1/8)		Mean	
	PM	TM	PM	TM	PM	TM	PM	TM
Team cohesion	3.5	4.1	3.3	3.6	4.5	3.9	3.8	3.9
Mutual relief	3.5	3.5	2.5	3.5	4.0	4.1	3.3	3.7
Support by management[1]	5.0	4.0	4.0	3.5	5. 0	4.3	4.7	3. 9
KSA-Homogeneity (knowledge, skills, abilities)	4.0	3.8	1.0	2.5	3.0	2.1	2.7	2. 3
VSA-Homogeneity (values, status, attitudes)	4.0	4.0	2.5	2.5	4.0	2.9	3.5	3. 1
Team engagement and valued norms	2.7	3.8	2.3	3.0	4.3	3.4	3.1	3. 4
Collaborative problem solving and decision making	3.0	2.8	3.0	3.5	4.0	3.9	4.1	4. 0
Communication (frequency of informal comm., exchange quality, balanced contribution, customers)	4.1	4.2	4.0	3.5	4.3	4.3	4.1	4. 0
Team management[1] (goal setting, feedback, task-coordination, team self-efficacy)	4.6	4.4	4.2	3.5	4.8	4.2	3.3	3. 4

PM=Project Manager, TM=Team member; number of PM/TM in parenthesis; scale values: 1= not important, 5=very important

Table 3: Significance ratings (means) of KSA critical to teamwork performance

Team/Task [1]KSA	Firm A PM	Actual state PM/TM	Firm B PM	Actual state PM/TM	Firm C PM	Actual state PM/TM	Mean PM/TM
Functional competence[1]	4.0	**4.0/ 4.1**	4.0	4.4/ 3.5	3.8	3.6/ 3.4	3.9/3.7
Methodological competence[1]	4.2	3.4/ 4.0	4.6	4.0/3.1	4.4	3.5/ 3.2	4.4./3.4
Social competence	3.8	3.8/3.9	4.0	4.2/ 3.3	4.3	3.5/ 3.6	4.0/3.6
Self-management competence[1]	3.2	3.6 /3.9	3.6	3.2/ 3.2	4.2	3.1/3.7	3.7/3.6
Preference for teamwork	4.0	3.0/ 3.6	4.0	3.0/3.5	5.0	4.0/4.0	4.3/3.7
Achievement motivation[1]	4.0	4.0/ 4.2	5.0	5.0/3.0	5.0	4.0/4.3	4.7/3.8
Identification with customer's needs[1]	5.0	5.0/ 4.2	3.5	3.5/ 3.0	5.0	3.5/ 3.9	4.5/3.7
Identification with firm[1]	3.0	3.0/ 3.6	4.0	4.5/ 3.5	3.0	3.0/2.9	3.3/3.7

VSA homogeneity (values, status, attitudes), Team engagement, and Collaborative problem solving.

Team members of the firms also disagree on KSA and VSA homogeneity and Collaborative problem solving. Results on the significance ratings of team- and taskwork KSA are depicted in table 3. With respect to the PM rankings, the most important factors are Achievement motivation, Identification with customer needs, and Methodological competence. PM and TM agree on the first two positions. These factors are taskwork KSA. The Teamwork KSA Preference for teamwork and Social competence rank at position four and five. Least important is Identification with the firm. Overall agreement on the significance of the KSA is r=.62. Both groups agree better on the importance of KSA as compared with the team- and taskwork factors. Differences (≥ 1 scale units) between firm ratings of project managers are Self-management competence, Preference for teamwork, Achievement motivation, Identification with customer's needs and with the firm. TM disagrees on Achievement motivation, Identification with customer's needs and with the firm.

The results of the ratings show the significance of the taskwork factors Support by management and Team management, and the team factor Communication and Team cohesion. We obtained similar results on the rankings of taskwork competencies. On the other hand we found substantial differences in the ratings between the firms. In the next step we will analyze those contextual and organizational factors derived from constraint-based teamwork analysis, which are correlated with team- and taskwork factors and team competencies.

References

Adolph, L. (2000). Soziale Konflike bei verschiedenen Formen industrieller Gruppenarbeit und ihre Auswirkungen. Frankfurt a. M.: Peter Lang.
Beyer, H., & Holtzblatt, K. (1998). Contextual design: Defining costumer-centered systems. San Francisco: Morgan Kaufmann.
Paris, C.R., Salas, E. & Cannon-Bowers, J.A. (2000). Teamwork in multi-person systems: a review and analysis. *Ergonomics, 43, 1052-1075.*
Stevens, M.J. & Campion, M.A. (1999). Staffing work teams: Development and validation of a selection test for teamwork settings. *Journal of Management, 25, 207-228.*
Vicente, K. J. (2002). Ecological Interface Design: Progress and Challenges. *Human Factors, 44(1), 62-78.*

Prognostic Work Analysis Using a Simulation Approach

Gert Zülch, Sascha Stowasser, Rainer Schwarz

ifab-Institute of Human and Industrial Engineering,
University of Karlsruhe
Kaiserstrasse 12, 76128 Karlsruhe, Germany
sascha.stowasser@ifab.uni-karlsruhe.de

Abstract

Procedures for work analysis usually reflect exclusively the actual state of a real work system at a given point in time. If one wants to examine e.g. future work situations with respect to their scope of action and work requirements, appropriate objective prognosis methods are then necessary. The high complexity of the prospective work analysis demands computer-supported aids, e.g. in the form of simulation procedures, which allow the work analyst or planner not only to explore the static, but also the dynamic correlations within the work system.
In the following article an overview of simulation applications in the industrial environment will first be provided. Subsequently, attention will be drawn to the necessity of prospective work analysis. The simulation procedure discussed here serves the prospective work analysis and is aimed at forecasting the repercussions of planned work systems to the employee in this work system. A focus of this article is the exemplary implementation of the computer-supported simulation for prognostic analyses of assembly work.

1 Introduction into Simulative Planning

1.1 Field of Application

In order to ensure the long-lasting competitiveness modern production enterprises strive to attain a higher degree of cost efficiency and customer-orientation. Furthermore in the wake of the globalization of markets, it becomes necessary to create ever more elaborate and exact plans of the production systems.
When analytical procedures fail due to the complexity of the work system, simulation can be used as a planning tool. Normally, simulation tools, which are specialized for the application areas of production and logistics, are used for this purpose. Since these material flow oriented procedures were at the beginning of their development, merely able to represent the technical components of a production system the holistic consideration of sub-systems, including the personnel, has gained in importance in the past few years.
The planning of future personnel assignment and the construction of the work system is carried out afterwards as a derived planning stage which is often merely comprised of rough determinations of the necessary workplaces. A detailed, prospective analysis of the operational and qualifications-oriented demands and the future physiological and psychological requirements is however mostly not executed (Zülch, Heel, Lunze, Hohendorf & Schweizer, 1998, p. 93). Thus. the usual planning procedure is too limited in its ability to prospectively determine good ergonomically personnel structures and human-oriented work systems in industrial production.

1.2 Approach to Simulation

The system behaviour of modern production systems cannot be illustrated exactly in an analytical manner due to complex interferences, stochastic influences and increasing demands for flexibility. For this reason simulation technology can be used as a tool in the planning of these systems. The relevant guideline of the German Association of Engineers (VDI 3633, sheet 1, 2000, p. 2) defines simulation as the "reproduction of a system with its dynamic processes in an experimentations model, for the purpose of gaining insights that are transferable to the reality". In accordance with the planned simulation objectives, a model of the real/planned system, limited to the essentials, is created. With this model the user can execute experiment series in order to either understand the real/planned system behaviour, to develop various strategies for system operation or to obtain decision support for his work. The principle approaches to the execution of a simulation study are shown in Figure 1.

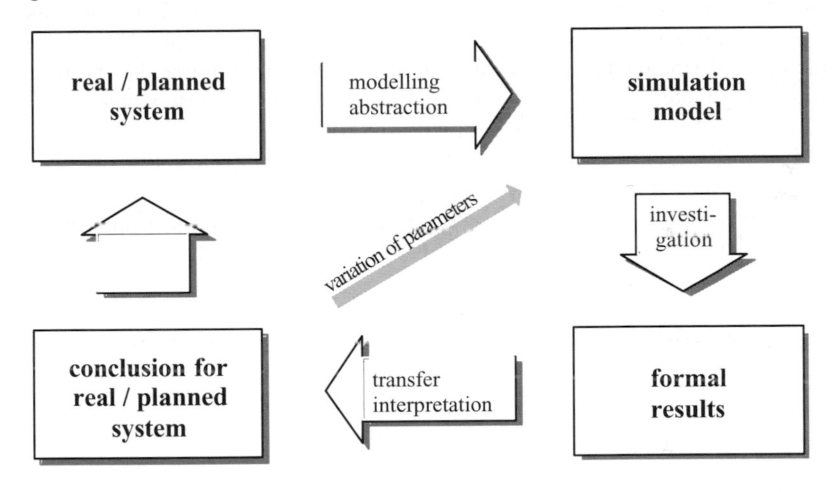

Figure 1: Procedure of a simulation study (following ASIM, 1997, p. 3)

The first experiments with the simulation model serve its validation. The simulation data obtained are compared with the data of the real system, whereby it must be ensured that the abstracted model represents the reality of the respective system with suitable exactness. If the necessary exactness is reached, empiric examinations of the system behaviour follow by systematically varying one or more system or process parameters. The results of a simulation run are composed of a large amount of data, which must then be compressed and interpreted, while keeping the surrounding conditions and simplifications from the model creation phase in mind. Prerequisites for the interpretation are e.g. results presented in the form of a graphs or animations. Finally, the solution approaches gained for the optimization of the system can be reviewed again in a simulation run and, in the case of positive results, transferred to the real/planned system.

2 Significance of Prospective Work Analysis

The analysis of work activities is focussed on work demands and work system attributes. The procedures which are used for this purpose usually have the disadvantage that they merely reflect the actual state of a real work system at a certain point in time. If one wants to examine e.g. future work situations with respect to their work demands and scope of actions, appropriate prognosis methods are then required. The high complexity of prospective work analysis demands computer-

supported aids, e.g. in the form of simulation procedures, which allow the analyst or planner to explore not only the static, rather also the dynamic correlations within the work system and thereby to support the planner in his search for a solution.
When dealing with such planning tools for the configuration of future work systems one can differentiate between static prognosis procedures, procedure for post-run-analysis and simulation procedures with integrated work analysis. While static prognosis procedures do not consider the dynamic coherences of a work system, thus not allowing for an exact reality approximating illustration of the work situation, post-run-analyses are well suited for the assessment of simulated work processes. Post-run-analysis procedures have already been developed for several application fields, such as e.g. for static and dynamic muscle work or noise load (see list of post-run-analysis procedures in Schindele, 1996).
In contrast, event simulation procedures allow for control strategies to react (e.g. work interruptions) to resulting work situations (e.g. to much stress for a worker) already during the simulation. The stress progression is hereby designated by the occurrence of events and reactions during the simulation. Such simulation procedures present the possibility to represent various statically operative factors influencing the worker e.g. dimate, noise or lighting. Furthermore, so-called dynamic influencing factors (work load, practise, time stress, monotony), which change over time or rather the course of the simulation and which are thus influenced by the specifications of the work demands, work organization and workplaces, are taken into account for the estimate of human reliability. There are only isolated approaches to simulation-integrated procedures in the area of work analysis, although they are of great importance as a prognosis and planning tool for work systems, as has been sufficiently shown in other application areas (e.g. productions logistics).

3 Illustration and Assessment of Future Assembly Work Activities

An simulation-integrated procedure for prospective stress analysis e.g. in the assembly area should allow for the temporal progression of stress to be determined over an arbitrarily long period of time (e.g. one shift, one week or one month) for each worker, dependent upon the activities executed (in simulation). Not only the objective, measurable stress factors should hereby be taken into account, rather also those stress factors, which are only indirectly ascertainable, for example due to work organization and work content. With respect to the development of the simulation procedure discussed here, this means that a suitable procedure or module for the determination of combined stress factors must be made available.

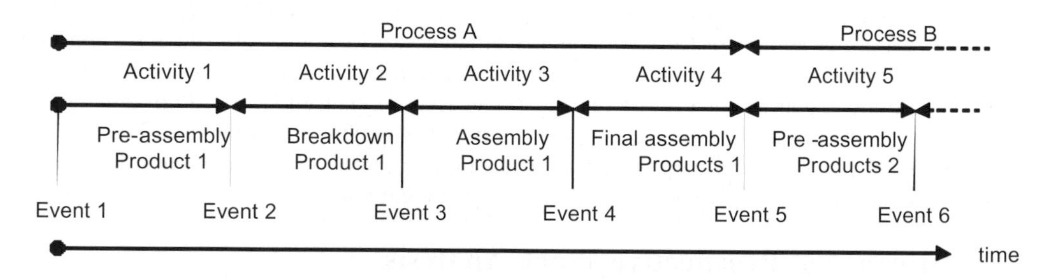

Figure 2: Representation of event-controlled activities in the assembly area (following Backes 1997, p. 119, modified)

The starting point of the simulation could be, for example certain, actual stress situations, which are determined with available or yet to be conceived work analysis procedures. The subsequent

simulative stress prognosis then considers, for example work organizational or technical changes to the work system as planning scenarios and the resulting stress situations can be investigated in a computer model. Not only the stress situation of an individual worker or workplace can be forecasted with such a procedure, rather also an entire work area or work group. In this manner overloading of individuals can be determined and offset with appropriate measures (e.g. with change of personnel assignments). The procedure could support the planner with hints as to which elements in the work system have been changed, and in what way, in order to avoid critical stress situations. A correlation between a simulated event and the executed activity can occur as illustrated in Figure 2. In an event-oriented system structure the system state remains constant between two events. For this reason it is know immediately after an event which events will occur next. These can be administered in a list, which assumes the timing. The activities 1-4 represented here, each terminated by two events, make up Process A, which is followed by Process B etc.
Each activity can then be assigned to a defined combination of stress factors. The resulting worker stress can be ascertained and transferred into an appropriated assessment scheme. A stress prognosis for an individual worker and a defined event timeframe can then be derived from this scheme (e.g. superposition, using a separate module). A subsequent assessment of the total stress of a worker can only be carried out in this manner to a limited degree since, for example the individual stress of a worker resulting from the stress factors and from personal and situational factors is not yet taken into account here.

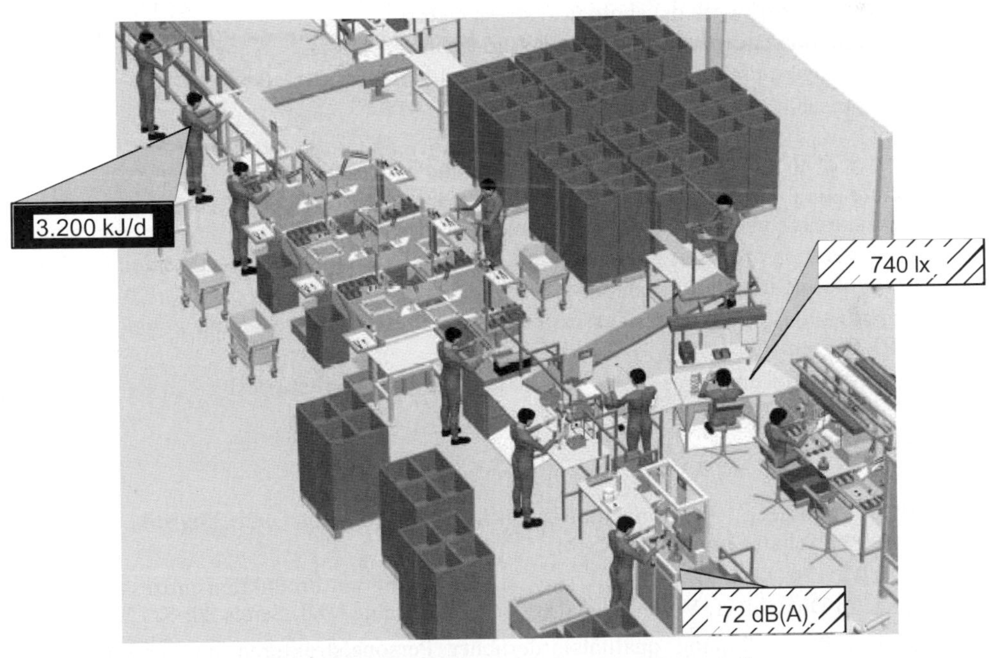

Figure 3: Stress levels in a 3D-layout-representation
(here ERGOMAS; following Zülch, 2000, p. 195)

The complete set of data required for a prospective simulation procedure is assuredly very complex, very difficult to acquire and must be continuously actualized after changes to the work and activity processes. The data material provided for this purpose can not only be used to describe the working conditions of the work processes, rather can also be prepared as an aid for the configuration of work systems or be made available as an information system. Generally, it seems feasible

that currently available logistic and plant management systems can be provided with this data material.
The visualization of this data requires a user-interface corresponding to communications ergonomic demands. One possible procedure is illustrated in Figure 3. As an example, the resulting stress for the workers from individual events, activities or even individual workplaces in connection with activities in an assembly area can be deposited in a 3D-layout of the area to be considered using a menu technique and, in an almost arbitrary manner, filled with further, relevant data. It must however be considered that, for example in case of the rearrangement of a machine in the shop floor, all measurement points need to be measured and entered anew. Experiences with computer-supported plant management systems show that the data administration associated with this type of data management entails large personnel expenditures. These expenditures are however justified when compared with the possible advantages from the examination of the simulation results.

4 Future Works

The solution elements of a simulation procedure presented in Figures 2 and 3 are only examples. Although this list is not exhaustive, a discrete event simulation procedure for worker stress assessment should support all of the following fields of functions:

- Simulation of work psychological stress factors,
- Simulation of organizations psychological stress factors,
- Planning-equitable visualization of the simulation results, and
- Representation of combined stress types.

The complexity of the models necessary for simulation-integrated prognostics increases dramatically with increasing number of stress types to be taken into account. Due to the complexity of the mentioned task areas some aspects of the following sub-modules should be developed in future:

- Consideration of pauses and operation time, simulation of the stress situation resulting from informational work content;
- Organizations psychological assessment with respect to the sequential completeness of work content;
- Assessment of time stress from tasks to be executed simultaneously.

5 References

ASIM (1997). Leitfaden für Simulationsbenutzer in Produktion und Logistik. Eigenverlag Arbeitsgemeinschaft Simulation (ASIM).

Backes, M. (1997). Simulationsunterstützung zur zielorientierten Produktionsprozessplanung und -regelung. In Düsseldorf: VDI Verlag. (Fortschritt-Berichte VDI, Series 20, Nr. 237)

Schindele, H. (1996). Planung qualitätsförderlicher Personalstrukturen im Fertigungsbereich. Karlsruhe, Uni Diss., 1996. (ifab-Forschungsberichte aus dem Institut für Arbeitswissenschaft und Betriebsorgansiation der Universität Karlsruhe, Band 12 - ISSN 0940-0559)

Zülch, G., Heel, J., Lunze, G., Hohendorf, Ch., & Schweizer, W. (1998). Simulationsunterstützte Personaleinsatzplanung. In A. Kuhn; M. Rabe (Eds.), *Simulation in Produktion und Logistik – Fallbeispielsammlung* (pp. 91-126). Berlin et al.: Springer-Verlag.

Zülch, G. (2000). Arbeitsschutz zwischen Umsetzungsdrang und Forschungsbedarf. In G. Zülch, B. Brinkmeier (Eds.), *Arbeitsschutz-Managementsysteme* (pp. 185-201). Aachen: Shaker-Verlag.

Section 6

Design & Visualisation

Mining Network Quality of Service with Neural Networks

Ajith Abraham and Johnson Thomas

Department of Computer Science
Oklahoma State University
USA
email: jpt@okstate.edu

George Ghinea

Department of Computing
Brunel University
UK
email: george.ghinea@brunel.ac.uk

Abstract

Recent research has been focused on how user satisfaction is impacted by varying multimedia Quality of Service (QoS). Some limited work has also been reported on the relationship between QoS and user understanding of multimedia presentations. However, the relationship between user understanding and satisfaction has not been addressed and in this paper we investigate this complex relationship. As a first step towards elucidating a formal model between understanding and satisfaction, we develop a neural network to model user understanding and satisfaction. A neural network based on the conjugate gradient algorithm providing high accuracy for prediction of user satisfaction and understanding is developed.

1 Introduction

Multimedia applications in entertainment, education, and business are increasingly being delivered over networks such as the Internet. It is widely recognized that the effectiveness of multimedia applications depends largely on the performance capabilities of networking protocols and communication delivery systems. However network congestion caused by limited bandwidth results in packet loss, bit errors and out-of-order arrivals result. Consequently, a great deal of research in this area has focused on the technical and networking aspects of delivering multimedia applications. The success of a particular application, however, is ultimately determined by the end-user's experience. Research into the end-user's perception of and satisfaction with multimedia applications delivered over networks has been relatively limited.

Quality of Service (QoS) refers to the quality of the network services that deliver a multimedia presentation. Traditional approaches of providing Quality of Service (QoS) to multimedia applications have focused on managing different technical parameters such as delay, jitter and packet loss over unreliable networks. Rather than technical parameters, the end user is more concerned with enjoying the overall multimedia display while at the same time assimilating its informational content. The quality of the presentation influences the satisfaction rating of the presentation from a user perspective and has an impact on user perception of the presentation. User perception of multimedia has been studied extensively in the educational psychology and Human Computer Interaction fields (Hapeshi et. al, 1992) (Reeves et al., 2000), where it has been assumed, that the underlying network is able to provide an excellent quality of multimedia presentation. Apteker showed that the dependency between human satisfaction and the required bandwidth of multimedia clips is non-linear (Aptekar et al., 1995). Most of these previous research focused on the satisfaction component of perception. Very little work has been reported on the effect of varying QoS on user assimilation and understanding.

We define a user-level defined *Quality of Perception* (QoP) which is a measure which encompasses not only a user's satisfaction with multimedia clips, but also his/her ability to perceive, synthesize and analyze the informational content of such presentations. In our study a number of experiments were conducted to determine the variation of user-level QoP with the Quality of Service associated with the multimedia presentation. We focused on QoS parameters at the application level, which can be broadly classified into temporal (interframe) and spatial (intraframe). In the case of video, the frame rate is a temporal parameter whereas the color depth is a spatial parameter. A questionnaire type of analysis was used to determine user QoP in terms of user satisfaction and information assimilation with varying QoS. The video clips ranged highly dynamic scenes such as a game of rugby to static scenes such as a chorus clip. ANOVA statistical analysis on the results obtained from the experiments (Ghinea et al., 1998) showed that

1. Reduced bandwidth caused by lower frame rates and reduced color depth does not significantly affect user perception
2. Although reducing the frame rate or changing the color depth individually does not significantly alter user satisfaction, the effect is significant when both parameters are modified simultaneously.
3. User satisfaction is dependent upon clip content.

The relationship between understanding (QoP) and satisfaction has not been studied. This relationship is complex and not well understood. Recently data mining has found lots of applications for nontrivial extraction of implicit, previously unknown, and potentially useful information (present knowledge in a form which is easily comprehensible to humans) from data by using several soft computing paradigms, statistical and visualization techniques. In this paper, our motivation is to improve the previous knowledge obtained using ANOVA analysis. We address the inter-relationship between understanding and satisfaction for the different video categories at a particular frame rate. Although our ultimate objective is to capture this relationship between understanding and satisfaction in the form of a function or *if-then* rules, in this paper as a first step we develop a neural network to model this relationship. The literature has reported very limited applications of neural networks and intelligent techniques including soft computing to the HCI field. For example, (Cheng et al., 1997) and (Geisler et al., 2001) model user knowledge and user preferences respectively using neural networks.

2 Artificial Neural Networks (ANN)

ANN is an information-processing paradigm inspired by the way the densely interconnected, parallel structure of the mammalian brain processes information. Learning in biological systems involves adjustments to the synaptic connections that exist between the neurons. Learning typically occurs by example through training, where the training algorithm iteratively adjusts the connection weights (synapses). These connection weights store the knowledge necessary to solve specific problems. Backpropagation (BP) is one of the most famous training algorithms for multilayer perceptrons. BP is a gradient descent technique to minimize the error E for a particular training pattern. For adjusting the weight (w_{ij}) from the i-th input unit to the j-th output, in the batched mode variant the descent is based on the gradient ∇E ($\frac{\delta E}{\delta w_{ij}}$) for the total training set:

$$\Delta w_{ij}(n) = -\varepsilon * \frac{\delta E}{\delta w_{ij}} + \alpha * \Delta w_{ij}(n-1) \quad (1)$$

The gradient gives the direction of error E. The parameters ε and α are the learning rate and momentum respectively.

In the Conjugate Gradient Algorithm (CGA) a search is performed along conjugate directions, which produces generally faster convergence than steepest descent directions. A search is made along the conjugate gradient direction to determine the step size, which will minimize the performance function along that line. A line search is performed to determine the optimal distance to move along the current search direction. Then the next search direction is determined so that it is conjugate to previous search direction. The general procedure for determining the new search direction is to combine the new steepest descent direction with the previous search direction. An important feature of the CGA is that the minimization performed in one step is not partially undone by the next, as it is the case with gradient descent methods. The key steps of the CGA is summarized as follows:

- Choose an initial weight vector w_i.
- Evaluate the gradient vector g_1, and set the initial search direction $d_1=-g_1$
- At step j, minimize $E(w_j + \alpha d_j)$ with respect to α to give $w_{j+1} = w_j + \alpha_{min} d_j)$
- Test to see if the stopping criterion is satisfied.
- Evaluate the new gradient vector g_{j+1}
- Evaluate the new search direction using $d_{j+1}= -g_{j+1} + \beta_j d_j$. The various versions of conjugate gradient are distinguished by the manner in which the constant β_j is computed.

An important drawback of CGA is the requirement of a line search, which is computationally expensive. The Scaled Conjugate Gradient Algorithm (SCGA) is basically designed to avoid the time-consuming line search at each iteration. SCGA combine the model-trust region approach, which is used in the Levenberg-Marquardt algorithm with the CGA. Detailed step-by-step descriptions of the algorithm can be found in Moller (Moller, 1993).

2.1 Experimental Setup Using Neural Networks and MARS

Over 70 subjects watched video clips at varying frame rates and color depths. The users were analyzed for QoP and satisfaction using appropriate questionnaires. In order to expand the data pool, we introduced simulated data interpolated from the real user data. The experiments reported here therefore are based on a combination of simulated data and real user data. Approximately 110 data items were used for training the neural network and the test set contained of 10 data items. The users viewed 12 different types of video clips at low QoS (8 frames per second, 5-bit color depth). The different clips had different characteristics. Some were highly dynamic (action movie for example) whereas others were static in nature (chorus). In some clips the video media was the primary conveyor of information, whereas in other the audio was of more importance. The resulting neural network therefore has 12 inputs and 12 outputs. Neural network training and testing were carried out on a Pentium IV 2 Ghz machine and the codes were executed using MATLAB. Test data was presented to the network and the output from the network was compared with the actual data.

- ANN – SCG algorithm

  We used a feedforward neural network with one hidden layers consisting of 30 hidden neurons. We used tan-sigmoidal activation function for the hidden neurons. The training was terminated after 1000 epochs.

3. Performance and Results

Figures 1 and 2 illustrate the training performance of the proposed neural networks. Figure 1 shows the training performance of a neural network modelling QoP (understanding) as input and satisfaction as output. Figure 2 shows the training performance of a neural network modelling satisfaction as input and QoP (understanding) as output. After training the root mean square error values are 0.0038 and 0.0032 respectively.

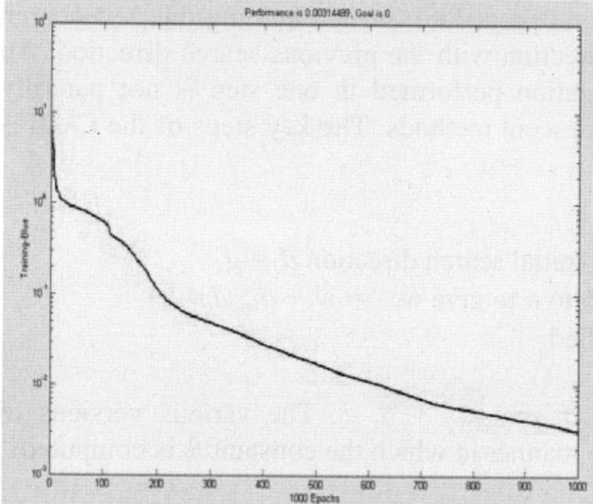

Figure 1: Training performance: modeling the understanding (input) and satisfaction level (output)

Figure 2: Training performance: modeling the satisfaction (input) and understanding (output)

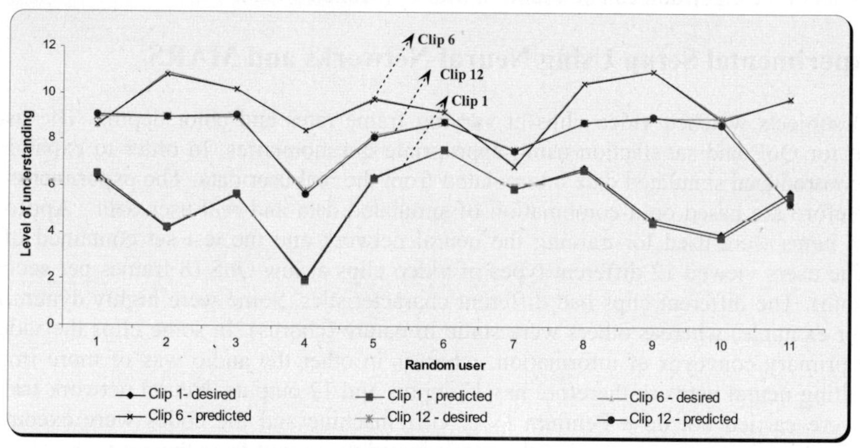

Figure 3: Predicted understanding and real understanding

Figures 3 and 4 shows the results on the test set. Given the user satisfaction level, the neural network model is able to accurately predict the user understanding level (figure 3). Figure 4 shows the predicted user satisfaction level given the understanding (QoP) level of a user. Moreover the neural network model is able to accurately predict user satisfaction (or understanding) across a wide variety of video clips. Irrespective of the dynamicity of the clip or the type of media that is the major conveyor of information (video, audio etc.) the neural network model provides a reliable and accurate prediction mechanism.

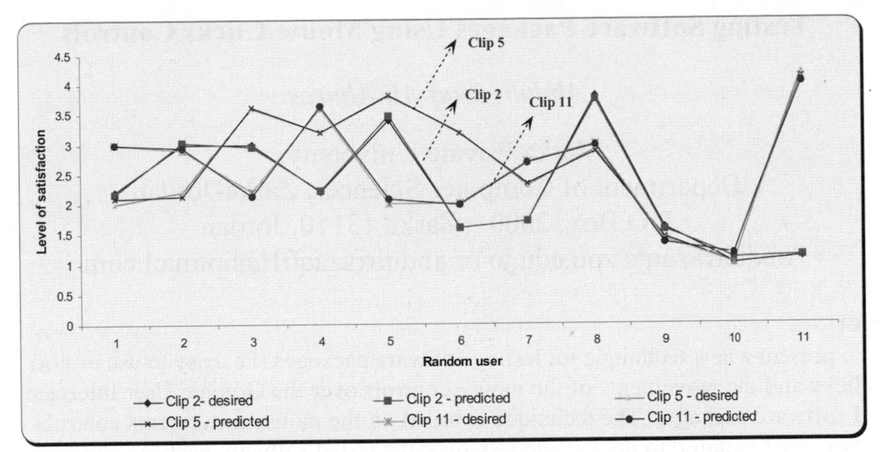

Figure 4: Predicted satisfaction and real satisfaction

4. Conclusions

In this paper we present a neural network model for predicting user satisfaction and user QoP. This is a first step towards providing a formal function for representing the QoP-satisfaction relationship of a human watching video. This relationship has significant repercussions for Internet based e-commerce, distance learning and entertainment. Of particular interest is the function between QoS and QoP/satisfaction. Predicting understanding and satisfaction at different levels of QoS is of great interest to network service providers. However our research is limited by the amount of real user data available. Our first step is to collect more user data. The main objective of our research is to model uncertainty and human perception and cognition (QoP, satisfaction etc) of multimedia by constructing computationally intelligent hybrid systems consisting of neural networks, fuzzy inference system, approximate reasoning and derivative free optimization techniques.

5. References

Apteker, R.T., Fisher, J.A., Kisimov, V.S., and Neishlos, H. (1995) Video Acceptability and Frame Rate, *IEEE Multimedia*, Vol. 2, No. 3, pp. 32-40.

Chen Q and Norcio A F. (1997). Modeling a User's Domain Knowledge with Neural Networks, *International Journal of Human-Computer Interaction* v.9, n.1, pp.25-40.

Geisler Ben, Vu Ha and Haddawy Peter. (2001). Modeling User Preferences via Theory Refinement *Proceedings of the 2001 International Conference on Intelligent User Interfaces*.

Ghinea G and Thomas J P. (1998). QoS Impact on User Perception and Understanding of Multimedia Video Clips, *Proc. ACM Multimedia Conference*

Hapeshi, K. and Jones, D. (1992). Interactive Multimedia for Instruction: A Cognitive Analysis of the Role of Audition and Vision", *International Journal of Human-Computer Interaction*, Vol. 4.

Moller A F. (1993). A Scaled Conjugate Gradient Algorithm for Fast Supervised Learning, *Neural Networks*, Volume 6, pp. 525-533.

Reeves, B. and Nass, C. (2000). Perceptual user interfaces: perceptual bandwidth, *Communications of the ACM*, Vol. 23, No. 3, pp. 65-70

Testing Software Packages Using Mouse Clicks Controls

Abdulrrazaq Ali Aburas

Zarka Private University,
Department of Computer Sciences, Zarka-Jordan
P.O.Box: 2000 – Zarka 13110, Jordan
abdulrrazaq@zpu.edu.jo or abdulrrazaq01@hotmail.com

Abstract

We present a new technique for testing software packages (i.e. easy to use or not) based on the left clicks and the movements of the mouse controls over the Graphic User Interface (GUI) of the tested software package. The technique is based on the mouse movements controls which are the left clicks that required to do a single task over the tested software package and computing the time that required to do the task which is the evaluation key for the testing. We provide the analysis for the behavior of the mouse controls over varity of users for known software package (Paintbrush program). The full operations of the new technique architecture are described. The conclusion of calculated results shows a clear relationship between the mouse clicks controls (i.e. user response) and the software package complexity is included.

1 Introduction:

When a new software package comes out to the real world from the producers, they are used several methods (software tools) to ensure that the software program, which are producing here, is very easy interactive Graphic User Interface (GUI) or it has good Man Machine Interface (MMI). Applying Fitt's Law[1][2], the interface must be user friendly and has very clear or unambiguous commands[3]. One of these common methods is by testing the new software package over different unknown randomly selected users which they have different computer knowledge then the producers take all the comments of how the users interact with it. Some other methods as by design and redesign over and over the GUI for that particular program[4].

We currently present a new approach "media" program that is taking in account the different users (i.e. mouse left clicks) and designing GUI (i.e. the movements) to do a single implementation task. The new approach we present in this paper is a software program taking advantage of the input mouse controls which are the movements of the curser (pointer) and the left click that comes from the users behaviour. The software provides a database for each user and it has default timer as the producers of the new package has set up the minimum time required to do the single implementation task using the Fitt's Law formula[1]. Our software program used for terminating the test user if the user time is set and for comparison purposes between users behaviours.

In the next section we introduce the new proposed approach "media" program that lay between the users and the tested new software package. In section 3, we provide the analysis of the experimentation to estimate the probability of successful and unsuccessful users of getting the single implementation task done, the probability of number of users did and did not do the task over several attempts and the number of users who did at least two successful tests. Section 4; the simulation test results and its method has been introduced. In section 5, shows the simple mathematical methods used and the distribution results over the experimentation tests. We provide concluding and remarks in section 6.

2 The New Proposed approach:

In this section we discuss the interesting and motivating factors in building up our new "media" program for evaluating software package[1][2][3]. In other word, all the new produced software package have Graphic User Interface (GUI) setting at the top of the functional code. The mouse controls is used to perform these functions via interface commands by clicking the left mouse bottom. Thus, the mouse is essential for each new software package[1] as a way of using it because the mouse provide easy movement, perform and control all over the interface software package and hence, it makes the process of our new approach "media" is valid and more effectively of testing the new produced software package[4].

For the above reason, we build up a program "media" such that the main input to this program is the mouse click controls, its movements and its left clicks locations above the tested software package the "media" simulated Paintbrush software package. We simulate the Paintbrush software package because it is very easy and well-known program. We used term "media" as better description for our new approach.

2.1 The New Proposed architecture:

We assume for simplicity and as matter of fact that the Paintbrush software package is tested program by wild range of users and to be easy tools for drawing, creating and displaying objects such as images, text, colour, etc.. . We simulate the Paintbrush software package (see Fig.1) and included with our new approach "media" as mask over it to prove it once again, it is an easy tools based on the mouse controls activity which is the only way of seeing how easy is the software package?. The "media" program for testing or evaluating software packages has three main functions as follows:

Function I: it is used the Fitt's Law as part of it to compute the number of the mouse left clicks when the user press it as a part of finishing his/her single implementation task over the tested package. This function will indicates that less numbers of left clicks means easy interfaced package and more numbers of left clicks means difficult interfaced package.

Function II: it is used to monitor the mouse user's movements over the tested software package. This function indicates that more movements over the tested package means that ambiguous GUI and difficult interfaced package, on the other hands, where less movements means that clear GUI and easy interfaced package.

Function III: it is used for monitoring the user time. The user are unknown and randomly selected. This time is set to finish his/her single implementation task (see Fig.2) over the tested software package, we set up time[1] called "timer" and the "media" program used it as a way of terminating the "media" program and it is used for efficiency as well.

By setting up these three functions in our "media" program we believe that it has been over controlled the mouse clicks controls. The "media" program have well structured data base which gives and produce a lot of calculated statistical results about the user behaviour to assist our analysis to determine how easy or difficult the new tested software package which is the aim of this paper.

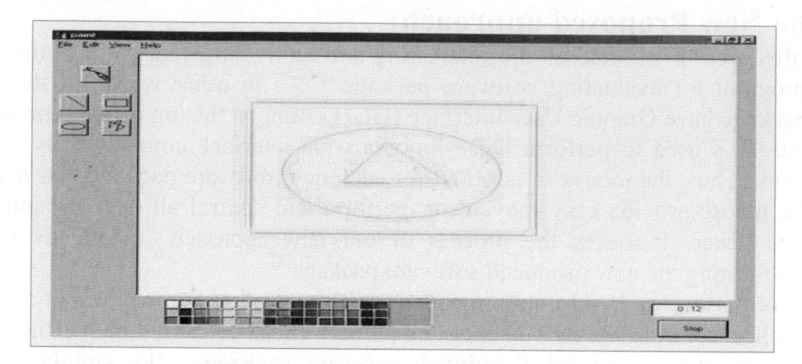

Fig.1: The Main window of the "media" program

In Fig.1, shows the "media" simulated Paintbrush program, main commands program, the working area and the timer for each experiment appear at the right bottom corner of the program included an example task to be preformed by the users, it is appear in the main working area.

3 The Experimentation test:

We provide single implementation task that consist of three attempts (tests) to estimate or determine how easy or difficult the new tested software package. We select random volunteers mail/female users used and unused Paintbrush software package to use our "media" program for testing a single implementation task (see Fig. 2) and must be done by the simulated Paintbrush program.

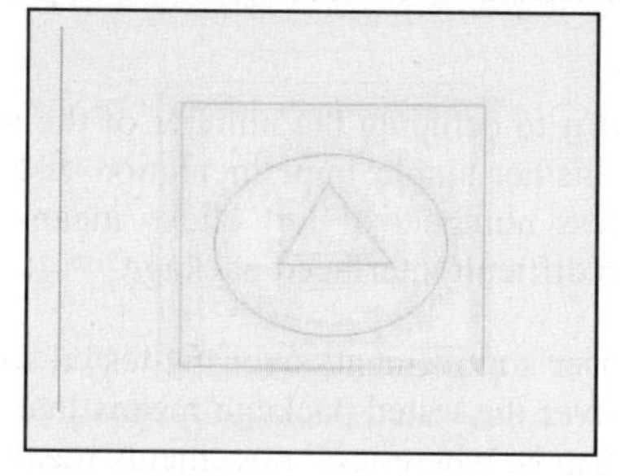

Fig. 2: The single implementation task

The single implementation task (see Fig. 2)is consists of three objects which are triangle inside circle inside a square respectively. The user is free to do the task in any order of drawing the objects as long as s/he finished before the timer of the "media" program is set.

4 The Simulation Test results:

Our "media" program creates a tables as database for every single user and its their experimentation results of how they behaved to finish the given single implementation task. The table (see Fig.3) shows a structure of the fields, some data comes from the tester and some data comes from the media program. The timer started as soon as the user uses the mouse and pushes the left click on the appropriate command on the interface of the "media" program.

	Test1			Test2			Test3			avr time	avr left click	Notes
	△	□	O	△	□	○	△	□	○			
1												
									Total Aver			

Fig.3: The table for recorded data.

The table for recorded data (see Fig.3) have fields such, as the data fields that come from the tester are test1, test2.test3 such as O, □, Δ and comments notes. The data fields that comes from the "media" program are test1, test2, test3 such as time, left click no., the avr_time and the avr_left click.

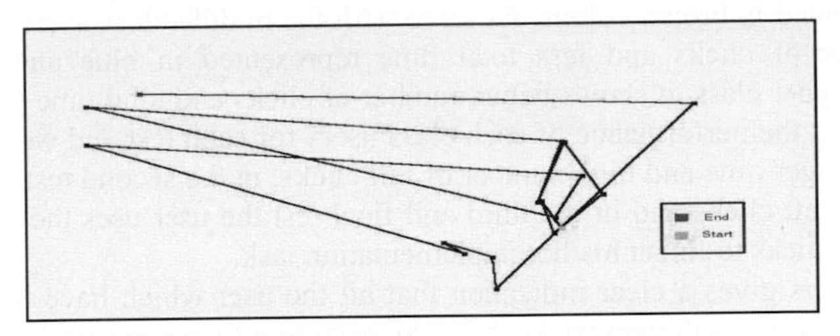

Fig.4: The user mouse movements

The user mouse movements (see Fig.4) shows the user mouse movements and left clicks positions over the GUI of the tested software package. The diagram shows also the user working area over the software package that indicates in undirected way of the position and location of the interface commands and menu which helps the analyst to determined the behaviours of the user during the testing time.

5 The Analysis of the simulation results:

We employed simple statistical method to analysis the results. Over many unknown randomly selected users, we divided the users into three different classes such as the beginner class, the middle class and the higher class depend on the knowledge of using the computer software packages, in general[5][6][7][8]. The methods are estimated different probability of presents that indicates the low and high complexity of the tested software package. The methods gives %44 of which user spend time to finish his/her test, it gives %54 of which user is left click the mouse to do the single implementation task. From the results of the previous, it is indicate that it is not difficult software package. Also it gives %52 of which user are finished test from first attempt, it gives %64 of which user are finished test from second attempt, and it gives %70 of which user are finished test from third and final attempt. As results of increasing the percents so that the indication it has a low complexity (easy) for the tested new software package and that is mean, it is easy package. The simulation results gives %70 of which user are used first time the computer and Paintbrush software program, also gives %36 of which users are finished all tests from first attempt, and it gives %66 of which users are finished two tests from first attempt. As results of increasing the percents, so that indicates how easy the tested new package because of high percents. Finally the simulation results indicates once again that it has a low complexity for the tested new software package.

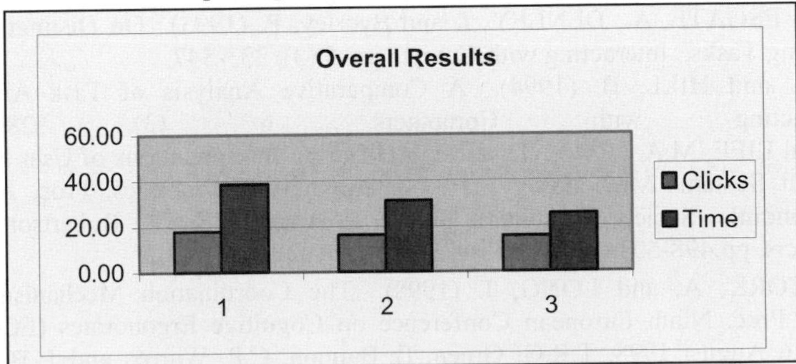

Fig.5: Over all Statistical Results.

From the diagram Over all Statistical Results (see Fig.5) which summarized over all the statistical results produced by our experimentation the "media" program using the simulation.
From the data base statistical results for all the users (Fig.5) shows as no. 1 in the Figure the beginner class of users where the number of clicks were high represented by blue colour and total time were high as well represented in brown colour. As same with the middle class users, no.2 in the figure shows less number of clicks and less total time represented in blue and brown respectively. In the no.3 and higher class, it shows better number of clicks and total time for users in this class. We have compared the performance of each class users for each test and we noticed that at first test the user take longer time and high number of left clicks, in the second test the user takes less time and number of left clicks and in the third and final test the user uses the smallest time and lowest number of left clicks to finish his/her implementation task.
This statistical results of all users gives a clear indication that all the user which have the same numbers of tests shows that the tested software package has an easy interface commands because the time and number of the mouse left clicks for each user was over all small and the number of users who did finished the single implementation task was high compared to the total number of users regardless of which user is used the Paintbrush software package or not.

6 Conclusion:

We introduced a new approach of "media" program used the mouse clicks controls to determine which software package is easy or not from the interface point of a view. The analysis of the results provided a sold answer that producers must redesign the GUI software package if the time and number of the mouse left clicks controls was high. Further work could be done is introducing a new strategies of designing GUI and possibility of introducing a new second mouse to the computer hardware for minimizing the mouse movements and number of clicks over the software package for efficiency. Possibility of using the number of the mouse left clicks as second password for users which could take it as one big advantage of this research work.

References

[1] Accot, J. & Zhai, S. (1997). Beyond Fitts' law: models for trajectory-based HCI tasks. Proceedings of ACM CHI'97 Conference on Human Factors in Computing Systems, pp 295 302

[2] Johnny Accot & Shumin Zhai (1999) "Performance Evaluation of Input Devices in Trajectory-based Tasks: An application of The Steering Law", Proceedings of ACM CHI'99 , pp 466 –472.

[3] The Human Computer Interaction; Jenny Preece, 1994, Addison-Wesley.

[4] Ethics in HCI, CHI2001, Panal Documentation.

[5] Developing Software for the user Interface; Len Bass & Joelle Coutaz, 1991, Addison-Wesley.

[6] DOWELL, J. (1995). Tabletop Crisis Training and the Requirements for Computer-Based Simulation. Journal of Intelligent Systems, 5 (2-4).

[7] LONG, J. (1996). Specifying Relations between Research and the Design of Human-Computer Interactions. Int. J. Human-Computer Studies, 44 (6), 875-920.

[8] WHITEFIELD, A., ESGATE, A., DENLEY, I. and Byerley, P. (1993). On Distinguishing Work Tasks and Enabling Tasks. Interacting with Computers, 5 (3), 333-347.

[9] WHITEFIELD, A. and HILL, B. (1994). A Comparative Analysis of Task Analysis Products. Interacting with Computers, 6 (3), 289-309.

[10] DUIGNAN, K. and LIFE, M.A. (1997). Disclosing Differing Interpretations of User Needs: An Application of Soft Systems Methodology. In Contemporary Ergonomics, Proc. Annual Conference of the Ergonomics Society, Grantham, UK, 15-17 April 1997, S.A. Robertson (ed). London: Taylor & Francis, pp.498-503.

[11] LAMBIE, T., STORK, A. and LONG, J. (1998). The Coordination Mechanism and Cooperative Work. In Proc. Ninth European Conference on Cognitive Ergonomics (ECCE9), Limerick, Ireland, 24-26 August 1998, T.R.G. Green, L. Bannon, C.P. Warren and J. Buckley (eds). France: EACE. 163-166.

Nine Principles for Actable Systems Design

Pär J. Ågerfalk

Dept. of Informatics (ESI), Örebro University
SE-701 82 Örebro, Sweden
pak@esi.oru.se

Abstract

Heuristic evaluation relies fundamentally on predefined criteria to direct attention to important aspects of the inspected phenomenon, such as Jacob Nielsen's ten usability heuristics. In this paper we present nine complementary principles explicitly focused on evaluation of IT-systems as tools for business action and communication. The principles are derived from the concept of actability, which embodies this view, and from an actability interpretation of the set of ten significant features of the casual face-to-face conversation suggested by Herbert H Clark.

1 Introduction

Heuristic evaluation relies fundamentally on predefined criteria (heuristics, principles) used to direct attention to important aspects of the inspected phenomenon. Several lists of criteria focusing user interface (UI) design and its relation to user and task characteristics have been proposed, including Jacob Nielsen's ten usability heuristics and Ben Shneiderman's eight golden rules (see, e.g., Keinonen, 1998). In this paper we present nine complementary principles explicitly focused on evaluation of IT-systems as tools for business action and communication, thus adopting the view that computers are essentially for communication – not computation (Flores, 1998). Along with the principles we discuss related implications for design. The principles are derived from the concept of actability (see, e.g., Ågerfalk, Sjöström, Eliason, Cronholm & Goldkuhl, 2002), which embodies the suggested view, together with an interpretation of the set of ten significant features of casual face-to-face conversations suggested by Clark (1996), from the perspective of actability. The principles stem from earlier work on actability heuristics (Ågerfalk et al., 2002) and actability quality ideals (Cronholm & Goldkuhl, 2002). These studies indicate that the suggested principles are often not upheld, which may be a source for many IT-system failures. To facilitate practical application, our presentation structures the principles differently and thereby stresses the coupling to underlying theoretical concepts. The nine principles aim to provide a thorough basis for the design of IT-systems acknowledging factors emerging when the IT-medium is introduced in social communication processes.

The ten features of the casual face-to-face situation suggested by Clark (1996) are as follows: *co-presence* (participants share the same physical environment), *visibility* and *audibility* (they see and hear each other), *instantaneity* (they recognize each other's action at no perceivable delay), *evanescence* (the medium fades immediately), *recordlessness* (actions do not leave any record or artefact), *simultaneity* (participants may receive and produce at once and simultaneously), *extemporaneity* (actions are formulated and executed in real time), *self-determination* (participants determine for themselves what actions to take when), *self-expression* (participants take action as themselves).

In the following section we discuss these features from the perspective of actability, stressing that IT-system use is the performance of social business action. In doing so, the nine actability principles are developed. The paper concludes with some experiences from a case study in which the actability principles were used.

2 Nine Principles for Actable Systems Design

2.1 The Principle of Action Elementariness

When using a business IT-system, communicators (speakers) and interpreters (listeners) are typically not co-present. When lacking co-presence, implicit references to anything outside of the system are likely to lead to confusion and misunderstandings. In the context of business IT-systems, there is typically neither visibility nor audibility. We say 'typically' since participants may see and hear each other with the adoption of multi-media type of interfaces, even though in typical business IT-systems, such means of communication are probably rare. When participants cannot see and hear each other, gestures, hawks, and other audiovisual clues are not possible to utilize. Taking these features into account *the principle of action elementariness* is founded on the notion that an IT-system is a system for handling action elementary messages (ae-messages) as results of elementary communication actions (e-actions), as suggested by actability. Ae-messages are regarded as elementary information units carrying a propositional content (what is talked about) and an associated action mode (what speaking does, an 'illocutionary force'). To represent the social context of e-actions, ae-messages also carry information about the communicator and intended interpreters of the message. Since participants cannot see, hear or by any other physical means recognize each other or what they are 'talking' about, these properties must be formalized and encoded in the system. From a design perspective the ae-message terms (propositional contents, signifier of action modes (such as illocutionary verbs), and information about communicators and intended interpreters) should be visible and kept together in the UI. The issue of making actors visible is not uncontroversial. CSCW research points at making work visible also open up for criticism, which may lead to more formalized work and reduction of sharing (Ackerman, 2000). On the other hand, Cronholm & Goldkuhl (2002) report on situations in which actor information is essential since interpreters need to get hold of communicators for clarification. The issue of visualizing actors must be treated in relation to the particular system and the important thing is to highlight the issue and to not fall into the trap of unreflective anonymization of information origin. Keeping ae-messages together means that it should be clear from the UI which action mode signifier belongs to which propositional content. It also means that there should be no propositional content that do not belong to any such signifier. The principle of action elementariness also means that separate messages should be kept separate (one thing at a time); it is recommended not to mix up messages directed towards different interpreters or forcing users to perform several e-actions by one UI manipulation (unless this is made clear in the UI and that it is the way things are usually done).

2.2 The Principle of Recorded Action

In contrast to the recordlessness and evanescence of face-to-face conversations, one of the advantages of IT-systems is that business actions may actually leave traces in the system's database. Utilizing this condition *the principle of recorded action* suggests that it is important to maintain an organizational memory of commitments and action prerequisites. This means that systems should be designed in such a way that information about performed actions, scheduled future actions and other action prerequisites are stored and easily accessible. Based on the principle of action elementariness, the action record should be considered as a record of ae-messages (i.e., including all ae-message terms, see above).

2.3 The Principle of Action Potentiality

From an actability perspective, IT-systems can be understood as the set of e-actions they afford and support – they provide an action potential (cf. Carroll, 1996). That is, action is only self-determined as far as the system design allows. Drawing on Habermas' (1984) notion of validity claims raised by communicators and presupposed to be accepted by interpreters, e-actions supported by IT-systems should be not only comprehensible, but refer to the true (or inter-subjectively believed) state of the world, represent sincere intentions, and be possible to perform in accordance with socially accepted norms. Therefore the system should assist users in raising and evaluating the validity of e-actions and corresponding ae-messages. To support such a design, *the principle of action potentiality* suggests that it is imperative that known and understandable effects of possible actions are clearly described and that their meaning and validity are understood and agreed upon. From a UI design perspective, these conditions stress the use of expressive interactive UI components (icons, labels, *etc*). They also stress that information the system requires from performers should be meaningful and easily provided to the system, and that information shown should be adequate (necessary and sufficient) so that actions can be based on it. A basic condition for achieving such designs is that the language used corresponds with the users' regular professional language. Finally, drawing on Habermas' (1984) notion of discursive action, if uncertainties arise concerning possible actions and their validity, systems should support justification by explanations, and preferably negotiation.

2.4 The Principle of Structured Action

While turn taking is self-determined in face-to-face settings and constrained by the very fact that any system restricts possible actions, and action sequences, to perform (see above), business rules and norms largely determine what actions to take when in a business context. E-actions are part of an 'action structure' and the set of possible e-actions to perform at any particular point in time changes as the interaction proceeds. Founded on these assumptions, *the principle of structured action* suggests that, in order for a system to be comprehensible, users should always know what they are doing, and what they are supposed to be doing, based on what is presented in the UI. That is, possible actions should not only be visible (as suggested by the principle of action potentiality), but choice of course of action to take should be informed by the system (the system could, e.g., help users to live up to previous commitments). To support users' participation in various overlapping activities, systems should admit focus and work task changes. Still, sequence restrictions should be enforced when necessary and desirable. It is also important that adopted principles for navigation in the system are made explicit and that the overall type of use situation at hand is communicated (e.g., whether it is possible to perform action through the system, or if the system only provides information for use outside of it).

2.5 The Principle of Irrevocable Action

When performing actions through and by means of an IT-system, the UI is typically not evanescent, but persistent for as long as participants desire. Unless an error occurs, a displayed message will stay displayed until it is deliberately turned off. This condition gives rise to the opportunity to formulate and execute e-actions reflectively during extended amounts of time. In order to facilitate this non-extemporaneity *the principle of irrevocable action* suggests that IT-systems should be explicit about when an action is actually performed and provide rollback (undo) as far as possible (i.e., until a new social fact has been established by use of the system).

2.6 The Principle of Remote Interpretation

The lack of co-presence (see above) also gives rise to the *principle of remote interpretation*, which suggests that systems should be designed as to take into account where and how messages are received; receipt and interpretation should be possible at desired places and in desirable ways.

2.7 The Principle of Delayed Interpretation

Even though IT-systems are interactive, participants may not perceive each other's actions instantaneously, but rather with considerable delay. Consider, for example, an order placed during the night. Such an order may not be displayed for an order-recipient until the next day. Therefore, relying on the principle of recorded action (see above), the *principle of delayed interpretation* suggests that keeping track of when an action is performed is important. It also suggests that care should be taken so that messages reach intended interpreters in due time.

2.8 The Principle of Delayed Feedback

When using an IT-system, participants either produce or receive – but not simultaneously, as in a face-to-face setting. This condition gives rise to *the principle of delayed feedback* stressing that users should understand that no feedback on business effects is given until a message has actually been delivered and interpreted. Obviously, the design should strive to minimize such delays.

2.9 The Principle of Delegated Action

All communication involves self-expression. Nonetheless, one aspect of business action stressed by actability is that people as well as IT systems may take action on behalf of other people and their organizations. This means that social obligations may be created through an IT-system without the responsible parties' direct interaction with the system. This condition leads to *the principle of delegated action* which suggests that performance of actions should be allocated to human actors and information systems so that users gain maximal support (e.g., in terms of decision support *vs.* automated actions), that description and explanation of the system's performed and scheduled future action(s) are readily available, and that all actors involved are aware of their action relationships, even though they may not be directly interacting with the system.

3 Some Experiences from Using the Actability Principles

The suggested actability principles have been used during an evaluation of a Swedish manufacturing company's intranet (referred to as The Intranet). The Intranet is, e.g., used for disseminating information and booking different resources. The study followed the evaluation approach described by Ågerfalk et al. (2002) and involved expert-based evaluation of The Intranet as well as a number of interviews with people using it in their daily work.

In the study, the actability principles helped to direct attention to phenomena not easily revealed at first. For example, the principle of action elementariness made us focus on actor visibility. At first we were pleased to see that information about resource owners (those who are responsible for particular resources) was displayed and that these people were easy to get in contact with (via links to contact information as well as mailto-links). Later on, the principle of delegated action made us investigate more thoroughly the actual responsibilities claimed by the system. This made us understand that the seemingly well-presented action responsibilities were not clear at all, but rather an important issue for future redesign of the system and its business context. It is, for example, necessary to sort out responsibilities and implement required consistency and integrity checks in the system. An example of problems related to this issue is that if a resource is deleted in the system altogether, it may still have bookings pending. This issue relates also to the principle of recorded action – the action of reporting deletion of resources was missing in the system. Furthermore, the principle of action potentiality made us focus on the information needed in performing different tasks. This revealed a problem with lack of search facilities regarding allocation of resource owners to resources owned. This proved to be a problem for both the resource manager (the one who allocates resources to resource owners), as well as to the resource owners. It was also interesting to see how the principles of delayed interpretation and delayed feedback helped to direct attention to a promise made in the system regarding when a response to a booking request was to be made by the resource owner (within five days). This was understood to be important information but it was suspected that the promise was not always kept. This turned out to be true, and also a major concern for the resource owner. The principle of remote interpretation made us suggest that requests might be handled via mobile devices when out of office. This turned out to be appealing to the resource owner who suggested it to be a redesign proposal to consider (including the integration of The Intranet with PDA organizers). Furthermore, the principles of structured action and irrevocable action directed our attention to a lack of action sequence restrictions needed to make sure that all bookings are cancelled before a booked resource may be deleted from the system. In this case, the deletion of a booked resource resulted also in cancellation of associated bookings, performed in the background by the system itself, and also resulting in inconsistent data. Obviously The Intranet could have warned the resource manager about pending resources, thus allowing the deletion action to be revoked as required.

4 References

Ackerman, M. S. (2000). The Intellectual Challenge of CSCW: The Gap between Social Requirements and Technical Feasibility, *Human-Computer Interaction,* 15(2/3), 179–203.

Ågerfalk, P. J., Sjöström, J., Eliason, E., Cronholm, S., & Goldkuhl, G. (2002). Setting the Scene for Actability Evaluation: Understanding Information Systems in Context, *Proceedings of ECITE 2002* (pp. 1–9), 15–16 July 2002, Paris, France.

Carroll, J. M. (1996). Becoming Social: Expanding Scenario-Based Approaches in HCI, *Behaviour & Information Technology,* 15(4), 266–275.

Clark, H. H. (1996). Using Language, Cambridge: Cambridge University Press.

Cronholm, S., & Goldkuhl, G. (2002). Actable Information Systems: Quality Ideals Put into Practice, *Proceedings of ISD 2002,* 12–14 September 2002, Riga.

Flores, F. (1998). Information Technology and the Institution of Identity: Reflections since Understanding Computers and Cognition, *Information Technology & People,* 11(4), 352–372.

Habermas, J. (1984). The Theory of Communicative Action, Cambridge: Polity Press.

Holistic Communication Modelling: Enhancing Human-Centred Design through Empowerment

Berki, Eleni eleberk@cc.jyu.fi

Isomäki, Hannakaisa[1] hannas@cc.jyu.fi

Jäkälä, Mikko mikko@cc.jyu.fi

Department of Computer Science and Information Systems
P.O.Box 35 (Agora)
FIN-40014 University of Jyväskylä
Finland

Abstract

The existing IS development methods are not communicative enough and therefore have not been widely used to achieve this purpose holistically and in an integrated manner. It is important for IS designers to use human-centred development methods, which also cater for the needs of the end-users. This paper focuses on strengthening the role of human involvement in IS design through *empowerment*. In so doing, communication aspects of IS development methods are examined, and the context and content of *holistic communication*[2] is developed. We furthermore suggest that in order to enhance human-centred IS design, we need to explore and apply the concepts and capabilities of holistic communication in the process of IS design.

1 Introduction

The selection of IS development methods is usually carried out by IS designers using scientific frameworks, or it is guided by ad-hoc knowledge and experience in a field. We might ask whether there is any communication-oriented method among the plethora of methods to adequately capture *all* the design requirements and organisation's lifecycle?

Modelling using design process models that are not communicative in their nature does not contribute to human-centred design and, thus, does not end-up to human-centred and communicative IS (Sölvberg 2000). Throughout the different stages of modelling, a successfully defined process needs to capture and communicate designers' interdisciplinary knowledge and analytical skills. Supporting the previous thoughts for co-operative concept modelling, Bubenko (1995) asserts that formal methods should be enhanced by developing user-designer communication. Design models should be improved by including IS designers' problem-solving intuitions and other cognitive needs related to conceptual modelling, such as own quality definitions concerning the holistic needs of the users. Eason (2001) restates that the design process

[1] Author for correspondence and off-print requests
[2] Holistic communication in ISD: Concept and term coined and defined by the authors of this paper

remains techno-centric; the ongoing development of virtual organisations points to the necessity of socio-technical approaches.

The traditional process of IS conceptualisation needs to be extended in order to capture the demands and interactions of users in the most effective way. The influence of multi-agents and multi-users of organisations in the design of IS and work processes is considered to be very powerful for the functionality and reliability of the IS functions. People need to be encouraged to identify relevant information and present it to others, taking into consideration the information content (Bubenko, 1995), different social, cultural, emotional (Isomäki, 2002), knowledge and linguistic backgrounds. The design process will benefit by communicating holistically these various needs.

2 Towards Communicative Process Models for IS Development

In general, the conceptual models for IS development are not utilised as thinking tools, because they do not represent people's views in a holistic manner. Moreover, they do not offer joint links to accommodate designers' knowledge, intuitions and creativity in an interactive and shared manner. Design processes need to be equipped with shared user models, and the people involved in them need to be connected through communicative ways of working. Yet we cannot model adequately human behaviour; however, we can model and communicate IS designers' different thoughts and concepts.

People have always been inventive in finding efficient ways of solving problems, even when they have not been acquainted with specialised knowledge. People are able to progressively develop and adopt creative and pleasant for them ways of work and act as multi-agents of knowledge under proper communication and thinking tools. For instance, the semantic and syntactic mechanisms that are needed to elicitate the linguistic patterns of stated problems can match expression, creativity and intuition. The interaction and contact with different knowledge disciplines gives insight to multiple viewpoints for many specialised situations.

The challenge for the IS designer is to represent the *semantics* of the stakeholders and the systems requirements in addition to *pragmatics* (correct understanding and individual usage) and *semiotics* (usage) in the real world (Berki, 2001). Nielsen's (1993) model for *usability engineering* emphasises in the predesign phase, that designers should know the users and define their individual characteristics, current and desired tasks besides performing functional analysis. The next challenge for the IS designer is to find an expressive way to integrate and communicate all these, and to extract the relevant and meaningful information that is to be used in IS design.

3 How Holistic Communication Can Enhance the IS Processes

Through the process of communication social relationships are being created, defined, modified, and dissolved. Theories of communication provide thinking tools to understand, describe, explain, evaluate, predict, and control human communication behaviour. They emphasise the communicative forces that prompt human action and enable to predict and control future communicative interactions. Theories of communication provide tools for observing communicators interacting over time in a medium (Cragan & Shields, 1998).

Littlejohn (1996) suggests a five-part model of communication theory. The model classifies communication theories into five genres: structural and functional, interactionist, cognitive and behavioural, interpretive, and critical theories. With the aid of these genres IS development may be enhanced to include holistic communication modelling. We hereby identify the absence of the later in the use and in the metaspecification of the ISD methods and from their associated tools. We furthermore support that IS metamodelling frameworks need to consider to integrate these communication features in their design philosophy (Berki, 2001).

Structural and functional theories include approaches of communication from the viewpoints of social and cultural structures and their attributes. Structuralism has its roots in linguistics and emphasises organisations of language and social systems. Hodge & Kress (1991) examine the notion of communication in society by viewing it essentially as a process, in which language has to be thoroughly assimilated into the social processes because through this empowerment function semantics is constituted and thus the ways of working are more effective.

Interactionist theories focus on communication and social life as a process, where interaction is seen as fundamental for the establishment of behaviour, meaning and language. Interpretive theories include the study of meaning in actions and texts describing how understanding occurs and the ways individuals perceive their own experience. This genre could broaden IS research and practice intellectually and give rise to new delineations of IS as social systems only technically implemented. In order to capture culturally mediated users' activity, interactionist theories could facilitate the interaction between users and ICT as a culturally mediated phenomenon (Kuutti, 1997). Organisational communication can be seen as the meaning generation process of interaction that creates, maintains, and changes the organisation (Pace & Faules, 1994).

Cognitive and behavioural theories focus on the individual and on the way people think. These theories have their roots in psychology. The inclusion of this genre is necessary for the IS designers to be aware of users' cognitive needs, particularly thinking errors which are due to unconscious cognitive biases which serve to reduce and combine mentally cumbersome quantities of information (Robillard, 1999). As Kirs *et al.* (2001) point out, IS designers should recognise cognitive biases in users in order to implement IS planning and design that aims at preventing humans from engaging in faulty actions during ICT use.

Critical theories focus on the quality of communication with special interest in inequality and oppression. These theories criticise the ways communication maintains domination of a group over another. Critical genre is needed in IS development for the recognition of users' emotional states, especially negative ones, which are often due to unequal or oppressing treatment. Brosnan & Davidson (1994) estimate that approximately one third of the population of the industrialised world suffers to some extent from technophobia, which refers to people's negative affective reactions to ICT. Empowerment inherent in holistic communication modelling reduces unequal or oppressing treatment in organisations, thus enabling all stakeholders – IS designers, managers and users - individual self-actualisation during IS development, enhancing the sharing of resources and decision making (Andrews & Herschel, 1996). Empowerment eliminates the emotional deficits at work.

4 Conclusions and Future Considerations

The aim of this work was to interconnect human-centred IS design and communication. We referred to the different philosophies and assumptions underlying IS development, which highlights the need for the integration of the functional and holistic issues that IS processes and stakeholders possess, and the need to communicate them adequately. We expressed our insights and arguments underpinning the need for an alternative way of systems thinking.

Holistic communication design models are required to be adopted in IS design process in order to facilitate and express people's thinking. In order to alleviate unwanted implications, communication issues must be considered in two ways: by facilitating thinking and communication among people involved in the design, and, by considering the capabilities of communication technology itself. A participatory IS design that incorporates multiple views in a holistic IS development process is the basis for establishing empowerment. This is a major contribution for the quality of people's life in learning communities and in IS. Future research will concentrate on further exploration of communication theories and modelling efforts in order to establish and finally simulate efficiently holistic communicative components and effective patterns of human-centred design process models.

References

Andrews, P.H. & Herschel, R.T. 1996. Organizational Communication: empowerment in a technological society. Boston: Houghton Mittlin Company.

Berki, E. 2001. Establishing a scientific discipline for capturing the entropy of systems process models, CDM-FILTERS: A computational and dynamic metamodel as a flexible and integrating language for the testing expression and re-engineering of systems. Ph.D. thesis, University of North London, Faculty of Science, Computing & Engineering. London: U.K.

Brosnan, M. & Davidson, M. 1994. Computerphobia: Is it a particularly female phenomenon? The Psychologist 7(2), 73-78.

Bubenko, J. A. jr. 1995. Challenges in Requirements Engineering. Second IEEE International Symposium on Requirements Engineering, York, U.K.

Cragan, J. F. & Shields, D. C. 1998. Understanding communication theory: the communicative forces for human action. Boston: Allyn and Bacon.

Eason, K. 2001. Changing perspectives on the organizational consequences of information technology. Behaviour & Information Technology 20 (5), 323-328.

Hodge, R. & Kress, G. 1991. Social Semiotics. Polity Press.

Isomäki, H.2002. The prevailing conceptions of the human being in information systems development: systems designers' reflections. Ph.D. thesis, University of Tampere, Dept. of Computer and Information Sciences. Tampere: Finland, A-2002-6.

Kirs, P.J., Pflughoeft, K. & Kroeck, G. 2001. A process model cognitive biasing effects in information systems development and usage. Information & Management 38, 153-165.

Kuutti, K. 1997. Activity theory as a potential framework for human-computer interaction research. In Nardi, B. (Ed.) Context and consciousness. Activity theory and human-computer interaction. Cambridge, MA: The MIT Press, 17-44.

Littlejohn, S. W. 1996. Theories of human communication. Wadsworth: Belmont, CA.

Nielsen, J. 1993. Usability Engineering. Cambridge, MA: Academic Press.

Pace, R. & Faules, D. 1994. Organizational communication. Englewood Cliffs: Prentice Hall.

Robillard, P.N. 1999. The role of knowledge in software development. Communications of the ACM 42 (1), 87-92.

Sölvberg, A. 2000. Co-operative concept modeling. In Brinkkemper, S., Lindencrona, F. & Sölvberg, A. (Eds.) Information Systems Engineering. State of the Art and Research Themes. London: Springer-Verlag.

The Context Quintet: Narrative Elements Applied to Context Awareness

Kevin Brooks

Motorola Human Interface Labs
210 Broadway St., 4th Floor
Cambridge, MA 02139 USA
kevin.brooks@motorola.com

Abstract

For a computational device to be context aware, it needs to be sensitive to physical, social, and task situations. While answering the five basic narrative questions of Who, What, When, Where and Why provides a tremendous amount of understanding and context to a story, making a computational or communications device sensitive to these same questions provide simple and powerful structural guidelines for context awareness in device interface design.

1 Introduction

In our drive to impart an ever-increasing level of intelligence in to our computational and communications devices, one promising subcategory of artificial intelligence is context awareness. If a device could be more aware of its own physical and task context, and aware of the user's social context, then it could potentially use this information to make better decisions on the user's behalf. The topic of context awareness is common within the research domains of pervasive computing, wearable computing and mobile computing. Devices that can be carried in the palm of the hand like a cell phone, or carried in a stylish shoulder bag like a laptop, are getting smaller, lighter, and far more powerful. As this happens, basic level communications and computational capabilities drop in price, further encouraging pervasiveness. If people are going to interact with devices that are plentiful, personal and small enough to always be with them, then these devices should also be aware of what is happening around them and what the user wants to do with them. It is a common metric of intelligence for both people and machines to quantify and qualify how much they know about the world around them. Another metric of intelligence is the ability to understand enough about the world to recount it in a story.

2 Story Context

Telling and understanding stories are intelligent acts, as is suggested by Marvin Minsky in his book *Society of Mind.* (Minsky, 1988). Storytelling is a natural and infinitely pervasive way for people to share information about the world. Most of our common daily communication with each other is in the form of stories. We thrive on stories. We express who we are through our personal stories. From an exceptionally early age we are told stories which begin the formation of our moral and cultural identity. Along with the chemical element carbon, stories are a basic element of our being. We are carbon and story-based beings. Given that story or narrative is so important to the way we think and live, it is reasonable, therefore, to look to narrative for clues about designing systems that communicate personally and cross culturally with people. It is in fact surprising that narrative is not a more prevalent component of interface design.

2.1 Relationship as Context

In storytelling, the storyteller establishes story context in part through the channel of *relationship* with the audience. (Martin, 1996) The words the storyteller uses in the story (assuming oral or written and not a visual story form) certainly go a long way toward establishing context as well. But it is the relationship with the audience that the teller establishes early on that is the conduit of the story context. That relationship is dependent on attributes of the situation – a situational context: whether the audience is composed of children, adults or mixed, whether the audience is composed of mostly men, women, or mixed, whether the audience is familiar or unfamiliar with the teller, and so on. Knowledge of the initial status of the relationship and an awareness of how that relationship changes over time is key to knowing what information the storyteller must relate and how to relate it. A woman telling a story to a group of women from her own culture does not need to specify many details about femininity, for instance, because both teller and audience share a common cultural definition.

One challenge for computational devices is that they were never children, nor gendered. Their inherent inhumanity makes them capable of only approximating the nature of this relationship as context conduit. As interface designers and researchers, we design the basic building blocks of these devices in a small enough granularity such that users can do the bulk of the work toward creating a relationship with our devices. That is, the devices provide the handles for relationship, in the form of customisable icons, reprioritizable interface elements, automatic linking of address book and calendar entries, etc., but it is the user who must work to grasp these handles. There is nothing unusual or wrong about this, but context awareness shifts the paradigm. If a device's awareness could allow it to actively improve the relationship between itself and its user, then it truly could become a personal device. As Karl Kroeber suggests, narrative represents a continuous potentiality for change. It folds and unfolds in on itself, "…transmuted by the very process of absorbing the meanings it has initiated." (Kroeber, 1992) Similarly, Lieberman & Selker refer to context as iterative and "…a state that is both input to and generated by the application persists over time and constitutes a feedback loop." (Lieberman & Selker, 2000)

2.2 Point of View as Context

If relationship is a conduit for context, then *point-of-view* is a result of context. Context means seeing from a point-of-view, usually somebody else's point-of-view. When an audience knows the context of a story, they have essentially adopted and accepted the point-of-view of the story-teller. They have accepted the world provided to them, the relationship is thoroughly established, and the conduit is open. From this point on, the storyteller has both power and responsibility to provide either what they expect/ask for or what they don't expect/ask for. What one asks for is not necessarily what one needs. When a user asks for contact information of a particular person, a context unaware device would simply do a database lookup and provide the information. A context aware device might provide the information and, according to an awareness of the situation based on current time, date, calendar information, and recently performed tasks, may also provide additional information or offer to perform a set of tasks for the user – order tickets, get the weather forecast, retrieve a traffic report, send someone a simple message, or all of the above.

This is more than a simple logical progression. If a user asks about the location of a theatre or where a particular movie is playing, it is a logical progression to assume that he wants to see that movie, therefore next offering to purchase tickets is a sort of first order logic. If someone asked the same question to another known person, a response of: *But I thought you wanted to see Greece?*, might seem nonsensical to a first order logic system. But to the receiver, such a response might not only make sense, but also make him reprioritize his evening such that instead of needing movie tickets, he goes home to work on his conference paper.[1]

3 Context Quintet

Given the context conduit of relationship and point-of-view as a result of communicated context, one method for approaching context awareness in intelligent devices is to employ the five often used narrative queries that go a long way toward establishing context in fiction or non-fiction stories. While this is nowhere near an exhaustive deconstruction of story context, these queries and their respective answers constitute an easily approachable structure for applying narrative context to context awareness research. The five queries are simply: *Who, What, When, Where,* and *Why*. When we listen to stories, we need most if not all of these questions answered in order to make sense of the story world, to make us feel comfortable with the characters, motivations, setting, etc. A personal communications and/or computation device that is aware of most, though not necessarily all of the answers to these questions would be acutely aware of the user's context. The questions below are listed in order from easiest or most straightforward to most challenging for a device. The ordering can also be interpreted as near-term to long-term technology delivery.

3.1 Straightforward

3.1.1 When

Acquiring and maintaining time and date information is common to all modern computational devices. Even cell phones store appointments like their larger computational counterparts, laptops and desktop workstations. Over-the-air time-date synchronization is also available in most phones and phone systems. Knowing time and date – a *temporal awareness* – is a fundamental component of context awareness because our culture is so dependent on specific time referencing.

3.1.2 Where

Having a physical *location awareness* means that it is possible to deduce distance, travel speed, and, to some extent, the location's personal meaning to the user, based on experience. This can be done on a cell phone on a rudimentary scale using cell tower triangulation, although this is information that consumers and, ironically, most developers do not usually have ready access to. When a device is not mobile, then simple caller ID can work for rudimentary location awareness. But when a device is out of doors, the best technique thus far for location awareness is global positioning systems (GPS). While plug-in GPS modules for PCs have existed for years, GPS functionality is just starting to appear in cell phones.[2] But since GPS uses satellite triangulation, it only works outdoors. For location awareness indoors, researchers and commercial companies are

[1] To Marvin Minsky and others (Lieberman, Liu, & Barry, 2003), such reasoning is called common sense reasoning, which is being actively applied to storytelling systems of various types. (Barry & Davenport, 2002)

[2] See Motorola phone model i88s, http://idenphones.motorola.com/iden/application?namespace=main

starting to look to 802.11b signal strength triangulation. Already, just by combining both indoor and outdoor location awareness and correlating that with calendar information, a device could learn locations of people and place names, and deduce if the user is where they said they would be or is on their way.

3.2 Under Development

3.2.1 What

What a user is doing, what task are they engaged in or trying to achieve is *task awareness*. Pattern analysis techniques or case-based reasoning could apply to this area of context awareness. Specifically in the domain of user interface, Cypher, Lieberman and others have written extensively on Programming by Demonstration (PBD), which works like a task context recorder/operator. (Cypher, 1994) (Lieberman & Selker, 2000) Instead of a user learning and employing a symbolic language like C++ or Java to represent context patterns and appropriate corresponding behaviours, PBD draws on the actions a user is already engaged in to learn what the user wants to achieve overall, such that the system can eventually take over the task for the user. A device with the ability to recognize user task patterns and then perform those tasks semi-autonomously would be a very valuable personal device. Coordinating that task awareness with location awareness means that a device could automatically perform simple tasks in the background whenever it is in a user or service provider-specified location. Coordinating task awareness and location awareness with temporal awareness means that a device could know when it will need to perform certain tasks and either begin them ahead of time or automatically coordinate its efforts with the user.

3.3 Challenging

3.3.1 Who

Determining and managing who – an *identity awareness* – is deceptively difficult. This, of course, is not simply the storage of names and categories that many address book software programs already provide. Nor is this determining what might be called "an absolute identity," with smart card technology, fingerprint readers, or retina scanners. Such identification systems already exist and networked devices can potentially acquire other personal data from internet sources (i.e. passport or social security numbers). More than just identification, identity awareness means managing a complex web of ever changing relationships. On a personal level, we identify each other not just by name, but (primarily) by relationship. Someone is not just "John," but also a family member, co-worker, friend, debtor, teacher, business partner, and so on. We identify people by what they mean to us. We know the world largely through the relationships we have with its elements. The more remote or disconnected something is, the less is seems we are likely to be interested in it.[3] In stories, the way a character's relationship changes over time is much more important than their name and some initial static classification. Identity awareness seeks to know, for example, what significance is associated with the user meeting with a particular person. Combined with time, location and task, knowing the meaning of *who* could provide insight into implementing a sophisticated model of device task prediction and resource management.

---

[3] One indicator of interest in this type of awareness is ongoing work on the semantic web, an extension of the world wide web that basically adds meanings to links. http://www.w3.org/2001/sw/

3.3.2 Why

Why is perhaps the hardest question of all. Interface design may never get to this point and perhaps should not try. We very often do not know the why of things ourselves. When someone gives us a *why*, we often counter with a different *why*. The why is so deeply rooted into who we are as human beings that it is absurd to believe that a machine could deduce or even guess a why answer of any real meaning. However, the why of the user's behavior is not nearly as informative from a device as the why of the device's behavior. The more complex and sophisticated the various algorithms of an interface and the more a device employs machine learning methods to absorb bits and pieces of a users life, the more it will need the ability to explain its actions when they are not expected by the user. The relationship between the user and the device would be strengthened through clear expressions of why the device is performing a particular task.

4 Conclusion

Combined with the other four awarenesses, the ability to provide a simple reasoning to the user is more than a safeguard, it also provides a window into the device's point-of-view. Narrative does not provide a comprehensive method of identifying and expressing context awareness, but for those of us whose lives are built on stories, which is all of us, it can provide a natural method for thinking about and managing the very complex computation associated with context awareness. Additionally, narrative offers a universally familiar domain for managing complexity. While the desktop interface metaphor may be particularly comforting to those of us chained to our traditional business desks, narrative metaphors cut across work environments, cultures and ages. Like a sonata, narrative is the music that turns complexity into sophistication.

5 References

Barry, B., & Davenport, G. (2002). *Why Common Sense for Video Production?* (Technical Report No. 02-01). Cambridge: MIT Media Lab.

Cypher, A. (Ed.). (1994). *Watch What I Do: Programming by Demonstration.* Cambridge: MIT Press.

Kroeber, K. (1992). *Retelling/Rereading - The Fate of Storytelling in Modern Times.* New Brunswick, New Jersey: Rutgers University Press.

Lieberman, H., Liu, H., & Barry, B. (2003). *Beating Some Common Sense into Interactive Applications.* Paper presented at the International Joint Conference on Artificial Intelligence, Acapulco, Mexico.

Lieberman, H., & Selker, T. (2000). Out of context: Computer systems that adapt to, and learn from, context. *IBM Systems Journal, 39*(3 & 4), 617-632.

Martin, R. (1996). Between Teller and Listener: The Reciprocity of Storytelling. In C. L. Birch & M. A. Heckler (Eds.), *Who Says? - Essays on Pivotal Issues in Contemporary Storytelling* (pp. 141-154). Little Rock, Arkansas: August House, Publishers.

Minsky, M. (1988). *Society of Mind.* New York: Simon & Schuster.

Software evaluation by the ergonomic assessment tool EKIDES

Heiner Bubb

Institute of Ergonomics, University of Technology Munich
Boltzmannstr 15. 85747 Garching
Germany
bubb@lfe.mw.tum.de

Iwona Jastrzebska-Fraczek

Institute of Ergonomics, University of Technology Munich
Boltzmannstr 15. 85747 Garching
Germany
fraczek@lfe.mw.tum.de

Abstract

The database system **EKIDES** (**E**rgonomics **K**nowledge and **I**ntelligent **De**sign **S**ystem) assists designers of technical systems, equipment, products and workplaces to meet ergonomic requirements for all system components and their interactions during the planning, development and subsequent design and blueprint processes.
The ergonomic tests can be carried out by using the *Basic* or *Consulting Modules of EKIDES.* Furthermore, if the required measurement equipment is not available, or if a qualitative task or product analysis is sufficient, the module *Checklist* may be used.

1 Introduction

EKIDES is composed of several modules (see Figure 1). Generic Design Module can be use for nonspecific analysis that utilizes generic ergonomics design considerations. Design Application Modules can be used for specific applications. In addition, Testing and Evaluation Modules can be used for analysis of workload, task analysis, product analysis, and general testing and evaluation. As user of EKIDES, sometimes you will be confronted with the question in which data folder you can find information about the specific facts relevant to your analysis. Two modules, i.e. the "Search for data folder" and "Keyword search" can be of assistance to you in such a situation.

1.1 Structure of Data

All ergonomic data in EKIDES are structured in the same way: For example data about Software Design contains 6 fields:

1. Remarks to software ergonomics (with 6 data folders),
2. User action and user guidance (with 9 data folders),
3. Text editing and depicting of text (with 2 data folders),
4. Processing and depiction of prescribed forms (with 2 data folders),
5. Processing and depiction of tables (with 2 data folders),
6. Processing and depiction of graphics (with 3 data folders).

The structure of data folders in all theme groups of database is similar (see Figure 2). The first column on the left side (c), with a color-coded sign, illustrates the meaning attached to the particular ergonomic requirement.

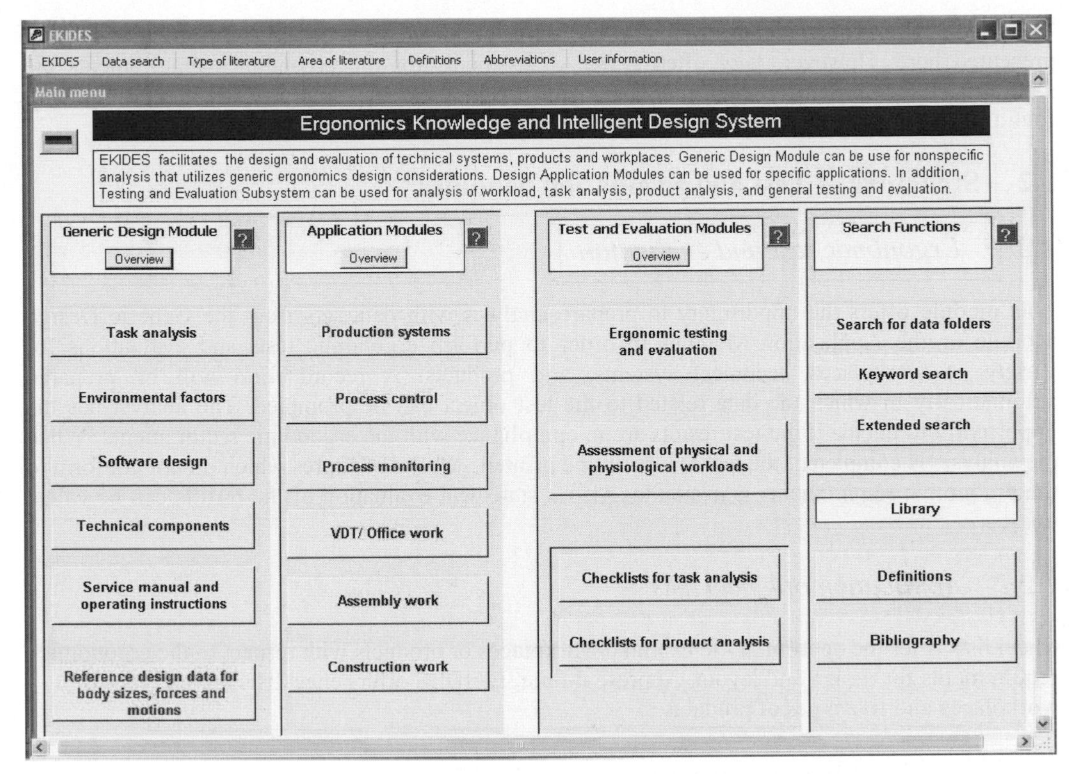

Figure 1: Main menu of EKIDES

Each requirement can be relevant to health, safety, performance, reliability/dependability, or comfort. For this classification, in many cases no defined guidelines from law, regulations, or standards are available. Where such guidelines are missed, an expert judgment was used. In the second column, a given item (component) is verbally described.

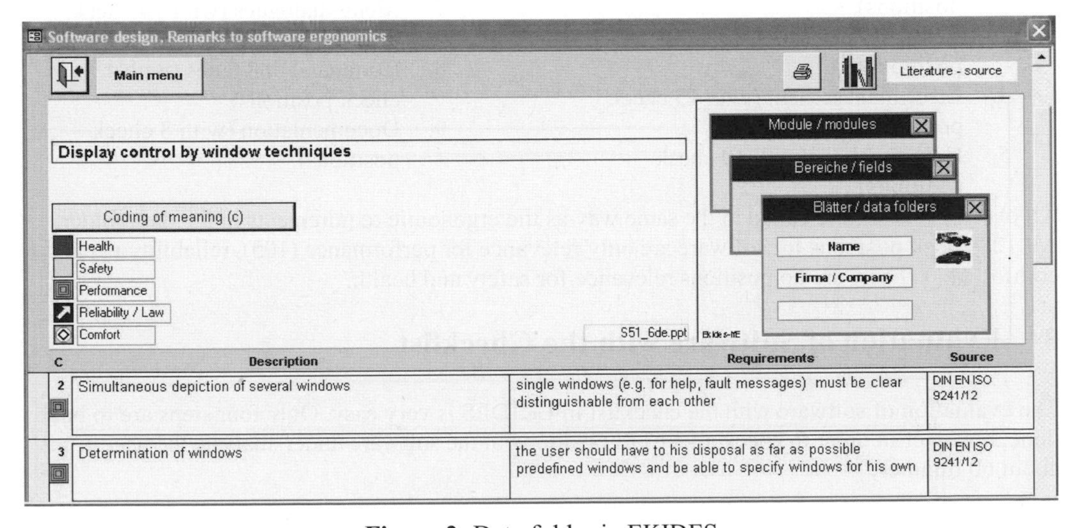

Figure 2: Data folder in EKIDES

The third column contains the ergonomic requirements. As much as possible, quantitative data are presented here. However, very often verbal assessments for the ergonomics requirements are indispensable. The fourth column contains information about the source if a reference has been found.

1.2 Structure of Test and Evaluation Modules

1.2.1 Ergonomic test and evaluation

This module offers the opportunity to prepare analysis with data sets from the Generic Design Module or the Application Modules in order to perform ergonomic tests and evaluations for variety of workplaces, technical systems, and products. A record form will be prepared automatically in which the data related to the test object can be prompted. The analyst has the opportunity to decide if the test results are in compliance with the ergonomic requirements. When the analysis is completed, the test report can be printed. A list of features which do not conform to the ergonomic requirements is available. Also a statistical evaluation of the results can be called up.

1.2.2 Application of Checklists

Checklists offer the opportunity to examine workplaces or products with respect to the ergonomic requirements for design and layout. At present time, EKIDES offers checklists for six types of workplaces and five types of products.

2 The Checklist for software in EKIDES

The Checklist for software in EKIDES is based on 9 fields:

1. Usability (with 12 check positions)
2. Information display (with 20 check positions)
3. User guidance (with 22 check positions)
4. Error management (with 15 check positions)
5. Help functions(with 10 check positions)
6. Dialog technique (with 11 check positions)
7. Menu dialogues (with 10 check positions)
8. Command dialogues (with 15 check positions)
9. Documentation (with 8 check positions)

All check positions are coded in the same way as the ergonomic requirements in the data folder. All 123 check positions for software are only relevance for performance (105), reliability (7) and comfort (11). There are no positions relevance for safety and health.

3 Evaluation of Software with the Checklist

The evaluation of software with the checklist in EKIDES is very easy. Only four steps are to be done, however in order to check about 123 positions in the software under analysis the user needs about 60 minutes.

3.1 Start a New Product Analysis

The user starts a new analysis with the selection of type of product. He must name the product and can add information about the company and he can add some comments for the analysis.

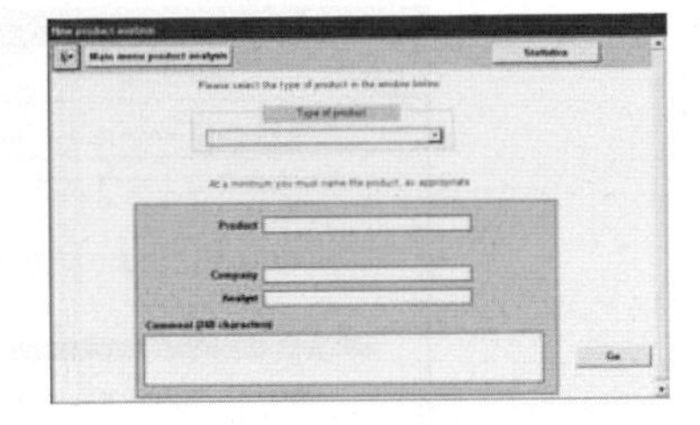

3.2 Selection of relevant Fields and Check positions

After start, the user must decide which the 6 fields are relevant for the analysis (i.e. "Are there any problems with the usability of the software?"). Within each field, he will find a list of problems which have to be answered with a "yes", "no" or "not applicable".

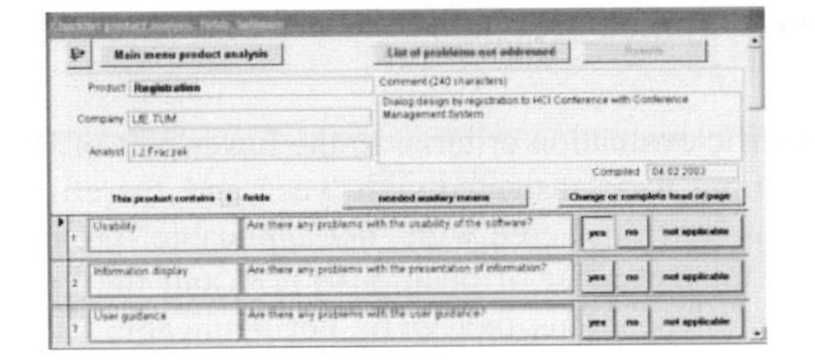

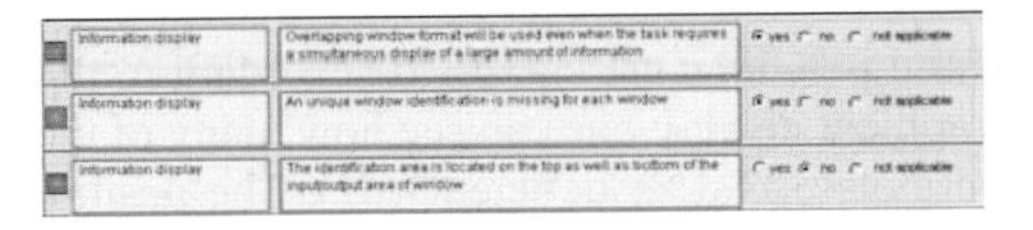

3.3 Result

After answer all the statements in the selected fields, the user can open a window with summary of results. He can see again the content of analysis and change the answer he has given during the evaluation session if necessary or he can view or print out only deficiencies in evaluated software.

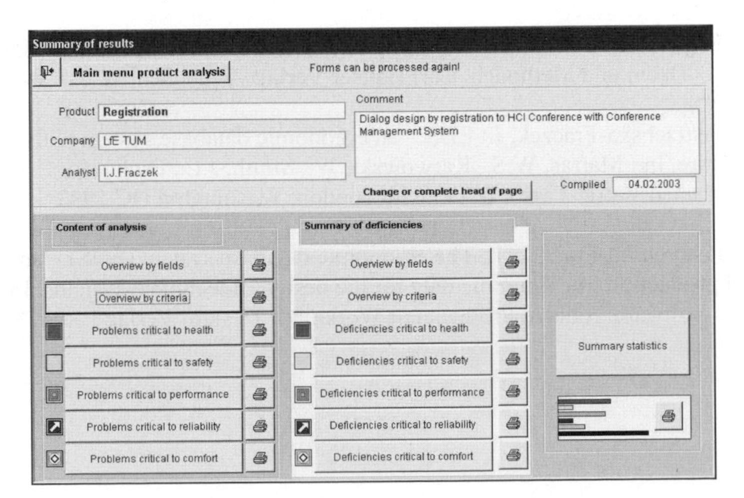

Figure 3: Result of the checklist

The button "Summary statistics" opens a graphic representation of the results.

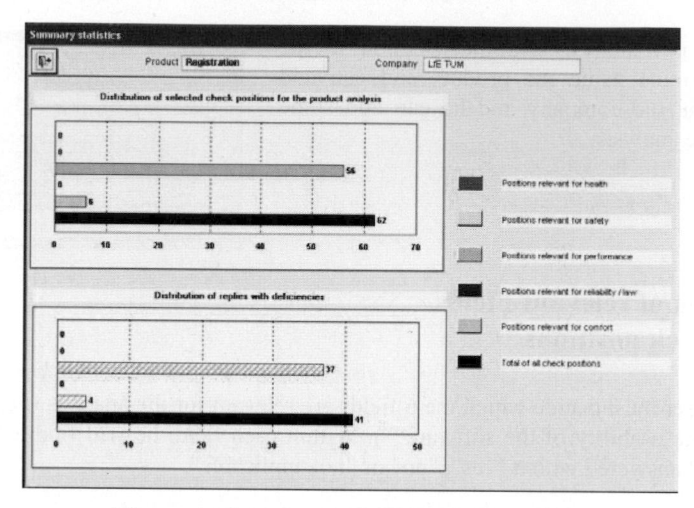

Figure 4: Graphic representation of results

These positions are classified according to the specific evaluation criteria. In the lower part of the graphical display one can see how many of the positions were answered with "yes", and, therefore, were classified as 'deficient' (41). The positions that were answered with "not applicable" are not included in the summary statistics. Therefore, the total number of positions (123) and the total number of answers (62) will differ by the number of positions that were 'not applicable'. The complete test results can be printed as a test report containing graphic illustration of the full scale evaluation. To compare the software with the checklist of EKIDES, the graphic form of analysis gives a first impression. The study of the check positions in each group (e.g. performance, comfort) allows the designers to identify all deficiencies of the software.

4 References

Schmidtke, H.: Ergonomische Prüfung von Technischen Komponenten, Umweltfaktoren und Arbeitsaufgaben – Daten und Methoden. Carl Hanser Verlag, München, Wien 1989

Schmidtke, H., Jastrzebska-Fraczek, I.: EDS – an ergonomic database system with computer-aided test procedure. In : Marras, W.S., Karwowski, W., Smith, J.L., Pacholski, L. (Ed.) The Ergonomics of Manual Work. Taylor & Francis, London, Washington DC 1993

Schmidtke, H., Jastrzebska-Fraczek, I.: The ergonomic database system (EDS) – an example of computer-aided production of ergonomic data for the design of technical systems. In: Landau, K. (Ed.) Ergonomic Software Tools in Product and Workplace Design. Verlag ERGON, Stuttgart 2000

Researches on Pen-Based User Interface

Dai Guozhong Tian Feng Li Jie Qin Yanyan Ao Xiang Wang Weixin

Laboratory of Human-Computer Interaction & Intelligent Information Processing, Institute of Software, Chinese Academy of Sciences
P.O.Box 8718, Beijing, 100080, P.R.China
guozhong@admin.iscas.ac.cn,{tf, lijie, qyy, ax, wwx}@iel.iscas.ac.cn

Abstract

Pen-based user interface is one of the main styles in Post-WIMP user interface. Continuous interaction and High bandwidth interaction are two important interaction features in PUI. Analysis and specification of these two features are declared at the beginning of this paper. Based on the research of two features, a new pen-based interaction paradigm named PIBG is presented. The main idea and advantages of the paradigm is declared in detail. At the end of this paper, some systems built on the paradigm are introduced.

1 Introduction

Pen-based user interface is one of the main styles in Post-WIMP user interface. It is designed on the Pen-Paper metaphor which is analogous to the user's real working environment. It's natural for people, especially fit for the Chinese culture. Pen-Paper metaphor is a universal and fundamental way for capturing daily experiences, communicating ideas, recording important events, conducting deep thinking and visual descriptions. Researches on PUI intend to make these traditional and ubiquitous activities computable with keeping the naturalness of them. Consequently, with the assistance of computing resources, people can achieve easier manipulations to information, such as maintenance, modification, retrieval, transferring, further processing and analysis. PUI is bringing many new features to these activities other than just simulating them.
Currently, many famous systems have built in PUI field, such as Tivoli (Pedersen, McCall, Moran & Halasz, 1993), LiveBoard (Elrod et al.,1992), SILK (Landay, 1996), DENIM (Lin, Newman, Hong & Landay, 2000), Cocktail Napkin (Gross, 1996), Flatland (Mynatt, Igarashi, Edwards & LaMarca, 1999), Classroom 2000 (Abowd et al., 1996), ASSIST (Alvarado & Davis, 2001), Teddy (Igarashi, Matsuoka, Tanaka, 1999). But most of these systems focus on specific application requirement. In this paper, we will focus on the common interaction features of PUI. Our research work will be discussed in three levels, including interaction information in PUI, interaction paradigm of PUI, and applications based on the interaction paradigm.

2 Interaction Information in PUI

Interaction information in PUI has two important characters. First, the information is continuous, instead of discrete event. So the flow of continuous interaction information from pen to PUI should be described. We use the Jacob's model (Jacob, 1996) to specify the information flowing form pen to PUI. This model is composed of two parts. One part is a graph of functional relationships among continuous variables. Only a few of these relationships are typically active at one moment. The other part is a set of discrete event handlers. These event handlers can cause specific continuous relationships to be activated or deactivated. The description is showed in Figure 1. Second, the interaction information created by pen is not only the 2D position in the coordination of paper, but also pressure, orientation, time, etc. So the structure of integrating various kinds of information should be set up. We apply a hierarchical structure to integrate multi-model information. In lexical level, the information integration is mostly applied in lexical feedback of stroke. We use an algorithm which gets information about pressure, orientation and time of pen, and then create stroke feedback which imitates the effect of Chinese Calligraphy. In syntactic level, the information integration is applied in the formation and classification of interaction primitive in PUI. These Information include pen information (integrated in Lexical level), the time span (start time and end time of the stroke), TimeUp (a value to specify if a user holds the pen on paper before he draws stroke), Distance (the length of the stroke). The information integration in semantic level can be applied in lots of fields. For example, in a pen-based mathematic expression editor, when a user inputs a stroke into an expression, information about the stroke isn't the only factor be considered, the position context and the interaction context of current stroke should also be considered. Figure 3 shows that when a user wants to extend the radical sign, he just draws a stroke near the current radical sign, and if the shape, the position context and time context of the new stroke fit for the special requirement, the new stroke will be recognized as extending the radical sign, instead of a new symbol.

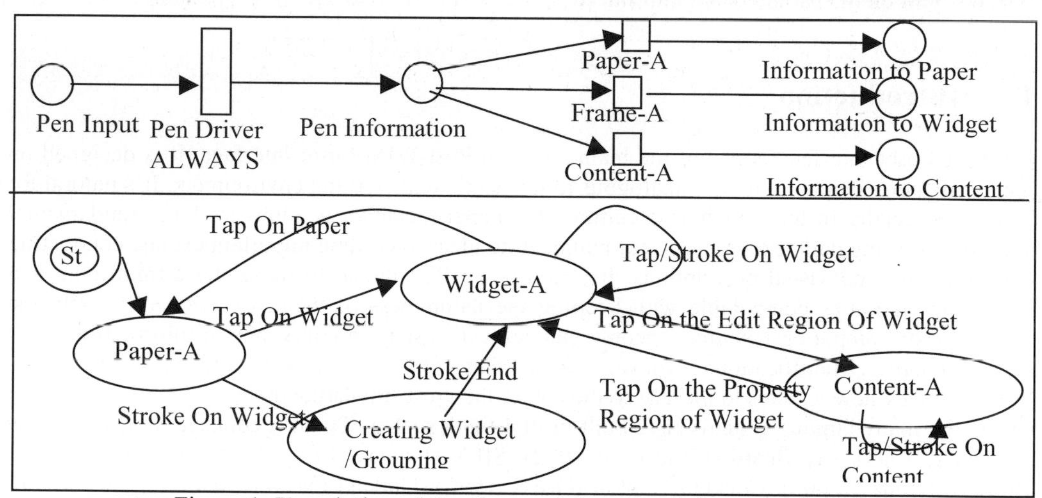

Figure 1: Description of Continuous Interaction Information in PUI

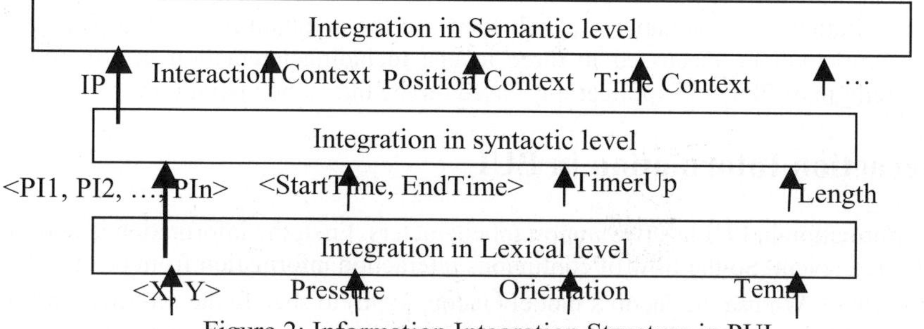

Figure 2: Information Integration Structure in PUI

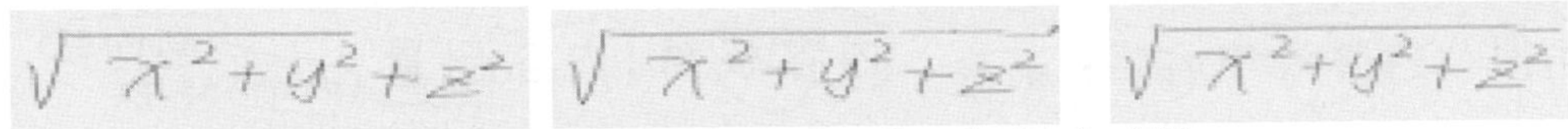

Figure 3: Process of Extending the Radical Sign

3 PIBG Paradigm

Based on the research of the interaction information in PUI, we present a new interaction paradigm of PUI, named PIBG. P means Physical Object. Two main interaction widgets in PIBG paradigm (paper and frame) belong to Physical Object. Paper is a kind of widget which serves as a container in PUI. It has two main responsibilities. First, it receives the information from pen and dispatches it to specific information receiver (a widget in the paper). Second, all widgets contained are grouped as a tree structure by Paper and managed by it. Frame is the most important widget in the PIBG paradigm. Its responsibilities are processing the interaction information and managing different type of data. Currently, there are seven kinds of frame in the system, showed in Figure 4. IB means Icon and Button. Menu is disappeared in PIBG paradigm. G is gesture. In PIBG paradigm, user action is changed from mouse pointing to pen gesture.

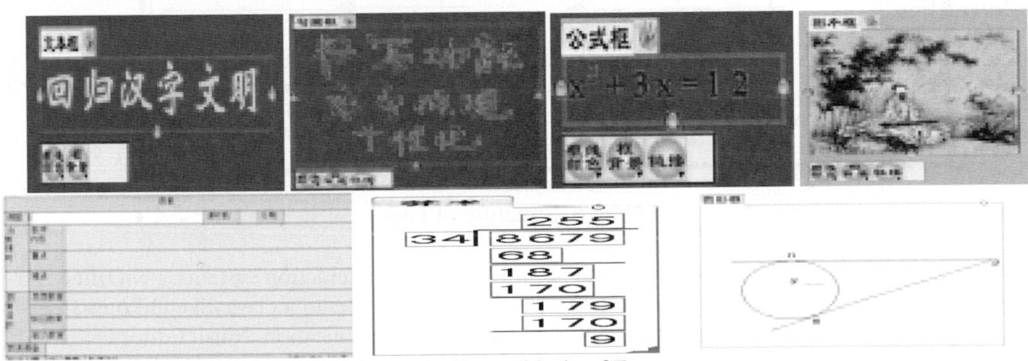

Figure 4: Seven kinds of Frames

Comparing with WIMP paradigm, the information presentation style and interaction style have changed. First, we discuss the information presentation style in PIBG paradigm. From the viewpoint of Cognition Psychology, when a user performs a task, his processing ability is limited by two factors: the processing resources available and the quality of data available (Norman & Bobrow, 1975). There is a trade-off between resources and data quality. Because lots of resources required in a task will cause heavy cognitive load. A primary design objective should be to minimize resource consumption by improving data quality. But the data quality improvement relies on training for the task. So lots of time will be spent. How to get high data quality without spending lots of time in training? PIBG paradigm gives us a good solution. The information presentation style in PIBG is analogous to natural working environment. It makes use of human's knowledge about the natural working environment which exists in human's mind for many years. So when a user interacts with PIBG paradigm, he will feel familiar with such interaction context. And the interaction efficiency will be increased. Second, the main interaction action in PIBG is pen gesture. Pen gesture has more advantages than mouse pointing. Gestures that we designed in PIBG paradigm are analogous to natural working style in paper. Figure 5 shows some gestures we designed in word processing task. At the same time, gesture can let a user focus on a task itself, instead of on how to interacting with some widgets which used to perform the task.

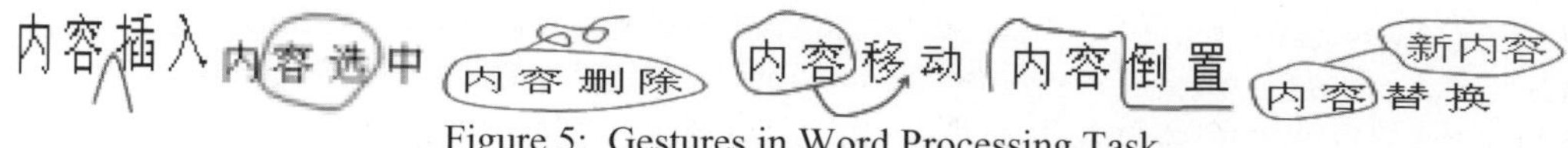

Figure 5: Gestures in Word Processing Task

The architecture of PIBG paradigm has three new features. It supports continuous interaction information, multi-model interaction information, and probabilistic interaction style. In PIBG, all widgets are built to support Continuous interaction and multi-model interaction according to the structure we introduced in Part 2. At the same time, we set up a structure which combines recognition, context-aware and User Mediation (Mankoff, Hudson, Abowd, 2000) techniques. Every widget has such structure to support this feature. Figure 6 shows the structure of Paper.

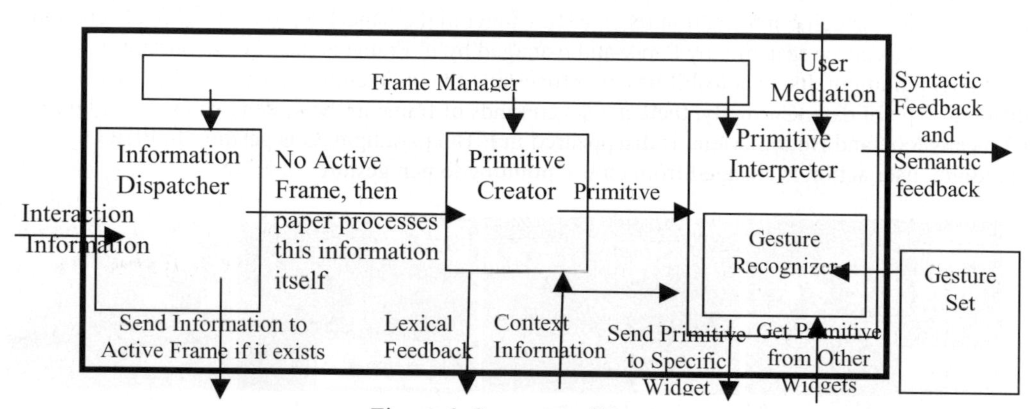

Figure 6: Structure of Paper

Based on the PIBG paradigm, we have built several pen-based interaction systems. PenOffice is a system for teacher preparing their presentations and deliver lectures to a class. SketchPoint(Li et al., 2002) is a prototype for experience capturing and idea communication. MusicEditor is a system that recognizes handwriting numbered musical notation, and exports it to audio result. CreativePen(Tian, Jie, & Dai, 2002) is a sketch system for children easily creating virtual worlds.

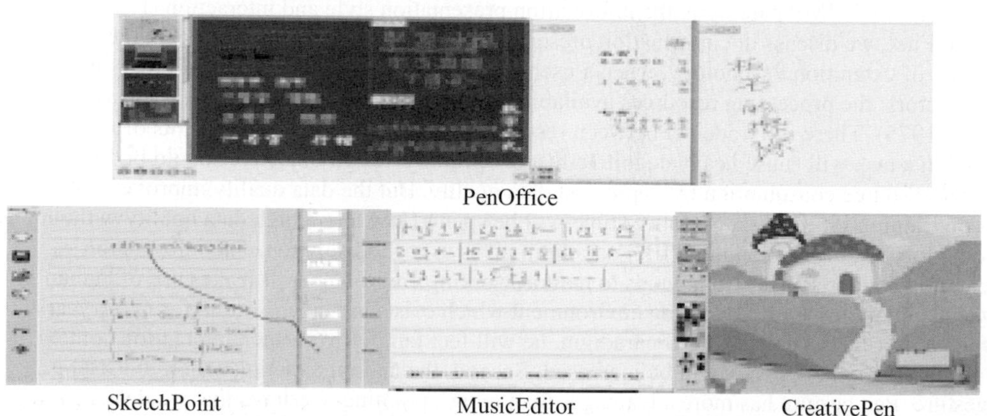

Figure 7: Four Pen-based Systems Based on PIBG Paradigm

4 Conclusion

In recent years, pen-based user interface has become a hot research field. But most of pen-base interaction systems built focus on specific application requirement. In this paper, the research focus is on the common features in PUI. We set up a new pen-based interaction paradigm based on these common features. And some pen-based interaction systems are built on it. In future, we will conduct a formal evaluation on the paradigm and enrich the content of the paradigm, such as interaction widget and pen gesture.

References

Abowd, G.D. et al. (1996). Teaching and learning as multimedia authoring: the classroom 2000 project, Proceedings of the fourth ACM international conference on Multimedia, p.187-198, November 18-22, 1996, Boston, Massachusetts, United States.

Alvarado, C. & Davis, R. (2001). Resolving ambiguities to create a natural sketch based interface. Proceedings of IJCAI-2001, August 2001.

Elrod, S. et al (1992) Liveboard: a large interactive display supporting group meetings, presentations, and remote collaboration. In Proceedings of the ACM Conference on Human Factors in Computing Systems: CHI'92, Monterey, CA, May 3 - 7, 1992, pp.599-607.

Gross, M.D. (1996). The Electronic Cocktail Napkin. A computational environment for working with design diagrams. Design Studies, 17(1):53--69, 1996.

Igarashi, T., Matsuoka, S. and Tanaka, H. (1999) Teddy: a sketching interface for 3D freeform design. Computer Graphics Proceedings, Annual Conference Series, 1999, ACM SIGGRAPH: 409-416.

Jacob, R.J.K. (1996). A Visual Language for Non-WIMP User Interfaces. Proceeding of IEEE Symposium on Visual Languages, 1996, 231~238.

Landay, J.A. (1996). "SILK: Sketching Interfaces Like Crazy ". Proceedings of Human Factors in Computing Systems (Conference Companion), ACM CHI '96, , 1996. pp. 398-399.

Li, Y., Guan, Z., Wang, H., Dai, G. & Ren, X. (2002). Structuralizing Freeform Notes by Implicit Sketch Understanding, in Proceedings of 2002 AAAI Spring Symposium: Sketch Understanding, pp.91-98.

Lin, J., Newman, M., Hong, J., & Landay, J. (2000). DENIM: Finding a Tighter Fit Between Tools and Practice for Web Site Design. CHI Letters: Human Factors in Computing Systems, CHI '2000,. 2(1): p. 510-517.

Mankoff, J., Hudson, S.E. & Abowd, G.D. (2000). Providing integrated toolkit-level support for ambiguity in recognition-based interfaces. In Proceedings of CHI'00 Human Factors in Computing Systems, ACM, pp.368–375.Canada. pp. 196 – 196.

Mynatt, E.D., Igarashi, T., Edwards, W.K. & LaMarca, A. (1999). Flatland: new dimensions in office whiteboards. In Proceedings of CHI'99 Human Factors in Computing Systems (May 15-20, Pittsburgh, PA), ACM, 1999, pp.346-353.

Norman, D., and Bobrow, D. On Data-Limited and Resource-Limited Processes, Cognitive Psychology 7, 1975, 44-64.

Pedersen, E.R., McCall, K., Moran, T.P. & Halasz, F.G. (1993). Tivoli: An Electronic Whiteboard for Informal Workgroup Meetings. In Proceedings of the ACM INTERCHI'93 Conference on Human in Computing Systems, 391-398.

Tian, F., Li, J. & Dai, G. (2002). A Pen-Based 3D System for Children. In Proceeding of APCHI'2002, Beijing, China, 612-623.

On the relevance of 3D shapes for use as interfaces to architectural heritage data

I.Dudek *J.Y Blaise* *P.Bénistant*

UMR MAP CNRS/MCC 694

EAML 184, av. de Luminy 13288 Marseille Cedex 09 France

idu@gamsau.map.archi.fr *jyb@gamsau.map.archi.fr* *pbe@gamsau.map.archi.fr*

Abstract

Documentation analysis and organisation are vital to the researcher when trying to understand architectural evolutions. Documentary sources provide partial evidences from which the researcher will infer possible scenarios on how an edifice or site may have evolved throughout the centuries. They provide *clues* that will need an *interpretation* from the researcher or conservator in order to *understand* the edifice or site's evolution. In this process, data interpretation is a critical step that base on the use 3D theoretical concepts originating from those who study edifices: architects, archaeologists, conservators, historians. What solutions can one base on in order to organise a documentation with regards to an architectural analysis of its content? How can we today provide architects, conservators or collection holders with a set of relevant architectural concepts that will serve as intermediates between the edifice's shape and its documentary sources? Our contribution introduces a methodological proposition that tries to cope with those questions, and evaluates possible uses of 3D scenes as interfaces for architectural heritage data visualisation. We describe the relations between documentation and architectural objects, and use the representation of 3D objects in order to access information but also to visualise the state of knowledge of each object.

1 Introduction

In writing hypertexts, establishing clear relationships between sources and destinations has been acknowledged as a vital question, and the same issue is raised when trying to attach 3D architectural shapes to documentary sources. But taking a closer look on what existing computer tools and formalisms offer when dealing with the architectural heritage shows that their relevance may not be optimal. Geometric modelling tools allow the construction of 3D models in which simulations of a morphology is possible. In parallel, GIS systems have proven useful in numerous site management experiences, particularly in archaeology. But whether there is a way in between those families of technologies remains to be proved. To put it more simply, can 3D models be efficient in data visualisation or retrieval? Can they offer a view on the data that other media forbid? Can they localise pieces of information with regards to a position in space and a moment in history? Can they inform the system's user on whether the proposed shape is original or reused, documented or hypothetical, etc.? We observe that although the edifice is not the information, the information is relative to the edifice. But the architectural documentation varies in type, precision and relevance. When representing the edifice as a set of 3D shapes for use as interfaces, we will therefore face a challenge to visualise shapes for which we have partial information, and with it a challenge to graphically notify it. Our methodology defines two complementary tasks:

- Concept modelling and instance documentation:
- Visualisation of the documentation's analysis and of elements of semantics

We consider that it is vital to use what the documentation says *as well as what it does not say*. 3D models, considered with regards to the above mentioned aspects, can be efficient in retrieving or visualising information about architectural evolutions. The paper introduces the research context, and then develop three aspects of our work : the methodology, the applications, and observations on the benefits and constraints connected to our vision of what 3D scenes can become.

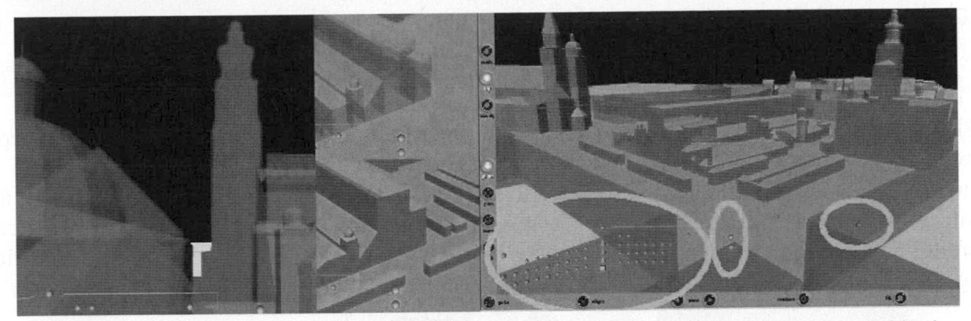

Figure 1 : Appearances (left) give indications on teach object's documentation analysis; interactive controls (right) nested in the VRML scene extend appearance modifications.

2 Research context

Architectural heritage is a domain in which both documentation and visualisation play essential roles. Moreover, ensuring their interdependence has clearly been acknowledged as a key issue (see for instance [Nak99]). Moreover, a physical object such as "an opening" can have been re-used several times during history, and often inside different edifices. Both the shapes reconstructed and the documentation relate to a moment in time. This introduces a level of complexity for which we lack adequate formalisms since such issues as dynamic data visualisation [Rus01] or time handling in GIS sytems [Bil97], although already addressed, do not bring operational breakthroughs in our domain. We propose to use the edifice's morphology as a support for data retrieval and documentation visualisation. Consequently, we need to isolate relevant architectural concepts and build out of them 3D models, as developed in [Dud01]. This idea is in fact nothing surprising for people involved in the analysis of edifices. The methodology used by historians of architecture and conservators in order to analyse evolutions of an architectural object is based on the interpretation and comparison of various types of documentation, as stated in the [Cra00] charter. Therefore the idea that different pieces of information are in relation to architectural elements (a building, a portal, etc.), is for them a natural (although often unspoken) part of their work methodology. But the documentation that serves as source of evidences is far from being exhaustive and non-ambiguous. What is more it is not structured with regards to the it documents. We therefore face several difficulties when implementing a document -> scene link :

1. We face a challenge to visualise shapes that in all cases are hypothetical. Consequently we will need to provide the scenes with graphical codes marking the evaluation of the hypothesis.
2. We face partial or contradictory evidence, lack of evidences, or rely on comparisons. We need to propose markings of the objects that correspond to their documentation .
3. Documentation about one element does not relate its sub-parts or to its super-parts : each concept should be documented independently from others. Scale, a notion oddly absent from 3D modelling, can act as this complementary filter in the information available on the edifice.
4. Inside an edifice that can be widely transformed, individual elements of architecture can, what is more, be reused or even moved somewhere else in the city, underlining another problem, this of localisation in time and space of architectural elements.

In [Hei00]'s experience, a 3D scene is used to navigate into a set of information about a city. The user may question the system on the localisation of services such as hotels, railway stations, etc.. Our experience differs in three main aspects. Elements supporting information are architectural elements (gates, arches, etc..), etc.. The information we deal with, as well as the shapes we represent, are in relation with a period in history. The documentation, and its avatar , a 3D morphology, need interpretation since information may be uncertain, incomplete, etc..

3 Methodological proposal

As shown in figure 2, three main elements have to be developed : a set of architectural concepts that will be represented in the scene, a documentation database and the scenes themselves.
In the first stage, concept modelling, we need to identify concepts that will be used as filters on the architectural documentation. The concepts are identified through an analysis of the morphological, structural and functional differences and similarities between the objects. They are then classified using the principle of heritage of properties (see [Con00]).

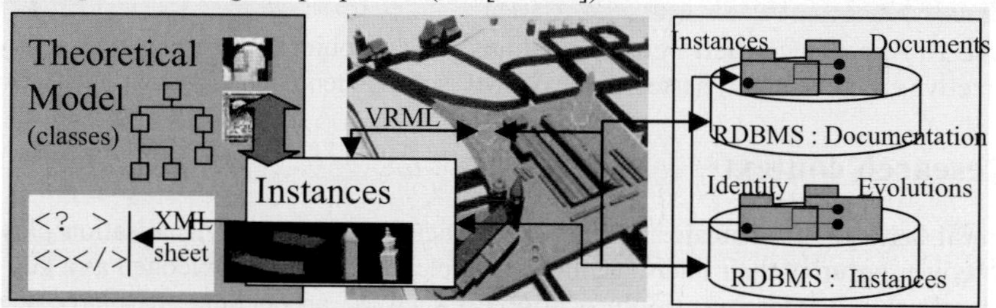

Figure 2 : A system architecture combining various formalisms

Once concepts are identified, and organised in a hierarchy, the making of 3D scenes results in the instanciation of the model's theoretical shapes and a call to the relevant representation method. Each concept is characterised by a morphology, i.e geometrical properties, but also by qualitative evaluators that support the documentation's analysis[1]. Each concept is implemented with a VRML representation method that graphically interprets the values of the qualitative evaluators. Finally, a bibliographical, iconographic and cartographic database stores the documentation itself in a traditional way (see [Ste91]). Each object inside a scene refers to the relevant pieces of documentation in the database using VRML's ANCHOR node. Each concept detains methods relevant for persistence handling in XML files and RDBMS context. It has to be stressed that autonomy and durability of the data sheets are of crucial importance in our application domain. We store the textual results (XML sheets) inside files that can be used independently from the system as a whole. We propose in line with [Wal02] a solution based on the idea that a unique input- the instance's XML sheet; has several outputs. Scenes are used as a query mode (predefined time-related scenes) by selecting an object or as a visualisation of the query's result, by instancing the objects corresponding to the search and calling their VRML representation method. Model and RDBMS platforms are chosen independent, we use Perl / PHP CGI Interfacing modules.

---

[1] Each object contains a group of attributes that are responsible for displaying the object with relevant graphical codes. They show inside a 3D scene the semantics associated to the source's analysis. For example the user of the scene can demand a visualisation of the level of certitude on an object's dating (represented in this example by a level of translucency). We distinguish justification attributes used to represent objects with a graphical code that indicates how credible the information we detain is with regards to specific themes (dating, shape, structure, function). And existence attributes used to represent objects with a graphical code that indicates whether or not we have documents about the object with regards to specific media types.

4 Application to Kraków's historical centre

The methodology we have briefly introduced is applied to the studying of the historical centre of Kraków , former capital of Poland. The city has a rich architectural heritage, and researchers deal with a growing critical mass of documentation, thereby underlining the necessity to investigate new methods of data management and visualisation. We have focused on the urban structure scale, corresponding roughly to the city's exteriors. More than six hundred evolutions have been identified, and with them the corresponding documentation. They cover a period of eight hundred years, with big differences in the number of evolutions between the various objects studied. Besides static VRML scenes at key periods, the web interface we have developed lets the user to query the system with regards to two main (and obligatory) parameters : the period he wishes to observe, the families of objects he wants to visualise (i.e the classes). In parallel, more parameters can be added in order to retrieve and visualise for instance only the objects for which we have inventories, or only the objects for which the dating is imprecise, etc.. VRML Scenes are calculated online. We use two types of information : appearance of the object (colouring / translucency) and anchoring (connection to a URL). Moreover, each scene is displayed with client-side interaction disposals (see fig1) that let the user to choose which DB should be queried or what document type is available on the object.

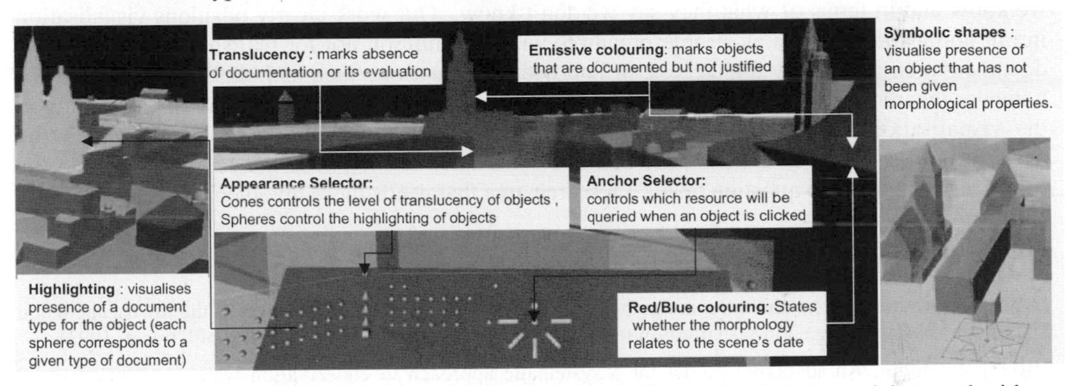

Each object is displayed natively with a translucency that indicates the dating precision, and with a colour that indicates whether the morphology proposed is this of the object *at the date of the scene* or whether it is a copy of a previous or later evolution of the object, thereby underlining needs for further investigation. Once the scene is displayed, other possibilities are left for the user, notably the highlighting of objects in relation to types of documents as described in the documentation database (plans, texts, illustrations, inventories, etc..). Finally, we need to cope with two possible incoherence : an object for which we have proposed a reconstruction but not yet established the documentation's analysis; and an object that we have fully documented but not yet reconstructed. The former are shown with an Emissive colouring that distinguishes them from the rest; the latter with a library of symbols that only localises the object. In both cases, the scene underlines a current state of knowledge, what is known and what is not, and therefore does remain consistent with the documentation.

5 Conclusion

Architectural historians and conservators face today two challenges when they have to study edifices or urban groups : they have to cope with a mass of documents about the city and its evolution, and deal with the constraints and logic of geometric modelling. What is more, there is a clear need to find a way to better capitalise the research efforts so that not only a seducing 3D

model but also the whole research process gets *persistent*. Finally, in our application field, research about *one* edifice or urban group strongly benefits from comparisons on the whole city.
Our position is that 3D models of the architectural shapes our documentation is about, are a natural and efficient filter for data visualisation and retrieval. A central improvement in the actual practice is the fact that architectural data finds its natural media, shapes, whereas it has traditionally up to now been centred on documentary descriptions (authors, editors, keywords, etc) On the documentation side, other benefits include the possibility given to reuse existing data sets, the possibility to visualise what a particular document *is about* (edifices quoted in it), the possibility to compare levels of information between various sectors or objects types inside the territory observed. On the virtual reconstruction side, the approach we defend helps the architects to build from his own words rather than from those of geometry, allows the author of the reconstruction to build an object *on which he has doubts* and to represent along with a morphology *the doubt itself*. It also fosters the emergence of a new vision of 3D models, a vision that says a 3D model can be a sustainable research tool if it reaches the readability of a geographical map.
But what our experience with this system has really revealed to us is its capacity to help us in putting the finger on questions that remain to be raised, or that are only partly addressed. In brief, the 3D scenes we provide help the researcher in clearly and synthetically visualise a state of knowledge on the city's evolution, and can be considered as key tools both in terms of what they we know and in terms of what they say we don't know. Our work clearly positions visualisation in our application domain as an interpretation, with an ambition not for realism but for the better documentation readability and access, in line with contributions such as [Alk93] or [Kan00]. We however regard our contribution as nothing more than a first step in trying to use 3D modelling in the visualisation of archival information. We believe that it s possible to greatly enrich the usefulness of 3D representations provided that some attention is put to the semantics behind the rendering, and that this question opens a research area that needs more involvement.

References

[Alk93] P. Alkhoven, The changing image of the city. A study of the transformation of the townscape using Computer assisted Design and visualisation techniques (Utrecht: PHD of the University of Utrecht, 1993).

[Bil97] A. Bilgin, An adaptable model for a systematic approach to conservation walks : an introductory study on GIS […], in Proceedings of. SFIIC 97 Conf., Chalon sur saone, France, 1997, 211-219

[Con00] D. Conway, Object Oriented Perl, (Manning Publications, Greenwich, Co, USA, 2000).

[Cra00] Cracow Charter 2000, in Proc Int. Conf on Conservation Krakow 2000, pp191-193, 2000.

[Dud01] I.Dudek, J.Y Blaise, Interpretative modelling as a tool in the investigation of the architectural heritage : information and visualisation issues, in Proc VIIP 2001, Marbella, SP, pp 48-54, 2001.

[Hei00] A.Heinonen, S.Pukkinen, I.Rakkolainen, An Information database for VRML Cities, in Proc. IV2000, London, UK, 2000.

[Kan00] J.Kantner, Realism vs Reality: creating virtual reconstructions of prehistoric architecture, in J.A Barcelo, M.Forte, D.H Sanders (Ed.) Virtual reality in archaeology, (Oxford: Archeopress 2000).

[Nak99] H. Nakamura, R. Homma, M. Morozumi, On the development of excavation support system, *in Proc. 17th eCAADe Conf. Turning to 2000*, Liverpool, UK, pp341-348,1999.

[Rus01] C. Russo Dos Santos, P.Gros, P.Abel, Dynamic Information visualization Using Metaphoric Worlds, in Proceedings. VIIP 2001, Marbella, SP, pp 60-65, 2001.

[Ste91] R. Stenvert, Constructing the past: computer-assisted Architectural-Historical Research PHD of the University of Utrecht,1991.

[Wal02] N.Walsh, XML: One Input – Many Outputs : a response to Hillesund, Journal of Digital Information, vol 3 issue 1, 2002.

Designing a box of inspirations: a story about intecreation from an information portal for puppetry

Kurt Englmeier

LemonLabs GmbH
Ickstattstr. 16, 80469 Munich, Germany
englmeier@lemon-labs.com

Abstract

This paper presents the rationale and the design of eStage – an information portal for puppetry that serves as a virtual box of inspirations besides the puppeteer's workbench. eStage is an information services that emerges from the ambience of stakeholders in the areas of puppetry. It is implemented using a turnkey information providing platform that supports intecreation, the interactive and cooperative development of content for an information ambience – a feature that supports in particular the integration of so far isolated content owners and creators.

1 Setting the scene

Perhaps one of the largest surprises in computing at the end of the 20th century was the extent to which entertainment became a major driving application. The power of the consumer market far outweighs in the meantime that of the scientific or even defence markets. This is a landmark indicating that computing found its way into a broad use of the society through entertainment. It seems only natural, then, that interactive entertainment should be considered as a valid application area for providing cultural as well as scientific information. This starts with intecreation when content owners, users, and experts rather than technology people are the major player in designing the presentation of an entertaining content.
eStage is an information provider platform for puppetry. Its user interface design emerges from the rationale of intecreation, the interactive and cooperative development of content for an information ambience – a feature fostering the creation of information repositories that lay ground for virtual communities and new ways of content and knowledge exchange.
eStage's data collection is built around material related to puppetry such as plays, descriptions of puppets, abstracts of plays, and useful literature like tales. The digital collection itself is available for free and lives from the free contributions of its community. This trait reflects the centre piece of eStage's objectives: It is a self-organising and self-sustained service established for and through the community active and interested in puppetry. The community's involvement addresses also the definition of the repository's organisational structure.

2 Intecreation

eStage bases on a platform technology that supports cross-media information provision and cross-owner content building. Its user interface design emerges from the paradigm of intecreation, the interactive and cooperative development of content for an information ambience – a feature fostering the creation of information repositories that lay ground for virtual communities and new ways of content and knowledge exchange. Intecreation takes into account that content building is much more a social process as it is conceived today in web computing. Social networks can create a self-organising structure of users, information, and expert communities. Such networks can play a crucial role in combining next-generation information ambiences from enterprise portals to web search engines with functions such as data mining and knowledge management to discover, analyse, and manage knowledge. (Zhong, Liu, & Yao, 2002; Nishida 2002). The underlying social process, however, challenges a system's capability to communicate with its users in a natural way, to learn from their interactions, and to adapt to their interests, capacities, and working environments. This requires advancing beyond the conventional information retrieval strategy of document access toward the automatic detection and adaptation of scenarios that arise from a suitable composition of multimedia sources.

3 The link of technical solutions and inspirational desire

The goal here is to be a source of inspirations for the ideas of puppeteers – including professionals and amateurs. The things puppeteers are after can be best clarified by the following example from a talk with the ensemble of a puppet theatre: "We were looking for a new play for children that should have some educational effect through a certain level of cruelty as it occurs likewise in many of Grimm's fairy tales. We looked for children's books in libraries, bookshops and even toy shops and run into a book in English telling the story of a mother and cannibal looking for a child to eat. The design idea for the necessary puppet came from a picture in a festival announcement showing a puppet representing a fierce devil."

This example reflects the cornerstones of the user interactions expected to be encountered in eStage:

- The access to a digital archive must enable the creative talents to go directly to the spot they are looking for while bypassing the vast majority of the collection.
- The interaction mode must cope with the requirement that users show up with a very vague idea that steers their searching and navigating. However, this idea is the only steering element.
- The structure of guidance provided by the system must be sensitive enough to let the users keep their vague idea as the steering element. Any too rigid structure may cause an unacceptable information overload.
- There is no such thing like a precise answer to this vague query. Anything that relates to it can be useful. Thus, any kind of information found by serendipity is welcomed.

The characteristic outlined so far demonstrates the expectations of a user community requiring a platform that supports actively their work - expectations that determine the rationale of an information platform in the specific context of fostering creative talents. Such a platform must be different from an application that digitises an archive or museum with their functions of information mediating as a leading design rationale. These users clearly do not want a portal realising a virtual tour through a puppetry museum. This may be expected by someone who wants to be informed of puppetry as a whole or at least of more comprehensive blocks of this theme. The visitors of a museum (a real or virtual one) expect to get an overview of the virtual collection she or he is entering, to be guided through the collections and to have learned something from this tour. Because of differing user expectations eStage did not align its user interaction mode to the museum metaphor. Nevertheless we tested a hybrid structure that links various documents much like traditional text- and guidebooks. Integrated into a retrieval environment like eStage it can be pretty helpful to guide the users to explanations or more details on a specific topic. They can broaden a certain topic or even compensate for missing knowledge by adding explanations to the pages a reader visits. Good examples for the application of the museum metaphor can be found at Fleck et al., 2002, Back et al., 2001, or Segbert & Tariffi, 2002.

4 Confidentiality and familiarity in information spaces

The design of a box of inspirations bears striking confidentiality and familiarity we associate with forms borrowed from everyday life. Through intecreation that extends the process of content building to the knowledge of the whole user community, this design idea pushes information architecture into a new age of IT-based sensing and opens new avenues for information mediating. The success of the architecture depends on this social process and as well as on the capability of its underlying approach to what extend it helps to reduce information overload in using the functional levels of the system. (Meisel & Sullivan, 2000) It can be pretty helpful if the system provides the users with semantic elements that help to describe concisely the things they expect to find while roaming an information spaces. These semantics shown in a suitable way enable the users to find easily the locations of the documents related to the context of the users' searching and navigating (Englmeier, 2000; Ramakrishnan & Grama, 1999). A semantic coordinate system endows the users with a concise as well as comprehensive vocabulary. Hierarchically arranged and grouped along major content facets this vocabulary acts as a stable coordinate system easy to comprehend and memorise. The users are thus much more in the position to localise themselves effortlessly. Successfully searching and navigating now means guided travelling from information to information just by changing the semantic coordinates, i.e. by pointing to relevant concepts. This structure on the other hand enables to pinpoint the semantic location of any kind of information.

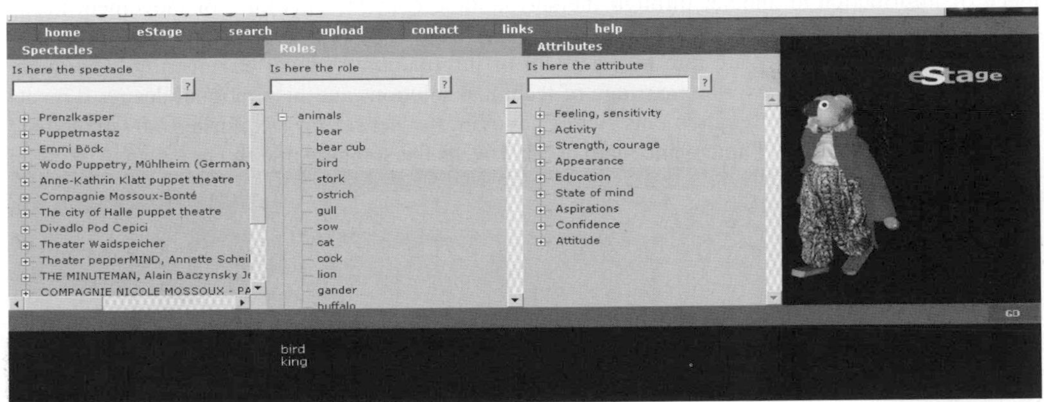

Figure 1: Searching and navigation in a semantic coordinate system: Selected concepts make up the initial query profile. While realising their retrieval strategy the users usually perform iterative steps of defining a query and analysing the retrieved results. In eStage, the documents are annotated solely with entries from the hierarchies.

Movement is a recurring theme for us, but it's an approach to movement that lets sense the network of information through entering plays, passing images, going through videos, crossing puppet characters and the like. Practicality isn't the primary goal, astonishment is the effect we are after. Information structures are intended to infuse data spaces with excitement, and creates a relationship to its users that has to do with curiosity.

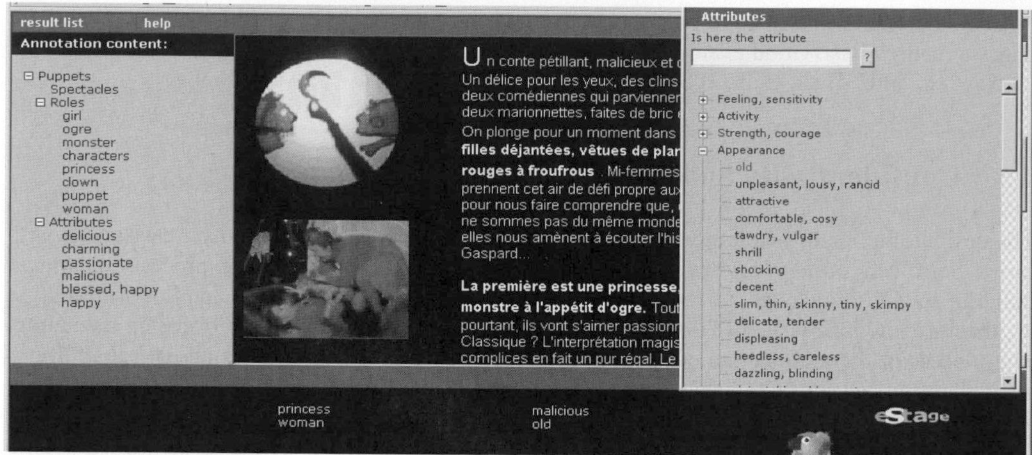

Figure 2: The document is presented together with its semantic surroundings. The interface shows the annotated concepts that are thus already familiar from the initial query formulation. Modifying the set of annotated concepts is thus tantamount with repetitive query formulation. The user can always invoke the respective concept window if concepts are required in addition to those annotated to the document.

With our design of information architecture the complex information space dissolves into an arrangement of familiar semantic structures. What we are seeing is the reflection of generic orientation. Roles, attributes, and spectacles define an information space and give it meaning. Our construction creates structures that transport people virtually from one point to another, form one piece of information to another through the surrounding context. This kind of movement keeps abreast of experience, revelation, serendipity, excitement. Overt expressions of serendipity combined with practicality are rare in the realm of information retrieval engineering. The query feature in eStage can achieve the marriage of technical solutions and inspirational desire. It meets a collection's structural requirements while evoking a sense for exploring an immediate information ambience. Its flamboyance inextricably weds the stunning and the practical.

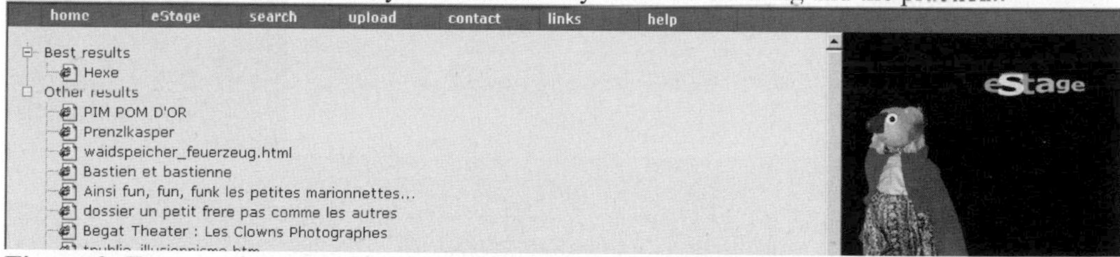

Figure 3. Exact retrieval results are marked as "Best results" in the workspace, while the others can be located more in the semantic surroundings of the best match.

5 Conclusion and perspectives

eStage is currently a collection of voluntary contributions from puppeteers, puppet theatres, and the like. Its content bases on intecreation, the new and inspiring metaphor drawn to the idea of mutual and collaborative design of future information ambiences.

From the puppeteers' point of view eStage is the implementation of a platform technology for "boxes of inspirations". Viewed from a more global angle it is a mining system. Unlike typical

data mining systems that restrict mining to numerical data, eStage's underlying technology links heterogeneous data types on a semantic level. So it enables mining simultaneously in templated data ("hard facts") as well textual data ("soft facts"). This characteristic of an information system is also required for other digital collections when it comes to providing access to huge collections that are made up by a number of heterogeneous databases.
The design of a unified access to large databases containing information on ancient coins, for instance, can benefit from this platform technology. It would be in the position to link the factual data (metrics of the coins) with the textual data (descriptions plus photographs and useful remarks) to provide a coherent view on the collection.
The integration of the context-oriented search feature for the domain experts and catalogue features or a museum metaphor opens way to information and learning providing services that emerge from the ambience of stakeholders in various areas, not only those of cultural information.

References

Back, M., Gold, R., Balsamo, A., Chow, M., Gorbet, M., Harrison, S., MacDonald, D., & Minneman, S. (2001). Designing Innovative Reading Experiences for a Museum Exhibition *IEEE Computer*, 34 (1), 80-87.

Englmeier, K. (1999). The link between data and tasks - the crucial challenge in designing user interfaces for information retrieval systems". In H.-J. Bullinger; J. Ziegler (Eds.), *Human-Computer Interaction. Communication, Cooperation, and Application Design. Vol. 2. Proceedings of HCI International '99*, (pp. 97-101).

Fleck, M., Frid, M., Kindberg, T., O'Brien-Strain, E., Rajani, R., & Spasojevic, M. (2002). From Informing to Remembering: Ubiquitous Systems in Interactive Museums. *IEEE Pervasive Computing*, 1 (2), 13-21.

Meisel, J.B. & Sullivan, T.S. (2000). Portals: the new media companies. *info – the journal of policy, regulation and strategy for telecommunications information and media.* 5 (2), 477-486.

Nishida, T. (2002). Social Intelligence Design for the Web. *IEEE Computer*, 35 (11), 37-41.

Ramakrishnan, N. & Grama, A.Y. (1999). Data mining: from serendipity to science. *IEEE Computer*. 32 (8), 34-37.

Segbert, M. & Tariffi, F. (2002). Cultural heritage trials: solutions and potential in a special project bouquet. Retrieved January 31, 2003, from www.trisweb.org/tris/trisportalpro/Papers/EVA_London_Paper-01.pdf.

Zhong, N., Liu, J., & Yao, Y. (2002). In Search of the Wisdom Web. *IEEE Computer*, 35 (11), 27-31.

Acknowledgement

Research outlined in this paper is part of the project eStage that was supported by the European Commission under the Fifth Framework Programme (IST-2000-28314). However views expressed herein are mine and do not necessarily correspond to the eStage consortium.

Comparing Presentation Styles of Help for Shoppers on the Web

Qin Gao,Pei-Luen Patrick Rau, Wei Zhang

Department of Industrial Engineering
Tsinghua University
Beijing 100084, China
gaoq02@mails.tsinghua.edu.cn, rpl@mail.tsinghua.edu.cn,
zhangwei@tsinghua.edu.cn

Abstract

This research investigates the effect of the presentation type of help information on enhancing the usability of E-Commerce Web sites. There are types of presentation, pop-up window type and pop-up message type, studied in this research. An experiment of eighteen participants in Tsinghua University was carried out. The results showed that the participants with pop-up message style of help took fewer steps than the participants with pop-up window style of help. However, no significant differences were found in performance time and satisfaction.

1 Introduction

Shopping on the Web is generating a lot of interest. Providing Web shoppers help information at the right time and in the right place can smooth the purchase process. Also, providing appropriate help information enhances both the usability of the Web site and customers' trust on the Web site. Many researchers have noticed that trust on E-Commerce Web sites is a critical concern, and a lot of studies have been done to find out how to improve the usability of the E-Commerce Web sites in order to enhance customer's trust. To offer the customer enough information to make them feel that they are controlling the transaction is an important consideration. Also, how to present help information to make it be percept in time, used effectively and considered friendly is a concerned problem. This study focuses on the presentation styles on designing help for shopping Web sites. The main objective of this study is to examine how the presentation style of help information might impact user's performance and perception of the E-Commerce Web site. This study provides an experimental evidence of the conclusion for Web designers to improve the design of the E-Commerce Web sites.

2 Customer's Trust on Shopping Web Sites

Usability is believed very important in E-Commerce. To ensure that end users are able to maximize what they can get out of it is absolutely the end purpose. Customers' trust is critical to the success of E-Commerce. The Faith-Trust-Confidence continuum, as defined by Arion et al. (1984), refers to the amount of available knowledge and cues on which to base one's belief. Trust acts as a mental mechanism, based on incomplete information, which helps reduce complexity to allow for decision making under uncertainty (Luhmann, 1988; Kahneman et al., 1982, Sisson, 1998).

Cheskin Research and Archetype Studio (1999) developed a model to understand trust consisting of four phases: unaware, build trust, confirm trust, and maintain trust. Proft (2003) suggested that the third phase, building trust, is the most important one. Building trust is a long process of several Web site specific activities, and can be achieved by good design strategies. Cheskin Research and Archetype Studio (1999) studied E-Commerce trust and suggested six primary components of trust: seals of approval, brand, navigation, fulfilment, presentation, and technology. Navigation reinforcement such as prompts, guides, instructions, and tutorials could aid and inform the user to complete the transaction or search on the Web site.

How to present the help information in a proper way ensuring the availability of knowledge but avoiding to be annoying might have effects on the perception quality of help information, and thus affects customers' trust.

3 Presentation Styles of Help

Two types of presentation style of help were compared in this study: pop-up window style and pop-up message style. The pop-up window style is illustrated in Figure 1. Help information is embedded in textual or graphical hyperlinks, and retrieved in a small new window while the user clicks the hyperlink.

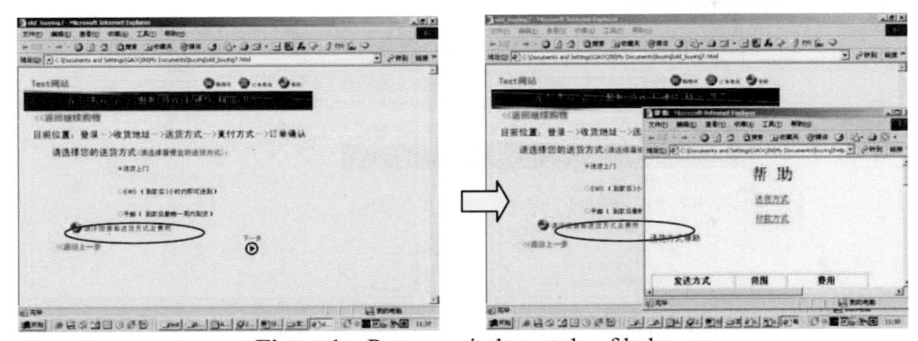

Figure1: Pop-up window style of help

The pop-up message style is illustrated in Figure 2. Help information will pop up if the user moves his mouse cursor on the object of a Web page. If the user feels confused, he/she may stop the mouse cursor on the text or the picture rather than performing any operation. The help information will pop up without any other further operation such as clicking so that the trust of users may be enhanced.

4 Methodology

An experiment was conducted to compare the two presentation styles of help. Two Web sites were created with all the same Web pages and the same help information except for the difference in presentation style of help. Eighteen participants, aged from 24 to 40, were recruited for participation at Tsinghua University, Beijing. The participants were randomly divided into two groups with nine participants in each group. The independent variable was presentation style of help (pop-up window and pop-up message). The dependent variables were performance time, number of steps, and satisfaction. The task asked the participant to shop two items on the testing Web site. The participant was told to find the lease expensive delivery and the fastest payment method.

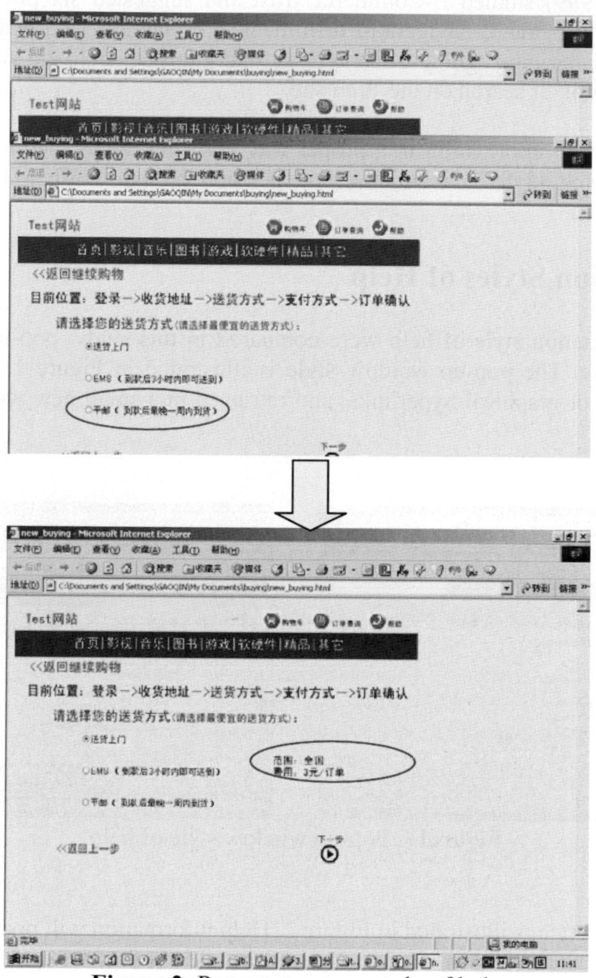

Figure 2: Pop-up message style of help

All participants began to fill out a general information questionnaire concerning their personal characteristics, including age, education, and past computer and Internet experience. Each participant was given on-screen instructions. A brief practice session was then conducted to help the participants understand the operation of the system and the tasks to be performed. Following the practice, each participant performed the shopping tasks. Participants were instructed to perform the tasks as quickly as possible without sacrificing accuracy. On the completion of the tasks, each participant was given questionnaire of satisfaction.

5 Results and Discussion

The data collected were checked for model adequacy. The data was transformed if model adequacy did not hold. The intention of the experiment was to examine how the presentation style of help might influence the browsing performance of Web shoppers. Significant difference in number of step ($t = -2.996$, $p = 0.008$) was found, as shown in Table 1. No significant differences were found in performance time and satisfaction. The participants with pop-up message style of help took fewer steps shopping than the participants with pop-up window style of help.

Table 1. Results for Comparing Two Presentation Styles of Help

Variables	Pop-up Window		Pop-up Message		t	p
	MEAN	SD	MEAN	SD		
Performance Time (sec)	285.52	186.125	325.54	129.136	-0.529	0.603
Number of Step	20.0	10.15	10.3	2.92	-2.996	0.008
Satisfaction	3.2	0.87	3.3	0.92	-0.663	0.510

5. Conclusion

The purpose of this study was to investigate the effectiveness of presentation style of help on browsing for Web shoppers. The results indicate that the pop-up message style was associated with fewer steps for participants. According to the results, it is suggested that providing pop-up message style of help information is expected to reduce the number of pages clicked for Web shoppers. However, the participants with pop-up message style of help did not perform the shopping task faster than the participants with pop-up window style of help. The possible reason could be that providing too many pop-up help messages may be annoyed and distracted for participants. Web designers who wish to enhance Web usability of E-commerce Web sites could utilize the results of this research. Pop-up message style of help information may reduce the number of steps and be retrieved easily for Web shoppers.

References

Arion, M, J.H. Numan, H. Pitariu & R. Jorna (1984). Placing Trust in Human-Computer Interaction. Proc. 7th European Conference on Cognitive Ergonomics (ECCE 7): 353-365.

Cheskin Research and Studio Archetype (1999). E-Commerce Trust Study. Retrieved November 15, 2002, from http://cheskin.com/think/studies/trustIIrpt.pdf.

Kahneman, D., Slovic, P. & A. Tversky (1982). Judgments Under Uncertainty. Cambridge Univ. Press.

Luhmann, N. (1988). Familiarity, Confidence, Trust: Problems and Alternatives. Gambetta, D. (Ed.) (1988). (1988). Trust: Making and Breaking Cooperative Relations, Basil Blackwell.

Proft, N. D. (2003). Building E-commerce Trust: Applying Usability Principles. Retrieved February 15, 2003, from http://user.online.be/cd01237/ecommercetrust/index.html.

Sisson, D. (1998). A thoughtful approach to web site quality. Retrieved November 15, 2002, from http://www.philosophe.com/commerce/trust.html

Smith, A. (2002). Improving web site usability. Retrieved November 15, 2002, from http://www.optimum-web.co.uk/improvewebsite.htm

Grid Transparent Windows

Antonio Gómez Lorente and Javier Rodeiro Iglesias

Informatics Department
University of Vigo
Vigo-Pontevedra-España
agomez@uvigo.es,jrodeiro@uvigo.es

Abstract

The main feature of the majority of recent user graphic interfaces is the inclusion of 2D windows. Another aspect deals with the existence of a visual hierarchy of those windows that leads to a well known increase from 2D to 2,5D. The purpose of this paper is to try to describe a user graphic environment paradigm that surpasses the 2,5D. A new interactive mechanism is proposed in which it is possible to see and interact with the contents of 2 simultaneous windows, in spite of the fact that these could belong to 2 different applications and occupy the same position on the screen. Although the perception of the contents of the overlapping and overlapped windows are seen hardly altered and blurred the simplicity and attractiveness of the paradigm reinforces its importance.
The underlying technique for this paradigm simply entails the elimination of a certain number of pixels from the modal frame window thus allowing the perception of the rest of the windows underneath it. The decision on the position and on the number of pixels to be cancelled is one of the most complex aspects of this technique. To select the pixels that are to be cancelled several techniques of the deterministic and probabilistic type have been tried. Once the pixels have been eliminated from the window, the user can not only perceive visually but also interact through them. In user graphic environments with X-Windows the shape extension library for nonrectangular windows is the one used for this technique.
The methods and difficulties of implementation of this paradigm are shortly analysed; its usage and usability are commented and as conclusion an image of a practical case is presented.

1 Aim

The interaction of users with computing systems through command lines constitutes a user working interface of a unique dimension. (1D) The succession and grouping of commands in scripts and menus can be considered an extension of that unique dimension (1,5D). The appearance of windows and graphic interfaces for windows introduces a new working dimension for the users (2D). The existence of a hierarchy of windows in working frames and the switching between windows and working frames leads to an increase of dimension from 2D to 2,5D. However the overlapping and covering/hiding of windows one on top of the other is a serious hindrance in the increase of dimension. The increase on the dimension level allowing the user to reach the more intuitive and user friendly 3D is an important goal of the paradigms of user graphic environments. The inconveniencies and obstacles found in this field are well known; a more rational use of windows and a more profitable distribution in working frames; overcoming the existence of one single modal window as foreground and a second working area as background that includes the rest of the windows is one of the means to surpass this 2,5D level. Grid transparent windows are an approximation to the intents of surpassing this 2,5D in user graphic environments, independently of the user interface paradigm that is used.

2 Introduction

The idea is simple. If objects in real life only hide or cover what is situated behind them but not what is beside them, why should the content of a user interface window hide that part of the other window that it overlaps? . The answer is obvious; it does so just as a sheet of paper hides other sheets of papers underneath it on the table. However there are also transparencies containing visible information that do not prevent seeing the rest of transparencies found underneath it. In computing we have something similar, transparent windows, as attractive and useful as the office acetates. Notwithstanding to be able to work with these windows (which is in itself quite a novelty) or working frames found on the bottom layers, as it occurs with acetates, one has to previously remove the top layers and thus stop using them. Would not it be possible that without having to remove the top layers and having these still visible one could access indistinctively top and bottom layers? That is, passing through the transparent areas of a modal working frame in order to reach the non transparent areas of the underlying working frames. Well, NO, it would not be possible. .As well as the information contained in an acetate is physically sustained by a flat body, that is, the acetate sheet in itself, the windows likewise have a flat dimension that can be transparent or as alfa transparent as one wishes but that still prevents going through them. Why could not this be feasible if it is possible to access a sheet of paper placed underneath another one by simply piercing a hole through it without having to remove the top sheet. After all, one does not need to go through a model frame with a pencil in order to be able to work on the other windows behind it. A pixel through which the mouse´s cursor could go through would suffice. That is the simple idea. It is possible. This is the most straightforward and intuitive description that one has thought of for grid transparent windows. We will have to ask the reader to be indulgent with this colloquial description. The pretension was to obtain a greater simplicity in the exposition of the matter.

3 Grid Transparent Window

Grid transparent windows are those windows that, independently of their level of transparency (or alfa transparency), enable interaction through them on any other window they may overlap. These windows not only allow visual perception of different applications (thus, different windows) at the same time- which in itself is quite important- but they also allow simultaneous interaction on these different applications (windows). One grid window suffices to have 2 simultaneously accessible overlapping windows. The switching does actually occur but to the user it appears as a transparent and automatic process. See snapshot.

4 Implementation

The implementation of this paradigm is also simple and intuitive. In a few words, it consists in *"adequately"* piercing or making a hole (metaphorically speaking) through the modal application window. These "holes" will allow seeing and interact with any overlapped window "*with hardly any loss of perception*". See snapshot 1.
The perforation technique is also formally simple. A set of "adequate" and "sufficient" pixels is chosen to be then cancelled on the desired window. Through these "*strategically*" chosen pixels

one can not only see the contents of any other underlying window (and this is important) but also (we insist here once more for this is still more important) introduce the mouse arrow and thus reach an overlapped window and work with it. This is done without losing sight of the modal or top window while we are working with the contents of the one lying underneath it. At any given moment one can go back to the pixels that were not cancelled and resume work on the initial window.

4.1 Design

The true suitability of this system lies in the optimum election of the pixels to be cancelled. The election of the pixels to be cancelled is what one can call the design of the grid. This design is not always as simple as it could seem. The difficulty lies in two aspects: the number and disposition of the pixels.
The number, or better still the percentage of pixels of a grid window is directly proportional to the relevant number of points and to the level of relevance of these points in the perception of the contents of a window before it undergoes the transparency process.
Generally speaking , windows with a graphic content with figures and images relatively big or without a high resolution admit a high perforation rate (up to 80%).
In application windows with abundant text and which were to be simultaneous with other windows which also had a lot of text , we were biased towards a 50% average perforation. In the majority of cases this is quite suitable.
As for the disposition of these points one has to take into account the following 2 principles: the cancellation probability of a pixel in a grid transparent window must be directly proportionate to the probability of that same pixel to be occupied by relevant image points in the contents of the window and the relevance level of these points.
The cancellation probability of a pixel in a grid transparent window must be inversely proportionate to the level of elimination probability of the pixels found in its proximity and in a higher proportion to those surrounding them.
If determining what percentage of pixels is to be cancelled is already a complex task, a higher complexity is attained when calculating the above probabilities and an even higher one when calculating the relevance of the pixels. In the queries made to localize references on the relevance of pixels in a specific window with respect to the contents of it, we have not found anything concrete that could help us in a specific way in the calculation and location of that desired number of pixels to be cancelled. Us authors would highly appreciate whatever reference sent, but we understand that the novelty of the matter (or our lack of knowledge) is the cause for this lack of bibliography on the subject. In that sense perhaps, grid transparent windows constitute a suitable tool in the study of pixel relevance of whatever type of information graphically represented. We believe that arriving at a certain point in the automatic design of the grids, simpler techniques of analysis and segmentation of images in the field of artificial perception could be used. See for example (Jäne).
The distribution of points to be cancelled has been done through a unique static grid using two very generic methods; a random or probabilistic approximation that consist in placing the points to be cancelled in the window so as to have a uniform distribution; a deterministic geometric approximation that consists in situating the points to be cancelled by the window in a homogeneously distributive manner through homogeneous geometrical segments. The probabilistic grid turns out to be more diffuse and gives out a weaker perception than that of the geometric grid. This comes to prove that the random grid introduces in the image a greater visual entropy and corroborates the Gestalt theory in the field of psychological perception. This interesting problem is well presented in (Arnheim).

4.2 Programming

The programming of grid transparent windows in X-Window is relatively easy thanks to the existence of the X-Widow Shape Extension Lib added in the Release 4 of Version 11 of the Standard X-Window. At first all the references to this library were made to supply nonrectangular borders (Shape) to the windows. Nowadays this library is mainly used to implement nonrectangular border images such as icons and application buttons using XPM library images. This technique has also proven useful in making irregular (cancelling pixels), not the border of the window, but the interior of it.

4.3 Performance

Unfortunately we must stress the weak performance of programs using X Shape extensions. As long as the algorithm used in the selection of pixels does not become excessively complex (in our case we managed to use none too complex, except in few very specific frustrating cases), the program executed using grid transparent windows does not undergo significant performance variations. Its features as far as performance is concerned and its use of computer resources as it executes as client are maintained. However, the Shape functions overload (we insist, not the client) in a significant manner the X-Window program server where they are being used. See reference [Lee].

5 Usability

Given on one side the inconveniences mentioned above and on the other side the novelty of the paradigm in which we have been working for a relatively short time, we were unable to provide to a great number of users (except to a reduced number of collaborators) applications using grid transparent windows. That is also the reason why it was not possible to present statistic field analysis on utility, usability and usage of the paradigm. We thereby can only make a rational and theoretical analysis of the given aspects.

5.1 Usage

The use of grid transparent windows is simple as long as we are able to anchor the grid window on the foreground and make it remain there. Luckily, for the case of our interest there are more newer X-Window managers that allow the positioning of a window "on top". As for the internal processing of applications to which grid windows have been added it would be desirable to append an internal method in these applications that would allow a change in the formal aspect of the contents of the windows. Thus it could prove useful the use of interactive facilities for size and colour changes on the fonts in the more "textual" applications and changes in the borderlines of the interior areas of colour on the more "graphic" applications.
On screenshot 2 a complete example of usage is presented.

5.2 Utilities

The theoretic advantage of this paradigm arises from the possibility and easiness of integration and interaction of two distinct windows with all the benefits of compensation, integration, extension and complementarities that go with it. It has been particularly useful to us authors in monitoring the data on the system while we were actually working on it. It has been also useful in the monitoring and in the tracking of execution of programs; consulting help tutorials while following its indications; writing documentation of programs while following their execution; consulting documentation on programming language functions while codifying those functions in a program; modifying a program code with an editor while seeing how it is being executed

5.3 Disadvantages

Grid transparent windows could be seen as unsightly, confusing, tiresome and non motivating. One has to realise that it is not the paradigm but the environment the cause of this situation. Contrasting, working with simultaneity and mixing various types of information originating at times from different sources requires a high level of mental and sensorial concentration. Our working level must improve, not due to the use of grid transparent windows but because of the goal aimed at. In certain cases, not so rare, due to the incompatibility existing between the outward appearances of the windows of the different applications being used at the same time, simultaneous perception and distinguishing between contents may result difficult. It is obvious that grid transparent windows are highly inadvisable for applications requiring high graphic precision and resolution.

6 Snapshot

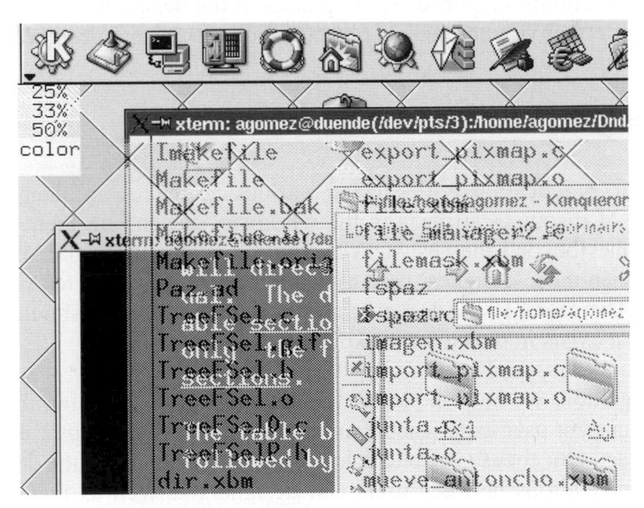

Snapshot 1. Grid Transparent xterm

References

Arnheim, Rudolf. 1971. Entropy and Art. University of California Press.

Jäne, B.. 1993. "Digital Image Processing: Concepts, Algorithms and Scientific Applications. Second Edition". Springer-Verlag Berlin Heidelberg.

Packard, Keith. "X Nonrectangular Window Shape Extensions Library. Version 1.0" MIT X Consortium.

Lee, Kenton. July 1994. "X Window System Application Performance Tuning". The X Journal.

Methods for exploring workplace activities and user contexts employing intermediate objects self-photos, personal view records, and skit performance

Kimitake Hasuike *Eriko Tamaru* *Mikio Tozaki*
Human Interface Design Development, Fuji Xerox Co., Ltd.
YBP E13F, 134 Goudo-cho, Hodogaya-ku,
Yokohama, Kanagawa, 240-0005, Japan
{kimitake.hasuike, eriko.tamaru, mikio.tozaki}@fujixerox.co.jp

Abstract

In this paper, we describe an outline and features of three methods we have developed for understanding user's work practices and their contexts. In the first method, we use self-photographs of users to share the context of the workplace between users and designers. In the second method, we use video records of the user's personal view to share user experiences. In the third method, we use skit performance as a prototype for evoking future activities and experiences. Although these three methods support different phases in the design process, their common focus is creation and use of effective intermediate objects among users and designers for the purpose of understanding user contexts.

1 Introduction

Users organize their practical activities using various resources in their workplaces. Their activities and usage of resources/artifacts in the workplace depend on their own context. These practices are sometimes beyond the designer's assumptions. In a particular situation, artifacts are used for aims other than those designers have assumed, are used unexpected way and combination with other artifacts. Therefore, In the design of a new system, designers should consider the broad range of activities that exists in the target workplaces.

Based on such recognition, we have explored methodologies for observing the work practices of users, for analysing current user experiences, and for prototyping future user experiences, in order to design a new artifact. Our focus is also to enhance communications in the design process, which includes both communications between users and designers as well as that among design team members. In our methods, we use self-photographs, users' personal view records, and skit performance, as mediums to help designers and users understand how users behave, the resources used, and the actual role of artifacts in a workplace.

2 Methods for exploring users and the contexts

2.1 Self-photo Study

Self-photo study is an observation method that employs the users' self-photos as a medium for understanding their backgrounds, workplaces, and practical activities (Tamaru et al., 2002).

In self-photo study, we give each participant a disposable camera and ask him/her to take some photographs during the course of his/her typical workday. They are instructed to take photos as records of their activities. These photos include their workplaces, environments, people they meet, and artifacts they use.

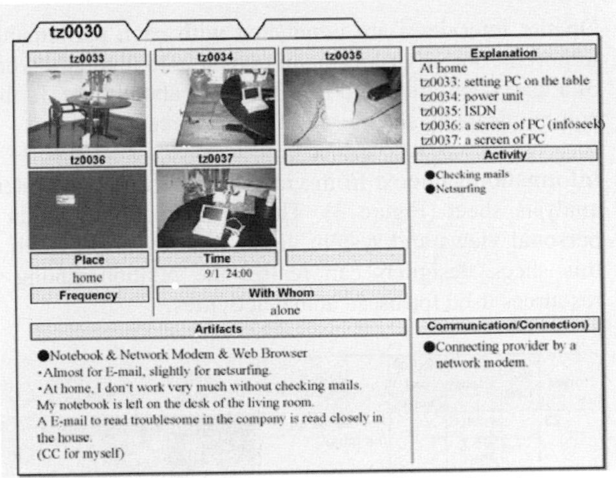

Figure 1. Workplace data sheet

An on-site interview is conducted with each participant after film development. The questions are about places they were, activities they did, tools and documents they used, the way they use these artifacts, people they communicate, the means for that communication, and so on.

Interviewing with the aid of self-photos makes it easier to ask the appropriate questions and to earn rich interview data. Self-photos function as tools for remembering the contexts of activities and for mediating the interaction between users and designers.

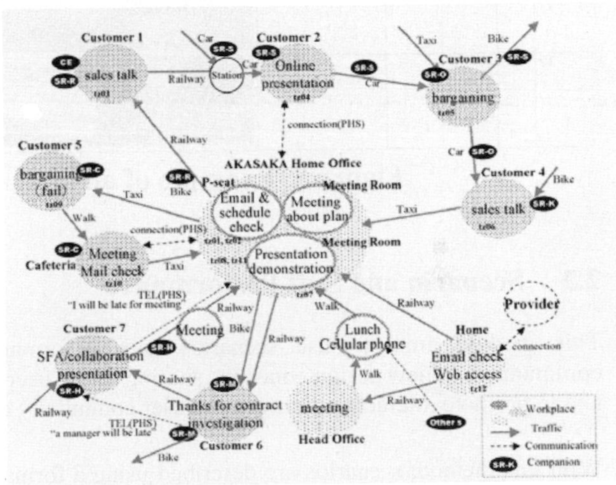

Figure 2. Example of an interaction map

Self-photographs and interview data are arranged on workplace data sheets. The format of this sheet has fields for self-photograph, place, time, activity, people, artifact, communication, and so on (Figure 1). Then an interaction map is created based on data sheets (Figure 2). This map visualizes the relationships among workplaces, workers, artifacts and activities. After the interviews, subjects and members of a design team hold a workshop to discuss their activities through the workplace data sheets and the interaction map.

2.2 Interaction Analysis Employing the User's Personal View

In direct observation, video records are commonly used to view and re-view user behavior. In our method, we use video records from user personal sight for exploring what user actually interacts, and what user actually experiences (Hasuike et al., 2001).

This method aims to observe user activity and experience through the user's personal view. An essential instrument for this is a small CCD camera located on a cap. Participants simply go about their usual activities while wearing this cap, and personal view video data is transmitted to a video recorder. We also record whole system view, audio, and describe observation notes.

On-site interviews are conducted with each participant after the observation. One of the most important objectives of this interview is to highlight which resources are effective in the sequence of a user's activities, and how they feel about these. In this interview, interviewers and participants communicate by sharing personal view records.

Information gathered from video/audio records and interview notes is integrated in an interaction analysis sheet (Figure 3). The format contains fields for scenes, places, situations, activities, personal views and system views captured from video, and interview protocols and notes. With this sheet, designers can realize the relation among user activities, personal views, and the resources used for user's actual activities.

Index	Scenes	Place	Situation/ Event	User's Activity	System View	User Personal View	Protocol	Interacted Design Element	Interviewer Protocol	Participants Protocol
23		In the Copy Area		Put the file on the table.				Table, File, Over all view of the copier, Location/Layout,	Did you select the copier in advance?	For stapling. I usually use another monochrome copier for more simple jobs not need staples.
									Did you know the machine by the location?	Yes.
27		Copy Area		Walking toward the target copier.				Indicator for Power On	What did you think about in this view?	Power is on or not.
									How did you know?	By the backlight of LCD panel.
35		In front of the Copier						LCD panel, ADF, Document	Now you put the removed staples at the clip Tray?	Unconsciousness. Always.

Figure 3. Example of an interaction analysis sheet

2.3 Scenario and Skit Performance

This method aims to use scenarios and performance with mock-ups as techniques for communicating new design concepts, sharing future user experiences between designers and users, and to promote interaction between the user community and the design team (Hasuike et al. 2002).

With this method, scenarios are described using a format, containing fields for scenes, situations, user activities, interaction elements, and design points. This format is basically the same as that for describing observed interaction, mentioned in Section 2.2.

We use skit-performance as a base technique that enables us to prototype a situation of "artifacts-in-use" (Bannon and Bodker, 1996) and "people at work" (Kuhn, 1996) regarding future artifacts in actual user contexts. There are some prior studies concerning performance, such as "Informance"(Burns et al., 1994), "Experience Prototyping" (Buchenau and Fulton, 2000), and the playful design approaches (Brandt et al., 2000). In our method, performances should take place in actual user environments. And there are two types of props: simple non-functional mock-ups as prototypes of future artifacts, and real artifacts existing in the use environment. Actors perform skits using these props as their resources for performance.

Index	Scene	...	Situation/ Event	User Activity	System View	User Personal View	...	Director	Actor
3.5	In the living room			She carries the e-pad to the kitchen counter.				"Where would you set the such communicators?"	
	In the living room			She places the e-pad on the kitchen counter.					A: "Here. Changes the direction to the Living room or the kitchen according to the situation."
									B: "May be same in my home."
									B: "I will changes the direction tc the kitchen while cooking."
				She views the lamp of the e-pad, which see supposes as to be at the top of it.				"OK. Please changes the direction."	B: " If the lamp is on, I will see the display."
	In the kitchen								A: "Oh, it will be useful."

Figure 4: Example of a performed skit in the interaction analysis sheet

During performing skit, the director suggests situations and facilitates user performance with prompts such as: "Now, you just got an e-mail from your colleague in the office. How would you become aware of that here?" The actor then responds to these questions by acting performance. The actor sometimes puts questions to the director. These include questions regarding the supposed situation, the features of the artifacts, and the functions and operational methods of the artifacts. Through these communications, skit is organized interactively.

Skits are observed and recorded with interaction analysis method employing users' personal views, and the results are described in the format mentioned in Section 2.2 (Figure 4). Thus the format is designed for both the design and analysis phases, and we can compare interaction scenarios considered by designers, with observed interaction sequences actually performed by user. By analyzing these differences, we can identify users' needs, ideas, and contexts, even if the users themselves are unaware of them.

3 Discussion

We have reviewed an outline of our three methods. In the first methods, we use self-photographs of users to share the context of the workplace between users and designers. In the second method, we use video records from the user's personal sight to share user experiences. In the third method, we use skit performing as a prototype for evoking future activities and experiences.

A common feature of these three methods is proactive use of visible intermediate objects, for sharing between users and designers as well as among design team members. Self-photos and users' personal view records are visual data simply captured in user workplaces. Skit performances with simple mock-ups are visible prototypes of future activities and experiences. Workplace data sheets and interaction analysis sheets are forms for arrangement of gathered data and visualization of observed activities and interviewed contexts. And an interaction map visualizes the design team's conception of the users' work practices.

In our method, there are some opportunities for communication employing these intermediate objects, on-site interviews, workshops, and skit dialogues. These communications include rich information about user themselves, user activities, backgrounds, problems, and ideas. From our experiences of studies employing these methods, intermediate objects do promote and facilitate deeper communication and understanding. With these settings, we can capture some unexpected user behaviors, unexpected usage of artifacts, unexpected users' assumptions for functions of future artifacts, and so on. These occurrences are resources for understanding problems and requirements in the actual context of use.

From the viewpoint of user participation, taking self-photos promotes users' proactive involvement in exploring workplace researches and design, and performing skits promotes users' proactive involvement in idea generation and design of the future work styles. From the viewpoint of design cycles, these visualized data can mediate the interaction between different design project teams. In a project, we utilized the work practice data of sales persons captured with self-photo method as a base for creating scenarios of future document handing system using electric-paper technology (Hasuike et al. 2002). These research-design cycles can promote deeper understanding of users work practices and can promote proper reflection of these understandings to the future design projects, from earlier phases of design process.

4 Conclusion

We have reviewed an outline of our three methods employing proactive use of visible intermediate objects. For developing effective databases of captured information with these methods, it is necessary to study more formal description method for describing workplace data sheets, interaction maps, interaction analysis sheets, and scenarios. It is also important to develop tools that support to handle these visible data.

References

(Tamaru et al., 2002) Tamaru, E., Hasuike, K., Tozaki, M., 2002, A Field Study Methodology Using Self-Photography of Workplace Activities, In the Design Research Society 2002 International Conference.

(Hasuike et al., 2001) Hasuike, K., Takada, R., Tamaru, E., Tozaki, M., 2001, A Method for Designing Physical Appearance Attributes from the Viewpoint of the Interaction Design, JP-061, Bulletin of the 5th Asian Design Conference, October, Seoul 2001.

(Bannon and Bodker, 1991) Bannon, J. and Bodker, S., 1991, Beyond the Interface: Encountering Artifacts in Use, In Carroll, J. M. (ed.), Designing Interaction: Psychology at the Human-Computer Interface, Cambridge, Cambridge University Press: 227-253.

(Kuhn, 1996) Kuhn, S., 1996, Design for People at Work, In Winograd, T. (ed.), Bringing Design to Software, NY, ACM Press: 223-289.

(Brandt et al., 2000) Brandt, Eva and Grunnet, Camilla, 2000, Evoking the Future: Drama and Props in User Centered Design, PDC'00, New York.

(Buchenau and Fulton, 2000) Buchenau, Marion and Fulton Suri, Jane, 2000, Experience Prototyping, Proceedings of DIS'00, Brooklyn, New York: 424-433.

(Burns et al., 1994) Burns, C., Dishman, E., Verplank, W., and Lassiter, B., 1994, Actors, Hairdos and Videotape - Informance Design, CHI94, Boston.

(Hasuike et al., 2002) Hasuike,K., Matsuo, T., Takeuchi,K., Tozaki, M., 2002, A Method for Designing and Analyzing Interaction Design at Earlier Phases of the Design Process -Use of the Scenario, Performance, and Description format. In the Design Research Society 2002 International Conference.

An examination method of human interface using physiological information

Yoshiaki Hayasaka[*, 1], *Tatsuhiro Kimura*[*, 2], *Shuhei Ogawa*[**, 3]
Norihisa Segawa[***, 4], *Kiyoyuki Yamazaki*[****, 5], *Masatoshi Miyazaki*[***, 6]

*Graduate School of Software and Information Science
Iwate Prefectural University
152-52, Sugo, Takizawa, Takizawa, Iwate, Japan
[1]g236z007@edu.soft.iwate-pu.ac.jp

Abstract

The authors have continued study of an objective guideline for development of a low-workload Human Interface (HI) utilizing psychophysiological measurement. An experiment was carried out in this study to assess interactive software users' responses to an irritating situation. The experimental task required the participants to respond to ten questions successively, then to wait for a time interval with a progress bar (PB), without a PB and with a blinking symbol-indicator. During task performance, electroencephalogram (EEG), heart rate, and plethysmogram were recorded and analyzed. Results showed that the high frequency component of the heart rate increased and the low frequency component decreased in the PB condition, even though alpha- and beta-wave EEG components showed no change. These results suggest that irritating situations stimulate the autonomic nervous system and induce changes in physiological indices. Possible application of the present method to development of a low-workload HI was discussed.

1 Introduction

Information terminals of computer network systems are widely spread, not only in industrial fields, but also in domestic use. Computer human interfaces (HI) of their systems have highly progressed through graphical processing technologies. Studies on ergonomic or bio-medical aspects of HI to reduce psychophysiological workload of users are sufficient compared with design, utility and/or efficacy of software (see Figure 1). Especially, there have been very few studies addressing emotional effects of HI use and their relation to users' mental and physical health condition.

The authors seek to investigate psychophysiological aspects of users who operate irritating HI through use of physiological measurements and analysis. As an irritating HI, we used interactive software without a progress bar (PB) that has the function of indicating processing status. Our hypothesis is that the PB has the effect of reducing mental stress by informing users of prospects for task termination. Irritating reactions may cause physiological changes in users through the autonomic nervous system and its changes, which can be detected by respiration, plethysmogram (PTG), heart rate (HR) and electroencephalogram (EEG) measurements.

** Bio-Medical Engineering, Graduate School of High-Technology for Human Welfare, Tokai University, 317, Nishino, Numazu, Shizuoka, Japan.

*** Faculty of Software and Information Science, Iwate Prefectural University, 152-52, Sugo, Takizawa, Takizawa, Iwate, Japan.

**** Bio-Medical Engineering, School of High-Technology for Human Welfare, Tokai University, 317, Nishino, Numazu, Shizuoka, Japan

Email: [2]kimuta@pop02.odn.ne.jp, [3]2afhm003@wing.ncc.u-tokai.ac.jp, [4]sega@acm.org
[5]ymzk@wing.ncc.u-tokai.ac.jp, [6]miyazaki@iwate-pu.ac.jp

A benchmark for development of low-workload HI with objective assessment of users' health condition will be required. This paper describes experimental results and the necessity of hygienic assessment of HI.

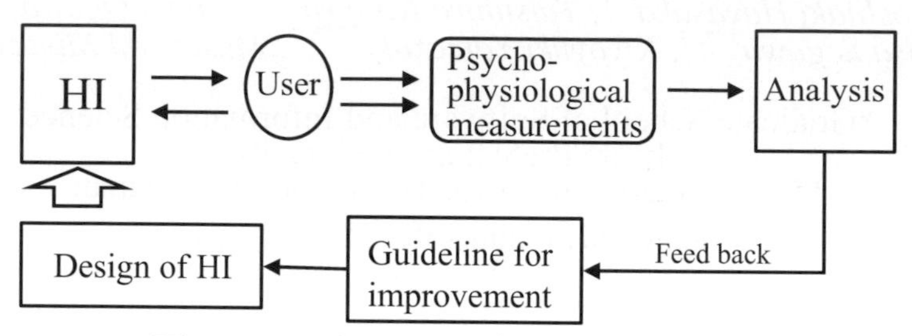

Figure 1: The concept of present study

2 Experimental method

Five university students (males-4, females-1; average age 23 years) participated in this experiment as volunteers. The experimental task consists of ten brief questions regarding geography, history, literature, science, and current topics. Questions were displayed successively on a personal computer CRT. Participants were instructed to press a key corresponding to the correct answer from among three alternatives. Interval times from the response to the start of the next question were set at three different times: 10 s, 20 s, or 30 s. The three intervals were arranged at random in a series of questions. In the interval time, three kinds of conditions were prepared for display: (1) a PB was shown, (2) no progress status indication was shown (irritating condition), or (3) a blinking symbol was shown. The PB is a quantitative indicator of processing, while the third condition is to provide qualitative information feedback of processing for the participants. As training for the experiment, all participants performed the task with easy questions and with a 10 s interval setting. Task software was programmed utilizing Microsoft Visual Basic ver.6 (see Figure 2).

During performance of the task above, physiological indices were measured in all participants. EEGs were recorded from three channels (Fz, Cz, and Pz) in a monopolar manner. Electrocardiogram (ECG, II derivation), finger PTG, and respiration curves were also recorded simultaneously. Measured data were recorded on a magneto-optical disk using a digital data recorder (TEAC, DR-M3) with sampling frequency of 200 Hz.

3 Analysis of physiological recordings

3.1 Heart Rate

Fluctuations of the HR were analysed during task performance to assess function of participants' autonomic nervous systems. The R-waves of the ECG were detected; then, averaged R-R intervals were calculated in each condition. The instantaneous R-R profile of 32 s segments were linearly interpolated and sampled in 1 Hz to obtain the time series R-R interval curves. Spectral analyses were made for these segments using FFT. Percentages of low frequency (0.04 to 0.15 Hz) and high frequency (0.15 to 0.4 Hz) spectral components were calculated (Akselrod et al., 1981).

3.2 RPA

The R-waves of the ECG to the onset of finger PTG asynchrony (RPA) were measured to estimate status of autonomic control to the vessels (see Figure 3). It is well known that the RPA indicates sympathetic nervous system activity through changes in vessel rigidity (Lane, Greenstadt, Shapiro & Rubinstein, 1983; Yamazaki et al., 1999).

3.3 EEG

The EEG recordings were analysed as follows to investigate attention and arousal of participants: EEGs recorded during task performance were divided into 2.56-s segments. Power spectra were calculated in each segment using FFT. Power spectra were averaged in each experimental condition. Percentages of alpha (8 to 14 Hz) and beta (14 to 20 Hz) components were obtained from the averaged power spectral density function.

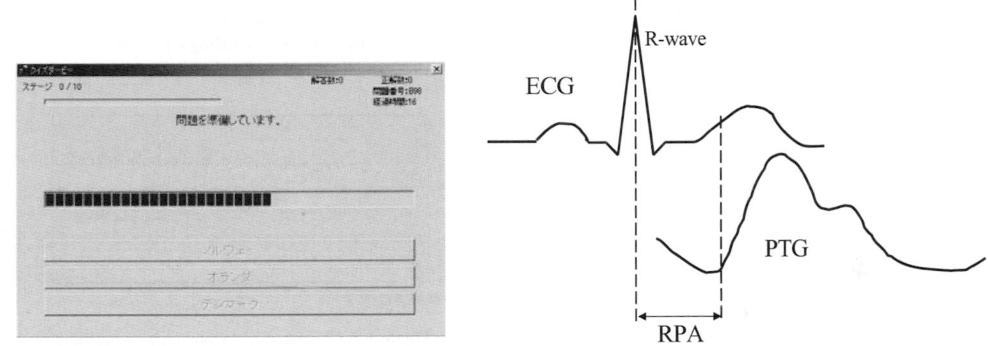

Figure 2: An example of experimental task

Figure 3: Schematic illustration of the RPA estimation

4 Results

HR recorded from a PB was shown, no progress status indication was shown (irritating condition) and a blinking symbol was shown no large change in any interval. On the other hand, low frequency components (LF) of heart rate fluctuation, as the sympathetic index obtained from power spectrum of R-R interval time series, increased in longer interval trials (see Figures 4-1 to 4-3). High frequency components (HF), as the parasympathetic index obtained from the same analysis, decreased in longer interval trials (see Figures 5-1 to 5-3). The RPA showed no large changes in any condition (see Figures 6-1 to 6-3). Relative alpha- and beta-wave amplitudes showed no large changes in any condition. One case showed decreased beta-wave amplitude in the condition with PB; that subject reported that the PB was effective to reduce irritation during a long wait for the next task trial.

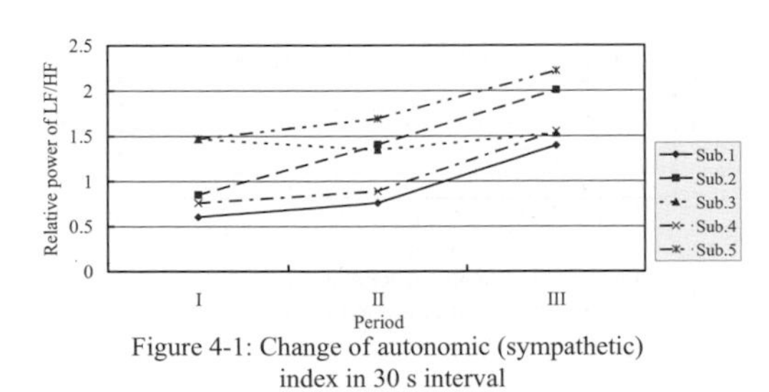

Figure 4-1: Change of autonomic (sympathetic) index in 30 s interval Conditon with PB

I : 1 to 10 s
II : 11 to 20 s in 30 s inter-trial interval
III : 21 to 30 s

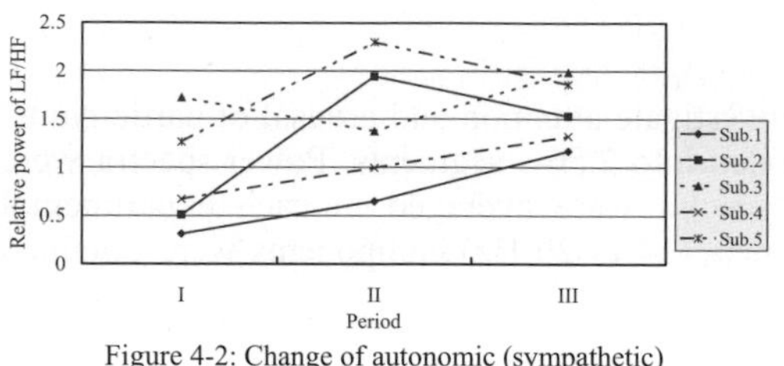

Figure 4-2: Change of autonomic (sympathetic) index in 30 s interval
Experimental condition without PB

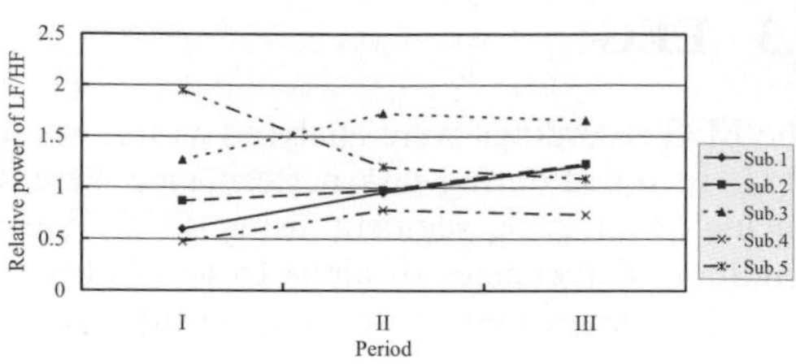

Figure 4-3: Change of autonomic (sympathetic) index in 30 s interval
Condtion with blinking indication

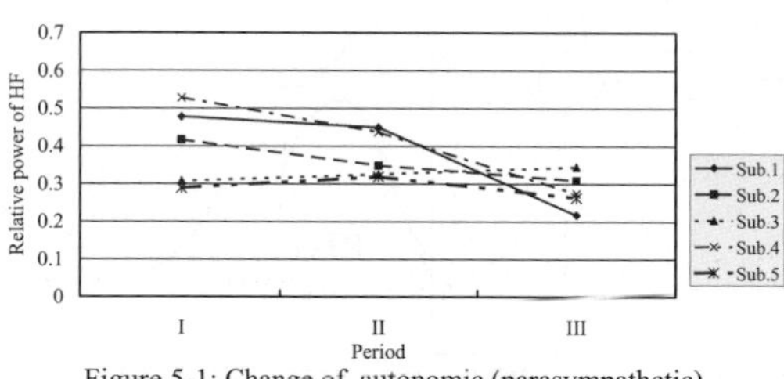

Figure 5-1: Change of autonomic (parasympathetic) index in 30 s interval
Condition with PB

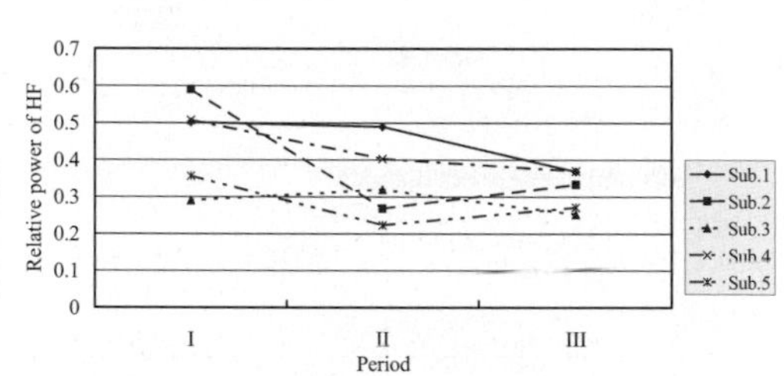

Figure 5-2: Change of autonomic (parasympathetic) index in 30 s interval
Experimental condition without PB

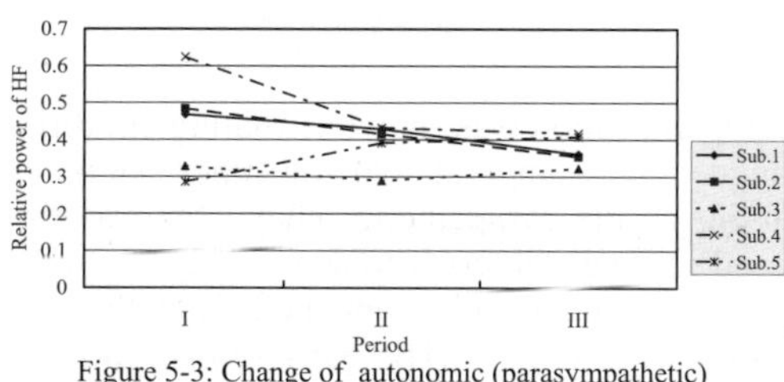

Figure 5-3: Change of autonomic (parasympathetic) index in 30 s interval
Condition with blinking indication

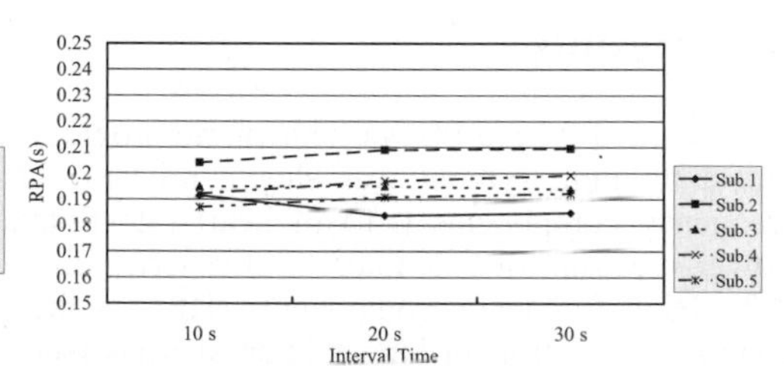

Figure 6-1: Change of RPA in each interval
Conditon with PB

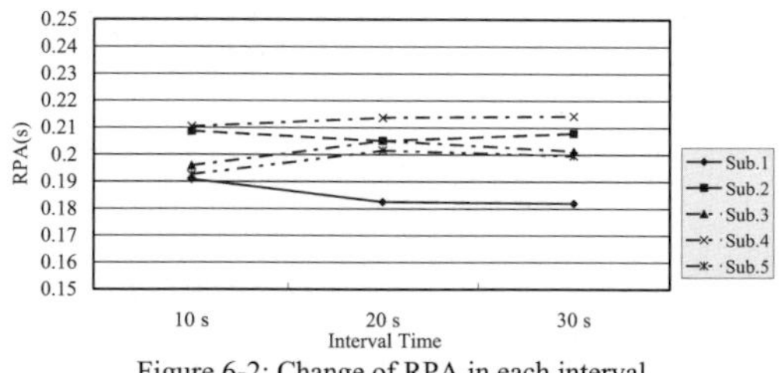

Figure 6-2: Change of RPA in each interval
Experimental condition without PB

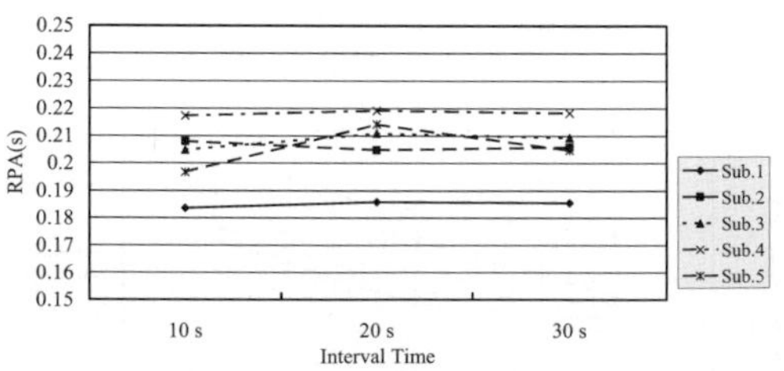

Figure 6-3: Change of RPA in each interval
Condtion with blinking indication

5 Discussion

The role of the PB is to provide users with information concerning processing progress. When the system's estimated time to complete a job is unknown, a user will have anxiety. In general, users prefer to have information whether the job is successfully processed or not. So, the PB is widely used in various software packages to reduce users' emotional stress. The authors used three experimental conditions in this study: a PB was shown, (2) no progress status indication was shown (irritating condition) and (3) a blinking symbol was shown. The third condition provides a comparative case with the PB to facilitate qualitative information feedback.

After the experiment, all participants reported that the PB was effective in waiting situations. There were some different subjective evaluations of the PB. A 'PB effective' participant showed an apparent decrease of EEG beta-wave amplitude. Other participants showed no EEG amplitude tendencies. This indicates that there are considerable idiosyncratic differences of subjective states in task performance using the PB.

On the other hand, autonomic nervous system function is more sensitive to irritating situations elicited by long waiting periods of the task interval. The LF / HF, which calculated from power spectrum of HR fluctuation, is well known as an index of the sympathetic state of the autonomic nervous system. The HF is calculated in the same way and is also well known as an index of the parasympathetic state. Results of the present experiment showed changes in the above indices, even though EEG, as the indicator of the central nervous system, showed no changes. This suggests that the irritating situation during task performance with a long waiting period generates relatively hidden and latent physiological changes.

The RPA is also an index of sympathetic nervous system based upon changes in small blood vessel rigidity caused by emotional stimulation. Experimental conditions of present study impose no emotional impacts to stimulate RPA change. The relation of latent physiological change to users' workloads must be investigated for low workload human interface design and development.

References

Akselrod S., Gordon D., Ubel F., Shannon D., Berger A., & Cohen R. (1981). Power spectrum analysis of heart rate fluctuation: a quantitative probe of beat-to-beat cardiovascular control. *Science*, 213:220-222.

Lane, D. J., Greenstadt, L., Shapiro, D., & Rubinstein, E. (1983). Pulse transit time and blood pressure: An intensive analysis: *Psychophysiology*, 20(1), 45-49.

Yamazaki, K., Suzuki, T., Hayasaka, Y., Nagashima, K., Okamoto, K., & Ikeda, K. (1999). Physiological and psychological assessment of hyper arousal utilizing a card sorting task: *Proceedings of the 1999 IEEE region 10 Conference*, vol. II, 1143-1145.

Interactive Maps on Mobile, Location-Based Systems: Design Solutions and Usability Testing

Fabian Hermann *Frank Heidmann*

Fraunhofer Institute for Industrial Engineering IAO,
Business Unit Usability Engineering
Nobelstr. 12, D-70569 Stuttgart, Germany
{Fabian.Hermann, Frank.Heidmann}@iao.fhg.de

Abstract

Map visualizations on mobile devices can display few details because of the small screens. This paper suggests design guidelines for interactive maps that were used for prototypes for a mobile fair guide. Abstract and simplified visualizations were combined with interactive linking to textual information, especially hidden labels for map objects that can be displayed in tooltips. Usability test results show this to be a promising strategy, if the hidden labels are combined with support of use cases that include search for known objects.

1 Introduction

Mobile systems help the user to use information in new contexts and location-aware systems can be used to adapt the required information to these contexts. In the project SAiMotion[1] a mobile, location-aware fair guide is being developed. A major requirement for such a system is the integration of spatial, context-related information and catalogue data like ontologies of exhibitors, or events. These information can be used e.g. when organizing visits or short parts during a trip, i.e. planning a tour. However, the display of detailed information on mobile devices is limited because of the small screens. Only few map objects can be visualized on a particular view, the usage of labels to describe the semantics of map objects is very restricted, and structural information like routes can only be displayed partly or in rough simplifications. The focus of this paper are design guidelines that were applied for the SAiMotion fair guide and its empirical testing. The process of interface conception, prototyping and usability engineering is described briefly as well as the used design principles and related results of user testing.

2 Interface Conception and Usability Engineering

To set up a user-centred design process, interface conception was strongly guided by usability engineering activities. The project started with a requirement analysis including empirical investigations with potential users. During the interface conception, GUI-prototypes were developed and tested in the usability lab.

---

[1] SAiMotion is funded by the "Bundesministerium für Bildung und Forschung" (BMBF, Funding-Key 01AK900A). The authors are responsible for the contents of this publication. The following Fraunhofer Institutes cooperate for SAiMotion: FIT (coordination), IAO, IGD, IIS, IPSI, IZM.

2.1 User Requirement Analysis

We followed a user-centred design approach, starting with the specification of usage scenarios (Carroll, 2000) and the formalization of these scenarios in a use case model. In order to get a basis for the task-oriented design of the user interface, *essential use cases* (Constantine & Lockwood, 1999) were specified which not only describe the set of system functionality required to perform tasks but also the sequence of interactions between user and system. This turned out to be an important and systematic input for the user interface conception guiding the use of visual variables for symbolization and derived principles for the interaction design. To participate users in the design process as early as possible, several empirical assessments of user requirements were performed in this phase (Hermann & Heidmann, 2002). An important outcome of this requirement analysis was the role of maps: planning and performing activities in mobile contexts make heavy demands on map visualizations. Besides the need of wayfinding maps for pedestrian navigation, interactive and adaptive overview maps play an crucial role. The strengths of a mobile, location-aware system on a business fair are usage of spatial information *combined* with content typically offered in a fair catalogue like lists of exhibitors, events, products etc.

2.2 Iterative Prototyping and Usability Testing

During the conception phase, two prototypes were developed and tested. With prototype 1, a laboratory test with 15 users was conducted. The participants were introduced to the general usage of the HTML-prototype running on an web-pad with stylus-interaction. The screen size of this prototype was assimilated to the Compaq ipaq. The participants had to solve several tasks focussing on navigating with different map views, changing displayed information, manipulating map objects, especially elements of a tour, reading and integrating information from list- and map-views. Specific tasks were given in order to test the usability of hidden labels (see below). To check the self-expressiveness of wording and iconic symbols, subjective and qualitative measures were recorded using the thinking-aloud method and semi-standardized interviews.

The results of the first test were used to change the interface specification and to develop a second HTML-prototype. This prototype was also tested with 11 users. Again, they were introduced to the basic interaction with the web-pad, and had to solve tasks.

3 Design Principles and Results from User Testing

3.1 Abstract and Simplified Visualization

To get a simple map visualization that can easily be displayed on a PDA, abstract objects must be used. An example is the abstract visualization of a tour path in a small scale that shows only the sequence of halls but not the detailed route within halls (see figure 1 b). Instead, additional information of parts of the tour within the halls could be retrieved by clicking on a symbol representing parts of the tour in each hall. This type of abstraction was easily understood by the participants of the user testing. However, the users wished to have the direction of the tour visualized especially in the big scales where the start and end points were not visible. In general, the visualizations of the second prototype was rated as very obvious.

An important requirement in the fair context was, to use the colour coding of the exhibition map. This restricted the colour design heavily, especially the one of the fair overview. A problem of the map of the chosen german fair was that some colours were to similar to each other for PDA colour displays, so that the users could not distinguish between these colours.

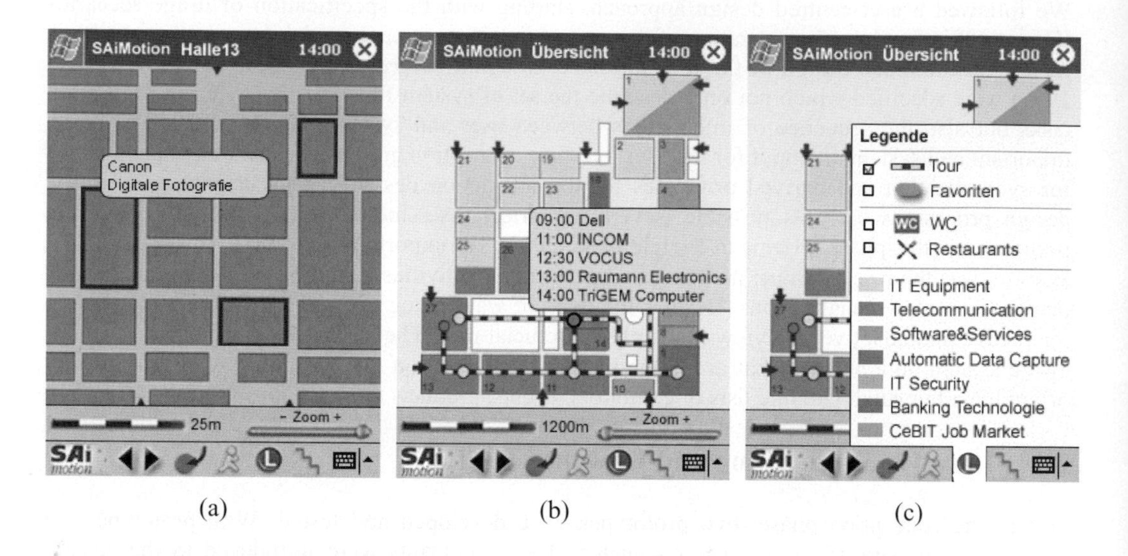

Figure 1: Prototype views of SAiMotion interactive maps, showing hidden labels in tooltips (a), tour elements and route together with additional information in a hidden label (b) and an interactive legend (c).

3.2 Interactive Manipulation of Map Objects

The user should be able not only to display spatial data in order to get an overview but also to use map visualizations to browse and manipulate data. E.g., a user should be able to select an object like an exhibitor on the fair from a list, to get a map on which the location of this object is highlighted, and to assign new attributes to this objects like being element of the user's tour or a point of interest. This requires the generation of the map view from a geographical database and the possibility to access this database by actions on the map view. The interaction design must provide intuitive means to perform the relevant actions on objects and to switch between list and map views. In our prototypes, this was realized by a context menu which offered actions on particular objects, like taking an exhibitor to a tour or displaying more detailed information, e.g. product lists or events on exhibition stands. Especially the linking between map objects, texts, and listings was highly appreciated by the users in both tests. Also the general navigation by back- and forward-buttons to switch between map-, text- and list-views was seen as very valuable.

3.3 Interactive and Dynamic Legend

The first prototype distinguished between two different types of legends. The first one explained the coding of information, e.g. colours or symbols used for particular categories of map objects. This legend was adapted to the information currently displayed on the map view. The second type of legend was interactive, i.e. it allowed the user to add particular layers, e.g. to display a planned tour on the map. This interactive legend always showed the same set of items which were considered as important attributes that should be directly accessible on any map view. However, in the first usability test the difference between those two types of legends was not clear to the users; they searched in both legends in order to read out codes as well as for changing displayed objects. Therefore, in the second prototype these two types of legend items were integrated in one legend, including stable items that can be ticked interactively as well as temporary items, that explain what currently can be seen on the map (see figure 1 b).

3.4 Hidden Labels for Map Objects on Small Screens

The standard solution to display objects and labels on cluttered maps is to give only an abstract overview in small scales and to force the user to zoom into big scales to retrieve more detailed map objects and labels. However, the user has to leave the overall small-scaled overview in order to retrieve labels in the bigger scales.

In contrast, labels can be hidden in tooltips that pop up when selecting a map object by clicking on it (see Figure 1 a). In the case of a tour visualization, not only names of tour elements, but also temporal information, like starting times of fixed appointments, can be presented in tooltips. This leads to much simpler map views that are more appropriate for small screens. The results of the two usability tests show, that the tooltips itself are a very easy-to-use interaction style that was immediately adopted by the users. They could retrieve object labels without leaving the small-scaled overview.

The results also indicate that hidden labels are sufficient for *some* tasks. In the usability tests, some tasks required to get a general overview of a spatial configuration, like identifying "problematic, unfavourable" parts of a route, or learning the route in order to sketch it afterwards. Most subjects used tooltips to complete these tasks, even when other possibilities were available. It seems that users preferred not to leave the general view and to read additional information in the tooltips to build up a mental model of the spatial and temporal structure of the tour.

For other tasks, the majority of subjects did not use tooltips, like e.g. to identify the distance between two stands in the same hall. I.e., when the users had to search for a known label, the tooltips where quite inconvenient to use. Therefore, hidden labels can be used to avoid cluttered screens. However, they must be augmented by interactive elements that prevent expensive search for known objects. One possibility is to provide a list of map objects displayed on a particular view that allows the user to select objects that should be highlighted and labelled. This solution was mentioned by several users in the second test. A text-based search may be used to find map objects. As search result, objects that match the search request should be highlighted and their labels should directly be displayed. In general, a design strategy could be to use the interaction history to decide which particular objects should directly be labelled on the map, and which labels could be hidden in tooltips. When it is likely that the user requested a map view to find the location of a set of objects the labels of these objects should directly be displayed. Examples are maps for exhibitors of a particular topic category, beginning and ending of routes, points of interests etc. Furthermore, important landmarks should be labelled directly.

3.5 Zooming by Clicking and Slider Control

Different solutions can be found for zooming: modes that are switched on by clicking on an icon, menu-commands, slider controls etc. We combined a horizontal slider control (see figure 1 a, b) with the possibility to click on halls or other map objects in the two small scales (fair overview and fair details). However, this produced an inconsistency in the effect of the user action "short click": while it evoked a zoom-into on the first two levels, in level four and three tooltips were displayed when clicking on map objects. So, in the third zoom scale (the hall overview) the users could not use direct clicks but must take the slider control to get into the biggest zoom level. Nevertheless, this inconsistency was accepted because it represents the most likely use cases in the different views: while no tooltip-information is necessary for recognizing the hall objects in the first two scales, the hidden labels are very important in the smaller scales. This design decision was tested in the second user test, where no disadvantage could be observed because of the inconsistency. Most of the users immediately understood the zoom slider as well as zooming by clicking, and used both commands. Only two of the eleven users did not use the biggest zoom level because they neglected the zoom slider after using zooming by clicking.

4 Conclusion and Further Steps

The experience with SAiMotion maps shows that abstraction and simplification are important design guidelines for interactive maps. The interactive linking of map objects with textual information, especially the usage of hidden labels, is a promising design solution for interactive maps on small displays in order to avoid cluttered screens. However, use cases like searching for known objects have to be supported. Our suggestion is to use the interaction history to select labels that probably are needed by the user. Furthermore, the user can search map objects by directly accessible lists of map objects for each single map view and with a text search for the whole set of map objects.

In the next step, an implementation of the map conception in the framework of the SAiMotion fair guide is realized. Important evaluation issues then are to test the proposed design guidelines in real usage settings including mobile situations on a business fair.

References

Carroll, J. M. (2000). Making Use: Scenario-Based Design of Human-Computer Interactions. MIT Press, Cambridge, MA

Constantine, L. L. & Lockwood, L. A. D. (1999). Software for use: a practical guide to the models and methods for usage centered design. Addison Wesley, Reading, MA

Hermann, F., & Heidmann, F. (2002). User Requirement Analysis and Interface Conception for a Mobile, Location-Based Fair Guide. In: Fabio Paternò (Ed.) *Human Computer Interaction with Mobile Devices. Proceedings of the 4th International Symposium Mobile HCI* (pp. 388-392). Berlin: Springer. Available at: http://link.springer.de/link/service/series/0558/tocs/t2411.htm

Searching for Patient Educational Information Using Electronic Resources: An Exploration of Nurses' Search Behavior

Josette F. Jones,

Indiana University
Indianapolis, IN – USA
jofjones@iupui.edu

Michael J. Smith,

University of Wisconsin
Madison, WI - USA
mjsmith@engr.wisc.edu

Abstract

Browser-based electronic resources are now a major source of information. Even though these computer-stored and displayed information resources seem to offer nurses what they like about information resources for patient care, their integration into nursing practice is rather inconsistent.
In order to deploy effective browser-based information systems for nursing practice, and to develop strategies and tools for its efficient use, the capture and analyses of the user's experiences while executing the tasks has been suggested as a method for system development.
The paper reports on an in-depth analysis of a small number of instances in which nurses in their work setting searched for information tailored to a patient's needs and preferences using browser-based electronic resources. Data collection occurred in vivo and consisted of multiple observational methods. Data analysis was based on content analysis using multiple reviewers. The theoretical framework for this research is based on the model of the information searching process developed by Kuhlthau (1996).

1 Introduction

Electronic resources available via browser-based interfaces on the World Wide Web (WWW) and Intranets are now a major source of information for most people. The interface is attractive, information can be presented in different forms and in different languages, it provides different organizational structures for information storage and access accommodating the user's preference and need. Literature review and experience has shown that creating technically excellent and advanced products does not matter as long as it does not meet the needs and experiences of the user.
Capturing and analyzing f user tasks and requirements are central within the development of tools and strategies for effective browser-based information resources. Therefore, we must begin its development by asking: Why do nurses use electronic resources? What stages do they follow, and within these stages what thoughts, feelings and actions are evident? And the last question to be answered is what are the perceived facilitators and barriers to the use of electronic resources in clinical nursing practice?

2 Background and Literature Review

Research on information searching in the context of work found that identification of information needs and the subsequent searching are influenced by perceived task complexity (Byström & Järvelin, 1995). Furthermore, the selection of the information source also depend on its perceived availability (Leckie, Pettigrew, & Sylvain, 1996), and also on the user's mental model of the information resource.

Add to this the fact that searching electronic resources is quite different from traditional information resources in which information is stored sequentially. Electronic resources such as the WWW and Intranet, consist of hyperlinked information (Allen, 1991, 1994). Although a hyperlinked information environment was intended to be ideal for browsing, it has been found to be potentially disorienting and to provide a heavier cognitive load for the user than traditional linear information resources (Marchionini, 1995). In addition, a user's performance in browser-based information systems is greatly affected by the user's previous experience and knowledge (Shneiderman, 1998).

Experience and knowledge generally can be defined as: the knowledge of the search topic and the domain, or the knowledge of the system used. Although it can be expected that the domain expert outperforms the novice with little domain knowledge, studies have found that domain knowledge begins to predict performance only after users have acquired some experience with the system (Egan, 1988).

Nurses have domain knowledge and most nurses have no difficulties operating the computer system where information is stored (Blythe & Royle, 1993; Royle, Blythe, DiCenso, Baumann, & Fitzgerald, 1997). In addition, electronic information resources provide nurses what they like about information resources for patient education: quantity and variety of real-time and up-to-date information about prevention and treatment of illness, information on the availability of health services and support groups. However, the integration of browser-based information resources in clinical practice is rather inconsistent. Access to the resources or its use is not the primary problem for nurses. Rather formulating search questions and choosing the most precise terms to locate relevant information arc (Royle, Blythe, Potvin, Oolup, & Chan, 1995). Also nurses' rigid work schedule and the limited extent to which information seeking in nursing is valued or encouraged create major obstacles to pursuing information seeking (Royle et al., 1995).

3 Theoretical Framework

Evaluating information searching effectiveness is not limited to assessing the precision and recall of the search results but also appraising the searcher's experiences during the search process. These ideas of user-oriented approach to information searching have been explored by Kuhlthau (Kuhlthau, 1996).

During a series of five research studies (Kuhlthau, 1996, p 77) developed and proposed a model of information searching based on the uncertainty principle; 2 further longitudinal studies over a period of 5 years confirmcd thc model. The uncertainty principle assumed information searching as an active, engaging process that begins with anxiety and uncertainty and in which all aspects of an experience are called into play to solve the uncertainty. The information searching process is described as a process of constructing meaning from the information encountered.

According to the Information Searching Model, the information searching process consists of six stages: initiation, selection, exploration, formulation, collection, and presentation. The theory incorporates aspects of the user (cognitive, affective, or physical) as well as external variables (social or cultural). Kuhlthau postulates and confirms by empirical research that information searching is a process of gradual refinement of the problem area, with information searching of one kind or another going on while that refinement takes place (Wilson, 1999). Thus, a successive search process is implicit in Kuhlthau's analysis of the search activity.

This research in part uses this theory of the information searching process as a guiding framework for data collection and data analysis.

4 Methods Summary

4.1 Design

The lack of current knowledge about how people, in this case nurses, search for information on browser-based electronic resources in the context of their work suggest the use of an exploratory approach (Marcus & Liehr, 1998). A non-experimental, observational study, with in-depth analysis of a small number of instances of searching for patient educational material using browser-based electronic resources was done.

4.2 Setting

Two local health care institutions - a teaching hospital, and an urban community hospital - granted permission to observe nurses in their work settings as they searched for patient educational material using browser-based electronic information resources. Human subject approval was obtained for observation of nurses in a medical-pediatric unit, a surgical unit, and a day-surgery unit in the community hospital; and for observation of nurses in the learning center of the teaching hospital.

4.3 Sampling Method

A convenience sample of eight nurses was approached. The nurses consented to be observed in the routine of their practice. Each nurse performed two separate instances of searching for patient educational information using electronic resources. Each search process was part of everyday patient education tasks and related to collecting and transferring information to the patient for health promotion, or disease prevention and management.

The nurses (3) from the community hospital used an institution-approved electronic information system, while the nurses (5) from the teaching hospital and clinics routinely used the WWW and the institution's intranet for locating and retrieving information for patient-centered education. Except for one nurse, both instances of searching per nurse were scheduled consecutively; with one session lasting an average of two hours.

4.4 Data Collection

Data collection occurred in vivo and consisted of multiple observational methods. It included audiotaping the nurse's verbal reflection about her or his experiences immediately after a need for patient-centered information was encountered and while searching electronic resources, recording of the researcher's observations during the search process, the capture of a detailed search log of the nurse's interactions with the electronic information sources, and nurses' answers to open ended questions after the search was completed. The open-ended questionnaire was built upon those used in Kuhlthau's (1996) and Belkin's research (2000). Belkin's research combines methods of cognitive task analysis and information searching strategies, and calls for the design of systems that support cooperative human-computer interaction (Belkin, 2000).

5 Data Analysis

Data analysis consisted of three distinct content analyses using multiple reviewers. Content analysis is a systematic and objective quantification of the observational data and the answers to the open ended questionnaires (McLaughlin & Marascuilo, 1990). In addition to the content analyses, the search paths followed by the nurses to locate the information related to the identified topic were reconstructed using content and process indicators. The collected data from the 16 observations were assembled in sets, one set for each search process and grouped by nurse participant. Subsequently, the data were broken up into an identifiable idea or opinion. Mutually exclusive and exhaustive categories were created or categories from the theoretical framework were used to sort data. An operational definition for each category was developed based on the

theoretical framework. Data units (i.e., ideas) were sorted into categories according to their semantic match to the operational definition.
The sorting was performed independently by two research-assistants, a graduate student in industrial engineering and a graduate student in nursing. These research assistants represented the two groups of people involved in information system design: the user and the interface designer.
For each analysis performed, intercoder reliability was calculated. The following formula was used to determine the percent agreement: $P = (N_A - N_D)/\text{ total}$
where N_A = number of agreements, N_D= number of disagreement, and total = number of identified distinct items (i.e. ideas) in the analysis. The percent agreement ranged from 78.23 to 92.21, which is indicative of high reliability (Neuendorf, 2002).

6 Results

The first analysis revealed the reasons why nurses use electronic information resources in clinical practice. Four clusters of reasons emerged: professional, personal, technological and organizational. The second content analysis served to identify the stages in nurses' information searching process. The stages in nurses' information searching process were similar to those identified in the Kuhlthau Model. However, differences from Kuhlthau's were observed. Nurses demonstrated some, but not all, of the patterns of thoughts, feelings, and actions Kuhlthau identified within each of the stages. Nurses also experienced some thoughts, feelings, and actions not identified by Kuhlthau. For example in the initiation stage, Kuhlthau observed four different thoughts, two patterns of feelings, and two possible actions. In the nurses' searching process three pattern of thoughts, two patterns of feelings, and three actions were identified. The second stage demonstrated in this sample no similar thoughts; some feelings and actions were similar, others identified by Kuhlthau though, were not observed. Additionally, nurses did not progress sequentially through the stages, but rather iterated back and forth between some stages, particularly the early stages. The third content analysis was aimed at the facilitators and barriers to the use of electronic information resources. The analysis revealed that nurses dislike long search processes. They appraised their search strategies inefficient, attributing this problem to the lack of knowledge regarding information technology and browser-based information systems, mechanisms employed by the search engines and by the constant changing information content and presentation. Nonetheless, nurses attempting to meet diverse patient educational needs perceived value in the diversity and variety of information available in electronic resources. Having access to the resources on the clinical unit via computers is considered a plus. Positive past experiences with searching for information in electronic resources help the nurses nurture confidence in self and in the resource, and allowed the development of more efficient search strategies. Perceived organizational policies disapproving electronic information access also influenced the use of electronic resources in practice.

7 Discussion

This study revealed that nurses are using electronic resources in their clinical practice. Factors affecting nurses' use of electronic resources are similar to those influencing information seeking in the context of work. It was also shown that nurses liked the variety of information available in electronic information sources but that the use of these resources was hampered by the way information systems operate. In addition, the findings also confirm earlier studies describing nurses' lack of information searching skills such as ill-focused search terms and rather heuristic search tactics and strategies. Nurses' information searching process, though, differed from the Information Searching Model as described by Kuhlthau. An explanation for this difference is suggested by the difference in context of searching, as well in the spectrum of the search. Kuhlthau observed students performing library searches for a research project over a semester; this study observed nurses searching for patient educational material during a restricted time period. During their search, students were focusing the topic of their information need, while nurses' information need mostly was well defined but were unable to locate the desired information. It is also noteworthy that the observational methods, especially the Think-Aloud technique, were not very sensitive in eliciting the affective aspects of the searching experience. Further exploration is advised. This study was undertaken as a preliminary step to deploy tools and strategies for nurses to improve the effectiveness of browser-based information searching. Understanding a process and identifying factors that intervene in the process of searching guide the building of interactive search interfaces. But this is not a one-step endeavor.

References

Allen, B. L. (1991). Cognitive research in information science: Implications for design. *Annual Review of Information Science and Technology, 26*, 3-37.

Allen, B. L. (1994, 1994). Perceptual speed, learning, and information retrieval performance. Paper presented at the Proceedings of the 17th SIGIR Conference.

Belkin, N. J. (2000). Helping people find what they don't know. Communications of the ACM, 43(8), 58-61.

Blythe, J., & Royle, J. A. (1993). Assessing Nurses' Information Needs in the Work Environment. *Bulletin of the Medical Library Association, 21*(4), 433-435.

Byström, K., & Järvelin, K. (1995). Task Complexity Affects Information Seeking and Use. *Information Processing and Management, 31*(2), 191-213.

Egan, D. E. (1988). Individual differences in human-computer interaction. In M. Helander (Ed.), Handbook of human-computer interaction (pp. 543-568). Amsterdam, Holland: Elsevier.

Kuhlthau, C. C. (1996). Seeking Meaning: A Process Approach to Library and Information Services. Norwood, New Yersey: Ablex Publishing Corporation.

Leckie, G. J., Pettigrew, K. E., & Sylvain, C. (1996). Modeling the Information Seeking of Professionals: a General Model derived from Research on Engineers, Health Care Professionals, and Lawyers. *Library Quarterly, 66*(2), 161-193.

Marchionini, G. (1995). Information seeking in electronic environments. Cambridge: Cambridge University Press.

Marcus, M. T., & Liehr, P. R. (1998). Qualitative Approaches to Research. In G. LoBiondo-Wood & J. Haber (Eds.), Nursing research: methods, critical appraisal, and utilization (4th ed., pp. 215-245). St. Louis, MO: Mosby.

McLaughlin, F. e., & Marascuilo, L. A. (1990). Advanced Nursing and Health Care Research. Philadelphia, PA: W.B. Saunders Company.

Neuendorf, K. A. (2002). The Content Analysis Guidebook. Thousand Oaks, CA: Sage Publications Inc.

Royle, J. A., Blythe, J., DiCenso, A., Baumann, A., & Fitzgerald, D. (1997). Do Nurses Have the Information Resources and Skills for Research Utilization? *Canadian Journal of Nursing Administration*, 9-30.

Royle, J. A., Blythe, J., Potvin, C., Oolup, P., & Chan, I. M. (1995). Literature Search and Retrieval in the Workplace. *Computers in Nursing, 13*(1), 25-31.

Shneiderman, B. (1998). Designing the User Interface: Strategies for Effective Human-Computer Interaction (3 ed.). Reading, Ma: Addison Wesley Longman, Inc.

Wilson, T. D. (1999). Models in Information Behaviour Research. *Journal of Documentation, 55*(3), 249-270.

InfoSky: Visual Exploration of Large Hierarchical Document Repositories

Frank Kappe[1], *Georg Droschl*[1], *Wolfgang Kienreich*[1] *Vedran Sabol*[2], *Jutta Becker*[2], *Keith Andrews*[3], *Michael Granitzer*[2], *Klaus Tochtermann*[2], and *Peter Auer*[3]

Abstract

InfoSky is a system enabling users to explore large, hierarchically structured document collections. Similar to a real-world telescope, InfoSky employs a planar graphical representation with variable magnification. Documents are assumed to have significant textual content, which can be extracted if necessary with specialised tools. Documents of similar content are placed close to each other and are visualised as stars, forming clusters with distinct shapes. For greater performance, the hierarchical structure is exploited and force-directed placement is applied recursively at each level on much fewer objects, rather than on the whole corpus.

1 InfoSky

InfoSky, enables users to explore large, hierarchically structured document collections. Similar to a real-world telescope, InfoSky employs a planar graphical representation with variable magnification and the metaphor of a zooming galaxy of stars, organised hierarchically into clusters. Documents of similar content are placed close to each other and are visualised as stars, forming clusters featuring distinct shapes, which are easy to recall.

InfoSky assumes that documents are already organised in a hierarchy of collections and sub-collections, called the *collection hierarchy*. Both documents and collections can be members of more than one parent collection, but cycles are explicitly disallowed. This structure is otherwise known as a directed acyclic graph. The collection hierarchy might, for example, be a classification scheme or taxonomy, manually maintained by editorial staff. The collection hierarchy could also be created or generated (semi-)automatically. Documents are assumed to have significant textual content, which can be extracted if necessary with specialised tools. Documents are typically plain text, PDF, HTML, or Word documents, but may also include spreadsheets and many other formats.

InfoSky combines both a traditional tree browser and a new telescope view of a galaxy, as shown in Figure 1. In the galaxy, documents are visualised as stars and similar documents form clusters of stars. Collections are visualised as polygons bounding clusters and stars, resembling the boundaries of constellations in the night sky. Collections featuring similar content are placed close to each other, as far as the hierarchical structure allows. Empty areas remain where documents are hidden due to access right restrictions, and resemble dark nebulae found quite frequently within real galaxies. The telescope is used as a metaphor for interaction with the visualisation. Users can pan the view point within the visualised galaxy, like an astronomer can point a telescope at any point of the sky.

[1]Hyperwave R&D, Albrechtgasse 9, A-8010 Graz, Austria, {fkappe|gdroschl|wkien}@hyperwave.com

[2]Know-Center, Inffeldgasse 16c, A-8010 Graz, Austria, {vsabol|jbecker|mgrani|ktochter}@know-center.at

[3]Graz University of Technology, Inffeldgasse 16c, A-8010 Graz, Austria, kandrews@iicm.edu and pauer@igi.tu-graz.ac.at

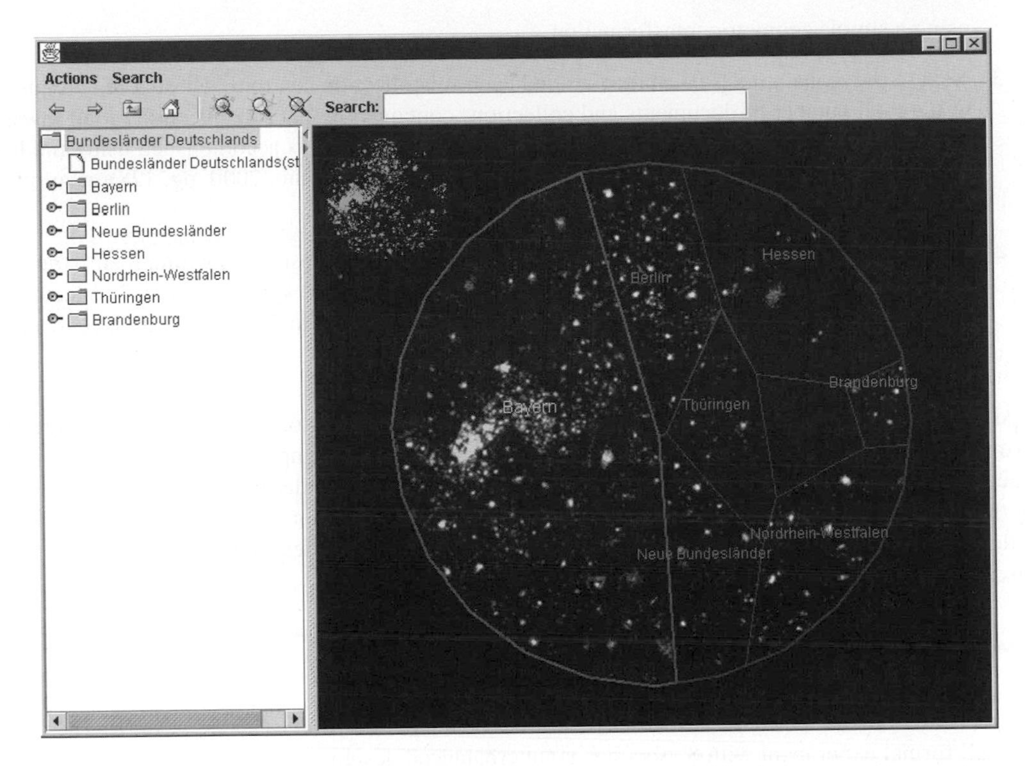

Figure 1: The original prototype of the InfoSky explorer, as used in the comparative study.

Magnification can be increased to reveal details of clusters and stars, or reduced to display the galaxy as a whole.

2 InfoSky Implementation

InfoSky is implemented as a client-server system in Java. On the server side, galaxy geometry is created and stored for a particular hierarchically structured document corpus. On the client side, the subset of the galaxy visible to a particular user is visualised and made explorable to the user.

The galactic geometry is generated from the underlying repository recursively from top to bottom in several steps:

1. First, at each level, the centroids of any subcollections are positioned in a normalised 2D plane according to their similarity with each other using a force-directed similarity placement algorithm. The similarities to their parent's sibling collection centroids are used as static influence factors to ensure that similar neighbouring subcollections across collection boundaries tend towards each other (they are not allowed to actually cross the boundary). The centroid of a synthetic subcollection called "Stars", which holds the documents at that level of the hierarchy, is also positioned.
2. The layout in normalised 2D space is transformed to the polygonal area of the parent collection

using a simple geometric transformation.

3. Next, a polygonal area is calculated around each subcollection centroid, whose size is related to the total number of documents and collections contained in that subcollection (at all lower levels). This polygonal partition of the parent collection's area is accomplished using modified, weighted Voronoi diagrams (Okabe, Boots, Sugihara, & Chiu, 2000, pg. 128), resulting in a recursive spider's web like subdivision of each area.

4. Finally, documents contained in the collection at this level are positioned within their parent's area using the similarity placement algorithm as points within the synthetic "Stars" collection, according to their inter-document similarity and their similarity to the subcollection centroids at this level, which are used as static influence factors.

Basing the layout on the underlying hierarchical structure of the repository has a major advantage in terms of performance. Similarity placement typically has a run-time complexity approaching $O(n^2)$, where n is the number of objects being positioned. However, since similarity placement is only used on one level of the hierarchy at a time, the value of n is generally quite small (the number of subcollection centroids plus the number of documents at that level). Full details of these algorithms are described in (Andrews et al., 2002).

3 User Testing

A small formal experiment with 8 users in a counterbalanced design was run to establish a baseline comparison between the InfoSky telescope browser and the InfoSky tree browser. Users were only allowed to use one or the other in isolation. The dataset used consisted of approximately 100,000 German language news articles provided by the Süddeutsche Zeitung. The articles are manually classified thematically by the newspaper's editorial staff into around 9,000 collections and subcollections upto 15 levels deep. Two sets of tasks were formulated (five pairs of equivalent tasks). The tasks were designed to be equivalent between the two sets in the sense that their solutions lay at the same level of the hierarchy and involved inspecting approximately the same number of choices at each level.

On average, the tree browser performed better than the prototype telescope browser for each of the tasks tested. The overall difference between tree browser and telescope browser was significant at $p < 0.05$ (paired samples t-test, 39 degrees of freedom, $t = 3.038$). The reasons for the slower performance of the telescope browser appear to be two-fold. Firstly, users typically have spent many, many hours using a traditional explorer-like tree browser and are very familiar with its metaphor and controls. Secondly, whereas the tree view component has already undergone many iterations of development, the telescope browser is a prototype at a fairly early stage of development.

When interviewed after the test, users indicated that they were very familiar with a tree browser and liked being able to use the mouse cursor as a visual aid when scanning lists. They liked the overview which the telescope browser provided and could imagine using it for exploring a corpus of documents. This study did not include a task asking users to find similar or related documents or subcollections, something which the telescope metaphor should support quite well. Users further indicated that a combination of both browsers and search functionality could be very powerful. This is something which InfoSky provides, but was not tested in this study.

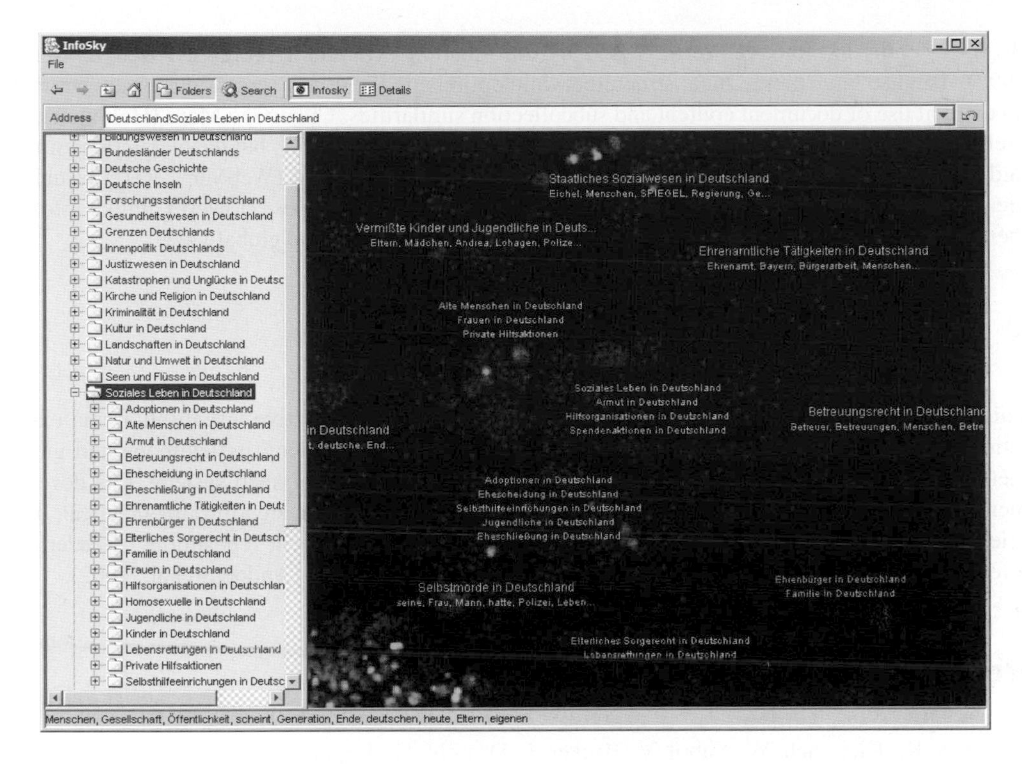

Figure 2: The revised version of InfoSky.

During the study, users complained that some of the Voronoi polygons were too small to see properly. Users were confused by the "Stars" collection used to hold documents at each level. When trying to access individual documents, users were also disturbed by the fact that the document labels appeared to jump around arbitrarily. In fact the prototype displayed only the titles of documents "near" to the cursor, but users were moving the cursor to step through what they perceived to be a list of documents. These problems were addressed in the new version of InfoSky shown in Figure 2.

We have not yet tested the complete InfoSky armoury of synchronised tree browser, telescope browser, and search in context against other methods of exploring large hierarchical document collections. Nor have we tested tasks involving finding related or similar documents or subcollections, something the telescope metaphor should be well-suited to. As development proceeds, we believe that the InfoSky prototype will constitute a step towards practical, user-oriented, visual exploration of large, hierarchically structured document repositories.

4 Related Work

Systems such as Bead (Chalmers, 1993) and SPIRE (Thomas et al., 2001) map documents from a high-dimensional term space to a lower dimensional display space, whilst preserving the high-dimensional distances as far as possible. In contrast to InfoSky, both operate on flat document repositories and do not take advantage of hierarchical structure. Systems such as the Hyperbolic

Browser (Lamping, Rao, & Pirolli, 1995) and Information Pyramids (Andrews, Wolte, & Pichler, 1997) visualise large hierarchical structures while optimising the use of screen real estate, but make no explicit use of document content and subcollection similarities. CyberGeo Maps (Holmquist, Fagrell, & Busso, 1998) use a stars and galaxy metaphor similar to InfoSky, but the hierarchy is simply laid out in concentric rings around the root. WebMap's InternetMap (WebMap, 2002) visualises hierarchically categories of web sites recursively as multi-faceted shapes. However, unlike InfoSky, there is no correspondence between the local view at each level and the global view.

5 Concluding Remarks

This paper presented InfoSky, a first prototype system for the interactive visualisation and exploration of large, hierarchically structured, document repositories. Using its telescope and galaxy metaphor, the InfoSky system addresses several key requirements for such systems. As development proceeds, we believe that the InfoSky prototype will constitute a step towards practical, user-oriented, visual exploration of large, hierarchically structured document repositories. Readers are referred to detailed descriptions of both InfoSky and the user study in (Andrews et al., 2002).

References

Andrews, K., Kienreich, W., Sabol, V., Becker, J., Droschl, G., Kappe, F., Granitzer, M., Auer, P., & Tochtermann, K. (2002). The infosky visual explorer: Exploiting hierarchical structure and document similarities. *Information Visualization*, *1*(3/4), 166–181.

Andrews, K., Wolte, J., & Pichler, M. (1997). Information pyramids: A new approach to visualising large hierarchies. In *Ieee visualization'97, late breaking hot topics proc.* (pp. 49–52). Phoenix, Arizona.

Chalmers, M. (1993). Using a landscape metaphor to represent a corpus of documents. In *Spatial information theory, proc. cosit'93* (pp. 377–390). Boston, Massachusetts.

Holmquist, L. E., Fagrell, H., & Busso, R. (1998). Navigating cyberspace with cybergeo maps. In *Proc. of information systems research seminar in scandinavia (iris 21).* Saeby, Denmark.

Lamping, J., Rao, R., & Pirolli, P. (1995). A focus+context technique based on hyperbolic geometry for visualizing large hierarchies. In *Proc. chi'95* (pp. 401–408). Denver, Colorado.

Okabe, A., Boots, B., Sugihara, K., & Chiu, S. N. (2000). *Spatial tesselations: Concepts and applications of voronoi diagrams* (Second ed.). Wiley.

Thomas, J., Cowley, P., Kuchar, O., Nowell, L., Thomson, J., & Wong, P. C. (2001). Discovering knowledge through visual analysis. *Journal of Universal Computer Science*, *7*(6), 517–529.

WebMap. (2002). *WebMap.* (http://www.webmap.com/)

VisJex – a Tool for Interactive Information Visualization

Silke Kleindienst and Christian Rathke

Hochschule der Medien
Wolframstraße 32, D-70191 Stuttgart
silke.kleindienst@gmx.de, rathke@hdm-stuttgart.de

Abstract

Making a precise decision becomes more and more complex due to the huge amount of data. We created a tool visualizing statistical data and giving the user the opportunity to discover playfully the meaning behind the data.

1 Introduction

In our times, in which information is growing exponentially, making a profound decision is becoming more and more complex. Managers are forced to come to decisions based on data, which need to be interpreted or which must be filtered and adapted to the task at hand.

Statistical data is mostly found in tables of some database. Accessing and using this data directly is difficult. Extraction tools are needed get to the data and to generate the requested diagrams i.e. tables or bar charts. Traditional software such as MS Excel offers integrated tools visualizing data extracted from tables (Figure 1).

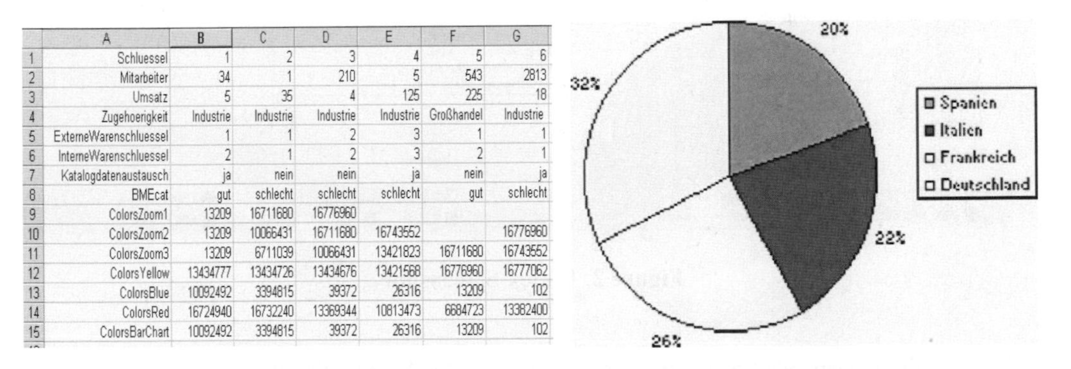

	A	B	C	D	E	F	G
1	Schluessel	1	2	3	4	5	6
2	Mitarbeiter	34	1	210	5	543	2813
3	Umsatz	5	35	4	125	225	18
4	Zugehoerigkeit	Industrie	Industrie	Industrie	Industrie	Großhandel	Industrie
5	ExterneWarenschluessel	1	1	2	3	1	1
6	InterneWarenschluessel	2	1	2	3	2	1
7	Katalogdatenaustausch	ja	nein	nein	ja	nein	ja
8	BMEcat	gut	schlecht	schlecht	schlecht	gut	schlecht
9	ColorsZoom1	13209	16711680	16776960			
10	ColorsZoom2	13209	10066431	16711680	16743552		16776960
11	ColorsZoom3	13209	6711039	10066431	13421823	16711680	16743552
12	ColorsYellow	13434777	13434726	13434676	13421568	16776960	16777062
13	ColorsBlue	10092492	3394815	39372	26316	13209	102
14	ColorsRed	16724940	16732240	13369344	10813473	6684723	13382400
15	ColorsBarChart	10092492	3394815	39372	26316	13209	102

Figure 1: Data in a table and as a pie chart

In addition to the commonly known commercial software packages there are some open source java libraries, for instance Chart2D, OpenThunderGraph, JFreeChart or JOpenChart with which various diagram types i.e. line charts, bar charts or pie charts can be generated. They offer a comfortable integration into java applications or applets.

All of these software libraries as well as the commercial software packages have one important drawback: the lack of being able to interact with them. Generating different diagrams is no problem, but it is impossible to dynamically change views by applying simple selection activities on the data being displayed. Interactively exploring what is behind the data is impossible with these mentioned tools.

2 Visualising Data with VisJex

We have developed a visualization tool called VisJex to serve these needs. As an example, VisJex is applied to data from a survey about standards in e-bussines carried out at the Fraunhofer IAO, Stuttgart. Figure 2 shows a screenshot of VisJex.

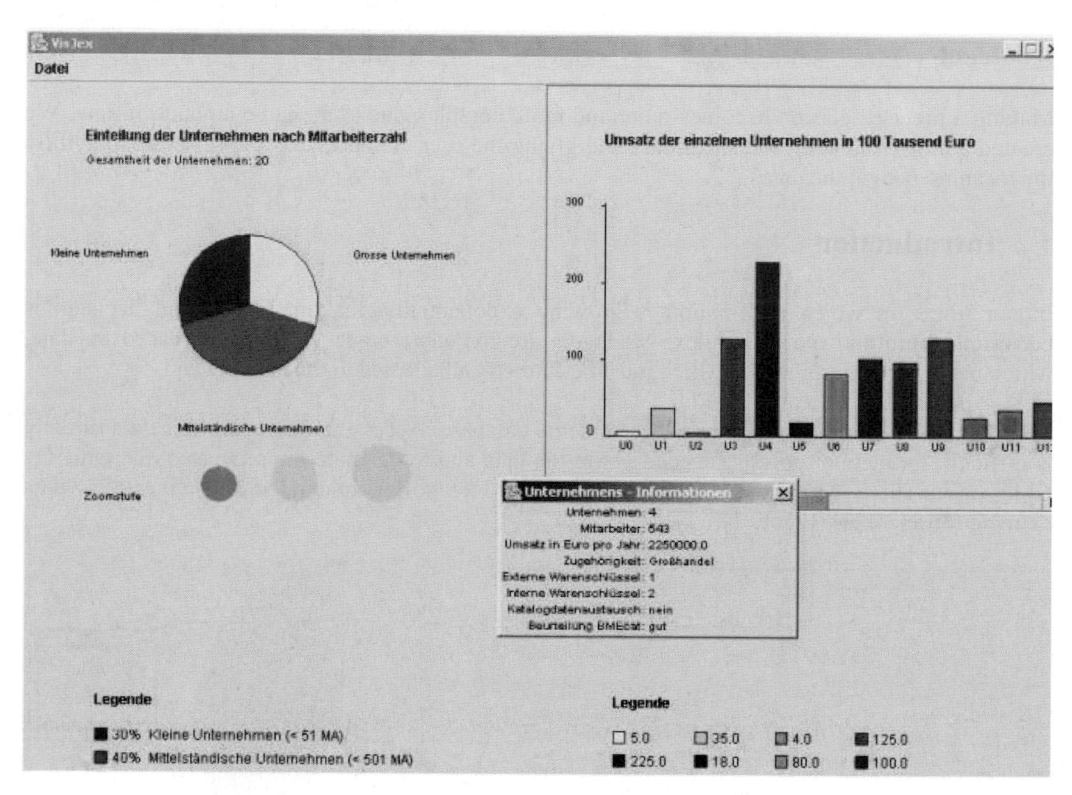

Figure 2: VisJex Screenshot

VisJex simultaneously shows various graphical formats. VisJex thereby combines traditional techniques of information presentation with an interactive approach to information visualization. Using familiar techniques helps inexperienced users to master the tool without effort. Being interactive provides the means to explore the huge amount of data, which traditional, passive techniques cannot match.

3 Augmenting traditional technologies with interaction

In the area of traditional technologies, pie charts and bar charts are the primary means to represent the different quantities of data.

- **Pie charts** are especially well suited to picture a data collection or a part of an aggregate, for instance, the number of employees of some companies. In our example application, a pie chart represents the segmentation of all surveyed companies in small, middle and large enterprises. But the diagram is not only for showing the visual representation of data. It is also a navigation tool. The segments of the pie chart are clickable. By choosing a segment the user gets the possibility to see the selected quantity of data from another perspective, in our example the volumes of sales. This information is displayed as a bar chart.

- **Bar charts** are especially well suited for representing time series and proportional numbers. Sales volumes are shown for the selected segment (small, medium, and large) as a bar chart. Using this representation it is easy to compare the numbers.

The interactive properties of the tool follow principles developed under the headings of: "Overview and Detail", "Zooming" and "Details-on-Demand".

- VisJex is primarily based on the idea of "**Overview and Detail**". To generate an overview about the entire data and to simultaneously gain insight into a particular feature of the data, both representations are shown simultaneously. The pie chart provides a summarization of the data, whereas the bar chart shows one possible detail view. See figure 3: VisJex Screenshot.

- **Zooming** is used within the pie chart itself. Zooming inside the pie chart simulates diving into the data. At the first zoom level, only three segments are shown representing the segmentation of companies into small, middle and large companies. While zooming in, the segmentation of the pie chart becomes more detailed, i.e. six respectively twelve segments are shown, representing subsets of the wider data sets. Figure 3 below show the different pie charts.

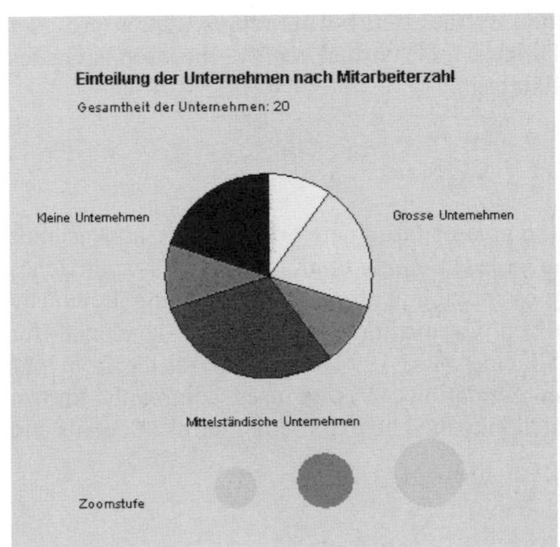

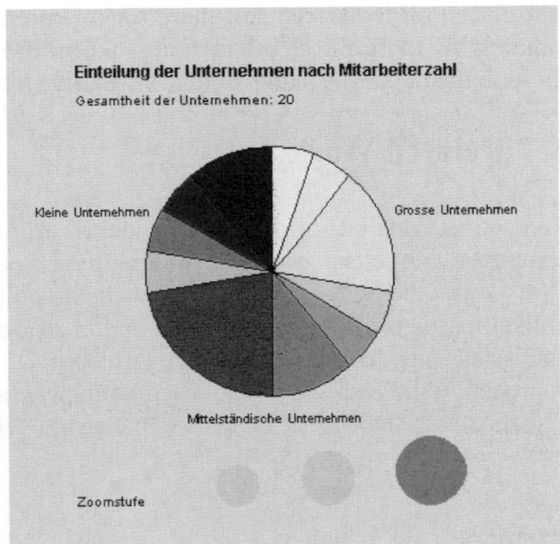

Figure 3: Zoom factors 2 and 4 on the same data

- **Details-on-Demand** is used to see very specific information about a single company without loosing the context of the overall information. Like the "clickable" and "zoomable" pie chart, the bar chart is a navigation tool as well. By clicking on one of the bar, representing the sales volume of one company, a dialog appears, showing detailed information about the chosen company (Figure 4).

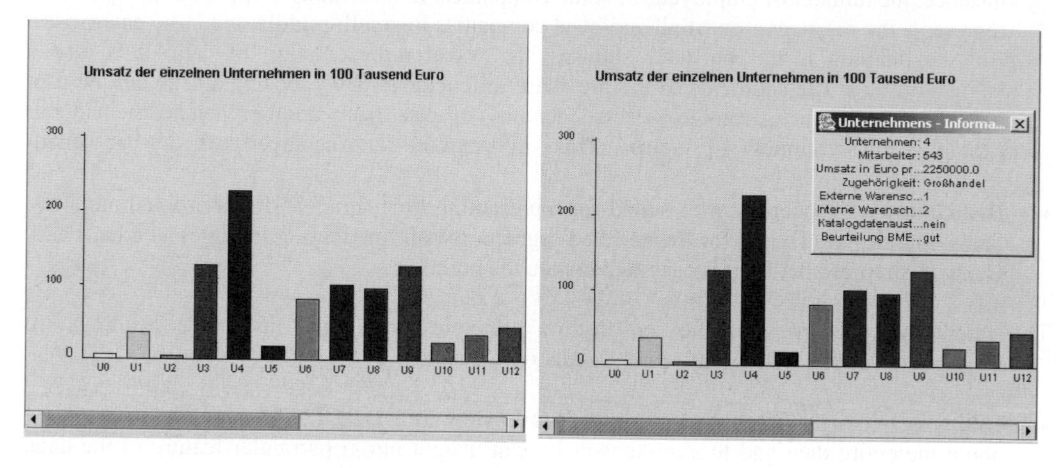

Figure 4: Bar Chart without and with details

4 Interactively looking behind the data

VisJex is based on the metaphor of "interactively looking behind the data". In VisJex, the 3D metaphor is applied to general, quantifiable data. VisJex is being developed into an interactive tool which takes graphically enhanced information as a starting pint for further exploration. Data is not just placed on the screen left there for interpretation by the user, but its representation provides handles for more direct interaction. Like in a 3D-Scene of a virtual reality, the user navigates through the data even using similar interactive techniques.

5 Related Work

Interactive visualizations have been looked at in the area of large networks or trees, for instance cone trees (Robertson et al., 1991) and hyperbolic trees (Lamping et al., 1995). Both approaches allow to interactively explore the network structure by moving the focus and expanding previously hidden parts of a large network. Also, specialized visualizations have been developed for databases, e.g., about movie data (Ahlberg & Shneiderman, 1994). They all differ from our approach with respect to the used graphical representations. VisJex uses commonly known visualizations such as bar charts and pie charts, which may then be explored in terms of details and views.

References

Robertson, G., Mackinlay, J., & Card. S. (1991). Cone Trees: Animated 3D Visualizations of Hierarchical Information. *Proceedings of the Conference on Human Factors in Computing Systems (CHI '91)*, pages 189-194.

Lamping, J., Rao, R. & Pirolli, P. (1995). A Focus + Context Technique Based on Hyperbolic Geometry for Visualizing Large Hierarchies. *ACM Conference on Human Factors in Computing Systems (CHI '95)*, Denver, Colorado.

Ahlberg, C. & Shneiderman, B. (1994) Visual information seeking: Tight coupling of dynamic query filters with starfield displays. In Proceedings of the ACM Conference on Human Factors in Computing Systems (CHI'94), 313-317.

Qualitative Evaluation of TT-Net Project

Jörn Krückeberg[1], Sigrun Goll[2], Marianne Behrends[1], Ingo Köster[1], Herbert Matthies[1]

[1]Institute of Medical Informatics, Hannover Medical School, 30623 Hannover, Germany

[2]Protestant University of Applied Sciences Hannover, 30625 Hannover, Germany

Abstract

This article describes the approach for qualitative evaluation of the TT-Net project, a multimedia learning environment for students of medicine. Based on the Kellog Foundation Framework Evaluation concept (Curnan et al., 1998), the evaluation places equal emphasis on the results of TT-Net as well as on the implementation of the project.

1 Description of the Project

The project 'Teaching and Training Network in Neurosurgery' (TT-Net), funded by the German Federal Ministry of Education and Research (BMBF-08NM150 A), under the auspices of Hannover Medical School aims at improving the education of students of medicine by composing web-based tutorial modules for 'Basic Procedures in Neurosurgery'. The multimedia learning modules are based on a collection of case studies and contain text, hyperlinks, video-clips, animation, virtual reality environments and digital radiograms (DICOM). This content is diagnosis-oriented and primarily based on the subject catalogue (GK) of the disciplines neurosurgery and neuropathology. The benefits of this projects will result from supporting 'learning by viewing', since this facilitates and improves the three-dimensional orientation of students, which is important for successful planning and procedures of operations.
An interdisciplinary editorial team (with employees of the Medicine, Media Education Science, Communication-Design and Medical Computer Science Faculties) is developing a didactic concept for TT-Net. It comprises three main aspects:

- web-based tutorial modules for self-study
- electronic communication
- conventional classes using the material of TT-Net.

The assignment of the Protestant University of Applied Sciences Hannover is to plan and realise a study for an external evaluation of the didactical concept of TT-Net.

2 Evaluation Concept

Following the methodological approach of Kellog Foundation Framework Evaluation (Curnan et al., 1998), the evaluation process is not only concerned with the web-based tutorial modules. A rather broader approach is chosen to analyse and evaluate contextual factors like motivations and perspectives of staff and stakeholders in the first instance, as well as experience and expectations of potential participants and the political circumstances in which the project is embedded. The main goal of this approach is to improve and to strengthen the process of project development. To achieve this goal, appropriate methods of research are chosen, such as group interviews, questionnaires and observation. Additionally, due to the specific product of the project, usability testing and knowledge assessment will be utilised as integrated parts of the framework with current methods. In general, this concept pursues the following aspects as evaluation guidelines:

- Consideration of contextual factors

Each phase of the evaluation process has political implications which will affect the issues focused on, the decisions made, how the outside world perceives the project, and whose interests are advanced and whose ignored. This understanding requires an ongoing dialogue with all groups involved in the project.

- Build capacity

Evaluation should be concerned not only with specific outcomes, but also with the skills, knowledge and perspectives acquired by the individuals involved in the project, in order to reach an increasingly sophisticated understanding of the project.

- Balancing the call to prove with the need to improve

Contrary to the conventional scientific research model based on hypothetical-deductive project legitimisation methodology, the chosen model for Framework Evaluation uses statistical analysis techniques, focusing on data gathering and analysis to improve the project development process (Curnan et al., 1998).

3 Implementing Evaluation of TT-Net

Within this framework, the process of evaluation focuses on project-level evaluation comprising three components. Within each component the method of questioning and the main focus concerning the content will be specified depending on the group to be asked.

3.1 Context Evaluation

The context evaluation component examines how the project functions within the economic, social and political environment and project setting.
As these contextual factors are different in each project, and since they are liable to change within a given project development phase, evaluation begins with a general question: "Which contextual factors have the greatest impact on success or obstruction in project work?"
The framework concept considers examination of the organizational structure of projects that have already progressed beyond the initial phase as an important part of context evaluation. With careful gathering and analysis, information concerning aspects such as management style, qualifications and the experience of staff or project partners can help to identify difficulties concerning project success or potential improvements. Combined with an analysis of the organizational context in which the project is embedded, this phase also involves an examination of working methods and verification of the way objectives are discussed within the project. According to the concept of Framework Evaluation, this process begins by identifying relevant subject areas for context evaluation. The following themes will be focused on:

- project goals (stated from multiple perspectives)
- success standards of project participants and of funding and higher institutions
- cooperative and administrative relations to higher institutions (MHH and BMBF), project partners and other medical multimedia projects
- characteristics of the situation of project staff and project management
- resources available to the project
- possibility for everyone to participate in evaluation process.

For this part of the concept, project management staff will be also asked about the research landscape concerning medical and general multimedia projects, future developments and research as well as trends in educational policy. As the second group, the developers (the editorial team) will be involved with statements about the tutorial modules, the planned virtual and non-virtual learning units including these modules, and the development of the project so far. This should identify their views and perspectives as well as strengths and weaknesses of the project.
Further participants include experts (e.g. physicians and teaching staff from clinics involved), users of TT-Net, students and medical specialists. All individuals in these groups will be interviewed about the tutorial modules and the learning units.
Table 1 below shows the individual groups to be surveyed during the context evaluation phase with their special topics and the surveying methods used. The first-named method in each case will be used in the context evaluation phase. The following methods will be used in the course of implementation and outcome evaluation.

Table 1: Group sampling and methods

GROUP	MAIN THEMES	METHOD
Project management (MHH) External developers (Berlin)	Objectives and core activities Tutorial modules and combined classroom courses Project organisation Communication and cooperation	Focus group, Questionnaire
Developers from: Media didactics, Communication design, Medicine, Medical informatics (MHH)	Objectives and core activities Project organisation Tutorial modules and combined classroom courses Communication and cooperation	Focus group, Questionnaire
Experts: (Project partners responsible for the contents)	Tutorial modules and combined classroom courses Project organisation Communication and cooperation	Interview, Questionnaire, Usability Tests
Participating teaching staff	Tutorial modules and combined classroom courses	Interview, Observation, Usability Tests
Users: Students, specialists	Tutorial modules and combined classroom courses	Observation Usability Tests, Questionnaire

3.1.1 *On the Choice of Focus Groups as a Survey Method*

The main aspects of context evaluation should be discussed by TT-Net project management and development staff. We consider focus group discussions to be a suitable method. Basically, the method of focus group can be described as a carefully planned discussion designed to obtain perception on a defined area of interest in a permissive, non-threatening environment. This method is also recommended in W.K. Kellog Foundation Evaluation Handbook: "One popular technique for conducting collective interviews is the focus group, where six to eight individuals meet for an hour or two to discuss a specific topic (...). Unlike a random sampling of the population, the participants in a focus group are generally selected because they share certain characteristics (...) which make their opinions particularly relevant to the study" (Curnan et al., p.77).
In this case, the group members share characteristics in being part of the TT-Net project, but they may and should also make statements from a personal point of view, for example concerning their experience with the progression of the project so far. "Role-specific experiences are stated by participants, discussed and commented on from various points of view, ensuring the maximum range of perspectives. This provides an opportunity to obtain a many-sided overall picture that should provide an optimal reflection of reality." (Lamnek, 1998, p.67).

3.2 Implementation Evaluation

Implementation evaluation examines the project in the phase following its structural set-up. This evaluation focuses on the main activities of the project in relation to its planned results. Successful project implementation often involves adapting the original project plan to local conditions, organisational dynamics and points of programmatic uncertainty.
The aim of this evaluation phase is to gather information concerning targets (or milestones) that are set and then not achieved. It also aims to increase the effectiveness of individual activities by making appropriate alterations where necessary. This should help the project to survive in the long-term. In addition, implementation evaluation should also generate documentation on the running and progress of the project. In this phase, each group will be questioned again with reference to their answers given during context evaluation.

3.3 Outcome Evaluation

Outcome evaluation will assess the short- and long-term results of the project. This phase will employ methods from the research field of usability testing as well as questionnaires and observations. The relevant groups for this phase are experts, teaching staff and users of TT-Net, such as students. The tests with these groups will be applied to the virtual learning room concept that has been realised with consideration of both systematic learning and problem-based learning. Participants are introduced to each of the various areas of the virtual learning room (illnesses, OP techniques, case studies and films) via a matrix. The matrix enables the learner to select from a range of learning paths. At the start of each session, the learner selects the materials s/he wishes to learn. The system records the materials already visited, giving each learner an overview of his/her knowledge. There will be a set of questions on each illness for final knowledge checking. The learner's answers are evaluated by the system and made visible by coloured markings in the matrix. Red stands for a poor result and green means that all questions were correctly answered (see Figure 1).
The structure of the tutorial modules is based on the various fields of everyday medicine. Besides clinical results, it also deals with aspects of both neuroradiology and neuropathology, as well as therapeutic procedures, especially operations. The most important OP techniques are not only

presented separately, but also linked to the illnesses. Various media are used to explain this complex medical material. Besides highly realistic illustrations of medical findings and video-clips, this means especially showing and elucidating OP techniques using modern 3D graphics.

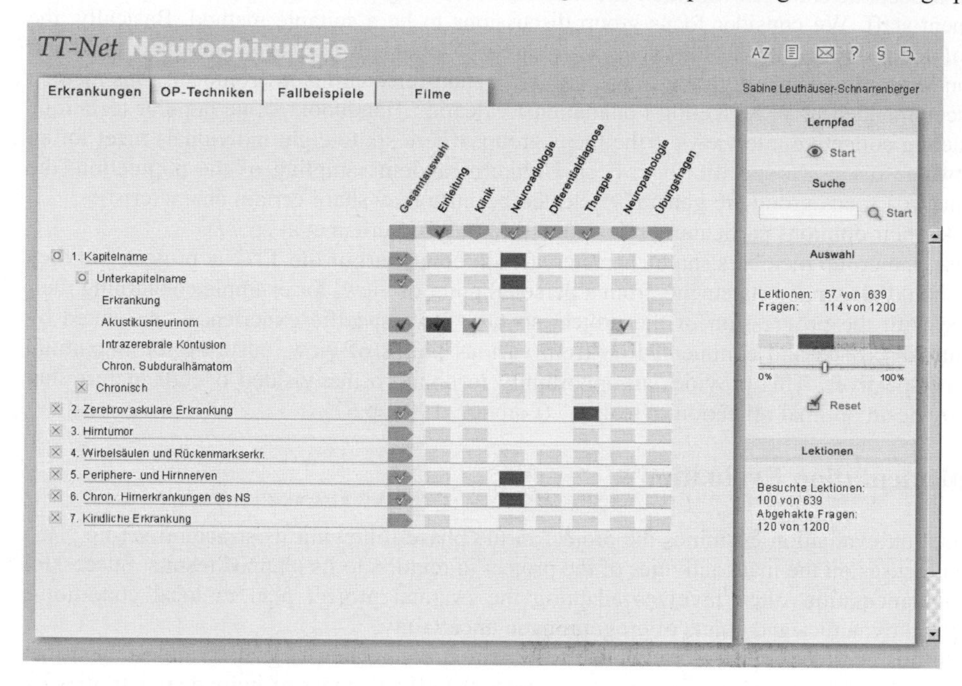

Figure 1: Matrix of TT-Net

4 Conclusions

In our opinion, the advantage of the chosen approach lies in the way evaluation is related to the project in question. This form of external evaluation running parallel to the project requires evaluators to be well informed but also to act at a certain distance. This can help to identify strengths and weaknesses in project implementation. Staff are then informed of these findings and can decide how use them for improving the projects results.

References

Curnan, S., LaCava, L., Langenburg, D., Lelle, M., Reece, M. (1998)
W.K. Kellog Foundation Evaluation Handbook. Michigan, W.K. Kellog Foundation
http://iol3.uibk.ac.at:8080/pub/evaluation/Kelloggs-Eval-Handbook.pdf

Hannover Medical School (MHH), Institute of Medical Informatics:
http://www.mh-hannover.de/institute/medinf/english_version/index.html

Lamnek, S. (1998) Gruppendiskussionen, Theorie und Praxis.
Weinheim, Psychologie Verlags Union

Computer-Supported Design Tools for Incorporating Multiple Levels of Cultural Context

Kun-Pyo Lee

Department of Industrial Design, Korea Advanced Institute of Science and Technology
373-1, Kusong-dong, Yusung-gu, Taejon, Korea
kplee@mail.kaist.ac.kr

Abstract

This paper aims to develop computer-supported tools for different levels of attributes of design problem for information appliances. Those include 'remote usability-testing through WWW' for functional attribute, 'internet-based conjoint analysis' for emotional attribute, and 'video annotating tool for video ethnography' for symbolic attribute. Prototypes of tools are introduced and the systematization of tools is explored for coping with dynamically changing design problems.

1 Changes of Design Methods in Different Levels of Cultural Context and Technological Paradigms

Design problems are so dynamically changing and unstable which makes it difficult to find appropriate design methods for them. Major forces for change include changes in technology itself and different levels of technology in its cultural context. In the view of culture, technology takes one of two domains in culture, which is overt and materialistic. The other one is more covert and cognitive part of culture. Many anthropologists and psychologists agreed on this view of dual nature of culture, some keeping the number of levels as two while some adding one more level producing three levels of model. Those include top layer of *artifact* which consists of the observable, objective and concrete such as technology, food, or housing; middle layer of *value* which includes something people know but cannot explicitly talk and elaborate; bottom layer of *basic assumptions* consisting of things in people's mind that are out of conscious awareness and taken for granted. (Hoft, 1996)

Similarly design is also known to consist of set of design attributes. Design attributes can be concrete and physical such as color, shape and texture, or they can be abstract like emotional feeling and symbolic meaning. Design attributes can be categorized into *functional attribute*, most fundamental attributes which design is supposed to fulfil as existential purpose and this can be measured objectively and easily quantified; *aesthetic attribute* is more subjective, and intangible which cannot be easily quantified; *symbolic attribute* which is most abstract and cannot be consciously evaluated. Comparing structures of culture and design reveals parallel similarities between elemental levels in terms of extent of their explicitness and locations in human mind: artefact and functional, value and aesthetic, and basic assumption and symbolic. Similarities between the structure of culture and design become even clearer when considering cyclical development of each level. At first, when new product introduced in real world it is perceived as functional artifact (e.g. car for driving), and some time later people get to form their individual value of like and dislike on it (e.g. 'sexy' car). Then, next, if the value is kept long enough and shared by society it would be gradually absorbed deep into subconscious level (e.g. car as society's icon). The reverse cycle can also happen. Out of unconscious level, some value is expressed, and if the value is salient enough, it can be manufactured as artefact.

Designers have been working on these three levels of culture and design no matter what paradigms of technology (manual tool, machine or information appliances) and no matter what levels of culture and design they designed for. Designers developed various tools and methods for them. A matrix of technological paradigm and levels of culture and design can disclose design methods applied for specific level of design problem under certain technological paradigm as shown Figure 1. (Lee 2002) Long-accumulated series of cyclical developments become evolutionary design history. As seen in diagonally expanding spiral arrow in Figure 1, important design method has been shifted from craftsmanship (how to make), to drawing (how to draw) and user-observation (how to understand users' contextual needs). We are currently living in the age of information appliances and major design methods of three levels of attributes include 'user-interface design', 'emotional engineering', and 'user-observation'.

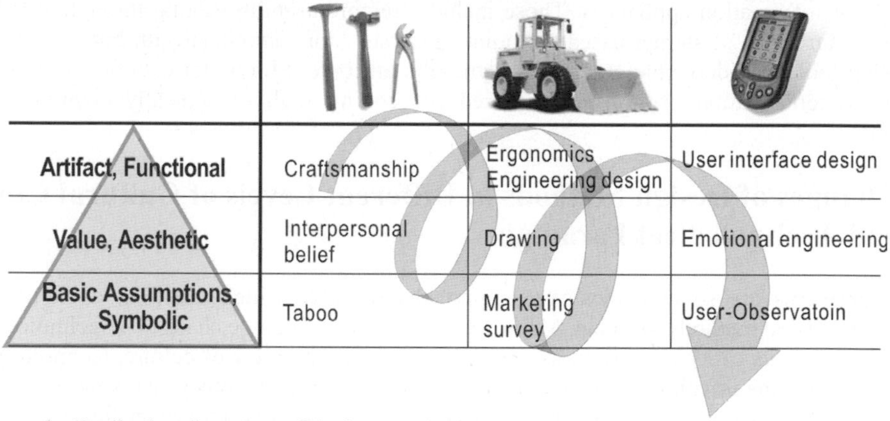

Artifact, Functional	Craftsmanship	Ergonomics Engineering design	User interface design
Value, Aesthetic	Interpersonal belief	Drawing	Emotional engineering
Basic Assumptions, Symbolic	Taboo	Marketing survey	User-Observatoin

Figure 1: Cyclic development of design methods over paradigms of technology and levels of culture and design.

2 Development of Computer-Aided Design Tools for Different Levels

Prototypes of computer-aided design tools for supporting three levels of design attributes for information appliances have been developed: namely 'remote usability testing tool using WWW for functional level', 'internet-based conjoint analysis for value level', 'video annotating tool for symbolic basic assumption level. They are introduced in more detail in the following sections.

2.1 Remote Usability Testing Tool for Functional Level

Usability testing becomes main method for user-interface design of functional level of information appliances for its advantages of direct observation of user's performance of task. However, despite of its advantages, there are some known limits of usability testing: high cost, effort and long time, and unnatural atmosphere of closed usability testing room. The prototype of tool called RUTIA (Remote Usability Testing for Information Appliances) was developed for solving major two problems mentioned above. RUTIA take advantage of WWW for allowing user to participate on usability testing in his own computer and collecting automatically usability testing data to server. (Lee 2002)

The tool, RUITA consists of three modules: module of testing, idea generation module, and analysis. In the module of testing, user is guided to a series of step-by-step usability testing process. Selected tasks are provided one by one on user's screen and he performs the tasks by operating the computer-simulated information appliances. User uses mouse to press control buttons, for which the product responses exactly same as real product. While performing tasks, user can refer user's manual for help or skip the task for difficult task at anytime. The sample screen of testing module is composed as shown in the left of Figure 2.

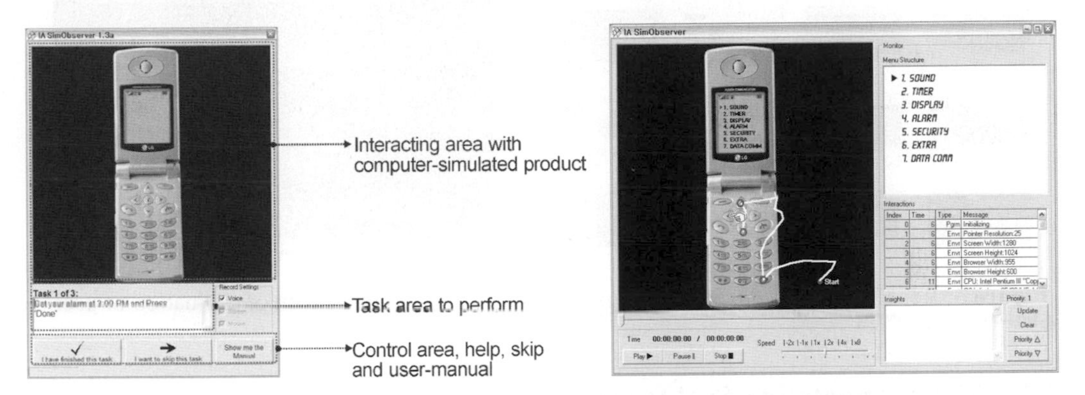

Figure 2: Sample screen of testing module (left) and analysis module (right)

After finishing all the given tasks, user is led to participate in the idea-generation session. In the session of idea-generation, user can actively generate her or his own ideas regarding layout of control buttons, grouping menus, and organizing interface structure. User can drag control buttons and configure his preferred way of layout. Finally, in the module of analysis, analysis is done with all the usability data including time taken, operational path and user's idea generated in the idea-generation session, which were transmitted to server. All the interacting processes by users while they were engaging in the usability testing are replayed with the exactly same operational paths, sequences and time. The operational traces are visualized in line over the product so that analyzer can easily see how user interacted, moved around, made errors and so forth. Besides, the data is summarized in the table: pressed buttons, time taken, user's action, and sequence. In this summarizing table, a researcher can sort out specific data for various purposes of analyses. The sample screen of analysis module is shown in the right of Figure 2.

2.2 Internet-Based Conjoint Analysis for Aesthetic Value Level

Currently emerging one of most promising design methods for aesthetic value level is 'emotional engineering' or '*Kansei* engineering' in Japanese term. Emotion is one of the most elusive terms which makes it difficult to deal with. For emotional engineering, statistical analysis called 'conjoint analysis' has been heavily used since the beginning of introduction of emotional engineering by Japanese originator, Nagamachi. (Nagamachi 1989) Conjoint Analysis provides designers with useful means to understand how much relative importance people put in features and levels of product.

A tool of Internet-based Conjoint Analysis was developed for easy collection of data and efficient analysis. User can participate in the experiment in his own computer-screen where different aesthetic images of product images are shown in turn. Users simply evaluate their subjective preference for displayed images and all the evaluated data is automatically transmitted to researcher's server for later analysis. (Lee 2000) Right part of Figure 3 shows the sample screen of the tool. Left part of Figure 3 shows 16 different images of product shown to users for evaluation.

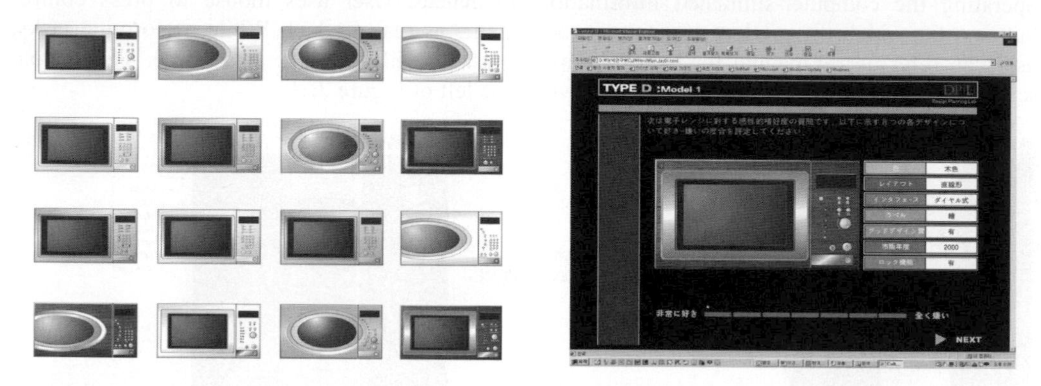

Figure 3: Representative set of products given to users and sample screen of evaluation of user's subjective preference

2.3 Video Annotating Tool for the Level of Symbolic Basic Assumption

As mentioned earlier, the design problems in the level of symbolic basic assumption is hard to understand because they are out of user's consciousness. Symbolic attributes can be dealt with only in relational context and through peoples' natural activities. It is this 'latency' of symbolic attributes that makes 'user-observation' valuable method for them. Particularly, video ethnography is one of most promising methods that applies video technology as a medium for getting insights on people's behaviours and understanding needs by experiential sampling and the cultural inventory. However, despite of its advantages, video ethnography is difficult to apply due to its exhaustive efforts and long time taken for analyzing considerable amount of video data.

The tool called VIDEOW was developed for annotating video data through computer. (Lee & Lee, 1998) VIDEOW allows designer to effectively watch, clip, mark, search, annotate, store, log, and edit video data right on the computer screen. This concurrency allows designer to have comprehensive contextual understanding of user behaviour. Interface of VIDEOW consists of five small windows for various functions: 'viewer', 'note pad', 'marker', 'note palette,' and 'activity chart'. (Figure 4) Viewer allows designer to play video data. While playing video, if designer finds some particularity to require further analysis, he can stop and press 'stamp' button to save the clip into Marker where he can draw any mark like an arrow or input any text on the part that needs attention and further explanation. Then, in Note Pad, designer can input more detail notes on insight or findings from video clip according to a set of frameworks. More frequently used notes such as 'press', 'rotate' or 'push' can be saved in Note Palette so that designer can just drag the word from and drop to Note Pad without need to input repeatedly. Finally Activity Chart includes all the notes written in Note Pad with video clip in sequence. He can easily search by just clicking relevant picture of video clip for re-analysis.

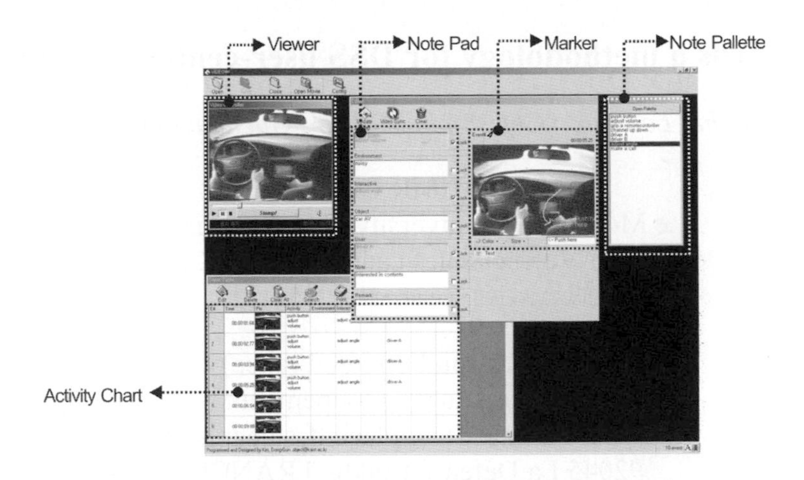

Figure 4: Interface of VIDEOW

3 Conclusion and Further Works

Design tools introduced so far are not entirely new concepts. The important point here is to consider all the different levels for design attributes in cultural context as a whole. These three tools can be integrated in system covering entire design process ranging from stage of planning to evaluation. Findings from one tool can be seamlessly handed over to the other tool. Besides, viewing design problems over the developmental cycle among different levels helps designer to flexibly program and to customize the design process. If a product has been developed long enough to reside in people's subconscious mind, designer may place more emphasis on observing users by heavily using video annotating tool. Or if product is regarded as aesthetic image in the level of emotion, designer may just jump to use tool for conjoint analysis. Furthermore database generated from the tool can bring many advantages. Database from video annotating tool can be used for patterning people's behavior and, similarly, database from the tool for conjoint analysis can be used to understand the people's emotional trend for various aesthetics.

4 References

Hoft, N. L. (1996). Developing a cultural model. In Elisa M. Del Galdo & J. Nielson. (Ed.). *International user interfaces*, (pp. 44-46). New York: John Wiley & Sons.

Lee, J. H. & Lee, K. P. (1998). A study on the user needs analysis based on observation methods in user-centred design. *Proceeding of 3rd Asian Design Conference*, 3, 347-349, Taichung, Taiwan.

Lee, K. P. (2002). Design methods and development of technology, *Form/Formdiskurs,* 183, 95-104.

Lee, K. P. (2002). Remote usability testing for information appliances through WWW: With the emphasis on the development of tools. *Proceeding of Common Ground*, 1, 631-639, Egham, United Kingdom.

Nagamach, M. (1989). Kansei engineering. Japan: Kaibundo.

Towards a methodology for DSS user-centered design

Sophie Lepreux[1], *Christophe Kolski*[1], *Guénhaël Queric*[2]

[1] LAMIH - UMR CNRS 8530
Le Mont-Houy, University of Valenciennes
59313 Valenciennes cedex 9, FRANCE
sophie.lepreux@univ-valenciennes.fr

[2] RFF (Réseau Ferré de France)
Direction du développement
Tour Pascal A – 6, place des degrés
92045 La Défense Cedex, FRANCE
guenhael.queric@rff.fr

Abstract

The paper describes a methodology for DSS user-centered design ; the DSS must respect the human decision making process. The proposed methodology integrates the development of components (in the component-based programming sense) in collaboration with the users. In this paper we explain the various phases of the methodology. This methodology is currently being applied in order to design a DSS for investments in the French railway infrastructure.

1 Introduction

Decision Support Systems (DSS) appeared thirty years ago. Much research has been performed in this area since then, notably in the areas of decision making, knowledge management, artificial intelligence, operational research, human-computer interaction and web based technologies. In this article, we concentrate on the human-computer interaction angle.

This type of system aims at helping the user to solve problems in order for him/her to make decisions. There are two types of DSS. The usual systems solve problems in order to help the user to make decisions. For a given problem, the systems provide the user with one or more solutions. The other systems help the decision-maker in the evaluation of solutions. Using the problem and its solutions, the system measures the various solutions according to criteria and then allows the user to select the best one.

In this paper, we concentrate on the second type of system. Indeed, we aim to design a decision support system for investments in the French railway infrastructure. The problems are known and recurrent and the solutions are also known. The difficulty of the investment is to find the best composition between solutions in order to satisfy the demand with minimal cost.

The paper is organised as follows. Section 2 shows the development methodology adopted for a general context. In Section 3, this methodology is applied to the design of a DSS for investments in the French railway infrastructure. Section 4 explains the preliminary results. We conclude and present future works in Section 5.

2 Methodology of design

Our research aims at proposing a methodology for designing a DSS. The method involves four phases: Analysis, Specification, Design and Validation, with an user-centered design approach (cf. Figure 1).
The analysis phase consists in the decomposition of the global problem into sub-problems. From these sub-problems, the solutions and their measurement tools are highlighted. Each of these tools (components) must be independent and autonomous. The methodology follows a component-based approach. Indeed, it is claimed in a global way that component-based engineering approaches provide the best possibilities of reuse during the project by using components according to the needs (Gamma, Hem R., Johnson, & Vlissides, 1994). The components themselves can be designed following classical design phases with a user-centred design approach. Therefore, from here, the design is separated in several ways : (1) the DSS design and (2) the design of each of these tools seen as components.
The specification phase, as usual, includes the study and consideration of functional requirements. In order to design a DSS, wc need to know the human decision process. Indeed, the DSS should follow the user's decision making process whilst supporting him/her at each step. For example, the process described by Simon (Simon H.A., 1960) is decomposed into four steps as follows: (1) environment study; (2) invention, development and analysis of different modes; (3) the choice of a particular action mode among the possible actions; (4) assessment of past decisions. These four steps must be respected by the designed system during the problem solving process. The specification phase makes it possible to know the form in which the problem will be presented by the user. From this representation, it is possible to know which decomposition of problem will enable the highlighting of the basic tools (components) to be used (or reused). At this stage, the designer must known how the system must behave.

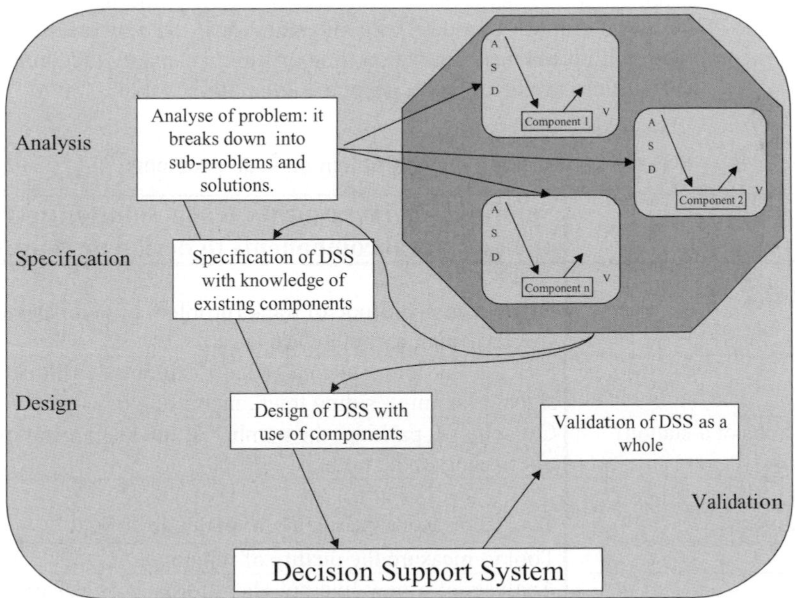

Figure 1 : DSS design methodology including component design

During the design phase, the system will be designed using knowledge of the specification (name, goal, attributes,...) and the state (under study, under construction, available) of the components. Indeed, here, we presume that the components are designed whereas they could still be at the development stage. Then, the designed DSS must be validated by the designers and the users using classical criteria (utility, usability...)(Nielsen, 1993). This stage should be done once the most of components have been designed, otherwise it will not be very valuable. It is at this stage that the component-programming language is selected (javabeans, EJB, CORBA...).
The interesting feature of our methodology is that the final DSS is not dependent on all the components (they are not essential to functioning of the system). Consequently, other components could be added to the system in the future without having to rethink the system's architecture. The future components will be designed in the same way as the others and integrated in the system. Only the capacities associated to the new components will be added to the capacities of the DSS.

3 Case study: DSS for investments in French railway infrastructure

The methodology was used for DSS design in the domain of railway infrastructure investments. The final system aims at supporting the RFF employees (who are, in fact, experts) in making investments decisions. RFF (Réseau Ferré de France) is the owner of the railway infrastructure in France. This company must ensure the maintenance and expansion of the railway infrastructure. In order to do this, the RFF employees have to evaluate the development proposed by railway operation. The RFF employees have to verify the usefulness of investments. They need a system which includes interactive measurement tools to analyse the various railway layouts. These tools will make it possible to formalise the experts' knowledge and to use it profitably.
In order to follow the alternative phases of the methodology, the type of problem (the needs) to be analysed is identified. Indeed, on Table 1 we can see several recurrent problems developed into sub-problems. For example, if the general problem is one of capacity, the user wants to know if the infrastructure allows the definition of a new path allocation or if the network is saturated. For that the user is confronted with problems of junction, line or station capacity (according to his/her problem). The potential solutions are listed in the second column of the table.

Table 1 : Breaking down problem into sub-problems

Problem	Solutions in investments (each solution is related to one or several components to design or reuse)
Capacity :	
- junction	Difference in altitude of track (problem of node capacity), speed of switch, put a third track
- line	Spacing system (manual block, automatic block of lights, signal planning), siding track, increasing of speed
- station (tracks of a station)	Capacity of parking, the number of tracks in a station, capacity of crossing in the node
Steadiness:	
- path allocation	Tool to measure the quality of path allocation
- running	Tool to measure the quality of running
"new plan"	Analysis of a new line but also of knock-on effect on the other lines
Station node :	
- crossing	nodal point, modify the speed, place the position

At each sub-problem or solution, measurement tools are identified; they will enable the assessment of solutions. The vital tools emerge from the first stage of the DSS design. For instance, the tools have to be able to calculate the residual capacity on a railway line (Lepreux, Abed, Kolski, Jung, & Legendre, 2001), to study the railway nodes, to simulate traffic in order to verify timetable robustness and to analyse railway circulation behaviour according to various signal layouts (block cutting), to support economical calculus, and so on. We can see an extract of these description tools in Table 2. For each tool which is considered as a component for the DSS, the characteristics (imperative and facultative input data, output data, form of results...) are known to the DSS.
In our industrial case, the difficulty of designing a DSS is that the users could be very different. Indeed, some users may use the DSS to have an idea about the railway capacity and to study globally if the demand will be satisfied, whereas other users may use the same DSS to perform precise calculus (for example to modify the plan of signals in the case of a particular demand in a specific infrastructure). Consequently the DSS must behave according to the requirements of the user. That is why the specification phase of DSS is very important. It is in this phase that the behaviour (adaptability) of the DSS is described.

Table 2: Two examples of tools to design

Tool	Definition	Imperative input data	Expert's input data	Facultative input data	results	needs
Simulation of timetable	Allows to visualise the place of train according to state signals	- Infrastructure (level of Remarkable Point) - Signals Place (kilometre Point) - Timetable	Length of trains	- Speed of signals - Coefficient of trains' acceleration and deceleration	- Time of release of signal - Graphic space-time	Robustness Spacing Capacity of line Modify the signal's place
Rate of use of station's infrastructure		- Map of station - Map of using of tracks	Length of trains	Coefficient of trains' acceleration and deceleration	Percentage, Bar graph indicating occupancy of each facility	Rating Spacing Explanation of investment

4 First Results

In order to design the DSS, we used the methodology. We performed the analysis stage, in which the tools to be designed are described (on Table 2) and we are now in the specification stage (concerning the DSS). In parallel, several components are under study or development, in collaboration with the experts. Evaluations and validations with the experts are being performed on mock-ups and prototypes. In particular we are evaluating how the human decision making process is respected. With this goals, we have defined a set of scenarii, in which the experts state the problem, and choose the tools to use; the system proposes the results coming from tools with several representation modes; these modes must be adapted to the user's typology (circulation railway expert, economical expert,...); the user can make the decision. During the scenarii, it is also possible to start again.

5 Conclusion and perspectives

In this paper, we have pointed to a methodology for designing a DSS. This methodology allows the design of a DSS using a centered-based approach and respecting the human decision making process. It allows the reuse of a set of basic components designed in collaboration with the users; it facilitates the evolution of the DSS by future integration of other components proposed by the experts. The tools have to correspond to the user typology, goals and needs.
In the future, personalization would make it possible to memorize the user's choices and recall them when other decisions are being made; among the visualisation tools, the system would sort the tools by preference order and by compatibility with results.

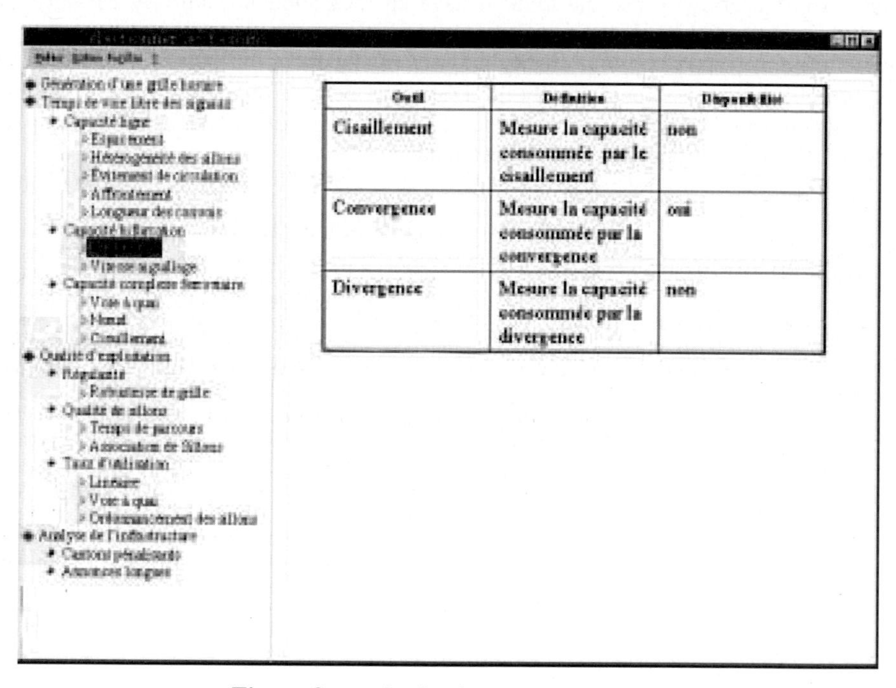

Figure 2: need selection screen

Acknowledgements.

The authors thank the RFF for its support, and also the Nord-Pas de Calais Region and the FEDER for supporting not only this research project but also our current reflections concerning new human-computer interactions (TACT NIPO Project).

References

Gamma, E., Hem R., Johnson, R., & Vlissides, J. (1994). Design Patterns: Elements of reusable Object-Oriented Software. Massachussetts: Addison Wesley.

Lepreux, S., Abed, M., Kolski, C., Jung, S., & Legendre, M. A methodology for decision support system design in railway capacity evaluation. In Morten Lind (Ed.). Danemark: DTU.

Nielsen, J. (1993). Usability engineering. Boston: Academic Press.

Simon H.A. (1960). The New Science of Management Decision. New York: Harper Brothers.

Concept Design of Mobile Phone for Chinese Deaf Mutes

Yan Li   *Wanli Yang*   *Rong Yang*   *Xuelian Li*

Legend Corporate Research & Development, Beijing P.R.China
liyani@legend.com

Abstract: The hearing impairment and language limitation impedes daily life and work of people with hearing and speech disability, and makes them unable to communicate with others effectively like normal people. With the popularization of the mobile phone, more and more deaf-mute persons begin to use the short text messaging services (SMS) of the mobile. But there is no mobile phone specially designed for them at the present market of China. Physical, psychological and communication requirements of them were studied by In-depth interview, focus group and questionnaire survey. Finally, the concept design aiming at improving the quality of communication for the hearing and speech disabilities will be provided in this paper.

1. Introduction

In 2001, a consortium of researchers from Aachen University of Technology (Germany), Ericsson Spain Research Department and Vodafone UK Limited, is working together on the project, called WISDOM (Wireless Information Services for Deaf People on the Move). Information in sign language from a video server via 3G (three generation) phone, give commands to their phones in sign language, and access a real time interpretation service to aid them in communicating with hearing people (Center for deaf studies). Nokia announced the release of the Nokia inductive loopset (LPS) in Taiwan in 2002. The device gives people who use a T-coil equipped hearing aid the freedom to talk on a digital mobile phone by transmitting speech from the phone to the hearing aid in the ear. At present, there are 20.57 million people with hearing and speech disability, occupying 34.3% of all the 60 million people with disability in China. The disability in hearing and speech greatly affects their daily life, making them difficult to get information, think and communicate with others like normal people. They also have to accept low income due to the limited choice of occupation. With the widespread use of the mobile phones and the short message service (SMS) in China, the deaf mutes have a brand new way of mobile communication. While it is a pity that the present mobile phones are not specially designed for them. Many functions that have nothing to do with them increase the cost of the mobile phones and some functions are not suitable for them, for example, ring for alert. To solve these problems, this paper works out the concept designs to optimize the mobile phone for deaf mute users.

Based on the initial analysis, the following hypotheses were proposed

(1) The alert mode of SMS should be vibration and flash mainly.

(2) The function of voice command is unnecessary.

(3) It must have the function to assist them to ask for help of the normal people more accurately and quicker.

(4) It should response to sound to hint them.

Later we further verify the above-mentioned hypotheses by means of questionnaire, in-depth interview, and focus group in China mainland area, and get some new conclusions.

2. Research Methods and Process

2.1 In-depth interview:

Twenty-six deaf mutes were involved in the in-depth interview, which focus on exploring their requirements of mobile communication. There were two kinds of interviewees: thirteen deaf-mutes with mobile phones and thirteen deaf mutes without mobile phones. The interview was focus on communication objects, communication problems and communication requirements

2.2 Questionnaire survey:

After the in-depth interview, all of the twenty-six deaf mutes were asked to fill in the questionnaire forms individually. Twenty-five valid samples had been collected. The questionnaire includes user's basic information, such as the Household and Family Characteristics, average income per month, average expense per month, the highest price that they would afford for a mobile phone etc. Deaf mutes with mobile phone provided information on their preferred functions and usage frequencies.

2.3 Focus group:

Eight normal people who live with deaf mute persons daily were invited to attend the Focus group, including two parents of deaf mute children, two adult children of older deaf mute persons, three teachers of school of deaf-mutes, one friend of the deaf mute person. From their point of views, more actual and deeper feelings and requirements of deaf mutes would be collected.

3. Research Result

3.1 Cognition characteristics:

From interviews with Chinese deaf mute persons, three kinds of cognition characteristics were found.

(1) Perception characteristics： Feel and understand objects outside mainly depend on sight, touch, taste and smell. People with hearing and speech disability are usually sensitive to light and flash.

(2) Memory characteristics: Have the strong visual memory ability. It's quick and easy to rehearse visual things

(3) Thinking characteristics: Have the obviously visual tendency. Thinking ability has little progress except the visual thinking ability

3.2 Psychology characteristics:

From interviews with Chinese deaf mute persons, three kinds of psychology characteristics were found.

(1) Inferiority complex. Due to the influence of the social circumstances, most of the deaf mutes were treated discriminately or abused in their childhood. Even the adult with hearing disability will be hurt unintentionally by children. Besides, their choice of occupation is so limited that they often get very low income and therefore live a poor life. All these factors make them feel severely inferior.

(2) Highly sensitive. Affected by the inferiority complex, they are much more sensitive and suspicious. They tend to put themselves in the adverse position and misunderstand others' behaviors.

(3) Monochromic orientation. They are eager to be treated by others as normal people. Therefore, their deepest desire is to communicate with the normal ones, and to use the products same with the normal people use. They hate to be regarded as abnormal and be laughed at for their physical disability.

3.3. Problems in communication process:

Usually the spouse of the deaf mute is also a deaf mute but his/her children or parents are often normal ones. It is inevitable that the deaf mutes to communicate with the normal people, especially with normal people outside their family. In the communication between the deaf mutes and normal ones, there are two important issues. One is that deaf mutes usually have weak ability to express themselves, for example, using wrong wording or reversed word order, making the normal ones confused. The other is they tend to have inferiority complex and often misunderstand the behavior of the partners they communicate with, which leads to communication failure. To help them communicate accurately and smoothly with normal people is one of the main purposes of the product. The main problems in the deaf mutes' communication were listed in table 1.

3.4、 The present usage of mobile phone:

After in-depth interview, we also investigate thirteen deaf-mute persons who have mobile phones by questionnaire and find out only two functions, i.e. SMS and phonebook are used by all of them and as for the other functions, only part of them use. The frequency of short messages sent daily is five times.

Table 1: The main problems in the deaf mutes' communication

Style	Method	No.	Problems	Count
Face -to- face	Sign language	01	Not all the people know sign language.	--
	Paper and pen	02	It is troublesome and a waste of time to communicate by pen and paper.	6
		03	They are poor at written expression. The wrong wording or reversed word order often results in misunderstanding.	5
		04	It is unsuitable on crowded or urgent occasions.	2
None face -to- face	Phone or beep pager by others	05	Being afraid of being troublesome, they are unwilling to trouble others.	6
		06	Normal people refuse to give a hand.	5
	Fax	07	Time-consuming and not applicable for urgent matters.	3
		08	Many of them don't have fax machines.	--
	SMS of mobile phone	09	They are difficult to be employed or have low income and most of them haven't mobile phones.	--
		10	The letters on the phone keys and the words on the display are too small to see clear.	10
		11	There are too many functions that are useless to or have nothing to do with them	4
		12	The vibration is too light in degree and too dull in ways to feel it.	3
		13	The display is too dim to see clearly.	3
		14	Some of the short messages are hard to understand.	3
		15	Some of them don't know how to spell.	3
		16	They have to take the mobile phones with them wherever they go otherwise they can't feel the vibration.	2
		17	The number of words permitted in one short message is so limited that they have to send many short messages sometimes to express one idea.	2
		18	The mobile phone couldn't send sign language.	1

4. Design and Conclusion

The sketch of the mobile phone for deaf mute users is showed as following

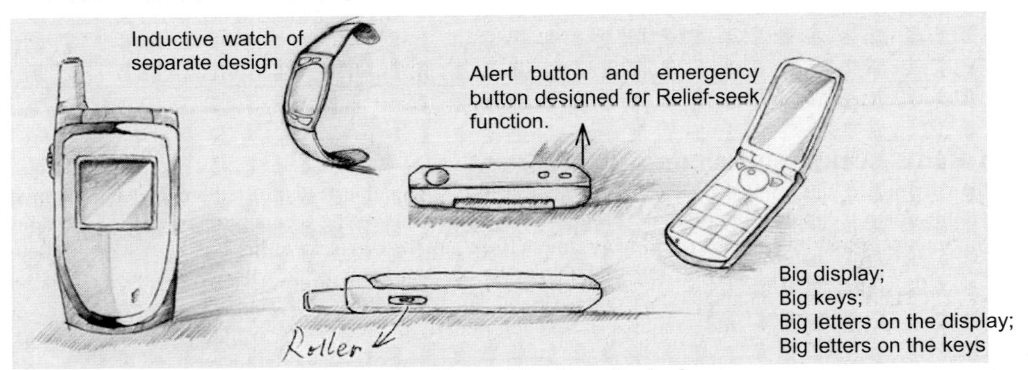

Figure 1: The sketch of the mobile phone for deaf mute users

4.1 Separate design

Now there are only two kinds of alert mode: ring and vibrate, among which ring is useless to them. We realize during the research that deaf mute users are not sensitive to the alert because the way of vibration is single or they don't take the phones with them and therefore they miss the short message or other information. The separate design was worked out for Problem 12, 16 in table 1. (see figure 1)

The separate design includes two aspects:

(1) Inductive wristwatch. Adds a new inductive wristwatch that is the same as the ordinary ones in appearance but has the function to respond to the short message of the phone or other alerted information. When the phone receives a short message, the wristwatch will vibrate to alert the user who then can operate the phone.

(2) Alert mode. Have varied modes of alert. The user can choose to turn on or off all the functions on the phone and the wristwatch.

4.2 Relief-seek design in emergency

During the preliminary study, we thought the traditional voice call was useless to them. While later we feel this function still need to be kept for the following reasons.

(1) Monochromic orientation. Based on it, they are unwilling to use any product that has the special mark of "deaf mute" and feel pleased to use the normal ones. So to cancel this function seems to be too "special".

(2) It is the dream of most of the deaf mute persons to regain the sense of hearing. The hearing aid becomes more and more popular. The new products aiming at regaining the sense of hearing have already come into being.

(3) The function of emergency call is much more important and practical to them. Due to the physical disability, the deaf mute people may face far more trouble or difficulty in the society. For example, when one faints (because of bus sickness or other diseases) in the bus, there must be certain delay if he couldn't inform his family immediately due to the mute, nor could he smoothly communicate with people surrounding him. The following figure 3 is designed for these problems.

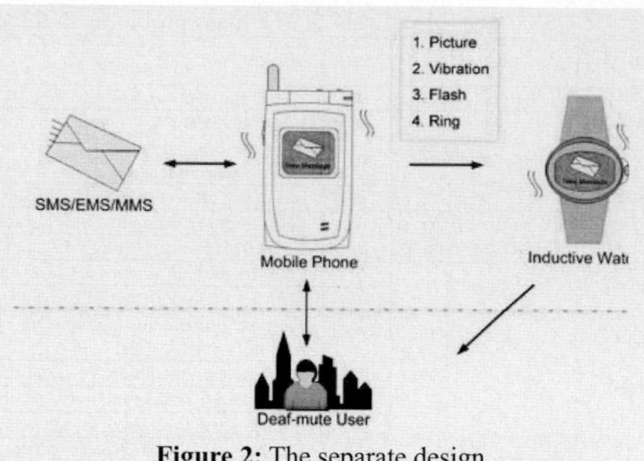

Figure 2: The separate design

Table 2: Alert mode for deaf mute users

Alert mode	Description
Picture	It is important to them according to the visual thinking characteristics of the deaf mute people.
Vibration	Varied to distinguish the short message from other information. Like the ring tone in the normal mobile phone, the swing, frequency and interval of the vibration can be defined.
Flash	Flash is an efficient way to alert deaf mute users because of their perception characteristics.
Ring	Remained. Deaf mute users can define whether it should be active.

4.3 Relief-seek design in daily life

In their daily life, the problem they most often meet is to communicate with the normal people, which is inevitable. Take shopping as an example. From selecting goods to settling accounts, the deaf mute person has to communicate with the salesmen for many times. If he has the company of his family or friends (normal ones), they can do all the deeds for him. Otherwise, he has to communicate with the salesman by pen and paper. Then Questions 02,03 and 04 in table 1 occur.

To solve these problems, we add a relief-seek function in daily life to the mobile phone, which includes two models:

(1) The pre-defined model. On the one hand, the daily expressions are already stored in the mobile phones before they are leave factory and the user can select them quickly. On the other hand, the user can add some new items of his own beforehand and use them in need. For example, when they take the bus, they set the destination at home and show it to the conductor when buying tickets.

(2) The temporary model. When needed, they can put them in at any time, which will be shown in the display. The deaf mute user may choose this function by sliding up and down the roller in the left side of the phone.

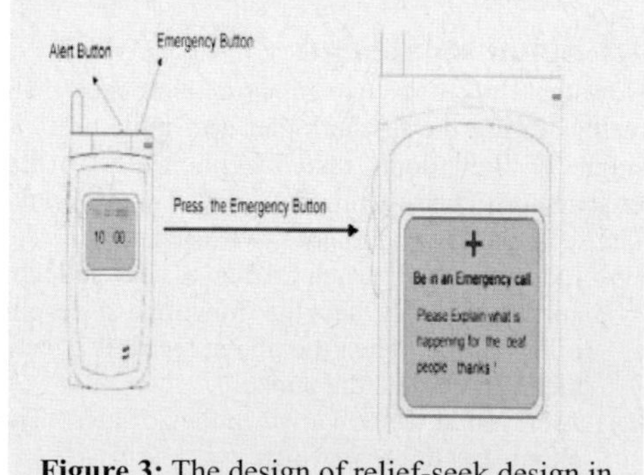

Figure 3: The design of relief-seek design in emergency

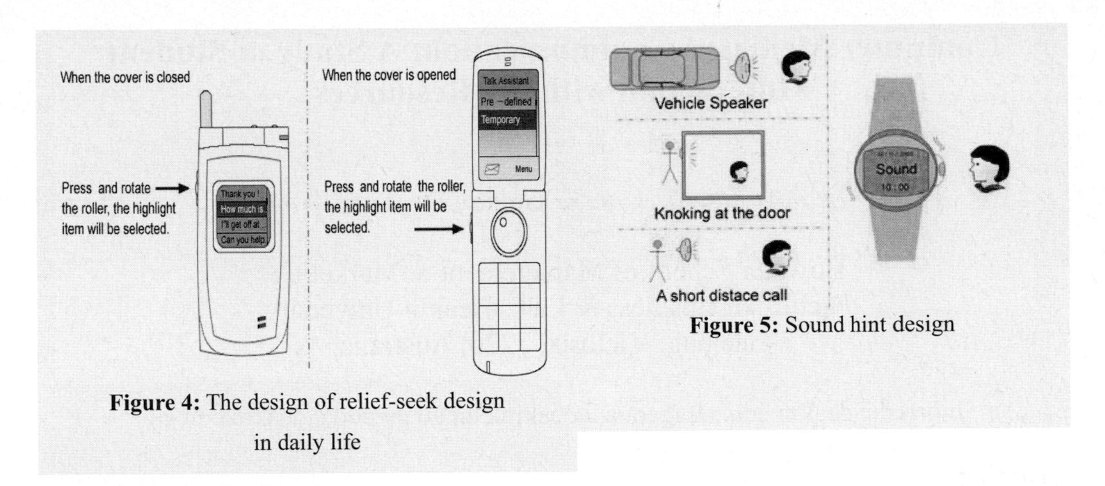

Figure 4: The design of relief-seek design in daily life

Figure 5: Sound hint design

4.4.Sound hint design

The percentage of deaf mute persons ranks first in that of all the people with disability, but the population is still very small. The normal people seldom meet a deaf mute in their daily life. Thus, when the deaf mutes need help, they may not obtain it timely or accurately or even be misunderstood. The following problems were found in the study. The sound hint design was worked out for these problems.

Table 3: Problems in the communication with normal people

No.	Problem
1	He couldn't dodge the vehicle behind him in time because of the incapability of hearing the vehicle loudspeaker and therefore may be blamed.
2	He can't hear his family knocking at the door and keeps them locked outside.
3	He may hear nothing when be called in a short distance until face-to-face.

5. Discussion

The above designs are special for the communication requirement of people with hearing and speech disability, provide solutions, such as the alert mode design, to the problems of the their communication on the move. The current is a first step towards designing mobile phone for deaf mute users. Future work will also deal with rules and differences of the sense on vibrations, the frequency and the preference of written language usage and the recognition accuracy of sign language.

Acknowledgement

Our Thanks to the colleagues of User Research Center of Legend Corporate Research & Development for supporting us and giving us a lot of valuable suggestions to carry out the study.

Reference and bibliography

1. Center for deaf studies, European Commission (2001). Introduction to WISDOM. Retrieved 20 Jan, 2003, from http://www.mobilewisdom.org
2. Nokia Press. (1998). PCS'98: Nokia debuts stylish new wearable cellular phone. Retrieved 8 Jan, 2003, from http://press.nokia.com/PR/200006783329_5.html
3. Qianlong Technology. (2002). Telecoil. Retrieved 21 Jan, 2003, from http://tech.21dnn.com/28/2002-11-18/71@527926.htm [2003]

Computer Mediated Communication: A Study of Student Interaction with the Resources

Leanne Morris, Lynette Genua, Dr Greg Wood

Bowater School of Management & Marketing,
Faculty of Business & Law, Deakin University,
Geelong, Victoria, 3220, Australia

morrisl@deakin.edu..au lgenua@deakin.edu.au gwood@deakin.edu.au

Abstract

This paper examines the student use of resources placed on the CMC system for them by the academics in a foundation Management unit of a Bachelor of Commerce degree at one of Australia's largest providers of off campus university education. The findings of this pilot study highlight that the students may not necessarily have the same appreciation of the value of the material that the academics perceive it to have. Also, it would appear that students are choosing to interact with the material at a sub-optimal rate, that in itself may be hampering their own learning experiences.

1. Introduction

When we read of 'virtual organisations' and 'electronic commerce' being the way of the future, it is not surprising to see organisations embracing technology with a fervour. They do this not just in an attempt to obtain a competitive advantage, but in many cases, just to stay competitive. The technological imperative that has driven many business organisations is no less evident in universities, particularly those engaged in distance education. As we live in a society in which knowledge is changing rapidly and the skills required by the individual are becoming more diverse, the traditional forms of distance education are no longer sufficient to meet the changing needs of today's students. No longer is there a single mass market for distance education. Increasingly there is a wider diversity of needs and smaller unique groups that require more individualised approaches to learning (Bates, 1995). Computer-mediated communication, (CMC), is one medium that can meet both the individualised needs of the student and the organisation's needs for flexible delivery.

This paper examines the student usage of on-line resources provided by one such tertiary CMC provider, in a unit that services in the vicinity of two thousand students per year, many of whom are new to the university sector.

2. The University

The university in this study is a major open and distance education university in Australia. It is dispersed geographically across 6 campuses in 3 cities in a mainland state. There are more than 31,000 students in the university proper with approximately 12,600 students classified as off campus. Seventy-eight percent of the students come from within Australia and eleven percent of the students come from Asia and the Middle East. The university's entrepreneurial arm has 45,000 students across Australia, Europe and Asia as it caters to the educational needs of large organisations, professional associations, unions and government instrumentalities.

3. What is CMC?

CMC evolved as organisations saw the need to network with individuals both inside and outside of the organisation and as a means of developing efficiency, productivity and coordination (Harasim, 1994; Kaye and Harasim, 1989). With pressure mounting on universities in Australia to increase their productivity and to be more efficient, it is not surprising therefore that many universities have embraced this form of technology as a potential alternative, or adjunct medium, for distance education.

CMC is one of the fastest growing technologies in terms of teachers using it and students accessing it (Bates, 1995; Lane and Shelton, 2001; Scagnoli, 2001). From the students' perspective, the use of networking in this way has been shown to be educationally advantageous (Presno, 1991). All students can contribute what and when they want and the asynchronous nature of CMC liberates the instruction from the constraints of time and distance. Students can download information and read it off-line and then respond when it suits.

4. Objectives and Methodology

This study focuses on the nexus between the perception of the academics involved in the unit as to the value of certain educational materials provided, and student interaction with these materials. The objectives of the study were to:

- Quantify the number of students who looked at, downloaded/saved the material
- Identify any patterns of material usage
- Identify the frequency of material usage
- Identify if students have different learning styles

The methodology was simple, but time consuming. Logs of off campus student interaction were compared by the academics conducting the course for each of the communications that occurred in respect to each one of the learning materials uploaded to the conferencing area. The interest was in student interaction and engagement with the available resources. The unit was an introductory Management unit with 154 off campus students enrolled. It is a core unit in a Bachelor of Commerce degree and all students in the Faculty take this unit at first year level.

5. Discussion of Results

The following section of this paper presents and discusses the results of the study.

5.1 Student Numbers

Initially there were 154 students enrolled in the unit, but by the end of the semester only 103 (66.9%) remained active. By active, one means that they were eligible to present for the examination in this subject. A mandated requirement in studying with this university at undergraduate level is that students must have the computer capacity to interact in all units of the course with the CMC initiatives of the university.

5.2 Lecture Note Outlines

Lecture note outlines were provided. In week one, 70 students downloaded the information, however by the final week only 36 students downloaded them. In the last week of the course, when revision notes were posted only a maximum of 46/103 (44.7%) of students actually read the message and a maximum of 38/103 (36.9%) downloaded the information. One can speculate that the students may not necessarily share the academics' enthusiasm for the provision of lecture notes. The 'popularity' of the material diminished as the semester progressed. The issue that needs to be addressed is: was this reduction in interest over the semester due to the lack of relevance of the lecture notes as perceived by students; or did the students have the information in other forms provided for within the course such as textbooks and /or study guides; or was it a feature of other, as yet undetermined factors.

5.3 Additional Resources provided for Students

Kaye and Harasim (1989) contend that conferencing areas are only empty rooms. They go on to suggest that what goes on in those rooms and what comes out of those rooms in many cases reflects the input that the teacher has put in to the development of the course. The success of CMC is dependent on the moderator and this form of tutoring is just as time consuming as other forms of tutoring. With this idea in mind, the academics designing this course placed on the website additional information, developed over many years of trial and error. Material that they believed would assist student performance in an academic arena.

A set of papers on academic writing skills was posted to the site and whilst 84/103 (81.6%) students opened the file only 34/103 (33.0%) actually downloaded it. Another set of papers on research skills was also posted to the site. The file was opened by 50/103 (48.5%) students and downloaded by 35/103 (34.0%) students. Once again, the expectations of student usage by the academics may not have matched activation by the student cohort.

Students appear to have a belief that they understand the basics of academic writing and research skills. It is a concern that their performance in these matters does not appear to match their own self-assessment. Historically, student errors, in the areas highlighted by the academic staff, are endemic amongst all cohorts doing this subject. The same errors occur from year to year. Hence, this information had been provided for perceived student benefit by academics who believed in its value to the students and the possible enhancement of the students' subsequent performance.

5.4 Assignment Material

The assignments were not compulsory and were subject to a 'roll over policy' that allowed the examination to be 100% of the assessment. Students were strenuously encouraged to do all pieces of assessment, but could opt to do only the examination.
The items provided for students in this area were again designed to enhance the students' performance. The message related to Assignment 1, the case study, was read by a maximum of 46/103 (44.7%) of students and downloaded by a maximum of 44/103 (42.7%) of students. It is of interest that only 67 students submitted the assignment. Hence, a maximum of 44/67 (66.7%) downloaded the information that would directly assist them with this piece of assessment.

With Assignment 2, which was an essay, a maximum of 74/103 (71.8%) read the message, but only a maximum of 19/103 (18.4%) downloaded the material for future reference. For this piece of assessment 53 students submitted the work. Hence, only a maximum of 19/53 (35.8%) downloaded information that should have been of vital use to them.

The point that amazes one about this lack of interaction to download this information is that the essay example provided for students is a model answer to a previous question of similar ilk, for which the same template could have been used for the assignment that the students were to consider.

Students do not appear to be extremely concerned about assignments that have past. The solutions to the assignments were only downloaded by a maximum of 21/67 (31.4%) for Assignment 1 and a maximum of 28/53 (52.8%) for Assignment 2. As this unit is the foundation unit for a major in this area of study, the lessons learnt from the feedback in this unit are ones that can be brought in to play throughout one's study of the major. Students do not seem to value, or even appear to be cognisant of this fact, or else for their own reasons they choose to ignore this information. This area needs further investigation to establish the reasons for responses that at first sight seem to suggest that students may be inadvertently short changing their future academic performance, by missing vital academic clues in the foundation unit, that could save them some 'possible grief' in subsequent units of the major.

5.5 Examination Material

For the academics administering this unit the lack of interaction with the examination material presents even more of a quandary, than the lack of interaction with the assignment material.

The examination in this unit was to some degree pre-seen. Ten days prior to the examination, the twelve essay questions that could appear on the examination were sent in the mail to all students. However, the examination multiple-choice questions were obviously not sent. These multiple-choice questions comprise 50% of the examination, yet only a maximum of 41/101[1] (40.6%) actually downloaded the sample multiple-choice questions provided for them prior to the examination.

Why would nearly 60% of students appear to choose to ignore such information at one of the most critical points of their semester of study? Like many of the questions raised by this study, answers are not easily or readily forthcoming.

6. Conclusion

CMC has the potential to open our universities to the world and the prospect of that is exciting. For distance education students, CMC offers a medium that can eliminate some of the problems associated with learning from a distance. At the same time, it can offer some of the features usually only associated with face-to-face teaching. As we are swept into the twenty-first century on the wave of the technological imperative, it appears that those organizations with the infrastructure to support the technology and their people using it effectively, will achieve a

---

[1] n did equal 103 students who were eligible to sit the examination, but 2 students submitted special consideration forms that precluded them from sitting the examination, hence the n is now 101.

competitive edge. However, the learning provided for students in this medium must be seen by them to be worthy of engaging, because the use of technology by itself will not transform this method of delivery into an effective educational tool (Laurillard, 1993).

The research findings in this study show that the value of the experience that the academics in this unit believed that they were providing may not be perceived as such by the students. It would appear that there is a definite under-utilisation of the resources at hand by the students engaged in this unit of study, yet this unit provides numerous resources that are more prevalent in number than in most other units in the relevant major or even in the rest of the degree.

This lack of engagement by students is a concern and answers are difficult to suggest let alone confirm. There may well be a paradigm difference between an academic's perception of the value of information and the perceived value of the same information by the student. Are we overloading the students with information in an attempt to enhance their learning: information between which they find it difficult to discriminate and therefore may just ignore? Are there gaps in our educative process for students about to engage with our CMC system? Do our students have different priorities to academics raised in earlier times and conditioned to the worth of constant improvement: i.e. are our students more mercenary than we may have been, in that they want to do the unit and exit it with minimal engagement? Many of the questions posed here our outside of the scope of this pilot study, but hopefully, more definitive answers will be forthcoming in subsequent investigations of this unit and method of study with future cohorts.

7. References

Bates, A.W.T. (1995). Technology, open learning and distance education. London: Routledge.

Harasim, L.M. (Ed.). (1994). Global Networks, Computers and International Communication. Cambridge Massachusetts: The MIT Press.

Kaye, T.M.R. & Harasim, L. (1989). Computer Conferencing in the Academic Environment Institute of Educational Technology. Open University: Milton Keyes.

Lane, D.R. & Shelton, M.W. (2001). The centrality of communication education in classroom computer-mediated-communication: Toward a practical and evaluative pedagogy. *Communication Education*, 50 (3), 241-255.

Laurillard, D. (1993). Rethinking University Teaching - a Framework for the Effective Use of Educational Technology. London: Routledge.

Presno, O.D. (1991). Children Help Change the Online World. *Matrix News*, 1 (5), 13-16.

Scagnoli, N.I. (2001). Student orientations for online programs. *Journal of Research on Technology in Education*, 34 (1), 19-27.

Text comprehension processes and hypertext design

Anja Naumann, Jacqueline Waniek, Angela Brunstein, & Josef F. Krems

Dept. of Psychology
Chemnitz University of Technology, 09107 Chemnitz, Germany
anja.naumann@phil.tu-chemnitz.de

Abstract

In hypertext, orientation and navigation problems occur frequently. Design principles, which improve navigation and knowledge acquisition of World Wide Web based documents are needed. First aim of this study was to identify factors, which have an influence on orientation and navigation in hypertext. For deriving these factors, text comprehension theories were used. It was argued that orientation problems occur when the mental representation of the content of a text and the representation of the text structure cannot be mapped onto each other. Therefore, the second aim was to develop and to test navigation aids. Results show a clear need for text structure information as orientation aid and for the facility of coherent navigation.

1 Introduction

A major advantage of hypertext is the individual and active acquisition of knowledge. Compared to books or other linear texts users have a much higher degree of freedom to choose their own path through the system. Nevertheless, in hypertext, orientation and navigation problems occur frequently (McDonald & Stevenson, 1998). This is reflected not only in a subjective sense of confusion but also in a measurable decrease of performance. Users do not know were to navigate next nor how to get there (Kim & Hirtle, 1995). When reading hypertext, less knowledge is acquired and more orientation problems occur than when reading electronic linear text (Gerdes, 2000; Naumann, Waniek, & Krems, 2001). Nevertheless, a lack of user's knowledge about the textual structure of the domain can be compensated by an adequate structure of the text (McDonald & Stevenson, 1998). Furthermore, structural aids and navigation tools can enhance performance in hypertext (Chen & Rada, 1996; McDonald & Stevenson, 1998). For this reason the aim of the present study is to find hypertext structures and navigation aids, which reduce the possibility of "getting lost" but preserves user's control over getting information. At first, factors were identified which have an influence on orientation and navigation. According to discourse comprehension theories (e.g. Graesser, Millis, & Zwaan, 1997) readers construct two mental representations when processing a text: one of the propositions given in the text (the text base) and one of the whole situation addressed by the text (the situation model). The situation model (e.g. van Dijk & Kintsch, 1983) is achieved by adding background knowledge to the information given in the text. By this means, a complete and coherent representation of the content is reached. While reader's situation model is essential for the understanding of the text content, the mental representation of the text structure is essential for orientation and navigation in hypertext (Calvi & De Bra, 1997). The processing of a incoherent (e.g. anachronous order of events) text is generally more complicated than the processing of a coherent text (e.g. Zwaan, Radvansky, Hilliard, & Curiel, 1998). This is particularly problematic in hypertext, because readers choose their own path through the document. If readers have to construct two mental structures while reading, then the

text structure should conform to the postulated situation model for improving the construction of both representations. When relevant dimensions of the two representations do not match, orientation problems are likely to occur.
The specific goal of this study was to test the effects of the text structure as an aid for orientation and navigation. It was assumed that using a coherent structure overview and linear navigation less orientation and navigation problems occur and more factual and structural knowledge is acquired. In contrast, the mental representation of the text structure should be better if participants could use this structure actively as a navigation tool, as opposed to simply using linear navigation via forward- and backward-buttons. Due to findings in previous studies (Chen & Rada, 1996; Naumann, Waniek, Krems, & Hudson-Ettle, 2001) that the differences in knowledge acquisition and navigational behaviour in hypertext were related to the complexity of the task (e.g. reading vs. information retrieval), reading and information retrieval were investigated separately. In addition, for high working memory load more orientation and navigation problems and less acquired knowledge were predicted than for low working memory load for the following reasons: working memory is important for text comprehension (Just & Carpenter, 1992); navigation in hypertext consumes working memory resources; and working memory capacity is limited (Baddeley, 1986). If processing demands are too high, processing and storage processes are impaired (Baddeley, 1986).

2 Method

2.1 Participants

In this experiment 128 students of Chemnitz University of Technology participated (age: M = 22.6, SD = 3.7; 79 % female) for course credit or monetary compensation.

2.2 Material

The material used consisted of two texts with different content. One text about the construction of the Trans-Siberian Railway and one about the Crusades. It was based on articles from a historical magazine. The texts contained 1810 to 1921 words, distributed over 16 text nodes. A trained experimenter analysed the text structure of the articles according to van Dijk & Kintsch (1983), using the three basic rules for gaining the macrostructure of a text (generalization, construction, and deletion). Four chapters, with three subchapters each, were extracted. In addition, macropropositions were derived for each chapter (e.g. “reasons for construction of the railway” and “start of construction”). These macropropositions were organized into an overview containing text structure information either according to the chronological order of events in a situation model (coherent) or according to the order of chapters in the original text (incoherent). The overview was shown in a frame on the left-hand side of the actual text and was either active or inactive. In the active overview text version (hypertext with active overview), participants could only navigate by clicking on the particular macroproposition within the overview. When participants navigated to one of these four chapters the macropropositions of the related subchapters were also presented. After clicking the subchapter, the related text was shown in the frame on the right hand side. In the inactive overview text version (hypertext with linear navigation), participants could only navigate within the text by using a back- and forward- button. Altogether, four different text conditions were constructed and compared:

a) coherent hypertext with linear navigation
b) incoherent hypertext with linear navigation

c) coherent hypertext with active overview
d) incoherent hypertext with active overview

Furthermore, one half of the participants had to solve a secondary task in order to increase working memory load. They had to memorize five-digit numerical orders during reading or information retrieval. These numerical orders popped up in erratic intervals in an additional window on the screen. Participants had to memorize them and close the window. After another erratic interval, participants had to chose the right number from four alternatives in another pop-up window.

2.3 Design and Procedure

The experiment was conducted as a between-subject design with four text conditions (see material). Participants were randomly assigned to one of these conditions. Moreover, one half of the participants had to perform the additional secondary task. The design was a 2x2x2 design with the factors navigation (linear navigation vs. active overview), coherence (coherent vs. incoherent), and working memory load (secondary task vs. no secondary task).
Dependent variables were problems with orientation and navigation in the text and problems with the text content (measured by means of a questionnaire), acquired factual knowledge (knowledge test with multiple choice questions, fill-in questions, and sentence recognition questions, and structural knowledge (card-sorting task with macropropositions written on the cards).
Participants were tested individually in two sessions, each lasting 90 to 120 min. At first, participants had to fill in questionnaires concerning their attitude and experience with computers, and prior knowledge about the text content domain. Then participants had to read the text about the Trans-Siberian-Railway carefully (first session) or to answer questions with the help of the text about the Crusades (second session). After finishing the task, participants were asked to write a short summary, and to answer multiple choice questions and fill-ins. A second questionnaire was concerned with orientation and navigation problems experienced in the text. Finally, participants had to complete a card-sorting task.

3 Results

For reading and information retrieval, groups did not differ significantly in attitudes towards computers, experience with computers, nor prior knowledge. Results are mostly based on multivariate analyses with three factors.

3.1 Reading

The results of reading show a significant influence of all three factors (navigation, coherence, and working memory load) on orientation (navigation: $F(2, 119) = 3.5; p < .05$; coherence: $F(2, 119) = 3.3; p < .05$; working memory load: $F(2, 119) = 16.9; p < .01$). There is no interaction between these factors. Participants reading the active hypertext experienced less orientation and navigation problems than did participants reading the text with linear navigation $F(1, 120) = 4.68; p < .05$. In contrast, coherence had an influence on learning problems; $F(1, 120) = 5.05; p < .05$. In reading the incoherent text, participants experienced significantly more problems with the text content than reading the coherent text. In addition, participants solving a secondary task report more orientation and navigation problems than participants without secondary task; $F(1, 120) = 6.69; p < .05$. However, they report less problems with the text content than participants with less working memory load; $F(1, 120) = 4.08; p < .05$.

Factual knowledge acquisition is influenced by navigation; $F(5, 112) = 3.5$; $p < .05$; and working memory load; $F(5, 112) = 4.6$; $p < .01$; but not by coherence of the text. No interaction was found between the factors. Participants reading the text with linear navigation identified significantly more sentences as not been in the text but correct in respect to content than participants reading the active hypertext; $F(1, 116) = 8.64$; $p < .01$. Likewise, participants who did not have to perform the secondary task identified significantly more sentences correctly than participants who performed the secondary task; $F(1, 116) = 10.34$; $p < .01$. Further, participants not performing the secondary task answered more fill-ins correctly than participants performing the secondary task; $F(1, 116) = 13.51$; $p < .01$.
In addition, a negative correlation between orientation problems and factual knowledge was shown. The number of correct answers in the multiple choice test was negatively correlated with orientation and navigation problems and orientation problems concerning the text content. The number of correctly answered fill-ins was also negatively correlated with orientation and navigation problems and orientation problems concerning the text content.
On structural knowledge, only the factor coherence had a significant influence. Having read the coherent text, participants were much more able to reproduce the original text structure than having read the incoherent text structure; Kruskal-Wallis, $H(1, N = 124) = 58.60$, $p < .01$).

3.2 Information retrieval

The results of information retrieval show a significant influence of navigation on orientation; $F(2, 117) = 4.37$; $p < .05$. The other factors had no influence on orientation, and there was no interaction between the three factors.
Factual knowledge was influenced by the factor coherence only; $F(5, 104) = 3.24$; $p < .01$. After answering questions with the help of the incoherent text, participants scored significantly higher in the multiple choice test than after answering questions with the help of the coherent text.
In addition, a negative correlation between orientation problems and factual knowledge was shown. Participants reporting more orientation problems concerning the text content answered also less fill-ins correctly.
As shown in reading, only the factor coherence had a significant influence on structural knowledge. Having read the coherent text, participants performed much better in the card sorting task than having read the incoherent text structure; Kruskal-Wallis, $H(1, N = 121) = 51.56$, $p < .01$).

4 Conclusions

As hypothesized, participants using a coherent structure overview reproduced the text structure much better than participants using an incoherent structure overview. As expected, in the reading task they also had less problems with the text content. Surprisingly, in information retrieval, they acquired less factual knowledge. This might be due to a more intensive search and thereby more incidental knowledge acquisition within the incoherent text.
As expected, participants reading the text with linear navigation acquired more factual knowledge than participants reading the text with active structure overview. In addition, participants with orientation problems acquired less factual knowledge. Both results support the findings of a previous study (Naumann, Waniek, & Krems, 2001). Contrary to the hypothesis, however, in reading the linear text, participants experience more orientation problems than in reading the hypertext version. One possible explanation could be the concentration on the navigation task and

the active and intense use of the structure overview in the hypertext version, which helps to keep orientation.

Results also confirm the hypothesis of limited working memory capacity. In reading, almost all dependent variables were impaired by the secondary task. However, the secondary task had no influence on information retrieval, which might be due to the high working memory load, inherent in the task.

To summarize, the data suggest that the construction of the reader's situation model from the text is best if coherent navigation is possible and an active overview of text structure for navigation is presented. Results also show the demand for variable hypertext features for diverse task, as shown before in previous studies (e.g. Naumann, Waniek, Krems, & Hudson-Ettle, 2001).

Acknowledgement

We would like to thank the German Research Council DFG, which provided support for this research.

References

Baddeley, A. D. (1986). *Working memory.* Oxford: Oxford University Press.

Calvi, L., & De Bra, P. (1997). Proficiency-adapted information browsing and filtering in hypermedia educational systems. *User Modeling & User-Adapted Interaction, 7*(4), 257-277.

Chen, C., & Rada, R. (1996). Interacting with hypertext: A metaanalysis of experimental studies. *Human-Computer Interaction, 11,* 125-156.

Gerdes, H. (2000). Hypertext. In: B. Batinic (Hrsg.), *Internet für Psychologen (Internet for Psychologists),* 2nd Edition (pp. 193-217). Göttingen: Hogrefe.

Graesser, A. C., Millis, K. K., & Zwaan, R. A. (1997). Discourse comprehension. *Annual Review of Psychology, 48,* 163-189.

Just, M. A. & Carpenter, P. A. (1992). A capacity theory of comprehension: Individual differences in working memory. *Psychological Review, 99,* 122-149.

Kim, H., & Hirtle, S. C. (1995). Spatial metaphors and disorientation in hypertext browsing. *Behaviour and Information Technology, 14,* 239-250.

McDonald, S. & Stevenson, R. J. (1998). Effects of Text Structure and Prior Knowledge of the Learner on Navigation in Hypertext. *Human Factors, 40,* 1, 18-27.

Naumann, A., Waniek, J. & Krems, J.F. (2001). Knowledge acquisition, navigation and eye movements from text and hypertext. In U.-D. Reips & M. Bosnjak (Eds.), *Dimensions of Internet Science* (pp. 293-304). Lengerich: Pabst.

Naumann, A., Waniek, J., Krems, J. F. & Hudson-Ettle, D. (2001). User behavior in hypertext based teaching systems. In: M. J. Smith, G. Salvendy, D. Harris & R. J. Koubek (Eds.), *Usability Evaluation and Interface Design: Cognitive Engineering, Intelligent Agents and Virtual Reality* (Volume 1 of the Proceedings of the HCI International, 2001, pp. 1165-1169). London: Lawrence Erlbaum.

Van Dijk, T.A., & Kintsch, W. (1983). *Strategies of discourse comprehension.* New York: Academic Press.

Zwaan, R. A., Radvansky, G. A., Hillard, A.E., & Curiel, J.M. (1998). Construction multidimensional situation models during reading. *Scientific studies of reading, 2*, 199-220.

A Method for Compression of Three Dimensional Bi-Level Image

Koji NISHIO and Ken-ichi KOBORI***

Department of Information Processing, Faculty of Information
Hirakata-City, Osaka, 573-0916, JAPAN
nishio@is.oit.ac.jp, kobori@is.oit.ac.jp

Abstract

We present a new method that reduces the information content of the three dimensional bi-level image. The process traces the surface of the boundary of the shape and generates a chain code extended to three dimension. To verify the effectiveness of the proposed method, we have experiments to transform some image into our code, and show its information content. Our method generates more compact representation than the DF representation that is the basic code for the octree representation. Applying some entropy compression technique, our code provides a compact representation.

1 Introduction

Until now, the boundary representation model has been used to define a three-dimensional shape model. However, the boundary representation model has a week point that it forces designers to handle of geometrical elements of a shape with exact knowledge. Recently, some approaches, applying a three-dimensional binary image to the definition of a three-dimensional shape, have been proposed. This representation is called bi-level voxel. These approaches enable intuitive operation of shapes. In the near future, three-dimensional bi-level images will be used in the three-dimensional shape design systems. However, the data size of the representation is lager than one of the boundary representation. In addition, it should be concerned that three dimensional shape data will be used on networks. So, in this paper, we propose a encoding method that reduces the information content of bi-level images.

2 Image compression

Currently, some encoding techniques are applied to a three-dimensional bi-level image compression.

2.1 Octree

Octree is the most popular encoding technique of three-dimensional bi-level image. This compact representation consists of a tree structure as shown in Fig.1. Its node is called an octant. An octant has 8 branches connected to smaller octants. Some octants represent larger space than a voxel. Each octant may be full, partially full, or empty (also called black, gray and white, respectively). A partially full octant is recursively subdivided into suboctants. Subdivision continues until all octants are either full or empty.

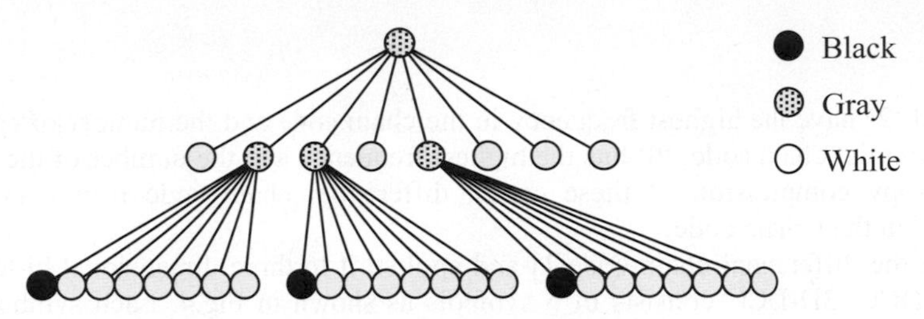

Fig.1 A tree structure of octree

DF representation is well known as a file format for octree. DF file consists of three kinds of symbols, such as '0', '1', and '('. Codes of DF representation are symbols ordered by depth first of the tree. In the case of Fig.1, the DF representation is represented as follows.

"(0 (1 0 0 0 0 0 0 0 (1 0 0 0 0 0 0 0 0 (1 0 0 0 0 0 0 0 0 0 0 "

In this case, the depth of the tree is 2 and we call it division level. Bi-level image takes 1 bit per cell and DF representation takes 2 bits per code. The image represented by the octree in Fig.1 takes 64 bits and DF representation takes 66 bits. In general, at the higher division level, DF representation is smaller than bi-level image.

2.2 Chain code

Chain code is one of data compression techniques for two dimensional bi-level images. Its symbol has 4 kinds of unit vector which traces the perimeter of the shape. Fig.2 shows symbols corresponding to each vector. The chain code of the shape shown in Fig.3 is represented as follows.

" 3 3 3 3 3 1 0 1 0 1 0 1 0 1 0 2 2 2 2 2 "

The two dimensional image shown in Fig.3 takes 64 bits. Correspondingly, the chain code takes 40 bits. Whereas the data size of bi-level image is in proportion to the area of the space, one of chain codes is in proportion to the length of the perimeter of the shape. To get more effective compression, we use the differential chain code. Defining cc_i as the i-th symbol of the chain code, the i-th symbol of the differential chain code dcc_i is represented as follows.

$$dcc_i = \mathrm{mod}(cc_{i+1} - cc_i + 4, 4)$$

Where mod(x, y) represents the remainder of x/y.
The differential chain code of the image shown in fig.3 is represented as follows.

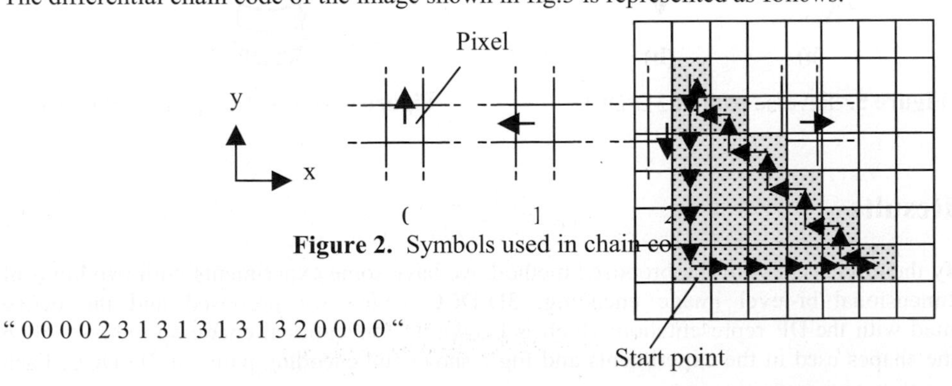

Figure 2. Symbols used in chain code

" 0 0 0 0 2 3 1 3 1 3 1 3 1 3 2 0 0 0 0 "

Figure 3. A path of chain code encoding

'3', '1' and '2' have the highest frequency in the chain code and the number of each symbol is 5. In the differential chain code, '0' has the highest frequency and the number of the symbol is 8. As far as entropy compression of these codes, differential chain code results in more compact representation than chain code.

We extend the differential chain code[1] and applied it to three dimensional bi-level images and call it 3D-DCC. 3D-DCC consists of 5 symbols as shown in Fig.4. Each symbol represents the relative position among vectors v_i, $v_{i-1}$and v_{i-2} which mean the connection of voxels placed on the surface of the shape. In general, chain code cannot trace the whole surface of a three dimensional shape by one stroke. Therefore, we extend 3D-DCC by adding symbol 'R' and 'J'. 'R' represents "reverse" and 'J' represents "jump". Fig.5(a) and (b) shows current set of three vectors and next set, respectively. In the case of turning back as shown in Fig.5(a), the encoding process inserts 'R' into the code and replaces v_{i-1} in Fig.5(b) with v_i in Fig.5(a). In order to trace the surface of the three dimensional shape, it is necessary to connect discontinuous paths. We insert a connection path shown in Fig.6, which is represented by 3D-DCC that has 'J's at its own terminals.

Encoding process starts at the black voxel located on the boundary of the shape (we call such voxel seed voxel), and ends when there is no black voxel on the boundary which has been unprocessed around the voxel in process. Encoding process starts again at a nearest seed voxel from the latest voxel. If there is no black voxel which is not processed, the encoding process ends. To get efficient encoding path, the process gives preference to '0', '2', '4', and 'R' over '1', '3', and 'J' in order to trace the boundary of the shape on the cross section parallel to X-Y plane and select a symbol which represents the vector parallel to X-Y plane after generating the connection path.

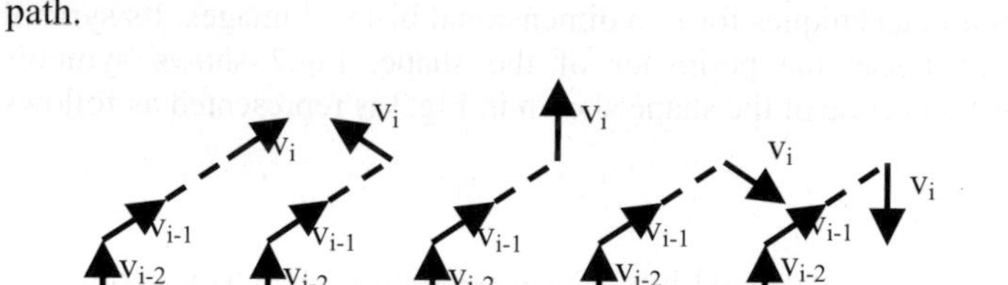

Figure 4. Symbols used in 3D-DCC

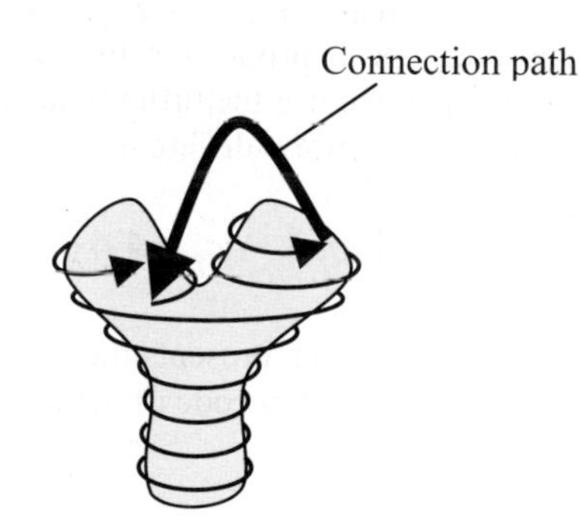

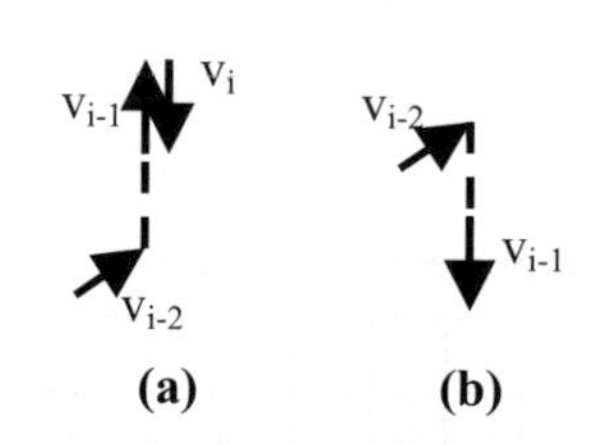

Figure 5. Reverse path of 3D-DCC

Figure 6. Connection path of 3D-DCC

3 Results

To verify the effectiveness of the proposed method, we have some experiments with two kinds of three dimensional bi-level image encoding, 3D-DCC which we proposed and the octree represented with the DF representation which is known as conventional encoding method. Fig.6 shows the shapes used in the experiments and Fig.7 shows the encoding paths of 3D-DCC. Each

shape is structured with a three dimensional bi-level image whose resolution along each axis is 128. Table1 and table2 show the number of symbol and information content of the DF representation and 3D-DCC, respectively.
In addition, we have the experiment with cross sections of 3D-DCC. Fig.8 shows encoding paths on three different cross sections and Table.3 shows the encoding results.

4 Conclusion

As mentioned above, our method is more effective than the DF representation of octree. The information contents of 3D-DCC are 27-40% to those of the DF representation. In addition, the compression ratios of 3D-DCC are different according to the direction of the cross sections. Our future work includes to decide the optimum direction of the cross sections.

References

[1] Bribiesca, E. "A chain code for representing 3D curves", Pattern Recognition, 33, pp.755-765, 2000.

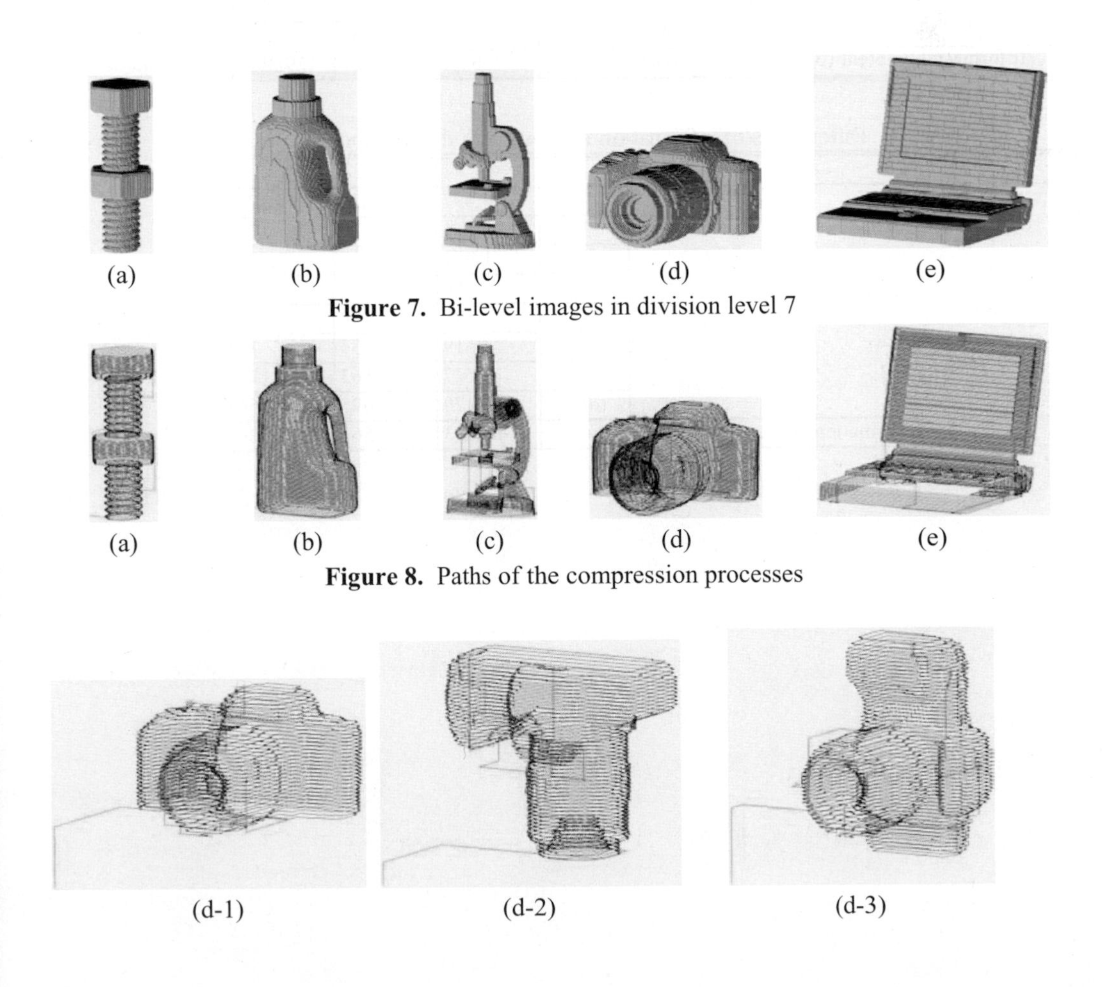

Figure 7. Bi-level images in division level 7

Figure 8. Paths of the compression processes

Figure 9. Paths and the direction of the shape

Tabel 1: Infoemation content of DF representation

		a	b	c	d	e
Number of symbol	'0'	17,283	18,555	37,787	11,838	33,609
	'1'	15,800	17,202	31,815	9,219	31,142
	'2'	4,726	5,108	9,943	3,008	9,250
Information content (bit)		53,586	57,932	112,469	33,902	104,907

Table 2: Information content of 3D-DCC

		a	b	c	d	e
Number of symbol	'0'	6,319	13,170	29575	8,367	19,810
	'1'	142	178	196	266	130
	'2'	7,106	7,327	1496	3,348	7,808
	'3'	154	149	149	237	136
	'4'	477	678	883	837	1,059
	'R'	22	4	22	24	78
	'J'	289	76	319	316	283
Information content (bit)		14,509	21582	32640	13,395	29,304

Table 3: Differences of information content with disposition

		d-1	d-2	d-3
Number of symbol	'0'	4,897	5,628	4,806
	'1'	46	79	104
	'2'	1,941	1,617	2,637
	'3'	50	104	89
	'4'	390	468	386
	'R'	19	36	17
	'J'	127	163	127
	total	7,470	8,095	8,166
Information content (bit)		10,043	10,562	11,526

Human-Human Collaborative Intentions in Learning Environment through Computer Graphics Interactions

César A. Osuna-Gómez, Leonid Sheremetov, Manuel Romero-Salcedo

Mexican Petroleum Institute
Eje Central Lázaro Cárdenas, 152
Col. San Bartolo Atepehuacan
07730 México DF
{cosuna, sher, mromeros}@imp.mx

Abstract

In this paper, the model APRI (Action-Perception-Reflection-Intention) and a software application in order to study intentions in collaborative learning situations based on graphic vocabulary are presented. The scenario is divided into five activities and its goal is the construction of a statement for the development of an information task. The communication media is a collaborative whiteboard that contains a set of icons, which represent different mental operations suitable for the debate. The intention measurement is carried out comparing the four Grice Maxims with a graphical conversational analysis. The first experiments showed that these Maxims have to be extended to give enough information about participant's intentions.

1 Introduction

Different collaborative applications exist which deal with the problem of the accomplishing a common task among the participants of a group. These applications have defined new research fields as CSCW (*Computer Supported Collaborative Work*) and CSCL (*Computer Supported Collaborative Learning*). In general, the idea of applying the collaboration principles has influence on the development of human-computer interfaces, specifically in two aspects. First, the collaboration is characterized by the fact that the participants' goal is common. Though the participants' actions can be different, these latter have to be oriented towards the successful task execution. In that sense, it is necessary to know the individual behaviour of the members in order to evaluate if they are carrying out collaborative actions or if the personal goals are interposed. Second, the will of participation plays a main role in the work group, since it allows to visualize results of a reflection through the performed actions when a member is interacting with the group.

However, to measure and to know the collaboration experience is a high complexity task because the elements which dictate the will of the individual are not completely known. There exist different contributions derived from the Conversational Pragmatic Theory and the Discourse Analysis, which allow to know the participation quality and direction of members of a group (Gallardo-Paúls, 1996). One of the strategies, in order to know the will of participation, is the study of the *intention*, which is defined as the determination oriented to reach a goal. That is to say, if the intention of a person is known then it is possible to determine the existence or not of the will of collaboration in an environment of group participation. Within this environment, the intention is represented by the actions that the participant carries out with an intrinsic purpose, which can agree or not with the goals established by a group. Some research works proposed how to measure the intention in the field of the conversational analysis (Stone, 2002). They also show

how to apply the Grice's collaboration based rules (Grice, 1975), through which qualitative signs about the direction of the will of a person in a transactional or conversation environment can be obtained. However, these research works have been carried out using pieces of natural language talk and they did not present cases where the language is a graphic-based vocabulary used in collaborative technological environments.

In this paper, a model and a software application in order to study the intention of the participants in a collaborative learning situation are presented, where a graphic-based vocabulary is used. First, a model relating intentions with the performed actions is described. Next, the Grice's collaboration based rules are shown. Then, the software application with metrics in order to study the intention is described. Finally, the conclusions derived from this work are presented.

2 Action-Perception-Reflection-Intention Model of Collaboration

The Action-Perception-Reflection-Intention (APRI) model is a representation of the cognitive process started with actions carried out by a user, in any environment, through computers or not. The model is derived from an analysis with CLE (Constructivist Learning Theory) and BDI (Belief, Desire and Intention) model (Bratman, 1987). First, CLE supports the process of learning based on a continuous reflection among previous knowledge, perception, intention and social situation. Second, BDI has proved that it can cope with the changing reality. APRI is a cycle configured by four human processes: Action, Perception, Reflection and Intention. *Action* refers to the activities done by the participant during a collaborative session. *Perception* tries to capture the change of the environment in relation with the personal experience. It is strongly related with Gestalt Theory. *Reflection* means the constructivist process, where a participant tries to learn about the changing environment and construct new knowledge. *Intention* means the result of the reflection as a "*set of executing threads*" (Georgeff et al, 1999) represented by actions.

In that sense, the intention provides the knowledge of changes in the cognitive state, since it is the subsequent result of an analysis made internally. An action is the physical expression of a voluntary state which results in a cognitive change (Jonassen & Land, 2002). In the learning case, the analysis of the actions allows to know if the reflection made by a user is oriented towards the cognitive activity for the construction of knowledge or not. Moreover, in a collaborative situation, users interact by means of actions in a shared collaborative space. Then, the effect to evaluate by each user comes from the actions carried out by him and by others.

During a conversational process, participants (playing sender and receiver roles) look for, in an implicit way, collaboration through the communication process. In that sense, Grice has proposed a non-normative principle that facilitates the detection of the collaboration or not of the dialogue participants (Grice, 1975). This principle is known as the "Cooperative Principle", which can be taken as a guideline to know the intentions of different members that compose a collaboration group. This principle is based on four maxims. 1. MAXIM OF RELEVANCE suggests that the sender contribution must adjust to the system of expectations and selective restrictions, which are ordered in relation with the collaborative goal of the group. This maxim defines weather the information given by the group members, is significant or not to the group goal. 2. MAXIM OF MANNER recommends that the participation in the conversation must be brief, ordered, clear and concise without ambiguity. Correct construction of each intervention in agreement with the grammar laws and the rhetoric is important. 3. MAXIM OF QUANTITY proposes that the members must offer a contribution, which is not more informative than necessary. 4. MAXIM OF QUALITY suggests the sender not to give his opinion about what s/he does not has evidence, what s/he considers false or what has not been verified. These maxims are applied in the APRI model in order to detect the information about the reflection of each one of the group participants.

3 Collaborative Debate Application

The developed application supports a constructivist learning situation. It was designed following the DELFOS model for the development of CSCL applications (Osuna et al, 2001). Its educative goal is the collaborative writing of an information task of the Big6 Model (Kuhlthau, 1993). The application includes five learning activities: GENERAL, it deals with the detection of subject matters, subjects, attributes, functions and factors. RELATION, it tries that students justify the relations that occur among attributes, functions and factors. ANSWER, it suggests that students provide a set of hypothesis in order to respond to the relations that were presented. DEBATE, it proposes a collaborative space to discuss and to argue on the different answers given by the members of the group. WRITING, it implies the information task writing. These five activities are carried out in a synchronous collaborative form. The first four phases are supported by the Collaborative Debate (CD) application, while the last one – by the Collaborative Editor of XML documents called ELXI (Sheremetov & Romero, 2003). Each participant has a CD-GUI (Collaborative Debate Graphics User Interface) using a set of icons to communicate with the others. Each one of the icons has a conceptual value within the activity. These icons represent the graphic vocabulary of a debate. As an example, Figure 1 depicts the GENERAL activity. In (a) we can observe a set of attributes, functions and factors made by students. Their subject matter was the petroleum. In (b) there are some icons of the graphic vocabulary and its meaning.

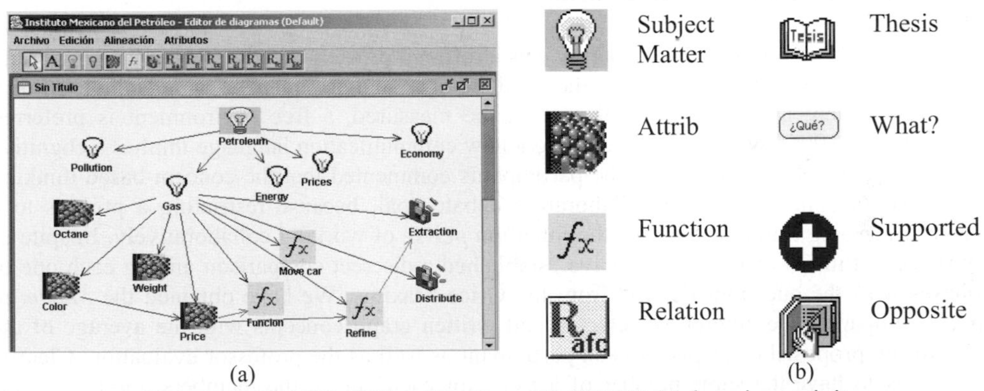

(a) (b)

Figure 1: The CD-GUI application. (a) GENERAL learning activity. (b) Graphic vocabulary and its meaning.

The CD-GUI has a component-based client-server architecture, in which components implement mechanisms for interaction control (replication, consistency and concurrency), shared workspace management and group awareness. The software component approach offers the possibility of developing open systems, which can be flexible and can be configurable by teachers (Dimitriadis et al, 2002). CD-GUI allows a workgroup to build the graphic structure of the debate of a particular topic using a basic set of elements. The clients are members of the group that make use of the editor's interface to modify the diagram and other elements of the debate. Each change is sent to the server that in turn resends them to the interfaces of the other members of the group. Communication between clients and server is carried out using Java RMI. Initially, each client contacts the server and subscribes to it to be able to send her contributions and to receive the changes of the other clients. Clients access the server connecting to the RMI registration service.

The CD-GUI interface of each member reflects the different phases of the debate, so for each phase there is an instance of a graphic component of interaction, by means of which the clients make their contributions. This interface reflects the active phase of the debate, presenting the graphic options in agreement with each particular phase. In particular, each component is the instance of a Java class parameterized with the options corresponding to each phase. This class in turn accesses another component class that maintains the storage of the complete structure of the debate. The server maintains a similar class, what allows it to conserve its own copy of the information of the debate. This allows a new member of the workgroup to request the server all the previous information of the debate and thus to incorporate it into the work session.

Grice Maxims have been added as non-normative references in order to detect the participant's intentions. With these four maxims, the actions performed by the participants were compared according to the context and goal of each phase. We have used the questions as follows. *Maxim of Quantity*: 1) Is the Information given complete? 2) Is the participation excessive? 3) Is the participation poor? *Maxim of Quality:* 1) Are logical elements given? 2) Are related contributions established? 3) Is there a contradiction between support and opposite arguments? *Maxim of Manner*: 1) Is the contribution clearly specified? 2) Are icons appropriately used? 3) Are there recurrent elements? *Maxim of Relevance*: 1) Does the debate go on? 2) Are context related data provided? 3) Is the main thread appropriated? 4) Is there continuity in the subject? In the following section, evaluation results are discussed.

4 Discussion and Conclusions

To study intention with graphics-vocabulary was a difficult process because it causes an important bias in the allowed interactions in the collaborative environments. In a conversational environment, where the intention is planned to be measured, a free environment is preferred (Gallardo-Paúls, 1996). Nevertheless, learning a new communication language implies a cognitive effort from the members that use it. The participants commented that the concept-based thinking process gives a concrete idea of collaborative debate goal, because restricting a process to a language describing it implies to have the intention *per se* of working collaboratively. In spite of the intrinsic definition restriction, we have established a discreet comparison among each one of the phases with the questions derived from the Grice Maxims. We have obtained the *Maxim of Quantity* comparing the amount of selected and written graph concepts with the average of the amount of the proposed concepts by each participant as well as the professor evaluation. Clearly, the idea was to have the same number of actions for each one of the members, indirectly, this allows us to visualize a well-established floor control. Nevertheless, the application does not handle the floor control and each student can have the number of participations that s/he wants. At least, the bias given by the new language is therefore restricted. *Maxim of Quality* has been implemented making logic comparisons among the contributions, that is to say, we compare if a concept is related to another one or not. The application allowed us to obtain the trajectory of participation given by each participant and to visualize if their opinions have been constant. *Maxim of Manner* was the most limited for its analysis, because the amount of actions allowed for interaction was restricted by the graphic language and the participants could hardly violate this principle. Finally, *Maxim of Relevance* has interested us because it allowed to visualize the importance of each action. In order to measure it, the help of the teacher was included. He evaluated if the participation of each one of the members are congruent with the problem solution. In that sense, the participation contents founded were: support contribution, opposite contribution, new main thread, changes to another main thread, incoherence and non participation.

The study of interactions allowed us to establish a new interface with more operations on the debate threads. The participants called icons the "debate language". This experience gave us an interesting guideline in order to evaluate the cognitive effects that the social knowledge has, from a communication media, as the debate icons. Finally, another issue to be included is the intention measurement and group supervision through personal assistant agent technology (Parra & Sheremetov, 2000). Nevertheless, it is necessary to establish more experiences in order to exactly define the different agent behaviours guided by intentions that the participants have.

Acknowledgements

We would like to thank to CONACyT-Mexico for partial support within the projects 31851-A and J32043A, and the Mexican Petroleum Institute within the project D.00006 (CDI). Special thanks to Manuel Chi and Miguel Contreras for their help in the programming of the CD-GUI software.

References

Bratman, M. (1987). Intentions, Plans, and Practical Reason. Cambridge. Harvard University.

Dimitriadis, Y., Asensio, J. Toquero, L. Estébanez, T. Martín, A., and Martínez, A. (2002). Toward a Component System for the Collaborative Learning Domain. In Proceeding of Symposium on Informatics and Telecommunications, Sevilla, Spain. (In Spanish).

Gallardo-Paúls, B. (1996). Conversational Analysis and Receptor Pragmatic. Valencia, Spain. University of Valencia. (In Spanish).

Georgeff, M., Pell, B., Pollack, M., Tambe, M., and Wooldridge, M. (1999). The Belief-Desire-Intention Model of Agency. In *Proceedings of Agents, Theories, Architectures and Languages Conference.* Paris, France.

Grice, H. (1975). Logic and Conversation. In Cole, P. and Organ, J. (Ed), *Syntax and Semantics III: Speech Acts* (pp. 41-58). Academic Press, New York: Academic Press.

Jonassen, D. & Land, S. (2002). Theoretical Foundations of Learning Environments. Mahwah, New Jersey: Lawrence Erlbaum Associates, Inc.

Kuhlthau, C. (1993). "Seeking meaning. A Process Approach to Library and Information Services". Greenwich, CT: Ablex.

Osuna C., Dimitriadis Y., and Martinez, A. (2001). Using a Theoretical Framework for the Evaluation of Sequentiability, Reusability and Complexity of Development in CSCL Applications. In *Proceedings of the European Computer Supported Collaborative Learning Conference*. Maastricht, the Netherlands.

Parra, B. & Sheremetov, L. (2000). SPAGAIA: A Simulation Model of Group Learning Process with Adaptive Intentional Agents. *In Proceedings of the Conference on Engineering and Computer Education.* São Paulo, Brazil.

Sheremetov L. and Romero-Salcedo, M. (2003). Telecommunication Technology Applied in the Virtual Corporate University Project at the Mexican Oil Institute, In Proceedings of the International Conference on Telecommunications. Papeete, Tahiti, French Polynesia.

Stone, M. (2002). Communicative Intentions and Conversational Processes in Human-Human and Human-Computer Dialog. To appear in J. Trueswell and M. Tanenhaus, (Ed), *World Situated Language Use: Psy-cholinguistic,Linguistic and Computational Perspectives on Bridging the Product and Action Tradition*s, MIT Press.

Experiments Using Combinations of Auditory Stimuli to Communicate E-mail Data

Dimitris Rigas and Dave Memery

School of Informatics, University of Bradford, Bradford BD7 1DP, United Kingdom
E-Mail: D.Rigas@bradford.ac.uk, D.Memery@mailgw1.leeds-lcot.ac.uk

Abstract
Experiments with a prototype version of a multimedia e-mail browser demonstrated that auditory stimuli could be used to communicate information in e-mail browsing. It was observed, however, that the graphical displays were still complex despite the fact some information was communicated aurally. These crowded displays could only be simplified further by hiding more visual information from the users. This is not often desirable as hidden information could be useful to users. The use of auditory stimuli has demonstrated that could help to communicate information that is not presented visually. In investigating this auditory approach, the paper describes four experiments in which four compositional sounds were used to communicate four e-mail categories. The compositional sounds used had a duration of 10 seconds (evaluated in the first experiment), 4 seconds (evaluated in the second experiment), 2 seconds (evaluated in the third experiment) and a mixture of 10, 4 and 2 seconds (evaluated in the fourth experiment). The experimental results indicated that short sequences of compositional sound could be easily identified and recognised by users. In most of the cases, a 2-second duration was sufficient for a user to identify an e-mail category. This finding is particularly useful as compositional sounds could be included in the audio-visual design of multimedia e-mail browsing. It is also particularly useful to auditory browsers for visually impaired users.

1. Introduction

When we browse e-mail data graphically, complex displays are often encountered. In some cases, the only solution to complex visual displays is to hide some information from the user and thus making the visual display more readable. There are a number of ways in which software engineers and designers could deal with this filtered or hidden information. For example, the filtered information could be presented in another visual instance but often at the cost of hiding some other graphical information and thus confusing the user. This will be particularly the case when the user needs to relate information seen in one visual instance with information seen in another. This is often the case as most of the graphical e-mail data connects implicitly or explicitly with other data. Another approach is to allow users to choose the parts of information that need to be filtered. In this case however a sufficient awareness of the content and importance of the information to be presented is needed. Alternatively, the filtered information could be presented in another medium (e.g., auditory). Thus no information will be hidden from the user. This paper describes experiments that investigated the possibility of using compositional sound to communicate e-mail categories and other relevant information for multimedia e-mail browsing.

2. Using Sound as a Metaphor

The use of audio as a communication metaphor includes the introduction of earcons. Earcons are short series of structured musical stimuli. There are one-element, compound, inherited and transformed earcons (Blattner et al. 1989). There are also guidelines for their design (Brewster 1994, Rigas 1996, and Rigas and Alty 1998) which aim to help multimedia designers to incorporate earcons in their systems. Sound, in the form of structured musical stimuli, was used to communicate graphical information to visually impaired users (Rigas 1996, Rigas and Alty 1997, and Alty and Rigas 1998). A word processor system was also developed for visually impaired users that used sound as a communication metaphor (Edwards 1989). There are also other examples of applications in which earcons were used to improve the usability of a graphics package for non-blind users (Brewster 1998). The contents of a software engineering database were also communicated using sound (Rigas 1993, and Rigas et al. 1997). The execution of an algorithm was also communicated successfully using sound (Rigas 1996).

The use of environmental sounds (auditory icons) is an alternative approach to the use of earcons. SonicFinder is a user interface application that used auditory icons. In this application, auditory icons communicated interface events in a way that each auditory icon implied the information communicated. Auditory icons were communicated in addition to the visual messages (Gaver 1986 and Gaver 1989). Every day sounds (e.g., tearing paper, hammering, walking) were also investigated (Vanderveer 1979). Users successfully matched sources of sounds and the sounds themselves particularly when the sounds derived from one source. There was however some confusion when similar sounds were produced from multiple sources. Bjork concluded that users interpreted the sources of sounds and not the pitch, timbre and other qualities (Bjork 1985). Sound sources have also being used to communicate environmental information such as physical events (e.g., a bottle breaks and smashes or bounces when dropped on the floor), events in space (e.g., an ambulance's siren approaching), dynamic changes (e.g., overflow can be detected when liquid is poured into a glass), abnormal structures (e.g., the sound of a car engine with a fault usually is different to the sound of the engine without the fault), and invisible changes (e.g., a hollow space in a wall can be identified by tapping blocks such as rhythm, timbre, register and dynamics (Mountford and Gaver 1990).

3. Multimedia E-mail Tool

A prototype multimedia system was developed to provide a platform to investigate the use of auditory communication metaphors in browsing e-mail data. The system was developed as an incremental working prototype. The prototype therefore provided the usual features and functions of an e-mail tool (including browsing techniques such as locating desired e-mails, and redirecting e-mails to WAP phones and pagers) but with the addition of audio-visual browsing. Two browsing techniques with auditory and visual metaphors were designed to investigate aspects of auditory information processing through two experimental stages. The first stage involved browsing with simple e-mail data (i.e., a few e-mails at a time) and the second stage involved browsing with complex e-mail data (i.e., many e-mails at a time). In these experiments, it was found that e-mail browsing could benefit from the use of tunes and short pitch sequences along with visual stimuli (Rigas et al. 2002).

This investigation is now extended to investigate additional types of auditory stimuli such as the use of compositional sounds in an attempt to reduce further complexity from visual displays often encounter in browsing graphically e-mail data.

3.1. Experiments using Variable Sequences of Composional Sound

Four experiments were performed with 60 users. In these experiments, four compositional sounds were used to communicate e-mail categories. Users were explained that sound A communicated e-mail category A, sound B communicated category B and so on to sound D that communicated e-mail category D. The four compositional sounds were the first 10 seconds of Vivaldi Four Seasons in G minor summer presto (referred to as sound A), Vivaldi The Four Seasons in E major spring allegro (referred to as sound B), Vivaldi The Four Seasons in C major sinfonia allegro molto (referred to as sound C) and Michael Nymann The Piano-The Embrace (referred to as sound D). Thus the hypothesis (two-tailed) was to examine whether users will recall four distinct e-mail categories using four distinct compositional sounds and whether the time duration of the sound affects the ability of users to recognise the e-mail categories. For this reason, different time durations of the sounds were chosen for the four experiments. The first experiment used 10 seconds of each compositional sound, the second experiment reduce the duration to 4 seconds, the third experiment reduced duration further to 2 seconds and the fourth experiment used a mixture of 10, 4 and 2 seconds. The method for all four experiments was the same. Users were presented with the four compositional sounds which communicated four e-mail categories. Each individual sound was repeated up to three times in order for users to establish some familiarity with the sounds. After this brief training session, users were presented these sounds in a random order. Each sound communicated a particular e-mail category. Each of the four compositional sounds was reviewed when users were requested to identify the matching e-mail category. The musical knowledge of all 60 users who participated in the experiments was assessed using a short questionnaire.

In the first experiment, 60 users were presented with 10 seconds of each compositional sound and they were requested to identify the e-mail category (A, B, C, or D) that the sound represented. Sound A and B were communicated once and sounds C and D were communicated twice for each user. All users successfully recognised e-mail category A and D as they were communicated with sound A and D and e-mail categories B and C were recognised by 59 users (98.3%). These results were highly significant[1]. The successful recognition of the 10 seconds sequences enable us to investigate whether a shorter sequence would also produce equally good results. In the second experiment, the same 60 users were presented with 4 seconds of each compositional sound and they were again requested to identify the e-mail category that the sound represented. E-mail category A was recognised by 58 users (96.6%), categories B and C were recognised by 59 users (98.3%), and category D was recognised by all users. These results were also highly significant[2].

These results encouraged us to reduce the duration of the compositional sound even further to 2 seconds. Therefore, the same 60 users were presented with 2 seconds of each compositional sound and they were again requested to identify the e-mail category that the sound represented. E-mail categories A, B and C were recognised by all users and category D was recognised by 59 users (98.3%). These results were also highly significant[3]. In the fourth experiment, the same 60 users were presented with either 10, 4 or 2 seconds of each compositional sound and they were again requested to identify the e-mail category that the sound represented. E-mail categories A, C and D

---

[1] χ^2=180 for e-mail category A, χ^2=172 for category B, χ^2 =352 for category C, χ^2=360 for category D. All χ^2 were calculated on a two-tailed hypothesis with dof=3 at $p<0.001$.
[2] χ^2 =329 for e-mail category A, χ^2=172 for category B, χ^2=172 for category C, χ^2 =360 for category D.
[3] χ^2 =360 for e-mail category A, χ^2=360 category B, χ^2=180 for category C, χ^2=172 for category D.

were recognised by all users and category B was recognised by 59 users (98.3%). Statistical results were similar to the previous experiments.

Further experiments were also performed to investigate the communication of other related e-mail data in addition to the relevant e-mail category. Initial results of those experiments demonstrated that additional information of an e-mail can be successfully communicated. For example, the information communicated included the priority status of the e-mail, whether the e-mail was read, unread or forwarded or the subject of the e-mail communicated using speech.

4. Discussion and Conclusions

Experimental results demonstrate that users could understand the auditory stimuli after a short training period. Even when the compositional sounds were only two seconds long, users could easily recognise the e-mail category that was communicated. When using compositional sounds, it is very important that users are trained with all relevant sounds prior to their interaction with the system. According to our experimental observations, training sessions often take 5 to 6 minutes for a small set of sounds (e.g., 4). The training session could also be included in the help section of the system. When choosing short sequences of compositional sounds, designers must take care to select those parts that pre-design trials show that they could be remembered easily. Therefore designers must first choose a selection of sounds and fully evaluate them with a small sample of users prior to their use in the system.

Furthermore, it is often the case that other type of auditory stimuli (e.g., speech, auditory icons) along with visual stimuli will be used. In this case, the combined perceptual effect must be investigated. Usually, there are issues of synchronisation of the information communicated and the perceptual context of the user. The information communicated visually could often help to provide a perceptual context to the user. The user would be expected to bridge the perceptual gaps of the information received visually with the information received aurally. Post-experimental interviews with the users indicated that once the users became aware of the four compositional sounds, the recognition was relatively easy. This is particularly important as users often quote problems with remembering the information communicated aurally.

The results of this work indicate that the use of short sequences of compositional sound could help users of e-mail tools to identify categories of e-mails. This is particularly useful for browsing complex displays of e-mail data. In fact, the experimental results suggest that a few seconds (typically 2 to 4) are sufficient for users to identify the category of an email. Most users were able to identify the e-mail category from the first 2 seconds. The next stage in our experiments is to combine short sequences of compositional sound, earcons (short sequences of structured musical stimuli), auditory icons (environmental auditory stimuli) and speech (synthesised and pre-recorded) in one integrated auditory message to complement the use of an audiovisual e-mail browser. This will involve the evaluation of each auditory stimulus and subsequently the addition of the visual stimuli. Experiments toward this direction are currently performed.

7. Authors

Dr Dimitrios Rigas is a Senior Lecturer in Computer Science at the University of Bradford, England, United Kingdom. Dr Rigas is also an Honorary Visiting Research Fellow at Loughborough University, United Kingdom. Mr David Memery is a research student at the Department of Computing at the University of Bradford.

References

Alty, J. L. and Rigas, D. I. (1998). Communicating graphical information to blind users using music: The role of context. In CHI-98, Human Factors in Computing Systems, pp. 574-581, Los Angeles, USA. ACM Press.

Bjork, E. A. (1985). The perceived quality of natural sounds. Acoustica, 57(3):185-188.

Blattner, M., Sumikawa, D. A., and Greenberg, R.M. (1989). Earcons and icons: Their structure and common design principles. Human-Computer Interaction, 4:11-44. Lawrence Erlbaum Associates, Inc.

Brewster, S. A. (1994). Providing a structured method for integrating non-speech audio into human-computer interfaces. PhD thesis, University of York, England, UK.

Brewster, S. A. (1998). Using earcons to improve the usability of a graphics package. In H. Johnson, L. Nigay, and C. Roast, editors, Proceedings of HCI'98: People and computers XIII, pp. 287-302, Sheffield, UK. Springer and British Computer Society.

Edwards, A. D. N. (1989). Soundtrack: An auditory interface for blind users. Human Computer Interaction, 4(1):45-66.

Gaver, W. (1986). Auditory Icons: Using sound in computer interfaces. Human-Computer Interaction, 2(2):167-177.

Gaver, W. (1989). The SonicFinder: An interface that uses auditory icons. Human-Computer Interaction, 4(1):67-94. Lawrence Erlbaum Associates, Inc.

Mountford, S.J., and Gaver, W. (1990). Talking and listening to computers. The art of human-computer interface design, pp. 319-314.

Rigas, D. I. (1993). A graphical browsing tool for the PCTE OMS. Master's thesis, University of Wales, Aberystwyth. MPhil.

Rigas, D. I. (1996). Guidelines for Auditory Interface Design: An Empirical Investigation. PhD thesis, Loughborough University, Leicestershire, UK.

Rigas, D. I. and Alty, J. L. (1997). The use of music in a graphical interface for the visually impaired. In INTERACT-97, International conference on Human-Computer Interaction, pp. 228-235, Sydney, Australia. Chapman and Hall.

Rigas, D. I. and Alty, J. L. and Long, F. W. (1997). Can music support interfaces to complex databases? In EUROMICRO-97, New Frontiers of Information Technology, pp 78-84, Budapest, Hungary. IEEE, Computer Society.

Rigas, D. I. and Alty, J. L. (1998). How can multimedia designers utilise timbre? In H. Johnson, L. Nigay, and C. Roast, editors, Proceedings of HCI'98: People and computers XIII, pp. 273-286, Sheffield, UK. Springer and British Computer Society.

Rigas, D. I. and Alty, J. L. (1998). Using sound to communicate program execution. In Proceedings of the 24th EUROMICRO Conference, volume 2, pp. 625-632, Vasteras, Sweden. IEEE, Computer Society.

Rigas, D., Memery, D., and Klearhou, K. (2002). Multimedia E-Mail Tool: An Empirical Case of Using Auditory Metaphors to Communicate and Browse Data. In Proceedings of the IASTED International Conference on Applied Informatics, pp. 199-203, ACTA Press.

Vanderveer, N. J. (1979). Ecological acoustics: Human perception of environmental sounds. PhD thesis, Dissertation Abstracts International, 40/09B, 4543.

Context Sensitive Interactive Systems Design: A Framework for Representation of contexts

Keiichi Sato

Illinois Institute of Technology
350 N. LaSalle Street Chicago, Illinois 60610 USA
sato@id.iit.edu

Abstract

The performance of interactive systems is determined in relation to the context in which the system and users interact for task execution. It is obvious that the significance and complexity of contextual information require systemic approaches for managing it in the design process. As the system complexity and diversity of types and ranges of use context increase, context sensitivity becomes a critical mechanism to enhance system performance and interactive qualities. This research attempts to introduce a representation framework for contextual information critical for developing a methodological foundation for user-centered design practice. First, this paper explains what compose contexts, how contexts are structured, and how they can be described. Then, strategies and methods to incorporate context sensitivity are discussed from the perspective of system development in relation to types of system architecture that can effectively implement context-sensitivity in design are discussed.

1 Introduction

The term "context" has been loosely defined and used in interactive systems design practice to represent various factors and conditions surrounding and influencing the use of the system. The performance of interactive systems is determined in relation to the context in which the system performs its intended roles. The system that performs well in one context may not necessarily perform well in other contexts. While the context dynamically changes, systems are usually designed to remain the same and to be operated within a very limited range of the context. For example, the user interface of an existing mobile information system cannot respond to a change from business use in the office to personal use in the automobile to achieve effectiveness of interaction and qualities of user experience. In order to maximize the system performance, therefore, the system needs to be sensitive to the change and range of the context.

Some contexts are composed of factors that reside in users such as cultural, social, chronological and cognitive factors; some are composed of external factors that reside in operational environments such as organizational, spatial and social factors; some contexts might be composed of combinations of factors that are distributed across users, systems and operational environments. These different types of contexts interact with each other and compose complex, underlying layers of operational conditions for the performance of interactive systems which are often referred as situations. This complex nature of contexts, points out the need for multidisciplinary viewpoints in developing frameworks for understanding contexts and for developing coherent mechanisms to incorporate those frameworks in interactive systems development.

This paper intends to discuss the following three areas of research: 1) The conceptual framework of the context in interactive systems design, 2) Design methodology for systematically incorporating the concept of contexts into interactive systems development, 3) Development strategies for context-sensitive interactive systems. It attempts to reveal the common foundation of Context Sensitive Design across different application domains and critical issues to be addressed in the next stages of research development.

The goal is to introduce a general foundation for context representation in interactive systems design methodology. This research effort particularly responds to the emerging needs for integrating physical and media systems with fast-growing network and embedded technologies that require management of complex contextual issues across many related domains. Only by addressing these issues and needs, users can be provided with living and work environments that most effectively integrate hardware, software, and communication technologies.

2 Contexts in Multiple Perspectives of Design

Much diversity can be found among definitions of contexts from different interests such as context aware computing, usability analysis, urban planning, and its academic origin in linguistics. In context aware computing, some descriptors of the domain such as location, identity, and time, describe conditions and environments of the system operation and are considered to be parameters of context (Dey et. al. 2001, Selker & Burleson 2000). In urban planning, community history is a part of context. In the area of office space planning, contexts of work include social, cultural and organizational aspects. Looking at the office workers' tasks level, information flow, project history, and daily activity patterns become important aspects of the context.

This indicates the characteristics of the concept of context: 1) Multiple aspects of context manifest based on the emergent relevance to the nature of actions and conditions, 2) The granularity of description varies depends on the focus of the viewpoints, 3) Contextual changes are evoked by triggers from different constituents of the domain, 4) Context evolves over time but some aspects change fast and others change slow. In order to understand the concept of context, it is useful to also define related concepts such as conditions, states, environments and situation.

Conditions are defined as individual variables in the domain of concern where the interaction is situated. They include environmental states, system states, and user states that include variables such as location, temperature, sound, users' emotion, and attention level.

Context is a pattern of behaviour or relations among variables that are outside of the subjects of design manipulation and potentially affect user behaviour and system performance. Example aspects of context for driving are chronological development of the user's activities for the day, a destination and purposes of driving, and a plan for intermediate activities before arriving at the destination. These are aspect models of contexts that all influence driving behavior. Simple terms such as "driving freeway" and "teenager" are also considered as contexts. These are examples of *ostensive* or *indexical use* of words implying particular contexts represented by typical patterns of conditions or characteristics associated with them instead of pointing to their immediate meanings. Some aspects of context take significant roles in forming situations for the current action; some aspects become irrelevant to the current action. We call the former *manifesting aspects of context* and the later *latent aspects of context*. Figure 1 shows how manifesting aspects of context and

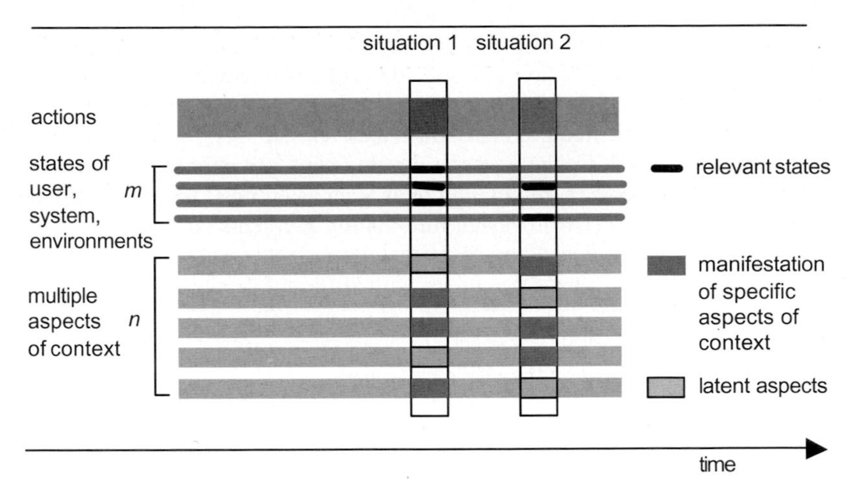

Figure 1: Formation of situations with actions, states, and contexts

conditions form the situation of the action. As a result of the action, some of the aspects of context change and evolve over time.

Situation is a collective condition at the scene of the interaction that is composed of relations among variables of conditions such as environmental states, contexts, systems and users states. Therefore the representation of situations needs to include description of relations as well as the set of descriptors as listed above. In some cases, actions in the situation consequentially become a part of the situation. In Figure 1, *m* state variables and *n* aspects of context are indicated, and a small number of states and aspects of context manifest relevance by having direct or collective effects to the action.

3 Representation of Contexts in Design

The key to the development of an effective methodology is to construct representation methods that enable reliable description, analysis, manipulation, and evaluation of contexts. Different representation methods will be introduced ranging from soft forms of description such as scenarios and story boards to more rigorous and formal description such as aspect-models for computational support of research and design. Different types of contexts also require different representation methods. The temporal dimension is also an important factor in the representation of the context, since some of contexts are constructed over time and interactions between contexts and contextual change take place over time. In order to capture contexts systematic approaches to record, document, and filter, field data is required at the scene of the field study.

Some of principles introduced in formal representation for contextual reasoning in AI are also useful for developing a framework of context representation for design. Examples include principle of locality (reasoning always happens in a context), principle of compatibility (there can be relationships between reasoning processes in different contexts), resolution of representing contexts, partiality of contextual information (stating no complete information, and no complete representation) (Bouquet, P. & Srafini, L. 2001).

The Design Information Framework (DIF) developed to provide structured guidelines and mechanisms to set formats for data inquiry and for the documentation, organization and interpretation of the collected data. DIF is a unified design information representation platform for bridging different viewpoints, activities and description methods involved in the system development process (Sato, K. 1991, Lim & Sato, 2001). It can be applied to represent contextual information by satisfying the requirements describes earlier. As shown in Figure 2, various concepts and variables used in the system development are structured with two levels of information elements. The basic representation units such as entities, acts, and attributes are called Design Information Primitives (DIP). Design Information Elements (DIL) represents higher levels of concepts such as functions, goals, and plans that can be represented by combination of DIP's. Once data is encoded by DIF, aspect models with different combinations of variables can be generated representing particular viewpoints for analysis, problem solving, evaluation communication (Lim & Sato, 2003). Aspects of context therefore can be represented as aspect models in appropriate forms built on the DIF mechanism.

4 Strategies for Incorporating Contexts in Design

In order to accommodate context sensitivity in interactive systems design, many different approaches at different aspects of the system must be considered. In the area of context aware computing, three models have been introduced and implemented for information processing level architecture; widget model, infrastructure-centered distributed or networked service model, and blackboard model (Winograd, 2002).

Implementation of comprehensive context sensitivity requires strategies for the three parts of overall systems as follows.

Sensing contextual changes: When relevant contexts are known and patterns of contextual changes are predicted, a sensing mechanism can be designed and embedded in the system to detect and capture indexes of changes, so that the system effectively changes its operational modes and characteristics to optimize its interfaces to users. This requires interpretive filters to detect signs of

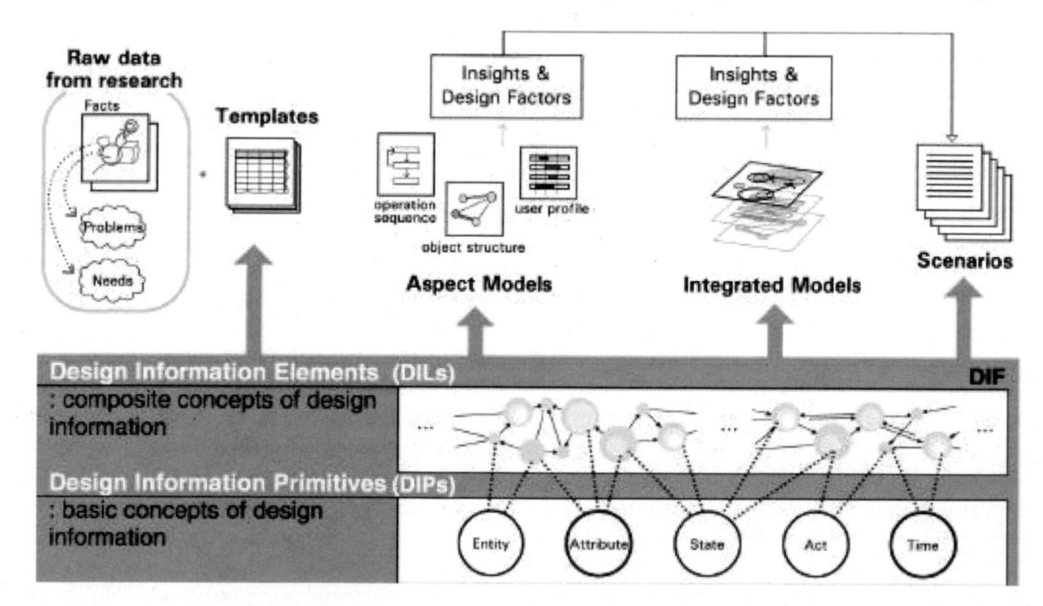

Figure 2: Structure of the Design Information Framework

contextual changes that need to be used as triggers for appropriate transformations of the system or interface configuration.

Re-configurable architecture: Use process of interactive systems is a learning process for developing more skills and knowledge about them. There are two approaches to this issue. One is to make a system re-configurable by its self adaptation mechanisms. The other is to assist users to re-configure the system by themselves.

Creating and managing contexts: In order to support particular user activities, the appropriate setting of contexts is necessary for the enhancement of user interactions with artifacts. Setting courses of cognitive activities leading to the particular actions, and setting a pattern of information distribution over the spatial environment for better user performance are examples of such effects.

5 Conclusion and perspectives

This paper demonstrated that the concept of context, which has been considered as soft and peripheral information in design, can and must become a critical resource for user-centered design practice. Although a general approach to the development of representational mechanisms for contextual information was explained, further formal and empirical research needs to be developed. Particularly use of context sensitivity in many different application domains opens a huge new space for research and system development including topics in architectural issues of physical and media spaces, interaction, and re-configurable interfaces as pointed out earlier.

6 Acknowledgement

A part of this research has been made possible by the grant from the Tangible Knowledge Research Consortium at the Institute of Design, IIT, sponsored by Motorola, Steelcase, SBI and Zebra technologies.

7 References

Bouquet, P. & Srafini, L. (2001). Comparing Formal Theories of Context in AI, *Human-Computer Interaction*, 16.

Dey, A. K., D., Abowd, G. D., & Salber, D. (2001). A conceptual framework and a toolkit for supporting the rapid prototyping of context-aware applications. *Human-Computer Interaction*, 16.

Lim, Y. & Sato, K. (2001). Development of Design Information Framework for Interactive Systems Design. *Proceedings of the 5th Asian International Symposium on Design Research*, Seoul.

Sato, K.(1991). Temporal aspects of user- interface design. *Proceedings of FRIEND 21 International Symposium on Next Generation Human Interface*, 5D. Tokyo, Japan

Selker, T. & Burleson, W.(2000). Context-aware design and interaction in computer systems. *IBM Systems Journal*, 39 (3&4)

Winograd, T. (2001). Architectures for context. *Human-Computer Interaction*, 16

Filter Effects of Mediating Technologies

Gunnvald B. Svendsen & Bente Evjemo

Telenor Research & Development
P. O. Box 1175, 9262 Tromsø, Norway
<gunnvald-bendix.svendsen><bente.evjemo>@telenor.com

Abstract

Mediation technologies like phones and IM can be viewed as filters on communication. The present study investigates which conversations are filtered and which are not by recording conversations in an office environment and later asking the workers to indicate which utterances would have been initiated over phone or e-mail if they were not co-located. The results indicate that 76% of all conversations would be filtered. It is argued that the filtering would be perceived as positive due to less disruptions in a telework setting, but that the long term effects would be negative. The filtering is explained by three mechanisms involving behavioral cost, memory and social cost respectively.

1 Introduction

Telework or distributed work has become increasingly widespread (Akselsen, 2001), moreover, recent research suggests that it increases both productivity and the workers' quality of life (Hopkinson, James & Maruyama, 2002a; Akselsen op cit). The main reason given for increased productivity is less disruption (Hopkinson, James & Maruyama, 2002b). This seems to be at odds with a long research tradition that points to the importance of informal office chat and informal meetings to productivity, knowledge distribution, innovation and the social well being of office workers (Wynn, 79; Suchman & Wynn, 84; Kraut, Fish, Root & Chalfonte, 90; Fish, Kraut, Root, & Rice, 93; Whittaker, Frohlich & Daly-Jones, 94; Isaacs, Whittaker, Frohlich & O'Conaill, 97).

Mediating technologies can be viewed as filters of communication. The cited telework studies suggest that filtering is positive (less disturbance), while the "office chat" studies argue that it is negative in the long run (less chat leads to less collaboration and innovation). The present study addresses this issue by investigating which conversations are filtered and which pass through mediating technologies.

2 Method

The study was done as naturalistic observations in an office environment divided into cubicles. Each cubicle was shielded with lightweight, movable walls about 1.5 m high. The work consisted of error recovery and planning of expansions of a telecom network. The workers belonged to different departments, but knew each other well and were experienced in their work. Five to eight workers occupied the office. All of them could overhear speech in a normal voice. The observations were done in intermittent sessions of two hours within a period of two weeks.

Altogether four sessions were conducted. The first phrase of all conversations was literally registered together with context information and time of occurrence. A total number of 79 utterances were registered; of these 11 were uttered by visitors and taken out of the analysis, leaving the total at 68.

After each session the workers were asked to identify which of the recorded conversations would not have taken place if the workers had not been co-located. Specifically, they were asked to state which utterances signified a conversation that would have resulted in a phone call, or other type of contact (non-filtered utterances), and which would not (filtered utterances). Phone and e-mail were explicitly mentioned as communication technologies that could have been used. Lastly, the utterances were analyzed and classified with regard to what effect, if any, they had on the work.

3 Results

According to the workers 52 of the 68 utterances (76%) would have been filtered in a telework setting. In a telework setting the 16 non-filtered utterances would have been directed to only one recipient. Assuming the utterances would have been distributed uniformly among the 6 workers, the number of conversations per worker would be about 2.5. This represents a 30-fold reduction of overheard conversations in a telework setting compared to a co-located one.

The 68 utterances fell into four main categories with respect to effect on work:

- Problem solving. This category includes utterances that give or ask for assistance related to specific tasks. Example: "What's the public code for this city?", "Does somebody know where the caps lock key is?"
- Information distribution. This category includes utterances related to personnel location and the status of key applications and office facilities. Examples: "The head quarter is jammed, I think.", "Mr. D has phoned you several times."
- Work coordination. This category encompasses task delegation and task prioritization. Examples: "Where is Peter – can somebody take care of this ...", "Did you manage to solve his problem?"
- No immediate effect. This category subsumes utterances that have no discernible effect on work. Example: "Look, it's raining."

A post hoc analysis indicates that different conversation-categories are filtered differently (chi square = 16.99, df = 3, p <=0.001). As shown in table 1, work coordination is filtered least with 44%, problem solving next with 59%, information giving is filtered 71%, while all conversations with discernible effect would be filtered. Of the 40 utterances with no immediate effect on work, 24 (60%) would have been filtered.

Table 1: Utterances by category and filtering

	problem solving	information giving	work coordination	no immediate effect	total
non-filtered	7	4	5	0	16
filtered	10	10	4	28	52
total	17	14	9	28	68

4 Discussion

The discussion will be focused around two themes, the effects of filtering and the reason why filtering occurs.

4.1 Effects of filtering

The results indicate that a telework setting would lead to a dramatic reduction of overheard conversations, and consequently a considerable reduction of disruptions. Thus they are in agreement with early studies of communication frequency as a function of distance (Kraut et al. 90) and with studies citing less disruption as a positive factor in making telework an effective way of work (Hopkinson et al. 2002b). However, a large amount (60%) of information *relevant to job performance* would also be filtered in a telework setting. It is prudent to ask what effect this would have on work performance.

Competent workers behave rationally. They know what is important to convey to others and they know when they need help. Thus the conversations that are filtered are probably the conversations the workers have deemed "unimportant" for the work at hand. This is indicated by the fact that all conversations without discernible effect on work would have been filtered, and a further analysis of the results supports this. For instance, in the "problem-solving" category the filtered questions could easily be replaced by other sources of information. This would require some extra effort by the problem solver, but this extra work would be measured in minutes and seconds. Consequently it would never show up as a significant factor in an analysis of work efficiency. The same holds, to even a larger degree, for the two other categories, "information distribution" and "work coordination". Thus it is hard to see an immediate negative impact on work performance as a result of the reduction in number of conversations.

The long-term effects of the reduced communication are harder to assess, but probably negative. This conclusion is based on an analysis of the "implied information" the conversations supplies. "Implied information" means information that can be assumed to be true given the utterance and the listeners knowledge of the context. The following filtered utterances exemplifies this:

- A question, overheard by everybody: "Tom, - Black Village?" Without hesitating Tom replies a two letters code. There is no introduction, no further explanation or closing comments. Apart from solving a problem for the questioner, this makes everybody in the office aware of the questioner's current tasks and field of work and Toms expertise in another.
- Bob tells Ann "I've just received the task you asked for". Earlier Ann, being part of another unit, had predicted that the task ought to come up. Bob's utterance shows that he understands that Ann was concerned, that he can be relied upon to offer this information and confirms their understanding of the workflow.
- Without any plausible occasion, Don says: "I really need to know more about the new services we are offering". Apart from saying just what it says, this utterance also makes the others aware that Don's work and the new services are related in some way, that Don doesn't feel quite up to it and probably shouldn't be given more new tasks at the moment.
- The utterance: " The head-quarter is jammed, I think", leads to an exchange on work routines and workflow strategies. This adds to their mutual understanding of how the company is functioning.
- Tim utters: "Nora, I really need a break now". The obvious message is that Nora must take the phone and that Tim is in the lounge if someone needs him. It also tells the others about

Tim's habits, when he needs a break, who he hands things over to and that he doesn't just sneak out.

As these examples show, the utterances provide implied information on a whole range of subjects; on workflow, on field of knowledge, on habits, on level of competence, on current tasks and so on. In other words, these conversations are both a means with which co-workers learn to know each other and a vehicle for expanding each other's knowledge of how the company works.

When work follows a pre-planned course, this type of knowledge is more or less redundant. It shows its importance as soon as work deviates from course. To do problem solving and diagnostics the worker needs a good model of the work possess, knowledge of who could be relied upon to have useful information, information on which resources are available where, and so on. Just the type of information the filtered utterances contain. The results don't shed light on this, but it seems a safe bet that implied information is the richest source of this type of information. Thus it is reasonable to assume that in a telework setting, office workers will omit to convey information that is important for the long-term conduct of work.

4.2 Why is communication filtered?

In the late eighties and early nineties much interest was devoted to video as a means of informal workplace communication (Kraut et al. 90). One argument for video was that the sight of others functions as a trigger for communication (op. cit). When the effect of video on communication frequency later was found to be less than predicted, it was argued that the behavioral cost associated with use of communication technology reduced its potential as a mediator of informal communication (Fish et al. 93) Echoing these ideas we propose that mediating technology filters communication by three different but intertwined mechanisms.

1. Behavioral cost. This hypothesis states that when the behavioral costs involved in communicating overshadows the perceived gains, the conversation will not be started. Thus, conversations about themes that are deemed as unimportant will occur only when the cost associated with talking is minimal. Under normal circumstances talking face-to-face has the least associated cost, while the cost associated with a phone call involves both finding and dialling the number, waiting for an answer, introduction of one self, etc. The question "Tom – Black valley" is filtered primarily by this mechanism.
2. Memory cue. This hypothesis states that initiating a conversation presupposes a reminder. In some circumstances this reminder is the physical sight of the other person, thus physical proximity has other filter characteristics than for instance e-mail or phone. The utterance: "Did you manage to solve his problem" is primarily filtered by this mechanism.
3. Social balance. This hypothesis states that there ought to be a balance between the importance of the theme introduced in a conversation and the task that is interrupted by the conversation. In a face-to-face setting the initiator of a conversation is able to gauge the listener's tasks. If the prospective conversational partners are physically separated, the initiator has to guess at the listener's task. Thus to place a phone call requires a theme of higher importance than an e-mail. The utterance: "I really need to know more about the new services we are offering", is probably filtered by this mechanism. In the co-located situation it is said into the air, and no one feels pressed to comment. As a start of a phone conversation or as the message of an e-mail, the utterance would have taken on a much higher significance.

The hypothesised mechanisms indicate that the speaker (consciously or unconsciously) relates the assumed cost and benefit of mediated communication to the assumed cost and benefit of non-communication. It is highly probable that it is the overt message that will be evaluated, not the implied information the message conveys. Thus, as long as implied information accompanies unimportant overt messages, it will be filtered whether it is important or not.

5 Summary and conclusion

The results indicate that mediating technology filters communication between co-workers to a large extent. This is positive in the sense that distributed work is a lot less prone to interruptions and disturbances. We have argued that the reduction in communication also is negative since the filtered information is important for the workers' ability to cope with unforeseen work circumstances. In order to explain why some, potential, important information is filtered, hypotheses concerning behavioral cost, memory and social cost have been put forth. Together with the implied nature of the filtered information, they can explain why filtering occur.

Clearly what we want is mediating technologies that filter disruption but let important implied information through. Such technology must both be able to convey information on colleagues, their work, field of expertise and habits, and at the same time do this in a manner that doesn't require or demand conscious attention. Specification of such mediating technologies requires a thorough understanding of both filter mechanisms and the role of implied information in the workplace.

References

Akselsen, S. (2001). Telework and life quality - Basic concepts and main results. EURESCOM Project Report, EDIN 0084-0904.

Fish, R. S., Kraut, R. E., Root, R. W. & Rice, R. E. (1993). Video as a Technology for Informal Communication. *Communications of the ACM*, 36(1), 48-61.

Hopkinson, P., James, P., & Maruyama, T. (2002a). Sustainable teleworking. Identifying and Evaluating the Economic, Environmental and Social Impacts of Teleworking. SUSTEL, IST–2001-33228

Hopkinson, P., James, P. & Maruyama, T. (2002b). Teleworking at BT. Report on survey results. SUSTEL, IST–2001-33228.

Isaacs, E. A., Whittaker, S., Frohlich, D., & O'Conaill, B. (1997). Informal communication reexamined: New functions for video in supporting opportunistic encounters. In: *Video-Mediated Communication*. K.E. Finn, A.J Sellen, & S.B Wilbur (eds); (pp 459-485) New Jersey: LEA.

Kraut, R.E., Fish, R. Root, R. Chalfonte, B. (1990). Informal communication in organizations: Form, function and technology. In: *Peoples reaction to Technology.* S. Oskamp & S. Spacapan (eds.) (pp. 145-199) Newbury Park: Sage.

Suchman, L., & Wynn, E. (1984). Procedures and problems in the office. *Office: Technology and People,* 2, 133-154.

Whittaker, S., Frohlich, D. & Daly-Jones, O. (1994). Informal workplace communication: What is it like and how might we support it? *CHI94.* (pp. 131-137). Boston: ACM.

Wynn, E. (1979). *Office chat as an information medium.* Unpublished PhD

Technical and Social Standards to support Appropriate Use of Digital Everyday Appliances

Hiroshi TAMURA

Hiroshi International University
Kurose, Hiroshima, Japan
tamura@mobilergo.com

Abstract

Digital technology introduced in work places at the last 20 years of 20th century, are now coming more and more familiar to everyday life at home and in public places. Digital appliances, like cellular, are carried in the pocket next to skin and stay by the bedside. Certain social and technical measures are in urgent need to facilitate good use of them, while preventing over use, forced use and ill use. This paper proposes standards to support appropriate use of digital everyday appliances at three levels, i.e. client, provider and technology.

1 Everyday Appliances

Automobile is an everyday product in many places. Drivers are not necessary to worry very much about engine and technical maintenance, or steering muddy road. Skills necessary for the city drivers are to keep proper speed depending on the cars ahead and road signs, to send proper turn signals to other traffic, to manoeuvre traffic jam, and to park in narrow space, etc. Skills are completely different from those to drive along lonely country roads, but are to drive in the socially acceptable manners. The socially acceptable manners might be matters of personal consciousness, but people need technical support to act as they think and to help acting in the social manner. Car turn signals are necessary equipment for the drivers oneself, ones behind and in the opposite lane, as well as street passengers. The turn signal might not always be shown correctly, still its public use is valuable.

2 Appropriate Use

Home, office, shop, hospital and other public spaces as well are full of digital everyday appliances furnished with automatic, information, and communication technology. The appliances are used for not only the functions intended by product manufactures, but also used to get services provided from the service operators. The value of the products is more or less determined by the service available from operators, but not the functions intended in original products (Tamura, 1995).

A user might be using more than two appliances at the same time. More than two users are using different appliances co-operatively or inconsistently. Some people are disturbed by the noise, ventilation, waste and traffic collisions. Certain social and technical measures to avoid overuse, forced use, ill use, and net-holic are necessary. An optimistic view will be to prevent inappropriate use of digital technology solely by education and good will of users. It is not like to happen in reality. Drivers run the car beyond speed limit not because of bad personality. They violate no parking sign, not because they are not well educated. Everybody is tempted to do, if others are doing such.

The appropriate use does not mean correct use in reference to some law, government rules, and intentions of manufacturers. It is appropriate as an understanding of community, thus appropriateness many vary by place, culture and community. Thus appropriateness measures have to be informed to users, or to appliances via digital network. With increase of digital appliances, rules and regulations for the appropriate use might increase tremendously, processing of regulations and rules to keep appropriate use become complex. Thus technical support becomes of essential importance to keep appropriateness while maintaining effectiveness and efficiency. Also, social or community rules will change, looking after how a certain application might accepted in public, appropriateness processors should be installed in advance to social rules, so that community might conveniently use the processor, when they think it necessary. Thus technology to install appropriateness processors, and to establish rules and regulations by making use of the technology, become key to wide social acceptance of digital everyday appliances.

3 Classification of Users

The people around the everyday appliances might be classified, active, passive and indirect users. Active users take initiative in using the appliance, while passive users receive direct effect of using the appliance. In case of fax machine, an active user is the one who sends the fax, and the passive user is the one who receives the fax. The active user has to be familiar with the operation of fax machine. Major target of usability studies in the past are mostly concerned with the active users. The passive users normally need not care about the machine operations. Occasionally, however if the fax paper is in need of supply, the passive user has to work out to set the fax paper even in the midnight. It is recommended not to force passive user to learn the operations unless he/she is willing to do.

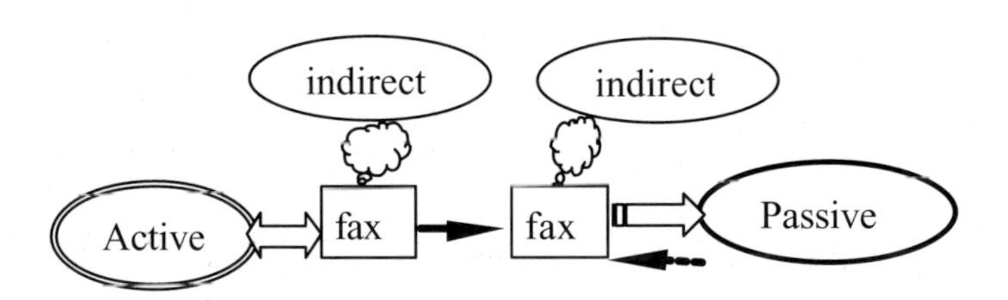

Fig 1: People around everyday appliances

The indirect user is those who do not get direct merit or demerit by operation of the system, but those who are influenced by sub-effect of using appliances, such as operation noise, alarm signal, ventilation outlet, smell and disposals. Each of such sub-effect might be decreased to some extent by technology, but increase in number of appliances and operation traffic might not allow the total sub-effect negligible. Thus measures to reduce total sub-effect have to be established.

For example, those who are not users of any technological systems, might be most nervous about the use of mobile phone. Those who are the owner of mobile phone, but not using it might also feel inconvenient when disturbed by calling sound of other users in a quiet space. A user of an appliance is a non-user of other appliances. Users of different appliances are influencing each other. When a user is using more than two appliances, one appliance might be influenced by others.

4 Matching the Individual Preference

People in everyday life are not in simple pursuit of goal and efficiency. They are in search of quality of lives, individually and sharing with others. Each has one's own preference, feeling and emotion, which change in time, place and occasion. Classical telephone had only one call signal. The tone is carefully selected, so that it may alert people effectively at low cost apparatus. When two phones are installed in a room, people are often confused to identify which bell is ringing. Nowadays in the mobile phones, user can assign different tones and melodies to identify the caller group and different functions of the receiver, such as phone call, mail arrival scheduler alert signals.

In the near future very many digital appliances are going to be in operation simultaneously, day an night, it is recommended for users to refrain from ill-use and over use, as well as to minimize indirect effects to non-users. Many social or community regulations will be proposed. However, social rules are often based on human understanding, which is varying by individual and situations.

Table 1: Appropriate use at Client level (Level 0._)
Self control by user to prevent ill-effects and over uses of appliances.

L0.0	Unified alerting signal, everywhere in the country. Easy to recognize an incidence. Difficult to differentiate events, locations, and urgency.
L0.1	Manual enhancement of call signals. Larger for elderly, lower for sleeping babies.
L0.2	Call signal suppression by user, introduction of vibration mode. Fully dependent on users conscious cooperation
L0.3	Choice of sound and volume, and manner mode selection.
L0.4	Setting the time patterns of alert signal, alarm signal pattern
L0.5	Setting sound quality, melody for call signal Extended user preferences, and choice options
L0.6	Assigning various sound and melodies to different callers and group
L0.7	Assigning various sound and melodies to different functions, such as call signal, alarm, mail, scheduler, etc.
L0.8	Combined use of optical, sound and vibration signals.
L0.9	Automatic release of manually suppressed sound signal at real emergency

5 Support for Appropriate Choice

Newspaper and broadcasting are the information services typically representing 20th century. They provided unified information at low cost to massive people. Information services for 21st century might provide varying information meeting to different users and needs. It might be easy to provide all the information at hand to every client equally, but this will enforce clients to look through all the data. Client may be in need of specific information applicable to specified time, place and preference. Information provider has to install such options to limit the information fitting to the user's need, referring to the history of use. The situation might occur when a client is driving car and in need of town information.

Information communication technology enable people to use various expressions, picture, image, colour and different fonts, etc. Also various media, like cellular, text mail, and internet are available(L1.1). People are in need of quick and short, detailed and exact, scientific and specialist report depending on the situations(L1.2). Information provider should care about the minimization of user load, such as watch and compare numbers, continuous monitoring of un-expected event, searching for a event through screen (L1.3).

Navigation log, retry and temporary records are also helpful for users (L1.4) for further use. Information provider should have certain methods to estimate the purpose (L1.5) of access of users and try to provide appropriate options to the menu. Information provider should meet the language, culture and history of using the systems by the user (L1.6). The history log will be used as user records of un/ favourable performances. The information should be screened or modified the expressions depending on the situations, like office, school, home, hotel, etc.

Table 2: Appropriate use at provider level (Level 1._)

Information/ Service provider should fulfil the requirements.

L1.0	Unified, mass distribution of information at low cost
L1.1	Providing variety of expressions, i.e. text, photo, coloured picture, movie, etc. Service available to various media, i.e. email, internet, mobile and on paper.
L1.2	Service available to various requirements, i.e. quick, detail, specific, high reliable, high quality, related, etc.
L1.3	User support to assimilate data. Load to compare, to watch continuously, search data. Decrease memory load, time and space constrain.
L1.4	User support to re-examine data, record, repeat operations.
L1.5	Adaptation to user purposes.
L1.6	Adaptation to language, culture and history
L1.7	Appropriate to situation of use, i.e. school, office, home, church. Appropriate to client, i.e. children
L1.8	Safe, harmless, beneficial
L1.9	Secure and reliable

6 Coordination among Appliances

People are often careless to forget to stop ringing call sound or alarm signal, where it is inadequate to ring. Where strict stop of ringing, or certain operations are essent ial, some wireless signal should be available, so that ringing is shut down by electronic signal. Phone receiver which are sensitive to the stop code is called elegant receiver. While receiver not accepting the code might be called primitive or egocentric receiver.

The decision to issue the stop code and the logic is made by social or community rules, the product manufacturer have to install such function so that the receiver might receive the signal effectively. Owners of egocentric receivers are fully responsible not to violate the social rules and should be constantly careful. Owners of elegant receiver may rely on the technical intelligence, might be regarded an elegant citizen.

The purpose of providing elegant receiver is not to let people forget about social rules and good manner of using mobile phone. When user consults the systems about the reasons for call stop signal, the elegant receiver should provide proper reasons for them.

If the reasons are dogmatic, people no longer hold on the rule. If the reasons are fitting to the actual situations, people will respect the rule. If announcement in crowded train say, "mobile phone is prohibited to avoid inconvenience to neighbours ...", people might agree. If trains were less populated, people will neglect the announcement.

Table 3: Appropriate use by Coordination (Level 2._)

Coordination functions among appliances, for effective, efficient and satisfactory results.

L2.0	Stand alone traditional appliances
L2.1	Sensor driven, timer controlled, programmed or AI driven appliances
L2.2	Environment sensitive, day/time sensitive, operation constrained appliances
L2.3	User adaptive, user history dependent, adaptive to people in room/house
L2.4	On site service(train guidance, way finding, helping people in need)
L2.5	Adaptive to user purpose, appliances available, network conditions
L2.6	User preference sensing (room temperature, lighting, channel selection) Preference based choice control. Priority rules, manner selection
L2.7	
L2.8	Environment support (waste classification, etc.)

Here L2.0 is dealing with traditional appliances operated by users. L2.1 is dealing with automatic appliances using sensors, timer, and programmed logic, which are common now in air conditioners. More elaborated air conditioning will be expected in L2.6. The coordination system will ask people sharing in a room, whether the air condition at the moment are favorable. Depending on the responses of people inside, the system will modify the temperature setting or the control program.

7. Conclusion

A three level structure to formulate standard for appropriate use of digital everyday appliances is proposed. Appliances so far were designed mostly as tool to perform certain functions. New types of appliances or services will be more dominant in new future. For the traditional appliances usability and utility were key importance. New appliances will be evaluated by relations and processes with users (Sawaragi, 2002). Standards for such new appliances and services have to be studies further.

References

Hiroshi TAMURA (1995) Information Technology in support of Living Oriented Society Postprint of 6th IFAC Symposium on Man Machine Systems, pp.157-162, Pergamon Press

Rui ZHANG, Hiroshi TAMURA, Yu SHIBUYA (1998) The Integration of Speech and Camera Control in Message Transfer TV Conferencing, International Journal of Human-Computer Interaction, vol. 10, pp. 327-341

Hiroshi TAMURA (1999) Communication beyond Reality Human Computer Interaction, vol. 2, pp. 1337-1340, ed. by: Bullinger, H., Ziegler, J., LEA

Hiroshi TAMURA (1999) Human Interface and Ergonomics in the Information Society, Proceedings of Ergonomics Symposium 2000, pp. 291-294, ESK

Tetsuo SAWARAGI(2002) Human-Automation Coordination mediated by Reciprocal Sociality, T-Fr-M06, IFAC 15th Triennial World congress, Barcelona

A Method of Volume Metamorphosis by Using Mathematical Morphology

Yuji Teshima and Ken-ichi Kobori

Osaka Institute of Technology,
Graduate School of Information Science and Technology
teshima@is.oit.ac.jp, kobori@is.oit.ac.jp

Abstract

Morphing technology which transforms one shape into another has been used to make special effects in movies. This paper presents a three-dimensional image morphing method without a source to a target object correspondence. This method applies to voxelized objects with full color texture. The algorithm consists of three parts, decision of the regions to be transformed, calculating distance images for those regions and interpolation by using morphology operations. First, expansion or contraction regions are decided from the position of two shapes. Secondly, the distance images of those regions are generated to interpolate the shapes. This approach utilizes Minkowski difference to compute the metamorphosis. As a result, automatic interpolation process is executed at needed frames. Some experimental results show that this method is more effective for volume-based metamorphosis.

1 Introduction

CG animation has been widely used in movies, CMs, and games, and it serves forceful images. Especially, morphing techniques which interpolate two different shapes and which animate the interpolation are remarkable. Conventional morphing methods apply to two-dimensional images. However, recent morphing methods apply to three-dimensional images because the viewpoint can be changed easily. Morphings are classified into boundary-based representation and volume-based representation by Lazarus[1)]. In general, the former uses triangle polygons, and the latter handles implicit function model and voxel data.

In boundary-based representation, there is a major restriction that the source and target objects should have the same topological structure. So, it is proposed that the morphing methods equalize topological structure between two shapes by using hamonic map[2),3)].

In volume-based representation, voxel data and implicit function data do not have topological structure, and the morphing process has a feature that computation time increases in higher resolution because the morphing quality depends on the resolution. There are many conventional methods by warping with manual correspondence between the source and target object[4),5)]. Those method generate good interpolation shapes by the correspondence. However, it is necessary to correspond to feature elements manually and it takes a few hours even if an expert designer works. Turk et al. proposed to generate to the interpolation using variational implicit functions[6)]. This method is difficult to control shape instinctively. Hughes proposed a method to generate interpolation automatically with Fourier transform[7)]. This method has two problems. One is that the method is time consuming, because the computation of each in-between image takes at least $O(n^3 logn)$, for a volume of size n^3. The other is that the method cannot control interpolations at all.

In this paper, we adopt volume data and propose a morphing method without manual correspondence. Also our method has ability to control interpolation easily by moving the position of the source and target image instead of corresponding. To transform from the source image to the target one we use mathematical morphology which is used in two-dimensional image processing. Especially, the method utilizes minkowski difference with ball structuring element.

2 Mathematical Morphology

Mathematical Morphology has two base operations which are Minkowski Sum and Minkowski Difference, and makes various filtering effects by combining them. In this paper, we introduce Minkowski Difference which removes noises mainly.

2.1 Minkowski Difference

Minkowski Difference operates on an object image and a structuring element. *Fg* is calculated with ***f*** which are position vectors in the object image *F* and ***g*** which is a position vector in the structuring element *G*.

$$F_g = \{z \in E : z = f + g, f \in F\} \quad (1)$$

In this paper, *E* is a three dimensional space though *E* means *N* dimensional Euclid space generally. Each *Fg* is generated with each ***g*** in *G*. Applying logical AND operation to those *Fg* is Minkowski Difference as

$$F \ominus G = \bigcap_{g \in G} Fg \quad (2)$$

In other words, a trace area results from moving the center of structuring element along the surface of the object. Minkowski Difference is the result image which is removed the trace area from the object image. Figure 1(a) shows a object image and Figure1(b) shows the result of Minkowski Difference. In this case, the shape of the structuring element is circle. Thus, Minkowski Difference object image shrinks. We can get various result shapes by changing the shape of the structuring element.

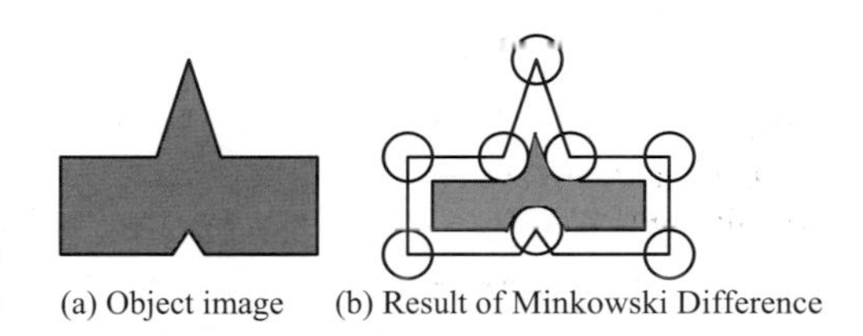

Figure1: Minkowski Difference

3 Morphing method

A proposed method generates interpolation shapes by using Mathematical Morphology. However Minkowski Difference is operated with single size structuring element in general, we use various size of ball structuring elements. In this paper, the source and target image has 24 bit(each R G B is 8bit) color and the surface of voxels has color data.

A proposed method consists of the pre-process and the interpolation process. We describe the morphing process in detail.

3.1 Pre-process

(1) After deciding the positions of source image and target image manually, we extract the regions to be transformed automatically; expansion regions, contraction regions and intersection regions. Figure2

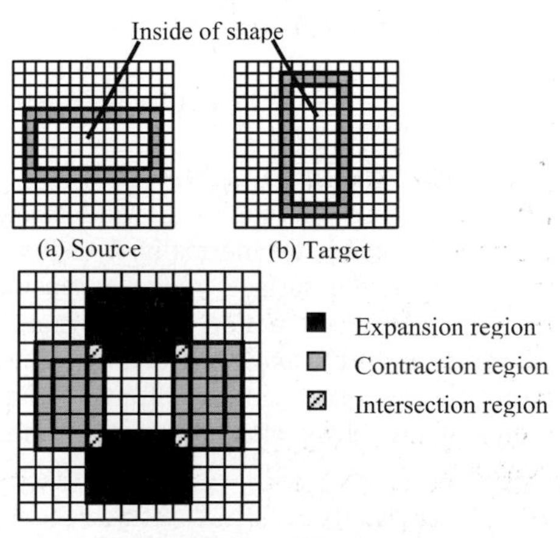

Figure2: Extraction of transformation

shows the extraction of transformation region. It is natural that the changing the position of the source and target image makes different interpolation.

(2) Distinguishing each closed region of expansion regions and contract regions. In figure2(c), closed regions are distinguished into four regions, two expansion regions and two contraction regions.

(3) Calculating distance value at each voxel with wave diffusion model for each distinguished expansion region and contraction region. At the same time, we give a label value each voxel in the regions. Label is 24bit colors corresponding to target voxel color. First, we decide start points of a wave. These points are surface voxels of the target image except the intersection region. Figure3 shows the start points of the wave. Secondly, we give distance value from these points in order by using distance addition table which is shown in Figure4. At same time, we give color of these points as label value in order. A result of propagation of distance and label value is shown in Figure5.

(4) Normalizing distance value in each closed region.

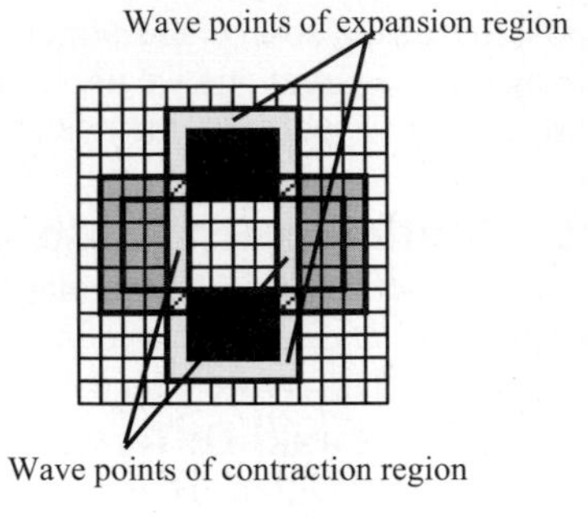

Figure3: Wave points

$\sqrt{2}$	1	$\sqrt{2}$
1	●	1
$\sqrt{2}$	1	$\sqrt{2}$

Figure4: Distance addition table

$$Dn = \frac{D \times Frame}{Dmax} \quad . \tag{3}$$

Where, a maximum distance of each closed region *Ai* is *Dmax*, a distance value in *Ai* is *D* and *Fame* is a number of morphing frame. A normalized distance is calculated as *Dn*. Applying **(3)** to all voxels in these region, each maximum distance is equal to *Frame*.

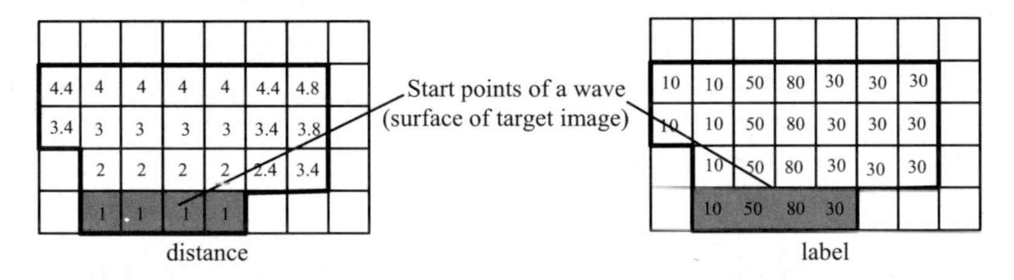

Figure5: Propagation of distance and label value

3.2 Interpolation process

(5) Calculating the radius of structuring element by using the normalized distance value *Dn* on the surface of the expansion region and the contraction region. Zigzag pattern in Figure6 shows the area of which radius is calculated in first frame. The radius of structuring element is given by the equation

$$radius = \frac{Dn}{Frame - currentFrame} \quad , \tag{4}$$

where *currentFrame* is current frame number.

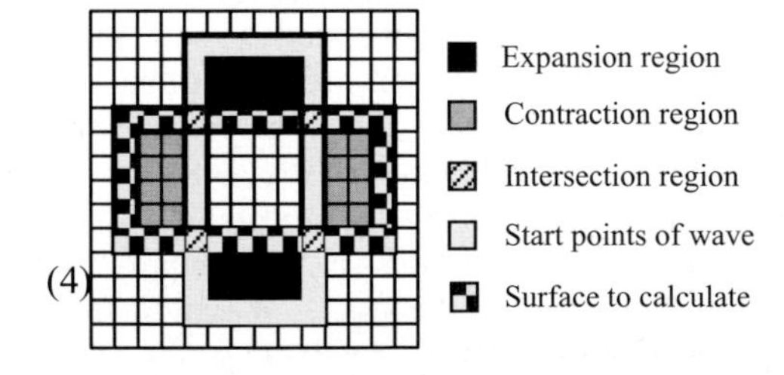

Figure6: Surface of transformation

(6) We generate *newShape* by giving Minkowski Difference to *currentShape* and interpolating the color of *currentShape*. Minkowski Difference is given to the surface of which *radius* is over 1. In

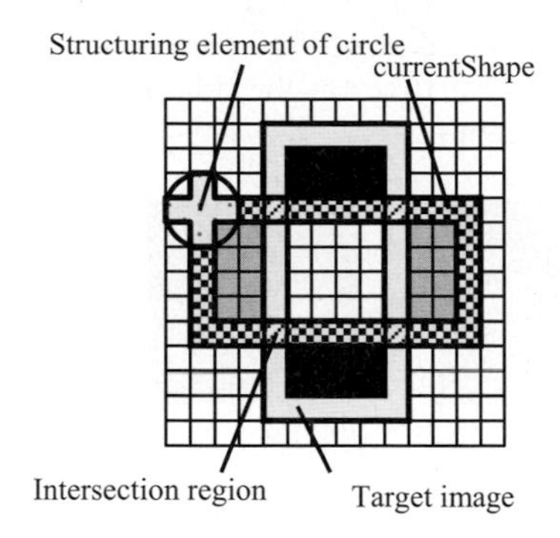

Figure7: Giving Minkowski Difference

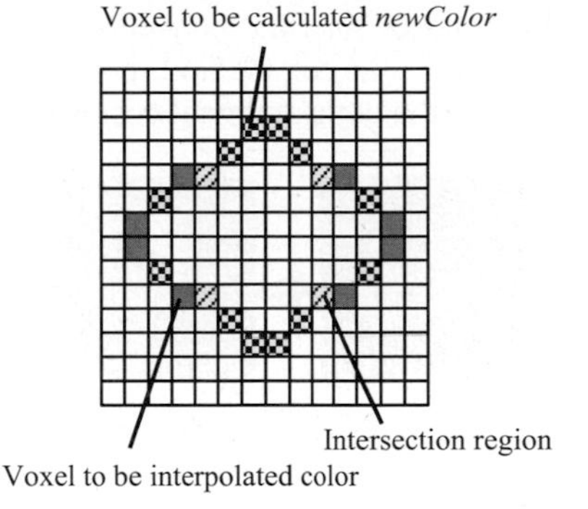

Figure8: Generation of newShape

first interpolation, *currentShape* is the source image. Minkowski Difference removes the voxels at which the center of the structuring element is located shown in Figure7. As a result, *newShape* shown in Figure8 is generated. The *newShape* consists of intersection region and new surface generated by Minkowski Difference except the start points of the wave. Color of new surface is defined as *newColor* and *newColor* is given by the equation

$$newColor = currentColor + \frac{(labelColor - currentColor) \times (radius + 1)}{D}, \tag{5}$$

where *labelColor* is label value and *currentColor* is color of *currentShape* corresponding to center of the structuring element. If the size of the structuring element is under 1, we interpolate the color by the equation

$$newColor = currentColor + \frac{labelColor - currentColor}{Frame - currentFrame} \tag{6}$$

without giving Minkowski Difference.

The only color is interpolated in the intersection region because the shape does not change. In the intersection region, *newColor* is given by the equation

$$newColor = currentColor + \frac{tColor - currentColor}{Frame - currentFrame}, \tag{7}$$

where *tColor* is the surface color of the target image.

(7) The process(5) and (6) are repeated and interpolation shapes are generated in order.

4 Experiment and consideration

We experimented to verify the method. Figure9 shows two pairs of source image and target image which are voxelized in 256^3 resolution. *Frame* of morphing is 50. Figure10 shows two morphing results between a clock and a mixer, a cannon and a bottle. In figure10, the interpolation shapes show smooth transformation from the source image to the target image. Minkowski difference with structuring element of different size brings us smooth interpolation shapes. Also, the interpolation shapes show that the color of source image changes into the color of the target image gradually according to the processing. This results from the process that makes the color data match the label during calculating distance values with wave diffusion model.

As an experiment of processing time, we compared our method with the conventional method with Fourier transform in volume-base. We measured average processing time of generating an interpolation shape per one frame for ten pairs of volume-data. The result shows that our method can process faster than the conventional method. Our method takes 48 % of the conventional processing time. In our method, processing time increases eight times at the double resolution. That is to say, the computational cost is $O(n^3)$ when the resolution is n.

Figure9: Source and target images

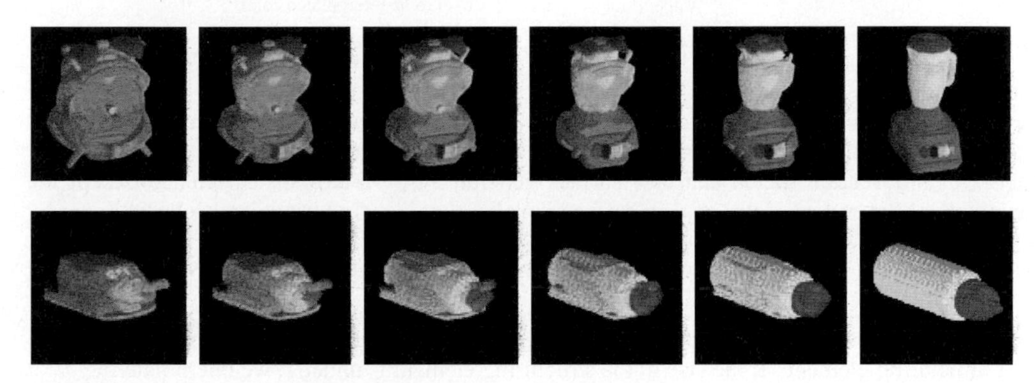

Figure10: Results of morphing

5 Conclusion and future work

In this paper, we proposed a morphing method for volume data without manual correspondence. Our method generates the interpolation shapes by giving Minkowski Difference to the transformation region extracted automatically. Also, changing the position of the source image and the target image can control the interpolation shapes. Some experiments show that our method is more effective than the conventional method. We will innovate volume rendering and gradient shading to get higher image quality. Future work also includes the verification of the effectiveness by the method using *Opening* operation and so on instead of Minkowski Difference.

References

1) F.Lazarus, A.Verroust : "Three-dimensional metamorphosis: A survey", The Visual Computer,14,pp373-389(1998)
2) T.Kanai, H.Suzuki, F.Kimura : "Metamorphosis of arbitrary triangular meshes", IEEE Comput. Graphics & Appl., 20,2,pp62-75(Apr.2000)
3) A.Gregory, A.State, M.C.Lin, D.Manocha, M.A.Livingston : "Interactive surface decomposition for polyhedral morphing", The Visual Computer, 15,9,pp453-470(1999)
4) A.Lerios, C.D.Garfinkle, M.Levoy : "Feature-based volume metamorphosis", SIGGRAPH95 Proceeding, pp449-456(Aug.1995)
5) D.Cohen-or, D.Levin, A.Solomovici : Three-dimensional distance field metamorphosis", ACM Trans Graph, 17,pp116-141(1998)
6) G.Turk, J.F.O'Brien : "Shape transformation using variational implicit functions", SIGGRAPH99 Proceeding, pp335-342(1999)
7) J.F.Hughes : "Scheduled Fourier volume morphing", SIGGRAPH92 Proceeding, 26,2,pp43-46(1992)

Designing Self-directed Learning Environments in Museum Settings: A Context Sensitive Approach

Karen Tichy

Interaction Design Consultant
44 Lothrop Street Beverly, MA 01915
ktichy@attbi.com

Abstract

Though user-centered approaches are commonly used in exhibit design today, exhibit messages and presentation formats are most often based on an average target visitor profile and result in a single, fixed visitor experience. The notion of contexts and context sensitive design appears to be a design method that can result in exhibits that better facilitate learning. This paper explores the issues of context for designing self-directed learning experiences and suggests possible foundations for future design methodology development that effectively incorporates context sensitivity.

1 Introduction

Thus far, the words "context" or "context sensitive" has primarily referred to the idea of providing visitors with information relevant to objects in the location they are at any point in time. Wireless handheld or wearable information systems such as HyperAudio, GUIDE, and HIPs, have been used to provide information based on visitor location and to a lesser extent rule-based inferences about visitor behaviour and/or explicitly input profile information. This information has been provided in various single and multi-modal forms such as audio and/or handheld screen display. So far, this technology has been applied in most cases to information retrieval systems for basic informational exhibits.

This view of context appears oversimplified when compared to the complex inter-relatedness of numerous, constantly changing conditions and attributes involved in the special conditions of self-directed educational environments. This paper begins to develop a contextual framework specific to the complex interactions between self-directed learners and informal learning environments and their affect on the resulting learning outcomes. First an overview of self-directed learning will be presented followed by an examination of conditions of learning in museum settings. Lastly, issues of context related to design of exhibits will be discussed.

2 Overview of Self-directed Learning in Museum Settings

What are the design elements and issues critical to designing self-directed learning environments? Which combinations of attributes and conditions define contexts? How to they change and interact throughout the learning process?

2.1 Design Components

A simple communication model has been readily adapted by visitor studies professionals to describe learning in self-directed environments (Screven, 1999). The primary components of these models include the intended exhibit message, exhibit media, and the viewer. Figure 1 details these elements from a user-centered design vantage point. Part of the media and message components have been joined together to form a fourth component called teaching strategies and activities to attempt to fully understand the instructional component in the learning process. Each of these components is comprised of over a dozen variables that may change during a visit making hundreds of possible combinations or conditions in which learning in museums might or might not take place.

Table 1: Design Components of Self-Directed Learning Environments

Visitor/ Self-directed learner	Exhibit interface/ Information design	Teaching strategies/ Activities	Exhibit Messages/ Content
Age	Media options:	Comparison	Purpose:
Social-economic/	audio, video, etc.	Matching	excite/peak interest,
Cultural aspects	Passive interpretive:	Repetition	change opinion,
Group member	text & graphics:	Practice	teach skill, teach
Paired	artifacts, objects,	Simulation	concept, teach fact,
Individual	diorama, models	Questions	environmental/
Family/group dynamics	Interactive interpretive	Diagram	health concern alert
Motivation level/ type	systems:	Analogy	Goals & objectives
internal/external	Computerized -	Metaphor	Quantity
Interests	kiosk, handheld	Games	Quality
Knowledge	Mechanical –	Role-playing	Complexity
Learning style:	flip labels,etc.	Model	Objectivity
primary intelligences/	Context sensitive	Motivational options:	Subjectivity
sensory modality	information systems	Intrinsic - challenge,	Passive/active
Independent learner/	Multi-modal presentation	positive feedback,	Associated intelligence:
joiner	Multiple points-of-entry	impart sense of	math, music, visual-
Visiting style	Wearable computing	accomplishment,	spatial, social...
Learning readiness/	Exhibit personality:	personalize content	Cognition type:
metacognitive skill	young/old	Extrinsic - prizes, etc.	skill, concepts, facts,
Preconceived ideas/	fun/serious	Pre-organizers	process, high/low
biases/misconceptions	easy/challenging	Post-organizers	conceptual
Time constraints	Environmental conditions/	Pre testing	Subject matter/
Physical comfort	Potential distractions:	Post testing	Museum type:
hunger, fatigue, crowds	traffic flow, lighting		historical/cultural,
	quality, acoustics,		science, botanical/
	climate control, exhibits		zoo/aquarium, etc.
	competing for attention		

Identifying attributes of these four components and understanding their relationships begins to lay a foundation for defining contexts in the exhibit domain. Analyzing changes in these relationships through out the learning process will further develop the framework.

2.2 The learning process in self-directed settings

What is learning? What are the required conditions for learning take place? What are the special conditions of self-directed environments?

Learning takes place when a visitor understands the concept the exhibit intends to convey and commits that information to memory. To understand conceptual messages the quality and quantity of a viewer's attention must last long enough to read and interact with the information and relationships that comprise such messages, be selective enough to bring key information into focus so it can be encoded, and be active enough to integrate new information with existing knowledge and attitudes. Not only must the quantity of time spent be sufficient, the quality of attention is also important. Attention can occur along a continuum from focused and active to casual and unsystematic and from mindful to mindless.

Conditions of visitors and public learning spaces are often at odds to optimal conditions for learning. Typically a visitor's time is limited and museum surroundings are typically hectic and distracting as exhibit and group members compete for attention. The extent a learning environment bridges these differences determines how well it facilitates learning.

Figure 1 shows the physical and cognitive paths a visitor might take in a typical goal-centered exhibit. Goal-centered exhibits, in addition to being informational, are more instructional by providing visitors with an achievable goal in the form of a task or activity. A goal-directed activity is intended to help focus attention on the exhibit message which begins to overcome the distractible nature of museums (Screven, 1999). Goal-centered activity also provides feedback to the visitor to help assess their own progress through the exhibit information.

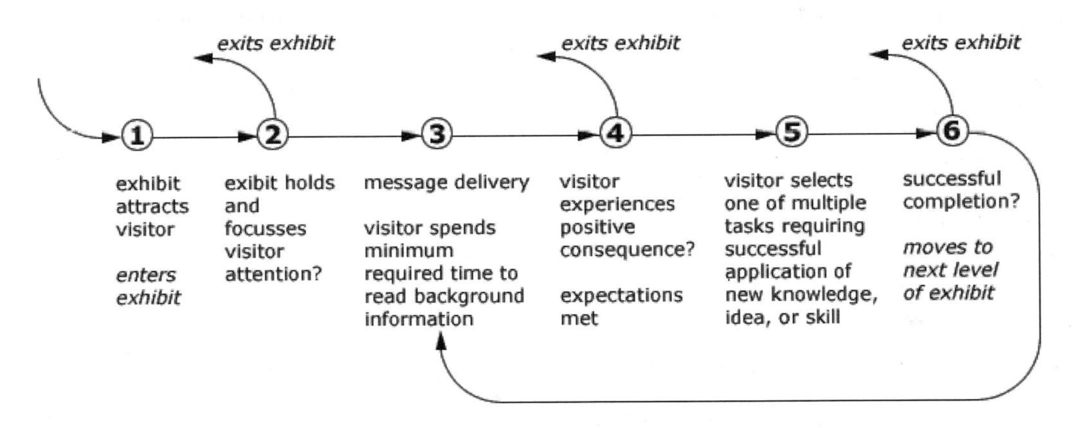

Figure 1: The learning process in self-directed settings

At the same time, these assessment and reflection mechanisms could also provide a method for capturing and interpreting contextual data that can be used to enhance the visitor's self-directed learning experience.

3 Context Issues of Exhibit Design

Which visitor attributes are most critical to the design problem? What combinations of attributes and conditions define contexts? Does the degree of stability and changeability of attributes affect contexts? How can instructional systems identify changes in contexts?

Figure 2 suggests a set of visitor attributes that are strongly associated with the learning process. These provide us with a base to define a contextual framework. Some of these user attributes are relatively stable, others can change dramatically during time spent at an exhibit. Some reside within the visitor, others are external. Metacognitive skill for example is relatively stable compared to level of attention. The combination of these attributes and conditions suggest ways the interface might change at critical points of the visitor-exhibit interaction to help make conditions for learning more optimal.

At critical points in the learning process qualitative and/or quantitative values can be assigned to these attributes and conditions. The combination of these attributes and conditions define contexts and suggest ways the interface might change to help make conditions for learning more optimal given that context. By sensing a visitor's level of attention in combination with knowing what best motivates the visitor (intrinsic or extrinsic incentive) the system could provide incentives as well as information and activities at the right time and at appropriate levels.

Inferences can be made about a visitor's level of attention by observing behaviours. Behavioral observation methodologies used by exhibit evaluators begin to provide qualitative and quantitative measures.

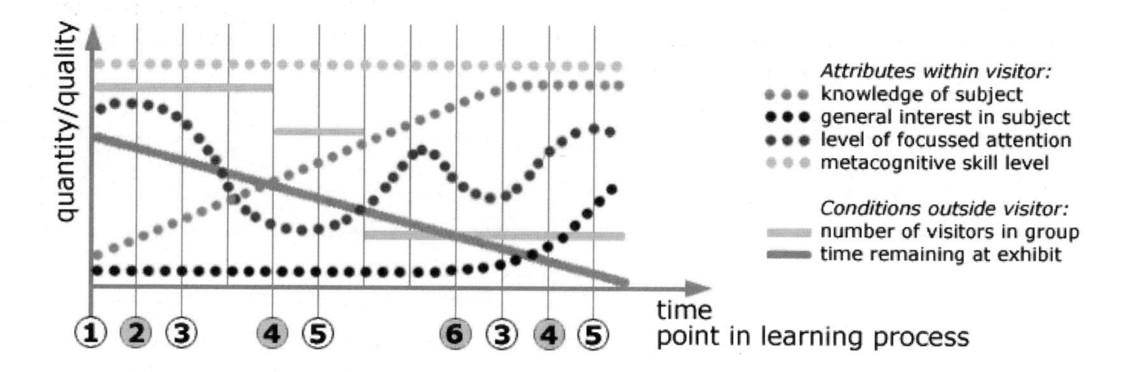

Figure 2: Change in visitor attributes over time

The set of critical attributes and conditions can be different for different exhibit messages, goals and content types. For example, an exhibit intended to alert the public to dangers of mercury contamination would need to identify special populations such as pregnant women to better insure that information was seen, read, and understood by populations at risk. Children at the same exhibit would experience much different information and activity options which are dynamically assembled to match contextual profiles. These profiles are defined by a combination of variables such as demographic information, point of entry behaviour, and interest profiles.

4 Summary and Perspective

This paper explored the issues of context related to the design of self-directed learning experiences and suggested conceptual foundations for future design methodology development that effectively incorporates context sensitivity. A distinction was made between purely informational exhibits and exhibits that attempt to provide instructional guidance. Exhibits with instructional orientation need to sense changes in contexts that most dramatically affect conditions of learning. Further investigation into ways to identify these changes is needed.

5 Acknowledgements

Thanks to Dr. Chandler Screven for sharing his vast knowledge and experience in visitor studies and exhibit design. Appreciation to The New England Aquarium Education Department for providing opportunity in and examples of effective exhibit design.

References

Aoki, P. M. and Woodruff, A. (2000). Improving electronic guidebook interfaces using a task-oriented design approach, *ACM Conference Proceedings on Designing Interactive Systems*, August 2000.

Bellotti, F., et. el. (2002). User testing a hypermedia tour guide, *Pervasive Computing*, April - June 2002.

Cheverst, K., et. el. (2000). Developing a context-aware electronic tourist guide: some issues and experiences, *Proceedings of the SIGCHI Conference on Human Factors in Computing*, April 2000.

Petrelli, D., et. el. (2001). Modelling and adapting to context, *Personal and Ubiquitous Computing*, 5(1).

Screven, C. (1999). Information design in informal settings: museums and other public spaces. In R. Jacobson (Ed.), *Information Design,* (pp. 131-192). Cambridge, MA: The MIT Press.

Semper, R. & Spasojevic, M. (2002). The electronic guidebook: using portable devices and a wireless web-based network to extend the museum experience. *Proceedings from Museum and the Web,* 2002.

Tichy, K. R., (1992). Information, media, and learning: designing adaptable learning tools through user controlled presentation styles, MS thesis, Institute of Design, Illinois Institute of Technology.

Vassilios, V., et. el. (2003). "Augmented reality touring of archaeological sites with the ARCHEOGUIDE system", *Cultivate Interactive*, 9(7).

Whitney, P. (1990). The electronic muse: matching information and media to audiences. *ILVS Review: A Journal of Visitor Behavior*, 1(2).

On the Statistical Distribution of Features in Content-Based Image Retrieval

G. A. Tsihrintzis and A. Theodossi

Department of Informatics, University of Piraeus
80 Karaoli & Dimitriou Street
Piraeus 18534, Greece

E-mail: {geoatsi, atheo}@unipi.gr

The problem of content-based image retrieval

Content-based image retrieval (CBIR) was proposed in the early 90's to overcome the difficulties involved with the traditional manually annotated (*text-based*) image retrieval systems of the 70's and 80's [1-3]. Specifically, in CBIR, images are indexed by their own visual content, that is by the distributions in them of features such as color, shape and texture (*general features*) or even human faces and fingerprints (*domain-specific features*).

These features are represented by vectors of real numbers of typical length on the order of 10^2 and may be defined either *globally* or *locally* in an image [4]. A feature may be represented in a variety of ways, each emphasizing a different aspect of the feature. Occasionally, the need arises for reduction of the dimensionality of the feature vector length in order to speed up the CBIR system performance.

To retrieve images from an image database using a CBIR system, a user submits a query in the form of a desired feature vector and the system returns images whose corresponding feature vector is "similar" (according to an appropriate similarity measure) to the query.

To define useful similarity measures for a feature, one needs to perform a study of the statistical distributions of the feature over a sufficiently broad collection of images and model them with useful statistical models. Statistical modeling is also useful in feature vector dimension reduction, as reduction algorithms usually rely on the second-order statistics of the feature vector [5-7].

Statistical signal modeling is mainly concerned with the extraction of information from observed data via application of models and methods of mathematical statistics. In particular, the procedure that generates the data is, in a first step, quantitatively characterized, either completely or partially, by appropriate probabilistic models and, in a second step, algorithms for processing the data are derived on the basis of the theory and techniques of mathematical statistics.

Clearly, the choice of good statistical models is crucial to the development of efficient algorithms, which, in the real world, will perform the task they are designed for at an acceptable performance level.

Traditionally, the signal and image processing literature has been dominated by Gaussianity assumptions for the data generation processes and the corresponding algorithms have been derived on the basis of the properties of Gaussian statistical models. The reason for this tradition is threefold:

(i) The well known Central Limit Theorem suggests that the Gaussian model is valid provided that the data generation process includes contributions from a large number of sources,

(ii) (ii) The Gaussian model has been extensively studied by probabilists and mathematicians and the design of algorithms on the basis of a Gaussianity assumption is a well understood procedure, and

(iii) (iii) The resulting algorithms are usually of a simple linear form which can be implemented in real time without requirements for particularly complicated or fast computer software or hardware.

However, these advantages of Gaussian signal processing come at the expense of reduced performance of the resulting algorithms. In almost all cases of non-Gaussian environments, a serious degradation in the performance of Gaussian signal processing algorithms is observed. In the past, such degradation might be tolerable due to lack of sufficiently fast computer software and hardware to run more complicated, non-Gaussian signal processing algorithms in real time.

With today's availability of inexpensive computer software and hardware, however, a loss in algorithmic performance, in exchange for simplicity and execution gains, is no longer tolerated. This fact has boosted the consideration of non-Gaussian models for statistical signal processing applications and the subsequent development of more complicated, yet significantly more efficient, nonlinear algorithms.

A class of statistical models that have attracted the interest of researchers in the signal and image processing field [eg., 8-12] are the, so-called, alpha-stable distributions and processes. These models generate time series (data) whose statistical distributions are heavy-tailed. In practical terms, the predicted probability that the data attain values that deviate from the average by a large margin is significantly higher than the probability predicted by Gaussian models.

This property of alpha-stable models has been found to reflect the distribution of real data better than Gaussian models. For example, this has been the case in modeling radiometrically uncalibrated infrared images [10], ultrawideband radar images [11], and medical ultrasound images [12].

In this paper, we perform an extended study of the statistical distributions of visual features in images. We show that the class of alpha-stable distributions provide a valid model for the distribution of these features and analyze the consequences in the design of similarity measures for use in retrieval algorithms and high-dimensional indexing.

References

1. Rui Yong, Huang S. Thomas, and Chang Shih-Fu, “Image retrieval: current techniques, promising directions, and open issues”, *Journal of Visual Communication and Image Representation,* vol.10, pp. 39-62, 1999.

2. IEEE Multimedia, Special Issue on Content-Based Multimedia Indexing and Retrieval, vol. 9, April-June 2002.

3. A.D. Theodossi and G.A. Tsihrintzis, ''Multimedia Indexing and Retrieval in Environmental Information Systems,'' in: *HELECO'03: An International Exhibition and Conference on Environmental Technology,* Athens, Greece, January 30 – February 2, 2003.

4. L. Amsaleg and P. Gros, ''A Robust Technique to Recognize Objects in Images and the Database Problems it Raises,'' in: *INRIA Technical Report No 4081*, 2000.

5. R. Ng and A. Sedighian, ''Evaluating Multi-dimensional Indexing Structures for Images Transformed by Principal Component Analysis,'' in: *Proceedings of SPIE Storage and Retrieval for Image and Video Databases*, 1996.

6. C. Faloutsos and K-I Lin, ''Fastmap: A Fast Algorithm for Indexing, Data-Mining and Visualization of Traditional and Multimedia Datasets,'' in: *Proceedings of SIGMOD,* pp. 163-174, 1995.

7. D. White and R. Jain, ''Algorithms and Strategies for Similarity Retrieval,'' in: *University of California-San Diego Technical Report VCL-96-101,* 1996.

8. G.A. Tsihrintzis, M. Shao, and C.L. Nikias, Recent Results in Applications and Processing of Alpha-Stable-Distributed Time Series, *Journal of the Franklin Institute*, vol. 333B, pp. 467-497, 1996.

9. G.A. Tsihrintzis and C.L. Nikias, Alpha-Stable Impulsive Interference: Canonical Statistical Models and Design and Analysis of Maximum Likelihood and Moment-Based Signal Detection Algorithms, in: *C.T. Leondes (ed.), Control and Dynamic Systems: Advances in Theory and Applications,* Volume 78*,* San Diego, CA: Academic Press, Inc., 1996, pp.341-388.

10. N. Nandhakumar, J. Michel, D.G. Arnold, G.A. Tsihrintzis, and V. Velten, A Robust Thermophysics-Based Interpretation of Radiometrically Uncalibrated IR Images for ATR and Site Change Detection, *IEEE Transactions on Image Processing: Special Issue on Automatic Target Recognition*, vol. IP-6, pp. 65-78, 1997.

11. R. Kapoor, A. Banerjee, G.A. Tsihrintzis, and N. Nandhakumar, Detection of Targets in Heavy-Tailed Foliage Clutter Using an Ultra-WideBand (UWB) Radar and Alpha-Stable Clutter Models, *IEEE Transactions on Aerospace and Electronic Systems*, vol. AES-35, pp. 819-834, 1999.

12. A. Achim, A. Bezerianos and P. Tsakalides, ''Novel Bayesian Multiscale Method for Speckle Removal in Medical Ultrasound Images,'' *IEEE Transactions on Medical Imaging,* vol. 20, pp. 772-783, 2001.

Dying Link

Koji Tsukada

Keio University Graduate School
of Media and Governance
5322 Endo, Fujisawa, Kanagawa
252-8520, Japan
tsuka@sfc.keio.ac.jp

Satoru Takabayashi *Toshiyuki Masui*

Sony Computer Science Laboratories, Inc.
Takanawa Muse Building 3-14-13 Higashi-Gotanda, Shinagawa, Tokyo 141-0022,
Japan
{satoru,masui}@csl.sony.co.jp

Abstract

We propose a new visualization technique called the *Dying Link*, which makes the links to Web pages look aged according to how old the linked pages are. Using our technique, users can easily tell how "fresh" the linked pages are, only by glancing at the appearance of the link string in a Web page. We describe the design, implementation, and evaluation of the Dying Link system.

1 Introduction

Although a number of attractive graphical Web pages exist on the Internet, we can find many outdated pages, maybe because it is difficult for people to keep those pages always up-to-date. One of the authors once made fancy Web pages with many graphical objects, but he had no chance to update them for three years, since they were too complicated for modification.

Although there exist many old Web pages, we cannot easily recognize how old they are, since their appearances do not change by their ages. In contrast, many real-world objects change their appearances as they get older, and we can vaguely see how old they are simply by glancing at them. For example, we can recognize how old a book is, from the color and shape of the book.

We propose a new visualization technique called the *Dying Link*, which makes the links to Web pages look aged according to how old the linked pages are, just like real-world objects. Using our technique, users can easily tell how "fresh" a Web page is, only by glancing at the appearance of the link string in a Web page.

2 Dying Link

Dying Link is a technique to display the "freshness" of Web pages pointed by hyperlinks in a Web page. When a user browses a Web page using the Dying Link system, the link to an old page is displayed in a blurred text or a vague figure, and he or she can easily see how old the target page is. Just like a user can tell visited Web pages from unvisited Web pages on standard Web browsers, he or she can tell how old the destination pages are, only by seeing how blurred the links to the pages are. Figure 1 illustrates how Dying Link can visualize the freshness of the linked pages.

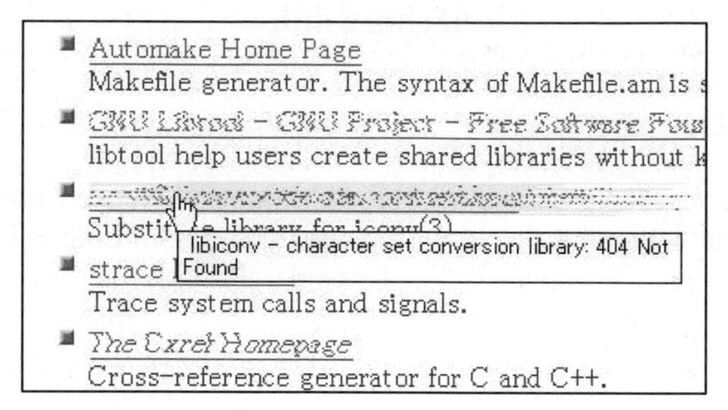

Figure 1: The visualization of Dying Link.

Since various font faces and colors are used in Web pages, it is difficult to visualize the freshness of the links only by changing those attributes of link strings. In contrast, almost all the texts and images in Web pages are designed to be displayed clearly, and blurred texts almost never appear in Web pages. For this reason, using blur is ideal for representing the freshness of link strings. According to Byrne et al. (Barrett, Maglio, & Kellem, 1997), the *locate* task to find desired information or links on a page is the second most time-consuming activity among WWW user tasks and they also argued that using different colors for visited Web pages helps users locate information. In the same way, Dying Link can help users locate fresh information efficiently.

3 Implementation

The Dying Link system is implemented as a Web proxy server. When it receives a request to display a Web page, it checks the modification time of the Web pages linked from the page, and simply adds special *dying tags* to the links according to the age of the linked pages. The dying tags use Cascading Style Sheets (CSS) to express the dying effects. For example, the following style sheet defines extreme effect for dead links and the effect can be applied by specifying "dead" to the *class* attribute in the HTML elements.

```
<style type ="text/css">
<!--  .dead
{ filter:wave(freq=5, lightstrength=0, phase=30, strength=50, add=false); }
 -->
</style>
...
<img class="dead" src="image.jpg">
```

Figure 2: Example of CSS to express the dying effects.

Since the ages of many linked pages should be checked before a Web page is displayed, all the operations are performed in parallel. If the system cannot determine how old the linked page is in less than eight seconds, the system will give up adding the effect to the link, since users usually cannot wait for more than eight seconds to view a Web page (Nielsen, 2000).

We developed three variations of Dying Link effects, namely *Distortion*, *Fade*, and *Blur*. Users can cycle these effects by hitting the "reload" button of standard Web browsers. Figure 3 shows the examples of the effects.

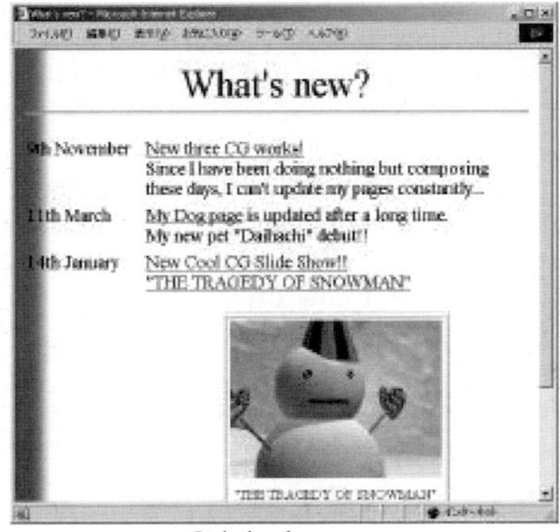

Original page

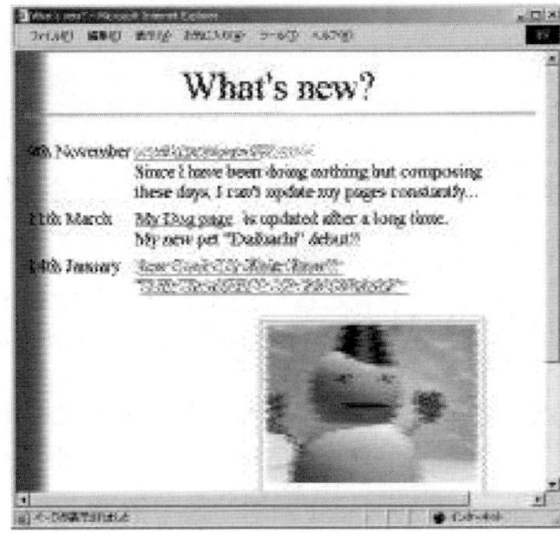

With Distortion effect

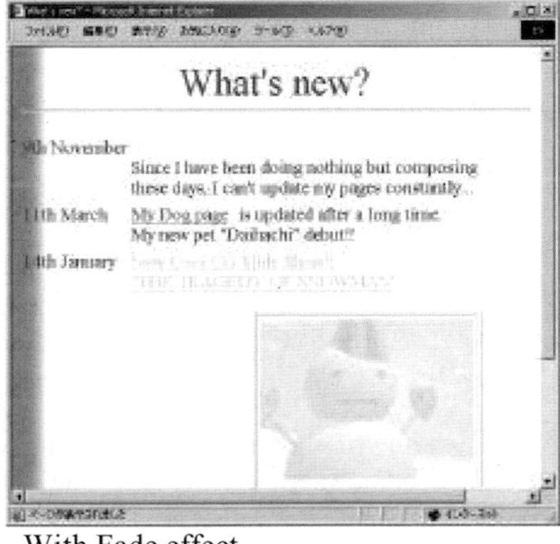

With Fade effect

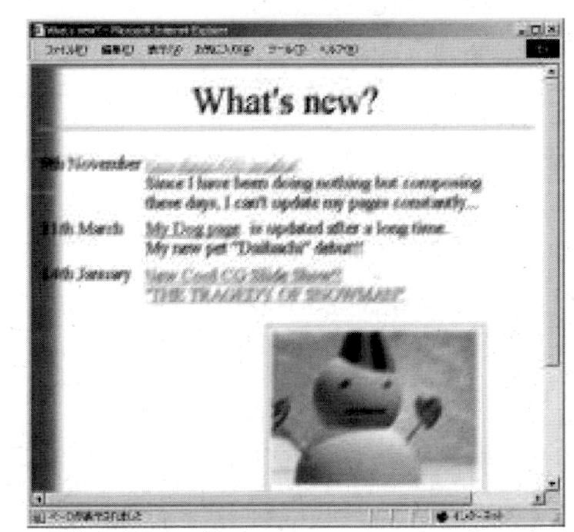

With Blur effect

Figure 3: Examples of Distortion, Fade, and Blur effects.

4 Discussions

Dying Link has been used by the authors for several months. Although a formal evaluation is not yet performed, we have found that Dying Link is especially useful in the following cases:

- Finding dead pages

According to Nielsen (Nielsen, 1998), 6% of the links from Web pages are "dead links" in average and 60% of users get frustrated about the dead links. Using the Dying Link system, users can easily recognize dead links, and they do not have to visit those pages.

- Finding active people

Some people in our laboratory update their Web pages frequently, and others do not. We can easily recognize the differences only by seeing the member list of our laboratory on the Web.

- Making people update their pages more frequently

Using the Dying Link system, users can very easily tell old pages from new pages. This fact can remind the Web page authors of how old their pages are, and make them feel like updating their pages more often than without using the Dying Link system.

Links to some Web pages are vaguely displayed even when they are actually very important, if they have not been modified for a long time. This is because current implementation of the Dying Link system determines the appearance of a link only from the modification time of the linked Web pages. Maybe it should determine the appearance also from how often the Web page is accessed, which may represent how important the Web page is.

At this moment, our system only works with the Internet Explorer, because of the restrictions of CSS. Also, it takes several seconds before displaying a Web page with hundreds of links. We are trying to eliminate these restrictions.

5 Related Works

Traffic Lights (Barrett, Maglio, & Kellem, 1997) (Maglio, & Barrett, 2000) annotates each link with simple colored icons around each hyperlink to indicate the speed of the network to those particular servers. While Traffic Lights focuses on the traffic speed of the network, Dying Link mainly focuses on the freshness of linked pages. Moreover, our approach is more intuitive because Dying Link uses aging effects instead of simple icons.

Visual Previews (Kopetzky, & Muhlhauser, 1999) shows visual thumbnails of linked pages. When the mouse pointer moved over a hyperlink, the preview window is opened by JavaScript. Unlike Visual Previews, Dying Link does not require users to move the mouse pointer to tell the freshness because our system adds visual effects to all hyperlinks directly.

Fluid Links (Zellweger, Chang, & Mackinlay, 1998) is a system for showing extra information attached to the displayed text string, when a user moves the mouse pointer over the text. The extra information should be prepared by the author, and users should move the mouse pointer over the string when they want to see that information. Although Dying Link only displays the freshness of the linked pages, the information is displayed without requiring special preparations.

Time-Machine Computing (Rekimoto, 1999) has a visualization technique using the modification time information. For example PostIt notes on the desktop gradually fade according to its modification time. While the main focus of Time-Machine Computing is personal desktop environment, our approach is optimized for efficient Web navigation.

Live Web Stationery (Seligmann, & Bugaj, 1997) employs a visualization technique similar to our system. It visualizes a Web page as if it were an old piece of paper. However, since the visual effect is generated as a background image, Live Web Stationery is difficult to be applied to pages with complex layouts, while Dying Link can be applied to arbitrary Web pages.

There are researches on the visualization of the users on WWW (Jung, & Lee, 2000) (Minar,

1999). We are planning to employ those techniques to reflect the popularity of Web pages to the link visualization in the future.

6 Conclusion

We proposed a new visualization technique called the Dying Link, which makes the links to Web pages look aged according to how old the pages are, just like real-world objects. We observed that our system can help users easily recognize the freshness of information pointed by hyperlinks, and it can urge people to think about updating their Web pages more frequently.

References

Barrett, R., Maglio, P. P., & Kellem, D. C. (1997). How to personalize the Web. In *Proceedings of the ACM Conference on Human Factors in Computing Systems (CHI '97)*, 75-82. Addison-Wesley.

Byrne, M. D., John, B. E., Wehrle, N. S., & Crow, D. C. (1999). The tangled Web we wove: A taskonomy of WWW use. In *Proceedings of the ACM Conference on Human Factors in Computing Systems (CHI '99)*, 544-551. Addison-Wesley.

Jung, Y., & Lee, A. (2000). Design of a social interaction environment for electronic marketplaces. In *Proceedings of Designing Interactive Systems (DIS '2000)*, 129-136.

Kopetzky, T., & Muhlhauser, M. (1999). Visual preview for link traversal on the WWW. In *Proceedings of the Eighth International World Wide Web Conference*, 447-454.

Maglio, P. P., & Barrett, R. (2000). Intermediaries personalize information streams. *Communications of the ACM*, 43(8), 96-101.

Minar, N. (1999). Visualizing the crowds at a Web site. In *CHI '99 Extended Abstracts*, 186-187.

Nielsen, J. (1998). Fighting Linkrot. Retrieved February 14, 2003, from http://www.useit.com/alertbox/980614.html

Nielsen, J. (2000). Designing Web Usability. New Riders Publishing.

Rekimoto, J. (1999). Time-machine computing: A time-centric approach for the information environment. In *Proceedings of the ACM Symposium on User Interface Software and Technology*, 45-54.

Seligmann, D., & Bugaj, S. (1997). Live Web stationery: virtual paper aging. In *Visual Proceedings: The art and interdisciplinary programs of SIGGRAPH '97*, 158.

Zellweger, P., Chang, B. & Mackinlay, J. D. (1998). Fluid links for informed and incremental link transitions. In *Proceedings of the ACM conference on Hypertext and Hypermedia'98*, 50-57.

Redesign the Data Dump – Statistical Vector Field

Philipp von Hellberg

Vorarlberg University of Applied Sciences
Achstr.1, 6850 Dornbirn, Austria – Europe
philipp.vonhellberg@fh-vorarlberg.ac.at

1 Abstract

A huge amount of statistical data e.g. demographic data is public available, but novices to statistical analysis cannot cope with large tables full of numbers or inappropriate static diagrams. The question is, how data-visualization can be improved with the help of time-based media elements. SVF is a prototype aimed to analyze time-based data in order to find trends, pattern, and correlations within statistical data sets. The target audience consists of non-computer experts and novices to statistical systems. The focus is set on visualization parameters based on the factor time. The most important are motion, sound and highly interactive elements, following the principle: "Overview first, zoom and filter, details on demand„ (Shneiderman 1996). The main focus is set on the perception of motion, which can help to observe and compare a large number of animated elements.

2 Background

The growing number of statistical data goes along with the progressive commercialization of public statistical data. Most federal census offices make the data available only in printed form as large tables of numbers, which is neither attractive to most citizens nor self-explanatory. This process can be seen antidemocratic. Attractive visualization and explorative tools can improve public knowledge about statistical facts and important social processes.

Standard chart creation software like Microsoft Excel does not provide visualizations in order to improve decision-making. In fact the decision of query creation has to be done before the visualization. Furthermore interactive media provide additional multimedia factors (e.g. motion, speed, flicker, sound, transparency, interactivity), which can be used to improve the density and clarity of information on the screen.

3 Statistical Vector Field

The huge amount of public statistical data makes it impossible to analyze even a small part in order to find new relationships and findings. Most statistical data sets are time-based and furthermore most of them get only relevant if analyzed over a certain period of time. This means that the amount of data grows with the number of recorded periods. But how to cope with the huge tables full of numbers? "Humans have no ′organ′ to perceive large quantitative information. Visualization and especially dynamic visualization can be a solution to similar problems„ (Kloberstein 1972). The unique functions and parameters of interactive media can give the designers of interactive visualization systems more opportunities to create tools for

“understanding„ multidimensional data. Statistical Vector Field (SVF) is a prototype that demonstrates some concepts of dynamic data visualization, which are discussed in the theoretical part of the project (German).
SVF is aimed to get a general grasp of a multidimensional data set. “A dynamic queries interface is useful for training and education (...) allow a wider range of people to explore the interaction„ (Shneiderman 1999) The visualization should enable the user to find outstanding parts and relationships within the data set in order to make a decision, which elements should be analyzed in more details.
Other modules are designed for a deeper analysis of the selected parameters. The “compare„ module has been already integrated into the given prototype; furthermore other visualization modules can be integrated.

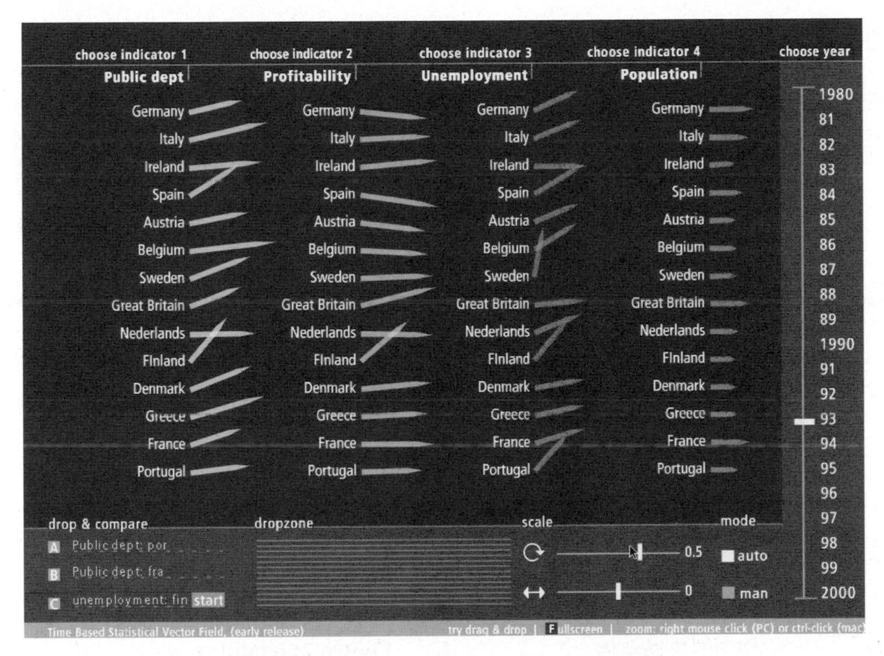

Figure 1: Interface: Statistical Vector Field

3.1 Approach

„Marks that move together will easily be selected from those with differing motions. (...) Objects with different motions are generally perceived as lying on different surfaces„ (Green, 1991). Based on these hypothesis it should be possible to identify without great effort graphical elements with differing or similar motion within a field of several elements. Human perception is very sensitive to stimuli related to motion. Thus motion can be used as an additional visualization variable, to strongly activate human attention.

3.2 Interface

The main graphical element should be understood as a vector. Vectors can convey two parameters: length and direction. The direction of each vector indicates the relative change (percentage) of an indicator’s value compared to the previous time period (e.g. year). The length of a vector indicates the exact value (scaled to fit in the grid), which can be observed with rollover-pop ups. The movement between the available values is interpolated. In that way the animation is smooth and without sharp steps. Otherwise it would cause an effect like closing the eyes, and a feeling like being in a new place. The animation lets keep the eyes open. While exploring a data-set one can follow the movement of the elements and doesn't have to compare different columns of numeric values.

If the animation is stopped (manual mode) the user can view and compare the actual value at a certain point of time, looking at the angle and the length of the different vectors. By dragging an element to the drop-zone users can build their own set of interest. Such a customized set can be saved as a standard comma-delimited file in order to be used it with other database systems or spreadsheet applications.

Besides the overall picture of the whole "vector field" the relative position of each element can also play a significant role.

"Humans can recognize the special configuration of elements in a picture and notice relationships among elements quickly. This highly developed visual system means people can grasp the content of a picture much faster than they can scan and understand text." (Shneidermann 1999, p.241)

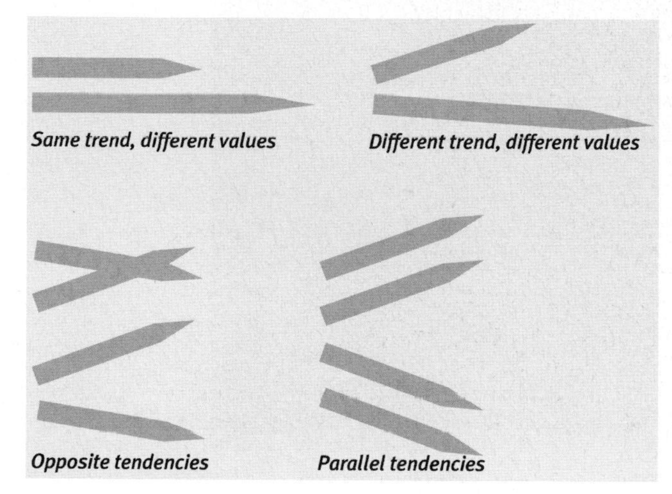

The relative position of the elements to each other visualizes the relationship of the chosen element´s values as shown in figure 2.

During animation (auto mode) the motion itself visualizes the trend. Within SVF motion is used like any other graphical variable like color, brightness or form. So every vector can communicate three parameters (table 1)

Figure 2: relative position of the vector elements

Table 1: Visual data mapping

Graphical variable	Parameter	Property
Length of vector	Absolute value	Dynamic / static (if stopped)
Motion of vector	Trend	Dynamic
Angle (direction) of vector	Relative change	Static (stopped)

3.3 Functionality

After selecting the category in the automatic mode it is possible to explore the tendencies of a large data base over a period of time. Figure 1 shows an example of demographic data over a period of 20 years. The rotation- and scale-factor can be adjusted. Small changes can be made visible too. There is a drop zone, which can be used for a more detailed analysis. Every element can either be dragged to the drop zone (to create custom sets of interest) or to the “drag & compare„ field, where the elements are recognized and made available for the compare function.

3.4 Plug-ins and extensibility

SVF also provides an additional optional view that can be used to guide detailed analysis of selected topics. Upcoming releases are intended to provide an open interface for supplementary visualization- and analysis plug-ins. By now two modules are available:

- The “compare window„ offers even more tools for a further analysis of the chosen indicators. A semitransparent multilayered area-plot displays the selected values. The colour provides additional information beside the graphical variable “form„. In doing so the blend of the different color shows correlations and exceptions.

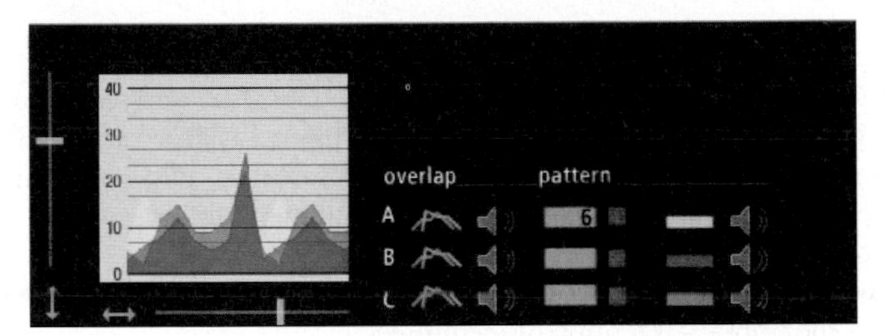

Figure 3: Plug-in: The Compare and Pattern Module

- The pattern module provides an input field for a threshold value. Because the auditory perception is more sensitive to rhythm and pattern “sonification can be used to recognize pattern within a set of data„ (Spence 2001). A short sound is played every time a value of the appropriate data row passes the threshold, while every element is distinguishable by the different pitch of the sound. “By using sound as unique coding the visual perception can be relieved„ (Shneiderman 1999). “Pitch and rhythm are the most distinguishing characteristics of a melody, and thus, can be considered more significant than, say, volume„ (Madhyastha & Reed 1992). Today there exist only a small number of models of rhythm perception. Still much research on this topic has to be done.

3.5 Technology

SVF was developed using the Macromedia Flash technology. Initially this solution was chosen for building different interface prototypes for rapid user testing. During this process it turned out that the technology would be suitable for the final prototype as well.
The macromedia generator technology allows a wide flexibility concerning the data base connection (JDBC/ODBC, URL [http, ftp, file] or Java Data Source). The system is ready to run on a web-server in order to provide best accessibility, but it will show his real potential only when there is a sophisticated backend. The coming release is developed by using only low-cost standard web technology based on FlashMX, MySQL and PHP. In that way it would be possible to provide search and filter functions leading to more “insight„.
Therefore SVF runs within every flash compatible browser. It offers platform independency, real-time “rendering„, a zoomable interface, easy maintenance and scalability.

4 Conclusions

The adequate use of dynamic elements has the potential to provide ease of use for non-expert statistical analysis tools.

- The use of motion as an addition graphical variable can help to easily recognize trends.
- Dynamic media-elements make it easier to analyze and "understand" time-based data.
- Sonification can relieve the visual perception; sound can be used as an additional unique coding.
- Further research has to be done on the suitability of sound for data visualization purposes.
- The project has shown that consumer authoring systems can act as a flexible tool for data visualization and for the development of platform independent prototypes to illustrate interface concepts.

5 Further work

Further work on the concept of straightforward access to large public databases has to be done. Some short user tests have shown a relevant improvement solving basic tasks by using SVF compared to the original data sheets from the European Statistical Yearbook. Further detailed usability tests and empirical studies will be conducted to evaluate SVF and the interface concept. New modules with filtering and sorting function are going to be developed.
To develop a highly interactive and customizable data exploration and visualization tool, accessible for most internet users is the primary intention of the project.

Acknowledgements

I would like to thank Karl-Heinz Weidmann, for his intellectual and motivating support. I also like to acknowledge my colleagues from the department of computer science and from the department of design.

References

Barrass, Stephen (1997). Thesis: Auditory Information Design thesis: CSIRO Australia http://viswiz.gmd.de/~barrass/thesis/

Green, M. (1991). Visual search, visual streams, and visual architectures. *Perception and Psychophysics*, 50

Kloberstein, H. (1973). Statistik in Bilder, Stuttgart, p.2

Madhyastha T.M and Reed D.A. (1992). A Framework for Sonification Design, in Kramer G. (ed) (1994) Auditory Display : Sonification, Audification and Auditory Interfaces, SFI Studies in the Sciences of Complexity, Proceedings Volume XVIII

Shneiderman, B. The eyes have it: A task by data type taxonomy for information visualizations. IEEE, Visual Languages '96, (September 1996) Boulder

Shneiderman, B. (1999). Dynamic Queries for Visual Information Seeking, in Stuard Card (eds.)Information Visualization, Morgan Kaufmann Publishers, p.254

Spence, Robert (2001). Information Visualization, Edinburgh, Pearson Education

Wilkinson, L. (1999).The Grammar of Graphics, NY

The Mars Exploration Rover / Collaborative Information Portal

Joan Walton	Robert E. Filman	John Schreiner
NASA	RIACS	NASA
Ames Research Center	Ames Research Center	Ames Research Center
Moffett Field, CA 94035	Moffett Field, CA 94035	Moffett Field, CA 94035
U.S.A.	U.S.A.	U.S.A.
Joan.D.Walton@nasa.gov	rfilman@arc.nasa.gov	John.A.Schreiner@nasa.gov

Abstract

We describe the architecture and interface of the Mars Exploration Rover/Collaborative Information Portal (MER/CIP), a system for integrating operational and scientific information for managing the 2003 Mars Exploration Rovers. MER/CIP displays schedules, notifies of events and the arrival of scientific data products, displays those products, and facilitates collaboration, all within the context of user-personalized access and interfaces. MER/CIP is an Internet Java application connecting to a web-services–based middleware and data- and meta-database back ends.

1 Managing Mars Rovers

Astrologers argue that the alignment of the planets governs human affairs. Scientists scoff at this idea. However, there is one important exception, when planetary alignment matters to science: launch windows for planetary exploration. In late May and early June, 2003, Mars will be in position. Two Mars Exploration Rovers (MER) (Figure 1) will rocket towards the red planet. The Rovers will perform a series of geological and meteorological experiments, examining geological evidence for water and conditions once favorable for life (Cornell/Athena 2003; Jet Propulsion Laboratory, 2003).

Back on earth, a small army of surface operations staff will work to keep the Rovers running, sending directions for each day's operations and receiving the files encoding the outputs of the Rover's six instruments. (Mars is twenty light minutes from Earth. The Rovers must be fairly autonomous.) The fundamental purpose of the project is, after all, Science. Scientists have experiments they want to run. Ideally, scientists want to be immediately notified when the data products of their experiments have been received, so that they can examine their data and (collaboratively) deduce results and plan the following experiments.

Figure 1: An artist's rendition of a Mars Exploration Rover on Mars

Mars is an unpredictable environment. We issue commands to the Rovers. However, we don't know if the commands will execute successfully, nor which of the objects sensed by the Rovers will be worthy of further examination. The steps of what, to a scientist, are conceptually individual experiments may be intermixed and scattered over a large number of activities. While the scientific staff has an overall strategic idea of what it would like to accomplish, concrete activities are planned daily. The data and surprises of the previous day need to be integrated into the negotiations for the next day's activities, all synchronized to a schedule of transmission windows. "Negotiate" is the operative term, as different scientists want the same resources to run possibly incompatible experiments. Many meetings plan each day's activities.

The Mars Exploration Rover/Collaborative Information Portal (MER/CIP) provides a centralized, one-stop delivery platform integrating science and engineering data from several distributed heterogeneous data sources. Key issues that MER/CIP addresses include

- *Scheduling and schedule reminders.* Operations planning is driven by meetings. Participants need dynamic information about where they need to be and cross-correlation to activities on Mars. Rather shortsightedly, all extant calendar tools presume a 24 hour day; A Martian Sol is 24.66 hours. Scheduling with respect to Mars time is critical, for the Rover is powered by sunlight. For scheduling, both time-scales must be visible.
- *Tracking the status of daily predicted outputs.* The outputs of experiments are radio-transmitted to earth daily. Experimenters want know what is planned to be done on a given Sol and what actually happened. They need to be informed when their data products have arrived. However, it is difficult to track the path from scientific command through the interleaving of command execution on to the data.
- *Finding and analyzing data products.* This includes searching through the data-products space and analyzing the data files found there. Such examinations can be as simple as viewing pictures or as complex as running scientist-created data analysis software. The data is stored in existing heterogeneous structures, developed independently and obliviously to the needs of the portal.
- *Collaboration.* Scientists and operations managers want to share information including not only data products, but also the results of analyses and annotations of these products and analyses.
- *Announcements.* MER/CIP serves as a primary mechanism for broadcasting to the staff announcements of events such as changes in schedules of meetings.
- *Personalization.* User interfaces, data product awareness and access rights to data must be personalized to the preferences and rights of each user. In particular, MER/CIP servers two very different communities: scientists, interested in the results of particular experiments, and operations staff, responsible for maintaining Rover health and safety.

2 Goals

Our goal in developing MER/CIP (and related projects, an emerging technology we call the *InfoCore Information Infrastructure*) (Walton, Filman, Knight, Korsmeyer & Lee, 2001; Walton, Filman, & Korsmeyer 2000) has been to create a generic information infrastructure for integrating scientific and engineering data. Key elements of this domain include:

- Integrating heterogeneous data sources
- Managing large amounts of data
- Supporting the use of unstructured data
- Controlling access to data in a distributed and possibly federated environment according to the rights and privileges of particular users
- Facilitating collaboration
- Providing tools for browsing and analyzing a range of data
- Presenting quality interfaces for the above tasks
- Doing all this in a familiar, easily installed and manipulated environment.

Figure 2 shows a screen-shot from the current version of MER/CIP. Key elements of this interface include the simultaneous access to a variety of different information sources using a variety of

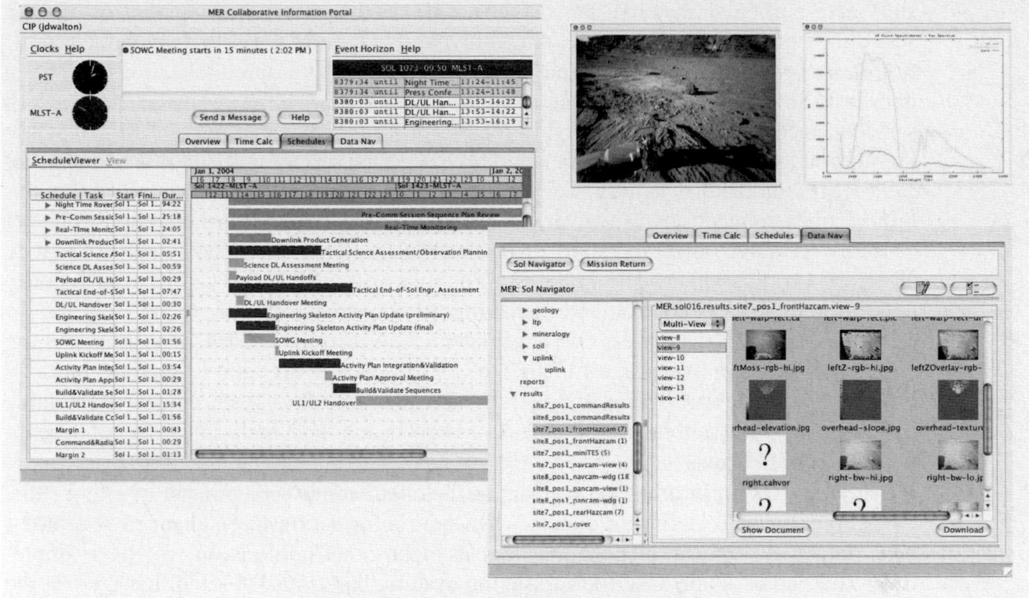

Figure 2: The MER/CIP user interface

GUI themes; the integration of numeric, structured, and photography information; scheduling tools based on Mars time, and the implementation of the system as a Java application.

3 MER/CIP Architecture

Architecturally, MER/CIP is a client-server web-services Java application (Figure 3). Key elements of this architecture are

- *Clients are applications.* The client program is a Java application, delivered over the net as a self-installing Jar file. The client application accesses services through web services and Java messaging. "Clients are applications" contrasts with our original design of "clients are applets in web browsers." Developers have more control over the organization and quality of interaction of applications than applets. We can skip debugging in each version of each browser. A Java application as a net-accessible program is enabled by recent implementations of executable jar files. However, running as an application requires us to manage our own security and encryption, and limits the variety of data types that are easily rendered. Java classes that render HTML are common; Java classes that substitute for exotic browser plug-ins are not. Fortunately, our application does not need the latter. In practice, the transition from applets to application has proven straightforward.
- *Tabbed pane GUI.* The dominant theme of the client GUI is a collection of sub-applications on tabbed panes beneath a "management area." The management area contains tools for quickly viewing the current time (Earth and Mars time zones), important announcements, and selected upcoming events. The tabbed panes contain more full-featured tools for navigating the space of files in the data repository, viewing event and staffing schedules, retrieving data products such as images, reading reports and summaries, and plotting telemetry data. The tools are interconnected and can call upon each other to provide functionality. For example, a report generated by the mission

staff at a particular meeting can be retrieved via the report-viewing tool or via the schedule-viewing tool. The user can then invoke the report viewer to display the report.

- *Middleware and web services.* The primary mechanism for sub-applications to obtain data is to invoke web services on a middleware server (Lea & Vinoski, 2003). Web services encode object structures and remote procedure invocations as XML, and use the normal HTTP protocols for transport. We use web services, in contrast to an earlier Java RMI architecture, as web services do not compromise firewall security by maintaining open connections. However, off-the-shelf XML/object translators don't present as richly recursive object model as RMI. The transition required modifying certain data structures to be simpler. The middleware is also responsible for vetting user identity and enforcing access control, and managing the movement of data to and from the repositories and CIP databases. Of particular relevance to the MER/CIP task is caching data: when an interesting data file arrives, 250 users may all want to see it simultaneously. The middleware server is driven by an Enterprise Java Bean model. It's notable in its use of both stateful and stateless session beans, and both container- and bean-managed entity beans.
- *Messaging.* Not all information flowing to the client comes as the result of client calls. One requirement is notifying clients when new information (e.g., changed schedules, new data files) appears. To handle this asynchronous notification, we have implemented a publish-and-subscribe messaging system, based on JMS and managed in the middleware. Clients subscribe to events of interest (like broadcast announcements and the appearance of new data files). Tools like the data acquisition monitor and broadcast announcer generate events that can match these subscriptions.
- *Data acquisition monitor.* The Rovers consume and produce a variety of data files, being presented with compilations of command sequences and returning the pictures and data from a variety of cameras and sensors. Scientists and operations staff want to know which commands have been or will be sent, and want to examine the data that has returned. This information resides on several different legacy MER Mission Data Servers. MER/CIP needs to know when things have arrived and where to find them. The data acquisition monitor process runs asynchronously, discovers new data files, inserts into the MER/CIP meta-database information about those files, and notifies the publish and subscribe system about them.
- *Meta-database.* MER/CIP (like the other systems in the InfoCore family) keeps two different kinds of data. Ordinary datasets (basically, data under control of MER/CIP, like user interface preferences) are kept in ordinary databases. Large data products (like Rover camera images) reside on the systems associated with the instruments that collected them. A meta-database keeps information about such files, including where to find them and searchable properties.. Dominating themes of the meta-database are (1) Scientific data is often naturally hierarchical. For example, a database may be composed of a series of experiments, each of which has a number of configurations. For each configuration, many identical steps may be performed (e.g., a series of photographs from a specific camera). The actual number of layers in this hierarchy varies among domains (and sometimes even varies within a domain) but the hierarchical nature is common. (2) The attributes of interest for any given experiment are numerous and vary from step to step. Thus, a conventional relational database organization of well-defined columns will prove inadequate. Instead, InfoCore systems use a relational database with mechanisms for expressing both "part-of" and dynamically defined relationships.

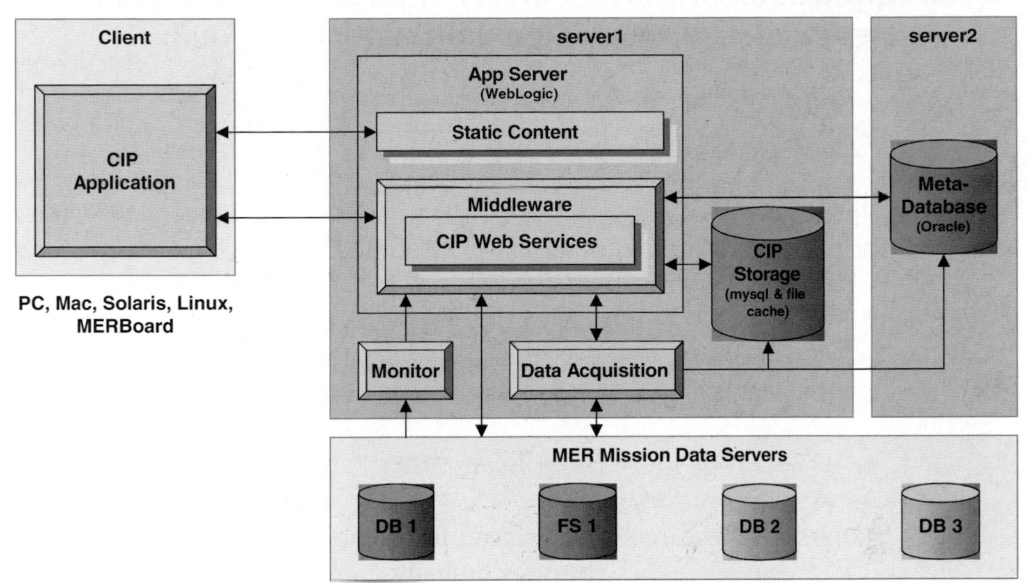

Figure 3: MER/CIP Software Architecture

4 Concluding Remarks

And astrology? Unlike most software projects, MER/CIP must be ready when the planets align. This hard deadline is proving to be an ambitious goal. At this writing, MER/CIP is nearing its version 1.1 release, consistently increasing its project management functionality and garnering enthusiasm from its prospective user base. In the complex mission operations environment, MER/CIP offers simple, easy-to-use tools that enable fast, secure access to critical information and afford MER/CIP users enhanced situational awareness. As the mission draws nearer, the need for an application that cuts across boundaries—between science and engineering data, Mars and Earth time zones, management and team member communications—to provide one-stop shopping for mission information is readily apparent. MER/CIP is rapidly moving beyond its initial role as mission enhancement and becoming an important contributor toward mission success.

References

Cornell/Athena Team (2003). Mission to Mars. http://athena.cornell.edu/the_mission/index.html

Jet Propulsion Laboratory (2003). 2003 Mars Exploration Rover Mission. http://mars.jpl.nasa.gov/missions/future/2003.html

Lea, D.; & Vinoski, S. (2003). Middleware for web services. *IEEE Internet Computing*, 7, 28–29/

Walton, J., Filman,R. E., Knight, C., Korsmeyer, D. J., & Lee, D. D. (2001). D3: A Collaborative Infrastructure for Aerospace Design. Workshop on Advanced Collaborative Environments, San Francisco, August 2001.

Walton, J., Filman,R. E., & Korsmeyer, D. J. (2000). The Evolution of the DARWIN System." *2000 ACM Symposium on Applied Computing* , Como, Italy, (pp. 971–977).

The Application of User-Centered Interaction Concepts to the Design of a Wireless Signal Strength Test Analyzer – A Case Study

Hong-Tien Wang

Center for Research of Advanced Information Technologies
Tatung Company
22 Chungshan North Road, Section 3
Taipei, 104 TAIWAN
avinw@tatung.com

Chien-Hsiung Chen

Graduate School of Design
National Taiwan University of Science and Technology
43 Keelung Road, Section 4
Taipei, 106 TAIWAN
cchen@mail.ntust.edu.tw

Hung-Liang Hsu

Center for Research of Advanced Information Technologies
Tatung Company
22 Chungshan North Road, Section 3, Taipei, 104 TAIWAN
dannyh@tatung.com

Abstract

The purpose of this design study is to apply the user-centered interaction concepts to the design of an innovative, user-centered wireless signal strength test analyzer that helps Manage Information System (MIS) personnel detect and analyze wireless signals. The user-centered interaction concepts were incorporated in the overall iterative interaction design process. Various user-centered interaction design techniques were explored in this design study, such as 1) the use of user requirements analysis among different MIS personnel based on various task objectives, 2) the application of rapid prototyping technique (both on-line and off-line) to the design of this innovative test analyzer, 3) The use of interface usability testing methods together with contextual inquiry and non-participant observation techniques for acquiring verbal and visual information pertaining to MIS personnel's interactions with this test analyzer. The results generated from this study demonstrate that MIS personnel can perceive and understand well the information presented on the proposed user interface and the application of user-centered interaction concepts in the design process can effectively enhance the functions and usability of this wireless signal strength test analyzer.

1 Introduction

As the progress of digital communication technology, humans from every corner of the world are able to interact with each other in a wireless manner. The quality of this type of wireless communication is determined by its wireless signal strength and distributed locations, i.e., the quantities and locations of access points. Because of this, an innovative test analyzer embedded within a Web Pad platform with a 10" touch-sensitive screen was used to help MIS personnel detect and analyze wireless signals. Because it is such an innovative device, user-centered interaction concepts were used to facilitate its overall iterative interaction design process. The purpose of adopting user-centered interaction concepts in this design study is to create useful user interface for this test analyzer based on the MIS personnel's viewpoints; i.e., to accommodate their interaction preferences and needs. By so doing will help MIS personnel establish a more stable and secure wireless environment within which users of various types of wireless devices (e.g., notebook computers, mobile phones, personal digital assistants, etc.) can enjoy the efficiency and effectiveness when communicating with each other.

2 User-Centered Interaction Design Concepts

To achieve the maximum usability of this test analyzer, the authors incorporated user-centered interaction concepts in the overall iterative interaction design process. The user-centered interaction concepts emphasize the placement of primary users at the center of all design considerations. These concepts have been taken into various discussions and applications by many researchers (Baecker, Grudin, Buxton, & Greenberg, 1995; Hewett & Meadow, 1986; Norman & Draper, 1986; Shneiderman, 1998) because understanding the basic principles of user-centered interaction concepts can facilitate interaction designers to create a device with better interface usability. In fact, the objective of applying user-centered interaction concepts to the design process is to help create interfaces with better utility to accommodate users' preferences and needs. Rubin (1994) claims that user-centered interaction design is an evolutionary process that requires interaction designers to understand that an easy-to-use user interface can be acquired through a process of trial and error, discovery, and refinement. The result of a user-centered interaction design process should be a higher quality interface that can enhance user's task performance while meeting their preferences and needs. User-centered interaction concepts not only represent the techniques, procedures, and methods for designing effective user interfaces, but also represent the techniques to be used in an iterative interaction design process. Therefore, discovering user preferences and needs is the first priority for an interaction designer to apply user-centered interaction concepts efficiently.

3 Design and Research Techniques

To achieve the objective of creating a useful wireless signal strength test analyzer, the authors first met with MIS personnel to determine the functions and specifications of this device. Basic task operating steps were also clarified during the initial meeting. After that, participatory design concepts were adopted in this design study. That is, the MIS personnel were asked to participate in the overall design and evaluation processes. In this design study, several major design and research techniques were used, which will be discussed in the following sections.

3.1 User Requirements Analysis

Generally speaking, user requirements are statements or constraints of a system (Kotonya and Sommerville, 1998). User requirements can be viewed as design specifications to facilitate the design of system functions shown on the user interface. After the user requirements are collected and analyzed, an interaction designer can start working on generating various design concepts based on these user requirements.

In this design study, various user interface sketches were first created based on the results from user requirements analysis. These sketches serve to help both the MIS personnel and the authors visualize the design concepts of this test analyzer. After that, a basic layout of required functions and information on the screen of this test analyzer was determined, and several refined ideas were then generated and represented by more computer-generated images. For example, the style of link quality and signal strength indicators were represented in a semicircular form together with numerical information shown below. It is agreed that the extra information provided by the indicator can help users read the measurements better (see Figure 1).

Figure 1: The Link Quality and Signal Strength Indicators were Designed with a Semicircular Form

Based on the results generated from user requirements analysis, the basic user interface features of this test analyzer were suggested. Specifically, the user interface was designed with a primary window to show all the wireless signals detected from various sources under that environment. The detailed signal source and strength information was designed in a group so that it could be easily read by the MIS personnel. A memo pad was also suggested to provide help information and allow users to take notes while using this device.

3.2 Rapid Prototyping

Rapid prototyping is a useful tool to help an interaction designer create and evaluate a design concept within a very short time. A user interface design prototype can be constructed by means of low-fidelity pencil and paper drawings or high-fidelity interactive computer simulations. Rapid prototyping is important because an interaction designer can quickly put his/her innovative ideas into concrete representation. It can also serve as a vehicle for potential users to interact with the design and evaluate its functions and usability. It can facilitate the communication between the interaction designer and potential users as well.

In this design study, after the initial screen layout and interface functions of this test analyzer were determined, an interactive type of simulation was created to help visualize the user interface. The interface functions of this test analyzer were evaluated by the MIS personnel in order to obtain their responses about the usability of this prototype. A task scenario was planned in advance for the evaluation purpose.

3.3 Interface Usability Testing

The concept of interface usability describes and measures how easy and effective a user interface is. In fact, research on interface usability should focus on both ease of learning and ease of use pertaining to users' interaction tasks (Lindgaard, 1994). Chapanis (1991) also argues that the usability of an interaction device can be measured by how easily and effectively the device can be used by its target users, given particular kinds of support to carry out a fixed set of tasks in a defined environment. An interaction designer can realize innovative concepts on useful devices by designing easy-to-use user interfaces. As a result, the device designed with better interface usability considerations can be used more easily and effectively by all its users.

In this design study, in order to ensure good interface usability, various user-centered interaction design principles with ergonomic considerations were incorporated in the interface design process. For example, this test analyzer adopted a graphic user interface design concept so that MIS personnel could interact with the test analyzer easily without memorizing any command languages. The touch-sensitive screen enables the MIS personnel to operate the test analyzer by tapping on the interface functions with a stylus.

In addition, Hackos and Redish (1998) argue that rapid, iterative prototyping with usability testing is an excellent means of testing various user interface design concepts. An interaction designer can easily tell which design concept is better based on the results of usability testing. In fact, to an interaction designer, the goal of conducting usability testing is to improve a user interface's usability by ensuring that it is not only easy to learn and use, but that it also can provide a high level of functionality to its target users at the same time. It is important for an interaction designer to understand that the process of usability testing is not just a single step. It is a series of interrelated, iterative testing process that can be planned and designed carefully to help generate a reliable result.

In this design study, interface usability was achieved by asking the MIS personnel to actually "walk through" the simulation prototype. Other useful research techniques, such as contextual inquiry that focuses on the open-ended interviewing and observing of the MIS personnel within their natural working environments and non-participant observation with detailed note-taking tasks were utilized to help collect both quantitative and qualitative data. More specifically, the contextual inquiry was conducted by interviewing the MIS personnel after they interacted with the simulation prototype. That is, by talking to them, it became clear what their interaction preferences and needs were regarding this test analyzer. In addition, in order to understand how the MIS personnel interact with the user interface, the authors went together with MIS personnel to several university campuses and observed and recorded how they used this test analyzer to detect wireless signals and other relevant activities. These data were then analyzed to help generate the patterns of MIS personnel's interaction styles and needs to make the user interface of this test analyzer more useful. This usability testing process was conducted several times until both the MIS personnel and the authors were satisfied with the final design outcome (see Figure 2).

4 Conclusions

This design study explored an iterative interaction process pertinent to the application of user-centered interaction concepts to the design of an innovative wireless signal strength test analyser. Several important design and research techniques (i.e., user requirements analysis, rapid prototyping, and interface usability testing methods together with contextual inquiry and non-participant observation techniques) were introduced in this design study. After several design iterations, the user interface of this test analyzer was implemented. It is hoped that by using the user-centered interaction concepts, MIS personnel who utilize this innovative wireless signal strength test analyser can detect and analyze wireless signal strength in a more efficient and effective manner.

5 Acknowledgements

The authors would like to express our special thanks to Dr. Alan Pan, Ms. Chiao-Tsu Chiang, Ms. Yueh-Chi Wang, Ms. Chieh-Yu Chan, and Mr. Shih-Chieh Chen for providing invaluable design recommendations and offering great help during the entire design process.

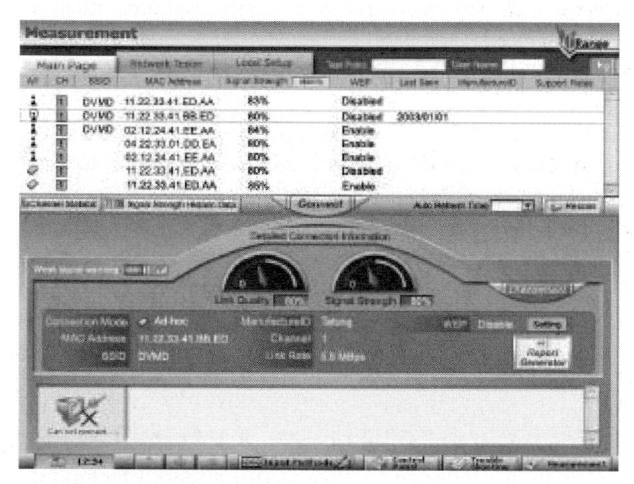

Figure 2: Final Design of the User Interface of the Wireless Signal Strength Test Analyzer

6 References

Baecker, R. M., Grudin, J., Buxton, W. A. S., & Greenberg, S. (1995). *Readings in Human-Computer Interaction: Toward the Year 2000*. San Francisco, CA: Morgan Kaufmann.

Chapanis, A. (1991). Evaluating usability. In B. Shackel & S. Richardson (Eds.), *Human Factors for Informatics Usability*. Cambridge: Cambridge University Press.

Hackos, J. & Redish, J. (1998). *User and Task Analysis for Interface Design*. New York: John Wiley & Sons.

Hewett, T. T., & Meadow, C. T. (1986). On designing for usability: An application of four key principles. *Proceedings of CHI '86*, 247-251.

Kotonya, G. & Sommerville. I. (1998). *Requirements Engineering: Processes and Techniques*. Chichester: John Wiley & Sons.

Lindgaard, G. (1994). *Usability Testing and System Evaluation: A Guide for Designing Useful Computer Systems*. London: Chapman & Hall.

Norman, D. A., & Draper, S. W. (1986). *User Centered System Design*. Hillsdale, NJ: Lawrence Erlbaum.

Rubin, J. (1994). *Handbook of Usability Testing: How to Plan, Design, and Conduct Effective Tests*. New York: John Wiley & Sons.

Shneiderman, B. (1998). *Designing the User Interface: Strategies for Effective Human-Computer Interaction*. (3rd ed.). Reading, MA: Addison-Wesley.

The WWW of Information Structures Design for Chinese Users

Chen Zhao   *Kan Zhang*

Human Factors and Engineering Psychology Laboratory

Chinese Academy of Sciences, Beijing, P. R. of China, 100101

Abstract

In this research, a Cognitive-Walkthrough experiment was conducted to explore whether there are the differences when Chinese users and American users organize the information structure. Moreover, the laboratory experiments and questionnaire were used to investigate how to design the information structure for Chinese users from the individual cognitive difference viewpoint. The results showed that American users prefer to use the network structure and Chinese users prefer to use the hierarchical structure while organizing information by drawing the mental map. For Chinese users, individuals' time cognitive style affects information structures on the efficiency of information retrieval. For the polychronic users, task complete time with a hierarchical structure was significantly lower than with a network structure. For the monochronic users, task complete time with a network structure of an information system was significantly lower than with a hierarchical structure.

1 Introduction

The Internet takes a more and more important role for accessing information. The World Wide Web has approximately 2.2 million Web sites offering publicly accessible content. These sites contain nearly 300 million Web pages. In China, there were 60,000 internet users in 1997 and by the April 2002, the users were 56.6 millions. However, as Internet users, we mostly met some of the following problems while surfing the Web: can't find what we want; can't get back; the content is unclear, feeling lost while surfing, the site is complicated, etc. Survey results show that users found the information they were looking for only 42% of the time; 51% of all sites are not organized according to simple, easy to understand concepts and 90% are not organized correctly. (Forrester, http://educorner.com/courses/ia/); 70% Enterprises are unsatisfied with their website designs; 62% shopper on-line gave up finally the commodities they were looking for. (Liu, Z. J. etc, 2001). The Internet is a wonderful tool for accessing information. But if all that content is not accompanied by context and organization it won't be of much help. Hence, how to build up information architecture to provide information efficiently in this complex net of information has been one of the most important fields in Human Computer Interaction.

Designing information architecture is the process of organizing, labeling, designing navigation and searching systems that helps people find and manage information more successfully. Organization systems are the ways content can be grouped and composed of organization schemes and organization structures. Information schemes define the shared characteristics of content items and influences the logical grouping of those items. The way we classify information is what we call an organizational scheme. Group items can be grouped alphabetically, geographically, by topics, task-oriented, etc. Information structures is the way different groups of information relate to each other is what defines the structure of the site (Rosenfeld. L. & Morville. P, 1998).

Mohageg M.A. (1992) did the research on the information structure. His results indicated that users of the hierarchical linking structure performed significantly better than those using network linking under searching task. Yee-Yin Choong (1996) did the cross culture study on designing information schemes for Chinese and American users. Her work showed that for the Chinese users, error rates with a thematic structure were significantly lower than with a functional structure. For the American users, error rate with a functional structure of an information system was 64% lower than with a thematic structure. Smith. P.A. (1996) reviewed the usability research applied to the World Wide Web on virtual hierarchies and networks. He concluded that the previous research has shown that exploratory tasks are best supported by a network or combination information structure, whilst search tasks are best supported by a hierarchical information structure. However, no single structure will ever be appropriate for all users at all times. Then what are the cross-culture challenges for information structure? How does or should information architecture address special populations and others with distinctive cognitive attributes? In this paper, the method of Cognitive Walk-through experiment was conducted to explore whether Chinese users have different information structure with American users while organizing information. Next the laboratory experiment was conducted to explore how to design the information structure for Chinese users. The purpose of this study is to probe the information structure design from the individual cognitive difference viewpoint and investigate how to design information structures for individual Chinese users.

2 Cognitive – Walkthrough Study

In the Cognitive-Walkthrough experiment, three American and two Chinese participants were given five tasks to finish by looking for the specific information in the user manual and operating on the Internet. Participants were asked to verbally work through problems which they encountered and how they thought the system should behave, where they expected to find information, and what they found confusing or difficult. After the walkthrough, the participants were asked to draw a map how they would organize the information in the Users' Manual. The mental maps showed that American users would like to use the network structure and Chinese users would like to use the hierarchical structure when they organized the information.

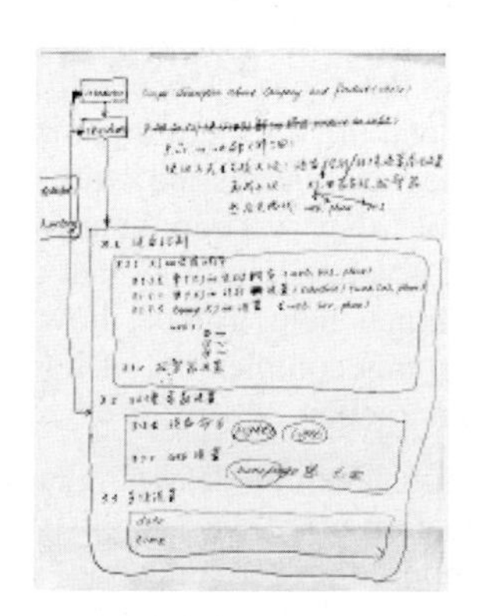

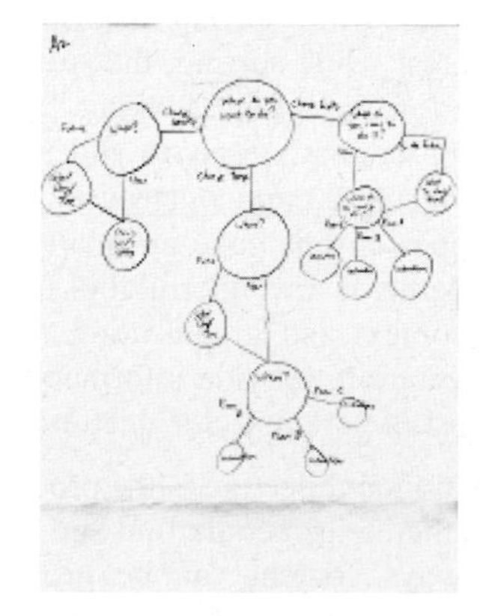

Figure 1 Mental maps of Chinese and American users' information structure

3 Experiment Study

The previous research indicated that no single structure could ever be appropriate for all users at all times. The purpose of this study was to explore how information structure should address populations with distinctive cognitive attributes. Cultural variables related to time perception have been studied extensively. One of the primary distinctions is between "monochronic" people and "polychronic" people. Cultures with a monochronic time orientation treat time in a linear manner. Monochronic people tend to organize activities sequentially and are considered to be good at doing one job at a time. Polychronic time orientation is characterized by many things happening at the same time. Polychronic people organize activities in parallel. People in cultures that follow polychronic time are considered to be good at multiple tasks. In general, North American and northern European cultures tend to be monochronic while countries from Asia, South American and Middle Eastern cultures tend to be polychronic (Hall, 1983, 1989). It is a natural assumption that hierarchical information structure matches the linear monochronic time orientation and the network information structure matches the multiple polychronic time cognitive style. However, according to the mental map from the cognitive walk-through study, Hall's assumption was not certified. In our other research, Chinese tend to be polychronic was not supported neither. Our previous research indicated that most Chinese participants were neutral in time orientation rather than polychronic as Hall had hypothesized (Zhao. C, Plocher, T, etc. 2002). Many scholars have been worked on the tools to measure time orientation. Of these, the MPAI3 (Modified Polychronic Attitude Index 3 by Lindquist et al. has shown high reliability. Therefore, MPAI3 questionnaire was used to categorize different time cognitive styles users. The reliability Alpha of MPAI3 questionnaire was 0.8486 in the survey for Chinese. So users' time cognitive styles were characterized into monochronic, polychronic and neutral.

32 Chinese undergraduates took part in this study. An on-line HyperCard-based astronomic database called AST was created specially for this study. Independent variables were Information structure (hierarchical vs. network) and Time Cognitive Style (polychronic, monochronic and neutral); Dependent variable was Task Complete Time. A 2×2 mixed design was used. The independent variable of information structure was within-subjects variable and the variable of cognitive style was between-subjects variable. Participants were given a period of time to explore

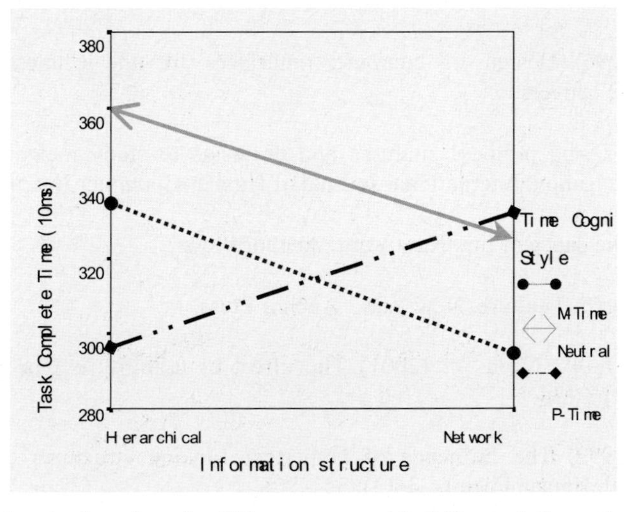

Figure 2. Task complete time for Chinese users with different information structures

the database and become familiar with hypertext environment. Then participants were given the experimental tasks to search for certain answers in the database. A repeat measure was performed on the task complete time data. The main effect of Information structure was not significant ($F(1,27)=1.706$, $p=0.203$); the main effect of Time Cognitive Style did not reach the statistic significance neither ($F(2,27)=1.398$, $p=0.264$). However, interaction between information structure and time cognitive style was significant ($F(2,27)=3.596$, $p=0.041$). For the polychronic users, task complete time with a hierarchical structure was significantly lower than with a network structure. For the monochronic users, task complete time with a network structure of an information system was significantly lower than with a hierarchical structure.

4 Discussion

The results showed that American participants would like to use the network structure and Chinese participants would like to use the hierarchical structure while organizing information by drawing the mental maps. For Chinese users, individuals' time cognitive style affects information structures on the efficiency of information retrieval. The hierarchical structure benefited the polychronic users and the network structure benefited the monochronic users on task complete time. These results indicate that it will benefit the efficiency of information retrieval to design the information structures based on user individual cognitive styles.

In the previous research on information structure, cross-culture comparison and dynamic study of time behavior were rarely found. This work was an exploratory study from individual cognitive perspective. In the cognitive walk-though study, American participants draw hierarchical structure and Chinese participants draw network structure while organizing information. However, we cannot conclude that since the sample of participants was very small. In the experimental study, only Chinese users participated the laboratory experiment. The comparison between American users and Chinese users should be addressed in order to investigate the information structure design for different culture users when the resources are available.

5 Reference

Choong, Y.Y. (1996) Design of computer interfaces for the Chinese population. Ph.D. Dissertation, Purdue University

Gray, S.H. (1990) Using protocol analyses and drawings to study mental model construction during hypertext navigation. International Journal of Human-Computer Interaction, 2:359-377

Hall, E.T. (1983) The dance of life. New York: Anchor Press.

Hall, E.T. (1989) Beyond culture. New York: Anchor Press.

Liu, z. J., Zhang. L. P., Chen. Y. (2001) The effect of usability engineering on IT product. ComputerWorld, Sept. 17th

Mohageg, M.F. (1992) The influence of Hypertext Linking structures on the efficiency of information retrieval. Human Factors, 34(3): 351-367

Rosenfeld. L. & Morville. P. (1998). Information Architecture for the World Wide Web. O'Reilly & Associates. Inc.

Smith, P. A. (1997) Virtual hierarchies and virtual networks: some lessons from hypermedia usability research applied to the World Wide Web. International Journal of Human-Computer Studies, 47: 67-95

Zhao, C., Plocher, T., Xu, Y.F., Zhou, R.G., Liu, X., Liang, M. & Zhang, K. (2002) Exploring Chinese Polychronicity Time Behavior. Proceedings of APCHI 2002, Beijing, Science Press

Section 7

Virtual Environments

Evaluation Consolidation of Virtual Reality Tools and Applications within VIEW project

Angelos Amditis, Ioannis Karaseitanidis, Niki Boutsikaki, Evangelos Bekiaris

Institute of Communication and Computer Systems of the National Technical University of Athens
9, Iroon Politechniou Str. Zografou GR-15773
angelos, gkara, niki@esd.ece.ntua.gr, ita@ita.gr

John Wilson

Institute for Occupational Ergonomics
University of Nottingham
Nottingham NG7 2RD

john.wilson@nottingham.ac.uk

Abstract

One of VIEW of the future project aims is the assessment of Virtual Environments (VE) when used to a work place. These will be evaluated from the Usability, Safety, Ergonomic and mostly Socio Economic point of view. The consolidation of all these evaluations will provide a useful tool to an industry interested in implementing Virtual Reality (VR) technology. Key stakeholders of the industries will become able to effectively utilize VR tools and applications, optimizing, at the same time, their benefits.

1 Introduction

Virtual Reality technology emerged since the late 70s but several technological barriers combined with high implementation costs prevented it from being widely accepted by the industry sector. Later research activities, however, have improved the cost effectiveness of VR technology by decreasing its cost and at the same time by overcoming serious technological barriers.

Thus, in recent years VR has increasingly started to be implemented for a vast variety of tasks, such as entertainment, manufacturing industries etc. Several case studies have indicated that VR tools and applications can be implemented to work places of an industry with positive results both to the companies and the end users. VR can be used to almost every sector of an industry varying from the designing face of a product, to prototyping, testing, manufacturing even marketing and management.

On the other hand, there are still a lot of inhibitory factors that do not allow the wide implementation of VR tools and applications to work places. Among these the still high implementation cost can be included, the lack of knowledge for the actual benefits of using VR and the lack of expertise that will enhance the efficiency of VR implementation to an industry's work place. Industries need to be informed of all possible implications of using Virtual Reality technology and also considering that the investment cost is relatively high they need to be able to foreseen the economic benefit of it.

VIEW of the future "Virtual and Interactive Environments for Workplaces of the Future" (IST-2000-26089) is a still ongoing European funded research project that aims on the development of VR products such as applications and guidelines to facilitate industries to successfully and effectively implement VR to their internal procedures. Innovative, user friendly, input devices will also be developed to further assist VR users to effectively work emerged in VR environments.

One major part of VIEW research activities is the evaluation of Virtual Reality technology used and developed within the project. The evaluation intents to enlighten the key stakeholders of industries that are interested in implementing VR to their companies in issues related to Usability, Safety, Applicability, Ergonomic and mostly Socio-Economic implications of VR systems. A Cost – Benefit Analysis of the VIEW products will further facilitate the investment decisions of industries.

2 Usability and Safety Evaluation

For the complete picture of a VR application or tool to be formed a Usability Evaluation need to take place. This is performed through a set of pilot tests where subjects are used to try various combinations of VR tools and applications. Both objective and subjective measures are used. Objective evaluation includes physiological measures in order to define the stress levels and general impacts of VR to the end-users health. Furthermore, subjective measures complement the usability evaluation in order to judge the acceptance rating of the end-user in terms of efficiency, satisfaction, helpfulness controllability and learn ability. Subjects define these parameters by filling usability questionnaires distributed after emerging to Virtual Environments.

One other very important factor than need to be assessed, when evaluating a VR system, is the impact on user's health and safety. The key concern from the literature is VR-induced sickness, experienced by a large proportion of VR participants, although the majority of these effects are mild and subside quickly (Nickols & Harshada, 2002). Physiological measures combined with questionnaires, where symptoms after VR immersion are recorded, are therefore used to this end.

3 Ergonomic Evaluation

Within ergonomic evaluation parameters such as Human Machine Interface (HMI) of VR, system complexity and interactivity are studied. This will help comparing and selecting the most appropriate VR tool each time for a specific task within the work place. Again both subjective and physiological measures are utilized. The impact and value of VR/VE will be assisted through careful application of ergonomics and real understanding of potential user needs (Wilson, 1999).

4 Applicability Evaluation

Major problems inhibitory of the expansion of Virtual Reality implementation to industries are the limited customizability, the lack of portability and the highly demanding hardware requirements of VR systems (Kan, Duffy & Su, 2001). Partly due to the lack of customer base, their functionality has been designed to focus on the industrial customers that manufacture high-end, high technology, and expensive products. Furthermore, most of the existing systems are generally difficult to customize. This makes it difficult for small-scale industrial companies to decide on the most cost - effective system to their specific needs.

Applicability evaluation within VIEW project will deal with practical advice that can be offered to user companies on how to utilise in a more efficient way VE-technology within their organisations. Counteracting measures for each problem that can be possibly acquainted during the implementation face of a VR system is reported.

For each device used to pilots and especially for the products developed within VIEW a set of guidelines for use is to be included to the consolidation addressing the several points such as; way of installation of the devices; the maintenance needed; the spaces they occupy; the specific purpose they will serve; the expertise knowledge and training that the devices will require from the user; the functions they will have; the other hardware and software they will require; the extensions they may have for future upgrade; the way they may interact with existing devices; the implications they may have to the working hours or procedures or work plan and other recommendations regarding the devices.

5 Socio Economic Evaluation

During Socio – Economic evaluation the effect of VR tools and applications on society, companies and end users is assessed. This is considered one of the most significant tasks since strict governmental laws and authorities may inhibit otherwise useful technologies if they suspect them to be harmful to the population. On the other hand a potential investor on VR technologies wishes to have a clear knowledge regarding the economic benefit and cost of the investment.

From society's point of view, Virtual Reality technology is a pollution free and has almost none harmful effects to the environment. On the contrary VR can help material savings of a company by reducing the need for prototyping and by optimizing the production procedures in a way that wastes are minimized. One problem usually acquainted during socio-economic analysis is the lack of actual data since Virtual Reality is not yet widely used so that it can influence society to a significant and measurable degree. This limitation is considered to VIEW project.

Additionally, from the company's point of view, assessed an initial survey within VIEW project has concluded that Virtual Environments implemented to a workplace can induce positive effects to almost every face of the product's development. Virtual Design tools help minimize the time to market a product, since the design is evaluated and later altered in significantly less time. Also, Virtual Design combined with Virtual Prototyping (VP) reduces the cost connected with the initial face of the development of each product. Using VP the cost of testing is much lower than when using "real" prototypes (Dai, 1998). At the same time VP provides a high level of accuracy of design analysis (Kan, 2001). Virtual Reality can also enhance the performance of the production procedures by monitoring and scheduling the material flow through the manufacturing chain. Virtual Manufacturing (VM) will optimize the production process in a way that cost and time are minimized. Management decisions can be further improved through virtual reality technology by the improvement of information quality available to the key stakeholders of each industry. For each potential change to manufacturing of a product, Virtual Reality can realize the depiction of possible problems before the beginning of manufacturing (Waly, Thabet & Beliveau, 1999). Industries implementing Virtual Technology are characterized by a significant degree of flexibility to changes and improvements and therefore have a competitive advantage on market.

From the end-user's point of view, geographically dispersed (virtual) collaborative work groups continue to become prevalent as electronic technologies expand organizational ability to link together resources that are separated in time and distance (Fulk & DeSanctis, 1995). According to a recent estimate made by the Department of Transportation of USA (DOT), at least 8.4 million US workers are currently members of dispersed teams or groups. The DOT study indicated that the number of virtual teams was over 30 million in the year 2000 (Horvath & Tobin, 1999). Thus, Virtual Collaboration technology has a significant impact on the structure of a work place and the way members of a virtual group interact with each other and socialize. Moreover, there is some evidence that the use of virtual reality may provide some benefits in improving safety and reducing risk in the physical work environment (Duffy, Wu & Parry, 2003). Virtual environment allows simulation of hazardous scenarios. However, the difficulty lies in how to quantify the risk reduction or reduction in severity of problems associated with the workplace due to the use of the new technology (Duffy et al, 1997).

6 Cost – Benefit Analysis

In order to provide information regarding the financial viability of a VR product a Cost – Benefit Analysis is to take place. Detailed cost data will be gathered during the development face of the VIEW project products and willingness to pay will be calculated from relative questionnaires distributed to the pilot subjects during the testing of the products. A market analysis and forecast of future demand will be performed by using expert questionnaires addressed to the most appropriate personnel of each of the companies that participate to VIEW consortium.

Lastly, all the formulas of Financial Analysis will be calculated as shown in Table 1. Financial Analysis will be accompanied by a Compatibility Analysis where the intangible aspects of the VIEW products, that are the ones that can not be expressed in monetary values, will be assessed and a quality profile will be formed by data taken from relative pilot questionnaires.

Table 1: Evaluation criteria for Financial Analysis of VIEW products

Indicator	Description	Measurement objective	Data required for the calculation
NPV	Net Present Value	If NPV>0, VIEW product will be a financially profitable product in the 5 years horizon	Estimated Annual sales of VIEW product in the next 5 years Total production costs of VIEW products Suggested price for VIEW product Initial investment cost
IRR	Internal Rate of Return	If IRR>r the introduction of VIEW product will be financially justifiable	
PBP	Pay-Back Period	To define the time period in which the monetary expressed benefits of VIEW product will outweigh its costs (investment and operational ones)	
EVC	Economic Value to the Costumer	To define the price of the VIEW product	The existing VIEW products total costs The Willingness to pay for VIEW product
TRP	Target Return Pricing	To determine the price that will yield the target rate of return on the investment	Estimated annual sales of VIEW product in the next 5 years Total Production costs of VIEW products Target Rate of Return Initial Investment Cost

7 Evaluation Consolidation Conclusion

In summary, Evaluation Consolidation of a VR tools or applications, within VIEW project, consists of Usability, Safety and Ergonomic Evaluation that will examine the VR system performance, HMI etc, the Applicability Evaluation that will give information to the potential user about the way VR can be installed and used and lastly the Socio Economic Evaluation and Cost - Benefit Analysis that will analyze the implications of VR to a national, organizational and individual level and assess the financial profitability of VIEW products.

The consolidation of all the above tasks will provide all valuable information regarding Virtual Reality technology. Industries that are interested in introducing Virtual Technology into their workplaces can benefit from VIEW project Evaluation Consolidation and make the optimal decision on the most appropriate VR system corresponding to their needs and the way to use it.

References

Dai F., Virtual Reality for Industrial Applications, Berlin, Springer, 1998.

Duffy Vincent G., Flora F. Wu and Parry P. W. Ng "Development of an Internet virtual layout system for improving workplace safety" Computers in Industry Volume 50, Issue 2, February 2003, Pages 207-230

Duffy V.G., C.J. Su, C.L. Hon, C.M. Finney, Safety implementation in manufacturing: implications for using virtual reality in the workplace. in: P. Seppala, T. Luopajarvi, C. Nygard, M. Mattila (Eds.), Proceedings of the 13th Triennial Conference of the International Ergonomics Association (IEA '97), Tampere, Finland, 29 June–4 July 1997, vol. 3, pp. 224–226

Fulk & DeSanctis, 1995. J. Fulk and G. DeSanctis , Electronic communication and changing organizational forms. Organization Science **6** 4 (1995), pp. 337–349.

Horvath, L., & Tobin, T. J. 1999. Twenty-first century teamwork: Defining competencies for virtual teams. Paper presented at the Academy of Management, Chicago, IL.

Kan H. Y., Vincent G. Duffy and Chuan-Jun Su "An Internet virtual reality collaborative environment for effective product design" Computers in Industry ,Volume 45, Issue 2, June 2001, Pages 197-213

Sarah Nichols & Harshada Patel "Health and safety implications of virtual reality: a review of empirical evidence" Applied Ergonomics, Volume 33, Issue 3, May 2002, Pages 251-271

Walyand Ahmed F. Walid Y. Thabet "A Virtual Construction Environment for preconstruction planning" Automation in Construction, Volume 12, Issue 2, March 2003, Pages 139-154

Waly, A., Thabet, W., Beliveau, Y. "Virtual construction for constructability improvement" Novel Design and Information Technology Applications for Civil and Structural Engineering 1999, Pages 71-76

Wilson John R. "Virtual environments applications and applied ergonomics" Applied Ergonomics, Volume 30, Issue 1, February 1999, Pages 3-9

Networked VR for Virtual Heritage

Makoto Ando

MVL Research Center,
Telecommunications Advancement Organization of Japan
4-6-1, Komaba, Meguro-ku, Tokyo, 153-8904 JAPAN
makoto.ando@toppan.co.jp

Abstract

This paper presents a case study of a VR application realized by applying Networked VR to support virtual heritage, which is one of the typical applications of VR. This Networked VR is based on our Scalable VR system, which integrates different types of VR systems. By using Scalable VR, the possibilities of the virtual heritage application can be extended to various fields. We present our preliminary implementation of Scalable VR content, a reconstruction of the Copan ruins of Honduras. Then, we introduce two usages of Networked VR for virtual heritage; a networked virtual guided tour of a museum exhibition, which can be used to convey knowledge of the Mayan civilization to a remote user, and an extracurricular class for children which supports both experiential and group study in a networked virtual Copan.

1 Introduction

Reconstruction of cultural and natural heritages using Virtual Reality (VR) technology has become an important and interesting area of research in archaeology (Forte, Siliotti & Renfrew, 1997). To archive our heritage in digital media for handing down to future generations, or to use VR technology as a tool to develop a procedure for the restoration of decaying heritage, are typical uses of virtual heritage. In addition, recently, virtual heritage presentations have also come into use for museum exhibitions and education, in addition to virtual archaeology ("IML", 2001).

Numerous types of VR systems, such as a fully immersive VR system, a haptic VR system, and a wearable VR system, are being researched and utilized in order to realize these new applications. However, in many cases these systems are used independently, or are interconnected only with the same kind of VR system. We are currently developing a Scalable VR system, a networked VR, which integrates different types of VR systems and enables different performances, and the use of different devices and different user environments (Tanikawa et al., 2002). The goal of the Scalable VR system is to realize a flexible VR application, with which users can select from among different usages and systems according to the environment, while sharing the same virtual space. With our Scalable VR system we aim to extend the virtual heritage applications to include interactive exhibitions in museums or historical education programs in schools.

In this paper, we describe our implementation of Scalable VR content, which is the reconstruction of the Copan ruins of Honduras, one of the ruins of the Mayan civilization. Also we introduce two case studies of the networked virtual Copan; a virtual guided tour for a museum exhibition, which can convey knowledge of the Mayan civilization to a remote user, and an extracurricular class for children which supports both experiential and group study in a networked virtual Copan.

2 The Virtual Copan

The Copan ruins are one of the typical ruins of the ancient Mayan civilization, and are located at the westernmost end of the Republic of Honduras, Central America. Many outstanding buildings constructed with advanced technology and many artistic stone carvings can still be seen in the Copan ruins, and the ruins are registered by UNESCO as a World Heritage in 1980. Many epitaphs and stone carving which serve as a key to understanding both of the Copan dynasties and the fate of the ancient city around the Copan, have been discovered in the Copan ruins, and thus the Copan ruins have become a center for the study of Mayan archaeology, and the subject of projects such as excavations and the deciphering of Mayan hieroglyphics. Therefore, the drawings and ether results of archaeological investigations are comprehensive, and therefore it is valuable to make a virtual representation of the ruins.

In addition to the free move in the virtual space, the ease of moving along a time-axis is also the feature of VR. The virtual Copan consists of two scenes; a faithfully reconstructed appearance of the present-day ruins, and a restored Copan as it was during the prosperous of the dynasties, based on the most recent research. Both scenes were produced under the supervision of the Honduran Institute of Anthropology and History. The area of ruins reconstructed is about 600m x 400m, including almost all key buildings and objects, such as a stone monument and an altar. It also has the background so that it may not become unnatural when looking down on the whole ruins from the sky (Figure1). To create the scene of the present-day ruins, over 2000 photographs taken with 6cm x 7cm film were scanned using a high-resolution drum scanner, and finally about 250 images processed from digitized pictures are used as textures. Figure 2 shows a stone object called "Altar-Q"' which forms the front of No. 16 shrine today. When experiencing the Copan ruins, visitors can observe the Mayan hieroglyphics engraved on the upper surface of the Altar-Q, and they can make the discovery that a Mayan character consists of two or more elements like a Chinese character, that each character has a phonographic role and an ideographical role, there is a strict rule for the arrangement of the characters, and so on. In order to make such observations possible by VR, the high-resolution texture (over 2048x2048) is applied to typical stone monuments, so that viewers can approach sufficiently close to the target without losing the detail.

For the scene of the past Copan, we aimed to restore the shape and color of buildings accurately based on the most recent research and archaeological investigation (Figure 3). The hues and color scheme of the shrine are examined by comparing a piece of mortar saved at the research institute with a standard color guide.

Figure 1: Snapshot of the virtual Copan ruins

Figure 2: A stone object "Altar-Q" (VR scene)

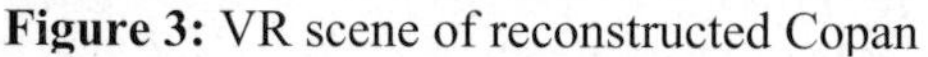

Figure 3: VR scene of reconstructed Copan

Figure 4: Old Copan at the summer solstice

Both of the scenes are include lighting simulation which render the sun in the position for the summer solstice. The summer solstice was one of the most significant days of the year for Mayan people, who had an advanced calendar, and viewers can experience this occasion spatially by means of VR (Figure 4).

3 An Implementation of Networked Virtual Heritage

3.1 A Virtual Guided Tour of the Copan Ruins

The presentation of virtual heritage in a museum is suitable for grasping the general situation of an exhibit, and also for grasping the presented ruins spatially. It allows a visitor to the museum to experience the ruins even through they cannot go to the actual ruins. In order to make it possible for people to join a guided tour not only from a theater but also from a remote place through the network, we propose a virtual guided tour system. The virtual guided tour system consists of two parts; a theater type VR environment with an immersive display system, and a PC based VR system connected through the network. By using the theater type VR environment and presenting a virtual heritage with very high-definition images, the audience can be as deeply impressed as if they have a real experience of the actual heritage. In addition, the tour guide can move around in the heritage, pointing out interesting objects to the audience interactively, according to the audience's reactions. The same views and explanations are provided simultaneously to the remote user, by real-time distributing of the guide's 3-dimensional position data in the virtual space and encoded audio data of the guide's explanation. In addition, each remote user can leave the guided tour temporarily and walk around in the virtual space to gain their own viewpoint, by operating a controller. At this time, the user can see the shadow of the guide as an avatar. When a certain period has elapsed after the user releases the controller, the viewpoint returns to the original position (the position of the guide) automatically. This sequence can be compared to the behavior of someone leaving the group just for a moment, and making their own explorations in an actual tour.

Figure 5 shows an experiment of a virtual guided tour. The theater type VR environment and PC are connected through a gigabit Ethernet, but this is not requirement. Currently, scene graph database of the virtual Copan is stored in each local machine, and thus the traffic is limited to real-time synchronization data and compressed audio data. Thus, the bandwidth is at most about 30kbps, and we believe that distribution of the tour to homes using the Internet is also possible.

The system also provides streaming of the VR scene. This function enables the contents of the guided tour to be distributed to the client machine, which has no 3D graphic capability. In the experiment, we used a 11Mbps wireless LAN to simulate mobile use. The resolution of the streaming image was 320x240. The average delay was 2-3 seconds, and the average frame rate was about 5 frames per second depending on the rapidity of scene movement. Although the image lacked the immersiveness of an experience, it was able to understand the explanation of the guide enough through the distributed VR scene.

Figure 5: The virtual guided tour system
Top: VR theater, Bottom: remote PC

Furthermore, the system provides the function of recording the entire sequence of the guided tour as a set of guide's motion path in the virtual space and his encoded audio data. Using this function, for example, it is possible to archive a lecture by a celebrated expert on the heritage site, and to open such archives to the public so that every user who has an interest in a heritage will be able to experience the lecture at any time, any location.

3.2 Experiential Study and Group Study in Virtual Copan

The use of virtual heritage as a tool for learning through cooperative work is more suitable for school education. In the educational field, it is well known that the discussion and the division of tasks within the group study are very important. VR media offer the advantages of spatial recognition, presence, and interactivity, compared with conventional teaching materials. A networked VR system supports interaction and communication between groups in the shared virtual space. Also by using our Scalable VR system, cooperative work and consensus decision making within the group are supported effectively, and the possibility of group study improves dramatically.

We installed a theater type VR environment and a number of PC based VR systems in the same location (Figure 6). All VR environments are connected through the network. Each group can share its own view as it is generated separately by an individual PC and a common view is displayed on the large immersive screen visible to all groups. The common view, the image of the theater type VR system, is operated by the teacher. In the virtual space, the teacher and all of the children are represented by avatars and they can be recognized by each other in virtual space. In this system, the members of each group cooperate among themselves, and they can work independently in the virtual space by controlling their controller. Moreover, the image on the large screen can be used to show objective view for all groups, to support to formation of an agreement in a group. The most significant feature of this system is that it is possible to present an objective view and a subjective view in the same context. The system is also able to support two important elements of educational learning simultaneously; experiential study by the theater type VR, and group study by the independent PC based VR.

The experiment was carried out in cooperation with an elementary school. We invited 11 children to our laboratory, and held a "Quiz Rally" in the virtual Copan. The children were divided into four groups, and a PC based VR system was assigned to each group. The rules of the Quiz Rally were as follows: a quiz about the Mayan civilization was hidden in the virtual Copan. Children explored the ruins, aiming to find the flag indicating a hint. When they discovered the flag, they decoded the Mayan character written on the flag as an identification number which was part of the quiz, and notified the teacher. They then received a question card. Finally, points can be obtained if their answer was correct.

Figure 6: Extracurricular class: The Quiz Rally in virtual Copan

Before the Quiz Rally, a tour of the virtual Copan using the large screen was performed in order to stimulate children's interest. Also during the rally, the large screen was used to display whole ruins, so that ever child can grasp his/her position in the virtual space at any time. This is very important for cooperative work within a group, because it is said that the member of the group have to recognize their work and situation each other from the objective view, in order to form division of tasks within the group.

From the analysis of the experimental results, we observed that the group uses the motion of their avatar displayed on the screen in order to discuss their present location. This indicates that the system functioned effectively to promote communication and agreement within a group. Also we observed the actions of pursuing the avatar of another group, or standing in front of a flag to interfere with another groups view. This indicates that the avatar functioned as a communication tool which helped participants to recognize each other.

4 Conclusion

This paper presented two implementations of the virtual heritage application using Networked VR. By introducing Networked VR, the usage of the virtual heritage was expanded extremely. As the case shows, it is especially effective for educational field, including the lifelong study. Our implementation is still an early stage, so we intend to improve user interface in the virtual space in order to make the system more flexible and effective for the interactive learning.

5 References

Forte, M., Siliotti, A. & Renfrew, C. (1997). Virtual Archaeology: Re-Creating Ancient Worlds. New York: Harry N Abrams, Inc.

Tanikawa, T., Yoshida, K., Ando, M., Wang, Y., Kuzuoka, H. & Hirose, M. (2002). A Study for Scalable VR by integrating heterogeneous VR Systems. *Proceedings of the Virtual Reality Society of Japan the Seventh Annual Conference*, 417-420 (Text in Japanese, Abstract in English)

TOPPAN PRINTING Co., Ltd. (2001). IML and Virtual Reality. Retrieved February 1, 2003, from http://www.toppan.co.jp/products+service/vr/intro_e.htm

Co-located interaction in virtual environments via de-coupled interfaces

Victor Bayon

VIRART
University Of Nottingham
vxb@cs.nott.ac.uk

Gareth Griffiths

VIRART
University Of Nottingham
Gareth.Griffiths@nottingham.ac.uk

Abstract

This paper describes an interaction style in virtual environments that allows co-located users to manipulate certain interactive features of virtual environments with 2D or non-inmersive types of input devices. The concept proposes to make interaction accessible to groups of users by mirroring some of the VE interaction characteristics and functionality and making them accessible from other types of input devices, such as Personal Digital Assistants (PDA), effectively de-coupling the interaction from the original forms and presenting the co-located users with a richer multi-modal multi-device interaction mechanism.

1 Introduction

Virtual reality multi-display multi-screen and CAVE types of systems are often used to visualise very large 3D environments and models converted from CAD to a virtual reality usable format (Bullinger, Blatch & Breining, 1999). These systems are traditionally used in conjunction with sophisticated 3D input, tracking devices and stereo projection to provide the participant with a "complete" immersive experience in the virtual environment (VE). In general terms, such 3D input devices are designed to perform tasks that can be described as navigation, selection, manipulation and system control (Bowman & Hodges, 1999).

There are different types of information that, when presented in three-dimensional form, can enhance the users' immsersive experience during the use of the system. However, there are instances where interaction and information can be hindered if represented in a three dimensional way (Lindeman, Sibert & Hahn, 1999). In such cases, the representation of such information can be achieved via screen artifacts of a more two dimensional nature, such as text or graphical widgets (Bowman, Kruijff, LaViola & Poupyrev 2000). Other approaches to enhance and facilitate the interaction during the immersive experience by restricting and constraining the 3D interaction to a 2D plane (Smith, Salzman & Stuerzlinger, 2001) make the manipulation of such 2D information with 3D input devices more effective.

As most of the 3D interaction techniques and interfaces can be difficult to implement for each specific input device (Eastgate, 2001), in this paper we propose to replicate some of the 2D functionality found in interactive VE, enhancing some of the manipulation tasks such as selection and task control by bringing them to the 2D interaction domain with 2D input devices with three main objectives:

- Provide a mechanism to allow the access of 2D interaction within the VE with 2D input devices.

- Support multi-modal forms of interaction and multi device and configurations to allow users to use the most convenient form of interaction in different circumstances.
- As the functionality can be replicated in more than one device, more than one (co-located) user could participate in different ways while using the virtual environment.

2 De-Coupling Interaction

The fundamental nature of 3D virtual environments relies on 2D structures such as scenegraphs and object geometry data. The files that form the virtual environment need to specify properties such as the geometry, position, location, rotation etc. The interaction capabilities that are described in the file format are structures of a 2D nature, where objects have attached methods that can be executed. All these structures are hidden from the user when the interactive VE is running, represented instead by the 3D graphics and the events that the input devices are capable of producing and the system is able to understand.

In de-coupling interaction, we propose to make explicitly visible some of the 2D structures that the 3D environment uses for its functioning, facilitating in turn the understanding of manipulation and implementation of 3D interactive VE. Rather than coupling all the interactions to specific 3D input devices and then using 2D input techniques to instigate the interactions, a common implementation approach in developing VE, the goal is to provide more generic methods that rely directly on the manipulation of the 2D structures and information via devices with 2D input capabilities, such as Personal Digital Assistants (PDA's). PDA's and other 2D capable devices have been already used as a means of providing access to 2D interaction features of VE (Hartling, Bierbaum & Cruz-Neira 2002).

3 Concept Demonstrator

This section describes an example of a VE that was created as a first prototype to introduce the concept to a group of VE users and car designers. The prototype VE consists of a car model that allowed a relatively high degree of interaction for the inspection of different parts of the car by groups users. We utilised a User Centred Design (UCD) methodology with the input of the expert VE users, developing and reviewing several interaction scenarios and the subsequent prototypes.

To provide the necessary information to de-couple the interaction from its original form, when the VE is first run, it creates an XML file describing different properties of all the entities present in the scenegraph of the VE (the representation of the 3D information in memory or a file). Among these properties, data to enable the external interaction with the entity is included.

Figure 1 (left side) depicts the VE created. The model allowed the users to open and close the doors, manipulate the bonnet and boot, navigate around the car in a first person perspective, select inside and away viewpoints and take screenshots of the current viewpoint. Figure 1 (right side) shows the XML representation of the VE that is produced after the VE is loaded. The different properties can be inspected directly on a XML compatible browser. As the scenegraph is almost replicated, it is possible to read and understand the structure of the interactions available in the VE.

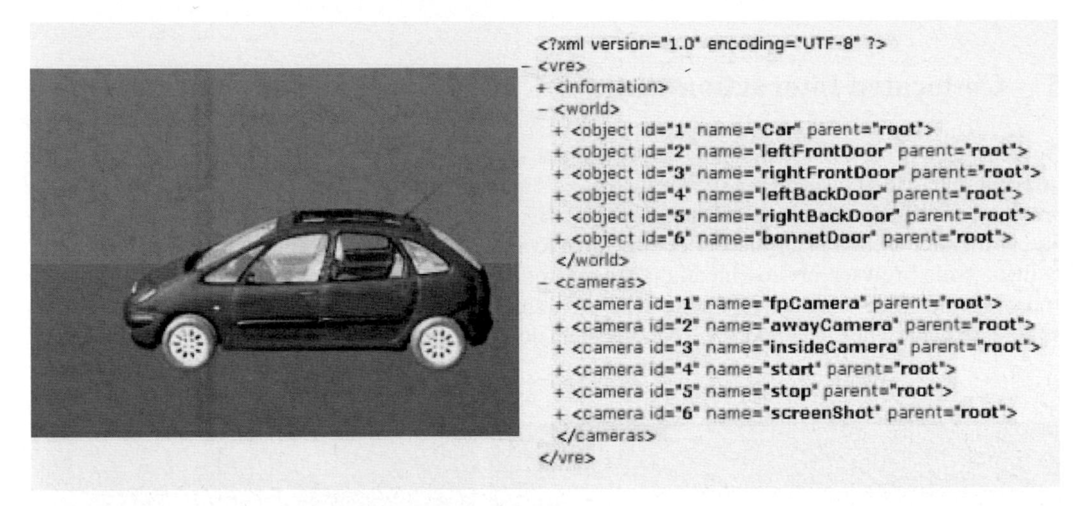

Figure 1: Car Model and XML Representation

4 Generating Interfaces

The XML per se, does not provide any functionality. To provide a way of querying properties of the VE, it can be transformed to a generic HTML page (Figure 2, left side) or a custom PDA application/interface (Figure 2, right side) that displays the properties and highlights the interaction capabilities. When a link is clicked on the web browser or on the custom interface, a message can be sent to the VE to take, for instance, a screenshot of the current viewpoint or select an object like a door in order to open it, de-coupling the selection/interaction from the original input device, such as a 3D mouse or a 3D wand selection tool that is traditionally used in the VE. The current implementation facilitates the VE and devices with communication via a network socket connection. The VE is capable of transforming the data received from the device and interpreting it as an interaction and executing it. In both cases, both representations are generated dynamically.

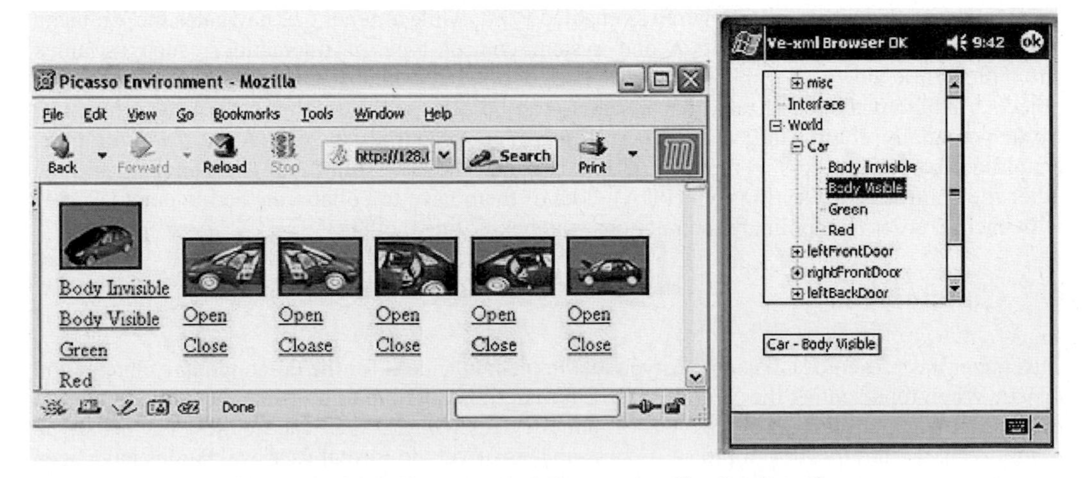

Figure 2: Web Browser And Custom Application/Interface

5 Co-located Interaction with the Environment

Devices such as a PDA with integrated wireless connections can be used ad-hoc to access the information as they incorporate web browsers as standard. As our VE could be accessed via a web browser, any PDA could be used straight away. By using the VE, it is possible to equip more than one participant with a PDA that will allow them to share the interaction and browse the VE via the built-in web browser or downloading the custom application. Compared with the traditional approach of only one active interactor with the immersive kit while the rest of the participants act as a passive audience, this approach can generate new group interaction modalities within the VE.

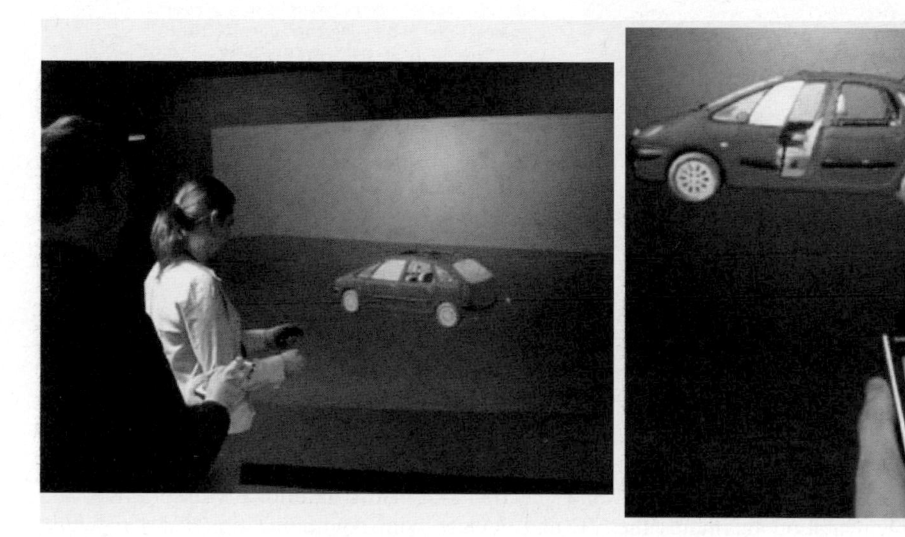

Figure 3: Co-located and PDA Interaction

Figure 3 shows two users during trial collaborating in the VE during an informal trial. A user is accessing the VE features with a wireless enabled PDA, while another one navigates the VE using a gyroscopic 2D mouse. Selection and system control type of interactions, such as quick navigation to a viewpoint, opening and closing doors and bonnets, changing colour, rendering objects visible/invisible and taking screenshots can be accessed through the PDA. Several users could potentially control the VE or query some of its properties using PDA at the same time, enabling co-located access to the VE. As one user had the navigation tool (gyro-mouse) and the other the control and selection tool (PDA), both of them have to collaborate and interact verbally with each in order to coordinate their actions while exploring the VE.

6 Conclusions

This paper has presented a basic prototype that implements ideas for the de-coupling of interaction in VE. When representing the VE in an XML format, it is possible to translate the interactions and make them compatible with other types of input devices with 2D GUI capabilities. The prototype is following an iterative development process and, as of yet, no formal user evaluations have been carried out to realize or validate the potential of this interfacing concept. As the development process involves expert users, we have received numerous ideas on how the system can be adapted to accommodate other situations and applications.

As the original interaction techniques supported by the VE could be substituted by other types of input devices (in this case by a web browser via a PDA), the interactions can be effectively decoupled from their original form, opening up new interaction modalities in interaction with VE different input devices and multiple users. Currently we are exploring the generation of other input modalities via the XML description speech processors, by means of turning the XML descriptive file into a speech grammar, and the generation of 2D menus inside the VE via the XML file controlled via a gesture interface.

7 Acknowledgements

This work has been conducted as part of the "Virtual and Interactive Environments for Workplaces of the Future" project funded by IST-2000-26089. The authors acknowledge support from project partners and the expert users that participated during the development of the interaction concepts.

8 References

Bowman, D. A. & Hodges, L. F. (1999). Formalizing the Design, Evaluation, and Application of Interaction Techniques for Immersive Virtual Environments. *Journal of Visual Languages and Computing*, 1999. 10 (1) 37-53.

Bowman, D. A., Kruijff, E., LaViola, J., & Poupyrev, I. (2000). The Art and Science of 3D Interaction. *IEEE Virtual Reality Tutorial*. 18 March 2000. Retrieved 14 February, 2003 from http://www.mic.atr.co.jp/~poup/3dui/TUT3DUI/

Bullinger, H-J., Blach, R, & Breining, R. (1999). Projection Technology Applications in Industry-Theses for Design and Use of Current Tools. *International Immersive Projection Technology Workshop*. Springer-Verlag.

Eastgate, E., (2001) *The Structured Development of Virtual Environments: Enhancing Functionality and Interactivity*. Unpublished Phd Thesis, University Of Nottingham. Retrieved 14 February 2003 from http://www.virart.nott.ac.uk/RMEPhD2001.pdf

Hartling, P., Bierbaum, A., & Cruz-Neira, C. (2002).Virtual Reality Interfaces Using Tweek. *ACM SIGGRAPH*. 21-26 July 2002. ACM Press.

Lindeman, R. W., Sibert, J. L.& Hahn J. K. (1999). Hand-held Windows:Towards Effective 2D Interaction in Immersive Virtual Environments. *IEEE Virtual Reality* . 13-17 Sep 1999. (pp. 205-212). IEEE Computer.

Smith, G., T. Salzman, & Stuerzlinger, W. (2001). 3D Scene Manipulation with 2D Devices and Constraints. *Graphics Interface*. Morgan Kaufman Publishers.

Building Virtual Environments using the Virtual Environment Development Structure: A Case Study

Mirabelle D'Cruz[ψ], *Alex W Stedmon*[ψ], *John R Wilson*[ψ],
Peter J Modern[γ] *& Graham J Shaples*[γ]

[ψ]VR Applications Research Team
University of Nottingham,
Nottingham, NG7 2RD, UK
mirabelle.dcruz@nottingham.ac.uk
alex.stedmon@nottingham.ac.uk

[γ]British Nuclear Fuels Plc.,
Springfields Works,
Preston, PR4 0XJ, UK
peter.j.modern@bnfl.com
graham.j.sharples@bnfl.com

Abstract

Project IRMA, virtual reality for manufacturing applications, aims to build, integrate, demonstrate and evaluate VR 'application demonstrators' which reflect different but complementary aspects of industrial manufacturing. Application demonstrators are being developed based on a fundamental design philosophy with generic software modules for handling messaging, object and communications management, between real world simulations and Virtual Environments (VEs). To assist developers, stakeholders and researchers alike, the Virtual Environment Development Structure (VEDS) has been developed which is a holistic, user-centred, approach for specifying, developing and evaluating VE applications. This paper provides a summary of VEDS and discusses its application to an industrial monitoring and control case study.

1 Introduction

In order that early decisions can be made about the design, usability and evaluation of new Virtual Environments (VEs), a clear progression taking user needs forward into system development is required. Whilst HCI and ergonomics guidelines exist, they have limited transfer to 3D, real-time interactive, VEs. To assist developers, stakeholders and researchers alike, the Virtual Environment Development Structure (VEDS) has been developed which is a holistic, user-centred, approach for specifying, developing and evaluating VE applications (Wilson, Eastgate & D'Cruz, 2002). This paper provides a summary of VEDS and discusses its application to an industrial case study: the British Nuclear Fuels Plc (BNFL) control and monitoring application (Modern, Stedmon, D'Cruz, Wilson & Sharples, 2003).

2 Project IRMA

Project IRMA, virtual reality for manufacturing applications, is an Intelligent Manufacturing Systems approved inter-regional Virtual Reality (VR) project with partners from the European Union, Japan, Switzerland, and Newly Associated States of Eastern Europe. The key aims of Project IRMA are to build, integrate, demonstrate and evaluate VR 'application demonstrators' which reflect different but complementary aspects of industrial manufacturing, and integrate different commercially available software simulation tools in VEs that support real-time, two-way, interaction (Modern, et al, 2003).

3 The Virtual Environment Development Structure

A summary of the main aspects of VEDS is illustrated in Figure 1, a more detailed explanation is presented in Wilson, et al, (2002).

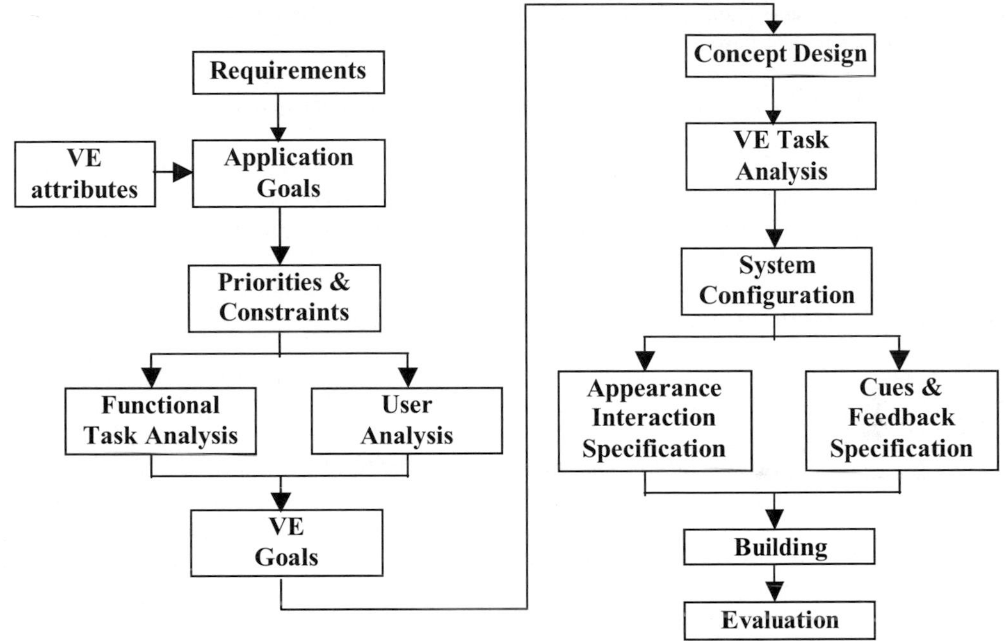

Figure 1: Summary of VEDS

At the beginning of any development process goals must be defined that provide the driving force of the VE building process. These goals may stem from a problem within the organisation, such as wastage, high costs or delays, or to increase competitiveness, support training, or reduce design life cycles (D'Cruz, Stedmon, Wilson & Eastgate, 2002). Through methods such as structured interviews, focus groups, and real world task analyses, the initial stages of VE development involve assessing which tasks and functions must be completed in the VE; determining user characteristics and needs; and allocating and dividing functions within the VE. As such, a VE can be specified in terms of its goals and expected user tasks, in relation to the complexity and balance between interactivity and exploration afforded. This specification should then be agreed by the VE development team and stakeholders (including end users). Further methods can be employed at this stage, such as storyboarding and virtual task analyses, specifying tasks to be performed by the participant within the VE. This is a critical component of the VE development process as there are trade-offs between the technical capabilities of VR applications and costs associated with VE complexity (related to sensory richness), update rate (related to sense of presence and any disturbing effects), and interactivity (numbers of objects that can be 'manipulated' in real-time, and how this is to be done).

Capturing user requirements can support the definition of expected (or unexpected) behaviours that underpin the successful use of a VE. Furthermore, understanding user requirements can also provide a basis for evaluation of a VE at a later stage of the design process. For all VE applications, evaluations should be made of both the environments themselves and also their use and usefulness (Wilson, et al, 2002). It is possible to divide such evaluations into examinations of validity, outcomes, user experience and process (D'Cruz, et al, 2003). This means that before VE building progresses too far, a more detailed examination can be made of how participants will respond to different elements of the VE; utilise its functionality; and comprehend all the interface elements to meet the application goals. As such, a basis can be set for encouraging participants to explore a VE and enable them to understand which elements may be interacted with, minimising dysfunctional participant behaviour and serious errors.

4 Applying VEDS to the BNFL application demonstrator

It is important to recognise that VR may not be the only or the best potential solution and that VR technologies are still relatively new. It is important, therefore, to consider from the outset what is already known about VR/VE attributes and consider how they match the application requirements and may compare to alternative technologies (D'Cruz, et al, 2002).

4.1 Application goals, priorities and constraints

The first stage of the development process was to decide on the application goals, priorities and constraints. As such, a number of scenarios were defined through meetings with the relevant personnel (eg. software engineers, process designers, control room operators and plant managers). The aim was to choose scenarios that the users (stakeholders and end-users) would recognise as key activities of the BNFL application demonstrator (and indeed any manufacturing process). The application goal was to investigate the use of the IRMA VE application demonstrator in comparison to the real world environment. Given this goal, the priorities were to ensure that the participants were given opportunities to explore key attributes of VR and simulation technologies (eg. 3D interactivity, multi-modal interfaces, presence). Constraints were faced in many forms but of particular importance was the safety critical issue of the nuclear industry. Defining goals at an early stage provide a major influence on VE design (and can also be used in evaluation as a criteria of success of the VE). The overall goal of the VE, as a research project, was to examine the use of VR technologies for a control and monitoring application. Therefore the goals were to provide a 'realistic' representation of scenarios, given chosen VR configurations, in terms of experience rather than visual fidelity, but also to explore what VR technology might offer in addition to a real world application.

4.2 User requirements

BNFL had a need to minimize the time in the initial stages of designing industrial plants by finding a method to test the accuracy of new process control software without the need for costly prototypes. Of particular interest was the use of digital solutions to investigate the use of VR technologies and simulation technologies. It was felt that a combined version of both virtual and simulation technologies, which represents the 'IRMA solution', would make a powerful and useful tool for industry (Modern, et al, 2003). In order to investigate this, the BNFL application demonstrator was developed for the flexible stacking and assembly of nuclear fuel pellets into fuel pins and fuel assemblies, which provided the opportunity to investigate various attributes of the technology in a safe and cost effective manner.

4.3 Analyses of the task and user

During the information gathering stages it is often useful to carry out task and user analyses. These involve a step-by-step 'walkthrough' of each scenario to note procedures of the task and what the user is doing and thinking, preferably by the real end users of the task to gather realistic data. This can be performed using techniques such as direct observation, photographic or video analysis, verbal protocol, interviews, and short questionnaires. In the case of the BNFL application demonstrator, access to the equipment was restricted which made it difficult to gather real world data and BNFL personnel therefore had to provide necessary information. This process can be problematic without a greater understanding of what type of information is required. Due to the limitations of current VR systems it is impossible and sometimes unnecessary to design a VE that reflects every interaction of the real world. Therefore only the information that affects the user's experience and addresses the goals of the application is necessary. For example, a desk maybe located in the working environment that is used for placing items on. If it has drawers, unless these are required to be opened to place things inside, there is no need to make them interactive. Therefore a desk can be either a simple non-interactive object or a more complex interactive object. Task and user analyses help to define the minimum type of interactivity and cues to interactivity which might be needed in the VE before it is built.

The stakeholders of an application often know too much information! They may be able to provide detailed schematics, manuals, and job specifications of an application that may not be necessary and may just prove cumbersome for the developers. In the development of the BNFL application demonstrator this was further compounded by the fact that the development team were located in three different countries, none of which were located near the user company. The exchange of information was therefore a learning process for all involved.

4.4 Concept design and VE task analysis

To provide a better understanding of the goals of the VE, storyboarding methods can be applied. A number of focus groups were held with key personnel from BNFL, HCI experts and developers to assess the application as it may be presented in the VE. Several iterations of the storyboard were made with particular attention to usability and user interface issues such as cues, feedback, levels of detail and interaction. Issues relating to usability of the VR system, for example, comfort, fit of equipment, temporal and spatial resolution limitations, field of view restrictions and general limitations implicit in using visual and auditory information, were discussed. Further to this, issues relating to VE design were considered, such as: representation of the participant and other participants, in the VE (eg. sizes, shapes, appearance, movements, facial expressions); supporting navigation and orientation (eg. interface tools and other aids, short-cuts, familiarisation routines, optimum world sizes); understanding and enhancing presence and involvement (eg. balancing pictorial realism, size and complexity, up date rules, sound, shadowing, enhancing interest); requirements for cues and feedback, minimising any side and after effects (eg. sickness, performance decrement, physiological change); and providing interface support and tools for interactivity (eg. mixed reality design, metaphors, interface elements, toolkits). For each usability issue it was important to ask whether: (a) is it significant for the safe and effective use of the VR/VE? (b) are the usability issues unique? and (c) is it possible to utilise data and criteria from other domains and from knowledge of performance of related tasks? (Eastgate, 2001).

5 Practical considerations

By careful, but flexible, use of a framework such as VEDS, the consequences of some of the barriers to VR application can be minimised. Technical limitations can be addressed, or their effects reduced by appropriate choice of technology, and careful specification of VEs to meet the user goals. Trade-offs, such as choices between visual fidelity and interactivity, can be resolved, and usability can be enhanced through selection of appropriate hardware and development of cognitive interface tools and aids. VE specification can take longer than people imagine, however given the choices of technologies and the possible designs of the VE, it is important to understand this process and consider it carefully. Information exchange without guidelines can be challenging as there is often a lot of information available which is not always relevant. Developers rely on an organisation to provide them with key information, but it is sometimes difficult to understand what this information is, especially if the opportunities and benefits afforded by VR/VEs are not fully understood. Furthermore, verbal and textual communication can be problematic for people working in different organisations, or even groups of people in different countries. VR applications can only be achieved through the careful consideration of VR/VE attributes and a thorough understanding of which constraints will apply and how to overcome them in specifying and building the VE. Finally, these applications will best show added value if a rational evaluation process is established at the outset, on the basis of realistic and achievable targets for use of VEs and for their incorporation into industrial/commercial practice, and/or education/training programmes (Stedmon, D'Cruz, Tromp, & Wilson, 2003) .

6 Acknowledgement

The work presented in this paper is supported by Intelligent Manufacturing Systems, Project 97007: Project IRMA, virtual reality for manufacturing applications and is sponsored within The EU FP5 Programme under Contract G1RD-CT-2000-00236.

7 References

D'Cruz, M., Stedmon, A.W., Wilson, J.R., & Eastgate, R. (2002). *From User Requirements to Functional and User Interface Specification: General Process.* University of Nottingham Report No.: IOE/VIRART/01/351.

Eastgate, R., (2001), *The Structured Development of Virtual Environments: Enhancing Functionality and Interactivity.* PhD Thesis, University of Nottingham.

Modern, P.J., Stedmon, A.W., D'Cruz, M., Wilson, J.R., & Sharples, G.J. (2003). The Factory of the Future? The Integration of Virtual Reality for Advanced Industrial Applications. In, *HCI International '03. Proceedings of the 10th International Conference on Human-Computer Interaction, Crete, June 22-27, 2003.* Lawrence Erlbaum Associates.

Stedmon, A., D'Cruz, M., Tromp, J., & Wilson, J.R. (2003) Two Methods and a Case Study: Human Factors Evaluations for Virtual Environments. In, *HCI International '03. Proceedings of the 10th International Conference on Human-Computer Interaction, Crete, June 22-27, 2003.* Lawrence Erlbaum Associates.

Wilson, J.R., Eastgate, R.M. & D'Cruz, M. (2002) Structured Development of Virtual Environments. In K. Stanney (ed). *Handbook of Virtual Environments.* Lawrence Erlbaum Associates.

Haptics in Museum Exhibitions

Koichi Hirota

RCAST, University of Tokyo
4-6-1 Komaba, Meguro-ku
Tokyo 153-8904 Japan
hirota@cyber.rcast.u-tokyo.ac.jp

Michitaka Hirose

RCAST, University of Tokyo
4-6-1 Komaba, Meguro-ku
Tokyo 153-8904 Japan
hirose@cyber.rcast.u-tokyo.ac.jp

Abstract

Haptic representation of virtual exhibits is a topic of interest in the application of virtual reality technology to museum exhibits. In this paper we discuss two fundamental techniques for the realization of such haptic representation. One is the simulation of haptic interaction with objects such as virtual exhibits. We propose a simulation method that enables realistic manipulation. Another is related to generating object models. We propose a method of generating object models from CT data for haptic interaction.

1 Introduction

Application of virtual reality and augmented reality technologies to exhibition presentations is a research topic that has been extensively investigated; it is expected that these technologies will provide a means of enriching the participants' experience of exhibitions by allowing multi-modal interaction with exhibits.

In addition to visual and auditory sensations, haptic sensation is an important part of the experience in such interaction with exhibitions; haptic sensation conveys various types of information such as weight, shape, and surface texture (Burdea, 1996). In most museums, exhibitions such as archaeological artifacts are so delicate that they must be displayed in showcases, however, if we can provide haptic sensations through virtual interaction with objects, participants will be able to experience the sensation of holding the artifacts in their own hands.

In this paper, we discuss our approach to creating haptic interaction for virtual exhibits; we propose a method of creating virtual object models for haptic interaction and a method for computing the forces in an interaction with object models.

2 Related Works

In comparison with visual and auditory displays, haptic devices are not widely used in exhibitions. This is not only because the technology to implement haptic devices is immature, but also because modeling and rendering methods for haptic interaction have not been established.

The development of haptic devices has been extensively pursued (Massie, 1996), and some devices such as PHANToM (SensAble Technologies) have been made commercially available. Also we have developed a wearable type device that can represent force in a relatively large working space (Hirose et al., 2001). Most of these devices have been designed successfully to simulate the interaction force that we receive through 'pen-type' tools. However, for the simulation of object manipulation using hands and fingers, the degrees of freedom of such devices are insufficient.

Regarding rendering algorithms to compute the output forces for haptic devices, previous works seem to have paid little attention to the simulation of manipulation considering the force distribution on finger pads and palms, partly because most have targeted 'pen-type' interactions. For interactions using these devices, the god-object method (Zilles & Salisbury, 1995) or other similar approach (Ruspini et al., 1997, and Ho et al., 1997) has commonly been used, where the force between a 'haptic interface point' and polygonal surfaces is simulated. One possible means of simulating the force distribution on pads and fingers is to define the plural haptic interface points and compute the force on each point. However, the previous algorithms are thought not to be suitable for this idea because they require a relatively large computation cost to find the initial contact for each haptic interface point. A common approach to object manipulation is to integrate the simulation of object dynamics (Yoshikawa et al., 1995).

For haptic representation of virtual exhibits of archaeological artifacts, it is necessary to build precise object models based on real objects. One approach is to obtain the 3D geometry of an object by using computerized tomography (CT). The most common method of generating a polygonal surface model based on CT data or other potential map data is known as the Marching Cubes method (Lorensen & Cline, 1987).

3 Manipulation of Object

We have previously proposed a fast algorithm for contact simulation. The algorithm computes the interaction force between a point and space that has been partitioned into tetrahedral cells. Since the algorithm accommodates plural haptic interface points, we approximately represent the user's hand by a cloud of haptic interface points and compute the discretely distributed forces during the interaction. In the computation, we also introduced the damping factors of pads and palms for stable manipulation; we assumed that each interaction point causes repulsive and damping forces. The repulsive force is computed as being proportional to the disparity between the haptic interface point and ideal the haptic interface point, which indicates the ideal position of the haptic interface point when it is constrained on an object surface, while the damping force is proportional to the relative velocity between them.

In addition, we simulated the dynamics of objects by solving the equations of motion, taking into account the force and torque caused by the repulsive and damping forces on all haptic interface points.

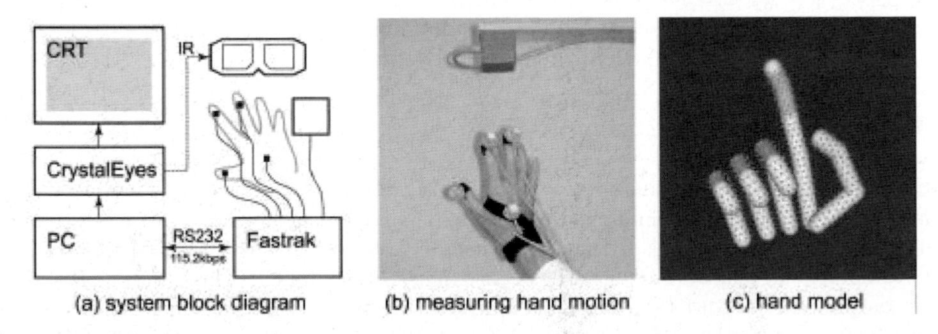

Figure 1: Experimental System

We implemented the algorithm in a prototype virtual environment (Figure 1(a)). The motion of fingers was measured by magnetic position sensors (Fastrak, Polhemous). All computation and visual rendering processes were performed by a PC (Dual Pentium4 2.2GHz) using a graphic card with an accelerator (GeForce3). We also realized stereoscopic viewing using liquid-crystal glasses (CrystalEyes, Stereo Graphics).
We used four sensors and measured the positions and orientations of the tips of the thumb, index and middle fingers, and palm (Figure 1(b)). From the sensor data, we estimated the position and orientation of all skeleton links; since we do not have sensors for the third and little fingers, we assumed that the orientation of each link of these fingers was the same as that of the links of the middle finger. We modeled the skin surface of the hand or fingers using a cylindrical surface over the link model, and arranged haptic interface points on the surface (Figure 1(c)).

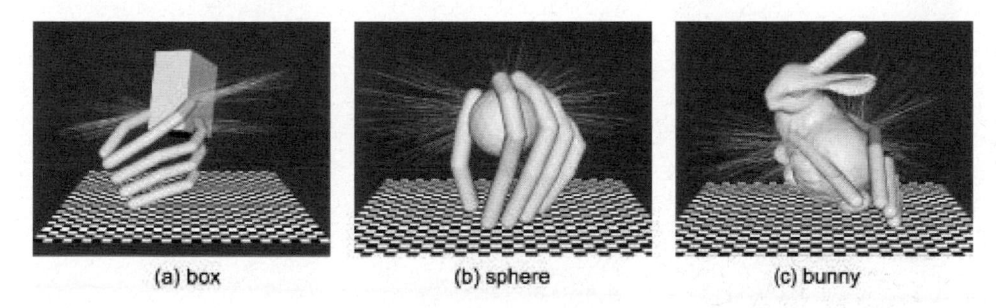

Figure 2: Examples of Object Manipulation

Examples of manipulations of several objects are shown in Figure 2. In the figure, the interaction force on each haptic interface point is presented as a vector. Through the experiment, we confirmed that our method works at a haptic rate; for example, the interaction between a hand model with 1260 haptic interface points and the bunny model (Figure 2(c)) was simulated at a rate of about 500Hz.

4 Generation of Object Model

We investigated a method for generating a tetrahedral cell mesh model from CT data that is applicable to our haptic interaction method. The mesh generation method is an expansion of the Marching Cubes method; in contrast to that method, which generates surface polygons from potential functions, our method generates a tetrahedral cell mesh that has an object surface inside it.
The problem with the direct expansion of the Marching Cubes method is that it generates a large number of tetrahedral cells inside and outside of the given isoplethic surface, and this results in the waste of memory and increase in computation time. In our approach, we propose a method, which we call 'adaptive mesh generation', that generates dense mesh only about the isoplethic surface. In the method, we iteratively divide the tetrahedral cells with which the isoplethic surface intersects. After dividing the mesh to a sufficient resolution, we divide the cells by the isoplethic surface and generate a polygonal boundary surface that approximates the isoplethic surface.
Furthermore, we attempted to reduce the number of tetrahedral cells in the mesh model by merging nodes. The model is still wasteful in that we can further reduce the number of cells by merging some neighboring cells, and the reduction of cells contributes to the reduction of computation time in the interaction that we discussed in the previous section.

In the operation of merging cells, we should not move the boundary nodes or nodes on the object surface, to avoid changing the shape of the boundary surface. Also, if we consider the characteristic of the haptic interaction method described above, the number of cells that is connected to the boundary nodes should not be increased to avoid increasing the computation time for haptic representation. We implemented an algorithm, which we call 'conditional node merging', that satisfies these restrictions on the merging of cells.

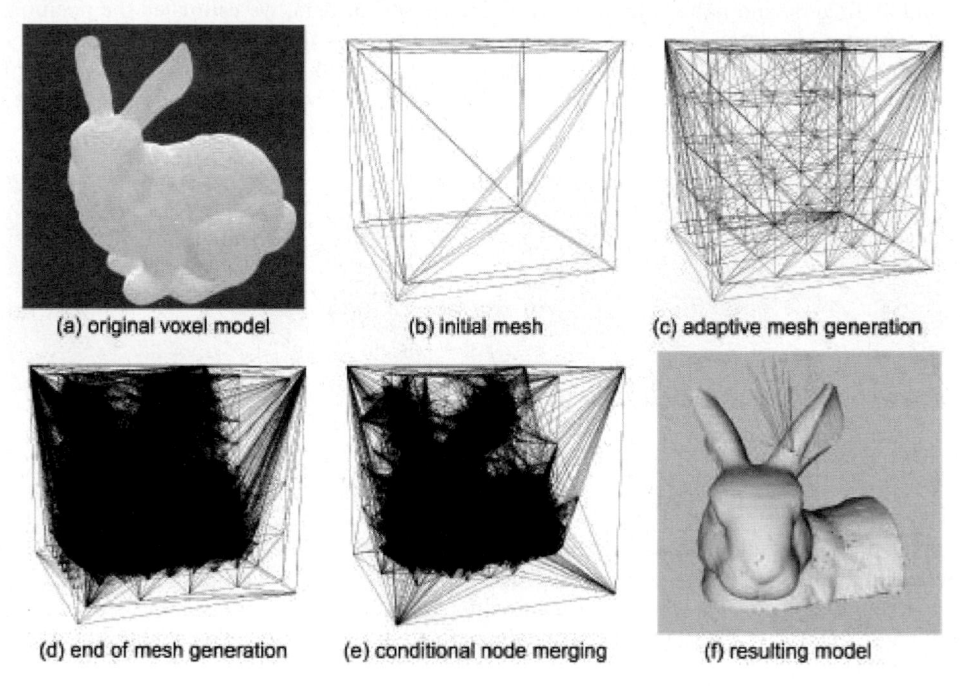

Figure 3: Process of Mesh Generation

We carried out experiments to evaluate the model generation method. Figure 3 shows an example of the mesh generation process by our method, where we are generating a cell mesh model from CT data. We used the CT data of a terra-cotta bunny from Stanford University (Levoy, 2000) in (a); started from the initial mesh in (b), applied the adaptive mesh generation algorithm in (c-d), made the boundary surface, and finally applied conditional node merging in (e). The model is immediately applicable to the haptic interaction algorithm discussed in the previous section. The bunny model being manipulated in Figure 2(c) was created by this process, and it consists of 59736 tetrahedral cells and 15548 boundary surface polygons.

5 Conclusion

We proposed methods of generating object models from CT data and interacting with the generated models. Through experiments we confirmed that our methods will provide a foundation for haptic interaction with objects, and we intend to apply the method to the interaction with virtual exhibits in museums. A problem remaining to be solved in our future work is to increase the visual reality of the object model by integrating visual modeling and rendering methods. Also, we are interested in developing haptic devices that are capable of providing distributed force across the surface of hands or fingers; we expect such a device will more effectively feed back force information computed by the proposed method.

References

Burdea, G. (1996). Force & Touch Feedback for Virtual Reality. A Wiley InterScience Publication, New York.

Massie, T. H. (1996). Initial Haptic Explorations with the Phantom: Virtual Touch Through Point Interaction. Master's Thesis, M.I.T.

Hirose, M., Hirota, K., Ogi, T., Yano, H., Kakehi, N., Saito, M., & Nakashige, M. (2001). HapticGEAR: The Development of a Wearable Force Display System for Immersive Projection Displays. Proc. VR 2001, 123-129.

Zilles C. B., & Salisbury J. K. (1995). A Constraint-Based God-Object Method for Haptic Display. Proc. IROS'95, 145-151.

Ruspini, D. C., Kolarov, K., & Khatib, O. (1997). The Haptic Display of Complex Graphical Environments. Proc. ACM SIGGRAPH'97, 345-352.

Ho, C., Basdogan, C., & Srinivasan, M. A. (1997). Haptic Rendering: Point- and Ray-Based Interactions. Proc. PHANToM Users' Group Meeting.

Yoshikawa, T., Yokokohji, Y., Matsumoto, T. & Zheng, X. Z. (1995). Display of Feel for the Manipulation of Dynamic Virtual Objects. Trans. ASME J. DSMC, 117(4), 554-558.

Lorensen, W. E., & Cline, H. E. (1987). Marching Cubes: a High Resolution 3D Surface Construction Algorithm. Computer Graphics, 21(4), 163-169.

Levoy, M. (2000). The Stanford Volume Data Archive. http://graphics.stanford.edu/data/voldata/

Wearable Computers and Field Museum

Atsushi Hiyama, Michitaka Hirose

Research Center for Advanced Science and Technology, The University of Tokyo
4-6-1, Komaba, Meguro-ku, Tokyo, Japan
{atsushi, hirose}@cyber.rcast.u-tokyo.ac.jp

Abstract

In this paper, we describe the application of wearable computers and augmented reality (AR) technologies for outdoor exhibition use. The developed system is designed to provide a proper exhibition content at the correct position in the field. In this research, in order to establish a portable, high-precision positioning system for AR exhibition, we embedded a number of radio-frequency identification (RFID) tags in the environment. Then we developed exhibition software for the purpose of achieving the exhibition easily in such systems.

1 Introduction

The expanding technology of digital archives has enabled museums to make the most use of information terminals. Generally, exhibitions at museums are constructed on the basis of results of research planned by curators. In fact, exhibitions are forced to be held inside the museum facilities, so that visitors can only minimally experience the exhibits' histories and cultural backgrounds with relation to the place of origin of the exhibits. In our field museum system, by adopting wearable computers and augmented reality technologies, we can post exhibits in an electronic medium and superimpose them on historical architecture or natural objects. In this paper, we describe the development of a Digital Field Museum system that can reconstruct the urban space from specific historic or cultural points of view.

2 Field Museum

2.1 Base Experiment

Our research group has carried out Field Museum experiments in public three times. In the third experiment in 2002, we developed a system in consideration of practical use at the International Exposition to be held in 2005. We made use of the prototype system that can supply image information to participants through a head-mounted display (HMD) or LCD panel embedded in the clothing. The information is supplied according to the participants' positional data taken from the global positioning system (GPS) (Hiyama, 2002). Through these experiments, we identified three characteristics of the Field Museum system.

- The exhibition data is in digital format, so we can present varied information about a place from various points of view.
- Participants can choose and edit exhibition contents according to their own interests.
- Exhibitors can collect feedback from participants in real time, since participants are experiencing the exhibits via the operation of wearable computers.

In order to develop a Field Museum system from the user's point of view, we especially paid particular attention to the first and second aspects of our system.

2.2 Design Plan

As a system that offers information according to the user's position, we actively implement augmented reality technology to convey the relationship between the exhibition site and the presented visual image accurately. Therefore, we must develop a high time-frequency resolution positioning system and detect the visual axis in a wide area. Additionally, we consider the principle of exhibition, which we could not do in past experiments.

3 Developed Systems

3.1 Experimental Field

We build an experimental field around our research building. The experimental field has an area of 1,700m^2,. There are 1349 RFID tags embedded in the ground. Considering the required resolution of position estimation for AR exhibition use and for existing structures, the tags were laid out at one tag per 1.2m gridiron interval (Fig. 1). The developed wearable computers are equipped with tag readers that can receive the ID signals transmitted from RFID tags. The wearable computer estimates its own position by retrieving the coordinate values from the ID list, input beforehand, each time it receives the ID signal from a tag. By using this positioning system, we can now identify the wearer's current position to an accuracy of 60 centimeters. In addition, the experimental field is covered with an 802.11b wireless network. Therefore, users are able to share information with one another.

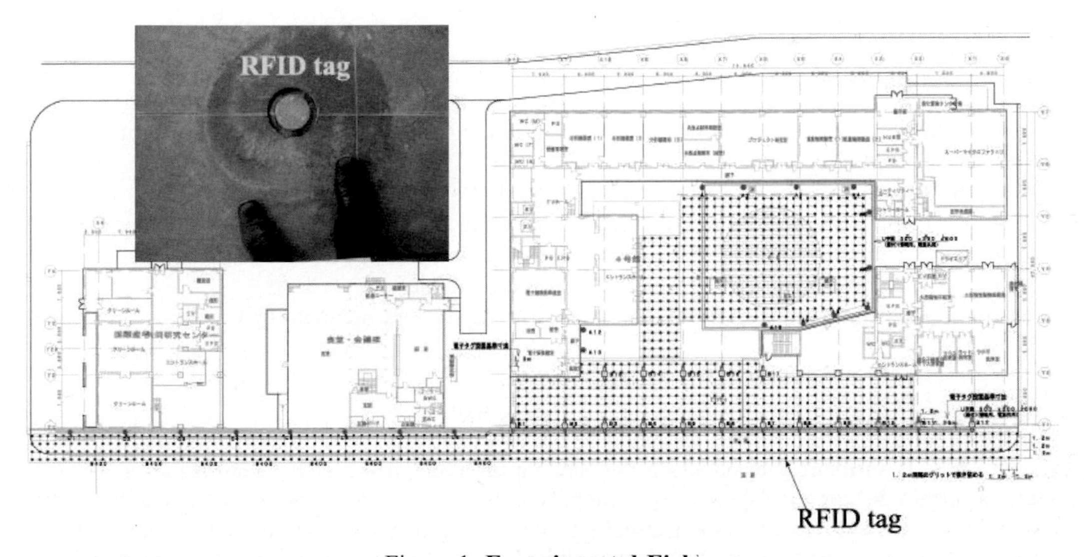

Figure 1: **Experimental Field**

3.2 Wearable Computer

As an information display device, we used a binocular see-through head-mounted display (HMD). There are an electromagnetic sensor, gyroscopic sensor and accelerometer on top of the HMD, for measuring the three-axis rotation of the observation point (Fig. 2). In combination with the developed positioning system, the user can enjoy visual information, superimposed on real scenery, concurrently with walking through the field.

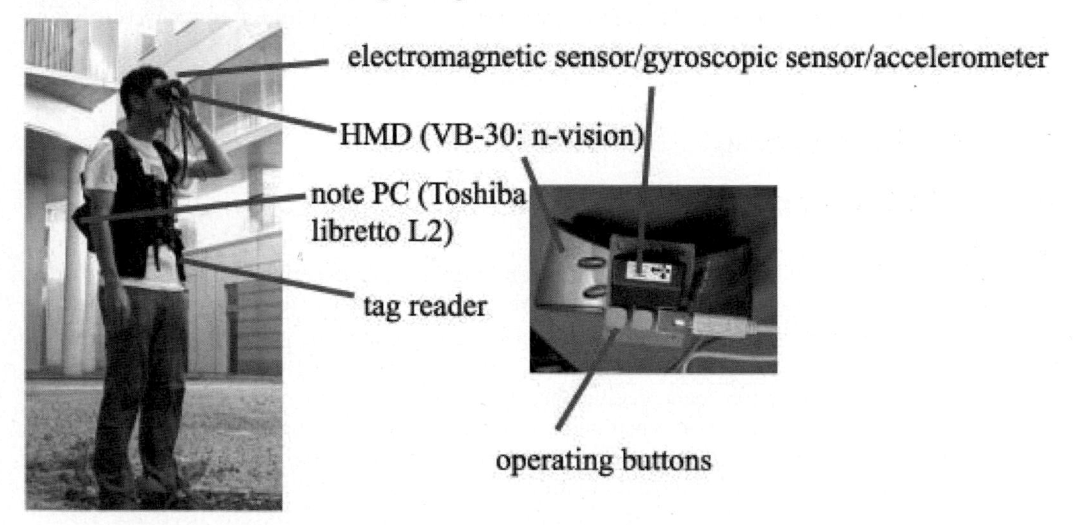

Figure 2: **Wearable computer**

4 Exhibition Software Design

Designing the exhibition software for the aforementioned hardware system is began with observing the behavior of museum visitors and building up a framework of software functions that museum visitors require.

At exhibition sites, visitors' behaviors can be divided into three modes.

- (1) Selecting an exhibition which one wants to observe. : Selection mode
- (2) Searching for an exhibit which one wants to observe next. : Navigation mode
- (3) Appreciation at an exhibit. : Appreciation mode

Moreover, according to the visitors' intentions, there are two manners of transition from one mode to another.

- (A) Observe exhibits in the order indicated.
- (B) Observe exhibits in an original preferred order.

For the second and third modes, we mounted both (A) and (B) types of appreciation algorithms. The (A) type is software automated. In the Navigation mode, if the visitor approaches an exhibit within a certain distance, the system automatically transits to the Appreciation mode, and when the visitor leaves an exhibit the system transits to the Navigation mode again. The visitor will go through this set of transitions until the last exhibit (Fig. 3). To appreciate exhibits in the manner of the (B) type, the visitor will select an exhibit, which he/she wants to observe next by operating two buttons mounted on the HMD (Fig.4).

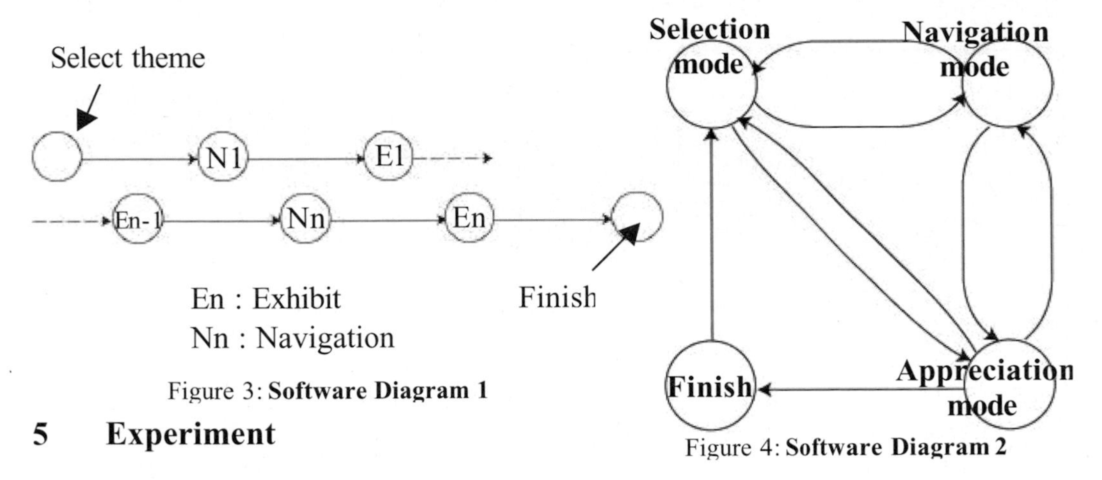

Figure 3: **Software Diagram 1**

Figure 4: **Software Diagram 2**

5 Experiment

5.1 Visual Interaction in Walkthrough

Before applying the developed system in the exhibition experiment, we specified an area between the Navigation mode and Appreciation mode, by examining visitors' walking paths in the exhibition field against three kinds of exhibits: small painting, large painting and 3D object. We measured 6 visitors' walking paths. Figs.5,6 and 7 show the results. Since the viewable angle of the HMD is 28 degrees horizontal and 21 degrees vertical, visitors are walking inside an area in which exhibits fit into the viewing field of the HMD. Therefore, we set up the boundary between the Navigation mode and the Appreciation mode as slightly larger than the area in which visitors can just glance at an exhibit.

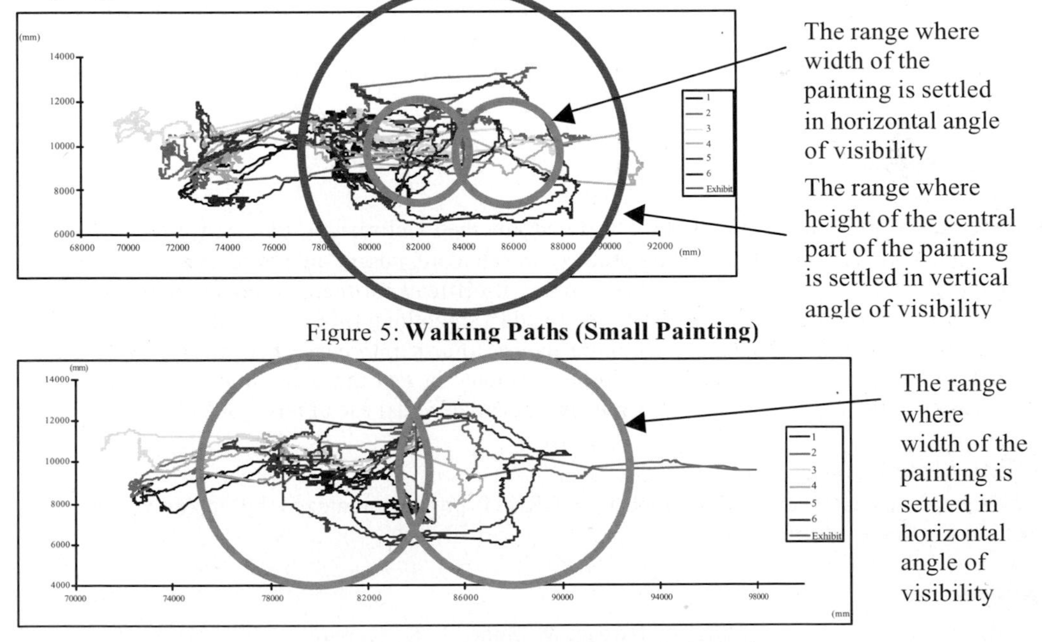

Figure 5: **Walking Paths (Small Painting)**

Figure 6: **Walking Paths (Large Painting)**

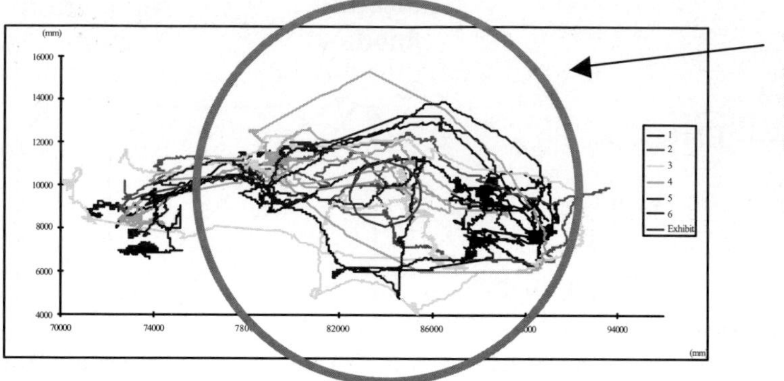

Figure 7: **Walking Paths (3D Model)**

5.2 Exhibition Experiment

In the developed Field Museum system, visitors are free to chose from a number of exhibition themes and appreciate exhibits. Visitors can change their preferred exhibition theme from one to another and learn the details of an exhibition site. For the exhibition, we selected "The Spread of Buddhist Art in Asia". The experimental was divided into 4 regions: Pakistan, India, China and Japan. Three exhibition themes were arranged according to the meaning of the work of art.

5.3 Result

The system must be designed for unspecified users so six participants were selected from a group without previous experience with augmented reality application. Four participants adapted easily to using the button interface at hand. The remaining two participants pointed out the difficulty of pressing the button because of its position. Five participants commented that the timing of automated transition between the Navigation mode and Appreciation mode was somewhat unsatisfactory.

6 Perspective

We are now continuing to demonstrate and upgrade the exhibition system. Since RFID tags are discretely positioned, there is a drifting problem in calculating users' positions even when they are standing still. This drift of users' positions makes it difficult for them to recognize the details of the exhibition contents. Thus we are planning to add a smoothing filter to the algorithm of position estimation to make the exhibit appear more natural. Furthermore, we plan to use the wireless network to provide and manage the exhibition contents in real time. We must also examine the characteristics of this exhibition system in order to realize actual use of this system.

Acknowledgements

This research has received the support of CREST project organized by Japan Science and Technology Corporation.

References

A. Hiyama, R. Ueoka, K. Hirota, M. Hirose, M. Sone, T. Kawamura. (2002). Development and Demonstration of Wearable Computer Based Interactive Nomadic Gallery. Proc. of the 6th International Symposium on Wearable Computers (ISWC2002),pp.129-130

Towards i-dove, an interactive support tool for building and using virtual environments with guidelines

P. Karampelas[1], *D. Grammenos*[1], *A. Mourouzis*[1], *C. Stephanidis*[2]

[1]Foundation for Research and Technology-Hellas
Institute of Computer Science
Science and Technology Park of Crete
Heraklion, Crete, GR-71110 Greece
{pkaramp, gramenos, mourouzi, cs}@ics.forth.gr

[2]University of Crete, Department of Computer Science

Abstract

The lack of appropriate guidelines and standards on best practice implementation and use of virtual reality technologies and virtual environments has driven our work towards the development of **i-do*ve***, a tool that shall integrate and exploit guidelines-related knowledge (e.g., guidelines, examples, reference materials). The level and type of guidance required by industrial users, as well as requirements for such a tool, have been elucidated and analyzed, inferring thereby the presented potential characteristics of the tool, in terms of objectives, target users, suggestive functionality, technical characteristics, etc. The evaluation plan of the system is also presented along with the preliminary results steaming from the evaluation of an explorative prototype of the system.

1 Introduction

Over the past few years, considerable efforts have been directed towards enriching the interface between humans and machines for the purpose of amplifying the creativity and productivity of people. This has led to the development of innovative user interfaces such as those that integrate Virtual Environments (VEs) and other emerging technologies (e.g., wireless networking, wearable devices). Despite significant and extensive research in both the VEs and the usability domain, until recently little attention was paid to the way VE-applications should be designed for usability, thus impeding VEs from living up to their full potential. Motivated by the numerous questions that arise from such lack of appropriate guidelines and standards on best practice implementation and use of VEs, the work presented here aims at the development of: (i) essential VE - related ergonomic knowledge; (ii) methodologies and tools for supporting the development and exploitation of such knowledge during all stages of the development life-cycle of industrial VE applications. In particular, this paper elaborates the requirements and design of **i-do*ve*** (**i**nteractive tool for the **d**evelopment **o**f **v**irtual **e**nvironments), an innovative tool for working with guidelines during the development and use of VE-applications. The main goal of the tool is to offer guidance and assistance in the appropriate format to VE development stakeholders.

2 Background & Related Work

Many of the tasks involved in developing VEs are difficult and generally characterized by lack of (structured) guidance, which often leads to hard to use, or even unsafe, end-products (Wilson, 2002). One of the main aims of the work presented in this paper is to provide active support (i.e. a tool) for the specification and building of VEs throughout the development life-cycle (e.g., design, implementation, evaluation). Previous related work in this field mainly focuses on two complementary research directions.

2.1 Guidelines - related Knowledge for VEs

In the domain of VE design and implementation, there are no established collections of design guidelines in any sense similar or analogous to the numerous collections of widely accepted or de facto standard guidelines that exist for conventional interfaces. Most guideline collections come from academic or collaborative research projects. One currently well-known example comes from the industry sector (IBM, 2000), while available examples of guidelines collections from academia include post-graduate research projects (Gabbard, 1997; Kaur, 1998). On the other hand, a significant number of documents exist, containing design knowledge from which guidelines can be extracted on the basis of research and best practice descriptions, but their presentation lies beyond the limits of this paper. A serious shortcoming of VE-related ergonomic sources is that they tend to become outdated, if not obsolete, in few years. This is mainly because VR technologies (hardware and software) have not yet reached a stable plateau, like for example PC technology did in the mid-90s, but is still rapidly and dramatically changing every few months.

2.2 Tools For Working With Guidelines (TFWWG)

Despite the indisputable value and importance of guidelines-related knowledge, several studies investigating the use of guidelines and existing knowledge by designers and developers of 2D applications have concluded that accumulated knowledge is frequently ignored. This is partly attributed to the fact that such knowledge is not easily exploitable by the user interface designers (Tetzlaff & Schwartz, 1991), and partly due to its traditional incarnation (i.e., paper based-manuals), which raises issues of ineffectiveness and lack of user-friendliness (Grammenos, Akoumianakis & Stephanidis, 1999). An approach that aims to overcome such limitations is the integration of guidelines into tools, which allow software designers to access design guidelines organised either as a database or hypertext. Nevertheless, it appears that such solutions are restricted to offer passive and predefined support, and this is the case with the only, currently known, software application that supports the design and development of VEs, a prototype hypertext-based tool (Kaur, 1998). Regarding 2D user interfaces, these shortcomings, among others, raised a compelling need for shifting from manual or passive and predefined support to active computer-supported use of guidelines (Grammenos et al., 1999) that may be achieved by monitoring software development activities and automatically providing guidance and hints as inferred necessary and relevant. In the novel field of VEs, no such tool is currently available.

However, significant research has been conducted on what is needed for an effective and useful guidance system for supporting the development of 2D user interfaces with outcomes that also apply to tools supporting VE development (e.g., Vanderdonckt, 1999). These requirements were taken into consideration while designing **i-do*ve***. Similarly, during the requirements specification and the design of the initial **i-do*ve*** prototype, the following were taken into account: requirements for a useful design support tool to be adopted in mainstream software development (Blatt &

Knutson, 1994); a methodology and tools for applying context-specific guidance (Henninger, 2000); an approach for balancing usability concerns for developers and end-users (Carter, 1999); and, a statistical approach for automatically generating cross-reference of guidelines (Goffinet & Noirhomme-Fraiture, 1999).

3 User Requirements Regarding Guidance and Tools

In order to acquire further insight into potential user profiles in terms of typical daily tasks, working conditions, use of computer systems and applications, etc, eleven VE designers, developers and users, members of the VIEW consortium, and five external potential users of the tool were interviewed. Interviewees also contributed to the improvement and fine-tuning of the initial interview forms, which were then utilized to produce questionnaires and conduct web-based surveys, aiming at capturing also the requirements of a wider population of potential users of the tool. Currently, feedback from twenty-one potential users has been received, both by means of interviews and surveys, providing a preliminary overview of the target population, as well as of the level and type of guidance required. The results of our preliminary study confirmed, among others, our initial hypothesis that a TFWWG simply offering guidance may not be sufficient; a number of participants in our study brought forward the need for integration between guidance support and other common tools such as schedulers and organisers, glossaries / dictionaries, indexes, notepads, instant messengers, etc, as well as the need for personal or shared data storage areas. Furthermore, a detailed description of potential user profiles in terms of user attitude / motivation for such a tool, and user knowledge or experience regarding guidelines and TFWWGs, was brought into focus elucidating among others: (i) the **user tasks and roles** that the tool needs to support along with their structure, frequency or priority; (ii) the **user goals** that should be accomplished while using the tool, including priorities among them; and (iii) the **user and context dimensions of diversity** that need to be taken into account during the design of the tool and supported by the tool interfaces.

4 Overview of i-dove Design

A number of about 70 high-level user tasks, covering a broad number of user roles, were carefully considered, aiming at identifying which of them can and shall be supported by the tool. More specifically, this inspection led to the definition of the functionality that will be offered by the tool, which includes:

- Context specific support and guidance, along with reminders, checklists and to do lists, workflow/Gantt charts, etc., including: (a) decision support for the specification of VEs, (b) VE design support and guidance, (c) VE elements repository populated with reusable design 'cases', patterns, interaction elements and techniques, (d) usability test battery, i.e., guidance and tools for measuring the usability and specifying proper operation of VEs and (e) guidelines of use, i.e., customizable checklists and guideline collections for the proper use of VEs.
- Alternative search and browsing mechanisms such as keyword search, index facility, predefined queries, catalogue hierarchies, advanced options allowing combined methods, along with facilities for sorting the results, e.g., sorting by the most recently created. Personalization of the above mechanisms is possible by employing user profiling.
- Personal / shared annotation and communication facilities for asynchronous or synchronous collaboration such as annotation mechanisms, message boards and chat.
- Resource scoring and statistics facility displaying, for example, the top rated or frequently used resources.

- Multiple content access levels, i.e., public, privileged, private and administrative facilities, for instance for: (a) adding, viewing, and removing resources or users; (b) manual or (semi-) automatic cross-referencing of related items; (c) extraction of administration-oriented content statistics, e.g., rate of use, number of users.

Potential users of the tool could be connected over the Internet using different web browsers, e.g., a desktop PC browser or a PDA/mobile web browser. Therefore, **i-do*ve*** will be a large-scale multi-tier web-based application and is now under development using ASP .NET technology, supported by the SQL Server database management system.

According to the aforementioned user goals and dimensions of diversity (including context diversity), the following user groups of the tool were identified:

- **Development stakeholders**, consisting of people who will be accessing the 'knowledge-base' of the tool mainly seeking information and guidance support for accomplishing the tasks assigned to them as part of the development life-cycle of a VE.
- **Academic users**, consisting of people who might be using **i-do*ve*** as a learning / teaching tool. These people appear to have additional requirements, such as, for example, access to lesson-like structured knowledge (e.g., tutorial sessions), evaluation of knowledge acquisition, provision of extensive reference material, etc.
- **VE application users**, i.e., people using (e.g., working in) a VE who will be referring to **i-do*ve*** seeking practical advice on healthy and proper operation of the VE in question, depending on the VE-technology used, the workplace environment and user typology (experience, age, etc.).
- **i-do*ve* administrators**, including people who act as system support staff (e.g., configure the tool and monitor its performance), but also those responsible for updating and maintaining the user and knowledge base of the tool.

These user groups represent distinct sets of behaviour patterns and goals -but not job descriptions- that are significant to the design of the user interfaces of **i-do*ve***. Each user group needs to be offered its own configurable and adaptable (e.g., personalized) interface.

5 Assessment / Evaluation

A usability evaluation plan was integrated within the development life-cycle of **i-do*ve***. First, a series of scenarios of use, representing the suggested functionality of the tool, were generated and used to design the system's user interfaces. Three usability experts inspected the scenarios and alternative paper-based prototypes and indicated potential usability problems (e.g., complicated screens, non-intuitive navigation model). The evaluation results provided feedback to the development of the first, mainly horizontal, software prototype. The evaluation process then continued using expert and experimental evaluation. Heuristics and cognitive walkthrough substantiated the opinions of experts, while thinking aloud and question asking protocol illustrated users feeling regarding the tool user interfaces. Three usability experts and five end users were invited to work, steered by the evaluators, with the tool in our usability laboratory and express their opinion to the development team. No major usability problems were identified in the first prototype. Nevertheless, it was suggested to reduce the complexity of some screens of the user interface in order to simplify their use. Another significant conclusion of the evaluation highlighted by experts was the need for improved accessibility of the tool.

6 Conclusions & Future Work

In conclusion, we argue that there is a significant need for developing an interactive guidelines support tool for the development of VEs such as **i-dove**. The development of **i-dove** will act catalytically towards: (a) avoidance of serious under-utilisation / wasteful regeneration of existing and future knowledge for creating usable, ergonomically designed, and safe VEs; (b) advancement of such existing knowledge and current practice; (c) detection of suitable ways for assessing and improving VE-applications, by measuring the usefulness of the knowledge and the success of the tool that will be created; (d) discovery of novel solutions and techniques towards advanced tools that will be able to affect the design of VE-applications in a positive way and, thereby, be adopted by the mainstream of software development. In the context of the current work, several areas were identified in which further research, as well as additional data collection and analysis, are needed, such as: (i) integration of VE-related ergonomic algorithms for (semi-) automated VE evaluation to the tool; (ii) refinement of the mechanism for (semi-) automatic cross-referencing of the tool's content; (iii) accommodation of users that aim to utilise the tool for educational/training purposes.

Acknowledgments

Part of the work reported in this paper has been carried out in the framework of the European Commission funded project "Virtual and Interactive Environments for Workplaces of the Future" - VIEW (IST-2000-26089). More information for the project and the consortium can be found at http://www.view.iao.fhg.de.

References

Blatt, L. A., & Knutson, J. F. (1994). Interface design guidance systems. In Nielsen, J., Mack, R., L., (Eds.), *Usability inspection methods* (pp. 351-384). New York: John Wiley & Sons, Inc.

Carter, J. (1999). Incorporating standards and guidelines in an approach that balances usability concerns for developers and end users. *Interacting with Computers*, 12, 179-206.

Gabbard, J. L. (1997). A Taxonomy of Usability Characteristics in Virtual Environments, MSc Thesis, Virginia Polytechnic Institute and State University.

Gofinet, L., & Noirhomme-Fraiture, M. (1999). Automatic cross- referencing of HCI guidelines by statistical methods. *Interacting with Computers*, 12, 166-177.

Grammenos, D., Akoumianakis, D., & Stephanidis, C. (1999). Support for Iterative User Interface Prototyping: The Sherlock Guideline Management System, In S. Chatty, P. Dewan (Eds.), *Engineering for Human-Computer Interaction* (pp.299-315). Kluwer Academic Publishers.

Henninger, S. (2000). A methodology and tools for applying context-specific usability guidelines to interface design. *Interacting with Computers* 12, 225-243.

IBM (2000). IBM RealPlaces Design Guide. Retrieved May 11, 2001, from http://www-3.ibm.com/ibm/easy/eou_ext.nsf/Publish/580.

Kaur, K. (1998). Designing Virtual Environments for Usability. PhD thesis, Centre for HCI Design, City University, London.

Tetzlaff, L., & Schwartz, D. (1991). The use of guidelines in Interface Design. In Proc. of CHI'91 (pp. 329-333). New Orleans, ACM Press.

Vanderdonckt, J. (1999). Development Milestones towards a Tool for Working with Guidelines. *Interacting with Computers*, 12, 81-118.

Wilson, J. R. (2002). From Potential to Practice: Virtual and Interactive Environments in Workplaces of the Future (VIEW). In Proc. of Virtual Reality International Conference (VRIC 2002).

GestureMan PS: Effect of a Head and a Pointing Stick on Robot Mediated Communication

Hideaki Kuzuoka

University of Tsukuba
1-1-1 Tennoudai, Tsukuba, Japan
kuzuoka@esys.tsukuba.ac.jp

Jun'ichi Kosaka

University of Tsukuba
1-1-1 Tennoudai, Tsukuba, Japan
kosaka_j@edu.esys.tsukuba.ac.jp

Shin'ya Oyama

Communications Research Lab.
2-2-2 Hikaridai, Seika-cho, Soraku-gun, Japan
oyamas@crl.go.jp

Keiichi Yamazaki

Saitama University
255 Shimo-ohkubo, Saitamashi, Japan
yamakei@post.saitama-u.ac.jp

Abstract

GestureMan PS is a robot that was designed as a video communication media. Especially we considered to provide the robot with the ability to express a remote person's pointing action as well as shifts in orientation and reference to the space and objects. For this purpose, we designed a robot's head and a pointing stick to be resources of remote person's orientation. Based on the experiment, we could confirm that the robot has the ability to support remote pointing.

1 Introduction

The aim of this paper is to show that a robot is an effective medium for communication when its head and arm (a pointing stick) are properly controlled to indicate remote person's orientation.
When a video mediated communication system is used, participants occasionally "encounter difficulties in making sense of each other's conduct even when undertaking seemingly simple actions such as pointing to objects within a particular environment (Heath, 2001, p.121)" In order to assist remote pointing, we have developed a remote control laser pointer named the GestureLaser (Yamazaki 1999) and it was mounted on a mobile robot named the GestureMan (Kuzuoka, 2000). However, the system could not perform pointing as well as people do in co-existing situation. One of the main reasons for this problem is lack of resources that clearly shows a remote person's orientation and frame of reference.
According to our study with the GestureMan (Heath, 2001), it was suggested that the following considerations might be relevant to develop an improved system;

- Provide participants with the ability to determine the location, orientation and frame of reference of others.
- Provide resources through which participants can discriminate the action of others which involve shifts in orientation and reference to the space and a range of objects, artefacts and features.

In the next section we introduce new GestureMan that was redesigned based on these suggestions. Then we explain an experiment to prove the effect of the system.

2 Gestureman PS

We designed the GestureMan with a pointing stick (GestureMan PS) so that it has multiple resources to help a co-present to determine a remote person's orientation and frame of reference.

2.1 Head and Body Design

GestureMan PS is shown in figure 1. A three-camera unit is mounted at the head of the robot. Each lens has 60 degrees of horizontal field of view thus the remote person can get 180 degrees of horizontal field of view in total.
Three-camera unit is covered with a white helmet. Also two ears and a visor are attached to a helmet. In this way the head gives a co-present clues to determine where the middle camera is orienting to. Pan and tilt motion of the head is made possible by two motors at the neck thus the robot can show both body orientation and head orientation independently. In order to clearly express body orientation shoulders are attached to the body.

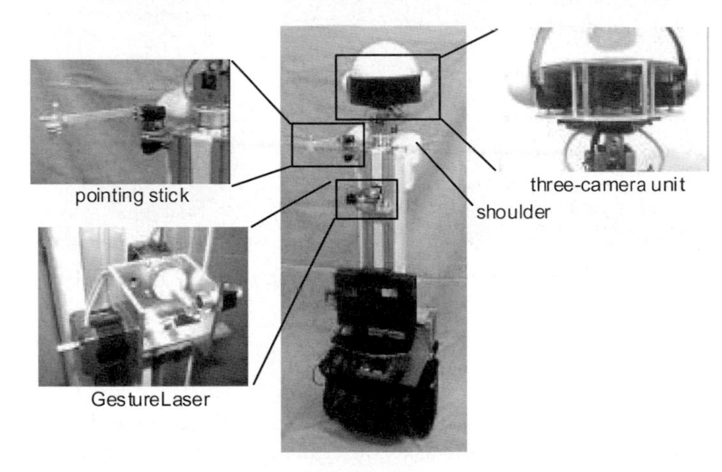

Figure 1: GestureMan PS

2.2 GestureLaser and Pointing Stick

GestureLaser is a remote control laser pointer that was developed by the authors (Yamazaki, 1999) and it is attached to the robot's body. A remote person controls the laser spot by a mouse. The laser beam is emitted only when a left mouse button is pressed. Otherwise the laser is turned off completely.
Because a red dot of the laser pointer appears apart from the robot, if there is not enough resource to determine which direction the laser is emitted to, a co-present occasionally has difficulty in finding the red dot (Heath, 2001). Therefore, in order to alleviate this problem, GestureMan PS is equipped with a pointing stick (Paulos, 1998) and it is controlled as to point to the same direction as the laser beam. Furthermore a small toy finger with a pointing posture is attached to the tip of the pointing stick in order to indicate that it is indicating direction. Soon after the left mouse button is pressed, a laser beam is turned on and the pointing stick starts to rise upward. When the laser is turned off the pointing stick gradually go back to downward position to indicate that nothing is pointed at.

2.3 User Interface for a Remote Person

Typically three liquid crystal displays (LCDs) and a joystick are placed in front of the remote person (Figure 2). Through the LCDs, the remote person can get 180 degrees of horizontal field of view around the robot. The robot's forward motion, backward motion, right turn, and left turn are controlled by leaning a joystick forward, backward, right, and left. The pan and tilt motion of the robot's head is controlled by a hat switch of the joystick. The laser pointer is controlled with a mouse. The laser is turned on only when the remote person pressed a left mouse button. Position of the laser pointer is controlled by mouse motion. Since the position of the laser spot could be monitored in LCDs, he/she can control the position of the laser spot just as controlling the mouse cursor. Pointing stick pointed downward when the laser pointer is turned off. Only when the laser pointer is turned on, the pointing stick is raised and starts to follow the red dot. A remote person and a co-present can talk to each other using wireless microphones and speakers.

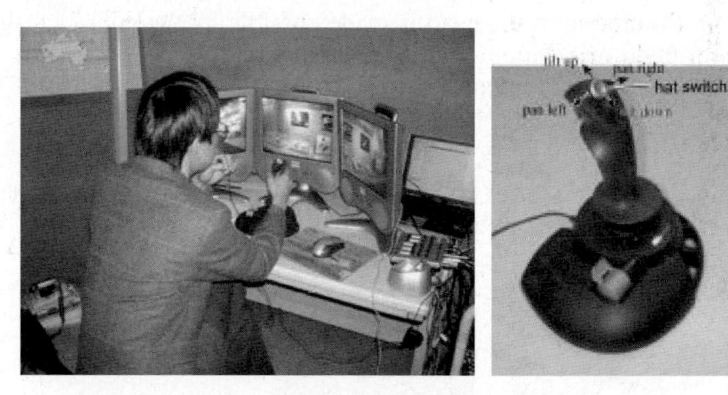

Figure 2: Typical user interface for a remote person

3 SIMPLE POINTING EXPERIMENT

We have conducted an experiment to examine the effect of the head and the pointing stick on indicating remote person's orientation and as a consequence it can assist finding the laser spot.

3.1 Study Design

Each subject was told to stand next to the GestureMan PS (Fig. 3). The robot was programmed to automatically point at 20 objects one by one with the laser pointer. Each object was attached with a number and a subject was asked to speak out the number when they found the pointed object. We measured the time to find each object. The distances from the robot to objects were ranged from 1 meter to 4.5 meters (Fig. 4). Each subject performed the task under the following three conditions;

condition 1: Only the laser pointer was used to point at objects and no other parts of the robot were moved.

condition 2: The laser spot and the head of the robot were used simultaneously. The head was controlled to turn to the same direction as the laser beam. Before each pointing was started the head oriented straight forward.

condition 3: The laser spot and the pointing stick of the robot were used simultaneously. The pointing stick was controlled to turn to the same direction as the laser beam. Before each pointing started the pointing stick pointed downward.

In order to eliminate the order effect, order of three cases was differentiated between subjects.

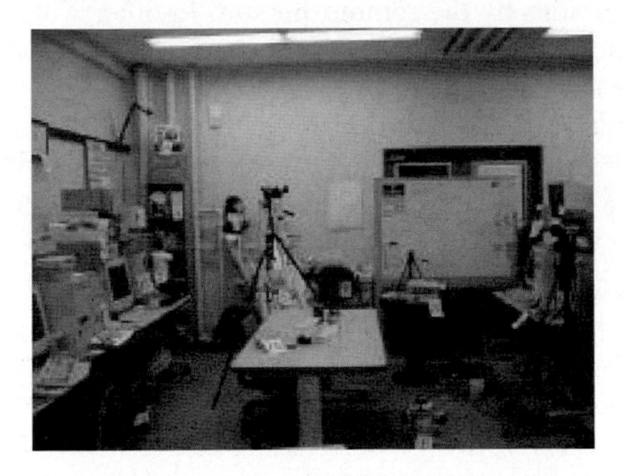

Figure 3: A room used for simple pointing experiment (from a subject's point of view)

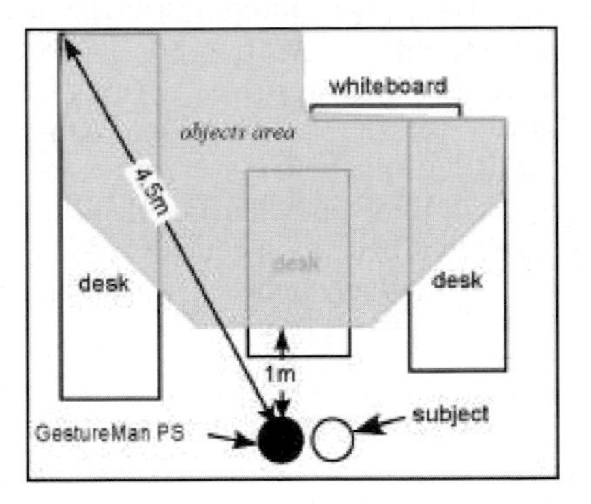

Figure 4: Positions of the GestureMan PS, a subject, and objects (grey area).

3.2 Time to Find Objects

Twelve students aged from 19 to 24 served as subjects. Figure 5 shows the average time to find one object. Since the number of subjects was limited we used nonparametric method for statistical test. As a result of Friedman's $?^2$ test, difference between three populations was significant ($p < 0.01$). Then Sheffe's method showed that both head orientation (condition 2) and pointing stick orientation (condition 3) significantly shorten the time to find an object ($p < 0.01$).
In condition 1 most of the subjects tried to find objects without looking at the robot. In condition 2 and 3, right after the robot started pointing, many subjects tried to find objects without looking at the robot. When subjects could not find the laser spot instantly, however, they turned to the robot to see the orientation of the robot's head or the pointing stick and then turned their heads back to the environment again (Fig. 6).

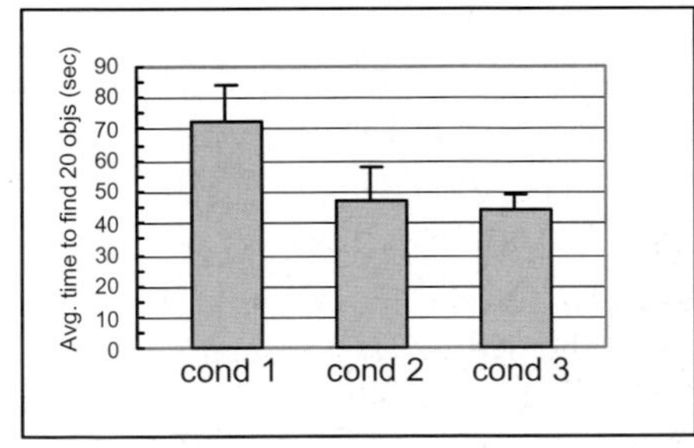

Figure 5: The average time to find objects

Figure 6: A subject saw a pointing stick to find an object.

From this result, it was confirmed that the GestureMan PS's head and pointing stick design can be resources for determining remote person's orientation.

4 DISCUSSIONS

Through the experiments we could show that the head and the pointing stick of GestureMan PS has the ability to express location, orientation, and frame of reference. It should be noted, however, that such ability was achieved not only by the existence of the head and the pointing

stick but also by the appropriate usage of these functions by the remote person. In other words, user interface for a remote person is also important in order to sequentially organize the robot mediated communication (Heath, 2000; Heath, 2001).
The current GestureMan PS system, however, still has drawbacks compare to communication in the co-existing situation. For example, the current user interface does not allow robot's head to be controlled as freely as a human. Also, the robot's head and the laser pointer cannot be controlled simultaneously because a remote person normally uses only his/her dominant hand so he/she switches alternatively between a joystick and a mouse. Therefore, further study on user interface to control the robot is necessary.

5 Conclusions

GestureMan PS is a robot that was designed as a video communication media. The system supports remote pointing to objects within the real environment. Especially we tried to provide the robot with the ability to express a remote person's orientations. For this purpose, GestureMan PS is equipped with the head and the pointing stick. Based on the experiments, we could confirm that these resources of the robot have the ability to support above mentioned considerations.

6 Acknowledgement

The research was supported by Telecommunications Advancement Organization of Japan, Japan Society for the Promotion of Science, Oki Electric Industry Co. Ltd., and Venture Business Laboratory (VBL) of Ministry of Education, Culture, Sports Science and Technology. We should thank the Communications Research Laboratory for supporting development of the GestureMan PS.

7 References

Heath, C., Luff, P. (2000). *Technology in Action*, Cambridge University Press, Cambridge UK.

Heath, C., Luff, P., Kuzuoka, H., Yamazaki, K., and Oyama, S. (2001). Creating Coherent Environments for Collaboration, *Proceedings of ECSCW 2001*, pp. 119-138.

Kuzuoka, H., Oyama, S., Yamazaki, K. and Suzuki, K. (2000). GestureMan: A Robot that Embodies a Remote Instructor's Actions, *Proceedings of CSCW 2000*, pp. 155-162.

Paulos, E. and Canny, J. (1998). ProP: Personal Roving Presence, *Proceedings of CHI' 98*, pp. 296-303.

Yamazaki, K., Yamazaki, A., Kuzuoka, H., Oyama, S., Kato, H., Suzuki, H., and Miki, H. (1999). GetsureLaser and GestureLaser Car: Development of an Embodied Space to Support Remote Instruction, *Proceedings of ECSCW '99*, pp. 239-258.

The Factory of the Future? The Integration of Virtual Reality for Advanced Industrial Applications

Peter J Modern[γ], Alex W Stedmon[ψ], Mirabelle D'Cruz[ψ], John R Wilson[ψ], & Graham J Sharples[γ]

[ψ]VR Applications Research Team,
University of Nottingham,
Nottingham, NG7 2RD, UK
mirabelle.dcruz@nottingham.ac.uk
alex.stedmon@nottingham.ac.uk

[γ]British Nuclear Fuels Plc.
Springfields Works,
Preston, PR4 0XJ, UK
peter.j.modern@bnfl.com
graham.j.sharples@bnfl.com

Abstract

Project IRMA, virtual reality for manufacturing applications, aims to build, integrate, demonstrate and evaluate VR 'application demonstrators' which reflect different but complementary aspects of industrial manufacturing. Application demonstrators are being developed based on a fundamental design philosophy that has produced generic software modules for handling messaging, object and communications management, between real simulated and Virtual Environments (VEs). The main VE application demonstrators within Project IRMA are a British Nuclear Fuels Ltd (BNFL) control and monitoring demonstrator and a Zanussi Electro-Mechanica (ZEM) industrial training demonstrator. Due to the integrated nature of Project IRMA and the drive towards a generic 'factory of the future' design philosophy, a number of technical and human factors related issues have been highlighted and these are discussed in the paper.

1 Project IRMA

Project IRMA, virtual reality for manufacturing applications, is an Intelligent Manufacturing Systems approved inter-regional Virtual Reality (VR) project with partners from the European Union, Japan, Switzerland, and Newly Associated States of Eastern Europe. The key aims of Project IRMA are to build, integrate, demonstrate and evaluate VR 'application demonstrators' which reflect different but complementary aspects of industrial manufacturing, and integrate different commercially available software simulation tools in VEs that support real-time, two-way, interaction. Whilst Saunders (2001) reviews the early development of Project IRMA, this paper details the integrated approach that Project IRMA is taking and reflects upon the underlying factors issues associated with VR applications of this nature.

2 The factory of the future

Application demonstrators are being developed based on a fundamental design philosophy that has produced generic software modules for handling messaging, object and communications management, between real simulated and VEs. Furthermore, the software modules integrate a number of commercial software simulation packages into an overall solution that has the facility to be further configured for other commercial software products that could be used in the 'factory of the future'.

The IRMA software solution, illustrated in Figure 1, is a generic solution for the integration of real world environment hardware (such as a factory process), commercial simulation software, and VR hardware and software.

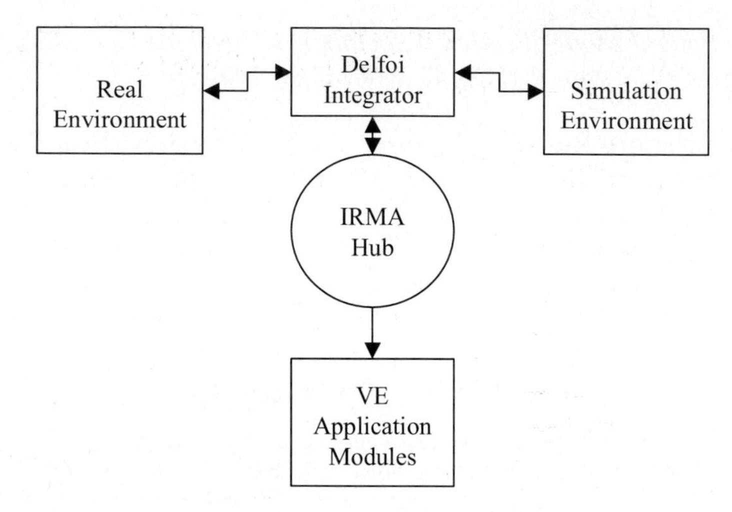

Figure 1: Schematic architecture of the Project IRMA solution

The Delfoi Integrator, IRMA Hub and VE application modules represent the overall IRMA technical solution. Communications modules link the Delfoi Integrator with the real environment and simulation environment. Information is processed through the 'IRMA Hub' and translated to the VE application demonstrators to support user interaction.

3 Project IRMA application demonstrators

The main VE application demonstrators within Project IRMA are a British Nuclear Fuels Ltd (BNFL) control and monitoring demonstrator and a Zanussi Electro-Mechanica (ZEM) industrial training demonstrator.

3.1 BNFL industrial control and monitoring demonstrator

This VE application demonstrator has been developed for the flexible stacking and assembly of nuclear fuel pellets into fuel pins and fuel assemblies. It represents a simple schematic fuel manufacturing plant simulation in Delmia QUEST software, together with a linked fuel handling and assembly machine simulated in Delmia IGRIP software. The VE application demonstrator is illustrated in Figures 2 and 3 below. A number of key mechanisms have been selected, that represent typical industrial manufacturing plant such as air actuated pistons, electric motors, robots, sensors, typical HMI units, and Programmable Logic Controller (PLC) systems. As such the components of the application demonstrator generalize to other industrial processes and environments, further supporting the generic nature of Project IRMA.

The tasks within the VE involve partially, or fully, immersed users, with tracked head and hand motion sensors. Users interact with the VE and conduct real world tasks such as setting machinery to datum and checking control system design specifications. The application is based on control residing in the simulation environment and linking real world hardware to the simulation environment so that changes in the VE can be translated back into the real world application.

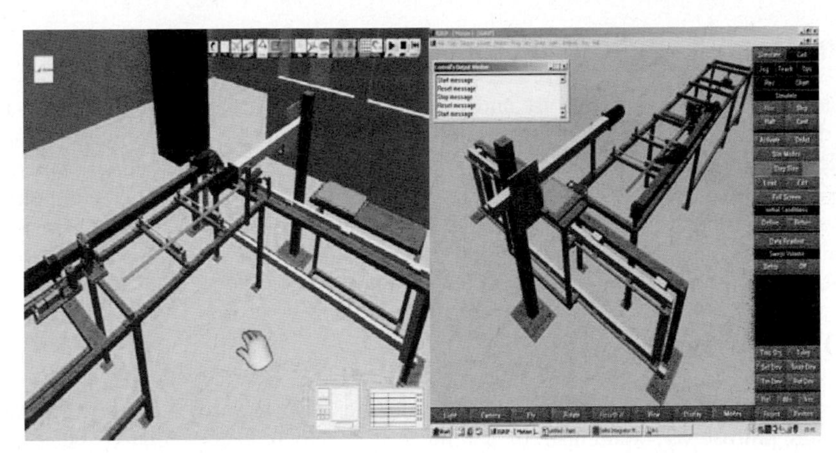

Figure 2: VE view of machine and interface icon graphics

Figure 3: IGRIP view of machine for fuel pellet stacking and assembly

3.2 ZEM industrial training demonstrator

The industrial training VE application demonstrator is set in the context of monitoring and fault rectification of a refrigerator compressor motor assembly and test line at ZEM. The ZEM VE application demonstrator comprises a refrigerator motor and stator line manufacturing plant modelled in Delmia-IGRIP software. The manufacturing line is modelled in varying levels of detail using the Delmia-IGRIP simulation tool and VEs created for operator or supervisor training utilising head-mounted displays and data-glove interaction. Figures 4 and 5 illustrate the IGRIP simulations of key machines and work-cells of the ZEM VE application demonstrator.

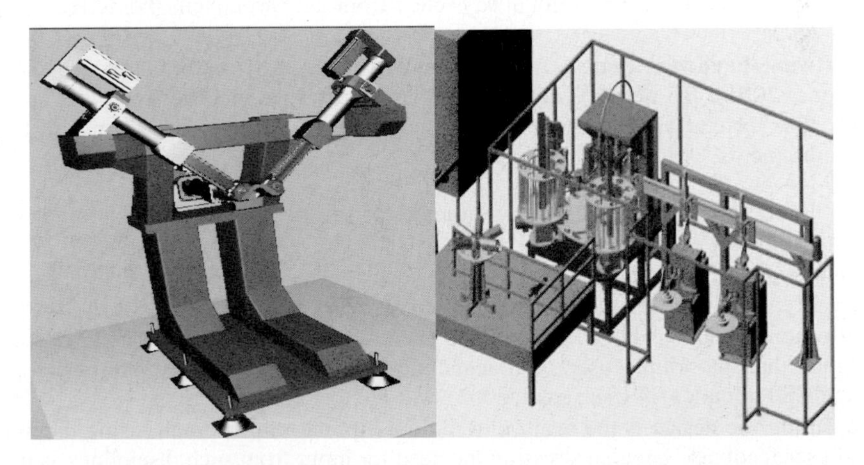

Figure 4: robotic coil winding machine

Figure 5: work cell layout

In accordance with the Virtual Environment Development Structure (D'Cruz, Stedmon, Wilson, Modern & Sharples, 2003) key machines in the line are modelled in higher detail than less important areas and it is the more detailed areas that relate to the training scenarios. In addition, the ZEM application demonstrator requires secure Virtual Private Network transmission of commercially sensitive VR communication data between two geographically different sites.

3.3 Further VE application demonstrators

Other VE application demonstrators are under development within Project IRMA which include: a free-form design which aims to use VR technologies to enable more intuitive working methods in product design; modelling of earthquakes using VR for the mathematical simulation and visualisation of built structures in simulated earthquake conditions; power plant components using VR for the modelling, simulation, prediction and visualisation of power plant component life; rapid prototyping of ship-building developing a demonstrator based on software for shipbuilding applications with the aim of reducing costs and timescales.

4 Development issues

Due to the integrated nature of Project IRMA and the drive towards a generic 'factory of the future' design philosophy, a number of technical and human factors related issues have been highlighted.

The process of software development and simulation model build within the Project has been an iterative cycle (D'Cruz, Stedmon, Wilson, & Eastgate, 2002). As such the design cycle has highlighted a number of issues:

- **distributed project team** - the EU software building group are located in four different countries working to specifications not in their native tongue. Furthermore this is compounded by the application demonstration sites being located in another two countries;
- **lack of guidelines** – for specifying VE development. This is a relatively new field of experience for industrial engineers and scientists (D'Cruz, et al, 2003);
- **level of detail** - required in the simulations and VEs, needs to be defined in relation to the user requirements which is not always clear from the participants themselves (D'Cruz, et al, 2002);
- **software/hardware issues** – with an emphasis on a user -centred approach (Stedmon & Stone, 2001), the iterative design cycle reveals further modifications that are required such as system architecture, specified software and hardware systems, further user requirements.

Underlying the technical development are a number of pertinent human factors issues which the project is investigating, such as:

- ensuring active involvement of users and a clear understanding of user and task requirements (including context of use) in VE development (D'Cruz, et al , 2003);
- allocating functions between users and technology (recognising that today's technology, rather than de-skilling users, can actually extend their capabilities into new applications and skill domains (D'Cruz, et al, 2002);
- ensuring the design is the result of a multidisciplinary input emphasising the importance of user feedback, but also stressing the need for input from such disciplines as marketing, ergonomics, software engineering, technical authors (D'Cruz, et al, 2003).

The utility of interactive VE visualisation for industrial applications depends on whether a VR application and the VE experience supports the user and is linked directly to the VE user requirements specification and the associated evaluation methodology (D'Cruz, et al, 2002). As such, the development of VR applications within Project IRMA is based on VEDS (D'Cruz, et al, 2003) and has highlight issues such as:

- the evaluation process should form an integral part of the VE user requirements specification. However the iterative nature of the software specification, build and testing process has shown that this is not straight-forward;
- the link between the written specification of user requirements and the actual VE visualisation and interaction experience are not clear at the outset since the latter is essentially visual – they need developing and refining via iterative feedback from the industrial user;
- a wide range of HMI options are possible. The most appropriate is not always known in advance and there is always a trade-off between VE design for reality vs. functional utility.

5 Conclusion

In addition to developing the integration software and VE application demonstrators, Project IRMA aims to evaluate the industrial application demonstrators paying particular attention to HCI issues that impact on usability and overall system performance (see, Stedmon, et al, 2003). Accordingly, development of an evaluation programme, linked with the specification of the application demonstrators, is underway in parallel with the building and testing of demonstrator software demonstrators.

6 Acknowledgement

The work presented in this paper is supported by Intelligent Manufacturing Systems, Project 97007: Project IRMA, virtual reality for manufacturing applications and is sponsored within The EU FP5 Programme under Contract G1RD-CT-2000-00236.

7 References

D'Cruz, M., Stedmon, A.W., Wilson, J.R., & Eastgate, R. (2002). *From User Requirements to Functional and User Interface Specification: General Process.* University of Nottingham Report No.: IOE/VIRART/01/351.

D'Cruz, M., Stedmon, A.W., Wilson, J.R., Modern, P.J., & Sharples, G.J. (2003). Building Virtual Environments using the Virtual Environment Development Structure: A Case Study. In, *HCI International '03. Proceedings of the 10th International Conference on Human-Computer Interaction, Crete, June 22-27, 2003.* Lawrence Erlbaum Associates.

Saunders, J.P. (2001). *IRMA: the Technology.* 1st International IMS Project Forum, Ascona, Switzerland, 8-10 October 2001. IMS International.

Stedmon, A.W., & Stone, R.J. (2001). Re-viewing Reality: Human Factors of Synthetic Training Environments. *International Journal of Human-Computer Studies.* 55, 675-698.

Virtual Prints: An Empowering Tool for Virtual Environments

A. Mourouzis[1], *D. Grammenos*[1], *M. Filou*[1], *P. Papadakos*[1], *C. Stephanidis*[1,2]

[1]Foundation for Research and Technology Hellas
Institute of Computer Science
Science and Technology Park of Crete
Heraklion, Crete, GR-71110 Greece
{mourouzi, gramenos, filou, papadako, cs}@ics.forth.gr

[2]University of Crete, Department of Computer Science

Abstract

The concept of Virtual Prints (ViPs), as digital counterparts of real-life tracks that people leave behind, has been introduced for supporting navigation, orientation and wayfinding in Virtual Environments (VEs) and has been explored using a prototype VE equipped with a simplified ViPs mechanism. This paper describes an elaboration of the ViPs mechanism with the aim to support a number of functions, popular, if not standard, in conventional applications, but also useful and required in VEs.

1 Introduction

In (Grammenos, Filou, Papadakos & Stephanidis, 2002), the concept of Virtual Prints (ViPs), as the digital counterparts of real-life tracks that people leave behind, was introduced for supporting navigation, orientation and wayfinding in Virtual Environments (VEs). ViPs can be manifested in three different types (Grammenos et al., 2002): (a) while 'inhabitants' of a virtual world are moving, they are leaving behind their Virtual Footprints (ViFoPs); (b) every time they are interacting ('contacting') with a VE their Virtual Fingerprints (ViFiPs) are 'imprinted' on it; and (c) Virtual Fossils (ViFossils) are special marks that can be permanently left within the virtual space, or on any object upon user request, and can be considered as a kind of personal landmark. Following the preliminary investigation that had been carried out using a prototype VE equipped with a simplified ViPs mechanism, this paper proposes a further elaborated ViPs concept and mechanism that can be employed to support not only navigation, orientation and wayfinding in VEs, but also an additional number of functions, popular, if not standard, in conventional applications, but potentially useful also in VEs, such as interaction shortcuts, bookmarks, help support, interaction history facility, back / forward facility, undo / redo and repeat facility, annotation facility, and facility for highlighting content or marking / identifying (non) visited areas.

2 Elaborated ViPs Properties & Characteristics

Just like footprints in the real world, ViFoPs can be visualised in VEs and thus provide a continuous (i.e., snail tracks like) or discontinuous (i.e., dashed like) three dimensional (3D) representation of the path followed by any user moving in the virtual space. ViFoPs can be depicted in various ways, depending mainly on the characteristics of the application and on the user's requirements and preferences. For a discontinuous representation of the user's path, ViPs can be depicted as simple 3D objects (e.g., cones which may also provide information regarding the user's orientation), or more realistic 3D objects (e.g., a 3D model of a shoe's sole). To make the path more distinguishable, *Connecting Lines* (e.g., thin 3D cylinders) between subsequent ViFoPs can be displayed upon user request (see figure 1), which is a similar technique to the one used when a continuous representation of the user path is desired. A continuous path can be depicted as a beeline using ViFoPs as edges and *Connecting Lines* as connectors. To increase the curvature of a continuous representation, and thus achieve a better graphical representation of 'continuity', the number of edges (i.e., the number of ViFoPs released per unit of time or space) needs to be increased. Alternatively, a number of intermediate *Nodes,* a simplified version of ViFoPs that can only store position co-ordinates x, y and z, can be used.

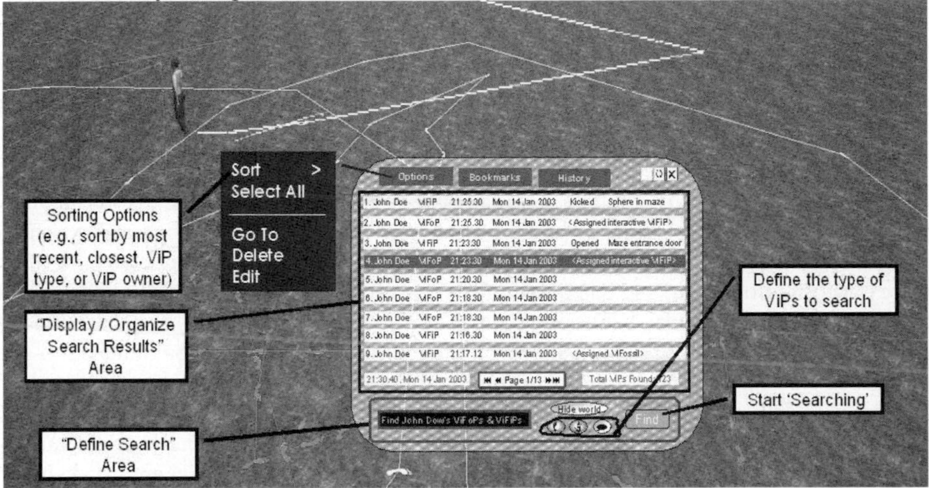

Figure 1: ***Search Console*** **(the Connection Lines between ViFoPs are also visible)**

Each ViFoP can store and therefore provide: (i) spatial information, i.e., position and orientation of the user in the virtual world; (ii) chronological information, e.g., creation, last accessed, or last modified time and date; (iii) personal data information, e.g., owner name or depiction of his/her ViFoPs; and (iv) information about assigned ViFiPs or ViFossils. Each ViFiP is always assigned to one and only ViFoP that is automatically released to record and represent the position and orientation of the user at the moment the interaction took place, just like each ViFossil is assigned to one and only ViFoP to record the position and orientation of the user at the moment the ViFossil was released. This way, a VE user may retrieve information regarding non-visited sites, including undiscovered options, through the ViFoPs of other user(s). Furthermore, a *Connecting Line* between the ViFoP and the assigned ViFiP / ViFossil can be drawn to make the relation visible. ViFoPs can be released anytime, upon user demand or automatically: (a) at specific time or / and space intervals; (b) each time a ViFiP is released; and (c) each time the user releases a ViFossil. Finally, the concept of Virtual Path (ViPath) can be introduced to group ViPs released by a single user that share common spatial (e.g., ViFoPs left in a specific virtual area), chronological (e.g., ViFoPs left during a specific day), or other (such as semantic, e.g., ViFoPs left while performing a specific task) characteristics. ViPs that belong to the same ViPath may 'inherit' part of the information stored in the 'parent' ViPath rather than store it themselves.

As mentioned above, the idea behind ViFiPs is that every time VE 'inhabitants' interact with a VE, their ViFiPs are 'imprinted' on it. Thus, ViFiPs are released automatically: (a) each time an interaction between the user and the VE is detected; and (b) each time the user collides with a component of the virtual world. Whenever possible, the ViFiP is 'imprinted' on the corresponding part of the virtual world. For example, in case of interaction using a pointing device, the user's ViFiPs can be 'imprinted' and thus visualised at the pointing area, e.g., as a cube or, more realistically, as a 3D model of a fingerprint. Non-visual ViFiPs (e.g., ViFiPs generated from speech-based interaction) can also be released, and it is the responsibility of the assigned ViFoP to make apparent the existence of the ViFiP in question. Thus, ViFiPs can be visual or non-visual and store: (i) spatial information when applicable, i.e., co-ordinate of the position and orientation of the interaction; (ii) personal data information, e.g., depiction of this ViFiP; (iii) elucidatory information about the interactive component, e.g., name of type of the object such as "Media Laboratory door"; (iv) descriptive information about the performed user action, e.g., "Left-click on the interactive device"; (v) other information about the VE reaction, such as semantic information e.g., "The Media Laboratory door opened".

ViFossils are special permanent marks that can be: (a) left anywhere within the virtual world; (b) 'pinned' (i.e., attached) to a specific virtual object; or (c) applied to a virtual object, e.g., by applying a specific texture on it. ViFossils can store: (i) spatial information, i.e., the coordinates of the ViFossil's position and orientation in the virtual world; (ii) personal data information, e.g., style and appearance of the ViFossil; (iii) a message in any digital form such as text, audio, or multimedia, e.g., voice-delivered instructions of use of an interacted component; and / or (iv) elucidatory information about the surrounding context of the ViFossil, e.g., "Left by the Media Laboratory door". As a result, ViFossils depending on the message may also be employed to provide or retrieve help support. For example, a number of ViFossils may be attached to specific components of a VE, providing related descriptions and / or guidance for inexperienced users. Just like ViFiPs, ViFossils can be visual or non-visual, and visual ViFossils can be depicted in various ways depending mainly on the characteristics of the application and on the user's requirements and preferences. Indicatively, ViFossils can be depicted as pins, yellow stickers, road signs, wall signs / posters, or pets. Furthermore, ViFossils can also be stored and retrieved through classic 'Favorites' like mechanism. Upon user's request, ViFossils can be grouped in folders. Each folder can be named with a significant title and hold a number of ViFossils and / or other sub-folders.

3 Interacting with ViPs

The ViPs mechanism (automatically) records a sequence of user actions by generating ViPs and thereby storing interaction history related information. A user interacting with a VE needs to be able, at any point in time, to activate or deactivate the ViPs mechanism aiming, for example, at memory space saving. Moreover, once the ViP mechanism is active, there are a number of options related to ViPs that VE users need to have access at any point in time, such as:

- Release a new ViP (e.g., a ViFossil).
- Start or delete a ViPath.
- Perform *ViPs-based Navigation & Interaction*. ViPs can support, among other things, synchronous and asynchronous *social navigation* in collaborative VEs (Grammenos et al, 2002). For example, it is possible to automatically follow another leading VE user / companion, which allows the user to focus on other tasks rather than on navigation. In addition, a VE user may 'take a shortcut' (i.e., be 'teletransported') to another of his / her ViP, such as to previous or next ViP (i.e., a 'Back & Forward' effect) or straight to

his first ViP (i.e., a 'Go to Home' effect). In a similar way, since ViFiPs hold information regarding the user interactions with components of VEs, ViPs may also support 'Undo', 'Redo', and 'Repeat' as well as *object-focused interaction* (Hindmarsh et al, 2000). It is also feasible to integrate such ViPs-based navigation and interaction functionality into a single component such as a *Navigation & Interaction Console* (e.g., see figure 2).

- Conduct *ViPs-based Search*. ViPs can provide, among other things, useful feedback regarding the position, time and nature of previous (inter-) actions of VE users. Such information can be visualised in a VE by the means of ViPs, but also accessed by the means of (2D) lists and catalogues. For instance, VE users can browse or search (for instance through keywords search or mixed search combining specific time or ViP type constrains) available lists of ViPs for specific information. This way, the ViPs mechanism may also support *navigation by query* (van Ballegooij & Eliëns, 2002) in VEs. This type of ViPs - related functionality can be integrated into a single component, such as a *Search Console* (e.g., see figure 1).
- Perform *ViPs-based Measurements*. ViPs hold information about their actual position in the virtual world, in terms of their co-ordinates in the three dimensions (x, y, z), and thus allow the automatic calculation of distances among them. In a similar way, since ViPs also store chronological information, such as creation or last accessed date and time, they allow, among others, the automatic calculation of the time required to navigate from one place to another or to perform a specific sequence of actions.

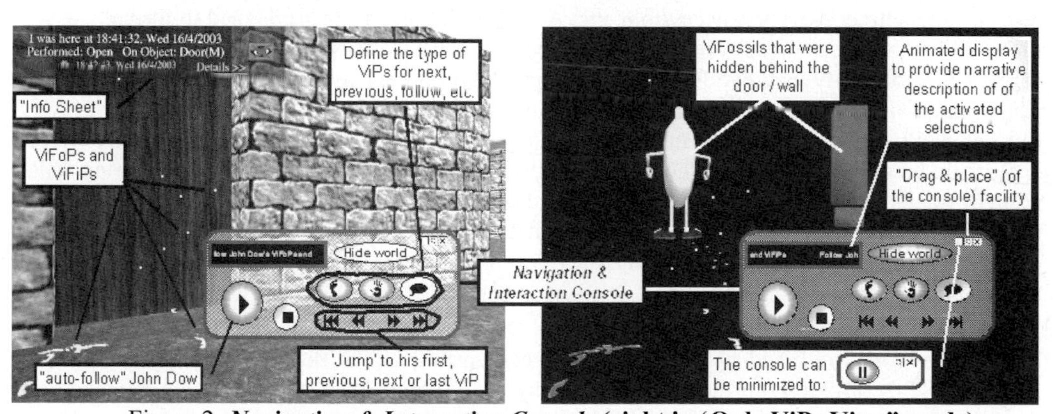

Figure 2: ***Navigation & Interaction Console* (right in 'Only ViPs View" mode)**

- Access *ViPs-based Interaction History*. Since the ViPs mechanism records information regarding the user's past navigation and interaction steps, it can also be utilized as an interaction history mechanism. For instance, a VE user may access a list of his / her previous actions performed during a specific day by revisiting the corresponding ViPath or by browsing / searching the corresponding lists of ViPs. It is feasible to incorporate such history-related functionality into the *Search Console* mentioned above.
- Access *ViPs-based Bookmarks*. Once again it is possible to incorporate any bookmarks- related functionality into the *Search Console* mentioned above.
- Access the *Configurations of ViPs*, such as: (i) modify the way ViPs are depicted; (ii) hide / display ViPs, while the ViPs mechanism is active; (iii) reduce or increase the number of displayed ViPs on a percentage basis, e.g., the user may choose to limit the

volume of ViPs by displaying only 60% of the recorded ViPs; (iv) hide or display the Connecting Lines; (v) interchange between 'Only ViPs' (i.e., ViPs 3D map, see figure 2, right part) and normal display (see figure 2, left part); (vi) scale ViPs up or down, i.e., an 'inflate' / 'deflate' effect; (vii) personalise or share own ViPs by specifying whether these are personal, 'read only', or 'read & write'; (viii) modify the way ViPs are depicted; or (ix) edit time – related configurations of ViPs, e.g., modify their time-sensitivity by increasing their lifespan.

According to (Grammenos et al, 2002) each ViP stores directly or indirectly significant user-related information (e.g., spatial, chronological, personal data) that can be displayed upon user request (e.g., when the user 'points' a specific ViP) through an *information sheet*. Furthermore, each ViP can be selected offering a number of alternative options, such as: (a) move, (re-) orient, bookmark, or delete the ViP; (b) 'select' the parent ViPath (i.e., the ViPath to which the selected ViP belongs) and thereafter copy & paste or bookmark its ViPs; (c) 'select' the owner, and thereafter access his / her *ViPs-based Interaction History* or perform a number of application dependent (i.e., ViPs independent) tasks such as block, talk to, or email, the owner, etc.

4 Conclusion & Future Work

ViPs, are 'inspired' from the real world, are easy to understand, and as digital entities, can be subject to processing, and can acquire a number of attributes, e.g., personal or shared, visible or hidden, interactive, dynamic, accessible, etc. A number of user tests are planed to formally assess the usability of the evolved ViPs mechanism (effectiveness, efficiency and user satisfaction) in supporting, not only navigation, orientation and wayfinding in VEs, but also a number of functions, commonly found in conventional applications, but also useful and required in VEs. Further work on ViPs currently under way concerns: (a) the identification of potential, e.g., in evaluation as a review tool; (b) the support of VE designers and developers by delivering ViPs-related guidelines for developing the ViPs mechanism; and (c) creating a re-usable ViPs mechanism to be used as a plug-in to VEs, so that ViPs can be easily adopted by the broader VE users and / or VE developers community.

Acknowledgments

Part of the work presented in this paper has been carried out in the context of "VIEW of the Future" (IST-2000-26089) project, and funded by the European Commission in the framework of the Information Society Technologies (IST) Programme.

References

Hindmarsh, J., Fraser, M., Heath, C., Benford, S., & Greenhalgh, C., (2000). Object-focused interaction in collaborative virtual environments. In *ACM Transactions on Computer-Human Interaction (TOCHI)*, v.7 n.4, p.477-509, Dec. 2000.

Grammenos, G., Filou, M., Papadakos, P., & Stephanidis, C., (2002). Virtual Prints: Leaving trails in Virtual Environments. In Proc. of the Eighth Eurographics Workshop on Virtual Environments, Barcelona, Spain, 30-31 May.

van Ballegooij, A.,& Eliëns, A., (2001). Navigation by Query in Virtual Worlds. In Proc. of Web3D 2001 Conference, Paderborn, Germany, 19-22 Feb 2001.

Avatar Communication: Virtual Instructor in the Demonstration Exhibit

Tetsuro Ogi[1, 2, 3], *Toshio Yamada*[1], *Takuro Kayahara*[1, 2], *Yuji Kurita*[1]

[1] Telecommunications Advancement Organization of Japan
[2] IML, The University of Tokyo
[3] Mitsubishi Research Institute, Inc.
2-11-16 Yayoi, Bunkyo-ku, Tokyo 113-8656, Japan
tetsu@iml.u-tokyo.ac.jp

Abstract

Video avatar is a technique that represents a realistic human image in the virtual space by using a live video, and it has been used to realize a high presence communication in the shared virtual world. In this technique, various kinds of video avatar models, such as the 2-dimensional plane model, the 2.5-dimensional depth model, and the 3-dimensional voxel model, have been proposed and used. Particularly, in this study, a video avatar studio and a video avatar server were developed in order to generate various kinds of video avatar data and use them in the application systems. These technologies were applied to the psychological demonstration exhibits in the shared immersive virtual world and they were used effectively.

1 Introduction

In the recent virtual reality systems, high presence virtual worlds can be displayed, and therefore representation of a realistic human image is also required. In particular, when the immersive virtual reality environment such as the CAVE is used, a real size image of the whole body expression is desired in order to realize the interaction as a first person experience. In addition, since the infrastructure of the broadband network has been constructed, a high presence human image can also be used as a natural communication tool in the share virtual world.

In this research, in order to meet these demands, video avatar technology has been studied. Video avatar is a technique that generates a high presence human image in the virtual world using a live video. In this method, various kinds of video avatar models, such as 2-dimensional, 2.5-dimensional and 3-dimensional video avatars, have been proposed and they are used according to the purposes of the application systems.

This paper describes the several kinds of video avatar technologies developed in this study, and the video avatar studio and video avatar server that generate various kinds of video avatar data and transmit them to the virtual reality application systems. Moreover, the video avatar was applied to the psychological demonstration exhibits, in which the psychologist was represented as a video avatar and he instructed the psychological experiment in the shared virtual world.

2 Video Avatar Technology

In general, video avatar is generated in the following process. First, the person's figure is captured by the video camera, and then only the person's image is segmented from the background as well as the geometry model of the person is created. By texture mapping the segmented person's image onto the geometry model, the video avatar is created. In this process, various kinds of video avatars can be generated due to the created geometry models and the segmentation methods of the person's figure.

The simplest video avatar method uses a 2-dimensional plane model. In this method, since the segmented person's image is placed as a 2-dimensional board in the virtual space, the video avatar itself does not have depth information. However, when the multiple cameras are used to film the person's image from various directions, 3-dimensional expression such as a motion parallax can be performed, by changing the selected video images according to the movement of the user's viewpoint. Figure 1 (a) shows the generation method of the 2-dimensional video avatar using the multi-viewpoint camera system.

On the other hand, when a stereo video camera is used, depth information for each pixel of the filmed image can be acquired using stereo matching algorithm. By arranging each pixel of the captured image in the virtual space according to the depth distance, a geometric model of the filmed image is created. Then, a stereo video avatar that has depth information can be generated, by texture mapping the segmented person's image onto the geometric model in real-time. However, since this model has depth information only for the front surface that faces the stereo camera, it is called a 2.5-dimensional video avatar (Ogi et al., 2001). Moreover, this model can be also used to perform 3-dimensional expression by changing two or more stereo cameras according to the movement of the user's viewpoint as shown in Figure 1 (b).

A 3-dimensional video avatar that has a complete surface model for of all directions can be generated, by using multiple cameras placed surrounding a person (Moezzi et al., 1996). In this method, the silhouette of the person's figure is calculated for each image filmed by the multiple cameras. By judging the position of each voxel in the three-dimensional space whether it is placed inside or outside the silhouette calculated by each camera, a voxel model of the person's figure is generated. Then, the 3-dimensional video avatar can be generated by texture mapping the segmented person's image onto the surface polygons created from the voxel data. Figure 1 (c) illustrates the generation method of the 3-dimensional video avatar.

Since these video avatar models have individual features about the calculation time and the image quality, the users should select the appropriate method according to the purposes of the application systems.

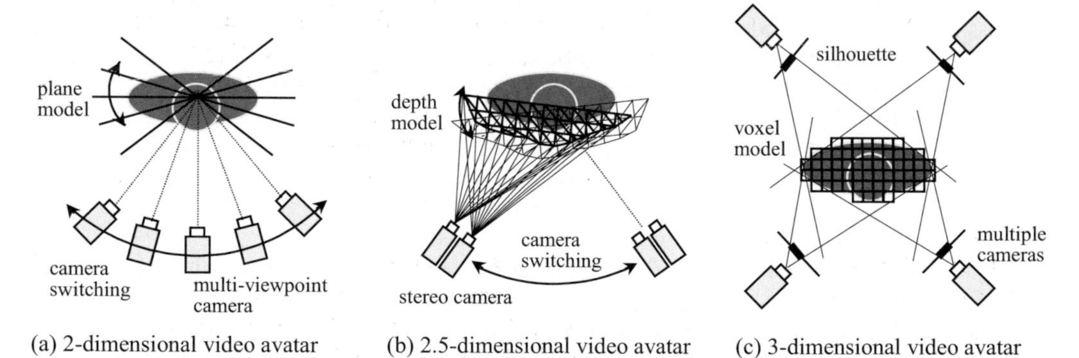

Figure 1: Generation method of various kinds of video avatar

3 Video Avatar Studio and Video Avatar Server

The video avatar technology can be used not only to represent a realistic human in the virtual world but also to realize a high presence communication, by transmitting the video avatar data mutually. In addition, the video avatar can be reproduced using the recorded data as well as it is transmitted in real-time. In this study, a video avatar studio and a video avatar server were developed in order to generate various kinds of video avatar data and use them in various methods. The video avatar studio is a cylindrical room with a diameter of 4000 mm and a height of 2570 mm. Figure 2 and Figure 3 show the system construction of the video avatar studio. In this system, eighteen CCD cameras (SONY DFW-X700) are mounted on the blue back wall at intervals of 20 degrees. The heights of the camera positions can be changed among 600 mm, 1200 mm, and 1575 mm from the floor. These cameras are attached on the outside of the wall and only the camera heads are stuck out into the inner side of the room through the holes. Since the background filmed by every camera is blue, the person's image can easily be segmented from the background using the chroma-key method.

Figure 2: Cylindrical video avatar studio

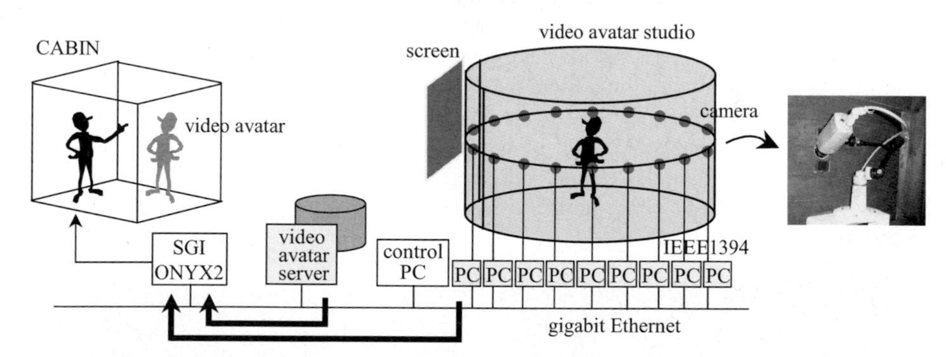

Figure 3: System construction of video avatar studio and video avatar server

The filmed images are transmitted to the PCs through IEEE 1394 connections, and they are used to generate various kinds of video avatar data. For example, the 2-dimensional video avatar using a plane model is created simply switching the images captured by eighteen cameras. When the 2.5-dimensional video avatar is used, two cameras placed in front of the user are used as a stereo camera to calculate the depth value. And the 3-dimensional video avatar is generated using all cameras to create a voxel model.

The PCs are connected to the graphics workstation through the gigabit Ethernet and the video avatar data are transmitted to the virtual reality system. These data can be stored in the database as well as they are superimposed on the virtual world in real-time. The stored data is transmitted to the virtual reality applications by the video avatar streaming server and they can be used repeatedly. The video avatar server communicates with the virtual reality applications and transmits the avatar data retrieved from the database according to the requests from the applications. In this case, the streaming server transmits the video avatar data at the same frame rate as the recording frame rate. Therefore, the application system can receive the video avatar data and integrate it in the virtual world without considering whether it is a real-time data or a recorded data.

4 Psychological Demonstration Exhibit

In this study, the video avatar technology was applied to the psychological demonstration in the immersive projection display CABIN (Hirose et al., 1999). CABIN is a CAVE-like multi-screen display that has five screens at the front on the left, right, ceiling and floor. In this system, the video avatar of the psychologist was superimposed on the demonstration of the virtual psychology laboratory, and the audiences experienced the psychological demonstrations such as the scene recognition and the pointing gesture recognition. In these demonstrations, the figure of the psychologist appeared in the virtual laboratory as a video avatar and he explained the psychological phenomenon and how the audiences can experience the demonstration in the shared virtual world.

In the demonstration of the scene recognition experiment, the audiences and the remote psychologist shared the virtual room through the network, and several objects were put on the table placed in it. In this experiment, the arrangement of the objects was changed and the table or the room was rotated. And then, the subjects were asked whether they could recognize the change of the positions of the objects.

Although the psychologist and the subjects are in the distant places, they were able to communicate with each other using the video avatar technology. In this demonstration, the psychologist's figure was represented as a 2.5-dimensional video avatar that has the front surface model, and he explained the contents of the experiment using his gesture in the virtual world as shown in Figure 4. The figure of the psychologist was also recorded as video avatar data and it was transmitted to the demonstration system using the video avatar server, so that it could be used in the demonstration exhibit repeatedly even when the psychologist was absent. Although the transmission of the psychologist's image was stopped during the experiment so that the subject carried out the experimental task alone, the psychologist was able to observe the subject's behaviour by his side without being noticed, because the subject's image was transmitted to the psychologist site.

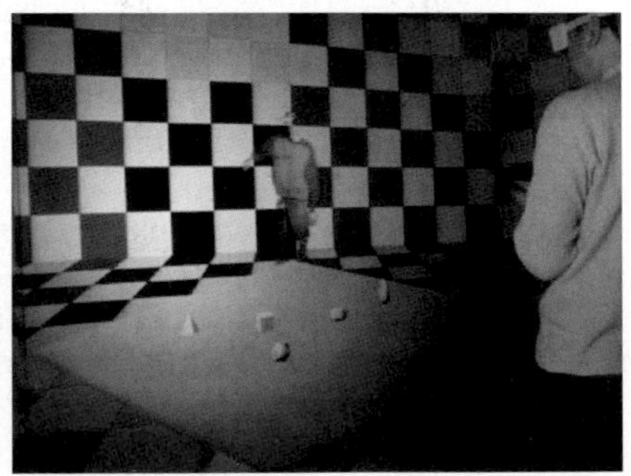

Figure 4: Demonstration of the scene recognition experiment

The demonstration of the pointing gesture recognition is a psychological experiment that shows how accurately the user can recognize the human's pointing gesture and which element of the person's body, such as the direction of the arm or the direction of the face, affects the user's recognition. Also in this demonstration, the psychologist explained the contents of the experiment to the remote subjects using the 2.5-dimensional video avatar. In addition, the figure of the person who was actually pointing at the objects in the real world was represented in the virtual world using the 2.5-dimensional or 3-dimensional video avatar. These video avatar data were recorded beforehand and were used in the experiment as well as they were transmitted from the remote site in real-time. Figure 5 (a) shows the video avatar of the psychologist who is explaining the contents of the experiment, and Figure 5 (b) shows the video avatar of the pointer who is actually pointing at the object in the real world.

In these demonstrations, several users experienced the psychological experiments in the shared virtual world, and the video avatar technology was effectively used for the instruction and the explanation of the demonstration exhibits.

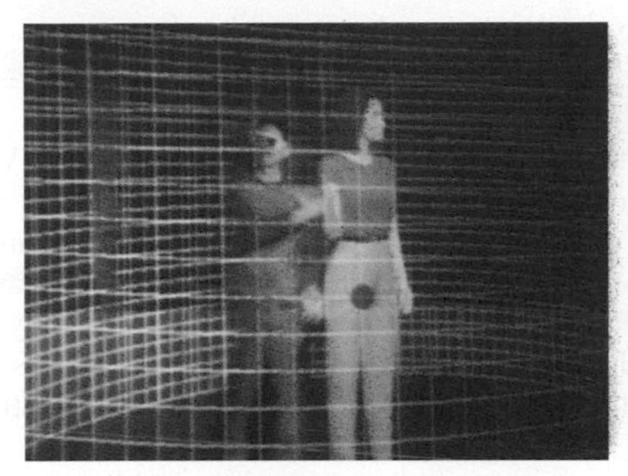

(a) Video avatar of psychologist

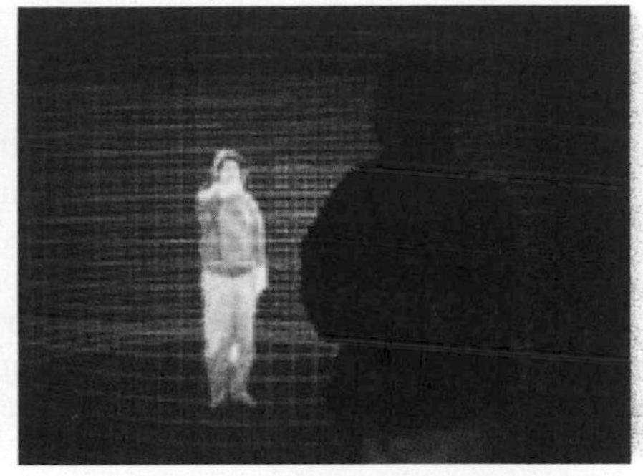

(b) Video avatar of pointer

Figure 5: Demonstration of the pointing gesture recognition experiment

5 Conclusions

In this study, various kinds of video avatar methods, such as using the 2-dimensional plane model, the 2.5-dimensional depth model and the 3-dimensional voxel model, were developed. These models can be easily generated and used using the video avatar generation studio and the video avatar streaming server technologies. In this study, these technologies were applied to the psychological demonstration exhibits in the shared virtual world and they were used effectively. Future work will include applying the video avatar communication technology to several practical exhibits and evaluating the effectiveness of this technology.

References

Ogi, T., Yamada, T., Kano, M., Hirose, M. (2001). Immersive Telecommunication Using Stereo Video Avatar, *Proceedings of Virtual Reality 2001 Conference*, 45-51.

Moezzi, S., Katkere, A., Kuramura, D.Y., Jain, R. (1996). Immersive Video, *Proceedings of VRAIS'96*, 17-24.

Hirose, M., Ogi, T., Ishiwata, S., Yamada, T. (1999). Development and Evaluation of Immersive Multiscreen Display "CABIN". *Systems and Computers in Japan*, 30 (1), 13-22.

Two Methods and a Case Study: Human Factors Evaluations for Virtual Environments

Alex W Stedmon, Mirabelle D'Cruz, Jolanda Tromp, & John R Wilson

VR Applications Research Team,
University of Nottingham,
Nottingham, NG7 2RD, UK
alex.stedmon@nottingham.ac.uk mirabelle.dcruz@nottingham.ac.uk
jolanda.tromp@nottingham.ac.uk john.wilson@nottingham.ac.uk

Abstract

In order that early decisions can be made about the design, usability and evaluation of new VEs, a clear progression taking user needs forward into system development is required. This paper presents an overview of two evaluation methods developed by the University of Nottingham, the Post-Immersion Evaluation Set and the VR Inspection Tool, in relation to a case study example, and assesses the relative merits of each approach. The methods are different in the way they elicit information from the target population. No single method is best suited to defining user, system and application needs. As such, a battery of methods provides a richer and deeper evaluation to take forward into VE design and usability.

1 Project IRMA

Project IRMA, virtual reality for manufacturing applications, is an Intelligent Manufacturing Systems approved inter-regional Virtual Reality (VR) project with partners from the European Union, Japan, Switzerland, and Newly Associated States of Eastern Europe. The key aims of Project IRMA are to build, integrate, demonstrate and evaluate VR 'application demonstrators' which reflect different but complementary aspects of industrial manufacturing, and integrate different commercially available software simulation tools in VEs that support real-time, two-way, interaction (Modern, Stedmon, D'Cruz, Wilson & Sharples, 2003).

2 From user requirements to system development

In order that early decisions can be made about the design, usability and evaluation of new VEs, a clear progression taking user needs forward into system development is required (D'Cruz, Stedmon, Wilson, Modern, & Sharples, 2003). As part of this process, a human factors evaluation has been conducted using two evaluation methods, developed by the University of Nottingham, the Post-Immersion Evaluation Set and the VR Inspection Tool. These methods have been applied to a VE application demonstrator that has been developed within Project IRMA for the flexible stacking and assembly of nuclear fuel pellets into fuel pins and fuel assemblies (see, Modern, et al, 2003. The evaluation was conducted in order to:

- evaluate the application and to feedback recommendations to developers and users;
- evaluate the choice of tools applied in view of revising them specifically for the overall project evaluation programme;
- show the user company the type and scope of information they can expect from such evaluation tools.

3 Human factors evaluation

An overview of the findings is presented below to illustrate the post-immersion evaluation set and the VR inspection tool.

The post-immersion evaluation set comprises of a number of questionnaires that have been designed to assess factors associated with the experience of Virtual Reality (VR) systems and Virtual Environments (VEs). These questionnaires are part of a wider set of pre- and post-immersion evaluation questionnaires, based on established research in relation to VR Induced Symptoms and Effects (Cobb, Nichols, Ramsay & Wilson, 1999). When used together the questionnaires offer an insight into the differences between user's VR experiences before and after using a VR application (Stedmon, Nichols, Patel & Wilson, 2003). In this instance only the post-immersion evaluation set was administered so that profiles could be drawn up based on the specific experience of using the BNFL application demonstrator. In the post-immersion evaluation set, the following questionnaires were used:

- **Simulator Sickness Questionnaire** (to evaluate user's immediate symptom level);
- **Stress Arousal Checklist** (to evaluate the user's levels of stress and arousal);
- **Presence Questionnaire** (to evaluate levels of perceived presence);
- **Usability Questionnaire** (to evaluate levels of perceived usability of the VR application);
- **Input Device Usability** Questionnaire (to evaluate levels of perceived usability for the input device used);
- **Behaviour During VE Viewing** (to assess behaviour patterns);
- **Post Immersion Assessment of Experience** (to assess the user's perception of their VR experience);
- **Enjoyment Questionnaire** (to measure overall enjoyment of the VR experience).

The VR Inspection tool is currently being developed as a method of self-assessment by application users and developers without the need for costly and time-consuming experimental trials. This tool has been created through widely accepted HCI design principles and expert assessments on presence and sickness (Tromp & Nichols, 2003). The tool consists of four main sections:

- **Section 1** – presents a description of the application being reviewed based on the technology of the VR system and VE description.
- **Section 2** – describes the user and task in more detail, based on specifying user groups and task analyses of the application (including length of immersion time). This section also highlights initial usability issues that have been reported by users or from informal observations.
- **Section 3** – contains a detailed structured list of usability issues, detailing ease of use as well as functionality.
- **Section 4** – this section presents 'the way forward' and describes and technical changes that can be directly recommended. It also provides an informal summary/expert rating based on the overall usability, with space for further comments or research questions/hypotheses that have not been highlighted elsewhere in the inspection.

The methodology involves visually assessing an interface and judging its compliance with utility heuristics and recognized general usability principles. A manual, which consists of forms, guides the assessor through the assessment and the depth of analysis is completely under control of the assessors (Tromp & Nichols, 2003).

4 General findings

The evaluation of the BNFL application demonstrator was carried out after a demonstration by two expert users. The responses in the post-immersion evaluation set are a reflection of the two user's experience of the application over a number of sessions including the demonstration session where data was collected. A more detailed account of the findings can be found in Stedmon, D'Cruz, Tromp & Eastgate (2002).

From the evaluation it is possible to make a number of general observations which are summarised below. From these, recommendations can be made that address issues that relate to usability, presence and sickness, as well as for the further development of the VE application more generally.

4.1 General observations

- the VE may promote symptoms associated with simulator sickness;
- the VE increased user arousal more than user stress;
- the VE generated moderate levels of presence;
- the VE was rated moderately for usability;
- the input device (mouse and joystick) scored average/moderate usability;
- there are potential physical factors concerning the VR equipment which might be alleviated by altering user behaviour;
- the users rated their experience close to what they expected from the application.

4.2 Usability recommendations

- improve interactivity so that users can move around and manipulate objects more easily;
- increase instructions, cues and feedback;
- decrease the level of clutter so that only salient features and objects are in the VE;
- the input device should be integrated more effectively to support the user's interaction with, and control over, the VE;
- the VR system response rate needs to be improved.

4.3 Presence recommendations

- improve the visual quality and realism of objects;
- increase interactivity to give the user a better sense of control;
- increase feedback for actions;
- include sound to increase realism;

4.4 Sickness recommendations

- reduce the temperature in room;
- keep the number of independent moving objects in the environment to a minimum;
- place the user nearer to objects which have to be manipulated;

4.5 General recommendations

From the observations a number of recommendations can be made:

- the visual quality of the VE and the realism of the objects could be improved to increase user's sense of feeling that they are actually 'in' the virtual world;
- more interactivity could be programmed into the VE so that users might reach out and believe they can touch objects;
- interaction needs to be considered so that it is easier to move around and manipulate objects with the VE, giving the user a better sense of control;
- instructions and feedback should be investigated in relation to the overall usability of the VE, to enhance presence, and to clarify user interaction with the VE;
- mistakes and errors should be investigated so that suitable instructions, designs and strategies for their recovery can be implemented;
- better integration of the input device should be made to support the user's interaction with, and control over, the VE;
- the physical aspects of the VR suite need to be considered and whist there is limited potential to make changes, findings could be incorporated into future instructions for use.

More generally, the users concluded that the current state of the application is insufficient, 'there is no interactive capability', 'the VE is too cluttered with unrecognisable objects/tools', 'not rendered well enough', and there are 'no interfaces, instructions or feedback for the user'. However, from the evaluation, it is apparent that the current prototype is 'usable and generates a sense of presence at a basic level and that both users enjoyed their experience'.

5 Comparing the methods

The methods are very different in the way they elicit information from the target population, one uses a questionnaire based methodology and the other employs a checklist and user driven evaluation process. Both the questionnaires and inspection tool are cost-effective methods. The inspection tool is a quicker method for gaining usable information as an evaluation can be carried out in a minimum of 2 hours, with a single expert or small expert focus group and without the need for recruiting large numbers of participants (Tromp & Nichols, 2003). The questionnaires provide a useful means of collecting information over a wide range of factors in a relatively short time, however, there is then the need for further collation and analysis for each participant which can be time consuming.

Both methods require some form of application to evaluate but the inspection tool lends itself to use at any stage of the development process, where initial findings can be fed back into the iterative VE development process such as the Virtual Environment Development Structure (D'Cruz, et al, 2003). As Tromp & Nichols (2003) state, the inspection tool can also be used to illustrate and quantify the usefulness of VEs for task-improvements in industry, and it can be used to provide an overview of incremental and iterative changes on an application under development. It has helped raise an awareness of utility and usability concerns amongst the non-usability expert employers of the method. It can be used at any time, from the very early design stages, through to the final product in use in the workplace (Tromp & Nichols, 2003).

6 Conclusion

From the evaluations it is apparent that there is not enough interaction and the scenarios are not fully developed to fulfil user or application needs. The findings highlight particular areas where develop effort might focus on reducing undesirable effects or increasing the usability of the application. The findings provide a basis for the iterative design process in the future development and evaluation of the VE under investigation, as well as serving to illustrate the complimentary nature of the methodologies employed. No single method is best suited to defining user, system and application needs, and, as such, a battery of methods provides a richer and deeper evaluation to take forward into VE design and usability.

7 Acknowledgement

The work presented in this paper is supported by Intelligent Manufacturing Systems, Project 97007: Project IRMA, virtual reality for manufacturing applications and is sponsored within The EU FP5 Programme under Contract G1RD-CT-2000-00236.

8 References

Cobb, S.V.G., Nichols, S.C., Ramsey, A.R. & Wilson, J.R. (1999). Virtual Reality – Induced Symptoms and effects (VRISE). *Presence: Teleoperators and Virtual Environments,* 8(2) pp.169-186

D'Cruz, M., Stedmon, A.W., Wilson, J.R., Modern, P.J., & Sharples, G.J. (2003). Building Virtual Environments using the Virtual Environment Development Structure: A Case Study. In, *HCI International '03. Proceedings of the 10th International Conference on Human-Computer Interaction, Crete, June 22-27, 2003.* Lawrence Erlbaum Associates.

Modern, P.J., Stedmon, A.W., D'Cruz, M., Wilson, J.R., & Sharples, G.J. (2003). The Factory of the Future? The Integration of Virtual Reality for Advanced Industrial Applications. In, *HCI International '03. Proceedings of the 10th International Conference on Human-Computer Interaction, Crete, June 22-27, 2003.* Lawrence Erlbaum Associates.

Stedmon, A.W., Nichols, S.C., Patel, H., & Wilson, J.R. (2003). Free Speech in a Virtual World: Speech Recognition as a Novel Interaction Device for Virtual Reality Applications. In, P. McCabe (ed). *Contemporary Ergonomics 2003. Proceedings of The Ergonomics Society Annual Conference, Edinburgh. UK. 15-17 April, 2003.* Taylor & Francis Ltd. London.

Stedmon, A.W., D'Cruz, M., Tromp, J., & Eastgate, R. (2002). *Evaluation of the Virtual Flexpin Rig.* University of Nottingham Report for Project IRMA.

Tromp, J., & Nichols, S.C. (2003). VIEW-IT: Usefulness Assessment Tool for CAD/VR Hybrids for use in Industry. In, *HCI International '03. Proceedings of the 10th International Conference on Human-Computer Interaction, Crete, June 22-27, 2003.* Lawrence Erlbaum Associates.

Design of Interaction Devices for Optical Tracking in Immersive Environments

Oliver Stefani, Hilko Hoffmann, Jörg Rauschenbach

Fraunhofer Institute Industrial Engineering
Nobelstrasse 12, 70569 Stuttgart, Germany
Oliver.Stefani; Hilko.Hoffmann; Jörg.Rauschenbach{@iao.fhg.de}

Abstract

In this paper we present an interaction concept and input devices for optical tracking in virtual immersive environments. Requirements to input devices for immersive environments will be based on a document, outlining major requirements for VR-systems. A preliminary testing of the usability will be discussed. We conclude this paper by presenting our interim findings.

1 Introduction

1.1 Input Devices for virtual environments

For the design of handheld devices, anthropometrical measurements of the human hand are essential (Bandera, Kern, Solf, 1986). There is a great variety of ergonomic grips comprising different kinds of handles, all of which have their advantages (Kapandji, 1992). For selection and manipulation in virtual environments a suitable alternative for a handle does not have to transmit high forces but has to provide a shape to allow for precise and well-aimed movements. It should also allow for a variation between different grip possibilities to avoid fatigue. At the same time the devices have to include the technological equipment, needed for real time tracking.

1.2 Spatial interaction

Exploration, evaluation and manipulation are currently the main tasks in immersive Virtual-Reality-Systems. Exploration is the walk through a virtual environment to look at virtual models like, for example, architectural models. Evaluation enables the movement of virtual objects to perform, for example, ergonomic evaluations with a digital human model. Manipulation interferes intensively with a virtual environment. An object itself may be transformed in its properties like colour or even shape. Using manipulation, objects may also be generated. One application which enables manipulation of objects is immersive modelling.

2 Requirements for Input Devices

In the framework of the European Union Project "VIEW of the Future"[1] users of virtual environments (Volvo, Alenia, PSA and John Deere) have produced a document which addresses

[1] IST-2000-26089

the users' needs regarding a virtual environment. The requirements regarding the interaction devices can be summarized as follows:

- Wireless
- Precise / high resolution
- Lightweight
- Quick response time
- 6 DOF
- Physical Ergonomics
- Performing main task directly with the input device
- Suitable for CAD-Evaluation

3 Interaction Concept for Optical Tracking

Humans are used to working with two hands. This can happen using both hands simultaneously (e.g. typing) or alternately (e.g. mouse and keyboard). Usually the two hands work together according to their capability. The dominant, skilful hand does the actual work (e.g. drawing) whereas the non-dominant hand assists (e.g. correcting the position of the paper). This shall be the basic idea for an interface: a smart distribution between both hands for a faster and more intuitive interaction with virtual environments.

Even though optically tracked systems provide high precision and fast response time, interaction with menus is still difficult. Pointing at a certain menu item is very difficult because one has to control 6 DOF. Therefore a tracking-decoupled menu concept has been integrated.

Figure 1 shows all the visible components of the interface: in the dominant hand the pointing device "dragonfly" with 6 DOF and in the non-dominant hand "Bug" with 3 DOF (only position). The concept is developed for dual-handed interaction. Yet, the "Bug" and the "Dragonfly" can also be used as stand-alone input devices, but with a different interaction concept. The determination of the position of the Bug enables the permanent possibility of visualizing a context sensitive menu. A button on the "Bug" in combination with a virtual ray, coming out of the dragonfly, serves as the selection tool and activates navigation/manipulation. The dragonfly does not have any buttons. The virtual object is connected to the "dragonfly" and follows directly its movements as long as the button on the "Bug" is pressed.

When evaluating a CAD model, the most important and most performed task is the positioning of the object relative to the user. The most natural way to do that is to actually play with the object in the hand. With an optically tracked system, the user has the possibility to interact with an object wireless and using lightweight interaction devices. The user is able to watch the object from all sides by just turning it in his fingers. With wired input devices one has to grab-release-grab and so on, to rotate and position the virtual object. Since there is no button on the pointing device, precise movements without losing control are possible.

The determination of the position of the "Bug" allows for the possibility of visualizing a context sensitive menu at all times. The menu will be used with a jog-dial. Pressing the jog-dial once will open the menu. By turning the jog-dial the user can scroll through the menu. Pressing a second time will select the menu item. In this way using the menu is de-coupled from the tracking system.

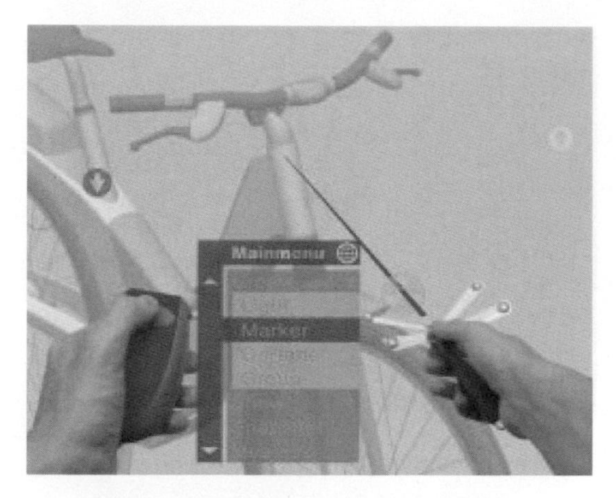

Figure 1: Components of the interface

4 Input Devices for Optical Tracking

4.1 Input device for the dominant hand: "Dragonfly"

The dragonfly is a pointing input device for real-time optical tracking systems. It has especially been designed to give the user a great variety of ergonomic grips. Depending on the projection set-up and the activity it is possible to change between many different grips to prevent fatigue of the hand (see figure 2 and figure 3). The device is designed for left- and right-handed persons. It does not have any buttons, electronics or batteries. This makes the device lightweight, thus eliminating fatigue of the hand. Six retro-reflective spheres in combination with the narrow form of the device allow for a precise identification of the device, its position and orientation in the 3D virtual environment, independent of the user's grip position. Having this basic structure the input device shows a dragonfly-like shape which is intensified with additional design elements. The dragonfly-metaphor is supposed to bring to mind a fast, precise, feathery and acrobatic flying input device.

Figure 2: Dragonfly in left hand with clench grip

Figure 3: Dragonfly in right hand with pinch grip

4.2 Input device for the non-dominant hand: "Bug"

The input device "Bug" comprises a jog-dial, identical to those in standard computer mice, and two additional buttons, which are well reachable for right- and left-handed persons (figure 4). In order to interfere with the user as little as possible, it is also wireless. Technically it is based on the standard components of a wireless mouse. Two retro-reflective spheres allow for the identification and position detection of the device. Since the Bug also offers the possibility to be used as a stand-alone device, it can be moved within one's hands intensively. The shape of the Bug is similar to a used piece of soap, which is intended, since it should be designed in a way to allow for a great variety of grip positions.

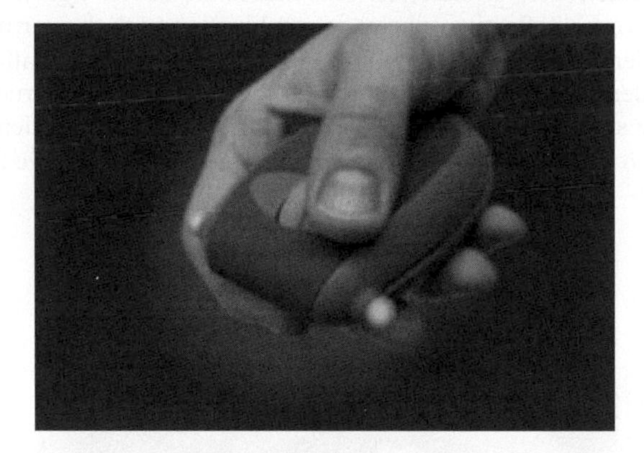

Figure 4: Non-dominant input device "Bug"

5 Evaluation

The new input devices have been tested by expert users at Fraunhofer IAO only. Further extensive testing in the context of VIEW is planned. So far, the practical experience has shown that use of tracking-decoupled menus makes the work in a virtual environment much more comfortable. Furthermore the ergonomic shape of the dragonfly is very much appreciated by the expert users. The opinions concerning the button-less concept are manifold. Although the ergonomic shape is appreciated by many users, having no button on the device can lead to difficulties. The Bug, although some users complained about its large size, seems to convince users due to its simple shape and of course due to its integrated jog-dial.

6 Conclusion

When designing input devices for virtual environments, consideration should be given to the technical limitations as well as to the possibilities a certain technology provides. The new devices for optical tracking feature an integration of the technology into a form which suits the ergonomic requirements for hand-held input devices, while making use of the advantages an optically tracked system can provide. Although extensive testing is not yet finished, we have found an improvement concerning comfort and usability. Preliminary results show that intuitive interaction concepts can only be designed if the concept is inherent in the input device and vice versa. Input devices should not be designed independently. Following our preliminary tests we assume that the new devices are very efficient for the evaluation of CAD models. As standard input devices for immersive virtual environments however, they will have a drawback for certain applications. The use of standard electronics as in wireless computer mice and mobile phones seems to be promising.

Acknowledgements

This work has been performed in the project "VIEW". Virtual and Interactive Environments for Workplaces of the Future (VIEW) is a three year long project funded by the European Union within the Information Society Technology program (IST).

References

Bandera J.E., P. Kern, J.J. Solf (1986). Leitfaden zur Auswahl, Anordnung und Gestaltung von kraftbetonten Stellteilen. *Schriftenreihe der Bundesanstalt für Arbeitsschutz, Dortmund,* Fb 494. Bremerhaven: Wirtschaftsverlag NW.

Kapandji I.A. (1992). Funktionelle Anatomie der Gelenke (2nd ed.), Bd. 1: Obere Extremitäten. Stuttgart: Ferdinand Enke Verlag.

Implementation of a Scalable Virtual Environment

Tomohiro Tanikawa

MVL Research Center,
Telecommunications Advancement Organization of Japan
4-6-1, Komaba, Meguro-ku, Tokyo, 153-8904 JAPAN
tani@cyber.rcast.u-tokyo.ac.jp

Abstract

In this paper, we present an implementation of a scalable virtual environment by integrating different types of virtual environments constructed with different types of input devices, rendering methods and display devices. By integrating numerous types of virtual environments, the VR system is able to realize a scalable VR application. To integrate these virtual environments, we propose an integration architecture that encapsulates each type of dataset, modeling and rendering algorithm as an independent virtual environment. In our experiment, we implement a photorealistic virtual environment to a high-end system such as the VR theater and several personal devices, such as the PC-based VR systems and networked Pocket PC-based PDA with a Wi-Fi card, and demonstrate the scalability of the proposed architecture.

1 Introduction

In this paper, we present a implimentation of scalable virtural environment by integrating different types of virtual environments constructed with different types of input devices, rendering methods and display devices. By integrating numerous types of virtual environments, VR system is able to realize a scalabile VR application. To integrate these virtual environments, we propose an integration architecture that encapsules each type of dataset, modeling and rendering algorithm, and the interaction model as an independent virtual environment. In our experiment, we implement a virtual environment to a high-end system such as we augment the VR theater and sevral personal devices, such as the PC-based VR systems and networked Pocket PC-based PDA with a Wi-Fi card, and demonstrate the scalability of this architecture.

2 Scalable VR Architecture

As mentioned above, different types of VR system have been proposed and developed, and each VR system has limitations. For example, immersive projection technology displays, such as the CAVE, CABIN and the VR theater, can provide realistic sensations to the users, as if he/she is in a ``real'' environment. However, even if the user can experience realistic sensations using such as this kind of display system, only one user at a time can interact in the virtual environment by using a special controller. Therefore, it is difficult for multiple users to obtain beneficial information simultaneously.

On the other hand, recently, PCs have been designed with high-performance processors, large memory spaces, high-resolution displays and broadband network connections, and therefore, PC-based VR systems can provide a high-quality virtual environment. Also, mobile devices such as the cellular phone and the PDA (personal data assistant) have become very popular, and we can easily access information from anywhere and at any time. By combining these different VR systems, we can overcome the limitation of each individual system. By integrating the PC-based or PDA-based VR system to the high-end VR system, all users can interact in the virtual environment and obtain beneficial information simultaneously.

2.1 Fundamental Architecture

To implement a scalable virtual environment using different types of VR systems (figure 1), we propose a fundamental architecture that encapsulates each type of dataset, modeling and rendering algorithms, along with their interaction models, as an independent environment. The fundamental architecture of the VR system is as shown in figure 2. In order to provide the feeling that the user is actually present in the virtual environment, the VR system presents a rendered image to the user according to his/her viewpoint in the virtual environment. Ordinarily, to construct a VR system, we construct the virtual environment using a single modeling and rendering methodology. By switching and combining separately constructed sub-environments, we can construct a scalable VR system, which supports the purposes, interactivity, and data handling of each constructed sub-environment (Tanikawa, Hirota & Hirose (2002)).

To support different types of display devices, modeling and rendering algorithms, interaction models, input devices, and network infrastructure, we modularize each sub-system of virtual environment, rendering, interaction and communication sub-systems as shown in figure 1. For encapsulating the difference in datasets, modeling and rendering algorithms, we propose a "rendering mechanism", for encapsulating the difference in interaction models and input devices, we propose a "user interface (UI) mechanism", and for encapsulating the difference in network infrastructure, we propose a "communication mechanism." By encapsulating each type of modeling and rendering algorithm along with the interaction model as an independent environments, we can construct a photorealistic environment based on optimal modeling and rendering methodology without interference from other environments.

(a) VR Theater (Cylindrical Screen) (b) Home VR Theater (c) PDA-based VR

Figure1: VR display systems

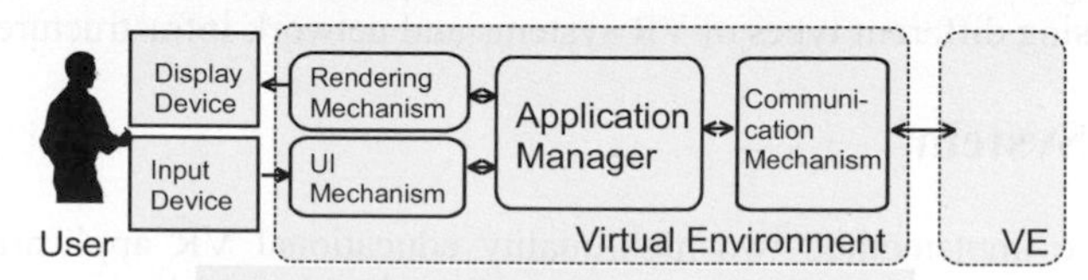

Figure2: Fundamental architecture

The use of architecture is similar to the approach of a networked virtual environment platform (A.Bierbaum et al (2001)) that provides an abstraction of the difference in hardware for portability. With this architecture, to support new users' requirements and systems, we only need to append the required new mechanism to be supported to the current virtual environment.

2.2 Rendering Mechanism and UI Mechanism

By integrating separately constructed virtual environments, we can implement a scalable virtual environment that supports all of the purposes, interactivity, and data handling requirements of ingredient virtual environment (Tanikawa, Hirota & Hirose (2002)). By encapsulating each type of dataset, modeling and rendering algorithms along with the interaction model as an independent environment with independent rendering and UI mechanism, we can focus on constructing each environment based on a single modeling and rendering methodology without interference from other environments. In this architecture, to support new users' requirements, we only need to append the required new rendering and UI mechanism to be supported to the current virtual environment.

To integrate different types of virtual environments, the integration of the interactions and rendered images of each sub-environment is indispensable. In order to provide the perception that the user is in one virtual environment although they are using different types of environments, the VR system should synchronize the user's viewpoints in each sub-environment and regulate any discordance between images and discontinuity of the image sequence on the display. Therefore, to generate the final rendered images, rendering mechanism compare not only depth information but also reliability, which is determined from data accuracy and resolution, display resolution and rendering errors from rendering methodologies. The reliability of each virtual environment is evaluated on the basis of rendered images and the angle of visibility from the user's viewpoint.

2.3 Communication Mechanism

By means of the rapidly growing network infrastructure, many types of VR systems have been connected with each other. In high-end VR systems, by using a broadband network, such as an Asynchronous Transfer Mode (ATM) network or a Gigabit Ether network, individual VR systems are able to transfer user's and scene information and thus share the virtual environment at an adequate refresh rate. Also, by using the rapidly growing wireless network infrastructure (Wi-Fi(IEEE802.11a/b) and 3G cellular phone services) and global positioning services, wireless devices such as notebook computers, PDAs, wearable computers and programmable cellular phones (i-mode) can share the virtual environment with other VR systems easily.

Of course, the available network infrastructure varies greatly in terms of bandwidth, latency time, packet loss rate and so forth. For example, most of the public network is not high quality, and data compression and verification are important considerations in sharing the virtual environment through such narrow-band networks. By switching and combining transmission and compression technologies according to the network conditions in communication mechanism, we can share a virtual environment using different types of VR systems and network infrastructure.

3 Prototype System

In our experiment, we implemented two high-quality educational VR applications on the VR theater (figure 1(a)): the Mayan ruins virtual environment for historical education, and the solar system environment for natural science education. A VR theater is a cylindrical screen with a 3-meter radius providing a 150 degrees horizontal by 45 degrees vertical field-of-view (TOPPAN (2001)). Also we implement the virtual environment to several personal devices, such as the PC-based VR systems (figure 1(b)) and networked Pocket PC-based PDA with a Wi-Fi card (figure 1(c)), and demonstrate the scalability of this architecture.

For Keio Youchisha Elementary School, we integrated these different types of VR systems. For the display of a photorealistic virtual environment, the VR theater is very useful. Using this kind of display, users can assimilate a large amount of information at one time. However, by accessing appropriate information according to the teacher's directions in the displayed virtual environment in the VR theater, students can analyze and handle the virtual environment individually.

For group studies, an educational specialist advised us that the VR system should provide interactivity for all users. Therefore, we implemented a scalable virtual environment that can fulfill such educational demands. As shown in figure 3, we installed several PC-based VR systems in front of the VR theater. Each VR system are connected through gigabit LAN by using the communication mechanism of each VR system, and managed by an application manager. The teacher operates the VR theater and guides his/her student groups in the Mayan ruins. With the teacher's guidance and through viewing the common view of the VR theater, the student groups can operate the PC-based VR system in front of them and experience the same virtual environment individually. In the virtual environment, the teacher and all of the students are represented by avatars, and they are able to recognize each other in the virtual environment (figure 4).

By using the proposed architecture, we can easily integrate different types of VR systems. The communication mechanism of each VR system switch shared information according to the system and network performance. And the rendering mechanism of the each VR system switch rendering

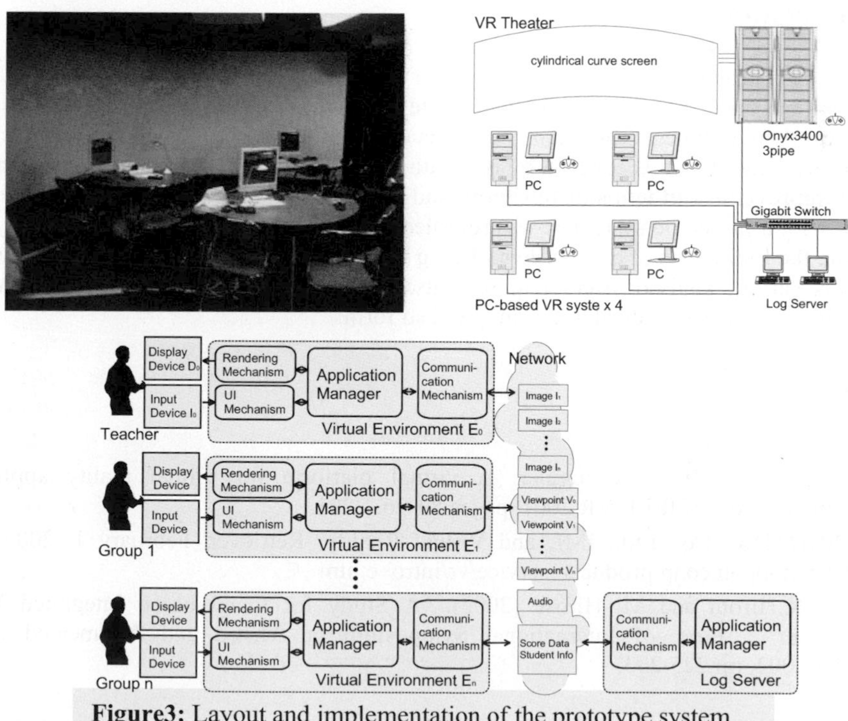

Figure3: Layout and implementation of the prototype system

Figure4: Scenes of group study

algorithm according to the system performance and display devices (figure 3). If the VR system has a sufficiently advanced enough 3D graphic capability to generate high-quality rendered images at an adequate frame rate, each communication mechanism can translate and receive the user's viewpoint data and event information and each rendering mechanism can render local scene graph database of the virtual environment at real time.

In the cases of the VR theater and the PC-based VR system, we can synchronize the teacher's and students' viewpoints. The PDA-based VR system does not have sufficiently advanced 3D graphic capability, and therefore the rendering and communication mechanisms of the system provide the virtual environment by receiving and displaying the streaming image compressed with MPEG-4 based compression technology according to the user's viewpoint. The resolution of the streaming image is 320x240, the average delay is 2-3 seconds, and the average frame rate is approximately 5 frames per second through an 11Mbps wireless LAN, and the user can share the virtual environment with the PDA-based VR system.

4 Conclusion

In this paper, we proposed a new architecture for building a scalable virtual environment by integrating various virtual environments. Our system is based on the concept that the sum of different types of environments can retain the merits of individual environments while overcoming their limitations. Based on this concept, by integrating different types of virtual environments, we can achieve improvement in terms of flexibility and accessibility. Through the development and experiments with a prototype system, it was revealed that a combination of different VR systems can overcome the limitations of each system. Using the proposed architecture, the VR system can easily support various kinds of users' requirements: datasets, purposes, rendering and modeling methodologies, input devices, display systems, and so forth.

References

A.Bierbaum et al (2001). VR Juggler: A virtual platform for virtual reality application development. Proc. of IEEE VR 2001, pages 89-96, 2001.

TOPPAN PRINTING Co., Ltd.. IML and Virtual Reality. Retrieved February 1, 2003, from http://www.toppan.co.jp/products+service/vr/intro_e.htm

T. Tanikawa, K.Hirota and M. Hirose (2002), "A Study for Image-based Integrated Virtual Environment", Proc. of International Symposium on Mixed and Augmented Reality (ISMAR) 2002, pp.225-233.

VIEW-IT:
A VR/CAD Inspection Tool for use in Industry

Jolanda G. Tromp *Sarah Nichols*
University of Nottingham
University park, NG7 2RD, Nottingham
Jola.Tromp@nottingham.ac.uk Sarah.Nichols@nottingham.ac.uk

Abstract

VIEW-IT is a rapid assessment method designed to identify the usefulness of virtual environments in terms of task-improvement and task-change through introduction of the new technology. This method has been developed for managers and end-users, such as CAD designers to help them decide whether a certain VR product will be useful to their product development tasks. The method has been newly created, based on two widely accepted HCI Inspection methods called Heuristic Evaluation and Cognitive Walkthrough.

1 Introduction

Companies are understandably reluctant to invest in a new technology until it has been proven to be cost-effective, reliable, compatible and useful, and it often proves difficult to convince the right people about the need or importance of introducing, changing or improving a system (Nielsen and Mack, 1994). Engineering market researcher Daratech, estimated the virtual prototyping & visualization market at 1.3 billion dollars in 2001 and the CAD market at 5.3 billion dollars (Briggs, 2002) and CyberEdge Information Services, Inc recently published that they expect the Visual Simulation/Virtual Reality marketplace will surge to $36.2 billion in sales in the near future (CyberEdge, 2002). VR is a natural progression from CAD, so there is in principle a sizable market for VR developments for use in the design and engineering industries.

The EU funded project Virtual and Interactive Environments for Workplaces of the Future (VIEW of the future, IST-2000-26089), with 12 partner institutions (Fraunhofer IAO, University of Stuttgart, John Deere, (D); VTT, WINTEC (FIN); Alenia (I); VOLVO (S); PSA Peugeot Citroen (F); COAT Basel (CH); ICS-FORTH, ICCS (EL); University of Nottingham (UK)), addresses these issues. It aims to develop best practice for industrial implementation and use of virtual environments, and integrate Virtual Environments (VEs) in product development, testing and training for workplaces of the future. During the project a rapid summative and descriptive usefulness VE assessment tool has been developed, called the VIEW inspection tool (VIEW-IT). VIEW-IT aims to assist people who have to make decisions about the usefulness of VR systems, by providing them with an awareness, a vocabulary, and tangible, comparable information, about the general usefulness of a VR system for a specified user group and task. The VIEW-IT assessment involves:

- One or more assessors
- visually assessing
- one or more elements of the interface
- by judging its compliance with newly created usefulness principles.

The method is based on widely accepted HCI design principles and two traditional HCI inspection methods: heuristic evaluation and cognitive walkthrough (Helander, Landauer and Prabhu, 1997; Preece, Rogers and Sharp, 2002) however, new assessment criteria have been compiled by the authors in order to address the novel concerns of VR/CAD users. The method consists of forms, which guide the assessor through the assessment (Tromp & Nichols, 2002). The depth of analysis is completely under control of the assessors. It takes a minimum of 2 hours to create a basic assessment report, which is converted into a 4 page summary document. This method has proved to be of help when establishing the added value of the new technology in terms of task improvement. It is also rapid and cost-effective, because it does not need experimental subjects.

Section 2 of this paper explains the development of VIEW-IT. Section 3 describes how to apply VIEW-IT, and finally section 4 summarizes the different ways in which the data generated by VIEW-IT can be used for industry and research.

2 Development of the VIEW Inspection Tool

VIEW-IT was developed in four stages. First a feasibility study was conducted to assess whether an assessment method could be made to fit the needs of the VIEW Project. Second, the method was created, adapted and extended based on the results of two focus-group meetings attended by seven leading VR usability experts. Third, the ease-of-use of the method was checked by means of three pilot-tests. And finally, the method was tested in-situ test by a team of four VOLVO developers - one of the user-groups of the VIEW Project. Since then the method is in use by the VIEW consortium to assess the experimental VEs developed during the project. We are still refining the method whilst testing it for reliability and validity. Additional plans exist to automate this method and to combine it with a VR/CAD systems design guidance tool.

2.1 Usefulness, Usability and the Virtual Experience

Within VIEW-IT, usefulness is assessed at two levels: usability and the virtual experience (see Figure 1). The virtual experience assessment is based on a traditional HCI method called heuristic evaluation, which is rapid and often educational for the assessors, as it assesses the VE with newly formulated heuristics about presence, general user friendliness, and VR-induced sickness – three components of the 'virtual experience' – a term used by Nichols (1999) to encompass the combination of effects experienced by the VR user.

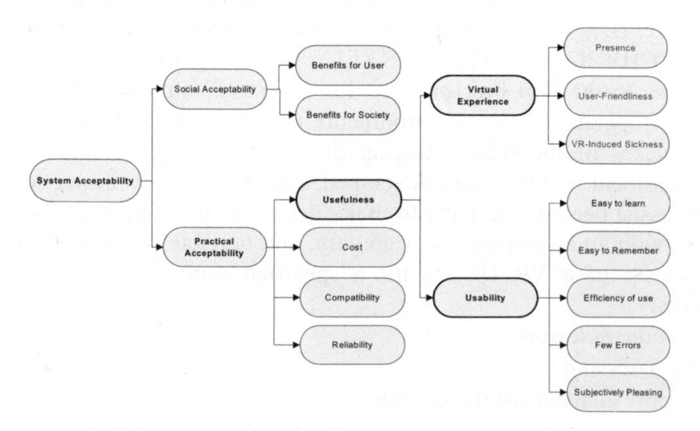

Figure 1: Usefulness for VEs, in the larger context of system acceptability.

The usability assessment is based on a traditional HCI method called cognitive walkthrough, which is extremely detailed and often educational for the assessors, making use of task independent interaction cycles (Kaur, 1998; Tromp, 2001) specifically formulated for VEs. During this assessment, all or specifically targeted elements of the total VE experience (such as input devices, output devices, 2D menus, 3D menus, navigation, object interaction, social interactions, etc), are judged and awarded points for visibility, feedback, and other usability principles. Together, these scores summarize and describe the VE. The final report is easily comparable with the reports of other assessed VEs and with an ideal score.

3 Applying the VIEW Inspection Tool

The assessors are guided through the assessment through a series of forms. Table 1 summarises these forms. The product summary forms are to be answered by the assessment team as a whole, because this has proved to help generate a common view of the product and consensus as to which elements of the system are most in need of attention and/or changing.

Table 1: Overview of forms used for VIEW-IT assessment.

Type of Form	Description	Acronyms
Demographic Summary	Information on the assessor.	DS Form
Product Summary	Information on the system	PS Forms (1-5)
Heuristic Evaluation	Virtual Experience Test	HE Forms (Presence, User-friendliness, Sickness)
Cognitive Walkthrough	Usability Test	CW Forms (a-j)
Assessment Summary	Summary of results	AS Forms: Four-pages (see Figure 2)

The Product Summary forms are used to identify which elements of the application are to be assessed for usability by means of the VIEW-IT Cognitive Walkthrough. Each CW form applies to one element of the complete VR system and assesses the task-flow and usability design of the element. For each problem that is found a rating (between 0 and 4) has to be given, and a unique reference number is allocated, for easy reference in future team discussions. The Heuristic Evaluation forms ask the assessors to express their agreement with three groups of heuristic statements on a five-point scale, with choices that range from Strongly Disagree, Disagree, Neutral, Agree, and Strongly Agree. Each assessor has to fill these forms in individually. The more assessors fill in this form the better, as the average score of multiple assessors will express a more accurate impression of these heuristics. Table 2 illustrates a typical extract from the heuristic evaluation within the VIEW-IT product summary document.

Table 2: A typical usability assessment summary of a 3D Menu.

Menu 1: 3D Menu 'Boule'		
3D Menu Interaction Cycle	**Potential problem and solution**	**Severity**
Can the user find the correct menu?	*One menu only. Correct icon selection occasionally difficult. Difficult to find correct option on menu.*	3
Can the user recognize the correct action on the menu?	*See above. Kept using incorrect icons.*	3
Can the other user(s) tell that user 1 is taking this action?	*No feedback to the active user.*	3
Can the user perceive that the action is taking effect?	*Only from the fact that they are applying it. No feedback.*	3
Can the other user(s) perceive that user 1 has executed the action?	*N/A. Single user*	0
Do the user(s) know what to do next?	*Users often had to be directed to function by experimentor.*	3

Table 3: A typical utility assessment summary of a VR/CAD environment in a 5-sided CAVE.

Rating from 0(unacceptable) to 4 (ideal):		
Presence	2	Level of presence could be improved by creating more functional details and making the interaction more predictable, including more consistent output from the VE to the user, the use of sound, and increased system response rate.
User-friendliness	2.25	Level of user-friendliness could be improved by improving the input device, improving the error-management, making the interaction more predictable, and improving the system response time.
Rating from 0(ideal) to 4 (unacceptable):		
VR-Induced Sickness	1.4	Level of VR-induced sickness could be lowered by making the room temperature more comfortable, using a smaller display, improving system response time, making the environment more realistic, have fewer independently moving objects.

The resulting final assessment report is compiled from all completed forms. The report is illustrated in Figure 2, and aims to provide an easily accessible overview and summary of the VR and its associated usefulness issues. The information is presented in highly condensed format so that there are sections 'Technical description', 'Virtual Environment description', 'User description', 'Task description', 'Initial usability issues', Cognitive Walkthrough results (see Table 2), Heuristic Evaluation results (see Table 3), 'Technical development recommendations', 'Research questions and hypotheses identified', and 'Additional comments'.

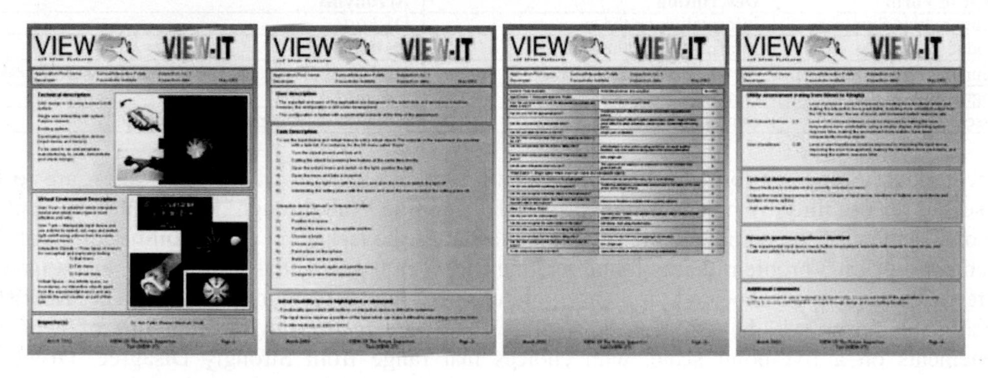

Figure 2: Four-page VIEW-IT final assessment summary report.

4 Conclusions

VIEW-IT is designed for both industry and research. On the one hand the method provides a summative report of the VEs developed for and during the VIEW Project, which can be used to establish an overview of issues relevant to use of CAD/VR in industrial organisations. On the other hand the method provides a descriptive evaluation report of Ves; especially useful when comparing different VEs. Finally, a scientific added value can be generated from the VIEW-IT database of assessed VEs, creating an opportunity for a meta-analysis of the type of design properties essential for VR when used for CAD purposes. Figure 3 shows the Heuristic Evaluation scores of 6 products to which VIEW-IT has so far been applied.

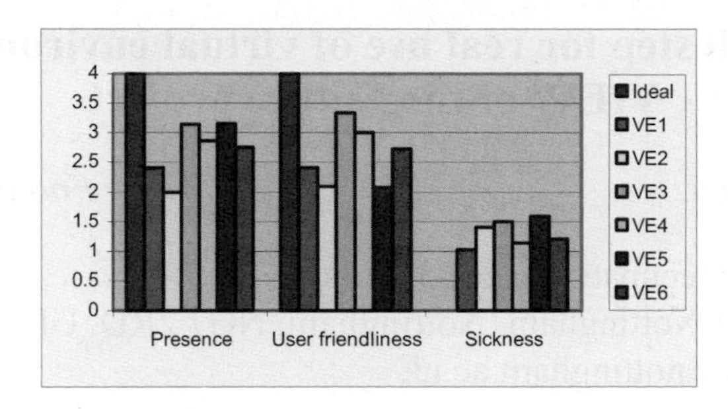

Figure 3: Virtual experience scores of the products assessed with VIEW-IT.

The VIEW-IT reports can be used in meetings to illustrate and quantify the usefulness of VEs for task-improvements in industry, and it can be used to provide an overview of incremental changes on a product under development. It has proved to help raise an awareness of the virtual experience and usability concerns amongst non-usability experts. It can be used at any time, from the very early design stages, through to the final product in use in the workplace. Additionally, the combined database of assessed VE products will give a long-term view on typical areas of VE design that are still insufficient for usefulness in the design and engineering industries.

5 References

Briggs, J.C., (2002). Virtual Reality is getting real: Prepare to meet your clone, in: The Futurist, May-June 2002, [electronic document at www.wfs.org].

CyberEdge, (2002). The Market for Visual Simulation/Virtual Reality Systems, Fifth Edition, [electronic document at www.cyberedge.com].

Helander, M., Landauer, T.K., Prabhu, P., (eds), (1997). Handbook of Human-Computer Interaction, 2nd ed, Elsevier Science B.V., The Netherlands.

Kaur, (1998). Designing Virtual Environments for Usability, PhD thesis, City University London, UK.

Nichols, S., (1999). Virtual Reality Induced Symptoms and Effects (VRISE): Methodological and Theoretical Issues. PhD Thesis: University of Nottingham.

Nielsen, J, Mack, R.L, (1994). Usability inspection methods, John Wiley and Sons, New York, NY.

Preece, J., Rogers, Y., Sharp, H., (2002). Interaction Design: Beyond Human-Computer Interaction, John Wiley and Sons, Inc., USA, pp. 420-425.

Tromp, J.G., (2001). Systematic Usability Design and Evaluation for Collaborative Virtual Environments, PhD. Thesis, University of Nottingham, UK.

Tromp, J.G., Nichols, S., (2002). VIEW-IT: Inspection Tool for CAD/VR Development and Acquisition, Technical Report VIRART/2002/103-JT01, University of Nottingham, UK.

Another small step for real use of virtual environments?: the VIEW of the Future project

John R Wilson [1]*(See Footnote 1)*

Institute for Occupational Ergonomics
University of Nottingham, Nottingham, NG7 2RD, UK
John.Wilson@nottingham.ac.uk

Abstract

During the past 10 years the nature of VR technology and of virtual environment experiences have changed markedly. Have the relevant most important human factors issues also changed? This paper draws upon a European Commission funded project, VIEW of the Future (IST 26089), to describe major human factors issues or problems and their relevance, in the context of particular technical challenges.

1 Introduction

VIEW of the Future (IST 2000 26089 - Virtual and Interactive Environments for Workplaces of the Future) is developing best practice for industrial use of virtual environments, for product development, testing and training. We will demonstrate application-independent VEs, develop strategy and guidance for building these, provide interactive VE design support tools, develop new interaction concepts and devices including a mobile VE system, and to provide better understanding of human factors to investigate VE participant behaviour and evaluate useability and utility. User needs are provided by three vehicle manufacturers (PSA, Volvo and John Deere) and the aerospace company Alenia Spazio, working with a variety of developer and human factors partners; there are twelve partners based in eight countries.

Work of the future, in industry and in commerce, will have an increasing array of technical systems available, of ever increasing sophistication and improving functionality. These new technical systems, however, will be of little value – and may even prove disruptive or harmful – if they are not implemented with proper understanding of their capabilities and of the human and organisational issues surrounding their use. By the end of the project it is envisaged that we will

---

[1] This paper is based on the work of the whole VIEW of the Future project. The partners are: University of Nottingham (Victor Bayon, Mirabelle D'Cruz, Gareth Griffiths, Sarah Nichols, Harshada Patel, Alex Stedmon, Jolanda Tromp, John Wilson); FhG-IAO (Hilko Hoffmann); University of Stuttgart (Oliver Stefani); VTT Information Technology (Raimo Launonen, Jukka Rönkkö); Suomen Wintec Oy (Seppo Laukkanen); John Deere (Paul Greif); Alenia Spazio (Valter Basso, Enrico Gaia, Marinella Ferrino); Volvo (Emma Johansson, Dennis Saluäär); PSA Peugeot Citroen (Séverine Letourneur, Jean Lorisson); COAT-Basel (Alex H. Bullinger, Karlheinz Estoppey, Ralph Mager); ICS-FORTH (Panos Karampelas, Alexandros Mourouzis, Constantine Stephanidis); ICCS (Angelos Amditis, Giannis Karaseitanidis)

be able to transfer the knowledge from VIEW of the Future to European industry, and provide concepts for the 'VE workplace' of the future.

2 The starting point

A number of the partners in VIEW of the Future have carried out state-of-the-art reviews and gaps analyses in virtual reality for national governments and the EU during the 1990s (e.g. Wilson et al, 1996). This work identified barriers as: Need for better technical development; Difficulties for technical integration between VR/VE and other commercial systems; A paucity of real working applications and poor evidence of added value; Concerns over possible participant health and safety problems and potential difficulties related to usability.

The general VIEW development system to frame the research needed is summarised in Figure 1. Two parallel streams of applications development (vehicle design and assembly training) will draw from a variety of VR platforms, and improved interaction specified for the applications as appropriate.

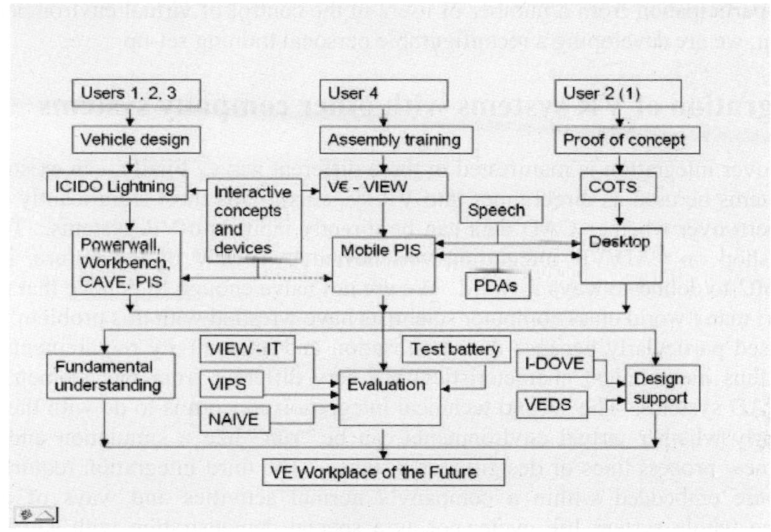

Figure 1:

3 Technical development

During the 1990s inadequacies identified in certain aspects of VR technical systems included: lags in the system; inadequate optics; and general worries about head mounted displays, with consequent debate on the merits of HMDs versus desktop VR. Into the 21st Century we are increasingly seeing improvements in technical characteristics of VR systems, and very much a move away from HMDs into projection systems, whether these be one-sided (projection wall or power wall), two-sided (PIS - personal immersive system), or three-, four-, five-, or six-sided systems (CAVEs). VIEW of the future is not developing technical systems per se, but rather is taking advantage of the state of the art technology developed previously or purchased by the partners. Amongst this are included CAVEs, powerwalls, personal immersive systems and holobenches. Mindful of the needs of many potential user companies, small wall projection systems and desktop systems are also explored. Contribution will be made to the further development of technical systems through the outcomes of the structured programme of user testing, and also through the relationship of the partners with the Fraunhofer spin off company, ICIDO. Generalisability of developments and findings will be aided by VIEW applications being structured around a Basic Applications Framework.

VIEW of the Future is carrying out substantial technical development in production of new interaction devices. Observations and structured experiments of people using our systems provide conceptual requirements for improved devices to control viewpoints and manipulate objects within virtual environments. Those being developed primarily include handheld devices, and new menu systems but also speech interfaces for various types of VR system.

In addition, one of the driving concepts behind VIEW, in terms of how we may work in the future, is the idea of mobile virtual environments. The concept of mobility itself has received much discussion – whether it is the individual, the system or the experience which is mobile. Technical development to date for a mobile user has been in personal digital assistants (PDAs), both to display virtual environments in a mobile fashion and also, perhaps more promisingly, to allow mobility and participation from a number of users in the control of virtual environments. For the mobile system, we are developing a reconfigurable personal training set-up.

4 Integration of VR systems with other company systems

The concern over integration is manifested in three different ways. Firstly, can existing company computer systems be used as direct input into VR systems? This most significantly comes in the form of concern over whether CAD data can be directly input into VR systems. To this end, a special workshop on CAD/VR integration was hosted by VIEW of the Future, in Athens in September 2002 to delineate ways forward. We are not naïve enough to believe that this will be a simple matter; many world class computer scientists have wrestled with this problem for a number of years, caused particularly because the visualisation and interactivity requirements from a VR system (and thus its technical characteristics) are very different from the geometric modelling required of CAD systems. The second technical integration concern is to do with the output from VR, particularly whether virtual environments can be "run" like a simulation and output data produced on new process lines or designs, for instance. The third integration requirement is that VR/VE become embedded within a company's normal activities and ways of doing things throughout the whole system life cycle, not as a special demonstration with something of the "wow factor". To this end, the project ensures the close involvement of user companies in the specification of VR technologies and interface elements. In comparison to other similar projects, this one has spent considerable time and other resources on producing and gaining general acceptance of a user requirements document, led by one of the user companies and providing description of user company operations which have potential for use of VR, a structured analysis of these in terms that can be provided to developers, and then translation into required functionalities and usability elements for the virtual environments which are produced.

5 Working applications and evidence of added value

The third barrier, which in some ways is a corollary of the first two, has been the absence of the real working applications of virtual environments that could be examined by other potential user companies and, partly as a consequence, no real evidence of the added value of VEs over alternative means of achieving the same tasks. VR technology configurations and VE applications are being built within the user companies by the companies themselves as well as by the developer partners, in order to meet specified user requirements and along the principles of the Virtual Environment Development System (Wilson et al, 2002). A further development which aids both the understanding of practical added value from virtual environments and also their integration within an organisation is the production within the project of a design support tool, I-Dove. The extensive user forum developed within VIEW has been asked to provide preliminary user requirements for such a design tool to support usable, well designed and safe virtual environments for workplace applications.

6 Human factors and usability in VIEW

Assessment of usability in virtual environments for a workplace of the future is being carried out within several broad streams: the effect of participating in the virtual environment on the participants, in terms of health and safety and in terms of their attitudes; the capability of participants to use the system effectively and productively, and the influence of system design on this; and how such participation in virtual environments may enhance the quality of performance of individuals and teams within workplace of the future.

Issues addressed include: 1) what are the needs of the user companies and how do these user requirements transfer into development aims and VE parameters, and what is the best structured development process to do this? 2) how should we best select different forms of VR platform and hardware and different types of VE application, on both human factors and organisational usability grounds? 3) how can we carry out application task analyses and virtual task analyses, to guide the development of the VEs and the related interface tools? 4) what user requirements (for such things as navigation, presence, interactivity, cues and feedback etc) are relevant to the design of VE interface tools and what is the best design of these? 5) what are the most appropriate methods for evaluating usability of the VE set ups and applications developed, including assessment of navigation, presence, interactivity, cues and feedback? 6) what are best methods for evaluating potential user side and after effects including health and safety consequences of using VE applications? 7) what are the best methods for evaluating organisational usability and integratability of VE systems and applications?

A number of special tools are being developed. VIEW-IT is a usability inspection tool (and associated inspectors' guidance document), which allows a structured walkthrough and assessment of virtual reality systems and virtual environments (see Tromp, 2001). A second tool identifies and records Virtual Prints (ViPs), digital counterparts, virtual footprints (ViFoPs), virtual fingerprints (ViFiPs) and virtual fossils (ViFossils) are respectively tracks of where the participants have moved in the virtual environment, of the virtual objects with which the participant has interacted, and of artefacts deliberately left by participants as some form of personal landmark within the environment (see Grammenos et al, 2002). NAÏVE (Griffiths, 2001) is a VE screening tool to evaluate how capable people are and how ready to participate within virtual environments in a skilled manner.

There are a very large number of possible psychological and psychophysiological tests to assess effects of virtual environments on participants, and early laboratory and field studies have been carried out to select an appropriate sub-set of potential methods and measures for incorporation in a usability test battery to assess side and after effects. Experiments have been carried out already to examine the comparative value and usability of different VR input devices, and to examine the preferred utilisation of speech as a means of accessing and participating in virtual environments. The partners involved in evaluation have also been visiting the VR facilities at the developer and user partner sites to utilise field test methods to evaluate the virtual environments. Current or future planned experiments include comparisons of interface concepts and devices and studies of virtual object recognition, internal level of guidance needed, orientation and size of environment, interaction hotspots, interaction metaphors, speech, stereo/mono, resolution and distancing and situation awareness.

The human factors of both development and evaluation in VIEW of the Future are many and varied. To handle these we have a large human factors team, drawn in inter-disciplinary fashion, but from such different backgrounds that they may have different priorities. For instance human factors experts at the user companies are particularly interested in supporting technical choice in the light of certain user requirements (e.g. resolution of the CAVE) and in developing user support tools (for instance to control mirror angles and views for vehicle occupant packaging). Within the developers there is particular interest in human factors support for iterative development of interaction devices and concepts, and investigating the mapping of control actions with menus appearing within the virtual environment. The human factors universities have perhaps more interest in formative evaluation than summative evaluation and, at some level, with further exploring the key questions of what it is to be present in, navigate around and interact with a virtual environment. All the human factors teams are interested in how the virtual environments will actually be used at user sites, and the value of such use.

7 Conclusions

It should be apparent from the description of this project that we could carry on investigating for another 10 or 15 years and not properly address, let alone solve, all the human factors issues and problems facing us. Nonetheless, through work on large scale projects like these as well as through smaller scale and perhaps more fundamental research, at times spread across many institutions in Europe, we are beginning to provide better answers to fundamental questions about what it is to participate in a virtual environment, and how VEs and VR technology might be best designed to support such participation. To do this is necessary to support the EU notion of the future workspace and design, appropriate work environments and tools.

References

Grammenos, D., Filou, M., Papadakos, P. and Stephanidis, C. (2002). Virtual Prints: Leaving trails in VEs. To appear in Proceedings of the Eighth Eurographics Workshop on Virtual Environments, Barcelona, May 2002.

Griffiths, G.D. (2001). *Virtual Environment Usability and User Competence: the Nottingham assessment of interaction within virtual environments (NAÏVE) tool*. Unpublished PhD Thesis, University of Nottingham.

Tromp, J.G. (2001). *Systematic usability design and evaluation for collaborative virtual environments*. PhD Thesis, University of Nottingham.

Wilson, J.R., Cobb, S.V.G., D'Cruz, M.D. and Eastgate, R.M. (1996). Virtual Reality for Industrial Application: Opportunities and Limitations. Nottingham: Nottingham University Press.

Wilson, J.R., Eastgate, R.M. and D'Cruz, M.D. (2002). Structured development of virtual environments. In K. Stanney (Ed.) *Handbook of Virtual Environments*. London: Lawrence Erlbaum Associates, p353-378.

Author Index

Subject Index